The Pesticide Encyclopedia

FSC
www.fsc.org
MIX
Paper from responsible sources
FSC® C013604

To all those who strive to keep the Earth green

The Pesticide Encyclopedia

Kalyani Paranjape

Vasant Gowariker

V.N. Krishnamurthy

Sudha Gowariker

www.cabi.org

CABI is a trading name of CAB International

CABI
Nosworthy Way
Wallingford
Oxfordshire OX10 8DE
UK

Tel: +44 (0)1491 832111
Fax: +44 (0)1491 833508
Email: info@cabi.org
Website: www.cabi.org

CABI
38 Chauncy Street
Suite 1002
Boston, MA 02111
USA

Tel: +1 800 552 3083 (toll free)
Email: cabi-nao@cabi.org

A catalogue record for this book is available from the British Library, London, UK.

Library of Congress Cataloging-in-Publication Data

The pesticide encyclopedia / Kalyani Paranjape … [et al.].
p. cm.
Includes bibliographical references and index.
ISBN 978-1-78064-014-3 (alk. paper)
1. Pesticides--Encyclopedias. I. Paranjape, Kalyani.

SB950.95.P47 2014
363.738′49803--dc23

2013040392

ISBN-13: 978 1 78064 014 3

Commissioning editor: Sreepat Jain
Editorial assistant: Alexandra Lainsbury
Production editor: Lauren Povey

Typeset by SPi, Pondicherry, India.
Printed and bound in the UK by CPI Group (UK) Ltd, Croydon, CR0 4YY.

Contents

About the Editors

Kalyani Paranjape is a management and enterprise consultant with a special interest in agricultural methodology. She is a postgraduate in business management, with work experience in multidisciplinary fields. Mailing address: 'Durvankur', 12 Shivparwati Housing Society, Near MIT College, Paud Road, Pune 411038, India.

Vasant Gowariker is a senior chemical technologist and a distinguished professor at the Indian Space Research Organisation (ISRO). Dr Gowariker was formerly Director of the Vikram Sarabhai Space Centre (VSSC); Secretary at the Department of Science and Technology, Government of India; Advisor on Science and Technology to four successive Indian prime ministers; and Chairman of the Rajiv Gandhi Science and Technology Commission (Cabinet rank), Government of Maharashtra. Mailing address: I – 101, Vanaraji Heights, Rambaug Colony, Near MIT College, Paud Road, Pune 411038, India.

V.N. Krishnamurthy is a senior chemist and industrial consultant. Dr Krishnamurthy was formerly Head of Physico-chemical-analytical Laboratories, R&D Complexes, dealing with chlorates, phosphates, nitrates, oxidizers, polymers, speciality chemicals and materials. Mailing address: 'Janaka', 38, PTP Nagar, Trivandrum 695038, India.

Sudha Gowariker is an active research scientist in plant and soil sciences and the author of several popular science books, essays and articles published since 1980 – some of which are included in high school textbooks in two major Indian states. Mailing address: I – 101, Vanaraji Heights, Rambaug Colony, Near MIT College, Paud Road, Pune 411038, India.

Utmost care has been taken to collect and study authentic information and document it as concisely as possible. Nevertheless, given the expanse of this subject, it is quite possible that some data, observations and/or descriptions given herein differ from what is available for or accessed by our readers. The editors/authors do not take responsibility for any error (technical or judgemental) that may be contained herein. However, they would be grateful if the same could be brought to their notice, for future reference. Email: encyclopediaoffice@gmail.com

Contributors

Mrs Leena Achari, MSc Life Sciences.
Dr Girish Bapat, Director Jnana Prabodhini.
Dr (Mrs) Ratnamala S. Bendre, Head, Department of Pesticides and Agrochemicals, School of Chemical Sciences, Jalgaon.
Dr B.C. Biswas, Formerly Additional Director – Fertilizer Association of India (FAI).
Miss Shubhangi Borse, MSc Organic Chemistry.
Kedar Champhekar, MSc Biodiversity, Currently Programme Officer at the Centre for Environment Education, Ahmedabad.
Mr Pradeep S. Chavan, Indian Agricultural Research Institute (IARI), New Delhi. Consultant for Agri-Horticultural and Irrigation Projects.
Dr T.K. Das, Indian Agricultural Research Institute (IARI), New Delhi.
Dr Manik Dhanorkar, Ashoka Buildcom (Agri Division), Nasik.
Dr (Ms) Prem Dureja, Fellow, National Academy of Agricultural Sciences of India (FNAAS), Ex. Head and Emeritus Scientist, Division of Agricultural Chemicals, Indian Agricultural Research Institute (IARI), New Delhi.
Dr Subhodh Mohanti, Vigyan Prasar and Author.
Shamika Mone, MSc Biodiversity.
Miss Monika Patankar, MSc Bio-informatics.
Dr Bishwajeet Paul, Indian Agricultural Research Institute (IARI).
Archana Pokale, MSc Physical Chemistry.
Dr Rajan Ramaswami, Consultant – Polyurethane (PU), Agro and Specialty Chemicals and Technical Head, McLube Asia Pvt. Ltd, Nashik.
Dr N.K. Roy, Formerly Head Chemical Division, IARI, New Delhi.
Dr Pratibha Sharma, IARI, New Delhi.
Ms Neha Shintre, MSc Biodiversity.
Dr Mahendra P. Srivastava, Ex-Professor and Head of Plant Pathology; Director of Planning, Monitoring and Evaluation, HAU (Chaudhary Charan Singh Haryana Agricultural University), Hissar; Chairperson of the International Working Group on Plant Health Clinic; and member of the German Phytomedical Society.
Dr T.S. Thind, Panjab Agricultural University, Ludhiana.
Dr Perumal Vidyasekaran, Fellow, Indian National Science Academy (FNA). Former Director of the Centre for Plant Protection Studies, Tamil Nadu Agricultural University.
Dr Sandhya Vyas, Pune University.
Dr Sudhir Vyas, Currently working in Albany Molecular Research Inc., Hyderabad.
Dr N.T. Yaduraju, International Crops Research Institute for the Semi-Arid Tropics (ICRISAT), India.

Foreword

The ecosystem on planet Earth supports many forms of life which are mutually dependent. When man learnt to farm and undertake planned agriculture, he learnt also that he had to compete with pests for the food he grew. Pesticides were discovered. In this relentless fight between man and pests, the tools of science and the power of technology became a weapon with which to combat pests. However, slowly it has become clear that these very tools employed for maximizing agricultural production have become a source of contamination of food.

Life cycle analysis and rational choice of pesticides have become necessary in agriculture. There are a large number of pesticides to select from for meeting the needs of agriculture. While cultural practices in agriculture promote the selection and choice of some pesticides over others, evolutionary biology of pests makes changes necessary. Pests develop resistance. Thus, indiscriminate use of pesticides is not in anyone's interest. Informed selection of pesticides based on scientific methodologies and tools call for authoritative expert systems. For these reasons the Department of Science and Technology considered it was essential to bring out an encyclopedia on pesticides.

Dr Vasant Gowariker and his ever-committed team, who had brought out an outstanding encyclopedia on fertilizers, were the ideal choice for creating *The Pesticide Encyclopedia*. I am delighted that they have created a masterly compilation of information on pesticides. *The Pesticide Encyclopedia* will help our pursuit of knowledge, since we need both food and the environment. This work serves as a rich source of unbiased information – not only on pesticides but also on toxicity, metabolism, worker safety, environmental aspects of pesticides, biocontrol of pests, integrated pest management (IPM) and much more. And the more we read this, the wider will be our perspective and the more prone we would be to explore the right mix for a brighter future for our children.

The Pesticide Encyclopedia is a monumental effort which combines commitment with professionalism. It links excellence with relevance. It is a high quality database. On the whole, *The Pesticide Encyclopedia* bridges an important knowledge gap and serves a purpose for the agriculture sector. I congratulate the authors for their outstanding effort of great value.

(T. Ramasami)

Honourable Secretary
Department of Science and Technology
New Delhi

Preface

We believe there are few things as awe-inspiring yet intimidating as attempting to write an encyclopedia. One gets to marvel at man's dogged pursuit of knowledge. Each branch of science (and its myriad sub-branches) is already widely developed and yet newer dimensions get unravelled. We believe that capturing the complete essence of a knowledge set is not only difficult but unimaginably impossible. And yet, new encyclopedias are constantly invading bookstores. We have bitten the bait too – a second time. The first was when we made *The Fertilizer Encyclopedia*.

We have modified our road map to *The Pesticide Encyclopedia* several times, in order to align its scope with the need for such a work. We hope the result meets the expectation of our readers and that they get to know most things about most things that have to do with the food that man grows and the organisms that he seeks to keep off the field … and most things in between!

In 2006, the Government of India (specifically, Dr T. Ramasami, the Honourable Secretary, Department of Science and Technology) invited us to create *The Pesticide Encyclopedia*, which would be a sequel to *The Fertilizer Encyclopedia*. It has taken us 6 years to put this information together. Although we still hold that it is impossible to capture all available information, time is of the essence and we feel it is necessary to present before the reader, *The Pesticide Encyclopedia*.

We now highlight some points that are important to keep in mind while reading the book.

- This work has attempted to bring out *only* the essence of three critical areas of interest: (i) crops; (ii) pests; and (iii) control agents. This encyclopedia is by no means the universe of these grave subjects, since each is a separate extensive subject of study by itself. What we have done is to *comprehensively cover each subject* to the extent considered relevant by the authors, reviewers and advisors.
- Solubility values of the active ingredients (a.i.) have been recorded in *The Pesticide Encyclopedia* only to indicate if the given a.i. is a danger to, for example, water sources or soils. In fact, some data expressed here in this volume may be totally different from that published elsewhere. For instance, in the case of sulfentrazone, the water solubility is given as 400 mg and 780 mg/l in two different sources. Both could be right, since it is possible to get different results at different temperatures and under the influence of other factors.
- In toxicity studies, LD/LC/EC values are calculated and measured by many relevant techniques, and interpreted from the stand point of safety of usage, for registration in a country. This volume contains a few such values for each a.i., indicating toxicity through the main routes of exposure. We feel that it is more important to indicate what each of these values means than to simply document values.
- Since the registration process of a pesticide a.i. is a long and arduous one, the status of any one ingredient in a country may change with time. *The Pesticide Encyclopedia* has tried to tap into relevant sources during the preparation of this work. It may not reflect the status at the time of publication. Similarly, this is the case for MRL/ADI/NOEL values.
- The inclusion of comprehensive notes on crops and important pests is to partially fill the gap between the academic and practical sides of agrochemicals. There are several other aspects of crop and pest management that have not found a place here. Also, pesticide application recommendations are based on practices prevalent in some regions, and may not be permitted in all parts of the world. In fact, pesticide recommendations may differ even when similar topics are being discussed. And since disease symptoms can be confusing due to their similarities, it may be necessary to consult a plant pathologist for a particular disease condition. However, we have attempted to uphold the spirit of being comprehensive.
- *The Pesticide Encyclopedia* has topics, which we refer to as 'entries'. There are over 1000 entries, covered in substantial detail and arranged alphabetically. If there is a concept/term embedded in an entry, we have treated it as a sub-entry. These sub-entries may be conveniently accessed by the reader (since they are listed in the Index). Entries and sub-entries are denoted in bold font. Entries for chemical pesticides mostly follow the pattern of mentioning the name of the a.i. first, followed by its chemical formula and then the IUPAC name in brackets. There is a vast amount of subsidiary information available in the entries that is not in bold and is not in the Index. This information includes names and brief details of many pests and several metabolites. The reader may thus benefit even more than expected from going through this work. Similarly, scientific names of species, as well as names derived from non-English terms are indicated in italics. When pesticide formulations are referred to, where possible, the international code used for pesticide formulations is included in brackets (see Appendix 2 for a full list of these codes). Besides entries and sub-entries, there is a Glossary that aims to explain terms and some concepts, in brief, and here at the front of the book there is a list of frequently used abbreviations.
- In noting trade names, we have tried to include at least one such commercial product but have not verified its existence today. These names are not meant to indicate endorsement of any commercial venture.

The Pesticide Encyclopedia would not have been possible if we did not lean on the works of great academics and researchers the world over, who have dedicated their lives to unravel the mysteries of life. Our heartfelt thanks go to all those thousands of great men and women, because any encyclopedia is a result of a gigantic churning of huge amounts of information. Our gratitude is evident from the voluminous bibliography at the end of the volume. Any missing reference is purely due to oversight.

Last but not least, we are proud to state that it was the scientific vision of Dr T. Ramasami that helped *The Pesticide Encyclopedia* see the light of day. As Secretary, Department of Science and Technology (DST), he allotted time from his extremely busy schedule to check on the progress of the work and to give his valuable inputs.

We would like to convey our gratitude to the funding agency, Technology Information Forecasting and Assessment Council (TIFAC), which has directly and indirectly assisted the smooth working of this project. We thank Jnana Prabodhini for managing the funds and conducting regular audits. We would also like to put on record our sincere thanks to all the reviewers, contributors and our project team mates, who have put in hard work to make it happen.

In spite of our earnest efforts, some errors could have crept in. We welcome our readers to kindly highlight the same, in order to make this encyclopedia even better.

Kalyani Paranjape   Vasant Gowariker   V.N. Krishnamurthy   Sudha Gowariker

Acknowledgements

The authors are grateful to: (i) the Technology Information Forecasting and Assessment Council (TIFAC) of the Government of India, for financing the project from September 2006 to March 2012; (ii) the Project Review Committee comprising Dr T. Ramasami (Chairman), Dr Vasant Gowariker, Dr Laxman Prasad, Dr Bishwajeet Paul, Dr B.C. Biswas, Dr N.T. Yaduraju, Dr M.P. Srivastava and Dr Girish Bapat, and Ms Kiran Abraham Zacharia (Member-Secretary), for reviewing the project road map and providing suggestions; (iii) the Review team consisting of Dr N.K. Roy, Dr Bishwajeet Paul, Dr T.K. Das, Dr N.J. Yaduraju, Dr M.P. Srivastava, Dr T.S. Thind, Dr (Mrs) Pratibha Sharma, Dr B.C. Biswas, Dr (Mrs) Ratnamala S. Bendre, Dr Rajan Ramaswamy, Dr Perumal Vidyasekharan, Dr Girish Bapat, Dr Manik Dhanorkar, Dr Subhodh Mahanti, Mr Pradeep S. Chavan, Dr Sandhya Vyas, Dr Sudhir Joshi and Dr (Ms) Prem Dureja, for going through the manuscript for gaps in the data; (iv) the technical and administrative personnel of the project, including Ms Shubangi Borse, Ms Shamika Mone, Ms Monika Patankar, Ms Leena Achari, Ms Archana Pokale, Mrs Vinaya Dhanorkar, Mr Kedar Champhekar, Ms Neha Shintre, Ms Kashmira Burande, Ms Anjali Sant, Mr Anay Gokhale, Mr Avinash Mehendale and Mr Shreyas Nakhare, for working steadfastly at ensuring accuracy and time lines; (v) Ms Ashwini Gowariker, Ms Manjusha Oak and Ms Vrinda Gupte for going through the manuscript and correcting it for English and logical flow; (vi) Dr T. Ramasami, Secretary, Department of Science and Technology (DST) for guiding the project team and writing the Foreword; (vii) the Director and administrative and accounts personnel of Jnana Prabodhini, for handling the administrative and financial matters of this project with TIFAC/DST; and (viii) all individuals and organizations from India, USA, Canada and New Zealand and some other regions that have spontaneously given permission to use their data and pictures for inclusion in *The Pesticide Encyclopedia*. For profiles of the reviewers and contributors see the Contributors section.

Frequently Used Abbreviations

ΔG	Gibb's free energy of vaporization (kJ/mol or kcal/mol)
ΔH_{fus}	enthalpy of fusion (kcal/mol)
ΔH_V	enthalpy of vaporization (kJ/mol or kcal/mol)
µg/g	micrograms per gram
µg/l	micrograms per litre
µg	microgram
µm	micrometre
^{14}C	radioactive labelled carbon-14 compound
2,4-D	2,4-dichlorophenoxyacetic acid
2,4-DB	2,4-dichlorophenoxybutyric acid
Å	Ångstrom (10×10^{-10} m)
AA	atomic absorption
AAPFCO	Association of American Plant Food Control Officials
ACCase	acetyl-coenzyme A (CoA) carboxylase
ACGIH	American Conference of Governmental Industrial Hygienists
ACOS	Advisory Committee on Organic Standards
ACP	Advisory Committee on Pesticides
ACRE	Advisory Committee on Releases to the Environment
ADI	acceptable daily intake
ADNT	acute delayed neurotoxicity
ADP	adenosine diphosphate
AEA	Agricultural Engineers' Association
AEC	anion exchange capacity
AHAS	acetohydroxy acid synthase
a.i.	active ingredient(s)
A_i	area of phase i (m^2)
AIC	Agricultural Industries Confederation
AIHA WEEL	American Industrial Hygiene Association's Workplace Environmental Exposure Level
ALPM	Automated Log-P Measurement
ALS	acetolactate synthase
AM	2-amino-4-chloro-6-methyl pyrimidine
ANSI	American National Standards Institute
AOEL	acceptable operator exposure level
AP	ammonium phosphate
aPAD	Acute Population Adjusted Dose
APHIS	Animal and Plant Health Inspection Service
APP	ammonium polyphosphate
aq.	aqueous
aRfD	acute reference dose
ARI	Aggregate Risk Index
Ark	activity ratio
AS	active substance
ATP	adenosine triphosphate
BAA	British Agrochemical Association
BCF	bioconcentration factor
BCPC	British Crop Protection Council
BHR	Beilstein Handbook Reference
BOD	biological oxygen demand (mg/l)
bp	boiling point
BP	British Pharmacopeia
BPAU	Biocides and Pesticides Assessment Unit
BPL	bone phosphate of lime (1% P_2O_5 = 2.185% BPL)
Bq	becquerel
BRC	British Retail Consortium
BRN	Beilstein Registry Number
BSI	British Standards Institute
b.w.	body weight
BWG	body weight gain
C	molar concentration (mol/l or mmol/m^3)
°C	degrees Celsius
C_A	concentration in air phase (mol/l or mmol/m^3)
CAC	Codex Alimentarius Commission
CAEPA	California Environmental Protection Agency
CAN	calcium ammonium nitrate
CAP	Common Agricultural Policy
CAS	Chemical Abstracts Service
CC	countercurrent chromatography
CCE	calcium carbonate equivalent
CCRIS	Chemical Carcinogenesis Research Information System
CE	capillary electrophoresis
CEC	cation exchange capacity (meq/100 g of soil)
CFR	Code of Federal Regulations
ChE	cholinesterase
ChEI	cholinesterase inhibition
CHEMTREC	Chemical Transportation Emergency Center
CI	colour index
CIBRC	Central Insecticide Board and Registration Committee
C_L	liquid or super-cooled liquid concentration (mol/l or mmol/m^3)
cm	centimetre
C/N	carbon:nitrogen ratio
CNC	critical nutrient concentration
CNS	central nervous system
Co.	company
COD	chemical oxygen demand
compd.	compound
Con	conductivity detector
COPA	Comité des Organisations Professionnelles Agricoles (of the European Union)

Corp.	corporation
cP	centipoise (measure of viscosity)
CPA	Crop Protection Association
cPAD	Chronic Population Adjusted Dose
CPC	centrifugal partition chromatography
CPM	Commission on Phytosanitary Measures
CRD	Chemicals Regulation Directorate of the Health and Safety Executive (HSE) of the UK
CRH	critical relative humidity
C^S	saturated aqueous concentration (mol/l or mmol/m^3)
C_S	solid molar concentration (mol/l or mmol/m^3)
CSF	Confidential Statement of Formulation
CSP	concentrated superphosphate
cSt	centistokes (measure of viscosity)
% CT	percentage of crop treated
CTFA	Cosmetics, Toiletries and Fragrance Association
C_W	concentration in water phase (mol/l or mmol/m^3)
CXL	Codex Maximum Residue Limit (Codex MRL)
D	density
D	D values (mol/Pa.h)
d-	dextrorotatory
D_A	D values for advection (mol/Pa.h)
D_{Ai}	D values for advective loss in phase i (mol/Pa.h)
DAP	diammonium phosphate
DAT	days after treatment
DC_{50}	designated concentration of a toxicant effect in 50% of the test organisms
DCD	dicyandiamide, nitrification inhibitor
DCI	data call-in
Dck	duck
DCP	dicalcium phosphate
DDT	1,1′-(2,2,2-trichloroethylidiene) bis(4-chlorobenzene)
DEA	diethanolamine
Dec	decompose, decomposition
DEEM	Dietary Exposure Evaluation Model
DEEM-FCID	Dietary Exposure Evaluation Model – Food Consumption Intake Database
der	dermal
Derive	derivative
DFR	Dislodgeable Foliar Residue
DIT	developmental immuno-toxicity
DMF	dimethylformamide
DMSO	dimethyl sulfoxide
DNA	deoxyribonucleic acid
DNT	developmental neurotoxicity
DOC	dissolved organic carbon
DOM	dissolved organic matter
DOT	Department of Transportation, USA
DP	dustable powder
D_{QW}	D value for total particle transport (dry and wet), (mol/Pa.h)
D_R	D value for reaction (mol/Pa.h)
D_{Ri}	D value for reaction loss in phase i (mol/Pa.h)
DRIS	Diagnosis and Recommendation Integrated System
D_{RS}	D value for rain dissolution (air–soil), (mol/Pa.h)
D_{RW}	D value for air–water dissolution (mol/Pa.h)
D_S	D value for air–soil boundary layer diffusion (mol/Pa.h)
DSC	differential scanning calorimetry
dS/m	deciSiemens per metre
D_{SW}	D value for water transport in soil (mol/Pa.h)
DT_{50}	Degradation time 50 is the time taken for half of a parent compound to disintegrate or disappear from water or soil
DTA	differential thermal analysis
D_{Ti}	total transport D value in bulk phase i (mol/Pa.h)
DTPA	diethylenetriaminepentaacetic acid
D_{VS}	D value for total soil–air transport (mol/Pa.h)
D_{VW}	D value for air–water diffusion (absorption), (mol/Pa.h)
DW	drainage water
DWLOC	drinking water level of comparison
E	emission rate (mol/h or kg/h)
EAS	existing active substance
EC	emulsifiable concentrate (formulation)
EC	European Commission
EC_{50}	effective concentration of a drug that produces the desired effect in 50% of the test organisms exposed
ECCA	European Crop Care Association
ECCO	European Community Co-ordination
ECD	electron capture detector, used in gas chromatography
ECJ	European Court of Justice (Luxembourg), for interpretation of European Community (EC) legislation
ECP	exchangeable cation percentage
ECPA	European Crop Protection Association
ED_{50}	effective dose that produces effect in 50% of the test organisms exposed
EDDHA	ethylenediaminedi-o-hydroxyphenylacetic acid
EDI	estimated daily intake
EDTA	ethylenediaminetetraacetic acid
EDWC	estimated drinking water concentration
EEC	estimated environmental concentration
EFSA	European Food Safety Authority
EINECS	European Inventory of Existing Commercial Chemical Substances
ELCD	electrochemical detector
ELISA	enzyme-linked immunosorbent assay
EMDI	estimated maximum daily intake
EO	ethylene oxide
EPA	Environmental Protection Agency of the USA
EPICS	equilibrium partitioning in closed system
EPPO	European and Mediterranean Plant Protection Organization

ERL	extraneous residue limit
ES	exchangeable sodium (meq/100 g soil)
ESP	exchangeable sodium percentage
ESR	exchangeable sodium ratio
Et	ethyl ($C_2H_5^-$)
ET	evapotranspiration
EU	European Union
EUP	end-use product
°F	degrees Fahrenheit
FAO	Food and Agriculture Organization of the United Nations
FC	field capacity
f-const	fragmental constants
FDA	Food and Drug Administration (USA)
FD&C	Food, Drug and Cosmetic (certified by the FDA)
FEMA	Federal Emergency Management Administration
FEPCA	Federal Environmental Pesticide Control Act of 1972
FFA	free fatty acid
FFDCA	Federal Food, Drug and Cosmetic Act
f_i	fugacity in pure phase i (Pa)
FID	flame ionization detector
FIFRA	Federal Insecticide, Fungicide, and Rodenticide Act
FLD	fluorescence detector
Fluo	fluorescence method
Fmus	female mouse
FOB	functional observation battery
Fp	freezing point
FPC	Fresh Produce Consortium
FPD	flame photometric detector for gas chromatography
FPD (S or P)	flame photometric detector (sulfur or phosphorus mode)
FQPA	Food Quality Protection Act
FRAC	Fungicide Resistance Action Committee
Frat	female rat
FSANZ	Food Standards Australia New Zealand
FTIR	Fourier Transform Infrared spectroscopy
FU	smoke generator
FYM	farmyard manure
G	advective inflow (m^3/h)
g	gram
GATT	General Agreement on Tariffs and Trade
G_B	advective inflow to bottom sediment (m^3/h)
GC	gas chromatography
GC/ECD	GC analysis with electron capture detector
GC/FID	GC analysis with flame ionization detector
GC/MS	Gas-liquid chromatograph coupled to a mass spectrometer
GC/NPD	Gas-liquid chromatograph coupled to a nitrogen phosphorus detector
GC-RT	GC retention time
GDD	growing degree days
gen.col	generator-column
GFAA	graphite atomic absorption spectrophotometer
GIS	geographical information system
GLC	gas liquid chromatography
GLN	guideline number
GPA	gallons per acre
GPC	gel permeation chromatography
Gpg	guinea pig
GPM	gallons per minute
GPS	global positioning system
GR	gypsum requirement
Gran	granular, granules
GRAS	generally recognized as safe
GUP	general use pesticide
GV	guideline value
h	hour
ha	hectare (10,000 m^2 or 2.47 acres)
HAL	health advisory level
HALL	Hall electrolytic conductivity detector
HDL	high-density lipoprotein
HDT	highest dose tested
HLC	Henry's Law constant ($Pa.m^3/mol$)
HPLC	high pressure liquid chromatography
HPLC/fluo	HPLC analysis with fluorescence detector
HPLC-k′	HPLC-capacity factor correlation
HPLC/MS	HPLC coupled with mass spectrometer
HPLC-RI	HPLC-retention time correlation
HPLC-RV	HPLC-retention volume correlation
HPLC/UV	HPLC analysis with UV detector
HPTLC	high performance thin layer chromatography
HRAC	Herbicide Resistance Action Committee
HSDB	Hazardous Substances Data Bank
IAA	indole-3-acetic acid
IARC	International Agency for Research on Cancer
IBA	indole-3-butyric acid
IC	ion chromatograph
IFA	International Fertilizer Association
IGR	insect growth regulator
Ihl	inhalation
IIBC	International Institute of Biological Control
im	intramuscular
Inc	incorporated
Inj	injection
IOBC	International Organization of Biological Control of noxious animals and plants
ip	intraperitoneal
IP	ionization potential
IPM	Integrated pest management
IPPC	International Plant Protection Convention
ISO	International Standardization Organization
IU	international units
IUPAC	International Union of Pure and Applied Chemistry
iv	intravenous
IW	irrigation water
J	Joule
JMPR	Joint Meeting on Pesticide Residues
K	Kjeldahl method
K	first-order rate constant, per hour (h^{-1} or /h)
K_1	uptake/accumulation rate constant, per day (day^{-1} or /day)
K_2	elimination/clearance/depuration rate constant, per day (day^{-1} or /day)
k_A	air–water mass transfer coefficient, air-side (m/h)

$K_{AR/W}$	aerosol–water partition coefficient
K_{AW}	dimensionless air–water partition coefficient
K_B	bioconcentration factor
K_b	biodegradation rate constant, per day (day^{-1} or /day)
kg	kilogram
K_h	association coefficient
K_h	hydrolysis rate constant, per day (day^{-1} or /day)
k_i	first-order rate constant in phase i, per hour (h^{-1} or /h)
K_{OC}	organic-carbon sorption partition coefficient
K_{OM}	organic-matter sorption partition coefficient
K_{OW}	octanol–water partition coefficient
K_P	sorption coefficient
K_p	photolysis rate constant, per day (day^{-1} or /day)
$K_{S/W}$	soil–water partition coefficient
$K_{SD/W}$	sediment–water partition coefficient
kV	kilovolt
k_W	air–water mass transfer coefficient, water-side (m/h)
kW	kilo watt
kWh	kilo watt hour
l	litre
l-	levorotatory
LAI	leaf area index
LC_{50}	concentration lethal to half the subjects tested
LD_{50}	dose (quantity) lethal to half the subjects tested
LDL	low-density lipoprotein
LE	larval equivalent, where 1 LE = 6×10^9 inclusion bodies
LED	light-emitting diode
LF	leaching fraction
LISA	low-input sustainable agriculture
LLC	limited liability company
LNG	liquefied natural gas
LOAEC	lowest observed adverse effect concentration
LOAEL	lowest observed adverse effect level
LOC	level of concern
LOD	limit of detection or limit of determination
logP	log of the octanol–water partition coefficient
LOQ	limit of quantitation
LPG	liquefied petroleum gas
LR	leaching requirement
LSC	liquid scintillation counter
LSD	least significant difference
Ltd	limited
M	million
MAC	maximum allowable concentration
MAP	monoammonium phosphate
max.	maximum, maxima
MCPA	(4-chloro-2-methylphenoxy) acetic acid
MCPB	4-(4-chloro-o-tolyloxy) butyric acid
Me	methyl (CH_3^-)
MEA	monoethanolamine
MEK	methyl ethyl ketone
mEq/l	milliequivalent per litre
µg	microgram
mg	milligram
mg/kg	milligram per kilogram
mg/kg/day	milligram per kilogram per day
mg/l	milligram per litre
min	minute(s)
min.	minimum
mix	mixture
mm	millimetre
mmho	millimho (a measure of electric conductance)
Mmus	male mouse
MOE	margin of exposure
mp	melting point
mph	miles per hour
Mrat	male rat
MRL	maximum residue level (or limit)
mRNA	messenger RNA
MS	mass spectrometer
MS (NCl)	mass spectrometer (negative chemical ionization mode)
MSDS	material safety data sheet
MSHA	Mine Safety and Health Administration
MTD	maximum tolerated dose
MUP	manufacturing-use product
Mus	mouse
mV	millivolt
MW	molecular weight
N	normal
NA	not applicable
NAD	nicotinamide adenine dinucleotide
NADPH	nicotinamide adenine dinucleotide phosphate
NBCI	National Biological Control Institute
NCI	National Cancer Institute
NEDI	national estimate of dietary intake
NF	National Formulary
NGO	non-governmental organization
NI	nitrification inhibitor
NIOSH	National Institute for Occupational Safety and Health
NMR	nuclear magnetic resonance spectroscopy
no.	number
NOAEC	no observed adverse effect concentration
NOAEL	no observed adverse effect level
NOEC	no observable effect concentration
NOEL	no observed effect level
NOHSC	National Occupational Health and Safety Commission of Australia (now the Office of the Australian Safety and Compensation Council)
NPDES	National Pollutant Discharge Elimination System
NPK	nitrogen, phosphorus and potassium
NRDC	Natural Resources Defence Council
NSF	National Science Foundation
NST	nitrogen selective thermionic detector
NUE	nitrogen use efficiency
O	ortho
o/w	oil in water
obs	observed
OC	organochlorine
OM	organic matter

OMRI	Organic Materials Review Institute (of the USA)
OP	organophosphate (the term organophosphorus is interchangeable with organophosphate)
OPP	EPA Office of Pesticide Programs
OPPTS	Office of Prevention, Pesticides and Toxic Substances of the EPA
OSHA	Occupational Safety and Health Administration in the United States Department of Labor
OSP	ordinary superphosphate
P	para
PAD	population adjusted dose
PAG	pesticide assessment guideline
PAM	pesticide analytical method
PAN	Pesticide Action Network
PANNA	Pesticide Action Network North America
PARP	partially acidulated rock phosphate
PCD	photoconductivity detector
PCR	polymerase chain reaction
PE	polyethylene
PEG	polyethylene glycol
PGR	plant growth regulator
Ph	phenyl ($C_6H_5^-$)
pH	negative log of the hydrogen ion concentration (potential hydrogen). It is also a measure of acidity or basicity of a liquid
PHED	pesticide handler's exposure data
PHI	pre-harvest interval
PIC	prior informed consent
pK	negative log of the dissociation constant
POE	polyoxyethylene
POP	persistent organic pollutant
ppb	parts per billion, 1/1000 of ppm
PPDB	Pesticide Properties Data Base
PPE	personal protective equipment
PPG	polypropylene glycol
ppm	parts per million (mg/kg or mg/l)
ppmv	parts per million in volume (ml/l)
ppmw	parts per million in weight (mg/kg)
PROTOX	protoporphyrinogen oxidase
PSD	Pesticides Safety Directorate (UK)
PSI	pounds per square inch
PVP	polyvinylpyrrolidone
QoI	quinone outside inhibitor
RAC	raw agriculture commodity
RADI	radio-assay
RBC	red blood cell
Rbt	rabbit
RCRA	Resource Conservation and Recovery Act (administered by EPA)
Ref	reference
REI	restricted entry interval
RfD	reference dose
RH	relative humidity
RIS	radiographic imaging system
RNA	ribonucleic acid
RQ	risk quotient
RSC	residual sodium carbonate (meq/l)
RUP	restricted use pesticide
s	second
SAFFIE	sustainable arable farming for an improved environment
SAR	sodium absorption ratio
SCBA	self-contained breathing apparatus
sc	subcutaneous
SC	GROW Tier I GROund Water computer model
SF	safety factor
SG	water-soluble granules
SLA	specific leaf area
SP	water-soluble powder
SPEC	spectrophotometer
SSP	single superphosphate
ST	2-sulfanilamido thiazole, a nitrification inhibitor
STEL	short-term exposure limit
t	ton
T&E	threatened and endangered species (from the Endangered Species Act)
TD	toxic dose
TDI	tolerable daily intake
TDS	total dissolved solids
temp	temperature
TGA	technical grade active ingredient
TLC	thin-layer chromatography
TLV	threshold limit value
TMDI	theoretical maximum daily intake
TSCA	Toxic Substances Control Act
TSD	thermionic specific detector
TSP	triple superphosphate
TVA	Tennessee Valley Authority
TWA	time weighted average
UAN	urea ammonium nitrate
UF	uncertainty factor
UKNPU	UK National Poisons Unit
ULV	ultra-low volume
UN	United Nations
USDA	United States Department of Agriculture
USG	urea super granule
US-OMRI	United States Organic Materials Review Institute
USP	United States Pharmacopeia
UV	ultraviolet
UV SPEC	ultraviolet spectrophotometer
v/v	volume-for-volume
VAM	vesicular arbuscular mycorrhizae
WG	water-dispersible granule
WHO	World Health Organization
WP	wettable powder (formulation)
WP	wilting point
WPS	Worker Protection Standard
WS	water-dispersible powder for slurry seed treatment
wt(s)	weight(s)
w/v	weight-for-volume
w/w	weight-for-weight
XSD	halogen specific detector

A

4-Allyl-2-methoxyphenol insecticide

4-Allyl-2-methoxyphenol (eugenol), $C_{10}H_{12}O_2$, is a plant-derived phenolic insecticide, present in plants such as cloves, cinnamon, nutmeg and holy basil (*Ocimum sanctum*) or *tulsi* in many Indian languages. Its CAS Registry Number is 97-53-0.

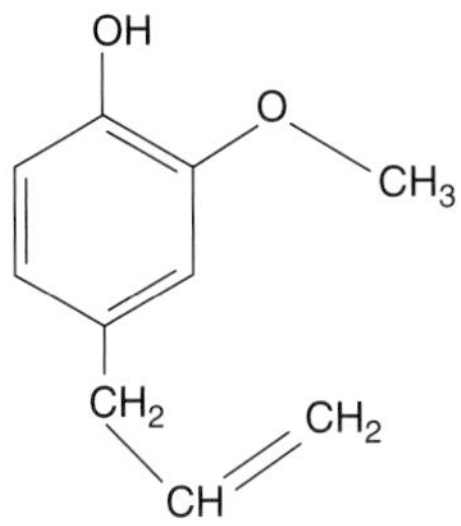

4-Allyl-2-methoxyphenol

4-Allyl-2-methoxyphenol is used to repel aphids, armyworms, beetles, cutworms and mites. The insecticide is also used as a herbicide (post-emergence) when used at high concentration and is usually used without other chemicals. One of the trade names it is available under is Matran. While the compound may be an irritant, there are no adverse effects on the environment.

4-Allyl-2-methoxyphenol is registered in Australia, South Africa, Vietnam and the USA.

4-Aminopyridine (4-AP) avicide

4-Aminopyridine (pyridin-4-amine) (4-AP), with the molecular formula $C_5H_6N_2$, is a widely used avicide. It is a colourless, stable solid, which is moderately soluble in water, acetone, methanol, ether and benzene. This avicide is prepared by the decarbonylation of the pyridine-4-carboxamide using sodium hypochlorite via the Hoffmann rearrangement. The CAS Registry Number of 4-aminopyridine is 504-24-5.

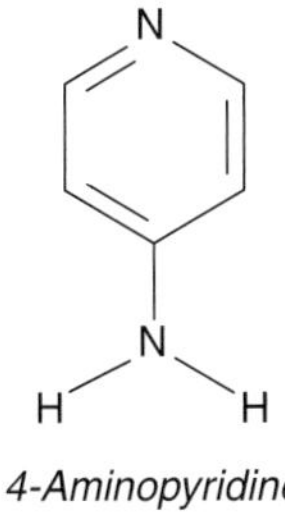

4-Aminopyridine

The avicide 4-AP is used against bird pests like blackbirds, grackles, pigeons and sparrows found on farms, public buildings and around animal pens. It is available as grain bait or powder concentrate under trade names such as Avitrol.

Some formulations of 4-AP are classified as a restricted use pesticide (RUP) by the EPA. The acute oral LD_{50} of this compound is 20 mg/kg in rats and 3.7 mg/kg in dogs. The LC_{50} in mallard ducks is 722 ppm. The NOEL is 200 ppm in dogs and 3 ppm in rats. The ADI is 0.0015 mg/kg/day. 4-AP is highly toxic to all mammals including humans. It is also an eye irritant. It can cause toxicity through the skin and result in methemoglobin (a condition where red blood cells lose the ability to carry oxygen, causing blue baby syndrome). 4-AP is considered a PAN Bad Actor chemical due to its acute toxicity. **Strychnine**, along with 4-AP, has been shown to cause toxicity to non-targeted species.

4-AP is rapidly absorbed into the bloodstream from the gastro-intestinal tract. It is readily broken down or metabolized in the liver and excreted from the body in the urine.

4-AP is registered for use in Australia and the USA.

Abamectin bioinsecticide

Abamectin, also known as Avermectin B_1, is a fermentation product of *Streptomyces avermitilis*, a soil bacterium (which is an actinomycete). A microorganism-derived insecticide, acaricide and nematicide, abamectin is an off-white crystalline powder which can cause a slight fire hazard.

The CAS Registry Number of abamectin is 71751-41-2 and the IUPAC name is: a mixture of ≥80% (10E,14E,16E)-(1R,4S,5′S,6S,6′R,8R,12S,13S,20R,21R,24S)-6′-[(S)-sec-butyl]-21,24-dihydroxy-5′,11,13,22-tetramethyl-2-oxo-(3,7,19-trioxatetracyclo [$15.6.1.1^{4,8}.0^{20,24}$] pentacosa-10,14,16,22-tetraene)-6-spiro-2′-(5′,6′-dihydro-2′H-pyran)-12-yl 2,6-dideoxy-4-O-(2,6-dideoxy-3-O-methyl-α-L-arabino-hexopyranosyl)-3-O-methyl-α-L-arabino-hexopyranoside and ≤20% (10E,14E,16E)-(1R,4S,5′S,6S,6′R,8R,12S,13S,20R,21R,24S)-21,24-dihydroxy-6′-isopropyl-5′,11,13,22-tetramethyl-2-oxo-(3,7,19-trioxatetracyclo[15.6.1.14,8.020,24]pentacosa-10,14,16,22-tetraene)-6-spiro-2′-(5′,6′-dihydro-2′H-pyran)-12-yl2,6-dideoxy-4-O-(2,6-dideoxy-3-O-methyl-α-L-arabino-hexopyranosyl)-3-O-methyl-α-L-arabino-hexopyranoside.

Streptomyces avermitilis generates different types of complex lactone derivatives called **avermectins**, and it is these which combine to produce abamectin (avermectin B_{1a} and B_{1b}). The molecular formulae of B_{1a} and B_{1b} are $C_{48}H_{72}O_{14}$ and $C_{47}H_{70}O_{14}$, respectively.

Abamectin acts on the pest's nervous system by stimulating and binding the γ-aminobutyric acid (GABA) at the nerve endings, thus blocking the electrical impulses to the nerves. The presence of GABA leads to the excessive flow of chloride ions into the cells. Abamectin is used to control mites, leaf miners, suckers, beetles and fire ants, in crops like cotton, citrus, pome fruit and vegetables.

Abamectin is available under trade names such as Agrimec, Avid, Affirm, Dynamac and Zephyr. It is formulated as emulsifiable concentrate (EC) and as a ready-to-use bait.

The application dose for various pests ranges from 5.6 g/ha to 28 g/ha, with its efficacy improving with paraffin oil. The avermectin mixture has a ratio of 80:20 of B_{1a} and B_{1b}. It is generally compatible with other crop protective agents.

Abamectin is highly toxic to mammals. The acute oral LD_{50} of abamectin is 10 mg/kg in rats and 30 mg/kg in mice. The acute oral LD_{50} in mallard ducks is 84.6 mg/kg, while the LC_{50} for a 96 h period in rainbow trout, is 3.2 µg/l. Abamectin is toxic to bees.

Abamectin is metabolized in plants and animals, forming metabolites 3′-demethylavermectin B_1 and 24-hydroxymethylavermectin B_1. In plants, a major metabolite is 8-9-(z)-Avermectin B_1. It binds tightly to soil but is degraded by soil microorganisms (see Abamectin at bottom of the page).

Abscisic acid as a PGR

Abscisic acid (ABA), $C_{15}H_{20}O_4$ ((2*Z*,4*E*)-5-[(1*S*)-1-hydroxy-2,6,6-trimethyl-4-oxocyclohex-2-en-1-yl]-3-methylpenta-2,4-dienoic acid), is a weak organic acid and plant growth regulator that belongs to the terpenoid family. It is a white crystalline odourless powder, soluble in water (3100 mg/l).

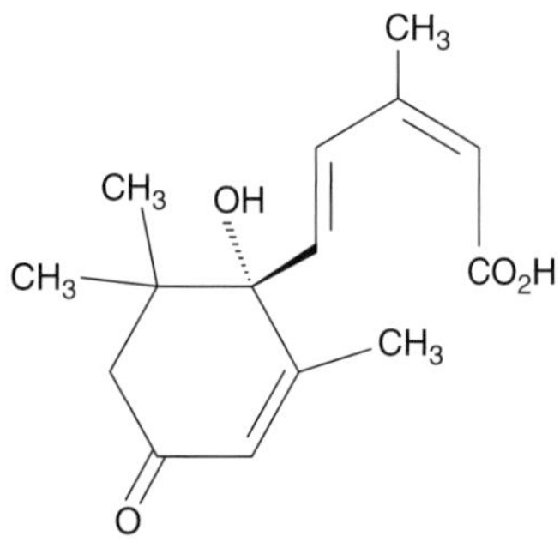

Abscisic acid

This acid is present in plants and has a role in the abscission of leaves, branches and flowers, and during stomatal movements. ABA is distributed throughout the plant body. ABA is synthesized mainly in chloroplasts. The rate of synthesis of ABA increases when the plant is under stress, like at the onset of winter or during water scarcity. Its production prevents fruits from ripening and seed from germinating.

The acute oral and dermal LD_{50} values in rats are > 5000 mg/kg. Inhalation LC_{50} (4 h) in rats is >5.1 mg/l. NOEL (90 days) in rats is 20,000 ppm/day. The LC_{50} in bobwhite quails is >2250 mg/kg. The acute LC_{50} (96 h) in rainbow trout is >121 mg/l. The LD_{50} in bees is >108 µg per bee.

Acaricide acrinathrin
See **Acrinathrin insecticide**

Acaricide aldicarb
See **Aldicarb insecticide**

Acaricide aldoxycarb
See **Aldoxycarb insecticide**

Acaricide azinphos-ethyl
See **Azinphos-ethyl insecticide**

Acaricide bifenthrin
See **Bifenthrin insecticide**

Acaricide butoxycarboxim
See **Butoxycarboxim insecticide**

Acaricide chlorfenvinphos
See **Chlorfenvinphos insecticide**

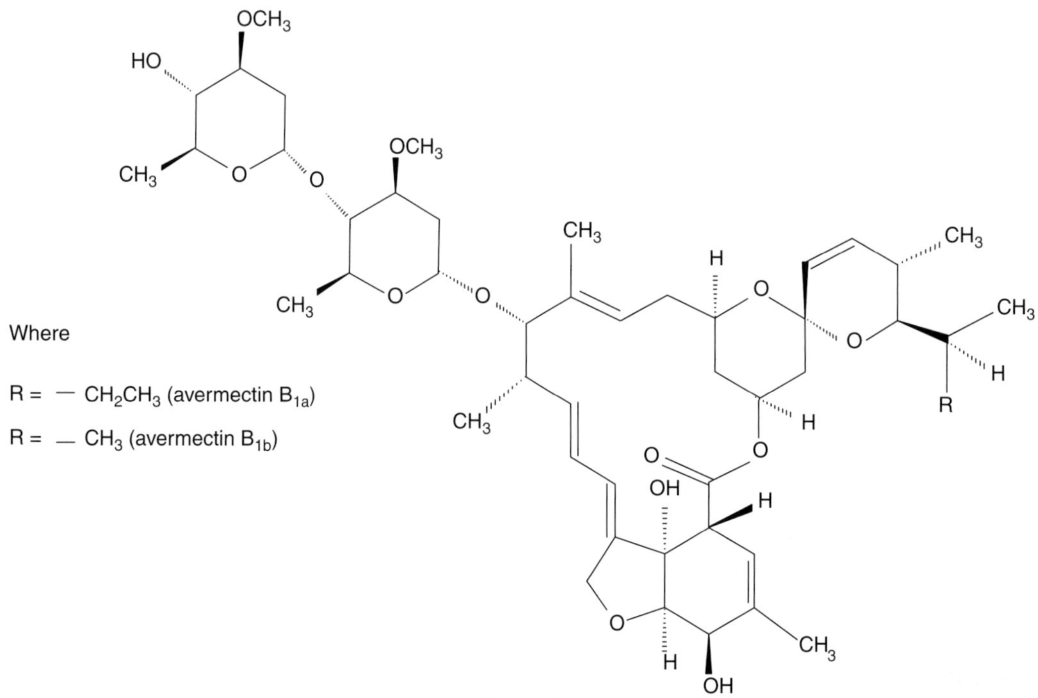

Abamectin

Acaricide chlorpyrifos-methyl
See Chlorpyrifos-methyl insecticide

Acaricide cyromazine
See Cyromazine insecticide

Acaricide DDT
See DDT insecticide

Acaricide demeton-*S*-methyl
See Demeton-*S*-methyl insecticide

Acaricide diazinon
See Diazinon insecticide

Acaricide dichlorvos
See Dichlorvos insecticide

Acaricide dicrotophos
See Dicrotophos insecticide

Acaricide dienochlor
See Dienochlor insecticide

Acaricide dimethoate
See Dimethoate insecticide

Acaricide disulfoton
See Disulfoton insecticide

Acaricide emamectin benzoate
See Emamectin benzoate insecticide

Acaricide endosulfan
See Endosulfan insecticide

Acaricide EPN
See EPN insecticide

Acaricide etofenprox
See Etofenprox insecticide

Acaricide fenamiphos
See Fenamiphos insecticide

Acaricide fenpropathrin
See Fenpropathrin insecticide

Acaricide fluazinam
See Fluazinam fungicide

Acaricide flucythrinate
See Flucythrinate insecticide

Acaricide formetanate
See Formetanate insecticide

Acaricide formothion
See Formothion insecticide

Acaricide methiocarb
See Methiocarb insecticide

Acaricide monocrotophos
See Monocrotophos insecticide

Acaricide oxamyl
See Oxamyl insecticide

Acaricide phenthoate
See Phenthoate insecticide

Acaricide quinalphos
See Quinalphos insecticide

Acaricide sulfotep
See Sulfotep insecticide

Acaricide sulfur
See Sulfur as a pesticide

Acaricide tetrachlorvinphos
See Tetrachlorvinphos insecticide

Acaricide thiometon
See Thiometon insecticide

Acaricide vamidothion
See Vamidothion insecticide

Acaricides (miticides)
An acaricide, also called miticide, is a type of pesticide that kills mites, ticks and small animals of the **class Arachnida** (of which, mites belong to the Acarina sub-class). But, the term is commonly used in reference to mites ('*acarus*' is mites in Latin). There are many important acaricidal compounds in use throughout the world, both of chemical and of biological origins. Crops most susceptible to mites include peas, beans, lentils, chickpeas, lupins and cassava.

Major pesticide groups such as pyrethroids, organophosphates (OP), organochlorines (OC), carbamates, antibiotics, growth regulators, nitrophenyl esters, sulfonic acid esters, organometallic compounds and halogenated benzyhydrol derivatives have acaricidal activity. Chemicals such as sulfur, nitrophenols, formamidines, **permethrin** and azobenzene have also found application as acaricides. **Diatomaceous earth** can also kill mites by cutting through their skin, which dries them out. **Methoprene**, hydroprene and permethrin are other common miticides. Most of these compounds are safe on non-target pests. Many acaricidal compounds have a broad-spectrum reach which even includes lice, cockroaches, fleas and mosquitoes.

ACCase inhibitors
ACCase (acetyl-coenzyme A (CoA) carboxylase) inhibitors form a herbicide family that inhibits plant lipid biosynthesis. They are specifically suited for grass control.

Aryloxyphenoxypropionic acids and cyclohexanediones are two herbicide groups that inhibit acetyl-CoA carboxylase, a key enzyme in lipid biosynthesis. Commonly, ACCase inhibitors are post-emergence herbicides. **Diclofop-methyl**, **fluazifop-butyl**,

quizalofop-ethyl, **haloxyfop-p** and **fenoxaprop-p-ethyl** are some group members.

Acceptable daily intake (ADI)

EPA defines ADI as 'the estimated amount of a substance that can be consumed every day for a lifetime by humans without presenting a significant risk to their health, based on current scientific evidence'. The ADI is the highest daily intake level of a chemical, which may be continued over the whole life of a person without appearing to pose a health risk and is thus a life-time exposure. The concept of the ADI was initially developed by the Joint FAO/WHO Expert Committee on Food Additives, or JEFCA, in 1961.

A toxin from a pesticide may be consumed via food and drinking water, and hence, acceptable levels of such consumption need to be established. ADI is usually expressed in milligrams per kilogram of body weight. It is calculated by dividing the threshold daily dose (mg/kg/day) for an appropriate species with a safety factor. This factor is often 100 but may vary from 10 to >1000. The term ADI is used interchangeably with the term **reference dose (RfD)**. The ADI is usually set 100 times lower than the no observed effect level (NOEL).

Acceptable risk

Acceptable risk is defined as the risk level at which a seriously adverse result is highly unlikely, but about which an assurance of 100% safety cannot be given. It is a compromise value, based on which a product or service can be accepted, with a full understanding of the risks involved.

Acceptable risk has become a commonly referred term ever since water and air pollution levels have started becoming a concern. It assumes importance today because of the heavy use of chemicals in crops, either as fertilizers or pesticides. It is common for chemicals to be retained on the crops and the soil, which find their way into the food chain.

Accumulator plants

Plants that concentrate large quantities of elements such as selenium and aluminium in their tissues are known as **accumulator plants**. Such plants can aid toxicity control as in **bioremediation**.

Acephate insecticide

Acephate, $C_4H_{10}NO_3PS$, ((*RS*)-(*O*, *S*-dimethyl acetylphosphoramidothioate)), is an organophosphorus insecticide (*N*-acetyl derivative of the insecticide methamidophos). It has a CAS Registry Number of 30560-19-1. It is a colourless, crystalline substance, highly soluble in water (790 g/l) but sparingly soluble in ethyl acetate, benzene and hexane. It is relatively stable to hydrolysis. Acephate is synthesized by the reaction of corresponding hydroxy compounds with dialkyl phosphorochloridothionates in the presence of a dehydrochlorinating agent. Phosphorothionate isomers convert into the more toxic phosphorothiolates when subjected to heat and UV light. Acephate is the *N*-acetyl derivative of methamidophos and is obtained by its acetylation.

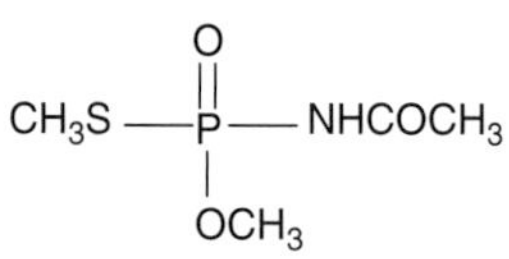

Acephate

Acephate, a systemic insecticide with contact and stomach action, is used to control a wide range of chewing and sucking insects such as moth larvae, thrips and aphids, in fruit, cotton, vines, soybeans, olives, groundnut, beet, brassicas, celery, potatoes and rice. Like any other OP insecticide, this chemical kills insects by interrupting the electrochemical processes that nerves use to communicate with muscles and other nerves.

The common formulations of acephate include wettable powder (WP), aerosol dispenser, encapsulated granule and water-soluble powder. Trade names include Orthene, Ortho12420 and Tornado. Acephate may be combined with many other pesticides.

Acephate is moderately toxic to rats and mice via the (acute) oral route. At high exposure levels, irritation to skin and eyes are common. It is toxic to bees with an LD_{50} of 1.2 μg per bee. The ADI for acephate is 0.03 mg/kg b.w. The NOEL (2 years) for acephate in dogs is 0.75 mg/kg b.w./day. This insecticide can injure some apple species, with symptoms such as marginal leaf burn. Symptoms of poisoning from acephate include excessive salivation, pinpoint pupils, sweating, nausea, vomiting, diarrhoea, weakness, tremor, incoordination, respiratory depression and tightness in the chest. WHO has classified this chemical as a Class III toxin (slightly hazardous). Acephate is a PAN Bad Actor chemical, since it is a cholinesterase inhibitor, suspected carcinogen and a potential groundwater contaminant.

The initial metabolic transformation of acephate (in plants and soils) is deacetylation into **methamidophos** (which constitutes about 10% of the applied acephate. Methamidophos in turn has a soil half-life of 2–6 days). The soil half-life of acephate is 3 days or less, although the residual activity lasts for about 2 weeks. It has a low potential to contaminate groundwater. Methamidophos is toxic to bees and birds. Another metabolic product is des-*O*-methylacephate (occurring in mammals).

Acephate is registered for use in countries such as India, Australia, New Zealand, the Philippines, South Africa, Tanzania, the USA and Canada. But it is banned or severely restricted throughout the EU.

Acetamiprid insecticide

Acetamiprid, $C_{10}H_{11}ClN_4$ ((*E*)-N^1-[(6-chloro-3-pyridyl)methyl]-N^2-cyano-N^1-methylacetamidine) is a neonicotinoid insecticide. It is a white, odourless powder soluble in water at 3660 mg/l at 25°C. This compound is stable to hydrolysis at normal temperatures. It has a CAS Registry Number of 135410-20-7.

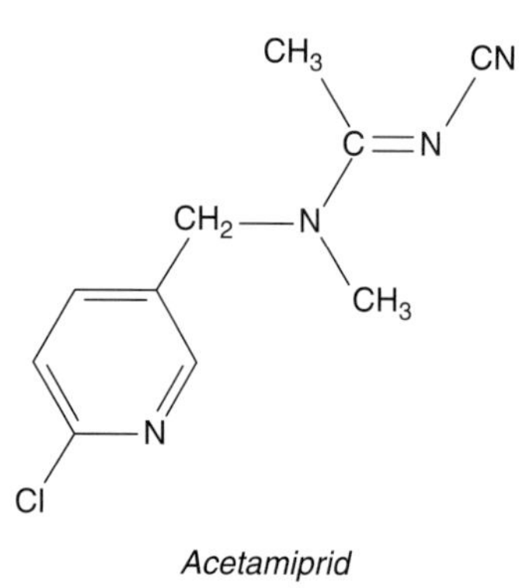

Acetamiprid

Acetamiprid, considered as a relatively safe insecticide, is effective on sucking insects that attack leafy vegetables, citrus and pome fruits, cotton, grapes and flowering plants.

Common formulations of acetamiprid are wettable powder (WP) and water-dispersible powder for slurry seed treatment (WS). The compound may be applied by ground equipment or by the aerial route. The application rates are very low at 0.62 kg/ha per season. Trade names of acetamiprid formulations include Assail, Pristine and Chipco.

Acetamiprid is considered slightly hazardous in dermal and inhalation tests on rats and rabbits. It is moderately toxic via the oral route among rats. The acute oral LD_{50} in male and female rats are 217 and 146 mg/kg, respectively. The acute percutaneous LD_{50} in rats is >2000 mg/kg. The compound is non-irritating to the skin and eyes. The LD_{50} in mallard ducks is 98 mg/kg and in fish (carp for 24–72 h) is >100 mg/l. It is slightly toxic to bees. The NOEL (2 years) of acetamiprid among rats is 701 mg/kg b.w. The ADI is 0.07 mg/kg b.w./day.

This insecticide rapidly degrades in aerobic soils, with a half-life between <1 and 8.2 days, while in aquatic systems, the anaerobic half-life is 45 days. However, since acetamiprid is mobile in soils, it may contaminate groundwater. It is hence necessary to ensure that spray drifts or runoffs do not occur. The degradation product is methyl (6-chloro-3-pyridyl) methyl amine, which is persistent in the environment.

Acetamiprid is registered in a large number of countries including India, Germany and the USA.

Acetic acid herbicide

Acetic acid, $C_2H_4O_2$ (ethanoic acid), also known as acetyl hydroxide, is a weak organic acid. It is used as a contact herbicide and in foods (especially for vinegar). It is produced industrially and by fungal fermentation. Acetic acid is miscible in water. Its CAS Registry Number is 64-19-7.

The use of acetic acid formulations controls broadleaved weeds and grasses in non-cropland, especially when sprayed on weeds at the early-growth stage. Its acidic property acts by drying out leaves.

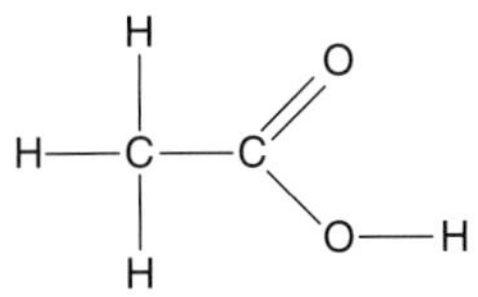

Acetic acid

Acetic acid is available as an aqueous solution in the market under trade names that include Grass killer. It breaks down into carbon dioxide and water and does not persist in the environment. It does not cause harm to non-target organisms and is safe when used in food.

Acetic acid is registered in a few countries of the EU such as Germany, Italy and Finland.

Acetochlor herbicide

Acetochlor, $C_{14}H_{20}ClNO_2$ (2-chloro-*N*-ethoxymethyl-6′-ethylacet-*o*-toluidide), is a **chloroacetanilide** herbicide. It has a CAS Registry Number of 34256-82-1. It is a pale yellow liquid, soluble in water (282 mg/l at 20°C), alcohol, acetone, toluene and carbon tetrachloride. Acetochlor is stable under normal temperatures and pressure. Thermal decomposition may result in the production of fumes of chlorine and oxides of nitrogen and carbon.

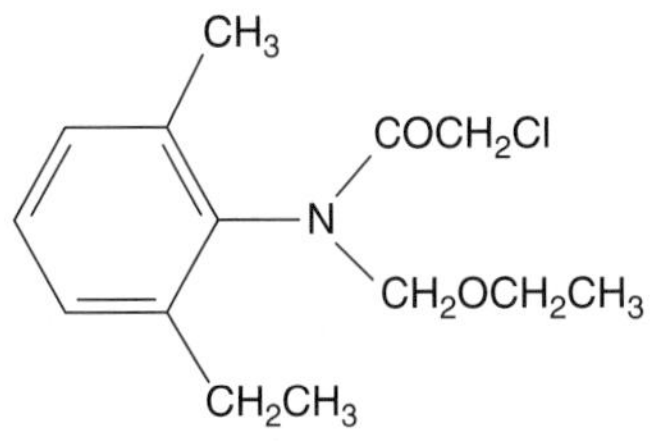

Acetochlor

As a selective herbicide, acetochlor acts as a lipid synthesis inhibitor. It inhibits the elongase and geranylgeranyl pyrophosphate (GGPP) cyclization enzymes, and hence affects the gibberellin pathway. It is used mainly as a pre-emergence compound to control annual grasses and broadleaved weeds in maize, cabbage, citrus, coffee, onion, potato, soybean and sugarbeet.

Acetochlor is compatible with most other pesticides and fluid fertilizers when used at recommended rates. Common formulations are emulsifiable concentrate (EC) or capsule suspension. It is soil incorporated or sprayed. Common dosage is 5 l/ha. Trade names of acetochlor include Guardian, Harness and Relay.

Acute oral and dermal LD_{50} values of acetochlor in rats are 1929 mg/kg and more than 2000 mg/kg b.w., respectively. The acute oral LD_{50} in bobwhite quails is 928 mg/kg. NOEL (2 years) in rats is 11 mg/kg b.w./day. Acetochlor is a probable human carcinogen and a well-established thyroid disruptor. It is highly toxic to fish (LC_{50} (96 h) in rainbow trout is 0.36 mg/l) and can also affect their development. This herbicide is moderately toxic to honey bees (acute 48 h oral LD_{50} being >100 µg per bee). Although WHO classifies the compound as a Class III toxin (slightly hazardous), it is classified as a PAN Bad Actor chemical.

Acetochlor is registered for use in Greece, Hungary, New Zealand, South Africa, Vietnam and the USA. The UK does not approve its use.

Many EU nations have fixed the MRL of acetochlor in maize grain as 0.05 mg/kg.

Acetohydroxy acid synthase (AHAS)

See **Herbicidal inhibition of ALS**

Acetolactate synthase (ALS)

See **Herbicidal inhibition of ALS**

Acetylcholine receptors

In the nervous system, messages are sent by nerve cells via neurotransmitter molecules which diffuse and bind to receptor proteins. These receptors open and allow transfusion of ions into the cells. Proper functioning of the nervous system depends on the communication between neurons at specific junctions called synapses.

Acetylcholine is one of the many neurotransmitters present in many organisms with functions in both the peripheral and the CNS. It is a part of the cholinergic system in the area of the brain. Its functions include a role in the organism's response to different stimuli and also a role in muscle contraction. Acetylcholine is an ester of acetic acid and choline. Acetylcholine is synthesized from acetyl-CoA and choline by the enzyme choline acetyltransferase.

Acetylcholine receptors (AChR) are proteins that respond to the binding of acetylcholine. AChR are found in the synapses between a nerve and muscle cells and are a crucial link between the brain and the muscles. Any malfunctioning of the AChR causes organ failure. AChR are of two types, namely, nicotinic acetylcholine receptors and muscarinic acetylcholine receptors.

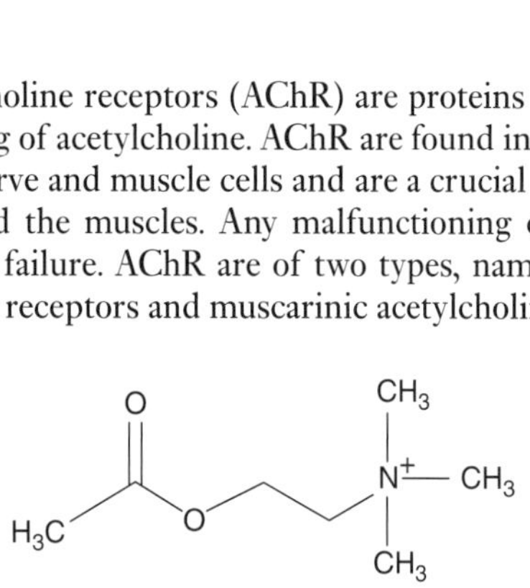

Acetylcholine

Many insecticides (both natural and synthetic) interfere with the functioning of the nervous system. The insect cholinergic system contains two proteins: (i) the enzyme **acetylcholinesterase**; and (ii) **nicotinic acetylcholine receptor (nAChR)**. Organophosphates (OP) and carbamates target the first of the proteins, namely, the enzyme acetylcholinesterase. **Neonicotinoids** and **spinosyns** target the nAChR. Synapses where acetylcholine is the neurotransmitter are known as cholinergic synapses. In animals, these are located in the CNS, peripheral ganglia and some nerve gland junctions. nAChRs are in the CNS of insects.

There are several compounds that interfere with nAChR in different ways. **Nicotine** (or its optically active form, (*S*)-nicotine) is one of them. It is an alkaloid in the tobacco plant that acts as an agonist of the nAChR, bringing about a change in acetylcholine, thus stimulating the ion channel. This leads to muscle twitching and convulsions in the target pests. Such an insecticidal property has made nicotine a popular choice as an insect control agent.

Methyllycaconitine, (with molecular formula $C_{37}H_{50}N_2O_{10}$ and CAS Registry Number 21019-30-7), an alkaloid isolated from *Delphinium* spp., is a highly active antagonist of insect nAChR. It is not very popular due to its limited insecticidal properties. Epibatidine (isolated from frog skin), **nithiazine** (nitromethylene family), **spinosad** (a macrolide antibiotic from actinomycetes) and some others are naturally occurring nAChR binders of varying efficiency.

Imidacloprid is an example of the neonicotinoid group. There are three other popular neonicotinoids in the market, namely **acetamiprid, nitenpyram** and **thiamethoxam**. This class of insecticides is a safer alternative to OPs and carbamates.

Acetylcholinesterase

The nervous system of vertebrates and invertebrates contains the enzyme acetylcholinesterase. The membrane-bound enzyme specifically hydrolyses the neurotransmitter acetylcholine and catalyses the hydrolysis of other choline esters. When acetylcholine is not degraded to choline and acetic acid in cholinergic synapses, then acetylcholine builds up and binds to the nicotinic acetylcholine receptor (nAChR), thus saturating the binding site. This results in the ion channel of the receptor remaining open, allowing the ions to pass through the synaptic membrane continuously. This results in membrane depolarization, continuous nerve cell firing, convulsions and death of say an insect. Sometimes, high concentrations of acetylcholine may desensitize nAChR, preventing transmission of nerve impulses, paralyzing the insect. Organophosphate (OP) and carbamate insecticides act on acetylcholinesterase.

The cholinergic nervous system is important in insects. **Choline**, found in almost all living cells, is a component of phospholipids and has a role in insect neurophysiology and fat metabolism. Membrane vesicles can accumulate choline in the nervous system. Insect cholinergic systems have assumed importance due to the predominance of OP and carbamate insecticides in crop protection. Insects are sensitive to anti-cholinesterase. Proper functioning of the nervous system depends on the communication between neurons at specific junctions called synapses. Acetylcholine is synthesized from acetyl-CoA and choline by the enzyme choline acetyltransferase. Acetylcholinesterase inactivates this choline acetyltransferase. Figure A.1 illustrates a typical cholinergic system.

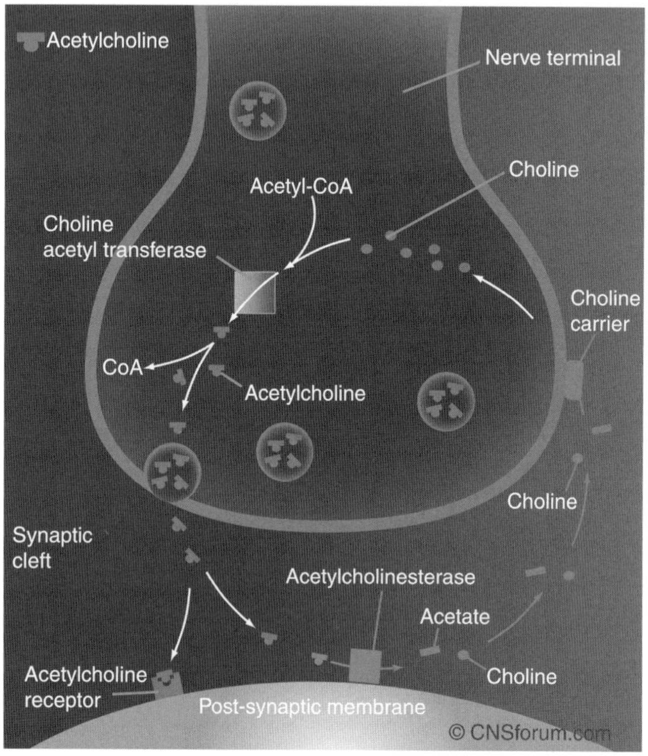

Fig. A.1. The mechanism of action of acetylcholinesterase. Courtesy of the Lundbeck Institute, with permission (http://www.cnsforum.com/educationalresources/imagebank).

Selectivity of commercial insecticides depends largely on differential metabolic detoxification. But, insensitivity of acetylcholinesterase to these chemical groups is the basis of insect resistance. The enzyme is the target site for the highly toxic nerve agent called sarin, from which OP insecticides were developed. Reversible inhibitors of acetylcholinesterase are useful in treating a number of medical conditions.

Carbamates (e.g. **carbofuran**) and carbamoyloximes (**methomyl**) also inhibit acetylcholinesterase by acylating (carbanoylating) the active-site serine but are reactivated through decarbamoylation at a much faster rate than OPs. Resistance to OPs and carbamates involves increased detoxification by glutathione *S*-transferases and a mutation acetylcholinesterase also termed as MACE (or modified acetylcholinesterase). This may make the enzyme less susceptible to inhibition by insecticides.

Acid

Acids are a major chemical class. Aqueous solutions of acids may have: (i) a sour taste; (ii) an ability to turn blue litmus paper red; (iii) an ability to react with metals to form salts; and (iv) an ability to react with bases or alkali forming salts. All acids contain hydrogen. Acids ionize in water and form hydronium ions (H_3O^+), usually written simply as hydrogen ions. In other solvents, acids ionize to give the positive ion of the solvent. The pH range of acids is 1–6.9.

Acids are classified as strong acids or weak acids, according to the concentration of hydrogen ions that results from ionization. Hydrochloric acid, sulfuric acid, nitric acid and perchloric acid are strong acids since they ionize almost completely in dilute aqueous solutions. Acetic acid and carbonic acid are weak acids. Acids are also classified as: (i) inorganic mineral acids such as hydrochloric acid and phosphoric acid; and (ii) organic acids such as acetic acid, phthalic acid, **oleic acid** and glycine. Organic acids are further divided into four major groups: mono carboxylic acids, dicarboxylic acids, fatty acids and amino acids. Organic acids can either be aliphatic or aromatic in nature.

G.N. Lewis strengthened the theory of acids by establishing the electron-pair concept. According to Lewis, any molecule or ion that can combine with another molecule by forming a covalent bond with two electrons from the second molecule or ion is an acid. Thus, an acid is an electron acceptor, for example boron trifluoride (BF_3).

The determination of the concentration of an acid in a solution is called acidimetry.

Acid equivalent

See **Active ingredient**

Acifluorfen herbicide

Acifluorfen, $C_{14}H_7ClF_3NO_5$ (5-(2-chloro-α, α, α-trifluoro-*p*-tolyloxy)-2-nitrobenzoic acid), is a herbicide of the ***p*-Nitrodiphenyl ether** group, with a CAS Registry Number of 50594-66-6. It is an off-white solid with solubility in water of >25 g/100 g. Its activity is enhanced by sunlight. It is produced by reaction of 2-chloro-4-trifluoromethylphenol with 2-nitro-5-chlorobenzoic acid.

Acifluorfen, a post-emergence herbicide, acts by blocking the production of pigments in plants (it is a **PROTOX inhibitor**). It controls certain weeds and grass species in groundnut and soybean crops. Examples of weeds controlled are velvetleaf (*Abutilon theophrasti*), **common lambsquarters** (*Chenopodium album*) and morning glory (*Ipomoea purpurea*).

Effective as a foliar spray, acifluorfen may be applied at pre- and post-crop emergence stages. The herbicide may be mixed with fertilizers and pesticides, as long as they do not contain oil and surfactants. It is available under the trade names Blazer and Tackle.

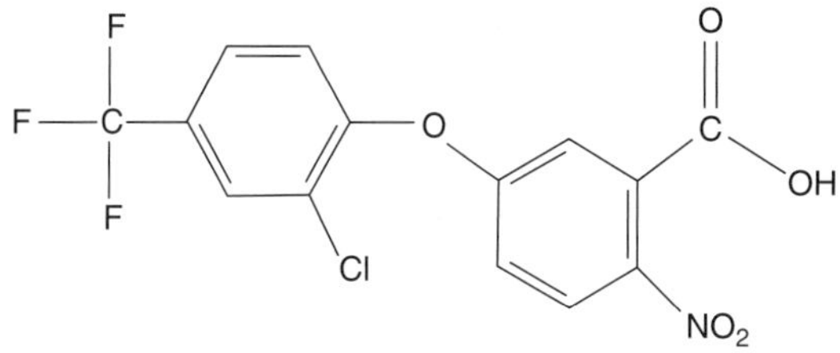

Acifluorfen

Acifluorfen is classified as a general use pesticide (GUP) by the EPA. WHO has classified it as a Class III toxin (slightly toxic). The acute oral LD_{50} of acifluorfen in rats is 1300 mg/kg, while the acute dermal LD_{50} is in the range of 450–2000 mg/kg in rabbits. It is slightly toxic to fish and non-toxic to bees, although it can cause serious eye injury. Products must display the signal word 'DANGER' on their labels. The ADI is 0.0125 mg/kg/day. The NOEL in rats (two generations) is 1.25 mg/kg b.w./day. Acifluorfen is classified as a PAN Bad Actor chemical because it has been found to be a carcinogen. Canada has registered the use of acifluorfen, while the USA has registered only its sodium salt.

Acrinathrin insecticide

Acrinathrin, $C_{26}H_{21}F_6NO_5$ ((S)-α-cyano-3-phenoxybenzyl (Z)-(1*R*, 3S)-2, 2-dimethyl-3-[2-(2, 2, 2-trifluoro-1-trifluoromethylethoxycarbonyl) vinyl] cyclopropanecarboxylate), is a pyrethroid insecticide. It has a CAS Registry Number of 101007-06-1. Acrinathrin is a colourless crystalline substance, which is sparingly soluble in water and readily soluble in organic solvents. It is stable in acids but hydrolyses at pH >7 (see Acrinathrin at bottom of the page).

Acrinathrin, a contact and stomach insecticide, is also an acaricide. It is effective against thrips and mites which attack vegetables, fruits, hops, vines and tobacco, among other crops. The action of this chemical is to interfere with the CNS.

The common formulations of acrinathrin are suspension concentrate (SC), wettable powder (WP) and emulsion. It is available in the market under trade names such as Rufast. The acute oral LD_{50} in rats and mice is >5000 mg/kg. Acute percutaneous LD_{50} in rats is >2000 mg/kg. Inhalation LC_{50} (4 h) in rats is 1.6 mg/l air. NOEL (90 day) in male rats is 2.4 mg/kg b.w. It is slightly toxic to birds and to fish.

According to WHO, acrinathrin comes under toxicity Class III (slightly hazardous). It does not irritate the eyes and may cause only mild irritation to the skin. The ADI is 0.016 mg/kg b.w./day. Exposure to large quantities (accidental consumption) can cause a burning sensation on the face, headache, nausea, vomiting and convulsions.

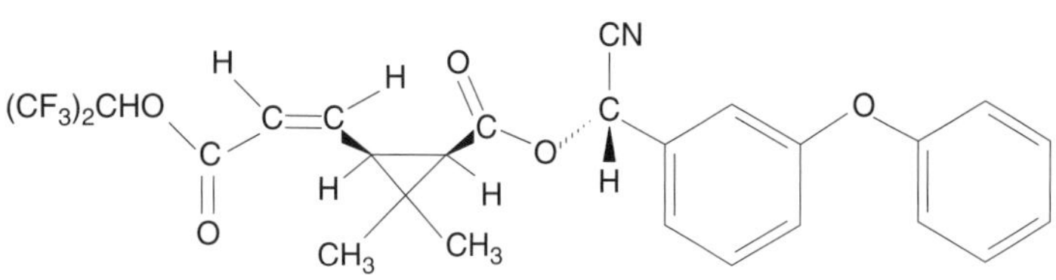

Acrinathrin

Animals eliminate acrinathrin either in its unchanged form or as metabolites. In soil, it gets strongly adsorbed onto soil particles.

Acrinathrin is registered for use in Portugal, South Africa and Vietnam. The Japan Food Chemical Research Foundation has fixed the MRL of mango as 1 ppm.

Acrolein slimicide

Acrolein, C_3H_4O (prop-2-enal), commonly known as acraldehyde or aqualin, is a slimicide and herbicide. It is the aldehyde derivative of acrylic acid and has a CAS Registry Number of 107-02-8. Acrolein is a colourless pungent liquid, soluble in water at 208 g/kg. It is a highly reactive chemical stable up to 80°C that polymerizes if exposed to light. This compound is produced as a combustion and oxidation product and is also used as an intermediate in industrial processes.

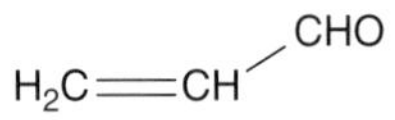

Acrolein

Acrolein reacts with the sulfhydryl groups of enzymes and helps in the breakdown of cell walls of slime-causing organisms. It is used to control submerged aquatic weeds, algae and floating weeds (*Pistia*, *Eichhornia* and *Jussiaea* spp.). Irrigation water is treated with 1–15 ppm of acrolein. Acrolein is available in liquid formulation, and a common trade name is Magnacideh. Acrolein is not mixed with other preparations.

An acute oral LD_{50} of acrolein in rats is 29 mg/kg and in mice it is 13.9 mg/kg. In bobwhite quails and mallard ducks, the acute oral LD_{50} values of acrolein are 19 and 30.2 mg/kg, respectively. Acrolein is toxic to aquatic animals, at concentrations above 5 mg/l. It is a skin irritant. NOEL (90 days) in rats is 5 mg/kg b.w./day. The ADI is 0.0005 mg/kg b.w./day. WHO and EPA classify acrolein as a highly hazardous chemical (Class Ib toxin).

The half-life of acrolein in water is 150 h (pH 5). In soil, it is easily metabolized and mineralized to carbon dioxide with DT_{50} up to 10.2 h. It may be detected in plants at harvest time.

Acropetal translocation

Translocation implies the transport of a dissolved substance within an organism. In plants, translocation takes place especially in the phloem or actively across a cell membrane. The movement of material absorbed by plants from the roots upwards is called acropetal translocation. The uptake of pesticides through irrigation is possible if there is acropetal translocation. This type of movement enables the uptake of pesticides via a rhizosphere drench.

Activator adjuvants

See **Adjuvants**

Active ingredient

Active ingredients (a.i.), in the context of pesticides, are the chemicals in a formulation that are responsible for pesticide activity. They are the technical material(s) in a formulation that kill or repel a pest. A pesticide compound may be formulated commercially in one of several ways, depending on the requirement, such as a wettable powder (WP), dust or emulsifiable concentrate (EC). The a.i. is mentioned on the label of the product.

The a.i. content in a dry formulation is expressed as a percentage by weight. For example, a WP may contain 80% a.i. and a granular product may contain 4% a.i. In a liquid formulation, it is expressed as grams (or pounds) per litre (or gallon). The a.i. is sometimes expressed as an **acid equivalent**, when the acid has to be converted into its salt or ester for penetrating the plant tissue, only to be reconverted to the acid form, once inside.

Acute inhalation toxicity

See **Acute toxicity**

Acute oral LD_{50}

See **Acute toxicity**

Acute toxicants

See **Anticoagulant rodenticides**

Acute toxicity

The acute toxicity of a chemical refers to its ability for quantifiable damage on the recipient, with a one-time dose of the chemical. A pesticide with high acute toxicity may be deadly when a very small amount is absorbed, with the exact results depending on: (i) the nature of the compound; (ii) the mode of exposure; and (iii) the susceptibility of the target. Acute toxicity thus refers to exposure of short duration (a single brief exposure).

A single short-term exposure is expected to show effects within 24 h. When a substance needs to be checked for acute toxicity, it is necessary to check its immediate effects on the body of test organisms. In such cases, **dermal tests** are useful.

The signal words on the label are based on the acute toxicity of the pesticide. Acute toxicity may be measured as acute oral (through the mouth), acute dermal (through the skin) and acute inhalation (through the lungs). Acute oral LD_{50} refers to the dose, given orally, that kills 50% of the test animals. The toxicity dose is usually measured in milligrams per kilogram of body weight (mg/kg b.w.) and such toxicity experiments are carried out in controlled conditions.

Acute toxicity is measured in many ways, depending on the nature of exposure. The commonly used term to describe acute toxicity is the lethal $dose_{50}$ (LD_{50}). It is the dose which kills 50% of the experimental animals on whom the chemical was administered under controlled laboratory conditions.

The acute dose is administered either by one oral or injectable dose, followed by observation for 14 days to record the number of deaths in one exposure. The dose is adjusted to arrive at the LD_{50} value. The lower the LD_{50}, the more acutely toxic is the pesticide. Therefore, a pesticide with an oral LD_{50} of 5000 would be much less toxic than a pesticide with an LD_{50} of 5. LD_{50} values are expressed as milligrams per kilogram (mg/kg), which means milligrams of chemical per kilogram of body weight of the animal. In some cases, the values are also expressed in ppm or ppb. The dose pertains to the a.i. of the compound. Hence, if a commercial formulation is a 50% emulsifiable concentrate (EC), then it will take two parts of the formulation to make one part of the a.i.

Acute inhalation toxicity is a single dose (or a short duration exposure) of a toxic compound inhaled to cause quantifiable damage to the test animal. It is expressed as lethal concentration$_{50}$ (LC_{50}). The concentration of an a.i. in the medium of air or any liquid (say, water) is considered during the measure of lethal concentration. The LC is thus the amount of pesticide inhaled from the air or absorbed (or consumed) through water. LC_{50} values, measured in milligrams per litre (mg/l), indicate the concentration of the lethal dose that can kill 50% of the test animals.

There are other lethal doses that indicate other measures of toxicity. There are measurements that indicate the minimum dose needed to kill the organism and for toxicity symptoms to be displayed. These are called Lethal dose low (LD_{Lo}) and **toxic dose low (TD_{Lo})**. Other measures that denote the extent of toxicity include lethal dose one (LD_{01}) and lethal dose hundred (LD_{100}). (See also **Toxicity**.)

Acylanilide herbicides

Acylanilides constitute a herbicidal group which inhibits photosystem II electron transport. Propanil is the major chemical in this group. Propanil is a contact, foliar herbicide that controls broadleaved weeds and *Echinochloa* spp. in cereals, mainly rice. Although it is a photosynthesis inhibitor, its variants have different modes of action. Its CAS Registry Number is 1918-16-7.

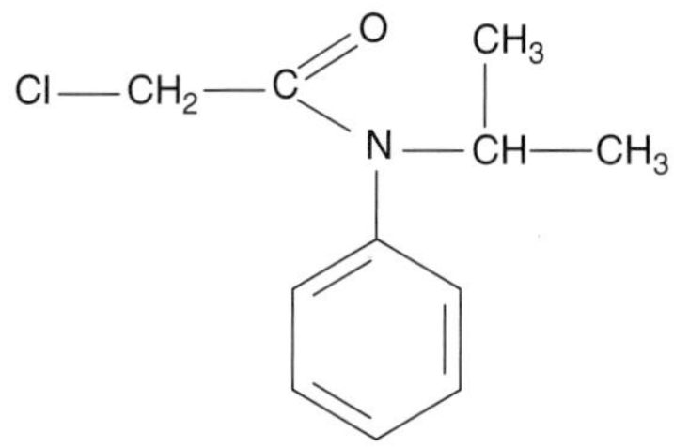

Propanil

Adalia bipunctata (ladybug) as a biocontrol agent

Adalia bipunctata, commonly known as the two-spotted ladybug or lady beetle, is a carnivorous **coccinellid beetle** of the family Coccinellidae. *A. bipunctata*, coloured orange or red with black spots, is used as a localized biological control agent against aphids, which can be serious pests on many crops. **Aphids** are known to feed on plant sap and cause deformity and disease in the plant. They excrete a sweet, sticky liquid called honeydew on which a black **mould (sooty mould)** develops. Aphids have several natural enemies like lady beetles, flower bugs, **green lacewings** and **syrphid flies**.

A. bipunctata is a predator of several aphid species and its larvae can feed on a total of up to 150 aphids/day. This character is exploited to render biocontrol of these pests on fruit and vegetable crops, greenhouse plants and ornamental and potted plants. An average of one to five larvae per flowering stem may be safely released for the desired result, in cases of roses, honeysuckle, strawberries and currants, while around 20 larvae may be ideal on fruit trees. Sunny and dry conditions favour the predatory habits of ladybug larvae, when they are released near aphid colonies. Periods of rain, wind and frost, as well as the presence of ants, can affect biocontrol.

Many recognized species in this family are pest predators and are reared as biocontrol agents.

Additives

Additives are substances added to another substance for improving some property of the main substance. Fertilizers, pesticides and foods often need additives for their intended use.

Additives include a group of carriers, solvents, dilutants and **adjuvants**. They are often added in small quantities and are inert materials. Additives enable better and uniform application of a pesticide, especially over a large area. They also serve to dilute the toxicity of the pesticide and make it safe for the user.

Kaolin or talc added to ammonium nitrate granules make the granules free flowing. Similarly, 0.3–0.4% **formaldehyde** is added to urea to make it abrasion resistant. Some additives such as adjuvants improve the functioning of the main ingredient.

The Food Protection Committee of the National Research Council, USA, defines a food additive as 'a substance or a mixture other than a basic food stuff that is present in food as a result of any aspect of production, processing, storage or packaging'. Such an addition of a substance is a conscious decision, aimed at enhancing food value or flavour, and is called an intentional additive.

While food additives are intentionally added to enhance food value and flavour, pesticide or fertilizer residues enter the food chain unintentionally. To avoid the harmful effects of such additives, most countries have guidelines laid down by specialized agencies, like the Food and Drug Administration (FDA) of the USA.

Solvents, a major component of additives, are of many types, depending on their chemical nature and their application. Some additives and inert ingredients can themselves cause harm to the applicator, making it necessary to have **personal protective equipment (PPE)** while applying pesticides/fertilizers. Some petroleum-based solvents are a potential hazard to humans via all the routes. Some additives can actually make the chemical stick to the PPE or the skin. A few types of additives are discussed below.

Aliphatic petroleum solvents such as diesel and aromatic petroleum solvents like xylene are ideal for granular formulations and some emulsifiable concentrates (EC). More examples of solvents are kerosene, mineral oils (aliphatic solvents), butanol and ethanol (alcohols), benzene and toluene (aromatic solvents), acetone and cyclohexanone (ketones), and many types of vegetable oils extracted from cottonseed and linseed. Properties such as polarity, **flash point**, surface activity and hydrophobicity are important when choosing the right solvent for a formulation. For instance, solvents with low evaporation rates are preferred for EC formulations. In case there is a need for more than one solvent for a formulation, Hensen's fractional solubility parameters are commonly used. In general, ideal characteristics of solvents include high solvency, flash point, low viscosity, volatility and phytotoxicity.

Dispersing agents are protective colloidal substances that are added to a pesticide formulation such that the a.i. is dispersed throughout the suspension, thus enabling a uniform spray without sedimentation. Additives are necessary since some ingredients do not dissolve in the solvent but remain suspended. As a result, agitation of the formulation may be required before application. Sodium carboxymethylcellulose and glue are examples of dispersing agents. Colloidal sulfur or copper are used in the form of paste, and when mixed with water, form a good suspension.

Water-dispersible formulations, liquid formulations, EC and suspensions have varying quantities of dispersing agents. Delayed sedimentation may occur due to increased viscosity of the liquid medium in which the rate of fall of the sediments is inversely proportional to the surface adsorption of the solid particles.

Wetting agents and spreaders are used when pesticidal a.i. are to be mixed in water. A good spreader must have good wetting, spreading and penetrating properties. It must not foam and must store well. Oils of cottonseed and castor seed, sulfonates and gelatin are examples.

When a crop is sprayed during rainy weather, some adhesive or binding material is added to the spray material, to ensure effectiveness. Adhesives, stickers and binders are compounds that increase the cohesive forces between solid particles, so that a film is formed between them, or by forming a bridge between the particles. Stickers tend to enhance tenacity of the deposit on the target. Some of them form insoluble derivatives which help in retaining the deposit. Polymeric surfactants such as lignin sulfonates and soluble polymers are examples of stickers. Binders increase the lattice energy of the solid matrix in a formulation. Binders must disintegrate when mixed with water. Maize (corn) syrup, molasses, PVC latexes, flour paste, polyethylene polysulfides and refined petroleum oil are some examples of binding material.

Additives are also commonly used in the process of formulating **baits**. Their role here is to enhance the taste/flavour of the bait. Common examples of additives used in baits are wheat flour, ground maize meal, groundnut meal and crushed grains mixed in vegetable oils.

Adjuvants

The term adjuvant is discussed here with respect to pesticides (especially herbicides). An adjuvant is any auxiliary, non-toxic material added to a herbicide formulation to enhance the activity and application of the formulation. Given that the performance of a pesticide or fertilizer is a factor of the effectiveness of its a.i., it is critical to choose the right adjuvant.

Attributes of good dispersal, wetting and absorbing are ideal for an adjuvant. An adjuvant may be needed in any step of pesticide uptake, such as bringing about an improvement in the spray tank mix or overcoming **pesticide antagonism**.

Adjuvants are commonly used in post-emergence applications. Preparing a herbicide formulation can sometimes pose a problem since either the carrier agent or the a.i. can react with each other and reduce the formulation efficacy. While the most common carrier agent is water, inter-molecular attraction, pH of the solute, mineral content and impurities in water can interfere with the performance of the herbicide. Unique features such as leaf scales, **epicuticular wax** structure and surface hairs can also interfere with the proper working of an adjuvant.

If a solute has characteristics that may not agree with a pesticide, an adjuvant can help bind the cations and prevent antagonistic interaction with the formulation.

Adjuvants can help reduce surface tension (the inward attraction of the surface molecules of a liquid, which causes them to be pulled inward). Surface tension prevents a spray from spreading on plant foliage because the surface of water acts like a barrier to prevent penetration, often resulting in the droplet remaining suspended or running off. Adjuvants increase the contact between the solution and the plant surface. A substance that helps reduce surface tension is called a surface active agent or **surfactant**. A surfactant interferes with the cohesion between water molecules at the water–air interface. As surface tension decreases with increasing concentration of the surfactant, weed control is made more effective. Many intervening concepts determine the actual effect of an adjuvant such as: (i) herbicide spray droplet size; (ii) herbicide spray retention; and (iii) herbicide spray coverage. Adjuvants that reduce the surface tension of the solution produce smaller droplets and have greater spray coverage. Smaller droplet size reduces surface tension and increased adjuvant concentration improves foliar herbicide penetration. Also, an increased concentration of a surfactant decreases the droplet size.

Spray modifiers are adjuvants that alter the physical characteristics of a solution, to result in improved retention and spread. Such modifiers include stickers and spreaders. Deposition agents are also called stickers. Stickers enhance spray retention, which can reduce runoff and increase rainfastness (help in adherence to the leaf surface). Spreaders help the spray to extend to a larger surface on the plant. Also called wetting agents, these are also used in spray mixtures to decrease the surface tension, causing the spray droplet to come into contact with the leaf surface.

Some adjuvants act to protect spray droplets from evaporating or volatilizing, thereby increasing absorption. These help to increase the effectiveness of the herbicide, although in some cases they may alter herbicide selectivity.

Activator adjuvants act to primarily increase the absorption, rainfastness and photo-transformation of herbicides. These adjuvants include many kinds of **wetting agents**, oils, surfactants and even some fertilizers. An activator adjuvant may perform one or many functions or can act differently on different target species. Emulsifiers enable the mixing of oil-based herbicides with water and increase the compatibility of the solution. An emulsion is a mixture of two liquids that are not miscible completely, such as in the case of oil in water. One of the components has to be dispersed in the other. The surrounding liquid is called the continuous phase, while the dispersed liquid is known as the discontinuous phase. **Emulsifying agents** promote the suspension of one liquid in another. In the case of water-in-oil, the emulsion is called invert emulsion. An invert emulsion is very viscous and is used to: (i) control drift; (ii) improve herbicide resistance to rain; and (iii) improve delivery. The character of the emulsifying agent is a large factor in determining the kind of emulsion that is formed.

Compatibility agents prevent clumps from forming in the solution, enabling easy spray. Dispersing agents make a spray solution more homogeneous, by reducing the attraction between like molecules.

Some adjuvants act as anti-foaming agents to control the excess foam formed as a result of mixing or agitation in the herbicide spray tank.

Adjuvants that reduce drift from herbicide sprays are called drift-control adjuvants. These adjuvants work by decreasing the size and number of fine droplets, while helping to produce larger-sized droplets, which do not drift due to their mass. Thickeners may also be added to increase the size of droplets.

Oil-based adjuvants are made from petroleum, vegetable seed oils or crop oils. Such adjuvants act as emulsifying agents and help increase spray retention. These are mostly non-phytotoxic oils. Oils made from petroleum sources need to be refined thoroughly before using. Crop-oil concentrates, which are the most popularly used oil-based adjuvants, contain around 20% emulsifiers. Crop-oil concentrates function to reduce surface tension between the leaf surface and the spray droplets. Crop-oil adjuvants allow the herbicide to penetrate the waxy leaf cuticle and prevent the spray from evaporating and the herbicide from crystallizing.

Adoxophyes orana granulosis virus as a biocontrol agent

Adoxophyes orana granulosis virus (family *Baculoviridae*), designated as AoGV, is a naturally occurring granulovirus. The larvae of the summer fruit tortrix moth (*Adoxophyes orana* F. von R.) act as hosts for the commercial production of this virus.

The virus is used for controlling summer fruit tortrix moths. The virus enters the moth's body while it feeds on apples and pears. Once inside, virus proteins are dissolved by the alkaline medium in the moth mid-gut, leading to virus particles being generated. These attack the moth, killing it. The virus is formulated as a concentrated suspension (having a pH range of 6–8).

Adsorption

The formation of a layer of gas, liquid or solid molecules on the surface of a solid or liquid, involving the evolution of heat, is known as adsorption. The two types of adsorption are: (i) chemisorption (involving a chemical bond); and (ii) physisorption (involving van der Waals forces). Finely divided or microporous materials with a large active surface (activated charcoal, activated alumina, silica gel) are strong adsorbents and are used for removing colour, odour and moisture. The quantity adsorbed depends on the forces of pressure and temperature.

The adsorptive properties of chemicals are the key to estimating their degradation and mobility. When molecules of many substances are present, those of one substance may be adsorbed more readily than those of others. This is called preferential adsorption.

Aerial spraying

Application of chemicals from the air is known as aerial spraying. Aircraft are customized for various agricultural uses. In aerial spraying, pesticides, fertilizers or seeds are sprayed from an aeroplane or helicopter, as shown in Fig. A.2.

Aerial spraying is particularly advantageous when the ground is inaccessible and the crop height or infestation is very great. Locust control and the control of malaria are carried out by aerial spraying.

However, due to environmental risks involved, aerial spraying is used only when necessary. Larger amounts of chemicals are consumed by this method, and the dangers of spray drift are significant. In addition, the high initial price of aircraft and high maintenance costs are other reasons that discourage use of this spraying method.

Aerosols

Aerosols constitute a type of pesticide formulation in which the a.i. is in a solvent, stored in a spray container under pressure, producing fine droplets or fog upon trigging the valve.

Most aerosols contain a low percentage of active chemical and are used as insect sprays. The two types of aerosol formulations are: (i) the ready-to-use type, which are available in pressurized sealed containers; and (ii) those in electrical or gasoline-powered aerosol generators that release the formulation as a smoke or fog. The larger aerosol applications are in greenhouses and outdoor locations, where the formulation can be re-filled as required. Smaller aerosol containers, useful in homes and gardens, for the control of mosquitoes and other biting insects, are portable and have a long shelf life.

The disposal of aerosols should be as per instructions on the label to prevent sudden inhalation.

Aflatoxins

Aflatoxins are a group of highly toxic polynuclear moulds (**mycotoxins**), designated as B_1 ($C_{17}H_{12}O_6$), B_2 ($C_{17}H_{14}O_6$),

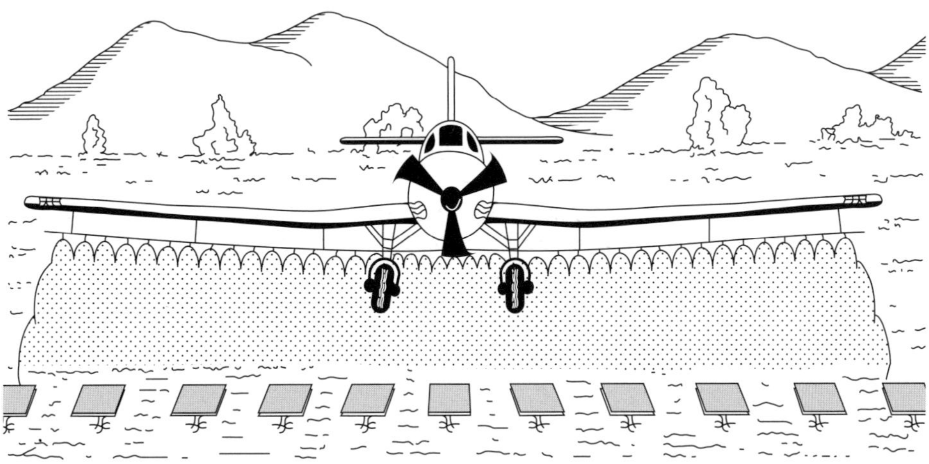

Fig. A.2. Aerial spraying. Courtesy of: www.stewardshipcommunity.com; 'Water-sensitive paper for monitoring spray distribution', Syngenta Crop Protection AG, CH-4002 Basel, Switzerland. With permission.

G_1 ($C_{17}H_{12}O_7$), G_2 ($C_{17}H_{14}O_6$), M_1 and M_2. They are crystalline substances soluble in, for example, acetone, methanol and chloroform but only slightly soluble in water (10–20 mg/l). Aflatoxins fluoresce strongly under UV light but are stable in cool and dark places. They become hydrolysed in alkaline medium. These toxins are produced mainly by the fungi *Aspergillus flavus* and *Aspergillus parasiticus*. Aflatoxins are carcinogenic (especially aflatoxins B_1, B_2 and G_1).

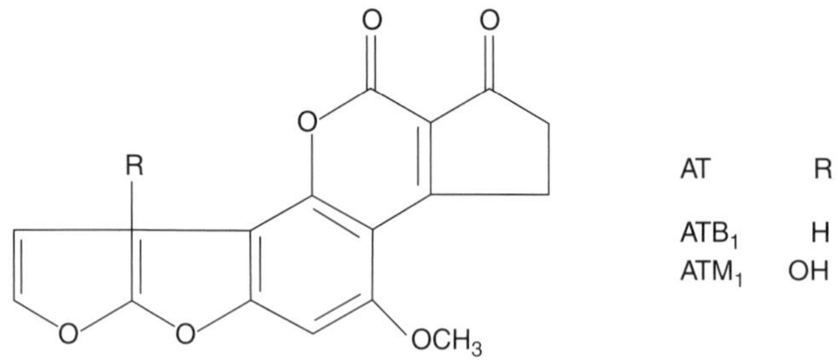

Aflatoxin

The *Aspergillus* species are active throughout the warmer climates in the world, in soils and in plant debris. They infect maize, cottonseed, groundnut and nut trees. They also contaminate stored grains, especially cereals and oilseeds. Infection is rampant when the moisture content is over 18%. The concentrations of such mycotoxins may vary even within a single infected grain, hence making it difficult to estimate the extent of toxicity.

Aflatoxins are toxic to fish, birds and mammals. High exposures can be hazardous to humans. Cattle will not eat contaminated grain. Ruminants metabolize B_1 and B_2 into M_1 and M_2 (which are the hydroxylated metabolites.)

In the USA, the FDA has set a limit of 20 ppb for aflatoxin B_1 in commercial grains and that of 0.5 ppb for aflatoxin M_1.

Agapeta zoegana (sulfur knapweed moth) as a biocontrol agent

Agapeta zoegana, a root-boring moth, is grown for its capability to feed on knapweed (genus *Centaurea*). This predator is a 10 mm long bright yellow moth with brown wing bands. Adults rest vertically on weed stems or under the leaves. The larvae feed on the weed root layers, damaging the vegetative stem bud and making the weed vulnerable to secondary infections. *A. zoegana* is active from mid-June to mid-August and an effective control is effected with 56 larvae per large root of knapweed.

Knapweeds are biennials that have long fibrous roots and propagate by seeds. They grow rapidly and compete with crops such as forage and forestry and are thus considered weeds. Knapweed species include spotted knapweed (*Centaurea maculosa*), Russian knapweed and diffuse knapweeds (*Centaurea diffusa*). *A. zoegana* is considered host specific to spotted and diffuse knapweed. Favourable conditions for knapweed growth are cool, humid weather and undisturbed soil. Their natural enemies are weevils, stem borers and root borers.

Agasicles hygrophila (alligator weed flea beetle) as a biocontrol agent

Agasicles hygrophila, also known as alligator weed flea beetle, is a recognized biocontrol agent for the alligator weed *Alternanthera philoxeroides*. The weed pest grows rapidly in water bodies and can invade crop lands too. Commonly found along the banks of water bodies and propagated by fragments, the weed is a pest in rice and wheat crops and ornamental plants that need irrigation for growth. This weed blocks waterways and canals, thus endangering navigation.

The adult beetle lays eggs on the leaves, roots and inside the stems of the alligator weed. Larvae feed on the weed while they grow, weakening the plant by hampering its normal biochemical functions. This keeps a check on the growth and proliferation of the weed. This natural course of weed control takes about 3 months, with the life of the adults being around 1.5 months.

Alligator weed flea beetles are a sound biocontrol method, especially when used with some species of stem borers. The beetle has been successful in halting the weed in aquatic situations but has little effect on terrestrial surroundings. Care has to be taken to avoid the application of pesticides, especially **toxaphene** and methyl parathion, since they can hurt the beetle.

Agricultural insect pests – broad types

Any animal or plant, in a broad sense, which causes damage to man and his possessions, is considered a pest. In the context of agriculture, an insect pest is any insect that causes damage to a crop by decreasing its yield or quality, and resulting in financial losses to the farmer. A pest becomes an economic liability when yield losses touch 5% or more.

At a practical level, insect pests in agriculture can be commonly classified into two major categories, based on their mode of feeding and the number of stages in their life cycle. These are: (i) borers and defoliators; and (ii) sucking pests.

Borers and defoliators usually feed on the leaves, roots, stems, flowers and fruits of crop plants. They generally have four stages in their life cycle: egg, larval, pupal and adult stages. Of these, the egg, pupal and adult stages are mostly harmless to crops, but the larval stage has potential to cause severe economic damage to farmers.

Sucking pests generally suck sap from the leaves, stem or fruits of crop plants. Their life cycle is divided into three stages: egg, nymph and adult stages. Of these, the egg stage is not dangerous, but the nymph and the adult stages can cause monetary losses to farmers.

Besides boring and sucking, these pests are carriers of viral diseases in certain crops.

Agrobacterium radiobacter (bacterium) as a biocontrol agent

Agrobacterium radiobacter, a Gram-negative rod bacterium, is recognized as a biocontrol agent for the control of the soil-borne pathogen *Agrobacterium tumefaciens*. *A. radiobacter* strains K84 and K1026 are prominently exploited for such use. This agent is more protective in nature than curative.

A. tumefaciens is responsible for the common crown gall disease that affects orchards of apples, pears, kiwi and grapes.

A. radiobacter formulations are made by growing their culture on peat substrates or on agar. Plant rootstocks and seedlings may be dipped in the diluted inoculation for necessary pest control. Wound sites are also sprayed with the inoculum. Control is brought about when the *A. radiobacter* strains colonize gall sites of the crops, thus preventing the growth of the pathogens. *A. radiobacter* isolates K84 and K1026 are registered for use in the USA, New Zealand, South Africa and Canada. These are observed to be safe to the environment.

A. radiobacter K84 produces an anti-agrobacterial compound called Agrocin 84, which inhibits pathogenic strains and species like *A. tumefaciens*. It also produces plasmids such as pAgK84 and pNOC, which retard the pathogenic strains from absorbing nutrients. pNOC also prevents some pathogenic strains from colonizing the crop root system. This biocontrol agent also has some residual effect.

A. radiobacter K1026 is the result of a genetic alteration of the strain K84. Strain K1026 is used to treat germinating seeds or roots and stems of many crops including cherries, plums and nut trees. *A. tumefaciens* and *Agrobacterium rhizogenes* can be controlled by this strain.

The use of the above-mentioned control agents are observed to be safe for the environment.

Agrobacterium radiobacter K1026

See *Agrobacterium radiobacter* **(bacterium) as a biocontrol agent**

Agrobacterium radiobacter K84

See *Agrobacterium radiobacter* **(bacterium) as a biocontrol agent**

Agrochemicals

Any chemical used in agriculture, including chemical fertilizers, pesticides, soil additives and amendments originating from plants, animals and minerals, is called an agrochemical. Most agrochemicals are mixtures of two or more chemicals.

Agrochemicals are designed to increase yield of food and maintain the health of domestic animals by controlling diseases, pests and predators. There are a.i. and inert ingredients in agrochemicals, the former providing the desired effects and the latter playing a stabilizing role.

The degree of use of agrochemicals in food and feed is a subject of regular debate.

Agroecosystem analysis (AESA)

Agroecosystem analysis (AESA) is an approach that is structured in field situations, to evolve a strategy for growing healthy crops. The various aspects involved are related to pests, moisture, soil pH, pest predators and climatic factors. This type of analysis is done by forming small groups of people and going on field trips, documenting and discussing observations, and evolving locally appropriate measures for plant health.

AIA as a bioPGR

See **Indole-3-acetic acid (IAA or AIA) as a bioPGR**

Air-blast sprayers

See **Application equipment – larger power-driven sprayers**

Alachlor herbicide

Alachlor, $C_{14}H_{20}ClNO_2$ (2-chloro-2′,6′-diethyl-N-methoxymethylacetanilide), is a **chloroacetanilide** herbicide. It is a yellowish to red, odourless solid. Alachlor has two CAS Registry Numbers, 15792-50-8 (used in the EU and Canada) and 15972-60-8.

This herbicide is a selective, systemic compound that acts by inhibiting lipid synthesis when absorbed by germinating weed shoots of grasses and broadleaved weeds. Alachlor is commonly used to protect soybean, maize and sorghum.

The acute oral LD_{50} of alachlor in rats ranges from 930 to 1350 mg/kg. Inhalation LC_{50} (4 h) in rats is >1.04 mg/l air. Acute oral LD_{50} in bobwhite quails is 1536 mg/kg and LC_{50} (96 h) in rainbow trout is 5.3 mg/l. LD_{50} in bees (48 h) is >100 µg per bee. This compound does not irritate the skin or the eyes in normal use. WHO classifies alachlor as a Class III toxin (slightly hazardous). It is a PAN Bad Actor chemical, because it is a cholinesterase inhibitor, a groundwater contaminant and a suspected endocrine disruptor. Its ADI is 0.01 mg/kg b.w./day.

The herbicide alachlor is registered in India, Cambodia, New Zealand, South Africa, Tanzania and Vietnam but not in the UK. In the USA, it is a restricted use pesticide (RUP) by the EPA.

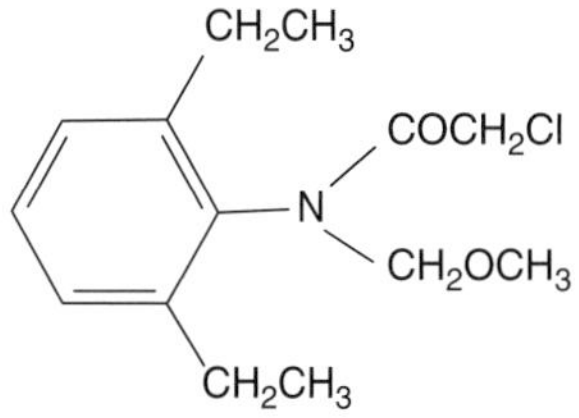

Alachlor

Alanycarb insecticide

Alanycarb, $C_{17}H_{25}N_3O_4S_2$ (ethyl (*Z*)-*N*-benzyl-*N*-[[methyl (1-methylthioethylideneamino-oxycarbonyl) amino] thio]-β-alaninate), is a carbamate insecticide. It is a colourless crystalline substance, sparingly soluble in water but readily soluble in acetone, methanol, benzene, ethyl acetate and dichloromethane. It is stable up to 100°C and decomposes at 195°C. Alanycarb is unstable in acidic and strongly basic media. It is produced by a reaction of *N*-methyl-(carbamoyloxy) thioacetimidate with *N*-benzyl-propylcarboxyethylatesulfenyl chloride. Its CAS Registry Number is 83130-01-2.

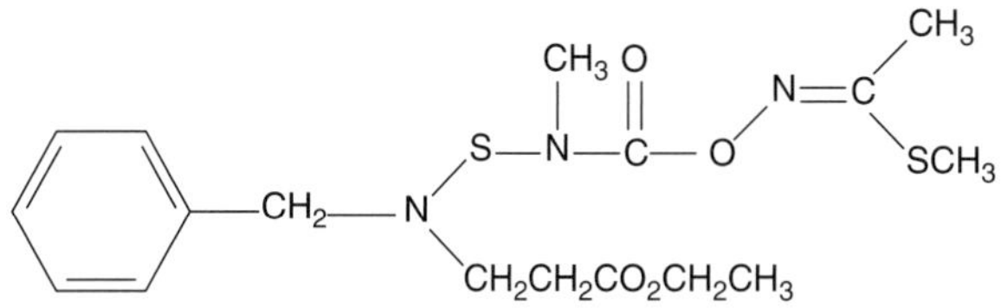

Alanycarb

Alanycarb, a broad-spectrum, cholinesterase inhibitor, is a contact and oral toxin. It is used as a foliar spray, and in soil or seed treatment. This insecticide is effective against many species of beetles, aphids, moths and thrips that attack vines, citrus and tobacco. *Polychrosis botrana* and *Helicoverpa assulta* are two examples of insect pests controlled. Orion is a common brand name.

The acute oral LD_{50} of alanycarb in male rats is 440 mg/kg while their acute percutaneous LD_{50} is >2000 mg/kg. The compound can irritate the eyes slightly but not the skin. The LC_{50} (8 days) in bobwhite quails is 3553 ppm, while the 48 h

LC_{50} in carp is 1.1 mg/l. The LD_{50} in bees is 0.674 µg per bee. According to WHO, alanycarb is a Class II toxin (moderately hazardous), and it is a PAN Bad Actor chemical because it inhibits cholinesterase.

In plants and soils, alanycarb is rapidly metabolized to methomyl oxime, which in turn metabolizes to acetonitrile and carbon dioxide. These are eliminated via urine and respiration. In plants, the N–S bond cleavage is the initial metabolic step, giving rise to methomyl, and finally degrading to carbon dioxide. The DT_{50} in soil is 1–2 days.

As per information available, alanycarb is registered for use in Vietnam.

Aldicarb insecticide

Aldicarb, $C_7H_{14}N_2O_2S$ (2-methyl-2-(methylthio)propionaldehyde*O*-methylcarbamoyloxime), is a carbamate insecticide. It is a colourless crystalline substance, soluble in water (4.93 g/l at pH 7), acetone, dichloromethane, benzene and xylene. It is hydrolysed by concentrated alkalis and decomposes above 100°C. The DT_{50} in water is 3240 days. Aldicarb has a CAS Registry Number of 116-06-3.

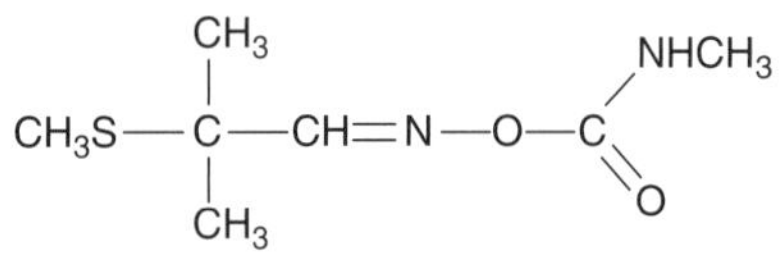

Aldicarb

A root-absorbed, systemic, cholinesterase-inhibiting insecticide, aldicarb is also an acaricide and nematicide. Structurally designed to resemble *O*-acetylcholine, aldicarb effectively controls aphids, whiteflies, leaf miners, soil-dwelling insects, worms and spider mites. Crops protected include potatoes, onions, groundnuts, sugarbeet and some fruits. One of the trade names of aldicarb is Termik10G.

Aldicarb is one of the most toxic pesticides in use today. It is a cholinesterase inhibitor (reversible), a groundwater contaminant, a carcinogen and a suspected endocrine disruptor. The compound is acutely toxic and is classified by WHO as a Class Ia toxin (extremely hazardous). The inhalation LD_{50} (4 h) in rats is 0.0039 mg/l air while the acute oral LD_{50} is 0.93 mg/kg. The acute dermal LD_{50} in male rabbits is 20 mg/kg. It is highly toxic to honey bees, some species of freshwater fish, invertebrates and birds (acute oral LD_{50} in mallard ducks is 1.0 mg/kg b.w.). The ADI is 0.003 mg/kg b.w./day. The NOEL, based on a human volunteer study, is 0.03 mg/kg b.w. Formulations with <10% a.i. may entitle it to be classified as a Class Ib toxin. It is both a PAN Bad Actor chemical and a member of PAN's **Dirty Dozen** chemicals.

The toxicity of aldicarb is evident in even small doses. Symptoms of aldicarb poisoning on handlers and mixers are dizziness, stomach cramping, diarrhoea, nausea and convulsions.

Degradation of aldicarb in soil and plants proceeds via hydrolytic, oxidative reactions, to produce aldicarb sulfoxide and then aldicarb sulfone. The metabolite sulfoxide is highly water soluble and more toxic than aldicarb. Another breakdown product of aldicarb is **aldoxycarb**. Mammals rapidly absorb and eliminate aldicarb with metabolic products like glucuronide or sulfate conjugates. Aldicarb residues range between 0.0011 and 0.06 mg/kg b.w.

Aldicarb is banned in many countries or its use is strictly restricted. India, Finland, Kuwait, Indonesia, Libya, Sweden and Tanzania have banned aldicarb. Argentina, Germany, Sudan, Belize and Togo have restricted its use. The USA has not registered its use. In some regions, its use on watermelons, cucumbers, cotton, ornamental plants and soybeans is registered.

Aldoxycarb insecticide

Aldoxycarb, $C_7H_{14}N_2O_4S$ (2-methyl-2-methylpropionaldehyde *O*-methylcarbamoyloxime) sulfone, is a carbamate insecticide. It is a white crystalline powder with sulfurous odour. Aldoxycarb is soluble in water (10 g/l at 25°C), and in solvents such as chloroform, acetonitrile, methanol, dichloromethane and acetone. It rapidly hydrolyses in acidic and alkaline media and decomposes above 140°C. Aldoxycarb is a known metabolite of **aldicarb**, produced by the oxidation of the thio-moiety. It has a CAS Registry Number of 1646-88-4.

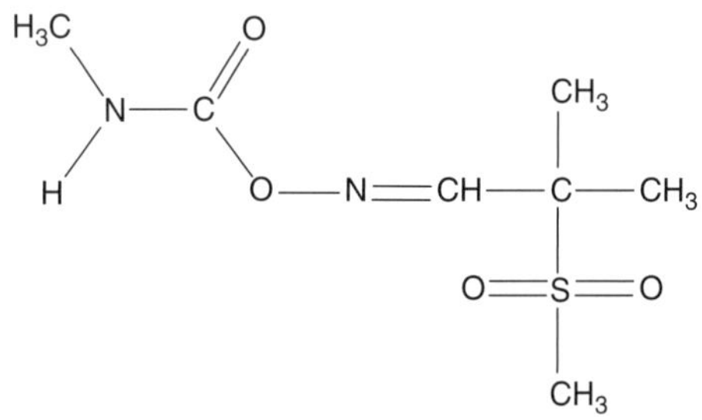

Aldoxycarb

Aldoxycarb, a systemic insecticide, acaricide and nematicide, causes reversible carbamalation of the acetocholinesterase enzyme of tissues. This allows the accumulation of acetylcholine at the cholinergic junction and affects the CNS. The chemical protects cotton, tobacco, potatoes and groundnuts against aphids, thrips, mites and nematodes.

Common formulations of aldoxycarb include wettable powder (WP), granules and suspension concentrate. It can be applied as a foliar spray, seed dressing or applied on the soil. A commercially available trade name is Standak.

Aldoxycarb is classified as a Class I toxin (highly toxic). Its acute oral dose in rats is 21.4 mg/kg, while its acute inhalation dose is 0.209 mg/l. The compound is highly toxic to mammals and to wildlife. Since it is a cholinesterase inhibitor and a highly toxic chemical, aldoxycarb has been classified as a PAN Bad Actor chemical. PPE is highly recommended for applicators and mixers. Symptoms of poisoning in humans include headache, dizziness, blurred or dark vision, muscle twitching, vomiting, diarrhoea, convulsions and death.

Aldoxycarb is metabolized in plants and animals, to finally yield aldoxycarb aldehyde, aldoxycarb alcohol, aldoxycarb acid and aldoxycarb nitrile, most of which turn to water-soluble glycosides and sulfates. In soil and water, it is reduced to methane sulfonic acid and 2-hydroxy isobutyraldehyde oxime. With a soil half-life of 2–4 weeks, aldoxycarb is mobile in certain soil types and contaminates groundwater under certain situations.

Argentina, Germany, Sudan, Belize and Togo have restricted the registration of aldoxycarb. Finland, Kuwait,

Indonesia, Libya, Sweden and Tanzania have banned it. Australia has registered it.

Aldrin insecticide

Aldrin, $C_{12}H_8Cl_6$ (1,2,3,4,10,10-hexachloro-1,4,4a,5,8,8a-hexahydro-exo-1,4-endo,-5,8-dimethanonaphthalene), is a cyclodiene organochlorine insecticide. It is a colourless solid soluble in water (0.027 μg/l) and petroleum hydrocarbons. Chemically stable, aldrin is made by the synthesis of hexachloro-1,3-cyclopentadiene with norbornadiene in a Diels–Alder reaction. It has a CAS Registry Number of 309-00-2.

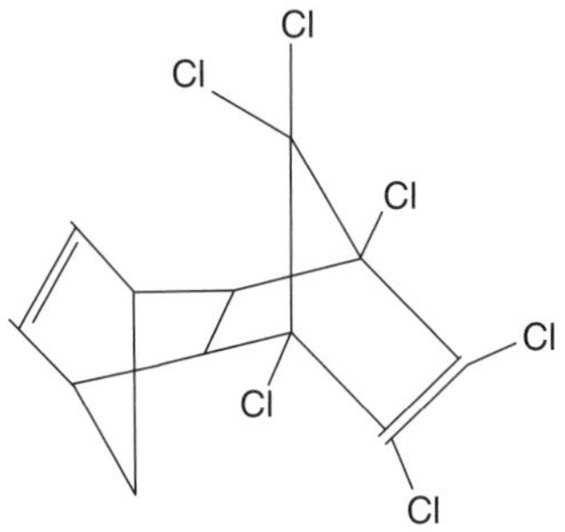

Aldrin

Aldrin is used to kill termites and **grasshoppers**, to protect maize and potatoes. It is also used in seed treatment.

The oral LD_{50} of aldrin in rats is in the range of 39–60 mg/kg. The chemical is highly toxic to birds and fish. It has been classified as a **persistent organic pollutant** (POP). Aldrin is a carcinogen and a suspected endocrine disruptor. This, along with its acute toxicity has led to it being classified as a PAN Bad Actor chemical and a PAN **Dirty Dozen** chemical.

In soil, on plant surfaces and in the digestive tracts of insects, aldrin oxidizes to the epoxide **dieldrin**, which is a very strong insecticide (neurotoxin). The chemical is difficult to metabolize.

Aldrin is banned in India and throughout the EU.

Algicide copper octanoate

See **Copper octanoate fungicide**

Algicide copper sulfate pentahydrate

See **Copper sulfate pentahydrate fungicide**

Algicide dichlone

See **Dichlone fungicide**

Algicide endothal

See **Endothal herbicide**

Algicide fentin acetate

See **Fentin fungicide**

Algicide monolinuron

See **Monolinuron herbicide**

Algicide nabam

See **Nabam fungicide**

Algicides (algaecides)

An algicide or algaecide is a substance used for killing or retarding the growth of algae. Algal growth can harm aquatic plants by suffocation. Weakened plants are thus prone to attacks by pathogens such as the *Pythium* species. Algicides help maintain the aesthetics of water bodies.

Chemical algicides include some fungicidal compounds like **cheshunt compound, Bordeaux mixture** and **copper sulfate pentahydrate** and very low concentrations of these a.i. are sufficient to kill algae. Copper compounds can kill fish and hence the dosage has to be strictly monitored. Polyquat is the common name for a class of algicides that control algae that attack pools and tubs (n-alkyl-dimethyl benzyl ammonium chloride, also known as benzalkonium chloride).

Organic compounds that act as effective algicides include barley straw and grapefruit seed extract. These can reduce algae without harming aquatic plants and fish.

Alkali

Any ionic salt of an alkali metal which is bitter when mixed in water, and more or less irritating or caustic to the skin, is an alkali. An alkali or a base turns red litmus, blue. It has a pH of more than 7. Sodium hydroxide, sodium carbonate, salt cake and sodium bicarbonate are examples. Both hydroxides and oxides of metals are classified as bases. Alkalis react with acids to yield salts. Sodium hydroxide and potassium hydroxide are very important industrial chemicals, used in the manufacture of, for example, soaps and glass. An alkali can combine with a proton to form a new compound.

The measurement of the concentration of alkali present is determined by titration with a standard acid, using an indicator.

Alkylating agents

See **Chemosterilant**

Allelochemicals

See **Weed control agents – biological origin**

Allelopathy

See **Weed control agents – biological origin**

Allergen

Any substance that acts like an antigen on coming into contact with an organism, by inhalation, ingestion or skin absorption, is an allergen. An allergen produces a specific reagin in the blood, which is similar to an antibody. People who are allergic to a substance produce reagins in response to a given allergen. The specificity of the allergen–reagin reaction and its dependence on molecular configuration is similar to the antigen–antibody reaction.

Allergies can occur to a host of elements in nature including plants, pollen, chemicals and foods.

Allethrin insecticide

Allethrin, $C_{19}H_{26}O_3$ ((*RS*)-3-allyl-2-methyl-4-oxocyclopent-2-enyl (1*RS*, 3RS; 1RS, 3SR)-2, 2-dimethyl-3-(2-methylprop-1-enyl) cyclopropanecarboxylate), is a non-systemic pyrethroid insecticide with contact, stomach and respiratory action. It has a CAS Registry Number of 584-79-2. Allethrin is almost insoluble in water but miscible with for example hexane, xylene and carbon tetrachloride. It is hydrolysed in alkaline media.

Each allethrin consists of eight possible stereoisomers. A variant of allethrin, consisting of only two stereoisomers, is called **bioallethrin**.

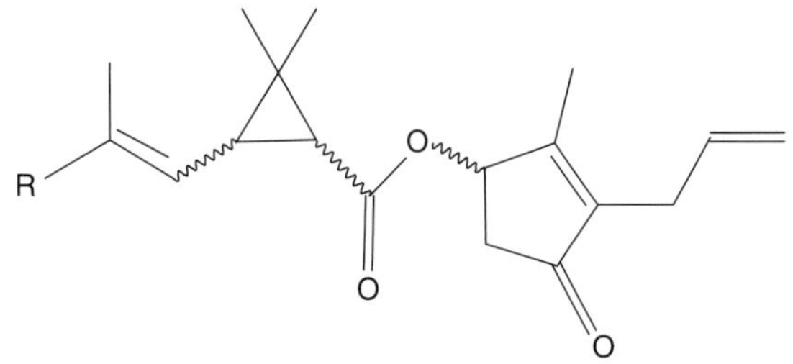

Allethrin I (R = $-CH_3$) Allethrin II (R = $-COOCH_3$)

Allethrin

Allethrin effectively controls flies, mosquitoes and ants in non-crop situations such as houses, gardens and public places. It is also used to control insects on farm animals. Purified *d*-trans-isomers of allethrin are often used for increased effectiveness. Some trade names are Alleviate and Pynamin; trade names of *d*-trans allethrin are *d*-trans and of bioallethrin, MGK 264.

The compound is formulated as wettable/dustable powder, and most often as coils, mats and oil formulations. It is also used as a ULV spray for outdoor mosquito control. Allethrin may be mixed with piperonyl butoxide.

Allethrin is slightly to moderately toxic by the dermal mode and by ingestion. Allethrin is very toxic to fish and invertebrates in water, with the LC_{50} being in the ppb range. Acute oral LD_{50} in male rats is 900 mg/kg. It is not toxic via skin in normal use (acute percutaneous LD_{50} in male rabbits is 4390 mg/kg). It is practically non-toxic to most bird species. The LC_{50} (96 h) in carp is 0.134 mg/l. Allethrin is slightly toxic to bees. However, since allethrin acts on the CNS, severe/long-term exposure to allethrin can cause symptoms such as nausea, convulsions, tremors, lack of coordination, incontinence and coma, and also can result in increased kidney and liver weight in some animals. The compounds have low toxicity for humans, hence its use in coils and mats to repel mosquitoes. Pesticide formulations containing allethrin are Class III toxins (slightly hazardous), and formulations must bear the signal word 'CAUTION' on the product label. Applicators and workers are advised to use PPE to avoid untoward exposure to allethrin formulations.

In surface waters, allethrin decomposes due to microbial action and photodegradation. In mammals, one methyl group is oxidized to an alcohol group and then to a carbonyl group.

India, Australia and South Africa have registered the use of allethrin, while the USA has not registered the use of this chemical.

d-Allethrin insecticide

d-Allethrin, $C_{19}H_{26}O_3$ ((*RS*)-3-allyl-2-methyl-4-oxocyclopent-2-enyl (1*R*)-*cis*, *trans*-chrysanthemate), is a synthetic pyrethroid insecticide. The technical material of *d*-Allethrin is a mixture of [1*R*,*trans*;1*R*] + [1*R*,*trans*;1*S*] + [1*R*,*cis*;1*R*] + [1*R*,*cis*:1*S*] (approximate ratio 4:4:1:1). This chemical is an oily liquid that is slightly soluble in water (5 mg/l). It hydrolyses in alkaline conditions and decomposes above 400°C. *d*-Allethrin is slightly volatile at room temperature. The CAS Registry Number is 42534-61-2. The racemic mixture of eight stereoisomers is known as allethrin (see *d*-Allethrin at bottom of the page).

d-Allethrin is used in both households and public places and is very effective on common non-crop pests like mosquitoes, houseflies and cockroaches.

d-Allethrin is commonly formulated as vaporizing mats and coils. It should not be combined with compounds of alkaline nature.

Although *d*-Allethrin is observed to be toxic to fish and bees (with a 96 h LC_{50} value ranging from 9.4 to 53 µg/l for fish species tested, and contact toxicity LD_{50} of 3.4 µg per bee), it is considered safe to use if label instructions are followed and as long as there is no accident/misuse. It is important to note that other derivatives like ***S*-bioallethrin** and allethrin may be toxic to fish and bees. *d*-Allethrin has low toxicity to mammals and birds. However, applicators and handlers are advised to wear protective clothing (especially if there is a spray application) and avoid eating, drinking or smoking in the area. Exposure to the chemical may cause irritation to the skin and eyes, numbness, dizziness and sometimes even seizures.

In mammals, one of the two methyl groups of the chrysanthemic acid moiety in *d*-Allethrin is oxidized to an alcohol group and then to a carboxyl group.

d-Allethrin is not a PAN Bad Actor chemical. It is registered by the USA, Canada, Australia, Greece, Tanzania and South Africa.

Alligator weed flea beetle

See ***Agasicles hygrophila* (alligator weed flea beetle) as a biocontrol agent**

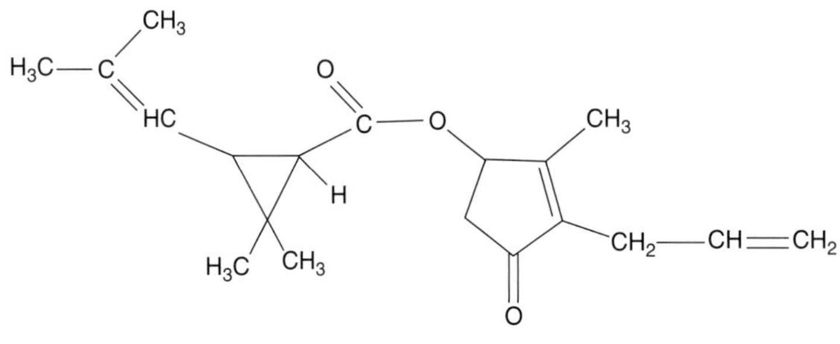

d-Allethrin

Alloxydim-sodium herbicide

Alloxydim-sodium, $C_{17}H_{24}NNaO_5$ (sodium (E)-(RS)-3-1 (allyloxyimino) butyl-4-hydroxy-6, 6-dimethyl-2-oxocyclohex-3-enecarboxylate), is a cyclohexanedione oxime herbicide. It is a colourless crystalline substance, soluble in water and other organic solvents. It is stable when dry but is very hygroscopic. Alloxydim-sodium is made by the reaction of mesityloxide with dimethyl malonate, the resultant product being reacted with *n*-butyryl chloride, followed by its reaction with hydroxyallylamine, in turn hydrolysing the ester to the acid. Alloxydim-sodium has a CAS Registry Number of 55635-13-7.

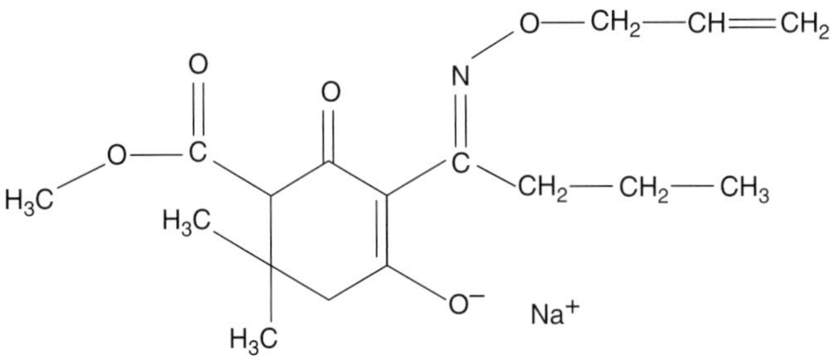

Alloxydim-sodium

Alloxydim-sodium, a selective, systemic, post-emergence herbicide, acts by inhibiting mitosis and also fatty acid synthesis. It is used to control grass weeds in sugarbeet and vegetables and broadleaved crops at 0.5–1.0 kg/ha.

Clout is one of the common trade names of alloxydim-sodium.

According to the WHO and EPA, alloxydim-sodium does not present acute hazard in normal use. The acute oral LD_{50} among rats and mice is between 2322 and 4000 mg/kg. Inhalation LC_{50} (4 h) among rats is >4.3 mg/l air. During 2 year NOEL feeding trials, rats and mice receiving 100 mg/kg diet showed no ill effects. Product labels are required to carry the signal word 'CAUTION'.

The high solubility of this compound in water makes it necessary to avoid any accidental spillage and leakages into water sources. In soil, DT_{50} is 2–10 days.

Alpha-cypermethrin insecticide

Alpha-cypermethrin, $C_{22}H_{19}Cl_2NO_3$ (recemate containing (*R*)-α-cyano-3-phenoxybenzyl (1*S*,3*S*)-3-(2,2-dichlorovinyl)-2,2-dimethylcyclopropanecarboxylate and (*S*)-α-cyano-3-phenoxybenzyl(1*R*,3*R*)-3-(2,2-dichlorovinyl)-2,2-dimethylcyclopropanecarboxylate), is a pyrethroid insecticide. It is a colourless crystal that is practically insoluble in water but soluble in many organic solvents. It is very stable in neutral and acidic media and also up to 220°C. Alpha-cypermethrin has a CAS Registry Number 67375-30-8.

Alpha-cypermethrin is a non-systemic, broad-spectrum insecticide that acts on the CNS and peripheral nervous system, even at a low dose. It is an ectoparasiticide in farms and in public places. It is used to control chewing and sucking insects on for example fruit, vines, cereals and oilseeds.

The formulations of alpha-cypermethrin include emulsifiable concentrate (EC), ULV liquid (UL) and tablet. It is available under trade names such as Fastac, Renegade and Dominex. Alpha-cypermethrin is applied at 10–15 g/ha and is compatible with most OP insecticides.

According to WHO, alpha-cypermethrin is a Class II toxin (moderately hazardous). Acute oral LD_{50} in rats is 57 mg/kg while inhalation LC_{50} (4 h) is >0.593 mg/l. The compound is non-toxic to birds but toxic to bees. It irritates the eyes and skin. The NOEL (1 year) for dogs is >60 mg/kg diet. The ADI is 0.02 mg/kg b.w./day. The parent chemical of this compound is **zeta-cypermethrin**, which is highly toxic (Class Ib, i.e. highly hazardous). Zeta-cypermethrin is a possible carcinogen and endocrine disruptor and thus a PAN Bad Actor chemical. If accidentally ingested, alpha-cypermethrin can cause headache, nausea, dizziness and diarrhoea.

This insecticide undergoes degradation in soil with DT_{50} of about 4 months in loamy soil.

Alpha-cypermethrin is registered for use in India, Australia and many countries of the African continent but not in the USA. Zeta-cypermethrin is registered in the USA and Australia (see Alpha-cypermethrin at top of the next page).

ALS inhibitors

See **Herbicidal inhibition of ALS**

Alternaria cassia (fungus) as bioherbicide

Alternaria cassia is an ascomycete fungus (family Pleosporaceae) and a biological herbicide.

Although *A. cassia* is a pest on soybean, groundnuts and cotton, its herbicidal properties are seen on sicklepod (*Senna obtusifolis*) and coffee senna (*Cassia occidentalis*).

The herbicidal product is made by a fermentation process. It is not recommended for combination with strong oxidizers, acids, bases or chlorinated water.

A. cassia is found to be non-toxic to humans, the environment and non-target organisms.

Alternaria destruens (fungus) isolate 059 as bioherbicide

Alternaria destruens isolate 059 is an ascomycete fungus (family Pleosporaceae) that is valued as a bioherbicide.

A. destruens controls the growth of many species of dodders (*Cuscuta* spp.). Infection takes place via active pathogenesis. This bioherbicide protects a variety of fruit and vegetables.

Common formulations of this herbicide, produced by way of fermentation, are granular formulation (GR) and wettable powder (WP). One of the commercially available trade names is Smolder G. Depending on the species, the compound may be applied at the early-emergence stage or at maturity.

A. destruens should not be combined with strong oxidizers, acids, bases and chlorinated water. *A. destruens* is safe for humans, mammals and the environment.

Alternaria leaf spot

See **Plant disease *Alternaria* leaf spot of cole crops**

Aluminium phosphide rodenticide

Aluminium phosphide, AlP, a compound of aluminium and phosphorus, is a rodenticide, fumigant and insecticide. It is a grey-green-yellow powder with a garlic odour. It reacts with moisture and acids to release **phosphine gas**. The CAS Registry Number is 20859-73-8.

(*R*)-alcohol (1*S*)-*cis*-acid

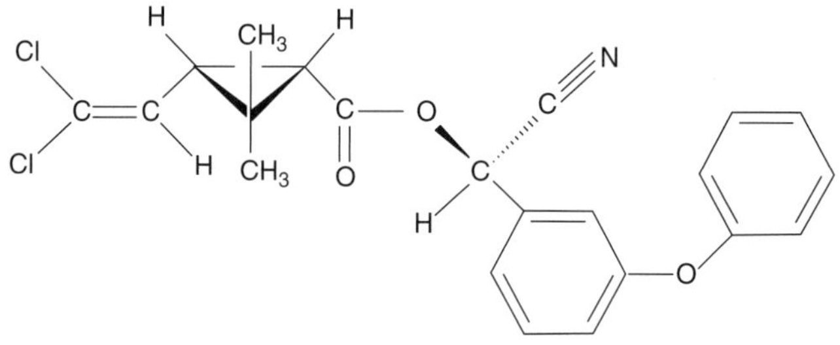

(*R*)-alcohol (1*R*)-*cis*-acid

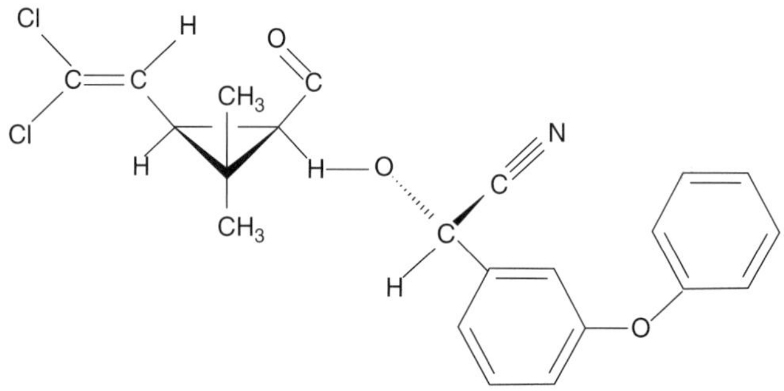

Alpha-cypermethrin

Al ≡ P

Aluminium phosphide

As a pre-harvest rodenticide, AlP is specifically used to control moles, rabbits and rodents. Once consumed, AlP reacts with digestive enzymes to generate the toxic phosphine gas. It is also used as a fumigant and insecticide for stored food grains in storehouses.

AlP is commonly formulated as pellets and baits (mixed with food). Rodents are attracted by the garlic odour. Trade names of AlP include Phostoxin, Phosphume and Phostek.

AlP is highly toxic on toxicity ratings. It is very dangerous to fish and other aquatic life.

There are many countries that register the use of aluminium phosphide, and the list includes India (restricted use), Germany, Australia and the USA.

Amblyseius mites as biocontrol agents

Mites form a widely known pest group that affects a wide range of crops like fruit orchards and vegetables. While **acaricides** are widely used to control mites, biological control is seen as an attempt to maintain the balance in the ecosystem.

The mite genus *Amblyseius* has been extensively studied and harnessed for the biological control of pest mites and thrips. Various species are useful in this regard.

Amblyseius californicus is an active predatory mite that helps in keeping pest mite species in control, thus rendering ecofriendly pest control. This mite species, also called the Cali mite, belongs to the family Tetranychidae. Targeted mite species include the common red spider mite or two-spot mite (*Tetranychus urticae*), carmine mite (*Tetranychus cinnabarinus*) and broad mite (*Polyphagotarsonemus latus*).

The motile cali mite is a robust organism that can survive even in hot, dry conditions. It is omnivorous in nature and can consume alternative food like pollen. The mode of operation of this predator is for the female to lay eggs in target mite colonies, allowing the six-legged larvae to feed on the larvae of target mites, thus effectively controlling their numbers. Not only the larvae but also adults attack and feed on pest mite species, by sucking out their fluids.

For effective biocontrol, a batch of 2000 mites, mixed with an inert carrier material, is released among pest colonies. In cases of serious infestation, acaricides such as fenbutatin oxide or hexithiazox and bifenazate may be used before the predator mites are released in the field. It is important to release these mites within 18 h of acquisition.

Other predator mites from the same genus include *Amblyseius newsami*, which are successful in the control of tea mites.

Amblyseius andersoni holds vast potential as a biocontrol agent of pest mites. This mite feeds on a vast range of mite species including red spider mite, fruit-tree red spider mite, gall mite, russet mite, two-spotted spider mite, European red mite and apple rust mite. In most cases, ornamental plants and fruit orchards are protected if these predators are released in a 1:2 ratio (predator:pest).

These mites should not be kept in storage for too long (over 18 h) or at temperatures lower than 10°C.

Amblyseius barkeri (or *Amblyseius mckenziei*) is another predator mite that feeds mainly on thrips but also on other mites. This species is considered robust as it can feed alternatively on pollen, fungi or spider mite eggs. Thrip species controlled include western flower thrips (*Frankliniella occidentalis*) and onion thrips (*Thrips tabaci*). These predators feed on the young thrip larvae on greenhouse crops.

The release of these predators at the pre-flowering stage can prevent the attack of mite pests such as strawberry mites (*Phytonemus pallidus*), bulb scale mites (*Steneotarsonemus laticeps*) and pineapple mites (*Steneotarsonemus ananas*). Release can be repeated if necessary.

A. barkeri may be conveniently grown in a mixture of bran and vermiculite (with each batch totalling up to 50,000 mites), to be released as required. Cooler temperatures affect these mites negatively, by inhibiting egg production.

Amblyseius cucumeris is a very small-sized, tan-coloured mite that lives on the underside of leaves or mature flowers. It is an effective biocontrol agent on thrip pests that also has alternate food sources like spider mites, strawberry mites, rust mites and pollen. This mite is used to control thrips on crops such as cucumbers, sweet pepper, melon and squash, tomatoes, and flowering plants like gerberas and roses.

The adult female lays eggs that hatch and grow within 10 days. Since the nymphs and adults are very small, they predate on thrip eggs and thrip larvae. They feed by piercing the victim and sucking out the body fluid. *A. cucumeris* should be released in large numbers since they are so small. Release rates range from 50 to 100 predators per plant. These predators may take up to 1 month to establish themselves, if dry and cold weather (<10°C) conditions are not present. Release time is preferably between March and November.

Target thrip species include *F. occidentalis*. *A. cucumeris* releases are assisted by another thrip predator, orius, *Amblyseius degenerans*.

A. degenerans is also used in the biological control of thrips on crops such as sweet pepper and aubergine. Biocontrol effects extend to western flower thrips (*F. occidentalis*), onion thrips (*T. tabaci*) and, to some extent, to tarsonemid mites.

Amensalism

Some organisms produce substances that are inhibitory to other organisms. The inhibitory effect of such substances is called amensalism (similar to **allelopathy**). For example, the fungus *Penicillium* inhibits bacteria.

American bollworms

See **Crop – cotton pest control**

Ametryn herbicide

See **Methyl- or ethyl-thiotriazine herbicides**

Aminoethoxyvinylglycine as a bioPGR

Aminoethoxyvinylglycine, $C_6H_{12}N_2O_3$ (*N*-(phenylmethyl)-1*H* purin-6-amine), is a microorganism-derived plant growth regulator (from a *Streptomyces* sp). Also called aviglycine or AVG, this compound inhibits the activity of 1-aminocyclopropane-1-carboxylic acid synthase (ACC synthase), thus blocking the production of ethylene (a plant hormone). Aminoethoxyvinylglycine's CAS Number is 49669-74-1.

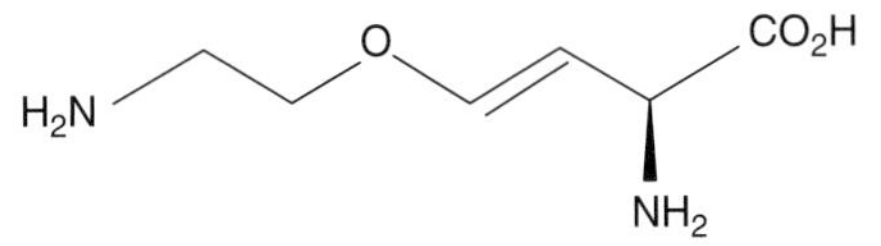

Aminoethoxyvinylglycine

A pre-harvest growth regulator, aminoethoxyvinylglycine is known to protect apples and pears from premature ripening and dropping. It acts by inhibiting the biosynthesis of **ethylene**, the plant hormone responsible for the maturation and ageing of fruit. This PGR is available as a water-soluble concentrate, at rates of up to 130 mg/l. A known trade name is Retain.

Aminoethoxyvinylglycine is not considered to be toxic to mammals, fish or honey bees. The acute oral LD_{50} in rats is >5000 mg/kg and acute dermal LD_{50} in rabbits is >2000 mg/kg. The compound is moderately toxic to birds, the acute oral LD_{50} being 121 mg/kg in bobwhite quails. The ADI is 0.002 mg/kg b.w./day. The NOEL (90 days) in rats is 2.2 mg/kg b.w. EPA classifies this PGR as a slightly toxic chemical. It degrades rapidly in the environment.

The USA, Canada, Australia and South Africa have registered the use of aminoethoxyvinylglycine.

Amisulbrom fungicide

Amisulbrom, $C_{13}H_{13}BrFN_5O_4S_2$ (3-(3-bromo-6-fluoro-2-methylindol-1-ylsulfonyl)-*N*, *N*-dimethyl-1*H*-1, 2, 4-triazole-1-sulfonamide), is a fungicide (sulfonamide group). It is a fine odourless powder, soluble in water (0.11 mg/l at pH 7). It has a CAS Registry Number of 348635-87-0.

A contact fungicide (with some penetration activity and rainfastness), amisulbrom is a quinone inside inhibitor (QiI) that inhibits fungal respiration. It protects potatoes, tomatoes and the cucurbit family from the serious tuber disease, **late blight**, and the foliar disease, **downy mildew**. *Phytophthora infestans* and *Pseudoperonospora cubensis* are the main target fungal species.

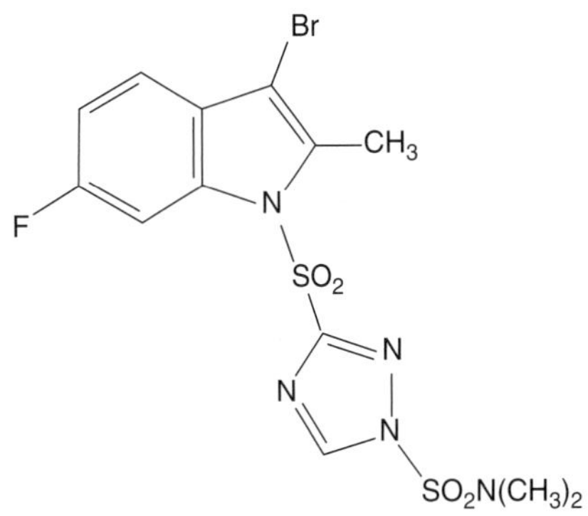

Amisulbrom

Amisulbrom is formulated as a suspension concentrate (SC) and flowable concentrate for seed treatment (FS). Common application doses are 100 g/ha, repeated up to six times per crop.

This compound is very toxic to aqueous organisms such as fish, *Daphnia* sp. and algae, and also for birds, bees and worms. Acute oral and percutaneous LD_{50} values in rats are both >5000 mg/kg. Acute oral LD_{50} in bobwhite quails is >2000 mg/kg. For carp fish, the 96 h LC_{50} is 22.9 µg/l. For bees, oral and contact LD_{50} values are >100 µg per bee.

Environmentally, the compound is readily degraded in soils, water and on plants. In some soils, it readily gets adsorbed. Soil DT_{50} is 3–13 days.

Amisulbrom is not registered in the USA but is registered for use in Germany and the UK.

Amitraz acaricide

Amitraz (*N*, *N*′-[(methylimino) dimethylidyne] di-2,4-xylidine), with molecular formula $C_{19}H_{23}N_3$, is a formamidine acaricide. It is a colourless needle-like substance with low solubility in water. It is non-corrosive and stable to heat and UV. Moisture may bring about slow decomposition. Amitraz is produced by the reaction of 2,4-xylidine, ethylorthoformate and methylamine. It has a CAS Registry Number of 33089-61-1 (see Amitraz at bottom of the page).

An insecticide and acaricide, amitraz is used to control red spider mites, leaf miners and **scale insects** on cotton. It is used to control bollworms, whitefly, leaf worms and aphids. It is used to control ticks, mites and lice on animals.

Amitraz is available in an emulsifiable concentrate (EC), wettable powder (WP) and a pour-on powder. Some trade names of amitraz are Taktic, Mitac, ByeBye, Edrizer and Sender.

WHO classifies amitraz as a Class III toxin (slightly hazardous). It is a reproductive toxin and can be carcinogenic. It is hence a PAN Bad Actor chemical (as per the PAN Pesticide Database). The acute oral LD_{50} in rats and rabbits are 600 mg/kg and >100 mg/kg, respectively. The ADI is 0.003 mg/kg/day. The substance is retained in apples, and hence apples treated with amitraz must not be processed.

Amitraz is eliminated almost unchanged via urine, upon oral administration. It does not enter water sources or the food chain in normal usage.

It is registered for use in Australia, Canada, South Africa, the USA and New Zealand.

Amitrole herbicide

Amitrole, $C_2H_4N_4$ (1*H*-1, 2, 4-triazol-3-ylamine), is a triazole foliar herbicide. It is a colourless crystalline substance, soluble in water but not in non-polar solvents. It is a powerful chelating agent, forming salts with most acids or bases. Amitrole is produced by the condensation of formic acid with amino guanidine. Structurally, amitrole is a five-membered heterocyclic ring. It has a CAS Registry Number of 61-82-5. Amitrole (or amino triazole) is one among a range of carotenoid (plant pigment) biosynthesis inhibitors, which includes **norflurazon**, **fluridone** and **isoxazolidinones**.

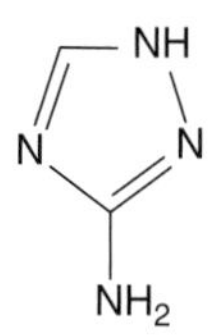

Amitrole

Amitrole, a **non-selective herbicide** applied at the pre-sowing or pre-planting stage, has good translocation properties. It acts by inhibiting carotenoid and histidine biosynthesis. This herbicide thus interferes with chlorophyll formation and re-growth from buds. Besides being used on fruit crops, barley, wheat, maize and peas, it is applied on non-crop sites (paths, parks and fallow) to control grasses and broadleaved weeds and aquatic weed plants. Examples of weeds controlled are dandelions, thistle species and foxtails.

This herbicide can be combined with many compounds (such as ammonium thiocyanate) and is formulated as soluble concentrate (SC) and water-soluble powder. Some known trade names are Weedazol, Azolan and Oriflam. For crops where amitrole is applied at post-emergence, an interval of 8 months is recommended before the next crop is planted.

Acute oral and dermal LD_{50} values for amitrole in rats are >5000 mg/kg. NOEL (2 years) in rats is 10 ppm. It is slightly toxic to aquatic species and non-toxic to birds and bees. Amitrole can irritate the eyes and the skin in case of exposure. It is classified as a Class U toxin (unlikely to be hazardous). However, at high doses the chemical is carcinogenic. Amitrole is a PAN Bad Actor chemical since it is a potential groundwater contaminant and endocrine disruptor, prompting an ADI of as low as 0.0005 mg/kg b.w./day.

Amitrole is eliminated unchanged within 2 days in animals. It readily metabolizes in plants. The compound has low soil persistence, with a half-life of 2 weeks. It degrades both chemically and microbially.

Amitus spiniferus (wasp) as a biocontrol agent

Amitus spiniferus, a small black wasp (1 mm long), parasitizes whitefly nymphs, predominantly the cottony whitefly *Aleurothrixus flocossus*.

The unique mode of action of this wasp is that it paralyses its host.

Ampelomyces quisqualis as a biofungicide

Ampelomyces quisqualis (AQ) is a mitosporic fungus (Class Coelomycetes) and is seen as a biofungicide to some extent. Isolate 10 of AQ is hyperparasitic on organisms in the genus *Erysiphe*, which cause **powdery mildew** diseases. The formulation is made by fermentation.

AQ-10 enters the hyphae and grows via fruiting bodies inside the host, affecting its growth and halting the progress of mildew. The biofungicide is used to protect grapes, apples, tomatoes and also members of the family Cucurbitaceae.

AQ-10 is available as water-dispersible granules, sprays and in a powdered form. This product can be combined with some bioinsecticides such as *Bt* (*Bacillus thuringiensis*) and also with some chemical fungicides.

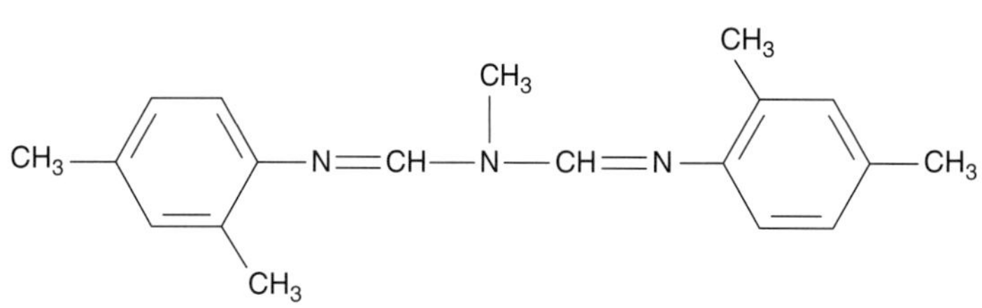

Amitraz

The presence of powdery mildews on a crop is a pre-condition for AQ-10 to be effective. Humidity and watery surroundings are also necessary.

Cyprus, Greece and the UK have registered the use of this fungicide, while the USA does not allow its use.

Ampholytic surfactants

See **Surfactants**

Amphotericin B

Amphotericin B is an antifungal antibiotic, suitable for the systemic therapy of both deep mycotic infections and superficial infections. It is produced by fermentation with a strain of *Streptomyces nodosus*.

Anaerobic infection

Infections caused by bacteria that are intolerant of oxygen (anaerobic) are termed anaerobic infections. Obligate anaerobes cannot tolerate the presence of oxygen and are actually harmed by it. Facultative anaerobes do not require oxygen for growth but may use it if it is present.

Anaerobes are prevalent throughout the body as indigenous flora, with all anaerobic infections arising endogenously. Commonly, anaerobic infections take place in an existing wound or when there has been a condition of short supply of oxygen. *Bacteroides fragilis*, *Peptostreptococcus* and *Clostridium* species are common anaerobic pathogens. Common diseases on man caused by anaerobic bacteria include tetanus and pneumonia. *Pseudomonas aeruginosa* is a facultative pathogen that causes symptoms such as soft rots in leafy vegetables, besides infecting animals and humans.

Anagrapha falcifera nucleopolyhedrovirus (AfNPV) as a bioinsecticide

Anagrapha falcifera nucleopolyhedrovirus (AfNPV) is a virus belonging to the family *Baculoviridae* and is known to be a bioinsecticide.

AfNPV is named after the celery looper (*A. falcifera*). The host range of this virus includes insect crop pests of *Heliothis*, *Helicoverpa* and *Lepidoptera* species such as the tobacco budworm and cotton bollworm. Crops protected by this virus include tomatoes, maize and fruit orchards.

Infection is initiated when a looper feeds on a virus-infected plant. The insect mid-gut fluids dissolve the virus protein matrix leading the virus to multiply into viral particles and a large number of occlusion bodies.

This bioinsecticide is available as a liquid concentrate and is sprayed on plant foliage. It may be used along with a variety of insecticides. The culture stores well if kept in a cool (<32°C), dry and dark place.

The virus is classified as slightly toxic, so caution is needed while handling this bioinsecticide. It is found to have no adverse effects on non-target organisms or the environment.

AfNPV is registered for use in the USA, among other countries.

Ancymidol as a PGR

Ancymidol, $C_{15}H_{16}N_2O_2$ (α-cyclopropyl-4-methoxy-α-(pyrimidin-5-yl) benzyl alcohol), is a plant growth regulator of the pyrimidine class. It is a white crystalline substance, soluble in water, acetone, methanol, hexane and ethyl acetate. An otherwise stable compound, ancymidol decomposes in strongly alkaline media (pH 11 and above). Ancymidol is produced by the metallation of 5-bromopyrimidine and a subsequent reaction with cyclopropyl 4-methoxy phenyl ketone. Ancymidol has a CAS Registry Number of 12771-68-5.

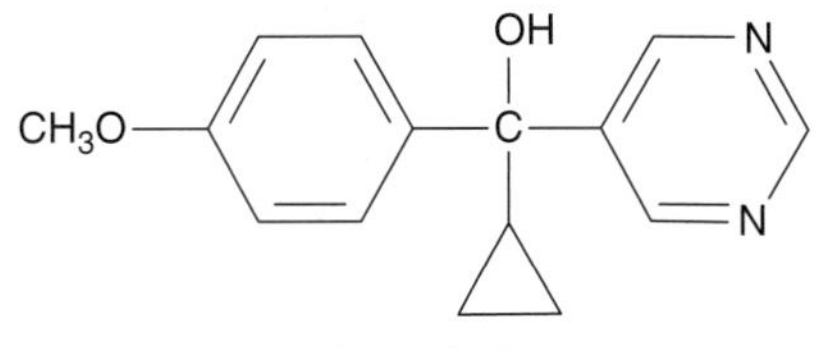

Ancymidol

Upon absorption by the plant (via translocation in the phloem), ancymidol reduces elongation, thus producing a more compact plant. It acts by inhibiting gibberellin synthesis. Ancymidol is used on ornamental and greenhouse plants.

Foliar application rate is 6–66 ppm for bedding, flowering and foliage plants, and 25–50 ppm for bulb crops. Ancymidol is formulated as a suspension concentrate and available under trade names including A-Rest.

Acute oral and percutaneous LD_{50} values of ancymidol in rats are 1721 mg/kg and >5000 mg/kg, respectively. Inhalation LC_{50} (4 h) in rats is 0.59 mg/l air. The compound is a mild eye irritant. WHO classifies ancymidol as non-hazardous (Class U), while EPA and the PAN Pesticide Database have classified it as being moderately toxic. It is slightly toxic to bees and some aquatic species. Ancymidol undergoes microbial degradation in soil.

Ancymidol is registered for use in the USA.

Angiosperms

An angiosperm is a plant of a large taxonomic group (subdivision Angiospermae) that includes most deciduous trees and flowering plants. The mature seed of angiosperms is enclosed in the ovule (within a fruit). Trees that are angiosperms are referred to as hardwoods.

Angiosperms can be divided into monocotyledons and dicotyledons. Monocotyledons or monocots are herbaceous plants with a single cotyledon or seed leaf, parallel-veined leaves and three-part flowers. Dicotyledons or dicots may be either herbaceous or woody plants, with two cotyledons or seed leaves, net-veined leaves and ring-shaped vascular tissues.

Anilofos herbicide

Anilofos, $C_{13}H_{19}ClNO_3PS_2$ (*S*-4-chloro-*N*-isopropylcarbaniloylmethyl *O*,*O*-dimethyl phosphorodithioate), is an OP herbicide. It is a crystalline solid, moderately soluble in water and soluble in organic solvents. Stable under normal storage conditions, it decomposes at 150°C. Anilofos is made by reacting *O*,*O*-dimethylphosphorodithioate with *p*-chlorobenzene *N*-isopropyl carbamoyl chloride. Anilofos has a CAS Registry Number of 64249-01-0.

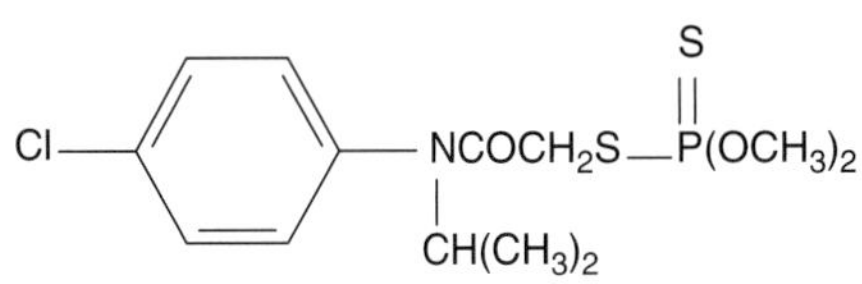

Anilofos

A selective, pre-emergence herbicide, anilofos is mainly used in rice crops, to control annual grasses and sedges in transplanted rice. Examples of common weeds that are inhibited by anilofos are Indian goosegrass (*Eleusine indica*), **barnyard grass** (*Echinochloa* spp.) and sprangletop.

Commonly available formulations of anilofos are emulsifiable concentrate (EC) and granule (GR) and it is available in the market under trade names such as Arozin and Rico. The chemical is usually soil-applied and is compatible with herbicides that control broadleaved weeds, like **2,4-D** and **propanil**.

Acute oral LD_{50} of anilofos in rats is between 830 and 472 mg/kg, respectively. Acute percutaneous LD_{50} is >2000 mg/kg. Acute oral LD_{50} in Japanese quails is from 2339 to 3360 mg/kg. The compound is moderately toxic to birds and slightly toxic to fish (96 h LC_{50} in goldfish is 4.6 mg/l) and bees. It is slightly irritating to the skin. According to WHO, anilofos is a Class II toxin (moderately hazardous). It is a PAN Bad Actor chemical because it inhibits cholinesterase. Symptoms of poisoning with anilofos are excessive salivation, sweating, headache, dizziness, diarrhoea and nausea.

In soil and water, anilofos produces a phosphoric acid compound, chloroaniline, and carbon dioxide as end products. Soil DT_{50} is up to 45 days.

India, the Philippines and Vietnam are some countries that have registered the use of anilofos, while the USA has not registered its use.

Animal Kingdom

The Animal Kingdom or Kingdom Animalia is a large group of multi-cellular, **eukaryotic organisms**. All the organisms of this kingdom are motile at least at some stage of life. Animal cells lack cell walls which distinguish them from plants, algae and fungi. Most animals digest food in an internal chamber with one or two openings. With the exception of sponges and placozoans, animal cells are differentiated into specialized tissues such as muscles for locomotion and nerves for sending and processing signals.

Animals show enormous diversity ranging from the simple corals and sponges, to insects, earthworms and nematodes, to higher forms such as fish, reptiles, amphibians, birds and mammals.

Animals are grouped into various hierarchical taxa with the species as the basic unit. An example of the scientific classification of a species of damselfly is illustrated in Fig. A.3.

Animals are also grouped as vertebrates and invertebrates. Vertebrates are animals with a vertebral column (or a backbone) and belong to phylum Chordata and sub-phylum vertebrata. Vertebrates include amphibians, reptiles, birds and mammals. All the animals in this phylum possess a cranium or skull.

Invertebrates are animals that do not have a vertebral column. This group includes all animal species except the ones included in the sub-phylum vertebrata of phylum Chordata. About 95% of the species in the animal kingdom are invertebrates.

In our context, a study of taxonomy is important in order to identify insect species that can cause a nuisance to man and his crops (called pests), or those that can act as a control measure against pests (called biocontrol agents). Table A.1 lists some important pest orders and their broad relevance.

Anionic surfactants

See **Surfactants**

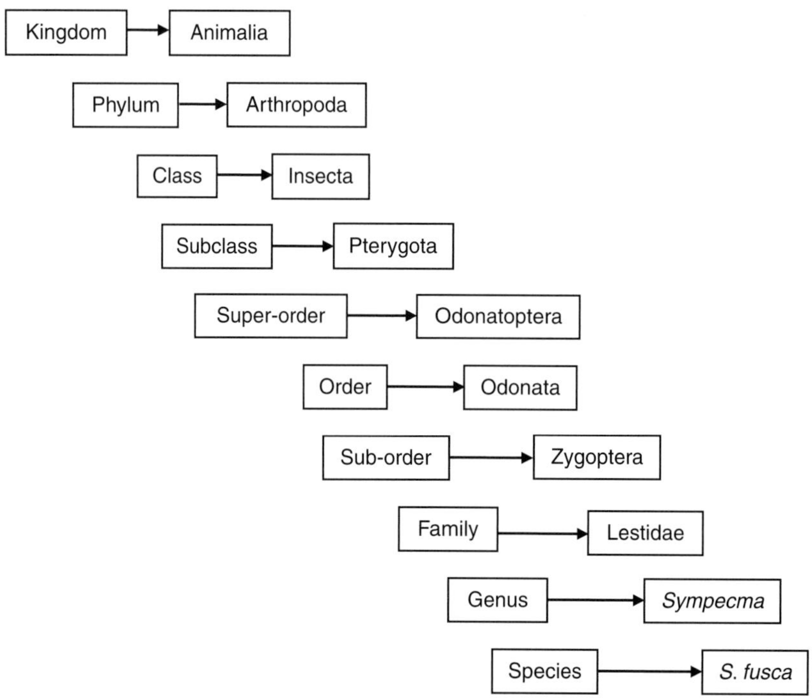

Fig. A.3. Scientific classification of a damsel fly. Courtesy of Kalyani Paranjape.

Table A.1. Illustration of some important pest orders and their significance. (Table courtesy of Kedar Champhekar.)

Phylum	Class	Order	Approx. no. of species recorded	Importance
Arthropoda	Insecta	Odonata (dragonflies, damselflies)	4,500	Nymphs eat mosquito larvae; adults are natural insect predators
		Blattaria (cockroaches)	3,500	Household pests
		Mantodea (mantids)	1,800	Natural predators of other insects
		Isoptera (termites)	1,900	Wood pests
		Orthoptera (grasshoppers, crickets)	10,000	Crop pests
		Phasmatodea (stick insects)	2,000	Crop pest of shade trees and oaks
		Mallophaga (bird lice)	2,600	Poultry pests
		Anoplura (sucking lice)	250	Human and livestock pests
		Thysanoptera (thrips)	5,000	Crop pests
		Hemiptera (bugs)	55,000	Crop and human pests, natural predators of other insects
		Neuroptera (lacewings and ant-lions)	4,000	Predators of crop insect pests
		Coleoptera (beetles)	300,000	Crop pests and predators of other insect pests of crops
		Lepidoptera (butterflies, moths)	120,000	Major crop pests
		Diptera (true flies and mosquitoes)	75,000	Human pests, pollinators
		Siphonaptera (fleas)	1,800	Human and livestock parasites
		Hymenoptera (ants, bees, wasps)	100,000	Household and crop pests, major pollinators; predators of insects
	Arachnida	Araneae (true spiders)	20,000	Natural control of insect pests
		Acarina (mites and ticks)	10,000	Crop, human and livestock pests
Nematoda	Secernentea	Tylenchida	>60 families	Crop pests
Mollusca	Gastropoda		35,000	Crop pests

Annuals

A plant that completes its life cycle from the stage of emergence to that of physiological maturity in 1 year or less is called an annual. In the present context of pests, there are many weeds that are annuals, either summer annuals (plant emergence in spring or early summer, growth during summer and seed production in autumn) or winter annuals (plant emergence in autumn, growth during winter and seed production by spring). These plants grow rapidly and produce a large number of seeds, which form their main source of propagation. The seeds germinate only if the weather conditions are favourable. Shepherd's purse (*Capsella bursa-pastoris* (L.)), pepper grass (*Lepidium sativum*) and corn cockle (*Agrostemma githago*) are examples.

Annual weeds are a nuisance to annual crops. Weed management techniques and tillage practices can be used to keep annual weeds in control. Mowing and **hoeing** during the early vegetative stage of an annual can control its rampant growth. Soil-applied herbicides usually control these weeds effectively by inhibiting seed germination.

Annuals are commonly managed by pre-emergence herbicides (preferably a fortnight before seed germination). The rule of thumb is summer annuals should be sprayed in the spring and winter annuals should be sprayed in the autumn. Simultaneously, it is important to clear the soil surface of mulch and leaves before applying herbicides for effective soil contact. Irrigating the land after application can lead to deeper penetration.

Antagonism

See **Pesticide antagonism**

Anthocorid bugs

See **Biocontrol of crop pests – predators**

Anthraquinone biorepellent

Anthraquinone, $C_{14}H_8O_2$ (IUPAC name also being anthraquinone), is a plant-derived bird repellent. It is also known as 9,10-dioxoanthracene. Although stable to hydrolysis in water, it is sensitive to light with DT_{50} in aqueous solution of about 9 min. Anthraquinone is found in a wide variety of plants including teak (*Tectona grandis*) and perennial rye grass (*Lolium perenne*). Its CAS Registry Number is 84-65-1.

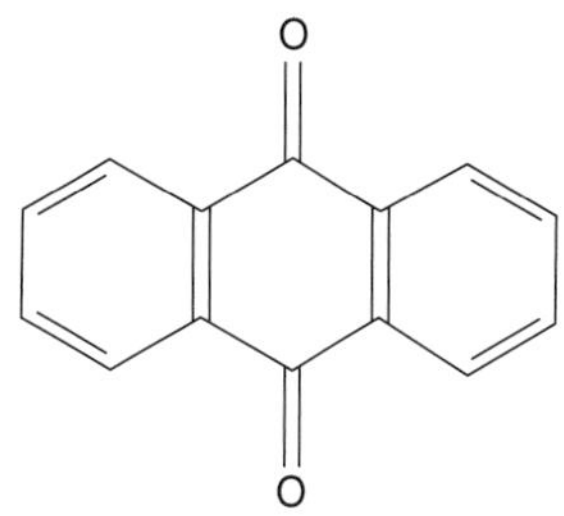

Anthraquinone

Commonly used on non-crop sites, anthraquinone finds application in keeping birds away from airstrips, nurseries and golf courses. This chemical is also used in seed treatment cases, where fungicidal and insecticidal compounds on seeds may prove lethal to birds.

Anthraquinone is available as dry and wettable powders (WP), water-dispersible powder for slurry seed treatment (WS) and as liquid concentrate. This chemical is available for example as Corbit and Flight Control. Spray applications may be repeated if required (50% concentrations are adequate).

Anthraquinone is considered non-toxic to mammals (WHO classifies it as a Class U toxin). The acute oral and acute dermal LD_{50} in rats is >5000 mg/kg. It is not irritating to the eyes and skin. The LD_{50} in Japanese quails is >2600 mg/kg. The LC_{50} in worms is >1000 mg/kg for dry soil. However, anthraquinone is observed to be a carcinogen, and is thus a PAN Bad Actor chemical.

Anthraquinone does not persist in the environment, with the compound degrading rapidly in soils, possibly due to microbial activity. DT_{50} is up to 10 days. It is eliminated within 48 h.

Anthraquinone is registered for use in the USA.

Antibiotics

The discovery of antibiotics was a major milestone in the development of medical science. It began with the discovery of penicillin (produced by strains of *Penicillium notatum* and *Penicillium chrysogenum*) in 1928 by Alexander Fleming.

An antibiotic is a chemical substance produced by microorganisms, with the capacity to inhibit the growth of, and even to destroy, bacteria and rickettsiae. Viruses are not controlled by antibiotics. Antibiotics may be broad spectrum (active against a wide range of pathogens) or specific (active against one or a few classes). Antibiotics used for clinical purposes are commonly produced by the large group of Gram-positive bacteria, called actinomycetes (streptomycetes). Some antibiotics commonly in use are penicillin, erythromycin, streptomycin, **tetracycline**, **amphotericin B**, nystatin, griseofulvin and mitomycin C.

Antibiotics interfere with the vital processes of targeted bacteria, such as protein synthesis, nucleic acid synthesis, cell wall formation and maintenance of the cell membrane.

Antibiotic compounds are widely used to control fungal pathogens, some examples being **blasticidin-S**, **kasugamycin**, cycloheximide and **validamycin**.

Repeated use of antibiotics on crops can lead to the development of resistance among pathogen species. Reasons for the development of resistance include the detoxification of the enzyme inhibited by the antibiotic. A typical case is that of resistance to the antibiotic **streptomycin**. Streptomycin is commonly used on apple, pear, tomato and pepper, against such diseases as fire blight, canker, dieback and spot diseases. Resistance can be managed by the use of alternative compounds: for example, in the case of streptomycin resistance, tetracycline may be used. The use of antibiotics may work against **beneficial microorganisms**. Handlers exposed to these compounds might be vulnerable to diarrhoea and allergies.

Anticarsia gemmatalis nucleopolyhedrovirus (MNPV or AgMNPV) as a bioinsecticide

Anticarsia gemmatalis nucleopolyhedrovirus (AgNPV) is a virus belonging to the family *Baculoviridae* and is a bioinsecticide. The culture is prepared on the velvet bean caterpillar (*Anticarsia gemmatalis* Hubner). This insecticide is effective on sugarcane borer and the velvet bean caterpillar and may, hence, be used on sugarcane and soybean crops.

Also known as MNPV or AgMNPV, the virus' mode of action begins when it is eaten by the caterpillar. Once inside, the guttural enzymes dissolve the protective protein matrix of the occlusion bodies, allowing the virus particles to enter the insect's haemolymph. These occlusion bodies are released upon the death of the caterpillar, thus spreading the infection.

MNPV is available as dried, ground powder as Polygen and Mutagen. The formulation is viable if stored in a cool and dry place. It may be used with other compounds, as long as they are not strong oxidizers, acids, bases or chlorinated water.

Anticoagulant rodenticides

Rodenticides are classified into anticoagulants and non-anticoagulants. Rodenticides that are made of chemical compounds that block the synthesis of vitamin K and inhibit clotting factors II, VII, IX, etc. (known as **prothrombin**, proconvertin, etc.), thus causing death by internal haemorrhaging are called anticoagulant rodenticides. Such compounds, discovered in the 1940s, include **warfarin**, **pindone**, **diphacinone**, **chlorophacinone**, **bromadiolone** and **brodifacoum**.

Anticoagulants, which are commonly classified as coumarins, 4-hydroxy coumarins and 1,3-indandiones, are chronic toxins. Upon consumption of a lethal dose, the pest starts to exhibit symptoms of toxicity like weakness and internal bleeding. These compounds have shorter elimination half-lives and have to be administered in higher concentrations (e.g. 0.005–0.1%), over a period of time and at repeated dosages. This group of anticoagulants is referred to as first generation, since these compounds are required to be consumed continually and show their effects only after 1 week (warfarin, **coumatetralyl** and **chlorophacinone**). While this delayed action removes the fear of bait, it also develops resistance in the rodents, as in warfarin's case.

The development of more potent anticoagulant rodenticides, effective on species resistant to first-generation compounds, came to be known as the second-generation anticoagulants. These rodenticides are derivatives of 4-hydroxycoumarins (e.g. 4-hydroxy-1-benzothiin-2-one). Second-generation anticoagulants can cause death even in one feed, if the lethal dose is consumed during the first feed. These compounds also display chronic effects (i.e. they bring upon death even after administration of multiple doses of very small quantities over a time period). **Brodifacoum**, **difenacoum**, **flocoumafen** and **bromadiolone** are a few examples of second-generation anticoagulants.

All anticoagulants are advised to be administered over a period of 15 days or so, to ensure the effective control of the entire population of a region. Anticoagulant baits are formulated as ready-to-use grain baits, pellets or wax baits for controlling Norway (brown) rats (*Rattus norvegicus*) and house mice (*Mus musculus*). Wax baits or paraffin-type baits are suitable for roof rats. Norway rats are susceptible to first-generation anticoagulants, while house mice are susceptible to all types of anticoagulants.

Anticoagulant compounds are toxic to all warm-blooded animals to varying degrees as a result of which, non-target animals such as dogs and cats may be killed.

The second classification of rodenticides, the non-anticoagulants, is also called acute toxicants. These are based on **arsenic**, **red squill** and phosphorus. **Zinc phosphide**, **cholecalciferol** (vitamin D_3) and **bromethalin** are currently three acute toxicant rodenticides that are registered and commercially available. Zinc phosphide, though requiring **pre-baiting**, is more fast-acting than the others. Acute toxicants are more effective than anticoagulants.

Anticoagulant rodenticides are generally combined with an antibiotic. This addition is used to suppress the symbiotic microflora in the animal gut or intestine, which can be a source of vitamin K.

Antidote

An antidote is a substance that reduces or counteracts the effect of poisoning. Pesticides contain several poisonous compounds. Treatment of victims can differ based on for example their age, gender and medical history. Antidotes are the first step towards reducing the toxic effects of a poison. Normally, mild acids, mild alkalis, salt water, milk and activated charcoal are used to nullify the effect of the poison and induce vomiting, and protect the throat and stomach lining.

Antidotes work in distinct ways. Chemical antidotes mix or interact with the poison and produce a harmless chemical. A mechanical antidote prevents absorption of the poison. A physiological antidote acts in three ways. It may bind to the poison through metabolic antagonism, combine with it to aid elimination, or compete with it for the receptor site.

Some examples of antidotes are given here. Activated charcoal, diazepam, iphenhydramine hydrochloride and sodium barbital help in allaying poisoning brought about by pyrethroids, chlorinated hydrocarbons and **nicotine** compounds. Calcium gluconate acts as an antidote against chlorinated hydrocarbons and mercurial compounds, while dimercaprol reduces toxicity caused by arsenic, copper, mercury and other heavy metals, as well as methyl and ethylene bromide.

It should be noted that nothing should be forced into the mouth of an unconscious person. Alcohol must be avoided as an antidote. An antidote is effective against only one kind of poison.

Anti-foaming agents

See **Adjuvants**

Antimetabolites

Antimetabolites are chemicals that inhibit and block a natural cell metabolite, thus disturbing normal metabolic functioning of a living system. An antimetabolite chemically and structurally resembles a normal cell metabolite, misleading the enzymatic substrate to act upon it, leading to abnormal compounds. Insect sterilants and many other pesticidal groups work as antimetabolites by inhibiting enzyme function. They interfere with the synthesis of nucleic acids, by competing with normal nucleotides thus preventing the formation of the correct sequence of nucleotides in nucleic acids (Table A.2).

Table A.2. Classes of antimetabolites including their major groups. (Table courtesy of Leena Achari.)

Antimetabolites	Major groups
Purine analogues	5-Bromouracil, 5-fluorouracil, 5-fluoroorotic acid, 2-thiouracil, 2,6-diaminopyrine
Pyrimidine analogues	2-Amino-4-methyl-pyrimidine, 1,2,3,6-tetrahydro-2,6-dioxo-4-pyrimidine acetic acid
Folic acid analogue	Aminopterin, amithopterin, methotrexate, folic acid
Pyrines, pyrimidines	Uracil, uracil 5-bromo, uracil 4,5-diamino-sulfate
Other	Methionine, thymine, tyrosine, methotrexate

Anti-sporulants

See **Fungicides**

Aphamide chemosterilant

Aphamide (N,N′-bis[bis-(aziridin-1-yl)phospharyl]-N,N′-dimethylethane-1,2-diamine) is a **chemosterilant**. The molecular formula of this chemosterilant is $C_{12}H_{26}N_6O_2P_2$. It is effective against the housefly, mosquito, fruit fly, German cockroach and citrus red mite.

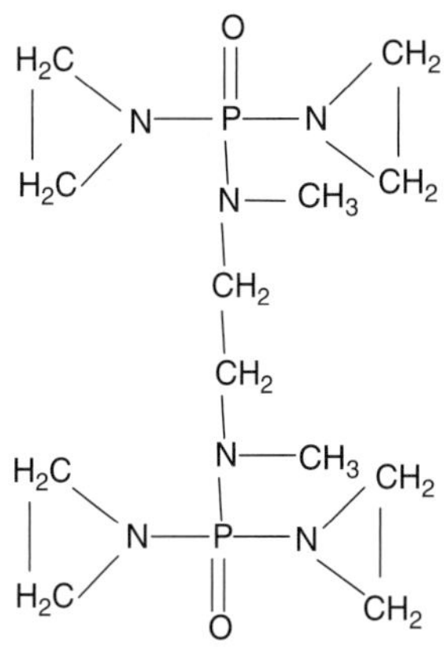

Aphamide

Aphid

See **Pest – aphid**

Aphid midge

See ***Aphidoletes aphidimyza*** **(aphid midge) as a biocontrol agent**

Aphid mummy

See ***Aphidius*** **wasps as biocontrol agents**

Aphidius colemani as biocontrol agent

See ***Aphidius*** **wasps as biocontrol agents**

Aphidius ervi as biocontrol agents

See ***Aphidius*** **wasps as biocontrol agents**

Aphidius matricariae as biocontrol agent

See ***Aphidius*** **wasps as biocontrol agents**

Aphidius wasps as biocontrol agents

The wasp genus *Aphidius* is a large entomopathogenic group, specific to aphids. Harnessing this property to bring about biocontrol of aphid pests has met with significant success among crops and landscape gardens alike. These wasps parasitize both nymph and adult aphids, by sniffing out honeydew from long distances. Important species demonstrate their pathogenic traits more or less along the lines mentioned below.

Aphidius wasp species are 2 mm long, dark/blackish slender insects with long and prominent antennae and wing venation. The female lays an egg inside an aphid in the aphid colony, causing the aphid to eat abnormally more and produce more honeydew. The larvae that hatch out of the eggs start eating the aphid from inside, weakening the aphid. While still inside the aphid, the larva starts producing a cocoon around itself, making the aphid swell and turn leathery and brownish from the outside, this form being termed an **aphid mummy**. Within 1 week, the adult wasp emerges from the carcass of the aphid by boring a hole through it.

Aphidius colemani is grown to control, for example: (i) the cotton aphid (*Aphis gossypii*); (ii) the peach potato aphid (*Myzus persicae persicae*); (iii) the melon aphid and the green peach aphid (*Myzus persicae*); and (iv) the tobacco aphid (*Myzus persicae* var. *nicotianae*). Cucumber, sweet pepper and greenhouse plants, including rose and chrysanthemum, are protected. Once released, these wasp cocoons take 1 week to settle in, with an effective introduction of one or two per square metre. Complementary biocontrol agents such as the gall midge (***Aphidoletes aphidimyza***) and ladybirds (***Adalia bipunctata***) may help, and also application of **pirimicarb**. Cucumber is most effectively protected.

Aphidius matricariae is very effective on green peach aphid. With each female capable of laying many eggs, control on aphids takes only up to 10 days. This is a robust species that can survive in cold weather; hence year-long protection to plants is possible. Chrysanthemums are best protected by these parasitic wasps. The release rate is around 50 wasps per 500 cuttings.

The ideal application rate *of A. matricariae* depends on the infestation and may vary from one or two per square metre. Wasp cocoons (inside mummified aphids), introduced in a potentially aphid-prone area, take about 7 days to establish.

Aphidius ervi protects greenhouse crops from many aphid species such as potato aphid and foxglove aphid. The aphid mummies in these cases are greyish brown. These wasps may be released at the rate of around 20–60 adults/m^2. Biocontrol is most effective when cocoons are introduced at the pre-aphid stage. **Green lacewings** and **aphid midges** can complement the task of biocontrol.

Aphidoletes aphidimyza (aphid midge) as a biocontrol agent

The aphid midge (*Aphidoletes aphidimyza*) is an insect whose larvae are predators of some aphid species.

The adult midge fly lays eggs in aphid colonies in crops and orchards. The larvae are bright orange, slug-like creatures. The larvae paralyse several aphids by injecting a toxin into them. The midge then feeds on the aphids (up to 65 aphids/day) by sucking out the body mass. Crops such as potatoes, cauliflowers, cabbage, greenhouse crops, orchard crops and berries are protected.

Release rate of the aphid midge is ten cocoons per square metre, after rearing them on aphid-infested plants along with peat and vermiculite. Release time should be planned in order to avoid cool weather, which does not suit aphid growth.

Pesticide sprays can directly affect midge larvae and adults. In addition, cool temperatures are unfavourable for aphid growth, making larvae release-time a critical factor. Midges can be released along with the wasp ***Aphidius matricariae***, for better results.

Apholate chemosterilant

Apholate, $C_{12}H_{24}N_9P_3$ (2,2,4,4,6,6-hexakis(aziridin-1-yl)-1,3,5,2λ^5,4λ^5,6λ^5-trizatriphosphinine), is a **chemosterilant**. It is a white-coloured substance, which is partially soluble in acetone, alcohol and chloroform. It remains stable in cool and dry conditions.

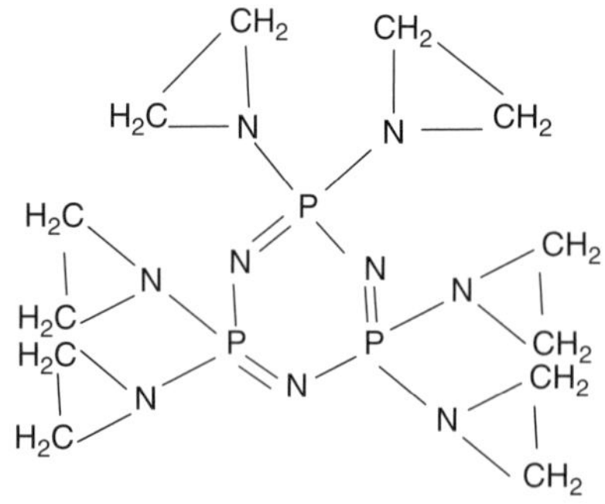

Apholate

Apholate is used against houseflies, mosquitoes, the stable fly, screw worm fly, Mexican bean beetle, German cockroach and citrus red mite. It is effective against houseflies at 0.5% when applied through food. Excessive use of apholate may act as sterilant to the **braconid** biocontrol agent (*Bracon phylacteophagus*), which is a parasite of some agricultural pests.

Apiculture

See **Bees**

Apophytes

Apophytes are the plants that have left their native homes and have settled on new sites. The migration of plants has been taking place ever since man started travelling to distant continents and the beginning of urbanization. Natural disasters, erosion and widespread cultivation are some reasons that have forced plants to migrate.

Having changed their surroundings, apophytes escape their natural enemies and thus have the potential to become invasive and noxious in the new location.

An ergasiophyte is a plant which has deliberately been introduced to a region for cultivation.

Application equipment – hand sprayers

Hand sprayers are used to apply small quantities of pesticides to a pest. These sprayers usually have an air pump which compresses air into the tanks and pressurizes the spray mixture. The pressure slowly drops as the liquid is sprayed. Hand-sprayer equipment is also called a push–pull hand-pump sprayer. It has a capacity of usually up to 10 l. Hand sprayers usually operate at 350 kPa (50 psi) or less. Hand sprayers are economical, simple to operate and easy to clean. Some types of hand sprayers are mentioned below.

Some of the tanks or cans are sealed with compressed gas or pesticide solutions and are not reusable. They are called pressurized can sprayers. In a trigger pump sprayer, the pesticide and diluents are forced through the nozzle when the trigger is pulled, and these kinds of sprayers are used to control household pests. These may have a capacity of up to 5 l. A hose-end sprayer causes a fixed rate of pesticide to mix with the water flowing through the hose to which it is attached. It is used for small pest control projects. A compressed-air sprayer is a hand-carried sprayer which operates under pressure, created by a self-contained manual pump. With capacity of 4.5–13.5 l, it is commonly used by homeowners and pest control operators. A backpack sprayer is a self-contained unit and is carried on the operator's back. It has a capacity of up to 20 l. Then, there are battery-operated sprayers and portable sprayers which have a connection to a larger spray tank, making it convenient for the handler to operate. A bucket or trombone sprayer is another hand sprayer that consists of a double-action hydraulic pump which is operated with a push–pull motion. The pesticide is sucked into the cylinder and pushed out through the hose and nozzle with the stroke. It has a capacity of less than 25 l.

Small motorized sprayers are semi-automated spraying equipment that may be of low or high pressure. These sprayers are used for smaller outdoor patches such as small orchards and nurseries. Small motorized sprayers have a larger capacity than hand sprayers. Small motorized sprayers are of two types: (i) estate sprayers; and (ii) power wheelbarrow sprayers. Estate sprayers are mounted on a two-wheel cart with handles for pushing and may have a capacity of about 200 l. A power wheelbarrow sprayer is a simple powered version of the manually operated sprayer, with a tank size of about 50 l. Some hand sprayers are shown in Fig. A.4.

Application equipment – larger power-driven sprayers

The larger power-driven sprayers are generally of three types: (i) low-pressure sprayers; (ii) high-pressure sprayers; and (iii) air-blast sprayers.

Low-pressure sprayers are designed to distribute diluted pesticides over large areas and are usually mounted on tractors or trucks (Fig. A.5). The capacity of these sprayers is >4000 l. They are used for pesticide application on huge tracts of agricultural land, forestry or parks.

High-pressure sprayers are used to spray dense foliage or the tops of tall trees. Such applicators are usually called 'hydraulic' sprayers with a tank capacity of around 4000 l.

Air-blast sprayers use a combination of air and liquid rather than liquid alone to deliver the pesticide to the surface being treated. These sprayers are built with a high-speed fan. The air blast shatters the drops, as they come out of the nozzle, into fine droplets and transports them to the target. The air blast is directed to one or both sides as the sprayer moves forward, or the spray is delivered through a movable nozzle (Fig. A.6).

Fig. A.4. A few types of hand sprayers. (a) Knapsack (backpack) sprayer; (b) mini-fogger; (c) portable sprayer; (d) turbo battery sprayer; (e) herbicide sprayer; (f) pressure sprayer. Courtesy of Padgilwar Corporation Ltd, India. With permission.

Fig. A.5. Field sprayer applying low-volume sprays. Courtesy of: www.stewardshipcommunity.com; 'Water-sensitive paper for monitoring spray distribution', Syngenta Crop Protection AG, CH-4002 Basel, Switzerland. With permission.

Fig. A.6. An air-blast sprayer used in orchards. Courtesy of: www.stewardshipcommunity.com; 'Water-sensitive paper for monitoring spray distribution', Syngenta Crop Protection AG, CH-4002 Basel, Switzerland. With permission.

Application equipment – sprayers

Pesticides are effective only when they are applied in a proper way. The point of contact and the size of the drop are very important to ensure prevention of harmful effects. The application of pesticide or a nutrient to a crop in the form of fine droplets is known as spray application.

Spraying is the most common means of pesticide application. While aerial spraying can successfully apply a small amount of pesticide to a large area, other types of sprayers (such as hand-held and motor-driven sprayers) are used to administer specific doses of pesticides to crop foliage or the soil. Important fungicides commonly sprayed are **Bordeaux mixture**, ethirmol, **copper oxychloride**, **zineb**, **maneb**, zinam, **captan**, **chloroneb**, **carbendazim**, benomyl, **tridemorph** and difolatan. Bacterial diseases may be taken care of by spraying copper compounds, but copper-sensitive crops are sprayed with **streptomycin** and aureofungin and other such bactericides. Spraying is advocated mainly for two reasons, namely: (i) when the action is protective in nature; and (ii) when uniform and thorough coverage is sought.

A sprayer is the most common type of equipment used to apply pesticides to crops. A sprayer consists of a tank from which the liquid is pumped through nozzles under pressure. The structure of the nozzle differs with the kind of spray desired. It may be fan shaped, which is used for low- and medium-volume spraying, or cone shaped, which is used for high-volume spraying.

Ultra-low volume (ULV) sprayers use special pesticide concentrates. They are used in agricultural, ornamental, turf and forestry situations, and some other pest control operations. Spinning disc sprayers use a special type of nozzle which spins at a high speed and breaks the liquid into uniformly sized droplets by centrifugal force. The droplets may be carried to the target by gravity or by an airstream created by a fan. Power to spin the nozzles is provided by an electric or hydraulic fan. The size of a spinning disc sprayer ranges from a small hand-held type (Fig. A.7) to large tractor-mounted and trailer-mounted units. Such a sprayer is suitable for herbicide applications. Recirculating sprayers are used to apply contact herbicides to weeds that are taller than the crop in which they are growing. Recirculating sprayers use small quantities of pesticides and protect susceptible non-target plants from injury. This type of sprayer is relatively costly. In the electrostatic sprayer,

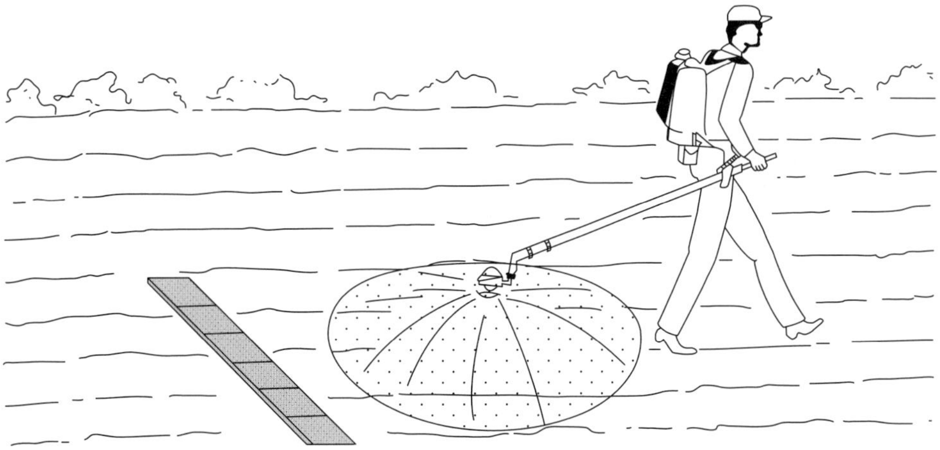

Fig. A.7. Spinning disk sprayer. Courtesy of: www.stewardshipcommunity.com; 'Water-sensitive paper for monitoring spray distribution', Syngenta Crop Protection AG, CH-4002 Basel, Switzerland. With permission.

systems are designed to reduce drift and apply less pesticide per unit area. The pesticide is charged with a positive electric charge as it leaves the nozzles. Plants have a natural negative charge, so the pesticides are attracted to the plants.

Sprayers are classified according to their application capacity (see Table A.3).

Table A.3. Broad classification of sprayers based on application capacity. (Table courtesy of Shubhangi Borse.)

Types of sprayer	Subtype of sprayer	Capacity/tank size
Hand sprayers	Pressurized cans	Less than 1 l
	Trigger pump sprayer	0.568–4.55 l
	Hose-end sprayer	1 l
	Push–pull hand pump sprayer	1 l or less
	Compressed air sprayer	4.5–13.65 l
	Bucket or trombone sprayer	22.5 l or less
Small motorized sprayers	Estate sprayers	Tank size 225 l
	Power wheelbarrow sprayer	Tank size 54–81 l
Larger power-driven sprayers	Low-pressure sprayers	Tank size <225–4450 l
	High-pressure sprayers	Tank size 90–2250 l
	Air-blast sprayers	Tank size 450–4500 l
Other sprayers	Ultra-low volume (ULV) sprayers	
	Spinning disc sprayers	
	Recirculating sprayers	

Calibration of crop protection equipment is absolutely necessary for any equipment. Calibration is a method of adjusting or correcting an instrument with a standard. This adjustment should be done from time to time, since valves and nozzles tend to become ineffective with use. Calibration ensures that the quantity of a toxin to be applied remains constant. Calibration precedes the spraying activity. Spray volume and application rates can be worked out by accurate sprayer calibration.

The volume of a spray depends on the discharge rate of the nozzle (spray liquid used), spraying or operating speed and spray swathe. Spray volume can be theoretically worked out using the formula shown in Eqn A.1 (see Eqn A.1 at bottom of the page).

However, there is another practical farmer-friendly formula too which is shown in Eqn A.2 (see Eqn A.2 at bottom of the page).

Aquatic weeds

A group of plants that have evolved and adapted themselves to survive and flourish in water are called aquatic plants or hydrophytes. Many such plants interfere with irrigation waters or fish culture and may promote disease. Many species of floating plants, immersed and submerged weeds and algae are aquatic weeds.

Algae form a major part of a water body and are known to increase the dissolved oxygen levels in water. The cyclical imbalance in the oxygen content and increased rates of feeding may cause uncontrolled algal blooms.

Species of duckweed (common and greater), water lettuce, mosquito ferns and common water hyacinth (*Eichhornia* spp.) live on the surface of water and are called floating hydrophytes or floaters. The size of these plants varies. They float on inflated air-filled hollow leaf stalks and the surface of the leaves are covered in a water-resistant substance that assists with floating.

Banker weeds or immersed weeds have long and hollow stems and they root themselves in the soil near the banks. They usually grow across banks of small water bodies. Examples are torpedo grass, water grass and pennywort.

$$\text{Application rate (l / ha)} = \frac{\text{Constant (600)} \times \text{Nozzle discharge rate (l/min)}}{\text{Effective swathe width (m)} \times \text{Spraying speed (km/h)}} \quad \textbf{(A.1)}$$

$$\text{Application rate (l / ha)} = \frac{\text{Plot area}}{\text{Marked area}} \times \text{Spray volume used for marked area} \quad \textbf{(A.2)}$$

Sinkers are rooted in the bed of the water body and are often fully submerged. However, they grow rapidly and slow down the normal flow of water. Hydrilla and Brazilian elodea are examples.

There are several subtypes of weeds besides the floaters, banker weeds or sinkers discussed above.

Several herbicides such as **endothal**, **diquat**, **glyphosate** and **2,4-D** control aquatic weeds.

Arsenic compounds as rodenticides

Arsenic compounds constitute a class of highly toxic rodenticides. Arsenic compounds are more toxic than elemental arsenic. Arsenites are often more toxic than arsenates. The acute minimal lethal dose of arsenic compound is between 50 and 200 mg/kg/day in humans. Arsenic trioxide is estimated at 500 times more toxic than pure arsenic.

Among pests, death can be brought about by the oral LD_{50} of less than 50 mg/kg b.w. Circulatory failure may occur due to the consumption of arsenic. Symptoms of poisoning are cramps, convulsion and bloody diarrhoea.

Arsenical inorganic insecticides

Arsenic salts were the first synthetic inorganic compounds used to control chewing insects such as potato beetle (*Leptinotarsa decemlineata*) and the cotton boll weevil (*Anthonomus grandis*). The As^{3+} ion of arsenic compounds, specifically, inhibits sulfhydryl groups in enzymes involved in cell respiration.

Arsenic compounds are toxic on ingestion, inhalation and through the skin. They react with the thiol groups in the tissues and enzymes and interfere with the phosphorylation process. These compounds affect, for example, the nervous system, blood vessels, liver and kidney. The pentavalent arsenic compounds like sodium arsenate (NaH_2AsO_4) and trivalent arsenic compounds such as copper arsenite and calcium arsenite are some examples. Copper arsenate ($Cu_3(AsO_4)_2.4H_2O$), a blueish-green powder, soluble in aqueous ammonia and diluted acids, is used as a herbicide, a fungicide and a rodenticide. It is formulated as wettable powder (WP).

Calcium arsenate ($Ca_3(AsO_4)_2$), formulated as flowable powder, was used to control boll weevil on cotton and fruit.

Lead hydrogen arsenate ($PbHAsO_4$) formulated as a water-soluble acid, is used against the potato beetle and the gypsy moth. It is relatively safe on plants, although toxic to applicators since it is absorbed through the mucous membrane.

Copper acetoarsenite ($Cu(CH_3COO)_2.3Cu(AsO_2)_2$), also known as Paris Green, a bright green pigment, first used as a wall paint, is a rodenticide and an insecticide.

Sodium arsenite ($NaAsO_2$) and arsenic trioxide are used in water dips to control cattle ticks.

Zinc arsenate is an insecticide, effective on potatoes and tomatoes (see Arsenical inorganic compounds at bottom of the page).

Aspergillus flavus isolates (fungus) as biocontrol agents

Aspergillus flavus (order Eurotiales) is a commonly occurring, soil-borne fungal species that causes moulds on stored grains and groundnuts. It is also mainly responsible for the production of **aflatoxins** that can have fatal effects on humans and cattle.

Two isolates of *A. flavus*, Link AF36 and NRRL 21882, are non-aflatoxin-producing strains. They are used to inoculate cotton and groundnut crops as they prevent the growth of other **mycotoxin**-producing strains by competing with them and dominating the field.

The inoculum is commercially available as Alfa-guard. A commonly acceptable dose is 11 kg/ha at the pre-blossom stage. It is important for moisture to be present during application. The product stores well in cool and dry containers. The product is not recommended to be combined with other fungicides. The isolates of *A. flavus* are not toxic to humans or the environment.

as-Triazine herbicides

The triazine family of herbicides is a large group of compounds, divided into *s*-triazines and asymmetrical triazines, also known as *as*-triazines or triazinones.

Triazine herbicides are selective photosystem II electron transport inhibitors. They are commonly selective herbicides that control broadleaved and grassy weeds in, for example, cereal crops, soybean and pineapple.

as-Triazine herbicides, which include metribuzin, are taken up better by plants at higher soil pH. They have a shorter soil life than symmetrical triazines and are relatively less volatile than triazines. **Metribuzin** selectively protects potato, soybean and lucerne, while ethiozin is effective against annual grasses.

as-Triazines can injure tolerant plants if applied in moist conditions, since detoxification of the compound becomes slower.

Asulam herbicide

Asulam, $C_8H_{10}N_2O_4S$ (methyl sulfanilylcarbamate), is a sulfonamide carbamate herbicide. Its action is similar to *N*-phenylcarbamate herbicides. It is a colourless substance, soluble in water (5 g/l), ethanol (120 g/l) and dimethylformamide (>800 g/l). It is stable at normal storage conditions. The CAS Registry Number of asulam is 3337-71-1. Asulam forms salts with sodium and potassium, giving asulam-potassium and asulam-sodium (with a molecular formula $C_8H_9N_2NaO_4S$, IUPAC name sodium methyl (EZ)-sulfanilylcarbonimidate and CAS Registry Number of 2302-17-2). Asulam is made by the reaction of 4-aminobenzenesulfonamide, after protecting the amino group, with ethyl chloroformate in the presence of a base.

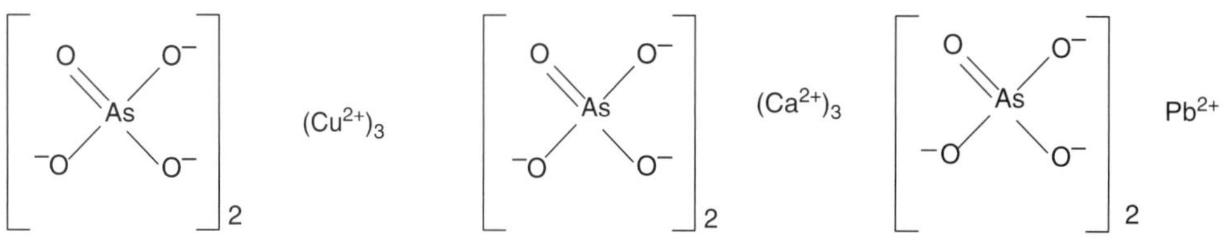

Copper arsenate Calcium arsenate Lead hydrogen arsenate

Some arsenic compounds

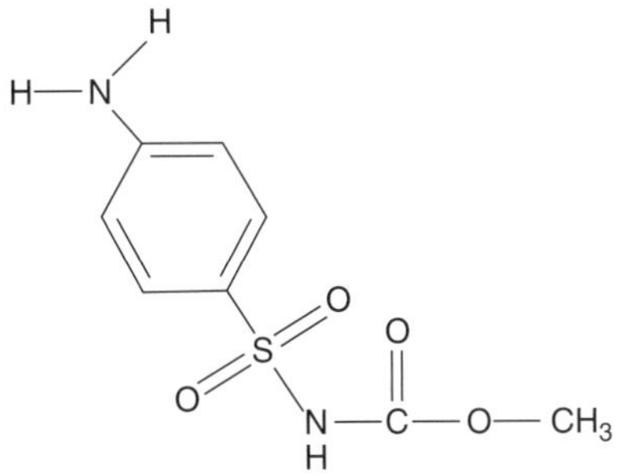

Asulam

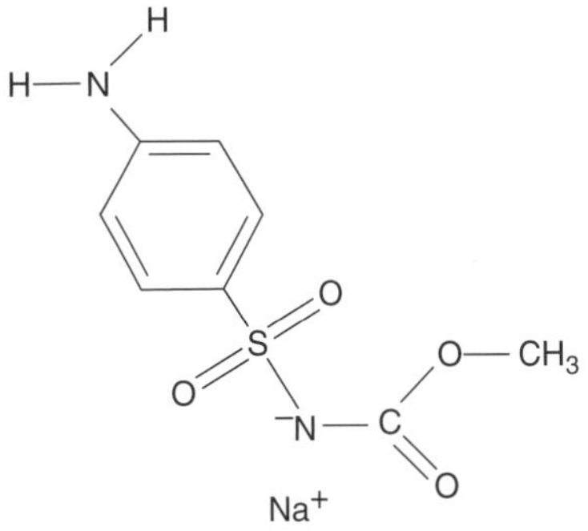

Asulam-sodium

Asulam is a systemic, selective herbicide that inhibits photosynthesis at photosystem II. It inhibits microtubule assembly as well as meristematic growth. Since it is translocated to the roots, it is ideal for perennial weeds. Asulam causes a slow **chlorosis** in susceptible plants. It controls many species of thistle and bracken fern in forests and pastures. It also controls wild oats and many perennial grasses and broadleaved weeds in spinach and lucerne. Depending on the crop, the application rate ranges from 1 to 10 kg/ha. It is formulated as a soluble concentrate (SL). Asulam is available as Asulox.

The toxicity of asulam towards all mammalian, avian and fish species is of low order of magnitude. The acute oral LD_{50} in rats, mice and rabbits is >4000 mg/kg. Inhalation LC_{50} (6 h) in rats is >1.8 mg/l air. The ADI is 0.36 mg/kg b.w./day. The acute oral LD_{50} in mallard ducks and pigeons is >4000 mg/kg. It is not toxic to bees. Asulam is a possible carcinogen, although it has been classified as a Class U toxin (unlikely to be hazardous) by WHO.

When orally administered, asulam is eliminated within 3 days. The half-life in field conditions is 1 week. It has low soil persistence and is commonly found in the top 15 cm of soil. Microbial degradation begins with the loss of amino group.

Asulam is registered for use in many countries in Africa and in Australia and the USA.

Asulam-sodium herbicide

See **Asulam herbicide**

Atomic absorption spectrometer (AAS)

Atomic absorption spectroscopy involves converting a sample into an atomic vapour and measuring the absorbance of this atomic vapour at a specific wavelength (λ), which characterizes a specific element. The measured absorbance is proportional to the concentration and is compared under the same experimental conditions to that of a reference sample of known composition.

An atomic absorption spectrometer (AAS) is used extensively in the analysis of trace elements in soils as well as elemental contaminants in polluted waters.

The advantages of an AAS include its high sensitivity for as many as 60 elements measurable in the ppm range. Besides, the measurement is simple, rapid, relatively low cost and highly specific (with very little interference).

An AAS (Fig. A.8) has four basic components. They are: (i) a light source, emitting the spectrum of the desired analyte elements; (ii) a sample atomization cell, such as a flame or graphite tube furnace; (iii) a monochromator to isolate the desired source emission line; and (iv) a detector/readout system, to allow measurement of change in the source line intensity by sample atom absorption. Many commercial instruments based on either a single- or double-beam design are available.

Fig. A.8. Atomic absorption spectrometer (AAS) 'novAA 350'. Courtesy of Analytik Jena AG, Germany. With permission from Corina Steidl.

Atrazine herbicide

Atrazine, $C_8H_{14}ClN_5$ (6-chloro-N^2-ethyl-N^4-isopropyl-1,3,5-triazine-2,4-diamine), is a **chlorotriazine** herbicide. It is a colourless powder, soluble in water (33 mg/l), methanol, ethylacetate, acetone, toluene and hexane. The compound is relatively stable in neutral to weakly acidic or weakly alkaline medium. It has a CAS Registry Number of 1912-24-9.

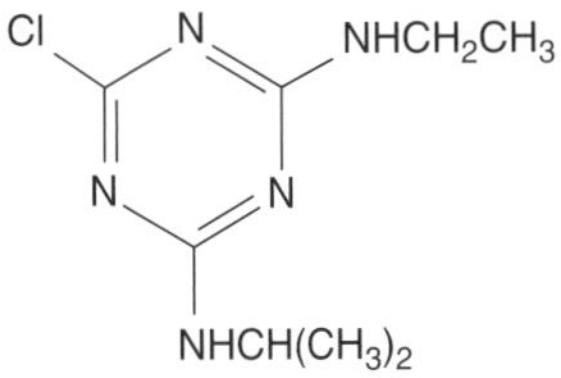

Atrazine

Atrazine, a systemic, selective herbicide, is a photosynthesis inhibitor that is quickly absorbed and translocated acropetally. It is used in the pre- and post-emergence control of a wide variety of weeds in maize, sorghum, sugarcane and grasslands.

The commonly available formulations of atrazine are flowable liquid and water-dispersible granules (WG). It can be combined with a variety of herbicides and even fertilizers, and may be applied at up to 4 kg/ha. Known trade names are Marksman, Atrazina and Atrazol.

The oral LD_{50} for atrazine is 1900–3090 mg/kg in rats and 1330–3990 mg/kg in mice. The dermal LD_{50} values in rabbits and rats are >3000 mg/kg. Atrazine is practically non-toxic to birds (LD_{50} >2000 mg/kg in mallard ducks). Atrazine is slightly toxic to fish and other aquatic life. Atrazine is not toxic to bees and slightly toxic to humans, with symptoms of poisoning being pain in the stomach, diarrhoea and vomiting, and irritation to the eyes, mucous membranes and skin. However, it is a groundwater contaminant, a carcinogen and a suspected endocrine disruptor. Thus, it has been classified as a PAN Bad Actor chemical. It is also phytotoxic to vegetables, potato, soybean and groundnut, if not used as per the label instructions. Atrazine has been classified as a restricted use pesticide (RUP) by the EPA due to its potential for groundwater contamination. It has maximum contaminant level (MCL) of 0.003 mg/l. It has a **reference dose (RfD)** of 0.018 mg/kg b.w./day. NOEL (2 years) in rats is 3.5 mg/kg/day. According to WHO toxicity classification, it is a Class U toxin (unlikely to be hazardous).

Although subject to microbial degradation (by *Pseudomonas* species and other bacterial species), atrazine has a soil half-life that ranges from 13 to 261 days, depending on the type of soil, and hence it is highly persistent. The degradation route begins with the hydrolysis of the C–Cl bond, finally yielding cyanuric acid (unstable with respect to ammonia and carbon dioxide). 2-Chloro-4-hydroxy-6-amino-1,3,5-triazine is another degradation product. In mammals, atrazine is metabolized rapidly and completely by oxidative dealkylation of the amino groups. Atrazine accumulates in the tissues of sensitive plants and results in **chlorosis** and death.

Atrazine is a restricted chemical in many countries including those of the EU. However, because of its effectiveness it is registered in many countries including India, the USA, Australia, Canada and many African countries. (See also **Chlorotriazine herbicides**.)

Attractants

Any device or means designed to lure an organism to a specific spot where it gets killed or repelled is called an attractant. Insect attractants are important for an environmentally acceptable pest control strategy. Attractants include food baits, pheromones, odours and simple sticky traps. For example, termites are attracted to the smell of decaying wood.

Attractants also help with monitoring insect populations in order to decide when control measures should be initiated. Sex pheromones are designed to create confusion during mating periods.

Chemical attractants can be used along with simple, inexpensive sticky traps. Two general types of attractants have been studied widely. These are: (i) sex pheromones; and (ii) plant kairomones. **Sex pheromone** attractants are generally volatile esters, alcohols and aldehydes that facilitate sexual communication and mating over long distances and can be perceived in nanogram to microgram quantities. These substances may be customized to attract male insects which can then be killed. Sex pheromones are used in intra-species chemical communication.

Kairomone attractants are volatile chemicals released by plants, which serve as major cues for phytophagous insects. Increasing numbers of kairomones have been identified and found to be useful insect attractants. The most successful mixtures are eugenol (with a molecular formula $C_{10}H_{12}O_2$, an IUPAC name of 4-allyl-2-methoxyphenol and a CAS Registry Number of 97-53-0) with **geraniol** (with a molecular formula of $C_{10}H_{18}O$ and a CAS Registry Number of 106-24-1) or phenyl ethyl propionate (with a molecular formula of $C_{11}H_{14}O_2$, an IUPAC name of 2-phenylethyl propanoate and a CAS Registry Number of 122-70-3). Kairomones are used in inter-species chemical communication.

Augmentative or inundative method of biological weed control

See **Biological weed control agents – application methods**

Aureofungin biofungicide

Aureofungin, an antibiotic fungicide, available as Aureofungin A ($C_{59}H_{86}N_2O_{19}$) and Aureofungin B ($C_{57}H_{85}NO_{19}$). It has a CAS Registry Number of 8065-41-6. It was discovered and developed by The Research Centre of Hindustan Antibiotics Limited (Pimpri, Pune, India).

Aureofungin, a systemic, broad-spectrum agent that quickly translocates in living plants, is isolated from the mycelia of *Streptoverticillium cinnamomeus* var. *terricola*. The antibiotic belongs to the polyene (heptaene) macrolide group of antibiotics. The fungicide can effectively control seed-borne diseases in many crops including sugarcane, sugarbeet, maize, tobacco, potato, tomato, wheat, citrus, coconut, cabbage, banana, apple, ginger, grapes, groundnut and cumin. Aureofungin is also an effective agent against post-harvest diseases (storage diseases) on vegetables and fruits.

Aureofungin is compatible with several insecticides, fungicides and plant hormones. It is needed in very low doses and displays rainfastness. Seeds remain viable while fending off saprophytic fungi during storage. This biofungicide is non-toxic and non-polluting.

Aureofungin is registered for use in India, although it is not registered by the EPA for use in the USA.

Autographa californica nucleopolyhedrovirus as a bioinsecticide

Autographa californica nucleopolyhedrovirus (AcNPV or AcMNPV) is a virus belonging to family *Baculoviridae*. This microorganism has been named after it was first isolated from the alfalfa (lucerne) looper *A. californica*. However, there are more than 30 lepidopteran species that the virus invades. The insecticidal properties of AcNPV are commonly used on maize, vegetables, fruit crops and ornamentals and any other crop the looper infests.

Being a **baculovirus**, the insecticidal action of AcNPV becomes evident only on its being ingested. The protective virus protein matrix of the polyhedral occlusion bodies dissolve in the insect larval mid-gut allowing the entry of the virus into the insect's circulatory system. The virus invades all the cell types of the larva and multiplies, thus releasing large numbers of occlusion bodies.

AcNPV is available as a liquid concentrate, available under the trade name Gusano. A foliar spray can keep the insect pest under control. The formulation stores well in cool, dry, dark and sealed containers (with temperatures below 32°C). AcNPV may be combined with non-repellent insecticides to increase efficacy.

The AcNPV virus is non-toxic to humans, non-target organisms and the environment.

Auxins

Auxins (originating from Greek 'auxein' meaning 'to increase') are a group of naturally occurring hormone-like substances essential for plant growth and regulation. Other major plant hormones include **cytokinins**, **gibberellins**, **ethylene** and **abscisic acid**. Auxins include **indole-3-acetic acid (IAA)**, 4-chloro-indoleacetic acid, phenylacetic acid (PAA) and indole-3-butyric acid (IBA). The most active endogenous auxins contain the indole ring system. Although IAA is very important, the non-indole analogue indole butanoic acid (IBA) has become very popular (especially as a root-inducing agent for cuttings).

While the most important function of auxins is perceived to be cell elongation (meristems), other functions include the growth of fruits and stems, root initiation, organ differentiation and abscission. Agricultural chemists have used this property to synthesize herbicides that induce so much abnormal growth in the plant that it becomes fatal. As the physiological action depends on the auxin structure and concentration, the biosynthesis, conjugation and degradation of the compounds can be monitored. The various constituents of cell walls, while keeping the cell wall intact, prevent the sudden growth or expansion of the cell volume. That is why auxins are necessary. For reference, a plant cell wall is shown in Fig. A.9. The plant cell wall is made up of cellulose, proteins and lignin and is very firm. This prevents any sudden expansion of cell volume, unless there is auxinic activity.

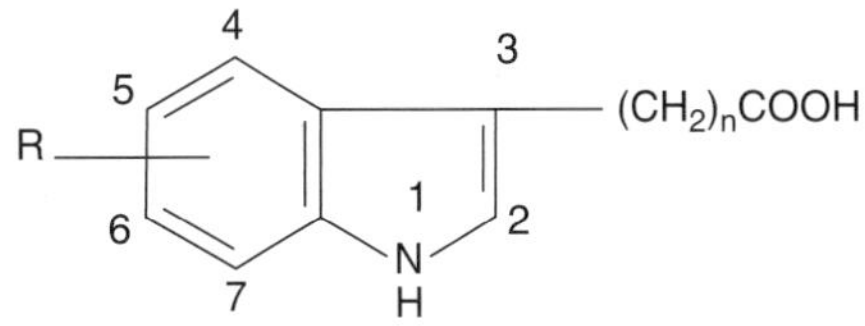

Indole auxins showing the indole ring and position of compounds: (1) n = 1, R = H – indole acetic acid; (2) n = 2, R = H – indole propionic acid; (3) n = 3, R = H – indole butanoic acid (IBA); (4) n = 1, R = 4-Cl – 4-chloro indole acetic acid; (5) n = 1, R = 7-Cl – 7-chloro indole acetic acid; (6) n = 1, R = 5-Br – 5-bromo indole acetic acid

Synthetic auxinic herbicides can be categorized into four classes, namely: (i) **phenoxy acids**; (ii) **benzoic acids**; (iii) pyridine acids; and (iv) quinolinecarboxylic acid. (The auxinic herbicide, benazolin, is a benzothiazolacetic acid compound.) The basic structure of auxinic herbicides includes a carboxylate, butyrate, acetate or propionate moiety bound to a ring structure by an ether linkage, and commonly halogenated with chlorine.

The effect of an auxin is generally felt at a site different from where it is produced. Auxinic herbicides are effective even at low concentrations.

Auxinic herbicides are effective on dicot species because these chemicals easily destroy their phloem. Grasses, however, are not so susceptible (translocation is not so rapid). The herbicidal effect in grasses is seen only in the early growth stages.

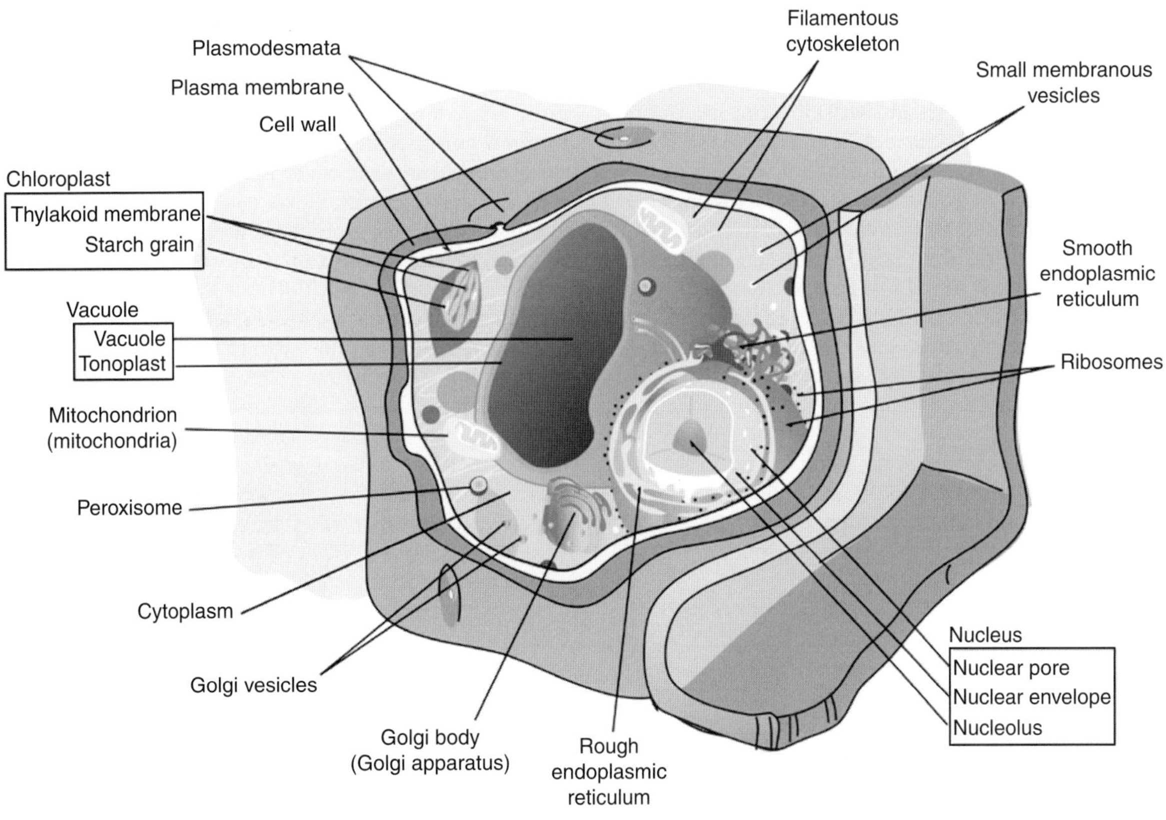

Fig. A.9. Diagram of a plant cell showing the cell wall. Courtesy of Mariana Ruiz Villarreal (http://en.wikipedia.org/wiki/Auxin). With permission.

The first auxinic herbicides used in agriculture were based on phenoxyacetic acids, and included **MCPA** ((4-chloro-2-methylphenoxy) acetic acid) and **2,4-D** ((2,4-dichlorophenoxy) acetic acid).

There is concern over persistence of auxinic herbicides in the soil. With repeated use over years, several weeds have also developed resistance to them. These include wild mustard (*Sinapis arvensis*), nodding thistle (*Carduus nutans*), yellow star thistle (*Centaurea solstitalis*), wild carrot (*Daucus carota*), kochia (*Kochia scoparia*), Italian thistle (*Carduus pycnocephalus*), Canada thistle (*Cirsium arvense*), **barnyard grass** (*Echinochloa crus-galli*), gooseweed (*Sphenoclea zeylanica*) and common chickweed (*Stellaria media*).

Avermectin

See **Abamectin bioinsecticide**

Avermectin B_{1a}

See **Abamectin bioinsecticide**

Avermectin B_{1b}

See **Abamectin bioinsecticide**

Avian repellents

Avian repellents are non-lethal substances that protect crops from birds. Scarecrows, loud sounds and other such visual or acoustic methods have been used for years to protect fields (Fig. A.10). A toxic plant *Veratrum* spp. (California false hellebore) protects seeds from birds.

Birds become a nuisance on a farm as they can access abundant food with minimum effort. Crows, pigeons, sparrows, doves and blackbirds are examples of bird pests.

Techniques to manage bird damage involve making it so time-consuming for the bird to peck off the crop that the farm ceases to be tempting. A repellent works by altering the perceived benefits of feeding on the crop.

The perceived benefits of a farm can be altered by applying a chemical repellent to a portion of the seed crop, mixing it with an equal amount of untreated seed and then broadcasting the mixture on the field. Alternatively, alternate rows can be sprayed with the desired chemical repellent. Another strategy involves offering a patch of favoured/cheaper fodder crop for birds, so that they are not attracted to the main crop.

The repellent treatment should make feeding as time-consuming as other activities such as nest building, feather maintenance and defence. At this point, the bird will move to other locations or search for other types of food.

Repellents become critical if the crop is economically valuable. Chemical repellents include **methiocarb**, **anthraquinone**, methylanthranilate, **copper oxychloride** and **4-aminopyridine**. Birds are averse to some seed fungicides including **thiram**, **ziram**, **copper oxychloride** and **copper hydroxide**. Tanin, thymol, **neem** extract, fish oil and chilli are also used as repellents. Many of them are toxic but do not always harm the birds since only small amounts are ingested. Some cause the birds pain on contact (primary repellents), while others act on the digestive system and so take longer to produce the desired effect (secondary repellents). Some repellents affect the nervous system and in these cases, birds 'learn' to avoid crop seeds after the bad experience.

The use of bird repellents depends on: (i) the acute toxicity; (ii) degradability of the compounds; (iii) phytotoxicity; and (iv) the regulations of the local environmental agency. Repellents are applied as granules or sprays.

Avicide chloralose

See **Chloralose rodenticide**

Avicide diazinon

See **Diazinon insecticide**

Avicide endrin

See **Endrin insecticide**

Avicides

An avicide is a substance that can be used to kill birds. Although not a favoured choice, avicides are used sometimes in serious cases where birds are a nuisance on crops and in public places. **Avian repellents** are normally used instead of avicides. Avicides act as a repellent to the whole flock by killing a few initial peckers and thus creating panic. These chemicals are sometimes sprayed by aircraft over bird nests. Most avicides are formulated as grain baits or powder concentrates. Most of them are also labelled restricted use pesticides (RUP) according to the EPA.

Commonly used avicides include **strychnine**, **parathion**, 3-chloro-4-methylaniline hydrochloride, starlicide or CPTH

(a) (b) (c)

Fig. A.10. Bird repellents made out of local material. (a) Bird repellent made out of a plastic bottle. (b) An old tin can and wires used to make a contraption to repel birds. (c) A scarecrow. Courtesy of Kalyani Paranjape (a and b) and Milind Paranjape (c).

(3-chloro-p-toluidine) and **4-aminopyridine**. **Chloralose** is also used as an avicide.

Starlicide, C_7H_8ClN (3-chloro-4-methylaniline) is a yellowish-brown liquid that is soluble in hot water, where the central benzene ring is modified by amine, chloro and methyl substituents. The CAS Registry Number of starlicide is 33240-95-8.

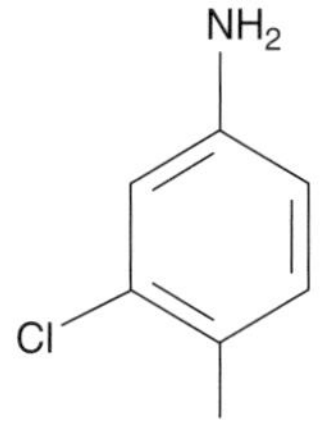

Starlicide

Starlicide is highly toxic to European starlings and gulls and kills by affecting the kidney. But the chemical does not harm non-target birds or pets and mammals. It is also known as CPTH or 3-chloro-p-toluidine. Of the avicide classes, starlicide and **fenthion** are the least hazardous.

Awareness of the local laws regarding killing of birds is necessary before using avicides. **Pre-baiting** is required before the use of avicides, so that birds are not averse to poisonous baits.

Azaconazole fungicide

Azaconazole, $C_{12}H_{11}Cl_2N_3O_2$ (1-{[2-(2, 4-dichlorophenyl)-1, 3-dioxolan-2-yl] methyl}-1*H*-1, 2, 4-triazole), is a fungicide of the triazole class (azole class). It is a beige-brown powder, soluble in water (0.3 g/l), acetone (160 g/l), hexane (0.8 g/l) and methanol (150 g/l) at 20°C. The compound is stable at about 220°C and to light under normal storage conditions. It has a CAS Registry Number of 60207-31-0.

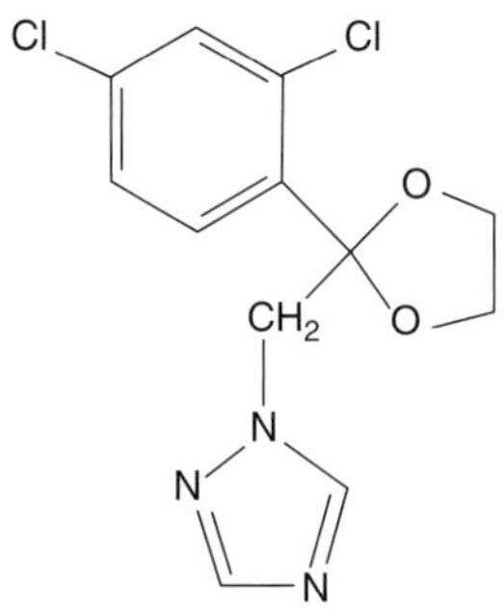

Azaconazole

Azaconazole acts by inhibiting sterol demethylation. It finds use as a storage protectant (to spray into storage boxes of fruits and vegetables) and in mushroom crops. It is also active against wood-destroying and sap stain fungi on tree wounds.

The fungicide is available as emulsifiable concentrate (EC) and oil-miscible liquid (OL) and may be combined with imazalil.

The acute oral LD_{50} among rats is 308 mg/kg, while in mice it is 1123 mg/kg. The dermal LD_{50} in rats is >2560 mg/kg. It is slightly irritating to the eyes and skin. The inhalation LC_{50} (4 h) in rats is >0.64 mg/l air. The NOEL in rats is found to be 2.5 mg/kg b.w./day. The LC_{50} in pheasant species tested is >5000 mg/kg and in rainbow trout, 42 mg/l. Azaconazole is moderately hazardous (a Class II toxin) according to WHO.

Azaconazole is registered for use in Australia and New Zealand. It is banned in the EU.

Azadirachtin bioinsecticide

Azadirachtin, $C_{35}H_{44}O_{16}$ (dimethyl (2a*R*,3*S*,4*S*,4a*R*,5*S*,7a*S*,8*S*,10*R*,10a*S*,10b*R*)-10-acetoxy-3,5-dihydroxy-4-[(1a*R*,2*S*,3a*S*,6a*S*,7*S*,7a*S*)-6a-hydroxy-7a-methyl-3a,6a,7,7a-tetrahydro-2,7-methanofuro[2,3-*b*]oxireno[*e*]oxepin-1a(2*H*)-yl]-4-methyl-8-{[(2*E*)-2-methylbut-2-enoyl]oxy}octahydro-1*H*-naphtho[1,8*a-c*:4,5-*b′c′*]difuran-5,10a(8*H*)-dicarboxylate), is a plant-derived insecticide with some nematicidal, fungicidal and anti-feedant properties. It is a greenish-yellow powder with a sulfur-garlic odour. It is practically insoluble in water. It has a CAS Registry Number of 11141-17-6 (see Azadirachtin at bottom of the page).

Azadirachtin, found in the seeds of the **neem** plant (*Azadirachta indica*, Fig. A.11) belongs to the tetranortriterpenoid class of chemicals. The seed also contains limonoids such as nimbolide, nimbin and salannin, which also show insecticidal properties. The salannin in the neem seed also acts by blocking **ecdysones**, the insect hormone responsible for metamorphosis, leading to incomplete insect life cycles. Azadirachtin targets a wide range of insects such as whitefly, leaf miners, caterpillars, aphids, thrips, beetles and soil pathogens. It is also effective against fungal pathogens that cause **powdery mildews**. Thus, it protects, for example, cabbage, potato, cotton, tea and coffee.

Using azadirachtin as an oil-mixed spray makes it work better. It can be sprayed often. Azadirachtin is compatible with other crop protective agents, except **fenvalerate**.

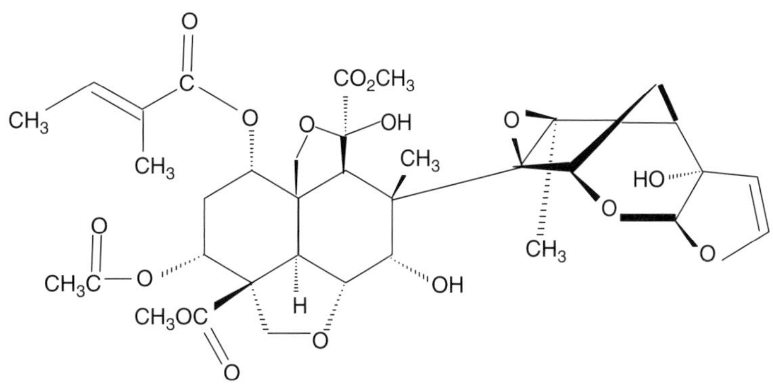

Azadirachtin

Fig. A.11. A neem tree. Courtesy of Kalyani Paranjape.

Azadirachtin is not considered toxic to mammals. The acute oral LD_{50} of azadirachtin in rats is >5000 mg/kg. The NOEL (daily) in rats is 5 mg/kg while that in mallard duck is 1–16 mg/kg. It causes slight eye irritation. The ADI is 0.1 mg/kg b.w./day. It belongs to toxicity Class IV (practically non-toxic), according to the EPA classification. The soil half-life is from 4 to 25 days. Azadirachtin degrades within 100 h when exposed to light and water.

Azadirachtin is registered for use in many countries including India, the USA, Canada, Australia and South Africa. (See also **Insect growth regulators**.)

Azafenidin herbicide

Azafenidin, $C_{15}H_{13}Cl_2N_3O_2$ (2-(2, 4-dichloro-5-prop-2-ynyloxy-phenyl)-5, 6, 7, 8-tetrahydro-1, 2, 4-triazolo [4, 3,α] pyridin-3 (2*H*)-one), is a triazolone herbicide. It is a rust-coloured solid, soluble in water (17 mg/l). It photolyses rapidly in an aqueous system. It has a CAS Registry Number of 68049-83-2. Azafenidin is made by reaction of *N*-(2-piperidyl)-*N′*-(2,4-dichloro-5-oxy-methylacetylene) hydrazone with methylchloroformate.

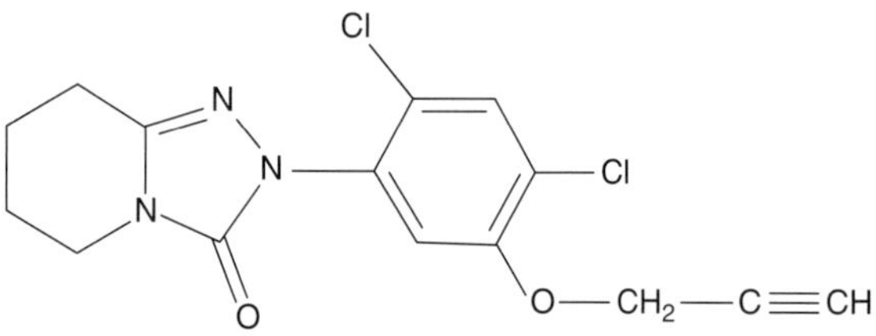

Azafenidin

Azafenidin is a selective pre- and post-emergence herbicide for broadleaved weeds and grasses (most important being **Bermuda grass**, nut sedges, knotweed and chickweed). It acts by inhibiting the porphyrin biosynthetic pathway, leading to accumulation of protoporphyrin IX, and thus cell membrane disruption. It is used on citrus, grapes and sugarbeet at the pre-emergence stage.

Azafenidin is used as a spray. The soil half-life ranges from 1 week to more than 6 months.

Azafenidin causes moderate eye irritation making it critical to wash exposed body parts thoroughly with soap and water after handling. The chemical should not be applied directly to or near water bodies. It may harm non-target plants. Azafenidin has an acute oral LD_{50} among rats of >5000 mg/kg and an acute dermal LD_{50} among rabbits of >5000 mg/kg. It is slightly toxic for rainbow trout (96 h acute LC_{50} is 33 mg/l). Ingestion of azafenidin may cause liver toxicity and anaemia in human beings. It is practically non-toxic for honey bees. However, as per the PAN Pesticide Database, azafenidin is a potential groundwater contaminant. It is also considered a reproductive/development toxin by the Pesticide Properties DataBase (PPDB) developed by the Agriculture and Environment Research Unit (AERU) at the University of Herefordshire, UK.

Azafenidin is registered for use in Australia but not in the USA.

Azamethiphos insecticide

Azamethiphos, $C_9H_{10}ClN_2O_5PS$ (*S*-6-chloro-2, 3-dihydro-2-oxo-1, 3- oxazolo [4, 5-b] pyridin-3-ylmethyl *O*, *O*-dimethyl phosphorothioate), is an organophosphorus insecticide. It is also known as an avian repellent. It is a colourless crystalline substance with solubility in water being 1.1g/l. It is soluble in dichloromethane, benzene, methanol and *n*-octanol, while being unstable in acids and alkalis. It is produced by the reaction of 3-chloromethyl-6-chlorobenzoxazol-2(3H)-one with the dimethyldithiophosphoric acid. It has a CAS Registry Number of 35575-96-3.

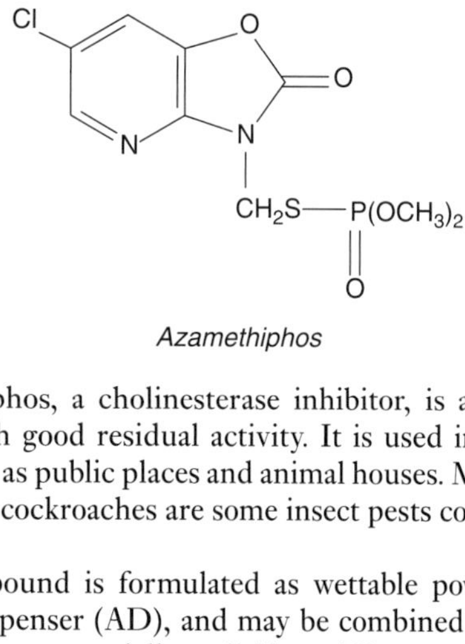

Azamethiphos

Azamethiphos, a cholinesterase inhibitor, is an effective insecticide with good residual activity. It is used in non-crop situations such as public places and animal houses. Mosquitoes, tsetse flies and cockroaches are some insect pests controlled by azamethiphos.

The compound is formulated as wettable powder (WP) and aerosol dispenser (AD), and may be combined with other pesticides. It is commercially available as Alfacron.

The insecticide is toxic to bees and highly toxic to birds and fish. However, sub-lethal doses have a repellent effect on birds, so the risk for birds is significantly reduced.

Azamethiphos metabolizes by the cleavage of phosphorothiolate C–S bond, followed by N-demethylation and opening of the oxazolinone ring. In mammals, the glucuronic acid conjugate of 2-amino-3-hydroxy-5-chloro-pyridine represents the major metabolite, followed by the corresponding sulfuric acid conjugate. In loamy sand, DT_{50} is 6 h.

Azamethiphos has an acute oral LD_{50} and a dermal LD_{50} in rats of 1180 mg/kg and >2150 mg/kg, respectively. The acute oral LD_{50} in bobwhite quails is 30.2 mg/kg. The ADI is 0.003 mg/kg b.w./day. According to WHO, it is a Class III toxin (slightly hazardous). But, since it is a cholinesterase inhibitor, azamethiphos has been classified as a PAN Bad Actor chemical.

Azamethiphos is registered in Canada, Denmark, South Africa and Tanzania.

Azinphos-ethyl insecticide

Azinphos-ethyl, $C_{12}H_{16}N_3O_3PS_2$ (*S*(3,4-dihydro-4-oxobenzo[d]-[1, 2, 3]-triazin-3-ylmethyl) *O*, *O*-diethyl phosphorodithioate), is an organophosphorus insecticide. It is a colourless needle-like substance with solubility in water being 4–5 mg/l. It is also soluble in *n*-hexane, isopropanol, dichloromethane and toluene. It is rapidly hydrolysed in alkaline media. It has a CAS Registry Number of 2642-71-9.

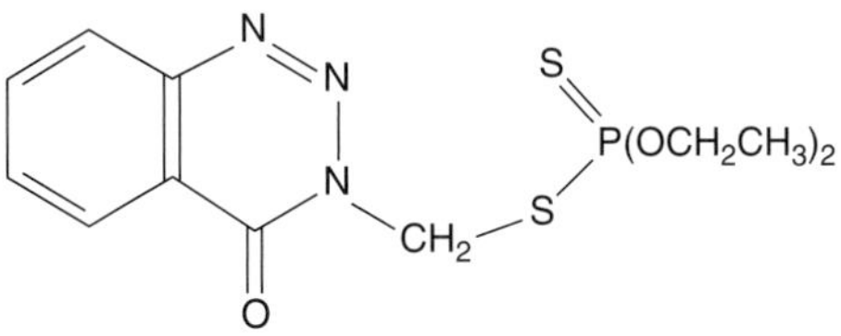

Azinphos-ethyl

Azinphos-ethyl, a cholinesterase inhibitor, is an effective broad-spectrum and non-systemic insecticide and acaricide. It is used to control aphids, caterpillars, potato bugs, beetles, boll weevils, whiteflies, boll worms, thrips and spider mites on crops like cereals, tobacco, cotton, coffee, potatoes, beet and other vegetables.

The insecticide is formulated as emulsifiable concentrate (EC), dustable/wettable powder (WP) and ULV liquids (UL). It is incompatible with alkaline materials. It is available under the trade names of Ethyl guthion and Gusathion A.

Azinphos-ethyl is highly toxic to mammals (acute oral LD_{50} in rats is 12 mg/kg) and is readily absorbed via the skin and by inhalation. WHO has classified this chemical as Class Ib (highly hazardous). It is highly toxic to some fish species and harmful to bees. Azinphos-ethyl is phytotoxic when formulated as an EC, the symptoms being russeting. Symptoms of poisoning include respiratory difficulties, gastrointestinal problems and CNS disturbances. Labels should display the signal words 'DANGER' and 'POISON' clearly.

Mammals eliminate almost all of azinphos-ethyl within 2 days via urine, with a monodesethyl compound and benzazimide being the major metabolites. In plants, metabolites include azinphos-ethyl-oxon, benzazimide, dimethylbenzazimide sulfide and dimethylbenzazimide disulfide. In soils, azinphos-ethyl is metabolized to form desethyl azinphos-ethyl, sulfonmethylbenzazimid, bis (benzazimidmethyl) ether, methylthiomethylsulfoxide and methylthiomethylsulfone. Although it does not leach in soil, it persists in water significantly. It is a PAN Bad Actor chemical since it is a cholinesterase inhibitor and a highly toxic compound.

India, Belize, South Africa and Thailand have banned the use of azinphos-ethyl. Cambodia has registered its use.

Azinphos-methyl insecticide

Azinphos-methyl, $C_{10}H_{12}N_3O_3PS_2$ (*S*-3,4-dihydro-4-oxo-1,2,3-benzotriazin-3-ylmethyl *O*,*O*-dimethyl phosphorodithioate), is an organophosphorus insecticide. It is colourless crystalline substance, soluble in water (28 mg/l), hexane and isopropanol. It has a CAS Registry Number of 86-50-0. It gets rapidly hydrolysed in alkaline and acidic media (with DT_{50} at pH 9 being 4 days). It photo degrades on soil surfaces and water, and decomposes above 200°C. It is obtained by the reaction of *N*-chloromethyl-benzazimide with ammonium dimethyldithiophosphate or directly from benzazimide by reaction with formaldehyde and dimethyldithiophosphonic acid in the presence of hydrochloric acid.

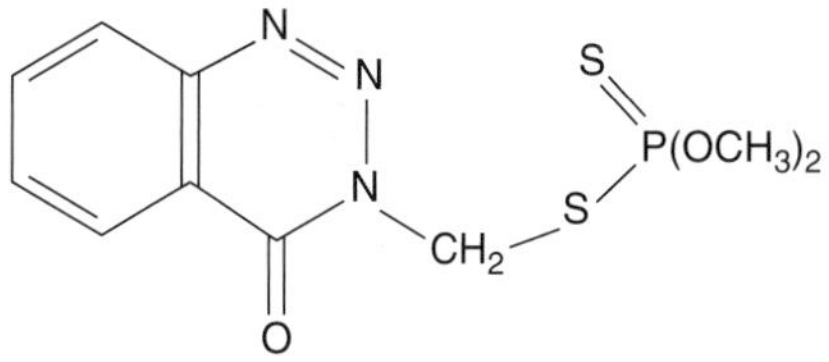

Azinphos-methyl

Azinphos-methyl, a very effective, non-systemic insecticide, is a cholinesterase inhibitor. It controls a variety of insects such as moths, a range of flies, mosquitoes, plant hoppers, aphids, thrips, weevils and beetles. The crops on which it is used include potatoes, cotton, soybeans, sugarcane, coffee, beet, and vegetables and fruits.

Common formulations of azinphos-methyl include wettable powder (WP), emulsifiable concentrate (EC) and dustable powder (DP). It is compatible with many pesticides, except those which are alkaline in nature. It is available as Gusathion-M, Cotoin-methyl and Guthion.

Azinphos-methyl is highly toxic by inhalation, dermal and ingestion routes. The acute oral LD_{50} in rats is 9 mg/kg. Dermal LD_{50} in rats is 150–200 mg/kg. It is a mild eye irritant. Inhalation LC_{50} (4 h) in rats is 0.15 mg/l air. NOEL (2 years) in rats and mice is 5 mg/kg diet. It can cause severe injury to the eyes. **Occupational Safety and Health Administration (OSHA)** has set an 8 h limit on exposure to azinphos-methyl. Azinphos-methyl is slightly to moderately toxic to birds and fish. WHO has classified azinphos-methyl as a Class Ib toxin (highly hazardous). The symptoms of toxicity among handlers include headache, irritability, vomiting, muscle cramps and dizziness. Any formulation containing above 13% of the active ingredient is classified as a restricted use pesticide (RUP) according to the EPA. Azinphos-methyl is also a potential groundwater contaminant and a cholinesterase inhibitor. These two properties along with its acute toxicity qualify this compound as a PAN Bad Actor chemical. All products must bear the signal words 'DANGER' and 'POISON'. The ADI is 0.003 mg/kg b.w./day. ECs of azinphos-methyl may cause russeting on some fruit varieties.

In mammals, more than 95% is eliminated and major metabolites are the monodesmethyl compound and benzazimide. In plants, major metabolites include azinphos-methyl oxon, benzazimide, mercaptomethyl benzazimide and cysteinmethyl benzazimide. In soil and water, degradation involves oxidation, demethylation and hydrolysis. The compound has a low mobility on the basis of the K_{OC} values and leaching studies. Soil half-life is several weeks.

Azinphos-methyl is registered in the USA, Canada, Australia, Indonesia, Thailand and South Africa. However, this chemical is banned throughout the EU and in India.

Azocyclotin acaricide

Azocyclotin (tri (cyclohexyl)-1*H*-1, 2, 4-triazol-1-yltin) with molecular formula $C_{20}H_{35}N_3Sn$ is an **organotin** acaricide. It is a colourless crystal, almost insoluble in water and slightly soluble in cyclohexane, toluene, dichloromethane and 2-propanol. Azocyclotin is produced by the reaction of tricyclohexyltin chloride with the sodium salt of 1,2,4-triazole or by the reaction of 1,2,4-triazole with bis-tricyclohexyltin oxide. It has a CAS Registry Number of 41083-11-8.

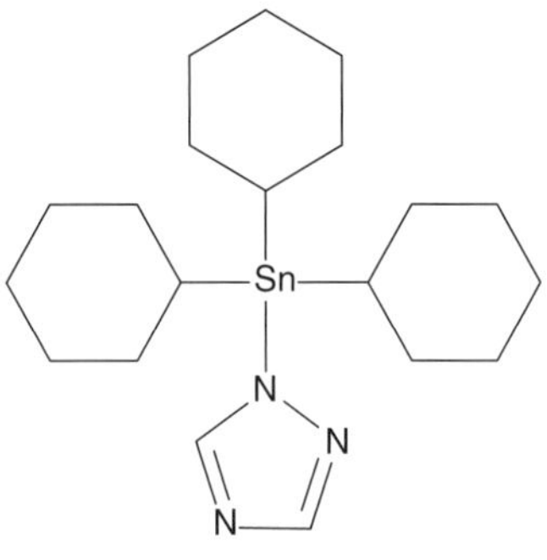

Azocyclotin

Azocyclotin is a contact, broad-spectrum acaricide, effective against phytophagous mites. It interferes with the lymphocyte function. It is used on fruit crops, especially apples.

The principal trade name of azocyclotin is Peropal. It is formulated as a soluble concentrate (SL) (at a dose of 500 g/l) and as a wettable powder (WP) (0.02 kg a.i./100 l).

The WHO classifies the acaricide as a Class II toxin (moderately hazardous). It is a skin and eye irritant. The acute oral LD_{50} in rats is >150 mg/kg. The ADI is 0.007 mg/kg/day. Azocyclotin is very rapidly hydrolysed into **cyhexatin**, whose residues are found on crops.

Azocyclotin is registered for use in Cambodia, Cyprus and Tanzania, but not in the USA.

Azoxystrobin biofungicide

Azoxystrobin, $C_{22}H_{17}N_3O_5$ (methyl(*E*)-2-{2-[6-(2-cyanophenoxy) pyrimidin-4-yloxy] phenyl}-3-methoxyacrylate), is a quinine outside inhibitor (**QoI**) fungicide (belonging to class β-methoxyacrylates extracted from **strobilurins**). It is a white solid, with solubility in water of 6 mg/l, ethyl acetate of 130 mg/l and dichloromethane 400 mg/l at 20°C. It is stable to hydrolysis at pH 5–7 and at room temperature. The CAS Registry Number of azoxystrobin is 131860-33-8. It was first discovered on fungi *Strobilurus tenacellus* and *Oudemansiella* spp. (see Azoxystrobin at top of the next column).

Azoxystrobin is a systemic and curative compound. Its QoI mode of action leads to the inhibition of mitochondrial respiration in fungi by blocking electron transfer between cytochrome b and cytochrome c_1, at the ubiquinol oxidizing site.

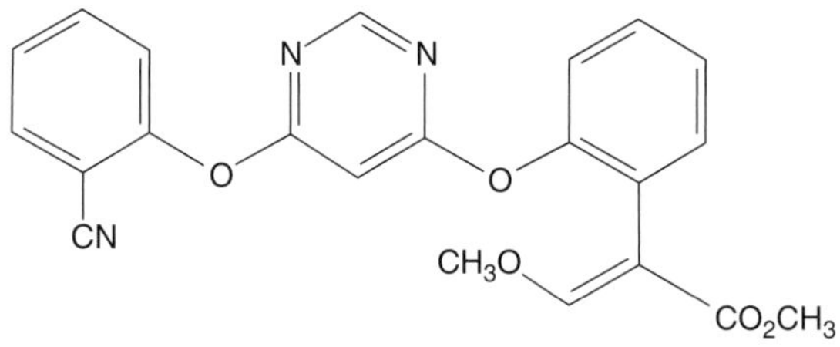

Azoxystrobin

This leads to reduced spore germination and mycelial growth. It also controls the fungal pathogenic strains that are resistant to the 14-demethylase inhibitors, **phenylamides, dicarboximides** or benzimidazoles. The fungicide bears protectant, curative, eradicant, translaminar and systemic properties. The fungicide exhibits anti-sporulant activity in pathogenic fungi and acts by inhibiting spore germination and mycelial growth.

Azoxystrobin is formulated as granules (GR), suspension concentrate (SC) and water-dispersible granules (WG). A proprietary product of azoxystrobin is ICIA 5504.

This biofungicide controls the growth of plant pathogenic fungi causing **powdery mildews** in cereals and grapes (*Erysiphe graminis* and *Uncinula necator*), rust diseases (*Puccinia* spp.), cereal leaf blotch (*Septoria tritici* and *Leptosphaeria nodorum*), rice blast (*Pyricularia oryzae*) and damping-off in rice (*Rhizoctonia solani*), **downy mildew** (*Plasmopara viticola*), **late blight** in potato (*Phytophthora infestans*), and **early blights** in tomatoes and potatoes (*Alternaria solani*). Citrus, cereals like rice, wheat and oats, potatoes, tomatoes, bananas and coffee are some of the crops that are protected by azoxystrobin.

WHO has categorized azoxystrobin as a Class U toxin (unlikely to be hazardous). The mammalian oral toxicity found in rats and mice is LD_{50} >5000 mg/kg. The oral toxicity in birds (mallard duck and bobwhite quail) is observed to be moderate, with the LD_{50} being >2000 mg/kg. It is very toxic to honey bees and to fish species. In earthworms the LC_{50} is 283 mg/kg and in various fish varieties the LC_{50} ranges from 0.47 to 1.6 mg/l. Phytotoxicity may be witnessed only in some varieties of apples. The NOEL in rats is found to be 18 mg/kg b.w./day. The UK ACP endorsed an ADI of 0.2 mg/kg b.w./day based on a **NOAEL** from a 13 week dietary study in the rat, with a 100-fold safety factor. The PAN Pesticide Database finds this compound to be a potential groundwater contaminant.

Azoxystrobin is metabolized in animals and poultry. Plants also metabolize the compound. In soils, azoxystrobin is known to produce up to six distinct metabolites. It is moderately mobile in soils, but the compound or the metabolites do not go deeper than 15 cm into the soil. In goats and hens, it is excreted rapidly, with low residues in milk, meat or eggs. Azoxystrobin degrades in the atmosphere by reacting with hydroxyl radicals.

India, Germany and the USA have registered the use of azoxystrobin.

1,4-Benzoxazin-3-ones

See **Insect growth regulators (IGRs)**

2-Butene-1,4-diyl, bis (bromoacetate) slimicide

2-Butene-1,4-diyl, bis (bromoacetate), $C_8H_{10}Br_2O_4$ (*bis*(bromoacetoxy)-2-butene), is a slimicide, bactericide and fungicide. It is a dark-brown oily liquid. It is practically insoluble in water. 2-Butene-1,4-diyl, bis (bromoacetate) is stable under ordinary conditions. Its CAS Registry Number is 20679-58-7.

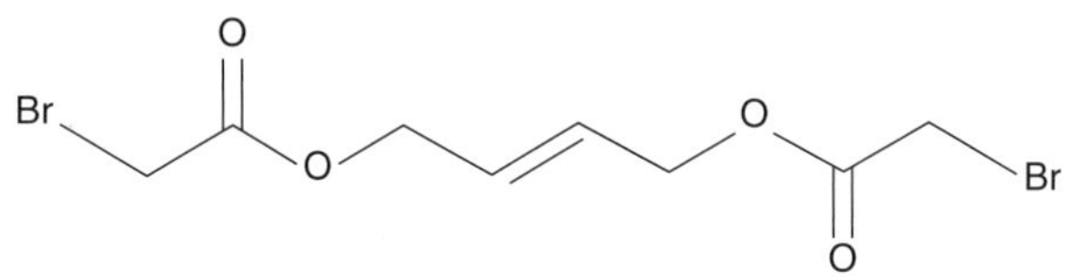

2-Butene-1,4-diyl, bis (bromoacetate)

The slimicide is used as a **biocide** to prevent slime mould, yeasts and fungi. Its key function is in water systems.

The acute oral LD_{50} of butene-1,4-diyl, bis (bromoacetate) in rats is 191 mg/kg.

6-Benzylaminopurine (BAP) as a PGR

6-Benzylaminopurine, $C_{12}H_{11}N_5$ (6-benzyladenine), also known as 6BA benzyl adenine or BAP, is a naturally occurring plant growth regulator in higher plants. It is an odourless, colourless, fine-needle-like substance, soluble in water (62.2 mg/l), methanol, acetone, *n*-hexane and toluene. It is stable in acidic, alkaline and neutral aqueous solutions. Its CAS Registry Number is 1214-39-7.

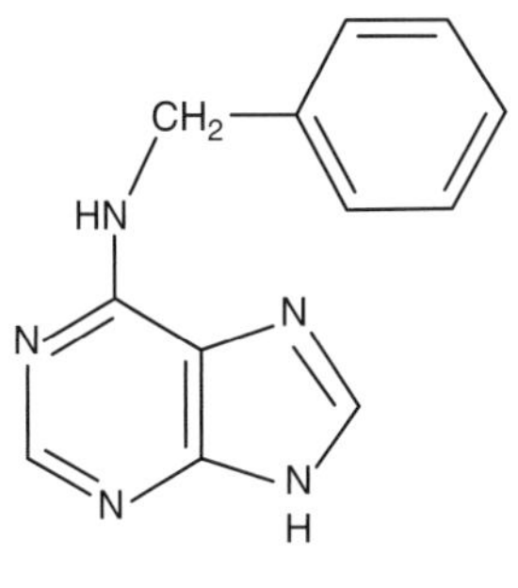

6-Benzylaminopurine

This PGR acts by stimulating RNA and protein biosynthesis, increasing cell division and lateral bud formation and delayed senescence in grapes, citrus and It also has post-harvest applications.

BAP is sprayed in large quantities, with doses ranging from 150 to 600 ppm. It may be mixed with several compounds. Important trade names include Accel.

BAP is not considered to be toxic. The acute oral LD_{50} in rats and mice are 2125 mg/kg and 1300 mg/kg, respectively. The acute percutaneous LD_{50} in rats is >5000 mg/kg. It is slightly toxic to some bird species (the 14 day LD_{50} in bobwhite quails is >2250 mg/kg) and not toxic to fish and bees. It is not a skin and eye irritant. Inhalation LC_{50} in male rats is >4.77 mg/l. The NOEL (2 year) in rats is between 5.2 and 6.5 mg/kg b.w./day. The ADI is 0.05 mg/kg/day. According to the WHO hazard classification, BAP is a Class U toxin (unlikely to be hazardous). BAP can cause irritation and even damage to eyes on prolonged or excessive contact. Hence, care has to be taken to ensure appropriate PPE is worn before handling the compound.

The PGR is metabolized and excreted through urine and faeces. Three metabolites are formed in mammals and nine metabolites in plants (soybeans, grapes, maize). It degrades in sandy loam soil and in clay loam soil in a fortnight.

Bacillus licheniformis (bacterium) as a biocontrol agent

Bacillus licheniformis is a Gram-positive thermophilic soil bacterium, belonging to the order Bacillales and family Bacillaceae. It is considered useful in biocontrol of pests. It can survive harsh weather conditions and will produce spores in favourable conditions. It grows on agar and meat extract. *B. licheniformis* isolate SB 3086, an important strain, is produced by fermentation.

The mode of action of *B. licheniformis* is the production of a small antibiotic substance along with anti-fungal hydrolytic enzymes, such as xylanase, mannase or protease. This secretion prevents the growth and development of many fungal pathogens including *Alternaria* spp., *Botrytis* spp., *Fusarium* spp., *Erysiphe* spp. and many more. Diseases such as leaf spot, **powdery mildews**, **scab** and rusts on non-food crops like ornamental plants, turf and lawns are controlled.

Isolate SB 3086 is commercially formulated as a liquid concentrate and sprayed on foliage or applied directly to the soil, after appropriate dilution. The common trade names include Green-Releaf. The product is applied at a rate of 1.5–7.5 l/ha, with fortnightly repetitions depending on the severity of the infection. This bacterium may be combined with other fungicides and insecticides. Its viable spores can remain active for a year, especially if not used with acids, alkalis or oxidizers.

B. licheniformis SB 3086 is non-toxic to humans, non-target organisms or the environment.

Bacillus mycoides (bacterium) as a biocontrol agent

Bacillus mycoides is a Gram-positive soil bacterium from the family Bacillaceae and order Bacillales. *B. mycoides* isolate J has been studied to act as a biological fungicide. With an OPP chemical code of 006516, the isolate is grown as a culture on sugarbeet.

B. mycoides isolate J controls the target fungal pest *Cercospora beticola* Sacc., which causes *Cercospora* leaf spot disease. This intervention on sugarbeet crops increases the systemic resistance to the disease. Although this pathogen is regularly treated by **mancozeb** or triphenyl tin hydroxide (TPTH), formulations of this bacterium are also used. Isolate J is applied as an aerial spray or applied directly after dilution, with fortnightly treatment, or more.

The above-mentioned bacterium is found to be non-toxic to humans and has no adverse effects on non-target organisms or the environment.

Bacillus pumilus (bacterium) as a biocontrol agent

Bacillus pumilus is a Gram-positive, rod-shaped, aerobic, endospore-forming, motile bacterium, which is found in soil and the aquatic environment. Some other common species of the genus *Bacillus* include *Bacillus subtilis*, *Bacillus sphaericus* and *Bacillus cereus*.

B. pumilus is of research and economic interest for various reasons, including: (i) its widespread presence throughout most soils and aquatic ecosystems; (ii) its resistance to harsh environmental conditions; (iii) the fungicidal and bactericidal potential of some isolates; (iv) the possible presence of toxigenic factors (presence of a toxin gene in at least one isolate); and (v) its contribution in the production of alkaline protease. In the present context of this book, *B. pumilus* is relevant for its pesticidal properties and because it is safe for humans and non-target organisms. Examples of two isolates are discussed in detail.

B. pumilus isolate GB34 effectively fends off the growth of *Pythium* spp., *Rhizoctonia* spp. and *Fusarium* spp. The isolate, found on soybean roots and in the rhizosphere, infects the fungal spores and thus inhibits their infection of the crop. Formulations of this isolate are made by fermentation.

Spores of isolate GB34 can be stored under dry and ambient conditions for 3 months, especially if they are not mixed with strong acids, bases or oxidizers. A commercial trade name is 'GB34 Concentrate Biological Fungicide'.

The fungicidal activity of *B. pumilus isolate* QST 2808 is evident in its control of *Alternaria* spp., *Erysiphe* spp., *Sclerotinia* spp. and *Botrytis* spp. which cause diseases like powdery mildews, **downy mildews**, many types of blights, **scab** and rust seen on cereals, bulb vegetables, leafy vegetables, vines and groundnuts. This soil bacterium, produced by fermentation, is also known to develop the immune system of the host plant. It is available either as a suspension or a wettable powder (WP), and is applied as a spray at the rate of approximately 8 l/ha. Isolate QST 2808 may be applied at fortnightly intervals if required.

Isolate QST 2808 may be combined with other fungicides and insecticides, except those with strong alkaline, acidic and oxidizing properties. This strain is found to be non-toxic to humans and has no adverse effects on non-target organisms or the environment.

Bacillus pumilus isolate GB34

See ***Bacillus pumilus*** **as a biocontrol agent**

Bacillus pumilus isolate QST 2808

See ***Bacillus pumilus*** **as a biocontrol agent**

Bacillus sphaericus (bacterium) as a biocontrol agent

Bacillus sphaericus is a naturally occurring, aerobic, obligate, motile, Gram-positive, endospore-forming bacterium from the family Bacillaceae and order Bacillales. It is exploited for its effective control of mosquito larvae. It has a CAS Registry Number of 143447-72-7. It is used as a biological insecticide and larvicide against the mosquito family Culicidae, especially *Culex* spp., *Aedes* spp., *Psorophora* spp. and *Anopheles* spp. It is rapidly hydrolysed under even mild alkaline conditions.

The mode of action of the *B. sphaericus* begins with the production of parasporal, proteinaceous, crystal inclusion bodies during sporulation. This material is toxic to the larvae, upon ingestion. Sustained results are seen if the bactericide is applied from the first instar up to early fourth instar of the mosquito life cycle.

The regular use of *B. sphaericus* in, for example, sewage waters and dams, can keep mosquito populations under control. It is formulated as water-soluble granules (SG) and applied by hand-held application equipment to water bodies. Trade names include SPIC Biomass and Sphericide. Compatible with other major pesticides except copper-based fungicides or algal control agents, *B. sphaericus* has shown no adverse effects in field use on non-target organisms (LD_{50} on rats >5000 mg/kg). *B. sphaericus* is registered for use in the USA and Canada.

Bacillus subtilis (bacterium) as a biocontrol agent

Bacillus subtilis is a Gram-positive, rod-shaped, aerobic, endospore-producing bacterium that is prevalent in soil, water and in rotting plant matter. It belongs to the order Bacillales and family Bacillaceae. The endospore keeps it safe from unfavourable environmental conditions. The CAS Registry Number is 68038-70-0.

B. subtilis is responsible for producing enzymes such as amylase and protease as well as antibiotics like bacillomyin B. These properties make this species useful in the textile and leather industries and for preparing medicines for skin infections. *B. subtilis* is by-and-large safe for humans, animals and the environment. A cross section of the organism is shown in Fig. B.1.

As already stated, *B. subtilis* is found to be environmentally safe and non-toxic to birds and rats. The acute oral LD_{50} in bobwhite quail is >5000 mg/kg, and the LC_{50} in bees and worms is 5663 ppm (5 days, oral) and >1000 mg formulation/kg dry soil, respectively. Some of the many isolates of this useful biocontrol agent are discussed here.

In the present context, *B. subtilis* has great significance because of the fungicidal and insecticidal properties of its isolates.

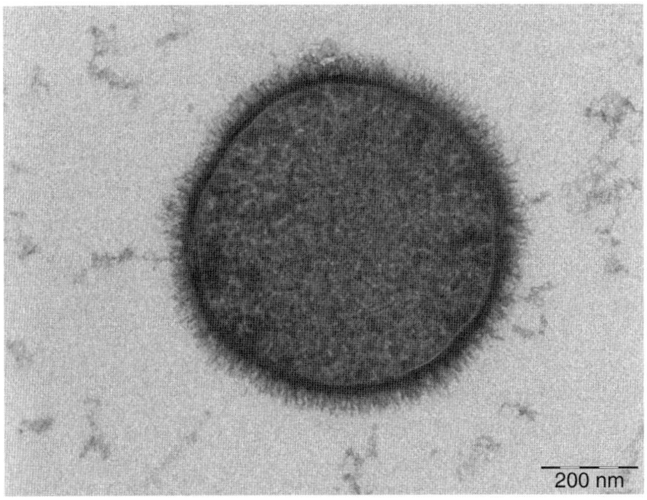

Fig. B.1. Transmission electron micrograph of a *Bacillus subtilis* cell in cross section (scale bar = 200 nm). Courtesy of Allon Weiner, The Weizmann Institute of Science, Rehovot, Israel. 2006. Wikimedia Commons released into the public domain by the author (transferred from en.wikipedia), with permission.

The general mode of action is the production of enzymes that are toxic to the pathogens. The lipopeptide in the bacterium disrupts the cell membranes of the pathogen. Then the bacterium colonizes the leaf surface and the rhizosphere, thus inhibiting the growth of the fungus.

Some strains of *B. subtilis* are particularly used against mosquitoes, while most control prominent fungal species including *Rhizoctonia solani*, *Fusarium* spp., *Alternaria* spp., *Aspergillus* spp. and *Botrytis cinerea* in cotton, legumes, aubergines, tomatoes and strawberries.

Common formulations of *B. subtilis* isolate products are water-dispersible granules (WG) and wettable powder (WP), seed-treatment formulations and liquid formulations. If stored in dry and sealed containers, they remain viable for 2 years at room temperature. The formulations can be used in combination with **metalaxyl** and **quintozene** but not with copper-based products or with bactericides like **streptomycin**. Highly alkaline or acidic media are detrimental to the growth of *B. subtilis* and its isolates. Most *B. subtilis* products may be applied as required (three to four times) with a definite degree of safety.

The soil bacterium *B. subtilis* var. *amyloliquefaciens* isolate FZB24 is, for instance, used to control *Rhizoctonia* and *Fusarium* among ornamentals.

Formulations of WG are sprayed, after adequate dilution, on the plants or on the soil. Rhizo-Plus is a known trade name of this isolate.

B. subtilis isolate GB03, another soil-dwelling strain, is exploited for its anti-fungal properties on many crops such as soybeans, cotton, groundnuts, wheat and barley. Some of the seedling fungal pathogens controlled thus include *Fusarium* spp., *Pythium* spp. and *Rhizoctonia* spp. Common trade names include Quantum 4000.

Produced by fermentation process, viable spores of isolate GB03 are available as seed treatments, liquid formulations and water-dispersible granules (WG). This isolate can be used with other fungicides except those which are strongly acidic or alkaline.

B. subtilis isolate HAI-0404 is another example of a biological fungicide that controls pathogens, particularly *B. cinerea*, on vegetables.

B. subtilis isolate QST 713 is used widely for the control of many fungal species such as *B. cinerea*, *Venturia* spp., *Phytophthora infestans*, *Uncinula necator*, *Podosphaera leucotricha* and *Erysiphe* spp. Bacterial species from the genus *Xanthomonas* are also controlled. Grapes, stone fruits, peppers, tomatoes, potatoes, onions, carrots, members of the cucurbit and cabbage families, and other vegetables are protected by this isolate. Many ornamentals are also covered by this product.

Bacillus thuringiensis (*Bt*) as a biocontrol agent

The bacterium *Bacillus thuringiensis* (*Bt*), named after the German town Thuringen where it was first discovered, is a Gram-positive, motile, soil microorganism that forms an endospore whenever necessary. It produces proteinaceous parasporal, crystal-inclusion bodies that are insecticidal in nature, when ingested by the pest. These bodies react with the proteases of the insect gut (mid-gut membranes), leading to production of a toxic substance called delta endotoxin, causing the death of the insect. Large numbers of such crystals are identified on several strains of *B. thuringiensis*, and this substance acts on coleopterans (beetles) and dipterans (flies) thus making this species a very suitable biological insecticide. Colorado potato beetle (*Leptinotarsa decemlineata*) occurring on potato crops is effectively controlled by *Bt*. *Bt*, unlike other insect pathogenic microbes, is a poor infectious agent.

Some *B. thuringiensis* products are prepared by fermentation. These products use genetically modified cells of *Pseudomonas fluorescens*, which die but carry the delta endotoxin in the microcapsule. Such products come either in capsule form or as granules and may be mixed with fungicides or **acaricides** that are not strongly acidic or alkaline in nature. A well-known trade name is M-Trak. The *Bt* encapsulated delta endotoxin is not toxic to humans or non-target organisms. (See also **Biocontrol of crop pests – entomopathogenic microorganisms**.)

Backpack sprayer

See **Application equipment – hand sprayers**

Bacteria

Bacteria are microscopic single-celled, mostly motile, prokaryotic organisms. They do not have a definite nucleus. These cells may vary in shape, for example they may be rod-shaped, spiral or filamentous. They reproduce by dividing in two. This asexual division may occur as often as once every 30 min. Bacterial species can be divided into two large groups depending on their cell wall structure and this grouping is done by the colour on the cell wall, on reaction with chemicals. Invented by Hans Christian Gram, bacteria are grouped as either Gram-negative or Gram-positive. The former is stained pinkish red while the latter becomes violet-blue.

Based on the source of food (energy) bacteria are referred to as autotrophic or heterotrophic bacteria, facultative or obligate bacteria, saprotrophic or parasitic bacteria.

Soil bacteria generally thrive best when they have: (i) adequate oxygen and a relative humidity of over 60%; (ii) temperatures ranging between 20 and 40°C; (iii) adequate organic matter and a large amount of exchangeable calcium ions (Ca^{2+}); and (iv) a pH of 6–8.

Bacteria are largely responsible for the decay and decomposition of organic matter as well as for recycling elements such as carbon, oxygen, nitrogen and sulfur. In general, the presence of bacteria in soil is the sign of a good agricultural soil.

Bacteria infect all animals, humans and plants, with many species succumbing to bacterial infections. Most infections on plants occur in humid or cool weather, especially when light intensity and duration are low. The tropical weather pattern is another environment conducive to bacterial infections. Most of these pathogens penetrate plant tissues when there are plant wounds (via injury, trimming, cutting). Common symptoms include leaf spots, fire blights, blights, blotch and wilting. Some important genera of bacterial pathogens include *Xanthomonas*, *Pseudomonas*, *Erwinia*, *Ralstonia* and *Spiroplasma*. Bacterial blight of rice, soybean and others, and bacterial **wilt** are some known diseases.

Bacterial infections are of a distinct type. Some species penetrate the plant tissue and destroy cell parts thus causing some cells to die. This leads to the rotting of surrounding plant parts. Some species produce thick polysaccharide sugars that block the xylem, leading to deprivation of fluids to plant parts. Sometimes, growth-promoting proteins are produced by some species, causing the formation of tumours.

Bactericides are the compounds that kill bacteria. Some common bacterial diseases like blight, leaf spot and blotch are controlled by antibacterial compounds, the oldest known of which are copper compounds. Some of the major bactericides are **bronopol**, **copper hydroxide**, hexachlorophene, acibenzolar-s-methyl, benzalkonium chloride, **kasugamycin**, **nitrapyrin**, oxylinic acid, **probenazole**, **streptomycin** and thiodiazole-copper.

There are a large number of copper-containing bactericides available including $CuSO_4$, $CuSO_4$ + lime, $Cu(OH)_2$ and $Cu(NH_3)_2 + CuCO_3$. Free copper ions (Cu^{2+}) are toxic to bacteria. Due to phytotoxicity, copper-based bactericides are mixed with copper-chelating fungicides such as **mancozeb** or **maneb** or with ferric chloride ($FeCl_3.6H_2O$). Antibiotics such as **streptomycin** control bacterial, algal and fungal attacks on fruit orchards, vegetable crops and in aquatic bodies – though copper compounds are more effective.

Crops that benefit most from copper-containing bactericides are apple, cucumber, wheat, rice, barley, pear, sweet potato, cotton and mushroom, although resistance to some compounds is now observed. Antibac and Phyton 27 are some commercial formulations.

Bacterial infection may be prevented by cultural control. The use of sterile tools, maintenance of crop rows, crop rotation and optimum irrigation are important measures.

Another type of bactericide is a natural toxin called bacteriocins secreted by some species of bacteria that kill other species of the same family. This heat-stable proteinaceous substance causes cell wall leakage of the target bacterial cells. It has a narrow range of specificity.

Bactericide 2-phenylphenol
See **2-Phenylphenol fungicide**

Bactericide 2-(thiocyanatomethylthio) benzothiazole
See **2-(Thiocyanatomethylthio) benzothiazole fungicide**

Bactericide Bordeaux mixture
See **Bordeaux mixture fungicide**

Bactericide copper hydroxide
See **Copper hydroxide fungicide**

Bactericide copper octanoate
See **Copper octanoate fungicide**

Bactericide copper oxychloride
See **Copper oxychloride fungicide**

Bactericide di-halo
See **Di-halo slimicide**

Bactericide formaldehyde
See **Formaldehyde fumigant**

Bactericide fosetyl-aluminium
See **Fosetyl-aluminium fungicide**

Bactericide kasugamycin
See **Kasugamycin biofungicide**

Bactericide nabam
See **Nabam fungicide**

Bactericide probenazole
See **Probenazole fungicide**

Bactericide thiadiazole-copper
See **Thiadiazole-copper fungicide**

Bactericides
See **Bacteria**

Bacteriocins
See **Bacteria**

Baculoviruses as biocontrol agents
See **Biocontrol of insects – baculoviruses**

Bait
Bait is any enticement (like food) used to attract a pest and kill it. Commonly, a toxin concealed in food serves as effective bait. **Arsenic** compounds and **Bordeaux mixture** are typically concealed in interesting feeds and make effective baits.

Additives like ground maize meal, cottonseed oil, groundnut oil and wheat flour are used to conceal the toxin and make it palatable and tempting to pests.

Bait selection and formulations of rodenticides
A rodenticide mixed with food to misguide a rodent into consuming the lethal dose is called rodent bait. Baits are used to control rodent species. Knowledge of rodent behaviour is helpful

in an effective rodent-control exercise. First, rats are suspicious of new foods and hence take a while to taste the baits. They also tend to move food from one place to another. Hence the choice of food and the form in which the food is offered affect the efficacy of the bait.

The choice of food favoured by rodents differs with species and individual members. However, they usually favour food that is familiar. They also prefer fresh food to stale, rancid and infested food. Fresh baits made of maize, wheat, oats or barley are a favourite.

Bait formulations include wax blocks, pellets, powders, pastes and dusts, to take care of a rodent's natural preference. Baits made as a loose meal or pellets are light-weight and easy to carry. A population may have a unique choice. This can be tested by placing non-toxic baits of different forms at fixed intervals in a large area and then checking consumption after 3–4 days.

Banker plants

Plants that are designed to be host to a pest parasite are known as banker plants. Such plants rear the parasite that feeds on the crop pest. Banker plants can even be the same crop plant that needs to be protected. In this natural system, natural enemies of a crop pest are grown in close proximity to the chosen hide-outs of the crop pest, in order to bring about their biocontrol.

Banker plants are chosen in such a way that they become the chosen hosts to **parasitoids** of crop pests. The banker plant system thus becomes sustainable.

It is important for a couple of generations of the parasitoids to get established before crop growth takes place. For example, parasitoids of green peach aphids are grown on wheat or oat plants. They may be fed parasitized aphids or other non-pest aphids till the specific crop pest aphid starts attacking. Once established, these parasitoids control green peach aphids. *Aphidius colemani* is often used as a parasitoid.

It is important to use mildew-resistant oats or rye plants in the banker plant system, lest the banker succumbs to infection. The weakness of this system is felt when the control agent cannot control all the species of the pest.

Banker weeds

See **Aquatic weeds**

Banned pesticides

Certain chemical pesticides are highly toxic to the users, and hence are banned. When used, they cause various **chronic** or **acute toxicity** effects. They can give rise to various genetic or immune disorders or fatal diseases. Some pesticides pollute groundwater and damage the environment.

Every country regulates the use and manufacture of chemical pesticides. These regulations and directives are laid down to control and manage environmental damage in their region.

The EPA maintains the list of banned or severely restricted pesticides as part of its participation in a voluntary international programme known as the Prior Informed Consent (PIC) procedure.

A banned pesticide is defined as one for which all registered uses have been prohibited by final government regulatory action or for which all requests for registration, re-registration or equivalent action for all uses have, for health or environmental reasons, not been granted.

A severely restricted pesticide is defined as a pesticide for which virtually all registered uses have been prohibited by final government regulatory action, but for which certain specific registered use or uses remain authorized. Following are a few examples of agencies that control and monitor pesticide usage strategy.

In India, pesticide regulation is monitored by the **Central Insecticide Board and Registration Committee (CIBRC)**. The **Environmental Protection Agency (EPA)** was established in the USA in 1970 and works for protection of human health and the environment. The **World Health Organization (WHO)** is the directing and coordinating authority for health within the United Nations (UN). It is responsible for: (i) providing leadership on global health matters; (ii) shaping the health research agenda; (iii) setting norms and standards; (iv) articulating evidence-based policy options; (v) providing technical support to countries; and (vi) monitoring and assessing health trends. **Pest Management Regulatory Agency (PMRA)** based in Canada is a branch of 'Health Canada'. It was established in 1995 and is responsible for pest management regulation amd its resources. The Food and Drug Administration (FDA) is an agency within the Department of Health and Human Services of the USA. It promotes and protects public health by monitoring products for continued safety. It also helps the public get accurate, scientific information needed to improve health. (See the **Pesticide legislation** entries in select countries.)

BAP

BAP is another term for **6-Benzylaminopurine**, a PGR.

Barium hexafluorosilicate insecticide

See **Fluoride inorganic insecticides**

Barnyard grass

See **Weed – barnyard grass**

Basipetal translocation

The movement of a dissolved substance in the plant is called translocation. When the translocation of a substance through a plant is downwards from the leaves to the roots, it is basipetal translocation.

Beauveria bassiana (fungus) as a biocontrol agent

Beauveria bassiana, found in many types of soils, is considered a useful broad-spectrum fungal species that provides biological control against a variety of insect pests in crop situations. The parasitic species belongs to the phylum Ascomycota and becomes distinctly evident by a whitish colour on affected hosts. Formulations are made by fermentation.

B. bassiana is known and grown for keeping a variety of pests such as whiteflies, weevils, thrips, aphids, termites, beetles and mosquitoes under control. Colorado potato beetles, black vine weevils, coffee borers, stem borers and codling moths are some of the many pests controlled by *B. bassiana*. Crops on which this biofungicide is used are cotton, maize, coffee and sugarcane.

Pest control is initiated only when fungal spores come in contact with the insect. The fungus infects the insect through its skin by the conidia which germinate. The hyphae penetrate the cuticle by secreting enzymes that dissolve the outer layer. To establish itself inside, the fungus produces toxic beauvericin that weakens the insect immune system and finally kills it. The fungus grows to fill up the body cavity and, when mature, comes out of the insect cadaver, producing a white substance; this symptom is referred to as white muscardine (Fig. B.2). The white mould produces more spores and conidia and the process continues, as the next target is infected. It can take up to 5 days to kill an insect.

Fig. B.2. White muscardine disease *Beauveria bassiana* (Bals.-Criv.) Vuill. Courtesy of Joseph O'Brien, USDA Forest Service, Bugwood.org, with permission.

B. bassiana is used as foliar sprays, wettable powder (WP), water-dispersible granule (WG) and oil-based emulsifiable suspensions. Typical application rates are 750–1000 ml/ha for indoor plants, depending on the crop and the pests to be controlled. It is necessary for the fungal spores to touch the pest for infection to begin.

Bees

Bees, belonging to the superfamily Apoidea, are known to exist in every continent except Antarctica. Bees, ants and wasps are closely related and form the order Hymenoptera under class Insecta. **Honey bees** are the most well-known species of bees and are reared by humans for extracting honey as well as for artificial pollination. Rearing of bees is called apiculture.

Bees bear two pairs of wings with the hind wings being shorter than the fore wings. Nectar and pollen from flowers form the diet of bees. Bees have a long proboscis for extracting nectar from flowers. They are hairy and carry an electrostatic charge on their bodies, making pollen stick to their bodies. Nectar serves as a source of energy while proteins and other nutrients are derived from pollen. Their legs or abdomen have a hairy mass called a scopa that packs the pollen collected by the bees.

Of the over 20,000 species of bees, a few are social and live in colonies. Examples are honey bees, bumble bees and stingless bees. But most are solitary creatures that live in forests.

Bees are extremely important pollinators of flowering plants throughout the world and are considered to be **beneficial organisms**. A significant portion of the farm produce depends on insect pollination, a majority of which is performed by bees. Examples of known pollinators are the European (or western) honey bee (*Apis mellifera*) and Asian hive bee (*Apis cerana*).

Bee populations are observed to have declined for over a decade, with entire colonies disappearing, the **phenomenon** being called colony collapse disorder. Pesticides are said to have caused this decline in bee populations all over the world. Pesticides such as **imidacloprid** and **clothianidin** which are used to treat seeds may negatively impact bees. Pesticides kill bees by direct poisoning and also by indirect poisoning (via contamination of pollen and nectar).

Benalaxyl fungicide

Benalaxyl, $C_{20}H_{23}NO_3$ (methyl *N*-(phenylacetyl)-*N*-(2,6-xylyl)-DL-alaninate), is a fungicide belonging to the **phenylamide** group. It has a white, odourless solid form. It is soluble in water (28.6 mg/l), acetone, methanol, ethyl acetate and xylene. The compound is stable at pH 4–9 (DT_{50} 86 days), to sunlight and up to 250°C. Benalaxyl is hydrolysed in concentrated alkaline media. Benalaxyl is produced by the reaction of the methyl ester of *N*-(2,6-xylyl)-DL-alanine with benzoyl chloride in the presence of a base. It has a CAS Registry Number of 71626-11-4.

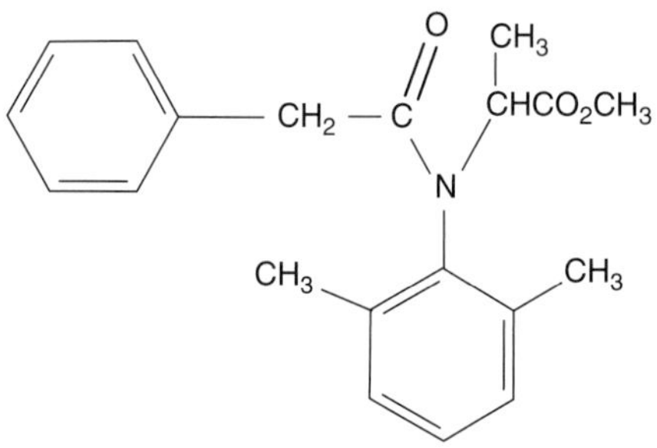

Benalaxyl

Benalaxyl is a systemic fungicide (with acropetal translocation). It disturbs the synthesis of rRNA by inhibiting RNA polymerase. It is used as both a curative and protective agent against oomycete fungi. It effectively prevents spore germination and controls growth of mycelia. **Late blights** of potatoes and tomatoes, and **downy mildews** of vines, curcubits, onions, soybeans and tobacco, are effectively controlled by this chemical. Species of *Phytophthora*, *Plasmopara*, *Pseudoperonospora*, *Sclerospora*, *Bremia* and *Pythium* are effectively controlled by benalaxyl.

Benalaxyl is formulated as emulsifiable concentrates (EC), granules (GR) and wettable powders (WP). It can be combined with other pesticides as long as they are not alkaline in nature. Application dosage ranges from 100 to 240 g/ha.

As per the WHO, benalaxyl is a Class U toxin (unlikely to be hazardous). It is non-carcinogenic, non-mutagenic and non-teratogenic in nature. The acute oral LD_{50} in rats is 4200 mg/kg while in mice it is 680 mg/kg. The 1.5 year NOEL in mice is found to be 250 mg/kg diet. It is non-irritating to the skin and eyes. Inhalation LC_{50} (4 h) in rats is >4.2 mg/l. It is practically non-toxic to birds and bees. It is moderately toxic to fish (among the fish species tested, the LC_{50} recorded ranges between 3.75 and 7.6 mg/l). The ADI is 0.07 mg/kg b.w./day. However, benalaxyl is to be applied and stored with care and as per label instructions, because it is toxic if inhaled.

Rats metabolize and eliminate benalaxyl via the urine and faeces within 2 days. Plants metabolize the chemical to glycosides, although residues are occasionally found. Benalaxyl is degraded by microorganisms to produce acidic metabolites. The DT_{50} in silt loam soil is 77 days.

The fungicide benalaxyl is registered for use in for example Australia, the EU and South Africa, although it is not registered by the USA.

A related compound, similar in properties to benalaxyl, is benalaxyl-M, $C_{20}H_{23}NO_3$ (methyl *N*-(phenylacetyl)-*N*- (2, 6-xylyl)-D-alaninate).

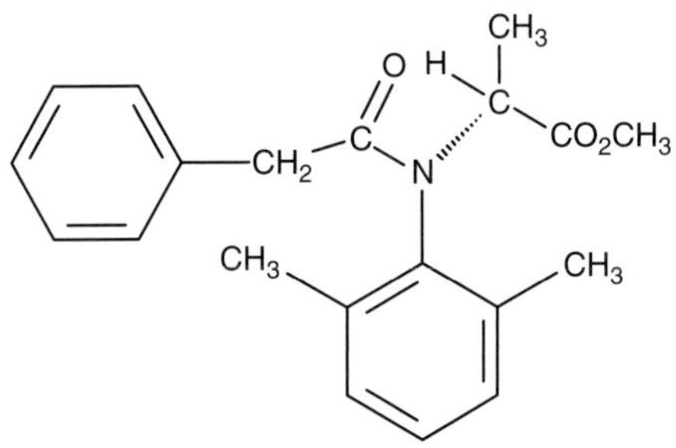

Benalaxyl-M

An odourless crystalline substance, with a CAS Registry Number of 98243-83-5, benalaxyl-M is soluble in water at 33.00 mg/l (pH 7) and some solvents like acetone and methanol.

This fungicide is a systemic with protective and curative action. Once absorbed, it translocates well throughout the plant.

Benalaxyl-M does not pose a major threat to small mammals, birds or bees. It is moderately toxic to fish (LC_{50} in rainbow trout is 4.9 mg/l). The ADI is 0.04 mg/kg b.w./day. In other aspects it is similar to benalaxyl.

In animals, following oral administration, benalaxyl-M metabolizes rapidly and is eliminated in the urine and faeces within 2 days. In plants it is slowly metabolized to glycosides. In soils, it slowly degrades microbially to various metabolites (DT_{50} ranging from 36 to 124 days).

Benalaxyl-M is registered for use in Germany, Greece and Portugal.

Benalaxyl-M fungicide

See **Benalaxyl fungicide**

Bendiocarb insecticide

Bendiocarb, $C_{11}H_{13}NO_4$ (2,2-dimethyl-1,3-benzodioxol-4-yl methylcarbamate), is a carbamate insecticide (*N*-methyl carbamate class), with a CAS Registry Number of 22781-23-3. It is a colourless, odourless, crystalline substance, soluble in dichloromethane, chloroform and acetone, but is hardly soluble in water. It hydrolyses rapidly in alkaline media. The DT_{50} at 25°C is 2 days (at pH 7), the degradation products being 2,3-isopropylidenedioxyphenol, methylamine and carbon dioxide. It is stable to light and heat. Bendiocarb is made by the reaction of 2,2-dimethyl-1,3-benzodioxol-4-yl with methylisocyanate.

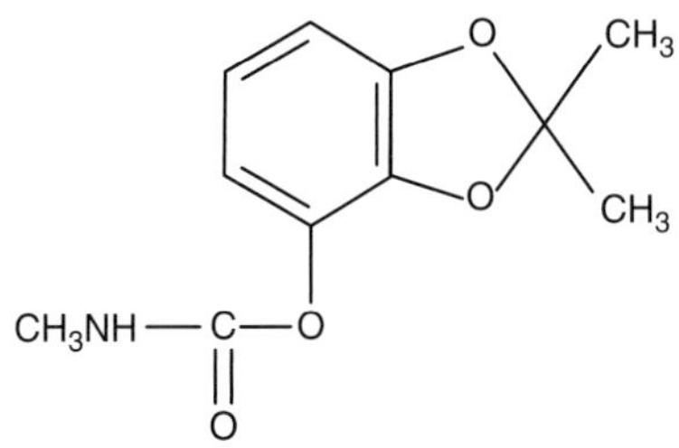

Bendiocarb

Bendiocarb, a cholinesterase inhibitor, is a systemic, broad-spectrum insecticide. An acutely toxic chemical, bendiocarb has a rapid residual effect. It acts by temporarily disrupting the nervous system by interfering with the enzyme acetyl cholinesterase, by binding to its active site leading to the accumulation of acetylcholine, which is required for nerve impulse transmission at nerve muscle sites. It is effectively used to control many members of the family Formicidae (ants), order Blattodea (cockroaches), family Culicidae (mosquitoes), family Muscidae (flies) and order Siphonaptera (fleas). Pests of public health and storage, and pests found in, for example, maize, sugarbeet and turf, are controlled by application of bendiocarb. Wireworm (*Agriotes* spp.), thrips (order **Thysanoptera**) and fruit fly (*Oscinella frit*) and many beetle varieties are some examples of pests commonly controlled.

Bendiocarb is formulated in various forms including ULV sprays and in granular forms, wettable powders (WP) and suspension concentrates (SC). It is compatible with alkaline materials. The WHO has recommended bendiocarb in malaria control. It is available in the market under the trade names of for example Ficam and Dycarb.

The insecticide has been categorized as a Class II toxin (moderately hazardous) by the WHO. It is a cholinesterase inhibitor and a developmental and reproductive toxin and hence a PAN Bad Actor chemical according to the PAN Pesticide Database. It is highly toxic to rabbits, guinea pigs, fish and rats. The acute oral LD_{50} in rats is 25–156 mg/kg. The acute contact LD_{50} in rats is 556–800 mg/kg. It is non-irritating to skin and eyes. The inhalation LC_{50} (4 h) in rats is 0.55 mg/l air. The acute oral LD_{50} in bobwhite quails is 19 mg/kg. The LC_{50} in rainbow trout fish is 155 mg/l. It is toxic to bees. The ADI is 0.004 mg/kg b.w./day. Symptoms of toxicity include muscle weakness, dizziness, vomiting, abdominal pain, seizures and cardio respiratory depression.

In mammals, bendiocarb is absorbed, metabolized and eliminated rapidly following oral administration or inhalation. The principal metabolite is 2,2-dimethyl-1,3-benzodioxol-4-ol, which forms conjugates and is detoxified. In soil, bendiocarb degrades by hydrolysis of the methyl carbamate and heterocyclic rings, the rate of breakdown being higher in alkaline conditions (soil DT_{50} being 3–21 days).

Bendiocarb is registered for example in India, Australia and South Africa but is not registered by the USA with manufacturing of formulations being stopped completely.

Bendiocarb technical analysis

Technical grade bendiocarb is an odourless, off-white solid, with a maximum of 5 g/kg of water. The minimum purity of bendiocarb content for analysis must be 960 g/kg of the sample. The special equipment required includes a liquid chromatograph with stainless steel columns (25 cm long and 4.6 mm in internal diameter, packed with 10 μm C-18 bonded silica gel (Partisil 10 ODS or equivalent)), equipped with a pumping system capable of maintaining a pressure of 17×10^3 kPa, and a UV spectrophotometer detector capable of measuring absorption at 254 nm. The liquid chromatography should have an ambient column temperature, a flow rate of 2 ml/min, a wavelength of 254 nm and an injection volume of 5 μl.

The retention time for bendiocarb is from 3 to 5 min and for the internal standard (propiophenone), 4.5–7.5 min.The liquid chromatography mobile phase is a mixture of 400 ml acetonitrile in water, which is degassed by applying reduced pressure for 10 min, until the mixture just boils.

The three special solutions are: (i) internal standard solution, made by dissolving 1 ml propiophenone in 1 l acetonitrile; (ii) bendiocarb standard solution, made by weighing about 0.5 g (to the nearest 0.1 mg) of standard analytical grade bendiocarb into a 100 ml conical flask with a stopper; and (iii) HPLC grade acetonitrile and water.

About 0.5 g of the bendiocarb sample to be analysed is weighed, to the nearest 0.1 mg, and added to a 100 ml conical flask; 25 ml internal standard solution is added by pipette and the bendiocarb is dissolved; 50 ml acetonitrile is pumped through the column, followed by 50 ml of a mixture of 750 ml/l acetonitrile in water. Change to the mobile phase mixture and pump until the system is equilibrated (flat baseline).

The standard (calibration) solution (5 μl) is injected until the response is stable and the ratios of the peak area or height of the bendiocarb peak to the internal standard peak agree to within 2% of their mean. The sample solution is injected. The peak area or height ratio for the sample solution must not differ by more than 10% from the peak area or height ratio for the calibration solution. If it does differ by more than 10%, the sample solution is reinjected or reweighed to match the calibration solution.

Equations B.1 and B.2 are used to calculate the bendiocarb content.

$$\text{The response ratio (r)} = \frac{\text{Area or height of bendiocarb peak}}{\text{Area or height of internal standard peak}} \quad \textbf{(B.1)}$$

$$\text{Bendiocarb content} = \frac{r_2 \times m_1 \times P}{r_1 \times m_2} \quad \textbf{(B.2)}$$

where

r_1 = Average response ratio for standard solution
r_2 = Average response ratio for sample solution
m_1 = Mass (g) of bendiocarb in the standard solution
m_2 = Mass (g) of sample taken
P = Purity (g/kg) of bendiocarb standard

The water-dispersible powder of bendiocarb is a homogenous mixture of technical bendiocarb along with fillers and necessary additives and is a fine, free-flowing powder that wets readily on stirring into water. To analyse such a sample, bendiocarb is extracted from the sample with acetonitrile. To this extract, the internal standard propiophenone is added. The bendiocarb is estimated by HPLC using a reverse-phase column and a mixture of acetonitrile/water as the mobile phase as was described above for technical grade bendiocarb.

Beneficial organisms

In the context of agriculture and crop protection, reference is often made to beneficial organisms. There are several organisms that are beneficial to man and his crops. These creatures are termed beneficial because of their role in keeping crop pests under control. This encyclopedia has many illustrations of pest control being brought about by natural enemies of pests (which can be predators, **parasitoids** and parasites), often described as biocontrol agents.

It is to be noted that this vast range of beneficial organisms remains beneficial as long as they are used or exploited by man wisely. The various species of lacewings, assassin bugs, spiders and parasitic and predator insects have a direct role in pest control, since they feed on crop pests such as aphids, mites, scale insects and many others. However, most beneficial organisms may turn to man's crops if they cannot find the food (i.e. the targeted crop pest) that they like to feed on. For example, the snail *Rumina decollata* is a predator of the small garden snail species, but can eat young vegetable plants in the absence of its preferred food.

It is important to note that any infusion of toxins in the environment of biocontrol agents can kill or maim them, thus disturbing the ecological balance.

There are a host of other organisms that are not reared as a community for pest management but have a role in maintaining crop yields, forests and a balance in nature. These are also included in the category of beneficial organisms. They include creatures, great and small, from bacteria and fungi to birds and mammals. Each has a function in nature.

Wild and domestic bees, honey bees, butterflies and some flies and moths are essential for pollination of crops, wild vegetation and forestry. Earthworms and soil microorganisms not only help in the process of decomposition but in breaking down chemicals that have been applied to plants. These creatures condition the soil and also recycle nutrients from organic matter.

Lizards and snakes control rodent and insect populations throughout the field. Bats and shrews are known to control flies (e.g. sawflies), mosquitoes and other pests. Toads eat slugs and soil insects. Birds of all types are effective insect control agents, especially on turf and lawns. Insect and reptile eggs, caterpillars and larvae are a common food for a large number of birds. Birds protect crops from insect pests. Chickens and ducks are control agents of soil insects. Fish prey on mosquito larvae and other water insects.

To encourage beneficial insects, pesticides should be used only when necessary. Toxic chemical sprays and soap sprays should be avoided in gardens or on meadows. To encourage populations of bees and butterflies, an attractive food supply in the form of flowering plants such as fennel, parsley, dill and members of the mustard family should be grown in and around crop fields. Most of these plants produce pollen and nectar, which are the favoured food for the adults of beneficial insects.

Soil **mulches** provide a safe place to hide for most lizards, shrews and small mammals, as do hedges and windbreaks. Maintaining small water bodies at convenient spots help mammals, birds and insects to quench their thirst, especially during dry and hot periods. Farmers can encourage birds to control pests by placing nesting boxes around their fields.

Many species of conventional biocontrol agents such as predatory mites, nematodes, spiders and snails are very efficient at controlling pest populations and have a beneficial role in nature. Details of the action of these biocontrol agents are presented throughout this encyclopedia.

Benomyl analysis

Benomyl (methyl 1-(butylcarbamoyl)-2-benzimidazolecarbamate) is a protective and eradicant fungicide, particularly in annual crops and is effective against **scab** and **powdery mildew**, particularly in apples. It is also effective against mites.

Analysis of benomyl residues in plant and animal tissues and in soil can be done by extraction, conversion to 2-aminobenzimidazole and measurement by direct fluorometry or by colorimetry after bromination. The analytical procedure is outlined as follows.

The sample (50 g) is blended with 150 ml ethyl acetate at high speed for 5–10 min. The sample is centrifuged at 2000 r.p.m. for 10–15 min and the ethyl acetate decanted through cotton into a 500 ml beaker. The beaker is rinsed out with portions of ethyl acetate and added back to the centrifuge bottle; 25 ml 0.1 N hydrochloric acid is added to the combined extracts and evaporated to a volume of 10–15 ml. The sides are washed with 15 ml hot water and transfered to a 125 ml separating funnel. When this mixture is cooled to room temperature, the phases are allowed to separate, by repeatedly washing with hexane or by further centrifuging and removing the hexane layer.

The aqueous phase is transfered to a 150 ml beaker, rinsed with small portions of water making the total volume to be about 40 ml. To this 15 ml 6.5 N sodium hydroxide solution is added, covered and boiled gently on a hot plate for 15 min. The contents of the beaker are transfered to a 250 ml separating funnel, after rinsing out the container with hot water. Then it is cooled to room temperature and the aqueous phase is extracted with four portions of 75 ml ethyl acetate, shaking for 2–3 min each time. After separating the ethyl acetate, it is filtered through a cloth into a 400 ml beaker. It is concentrated (reduced) to about 50 ml by evaporation at room temperature. The contents of the beaker are transfered to a 125 ml separating funnel, rinsing with small portions of ethyl acetate. Finally the beaker is rinsed with 10 ml 0.1 N hydrochloric acid, and the solution added to the separating funnel and shaken for 2 min. After the phases separate, the aqueous phase is drained into a second 125 ml separating funnel. It has to be ensured that the aqueous phase is acidic, adding 1 ml 0.1 N hydrochloric acid, if required. The ethyl acetate is extracted with a second 10 ml portion of 0.1 N hydrochloric acid and the acid phases combined. Then 2 ml 6.5 N sodium hydroxide solution is added to the combined acid extract to make the solution strongly basic. The solution is extracted with four 50 ml portions of water-saturated ethyl acetate, shaking for 2–3 min each time. After phase separation, the ethyl acetate extract is filtered through a cloth into a 250 ml beaker and evaporated to a volume of about 20 ml. This extract is transfered to a 50 ml beaker and evaporated to dryness. The sample is now ready for determination either by fluorometric or colorimetric methods.

For the fluorometric determination of the sample, the residue is first dissolved in a 9:1 methyl alcohol-aqueous 0.1 N sodium hydroxide solution. It is quantitatively transfered to a 10 ml volumetric flask, washing or rinsing with small quantities of methanol-sodium hydroxide solution mixture until the volume is made up. The fluorescence of the solution is determined at 335 nm, with an excitation wavelength of 285 nm. The reading is compared with a calibration curve prepared by measuring the fluorescence of 5, 10, 20, 30 and 50 μg of 2-amino-benzimidazole in 10 ml of 9:1 methanol–0.1 N sodium hydroxide solution. The quenching factor on one or more control samples is determined from each sample series by adding a known amount of 2-aminobenzimidazole to the untreated sample extract and comparing the relative intensity with the value obtained on a standard solution of the same apparent concentration. The benomyl content is calculated using Eqn B.3.

$$\text{Benomyl (ppm)} = \frac{(c)(2.18)}{(q)(r)(w)} \qquad \textbf{(B.3)}$$

where
c = 2-Aminobenzimidazole content in micrograms (μg) obtained from the calibration curve
q = Quenching factor
r = Recovery factor
w = Sample weight, in grams (g)

In the colorimetric determination of benomyl, 10 ml saturated bromine water is added to the residue and this is placed in a steam bath for 30 min. The steam bath is uncovered and left open until excess bromine is completely eliminated as determined by the absence of odour. All contents (rinsing out with hot water) are quantitatively transfered to a 125 ml separating funnel, cooled and diluted to 30 ml with water. The residue beaker is rinsed with 25 ml water, saturated with ethyl acetate and added to the separating funnel, shaking for 3–4 min vigorously. After layer separation, the ethyl acetate is filtered through cotton into a 50 ml beaker. The entire process is repeated with a second portion of 25 ml water saturated with ethyl acetate. The combined extracts are concentrated to 1 ml and transfered to a 2 ml volumetric flask and the beaker rinsed with ethyl acetate and diluted to the mark and mixed. The absorbance is measured at 445 nm, using semi-microquartz cells having a 10 mm light path against ethyl acetate as a reference. A calibration curve is prepared by carrying 5, 10, 20, 30 and 50 μg of the standard through the bromination method as described above. The 2-aminobenzimidazole (the primary degradation product of benomyl) content in the sample is determined and the benomyl content is calculated using Eqn B.4.

$$\text{Benomyl (ppm)} = \frac{(c)(2.18)}{(r)(w)} \quad \textbf{(B.4)}$$

where c, r and w have the same significance as in the flurometric determination.

Benomyl fungicide

Benomyl, $C_{14}H_{18}N_4O_3$ (methyl 1-(butylcarbamoyl) benzimidazol-2-ylcarbamate), is a **benzimidazole fungicide**, with acaricidal properties. Benomyl is a whitish crystalline substance, soluble in water (2.9 ml/l), chloroform (94 g/kg) and xylene (10 g/kg). Benomyl is stable to light but decomposes on contact with water and in moist conditions in soil. It also dissociates to **carbendazim** and butyl isocyanate in some solvents. Its half-life is 3.5 h at pH 5 which decreases with increasing pH. Benomyl is produced by the reaction of carbendazim with butyl isocyanate. The CAS Registry Number of benomyl is 17804-35-2.

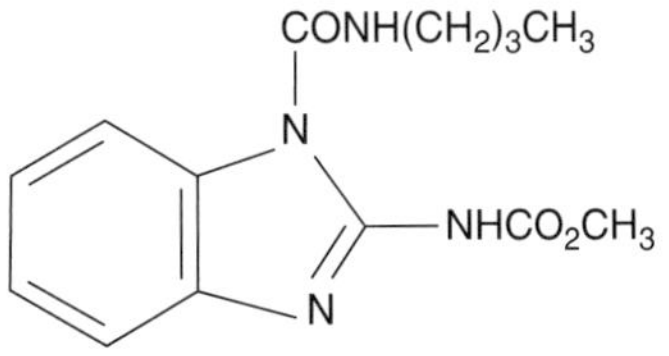

Benomyl

Benomyl used as a systemic, protective, pre- or post-harvest fungicide, acts on ascomycetes, deuteromycetes and basidiomycetes. The chemical inhibits mitosis by binding to beta-tubulin. As an acaricide, it has ovicidal action. The fungicide controls diseases of cereals, legumes, groundnuts and fruits, including apple, grapes and berries. Leaf spot, rice blast, rot, **powdery mildew** on cereals and legumes, **scab** on apple and groundnut tikka diseases are controlled. It can also be used on stored fruits.

Benomyl is formulated as a wettable powder (WP). Common application rates of benomyl are 140–550 g a.i./ha on field and vegetable crops. The chemical can be used with many non-alkaline fungicides. For post-harvest storage use, benomyl is applied at lower rates. It is marketed in India as Benlate-50 WP. Other trade products are for example Benofit, Fundazd and Viben.

Benomyl is classified under toxicity Class U (unlikely to be hazardous) according to the WHO. Its toxicity in rats is slight, with acute oral LD_{50} >5000 mg a.i./kg. It causes slight irritation to the eyes but is non-irritating to the skin (acute dermal LD_{50} is >5000 mg/kg). Inhalation LC_{50} (4 h) in rats is >2 mg/l air. The LC_{50} in rainbow trout is 0.27 mg/l. Benomyl is practically safe on bird species tested (dietary LC_{50} being >10000 mg/kg diet) and beneficial insects like worms and bees. NOEL for 2 years among rats is found to be >2500 mg/kg of diet. Benomyl is reported to be a reproductive toxic, suspected endocrine disruptor and a possible carcinogen, hence it is a PAN Bad Actor chemical and its production has been stopped.

In animals and plants, benomyl degrades by the removal of butylcarbamoyl group to give the relatively stable compound, carbendazim. Metabolites include 2-aminobenzimidazole, 5-hydroxybenzimidazole carbamate, and 4-hydroxy-2-benzimidazolemethylcarbamate. In soils and/or the environment, it is rapidly converted to carbendazim. The DT_{50} is 2 h in water and 19 h in soil.

Benomyl is registered for use in India, South Africa and many other African countries, Cambodia, the Philippines and Vietnam. It is not registered by the USA.

Benoxacor herbicide safener

Benoxacor, $C_{11}H_{11}Cl_2NO_2$ ((RS)-4-dichloroacetyl-3,4-dihydro-3-methyl-2*H*-1,4-benzoxazine), is a herbicide safener. It is a colourless, crystalline solid substance, which is not easily soluble in water but readily so in for example methanol, cyclohexanone and acetone. Benoxacor hydrolyses in alkaline media (DT_{50} about 50 days) and shows rapid photolytic degradation in aqueous solution. But it is otherwise stable in acidic media and up to 200°C. It has a CAS Registry Number of 98730-04-2.

While benoxacor does not have any herbicidal property, it gets absorbed into the crop where it enhances crop tolerance for pesticides. Increasing the tolerance of maize to **metolachlor** (without affecting herbicide activity) is its main property. Benoxacor is normally maintained in the soil for 10–14 weeks. Benoxacor is available under trade names such as Dual.

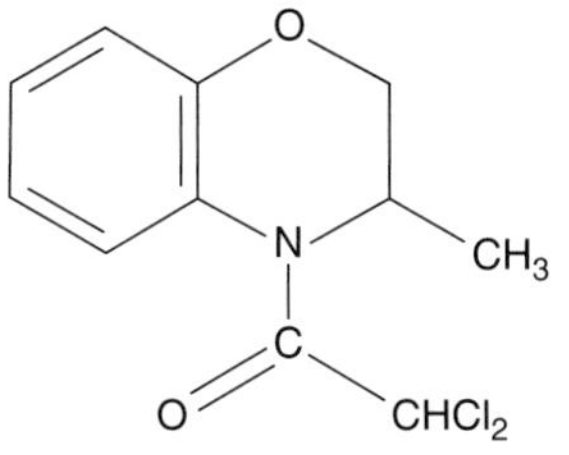

Benoxacor

Benoxacor is classified as a Class U toxin (unlikely to be hazardous) according to the WHO. In toxicological studies, its acute oral LD_{50} among rats is >5000 mg/kg, while among tested bird species (mallard duck and bobwhite quail) it is >2150 mg/kg. This chemical is practically safe for bees and earthworms, while in fish, the 96 h LC_{50} is in the range of 3.5–10 mg/l. It is non-irritating to skin and eyes. The NOEL (2 years) in rats is 0.5 mg/kg b.w. The ADI is 0.004 mg/kg b.w./day.

In animals and plants, benoxacor metabolizes to water-soluble conjugation products following a series of reactions like ring hydroxylation and deacetylation. Benoxacor is registered for use in Portugal but not in the USA.

Bensulide herbicide

Bensulide, $C_{14}H_{24}NO_4PS_3$ (*O,O*-diisopropyl *S*-2-phenylsulfonylaminoethyl phosphorodithioate), is an organophosphate herbicide. It is a colourless solid with a CAS Registry Number of 741-58-2. It is soluble in water and in organic solvents. It is relatively stable to acids and alkalis. It decomposes slowly in light. Bensulide is made by reaction of *O,O*-di-isopropylphosphorodithioate with *N*-chloroethylbenzene sulfonamide.

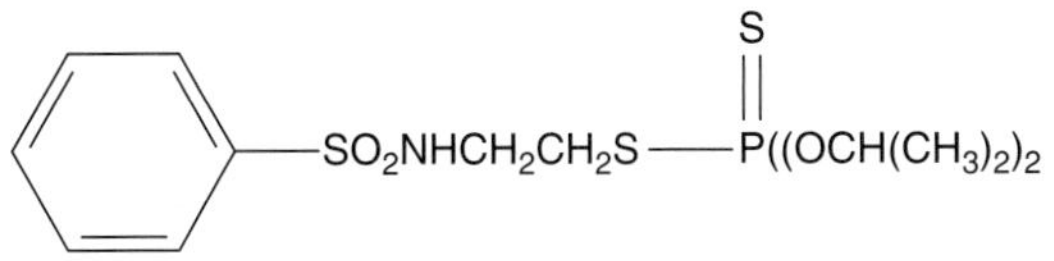

Bensulide

Bensulide, a pre-emergence, selective herbicide, is absorbed by roots minimally, the major portion getting adsorbed to roots. Translocation via the roots is limited to metabolites.

The mode of action is lipid synthesis inhibition, resulting in poor seed germination of annual grasses (e.g. bluegrass and crabgrass) and broadleaved weeds. Crops that benefit from the use of bensulide are cotton, carrots, cucumbers, peppers and melons.

The herbicide is formulated as emulsifiable concentrate (EC) and as granules (GR). It is available in the market under the trade name of Betasan.

Bensulide is classified as a Class II toxin (moderately hazardous), according to the WHO and EPA and this classification is evident in the acute oral LD_{50} for rats being in the range of 271–360 mg/kg. It also causes mild irritation to the eyes and skin. Inhalation LC_{50} (4 h) in rats is >1.75 mg/l. The ADI is 0.005 mg/kg b.w./day. Bensulide is slightly toxic to birds, and is slightly-to-highly toxic to aquatic organisms and bees. The acute oral LD_{50} in bobwhite quails is 1386 mg/kg. The LC_{50} in rainbow trout is 1.1 mg/l, while the LD_{50} in bees is 0.0016 mg per bee.

The chemical bensulide is classified as a general use pesticide (GUP) by the EPA. The WHO classifies it as a moderately toxic chemical (Class II toxin). Products containing bensulide have to bear the signal word 'CAUTION'. Handlers and applicators are cautioned against excessive exposure (via ingestion and inhaling of dust) to bensulide. Symptoms of poisoning are typical of organophosphates like salivation, sweating, incontinence, tremors, muscle twitching, depression and seizures. This leads to the strong recommendation to handlers to be equipped with protective gear at all times.

Bensulide persists in both plants and soil due to its high adsorption properties. Degradation occurs as a result of rising temperature and presence of organic matter and also the action of microorganisms. It is also considered a potential groundwater contaminant.

Bensulide is registered in Australia, Canada and the USA.

Bensultap insecticide

Bensultap, $C_{17}H_{21}NO_4S_4$ (*S*,*S*′-2-dimethylaminotrimethylene di (benzenethiosulfonate)), is a carbamate insecticide. It is a pale yellowish crystalline powder, hardly soluble in water (0.7–0.8 mg/kg) but soluble in acetone, acetonitrile and *N*,*N*-dimethyl formamide. Bensultap is hydrolysed in neutral or alkaline solutions and is stable up to 150°C under diffused light. Its CAS Registry Number is 17606-31-4.

Bensultap, a nerve toxin, is a derivative of **nereistoxin**,

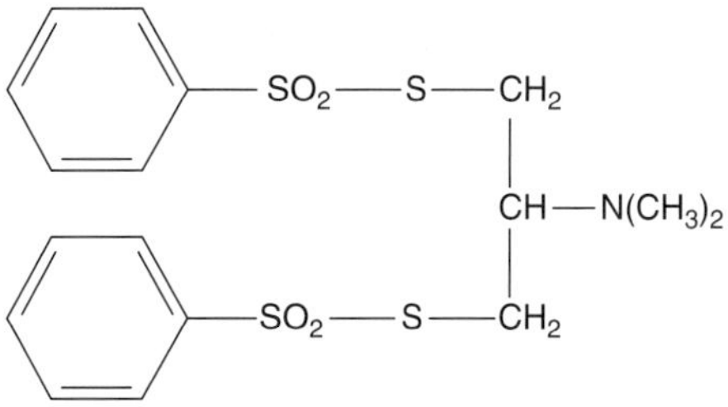

Bensultap

which was found in the marine annelid or segmented worm, *Lumbrineris brevicirra*. The mode of action is the blocking of the synaptic neuron transmissions in the CNS of beetles.

The use of bensultap can effectively control many types of beetles, leaf rollers (*Caloptilia theivora*), leaf miners (*Hydrellia griseola*), borers, boll and root weevils and various moths such as the diamondback (*Plutella xylostella*). Crops like rice, maize, cotton, tea, the brassica group and fruits, including apples and pears, are protected.

Popular formulations of bensultap are wettable powders (WP), granules (GR) and dustable powders (DP). This insecticide may be combined with other non-alkaline compounds to enhance the pesticidal effect. Known brand names include Bancol and Victenon.

The WHO and EPA classify bensultap as slightly hazardous. The acute oral LD_{50} in rats is 1105 mg/kg, while its acute inhalation LC_{50} is 0.47 mg/l. The acute dermal LD_{50} in rabbits is >2000 mg/kg. It is a slight eye irritant. The NOEL (90 days) in rats is 250 mg/kg diet. The acute oral LD_{50} in bobwhite quail is 311 mg/kg. LC_{50} (48 h) in carp is 15 mg/l. The acute 48 h LD_{50} in honey bees is 25.9 μg per bee. The acute 14 day LC_{50} in earthworms is 30 mg/kg.

Bensultap is not registered by the USA or countries of the EU.

Benthiavalicarb fungicide

Benthiavalicarb, $C_{15}H_{18}FN_3O_3S$ ([(*S*)-1-{[(1*R*)-1-(6-fluoro-1,3-benzothiazol-2-yl)ethyl]carbamoyl}-2-methylpropyl]carbamic acid), belongs to the amino acid carbamate class. The CAS Registry Number is 413615-35-7 (see Benthiavalicarb at bottom of the page).

It is effective against fungal (oomycetes) pathogens, inhibiting phospholipid biosynthesis and cell wall synthesis. It controls infections and also protects crops by preventing or

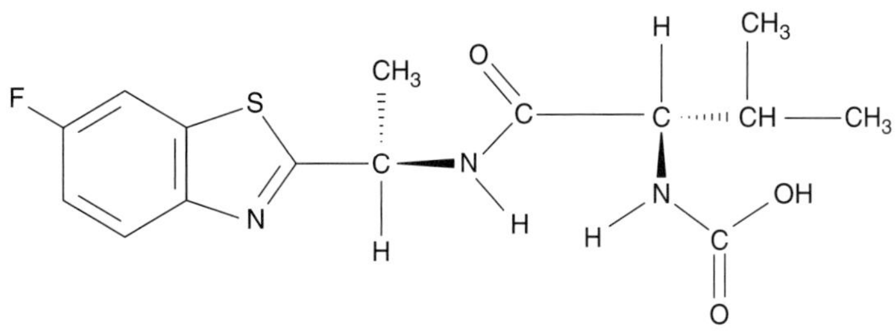

Benthiavalicarb

inhibiting spore germination and mycelial growth. With good residual activity, this chemical effectively controls: (i) **downy mildews** in vegetables and grapes caused by *Plasmopara viticola* and *Pseudoperonospora cubensis*; and (ii) **late blights** in tomatoes and potatoes caused by *Phytophthora infestans*.

Benthiavalicarb is formulated as a flowable concentrate (SC) and as water-dispersible granules (WG). A rate of up to 75 g/ha is generally enough to bring effective control of the infection. One of the trade names is KIF-230.

Benthiavalicarb is not very toxic to rats, mice and birds (with the acute oral LD_{50} >5000 mg/kg). The NOEL for 2 years in rats is found to be between 9.9 mg/kg and 12.5 mg/kg b.w. The LC_{50} in fish is >10 mg/l. Benthiavalicarb is safe for beneficial organisms inhabiting the surrounding environment.

The half-life of benthiavalicarb in soil ranges from 11 to 40 days (the period increasing with anaerobic conditions). The compound is metabolized faster in animals than in plants and includes reactions such as glutathione conjugation or hydroxylation of the benzene moieties.

Benthiavalicarb is not a PAN Bad Actor, unlike its parent, benthiavalicarb-isopropyl. It is registered in many countries in the EU.

Benthiavalicarb-isopropyl, $C_{18}H_{24}FN_3O_3S$ (isopropyl [(*S*)-1-{[(1*R*)-1-(6-fluoro-1,3-benzothiazol-2-yl)ethyl]carbamoyl}-2-methylpropyl]carbamate), is a carbamate fungicide (the valinamide class). It has a CAS Registry Number of 177406-68-7. This fungicide is a white powder, soluble in water (13–14 mg/l depending on pH), methanol (41.7 mg/l) and xylene. It is stable to hydrolysis and undergoes photolysis in water under different parameters (see Benthiavalicarb-isopropyl at bottom of the page).

Benthiavalicarb-isopropyl is a protective and curative fungicide and acts by inhibiting mycelial growth and germination of sporangia. It can be used with other fungicides such as **mancozeb** and **folpet**. It is active against downy mildews and late blight on vegetables, potatoes and grapes at 25–75 g/ha. It is formulated as soluble concentrates (SL) and water-dispersible granules (WG). It is available under trade names such as Valbon and Vincare.

This fungicide is a carcinogen and according to the PAN Pesticide Database, it is a PAN Bad Actor chemical. This chemical is registered in Germany, Greece and the Netherlands. Benthiavalicarb-isopropyl is not acutely toxic to rats on ingestion or via the inhalation route. The oral LD_{50} in male and female rats is >5000 mg/kg. The inhalation LC_{50} (4 h) in male and female rats is >4.6 mg/l air. It does not irritate skin but is a mild eye irritant. The ADI is 0.1 mg/kg b.w./day. It is non-toxic to bees and worms and practically so to birds and fish.

Following administration, excretion of the chemical takes place within 7 days in rats. The metabolism is complex and proceeds slowly in plants. The fungicide is readily degraded in soil (DT_{50} 11–19 days) in aerobic conditions.

Benthiavalicarb-isopropyl fungicide

See **Benthiavalicarb fungicide**

Benzimidazole fungicides

Benzimidazoles are a group of compounds that have fungicidal properties. As systemic, broad-spectrum compounds, their mode of action is to bind to the β-subunit of the protein tubulin, making it difficult for nuclear division to take place in the fungus. This mode of action also involves interference with microtubule assembly. This action is seen in the inhibition of spore tube formation and mycelia. These fungicides, in spite of resistance being developed, are used for their protective and curative activity against a number of crop pathogens. Some benzimidazoles are also **acaricides**.

Benzimidazole fungicides include **benomyl, carbendazim, thiophanate-methyl, thiabendazole** and **fuberidazole**. *N*-Phenylcarbamates, *N*-phenylformamidoximes and other classes of compounds have a similar mode of action to benzimidazoles. Most have residual activity and produce methyl benzimidazol carbamate as a degradation product.

Benzimidazoles, required in low doses, have proved extremely efficient on cereal crops, vines, groundnuts and vegetable crops. The diseases controlled mainly include powdery mildew, anthracnose, brown rot, leaf spot and blossom rot caused by *Fusarium* spp., *Botrytis* spp., *Venturia* spp. and *Rhizoctonia* spp.

Since resistance has developed on a large scale as far as most of the members of the benzimidazole group are concerned, combining these compounds with those with a different site of action and restricting the number of applications per season become important **resistance management** techniques. Benzimidazoles, being curative in nature, can be combined with protective compounds like **thiram, captan** or **mancozeb**.

Belgium Agriculture Ministry has imposed restrictions on the benzimidazole fungicides, **carbendazim**, benomyl and thiophanate-methyl.

Some management guidelines for benzimidazole fungicides followed in New Zealand are given in Table B.1.

Benzobicyclon herbicide

Benzobicyclon, $C_{22}H_{19}ClO_4S_2$ (3-(2-chloro-4-mesylbenzoyl)-2-phenylthiobicyclo [3.2.1] oct-2-en-4-one), is a herbicide. It is a slightly yellow, odourless crystalline substance, almost

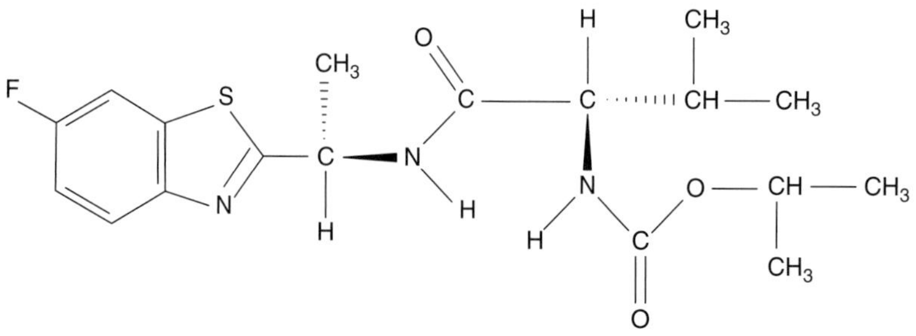

Benthiavalicarb-isopropyl

Table B.1. Benzimidazole use guidelines for specific crops. (From Martin *et al.*, 2005; with permission.)

Crop	Disease	Recommendations
Stone fruit	Brown rot	Confine use to the blossom period only
Pip fruit	Fruit rots, *Botrytis* dry eye rot	Confine use to the blossom period and use a fungicide from another group for black spot control. Apply only in mixture with a protectant fungicide (e.g. thiram, captan, manocozeb or metiram)
Grapes, strawberries, raspberries, brambles, tomatoes and beans	Grey mould (*Botrytis* sp.)	Avoid use as resistance is known to be present, otherwise do not exceed one application per season
Cucurbits	Powdery mildew	Use in mixture or alternation with alternative products
Tomatoes, beans and lettuce	*Sclerotinia*	Limit use to early season (blossom to 14 days post-blossom on flowering crops)
Wheat	Eyespot	Apply only in second-year crop. If control was below expectation, use an alternative in the third year
Wheat	Speckled leaf blotch	Comply strictly with recommended timing (last week of August)
Barley	Scald	Comply with label rates and timing

insoluble in water (0.052 mg/l), and thermally stable up to 150°C and rapidly hydrolysed. The CAS Registry Number of benzobicyclon is 156963-66-5.

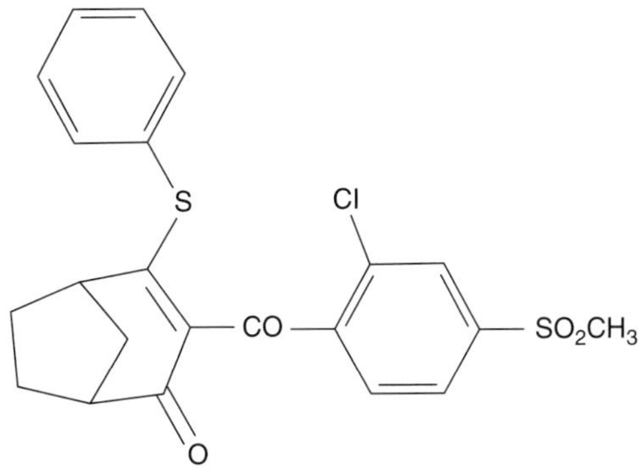

Benzobicyclon

Benzobicyclon is a selective, pre-emergence herbicide. It is a *p*-hydroxyphenyl pyruvate dioxygenase inhibitor, which leads to bleached leaves of weeds such as grass, sedge and annual broadleaved weeds that are commonly seen in rice crops.

The herbicide benzobicyclon is very effective at doses of 200–300 g/ha. It is available as granules and suspension concentrates, with trade names including ShowAce. It can be combined with other herbicides like **tembotrione**.

The acute oral LD_{50} of benzobicyclon in rats and mice is >5000 mg/kg. It is not a skin irritant. The LD_{50} in bird species (bobwhite quail) tested is >2250 mg/kg, while the 48 h LC_{50} in carp is >10 ppm. The oral and contact LD_{50} of benzobicyclon in the honey bee is > 200 µg per bee.

Benzofenap herbicide

See **4-Hydroxyphenylpyruvate dioxygenase (HPPD)**

Benzoic acid herbicides

Benzoic acid ($C_7H_6O_2$) is a weak carboxylic acid that exists as a crystalline solid substance. With a CAS Registry Number of 65-85-0, benzoic acid is soluble in water (to the extent of 0.34 g/l) and also in ethanol and methanol.

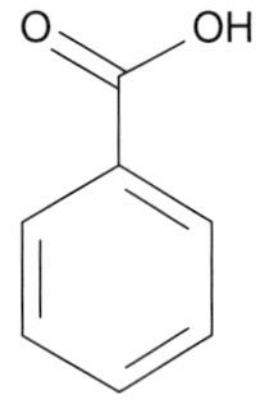

Benzoic acid

The benzoic acid class of chemicals is a key ingredient in the making of several organic compounds as well as in the preparation of insecticides (**benzyl benzoate**) and herbicides (auxin mimics). Benzoic acid is only slightly toxic to mammals and other non-target organisms, and is registered in the USA, while some of the derivatives of benzoic acid like benzyl alcohol and sodium benzoate are not registered by EPA.

As auxinic herbicides (which are chlorinated), benzoic acid derivatives act as growth regulators as well as effective control agents of perennials and broadleaved weeds in cereal crops. **TBA**, 2,3,5-triiodobenzoic acid (TIBA), **chloramben**, **dicamba**, tricamba and dinoben are some of the herbicides developed in this class. Some information on benzoic acid herbicides is presented for easy reference in Table B.2.

Benzoximate acaricide

Benzoximate (3-chloro-α-(*EZ*)-ethoxyimino-2,6-dimethoxybenzyl benzoate), with the molecular formula $C_{18}H_{18}ClNO_5$, is a bridged diphenyl acaricide. It is a colourless crystalline solid that is almost insoluble in water but soluble in benzene, xylene and dimethylformamide.

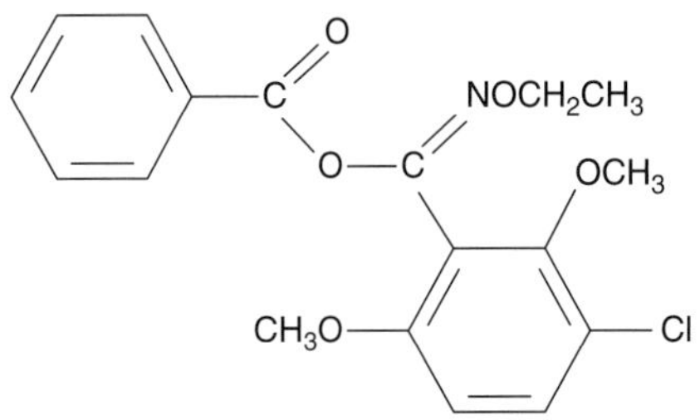

Benzoximate

Acaricide benzoximate is a non-systemic compound with contact and stomach action. It is used against mites on apples

Table B.2. General and toxicological properties of chloramben, dicamba and TBA. (Table courtesy of Shubhangi Borse.)

Properties	Chloramben	Dicamba	TBA
IUPAC	3-Amino-2,5-dichlorobenzoic acid	3,6-Dichloro-*o*-anisic acid	2,3,6-Trichlorobenzoic acid
Molecular formula	$C_7H_5Cl_2NO_2$	$C_8H_6Cl_2O_3$	$C_7H_3Cl_3O_2$
Solubility in water	1.96 g/ml (20°C)	4.5 g/l (25°C)	7.7 g/l (22°C)
Acute toxicity oral LD_{50} (mg/kg)	>5000 (rat)	>1707 (rat)	1500 (rat), 1000 (mice), 600 (rabbit)
Fish LC_{50} 96 h (mg/l)	Not toxic	135 (rainbow trout, bluegill sunfish)	Data not available
Bird LD_{50} (mg/kg)	4640 (mallard duck)	2000 (mallard duck)	>1500 (hens)
Honey bee LD_{50} (µg per bee)	Not toxic	Not toxic (>110)	Not toxic
Eye sensitivity (rabbit)	Mild irritant	Extremely corrosive	Data not available

and citrus fruits, especially *Panonychus* spp. and *Eotetranychus* spp. It is produced by the reaction of ethyl-3-chloro-2,6-dimethoxy hydroximate with benzoyl chloride in the presence of a base. The trade names of benzoximate are Acarmate, Artaban and Citrazon. The CAS Registry Number is 29104-30-1.

The WHO classifies benzoximate as a Class U toxin (unlikely to be hazardous). The acute oral LD_{50} in rodent species tested is in the range of 5000–14,500 mg/kg. The NOEL in rats (2 year feed) is 20 mg/kg/day.

Benzoximate is registered in South Africa and is limited to citrus fruits, with an MRL of 2 ppm.

Benzyl adenine as a PGR

Benzyl adenine is another term for **6-Benzylaminopurine**, a PGR.

Benzyl benzoate acaricide

Benzyl benzoate (benzyl benzoate), with molecular formula $C_{14}H_{12}O_2$, is a bridged diphenyl acaricide. It is a colourless, oily substance with a faint aromatic odour and is insoluble in water. It should be stored in a cool, dry place in tightly sealed containers, protected from heat and light. It is incompatible with strong oxidizers, acids, bases and reducing agents.

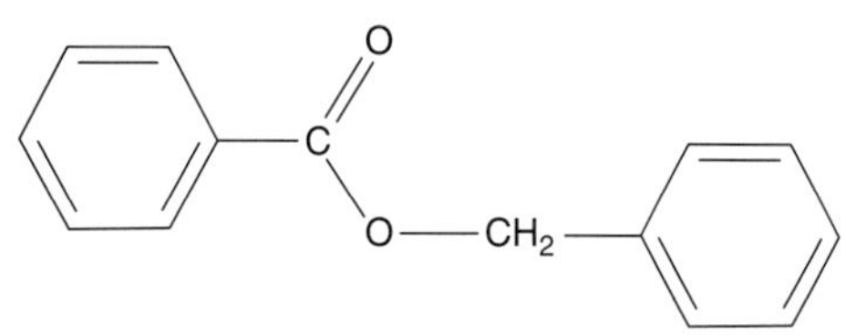

Benzyl benzoate

Benzyl benzoate, the ester of benzyl alcohol and **benzoic acid**, is used as an insecticide. It is also used to cure the skin condition scabies.

The acute oral LD_{50} in rats and mice is between 1400 and 1700 mg/kg. The acute dermal LD_{50} in rabbit is 4000 mg/kg. The substance irritates the eyes, skin and the respiratory tract.

Benzyl benzoate is registered for use in Australia, Canada, South Africa and the USA.

Bermuda grass

See **Weed – Bermuda grass**

Best management practices (BMP)

The activities related to management practices that take habitat and environment into consideration are referred to as best management practices (BMPs). The practices such as strip-cropping, cover crops, green manure crops, soil testing, liming and appropriate application of chemicals for both crop nutrition and protection are necessary for good food production. If managed properly, they would prevent or control the movement of sediment, nutrients, pesticides and other pollutants from the land surface or groundwater.

BMPs are useful in maintaining water quality, watershed protection and management, and soil conservation. Many agencies concerned with environment protection worldwide acknowledge the usefulness of BMPs in reducing pollution.

Beta-cypermethrin insecticide

Beta-cypermethrin, $C_{22}H_{19}Cl_2NO_3$ (reaction mixture comprising the enantiomeric pair (*R*)-α-cyano-3-phenoxybenzyl(1*S*,3*S*)-3-(2,2-dichlorovinyl)-2,2-dimethylcyclopropanecarboxylate and (*S*)-α-cyano-3-phenoxybenzyl(1*R*,3*R*)-3-(2,2-dichlorovinyl)-2,2-dimethylcyclopropanecarboxylate in the approximate ratio 2:3 with the enantiomeric pair (*R*)-α-cyano-3-phenoxybenzyl(1*S*,3*R*)-3-(2,2-dichlorovinyl)-2,2-dimethylcyclopropanecarboxylate and (*S*)-α-cyano-3-phenoxybenzyl(1*R*,3*S*)-3-(2,2-dichlorovinyl)-2,2-dimethylcyclopropanecarboxylate), is a pyrethroid insecticide. It has a CAS Registry Number of 52315-07-8. It is a creamish crystalline substance, soluble in water, isopropanol, xylene, dichloromethane, acetone, ethyl acetate and ether. It is stable at 150°C in air and sunlight, and in neutral to slightly acidic media. It is hydrolysed in strongly alkaline solution.

Beta-cypermethrin is a non-systemic insecticide that is a nerve toxin. It acts by disturbing the function of neurons by interacting with the sodium channel. As a crop protection agent, it is effective on leaf borers, weevils, armyworms and others that affect vegetables, mustard, fruit, rice and cotton. The compound may also be applied on animals and in public places (see Beta-cypermethrin at top of the next page).

Beta-cypermethrin is popularly formulated as an emulsifiable concentrate (EC), ULV liquid (UL) and suspension concentrate (SC). The insecticide may be sprayed by diluting 15 ml of the active ingredient in a 15 l tank. A known brand name is Stronger.

The WHO and EPA classify beta-cypermethrin as moderately hazardous (Class II toxin). The acute oral LD_{50} among female rats is 166 mg/kg, while the inhalation LC_{50} value stands at >1.97 g/l. The acute dermal LD_{50} in rats is >5000 mg/kg. The NOEL (2 years) in rats is 250 mg/kg diet. The acute oral LD_{50} in bobwhite quails is 8030 mg/kg. Fish species are not harmed under normal field conditions. The LD_{50} (48 h, oral) in bees is 0.0018 mg per bee. Labels should have the signal word 'WARNING'. The ADI for beta-cypermethrin is 0.05 mg/kg/day.

(*R*)- alchohol (1*S*)-*trans*-acid

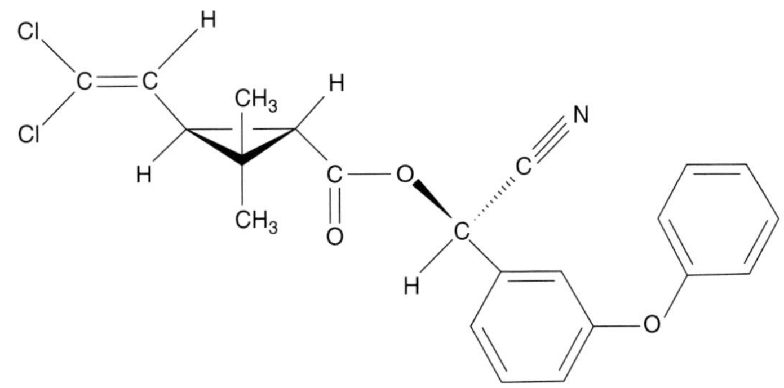

Beta-cypermethrin

It is a possible carcinogenic compound and a suspected endocrine disruptor. Symptoms of toxicity include burning, irritation of the face and respiratory tract, vomiting, fluid in lungs and convulsions.

The half-life of beta-cypermethrin in water is 1.2 days while in the soil it is 10 days. Beta-cypermethrin is registered in Australia, Cameroon, the Philippines, South Africa and the USA.

Bialaphos herbicide

See **Herbicidal inhibition of glutamine synthetase (GS); Bilanaphos bioherbicide**

Biennials

Biennials are plants that live for up to 2 years. The critical flowering stage marks the end of their life cycle. Commonly seen in cool temperate climates, these plants grow vegetatively and form rosettes (a circular arrangement of leaves of a single height) and a sturdy root system in the first year. Biennials form a significant part of the weed population. Wild carrot and dandelion are examples of biennial plants.

Bifenox herbicide

Bifenox, $C_{14}H_9Cl_2NO_5$ (methyl 5-(2, 4-dichlorophenoxy)-2-nitrobenzoate), is a nitrophenyl ether herbicide. With a CAS Registry Number of 42576-02-3, bifenox exists as light yellow crystals and is very highly soluble in xylene, methanol and dichloromethane, but hardly so in water (0.1 mg/l).

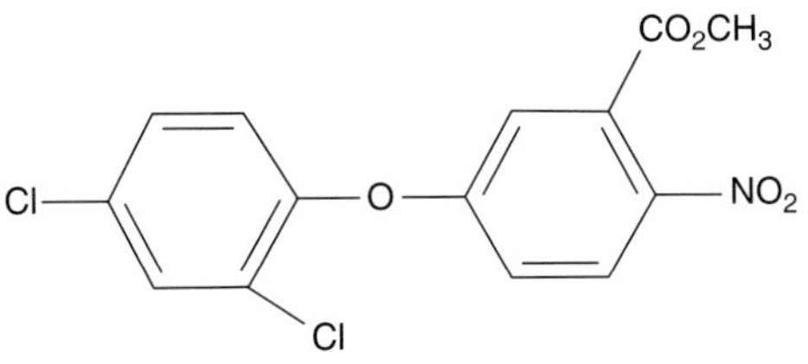

Bifenox

Bifenox, a selective, pre- and post-emergence herbicide, is absorbed by plant parts and acts by interfering with haem production (the inhibition of protoporphyrinogen oxidase). With such a mode of action, bifenox controls broadleaved weeds and grass species. Bifenox is similar in activity to many of the other members of the ***p*-nitrodiphenyl ether** group (including **acifluorfen** and **fomesafen**). Bifenox is applied to soybeans, rice and some small-grained cereals.

Bifenox is available as a soluble concentrate (SL), for easy spraying with water. It may be used with other herbicides for enhanced effect. Common trade names are Modown and Foxpro.

The herbicide is classified under Class U (unlikely to be hazardous) according to the WHO. The acute oral LD_{50} in rats is >5000 mg/kg. It is not irritating to the skin. The NOEL in rats (2 years) is >80 mg/kg b.w./day. The chemical is not toxic to birds, honey bees or earthworms. The inhalation LC_{50} in rats is 0.91 mg/l. The LC_{50} (96 h) in rainbow trout is >0.67 mg/l. The ADI is 0.3 mg/kg b.w./day. The EPA rates this chemical as moderately toxic.

Bifenox is rapidly absorbed and eliminated from the body, mainly as 5-(2,4-dichlorophenyl)-2-nitrobenzoic acid (metabolite of bifenox). In soil, the activity is present for 7–8 weeks. Degradation is mainly chemical and microbial.

In animals, bifenox is de-esterified to form 5-(2,4-dichlorophenoxy)-2-nitrobenzoic acid. Bifenox is registered for use in many EU countries.

Bifenthrin insecticide

Bifenthrin, $C_{23}H_{22}ClF_3O_2$ (2-methylbiphenyl-3-ylmethyl (Z)-(1*RS*, 3*RS*)-3-(2-chloro-3, 3, 3-trifluoroprop-1-enyl)-2, 2-dimethylcyclopropanecarboxylate), is a pyrethroid insecticide. It is a viscous liquid, sparingly soluble in water, heptane and methanol, but soluble in acetone, chloroform and toluene. It should be stored in a cool, dry and ventilated place in temperatures not exceeding 50°C. It is produced by the reaction of 3-(2-chloro-3,3,3-trifluoro-1-propenyl)-2,2-dimethylcyclopropanecarboxylic acid with 3-(bromomethylene-2-methyl) biphenyl in the presence of potassium hydroxide. It has a CAS Registry Number 82657-04-3.

Bifenthrin, a contact, broad spectrum insecticide, is also an acaricide and termiticide. It affects the pest nervous system. A variety of moths, flies, bugs, leaf and stem hoppers, mites, aphids and locusts are controlled, giving protection to cereals, citrus, cotton and vegetables. It also controls termites and has the longest residual activity against termites.

Insecticide bifenthrin is formulated as an emulsifiable and suspension concentrate, wettable powder (WP) or as granules. Trade names include Talstar. It is compatible with neutral insecticides and fungicides.

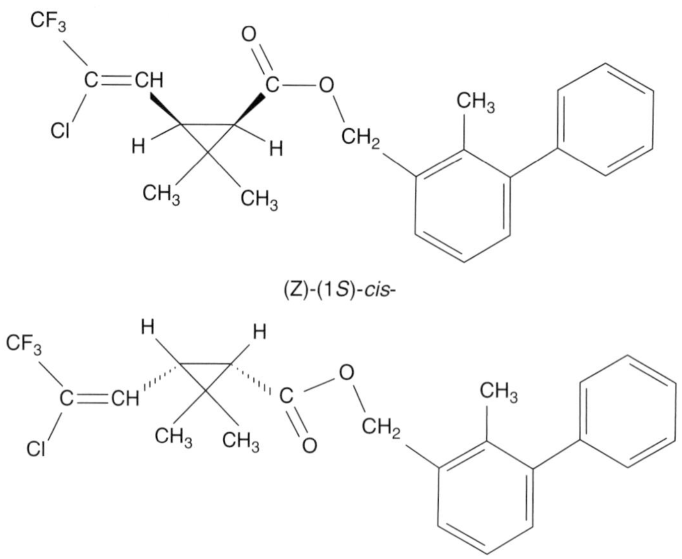

Bifenthrin

The WHO and EPA classify bifenthrin as moderately hazardous. This product has moderate inhalation and low dermal toxicity. It is extremely toxic to fish and other aquatic animals (LC_{50} (96 h) being as low as 0.00015 mg/l) and bees (oral LD_{50} 0.1 μg per bee). The ADI is 0.015 mg/kg/day. The acute oral LD_{50} in rats is 54.5 mg/kg, while the acute dermal LD_{50} in rabbits >2000 mg/kg. The acute oral LD_{50} in bobwhite quails is 2150 mg/kg. The symptoms of poisoning include increased sensitivity to sound and touch (like tingling, prickling and itching), headache, dizziness, nausea, vomiting and muscle twitching. Applicators and handlers are cautioned against any exposure to the chemical. It is also a suspected endocrine disruptor and a possible carcinogen. Bifenthrin is classified as a PAN Bad Actor chemical by the PAN Pesticide Database.

Bifenthrin persists in soil for a long period of time, evident from a half-life of up to 8 months. It has a long residual activity. In animals, the administered bifenthrin is excreted within 48 h. In plants, it metabolizes by hydrolysis.

Bifenthrin is registered in India, Australia, Canada, Cyprus, Denmark, Germany, Greece, Hungary, the Netherlands, New Zealand, the Philippines, South Africa, Tanzania, Uganda and the USA.

Bilanaphos bioherbicide

Bilanaphos, $C_{11}H_{22}N_3O_6P$ (4-[hydroxyl(methyl)phosphinoyl]-L-homoalanyl-L-alanyl-L-alanine), is an organophosphate herbicide. Its CAS Registry Number is 71048-99-2. It is soluble in water but insoluble in organic solvents. It is unstable in strong acids and strong alkalis. Bilanaphos is made by reaction of 2-nitro-5-phenol with phosphorus trichloride, followed by treatment with sulfur, and then with ethanol and butylamine. Bilanaphos is commonly present as bilanaphos-sodium. Bilanaphos is isolated from the soil-inhibiting actinomycete, *Streptomyces hygroscopicus* and produced as bilanaphos-sodium during fermentation.

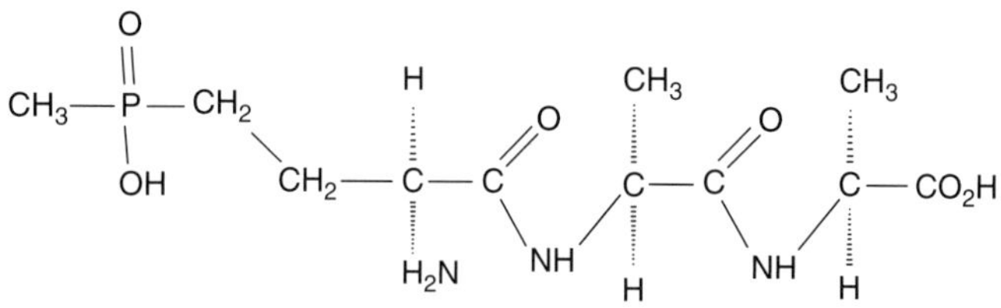

Bilanaphos

Herbicide bilanaphos is a **non-selective**, post-emergence, systemic herbicide that works by inhibiting photosynthesis. The mode of action includes irreversible inhibition of enzyme glutamine synthatase, and photophosphorylation, causing photodynamic damage. Supply of glutamine can reverse the herbicidal trend. The crops protected against many annual and perennial weed species include vine, apple, cucurbits, rubber and mulberries.

Bilanaphos is available as soluble powder and a water-soluble powder. Bilanaphos is compatible with other crop protective agents. It is available in the market under the trade name Meiji and Herbiace. A post-emergence dose is in the range of 0.5–1.0 kg/ha.

According to the WHO, bilanaphos is classified as a Class II (moderately hazardous) toxin. The acute oral LD_{50} among rats is between 268 and 404 mg sodium salt/kg. This chemical also has moderate inhalation toxicity. The acute percutaneous LD_{50} in rats is >3000 mg/kg. It is non-irritating to eyes and non-toxic to soil microorganisms like earthworms. Its LC_{50} in a 48 h period among carp is 1000 mg/l. It is a cholinesterase inhibitor and the PAN Pesticide Database has classified this compound as a PAN Bad Actor chemical. Some of the symptoms of poisoning include weakness, headache, excessive salivation, sweating and tremors.

In plants, bilanaphos is converted to **phosphinothricin** (L-isomer of glufosinate) which cannot be further degraded.

When made biologically, the presence of the enzyme phosphinothricin acetyl transferase makes bilanaphos non-inhibitory to glutamate synthesis and is thus non-phytotoxic to crops. In the case of mice, after metabolism, 2-amino-4-[(hydroxyl) (methyl) phosphinyl] butyric acid is formed. Bilanaphos is degraded in soil and water. Bilanaphos is not registered by the USA.

Bioacaricide karanjin

See **Karanjin bioinsecticide**

Bioaccumulation

Bioaccumulation refers to the accumulation of organic or inorganic substances in living tissues. It is seen when the rate of intake is faster than the rate of excretion in an organism over a period of time.

Bioaccumulation occurs when a pesticide flows with water to join a lake or any other water body and is ingested by aquatic organisms. Such organisms may die of an overdose. Chemical groups like the organochlorines are infamous for accumulating in the tissues of animals, which then enter the food chain.

Bioallethrin insecticide

Bioallethrin, $C_{19}H_{26}O_3$ ((*RS*)-3-allyl-2-methyl-4-oxocyclopent-2-enyl (1*R*,3*R*)-2,2-dimethyl-3-(2-methylprop-1-enyl)cyclopropanecarboxylate), is an orange-yellow viscous liquid, which is a potent, non-systemic, non-residual pyrethroid insecticide that produces a rapid knock-down effect. It has the CAS Registry Number of 584-79-2. Bioallethrin is an ectoparasiticide. It consists of two of the eight stereoisomers of allethrin. It is soluble in water (4.6 mg/l) and completely miscible in for example acetone, chloroform, ethanol and toluene. It degrades by UV light (DT_{50} 48.8 h) and in alkaline conditions (DT_{50} in aqueous media being 4.3 days at pH 9).

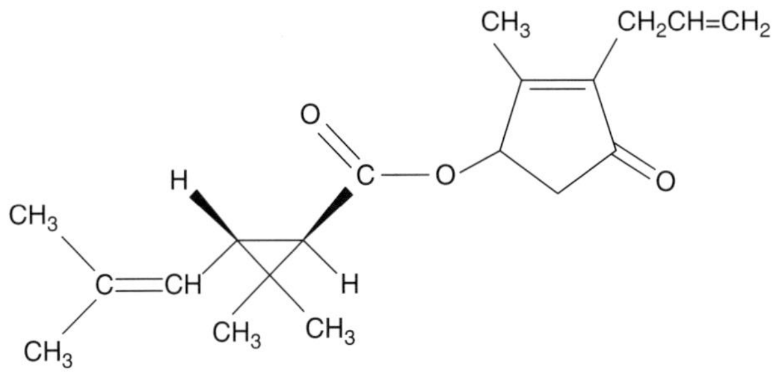

Bioallethrin

Bioallethrin, a contact, non-systemic insecticide, is a nerve toxin. It is effective against household pests such as cockroaches (order Blattodea), mosquitoes (family Culicidae) and flies (family Muscidae).

This insecticide can be combined with other non-alkaline pesticides (with the addition of **piperonyl butoxide**). Principal trade names are Bioallethrin and D-Trans. Bioallethrin is formulated as mats and coils.

The WHO and EPA categorize bioallethrin under Class II (moderately hazardous), with acute oral LD_{50} among rats being between 700 and 1042 mg/kg. The chemical is toxic to fish. The acute percutaneous LD_{50} in rabbits is >3000 mg/kg. Inhalation LC_{50} (4 h) in rats is 2.5 mg/l air. The NOEL (90 days) in rats is 750 mg/kg diet. The ADI is 0.005 mg/kg b.w./day.

Bioallethrin administered to rats at 100–200 mg/kg b.w. is eliminated in the urine and faeces within 2 and 3 days. In soil, bioallethrin degrades at pH 5.

Bioassay

Bioassays are experiments that estimate the potency of a chemical by analysing its reaction with a living organism. A quantitative bioassay can be defined as an assay in which an unknown amount of a phytotoxic pesticide is calculated by comparing its reaction with a target site with the reaction caused by known standard rates of the same compound. Such testing indicates possible hazards in agriculture or the environment.

Bioassay, an indirect method to determine herbicide concentration, has some limitations. Only phytotoxic compounds can be detected. These are sufficiently sensitive and highly suitable for screening new compounds where no standards have been established. No indication of hazards can be known by bioassays. Such a study can be useful only if sufficient time is allowed for metabolism of a pesticidal compound to be completed. But bioassays have only a limited range of sensitivity because a close relationship between dose and response occurs only within a comparatively small range of concentrations.

Despite their weakness, bioassays can provide low-cost and sensitive analyses, especially for highly biologically active compounds such as sulfonyl ureas or **imidazolinones** for screening purposes. For highly biologically active groups, bioassays are still a standard method of quantitative determination.

Biobactericide kasugamycin

See **Kasugamycin biofungicide**

Biocide

A synthetic toxic substance that can destroy microorganisms like bacteria, moulds and other fungi and insects is called a biocide. Biocides may be herbicides or fungicides or any other pesticide. Biocides may also sometimes be toxic to humans and higher animals.

Biocide chemicals are used in all kinds of pesticidal formulations, and find use as anti-fouling agents and as a wood preservative. Biocides include **terbutryn**, chlorinated hydrocarbons, organic sulfur compounds, metallic salts, quaternary ammonium compounds and phenolics.

Bioconcentration factor (BCF)

Bioconcentration is the disproportionate increase of chemical concentration in an organism (through tissue absorption) with respect to its rate of metabolism and excretion.

The bioconcentration factor (BCF) assumes importance since a higher uptake of chemical compounds causes toxins to move up the food chain. Chemical pesticides, chlorinated compounds and nitrogen fertilizers cause bioconcentration, which become lethal to many birds, frogs, fish and aquatic life.

Chemical uptake also takes place in plants, which can absorb contaminants from both the soil and the air. Such an unacceptable and uncontrolled uptake of nutrient chemical in plants can cause its concentrations to increase in meat, milk and other dairy products. Many methods have been devised to calculate and assess bioconcentrations (often it is calculated as a function of logP). However, one cannot rely on generalized BCFs because a BCF varies greatly with biotic and physical factors such as species, season, pH, temperature and salinity.

In the context of plant life, the BCF is the ratio of the plant shoot metal concentration to the soil metal concentration. Most **accumulator plants** have a BCF ratio of <1. In the context of applying the technique of **phytoremediation**, to restore land after chemical excesses, it is estimated that it would take 10 years for a crop producing 10 t biomass/ha with a BCF of >20 to reduce soil metals by half. These figures put the potential environment hazards in perspective.

Biocontrol agents

Any living creature used in the control of insect pests or disease-causing plant pathogens is called a biocontrol agent. Such an agent could be a predatory/parasitic insect or a microorganism such as a fungus or bacterium.

The following are some examples of biocontrol agents: (i) ***Bacillus thuringiensis (Bt)*** is used in the control of larvae of moths infesting crops like cabbage, tomato, aubergine (brinjal), cotton, soybean, okra and sunflower; (ii) ***Bacillus subtilis*** and some of its strains and isolates control the stem rot disease of carnation and groundnut; and (iii) ***Agrobacterium radiobacter*** strains are developed to control the soft-rot disease of vegetables, caused by *Agrobacterium tumefaciens*.

In fungi, various species of the genera *Trichoderma* (e.g. *Trichoderma viride*, *Trichoderma harzianum*, *Trichoderma hamatum* and *Trichoderma virens*), *Paecilomyces*, *Verticillium*, *Nomuraea*, *Metarhizium* and ***Beauveria*** are identified as biocontrol agents.

Paecilomyces lilacinus is used to control root-knot forming nematodes in all vegetable crops. *Verticillium lecanii* attacks aphids, whitefly, **scale insects** and mites, and is used as a biocontrol agent in fruit crops such as grapes, pomogranate, guava and custard apple.

Helicoverpa **nucleopolyhedrovirus** (HNPV) is a virus that kills the larvae of *Helicoverpa* by infecting them. These larvae destroy the crops of sorghum, gram, pigeon pea, cotton, sunflower, maize, tomato and safflower, and can be controlled by the use of HNPV.

Similarly, ***Trichogramma*** **sp.** is an example of a biocontrol insect which acts against insect pests like bollworm, stem borers and fruit borers in crops such as sugarcane, sorghum, maize, rice, cotton and okra.

There are many instances of pest control being brought about by the use of predators.

Biocontrol of crop pests – entomopathogenic microorganisms

Entomopathogens are disease-causing organisms (such as bacteria, viruses, fungi and protozoa) that kill their hosts (generally insects) or debilitate their future generations. The infected insects are unable to feed properly, remain stunted or become immobile or paralysed. Entomopathogens enter the body cavity through contaminated feed, and multiply. Bacteria can also invade through the integument of the insect, which may have been injured or pierced by parasitoids and infections caused by protozoa or nematodes. Death of the insect may either occur due to intoxication, sudden lack of oxygen, chemical changes in the gut or by the toxins or crystals produced in the bacterial cells.

Bacterium ***Bacillus thuringiensis*** (***Bt***) and its subspecies are some of the most important entomopathogenic microorganisms used in agricultural situations. They are 3–6 μm long, Gram-positive rods with an irregular chromatic network and are considered true insect bacteria. The subspecies of *Bt* (e.g. *israelensis*, *kurstaki* and *sphaericus*) have been successfully exploited to control many Lepidopteran, Dipteran and Coleopteran pests such as cotton bollworm, beetles, flies and borers that affect tobacco, sunflower, beans, pulses and other vegetables. The pod borer (*Adisura atkinsoni*), cotton bollworm (*Helicoverpa armigera*), **pink bollworm** (*Pectinophora gossypiella*) and *Spodoptera litura* (leafworm moth) are just some of the insect pests that are controlled by *Bt* formulations.

Formulations of *Bt* are made by the fermentation process. Common formulations include powder or liquid forms, with application doses of 0.3–0.5 kg/ha, especially for the suppression of *H. armigera*.

Several crops like cotton, maize and aubergine (brinjal) are genetically engineered to carry the **endotoxin** gene from *Bt*, to provide the crops with the necessary protection.

Paenibacillus popilliae, the soil-dwelling, Gram-positive, rod-shaped bacterium is an important biocontrol agent that can effectively control white grubs. During the mating season in August when grubs eat vociferously, the bacterium is ingested by them and it starts to reproduce rapidly, destroying the grub's tissues and finally killing it. This measure protects groundnut and sugarcane crops. *P. popilliae* infection is referred to as 'milky spore'. Grub species controlled by this method include *Leucopholis coneophora* and *Leucopholis lepidophora*.

P. popilliae formulations are applied to the soil at a rate of 0.5–1.0 kg/ha at the time of planting. For proper distribution of spores, 2 g spore dust (containing 200 million spores) is deposited with a spacing of 3.05 m. (See also the several entries on ***Bacillus*** species.)

Biocontrol of crop pests – parasitoids

Parasitoids are organisms that are parasitic on a host for one stage of their life cycle, resulting in the eventual death of the host. The relationship commonly sees the host being unable to reproduce. Having disabled and killed the host, the adult parasitoid develops into a free-living creature. A parasitoid may be specific or polyphagous.

Parasitoids are exploited for their entomopathogenic properties. The female parasitoid lays eggs inside or on an insect host. The eggs hatch into legless larvae which feed on the host internally (endoparasitoid) or externally (ectoparasitoid). This feeding weakens and kills the host. Adult parasitoids may also feed on nectar and pollen, thus aiding in pollination.

Wasps belonging to the genus ***Trichogramma*** are known to be very effective parasitoids. The main species include *Trichogramma chilonis*, *Trichogramma pretiosum*, *Trichogramma exiguum*, *Trichogramma brasiliensis* and *Trichogramma japonicum*, which have attained varying degrees of success in the control of, for example, loopers, borers, bollworms (*H. armigera* and *Pectinophora gossypiella*) and codling moths on a host of crops like sugarcane, groundnut, maize, sunflower, sorghum and rice.

Similarly, tachinid flies or tachinids (family Tachinidae and order Diptera) are a large family of parasitoids that are grown to control various lepidopteran pests on crops. The female lays many small white eggs on the pest host plant, which are eaten by pest host organisms like the caterpillars or larvae of sawflies, moths, beetles and **grasshoppers**. The grubs hatch in the digestive canal, and bore into the body cavity to feed and develop. Sometimes, the female lays eggs on the pest larvae.

The larvae of the moth *Epiricania melanoleuca* (family Epipyropidae and order Lepidoptera) have been used for control of the **leafhopper** *Pyrilla perpusilla* on sugarcane. This pest, which sucks sugarcane sap, can overwinter, growing into adults on ratoon crops. The woolly and cottony caterpillars of *E. melanoleuca* moths are ectoparasitic in nature and suck on the sap of the leafhopper, making them unable to reproduce. Cultivating these moths and spreading around 500,000 eggs/ha can bring about good control. A smaller number of adults (around five per leaf) are sufficient to prevent an outbreak of an infestation. They are commonly supplied as cocoons so that they can establish themselves in the field comfortably.

Biocontrol of crop pests – predators

Animals that feed on other animals are predators. Predators require several prey to complete their life cycle. Almost all insects, beetles, spiders, mites, rodents, toads, frogs and birds are predators. Some of them are used as pest controlling agents by man.

Coccinellid beetles (family Coccinellidae), commonly known as **ladybirds**, feed on hemipteran pests like aphids, scale insects, **mealybugs** and mites. Vidalia beetle and the mealybug ladybird (*Cryptolaemus montrouzieri*) are voracious predators of mealybugs, **scale insects** and psyllids and thus protect horticultural and fodder crops. The presence of 10–15 beetles per plant normally keeps the pest under control.

The family Chrysopidae of the order Neuroptera has a large number of species of insects that are biocontrol agents. ***Chrysoperla** carnea*, *Mallada boninensis*, *Apertochrysa crassinervis* and *Mallada astur* are the most common chrysopid predators. The first two are useful for protection from aphids, fruitfly maggots, eggs of **leafhoppers** and many soft-bodied insects. *C. carnea* has been found useful on cotton, green gram, sorghum, maize, safflower, sunflower and pigeon pea.

Anthocorid bugs of the order Hemiptera are called flower bugs or **minute pirate bugs**. *Orius tristicolor* and *Anthocoris nemoralis* are common examples. They are used as agents in the management of thrips, aphids and the eggs and young larvae of moths. They feed on these insect pests that usually attack and destroy sugarcane crops and fruit orchards.

Syrphids or hover flies have been recorded as important predators of aphids on several crops. The female lays eggs in aphid colonies and must feed on **honeydew** for healthy eggs. The emerging larvae feed on the aphids and thus control their numbers. *Syrphus rectus* and *Allograpta obliqua* are two common examples of syrphids.

There are many other predators that are reared for their effective control of crop pests. *Cyrtorhinus lividipennis* (control of brown plant hopper) and ***Amblyseius*** spp. are predatory mite species that control phytophagous mites in several crop ecosystems.

Biocontrol of insects – baculoviruses

Baculoviruses are viral pathogens that attack insects. The family *Baculoviridae* includes large, rod-shaped viruses, which are divided into two genera, namely, nucleopolyhedroviruses (NPVs) and granuloviruses (GVs). The majority of baculoviruses used as biocontrol agents are NPVs. These are species specific and have a narrow-spectrum insecticidal application. Besides, they show no negative effect on plants, mammals, birds or fish. Because of these two factors they are desirable in the IPM programme.

Baculoviruses are unable to reproduce without a host, a trait that makes them obligate parasites. They are often found wherever insects exist. When an insect consumes a virus particle, the infection starts. Infective baculovirus particles (virions) carry their genetic material to all the cells of the host body, creating so many virus cells that the insect finally dies.

The nucleopolyhedroviruses (NPVs) are irregular, spherical, cubic or hexagonal crystals with blunt or 'melted' edges, which are deposited in the nuclei of the fat body cells. Each polyhedron contains many particles or virus rods, which are alkali sensitive.

The NPVs have been successful in the control of many moths and insects on, for example, pigeon pea, groundnut, tobacco, cotton and sunflower. The field-bean pod borer (*Adisura atkinsoni*), the caterpillar of the leafworm moth (*Spodoptera litura*), the caterpillar of the cotton bollworm (*Helicoverpa armigera*) and the red hairy caterpillar of moth *Amsacta albistriga* are some examples of insect pests that are controlled by NPVs.

NPV suspensions are compatible with low doses of insecticides and increase their efficacy when combined with boric acid, jaggery, neem seed-kernel extract (NSKE) and endosulfan. It can also be combined with wasps belonging to the genus ***Trichogramma*** and the wasp ***Chrysoperla** carnea*.

Recommended doses of NPV are $1.5–3.0 \times 10^{12}$ proteionic occlusion bodies (POB)/ha (250–500 larval equivalent (LE)/ha where 1 LE = 6×10^9 inclusion bodies) for pigeon pea and 3.0×10^{12} POB/ha (500 LE/ha) for cotton.

GVs have also demonstrated their lepidopteran pest control capability, especially against the sugarcane shoot-borer (*Chilo infuscatellus*). GVs contain oval granules, with each containing one virus rod in an occlusion body. The occlusion bodies have a granular appearance and are relatively small sized, as seen in infected cells. The occlusion body, made of a single viral protein called granulin, envelopes the virion. GVs are exploited for their biocontrol properties, for example in the case of the codling moth (*Cydia pomonella*), a common insect pest of pear and apples. The virus is called codling moth granulosis virus (CMGV).

Baculoviruses have been shown to have no negative impact on plants, mammals, birds, fish or even on non-target insects. Table B.3 provides details of some important baculoviruses used against crop pests. (See also **Biocontrol agents**.)

Biocontrol of insects – entomofungi

Fungal species that attack insects belong to the phycomycetes, ascomycetes, basidiomycetes and deuteromycetes. Twenty such entomofungal species are known and are grown for their biocontrol properties.

The fungal pathogen penetrates the cuticle of the insect. It secretes toxins in the insect, while covering the host with mycelium and spores of a particular colour, and this condition is commonly referred to as muscardine disease. The attack makes the insect stop feeding and become weak, before dying.

Table B.3. Some important baculoviruses and their role in biocontrol. (Table courtesy of Shubhangi Borse and Leena Achari.)

Name of virus	Insect controlled	Crop protected
Spodoptera littoralis nucleopolyhedrovirus	African cotton leaf worm (*Spodoptera littoralis*)	Cotton, maize and tomato
Codling moth granulosis virus	Codling moth (*Cydia pomonella*)	Apples, pear and plum
Cabbage armyworm nucleopolyhedrovirus	Cabbage moth, American bollworm, diamondback moth, potato tuber moth and grape berry moth	Cotton, cabbage and tomato
Anagrapha falcifera nucleopolyhedrovirus	Celery looper (*Anagrapha falcifera*)	Vegetables

The genus *Beauveria* has many species that are exploited for their entomofungal properties. Often found in soil, *Beauveria* species can effectively kill termites, whiteflies, many types of beetles (including the Colorado potato beetle), borers, weevils and many others. ***Beauveria bassiana*** is a prominent species used for biocontrol protecting, for example, cotton, maize, lucerne, safflower and sunflower. *Hirsutella thompsonii*, *Metarhizium* spp., *Verticillium lecanii*, *Nomurea rileyi* and *Entomophthora muscae* also infect many species of flies, maggots, aphids, mites and thrips. These biocontrol agents are used on sugarcane, castor, groundnut, pigeon pea and cotton.

Most entomofungi pathogens can be grown on a variety of substrates like coconut water, rice bran, wheat bran and maize meal. Climatic conditions suitable for their proper growth are humid weather and low temperatures. Exposure to pesticides, especially herbicides, can severely restrict the germination and growth of these biocontrol agents. The fungal formulations are commonly available in powder and water suspension forms. A common dosage is a culture of 10^9–10^{11} spores mixed with 1 l of water and sprinkled in a specified area, for effective grub control.

There are hundreds of protozoans and microsporidia too that attack insects, fish and crustaceans and hold some potential as biocontrol agents. **Microsporidia** are unicellular parasites that form spores and are classified under fungi. *Mattesia oryzaephili*, *Nosema* sp. and *Paranosema* sp. are examples. These provide natural control of lepidopteran members, especially among stored grains.

Biocontrol of insects – entomopathogenic nematodes

Entomopathogenic nematodes are those species that feed on insects, making them important as biocontrol agents. Soil-borne nematode species of *Steinernema* (e.g. *Steinernema feltiae*) and *Heterorhabditis* (order Rhabditida) are exploited for their parasitism on insects, in conjunction with some bacterial species like *Xenorhabdus* sp. or *Photorhabdus* spp. (*Photorhabdus luminescens*). This method of biocontrol has limited scope, since nematodes cannot move freely in the absence of soil water.

The insect–pathogen pathway proceeds with the nematode nestling the bacterium in its body, entering the insect gut via the haemocoel. There, the bacterium reproduces and feeds on the insect tissues, while producing a toxin that kills the insect within 2 days. Biostatic substances in the bacteria prevent decay of the insect body, thus making food available for both the nematode and the bacteria.

Entomopathogenic nematodes are effective in greenhouse situations, potted plants and in crops with sufficient irrigation.

Biocontrol of pests, age-old methods

Insect pests have been in existence from time immemorial and farmers have to control them by all kinds of ways. Before chemical pesticides were invented, farmers the world over used materials of biological origin for protecting their crops. The following passages illustrate just a few of the biocontrol methods practised in the past.

In ancient Indian literature, one comes across many prescriptions for insect pests. They are applicable even today. The 11th century AD historic treatise *Vrikshayurveda* by Surpala and the 16th century AD document *Vishvavallabha* by Chakrapani Mishra have recommended several herbs, plant parts and materials of animal origin for controlling insect pests. Out of them *karanja* (*Pongamia pinnata*), *nimba* or **neem** (*Azadirachta indica*), *shatapushpa* (*Anethum sowa*) and *vasa* (*Adhatoda vasica*) had been known to have insecticidal and antimicrobial properties. These plants along with cow urine and ash were made into pastes and used in several ways for killing or repelling insect pests (for instance, it is applied to the affected trees or used for fumigation). *Vishvavallabha* also recommends: (i) planting of *shatapushpa* (*Anethum sowa*) and *kuberakshi* (*Caesapinia crista*) densely between tree rows for controlling insect pests; and (ii) planting of *ajagandha* (*Cleome gynandra*) to repel termites.

In the traditional Sri Lankan agriculture scenario, the treatises reveal the planting of poisonous plants for controlling pests. Also, branches, stems and leaves of a poisonous creeper, the glory lily (*Gloriosa superba*), were spread over the field to kill pests on the rice roots. In paddy fields, a crude paste of neem (*A. indica*) and *mahapatta* leaves, wrapped in a cloth, were placed at the entry point of the water canal. These carried the pesticidal element right through the fields. Leaves of *mahapatta* and of neem soaked in water for several days and then sprayed over bean plants was another innovative method for repelling insects. Farmers in some countries grew trees along the borders of the fields or kept branches of trees in the area so that birds like owls could perch and prey on insects.

Dusting plants with wood ash in the morning when leaves are wet with dew was a usual ancient practice for repelling insects.

Many of the above-mentioned non-invasive techniques are seen to be in use in many Asian farming communities even today.

Biodegradability

Biodegradation is the process of reducing a substance to a simpler molecular structure by microbial processes. Biodegradability is the susceptibility of a substance to such reduction. Soluble sugars formed from cellulose are an example of biodegradation. The carbohydrates and proteins are converted into carbon dioxide (CO_2), and ammonium (NH_4^+) and nitrate (NO_3^-) ions, respectively, in the final stage of biodegradation.

Substances have their own resistance or susceptibility to degradation. Branched alkylbenzene sulfonates are much more resistant to biodegradation than are linear alkylbenzene sulfonates.

A long, straight alkyl chain is susceptible to bacterial attack. Organophosphorus insecticides are more easily degraded than chemicals like **DDT**.

Microorganisms attack substances, using enzymes, to get energy, carbon, nitrogen or other nutrients. The enzymes completely destroy or simplify the molecular structure of the substance. By-products of microbial physiology can bring about further changes (fermentation by-products, change in pH, etc.).

Biodegradation can be measured by applying the chemical and physiological assays of pure or mixed cultures or samples of soil, water or industrial sludge. The amount of carbon dioxide evolved in aerobic soils or the amount of ^{14}carbon released from a radioactive substrate are common measures of biodegradability. Biodegradation and its monitoring are relevant in view of recycling of limited natural resources.

Biofungicide carvone

See **Carvone as a bioPGR**

Bioherbicides

Bioherbicides are biological weed control agents, with a living microorganism as the active ingredient. Most agricultural pests have natural enemies, which help control pests. When such naturally occurring enemies are used as weed control agents, they are called bioherbicides.

Bioherbicidal agents include fungi, protozoa and bacteria, which have compounds that may be phytotoxic to plant residues. The bioherbicide in which a fungus is the control agent is called a **mycoherbicide**. Many mycoherbicides are persistent in the environment.

Although the specificity of bioherbicides is similar to that of chemical herbicides, environmentally, the agents of biological control seem to be safer than synthetic chemicals.

BioIGR karanjin

See **Karanjin bioinsecticide**

Biological control defined

Professor Henry S. Smith, who was an entomologist from the University of California, is credited with coining the term 'biological control' in 1919. Biological control is defined as: 'the actions of parasites, predators and pathogens in maintaining another organism's density at a lower average than would occur in their absence' (http://nature.berkeley.edu/biocon/What%20is%20Biological%20Control.htm).

Biocontrol measures comprise the use of microbial, viral and fungal organisms that attack or infect the pest, for food or as a host. Entomophagous predators like bigger insects, fungi, nematodes or reptiles, form another natural biocontrol system for pests. Use of plant-derived pesticides, **pheromones**, repellents and genetically modified crops that can resist pathogen attacks are other methods to implement biological pest control.

In spite of successful results, especially with ***Bacillus thuringiensis*** (***Bt***)-based insect control products and a few biofungicide products, **biopesticides** have not occupied an important place in the world pesticide market. Biologically prepared formulations comprise around 1% of the world market. Reasons for this low take up include: (i) limited formulations; (ii) inconsistent field results; and (iii) not as remarkable an improvement in yield when compared with chemical pesticides.

However, the overall benefits accruing from biological pest control methods are expected to help increase acceptance of biopesticides and biological control measures more widely.

Biological control methodology

Biological control consists of the use of natural enemies to control pests. These natural enemies include predators and parasitoids. A vast number of insects and nematodes predate upon and parasitize agricultural pests. These control agents are discussed below.

A predator is an organism that preys on another and assumes importance in the present context for their biological control of pests. Predators catch and consume large numbers of crop pests during all or some stages of their lives. Although birds and reptiles do contribute towards biological control, a vast number of insect species become important from the point of view of biological control. The more important orders under **phylum Arthropoda** are hemiptera (bugs), coleoptera (beetles), diptera (flies), neuroptera **(lacewings)**, and acarina and araneae (spiders and mites).

Predators may either be monophagous, polyphagous or oligophagous, each type characterized by the number of species the predator can feed on. Oligophagous predators, which prey on a limited number of species, are considered ideal as these can maintain themselves by shifting to available prey if the most desirable target organism is not present.

Parasitoids are organisms that complete their life cycle inside or on a host, consuming the host alive and eventually killing it. It is a form of parasitism which kills the host. Parasitoids are often host specific and have a good ability to search and locate their host individuals. Often they are free-living organisms for the better part of their lives and parasitize a host for a particular stage, usually for laying eggs. Endoparasites lay eggs inside the host and complete their development inside whereas ectoparasites lay eggs on the host and the larvae remain outside usually and feed by inserting their mouthparts into the host body. Many species of insect in the orders Hymenoptera, Diptera and Coleoptera are predators of crop pests and are hence exploited.

A suitable control agent has to be carefully chosen from the options available, and how they are applied (or released) on to a field has to be decided. Three application techniques are: (i) inoculation technique; (ii) inundation technique; and (iii) conservation technique.

In the inoculation technique, it is well known that a natural enemy may not be able to control an outburst of pest in its new home. Selection of a voracious predator might be in order. Before introducing a new control agent, it is important to confirm that it is quarantined and checked for any infection. A high reproductive capacity and its capacity to hunt for prey are important considerations while choosing a natural enemy.

The inundation technique involves heavy release of a biocontrol agent in a location that has a pest on whom the agent depends heavily. A decline in pest numbers brings about the decline of agent population.

The conservation technique of activating biocontrol agents focuses on naturally growing the population of natural enemies of crop pests. The conservative and wise use of insecticides and the creation of comfortable conditions for growth of biocontrol agents are two approaches for applying biocontrol techniques conservatively. Examples of conservation include: (i) keeping small parts of the field with host weeds for harbouring these natural enemies; (ii) maintaining desired humidity within the crop area for their growth; and (iii) making alternate hosts available for the pest natural enemies. (See also **Biological weed control agents – application methods**.)

Biological control of locusts

Desert locusts have long been known to cause tremendous damage to crops. Locusts lay eggs in the desert, from which wingless nymphs, called hoppers, hatch and develop into adults (Fig. B.3). These desert locusts attack crops in swarms. The African continent has often been affected by locust attacks.

Fig. B.3. An adult locust. Sketch courtesy of Kedar Champhekar.

Biological control of these pests has been made possible by the use of a new variety of an entomopathogenic fungal species, namely *Metarhizium anisopliae* var. *acridum*, since it specifically targets the family Acrididae, to which major locusts and grasshopper species belong. This was possible by the support of a French project called LUBILOSA, which is the acronym of 'Biological control of locusts and grasshoppers' in French. This biofungicide is effective in an oil suspension, which ensures that the spores do not blow away in the desert winds. A known trade name of this formulation is Green Muscle. This biocontrol method stops locust encroachment on to crops, especially if the fungicide is sprayed on the wingless nymphs. The fungus persists in the environment and the cycle of biological control continues.

Biological control of nematodes

Although there are many methods by which nematode populations can be checked chemically, most of these nematicides are costly, harmful and sometimes counterproductive (e.g. development of resistance to these chemicals). However, use of natural enemies of nematodes can inhibit pest growth.

Tardigrades, enchytracids, mites, larger insects, fungi, bacteria and predator nematodes are some nematode predators. Tardigrades (phylum Tardigrada) are small creatures which are similar to arthopods as well as to nematodes because they have a triradiate pharynx with stylets. Some species are carnivorous and are known to be nematode predators. Another group of organisms that are predators of nematodes are members of the Enchytracidae which have species that live on land and in ice fields (potworms). Some nematodes (genus *Mononchus*) predate on smaller nematodes.

There are many species of fungi that feed on nematodes. Extensive study of the trapping mechanisms of fungi reveals that nematophagous fungi use various methods to capture their prey. The nematode-trapping fungi use adhesive knobs, branches, networks or conidia, or often, constricting rings to trap their prey on contact. These mechanisms become activated the instant the nematode touches the waiting fungus, while moving about. In the case of rings, by increasing the ring cell volume and increasing the osmotic pressure of the ring cell more than that of the nematode cells, the fungus makes the nematode cell wall collapse. The fungus then injects its invasive hyphae into the live nematode and feeds on its body, enabling the fungus to grow reproductively and vegetatively. Examples of such fungi using various mechanical or adhesive devices to trap nematodes are: *Arthrobotrys oligospora*, *Dactylaria candida*, *Arthrobotrys dactyloides* (constricting rings), *Nematoctonus leiosporus* and *Dactylaria brochopaga*. A particular bacterium, *Pasteuria penetrans*, has the potential to be a biological control agent for nematodes especially the *Meloidogyne* spp. (*Meloidogyne incognita*, *Meloidogyne exigua* and *Meloidogyne coffeicola*) responsible for root galls and root necrosis in coffee plantations. Biocontrol is brought about by the bacterium attaching itself to the nematode, developing a penetration tube and penetrating the nematode cuticle via its spores. Extensive reproduction in the nematode body weakens the host, resulting in reduced reproduction.

Another bacterium, *Bacillus firmus*, is also being commercially exploited (e.g. available as BioNem) to protect against nematode attacks on tomatoes, cucumber, pepper and basil.

There are many small organisms like rotifers, predatory nematodes and protozoans which play an important part in controlling nematodes in natural settings.

The biological control properties of mustard (the seed and leaves) have been a refreshing discovery for farmers cultivating potatoes. Mustard is planted in rotation between the wheat and potato crops. It makes useful green manure especially when incorporated after 6 weeks. The decomposition leads to the release of the fumigant **methyl isothiocyanate** into the soil, in a deferred manner. This renders the required nematicidal effect.

Sustained use of mustard and other biologically active agents can help prevent groundwater contamination caused by **ethylene dibromide** and other chemicals.

Some biological nematicides have been developed that act as control agents. A proprietary formulation of chitin and urea (the trade name being Clandosan) contains biological matter from crab and crawfish exoskeletons. The active ingredients therein encourage the growth of soil microorganisms that degrade the **chitin** in the nematodes and their eggs. Thus, the formulation acts as a biological agent. There is another biological formulation (Di-Tera), produced by antibiotic fungal metabolites from *Myrothecium verrucaria*. Organic matter, with a low carbon to nitrogen ratio, is believed to be suitable as a nematicidal agent. Extracts of seeds of the **neem** plant (*Azadirachta indica*) are also used as nematicides.

Biological half-life

Compounds like heavy metals, radioactive substances or large organic molecules do not get metabolized easily on entering the body. They instead accumulate in different body parts. They take many days or even months to get completely eliminated from the living system. During this period, the substances undergo a slow process of decomposition either due to radioactivity or due to metabolic processes of the consumer.

The time required by a substance (metabolite, drug, signalling molecule or a radioactive nuclide) to reduce to half of its original quantity is known as the biological half-life. In other words, it is the time taken by a substance to lose half of its radiological, chemical or physiological activity. It is generally denoted as $t_{1/2}$. For example, the average biological half-life of **malathion** is 6 days. It is very important to consider the half-life of a chemical when using it as a pesticide.

Biological weed control

Techniques used to control weed pests in a natural way are comprehensively known as biological weed control. Biological weed control is a challenging task and may completely fail (if not monitored in the initial stages), compelling the use of chemical and cultural methods. However, use of natural predators to control noxious weeds is an old practice.

A weed normally becomes a problem when it leaves its natural surroundings, since it leaves its natural enemies behind. Thus, an introduced weed normally thrives in its new home. Orchestrating an attack by its natural predators (if known), by one of the application methods, is a good way to bring about a check on the weed. Such control agents could be insects, fungi and other microbes. It is important to note that biological control is just one of the many techniques used to control and manage weeds.

Advantages of biocontrol of weeds are self-perpetuation, specificity and cost-effectiveness.

It is to be noted that the introduction of a new organism as a control agent may have far-reaching effects on the environment as it disturbs the existing balance of flora and fauna, sometimes even creating new enemies out of innocuous organisms. For example, the Argentinian cactus moth (*Cactoblastis cactorum*) was introduced to Australia to control the prickly pear cactus but started targeting non-weedy prickly pear cactus.

Biological weed control agents – application methods

Biological agents are increasingly playing an important role in pest management. Crop protection techniques include rearing natural enemies of pests, so that they may be released on demand, to nip pest damage in the bud. The menace of weeds is a major impediment in crop yields and requires extensive quantities of herbicides. There are four main approaches to bringing about biological control of pests, namely: (i) the classical method (also called importation or inoculative approaches); (ii) the augmentative (or inundative) method; (iii) the conservation method; and (iv) the broad-spectrum method. Examples are provided here for weed pests although they are similar for other pests.

The classical (or inoculative or importation) method of biological weed control is an ecological approach. The control comes about by introducing host-specific natural enemies of the target weed which have adapted to their new homes. Often, in the absence of natural enemies, weeds become noxious in the new surroundings. Identifying the original home of a weed and its natural enemies (with their specificity) are the two steps of the classical method. The classical method is suitable for fungal and arthropod agents because of their specificity. Rusts are often used as biocontrol agents against certain weeds.

The augmentative or inundative method of biological weed control is based on the release of a large number of control agents in quick succession so that the area is inundated with the control agents. This method gives effective results; however, it requires that both the target and the control agents are natives of the land. It also demands precise knowledge of specificity of the agent and its behaviour in mass captivity. This constraint makes this technique less practical. Nematodes are an appropriate choice.

The conservation method of application of biological weed control is a theoretical concept that advocates the natural conservation and/or increase in population of identified natural enemies and predators. It is believed that these pests would be naturally controlled.

The broad-spectrum method of weed biocontrol involves manipulating both the area to be treated and the control agent so that control is brought about swiftly. Selective grazing is an example of this technique. Table B.4 illustrates some biological control agents and their mode of application.

Bionematicide dihydroazadirachtin

See **Dihydroazadirachtin bioinsecticide**

Biopesticides

Pesticidal formulations synthesized out of biologically present substances or microorganisms are called biopesticides. These encompass microbial organisms, plant-derived pesticidal compounds and insect **pheromones**. Biopesticides are environmentally friendly and safe. Today, biopesticides are used for niche crops. ***Bacillus thuringiensis*** (***Bt***) is an example. Their low toxicity, environmental safety and ease of registration make biopesticides an attractive option.

However, compared with the potential that exists for biopesticides, there has not been significant progress in this direction. Currently, the number of products registered as biopesticides is fewer than 200.

The present low acceptance rate of biopesticides is largely, among other reasons, a result of: (i) inconsistent field results; (ii) inadequate formulations; and (iii) extreme pest specificity. Use of biopesticides could be increased if they were used in combination with reduced rates of chemical pesticides.

Bio-purification factor

The bio-purification factor (BPF) is the inherent capacity of a biological system to resist the uptake of toxic elements. The BPF of a toxic element E is determined in relation to the nutrient element calcium present in food and consumed.

Since the BPF is functional, the toxic element consumed through food is seen to decrease in the food chain while the proportion of calcium increases. For example, the toxic element in grass automatically gets transferred to the animal at a reduced rate.

Table B.4. Some examples of biocontrol agents used against weeds with their methods of application. (Table courtesy of Kalyani Paranjape.)

Biocontrol agent	Type	Weed	Application[a]	Remarks
Dactylopius ceylonicus, *Dactylopius opuntiae*	Cochineal insect	Prickly pear cactus	C	
Chrysolina quadrigemina	Leaf beetle	St John's wort	C	
Hyles euphorbiae	Leafy spurge hawkmoth	Leafy spurge (*Euphorbia esula* L)	C	
Oberea erythrocephala	Root and stem boring beetle	Leafy spurge (*E. esula* L)	C	
Rhinocyllus conicus	Weevil	Musk thistle	C	
Phytophthora palmivora	Fungus	Stranglevine on citrus orchards	I/A	A mycoherbicide that can damage sensitive crops
Fusarium oxysporum f. sp. *cannabis*	Fungus	Unwanted marijuana plants	I/A	Inoculum on barley straw or soybean meal
F. oxysporum	Fungus	Witchweed among sorghum	I/A	Inoculum grown on sorghum straw. Increases yield
Puccinia chondrilla	Rust fungus	Rush skeletonweed	I/A	
Puccinia punctiformis	Rust fungus	Canada thistle	I/A	
Pseudomonas syringae	Bacteria	Canada thistle	I/A	Sprayed with a surfactant
Xanthomonas campestris pv.	Bacteria	Bluegrass varieties	I/A	Injures and causes wilt
Trichechus spp.	Sea manatee	Cattails and water hyacinth	B	
Sheep and goats	Mammal	Leafy spurge, gamble oak	B	Eat fast; destroy crops
Geese and ducks	Bird	Grasses, small broadleaved weeds	B	Geese selectively control nutsedge

[a]B, Broad spectrum; C, classical; I/A, inundative/augmentative.

Biorational products

Biorational products are considered third-generation pesticides that are environmentally sound, non-toxic and structurally similar to chemicals produced by insects and plants.

Most biorational insecticides are as effective (if not more effective) on insect pests as synthetic chemical pesticides. But, since they have a short residual life, more intense scouting for target pests is required on the field. Cultural methods need to complement the effects of these compounds. For instance these products should be applied at the time in the pest's life cycle when the pest is most vulnerable.

Biorational insecticides are classified into two distinct groups: (i) biochemical; and (ii) microbial. Biochemical products include hormones, enzymes, **pheromones** and natural growth regulators. Microbial products originate from biological organisms such as bacteria, fungi, nematodes, protozoa and viruses. Such a categorization overlaps with biological pesticides. But whereas **nicotine** would be a bioinsecticide, it cannot be considered a biorational product due to its high mammalian toxicity. Biorational products also include low-toxicity soaps, oils, sulfur, white-clay-based products and potassium bicarbonate. Most of these products are effective against insects, fungi, mites and bacteria. Their phytotoxicity has to be ascertained before use.

Bioremediation or bio-treatment

Bioremediation or bio-treatment is treating contaminated land with naturally occurring or genetically modified microbes that degrade complex organic compounds (e.g. pesticides, insecticides, polychlorobenzenes, radioactive substances and heavy metals) into less harmful and less mobile substances, making the soil suitable for cultivation.

Some plants naturally accumulate large quantities of heavy metals (lead, cadmium and mercury). These are called hyperaccumulator plants. Higher plants like common ragweed (*Ambrosia artemisifolia*), hemp dogbane (*Apocynum cannabinum*) or nodding thistle (*Carduus nutans*) have been shown to absorb from 400 to 1250 mg lead (Pb)/kg shoots. After phytoextraction, the above-ground portion with the metal content of the plant is harvested and removed.

The success of phytoextraction depends on: (i) the extent of soil contamination; (ii) metal availability for root uptake; and (iii) the ability of plants to intercept and absorb the metal for shoot accumulation. It also depends on the interactions of soils, metals and plants. Metal adsorption impedes the process of phytoextraction, while the addition of chelates stimulates the release of metals in solution and enhances root uptake. Some forage crops and halophytes are able to take up a significant quantity of salt. There are also some microorganisms that perform this task, contributing to bioremediation. (See also **Phytoremediation**.)

Biorepellent capsaicin

See **Capsaicin bioinsecticide**; **Black pepper oil as a repellent**

Bioresmethrin insecticide

Bioresmethrin, $C_{22}H_{26}O_3$ (5-benzyl-3-furylmethyl (1*R*,3*R*)-2,2-dimethyl-3-(2-methylprop-1-enyl) cyclopropanecarboxylate), is a viscous yellowish liquid pyrethroid insecticide. It has a CAS Registry Number of 28434-01-7. It is soluble in for example water, ethanol, acetone and chloroform. It decomposes above 180°C and under UV light. It is readily hydrolysed in alkaline media.

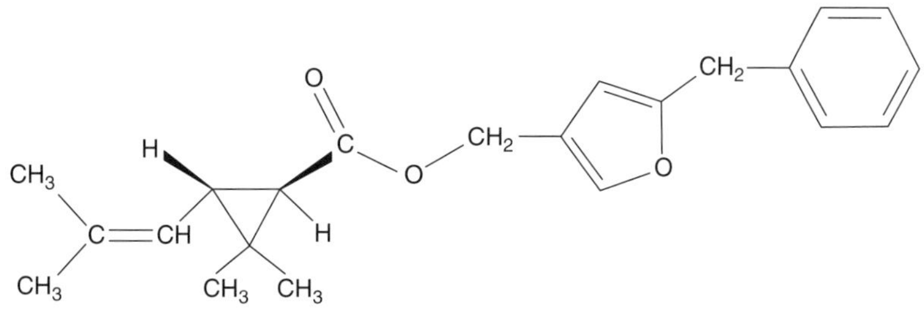

Bioresmethrin

Bioresmethrin, a contact insecticide, acts on the nervous system of crop pests, and also of mosquitoes, cockroaches and flies.

Bioresmethrin is available in the market under the trade name of Isothrin. It is usually formulated as an aerosol dispenser and oil miscible liquid.

According to the WHO, bioresmethrin comes under the toxicity Class III (product that is slightly hazardous). It has an oral LD_{50} of 8600 mg/kg and dermal LD_{50} of 10000 mg/kg in rats. The inhalation LC_{50} (4 h) in rats is 5.28 mg/l air. The NOEL (90 days) in rats is 1200 mg/kg diet. The 96 h LC_{50} in rainbow trout is 0.00062 mg/l. The ADI is 0.03 mg/kg b.w./day. It is very toxic to honey bees and fish. It is a PAN Bad Actor chemical.

Symptoms of over-exposure include irritability to sound or touch (like prickling, itching and burning), headache, dizziness, vomiting, diarrhoea and fatigue.

In animals, bioresmethrin is metabolized by the oxidation reaction in the isobutenyl methyl group by ester cleavage and conjugation reaction. It is strongly adsorbed on soil.

Bioresmethrin is registered in the USA but not registered in the EU. It is registered for use in homes and gardens in Australia.

Biphenyl fungicide

Biphenyl, $C_{12}H_{10}$ (1,1′-biphenyl), is an aromatic hydrocarbon compound used as a fungicide and a **fungistatic** (anti-sporulation). Also known as **diphenyl** and **phenylbenzene**, it has a CAS Registry Number of 92-52-4. It is a colourless, crystalline substance that is readily soluble in ethanol, diethyl ether and soluble in water at (4.45 mg/l). The compound is stable in acidic and alkaline media. Biphenyl is produced by the reaction of phenylmagnesium bromide with 1,4-dimethyl-2,3-dibromobutane in an inert solvent.

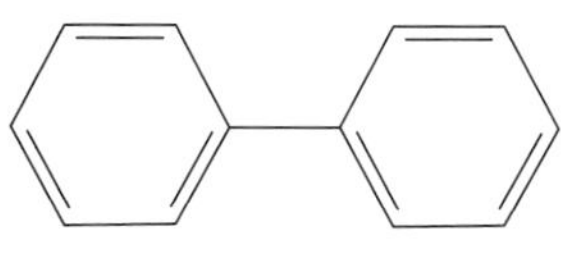

Biphenyl

Biphenyl, mainly used in post-harvest treatment, finds fungistatic application during the storage and transportation of citrus fruits. The chemical acts on the CNS of the pest, causing convulsions followed by death.

Biphenyl is classified as a Class U toxin (product unlikely to present acute hazard in normal use) by the WHO and as a Class III toxin (slightly toxic) by the EPA. The oral LD_{50} doses in rats and rabbits have been recorded as 3280 and 2400 mg/kg, respectively. The LC_{50} (96 h) values for some fish species are between 1.5 and 7 mg/l. The vapour or dust of the compound causes respiratory discomfort and prolonged exposure can cause serious harm to humans too (especially when the vapour concentration is in excess of 0.005 mg/l air). That is why it is necessary to store biphenyl in a tightly closed container, in a cool, dry area. Symptoms of toxicity include aches in the body, stomach and head, and also nausea. Biphenyl is also suspected to be an endocrine disruptor, although it is not a PAN Bad Actor chemical.

On ingestion by animals, biphenyl forms 4-hydroxybiphenyl and other phenolic compounds. Residues of biphenyl are found in plants.

Biphenyl is not registered in the USA, but Australia registers its use.

Biphenyl technical analysis

Biphenyl, a fungicide with post-harvest applications, is commonly used during wrapping of citrus fruits to inhibit the mycelial growth and spore formation of citrus fruit mould. The biphenyl content of fruit peel is determined by steam distillation followed by thin-layer chromatography (TLC) and spectrophotometric measurement. It is analysed in the following manner.

Blend the peel of the citrus sample in the high-speed blender for 5 min after determining the peel:fruit ratio by weight. Place 300 g of the peel in a 1 l round-bottomed flask and add water to half the volume of the vessel (500 ml). Fill the moisture test apparatus head with water to the overflow point and add 3 ml *n*-heptane on the water layer. Wrap the exposed portion of the flask and the extraction unit with aluminium foil. Gradually heat the flask evenly until the contents start to boil vigorously. Extract for 3 h and drain the contents into a separating funnel. Discard the lower aqueous layer and drain the heptane extract through a glass column (with an outer diameter of 2.5 cm and an internal diameter of 8–10 mm) containing granular anhydrous sodium sulfate (to absorb the water in the liquid), into a 10 ml volumetric flask. Rinse the column with 1 ml heptane. Rinse the extraction unit with five 2 ml portions of ethanol and collect each rinse in the separating funnel. Add 5 ml *n*-heptane and shake vigorously for a few seconds. Add 50–75 ml water, shake and allow the layers to separate. Discard the lower aqueous layer and pass the heptane layer through the same sodium sulfate column. Rinse the separating funnel and column with enough *n*-heptane to dilute to 10 ml volume.

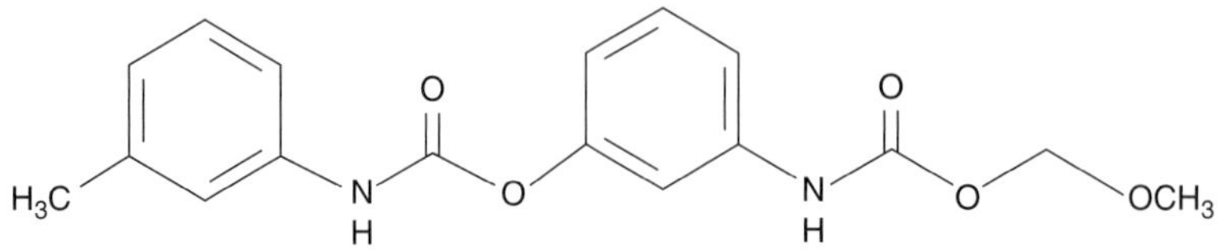

Phenmedipham

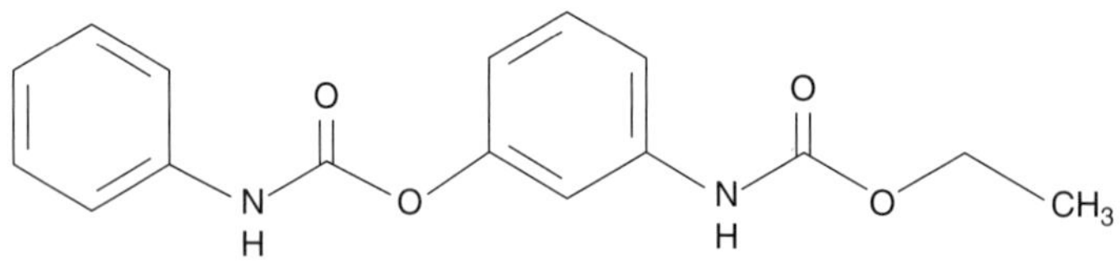

Desmedipham

Prepare a slurry of 40 g of silica gel and 80 ml of water by shaking vigorously for 5 s and eliminate air bubbles. Coat 20 × 20 cm glass plates to a thickness of 300 μm. Air dry and activate the absorbent by heating for 2 min at 100°C and store in a desiccator. Condition the developing tank with *n*-heptane for 1 h before use. Spot approximately 100 μl samples 3 cm from the bottom edge of the plate. Pour 10–15 ml heptane into the tank, insert the plate and seal the tank. Develop until the solvent front is approximately 30 mm from the top of the plate. Air dry and view under UV light. Diphenyl appears as a bright blue spot on a yellow background. Mark the area containing biphenyl and an equal area containing no diphenyl for reference. Remove the spots by scraping and transfer to 10 ml volumetric flask. Dilute to volume with ethanol and filter through Whatman Grade no. 44 filter paper and measure the absorbance at 248 and 300 nm in 1 cm cells. Determine the diphenyl content from a calibration curve. The absorbance at 300 nm is used to correct for interferences.

Bipyridiliums

See **Herbicide action – contact membrane disrupters**

Bis-carbamate herbicides

Bis-carbamates are a herbicidal group that inhibits photosystem II electron transport. **Phenmedipham** and **desmedipham** are the two members of the *bis*-carbamate group, which render post-emergence control of weeds in sugarbeet crops (see Phenmedipham and Desmedipham at top of the page).

Phenmedipham, $C_{16}H_{16}N_2O_4$ (methyl 3-(3-methylcarbaniloyloxy)carbanilate, or 3-methoxycarbonylaminophenyl 3-methylcarbanilate), is a carbanilate herbicide. The CAS Registry Number is 13684-63-4.

This herbicide is used post-emergence, commonly sprayed on the leaves of broadleaved weeds and grasses that invade sugarbeet crops, peas and spinach. The herbicide is formulated as emulsifiable concentrates (EC). Kemifam and Herbasan are two trade names.

The WHO classifies phenmedipham as a Class U toxin (unlikely to be hazardous), although it is considered as moderately toxic as per acute toxicity studies. The herbicide is a potential groundwater contaminant as per PAN guidelines. (See also **Desmedipham herbicide**.)

Bis-(dimethylthiocarbamyl) disulfide fungicide

Bis-(dimethylthiocarbamyl) disulfide is another term for thiuram disulfide or thiram. (See also **Thiuram disulfide fungicide**.)

Bismerthiazol bactericide

Bismerthiazol, $C_5H_6N_6S_4$ (5,5′-(methylenediimino)*bis*(1,3,4-thiadiazole-2-thiol)), also known as bismerthianol, is a bactericide. It is a white crystal or light yellow powder. It is soluble in 2-methyl-formamide, pyridine, ethyl alcohol and methanol, but sparingly so in water. The copper salt is also known and used as a fungicide. It has a CAS Registry Number of 79319-85-0. The copper salt has a molecular formula $C_5H_4CuN_6S_4$.

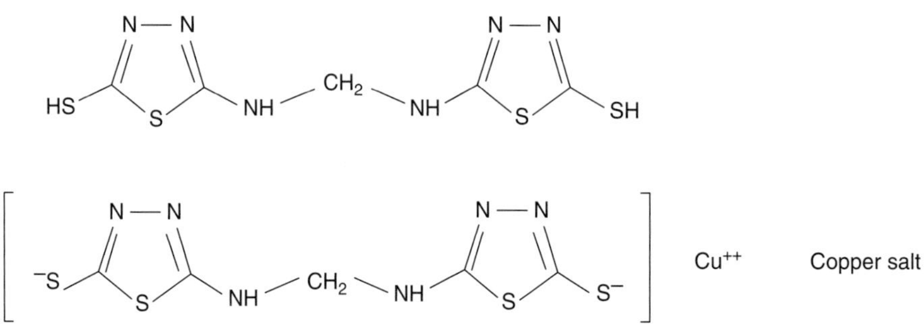

Bismerthiazol

Bismerthiazol, a preventive and curative bactericide, is seen to control species of *Xanthomonas* spp. like *Xanthomonas citri* and *Xanthomonas oryzae* pv. *oryzae* that cause rice leaf blight and **citrus canker** of orange. Thus, it protects rice and citrus fruits.

Bismerthiazol is available as a wettable powder (WP) and suspension concentrate (SC). Toxicologically, the oral LD_{50} of bismerthiazol in rats is 3160–8250 mg/kg, whereas that of the copper salt is >2000 mg/kg. The acute dermal LD_{50} for copper salt is >5000 mg/kg in rats. The copper salt is not a skin irritant but an eye irritant.

Bispyribac-sodium herbicide

Bispyribac-sodium, $C_{19}H_{17}N_4NaO_8$ (sodium 2,6-*bis*(dimethoxy-pyrimidin-2-yloxy)-benzoate), is a herbicide. It is a colourless powder readily soluble in water (73,000 mg/l) and soluble in methanol (at 26.3 g/l). It has a CAS Registry Number of 125401-75-4. Bispyribac-sodium is made by reacting 2,6-dihydrobenzoic acid with 2-chloro-4,6-dimethoxypyrimidine.

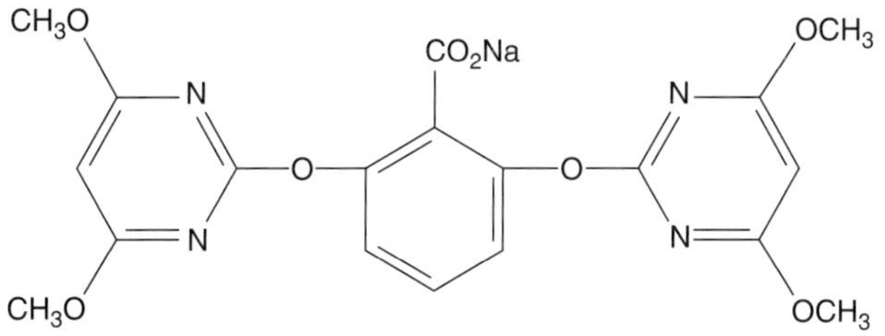

Bispyribac-sodium

Bispyribac-sodium, a selective, post-emergence, systemic herbicide, is absorbed both by the roots and by the leaves. This chemical is used mainly for weed control in directly seeded rice. This chemical is also suitable for weed control on non-crop sites. The mode of action is the inhibition of branched-chain amino acid synthesis (ALS or AHAS). The target weeds are grasses, sedges and broadleaved weeds, especially **barnyard grass** (*Echinochloa* spp.).

Formulations of bispyribac-sodium include water-soluble liquid (SC). Rates in the range of 15–45 g/ha are common. Nominee is a common trade name.

Bispyribac-sodium has an acute oral LD_{50} among rats and mice between 2635 and 4111 mg/kg. It has an acute percutaneous LD_{50} in rats of >2000 mg/kg. It has an acute oral LD_{50} in bobwhite quails of >2250 mg/kg. The ADI is 0.011 mg/kg. The 96 h acute LC_{50} in rainbow trout is >100 ppm. It is non-toxic to bees (LD_{50} >200 μg per bee).

In animals, bispyribac-sodium is found in urine and faeces within 7 days. It is a potential ground contaminant.

Bispyribac-sodium is registered in India, the USA, Vietnam, Portugal, the Philippines and Canada.

Bis (trichloromethyl) sulfone slimicide

Bis (trichloromethyl) sulfone (sulfonylbis (trichloromethane)) is a slimicide and fungicide with molecular formula $C_2C_6O_2S$. The CAS Registry Number is 3064-70-8. It is a whitish powder with a strong odour.

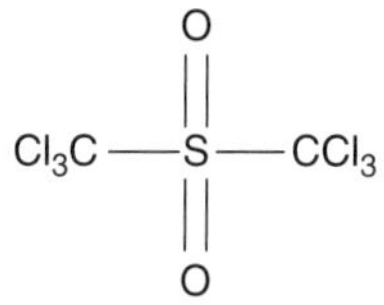

Bis (trichloromethyl) sulfone

This is a broad-spectrum compound and is registered by the USA as an algicide, fungicide, herbicide and antimicrobial. It is formulated as liquid concentrates.

Bitertanol fungicide

Bitertanol, $C_{20}H_{23}N_3O_2$ (1-(biphenyl-4-yloxy)-3,3-dimethyl-1-(1*H*-1,2,4-triazol-1-yl)butan-2-ol), is a triazole fungicide. It consists of a 20:80 ratio of (1*RS*,2*RS*) and (1*RS*,2*SR*) isomers and has a CAS Registry Number of 70585-36-3. Bitertanol is a whitish crystalline powder, very poorly soluble in water (0.004 g/l), and soluble in xylene (18 g/l) and dichloromethane (>250 g/l). The fungicide is stable in neutral, acidic and alkaline media (DT_{50} >1 year at 25°C).

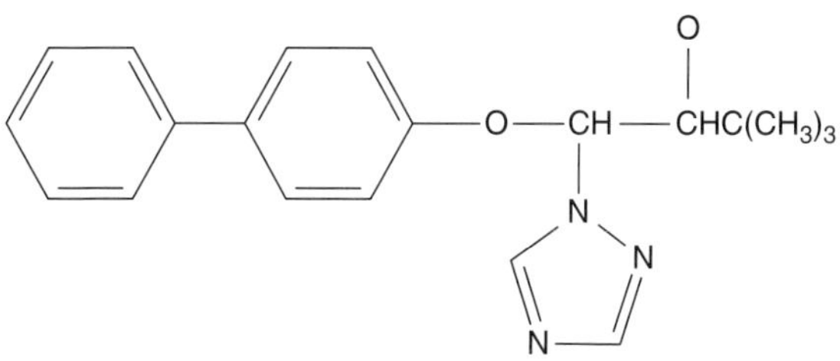

Bitertanol

Bitertanol, a fungicide used mainly on crop foliage and for treating seeds, acts by inhibiting sterol demethylation and acts to prevent fungal spore germination. Bitertanol prevents formation of spores and further inhibits mycelial development and the biosynthesis of **ergosterol**. The chemical controls **smut** and bunt diseases of wheat, sigatoka of bananas and several rusts, powdery mildews and leaf spots on vegetables, cereals, legumes, tea and ornamental plants.

The fungicide bitertanol is available as paste (PA), wettable powder (WP), emulsifiable concentrate (EC), suspension concentrate (SC) and as a water-dispersible powder (WS) especially for seed treatments. For seed dressing of wheat, a common application dose is 4–8 g/ha, while on bananas, a dosage from 105 to 195 g/ha is sufficient. In the case of rust and **powdery mildew** a larger dose of up to 0.5 kg/ha is commonly applied for effective control. Fruit crops are more tolerant of the wettable powder than the emulsifiable concentrate formulation.

Bitertanol does not pose a serious threat to mice, rats, birds and worms. The acute oral LD_{50} values in rats and mice are >5000 and 4300 mg/kg, respectively, while those for bird species tested are in the range of 776 to >2000 mg/kg. In bees, the oral LD_{50} is >104.4 μg per bee. The 2 year NOEL in rats and mice is found to be 100 mg/kg diet. It is slightly to moderately toxic to fish (with the 96 day LC_{50} being 2.14–3.54 mg/l). Bitertanol is a suspected endocrine disruptor, according to the PAN Pesticide Database.

When ingested by mammals, bitertanol is eliminated rapidly, either in its unchanged form or as a degradation product. It photodegrades in water and degrades in soil rapidly. Accumulation is minimal and easily traceable in plants. Triazolyl acetic acid is a metabolite of bitertanol.

Bitertanol is registered in India, Australia, South Africa, the Philippines, Finland and Denmark.

Black arm stem blight *Karpa*

See **Crop – cotton pest control**

Black pepper oil as a repellent

Black pepper oil, obtained from the seeds of black pepper plant (*Piper nigrum* L.), is an animal repellent. It is obtained by steam distillation of unripe, dried fruits of the black pepper plant, which has a green or pale yellow colour and a sharp odour.

Black pepper oil contains piperine (molecular formula $C_{17}H_{19}NO_3$, with an IUPAC name 1-[5-(1,3-benzodioxol- 5-yl)-1-oxo-2,4-pentadienyl]piperidine) and **capsaicin** ($C_{18}H_{27}NO_3$, with an IUPAC name 8-methyl-*N*-vanillyl-*trans*-6-nonenamide), which together repel small mammals such as dogs and cats. It can be used in industries and homes in storage situations.

Black pepper oil is available in granular formulations and may be applied by hand or with a spreader. A trade name is Animal Repellent Granular. This compound is safe for handlers and also to mammals, birds and fish. This oil is used inside in for example basements, sheds and garbage stores and also outdoors on lawns and gardens. It is applied manually or with a spreader and has other uses such as aromatherapy and in skin products.

Black pepper oil has no adverse effect on human beings except its repellent effect. It is generally recognized as safe for use in food (i.e. GRAS) by the FDA and may be used as a flavouring agent in food. Black pepper oil is not toxic to mammals, birds and fish or to the environment.

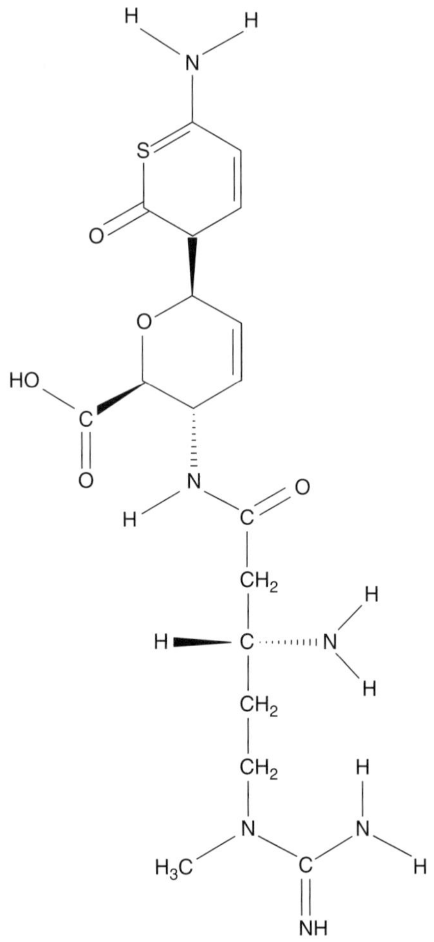

Blasticidin-S

Black sooty mould

See **Mould (sooty mould)**

Blast

See **Plant disease blast**

Blasticidin-S antibiotic

Blasticidin-S, $C_{17}H_{26}N_8O_5$ (1-(4-amino-1,2-dihydro-2-oxopyrimidin-1-yl)-4-[(S)-3-amino-5-(1-methylguanidino) valeramido]-1,2,3,4-tetradeoxy-β-D-erythro-hex-2-enopyranuronic acid), is a fungicide derived by fermentation from the soil actinomycete, *Streptomyces griseochromogenes*. The white powder has a CAS Registry Number of 2079-00-7 (see Blasticidin-S at top of the second column).

Blasticidin-S, a peptidyl nucleoside antibiotic compound, is a fungicidal and antibacterial agent that prevents the growth of both eukaryotic and prokaryotic cells. It has protective and curative properties. It is available as a benzylaminobenzenesulfonate salt and is used in the control of rice blast caused by *Pyricularia oryzae*. The mode of action is the inhibition of peptidyl transfer and protein chain elongation, thus disturbing protein biosynthesis and inhibiting growth of mycelia.

Blasticidin-S is normally sprayed on the leaves strictly in the range of 100–300 g a.i./ha (or 10–40 ppm). It can be formulated as dispersible powders (DP), emulsifiable concentrates (EC) and wettable powders (WP). This compound may be combined with other non-alkaline pesticides like **fenitrothion**. Bal-S is a known trade name.

The WHO has categorized blasticidin-S as highly hazardous (Class Ib toxin). The LD_{50} of blasticidin-S in rats is 55.9 mg/kg. In carp the LC_{50} for 48 h is >40 mg/l. It is phytotoxic to potatoes, tomatoes, tobacco and even to rice at higher doses. Exposure of the eyes and the skin to the compound can cause irritation and also conjunctivitis (eyes). Handlers are advised to ensure that calcium acetate is mixed with the antibiotic to minimise injury to the eyes, in case of an accident.

Enzyme BS deaminase, isolated from *Bacillus cereus* and *Aspergillus terreus*, has been shown to hasten the degradation of blasticidin-S to form the non-toxic deaminohydroxy derivative.

This antibiotic is excreted through urine and faeces within 24 h in mice. The soil DT_{50} is 2 days. In rice, cytomycin and deaminohydroxy products are the two metabolites formed.

Blasticidin-S is registered for use in Cambodia.

Blight

See **Plant disease bacterial blight**

Blight on onion

See **Crop – onion pest control**

Boom height

See **Spray application concepts**

Borax herbicide

Borax, $Na_2B_4O_7.10H_2O$ (disodium tetraborate decahydrate), is a naturally occurring herbicide as well as an insecticide. Borax or tetraborate decahydrate is a salt of boric acid and has a CAS Registry Number of 1330-43-4. It is a colourless crystalline solid, highly soluble in water, glycerol and ethylene glycol but insoluble in ethanol. Above 62°C it loses water of crystallization, forming the pentahydrate followed by anhydrous tetraborate.

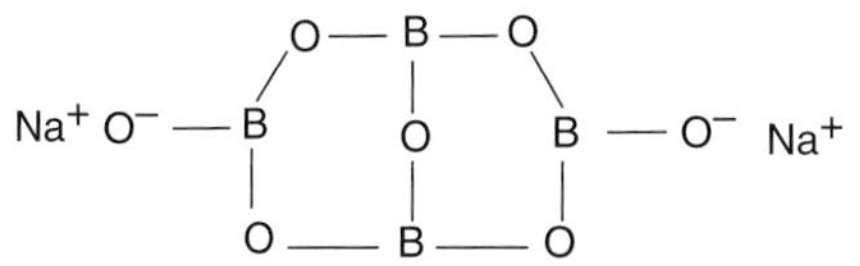

Borax

Borax, a **non-selective herbicide**, also acts as an insecticide (insect bait). Perennial weeds such as creeping Charlie (*Glechoma hederacea*, also known as ground ivy) are controlled. It is especially effective on shallow-rooted weeds. Being a micronutrient, its excess causes toxicity, which gives borax its herbicidal effect. It causes necrosis and dessication of leaves. As an insecticide, it can control the members of the family Formicidae (which includes ants).

Borax is used in combination with for example **atrazine**, **simazine** and sodium chloride available in the market under the trade names Dehybor, Neobor and Pyrobor. It is also an essential micronutrient that aids in sugar transportation in plants.

Borax is categorized as a chemical that is safe to use under normal conditions. The WHO categorizes it as a Class U toxin. The LD_{50} in rats is in the range of 4500–6000 mg/kg, and an acute percutaneous LD_{50} in rabbits is >10,000 mg/kg. In fish species, the LC_{50} is 1320 mg/l. It is non-toxic to bees. Irritation of skin and respiratory tract, nausea and vomiting are some of the symptoms of borax poisoning. The NOEL (2 years) in rats is 154 mg/kg b.w.

Borax is adsorbed to inorganic particles in the soil. Thus, it can persist in nature for up to 1 year, unless watery or acidic conditions prevail. Borax salts, like sodium borate, are also observed to persist.

Countries like the USA, Canada, South Africa and Australia have registered borax use.

Bordeaux mixture fungicide

Bordeaux mixture, $Ca_3Cu_4H_6O_{19}S_4$, is an inorganic copper fungicide as well as a bactericide. A light green, fine powder, Bordeaux mixture is slowly soluble in water, but more readily so in for example acetic acid, methanol and acetone. Being non-free-flowing, this compound needs a stabilizing agent. Bordeaux mixture is also called bordo mixture, neutral Bordeaux and bouillie bordelaise. It is prepared by combining fresh calcium hydroxide (hydrated lime) and **copper sulfate** (in equal measure) with a hundred times as many litres of water. Bordeaux mixture was developed and introduced by a French scientist, A. Millardet, in 1882. He is said to have developed this mixture, taking a cue from vineyard owners in Bordeaux, France. Bordeaux farmers sprayed a solution of hydrated lime and copper sulfate on the leaves and grapes of plants on the periphery of their farms in order to repel trespassers and workers from picking the grapes. However, this potent solution also unwittingly cured mildew on the treated plants.

Ever since it became popular, Bordeaux mixture, a contact protectant fungicide and bactericide, has been effectively used to control fungal diseases like downy mildew of grape vines caused by *Plasmopara viticola*, **scab** of apple and potato blight caused by *Phyophthora infestans*. It also controls bacterial diseases such as **citrus canker**, caused by *Xanthomonas citri*, and also several other leaf spot diseases. Some other diseases that are controlled include anthracnose, **downy mildew** and cankers. The mode of action is the interference in normal cell and spore activity due to the toxicity of copper ions.

Bordeaux mixture is formulated as a soluble concentrate (SL), wettable powder (WP) and as water-dispersible granules (WG). It can be combined with neutral pesticides. It is important to apply this mixture while fresh, lest it reduce its fungicidal nature.

Other formulations of Bordeaux mixture include the Bordeaux paste, made by combining copper sulfate, quick lime and water in the ratio of 1:1.5:15. Bordeaux paint is made by combining monohydrated copper sulfate, lime and linseed oil in the ratio 1:2:3. Pruning tips are applied with one of these formulations, to prevent fungal attacks on open wounds. Bordeaux mixture is available commercially under trade names such as Bordocop and Z-Bordeaux.

Bordeaux mixture is slightly toxic to man and animals, with the acute oral LD_{50} in rats being >2302 mg/kg. The acute percutaneous LD_{50} in rats is >2000 mg/kg, while the inhalation LC_{50} (4 h) in male and female rats is <5 mg/l. The ADI is 0.15 mg Cu/kg. b.w./day. From the ecological toxicity point of view, the toxicity limits defined in bobwhite quail in terms of the LD_{50} is 616 mg Cu/kg. In rainbow trout the LC_{50} for 96 h is >21.39 mg Cu/l, and in bees the LD_{50} is 23.3 μg Cu per bee. The NOEL is optimized at 17 mg Cu/kg b.w./day. This mixture may be toxic to some copper-sensitive crops and will harm foliage if the copper concentration is high.

Copper, an essential plant element, strongly binds to soil particles, hence limiting the availability of free copper ions in the soil solution. The tolerance limit in fruits and vegetables is 5 mg/kg and in meats and eggs 2 mg/kg.

Bordeaux paint

See **Bordeaux mixture fungicide**

Bordeaux paste

See **Bordeaux mixture fungicide**

Bordo mixture

See **Bordeaux mixture fungicide**

Boron as a plant nutrient

See **Plant nutrient excesses and deficiencies**

Boscalid fungicide

Boscalid, $C_{18}H_{12}Cl_2N_2O$, (2-chloro-*N*-(4′-chlorobiphenyl-2-yl) nicotinamide), also known as **nicobifen**, is a pyridine or pyridinecarboxamide fungicide. With a CAS Registry Number of 188425-85-6, boscalid is a white, odourless crystalline substance, soluble in water (4.6 mg/l), methanol 40-50 g/l and acetone 160–200 g/l. The compound is stable to hydrolysis at pH 4, 5 and 9. It is stable to photolysis in aqueous solution.

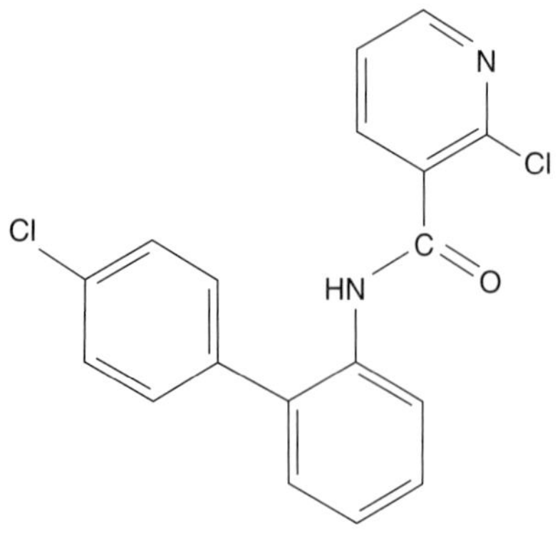

Boscalid

Boscalid, with a mainly foliar application, is a systemic compound that affects almost every stage of fungal growth (e.g. prevention of spore germination and germ tube elongation), although its main mode of action is the inhibition of the enzyme succinate ubiquinone reductase (which interferes with the mitochondrial electron transport chain). This fungicide controls diseases caused by *Alternaria* spp., *Sclerotinia* spp., *Botrytis* spp. and many others, and covers crops including cereals, groundnuts, potatoes, grapes and turf.

Boscalid is available as a suspension concentrate (SC) and as water-dispersible granules (WG). This chemical can be combined with other compounds to widen its control range.

Boscalid is practically safe on rats, (LD_{50} >5000 mg/kg) and also on worms. The inhalation LC_{50} (4 h) in rats is >6.7 mg/l. But it is slightly to moderately toxic to birds (with an LD_{50} in bobwhite quails being >2000 mg/kg b.w.) and toxic to fish (with the LC_{50} (96 h) in rainbow trout being 2.7 mg/l). It does not irritate the eyes or the skin in normal use. The NOEL in rats is 5 mg/kg. b.w. However, it is a possible carcinogen. The ADI for boscalid is 0.04 mg/kg b.w./day.

Boscalid is rapidly eliminated after ingestion by small mammals, with several metabolic reactions taking place including hydroxylation of the biphenyl ring. It is rapidly excreted mainly via faeces. Similar reactions take place in plants. Unchanged boscalid is also present in both plants and animals. It degrades both in water and in soil quite well (DT_{50} is 108 days to more than 1 year).

Boscalid is registered for use in the USA, Australia, South Africa and many EU countries.

Bouillie bordelaise

See **Bordeaux mixture fungicide**

Braconid (*Bracon* spp.) as a biocontrol agent

The *Bracon* spp. (from order Hymenoptera and family Braconidae) is a minute parasitic wasp that is known for parasitizing many crop pests.

The 2.5 mm adult wasp, which is a thin, brownish, free-living, nectar-feeding organism, parasitizes ants, aphids, bollworms, caterpillars, leaf miners, **leafhoppers** and maggots of various crops, by laying eggs inside the host body. The hatched eggs feed and often pupate inside the host, thus inhibiting the latter's life cycle (larval, pupal and nymphal stages) leading to death. The mode of parasitism may differ with species.

Many crops including cotton, wheat, maize, tobacco and sunflower are protected due to the wasp's parasitism on, say, the larval stage of the gram caterpillar or the **spotted bollworm**.

Major pests are controlled, in their various immature stages, by various species of *Bracon* such as: (i) the spotted bollworm, American bollworm and **pink bollworm** of cotton; (ii) the stem borer and pink stem borer of sorghum; (iii) caterpillars of gram and tobacco; and (iv) the wheat stem sawfly and stem or internode borer of sugarcane. Examples of *Bracon* spp. used for these control measures are *Bracon brevicornis*, *Bracon greeni*, *Bracon gelechidophagus*, *Bracon chinensis*, *Bracon drinensis*, *Bracon meromyzae*, *Bracon cephi* and *Bracon lissogaster*.

Breadth of activity

See **Fungicides**

Broad-spectrum method of weed biocontrol

See **Biological weed control agents – application methods**

Brodifacoum rodenticide

Brodifacoum, $C_{31}H_{23}BrO_3$ (3-[3-(4′-bromobiphenyl-4-yl)-1,2,3,4-tetrahydro-1-naphthyl]-4-hydroxycoumarin), is a rodenticide of the coumarin class. This chemical is a tasteless, odourless, whitish powder, soluble in water (<10 mg/l) but more so in acetone. It is also thermally and photolytically stable. It is degraded by UV light when in solution. Its CAS Registry Number is 56073-10-0.

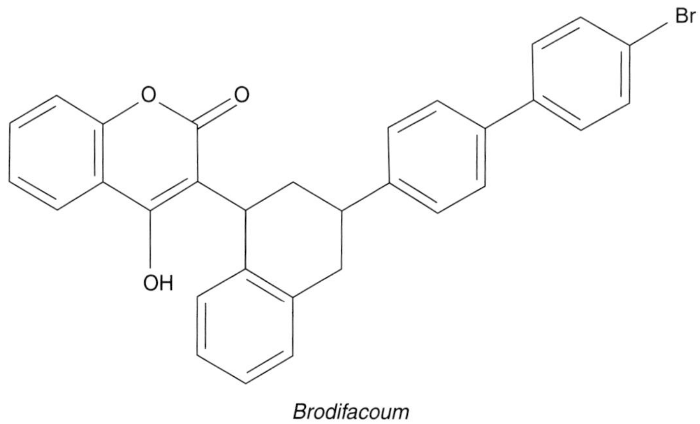

Brodifacoum

Brodifacoum, a second-generation anticoagulant, is effective against many types of rodents, commonly for example rats (e.g. *Rattus mindanensis*, *Rattus argentiventer*) and voles.

The mode of action of brodifacoum is the inhibition of the enzyme vitamin K epoxide reductase, which synthesizes vitamin K, responsible for clotting the blood. Brodifacoum is important since it is effective on pests that have developed resistance to **warfarin**.

The common formulations of brodifacoum are pellets, wax and cereal baits, while common trade names include Klerat, Talon, Sorex and Havoc. A single low dose of this chemical (of up to 50 mg/kg) is sufficient to kill the rodent.

The WHO classifies brodifacoum under the toxicity Class Ia (extremely hazardous). The acute oral LD_{50} values among rats and mice range from 0.27 to 0.4 mg/kg, while those in dogs, chickens and mallard ducks are <5 mg/kg. It is a mild skin and eye irritant (acute percutaneous LD_{50} in rats are up to 5.21 mg/kg). The LC_{50} in worms (14 days) is >994 mg/kg dry soil. The inhalation LC_{50} for a 4 h period ranges from 0.0005 to 0.005 mg/l air. Due to acute toxicity, brodifacoum is considered a PAN Bad Actor chemical according to the PAN Pesticide Database. As a result of its anti-clotting manifestations, typical symptoms of over-exposure to brodifacoum are bleeding of the gums, nose and via urine.

This rodenticide degrades in soil and water in the pH range of 5.5–8. In mammals, many hydroxycoumarins are formed.

Brodifacoum is registered for use in the USA, Canada, Australia, Germany and the Netherlands. (See also **Anticoagulant rodenticides**.)

Bromacil herbicide

Bromacil, $C_9H_{13}BrN_2O_2$ (5-bromo-3-sec-butyl-6-methyluracil), is a uracil herbicide. It is an odourless, whitish crystalline solid, soluble in water and organic solvents. While being stable to normal storage conditions, this compound can cause a hazard when exposed to heat, flame or strong oxidizers. It slowly decomposes in the presence of strong acids. Bromacil is made by the reaction of isopropyl isocyanate with methyl 3-aminocrotonate, followed by cyclization in sodium hydroxide at elevated temperature and bromination in acetic acid. It has a CAS Registry Number of 314-40-9.

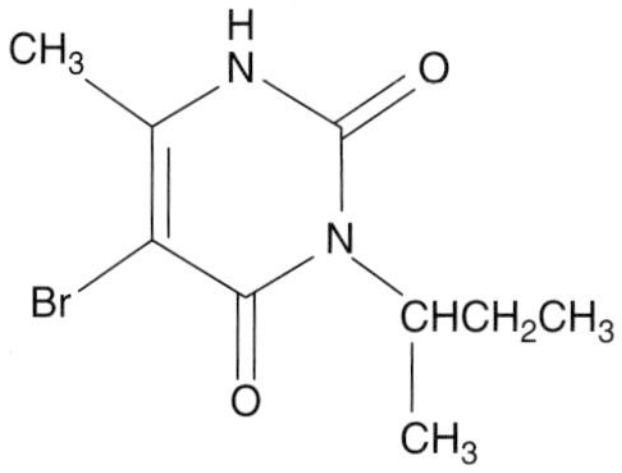

Bromacil

Bromacil, a general use herbicide (as per the EPA), is used on brushes and weeds on unfertilized land. It may selectively control many weeds in citrus and pineapple crops at rates of 1.5–8 kg/ha.

The commonly available formulations of bromacil are wettable powder (WP), soluble concentrates (SL) and granules (GR). It may be combined with other pesticidal compounds (except water-based formulations with ammonium sulfamate and liquid preparations of **amitrole**). It is available in the market under the trade names of Hyvar X-L, Rokar XL and Uragan.

According to the WHO and EPA, bromacil is classified as practically non-toxic in acute toxicity studies (toxicity Class U and Class IV, respectively). It has an oral LD_{50} in rats of 2000 mg/kg and in bobwhite quail of 2250 mg/kg. It can cause slight to moderate irritation to the skin and the eyes. Inhalation LC_{50} (4 h) in rats is >5.6 mg/l air. The ADI is 0.1 mg/kg b.w./day. The LC_{50} in rainbow trout is 36 ppm. It is non-toxic to bees. Bromacil is a PAN Bad Actor chemical since it is believed to affect the endocrine system and may be a carcinogen.

Since liquid formulations of bromacil (when compared with dry formulations) are moderately toxic, labels should clearly display the signal word 'WARNING'. Bromacil causes irritation to eyes, skin and the respiratory tract, and symptoms of poisoning include dilation of pupils, muscular weakness and diarrhoea. Care has to be taken to prevent contact with skin or eyes especially while applying as a spray mist. Washing of hands, face, clothes, mixers and tools is highly recommended. Disposal of unused bromacil and water used for rinsing should be carried out as per the recommendation of the local regulating agency.

In animals and plants, the principal metabolite of bromacil is 5-bromo-3-*sec*-butyl-hydroxymethyluracil. It has a soil half-life of about 2 months, besides being mobile in soil, thus increasing the risk of groundwater contamination.

Bromacil is registristed for the use in the USA, Vietnam, Tanzania, South Africa, the Philippines, New Zealand, Canada, Cambodia and Australia.

Bromadiolone rodenticide

Bromadiolone, $C_{30}H_{23}BrO_4$ (3-[3-(4′-bromobiphenyl-4-yl)-3-hydroxy-1-phenylpropyl]-4-hydroxycoumarin), is a rodenticide that is a 4-hydroxycoumarin derivative. It has a CAS Registry Number of 28772-56-7. It is a yellowish powder that is soluble in water (19 mg/l), dimethylformamide (730 g/l) and ethyl acetate but sparingly soluble in diethyl ether and hexane. Bromadiolone is a mixture of four stereoisomers, out of which two stereoisomeric centres are at the phenyl- and the hydroxyl-carbons in the carbon chain of the substituent at the 3 position of the coumarin. It is thermally stable below 150°C (see Bromadiolone at top of the next page).

Bromadiolone, a second-generation rodenticide, is an anticoagulant used in rodent baits, thus protecting crops, industrial areas and stored grains. Bromadiolone acts in a way similar to warfarin but is much more effective. This is manifested in immediate death rates at even very small doses of 0.005% concentration.

Common formulations of bromadiolone are cereal-based baits, tracking dusts and powder concentrates, and common trade names are Super Caid, Lanirat and Rato ban.

The WHO and EPA have categorized bromadiolone as extremely hazardous (Class Ia). The acute oral LD_{50} values in rats, mice and rabbits are between 1 and 1.75 mg/kg. It causes skin and eye irritation and may lead to reduction in prothrombin among cats and dogs. The inhalation LC_{50} in rats is <0.02 mg/l. The acute oral LD_{50} in Japanese quail is

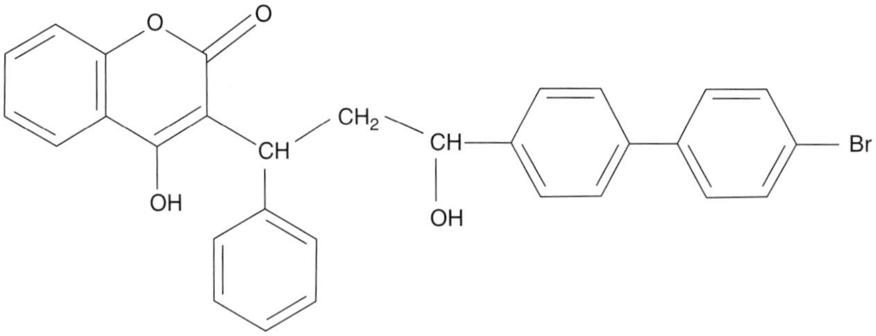

Bromadiolone

134 mg/kg b.w., and the LC_{50} (96 h) in fish (rainbow trout) is 2.89 mg/l. It is not toxic to bees. Acute exposure causes bleeding of gums, nose and through urine. Due to acute toxicity, bromadiolone is listed as a PAN Bad Actor chemical.

Leaching behaviour is dependent on soil type. Bromadiolone is registered for use in India, Australia, New Zealand, the USA, Canada, South Africa, Cambodia, Tanzania and Vietnam. (See also **Anticoagulant rodenticides**.)

Bromeliads

See **Epiphytes**

Bromethalin rodenticide

Bromethalin, $C_{14}H_7Br_3F_3N_3O_4$ (α,α,α-trifluoro-*N*-methyl-4,6-dinitro-*N*-(2,4,6-tribromophenyl) -*o*-toluidine), is a diphenylamine rodenticide. It is a pale yellow crystalline substance that is soluble in water (<0.01 mg/l), dichloromethane (300–400 g/l) and chloroform (200–300 g/l). Bromethalin is made by reacting 2,4-dinitro-6-trifluoromethyl methylaniline and 1,2,3,5-tetrabromobenzene. The chemical decomposes under UV light. It has good storage stability under normal conditions. Its CAS Registry Number is 63333-35-7.

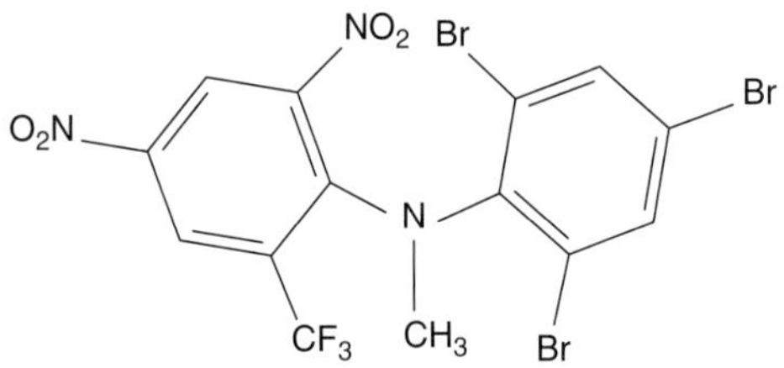

Bromethalin

Bromethalin, an acutely toxic rodenticide, is a neurotoxin. Its mode of action includes uncoupling the phosphorylation in the cells of the CNS. It is a good alternative to pests that have developed resistance to anticoagulant rodenticides. Bromethalin does not induce bait shyness. This chemical is available as ready-to-use baits. Common trade names include Assault, Vengence and Trounce.

The WHO lists bromethalin as a Class Ia toxin (extremely hazardous). The acute oral LD_{50} among rats, mice, dogs and cats is up to 5 mg/kg b.w., while the inhalation LC_{50} in rats is 0.024 mg/l air. The NOEL (90 days) in rats is 0.025 mg/kg/day. Bromethalin is a PAN Bad Actor chemical. Symptoms of poisoning include paralysis of the hind limbs and respiratory paralysis.

In animals, this rodenticide follows the metabolism route of *N*-demethylation. (See also **Anticoagulant rodenticides**.)

Bromethalin is registered for use in Canada and the USA.

Bromofenoxim herbicide

Bromofenoxim, $C_{13}H_7Br_2N_3O_6$ (3,5-dibromo-4-hydroxybenzaldehyde-2,4-dinitrophenyloxime), is a phenoxy herbicide. It has a CAS Registry Number of 13181-17-4. This herbicide is a cream-coloured, crystalline substance, poorly soluble in water but soluble in acetone. Its hydrolysis DT_{50} is 41.4 h (at pH 1) and 9.6 h (at pH 5). Bromofenoxim, a precursor of **bromoxynil**, is made by reacting 3,5-dibromo-anisaldehyde oxime with 2,4-dinitro-chlorobenzene in the presence of sodium ethoxide. It is categorized under HRAC Group C3.

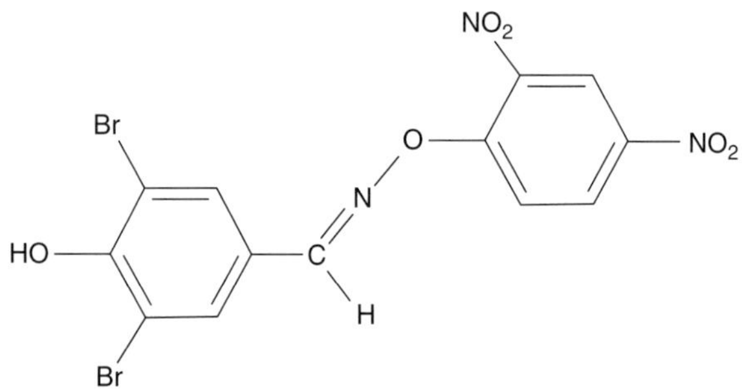

Bromofenoxim

Bromofenoxim is a selective, post-emergence, contact herbicide, which is absorbed by the leaves and shoots. It acts by inhibiting photosynthesis at photosystem II and this way controls annual broadleaved weeds in cereals such as maize and also on grasslands.

Herbicide bromofenoxim is available as wettable powder (WP) and soluble concentrate (SL). It is available in the market under the trade name of Faneron.

According to the WHO and EPA, bromofenoxim comes under the toxicity Class III (slightly hazardous). The oral LD_{50} among rats is 1217 mg/kg. It is highly toxic to fish (with the LC_{50} values for a variety of fish species ranging from 0.056 to 0.24 mg/l).

In mammals, dinitrophenol and hydroxydibromobenzonitrile are metabolites of bromofenoxim. In plants, the metabolites are 3,5-dibromo-4-hydroxybenzoic acids.

Bromofenoxim is classified as obsolete by the EU and is not registered in the USA.

Bromophos-ethyl acaricide

Bromophos-ethyl, $C_{10}H_{12}BrCl_2O_3PS$ (*O*-4-bromo-2, 5-dichlorophenyl *O,O*-diethyl phosphorothioate), is a halogen-containing organothiophosphate acaricide. It is a yellow fluid, practically insoluble in water. Although stable in an aqueous suspension, it is unstable in alkaline medium. The CAS Registry Number is 4824-78-6.

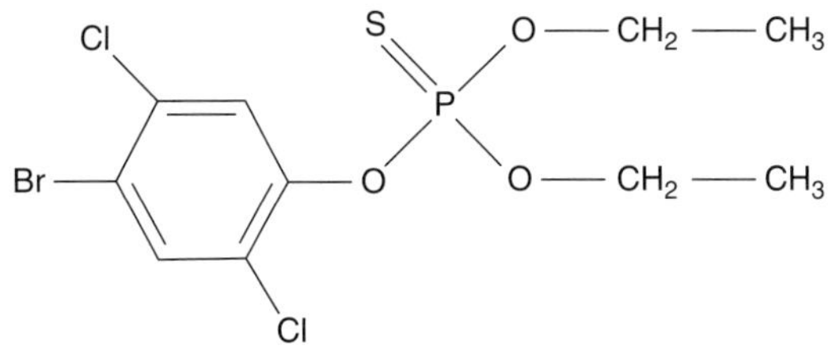

Bromophos-ethyl

Bromophos-ethyl is a non-systemic, broad-spectrum insecticide and acaricide. It acts as a stomach poison and is effective against many chewing and sucking pests like caterpillars, root maggots, aphids, flies and beetles. As an acaricide, it works against ticks on animals. Bromophos-ethyl is also effective against mosquito larvae.

Acaricide-insecticide bromophos-ethyl is available as an emulsifiable concentrate (EC), ULV concentrate, a wettable powder (WP) and dip formulation. It is compatible with insecticides and fungicides. Typical dosage rates are between 0.2 and 1.2 kg/ha. Due to its toxicity, application should be carried out with care in order not to harm bees and other non-target organisms. It is considered a PAN Bad Actor chemical since it inhibits cholinesterase.

Bromophos-ethyl is absorbed in the gastrointestinal tract and excreted in urine and faeces. Principal metabolites formed in mammals, plants and soil include dichloro-bromophenol, 2,5-dichloro-4-bromophenol and bromoxon-ethyl. Residues have been found in apples and lettuce.

Bromophos-ethyl is not registered in the USA or the EU.

Bromopropylate acaricide

Bromopropylate, $C_{17}H_{16}Br_2O_3$ (isopropyl-4, 4′-dibromobenzilate), is a bridged diphenyl acaricide. It is a colourless crystal that is almost insoluble in water but soluble in most organic solvents. Bromopropylate is produced by esterification of 4, 4′-dibromobenzilic acid with 2- propanol.

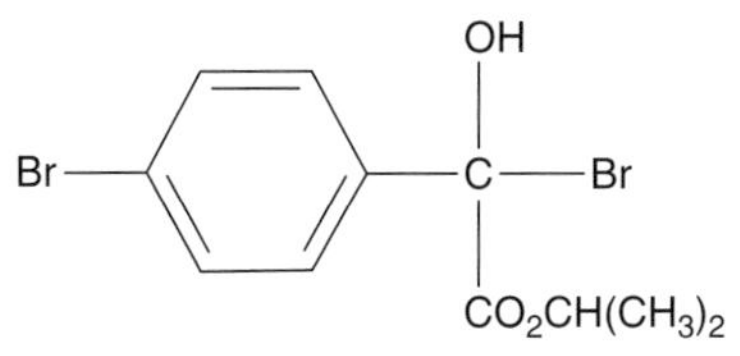

Bromopropylate

The major crops that are protected by bromopropylate are citrus, vines, pome and stone fruits, sugarbeet, sugarcane and tea. The target mite species include European red mite, two-spotted mite, carmine mite and apple rust mite.

The acaricide is sprayed by either hand-held or motorized sprayers. It is formulated as an emulsifiable concentrate (EC). A common trade name of bromopropylate is Neoron.

The WHO classifies bromopropylate as a Class U toxin (unlikely to be hazardous). It is non-mutagenic and is not irritating to eyes and skin. The acute oral LD_{50} in rats is >5000 mg/kg. The NOEL in rats (2 years) is 3.7 mg/kg/day. The ADI is 0.03 mg/kg/day.

In animals, bromopropylate accumulates in the fat tissues from which it is eliminated. In studies on soybeans and apples, it was found not to penetrate the tissues. Residues include the unchanged compound and the metabolite 4, 4′-dibromobenzilic acid.

Bromopropylate is registered for use in Australia, New Zealand and South Africa but not in the USA. The MRL values for banana (whole) and citrus juice in Germany are 3 mg/kg and 0.2 mg/kg, respectively.

Bromoxynil herbicide

Bromoxynil, $C_7H_3Br_2NO$ (3,5-dibromo-4-hydroxybenzonitrile), is a **hydroxybenzonitrile** herbicide. Its CAS Registry Number is 1689-84-5. This herbicide exists as a colourless solid, sparingly soluble in water and readily soluble in polar organic solvents. It is thermally stable and stable in dilute alkalis and acids. Bromoxynil is made by the reaction of sodium hypobromide with 4-hydroxybenzonitrile.

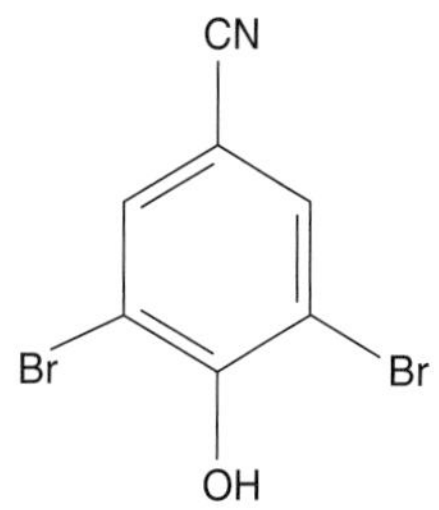

Bromoxynil

Bromoxynil is a selective, post-emergence, contact herbicide with some systemic activity (it is absorbed by the leaves). It acts by inhibiting photosynthesis and uncoupling oxidative phosphorylation. Annual broadleaved weeds in maize, sorghum, onions, garlic and turf are controlled with this herbicide.

The available formulations of bromoxynil are emulsifiable concentrates (EC), wettable powders (WP) and soluble concentrates (SL). It is available in the market as for example Brominal, Buctril and Bromotril. The application dose commonly applied to cereal crops ranges from 300 to 560 g/ha.

Both the WHO and EPA have classified bromoxynil as moderately hazardous (toxicity Class II). The oral LD_{50} among rats ranges from 80 to 180 mg/kg in rats, while the same in rabbits is 260 mg/kg. Dermal LD_{50} of 3600 mg/kg in rabbits suggests only slight irritation. The compound may cause slight irritation to the eyes. The LC_{50} (4 h) in rats is 0.15–0.38 mg/ml. The ADI is 0.01 mg/kg/day. Bromoxynil is highly toxic to birds (with the LD_{50} of several tested species ranging from 50 to 240 mg/kg). Similarly, it is highly toxic to many fish species, but it is not toxic to bees. The chemical is considered to be a restricted use pesticide (RUP) by the EPA, indicating use of this chemical only with formal consent. Product labels should carry the signal word 'WARNING'. Fields applied with bromoxynil should not be harvested for livestock or grain for a period of 14–60 days, depending on the dose applied.

Bromoxynil is degraded by hydrolysis and debromination to less toxic substances such as hydroxybenzoic acid. Bromoxynil has low persistence in soil with soil DT_{50} of about 10 days.

The MRL of bromoxynil (residue being 3,5-dibromo-4-hydroxybenzonitrile) in Canada (as per Health Canada) is fixed at 0.1 ppm for cereals such as millet and sorghum. This figure is similar to the MRL values in the USA and some other countries. (See also **Hydroxybenzonitrile herbicides**.)

Bromuconazol fungicide

Bromuconazol, $C_{13}H_{12}BrCl_2N_3O$ (1-[4-bromo-2-(2,4-dichlorophenyle) tetrahydrofurfuryl]= 1*H*-1,2,4-triazole), is a triazole (conazole) fungicide. This colourless powdered compound is a mixture of two diastereoisomers, LS 850646 and LS 850647 (commonly in the ratio 54:46). Bromuconazol is soluble in water at 0.05 g/l and moderately soluble in most organic solvents. The CAS Registry Number of bromuconazol is 116255-48-2.

Bromuconazole, a sterol dimethylation (**ergosterol** biosynthesis) inhibitor (DMI), is a systemic compound and is used to control pathogenic species of *Alternaria*, *Fusarium* and *Pseudocercosporella* that infect cereals, fruits, vegetables, vines, turf and ornamentals.

Commonly available as granules, suspension concentrate (SC) and emulsifiable concentrate (EC), bromuconazole is applied as a foliar spray at a maximum rate of 300 g/ha.

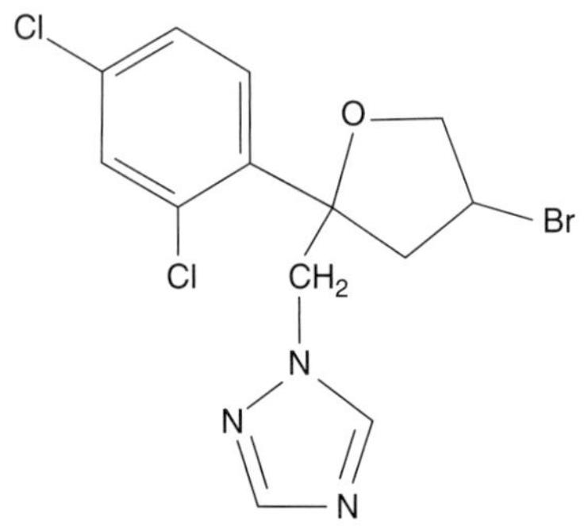

Bromuconazol

The WHO has classified bromuconazole as a Class II toxin (moderately hazardous). The LD_{50} in rats is 365 mg/kg and in mice, 1151 mg/kg. The fungicide is also considered toxic to birds like the bobwhite quail and mallard duck (with oral LD_{50} >2150 mg/kg b.w.) and to rainbow trout (with LC_{50} (96 h) of 1.7 mg/l). However, it is not toxic to bees and earthworms, if used as per label instructions. The NOEL in rats (2 years) is 20 ppm (1.1 mg/kg b.w./day).

This fungicidal compound is extensively metabolized in animals, although it does not accumulate in organs or tissues. In most plants, polar metabolites and conjugates are formed. Bromuconzole is not very mobile in soil but is very persistent, with a half-life of 123–600 days.

The fungicide bromuconazole is registered for use in the USA, Australia and South Africa, but it is not permitted in EU countries. India too has not registered this compound.

Bronopol bactericide

Bronopol, $C_3H_6BrNO_4$ (2-bromo-2-nitropropane-1,3-diol), is a bactericide, a bacteriostat and a fungicide. With a CAS Registry Number of 52-51-7, the hygroscopic bronopol is a pale brown-coloured solid, readily soluble in water (from 250 to 500 g/l), acetone and ethyl acetate, and slightly soluble in chloroform, diethyl ether and benzene. The compound is unstable in aluminium containers.

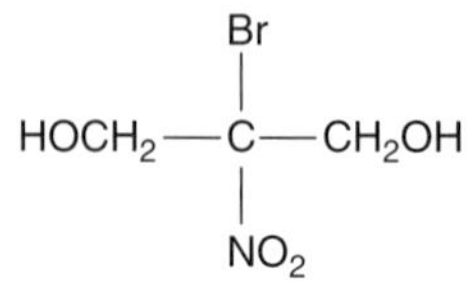

Bronopol

Bronopol, a seed-treatment compound, inhibits the enzyme dehydrogenase, thus damaging cell membranes of bacterial pathogens like *Xanthomonas campestris* pv. *malvacearum* causing black arm disease of cotton. This compound is used as preservative in the pharmaceutical industry.

The bactericide is available as water-soluble powder (SP), wettable powder (WP) and as a powder for dry seed treatment. The commonly used commercial product is named Bronotak.

Bronopol is classified under Class II (moderately hazardous) by the WHO and EPA. The acute oral LD_{50} toxicity in rats and mice is in the range of 180–500 mg/kg. The same in mallard ducks is observed at 510 mg/kg. The LC_{50} in trout is 20 mg/l. It is a moderate skin and eye irritant. The inhalation LC_{50} (6 h) in rats is >5 mg/l air. The ADI is 0.02 mg/kg b.w./day. The NOEL (72 days) in rats is recorded to be 1000 mg/kg diet. Bronopol is a PAN Bad Actor chemical.

This bactericide is absorbed and excreted rapidly when consumed by animals. In both animals and plants, metabolite 2-nitropropane-1,3-diol is formed.

Bronopol is registered for use in the USA, Tanzania, New Zealand, the Netherlands and Canada, but not in the EU.

Bucket or trombone sprayer

See **Application equipment – hand sprayers**

Buffer zone

See **Spray application concepts**

Bunyaviridae viruses

Members of the family Bunyaviridae are negative-strand RNA viruses. These viruses usually infect arthropods and rodents and sometimes even humans. Only members of the genus *Tospovirus* are plant-infecting viruses. Except for members of the genus *Hantavirus* (serious animal pathogens), all other members of the family are transmitted by arthropod vectors.

Bupirimate fungicide

Bupirimate, $C_{13}H_{24}N_4O_3S$ (5-butyl-2-ethylamino-6-methylpyrimidin-4-yl dimethylsulfamate), is a pyrimidine or aminopyrimidinol fungicide. With a CAS Registry Number of 41483-43-6, bupirimate is a pale, waxy solid that is soluble in water (13.06 mg/l) and many common organic solvents. It is rapidly hydrolysed in acidic media and unstable to storage above 37°C. Bupirimate is produced by refluxing the sodium salt of 2-ethylamino-4-methyl-5-butyl-6-hydroxypyrimidine with dimethylsulfamoyl chloride in toluene.

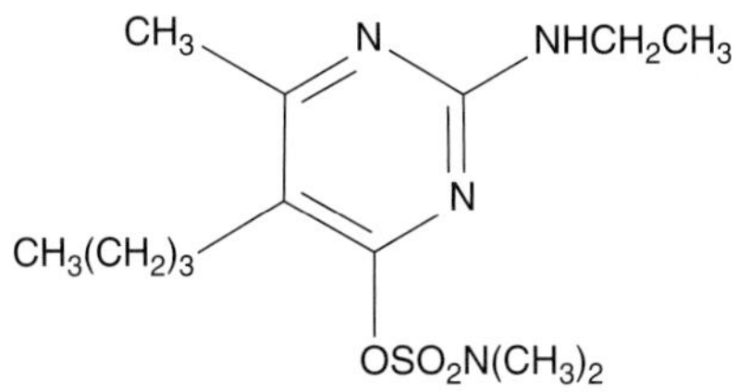

Bupirimate

The systemic fungicide bupirimate, absorbed by the leaves, acts by inhibiting the enzyme adenosine deaminase, thus disturbing nucleic acid synthesis. It leads to poor sporulation.

Bupirimate is effectively used as emulsifiable concentrates (EC), against fungi causing **powdery mildew** of apples, pears, peaches, berries, grapes, roses and cucurbits. Dosage ranges from 150 to 750 g/ha. Trade names include Nimrod.

Bupirimate is classified as a Class U toxin (unlikely to be hazardous) by the WHO and under Class III by the EPA. As per the acute oral and percutaneous LD_{50} studies of bupirimate, the fungicide is not very toxic (>4000 mg/kg among rats, mice, rabbits and guinea pigs). It is practically non-toxic to birds (acute oral LD_{50} being >5200 mg/kg b.w.) and non-irritating to skin and eyes. The NOEL (2 years) in rats is recorded to be 100 mg/kg diet. It has an ADI of 0.05 mg/kg b.w/day. The fungicide bupirimate is toxic to fish (LC_{50} in rainbow trout is 1.4 mg/l).

Once ingested, mammals excrete bupirimate within 10 days via the urine. The soil half-life is up to 3 months and it degrades to form erithrimol, among others.

Bupirimate is registered in Australia, South Africa, New Zealand, the Netherlands, Hungary, Greece and Cyprus, but not in the USA.

Buprofezin as an IGR

Buprofezin, $C_{16}H_{23}N_3OS$ ((*Z*)-2-tert-butylimino-3-isopropyl-5-phenyl-1,3,5-thiadiazinan-4-one), is an insect growth regulator. It is a white crystalline substance, sparingly soluble in water but soluble in organic solvents. It is stable in acidic and alkaline media and also to heat and light. The compound is prepared by cyclocondensation of *N*-phenyl-*N*′-tert-butylthiourea with *N*-methylene chloride *N*-carbonylchloride-2-propylamine.

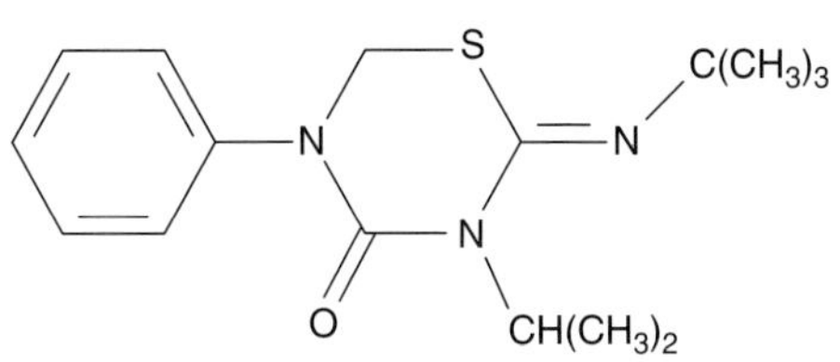

Buprofezin

Buprofezin, a persistent compound, acts by inhibiting **chitin** synthesis. It is both an insecticide and an acaricide with contact action. It inhibits **moulting** of nymphs and larvae and also suppresses oviposition by adults. It is effective against many species of **leafhoppers**, plant hoppers, cicadas, aphids, whiteflies and **scale insects** on for example rice, citrus and cotton.

The typical formulations are soluble concentrate (SL), emulsifying concentrate (EC) and wettable powder (WP). The principal trade name of buprofezin is Applaud.

Insecticide-acaricide buprofezin is classified as unlikely to be hazardous by the WHO. It is a possible carcinogen. It is not irritating to the skin and eyes of rabbits. The acute oral LD_{50} in rats is between 2198 and 2355 mg/kg. The acute percutaneous LD_{50} in rats is >5000 mg/kg. The NOEL in rats is between 0.09 and 1.12 mg/kg/day. The ADI is 0.01 mg/kg b.w./day.

Buprofezin is registered in Australia, the USA, India, Mali, New Zealand and South Africa.

Burgundy mixture fungicide

Burgundy mixture, an old fungicidal formulation invented in France, is a combination of dissolved **copper sulfate** ($CuSO_4$) and sodium carbonate (Na_2CO_3) in ratios of up to 1:18, to keep the copper concentration very low. The CAS Registry Number of this mixture is 11125-96-5.

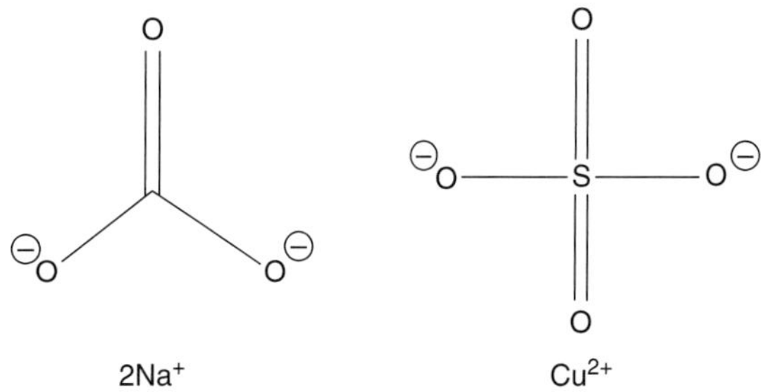

Burgundy mixture

Burgundy mixture is a fungicide that gives long-term protection against fungal infections including blight disease caused by *Phytophthora infestans*. The Cu (II) ions inhibit spores from germinating. But since copper can be phytotoxic, chelated and non-reactive forms of copper are used. Spraying the underside of leaves is considered the best method of controlling blight on potatoes.

Burning of weeds

See **Weed management by mechanical methods**

Butachlor herbicide

Butachlor, $C_{17}H_{26}ClNO_2$ (*N*-butoxymethyl-2-chloro-2′,6′-diethylacetanilide), is a chloroacetyl anilide herbicide. It is a yellowish oily liquid with a faint sweet odour. Its CAS Registry Number is 23184-66-9.

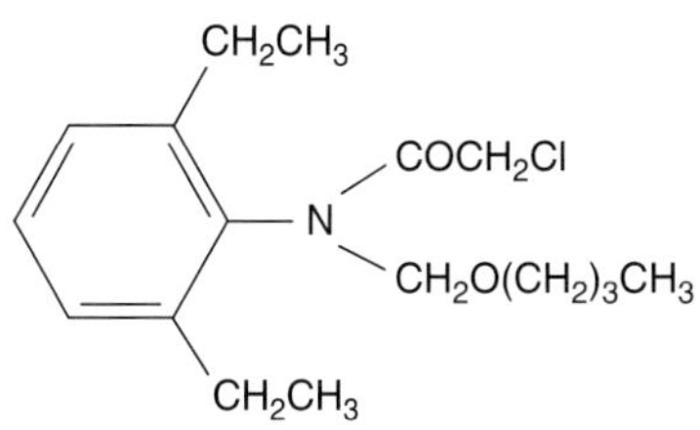

Butachlor

A selective, systemic rice herbicide, butachlor is commonly used as a pre-emergent/just post-emergent herbicide against broadleaved weeds and annual grasses in paddy. **Barnyard grass** (*Echinochloa crus-galli*), hairy crabgrass (*Digitaria sanguinalis*), foxtail millet (*Setaria* spp.), witch grass or torpedo grass (*Panicum* spp.) and species of wild rice and red rice such as *Oryza longistaminata*, *Oryza breviligulata* and *Oryza rufipogon* are some examples of weeds that bother rice crops. Other crops that benefit from butachlor applications include wheat, barley, groundnut, brassicas and cotton. It is reported to inhibit the synthesis of very long chain fatty acids.

The effective dose of butachlor has a wide range of 750–4500 g/ha. This herbicide may be used with other compounds like **pretilachlor**, **anilofos** and **thiobencarb**. It is normally applied on rice during seedbed preparation, and granular formulations and emulsifiable concentrates (EC) are preferred. It is important to take care that the herbicide does not injure the seed. It is commercially available as Machete, Butanex and Butataf.

The WHO classifies butachlor as a Class U toxin (unlikely to be hazardous). The acute oral LD_{50} values among rats, mice and rabbits are 2000, 4747 and 5010 mg/kg, respectively. The inhalation LC_{50} (4 h) in rats is >3.34 mg/l air. It does not irritate the eyes or the skin. It is non-toxic to bees (contact LD_{50} is >100 μg per bee). It is practically not toxic to birds in normal, recommended usage. The ADI is 0.037 mg/kg b.w./day. Butachlor is toxic to fish (LC_{50} (96 h) for rainbow trout is 0.52 mg/l). Also, its application to flooded alluvial paddy field is observed to inhibit methane (CH_4) production and bring down the number of methanogenic bacteria. So, the application dose and frequency are important. It is also observed to be a carcinogen, and so it is classified as a PAN Bad Actor chemical.

Butachlor is absorbed and eliminated rapidly in mammals with the production of water-soluble metabolites. In plants, it is similar. In soil, microbial degradation is prominent and rapid in alluvial soils, waterlogged conditions and organic soils. In soil, it persists for 6–10 weeks. It is converted to water-soluble derivatives with a slow evolution of CO_2.

Butachlor is registered for use in India, the Philippines, Cambodia and Vietnam.

Butam herbicide

See **Tebutam herbicide**

Butamifos herbicide

Butamifos, $C_{13}H_{21}N_2O_4PS$ (*O*-ethyl *O*-6-nitro-m-tolyl sec-butylphosphoramidothioate), is an organophosphate herbicide and has a CAS Registry Number of 36335-67-8. It is a yellowish-brown liquid, readily soluble in acetone, methanol and xylene but not so much in water. It is made by reacting 2-nitro-5-methylphenol with *O*-ethyl-*N*-butyl phosphamido thinochloride.

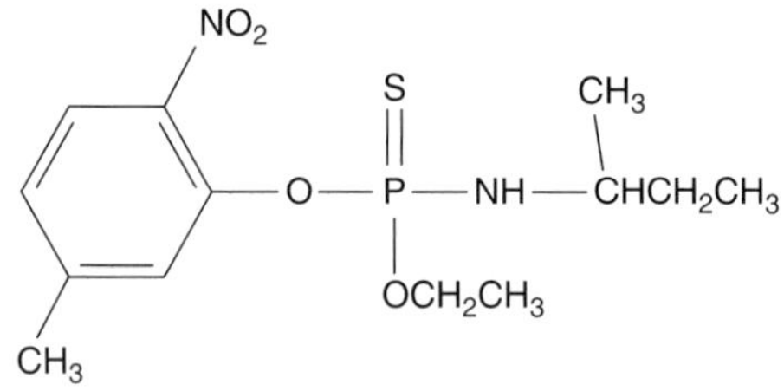

Butamifos

Butamifos is a pre-emergence, selective, non-systemic herbicide, used to control annual weeds in for example beans, rice and vegetables. The chemical works by inhibiting assembly of microtubules.

The herbicide is formulated as emulsifiable concentrates (EC) and as granules. The herbicide is available under the trade name of Cremart and Hercules 26905. It is classified as a K1 chemical under the **Herbicide Resistance Action Committee (HRAC)**. Butamifos is applied at 0.9–1.0 kg/ha for rice and 1.0–2.0 kg/ha for vegetables.

According to the WHO, butamifos comes under the toxicity Class II (moderately hazardous). Among rats, the acute oral LD_{50} is 1070 mg/kg, while the dermal LD_{50} is >5000 mg/kg. It is moderately toxic to fish (LC_{50} in carp is 2.4 mg/l). Butamifos is a PAN Bad Actor chemical, since it is a cholinesterase inhibitor.

The herbicide butamifos is not registered in the USA or any of the EU countries.

Butenachlor herbicide

See **Chloroacetanilide herbicides**

Butocarboxim insecticide

Butocarboxim, $C_7H_{14}N_2O_2S$ (3-(methylthio) butanone *O*-methylcarbamoyloxime), is an oxime carbamate (N-methyl carbamate) insecticide. It has a CAS Registry Number 34681-10-2. It is a brown viscous liquid soluble in water (35 g/l) but is miscible with aromatic hydrocarbons, esters and ketones. It is hydrolysed by strong acids and alkalis. Otherwise, it is stable to normal storage conditions, sunlight and up to 100°C. Butocarboxim is produced by treatment of 3-methylthio-2-butanone with hydroxylamine, followed by a reaction with methylisocyanate.

Butocarboxim, a systemic, broad-spectrum insecticide, with additional acaricidal properties, acts on the nervous system of the pest by inhibiting cholinesterase. Its residual activity gives effective control over mites. Pests controlled include aphids, thrips, whitefly, **mealybugs** and spider mites. Fruit trees, vegetables, cereals, cotton and tobacco are some of the crops protected.

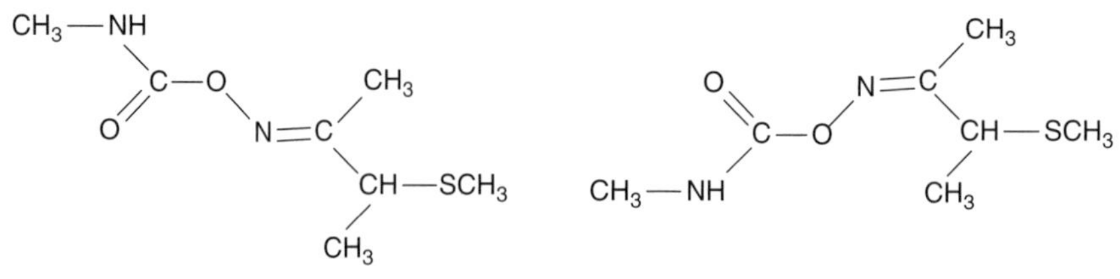

(*E*)- (*Z*)-

Butocarboxim

Available formulations of butocarboxim are emulsifiable/soluble concentrates and aerosol.

Butocarboxim is highly toxic to small mammals and honey bees. The WHO classifies it as highly hazardous (Class Ib toxin).The acute oral LD_{50} in rats is from 153 to 215 mg/kg, while its dermal LD_{50} is 360 mg/kg b.w. and inhalation LC_{50}, 1.0 mg/l. The NOEL (2 years) in rats is 100 mg/kg diet. The ADI is 0.02 mg/kg b.w./day. Since the chemical is a cholinesterase inhibitor and is acutely toxic, it is considered a PAN Bad Actor chemical according to the PAN Pesticide Database. Symptoms of poisoning include sweating, headache, muscle weakness, dizziness, vomiting and diarrhoea.

In plants and in the soil, butocarboxim degrades sulfoxide and sulfone metabolites. With a soil DT_{50} being a maximum of 8 days, the residues of the chemical are also negligible.

Butocarboxim is registered for use in Cambodia.

Butoxycarboxim insecticide

Butoxycarboxim, $C_7H_{14}N_2O_4S$ (3-mesylbutanone *O*-methylcarbamoyloxime), is an oxime carbamate insecticide. It has a CAS Registry Number 34681-23-7. It is a colourless, crystalline substance that is readily soluble in water and polar organic solvents. Technical butoxycarboxim contains the (*E*)- and (*Z*)- isomers in a ratio of 85–90:15–10. It has an aqueous hydrolysis DT_{50} that decreases with alkalinity (501 days at pH 5 and 16 at pH 9). It is stable to UV light and up to 100°C. It is obtained by the reaction of 3-methylsulfon-2-butanone with hydroxylamine, followed by treatment with methylisocyanate. Its CAS Registry Number is 34681-23-7 (see Butoxycarboxim at bottom of the page).

Butoxycarboxim, a cholinesterase inhibitor, is a systemic insecticide and acaricide used in the control of sucking insects like aphids and spider mites.

The insecticide-acaricide is available as a ready-to-use formulation on a cardboard stick. It is available in the market under the brand name of Plant Pin.

According to the WHO, butoxycarboxim comes under toxicity Class Ib (highly hazardous). It has an acute oral LD_{50} of 458 mg/kg and a dermal LD_{50} of >2000 mg/kg b.w. in rats. It is non-toxic to fish and bees. Considering the toxicity as well as the property of cholinesterase inhibition, butoxycarboxim is a PAN Bad Actor chemical. Symptoms of poisoning with butoxycarboxim include muscle weakness, sweating, headache, dizziness and diarrhoea.

In mammals, butoxycarboxim is eliminated both in the unchanged form and as metabolites. Soil DT_{50} of the chemical is around 44 days (at 20°C).

Butroxydim herbicide

Butroxydim, $C_{24}H_{33}NO_4$ ((5*RS*)-5-(3-butyryl-2,4,6-trimethylphenyl)-2-[(*EZ*)-1-(ethoxyimino)propyl]-3-hydroxy-cyclohex-2-en-1-one), is a cyclohexene oxime herbicide. It is a colourless powdery solid, sparingly soluble in water (6.9 mg/l). It has a CAS Registry Number of 138164-12-2. Butroxydim is produced by reaction of 5-(2,4,6-trimethylphenyl)-2-(1-ethoxyimino)-3-hydroxy-cyclohex-2-enone with butanoic acid chloride in the presence of $AlCl_3$.

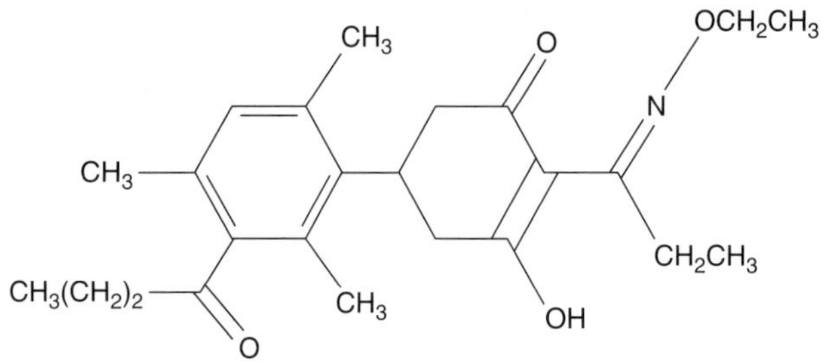

Butroxydim

Butroxydim is a selective, systemic, post-emergence herbicide, used to control annual ryegrass that hinders the growth of broadleaved crops such as lucerne, soybeans, beans and lentils. It inhibits lipid biosynthesis (Class A of HRAC). Factor is a known trade name for butroxydim.

According to the WHO, butroxydim comes under the toxicity Class III (slightly hazardous). It has an acute oral LD_{50} in male rats of 3476 mg/kg and in female rats of 1635 mg/kg. The dermal LD_{50} is >2000 mg/kg in rats. It has an inhalation LC_{50} of >2.99 mg/l in rats. It is a moderate skin and eye irritant. It is not hazardous to birds, honey bees or fish under normal use. It has an ADI of 0.025 mg/kg b.w./day. Butroxydim is not a PAN Bad Actor chemical.

In animals, following oral administration, more than 90% is excreted within 7 days. In plants it is metabolized rapidly. The butyryl chain is easily oxidized. It degrades rapidly in aerobic soil with a half-life of around 9 days.

Butroxydim is registered for use in Australia but not in the USA and the UK. The MRL of butroxydim on soybeans, beans, groundnuts, sunflower, cotton and many others has been fixed at 0.01 ppm by The Japan Food Chemical Research Foundation.

Butylate herbicide

Butylate, $C_{11}H_{23}NOS$ (*S*-ethyl di-isobuylthiocarbamate), is a thiocarbamate herbicide. It is a colourless liquid, moderately soluble in water (44 mg/l) and miscible with organic solvents. Butylate is hydrolysed by strong acids, alkalis and by sunlight. It is thermally stable up to 200°C. Butylate is made by reaction of di-isobutylcarbamoyl chloride with ethylmercaptan. It has a CAS Registry Number of 2008-41-5.

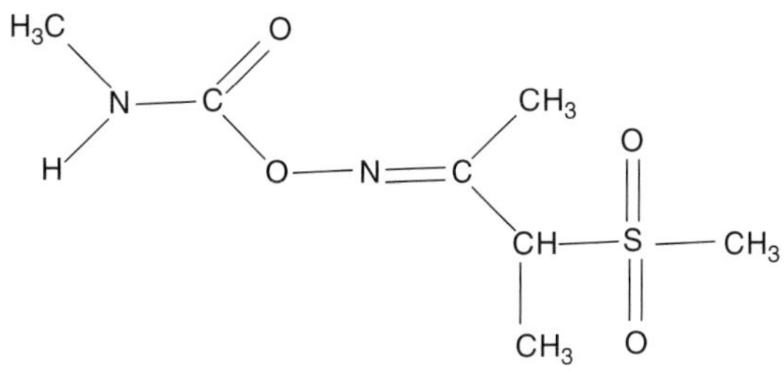

Butoxycarboxim

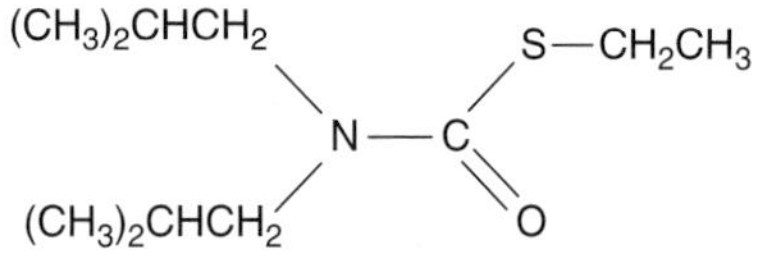

Butylate

Butylate is a selective, pre-plant, systemic herbicide, absorbed by the roots and tips of the emerging shoots in the soil. This herbicide is mainly used on maize for the control of nut grass and millet grass and also broadleaved weeds. It acts by inhibiting lipid biosynthesis.

Butylate is formulated as an emulsifiable concentrate (EC) and as granules. These formulations are available as for example Topanol, Sutan Plus and Dalpac. Application of this herbicide before planting maize helps to inhibit the germination of weed seeds. Control is enhanced when butylate is combined with **atrazine** and cyanazine.

The WHO classifies butylate as a Class U toxin (unlikely to be hazardous in normal use). The oral LD_{50} value for butylate is >3500 mg/kg in rats, and the dermal LD_{50} in rabbits is >5000 mg/kg. It does not irritate the skin and eyes if proper label instructions are followed. The ADI is 0.05 mg/kg b.w./day. The herbicide is non-toxic to birds and bees but is moderately toxic to fish, with an LC_{50} for a 96 h exposure being between 4.2 ppm and 6.9 ppm.

Butylate is a cholinesterase inhibitor, with potential for **bioaccumulation** and groundwater contamination. Thus, it is classified as a PAN Bad Actor chemical. Products must bear the signal word 'CAUTION'. Symptoms of poisoning can include irritation to the throat and the respiratory tract, and also to the skin and eyes. However, the EPA has classified it as a general use pesticide (GUP) and has declared as permissible, a maximum of 700 μg butylate/l drinking water.

This herbicide degenerates in mammals and plants to produce *N,N*-dialklylcarbamoyl conjugates, carbon dioxide, diisobutylamine and fatty acids. The herbicide breaks down in soil to form sulfoxides, with a half-life in the range of 2–2.5 months. Microbial action on butylate is also notable and causes degradation. However, butylate is also believed to produce ethanethiol (which acts like a rodenticide) and diisobutylamine in soil, that may be a threat to the environment. Ethanethiol (C_2H_6S) has a CAS Registry Number of 75-08-1 and its structure is presented below. Diisobutylamine ($C_8H_{19}N$) has a CAS Registry Number of 110-96-3.

Butylate is registered for use in the USA, but according to the Local Environment Risk Assessment for Pesticides, it is disallowed in the UK.

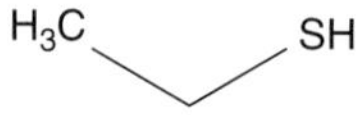

Ethanethiol

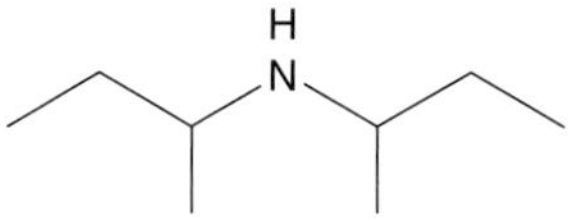

Diisobutylamine

C

4-Chlorophenoxyacetic acid (4-CPA) as a PGR

4-CPA, $C_8H_7ClO_3$ (4-chlorophenoxy acetic acid), is a plant growth regulator (PGR) and herbicide. It is a colourless crystalline substance, soluble in most organic solvents. 4-CPA is produced by the reaction of p-chlorophenol with chloroacetic acid. The CAS Registry Number is 122-88-3.

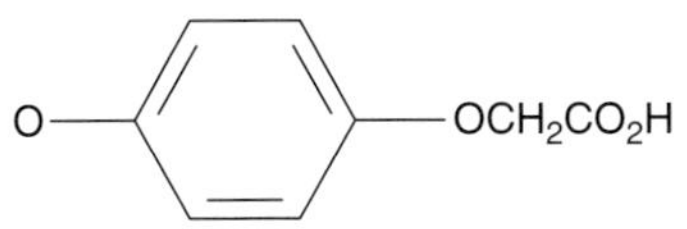

4-CPA

4-CPA is used to improve fruit setting, inhibit sprouts and regulate fruiting in tomato, beans and peaches. It can also be used as a herbicide. It is available as Zomatotone. 4-CPA is formulated as a soluble concentrates and tablets. The WHO rates it as a Class III toxin, indicating slight hazard.

Cadusafos insecticide

Cadusafos, $C_{10}H_{23}O_2PS_2$ (*S*,*S*-di-*sec*-butyl *O*-ethyl phosphorodithioate), is an aliphatic organophosphorus insecticide. It has a CAS Registry Number of 95465-99-9. It is a yellowish liquid that is highly soluble in water (248 mg/l) and completely miscible with acetone, toluene, acetonitrile, dichloromethane and ethyl acetate. It is stable at 50°C. Its half-life in light is more than 115 days. Cadusafos is prepared by reacting ethyl dichlorophosphate with *sec*-butanethiol in the presence of a base, under anaerobic conditions.

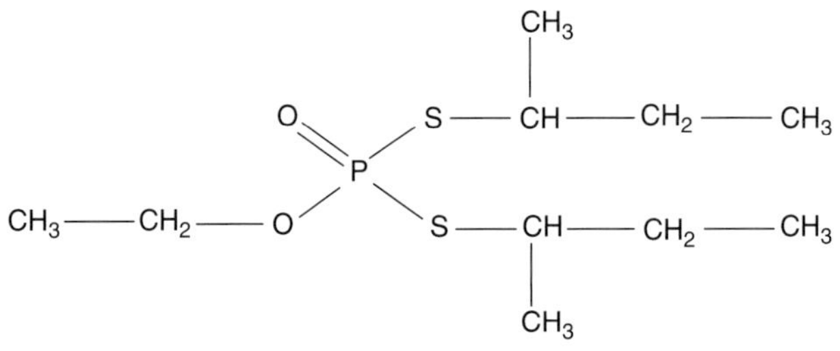

Cadusafos

Cadusafos, a broadcast insecticide with nematicidal properties, is an effective cholinesterase inhibitor. It controls soil-borne insects and nematodes, partly because it does not translocate too much in the upper portions of the plant. It is applied on bananas, citrus, maize, sugarcane, vegetables, potatoes and tobacco, against beetles and moths (e.g. members of the family Noctuidae, *Agriotes* spp. and *Melolontha* spp.), as well as nematodes such as *Meloidogyne* spp. and *Globodera* spp.

Cadusafos is commonly formulated either as granules or an emulsifiable concentrate (EC). In banana plantations, 4 g per plant gives adequate control with a residual effect of 6 months, while on potato and tobacco, a dose of 3–10 kg/ha is common. Trade names include Rugby and Apache.

The acute dermal LD_{50} in rabbits is 12 mg/kg. It does not irritate the skin or eyes of rabbits. This compound is moderately toxic to aquatic life as per acute toxicity tests. The WHO has classified cadusafos as a Class Ib (highly hazardous) toxin. It is a PAN Bad Actor chemical.

Cadusafos is eliminated rapidly by animals with major metabolites being hydroxy sulfone followed by phosphorothioic and sulfonic acids. Its DT_{50} in silty clay and sandy loam soils is 45 days. It has been observed that cadusafos undergoes biodegradation by the action of Gram-positive bacteria.

The ADI of cadusafos was set as 0.0004 mg/kg/day in the European Food Safety Authority's (EFSA) conclusion of the peer review of the pesticide risk assessment of the active substance cadusafos (EFSA, 2009). According to the FAO, the NOEL has been fixed at 0.03 mg/kg b.w./day. The MRL for banana was proposed at 0.01 mg/kg by the EFSA.

Cadusafos is registered only in Australia, South Africa and Tanzania.

Cafenstrole herbicide

Cafenstrole, $C_{16}H_{22}N_4O_3S$ (*N*,*N*-diethyl-3- mesitylsulfonyl-1*H*-1,2,4-triazole-1-carboxamide), is a crystalline substance,

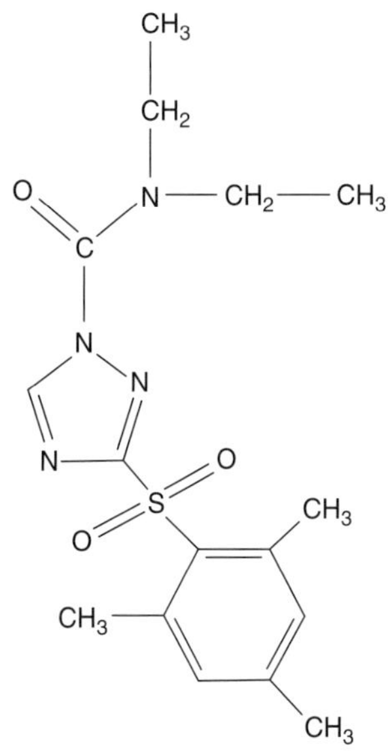

Cafenstrole

sparingly soluble in water and an amide herbicide. Cafenstrole is synthesized by the reaction of 2-sulfonyl-(2′,4′,6′-trimethylphenyl)-1,3,4-triazole with *N*-diethylcarbamoyl chloride. Its CAS Registry Number is 125306-83-4.

Cafenstrole is used both pre- and post-emergence (although it is more effective when used pre-emergence) for the control of grassy weeds like crabgrass (*Digitaria ciliaris*), green bristlegrass (*Setaria viridis*), annual bluegrass (*Poa annua*) and Manila grass (*Zoysia matrella*) in turf and rice crops.

Cafenstrole is applied mostly at around 40 g/ha and is commonly applied on leaves.

The 48 h LC_{50} in fish is >1.2 mg/l, while in bird species the LD_{50} is >2000 mg/kg. Cafenstrole is not mobile in the soil and often remains at the depth of 2 cm. The Japan Food Chemical Research Foundation has fixed an MRL of 0.02 ppm for brown rice.

Calcium arsenate insecticide

See **Arsenical inorganic insecticides**

Calcium as a plant nutrient

See **Plant nutrient excesses and deficiencies**

Calcium cyanamide as a molluscicide

Calcium cyanamide, $CaCN_2$, used as a nitrogen fertilizer, has molluscicidal properties. Calcium cyanamide occurs as hexagonal crystals. Pure calcium cyanamide is non-volatile and non-combustible. It is essentially insoluble in water but undergoes partial hydrolysis to form soluble calcium hydrogen cyanamid. Its CAS Registry Number is 156-62-7.

It is effective against green apple snails in directly sown rice. It is formed when calcium carbide reacts with nitrogen. It is commercially known as Nitrolime.

Calcium cyanamide has to be applied 10 days before sowing rice. It is safe if used as recommended. It can be broadcast on the field at a rate of as much as 100 kg/ha for an effective application. To prevent the entry of snails and slugs in the field, calcium cyanamide is used along the edges of the field.

Calcium cyanide rodenticide

See **Hydrogen cyanide rodenticide**

Calcium oxide as a molluscicide

*S*ee **Molluscicides**

Calcium phosphide rodenticide

Calcium phosphide, with molecular formula Ca_3P_2, is used as a rodenticide. It is a red-brown, combustible crystalline powder, which releases phosphine in water and acids. Metal phosphides include **aluminium phosphide, zinc phosphide,** copper phosphide and magnesium phosphide. Calcium phosphide has a CAS Registry Number of 1305-99-3. Its trade name is Photophor for the incendiary use or Polytanol for the use as rodenticide.

Calcium phosphide is formulated as bait, wherein it is mixed with food. Upon ingestion, the digestive juices react with the phosphide to generate **phosphine gas**. Calcium phosphide is a suitable agent when rodents are resistant to other rodenticides. Common trade names of calcium phosphide rodenticides include Polythanol.

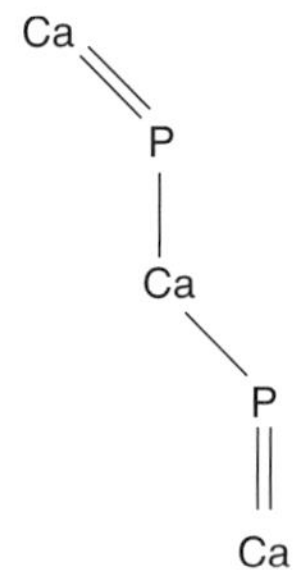

Calcium phosphide

Calcium polysulfide fungicide

Calcium polysulfide (also called **lime sulfur**), with molecular formula CaS_x, is a polysulfide fungicide with secondary insecticidal and acaricidal activity. It is a deep-orange liquid, soluble in water. It has residual action. It can be prepared by boiling calcium hydroxide and sulfur. It decomposes with CO_2, metal salts and acids to form insoluble sulfides, sulfur dioxide and hydrogen sulfide. Calcium polysulfide has a CAS Registry Number of 1344-81-6.

Calcium polysulfide is a protective, systemic fungicide that controls **powdery mildews**, anthracnose, **scab** and other diseases on lucerne, beans, clover and fruit. On oranges and limes, it can effectively control green mould caused by *Penicillium digitatum*. It acts on **scale insects** and spider mite eggs on fruit trees. It is also used on animals against ringworm.

The principal trade names of calcium polysulfide are Policalcio and Sulka. It is formulated as a soluble concentrate (SL) and an emulsifiable concentrate (EC). It is applied at the rate of 9–12 l/ha. It acts by disrupting metabolism. It also utilizes the fungicidal properties of **elemental sulfur** during its degradation and so acts as a 'protectant'.

The EPA classifies calcium polysulfide as a Class I toxin. It causes eye damage and skin irritation due to its high alkalinity. The acute oral LD_{50} for male and female rats is 820 mg/kg. The acute dermal LD_{50} in rats is >2000 mg/kg. Its reaction products inside organisms include the toxic hydrogen sulfide. It is phytotoxic to sulfur-sensitive apricots, raspberries and cucurbits.

Since calcium polysulfide is combustible, safety requirements need to be adhered to while applying or handling the substance.

Calcium polysulfide rapidly degrades to calcium hydroxide, also called hydrated lime (which is slightly toxic), and sulfur in the environment and the human body.

It is registered for use in for example India, South Africa, Australia, the USA and Canada.

Calibration of crop protection equipment

See **Application equipment – sprayers**

Canola oil bioinsecticide

Canola oil is a plant-derived insecticide, a repellent and an adjuvant. Canola oil is obtained from seeds of *Brassica napus* L. and *Brassica campestris* that belong to the family *Cruciferae* (mustard family). It is insoluble in water.

Apart from being an edible oil, canola oil is effective against insect pests, aphids, beetles, beetle larvae, cabbage worms, earwigs, gnats, gypsy moths, **leafhoppers**, leaf miners, leaf rollers,

mealybugs, mites, psyllids, scales, stinkbugs, tent caterpillars, thrips and whiteflies. This compound is especially used on citrus, maize, nut trees, sugarbeet, vegetables, melons, olives and house plants. Canola oil is applied through the irrigation system. A mixture of canola oil prepared with organic repellents is sprayed on foliage. Canola oil has a non-toxic mode of action and is believed to repel insects by altering the outer layer of the leaf surface or by acting as an insect irritant.

Canola oil is not likely to mix with water and leach into soil and groundwater. According to the EPA, canola oil is slightly toxic. However, it quickly decomposes in the environment, thus limiting its toxic effect. Canola oil is approved by the Organic Materials Review Institute of the USA (US OMRI) for use in organic farming. It is registered in Australia, New Zealand and South Africa.

Capsaicin bioinsecticide

Capsaicin, $C_{18}H_{27}NO_3$ (8-methyl-*N*-vanillyl-*trans*-6-nonenamide), a stable alkaloid, is an animal repellent that has also been registered for use as an insecticide, nematicide, fungicide, miticide, rodenticide and feeding depressant. It is mainly present in the solanaceous plants of the genus *Capsicum*, the compounds present therein being called capsaicinoids (hot pepper extracts). Capsaicin and dihydrocapsaicin (DHC) are the most prominent ones of these capsaicinoids. Capsaicin is obtained by making a fine powder of dried, ripe chilli pepper (*Capsicum frutescens* L.). A highly concentrated liquid can also be made by heating in a solvent.

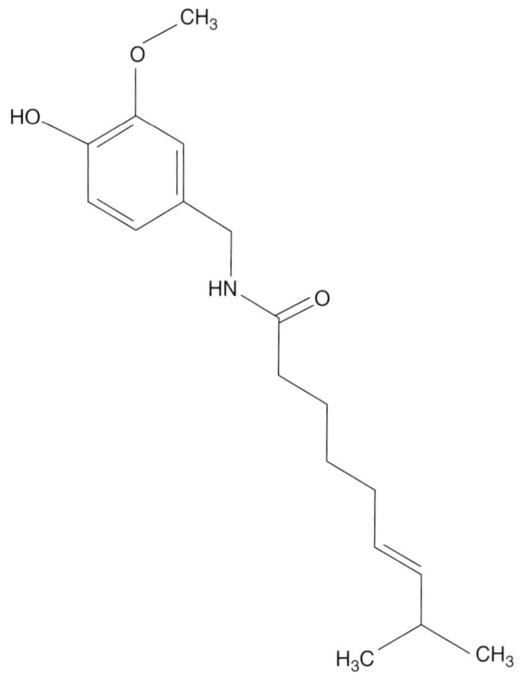

Capsaicin

Capsaicin serves as an insect repellent for mites, birds and mammals such as rabbits, squirrels, deer, voles, raccoons, cats and dogs. It also reduces transpiration in treated crops. It is used on crops, trees, buildings and public places. When applied to foliage of plants, it deters feeding by insects such as spider mites, lace bugs and other invertebrates.

Capsaicin may be combined with extracts of organic control agents like garlic and mustard. It is available as a paraffinic concentrate, liquid emulsifiable concentrate (EC) and soil drench. It may be sprayed on foliage two to three times for effective control against pests.

The acute oral LD_{50} values were determined to be 97.4 mg/kg and 118.8 mg/kg in female and male mice, respectively, and 148.1 mg/kg and 161.2 mg/kg in female and male rats, respectively.

The trade names for capsaicin include Armorex and Nemastroy. Capsaicin is only toxic to beneficial insects like honey bees. Capsaicin was first registered for use in the USA in 1962. It is classified as a biochemical pesticide due to its mode of action and the fact that it is a naturally occurring substance.

Captafol fungicide

Captafol, $C_{10}H_9Cl_4NO_2S$ (*N*-(1,1,2,2-tetrachloroethylthio) cyclohex-4-ene-1,2-dicarboximide), is a **phthalimide** fungicide. It has a CAS Registry Number of 2425-06-1. It is an almost colourless crystalline substance, soluble in water (1.4 mg/l), benzene (25 g/kg), acetone (43 g/kg) and toluene (17 g/kg). It slowly hydrolyses in aqueous emulsion but rapidly hydrolyses in acidic and alkaline media. Captafol is obtained by the reaction of the sodium salt of tetrahydrophthalimide in benzene with tetrachloroethylsulfenyl chloride.

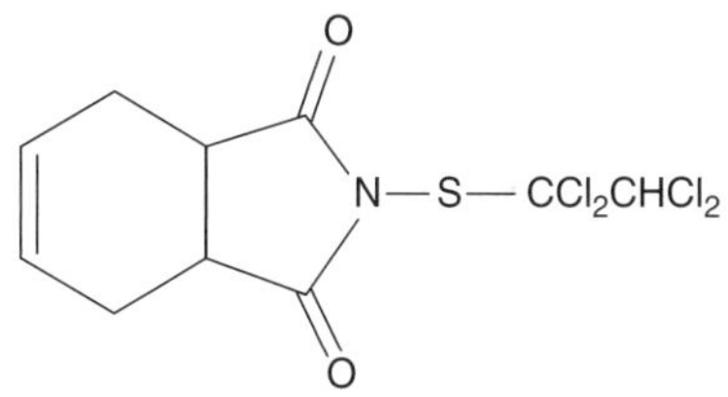

Captafol

Captafol, a sulfanilamide compound, is a contact, broad-spectrum, protective fungicide that inhibits respiration. It is also a sporicide. The fungicide is used to control diseases of leaves, fruits and seeds including **downy mildew**, black rot, shot-hole and seed-borne diseases caused by *Pythium* and *Phoma* spp. Crops thus protected include tomatoes, coffee, potato, groundnuts, onions, apples, citrus, pome, vine, stone fruit, wheat, barley, maize, rice and sugarbeet. Captafol also acts as a wound protectant during pruning and grafting. Besides, it helps to prevent lumber from rotting.

Some formulations of captafol include dustable powder (DP), wettable powder (WP) and powder for dry seed treatment (DS). Trade names include Crisfolatan, Difolatan and Difosan. Captafol is incompatible with highly alkaline materials. With the exception of alkaline preparations, captafol can be mixed with active ingredients like triadimefon, **ethirimol**, **folpet**, halacrinate, **propiconazole** and **pyrazophos**. Safety goggles, gloves and eye fountains are a prerequisite for handlers of captafol.

Captafol is categorized as a general use pesticide (GUP) by the EPA and has a toxicity classifcation of IV (relatively non-toxic). It is corrosive to eyes and may cause mild skin irritation or allergies, all of which may lead to harmful effects on prolonged exposure. The acute oral and percutaneous LD_{50} in rats and rabbits, respectively, range from 5000 to >15,400 mg/kg. The NOEL (2 years) in rats and dogs is 500 mg/kg diet and 10 mg/kg/day, respectively. Captafol is considered moderately to very highly toxic to fish and freshwater invertebrates, although it is non-toxic to bees. However, it is a potential carcinogen. It may become phytotoxic in grapes, apples and roses. Captafol is not listed as a PAN Bad Actor chemical, but its *cis* isomer is a PAN Bad Actor. Product labels must carry the signal word 'WARNING'. The WHO rates the unspecified isomer of captafol as a Class Ia toxin (extremely hazardous).

In animals, captafol undergoes hydrolytic cleavage to form the metabolites, dichloroacetic acid and tetrahydrophthalimide (THPI), which is degraded to tetrahydrophthalimidic acid and further to phthalic acid and ammonia and other soluble metabolites. In plants, the metabolism is the same as in animals. The half-life of captafol applied to most crops is 5 days.

Captafol has a half-life of 8 days in all types of soils. Leaching is low, especially from basic soils. Captafol and its metabolites are absorbed by roots and shoots of plants. It is also translocated in plant tissue via seed/soil/foliar treatment.

Captafol is not registered in India, the EU or the USA. In India, however, the compound and its formulations are allowed for export. Its generally accepted MRL on tomato is 5 mg/kg.

Captan fungicide

Captan, $C_9H_8Cl_3NO_2S$ (*N*-(trichloromethylthio) cyclohex-4-ene-1,2-dicarboximide), belongs to the phthalamide (dicarboximide) class of fungicides. It is also known as Kittleson's killer. The compound is a beige, amorphous crystalline solid, soluble in water (3.3 mg/l), xylene (20 g/l), chloroform (70 g/l), benzene (21 g/l), ethanol (2.0 g/l) and in many other organic solvents. Captan has a CAS Registry Number of 133-06-2. It hydrolyses rapidly in alkaline pH, with a half-life (DT_{50}) of <2 min at pH 10.

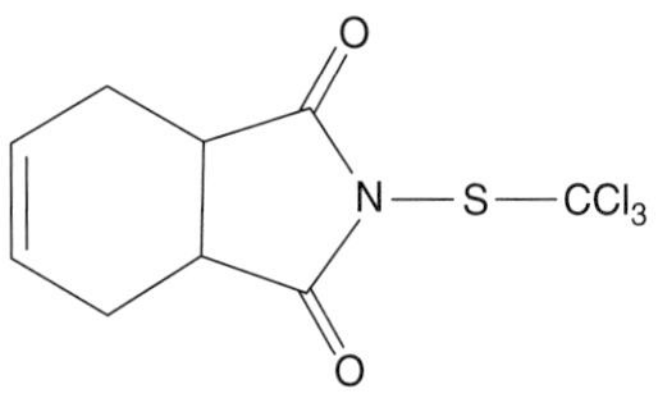

Captan

Captan is a **non-systemic fungicide** and provides protective and curative action on a wide range of fungal diseases. It is a non-specific thiol reactant that inhibits respiration. It protects against a wide variety of fungal infections in crops both at the seedling stage and in mature plants. It controls **scab**, rot diseases (black, brown and bull's eye), leaf curl, leaf spots, blights, anthracnose and shot-hole of stone fruits. It therefore serves both as a seed-dresser and as a foliar spray to control seed-borne diseases, soil-inhibiting fungi and foliar/fruit diseases. Captan technical (i.e. technical grade) is available as a 92% w/w. Crops that are protected include tomatoes, pome, peach, turf and vegetables. Seed treatment is effectively carried out against pathogens like *Pythium* spp., *Phoma* spp. and *Rhizoctonia* spp., by the use of captan on crops such as maize and canola.

Trade names for captan include Orthocide, Clomitane, Vancide 89, Agrox and Merpan. It is formulated as a dustable and wettable powder, soluble concentrate, dry powder for seed treatment and water-dispersible powder in slurry treatment. A rough dose is 1 tbs/l water.

The acute oral LD_{50} values in rats and mallard ducks are 9000 mg/kg and >5000 mg/kg, respectively, while in bees it is recorded to be 91 μg per bee.The compound is corrosive to eyes and skin. Captan is classified as a Class U toxin (unlikely to pose hazard) by the WHO and as a Class IV toxin by the EPA. However, it is a potential carcinogen and hence a PAN Bad Actor chemical. It is considered to be highly toxic as per the EPA's acute toxicity rating. The NOEL in rats (for a 3 year period) is recorded to be 12.5 mg/kg/day.

In animals, captan metabolizes mainly to tetrahydrophthalimide (THPI). THPI has a half-life of up to 20 days. In soil, where captan is only slightly mobile, the metabolites are THPI and thiophosgene. The soil DT_{50} is about 1 day at pH 7.2, with temperature around 25°C.

Captan is registered for use in India, Canada, Australia, South Africa, New Zealand, the USA and several EU countries such as the UK, Germany, Greece, France, Malta, Spain and Italy. The MRL value fixed for grapes in the UK is 0.02 mg/kg.

Caraway oil bioinsecticide

Caraway (*Carum carvi*, belonging to the Umbelliferae family) is a biennial plant with origins in Asia Minor. It is the source of an essential oil called caraway oil and is known to have medicinal and insecticidal properties, because of components like **carvone** (an optical isomer), aldehyde, ketone, terpene and D-limonene. Caraway oil is obtained by steam distillation of the crushed fruits. It is an almost colourless substance with a strong, warm, sweet-spicy odour. Carvone, the main constituent of caraway essential oil, is also classified as an antiseptic, CNS stimulant, insecticide and vermicide (dewormer).

Carbamate as herbicides

See **Herbicide action – meristematic mitotic inhibitors**

Carbamate insecticides

Carbamate insecticides, derivatives of carbamic acid (NH_2COOH) developed from the alkaloid physostigmine, are synthetic organic insecticides (organochlorine and organophosphorus compounds). Most carbamates are derivatives of methyl carbamic acid. Methylcarbamates are esters of substituted phenols and oximes.

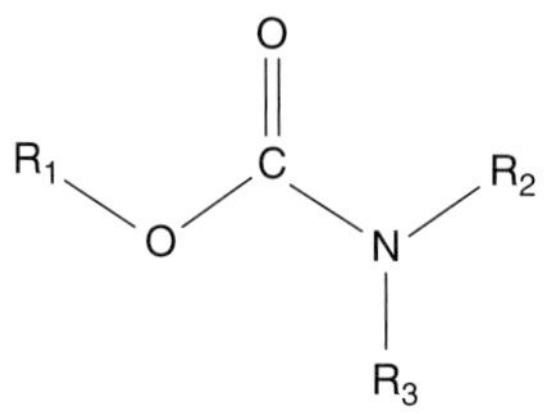

where, R_1, R_2 and R_3 are alkyl or aryl groups

Carbamate

Carbamate insecticides act on the nervous system by reversibly inhibiting **acetylcholinesterase**, making them a safe option.

This feature, along with their **biodegradability** (hence their non-persistence), has made this group of insecticides popular with crop-care specialists. These are used against soil and leaf insects on vegetable crops.

A nerve impulse stimulates the generation of several neurotransmitters, the most important of which being acetylcholine (ACh), with molecular formula $CH_3COOCH_2CH_2N^+(CH_3)_3$. ACh functions in the CNS and peripheral nervous system and activates muscles and has a role in enabling sensory perception. Muscle contraction takes place after a series of steps that begins with ACh binding to ACh receptors that open up the sodium channels. The enzyme acetylcholinesterase (AChE) functions as a degenerator of ACh (by hydrolysing it into choline and acetic acid). The function of the AChE is critical since it keeps the concentration of the ACh in check preventing over-excitation.

Inhibition of the enzyme AChE begins by the formation of an inhibitor–enzyme complex where serine hydroxyl gets carbamylated. But the carbamylated enzyme can spontaneously get regenerated to the original form.

Inhalation and ingestion are two routes of entry of carbamate insecticides, with the symptoms of poisoning being muscle weakness, dizziness, physical discomfort and vomiting. However, the effect is normally for a short duration only.

There are four classes of carbamate insecticides based on the type of constituent compounds. They are: (i) the dimethyl carbamates of heterocylic enol (**pirimicarb**); (ii) the methyl carbamates of phenols (**carbaryl, carbofuran, carbosulfan, bendiocarb, ethiofencarb, fenobucarb, formetanate, furathiocarb, isoprocarb, methiocarb, metolcarb, propoxur, trimethacarb** and **xylylcarb**); (iii) the methyl carbamates of oximes (**alanycarb, aldicarb, aldoxycarb, butocarboxim, butoxycarboxim, methomyl, oxamyl, thiodicarb** and **thiofanox**); and (iv) the **nereistoxin** precursors (**cartap, thiocyclam** and **bensultap**).

Carbamothioate

See **Herbicide action – meristematic mitotic inhibitors**

Carbaryl insecticide

Carbaryl, $C_{12}H_{11}NO_2$ (1-naphthyl methyl carbamate), is a methyl carbamate insecticide of phenol class. It is a pale-coloured crystalline substance, slightly soluble in water and soluble in polar organic solvents. It is stable under slightly acidic conditions but hydrolyses in an alkaline medium (producing 1-naphthol). Carbaryl is made by the reaction of α-naphthol with methylisocyanate in the presence of a catalytic amount of triethylamine. Carbaryl has a CAS Registry Number of 63-25-2.

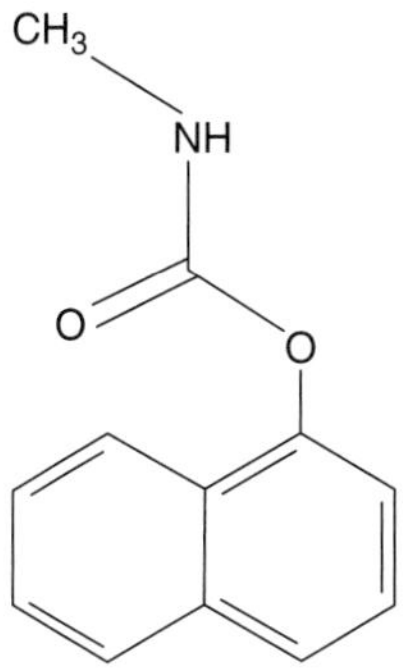

Carbaryl

Carbaryl is a broad-spectrum, systemic insecticide (of chewing and sucking insects) and a PGR. It inhibits cholinesterase and targets members of the orders Lepidoptera and Coleoptera, protecting cotton, maize, soybean, sorghum, rice, potatoes, groundnuts, vegetables and citrus.

Carbaryl can also be mixed with many insecticidal, fungicidal, acaricidal formulations as long as they are not alkaline in nature. It is available as bait, dusts, wettable powders, granules, dispersions and suspensions.

The WHO and EPA classify carbaryl as moderately toxic (Class II). It is reported to be toxic to rabbits (LD_{50} >2000 mg/kg) and rats (LD_{50} >4000 mg/kg). It is also toxic to pheasants, ducks, fish and bees. It also inhibits growth of earthworms. It may contaminate groundwater. It is categorized as a PAN Bad Actor chemical.

In sandy loam and clay loam, carbaryl degrades with a half-life of 7–14 days and 14–28 days, respectively. Sunlight and bacterial action are responsible for degradation in soil. It does not accumulate in body tissue and is instead metabolized into non-toxic substances.

Carbaryl is registered for use in India, South Africa, Australia, the USA and many other countries. Its use is not permitted in the EU. Joint FAO/WHO Expert Committee on Food Additives has fixed an MRL for chilli peppers at 2 mg/kg and cranberries at 5 mg/kg.

Carbendazim fungicide

Carbendazim, $C_9H_9N_3O_2$ (methyl benzimidazole-2-ylcarbamate), is a fungicide in the **benzimidazole** or benzimidazolylcarbamate group. It is a crystalline powder, soluble in water (8 mg/l, pH 7), and stable in acids but decomposes slowly in alkaline solutions and at 307°C. It is produced by the reaction of *o*-phenylenediamine with methoxycarbonylcyanamide or with methoxycarbonyl isocyanide dichloride (formed from cyanogens chloride, phosgene and methanol) in the presence of a base. A related product in this group is carbendazim phosphate ($C_9H_{12}N_3O_6P$), which is soluble in water at 9 g/l (pH 2). The CAS Registry Number is 10605-21-7.

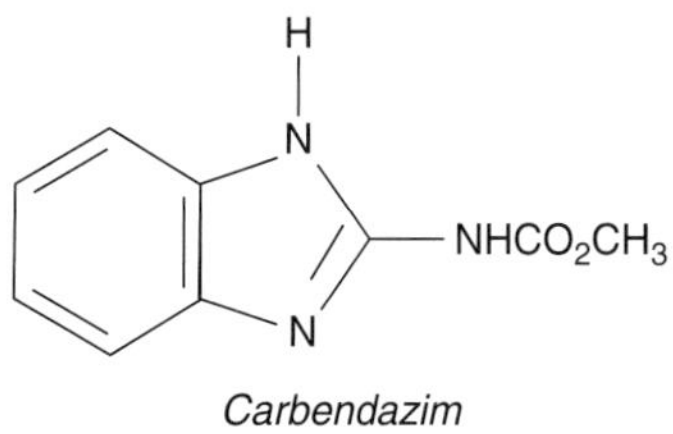

Carbendazim

Carbendazim, being a systemic and protective fungicide, inhibits mitotic microtubule formation, after being translocated acropetally. Its action inhibits the development of germ tubes, appresoria and mycelial growth. Its reach extends to protection from storage diseases (post-harvest) and seed treatment (preplant). The pathogens controlled include *Septoria*, *Fusarium*, *Erysiphe*, *Pseudocercosporella*, *Sclerotinia*, *Alternaria*, *Cercospora*, *Botrytis*, *Cladosporium*, *Tilletia*, *Venturia*, *Podosphaera* and *Monilinia* in cereals, canola, sugarbeet, grapes, tomato, cotton, pome and stone fruit.

Both carbendazim and carbendazim phosphate are available in the form of flowable concentrate, oil-dispersible powder, soluble concentrate (EC), suspension concentrate (SC), wettable powder (WP), etc. Trade names include Eertavas. The application rates vary in the range of 120–600 g/ha, and 0.6–0.8 g/kg for seed treatment (cotton and cereals) and 0.3–0.5 g/l as a fruit dip. This compound can combine with triazoles and **dithiocarbamates**.

Carbendazim is classified as a Class U toxin by the WHO and is found to be very slightly toxic in mammals as well as birds. However, it is highly toxic to aquatic organisms (in rainbow trout, the LC_{50} is 0.83 mg/l). Its acute oral LD_{50} in rats is determined to be 6400 mg/kg, for birds (quail) in the range of 5826–15595 mg/kg and in bees (contact) >50 μg per bee. Carbendazim adsorbs to soil, harming earthworms (LC_{50} of *Eisenia foetida* is 6 mg/kg soil). Carbendazim is suspected to disrupt endocrine function and is a possible carcinogen.

It is necessary for carbendazim containers and application equipment to be disposed of either by incineration or in a landfill, or according to the local environment regulating agency. All care should be taken to ensure that no water source is contaminated in the process. Protective equipment is a must while applying and cleaning up.

The said fungicide is decomposed in the environment with a half-life (DT_{50}) of 6–12 months in soil, and 5–25 months in water, under both aerobic and anaerobic conditions. Plants readily absorb carbendazim that degrades to produce a minor metabolite called 2-aminobenzimidazole. The half-life of the metabolite is 8–32 days in field conditions.

Carbendazim is registered for use in India, New Zealand, Australia, the USA and several countries of the EU including the Netherlands, the UK, Spain and Germany. The MRL fixed by the Pesticide Residues Committee of the UK on pears and apples is 0.2 mg/kg.

Carbetamide herbicide

Carbetamide, $C_{12}H_{16}N_2O_3$ ((*R*)-1-(ethylcarbamoyl) ethyl carbanilate), is a carbanilate (carbamate) herbicide. It is a colourless crystalline substance, which is fairly soluble in water and readily soluble in polar organic solvents. It is stable under normal storage conditions. It is made by condensation of phenyl isocyanate with *N*-ethyl lactamide. Carbetamide has a CAS Registry Number of 16118-49-3.

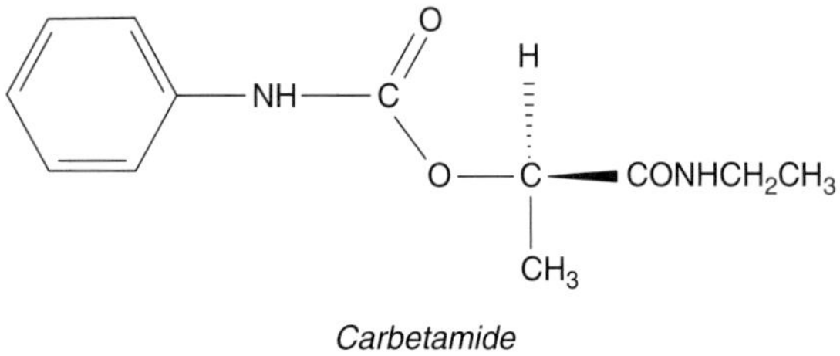

Carbetamide

Carbetamide, a selective pre-emergence (or early post-emergence) herbicide absorbed by roots and leaves, acts by inhibiting cell division in young shoots and roots. It is used to control annual grasses and some broadleaved weeds in sugarbeet, clover, vegetables, oilseed and fruit orchards.

The herbicide carbetamide is available in the market under trade names such as Legurame, Riot and Carbatamex. Applications of around 2 kg a.i./ha, 6 weeks before harvest is recommended. It is usually applied as a spray and formulated as wettable granules or wettable powders.

According to the WHO and EPA, carbetamide comes under the toxicity Class U and Class IV (practically non-toxic), respectively. It is not a PAN Bad Actor chemical. It is an eye irritant and is suspected to be a possible carcinogen.

Metabolites of carbetamide include 2-(4-hydroxyphenylcarbamoyloxy)-propionic acid, which is formed in plants via hydrolysis.

Carbetamide is registered in Australia, the UK, Belgium, the Netherlands and many other countries, but not in the USA. As per the Atlanta MRL database (version 2), the MRL fixed by Germany for table grapes, small berries and fruits was 0.05 mg/kg.

Carbodiimide CGA 140408 acaricide

See **Diafenthiuron acaricide**

Carbofuran insecticide

Carbofuran, $C_{12}H_{15}NO_3$ (2,3-dihydro-2,2-dimethylbenzofuran-7-yl methylcarbamate), is a carbamate insecticide of the benzofuranyl methylcarbamate group. It has a CAS Registry Number of 1563-66-2. It is a colourless crystalline substance, soluble in water (0.32g/l), dichloromethane, isopropanol and toluene. It is hydrolysed in alkaline medium. Carbofuran is prepared by the claisen rearrangement and cyclization of 2-methallyloxy-1-nitrobenzene to 2,3-dihydro-2,2-dimethyl-7-nitrobenzofuran. Reduction of the nitro compound gives the corresponding aniline, which is diazotized and concentrated by boiling to form the phenol. The phenol thus obtained reacts with methyl isocyanate (MIC) to give carbofuran.

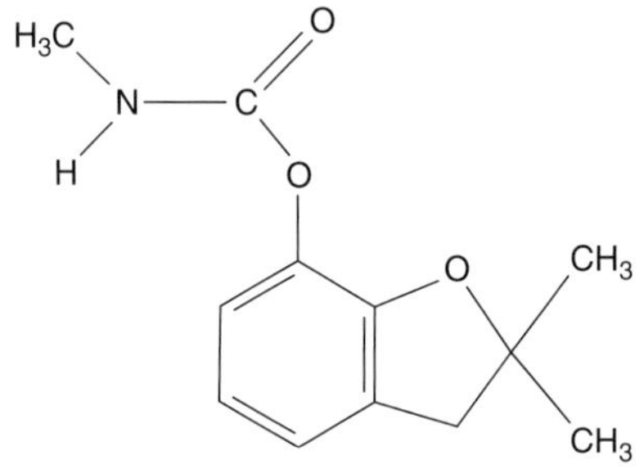

Carbofuran

Carbofuran, a cholinesterase inhibitor, is a systemic insecticide, with nematicidal and acaricidal properties. It is used to control soil and leaf insects mainly in rice, bananas, soybeans, sunflower, coffee and sugarcane. Other crops that are covered include vegetables, ornamentals, beet crops, maize, sorghum, lucerne, groundnut, cotton, citrus and strawberries. Insects like wireworms, grubs, millipedes, many types of flies (fruit, bean seed, sciarid and root), flea beetles, weevils, aphids, thrips, as well as nematodes are thus controlled.

Carbofuran is available in the market under the brand names of Furadan, Agrofuran, Curater and Carbodan. It is compatible with many non-alkaline pesticides. Carbofuran is formulated in forms such as granules, wettable powder and flowable concentrate for seed treatment.

The ADI of carbofuran is 0.01mg/kg/day. It is toxic to bees. Carbofuran is highly toxic to birds (especially in its granular form) and fish (LD_{50} in bluegill sunfish is 0.24 mg/l). It is highly toxic by inhalation and ingestion and moderately toxic by dermal absorption. Death may result at high doses. It is a potential groundwater contaminant and a suspected endocrine disrupter. It is a PAN Bad Actor chemical. Signal words 'WARNING' and 'DANGER' are a prerequisite on product labels.

In plants as well as in soils, carbofuran metabolizes into 3-hydroxycarbofuran, phenol compounds, CO_2 and keto-carbofuran, which also have insecticidal properties (although it further hydrolyses into 3-ketocarbofuran-7-phenol). Carbofuran is degraded in soil by hydrolysis, oxidation, sunlight and microbes. It is also mobile in many types of soils and persists for quite long with a soil DT_{50} of 30–60 days.

Many countries have either banned or severely restricted the use of carbofuran, for example in the case of Canada, Kuwait, the USA, Belize and Libya. India, Australia, Vietnam, South Africa and the USA have registered it, although there is much debate on the fate of its granular formulation. A reference MRL for grapes and berries is 0.05 mg/kg, based on the Atlanta MRL database (version 2).

Carbon disulfide

See **Fumigants**

Carbophenothion acaricide

Carbophenothion, $C_{11}H_{16}ClO_2PS_3$ (*S*-4-chlorophenylthiomethyl *O*,*O*-diethyl phosphorodithioate), is an organothiophosphate acaricide. It is a yellow-brown liquid with a mild mercaptan-like odour. It is non-corrosive and relatively stable to hydrolysis and to temperatures up to 80°C. The solubility in water is 0.34 ppm at 20°C. It is miscible in solvents such as alcohols, ethers, toluene, mineral and vegetable oils. Its CAS Registry Number is 786-19-6.

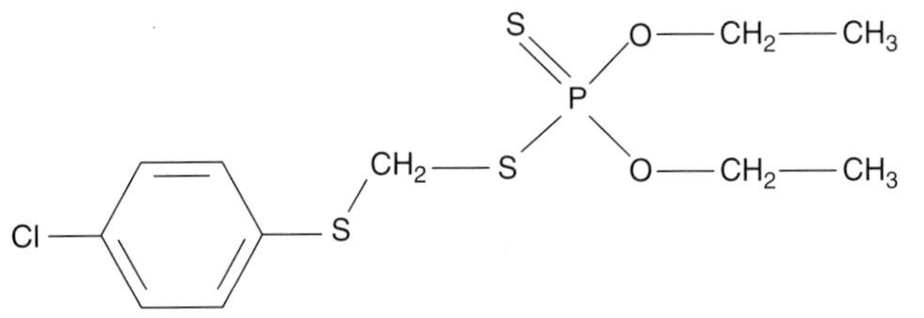

Carbophenothion

Carbophenothion is an insecticide and acaricide. The compound has a long residual activity that can control growth of eggs and neonates of many species. It is used on fruit and nut crops, vegetables, sorghum, maize and cotton to control aphids, scale insects and spider mites. Mixing with petroleum oil broadens its control spectrum to include parasites like ticks and suckers.

Acaricide-insecticide carbophenothion is available as an emulsifiable concentrate, wettable powder, granules and a seed treatment formulation. The trade names for products containing carbophenothion include Trithion, Garrathion, Dagadip, Endyl and Lethox.

The EPA has classified carbophenothion under Category I (i.e. highly toxic). The products containing the a.i. must bear the signal word 'DANGER'. The acute oral LD_{50} is in between 10 and 30 mg/kg in rats and 218 mg/kg in mice. The dermal LD_{50} of carbophenothion ranges from 27 to 1850 mg/kg in rats and rabbits, with the NOEL being 1.1 mg/kg/day. The LC_{50} is 0.002 mg/l of air. It is highly toxic to birds, with the LD_{50} being 5.6–35 mg/kg for starlings and geese. It is highly toxic to fish, crustaceans and amphibians. The ADI is 0.0125 mg/kg/day. Carbophenothion is a PAN Bad Actor chemical.

Carbophenothion is absorbed through the skin very quickly (in <1 h in rats). It is excreted from the body rapidly, mainly via urine and faeces within 72 h. Carbophenothion persists in soils for longer than 6 months indicating relative immobility in sandy loam soils.

Carbophenothion is not registered in the USA or in the EU. It is considered an obsolete pesticide.

Carbosulfan insecticide

Carbosulfan, $C_{20}H_{32}N_2O_3S$ (2,3,-dihydro-2,2-dimethylbenzofuran-7-yl(dibutylaminothio)methylcarbamate), is a carbamate insecticide. It has a CAS Registry Number of 55285-14-8. It is a slightly orange clear viscous liquid with solubility in water (0.3 ppm). It is miscible with xylene, hexane, chloroform, methanol and ethanol. Carbosulfan is produced by the reaction of sulfenyl chloride with dibutylamine, followed by treatment with carbofuran.

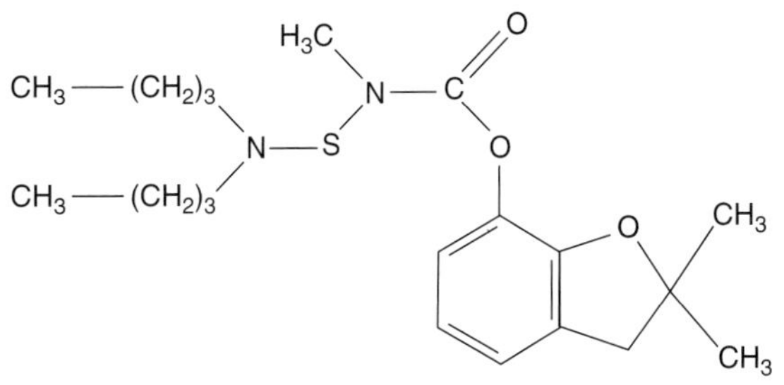

Carbosulfan

The insecticide, an *N*-sulfenyl-*N*-methylcarbamate compound, is a pro-insecticide of **carbofuran** (a metabolite of carbosulfan in plants). It is a cholinesterase inhibitor, a broad-spectrum carbamate insecticide used to control insects, mites and nematodes via soil, foliar and in seed treatments. Lucerne, apple, potato, sugarbeet, rice, maize and citrus are thus protected. In plants, carbosulfan is not translocated and its systemic activity is due to metabolites.

Carbosulfan is formulated as emulsifiable concentrate, dust and granules. A known brand name is Marshal. It is either broadcast or side-dressed, or applied in furrows.

According to the WHO and EPA, carbosulfan is a moderately hazardous toxin (Class II toxin) and extremely toxic (Class I), respectively. Due to its acute toxicity rating and cholinesterase inhibiting property, it is classified as a PAN Bad Actor chemical.

The main metabolic routes for carbosulfan in plants and animals are N–S bond cleavage, oxidation, conjugation and hydrolysis. The products of metabolism in plants are carbofuran and 3-hydroxycarbofuran. In mammals, it rapidly undergoes hydrolytic and oxidative processes followed by conjugation. It is not persistent in soils, with DT_{50} of 2–5 days. Carbosulfan gets rapidly degraded to carbofuran in sandy loam soil.

Carbosulfan is registered for use in many countries including India, Australia, the Philippines, South Africa and many other African countries. It is not registered for use in the USA. The MRL of potatoes, among other crops, is commonly fixed at 0.05 mg/kg.

Carboxin fungicide

Carboxin, $C_{12}H_{13}NO_2S$ (5,6-dihydro-2-methyl-1,4-oxathiine-3-carboxanilide), is a fungicide belonging to the fungicidal class oxathiincarboxamide. Its CAS Registry Number is 5234-68-4. Carboxin is a yellowish white crystalline substance, which is soluble in water (195 mg/l), benzene, methanol and acetone (600 g/l). It is stable to hydrolysis in alkaline media. But if exposed to sunlight, its half-life is 1.54 h, in the solution form. It is produced by the reaction of 2-chloro-3-oxobutyricanilide with 2-mercaptoethanol, followed by acid-catalysed ring closure with elimination of water.

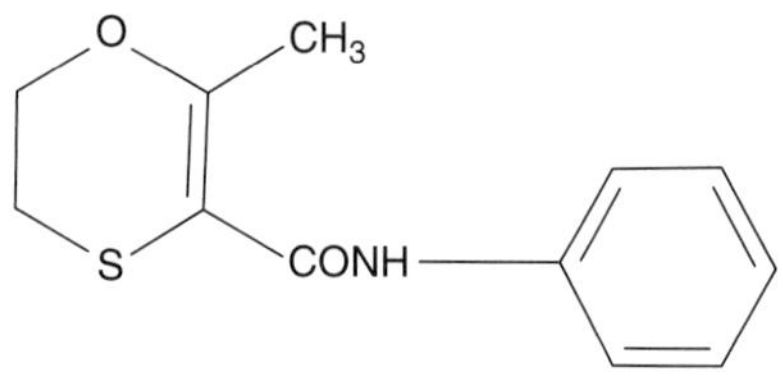

Carboxin

Carboxin, a systemic fungicide, is a succinase dehydrogenase inhibitor that inhibits the mitochondrial function by disrupting complex II (succinase dehydrogenase) in the respiratory electron transport chain. This chemical is both preventive and curative in nature. It controls seed-borne diseases like **smut** (caused by *Ustilago* spp.), bunt diseases, blight and seedling rot (caused by *Rhizoctonia* spp.) as well as turf diseases. Crops on which this fungicide is used are barley, oats, maize, wheat, rice, cotton and vegetables.

The fungicide is commonly applied at the rate of 50–200 g/100 kg seeds. The chemical is available as soluble concentrate (SL), flowable concentrate for seed treatment (FS), wettable powder (WP) and water-dispersible powder for slurry seed treatment (WS). Some common trade names are Vitavax and Sunboxin. Carboxin can be combined with pesticides that are neither highly alkaline nor acidic.

Carboxin is classified as a Class U toxin by the WHO and as Class III by the EPA. It is moderately toxic to mammals as well as birds, but highly toxic to aquatic organisms. Its acute oral LD_{50} in rats is 2864 mg/kg, for birds (bobwhite quails) around 3302 mg/kg and in bees >100 μg per bee. The LC_{50} in rainbow trout is 2.3 mg/l and in earthworms it is 500–1000 ppm. Carboxin formulations have to carry the signal word 'CAUTION' on the label. However, carboxin is also observed to produce developmental/reproductive defects in some tested animals. This chemical is also a PAN Bad Actor chemical.

Upon ingestion by rats, carboxin is metabolized and eliminated as 4-hydroxycarboxin sulfoxide and its glucuronide. In plants, it undergoes oxidation to carboxin sulfoxide and carboxin sulfone. In soil, carboxin and carboxin sulfoxide are mobile, making both parent and the metabolite potential groundwater contaminants. The soil half-life is <1 day at 20°C.

Carboxin is registered in India, Australia, New Zealand, Mali, South Africa, the USA and most countries of the EU. However, the EU has banned carboxin sulfoxide. The MRL values set out for various forms of harvest products of oats, rice, wheat and groundnuts (like the grain, forage and straw) are being proposed at around 0.2 ppm by the Reregistration Division of the Office of Prevention, Pesticides and Toxic Substances (OPPTS) of the EPA, USA.

Carcinogen

A substance that causes the development of cancerous growth in living tissues is a carcinogen.

Exposure to such a carcinogenic substance can result from exposure to toxic chemicals such as asbestos, benzopyrene, nickel carbonyl, chloromethylether and a large number of pesticidal ingredients. Such exposure is evident while handling carcinogenic chemicals in the field, laboratory or a manufacturing plant. Cyclamate compounds and 4-dimethylaminoazobenzene are carcinogenic chemicals.

There are several hundred known and suspected carcinogenic compounds and new ones are constantly being discovered. In 1978, the **Occupational Safety and Health Administration** (OSHA) of the USA issued a tentative list of carcinogenic substances, subject to periodic revision.

Many agencies throughout the world categorize chemicals as carcinogens in order to caution users about its application. For example, the **International Agency for Research on Cancer** (**IARC**) and USA National Toxicology Program (NTP) evaluate chemicals and rank them on their toxicity. IARC, which is a part of the WHO, aims to conduct research on causes of cancer in humans and its control. It also works towards spreading awareness and information on the subject. It ranks carcinogens into Groups 1, 2a, 2b, 3 and 4, which indicate the extent of carcinogenicity of a chemical. Group 1 ranks known carcinogens, Group 2a indicates probable carcinogens, Group 2b stands for possible carcinogens, Group 3 indicates the group of chemicals which cannot be classified as carcinogenic for want of complete and conclusive data, while Group 4 indicates those chemicals that are probably not carcinogens.

More information about carcinogenic chemicals may be obtained from www. pesticide.org/.../corrected-2000-cancer-report-public.pdf, included in Appendix 13. This information is courtesy US **Environmental Protection Agency (EPA)**, Office of Pesticide Programs, Health Effects Division, after correspondence with Mary-Jean Lormand.

Carfentrazone

See **Carfentrazone-ethyl herbicide**

Carfentrazone-ethyl herbicide

Carfentrazone-ethyl, $C_{15}H_{14}Cl_2F_3N_3O_3$ (ethyl (*RS*)-2-chloro-3-{2-chloro-5-[4 (difluoromethyl)-4,5-dihydro-3-methyl-5-oxo-1*H*-1,2,4-triazol-1-yl]-4-fluorophenyl}propionate), is a triazolone herbicide. It is a viscous liquid with a faint petroleum odour. With solubility in water being 23 mg/ml, the compound is soluble in toluene and hexane and also miscible with acetone, ethanol, ethyl acetate and methylene chloride. The CAS Registry Number is 128639-02-1. This herbicide is a derivative of carfentrazone.

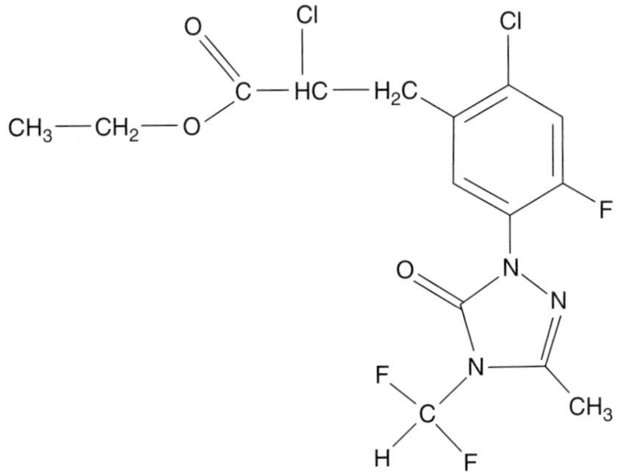

Carfentrazone-ethyl

Carfentrazone-ethyl is a post-emergence herbicide that controls broadleaved weeds and several mustard species in wheat, maize and soybean. The chemical control of weeds proceeds via membrane disruption which is initiated by the inhibition of the enzyme protoporphyrinogen oxidase. The compound inhibits the chlorophyll biosynthetic pathway.

The application dose of carfentrazone-ethyl to control wild mustard is in the range of 9–35 g/ha. The compound is sprayed by using boom sprayers. Trade names include Shark.

Carfentrazone-ethyl is practically a safe compound in normal use, for example on mallard duck and the bobwhite quail (acute LD_{50} is >2250 mg/kg). It is not carcinogenic, neurotoxic or mutagenic. The NOEL for carfentrazone-ethyl in rats is 3 mg/kg/day. However, the compound is highly toxic to aquatic plants and moderately toxic to fish. The EC_{50} for this compound is 9.8 mg/l. According to the European Commission (EC) carfentrazone-ethyl has an ADI of 0.03 mg/kg b.w./day. The WHO classifies it as slightly hazardous.

It breaks down rapidly in the environment, while its degradation products persist in water. In plants, it is converted to the acid, which is hydroxylated and then oxidized to form the dibasic acid. Because of its low application rate, carfentrazone-ethyl residues are expected to occur at low levels in surface water and groundwater.

Carfentrazone-ethyl is registered for use in Australia, the USA, Canada, South Africa and many EU nations like the UK, Finland, Poland, Denmark, Germany and France.

The herbicide carfentrazone-ethyl is derived from carfentrazone. With molecular formula $C_{13}H_{10}Cl_2F_3N_3O_3$ and IUPAC name (*RS*)-2-chloro-3-{2-chloro-5-[4-(difluoromethyl)-4,5-dihydro-3-methyl-5-oxo-1*H*-1,2,4-triazol-1-yl]-4-fluorophenyl}propionic acid, carfentrazone is a triazolone herbicide. It is a viscous liquid that is a cell membrane disruptor. It is not registered in the USA or in the EU.

Carotenoids

See **Herbicide action – carotenoid synthesis inhibition**

Carpropamid fungicide

Carpropamid, $C_{15}H_{18}Cl_3NO$, is a mixture of (1*R*,3*S*)-2,2-dichloro-*N*-[(*R*)-1-(4-chlorophenyl)ethyl]-1-ethyl-3-methylcyclopropanecarboxamide, (1*S*,3*R*)-2,2-dichloro-*N*-[(*R*)-1-(4-chlorophenyl)ethyl]-1-ethyl-3-methylcyclopropanecarboxamide, (1*S*,3*R*)-2,2-dichloro-*N*-[(*S*)-1-(4-chlorophenyl)ethyl]-1-ethyl-3-methylcyclopropanecarboxamide and (1*R*,3*S*)-2,2-dichloro-*N*-[(*S*)-1-(4-chlorophenyl)ethyl]-1-ethyl-3-methylcyclopropanecarboxamide, where the first two cited comprise at least 95% of the total. It has a CAS Registry Number of 104030-54-8. Carpropamid is a colourless crystalline solid that has low solubility in water.

Carpropamid is a systemic, long-term protective fungicide which is specifically used in the control of rice blast disease, caused by *Pyricularia oryzae*. It can also control some bacterial diseases. It is also believed to increase phytoalexin production, which increases resistance. It acts by inhibiting melanin biosynthesis by inhibiting dehydration reactions (MBI dehydratase).

Carpropamid is formulated as granules (GR), emulsifiable concentrate (EC), suspension concentrate (SC), flowable concentrate for seed treatment (FS) and water-dispersible powder for slurry seed treatment (WS). Very low quantities are required to be applied to the crops. It can also be combined with for example **imidacloprid**, **clothianidin** and **carbosulfan**.

The fungicide is classified as a Class U toxin by the WHO. Its acute oral LD_{50} in rats is quite high (>5000 mg/kg), and also in birds too, as for Japanese quails it is >2000 mg/kg. The LC_{50} (48 h) in fish is determined to be 5.6 mg/l, while in earthworms it is >1000 mg/kg. The NOEL (2 years) in rats and mice is observed to be 400 mg/kg diet. This chemical is not a PAN Bad Actor chemical.

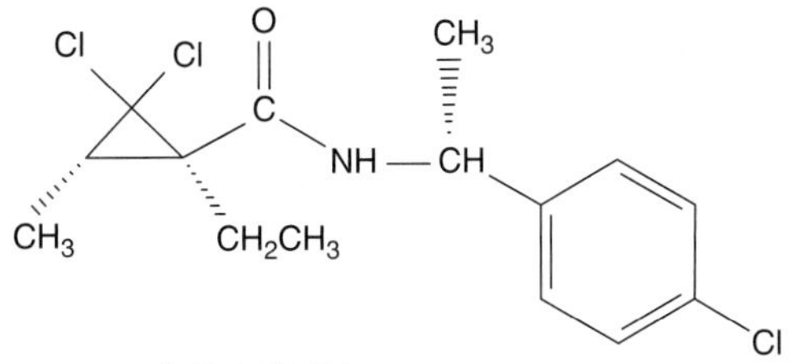

(1*R*, 3*S*)1*R*-isomer

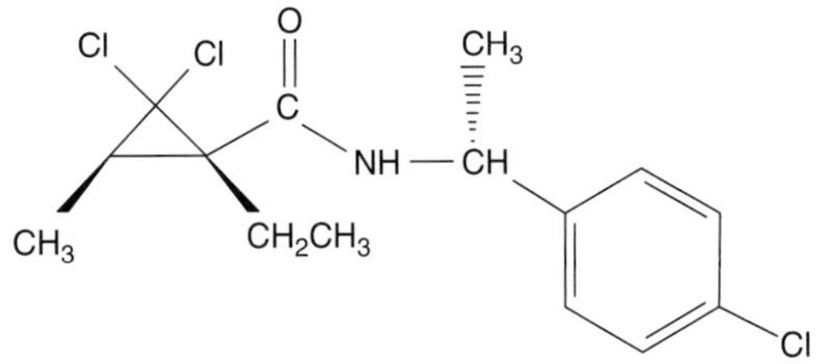

(1*S*,3*R*)1*R*-isomer

Carpropamid

Carpropamid is stable in sterile aqueous buffer solutions. Its degradation is significantly faster in pH neutral water. If orally administered in rats, it gets metabolized oxidatively in the liver. Similar metabolism takes place in paddy soil conditions. The parent compound itself is found as a major residue in rice. The soil DT_{50} of the chemical can last up to several months.

The countries that register carpropamid include India and Vietnam. It is not registered by the EPA for use in the USA.

Cartap hydrochloride insecticide

See **Cartap insecticide**

Cartap insecticide

Cartap, $C_7H_{15}N_3O_2S_2$ (*S*,*S*′-(2-dimethylaminotrimethylene) *bis*(thiocarbamate), is a carbamate insecticide and a **nereistoxin** analogue. Cartap salts include cartap hydrochloride ($C_7H_{16}ClN_3O_2S_2$). It is a colourless crystalline, slightly hygroscopic solid with a slight odour. It is soluble in water (200 g/l) and slightly soluble in methanol and ethanol. It is hydrolysed in neutral or alkaline media to the dihydronereistoxin, which is oxidized to the insecticide, **nereistoxin**. Cartap is almost always used as cartap hydrochloride. Cartap is obtained by hydrolysing 1,1-dihydrocyanato-2-dimethylaminopropane with hydrochloric acid. The CAS Registry Number of cartap is 15263-53-3, while that of cartap hydrochloride is 22042-59-7.

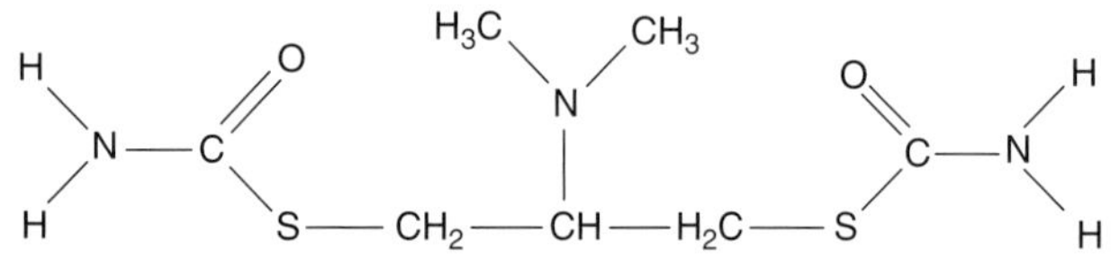

Cartap

Cartap, a pro-insecticide of the natural toxin nereistoxin, traced back to marine worms *Lumbrinereis heteropoda* and *Lumbrinereis brevicirra*, is a broad-spectrum, systemic, pre-harvest insecticide. It antagonizes the nicotinic acetylcholine receptor, affecting the cholinergic transmissions in the nervous system, leading to paralysis in the insects. Nereistoxin does not inhibit cholinesterase. Cartap is used for the control of chewing and sucking insects mainly insects in the orders Lepidoptera and Coleoptera on rice, potatoes, vegetables, soybeans, groundnuts, sunflowers, maize, sugarbeet, wheat, pome/stone/citrus fruit, ginger, tea, cotton and sugarcane. Pests include *Chilo suppressalis*, *Cnaphalocrocis medinalis*, *Lissorhoptrus oryzophilus* and rice-leaf beetle (*Oulema oryzae*).

The available formulations of cartap are dustable powder, granule and water-soluble powder: 25% and 50% water-soluble powder or 2% dust are two common practical formulations used. Applications are made up to 21 days before harvest. It may sometimes be phytotoxic to cotton, tobacco and apples. Cartap is available in the market as Padan. It is not compatible with alkaline pesticides. Other nereistoxin analogue insecticides include **bensultap** and thiosultap.

According to the WHO and EPA, cartap is rated as moderately hazardous (Class II toxin). In mammals, the carbonyl is hydrolysed, and the sulfur oxidized, with *N*-demethylation of thiomethyl derivatives. No accumulation occurs in tissues. Cartap is rapidly excreted via urine. The soil DT_{50} is 3 days. Cartap and its salts are not PAN Bad Actor chemicals.

Cartap hydrochloride is registered in India, Vietnam and Cambodia, whereas cartap monohydrochloride is registered in South Africa, the Philippines and Uganda.

Carvone as a bioPGR

Carvone, $C_{10}H_{14}O$ ((*RS*)-5-isopropenyl-2-methylcyclohex-2-en-1-one), is a botanical fungicide and an unclassified plant growth regulator. It is also referred to as d-carvone and its CAS number is 99-49-0. It is a racemic mixture of two optical isomers, namely (*S*)-carvone and (*R*)-carvone. (*S*)-carvone is a single isomer. Carvone is a pale yellow to colourless liquid, with a characteristic odour. It is insoluble in water and glycerol but soluble in alcohol, ether, chloroform, propylene glycol and mineral oils. The isomer (*S*)-carvone is extracted from **caraway oil**.

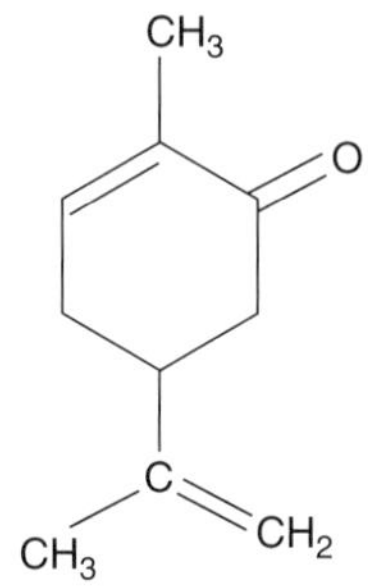

Carvone

Carvone is effective against storage diseases of potatoes like silver scurf and dry rot caused by *Helminthosporium solanii* and *Fusarium sulfureum*. As a PGR, it is used to inhibit potatoes from sprouting. The most commonly selected product is Talent.

The acute oral LD_{50} dose in rats is 1640 mg/kg whereas that of isomer (*S*)-carvone is 3710 mg/kg. The NOEL for 3 months in rats is 93 mg/kg b.w. Carvone has been used in mouth fresheners and aromatherapy. Metabolites of carvone include dihydrocarvonic acid and carvonic acid.

As a PGR, d-carvone is registered in the EU but not in the USA.

Catch crops

In agriculture, a catch crop is a fast-growing crop that is grown simultaneously with, or between, successive plantings of a main crop. Crops planted between or along with two or more principal crops are called catch crops. Turnips, ryegrass, spinach, radish and millet are examples of catch crops. Rye, clovers, radish and cowpeas are also examples of catch crops, depending on the season.

Catch crops are planted to provide fresh feed to livestock. They are also a quick source of livelihood and also of green manure for the soil. At other times, a catch crop is part of the crop rotation cycle or a catch crop is planted to retain the soil nutrients and prevent them from leaching out of or running-off the soil. All these options are possible since catch crops grow rapidly.

Cationic surfactants

See **Surfactants**

Central Insecticide Board and Registration Committee (CIBRC)

The issue of pesticide use and legislation was studied in 1964–1967 by an expert committee of the Indian Council of Agricultural Research headed by Professor M.S. Thacker.

Based on the recommendations of the Expert Committee, a comprehensive Insecticides Act was passed in 1968 to regulate the import, manufacture, sale, transport, distribution and use of insecticides with a view to preventing risks to human beings and animals and for other related matters.

While the Ministry of Agriculture oversees the enforcement of the Insecticides Act, the Department of Agriculture frames the rules. The department was responsible for forming the Central Insecticides Board and Registration Committee (CIBRC), with necessary machinery and functionaries at state levels. All the provisions of the Insecticides Act came into force by 1971. As per the Act, all pesticides have to be compulsorily registered at the central government level, while licences for their manufacture, formulation and sale are dealt with at the state level.

The Central Insecticide Board (CIB) fixes tolerance limits for insecticide residues, safety period and shelf life, based on the data provided by the manufacturers and research findings of scientists in the country. The CIB may advise the central and state governments regarding: (i) the risks, to human beings or animals, involved in the use of insecticides and the safety measures necessary to prevent such risk; and (ii) the manufacture, sale, storage, transport and distribution of insecticides with a view to ensure safety to human beings or animals. It also has a role in fixing the chemical classification based on toxicity, shelf life of each product, and so on.

The Registration Committee (RC), after due diligence, grants registrations to individual people who wish to import or manufacture insecticides. The RC scrutinizes the formulae of such compounds and verifies claims to their bio-efficacy and safety to human beings and animals.

More information on the agencies may be obtained from http: //cibrc.nic.in.

Chemigation

When a chemical is added to irrigation water, it is called chemigation. This input may be either a pesticide or fertilizer. Many pre-emergence herbicides are applied by chemigation.

The equipment used for such application through irrigation waters is varied and specialized since a large volume of a pesticide has to reach a vast area of the field.

The pesticide is injected into the main irrigation stream with a positive displacement pump. It is very important that the calibration is accurate. Anti-siphon and switch valves prevent contamination of the irrigation water source or overflow into the slurry feed tank.

Although chemigation is an effective means of pesticide application, the method has some disadvantages. These include: (i) as water is the carrier, some chemicals may move away from the targeted site or leach into the soil zone, which may also lead to the application of chemicals to sites that do not need any treatment; and (ii) there is a need for constant agitation, which is an impediment in the work, because if sedimentation occurs, the pesticide may be unevenly distributed.

Chemosterilant

A chemosterilant is any chemical compound used to control economically destructive or disease-causing pests (usually insects) by causing temporary or permanent sterility of one or both of the sexes or preventing maturation of the young to a sexually functional adult stage.

The use of chemosterilants is a pest management technique, where the emphasis is on management not control. Curbing the pest's reproductive capability automatically restricts their numbers. Chemosterilants can affect the targeted pest temporarily or permanently.

Chemosterilants are used on insects, rodents and birds. Once having sterilized a percentage of the population, the normal mating process is allowed to continue, resulting in a manageable offspring population. Care must be exercised in the direct application of a chemosterilant to a natural population, because most chemosterilants cause genetic changes and are thus dangerous both to beneficial animals and to man. And yet, this method of pest control is considered more humane than the use of pesticides. The hormone mestranol, antimetabolites, triethylenemelamine, alkylating agents and terpenes are some chemicals (or groups) that act as sterilants.

Insects reared for laboratory testing are sterilized by feeding them treated foods or by placing them in contact with treated materials.

Antimetabolites are substances that resemble naturally occurring cell constituents or metabolites that the body attempts to utilize in its normal functioning. However, this mismatch leads to halting of normal body functions. Amethopterin and aminopterin cause sterility in female insects thus preventing egg formation or hatching activity.

Alkylating agents are extremely reactive agents that damage chromosomes, destroy cells and induce mutation. These include **TEPA**, **METEPA** and **apholate**. Alkylating agents alter genetic material in both male and female reproductive cells.

Exposure of a pest to terpene signals the halting of moulting in insects since the chemical is similar to a **juvenile hormone (JH)**. Terpenes are organic compounds that are naturally present in some trees and some insects and are a constituent of resins. Terpenes are considered essential constituents of all living organisms. JH holds back insects in their juvenile stage (e.g. the life cycle stages of eggs or larvae). Natural development into adulthood can take place only in the absence of these hormones. The mimicking of JH by terpenes is another chemosterilant action. In yet other cases, some chemicals hasten the development so much that the organism develops abnormal reproductive organs.

Chenopodium ambrosioides bioinsecticide

An extract of *Chenopodium ambrosioides* (commonly known as American wormseed), which contains lipophilic constituents, is an insecticide. This oil destroys the exoskeleton and joints of insects, leading to leaking of fluid from the body and crippling them.

C. ambrosioides controls aphids, mites, whiteflies, thrips and leaf miners on brassicas, bulbs, leafy-fruit vegetables, citrus and nut trees.

The common formulation of *C. ambrosioides* is an emulsifiable concentrate, applied at the rate of 0.5–1% emulsion in a high volume spray. This insecticide can be combined with several other naturally derived control agents. A known trade name is Requiem.

C. ambrosioides is not toxic to mammals and birds, but may be toxic to fish. The acute oral LD_{50} of this insecticide in rats is >5000 mg/kg, while the acute oral LD_{50} in bobwhite quails is

>2250 mg/kg. This compound is classified as slightly toxic by the EPA and WHO (Class III). It is approved for use in organic farming by the US OMRI.

Cheshunt compound fungicide

Cheshunt compound, a soil fungicide, is made of a dilute solution of **copper sulfate** and ammonium carbonate. Suitable for organic farming, cheshunt compound, diluted in water, is an ideal pre-sowing spray application. The compound is used for preventing damping-off in seedlings.

Besides being a fungicide, cheshunt compound is used to control or inhibit aquatic plants and algae. It controls ectoparasites on fish. Copper sulfate also inhibits the bacteria *Escherichia coli*.

However, since copper is toxic to fish, care has to be taken about its dosage.

Chinomethionat fungicide

Chinomethionat, $C_{10}H_6N_2OS_2$ (6-methyl-1,3-dithiolo[4,5-*b*] quinoxalin-2-one), is a quinoxaline fungicide with fumigant and acaricidal (spider mites) properties. Some common names of the chemical are **oxythioquinox**, quinomethionate and chinomethionate. It exists in the form of yellow crystals, which are soluble in water (1 mg/l), toluene (25 g/l), hexane (1.8 g/l), DMF (10 g/l), as well as hot benzene and dioxane. It hydrolyses in alkaline media and is incompatible with mineral oils and **thiram**-based formulations. Chinomethionat is produced by the reaction of 7-methyl-2,3-quinoxaline thiol with carbonyl chloride. Its CAS Registry Number is 2439-01-2.

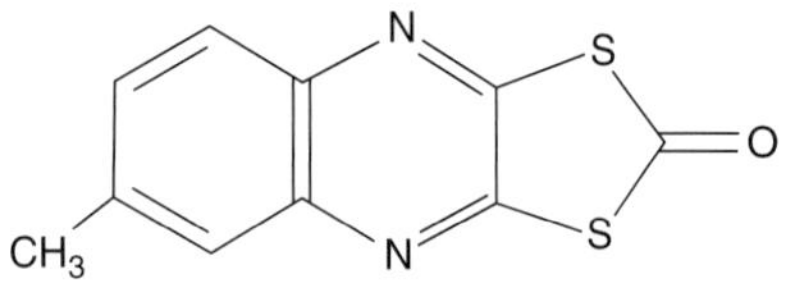

Chinomethionat

Chinomethionat is a selective, non-systemic, contact fungicide and a fumigant. It provides a protective cover to crops including cucurbits, cotton, coffee, tea and vegetables by controlling **powdery mildews** and spider mites.

Some common formulations of chinomethionat are dustable powder (DP), smoke generator (FU) and wettable powder (WP). Trade names include Joust and Morestan.

This compound is classified by the WHO as slightly toxic. It is a severe eye irritant as tested on rabbits. The acute percutaneous LD_{50} in rats is >5000 mg/kg. The acute oral LD_{50} values in male rats, female rats and bobwhite quails are 2541, 1095 and 196 mg/kg, respectively. The NOEL (2 year) in rats is 40 mg/kg diet, while in male mice, it is 270 mg/kg diet. The ADI is 0.006 mg/kg b.w./day. It is not toxic to bees. It is phytotoxic to certain varieties of apples, pears and roses. Chinomethionat is a PAN Bad Actor chemical because it is a proven developmental/reproductive toxin and a carcinogen.

In rats, chinomethionat is metabolized rapidly and eliminated within 3 days through the faeces and urine. The principal metabolite is chinomethionat acid. The metabolite detected in plant parts is dihydromethylquinoxalinedithiol.

The fungicide-acaricide is registered only in Australia, not in the USA or in the EU. India has refused registration to certain products with this compound. The MRL for apples and pears in Australia is 0.5 mg/kg, while the provisional MRL for mangoes as established by the Japan Food Chemical Research Foundation is 0.3 ppm.

Chirality of pesticides

Chirality is a property of molecules having a non-superimposable mirror image. All organisms constitute a chiral environment. When synthetic agrochemicals, which are chiral and prochiral compounds, interact with biological enzymes in a plant or an organism, enantiomeric and enantiotopic selectivity is demonstrated while causing their effect on the target.

Generally one enantiopure isomer is more active than its counterpart. Insecticides derived from natural products, like **pyrethroids** or **nicotine**, have chiral structures. One or more individual enantiomer of the compound having an insecticidal property, reacts with the site of action of the pest nervous system, since the enzyme present at the site also has a chiral nature.

Chitin

Chitin is a polysaccharide derivative of glucose, comprising 2-(acetylamino)-2-deoxy-D-glucose. It contains nitrogen and is structurally similar to cellulose. It is a component of fungal cell walls and the exoskeleton of insects. Its molecular formula is $(C_8H_{13}O_5N)_n$. Chitin is composed of β(1-4)-linked units of the amino sugar *N*-acetyl-glucosamine and is the main source of production of chitosan, which is used in a number of applications. Chitin serves a protective function in insects and fungi. In insects, the synthesis of chitin is also responsible for insect **moulting**, until adulthood. This property of chitin has led to the development and synthesis of a range of chitin synthesis inhibitors, which play a role in pest control. Benzoylphenylurea-based compounds, which include **flucycloxuron**, hexaflumuron, chlorbenzuron, penfluron, **teflubenzuron** and **triflumuron**, inhibit the synthesis of chitin.

Inhibitors of chitin production are a part of a bigger group called **insect growth regulators (IGRs)**, which also include **juvenile hormones**, juvenile hormone mimics and moulting hormones. Other IGRs include fenoxycarb, **hydroprene**, **ecdysterone**, α-ecdysone and precocenes. These active ingredients block the synthesis of chitin, thus hampering the production of the new exoskeleton required in the moulting process.

Chitin also finds application in medicine.

Chitin synthesis inhibitors

See **Insect growth regulators (IGRs)**; **Chitin**

Chloralose rodenticide

Chloralose, $C_8H_{11}Cl_3O_6$ ((*R*)-1,2-*O*-(2,2,2-trichloroethylidene)-α-D-glucofuranose), is an avicide, avian repellent and rodenticide. Its CAS Registry Number is 15879-93-3. It is a crystalline powder soluble in water (4.44 g/l), alcohol and diethyl ether in varying degrees. But chloralose is practically insoluble in petroleum ether. The beta-isomer of chloralose is less soluble in these substances than the alpha-isomer. Chloralose is produced by reacting glucose and water-free chloral, under heating.

The toxic action of chloralose is initiated when chloralose is reduced to trichloroethanol (2,2,2-trichloroethanol). Trichloroethanol slows down the metabolic processes in the

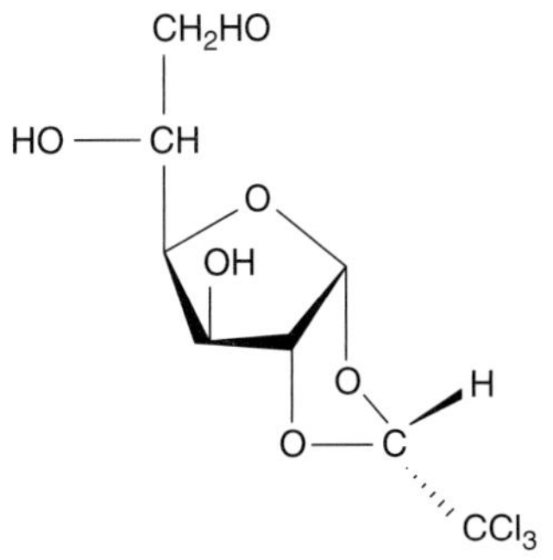

Chloralose

target pest, sometimes causing hypothermia and death. Small rodents like mice are most effectively controlled by chloralose, even at temperatures below 15°C.

Chloralose is commonly formulated as a bait and bait concentrate. As rodent bait, the common concentration dosage of chloralose is 2.8%.

The WHO has classified chloralose as a Class II toxin (moderately toxic). It is classified as moderate on the acute toxicity test, although it is not a PAN Bad Actor chemical. The acute oral LD_{50} in mice is 32 mg/kg, but in rats it is much higher at 400 mg/kg. The same for birds is in a range of 32–178 mg/kg. In animals, chloralose metabolizes to chloral and oxidizes to trichloracetic acid and reduces to trichloroethanol. This compound is not permitted for use in the EU, but is permitted for use in Australia and New Zealand.

Chloramben herbicide

Chloramben, $C_7H_5Cl_2NO_2$ (3-amino-2,5-dichlorobenzoic acid), is a **benzoic acid herbicide**. It is a colourless, crystalline solid, soluble in water (799 mg/l) and to a much lesser degree in acetone, ether and ethanol. It degrades in water exposed to sunlight. Its salts include chloramben-methyl and chloramben-ammonium. The CAS Registry Number of chloramben is 133-90-4.

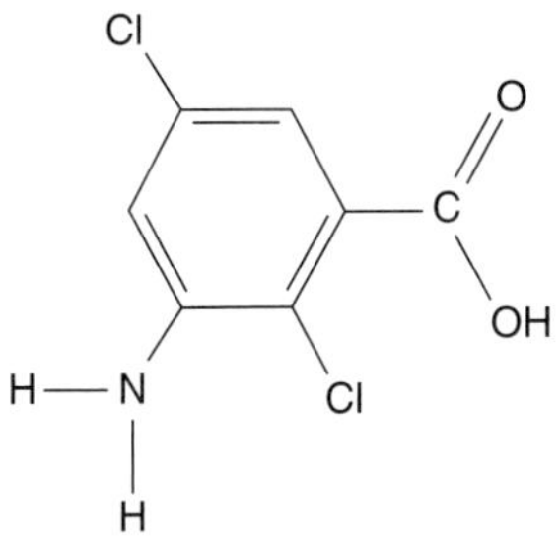

Chloramben

Chloramben, a selective, pre-emergence, systemic benzoic acid herbicide, belongs to the group of **auxin**-like herbicides. It selectively controls annual grasses and broadleaved weeds in several broadleaved crops, including soybean, maize, lettuce, dry beans, groundnut, carrot and tomato. *Amaranth* spp. and oakleaf goosefoot (*Chenopodium glaucum*) are some of the weeds that are controlled by chloramben. Since chloramben has poor translocation, it is incorporated in the soil, to protect seeds and seedlings.

Other herbicide groups that mimic auxins are **phenoxy acid-**based herbicides, **picolinic acid-**based herbicides and others like **quinclorac**.

Chloramben has low toxicity to birds (with LC_{50} of mallard birds being 4640 mg/kg) and fish. It is not toxic to bees. The oral LD_{50} for chloramben in rats is 3500 mg/kg and in mice is 3725 mg/kg. The herbicide chloramben is sensitive to microbial action and photolytic degradation. While dechlorination and deamination take place in moist soil, only deamination reactions take place in dry soil. Major degradation products include 3,5-dihydroxybenzoic acid, 5-chlorosalicyclic acid and 2,5-dichlorobenzoic acid. In plants, chloramben is directly conjugated to glucose forming *N*-glucosylchloramben [*N*-(3-carboxy-2,5-dichlorophenyl) glucosylamine]. There is no accumulation in animal tissues. Since degradation is rapid in sandy soils, moist soils and intense sunlight, persistence is insignificant. In soils rich in organic matter, and kaolinite soils, chloramben is adsorbed and persistence is high.

Chloramben is not registered in the USA and is banned in the EU.

Chlordane insecticide

Chlordane (octachloro-4,7-methanohydroindane) is an organochlorine (chlorinated cyclodiene) insecticide. It is a mixture of *cis*-isomer (1α, 2α, 3aα, 4β, 7β, 7aα) and *trans*- (or gamma- or beta-) isomer (1α, 2β, 3aα, 4β, 7β, 7aα). It is an amber-coloured liquid that is insoluble in water but soluble in organic solvents and unstable in alkaline medium. It is made by condensing hexachlorocyclopentadiene with cyclopentadiene yielding chlordane which is chlorinated to the octachloro derivative. The CAS Registry Number of chlordane is 57-74-9.

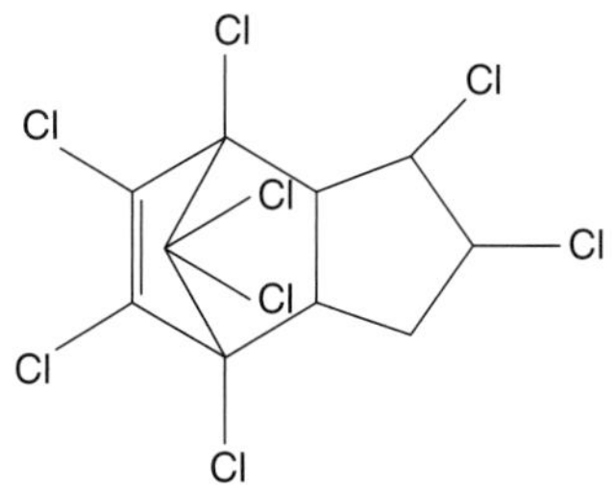

Chlordane

Chlordane, a contact insecticide with fumigant properties, is used extensively for the control of ants, termites and insects found in houses and soil. It is used as a wood preservative for protective treatment of underground cables, lawns, foliage and seeds. It is used against members of the Formicidae family (ants), the order Coleoptera (beetles), Noctuidae larva (i.e. moths) and subterranean termites. Chlordane inhibits the GABA (neurotransmitter γ-aminobutyric acid) receptor chloride channel complex by mimicking the nerve excitant picrotoxin. The action of the chemical is also seen in the inhibition of the enzyme ATPase, essential in calcium transportation. This action leads to a state of uncontrolled excitation in the insect pest.

Commonly used formulations of chlordane include emulsifiable concentrates, wettable powders and granules.

Chlordane has been classified as a Class II toxin (moderately hazardous). It is absorbed by the skin and causes irritation to the eyes and skin. The acute oral LD_{50} in rats ranges from 150 to 700 mg/kg b.w. The dermal LD_{50} in rats is 590–840 mg/kg. It is moderately toxic to birds (LD_{50} is 83 mg/kg) and highly toxic to fish (LC_{50} of 0.07–0.09 mg/l) and bees. Since chlordane is feared to be an endocrine disruptor, a cholinesterase inhibitor and a groundwater pollutant, it is considered a PAN Bad Actor chemical. Thus, many applications of the chemical have been cancelled. It is included in PAN's list of **Dirty Dozen** chemicals.

Chlordane is a persistent insecticide, with a soil half-life being around 350 days. The USA has banned all uses of chlordane and its derivatives (isomers). Similarly, many countries such as India, Korea, Japan and Singapore and the entire EU have banned or severely restricted the use of the chemical. As per the Agency for Toxic Substances and Disease Registry, USA, the MRL for chlordane is fixed at 0.001 mg/kg/day (acute oral test).

Chlordene insecticide

Chlordene, $C_{10}H_6Cl_6$, or *cis*-chlordene is an organochlorine insecticide. It is produced by reacting cyclopentadiene with hexachlorocyclopentadiene at around 90°C. The CAS Registry Number of chlordene is 3734-48-3.

Chlordene is directly applied to the soil for pest control. However, this compound is not being produced in a major way, although it is available as dusts, emulsifiable concentrates, granules, oil solutions and wettable powder formulations.

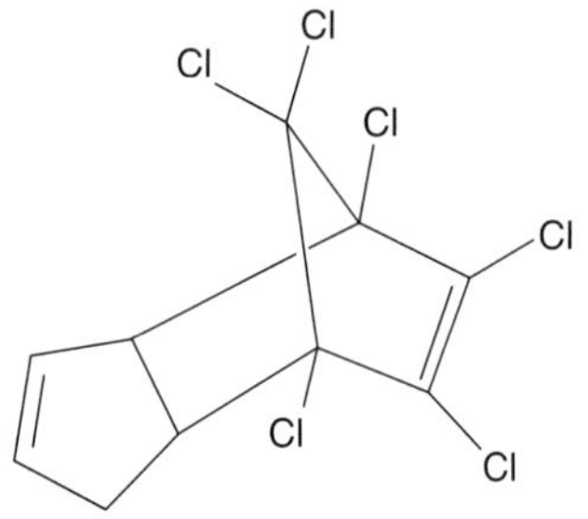

Chlordene

Chlordimeform hydrochloride acaricide

Chlordimeform hydrochloride, $C_{10}H_{14}Cl_2N_2$ ((*EZ*)-N^2-(4-chloro-*o*-tolyl)-N^1,N^1-dimethylformamidine hydrochloride), is a formamidine acaricide. It exists as a colourless crystalline substance, soluble in water at 20°C. It is rather stable in strong acids but is hydrolysed in weakly acidic to alkaline solutions. The CAS Registry Number is 19750-95-9.

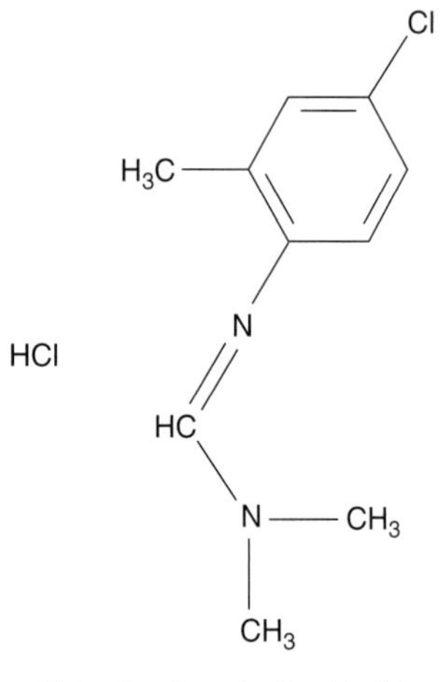

Chlordimeform hydrochloride

Chlordimeform hydrochloride is a broad-spectrum acaricide and insecticide which has contact, stomach and vapour activity. When applied to the foliage of fruit, vegetables and field crops, it effectively controls all stages of insects and mites. Chlordimeform hydrochloride is effective against mites that have developed resistance against OP insecticides. This compound also has anti-feedant and animal parasiticidal uses, for example against cattle ticks. The crops protected include fruit trees, grapes, vegetables, cotton and hops.

The acaricide is used as a water-soluble powder and as dips for farm animals. Some known trade names are Galecron, Acaron, Fundal and Spike. The dose ranges from 0.01 to 0.1%.

The substance can be absorbed into the body by inhalation and by ingestion. Repeated or prolonged contact with skin may cause dermatitis. It is a highly toxic substance. The acute oral LD_{50} in mice is 220 mg/kg and in rats, 265 mg/kg. The NOEL in rats is 5 mg/kg b.w./day. The ADI in humans is 0.01 mg/kg. It is also a carcinogen. It is thus a PAN Bad Actor chemical.

Chlordimeform hydrochloride appears to be fairly rapidly excreted in animals and does not appear to be stored in body tissue. The compound is not EPA registered.

Chlorethoxyfos insecticide

Chlorethoxyfos, $C_6H_{11}Cl_4O_3PS$ (*O*,*O*-diethyl *O*-(1,2,2,2,-tetrachloroethyl) phosphorothioates), is an aliphatic organophosphorus insecticide. It has a CAS Registry Number of 54593-83-8. It is a colourless crystalline powder, which is soluble in water (<1 mg/l), acetonitrile, chloroform, ethanol, hexane and xylene. It can be stored well at room temperature. The hydrolytic half-life values at pH 5 and 9 for this compound are 4.3 and 72 days, respectively. It is produced by the reaction of phosphorus trichloride with acetonitrile and chloral, followed by treatment with hydrogen sulfide and ethanol.

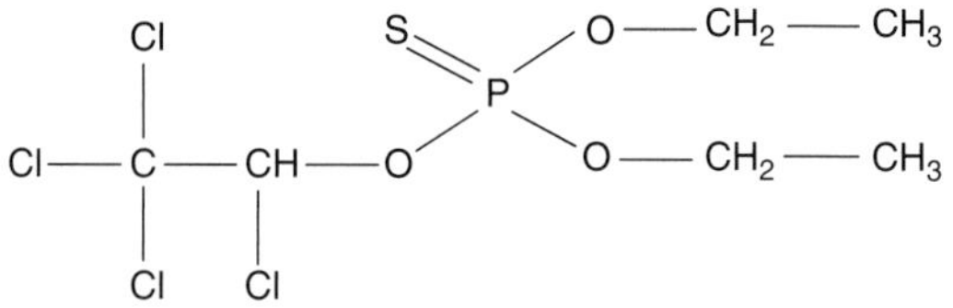

Chlorethoxyfos

Chlorethoxyfos is a soil insecticide especially effective on maize pests such as rootworms, cutworms, maggots and white grubs. Chlorethoxyfos is formulated as granules and has a vapour action, and hence it is ideal for soil incorporation. A single, pre-plant band application is found sufficient. Common trade names include Fortress. Care has to be taken while loading or applying the chemical (a closed application system is recommended) in order not to cause exposure to handlers.

The insecticide is very toxic to rats, mice, rabbits and fish. The compound causes mild irritation to the skin and eyes. The chemical is a cholinesterase inhibitor. This, and the fact that it is very toxic as per acute toxicity tests, has lead to chlorethoxyfos being classified as a PAN Bad Actor chemical. It is a Class Ia toxin (extremely hazardous) according to the WHO.

In mammals, chlorethoxyfos is metabolized into trichloroacetic acid, dichloroacetic acid, chloral hydrate and **trichloroethanol** by the cleavage of the P–*O*-tetrachloroethyl bond. Many of the metabolites are mobile but degradable. Carbon dioxide, serin, glycine and glycine conjugates are some of the final metabolites. In plants, it degrades into trichloroacetic acid followed by oxalic acid. The soil DT_{50} is within 1 week.

Chlorethoxyfos is registered in the USA only as an in-furrow application.

Chlorfenapyr acaricide

Chlorfenapyr, $C_{15}H_{11}BrClF_3N_2O$ (4-bromo-2-(4-chlorophenyl)-1-ethoxymethyl-5-trifluoromethyl-1*H*-pyrrole-3-carbonitrile), is a pyrazole acaricide and specifically a pro-insecticide (meaning it is metabolized into an active insecticide after entering the host). It is a white solid, which is practically insoluble in water, and soluble in acetone, diethyl ether, dimethyl sulfoxide, tetrahydrofuran, acetonitrile and alcohols. It has a CAS Registry Number of 122453-73-0.

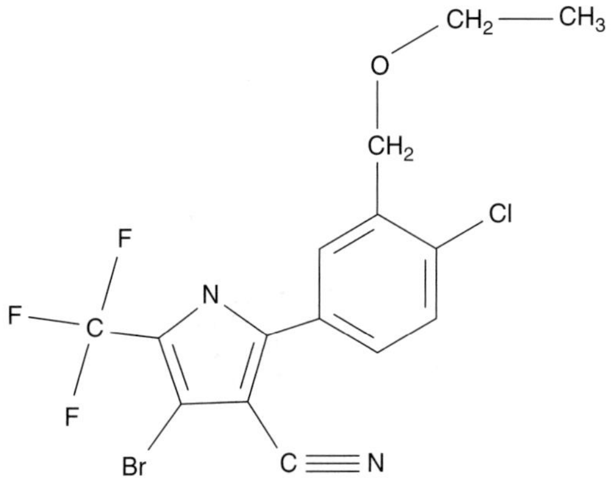

Chlorfenapyr

The acaricide is produced by the reaction of 4-bromo-2-(*p*-chlorophenyl)-5-(trifluoromethyl) pyrrole-3-carbonitrile with sodium hydride in tetrahydrofuran, followed by reaction with chlorobromomethane, and subsequently with sodium ethoxide.

Chlorfenapyr is an insecticide and acaricide with mainly stomach and some contact action. It exhibits good translaminar but limited systemic activity in plants. The main activity is the inhibition of ATP synthesis. The compound controls many species of insects and mites, including those resistant to carbamate, organophosphate and pyrethroid insecticides, and also acts as **chitin synthesis inhibitors** in cotton, vegetables, citrus, top fruit, vines and soybeans. Among pests resistant to conventional products, which are controlled by chlorfenapyr, are *Brevipalpus phoenicis* (leprosis mite), *Leptinotarsa decemlineata* (Colorado potato beetle), *Helicoverpa* spp., *Heliothis* spp., *Plutella xylostella* (diamondback moth) and *Tetranychus* spp. Its use in **resistance management** programmes for control of various cotton pests is under evaluation.

Trade names of chlorfenapyr include Pirate, Stalker and Alert. The formulation types of chlorfenapyr are emulsifying concentrates (EC) and soluble concentrates (SL).

The acute oral LD_{50} in male rats is 441 mg/kg, while in female rats it is 1152 mg/kg. The acute percutaneous LD_{50} for rabbits is >2000 mg/kg. It is moderately irritating to the eye but non-irritating to skin. The LC_{50} in rats is 1.9 mg/l air. The WHO classifies the compound as a Class II toxin (moderately hazardous). The acute oral LD_{50} in mallard ducks is 10 mg/kg. The LD_{50} in bees is 0.2 mg per bee.

Chlorfenapyr is registered in India, Australia, Cambodia, the Philippines, South Africa, Tanzania, Vietnam and the USA.

Chlorfenvinphos insecticide

Chlorfenvinphos, $C_{12}H_{14}Cl_3O_4P$ ((*EZ*)-2-chloro-1-(2,4-dichlorophenyl)vinyl diethyl phosphate), is an organophosphorus insecticide-acaricide (vinyl phosphate class). It is a colourless liquid, moderately soluble in water (145 mg/l) and miscible with ethanol, acetone, hexane and kerosene. It hydrolyses rapidly in strongly alkaline solutions. It is produced by the reaction of 2,4-dichoroacetophenone with triethyl phosphate. Its CAS Registry Number is 470-90-6 (see Chlorfenvinphos at bottom of the page).

Chorfenvinphos, being a cholinesterase inhibitor (affecting the nervous system), has a long residual activity. It effectively controls cutworms, rootworms, stem borers, **leafhoppers**, phorid flies, bulb flies, mites and soil insects. The crops covered include maize, rice, wheat, cotton, citrus, sugarcane and potatoes. Chlorfenvinphos is also effective on ectoparasites on farm animals and on mosquito larvae.

The insecticide-acaricide is compatible with many non-alkaline pesticides. It can be formulated as granules, emulsifiable concentrates and wettable powder. It is applied on soils, seeds and on leaves. Some trade names include Birlane, Supona, Sapecron and Vinyphate. Typical application doses for cutworms and rootworms are in the range of 2–4 kg/ha, while mite infestation on citrus is controlled with a dose of 200–400 g/ha.

The WHO has classified chlorfenvinphos as an extremely hazardous chemical (Class Ia) due to its high toxicity in acute

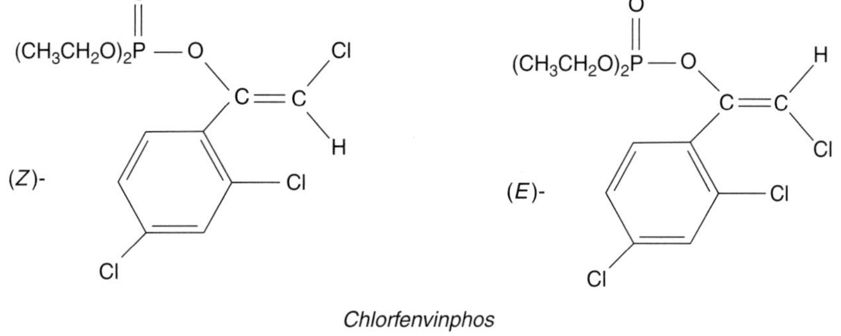

Chlorfenvinphos

toxicity tests. It can also damage seeds during direct seed application. It is also a suspected endocrine disruptor as well as a cholinesterase inhibitor and so is classified as a PAN Bad Actor chemical. It is slightly toxic to bees, but moderately toxic to fish and molluscs.

The poor degradation activity of rat liver microsomal enzymes makes chlorfenvinphos selectively toxic to rats. Other mammals metabolize chlorfenvinphos rapidly and completely, when administered orally, and the metabolites are rapidly excreted in the urine. In soils, chlorfenvinphos decomposes gradually to form α-dichlorophenyl-ethanol, dichloroacetophenone, diethyl hydrogen phosphate and desethyl chlorfenvinphos.

India has banned the use, manufacture and sale of chlorfenvinphos. It is only registered in Australia, New Zealand and Tanzania. (See also **Organophosphorus insecticides**.)

Chlorfluazuron insecticide

Chlorfluazuron, $C_{20}H_9Cl_3F_5N_3O_3$ (1-[3,5-dichloro-4-(3-chloro-5-trifluoromethyl-2-pyridyloxy) phenyl]-3-(2,6-difluorobenzoyl) urea), is an insecticide belonging to the benzoylphenylurea class. It is a white, crystalline substance, soluble in water (<0.01 mg/l), xylene, methanol, isopropanol, dichloromethane, acetone and cyclohexanone. It is stable to heat, light and hydrolysis. It has a CAS Registry Number of 71422-67-8 (see Chlorfluazuron at bottom of the page).

Chlorfluazuron is an insect growth regulator (IGR) and an anti-moulting agent that blocks the synthesis of the critical insect cuticle component, **chitin**, leading to the death of the larvae and pupae of insect pests belonging to a broad range of members of the order Lepidoptera (especially species of the genera *Heliothis* and *Spodoptera* and the diamondback moth *Plutella xylostella*), the whitefly *Bemisia tabaci*, termites and thrips species (family Thripidae). Chlorfluazuron is used on for example cotton, potatoes and tea.

The insecticide-IGR is available in formulations such as emulsifiable concentrates (EC) and soluble concentrates (SL). The chemical is effective in low doses from 10 to 50 g/ha. It is available under trade names like Aim, Atabron, Ishipron and Jupiter.

According to the WHO, chlorfluazuron comes under the toxicity Class U (i.e. unlikely to be hazardous). It is not acutely toxic by the oral, dermal or the inhalation routes, although it may cause slight irritation to the eyes.

Chlorfluazuron is metabolized in rats by cleavage of the urea bridge. It is slowly degraded in plants. In soil the half-life varies from 6 weeks to a few months. It is also shown to bio-accumulate in animal tissues of rats and goats.

Chlorfluazuron is registered for use in Australia, Vietnam and the Philippines, but not in the USA or the EU. The Japan Food Chemical Research Foundation has fixed an MRL on mangoes at 2 ppm.

Chlorflurenol as a PGR

Chlorflurenol, $C_{14}H_9ClO_3$ ((*RS*)-2-chloro-9-hydroxyfluorene-9-carboxylic acid), is a morphactin type of herbicide and a plant growth regulator. It has a CAS Registry Number of 2464-37-1. It is a cream-coloured crystalline substance soluble in water, acetone, ethanol, benzene and cyclohexane. It undergoes rapid photolysis in light and loses efficacy in alkaline medium.

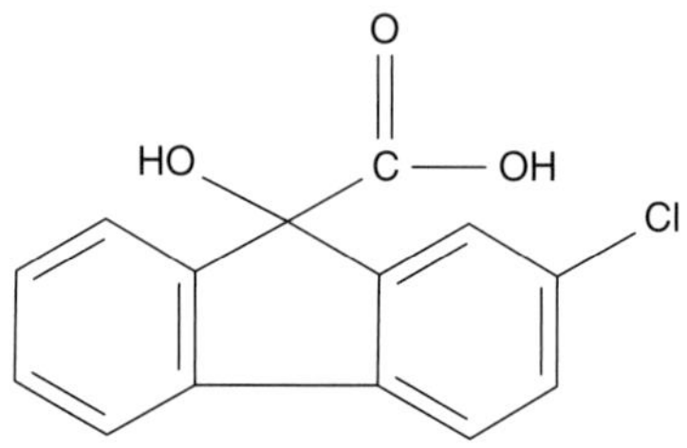

Chlorflurenol

Chlorflurenol is a post-emergence herbicide and a PGR that retards or kills broadleaved weeds as well as grasses on non-farmland like turf, golf courses and fences. It inhibits growth tips after being absorbed and translocated both acropetally and basipetally. The only agricultural application is on pineapples.

The herbicide is compatible with many other PGRs and herbicides, except those that are alkaline in nature. Chlorflurenol is commonly formulated as an emulsion concentrate, granule or a liquid formulation. It can be sprayed by an air blast, hand gun or any other means, as long as spray drift is minimized. It is commonly applied at rates from >1 to 3.25 kg/ha. It is available in the market under trade names like Curbiset and Multiprop.

Chlorflurenol carries a risk of groundwater pollution. Terrestrial and semi-aquatic plants, insects and fish are affected. Since there have been accidents on freshly treated turf, the application dose is restricted to two applications of 250 g/ha. There is no acute toxicity to birds, reptiles and mammals. Those involved in applying and handling the chemical must

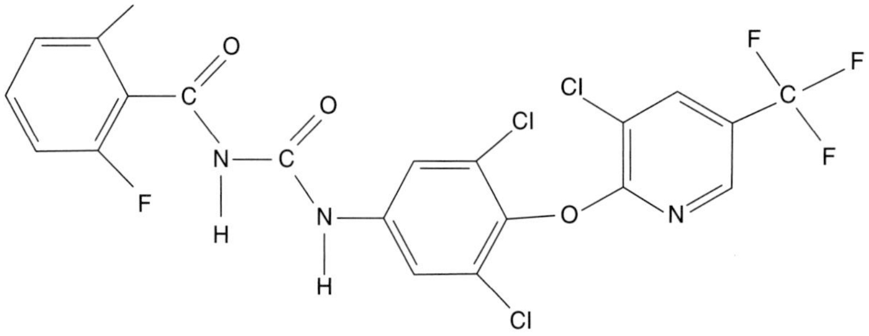

Chlorfluazuron

ensure that they wear chemical-resistant gloves, head gear and clothes (full sleeves and long trousers).

In animals, almost complete elimination of chlorflurenol occurs within 24 h. The chemical is observed to be highly mobile, though it degrades rapidly.

Chlorflurenol methyl ester, $C_{15}H_{11}ClO_3$ (methyl (*RS*)-2-chloro-9-hydroxyfluorene-9-carboxylate), is a derivative of chlorflurenol. It is soluble in water (18 mg/l), cyclohexane, methanol and isopropanol, to varying degrees.

Chlorflurenol methyl ester is a PGR, herbicide and a pruning aid used on for example turf and lawns. It is practically non-toxic to birds and mammals.

The USA has registered the use of chlorflurenol and its parent compound flurecol.

Chlorflurenol methyl ester

See **Chlorflurenol as a PGR**

Chlorine as a plant nutrient

See **Plant nutrient excesses and deficiencies**

Chlormephos insecticide

Chlormephos, $C_5H_{12}ClO_2PS_2$ (*S*-chloromethyl *O,O*-diethyl phosphorodithioate), is an **organophosphorus (OP)** aliphatic insecticide. It is a colourless liquid with solubility in water being 60 mg/l. It is miscible with most organic solvents. It is hydrolysed at 80°C and in alkaline medium. It is obtained by the reaction of *O, O*-diethyl thionothiolphosphate with bromo-chloro-methylene. The CAS Registry Number is 24934-91-6.

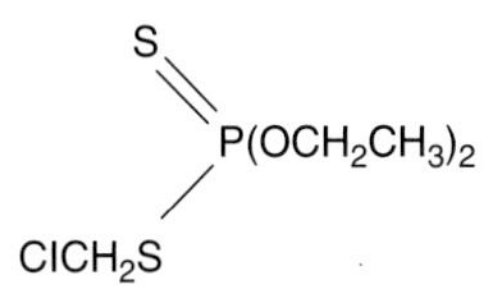

Chlormephos

The insecticide is a non-systemic, cholinesterase inhibitor. It is used to control soil insects like white grubs, wireworms and millipedes in maize, sorghum, soybean, sugarbeet, fodder beet, sugarcane, potatoes, tobacco and cereals.

Chlormephos is formulated as granules. It is combined with systemic OP insecticides. A dose of 2 kg/ha may be broadcast or incorporated into the soil. It is available as Dotan, among other trade names.

The WHO has categorized chlormephos as extremely hazardous (Class Ia). It is moderately to very highly toxic to fish, crustaceans and bees. Chlormephos may injure sorghum and soybean crops. The acute oral LD_{50} in rats is 7 mg/kg. In the presence of certain substances like hydrides, chlormephos can release toxic **phosphine gas**. Since the chemical is a cholinesterase inhibitor and rated very high on the acute toxicity scale, it is categorized as a PAN Bad Actor chemical. Care must be taken by applicators and handlers of chlormephos to ensure their safety. Self-contained breathing apparatus and special protective clothing are recommended.

In mammals, diethyl phosphate and diethyl phosphorothioate are produced upon degradation of chlormephos within a day. Elimination takes place via urine. In soil, chlormephos is converted to **ethion** by the hydrolytic product, *O,O*-diethyl hydrogen phosphorothioate. Persistence in soil is likely to be longer in cool months.

The USA and the EU have banned the use of this chemical.

Chlormequat as a PGR

Chlormequat, $C_5H_{13}ClN$ (2-chloroethyl trimethylammonium), is a plant growth retardant. It is an extremely hygroscopic crystalline, colourless substance. It is soluble in water, ethanol, dichloromethane, ethyl acetate, *n*-hexane, acetone and chloroform. The aqueous solution is quite stable up to 245°C. The CAS Registry Number of chlormequat is 7003-89-6.

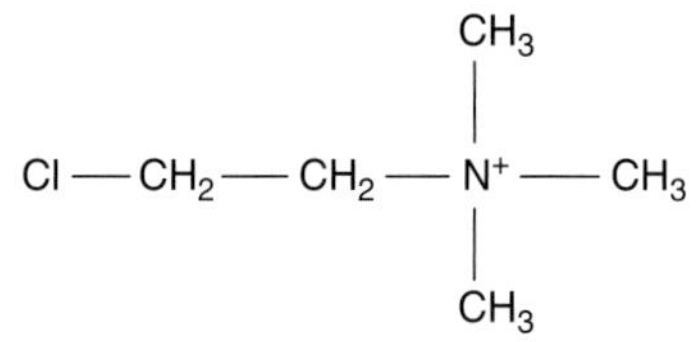

Chlormequat

Chlormequat, a growth regulator, inhibits cell elongation. In cereal crops like wheat, rye and oats, yields are seen to increase while lodging is reduced. The product that is generally used in formulations is chlormequat chloride.

Chlormequat is formulated as a soluble concentrate (SL) and dustable powder and is available in the market under the trade names of Cycocel, CeCeCe, CCC700 and Quintacel. It is compatible with growth regulators and pesticides, except with contact herbicides. Chlormequat chloride is generally sprayed at rates of 750 gm/l as a soluble concentrate. It also controls soil insects, foliar fungi and acts like a weedicide.

According to the WHO, chlormequat is a Class III toxin (i.e. a product that is slightly hazardous). While the compound chlormequat is not considered a PAN Bad Actor chemical, its salt chlormequat chloride is classified as a PAN Bad Actor since it is slightly toxic in toxicity tests (according to the EPA).

The ADI for chlormequat is 0.05 mg/kg/day. In animals, it is eliminated in 24 h and in plants it is converted to choline chloride. In soil it is rapidly degraded by microbial activity.

Chlormequat is not registered in the USA. Two of its derivatives, phenyl 2-trimethylammonium ethanesulfonate methyl sulfate and choline chloride, are also not registered. India, Australia, Denmark, Belgium, Canada and the USA are some of the countries that have registered the use of chormequat chloride. According to Health Canada, the MRL of chlormequat chloride on wheat is 0.1 ppm.

Chlormequat chloride as a PGR

See **Chlormequat as a PGR**

Chloroacetamide herbicides

See **Herbicide action – others**

Chloroacetanilide herbicides

Chloroacetanilides are organic compounds that are among the most frequently used herbicides in the world, because of their wide use in crop protection. Chloroacetanilides are water-soluble, pre-emergence, soil-applied herbicides that are environmentally safe since they degrade easily. But they are usually used with safeners (e.g. flurazole, **oxabetrinil**) since they are toxic to plants. This family of herbicides targets grass weeds among rice, soybean and wheat. The mode of action is the inhibition of protein synthesis. They inhibit the growth of new and young weeds. Seedlings grow abnormally, if they emerge at all.

Chloroacetanilide herbicides have many active ingredients, such as **acetochlor**, alachlor, **butachlor**, butenachlor, delachlor, diethatyl, dimethachlor, ethachlor, ethaprochlor, **metazachlor, metolachlor, S-metolachlor, pretilachlor, propachlor, propisochlor**, prynachlor, terbuchlor, thenylchlor, xylachlor, most of which are frequently soil-applied, to prevent or inhibit new growth in crops such as sorghum, maize, rice and groundnuts. These compounds are not as effective on older and established weed plants.

However, caution has to be exercised while using these chemicals in large quantities. These chemicals are toxic to fish and other aquatic species, and are considered to be carcinogenic. Exposure may lead to organ dysfunction and abnormalities. They can also be toxic by the inhalation route. Residues of chloroacetanilides are also found in groundwater sources.

Only a few active ingredients of the chloroacetanilide family are highlighted here. Other members find independent places alphabetically.

Alachlor, similar to metolachlor, is a pre-emergence herbicide. **Yellow nutsedge** and annual grasses and broadleaved weeds of maize, sorghum, groundnut, soybean and dry beans are controlled by alachlor. A common trade name is Lasso. Metolachlor, while similar to alachlor, has longer soil persistence. Propachlor lends grass control in maize and soybean. It is commonly combined with triazines. Trade names include Ramrod. Acetachlor is effective against particular small-seeded broadleaved weeds in soybean and maize crops. A common trade name is Harness. Butachlor (trade name Machete) is not only effective against many annual grasses and broadleaved weeds but also some aquatic species common in rice fields. Dimethanamid, with a five-member thiophene ring, is also a typical maize and soybean herbicide controlling yellow nutsedge, annual grasses and broadleaved weeds. A known trade name is Frontier.

Butenachlor, $C_{17}H_{24}ClNO_2$ ((Z)-N-but-2-enyloxymethyl-2-chloro-2′,6′-diethylacetanilide), is another member of the chloroacetanilide group, with a CAS Registry Number of 87310-56-3. The compound or its metabolite 2-chloro-2′,6′-diethylacetanilide are not registered in the USA.

Delachlor, $C_{15}H_{22}ClNO_2$ (2-chloro-N-(isobutoxymethyl) acet-2′,6′-xylidide), is a chloroacetanilide herbicide. It has a CAS Registry Number 24353-58-0 and is soluble in water at 59 ppm at 20°C. It is used on grasses in sugarbeet. Delachlor is not a PAN Bad Actor chemical. It is not registered in the USA.

Diethatyl, $C_{14}H_{18}ClNO_3$ (N-chloroacetyl-N-(2,6-diethylphenyl)glycine), a chloroacetanilide herbicide is generally available as diethatyl-ethyl (CAS Registry Number 38727-55-8). An obsolete pesticide (according to the WHO), it controls grasses in some crops. However, it is a potential groundwater contaminant (solubility in water is 120 mg/l). It is not registered in the USA (see Diethatyl at top of the next column).

Dimethachlor, $C_{13}H_{18}ClNO_2$ (2-chloro-N-(2-methoxyethyl) aceto-2′,6′-xylidide), is a chloroacetanilide herbicide. Its CAS

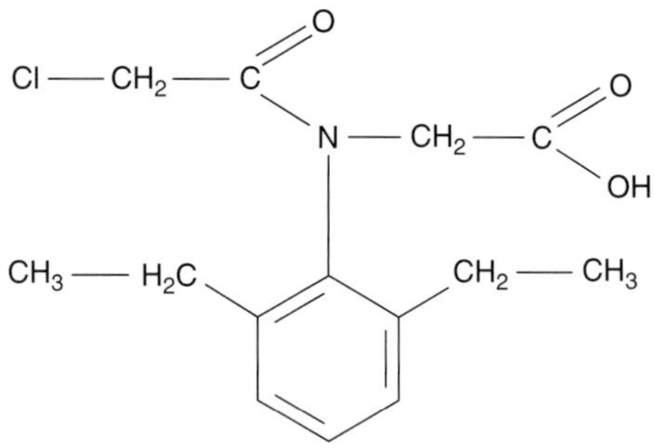

Diethatyl

Registry Number is 50563-36-5. It is a selective herbicide absorbed by new shoots and roots.

According to the WHO, dimethachlor comes under toxicity Class III (slightly hazardous). The ADI of dimethachlor in rats is 0.1 mg/kg b.w./day.

Dimethachlor is registered in some EU countries like Germany, France, Poland and Hungary but not in the UK. According to the EU Pesticide Database, the MRL for rapeseed is 0.05 mg/kg.

Prynachlor, $C_{12}H_{12}ClNO$ ((RS)-2-chloro-N-(1-methylprop-2-ynyl) acetanilide), is a chloroacetanilide herbicide. It has a CAS Registry Number of 21267-72-1. It has an oral LC_{50} of 1177 mg/kg and a dermal LD_{50} of 1926 mg/kg in rats. It is not a PAN Bad Actor chemical. It is registered for use in Hungary but not in the USA.

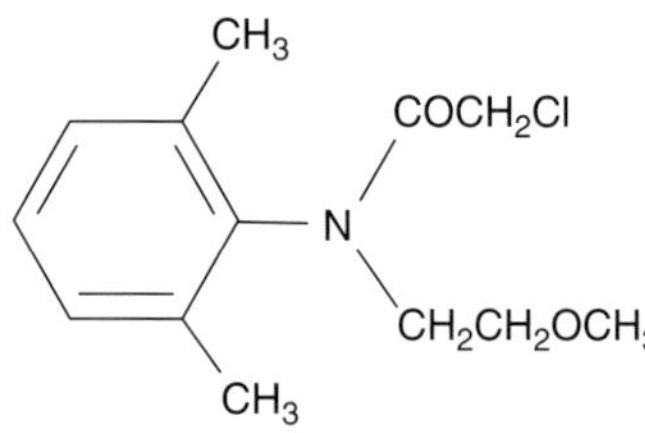

Dimethachlor

Thenylchlor, (2-chloro-N-(3-methoxy-2-thenyl)-2′,6′-dimethylacetanilide), another chloroacetanilide herbicide, is absorbed through the roots, from where it blocks protein synthesis and, hence, cell division. Its CAS Registry Number is 96491-05-3 and molecular formula, $C_{16}H_{18}ClNO_2S$.

Chlorobenzilate acaricide

Chlorobenzilate, $C_{16}H_{14}Cl_2O_3$ (ethyl 4,4′-dichlorobenzilate), is a bridged diphenyl acaricide. It is a brownish-yellow liquid. It is practically insoluble in water but soluble in benzene, acetone, methyl alcohol, toluene, hexane, octan-1-ol and alcohol. It is stable under normal temperatures and pressures, though it may burn (but not ignite). The breakdown of chlorobenzilate by heat may create corrosive fumes of hydrogen chloride and toxic oxides of carbon. The exposure to alkalis or strong acids may cause hydrolysis. Its CAS Registry Number is 510-15-6 (see Chlorobenzilate at top of the next page).

Chlorobenzilate, a non-systemic compound, has contact action. Being a neurotoxin, it disrupts the functioning of the nervous system of mites on citrus crops and in beehives. It has little insecticidal action, killing only ticks and mites.

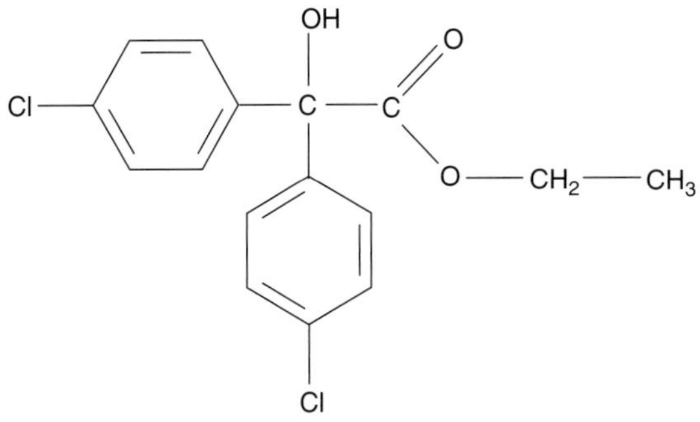

Chlorobenzilate

The acaricide is available as emulsifiable concentrate (EC) and as wettable powder (WP) formulations. Product mixtures are foliar sprayed strictly as per label instructions. Trade names include Acaraben, Folbex, Benzilan, ECB, Geigy 338 and Benz-o-chlor.

Chlorobenzilate is slightly to moderately toxic to humans. It is a mild skin and eye irritant. The acute oral LD_{50} in rats and hamsters is 700 mg/kg, while in mice it is 729 mg/kg. The 7 day LC_{50} is 3375 ppm in bobwhite quails. It is a suspected carcinogen in animals and humans. It has produced liver tumours in mice, but the evidence for carcinogenicity in rats is uncertain. Before being banned for use in the USA, the compound was classified by the EPA as a restricted use pesticide (RUP). Chlorobenzilate may be phytotoxic.

Humans rapidly excrete chlorobenzilate, usually within 3–4 days. In dogs, portions of the compound have been shown to remain unchanged for up to 9 months. Chlorobenzilate is practically insoluble in water and adsorbs strongly to soil particles. Hence, it is unlikely to leach into groundwater. It is fairly persistent on plant foliage.

Chloroneb fungicide

Chloroneb, $C_8H_8Cl_2O_2$ (1,4-dichloro-2,5-dimethoxybenzene), is an aromatic hydrocarbon fungicide (of the chlorophenyl/nitroaniline class). It is a colourless, crystalline solid, soluble in water (8 mg/l), acetone (115 g/kg), xylene (89 g/kg) and dichloromethane (133 g/kg) at 21°C. The compound is stable up to its boiling point (268°C/760 mm Hg). Chloroneb has a CAS Registry Number of 2675-77-6.

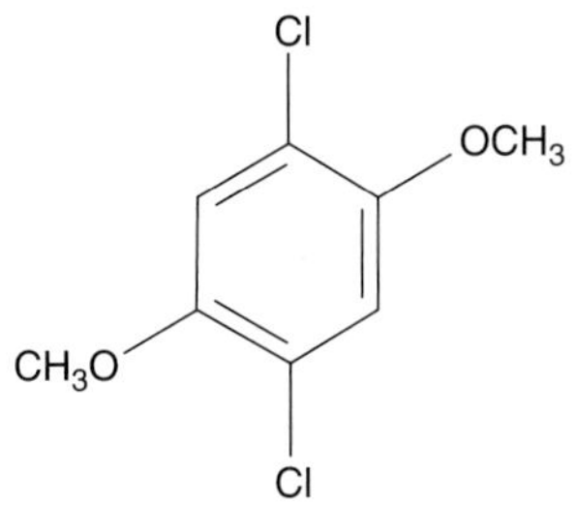

Chloroneb

Chloroneb, a systemic compound, which inhibits lipid peroxidation, has a soil and seed application. It remains concentrated near the roots and lower stems, and it acts as a fungistatic. Seedling diseases of cotton, beans, sugarbeet and soybeans caused by *Rhizoctonia*, *Typhula*, *Pythium* and *Sclerotium* pathogens (causing diseases such as snow mould, damping-off and blight) are well controlled. Turf and ornamentals are also protected by the use of chloroneb.

The fungicide chloroneb is available as a wettable powder (WP) and other formulations for seed treatment. Soil drench application and **chemigation** are also commonly used.

The fungicide has been classified by the EPA and WHO as a slightly hazardous compound (Class III toxin). All labels should display the signal word 'CAUTION'. It is not a PAN Bad Actor chemical. The acute oral LD_{50} dose in rats is found to be >11000 mg/kg whereas for some bird species tested it is found to be >5000 mg/kg. The LC_{50} (46 h) in some fish species is >4200 mg/l. The NOEL for 2 years in rats is 2500 mg/kg diet. In the mammalian toxicological studies, chloroneb is not found to be a threat to animals, fish and birds.

The metabolism of chloroneb in animals is observed to proceed by the formation of metabolite, 2,5-dichloro-4-methoxyphenol. In soil, it breaks down by microbial action. The fungicide leaches from sandy soils and may enter groundwater sources, as per the Re-registration Eligibility Decision (RED) document of the EPA (September 2005). But it is highly adsorbed in turf soil. The soil half-life of chloroneb is in the range of 10–180 days.

The USA has registered chloroneb. Health Canada's **Pest Management Regulatory Agency (PMRA)** has discontinued the use of chloroneb in agriculture. The MRL values set for beans, sugarbeet and the meat and fat of hog (large pig), goat and horse are 0.2 ppm, while for forage and hay of cowpea and soybean, it is 2 ppm.

Chloroneb technical analysis

The fungicide chloroneb is highly effective against seedling diseases caused by *Rhizoctonia*, *Pythium* and others. Chloroneb and its metabolite 2,5-dichloro-4-methoxyphenol can be determined by separating them from soil or organic tissues by processes including steam distillation, extraction with hexane, and determination by gas chromatography (GC) using programmed column temperature and a selective microcoulometric detector (described as follows).

Place up to 100 g of the sample in a 2 l round-bottomed flask, along with porcelain boiling chips, minute quantities of antifoaming agent (e.g. Dow Corning Antifoam A) and 1 l of 5 N phosphoric acid. Set up the apparatus for collecting the extract by heating, with the help of a distillation extraction head, a water-cooled condenser and 400 ml of *n*-hexane in a 1 l flask. Apply heat to both the flasks at such a rate that condensed hexane and water pass through the capillary in the form of small droplets. Extraction may be done for 12 h.

Upon extraction, transfer the hexane from the flask (along with rinsates, as a result of washing with hexane) to a 600 ml beaker. Evaporate the solution at room temperature to approximately 4 ml, and then transfer this to a 5 ml volumetric flask, dilute to the mark with hexane and shake. Analyse the solution by gas chromatography using a GC column tube that is 120 cm long with an internal diameter of 10 mm. This is packed with 80–100 mesh acid-washed and dimethyldichlorosilane-treated Chromosorb W, coated with 10% by weight DC-560 phenyl methyl silicone oil plus 0.2% by weight Epon1001epoxl resin. The temperature of the injection port must be 240°C, and that of the column, 100°C (for 2 min). Programme from 10°C/min to 180°C/min, and hold it there isothermally until the end of the analysis. The carrier gas is helium, the flow rate through the column is 80 ml/min and the purge flow rate is 50 ml/min. The detector is microcoulometer, the furnace temperature must be 880°C and the oxygen flow 50 l/min. The retention time of chloroneb is 10.7 min and that of the metabolite is 9.3 min.

Chlorophacinone rodenticide

Chlorophacinone, $C_{23}H_{15}ClO_3$ (2-[(2*RS*)-2-(4-chlorophenyl)-2-phenylacetyl]indan-1,3-dione), is a rodenticide belonging to the chemical class 1,3-indandione. It is a pale yellow crystalline substance, soluble in water (100 mg/l), methanol, ethanol, acetone and acetic acid. Chlorophacinone is produced by condensing 1-phenyl-1-(4-chlorophenyl) acetone with dimethylphthalate in the presence of sodium methoxide. Its CAS Registry Number is 3691-35-8.

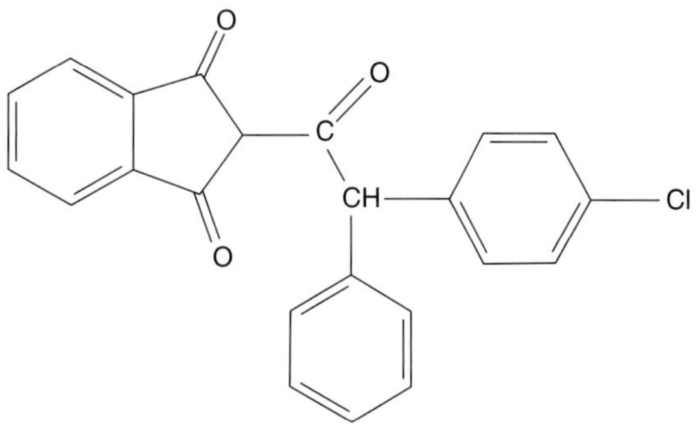

Chlorophacinone

Chlorophacinone, with anticoagulant properties, acts on rodents by blocking the formation of **prothrombin**, causing uncoupling of oxidative phosphorylation. This causes internal haemorrhaging in the animal, causing death.

Common trade names of chlorophacinone are Caid, Raviac and Liphadione, with common formulation types being block or grain bait, bait concentration and tracking powder. It is commonly mixed with sulfaquinoxaline. It is a single-dose killer (so there is no bait shyness), with a concentration dose range of 0.005–0.01%.

The acute oral LD_{50} in rats is 20.5 mg/kg while the 1 h inhalation LC_{50} is >3.0 mg/l air. The EPA has classified the technical product as extremely toxic and the tracking powder as moderately hazardous. The WHO has classified this chemical as an extremely hazardous one (Class Ia). It is considered a PAN Bad Actor chemical since it is shown to be highly toxic according to acute toxicity studies. Chlorophacinone has low toxicity to birds and is not toxic to bees, if used as directed. In mammals, it is eliminated within 2 days as metabolites.

It is important that chlorophacinone is transported in leak-proof containers that are clearly labelled. The chemical should be kept out of reach of children and people who are bleeding. Vitamin K (phytomenadione) acts as an **antidote**, administered either intravenously or orally, generally not exceeding 10 mg twice daily, or as advised by a doctor.

Degradation studies of chlorophacinone show that a portion is excreted as the unchanged chemical, whereas the remaining is metabolized as hydroxylated products, which are not toxic.

The USA, Canada, Germany and Tanzania have registered the use of chlorophacinone.

Chloropicrin analysis

Chloropicrin (CCl_3NO_2) is an insecticidal soil fumigant, nematicide and a fumigant for stored grains and cereal products. It is also effective against soil-borne fungi. The following passage describes the method of analysing chloropicrin.

An absorber tube containing 2.5 ml 2% sodium peroxide solution in *N*-sodium hydroxide and 2.5 ml isopropyl alcohol is prepared, as part of the apparatus to isolate chloropicrin. The absorber tube is placed in an apparatus train, without the aeration flask being attached. A temperature of 85°C is maintained in the condenser by circulating hot water through the condenser. The following are placed in the aeration flask (with the stopcock closed): 100 ml water; 10 ml 6 N sulfuric acid; 16 ml phosphotungstic acid solution (with 20 g $P_2O_5.24WO_3.25H_2O$ in 80 ml water); and about 0.5 g of a defoaming agent.

The sample (200 g) is blended with 200 ml water to form a slurry. The slurry is transferred to the flask, and air or nitrogen is supplied to the condenser to permit a slow stream to pass through the slurry. The solutions are kept boiling by heating and by increasing the flow rate of gas to approximately 40 ml/min. The distillate in the absorber tubes are then collected and transferred to a 100 ml volumetric flask. Each absorber is washed with 10 ml water and 15 ml isopropyl alcohol and the washings are added to the flask. Isopropyl alcohol is then added to dilute the solution to 100 ml.

A suitable aliquot of the distillate is transferred to a 125 ml boiling flask. The water, sodium peroxide and isopropyl alcohol in the mixture are calculated and adjusted to 3.5 ml water, 5 ml 2% sodium peroxide solution and 11.5 ml isopropyl alcohol. A ratio of 11.5 of nopropyl alcohol to 8.5 ml total water is critical. Aliquots of diluted standard solution (0, 1.0, 2.0, 3.0, 4.0, 5.0, 6.0 and 7.0 ml) are placed into 125 ml boiling flasks containing 11.5, 10.5, 9.5, 8.5, 7.5, 6.5, 5.5 and 4.5 ml isopropyl alcohol, respectively, to which 5 ml 2% sodium peroxide and 3.5 ml water are added. A reagent blank and a reflux condenser are prepared. The boiling rate for the vapours is adjusted to reach well up in the condensers, with a reflux for 30 min. Isopropyl (5 ml) is used to rinse the condensers, which are then detached from the flask. The cooled contents are transferred to 50 ml volumetric flasks, keeping the total volume below 45 ml. A drop of phenolphthalein indicator is added, to be neutralized with concentrated hydrochloric acid. To this solution 1 ml 1% aqueous sulfanilic acid solution and 1 ml 0.1% N-(naphthyl) ethylenediaminidihydrochloride are added one after the other with a little time interval, while diluting with water and mixing thoroughly. Absorbance can be measured at 550 nm versus water. Calibration can be prepared with the values obtained for the standard solutions and chloropicrin content determined from the standard curve.

Chloropicrin insecticide

Chloropicrin, CCl_3NO_2 (trichloro (nitro)methane), is a fumigant insecticide and also a nematicide. It is a colourless liquid, soluble in water at 1.62 g/l at 25°C and miscible with acetone, benzene and carbon tetrachloride. Chloropicrin is unstable in alkaline media. It is made by chlorinating nitro-methane with alkaline metal hypochlorites. It photodegrades to yield phosgene and nitrosyl chloride. Its CAS Registry Number is 76-06-2.

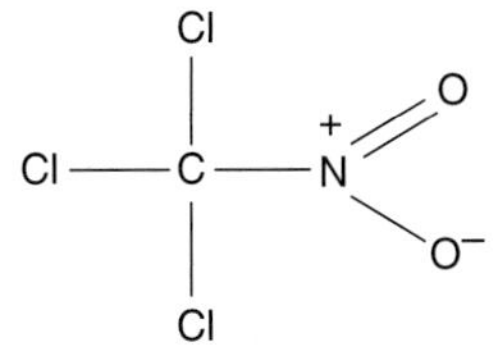

Chloropicrin

Chloropicrin is an established nematicide and a soil insecticide. It also has fumigant properties that make it suitable as a protective agent for stored grain.

It is commonly formulated as an oil miscible liquid or a tablet, and used in combination with other fumigants such as **methyl bromide** or **1,3-dichloropropene** or methylisothio-cyanate.

Chloropicrin is highly phytotoxic in nature and has a lachrymatory action. It is classified by the EPA as highly toxic. The acute oral LD_{50} in rats is 250 mg/kg, while the inhalation LC_{50} is 0.008 mg/l air. It is a PAN Bad Actor chemical and is also a potential groundwater contaminant. As a highly toxic and dangerous chemical, chloropicrin has to be stored in a cool, dry, leak-proof original container with clear labels on it. The use of the chemical must be strictly as per the label instructions, and only where appropriate PPE is used. Chloropicrin is degraded in soils by *Pseudomonas* spp.

Chloropicrin is registered in the USA, Canada, Australia, New Zealand and Greece. According to the FSANZ Joint Food Standards Code of Australia, a MRL of 0.10 mg/kg has been fixed for cereal grains. According to the Pest Control Products Act of Health Canada (PMRA) the MRL of the chemical for broccoli, cucumbers, onions, carrots, potatoes and fruits (e.g. apples, pears, apricots, peaches and many others) is fixed at 0.025 mg/kg.

Chlorosis

Chlorosis is a plant condition where green plants lose their colour due to lower chlorophyll levels, leading to the plant becoming unhealthy, yellow and pale. Nutrient deficiency (mainly of nitrogen, iron, **magnesium** and copper), lack of proper sunlight and certain viral attacks make it difficult to produce chlorophyll, causing chlorotic conditions in plants. Chlorosis caused by lack of light is called etiolation.

As in the case of **necrosis**, chlorotic conditions leave plants open to pathogen attacks. Attention to the overall health of a plant is an ideal strategy towards healthy, pest-free crop life.

Symptoms of chlorosis are seen as even or uneven pale green or yellow-green colouration of the foliage, with the veins remaining green. The more yellow the colour, the more severe the chlorosis. Patches of light-coloured crops in field crops indicate the presence of chlorosis.

Nutritional deficiency is the major reason for chlorotic plant growth. However, this may be misleading, since a deficiency does not necessarily mean the absence of that nutrient to the plants. The nutrient may well be present in the soil but may be 'unavailable' to the plant (e.g. it may be in a form that cannot be taken up). Sometimes, unaerated conditions and a high soil pH can hinder the uptake of the nutrients that are present. Waterlogged, damaged or compacted roots can prevent nutrition uptake in many crops. Uptake of iron and **manganese** is hindered by excessive presence of potassium, magnesium and phosphorus. In highly alkaline soils too, nutrients become unavailable to plants. Pesticides and particularly herbicides can cause chlorosis on crops, as well as targeted weeds. Exposure to sulfur dioxide can also cause chlorosis.

Some pathogens cause chlorosis in plants. For instance, the tomato mosaic virus infects various crops including tomato, tobacco, spinach, cherries, peppers, cucumbers, squashes and many ornamental plants, cause varying degrees of chlorosis in plants, apart from causing green-yellow patterns on leaves and dwarfing.

Providing the nutrients that are deficient, especially micronutrients, is the first measure for correcting chlorosis. Soil incorporation of chelated iron compounds (and perhaps, as a quick solution, foliar application of iron solution), nitrogen compounds, and iron or manganese sulfate are some examples of nutrient supplements. Adjusting the soil pH after testing water and soil can help in correcting chlorosis. Care has to be taken that applications are carried out when temperatures are not too high and that enough irrigation is used during such operations.

Simultaneously, use of good quality, insect-resistant seeds, and control of **vectors** of viruses (e.g. aphids), can keep a good control over pathogen attacks and, hence, chlorosis.

Chlorothalonil fungicide

Chlorothalonil, $C_8Cl_4N_2$ (tetrachloro-isophthalonitrile), is an aromatic, chloronitrile fungicide. It is a colourless, odourless, crystalline solid, soluble in water (0.81 mg/l), acetone (20.9 g/l), ethyl acetate (13.8 g/l), xylene (74.4 g/l) and in 1,2-dichloroethane (22.4 g/l). It is stable at normal temperatures, UV and in moderately alkaline aqueous solutions. However, hydrolysis occurs above pH 9. It is produced by the reaction of tetrachloroisophthalyl chloride with ammonia to give diamide, which, on treatment with phosphorus oxychloride, yields the dinitrile. The CAS Registry Number is 1897-45-6.

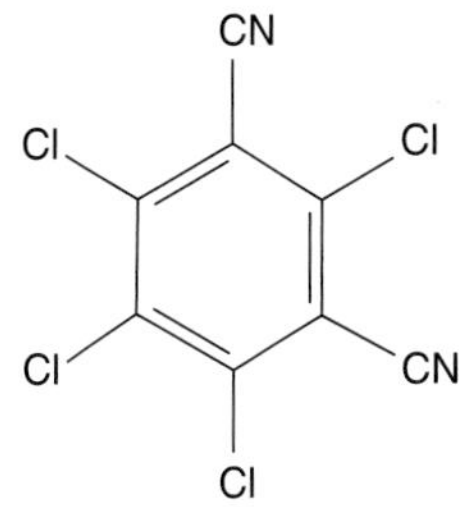

Chlorothalonil

Chlorothalonil, a multi-site, foliar, non-systemic fungicide and **fungistatic** agent, depletes the glutathione (thiol group) in germinating fungal cells and disrupts glycolysis and energy production in cells. The chemical is used in controlling fungal diseases in for example fruits, vegetables, cereals, oilseeds, tuber crops and mushrooms. This chemical is also used for protection of wood against mould, **sap stain** and decay.

The fungicide is available as a soluble concentrate (SL), water-dispersible granule (WG) and a wettable powder (WP). Some popular brands are Bravo, Amconil, Chloronil and Kavach.

Phytotoxicity of chlorothalonil can result in russeting and injury in ornamentals, apples and grapes. The foliage of *Pittosporum* (Japanese mockorange/pittosporum) or *Schefflera* (octopus tree) is particularly sensitive to chlorothalonil. The fungicide is incompatible with oils and will cause phytotoxicity when mixed with oils or oil-containing substances.

Chlorothalonil causes irritation to the human eye and skin. The chemical is moderately toxic to birds and bees. The acute oral LD_{50} dose in bobwhite quails is >2000 mg/kg b.w. The LC_{50} concentration in bees is >63 μg per bee. But the chemical is highly toxic to aquatic organisms (with the 96 h LC_{50} in bluegill sunfish being 59 μg/l and in rainbow trout, 39 μg/l). The acute oral LD_{50} dose in rats is found to be >5000 mg/kg. It is also shown to cause kidney disorders in rats and mice. For chronic effects, a NOEL of 2 mg/kg b.w. is fixed in rats. Chlorothalonil is shown to be carcinogenic in nature and a potential groundwater contaminant and hence a PAN Bad Actor chemical. The WHO categorizes this chemical under Class U (unlikely to be hazardous).

The metabolism of chlorothalonil in animals (ruminants) leads to formation of a 4-hydroxy derivative. It reacts with glutathione in the gut to yield mono-, di- or tri-glutathione conjugates and further to di- and tri-thiol or mercapturic acid derivatives that are excreted through urine and faeces. In plants, however, the residues remain as the parent compound.

Chlorothalonil degradation is fast in biotic aquatic systems. It is less mobile in clayey, silty loamy soil since it binds to these soils. It is mobile in sandy soil. The soil DT_{50} (aerobic) is 3 months.

The fungicide is registered in Australia, New Zealand, the USA, Canada and most countries of the EU (e.g. the UK, the Netherlands and Germany). According to the Prevention of Food Adulteration Rules 1955 of India, the MRL for potatoes is set at 0.1 mg/kg. It is the same in Australia. According to the Pest Control Products Act of Health Canada (PMRA) the MRL of the chemical for cabbage, cauliflower, garlic, onions and many others is set at 5 mg/kg.

Chlorotriazine herbicides

Chlorotriazines are a type of triazine herbicides with a chloro substituent in their structure. The chlorotriazine group includes **atrazine**, **simazine**, cyanazine and **hexazinone** (e.g. of trade names being Aatrex, Bladex and Velpar), which target different weeds on a wide range of crops.

Chlorphonium as a PGR

Chlorphonium, $C_{19}H_{32}Cl_2P$ (tributyl(2,4-dichlorobenzyl phosphonium)), is a plant growth regulator/inhibitor. Chlorphonium is colourless and has a crystalline property. It is soluble in water, methanol and acetone but insoluble in diethyl ether and hexane. It is stable under normal storage conditions. Chlorphonium is produced from 2,4-dichlorobenzyl chloride and tributyl phosphine. The CAS Registry Number of chlorphonium is 115-78-6 (see Chlorphonium at top of next column).

Chlorphonium chloride is a salt of chlorphonium, and an organophosphate systemic herbicide. Chlorphonium and its derivatives reduce internodal length and increase the number of buds in ornamentals like chrysanthemums, geraniums and petunias.

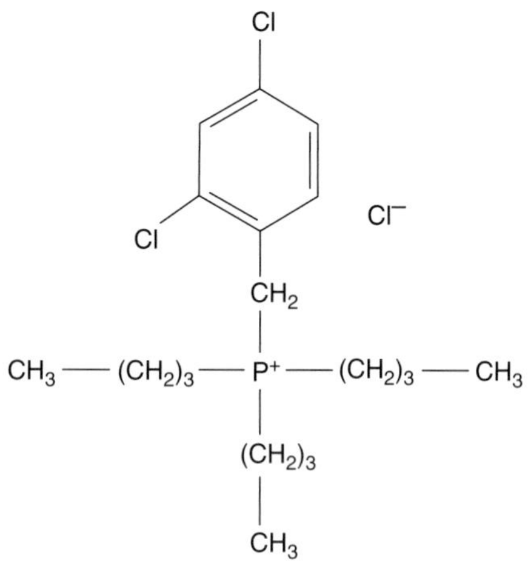

Chlorphonium

The compound, also known as phoston, is available as for example Phosfleur, Phosfon and Phosfon D. It is formulated as a granular product or as a soluble concentrate.

According to the WHO, chlorphonium is moderately hazardous. The oral LD_{50} in rats is 178 mg/kg, while the dermal LD_{50} in rabbits is 750 mg/kg. It is non-toxic to bees.

It is not registered in the USA and EU countries.

Chlorpropham herbicide

See **Herbicide action – meristematic mitotic inhibitors**

Chlorpyrifos insecticide

Chlorpyrifos, $C_9H_{11}Cl_3NO_3PS$ (*O,O*-diethyl *O*-(3,5,6-trichloro-2-pyridyl) phosphorothiate), is an organophosphorus insecticide (organothiophosphate or OP). It comprises colourless crystals with a mild mercaptan odour, and is soluble in water (1.4 mg/l) but highly soluble in organic solvents such as benzene, acetone, chloroform, carbon disulfide, diethyl ether, xylene, iso-octanol and methanol. Chlorpyrifos hydrolyses at higher pH and in the presence of chelating metals (copper). It is produced by the reaction of 3,5,6-trichloro-2-pyridinol with diethylthiophosphoryl chloride. The CAS Registry Number is 2921-88-2.

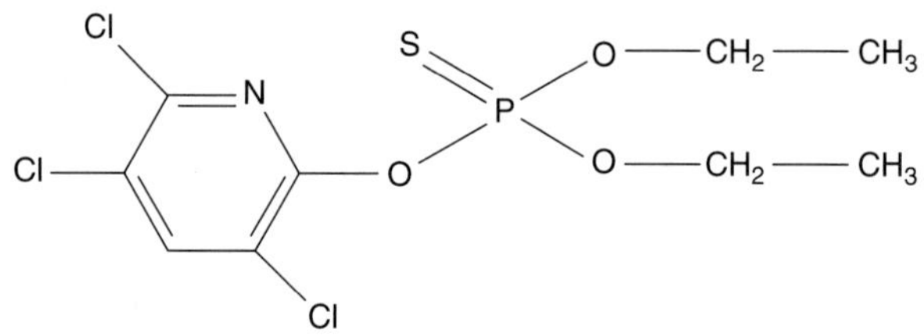

Chlorpyrifos

Chlorpyrifos, a cholinesterase inhibitor, is a non-systemic soil and foliar insecticide. Its multifarious actions (e.g. on stomach and respiration) render it effective against pests in households (mosquitoes, ants, rodents and termites), warehouses and animal dwellings on farms and at home. It is used to control many members of the orders Coleoptera, Diptera, Homopetera and Lepidoptera in maize, cotton, almond, fruits (orange, apple, strawberries, figs and bananas), vine, potatoes, beet, tobacco, soybeans, lucerne, sunflowers, sweet potatoes, groundnuts, rice, maize, sorghum, asparagus and mushrooms. It has additional acaricidal and nematicidal properties. Non-food uses include turf, forestry and ornamentals.

The insecticide is formulated in forms such as granule (GR), emulsifiable concentrate (EC), wettable powder (WP), dustable powder (DP), ULV liquid (UL) and as a microcapsule. It should not be brought in contact with alkaline materials. It is available as trade names such as Borodan and Dursban.

The OP insecticide metabolizes rapidly in mammals and plants into 3,5,6-trichloro-2-pyridinol, which is excreted in the urine. In soil, it degrades slowly (with a half-life of up to 120 days) to 3,5,6-trichloropyridin-2-ol, organochlorine compounds and carbon dioxide. Thus, persistence of chlorpyrifos and its metabolites is an environmental concern.

Chlorpyrifos is shown to cause poisoning and detrimental effects on the nervous, cardiovascular and respiratory systems. It has also shown to affect **beneficial organisms**, birds and fish. Chlorpyrifos may also injure roses, poinsettias and camellias. Due to its toxicity, the use of chlorpyrifos on a wide scale is avoided.

The metabolite 3,5,6-trichloro-2-pyridinol (CAS Registry Number of 6515-38-4) does not have a significant bioaccumulation potential. But, with a moderately high solubility in water, it is toxic to algae, fish and earthworms. It has a potential for groundwater contamination, and so is classified as a PAN Bad Actor chemical.

Chlorpyrifos is registered by the EPA for agricultural use and as an ant and rodent bait, but the USA does not allow the use of any of the metabolites of this chemical (except for **chlorpyrifos-methyl**). Prevention of Food Adulteration Rules 1955 of India has fixed 0.01 mg/kg as the MRL of chlorpyrifos for onions, potatoes, cabbages and cauliflowers, whereas it is 0.05 mg/kg for cereal grains. FSANZ's Food Standards Code has fixed the MRL at 0.1 mg/kg for cereal grains (temporarily), while it is 0.05 mg/kg for crops such as potatoes, strawberries and passion fruit.

The insecticide is registered in Australia, Burkina Faso, Cambodia, Canada, Cape Verde, Chad, Guinea-Bissau, India, Madagascar, Mali, Mauritania, New Zealand, Niger, the Philippines, Senegal, South Africa, Tanzania, Uganda, Vietnam, the USA and much of the EU.

Chlorpyrifos-methyl insecticide

Chlorpyrifos-methyl, $C_7H_7Cl_3NO_3PS$ (*O,O*-dimethyl *O*-3,5,6-trichloro-2-pyridyl phosphorothioates), is an organophosphorus insecticide (OP). It is a colourless crystalline substance, with a slight mercaptan odour, soluble in water (4 mg/l) and highly soluble in organic solvents such as acetone, benzene, diethyl ether, chloroform, methanol and hexane. It is relatively stable under neutral conditions, but hydrolysed by acids (pH 4–6) and, more readily, by alkalis (pH 8–10). The half-life of chlorpyrifos-methyl is 3 days. It is produced by the reaction of **3,5,6-trichloro-2-pyridinol** with dimethylthiophosphoryl chloride. It has a CAS Registry Number of 5598-13-0 (see Chlorpyrifos-methyl at top of next column).

Chlorpyrifos-methyl, a non-systemic insecticide, is a cholinesterase inhibitor. It also has acaricidal properties. It is used to control insects belonging to the orders Coleoptera (beetles), Diptera (flies, especially of family Muscidae), Homoptera (sucking pests e.g. plant hoppers, **leafhoppers**, aphids, cutworm, cricket and whiteflies) and Lepidoptera (butterflies and moths)

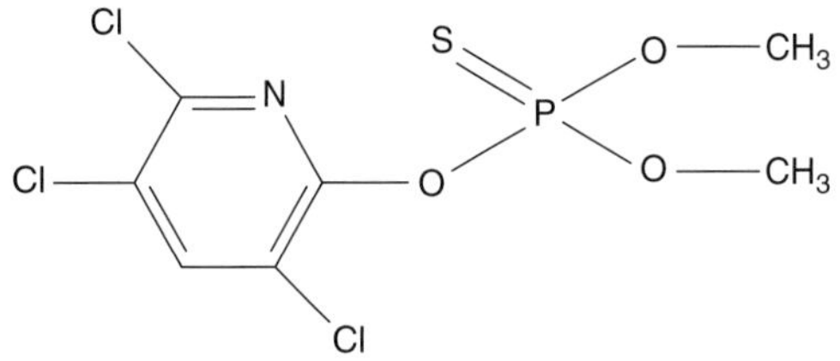

Chlorpyrifos-methyl

in cereals including wheat, rice and sorghum, a variety of fruit (pome, stone and citrus), tomatoes, tea, cotton and other crops.

This OP compound is marketed as an emulsifiable concentrate (EC), ULV liquids (UL) and a fogging concentrate. It is incompatible with alkaline and strongly acidic materials. It is available under the trade names of Reldan and Pyriban-M. Specifically, for the control of **rice stem borer**, it is used in France, Korea and Japan at up to 800 g/ha. Field conditions demand a dosage of up to 2 kg/ha, while storage needs up to 10 mg/kg of grain. It is also used in industrial and public health sites and stored-grain houses.

Since chlorpyrifos-methyl is a cholinesterase inhibitor, it has been categorized as a PAN Bad Actor chemical. The WHO has categorized it as a Class U toxin. It is a concern for many countries to observe chlorpyrifos-methyl residues in food samples. It is toxic to crustaceans. The LD_{50} for skin absorption in rabbits is >5000 mg/kg. The oral LD_{50} in rats is between 2000 and 3000 mg/kg. Prolonged exposure in humans may cause cataracts in eyes. Use of PPE such as eye wear, full-sleeved torso and full-length trousers with boots is recommended.

Chlorpyrifos-methyl is metabolized into 3,5,6-trichloro-2-pyridinol in plants and soils as well as animals. In animals, it is excreted in urine. In soils and water, it undergoes microbial degradation to 3,5,6-trichloropyridin-2-ol, which is subsequently degraded to organochlorine compounds and carbon dioxide. The DT_{50} in soil varies between 1.5 and 33 days, depending upon soil type and microbial activity. The half-life of chlorpyrifos-methyl in stored wheat is 3–5 months, but it decomposes faster at higher temperature and humidity levels.

India, most countries of the EU, the USA, Australia, Cameroon, Tanzania, the Philippines and New Zealand have registered the compound. There are set standards of residue limits that guide what is safe to be consumed by animals and humans. The MRL in milk is 0.01 mg/kg, rice 0.1 mg/kg, white bread 0.5 mg/kg, wholewheat bread 2 mg/kg, raw wheat, maize and sorghum 10 mg/kg and bran 20 mg/kg.

Chlorpyrifos technical analysis

The insecticide chlorpyrifos is available as a crystalline solid. The technical grade chlorpyrifos sample must be free from extraneous impurities or modifying agents for proper analysis. The sample must contain chlorpyrifos of 94% purity, an acid content expressed by sulfuric acid of up to 3 g/kg and a maximum water content of 0.4 g/kg.

The special apparatus used for technical analysis includes liquid chromatography equipment with a stainless steel column, a pumping system capable of maintaining a pressure of 11×10^3 kPa and a UV spectrophotometer detector capable of measuring UV absorption at 300 nm. The stainless steel column is 25 cm long with a 4.6 mm internal diameter packed with C-18 bonded silica gel (zorbax ODS or equivalent) and protected with a 2 μm inlet filter. The liquid chromatography mobile phase is a degassed mixture of 82.5 ml acetonitrile, 17.5 ml water and 0.5 ml glacial acetic acid. The operating conditions for the HPLC are ambient column temperature, a flow rate of 2 ml/min, a detector sensitivity of 1.0 AUFS and an injection volume of 10 μl. The retention time of chlorpyrifos is 3.55 min, of the internal standard, 7.10 min and of the alternative standard, 18.0 min.

Special reagents used are chlorpyrifos standard solution, internal standard solution and acetonitrile of HPLC grade. The chlorpyrifos standard solution is made by weighing about 80 mg of analytical grade chlorpyrifos of known purity into a glass vial. The internal standard solution of 1,4-dibromonaphthalene is made by weighing about 1.5 g of the compound to the nearest 0.1 g into a volumetric flask, diluted to volume with acetonitrile and mixed. This internal standard solution (25 ml) is pipetted into the standard solution and mixed well.

The standard chlorpyrifos calibration solution (10 μl) is injected into the HPLC port and attenuation adjusted to give the largest possible on-scale peaks. Injections are repeated until the ratios between the chlorpyrifos peak areas or height and the internal standard peak areas or height agree within ± 0.5%. Without changing the conditions, inject 10 μl aliquots of the sample solution until the response ratios agree within ± 0.5 %. Average the last two response ratios of the sample solution. Using Eqn C.1 and Eqn C.2 the chlorpyrifos content can be calculated.

$$\text{The response ratio}(r) = \frac{\text{Area or height of chlorpyrifos peak}}{\text{Area or height of internal standard peak}} \qquad (C.1)$$

$$\text{Chlorpyrifos content (g / kg)} = \frac{r_2 \times m_1 \times P}{r_1 \times m_2} \qquad (C.2)$$

where
r_1 = Average response ratio for standard solution
r_2 = Average response ratio for sample solution
m_1 = Mass (g) of chlorpyrifos in the standard in solution
m_2 = Mass (g) of sample taken
P = Purity (g/kg) of chlorpyrifos standard

The emulsifiable concentrate of chlorpyrifos is analysed by dissolving the sample in acetonitrile along with 1,4-dibromonaphthalene added as an internal standard. The chlorpyrifos content is estimated by HPLC, using a reverse-phase column and a mixture of acetonitrile, water and acetic acid as the mobile phase, as described above for technical grade chlorpyrifos.

Chlorsulfuron herbicide

Chlorsulfuron, $C_{12}H_{12}ClN_5O_4S$ (1-(2-chlorophenylsulfonyl)-3-(4-methoxy-6-methyl-1,3,5-triazin-2-yl)urea), belongs to the triazinylsulfonylurea family of herbicides.

Chlorsulfuron is a systemic sulfonylurea herbicide for the selective pre- and post-emergence control of broadleaved and grassy weeds in cereal crops. It is available as water-dispersible granules (WG). It is highly soluble in water. The CAS Registry Number is 64902-72-3. A known trade name is Glean (see Chlorsulfuron at top of next column).

Chlorsulfuron is active on annual weeds and grasses, showing the best results on for example velvetleaf, pigweed, mustards, lambsquarters and Canada thistle.

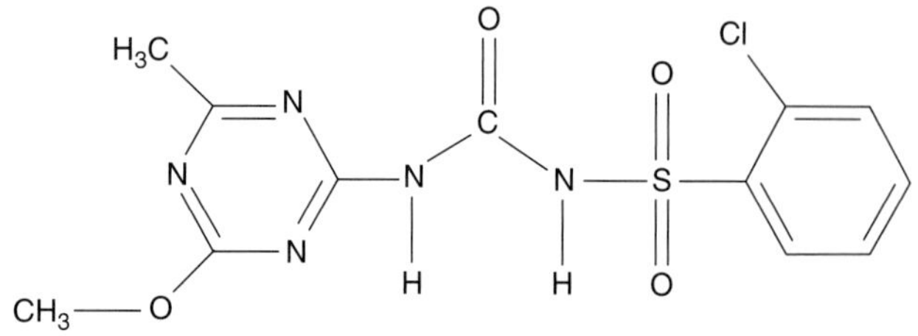

Chlorsulfuron

The WHO has classified chlorsulfuron as a Class U toxin (unlikely to be hazardous). It does not pose a hazard to fish, birds and non-target creatures. It is said to be a potential groundwater contaminant. Chlorsulfuron is rapidly eliminated when ingested by mammals. It undergoes *O*-demethylation and the cleavage of sulfonylurea linkage.

The herbicide chlorsulfuron is registered in the USA, Canada, Australia, New Zealand and Tanzania.

Chlorthal-dimethyl herbicide

See **Chlorthal herbicide**

Chlorthal herbicide

Chlorthal, $C_8H_2Cl_4O_4$ (tetrachloroterephthalic acid), is a phthalic acid herbicide. Its solubility in water is 0.5 mg/l, while the solubility in benzene, toluene, xylene, dioxane and carbon tetrachloride is higher. It exists as chlorthal-dimethyl ($C_{10}H_6Cl_4O_4$) and has a CAS Registry Number of 1861-32-1. It is also known as DCPA.

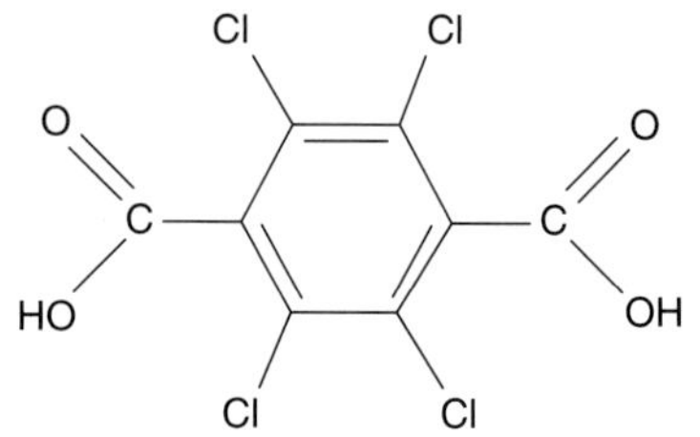

Chlorthal

Chlorthal is a pre-emergence herbicide used on annual grasses and annual broadleaved weed species in a wide range of vegetable crops.

The available formulations of chlorthal are wettable powders (WP), granules (GR) or as suspension concentrate (SC). It is available as Dacthal and Dacthalor.

Chlorthal has very low toxicity to mammals, with both oral and dermal LD_{50} values in rats, birds and rabbits being between 2000 and 12,000 mg/kg. It is a mild eye irritant and is slightly toxic to bees. It is classified as a general use pesticide (GUP) by the EPA, with a prerequisite of the signal word ‘CAUTION’ on labels.

The herbicide is moderately persistent in soil. However, moisture is essential for degradation of chlorthal in soil. The chlorthal metabolite, tetrachloro-terephthalic acid (TTA or diacid), is much more water soluble than the parent compound and is subject to leaching in some soils.

Chlorthiamid herbicide

Chlorthiamid, $C_7H_5Cl_2NS$ (2,6-dichlorothiobenzamide), belongs to the family of thioamide herbicides. It is an off-white solid with solubility in water of 950 mg/l. It is also soluble in aromatic and chlorinated hydrocarbons. It is thermally stable (up to 90°C) and stable in acidic solutions. In alkaline solutions, chlorthiamid is converted to **dichlobenil**, of which chlorthiamid is a precursor.

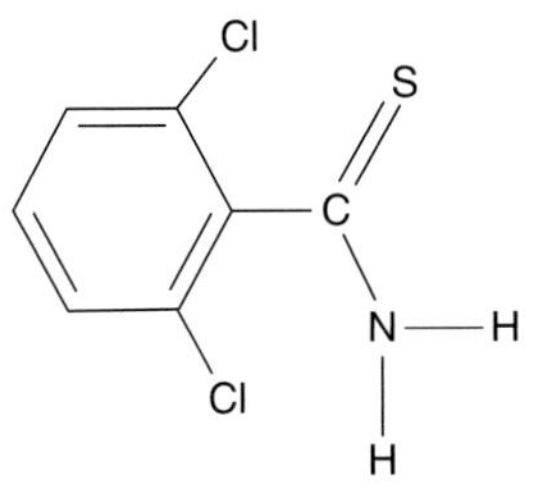

Chlorthiamid

Chlorthiamid is a systemic herbicide, which is mainly absorbed by the roots. It inhibits cell wall (cellulose) synthesis in weed seeds, making germination of seeds slow and difficult. Chlorthiamid is thus successful in controlling growth of grass and broadleaved weeds in pome and citrus fruits, berries and in pastures.

The herbicide chlorthiamid is usually formulated as granule (GR). It is not to be mixed with alkaline material. It is available under the trade name Prefix.

Chlorthiamid has an acute oral LD_{50} of 757 mg/kg in rats and of 500 mg/kg in mice. It has an acute percutaneous LD_{50} in rats of >1000 mg/kg. The NOEL in rats is 100 mg/kg. It is not toxic to non-target organisms and fish (LC_{50} in harlequin fish is 41 mg/l). According to the WHO, chlorthiamid comes under the toxicity Class III (slightly hazardous).

In soil, chlorthiamid is converted to dichlobenil, which then gradually undergoes microbial degradation to give 2,6-dichlorobenzamide and 2,6-dichlorobenzoic acid. Dichlobenil is a major metabolite of chlorthiamid.

Cholecalciferol rodenticide

Cholecalciferol is a white needle-like crystalline substance with molecular formula of $C_{27}H_{44}O$. It is almost insoluble in water but soluble in organic solvents. The CAS Registry Number of cholecalciferol is 67-97-0.

Calciferol, cholecalciferol and **ergocalciferol**, which are the forms of vitamin D, are required in minute quantities to maintain calcium and phosphate levels.

Cholecalciferol is an acute and chronic rodenticide which can effectively kill house rats, roof rats, Norway (brown) rats, either in a single dose or via multiple doses. It acts by increasing the concentration of blood calcium (vitamin D toxicosis) and increasing serum calcium (by converting cholecalciferol to calcitrol). Several hormones regulate blood calcium levels, and its sudden rise can lead to fatal effects since calcification affects kidneys, liver and heart (see Cholecalciferol at bottom of the page).

Since cholecalciferol is tasteless, bait shyness does not occur in rodent pests. A concentration of 0.075% in baits is sufficient to have a lethal effect. Cholecalciferol baits are safer than second-generation anticoagulants or **acute toxicants**. The commonly known brand names of this rodenticide include Quintox, Rampage, Hyperkil and Sorexa D.

The oral acute LD_{50} in rats ranges from 42.5 mg/kg to 43.6 mg/kg, while in bird species it is 2000 mg/kg and in dogs, 88 mg/kg.

The USA has registered the use of cholecalciferol but not its degradation products like 1-alpha-hydroxycholecalciferol and vitamin D2. Australia and New Zealand have also registered this chemical. The rodenticide is not allowed for use throughout the EU countries.

Choline

See **Acetylcholinesterase**

Chromatography

See **Gas chromatography**

Chronic toxicity

The effect of a long-term exposure to a toxin, at low levels, is called chronic toxicity. Repeated exposure at low levels or accumulation of different chemicals in body tissues can have chronic and serious effects which may not be immediately evident. Persons routinely exposed to small amounts of pesticides while mixing, loading and applying them or by working in fields, post-pesticide application, are subject to chronic toxicity.

To test chronic toxicity, the concerned substance needs to be orally administered or inhaled for a specific duration of time.

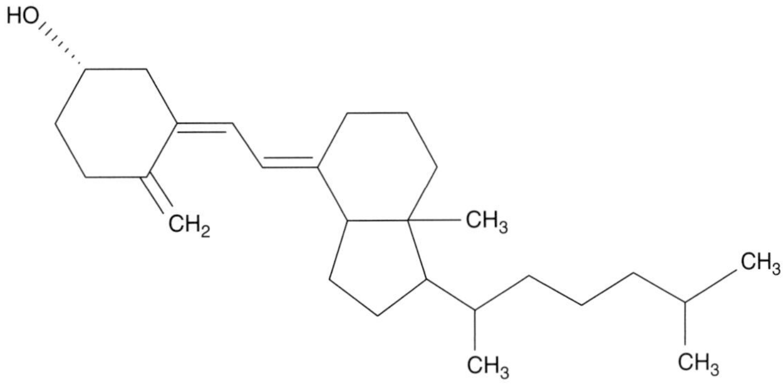

Cholecalciferol

For purposes of testing the toxic effects of pesticides, exposure is often recorded for half the life time of the population of organisms. However, there is no standard measure of chronic toxicity like LD_{50} or LC_{50}.

The effects of acute and chronic toxicity are often quite different. Chronic toxicity is often seen in organs such as the kidney or liver getting damaged, carcinogenic effects developing, mutagenesis and endocrine disruption. A chemical with high acute toxicity might not have high chronic toxicity, with an acute toxic effect being proportional to the dose amount. For example, if rats eat **demeton-*S*-methyl**, it quickly passes through the intestinal tract causing hardly any harm. However, if **daminozide** and **DDT** are fed every day in small amounts, rats gradually develop tumours (since these chemicals are carcinogens). Another example of chronic toxicity is the teratogenicity of **dichlorprop** (which is a reproductive toxin).

The risk of harm from pesticide exposure is equal to toxicity multiplied by the amount and route of exposure of the pesticide (Eqn C.3).

$$\text{Risk} = \text{Toxicity} \times \text{Exposure} \qquad \text{(C.3)}$$

Chrysopa as biocontrol agents

See **Green lacewings as biocontrol agents**

Chrysoperla as biocontrol agents

The lacewing *Chrysoperla carnea* (family Chrysopidae) is a predatory insect on many crop pests including aphids, thrips, **mealybugs**, whiteflies, bollworms, beetle larvae, spider mites, leaf miner nymphs and the eggs of **leafhoppers** and **scale insects**. Adult lacewings, 18 mm long, are yellowish green in colour and possess four translucent, membranous wings with many veins and cross veins. They are free-living and survive on nectar and the pollen of flowers. They can adjust to unfavourable environmental conditions. The adult female lays oblong green eggs on the lower side of the leaves, attached with a filament. The eggs, which hatch in 4 days, give pale-coloured larvae that have a reptilian appearance (Fig. C.1a). These predatory larvae captivate the host insect by their sucking jaws, inject poison into their body and suck out the body fluids. Then they pupate for 5 days, after which they are transformed into adult lacewings (Fig. C.1b).

Green lacewings are commercially available as eggs, young larvae, pupae and adults. Generally, populations resistant to pyrethroids, organophosphates and carbaryls are controlled biologically by these lacewings. *Chrysoperla* spp. serve as important predators of aphids on potatoes, pepper, tomato, sugarbeet, aubergine, maize, peas, cabbage, cotton and apples. (See also **Green lacewings as biocontrol agents**.)

Cinidon-ethyl herbicide

Cinidon-ethyl, $C_{19}H_{17}Cl_2NO_4$ (ethyl (*Z*)-2-chloro-3-[2-chloro-5-(cyclohex-1-ene-1,2-dicarboximido)phenyl] acrylate), belongs to the herbicide family **dicarboximide**. The CAS Registry Number is 142891-20-1. It is a white, crystalline, odourless powder, soluble in water (0.057 mg/l), acetone, methanol and toluene. It is rapidly hydrolysed (DT_{50} of 5 days) and photolysed.

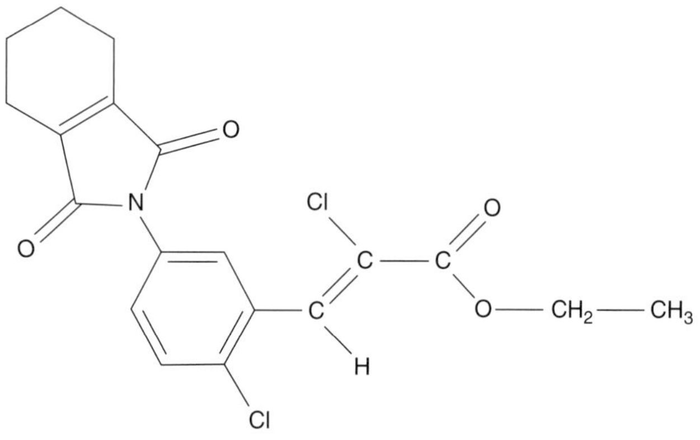

Cinidon-ethyl

The compound, a contact, selective, post-emergence herbicide, controls annual broadleaved weeds in cereal crops like wheat. It is a protoporphyrinogen oxidase **(PROTOX)** inhibitor. Cinidon-ethyl is available as emulsifiable concentrate. Prominent trade names include Lotus and Solar.

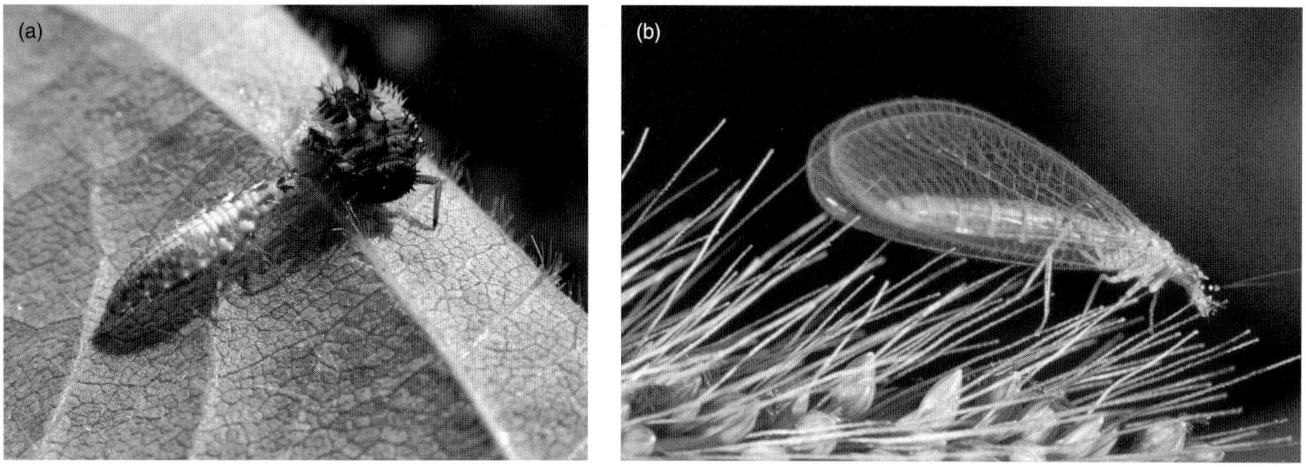

Fig. C.1. Common lacewing larva (a) and adult (b). Courtesy of John Obermeyer (a) and Brian Christine (b), Purdue Extension Entomology. With permission.

Cinidon-ethyl is not a major hazard when used as per label instructions. The dermal and acute oral LD_{50} in rats is >2000 mg/kg, while is not toxic to earthworms, fish and honey bees. The ADI is 0.01 mg/kg b.w./day.

Upon ingestion, cinidon-ethyl is rapidly absorbed and eliminated. It readily biodegrades in soil and water and leaves no traces of residue on grain. The DT_{50} in soil is less than 2 days, with the compound getting mineralized.

Cinidon-ethyl is registered for use in some EU countries such as Germany, France, the Netherlands, Hungary and the UK. It is not registered in the USA.

Cinmethylin herbicide

Cinmethylin, $C_{18}H_{26}O_2$ ((1*RS*, 2*SR*, 4*SR*)-1, 4-epoxy-*p*-menth-2-yl 2-methylbenzyl ether), is an unclassified herbicide. The CAS Registry Number is 87818-31-3. It is a dark orange liquid, stable between pH 3 and pH 11. It is soluble in water at 63 ppm (see Cinmethylin at bottom of the page).

Cinmethylin, a selective, pre-emergence, systemic herbicide, is absorbed by the roots and used to control annual grass weeds, mainly in rice. This herbicide is available as an emulsifiable concentrate (EC) and as granules (GR). Argold is one of the trade names containing cinmethylin.

According to the WHO, cinmethylin is unlikely to be hazardous in normal use (Class U toxin). The acute oral and dermal LD_{50} among rats is 4553 mg/kg and >2000 mg/kg, respectively. It is safe on birds, fish and *Daphnia magna*, when used as directed. The ADI is 0.3 mg/kg b.w./day. Product labels should display the signal word 'CAUTION'.

Cinmethylin is registered for use in India, Australia, Cambodia, Tanzania and Vietnam, but not in any of the EU countries.

Cinnamaldehyde biofungicide

Cinnamaldehyde (2E)-3-phenylprop-2-enal) is an organic plant-derived compound. It is also called cinnamic aldehyde, cassia aldehyde and *trans*-cinnamaldehyde. It is derived from the bark of the cinnamon tree (*Cinnamomum zeylanicum*; Fig. C.2), which owes its characteristic flavour to cinnamaldehyde. It is a yellow volatile oil that is obtained from distilling the leaves (Fig. C.3) and twigs of the cinnamon tree, with steam. Cinnamon is commonly used as a food additive and is a well-known spice (Fig. C.4). An extract of *Cassia tora* also contains similar plant components. The CAS Registry Number of cinnamaldehyde is 104-55-2.

Cinnamaldehyde acts as a fungicide, insect attractant and strong animal repellent (cats and dogs). As a fungicide, it disrupts

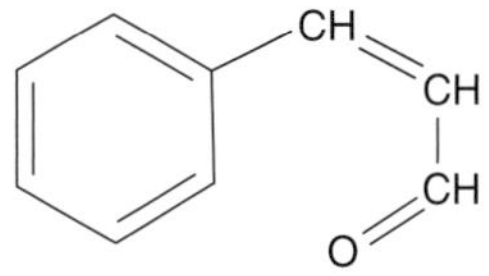

Cinnamaldehyde

the fungal membrane. It is used for the control of dry bubble (*Lecanicillium fungicola*), pitch canker disease and diseases caused by *Verticillium*, *Rhizoctonia* and *Pythium*. The crops that cinnamaldehyde covers are mushrooms, maize, horticultural crops, turf and pine forests. Additionally, it is also believed to kill mosquito larvae.

Cinnamaldehyde is available as a 30% wettable powder formulation. It is applied at the rate of 0.2–0.5%, when being

Fig. C.2. The cinnamon tree. Courtesy of Kalyani Paranjape.

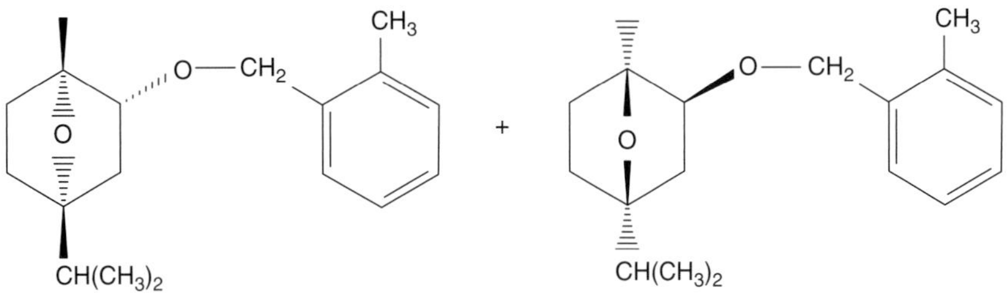

Cinmethylin

Fig. C.3. A closer view of the leaves of the cinnamon tree. Courtesy of Kalyani Paranjape.

Fig. C.4. Cinnamon sticks. Courtesy of Leena Achari.

used as a preventive. Generally cinnamaldehyde is not used with other crop protective agents. Trade names include Vertigo and Cinnacure A3005.

This biofungicide is not acutely toxic to mammals. Its acute oral LD_{50} in rats is 2.25–3.35 g/kg, while its acute dermal LD_{50} in rats is >1.2 g/kg. It may cause irritation to the eyes and skin.

Cinnamaldehyde is rapidly degraded in soil and does not cause harm to the environment or non-targeted animals. It is registered in the USA as a repellent, fungicide and insecticide.

Cinosulfuron herbicide

Cinosulfuron, $C_{15}H_{19}N_5O_7S$ (1-(4,6-dimethoxy-1,3,5-triazin-2-yl)-3-[2-(2-methoxyethoxy) phenylsulfonyl]urea), is a triazinylsulfonylurea herbicide. It is a colourless crystalline powder, soluble in water at 4000 mg/l (pH 6.7). The CAS Registry Number is 94593-91-6.

Cinosulfuron is a post-emergence herbicide that controls broadleaved weeds such as morning glory, cocklebur, pigweed and sunflower in soybeans and groundnuts. It acts by inhibiting branched-chain amino acid synthesis (i.e. is an **ALS inhibitor**), especially valine and isoleucine, in turn stopping cell division.

The herbicide cinosulfuron is available as granules, both water dispersible and dry. It is available in the market under the trade names of Classic and Darban.

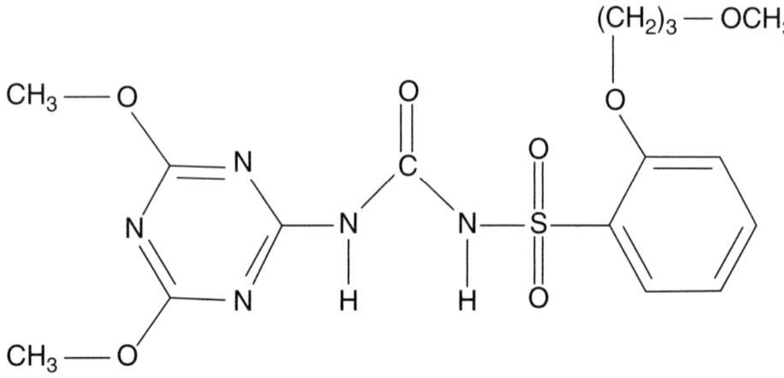

Cinosulfuron

The WHO has classified this herbicide as a Class II toxin (moderately hazardous). The acute oral LD_{50} in rats and birds is 4102 mg/kg and >2510 mg/kg, respectively. In normal use, it does not cause irritation to the eyes or skin. It is not toxic to bees. It has an ADI of 0.020 mg/kg b.w./day. Over-exposure to the compound may cause symptoms like vomiting, headache and breathlessness.

In animals, cinosulfuron is rapidly and extensively metabolized and produces up to 18 metabolites. The soil half-life is 20 days. The herbicide is considered mobile and highly leachable.

Cinosulfuron is registered for use in Vietnam.

Citral repellent

See **Citronella oil biorepellent**

Citric acid bioinsecticide

Citric acid, $C_6H_8O_7$ (3-carboxy-3-hydroxypentanedioic acid), is an insecticide derived from citrus fruits. The CAS Registry Number is 77-92-9.

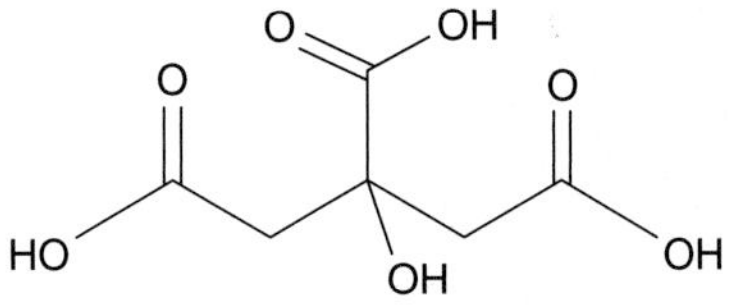

Citric acid

Citric acid is used against a variety of insects including ants, beetles, aphids, caterpillars, mites, whiteflies and leafhoppers. Citric acid is applied on ornamental plants, vegetables, shrubs and fruit trees. It may be applied, mostly on the foliage, based on individual judgement, at the sight of the pest, as often as needed. It is available as an aqueous concentrate solution. Spraying in the sun should be avoided. A popular trade name Sharp Shooter may be used as often as required even up to the day of harvest, and at the first sight of insects. This agent may also be combined with some other insecticides and with **mint oil**. The combination of citric acid and mint oil is an effective fungicide and bactericide. Fungi like *Phytophthora* spp., **downy mildews**, *Sclerotinia* spp. and *Botrytis* spp., and bacterial pathogens such as *Pseudomonas*, are effectively controlled. The oil and acid combination inhibits respiration. A known trade name is Fungastop.

The use of citric acid is safe for crops, non-target organisms, mammals and the environment. It may also be used in organic farming as it is approved by the US OMRI, provided it is produced by microbial fermentation of carbohydrate substrates and not derived from genetically modified microorganisms.

The USA, Canada and Australia have registered the use of citric acid, but calcium citrate and sodium citrate are not registered by the EPA. However, it is a PAN Bad Actor chemical since it is rated acute in toxicity studies.

Citronella oil biorepellent

Citronella (*Cymbopogon nardus* and *Cymbopogon winterianus*) belongs to a group of grassy trees and plants in the family Cardiopteridaceae that grow in the tropics (Fig. C.5). The plant grows well in regions of bright sunlight, plenty of water and well-drained soil. The oil of citronella, an **essential oil**, extracted by distillation, is a plant-derived insect repellent. The CAS Registry Number is 8000-29-1. Citronella oil is used as an insect and animal repellent and an anti-feedant.

Citronella oil is a derivative of citral (CAS Registry Number of 5392-40-5). Citral is said to be slightly toxic and is a pheromone that works best as an insect and animal repellent.

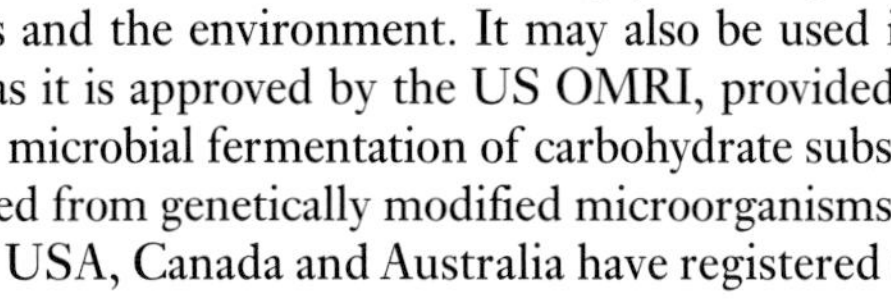

Citronella

Citral is not a PAN Bad Actor chemical but is not registered in the USA.

The odour of citronella oil makes it difficult for insects to locate their host and thus repels them. Target pests include mosquitoes, gnats, flies and fleas. The plant extract also contains components such as alpha-citronellal and citronellol.

This oil is not commonly used with other chemical insecticides. It is available as a candle, spray and as a skin lotion. Bio Bug Insect Repellent and Green Ban are two common trade names. It is an ingredient in many repellent formulations.

Citronella is naturally present in the environment and does not show any harmful effects on mammals or on non-target organisms. It is registered in the USA, the UK, Australia and South Africa.

Citrus canker

See **Crop – orange orchard pest control**

Class Arachnida

Class Arachnida is a class of invertebrate organisms that are commonly eight legged and wingless. They normally breathe through gill-like structures. Their bodies are divided into two main parts, one bearing the head and the thorax and the other, the abdomen. In addition, they have two pairs of appendages which are used for feeding and sensing. The members, called arachnids, are generally aggressive, predatory creatures preying on small arthropods. This class belongs to the phylum Arthropoda and sub-phylum Chelicerata. This class includes spiders, mites and ticks.

The order Araneae includes 20,000 species of spiders. These abundantly numbered insects range in length from less than 0.76 mm to over 12 cm. The mouthparts are designed to grasp food and these distinguish spiders from other arthropods' mandibles. These mouthparts are called chelicerae. Sometimes, these parts secrete poisonous juices that they inject into their prey. The abdomen has silk glands.

Mites and ticks constitute the order Acarina. There are 10,000 species of diverse tiny members of mites and ticks. The division of the body in mites and ticks is very unclear with the mouthparts fused with rest of the body and separated by a flexible

Fig. C.5. Citronella. (a) Leaves. (b) Whole plant. Photos courtesy of Kalyani Paranjape.

cuticle. Most lack sharp eyesight. Their mouthparts are adapted to biting, sucking and chewing. Many parasitic and free-living species inhabit leaf litter. Larval stages normally have three pairs of legs, while nymphal and adult stages have four pairs.

Class Gastropoda

The class Gastropoda belongs to phylum Mollusca and consists of snails and slugs. Snails are small, muscular creatures that have a hard protective shell from above. Slugs are usually grey, worm-like and without an external spiral shell. Snails and slugs move over the surface on a muscular foot that secretes mucous, which dries up into a glossy trail. These pests are commonly hermaphrodites and can lay eggs up to six times a year. Snails and slugs are voracious eaters and are active at night and on humid days. Their diet includes live plants (young bark, fruits and leaves) and decaying organic matter. Slugs prefer temperate humid climates to warmer and dry areas. Snails survive in the dry season by covering their shell opening with a membrane. Some species of snails have both a gill and a lung, allowing them to be amphibious.

Some species of snails and slugs are considered crop and garden pests. The golden apple snail (*Pomacea canaliculata*; Fig. C.6) is a specific example. Others include *Helix aspersa* and *Tympanotonus fuscatus*, *Milax gagates* and *Agriolimax reticulatus*.

Fig. C.6. *Pomacea canaliculata* laying eggs. Photo courtesy of Katasha Cornwell, Withlacoochee River, Florida, USA. With permission.

Besides causing direct damage to crops, some mollusc species are host to pathogenic organisms, which infect animals and humans (e.g. *Angiostrongylus cantonensis*, *Catadiscus pomacea*, *Fasciolopsis buski* and *Dicrocoelium dendriticum*). Molluscicides inhibit or kill molluscs.

Mollusc pests prefer succulent vegetation and fruits, and so for example attack strawberries, artichokes, tomatoes, lettuce, marigolds and direct-seeded rice.

Snails and slugs have many natural enemies including ducks, geese, beetles, snakes and toads and the predator land snail, the decollate snail (*Rumina decollata*).

Chemicals such as **copper sulfate**, **calcium cyanamide**, **organotin** compounds, **metaldehyde**, **methiocarb**, iron phosphate, **endosulfan** and **niclosamide** are used to control molluscs. Extracts of *Derris elliptica*, *Azadirachta indica*, vulgarone B (from *Artemisia douglasiana*), camellia seed cake and tobacco dust from **nicotine** ($C_{10}H_{14}N_2$) have also proved to be effective molluscicides. A combination of natural and chemical agents is also used, along with hand picking and trapping of these pests (e.g. a beer can placed slightly below ground level is an effective trap as snails and slugs can fall into it).

Classical (or inoculative or importation) method of biological weed control

See **Biological weed control agents – application methods**

Classification of herbicides – a general view

Although chemical herbicides are commonly classified by chemical group or mode of action, some other criteria are also used.

Classification can also be done based on the crop to be protected. For example, **acetochlor** (chloroacetalinide group), **paraquat** (foliar contact and systemic) and **trifluralin** (dinitroanilines-toludines group) can be classified under one group, namely one that ensures soybean crop protection. Such a grouping provides farmers with essential practical knowledge.

Herbicides are also classified based on observed effects, but this may be misleading because visual effects on the weed (e.g. burning, growth anomalies, **chlorosis** and necrosis) may not always reflect the actual effects on growth.

Classification by site of uptake has practical advantages for the farming sector. Leaves and roots are the means of uptake. Foliar-active groups are generally contact herbicides and include **diphenyl ether** herbicides, phenoxyacid (with easy translocation), **bipyridilium** and arsenical herbicides (also with easy translocation) and selective oils. Soil-active herbicide groups include **triazines**, **sulfonylureas**, **imidazolinones**, **carbamothioates**, **dinitroanilines** and **chloroacetamides**. Here the uptake is via roots. Some formulations are designed to be soil and foliar active.

A herbicide may also be broadly classified on the basis of its immediate activity – contact or systemic. If a herbicide is translocated in the plant, it is said to have systemic activity. This contrasts with contact herbicides which have an external activity.

Selectivity is another broad basis of classifying herbicides. Selectivity depends, for example, on plant age, growth stage, absorption and translocation. It is used as a rule of thumb for administering herbicides and depends on concentration and rate of application.

There is yet another classification based on time of application, namely, for example, pre-planting, pre-emergence or post-emergence. This has practical significance but is too broad a division to justify a formal classification.

Clethodim herbicide

Clethodim, $C_{17}H_{26}ClNO_3S$ ((5*RS*)-2-{(1*EZ*)-1-[(2*E*)-3-chloroallyloxyimino]propyl}-5-[(2*RS*)-2-(ethylthio)propyl]-3-hydroxycyclohex-2-en-1-one), is a cyclohexene oxime compound. It is an amber-coloured, viscous liquid, soluble in water at 6630 mg/kg but poorly soluble in most organic solvents. It is

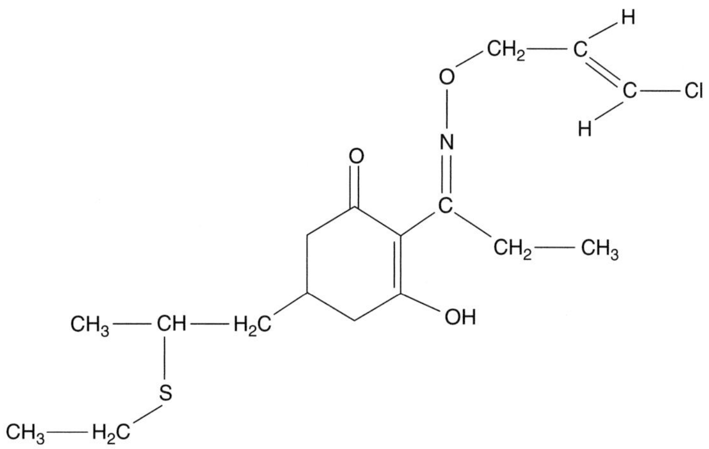

Clethodim

unstable to UV exposure. Clethodim is produced by reacting ethylmercaptan with crotonal, followed by treatment with acetone, dimethylmalonate, hydrolysis reagents, propionic acid chloride and α-chloro-hydroxyallylamine. It has a CAS Registry Number of 99129-21-2 (see Clethodim at top of the page).

Clethodim is a systemic, post-emergence herbicide, rapidly absorbed and translocated from foliage to the growing parts of the plant. It acts as a fatty acid synthesis inhibitor. It is used for the control of annual and perennial grasses (e.g. signalgrass and **barnyard grass**) in a wide range of broadleaved crops including lucerne, flax, soybean, cotton, groundnuts, sunflower and vegetables.

Clethodim is available in the market as Select and Cletodime, among others. Clethodim is formulated as an emulsifiable concentrate (EC). It may be combined with 2,4-DB, among other chemicals.

According to the WHO and EPA, clethodim is a slightly hazardous chemical (Class III toxin). The oral LD_{50} values in male and female rats are 1630 mg/kg and 1360 mg/kg, respectively. The technical product does not harm the skin, but the formulations cause moderate irritation to the eyes and the skin. Thus, labels must carry the signal word 'WARNING'. The chemical is practically non-toxic to birds and honey bees and does not harm non-target species. It is slightly toxic to fish and aquatic organisms. It is also a possible groundwater contaminant.

In animals and plants, clethodim is metabolized to its sulfoxide (mainly by photolysis in plants). In soils, clethodim does not persist for long but is slightly mobile. Degradation takes place by aerobic processes and photolysis, with the DT_{50} being 3 days. The chemical breaks down into sulfoxide, sulfone and oxazole sulfone. But in an aquatic environment clethodim is highly persistent, with a DT_{50} of at least 128 days (much more at higher pH).

FSANZ's Joint Food Standards Code 1.4.2 to v116 (20 May 2010) has fixed the MRL of clethodim for onions and garlic as 0.30 mg/kg, while for lettuce, wheat grain and peas it is 0.1 mg/kg. The EPA (40CFR180 (e-CFR)) (the electronic code for Federal Regulation, with 40 referring to protection of the environment) has fixed the MRL for onions and garlic as 0.20 mg/kg.

Clethodim is registered in the USA, Australia, New Zealand, Canada, Vietnam, Uganda and some countries of the EU such as Germany, Poland, France and Italy, but not the UK.

Clodinafop herbicide

See **Clodinafop-propargyl herbicide**

Clodinafop-propargyl herbicide

Clodinafop-propargyl, $C_{17}H_{13}ClFNO_4$ (prop-2-ynyl (*R*)-2-[4-(5-chloro-3-fluoro-2-pyridyloxy) phenoxy] propionate), belongs to the herbicide family, aryloxyphenoxy propionates. It is a cream-coloured powder that is almost insoluble in water (4 ppm). It stores well in its original container in a cool, dry, well ventilated room. The compound has a CAS Registry Number of 105512-06-9 (see Clodinafop-propargyl at bottom of the page).

This herbicide is a systemic, post-emergence herbicide and also a plant growth regulator that controls grass weeds like foxtail, **barnyard grass** and volunteer crops (it is ineffective on broadleaved weeds). It is rapidly absorbed and translocated by the foliage. It interferes with the production of fatty acids in the targeted weeds.

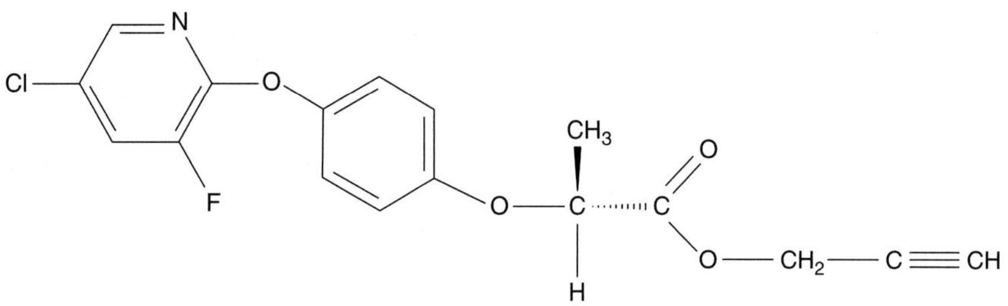

Clodinafop-propargyl

Common formulations of clodinafop-propargyl include liquid and emulsifiable concentrate (EC). An early application kills competition from weeds and thus improves yields. A dose of 56–70 g/ha of clodinafop-propargyl is recommended on spring wheat.

The herbicide clodinafop-propargyl is moderately toxic to mammals and birds (the oral LD_{50} among rats and quails is 1828 and >1455 mg/kg, respectively). Similarly, the herbicide is toxic to fish, with the LC_{50} in trout being 0.39 mg/l. However, it is practically non-toxic to honey bees. According to Cancer Assessment Review Committee (CARC), clodinafop-propargyl is classified as 'likely to be carcinogenic to humans'.

This herbicide breaks down rapidly in soil. It is mobile in soil. It converts from the ester form to the active acid and then to biologically inactive compounds.

Clodinafop-propargyl is registered by India, the USA, Australia, Canada, Germany, France, Italy, Denmark, the UK and some other countries.

This chemical is derived from clodinafop. With a CAS Registry Number of 114420-56-3 and molecular formula $C_{14}H_{11}ClFNO_4$, this compound is an aryloxyphenoxypropionic herbicide. It is not a PAN Bad Actor chemical and is registered in many countries of the EU including Denmark, Germany, France, Italy and the UK.

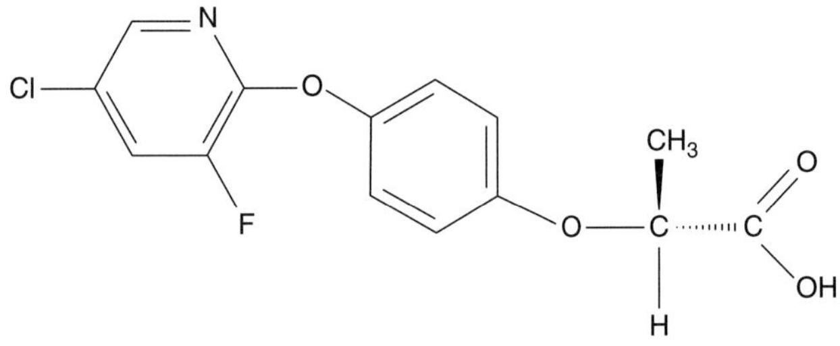

Clodinafop

Cloethocarb insecticide

Cloethocarb, $C_{11}H_{14}ClNO_4$ (2-[(*RS*)-2-chloro-1-methoxyethoxy] phenyl methylcarbamate), is a non-hygroscopic, colourless, crystalline substance that is an insecticide (of the phenyl methylcarbamate class), molluscicide and a carbamate nematicide. Its solubility in acetone and chloroform is over 1 kg/kg at 20°C, whereas in water it is 1.3 g/kg. Cloethocarb is hydrolysed by strong acidic and alkaline material. It has a CAS Registry Number of 51487-69-5.

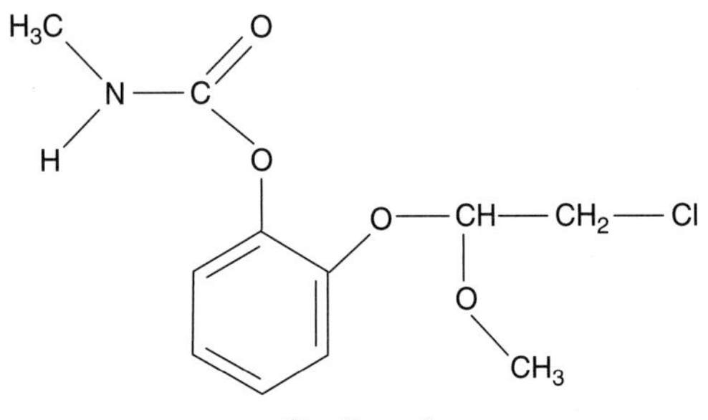

Cloethocarb

Cloethocarb is a cholinesterase inhibitor. Its systemic nature works well on soil pathogens and thus controls seed diseases. Potatoes, sugarbeet, tobacco, maize, rice, wheat and sorghum are protected from rootworms, caterpillars, aphids and nematodes (e.g. *Heterodera schachtii*).

Granules and pellets are common formulations, with a known trade name being Lance.

Cloethocarb is toxic to both fish and bees and is classified as a Class III toxin (slightly hazardous) by the EPA. The acute LD_{50} in rats is 3.4 mg/kg. It has a residual activity of up to 7 weeks in soil and water and it is non-phytotoxic when used as instructed. It is declared a PAN Bad Actor chemical and is considered to be obsolete by the WHO. The USA has not registered this chemical.

Clomazone herbicide

Clomazone, $C_{12}H_{14}ClNO_2$ (2-(2-chlorophenyl) methyl-4,4-dimethyl-3-isoxazolidinone), is a herbicide. It is a viscous liquid above room temperature and a white crystalline solid when cooled. It is soluble in water (1100 ppm), acetone, chloroform, cyclohexanone, hexane and methanol. It is relatively stable to UV light. The CAS Registry Number is 81777-89-1.

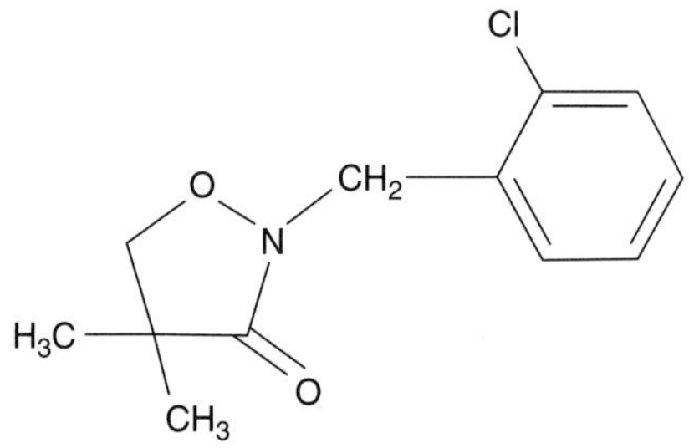

Clomazone

Clomazone is a broad-spectrum, pre-plant, pre-emergence herbicide, used for controlling annual grasses and broadleaved weeds in cotton, soybeans, tobacco and many vegetables. It inhibits production of chlorophyll and **carotenoids** in plants, thus cutting off nutrition.

The herbicide is formulated as emulsifiable concentrate or capsule suspensions, appropriately diluted. It is sprayed on the foliage. It is available as trade names Command, Commence and Gamit. Since the chemical is volatile, care has to be taken to prevent drift or vapour affecting the surrounding areas.

The herbicide is slightly toxic via the oral, inhalation and dermal routes. The oral LD_{50} values among rats, rabbits and birds range from 1369 mg/kg to 2510 mg/kg, while the inhalation LC_{50} among rats is between 4.23 and 6.52 mg/l. The 96 h LC_{50} values among fish and *Daphnia magna* are between 19 ppm and 40.6 mg/l. Products containing clomazone must bear the signal word 'CAUTION' on the label. The ADI for clomazone is 0.133 mg/kg/day. Exposure to the herbicide causes irritation to the skin and eyes. High volatility makes clomazone phytotoxic to roses, small grain crops and vegetables. Handlers are advised to equip themselves with face shields and goggles. The WHO classifies clomazone as moderately toxic (Class II).

Though highly soluble in water, clomazone tends to adsorb to soil. Hence it has a low to moderate potential to contaminate groundwater. The herbicide is persistent, with soil DT_{50} ranging from 28 to 84 days. But degradation is hastened by warm temperatures, water, microbial action and sandy loams.

According to the EU Pesticide Database, the MRL in potatoes and rapeseed is 0.01 mg/kg. It is registered in India, Australia, the Philippines, New Zealand, the UK, the USA and Canada.

Clomeprop herbicide

Clomeprop, $C_{16}H_{15}Cl_2NO_2$ [(*RS*)-2-(2,4-dichloro-m-tolyloxy) propionanilide], is a phenoxy-carboxylic acid herbicide. It is soluble in water (0.032 mg/l) and in organic solvents. The CAS Registry Number is 84496-56-0.

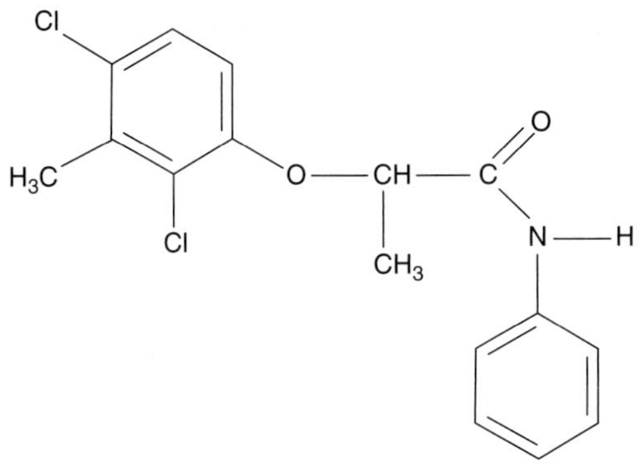

Clomeprop

Clomeprop, a rice herbicide, is a selective, post-emergence compound, used to control broadleaved weeds and sedges. It mimics the growth promoter, indole acetic acid.

Formulated as an emulsifiable concentrate and as granules, clomeprop is classified as a Class U herbicide (unlikely to be hazardous) with an acute oral and dermal LD_{50} among rats being >5000 mg/kg and the 48 h LC_{50} in carp being >10 mg/l.

Cloprop as a PGR

Cloprop, $C_9H_9ClO_3$ ((*RS*)-2-(3-chlorophenoxy) propionic acid), is a plant growth regulator. It is also a herbicide (phenoxypropionic class). Cloprop, a colourless crystalline substance, is soluble in water, acetone, dimethyl sulfoxide, ethanol, methanol, iso-octanol, benzene, toluene, diethylene glycol, dimethyl formamide and dioxane, and is highly stable. It is produced by the reaction of sodium m-chlorophenolate in absolute ethanol with ethyl bromopropionate, followed by hydrolysis of the ester of the corresponding acid. The CAS Registry Number of cloprop is 101-10-0.

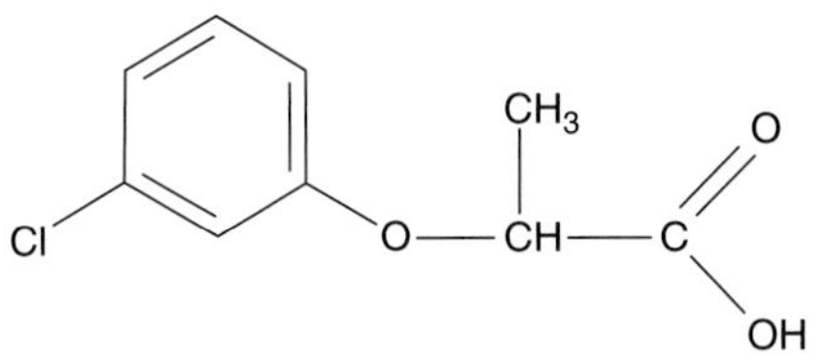

Cloprop

As a growth regulator, cloprop is absorbed by the foliage but is not readily translocated. It is used on fruits like peaches, pineapples (reduces crown growth) and plums (fruit thinning). It also delays fruit ripening and increases the size of fruit.

Cloprop is formulated as a soluble concentrate (SL). It is available in the market under trade names like Fruitone, Swellpine and Fruitone CPA. A common application dose is 80 g/l.

According to the EPA, cloprop is a slightly toxic chemical.

Clopyralid herbicide

Clopyralid, $C_6H_3Cl_2NO_2$ (3,6-dichloropyridine-2-carboxylic acid), is a pyridine carboxylic acid herbicide. It exists as colourless crystals, soluble in water and organic solvents. It is stable to light and acidic medium. Its CAS Registry Number is 1702-17-6. It is produced by photochlorination of picoline, followed by oxidation.

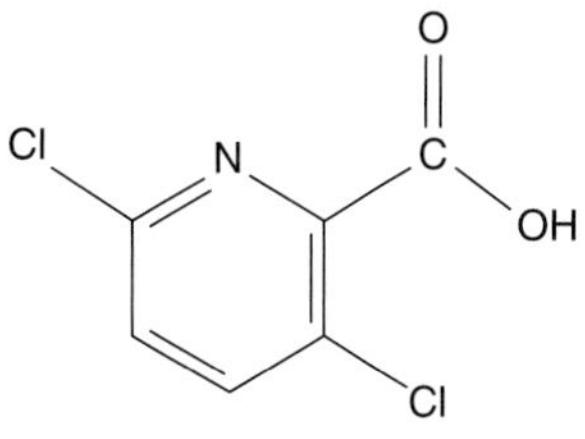

Clopyralid

The herbicide, a selective, post-emergence, systemic compound, is translocated both acropetally and basipetally. It mimics the action of auxins, like indole acetic acid, and acts on cell elongation and respiration. The chemical controls annual and perennial broadleaved weeds in for example maize, sugarbeet, sorghum and turf. Thistles and clovers are some target weeds.

Clopyralid is available as a soluble concentrate (SL). Trade names include Lontrel and Stinger. It is compatible with **2,4-D**, **MCPA**, benazolin, **dichlorprop**, **fluroxypyr** and **triclopyr**.

According to the WHO and EPA, clopyralid is unlikely to present acute hazard in normal use. It causes irritation to the skin and eyes. It is also observed to be harmful to beneficial insects and non-target plants. Potato is very sensitive to the use of clopyralid. Clopyralid is a potential groundwater contaminant and hence is a PAN Bad Actor chemical.

This chemical is persistent in soil for as long as 14 months. Microbial action via decarboxylation takes place in the soil. In tolerant plants, clopyralid gets metabolized and inactivated. Among mammals, it gets quickly eliminated via the urine, with the acute oral LD_{50} in male rats being 3738 mg/kg.

The USA, Australia, Denmark, the UK and many other countries have registered the use of clopyralid. The copper, potassium or isopropanolamine salts, which are related to clopyralid, are not PAN Bad Actor chemicals, although some of them may be classified as restricted use pesticides (RUP) by the EPA or are not registered in some countries (e.g. the isopropanolamine salt is not registered in the USA).

Cloquintocet herbicide

See **Cloquintocet-mexyl herbicide safener**

Cloquintocet-mexyl herbicide safener

Cloquintocet-mexyl, $C_{18}H_{22}ClNO_3$ ((*RS*)-1-methylhexyl (5-chloroquinolin-8-yloxy) acetate), is a herbicide safener. It is a colourless crystalline substance with solubility in water of 0.59 mg/l (at 25°C). It is soluble in ethanol, acetone, toluene, *n*-hexane and *n*-octanol. It is hydrolysed in alkaline media. The CAS Registry Number of this compound is 99607-70-2.

As a safener, the compound protects small cereal crops against the toxicity of **clodinafop-propargyl**, so as to enable the herbicide to control annual grasses such as *Avena* spp. and *Setaria* spp.

This safener is available as an emulsifiable concentrate (EC) or wettable powder (WP). It is compatible with herbicides such as clodinafop-propargyl, **pinoxaden** and propargyl. It is sprayed with the help of **wetting agents** and surfactants like lignin sulfonates. Trade names include Topik and Rescue.

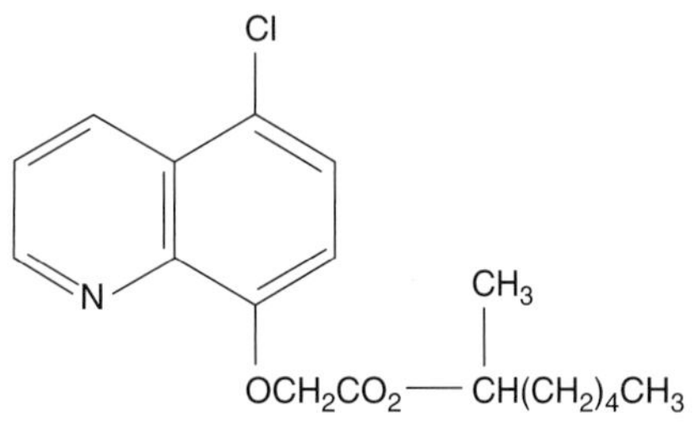

Cloquintocet-mexyl

The WHO has categorized cloquintocet-mexyl as a Class III toxin (slightly hazardous). It has an oral LD_{50} among rats and birds of >2000 mg/kg and a 4 h inhalation LC_{50} among rats of >0.935 mg/l air. It is non-toxic to honey bees. An ADI is 0.04 mg/kg b.w./day.

Cloquintocet-mexyl is degraded to the free acid in plants. In soils, it adsorbs to soil particles. The herbicide safener is registered in Australia and Canada and is known to be used in Hungary, Portugal, France and Germany.

The herbicide safener is a derivative of **cloquintocet**, which is also a herbicide safener. It has the molecular formula $C_{11}H_8ClNO_3$, an IUPAC name of (5-chloroquinolin-8-yloxy) acetic acid and a CAS Registry Number of 88349-88-6.

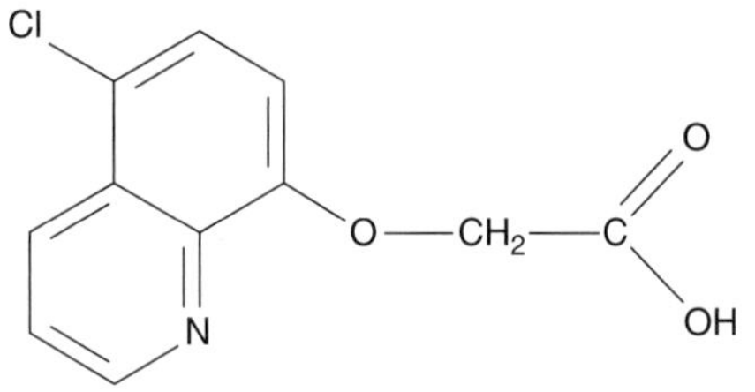

Cloquintocet

Cloransulam

See **Cloransulam-methyl herbicide**

Cloransulam-methyl herbicide

Cloransulam-methyl, $C_{15}H_{13}ClFN_5O_5S$ (methyl 3-chloro-2-(5-ethoxy-7-fluoro[1,2,4]triazolo[1,5-*c*]pyrimidin-2-ylsulfonamido) benzoate), is a **triazolopyrimidine** herbicide. It is an off-white powder, soluble in water at 16 ppm, and also in acetone, acetonitrile and ethylacetate. It remains stable at room temperature. The CAS Registry Number is 147150-35-4. Cloransulam-methyl is a derivative of cloransulam, $C_{14}H_{11}ClFN_5O_5S$ (3-chloro-2-(5-ethoxy-7-fluoro[1,2,4]triazolo[1,5-c]pyrimidin-2-ylsulfonamido) benzoic acid or 3-chloro-N-(5-ethoxy-7-fluoro[1,2,4]triazolo[1,5-c]pyrimidin-2-ylsulfonyl) anthranilic acid). The CAS Registry Number is 159518-97-5. Cloransulam is a triazolopyrimidine or sulfonanilide herbicide (see Cloransulam-methyl and Cloransulam at top of the next column).

Cloransulam-methyl is a pre- and post-emergence herbicide applied to control broadleaved weeds in soybean crops.

It is usually formulated as water-dispersible granules (WG) and is applied on or into the soil. A known trade name is FirstRate. Ground application doses are in the range of 42–52 g/ha.

The WHO classifies cloransulam-methyl as a Class U toxin (unlikely to be hazardous). The LD_{50} values among rats (acute oral), rabbits (dermal) and bobwhite quails (oral) are

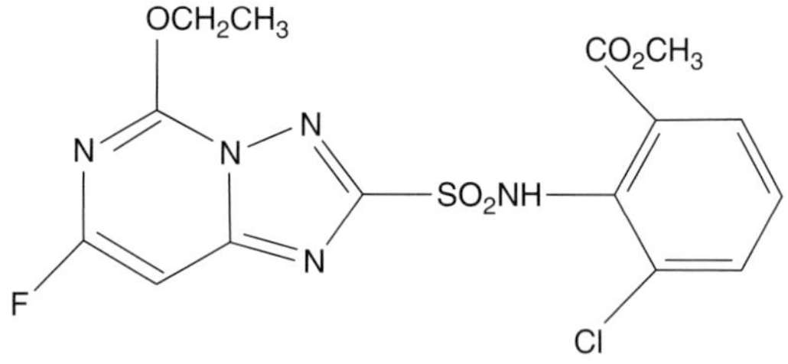

Cloransulam-methyl

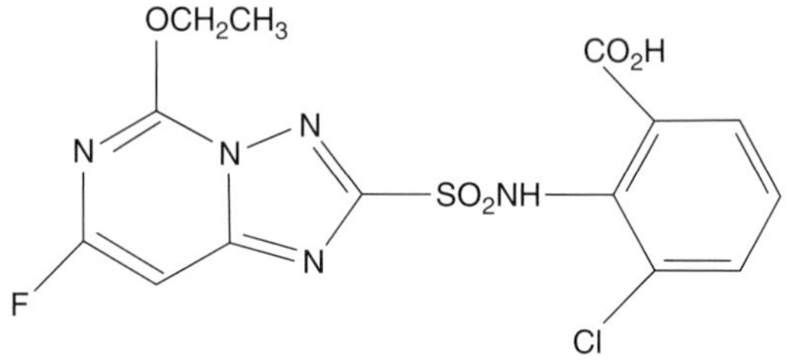

Cloransulam

>5000 mg/kg, >2000 mg/kg and >2250 mg/kg b.w. It is practically non-toxic to fish species and honey bees. The RfD for cloransulam-methyl is 0.1 mg/kg/day. In cases of excessive exposure, breathlessness and skin and eye irritation can occur. Moving the patient outdoors to the open, flushing the eyes and drinking water to dilute the toxin can help.

Cloransulam-methyl is registered in the USA and Canada.

Clothianidin insecticide

Clothianidin, $C_6H_8ClN_5O_2S$ (*E*)-1-(2-chloro-1,3-thiazol-5-ylmethyl)-3-methyl-2-nitroguanidine), is a neonicotinoid insecticide (nitroguanidine subgroup of nicotinoid). It is soluble in water at 0.3 g/l and is stable to hydrolysis. Its CAS Registry Number is 210880-92-5.

Clothianidin is a major metabolite of a neonicotinoid precursor, **thiamethoxam**. Clothianidin, a systemic insecticide, is used in the seed treatment of maize and canola and is recommended only in commercial slurry treatment. It acts by antagonizing the neurotransmitter, acetylcholine. Clothianidin is used along with dinotefuran to control black vine weevils on ornamentals.

The rate of application is up to a 100 g/ha, while for seed treatment the rate of application is 150–450 g a.i. per 100 kg of seed. The widely used brand of clothianidin formulation is Poncho 600.

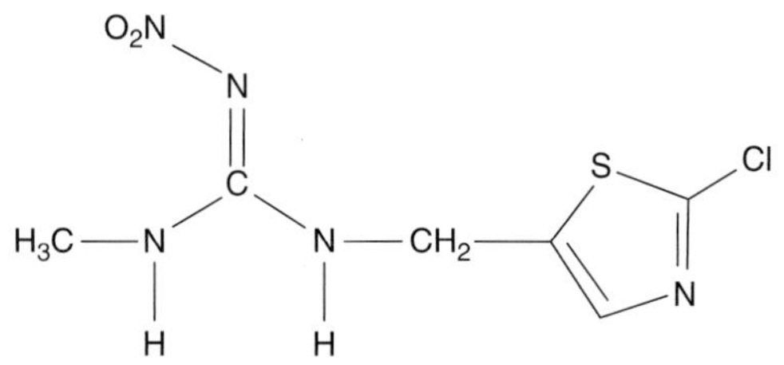

Clothianidin

This insecticide is not considered to be a threat to the environment and non-target creatures if used as per label instructions.

Clothianidin is non-toxic to most freshwater fish species and practically non-toxic to birds (LD_{50} in bobwhite quails being >2000 mg/kg). It is very highly toxic to the midge species *Chironomus riparius*. To rats, mice, rabbits and guinea pigs, the lowest acute oral LD_{50} is 425 mg/kg in mice. But clothianidin is toxic to endangered birds and small mammals via the ingestion of treated seeds (possibly due to contaminated pollen and nectar). It has been observed to be highly toxic to bees (LD_{50} of >0.044 μg per bee). Clothianidin and **imidacloprid** are said to have contributed to mass die-offs and colony collapse disorder in bees.

Clothianidin is also said to be mobile (to very mobile) in soil, threatening contamination of groundwater and thus affecting drinking water. It also persists in soil. Clothianidin is thus recommended under very strict guidelines. The most important one is that the compound is not to be used in on-farm seed treatment applicators (e.g. slurry boxes). All treated seeds have to be coloured in distinctive colours so as to avoid use as food or feed. Treated seeds have to be covered to avoid being a hazard to birds (with all spilled seeds being buried, away from water bodies). Farms planted with such seeds should not be sown with any other crop (except for non-grass animal feeds, soybeans and some others) for at least a year. Applicators are recommended to use protective gear (e.g. gloves and long-covering garments).

Clothianidin is registered for use in India, Australia, the USA, Canada, New Zealand, the Philippines and many EU members including the UK, Germany, Finland and Denmark.

Cloxyfonac as a PGR

Cloxyfonac, $C_9H_9ClO_4$ (4-chloro-α-hydroxy-*O*-tolyloxyacetic acid), is a plant growth regulator. It is a colourless crystalline substance, soluble in water, acetone, dioxane, ethanol and methanol but insoluble in benzene and chloroform. Cloxyfonac has a CAS Registry Number of 6386-63-6.

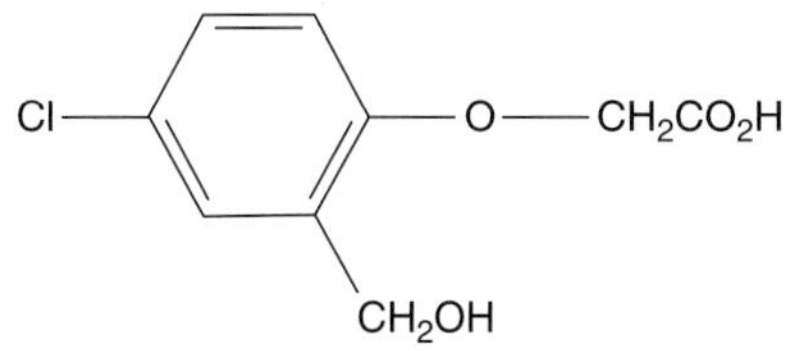

Cloxyfonac

Cloxyfonac is used to encourage the setting of fruit in tomatoes and aubergines, so that the fruit is of a uniform size. The PGR is formulated as soluble concentrate (SL). It is available in the market under the trade name of Tomatlane, among others.

The WHO has categorized cloxyfonac as a Class U toxin (unlikely to be hazardous). It is not a PAN Bad Actor chemical.

Coccinellid beetle

See ***Adalia bipunctata*** **(ladybug) as a biocontrol agent**

Coenzymes

See **Enzymes**

Cofactors

See **Enzymes**

Colonization potential

The ability of microorganisms to inhabit and colonize the rhizosphere-rhizoplane environment is known as **colonization potential** or **rhizosphere competence**. (See **Rhizosphere**.)

Common lambsquarters

See **Weed – common lambsquarters**

Common viral plant diseases

Mosaic, wilt, blight and mottle are some of the common viral infections found in plants.

Mosaic disease is a very common viral disease, affecting most plant families. A mottled appearance, stunted and distorted growth, discoloration and upward rolling of leaves are the symptoms. The taste of the fruit turns bitter. Examples of mosaic diseases are sugarcane mosaic, rose mosaic, wheat streak mosaic and cowpea mosaic.

There is no chemical available to control the spread of this disease. But since it is transmitted by **vectors** such as aphids or beetles, an effective insecticide can keep the pathogen in control.

Wilt (virus) disease gets its name because it causes the plant to lose rigidity and the ability to transpire. The symptoms include downward curling of leaves, stunted growth, spots on leaves and fruits, and long streaks on stems. An example is the tomato spotted wilt. Wilt disease virus spreads via the insect **thrip**. Controlling infestation by thrips can help control the spread of wilt.

Blight (virus) is a disease in which there is **chlorosis** and browning followed by the death of leaves or the stem. It is caused by pathogenic organisms such as viruses and fungi. An example of blight caused by a virus is head blight of sorghum.

Mottling is a type of disease in which uneven yellow spots occur on plant parts. Examples include rusty mottle on cherry and chlorotic mottle in cowpea.

Some symptoms of viral diseases like mottling and blight also occur due to malnutrition, posing a problem for accurate diagnosis. Many other symptoms are also common among infections caused by fungal or bacterial pathogens, or due to insect infestation. These include necrosis, chlorosis, hyperplasia and hypoplasia. Hypoplasia is a condition where new plant cells are below normal in size. This affects the development of the tissues and results in stunted growth of the plant. Hyperplasia is the reverse of hypoplasia. Here the rate of cell production abnormally increases and cell division becomes excessive. This leads to abnormal growth of plant tissues and the production of tumours or swollen parts on plants.

Compound 1080 rodenticide

Sodium fluoroacetate or sodium monofluoroacetate is popularly known as Compound 1080. This compound is a rodenticide. (See **Sodium fluoroacetate rodenticide**.)

Compressed-air sprayer

See **Application equipment – hand sprayers**

Conjugation

Conjugation is a process in which a pesticide or a **metabolite** is linked to a natural product of the metabolizing organism, such as sugars and amino acids. This process occurs in higher organisms such as plants, birds and mammals, and their products are known as conjugates. Conjugates are generally more water soluble than the parent product and are therefore more readily excreted. This metabolic process is important in determining the biological activity of a molecule.

The presence of two or more double bonds that alternate with single bonds is an unsaturated compound, and is also known as conjugation. An example is 1,3-butadiene (CH_2=CH–CH=CH_2) or maleic acid.

Contact angle of a droplet

See **Spray application concepts**

Contact pesticides

Any compound that causes death or injury to an organism when it comes into contact with it is called a contact pesticide. It does not have to be ingested. The important feature of a contact pesticide is that it must penetrate the outer covering or the external body surface of the pest.

For instance, such a compound can damage exposed plant tissue and leaf tissues and hence inhibit photosynthesis and respiratory metabolism (as in the case of **paraquat** and **bipyridilium** herbicides), or in the case of insects enter the body through the integument and wax coat or enter the respiratory system through the spiracles (as in the case of **tetramethrin, chlorpyrifos, thiamethoxam** and **methamidophos**). In the case of fungal pathogens, many contact fungicides inhibit respiration, thus killing or inhibiting them. Contact fungicides may inhibit multiple sites (e.g. **captan, mancozeb**).

The contact pesticidal compound, also known as non-systemic pesticide, forms a layer over the plant, protecting it from attack. A spray application is common among contact pesticides.

However, contact pesticides are often non-selective and can harm other species of wildlife and even crops.

Some systemic pesticidal compounds have some contact action too.

Container dusters

See **Dusting machinery**

Contaminant

A contaminant is an undesirable foreign material present in a product. Contaminants, when present in air, water or food products, make the medium toxic or otherwise harmful. Sulfur dioxide resulting from the combustion of high sulfur fuels, pesticide residue in vegetables, fish and other foods and industrial dusts are some examples of contaminants.

Contamination of the food chain

See **Pesticides and the environment**

Contamination of groundwater

See **Pesticides and the environment**

Continuous phase

See **Adjuvant**

Controlled release formulations of pesticides

Pesticides come in various forms, for example as emulsifiable concentrates (EC), wettable powders (WP), water-dispersible granules (WG) and dustable powders (DP). The agent or the active ingredient (a.i.) in a formulation is available completely and immediately to the environment, after its application.

To avoid rapid availability of the a.i., release is controlled at a desired rate or stalled for a desired period in the future. The controlled release rate is fixed not only to render effective pest control but also to prevent harm to current or subsequent crops, applicators and non-target organisms. It also ensures activity over a longer period.

Controlled release formulations constitute a mix of the a.i. and an inert material. Inert materials are commonly degradable polymeric macromolecules that play a role in enhancing the physical characteristics of the formulation. Such substances do not hamper the activity of the pesticide or cause any harm to the environment. Controlled-release pesticide formulations may be used in IPM strategies.

The mechanism and kinetics of controlled-release pesticide formulations can be brought about chemically, physically and biologically. Formulations made by backbone linking, side-chain bonding and matrix degradation are chemical interventions. When formulations are designed to be activated through micro- and macro-encapsulation, a coated solid, or via hollow fibres and foams, the active compound is said to be released by physical manoeuvring. Biologically controlled release formulations are brought about by using dead or living cells as delivery mechanisms.

There may be several variations and innovations to bring about controlled release.

Convergent lady beetles as a biocontrol agent

See **Crop – cotton pest control**

Copper 8-quinolinoleate fungicide

See **Oxine-copper fungicide**

Copper acetoarsenite

See **Arsenical inorganic insecticides**

Copper arsenate

See **Arsenical inorganic insecticides**

Copper as a plant nutrient

See **Plant nutrient excesses and deficiencies**

Copper hydroxide fungicide

Copper hydroxide, CuH_2O_2 (copper (II) hydroxide), is an inorganic fungicide. It is a blue powder, soluble in water (5.06×10^{-4} g/l), *n*-heptane (7010 μg/l), acetone (5000 μg/l) and ethyl acetate (2570 μg/l). It is dehydrated above 50°C and decomposed above 100°C. The fungicide is often available as a mixture of copper (II) carbonate and hydroxide. The CAS Registry Number is 20427-59-2.

$$HO^- \quad Cu^{2+} \quad HO^-$$

Copper hydroxide

Copper hydroxide is a protectant fungicide and bactericide and, to a certain degree, a nematicide. Its role is similar to that of **Bordeaux mixture**. It inhibits germination of fungal spores. Copper (II) hydroxide is used to control root growth, when combined with latex paint. It is also used as a ceramic colourant.

Copper hydroxide is commercially available under trade names Champ, Coprodate and Coproxide. It is also used in combination with other fungicides such as ipconazole. Some of the common formulations are water-dispersible granule (WG) and wettable powder (WP). Copper hydroxide is incompatible with acids, dicloran or **calcium polysulfide**.

The compound is classified by both the WHO and EPA as a slightly toxic pesticide. It is very irritating and corrosive to eyes. The acute oral LD_{50} in rats ranges from 489 to 1280 mg/kg, while in mallard ducks it is 223 mg Cu/kg. At low temperatures, it may injure plum and peach foliage. The NOEL is 16–17 mg Cu/kg b.w./day. The ADI is 0.5 mg/kg b.w./day.

Copper, an essential element in animals and plants, does not degrade or metabolize. In soil, copper is strongly bound to a wide range of soil substances hence limiting the amount of free copper ions in soil. In acid soils, copper ions are present in greater concentration than at neutral or alkaline pH, limiting the presence of free copper ions. In water, copper is not hydrolysed. It binds to mineral particles so as to precipitate as insoluble inorganic salts.

Copper hydroxide is not a PAN Bad Actor chemical and is registered for example in India, Canada, South Africa, New Zealand, the UK and the USA. (See also **Inorganic pesticides**.)

Copper octanoate fungicide

Copper octanoate, $C_{16}H_{30}CuO_4$ (copper dioctanoate), is an inorganic compound. It is a solid with a CAS Registry Number of 20543-04-8.

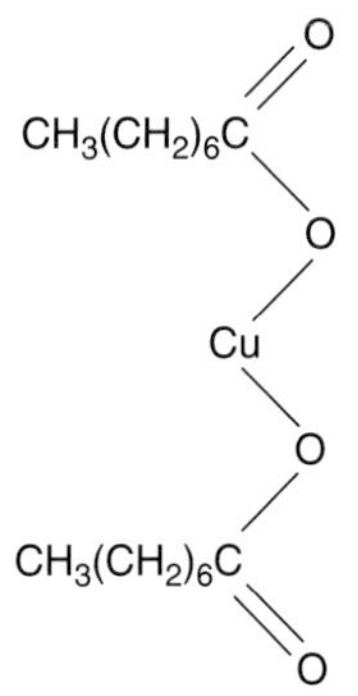

Copper octanoate

A fungicide and bactericide with additional algicidal property, copper octanoate controls **powdery mildews** (*Erysiphe* spp.), **downy mildews** (*Plasmopara* spp.), leaf and fruit spots, blights (*Phytophthora* spp.) and rusts (*Puccinia* spp.).

The EPA classifies copper octanoate as slightly hazardous (Class III). It does not irritate the skin or eyes. The acute oral LD_{50} in rats is >2000 mg/kg. The acute dermal LD_{50} in rats is >2000 mg/kg. It is toxic to fish but non-toxic to bees and earthworms. Copper octanoate is incompatible with oils. It may be phytotoxic to some varieties of roses and apples. It is not a PAN Bad Actor chemical.

In soil, copper octanoate degrades to free copper and octanoic acid. It has to be noted that the total soil copper concentration increases with the addition of copper pesticides since copper does not degrade.

Copper octanoate is registered for use in Australia and the USA.

Copper oxychloride fungicide

Copper oxychloride, $ClCu_2H_3O_3$ (dicopper chloride trihydroxide), is an inorganic fungicide. It is an odourless bluish-green powder, which is almost insoluble in water ($<10^{-5}$ mg/l) and organic solvents. It is soluble in dilute acids and ammonium hydroxide forming Cu (II) salts and complex ions, respectively. Copper oxychloride has a CAS Registry Number of 1332-40-7. It is incompatible in acidic media, mercury-containing compounds and **dithiocarbamates**.

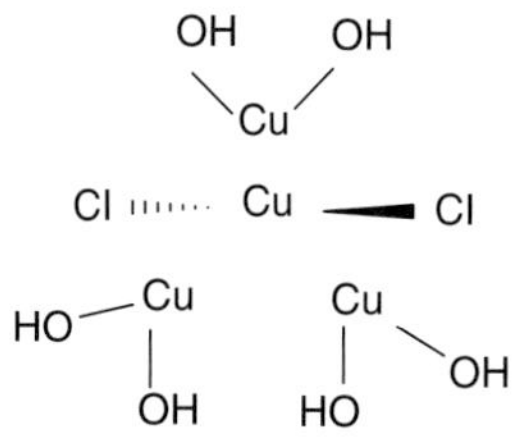

Copper oxychloride

A foliar fungicide with bactericidal properties, copper oxychloride controls **late blight** of potatoes and tomatoes. Leaf spot, **downy mildews** and shot hole are controlled among beet crops, olives, vines and spinach. The fungicide also controls bacterial diseases in lettuce.

The principal trade names of copper oxychloride are Cekuper, Blitox, Cobox and Funguran. It is used with fungicides such as **benalaxyl**, cymoxanil, **metiram** and **mancozeb**. Some of the common formulation types are wettable powder (WP), paste (PA) and dustable powder (DP).

Acute oral and percutaneous LD_{50} values in rats are 700 mg/kg and >2000 mg/kg, respectively. In fish, the LC_{50} in carp is 2.2 mg/l. It is not toxic to bees. Contact with eyes causes watering and redness, while with skin, it can cause redness, scaling and itching. The use of protective equipment and safe industrial practices are necessary. The WHO and EPA classify it as slightly hazardous (Class III).

In animals, copper oxychloride is excreted via faeces. The compound is registered in countries such as India, New Zealand, Cambodia, the UK and the USA.

Copper sulfate fungicide

See **Copper sulfate pentahydrate fungicide**

Copper sulfate pentahydrate fungicide

Copper sulfate ($CuSO_4$), an inorganic fungicide, exists in nature with varying degrees of hydration, one of them being copper sulfate pentahydrate ($CuSO_4.5H_2O$). Copper sulfate pentahydrate is soluble in water (230 mg/kg at 25°C) and insoluble in organic solvents. It decomposes at increasing temperatures. Copper sulfate is registered in India and most EU countries.

Copper sulfate pentahydrate ($CuSO_4.5H_2O$) is used as an algicide, fungicide and a molluscicide. But, its use as a fungicide is prominent. The compound is mixed with, for example, ammonium carbonate or lime and applied as a crop protection agent. It may also be mixed with fertilizers in a ratio of two parts copper sulfate pentahydrate:40 parts fertilizer.

For use against terrestrial gastropods, the powdered form is applied at a rate of 10–20 kg/ha. An aqueous solution at 3 kg/100 l is also recommended. Snails and slugs must come in contact with the copper sulfate for it to be effective. The compound damages and dries up the mucous of these creatures, thus killing them.

Copper sulfate pentahydrate is toxic to fish and it is not registered in some countries (e.g. Japan) because of its negative effects on the environment. The WHO has classified it as a

Class II toxin (moderately hazardous). Due to its toxicity, it has to be strictly used as per the recommendations. Standing crops are not treated with copper sulfate as it can damage them. It is not a PAN Bad Actor chemical.

The USA, Canada, New Zealand, Australia, South Africa and some other countries have registered the use of copper sulfate pentahydrate. (See also **Inorganic pesticides**.)

Cornexistin bioherbicide

Cornexistin, $C_{16}H_{20}O_6$ ((4R,5R,8R,9E)-9-ethylidene-5,8-dihydroxy-4-propyl-5,7,8,10-tetrahydro-4H-cyclonona[c]furan-1,3,6-trione), with a CAS Registry Number of 123068-35-9, is a fungus-derived potent herbicide used on maize crops. It is derived from *Paecilomyces variotii*. The result of an extraction process forms a needle-like substance that is insoluble in water.

Cornexistin is a selective, post-emergence herbicide that controls many weeds in the maize field. It is found to be non-toxic to mice.

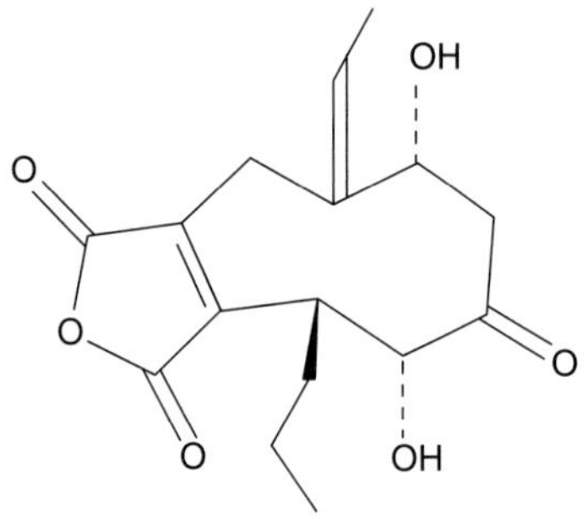

Cornexistin

Corn (maize) oil bioinsecticide

Corn (maize) oil (CAS Registry Number 8001-30-7) is used in some crop situations to control insects. It is not acutely toxic and is used in combination with biological ingredients, as pesticidal agents.

Corn oil is also an **adjuvant** used in herbicide sprays. It improves wetting, reducing evaporation, increasing penetration and enhancing deposition and compatibility. Corn oil is added for better penetration in plant leaves.

It is registered for use in Vietnam and the Philippines.

Cotton *Dahiya*

See **Crop – cotton pest control**

Cottonseed oil bioinsecticide

Cottonseed oil (CAS Registry Number 8001-29-4) is an **essential oil** that is an ingredient in many processed foods and salads, while being one of the most unsaturated vegetable cooking oils available. The oil, extracted from crushing the seeds, is also an effective insecticide, and together with other oils can control mites, aphids and whiteflies, especially in greenhouse sites and on ornamentals, vines, lucerne, maize and tomatoes, at a rate of around 900 g–1.3 kg/ha. It has slightly acute toxicity but is not a PAN Bad Actor chemical, and is registered in many countries, for example Vietnam.

Cottonseed oil plus maize oil plus garlic oil biofungicide

The combination of essential oils is commonly used to protect crops from specific pests such as insects or fungal pathogens. One such combination is the oil of three crops – cotton, maize and garlic. This combination is an effective fungicide, controlling **powdery mildew** by disrupting the membranes of germinating spores. A common formulation is the aqueous concentrate sprayed on the leaf surface of infected plants.

The oil combination is non-toxic to mammals and the environment. It is categorized as a GRAS compound. This combination has US OMRI approval for organic farming.

Coumaphos insecticide

Coumaphos, $C_{14}H_{16}ClO_5PS$ (*O*-3-chloro-4-methyl-2-oxo-2H-chromen-7-yl *O*,*O*-diethyl phosphorothioates), is an organophosphorus insecticide (heterocyclic organothiophosphate class). It is also considered as an organothiophosphate acaricide for its treatment in the control of scabies mites. It is a colourless crystalline substance soluble in water at 1.5 mg/l but more soluble in organic solvents. It is stable to hydrolysis in aqueous media. It is obtained by condensation of resorcinol with α-chloroethylacetone in the presence of an acid, followed by reaction with diethylthiophosphoryl chloride. The CAS Registry Number is 56-72-4.

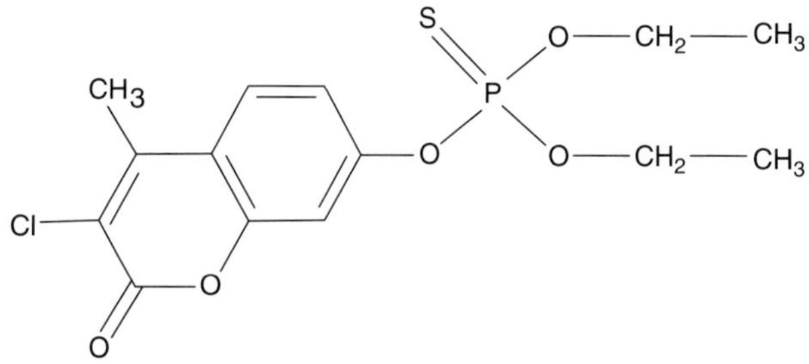

Coumaphos

Coumaphos, a veterinary and non-systemic insecticide, is a cholinesterase inhibitor. It controls ectoparasites such as cattle grubs, lice, horseflies, face flies and ticks on cattle, sheep, pigs, poultry, goats and horses, generally at the outset of the fly season in early spring. It controls larval infestation, when added to animal feed (commonly applied at 2 mg (a.i.)/kg b.w./day for 1 week). Coumaphos also controls parasitic mites (*Varroa jacobsoni*) on bees.

The insecticide is formulated as crumbles, dusts, powder, feed premix, spray or liquid. It is available in the market under the trade names of Asuntol and Perizin.

The WHO has categorized coumaphos as a Class Ib toxic (highly hazardous). Since coumaphos is a cholinesterase inhibitor and a highly toxic compound, it is considered a PAN Bad Actor chemical. Coumaphos is toxic to aquatic invertebrates and fish and hazardous to birds (on ingestion of cattle hair and other contaminated material). However, environmental protection agencies are satisfied that alerting users of coumaphos (through warnings on labels) would be the best way to minimize any threat to life and the ecosystem. Handlers, mixers and applicators of coumaphos, who are also under threat from exposure, are advised to use necessary equipment and clothing to avoid any danger. The LD_{50} values for male and female rats are 1 and 16 mg/kg, respectively. However, some rats can tolerate quantities of up to 100 mg/kg diet.

Cows and rats are able to degrade coumaphos in the liver rapidly, excreting the principal metabolites such as diethyl hydrogen phosphorothioate and some deethylation products in the urine. The photolytic DT_{50} of coumophos on the soil surface is 23.8 days. Coumaphos is not mobile in soil but adsorbs to soil particles and hence holds no potential for contaminating groundwater sources. Microbial degradation also takes place by *Pseudomonas monteilii* and a few more strains, breaking coumaphos into chlorferon and diethylthiophosphoric acid.

The USA, Canada and New Zealand have registered the use of coumaphos. Health Canada has established an MRL for coumaphos to be 1 ppm in bee wax.

Coumarin rodenticide

Coumarin, $C_9H_6O_2$ (2H-chromen-2-one), is a compound in the phenylpropenoid group of chemicals. It is a rodenticide with additional fungicidal properties. The CAS Registry Number is 91-64-5. It has a vanilla fragrance and hence is used as a food additive. Coumarin can be prepared by reacting salicylaldehyde and acetic anhydride by the Perkin reaction method.

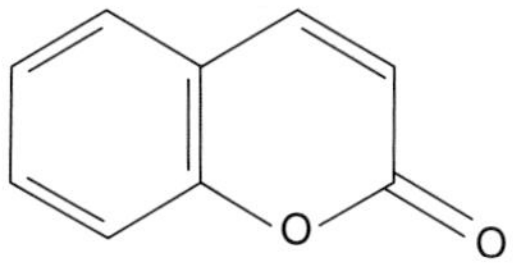

Coumarin

To be a rodenticide, coumarin, a pro-pesticide, has to be converted into dicoumarol. Upon ingestion, the toxic metabolite formed is 3,4-coumarin epoxide. This compound causes internal haemorrhage leading to death. It blocks vitamin K from being synthesized hence preventing clotting.

Rodent species and humans both convert coumarin into toxic metabolites. However, in the case of humans, the metabolite 7-hydroxycoumarin is formed, which is of low toxicity.

Coumarin is also used in medical practice as a blood thinner and for treating tumours. It is a naturally present toxin in many plants, such as tonka beans (*Dipteryx odorata*), sweet clover, strawberries, cassia bark (cinnamon variety) and apricots. Its bitter taste acts as a feed-repellent. **Brodifacoum**, **bromadiolone**, ensaculin coumafuryl, **difenacoum** and **warfarin** belong to the coumarin family.

Coumarin is a potential kidney and liver toxin, so its inclusion in food has to be strictly monitored. Natural derivatives include umbelliferone (7-hydroxycoumarin) and aesculetin (6,7-dihydroxycoumarin).

Coumatetralyl rodenticide

Coumatetralyl, $C_{19}H_{16}O_3$ (4-hydroxy-3-(1,2,3,4-tetrahydro-1-naphthyl)coumarin), is a rodenticide. It is a colourless to yellowish crystalline substance, which is soluble in water at increasing pH levels (i.e. 4 mg/l at pH 4.2 and 100–200 g/l at pH 9), and also in dimethyl formamide, dichloromethane (50–100 g/l) and isopropanol (20–50 g/l). It remains stable up to 150°C. It is not hydrolysed by water for up to 5 days and has a DT_{50} of >1 year. Its aqueous solution decomposes rapidly under UV light with a DT_{50} of about 1 h. Coumatetralyl is produced by the condensation of 4-hydroxycoumarin with 1,2,3,4-tetrahydro-1-naphthol. The CAS Registry Number is 5836-29-3.

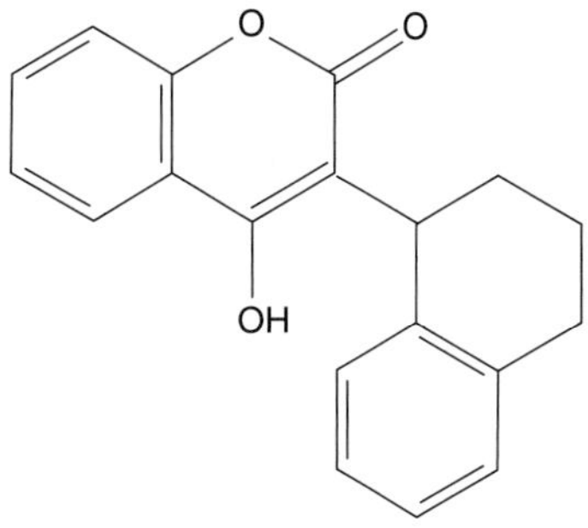

Coumatetralyl

Coumatetralyl, a **coumarin** anticoagulant, is known and widely used as a rodenticide and can effectively control Norway (brown) rats, but it cannot control mice to the same extent. Several doses are needed for the lethal effects to be displayed. It does not induce bait shyness.

The common formulations of coumatetralyl are grain baits, ready-to-use baits, bait concentrate and tracking powder. A common trade name is Racumin. Baits commonly contain 0.375% concentration. More than half of coumatetralyl gets mineralized within 3 months.

In rats, the acute oral LD_{50} is 16.5 mg/kg, while in mice it is >1000 mg/kg, for rabbits it is >500 mg/kg and for Japanese quail it is >2000 mg/kg.

The WHO has categorized coumatetralyl as a Class Ib toxic (highly hazardous) and it is classified as a PAN Bad Actor chemical. India, Australia, New Zealand, the Philippines and South Africa have registered the use of coumatetralyl, while the USA has not registered its use.

Cover crop

A cover crop is any crop that is grown usually in a non-crop period, to manage soil fertility, erosion, water, weeds and pests in an agro-ecosystem. They are an important part of sustainable agriculture as many cover crops are annual grasses, legumes and cereals.

The advantage of a cover crop is that it can be tilled back into the soil to become green manure. Its rapid growth controls weeds before the main crop-sowing activity can start. Cover crops retain moisture, prevent runoff and add phosphorus and nitrogen to the soil.

Examples of cover crops are forage plants and legumes such as joint vetch, clovers, butterfly pea and hairy indigo. Palm trees and rubber plants are two out of many crops that benefit from added nutrition from cover crops.

Critical micelle concentration (CMC)

See **Surfactants**

Crop – castor pest control

Castor (*Ricinus communis*) has two centres of origin, namely the Indian subcontinent and Africa. India is a major castor-producing country in the world. The plant can grow to a height of up to 10 m. The leaves are large (20–40 cm) with a long petiole. The stem, leaves and the petiole are covered with wax. The fruit is a capsule that bears thorns. The seed is about 2 cm long and 1.5 cm broad. The crop has multiple uses including oil, manure (oil cake), fodder (for goats), lubricant, biodiesel, medicine (for skin, hair and stomach), and as an input in printing-ink, paper-pulp and soaps. It is also being explored as an alternative to mulberry plants in the sericulture industry. Since castor is an important crop, its pest management is an important subject of study.

Castor is best grown in soils that are deep, well-drained, slightly acidic and not very fertile. Waterlogged conditions and castor monoculture are to be avoided at all times.

Castor can also be an intercrop or a windbreak around vegetable crops. Ukraine, India, China and Russia are some countries that extensively cultivate castor with high yields per hectare.

High-yielding hybrid castor is dwarf and annual in nature and can mature within 5–10 months. Some of the high-yielding draft varieties are UC53, Baker 44, AKC-1, GSH-1, GSH-4, RG 2659 and RG 266 which are also wilt and insect resistant (hopper).

Chelated micronutrient preparations with emphasis on zinc, **magnesium**, **boron**, **manganese**, **molybdenum** and **copper** are applied either through the drip system or as foliar spray to correct deficiencies in high-yielding varieties. In the case of a **wilt** disease, a 3 year gap is recommended before rotating the crop.

Many types of pests attack the castor crop. A few are grasshoppers, bugs (*Helopeltis* spp., also known as tea mosquito or mirid bug), castor semilooper (Fig. C.7), caterpillars, whiteflies (*Trialeurodes ricini*), castor shoot and capsule borer (*Conogethes punctiferalis*; Fig. C.8), armyworms, tussock hairy caterpillar (in the family Lymantriidae), leaf spot (*Cercospora reicinella*), jassids (*Empoasca flavescens*, otherwise called **leafhoppers**), rust (*Melampsora ricini*) and *Alternaria* leaf spot.

Pest control is carried out by several means. Use of light, water and sticky traps, manual removal of pests and monitoring through use of **pheromones** are steps recommended in IPM. *Trichoderma*-based biofungicides are commonly used in place of chemical fungicides to fight wilt disease. Extracts of garlic and chillies are also used for the control of armyworm. In place of urea, the use of nitrogenous fertilizers such as calcium nitrate/calcium ammonium nitrate is recommended. Since the castor crop takes about a month to reach the four-leaf stage, there is competition from weeds. Sufficient row space to enable efficient weeding is always recommended.

Seed treatment of 3 g **thiram**/kg of seed is recommended for the control of leaf spot. Seed treatment with **imidacloprid** or **carbosulfan** is also recommended. **Monocrotophos** (0.05%) or **dimethoate** (0.05%) can give overall protection against sucking pests, including jassids.

Quinalphos, chlorpyrifos, carbaryl, monocrotophos, dimethoate and **parathion-methyl** are some of the chemicals used to control castor pests. The herbicide **fluchloralin** at 1 kg/ha should be sprayed into the soil prior to sowing.

Crop – cotton pest control

Cotton (*Gossypium hirsutum*, family Malvaceae) is an important cash crop in most countries and particularly in China, the USA, India, Pakistan and Uzbekistan. Cotton is harvested for its seed and fibre. It grows well in tropical climates with temperatures of up to 25°C. The plant needs moist and fertile soil, especially during seeding and seedling emergence.

Fig. C.7. Castor damage caused by the castor semilooper. Courtesy of: http://takingroots.in/ipm_castor_semilooper (licensed under creative commons). With permission from Ramanjaneyulu GV, Executive Director, Centre for Sustainable Agriculture, Secunderabad, India.

Although rain-fed cotton is also common, the yields are low. The roots grow deep downwards and leaves, about 15 cm long and wide, have a spiral arrangement. A long flowering period leads to fruiting (the fruit is called a cotton boll). The boll contains the fibre, which opens after the full growth of the fruit.

Successful cotton cultivation is dependent on: (i) protection against pest attacks (at almost every stage); (ii) fertile soil; and (iii) abundant water. The cotton crop is vulnerable to many pests like pink bollworm, American bollworm, aphids, thrips, mirid bug, nematodes, and **wilt** and seedling diseases. So, crop management via cultural and chemical methods has become imperative in proper cotton cultivation.

Cultural management techniques include: (i) deep ploughing; (ii) precision tillage; (iii) removal and destruction of crop residues; (iv) prevention of waterlogged soil; (v) crop rotation (with e.g. legumes, sorghum, lucerne); (vi) trap cropping (e.g. castor, marigold, lucerne); (vii) intercropping (e.g. chillies); (viii) use of optimum spacing and density; and (ix) the use of pheromone traps. The use of resistant varieties of cotton also helps in keeping diseases away.

Fig. C.8. Castor damage caused by the castor shoot and capsule borer. (a) Damaged branch; (b) bored capsules. Courtesy of: http://takingroots.in/ipm_cscb. (licensed under creative commons). With permission from Ramanjaneyulu GV, Executive Director, Centre for Sustainable Agriculture, Secunderabad, India.

Pink bollworm (*Pectinophora gossypiella*) is a serious pest in cotton crops. The larvae of the bollworm moth burrow into the boll, staining and damaging the fibre within (Fig. C.9). Use of **chlorpyrifos**, **cypermethrin**, **spinosad** and lambda cyhalothrin (25 g/ha) can chemically control the pink bollworm.

The American bollworm or cotton bollworm (*Helicoverpa armigera*; Fig. C.10) is a polyphagous larva, which prefers cotton, cereals like sorghum and maize, cowpea, pigeon pea, tobacco and many vegetables. Also called the corn earworm, the life cycle of this pest can be completed in 1 month. Three to five larvae in a patch of the field are sufficient to bring down the yields of cotton, the symptoms including premature opening of blooms, non-fruiting of blooms and boll drops. Similar attacks are seen by the **spotted bollworm** (*Earias vittella*). In both the cases, 10% emulsifiable concentrate **bifenthrin** (80 g/ha) and similar spray of **quinalphos** bring about effective control. The eggs

Fig. C.9. Damage (images show half-opened bolls) caused on cotton by pink bollworm infestation. Courtesy of: http://takingroots.in/ipm_pink_bollworm (licensed under creative commons). With permission from Ramanjaneyulu GV, Executive Director, Centre for Sustainable Agriculture, Secunderabad, India.

of ***Trichogramma** chilonis* released twice within 2 months of planting can control the bollworm.

The larval stage of the greyish moth, the beet armyworm (*Spodoptera exigua*), is very harmful since it feeds on cotton seedlings, terminal buds and young bolls. Apart from cotton, the beet armyworm attacks many other crops like lucerne and vegetables, and some weeds (e.g. *Amaranthus* spp. and *Chenopodium* spp.). Biological control can be brought about by introducing predators such as lacewings, **minute pirate bugs** and spiders. Application of pyrethroids, carbamates and the use of ***Bacillus thuringiensis*** are considered useful in the control of this insect pest. Methoxyfenozide, flubendiamide, **esfenvalerate**, **diflubenzuron** and prophenofos are some of the chemicals used.

It is critical to protect cotton during the first 3 months of growth against the sucking type of pests like aphids, hoppers, whiteflies, **mealybugs**, mites and thrips, and hence seed treatment assumes great importance. Application of **carbendazim** (1.5 g/kg seed) followed by imidacloprid (7 g/kg) can provide protection against wilt disease caused by *Fusarium* spp. Application of carbendazim, followed by imidacloprid and then **phorate** granules, prior to sowing can control sucking pests in the early stages of plant growth. The biological control of the seed disease is also brought about by *Azospirillum* (25 g/kg).

There are aphid species that attack cotton, mostly at the seedling stage. The cotton aphid (*Aphis gossypii*) is an important pest. These are soft-bodied yellow, black and green nymphs that grow into winged/wingless adults. They reside on the lower side of the leaves and young apical shoots and feed on the sap, excreting the sugary **honeydew** (which attracts the **sooty mould**, *Capnodium* spp.). The honeydew affects the photosynthesis of plants, evident in curled up leaves, unopened leaves, sticky honeydew accumulations and small-sized bolls (Fig. C.11). If 20% of the crop is affected, the damage may cause economic losses.

For aphid control, a balance between biocontrol measures and chemical intervention is necessary to keep resistance to insecticides (organophosphates, pyrethroids and carbamates) at bay. Planting of cotton in the early part of the season can help the plant take root and establish itself before aphids attack them. Selecting resistant cultivars is an essential part of aphid management.

Natural predators of aphids include bugs, wasps and beetles. *Lysiphlebus testaceipes* is a tiny wasp that lays its eggs inside the aphid and has its young ones feed on the aphid, thus killing them. The convergent lady beetle (*Hippodamia convergens*) needs to consume aphids in order to reproduce. The beetle larvae may eat up to 50 aphids a day. The larvae and the adults both pierce a hole in the aphid body and suck out its sap. The big-eyed bug, the pirate bugs, **green lacewings** (***Chrysoperla*** and *Chrysopa* spp.) and a fungus (*Entomophthora* sp.) feed on aphid populations and may be used to control their infestation. **Acetamiprid, thiamethoxam**, imidacloprid, chlopyrifos and aldicarb are some of the chemicals used in the control of aphids. Most of them have high specificity and have a season-long effect.

There are many species of hoppers that attack the cotton crop like the potato leafhopper (*Empoasca fabae*) and southern garden leafhopper (*Empoasca solana*). These greenish chisel-shaped insects suck the sap out of the lower side of the leaf, distorting and discolouring it. Cotton yields are reduced

Fig. C.10. Common pests of cotton. (a) Old World bollworm (*Helicoverpa armigera*). (b) Cotton bollworm *Helicoverpa armigera* (Hübner). (c) Bollworm adult (*Helicoverpa zea*). Courtesy of (a) Gyorgy Csoka, Hungary Forest Research Institute, Bugwood.org. (b) Paolo Mazzei, Bugwood.org. (c) Ronald Smith, Auburn University, Bugwood.org. With permission.

Fig. C.11. Damage caused on cotton by aphid infestation. Courtesy of: http://takingroots.in/ipm_aphids. (licensed under creative commons) with permission from Ramanjaneyulu GV, Executive Director, Centre for Sustainable Agriculture, Secunderabad, India.

because of this attack. Even as few as three nymphs are enough to cause losses.

On rain-fed cotton, thrip species like **western flower thrips** (*Frankliniella occidentalis*) are observed in mid-August. These yellow, orange and brown small insects (1 mm in length and whose adults are winged creatures) use rasping and sucking mouthparts to suck the sap from leaves. The affected part turns whitish and later turns brown. The **economic threshold level (ETL)** is ten thrips per leaf. Although there are many chemicals (e.g. **acephate**) that may be applied to control thrips, it is recommended that cultural methods (e.g. weeding out alternate hosts such as morning glory) are used in order to retain a small population of thrips, since thrips can control the mealybug population.

The nymphs of the yellowish adult whitefly (2–3 mm) with grey or white wings suck sap from the lower side of the leaves and excrete a sugary substance (which attracts the black sooty mould). The silverleaf whitefly (*Bemisia argentifolii*) is one of the prominent pest species. Damage may lead to losses when there are about eight to ten adults per leaf or 20 nymphs per leaf (which normally happens in November). Emusifiable concentrates or water-dispersible granules of imidacloprid, **profenofos** and **diafenthiuron** may be applied to control the pest. In addition, beetles, spiders and green lacewings (*Chrysopa* spp.) are used to control whiteflies in the natural way. Spraying **neem** oil, or a combination of **triazophos** and a detergent, are other measures, especially if low chemical intervention is needed.

Phenococcus solenopsis is the most important species among mealybugs. These blackish green, oval-bodied pests feed on buds, flowers, bolls and leaf petioles of cotton, aubergine (also called brinjal or eggplant) and tomatoes and, in the process, excrete a sticky substance (attracting sooty mould). Plant growth is affected and the bolls get stunted and deformed. An emulsion of **buprofezin**, or a dose of acephate in combination with a detergent, is recommended for the control of mealybugs. Cultural methods include removal of parthenium weeds, weeds of the hallucinogenic genus *Datura* and weeds of *Chenopodium* spp. (e.g. lambsquarters, goosefoot, pigweed).

There are many regional diseases that affect the cotton crop. *Kawadi* disease is one such. The causal agent is the fungus *Colletotrichum gossypii*, which spreads through the infected seed. The fungus also affects the stem and the bolls resulting in stunted growth. The leaves turn grey black, the bark starts to peel off, and the lint becomes yellow and does not detach easily. Management techniques include the use of disease-resistant, clean and certified seed varieties, treated with *Trichoderma* and carbendazim, and the spray application of a mixture of carbendazim 10 g and **streptomycin** 1 g/10 l water.

Black arm stem blight *Karpa* (Zulsa) is a seed-borne disease caused by the bacterium *Xanthomonas axonopodis* pv. *malvacearum*. The lint in the bolls turns yellowish affecting the quality of the fibre and causing significant loss. The bacteria interfere with the functioning of photosynthesis. The infected seed does not germinate well. The lower surface of the leaf exhibits small grey-black triangular spots. The leaf turns yellow and dries up. In severe cases, black spots appear on the branches and stems, hence the name. The use of disease-resistant, acid delinted certified seed varieties, treated with *Trichoderma* prior to sowing, is recommended. **Copper oxychloride** (25 g) + **streptomycin** (1 g/10 l water), sprayed frequently, can give the desired control. Soaking delinted seeds in 100 ppm solution of **streptocycline** for 2–10 h is another way to control the disease.

The causal agent of the cotton *Dahiya* is the fungal species of genus *Ramularia*. High humidity levels favour the disease. Whitish spots appear on both sides of leaves, the lower sides first, to give the effect of yoghurt (curd or the Indian word, *dahi*) sprayed on the foliage. This infection leads to leaf drop and poor fruit set. Dusting the plant in the early dew-laden hours of the day with 300-mesh sulfur at 20 kg/ha renders control.

Alternaria leaf spot caused by the fungal agent *Alternaria microspora* causes grey-coloured spots with yellow borders on the leaves. The infection leads to leaf drop and yellow stain on the bolls. Management measures include the use of disease-resistant varieties and treating certified disease-free seeds with *Trichoderma* spp. An additional spray application of *Trichoderma* spp. or a water solution of phytolan (25 g) and streptomycin (1 g) is common. These can be applied up to four times during crop life.

Root rot, caused by the *Rhizoctonia* fungus, leads to plant wilting and the roots having a peeled appearance. Again, seed treatment and root-zone drench with *Trichoderma* spp. help control the disease. **Carbendazim** 0.3% solution applied near the stem may cure the condition. Crop rotation is recommended to prevent root rot.

A nutrient deficiency compounded by anaerobic conditions also affects the cotton crop. *Lallya* is a condition that occurs due to waterlogged condition and the deficiency of nitrogen and **magnesium**. A foliar application of 0.2% $MgSO_4$ and 2% urea is effective.

Weedicides may be applied at various growth stages of the crop growth. Diuron 80 WP at 0.5 l/ha (500 l solution) is effective at the germination stage. **Fluchloralin** may be sprayed pre-germination to be followed by another within a month. **Pendimethalin** 1.5 l/ha could be used prior to germination as a single spray.

Crop – grape pest control

Grapes (*Vitis vinifera*) are berries of the perennial deciduous vines of the genus *Vitis*. Grapes produced in the world are used mainly for wine making, the remaining being used as table grapes and for producing raisins. Grapes grow the best in the subtropical, mild or hot tropical climate, with long sunny days. Soil must be fertile and well drained so that the roots can penetrate deep. Italy, Spain, France, the USA and China are five top grape-producing countries.

The annual cycle of the grape vine in the northern hemisphere begins in March and the harvest is around the end of September, whereas in the southern hemisphere, the two periods are September and from February to April. Deficiency of nitrogen and phosphorus combined with low intensity sunlight can lead to unhealthy growth of berries and economic losses. Lack of sufficient **boron** may lead to green berries. Potassium deficiency leads to the reduction of the bunch size and white patches on the berries.

Given below are measures that help control some major diseases that affect grape crops. The selection of disease-resistant plants and rootstocks, balanced nutrition, maintenance of soil structure and drip irrigation keep the plant healthy and disease-free as far as possible.

Powdery mildew of grapes, caused by the fungus *Uncinula necator*, is prevalent all over the world. The disease affects canes, leaves, lateral shoots, flowers and the berries. White patches develop on the lower surface of the leaves, followed by mottling, curling and withering. Berries may show white patches and reduced size, and in severe cases, cracking and withering. The spores survive in the flowers.

Cultural measures for the management of powdery mildew include: (i) allowing sunlight and aeration by maintaining adequate spacing between vines; (ii) ensuring hygienic conditions; and (iii) the general measures mentioned above.

Damage to the berries from powdery mildew is seen immediately, if infection occurs post-bloom. Hence, timing of fungicide application is very important. **Quinoxyfen, metrafenon, myclobutanil** and **trifloxystrobin** are some of the fungicides that provide effective control over powdery mildew, with a maximum of two applications per season. Dosage range is 100–750 g/ha. **Lime sulfur** and potassium bicarbonate formulations are also good for disease control. ***Bacillus thuringiensis* (*Bt*)**, the biofungicide, is also applied for early disease control. Label instructions with respect to mixing and dosing are to be carefully followed.

Fungi such as *Ampelomyces* sp. are used in biological control of fruit rot. Spraying vines with *Trichoderma* spp. culture (5 g/l water) from September until the time of pruning in October helps prevent **downy mildew**, powdery mildew and *Alternaria* on the leaves. Spraying with 0.2% **hexaconazole** or **flusilazole** in the second week of May and 10 days thereafter is also found to be effective.

Downy mildew of grapes, caused by the fungus *Plasmopara viticola*, is known to be one of the major diseases in North America and Europe, the affected parts being the leaves and fruit. Young leaves develop yellowish-green spots on them, and whitish spots underneath. Young berries turn dark and do not ripen. The disease favours warm and very humid conditions. Proper row spacing, adequate drainage and removal of excess foliage are specifically beneficial for control, besides the measures mentioned above. Wettable granule formulation of iprovalicarb and **propineb**, singly or in combination, are effective in disease control.

Leaf spot, caused by *Gleosporium* spp., is observed in all grape-growing areas. The symptoms of the infection are brownish spots on young leaves, newly developing berries and tender stems, which may expand over the entire surface. In severe infections, the pathogen can enter the stem and carry on into the next life cycle. For effective control of leaf spot, 5 g of *Trichoderma* spp. culture dissolved in 5 l water can be sprayed on each vine in June and July.

Anthracnose of grapes, caused by *Elsinoe ampelina*, can affect all above-ground parts. The symptoms are greyish spots on leaves and berries. Affected leaves develop dark green margins, get distorted and curl backwards. Berries develop a reddish border around the grey spots. Control of the disease includes removal of infected plant parts followed by the application of the **Bordeaux paste** to the pruned surfaces. Spray application of vines with 0.3% **mancozeb** or **copper oxychloride** or 0.1% **carbendazim** is also effective.

Grape crops are susceptible to a number of pests, some of which are covered below, along with remedial or preventive measures. The pests include beetles, **mealybugs**, thrips and hoppers. Spray application of **thiamethoxam** (25 g wettable granules/ha), and a soil drench of the same ingredient, can help in controlling most of the insect pests. **Imidacloprid** (spray solution of 3 mg/l water) also helps control a large number of insect pests. This effect is seen without affecting the pests' natural enemies.

Flea beetles (*Scelodonta strigicollis*) feed on the tender top growth and buds, as a result of which new growth withers and buds fail to sprout. Mature leaves get holes in them. The infestation is particularly severe after each pruning. Removal of the affected part followed by the application of paste of copper oxychloride and **carbaryl** to the pruned surfaces is a popular control method. Post-pruning, spray application of **quinalphos** or even carbaryl, once or twice a week, helps to render protection to newer emerging leaves and buds.

The adults and the nymphs of the thrip *Rhipiphorothrips cruentatus*, attack flowers, young berries and leaves, causing spots on the parts, leading to leaf drop. Emamectin (11 g/ha) helps control thrips. The adults and the nymphs of the sap-sucking hopper *Arboridia viniferata* feed on the sap of the lower sides of leaves, causing white spots which develop into grey bands, causing leaf drop. Sprays of quinalphos (0.05%) and **monocrotophos** (0.1%) at early stages of infestation help control hoppers.

Mealybug species (*Maconellicoccus hirsutus*) feed on young leaves and berries, apart from producing **honeydew**, attracting the growth of **sooty mould** (*Capnodium* spp.) Since the female lays several eggs in a capsule-like sac, inside peeling barks, and under leaves and lower portions of stems, it is difficult to target and kill the eggs. A mixture of 400 g red clay, 5 ml **dichlorvos** and 2.5 g copper oxychloride/l water is effectively used to control mealybugs. **Parathion-methyl** dust (2% at 20 g per plant) applied twice a year to the soil helps keep away ants who share a symbiotic relationship with mealybugs. Such a treatment also controls stem borers. Biocontrol fungal agents *Verticillium lecanii* or ***Beauveria bassiana*** and lady beetles *Cryptolaemus montrouzieri* also help to control mealybugs.

The tobacco caterpillar (*Spodoptera litura*) is another pest of the grape crop. The larva of the moth hatches out on the lower side of leaves, feeding on leaves and flowers, making the leaf surface papery. The use of chlorpyrifos, carbaryl, dichlorvos, **methomyl** and wettable sulfur in the concentration range of 0.05–0.125% is found to be effective.

Crop – groundnut pest control

Groundnut (*Arachis hypogaea*) is one of the top five oilseeds grown all over the world. It is a member of the family Fabaceae. Groundnut is an annual, self-pollinating, leguminous plant, with yellow, five-petalled flowers and pinnate leaves that grow around 61 cm in height. Post-blossom, the pedicel elongates to enter the soil, where pods of the groundnut grow and develop. Groundnut (or peanut) grows well in sandy loamy soil with good drainage. It is commonly grown as a ***kharif*** crop, in places that receive >1000 mm of rainfall. It can grow in well irrigated areas too.

Groundnut is an important crop (Fig. C.12). It yields high quality edible oil. Livestock is fed with oil cake and the plant parts after harvest. The plant is also a source of manure. The plant is a commonly accepted partner in a rotation system, due to its nitrogen-fixing property. It is commonly rotated with wheat, sorghum (*Sorghum bicolor* and called *jowar* in India), onions, potatoes and garlic. Fertilization programmes proceed with a base dose of a nitrogen-phosphorus-potassium (NPK) in the ratio of 10:20:20. Micronutrients like zinc and **boron** help prevent diseases right from the sowing stage.

The countries that produce groundnuts include India, China, the USA and African countries such as Sudan, Nigeria and Senegal.

The groundnut crop is attacked by many pests that affect its productivity. They include weeds, insects and fungal species. Use of pest control measures, along with good cultural practices, hence becomes necessary for an environmentally safe cultivation of this crop.

The use of pheromone traps, light or sticky traps, removal of infected plants and maintenance of a healthy crop row space (at least 30 cm) are basic IPM measures. Application of **neem** or castor cakes keeps nematodes and insect pests away. Solarization and deep ploughing remove soil pests effectively.

Seed treatment is essential for prevention of seed diseases. **Thiram** and **carbendazim** applied at 2.5–5 g/kg seeds are two chemical options. Cultures of *Rhizobium*, *Pseudomonas fluorescens* and *Trichoderma* (10–15 g/kg) on well-dried seeds help prevent fungal seed rot caused by *Sclerotium rolfsii* and some others. Seed treatment with **chlorpyrifos** 20% at 2 l for 80 kg seed can keep **white grubs** from attacking the groundnut tap root. Dust formulations of some of these chemicals can keep many soil pests in control. Seedling rot caused by *Aspergillus niger* is kept under control by the use of resistant varieties (e.g. J-11, JCG-88).

Weeds are difficult to control and so pre-emergence herbicides or biological agents are ideal in weed control. Hand weeding or mechanized weeding are equally practised. The noxious pearl millet varieties, **Johnson grass** (*Sorghum halepense*), parthenium (*Parthenium hysterophorus*) and Sudan grass (*Sorghum bicolor drummondii*) may come into the field while the sorghum crop is rotated with groundnut. **Bermuda grass** (*Cynodon dactylon*) and nutgrass (*Cyperus rotundus* L) are some of the weeds that affect groundnut productivity. **Fluchloralin** (up to 1.5 kg/ha) at the pre-sowing stage and **alachlor** (up to 2 kg/ha) and **pendimethalin** 30% EC (1 kg/500 l water) at the pre-emergence stage are some herbicides used to keep weeds at bay.

Fungal leaf spot (called *Tikka* disease in India) is a common disease that affects the leaves of the groundnut plant. Caused by the *Cercospora* fungus (from the family Mycosphaerellaceae), the infection is evident by the brown-black spots appearing on leaf surfaces (first upper and then lower). Infected plants from a previous crop harbour this pathogen which then spreads with the breeze. The use of resistant varieties (e.g. Girnar-1, NC-4 and ALR-2) is a good way to prevent crops from becoming infected. Groundnut should be rotated with onion or garlic crops, and intercropped with pigeon peas (*Cajanus cajan*, called *Toor* in India), black gram and pearl millet (*Pennisetum americanum*, called *Bajra* in India) to reduce the spread of both tikka and rust. Advancing sowing dates helps escape the disease. Application of a liquid formulation of carbendazim or **mancozeb** is helpful.

Rust is another disease that is apparent from the orange-brown spots on the underside of leaves of the groundnut. The fungus *Puccinia* spp. (from the family Pucciniaceae) spreads widely with the wind, especially in humid conditions and ambient temperatures, such as 20°C. Liquid formulations of **mancozeb** and **tridemorph** give a reasonable cover against rust disease, repeated as per the severity. At the same time,

Fig. C.12. A groundnut crop. Courtesy of Sudha Gowariker, with permission.

Trichoderma harzianum and *Verticillium lecanii* are biologically active agents for rust control.

Viruses also attack groundnut plants, causing necrosis of the bud and the stem, and mottling. Typical symptoms are yellowing of the plant along with stunting. Parthenium weed may become the host and thrips are vectors of these attacks. Infectious cell-free virus and panicum mosaic virus are the viral pathogens bringing about such infections.

When it comes to insect pests, it is necessary to control the insect complex of aphids, thrips and hoppers, since their excreta (honeydew) attract *Capnodium* fungus. **Dimethoate**, **phosphamidon** or **quinalphos** are applied to control thrips, aphids and hoppers. Applications may be repeated, as required. Similarly, the armyworm (*Spodoptera litura*), red hairy caterpillar (*Amsacta* sp.) and leaf miner (*Stomopteryx subsecivella*) cause substantial losses (Fig. C.13).

Many of these insect pests are controlled by **quinalphos**, **endosulfan**, **phosphamidon**, **carbaryl**, **dichlorvos** and **fenvalerate**. Use of **trap plants** (cowpea) and border crops of sorghum are useful methods. In case of insect infestation, soybean should be avoided from any rotation. Introducing eggs of *Telenemus remus*, a preparation from ***Bacillus thuringiensis*** (***Bt***) (to control armyworm) and those of ***Trichogramma*** *chilonis* for leaf miners are some of the biological control remedies for insects.

There are many other pests like termites, rodents and nematodes. Suitable combinations of the above chemical agents along with biocontrol programmes can help control these pests.

Fig. C.13. Damage to groundnut crop as a result of leaf miner causing leaves to shrivel. Courtesy of: http://takingroots.in/ipm_aphids (licensed under creative commons) with permission from Dr Ramanjaneyulu GV, Executive Director, Centre for Sustainable Agriculture, Secunderabad, India.

Crop – lucerne (alfalfa) pest control

Lucerne (*Medicago sativa* or alfalfa) is considered one of the best forage crops used by man. Believed to have originated in Iran, it is rich in proteins, calcium and phosphorus and is capable of fixing atmospheric nitrogen. It is capable of improving soil. It is a good crop in rotation cycles. It has a high fodder value for horses, cattle, rabbits and poultry, in combination with some cereals. It belongs to the order Fabales and family Fabaceae.

Lucerne is a small-seeded, perennial legume plant that grows in dry and cool weather with well-drained soils. It can tolerate temperatures up to 40°C and pH up to 8, but it cannot tolerate waterlogged or acidic soils. It survives grazing or heat via underground vegetative growth.

Many types of pests attack the lucerne plant. IPM strategies auger well for the healthy growth of lucerne, especially since IPM can keep a check on pesticide residues in the fodder, and thereby in the food chain. **Agroecosystem analysis** (**AESA**) and personal monitoring through pheromone, light, water and sticky traps form the foundation of IPM. Seed treatment with *Rhizobium* culture encourages healthy seedlings.

Root rot, bacterial **wilt**, seedling blights and damping-off are some of the fungal diseases that infect the lucerne plant. Nematodes also attack the roots, while leaf spot and other foliar diseases can take their toll on the growth of the plant.

Phytophthora megasperma and *Aphanomyces euteiches* are examples of two fungal pathogens that affect the roots, seeds and the seedlings. The former commonly infects the seeds even before they germinate, whereby the seedling growth is instantly affected, sometimes leading to plant death. *Phytophthora* root rot, stem rot, crown rot, damping-off and seedling blight are caused by this pathogen. *A. euteiches* infection allows the seedling to emerge, but leads to stunted growth with reddish-brown leaves. Since the growth is slow, the plant faces disproportionate competition with weeds. Root diseases are severe in wet and highly clayey soils. These are accompanied with **chlorosis** and necrosis of above-ground plant parts. Young plants commonly die and older plants develop cankers on the sides of the roots. In crown rot, the entire crown area is damaged.

It is advisable to use resistant varieties. Root rot can be controlled with soil treatment with **metalaxyl**, which can control the disease at the seedling stage itself.

Bacterial leaf spot, caused by *Xanthomonas campestris* subspecies *alfalfae*, causes severe defoliation and kills seedlings, with the disease conditions becoming worse during extended periods of hot, rainy, windy weather. The bacterium, which overwinters in crop debris and seed, spreads with rain, wind and infected tools or plants.

Bacterial wilts, verticillium wilts and *Fusarium* wilts are some wilt diseases that affect lucerne. These diseases are caused by *Corynebacterium insidiosum*, *Verticillium albo-atrum* and *Fusarium oxysporum*, which attack the roots and stems. The affected plants show stunted growth before they die. The bacteria survive for a period of 3 years in the soil. Disease resistant varieties are strictly recommended for use.

An insect pest called clover root curculio (or sitona weevil) eats at the roots of the plant and causes injury to the plant, making the plant vulnerable to secondary infections. A threatening insect pest of lucerne is the alfalfa weevil (*Hypera postica*). Adults chew on leaves and, after several weeks, make cavities in the stems and lay eggs in the cavities. Within a fortnight, the larvae hatch and begin to feed on the inside of the stem, slowly moving to the leaf bud. This leads to distorted growth of the plant. The larvae then pupate on the ground, to emerge as adults in around 10 days, and the cycle continues. Immediate cutting of the plant after harvest and destroying the crop residue can help control the insect. Ladybug beetles, **lacewings**, damsel bugs and some wasp species prey on the larvae, thus controlling the infestation biologically.

There are many leaf diseases of lucerne. They are mostly fungal diseases and include common leaf spot, *Stemphylium* leaf spot and **downy mildew**. The common leaf spot caused by *Pseudopeziza medicaginis* creates small bumps on the leaves. The *Stemphylium* leaf spot caused by *Stemphylium botryosum* causes oblong spots on leaves surrounded by a yellow halo. The downy mildew caused by *Peronospora trifoliorum* causes yellow patches on the leaves. The pathogen produces spores and mycelia, which are visible under the leaf. The black stem leaf spot is common during warm, moist conditions and is evident as small brown spots. Use of resistant varieties and early harvest of stands are two management strategies.

Aphids transmit a viral disease caused by the alfalfa mosaic virus called alfalfa mosaic. It may look similar to downy mildew. Here, yellow streaks form on the leaf. Use of aphid-resistant varieties and controlling aphid growth are two ways to manage this disease.

For weed control, it is advisable to use certified, clean, weed-free seed. Since lucerne grows fast, hand weeding is possible only in the first month. The main weed species that bother lucerne are sweet clover (*Melilotus* spp.), yellow sweet clover (*Melilotus officinalis*), dodder (*Cuscuta* spp.), star thistles and **yellow nutsedge**. Pre-plant herbicides during seedbed preparation are important in lucerne cultivation, since these agents can kill weeds yet to emerge before the sowing of lucerne can begin; for example, **fluchloralin** (at 2 l/ha/500 l water may be sprayed at the pre-sowing stage), **dalapon** (pre-sowing spray at 4 kg/ha, followed by disc harrow action to penetrate the soil) or **oxyfluorfen** (0.4 l/ha/500 l water). **2,4-DB**, **imazethapyr**, **trifluralin** and **imazamox** can control some grass weeds and broadleaved weeds, including the parasitic weed dodder.

Crop – maize pest control

Maize (*Zea mays*), a native of the American continents, is a member of the grass family. It was called maize by the native Indians. It is believed by some that today's maize originated from a Mexican wild grass called teosinte, *Zea mexicana*. Maize is also referred to as corn.

The maize plant grows to around 2 m and bears separate staminate inflorescence (the tassels on the stalk) and pistillate inflorescence (which forms the ear of maize) (Fig. C.14). The ear, containing the kernels, is tightly covered with a husk or bract, making it hard for seeds to disperse. The maize plant grows on all soils, except very clayey or very sandy soils. It cannot tolerate water logging and is moderately sensitive to salinity. Maize grows best in temperate to tropical climates, with daily average temperatures above 20°C.

Maize contains large amounts of carbohydrate but low amounts of proteins. According to the nature and composition of starch in the endosperm, maize can be of various types, such as sweetcorn, popcorn and flour corn.

Maize is used as animal feed and as an ingredient of many processed foods and industrial products such as corn (maize) starch, laundry starch and glue.

Fig. C.14. A maize field. Courtesy of G.E. Cardon, Bugwood.org. With permission.

Effective management practices for maize cultivation involve common strategies that help optimize yields, reduce the risk of pathogens and maintain ideal soil conditions. It is a common practice to cultivate hybrid maize varieties that have a robust root system and produce better yields. Early planting is an accepted practice to minimize rootworm attack. The eggs hatch in spring, and the larvae feed on young roots while the adults feed on leaves and pollen. If the plant completes the growth and pollination stages before the worms grow into adults, the loss caused by these pathogens can be minimized.

Crop rotation is considered a useful management practice in maize production. Maize is usually rotated with soybean to fight pathogen attack, considering the nature of residues left behind by either crop. Maize generates more residues than soybean. Maize residues breakdown gradually while soybean residues disintegrate easily. A good combination of these residues and with the help of good tillage, waterlogged conditions may be kept at bay, thus aiding the maize crop.

Generally accepted tillage practices for maize in low rainfall areas are ridge tillage and no-tillage. A one-pass field cultivation, or autumn chisel ploughing and spring cultivation are considered ideal on poorly drained soils. However, the maize yield decreases with no-till practice.

Considering that poor soils lead to poor crops, the management of nutrients is another crucial task for optimum maize yields. Nitrogen (N) may be supplied through anhydrous ammonia and injecting UAN (urea-ammonium nitrate) into the soil. Phosphorus and potassium may also be incorporated, keeping in mind not to overdo phosphorus application, since it leads to zinc deficiency. Incorporation of manure and soybean residues is also useful. Row cleaning devices are useful in reduced tillage conditions. This aids stand establishment and early growth.

Maize, like any other crop, is susceptible to attacks, mostly from insects and fungi. *Diabrotica virgifera*, the western corn rootworm, and *Diabrotica barbari*, the northern corn rootworm, are two beetles that pose a threat to maize crops. Western corn rootworm, which can create major losses, has a black wing cover, while the northern corn rootworm is greenish. Western corn rootworm adults lay eggs in the soil in late summer months, which overwinter and then they hatch by May the following year. The larvae (that develop via many instars) feed on the root system (root hairs, small roots and even primary roots) causing injury to the plant and resulting in lodging. The larvae pupate and then emerge as adults that feed on leaves and pollen for over a month, causing further damage. They then mate and lay eggs, continuing the cycle. These pests remain viable even when maize is rotated with soybean.

Manual inspection, use of yellow sticky traps and rootworm-resistant varieties, seed treatment and foliar insecticidal sprays are all useful in managing this insect pest. Seed treatment with **clothianidin** and **thiamethoxam** are considered useful in the control of corn rootworms. Soil insecticides such as **chlorpyrifos**, **terbufos** and **carbofuran** are also effective in larval control, while keeping resistance development in check. Early planting is an accepted practice to minimize rootworm attack. As already explained, the eggs hatch in spring, and the larvae feed on young roots while the adults feed on leaves and pollen. If the plant completes the growth and pollination stages before the worms grow into adults, the loss caused by these pathogens can be minimized.

The European corn borer (*Ostrinia nubilalis*; Fig. C.15a) that belongs to the order Lepidoptera, is another serious pest of maize. These insects see two generations per year. In humid June, the 2.5 cm adult moths lay many white eggs under maize leaves, along the midrib, which hatch in a week's time. The five-stage larval development begins with the first two instars feeding on the leaves (causing holes in the leaf tissue), the subsequent instars boring into the stalk to pupate and emerge as moths in early August. These moths mate again and lay eggs that hatch into larvae that feed on leaves, leaf sheaths, ear tips, ear shank and ear.

The use of hybrid varieties, light traps and pheromone traps are some ways to manage these borers. Heavy rain or irrigation can drown young larvae, just as dry windy conditions can dry up egg masses. Natural control by the use of fungal agent ***Beauveria bassiana*** and wasp ***Trichogramma*** spp. can control these pests. Broad-spectrum insecticides like lambda cyhalothrin can control most larvae. **Parathion-methyl**, **fonofos** and **chlorpyrifos** may also be used for control.

Similarly, the polyphagous larva of moths such as *Helicoverpa zea* (and often *Helicoverpa armigera*) is called the corn earworm when it feeds on maize (Fig. C.15b). The yellowish-brown female moth of *H. zea*, which can lay up to 3000 eggs in its lifetime, lays single eggs on fresh and young maize silks. The larva hatches out and feeds on the silk, the kernel and the ear tips. This damages the produce significantly. Within 2 weeks, they pupate in the soil, overwintering until the adult emerges in

Fig. C.15. Two common pests of maize. (a) European corn borer. (b) Corn earworm. Courtesy of John Obermeyer, Purdue Extension Entomology. With permission.

another fortnight. It is difficult to control the larva since it is inside the maize for the majority of its lifespan.

Pyrethroids can control earworm damage. But they have to be used in moderation, lest the pest develops resistance.

Maize rust, caused by *Puccinia sorghi*, is a fungal disease that affects sweetcorn and is evident by brownish pustules on leaves that rupture and reveal reddish spores. In dry and warm weather, the spores get blown on to different locations of the field. Applications of **mancozeb** can prevent the epidemic spread of the disease.

Another disease is the southern maize leaf blight caused by the fungus *Bipolaris maydis*. This pathogen forms lesions in maize leaves. The fungus overwinters as mycelia in infected plants, and begins sporulating as warm and humid conditions occur. The conidia are transported by wind and rain on to new plants, where the fungus infects via the stomata. The symptoms are seen as lesions on the leaves that produce conidia, which becomes the base for secondary infection. Maize (corn) blight can significantly damage maize crops if not controlled. The plant is unable to carry out photosynthesis, affecting grain filling.

Disciplined crop debris management and manipulating planting time can help control the initiation of the disease. The use of broad-spectrum fungicides such as pyraclostrobin and **azoxystrobin** can achieve effective control of infection.

Corn (maize) **smut** is a major fungal disease of maize caused by *Ustilago maydis*. This fungus has two spore stages, namely teliospores and basidiospores. The pathogen overwinters on plant residues or on the soil in the teliospore stage. It germinates in spring as the mean temperatures start to increase and the basidiospore stage begins. The fungus reaches young and growing tissues of maize plants and produces galls or tumours of brownish-black spores in the ears, stalk and the tassels which then rupture through the membranes (Fig. C.16). In this way, propagation of the fungus is carried on. Corn smut is commonly present, throughout all maize producing regions, at least to some extent.

Manual removal of galls while they are still young (greyish colour), and use of resistant varieties, are two important measures of managing this disease. Use of copper and sulfur fungicides is also effective, especially in the initial stages of the infection. Use of for example **fonofos** and chlorpyrifos can help in controlling the spread of the disease.

Crop – onion pest control

Allium cepa or onion is a pungent vegetable believed to have wide-ranging medicinal properties. In India and many Asian countries, onions form an essential part of the daily diet.

A. cepa grows in both temperate and subtropical conditions. They grow well in raised, well-drained, medium clay soils, but can adapt to a range of soil and climatic conditions. In rain-fed areas with erratic monsoons, ready seedlings can be transplanted between seasons to take advantage of the rains.

A number of fungal, bacterial, viral and nematodal infections attack the onion crop. The intensity of attacks varies with the pathogen and farm management techniques used. The following measures form the foundation of IPM in onion farming: (i) **agroecosystem analysis (AESA)**; (ii) field surveys; (iii) monitoring through the pheromone, light, water and sticky traps; (iv) **crop rotation** (with soybean and cereals); (v) land preparation; (vi) use of clean, certified disease-free seed; (vii) use of resistant varieties (e.g. N 2-4-1 and ARKA Niketan); and (viii) seed treatment. **Soil solarization** prior to sowing and soil treatment with **copper oxychloride**, **thiram** and **phorate** (at 20–30 g copper oxychloride per bed of 3 × 1 m) are found to be helpful. Maintaining planting discipline, such as proper row spaces, uprooting volunteer onions, removing remains of harvests from the field and avoiding excessive irrigation, are very useful practices. Rotating crops and maintaining a clean, disciplined cropping practice can reduce the severity of most infections. Seed treatment includes thiram or captan (3 g/kg seed) or *Trichoderma* spp. (5 g/kg). **Neem** oil cake at 200 kg/ha helps in preventing nematodes and fungal diseases. Green manuring helps to enrich the soil and control unnecessary vegetation that may become hosts to pathogens.

Weeds are a problem for the major part of the growing period of onions. To keep weeds in check, **fluchloralin** (670–900 g a.i./ha) can be applied fortnightly, starting at the preplant stage. A spray application of **demeton-*S*-methyl** and dithane a month after sowing, followed by application of urea

Fig. C.16. Smut gall on stalk. Courtesy of Gary Munkvold, Department of Plant Pathology, Iowa State University (www.plantpath.iastate.edu). With permission.

and **phorate** granules a fortnight after germination, is recommended. If the young plants are to be transplanted, this should be done 2 months after sowing.

Blight on onion is caused by different agents such as *Alternaria* spp., *Stemphylium* spp. and *Botrytis* spp. The disease spreads rapidly in the ***kharif*** season, in the midst of humid, cloudy weather. The first symptom of blight is the appearance of small sunken white patches on leaves, which turn purple red with yellow edges. The patches start from the tips and progress downwards. The plant gets a burnt appearance and the disease affects the production of seeds and bulb. Onions affected by blight do not store well; in fact they cause secondary infection, which spreads rapidly via air. Blight-causing *Stemphylium vesicarium* fungus attacks leaves, forming yellowish-brown patches, which spread from the base to the tip. Leaves shrivel up and get burnt tips. Blight caused by *Botrytis cinerea* appears in the form of several white spots and burnt leaf tips. A thrip attack encourages *B. cinerea*, which in turn intensifies the attack of *Alternaria* blight.

Deep ploughing, harrowing, solarization and rotation help in managing blight. Seeds are treated with **captan** and **carbendazim** (2 g/kg seed). Post-germination, a spray of a mixture of **mancozeb, dimethoate** and water helps to prevent the disease. If transplantation is chosen, roots of seedlings are dipped in a solution of **carbendazim** 50 WP 10 ml/l water and later in an *Azotobacter* culture (500 g in 5 l water). Post-transplantation, an EC of mancozeb may be alternated with **oxydemeton** fortnightly (**carbendazim** and **cypermethrin** may be used instead), with an insect trapping sticker.

Onion leaf or silver shoot (i.e. elongation of leaf sheaths and tubular gall at the base of a tiller) is caused by the rice gall midge (*Orseolia oryzae*) (Fig. C.17). Maggot-like larvae feed inside developing buds and form elongated eggs. The Asian rice gall midge is found in irrigated or rain-fed wetland environments during the tillering stage of the rice crop. It is also common in upland and deep-water rice. The adults are nocturnal and they can easily be trapped using light traps. During the dry season, the insect remains dormant in the pupal stage. They become active again when the buds start growing after the rains. The population density of the Asian rice gall midge is favoured mainly by: (i) cloudy or rainy weather; (ii) cultivation of high-tillering varieties; (iii) intensive management practices; and (iv) low parasitization.

Fig. C.17. Damage caused by the rice gall midge on onion shoots. Courtesy of: http://takingroots.in/ipm_rgm (licensed under creative commons). With permission from Ramanjaneyulu GV, Executive Director, Centre for Sustainable Agriculture, Secunderabad, India.

Leaf **smut** is caused by the soil-borne fungus *Urocystis sepulae*. Immediately after germination, cotyledons, new leaves and shoots exhibit black streaks or patches, leading to stunted growth. Older leaves and bulbs have patches, from which the black, dust-like fungus comes out, upon harvesting. Cool, wet springs intensify smut infection.

Crop rotation, seed treatment with WP of carbendazim or captan and the soil application of 30 g copper oxychloride (for every 3 × 1 m bed) while preparing seedbeds are two useful control measures. Transplanting only healthy seedlings from the nursery to the field is a good strategy.

Downy mildew in onion is caused by the air-borne fungus *Peronospora destructor*. The primary source of infection is the seed, while it is transported by air and water. The disease causes yellow-green patches (lesions) on leaves, which ultimately collapse and die. The disease covers the entire leaf, leading to burnt tips and deformed bulbs. Infection spreads with humidity. Downy mildew also leads to other fungal and bacterial infections.

Downy mildew management techniques include the selection of well-drained soil, use of certified disease-free seeds and crop rotation. At the initial stages, mancozeb and copper oxychloride should be alternately applied every 10 days, for three such repetitions. **Metalaxyl** (at 25 g/10 l water) sprayed twice every 10 days stops severe infection.

The onion crop also suffers from post-harvest and storage diseases. Some pathogens begin the infection process at the time of harvest and affect the crop in storage. Black mould and a bacterial soft rot are examples.

Black mould on the bulb is caused by the fungus *Aspergillus* spp. The fungus first infects the parts where the leaves are cut while harvesting, and then spreads to the bulbs. Infected bulbs exhibit black mould on the outer membrane which can be wiped off with a finger. This infection spreads to other onions in storage. Since higher temperatures and the presence of free water on the yield encourage sporulation, it is important to maintain dry and well-ventilated conditions in the warehouse at all times.

Bacterial soft rot of onion, a post-harvest disease caused by *Erwinia carotovora*, *Pectobacterium carotovorum* and *Dickeya chrystanthemi* and *Pseudomonas alliicola*, infects bulbs at the time of maturity, turning the outer membranes wet and yellowish. The skin becomes slippery and sour, and when squeezed, releases a peculiar smelling yellow discharge. This discharge forms the primary source of infection, which then spreads to injuries on the plant. Chemical intervention is not as useful as preventive methods. Effective preventive measures are: (i) avoiding injuries while harvesting or storage; (ii) mechanical drying of the onions harvested; and (iii) storing in well-ventilated warehouses.

Botrytis allii fungus causes neck rot in onions. This disease is initiated by an injury to leaves, resulting in the browning and rotting of bulbs. The surface turns rough, membranes shrivel and grey mould becomes visible. The disease spreads through seeds and the soil.

The choice of well-drained soil, disciplined crop rotation and moderate use of chemical fertilizers are important in rot disease management. At the first sight of disease symptoms in crops meant for seed production, infected areas should be drenched with **streptomycin** (3 g) and copper oxychloride (30 g) in 10 l of water. While being careful to avoid injury to bulbs or leaves, bulbs meant for storage should be sprayed with carbendazim or mancozeb along with a sticker, a fortnight before harvest. Dry leaves should be cut 6 cm from healthy bulbs meant for collection. They should then be dried for a fortnight before being stored in well-ventilated storehouses, pre-sprayed with **mancozeb** (25 g), carbendazim (10 g) and **dichlorvos** (20 ml) in 10 l of water.

Some of the insect pests that affect onion crops are given below.

Thrips (*Thrips tabaci*) lacerate the host tissue and suck sap. This causes blotching, wherein affected leaves turn brown and roll up. Indirect damage is in the form of the transmission of viral diseases such as spotted wilt virus of onion and blight caused by *B. cinerea*. The **economic threshold level (ETL)** is usually about 10% affected plants.

If introduced early, **ladybird beetles** and ***Chrysoperla*** spp. can adequately control thrips, as can sticky traps. Use of both systemic and contact pesticides like endosulfan (0.05%) and **monocrotophos** every 15 days, until some control is effected, is recommended.

The stem and bulb nematode (*Ditylenchus dipsaci*) feeds on parenchymal tissues, as a result of which seedlings look pale green, young bulbs become soft, swollen and deformed and leaf tips exhibit dieback symptoms. Severe infestation can cause damping-off. The entire life cycle of this pest can be completed in 25 days. **Isazofos, aldicarb** and benfuracarb are applied to control this species of nematodes.

Root maggots are another pest that affects the crop. The injury caused by these pests often leads to basal rot (caused by *Fusarium oxysporum*). Chlorpyrifos is commonly used to control these pests. *F. oxysporum* can infect the plant at any stage of growth. The symptoms are seen as leaves turning yellow or brown, from the tip to the base. Rot appears at the basal plate of the root region, spreading to the bulb. Such infected bulbs become watery and brown from inside, and while they may appear normal, the rot continues to spread, even in store houses.

The tobacco caterpillar (*Spodoptera* spp.) can be mechanically removed at different stages of its development and destroyed. Alternatively, *Telenomous remus* may be released at weekly intervals over 4 (consecutive) weeks to manage the infestation of the tobacco caterpillar and other insect pests. **Neem** extracts are also helpful in controlling tobacco caterpillars. Dust applications of **carbaryl** or spray applications of **lindane** or **endosulfan**, with a mandatory pre-harvest interval of a fortnight, are useful for the control of these caterpillars.

The damage caused by the *Eriophyes* spp. of mites reduces the plant stands, thus stunting their growth. These pests directly damage bulbs on the plant and in storage. Additionally, mites also help transmit viruses such as the mite-borne latent virus. Applying dicofol (450 g/ha) on first sight of infestation is a control measure.

Crop – orange orchard pest control

The sweet orange *Citrus sinensis* is a small flowering tree growing about 10 m in height with evergreen leaves (Fig. C.18). It originated in South-east Asia and is a subtropical crop requiring dry and warm temperatures (40°C and more), with low rainfall during its growing period. Well-drained, lightly loamy soils are suitable for good root development. Favourable annual precipitation varies from 12.5 to 50 cm, though oranges are frequently grown in high rainfall areas. Many orange species have adapted to cooler regions.

Fig. C.18. The orange plant. Courtesy of Kedar Champhekar. With permission.

Adoption of timely and effective IPM programmes is a necessity, as pest control is expensive and time consuming. Some methods and practices that control insects and other pathogens, while ensuring a good crop yield, are: (i) **Agroecosystem analysis** (AESA); (ii) field surveys; (iii) use of pheromone, light and water traps; (iv) deep ploughing; (v) solarization in the summer months; (vi) the use of a mixture of *Trichoderma* spp. and organic matter; (vii) disciplined pruning; (viii) drip irrigation; and (ix) maintaining a minimum planting distance of 6 × 6 m. Intercropping is suggested with, for example, cowpea, soybean and peas, to encourage the numbers of natural predators such as the **ladybird beetle** and ***Chrysoperla*** spp. during the initial years. Seeds used for growing the rootstocks should be treated with fungicides to control fungal diseases. Sufficient sunlight and ventilation keeps humidity to its optimum, making pests less comfortable. These form the foundation of IPM.

And yet, the orange plant gets attacked by many pathogens like insects and fungi, some of which are as follows.

Gummosis of orange or foot rot is a common fungal disease of citrus crops. The agents *Phytophthora parasitica* and *Phytophthora citrophthora* cause root rot, gumming of bark and discoloration of the lower part of the trunk. With increased moisture levels, the pathogens produce zoospores that penetrate the roots and germinate there. The plant is finally killed. Application of **Bordeaux paste** or copper oxifluoride to the pruned surfaces, prior to the rainy season, proves effective.

Fruit rot or brown rot caused by *Phytophthora palmivora* is also prevalent, but may be managed by a mixture of copper and potassium phosphate a month before harvesting.

Anthracnose diseases are caused by many species of ascomycetous fungi and affect foliage, flowers and stems of many plants by producing dark-coloured lesions on the affected part. One of the fungal agents causing anthracnose is *Colletotrichum gloeosporioides*. Post-bloom fruit drop and petal discoloration are common manifestations in citrus. A spray of **carbendazim** or **copper oxychloride** must be applied and repeated fortnightly, if required.

Citrus trees are also affected by the **sooty mould**, caused by *Capnodium citri*. Spraying of starch is one of the ways to detach the mould from the leaves.

Black fly (*Aleurocanthus woglumi*) is a major pest of orange. The blackish-grey winged adult lays eggs on the lower side of the leaves. Nymphs hatch, suck sap (along with the adults), proceed through four instars and pupate on the leaf itself. Both nymphs and adults excrete a sugary substance on which the black sooty mould grows, spreading to branches and the fruits. This disease condition, called ***kolshi*** in a local Indian language, is characterized by loss of sap and reduced photosynthesis. This affects the quality and quantity of the fruit. Application of 15–25 ml **monocrotophos**, **quinalphos** or demeton methyl, diluted in 10 l of water, is an effective control strategy, especially when the eggs start hatching. A mix of **neem** oil and a detergent powder dissolved in water may be sprayed at the hatching stage, and repeated if required. These measures reduce the mould as well as the direct damage caused by the pest.

Another fly, the citrus whitefly (*Dialeurodes citri*) creates notable damage to the orange plant. The 1.5 mm long adult moth lays oval eggs on the underside of the leaves. Eggs hatch into yellowish-green nymphs. Besides feeding on the sap of the plants, these insects excrete the characteristic **honeydew** that attracts the fungal sooty mould. If over ten nymphs are found per leaf (the **economic threshold level, ETL**), the release of ten larvae of the parasitic **green lacewing** (*Chrysoperla carnea* and *Chrysoperla rufilabris*) is recommended to control the growth of the pest. The release of the ladybird *Clitosthetus arcuatus* is also effective for biocontrol. These and other predator insects also control aphids, **mealybugs** and thrips. Two fungal species of the genus *Aschersonia* also help in controlling citrus whitefly.

The lemon butterfly and the fruit sucking moth are two other important pests of the orange tree. The caterpillar of the lemon butterfly (*Papilio demoleus*) eats up all the leaves of the young plant, with the most damage being caused between July and September. The second pest, the adult fruit sucking moth (*Achaea janata*), is a large, colourful lepidopteran pest. It attacks the ripened fruit at night, while the larvae survive on alternate weeds *Tinospora cordifolia* (called *gulwel* in the Indian language of Marathi) and *Psoralea corylifolia* (called *babchi* in Indian languages Hindi and Marathi). Control measures include: (i) physical removal of the pest; (ii) elimination of alternate host weeds; (iii) application of ***Bacillus thuringiensis*** preparations; and (iv) a spray of 35% **endosulfan** or 50% **carbaryl** diluted in 10 l of water. Application of monocrotophos and quinalphos may also be considered. Light traps or pest baits of jaggery and **malathion**, mixed in water, may be hung from branches to capture the fruit sucking moth. Also, burning wet garbage around the trees can kill or repel these pests.

Orchards that are older and not so well tended attract insect pests like the bark-eating caterpillar (*Indarbela quadrinotata*) that lays eggs in the bark of the tree, thus injuring the plant. It makes a web out of saliva and excreta, causing the plant to wilt. Manually removing the web and applying monocrotophos into the holes created in the bark can manage this disease.

Mite pests *Tetranychus* spp. and *Typhlodromus* spp. are microscopic mite pests that attack the leaves and fruit by lacerating them and then sucking the sap, causing white patches on the leaves and reddish patches on the fruit. At an ETL of 2% infested fruit, the quality of the pulp is significantly affected. The mites can be controlled by 20 g sulfur, 7.2 ml **phosalone** and 13.5 ml dicofol sprayed with 10 l water. *Chrysoperla* larvae are also an effective control agent.

Aphids are controlled by applying **imidacloprid** (10 g/ha) and using healthy seedlings.

The leaf miner fly *Phyllocnistis* spp. is a 1.5 mm long, foliar pest that punctures the leaf and sucks the sap, and lays eggs in the hole that has been created. The larvae that hatch out feed on the leaf, causing them to curl up and break. At an ETL of 10% of affected leaves, an application of quinalphos (20 ml/10 l water), or a combination of 20 ml methyl demeton and 10 g detergent powder/10 l water, can be applied. Imidacloprid sprays are also recommended for its control. A pre-sowing soil treatment with **phorate** (10 kg /ha) can help prevent the occurrence of the pest.

The rapidly breeding mealybug *Pseudococcus filamentosus* can cause great damage to leaves and branches, causing twisting of leaves and stunted growth. Both larvae and adults feed on the sap, thereby weakening the plant. Since the adult lays eggs in the soil, treating the soil before sowing can prevent infestation. Applying grease to the stem prevents the pest from climbing up the tree. Pruning to allow sunlight to filter through is advisable. At an ETL of 10% infested fruit, **dimethoate** or **carbaryl** mixed with kerosene and diluted in water may be applied. The predator beetle *Cryptolaemus montrouzieri* and the parasitic wasp *Leptomastix dactylopii* can also control the citrus mealybug.

Citrus psylla (*Diaphorina citri*), a kind of plant louse, is a sap-sucking pest, laying around 900 eggs in half-opened leaves. Both adults and nymphs that feed on the sap of leaves can be controlled by spray application of monocrotophos and **neem** formulations. Control of alternate hosts such as curry-leaf plants can also control the pest.

For damage caused by root-knot nematodes (*Meloidogyne* spp.) apply neem seed cake and use resistant rootstocks. The application of cultures of *Pseudomonas fluorescens*, which are Gram-negative, rod-shaped bacteria, exploited for the control of root diseases like *Fusarium* spp. and *Pythium* spp., may be used to control nematodes. The modes of action are the production of 2,4-diacetylphloroglucinol (2,4-DAPG) and competition with nematode pathogens for colonizing plant roots. Severe infestation is addressed by the application of 1 kg carbofuran granules/ha.

Insect infestations lead to vector-borne diseases caused by viral pathogens. Greening of citrus is a disease caused by *Citrus tristeza* virus (CTV), a thread-like rod virus, whose particles live in the plant's phloem. The transmission of this infection is brought about by aphid species such as the citrus brown aphid and melon aphid. The use of certified bud-stocks and resistant rootstocks help to prevent the spread of disease. Innoculation with a mild strain of CTV is also done to prevent infection.

Citrus canker is a bacterial disease caused by *Xanthomonas citri*. The pathogen enters through the stomata and through injuries and wounds caused by pruning and insect infestations. The symptoms of infection are tiny, blister-like lesions and early fruit and leaf drop. The presence of free moisture induces the bacteria to ooze out and disperse. Optimum irrigation, along with the application of **streptocycline**, **Bordeaux mixture** and any other copper fungicide helps in controlling the canker. **Mancozeb** may also be applied.

To control soil-borne diseases such as *Phytophthora* and gummosis, 50 g of *Trichoderma* spp. culture in 10 l of water may be applied as a soil drench or through a drip irrigation system, at the time of planting. As mentioned above, soil treatment with phorate (10 kg/ha) at the pre-plant stage can help prevent the occurrence of many pests.

Crop – pigeon pea pest control

Cajanus cajan, pigeon pea or red gram (also called *arhar* or *toor* in India) is a legume crop grown mainly in Asia, East Africa and the American continent. Well-drained, non-acidic soil and a temperature of 30–35°C during the early stages are the conditions in which pigeon pea grows well. Moisture must be made available in the early growth stage, since this is a ***kharif*** crop (crops sown just before the onset of monsoons, i.e. June–July). Since it is a deep-rooted plant, proper seed preparation goes a long way to having a healthy crop. Like all legumes, the pigeon pea fixes nitrogen in the soil, while being a source of protein. Normally, it is grown as an intercrop with cereals such as sorghum, maize, groundnut, cotton and pearl millet. There are many drought-resistant varieties available, making *toor* cultivation possible in many areas of the continents mentioned above.

Several pests and diseases attack pigeon pea, hence prophylactic measures are crucial. Cultural practices that are essential for an environmentally sound pest management strategy, while ensuring high yields, include: (i) manual inspection; (ii) use of light and pheromone traps; (iii) crop rotation; (iv) use of good, healthy, clean and disease-resistant seeds; (v) planting of trap crops; and (vi) **soil solarization**.

Since the pigeon pea crop grows slowly in the first 2 months, weeds can outpace them very quickly. Herbicides such as alachlor, **prometryne**, **fluchloralin** and **paraquat** are some compounds that may be used singly or in a recommended combination to keep broadleaved and grass weeds at bay. **Pendimethalin** (at 1.5 kg/ha) may be sprayed when the soil is wet, after sowing but before germination. Spraying of **glyphosate** (100 ml in 10 l water) on existing weeds but before sowing of the crop seed is a common practice. Use of glyphosate-resistant seeds is another measure to prevent a weed problem. Manual weeding between rows, with hoes, is essential at regular intervals in these 2 months.

The green larva of the pod borer *Helicoverpa armigera* is a significant pest of the pigeon pea crop. The larvae that hatch out of the light brown eggs laid on flowers start feeding on leaves and pods, leaving behind holes in the pods. Bird perches at a low level and pheromone traps are means of controlling these pod borers. **Monocrotophos** sprayed as an emulsifiable concentrate (EC) or an application of **fenvalerate** or **cypermethrin** are very effective for the control of the pod borer larvae. **Endosulfan** (EC 20 ml) may also be applied on the underside of leaves, if necessary. As a prophylactic measure, ***Helicoverpa* nucleopolyhedrovirus** (HNPV 250 LE 550 ml/ha) is sprayed on the crop. Besides *H. armigera*, there are many species of pod borers that cause varying degrees of damage to the crop. Other insect pests include bugs, beetles and flies such as the blister beetle (*Mylabris pustulata*) and pod fly (*Melangromyza obtusa*). The pod fly is controlled by an EC of **dimethoate** (15 ml per 10 l spray). The plume moth or *Pisari patang* (*Exelastis atomosa*) is controlled by endosulfan (as EC), **phenthoate**, **quinalphos** or **phosalone**. The nymphs and the adults of aphid species *Aphis* *craccivora* suck the sap of the stem, leaves, flowers and pods, excreting **honeydew** (Fig. C.19). This results in secondary fungal infection. The whole plant can die as a result. The natural enemy **coccinellids** can keep the aphid population under control. The EC of methyl dematon (25% EC spray solution in 300 l water/ha) and phosphomidon are useful chemical agents for aphid control.

Fig. C.19. Damage caused on pigeon pea by aphid infestation. Courtesy of: http://takingroots.in/ipm_aphids (licensed under creative commons). With permission from Ramanjaneyulu GV, Executive Director, Centre for Sustainable Agriculture, Secunderabad, India.

Fungal pathogens also infect the pigeon pea crop and may damage portions of it as a result. Seed treatment with **carbendazim** and **thiram**, and use of agents of biological origin like *Trichoderma* and *Pseudomonas* spp., are crucial in avoiding many fungal diseases and ensuring a healthy crop. **Wilt**, caused by soil-borne *Fusarium* spp., can lead to damage during flowering and the podding time, because of the tap root and the lateral roots getting infected. Crop rotation with crops such as sorghum or tobacco and the use of resistant varieties (e.g. BSMR-736 and C-11) are an effective method to counter this disease. The **powdery mildew** (*Erisyphe polygoni*), which attacks leaves and flowers with a powdery white growth, leads to defoliation and stunted growth. This disease is controlled with 300-mesh sulfur dust (20 kg/ha). Alternatively, a water spray of **sulfur** may also be used. Triadimefon 25 at 0.03% or **carbendazim** (0.5–1.0 g/l) effectively controls the disease. Small necrotic spots that form lesions on the pigeon pea foliage with brownish concentric rings is known as ***Alternaria* leaf spot**. This leads to leaf drop and drying up of branches. Use of resistant varieties like ICPL 366 and DA 2 is helpful. Chemical intervention includes **mancozeb** (3 g/l).

Bacterial leaf spot and stem canker are caused by *Xanthomonas campestris*. The disease forms minute brown lesions with a yellow halo on the leaves. As the infection spreads, lower leaves drop and branches dry up. Seed treatment with **streptocycline** followed by spraying with the same antibiotic at 100 ppm (if required) controls the disease.

Sterility mosaic, a viral disease, carried by mites (the eriophyid mite) affects the pigeon pea crop to a large extent (Fig. C.20). The infection leads to yellowing of the younger leaves and stunted appearance of leaves and branches. Even one vector mite can spread the disease, which is incurable. It is easy to spot the extent of the disease by observing patches of stunted, pod-less, pale green plants. In severe cases, pods and seed set are almost negligible. Humid and wet conditions favour the spread of this disease.

Fig. C.20. Damage caused by viral disease in pigeon pea transmitted by mites. Courtesy of: http://takingroots.in/ipm_mites (licensed under creative commons). With permission from Ramanjaneyulu GV, Executive Director, Centre for Sustainable Agriculture, Secunderabad, India.

Control over mites is a preventive measure, and hence a spray of a combination of wettable sulfur and monocrotophos is recommended. Dicofol (3 ml/l water) or metasystox (0.1% in 1 l water) can also help. Roguing of the affected plants and practising rotation are the cultural control methods. The use of resistant varieties such as BDN 736 is recommended. Removing crop residues and volunteer crop plants deprives mites of their home, thus helping to prevent the incidence of mites.

Among many nematode species associated with the crop, root-knot, spiral, cyst, reniform and lesion-forming species can cause disease in pigeon pea plants. The use of nematode-resistant, healthy, clean seeds (e.g. Basant), crop rotation, ploughing and solarization with plastic sheets reduces the occurrence of nematodes. Application of **phorate** granules during soil preparation is the chemical measure to control nematodes.

Crop – potato pest control

Potato (*Solanum tuberosum*) originated in South America and has today spread to the rest of the world. The potato is a member of the *Solanaceae* family, which also includes tomato, tobacco and aubergine (brinjal). The crop, with high calorific value, is one of the most widely consumed foods of most people in the world. Major producers include China, Russia, the USA, countries in the EU and India.

The potato plant is a short duration crop and within 4 months it is ready to be harvested. It needs cool weather and is thus sown by early spring in temperate regions and by late winter in warmer regions. The ideal soil is loose, aerated, well drained and rich in organic matter. Non-saline and non-alkaline soils are unsuitable for growth. Sandy and loamy soils and soil with a pH of 5.2–6.4 are considered ideal. Propagation takes place via tubers (or parts thereof) or by the seeds present in the fruits of the plants.

The *Solanum* spp. produce toxins called glycoalkaloids (solanine and chaconine) on sprouts, leaves and stems. While for the plant this toxin may be produced for self-protection, consumption can cause headaches, confusion and stomach cramps.

Potato, like any other crop, is attacked by several pests, which results in low yields and losses. While selecting disease-resistant varieties is important, time of planting (region-wise), irrigation schedule, and organic and chemical fertilizer/pesticide application are equally important. Similarly, to maintain the crop, cultural practices and pest management techniques, interspersed with pesticide application, are important. The main IPM techniques are: (i) monitoring through the use of pheromone, light and water traps; (ii) crop rotation (with e.g. cucumber, pumpkins, maize, lettuce); (iii) solarization; and (iv) the use of natural enemies.

Early blight of potato, caused by fungal pathogen *Alternaria solani*, is one of the most widely known foliar diseases of potatoes. Moisture on the foliage is conducive to the spread of the pathogen. The spores and mycelia of the pathogen overwinter and grow on infected plant debris or infected tubers in a field, and, in the midst of rains and wind, the air-borne spores move to newer plants. Large portions of newly sown potatoes can be affected by this infection. The symptoms are circular or irregular brown spots that appear between the leaf veins on both sides of mature leaves. As these lesions increase, the leaves become chlorotic and start to wither and die (while still on the branch) giving them a burnt look. When lesions appear on stems, it leads to smaller-sized tubers. The symptoms of early blight on tubers are seen as circular, brown, sunken lesions. The pathogen enters the tuber during harvest, where they may be prone to injury during the harvest procedure or if the soil is coarse. The early blight pathogen completes many life cycles of spore production in just one attack; spores from the leaf lesions spreading secondary infection.

Protectant fungicides such as **maneb**, **mancozeb** and triphenyltin hydroxide may be foliar sprayed on the potato plant (from 175 g to 2.25 kg/ha) along with **strobilurin** applications. However, it is important to be aware, at all times, that *A. solani* and many other fungal species have developed resistance to the strobilurin group. **Azoxystrobin**, **trifloxystrobin**, **famoxodone**, pyrethamil, **fenamidone** and **boscalid** are some of the systemic agents that must be mixed with the above-mentioned protectant fungicides. A spray of **copper oxychloride** (0.3% a.i.) and **chlorothalonil** (15 g/10 l) spray may also be applied. The timing and the frequency of application is crucial to the control of early blight of potatoes. Balanced fertilization, especially with a spray of urea on the 45th day of planting, is also helpful.

Late blight is caused by the fungus *Phytophthora infestans* (an oomycete water mould), which is a soil-borne pathogen, observed all over the world. The disease may attack at any stage of the crop, and severe infection may destroy the entire crop. The symptoms include green or brown spots on lower leaves that transcend leaf veins. The leaves later appear as sunken and water-soaked lesions. In severe cases, the entire leaf withers and dies. Again, wet and humid conditions favour the initiation of infection, and rain and wind can carry the spores to neighbouring plants. During cool conditions, the leaf has a velvety look. The infection can spread to the roots and to the stem, with the stem appearing reddish. The spores reach the tuber with the flow of water. If left unchecked, this infection can spread rapidly, with many cycles of spore production in one spate of infection. The infection on the tuber is evident by brown patches and spots on the skin, rot-like conditions inside and finger-like extensions appearing in the healthy tissue. However, the pathogen cannot survive outside living tissue, and hence, if the plant freezes or is buried deep inside the soil, the disease cycle can break.

Uprooting and destroying (burning) the infected plants can control the disease. Other ways to control the occurrence of the disease are: (i) controlling irrigation so that leaves remain dry for a substantial period of the day; and (ii) allowing infected tubers or plants to freeze. On sighting the disease, application of **metalaxyl** and mancozeb at 0.25% a.i. is recommended with fortnightly repetitions.

Fusarium solani and also *Fusarium oxysporum* fungi cause fungal wilt. The infected plants turn yellow and die. The stem and its internal parts turn brown. *Sclerotium rolfsii* fungus also causes wilting of the potato plants. Formation of knot-like bodies resembling the mustard seed near the stem is the symptom of the disease.

Soaking the tubers in 0.2% mancozeb solution for 10 min and drying them in the shade before planting lessens the possibility of disease. If the planted tuber seed is not covered well by soil, the tuber is easily susceptible.

Bacterial diseases like **wilt** and rot are also common on the potato crop. The bacterium *Ralstonia solanacearum* causes potato bacterial wilt. Symptoms are seen as a brown milky liquid oozing out of the infected stem or tuber, or even the buds of the tubers in severe cases of infection. Use of disease-free seed and crop rotation with onion, garlic, cabbage or cereals are recommended. Keeping the land fallow is another option. Sterilizing knives and other implements with 0.1% solution of mercury chloride and soaking the tubers in 200 ppm **streptomycin** solution for 1 h prior to planting are also practised. Application of stable bleaching powder at 12 kg/ha is recommended.

The causal agent of the potato bacterial ring rot is *Clavibacter michiganensis*. The bacterium, that spreads through soil and water, causes yellowing and stunting of plants. The tuber develops a ring-like brown spot when cut, and a yellowish secretion exudes from the flesh. Preventive measures are important in the management of this disease. The seed potato can be soaked for 30 min in a solution of **streptocycline** (1 g/10 l water). Streptocycline may be applied twice in the season.

Soaking the seed tubers in the solution of **captan** (30 g) and **carbendazim** (10 g) in 10 l of water prior to planting is recommended. Also an application of 5 kg *Trichoderma* culture/ha and **neem** oil cake at 2.5 t/ha is recommended.

Dry rot, caused by *Macrophomina phaseolina*, spreads through the soil and infects potatoes in storage. Black spots appear near the buds and spread to cover the entire tuber. The tubers rot in high humidity and dry up in dry weather. The fungus fills up the tuber from inside.

Use of disease-free seed potatoes and organic manure help prevent the disease. Select early varieties, so the crop can be harvested before the temperatures rise. Store the seed potatoes after soaking them in the solution of 20 g carbendazim/10 l water for 30 min.

There are about 25 viral diseases identified on the potato crop in India. Leaf roll virus, purple top roll virus, potato virus X and potato virus Y are the major ones. The leaf roll virus causes the plant leaves to curl up and turn yellow and other parts to appear burnt. The infected plants are stunted thus affecting the yield of tubers. The **vectors** of this virus are aphids and hoppers. The burnt appearance is due to the aphids. The purple top roll virus is caused by a phytoplasma spread by hoppers. Here, the leaves dry up and the plant is stunted. Potato virus X causes light green spots on the leaves, with the veins turning yellow. The disease spreads through contaminated equipment.

Effective measures to prevent the occurrence of this disease include: (i) use of disease-free seed potatoes; (ii) roguing out and destroying infected plants; (iii) practising crop rotation; and (iv) controlling aphids and hoppers.

Many insect pests too create damage to the crop. The potato tuber moth or *Pthorimaea operculella* causes two distinct types of damage. The larvae, when small, attack the inner portions of the leaves and are hence called leaf miners. The same larvae grow to be adults and attack and damage tubers in storage, and are then known as tuber moth. The adult is brown with greyish wings and has a shimmery look because of the fine hair on the body. The female lays over 100 eggs on the plant, which hatch into larvae and start feeding inside the leaves. After 2 weeks, the larvae pupate into a silky cocoon to emerge as adults in a week.

The damage caused by the larvae and the moths can lead to major losses, especially in subtropical regions, because they produce 12–14 generations in a year. The **economic threshold level (ETL)** is 15–20 moths caught consecutively for 3 days in the pheromone traps.

Some pest management techniques are: (i) crop rotation; (ii) intense soil cultivation; (iii) use of pheromone traps (during the leaf senescence stage); and (iv) use of organic manure to prevent cracking of the soil surface. Prompt harvesting of the tubers and subsequent cold storage, and fumigation of the warehouse before stocking the harvest, are other ways. Granulosis virus and the **parasitoid** wasp *Copidosoma koehleri* are also known to control the pest. As a last resort systemic insecticides may be used to control the leaf miner.

There are several other pests that also affect the potato crop, namely, aphids, hoppers, red mites, beetles, whiteflies and thrips. The Colorado potato beetle is a defoliating pest that attacks potatoes and also tobacco, tomatoes and aubergine. If unchecked, plant leaves can be badly eaten by the larvae.

All such infestations weaken the plant and promote the spread of viral diseases. The following treatment is found to be helpful: 10 kg of **phorate** granules/ha, 10 days after planting, followed by a spray of **endosulfan** (15 ml) or **dimethoate** (10 ml) or methyl demeton (10 ml) in 10 l of water. The menace of white grub larvae can be controlled by the application of carbofuran granules at 1 kg a.i./ha. A higher dose can also help control cyst nematode infection. This, along with the use of light traps, is useful. Cut worms can be controlled with **chlorpyrifos**. Rodents are also a nuisance and may be controlled by the use of **bromadiolone** at 10–15 g per burrow, or **zinc phosphide** (2–2.5 % baits at 10 g per burrow after baiting with plain bait to mislead the pest). Weeds should be removed from the bund and the bund cleared of crop residues so that rodents do not hide or burrow in them.

Crop – rice pest control

Rice (*Oryza sativa*) is a very widely consumed cereal, grown in over 100 countries. Rice is believed to have been in existence for over 7000 years.

Rice is an annual multi-stalked plant, growing to around 1 m in height (Fig. C.21). It is best suited to a watery ecosystem and is mostly grown in flooded fields. The grains, borne by a panicle, are surrounded by bracts. Air columns in the stalk enable the diffusion of air through the stem down to the roots. This way, they are able to resist decay even while they remain submerged for long periods. The processes of threshing and winnowing, followed by milling and polishing, convert the harvest into the product commonly consumed. Many varieties are distinct in their appearance and taste, such as long-/medium-/short-grained rice, black rice and glutinous rice. Rice is a good source of thiamine and niacin (both vitamin B). It also provides iron, phosphorus and magnesium. It has higher levels of the amino acid lysine compared with other cereal grains. Optimal levels of rice production are achieved by a balanced addition of sulfur, iron and zinc, as well as NPK.

Fig. C.21. Rice crop, variety Lemont, which is a high-yielding semi-dwarf and an early maturing variety. Courtesy of David Nance, USDA Agricultural Research Service, Bugwood.org. With permission.

Rice has adapted to grow in difficult situations, for example under conditions of drought, floods or nutrient stress, and can grow in rainfed lowlands, deepwater, tidal wetlands, rainfed uplands and irrigated areas. Upland rice grows without ponding.

Rice is also categorized on the basis of the nature and treatment of the grain (i.e. size, variety and processing). The nutrient value decreases when rice is milled and polished.

Methods for increasing the efficiency of rice cultivation include: (i) developing disease-resistant rice plants; (ii) modifying the life cycle of rice by delaying crop senescence and inducing early flowering, for higher yields; and (iii) controlling the menace of weeds.

Like all crops, rice is attacked by several diseases. The yield of rice is affected all over the world by many bacterial and fungal diseases. These pathogens attack various parts of the plant such as the leaf, stem, panicle and blossom. Although the severity and extent of damage varies with the geographic location or the variety of rice, the major diseases include bacterial leaf blight, leaf blast, stem rot, bacterial leaf streak, sheath blight, seedling blight, brown leaf spot, kernel **smut** and sheath spot. In addition to these, there are many other diseases which are specific to certain locations and seasons. Some of the ways to control or prevent infection or pest attacks are: (i) destroying the crop residues; (ii) avoiding over use of N fertilizers; (iii) early planting; (iv) proper flood management; and (v) the use of resistant varieties.

Xanthomonas oryzae, a Gram-negative pathogen, is a member of phylum Proteobacteria. It is responsible for bacterial leaf blight, a disease which is very common among rice-growing countries in South Asia. The pathogen attacks the plant at the seedling stage, which first makes the leaves turn greyish brown and then wilt and die. The leaf develops water-soaked lesions at the margin of its leaf blade, which then coalesce. The leaf discharge appears as white droplets. The disease becomes severe and contagious when there is heavy rain and wind.

The use of resistant varieties of rice and effective bactericides/antibiotics like **streptocycline** can control the spread of the disease. In some areas, rolling of stubble to encourage decay controls the disease. The use of mint and **citronella** extracts in water as well as **copper oxychloride** is found to control bacterial diseases in rice.

Rice leaf blast is caused by the selective fungal pathogen *Pyricularia grisea* (*Magnaporthe grisea* or *Pyricularia oryzae*). Leaf blast in rice affects the rice yield in both flooded and non-flooded areas. The plant is affected in two steps – the leaf and the panicle. Thus, this disease is also commonly known as panicle blast and rotten neck because the fungus prevents the maturation of the rice grain and breaks the seed head of the rice plant.

Another disease seen in rice is rice stem rot, caused by the fungus *Sclerotium oryzae*. The fungus attacks the leaf sheath near the water line at the reproductive stages, causing rectangular black lesions to expand, making way for the pathogen to enter the stems. The panicles are not filled, and the grain is minimal. The stems are weakened via lodging, thus killing the plant before harvest time. The inside of the stem and sheaf is filled with numerous dark spores and mycelia.

Stem rot may be managed by: (i) crop rotation; (ii) varying the flood levels; (iii) avoiding excess nitrogen fertilization; and (iv) burning the stubble to break the cycle. The single application of triphenyltin hydroxide (around 1.12 kg/ha) at the mid-tillering stage controls the spread of the disease.

Rice is also attacked by insects like stem borers, leafhoppers and plant hoppers.

The rice stem borer (*Chilo plejadellus* Zincken) is a serious pest at its larval stage. It is a member of the family Crambidae of class Insecta. Adult moths, which are whitish brown and 4 cm long, lay egg masses on leaves or behind leaf sheaths. Upon hatching in 5 days, the larvae bore into the sheath or bottom of the panicle, thus entering the stem. They feed on the inner tissues of the stem, thus depriving the plant of its nutrients and killing the central shoot. So panicles do not emerge, and if they do, they turn white and have no grain. The sugarcane borer *Diatraea saccharalis* also attacks the rice plant.

Termites and rice water weevils from the orders **Isoptera** and Coleoptera, respectively, attack the roots and damage the growing plant. Weevil larvae (*Lissorhoptrus oryzophilus*) are especially dangerous since they reduce the root system severely. Similarly, the armyworm *Pseudaletia unipuncta* (order Lepidoptera) may also sometimes act as a defoliator of rice.

Rice brown plant hoppers from the family Delphacidae are examples of sucking insects which feed on plant sap from the plant xylem and phloem. The plant dries up and looks brown and burnt, and the condition is called **hopper burn**. These insects are rampant in Asia's rice-growing countries. Not only do they damage plants directly, but some species (e.g. *Nilaparvata lugens*) are **vectors** of grassy stunt virus. The fungal pathogen *Pandora delphacis* can control the growth and spread of the brown plant hoppers. **Leafhoppers** from the family Cicadellidae attack the above-water parts of the rice plant and can damage much of the field, if left unchecked. Then there are green leafhoppers that affect rice seriously (Fig. C.22).

Fig. C.22. Damage caused by rice tungo, a viral disease in rice transmitted by the green leafhopper. Courtesy of: http://takingroots.in/ipm_rglh (licensed under Creative Commons). With permission from Ramanjaneyulu GV, Executive Director, Centre for Sustainable Agriculture, Secunderabad, India.

The Asian rice gall midge is a common pest of rice in all environments including wetland, irrigation and highland fields. Adult midges are nocturnal and become active with the onset of rains, when the buds of the rice plant emerge. The larvae, which are the pestilence stage of the midge, feed inside the bud of the emerging tiller by injecting saliva into the bud. This leads to abnormal growth of the tiller, causing a hollow cavity to develop at the base of the affected tiller and the emergence of the gall. Plant growth is stunted and the leaves become wilted and rolled. Tillers do not produce panicles.

The nymphs and adults of the rice gundhi bug (*Leptocorisa varicornis*) suck the milk from the developing grains and stem in the early stage of grain formation. Panicles become discoloured and some of them are empty or poorly filled. Infested paddy straw has a foul smell.

The main weeds that occur in rice include *Echinochloa crus-galli*, an annual grass that grows several tillers producing seeds with which they proliferate. *Echinochloa colona* is another annual that is difficult to weed out. Sprangletop (*Leptochloa fascicularis*) and watergrass are some of the other weeds that affect normal rice growth.

Proper seedbed and controlled irrigation are some of the cultural methods of controlling weeds in rice. A pre-plant application of **thiobencarb** gives good control of most of these weeds. Application of bensulfuron methyl at the one- to three-leaf stage of rice (or when it is well rooted) and when the weeds are still emersed in water is another chemical agent that can control rice weeds. **Propanil**, **cyanophos**, **cyhalofop-butyl** and **fenoxaprop-p-ethyl** are also alternatives that may be considered for insect and weed control.

Crop – sorghum pest control

The cereal *Sorghum bicolor* (family Poaceae and order Poales) is the main species of the genus that is used for direct consumption or in processing, in many parts of almost all continents (Fig. C.23). The genus has many grasses that either have a fodder value or a nuisance value (as in the troublesome perennial weed *Sorghum halapense* or **Johnson grass**). *S. bicolor* is the major crop of people in many drought-prone regions of Africa and Asia and is thus an important world crop.

Fig. C.23. A sorghum crop. Courtesy of Howard F. Schwartz, Colorado State University, Bugwood.org. With permission.

Growing in a variety of altitudes, sorghum grows best in clayey/loamy soils with high water retention. It tolerates high temperatures throughout its life, and alkaline soils. It is sown towards the cooler months (in India, it may be grown as a ***rabi*** crop in the south but may be a ***kharif*** crop in the north). Its planting may be considered between September and January. But in many cases, the crop is sown before 15 June to escape the shoot fly. In general, early planting with sufficient row spaces help keep away autumn armyworms, earworms, midges and borers. Early planting helps avoid the occurrence of leaf spot but places the plant under threat from **smut** disease. Fertilization with NPK (80-40-40) is ideal for its quick and healthy growth. Pre-planting activity may only constitute removal of weeds, since sorghum can grow as a zero tillage crop. Dense planting helps in the generation of high biomass, although the costs of insect infestation should be considered before deciding on plant density. Key points to be noted are that varieties that mature late (approximately 60 days) yield higher harvest. In situations of low moisture, an early maturing variety is preferred. Taller varieties are high yielding but face problems of lodging.

Cultural practices (e.g. destroying volunteer sorghum, post-harvest stubble management and elimination of host plants like Johnson grass) are very important to ensure pest-free cultivation of sweet sorghum. Rotating sorghum with non-grass crops like soybean and cotton helps in breaking the life cycles of some insects like wireworms, cutworms and white grubs. Use of seeds resistant to insect attacks and the use of pre-treated seeds are two other methods of keeping the crop healthy until harvest. Such seeds should germinate early so that they escape attacks from worms, aphids and green bugs, besides the earlier mentioned pests.

As is the case with all crops, sweet sorghum is attacked by a plethora of diseases such as rots, leaf spots, blights and rust. There are also many insect pests that infest the plant causing a strain on the value of the harvest.

Sorghum leaf rust caused by *Puccinia purpurea* is common in regions of high humidity. Young plants develop pustules on the leaf surfaces which break out as reddish spore masses. This disease can take on epidemic proportions if not checked in time. **Mancozeb** applications are commonly used to control the spread of the disease.

Many types of spot diseases affect the sorghum crop. Some examples are grey leaf spot, oval leaf spot, zonate leaf spot and rough leaf spot. Some of the causal agents are *Cercospora sorghi*, *Cercospora fusimaculans*, *Gleocercospora sorghi* and *Ascochyta sorghi*. In all cases, angular or non-definite lesions appear on the lower leaves, which then grow in size and join together, affecting the whole leaf. Humidity favours its spread. Cultural practices of optimum irrigation and sufficient row spaces can control the spread of the disease.

Downy mildew caused by *Sclerospora sorghi* results in green and white stripes on the lower surfaces of leaves, which finally tear along the stripes. Besides roguing out the infected plants, seed treatment with **metalaxyl** at 4 g/kg seed is recommended. A spray of metalaxyl or mancozeb also helps control the disease.

Leaf blight caused by *Helminthosporium turcicum* occurs during humid and cloudy conditions forming 2.5 cm spots with straw-coloured centres. Seed treatment with **thiram** at 2 g/kg seed and spray application with mancozeb or copper oxychloride are the recommended control measures. **Ergot** disease is common among sorghum.

Bacterial diseases caused by *Pseudomonas syringae* and *Xanthomonas campestris* can result in significant damage if ignored, leading to destruction of grain heads. Long wet spells or over-irrigation encourages the spread of bacterial blights. Round-to-oval lesions are formed on leaf blades, which enlarge and become purplish brown. Bacterial ooze appears as shiny granules on the leaf blades which wither and die. These pathogens and their subspecies also cause leaf stripes and bacterial leaf spot on sorghum. The bacteria overwinter in the leaf residues and infect the young plant tissue as it germinates by entering via stomata. Secondary infections occur by wind-spread and infected plant parts. Use of appropriate chemicals in moderation, combined with use of resistant seeds, clean tools and disciplined irrigation, help to avoid infection.

Common armyworms, cutworms, locusts, midges and many others can be controlled by a spray application of 700 ml **chlorpyrifos** in 100 ml water. Seed treatment with thiamethoxam provides systemic, broad-spectrum control of a wide range of insects like aphids, leaf miners and beetles. **Cyfluthrin, zeta-cypermethrin, dimethoate** and malathion are also used to control sorghum insect pests.

The sorghum shoot fly (*Atherigona soccata*) attacks the plant in the first month of germination. The fly lays eggs on the lower side of the leaves, which hatch within 3 days. The larvae start feeding on the growing parts of the plant, leading to wilting of the plant. The plant starts to tiller unusually.

The stem borer (*Chilo partellus*) attacks 40 days after sowing. The young growing shoots are eaten by the borer larvae, which causes stunted growth of the plant. The stem quality suffers since the larvae move into the stem.

Endosulfan 35EC (1.5–2.0 ml/l water) twice at an interval of 1 week on day 7 and day 14 of sowing is considered effective on both the shoot fly and the stem borer. Application of carbofuran granules to the top shoot whorl controls the damage of the borer. The use of ***Bacillus thuringiensis*** (*Bt*) (1 g/l) controls leaf damage caused by the borer. A biopesticide made of **neem** and other limosides, and called Nimbecidine, is used to control the growth of the borer. The fungus *Metarhizium anisopliae* acts as a biocontrol agent to control the pests.

The larvae of the midge fly (*Contarinia sorghicola*) feed on the ovaries and the grain in the ear, leading to empty ears. Endosulfan, **carbaryl** and **malathion** may be used to control midge fly.

The sorghum flea beetle feeds on the leaves, turning them yellow. This slows plant growth. An application of 3% malathion dust is effective in controlling the beetle. Similarly, the shoot bug (*Peregrinus midis*) attacks in wet conditions. They can affect the growth of the plant. Here too, the application of endosulfan granules or carbaryl can control the pest. Aphid attack (*Rhopalosiphum maidis*) results in poor emergence of the panicle, causing blotches and necrosis on the plant. **Oxydemeton-methyl** or **carbofuran** may be applied to control the aphid.

Crop – sugarbeet pest control

Sugarbeet (*Beta vulgaris*) is a hardy, temperate tuber crop, whose roots store sugar. With many parts of the EU, USA and Russia being major producers of sugarbeet, this crop accounts for nearly 25% of the world's total sugar production. Modern biotechnological research has developed varieties that can be grown in subtropical regions as a winter crop. This crop is mainly planted in the spring and harvested in autumn. Besides being a source of sugar, sugarbeet is used as a fodder crop and can be rotated with other crops like soybean. The sugarbeet plant grows short and has large leaves. It grows in a variety of soils and flourishes if the days are bright and sunny and the soil not acidic (ideal pH is between 6 and 8). Following are a few obstacles and challenges that a sugarbeet crop has to face.

The main threat to a healthy sugarbeet crop is weeds. The crop has poor tolerance to weeds, especially for the first 2 months of its growth. A wide variety of pre-emergence and post-emergence herbicides are used to control weed problems. The main weeds are **barnyard grass, common lambsquarters**, Canada thistle, Russian thistle, wild oats and wild mustard. Extensive use of glyphosate-resistant sugarbeet crops is a common practice. Specifically, **diethatyl, EPTC, clopyralid** and **desmedipham** are some pre- and post-emergence herbicides that control weed species.

Cutworms, the larval stage of some moth species (Lepidoptera), pose a major problem not only to sugarbeet but also to a variety of vegetables and cereals. Cutworms can damage nearly three-quarters of a crop, if left unchecked. These either cut off the sapling at the soil level or feed on the root under the soil, or climb the plant and eat the flowers and leaves. The black cutworm (*Agrotis ipsilon*) cuts off the sapling of sugarbeet and must be controlled early. **Heptachlor** (soil drench), used around the plant, is a common agent used for cutworm damage. The predator nematode *Steinernema* spp. and a ***Bacillus thuringiensis*** formulation are also useful in cutworm control. Other insect pests that attack sugarbeet include the flea beetle (*Monolepta signata*), weevils, borers, **leafhoppers** and hawk moths. These can be controlled by manually picking and destroying the eggs and caterpillars. The use of pheromone traps and also ***Trichogramma*** spp. is also known. **Monocrotophos** and **quinalphos** and dust applications of **carbaryl** are also useful.

Sugarbeet is also susceptible to diseases caused by fungi, bacteria, viruses and nematodes. The more important diseases of the crop include seedling blights and root rots caused by the fungi *Pythium* spp., *Sclerotium rolfsii* and *Rhizoctonia solani*. The diseases can be controlled partly through seed-polishing and seed treatment. Root rot, caused by *Sclerotium rolfsii*, can be controlled by **chlorpyrifos** in a soil drench application before appearance of disease. Even application of the bioagent *Trichoderma viride* has been useful in controlling the fungus in various crops. *Cercospora* leafspot caused by the fungus *Cercospora beticola* is a serious fungal foliage disease of sugarbeet. **Mancozeb** and triphenyl tin hydroxide are two effective agents against this disease. A 3 year crop rotation helps to reduce the inocula.

Most of the pest attacks mentioned above may be controlled by suitable cultural practices. Optimum irrigation and avoidance of excess moisture on the canopy helps keep the fungal pathogens under control. Tilling so that cutworm moths do not lay eggs around the plant, and the removal of crop residues and volunteers, discourages overwintering of larvae. Use of light traps, pheromone traps, manual inspection and introduction of natural enemies are all part of the IPM strategy.

Crop – sugarcane pest control

The term sugarcane includes up to 35 species of tall perennial grasses of the genus *Saccharum*. Known to be the world's chief sugar-producing crop, sugarcane contributes nearly 75% to the global sugar pool. In India, sugarcane has significance not only as the prime source of sugar but as a major agro-industrial crop.

The cultivation of sugarcane (Fig. C.24) requires a tropical or subtropical climate, and needs well-drained heavy soils and rainfall of up to 1200 mm. Planting is normally done in October or November, for best yields.

Fig. C.24. The lower portion of a sugarcane plant. Courtesy of Kalyani Paranjape.

Sugarcane is vulnerable to several insect pests and diseases, especially in the subtropical regions. Major insect pests include moths, borers, scales, white grubs, termites, whiteflies, black bugs, **mealybugs** and white woolly aphids. Diseases that affect the yield of sugarcane include red rot, **smut**, rust (fungi), grassy-shoot (mycoplasma) and a ratoon-stunting disease (bacteria). The disease **sugarcane whip smut** caused by the fungus *Ustilago scitaminea* is easy to identify, as it produces a long whip-like dusty black shoot from the growing tip. Sugarcane red rot caused by *Colletotrichum falcatum* Went is a major fungal disease of sugarcane in India. The symptoms are seen on leaves, as they discolour and wither from the tip to the base, with a few lesions on them. The tissues are reddened and the plant may die in the matter of a week, in cases of severe infections. **Thiophanate-methyl**, a systemic fungicide, provides long-term protection against red rot. Sugarcane is also at risk with weeds, as they harbour pests and disease-causing agents.

Means of controlling pests include: (i) mechanical methods (removing egg masses or adults, and roguing infested plants); (ii) cultural means (**hoeing** and weeding in the early growth stages, avoiding late planting, destroying old cane stalks); (iii) biological methods; and (iv) the use of resistant varieties. It is also important to add a few inches of soil near the main stem, a process called hilling or earthing-up, for healthier growth of cane. The planting of vigorous and disease-free seed canes from a well-nourished nursery crop, followed by the prevention of secondary infection, helps eliminate diseases. If seed is raised from heat-treated canes, the plant has a good chance to be disease free.

Pesticide formulations of chlorinated hydrocarbons, organophosphates and carbamates have been found to be effective in reducing damage caused by a variety of insect pests. Weed control can be brought about through a pre-emergence application of **atrazine** at 2 kg a.i./ha, followed by **2,4-D** spray at 1 kg/ha (60 days after planting).

Crop – tea pest control

Camellia sinensis or the tea bush is considered the world's most consumed beverage after water. Known for its stimulating and antioxidant qualities, this evergreen is grown in over 50 countries. *C. sinensis* is cultivated at high altitudes in tropical climates. The plant prefers moist organic soil and heavy rainfall.

Out of various pests that are known to attack the tea bush, a few are highlighted here.

The tea mosquito bug (*Helopeltis theivora*) is a major pest of the tea crop. The adults and the nymphs that hatch out of eggs laid inside the stem, suck the sap causing brown leaf spots and curled and swollen buds and shoots. Frequent plucking, weeding and exposure to sunlight help reduce the population of *H. theivora*. Biological control includes releasing ***Chrysoperla** carnea* (up to 1000 larvae/ha). Chemical control measures include spraying of **endosulfan**, **quinalphos** or **chlorpyrifos** (all in EC and 750 ml/ha), metasystox, ethofenoprox and alphamethrin.

Thrips (*Scirtothrips* spp.) attack tender leaves and buds, thus damaging them. Control may be brought about by releasing **anthocorids**, predatory thrips (*Aeolothrips*, *Mymarothrips* spp.) or *Chrysoperla carnea* (up to 1000 larvae/ha). The threshold limit is three thrips per shoot. These pests may be controlled by chlorpyrifos, quinalphos and endosulfan.

Tea green hoppers (*Empoasca flavescens*) are serious pests, as they hide and feed on the lower surface of tea leaves, resulting in withering and burning of young leaves. Females lay eggs in the soft tissue of buds, and nymphs that hatch grow to adulthood through four instars. Nymphs and adults pierce the leaf to suck its sap. Since females can lay as many as 100 eggs throughout their lives and the eggs are laid inside the tissues, early control can benefit the crop. Manual removal and spraying with endosulfan or quinalphos are recommended.

The larval stages of lepidopteran and others like aphids, **mealybugs** and scales attack the tea plant and are controlled by ***Bacillus thuringiensis***, **fenvalerate**, **deltamethrin** and chlorpyrifos.

Tea tortrix, which is a leaf roller that attacks both tea and coffee plantations, can be parasitized by *Phytodietus spinipes*.

The shot hole borer (*Xyleborus fornicatus*), named after the characteristic shot holes it creates in infested branches, makes tunnels within stems and causes bud mortality and dieback symptoms in branches. Its **economic threshold level (ETL)** is 15% of the infested shoots. A culture of ***Beauveria bassiana***, applied at the end of May and then again in October, has been found to be effective against this borer. Pesticides include fenvalerate, deltamethrin and **cypermethrin**, sprayed in April, May, October and December.

Aphid species *Toxoptera aurantii*, which attack tea plants soon after pruning causing leaf curl and stunting of growth, can be controlled by the removal of infested shoots and spraying of **neem**-based pesticides.

Termites (*Microcerotermes* spp.) may attack the soft-wooded tea plants. Emulsifiable concentrates (EC) (approximately

20 ml/10 l water) of endosulfan or chlorpyrifos are used. Spraying with **imidacloprid** 200 SL (4 ml in 10 l water) is also effective.

Mites, especially red spider mites, are serious pests of the tea crop. Other mite pests are the pink mite (*Acaphylla theae*), scarlet mite (*Brevipalpus* spp.) and yellow mite (*Polyphagotarsonemus latus*) that cause varying degrees of damage. Mites suck the sap of leaves, starting at the midrib and covering the entire leaf. Spray oils with a paraffin base and sulfur and **lime sulfur** are used in the control of mites. In addition, neem formulations are recommended. Ethion or wettable sulfur helps control mites (of the Eriophyidae family).

Females of the red spider mite (*Oligonychus coffeae*) lay eggs on the upper portion of mature leaves. The larvae, the nymphs and the adults feed on the sap of the leaves, which cause brown patches on the leaf surface. Cultures of entomopathogenic fungi such as *Verticillium lecanii*, *Paecilomyces* spp. and the *Hirsutella* spp. are also seen to be effective. **Ethion**, quinalphos and **fenpropathrin** (preferably EC formulation, up to 500 ml) and 1 kg sulfur in the form of wettable powder (WP) are recommended pesticides.

Cutworm (*Spodoptera litura*) is monitored by placing five pheromone traps/ha. Root-knot nematodes (*Meloidogyne* spp.) which attack tea roots can be checked by applying 2.5 g of granular pesticides such as **carbofuran** and **phorate** per young plant, 5 g per older plant. Carbofuran can be applied to mature bushes after pruning. These measures can be repeated after 3 months if needed.

Some blight diseases also affect the tea plant. For instance, grey blight (*Pestalotiopsis* spp.) and blister blight (*Exobasidium vexans*) are controlled with dithane or **carbendazim** diluted in water. A combined fortnightly application of **hexaconazole**, copper oxychloride and triton may be made to control blight. Regular mulching and manual weeding during the early growing stages of tea can help keep tea weeds under control. Spray applications of **2-4,D** and a mixture of 2-4,D with **glyphosate** are recommended, making sure the manual knapsack sprayers are equipped with special regulator nozzles to avoid drift to the crop.

Crop – wheat pest control

Wheat (*Triticale* sp.), like rice and maize, forms one of the most extensively available foods grown almost the world over. This crop is from the grass group (family Poaceae and order Poales). The cereal grain is believed to have originated in and around Iran, Iraq and Turkey a few thousand years ago. The plant grows to a height of around 1 m and has narrow green leaves. Upon germination, it develops a seminal root system. Each bud at the leaf base can develop into a tiller that can mature to hold the wheat head. The head holds edible seeds (kernels). Wheat grows well in temperate grasslands with 30–90 cm of rain and cool temperatures during its early growing stages. During the later stages, the kernel growth is encouraged by sunny days. Based on cropping time, there are two types of wheat: winter wheat (September–October) and the spring wheat (mid-May). Rain-scarce areas produce hard wheat, while rain-abundant areas produce soft wheat. Common wheat species of today are bread wheat (*Triticum aestivum*) and durum wheat (*Triticum turgidum*).

The grain has three parts: (i) the endosperm; (ii) the bran; and (iii) the germ (embryo). Nutrition through wheat foods differs with varying compositions of these three constituents, with bran and germ accounting for necessary supplements like vitamin B, manganese, zinc, **magnesium** and others. Wheat is an ingredient in breads, spaghetti, noodles and sauces, and contributes to the fermentation process of alcohol.

Wheat is prone to many foliar-, soil- and seed-borne diseases throughout its growth. Many fungal, viral and bacterial diseases affect wheat yields worldwide. Rust diseases, loose **smut**, **powdery mildew** and take-all affect almost all areas.

Rust, a fungal disease, is of two kinds, namely leaf rust and stem rust. Leaf rust is caused by *Puccinia recondita tritici* and is more common and destructive than stem rust, caused by *Puccinia graminis tritici*. Rust prospers in conditions of moist and warm weather. Wheat leaf rust, which is a common fungal disease in wheat, caused by *P. recondita tritici*, is evidenced by the appearance of circular pustules with red-orange spores on the upper leaf surfaces and leaf sheaths. These pustules appear first during tillering. The pustules grow to cover the entire leaf stage of wheat growth. They grow in large numbers covering the whole leaf, turning the leaf yellow and eventually killing it. The pustules turn a rust shade and then black. Leaf rust results in the reduction of tillers, kernels and grain weight. The pathogen has both sexual and asexual life cycles and needs an alternate host to complete its life cycle.

Wheat stem rust, caused by the fungus *P. graminis tritici*, is seen to occur when dark, reddish-brown pustules (containing spores) appear on the leaf surfaces, sheaths and stems. The pustules are larger in stem rust than in leaf rust and are spindle-shaped with ragged edges. These pustules appear black as the plant matures and spores turn dark brown. Stem rust weakens the stems resulting in lodging and reduced grain weight.

There is another rust disease called the stripe rust or the yellow rust characteristic of yellow pustules on leaves that affect the head. This is caused by the pathogen *Puccinia striiformis* f. sp. *tritici*.

Rust diseases in wheat are usually controlled by the use of triazole fungicides.

Wheat loose smut is caused by the fungus *Ustilago tritici*, whose spores settle on wheat flowers, grow mycelium and infect the seed embryo. Humid weather and moderate temperatures favour the disease. Infected wheat heads emerge slightly earlier than the ones from healthy plants and are not distinguishable. The seed is completely destroyed and the entire head is full of a dark mass of spores. If such kernels are planted, then the fungus becomes active and grows systemically in the plant, finally killing it.

The use of disease-resistant varieties and healthy seeds can control the disease. Fungicides such as **carboxin**, **triadimenol** or imazilil and seed treatment with some protectant fungicides are effective against smut. Seed treatment fungicide should be applied at 2.5 g/kg seed. Heat or solar treatment can also be used to control the disease.

Powdery mildew of wheat is another disease that occurs on wheat fields. It is caused by the ascomycete, *Erysiphe graminis* f. sp. *tritici*. It overwinters as cleistothecium, and in humid conditions gets blown by the wind on to healthy plants. There, it produces chain-like conidia. The symptoms are greenish patches on leaves and sheaths.

Wheat streak mosaic or yellow mosaic is caused by the mosaic virus that is transmitted by an eriophyid mite species,

Acerina tulipae, which causes leaves to curl. This mite produces a web, with which it gets carried by the wind. Both the nymphs and the adults carry the disease. It feeds on the leaves. The symptoms of the viral infection are mottled leaves that have yellow streaks. The growth is stunted and infected plants bear little grain. Cultural practices such as prevention of voluntary wheat and use of **acaricides** can help control the disease.

In general, crop cultural practices are crucial in managing fungal and bacterial infections on wheat, including: (i) crop rotation (with an oilseed crop or barley); (ii) removal of volunteer wheat; and (iii) removal of weeds, not only for their own sake but also to deprive pathogens of alternate hosts. Carrying out cultivation of the fields is also essential. Over-fertilization, extended rainy spells and monoculture help establish the disease. Disease-resistant varieties can prevent disease. Planting winter wheat in early September helps plants to flower and mature before the weather warms up to attract infection.

Major weeds of wheat include Mexican prickly poppy, lambsquarters (*Chenopodium album*), little seed canary grass (*Phalaris minor*) and wild oats (*Avena luoviciana*). Pre-emergence applications of **pendimethalin**, and post-emergence application of sulfosulfuran (25 g/ha) and **metribuzin** (175 g/ha), are suitable for both grassy and broadleaved weeds. **Clodinafop** and fenoxaprop are also used for grassy weeds.

Insect pests like aphids and termites can attack the wheat crop. Nematodes can also cause harm to the plant. Cultural practices of repeated summer ploughing and rotation with radish, carrot and gram break pest cycles. A spray of **dimethoate**, formathion, **oxydemeton methyl** or **monocrotophos** can control the growth of aphid and other pests, while the use of biocontrol agents such as ***Chrysoperla*** or *Apertochrysa*, *Metarhizium anisopliae*, *Verticillium lecanii* and ***Beauveria bassiana*** can also be helpful.

Crop protection

Crop protection is an important step in agricultural production and includes a set of measures that aims to protect crops from diseases. Protecting yields, either as standing crops or post-harvest, is as important as growing them. Crop protection systems encompass practices that attempt to improve crop quantity and quality. On closer study, it is seen that all agricultural disciplines, including crop protection, are interdependent.

The choice of crop protection measures reflects a balance between the potential damage and expected profits. If the costs of pest control exceed the expected profit, the damage is tolerated. Damage to crops, especially cash crops, can upset the foreign exchange position of a country. Examples are: (i) the collapse of the coffee industry in Sri Lanka caused by coffee rust; and (ii) the destruction of cocoa in Nigeria due to a combination of viral and fungal infection.

Crop protection measures are generally classified as: (i) chemical; (ii) biological; (iii) physical and mechanical; (iv) biotechnological; and (v) genetic engineering methods.

Chemical agents are characterized as insecticides, acaricides, fungicides, herbicides, nematicides, rodenticides and others. Insecticides and herbicides account for nearly three-quarters of the crop protection products. Rice and cotton are among a few crops which require an extensive amount of crop control agents. Chemical agents have been developed to target a wide range of pests, resulting in the availability of pesticides with the required selectivity, persistence and non-toxicity to non-target pests yet that only need be applied in the smallest dose.

Biological methods of crop protection against insects include the use of pest management, seen in the use of repellents, attractants, **pheromones** and growth hormones. It is also seen in the appropriate use of natural enemies to achieve sustainable pest numbers. For instance, weed control is brought about by the use of insects that need specific host plants for the growth cycle (monophagous and oligophagous).

With biotechnological inputs, microorganisms and their metabolized products can be used as pathogens or pathogen antagonists. Examples include: (i) some highly specific products of some microorganisms that can be used as antibiotics such as **blasticidin**, **kasugamycin** and **polyoxin**; (ii) spore preparations from *Verticillium lecanii* (a fungus that affects aphids and whiteflies); and (iii) the antagonistic effect of *Trichoderma harzianum* against the phytopathogenic fungus *Stereum purpureum*. Genetic engineering has a role in developing varieties that are resistant to pathogen attacks and the unintended exposure to pesticides.

Crop protection can also be brought about by devising suitable farming techniques (cultivation methods, rotation, mechanical methods and time scheduling such as tillage, planting dates and seedbed preparation). Other important physical crop protection methods are: (i) **hoeing** and harrowing; (ii) fencing of crops to protect against grazing animals; (iii) flooding of the soil to kill nematodes and even field mice; (iv) trimming of branches to control fungal diseases; (v) soil steaming; and (vi) burning crop residues. A balance between fertilization and pesticide application can ensure crop growth and protection. Tillage practices, correct selection of crops for rotation, suitable seed dressing and quarantine measures are used against infestation of pests and infection of pathogens.

Selective action in a crop protection agent changes with the kind of control agent. Herbicides should target all plants in the area with the exception of the crop itself. Insecticides and fungicides, on the other hand, should target only the one specific species which is the target organism, although insecticides can be monotoxic (acting on a specific species), oligotoxic (group specific) or polytoxic (broad-spectrum effect). To prevent ecological imbalance, high specificity is preferred, in spite of the risk of pest resistance.

The ideal property of a crop protection compound is to persist in the environment until it achieves the necessary control, after which it should disintegrate to natural substances.

Crop rotation

Crop rotation refers to the growing of crops in a regularly scheduled sequence, in contrast to growing one crop continuously or different crops haphazardly on the same land or farm. In crop rotation, two or more crops are grown in a field in a fixed sequence. A traditional farmer usually resorts to crop rotation to produce several crops in a season. Rotation facilitates stable production, diversity in crop availability, and income and labour distribution. Maize, in rotation with legume sod, outperforms monoculture maize.

One of the earliest agriculture systems practised was the four-course rotation of Norfolk, UK. This rotation follows a root crop-cereal-ley-cereal sequence. As the needs changed, sugarbeet replaced other root crops. Potato was introduced in a still widely used five-course rotation which comprises a sugarbeet-barley-ley-potato-wheat sequence. In some areas, a six-course rotation (East Lothian rotation) is followed, whereas elsewhere, other course systems have evolved (e.g. Fenland rotation). There is another popular rotation involving two-third cereals and one-third dicotyledons (clover, potatoes and beets). The types that evolve also depend on livestock production, thus leading to ley farming or alternate farming.

In comparison to monoculture practices, crop rotation enables: (i) reduced chances of diseases and pests; (ii) better weed control; (iii) improved fertility; and (iv) a more even distribution of annual labour requirement.

Crop rotation is greatly beneficial in reducing the build-up of species-specific pests because a change in the crop species breaks the life cycle of the pest. A non-host plant in rotation prevents the pest from increasing its numbers on the crop plant. While crop rotation does not totally eliminate the pest, it helps control pest numbers. For example, corn rootworm which thrives only on maize root is reduced to a manageable level by crop rotation. When crop rotation is not adequate, insecticides should be used on the soil.

Crop rotation, with suitable associated factors, has also a beneficial effect on crop yields. Cereals grown after legume-grass sod crops, with good drainage, aeration and organic matter, may lead to a higher and more tolerant yield. Many fungal diseases that need a cereal crop to cause take-all or eye-spot (also called *Cercosporella* foot rot) can be controlled by crop rotation with a non-grass crop, as these diseases do not survive in the soil in the absence of cereal plants.

Established soil pests may require longer rotations.

Crop-oil concentrates

See **Adjuvant**

Cryolite

See **Fluoride inorganic insecticides**

Cultural practices for nematode management

While the damage caused by a nematode attack can be extensive, the use of nematicides and fumigants may not always be the answer. This is due to their adverse effects on the environment and the high costs involved. Thus, a population of nematodes is managed using cultural practices.

Some examples of possible crop sequences that have yielded good results in root-knot nematode suppression in the Mid-Atlantic region of the USA are illustrated in Table C.1. Depending on local parameters such as humidity, temperature and soil conditions, these might vary with region.

Crop rotation with non-host crops helps to control populations of nematodes. Since most nematode species have a wide host base, crop rotation becomes a challenge. However, winter cereals, cotton and groundnuts are non-hosts to *Meloidogyne incognita*, *Meloidogyne javanica* and *Meloidogyne arenaria*.

Some plants help break the cycle of specific nematode species. For instance, a summer cover crop called hairy indigo (*Indigofera hirsuta*) can control sting and root-knot nematodes.

Planting of an intermediate crop and a fodder source, sorghum sudangrass, in the second half of July after an early season potato, cucumber or pea crop helps suppress/reduce root-knot nematode populations. Subsequently, if the biomass of sorghum sudangrass can be incorporated as a green manure in 2 months, it can suppress root-knot and root-lesion nematodes.

Castor bean can be used similarly against the root-knot nematode, since it is a poor host just like oat or rapeseed. Pineapple is rotated with sugarcane to manage the reniform nematode (*Rotylenchulus reniformis*). Similarly, sorghum and maize are rotated with reniform-resistant varieties of soybean.

Weed control is critical since nematodes reside in them as alternate hosts. Similarly, keeping tracts of land fallow in periods of high temperatures helps by depriving the nematodes of fresh roots.

Flooding fields wherever possible can be effective against a few species like root-knot nematodes (but not soybean cyst nematode). Flooding should be alternated with periods of dry spells so that host plants do not survive.

Planting time can be modified, whenever possible, to escape the major part of nematode attack. Sowing in late autumn gives the crop enough time to grow before nematodes can damage roots (nematodes prefer a warmer climate), although success cannot always be guaranteed.

Optimum use of soil amendments is helpful in allaying nematode attacks. Sawdust, chicken manure, oil cakes, leaves of **neem**, castor, mustard and rice incorporated into the infested

Table C.1. Potential cropping sequences to minimize damage from root-knot nematode in the Mid-Atlantic, USA. (From Sustainable Agriculture and Research Education (SARE), www.sare.org, with permission.)

Year 1				Year 2			
Crop 1	Crop 2	Amendment	Winter cover crop	Crop 1	Crop 2	Amendment	Winter cover crop
Potato	Soybean 'Manokin'	None	Oat	Cucumber	Soybean 'Manokin'	None	Oat
Potato	Castor bean	None	Oat	Cucumber	Castor bean	None	Oat
Potato	Sorghum sudangrass	None	Rapeseed	Cucumber	Sorghum sudangrass	None	Rapeseed
Potato	Sorghum sudangrass	7,000 lb/acre poultry litter	Oat	Cucumber	Sorghum sudangrass	7,000 lb/acre poultry litter	Oat
Potato	Sorghum sudangrass	10,500–31,000 lb/acre poultry litter compost	Oat	Cucumber	Sorghum sudangrass	10,500–31,000 lb/acre poultry litter compost	Oat

soils have shown control of stem damage caused by *R. reniformis*, *M. incognita* and some others.

If seeds are treated with extracts of neem seed kernels they escape nematode attacks, especially root-knot nematodes. In turf crops of **Bermuda grass**, lawns and sport turf, deep and infrequent watering, a balanced fertilization programme and adequate mowing height are considered to be ideal management practices.

Cover crops have to be planted wisely such that they starve the nematodes that are most problematic to the crop that is planned for sowing. For example, Bermuda grass (*Cynodon dactylon*) grown as a cover crop can effectively control the root-knot nematode, while sunn hemp (*Crotalaria juncea*) effectively controls both sting and root-knot nematodes. Table C.2 shows a few planting dates for nematode-suppressive cover crops in Maryland, USA.

Cyanamide herbicide

Cyanamide, H_2NCN, is a herbicide which primarily acts as a plant growth regulator and herbicide. It is a colourless, crystalline solid, which is hygroscopic in nature (solubility in water is 4.59 kg/l). It is also soluble in organic solvents such as alcohols, phenols and ethers. But it has poor solubility in benzene and halogenated hydrocarbons. It is insoluble in cyclohexane. It is stable to light and temperature (180°C). Upon contact with alkalis, it polymerizes. Similarly, it is decomposed by acids to give urea.

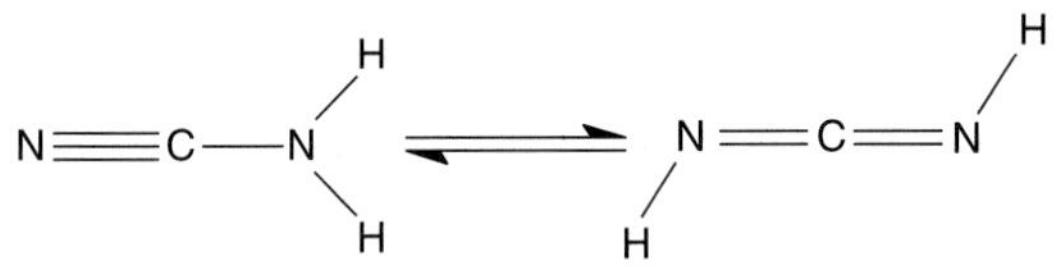

Cyanamide

Cyanamide is a contact herbicide and PGR. It inhibits side shoots in plants and sprouting in hops and vines. As a herbicide, it controls broadleaved weeds in crops such as onions, leeks, shallots, chives and garlic as well as bulb flowers.

The herbicide-PGR is a severe skin and eye irritant. It has an acute oral LD_{50} in rats of 223 mg/kg and acute inhalation LC_{50} of 1 mg/l. It is toxic to fish as well as bees. It has an oral LD_{50} in birds (bobwhite quail) of 350 mg/kg. When inhaled, it causes a cough and shortness of breath. It is commonly available under the trade name of Alzodef.

Acetylcyanamide is the main metabolite produced in rats. In the process of cyanamide degradation, it is easily converted into nitrogenous compounds and is rapidly taken up by plants. Therefore, cyanamide does not have residual activity.

Cyanophos insecticide

Cyanophos, $C_9H_{10}NO_3PS$ (*O*-4-cyanophenyl *O,O*-dimethyl phosphorothioates), is an organophosphorus insecticide (phenyl organothiophosphate). It is a yellow-red liquid with a faint odour, soluble in water (46 mg/l), methanol, acetone and chloroform. Although it is stable under normal storage conditions, it rapidly decomposes under alkaline conditions, exposure to light and at temperatures of 120°C. It is produced by the reaction of dimethylthiophosphoryl chloride with sodium *p*-cyanophenolate. Cyanophos has a CAS Registry Number of 2636-26-2.

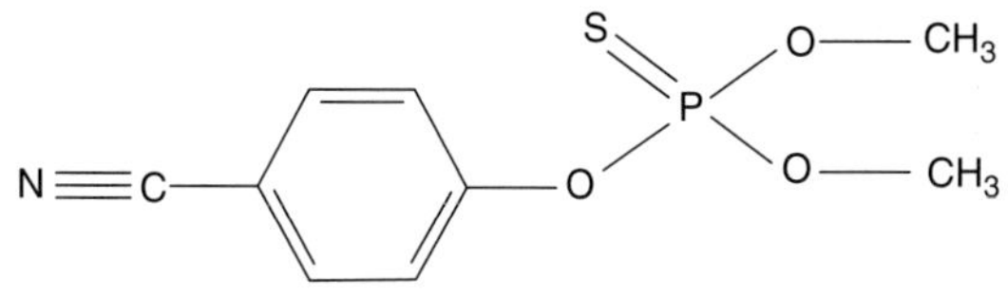

Cyanophos

Cyanophos, a cholinesterase inhibitor, protects cotton crops and many fruits and vegetables against aphids (Aphididae), **scale insects, mealybugs** (Coccidae, Margarodidae) and moths and butterflies (Lepidoptera). It is also used to control many insects in rice such as the **rice stem borer**, yellow rice borer, leaf roller and **leafhopper**. It is also effective against locusts and pests in public places (e.g. cockroaches belonging to the order Blattodea and mosquitoes in the family Culicidae).

The insecticide is formulated in forms such as dustable powder (DP), emulsifiable concentrate (EC), ULV liquids (UL) and wettable powder (WP). Common application rates are up to 50 g a.i./ha. It is commonly available under the trade name of Cyanox.

Table C.2. Production practices for nematode-suppressive crops in Maryland, USA. (From Sustainable Agriculture and Research Education (SARE), www.sare.org, with permission.)

Crop	Planting date[a]	Seeds/acre	Incorporation date	Comments
Resistant soybean	10 July–2 August	250,000	N/A	'Manokin'[b] is resistant to root knot nematode and soybean cyst nematode. However, it is susceptible to root-lesion nematode. Therefore, use this practice where root-lesion nematode is not a problem
Castor bean	10 July–2 August	11,500–24,500	25 September–19 October	Nematode suppressive effect was observed in some, but not all environments. This may be a difficult crop for many farmers to incorporate into production
Sorghum sudangrass	10 July–2 August	200,000	25 September–19 October	Nematode suppression is not observed every year. Suppression is enhanced by addition of poultry litter compost
Rapeseed	9–27 October	861,400	12 March–18 April	This crop is planted in the 'off-season' and because of timing may not replace a cash crop

[a]Planting dates may need to be modified in other areas in the Mid-Atlantic region.
[b]'Manokin' is a nematode-resistant variety of soybean.

The insecticide is rated as moderate on an acute toxicity rating. The WHO categorizes it as a Class II toxin (moderately hazardous). Since it is a cholinesterase inhibitor, it is considered a PAN Bad Actor chemical. The compound is toxic to wild bird species (LD_{50} 3 mg/kg), and the LD_{50} in rats is 215 mg/kg. The compound is also toxic to bees.

In mammals, the main biodegradation pathways are demethylation and aryl ester bond cleavage; the liberated cynophenol is excreted in the form of the sulfate ester, formed by conjugation. Desmethylcyanophos, cyanophos oxon (which in turn converts to desmethylcyanophos oxon) and 4-cyanophenol are the main metabolites.

The compound is not registered in the USA and countries of the EU, but is allowed in South Africa, Tanzania and some other African countries.

Cycloate herbicide

Cycloate, $C_{11}H_{21}NOS$ (*S*-ethyl *N*-cyclohexyl(*N*-ethyl)thiocarbamate), is a thiocarbamate herbicide. With a CAS Registry Number of 1134-23-2, it is readily soluble in water (95 mg/l).

A selective, systemic, pre-emergence herbicide, cycloate is effective on spinach and beet crops, for the control of perennial grasses and broadleaved weeds.

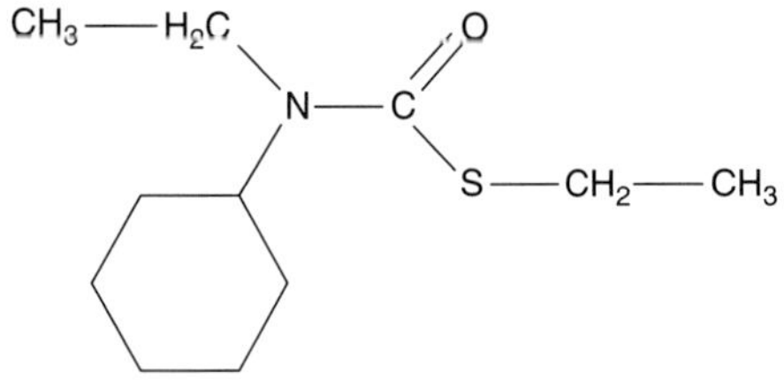

Cycloate

The WHO classifies cycloate as slightly hazardous (Class III toxin), since it tests low on acute toxicity. However, it is a potential groundwater contaminant, a reproductive/development toxin and a cholinesterase inhibitor. It also causes irritation to the skin and eyes.

The soil half-life ranges from 43 to 109 days and cycloate is thus considered moderately persistent. Cycloate is registered for use in the USA but disallowed throughout the EU.

Cycloheximide as a bioPGR

Cycloheximide, $C_{15}H_{23}NO_4$(4-{(2*R*)-2-[(1*S*,3*S*,5*S*)-3,5-dimethyl-2-oxocyclohexyl]-2-hydroxyethyl}piperidine-2,6-dione), is an antibiotic produced by ***Streptomyces griseus*** and is used as a plant growth regulator. It is also used as a fungicide but is not allowed for wide use because of its **chronic toxicity**. The CAS Registry Number is 66-81-9.

PGR cycloheximide is an odourless, colourless crystalline powder, slightly soluble in water but readily so in chloroform, ethanol and methanol. It stores well in solutions but degrades in alkaline media. It acts like a protein synthesis inhibitor in all eukaryotes, by interfering with the translocation step. Cycloheximide can also stimulate the production of **ethylene** and hence promote growth. It is also used as a rodenticide and a toxin to many other pests.

This PGR is formulated as a powder or a dust. Trade names include Acti-aid, Naramycin and Actispray.

Cycloheximide is toxic by inhalation and via the skin. It is moderately to highly toxic to tested animals. It causes irritation to the eyes. It causes adverse effects on reproductive functions and can be fatal to the fetus. It is thus a PAN Bad Actor chemical and is not registered in the USA.

S. griseus is a Gram-positive soil bacterium, which is known to produce antibiotics that have many uses. *Streptomyces* is the largest genus in the phylum Actinobacteria, and many of its species have been exploited for their antibacterial properties in medical science.

Cycloprothrin insecticide

Cycloprothrin, $C_{26}H_{21}Cl_2NO_4$ ((*RS*)-α-cyano-3-phenoxybenzyl (*RS*)-2, 2-dichloro-1- (4-ethoxyphenyl) cyclopropanecarboxylate), is a pyrethroid ester insecticide. It is a yellow-to-brown viscous liquid, soluble in water and readily soluble in common organic solvents but only moderately soluble in aliphatic hydrocarbons. Cycloprothrin is produced by esterification of 3-phenoxy-α-(*RS*)-cyanobenzyl alcohol with 1-ethoxy phenyl-2, 2-dichloro cyclopropionic acid. It has a CAS Registry Number of 63935-38-6 (see Cycloprothrin at bottom of the page).

Cycloprothrin is an insecticide with contact and stomach action. It also possesses anti-feedant and repellent activity. It is used to control insects in rice (e.g. rice water weevils), fruits, vegetables, tea, cotton, soybeans and forestry. It is non-phytotoxic when used as directed.

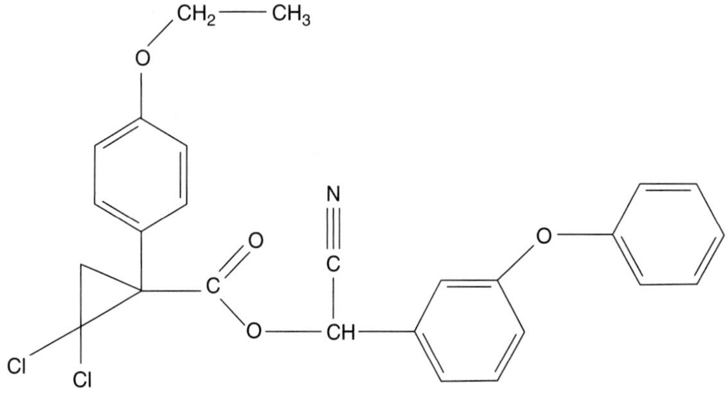

Cycloprothrin

The compound is commonly formulated either as a granule (GR), dustable powder (DP), wettable powder (WP) or emulsifiable concentrate (EC). It is available in the market under the trade name Cyclosal.

The insecticide is categorized as a Class U toxin (unlikely to cause hazard). It is not considered a PAN Bad Actor chemical. It is not teratogenic or carcinogenic but it is toxic to bees. Aerial spraying has been found to be hazardous to wild bird species.

In animals, cycloprothrin is rapidly and completely excreted via urine and faeces. When it is applied to simulated paddy water, cycloprothrin is absorbed by rice plants, especially with time.

The insecticide is not registered by the EPA.

Cyclosulfamuron herbicide

Cyclosulfamuron, $C_{17}H_{19}N_5O_6S$ (1-[2-(cyclopropylcarbonyl) anilinosulfonyl]-3-(4,6-dimethoxypyrimidin-2-yl) urea), is a pyrimidinylsulfonylurea herbicide. This herbicidal compound is a whitish, solid substance soluble in water (6.52 ppm in pH 7 at 25°C). The CAS Registry Number is 136849-15-5.

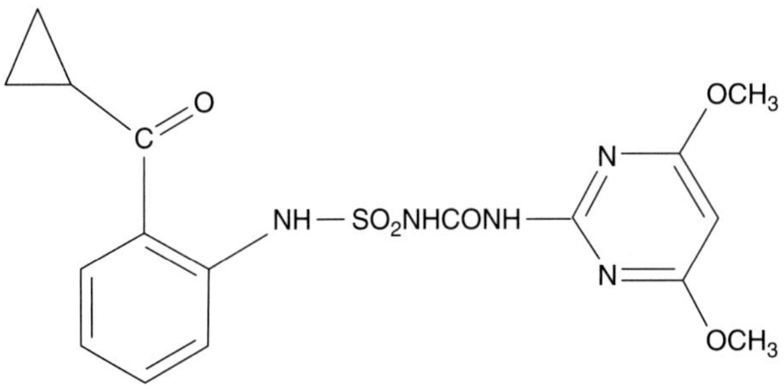

Cyclosulfamuron

The herbicide cyclosulfamuron, an inhibitor of acetolactate synthase (ALS), is effective on weeds in cereal crops like rice and wheat. It acts by inhibiting the synthesis of essential amino acids, thus halting cell division. It is used effectively on rice, wheat and barley against a variety of weeds including spike rush, mustard and flatsedge.

Cyclosulfamuron is available in wettable granules and powder formulations, with Invest being a known trade name. Some weed species have developed resistance to this compound.

This herbicide is not toxic in normal use, with the acute oral LD_{50} in rats being >5000 mg/kg and inhalation LC_{50} being 5.2 mg/l. The dermal LD_{50} in bobwhite quails is 1880 mg/kg. It does not cause irritation to the skin and is practically non-toxic to honey bees, earthworms and fish. The WHO has classified cyclosulfamuron as a Class U toxin (unlikely to be hazardous).

The herbicide is rapidly metabolized upon absorption. Animals eliminate the substance and its inactive metabolites. In plants, it is metabolized to non-toxic compounds.

Cyclosulfamuron is registered for use in the USA and Vietnam.

Cycloxydim herbicide

Cycloxydim, $C_{17}H_{27}NO_3S$ ((5*RS*)-2-[(*EZ*)-1-(ethoxyimino) butyl]-3-hydroxy-5-[(3*RS*)-thian-3-yl]cyclohex-2-en-1-one), is a cyclohexene oxime herbicide. It has a CAS Registry Number of 101205-02-1. It is soluble in water (53 mg/l) and very highly soluble in acetone and hexane.

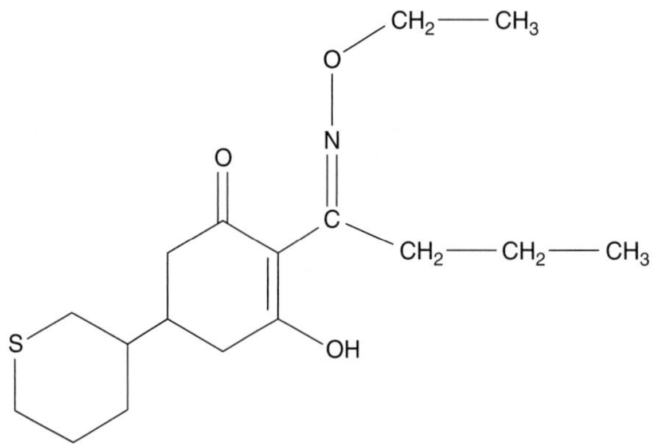

Cycloxydim

The herbicide is a selective, systemic compound, absorbed by the leaves. It is a fatty acid synthesis inhibitor on grass weeds in various crops.

Herbicide cycloxydim is normally formulated as an emulsifiable concentrate. Focus, Ultra and Laser are trade names containing this compound.

The WHO classifies cycloxydim as a Class U toxin (unlikely to be hazardous). The acute oral LD_{50} values among birds and rats are in the range of 2000–3940 mg/kg.

In soils, the herbicide cycloxydim degrades extremely rapidly with a half-life of half a day. Cycloxydim sulfoxide and cycloxydim sulfone are two metabolites of cycloxydim and they are known to harm the environment.

Cycloxydim is registered in many countries of Europe, South Africa and Tanzania.

Cyfluthrin insecticide

Cyfluthrin, $C_{22}H_{18}Cl_2FNO_3$ ((*RS*)-α-cyano-4-fluoro-3-phenoxybenzyl (1*RS*,3*RS*;1*RS*,3*SR*)-3-(2,2-dichlorovinyl)-2,2-dimethylcyclop), is a mixture of diastereoisomeric pairs of enantiomers. It is soluble in water (0.002 mg/ml at 20°C) and also in dichloromethane and toluene (both >200 g/l) and hexane (20–50 g/l). In water, cyfluthrin hydrolyses in alkaline conditions. It has a CAS Registry Number of 68359-37-5 (see Cyfluthrin at top of the next page).

Cyfluthrin is a non-systemic, broad-spectrum pyrethroid insecticide that protects cotton, maize, groundnuts, turf, potatoes and vegetables from beetles, termites, worms (e.g. maize worms, cabbage worms, armyworms), flies, fleas and mosquitoes. Sucking and chewing insects in the orders Lepidoptera, Coleoptera and Hemiptera are also controlled. The mode of action is the blocking of sodium channels of the insect, thus interfering with nervous transmissions. Cyfluthrin also controls pests of public places. It also has acaricidal properties.

The compound is formulated as wettable powders, aerosols and granules. Most formulations contain a chemical synergist called **piperonyl butoxide**. The common trade names are Baythroid, Tempo, Solfac and Contur. A common application dose is between 10 and 55 g/ha.

Cyfluthrin is considered moderately toxic to sheep, rats, mice and rabbits. In humans, exposure causes skin and eye

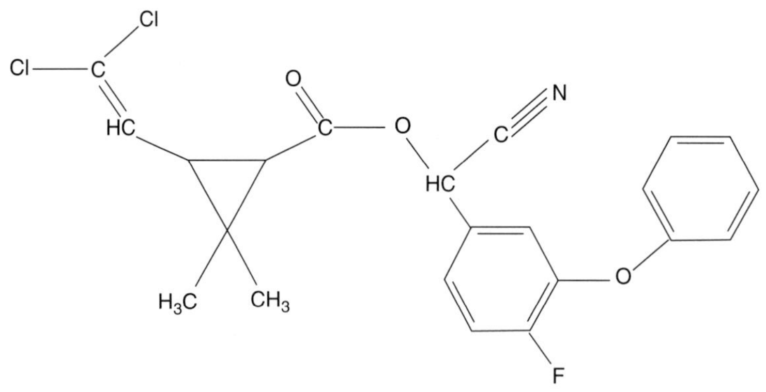

Cyfluthrin

irritation, burning and stinging. Cyfluthrin is highly toxic to marine and freshwater fish (LC_{50} is in the range of 0.000068–0.022 mg/l) and also bees (LC_{50} is 0.037 mg per bee). It is also hazardous to **beneficial organisms**. The NOEL is 2.5 mg/kg/day. The WHO has categorized this chemical as a Class II toxin (moderately hazardous). But it is not included as a PAN Bad Actor chemical.

In humans and mammals, more than half of cyfluthrin is eliminated via urine within 24 h. Cyfluthrin breaks down in sunlight. On surface soils, its half-life is up to 4 days. The primary degradation products are CO_2, 4-fluoro-3-phenyl-benzaldehyde and 4-fluoro-3-phenoxybenzoic acid.

India, Australia, the USA, Canada, New Zealand and some EU countries like Spain, Portugal, Belgium, France and Greece have registered the use of cyfluthrin. However, a breakdown product, 4-fluoro-3-phenoxybenzoic acid is not registered.

FSANZ's Joint Food Standards Code 1.4.2 to v116 (20 May 2010) has fixed the MRL of cyfluthrin as 0.2 mg/kg for aubergines and tomatoes but 2 mg/kg for cereal crops. On the grains of barley, oats, wheat and rye, the EPA (EPA 40CFR180 (e-CFR) to 19 May 2010) has fixed the MRL as 0.15 mg/kg, while on maize, it is 0.05 mg/kg.

Cyhalofop-butyl herbicide

Cyhalofop-butyl, $C_{20}H_{20}FNO_4$ (butyl (*R*)-2-[4-(4-cyano-2-fluorophenoxy)phenoxy]propionate), is an aryloxyphenoxy-propionate herbicide. It is an amber clear liquid, soluble in water and stable to metals and metal ions when stored at 50°C. Its CAS Registry Number is 122008-85-9. This substance is a derivative of cyhalofop, $C_{16}H_{12}FNO_4$ ((*R*)-2-[4-(4-cyano-2-fluorophenoxy)phenoxy]propionic acid), with a CAS Registry Number of 122008-78-0.

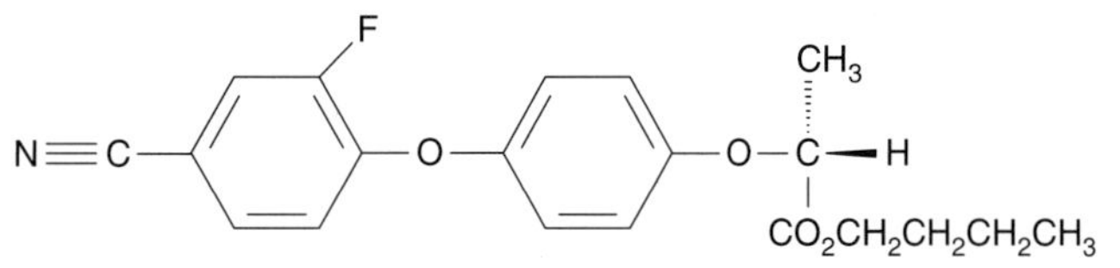

Cyhalofop-butyl

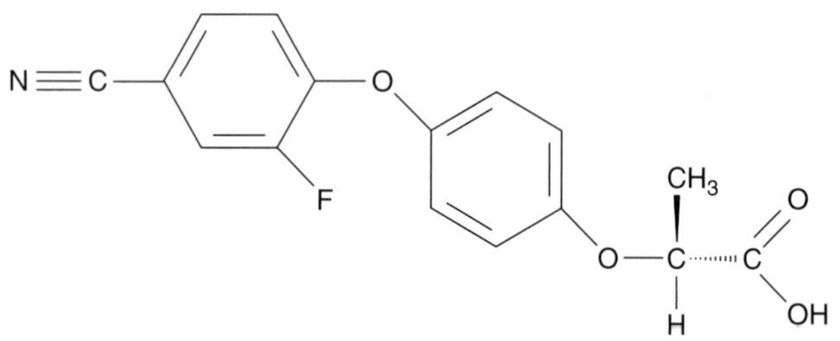

Cyhalofop

Cyhalofop-butyl is a post-emergence grass weed herbicide of rice. It acts by inhibiting acetyl coenzyme-A carboxylase and fatty acid biosynthesis. This results in the death of growing points. It is effective on **barnyard grass, Johnson grass** and red sprangletop.

The herbicide is usually formulated as an emulsifiable concentrate (EC). It can be sprayed aerially or by ground equipment (at 200–300 g/ha), with a maximum of two applications per season.

The WHO classifies the herbicide cyhalofop-butyl as a Class U toxin (unlikely to be hazardous), with an acute oral and dermal LD_{50} among rats being >5000 mg/kg and >2000 mg/kg, respectively. It may cause eye and skin irritation. The acute LC_{50} among bluegill and *Daphnia magna* are 0.79 mg/l and >2.7 mg/l, respectively. It is not toxic to birds, honey bees and earthworms. The ADI of cyhalofop-butyl is 0.003 mg/kg b.w./day. But this rice herbicide is highly toxic to terrestrial plants and hence demands planning if paddy waters are to be used outside the field. It may also contaminate groundwater supplies.

Cyhalofop-butyl is registered in India, Australia, EU members like Spain, France, Italy, Portugal and Greece, the Philippines, Vietnam and the USA.

Cyhalothrin insecticide

Cyhalothrin, $C_{23}H_{19}ClF_3NO_3$ ((*RS*)-α-cyano-3-phenoxybenzyl (1*RS*,3*RS*)-3-[(*Z*)-2-chloro-3,3,3-trifluoropropenyl]-2,2-dimethylcyclopropanecarboxylate), is a pyrethroid insecticide. It is a yellow-to-brown viscous oil, insoluble in water and readily soluble in acetone, dichloromethane, methanol and ethyl acetate.

It is stable to light and also to *cis/trans* isomerization for at least 4 years in the dark at 50°C. Cyhalothrin is produced by esterification of 3-(2-chloro-2-trifluoro methylcyclopropanecarboxylic acid) with α-cyano-3-phenoxybenzyl alcohol. It has a CAS Registry Number of 68085-85-8.

Cyhalothrin is a non-systemic, pyrethroid insecticide with contact and stomach action. It also has repellent, ectoparasiticidal and acaricidal properties and also a long residual activity. As animal ectoparasiticides, it is used as an animal dip or as a spray, especially against the tick species *Rhipicephalus microplus*.

This insecticide is usually formulated as an emulsifiable concentrate (EC). Cyhalothrin is available in the market under trade names Cyhalon and Grenade (see Cyhalothrin at bottom of the page).

According to the WHO, cyhalothrin is classified as a Class II toxin (moderately hazardous). The ADI for cyhalothrin is 0.02 mg/kg/day. It is a suspected endocrine disruptor. It is toxic to aquatic organisms and may cause harm to **beneficial organisms**.

In animals, cyhalothrin is rapidly eliminated via urine and faeces. In soil and water, leaching of cyhalothrin and its degradation products is negligible.

India, the USA and the EU have not registered the use of cyhalothrin. However, all three have registered the use of lambda cyhalothrin, which is a subset of cyhalothrin.

Cyhexatin acaricide

Cyhexatin (tricyclohexyltin hydroxide) with molecular formula $C_{18}H_{34}OSn$ is an **organotin** acarcide. It is a colourless crystal, which is almost insoluble in water (i.e <1 mg/l). It is soluble in acetone, chloroform and methanol but less soluble in most organic solvents. It is incompatible with strong oxidizers and UV light. The Grignard reaction of cyclohexylmagnesium chloride with tin tetrachloride followed by hydrolysis with aqueous sodium hydroxide produces cyhexatin. It has a CAS Registry Number of 13121-70-5.

Cyhexatin is a broad-spectrum, contact acaricide that has strong action. It is an effective acaricide used to control a wide range of mites, mainly in fruit such as apple, orange, grapes, pear, hawthorn, peach, tea, flowers and cotton. Trade names of cyhexatin include Acarotine, Pennstyl and Aramol. It is usually supplied as a flowable, wettable powder (WP).

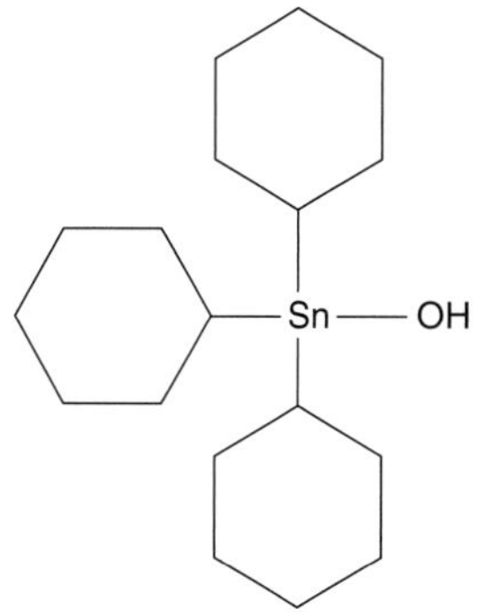

Cyhexatin

The WHO classifies cyhexatin as a Class III toxin (slightly hazardous). It is irritating to the eyes. The acute oral LD_{50} in rats is 540 mg/kg, while in rabbits it ranges from 500 to

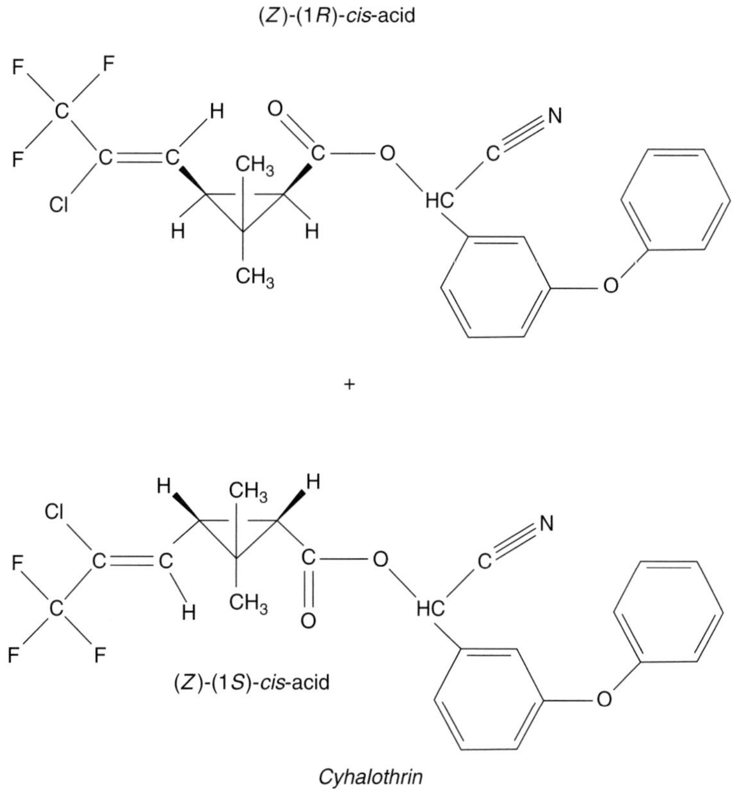

Cyhalothrin

1000 mg/kg. The NOEL in rats is 0.1 mg/kg/day. The ADI is 0.007 mg/kg/day.

Cyhexatin is a PAN Bad Actor chemical and is not registered in the USA or the EU. It is used in South Africa.

Cypermethrin insecticide

Cypermethrin, $C_{22}H_{19}Cl_2NO_3$ (α-cyano-3-phenoxybenzyl (1*RS*,3*RS*;1*RS*,3*SR*)-3-(2,2-dichlorovinyl-)-2,2-dimethylcyclopropanecarboxylate), is a pyrethroid insecticide. It comprises yellowish, odourless crystals that exist as a viscous semi-solid at ambient temperatures. It is poorly soluble in water (0.004 mg/l at pH 7), but is soluble in for example acetone and chloroform. It is relatively stable in weak acidic media. Cypermethrin is part of a group of pyrethroids that are cost-effective substances and which are more effective and more toxic than pyrethrins. The chemical is obtained by the esterification of permethrin acid with α-cyano-3-phenoxybenzyl alcohol. The CAS Registry Number of this insecticide is 52315-07-8 (see Cypermethrin at bottom of the page).

Cypermethrin, a non-systemic synthetic pyrethroid, has a good residual activity on treated plants, and is commonly used to control attacks on agricultural pests from members of the orders Lepidoptera, Coleoptera, Diptera, Homoptera and others. The crops that are thus protected include potatoes, tomatoes, cereals, coffee, cotton, fruits and soybeans, among others. Cypermethrin is also used to control non-crop pests like mosquitoes, cockroaches, wasps, spiders and flies. It also has termiticidal properties, besides being an animal ectoparasiticide.

Except with alkaline materials, cypermethrin is compatible with many insecticides and fungicides. Cypermethrin is formulated either as an emulsion concentrate (EC), a wettable powder (WP), an ULV liquid (UL) or as granules (GR). Trade names include Cybush, Cynoff and Demon.

Cypermethrin is highly toxic to fish, bees and water insects, and mildly toxic to rats and rabbits, although birds are not so vulnerable. Cypermethrin is believed to damage the endocrine system. There are newer studies that link the use of pyrethroids with the occurrence of cancer and reproductive disorders.

Applicators and loaders of cypermethrin are cautioned against symptoms of burning, dizziness and itching, as a result of excessive exposure.

The WHO has categorized cypermethrin zeta as an extremely hazardous chemical (Class Ib toxin). This derivative is a PAN Bad Actor chemical.

Rats and mice eliminate more than half of cypermethrin within 2 days via urine (and to some extent, via faeces). Humans also eliminate the chemical rapidly. In soil and water, the ester bond of cypermethrin is cleaved, followed by hydrolysis in about 16 weeks. Soil degradation is brought about mainly by microbes. On foliage, the half-life is 5 days.

Cypermethrin is not mobile. It does not leach downward because it binds to the soil.

In India (Prevention of Food Adulteration Rules 1955 amended to GSR 367E, 22 May 2007) the MRL for cypermethrin for cabbage head is fixed at 2 mg/kg and for wheat grain at 0.05 mg/kg. In Australia (FSANZ Joint Food Standards Code 1.4.2 to v116, 20 May 2010), for grains of barley, oat, maize and rye, the MRL is 1 mg/kg, while for tomatoes it is 0.05 mg/kg.

Cypermethrin is registered for use in many countries like India, Australia, Canada, most of the EU (e.g. Hungary, Cyprus, Denmark, Germany and the UK), Burkina Faso, Cambodia, Madagascar, Mauritania, New Zealand, the Philippines, Senegal, South Africa, Tanzania, Uganda and Vietnam.

Cyromazine insecticide

Cyromazine, $C_6H_{10}N_6$ (*N*-cyclopropyl-1,3,5-triazine-2,4,6-triamine), is a triazine insecticide and an acaricide. It is a colourless crystalline substance soluble in water (13,600 mg/l) and organic solvents such as methanol, isopropanol, acetone, *n*-octanol, methylene chloride, toluene and hexane. It is stable at 310°C. It has a CAS Registry Number of 66215-27-8.

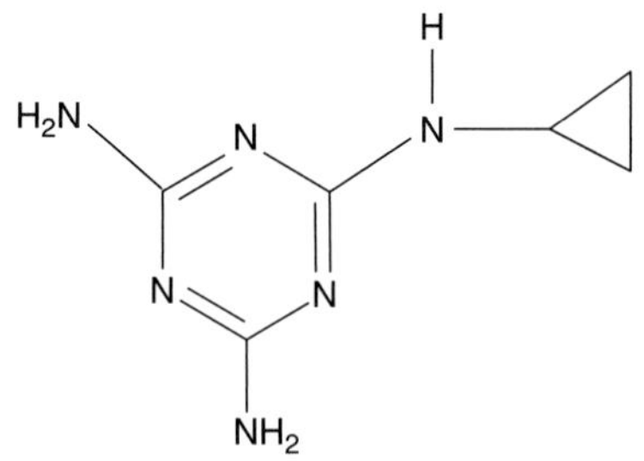

Cyromazine

Cyromazine is an **insect growth regulator** with contact action. It harms the nervous system of the immature larval stages of certain insects. It interferes with **moulting** and pupation. When it is applied to the leaves, it exhibits a strong translaminar effect. It is used as a foliar spray to control leaf miners (*Liriomyza* spp.) in vegetables, potatoes and ornamentals, and also to control diptera larvae. It controls flies on animals. Cyromazine is formulated as a wettable powder (WP).

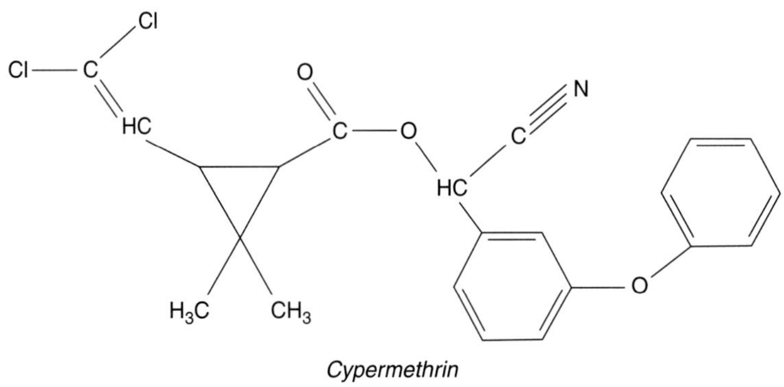

Cypermethrin

According to the WHO, cyromazine is a Class III toxin (slightly hazardous). The acute oral LD_{50} in rats is 3387 mg/kg. It is a groundwater contaminant and hence a PAN Bad Actor chemical.

Cyromazine is rapidly metabolized in plants with the principal metabolite being melamine. Its hydrolysis half-life is 28 days, while the aerobic soil half-life is 63 days.

The USA, Australia, Canada, New Zealand and a few countries of the EU such as France, Greece, Cyprus and Belgium have registered cyromazine but not the breakdown product melamine.

Cytokinins as PGRs

Cytokinins are plant-derived compounds that act as **plant growth regulators** and nematicides. Cytokinins are naturally present in plants and take part in initiation of plant cell division. Cytokinins have a CAS Registry Number of 308064-23-5. The products of cytokinins are extracted from seaweed (see Cytokinins at bottom of the page).

As natural PGRs, cytokinins are associated with cell division which leads to increased plant growth and delay of senescence in treated organs. As nematicides, the target crops for control of nematodes are avocados, berries, citrus, coffee, herbs, kiwi fruit, tree nuts, tropical fruit and vegetables.

Cytokinins are available as aqueous concentrates and granules. Some trade names are Suppress (for nematodes), Foliar Triggrr MFG, Organic Triggrr and Maxon. Formulations are mostly sprayed on crops at an early growth stage. Products of cytokinins are compatible with several other products.

Cytokinins are not acutely toxic to mammals. The acute oral LD_{50} of cytokinins in rats is >5000 mg/kg and the acute dermal LD_{50} in rabbits is > 2000 mg/kg. Cytokinins cause slight eye and skin irritation in rabbits. Cytokinins are rapidly degraded in soil, and hence are not harmful to the environment or non-target organisms. Cytokinins are approved for use in organic farming. They are registered in the USA and Vietnam.

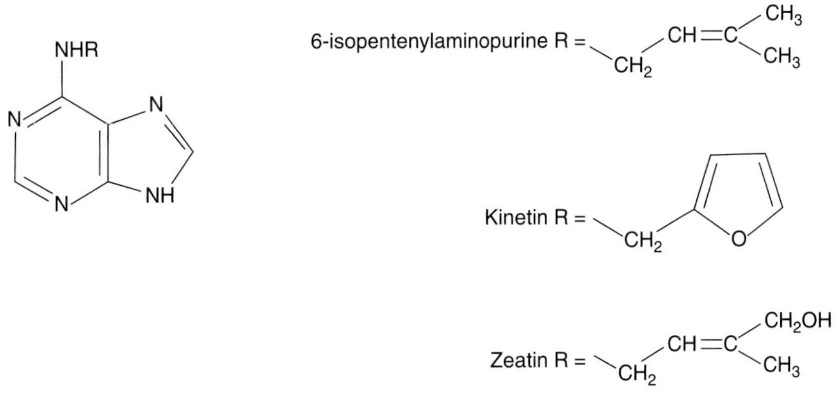

Cytokinins

D

1,3-Dichloropropene nematicide

1,3-Dichloropropene, $C_3H_4Cl_2$ ((*EZ*)-1,3-Dichloropropene), is a colourless to amber liquid with a sweet odour. It is soluble in water (2 g/l) and miscible with hydrocarbons, esters and ketones. It has a DT_{50} of 11.3 days in the pH range of 5–9. It is made by chlorination of propene. 1,3-Dichloropropene has a CAS Registry Number of 542-75-6.

$$Cl—CH_2—CH{=}CH—Cl$$

1,3-Dichloropropene

The compound is used as a soil fumigant and nematicide and has a pre-plant application. It can control most of the nematode species among citrus crops, nut trees, vines and vegetables. Soil insects and some fungal pathogens are also controlled.

The nematicide can be applied in combination with 1,2-Dichloropropene and is usually formulated as a liquid. Telone, D-D92 and D-D95 are common trade names.

1,3-Dichloropropene is an acute toxin with very low LC_{50} values in rats and birds being 110 and 140 mg/kg, respectively. The inhalation LC_{50} (4 h) among rats and mice is 2.7 and 3.7 mg/l air, respectively. It is not toxic to bees, when used as directed. It is a carcinogen and a groundwater contaminant, and hence a PAN Bad Actor chemical.

In soil and water, 1,3-Dichloropropene hydrolyses to the 3-chloroallyl alcohols and is thus non-persistent in animals. 1,3-Dichloropropene is excreted rapidly as mercapturic acid conjugates. It is metabolized rapidly in plants. This compound is registered in the USA, Australia, New Zealand and South Africa. (See also **Nematicides**.)

2,4-DB herbicide

2,4-DB (4-(2,4-dichlorophenoxy) butyric acid) is a phenoxybutyric herbicide with a molecular formula $C_{10}H_{10}Cl_2O_3$. The solubility of 2,4-DB in water is 46 mg/l, while the solubility in acetone is very high. 2,4-DB and its salts are very stable, but esters are hydrolysed under acidic and basic conditions. It is produced by reaction of 2,4-dichlorophenol with trimethylene chlorobromide, followed by reaction with sodium cyanide and hydrolysis of the resulting cyano compound to the acid. The CAS Registry No is 94-82-6.

Upon absorption into plants, it undergoes β-oxidation, leading to its transformation to the herbicide **2,4-D**. 2,4-DB is thus a pro-herbicide, giving it more selectivity than 2,4-D, evident in its non-phytotoxicity to lucerne, clover, soybean, groundnut and forage legumes.

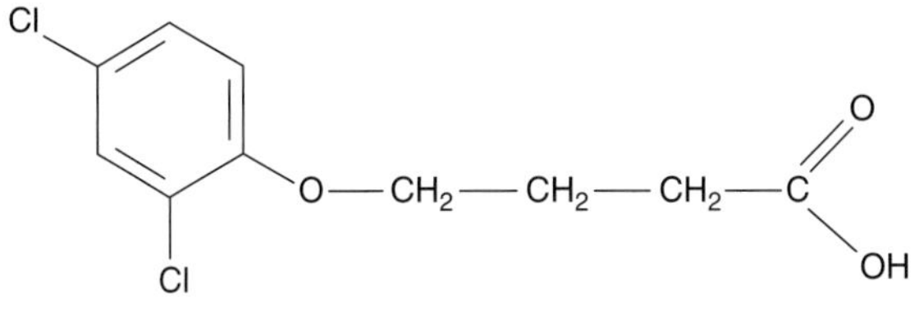

2,4-DB

2,4-DB is a reproductive toxin, a possible carcinogen and a suspected endocrine toxin, and hence is a PAN Bad Actor chemical. The field half-life is 5 days for 2,4-DB, and 10 days for the dimethylamine salt formulation.

In soils, microbial oxidation forms 2,4-D, which is susceptible to further degradation. 2,4-DB fed to dairy cows is excreted in the urine.

2,4-D herbicide

2,4-D (2,4-dichlorophenoxy)acetic acid), with a molecular formula $C_8H_6Cl_2O_3$, is a very efficient colourless powder and a **phenoxy acid herbicide**. Its solubility in water is 311 mg/l and in ethanol is 1250 g/kg. 2,4-D and its salts are very stable, but esters are sensitive to hydrolysis under acidic and basic conditions. 2,4-D is produced by interaction of 2,4-dichlorophenolate with monochloroacetic acid. The CAS Registry Number of 2,4-D is 94-75-7.

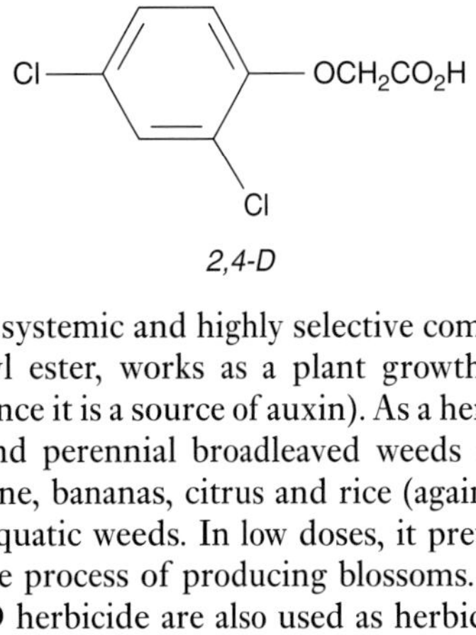

2,4-D

2,4-D is a systemic and highly selective compound, which as its isopropyl ester, works as a plant growth regulator on citrus crops (since it is a source of auxin). As a herbicide, it controls annual and perennial broadleaved weeds in maize, sorghum, sugarcane, bananas, citrus and rice (against grasses). It also controls aquatic weeds. In low doses, it prevents fruit fall and hastens the process of producing blossoms. The salts and esters of 2,4-D herbicide are also used as herbicides. Some of the commercial products are Damine, Deferon and Kay-D.

The herbicide is classified as a moderately toxic chemical (Class II) and is highly toxic to some fish species. It has an acute oral LD_{50} of 375 mg/kg. Upon exposure, it can irritate the eyes and skin, while causing symptoms like nausea and some nerve-related problems. The acute dermal LD_{50} in rats is >1600 mg/kg, while inhalation LC_{50} (24 h) in rats is 1.79 mg/l. It may be phytotoxic for up to 28 days. It is non-toxic to bees but toxic to fish. The LC_{50} in Japanese quails is >100 mg/l. The ADI is 0.01 mg/kg b.w./day. It is classified as a PAN Bad Actor chemical, since it is a potential carcinogen, potential groundwater contaminant and suspected endocrine disruptor.

Plants detoxify 2,4-D slowly by dechlorination, decarboxylation, hydroxylation and dealkylation and also via glucose conjugation. 2,4-D is excreted unchanged in the urine of humans and rats. 2,4-D is mobile, with a soil half-life of 10 days, which can be shortened in case of microbial action, thus reducing the potential for leaching, although traces are sometimes found. In animals, it is eliminated within 24 h.

2,4-D is registered in a large number of countries including India, Australia, New Zealand, the USA, many African nations and all except two countries of the EU.

Daimuron herbicide

Daimuron, $C_{17}H_{20}N_2O$ (1-(α,α-dimethylbenzyl)-3-*p*-tolylurea), is a **phenylurea** herbicide. It has a CAS Registry Number of 42609-52-9. It is an odourless, colourless needle-like substance, soluble in water at 1.2 mg/l, and also in for example acetone and benzene. It is stable to storage conditions and to heat and light.

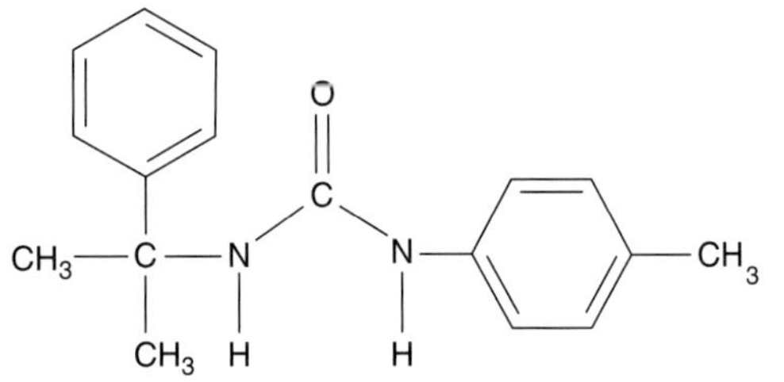

Daimuron

Daimuron is a selective, pre- and post-emergence compound that inhibits cell division after being absorbed by the roots of the target weeds. It is used in rice to control annual grasses and Cyperaceous weeds, the group to which **sedges** belong.

This compound is formulated as granules (GR) and soluble concentrates (SL) and available as Showrone.

According to the WHO, it belongs to the toxicity Class U (unlikely to be hazardous). It is not a PAN Bad Actor chemical. However, daimuron degrades slowly in paddy (half-life is around 50 days). Daimuron is not registered by the EPA as a herbicide.

Dalapon herbicide

Dalapon, $C_3H_4Cl_2O_2$ (2,2dichloropropionic acid), is an organochlorine herbicide. It is a colourless liquid with an acrid odour, highly soluble in water and soluble in ethanol, acetone, benzene and methanol. It is stable under normal temperatures and pressures, except in the presence of heat, flame or strong oxidizers, when there may be risk of explosion. Dalapon is usually formulated with sodium and magnesium salts and thus dalapon sodium and dalapon magnesium are available. Dalapon is produced by chlorination of propionic acid. The CAS Registry Number is 75-99-0.

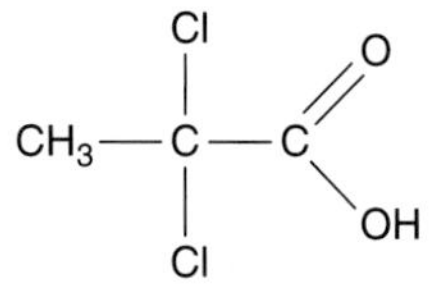

Dalapon

Dalapon, a selective, systemic, pre- and post-emergence herbicide, is also a plant growth regulator. It is used to control specific annual and perennial grasses, such as quack grass, **Bermuda grass** and **Johnson grass**. The compound is mainly used on sugarcane and sugarbeet, but is also used on potatoes and carrots.

Dalapon is compatible with several herbicides and fertilizers. It is available as a water-soluble powder that may be sprayed on the foliage. It is available in the market under the trade names of Dalacide and Diserbo Canali.

Dalapon is classified as a Class U toxin (i.e. it is unlikely to be hazardous). The oral LD_{50} values among male and female rats are 9330 mg/kg and 7570 mg/kg, respectively. Similarly, it does not pose a hazard to rabbits and guinea pigs, in normal use. It is non-toxic to fish and birds. However, it can cause moderate irritation to the eyes, skin and the respiratory tract. The RfD for dalapon is 0.03 mg/kg/day.

Dalapon has a low to moderate persistence in soil (the soil DT_{50} being 2–8 weeks). Degradation proceeds via microbial action, hydrolysis and photolysis. In mammals, dalapon is rapidly eliminated.

Dalapon is approved for use in Australia and the USA. In India the registration has been withdrawn.

Daminozide as a PGR

Daminozide, $C_6H_{12}N_2O_3$ (*N*-(dimethylamino)succinamic acid), is a plant growth regulator (growth retardant). It is a white substance and has an amine-like odour. It is readily soluble in distilled water, methanol and acetone but is insoluble in hydrocarbons. It is hydrolysed by acids, alkalis and heat and is slowly decomposed in light. Its CAS Registry Number is 1596-84-5. It is produced from succinic anhydride and 1-dimethyl hydrazine.

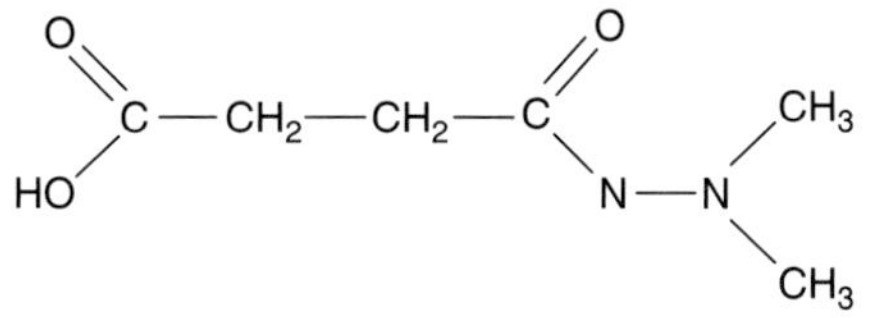

Daminozide

Daminozide is used as a PGR and is responsible for the production of compact plants by inhibiting elongation. It translocates throughout the plant via the leaves. It is used on for example chrysanthemums, hydrangeas and poinsettias.

The PGR is available as a water-soluble powder (SP). It is available in the market under the trade names of Alar and B-Nine.

Daminozide is relatively non-toxic to mammals in acute toxicity studies (the acute oral LD_{50} in rats is 8400 mg/kg, and in mice it is 6300 mg/kg). Similarly, the acute dermal LD_{50} in rabbits is >1600 mg/kg, while their inhalation LC_{50} is >147 mg/l.

It is not toxic to birds (96 h LC_{50} in quail being 5620 mg/kg). It is of low toxicity to fish, terrestrial wildlife and to bees. It causes irritation to eyes and the skin. It has been classified by the WHO as a Class U toxin (unlikely to be hazardous.) The 2 year NOEL in rats was found to be 5 mg/kg. However, daminozide is found to be a carcinogen and is classified as a PAN Bad Actor chemical. It is also observed to cause defects in reproductive organs and liver. Crop use of the chemical has hence been stopped by most countries.

Daminozide is soluble in water and very mobile in soils. It is unlikely to contaminate groundwater, since it breaks down very rapidly into metabolites like formaldehyde and 1,1-dimethyl hydrazine, both of which are considered an environmental hazard. On oral administration in animals, the compound is completely metabolized and eliminated through urine and faeces.

The countries that have registered the use of daminozide include Australia, New Zealand, Canada, South Africa, the USA and countries of the EU such as Cyprus, Denmark, Finland, Greece, Hungary, the Netherlands and Portugal.

Dazomet fungicide

Dazomet, $C_5H_{10}N_2S_2$ (3,5-dimethyl-1,3,5-thiadiazinane-2-thione or tetrahydro-3,5-dimethyl-1,3,5-thiadiazine-2-thione), is a fungicide, herbicide, insecticide and nematicide. Its CAS Registry Number is 533-74-4. It is a colourless crystal, soluble in water (3500 mg/kg) and cyclohexane (89,700 mg/kg). Dazomet is hydrolysed to carbon disulfide in acidic conditions. It is made by reacting methylammonium *N*-methyldithiocarbamate and formaldehyde. It is to be stored in a cool and dry place.

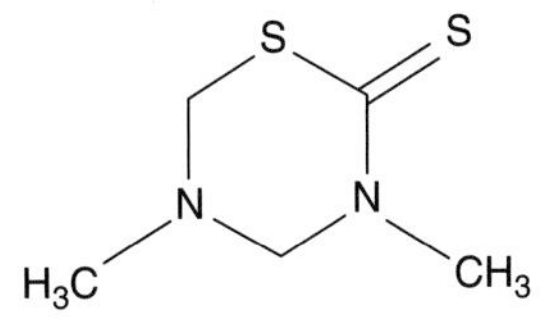

Dazomet

Dazomet, a pre-plant control agent, is a precursor to the fumigant agent **methyl isothiocyanate** and decomposes into it when applied. Thus, fungal species (e.g. *Fusarium*, *Rhizoctonia* and *Pythium*), weed seeds, soil insects and nematodes are controlled.

Dazomet is classified as a Category III toxin by the EPA and WHO (i.e. slightly toxic). The acute dermal LD_{50} in rats is >2000 mg/kg b.w. The acute oral LD_{50} in rats is 519 mg/kg. This chemical tends to cause irritation to the eyes. It is toxic to fish and other aquatic organisms, but not to bees if used as directed. The ADI is 0.003 mg/kg b.w./day. The NOEL in rats is 20 ppm. It is quite phytotoxic to plants and hence may be applied only at the pre-plant stage. There is a risk of application in the dust form and so labels should clearly mention the signal word 'CAUTION' on aerosol/dust products. The substance can be absorbed into the body by inhalation if it is applied as an aerosol, and by ingestion. Recommended care for handlers includes wearing of safety goggles along with breathing protection equipment, washing of hands, eyes, skin and clothes immediately after using/mixing/applying it, and keeping the chemical away from food and drinks. Labels on formulations caution against dispersion of dust.

The compound degrades in wet or humid soil to simple compounds like methylisocynate, hydrogen sulfide, formaldehyde and methylamine.

For most categories of citrus/stone/pome fruits, the MRL of dazomet as fixed by the European Food Safety Authority is 0.02 mg/kg.

Dazomet is registered in India, the USA, Australia, Canada, New Zealand, the Philippines, South Africa, Tanzania and most countries of the EU.

DCIP nematicide

DCIP, $C_6H_{12}Cl_2O$ (*bis* (2-chloro-1-methylethyl) ether), is an unclassified nematicide. It is a brownish oil. The CAS Registry Number is 108-60-1. It is soluble in water (1.7 g/l) and stable to heat and light.

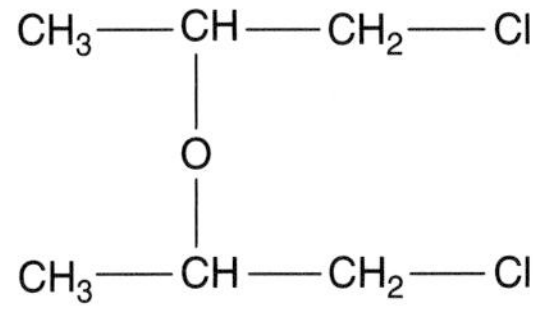

DCIP

DCIP controls nematodes in vegetables, fruits, tobacco and berries. It also controls ground **mealybugs** on citrus trees.

DCIP is often formulated as emulsifiable concentrate (EC) or as granules (GR). DCIP is recommended for application during the plant growth stage. The products usually include **chloropicrin**. Trade names include Nemamort and Root Guard.

The acute oral LD_{50} in rats and mice is up to 536 mg/kg. The acute dermal LD_{50} in rats is >2000 mg/kg. The inhalation LC_{50} for a 4 h period is 12.8 mg/l air. The ADI is 0.13 mg/kg/day.

Japan, Taiwan and the USA have registered the use of DCIP.

DCPA herbicide

See **Chlorthal herbicide**

DDT insecticide

Dichlorodiphenyltrichloroethane or DDT, $C_{14}H_9Cl_5$ (1,1,1-trichloro-2, 2-*bis*-(4-chlorophenyl) ethane), is an organochlorine contact insecticide, and also an acaricide (bridged diphenyl acaricide). It is a white crystalline solid, insoluble in water and soluble in hydrocarbons. DDT is readily dehydrochlorinated in alkaline solutions. DDT is produced by the condensation of chlorobenzene with chloral in the presence of strong sulfuric acid. The CAS Registry Numbers for DDT are 50-29-3 (for p,p′-DDT) and 789-02-6 (for o,p′-DDT).

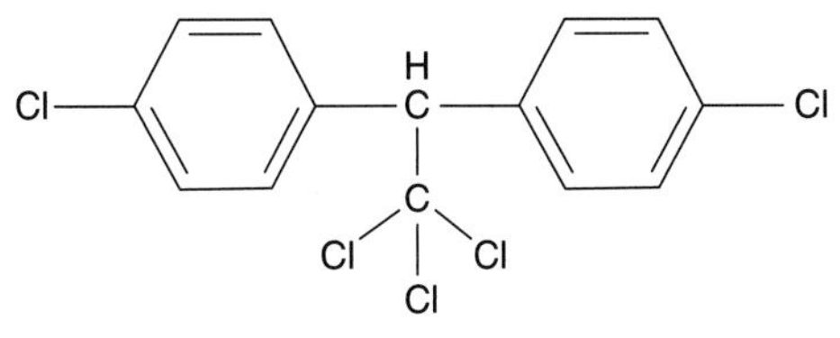

DDT

DDT is a non-systemic, broad-spectrum, organochlorine insecticide with contact action and a long residual effect. Although it acts by hindering the nervous system (central and peripheral systems, mainly interfering with axonal and synaptic transmission) and the liver, it has chronic effects since it accumulates in body tissues. The intended reactions in insects include hyperexcitability, convulsions or paralysis and death. Although it is used as a control agent against mosquito species (to control malaria) and a bacterial disease called typhus (agents *Rickettsia typhi* or *Rickettsia prowazekii* carried by lice and some flea species), DDT has also been used as a general insecticide, especially against gypsy moths.

DDT is variously formulated as either emulsifiable concentrate (EC), wettable powder (WP), dustable powder (DP), granule (GR) or aerosol (AE). It may be combined with non-alkaline pesticides. Some trade names are Anofex and Cesarex.

The WHO and EPA have classified DDT as a Class II toxin (moderately hazardous). It is moderately to slightly toxic to mammals. The acute oral LD_{50} ranges from 113 to 300 mg/kg in mice, rats and rabbits. The acute dermal LD_{50} in female rats is 2510 mg/kg. It is relatively non-toxic on inhalation. The ADI is 0.01 mg/kg b.w./day. DDT is highly toxic to fish and moderately toxic to birds. It accumulates in body tissues and fat cells and is a persistent organic pollutant (POP). It is considered a PAN Bad Actor chemical since it is a reproductive toxin, a carcinogen and a suspected endocrine disruptor. It is also on the IARC Carcinogen List. It is phytotoxic to cucurbits, beans, young tomato plants and some varieties of barley.

Symptoms of DDT poisoning in humans include nausea, diarrhoea, dizziness and sometimes tremors and convulsions.

DDT degrades very slowly in soils (with a half-life of up to 15 years), degradation taking place mostly by microbial action. Degradation is also slow in sunlight. The principal metabolites of DDT in animals and soils are DDD (1,1-dichloro-2, 2-*bis*-(4-chlorophenyl) ethane) and DDE (1,1-dichloro-2, 2-*bis*-(4-chlorophenyl) ethylene). DDT and its breakdown products are persistent in nature mostly due to its low water solubility and high fat solubility.

Canada, Singapore, Cuba, Sweden, Japan and the USA have totally banned the use of DDT. But in many countries including India, Sri Lanka, Ethiopia and the Philippines, it is allowed to be used (imported) by government agencies for managing public health diseases, mainly malaria.

DEET as an IGR

DEET or diethyltoluamide, $C_{12}H_{17}NO$ (*N*,*N*-diethyl-*m*-toluamide), is an insect repellent. It is a light amber liquid, practically insoluble in water but easily soluble in benzene, ethyl ether and ethanol. It has a CAS Registry Number of 134-62-3.

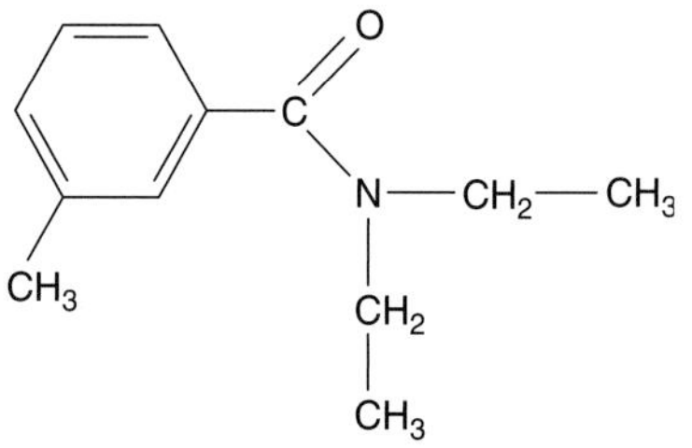

DEET

DEET is primarily used to repel mosquitoes and ticks and is one of the most widely used liquid insect repellents today. The mode of action is the inhibition of insect olfactory receptors, bringing about repellence.

DEET is formulated as a pressurized liquid spray (both pump and aerosol), and also as sticks and creams. The repellent insecticide, sold as Detamide, Metadelphene, OFF and Repel, is often combined with ethyl or isopropyl alcohol or ethanol.

The WHO has categorized DEET as a Class III toxin (slightly hazardous). It has a low toxicity in mammals, and a slight to moderate toxicity to fish. Although skin contact with DEET formulations of even 100% concentration is considered safe, it may cause irritation in people with psoriasis and acne, and in children. It may not be safe to use near water sources. Many natural alternatives to DEET have been used to repel insects and these include eucalyptus-based products (with eucalyptol), **neem** and lemon grass.

DEET is registered for use in the USA, Canada, Australia and New Zealand.

DEET technical

DEET (*N*,*N*-diethyl-*m*-toluamide) is a clear liquid, commonly used as an ingredient in many insect repellent compositions.

The requirements for analysing DEET from a substance are: (i) a minimum purity of 950 g/kg; (ii) acidity as H_2SO_4, a maximum of 0.3 g/kg; (iii) water content, a maximum of 3 g/kg; (iv) relative density 25°/25°C (0.992–0.999); and (v) Relative Index 25°C (1.520–1.524). The apparatus required for this analysis are a double-beam infrared spectrophotometer (Perkin Elmer model) and two equivalent infrared absorption cells, with sodium chloride windows and a path length of approximately 0.4 mm.

Weigh (to the nearest 0.1 mg) standard DEET into four 50 ml volumetric flasks, to give approximately 20, 40, 60 and 80 g/l concentrated solutions in carbon disulfide. Fill the reference cell with carbon disulfide and the sample cell with each of the standard solutions in turn and record the spectra.

Measure the absorption at 14.18 μm and at 14.48 μm and calculate the difference between these values, ΔA, for each of the solutions. Plot the values of ΔA against concentration (g/l) of the meta-isomer. The value of ΔA is determined by Eqn D.1.

$$\Delta A = [A_{14.18} - A_{14.48}]_{\text{reference}} - [A_{14.18} - A_{14.48}]_{\text{blank}} \qquad \textbf{(D.1)}$$

Weigh (to the nearest 0.1 mg) about 0.5 g of the sample and transfer it quantitatively to a 10 ml volumetric flask, and make up to the mark with carbon disulfide. Measure the infrared absorption at 14.18 μm and 14.48 μm using the same conditions. Determine the concentration of meta-isomer by comparing this value with the calibration curve. The DEET content (g/kg) is given by Eqn D.2.

$$\frac{C_1 \times P}{C_2} \qquad \textbf{(D.2)}$$

where

C_1 = Concentration (g/l) of standard DEET found from the calibration curve

C_2 = Concentration (g/l) of the sample taken

P = Purity (g/kg) of the reference standard

Defoliants

A chemical that desiccates leaves from trees and growing plants is called a defoliant. The agent may be either organic or inorganic. Some examples of organic compounds are phenoxyacetic acid, trichloro-picolinic acid, carbamates and nitro compounds. Specifically, **thidiazuron**, **pyraflufen ethyl**, **tribufos** and diuron are used as defoliants. Examples of inorganic compounds are arsenic compounds, cyanides, thiocyanates and chlorates.

Defoliants and **desiccants** are commonly known as harvest aids and help in more efficient and timely harvesting operations. Removal of excess foliage or parasitic growth on some plants allows air movement and prevents insect infestation. Cotton is one crop that needs harvesting aids.

Many persistent defoliants have been used in military operations. For example in the Vietnam War, Agent Orange was used by the Americans to thin out Vietnamese jungles used as camouflage by the Vietnamese. Many defoliants are toxic.

Defoliating insect pests

A large group of insects belonging to several insect orders (e.g. Coleoptera, Lepidoptera and **Orthoptera**) feed on crop leaves at some point through their life cycle. Occurrence of eggs and larvae on stems and leaves, holes in foliage and skeletonized leaves are some of the symptoms of attacks by defoliating insect pests.

Defoliation reduces the photosynthetic capacity of a plant and thereby decreases yields. Often, if severe infestation occurs when the plant is mature, growth is affected, with a possibility of death. At an early growth stage, crops can recuperate, by producing more tillers. Tobacco budworm, cotton bollworm, armyworm, blister beetles, budworms, **grasshoppers**, weevils, clover worms and corn rootworms are just a few of the several species of pests that defoliate plants. Examples of crops that are affected are cotton, rice, soybean and ornamentals.

Estimation of crop damage by defoliators may be made by sampling populations of larvae sighted in a certain area. When pest life-cycle stages are plotted against damage, defoliation follows an 'inverted U'pattern, with very little damage during the early life cycle of pests, such as at the onset of favourable climate, in early spring and during the mating season. The damage is most severe at the second or third instar larval stages, before pupation, when feeding is crucial. The damage decreases when for example insect numbers decline, insects move on to other areas, when winters set in and when crops are harvested.

Degradation of pesticides

Chemical pesticides, once applied, are subject to several routes of degradation. Degradation of a pesticide is the breakdown of its structure, into simpler substances, leading to reduced efficacy, since degradation products are commonly less toxic. Pesticide compounds are degraded by the action of for example soil pH, water, microbes, sunlight and UV radiation.

Microbial degradation in soil occurs when fungi and bacteria digest these compounds as food. Conditions like warm temperatures, favourable pH levels, adequate soil moisture, aeration and fertility that favour microbial growth may result in faster degradation of pesticides. This often necessitates frequent and higher application of some pesticides. However, adsorption of pesticides to soil colloids resists degradation by microbes, since these chemicals become unavailable to them.

Chemical degradation is the breakdown of a pesticide by a process not involving a living organism. Factors like adsorption of pesticides to the soil, soil pH level and soil moisture contribute to the rate and type of chemical reactions that occur. Many pesticides, especially organophosphorus insecticides, are susceptible to degradation by fluid decomposition in highly acidic soils or spray mixes.

Photodegradation is the breakdown of pesticides by sunlight. Pesticides applied to foliage or soil vary considerably in their stability when exposed to sunlight. Like other breakdown processes, photodegradation reduces the potency of the chemicals applied.

Degree days for pest management

A temperature-based planning model has been developed to predict the local appearance of insects by comparing average temperatures in a period of time with temperatures favourable for insect growth and development.

Since temperatures regulate pest development, scientists use degree day (or heat unit) measurements to track heat accumulation over time and predict the local appearance of various pests. The degree day formula assigns a value to each day's maximum and minimum temperature and compares that value to the temperature at which a particular pest begins to develop. Units accumulate daily, with a predetermined starting point in the pest's life cycle. If the pest is a caterpillar, for example, degree day units might be calculated from the first day of its adulthood. Using this information, entomologists predict when each pest will emerge from hibernation and become active. Such predictions may be inaccurate if there is sudden spike or fall in temperature, which can lead to rapid growth or slow down in insect development time. Suffice to say that this prediction model has an important role in scheduling planting, harvesting and pest control treatment.

Delachlor herbicide

See **Chloroacetanilide herbicides**

Delta endotoxin

See ***Bacillus thuringiensis* (*Bt*) as a biocontrol agent**

Deltamethrin analysis

Deltamethrin, $C_{22}H_{19}Br_2NO_3$ ((*S*)-α-cyano-3-phenoxybenzyl (1*R*,3*R*)-3-(2,2-dibromovinyl)-2,2-dimethylcyclopropanecarb oxylate), is a pyrethroid insecticide. The mixture of deltamethrin and other additives can be analysed as below, with the following prerequisites: (i) the deltamethrin content should be a minimum of 980 g/kg (or a purity of 98%), with the deltamethrin *R*-isomer being 10 g/kg; (ii) acid chloride corresponding to deltamethrin should be a maximum of 2 g/kg; and (iii) the acid + anhydride corresponding to deltamethrin should be 10 g/kg.

The apparatus required for this analysis includes liquid chromatography apparatus with stainless steel columns, equipped with a pumping system capable of maintaining a pressure of 15×10^3kPa, a UV spectrophotometer detector capable of measuring UV absorption at 254 nm and a loop-type injector. The column tube is 15–18 cm long and 4.6 mm in internal diameter packed with silica. The ideal conditions for the liquid chromatograph are ambient column temperature, flow rate of 1–1.7 ml/min, wavelength of 254 nm, injection volume of 20 μl and retention time of deltamethrin of about 9 min. The liquid chromatography mobile phase is a degassed mixture of 5 ml dioxane and 25 ml iso-octane.

Weigh two samples of about 50 mg each (to the nearest 0.1 mg) of deltamethrin into two 50 ml volumetric flasks (B_1 and B_2). Add a mixture of 20 ml dioxane and 80 ml iso-octane and shake to dissolve the samples. Dilute further to the mark with the same solution.

Inject the calibration solution from one of the two flasks (A_1 or A_2) until the peak areas/heights of two successive injections agree within 2%. Inject the calibration solutions A_1 and A_2 and sample solutions B_1 and B_2 in succession in the following sequence: A_1B_1, A_2B_2, A_1B_2 and A_2B_1. For each of these four groups, calculate the deltamethrin content according to Eqn D.3.

$$\text{Deltamethrin (g/kg)} = \frac{a_2 \times m_1 \times P}{a_1 \times m_2} \quad \text{(D.3)}$$

where
a_1 = Deltamethrin area/height of the calibration solution
a_2 = Deltamethrin area/height of the sample solution
m_1 = Mass (g) of deltamethrin standard in the calibration solution
m_2 = Mass (g) of sample taken
P = Purity of deltamethrin standard (g/kg)
The four results should agree within ±3% of their mean value.

For analysing deltamethrin *R*-isomer content, weigh about 200 mg of the sample (to the nearest 0.1 mg) in a 50 ml volumetric flask. Add 2.5 ml dioxane to dissolve and make up to the mark with iso-octane. The *R*-isomer content is determined by comparing the response of the sample with that of an *R*-isomer standard by HPLC, using a silica column and a mixture of dioxane and iso-octane as the mobile phase. The retention times are 8 min for the *R*-isomer and 9 min for deltamethrin.

The deltamethrin wettable/dustable powder (WP/DP), a fine, free-flowing whitish powder that wets readily, is extracted with the dioxane and iso-octane mixture, and estimated using HPLC as described above.

The emulsifiable concentrate (EC) of deltamethrin is the chemical, dissolved in a solvent and free from suspended impurities and sediments. Here too, the estimation is done by HPLC, by comparing the response of the sample with that of a standard deltamethrin, using a silica column and the dioxane-iso-octane mixture as the mobile phase as described above. Similarly, the dustable powder (DP) of deltamethrin is extracted from the sample with a mixture of dioxane and iso-octane and is estimated using HPLC.

Deltamethrin insecticide

Deltamethrin, $C_{22}H_{19}Br_2NO_3$ ((*S*)-α-cyano-3-phenoxybenzyl-(1*R*,3*R*)-3-(2,2-dibromovinyl)-2,2-dimethylcyclopropanecarbox ylate), is an insecticide (synthetic pyrethroid). It is a colourless crystalline substance that is almost insoluble in water but highly soluble in dioxane, cyclohexanone, dichloromethane and acetone. While otherwise stable in normal storage conditions, it is unstable in alkaline media. Deltamethrin is produced by esterification of dibromopermethrin acid with α-cyano-3-phenoxybenzyl alcohol. It has a CAS Registry Number of 52918-63-5.

(*S*)-alchohol 1(*R*)-*cis*-acid

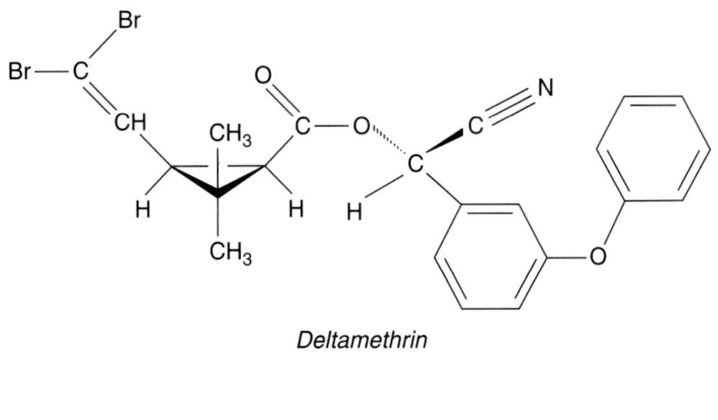

Deltamethrin

Deltamethrin, a non-systemic, broad-spectrum, fast-action insecticide, is also a cost-effective agent that is mostly used on insects in the orders Coleoptera, Heteroptera, Homoptera and Lepidoptera, and includes stored grain pests, crop pests, public health pests and farm pests (spiders, fleas, ticks, moths, carpenter ants, bees, bedbugs, termites, cockroaches, locusts, mosquitoes, the tsetse fly and crawling insects). The crops that are covered include maize, cotton, soybean and vegetables. The mode of action of deltamethrin is the inhibition of the nervous system.

Deltamethrin is formulated as an emulsifiable concentrate (EC), wettable/dustable powder (WP/DP) and as granules (GR). Deltamethrin may be used with insecticides and fungicides. Normally a small dose of 5–25 g/ha of deltamethrin is sufficient to control an infestation. Delta dust is a known trade name.

The WHO has classified deltamethrin as a moderately hazardous toxin (Class II). Deltamethrin is slightly toxic to rats, dogs, most birds and rabbits and is non-irritating to the skin. But it is toxic to fish, other aquatic species, earthworms and bees. Poisoning can occur to handlers/users if ingested and may lead to nausea, dizziness and even coma. The acute oral LD_{50} in rats ranges from 87 to 5000 mg/kg depending on conditions, while the inhalation LC_{50} (6 h) in rats is 0.6 mg/l air. The ADI is 0.01 mg/kg/day.

Degradation of deltamethrin occurs in soil and water by the action of microbes (the DT_{50} in field conditions is up to 23 days). Elimination of deltamethrin is quick in the case of rats and some other animals, with the hydroxylation of the phenyl ring and hydrolysation of the ester bond. Its metabolites include *trans*-deltamethrin. Deltamethrin does not leach since it is poorly soluble in water and most of it adsorbs to soil colloids. So it is not a threat to water sources. While it is not absorbed by leaves or roots, it easily photodegrades into non-harmful compounds.

Deltamethrin is registered for use in India, the USA, Canada, Australia, Cambodia, South Africa and many other countries in Europe and Africa.

Demethylation inhibitors (DMIs)

See **Sterol biosynthesis inhibitor fungicides (SBIs)**

Demeton-*S*-methyl insecticide

Demeton-*S*-methyl, $C_6H_{15}O_3PS_2$ (*S*-2-ethylthioethyl *O*,*O*-dimethyl phosphorothioate), is an organophosphorus insecticide. It is a pale yellow oil, soluble in water (22 g/l) and readily so in alcohols, ketones, dichloromethane and isopropanol. It hydrolyses rapidly in alkaline media (with a DT_{50} of 1 week at pH 9). Demeton-*S*-methyl has a CAS Registry Number of 919-86-8, while that of demeton-*O*-methyl is 867-27-6. It is prepared by the reaction of potassium *O*,*O*-dimethylphosphorothioate and 2-chlorodiethyyl sulfide.

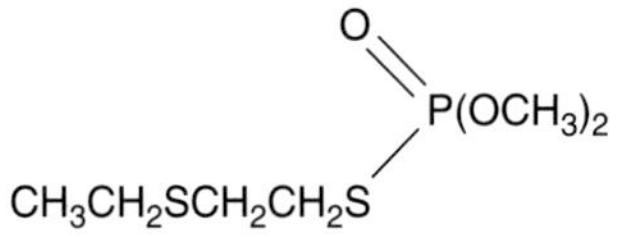

Demeton-S-methyl

Demeton-*S*-methyl is a systemic insecticide with good acaricidal properties. Being an organophosphate, cholinesterase inhibition is the mode of action. It has a rapid contact action. Due to good translocation, it can effectively control infestation of aphids, mites and other sucking insects. The crops that are covered include potatoes, fruits and vegetables. Aphid species, such as *Myzus persicae*, are also controlled.

Demeton-*S*-methyl is available in the market under the trade names Bayer 8169, Demeton-S, Mifatox and Mercaptophos. Formulations are generally in emulsifiable concentrates (EC). It is compatible with many non-alkaline pesticides.

This insecticide is a Class 1b toxin (extremely hazardous) according to the WHO. It is toxic to bees and fish species and also moderately toxic to mallard ducks. The acute oral LD_{50} in rats is 30 mg/kg, while the inhalation LC_{50} (4 h) is 0.12 mg/l air. It can be phytotoxic to some varieties of ornamental plants. This chemical is a PAN Bad Actor chemical since it rates high in acute toxicity tests. It is also a cholinesterase inhibitor and a suspected endocrine disruptor. The ADI is 0.0003 mg/kg/day.

Mammals eliminate almost all of the demeton-*S*-methyl in the urine within 1 day. In both mammals and plants, it metabolizes via many reactions such as oxidation of the thioethyl group to give sulfoxide (**oxydemeton-methyl**) and sulfone (demeton-*S*-methylsulfon). In soils and water, demeton-*S*-methyl degrades very rapidly.

Demeton-*S*-methyl is registered for use in South Africa but is not registered in the USA. Similarly, some of the degradation products like demeton-*S*-sulfone, demeton-*S*-methyl sulfone, demeton-*O*-methyl are classified as PAN Bad Actor chemicals and are thus not registered in countries like the USA and in the EU.

Deoxynivalenol (DON)

See **Vomitoxin**

Dermal test

See **Toxicity testing**

Dermal toxicity

Exposure to a toxic compound (commonly a chemical pesticide) through the skin is known as dermal exposure. Some toxic compounds can be absorbed through the skin, not only causing irritation and burning on the skin surface but also leading to disruption of the internal body system.

The frequency of dermal exposure is very high, because pesticides have often to be mixed, handled, applied and disposed of. Formulations such as liquids, dusts and wettable powders can be absorbed dermally.

The extent of damage caused because of dermal exposure depends on: (i) quantity; (ii) contamination area; (iii) time of the exposure; (iv) toxicity of the chemical; and (v) rate of absorption via the skin. The injury is very serious if exposure is in the lower groin area or the genitals.

Desiccants

Desiccants are organic/inorganic broad-spectrum compounds that aid harvest management. They bring about quick control of crop over-growth and the nuisance of weeds. Application of desiccants also allows for removal of potential weed hosts of pests and post-harvest crop growth. **Paraquat** and **diquat dibromide** are two examples of commonly used desiccants.

Desiccants are often used instead of, or along with **defoliants**. The mode of action of desiccants is the dehydration of foliage, leading to leaf drop. The results of a desiccant last for a short time and subsequent moisture supply can revive weed or crop growth. These compounds also facilitate even growth of crops for an efficient harvest. An even growth of crops ensures good quality yields, uniform moisture content in the harvest and thus more efficient storage. A desiccant is used especially when a crop is grown for seed production, as with the pea crop (application of the agent should be carried out closer to maturity). In the case of potatoes, the above-ground plant is killed in order to maintain the high quality of the tuber. Desiccants such as **endothal**, **pyraflufen**, pelargonic acid and **carfentrazone** are used. Applications may be split up if required for a better vine desiccation and for faster tuber maturity. Peas, pulses and some cereals also benefit from the use of desiccants. It is important to maintain the required pre-harvest interval, usually of up to 2 weeks. Normally, desiccants are ground applied or aerially applied, but care must be taken to avoid spray drift. Reglone is a common trade name.

Desmedipham herbicide

Desmedipham, $C_{16}H_{16}N_2O_4$ (ethyl 3-phenylcarbamoyloxyphenylcarbamate), is a carbanilate (***bis*-carbamate**) herbicide. These colourless crystals, poorly soluble in water (7 mg/l at lower pH) but readily soluble in dichloromethane, ethyl acetate and methanol, hydrolyse in neutral and alkaline media. Desmedipham is made by the reaction of *m*-aminophenol with ethylchloroformate, followed by treatment with phenylisocynate. Desmedipham has a CAS Registry Number of 13684-56-5.

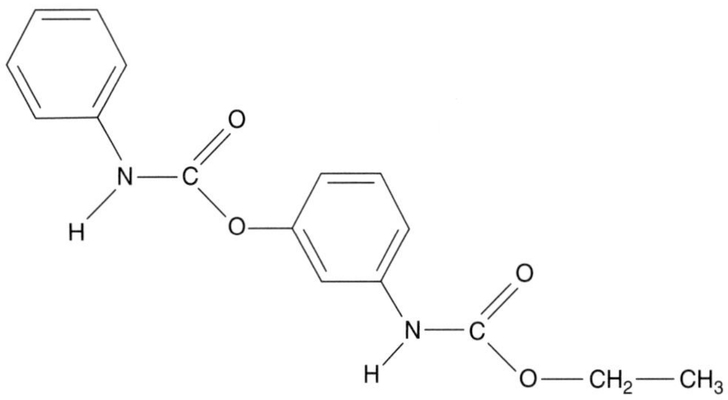

Desmedipham

Desmedipham is a selective, systemic, post-emergence herbicide, absorbed by the leaves and roots, with translocation primarily in the apoplast. It is a photosynthetic electron transport inhibitor. It is applied to control broadleaved weeds (e.g. **common lambsquarters**, chickweed, ragweed and yellow foxtail) in some leafy vegetables, sugarbeet and table beet (beetroot). The mode of action is the inhibition of photosynthesis.

Desmedipham is available under trade names including Betanal and Kemifam. It can be broadcast or sprayed aerially, or by a ground sprayer. It is compatible with non-alkaline preplant herbicides.

According to the WHO, desmedipham comes under the toxicity Class U (unlikely to be hazardous in normal use). The oral LD_{50} among rats is >10250 mg/kg, while the inhalation LC_{50} (4 h) is >7.4 mg/l. Based on a 2 year study of rats, the ADI has been determined as 0.03 mg/kg b.w./day.

In mammals, most of desmedipham along with its metabolites is eliminated via urine. Desmedipham is hydrolysed to ethyl *N*-(3-hydroxyphenyl) carbamate (EHPC). Hydrolysis and conjugation (to glucuronides and ethereal sulfates) are the major steps in metabolism. In sugarbeet, EHPC is the major metabolite, with *m*-aminophenol as a further metabolite. The DT_{50} in soil is about 34 days, with formation of ethyl 3-hydroxycarbanilate, which undergoes further degradation.

Desmedipham is registered in the USA, Canada, New Zealand and all EU nations except three. (See also **bis-carbamate herbicides**.)

Desmetryn herbicide

Desmetryn, $C_8H_{15}N_5S$ (N^2-isopropyl-N^4-methyl-6-methylthio-1,3,5-triazine-2,4-diamine), is a **methylthiotriazine herbicide**. It is a white, crystalline compound, soluble in water at 580 ppm. The CAS Registry Number is 1014-69-3.

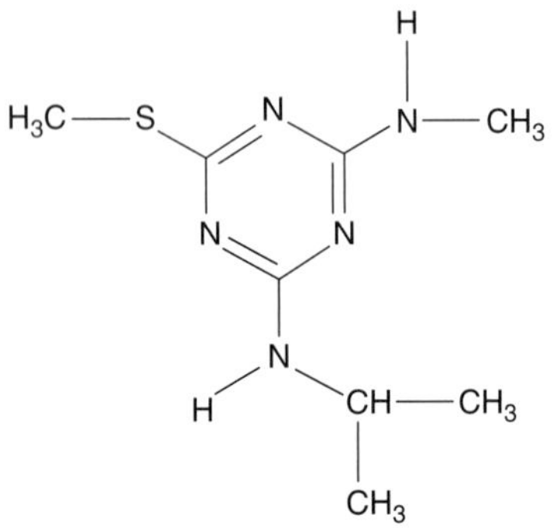

Desmetryn

Desmetryn is used as a post-emergence foliar herbicide for broadleaved and grassy weeds. It is also used on annual broadleaved weeds and grasses in cabbage and onion crops. It acts as a photosynthetic electron transport inhibitor. It is available in the market under trade name Semeron.

The WHO has categorized desmetryn as a slightly hazardous chemical (Class III). Testing on rats and mice indicates that there could be adverse effects on important internal organs, although it is not carcinogenic.

Diafenthiuron acaricide

Diafenthiuron, $C_{23}H_{32}N_2OS$ (1-tert-butyl-3-(2,6-di-isopropyl-4-phenoxyphenyl) thiourea), is a thiourea acaricide. Diafenthiuron is converted to the corresponding carbodiimide (carbodiimide CGA 140408) in sunlight and in aqueous solutions. So diafenthiuron may be termed a **pro-pesticide**. The CAS Registry Number of diafenthiuron is 80060-09-9.

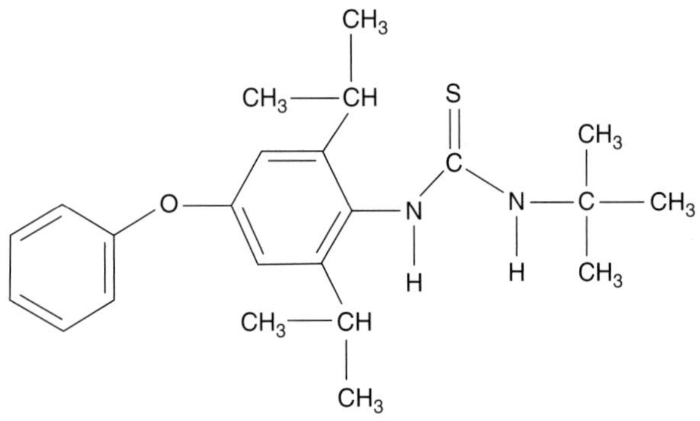

Diafenthiuron

Carbodiimides are a functional group formed from thioureas. Carbodiimide CGA 140408, $C_8H_{17}N_3HCl$ (1-tert-butyl-3-(2,6-di-isopropyl-4-phenoxyphenyl)-carbodiimide), is an inhibitor of mitochondrial respiration. It is an acaricide, insecticide and an ovicide. Phytophagous mites of the families Tetranychidae and Tarsonemidae (including species of spider mites and white mites) are well controlled by this chemical. Among insect pests, aphids, whiteflies, jassids and leaf-eating insects are controlled by diafenthiuron. Crops on which this acaricide-insecticide may be applied include cotton, soybean, vegetables and some ornamental plants.

Diafenthiuron is applied at the rate of 300–500 g/ha, and formulated as soluble concentrate (SC) and wettable powder (WP). The selected trade names include Pegasus and Polo. It may be combined with select pesticides like fenoxycarb.

The WHO has classified diafenthiuron as a Class U toxin. The acute oral LD_{50} in rats is 2068 mg/kg. The acute percutaneous LD_{50} in rats is >2000 mg/kg. It is non-irritating to the skin and eyes, in normal use. The LC_{50} (4 h) in rats is 0.558 mg/l air. The NOEL (90 days) in rats is 4 mg/kg b.w./day and for dogs is 1.5 mg/kg b.w./day. The ADI is 0.003 mg/kg b.w./day. The acute oral LD_{50} in bobwhite quails and mallard ducks is >1500 mg/kg. The LC_{50} (96 h) in carp is 0.0038 mg/l, rainbow trout is 0.0007 mg/l and bluegill sunfish is 0.0013 mg/l. It is toxic to honey bees. The LD_{50} (48 h) (oral) in honey bees is 2.1 mg per bee and (contact) 1.5 mg per bee. Diafenthiuron is safe on the adult and immature stages of many biocontrol agents such as predatory mites (*Amblyseius andersoni*, *Typhlodromus pyri*), spiders (Erigonidae, Lycosidae) and *Chrysopa carnea*. However, carbodiimide is known to cause irritation to the eyes and skin and be a hazard on exposure to those who apply the pesticide.

In animals, studies of the absorption, distribution and excretion in rats demonstrate that the major portion of the dose is excreted with the faeces. The compound is degraded to yield its corresponding carbodiimide, which in turn reacts with nucleophiles like water and fatty acids to form urea and fatty acid derivatives. Diafenthiuron and its main metabolites show a strong sorptivity to soil particles. The degradation in soils proceeds rapidly (DT_{50} is from <1 h to 1.4 days).

Diafenthiuron is registered in Australia, Cambodia and Vietnam but not in the USA.

Diallate

See **Thiocarbamate herbicides**

Diatomaceous earth as a biopesticide

Diatomaceous earth is mined from the fossilized silica shell of unicellular or colonial algae in the class Bacillariophyceae, known as diatoms. It has both abrasive and sorptive qualities. Like the silica aero gels, diatomaceous earth absorbs the waxy layer of the skin surface of insects, causing them to dry out. In addition to its **desiccant** action, it works abrasively to rupture insect cuticles, allowing cell fluid to leak out.

This substance is formulated as dust either alone or in combination with pyrethrin.

The biocompound is non-toxic to mammals. The oral LD_{50} in rats is between 3160 and 8000 mg/kg, depending on the formulation. It is safe for human consumption.

Of the two grades of diatomaceous earth – swimming pool grade and natural grade – only the natural grade is used for insect control as the swimming pool grade contains crystalline silica, which is a respiratory hazard to handlers of diatomaceous earth.

This substance may be used in homes for filling crevices and cracks with, in order to kill insects such as cockroaches, ants and termites. It may be placed in attics to kill passing insects. It can be fed to animals, where it has a scouring effect, to control intestinal parasites. It is applied around crop/non-crop beds to discourage slugs and snails.

Diazinon analysis

Diazion (technical grade), $C_{12}H_{21}N_2O_3PS$, a light brown liquid, is an acaricide and a repellent for birds and insects. The requirements for analysing diazinon are a purity of at least 95%, acidity (as H_2SO_4) of 0.3 g/kg, acetone (1.5 g/kg) and water content of 0.6 g/kg. Gas-liquid chromatography equipment with a flame ionization detector is essential. The chromatographic column should be made of a borosilicate glass tube (180 cm long, 4 mm in internal diameter and 6 mm in outside diameter), and the column packing material should be treated with 10% silicone DC 200. Operating conditions include an oven temperature of 190°C (±10°C), the temperature of the injection port, 240°C, and that of the flame-ionization detector, 240°C. The flow rate of hydrogen should be adjusted to fit the requirements of the detector and the flow rate of the carrier gas, nitrogen, should be 80–100 ml/min. The retention time of the diazinon peak is 5–6 min.

To prepare the diazinon calibration solution, weigh about 125 mg diazinon standard into a standard bottle with an aluminium-lined screw cap. Pipette 50 ml internal standard solution in acetone and shake mechanically for 30 min. To prepare the internal standard, weigh 4 g technical aldrin (to the nearest 0.1 g) into a 600 ml beaker. Add 400 ml acetone to dissolve. Filter through Whatman filter paper into a 1 l volumetric flask, washing with several small quantities of acetone and make up to 1000 ml.

Weigh 125 mg (to the nearest 0.1 mg) of the diazinon sample into a 125 ml round-bottomed bottle with an aluminium-lined screw cap. Pipette 50 ml internal standard of the diazinon and calibration standard and shake mechanically for 30 min. Allow to stand to let the insoluble materials settle. Inject portions of 3 ml calibration solution until the response is stable and the ratios of the peak area or height of the diazinon peak to the internal standard peak agree to within 2% of their mean. Then make duplicate injections of sample solution followed by duplicate injections of calibration solution. The peak area or height ratios of the calibration solution injections must agree to within ±1% of the first accepted calibration solution values.

Calculate the peak area or height ratios for both duplicate calibration solution injections preceding and following the sample solution injections. Average the four values. Similarly, calculate and average the peak area/height ratios of the two sample solution injections.

The response ratio (r) for each injection is given by Eqn D.4.

$$r = \frac{\text{Area or height of diazinon peak}}{\text{Area or height of internal standard peak}} \tag{D.4}$$

The diazinon content (g/kg) is given by the formula in Eqn D.5.

$$\text{Diazinon content} = \frac{r_2 \times m_1 \times P}{r_1 \times m_2} \tag{D.5}$$

where
r_1 = Average of response ratio for calibration solution
r_2 = Average of response ratio for sample solution
m_1 = Mass (g) of diazinon standard in the calibration solution
m_2 = Mass (g) of sample taken
P = Purity of diazinon standard (g/kg)

The water-dispersible powder of diazinon is a homogeneous mixture of technical grade diazinon with filler(s) and other formulation additives, in the form of a fine, free-flowing powder that wets readily. The diazinon is extracted from the sample with acetone, to which aldrin is added as an internal standard. The extracted solution is estimated by gas-liquid chromatography as described for technical grade diazinon. Similarly, an emulsifiable concentrate of this compound has diazinon in a suitable solvent with other necessary additives. The sample dissolved in acetone to which aldrin is added as an internal standard is estimated by gas-liquid chromatography as described for technical grade diazinon.

Diazinon insecticide

Diazinon, $C_{12}H_{21}N_2O_3PS$ (*O,O*-diethyl *O*-2-isopropyl-6-methylpyrimidin-4-yl phosphorothioate), is an organophosphorus insecticide. It is a yellowish liquid, soluble in water (40–60 mg/l) and miscible with ethers, alcohols, benzene, toluene, hexane, cyclohexane, dichloromethane, acetone and petroleum oils. It oxidizes above 100°C and decomposes above 120°C. It hydrolyses in acidic medium. It is produced by many processes including condensation of isobutyramidine with acetoacetate to yield the intermediate 2-isopropyl-4-methylpyrimidine, which in turn is treated with diethylthiophosphoric acid. It has a CAS Registry Number of 333-41-5.

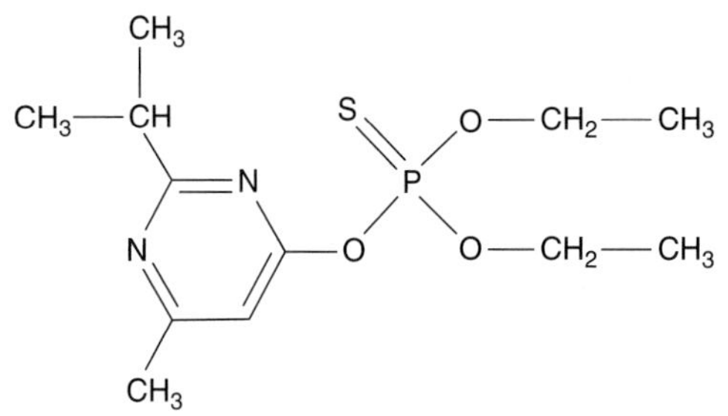

Diazinon

Diazinon, a cholinesterase inhibitor, is a non-systemic, broad-spectrum insecticide, which is also a veterinary ectoparasiticide, bird-repellent, avicide and acaricide with contact, stomach and respiratory action. It acts by disrupting normal functioning of the nervous system. It is absorbed by roots when applied to the soil and the foliage. As a crop protection agent, it is used in the control of soil insects, sucking and chewing insects, and mites. It is used in dry seed treatment. Crops that are protected include fruits, vegetables, potatoes, cereal crops, sugarcane, coffee, cocoa, tea, tobacco, lucerne and cotton. As an ectoparasiticide, it keeps ticks and fleas away from farm animals. It can also be used against pests in the home such as ants, cockroaches and bedbugs. It is also used on forage.

The insecticide is formulated in forms such as granule (GR), wettable/dustable powder (WP/DP), emulsifiable concentrate (EC) and as an aerosol. It may be used with other pesticides, except copper-containing compounds. It is available on the market under the trade names Alfa-Tox,, Bazuden, Ciazinon, Dazzle, Dizinon, ENT 19507, Gardentox, Geigy 24480, Nucidol, Saraolex and Spectracide.

Diazinon is classified as a Class II toxin (moderately hazardous) by the WHO. It is classified as a PAN Bad Actor chemical since it is a cholinesterase inhibitor and a developmental/reproductive toxin. It is also a suspected endocrine disruptor and a potential groundwater contaminant. The LD_{50} among rats is around 1250 mg/kg. The dermal LD_{50} among rabbits is 540–650 mg/kg. Labels should clearly state the appropriate signal word, 'WARNING'. Diazinon is highly toxic to birds (the LD_{50} in mallard ducks being between 2.75 and 40.8 mg/kg), fish and bees. The symptoms of diazinon poisoning include nausea, vomiting and even coma.

In animals, the principal metabolites are diethyl thiophosphate and diethyl phosphate, with a very short half-life (half a day). In plants too, rapid absorption and translocation takes place. Metabolism occurs via transformation and degradation of the hydroxypirimidine derivatives to carbon dioxide. In soil and water, degradation involves oxidation to the phosphate (diazoxon) and hydrolysis. Diazinon is fairly strongly adsorbed on to soil, with half-lives of up to 1 month.

Diazinon has been banned by Denmark. There are severe restrictions on its use on lawns due to its risk to birds and potential to contaminate groundwater. It is registered for use in India, the USA, Canada, Australia, South Africa and many other countries.

Dibber

See **Weed management by mechanical methods**

Dicamba herbicide

Dicamba, $C_8H_6Cl_2O_3$ (3,6-dichloro-*o*-anisic acid), is a **benzoic acid herbicide**. It forms colourless crystals, moderately soluble in water and readily soluble in organic solvents. It is stable in acidic and alkaline media and to photolytic degradation. It is highly soluble in water (27,200 mg/l). It is produced by hydrolysis of 1,2,4-trichlorobenzene with sodium hydroxide, followed by carboxylation, esterification with methanol, methylation of the hydroxyl group and hydrolysis of the ester to the acid. Dicamba has a CAS Registry Number of 1918-00-9. It forms salts with bases and amines.

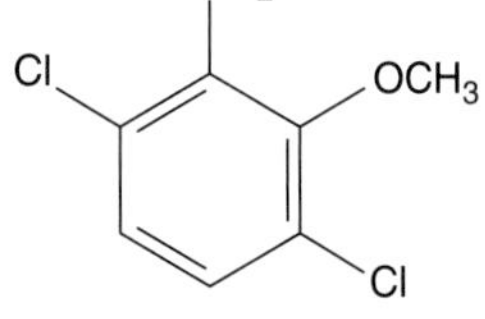

Dicamba

The herbicide dicamba is a selective, systemic herbicide, which also acts like a plant growth regulator (it is auxin-like). It acts by promoting plant growth (thus, tweaking the quantity can have either regulatory or herbicidal effects). It is easily absorbed by plant parts and translocated throughout (via both symplastic and apoplastic systems). It controls annual and perennial broadleaved weeds and brush species in maize, sugarcane, turf and public places.

The herbicide is available in granule (GR) formulations. Dicamba is often used in combination with one or more of the phenoxy acid-based herbicides. It is a popular compound, used widely, since it is required in small doses. And thus, as a logical extension, some weeds like common hemp nettle (*Galeopsis tetrahit*), kochia (*Kochia scoparia*) and wild mustard (*Sinapis arvensis*) have developed resistance to it. It is available in the market under trade names like Banvel, Camba and Diptyl.

According to the WHO and EPA, dicamba is a slightly hazardous chemical (toxicity Class III). The oral LD_{50} of this compound among rats is 757 mg/kg and the dermal LD_{50} is 2000 mg/kg. It is slightly toxic to some fish species. It is caustic to the eyes and can cause serious injury. It is phytotoxic to some cacti species. However, if ingested, dicamba can cause vomiting, headaches and muscle weakness, requiring medical attention. Applicators, handlers and others that come in contact with dicamba are well cautioned against eating/handling food and drink nearby or against eye contact. It is a PAN Bad Actor chemical since it is feared to contaminate underground water sources.

In mammals, dicamba does not change much, and is eliminated via the urine. In aerobic soils, microbial action produces 3,6-dichlorosalicylic acid. Otherwise, adsorption is significant in organic soils. Dicamba has a soil half-life of up to 2 weeks (aerobic), with a hydrolysis half-life of 30 days. In plants, dicamba metabolizes into 5-hydroxy-2-methoxy-3,6-dichlorobenzoic acid, with 3,6-dichlorosalicylic acid being another metabolite.

Dicamba is registered for use in the USA, Canada, Australia and South Africa. However, many restrictions are placed on its use on lawns and public places, due to the risk of groundwater contamination.

Dicarboximide fungicides

The dicarboximide fungicide group consists of cyclic imides belonging to three classes: (i) oxazolidinedione; (ii) imidazolinedione; and (iii) pyrrolidinedione. The known compounds in this group include dichlozoline, **vinclozolin**, chlozolinate, iprodione and **procymidone**. These compounds commonly have a low vapour pressure, solubility in water between 2 and 13 mg/l and solubility of >100 g/l in acetone and dichloromethane. These compounds are unstable in alkaline solution, with DT_{50} values below 1 h at pH above 8. Their preparation involves many chemical processes including cyclization of *N*-(3,5-dichlorophenyl) carbamates under basic conditions or dehydration of *N*-(3,5-dichlorophenyl) dicarboxylic acid amides.

The dicarboximides are protectant fungicides that occasionally have some degree of systemicity. They are mainly used against *Botrytis*, *Monilinia*, *Alternaria*, *Cercospora* and *Fusarium* on crops such as grapes, cereals, sunflowers, strawberries, groundnuts and the cucumber family. The mode of action is understood to be interference with the osmotic signal transduction pathway. Thus, conidial germination and mycelial growth are severely inhibited.

Dicarboximides are used as foliar sprays at 600–1000 g a.i./ha. The common formulations include wettable powders (WP), suspension concentrates (SG) or water dispersible granules (WG). The compounds can be drenched in soils and also used as seed treatment agents.

Since resistance has been observed among some isolates of *Botrytis cinerea*, dicarboximates are combined with other fungicidal group members such as **chlorothalonil, thiram, maneb,** imazalil and **carbendazim**. It is also recommended that dicarboximide active ingredients be used only when infection is severe and for a maximum of two to three times a season, with sufficient time intervals between applications.

Dicarboximides are rapidly eliminated in the urine and faeces, after oral administration in rats, with pathways like opening of the cyclic imide, alkyl or aryl hydroxylations.

Dicarboximdes are slightly mobile and not very persistent (except for procymidione which persists for several months). 3,5-Dichloroaniline is formed as a metabolite (due to cleavage of cyclic imide). Photodegradation leads to dechlorination. Reaction with hydroxide ions yields anilides.

Dichlobenil herbicide

Dichlobenil, $C_7H_3Cl_2N$ (2,6-dichlorobenzonitrile), is a benzonitrile herbicide (substituted benzene). It forms colourless crystals that are moderately soluble in water and soluble in polar organic solvents. Dichlobenil is stable in acids but rapidly hydrolysed by strong alkalis to 2,6-dichlorobenzamide. It has a CAS Registry Number of 1194-65-6. It is soluble in water at 21 mg/l. Dichlobenil is made with the starting material as 2,6-dichlorotoulene which goes through various conversions, with one of them being dichlobenil. It is produced by conversion of 2,6-dichlorotoluene into 2,6-dichlorobenzal chloride, which is converted into 2,6-dichlorobenzaldehyde, 2,6-dichlorobenzaldoxime and 2,6-dichlorobenzonitrile.

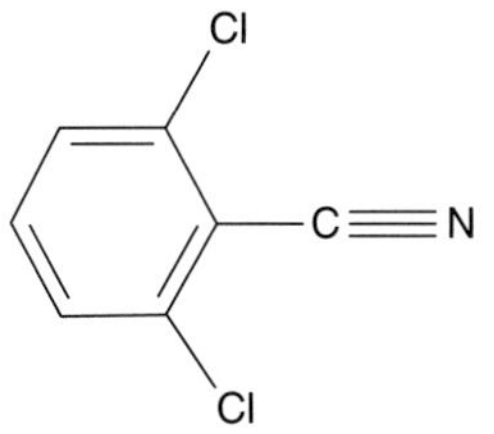

Dichlobenil

Dichlobenil is a systemic, pre- and post-emergence herbicide, which acts by inhibition of plant cell division among plant tissue. It thus affects seed germination and the plant root zone. It is used for controlling annual and perennial weeds in berry crops, fruit trees, vineyards and forest plantations. Weeds such as horsetail, bluegrass, Canada thistle and foxtail are effectively controlled.

Dichlobenil is available as granules (GR) and wettable powder (WP) formulations. It is often incorporated into the soil, where it affects the weed roots due to volatilization. It volatilizes rapidly at higher soil temperatures and wet soils (thus necessitating applying it under the soil surface). It is available in the market under the trade names of Casoron and Barrier.

According to the WHO, dichlobenil comes under the toxicity Class U (unlikely to be hazardous in normal use) and the EPA classifies it as a Class III toxic (slightly toxic). The oral LD_{50} is 3160 mg/kg among rats. It is toxic to fish (LC_{50} is 5 ppm for a 4 day exposure) and is observed to accumulate in their tissues. It is also a threat to beneficial soil organisms. Dichlobenil is also persistent in soil (half-life up to 6 months).

In plants, dichlobenil metabolism involves ring hydroxylation (at the 3-position and, to a lesser extent, at the 4-position) of both dichlobenil and 2,6-dichlorobenzamide, and subsequently conjugation with a sugar. In soil, dichlobenil gradually undergoes microbial degradation to 2,6-dichlorobenzamide, which is slowly broken down to 2,6-dichlorobenzoic acid.

Dichlobenil is registered in a few countries such as the USA, Canada and Australia.

Dichlone analysis

Dichlone (2,3-dichloro-1,4-naphthaquinone), is a seed protectant and a foliar fungicide. Dichlone is extracted by benzene and reacted with dimethylamine to a red colour, specific for dichlone. The following is an account of dichlone analysis which has to be done on the same day to avoid its decomposition.

Add 1 kg of the crop sample to 500 ml benzene in a closed glass jar and mix for 15 min (this may be done by shaking mechanically). Decant the benzene and dry using anhydrous sodium sulfate for 10 min. Filter through a fast filter paper (e.g. Whatman 40 paper). Place 38 ml benzene extract in a 50 ml cylinder with a ground-glass stopper. Dilute to 40 ml with anhydrous dimethylamine and mix thoroughly. Dilute a second 38 ml benzene extract to 40 ml with benzene and mix thoroughly. Using benzene as reference, measure the absorbance of both samples at 495 nm in 10 cm cells. Colour measurements should be made 5–20 min after the addition of diphenylamine. Determine the dichlone content from a standard curve. Interference from natural plant components may be expected around 0.02 ppm for untreated samples.

Dichlone fungicide

Dichlone, $C_{10}H_4Cl_2O_2$ (2,3-dichloro-1,4-naphthoquinone), is a quinone fungicide, with algicidal activity. It is a yellow crystalline substance, insoluble in alcohol, poorly soluble in water (0.1 mg/l), but easily soluble in xylene, 1,2-dichlorobenzene, acetone and acetic acid. It is stable to light and heat in dry conditions but is hydrolysed by alkalis. The CAS Registry Number of dichlone is 117-80-6. It is stable to acidic medium but slowly decomposes by light in its solution form.

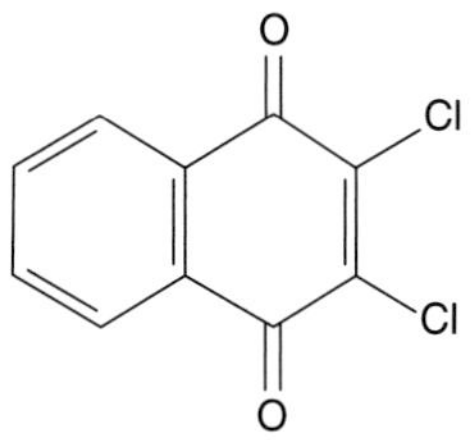

Dichlone

Dichlone, a foliar fungicide, is used to control brown rot, **scab** and blossom blights on apples, peaches and pears. It also effectively controls blue-green algae in ponds, lakes and swimming pools.

The fungicide is commonly formulated as a wettable powder (WP). It is compatible with fungicides and insecticides (**benomyl** and **captan**), as long as it does not contain copper, lime, calcium arsenate, sulfur or oil. The principal trade name is Kolo.

Dichlone is a slightly toxic compound. It irritates the skin and eyes and is particularly injurious to the mucous membrane when inhaled. The acute oral LD_{50} of dichlone in rats is 1300 mg/kg, while the acute percutaneous LD_{50} among rabbits is 5000 mg/kg. Applicators must wear protective clothing and suitable skin creams (glycerol or silicone-based). The NOEL (for 2 years) in rats receiving 1500 mg/kg diet showed no ill effects. It is not toxic to bees but toxic to fish.

In plant cells, dichlone forms a dimercapto compound by replacing chlorine atoms with a sulfhydryl group. It easily penetrates soil and contaminates groundwater and water bodies. It degrades in aerobic soil in 10 days.

Dichlormid herbicide safener

Dichlormid, $C_8H_{11}Cl_2NO$ (*N*, *N*-diallyl-2, 2-dichloroacetamide), is a herbicide safener. The compound is a clear-to-amber, viscous liquid. Dichlormid is stable to light and is not readily volatile. It becomes unstable when heated above 100°C and when heated with iron. Dichlormid has a CAS Registry Number of 37764-25-3.

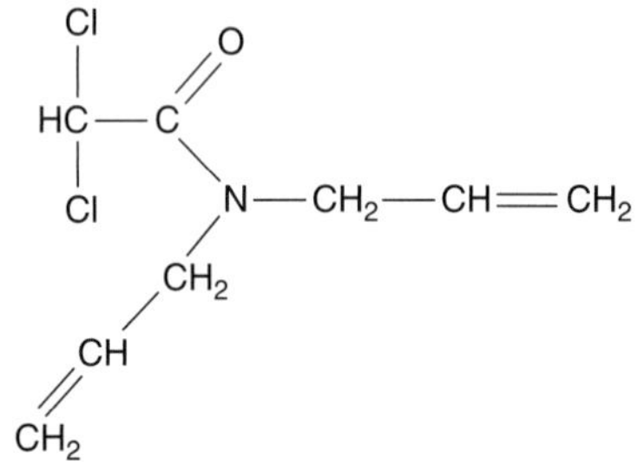

Dichlormid

Dichlormid, a systemic compound, is used to increase the tolerance of maize, wheat and rice against **thiocarbamate, chloroacetanilide, butachlor** and **metolachlor**. It may also be used in seed dressing. A principal trade name is Eradicane.

According to the WHO, dichlormid falls under the toxicity Class III (slightly toxic). The acute oral LD_{50} in male rats is 2816 mg/kg and in female rats is 2146 mg/kg. The NOEL (90 days) in rats is 20 mg/kg diet. Its 5 day dietary LC_{50} in mallard ducks is 14,500 mg/kg and in fish it is 141 mg/l.

Dichlormid is registered for use in countries including Hungary.

Dichlorprop herbicide

Dichlorprop, $C_9H_8Cl_2O_3$ ((*RS*)-2-(2,4-dichlorophenoxy) propionic acid), is a phenoxypropionic acid herbicide. It is a colourless crystalline substance, soluble in water (729–864.8 mg/l) and organic solvents. Dichlorprop forms salts with heavy metals. Dichlorprop is hydrolysed under acidic and basic conditions. Dichlorprop is made by reaction of 2-chloropropionic acid with 2,4-dichlorophenol. Only the resolved form (+)-isomer has herbicidal properties. It is produced by the reaction of 2-chloropropionic acid with 2,4-dichlorophenol. Dichlorprop has a CAS Registry Number of 120-36-5.

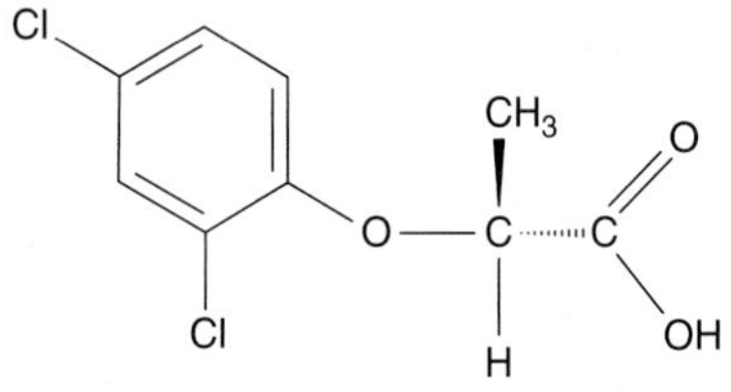

Dichlorprop

Dichlorprop is a selective, systemic, post-emergence, hormone-type herbicide. It acts by inhibiting formation of the abscission zone. It is absorbed by the foliage and is translocated to the roots. Dichlorprop controls annual and perennial broadleaved weeds in cereals and public places. It also acts against aquatic weeds.

The herbicide is available in the market as Basagran, Dicopur, Polymone, Redipon and Weedone.

According to the WHO and EPA, dichlorprop comes under the toxicity Class III (slightly hazardous). However, it is categorized as a PAN Bad Actor chemical since it is a reproductive toxin and a possible carcinogen. A related product dichlorprop-P is also a PAN Bad Actor chemical.

The average half-life of dichlorprop is 10 days. Both in soil and in plants, dichlorprop degrades to form 2-4-dichlorophenol. It may conjugate to form mono-, di- and triglucosides. Microbial degradation and plant uptake prevent leaching losses.

Dichlorprop is registered for use in South Africa and Australia but not in the USA.

Dichlorprop-P herbicide

Dichlorprop-P, $C_9H_8Cl_2O_3$ ((2*R*)-2-(2,4-dichlorophenoxy)propionic acid), is a phenoxypropionic herbicide (chlorophenoy acid or ester). It is a colourless crystalline substance, sparingly soluble in water but readily soluble in organic solvents. It is stable to light. Its CAS Registry Number is 15165-67-0. Dichlorprop-P is made in the same way as dichlorprop, with the (+) stereoisomer being obtained by fermentation.

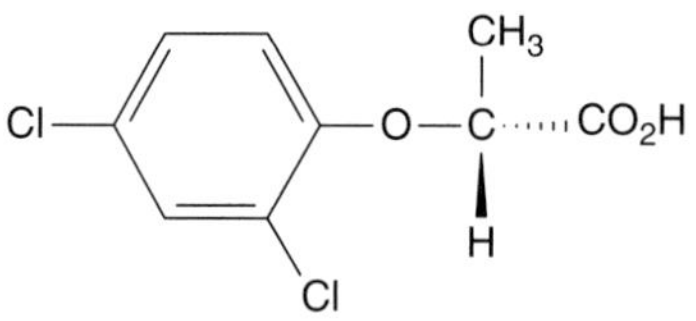

Dichlorprop-P

Dichlorprop-P is a selective, systemic, post-emergence, hormone-type herbicide, absorbed by the leaves, with translocation to the roots. The compound is used on cereal crops and on apples. It is particularly effective in controlling types of twining annuals such as *Bilderdykia convolvulus*, spotted ladysthumb and curlytop knotweed.

Dichlorprop-P may be combined with other herbicides. It is available in the market under the trade name of Duplosan. It is formulated as emulsifiable concentrate (EC), dustable powder (DP) and soluble concentrate (SL). According to the EPA, dichlorprop-P is rated as highly toxic as per acute toxicity tests. It is also a possible carcinogen and a potential groundwater contaminant.

Dichlorprop-P is registered in the USA, Canada, Australia and the UK, among other countries.

Dichlorvos analysis

Dichlorvos, $C_4H_7Cl_2O_4P$ (2,2-dichlorovinyl dimethyl phosphate), is an acaricide and an insecticide. This pale amber liquid may be analysed from a given sample with the following prerequisites. There must be a purity of 97% per kilogram of the sample, H_2SO_4 of 2 g/kg for acidity and water content of 0.5 g/kg.

The apparatus includes gas-liquid chromatography equipment with a glass column (a borosilicate glass tube 3 mm in internal diameter, and 6 mm in external diameter, bent to fit the chromatograph) and equipped with an on-column detection system, a high-sensitivity flame-ionization detector and a suitable recorder or integrator. The column-packing material is Gas-chrom Q (125–150 µm) or chromosorb W-HP (100–120 mesh) treated with 3% OV-225.

The operating conditions for gas-liquid chromatography include the temperature of the oven at 140°C, of the injection port at 160°C and of the flame-ionization detector at 250°C. The flow rate of the carrier gas nitrogen is 40 ml/min. The retention times of the dichlorvos peak and of the internal standard peak are 5 and 9.5 min, respectively.

The reagent, the internal standard solution, is an acetone solution containing diethyl pimelate at a concentration of 4 mg/ml. The dichlorvos calibration solution is made in the following way. Add about 0.20 and 0.25 g of analytical grade dichlorvos and 25 ml of the internal standard solution into two separate 50 ml volumetric flasks. Dilute with acetone. Inject 2–6 µl of the calibration solution and adjust the instrument controls so that maximum of the large peak gives a recorder deflection of 80–90%. The volume of liquid injected may be y µl. Now, inject duplicate volumes of y µl of each of the calibration solutions. Measure the dichlorvos and diethyl pimelate peak areas.

For each injection, calculate the calibration factor F using Eqn D.6.

$$F = \frac{m_1 \times P \times \text{Area of diethyl pimelate peak}}{1000 \times \text{Area of dichlorvos peak}} \quad \text{(D.6)}$$

where
m_1 = Mass of dichlorvos (g) in the calibration solution
P = Purity (g/kg) of dichlorvos standard
Repeat the injections until the F values for each calibration solution differ by not more than ±0.5% from each other.

Inject duplicate volumes of y µl of each sample solution. Calibration solution A is injected at the beginning and end of the sample injection solutions in duplicate. Determine the calibration factors F_1 and F_2 for each of the calibration solution injections.

The mean of the calibration factors F_c is given by Eqn D.7.

$$F_c = \frac{F_1 + F_2}{2} \quad \text{(D.7)}$$

The dichlorvos content (g/kg) can be estimated by Eqn D.8.

$$\text{Dichlorvos content} = \frac{F_c \times A_1}{m_2 \times A_2 \times 1000} \quad \text{(D.8)}$$

where
F_c = Mean calibration factor for dichlorvos
m_2 = Mass (g) of sample taken
A_1 = Area of dichlorvos peak in the sample solution
A_2 = Area of diethyl pimelate peak in the sample solution

To analyse an emusifiable concentrate sample of dichlorvos, dissolve the sample in acetone along with diethyl pimelate as the internal standard and estimate the dichlorvos content by gas-liquid chromatography using a 3% OV-225 on Gas-chrom Q or Chromosorb W-HP column as described for dichlorvos technical grade.

Dichlorvos insecticide

Dichlorvos, $C_4H_7Cl_2O_4P$ (2,2-dichlorovinyl dimethyl phosphate), is a subgroup of the vinyl phosphate class of organophosphate insecticides. It is a colourless-to-amber oily liquid, soluble in water (8 g/l), diesel oil, kerosene and iso-paraffinic hydrocarbons. It is hydrolysed by alkalis to form dimethyl hydrogen phosphate and dichloroacetaldehyde. Dichlorvos has a CAS Registry Number of 62-73-7. Dichlorvos is prepared by reacting trimethyl phosphite with chloral (the Perkow reaction).

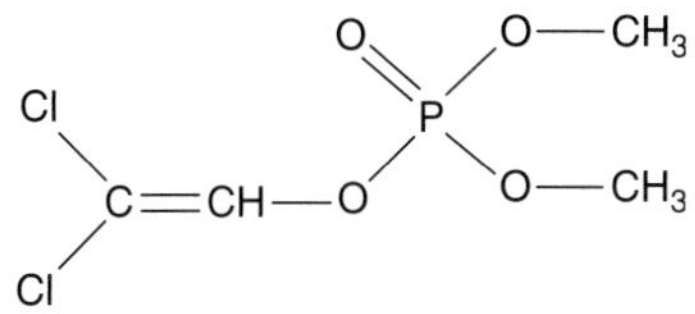

Dichlorvos

Dichlorvos, a cholinesterase inhibitor, is a broad-spectrum insecticide, with acaricidal properties. It affects the normal working of the nervous system and has stomach and respiratory action. Mushrooms, vines, rice, tea and cotton are protected by this insecticide from spider mites, sciarid flies, mushroom flies and sucking and chewing insects. It is also effective against household pests including flies, bed bugs, mosquitoes, cockroaches and ants.

Dichlorvos is formulated in forms such as granules (GR), an emulsifiable concentrate (EC) and aerosol dispenser. It is available in the market under the trade names of Vapona, Nuvan, Phosvit, among others. Dichlorvos is compatible with non-alkaline pesticides. It is also not to be used with **chinomethionat** and dichlofluanid. It is phytotoxic to some varieties of chrysanthemum.

The WHO and EPA have classified dichlorvos as a toxin of Class Ib (extremely hazardous). It is toxic to bees, birds and fish. It is very irritating to the eyes and skin. The oral LD_{50} among rats is between 56 and 108 mg/kg. It is also toxic if inhaled. Fatalities have taken place as a result of over-exposure to dichlorvos via touch and breath (as recorded by the UK National Poisons Unit or NPU). It is also considered to be a possible carcinogen according to the EPA and the IARC. The reference concentration (RfC) for dichlorvos is 0.0005 mg/m^3.

In mammals, dichlorvos rapidly degrades by hydrolysis and *O*-demethylation, with a half-life of 25 min. It is rapidly decomposed in plants and in the atmosphere and thus does not persist in the environment. The organic soil half-life is <1 day. The various metabolites of dichlorvos are phosphoric acid, CO_2 and a hydrolytic product, dichloro-acetaldehyde (which further breaks down into dichlorethanol and glycolic acid.)

Angola, Fiji, the USA and the EU have banned dichlorvos, while Indonesia, South Korea and Kuwait have restricted its use.

Diclofop-methyl herbicide

Diclofop-methyl, $C_{16}H_{14}Cl_2O_4$ (methyl (*RS*)-2-[4-(2,4-dichlorophenoxy)phenoxy]propionate), is a miscellaneous herbicide. It is a colourless crystalline substance, soluble in water (50 mg/l at 22°C), and in acetone and methanol (each 40 g/l). It is stable to light. Dichlofop-methyl has a CAS Registry Number of 51338-27-3. It is produced by reaction of 2,4-dichlorophenol with methyl 4-bromophenoxy- 2-propionate.

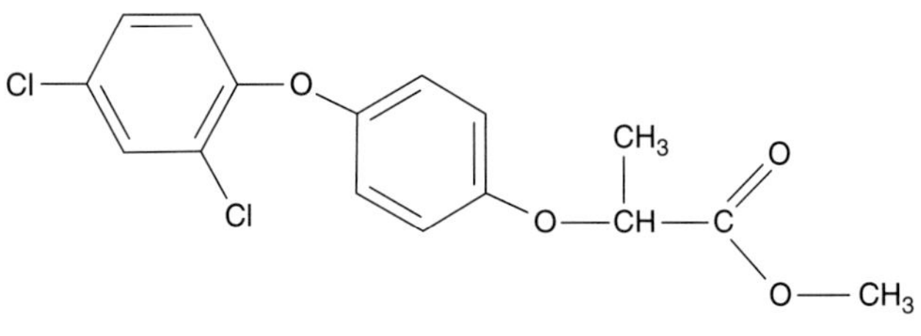

Diclofop-methyl

Diclofop-methyl, a selective, systemic, pre- and post-emergence compound, becomes effective when it converts to diclofop. It is absorbed by the leaves and translocated throughout the plant. Inhibiting fatty acid synthesis, destroying cell membranes, affecting photosynthesis and starving the roots of assimilates are the major herbicidal effects of this compound. The herbicide controls wild oats, wild millet and some annual grasses in cereals, beans, brassicas, onions and sugarcane.

This chemical is available as an emulsifiable concentrate (EC). It is available in the market as Hoegrass. It may be combined with urea-based herbicides.

According to the WHO and EPA, diclofop-methyl comes under toxicity Class III (slightly hazardous). It is categorized as a restricted use pesticide (RUP) by the EPA and must have the signal word 'DANGER' or 'CAUTION' printed on the labels of its formulations. The acute oral LD_{50} among rats is between 563 and 693 mg/kg and among birds from 4400 to 10,000 mg/kg. Since diclofop-methyl is moderately toxic and is a reproductive toxin and carcinogen, it is classified as a PAN Bad Actor chemical.

In plants, diclofop-methyl is taken up rapidly and almost completely. In soil, it is metabolized to diclofop, which then undergoes further degradation to 4-(2,4-dichloro-phenoxy) phenol, carbon dioxide and hydroxylated free acids. India, the USA, Australia, South Africa and a few EU members such as the UK, Spain, Italy, France and Portugal have registered the use of diclofop-methyl. (See also **ACCase inhibitors**.)

Diclosulam herbicide

Diclosulam, $C_{13}H_{10}Cl_2FN_5O_3S$ (2′,6′-dichloro-5-ethoxy-7-fluoro[1,2,4] triazolo [1,5*c*]pyrimidine-2-sulfonanilide), is a triazolopyrimidine sulfonamide herbicide. It is an off-white powder, soluble in water at 12.4 mg/100ml. It is stable under normal temperature and storage conditions. The CAS Registry Number is 145701-21-9.

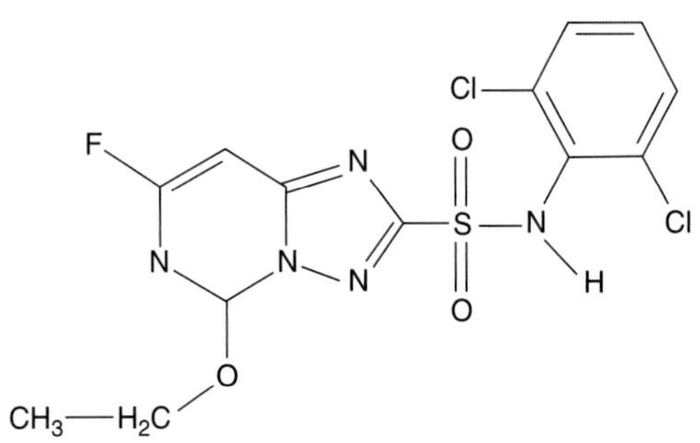

Diclosulam

Diclosulam, mainly a pre-emergence herbicide, is used on groundnuts and soybeans. It acts by inhibition of acetolactate synthase (ALS), and thus affects the biosynthesis of branched-chain amino acids.

The herbicide is formulated as wettable granules (WG). It is available in the market as Strongarm. An accepted application rate is up to 25 g/ha for groundnuts.

The WHO has classified diclosulam as a Class U toxin (unlikely to be hazardous). The acute oral LD_{50} in rats is >5000 mg/kg and the acute dermal toxicity LD_{50} is >2000 mg/kg. It can cause slight irritation to the eyes. It is practically non-toxic to birds, mammals, insects, earthworms, freshwater fish and invertebrates, and estuarine/marine fish and invertebrates, although it is important to dispose of diclosulam with care and according to the label instructions.

Diclosulam is registered for use in the USA.

Dicrotophos insecticide

Dicrotophos, $C_8H_{16}NO_5P$ ((*E*)-2-dimethylcarbamoyl-1-methylvinyl dimethyl phosphate), is an organophosphorus insecticide. It is a yellowish liquid, totally miscible with water and other organic solvents and slightly soluble in diesel, oil and kerosene. The compound decomposes on heating. The half-life of dicrotophos at pH 5 is 88 days and at pH 8 is 28 days. Dicrotophos is synthesized by the Perkow reaction from trimethyl phosphite and *N*,*N*-dimethyl-α-chloroacetoacetamide, consisting mainly of the (*E*)-form. Dicrotophos has a CAS Registry Number of 141-66-2.

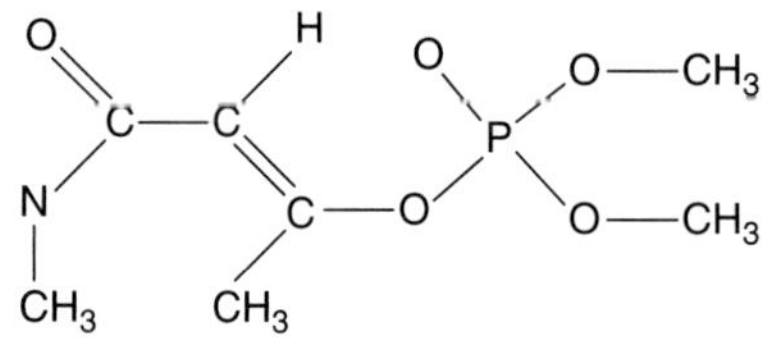

Dicrotophos

Dicrotophos, a cholinesterase inhibitor, is a systemic insecticide and acaricide with ectoparasiticidal properties. It has contact and stomach action and is moderately persistent. It is used to control sucking-chewing-boring insects and mites in cotton, coffee, rice, pecans, sugarcane, citrus fruit, tobacco, cereals, potatoes and palms. Dicrotophos is formulated in forms such as emulsifiable concentrate (EC), ULV liquids (UL) and soluble concentrate (SL). It is non-phytotoxic if used as directed, but may harm some varieties of apple and pear. It is compatible with most other pesticides. It is available under the trade names of Bidrin and Carbicron.

Dicrotophos is categorized by the WHO as a Class Ib toxin (highly hazardous). The LD_{50} among small mammals is 16–25 mg/kg and among birds is below 5 mg/kg. It is a possible carcinogen (as per the EPA carcinogen list). It is so very mobile in soils that it is considered a potential groundwater contaminant. It is thus considered a PAN Bad Actor chemical.

Due to the various toxicological findings of dicrotophos, this compound has been banned or its use severely restricted by the EPA and several countries including India and Pakistan. It is registered in the USA, Vietnam and Cambodia.

Dicyclanil as an IGR

Dicyclanil, $C_8H_{10}N_6$ (4,6-diamino-2-cyclopropylaminopyrimidine-5-carbonitrile), is a pyrimidine derivative that acts as an insect growth regulator. This white powder is made by reaction of 2-chloro-4,6-diamino-5-cyanopyrimidine with cyclopropylamine. It has a CAS Registry Number of 112636-83-6.

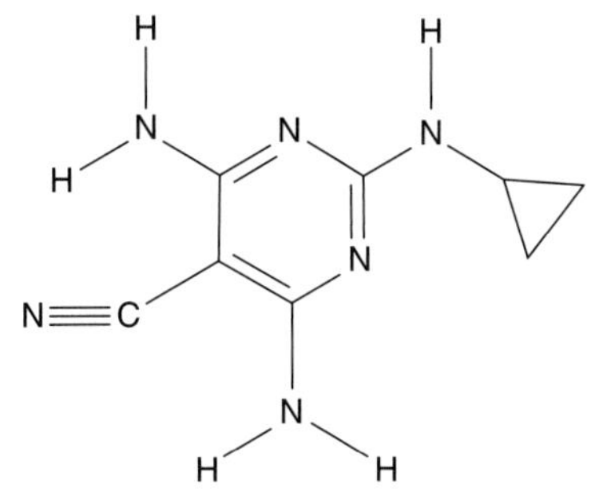

Dicyclanil

Dicyclanil, a compound with high-specificity, is commonly known as a veterinarian prescription that is used on sheep and cattle, with a common formulation being a spray. With a preventive property, dicyclanil prevents eggs from hatching and effectively controls the young growth stages of Diptera larvae, which constitute flies and midges.

Dicyclanil is registered in Australia and New Zealand but not registered in the USA.

Dieldrin insecticide

Dieldrin, $C_{12}H_{10}OC_{16}$ (1*R*,4*S*,4a*S*,5*R*,6*R*,7*S*,8*S*,8a*R*)-1,2,3,4,10,10-hexachloro-1,4,4a,5,6,7,8,8a-octahydro-6,7-epoxy-1,4:5,8-dimethanonaphthalene) or (1,2,3,4,10,10-hexachloro-6,7-epoxy-1,4,4a,5,6,7,8,8a-octahydro-*endo*-1,4-*exo*-5,8-dimethanonaphthalene), is a cyclodiene organochlorine insecticide. The product is insoluble in water and aliphatic hydrocarbons, but soluble in acetone and benzene. Dieldrin is made from aldrin by epoxidation with peracetic or perbenzoic acid. Aldrin by itself too breaks down to form dieldrin. The CAS Registry Number of dieldrin is 60-57-1. **Endrin** is a stereoisomer of dieldrin.

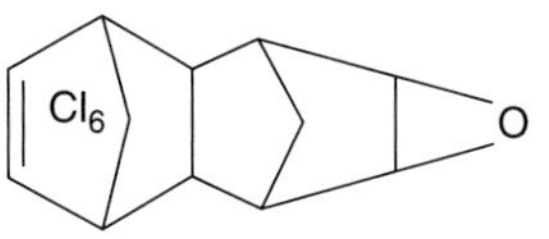

Dieldrin

Dieldrin and **aldrin** have been used extensively as soil insecticides and for seed treatments. Dieldrin, which is very persistent, can control migratory locusts. The spray residue can control the *Anopheles* vectors of malaria and the tsetse fly.

However, dieldrin is an extremely **persistent organic pollutant** (POP) and is also said to bio-magnify along the food chain. It is highly toxic to fish and other aquatic organisms. The US National Toxicological Program (NTP) ranks dieldrin as being moderately to highly toxic. It is shown to be carcinogenic, with reproductive, respiratory and other ill effects. It is also a suspected endocrine disruptor. Like aldrin, dieldrin is not only a PAN Bad Actor chemical but is also included in the PAN **Dirty Dozen** list.

Long-term exposure to dieldrin has proved to be toxic to insects, animals and humans, hence dieldrin has been banned or has severely restricted use in most countries. India has banned the use of endrin, aldrin and dieldrin from being manufactured or imported.

Dienochlor insecticide

Dienochlor, $C_{10}Cl_{10}$ (perchloro 1,1′-bicyclopenta-2,4-diene), is an organochlorine insecticide and an acaricide. Although dienochlor is stable under normal storage conditions, it decomposes in sunlight (with the DT_{50} being 1.6 min) and undergoes hydrolysis (with the DT_{50} being 30.5 days at pH 9 and 184 days at pH 5). It has a CAS Registry Number of 2227-17-0.

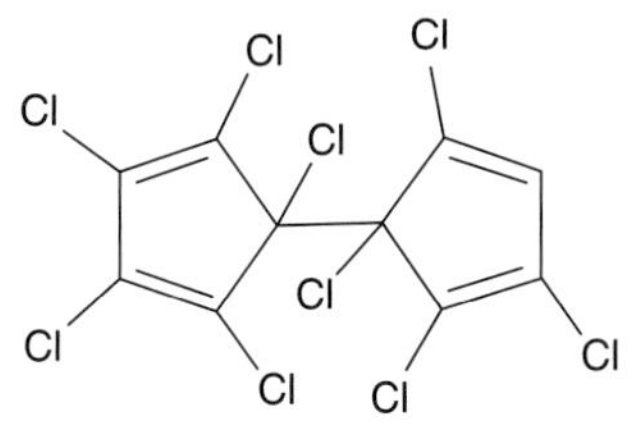

Dienochlor

Dienochlor is used as a selective acaricide to control mites on crops and ornamental plants. It has mainly a contact action and interferes with oviposition.

Dienochlor is rated as a Class III toxin (slightly hazardous). The oral LD_{50} among male rats is >3160 mg/kg. However, as dienochlor is an organochlorine, symptoms of acute exposure to it may result in nausea, vomiting, tremors and convulsions.

Degradation of dienochlor is rapid, especially due to photochemical breakdown. It decomposes in soils (with a DT_{50} of 3.1 days). The major degradation products of dienochlor in plants are perchloroketones. This compound is not permitted for use anywhere in the EU.

Diethatyl herbicide

See **Chloroacetanilide herbicides**

Diethatyl-ethyl herbicide

See **Chloroacetanilide herbicides**

Difenacoum rodenticide

Difenacoum, $C_{13}H_{24}O_3$ (3-[(1*RS*,3*RS*;1*RS*,3*SR*)-3-biphenyl-4-yl-1,2,3,4-tetrahydro-1-naphthyl]-4-hydroxycoumarin), is a **coumarin** rodenticide. It is a colourless crystalline substance, soluble in water (<10 mg/l), acetone and chloroform (both >50 mg/l). It is stable up to 100°C and to light. Condensation of 4-hydroxycoumarin and 3-biphenyl-4-yl-1,2,3,4-tetrahydro-1-naphthol results in difenacoum. Difenacoum has a CAS Registry Number of 56073-07-5.

The rodenticide, a second-generation compound, is a coumarin anticoagulant effective on warfarin-resistant mice and rats. It acts by blocking the clotting process by preventing the functioning of clotting factors II, VII, IX and X.

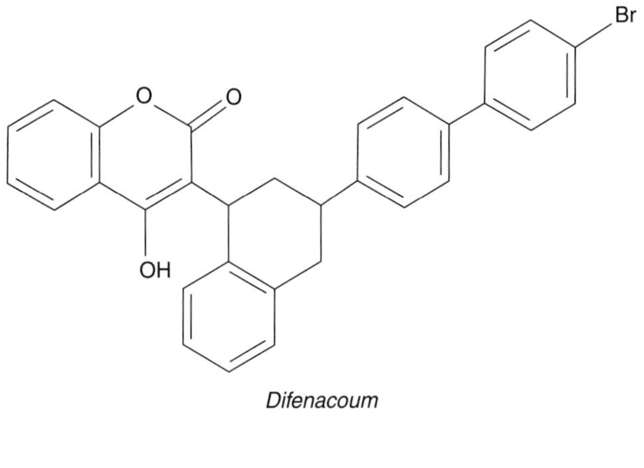

Difenacoum

Commonly known formulations of difenacoum are grain bait, block bait, bait concentrate and ready-to-use baits. Difenacoum formulations are used in mixtures with calciferol. Difenacoum, which is highly target-specific, is generally effective in very low doses (even 0.005%).

The WHO and EPA have categorized difenacoum as a Class Ia toxin (extremely hazardous). The acute oral LD_{50} in rats is 1.8–2.45 mg/kg and in mice is 0.8 mg/kg. Toxicity symptoms can be relieved by an injection of vitamin K.

Difenacoum is registered in the USA, Australia, the UK, Germany and South Africa. (See also **Anticoagulant rodenticides**.)

Difethialone rodenticide

Difethialone, $C_{31}H_{23}BrO_2S$ (3-[(1*RS*,3*RS*;1*RS*,3*SR*)-3-(4′-bromobiphenyl-4-yl)-1,2,3,4-tetrahydro-1-naphthyl]-4-hydroxy-1-benzothiin-2-one where the ratios of the racemates (1*RS*,3*RS*) to (1*RS*,3*SR*) lie within the ranges of 0–15 to 85–100, respectively), is a **coumarin** rodenticide. It is a white-yellowish crystalline powder, soluble in water (0.39 mg/l), ethanol (0.7 g/l) and dimethylformamide (332.7 g/l). Difethialone has a CAS Registry Number of 104653-34-1. Difethialone is prepared by condensing 4-hydroxy-2*H*-1-benzothiapyran-2-one with 3-(4′-bromo-4-biphenyl)-1,2,3,4-tetrahydro-1-naphthol in acetic acid and concentrated sulfuric acid.

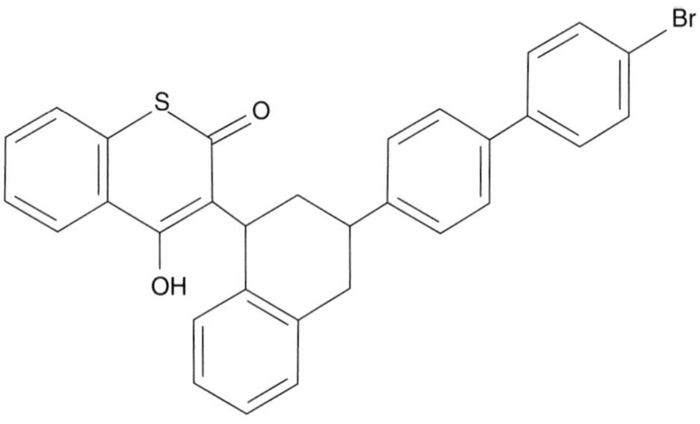

Difethialone

Difethialone, with anticoagulant properties, blocks the production of vitamin K. It is effective on rats, mice and voles, especially those that are resistant to **warfarin**.

The chemical is formulated commonly as grain and block baits, usually, wheat. As little as 0.0025% difethialone may be enough to administer the ideal dose. Commonly known trade names containing difethialone are Frap and Baraki.

The WHO has categorized difethialone as a Class Ia toxin (extremely hazardous). The acute oral LD_{50} of difethialone in rats is 0.56 mg/kg, in mice 1.29 mg/kg, in dogs 4 mg/kg and in pigs between 2 and 3 mg/kg. The inhalation LC_{50} for (a 4 h period) in rats is between 5 and 19.3 mg/l. The chemical is thus classified as a PAN Bad Actor chemical. Handlers and applicators are vulnerable to difethialone poisoning. Accidental consumption and skin contact can result in poisoning, the symptoms of which include bleeding gums, haemorrhaging, irritation of the eyes and skin.

Difethialone is registered for use in the USA, Canada, Australia, South Africa and in some countries of the EU, but most countries have rules about usage of this compound.

Diflubenzuron insecticide

Diflubenzuron, $C_{14}H_9ClF_2N_2O_2$ (1-(4-chlorophenyl)-3-(2,6-difluorobenzoyl)urea), is an insecticide of the benzoylurea class. It is a colourless crystalline substance, insoluble in water, soluble in dimethylformamide and dioxane. It is light sensitive when in solution. It is obtained by reaction of 2,6-difluorobenzamide with oxalyl chloride, followed by reaction of the obtained product with *p*-chloroaniline. It has a CAS Registry Number of 35367-38-5.

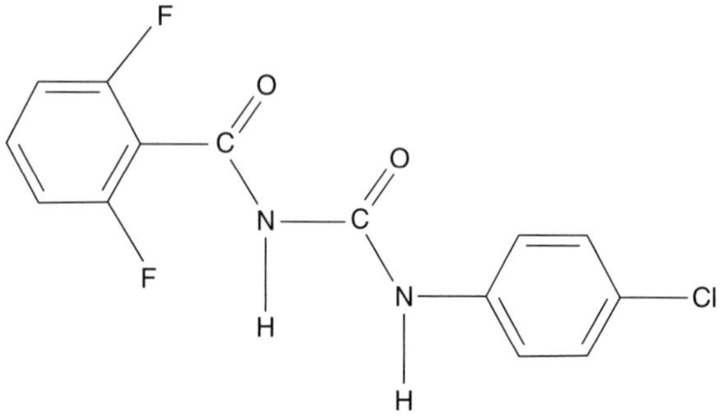

Diflubenzuron

Diflubenzuron, a halogenated benzoylphenyl urea compound, is a contact insecticide. It acts by interfering with the synthesis of **chitin** in insects, thus disallowing the proper formation of insect cuticle. Diflubenzuron can effectively control insect growth, even at the larval stage. It is applied to control mosquitoes (larvicidal), rust mites, lice, fleas and ticks (as an animal ectoparasiticide). Also it is used to control grasshoppers, flies and locusts in cotton, soybean, citrus, tea, vegetables and mushrooms.

Diflubenzuron is available as wettable powder (WP) and is available as the trade name Dimilin.

According to the EPA and WHO, diflubenzuron comes under the toxicity Class U (unlikely to be hazardous). The oral LD_{50} in rats and mice is >4640 mg/kg and the rat dermal LD_{50} is >10000 mg/kg. Diflubenzuron does not penetrate plant tissues and does not accumulate in the body tissues. This chemical is practically non-toxic to various bird species (bobwhite quails and mallard ducks both have an 8 day dietary LC_{50} of >4640 ppm). Labels of formulations with diflubenzuron should display the signal word 'CAUTION'. Diflubenzuron is also a suspected endocrine disruptor.

The chemical is rapidly degraded into 4-chlorophenylurea and 2,6-diflurobenzoic acid. The metabolites strongly adsorb to soil and so are not mobile. Degradation takes from a few days to 4 months. Its selectivity and rapid degradation in soil and water make it safe for **beneficial organisms** and natural enemies.

Diflubenzuron is registered for use in many countries including India, the USA, Australia, Canada, Germany, Denmark, South Africa and many other countries in Africa.

Diflufenican herbicide

Diflufenican, $C_{19}H_{11}F_5N_2O_2$ (2′,4′-difluoro-2(α,α,α-trifluoro-*m*-tolyoxy) icotinanilide), is an anilide and pyridine herbicide. It exists as colourless crystals, and is insoluble in water but soluble in organic solvents. Diflufenican has a

CAS Registry Number of 83164-33-4. Diflufenican is produced by the reaction of 2-(3-trifluoromethyl phenoxy) nicotinic acid with 2,4-difluoroaniline and sulfonyl chloride.

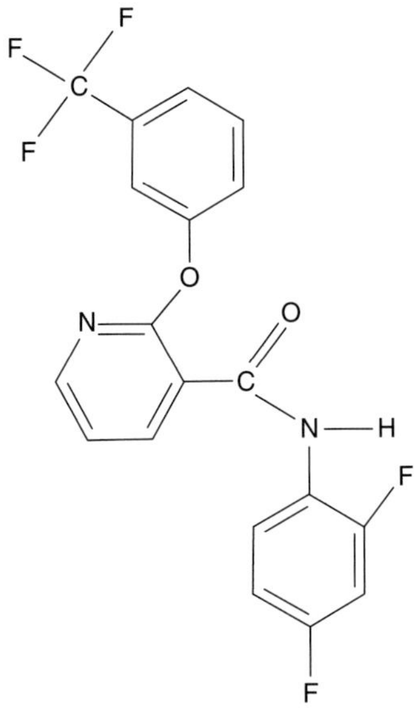

Diflufenican

Diflufenican, a selective, contact, pre- and early post-emergence herbicide, acts by inhibiting carotenoid synthesis and photosynthesis. It is absorbed by the very young shoots and has residual properties. Diflufenican is used to control grass and broadleaved weeds (e.g. *Veronica* and *Viola* spp.) in wheat, rye and barley.

Diflufenican is available as wettable powder (WP) and soluble concentrate (SL). It is available in the market under the trade name of Javelin, Alister and Quartz. This herbicide may be combined with a number of herbicides like **isoproturon**, **bromoxynil** and **mesosulfuron-methyl**, in various formulations.

According to the WHO, diflufenican comes under the toxicity Class U (i.e. product unlikely to present acute hazard in normal use).

In cereals, diflufenican is rapidly metabolized via nicotinamide and nicotinic acid to carbon dioxide. In soil, degradation of diflufenican proceeds via the metabolites 2-(3-trifluoromethyl-phenoxy) nicotinamide and 2-(3-trifluoromethyl-phenoxy) nicotinic acid to bound residues and carbon dioxide.

Diflufenican is registered in Australia, Denmark, Tanzania and several countries in the EU. The USA has not registered this compound.

Diflufenzopyr herbicide

Diflufenzopyr, $C_{15}H_{12}F_2N_4O_3$ (2-{(*E*)-1-[4-(3,5-difluorophenyl) semicarbazono]ethyl} nicotinic acid), is a urea herbicide (semicarbazone/benzoic acid). It is a tan solid, sparingly soluble in water. It is produced by reaction of 2-acetylnicotinic acid with 3,5-difluorobezene-*N*-2-methylsemicarbazone in the presence of pyridyl tosylate. Diflufenzopyr has a CAS Registry Number of 109293-97-2.

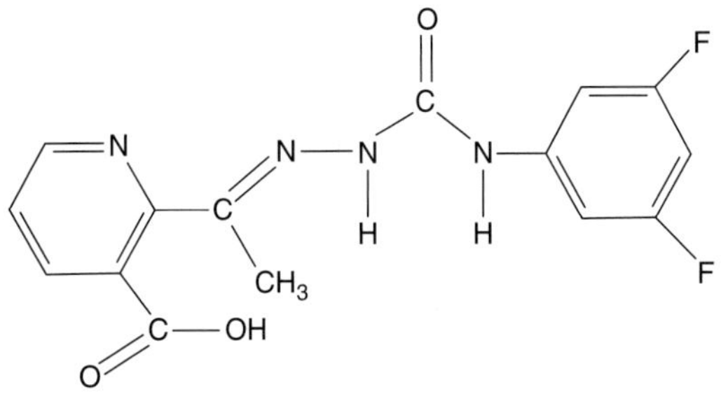

Diflufenzopyr

Diflufenzopyr, a polar auxin transport inhibitor, is a systemic, post-emergence herbicide used to control perennial weeds and annual broadleaved weeds in maize. Weeds include horse nettle, redroot pigweed and lambsquarters. This chemical is easily absorbed by the plant. Lower doses are sufficient to exercise weed control.

The herbicide may be combined with **2,4-D** and **clopyralid**. Spray application of this herbicide is to be carried out only at ambient temperature and humidity.

The acute rating from the EPA product label for diflufenzopyr is slightly toxic. Though diflufenzopyr does not pollute groundwater, the parent compound diflufenzopyr-sodium (CAS Registry Number of 109293-98-3) is a potential water contaminant (solubility in water is 4200 mg/l and soil half-life being between 55 and 164 days).

Both diflufenzopyr and diflufenzopyr-sodium are registered in the USA and Canada.

Di-halo slimicide

Di-halo, $C_5H_6BrClN_2O_2$ (1-bromo-3-chloro-5, 5-dimethyl-imidazolidine-2.4-dione), is a slimicide, bactericide and a fungicide. Structurally similar to hydantoin, di-halo is a white crystalline compound with a slight acetone odour, insoluble in water (0.15 g/100 ml at 25°C) but soluble in acetone. It is incompatible with acids and may cause a fire hazard. The CAS Registry Number of di-halo is 16079-88-2.

Di-halo has non-crop applications and is used in pools, tap water and in industries with large water use (e.g. involving pulp, paper and dyeing) and water cooling systems. It works well as a disinfectant and sanitizer besides being able to control bacterial and fungicidal pathogens and slimes. It has the properties of bromine and chlorine.

The acute oral LD_{50} of di-halo in rats is 1390 mg/kg. The dermal LD_{50} of di-halo in rabbits is >2000 mg/kg. Hence, this compound is moderately toxic.

Dihydroazadirachtin bioinsecticide

Dihydroazadirachtin (DAZA) is an insecticide derived from **azadirachtin** (AZA) which occurs naturally in the seeds and kernels of the **neem** tree, *Azadirachta indica*. This compound is formed by the hydrogenation of AZA. The CAS Registry Number is 108189-58-8 (see Dihydroazadirachtin at top of the next page).

A broad-spectrum nematicide as well as an insecticide, DAZA is used to control insects and soil-borne pests on ornamental plants, shrubs, trees and some food crops. DAZA, like AZA, interferes with the **moulting** process of insects and also acts like an anti-feedant.

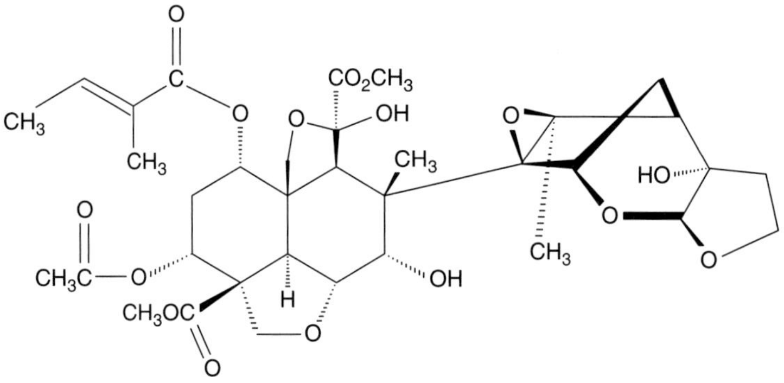

Dihydroazadirachtin

DAZA is used as a foliar application, which may normally be applied frequently during the same season (at the rate of 50 g/ha). DAZA is available under the trade name of DAZA, among others.

DAZA is not acutely toxic to mammals. The acute oral LD_{50} among rats is >5000 mg/kg and inhalation LC_{50} of DAZA in rats is >2.9 mg/l, while the acute dermal LD_{50} among rabbits is >2000 mg/kg. DAZA does not cause any adverse effect on the environment because it is metabolized by microorganisms in soil and the aquatic environment. DAZA is slightly toxic to birds (the acute oral LD_{50} for bobwhite quail being >816 mg/kg) and aquatic organisms (the LC_{50} for rainbow trout being 17.65 mg/l). DAZA is non-toxic to honey bees, although some other beneficial insects such as **green lacewing**, mites, whitefly **parasitoids** and **ladybirds** may be harmed by the application of DAZA.

Although most neem-derived insecticides are registered by many countries including India and the USA, DAZA is not registered in the USA.

Diisobutylamine

See **Butylate herbicide**

Dimepiperate herbicide

Dimepiperate, $C_{15}H_{21}NOS$ (*S*-1-methyl-1-phylethylpiperidine-1-carbothioate), is a thiocarbamate herbicide. It is a waxy solid, soluble in water (20 mg/l) and fairly soluble in organic solvents. It is stable in normal, dry conditions for a few months. Dimepiperate is produced by reaction of benzene isopropylmercaptan with piperidine carbamoyl chloride. Its CAS Registry Number is 61432-55-1.

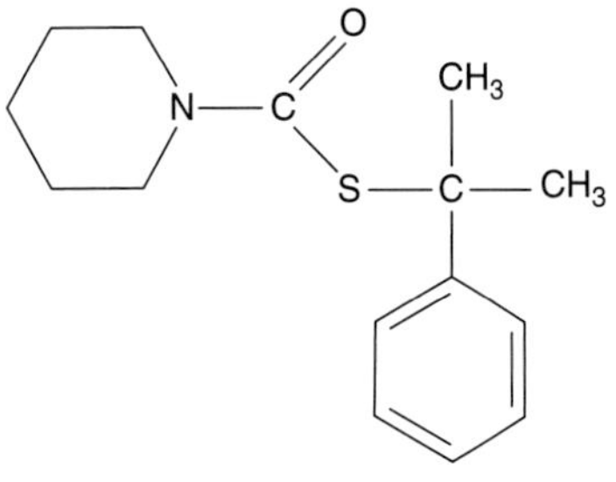

Dimepiperate

Dimepiperate, a systemic herbicide, is rapidly absorbed and translocated by the plant parts. It is a lipid-metabolism inhibitor. It is used for the control of **barnyard grass** (*Echinochloa crus-galli*) in flooded rice fields.

Dimepiperate is available in the market under the trade name of Yukamate.

According to the EPA, dimepiperate comes under the toxicity Class III (slightly hazardous). Since it is a cholinesterase inhibitor, dimepiperate is classified as a PAN Bad Actor chemical.

The DT_{50} of dimepiperate in field conditions is 7 days.

Dimethachlor herbicide

See **Chloroacetanilide herbicides**

Dimethenamid herbicide

Dimethenamid, $C_{12}H_{18}ClNO_2S$ ((*RS*)-2-chloro-*N*-(2,4-dimethyl-3-thienyl)-N-(2-methoxy-1-methylethyl) acetamide), is an amide herbicide. It is a yellow-brown liquid, which is viscous in nature, fairly soluble in water and organic solvents. It is stable to normal storage conditions for about 1 month. Dimethenamid is produced by reaction of 2,4-dimethyl-3-aminothiene with 2-chloro-3-methoxypropane, followed by treatment with chloroacetyl chloride. It has a CAS Registry Number of 87674-68-8.

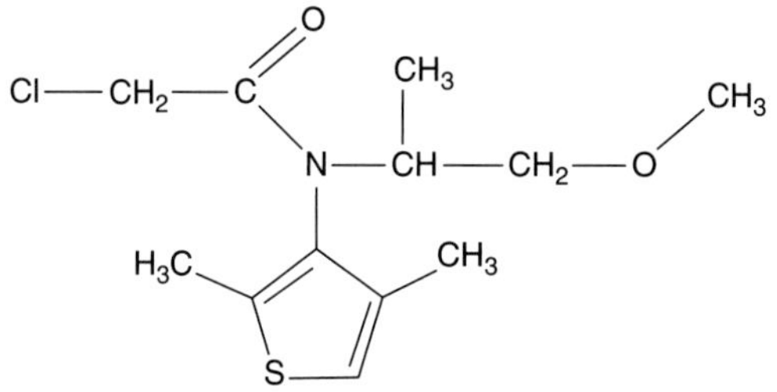

Dimethenamid

Dimethenamid, a systemic, pre-emergence herbicide, is absorbed by shoot tips. It is used for the control of annual grasses and many broadleaved weeds in maize, soybean and other crops.

Herbicide dimethenamid is available as an emulsifiable concentrate (EC). It is available in the market under the trade name of Frontier.

The EPA's acute rating for dimethenamid is that of moderate toxicity. It is also a possible carcinogen although it is not a PAN Bad Actor chemical. It is moderately toxic to fish and slightly toxic to bees.

Dimethenamid metabolizes to glutathione, cysteine, thiolactic acid and thioglycolic acid. In soil, it undergoes microbial action, and the DT_{50} is 8–30 days.

The herbicide dimethenamid is registered for use in the USA, Canada and South Africa. (See also **Chloroacetanilide herbicides**.)

Dimethipin as a PGR

Dimethipin, $C_6H_{10}O_4S_2$ (2,3-dihydro-5, 6-dimethyl-1,4-dithi-ine 1,1,4,4-tetraoxide), is a plant growth regulator. It is a white, crystalline solid, soluble in water (4.6 g/l), acetonitrile and methanol. It is stable to normal conditions for up to 1 year, but for 14 days at 55°C. It is photostable for 7 days. The CAS Registry Number is 55290-64-7.

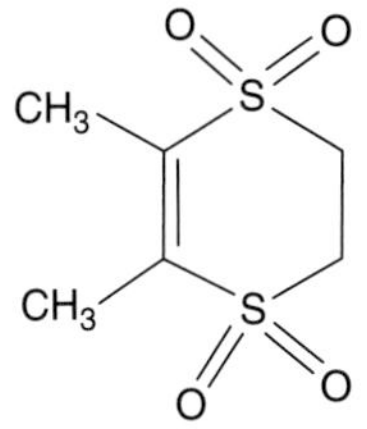

Dimethipin

Dimethipin is used as a **defoliant** as well as a **desiccant**. It defoliates cotton, rubber trees and potatoes. As a desiccant, it dehydrates seed moisture.

Harvade is one of the trade names of dimethipin.

Dimethipin is considered a slightly hazardous chemical by the WHO. It has the potential to enter the food chain through cottonseed-based cattle feed. It is a possible carcinogen and a potential groundwater contaminant. It is not a PAN Bad Actor chemical. This compound has low toxicity in fish.

Dimethipin is registered for use in Australia and South Africa but is not registered in the USA.

Dimethoate insecticide

Dimethoate, $C_5H_{12}NO_3PS_2$ (*O,O*-dimethyl *S*-methylcarbamoylmethyl phosphorodithioate), is an organophosphorus insecticide. It is a whitish crystalline substance, soluble in water (39,800 mg/kg) and in alcohols, ketones, benzene, toluene, chloroform, saturated hydrocarbons and *n*-octanol. It is hydrolysed in alkaline solutions and decomposed on heating, forming the *O, S*-dimethyl analogue. Dimethoate is produced by reaction of the sodium salt of *O,O*-dimethyldithiophosphoric acid with *N*-methylchloroacetamide. Its CAS Registry Number is 60-51-5.

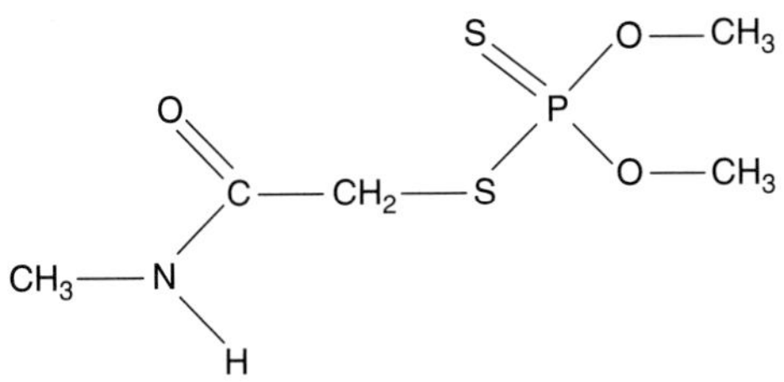

Dimethoate

Dimethoate, a cholinesterase inhibitor, is a systemic, broad-spectrum insecticide, and also an acaricide. It also has contact action. The wide range of pests controlled by this insecticide includes those belonging to the orders Coleoptera, Lepidoptera, Hemiptera (family Aphididae), and sub-classes Collembola and Acari. Whitefly, **mealybugs** and **scale insects** are some of the pests controlled by this chemical. Cereal crops, fruits and vegetables, tea, coffee and cotton are some of the crops that are efficiently protected. It is also used as an ectoparasite control agent (against mites and ticks).

Dimethoate is formulated as an emulsifiable concentrate (EC), wettable powder (WP), ULV liquid (UL) and aerosol (AE). It is incompatible with alkaline materials and sulfur-based formulations. It is available in the market under the trade names of De-fend, Perfekthion, Rogodial and Cygon. The pre-harvest interval is commonly fixed at 3 weeks.

Dimethoate is a moderately to highly toxic chemical. The WHO categorizes it as a Class II toxin (moderately toxic). The oral LD_{50} for dimethoate in rats is 60–387 mg/kg, while the values in cats, dogs, rabbits and pigs are in the range of 100–400 mg/kg. It is highly toxic to bees. It may be phytotoxic to fruits (e.g. peach, lemon and apple), cotton and vegetables (e.g. tomato and beans). Dimethoate is also a possible carcinogen, potential groundwater contaminant and a suspected endocrine disruptor, while it is a confirmed cholinesterase inhibitor and a reproductive toxin. Since dimethoate is injurious to the eyes and is absorbed through the skin, chemical spills should be taken care of as a matter of urgency. Contaminated clothes, containers and equipment should be removed, washed or disposed off as per local regulations. Symptoms of poisoning are many and include blurred vision, headaches, slurred speech, nausea and convulsions.

In mammals and plants, dimethoate metabolizes and oxidizes to phosphorothioate, which gives the corresponding oxone. This oxone is highly toxic and a cholinesterase inhibitor, which appears to be more persistent than dimethoate. Hydrolysis leads to *O,O*-dimethyl–phosphorodilthioate, –phoshorothioate, and –phosphate. It adsorbs very weakly in soil and is subject to leaching. The DT_{50} of aerobic soils is 2–4 days, whereas the photolytic soil DT_{50} is 7–16 days.

Many countries have banned the use of some forms of dimethoate. It is registered for use (for specific applications) in India, the USA, Canada, Australia, South Africa, the Philippines, New Zealand and other countries.

Dimethyl phthalate repellent

Dimethyl phthalate, $C_{10}H_{10}O_4$, is an insect repellent (of class alkyl phthalate). It is a colourless, viscous liquid, slightly soluble in water and readily soluble in alcohols, esters, ketones and chlorinated hydrocarbons. It is hydrolysed by alkalis. Dimethyl phthalate is made from phthalic acid and methanol by esterification. The CAS Registry Number of dimethyl phthalate is 131-11-3.

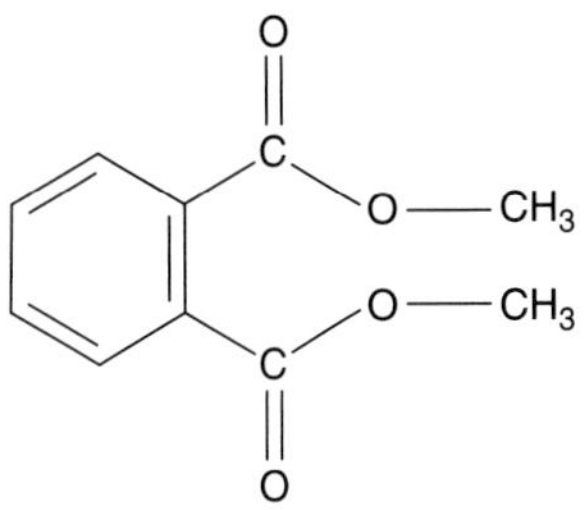

Dimethyl phthalate

Dimethyl phthalate has a broad spectrum of effectiveness. It provides effective control against the human lymphatic filarial nematodal parasite *Wuchereria bancrofti* by controlling the vector, the *Culex* spp. of mosquitoes. The *Culex* mosquitoes are vectors for other diseases such as avian malaria and filaria.

Dimethyl phthalate also repels the mosquito *Aedes aegypti*, which causes yellow fever, as well as leeches such as *Haemadipsa sylvestris*.

Dimethyl phthalate is commercially available as DMP. The WHO has classified dimethyl phthalate as a chemical that is unlikely to cause hazard (Class U). The LD_{50} in rats is 8200 mg/kg. Dimethyl phthalate is strongly irritating to the eyes, mucous membranes and skin. It is a suspected endocrine disruptor. Over-exposure to this chemical can cause stomach irritation and depression of the nervous system.

Dimethylvinphos insecticide

Dimethylvinphos, $C_{10}H_{10}Cl_3O_4P$ ((*Z*)-2-chloro-1-(2,4-dichlorophenyl)vinyl dimethyl phosphate), is an organophosphorus insecticide. It is a pale white crystalline solid with solubility in water being 130 ppm. It is soluble in organic solvents such as xylene, acetone and cyclohexane. It is unstable in sunlight and by hydrolysis (the DT_{50} is 40 days). Its CAS Registry Number is 2274-67-1.

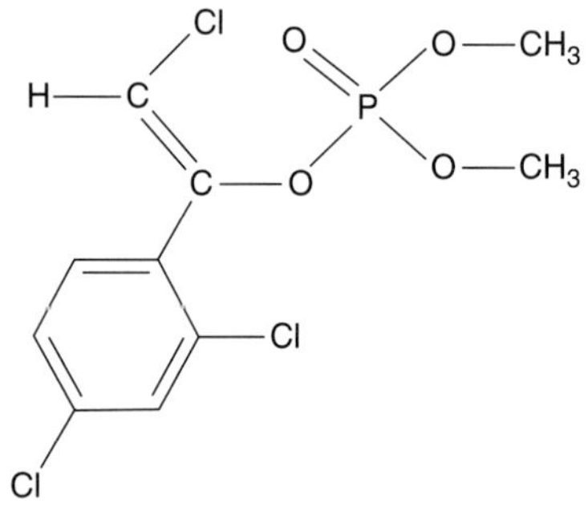

Dimethylvinphos

Dimethylvinphos, a cholinesterase inhibitor, is a moderately persistent stomach insecticide with contact and stomach action. It is effective against the leaf roller and stem borer of the rice crop at 0.6–0.8 kg a.i/ha.

The insecticide is formulated commonly as dustable powder (DP) (2%). It may damage peach leaves. It is available in the market under the trade name of Rangado.

The main degradation route in mammals is *O*-demethylation to desmethyl dimethylvinphos, which is hydrolysed to 2,2′,4-trichloroacetophenone.

Dinitramine herbicide

Dinitramine, $C_{11}H_{13}F_3N_4O_4$ (N^1,N^1-diethyl-2,6-dinitro-4-trifluoromethyl-*m*-phenylenediamine), is a dinitroaniline herbicide. It is a yellow crystalline substance, soluble in water at 1.1 mg/kg. The CAS Registry Number is 29091-05-2.

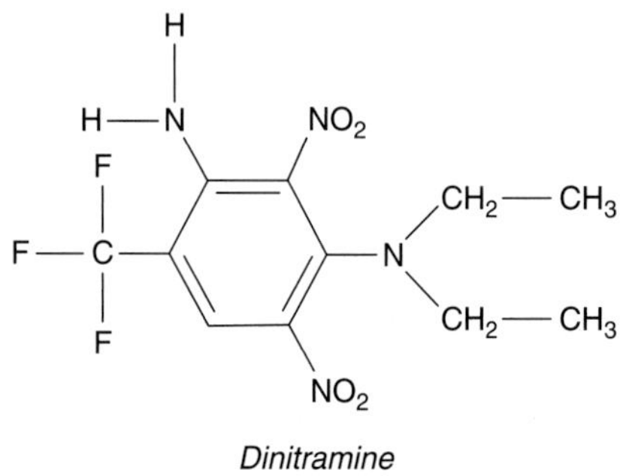

Dinitramine

A selective, pre-plant, soil herbicide, dinitramine is used to protect soybeans, cotton and groundnuts against many types of weeds. The acute oral LD_{50} among mammals is >3000 mg/kg. The half-life in soil is 38 days and it is considered to be moderately persistent.

Dinitroanilines

See **Herbicide action – meristematic mitotic inhibitors**

Dinocap analysis

Dinocap, $C_{18}H_{24}N_2O_6$ ((*RS*)-2,6-dinitro-4-octylphenyl crotonates and (*RS*)-2,4-dinitro-6-octylphenyl crotonates in which ‘octyl’ is a mixture of 1-methylheptyl, 1-ethylhexyl and 1-propylpentyl groups), is an acaricide and fungicide.

Dinocap is extracted with mixed hexanes and purified either by washing with sulfuric acid or by liquid chromatography and finally determined colorimetrically in *N*,*N*-dimethylformamide solution. The analysis of dinocap is given below.

Extract in a rolling tin apparatus (3.78 l (1 gallon) capacity) containing 500 g of macerated tissue and 1 l mixed hexanes (boiling point 60–80°C). Decant the extract into a beaker containing 200 g anhydrous sodium sulfate and mix completely. Filter the mixture through a fluted filter paper. Purify the extract by either of the following methods (a and b).

- Method a – Add 20 ml concentrated sulfuric acid to a 100–200 ml portion of the sample extract in a 500 ml separating funnel. Shake gently for 3 min and allow the phases to separate and discard the lower acid portion. Wash the upper portion with four 50 ml portions of water, discarding the water portion every time. Pass the remaining solution through a column (25 × 2.5 cm) containing 100 g anhydrous sodium sulfate. Rinse the separating funnel with 100 ml hexane in two portions and add to the column. Finally, rinse the column with 50 ml hexane and add this to the column effluent.
- Method b – Prepare the column containing 20 g florisil and pre-rinse with 50 ml 6% ethyl ether in mixed hexanes. Pass 100 ml portions of the sample extract through the column and follow by rinses of 50, 100 and 150 ml portions of 6% ethyl ether in mixed hexanes. Discard the 50 ml rinse and collect the other two.

After purification, evaporate under reduced pressure in a rotary evaporator with the flask immersed in a water bath at 40–50°C. Dissolve the residue in 4 ml dimethylformamide. Allow to stand for 20 min and filter through glass wool into a cuvet. Measure the absorbance at 440 nm versus that of *N*,*N*-dimethylformamide.

Prepare a calibration curve using solutions of dinocap in mixed hexanes containing 5–50 ppm. Evaporate and add 4 ml dimethyl formamide as for the sample. Beer’s Law is followed up to 50 μg of the sample.

Dinocap fungicide

Dinocap, $C_{18}H_{24}N_2O_6$ ((*RS*)-2,6-dinitro-4-octylphenyl crotonate and (*RS*)-2,4-dinitro-6-octylphenyl crotonate in which ‘octyl’ is a mixture of 1-methylheptyl, 1-ethylhexyl and 1-propylpentyl groups), is a dinitrophenyl crotonate fungicide. Its CAS Registry Number is 39300-45-3. It is a reddish-brown viscous liquid, which is soluble in water (0.151 mg/l) and in most

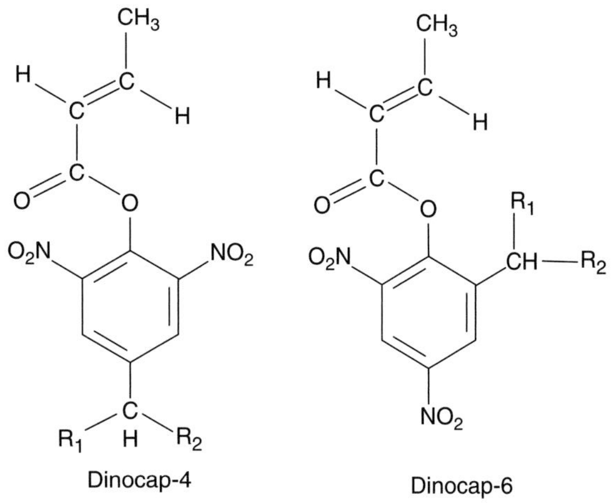

R1 = methyl, ethyl or propargyl R2 = hexyl, pentyl or butyl

Dinocap

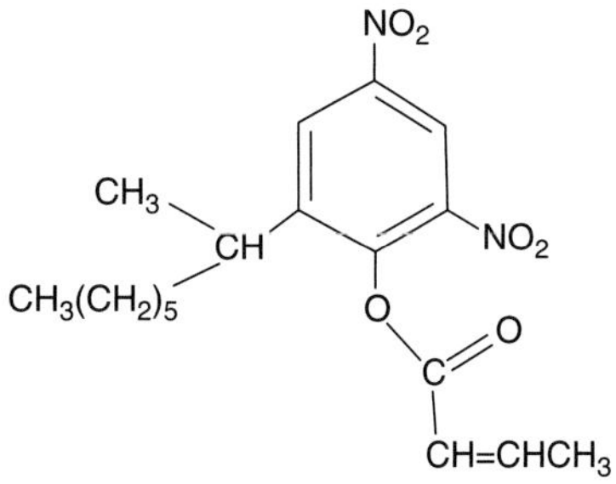

Meptyldinocap

organic solvents. It rapidly decomposes in light and is incompatible with oils and oil-based sprays and alkaline materials.

Dinocap is produced by nitration of *o*-(1-methylheptyl) phenol followed by esterification with crotonic acid. The industrial product is a mixture of isomers and is analysed by the Kjeldahl method.

A contact fungicide which has both protective and curative actions, dinocap is also a contact acaricide. As a fungicide, it controls **powdery mildews** on peaches, apples and cherries (causal agents include *Podosphaera leucotricha* and *Sphaerotheca pannosa*), on vines (*Uncinula necator*) and on members of the family Cucurbitaceae (e.g. cucumbers, melons and gourds) (*Erysiphe cichoracearum*). As an acaricide, it controls mite species such as spider mites, European red mite (*Panonychus ulmi*) and yellow tea mite (*Hemitarsonemus latus*) in apples and other fruit trees. Dinocap uncouples mitochondrial oxidative phosphorylation.

Some of the common formulation types are dustable powder (DP), emulsion concentrate (EC; 350 g a.i./l) and wettable powder (WP; 185 g a.i./kg). Karathane is a popular brand. The common application dose of dinocap on grapes is at the rate of 210 g/ha. Dinocap is combined with **myclobutanil, fenbuconazole** and **endosulfan**.

The EPA and WHO classify dinocap as a Class III toxin (slightly toxic). It is absorbed by the skin and is a skin irritant. The acute oral LD_{50} in male and female rats and bobwhite quails are 990 mg/kg, 1212 mg/kg and >2150 mg/kg, respectively. It is moderately toxic in mammals. The LD_{50} among mice is 53 mg/kg. Acute percutaneous LD_{50} in rabbits is >2000 mg/kg. The NOEL (2 years) in rats is 6–8 mg/kg b.w./day. The ADI is 0.008 mg/kg b.w./day. Dinocap is a PAN Bad Actor chemical because it is a developmental toxin and is moderately hazardous.

Dinocap is degraded in plants, animals and the environment to give 2,4- and 2,6-dinitrophenol (DNOP). In animals (rats), it is eliminated through urine and faeces, with residues seen only in the digestive tract. In soil, dinocap is strongly adsorbed, depending on the nature of the soil, and can be classified as having low mobility, with no leaching potential. Adsorption is observed to be lesser in its degradation products.

Dinocap is registered in India, South Africa, Australia, Czechoslovakia and Greece, but not in the USA. Recommended MRLs are 0.2 ppm for apples and 1 ppm for grapes.

Meptyldinocap, $C_{18}H_{24}N_2O_6$ (2-(1-methylheptyl)-4, 6-dinitrophenyl crotonate), is a single isomer of dinocap. It is a yellow-brown liquid with solubility in water being 0.248 mg/l (pH 7, 20°C). Like dinocap, meptyldinocap is also a contact fungicide, which acts by disturbing metabolism in the cell, and inhibiting spore germination and respiration. It controls powdery mildews on grapes, strawberries and cucurbits.

The fungicide meptyldinocap is irritating to the skin and eyes, but it is low on acute toxicity. Acute oral LD_{50} values in rats and mice are >2000 mg/kg. The acute percutaneous LD_{50} in rabbits is >2000 mg/kg. It is highly toxic to fish and invertebrates, but moderately toxic to algae.

In soil, meptyldinocap is degraded through hydrolysis and microbial degradation. Meptyldinocap is registered in Canada, Greece and Hungary, but not in the USA.

Dinoterb herbicide

Dinoterb, $C_{10}H_{12}N_2O_5$ (2-*tert*-butyl-4, 6-dinitrophenol), is a dinitrophenol herbicide. It is a pale yellow solid, with a phenol-like odour, soluble in water (4.5 mg/l) at pH 5, cyclohexanone, dimethyl sulfoxide and alcohols. In the form of an ester or a salt, dinoterb may exist as dinoterb acetate. It is stable for at least 1 month at a pH range of 5–9 in normal storage conditions.

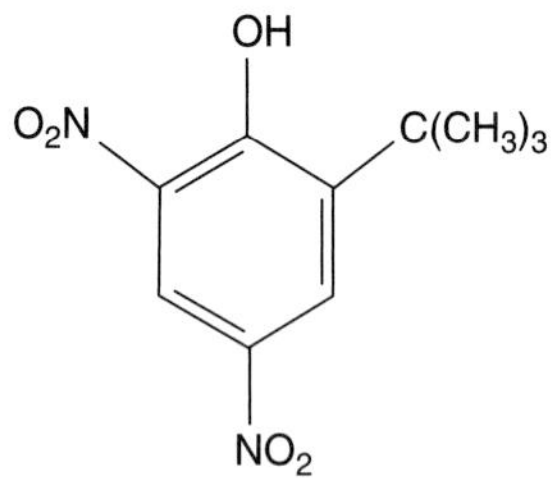

Dinoterb

Dinoterb is a selective, non-systemic, post- and pre-emergence herbicide with contact action. The mode of action is the process of oxidative phosphorylation. It is used for control of annual broadleaved weeds in maize, lucerne, beet, peas and beans.

The herbicide is available in emulsifiable concentrate (EC) and suspension concentrate (SC) formulations.

Dinoterb is highly acutely toxic, with an acute oral LD_{50} among rabbits of 100 mg/kg and that among rats of 62 mg/kg. The short-term dietary NOEL of dinoterb in rats for 2 years is 0.375 ppm diet. It is extremely toxic to fish, with an acute 96 h LC_{50} in rainbow trout of 0.0034 mg/l. The WHO has classified dinoterb as a Ib toxin (highly hazardous), and it is thus classified as a PAN Bad Actor chemical. It is dermally toxic. It is also observed to cause fetal harm.

Most of dinoterb is excreted in the faeces and urine within 7 days. Dinoterb may cause harm to the unborn child. It is toxic in contact with skin and very toxic if swallowed. Dinoterb is very toxic to aquatic organisms and may cause long-term adverse effects in the aquatic environment.

The EU Directives 76/895/EEC, 86/362/EEC, 86/363/EEC and 90/642/EEC has fixed the MRL of dinoterb for citrus fruits, almonds, cashew nuts, coconuts, apples, barley and many others as 0.05 mg/kg. The USA has not registered this compound.

Diofenolan as an IGR

Diofenolan, $C_{18}H_{20}O_4$, is a reaction mixture of (2*RS*,4*SR*)-4-(2-ethyl-1,3-dioxolan-4-ylmethoxy)phenyl phenyl ether (50%–80%) and (2*RS*,4*RS*)-4-(2-ethyl-1,3-dioxolan-4-ylmethoxy) phenyl phenyl ether (50%–20%). It is an insect growth regulator of the **diphenyl ether** class. It is a colourless liquid, which is partially soluble (4.9 mg/l at 25°C) and miscible with methanol, acetone, toluene and hexane. It is rapidly degraded by light. Diofenolan has a CAS Registry Number of 63837-33-2 (see Diofenolan at bottom of the page).

Diofenolan, an IGR with **juvenile hormone** activity, is effective against pests in the order Lepidoptera and also **scale insects** (order Hemiptera) and weevils (order Coleoptera) in citrus trees, mango, olives, nuts, tea and ornamentals.

The common formulation of diofenolan is as an emulsifiable concentrate (EC). It is available in the market under the trade name of Aware.

According to the WHO, diofenolan comes under the toxicity Class III (slightly hazardous). It is not registered in the USA but is not a PAN Bad Actor chemical.

Dioxapyrrolomycin bioinsecticide

Dioxapyrrolomycin is an antibiotic isolated from *Streptomyces fumanus* Sveshnikova. It is a tricyclic pyrrole compound that has insecticidal and acaricidal activity. It is commonly used as an antiparasitic for cattle, sheep and poultry. The compound and its derivatives are uncouplers of mitochondrial oxidative phosphorylation.

Dioxathion acaricide

Dioxathion, $C_{12}H_{26}O_6P_2S_4$ (*S*,*S*'-1,4-dioxane-2,3-diyl *O*,*O*,*O*',*O*'-tetraethyl *bis* (dithiophosphate)), is an organophosphate acaricide. It is a thick reddish-brown coloured liquid or powder with a garlic odour. It is insoluble in water. It is unstable in the presence of alkalis, iron, tin and strong acids. The CAS Registry Number of dioxathion is 78-34-2.

Besides being an acaricide on nut trees and citrus fruit crops, dioxathion is also used as an insecticide on ectoparasites like ticks, horn flies and lice on farm animals.

Dioxathion is applied as a spray and as a dip. A common trade name is Deltic.

Acaricide dioxathion is an acutely toxic compound, considered to be a restricted use pesticide (RUP) by the EPA and by most countries. The acute oral LD_{50} of dioxathion among rats is 23 mg/kg, while the dermal LD_{50} is 235 mg/kg b.w. It is slightly toxic to birds but very toxic to fish and *Daphnia magna*. The ADI of dioxathion is 0.0015 mg/kg b.w./day. It is a cholinesterase inhibitor and is thus declared a PAN Bad Actor chemical. Exposure to this chemical may lead to irritation of the eyes, pupil constriction and blurring of vision, abdominal cramps, laboured breathing, nausea, vomiting, diarrhoea, muscle cramps

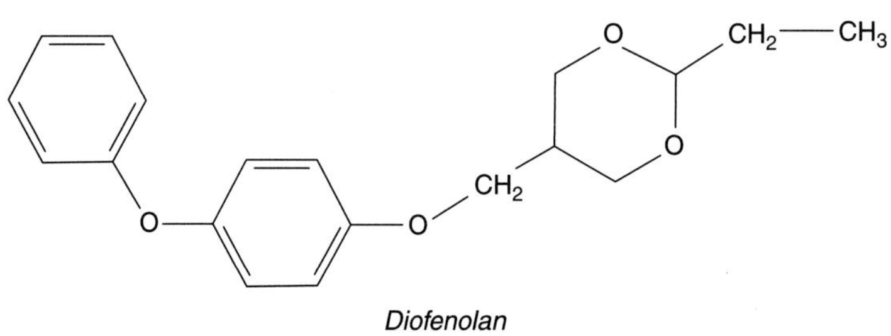

Diofenolan

and excess salivation. Dioxathion contact can be avoided or reduced by using appropriate personal protective clothing.

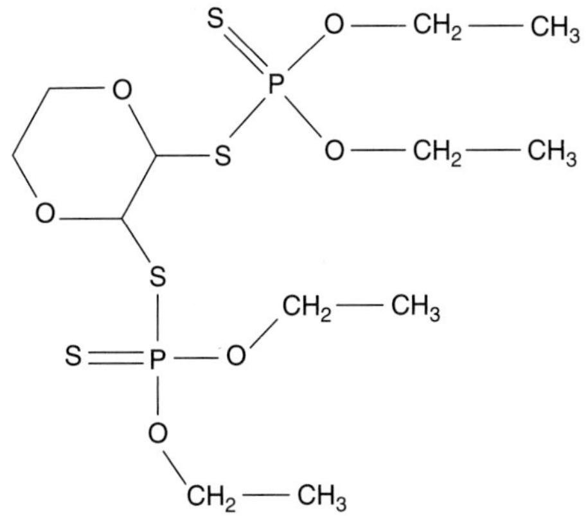

Dioxathion

Diphacinone rodenticide

Diphacinone, $C_{25}H_{16}O_3$ (2-(diphenylacetyl) indan-1,3-dione), is a 1-3,indandione rodenticide. It is a yellow crystalline substance soluble in water at 0.3 mg/kg, in chloroform at 204 g/kg and in toluene at 73 g/kg. It forms salts in alkaline medium. Diphacinone is not very stable in a pH range of 6–9, or beyond 338°C. It hydrolyses rapidly at pH 4 and decomposes in water. Diphacinone has a CAS Registry Number of 82-66-6. It is made by the condensation of 1,1-diphenyl acetone with dimethyl phthalate in the presence of sodium methoxide.

Diphacinone is an anticoagulant indandione rodenticide which requires repeated doses to be administered before there is lethal effect. It acts by inhibiting vitamin K-dependent steps in the synthesis of coagulation factors II, VII and X. It also uncouples oxidative phosphorylation (energy generation) in mammals. Diphacinone is used to control rats, mice, ground squirrels, voles and other rodent species.

The rodenticide is formulated as bait concentrate, powder, pellets and wax blocks. Diphacinone can be used with other rodenticides. Usually 0.1–0.5% powder concentrates or 0.005–0.05% in pellets and wax blocks are effective.

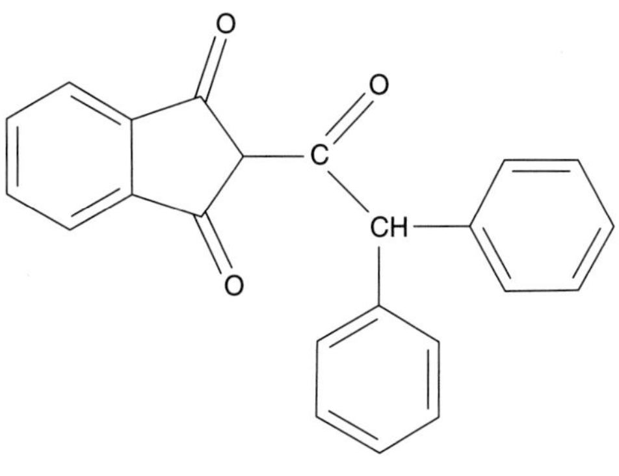

Diphacinone

The WHO categorizes diphacinone as a Class Ia toxin (extremely hazardous) and it is a PAN Bad Actor chemical. It has been classified by the EPA too as extremely toxic (Category I). The acute oral LD_{50} in rats is 2.3 mg/kg, in mice 340 mg/kg, in rabbits 35 mg/kg, and in dogs between 3 and 7.5 mg/kg. The inhalation LC_{50} in rats (for a 4 h period) is <2 mg/l air.

Since diphacinone works by inhibiting the process of blood clotting, sufficient doses of vitamin K_1 can help to revive victims accidentally exposed to this chemical. Labels of formulations should display the signal word 'WARNING'.

Diphacinone is metabolized by hydroxylation and conjugation.

Diphenyl ether

See **Herbicide action – protoporphyrinogen oxidase (PROTOX) inhibition**

Diphenyl fungicide

Diphenyl is another name for **biphenyl** fungicide.

Diquat dibromide herbicide

Diquat dibromide, $C_{12}H_{12}Br_2N_2$ (9,10-dihydro-8a,10a-diazonia-phenanthrene dibromide or 6,7-dihydrodipyrido[1,2-*a*:2′,1′-*c*] pyrazine-5,8-diium dibromide or 1,1′-ethylene-2,2′-bipyridyldiylium dibromide), is a herbicide of the **bipyridilium** class. It is commonly known as diquat. This chemical is highly soluble in water (>677,000 mg/l) and is decomposed in alkaline medium and by UV radiation. The CAS Registry Number of diquat dibromide is 85-00-7.

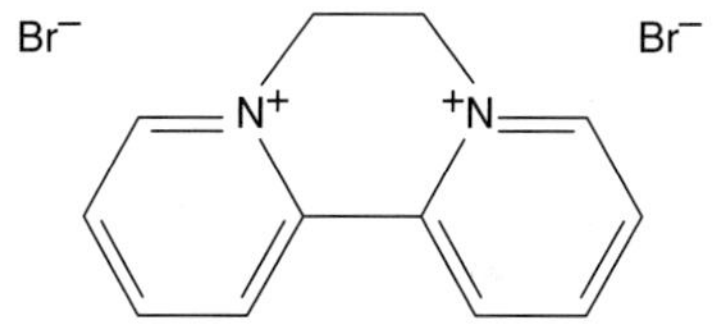

Diquat dibromide

Diquat dibromide, a fast-acting, **non-selective**, contact herbicide, is a **desiccant**, plant growth regulator and a herbicide. As a desiccant, it is used on potatoes for an effective harvesting activity, while as a PGR, sugarcane is applied with diquat dibromide to prevent flowering. Its activity is seen in its control of aquatic weeds, for example curly waterweed (*Lagarosiphon major*) and Brazilian waterweed (*Egeria densa*).

The herbicide is a moderately toxic chemical and is categorized under Class II (moderately toxic) by the WHO. The oral LD_{50} in rats is 120 mg/kg while it is 188 mg/kg for rabbits and guinea pigs. Diquat dibromide is a PAN Bad Actor chemical. There are post-application restrictions of 1 day before entering fields and golf courses that have been sprayed. It is toxic to humans if swallowed.

Diquat dibromide decomposes rapidly in soil and so has a low persistence. Its soil half-life ranges from 1060 to 3450 days. This chemical is a potential groundwater contaminant. It contains a toxic impurity called **ethylene dibromide**. Other degradation products include diquat bistribromide, which is not registered by the USA.

Diquat dibromide is registered in the USA, Canada, Australia, Germany, South Africa and Tanzania.

Diquat herbicide

See **Diquat dibromide herbicide**

Dirty Dozen chemicals
See **PAN International**

Discontinuous phase
See **Adjuvants**

Disease
See all entries for **Plant disease.**

Disinfection
Disinfection (eradicative or curative) is the killing or inactivation of disease-producing microorganisms in animals, seeds or other plant parts. *Infectus* in Latin means spoiled or infected. Often, disinfection is mistaken for sterilizing (which is **disinfestation**). In plant pathology, disinfection is usually restricted to the destruction or inactivation of plant pathogens that have the potential to cause disease. Curative treatments occur only when the plant shows symptoms of disease. The basic difference is one of degree of pathogen involvement with the host.

Disinfection is usually achieved by the application of chemicals, heat or a suitable combination. Immersing plant bulbs in formalin solution is an example of disinfection to eliminate fungi and nematodes.

Disinfectants are usually milder (as low a concentration range of 0.001–2%), in order to prevent killing/injuring plants while killing pathogens. However, the efficacy of a disinfectant is measured by comparing a given disinfectant with phenol, which is the standard. The efficiency rating system, called the phenol coefficient, tests the given disinfectant and phenol on a standard microbe, usually *Salmonella typhi* or *Staphlococcus*. The given compound is rated as <1, >1 or $=1$. This value is arrived at by setting up the given dilutions of both phenol and the compound and incubating the microbe and reading the mortality.

Systemic fungicides such as **benzimidazole**, oxathins (**carboxin**) and **ergosterol** biosynthesis-inhibiting fungicides, as well as formalin and **mercuric chloride** are some of the chemicals that are generally used as disinfectants, due to their non-phytotoxic nature.

Suffice to say prudent use can ensure resistance to disinfectants does not arise.

Disinfestation
Disinfestation is the use of agents that destroy or inactivate organisms that are in non-vital association with host plant parts such as seed surfaces and plant debris. These organisms are also found in irrigation water, on soil surfaces, farm tools, pots, hydroponic solutions and buildings. These agents serve as protectants.

Disinfestation removes pathogens from plant surfaces before infection. The process of disinfestation is a precondition for plant quarantine.

Disinfestation of surfaces may be carried out by: (i) fumigation; (ii) solarization; (iii) dusting; (iv) drenching; (v) the use of steam and hot water; and (vi) UV radiation. A variety of chemicals including formalin, mercuric and mercurous chlorides, organic mercury, copper salts, disulfide and methyl bromide are used in higher concentration to serve as effective disinfestants. These chemicals cleanse pruning tools used for removing plant parts infected with for example blight and rot. Copper compounds are used to kill zoospores of phytophagous fungal species.

Chlorination of irrigation water is also part of disinfestation.

Disk
See **Weed management by mechanical methods**

Dispersing agents
See **Adjuvants**

Disposal of pesticides
See **Pesticide disposal**

Distribution coefficient (K_d)
The ratio of a chemical's sorption in soil or sediment to its dissolved concentration at equilibrium is the distribution coefficient. It is the measure of a pesticide's affinity for adsorption as against its potential for leaching through the soil. The sorption concentration is measured in milligrams per kilogram (mg/kg), while the dissolved concentration is measured in milligrams per litre (mg/l).

Disulfoton insecticide
Disulfoton, $C_8H_{19}O_2PS_3$ (*O*,*O*-diethyl *S*-2-ethylthioethyl phosphorodithioate), is an organophosphorus insecticide (aliphatic organothiophosphate). It is a yellowish oil with an odour and is soluble in water (12 mg/l) and readily miscible with *n*-hexane, dichloro-methane, isopropanol and toluene. It is hydrolysed in alkaline media. It is produced by the reaction of the sodium salt of diethyldithiophosphoric acid with ethylmercaptoethyl chloride. The CAS Registry Number of disulfoton is 298-04-4.

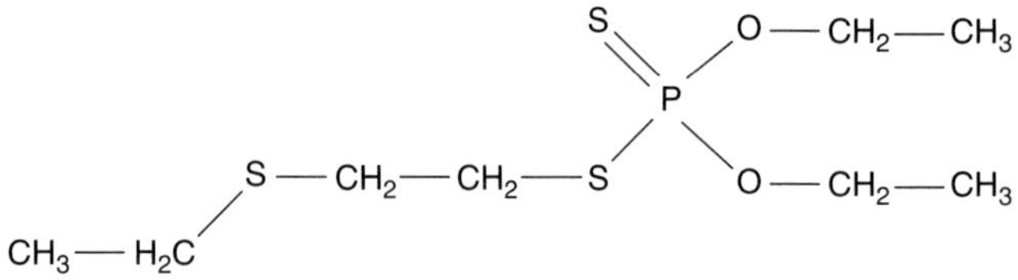

Disulfoton

Disulfoton, a cholinesterase inhibitor, is a systemic insecticide and acaricide (organothiophosphate). Upon being absorbed by plant parts and translocated throughout, disulfoton controls aphids, thrips, **mealybugs** and spider mites in potatoes, tobacco, cereal crops and fruit crops. Some virus infections (e.g. caused by cucumber mosaic virus and potato leaf roll virus) are also prevented.

Disulfoton is formulated as emulsifiable concentrate (EC), powder and granule (GR). In its powder form, and in the prescribed quantities, it is effectively used in seed treatment. It is available in the market as Di-syston, Disystox and Dithiosystox.

The main metabolites of disulfoton in mammals are disulfoton sulfoxide and sulfone, their oxygen analogues and diethylthiophosphate. In plants, disulfon is rapidly metabolized. The soil half-life is 9 days and it exhibits medium to low mobility.

The WHO has classified disulfoton as a Class Ia toxin (extremely hazardous). It is a PAN Bad Actor chemical on the basis of its acute toxicity and its being a cholinesterase inhibitor. This chemical is also a potential groundwater contaminant. Disulfoton is toxic to bees and fish depending on its application mode.

Belize has restricted the use of disulfoton. India, Pakistan and Thailand have banned disulfoton. Many countries in Asia, Europe and Africa have not registered its use, but the USA, Australia and South Africa allow its use.

Dithiocarbamate fungicides

Dithiocarbamate fungicides are a combination of carbon and metal complexes and have become a major part of fungicidal strategy. The credit for the development of this group goes to Tisdale and Williams. Dithiocarbamates are developed from the derivatives of *N*,*N* dimethyldithiocarbamic acid (DDC) and *N*,*N*-ethylenebisdithiocarbamic acid (EBDC), with each of these acids combining with specific metals (or their complexes) to form the active ingredients. The mode of action of this group of fungicides is the inhibition of fungal growth by excessive inputs of heavy metal into the cell or by letting functional groups (e.g. sulfydryl or amino groups) react with the chemicals.

Dithiocarbamates are non-selective, broad-spectrum contact fungicides, compatible with many a systemic fungicide. The compounds that are included in the dithiocarbamate group of fungicides are **ziram**, **ferbam**, **thiram**, **maneb**, **zineb**, **mancozeb**, methyldithiocarbamic acid (**metam**) and **metam-sodium** (sodium salt of metam). While ziram, ferbam and thiram are formed by the combination of DDC with iron and zinc, maneb, zineb and mancozeb are the complexes of EBDC with manganese or zinc or their combination.

The main target pathogens are from the genera *Alternaria*, *Aphanomyces*, *Pythium*, *Puccinia*, *Septoria*, *Botrytis*, *Peronospora* and several others. Crops protected thus include cereal crops, grapes, bananas, tomatoes, berries, pome and stone fruits, potatoes and several others. Common diseases include **early** and **late blights**, rust, **downy mildew** and *Cercospora* leaf spot.

These are effective fungicides that are much in demand, although as they are **non-systemic** repeated applications are required.

Dithiocarbamic acid

See **Sulfur as a pesticide**

Dithiopyr herbicide

Dithiopyr, $C_{15}H_{16}F_5NO_2S_2$ (*S*,*S*'-dimethyl 2-difluoromethyl-4-isobutyl-6-trifluoromethylpyridine-3,5-dicarbothioate), is a pyridine herbicide. The CAS Registry Number of dithiopyr is 97886-45-8. This colourless crystalline solid is soluble in water at 1 mg/kg. It is stable to hydrolysis and photolysis.

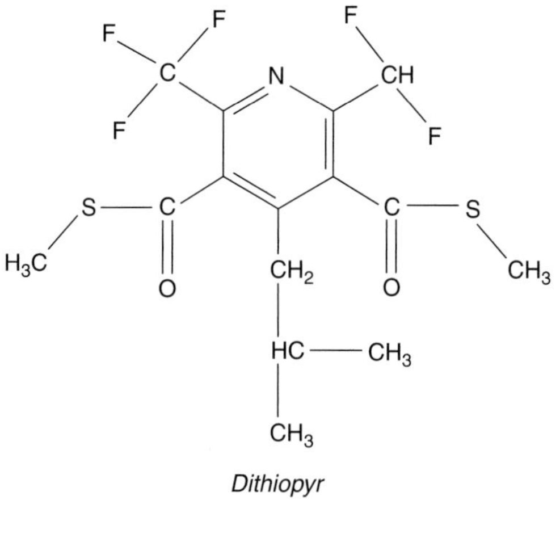

Dithiopyr

Dithiopyr is a pre- and post-emergence herbicide that effectively controls crabgrass, foxtail, **barnyard grass**, chickweed and many other grassy weeds and broadleaved weeds in turf and rice. It acts by inhibiting microtubule assembly.

Water-soluble formulations are common with dithiopyr. It may be mixed with some broad-spectrum herbicides. Dimension is one important trade name.

According to the WHO, dithiopyr is a Class U toxin (unlikely to be hazardous). The ADI/RfD as fixed by the EPA is 0.0036 mg/kg.

Dithiopyr is registered for use in the USA, Canada and Australia.

DMDP bionematicide

DMDP, $C_6H_{13}NO_4$ ((2*R*, 5*R*)-dihydroxymethyl-(3*R*, 4*R*) dihydroxypyrrolidine), is a plant-derived nematicide isolated from tropical legume plants called lance pods (*Lonchocarpus*) and Indian beech or *Milletia pinnata* (*Karanj* in Hindi). The CAS Number is 59920-31-9. These plants (lance pods and Indian beech) contain traces of **rotenone**, which is insecticidal/nematicidal in nature. The sugar (fructose) analogue in DMDP disrupts the functioning of the insect gustatory nerves. The rotenone and the fructose together make DMDP an effective nematode control agent.

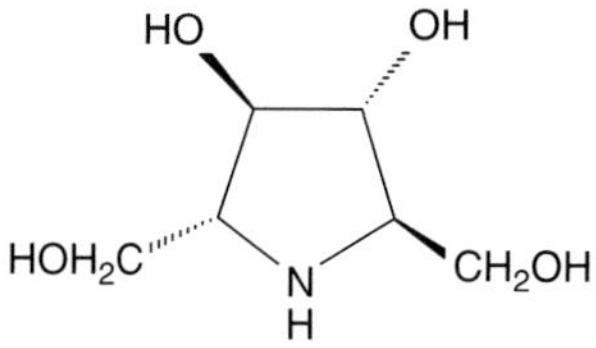

DMDP

DMDP, when sprayed on foliage, is translocated to the roots. The agent controls root attacks by soil insects. Crops that are sprayed with DMDP include vegetables, bananas, coffee, tobacco and cotton.

The toxic effect of DMDP on mammals and on the environment is low.

Dodemorph acetate fungicide

See **Dodemorph fungicide**

Dodemorph fungicide

Dodemorph, $C_{18}H_{35}NO$ (4-cyclododecyl-2,6-dimethylmorpholine), is a morpholine fungicide. This stable compound, being a mixture of two isomers (*cis* and *trans*), exists either as a colourless oil or as a solid, soluble in water up to 100 mg/l and soluble in chloroform (>1000 mg/l), ethanol and acetone. Dodemorph can exist as a salt, as in dodemorph acetate.

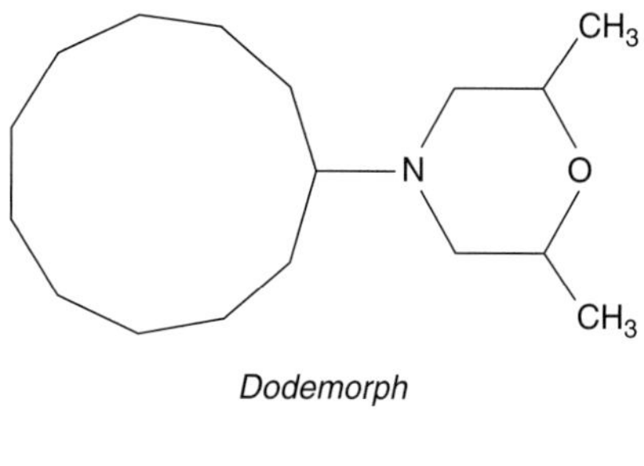

Dodemorph

Dodemorph is an **ergosterol** biosynthesis inhibitor (a **sterol biosynthesis inhibitor (SBI)**). It is a systemic fungicide which has both protective and curative properties. It translocates well in plants. It is a fungicide mainly used on ornamentals (e.g. roses) for the control of **powdery mildew**.

This fungicide is generally formulated as an emulsifiable concentrate (EC). Common application rates include 100 g/100 l of water. It is marketed under the trade name of Meltatox, among others.

The fungicide dodemorph is classified as a Class U toxin (unlikely to be hazardous) by the WHO. It is not acutely toxic to mammals and most other non-target organisms, but irritates the skin and eyes. But, it is considered moderately toxic to fish and to *Daphnia* sp. The ADI is optimized at 0.082 mg/kg b.w./day. However, it is feared to have chronic ill effects on the reproductive system.

Dodemorph is considered moderately persistent due to its high adsorptive properties. Its stability in most soils also strengthens its persistence. The half-life in most soils is up to 41 days but it undergoes aqueous photolysis within 4 days.

Dodemorph acetate, $C_{20}H_{39}NO_3$, is a colourless solid, highly soluble in water (520 mg/l at 25°C), benzene and ethanol. It is used to control powdery mildews and performs in a similar way to dodemorph. It is a skin and eye irritant. The acute dermal LD_{50} in rats is >4000 mg/kg.

Both dodemorph and dodemorph acetate are registered in a few countries of the EU and South Africa, but not in the USA.

Downy mildew
See **Plant disease downy mildew**

Downy mildew of grapes
See **Crop – grape pest control**; **Plant disease downy mildew**

Drift-control
See **Adjuvants**

Dry flowable formulations (DF)
Water-dispersible granules (WG) are also known as dry flowables because they can be measured volumetrically, as a suspension concentrate or 'flowable'. DF exhibits the shelf stability properties of a solid product. WG and DF vary widely in appearance and bulk behaviour, which depends on the specific method of manufacturing.

Dry formulations
Pesticides can be formulated broadly into wet or dry forms. A dry formulation is a combination of ingredients, either as a slurry or converted into a dry form that may be applied to the target pest. Dry formulations are divided into those for: (i) direct use; (ii) dispersion; and (iii) dissolution. The direct-use type includes dustable powders, powders for seed treatment, dry applied granules and tablets for direct application. The dispersion type includes wettable powders, powders in water-soluble packaging, water-dispersible powders for slurry seed treatment and water-dispersible granules and tablets. The dissolution type includes water-soluble powders, water-soluble powders for seed treatment, water-soluble granules and water-soluble tablets.

Just as in any formulation, the ingredients in a dry formulation must be compatible, safe and storable. The components should be of uniform size with a high critical relative humidity. Bulk blending is a form of dry mixing in which all nutrients are granular (1–3.3 mm diameter). Granules must be stable, hard and free of dust. (See also **Pesticide formulations**.)

Dry-applied granules (GR)
See **Granules**

Dust
A finely divided solid particle that may have damaging effects on human beings on inhalation or that constitute a fire hazard is called dust. Dust includes: (i) metallic particles; (ii) silica, mica, talc, graphite, clays, calcium carbonate and asbestos; and (iii) organic materials such as chemical pesticides, cereal flours, cellulose and coal.

Dusts have many uses in industrial, agricultural and household situations. Coating a pesticidal or fertilizer formulation with a fine dust powder ensures a longer life for the compound and its slow release. Pesticide dusts are applied directly as spot treatment for insect control, and can be applied into cracks to eliminate insect colonies.

Dusts constitute particles 10 μm or less in diameter. Dust sizes ranging from 0.5 to 2 μm can damage the lungs. Silicosis is a respiratory disease caused by continuous exposure to uncombined crystalline silica dust in mines and quarries. The tolerance values for various dusts are available in the literature and the safety regulations of the **Occupational Safety and Health Administration (OSHA)**.

Dust suspensions in enclosed areas are a fire hazard regardless of the chemical nature of the dust, with a possibility of an explosion by static charge or open flame.

The formulae shown in Eqn D.9 and Eqn D.10 provide a guideline for finalizing a dust formulation.

$$\text{Percentage of a.i. in a dust} = \frac{\text{Quantity of formulated pesticide (in kg)} \times \text{Percentage of a.i. in formulation}}{\text{Quantity of dust prepared (in kg)}} \qquad \textbf{(D.9)}$$

$$\begin{array}{l}\text{Quantity of formulated pesticide required to prepare a dust in}\\ \text{a certain concentration}\end{array} = \frac{\text{Percentage a.i. required} \times \text{Quantity (in kg) of dust required}}{\text{Percentage of a.i. in formulation}} \qquad \textbf{(D.10)}$$

Dust treaters for seeds

See **Seed protection devices**

Dustable powder (DP)

Dustable powder (DP) is one of the dry formulations of pesticides. It consists of the active ingredient (a.i.) in a powder carrier. DPs are manufactured very simply by pre-blending an a.i., powder carrier and accompanying additives in an efficient blender, followed by size reduction using a mechanical energy mill. The equipment required is inexpensive and readily available.

Since a DP is meant for mechanically sticking to the target, efficient adhesion of the a.i. to the powder carrier is essential. In order for the formulation to remain stable and effective, additives such as anti-caking agents and UV protectants are added.

The benefits of DPs are: (i) the ease of manufacture; (ii) low cost; (iii) ease of use (no prior mixing needed); and (iv) positive shelf-life attributes.

In spite of all these advantages, there are some serious drawbacks of DPs. The most important one is that DPs tend to drift to non-target organisms. Other disadvantages include: (i) they are harmful to the applicator and the operator involved in their manufacture; and (ii) the use of DP in restricted areas may be an explosion hazard. In spite of these drawbacks, DPs are still used in many parts of the world.

Some special dusts known as tracking powders are used for monitoring and controlling rodents and insects. This finely ground dust contains stomach poison which sticks to the feet or fur of the pest and is ingested during grooming. Tracking powders are useful when baits are refused due to abundant food supply. (See also **Pesticide formulations**.)

Dusting machinery

For applying dust formulations, there are specially designed equipment called dusters and dusting machinery. These machines consist of a hopper (container) that holds the agitated dust particles that are to be fed into a current of air so that the particles are ultimately discharged in the form of fine dusts dispersed over a large area or in the form of a cloud. This equipment may be powered by hand or an engine, and accordingly there are: (i) hand-operated dusters; and (ii) power dusters.

Hand-operated dusters are of three types: (i) container dusters; (ii) hand dusters; and (iii) rotary hand dusters. Container dusters are of a basic type in which the dust from a plastic, rubber or leather container is discharged by squeezing or via a valve, in a small stream of air. These are useful over very small areas. Hand dusters are usually small and consist of an airtight container with a controlled inlet and outlet. The dust-filled container is operated with a plunger pump. These dusters are only used as spot treatment on animals or over a small kitchen garden. An adaptation to the hand duster is the rotary hand duster. Here a current of air is produced by a fan, driven by a geared handle. Dust is fed via a rotating brush into the inlet vent either directly from the chambers or into the stream of air. Spoon-shaped nozzles are useful for effective dispersal.

Power dusters run on the power of an engine on which they are mounted. The hopper capacity is around 15 kg and such dusters may have from one to eight outlets with flexible pipes and filled with a boom with multiple nozzles. Discharge rates are high and so is the coverage (commonly around 12 ha/h).

E

(*E*)-3,7-dimethyl-2,6-octadien-1-ol bioinsecticide
See **Geraniol bioinsecticide**

Early blight
See **Plant disease early blight**

Ecdysone
See **Insect growth regulators (IGRs)**

Ecdysteroids
See **Insect growth regulators (IGRs)**

Economic entomology
See **Entomology**

Economic threshold level (ETL)
See **Entomology**

Ecotoxicology
Ecotoxicology is the science that studies the action of physical and chemical agents on populations of organisms (e.g. plants, animals), soil, water and other factors within a defined ecosystem. It includes transfer of substances and interactions with the environment. Ecotoxicology is also referred to as environmental toxicology.

Ectoparasitic nematodes
See **Nematodes**

Edifenphos fungicide
Edifenphos, $C_{14}H_{15}O_2PS_2$ (*O*-ethyl *S*,*S*-diphenyl phosphorodithioate), is an organophosphorus fungicide. It is a clear, yellowish liquid, which is combustible in nature. It is soluble in water (0.056 g/l), *n*-hexane, dichloromethane, isopropanol, toluene, acetone and xylene. Edifenphos is hydrolysed by strong acids and alkalis. Its half-life varies from 29 min to 11.5 days depending on pH. Edifenphos is produced by a reaction of *O*-ethylphosphoryl dichloride with thiophenol in the presence of a base. It has a CAS Registry Number of 17109-49-8.

Edifenphos is a selective and protective foliar fungicide, best known for its use in rice for the control of: (i) *Rhizoctonia solani*, which causes sheath blight; (ii) *Pyricularia oryzae*, which causes blast; and (iii) *Cochliobolus miyabeanus*, which causes brown spot. The mode of action is believed to be the inhibition of **chitin** biosynthesis in the fungal cell wall.

Available formulations of edifenphos are emulsifiable concentrate (EC), dustable powder (DP) and wettable powder (WP). Up to 1 kg/ha is commonly applied for control of rice blast. Multiple applications are permitted depending on the intensity of the infection. A pre-harvest interval of 3 weeks to 2 months is maintained. For instance, it is 21 days in Japan and many rice-producing nations, but is 60 days in Italy.

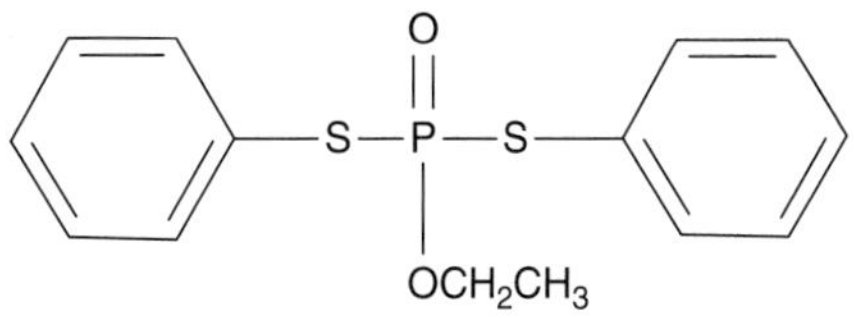

Edifenphos

p-Hydroxylated edifenphos is an intermediary metabolite in mycelial cells of *P. oryzae*. Edifenphos shows no delayed neuropathic symptoms.

The WHO has classified edifenphos as a Class Ib toxin (highly hazardous). Heating may produce irritating, corrosive and toxic gases. The acute LD_{50} in rats is 100–260 mg/kg, while in guinea pigs and rabbits it ranges from 350 to 400 mg/kg. Percutaneous LD_{50} of edifenphos in rats is 615 mg/kg. The 2 year NOEL in male rats is 5 mg/kg. Edifenphos has an acute oral LD_{50} in bobwhite quails ranging from 281 to 293 mg/kg. It is moderately to highly toxic to aquatic animals (the LC_{50} in carp is 1.3–1.55 ppm/48 h). The ADI for humans is up to 0.003 mg/kg b.w./day. Edifenphos is a PAN Bad Actor chemical. Symptoms of toxicity can include excessive salivation, sweating, respiratory depression and loss of consciousness. So, appropriate PPE such as gloves, special clothing, goggles and positive pressure self-contained breathing apparatus (SCBA) are recommended.

Edifenphos is no longer produced or registered in the USA and most EU members. It is registered for use in India, the Philippines, Vietnam, Japan and Korea.

Japan has fixed 0.2 mg/kg as the MRL for rice.

Electrostatic sprayer
See **Application equipment – sprayers**

Elemental sulfur fungicide
Elemental sulfur is one of the forms of naturally occurring sulfur, and the only inorganic **phytoalexin**. It is also naturally produced by some disease-resistant species of tomatoes, cocoa, beans and cotton. Elemental sulfur is stable to heat and moisture, but oxidizes to the sulfate when it comes in contact with soil. It is absorbed by plants in its sulfate form.

Sulfur has long been known as a fertilizer as well as a pesticide (especially as a fungicide). As a fungicide, elemental sulfur has been used mainly to control **powdery mildew** diseases. Fungal species that are ascomycetes, basidiomycetes and deuteromycetes are controlled by this compound.

Elemental sulfur is commonly available as sulfur dust and wettable sulfur. Sulfur is ground, mechanically or by an air-blast separation process, and mixed with clay-like kaolin or talc to form a fine free-flowing dust, which is very effective for seed treatment. The particle size of the dust must be of 6 μm or less. The finer the particles the more effective elemental sulfur is. Wettable sulfur is produced by the addition of a wettable/dispersing agent (and sometimes bentonite clay) to sulfur dust, so that further addition of water may make the formulation uniform and homogenous and conducive to spraying. Wettable sulfur is also available as a paste or powder. Pastes and wettable powders normally have 70% sulfur content with a particle size of 6 μm or more. Micronized sulfur, made by grinding sulfur in a micronizer mill by high-pressure jets, is a common form of wettable sulfur.

Many sulfur formulations are registered in many countries, including India.

Emamectin benzoate bioinsecticide

Emamectin benzoate (a mixture of ≥90% (10*E*,14*E*,16*E*)-(1*R*,4*S*,5′*S*,6*S*,6′*R*,8*R*,12*S*,13*S*,20*R*,21*R*,24*S*)-6′-[(*S*)-*sec*-butyl]-21,24-dihydroxy-5′,11,13,22-tetramethyl-2-oxo-(3,7,19-trioxatetracyclo[15.6.1.1^{4,8}.0^{20,24}]pentacosa-10,14,16,22-tetraene)-6-spiro-2′-(5′,6′-dihydro-2′*H*-pyran)-12-yl 2,6-dideoxy-3-*O*-methyl-4-*O*-(2,4,6-trideoxy-3-*O*-methyl-4-methylamino-α-L-*lyxo*-hexapyranosyl)-α-L-*arabino*-hexapyranoside benzoate and ≤10% (10*E*,14*E*,16*E*)-(1*R*,4*S*,5′*S*,6*S*,6′*R*,8*R*,12*S*,13*S*,20*R*,21*R*,24*S*)-21,24-dihydroxy-6′-isopropyl-5′,11,13,22-tetramethyl-2-oxo-(3,7,19-trioxatetracyclo[15.6.1.1^{4,8}.0^{20,24}]pentacosa-10,14,16,22-tetraene)-6-spiro-2′-(5′,6′-dihydro-2′*H*-pyran)-12-yl2,6-dideoxy-3-*O*-methyl-4-*O*-(2,4,6-trideoxy-3-*O*-methyl-4-methylamino-α-L-*lyxo*-hexapyranosyl)-α-L-*arabino*-hexapyranoside benzoate) is a microorganism-derived insecticide, acaricide and nematicide of the **avermectin** class. It contains two homologue constituents, namely emamectin B_{1a} ($C_{56}H_{81}NO_{15}$ (B_{1a})) and emamectin B_{1b} ($C_{55}H_{79}NO_{15}$ (B_{1b})).

Emamectin benzoate, a photosensitive, creamish crystalline powder, is an avermectin compound, soluble in water (110 mg/l), methanol and acetone. It has a CAS Registry Number of 155569-91-8. It is incompatible to alkaline media (see Emamectin benzoate at bottom of the page).

Emamectin benzoate acts on the nervous system by stimulating the release of γ-aminobutyric acid (GABA) from nerve endings, thus leading to excess chloride ions in the cells. This inhibits the proper functioning of the nerves and paralyses and kills pests such as nematodes, insects and mites. Leaf miners, moths, caterpillars, mites, bollworms, cutworms and armyworms are effectively controlled by this compound. Crops that are protected include tobacco, cotton, tea, brassicas, tomatoes, lettuce, maize, soybeans and pines.

Emamectin benzoate is available in formulations such as emulsifiable concentrate (EC) and soluble granule (SG). Common dosage for cotton and brassicas is 16 g/ha. This compound can be combined with many other pesticides, as long as they are not alkaline in nature. It is best to store emamectin benzoate in tightly sealed containers in a cool, dry place. The trade names of emamectin benzoate include Proclaim, Affirm, Shot-one, Arise and Denim.

The EPA categorizes emamectin benzoate as highly toxic in its acute toxicity ratings. The compound is an eye and skin

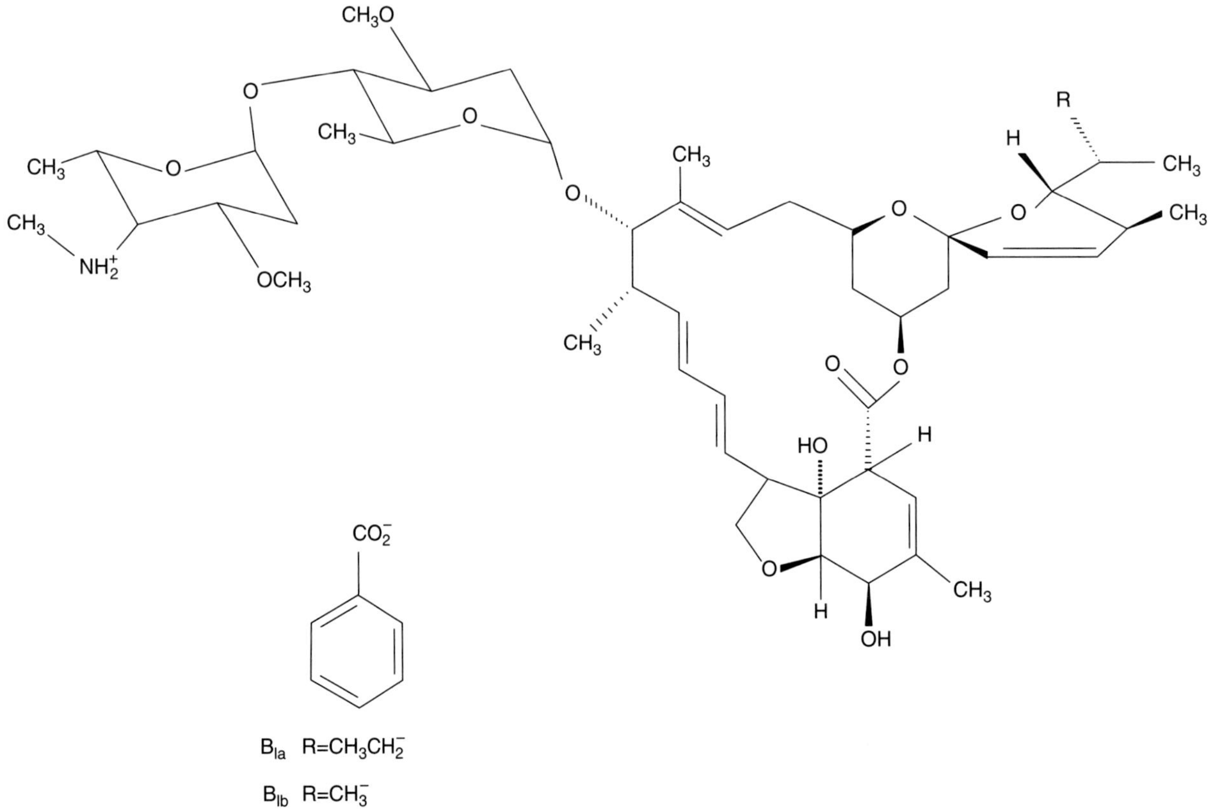

Emamectin benzoate

irritant, and is said to cause harm to **beneficial organisms** and **bees**. It has been declared a PAN Bad Actor chemical. Its acute oral LD_{50} in rats is 70 mg/kg, while the inhalation LC_{50} (4 h) is 2.12–4.44 mg/m^3. The LC_{50} in mallard ducks is 46 mg/kg. Emamectin benzoate is also a potential groundwater contaminant. It has a half-life of 211 days in aerobic soil and of 427 days in anaerobic soil. The ADI is 0.0025 mg/kg/day. Emamectin benzoate products are permitted for use in some countries as fish feed, to prevent certain infections.

Animals partially metabolize emamectin benzoate but eliminate the compound within 51 h. In plants, the compound is degraded by the sun into residual products, the most significant being the unchanged compound itself.

Emamectin benzoate is registered for use in India, Australia, New Zealand, South Africa, Vietnam and the USA.

Emulsifiable concentrates (EC)

Emulsifiable concentrates (EC) are among the most commonly used pesticide formulations. ECs are recommended in situations where the active ingredient (a.i.) is water insoluble or water miscible but soluble in oil. In such cases, the solvent and the a.i. are mixed with an emulsifier so that further addition of water can make a suitable spray solution. The addition of surfactants or emulsifying agents ensures the formation of homogenous, non-crystallizing stable emulsions. The suspension that is formed is milky in appearance and called an **emulsion**.

The advantage of an EC is that it is economical and easy to use. An organic carrier facilitates better uptake and translocation through penetration and cuticular solubilization. There is uniform distribution of toxic material on the target since the EC has a higher fraction of fine particles.

Major disadvantages of ECs are that the organic carriers may be volatile, inflammable and phytotoxic. The organic carriers may harm the environment.

Emulsifiers

See **Adjuvants**; **Emulsifying agents**

Emulsifying agents

Emulsifying agents are materials that stabilize the emulsion of two liquids which are not miscible or which tend to separate out from each other. Emulsifying agents maintain the stability of the concentrate within limits. Products such as polyoxyethylene and ethylcellulose are used as agents, for either uniform distribution or a controlled release of the active ingredients, or both. (See also **Pesticide formulations**; **Adjuvants**.)

Emulsion

A mixture of two or more immiscible liquids, in which one is dispersed in the other as microscopic or ultramicroscopic particles, is an emulsion. An example is oil in water or water in oil. Products like milk, butter, latex paints and floor wax are all emulsions.

Emulsions are stabilized by agents/**emulsifiers** that form films on the droplet surface that make them physically stable (e.g. in the case of soap molecules, milk, bentonite clay or proteins). They also provide the emulsion homogeneity. Less stable emulsions eventually separate spontaneously into two liquid layers.

Stable emulsions may be broken by adding multivalent ions of the charge opposite to the emulsion droplet, such as the addition of acids.

Fertilizers and pesticides that cannot be dissolved directly in water are dissolved in an organic solvent which can then be mixed with water to form an emulsion using an emulsifier. For example, a sulfur emulsion for foliar spraying is made by dissolving sulfur in carbon disulfide and mixing with water.

Oil-in-water emulsions are suitable for oil-soluble pesticidal compounds, emulsified into a water carrier, along with specific surface-active agents. Oil-in-water emulsions exhibit good shelf life, though the oil may be inflammable or harmful to the skin. Such emulsions are used for seed treatment.

When a group of pesticidal ingredients that are water insoluble or water soluble are mixed in water along with suitable surfactants and surface active agents, they form a kind of a homogeneous and clear emulsion called the **micro-emulsion**. These emulsions are stable in a specific temperature range and disperse well in spray water.

An invert emulsion contains a water-soluble pesticide dispersed in an oil carrier. It requires a special kind of emulsifier that allows the pesticide to be mixed with a large volume of petroleum-based carrier, usually fuel oil. The droplets in invert emulsions are heavier and aid in reducing drift in windy conditions. It also facilitates better absorption, even in the rains. Invert emulsions are commonly used to control vast tracts of weeds.

Emulsions for seed treatment are prepared in a tumbling device. Seed treatment emulsions are manufactured as oil-in-water emulsions and the water component of the emulsion may incorporate water-soluble active ingredients for co-application to the seed. The ingredients need to be non-phytotoxic to the seeds. The seeds need to be dried carefully, to ensure they do not germinate but remain viable.

Emulsions for seed treatment

See **Emulsion**

Encapsulation

Encapsulation in the context of pesticides is the process of enclosing, coating or embedding a material (as a whole or in parts) in a molten film, sheath or foam. Pesticide and fertilizer granules are encapsulated with substances that control the release or dissolution rate of the active ingredient.

When a material is enclosed in very small capsules (e.g. 20 μm diameter) such that it is not readily permeable through the capsule, the process is termed **micro-encapsulation**. Fertilizers, nutrients, pesticides, **pheromones** and pharmaceuticals are micro-encapsulated. The covering layer is made of substances like polyamides, elastomer and gelatine.

In organisms too, enzymes are micro-encapsulated till they reach the desired site of action.

Enchytracidae

See **Biological control of nematodes**

Endod

See **Molluscicides**

Endoparasitic nematodes

See **Nematodes**

Endosulfan insecticide

Endosulfan, $C_9H_6Cl_6O_3S$ (6,7,8,9,10,10-hexachloro-1,5,5a,6,9,9a-hexahydro-6,9-methano-2,4,3-benzodioxathiepine-3-oxide), is a cyclodiene organochlorine insecticide with additional acaricidal properties. With a CAS Registry Number of 115-29-7, endosulfan is a brownish solid, soluble in petroleum solvents but almost insoluble in water. Endosulfan is made by mixing α and β stereoisomers, in a 7:3 ratio. At high temperatures the α-isomer is converted to the β-isomer.

The chemical is produced by reacting hexachlorocyclopentadiene with *cis*-butene-1,4-diol and then with thionyl chloride. It is available in the market under trade names Thionex, Endocil, Phaser and Benzoepin.

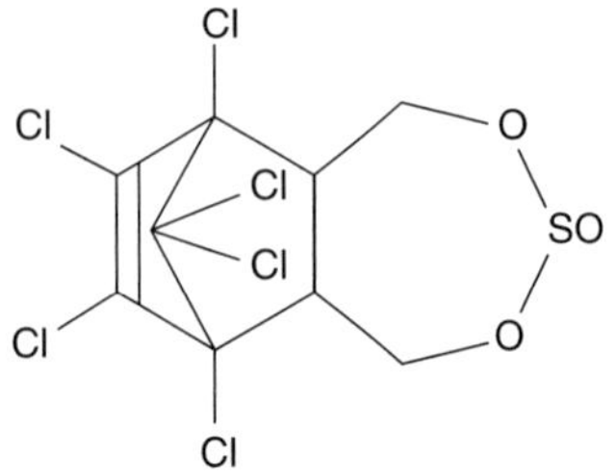

Endosulfan

Endosulfan acts on the nervous system by antagonizing the GABA receptors (γ-aminobutyric acid) that disturb chloride channels. Target pests include whiteflies, **leafhoppers** and many types of beetles, worms and aphids. Important crops protected by endosulfan are vegetables, cereals (e.g. rice, maize and sorghum), tea, coffee, tobacco and cotton. A non-farm use is as a wood preservative.

Endosulfan is formulated as an emulsifiable concentrate and wettable powder. It is compatible with other non-alkaline pesticidal materials such as **dimethoate**, **malathion**, **methomyl**, **monocrotophos**, **pirimicarb**, **triazophos**, fenoprop, **parathion** and **oxine-copper**. Application rates range from 0.45 kg a.i./ha to 1.4 kg a.i./ha. Pre-harvest time period is prescribed as 42 days.

Endosulfan is classified as a Class II toxin (moderately toxic) by the WHO, and as a moderately to highly toxic chemical by the EPA. The oral LD_{50} is 80 mg/kg in rats and 7.36 mg/kg in mice. It does not cause significant irritation to the eyes or skin. Endosulfan can be highly toxic to birds (with the lowest LD_{50} rate in some species being 31 mg/kg). It is extremely toxic to fish (96 h LC_{50} for rainbow trout being 1.5 μg/l) and moderately toxic to honey bees. Adverse effects have not been observed on beneficial insects. The compound is a suspected endocrine disruptor and is classified as a PAN Bad Actor chemical. The ADI of endosulfan is 0.006 mg/kg/day and the RfD is 0.00005 mg/kg/day.

Extreme care has to be taken to prevent exposure to endosulfan. The symptoms of toxicity can include lips or fingernails turning blue, breathlessness, confusion, convulsions, diarrhoea and vomiting. Removal of contaminated clothes, exposure to fresh air and administration of gastric lavage are the main first-aid measures before professional medical attention is provided.

Endosulfan mainly breaks down into endosulfan sulfate (CAS Registry Number 1031-07-8). Endosulfan moderately persists in soil (half-life up to 50 days). Its half-life is 5 weeks in neutral water, and 1 day in alkaline media. In soils, breakdown is helped by fungal and bacterial action. Endosulfan is seen to accumulate in body tissues. It is present in surface waters in large proportions although the half-life on vegetable residues is about 1 week.

Endosulfan is banned in India, following reports of reproductive deformities that initiated public pressure against the use of this chemical. It is registered in Australia, the Philippines and many African countries. Endosulfan is banned in the EU and many other countries.

The prescribed US FDA upper limit for endosulfan residues is 2 ppm on most food products.

Endothal herbicide

Endothal, $C_8H_{10}O_5$ (7-oxabicyclo [2.2.1] heptane-2,3-dicarboxylic acid), is a herbicide, algicide and a plant growth regulator (defoliator) of the dicarboxylic acid class. It is a colourless crystalline substance, soluble in water, benzene, isopropanol, acetone, dioxane, diethyl ether and methanol. It is stable to light and temperature up to 90°C, above which it undergoes slow conversion to the anhydride. Endothal has a CAS Registry Number of 145-73-3. Liquid formulations are required to be stored at 0°C or above. The compound is a dibasic salt and forms water-soluble amine and alkali metal salts.

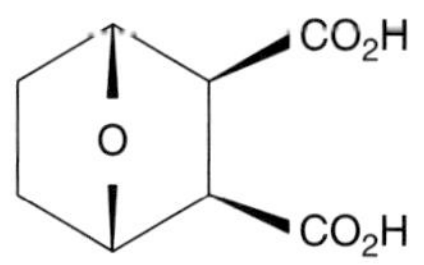

Endothal

Endothal is a selective, contact, pre- and post-emergence herbicide with limited translocation. It controls annual grass and broadleaved weeds in sugarbeet, fodder beet, beetroot, spinach and turf. As a PGR, endothal acts as a **defoliant** (cotton) and **desiccant** (in lucerne, clover and hops). It is also an algicide and controls aquatic weeds like plankton and pondweed.

This herbicide is available as granules (GR) and as a soluble concentrate (SL). Known trade names are Accelerate and Herbicide 273, among others.

The potassium and amine salts of endothal are used as aquatic herbicides to control a variety of plants such as plankton, pondweed, milfoil, elodea and algae in water bodies and rice fields. Endothal reduces sucker branch growth in hops, and is used as a desiccant to aid the harvest of lucerne, potatoes, clover and cotton.

The EPA has classified endothal under toxicity Class II (moderately toxic). The LD_{50} values for rats, guinea pigs and rabbits are 51, 250 and 100 mg/kg, respectively. It is not toxic to bees but toxic to some fish species. Endothal is toxic to humans and can cause neurotoxicity, carcinogenicity and reproductive and developmental toxicity. Its ADI and RfD for humans is 0.02 mg/kg/day. Endothal is not a PAN Bad Actor chemical.

Exposure to endothal can cause symptoms of toxicity like irritation to skin, eyes, respiratory and gastrointestinal tracts, and convulsions. Immediate medical attention is a must after administering first-aid measures like washing the affected area with plenty of water, drinking cold milk or aluminium hydroxide gel and removal of contaminated clothes.

Although endothal is mobile in soil, leaching is not a problem since the compound degrades, partly by microbial degradation, in a maximum of 21 days. It also degrades easily in water under aerobic conditions.

Many salts like dipotassium, disodium, diammonium and di-(*N*,*N*-dimethylalkylamine) and glutamic acid are produced as breakdown products, but none of them is a PAN Bad Actor chemical.

Endothal is registered in the USA, Australia, Canada and New Zealand. Its MRL on maize, wheat, barley, sorghum, rice, apples and milk is generally accepted to be 0.02 ppm.

Endotoxin

Endotoxin is a component of the outer cell membrane of certain species of Gram-negative bacteria. It is made of large molecules of lipids and polysaccharides (known as lipopolysaccharides) and is believed to be important for the following reasons: (i) as a permeability barrier; (ii) as a protection from external threats; and (iii) for the development of immunity in the microorganism.

Escherichia coli, *Salmonella* spp., *Pseudomonas* spp. and *Haemophilus influenzae* are some of the bacterial species in which endotoxins occur.

Endotoxins cause many types of reactions in their hosts, such as fever, haemorrhage and alterations in immune response.

Endrin insecticide

Endrin, $C_{12}H_8Cl_6O$ (1*R*,4*S*,4a*S*,5*S*,6*S*,7*R*,8*R*,8a*R*)-(1,2,3,4,10,10-hexachloro-1,4,4a,5,8,8a-hexahydro-6,7-epoxy-1,4-endo, endo-5,8-dimethanonapthalene), is a cyclodiene organochlorine insecticide as well as a rodenticide and an avicide. It is an almost odourless, whitish-cream solid, nearly insoluble in water (23 μg/l) and moderately soluble in benzene and acetone. As an *endo* isomer of **dieldrin**, endrin transforms into an insecticidal substance when mixed with strong acids or exposed to sunlight. The multi-stage production of endrin involves many chemicals such as hexachlorocyclopentadiene and cyclopentadiene. Endrin has a CAS Registry Number of 72-20-8.

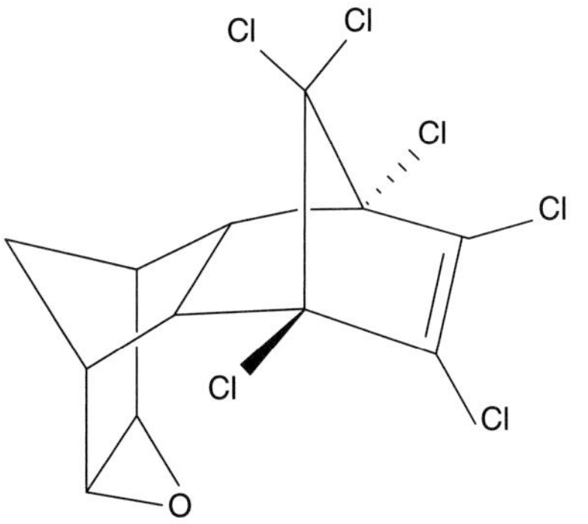

Endrin

Endrin is used on rice, wheat, maize, sugarcane, beans, cole and beet crops, tobacco and cotton, to control insect pests such as cotton bollworms, corn borers, cutworms and **leafhoppers**. Spray applications and seed treatment are often used on crops. Ground applications are common for the control of mice and voles (when used as a rodenticide).

Endrin is compatible with many pesticides and is available in forms such as emulsifiable concentrate (EC), wettable powder (WP), dust (DP) and granules (GR).

The acute oral LD_{50} of endrin in rats is in the range of 7.5–18 mg/kg, while the dermal LD_{50} is in the range of 15–18 mg/kg. The IRAC has classified endrin as not being carcinogenic. The compound is very toxic to fish, birds and bees. It is a PAN Bad Actor chemical since it is rated highly toxic and is a suspected endocrine disruptor.

Symptoms of endrin poisoning include headaches, mental confusion and tremors. Victims of poisoning may require gastric lavage with water and purgatives and respiratory assistance. PPE must be used, which must include safety glasses and gloves.

Orally ingested endrin is eliminated mostly in the faeces. The soil half-life of endrin is over 10 years. Since endrin is lipophilic, it dissolves in the fat of body tissues and thus accumulates and persists there. In comparison with dieldrin, the degree of persistence of endrin in organisms is slightly lower, probably due to its somewhat faster excretion in the bile. Endrin absorbed through the skin or fatty tissues or inhaled as dust or vapour is less likely to be excreted easily, and can remain there for a long time. Thus, it is classified as a persistent organic pollutant (POP) and is banned in many countries including India. The maximum contaminant level for drinking water is set at 2 ppb.

Enestroburin fungicide

Enestroburin, $C_{22}H_{22}ClNO_4$ (methyl 2-{2-[3-(4-chlorophenyl)-1-methylallylideneaminaoxymethyl] phenyl}-3-methoxyacrylate), is a **strobilurin** type of fungicide (methoxyacrylate). Also known as enestrobin or SYP-Z071, it is also insecticidal towards members of the order Lepidoptera. Enestroburin is a white crystalline substance, insoluble in water but easily soluble in acetone, ether and chloroform.

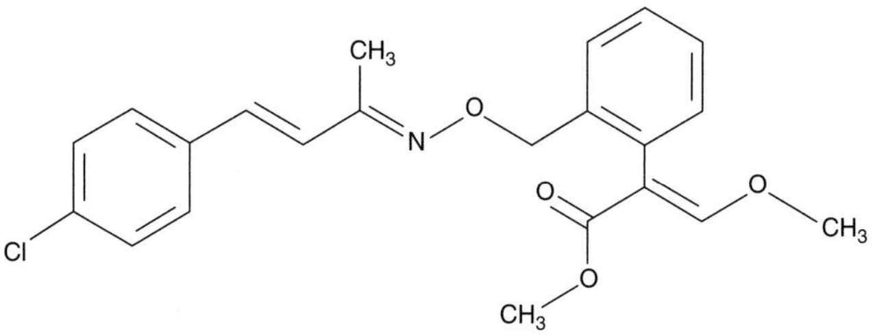

Enestroburin

Enestroburin is a systemic, broad-spectrum, protective and curative fungicide. It is a **QoI** inhibitor (quinone outside inhibitor), which inhibits respiratory electron transport. Generally formulated as an emulsifiable concentrate (EC), this compound effectively controls cucumber diseases like **powdery mildew**, **downy mildew** and grey mould (caused by *Pseudoperonospora cubensis*, *Sphaerotheca fuliginea*, *Botrytis cinerea*). SYP-Z071 also protects wheat from powdery mildew.

Toxicologically, the acute oral LD_{50} of enestroburin in rats is in the range of 749–926 mg/kg. It is not a skin irritant but a slight eye irritant.

Entomology

The field of biological sciences that deals with insects is called entomology. This science has assumed importance, since there are large numbers of insect species with either beneficial or destructive characteristics. While numerous insect pests can cause havoc on the field, numerous other species benefit the ecosystem; for example, pollinating insects promote fruit crops.

Medical entomology is concerned with insect vectors that can cause diseases in humans and animals. Many plants are attacked by fungal, viral and bacterial diseases that are transmitted by insects. Insect parasites have been studied from the point of view of controlling destructive insects. Such studies form the basis of economic entomology. The study of insects that have a direct influence on man's financial stability is called economic entomology. Since insect pests have a tangible effect on the lives of people (i.e. causing reduced yield or monetary loss), this branch of entomology focuses on harmful insects rather on beneficial insects. From this emerges the concept of the economic threshold level (ETL), when control measures become critical. When the expectation is of a high crop yield, the economic threshold would be very low. It is important to note that allowing a pest population to remain below the ETL is an ideal tool of IPM.

Environmental concerns during pesticidal application

The application of chemical pesticides, however necessary, has to take into account the effect it would inadvertently have on the soil, water, air and living organisms in the neighbouring areas. To minimize the ill effects on the environment, applications should be made only during favourable physical and natural conditions. The following passage enumerates how variations in temperature, wind and moisture witness changes in the mode of pesticide application. Suitable climatic conditions support optimization of pesticide use.

Temperature influences the effectiveness of pesticide applications. Low temperature may slow down pesticide activity, while higher temperatures and direct sunlight may break down the chemical prematurely. Additionally, high temperatures with low humidity increase pesticide vaporization and drift. Thus, the ideal temperatures for pesticide application may be between 10°C and 40°C, although precise temperatures for application should be confirmed from the product label.

The presence of moisture in a field, as observed during irrigation or rain, affects the efficacy of pesticides sprayed on plants or the soil. The effect of a few chemicals will be enhanced, while in other cases the application may be wasted. Label instructions precisely indicate if moisture is useful for a specific formulation and if so, how much. Some protectant fungicides are designed to be applied during or just before expected rain or watering. Soil insecticides need to seep into the soil and hence need plenty of rain for best results. However, most contact herbicides, insecticides and some fumigants lose their pest control action in a rain storm, since most of the compounds get washed out, necessitating repeated applications. In some cases, there may be a need to wait for a while before repeating an application, in order to avoid its excessive presence in a single location. Such words of caution are reflected on product labels. Before application of any pesticide, it is advisable to check the weather forecast for at least up to a week.

Air movement from wind or ventilation can alter pesticide effectiveness. Too much air movement can blow pesticides off target and result in inadequate control. The amount of acceptable air movement depends on the type of formulation and the application technique used. Drift has to be anticipated and application of broad-spectrum pesticides planned accordingly. Aerial equipment may not be the right choice at higher wind speeds (>15 km/h). Sometimes, to offset air movement, air is blown towards the area to be protected.

Safety precautions should also be considered while applying pesticides. If possible, application should be carried out during the early morning hours, late evening, or even at night, to take advantage of cooler temperatures, lower volatilization and fewer unsuspecting passers-by. Such operations should be carried out with artificial lights and with the help of colleagues or assistants, to avoid accidents.

Environmental contamination by pesticides

See **Pesticides and the environment**

Environmental factors for PPE

See **Worker safety norms and first-aid techniques**

Environmental impact assessment (EIA)

The assessment of the impact of development or industrial projects on the environment is stated in an environmental impact assessment (EIA) notification. EIAs note negative as well as positive effects on the environment. The information in an EIA is seen as an influencing factor in decision making. EIAs exist in some form or other in most countries. The EIA, as well as an environmental impact statement (EIS), together provide guidelines towards risk management in any major project, such as a mine, dam, power plant or chemical plant. Agencies drafting an EIA are also responsible for analysing critical cases and ensuring enforcement of policies and laws in the direction of safe implementation.

The International Association for Impact Assessment (IAIA) defines an EIA as 'the process of identifying, predicting, evaluating and mitigating the biophysical, social and other relevant effects of development proposals prior to major decisions being taken and commitments made.' Australia, the EU, India, Sri Lanka, Nepal, Canada and the USA are members of the IAIA.

Any EIA, to be effective in an area, needs abundant primary and secondary data collected over the years, which can give a definite picture of, for example, the quality of air, amount of green cover, soil quality and status of erosion in that area. There is a need for information to be collected from all sources, and made available at one place for better impact assessment and policy implementation.

From the point of view of project implementers, a group, task force or a separate environmental department needs to be at the forefront of the EIA team. The involvement of the local staff is critical in making a realistic assessment of the environment.

The ISO 14000 series relate to 'Environmental Management Standards'. All standards except 14001 are guidance documents and are voluntary. These standards help companies improve their environmental performance and prevent issues from getting in the way of trade. However, the assessment and subsequent statement about a particular project may not stop the project from being implemented.

In India, the Ministry of Environment and Forests, which is responsible for enacting laws for the safety of environment, undertakes EIA. Acts relating to the environment include

The Environment (Protection) Act (1986), The Air (Prevention and Control of Pollution) Act (1981) and others concerning water use and wildlife protection. Another body active in this area is the Central Pollution Control Board. The implementation of EIA faces challenges of lack of information within a stipulated time interval. The name EIA comes from Sec. 102 (2) of the National Environmental Policy Act (NEPA), 1969, USA.

Laws under the NEPA require the EIA to be undertaken, leading to the issue of the EIS or the declaration of a 'Finding of No Significant Impact' (FONSI) document. The assessment, once done, is open for public comment, in the form of a draft EIS, before becoming a final EIS.

Environmental protection

The system of procedures that limit damage to the quality of water, air and land is termed environmental protection. Protective measures include restoring the balance in the ecosystem and preserving natural resources.

Multifarious human activities produce harmful waste products including vapours, gases, solids, liquids or radiation. All living systems are affected, to at least some extent, by the introduction of such effluents. Thoughtless disposal of industrial waste, for instance, creates toxicity in water bodies, which immediately disturbs the biological systems in the aquatic environment. Similarly, dumping plastic and other rubbish in a haphazard manner can also pollute the environment in several ways.

In the context of pesticides, soil, water and living things are constantly being subjected to exposure to chemicals. In the process of eliminating pathogens, the ecosystem and **beneficial organisms** are often harmed. If such chemicals do not degrade in a short period of time, they can threaten the balance of nature in the area. Examples of persistent chemicals are chlorinated hydrocarbons.

Environmental protection thus aims to protect the environment from excesses such as toxins and radiation, in order to reduce the ill effects on living things and nature.

There are many agencies in most countries that are responsible for the upkeep of the environment. Most of these agencies are responsible for enacting laws, often with the help of relevant government ministries, to prevent the misuse of natural resources, reduce pollution and ensure optimal use of pest control agents, thus maintaining the balance of ecosystems. Concepts and terms relating to toxicity of the soil, residual remains of chemicals and specific crop protection agencies are noted throughout this book.

Environmental Protection Agency (EPA) of the USA

The Environmental Protection Agency (EPA) of the USA was established on 12 December 1970 with a mission to protect human health and the environment. It has the responsibility of consolidating 'a variety of federal research, monitoring, standard-setting and enforcement activities to ensure environmental protection'.

The stated purpose of EPA's tasks is to ensure that Americans live in a safe environment, with federal laws tailored to ensure protection of the health of man and the environment. The EPA facilitates communities and individuals to have access to accurate information regarding policies and decisions about environment protection. As a complement to its other activities, the EPA coordinates and supports research and anti-pollution activities by state and local governments, private and public groups, individuals and educational institutions. The EPA also strengthens efforts among other federal agencies with respect to the impact of their operations on the environment. In all, the EPA is designed to serve as the public's advocate for a liveable environment.

Some of the federal regulations that strengthen EPA functioning are 40 CFR (Code of Federal Regulations), the Clean Air Act, the Clean Water Act and the Toxic Substances Control Act.

The EPA has been working for a cleaner, healthier environment for the American people. For more details on the regulating agency, visit http://www.epa.gov/lawsregs/index.html and other related sites.

Environmental protection agencies of other countries work on similar lines.

Enzymes

An enzyme is a protein that catalyses chemical reactions in living cells or substrates. Like any catalyst, enzymes speed up cell reactions without taking part in the reaction. They speed up reactions such as redox reactions and transfer/modification of a chemical group. There are a few thousand enzymes involved in various life processes. In its action, a substrate may form a complex with the active sites of the enzyme (Fig. E.1).

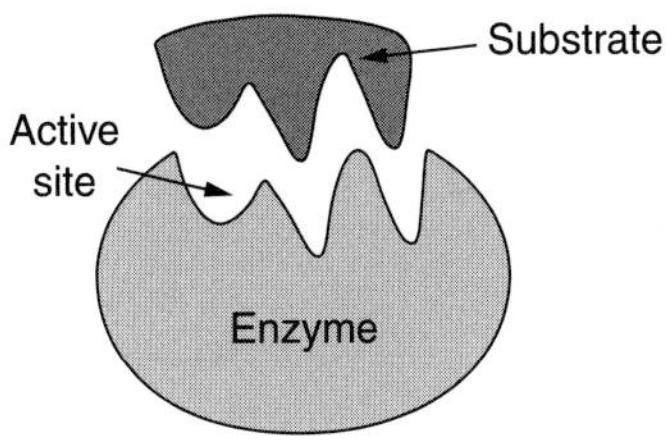

Fig. E.1. Schematic diagram showing the active site of an enzyme. Courtesy of: Drug Strategies to Target HIV: Enzyme Kinetics and Enzyme Inhibitors, Chemical Kinetics Experiment, Authors: Rachel Casiday and Regina Frey, Department of Chemistry, Washington University, with permission from R. Frey.

Cofactors are non-protein accessory molecules (and could be either organic or inorganic) that may be needed to activate enzymes. Organic cofactors are either bound to the protein (in which case they are called prosthetic groups) or are released at the site of the activity (when they are called coenzymes). Examples of coenzymes include nicotinamide adenine dinucleotide (NADH) and adenosine-5′-triphosphate.

Enzymes are specific in their catalytic behaviour on the substrate, for instance an enzyme will only speed up a specific reaction or set of reactions in a cell's metabolic pathway. Increase in stress levels, temperatures or disease conditions affects enzyme activity.

In the context of pesticide mode of action, there are many molecules that interfere with the activity of enzymes, thus either inhibiting their activity or increasing it. Inhibition of enzymes can be temporary (reversible) or permanent (irreversible). When inhibitor molecules compete with the substrate to bind to the enzyme active site in a non-covalent bond (i.e. in a bond that does not involve electron sharing), the inhibition is reversible. Some enzymes that are inhibited by drugs are: (i) **acetylcholinesterase** or AChE (in humans and insects for neurotransmitter function); (ii) protease; and (iii) 3-phosphoshikimate 1-carboxyvinyltransferase (in plants). Examples of reversible reactions are those caused by **carbamate insecticides**. Examples of those that permanently inhibit enzyme activity are those caused by **organophosphorus insecticides** (e.g. **malathion**). When a covalent bond is formed between the active site and the inhibitor, then the inhibition is irreversible.

EPA

EPA is the abbreviation of **Environmental Protection Agency (EPA) of the USA**.

Epicuticular wax

The wax formed on leaf cuticles of certain plants such as peas, cabbage and plantain is called epicuticular wax. This wax is formed out of straight-chain aliphatic hydrocarbons made of substituted groups. Epicuticular wax is an important part of the cuticle and helps to reduce water loss as well as to reflect light.

The amount of wax increases with the amount of stress a plant is under (caused by for example increased sunlight, UV radiation and water shortage). Wax can also be formed inside plant cuticles (cuticular wax).

Epidemiology

The study of the epidemic nature of disease is known as epidemiology. An epidemic is the phenomenon of an infectious disease spreading through an entire community. Historically, epidemiology involves the study of the nature, spread and prevention of human diseases such as cholera, influenza and jaundice. The science has now expanded to include the study of non-contagious diseases such as mental illness, arthritis, cancer and heart diseases. Similarly, epidemiology is becoming relevant in crop situations as well as among poultry and other farmed animals.

Plant epidemiology studies the causes of plant diseases and looks at ways to prevent them. It involves the study of related sciences such as biology, microbiology, agriculture, mathematics and ecology.

Epiphytes

Plants that live for most or some part of their lives on other plants and have no parasitic characteristics are called epiphytes. The host plant provides the epiphyte with physical support. In the absence of a host plant, the epiphyte can use a pole or a wall for such support. Some of the 30,000 species of epiphytes are weed species, which makes this topic significant to the present work.

Epiphytes are more abundant in the tropics, rainforests and the northern temperate regions. Orchids, lichens, cacti, bromeliads and mosses are some types of epiphytes. The roots of such plants are aerial.

Epiphytes play an important part in biodiversity. Epiphytes are known to be a microhabitat. Non-vascular epiphytes such as bryophytes and mosses are safe havens for arthropods such as mites, ants, moth larvae and **thrips**, especially when dense epiphytic growth forms a thick mat in forest canopies. Vascular epiphytes such as bromeliads are hosts to frogs, mosquitoes, insects, snails and stinging ants because water collects in their upturned leaves.

True epiphytes live on a host plant throughout their lives. *Drymoglossum heterophyllum* (bird's nest fern) is a stubborn creeping fern that is found on rubber, coconut and fruit trees. Hemi-epiphytes or half-epiphytes complete one stage of their life as epiphytes (by having adventitious roots) and the other as ordinary plants (with the roots reaching the soil). Some species of cacti (e.g. *Opuntia ficus-indica*) are examples. If the plant is epiphytic in the latter stage of its life, it is called a secondary epiphyte.

Bromeliads are an advanced type of epiphyte considered harmful to beneficial trees and humans. These broadleaved plants bear special epidermal cellular structures called trichomes. The leaves are structured such that they form a watertight spiral or a container in which a large amount of water can collect. This stagnant water becomes home to mosquitoes, which act as vectors to diseases such as malaria. Bromeliads can grow in the ground as well, as in the case of pineapple, which is a cultivated bromeliad.

Epiricania melanoleuca caterpillars

See **Biocontrol of crop pests – parasitoids**

EPN insecticide

EPN, $C_{14}H_{14}NO_4PS$ ((*RS*)-(*O*-ethyl *O*-4-nitrophenyl phenylphosphonothioate)), is an organophosphorus insecticide of the phenyl phenylphosphonothioate class. It has a CAS Registry Number of 2104-64-5. EPN is a yellow, crystalline substance with an aromatic odour. It is insoluble in water but soluble in benzene, toluene, xylene, acetone, isopropanol and methanol. It is stable in neutral and acidic media but is hydrolysed by alkaline media. Hydrolysis liberates *p*-nitrophenol. EPN is known for forming the toxic **phosphine gas** (in the presence of hydrides) and phosphorus oxides on partial oxidation. Heating also leads to the formation of the *S*-ethyl isomer.

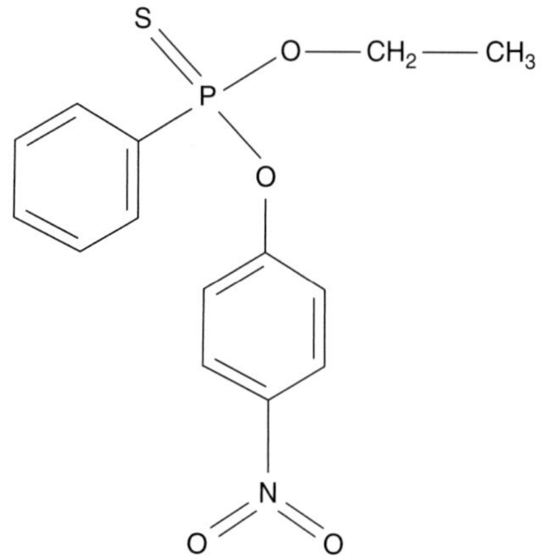

EPN

EPN, a cholinesterase inhibitor, is a non-systemic insecticide with acaricidal properties. It is effective against a wide range of Lepidopteran larvae, especially bollworms in cotton (e.g. *Heliothis* spp. and *Pectinophora gossypiella*), stem and stalk borers in rice (*Chilo* spp.) and leaf-eating larvae in fruit and vegetables.

EPN is formulated as wettable powder (WP) and emulsifiable concentrate (EC). It is compatible with many pesticides such as **Bordeaux mixture** and zinc-lime except those which are alkaline in nature. EPN may be phytotoxic to some varieties of apple. It is available under trade names such as EPN (Nissan).

The WHO has categorized EPN as a Class II toxin (moderately hazardous). EPN is toxic to rats (the acute oral LD_{50} is 14 mg/kg) and very toxic to bees (the LD_{50} is 0.011 mg per bee). The dermal LD_{50} in rats is 538 mg/kg b.w., while the inhalation LC_{50} in rats is 0.106 mg/l for 1 h. EPN is also very toxic to fish (with an acute 96 h LC_{50} for rainbow trout being 0.110 mg/l) and *Daphnia magna*. It can be fatal if swallowed (the fatal dose is 0.3 g for an adult of 70 kg). EPN is classified as a PAN Bad Actor chemical. Symptoms of poisoning can include headaches, nausea, dizziness and convulsions. It is necessary to wear PPE, such as safety glasses and gloves, during handling and application. The chemical should not be inhaled or touched. Containers should be kept far away from fires, as they can explode.

The principal biodegradation pathway of EPN in mammals is oxidative dearylation to produce nitrophenol and *O*-ethyl hydrogen phenylphosphonothioate. In mammals, metabolism is carried out via desulfuration and removal of *p*-nitrophenol, and also by reduction of the nitro group to an amino group. The nitro group reduces to an amino group in the environment. The major metabolite in plants is ethyl hydrogen phenylphosphonate. The DT_{50} in loamy soil is 15–30 days.

The USA and many other countries have restricted the use of EPN. India has refused to register the use of EPN and the EU has banned its use.

EPSPS

See **Herbicidal inhibition of EPSPS**

EPTC herbicide

EPTC, $C_9H_{19}NOS$ (*S*-ethyl dipropyl(thiocarbamate)), is a thiocarbamate herbicide, with a CAS Registry Number of 759-94-4. It is a colourless liquid, moderately soluble in water (245 mg/l) and miscible with acetone, ethyl alcohol, kerosene and xylene. It is stable up to 200°C, but is hydrolysed by strong acids on heating. EPTC is made by reaction of di-*n*-propylamine with phosgene, followed by treatment with ethyl mercaptan.

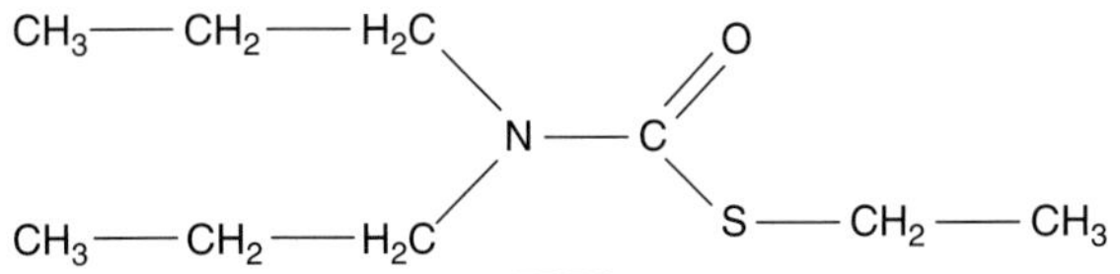

EPTC

EPTC is a selective, pre-plant or pre-emergence, systemic herbicide, absorbed by the roots and shoots and translocated upwards. It is a lipid metabolism inhibitor. When mixed into soil, it controls annual and perennial grasses and broadleaved weeds in beans, forage legumes, potatoes, maize and sweet potatoes.

Available as emulsifiable concentrates (EC) and granular (GR) formulations, under trade names Eptam, Eradicane and Alirox, EPTC may be combined with most pesticides.

According to the WHO, EPTC comes under the toxicity Class II (moderately toxic). This compound is a cholinesterase inhibitor, and is irritating to the skin and eyes (as tested on rabbits). The oral LD_{50} values in rats, mice and cats are 1632, 3160 and 112 mg/kg b.w., respectively, while the 1 h inhalation LC_{50} of EPTC in rats is 31.56 mg/l. Although it may be considered relatively safe for birds, EPTC is slightly toxic to aquatic organisms (the 96 h LC_{50} of EPTC in rainbow trout is 19 mg/l).

Symptoms of poisoning by EPTC include irritation, redness of eyes and skin, discomfort in breathing and dizziness. People that come in contact with the chemical should remove contaminated clothing and food, wash the affected parts and move outdoors into fresh air. The signal word 'CAUTION' must be prominently displayed on EPTC labels. It is feared that it can contaminate groundwater and it is a PAN Bad Actor chemical.

In rats, only limited amounts of EPTC are quickly eliminated via urine and faeces. In plants, EPTC is rapidly metabolized to carbon dioxide and other natural metabolites. The soil half-life of this chemical is between 6 and 42 days (aerobic conditions) depending on the soil. The hydrolysis half-life is about 1 month. In soil, it rapidly degrades microbially to a mercaptan residue, an amino residue and carbon dioxide.

EPTC is registered for use in the USA, Australia, Canada and South Africa. (See also **Herbicide action – by inhibition of fatty acid elongation**.)

Eradicants

See **Fungicides**

Ergasiophytes

See **Weeds – classification**

Ergocalciferol biorodenticide

Ergocalciferol ((5*Z*,7*E*,22*E*)-(3*S*)-9,10-secoergosta-5,7,10(19),22-tetraen-3-01) or vitamin D_2 is a natural vitamin. It is a colourless crystalline substance, soluble in water at 50 mg/l and in acetone at 69.5 g/l. This compound deteriorates in air, at high temperatures (>120°C) and in acidic media. Ergocalciferol is produced by irradiating **ergosterol** obtained from yeast.

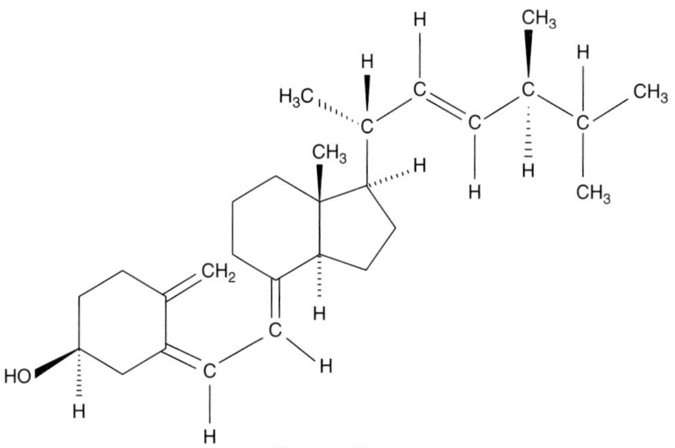

Ergocalciferol

Naturally occurring ergocalciferol is used for rodent control, either by itself or in combination with very small doses of **warfarin** and **difenacoum** (0.005–0.025%). In high doses, ergocalciferol induces mobilization of skeletal calcium and its absorption into the intestine. This activity causes hypercalcaemia and an increase in serum cholesterol.

Common formulations are grain and block baits, which are sold under trade names such as Sorexa. Ergocalciferol may be compatible with other rodenticides.

The acute oral LD_{50} of ergocalciferol in rats is 5 mg/kg. It is not dangerous to bees or domestic animals if used as recommended. In mammals, ergocalciferol is hydroxylated to 1∝, 25-dihydroxy derivative in the liver, and 24(*R*), 25-di-hydroxy derivative in the kidneys.

Ergosterol

Sterols are crucial components of eukaryotic organisms, responsible for structural and biological processes. Ergosterol is a sterol present in fungi. Ergosterol is relevant to our study since it is used as a rodenticide.

Ergosterol, $C_{28}H_{44}O$ (ergosta-5,7,22-trien-3β-ol), is a colourless crystalline substance. A biological precursor to vitamin D_2, it belongs to the class of unsaponifiable lipids. Ergosterol is insoluble in water but soluble in organic solvents. Upon irradiation, it is converted into **ergocalciferol**. Ergosterol has a CAS Registry Number of 57-87-4.

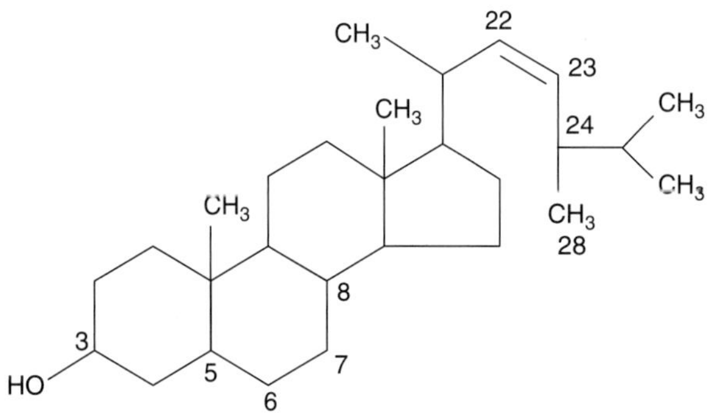

Ergosterol

Ergosterol, a component of fungal cell membranes, was first isolated from fungal species in the genus *Claviceps*, which are called ergots and grow on, for example, millet, sorghum, rye and barley. These organisms produce an alkaloid that is harmful to mammals (including humans). Research into ergosterol functioning has led to many anti-fungal drugs (such as **amphotericin B**) which inhibit the working of ergosterol.

Excessive doses of ergosterol can cause hypocalcaemia, which increases the level of calcium in the blood. This state can affect the kidneys and liver, leading to the death of rodents.

Ergosterol differs from cholesterol by having three positions of unsaturation, at carbon atoms 5-6, 7-8, and 22-23, and containing a methyl group (carbon atom 28) substituted for a hydrogen atom at carbon atom 24.

Ergot fungi

These fungi belong to the genus *Claviceps*. Ergot fungi such as *Claviceps purpurea* and *Claviceps paspali* grow on grasses such as rye, wheat, millet, barley, rushes and sedges (Fig. E.2).

When ergot fungal spores infect their host, the pathogen develops a hardened mass of mycelium which contains a toxic alkaloid of the ergotamine type (a large, tetracyclic peptide linked to lysergic acid) and other lysergic acid derivatives such as ergotoxine and dihydroergotoxine. Upon inadvertent consumption, these fungi can cause a disease called ergotism in humans and animals. The disease is characterized by the malfunction of the nervous system. The disease may have three types of manifestations: (i) gangrenous; (ii) hallucinogenic; and (iii) convulsive.

Fig. E.2. *Claviceps purpurea*, the 'ergot fungus', on wheat. Courtesy of Doug Waylett, with permission.

Ergotism

See **Ergot fungi**

Esfenvalerate insecticide

Esfenvalerate, $C_{25}H_{22}ClNO_3$ ((α*S*)-α-cyano-3-phenoxybenzyl (2*S*)-2-(4-chlorophenyl)-3-methylbutyrate), is a synthetic pyrethroid insecticide. With a CAS Registry Number of 66230-04-4, esfenvalerate is a colourless, crystalline substance soluble in water, xylene, acetone, chloroform, ethyl acetate, dimethylformamide and dimethyl sulfoxide. Esfenvalerate decomposes thermally and also in alkaline media, giving rise to hydrogen cyanide. Structurally, esfenvalerate is identical to the naturally occurring **fenvalerate**, except in the proportions of the four isomers. It is the fully resolved (*S*,*S*)-isomer of fenvalerate.

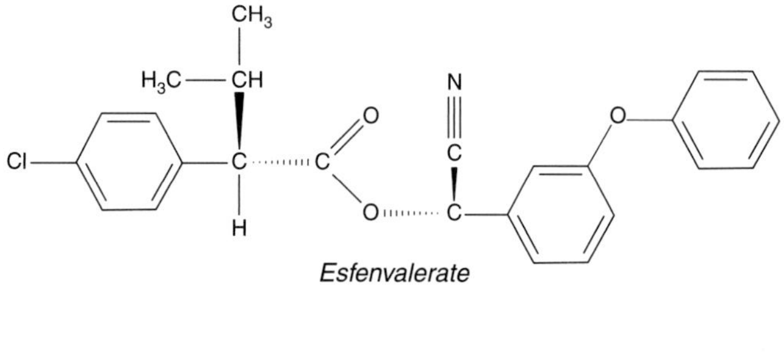

Esfenvalerate

Esfenvalerate is a contact insecticide that attacks the insect nervous system, resulting in the sodium channels remaining open. This leads to symptoms such as convulsions and paralysis. This chemical protects cotton and numerous vegetables against a broad range of insect pests including members of the Coleoptera (beetles and weevils), Diptera (flies and midges), Heteroptera (bugs), Lepidoptera (moths and butterflies) and **Orthoptera**. Affected insects die after hyperexcitation, convulsions and paralysis.

Esfenvalerate is formulated as an emulsifiable concentrate (EC), suspension concentrate (SL) and also ULV liquid (UL), under the trade names of Sumi-alpha and Asana, among others. It is commonly mixed with carbamates, organophosphates and some herbicides, since it is effective on insect strains that are resistant to these chemical classes. Esfenvalerate should, however, not be mixed with strong oxidizing agents and alkaline substances.

The WHO classifies esfenvalerate as a Class II toxin (moderately hazardous). It can cause irritation to the skin and eyes of handlers and applicators. The acute oral LD_{50} in rats is 458 mg/kg, while it is only slightly toxic to rats and rabbits via dermal and inhalation routes. The chemical is moderately to slightly toxic to birds (with the oral LD_{50} for bird species tested being between 1312 and 2250 mg/kg). It is highly toxic to bees. The 96 h LC_{50} for rainbow trout is 0.0003 mg/l. The emulsifiable concentrate is a restricted use pesticide (RUP) according to the EPA, since it may adversely affect aquatic organisms. Esfenvalerate product labels are required to display the signal word 'WARNING'.

Symptoms of esfenvalerate poisoning include headache, dizziness and muscle twitching. Use of protective gear (e.g. long-sleeved shirt and long trousers, chemical-resistant gloves made of material such as neoprene or nitrile rubber, shoes and socks), breathing apparatus and safety goggles is critical. In the case of accidents, first-aid measures include: (i) washing the eyes and body; (ii) taking the person outdoors for some fresh air; and (iii) application of vitamin E cream or oil on the affected skin. In case of a spill or leak, the chemical should be soaked up by soil or any man-made absorbent material.

Animals metabolize esfenvalerate via many processes, including hydroxylation of 2′- and 4′-hydroxyl moieties and ester cleavage, leading to the conjugation of the acidic metabolites with sulfate. In plants, major metabolites found include decarboxylated fenvalerate. Esfenvalerate is moderately persistent with a half-life of up to 3 months, depending on the soil. The degradation products, as well as the unchanged esfenvalerate, are relatively immobile in soil. In sunlight, the half-life in water is about 3 weeks.

Esfenvalerate is registered for use in Australia, New Zealand, Denmark, the USA, South Africa and many countries of the EU.

Esprocarb herbicide

Esprocarb, $C_{15}H_{23}NOS$ (*S*-benzyl-1,2-dimethylpropyl (ethyl) thiocarbamate), is a thiocarbamate herbicide, with a CAS Registry Number of 85785-20-2. It is a liquid, moderately soluble in water and soluble in organic solvents. Esprocarb is made by reaction of *N*-ethyl-*N*-(methyl) isobutylcarbamoyl chloride with mercaptotoluene.

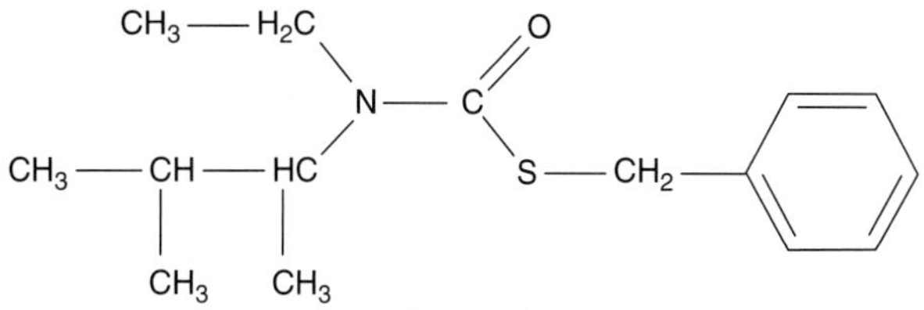

Esprocarb

Esprocarb, a pre- and post-emergence herbicide, inhibits lipid metabolism and is used to control annual weeds and **barnyard grass** (*Echinochloa* spp.) in rice.

The WHO classifies esprocarb as a Class III toxin (i.e. slightly hazardous). However, since it poses a risk to handlers by way of inhibiting cholinesterase, esprocarb is classified as a PAN Bad Actor chemical. Esprocarb can cause dizziness, nausea and irritation to the skin, eyes and respiratory tract.

Essential oils

See **Insecticides**

Estate sprayers

See **Application equipment – hand sprayers**

Estimation of pesticide residues in soil

The following method helps treated samples to be analysed for the presence of pyrethroids, organochlorines, organophosphates and carbamates. Such an analysis makes use of gas chromatography (GC).

About 15 g of a representative sample of soil is well ground and air dried in a 100 ml beaker. Two drops of ammonia solution (0.25 ml) (specific gravity 0.91) are added and mixed thoroughly with a glass rod and set aside for 30 min until the ammonia evaporates. To this soil mixture 0.3 g activated charcoal, 0.3 g **florisil** (prewashed and activated) and 10 g anhydrous sodium sulfate are added and mixed thoroughly. The above mixture is compacted in a glass column (60 cm × 22 mm, fitted with a teflon stop cock) in between the two layers of anhydrous sodium sulfate.

The column is eluted with 125 ml acetone-hexane mixture (1:9 v/v), ensuring a flow rate of 4 ml/min. The extracts are collected and used for GC analysis directly without purification or cleaning. A drop of mineral oil is added and the material is nearly dried using a rotary flash evaporator. The total volume is made to 2 ml with hexane and injected into a GC/ECD system for estimation of synthetic pyrethroids and organochlorines and into a GC/NPD-capillary system for carbamates and organophosphates.

Estimation of pesticide residues in vegetables and fruits

In order to estimate the amount of pesticide residue in food, small slices of vegetables or fruits totalling about 2 kg are mixed thoroughly. Of this sample (post-slicing), 20 g is macerated with 4–5 g anhydrous sodium sulphate and to this 100 ml acetone is added and the mixture is mechanically shaken for 1 h. After this the extract is filtered through 2–3 cm layers of anhydrous sodium sulfate.

After adding a drop of mineral oil to this mixture, the acetone extract is concentrated to 40 ml in a rotary flash evaporator.

About 250 ml of a 10% aqueous sodium chloride solution is added to the extract and the solution is extracted three times with ethyl acetate (50, 30, 30 ml) in a separatory funnel, by shaking away the organic phase. The organic extracts are filtered through anhydrous sodium sulfate. The concentrated extract is divided into two parts for separate estimation of organochlorines and synthetic pyrethroids on the one hand, and for organophosphates and carbamates on the other.

The following steps are generally followed for organochlorine and synthetic pyrethroid estimation. A glass column with an internal diameter of 60 cm × 2.2 cm is packed with 5 g adsorbent mixture (**florisil**: activated charcoal in the ratio of 5:1 w/w) in between two layers of anhydrous sodium sulfate. The column is pre-wet with 50 ml hexane and the concentrated extract of ethyl acetate (from above) is transferred into it. The column is eluted with 125 ml solution of ethyl acetate and hexane (3:7 v/v). The eluate is concentrated to near dryness using a rotary flash evaporator after adding a drop of paraffin oil and concentrated using a gas manifold evaporator. The final volume is made up to 2 ml in ethylacetate: *n*-hexane (3:7 v/v). Analysis by gas chromatography (GC) is then carried out.

The GC parameters for estimation of organochlorines and pyrethroids are as follows: (i) GC model – HP 5890 A or equivalent; (ii) detector – electron capture detector (ECD) Ni 63; and (iii) column – capillary column, HP-5, 30 m × 0.32 mm id × 0.25 nm in film thickness of 5% diphenyl and 95% dimethyl polysiloxane.

The temperature programming is as follows: (i) for the oven – 150°C for 5 min, followed by 190°C for 2 min, followed by 280°C for 10 min; (ii) the injection port temperature – 280°C; and (iii) the detector temperature – 300°C. The nitrogen flow rate is 2 ml/min through the column, increasing to 60 ml/min with a split ratio of 1:10.

For an estimation of residues of **organophosphorus** and **carbamates**, a glass column with an internal diameter of 60 cm × 2.2 cm, uniformly packed with 5 g adsorbent mixture of silica gel (60–120 mesh) and activated charcoal (5:1 w/w) in between three to four layers of anhydrous sodium sulfate is used. The column is pre-wet with 50 ml hexane and the concentrated extract from above is loaded into the column. The column is eluted with a 125 ml mixture of acetone and hexane (3:7 v/v). A rotary flash evaporator and a gas manifold evaporator is used to dry the liquid, to make a 2 ml sample in the acetone: or-hexane ratio of 3:7 (v/v). This sample can be taken for GC analysis.

The GC parameters for estimation of organophosphate and carbamate residues are: (i) GC model – HP 5890 A or equivalent; (ii) detector – nitrogen phosphorus detector (NPD); and (iii) column – capillary column, HP-1, 10 m × 0.53 mm id × 0.65 cm film thickness of 100% diphenyl polysiloxane.

The temperature programming is as follows: (i) for the oven – 100°C for 1 min, followed by 240°C for 3 min; (ii) the injection port temperature – 250°C; and (iii) the detector temperature – 275°C. The gas flow rate of H_2 is 1.8 ml/min, O_2 is 130 ml/min and N_2 is 18 ml/min.

A stock solution of 1–2 ppm analytical grade of standard is injected and the retention time is recorded. The cleaned test sample (2–5 µl) is injected into the GC column. The residues are identified by comparing the retention times of samples with the standard and the amount of residues recorded in the integrated chart. The amount of residue in ppm is calculated using Eqn E.1.

$$\text{Residue in ppm} = \frac{A_1 \times V_1 \times C \times R_f}{A_2 \times W} \quad \textbf{(E.1)}$$

where
A_1 = Area of sample in the chromatogram
A_2 = Area of standard in the chromatogram
V_1 = Total volume of the sample in millilitres (ml)
C = Concentration of analytic standard in ppm
W = Weight of sample in grams (g)
R_f = Recovery factor

Estimation of pesticide residues in water

The liquid extraction method is used for estimation of pesticide residues in water samples, for example from drinking water, river water, pond water or rain water. Even very minute quantities (e.g. µg/l) can be detected using this method.

The water sample (500 ml) is placed in a 1 l separating funnel. About 10–15 g pure sodium chloride is dissolved in the water by gently shaking the flask and 50 ml of 15% dichloromethane in hexane is mixed into the solution by slowly shaking to release the pressure. Once the layers start to separate, the water layer is transferred to another 1 l funnel and the extraction is repeated twice, using fresh batches of 50 ml 15% dichloromethane in hexane. The combined extract (organic phase) is filtered through a glass funnel on layers of anhydrous sodium sulfate, with a small ball of glass wool or cotton at the bottom.

The extract is reduced to about 10 ml in a rotary flash evaporator at 45–50°C, after adding a drop of liquid paraffin oil. To re-dissolve the residue, 5 ml hexane is added and the evaporation process is repeated two or three times to ensure the complete removal of dichloromethane. The residue is dissolved in hexane for analysis using gas chromatography (GC). The detection limit is 5–10 µg/l.

The solid phase extraction method of the water sample is more sensitive than liquid–liquid extraction and has a detection limit of 1 ng/l or lower. Here, the water sample is passed through a solid phase extraction cartridge containing (octadecyl) C_{18} group, chemically bonded to silica.

The cartridge, made of glass/polypropylene with teflon frits, is initially washed with 10 ml eluting solvent (acetonitrite/methanol). One litre of water is passed through the cartridge at 25–30 ml/min followed by 10 ml eluting solvent (ethyl acetate). The eluate is collected while removing excess moisture (by treating with anhydrous sodium sulfate) and with the help of a manifold evaporator and dry nitrogen flow. The eluate is analysed by GC using a capillary column.

Ethanethiol

See **Butylate herbicide**

Ethephon as a PGR

Ethephon, $C_2H_6ClO_3P$ (2-chloroethyl phosphonic acid), is a plant growth regulator (**defoliants**, ethylene releasers). With a CAS Registry Number of 16672-87-0, ethephon is a colourless solid substance, readily soluble in water, methanol, ethanol, isopropanol, acetone, diethyl ether, benzene and toluene but not in kerosene and diesel. It is stable in aqueous solution having a pH of <3.5. Ethephon is sensitive to UV irradiation and is unstable above 75°C.

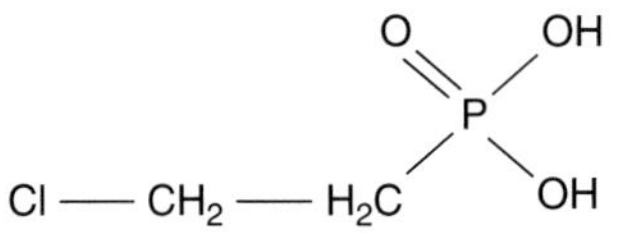

Ethephon

Ethephon has growth-promoting properties that stimulate flowering, lateral branching and production of latex. The chemical penetrates plant tissues and breaks down into carbon dioxide, phosphate and **ethylene**, the latter hastening plant growth. It has both pre-harvest and post-harvest applications. It is used on cotton, rubber, sugarcane, tobacco, cereals (wheat, rye and barley), fruits (pineapple, apples, grapes, guava, blackberries and cherries) and tomatoes. To hasten post-harvest ripening, ethephon is used on bananas and mangoes.

Ethephon is formulated as solution concentrate (SC) and emulsifiable concentrate (EC). It may also be used in combination with two **plant growth regulators, mepiquat** chloride and **chlormequat chloride**. Ethephon is incompatible with alkaline materials and solutions that contain metal ions. It is commercially available as Ethrel, Cerone, Sierra, Terpal, Bromeflor, Etheverse and others.

According to the WHO, ethephon is classified under Class U (unlikely to present acute hazard in normal use), while the EPA categorizes it as slightly toxic. The acute oral LD_{50} values of ethephon in rats, mice and rabbits are 1.6 g/kg, 2850 mg/kg and 5000 mg/kg, respectively. The acute oral LC_{50} among different bird species ranges from 596 to 3750 mg/kg. It is non-toxic to bees and earthworms.

However, since there is a risk of cholinesterase inhibition to handlers and applicators, it is considered a PAN Bad Actor chemical. Hence, special attention must be paid to symptoms of poisoning such as excessive salivation, sweating, respiratory depression, headache, dizziness, nausea and seizures.

The ADI for ethephon is 0.05 mg/kg/day. The RfD of ethephon is 0.005 mg/kg/day.

In animals, ethephon is rapidly excreted via urine and as ethylene via the exhaled air. In plants too, ethephon and ethylene are the major metabolites. In soils, ethephon degrades to phosphoric acid, ethylene and chloride ions, and the unchanged ethephon is not very mobile and so does not threaten groundwater. The monochloroethyl ester of (2-chloroethyl)-phosphonic acid, a common impurity in ethephon, degrades into monochloroacetic acid.

Ethephon is registered for use in India, Australia, New Zealand, the Philippines, Vietnam, many EU nations such as the UK, Denmark, Germany, South Africa and some African nations, Canada and the USA.

Ethiofencarb insecticide

Ethiofencarb, $C_{11}H_{15}NO_2S$ (α-ethylthio *o*-toyl methyl-carbamate), is an *N*-methyl carbamate insecticide. This chemical has a CAS Registry Number of 29973-13-5. It occurs as colourless crystals, moderately soluble in water, fairly soluble in hexane, dichloromethane, isopropanol and toluene. It hydrolyses in alkaline solution and photodegrades in its aqueous form. Ethiofencarb is prepared by the reaction of 2-chloromethyl phenol with sodium ethylmercaptide to form 2-ethylthiomethylphenol, which in turn is reacted with methylisocyanate.

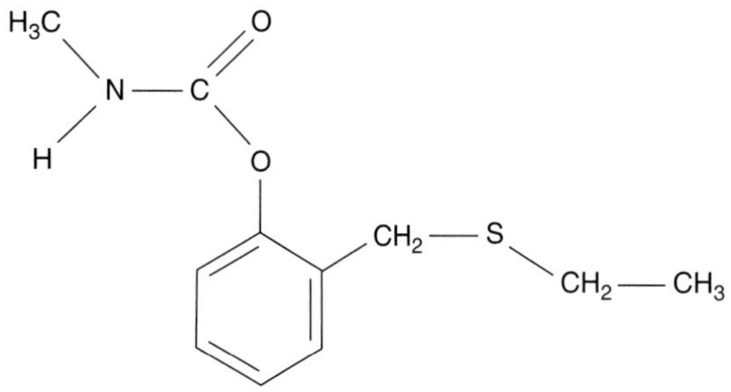

Ethiofencarb

Ethiofencarb is a systemic insecticide having cholinesterase inhibition activity. It shows contact and stomach action. It is used in the control of aphids on fruit (pome and stone), vegetables and sugarbeet.

Formulated as granules (GR) and emulsifiable concentrates (EC), ethiofencarb is available in the market under the brand name of Croneton.

The WHO classifies ethiofencarb under toxicity Class Ib (highly hazardous). It is toxic to bees. Ethiofencarb may be phytotoxic to some ornamental plants. PAN lists ethiofencarb as a Bad Actor chemical, while the EU's Nordic Council of Ministers lists ethiofencarb as 'dangerous for the environment'.

Exposure to ethiofencarb may lead to any of many symptoms such as muscle weakness, muscle twitching and slurred speech, seizures, miosis with blurred vision, nausea, abdominal pain and nervous system depression.

In mammals, ethiofencarb metabolizes into ethiofencarb sulfoxide and sulfone, ethiofencarb phenol and the corresponding sulfoxide and sulfone. In plants, ethiofencarb sulfoxide and sulfone, ethiofencarb phenol-sulfoxide and the conjugates are formed. While ethiofencarb has a relatively high mobility in soil and water, it is very rapidly degraded to its sulfoxide and sulfone and hydrolysed to the corresponding phenolic metabolites.

The EU has cancelled the registration of ethiofencarb as an insecticide, while India has registered its use.

Ethion insecticide

Ethion, $C_9H_{22}O_4P_2S_4$ (*O,O,O′,O′*-tetraethyl *S,S′*-methylene bis(phosphorodithioate)), is an organophosphorus insecticide. With a CAS Registry Number of 563-12-2, ethion is a yellow-amber liquid, soluble in water (2 mg/l) but miscible with acetone, ethanol, xylene, kerosene and petroleum oils. It is hydrolysed by aqueous acids and alkalis and is oxidized by air. Photo-oxidation may occur in sunlight. Ethion is synthesized by reacting diethyl hydrogen phosphorodithioate (2 moles) with bromochloromethane or dibromomethane (1 mole) (see Ethion at top of the next page first column).

Ethion, a non-systemic insecticide, has additional acaricidal and ectoparasiticidal properties. The chemical acts as a cholinesterase inhibitor. Aphids, spider mites, scales, thrips, sucking insects and lepidopteran larvae are effectively controlled at

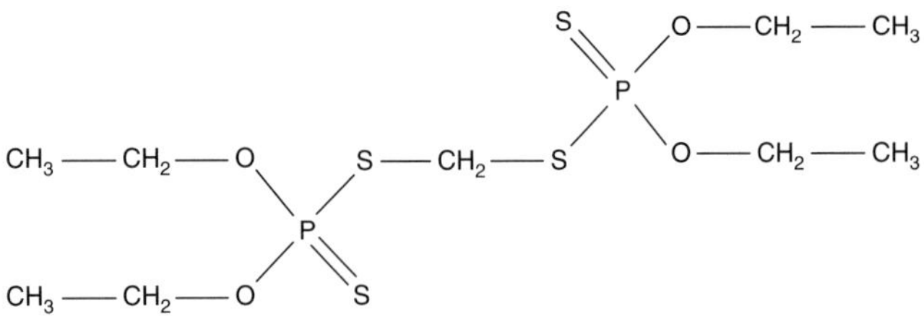

Ethion

different life stages. As an ectoparasitic insecticide, it protects cattle from ticks and other insects. The crops that are protected are citrus fruits, apples, onions, cotton and cereals.

Ethion is commonly available as an emulsifiable concentrate (EC) or dustable powder (DP). A spray application of ethion mixed with oil is ideal for trees. It may be used with a range of pesticides, except those that are alkaline in nature. High temperatures and UV exposure can make ethion more toxic. It is available under trade names like Bladan, Drithion, Embathion, Niagara 1240 and many more.

The WHO classifies ethion as a Class II toxin (moderately toxic), with the acute oral LD_{50} among rats being 208 mg/kg, and among mice and guinea pigs being 40 mg/kg. The chemical is also moderately toxic via inhalation, with 4 h LC_{50} in rats being 0.864 mg/l. It is moderately to highly toxic to fish and other aquatic organisms (the 96 h LC_{50} ranges from 0.210 to 0.72 mg/l among the species tested). Ethion is highly toxic to some bird species and extremely toxic to honey bees (LD_{50} is 20.55 μg per bee). On contact with skin, it can cause burns. Since this chemical scores high on acute toxicity and is a cholinesterase inhibitor, ethion is a PAN Bad Actor chemical. Thus it is mandatory for ethion products to have the signal word 'WARNING' on the label. The RfD of ethion is 0.0005 mg/kg/day.

Symptoms of poisoning from ethion may include fatigue, nausea, blurred vision, cramps, diarrhoea, excessive salivation, headache, tightness in chest and convulsions.

Upon ingestion by animals, the P-S and C-S linkages of ethion are cleaved to give *O,O*-diethyl hydrogen phosphorothioate and *O,O*-diethyl phosphorodithioate, respectively. Mono- and dioxon products also occur. In soil, the half-life can be up to 2 months and almost 1 year in greenhouse conditions, showing its high persistence in soil. Since it is almost insoluble in water but adsorbs strongly to soil particles, ethion does not leach, but is said to accumulate in plant and animal tissues. Ethion is degraded by microbial action. Its hydrolysis half-life at 25°C are 63, 58, 25 and 8.4 weeks at pH 5, 6, 7 and 8, respectively.

Ethion is banned in numerous countries including the USA, the EU and New Zealand, but is registered in India, Australia and Tanzania.

Ethirimol fungicide

Ethirimol, $C_{11}H_{19}N_3O$ (5-butyl-2-ethylamino-6-methylpyrimidin-4-ol), is a pyrimidine fungicide. Ethirimol has a CAS Registry Number of 23947-60-6. It is a colourless crystalline substance, soluble in water and chloroform, ethanol and acetone. It is stable to heat, and in acidic and alkaline solutions in the dark. Ethirimol is produced by reaction of ethyl 2-butylacetoacetate and *N*-ethylguanidine.

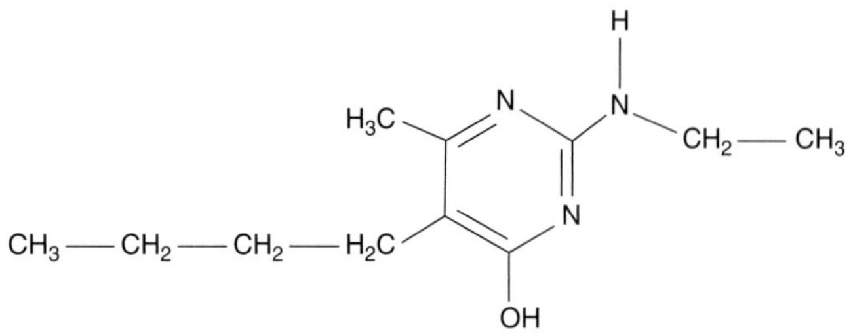

Ethirimol

Ethirimol is a systemic fungicide (via both the roots and the stems) but translocates only through the xylem. With both protective and curative action, it is used for controlling powdery mildew, mainly on cereal crops like barley, wheat and oats.

Ethirimol is available as suspension concentrate (SC), emulsifiable concentrate (EC) and as a liquid seed treatment agent, and, as such, has varied applications (i.e. on seed, foliage and into the soil). It is available under the trade name Mligo, among others.

Ethirimol has an acute LD_{50} in rats of 6340 mg/kg and mice 4000 mg/kg. The acute percutaneous LD_{50} in rats is >200 mg/kg. It is a mild eye irritant. The acute LD_{50} among partridge species is higher than 3000 mg/kg, while the acute 96 h LC_{50} in rainbow trout is 60.8 mg/l. It is very harmful to honey bees (acute 48 h LD_{50} being 1.6 μg per bee). According to the WHO, ethirimol is a Class U toxin (unlikely to be hazardous). Its ADI is 0.05 mg/kg b.w./day.

Animals eliminate ethirimol via the urine, with the metabolism involving hydroxylation of the butyl group. In plants, ethirimol undergoes rapid degradation to produce metabolite 2-amino-5-butyl-6-hydroxy-4-methylprimidine.

This compound is banned throughout the EU and in the USA.

Ethofumesate herbicide

Ethofumesate, $C_{13}H_{18}O_5S$ ((*RS*)-2-ethoxy-2,3-dihydro-3,3-dimethylbenzofuran-5-ylmethanesulfonate), is a benzofuranyl alkylsulfonate herbicide. It is a white suspension, soluble in water at 50 ml/l. It is stable in normal storage conditions but incompatible with acids and oxides. The CAS Registry Number is 26225-79-6.

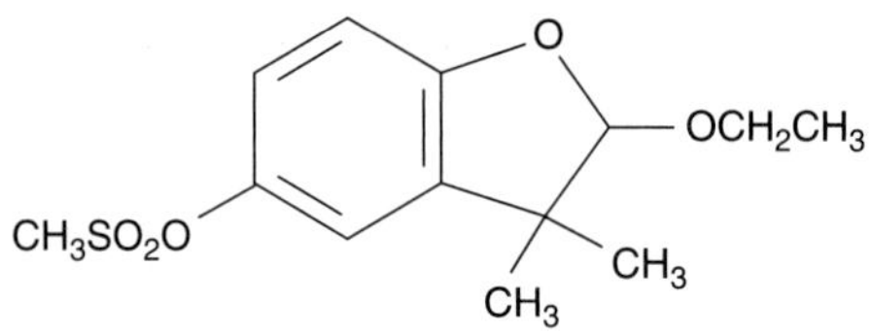

Ethofumesate

Ethofumesate is a selective, pre-and post-emergence herbicide that acts by inhibiting mitosis, photosynthesis and respiration in broadleaved and grass weeds like meadow grass and black grass in crops of carrots, spinach and sugarbeet. The herbicide also helps to control weeds during seed production of sod, turf and grass.

Flowable concentrate (SC) and emulsifiable concentrates (EC) are common formulations of this herbicide. The acceptable dosage is 2 l/ha, with a cap of 1 kg of ethofumesate per

3 year period. It is commonly incorporated into the soil at the pre-plant or pre-emergence stage. If application is necessary at the post-emergence stage, spraying must be done in moist conditions, before weeds are tillered. Grazing should not be allowed up to a fortnight after application.

The WHO has classified this herbicide as unlikely to be hazardous, although it is toxic to fish (the LC_{50} being 38.50 mg/l), algae and *Daphnia*. It is not harmful to honey bees and earthworms, but is a potential groundwater contaminant. It may be phytotoxic to lettuce and tomatoes. Herbicide handling and disposal must be such that it does not drain into water sources.

Anaerobic degradation of ethofumesate in soil takes a long time (half-life >90 days), with microbial action aiding the process. However, it photodegrades on the soil surface. The hydrolysis DT_{50} is 2900 days.

Ethofumesate is registered for use in the USA, Australia, Hungary, Germany, France, Italy, Belgium, the UK and New Zealand.

Ethoprophos insecticide

Ethoprophos, $C_8H_{19}O_2PS_2$ (*O*-ethyl *S*,*S*-dipropyl phosphorodithioate), is an organothiophosphate insecticide and an organothiophosphate nematicide. This pale yellow liquid chemical has a CAS Registry Number of 13194-48-4. It is soluble in water (700 mg/l), acetone, ethanol, xylene and 1-2 dichloroethane. It is rapidly hydrolysed in alkaline to weakly acidic media but is stable in water up to 100°C.

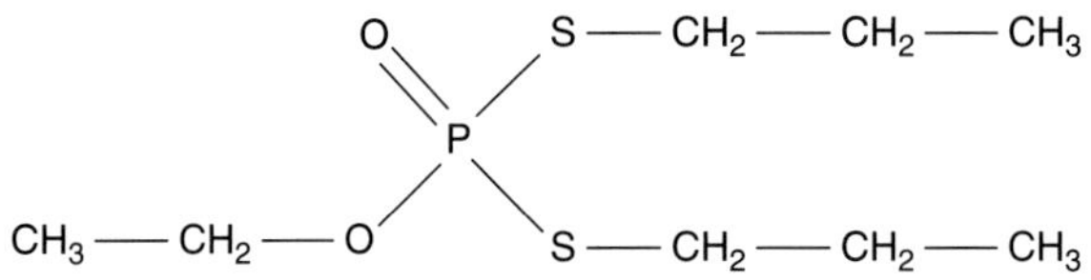

Ethoprophos

Ethoprophos, a broad-spectrum, pre-plant agent, is a non-systemic insecticide and nematicide. It is absorbed into the roots and has negligible translocation. It inhibits cholinesterase, leading to acetylcholine being accumulated at the nerve synapses followed by death of target pests. It controls plant-parasitic nematodes and soil insects, examples being rootworms, wire worms and soil arthropods. Ethoprophos is used on a wide range of crops including potatoes, tomatoes, groundnuts, sugarcane, tobacco, cereals, turf grass and fruits (e.g. citrus, bananas).

This insecticide is formulated as granules (GR) or an emulsifiable concentrate (EC). It is not to be used with alkaline pesticides. It can be applied up to a dose of 12 kg/ha. It can be broadcast on the soil at the pre-plant stage by equipment that would be appropriate to prevent inhaling the product. Ethoprophos is also incorporated up to a depth of 15 cm into the ground. It is available under trade names like Mocap and Sanimul.

The acute oral LD_{50} values of ethoprophos in rats, mice and rabbits range from 31 to 62 mg/kg b.w., making it a very toxic chemical. It is also very toxic via the dermal route for mice and rabbits (the dermal LD_{50} in rabbits and mice is 26 mg/kg b.w. and 18 mg/kg b.w., respectively). The inhalation route also causes toxicity. As per the Pesticide Properties DataBase (PPDB) developed by the Agriculture and Environment Research Unit (AERU) at the University of Hertfordshire, UK, ethoprophos is moderately toxic to fish, aquatic invertebrates, honey bees and earthworms, while being highly toxic to birds, mammals and crustaceans (although it is believed that the chemical is not dangerous to bees if used as directed). The WHO classifies ethoprophos as a Class Ia toxin (extremely hazardous).

Symptoms of poisoning commonly include nausea, vomiting, excessive sweating and salivation, slurred speech and muscle twitching and may include convulsions and coma. Precautions necessary during handling, mixing and application are the wearing of protective clothes, impermeable boots and gloves, a hat and a respirator. Disposal of ethoprophos and its containers should be carried out as per label instructions and local regulations.

The chemical should be transported separately in sealed containers, away from food, drink and unprotected persons. Accidental skin contact should be immediately treated with thorough washing with soap and water. In case of the chemical having entered the eyes, washing with water or isotonic saline water is the ideal first-aid measure. In case ethoprophos is accidentally consumed, gastric lavage is recommended, before formal medical treatment can be made available.

Mammals absorb and eliminate ethoprophos soon after consumption, hydrolysing to produce the metabolite *O*-ethyl-*S*-propylphosphothioic acid. Plants also rapidly break down ethoprophos into non-toxic metabolites, methyl propyl sulfide and methyl propyl sulfone. The DT_{50} in organic soil is 87 days at pH 4.5. Residues are not observed on treated foliage.

Korea, South Africa, the USA, Canada and some EU members such as Belgium, France and the UK have registered the use of ethoprophos. Germany, Denmark and Hungary are some countries that have disallowed its use.

Ethoxysulfuron herbicide

Ethoxysulfuron, $C_{15}H_{18}N_4O_7S$ (1-(4,6-dimethoxypyrimidin-2-yl)-3-(2-ethoxyphenoxysulfonyl) urea), is a member of the pyrimidinylsulfonylurea class of herbicides. It is a whitish beige powder that hydrolyses in slightly acidic conditions. The CAS Registry Number is 126801-58-9.

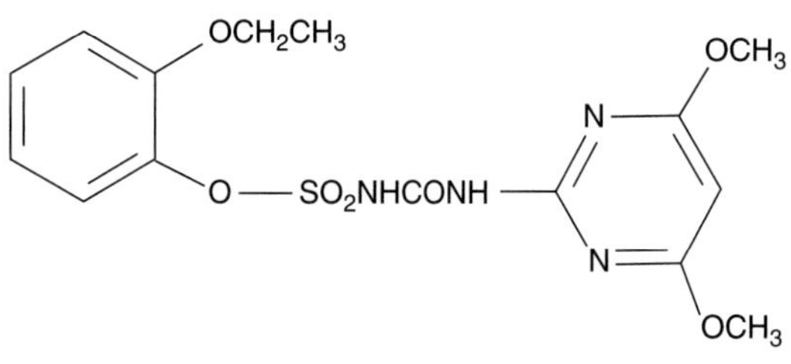

Ethoxysulfuron

The post-emergence herbicide is a branched-chain amino acid synthesis inhibitor (an **ALS inhibitor**) that halts the production of valine, leucine and isoleucine, hence stopping cell division and plant growth. It specifically inhibits the enzyme acetohydroxy acid. It is used on grains and sugarcane.

Ethoxysulfuron has an acute oral LD_{50} in rats of >3270 mg/kg and an acute percutaneous LD_{50} of >4000 mg/kg.

This herbicide is not persistent in aerobic soil, with the DT_{50} being 20 days or fewer. However, aqueous hydrolysis extends to 250 days, making it persistent in water.

The herbicide ethoxysulfuron is registered in India, Australia and Italy, but not in the USA, the UK and many other EU countries. The generally accepted MRL value for most citrus/pome/stone fruits is 0.05 mg/kg.

Ethychlozate as a PGR

Ethychlozate, $C_{11}H_{11}ClN_2O_2$ (ethyl 5-chloro-3(1*H*)-indazolylacetate), is a plant growth regulator of the unclassified class. It is a yellow, crystalline substance, with a CAS Registry Number of 27512-72-7. It is soluble in water, acetone, ethyl acetate, ethyl alcohol, hexane, kerosene, methanol and isopropanol. It decomposes above 250°C.

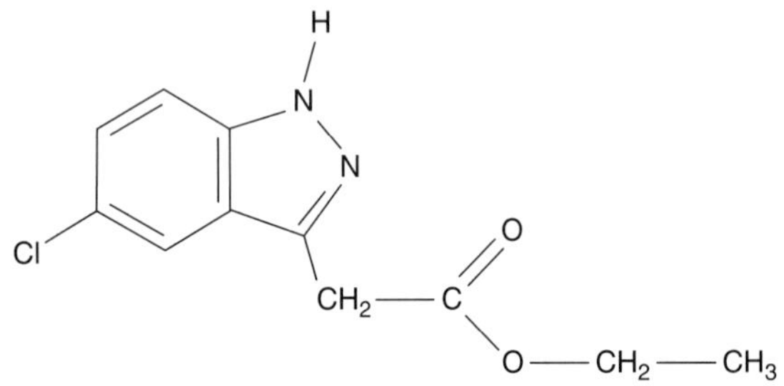

Ethychlozate

Ethychlozate, with an auxin-like activity, stimulates the production of ethylene that enhances growth and the abscission layer in fruits. This chemical is commonly used for improving fruit colour, quality and size, especially of citrus fruit.

This PGR is formulated as an emulsifiable concentrate (EC). A common trade name is Figaron. Ethychlozate is rapidly excreted via urine of animals within 24 h. It decomposes in the soil within 1 week.

Ethylene as a PGR

Ethylene, C_2H_4 (ethene), is a colourless, flammable gaseous hydrocarbon of the alkene series occurring in petroleum and natural gas. Ethylene is also a plant growth regulating hormone. It is used for fruit-ripening in bananas, although excess use can result in premature opening of buds, senescence and abscission.

Ethylene can be produced by heating higher hydrocarbons. However, at higher temperatures and pressure, ethylene is converted into polyethylene (polymerization). Ethylene reacts with many chemicals to produce ethanol, gasoline, antifreezes and plastics. Ethylene manufacturing plants pollute the surroundings.

Tomato plants are excellent biological indicators of ethylene, with symptoms such as epinasty of the petiole, foliar **chlorosis** and formation of adventitious roots on the stem above the soil layer. Ethylene can cause developmental abnormalities in chronic use.

This PGR is registered in the USA, Canada and Australia.

Ethylene dibromide insecticide

Ethylene dibromide, $C_2H_4Br_2$ (1,2-dibromoethane), is a halogenated organic compound that is a fumigant insecticide as well as a nematicide. It is a colourless, chloroform-odoured liquid, soluble in water (4.3 g/kg) and quite soluble in diethyl ether, ethanol and many others. It is decomposed by alkaline materials and light. Ethylene dibromide has a CAS Registry Number of 106-93-4. It is made by brominating ethylene.

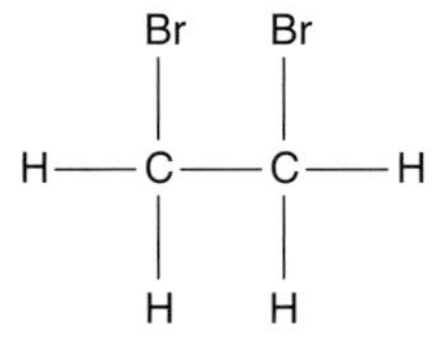

Ethylene dibromide

Ethylene dibromide, as a soil sterilant, fumigant and a nematicide, effectively controls soil insects such as wire worms, beetles, termites and stored grain insects (when used as a fumigant). With such properties, ethylene dibromide protects cereals, citrus and turf. It is additionally used to treat wood and protect against wax moths in beehives.

Ethylene dibromide is formulated as oil. Compositions constitute a mixture of ethylene dibromide with **methyl bromide** and **chloropicrin** in different proportions. It is incompatible with strong oxidizers, magnesium, alkali metals and liquid ammonia. Trade names include Bromofume, Soilfume and Celmide.

The acute oral LD_{50} in rats ranges from 146 to 420 mg/kg, while the oral LD_{50} in rabbits is 55 mg/kg. It is absorbed into the skin and causes skin and eyes to burn. It is harmful to inhale the fumigated chemical. It is also phytotoxic.

Poisoning can take place either by consumption, dermal absorption or inhalation and symptoms of poisoning from ethylene dibromide include vomiting, diarrhoea, drowsiness, CNS depression, a weak pulse, liver and kidney failure and even death (blisters, burning and redness are common on the skin). Hence, much care has to be taken by applicators using this pesticide, for example wearing fully-encapsulating, vapour-protective clothing. Eating, drinking and smoking should not take place in the vicinity of the chemical. The chemical should be stored in cool, well-ventilated, sealed containers. It should not be brought in contact with sodium, potassium, calcium, zinc and magnesium, liquid ammonia and strong oxidizers.

If exposed, the person concerned should remove contaminated clothes and wash the affected part with soap and water immediately. The EPA has calculated a provisional RfD of 0.0002 mg/m^3 for ethylene dibromide based on reproductive effects in humans. Since ethylene dibromide is an acute toxic, a carcinogen, a suspected endocrine disruptor and a reproductive toxic, it is declared a PAN Bad Actor chemical and is on the PAN **Dirty Dozen** list.

If ingested, ethylene dibromide is eliminated via the urine and liver within 48 h. Metabolism is by conjugation with glutathione, with inorganic bromide being formed. In the air, ethylene dibromide has a half-life of 32 days and will degrade by reacting with hydroxyl radicals. The half-life of ethylene dibromide in water is 1–5 days. It is found to contaminate groundwater sources. In soil, the chemical degrades by the action of microorganisms.

Ethylene dibromide is not registered for use in the USA. India, South Africa and Tanzania have registered its use. In India, a mixture of ethylene dibromide and carbon tetrachloride in the proportion of 3:1 is registered and is called EDCT.

Ethylenebisdithiocarbamate (EBDC) fungicides

The ethylenebisdithiocarbamate (EBDC) group of sulfur fungicides includes nabam, zineb, maneb and vapam. EBDCs are commonly used for treating **early** and **late blights** of potatoes and tomatoes. Many fungal diseases of vegetables, fruits and commercial sod are also controlled by this group of chemicals.

Although care and precaution is highly recommended while applying and handling these chemicals, they are considered safe if used as instructed, because of their low residual levels.

Nabam, $C_4H_6N_2Na_2S_4$ (disodium ethylenebisdithiocarbamate), is a generic name for sodium ethylenebisdithiocarbamate. It is used mainly as an algicide and a fungicide. It also has slimicidal and bactericidal effects.

Zineb, $C_4H_6N_2S_4Zn$ (polymeric), also called zinc ethylenebisdithiocarbamate, is used as a fungicide and a nematicide. It is commonly used for controlling the blight of potatoes and damping-off disease. *Meloidogyne* spp. is a typical nematode that is controlled by zineb. Use of zineb thus protects a range of fruit and vegetable crops from soil and seed pests.

Maneb, $C_4H_6MnN_2S_4$ (manganese ethylenebisdithiocarbamate) (polymeric), is an organic sulfur fungicide. When sprayed, maneb controls blights of potato and tomato as well as many diseases in fruits, vegetables, cereals and ornamentals.

Vapam, $C_2H_4NNaS_2$ or **metam-sodium** is **sodium N-methyl-dithiocarbamate**. This chemical is a fungicide with additional nematicidal and herbicidal properties. It has fumigant properties, all of which help to control soil insects, nematodes (e.g. root-knot nematodes) and weed seeds. Also fungal diseases such as cotton wilt, damping-off and root rot are controlled by use of this chemical.

Metiram, $C_{12}H_{12}N_6S_{16}Zn$ (zinc ammoniate ethylenebisdithiocarbamate–poly(ethylenethiuram disulfide)), is another ethylenebisdithiocarbamate chemical. This is a light yellow solid that acts as a contact poison, it being moderately toxic in acute toxicity tests. It is a formulation containing zinc ethylenebisdithiocarbamate and ethylene bisthiuram polysulfide. Its LD_{50} in rats is 6100 mg/kg and the tolerance limit in food products of vegetable origin is 1 mg/kg.

Mancozeb or manzeb and cuprosan are two more important ethylenebisdithiocarbamate chemicals that are valued as fungicidal agents.

Ethylenethiourea (ETU)

See **Mancozeb fungicide**

Etobenzanid herbicide

Etobenzanid, $C_{16}H_{15}Cl_2NO_3$ (2′,3′-dichloro-4-ethoxymethoxybenzanilide), is an anilide herbicide. It is a crystalline, odourless substance that is slightly soluble in acetone, acetonitriles and xylene. Its solubility in water is 0.92 ppm. The CAS Registry Number is 79540-50-4.

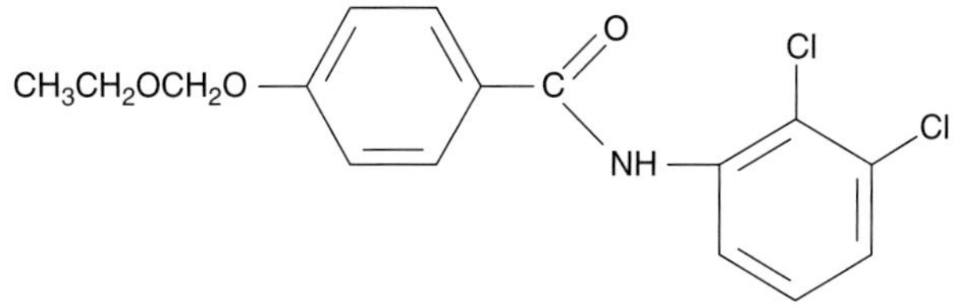

Etobenzanid

Etobenzanid is a rice herbicide, used for both pre- and post-emergence paddy weeds. The compound is used mainly as a wettable powder (WP) or as granules (GR).

Besides being slightly irritating to the skin and eyes, etobenzanid is practically non-toxic to rats (LD_{50} >5000 mg), although it can be fairly toxic to some species of birds and fish. Honey bees are safe to the exposure of this compound.

The MRL of etobenzanid, fixed by Japan for rice, is 0.1 ppm.

Etofenprox insecticide

Etofenprox, $C_{25}H_{28}O_3$ (2-(4-ethoxyphenyl)-2-methyl propyl-3-phenoxy benzyl ether), is a pyrethroid ether insecticide with acaricidal properties. Etofenprox has a CAS Registry Number of 80844-07-1. It is a colourless crystalline substance, almost insoluble in water but soluble in acetone, chloroform, ethyl acetate, xylene and methanol. It is stable in acidic medium at 80°C and to light. Etofenprox is produced by the esterification of the *p*-ethoxy-*tert*-butyl phenylalcohol with (3-phenoxy) benzyl chloride.

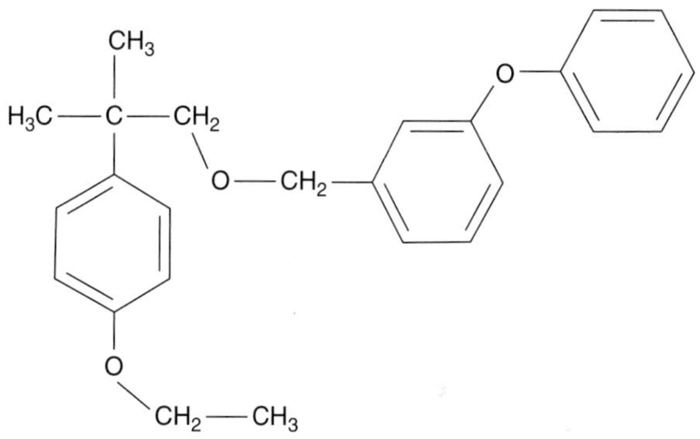

Etofenprox

As a contact insecticide, etofenprox targets many types of insects belonging to the insect orders Lepidoptera, Homoptera, Coleoptera, Diptera and **Thysanoptera**. It is used to control rice water **weevils**, leaf beetles, **leafhoppers**, plant hoppers, **aphids**, moths, **whitefly**, leaf miners and leaf rollers, **thrips** and borers that are seen attacking crops such as rice, fruit (pome/stone/citrus), tea, soybean, sugarbeet and vegetables (e.g. brassicas and cucumbers). Etofenprox is also used to control pests on livestock (e.g. ticks, fleas and earwigs) and household pests (e.g. cockroaches, silverfish and spiders).

Etofenprox is variously formulated as emulsifiable concentrate (EC), wettable powder (WP) and dustable powder (DP), oil in water emulsion (EW), capsule suspension (CS), soluble concentrate (SL) and ULV liquid (UL). A maximum of three applications per season, each application being of 0.15–0.4 kg/ha, is common on rice. Etofenprox proves effective on some rice pests that have developed resistance to **organophosphorus insecticides**. Etofenprox is compatible with many other insecticides and fungicides. A common trade name is Trebon.

The acute oral LD_{50} tests on mammals and birds show that etofenprox does not pose a threat to them in normal use (the acute oral LD_{50} in rats is >42,880 mg/kg, for male mice >107,200 mg/kg, dogs >5000 mg/kg and in mallard ducks >2000 mg/kg.) It is non-irritating to the skin and eyes in rabbits in normal use, although applicators and handlers may experience some mild skin irritation. Etofenprox is toxic to fish and earthworms. The ADI is 0.03 mg/kg/day. According to the WHO and EPA, etofenprox is a Class III and a Class IV toxin, respectively (slightly hazardous).

Etofenprox is mainly metabolized via a range of reactions including desethylation, hydroxylation and oxidation, leading to formation of polar compounds. In animals, etofenprox metabolizes into 2-(4-hydroxy phenyl)-2-methyl propyl 3-(phenoxy benzyl) ether and other products. In plants, the main metabolite of etofenprox is 2-(4-ethoxy phenyl)-2-methyl propyl 3-phenoxy benzoate. The DT_{50} of etofenprox in soil is 6 days.

Etofenprox is registered in Cambodia, Chad, Canada, Gambia, Germany, Hungary, India, the USA, Vietnam and many other countries.

Etrimfos insecticide

Etrimfos, $C_{10}H_{17}N_2O_4PS$ (*O*-6-ethoxy-2-ethylpyrimidin-4-yl *O*,*O*-dimethyl phosphorothioate), is a pyrimidine organothiophosphate insecticide. Etrimfos has a CAS Registry Number of 38260-54-7. It is a colourless liquid, soluble in water (40 mg/l) and miscible with acetone, ethanol, diethyl ether, chloroform, dimethylsulfoxide, ethyl-acetate, hexane and kerosene. It is hydrolysed in aqueous media, with the DT_{50} at pH 9 being 14 days.

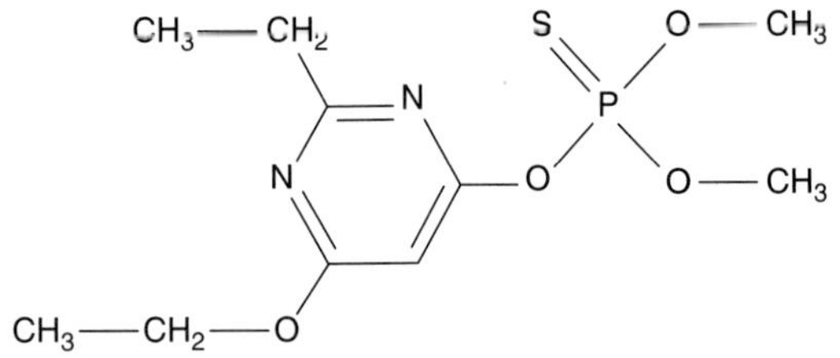

Etrimfos

Etrimfos is a non-systemic insecticide and acaricide. The mode of action is the inhibition of cholinesterase. It effectively controls members of the Lepidoptera and Coleoptera. Mites are also kept in check. Thus a number of crops and stored grains are protected.

This insecticide is formulated as dustable powder (DP), granules (GR), emulsifiable concentrate (EC) and ULV liquid (UL). Etrimfos may be applied as a dust or spray for seed treatment, at up to 15 mg/kg, depending on the severity of the infestation and storage needs. It is available in the market as Ekamet and Satisfar, among others.

The acute oral LD_{50} of etrimfos in mice is 470 mg/kg, while in male and female rats it is 1800 and 2354 mg/kg, respectively. The WHO classifies etrimphos as a Class II toxin (moderately hazardous). Etrimfos is toxic to bees. Cholinesterase inhibition and moderately acute toxicity have resulted in etrimfos being classified as a PAN Bad Actor chemical.

Excessive salivation, headaches, sweating, vomiting, respiratory depression and seizures may occur in cases of poisoning. The ADI in man is 0–0.003 mg/kg b.w./day.

The principal degradation pathway in mammals, plants and soils is via the cleavage of the pyrimidinyl ester bond, the major metabolite being 6-ethoxy-4-hydroxy-2-ethylpyrimidine. In mammals, etrimfos degrades to demethyl etrimfos. The DT_{50} in soil at pH 6.8 is under 10 days. It hydrolyses in soil to 6-ethoxy-2-ethyl-4-hydroxy-pyrimidine (EEHP). The MRL of etrimfos in bran of wheat (unprocessed) is 20 mg/kg, while for barley, maize, wheat and rapeseed it is 10 mg/kg.

Eukaryotic organisms

Any organism whose cells contain structures bound by membranes is a eukaryote. The main difference between eukaryotes and prokaryotes is the presence, in eukaryotes, of a nucleus which contains the genetic material. The main constituents of a eukaryotic cell are nucleus, mitochondria, Golgi, ribosomes and plasma.

Eukaryotes belong to the taxa Eukaryota and constitute a very wide domain of living organisms including unicellular organisms, fungi, plants and animals, although there are very clear distinctions between the cells of various eukaryotes.

Fagarol

See **Insect growth regulators (IGRs)**

False seedbed

False seedbed is another term for **stale seedbed**. This is an organic weed control technique. A seedbed is created before seeds are sown. Weed seeds embedded in the soil germinate in response to the moisture added during seedbed preparation. After the weed is established, it can be destroyed by chemical or non-chemical means. The crop seeds can then be sown.

Famoxadone fungicide

Famoxadone, $C_{22}H_{18}N_2O_4$ (3-anilino-5-methyl-5-(4-phenoxyphenyl)-1,3-oxazolidine-2,4-dione), is an oxazolidinedione fungicide. A creamish powder, famoxadone is soluble in water (52 μg/l), acetone, toluene, dichloromethane and hexane. It is unstable in neutral to basic conditions. It is stable in cool, dark conditions and in acidic aqueous solution. It is produced by the reaction of 2-(4-phenoxyphenyl) lactate with a mixture of phenoxyphenyl and 1,1′-carbonyldiimidazole in dichloromethane, followed by treatment with phenylhydrazine. The CAS Registry Number is 131807-57-3.

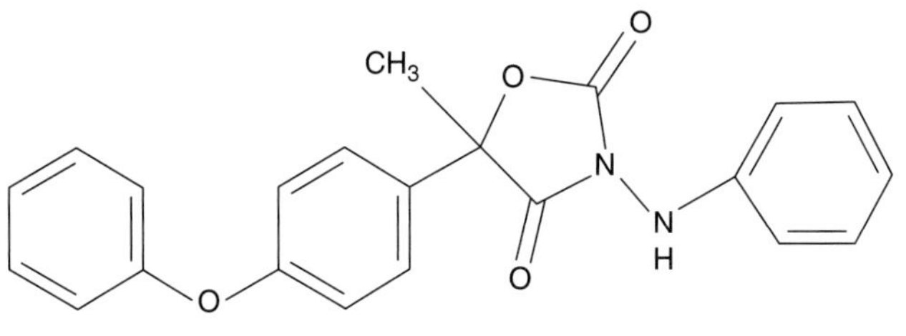

Famoxadone

Famoxadone, a quinone outside inhibitor (QoI), is highly active against spore germination and mycelial growth, by inhibiting mitochondrial electron transport by blocking ubiquinol, thus decreasing production of ATP by the fungal cell. It has protectant and residual properties. It is effective against **downy mildew** (*Plasmopara viticola*), **late** and **early blight** (*Phytophthora infestans* and *Alternaria solani*) and the *Septoria* complex. Target crops include vines, potatoes, tomatoes, cereals, sugarbeet, canola and cucumber.

The fungicide may be mixed with triazoles. The chemical is safe on crops, while being rainfast. Commonly, this compound is applied at the rate of 50–200 g/ha.

Famoxadone is a Class U toxin (unlikely to be hazardous) according to the WHO. It is also not a skin or eye irritant in formal use. An ADI for humans is 0.006 mg/kg b.w./day.

While famoxadone degrades rapidly (within 2 h) in alkaline conditions, the half-life is >1 month in acidic conditions and dark surroundings. It is non-mobile and is said to have a good ecotoxicological profile.

Fungicide famoxadone is registered in India, the USA, Canada and many EU members.

Famphur insecticide

Famphur, $C_{10}H_{16}NO_5PS_2$ (*O*-4-dimethylsulfamoylphenyl *O*,*O*-dimethyl phosphorothioate), is an organophosphorus insecticide. It is a colourless crystalline powder, sparingly soluble in water and aliphatic hydrocarbons and soluble in isopropanol, xylene, acetone, carbon tetrachloride, chloroform and toluene. It is stable at ambient temperature. Famphur is obtained by reaction of 4-hydroxy-*N*,*N*-dimethylbenzenesulfonamide with dimethylthionophosphoric acid chloride. The CAS Registry Number of famphur is 52-85-7.

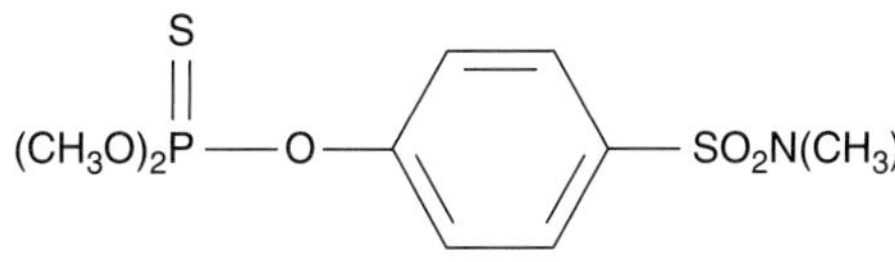

Famphur

Famphur, a cholinesterase inhibitor, is a veterinary ectoparasiticide. It controls grubs, nose flies, lice and other pests that attack farm animals. Specific examples of pests include *Haematabia irritans*, *Hypoderma* spp., *Bovicola bovis* and *Linognathus vituli*. The insecticide is formulated as a pour-on and dustable powder (DP). Trade names include Bo-Ana and Warbexol.

The compound is considered toxic with chronic effects. It is toxic to humans and is reported to be a carcinogen and a developmental toxic. Its solubility also makes it a possible groundwater contaminant.

The principal degradation routes in mammals are P-O-phenyl bond cleavage and *O*-demethylation. Oxidative desulfuration and *N*-demethylation also form toxic metabolites.

Fatty acid

See **Oleic acid herbicide**

Fatty acid elongation inhibitors

See **Herbicide action – lipid synthesis inhibitors**

Federal Insecticide, Fungicide, and Rodenticide Act (FIFRA)

The Federal Insecticide, Fungicide, and Rodenticide Act (FIFRA) provides for federal regulation of pesticide distribution, sale and use in the USA. All pesticide active ingredients (a.i.), distributed or sold in the USA, must be registered (licensed) by the EPA. A pesticide may be registered under FIFRA prior to the EPA. An applicant has to prove, apart from specific benefits of the a.i., that it 'will not generally cause unreasonable adverse effects on the environment'.

Here, 'unreasonable adverse effects on the environment' mean: (i) any unreasonable risk to man or the environment, taking into account the economic, social and environmental costs and benefits of the use of any pesticide; and (ii) a human dietary risk from residues that result from the use of a pesticide in or on any food, inconsistent with the standard under section 408 of the Federal Food, Drug and Cosmetic Act. More information may be accessed from http://www.epa.gov/agriculture/lfra.html and related websites.

Fenamidone fungicide

Fenamidone, $C_{17}H_{17}N_3OS$ ((*S*)-1-anilino-4-methyl-2-methylthio-4-phenylimidazolin-5-one), is an imidazole fungicide. It is a white powder, soluble in water (0.0078 g/l), acetonitrile (86.1 g/l), acetone, dichloromethane (330 g/l) and methanol. The CAS Registry Number is 161326-34-7.

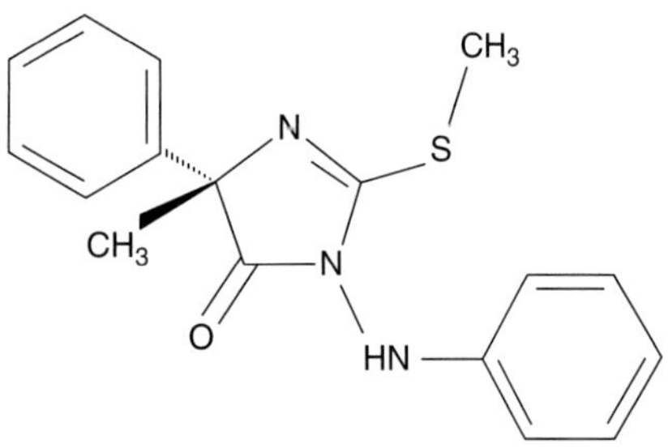

Fenamidone

The fungicide fenamidone, a protectant and curative compound with a foliar application, inhibits mitochondrial respiration by blocking electron transport at a level of the enzyme ubihydroquinone: cytochrome c oxidoreductase (complex III). It controls infections caused by *Alternaria*, *Phytophthora*, and *Peronospora* on citrus fruits, potatoes and vegetables, and by *Mycosphaerella* on bananas.

To combat issues of resistance, it is common for fenamidone to be combined with fungicidal compounds that have multiple modes of action, such as **fosetyl-aluminium** or **chlorothalonil**. It is formulated as a suspension concentration (SC) or as water-dispersible granules (WG), in doses of 500 g/ha. It is commercially available as Reason and Censor.

The fungicide in question is low on acute toxicity (with an acute oral LD_{50} of >5000 mg/kg and an acute dermal LD_{50} >2000 mg/kg in rats). It is practically non-toxic to birds and does not irritate the skin or the eyes. The ADI of fenamidone is 0.03 mg/kg b.w./day.

Fenamidone is readily photodegraded under aqueous conditions. One of the only significant metabolites is RPA 405862, formed by hydrolysis of the thiomethyl side-chain.

Fenamidone is registered in India, the USA, the UK, Canada, Denmark, Finland, Germany, Greece and South Africa.

Fenamiphos insecticide

Fenamiphos, $C_{13}H_{22}NO_3PS$ (ethyl 4-methylthio-*m*-toyl isopropylphosphoramidate), is an organophosphorus insecticide. It is a colourless, crystalline substance, soluble in water at 400 mg/l. It is also soluble in organic solvents such as dichloromethane, isopropanol, toluene and hexane. It is stable in aqueous media, with the DT_{50} being 1 year under the following conditions: 1 day at pH 4 and 3 years at pH 9. Fenamiphos is prepared by condensing 4-methylyhio-*m*-cresol with *O*-ethyl *N*-isopropyl phosphoramidochloride, or with ethyl dichlorophosphate followed by reaction with isopropylamine. The CAS Registry Number of fenamiphos insecticide is 22224-92-6.

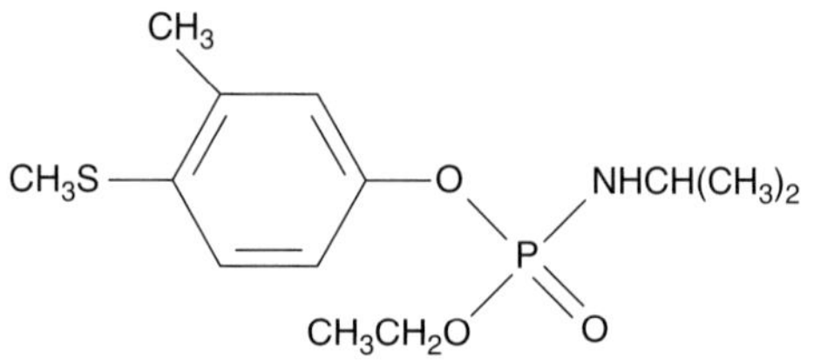

Fenamiphos

The insecticide fenamiphos, a cholinesterase inhibitor, is also a systemic nematicide with acaricidal properties. It moves throughout the plant and demonstrates contact, protective and residual action. It controls a variety of nematodes, especially those living near tomato crops. It also controls spider mites and aphids. Fenamiphos is used on many stone and citrus fruits, bananas, cotton, coffee, tomatoes, potatoes, sugarbeet, vegetables and tobacco.

Fenamiphos is formulated as an emulsifiable concentrate (EC) and as granules (GR). It is commonly broadcast or applied in rows or bands before or at planting time. It is available in the market under the trade name of Nemacur.

When ingested by mammals, fenamiphos metabolizes to its sulfoxide and sulfone analogues, followed by several reactions before being excreted through the urine. In plants, degradation is by thio-oxidation and hydrolysis. The major metabolites are fenamiphos sulfoxide and fenamiphos sulfone. The oxidation of the thiomethyl group to sulfoxide and sulfide is required for nematicidal activity in plants.

Fenamiphos degrades in water and soils and is not mobile. The soil DT_{50} is many weeks.

The EPA has made fenamiphos a restricted use pesticide (RUP) due to high acute toxicity and toxicity to wildlife. The WHO rating is Class Ib (highly hazardous). Its acute toxicity, cholinesterase inhibiting property and potential to contaminate groundwater have made it a PAN Bad Actor chemical.

Fenamiphos is banned in many countries including the UK. But it is registered in Australia, the USA, South Africa, the Philippines, France, Cyprus and Germany.

Fenarimol fungicide

Fenarimol, $C_{17}H_{12}Cl_2N_2O$ ((*RS*)-2,4′-dichloro-α-(pyrimidin-5-yl)benzhydryl alcohol), is a pyrimidine or pyrimidinyl carbinol fungicide. It is an off-white crystalline substance soluble in water (13.7 mg/l), acetone and methanol. It is decomposed rapidly by sunlight. Fenarimol is produced by reaction of 2, 4- dichlorobenzophenone with 5-lithiopyrimidine at −125°C. The CAS Registry Number of fenarimol fungicide is 60168-88-9.

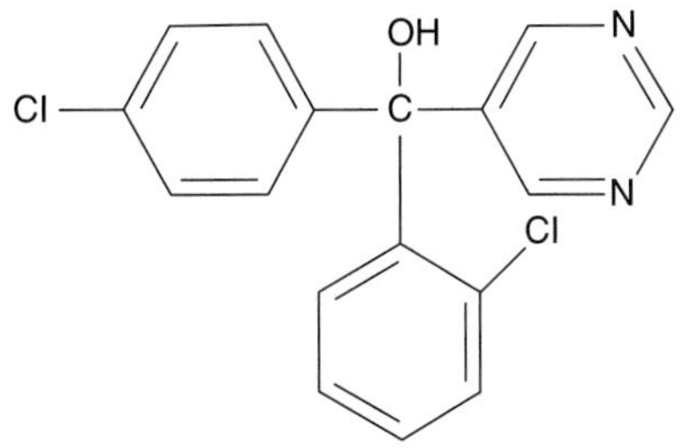

Fenarimol

Fenarimol, a systemic fungicide with protective and curative properties, is an **ergosterol** biosynthesis inhibitor. It is used to control **powdery mildews** in fruits and vegetables.

The fungicide is formulated for example as a wettable powder (WP) and an emulsifiable concentrate (EC) and is available under trade names Rubigan and Osadan.

Fenarimol is classified as a Class U toxin (unlikely to be hazardous), as the acute LD_{50} in rats is 2500 mg/kg b.w. But it is a potential groundwater contaminant and a suspected endocrine disruptor (according to test results documented by the PAN Pesticide Database).

The fungicide forms many photodegradation products in plants and is excreted rapidly upon ingestion by animals. Fenarimol is registered in India, Australia, Hungary, Madagascar, South Africa and the USA.

Fenazaquin acaricide

Fenazaquin, $C_{20}H_{22}N_2O$ (4-tert-butylphenethylquinazolin-4-yl ether), is a quinazoline acaricide. It consists of colourless crystals, almost insoluble in water (0.102 mg/l at pH 5 and 7) but soluble in many organic solvents such as acetonitrile (33–50 g/l) and acetone (400–500 g/l). Fenazaquin is produced by reaction of 4-hydroxyquinazoline with phosphate and chlorine and treating the resulting haloquinazoline with 4-tert-butylphenylethanol. Its CAS Registry Number is 120928-09-8.

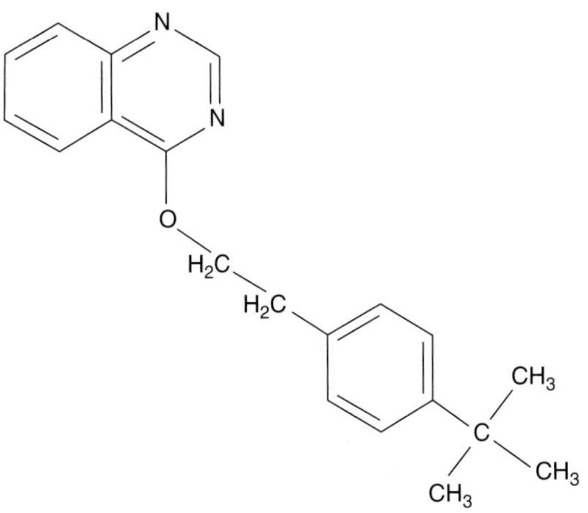

Fenazaquin

Fenazaquin targets mite and insect pests, with the ingestion and dermal routes disrupting insect mitochondria. It even affects the eggs of mites. It is used against mite species of *Eutetranychus*, *Panonychus*, *Tetranychus* and *Brevipalpus* in almonds, apples, citrus, cotton, grapes and ornamentals. The emulsifiable concentrate (EC) of fenazaquin 100 is commonly used on most crops, while the suspension concentrate (SC) diluted as required is also used for spray applications. The commonly accepted application rates are between 100 and 150 ml/100 l. The trade names of fenazaquin include Magister.

The WHO classified fenazaquin as a Class II toxin (i.e. moderately hazardous). It is acutely toxic to rats (the acute oral LD_{50} is 134 mg/kg). Fenazaquin is registered in India, Greece, Portugal and other countries, but not in the USA.

Fenbuconazole fungicide

Fenbuconazole, $C_{19}H_{17}ClN_4$ ((*RS*)-4-(4-chlorophenyl)-2-phenyl-2-(1*H*-1,2,4-triazol-1-ylmethyl)butyronitrile), is an conazole fungicide. It is a colourless crystalline substance, soluble in water (13.7 mg/l), benzene and dichloromethane. It is extremely stable to heat, light and atmospheric oxygen. The CAS Registry Number of fenbuconazole fungicide is 114369-43-6.

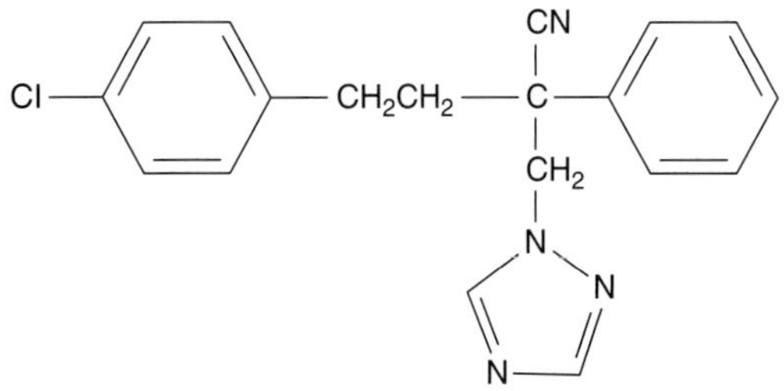

Fenbuconazole

Fenbuconazole is produced by a series of reactions involving benzylnitrile, potassium hydride, bromomethane and other compounds. Fenbuconazole, a systemic fungicide with protectant and curative properties, is used to control rusts, **powdery mildews**, bunt and **smut** on cereal crops and on fruits such as apples, peaches and apricots. *Septoria* and *Puccinia* spp. are among specific fungal pathogens controlled by this fungicide. It is used on foliage and on seeds. It also has a post-harvest application.

The fungicide fenbuconazole is available in the market as Indar 5EW and Enable. Spray applications are common. It is compatible with many pesticides.

The fungicide has a low acute toxicity and is classified as a Class U toxin by the WHO (i.e. unlikely to be hazardous). The oral and dermal LD_{50} in rats are 2631 mg/kg and >5000 mg/kg, respectively. An ADI in humans is 0–0.03 mg/kg b.w./day. However, it is phytotoxic to tangerines and some grapefruit varieties and is moderately toxic to fish species. It is also observed to be a possible carcinogen and endocrine disruptor.

Fenbuconazole is registered in the USA, Australia, Canada, Greece, Portugal and South Africa.

Fenchlorazole herbicide safener

Fenchlorazole, $C_{10}H_4Cl_5N_3O_2$ (1-(2,4-dichlorophenyl)-5-trichloromethyl-1H-1, 2,4-triazole-3-carboxylic acid), is a herbicide safener. It is a white, odourless solid, soluble in water (0.9 mg/l at pH 4.5), *n*-hexane, methanol, toluene and dichloromethane. It can take the form of a salt or an ester, such as fenchlorazole-ethyl. The CAS Registry Number of this safener is 103112-35-2. The aqueous hydrolysis DT_{50} is 5.5 days (pH 7).

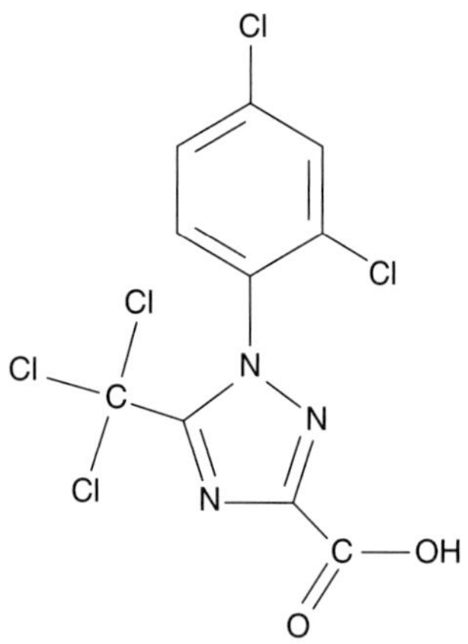

Fenchlorazole

Fenchlorazole plays an important part in breaking down herbicides like fenoxaprop-ethyl or **fenoxaprop-p-ethyl**, into safer metabolites, by which crops do not suffer from leaf discolouration and **chlorosis**. This safener commonly sees usage in wheat crops.

The acute oral and percutaneous LD_{50} values among mammals and birds suggest the absence of threat to these species, under normal usage. The chemical does not harm bees and does not irritate the skin or the eyes, if used as instructed. However, it is toxic to fish and aquatic organisms (with the LC_{50} among rainbow trout (96 h) being 0.08 mg/l). The WHO classifies fenchlorazole as a Class U toxin (unlikely to be hazardous). It is not registered in the USA.

Fenclorim herbicide safener

Fenclorim, $C_{10}H_6Cl_2N_2$ (4,6-dichloro-2-phenylpyrimidine), is a herbicide safener of the pyrimidine class. It is a colourless crystalline powder, soluble in water (2.5 mg/l), acetone (140,000 mg/l), cyclohexanone, toluene, xylene, hexane, *n*-octanol and isopropanol. It releases toxic oxides of carbon and nitrogen, upon thermal decomposition. It stores well in the original container in a cool, dry, ventilated place away from direct heat or sunlight. The CAS Registry Number of fenclorim is 3740-92-9.

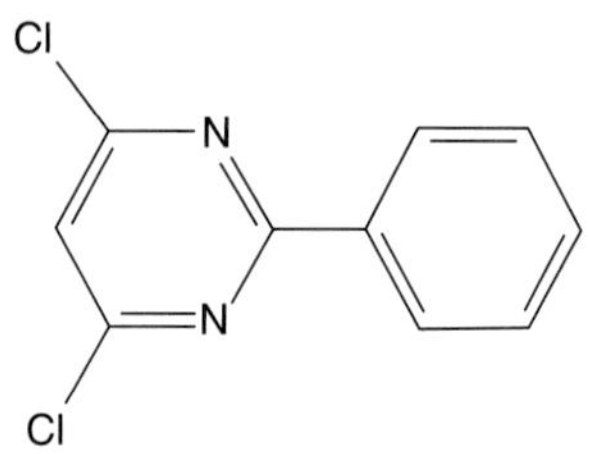

Fenclorim

Fenclorim acts by metabolizing **pretilachlor** in direct-seeded rice. The safener thus protects the crop from seedling damage. The compound is formulated as an emulsifiable concentrate (EC) or as granules (GR). It is compatible with many compounds.Trade names include Sofit.

The safener is not acutely toxic to mammals, earthworms and bees, in normal use. It is moderately toxic to birds (the LD_{50} in Japanese quails is >500 mg/kg), while it is toxic to fish (with the acute LC_{50} in rainbow trout for 96 h being 0.6 mg/l).

Personal protection is advised at the time of handling fenclorim, like the use of a chemical protection suit, safety goggles, boots and self-contained breathing apparatus (SCBA), approved by NIOSH. It is advised that fenclorim be kept away from fire, light, food, feed and drink. Handlers are advised to wash themselves and their clothing thoroughly after contact with fenclorim.

Fenclorim is rapidly absorbed and metabolized to polar compounds and excreted. In plants too, it has the same route. The metabolites are strongly adsorbed. It does not accumulate in body tissues or leave residues on crops. The soil DT_{50} is 26 days. Aqueous hydrolysis DT_{50} of fenclorim is 190 days, demonstrating its persistence.

Fenclorim is registered for use in the Philippines and Vietnam.

Fenfuram fungicide

Fenfuram, $C_{12}H_{11}NO_2$ (2-methyl-3-furanilide), is a phenylamide fungicide. It is a colourless crystalline substance, soluble in water, acetone, methanol and xylene. It is stable to heat, light and neutral media, but is hydrolysed by strong acids and alkalis. Fenfuram is produced by heating 2-methyl-4, 5-dihydrofuran-3-carboxylic acid with sulfonyl chloride, followed by reaction with aniline. The CAS Registry Number of fenfuram is 24691-80-3.

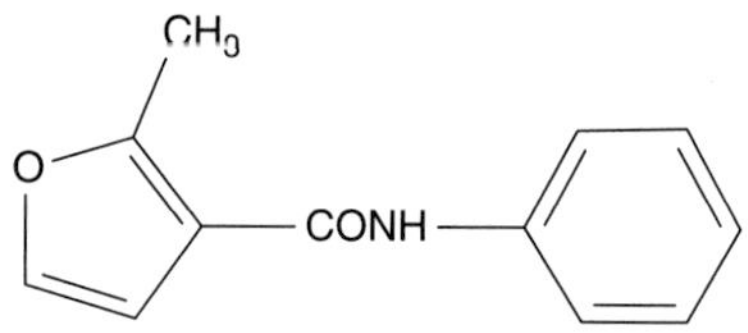

Fenfuram

The fungicide controls bunts and **smuts**, in cereal crops. Seed dressing is the preferred application method. It is commercially available as Panoram, Furadan and Nuvan.

This compound has an LD_{50} in rats of 12,900 mg/kg. It has the NOEL in rats of 10 mg/kg. The WHO categorizes fenfuram as a Class U toxin (unlikely to be hazardous). Fenfuram is not toxic to bees when used as directed.

Fenfuram is not registered in the USA.

Fenitrothion insecticide

Fenitrothion, $C_9H_{12}NO_5PS$ (*O*,*O*-dimethyl *O*-4-nitro-*m*-toyl phosphorothioate), is an ester of phenol, and in the class of organophosphate insecticides. It is a pale reddish-brown oily liquid, with a slight odour. It is almost insoluble in water, but is readily soluble in alcohols, aromatic and chlorinated hydrocarbons, and others. It is relatively stable to hydrolysis in normal conditions. Fenitrothion is prepared by the reaction of sodium 3-methyl-4-nitrophenolate with dimethylthiophosphoric acid chloride. It has a CAS Registry Number of 122-14-5.

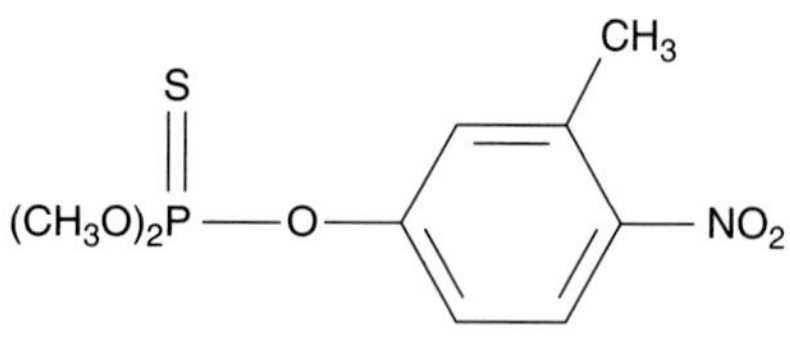

Fenitrothion

Fenitrothion, a non-systemic insecticide and a selective acaricide with contact and stomach action, is a cholinesterase inhibitor. The compound controls pests on a wide range of crops and in forests and non-farm situations. Cereals, sugarcane and a variety of fruits and vegetables are protected from grain beetles, grain weevils, locusts and **grasshoppers**. Similarly, forest vegetation like spruce and conifers are protected from mites and insects. In non-farm situations, flies, cockroaches and mosquitoes are controlled by fenitrothion.

Typical application doses include 200 mg a.i./l. Fenitrothion may be mixed with non-alkaline compounds like tricyclozole or **ferimzone**, along with oxalic acid or boron salts, to increase shelf life. Fenitrothion may be formulated as an emulsifiable concentrate (EC), granules (GR), ULV liquid (UL) or dust (DP). Some established trade names include Folithion, Sumithion and Dicofen.

Fenitrothion is a moderately toxic compound. It is toxic to bees, with contact LD_{50} being 0.383 μg per bee). The ADI is up to 0.005 mg/kg b.w./day.

This compound is metabolized into dimethylfenitrooxon and 3-methyl-4-nitrophenol (the latter, especially in upland conditions). In anaerobic conditions, aminofenitrothion is formed. Fenitrothion undergoes photodegradation.

India, the USA, Australia and many African countries have registered fenitrothion. Canada has severely restricted fenitrothion, while it has been totally banned from the EU.

Fenobucarb insecticide

Fenobucarb, $C_{12}H_{17}NO_2$ (2-*sec*-butylphenylmethylcarbamate), is a carbamate insecticide. It is a colourless solid with solubility in water being 420 mg/l at (20°C). It is also readily soluble in acetone, benzene, chloroform, xylene and toluene. It is stable under normal storage conditions and to light. It is hydrolysed by acids and alkalis (DT_{50} of 2.06 days at pH 10). Fenobucarb is prepared by reacting 2-*sec*-butylphenol and methylisocyanate. The CAS Registry Number of fenobucarb is 3766-81-2.

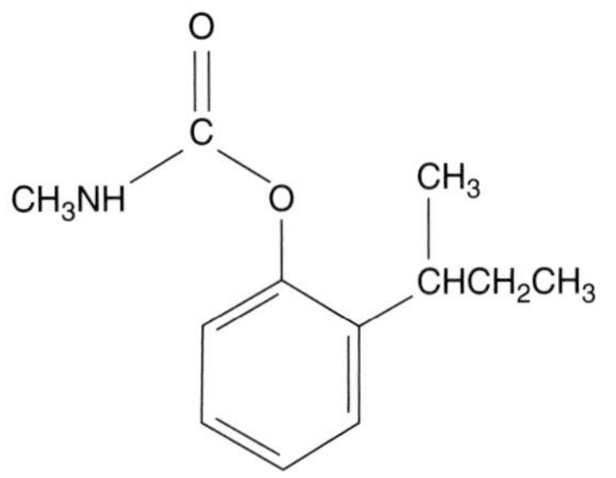

Fenobucarb

Fenobucarb is a cholinesterase inhibitor. It is a non-systemic insecticide with contact action. It is used in the control of **leafhoppers**, **aphids**, plant hoppers, **thrips** (Thysanoptera) and weevils on rice, tea, sugarcane, **cotton**, wheat, cucurbits, aubergines and capsicums. It is used in the control of bollworms and aphids on cotton.

Available formulations of fenobucarb are **emulsifiable concentrate (EC)**, dustable powder (DP), ULV liquid (UL) and microgranules. It is available under brand names like Bassa.

According to the WHO, fenobucarb comes under toxicity Class II (moderately hazardous). And since it is a cholinesterase inhibitor, it is classified as a PAN Bad Actor chemical.

In animals and plants, fenobucarb forms 2-(2-hydroxy-1-methylpropyl)-phenyl *N*-methylcarbamates, as the main metabolite. In soil, the DT_{50} ranges from 6 to 30 days and from 6 to 14 days under paddy and upland conditions, respectively.

Fenothiocarb acaricide

Fenothiocarb, $C_{13}H_{19}NO_2S$ (*S*-4-phenoxybutyl dimethyl (thiocarbamate)), is a thiocarbamate acaricide. It is a colourless crystalline substance, sparingly soluble in water but readily soluble in cyclohexanone, acetonitrile, acetone, xylene, methanol and hexane. Fenothiocarb is produced by reaction of the sodium salt of *N*,*N*-dimethyl carbamothioate with phenoxybutylchloride. Its CAS Registry Number is 62850-32-2 (see Fenothiocarb at bottom of the page).

Fenothiocarb is a non-systemic compound, used to control eggs and young stages of citrus red spider mites and the European red spider mites and other species in the genus *Panonychus* (family Tetranychidae). A common trade name is Panocan.

The WHO classifies fenthiocarb as a Class III toxin (slightly hazardous). The acute oral LD_{50} in rats and mice is 1150–1200 mg/kg and 4875–7000 mg/kg, respectively. The NOEL in rats (2 years feed) is 1.86–1.94 mg/kg/day. The ADI is 0.0075 mg/kg/day.

Fenthiocarb is not registered in the USA.

Fenoxaprop-p-ethyl herbicide

Fenoxaprop-p-ethyl, $C_{18}H_{16}ClNO_5$ ((*R*)-2-[4-(6-chloro-1,3-benzoxazol-2-yloxy) phenoxy] propionic acid; (*R*)-2-[4-(6-chlorobenzoxazol-2-yloxy) phenoxy]propionic acid), belongs to the aryloxyphenoxypropionate family of herbicides. It is a white powder with solubility in water being 0.9 mg/l, in acetone and toluene being 200 g/l and in ethanol, 24 g/l. Its CAS Registry Number is 71283-80-2 (see Fenoxaprop-p-ethyl herbicide at top of the next page).

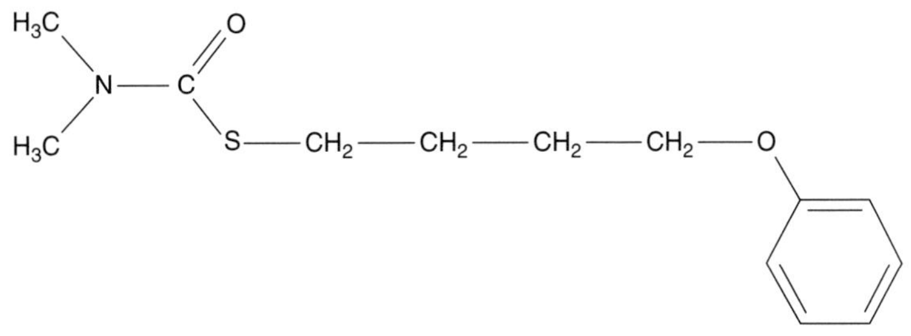

Fenothiocarb

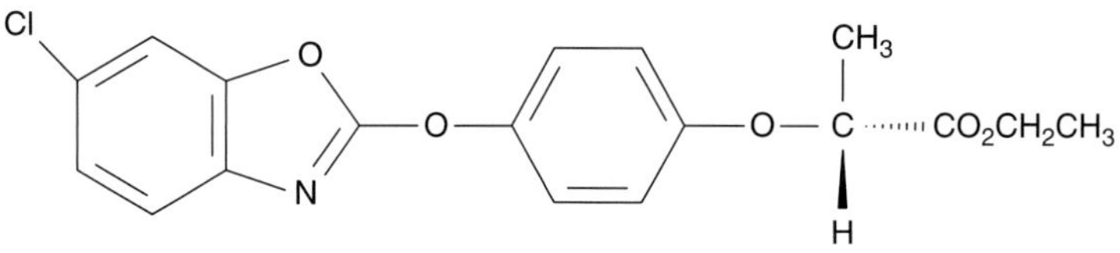

Fenoxaprop-p-ethyl

Fenoxaprop-p-ethyl, a fatty acid synthesis inhibitor acting on the acetyl CoA carboxylase (**ACCase**), is a selective, post-emergence herbicide, absorbed mainly by the leaves of susceptible weeds in potatoes, beans, soybeans, vegetables, groundnuts, flax, cotton and cereals (e.g. wheat, rye and barley). Weeds include wild oats and black grass.

This herbicide is formulated as an emulsifiable concentrate (EC) among others. It is commercially supplied as Furore Super, Puma, Triumph and Whip Super.

Herbicide fenoxaprop-P-ethyl is not acutely toxic to mammals, since its acute oral LD_{50} in rats and mice ranges from 3150 to >5000 mg/kg. It does not pose a threat to bees or birds either. But is highly toxic to fish and other aquatic organisms (the LC_{50} in tested fish species ranges between 0.46 and 0.58 mg/l). It causes irritation to the skin and eyes. It can also cause harm to non-target plants, in case of spray drift.

In plants, fenoxaprop-p-ethyl is metabolized via fenoxaprop to 6-chloro-2,3-dihydrobenzoxazol-2-one. In soil, fenoxaprop-p-ethyl is rapidly hydrolysed to fenoxaprop.

This herbicide is registered in over 75 countries worldwide such as India, Argentina, Australia, Brazil, Canada, China, Columbia, Pakistan, Russia, Turkey, the USA, and all (but two) countries of the EU.

Fenpiclonil biofungicide

Fenpiclonil, $C_{11}H_6Cl_2N_2$ (4-(2,3-dichlorophenyl)-1*H*-pyrrole-3-carbonitrile), is a cynopyrrole fungicide. It is a white crystalline substance soluble in water (4.8 mg/l), and highly soluble in acetone (360,000 mg/l), *n*-hexane and *n*-octanol. It is stable up to 250°C. Fenpiclonil can be produced by the reaction of either 2-(2, 3-dichlorophenyl) acrylonitrile or 1-cyano-2(2, 3-dichlorophenyl) ethyl cinnamate with *p*-toluenesulfonylmethyl isocyanate in dimethoxyethane in the presence of a base such as sodium hydride or sodium methoxide. Its CAS Registry Number is 74738-17-3. The origin of this fungicide is traced to *Pseudomonas pyrrocinia* (from the natural antibiotic pyrrolnitrin).

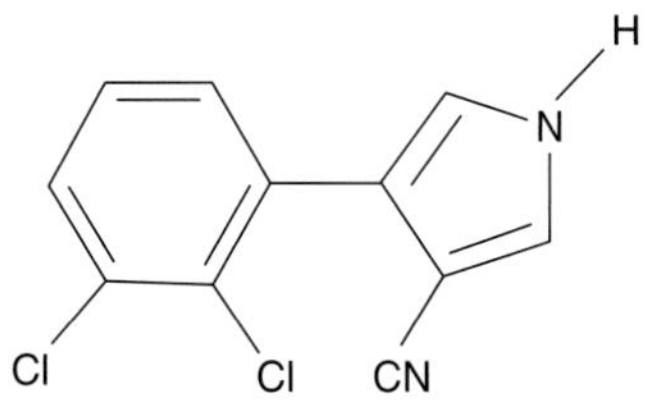

Fenpiclonil

Fenpiclonil, a systemic fungicide, with some contact action too, is mainly used on cereal crops, against seed-borne fungal diseases, caused by *Fusarium* sp. and *Tilletia* sp. With multiple modes of action and long-lasting activity, this fungicide targets many robust isolates of *Fusarium nivale*, especially those resistant to **carbendazim**.

The common formulations of fenpiclonil are flowable concentrates (SC) and powder for seed treatment, and the commercial trade names include Beret and Electer.

Fenpiclonil has an LD_{50} in rats of >5000 mg/kg. Fenpiclonil has the NOEL in rats of 25 ppm. According to the WHO, fenopiclonil is a Class U toxin (unlikely to be hazardous). It is non-toxic to honey bees. Labels, however, carry the signal word 'CAUTION'.

Degradation of fenpiclonil proceeds via many reactions like oxidation of the pyrrole ring followed by hydrolysis of the nitrile group. However, the DT_{50} of this compound in soil is 242 days, making it persistent in nature, although it is not mobile in soil.

Fenpropathrin insecticide

Fenpropathrin, $C_{22}H_{23}NO_3$ ((*R*,*S*)-α-cyano-3-phenoxybenzyl 2,2,3,3-tetramethylcyclopropane carboxylate), is a pyrethroid insecticide and acaricide. It is a brownish solid, sparingly soluble in water and also in xylene, cyclohexanone and methanol. It is decomposed in alkaline solution and is oxidized by light and air. It may be stored at room temperature in cool, dark and dry conditions for >2 years. Fenpropathrin is produced by the reaction of α-cyano-3-phenoxybenzyl alcohol with 2,2,3,3-tetramethylcyclopropane carboxylic acid chlorides. The CAS Registry Number of fenpropathrin insecticide is 39515-41-8.

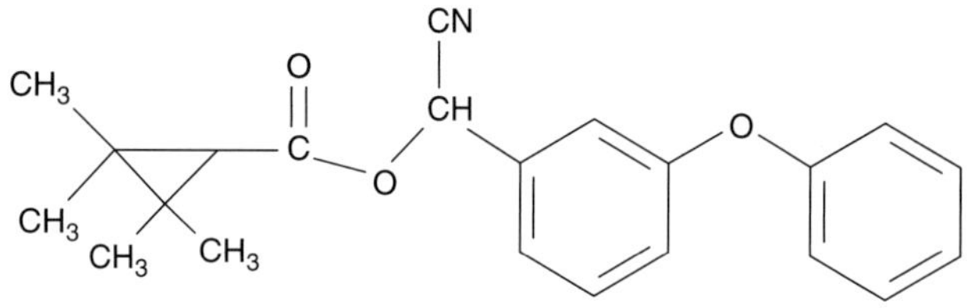

Fenpropathrin

Fenpropathrin, with both contact and stomach action, is used as a repellent, insecticide and acaricide. It effectively controls many species of mites, insects (e.g. whitefly, leaf miners, leafworms, bollworms) and lepidopterous larvae. It is formulated as emulsifiable concentrate (EC). It is available under trade names like Danitol, Rody, Herald and Meothrin.

Fenpropathrin is a moderately hazardous chemical (Class II toxin). It has an acute oral LD_{50} in the range of 66.7–70.6 mg/kg among rats. The acute oral LD_{50} in mallard ducks is 1089 mg/kg. It is highly to very highly toxic to fish and amphibians (with the LC_{50} for bluegill sunfish being 1.95 μg/l). Due to its acute toxicity, it is a PAN Bad Actor chemical. It does not irritate the skin during normal usage, but might irritate the eyes. The ADI in man is 0.03 mg/kg b.w./day.

Fenpropathrin is degraded principally by photolysis. It is a potential water contaminant.

The insecticide is registered in India, the USA, Canada, Cambodia and South Africa.

Fenpropidin fungicide

Fenpropidin, $C_{19}H_{31}N$ (1-[(*RS*)-3-(4-*tert*-butylphenyl)-2-methylpropyl] piperidine), is a morpholine analogue fungicide. It is a pale yellow, odourless, slightly viscous liquid, its solubility in water being 0.53 g/l. It is completely miscible in acetone, ethanol, toluene, *n*-octanol and *n*-hexane. It is stable to storage, to UV radiation, when in an aqueous solution, and to hydrolysis at 80°C at pH 4, 7 and 10. Fenpropidin is produced by the condensation of 4-*tert*-butylbenzaldehyde with propanal to give 4-*tert*-butylphenyl-2-methylpropylinealdehyde, the hydrogenation product of which is condensed with piperidine, followed by hydrogenation. The CAS Registry Number of fenpropidin is 67306-00-7.

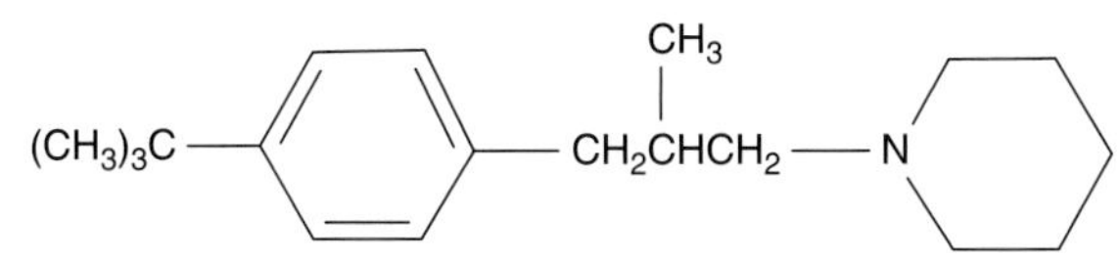

Fenpropidin

Fenpropidin, a systemic fungicide with protective and curative action, is an ergosterol biosynthesis inhibitor, used to control **powdery mildew** and rust in cereal crops. *Erysiphe graminis*, *Puccinia* spp. and *Rhynchosporium secalis* are some target pests.

The fungicide is formulated as an emulsifiable concentrate (EC) and is commercially available as Tern, among others.

The WHO has categorized fenpropidin as a Class II toxin (moderately hazardous). It has an LD_{50} in rats of >1400 mg/kg. It causes irritation to the skin and eyes. The ADI is 0.02 mg/kg/day. It is also observed to be a possible carcinogen. When equipped with PPE such as gloves and coveralls, applicators are not at risk in normal usage of this fungicide.

In rats, following oral administration, fenpropidin is rapidly absorbed, distributed, metabolized and excreted in the urine and faeces. In wheat, the principal metabolic pathway involves hydroxylation of the piperidine ring and oxidation of the tertiary butyl group of fenpropidin. Fenpropidin is strongly adsorbed and extensively degraded in soil. So the risk of the compound or its metabolites leaching is practically absent.

Fenpropidin is registered for use in many countries of the EU including Portugal, Finland, France, Germany, Greece, Hungary, Spain and the Netherlands, and in New Zealand.

Fenpropimorph fungicide

Fenpropimorph, $C_{20}H_{33}NO$ ((±)-*cis*-4-[3-(4-*tert*-butylphenyl) -2-methylpropyl]-2,6-dimethylmorpholine), is a morpholine fungicide. It is colourless, odourless oil, soluble in water (4.3 mg/l), acetone, chloroform, toluene and ethanol. It is stable to storage at room temperature, to light and to hydrolysis from pH 3 to 9. It is produced by reductive amination of 3-(4-*tert*-butylphenyl)-2-methyl-propanal and *cis*-2,6-dimethylmorpholine. The CAS Registry Number of fenpropimorph is 67564-91-4 (see Fenpropimorph at bottom of the page).

Fenpropimorph, a systemic fungicide, absorbed through the leaves and roots, has protective and curative properties. It is mostly used in the control of fungal diseases, including rusts (brown and yellow) and **powdery mildew**, on wheat, barley, oats, sugarbeet and beans. An **ergosterol** biosynthesis inhibitor, it is used in the control of, among others, *Uromyces* sp. in all the above-mentioned crops as well as *Puccinia* sp. and *Erysiphe* sp.

The fungicide fenpropimorph is formulated as an emulsifiable concentrate (EC) and may be combined with many pesticides like **azoxystrobin** and cyclohexanone. The application rates are up to 1 l/ha with a maximum of two applications per treatment. A pre-harvest interval of 5 weeks is strongly recommended. Fenpropimorph is commercially available as Corbel, Mantra and Jackdraw, among others.

Fenpropimorph has an acute oral LD_{50} in rats of >1400 mg/kg. It is not an eye irritant (as tested on rabbits). An ADI in humans is 0.003 mg/kg/day. It is practically safe on birds, earthworms and honey bees. It may be toxic to fish.

In animals, plants and soil, fenpropimorph is degraded by a series of reactions that include the oxidation of the tertiary butyl group. The half-life of fenpropimorph in soil ranges from 15 days in moderately humus-rich loamy sand to 93 days in very humus-rich loamy sand. A major metabolite formed in soil is fenpropimorph carboxylic acid.

Fenpropimorph is registered in most countries of the EU, New Zealand and the USA.

Fenthion insecticide

Fenthion, $C_{10}H_{15}O_3PS_2$ (*O*,*O*-dimethyl *O*-4-methylthio-*m*-toyl phosphorothioate), is an organophosphorus insecticide. It is a brown oily liquid, soluble in water (4.2 mg/l), dichloromethane, toluene, isopropanol and hexane. It is stable to light (up to 210°C) and acidic conditions, and moderately stable in alkaline conditions. Fenthion is prepared by the reaction between the sodium salt of 3-methyl-4-methylmercaptophenol and dimethylthionophosphoric acid chloride. It has a CAS Registry Number of 55-38-9.

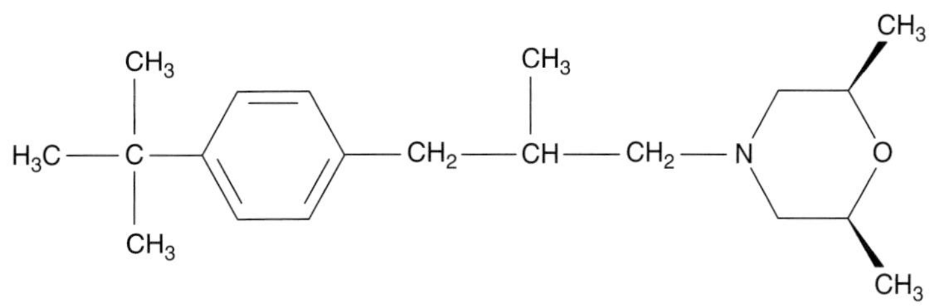

Fenpropimorph

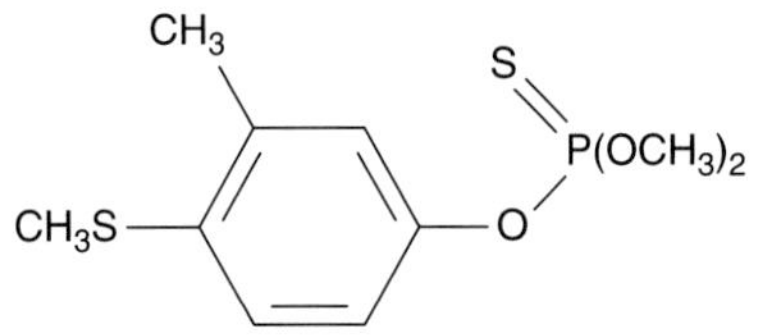

Fenthion

Fenthion, a cholinesterase inhibitor, is a non-systemic, broad-spectrum insecticide, used to control stem borers, bugs, fruit flies and **leafhoppers**. Fruits, vines, olives, vegetables, cotton, tea, sugarcane, rice, tobacco and ornamentals are protected using this insecticide. Fenthion is also an animal ectoparasiticide and household insecticide (against e.g. flies, ants and mosquitoes).

Fenthion is formulated as wettable and dustable powder, **emulsifiable concentrate (EC)**, granule (GR), ULV liquid (UL) and a pour-on. Some varieties of apples and cotton may be injured. It is compatible with non-alkaline insecticides and fungicides. It is available under trade names of Baycid, Bayer 9007, DMTP, Entex, Lebacid and Talodex.

In animals, elimination is mainly via hydrolysis with major metabolites being fenthion sulfoxide and fenthion sulfone and their oxygen analogues. In plants, fenthion is finally degraded to their monodesmethyl compounds. In soil and water, fenthion degrades into fenthion sulfoxide and fenthion sulfone. All degradation products hydrolyse into their phenol compounds. Fenthion is rapidly degraded in water with a DT_{50} of 1.5 days.

Fenthion is very highly toxic to birds, with reported LD_{50} values ranging from <4 to 26 mg/kg in ducks. Fenthion is moderately toxic via the oral and dermal routes (oral LD_{50} values of 180–298 mg/kg in rats and 150 mg/kg in rabbits). It is also said to be a potential groundwater contaminant and a suspected endocrine disruptor and hence is a PAN Bad Actor chemical. Excessive exposure causes nausea, cramps and excessive salivation.

The USA and the EU have banned the use of fenthion. A large number of countries like India, Australia, Cambodia, the Philippines, New Zealand and South Africa have registered its use.

Fenthion technical

The technical analysis of fenthion is presented here. The sample for analysis should contain a minimum purity of 940 g fenthion/kg sample, up to 3 g H_2SO_4 for acidity and up to 2 g water/kg of sample. The special apparatus needed for the analysis is a spectrophotometer or photoelectric colorimeter capable of measuring absorbance at 530 nm. The special reagents include: (i) 3-methyl-4-(methylthio) phenol (with a melting point of 56°C ±0.5°C); (ii) ethylene glycol monomethyl ether; (iii) a buffer solution; and (iv) a diazo solution.

To prepare a fraction of ethylene glycol monomethyl ether, reflux 1 l of the pure reagent with 50 g potassium hydroxide pellets for 1 h. Distill and collect the fraction, while boiling between 130 and 140°C. To purify the reagent 3-nitraniline-4-sulfonic acid, dissolve 10 g of the reagent in 1 l hot distilled water. Add activated charcoal, stir, filter and cool the filtrate in ice. Filter off the crystalline acid, wash with water and dry at 50°C. The buffer solution is made by first dissolving 106 g Na_2CO_3 in distilled water to make up to 1000 ml of solution I, then dissolving 50 g sodium bicarbonate in distilled water, to make up to 1000 ml of solution II. Mix 350 ml solution I and 130 ml solution II, ensuring a pH of 10.4 at 20°C. To make the diazo solution, dissolve 0.35 g of the purified 3-nitraniline-4-sulfonic acid in 5 ml 1N sodium hydroxide. Add to it 350 ml of a mixture of ice and water, followed by 5 ml concentrated hydrochloric acid. Diazotize at 0.5°C, with 0.1 M sodium nitrite (~16.5 ml). Determine the end point of diazatization by the spot test on starch potassium iodide paper. Dilute the solution to 500 ml in a volumetric flask and use immediately.

The experimental procedure begins with the determination of free 3-methyl-4-(methylthio) phenol, which involves the preparation of a calibration curve. Dissolve 400 ml pure 3-methyl-4-(methylthio) phenol in 5 ml 1N sodium hydroxide and 10 ml isopropanol and make up to the mark in a 1000 ml volumetric flask with distilled water. Transfer 50, 75, 100, 125, 150 and 175 ml of the completely mixed solution into six 500 ml volumetric flasks and dilute to the volume with distilled water. The stock solution thus obtained contains 20, 30, 40, 50, 60 and 70 mg of 3-methyl-4-(methylthio) phenol in 500 ml flasks. Transfer from each of the solutions, 5 ml (corresponding to 0.2–0.7 mg 3-methyl-4-(methylthio) phenol) into six 50 ml volumetric flasks. Add to each, 10 ml isopropanol, 20 ml buffer solution and 10 ml diazo solution and make up to the mark with water. After mixing thoroughly, the flasks are kept at 20°C in a water bath for 2 h. Measure the absorbance at 530 nm using a 1 cm cell in a spectrophotometer, using reagents as reference. Plot graphically the measured absorbance corresponding to 0.2–0.7 mg of 3-methyl-4-(methylthio) phenol. Since the standard solution of 3-methyl-4-(methylthio) phenol deteriorates on standing, a fresh solution is to be prepared each day.

The procedure involves weighing an amount of the sample with about 5 mg 3-methyl-4-(methylthio) phenol into a 50 ml volumetric flask and dissolving 2-propanol into it, making it up to the mark. After shaking and filtering, transfer 5 ml of the clear solution to a 50 ml volumetric flask and add the first 10 ml of the diazo solution. Make up to the mark with water. After shaking thoroughly, set aside the flask in a thermostatically controlled water bath at 20°C for 2 h. Measure the absorbance in a 1 cm cell at 530 nm using a spectrophotometer. Read from the calibration curve the amount of 3-methyl-4-(methylthio) phenol corresponding to the absorbance. If the amount is not in the range 0.2–0.7 mg, repeat the determination with a smaller or larger sample.

For determination of fenthion, prepare the calibration curve, by transferring 5 ml of each of the stock solutions (as prepared above) into 50 ml volumetric flasks; add successively 3 ml isopropanol from a burette, 30 ml buffer solution and 7.5 ml diazo solution and make up to the mark with the buffer solution. After mixing thoroughly, keep the flasks for 2 h in a thermostatically controlled water bath at 20°C. Measure the absorbance in a 1 cm cell at 530 nm using a spectrophotometer. The reagents are used as reference.

The procedure involves weighing about 0.08 g (to the nearest 0.1 mg) of the sample and rinsing with 25 ml ethylene glycol monoethyl ether into a 100 ml flask. Add 1.65 g potassium hydroxide pellets (about 27 pellets) and reflux the contents for 7 h using a reflux condenser. Cool, transfer to a 500 ml volumetric flask, dilute to the mark with water, and mix thoroughly. Transfer 5 ml of this solution into a 50 ml volumetric flask and add successively 3 ml 2-propanol from a burette,

30 ml buffer solution and 7.5 ml diazo solution. Dilute to volume using buffer solution. After mixing thoroughly, keep the flask in a thermostatically controlled water bath at 20°C for 2 h. Measure the absorbance in a 1 cm cell at 530 nm in a spectrophotometer using the reagents as reference. Read from the standard curve the amount of 3-methyl-4-(methylthio) phenol corresponding to the absorbance. Use Eqn F.1 to calculate the fenthion content.

$$\text{The fenthion content (g/kg)} = \left(\frac{m_1 \times 10}{m_3}\right) - \left(\frac{m_2}{m_4}\right) \times 18.05 \quad \textbf{(F.1)}$$

where

m_1 = Total amount (mg) of 3-methyl-4-(methylthio) phenol found after hydrolysis

m_2 = Amount (mg) of free 3-methyl-4-(methylthio) phenol

m_3 = Mass of sample used for determination of the total 3-methyl-4-(methylthio) phenol

m_4 = Mass of sample used for determination of 3-methyl-4-(methylthio) phenol

A sample material, containing water-dispersible powder (WS) of fenthion, is commonly a homogeneous mixture of fenthion (technical) with fillers and other additives; it is free flowing in nature, and readily wettable on stirring into water. The determination of fenthion is the same as described under fenthion (technical).

The active ingredient is hydrolysed with a solution of potassium hydroxide in ethylene glycol monoether and the resulting 3-methyl-4-(methylthio) phenol is coupled (at a pH of around 10.4) with diazo-3-nitraniline 4-sulfonic acid to form a red-violet azo dye, the absorbance of which is measured at 530 nm. The free 3-methyl-4-(methylthio) phenol, the most likely impurity, is also determined by the same method but on a sample that has not been hydrolysed. The difference between these values is taken as the fenthion.

Fentin acetate fungicide

See **Fentin fungicide**

Fentin chloride

See **Organotin fungicides**

Fentin fungicide

Fentin, $C_{18}H_{15}Sn$ (triphenyltin), is an **organotin** fungicide with algicidal and molluscicidal properties. It also has antifeedant properties. The CAS Registry Number of fentin is 668-34-8. The compounds fentin acetate and fentin hydroxide are derivatives of fentin.

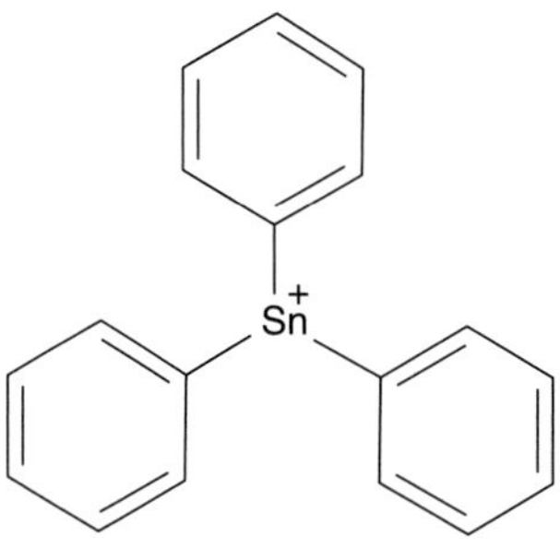

Fentin

Fentin acetate, $C_{20}H_{18}O_2Sn$ (triphenyltin (IV) acetate; triphenyltin acetate), is a colourless crystal, soluble in water (9 mg/l), ethanol (22 g/l), hexane (5 g/l) and toluene (89 g/l). It is stable when dry but is unstable in acids and alkalis and in water (where it converts to fentin hydroxide). Fentin acetate is produced by the treatment of triphenyltin chloride with sodium acetate. The CAS Registry Number is 900-95-8.

Fentin acetate is a contact fungicide, an algicide and a molluscicide, with protective action. By inhibiting oxidative phosphorylation (ATP synthase), it controls **early blight** and **late blight** of potatoes, **leaf spot** diseases of sugarbeet, onions and groundnuts and anthracnose of beans. The chemical may also be used on citrus fruits. It controls snails and algae in rice fields and fish ponds.

The trade names of this compound include Super-Tin, Suzu and Brestan, while it is applied as a wettable powder (WP), at a dose up to 300 g/ha. It is incompatible with oil emulsions and emulsifiable concentrate (EC) formulations.

The WHO and EPA classify fentin acetate as a Class II toxin (moderately hazardous). The acute oral LD_{50} in rats is 140–298 mg/kg, while the acute percutaneous LD_{50} in rabbits is 127 mg/kg. It is a skin and mucous membrane irritant. It is not toxic to bees. It may harm fruit crops and wet-seeded rice saplings. The ADI is 0.0004 mg/kg b.w./day. But it is a suspected endocrine disruptor.

In soil, fentin acetate is degraded to inorganic tin via di- and mono-phenyltin compounds.

It is not registered in the USA, banned throughout the EU and refused registration in India. But, it is used in South Africa. The MRL on citrus fruits (e.g. lemon and orange) is 0.05 mg/kg.

Fentin hydroxide, $C_{18}H_{16}OSn$ (triphenyltin (IV) hydroxide; triphenyltin hydroxide), is a colourless crystal, soluble in water (1 mg/l), ethanol (32 g/l) and polyethylene glycol (41 g/l). It is stable in the dark at room temperature, but slowly decomposes in sunlight and UV to inorganic tin. It is produced by the treatment of triphenyltin chloride with potassium hydroxide and has a CAS Registry Number of 76-87-9.

Fentin hydroxide is also a contact fungicide with protective action, with most of the pest control pattern and mode of action being similar to fentin acetate. It is formulated also as a soluble concentrate (SL). Fentin hydroxide is incompatible with strongly acidic compounds, oils and liquid formulations. Trade names of the compound include Suzu and Tubotin.

The WHO and EPA classify fentin hydroxide as a Class II toxin (moderately hazardous). The acute oral LD_{50} in rats ranges from 150 to165 mg/kg. The acute percutaneous LD_{50} in rabbits is 127 mg/kg. It is very toxic to fish and aquatic organisms. It is a skin and mucous membrane irritant. It is not toxic to bees, but may be phytotoxic to tomatoes and apples, particularly when used as surfactants and spreaders. The ADI is 0.0004 mg/kg b.w./day. This chemical is carcinogenic and a reproductive toxin, and it is a suspected endocrine disruptor. The chemical is classified as a PAN Bad Actor chemical.

In soil, fentin hydroxide degrades to inorganic tin via di- and mono-phenyltin compounds. The compound is refused registration in India, and banned throughout the EU. This compound is registered for use in the USA. The MRL on fruits such as mangoes and bananas is 0.05 mg/kg.

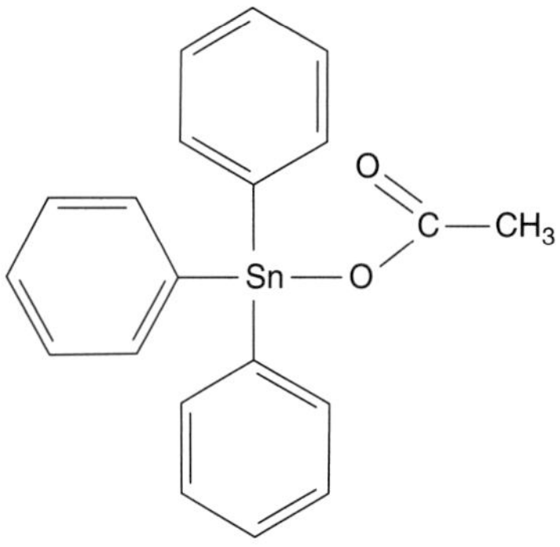

Fentin acetate

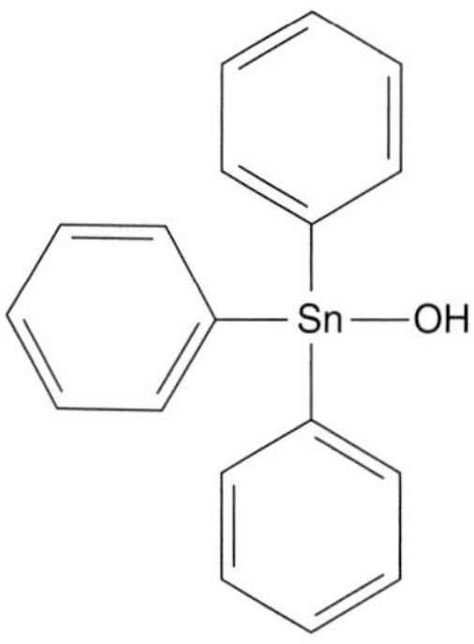

Fentin hydroxide

Fentin hydroxide fungicide

See **Fentin fungicide**

Fentrazamide herbicide

Fentrazamide, $C_{16}H_{20}ClN_5O_2$ (4-(2-chlorophenyl)-*N*-cyclohexyl-*N*-ethyl-4, 5-dihydro-5-oxo-1*H*-tetrazole-1-carboxamide), is an amide herbicide. It is a colourless crystalline solid, soluble in water and organic solvents like heptane, xylene, isopropanol and dichloromethane to different degrees. The CAS Registry Number of fentrazamide is 158237-07-1.

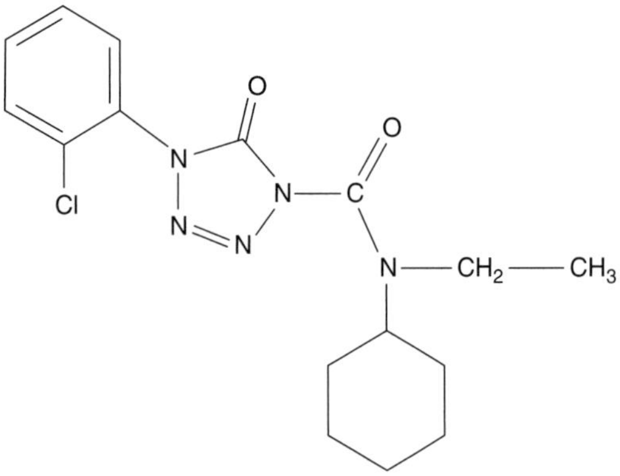

Fentrazamide

Fentrazamide, a selective compound, inhibits cell division by interfering with very long chain fatty acid synthesis. Such chemical activity is seen in suppressed weed growth and leaf emergence. Fentrazamide is mainly used on rice and controls **barnyard grass** (*Echinochloa crus-galli*), sedge grass and wild millets.

Fentrazamide formulations include granules (GR) and water-dispersible granules (WG) and can be combined with other compounds such as **propanil** or bensulfuron-methyl.

The acute oral LD_{50} of the compound among rats is >5000 mg/kg and among birds is 2000 mg/kg. For all practical purposes, it is not toxic to honey bees, earthworms, *Daphnia magna* and fish species. The DT_{50} in water is between 20 and 30 days.

Fentrazamide is not approved for use in the EU or the USA but is registered in Vietnam, Korea, Japan and the Philippines.

Fenuron herbicide

Fenuron, $C_9H_{12}N_2O$ (1,1-dimethyl-3-phenylurea), is a phenylurea herbicide. It is a colourless crystalline substance, soluble in water (3.85 g/l), ethanol, acetone, benzene, chloroform, groundnut oil, *n*-hexane, diethylether and hydrocarbons. It is produced by the reaction of phenyl isocyanate with dimethylamine. The CAS Registry Number is 101-42-8.

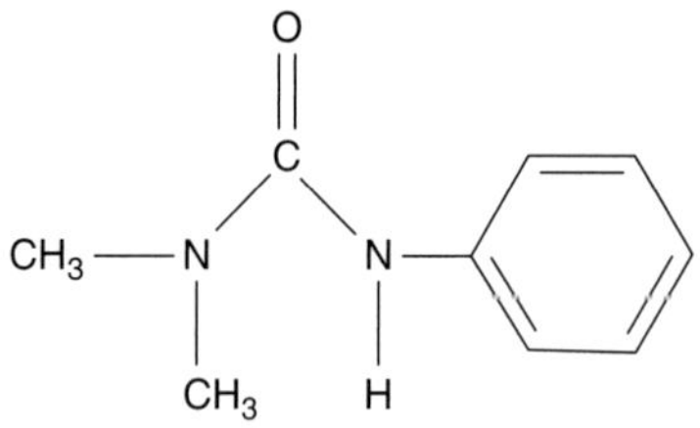

Fenuron

Fenuron is a **non-selective**, systemic herbicide, absorbed mainly by the roots. It is a photosynthesis inhibitor. It effectively controls perennial weeds, broadleaved weeds and woody plants in beet crops.

It is compatible with other wettable powder herbicides to varying degrees. It is commonly used with chloropropham. Quintex and Dybar are some known trade names.

The acute oral LD_{50} of fenuron in rats is 4000–5700 mg/kg. It is categorized as unlikely to be hazardous (Class U).

In animals, metabolism proceeds with dealkylation and hydroxylation of the aromatic ring of phenylurea. In the atmosphere, degradation of vapour-phase fenuron produces hydroxyl radicals. In soil, enzymatic and microbial degradation involves step-by-step *N*-demethylation. The soil half-life is up to 4.5 months. In plants, *N*-demethylation occurs step by step. This chemical is considered obsolete.

Fenvalerate insecticide

Fenvalerate, $C_{25}H_{22}ClNO_3$ ((*RS*)-α-cyano-3-phenoxybenzyl (*RS*)-2-(4-chlorophenyl)-3-methylbutyrate), is an insecticide belonging to the pyrethroid class. It is a yellow-brown viscous liquid, almost insoluble in water but readily soluble in organic solvents. It is stable to moisture and temperature but unstable in alkaline media. Fenvalerate is a mixture of four optical isomers that make it an effective insecticide. It is the first pyrethroid insecticide with no cyclopropane ring in its molecule. Fenvalerate is produced by the reaction of α-cyano-3-phenoxybenzyl alcohol with 2-(4-chlorophenyl)-3-methylbutyric acid chloride. Its CAS Registry Number is 51630-58-1.

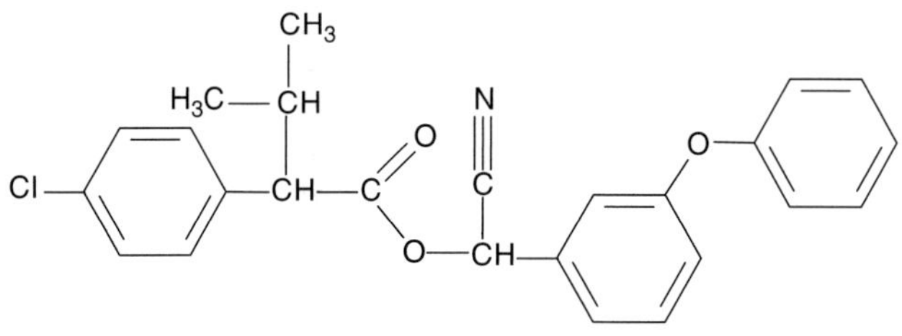

Fenvalerate

This compound, a nerve toxin, is effective on insects that attack fruits and vegetables such as apple, peach, pear, cabbage, cauliflower and others in the brassica group. In addition, fenvalerate also controls flies and ticks in animal shelters.

Common brand names are Belmark, Pydrin, S-5602, SD43775 and Sumicidin. Fenvalerate products come as emulsifiable concentrates (EC) and wettable powders (WP).

Fenvalerate is toxic to rats via the oral and inhalation routes. The acute oral LD_{50} in rats is 451mg/kg, while the inhalation LC_{50} is >0.101 mg/l. The dermal LD_{50} in rabbits is >1000 mg/kg b.w. The ADI of fenvalerate is 0.02 mg/kg b.w./day. This compound is practically non-toxic to birds (with the acute LD_{50} in mallard ducks being 9932 mg/kg). It is extremely toxic to fish and honey bees (the acute 96 h LC_{50} in rainbow trout is 0.0036 mg/l and the acute 48 h contact LD_{50} in honey bees is 0.23 µg per bee). Keeping the application rates to a minimum is the only way to prevent harm to non-target organisms.

Fenvalerate degrades via the ester and diphenyl ether cleavage, followed by the hydration of the cyano group to finally form carbon dioxide. In plants, degradation results in decarboxy-fenvalerate. In plants, soils and water, the half-life is around 2 weeks.

The insecticide is registered in India, Australia, the Philippines, South Africa, Tanzania, Uganda and many other countries.

Ferbam fungicide

Ferbam, $C_9H_{18}FeN_3S_6$ (iron (III) dimethyldithiocarbamate), is a dimethyldithiocarbamate fungicide. It is a black powder soluble in water (130 mg/l), chloroform, pyridine, acetonitrile and acetone. It decomposes on exposure to moisture and heat. Ferbam is produced by the precipitation from an aqueous solution of sodium dimethyldithiocarbamate in the presence of a water-soluble iron salt. The CAS Registry Number of ferbam fungicide is 14484-64-1.

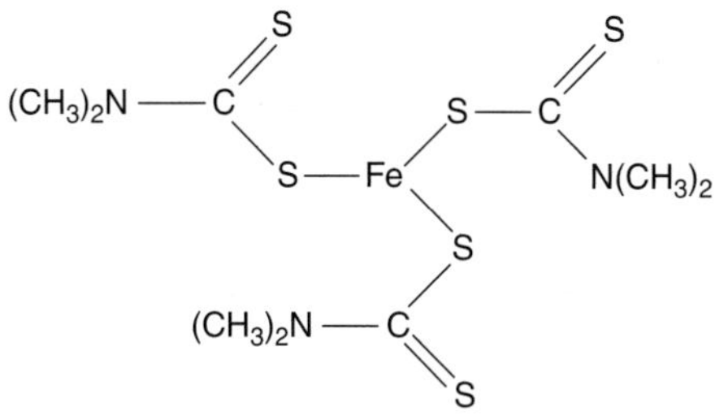

Ferbam

Ferbam, a foliar fungicide with protective action, controls **scab**, leaf curl and other diseases on a variety of fruits and tobacco.

This fungicide is used along with dithiocarbamates to expand the spectrum of activity. Ferbam is available as a wettable powder (WP) and is commonly offered under the trade name Trifungol.

Ferbam is classified as a Class U toxin (unlikely to be hazardous). Its acute oral LD_{50} in rats is >4000 mg/kg. It is irritating to skin and mucous membranes. It is moderately toxic to fish, but not toxic to bees. Ferbam may cause russeting in some apple varieties. An ADI of ferbam for humans is 0.2 mg/kg/day. It is a PAN Bad Actor chemical since it is a confirmed carcinogen.

The metabolites of ferbam include the dimethylamine salt of dimethyldithiocarbamic acid, tetramethylthiuram mono- and di-sulfide, carbon disulfide and sulfur.

Ferbam is registered in Canada and the USA although India has withdrawn this compound from the market.

Ferimzone fungicide

Ferimzone, $C_{15}H_{18}N_4$ ((*Z*)-2′-methylacetophenone 4,6-dimethylpyrimidin-2-ylhydrazone), is a fungicide. The colourless, crystalline compound has a good solubility in water (162 mg/l). It is soluble in ethanol, ethyl acetate, chloroform, xylene and acetonitrile to varying degrees. It is stable to sunlight and to neutral and alkaline aqueous solutions. It is produced by the condensation of 2-methylacetophenone with 4,6-dimethyl-2-hydrazinopyrimdine. The CAS Registry Number of ferimzone fungicide is 89269-64-7.

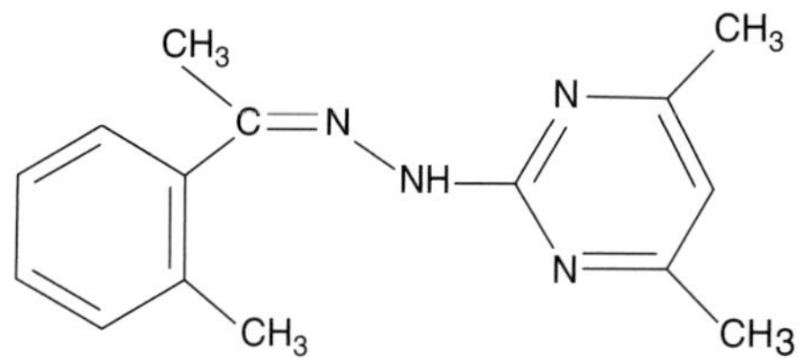

Ferimzone

Ferimzone, a rice fungicide, is specifically used in the control of diseases of rice such as leaf spot disease (caused by *Helminthosporium oryzae* and *Cercospora oryzae*) and rice blast (caused by *Pyricularia oryzae*).

The fungicide ferimzone, formulated as a wettable powder (WP), is classified as slightly toxic in acute toxicity studies. It has an acute oral LD_{50} in rats of 725 mg/kg and in bobwhite quails of 2250 mg/kg. It has an LC_{50} in fish (72 h) of 10 mg/l. Eye or skin contact is harmful. It is advisable to keep cans and tools with ferimzone away from food and feed. This chemical is not registered by the USA.

Ferrous sulfate as herbicide

Ferrous sulfate, FeO_4S (iron (II) tetraoxosulfate), is the most common source of iron, used both as a fertilizer and as a pesticide. This inorganic compound, an odourless, blue-green crystalline substance is used as a contact herbicide and also as molluscicide. It is highly soluble in water (48.6 g/100 g) at 50°C. It stores well in cool and dry conditions, since it oxidizes in moist conditions, producing the corrosive ferric sulfate. The CAS Registry Number of ferrous sulfate is 7720-78-7.

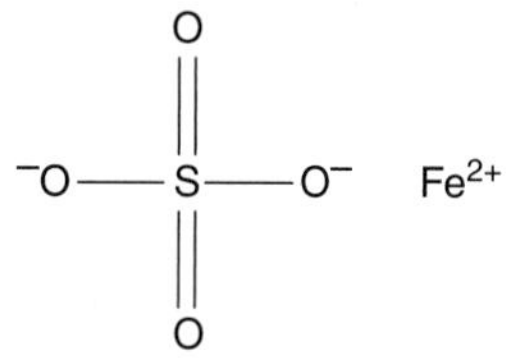

Ferrous sulfate

The herbicide ferrous sulfate controls moss and lawn weeds very effectively and keeps mollusc species under control. It is used on maize, grapes and cotton.

Ferrous sulfate, commonly formulated as granules (GR) and water-soluble powder (SP), may be used along with other compounds like **2,4-D**, and **mecoprop-P**, but is incompatible with alkalis, soluble carbonates and oxidizing materials. It also supplies iron at micronutrient doses.

The acute LD_{50} of ferrous sulfate among birds is >2510 mg/kg, while among rats it is 319 mg/kg. The acute LC_{50} for crustaceans is >7.1 ppm. It can cause irritation to the respiratory tract and skin, with symptoms being redness, itching and pain. It is necessary to take proper care while handling the compound (i.e. wearing protective clothing, boots, etc.) due to the corrosive nature of ferric sulfate.

Ferrous sulfate is abundantly available in nature and does not cause any serious toxicity issues, in normal use. It does not bioaccumulate in body tissue.

Ferrous sulfate is registered for use in Germany, Portugal and the USA.

Fire diamond

See **Fire hazards and prevention**

Fire hazards and prevention

The process of producing, refining, storing, transporting and using chemicals is always exposed to risks and fire hazards, since many solid or liquid chemicals or chemical mixtures are highly reactive or combustive even at normal temperatures. Any negligence during storage or transportation can cause emergency situations, where loss of life and property cannot be ruled out.

Since a large number of personnel come into contact with chemical compounds – from the synthesis/refining stage to their final consumption – there is a need for a common code for signalling danger that is easily understood by all. The National Fire Protection Association (NFPA) in its Section No. 704 of the National Fire Code has developed a diamond-shaped hazard identification signal that has found acceptance in most countries, although countries may add their own locally recognizable symbols if necessary.

The fire diamond has four colours: red, blue, yellow and white. Each of these colours indicates a specific hazard warning: (i) the red colour indicates flammability of the material; (ii) blue indicates a health hazard on exposure to the substance; (iii) yellow indicates reactivity; and (iv) white indicates that precautions may be required.

Each colour has numbers 0 to 4 indicating the kind and grade of hazard.

Two symbols, ~~W~~ and OX, are used to indicate special characteristics of materials. ~~W~~ indicates the unusual reaction of the material with water, as in the case of sodium or caesium. OX indicates the oxidizing properties of the substance, as in the case of ammonium nitrate and hydrogen peroxide. Other commonly used symbols in fire safety include ACID, ALK and COR, indicating the acidic, alkaline and corrosive properties of the substance in question, respectively.

Besides the hazard recognition colour code, the NFPA has classified fire into four classes, A, B, C and D, to recognize the nature of fire. Class A fires are those involving for example wood, paper or cloth that can be controlled by water. Substances like combustible liquids and flammable gases that are controlled by breaking the chain reaction and in the absence of oxygen are Class B fires. Fires involving energized electrical equipment, which need non-conductive extinguishing agents to control them, are classified as Class C. Class D fires are those involving combustible metals such as sodium and potassium and that require specialized extinguishers. More information may be received from http://www.nfpa.org and related links.

Flamprop herbicide

Flamprop, $C_{16}H_{13}ClFNO_3$ (*N*-benzoyl-*N*-(3-chloro-4-fluorophenyl)-DL-alanine), is an arylalanine herbicide. This compound is a colourless, crystalline substance, soluble in water (35 mg/l), acetone, cyclohexanone, carbon tetrachloride, ethanol and hexane. Flamprop may exist as an ester or a salt, for example flamprop-isopropyl or flamprop-methyl. It hydrolyses in alkaline medium. The CAS Registry Number of flamprop herbicide is 58667-63-3.

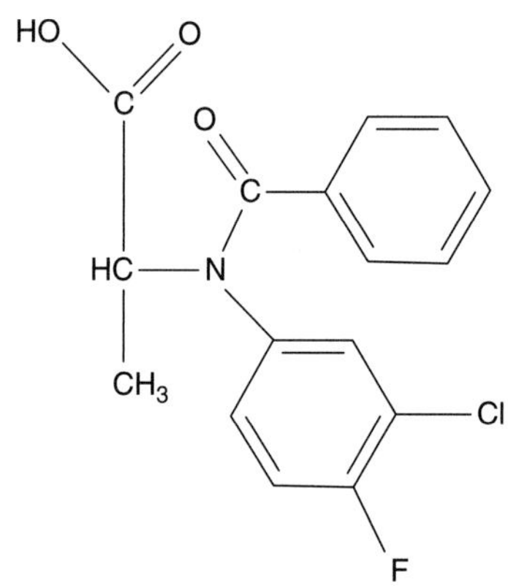

Flamprop

This herbicide is used in post-emergence control of wild oats (*Avena* spp.) in wheat. It also controls *Alopecurus myosuroides* (black grass/slender foxtail).

Flamprop is miscible with fungicides, **chlormequat chloride** and foliar nutrients but ought not to be mixed with phenoxy herbicides, for the risk of reducing its efficacy. It is commonly formulated as an emulsifiable concentrate (EC).

Flamprop is moderately toxic to mammals and birds (the acute oral LD_{50} in rats, mice and bobwhite quails are 1210, 720 and 4640 mg/kg, respectively). It is non-irritating to the skin. The LC_{50} of flamprop in rainbow trout is 4.7 mg/l. Flamprop is non-toxic to bees, worms and soil arthropods. The NOEL (2 years) of flamprop in rats is 2.5 mg a.i./kg of diet.

Flamprop is registered for use in Australia but disallowed throughout the EU and the USA.

Flamprop-M-methyl, also systemic in nature, is absorbed by the leaves. It inhibits plant growth at the meristem. It is hydrolysed to the biologically active flamprop. The CAS Registry

Number of flamprop-M-methyl is 63729-98-6. The MRL value of flamprop-methyl on wheat straw and fodder is commonly fixed at 0.1 mg/kg. Similarly, flamprop-M-isopropyl is a selective systemic herbicide, absorbed by the foliage. It undergoes hydrolysis to flamprop-M, and further reaches the meristem, where it inhibits elongation, division and growth of cells. It then undergoes conversion to a biologically-inactive conjugate. Mammals metabolize and eliminate flamprop-M-isopropyl within 4 days. The major soil degradate from both esters is free acid. Its CAS Registry Number is 63782-90-1.

Flamprop-M-isopropyl herbicide

See **Flamprop herbicide**

Flamprop-M-methyl herbicide

See **Flamprop herbicide**

Flash point

The temperature at which a volatile solid or liquid gives off vapour, which when mixes with the air around it and becomes combustible, is known as the flash point of the substance. There are methods developed to determine flash points, for example open cup or closed cup methods. Generally, liquids with a flash point below 37.7°C are considered flammable.

Flazasulfuron herbicide

Flazasulfuron, $C_{13}H_{12}F_3N_5O_5S$ (1-(4,6-dimethoxypyrimidin-2-yl)-3-(3-trifluoromethyl-2-pyridylsulfonyl) urea), is a pyrimidinylsulfonylurea herbicide. It is available as an odourless, white, crystalline powder. It is soluble in water (16.1 mg/l), acetone, acetic acid and toluene. It is produced by the reaction of 2-sulfamoylchloride-3-trifluoromethylpyridine with 2-amino-4,6-dimethoxypyrimidin. The CAS Registry Number of flazasulfuron is 104040-78-0 (see Flazasulfuron at bottom of the page).

Flazasulfuron, a systemic, pre- and post-emergence broad-spectrum herbicide, is an inhibitor of acetolactate synthase (ALS), a key enzyme for branched-chain amino acid synthesis which stops cell division and plant growth. When applied in sugarcane, grape and non-crop areas, this herbicide controls annual and perennial weeds, sedges, broadleaved weeds and turf (e.g. *Zoysia* and *Cynodon* spp.). It also causes discoloration, leaf necrosis and desiccation before killing the herbs.

The herbicide flazasulfuron is formulated as water-dispersible granules (WG) and wettable powder (WP). Foliar sprays from the ground level or spot treatment are recommended when the required dose is between 25 and 100 g a.i./ha.

The herbicide is not toxic to mammals, birds and fish at a practical level. The acute oral LD_{50} in rats and mice is >5000 mg/kg, while their dermal LD_{50} is >2000 mg/kg. It is non-irritating to the skin and eyes of rabbits. The chemical may be slightly toxic to honey bees in the case of prolonged exposure. According to European Commission (EC), the ADI is 0.013 mg/kg b.w./day and according to the USA a RfD is 0.5 mg/kg b.w.

Flazasulfuron is absorbed, metabolized and completely excreted within 7 days in animals. It is safe for birds, fish, bees and other **beneficial organisms**. The DT_{50} in water is 17.4 h at pH 4. The soil DT_{50} is <7 days.

Flazasulfuron is registered for use in Australia, Canada, the USA, Japan, the UK, Denmark, Finland, Germany, Greece, South Africa and Tanzania. The Japan Food Chemical Foundation has provisionally fixed 0.02 mg/kg as the MRL for mangoes.

Floaters

See **Aquatic weeds**

Floating hydrophytes

See **Aquatic weeds**

Flocoumafen rodenticide

Flocoumafen, $C_{33}H_{25}F_3O_4$, with IUPAC name 4-hydroxy-3-[1,2,3,4-tetrahydro-3-[4-(4-trifluoromethylbenzyloxy) phenyl]-1-naphtyl]coumarin (a mixture of *cis*- to *trans*-isomers in the ratio range of 60:40 to 40:60), is used as a rodenticide. It is a solid, powdery, whitish substance soluble in water (1.1 mg/l), and readily so in acetone (>600 g/l), ethanol (34 g/l) and octanol (44 g/l). Flocoumafen is stable up to 250°C and to hydrolysis from pH 7 to 9. Its CAS Registry Number is 90035-08-8 (see Flocoumafen at top of the next page).

A **coumarin** anticoagulant, flocoumafen is one of the most potent second-generation compounds. It works by inhibiting vitamin K_1 and blocking the production of **prothrombin**. It has applications in industry, fields, plantations and buildings, and works well even on resistant rodent strains. Target species include *Rattus norvegicus*, *Rattus rattus* and *Rattus argiventer*.

Common formulations are ready-to-use baits, grain baits, wax briquettes and pellets. A well known trade name is Storm.

The acute oral LD_{50} of flocoumafen among rats, mice, dogs and rabbits is 0.8 mg/kg or below. This compound is very toxic to dogs (with an oral LD_{50} of 0.25 to 0.75 mg/kg). The inhalation LC_{50} in rats is between 0.16 and 1.4 mg/l. A 0.005% bait concentrate is generally sufficient to kill the pest. The EPA and WHO have categorized flocoumafen as an extremely toxic substance (Class Ia).

Flocoumafen is metabolized to a number of hydroxycoumarins. (See also **Anticoagulant rodenticides**.)

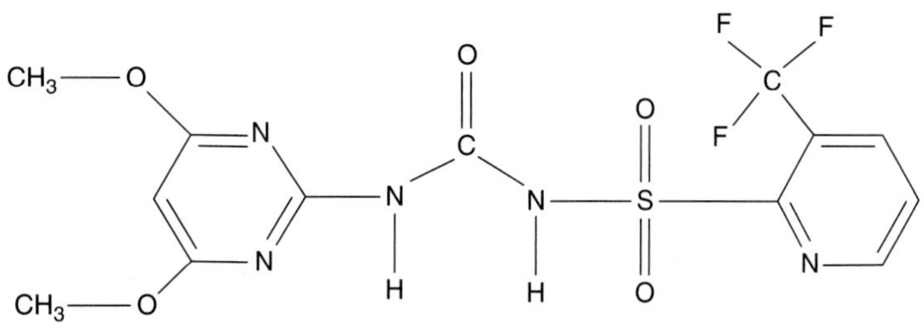

Flazasulfuron

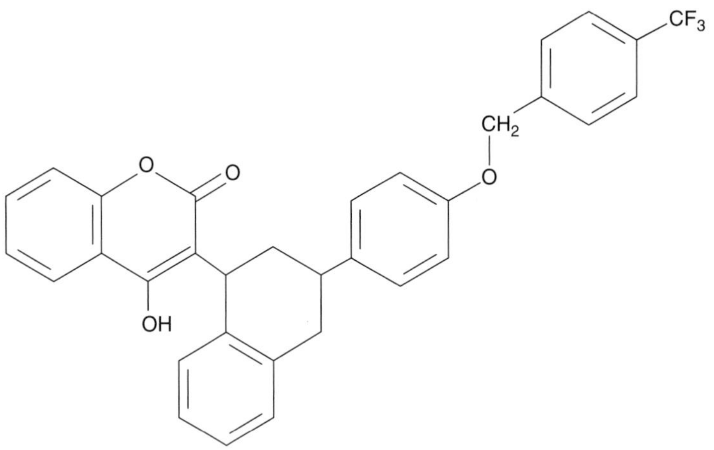

Flocoumafen

Florasulam herbicide

Florasulam, $C_{12}H_8F_3N_5O_3S$ (2,6,8-trifluoro-5-methoxy [1,2,4] triazolo[1,5-c]pyrimidine-2-sulfonanilide), is a triazolopyrimide sulfonanilide herbicide. Its solubility in water is 0.121 g/l, in acetone 123 g/l, and in acetonitrile 72.1 g/l. Florasulam is stable under normal storage conditions, away from moisture and extreme temperatures and also for 30 days in the pH range of 5–7. It is incompatible with strong bases, acids and oxidizing materials. The CAS Registry Number is 145701-23-1.

Florasulam is a selective, post-emergence herbicide that acts by inhibiting the plant enzyme acetolactate synthase (ALS), thus blocking growth processes leading to death of the plant. It controls broadleaved weeds, such as the annual cleavers or stickywilly (*Galium aparine*), chickweed (*Stellaria media*), buckwheat and mayweed. The crops protected include cereals such as maize, wheat, barley, oats and rye.

Florasulam is commonly applied at rates from 4 to 7.5 g/ha for wheat, barley, oats and maize, with applications on maize being at the higher range. The compound is effective when applied as a single application during weed growth stage (between two leaf and flag leaf emergence stage). The recommended formulation type is the suspension concentrate (SC). Ground and aerial equipment may be used.

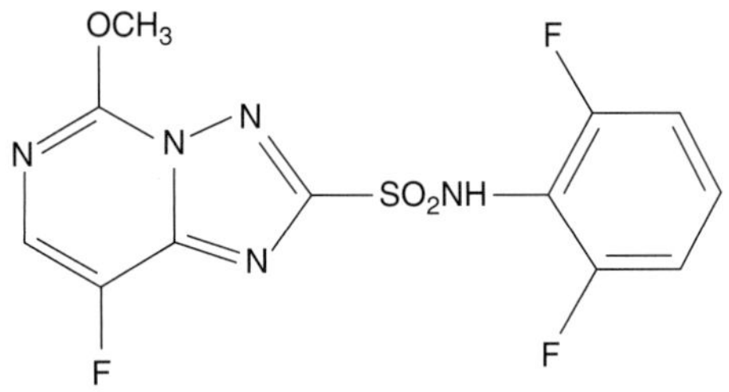

Florasulam

The acute oral LD_{50} values of florasulam for rats, mice and quails are >5000 mg/kg (former two) and 1046 mg/kg, respectively. The acute dermal LD_{50} of florasulam for rat is >2000 mg/kg. The acute LC_{50} at 96 h in bluegill sunfish and rainbow trout is >98 mg/l and >96 mg/l, respectively. The LC_{50} of florasulam in *Daphnia* sp. at 48 h is >292 mg/l. The LD_{50} of florasulam in bees at 48 h is >100 μg per bee (oral and contact). The LC_{50} of florasulam for earthworm for 14 days is >1320 mg/kg. Florasulam is not toxic to bees, fish and birds on an acute basis, but is highly toxic to algae. The NOEL of florasulam for 90 days in rats and mice is 100 mg/kg b.w. According to European Commission, the ADI of florasulam is 0.05 mg/kg b.w./day. Applicators and handlers are recommended to avoid skin and eye contact and advised to bathe, and clean/wash garments and accessories after handling the compound.

Upon oral administration in test mammals, florasulam is rapidly eliminated as the unchanged product. In plants, florasulam is rapidly metabolized via hydroxylation in the 4-position of the phenyl ring followed by other reactions, leading to small polar components. Florasulam degrades rapidly in soil, with a half-life ranging from 2 to 18 days. Biomagnification potential for florasulam is low.

This herbicide is registered in most of the countries of the EU including the UK, France, Germany, Greece, Spain and Belgium. It is also registered in the USA, Canada, Australia, South Africa and Tanzania.

Florisil

Florisil is an extremely polar and weakly alkaline sorbent used in pesticide analysis. It is made of magnesium oxide and silica gel.

Flowables

See **Suspension concentrates (SC)**

Fluazifop herbicide

Fluazifop, $C_{15}H_{12}F_3NO_4$ ((*RS*)-2-{4-[5-(trifluoromethyl)-2-pyridyloxy] phenoxy} propionic acid), is an aryloxyphenoxypropionic herbicide. In its salt or ester form, it may exist as fluazifop-butyl or fluazifop-methyl or fluazifop-p. The (*R*)-isomer of this substance has the ISO common name of fluazifop-p. The CAS Registry Number of fluazifop is 69335-91-7. Fluazifop-p-butyl is the *n*-butyl ester of the acid fluazifop-p. Fluazifop-p is the *R*-enantiomer of the acid, while fluazifop typically contains equal proportions of both the *R*- and the *S*-enantiomers. Fluazifop-p is the herbicidally active portion of the mixture.

Fluazifop, a selective, systemic, post-emergence compound, is absorbed by the leaves of susceptible grass weeds (e.g. quack grass (*Elytrigia repens*) and proso millet) and interferes with their ATP production. It is used in cotton, soybeans, citrus/pome fruit, groundnuts, potatoes and bananas. Broadleaved weeds break this herbicide down and are not controlled.

Common application rates range from 0.14 to 0.36 kg/ha, depending on the severity of the situation. Fluazifop is not registered in the USA.

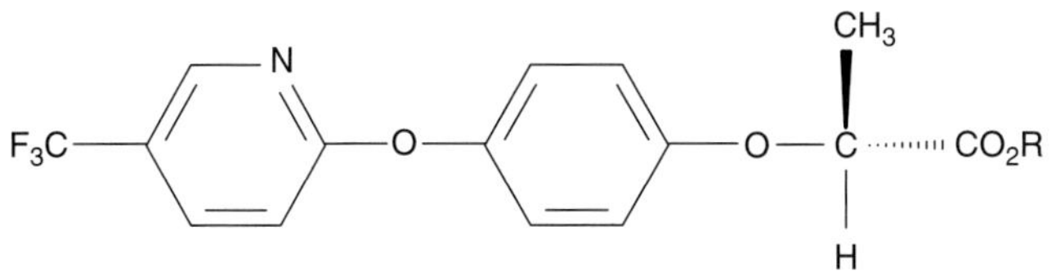

R = H (fluazifop; *R*-enantiomer only is fluazifop-p)

R = —(CH$_2$)$_3$ CH$_3$ (fluazifop-butyl; *R*-enantiomer only is fluazifop-p-butyl)

Fluazifop

Fluazifop-butyl, $C_{19}H_{20}F_3NO_4$ ((*RS*)-2-{4-[5-(trifluoromethyl)-2-pyridyloxy] phenoxy} propionate), is an aryloxyphenoxypropionic herbicide. It is a creamish, odourless liquid, soluble in water (1 mg/l) and miscible with acetone, cyclohexanone, hexane, methanol, dichloromethane and xylene. It rapidly hydrolyses in alkaline medium, and hydrolyses to fluazifop. It is applied at between 175 and 350 g/ha in field crops. Fluazifop-butyl is compatible with many herbicides, but this combination is not without the risk to the crop and reduced efficacy of fluazifop-butyl. Fluazifop-butyl is available in the form of emulsifiable concentrates (EC) and wettable powder (WP). Common trade names include Fusilade 2000 and Onecide.

Fluazifop-butyl, like fluazifop does not pose a serious threat to mammals, bees and birds, under normal use (the acute oral LD_{50} values in male and female rats are >2000 mg/kg and >3600 mg/kg, respectively). According to the EPA, its ADI value is 0.01 mg/kg b.w./day. Handling or mixing the compound should proceed with a full-face shield, protective clothing and PVC gloves to avoid accidents of a serious nature. According to the PAN Pesticide Database, fluazifop-butyl is a potential groundwater contaminant and a suspected endocrine disruptor.

In plants and in rats, fluazifop-butyl is rapidly hydrolysed to fluazifop which is not further metabolized. Fluazifop-butyl is rapidly degraded to fluazifop in moist soil, the action being accelerated by microbes. With a half-life of less than 1 week, the degradation product is fluazifop. Post-harvest residues on some crops like potatoes were found to be <0.05 μg/g.

Fluazifop-butyl is not registered in the USA, but it is registered in Tanzania and Vietnam.

Fluazifop-p, $C_{15}H_{12}F_3NO_4$ ((R)-2-(4-((5-(trifluoromethyl)-2-pyridyloxy) phenoxy) propionic acid), is also a grass-selective herbicide. The CAS Registry Number is 83066-88-0.

Fluazifop-p harms most grasses, including pasture and bent grass species, but not broadleaved plants such as clovers. The application rates are 0.4 kg/ha on post-emergence weeds. However, they may be phytotoxic to lilies and iris.

The half-life of fluazifop-p is around 3 weeks, but it is not mobile, and thus not a threat to groundwater.

Fluazifop-p-butyl, $C_{19}H_{20}F_3NO_4$ ((*R*)-2-(4-((5-(trifluoromethyl)-2-pyridyloxy) phenoxy) propionic acid), is also an aryloxyphenoxypropionic herbicide. It is a pale straw-coloured liquid, soluble in water (1 mg/l) and miscible with acetone, hexane, methanol, dichloromethane, ethyl acetate, toluene and xylene. Fluazifop-p-butyl is stable to UV light for over a year, but decomposes at 210°C. It hydrolyses to fluazifop-p, once absorbed by the leaf, before being stored in the meristem.

Via its mode of action, fluazifop-p-butyl can control *Bromus* spp., *Elytrigia repens* (quack grass) and *Panicum* spp. (witch-grasses). It is a fatty acid synthesis inhibitor, which disrupts the production of acetyl CoA carboxylase.

However, it does not affect broadleaved species. It is available as an emulsifiable concentrate (EC). The addition of an oil adjuvant or non-ionic surfactant increases efficiency. The formulations may contain fluazifop-butyl. Trade names include Horizon and Fusion.

As with others of this group, fluazifop-p-butyl is not very toxic in normal use. The WHO classifies it as a slightly hazardous chemical, with acute oral LD_{50} values for female and male rats being 2451 mg/kg and 3680 mg/kg, respectively. The dermal LD_{50} for the compound is >2450 mg/kg for rabbits. Though it is practically non-toxic to birds, it may become toxic to fish, aquatic invertebrates and bees. The acute LC_{50} of fluazifop-p-butyl in rainbow trout is 1.07 mg/l, while in bees the LD_{50} is >0.2 mg per bee (oral and contact). It is only a mild skin and eye irritant and causes no skin sensitization. Labels must carry the signal word 'CAUTION'. Fluazifop-p-butyl is non-phytotoxic to broadleaved crops.

In plants, fluazifop-p-butyl is rapidly broken down in the presence of water. It breaks down rapidly in moist soils, with a half-life >1 week. It has low mobility in soils and hence does not contaminate groundwater. The main breakdown product is fluazifop acid.

Fluazifop-butyl herbicide

See **Fluazifop herbicide**

Fluazifop-p herbicide

See **Fluazifop herbicide**

Fluazifop-p-butyl herbicide

See **Fluazifop herbicide**

Fluazinam fungicide

Fluazinam, $C_{13}H_4C_{12}F_6N_4O_4$ (3-chloro-*N*-(3-chloro-5-trifluoromethyl-2-pyridyl)-α,α,α-trifluoro-2,5-dinitro-*p*-toluidine), is a fungicide with specific acaricidal activity. It shares some structural similarities with **dinocap**, in that it is a dinitrophenyl derivative, although, in this case, a derivative of 2,6-dinitroaniline. It is a light yellow, crystalline substance soluble in water (1.7 mg/l), ethanol and dichloromethane. It is stable to acidic and alkaline media and to heat. The CAS Registry Number of fluazinam is 79622-59-6.

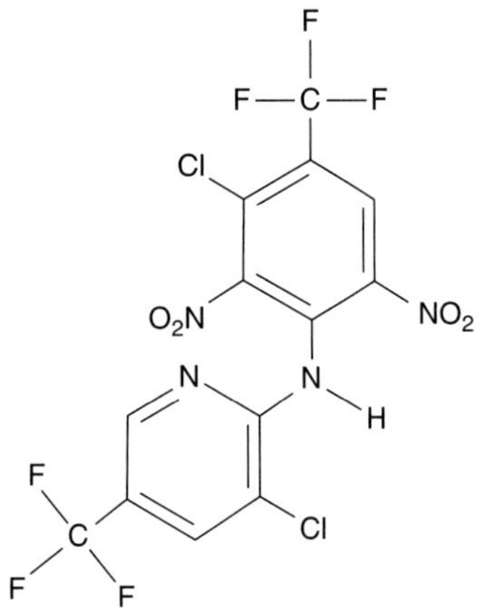

Fluazinam

Fluazinam, a protective, broad-spectrum, non-systemic fungicide, uncouples the process of mitochondrial oxidative phosphorylation, thus inhibiting spore germination and appressorial formation and in turn, preventing host penetration.

The compound controls **downy mildew**, mould conditions and certain types of blights on vines, apples, groundnuts and potatoes. It is also used in the control of mites in apples. It has a good residual effect and rainfastness, which gives long-term protection.

Fluazinam is formulated as a suspension concentrate (SC), wettable powder (WP) or a dustable powder (DP), and is available under the trade name of Shirlan, among others.

Fluazinam shows a relatively low mammalian toxicity with an acute oral LD_{50} of >5000 mg/kg and an acute dermal LD_{50} of >2000 mg/kg in rats. It is toxic to fish.

The USA and all (but two) members of the EU have registered this compound.

Fluazuron acaricide

Fluazuron, $C_{20}H_{10}Cl_2F_5N_3O_3$ (3-[3-(3-chloro-5-trifluoromethyl-2-pyridinyloxy)-4-chlorophenyl]-1-(2,6-difluorobenzoyl)-urea), is an acaricide on animals. It is a crystalline, pinkish powder, insoluble in water and soluble in organic solvents such as methylene chloride, cyclohexanone and isopropanol. It is stable up to 219°C. The CAS Registry Number of fluazuron is 86811-58-7 (see Fluazuron at bottom of the page).

Fluazuron is used in cattle for the control of a cattle tick species called *Boophilus microplus*. This acaricide acts by inhibiting **chitin** incorporation into tick cuticle, thus interrupting the life cycle of immature ticks. This compound is formulated as a pour-on solution. It is available in the market under trade names such as Acatak.

According to the WHO, fluazuron comes under the toxicity Class III (slightly hazardous).

Flucarbazone-sodium herbicide

Flucarbazone-sodium, $C_{12}H_{10}F_3N_4NaO_6S$ (sodium 4,5-dihydro-3-methoxy-4-methyl-5-oxo-*N*-{[2-(trifluoromethoxy) phenyl] sulfonyl}-1*H*-1, 2,4-triazole-1-carboximidate), is a sulfonylaminocarbonyltriazolinone herbicide. Flucarbazone-sodium, a colourless, odourless crystalline powder, is soluble in water (44,000 mg/l). Its CAS Registry Number is 181274-17-9.

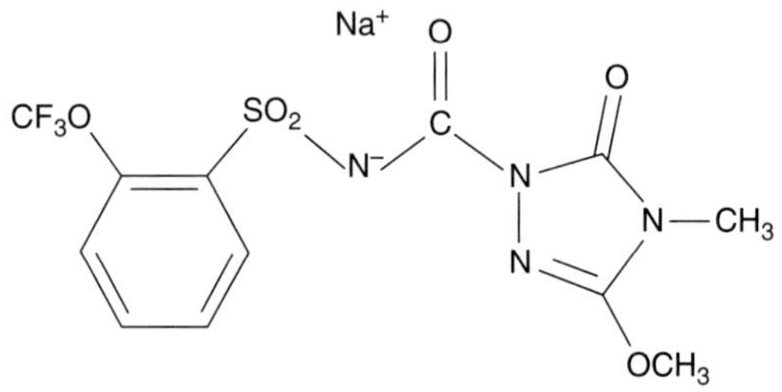

Flucarbazone-sodium

Flucarbazone-sodium, a post-emergence herbicide, absorbed through the roots and foliage, inhibits amino acid synthesis (**ACCase** inhibition). Its targets are wild oats (*Avena fatua*), foxtail (*Setaria viridis*) and *Amaranthus retroflexus*. The action results in stunting and discoloration.

The compound comes in water-dispersible granules (WG), applications of which are to be restricted to one a season with a pre-harvest interval of at least 2 months, especially when rotated with oats and mustard. It may be applied at 21 g/ha. Trade names include Everest.

This herbicide is practically non-toxic to rats, birds, earthworms and fish (with the acute oral LD_{50} and acute percutaneous LD_{50} in rats being >5000 mg/kg, the acute LD_{50} in bobwhite quails being >2000 mg/kg and the acute LC_{50} in earthworms being 1000 mg/kg). The WHO has categorized the compound as unlikely to be hazardous. The acute **reference dose (aRfD)** is 3.0 mg/kg b.w.

Flucarbazone-sodium is almost completely excreted via faeces and urine within 48 h of oral administration to rats. The main metabolite is *N*-desmethyl. The average soil DT_{50} is 17 days while the photolytic DT_{50} is >500 days. The compound does not leach below 30 cm into the soil. The soil half-life is up to 24 days.

Flucarbazone-sodium is registered for use in Canada and the USA.

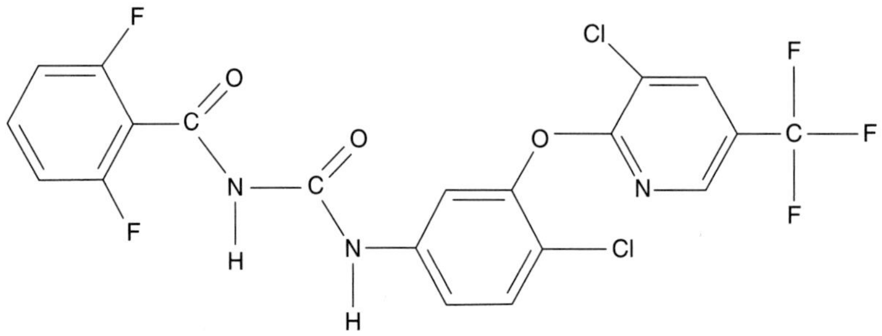

Fluazuron

Flucetosulfuron herbicide

Flucetosulfuron, $C_{18}H_{22}FN_5O_8S$ ((1*RS*, 2RS; 1RS, 2SR)-1-{3-[(4,6-dimethoxypyrimidin-2-ylcarbamoyl) sulfamoyl]-2-pyridyl}-2-fluoropropyl methoxyacetate), is a pyrimidinylsulfonylurea herbicide. It is an odourless white solid, soluble in water (at 114.0 mg/l at 20°C), acetone, ethyl acetate, *n*-hexane and methanol. The CAS Registry Number is 412928-75-7.

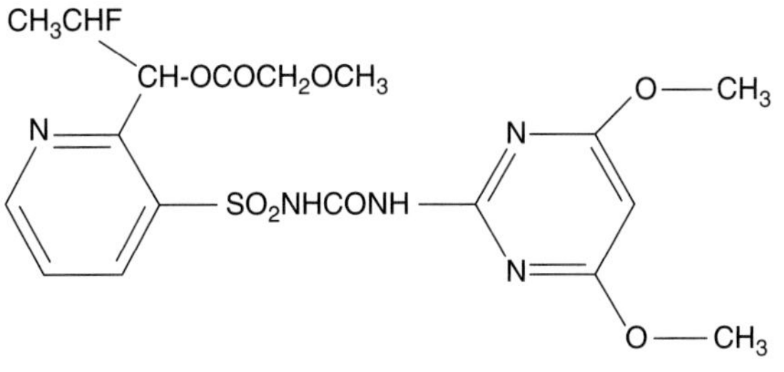

Flucetosulfuron

Flucetosulfuron, a broad-spectrum, systemic herbicide, inhibits acetolactase synthase (ALS), thus causing **chlorosis** of the plant, leading to death of apical meristems. It is mainly used on cereals, controlling a wide range of broadleaved weeds, grass weeds and sedges, like **barnyard grass** (*Echinochloa crus-galli*), cleavers (*Galium aparine*), chamomile and poppy.

Common formulations include granules (GR) and water-dispersible granules (WG), applied to both soil and foliage at 15–30 g/ha, especially on rice.

The herbicide is practically non-toxic to rats, mice and dogs (acute oral LD_{50} in rats, male and female mice is >5000 mg/kg, while in dogs it is >2000 mg/kg). The acute LC_{50} of flucetosulfuron for carp and *Daphnia* is 10 ppm.

Flucetosulfuron is registered for use in Vietnam.

Fluchloralin herbicide

Fluchloralin, $C_{12}H_{13}ClF_3N_3O_4$ (*N*-(2-chloroethyl)-2, 6-dinitro-*N*-propyl-4-(trifluoromethyl) aniline), is a 2,6-dinitroaniline herbicide. Fluchloralin is an orangish-yellow solid, soluble in water (<1 mg/l) and highly soluble in acetone, benzene and cyclohexane (1 kg/l). It is decomposed by UV light but is otherwise stable when stored in original containers in dry conditions at room temperature. It is produced by the reaction of 2,6-dinitro-4-trifluoromethylbenzene with *N*-chloroethyl-*N*-isopropylamine. The CAS Registry Number is 33245-39-5.

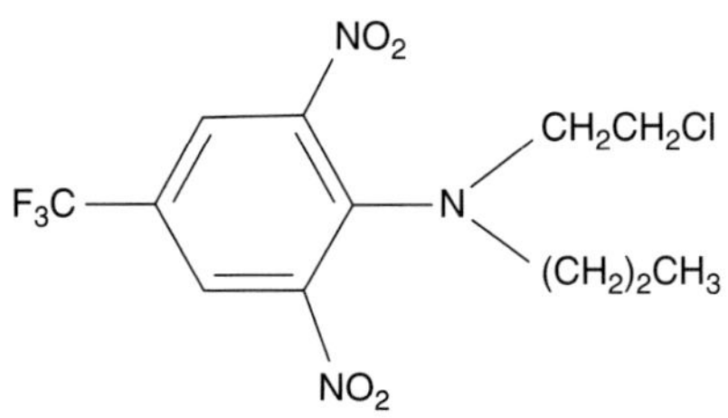

Fluchloralin

Fluchloralin, a selective, pre-emergence, pre-plant soil herbicide, inhibits seed germination and growth processes. Fluchloralin is used to control annual grasses and some broad-leaved weeds in cotton, transplanted rice, soybeans, onions, groundnuts, potatoes, jute, sunflowers and vegetables (e.g. chillies, bean varieties, tomatoes, cabbage and lady's finger).

Commonly accepted formulations of fluchloralin are emulsifiable concentrates, with application dosage ranging from 0.5 to 1.5 kg/ha.

The WHO has classified the compound as a slightly hazardous chemical. The acute oral LD_{50} values in rats, rabbits, mice, dogs, mallard ducks and bobwhite quails are 1550 mg/kg, 8000 mg/kg, 730 mg/kg, 6400 mg/kg, 13,000 mg/kg and 7000 mg/kg, respectively. It is a mild eye and skin irritant. Care has to be taken to avoid exposure via dermal contact or inhalation. It is not toxic to bees, however is toxic to fish (with the LC_{50} in rainbow trout being 0.012 mg/l). It may be phytotoxic to some finger millet and rice varieties if applied without sufficient care (and especially if mixed with DBCP (dibromochloropropane), a fungicide and nematicide. Proprietary formulations must warn against discharging effluents into water bodies, without the permission of the concerned authority.

Fluchloralin degrades rapidly (DT_{50} of 5 h) in the vapour-phase by reacting with photochemically produced hydroxyl radicals, when released to the atmosphere. In soil or water, fluchloralin dissipates rapidly through photodegradation, volatilization and microbial action, with a DT_{50} of up to 4 days. It gets adsorbed to clay colloids and does not leach.

The herbicide is registered in India, although the USA does not permit its use.

Flucycloxuron acaricide

Flucycloxuron, $C_{25}H_{20}ClF_2N_3O_3$ (1-{α-[(*EZ*)-4-chloro-α-cyclopropylbenzylideneamino-oxy]-*p*-tolyl}-3-(2,6-difluorobenzoyl) urea), is a benzoylurea acaricide and an insect growth regulator. It exists as off-white to yellow crystals, insoluble in water, sparingly soluble in cyclohexane, xylene, ethanol and soluble in *N*-methylpyrrolidone. Flucycloxuron is produced by the reaction of 4-chlorocyclopropylphenone with hydroxylamine to yield the corresponding oxime, which is treated with 4-nitrobenzylbromide, followed by reduction to give the corresponding amine that is further treated with 2,6-difluorobenzoyl isocyanate. The CAS Registry Number is 113036-88-7 (see Flucycloxuron at top of the next page).

Flucycloxuron is a **chitin synthesis inhibitor** and a mite growth regulator, active against pest egg and larval stages, on a variety of crops. Trade names include Andalin.

The WHO classifies flucycloxuron as a Class U toxin (unlikely to be hazardous). It is non-irritating to skin but mildly irritating to eyes. It is non-mutagenic, non-teratogenic and non-carcinogenic. The acute oral LD_{50} in rats is >5000 mg/kg. The acute dermal LD_{50} in rats is >2000 mg/kg. The NOEL in rats (2 year feed) is about 6 mg/kg/day.

Flucycloxuron is not a PAN Bad Actor chemical. It is not registered by the EPA for use in the USA.

Flucythrinate insecticide

Flucythrinate, $C_{26}H_{23}F_2NO_4$ ((*RS*)-α-cyano-3-phenoxybenzyl (*S*)-2-(4-difluoromethoxyphenyl)-3-methylbutyrate), is a pyrethroid insecticide. It is an amber liquid, almost insoluble in water (0.5 mg/l) but highly soluble in acetone, xylene and corn (maize)/soybean oil. It hydrolyses in water under alkaline conditions with a DT_{50} at pH 9 being 6.3 days. It degrades in sunlight. It is produced by the reaction of 1-(4-difluoromethoxyphenyl-2-methyl) butyric acid with α-cyano-3-phenoxybenzyl chloride. The CAS Registry Number of flucythrinate insecticide is 70124-77-5 (see Flucythrinate at bottom of the next page).

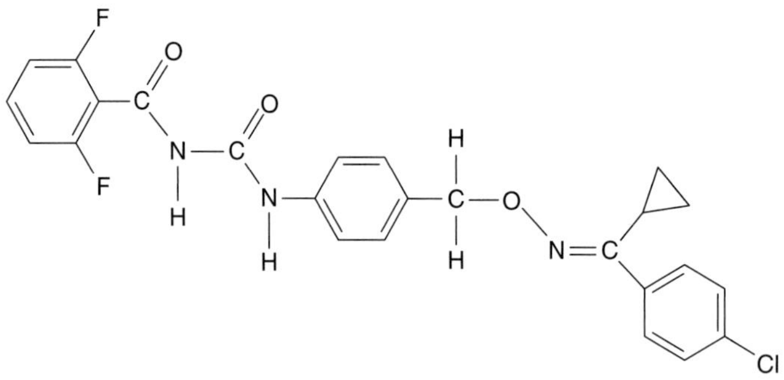

(*E*)-isomer

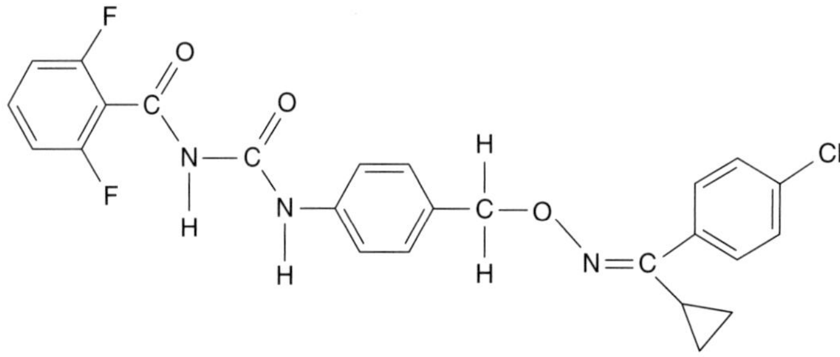

(*Z*)-isomer

Flucycloxuron

A non-systemic, broad-spectrum compound, flucythrinate has contact and stomach action, is mainly used on cotton, but is also suited for a variety of fruits, cereal crops, coffee, cocoa and tobacco. A wide variety of lepidopteran, homopteran and coleopteran pests are controlled, especially *Heliothis* spp., bollworms, leafworms, whitefly and mite species.

Flucythrinate is commonly formulated as an emulsifiable concentrate (EC) or as a wettable powder (WP). It may be mixed with **chlorpyrifos**, **phenthoate** and **dimethoate**. Commonly known trade names are Cybolt, Cythrin, AASTAR and Pay-off.

Flucythrinate is a Class Ib toxin as classified by the WHO (which indicates a highly hazardous pesticide). Label instructions are to be strictly followed. Some formulations are classified as restricted use. On mammals, it is highly toxic via the oral route (LD_{50} in rats is between 76 and 81 mg/kg). Via the inhalation route, flucythrinate is moderately toxic. Although flucythrinate is almost non-toxic to birds, it is very highly toxic to fish species and bees (LD_{50} is 0.078 μg per bee). Due to its acute toxicity, flucythrinate is considered a PAN Bad Actor chemical. The ADI of flucythrinate is 0.02 mg/kg b.w./day.

Mammals and humans eliminate most of flucythrinate within 24 h via the urine and faeces. Although most of the parent compound is eliminated unchanged, hydrolysis is the route of degradation. In soils and water, it shows low mobility and leaching.

Cambodia has registered the use of this pesticide. The USA does not permit its use.

Fludioxonil biofungicide

Fludioxonil, $C_{12}H_6F_2N_2O_2$ (4-(2,2-difluoro-1, 3-benzodioxol-4-yl)-1*H*-pyrrole-3-carbonitrile), is a cynopyrrole fungicide (**phenylpyrrole**). Originally synthesized out of the natural antibiotic pyrrolnitrin of *Pseudomonas pyrrocinia*, fludioxonil is a colourless crystalline substance, soluble in water (1.8 mg/l), acetone, ethanol, *n*-octanol and hexane. The CAS Registry Number of fludioxonil is 131341-86-1.

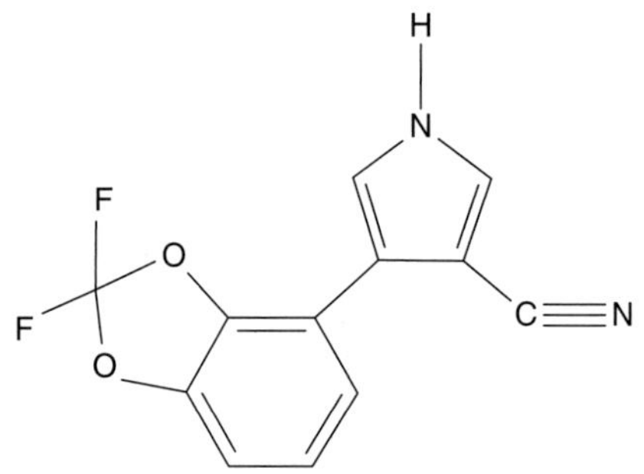

Fludioxonil

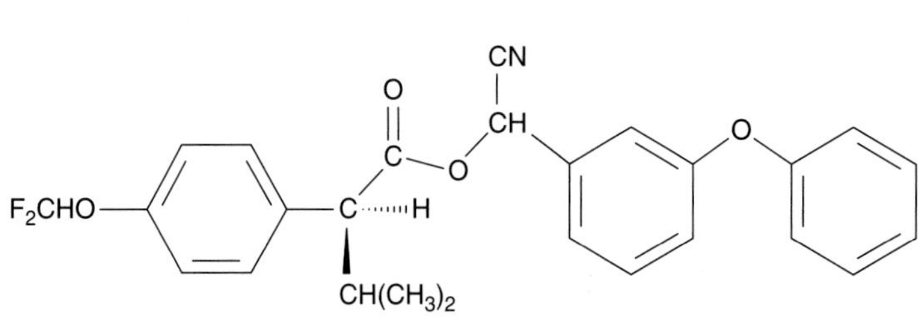

Flucythrinate

Fludioxonil, a contact compound, is used to control fungal pathogens mainly infecting seeds and foliage of cereal crops, vines, stone fruit and vegetables. As a seed treatment agent, species of *Gibberella* (in rice), *Fusarium*, *Tilletia*, *Helminthosporium* and *Septoria* are controlled. As a foliar fungicide, crops are protected against *Botrytis*, *Sclerotinia*, *Rhizoctonia* and *Alternaria*.

The fungicide is formulated as a wettable powder (WP), a flowable concentrate (SC) and as water-dispersible granules (WG). It may be combined with **tefluthrin** and **metalaxyl-M**. Fludioxonil is available in the market under the trade name of Celest.

Fludioxonil is known to be persistent in soil, with a typical half-life of over 125 days, while it is stable to aqueous hydrolysis. An ADI in humans is 0.37 mg/kg b.w./day. It has relatively high toxicity to fish, aquatic invertebrates and aquatic plants and slight toxicity to bees. Some studies show that it could contaminate groundwater sources.

Fludioxonil is registered in Australia, Canada, the USA, Japan and throughout the EU. The MRL for strawberry and wine grapes are 2 ppm.

Flufenacet herbicide

Flufenacet, $C_{14}H_{13}F_4N_3O_2S$ (4′-fluoro-*N*-isopropyl-2-[5-(trifluoromethyl)-1, 3,4-thiadiazol-2-yloxy] acetanilide), is a herbicide. Formerly known as fluthiamide, it is a whitish solid, soluble in water at pH 4 (56 mg/l), acetone, dichloromethane, dimethylformamide, toluene and dimethyl sulfoxide, isopropanol and hexane. Flufenacet is stable to hydrolysis and also photolysis at pH 5 and at normal storage temperatures. Flufenacet is moderately stable in both aerobic and anaerobic soil environment. The CAS Registry Number is 142459-58-3.

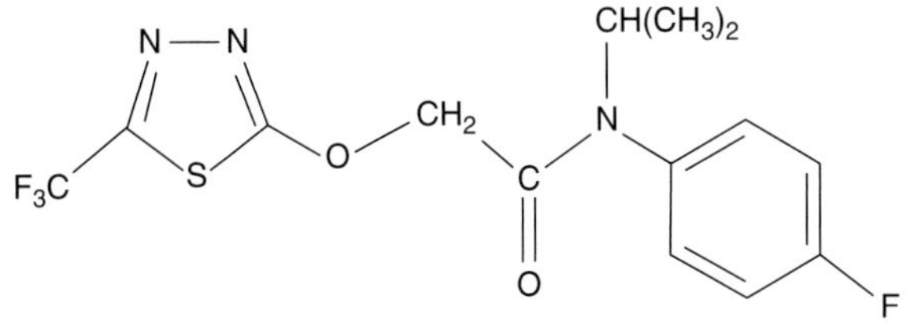

Flufenacet

Flufenacet is a pre- and post-emergence, systemic herbicide with meristematic activity. It acts by inhibiting cell division and cell growth, by inhibiting fatty acid metabolism. Black grass (*Alopecurus myosuroides*), wind bentgrass (*Apera spica-venti*) and **barnyard grass** (*Echinochloa crus-galli*) are some of the weeds controlled in potato, soybean and maize fields at the pre-emergence stage, while it is applied to maize, wheat and rice at the post-emergence stage. In the case of tomatoes, flufenacet is applied at the pre-plant stage.

There are many formulations of flufenacet and they include emulsifiable concentrate (EC), granule (GR), suspension concentrate (SC), water-dispersible granule (WG) and wettable powder (WP). Trade names include FOE 5043 DF Herbicide, Axiom DF Herbicide and Cadou. Flufenacet can be applied with **terbuthylazine** (selective herbicide), **diflufenican** (foliar pre- and post-emergence herbicide), **isoxaflutole**, **metosulam** and **metribuzin**, to increase the control spectrum. They are applied at 0.5–1.5 kg/ha. Italian ryegrass can detoxify the herbicide and is thus not affected by flufenacet.

The WHO has categorized flufenacet as a slightly hazardous chemical. The acute oral LD_{50} in rats ranges from 589 to 1617 mg/kg, and for mice, from 1331 to 1756 mg/kg. The herbicide does not cause skin or eye irritation. The acute LD_{50} in bobwhite quails is 1608 mg/kg, while the acute LC_{50} in fish is 2.4–3.49 ppm. The RfD is 0.004 mg/kg/day.

Upon intake, flufenacet is rapidly excreted by animals and plants, with none of the parent material being detected. The soil DT_{50} is 10–54 days, carbon dioxide being the by-product. This herbicide is registered in India, the USA and many countries in the EU.

Flufenoxuron acaricide

Flufenoxuron, $C_{21}H_{11}ClF_6N_2O_3$ (1-[4-(2-chloro-α,α,α-trifluoro-*p*-tolyloxy)-2-fluorophenyl]-3-(2,6-difluorobenzoyl) urea), is a benzoylurea acaricide. It is a colourless, crystalline solid that is almost insoluble in water and moderately soluble in acetone, xylene and dichloromethane. Flufenoxuron is produced by a series of reactions involving 2-fluoro-4-hydroxyaniline, dimethyl sulfoxide and 3,4-dichlorotrifluoromethyltoluene. Its CAS Registry Number is 101463-69-8 (see Flufenoxuron at bottom of the page).

Flufenoxuron is a **chitin synthesis inhibitor** and a mite and insect growth regulator, with contact and stomach action. It controls the immature stages of many phytophagous mites and insect pests on pome fruits, vines, citrus fruits, tea and ornamentals.

The acaricide flufenoxuron is available in a variety of formulations including dispersible powders, emulsifiable concentrates and suspension concentrates. Trade names of flufenoxuron include Cascade.

The WHO classifies flufenoxuron as unlikely to be hazardous. It is non-irritating to the skin and eyes. The acute oral LD_{50} in rats is >3000 mg/kg, while the acute dermal LD_{50} in rats is >2000 mg/kg. It is extremely toxic to *Daphnia* sp.

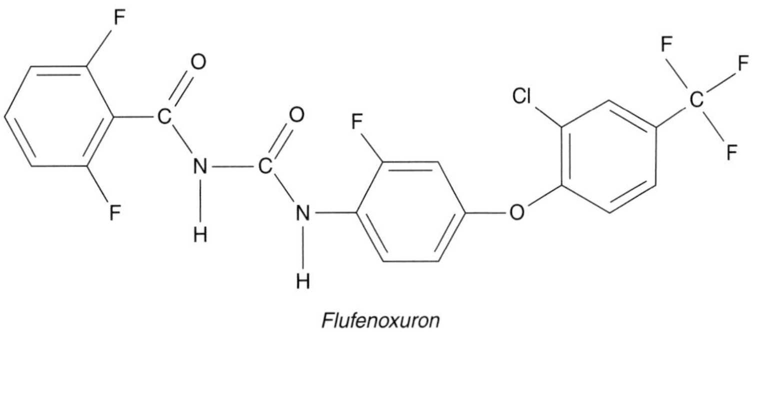

Flufenoxuron

Flufenoxuron persists in soil. It bio-accumulates in fish and aquatic gastropods. The compound is registered in Australia, India, Canada and South Africa but not in the USA.

Flufenpyr-ethyl herbicide

Flufenpyr-ethyl, $C_{16}H_{13}ClF_4N_2O_4$ (ethyl 2-chloro-5-[1,6-dihydro-5-methyl-6-oxo-4-(trifluoromethyl)pyridazin-1-yl]-4-fluorophenoxyacetate), is a **pyridazinone** herbicide. It is the derivative of flufenpyr. The CAS Registry Number is 188489-07-8.

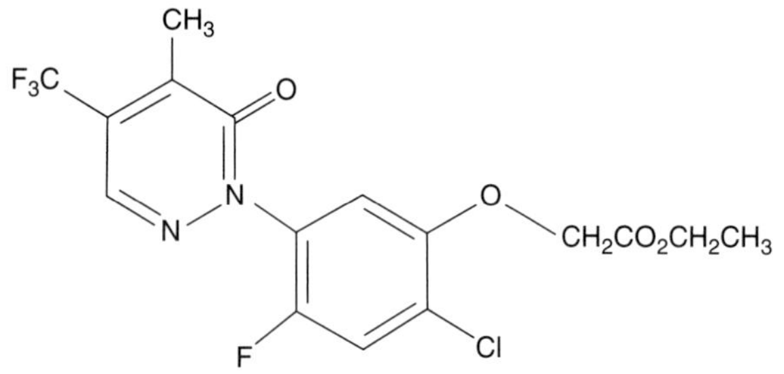

Flufenpyr-ethyl

Flufenpyr-ethyl acts as a protoporphyrinogen oxidase inhibitor causing protoporphyrins to accumulate, damaging the membrane structure and cellular function. It controls velvetleaf and morning glory in maize, soybean and sugarcane.

The compound is commonly formulated as water-dispersible granules (WG). The rate of application is 30 g/ha. The 90 day oral **NOAEL** in male mice is 395 mg/kg/day.

Flufenpyr-ethyl is registered in the USA, although flufenpyr is not registered.

Flumetralin as a PGR

Flumetralin, $C_{16}H_{12}ClF_4N_3O_4$ (*N*-(2-chloro-6-fluorobenzyl)-*N*-ethyl-α,α,α-trifluoro-2, 6-dinitro-p-toluidine), is a plant growth regulator. It is a yellowish orange, odourless, crystalline substance, soluble in water, acetone, toluene, ethanol, *n*-hexane and *n*-octanol. It decomposes exothermally above 250°C and is stable between pH 5 and 9. The CAS Registry Number of flumetralin is 62924-70-3.

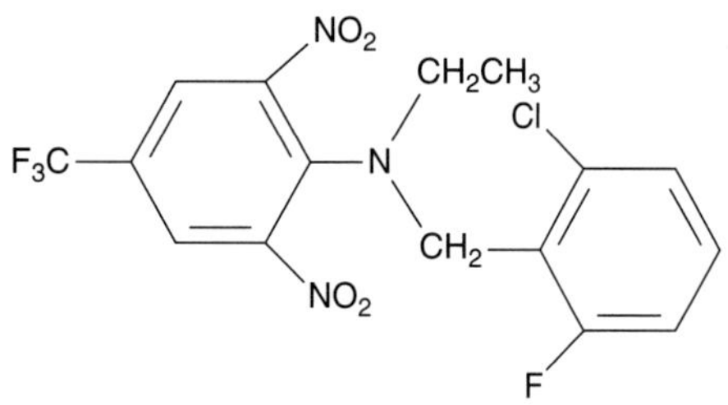

Flumetralin

Flumetralin primarily controls the growth of auxiliary buds in tobacco crops. It is formulated as an emulsifiable concentrate (EC). It is available in the market under trade names such as Prime. It has a long residual activity which renders season-long control.

According to the WHO, flumetralin is unlikely to present acute hazard in normal use. It is toxic to fish but not toxic to bees. It may injure immature plants, if not used as per label directions.

In animals, metabolic reactions include reduction of nitro groups, acetylation of the amino groups and hydroxylation of the phenyl ring. It is registered in the USA, but disallowed throughout the EU.

Flumetsulam herbicide

Flumetsulam, $C_{12}H_9F_2N_5O_2S$ (2′,6′-difluoro-5-methyl [1,2,4] triazolo [1,5-*a*] pyrimidine-2-sulfonanilide), is a triazolopyrimidine herbicide. It is an off-white odourless solid, soluble in water (49 mg/l) at pH 2.5, the solubility increasing with pH. It is very slightly soluble in acetone and methanol and insoluble in hexane and xylene. It is stable in a safe, airtight container up to 50°C. The CAS Registry Number is 98967-40-9.

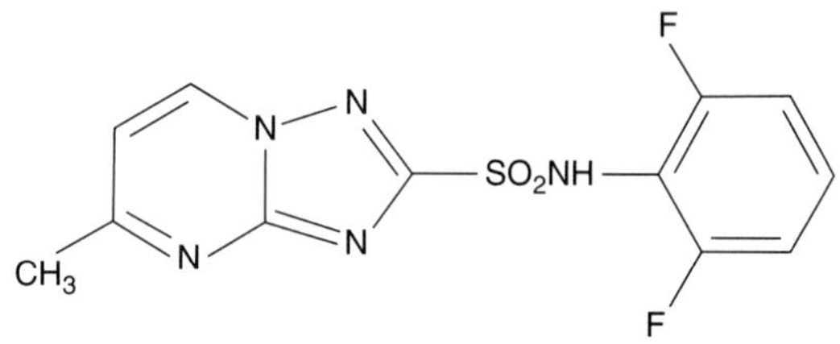

Flumetsulam

Flumetsulam, which is both a pre- and post-emergence systemic herbicide, gets absorbed through the foliage and roots and inhibits enzyme acetolactate synthase (ALS) essential for the synthesis of amino acids leucine, isoleucine and valine. Broadleaved weeds and grasses growing in soybeans, field peas and maize are targets of flumetsulam.

Flumetsulam is used in combination with other agents such as **trifluralin, metolachlor, acetochlor** and **clopyralid**. It is available in oil-miscible flowable concentrate (OF), suspension concentrate (SC) and water-dispersible granule (WG) formulations. Trade names include Broadstrike, Hornet and Sureshot. A single application per year is recommended, and this should be applied in normal moist conditions. Herbicidal activity of flumetsulam increases with pH but decreases with organic matter content.

The acute oral LD_{50} of flumetsulam in rats and birds is >5000 mg/kg and >2250 mg/kg, respectively. Though not so toxic to the skin, this chemical may irritate the eyes. The inhalation LC_{50} of flumetsulam (4 h) is >1.2 mg/l. The NOEL in mice is >1000 mg/kg. The LC_{50} in honey bees is >100 μg per bee. Unscrupulous use of the chemical can harm sugarbeet, cotton and tomatoes and also rotational crops. Handlers are advised to wear chemical-resistant gloves, shoes with socks and protective eyewear. Garments that have come in contact with flumetsulam should be discarded and disposed of.

Animals rapidly eliminate flumetsulam. Metabolites include 5-hydroxy or 5-methoxy derivatives. In plants, the DT_{50} varies from 2 h in maize to 131 h in *Chenopodium* spp. In water, the photolysis half-life is 1 year. The soil photolysis DT_{50} is 3 months.

The herbicide is registered in the USA, Australia, Canada and New Zealand, but not in the EU.

Flumiclorac herbicide

See **Flumiclorac-pentyl herbicide**

Flumiclorac-pentyl herbicide

Flumiclorac-pentyl, $C_{21}H_{23}ClFNO_5$ (pentyl [2-chloro-5-(cyclohex-1-ene-1, 2-dicarboximido)-4-fluororophenoxy] acetate), is a herbicide. It is a beige solid, emulsifiable in water at 0.189 mg/l, while it dissolves in methanol, hexane, *n*-octanol and acetone. It stores well in well-ventilated areas, in airtight containers and away from heat and flame. The CAS Registry Number is 87546-18-7.

This herbicide is a derivative of flumiclorac. With an IUPAC name [2-chloro-5-(cyclohex-1-ene-1,2-dicarboximido)-4-fluororophenoxy]acetic acid, and a CAS Registry Number of 87547-04-4, flumiclorac is a post-emergence **dicarboximide** herbicide with molecular formula $C_{16}H_{13}ClFNO_5$.

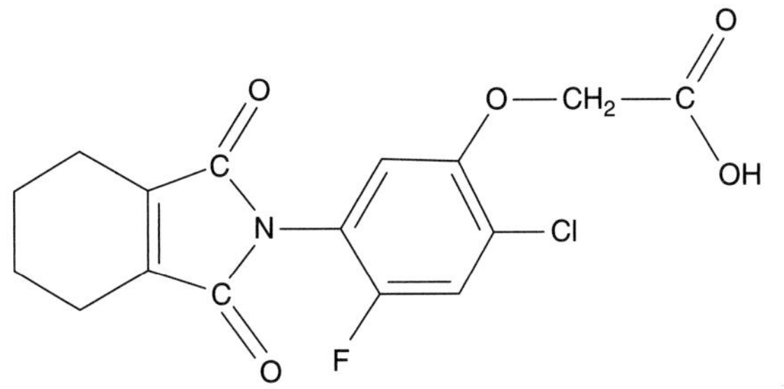

Flumiclorac

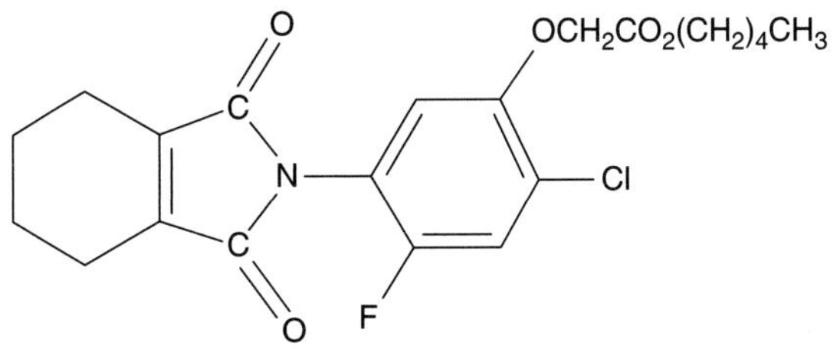

Flumiclorac-pentyl

The herbicide flumiclorac-pentyl, a fast-acting, contact, post-emergence herbicide, acts by inhibiting protoporphyrinogen oxidase, thus disrupting the cell membrane. Flumiclorac-pentyl induces porphyrin accumulation which causes membrane lipid peroxidation, damaging the membrane function and structure irreversibly. It causes desiccation, wilting, bleaching, browning and necrosis. It is applied commonly to broadleaved weeds in maize and soybean, including *Chenopodium album* (lambsquarters), *Amaranthus* spp. (pigweeds), *Datura stramonium* (jimson weed), *Abutilon theophrasti* (velvetleaf), cocklebur and common ragweed.

Flumiclorac-pentyl is available in emulsifiable concentrate (EC) formulation and is applied at the rate of 0.030–60 g/ha.

This herbicide is considered slightly hazardous, with an acute oral LD_{50} in rats of 3.2 g/kg. The dermal LD_{50} in rabbits is >2 g/kg. The inhalation LC_{50} in rats for a 4 h period is 5.51 mg/l. The compound is moderately toxic to birds (with the acute oral LD_{50} in bobwhite quails being >2250 mg/kg). It is highly toxic to fish (with the acute LC_{50} in rainbow trout being 1.1 mg/l) and honey bees (the acute contact LD_{50} is >106 μg per bee). The NOEL of the compound is 300 ppm in mice. In case of accidental exposure, flushing with water and soap is recommended, while giving milk or water to the affected person.

Flumiclorac-pentyl rapidly degrades in soil, the DT_{50} in loamy-sandy soil being 0.48–4.4 days. The compound is immobile in soil and the metabolites have low to medium mobility. It hydrolyses at pH 5 with a DT_{50} of 4.2 days. Aqueous photolysis also occurs.

The herbicide flumiclorac-pentyl is registered in the USA and Australia. It is banned in the EU.

Fluometuron herbicide

Fluometuron, $C_{10}H_{11}F_3N_2O$ (1,1-dimethyl-3-(α,α,α-trifluoro-*m*-tolyl) urea), is a substituted urea herbicide. It is a white crystalline substance with an amine-like odour. It is neither flammable nor corrosive. Fluometuron is soluble in water (110 mg/l), methanol, acetone, dichloromethane, *n*-octanol and hexane. It is stable in acidic, neutral and alkaline media (at 20°C) but may explode in the presence of fire. It is decomposed by UV irradiation and thermal exposure, the latter releasing toxic fumes of fluorides and oxides of nitrogen and carbon. It is produced by the reaction of 3-trifluoromethylphenyl isocyanate with dimethylamine. The CAS Registry Number is 2164-17-2.

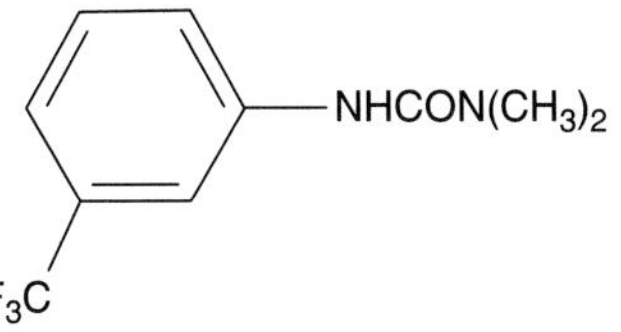

Fluometuron

Fluometuron is a selective, pre- and post-emergence, systemic, long-lasting (2–5 months) cotton herbicide that is absorbed by the roots and acts by inhibiting photosynthesis.

Fluometuron is available as liquid, dry flowable and wettable powder formulations. Trade names include Cotoran, Cotorex, Higalcoton and Pakhtaran. The pre-harvest interval (i.e. the time after the last application up to the time of harvest) for cotton is 1 week, while for sugarcane it is a few months.

This herbicide is not very toxic to rats in normal use. The acute oral LD_{50} among rats is between 1515 and 8900 mg/kg, and in rabbits it is 2500 mg/kg. It is slightly toxic to birds and is moderately toxic to fish, *Daphnia* spp., earthworms and honey bees (the oral LD_{50} is >155 μg per bee). Fluometuron causes mild skin and eye irritation to humans. Its dust can irritate the lungs too. In view of this fact, loaders and handlers are advised to wear protective clothing like coveralls, long-sleeved shirt, shoes, impermeable gloves and eye protection. The ADI of fluometuron is 0.00025 mg/kg/day. It is a possible carcinogen, and may contaminate groundwater. The WHO has rated it as a Class U toxin.

In animals (rats), almost all of fluometuron is eliminated within 1 week, with metabolites being demethylated compounds and gluconuronic acid conjugates. In plants, fluometuron is degraded to form an aniline derivative. Fluometuron degrades in soil by the action of soil microbes but its half-life widely varies between 30 days and 340 days, hence making it a persistent compound. The half-life of fluometuron in water (hydrolysis half-life) ranges from 110 to 144 weeks. In an aqueous solution and exposed to natural sunlight, most of fluometuron decomposes in 3 days.

Fluometuron is registered in the USA for use on cotton and sugarcane while it is also registered in Australia, Vietnam, Belgium, Greece, Spain and several African countries.

Fluoride inorganic insecticides

A number of fluoride insecticides such as sodium fluoride, cryolite, sodium hexafluorosilicate and barium fluorosilicate have been developed as safer stomach poisons, compared with other toxic alternatives. Inorganic fluoride compounds act by inhibiting metal-containing enzymes such as phosphatase, phosphorylases and others like enolase (or phosphopyruvate dehydratase), by forming complexes with them (like a complex with magnesium ions forms magnesium fluorophosphates). Thus these enzyme-complexes throw all normal biochemical processes out of gear, thus killing the pest. An advantage of fluorides is their high activity that does not leave significant amounts of residues on the crops.

Sodium fluoride, when combined with pyrethroids, becomes more effective as an insecticide against household crawling insects and larvae of ants, cockroaches and lice, as well as those in warehouses. Sodium fluoride is highly toxic and should be carefully handled to avoid loss of life to animals, plants and humans. Commercial products include Florocid. The CAS Registry Number of sodium fluoride is 7681-49-4.

Naturally occuring cryolite (trisodium hexafluoroaluminate), with molecular formula AlF_6Na_3, is an insecticide used against leaf-eating pests on grapes, potatoes, lettuce and citrus. It may be applied as a dust or a wettable powder, from the ground, or from the air. Application doses depend on the need (a common dose is 33.6 kg/ha on citrus). Cryolite causes irritation to the eyes, but is not a serious threat to handlers or the environment. Registered in the USA, it has a CAS Registry Number of 15096-52-3.

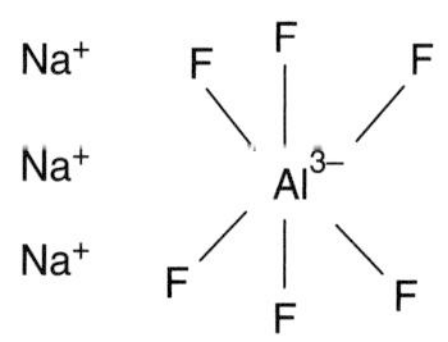

Cryolite

Sodium hexafluorosilicate (disodium hexafluorosilicate), with molecular formula F_6Na_2Si, is a soil insecticide of horticultural crops. It is a colourless crystalline substance soluble in water. But since it is phytotoxic, it is used as a non-farm pest control agent for cockroaches, lice, weevils and cutworms. Commercial products include Satsan and Prodan. The CAS Registry Number of sodium hexafluorosilicate is 16893-85-9.

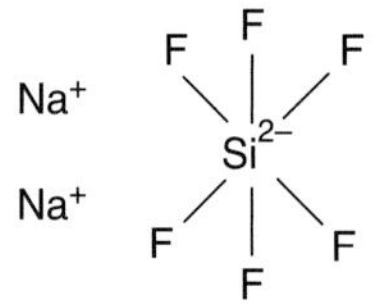

Sodium hexafluorosilicate

Barium hexafluorosilicate, with molecular formula BaF_6Si, is used as poison for plant-feeding insects. Its CAS Registry Number is 17125-80-3.

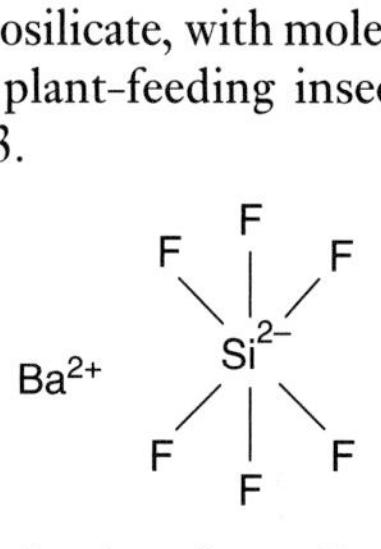

Barium hexafluorosilicate

Fluoroacetamide rodenticide

Fluoroacetamide, C_2H_4FNO (2-fluoracetamide), is a colourless crystalline powder soluble in water, acetone and ethanol. Fluoroacetamide is produced by the reaction of α-chloroacetamide with potassium fluoride in tetrachloroethylene at high temperatures. Its CAS Registry Number is 640-19-7.

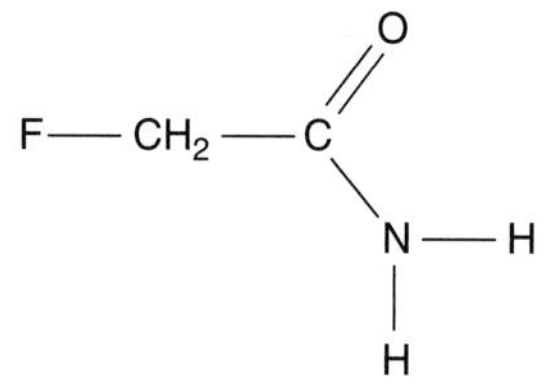

Fluoroacetamide

This rodenticide, a relatively fast-acting compound, does not induce bait shyness. It acts on the heart of rodents and subsequently on the CNS and provides effective control over rodent pests. Fluoroacetamide is required at concentrations of 1–2% in baits. The common trade names include Rodex.

The WHO classifies fluoroacetamide as an extremely hazardous chemical (Class Ib toxin). The acute oral LD_{50} is 13 mg/kg for Norway (brown) rats. So, it is considered to be a PAN Bad Actor chemical.

This rodenticide is not registered in the USA.

Fluoroglycofen-ethyl herbicide

Fluoroglycofen-ethyl, $C_{18}H_{13}ClF_3NO_7$ (ethyl *O*-[5-(2-chloro-α,α,α-trifluoro-*p*-tolyloxy)-2-nitrobenzoyl] glycolate), is a selective *p*-nitrophenyl ether herbicide. It is a dark amber solid, soluble in water (0.6 mg/l) at 25°C and readily soluble in organic solvents, except hexane. In aqueous solutions, fluoroglycofen-ethyl is stable at 22°C, with its DT_{50} being 231 days at pH 5. It is produced by the reaction of 5-[2-chloro-4-(trifluoromethyl) phenyl]-2-nitrobenzoyl chloride with ethyl hydroxyacetate in the presence of trimethylamine. Its CAS Registry Number is 77501-90-7.

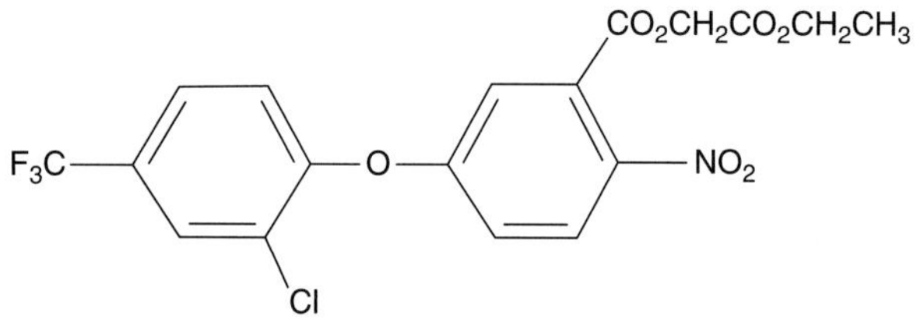

Fluoroglycofen-ethyl

Fluoroglycofen-ethyl is a selective leaf and root herbicide, which acts by inhibiting the enzyme protoporphyrinogen oxidase, leading to irreversible damage to the structure and function of cell membranes. It controls broadleaved weeds and grasses particularly *Galium* spp. (also known as cleavers), *Viola* and *Veronica* spp. in wheat, barley, oats, groundnuts, rice and soybeans. A trade name is Compete. It is incompatible with suspension concentrates.

This herbicide is not a big threat to most of the tested species of mammals and birds. The acute oral LD_{50} among rats is 1500 mg/kg, while the acute percutaneous LD_{50} in rabbits is >5000 mg/kg. The acute oral LD_{50} among birds is >3160 mg/kg, with an 8 day dietary LC_{50} in mallard ducks and bobwhite quails being >5000 mg a.i./kg. It is slightly irritating to the skin and eyes of rabbit. The (96 h) LC_{50} in fish ranges from 1.6 to 23 mg/l. The contact LD_{50} in honey bees is >100 μg per bee. The NOEL in dogs is 320 mg/kg diet for 1 year. The ADI is 0.009 mg/kg/day.

In both animals and plants, fluoroglycofen-ethyl is degraded by hydrolysis of the ester and reduction of the nitro group. In soil and water, fluoroglycofen-ethyl is very rapidly hydrolysed to the corresponding acid, with microbial degradation in soil being significant too. The DT_{50} of fluoroglycofen-ethyl in soil is 11 h.

Fluoroimide fungicide

Fluoroimide, $C_{10}H_4Cl_2FNO_2$ (2,3-dichloro-*N*-4-fluorophenylmaleimide), is a fungicide. It is a pale yellow crystal soluble in water (5.9 mg/l) and methanol. It is stable under UV and at 120°C. This fungicide is produced by the reaction of 2,3-dichloromaleic acid in aqueous solution with *p*-fluoroaniline. The CAS Registry Number is 41205-21-4.

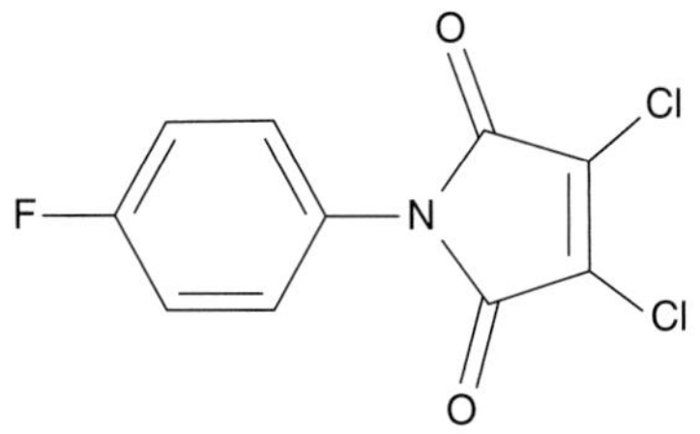

Fluoroimide

Fluoroimide is a foliar fungicide with protective action. It inhibits spore germination and is used in the control of blossom blights in apples, citrus, rubber and pine trees. *Monilinia mali*, *Diaporthe citri* and *Corticium* sp. are some of the pathogens that are controlled.

Fluoroimide is formulated as a wettable powder (WP) and is commonly available as the trade name Sparticide. It has an acute oral LD_{50} in rats of >15000 mg/kg, and is not considered very toxic to the environment at large. According to the WHO, fluoroimide is unlikely to present acute hazard.

Flupoxam herbicide

Flupoxam, $C_{19}H_{14}ClF_5N_4O_2$ (1-[4-chloro-3-(2,2,3,3,3-pentafluoropropoxymethyl)phenyl-5-phenyl]-1*H*-1,2,4_triazole-3-carboxamide), is an amide herbicide. It is a beige crystalline substance, sparingly soluble in water and readily soluble in polar organic solvents. Flupoxam is made by the reduction of the nitro group of 2-chloro-5-nitrobenzene followed by diazotization, addition of benzylamino ethylcarboxylate and cyclization of the product. The CAS Registry Number is 119126-15-7.

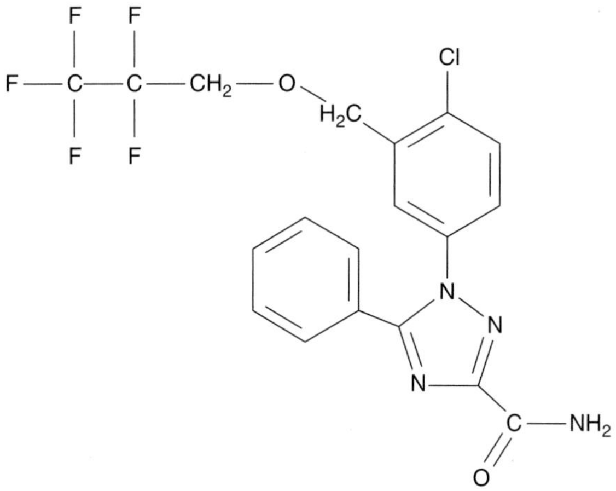

Flupoxam

Flupoxam, mainly a pre-and post-emergence cereal herbicide, is a cell elongation inhibitor, and mainly acts on the actively growing meristematic tissues (both root and foliar) of broadleaved weeds in winter wheat and winter barley crops. It is available under trade names like Ovation. Flupoxam is available as a suspension concentrate (SC).

Flupropanate herbicide

Flupropanate, $C_3H_2F_4O_2$ (2,2,3,3,-tetrafluoropropionic acid), is a halogenated aliphatic herbicide. It is present as colourless crystals, soluble in non-polar organic solvents. It is stable to heat, light and water. It is made by the oxidation of 2,2,3,3-tetrafluoropropanol with chromium trioxide. The CAS Registry Number is 756-09-2.

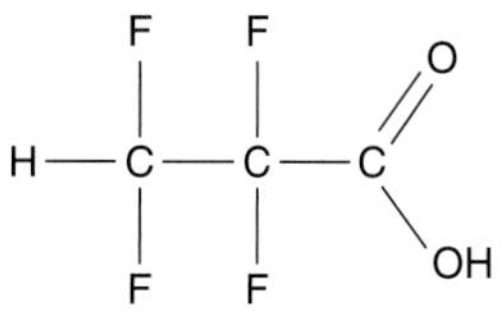

Flupropanate

Flupropanate is a systemic herbicide with low contact activity. It controls annual and perennial grasses, after being taken up by the roots. Applied to pastures and uncultivated land, flupropanate is commonly available as soluble concentrate (SL) and granules (GR). A common trade is Frenock. The WHO classifies it as a Class U toxin (unlikely to be hazardous).

Flupyrazofos insecticide

Flupyrazofos, $C_{14}H_{16}F_3N_2O_3PS$ (*O,O*-diethyl *O*-(1-phenyl-3-trifluoromethylpyrazol-5-yl) phosphorothiote), is an organophosphorus insecticide. It is soluble in water (0.8 mg/l) and is slowly hydrolysed in aqueous media. Its DT_{50} is 121 h at pH 9. The CAS Registry Number is 122431-24-7.

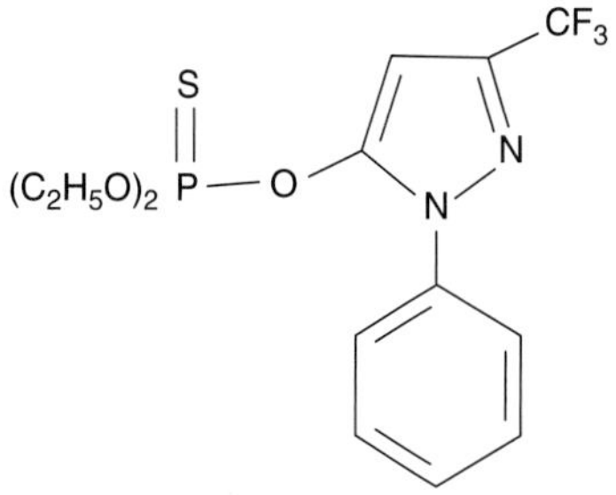

Flupyrazofos

This insecticide, with contact action, is used for controlling caterpillars, especially of the diamondback moth (*Plutella xylostella*).

This insecticide is an acutely toxic chemical, with the acute oral LD_{50} in rats being 372–605 mg/kg. The principal route of metabolism of flupyrazofos is by oxidative desulfuration, followed by hydrolysis to 1-hydrogen phosphate. Its soil DT_{50} is about 14 days.

Fluquinconazole fungicide

Fluquinconazole, $C_{16}H_8Cl_2FN_5O$ (3-(2,4-dichlorophenyl)-6-fluoro-2-(1*H*-1,2,4-triazol-1-yl) quinnazolin4 (3*H*)-one), is an azole fungicide. It is a pale brown crystalline solid, soluble in water (1.15 mg/l), acetone, xylene and ethanol, and stable to light. It is produced by the reaction of 2-chloro-6-fluoro-*N*-3-*p*-chlorophenylquinazoline-4(3*H*)-one with 1,2,4-triazole in an inert solvent in the presence of potassium carbonate. The CAS Registry Number is 136426-54-5.

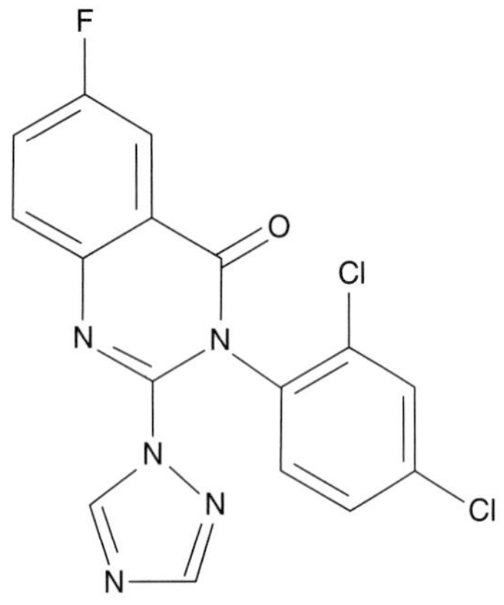

Fluquinconazole

Foliar-applied fluquinconazole, transported through the xylem, inhibits **ergosterol** biosynthesis for ascomycetes, deuteromycetes and basidiomycetes that attack cereal crops and fruits. It has protectant and eradicant properties. *Venturia inaequalis* and *Podosphaera leucotricha* are two pathogens controlled by fluquinconazole.

Common formulations of this fungicide include suspension concentrates (SC) and wettable powders (WP). The chemical is commonly available as Castellan.

The fungicide is extremely toxic to rats (the acute oral LD_{50} is 112 mg/kg), while in birds and fish, it is practically non-toxic, if label instructions are followed. It is non-irritating to skin and eyes. The ADI is fixed at 0.0003 mg/kg/day. In plants, fluquinconazole is metabolized into 1,2,4-triazole.

Fluquinconazole is registered in Australia, many members of the EU (e.g. Germany and the UK) and South Africa. The European MRL for this fungicide on wheat grain is 0.1 mg/kg.

Fluridone herbicide

Fluridone, $C_{19}H_{14}F_3NO$ (1-methyl-3-phenyl-5-(α,α,α-trifluoro-*m*-tolyl)-4-pyridone), is a 4-pyridone aquatic herbicide. It is a white, odourless, crystalline solid, soluble in water (at 12 mg/l at pH 7 and 25°C), methanol, chloroform, diethyl ether ethyl acetate and hexane. It is stable to hydrolysis from pH 3 to 9. Fluridone decomposes at around 200°C and also by UV irradiation. The CAS Registry Number of fluridone herbicide is 59756-60-4.

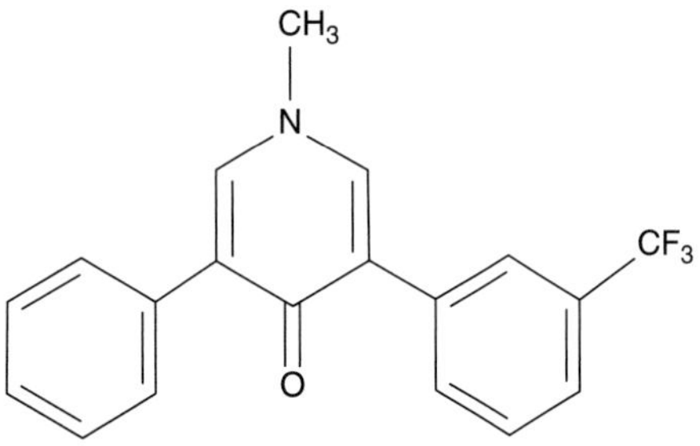

Fluridone

Fluridone is a systemic, broad-spectrum herbicide, absorbed by plant shoots and roots. It inhibits carotenoid synthesis, which is responsible for the production of chlorophyll. The compound degrades chlorophyll, thus producing white, chlorotic growing points in plants. It is applied in freshwater ponds, lakes, reservoirs, drainage canals, irrigation canals and rivers and controls submerged and emerged aquatic plants like Carolina fanwort (*Cabomba caroliniana*), coon's tail (*Ceratophyllum demersum*), elodea (*Elodea canadensis*) and many others. Fluridone does not affect algae adversely. The compound also controls **barnyard grass**, **Bermuda grass**, **common lambsquarters**, crabgrass and many more.

The herbicide fluridone is available in aqueous suspension and pellet forms, and may be applied by surface spray. Just a little moisture is enough to activate it. Rates of 4.5 kg/surface ha usually suffice. The chemical may be mixed with other herbicides.

The herbicide is not toxic to rats and mice via the oral or inhalation routes (the acute oral LD_{50} in rats and mice is >10,000 mg/kg while the inhalation LC_{50} in rats is >4012 mg/l air). It is not a hazard to bird species tested. It may be slightly to moderately toxic to fish, crustaceans, molluscs and insects. Handlers are advised to be equipped with breathing apparatus and protective clothing to avoid any harm to skin and eyes. It is also critical to dispose of the waste material in a manner suggested on the label. The ADI of fluridone is 0.08 mg/kg/day, with a maximum permissible intake (MPI) of 4.8 mg/day for an adult.

The DT_{50} of fluridone in water is 23 h. It photodegrades with a half-life of 34 h in neutral pond water. It is persistent under anaerobic aquatic conditions, with a half-life of 9 months. Normally, the half-life in water is 20 days.

Fluridone is registered in the USA.

Flurochloridone herbicide

Flurochloridone, $C_{12}H_{10}C_{12}F_3NO$ ((3*RS*, 4*RS*; 3*RS*, 4*SR*)-3-chloro-4-chloromethyl-1-(α,α,α-trifluoro-*m*-tolyl)-2-pyrrolidone (in a ratio 3:1)), is a herbicide. It is a brown-beige waxy solid, soluble in water (28 mg/l), acetone, chlorobenzene and xylene. Flurochloridone is stable to hydrolysis in acidic media but decomposes in 138 days (at 100°C). Aqueous photolysis of flurochloridone occurs in 6.5 days at pH 7. The CAS Registry Number of flurochloridone herbicide is 61213-25-0.

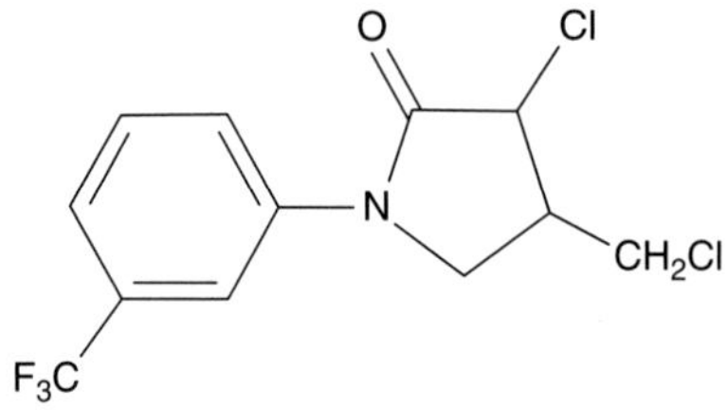

Flurochloridone

Flurochloridone, a selective, broad-spectrum, pre-emergence herbicide, inhibits carotenoid biosynthesis, due to which the chlorophyll does not undergo photo-oxidation. Flurochloridone is used to control chickweed (*Stellaria media*), ivyleaf speedwell (*Veronica hederifolia*), redroot pigweed (*Amaranthus retroflexus*),

Portulaca oleracea and black nightshade (*Solanum nigrum*), *Galium aparine* and *Veronica persica* in a wide range of crops such as sunflowers, winter wheat, winter rye, potatoes and cotton.

The herbicide, available as a capsule suspension (CS) and emulsifiable concentrate (EC), is sprayed after mixing with water. It is sprayed at 250–750 g/ha.

Besides being a slight skin irritant and being slightly hazardous to fish, this herbicide is not toxic to mammals, birds, bees or earthworms. The ADI of flurochloridone in rats is 0.02 mg/kg b.w./day and the aRfD is 0.025 mg/kg b.w./day.

Flurochloridone is extensively metabolized and rapidly excreted in rats (>95% in 90 h), following a series of reactions like oxidation, hydrolysis and glutathione conjugation. The residue level of flurochloridone in crops is generally <0.05 mg/kg. Flurochloridone does not leach because it is adsorbed and readily degraded in soil (soil DT_{50} is 9–70 days).

The herbicide is registered in France, Greece, Hungary, Poland and South Africa. The EU Pesticide Database has established 0.01 mg/kg as the MRL for sunflowers and potatoes.

Fluroxypyr herbicide

Fluroxypyr, $C_7H_5Cl_2FN_2O_3$ (4-amino-3,5-dichloro-6-fluoro-2-pyridyloxyacetic acid), is a pyridine herbicide. It is a colourless, crystalline solid, moderately soluble in water and in organic solvents. Fluroxypyr, being a stable acidic compound, reacts with alkalis to form salts like fluroxypyr-meptyl. Fluroxypyr is made by the reaction of 4-amino-3,5-dichoro-2-(methylsulfonyl) pyridine with methylglycolate. The CAS Registry Number is 69377-81-7 (see Fluroxypyr at bottom of the page).

Fluroxypyr, a post-emergence herbicide, is absorbed by the foliage and is herbicidal in the acid form. This form is translocated rapidly in the target plant. It controls a number of broadleaved weeds in wheat, barley, oats and orchards. It induces auxin-type responses, leading to leaf curling.

In soils, microbial action on fluroxypyr removes the carboxyl group or acetic acid group and metabolizes it within 9 days. Mammals excrete it rapidly without change. This stable compound has a DT_{50} of 185 days. The WHO has classified this compound as a Class U toxin (unlikely to be hazardous).

Flurprimidol as a PGR

Flurprimidol, $C_{15}H_{15}F_3N_2O_2$ ((RS)-2-methyl-1-pyrimidin-5-yl-1-(4-trifluoromethoxyphenyl) propan-1-ol), is a pyrimidine plant growth retardant. It is a colourless, crystalline substance, soluble in water (127 mg/kg), hexane, toluene, methanol and acetone. It is photolytically decomposed in water with a half-life of 3 h. The CAS Registry Number is 56425-91-3.

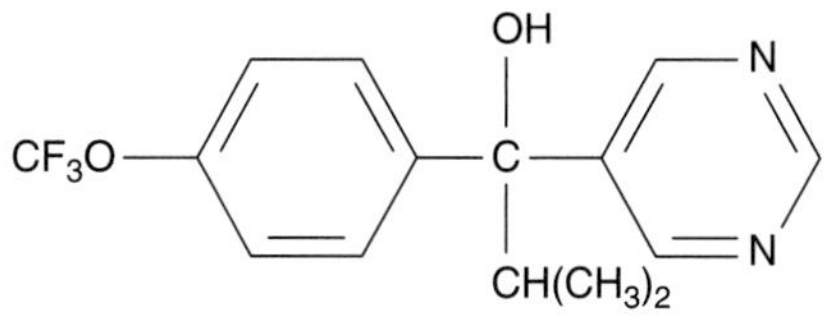

Flurprimidol

The PGR compound, well absorbed and translocated throughout the plant, interrupts the production of gibberellic acid, thus reducing internode elongation of weed species, turf grasses and other non-crop situations. It is formulated as wettable powder (WP), emulsifiable concentrate (EC) and suspension concentrate (SC). A common flurprimidol trade name is Cutless.

The PGR flurprimidol is classified as a Class III toxin, indicating slight toxicity. It is, however, a potential groundwater contaminant. Upon oral administration, it is excreted via urine and faeces within 48 h. It is degraded in soil under aerobic conditions. The USA has registered the use of flurprimidol.

Flurtamone herbicide

Flurtamone, $C_{18}H_{14}F_3NO_2$ ((*RS*)-5-methylamino-2-phenyl-4-(α,α,α-trifluoro-*m*-tolyl) furan-3(2*H*)-one), is a herbicide that exists as a yellowish powder. It is soluble in water (at 35 mg/l at 20°C), acetone, dichloromethane, methanol and isopropanol. Flurtamone degrades rapidly in sunlight (the DT_{50} is 13.0–16.8 h) and may lose efficacy in concentrated acids and bases. It stores well in original, airtight containers, kept in cool, dark and ventilated locations. The CAS Registry Number of flurtamone herbicide is 96525-23-4.

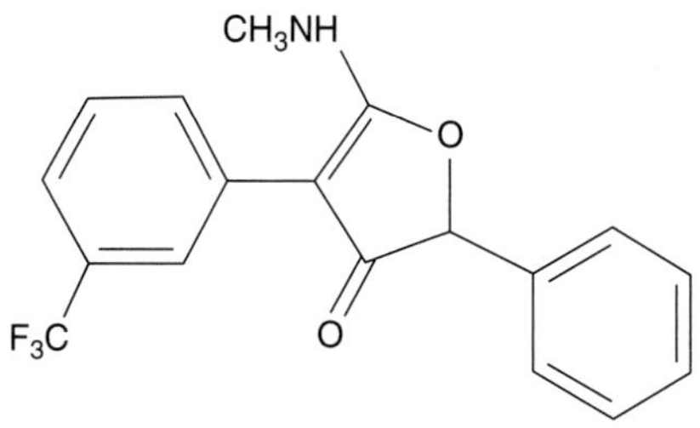

Flurtamone

Flurtamone, a pre- or post-emergence herbicide, blocks carotenoid biosynthesis (for chlorophyll) by inhibition of phytoene desaturase. It has a transitory bleaching effect. It may be used at the pre-plant stage on soybeans, groundnuts, cotton, peas and sunflowers, for the control of broadleaved weeds and grass weeds.

Flurtamone is available as suspension concentrates (SC), water-dispersible granules (WG) and as wettable powders (WP). It can be applied at a range of 250–375 g/ha.

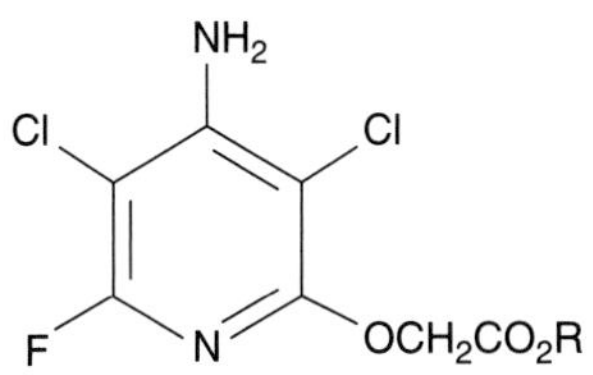

Fluroxypyr	R = H
2-Butoxy-1-methylethyl	R = $CH_3(CH_2)_3OCH_2CH(CH_3)$-
Meptyl (1-methylheptyl)	R = $CH_3(CH_2)_5CH(CH_3)$-

Fluroxypyr

Flurtamone is irritating to the skin and eyes, as seen in tests on rabbits. Otherwise, it is not very acutely toxic to rats and birds. The short-term dietary NOEL values among rats, mice and dogs are in the range of 5.6–200 mg/kg. The acute LD_{50} of flurtamone in honey bees is >100 μg per bee (contact). The acute LC_{50} (14 days) in earthworms is >1800 mg/kg. The ADI of flurtamone is 0.03 mg/kg b.w./day. Normal precautions like avoiding contact with skin and eyes, and following label instructions are advised for handlers.

The herbicide is registered in about ten member countries of the EU.

Flusilazole fungicide

Flusilazole, $C_{16}H_{15}F_2N_3Si$ (bis(4-fluorophenyl)(methyl)(1H-1,2,4-triazol-1-ylmethyl)silane), is a triazole (conazole) fungicide. It is a colourless crystalline substance, soluble in water and organic solvents. It is stable under normal storage conditions, to light and up to 310°C. It can be produced by addition of butyllithium to a solution of chloromethyldichlorosilane and 1-bromo-4-fluorobenzene and subsequent reaction with triazole. The CAS Registry Number is 85509-19-9.

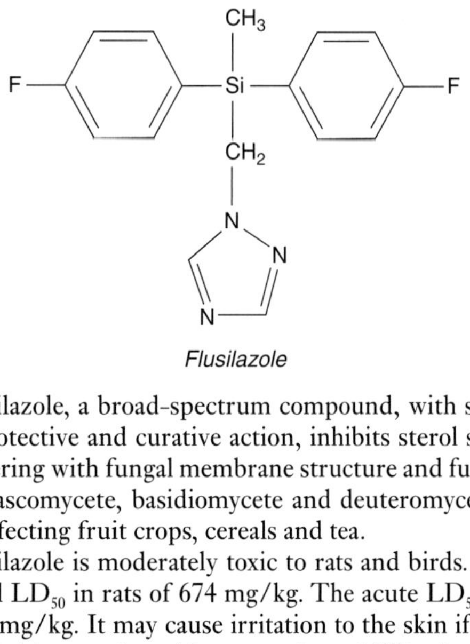

Flusilazole

Flusilazole, a broad-spectrum compound, with some systemic, protective and curative action, inhibits sterol synthesis, by interfering with fungal membrane structure and function. It controls ascomycete, basidiomycete and deuteromycete membranes infecting fruit crops, cereals and tea.

Flusilazole is moderately toxic to rats and birds. It has an acute oral LD_{50} in rats of 674 mg/kg. The acute LD_{50} in birds is >1590 mg/kg. It may cause irritation to the skin if not used as per label instructions (the acute dermal LD_{50} in rabbits is >2000 mg/kg). The acute LD_{50} for bees is >150 μg per bee, indicating toxicity. The WHO has categorized this fungicide as slightly hazardous (Class III toxin). It is available under trade names like Punch.

The fungicide is registered in India, the USA, Canada, Australia, Vietnam, South Africa and many EU nations. The typically accepted MRL for bananas, grapes and raisins is 0.5 ppm.

Fluthiacet-methyl herbicide

Fluthiacet-methyl, $C_{15}H_{15}ClFN_3O_3S_2$ (methyl {2-chloro-4-fluoro-5-[(*EZ*)-5,6,7,8-tetrahydro-3-oxo-1*H*,3*H*-[1,3,4] thiadiazolo[3,4-*a*]pyridazin-1-ylideneamino] phenylthio}acetate), is a herbicide. A derivative of fluthiacet, it is a white powder, soluble in water (0.85 mg/l at 25°C), acetone, acetonitrile, dichloromethane, ethyl acetate, *n*-hexane, *n*-octanol, methanol and toluene. While it is stable at 150°C, its aqueous photolysis DT_{50} in natural water is 5.88 h. At lower pH levels the DT_{50} in water is >1year. Photolysis in UV light takes 2 h. It is produced by the reaction of 1-[(2-fluoro-4-chloro-5-thiomethylacetate)-phenylthiocarbamoyl]-2-methoxycarbonylhexahydropyridazine with methanolic potassium hydroxide, followed by treatment with trichloromethyl chloroformate. The CAS Registry Number of fluthiacet-methyl herbicide is 117337-19-6.

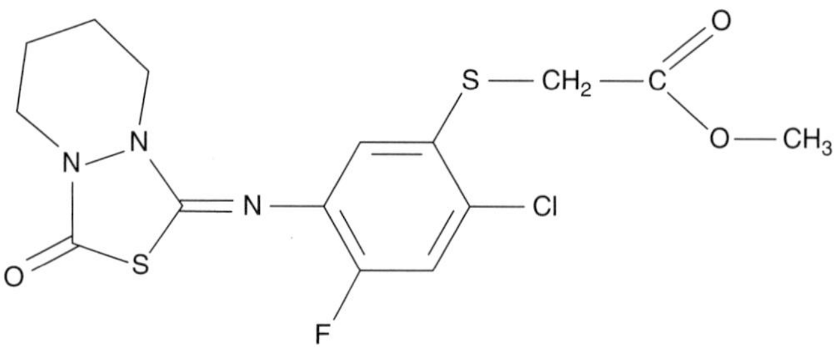

Fluthiacet-methyl

Fluthiacet-methyl is a selective, contact, post-emergence herbicide which requires light for activity. It is a protoporphyrinogen oxidase inhibitor causing accumulation of protoporphyrins, enhancing peroxidation of membrane lipids, and leading to irreversible damage to membrane structure and cellular function. It is applied on annual broadleaved weeds like velvetleaf (*Abutilon theophrasti*) that grow in crops like soybeans.

The compound is formulated as a wettable powder (WP) (4.75%) and an emulsifiable concentrate (EC). It is normally applied at 4–15 g/ha.

The acute oral LD_{50} of fluthiacet-methyl among rats and birds is >5000 mg/kg and >2250 mg/kg. It is not irritating to the skin, based on dermal tests on rabbits. It is practically non-toxic to honey bees, small mammals, birds, earthworms and many other soil organisms. But it is highly toxic to fish, toxic to freshwater invertebrates and also to duckweed and some alga species. This fact has to be mentioned clearly on the label, and hence it is necessary to avoid the compound flowing into water bodies. Disposal of effluent has to be in accordance with the respective national agency responsible for waste disposal. It is also toxic to onions and cucumber. According to the EPA, the ADI of fluthiacet-methyl is 0.001 mg/kg b.w./day. According to the PAN Pesticide Database, this compound is a carcinogen, and hence a PAN Bad Actor chemical.

Most of the fluthiacet-methyl is eliminated from rats via the faeces, after it is metabolized via hydrolysis of the methyl ester, isomerization at the thiadiazole ring and hydroxylation of the tetrahydropyridazine moiety. In beans, <0.01 ppm of the herbicide residue is found. Photolysis in soil occurs in 21 days, with the DT_{50} in loamy soil being 1.2 days at 25°C.

Fluthiacet-methyl is registered for use in the USA, but not registered throughout the EU.

Fluthiamide herbicide

See **Flufenacet herbicide**

Flutolanil fungicide

Flutolanil, $C_{17}H_{16}F_3NO_2$ (α,α,α-trifluoro-3′-isopropoxy-*o*-toluanilide), is a phenylbenzamide fungicide. It is a colourless crystalline substance, soluble in water (10 mg/l), acetone, acetonitrile, dichloromethane, ethyl acetate, *n*-hexane, *n*-octanol and toluene. It is stable in both acidic and alkaline media. Flutolanil is produced by the reaction of 2-(trifluromethyl) benzoyl chloride with 3-iso-propoxyaniline. Its CAS Registry Number is 66332-96-5.

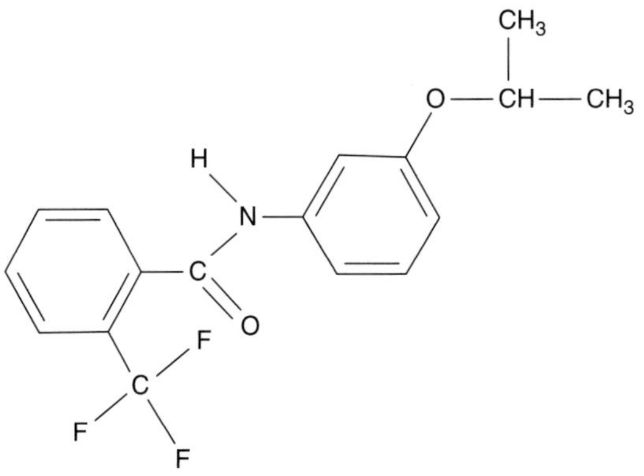

Flutolanil

A systemic fungicide with protective and curative action, flutolanil acts as a succinate dehydrogenase (complex II) inhibitor, in the respiratory electron transport chain, leading to inhibition of aspartate and glutamate synthesis. It prevents growth and host penetration of for example *Rhizoctonia solani*, *Rhizoctonia cerealis* and *Corticium rolfsii* on rice, sugarbeet, potato and cereals.

Flutolanil is slightly toxic. The acute oral LD_{50} in rats is >5000 mg/kg, while that for bird species, is >2250 mg/kg. The ADI value in humans is 0.001 mg/kg/day. It may irritate the eyes. The WHO classifies the compound as a Class U toxin (unlikely to be an acute hazard in normal use). A common trade name is Moncut.

In rats, flutolanil is eliminated via the faeces and urine within 48 h. The metabolism proceeds via hydrolysis of the methyl ester, isomerization at the thiadiazole ring and hydroxylation of the tetrahydropyridazine moiety.

Flutolanil is registered in the USA, Australia and some EU members including Finland, the UK, France, Greece and the Netherlands.

Fluvalinate acaricide

Fluvalinate, $C_{26}H_{22}ClF_3N_2O_3$ ((*RS*)-α-cyano-3-phenoxybenzyl N-(2-chloro-α,α,α-trifluoro-p-tolyl)-DL-alinate), is a pyrethroid ester acaricide and an insecticide. It is a viscous, yellow oil that is insoluble in water (0.005 μg/ml), but very soluble in organic solvents and aromatic hydrocarbons. It is a stable, non-volatile, fat-soluble compound. Its CAS Registry Number is 69409-94-5 (see Fluvalinate at bottom of the page).

Fluvalinate, a broad-spectrum compound, with contact and stomach action, is used to control *Varroa* mites in honey bee colonies. It is also used against moths, beetles and Hemipteran insects of cotton, cereals, grapes, potatoes, fruits, vegetables and turf.

The compound is available in emulsifiable concentrate (EC) and flowable formulations. Trade names include Apistan, Klartan, Minadox, Spur and Yardex.

Fluvalinate is a moderately toxic material. The oral LD_{50} in rats ranges from 261 to 282 mg/kg. It is slightly toxic to birds, with the acute oral LD_{50} in bobwhite quails being >2510 mg/kg. It is highly toxic to fish. The compound causes irritation to the eyes and the skin. Other symptoms include coughing, throat irritation, headache, rashes and nausea. Hence, applicators are advised to be equipped with goggles, face shield, gloves and full-sleeved coats and long trousers, while handling this chemical. The ADI is 0.01 mg/kg/day (based on a 2 year rat feeding study). Products containing fluvalinate must bear the signal word 'DANGER' on the label. It is classified as a restricted use pesticide (RUP) according to the EPA because of its high toxicity to fish and aquatic invertebrates. Fluvalinate is observed to be a developmental toxin. The WHO classifies fluvalinate as unlikely to be hazardous.

Fluvalinate is nearly insoluble in water and has a strong tendency to bind to soil particles. It is therefore unlikely to contaminate groundwater; however, metabolites of fluvalinate may leach. It is registered for use in India and Hungary.

Folpet fungicide

Folpet, $C_9H_4Cl_3NO_2S$ (*N*-[(trichloromethyl) thio] phthalimide), is a thiophthalimide fungicide. It is a colourless, crystalline substance, hardly soluble in water (1 mg/l) and slightly soluble in organic solvents. It is produced by the reaction of the

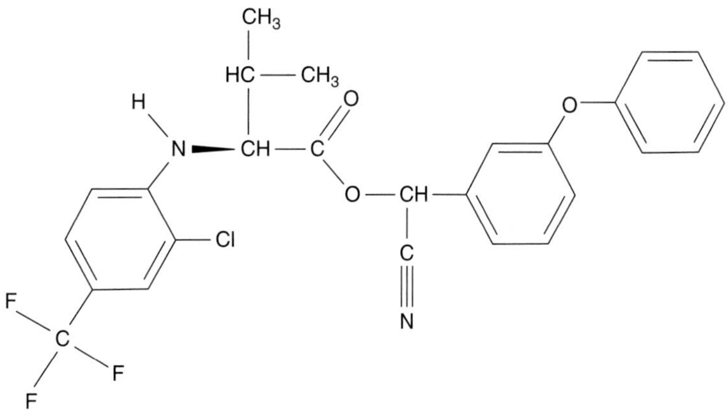

Fluvalinate

potassium salt of phthalimide with dichloromethane sulfenyl chloride in an organic solvent. The CAS Registry Number is 133-07-3.

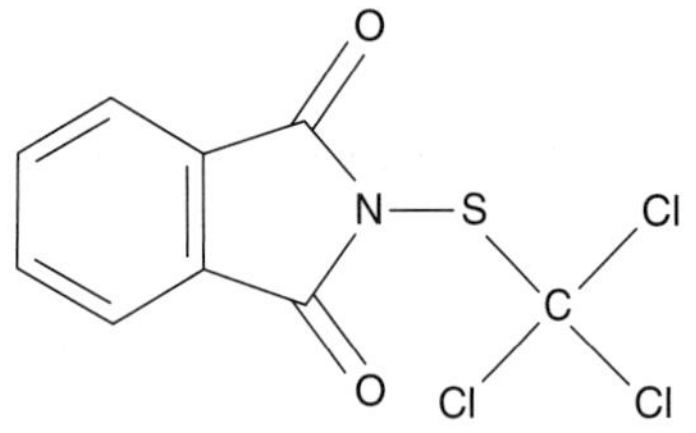

Folpet

The fungicide folpet, a protective, broad-spectrum foliar fungicide, inhibits normal cell division in pathogens causing leaf spot, mildew and **scab** on fruits, vegetables and ornamental plants. It also has industrial applications in the paint, plastic and construction industries.

Folpet is incompatible with strongly alkaline preparations. It is formulated as wettable powders (WP) and dusts (DP) and is available as trade names Faltan, Folpel and Fungitrol 11.

Apart from being a skin and mucous membrane irritant, folpet is not considered hazardous to mammals via oral or dermal exposure. The acute oral toxicity in rats >10 g/kg and an acute dermal toxicity in rabbits is >5 g/kg. It is also not considered an eye irritant. It is slightly toxic to some bird species upon ingestion. However, it is highly toxic to fish and many other aquatic organisms. It is a cholinesterase inhibitor and hence a PAN Bad Actor chemical. The signal word required to be displayed on product labels is 'WARNING'. The WHO classifies it as a Class U toxin (unlikely to be hazardous).

In soil and groundwater, degradation follows the removal of three chlorine atoms due to the thiol group, forming trithiocarbonate, thiophosgene and **phthalimide**.

Folpet is registered in the USA, Canada, South Africa and many members of the EU.

Fomesafen herbicide

Fomesafen, $C_{15}H_{10}ClF_3N_2O_6S$ (5-(2-chloro-α,α,α-trifluoro-*p*-tolyloxy)-*N*-mesyl-2-nitrobenzamide), is a selective herbicide belonging to the *p*-nitrophenyl ether group. It is a white crystalline solid, soluble in pure water at >600 mg/l. It is stable to hydrolysis under both acidic and alkaline conditions, and stores well in a cool, dry and well-ventilated area up to 50°C. It is stable to aqueous photolysis (the DT_{50} is 32 days). It is manufactured by the reaction of 2-chloro-4-trifluoromethylphenolate with 4-chloro-*N*-methylsulfonyl-2-nitrobenzamide. The CAS Registry Number of fomesafen herbicide is 72178-02-0.

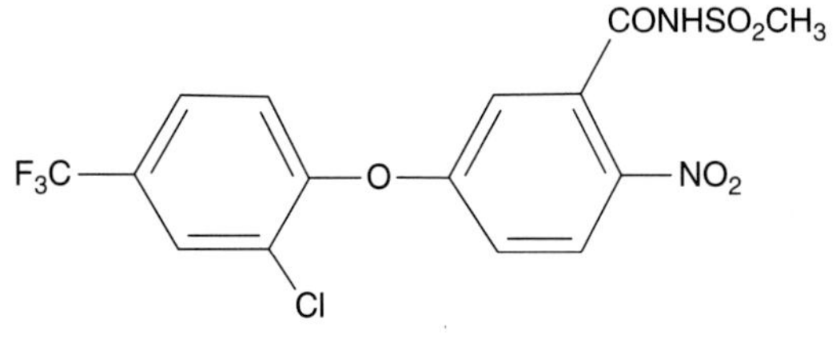

Fomesafen

Fomesafen, a selective, post-emergence herbicide, disrupts photosynthesis by generating superoxide radicals in chloroplasts, and in this way, controls broadleaved weeds and grasses including velvetleaf (*Abutilon theophrasti*) and ragweed (*Ambrosia artemisiifolia*), mainly in soybean crops.

The herbicide is available in a soluble concentrate (SL) formulation, with a typical application being between 200 and 400 g/ha. It is more effective if used with **adjuvants**. It is also not ideal to mix it with acids. The trade names of fomesafen include Flex and Reflex.

Fomesafen is a slightly hazardous toxin, according to the WHO. The acute oral LD_{50} of fomesafen in rats and in bobwhite quails is >2000 mg/kg and >20,000 mg/kg, respectively. But for the mallard duck, the value is >5000 ppm. The 4 h inhalation LD_{50} is 5.48 mg/l. It is not very toxic to all fish species, but is toxic to *Daphnia magna* (with the acute LC_{50} being 0.33 mg/l). Fomesafen is corrosive to eyes and can cause irreversible damage. Caution is advised to ensure that the compound is not ingested, inhaled and mixed with food or drink. The NOEL (2 year) of fomesafen in rats is 5 mg/kg b.w./day and in mice, it is 1 mg/kg b.w./day. The EPA has fixed the ADI as 0.0025 mg/kg b.w./day. This chemical is known to leach and may contaminate groundwater, especially if label instructions are not followed. It is considered a possible carcinogen according to the PAN Pesticide Database and the EPA OPP carcinogen list.

Fomesafen is persistent in soil, with a soil DT_{50} of >6 months under aerobic conditions and 1–2 months in anaerobic conditions. The photodegradation DT_{50} on the soil surface is up to 104 days.

Use of fomesafen was allowed on a restricted basis in many countries in the EU, for example in France and Italy where it was allowed for soybean and beans. It is now banned in the EU. It is registered in Canada, Hungary, South Africa, Tanzania and the USA. (See also ***p*-Nitrodiphenyl ether herbicides**.)

Fonofos insecticide

Fonofos, $C_{10}H_{15}OPS_2$ (*O*-ethyl *S*-phenyl (*RS*)-ethylphosphonodithioate), is an organophosphorus insecticide. It is a clear, aromatic liquid, soluble in water (17 mg/l) and is miscible with acetone, ethanol, xylene and kerosene. It is hydrolysed in acidic media and unstable above 100°C. The optical rotations of the chiral forms are reversed in solution with carbon tetrachloride, cyclohexane and methanol. The (*R*)-isomer is biologically more active than the (*S*)-isomer. It is produced by the reaction of thiophenol with *O*-ethylethane thiophosphoryl chloride. The CAS Registry Number of fonofos is 944-22-9.

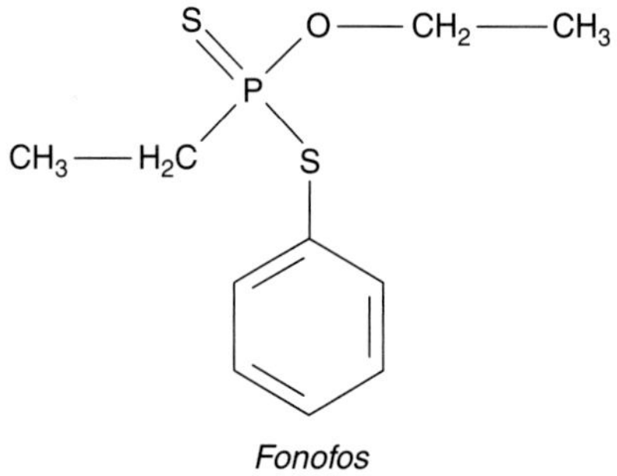

Fonofos

Fonofos, a cholinesterase inhibitor, is a broad-spectrum, soil insecticide with contact and stomach action. Flies and worms controlled include corn rootworms, cabbage root fly, onion fly, carrot fly and weevil larvae on cereals, vegetables and fruits.

The insecticide is formulated as micro-granules, emulsifiable/capsule suspension (CS) and commonly applied for treating seed. Known trade names are Dyfonate and Capfos.

It is considered extremely hazardous by the WHO. It is highly toxic to bees and fish. It is a cholinesterase inhibitor and a potential groundwater contaminant. Thus it is classified as a PAN Bad Actor chemical.

In plants, metabolism involves oxidation to the phosphonothioate and hydrolytic cleavage of the thiophenol residue. The soil DT_{50} is in the range of 3–16 weeks, depending on the kind of soil. Degradation products include *O*-ethyl ethanephosphonothioic acid fonofos oxon, *O*-ethyl ethanephosphonic acid, diphenyl sulfide and methylphenyl sulfoxide.

Many countries including Belize, Colombia and Thailand have banned the use of fonofos. Canada and India are among the few countries that have registered its use.

Food additives

See **Additives**

Food contamination by pesticides

See **Pesticides and the environment**

Foramsulfuron herbicide

Foramsulfuron, $C_{17}H_{20}N_6O_7S$ (2-[[[[(4,6-dimethoxy-2-pyrimidinyl)-amino] carbonyl] amino] sulfonyl]-4 (formylamino)-*N*, *N*-dimethylbenzamide), is a sulfonylurea herbicide. It is a beige powder with slightly acidic odour, soluble in water (3290 mg/l) at 20°C and pH 7. Although stable to photolysis, it undergoes abiotic hydrolysis with a DT_{50} of 10 days at pH 5. The CAS Registry Number of foramsulfuron is 173159-57-4.

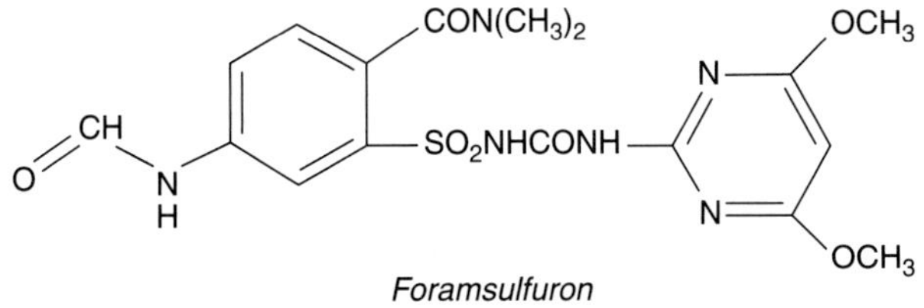

Foramsulfuron

Foramsulfuron, a post-emergence herbicide, is taken up by the leaves and is translocated throughout the plant. It interferes with amino acid synthesis through inhibition of acetolactate synthase (ALS). This action affects the meristematic regions, causing chlorotic and necrotic growth, and ultimately, death. It is used to control grass weeds, **barnyard grass** (*Echinochloa crus-galli*), yellow foxtail (*Setaria glauca*) and **Johnson grass** (*Sorghum halepense*) and broadleaved weeds in maize. When foramsulfuron is mixed with **iodosulfuron-methyl-sodium**, the broadleaved weed spectrum is enhanced to include weeds such as *Abutilon theophrasti*, *Chenopodium album*, *Xanthium strumarium* and *Ipomoea* spp.

Foramsulfuron is available in oil-miscible flowable concentrate (OF), suspension concentrate (SC) and water-dispersible granules (WG), commonly at 35%. It is available under trade names like Equip and Option. Common application doses range from 30 to 40 g/ha, with a maximum of two applications per season. Since it can be phytotoxic to crops, spray drift should be avoided. Sensitive crops can be safely planted only 2 months after the application of particular foramsulfuron formulations. A field sprayed with foramsulfuron can be planted with maize and soybean, after a period of 2 weeks.

Among rats, the acute oral LD_{50} of foramsulfuron is >5000 mg/kg, while the acute LD_{50} of foramsulfuron in mallard ducks is between 2788 and 5000 mg/kg. It is slightly irritating to the eyes. The ADI of foramsulfuron is 849 mg/kg b.w./day. Foramsulfuron is toxic to non-target plants. It is a potential groundwater contaminant.

In rats, following ingestion, most of the compound is eliminated within 24 h as unchanged foramsulfuron. In plants, residue levels are low with metabolism occurring by hydrolysis of the sulfonylurea bridge and formamide moiety on the phenyl ring, leading to a product that is highly polar and water soluble. The DT_{50} in soil is 1.5–12.7 days, whereas the aerobic soil metabolism of foramsulfuron takes 40 days. It is not persistent in a water/sediment system.

Foramsulfuron is registered in Argentina, Brazil, Canada, the USA, New Zealand and many countries in the EU including France, Germany, Hungary, Austria, Poland and Romania. The MRL proposed by New Zealand (New Zealand Food Safety Authority) on maize is 0.01 mg/kg.

Forchlorfenuron as a PGR

Forchlorfenuron, $C_{12}H_{10}ClN_3O$ (1-(2-chloro-4-pyridyl)-3-phenylurea), is a plant growth regulator. It is a whitish crystalline powder soluble only in water. It is hydrolytically stable at pH 5, 7 and 9 in aqueous buffer solutions, and photolytically stable at pH 5 in the same.

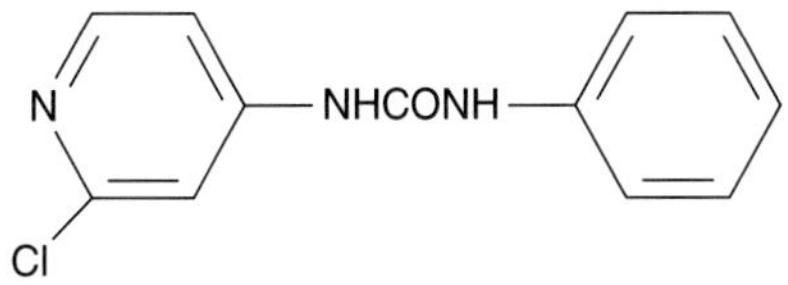

Forchlorfenuron

Forchlorfenuron is absorbed by most plant parts, where it acts synergistically with natural **auxins** to promote cell division and growth. It is also used to improve fruit size, fruit set and cluster weight mainly for kiwifruit and grapes. It is non-carcinogenic and does not create any neurotoxicity. It is available in the market under trade names like Sitofex.

In animals, forchlorfenuron is rapidly absorbed and metabolized. It is registered in India, the USA and EU members, Cyprus, Spain, Italy and Greece.

Formaldehyde fumigant

Formaldehyde (or methanal) with molecular formula CH_2O, is a colourless, pungent, light-sensitive, naturally occurring flammable gas (also available as a liquid). It is a fumigant and a bactericide. Made up of carbon, hydrogen and oxygen, it is soluble in water and miscible with acetone, alcohols and ethers. Being a powerful reducing agent, it reacts explosively with nitrogen dioxide and perchloric acid, among others. It is also a powerful reducing agent, incompatible with oxidizing agents and alkalis. The CAS Registry Number is 50-00-0.

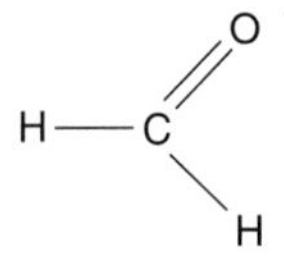

Formaldehyde

Formaldehyde is used in glasshouses, mushroom houses and as a silage preservative. It is also a soil sterilant. It is formulated as a hot fogging concentrate and a soluble concentrate (SL).

The fumigant is a toxic chemical, with its vapours being hazardous to animals and man. It is very irritating to the eyes, skin and the mucous membranes.The acute oral LD_{50} in rats ranges from 550 to 800 mg/kg and the acute percutaneous LD_{50} in rabbits is 270 mg/kg. It is toxic to fish and is also extremely phytotoxic. Formaldehyde, along with its by-product bis-chloromethyl ether, is carcinogenic in nature and hence a PAN Bad Actor chemical.

In animals, formaldehyde is absorbed rapidly via respiratory and gastrointestinal routes. In the human body and in the environment, it metabolizes quickly and so it does not accumulate in the food chain. In the human body, it is converted to **formic acid**. It also breaks down by the action of bacteria and by sunlight in soil or water.

Formaldehyde is registered in the USA, Canada, Australia and New Zealand, but not in the EU.

Formetanate hydrochloride acaricide

Formetanate hydrochloride, $C_{11}H_{16}ClN_3O_2$ (3-dimethylaminomethyleneaminophenyl methylcarbamate hydrochloride), is a carbamate compound that is used as an acaricide and insecticide. It is a colourless, crystalline powder, soluble in water (822 g/l at 25°C), methanol (283 g/l) and acetone (0.074 g/l). It is stable for at least 8 years at room temperature. It decomposes at about 200°C. It is not compatible with alkaline materials. Its CAS Registry Number is 23422-53-9.

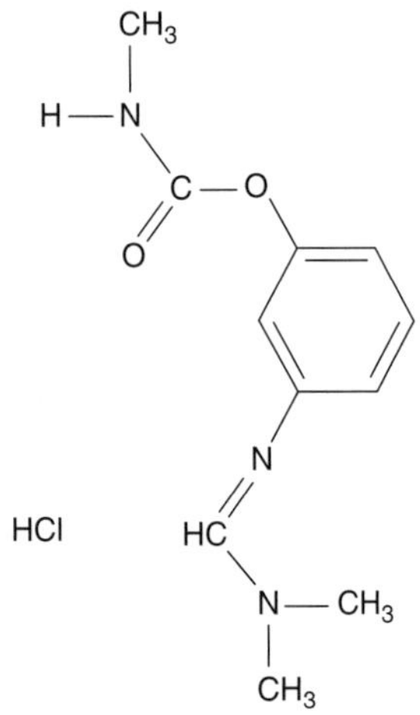

Formetanate hydrochloride

Formetanate hydrochloride is an acaricide and insecticide with contact and stomach action, inhibiting the enzyme cholinesterase of spider mites and some insects in the orders Diptera and Hemiptera. It is used to protect ornamentals, lucerne, vegetables and fruit. The formulation type is water-soluble powder (SP).

The acaricide is classified by the WHO and EPA as extremely hazardous. It is an eye irritant. The acute oral LD_{50} in rats is 14.8 mg/kg, while in dogs it is 19 mg/kg. In normal use, the compound does not irritate the skin (the acute percutaneous LD_{50} in rats is >5600 mg/kg). It is carcinogenic by nature and is feared to be a groundwater contaminant. It is very toxic to bees (with the LD_{50} being14 μg per bee) and at high levels is phytotoxic to peas, beans, cucumbers and some rose species. The ADI is 0.004 mg/kg b.w./day.

Formetanate hydrochloride is registered for use in Canada, Portugal and the USA.

Formetanate insecticide

Formetanate, $C_{11}H_{15}N_3O_2$ (3-dimethylaminomethyleneaminophenyl methylcarbamate), is a carbamate insecticide and an acaricide. It is a colourless, crystalline powder, soluble in water, methanol, acetone, toluene, dichloromethane, acetate and *n*-hexane. Though stable at room temperature, it decomposes at 200°C and hydrolyses in 2 h at pH 9. It is produced by the reaction of 3-dimethylaminomethyleneaminophenol with methyl isocyanate. The CAS Registry Number is 22259-30-9.

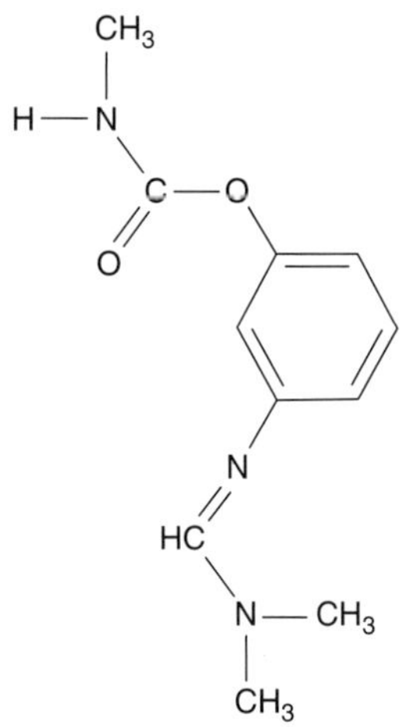

Formetanate

A cholinesterase inhibitor and thus a neurotoxin, formetanate is used on apples, pears, citrus fruits and luceme (when grown for seed) to control thrips and mites. The chemical is absorbed into the leaves but is not translocated.

This insecticide is available as a wettable powder (WP), sprayed aerially or from the ground, at a maximum of 1.3 kg/ha. It is commercially available commonly as Carzol.

Formetanate has high acute toxicity via the oral route. It is highly toxic to birds and mammals and moderately toxic to honey bees. It is slightly to moderately toxic to marine organisms. It is a PAN Bad Actor chemical.

In animals, formetanate metabolizes to form the glucuronide and ethereal sulfate conjugates of *m*-acetamidophenol. It rapidly degrades in soil under aerobic conditions, but appears to leach and hence there is a potential for groundwater contamination.

The insecticide is registered in South Africa, Cambodia, New Zealand and a few EU members including Greece, France and Spain.

Formic acid acaricide

Formic acid, CH_2O_2 (other names being methanoic acid, formylic acid or aminic acid), is a naturally occurring carboxylic-acid-based acaricide. It is an irritating, pungent liquid produced by bees and ants. The CAS Registry Number is 64-18-6.

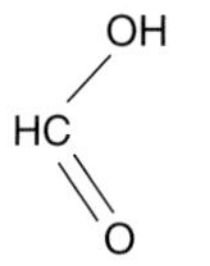

Formic acid

Formic acid finds application in controlling mite populations (e.g. *Varroa* and Tracheal mites) in bee hives, especially between 10 and 25°C. It can form a part of crop IPM, especially in cases of excessive mite infestation. The chemical irritates and kills mites without harming the bees or the crops.

Formic acid is available as a gel, ensuring slow release of the substance. The trade names of formic acid include Formic Acid Gel, For-Mite and Mite Away II. The acid is used often as a bactericide on silage and fodder, to extend its storage.

Formic acid is mildly toxic in the case of oral and inhalation routes, highly irritating to the respiratory tract and corrosive to the skin. However, it does not harm the environment and is not considered a hazard in normal use.

This acaricide is registered for use in Australia, Canada and the USA.

Formothion insecticide

Formothion, $C_6H_{12}NO_4PS_2$ (*S*-[formyl(methyl)carbamoylmethyl]*O*,*O*-dimethyl phosphorodithioate), is an organophosphorus insecticide and an acaricide. It is an odourless, pale yellow, viscous liquid, soluble in water at 2.6 g/l and completely miscible with alcohols, ethers, ketones, chloroform and xylene. It is hydrolysed by water to **dimethoate** and dimethoxyphosphinothioylthio acetic acid, and more rapidly so in alkaline media. Its dilute solutions in non-polar organic solvents are stable. Formothion is produced by the reaction of the sodium salt of *O*,*O*-dimethyldithiophosphoric acid with *N*-formyl carbamoylmethyl chloride. The CAS Registry Number is 2540-82-1.

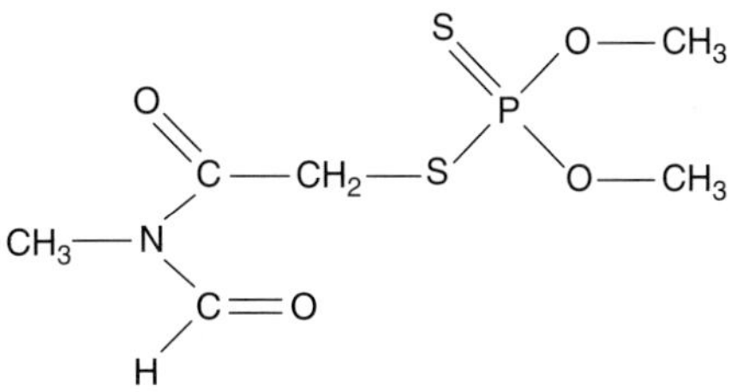

Formothion

Formothion, a cholinesterase inhibitor, is a systemic compound with contact and stomach action that controls many sucking and chewing insects such as aphids, **leafhoppers**, plant lice, thrips and spider mites. It is used on cotton, rice, sugarcane, tobacco and fruit crops.

This insecticide is formulated as emulsifiable concentrate (EC) and ULV liquid (UL). Combination with alkaline substances has to be avoided. It is available commercially as Anthio. For mite control, an accepted dose is 250–500 g a.i./ha.

Formothion is toxic to bees and rats and moderately toxic to fish and birds. The oral acute LD_{50} among rats is 365 mg/kg. It may be phytotoxic to some varieties of peach, apricot and cherry. The WHO classifies it as a Class II toxin (moderately hazardous). It is non-irritating to skin of mammals in normal use. Besides the toxicity, since it is a cholinesterase inhibitor and a suspected endocrine disruptor, it is considered to be a PAN Bad Actor chemical.

In mammals, upon ingestion, formothion is metabolized to (dimethoxyphosphinthinthioylthio) acetic acid and polar metabolites, and eliminated quickly. It metabolizes in plants to dimethoate (dimethoxyphosphinthinthioylthio) acetic acid, bis (dimethylthiophosphoryl) disulfide, **omethoate** and *O*,*O*-dimethyl hydrogen phosphorodithioate. The loamy soil DT_{50} is <1 day.

India has now withdrawn registration of formothion.

Formulations

See **Pesticide formulations**

Fosamine herbicide

Fosamine, $C_3H_8NO_4P$ (ethyl hydrogen carbamoylphosphonate), is an organophosphorus herbicide and a plant growth regulator. It is a colourless crystal, soluble in water (>2.5 kg/l at 25°C), methanol, ethanol, dimethylformamide, benzene, chloroform, acetone and hexane. Fosamine is stable in neutral and alkaline media and it decomposes in dilute acidic medium. The ammonium salt, fosamine ammonium, functions as a PGR. The CAS Registry Number of fosamine herbicide is 25954-13-6.

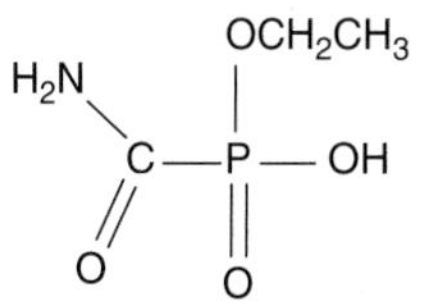

Fosamine

Fosamine, a post-emergence herbicide with both systemic and contact activity, inhibits mitosis, thus preventing dormant tissues from re-growing. A non-crop herbicide, it controls brush growth as well as woody growth in parks, industrial sites, forestry and plantations (e.g. in pines, maple and vine maple).

This herbicide, used either alone or with fosamine ammonium, and with a surfactant, is available in soluble concentrate (SL) formulations, with spray applications made on a dry day, avoiding spray drift. Trade names include Krenite UT. The maximum application rate for low shrubs/brush is up to 20 kg/ha. Degradation rates are fast in acidic medium.

Fosamine is not acutely toxic to most tested species. For instance, the oral LD_{50} of fosamine among rats is between 5000 and 24,400 mg/kg, while the dermal LD_{50} in rabbits is >1683 mg/kg. It is practically non-toxic to birds, bees and fish. The acute topical LD_{50} for honey bees is >200 μg per bee. The WHO classifies the fosamine salt as a Class U toxin (unlikely to be hazardous), but due to moderate skin and eye irritation that can be caused, the signal word of 'WARNING' is displayed. Handlers are advised to wear necessary protective garments and wash hands and clothing before eating and drinking.

The typical soil half-life for fosamine is about 8 days, while that of fosamine ammonium is up to 2 weeks. Fosamine is mobile in some soils, but since it is subject to microbial degradation and does not bioaccumulate, it is not a threat to groundwater resources and the environment. Fosamine ammonium too has the potential to be mobile in the environment.

The USA and Canada have registered the use of fosamine and its salt, but it is not permitted in the EU.

Fosetyl-aluminium fungicide

Fosetyl-aluminium, $C_6H_{18}AlO_9P_3$ (aluminium tris(ethyl phosphonate)), is an organophosphate fungicide, with some antibacterial properties. It is a colourless powder, soluble in water (110 g/l), acetone, ethyl acetate and methanol (>800 mg/l). It hydrolyses under extreme acidic or alkaline conditions and must be stored in airtight containers in dry, ventilated areas away from food, feed, open flame or heat. It is stable to aqueous hydrolysis. The CAS Registry Number of fosetyl-aluminium is 39148-24-8.

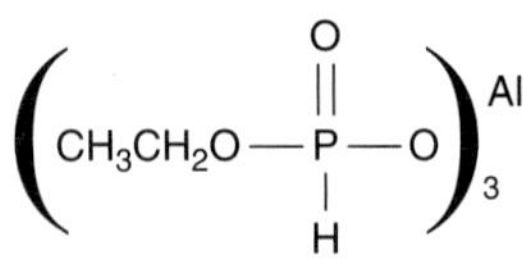

Fosetyl-aluminium

Fosetyl-aluminium, a systemic, protective fungicide, is readily translocated throughout the plant. It inhibits spore germination and host penetration, while enhancing the plant's own natural defence systems. It controls infections of *Phytophthora*, *Pythium* and *Alternaria*, and can also halt the advance of bacterial *Pseudomonas* sp., thus preventing blister and purple spot, crown rots and root rots.

The fungicidal compound is formulated as water-dispersible granule (WG) and wettable powder (WP). It is commonly available as Aliett.

Fungicide fosetyl-aluminium has an acute oral LD_{50} in rats of >7080 mg/kg and acute percutaneous LD_{50} in rabbits of >2000 mg/kg. It irritates the respiratory tract. According to the WHO and EPA, fosetyl-aluminium is a slightly toxic compound (Class III). It has low toxicity to fish but is moderately toxic to aquatic invertebrates. It is practically non-toxic to honey bees. However, according to the PAN Pesticide Database, this chemical is high on acute toxicity and holds the potential to contaminate groundwater. Hence, it is a PAN Bad Actor chemical.

Fosetyl-aluminium undergoes extensive metabolic transformation, to form CO_2 and phosphorus acid, in animals. In plants, the metabolism of fosetyl-aluminium is through the hydrolytic cleavage of the ethyl ester bond. In soil, fosetyl-aluminium has an extremely short half-life under both aerobic and anaerobic conditions, with rapid dissipation and metabolism.

This fungicide is registered in countries including India, the USA and all countries of the EU, Australia, New Zealand, Tanzania and Vietnam.

Fosthiazate nematicide

Fosthiazate, $C_9H_{18}NO_3PS_2$ ((RS)-*S*-sec-butyl *O*-ethyl 2-oxo-1,3-thiazolidin-3-ylphosphonothioate), is an organophosphorus pesticide. It is a light-brown liquid with solubility in water being 9.85 g/l. It is hydrolysed in aqueous media. It is produced by treating 2-oxo-3-thiazolidine with *S*-sec-butyl *O*-ethyl phosphorochloridothioate in toluene. Its CAS Registry Number is 98886-44-3.

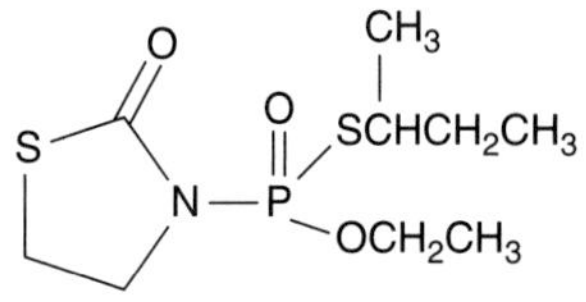

Fosthiazate

The compound, a broad-spectrum, cholinesterase inhibitor, is primarily a nematicide. But it is also used to control aphids and thrips. Fosthiazate is used against root-knot nematodes, specifically *Meloidogyne* sp. and *Heterodera* sp. It protects tomatoes, potatoes and vegetables. It is an effective alternative to **methyl bromide**, which is banned or restricted in many countries.

Nematicide fosthiazate is formulated as granules (GR) or as an emulsifiable concentrate (EC). Soil incorporation is a common way of application at the pre-plant stage. The maximum limit of application is set as 1.1–1.70 kg/ha, with only one application a year. It is commonly available as the trade name Nemathorin.

Fosthiazate is categorized as highly toxic to birds (in the case of the bobwhite quail, the LD_{50} is 16.6 mg/kg), honey bees and crustaceans. It is moderately toxic to mammals. Fosthiazate has been observed to be moderately toxic via the dermal and oral routes, slightly toxic on inhalation and moderately irritating to the eyes and skin. Hence applicators are advised to handle the chemical with proper PPE. Since fosthiazate is a cholinesterase inhibitor, it is listed as a PAN Bad Actor chemical. **Chemigation** should be monitored to avoid harm to birds.

The nematicide degrades faster in alkaline conditions (half-life of 1 week). This chemical is not mobile, with the residue products (methyl sec-butyl sulfone and 2-butanesulfonic acid) penetrating only to depths of up to 30 cm into the soil. The DT_{50} at pH 9 is 3 days.

Fosthiazate is considered a viable alternative to methyl bromide used for the tomato crop. Use of methyl bromide has been banned or restricted, since it is known to threaten the ozone layer.

The compound is registered in the USA, Japan, the Philippines, South Africa and about half of the EU including Cyprus, Germany, Greece, Hungary, Italy, the Netherlands and the UK.

Fouling

See **Organotin fungicides**

FRAC Code List

Fungicides, like any other pesticide, work by inhibiting certain biochemical processes or biosynthetic pathways, thus retarding stages of fungal growth or killing the fungi. The action of these chemical agents, blocking fungal biosynthetic pathways is known as a mode of action (MOA). The **Fungicide Resistance Action Committee (FRAC)** maintains a list of commercial fungicides with their common names, chemical classification, target site and MOA, with a view to manage resistance better. Please refer to Appendix 10 for the FRAC Code List which provides these details of fungicides, sorted by their MOA. The FRAC Code List is reproduced with permission from Dr Randall Gold, European Technical Management, Speciality Crop Fungicides, BASF –The Chemical Company (www.frac.info.).

Fuberidazole fungicide

Fuberidazole, $C_{11}H_8N_2O$ (2-(2′-furyl) benzimidazole), is a **benzimidazole fungicide**. It is a fine crystalline, odourless powder with solubility in water of 71 mg/l. It is soluble in isopropanol, toluene and dichloromethane. Being unstable in light, it should be kept in a cool, dry, dark location in an airtight container. The CAS Registry Number is 3878-19-1.

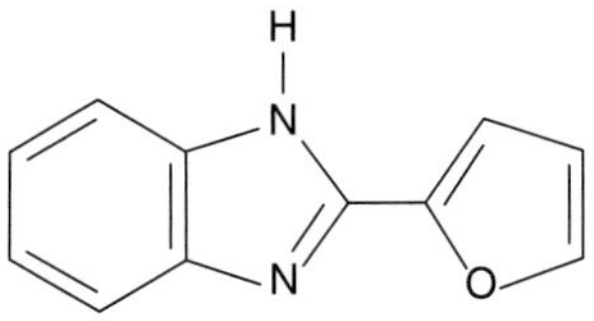

Fuberidazole

A mitosis inhibitor, fuberidazole is a systemic fungicide, used to control *Fusarium* spp. in cereals. It is used in treating seeds.

The fungicide fuberidazole is formulated as a powder for dry seed treatment (DS), a flowable concentrate (SC) or a solution, all of which cater for dry and wet seed treatment. It is incompatible with isocyanates, halogenated organics, peroxides, phenols (acidic), epoxides, anhydrides and acid halides. The fungicide is available in brand names like Voronit.

Fuberidazole has an LD_{50} in rats of 500 mg/kg and is classified as moderately hazardous (Class II toxin) by the WHO. It is non-toxic to bees, but is moderately toxic to some aquatic organisms. The ADI in man is 0.0025 mg/kg b.w./day.

In animals, metabolism of fuberidazole is almost completely via hydroxylation, followed by conjugation and cleavage of the furan ring. Mobility in soil is low and degradation, rapid.

This fungicide is registered in the USA and EU member states (e.g. the UK, Germany, Finland, Denmark and some others).

Fumigant nematicides

See **Nematicides**

Fumigants

Fumigants are volatile toxic chemicals that are applied to kill pests. The application is commonly made on or in soil, in grain store houses, in hospitals, or any infected area which needs to be sterilized.

In agriculture, fumigants effectively destroy insects, microorganisms, weeds, weed seeds and nematodes. Fumigation, a pre-plant treatment, is undertaken on stored grains to control pests. Fumigants are generally sprayed. Fumigation compounds are also applied as **soil fumigants** or as **soil sterilants**. Halogenated alkanes form the base of soil fumigants. It is best to apply these compounds in cooler temperatures.

Chloropicrin (Cl_3CNO_2), carbon disulfide (CS_2), methyl bromide (CH_3Br), hydrogen phosphide gas (H_3P), **hydrogen cyanide** (HCN) and carbon tetrachloride (CCl_4) are generally used for fumigation. Chloropicrin is very effective on stored grain. It is available as a liquid, which upon exposure to air turns into gas. It is persistent, hence effective for the long term, but since it persists on the seed as well, several countries have restricted its use to empty storehouses and containers. Because of high toxicity, applicators should be equipped with masks, breathing apparatus and full-body protective clothing. The EPA has given some guidelines regarding the use of fumigants. **Methyl bromide**, alone or in combination with chloropicrin, had become the standard against which all fumigation was measured. It is a colourless, odourless gas available in a compressed form. It quickly spreads through the air and becomes effective. It is effective in the control of stored grain pests. However, it is found to be toxic, acting on the respiratory system. Since it destroys the stratospheric ozone, it is scheduled to be banned all over the world by the year 2015.

Soil fumigation is carried out by injecting the soil up to a certain depth with the compound, or by applying through irrigation channels. Only mobile (in water) compounds are suitable for being soil fumigants. Halogen and isothiocynate, propylene oxide, **1,3-dichloropropene** and methyl iodide are suitable soil fumigants, while **propargyl bromide** is applied via drip irrigation with an emulsifying agent.

Fumigants in grain and flour – technical analysis

Methyl bromide, ethylene oxide, ethylene dibromide and ethylene dichloride are efficiently extracted from grain and flour with a mixture of acetone and water in the ratio 5:1.

Weigh about 10 g wheat or 5 g flour and transfer to a stoppered flask containing 30 ml of the mixture of acetone and water mentioned above. Allow to stand at 20°C, for different durations, depending on the fumigant and the sample (e.g. for methyl bromide or ethylene oxide, allow 1 h for flour and 6 h for wheat, while for ethylene dibromide, allow much longer for flour or wheat).

Inject 2 μl of the supernatant solution into the gas chromatograph, allowing 12 min between injections for complete removal of water from the column, so that the sensitivity is not lowered. The conditions of the experiments differ with the fumigant and are given below.

For methyl bromide and ethylene oxide, the column should be the size of 2 m × 4.6 mm (internal diameter). The temperature of the column should be 85°C, while that of the injector, 125°C. The carrier gas is taken to be helium, with the flow rate being 80 ml/min. The absolute retention time is 42 s for ethylene oxide and 45 s for methyl bromide. The detection limit for ethylene oxide is 2×10^{-9} g and for methyl bromide, 5×10^{-11} g. For ethylene dichloride and ethylene dibromide, use the column measuring 4 m × 4 mm (outer diameter), packed with the column packing of 10% by weight carbowax 1540 on teflon. The temperature of the column in the case of ethylene dichloride is 65°C, and for ethylene dibromide, it is 120°C. The injector temperature is 130°C. Nitrogen is used as the carrier gas and the flow rate is 30 ml/min. The absolute retention time for ethylene dichloride is 12.5 min, while for ethylene dibromide it is 8 min. The detection limit for ethylene dichloride is 1×10^{-9} g, while that of ethylene dibromide, it is 1×10^{-11} g.

Fumonisins

Fumonisins are a group of toxic metabolites (**mycotoxins**) produced primarily by the fungi *Fusarium moniliforme* and *Fusarium proliferatum* which infect maize, grasses and herbaceous plants. Other *Fusarium* species that produce fumonisins are *Fusarium napiforme* and *Fusarium nygamai*. Fumonisin B1, the most common toxin, is generally produced on mouldy maize and in silage by certain strains of *F. moniliforme* and is seen in animal feeds.

These fungi grow over a wide range of conditions, with some *F. moniliforme* strains producing more of these toxins than others. The incidence of *F. moniliforme* appears to be related to dry conditions. Early season drought stress increases infection by *F. moniliforme* and its colonization in maize. Low concentrations of fumonisins are present in good quality maize and in most processed, commercially available maize products.

Upon ingestion by animals, fumonisins destroy areas of their brain, while it may cause cancer in human beings. The levels of feed recommended are to be no higher than 5 ppm for horses, 10 ppm for swine (pigs) and 50 ppm for cattle.

Fungi

Fungi are an important group of microflora with generally filamentous and branched somatic (physical) structures. They are **eukaryotic**, achlorophyllous thallophytes that have cell walls and prominent nuclei. They reproduce by both sexual and asexual means.

Biochemical and molecular biology have now classified microorganisms similar to fungi, in the Kingdoms Protista, Chromista, Protozoa and Fungi. Kingdom Protista, divided into myxomycota and oomycota, has heterotrophic or autotrophic modes of nutrition and includes slime moulds. The Kingdoms Chromista and Protozoa mainly consist of non-fungal phyla.

Kingdom Fungi includes the fungi (mycelial or yeast) with absorptive and heterotrophic nutrition. Kingdom Fungi is divided into: (i) the Chytridiomycota, constituting aquatic fungi with sexual and asexual spores, with posterior flagella; (ii) the Zygomycota (bread moulds), with aseptate hyphae, sexual resting spores called zygospores and asexual spores formed in sporangia; (iii) the Ascomycota (mycelial fungi or yeasts), with sexual spores called ascospores formed in asci and asexual spores that grow as conidia; and (iv) the Basidiomycota with sexual spores called basidiopores, formed on a basidium.

In soils, fungi exist in great numbers (about a million per gram of soil) and perform many functions. The size of fungi varies from microscopic (lower fungi) to visible specimens (higher fungi, such as mushrooms) and examples include yeasts, moulds and mildews. The presence of **chitin** in the cell wall is one of the major distinguishing features of the organisms in the Kingdom Fungi. Fungal spores help them survive in adverse conditions. Fungi are heterotrophic and aerobic and may be parasites or saprophytes, but they play an important part in releasing nutrients to the soil from dead plants and animals, especially in acid soils. But as the decomposable material declines, the fungi too decline.

In neutral or slightly alkaline conditions, fungi produce a filamentous growth known as mycelium, which may extend up to several centimetres in undisturbed ecosystems. Fungal mycelium spreads over surfaces and helps bind mineral particles, turning them into aggregates. They are sensitive to wet soil conditions. They can convert nearly half of the carbon into cell material, while releasing the remaining as carbon dioxide. Thus fungi have a role in humus formation and nutrient recycling (compared with bacterial efficiency of only 20%).

Fungi are one of the most economically important groups in both, positive and negative ways. In addition to edible mushrooms, fungi are used in fermentation processes for the production of bread, alcohol, cheese and antibiotics (e.g. *Penicillium chrysogenum* in the production of penicillin). Fungi are also useful as biocontrol agents on crops.

On the other hand, pathogenic fungi are responsible for destroying crops that have impacted human history, and for debilitating diseases of humans and domestic animals. Rust fungi and **smut** fungi are examples of damaging pathogens harmful to cereal crops. Most of the parasitic fungi cause crop diseases.

Fungicidal antibiotics

Antibiotics are substances produced by microorganisms that are toxic to bacteria and other microorganisms. Antibiotics show definite growth-retarding properties and are often toxic to the target microorganism. Streptomycin, oxytetracycline, chloramphenicol and griseofulvin were antibiotic compounds developed initially. Most of these are unstable in alkaline media. Antibiotics may sometimes be mixed with chemical compounds for specific desired effects.

Streptomycin, a fermentation product of *Streptomyces griseus*, is soluble in water and methanol, but is unstable in strong acids and alkaline material. It is used against bacterial diseases such as canker, blights, rots and **wilts** on citrus, pome fruits, cotton, potato and tobacco.

Griseofulvin is a systemic fungicide that acts against *Botrytis* spp. and **powdery mildew** diseases. It is produced from the fungal mould *Penicillium griseofulvum*. **Cycloheximide**, extracted from a *Streptomycin* species, controls **downy mildew** and rusts but is phytotoxic, hence has a limited use. **Blasticidin-S**, a non-systemic and preventive fungicide, is effective against rice blast (caused by *Pyricularia oryzae*) and other rice diseases. It is a nucleoside antibiotic made out of a *Streptomyces* species and is soluble in water and acetic acid. A foliar application of 5.40 ppm is common. **Kasugamycin** acts by inhibiting protein synthesis and controls rice blast and many rice diseases. This agent is a fermentation product of *Streptomyces kasugaensis* and is considered an aminoglycoside antibiotic.

Polyoxin B, $C_{17}H_{25}N_5O_{13}$, is an amorphous powder, soluble in water but insoluble in most organic solvents. It acts by inhibiting **chitin** synthesis. It is not phytotoxic and is safe on animals and humans. Polyoxin B, produced by fermentation of *Streptomyces cacaoi* var. *asoensis*, is a systemic foliar antibiotic fungicide. It is used to control black spot in pears and also rice blast. Polyoxins act by inhibiting chitin synthesis.

Validamycin, $C_{20}H_{35}NO_{13}$, is a colourless powder, which is soluble in water. It is produced by fermentation of *Streptomyces hygroscopicus* var. *limoneus*. Validamycin is a non-systemic antibiotic with fungicidal properties against basidiomycetes. It is effective against sheath blight of rice and black scurf of potatoes. Validamycin has a short half-life (<4 h) and hence is environmentally safe. It is non-phytotoxic, and safe if used in the presence of animals and humans. Its available formulations are dust and liquid. It is more often used as a curative rather than as a protective agent.

Mildiomycin is a nucleoside antibiotic that works by inhibiting protein synthesis. It controls powdery mildew on roses. It is made from *Streptoverticillium rimofaciens*.

Fungicidal foliage treatment

See **Fungicide application methods**

Fungicidal modes of action

Fungicides are developed bearing in mind the physiology of the fungal species they are to control and the crucial biochemical processes required by the organism. Fungicides kill or control fungi by attacking a certain biochemical process or group of processes. This targeted activity is called their mode of action (MOA). The following passage illustrates the major MOA of fungicides.

Some fungicides disrupt the cell functioning in the organism. These agents are cell toxicants that usually have protective, multi-site activity. Normally, these are broad-spectrum compounds that cover a wide range of species. The compounds of sulfur, copper, mercury and tin in combination with lime work by disrupting fungal cell function. Examples of organic compounds include **dithiocarbamate** (e.g. **ziram, thiram** and **zineb**), **phthalimide** (**captan, captafol** and **folpet**), sulfamide (dichlofluanid), **triazine** (anilazine), chlorophenyl (**chlorothalonil**), quinine (dithianon), nitroparaffin (fenitropan) and quinozaline (**chinomethionat**).

Some fungicides work by disrupting fungal membrane function. Sterols help in keeping membrane integrity among all **eukaryotes**, while glycerophospholipids prevent permeability for ion movement outside the cell membrane. **Demethylation inhibitor (DMI)** fungicides are site-specific fungicides that disrupt the synthesis of sterols. These act on higher orders of fungi such as ascomycetes, deuteromycetes and basidiomycetes, since these pathogens synthesize **ergosterol**. Sterol inhibitors are divided into C14-demethylation sterol inhibitors (imazalil, **prochloraz, triflumizole, nuarimol**) and $\Delta^{8,7}$ isomerase Δ^{14} reductase inhibitors (**fenpropimorph, tridemorph**). Inhibition of glycerophospholipid biosynthesis brings about general disruption of cell functions. Inhibition of phosphatidyl-choline synthesis (an essential part of cell growth) is brought about by **iprobenfos** and **edifenphos**. Similarly, phosphatidylinositol (which is critical for cell growth) is inhibited by **validamycin A**.

Disruption of fungal nuclear processes works against nucleic acid metabolism and tubulin biosynthesis. Nucleic acids, such as RNA and DNA, play a key role in many biochemical reactions involving pyrimidine precursors. The primary sites of action of the chemical classes of phenylamides and **hydroxypyrimidines** are nucleic acid pathways. **Phenylamide fungicides** specifically act against oomycete fungi and inhibit specific fungal growth stages. Hydroxypyrimidines inhibit adenosine deaminase, which is found in a wide range of fungi, but they are highly specific for the control of the fungal family Erysiphaceae. This family includes **powdery mildew**-causing pathogens. **Benzimidazoles** inhibit tubulin synthesis, and thus can control ascomycetes, basidiomycetes and deuteromycetes, but not oomycetes. The systemic fungicidal compounds in the benzimidazole group are **benomyl, carbendazim, thiophanate-methyl, fuberidazole** and **thiabendazole**. There are some other tubulin synthesis inhibitor fungicides, like phenylcarbamates and benzamides.

Disrupting the cell wall of fungi is another MOA of fungicides. **Chitin**, synthesized by the help of enzyme chitin-UDP-*N*-acetylglucosaminyl transferase, is an important constituent of the cell wall and it has a structural role. Chitin is found in ascomycetes, deuteromycetes and basidiomycetes. **Polyoxins** are selective agents that disrupt the cell wall function of chitin causing the swelling of the germ tube and hyphal tips, thus killing the fungus. Melanin synthesis is the other function of the fungal cell wall and helps the appressorial wall to penetrate the host epidermis by infection hyphae. **Tricyclazole fungicide** inhibits the melanization of appressoria and the infection hyphae. Melanin synthesis inhibitors are effective against ascomycetes and deuteromycetes.

Fungicides can also kill by targeting protein synthesis. **Blasticidin-S** and **kasugamycin**, two prominent systemic agents, prevent the elongation of the protein chain.

Inhibition of fungal respiration is another MOA of fungicides. Such compounds disrupt oxidative phosphorylation and inhibit complex II and III. Such inhibitors have a broad spectrum of activity, because they uncouple phosphorylation by permitting electron transfer, without ATP production. Nitrophenol derivatives, drazoxolon, **fluazinam** and **fentin** inhibit oxidative phosphorylation. Some fungicidal compounds inhibit important pathways in respiration – succinate dehydrogenase and complex II. These compounds, specific to basidiomycetes, block electron transfer. Carboxamides (**fenfuram, thifluzamide**, mepronil, methfuroxam) inhibit complex II.

A few compounds block electron transfer of the mitochondrial electron transport chain and thereby inhibit complex III. Such compounds include **strobilurin** and **famoxadone**.

Guanidine fungicides and **guazatine** work by altering cell membrane permeability. This reduces nutrient uptake.

Some compounds work by an unspecified MOA. **Organophosphorus (OP) fungicides**, anilinopyrimidines, aromatic hydrocarbons **fosetyl-aluminium, phenylpyrroles, quinoxyfen**, cymoxanil, prothiocarb, **propamocarb hydrochloride** and dimethomorph are such chemical classes or active ingredients (a.i.). While anilinopyrimidines focus on reducing the production of lytic enzymes, the OP fungicides inhibit conidial germination, and some others inhibit methionine biosynthesis, fungal mutagenic properties or disrupt lipid peroxidation (aromatic hydrocarbons).

Table F.1 illustrates the major fungicidal chemical classes with examples of a.i.s, their MOA and the diseases they control.

Fungicide application methods

Fungal infections on crops are widespread and can become chronic, if left unchecked. This explains why fungicides account for a large portion of pesticides used worldwide. An understanding of the need to apply fungicides, the mode and time of application, and the precise formulations are necessary for an efficient pest management strategy. Following are some modes of fungicide treatment on plants, although it should be noted that the methods are also common to other pesticide types such as insecticides, nematicides and bactericides.

Seeds are frequently attacked by fungal pathogens, resulting in abnormal germination or rotting of the seed. Pathogens can either attack the surface of the seed coat or penetrate the seed and infect from within. So, fungicides are developed to treat seeds. Seed treatment is a method of applying fungicide to seeds, so that both soil-borne and seed-borne fungi and bacteria that cause diseases like seedling **blights**, seed decays, **smuts** and bunts are destroyed. This takes place as either wet seed treatment or dry seed dressing. **Rotary seed dressers** and slurry treaters are the main tools used for such treatments. The former, suitable for a small seed lot, is for dry seed dressing, while for the latter, a larger seed quantity is treated with fungicide slurry, so as to coat the

Table F.1. An overview of fungicidal compounds with their active ingredients and properties. (Table courtesy of Kalyani Paranjape.)

Active ingredient (a.i.)/chemical group/toxicity	a.i. with mode of action (MOA)/treatment/ symptoms/others	Disease/target species
Brine (sodium chloride solution) and lime	Sodium and calcium. Non-systemic application used for seed treatment	Bunt diseases on wheat, *Suillus luteus* at high NaCl concentrations
Copper causes general disruption of cell metabolism. Very toxic, non-systemic, immobile protectant	Copper sulfate, copper carbamate and copper hydroxide. MOA – reacts with carboxy, sulfydryl and amino groups in the cell	Bunt diseases on wheat, mildew, leaf spots, blights and apple scab
Phenyl- and alkyl mercury	Safe for seed treatment. Cheaper than inorganics	Wheat bunt, seed-borne diseases on barley, tomatoes
Elemental sulfur, lime sulfur – disrupt biochemical sites such as respiration and proteins	Sulfur. MOA – S oxidizes to sulfur dioxide and sulfur trioxide or reduces to hydrogen sulfide. Foliage sprays on fruit trees	Powdery mildew in grape vines and hops; apple scab and peach leaf curl
Mercury – inorganic. Very toxic to mammals	Phenyl mercury chloride and methoxy-ethyl mercury chloride. MOA – disrupts cell functions. Used in seed and foliar treatments and on turf grass and orchard trees	*Plasmodiophora*, *Nectria galligena* for club root disease and apple canker
Tin – biocidal property	Tributyl tin, fentin acetate and fentin hydroxide. MOA – general disruption	Wood treatment/preservatives
Bordeaux mixture (lime and copper sulfate) and combinations of lime and copper oxychloride, cuprous oxide, calcium polysulfide. Immobile protectant. Toxicity of copper sulfate decreases with lime. Low in human toxicity and are economical	Cu and S. MOA – forms complexes with sulfydryl, hydroxyl, amino and other enzymes. Use of Bordeaux mixture is permitted in organic farming. Fungi commonly do not develop resistance against S products	Downy mildew of grapes and potato blights
Carboxamides – systemic and safe fungicide. Inhibitor of mitochondrial energy production and complex II, involving related proteins (respiration activity). Structure similar to *N*-phenyl-2-butenamide	Fenfuram, mepronil, methfuroxam, flutolanil and thiflumazid. MOA – inhibition of electron transfer from non-haem iron-sulfur protein to ubiquinone. Used as wettable powder or for seed dressing with/without other fungicides (e.g. thiram and imazalil) to target non-basidiomycetes. Complex II – a membrane-bound enzyme that catalyses oxidation of succinic acid to fumaric acid	Basidiomycetes, which causes smuts, bunts and rusts. Also *Rhizoctonia solani*, *Helminthosporium* spp., *Ustilago* spp. Maize, canola, legumes and cotton are also protected
Antibiotics – from naturally occurring products. Systemic and specific action. Protective and curative properties. Inhibition of protein synthesis	Blasticidin S, kasugamycin, polyoxins, validamycin A, streptomycin, oxytetracycline, chloramphenicol. MOA – inhibits protein synthesis while interacting with ribosomal RNA complex, inhibits hyphal growth. Emulsifiable concentrate and wettable powder used as foliar application. If combined with other fungicides, it increases effectiveness	Controls rice blast caused by *Pyricularia oryae*. Controls downy mildew on onion, powdery mildew on roses, apples and pears. *R. solani*, *Alternaria* spp. and *Botrytis cinerea* in rice, vines, potatoes, vegetables and ginger
2-Amino-pyrimidine – part of 2-amino-4-hydroxypyrimidine chemicals. Systemic and protectant activity	Ethirimol, dimethirimol, buprirmate. MOA – affects appressorium development. Soil drench, seed treatment. Ethirimol is mixed with sterol demethylase inhibitor, flutriafol and thiabendazole	Powdery mildews on cucurbits and orchards
Dicarboxymides – inhibits ribosomal RNA synthesis (lipid and membrane activity). Used with multi-site and sterol biosynthesis inhibitors	Three classes: (i) oxazolidinedione (vinclozolin, chlozolinate); (ii) hydantoin or imidazolinedione (iprodione); and (iii) succinimide or pyrrolidinedone (procymidione). MOA – inhibit mycelial growth, conidia and hyphal branching. Foliar sprays, soil drenches, seed treatments and dips	*Botrytis cinerea*, *Monilia* spp., *Sclerotinia sclerotiorum*, *Septoria*, *Helminthosporium* and *Alternaria* in grape vine, canola, fruit, cereals, tomatoes and beans
Sulfamides – broad spectrum, surface protectant. Disrupts cell functions	Dichloflanid (similar to a phthalimide, has fluorine instead of chlorine in structure), tolyfluanid. MOA – degradation of dichloro-fluoro-methylthio moiety	*Venturia* spp. in apples and pears; *B. cinerea* and *Plasmopara viticola* in grapes; and *Alternaria solani*

Continued

Table F.1. Continued

Active ingredient (a.i.)/chemical group/toxicity	a.i. with mode of action (MOA)/treatment/ symptoms/others	Disease/target species
Dithiocarbamates (non-selective, broad-spectrum, multi-site inhibitors which are immobile protectants). Safe on plants. They are used with fungicides like acylalanines	Two types of derivatives: (i) **dimethyldithiocarbamic acid** (DDC), i.e. the *N,N*-disubstituted **dithiocarbamate group (ziram, thiram, feram)**; and (ii) **ethylenebisdithiocarbamic acid** (EBOC), i.e. the *N*-monosubstituted dithiocarbamate group (**maneb, zineb, mancozeb** and **propineb**). MOA – binding of the dithiocarbamic acid/carbamic acid to sulfydryl groups, and depriving fungal cells of essential heavy metals. Wettable powders, granules, suspension concentrate for seed (thiram), soil (metam, thiram) and foliar sprays	*Aphanomyces* spp., *Alternaria* spp., *Pythium* spp., *Venturia* spp., *Botrytis* spp., *Puccinia* spp. and *Fusarium* spp. on banana, potatoes, cereals, tobacco, vegetables and vines
Phthalimides (N-(chlorinated alkyl) thiosubstituted derivatives of tetra hydrophthalamide or phthalimide), multi-site broad-spectrum, surface protectant. Disrupts cell functioning	Captan, captafol and folpet. MOA – the oxidation of sulfydryl groups (captan) in fungal cells and oxidation of gluthione (both captan and folpet) and amino groups. Used in sprays, root dips and for seed and soil treatments	*Alternaria*, *Botrytis*, *Cercospora*, *Pythium*, *Rhizoctonia*, *Venturia* and *Septoria* in cereals, citrus fruits, turf, apple, pear, vegetables and grapevines
Chlorothalonil – has a chlorinated isophthalonitrile structure and is from the chlorophenyl group. A multi-site inhibitor and surface protectant. Disrupts cell function	MOA – disturbing physiological activities of fungal cell constituents by binding reactions to sulfydryl and mercapto groups. Foliar fungicide	Banana, beans, coffee, coconut, potato, rice, fruit, rubber, tobacco against species like *Aphanomyces*, *Rhizopus*, *Phytophthora*, *Alternaria* and *Botrytis*
Triazines – organic immobile protectant. Disrupts cell function	Anilazine (1,3,5 triazine). MOA – interacts with amino and sulfydryl groups in cell enzymes. Foliar spray	*A. solani*, *B. cinerea* in vegetables, *Stagonospora nodorum* in turf and coffee
Benzimidazoles – a group of tubulin-binding compounds which belong to the 2-aminobenzimidazole class	Benomyl, carbendazim, thiophanate-methyl thiabendazole and fuberidazole zoxamide. MOA – disrupt mitosis during cell division at metaphase. Wettable powders, dust and paste form are sprayed on leaves, soil and seeds. These are also mixed with other fungicides like thiram	Ascomycetes, basidimycetes and deuteromycetes. *B. cinerea* in citrus fruits, grape, lettuce and tomato. *Diplocarpon mali* on apple, *S. sclerotiorum* on cabbage and many more
Phenylamides (acylalanines) – disruption of nuclear processes by inhibiting RNA synthesis. Safe systemic fungicides with long-lasting, preventive and curative properties as well as different degrees of specificity	Metalaxyl, metalaxyl-M, furalaxyl, benalaxyl, ofurace and oxadixyl, the most popular being the first two. MOA – inhibits later developmental stages by inhibiting synthesis of RNA. Seed treatments and root and foliage applications. These are mixed with contact multi-site inhibitors like mancozeb and folpet	Controls oomycetes in orders Peronosporales, Sclerosporales and Pythiales. Downy mildews of *Albugo*, *Bremia*, *Peronospora*, *Sclerophthora* and *Pythium* in grapes, peas, beans, maize, canola and cotton; late blight of potato, blue mould of tobacco and soil and seed-borne diseases
Phenylpyrroles – active broad-spectrum systemic fungicide, synthesized from a natural antibiotic pyrrolnitrin from *Pseudomonas pyrrocinia* after making it more photo-stable. Safe for humans, rats, dogs, rabbits, bees, birds and earthworm species if used as directed, but toxic to fish	Fenpiclonil and fludioxonil. MOA includes inhibition of membrane-dependent transport processes, prevention of glycerol synthesis and inhibiting phosporylation of glucose. This action finally leads to inhibition of spore germination and elongation of the germ tube. Available as wettable powders and water-dispersible granules. These are also mixed with other fungicides (e.g. folpet and mancozeb). Used in seed treatment and as a foliar application both with and without other fungicides	Protects against many seed- and soil-borne diseases of cereal crops, caused by members of the genera *Alternaria*, *Aspergillus*, *Fusarium*, *Rhizoctonia* and *Penicillin*
Fosetyl aluminium, a breakdown product of H_3PO_4. A phloem-mobile organophosphorus fungicide. Low toxicity to non-target species	MOA – directly inhibits spore germination and mycelial development. Water-dispersible granules. Mixed with folpet, chlorothalonil, mancozeb	*Phomopsis viticola*, *Phytophthora citrophthora*, *Phytophthora parasitica*, *Phytophthora citricola*, *Pythium* and *Peronospora* on grapevine, hops, citrus
Anilinopyrimidines – these fungicides do not have any substituent on the phenyl ring. Low mammalian toxicity	Cyprodinyl, mepanipyrim, pyrimethanil. MOA – interferes with the biosynthesis of methionine (amino acid) and blocks the production of hydrolytic enzymes and proteins. Formulated as wettable powder or suspension and used as foliar sprays	Grey mould on grapevine, powdery mildew, eye spot and leaf blotch. Ascomycetes and basidiomycetes. *Phytophthora*, *Pythium*, *Peronospora*, *B. cinerea*, *Venturia* spp., *Alternaria*, *Erysiphe graminis* and *Pyrenophora* spp. On wheat, barley, grape, pea, pome fruit and vegetables

Continued

Table F.1. Continued

Active ingredient (a.i.)/chemical group/toxicity	a.i. with mode of action (MOA)/treatment/ symptoms/others	Disease/target species
Strobilurins – natural product derived from *Strobilurus tenacellus*, a wood-decaying basidiomycete. This is a complex III inhibitor. There are 16 variants of strobilurins like kresoxim-methyl, azoxystrobin and trifloxystrobin. Slow-acting systemic, broad-spectrum protectant fungicides with relatively good translocation within the plant. All are extremely safe to humans and mammals if used as directed	Kresomix-methyl, azoxystrobin, trifloxystrobin, metaminostrobin, picoxystrobin, BAS 500F (pyraclostrobin). MOA – blocking critical reactions at Q_o site of mitochondrial energy production. These are formulated as suspension concentrates, emulsions or wettable granules	Controls against ascomycetes, basidiomycetes and oomycetes against a broad range of diseases such as scab, powdery mildew on apple, cucurbits, sugarbeet and cereals, grapevines; rusts, scald, brown patch, blights on vegetables, cereals, potatoes and rice. *Venturia inaequalis, Podosphera leucotricha, Puccinia recondita, Puccinia arachidis, Septoria nodorum, Septoria tritici, Rhynchosporium secalis, R. solani* and *Magnaporthe grisea*
Quinoxyfen – quinoline-based fungicide with high specificity and protectant cover (up to 42 days for cereals and up to 21 days for grapevines). Safe on birds, bees, earthworms and humans	MOA – unidentified. Inhibits biochemical processes thus stopping spore germination and appressorium formation	Controls powdery mildew on cereal crops and grapevines, caused by *Erysiphe graminis, Erysiphe betae, Erysiphe cichoracearum, Uncinula necator*
Famoxadone – oxazolidinedione fungicide, broad spectrum, protective action and residual effect. Low toxicity concerns for animals and humans	MOA – inhibition of enzymes at the Q_o site of complex III (shares similar MOA to strobilurin)	Against downy mildew and late and early blights caused by *Phytophthora viticola, Phytophthora infestans* and *A. solani* on crops like vines, potatoes/tomatoes, cereals, sugarbeet
Acibenzolar-*S*-methyl (ASM) – benzothiadiazole class, which has SAR (systemic acquired resistance) activity. Lends long-term broad spectrum SAR to crops	MOA – activates host defence mechanism. Soil treatment	Covers a variety of diseases of viral, bacterial and insecticidal as well as fungal nature. Powdery mildew and leaf rust in wheat. Protects cereals, tobacco, vegetables, pome fruit, grapes and bananas
Probenazole – inhibits early fungal development stages, appressorium formation	MOA – increases the activity of phenylalanine ammonia lyase and peroxidase. Spray application on rice	*Pyricularia grisea* causing rice blast in Japan

seeds, before they are packed. Yet another method of treating seeds is known as the seed-dip method, in which seeds are kept dipped in a fungicidal-water suspension for a specified period of time before planting. Dust treatment involves mixing seed with fungicide dust in the planter or hopper-box of a seed drill while sowing. Hot water treatment is effectively carried out against black rot of crucifers and moist hot air treatment against seed or set-borne diseases of sugarcane.

Similarly, fungicidal foliage treatment is also a necessary practice which requires the use of wettable powders, emulsifiable concentrates, suspensions and water-dispersible granules. Solutions of finely dispersed fungicidal particles are applied to foliage. The fungicides may be systemic or **non-systemic** in nature. Non-systemic sprays, being protective, have to be applied before any infection begins and may have to be used in larger quantities to account for loss of the ingredient, via rain or weathering. Fungicides such as **metalaxyl**, etridiazole and propamocarb are formulated as granules as well as spray formulations for ease of application on perennial crops. Similar treatment is given for fruit protection.

Soil harbours a wide variety of microorganims, many of which are phytopathogens. Soil treatment is carried out either by physical or chemical methods. Physical methods include the use of steam, fire or electricity to eliminate pathogens, while chemical methods use both volatile and non-volatile chemicals. Soil drenching is a treatment, involving application of mobile, aqueous fungicide solutions, before or after crop emergence. Such fungicidal active ingredients need to penetrate the soil up to 15 cm. Soil drenching controls soil-borne diseases such as damping-off, root rots or infections at the soil surface. Non-volatile fungicides may be broadcast in the field or mixed with soil and spread on the field. **Fumigants**, which are volatile compounds, act as **soil sterilants** and control losses by **fungi,** weed seeds, insects and **nematodes**. Since most of the known fumigants are toxic, their use is restricted.

Root diseases of fruit trees are controlled by trunk injections. Fungicidal compounds like **fosetyl-aluminium** are injected into trunks by boring many holes into the trunk and applying the fungicide under pressure.

Post-harvest treatment is equally critical during the protection of harvested crops and fruits, against insects, bacteria and fungi. Normally it is a pest complex that destroys produce. While fungi may be initial infectors, secondary bacterial infections may hasten destruction by causing tissue breakdown and rotting. Fungicides used in treating harvested fruit (e.g. banana and citrus) include **benzimidazole, dicarboximides** and triazoles. Formulations include an aqueous dip or spraying.

Fungicide azadirachtin

See **Azadirachtin insecticide**

Fungicide development

Man has suffered from fungal diseases since ancient times, and has been trying to get rid of them using various methods.

However, real progress in fungicide development started to take place in the 20th century.

In the 18th century, sulfur, copper, chlorine, brine and lime were used as fungicidal compounds in some form or other. As science progressed, these chemicals were supplemented with phenyl and alkyl-mercury compounds, **Bordeaux mixture** and tin compounds. These active ingredients (a.i.) were effective surface agents but they injured crops if not used carefully. They were used on wheat, barley, grapes and apples for treating seed and foliar diseases. They acted by disrupting cell membranes and affecting respiration, for example. All these compounds are grouped as first generation fungicides.

Use of **dithiocarbamates** (the class that includes **thiram, zineb** and **maneb**) marked the beginning of the second generation fungicides. These were followed by **phthalimides** (the class that includes a.i. like **captan, captafol** and **folpet**) and **chlorothalonil** (a class in itself) and phosphorylation uncouplers such as **dinocap,** which is a substituted dinitrophenol. This generation of fungicides was far more effective to a broader spectrum of diseases, but less injurious to plants. However, these were still only surface fungicides and required repeated applications.

The development of systemic fungicides marked the beginning of the third generation fungicides. These included benzimidazoles (**carbendazim**), carboxamides, organophosphorus, morpholines, **phenylamides**, triazoles, **fosetyl-aluminium** and **dicarboximides**. These classes of fungicides are uniquely efficient in low doses with very low mammalian toxicity. These are highly specific and thus relatively safe to the environment.

The melanin biosynthesis inhibitors (e.g. pyroquilon), **phenylpyrroles**, sterol biosynthesis inhibitors and **strobilurins** are termed as the fourth generation fungicides. These compounds offer effective control over a diverse group of fungi, and in some cases (e.g. strobilurins) they cover all classes of plant pathogenic fungi, namely, ascomycetes, basidiomycetes, deuteromycetes and oomycetes. These are effective in their control but suffer from the threat of pest resistance. Table F.2 shows the list of commonly used a.i. with their chemical fungicide groups and their year of discovery.

Fungicide di-halo

See **Di-halo slimicide**

Fungicide oleic acid

See **Oleic acid herbicide**

Fungicide oxythioquinox

See **Oxythioquinox acaricide**

Fungicide Resistance Action Committee (FRAC)

As pathogens start developing resistance to various fungicides, there arises the need for some management strategy for their application. Frequent occurrence of resistance resulted in the formation of the Fungicide Resistance Action Committee (FRAC), a specialist technical group of Crop Life International. The main aims of FRAC are to identify, collate and distribute information about potential resistance problems and management guidelines.

FRAC has identified a few plant pathogens that have shown a high risk of development of resistance to fungicides. Table F.3 lists the names of some such pathogens with the diseases they cause on crops.

Fungicide uniconazole

See **Uniconazole as a PGR**

Fungicides

Any chemical compound, natural or synthetic, designed or developed to kill or control the growth of fungus is known as a fungicide. Fungicides occupy a large part of the pesticide world, their use extending from industries to agriculture to one's home, for a number of purposes, including: (i) protection of seed grain during storage and/or shipment; (ii) protection of mature crops, seedlings and flowers in fields as well as in storage; (iii) suppression of mildews that attack painted surfaces; (iv) control of slime in many water-based industries (e.g. paper pulp); and (v) protection of wood and fabric in homes. Much of the supply of fungicides goes to prevent or minimize crop losses caused by phytopathogenic fungi. Most infections develop in humid conditions at moderate temperatures, making the consumption of all pesticide types increase at the onset of the period of rains.

Fungicides must be sufficiently toxic to the target fungi at dose concentrations that the plant can tolerate. Fungicidal activity can also be **fungistatic** in nature. Fungistatic agents do not kill the fungi but inhibit fungal growth temporarily. Some fungistatic chemicals like phenanthrene derivatives and Bordeaux mixture inhibit spore production without affecting the growth of vegetative hyphae. These chemicals are anti-sporulants. They cannot stop the disease but reduce the available inoculum that can spread the infection.

Table F.2. Development of fungicide classes over the years. (Adapted from Srivastava (2003), with permission.)

Common name	Chemical group/inhibitors	Year of discovery
Dinocap, PCNB and chlorothalonil	Aromatics	1946, 1930, 1964
Carboxins, oxycarboxins	Oxathiins	1966
Carbendazim, benomyl	Benzimidazoles	1968
Edifenphos, tolcophos methyl	Organophosphates	1970s
Acylalanines, oxazolidinones	Phenylamides	1970s–1980s
Prothiocarb, propamocarb	Carbamates	1974, 1981
Tridemefon, propiconazole, penconazole	Triazole	1973, 1979, 1983
Pyroquilon, tricyclazole	Melanin biosynthesis	1980s
Blasticidin S, kasugamycin	Protein synthesis	1995, 1965
Validamycin	Phosphatidylinositol synthesis	1970
Phenyl pyrroles, strobilurin	Novel fungicide	1990s

Table F.3. Fungal pathogens that pose a high risk to development of resistance. (From Russel (2003) (FRAC Risk List December 2005), with permission from Dr Randall Gold, European Technical Management, Speciality Crop Fungicides, BASF –The Chemical Company (www.frac.info).)

Pathogen	Crop	Disease
Botryotinia fuckeliana (*Botrytis cinerea*)	Various, especially grapevine	Grey mould
Erysiphe (=*Blumeria*) *graminis*	Wheat/barley	Powdery mildew
Mycosphaerella fijiensis	Banana	Black sigatoka
Penicillium spp.	Citrus, various	Post-harvest rot
Phytophthora infestans (RNA polymerase)	Potato/tomato	Late blight
Plasmopara viticola	Grapevine	Downy mildew
Pseudoperonospora cubensis and related species	Cucurbits, various	Downy mildews
Pyricularia spp.	Rice, turf	Rice blast, leaf spot
Sphaerotheca fuliginea and related species	Cucurbits, various	Powdery mildews
Venturia spp.	Apple, pear	Scab

Fungicides can be classified in many ways, for example: (i) mobility in the plant; (ii) type of protection, to the plant; (iii) breadth of activity; (iv) mode of action (MOA); and (v) chemical group.

Fungicidal mobility can be further grouped into contact fungicides and residual fungicides (**non-systemic fungicides**), and systemic fungicides.

A contact fungicide is one that remains on the surface where it is applied but does not penetrate into the plant system. They form a protective layer over the surface and prevent the fungus from entering the host, thereby inhibiting spore germination or mycelial growth, thus displaying a protective function. The activity on the plant surface prevents infection from occurring (e.g. such fungicides include **zineb**, sulfur, 1,4-oxathiin, imidazole and pyridines). These fungicides react with the thiol groups present in the enzymes of the fungus and inhibit metabolic processes. They are multi-site inhibitors (i.e. they act on different sites in a metabolic process). Repeated applications are needed to protect new growth of the plant and to replace material that has been washed off by water, or degraded by environmental factors like sunlight. Some examples of contact fungicides are **sulfur**, **copper oxychloride** and **dithiocarbamates**.

Systemic fungicides are absorbed into plant tissue and may offer some post-infection activity. Very few fungicides are truly systemic, that is very few move freely throughout the plant. Some move only upwards in the plant through the xylem tissue, and some move locally into treated leaves and around the treated area, while still others move downwards in the plant from the site of application.

Systemic compounds have curative properties and can arrest the pathogen post-infection. They are commonly site specific and act only on select species. They interfere with biosynthetic processes such as respiration, protein synthesis and membrane functions. They inhibit fungal growth and spore production more effectively than spore germination. Systemic fungicides are needed in significantly smaller amounts per hectare. Examples of systemics include **carbendazim** and **benomyl** (both **benzimidazoles**), **ofurace**, **metalaxyl**, **benalaxyl** (all phenylamides), **iprobenfos** and **edifenphos** (both organophosphates) and **penconazole**, **triadimenol**, **bitertanol** (all triazoles). Chemical classes, for example pyridines, pyrimidines and imidazoles, also include active ingredients (a.i.) that are systemic in nature. Fungicides that are capable of eradicating a fungus after it has caused infection are called therapeutants. The a.i. in this group can penetrate the plant and stop the growth of the pathogen in the plant tissue; they are usually most effective 24–72 h after the occurrence of infection, depending on the fungicide. The fungicides that not only stop disease development after symptoms have developed but provide protection to the host are called **eradicants**. For example, organic mercurials, **lime sulfur**, **phenylamides** and **strobilurins** are eradicants. Anti-sporulant fungicides are those that contain a.i.s that can prevent sporulation or spore production.

Based on the breadth of activity, fungicides can act on a single site (acting only on one point in one metabolic pathway, or against a single critical enzyme or protein needed by the fungus), or have multiple sites (different metabolic pathways). Single-site fungicides are less toxic to plants.

Fungicides act by interfering with key processes, or by inactivating critical enzymes or proteins. These are known as the compound's MOA. Interfering with energy production or respiration are some examples of MOA. Fungicides can be classified under particular chemical groups, indicating a common biochemical MOA. These groups may or may not have similar chemical structures.

On the basis of protection type, fungicides can be: (i) seed protectants (e.g. **captan**, **thiram**, organic mercurials); (ii) soil fumigants (e.g. **chloropicrin**, **formaldehyde**, **vapam**); or (iii) tree-wound dressing fungicides (e.g. **Bordeaux paste** or spray).

Fungigation

Fungigation is the application of fungicides through irrigation waters in an open or closed system.

Fungistatic

A substance or a compound, which temporarily inhibits fungal growth, is known as a fungistatic. (See also **Fungicides**.)

Fungitoxic

A substance or a compound, which is toxic to a fungus, is said to be fungitoxic.

Furathiocarb insecticide

Furathiocarb, $C_{18}H_{26}N_2O_5S$ (butyl 2,3-dihydro-2,2-dimethylbenzofuranyl *N*,*N*′-thiodicarbamate), is a thiocarbamate insecticide. It is a yellow viscous liquid, which is sparingly soluble in water and readily miscible with most common organic solvents including acetone, methanol, isopropanol, hexane and toluene. Furathiocarb is obtained by the reaction of *N*-*n*-butyl-methylcarbamate with sulfenyl chloride, followed by treatment of the product with 2,3-dihydro-2,2-dimethyl-7-benzofuranyl *N*-methylcarbamate. It is stable up to 400°C. The CAS Registry Number is 65907-30-4.

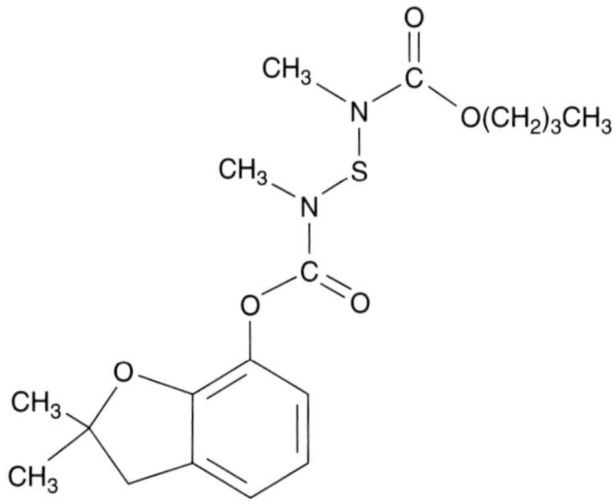

Furathiocarb

Furathiocarb, a systemic, cholinesterase-inhibiting soil insecticide has contact and stomach action. It controls insects in maize, canola, sorghum, sugarbeet, sunflower and vegetables.

This compound is formulated as a dustable powder (DP), granule (GR) and emulsifiable concentrate (EC). It is also applied to foliage or seed. It is compatible with most other pesticides with a commonly known trade name being Deltanek.

According to the WHO and EPA, furathiocarb is a highly hazardous chemical (Class Ib). It is considered a PAN Bad Actor chemical due to its cholinesterase inhibiting property and acute toxicity.

In animals, the metabolic transformation of furathiocarb in rats proceeds via rapid and complete hydrolysis, followed by oxidation and conjugation. In plants, furathiocarb is metabolized to **carbofuran** and its hydroxy and keto derivatives. In soils, furathiocarb is rapidly decomposed to carbofuran, followed by complete metabolism.

The insecticide furathiocarb is registered in many countries in Africa, Australia and New Zealand but not in the USA.

Furrow openers

See **Weed management by mechanical methods**

Fusarium species as biocontrol agents

Fusarium species (family Nectriaceae) are well known for their phytopathogenic properties. Most soil-borne diseases of plants such as root rots, damping-off and seed **blights** are associated with this genus. *Fusarium* species may be pathogens or endophytes (those that complete a part of their life cycle in a plant, without causing harm to/death of the host) and they also have a wide animal host range. With regards to host plants, they usually colonize the roots and are also present in seed. Fusarium wilt is a common wilt disease that affects tomatoes, tobacco and members of the cucurbit group. Causal agents are *Fusarium oxysporum* and *Fusarium moniliforme*.

However, there are many strains or isolates of *Fusarium* species that have been identified as biological pest control agents. They are abundantly present in almost all soil types. But, most of the isolates have a narrow host range. They colonize the rhizosphere of the crop and then invade the crop root surface. This makes it difficult for the pathogenic isolates to grow nearby. In addition, the invasion of the crop root activates the immune system of the plant, leading to the production of **phytoalexins**. To that extent, some isolates may be used as biocontrol agents.

Such isolates, code-named Fo47 and CS-20 for example, are normally cultivated by growing them on agar plates or oatmeal and are formulated as micro-granules via solid/liquid fermentation for viable storage. Trade names of isolate Fo47 include Biofox C. Chlorinated water, strong oxidizers and acidic fungicides are to be avoided, to conserve these beneficial isolates.

Fusarium isolates with biological control properties are safe for use, in normal doses, in the presence of mammals, birds and fish and are approved for use in organic farming.

G

Gametocide

A gametocide is a substance which can control pollination of plants by selectively killing plant sex cells (gametes). Some suggested gametocides are **maleic hydrazide**, sodium dichloroacetate (dichloroacetic acid) and beta-dichloroisobutyrate. Male sterility has been brought about in experiments with gametocides and certain plants like the common snap dragon (*Antirrhinum majus nanum* L).

The use of gametocides is, however, not always viable due to uncertainty about application doses.

Gamma-HCH insecticide

Gamma-HCH is an isomer of lindane. (See also **Hexachlorocyclohexane insecticide**)

Garlic oil as fungicide

Garlic oil (CAS Registry Number is 8008-99-9) possesses antibacterial, fungicidal and insecticidal properties. It contains compounds like allicine and diallyl sulfide. Minced garlic may also be combined with mineral oil or fish emulsion or any other biological ingredient for effective control of aphids, bugs, leafhoppers and whiteflies.

However, along with pathogens, garlic oil may be phytotoxic to crops too, and hence direct spraying is not advised. Temperatures above 26°C are also not conducive to garlic spray application.

Gas chromatography

Chromatography (literally, colour graph) is an advanced technique of analytical chemistry, which is used to identify the constituents of a compound or mixture. The chromatograph, which works on the principle of partition coefficient, involves passing the mixture (or the sample) through a mobile phase and a stationary stage. The fluid of the mobile phase is termed the eluate and the substance to be separated is the analyte. The mobile phase is commonly an organic solvent in which the sample can dissolve. The analyte passes through the columns of the instrument at various temperatures, speeds and other factors, and while travelling, the parts of the mixture react with the stationary phase differently, and finally separate out.

Chromatography is undertaken to identify for example: (i) adulteration; (ii) purity of a compound; and/or (iii) the concentration of the various parts of a mixture. There are many types of chromatographic techniques depending on the phases and the detectors used. Sometimes the stationary phase is placed in a tube or the column bed (column chromatography) and otherwise, it is on a plane (planar chromatography). When the mobile phase is a liquid, the separation technique is called liquid chromatography (LC).

Gas chromatography (GC) is a technique invented by Antony James and Archer Martin. GC is usually column-bound. The mobile phase is a gas (usually nitrogen or helium). The stationary phase is a liquid that adsorbs on the surface of an inert substance like **diatomaceous earth**. The gas percolates through the stationary phase to help the separation.

GC can analyse about 80% of the total pesticides used today, including gaseous and volatile substances. For comparison, non-volatile chemicals which may have high molecular weight are generally analysed by LC or even by thin layer chromatography (TLC). The major advantages of GC are: (i) the ease of operation; (ii) the short time taken to obtain results; (iii) the quantitative and qualitative accuracy of analysis; and (iv) the acceptance of solid, liquid and gaseous samples.

The basic components of the GC instrument are a carrier gas, injection port, over/column, detector, and recorder or database unit (Fig. G.1). GC is also called gas-liquid chromatography.

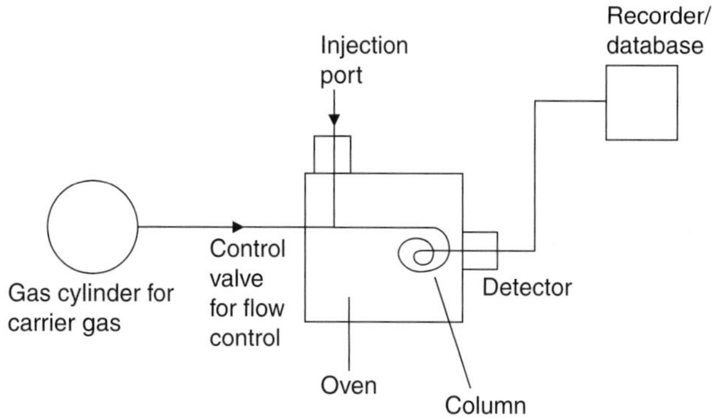

Fig. G.1. Schematic diagram of the basic components of the gas chromatography (GC) equipment.

The carrier gas (usually nitrogen, helium, argon or carbon dioxide – most commonly nitrogen) helps in burning and carrying the analyte vapours from the injection port to the detector through the columns. The gas must be devoid of moisture and inert in nature. It must be suitable for the detector used. The flow of the gas is controlled by needle valves or a flow controller. The injection port is the device from where the sample, dissolved in solvents such as hexane or ethyl acetate, is introduced into the carrier gas stream. It is important for the sample not to be too large. The sample is either injected with a microlitre syringe or with a split/splitless injection system (the latter is suitable for capillary columns).

The process of separation takes place in the chromatographic column. The column, which can be straight, bent or coiled, varies in length from 91 cm to more than 3 m, while it may range from 10 to 100 m in the case of capillary columns. The efficiency of the column is inversely proportional to the column diameter. The diameter of the column ranges from 3 to 4 mm for packed columns and is 320 μm for capillary and megabore columns. The film thickness of the stationary phase in a capillary column is 0.25 μm and 2.65 μm for megabore columns.

There are three types of columns: (i) packed; (ii) capillary; and (iii) fused silica open tubular column. Packed columns contain a finely divided, inert, solid support material coated with liquid stationary phase. These are suitable for analysing gases of low molecular weight. Capillary columns are made of fused silica and synthetic quartz, coated on the outside with polyamide. These flexible columns are of two types, namely, wall-coated open tubular and support-coated open tubular. The support-coated open tubular columns are generally more efficient. The fused silica open tubular column is used for pesticide analysis. The solid support provides a large uniform inert surface area for the distribution of the liquid phase.

The stationary phase of the column system is selected after understanding the nature of the analyte (i.e. its polar character, its volatility range and the temperature of the column). The two main solid supporters are Chromosorb-P and Chromosorb-W, the latter being preferred for being more inert and suitable for polar compounds. The solid support must be inert (in order to avoid adsorption), mechanically strong and thermally stable. It must have a large surface area (1–20 m^2/g) and exist as small particles to afford column efficiency and uniform packing.

Liquid phases, which may include high-melting-point silicone greases such as SE-30 or OV-101 (dimethylsilicone), OV-17 (50% phenyl methyl silicone), carbowax 20M (polyethylene glycol) and DEGS (polydiethylene glycol succinate), help in the separation process through differential solubility to components of the sample. A liquid phase must be chemically inert, thermally stable, non-volatile and be able to dissolve the sample to a reasonable degree.

A detector is a device that senses the presence of components that are different from the carrier gas, and converts that information into an electrical signal. Ideally, a detector must be highly selective, sensitive and must respond to different types of compounds. Commonly used detectors for pesticide residue analysis are: (i) the thermal conductivity detector (TCD); (ii) the flame ionization detector (FID); (iii) electron capture detectors (ECD); (iv) the nitrogen phosphorus detector (NPD); and (v) the alkali-flame ionization detector (AIFD).

The data acquisition system amplifies the signal from the detector and records its peak, called the chromatogram. The area of the peak represents the concentration of the component. Devices used for generating a chromatogram and the report are the integrator and personal computer (PC). Quantification of a solute is done by comparing the height or area of each peak of unknown solutes with that of peaks of known standards. The R_f value is a function of column, temperature, carrier gas flow rate, and affinity of the solute for the mobile and stationary phases. Under given conditions, the R_f value of the compound remains constant. The partition coefficient (K_1) can be calculated using Eqn G.1.

$$\text{The partition coefficient } (K_1) = \frac{Cs}{Cg} \qquad \textbf{(G.1)}$$

where
Cs = Concentration in stationary phase
Cg = Concentration in gaseous phase

A narrow ratio means faster separation and vice versa. The number of theoretical plates $N = 16\,(x/y)^2$, where 'y' is the length of the baseline cut by two tangents (width of the peak) and 'x' is the distance from the injection peak maximum (retention time).

Resolution (R) is a measure of both the column and the solvent efficiencies. It accounts for both the narrowness and separation between maxima. If R = 1, the resolution of two equal-area peaks is approximately 98% complete. If R = 1.5, baseline separation is achieved.

Genetic manipulation for pest control

Genetic control refers to a variety of methods for suppressing insect pests through manipulation of the inheritance mechanisms of the population.

The many common methods used to control pests include: (i) the mass release of sterilized males; (ii) the use of **chemosterilants**; and (iii) the mass introduction of deleterious mutations. Gamma radiation from Cobalt-60 over a large area and for a sustained period has made it possible for eradication of some pests from certain locations, such as eradication of the screwworm fly from the southern USA.

Similar manipulation in crop plants increases their tolerance to attacks by microorganisms and insects. Genetic host plant resistance is produced by antixenosis, antibiosis or by tolerance.

Genetic engineering provides opportunities to produce host plant resistance through transfer of genes for the production of antibiotic factors such as ***Bacillus thuringiensis*** (***Bt***) toxin into the genomes of crop plants such as tobacco, tomato, potato, maize and soybean. The insecticidal properties of the toxin protect the crop plant from damage by insect herbivores. However, unrestrained use of this technology has led to resistance to *Bt*.

Geraniol bioinsecticide

Geraniol bioinsecticide, $C_{10}H_{18}O$ ((*E*)-3,7-dimethyl-2,6-octadien-1-ol), also known as guaniol, lemonol and geranyl alcohol, is a plant-derived insecticide and insect repellent. The compound is a mixture of citronellal (3,7-dimethyl-6-octenal) + geraniol ((*E*)-3,7-dimethyl-2,6-octadien-1-ol) + citronellol (3,7-dimethyl-6-octan-2-ol) + geranyl acetate (3,7-dimethyl-6-octen-1yl acetate). The source of this repellent-insecticide is the **citronella** plant (*Cymbopogon nardus*). The CAS Registry Number is 9007-92-5.

Geraniol and citronellol are the main constituents of this bioinsecticide. Geraniol is an essential oil like geranium oil, lavender oil and lemon grass oil. It is effective in repelling mosquitoes, houseflies, stable flies, horn flies, head lice and ants and is used to maintain public and animal health and hygiene.

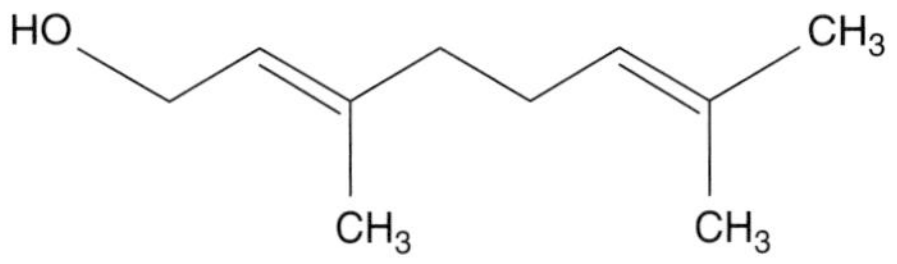

Geraniol

Sprays and candles are some of the formulations of (*E*)-3,7-dimethyl-2,6-octadien-1-ol, and trade names include Biomite and Japanese Beetle Combo Bait. This insect repellent is compatible with strong oxidizing agents.

This insect repellent does not cause harm to mammals, non-target organisms or the environment. It is permitted for organic farming.

Geranyl alcohol

See **Geraniol bioinsecticide**

Germicides

Germicides, also called bactericides, are substances that destroy harmful microorganisms. Germicides vary greatly in their potency and specificity. Germicidal compounds include a large range of control agents that include biologically derived substances (like bacteriophages), chemical compounds and even short-wave radiation (e.g. gamma radiation).

Giant ragweed

See **Weed – giant ragweed**

Gibberellin A_4 and gibberellin A_7 as a PGR

See **Gibberellins**

Gibberellins

Gibberellins are derivatives of terpenoids (tetracyclic diterpene acids). There are more than 100 natural gibberellins, found in plants, fungi and bacteria. The gibberellins isolated and studied are named as gibberellin A_1, A_2, and so on.

Gibberellins, which are variously arranged 19 carbon units, are weak acids, and have a role to play in promoting growth and development in living systems. In plants, they are found in the seed and help hydrolyse starch in the endosperm, inducing germination. They reverse the inhibiting effect of **abscisic acid**. However, the presence of an **auxin** may increase their efficiency. Gibberellic acid (molecular formula, $C_{19}H_{22}O_6$) has a CAS Registry Number of 77-06-5 and is a dihydroxylated gibberellin.

Gibberellins have been classified as a Class U toxin (not likely to be hazardous). They are registered as growth promoters in India, the USA, Canada, Australia, New Zealand and South Africa. Gibberellic acid (GA) is registered in India.

The mixture of gibberellin A_4 and gibberellin A_7,$C_{19}H_{24}O_5$ ((3S,3aR,4S,4aR,7R,9aR,9bR,12S)-12-hydroxy-3-methyl-6-methylene-2-oxoperhydro-4a,7-methano-3,9b-propanoazuleno[1,2-b]furan-4-carboxylic acid), is a plant growth regulator. Its CAS Registry Number is 8030-50-3. It is a yellow-white crystalline powder, slightly soluble in water and cyclohexane, but readily soluble in acetone and ethyl acetate. It is stable in a cool, dry place in airtight containers (see Gibberellin A_4 and Gibberellin A_7 at bottom of page).

Gibberellin A_4 with gibberellin A_7 is used to reduce russeting in apples, improve the length and quality of fruit and increase seed germination.

This mixture of GA_4 and GA_7 is available as soluble concentrate (SL), under trade names like Provide, Berelex and Procone.

The GA_4-GA_7 mixture is registered in the USA, Canada, Australia, New Zealand, Greece and many other countries.

Gliocladium catenulatum (fungus) as a biocontrol agent

Gliocladium catenulatum is a widespread, soil-borne fungal saprophyte. It belongs to the Phylum Ascomyota order Hyprocreales. It secretes a secondary metabolite that inhibits plant pathogenic fungal strains.

Some strains are thus grown on artificial media and exploited for their biocontrol properties. J1446 is a prominent example.

G. catenulatum strains are formulated as powders that can be sprayed as a solution on foliage, into soil and as a seed dip. The formulation controls diseases such as damping-off, rot diseases and **wilt** on vegetables and turf.

The biocontrol agent does not harm the environment and is not toxic to man and animals. It is registered in the USA and the EU for specific applications.

Another species which is important is *Gliocladium virens* (strain GL21). With its preventive properties against many fungal species such as *Pythium* spp., *Rhizoctonia solani* and *Sclerotium rolfsii*, it protects cotton, vegetables and some other crops against rot diseases and damping-off.

Cultures are grown, mixed with water and applied to the soil so that the roots are protected at sowing or planting

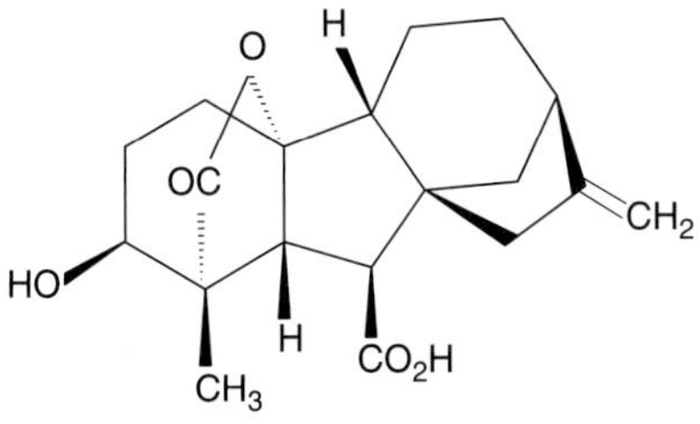

Gibberellin A_4

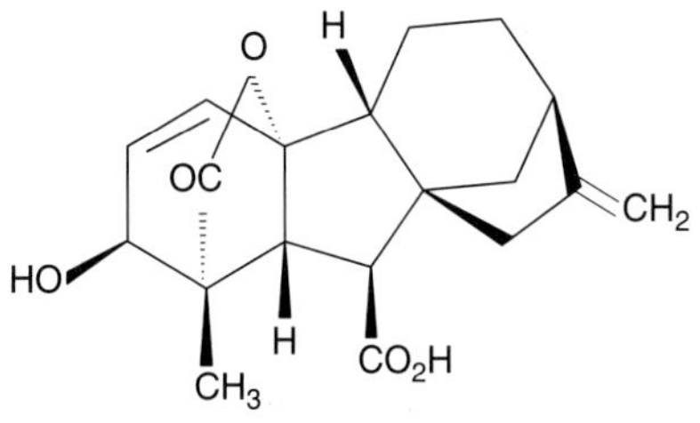

Gibberellin A_7

time. Again, this agent is not harmful to man, animals or the environment.

Gliocladium virens

See *Gliocladium catenulatum* **(fungus) as a biocontrol agent**

Glufosinate herbicide

See **Glufosinate-ammonium herbicide**

Glufosinate-ammonium herbicide

Glufosinate-ammonium, $C_5H_{15}N_2O_4P$ (ammonium-4-[hydroxy(methyl)phosphinoyl]-DL-homoalaninate), is an organophosphate herbicide. Its CAS Registry Number is 77182-82-2. It is a colourless crystalline substance, readily soluble in water (1.37 kg/l) and moderately soluble in inorganic solvents. It is stable to light. This herbicide is produced by reaction of *O,O*-diethyl methyl phosphonate with 2-amino-4-bromobutanoic acid. It is the ammonium salt of glufosinate. Glufosinate-ammonium is also referred to as phosphinothricin.

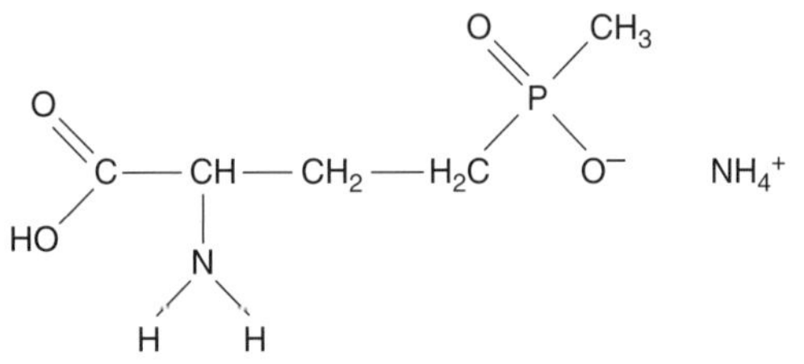

Glufosinate-ammonium

Glufosinate-ammonium is a **non-selective**, post-emergence, contact, broad-spectrum herbicide. The chemical is translocated within the leaves, mainly up to the leaf tip. It inhibits the enzyme **glutamine synthetase**, leading to less glutamine and over-accumulation of ammonium ions. The plant dies because photosynthesis does not take place properly. A wide range of annual and perennial broadleaved weeds and grasses found among vineyards, rubber and fruit trees are controlled.

Glufosinate-ammonium is available as water-soluble powder (SP). It is available in the market under the trade names of Basta and Liberty.

According to the WHO and EPA, glufosinate-ammonium is a slightly hazardous chemical (Class III toxin). The ADI in man is 0–0.02 mg/kg b.w./day. The oral LD_{50} is 436–464 mg/kg in mice and the dermal LD_{50} is 1510–1660 mg/kg in rats. The acute oral LD_{50} in birds is 2000 mg/kg. Excessive exposure may cause convulsions, salivation and hypersensitivity.

In animals, glufosinate-ammonium breaks down to 3-[hydroxy(methyl)phosphinoyl] propionic acid and 2-methylphosphinylacetic acid (MPAA-2). In plants, the final product is carbon dioxide. In soil, it is degraded by microbes, with the soil half-life being in the range of 3–70 days (with a shorter half-life in clayey and organic soil). It is categorized by the EPA as persistent and mobile. MPAA-2 is a neurotoxin.

Glufosinate-ammonium is registered in India, the USA, Canada, Australia, the Philippines, Vietnam, Tanzania, South Africa and throughout the EU (except Lithuania and Germany).

Glufosinate, $C_5H_{12}NO_4P$ ((2*RS*)-2-amino-4-[hydroxy (methyl)phosphinoyl]butyric acid), the parent compound of glufosinate-ammonium, is an organophosphate herbicide. It has a CAS Registry Number of 51276-47-2. It is a suspected endocrine disruptor and is not registered in the USA, but is registered for use in the EU and Cambodia. Glufosinate is classified as a Class III toxin (unlikely to be hazardous) by the WHO.

Glutamine synthetase (GS)

See **Herbicidal inhibition of glutamine synthetase (GS)**

Glyphosate herbicide

Glyphosate, $C_3H_8NO_5P$ (*N*-(phosphonomethyl) glycine), is an organophosphate herbicide. It is a colourless crystalline substance, fairly soluble in water, insoluble in organic solvents and stable up to 60°C. Glyphosate is synthesized by heating a mixture of phosphorus acid and α-amino acetic acid and then adding formaldehyde. The CAS Registry Number is 1071-83-6.

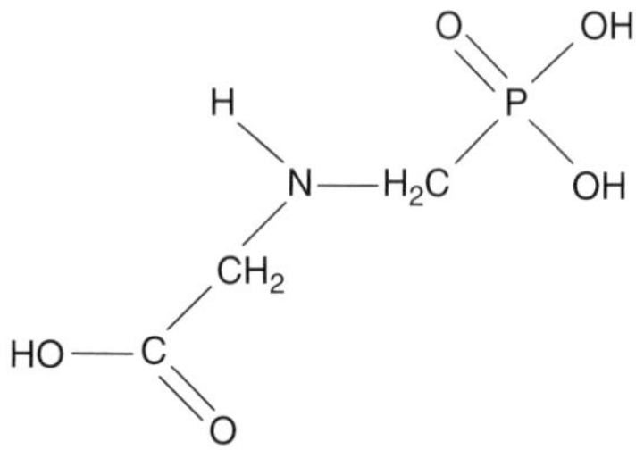

Glyphosate

Glyphosate is a broad-spectrum, **non-selective**, systemic herbicide that spreads all through the plant, once absorbed by the foliage and shoots. It acts on the amino acid formation (by inhibiting the enzyme 3-enoyl pyruvylshikimate-3-phosphate synthase), and thus affects many enzymatic reactions. Glyphosate is very widely used to control many types of grasses and broadleaved weeds, of both the annual and the perennial type, commonly seen in cereals, peas, beans, canola, flax and mustard. Glyphosate is available as a water-soluble powder (SP). It is available in the market under trade names like Rodeo and Roundup.

According to the WHO, glyphosate is a Class U toxin (product unlikely to present acute hazard in normal use) and the EPA places it under Class III (slightly toxic). The oral LD_{50} in rats of pure glyphosate is 4230–5600 mg/kg. But, formulations prepared with the use of surfactants like polyoxyethylene amine (POEA), become toxic due to the nature of the surfactant, and such formulations are toxic to fish and tadpoles.

Glyphosate is registered by India, the USA, Australia, New Zealand and many EU countries. Some countries have restricted the use of certain formulations with surfactants that are deemed to be toxic. (See also **Herbicidal inhibition of EPSPS**.)

Good agricultural practices (GAP)

See **Pesticide deposits and residues**

Goosegrass

See **Weed – goosegrass**

Grain-treating machines

See **Seed protection devices**

Gram-negative bacteria

Gram-negative bacteria, named after a Danish physician Hans Christian Gram, are bacteria that lose the initial colour of the Gram stain and then take on the colour of the final stain (light pink). *Rhizobium* spp. are Gram-negative bacteria. Gram-positive bacteria are bacteria that retain the Gram stain. *Bacillus* spp. are Gram-positive organisms.

Gram-positive bacteria

See **Gram-negative bacteria**

Granules

A pesticide in a granular form is an example of a solid formulation. However, at the time of application, granules may be dry-applied or dispersed in water. Dry-applied granules (GR) are the most economical product form available to the farmer. The product mainly consists of an active ingredient (a.i.) and a solid carrier, either of mineral or of biological origin, and may also contain formulants to ensure bioavailability and physical compatibility of the a.i.

Pesticide granules, commonly dispersed within furrows or on the soil surface, are applied either mechanically or by hand. Granules store well, are transported easily, economical and easily available. Blank carrier granules are tumble-mixed in a rotating hopper with the a.i. in the form of a solution or slurry, for efficient coating.

But GR products may be dusty and may contain toxic a.i.s, making it necessary to protect handlers from exposure.

Water-dispersible granules (WG) are any agglomerated composition designed to break up and disperse in water to form uniform suspensions for application. Due to their relative uniform bulk density, they can be conveniently measured for application. They can be handled easily and have a greater degree of safety due to lowered dust levels. WG products are made by fluid-bed granulation followed by spray drying and high-speed mixing technologies.

Other preparation methods include pan granulation leading to smooth, spherical granules, resulting in small uniform cylindrical particles and intensive mixing, producing small irregular agglomerates.

Volumetric measurement and an excellent shelf life are two other advantages of WG. However, WG products are an expensive choice, since they need specialized technology for packing and transportation. In addition, they normally need pneumatic applicators. WG are also referred to as **dry flowable formulations (DF)**.

Water-soluble granules (SG) are another granular formulation. Using the same agglomeration techniques as WG products, SG should be made with attention to processing of the granules, to avoid partial solubilization and re-crystallization of the a.i. during drying, resulting in granules not behaving properly on dissolution.

Grasshopper

See **Pest – grasshopper**

Gravity seed dressers

See **Seed protection devices**

Green lacewings as biocontrol agents

Green lacewings (or lacewings) are insects that belong to the family Chrysopidae in the order Neuroptera. These creatures vary in length from 6 mm to 65 mm and are found in a variety of regions such as the USA, Europe and the tropical regions. They are usually green to brownish in colour with translucent wings and are valued for their pest control properties.

Females lay hundreds of eggs on the underside of leaves, near aphid colonies. As soon as larvae hatch, they move to the leaf and start looking for food, which includes **honeydew**, pollen, nectar and aphids and other insect pests. When they grasp their prey, especially an aphid, they spray a digestive juice over them through their maxillae which kills the pest. Eating voraciously for about 3 weeks they pupate in a cocoon and emerge as adults.

The two well-studied genera are *Chrysopa* and ***Chrysoperla***. The species of these genera are similar and are called common lacewings. *Chrysoperla* species include *Chrysoperla carnea* (which is called the common green lacewing; Fig. G.2) and *Chrysoperla lucasina*, while the genus *Chrysopa* includes *Chrysopa pallens* and *Chrysopa perla*.

Fig. G.2. *Chrysoperla carnea*. Courtesy: Picture copyright 2006 J.K. Lindsey. Reproduced with permission.

Common lacewings (lacewings/green lacewings) are valued for their predatory habits, as many of their prey are pests (e.g. aphids, thrips and **mealybugs**). Lacewings are intentionally grown so that the larvae may be introduced into fields afflicted with such insect pests. Their favourite plants (e.g. sunflowers, daisies, asters) are grown in the vicinity of the crop to attract them to the field and exploit their predatory nature. (See also ***Chrysoperla* as biocontrol agents**.)

Griseofulvin antibiotics

See **Fungicidal antibiotics**

Guanidine fungicide

Guanidine, CH_5N_3, is a crystal formed by the oxidation of naturally occurring guanine. The derivative compounds that are synthesized from this water-soluble, alkaline substance are fungicidal in nature.

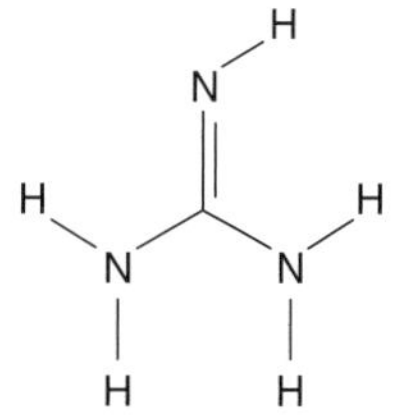

Guanidine

The two derivatives of guanidine, dodine and iminoctadine triacetate have one or more guanidine group attached to the long-chain alkyl group. Fungitoxicity increases with the length of the alkyl chain up to a point, beyond which it reduces. These fungicides are broad-spectrum contact compounds that act by binding to the anionic sites on the cell membranes, causing cell constituents to leak out, thus killing the pathogen. Guanidine fungicides control **scab** and canker disease, seen mostly on apples and pears. The common pathogens that are kept under check are *Alternaria*, *Venturia*, *Mycosphaerella*, *Fusarium*, *Cladosporium* and *Sclerotinia*. Dodine and iminoctadine triacetate are generally used to protect stored products (e.g. fruit), since they are phytotoxic in nature.

Guaniol

See **Geraniol bioinsecticide**

Guazatine acetate fungicide

See **Guazatine fungicide**

Guazatine fungicide

Guazatine, $C_{18}H_{41}N_7$ (a mixture of the reaction products from polyamines, comprising mainly octamethylenediamine, iminodi(octamethylene)diamine, octamethylenebis(imino-octamethylene)diamine, and carbamonitrile), is an aliphatic nitrogen fungicide, and also a post-harvest protective agent, an anti feedant and a bird repellent. Guazatine comprises a mixture of guanidated di-, tri- and poly-amines derived from octamethylenediamine. The product has been prepared with C14 labelling of the terminal positions of the octamethylene chains. It has a CAS Registry Number of 108173-90-6.

Guazatine, a **non-systemic** contact fungicide, is mainly an agent for the control of seed-borne diseases of cereals. As a post-harvest protective agent against sour rot and blue mould, it is used on citrus fruit (lemons and oranges). Guazatine acts on the membrane function of fungal pathogens. Seedling blight (*Fusarium* spp.), blotch (*Septoria* spp.), **smut** (*Ustilago* spp.) and common bunt (*Tilletia* spp.), etc., are controlled by guazatine.

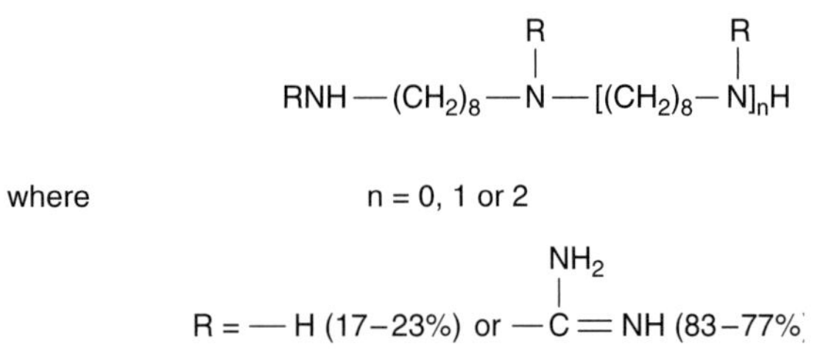

Guazatine

The WHO classifies guazatine as a Class II (moderately hazardous) toxin. The acute oral LD_{50} in rats is 300 mg/kg. The NOEL (2 years) in rats is 2.8 mg/kg b.w./day. Animals rapidly eliminate most of the same, mostly unchanged. The half-life on loamy soil was around 60 days, but it ranges up to 176 days, depending on the soil type. The seed-treated guazatine degrades into carbon dioxide in the presence of moisture.

Guazatine is registered in South Africa, Hungary, Greece, New Zealand and the UK but not in the USA.

Guazatine acetate, $C_{18}H_{41}N_7$ (1,1'-(iminodioctamethylene) diguanidine triacetate), is a salt of guazatine which is a yellowish-brown liquid that is soluble in water (>3 kg/l), methanol (510 g/l) and *n*-hexane (0.1 g/l). It is sparingly soluble in xylene and other hydrocarbons. Its CAS Registry Number is 115044-19-4.

Guazatine acetate is used in seed treatment, post-harvest dips and as a wood protectant. A wide range of fungal species including *Septoria*, *Tilletia*, *Fusarium*, *Pyricularia* and *Cercospora* are controlled. Rice, maize, groundnut, soybean, potatoes and sugarcane are effectively protected by this fungicide.

The formulation types of guazatine acetate are a powder for dry seed treatment (DS), a solution for seed treatment (LS) and soluble concentrate (SL). A common trade name is Panoctine.

The acute oral LD_{50} among rats of guazatine acetate is 360 mg/kg. The acute percutaneous LD_{50} in rats is >1000 mg/kg and in rabbits is 1176 mg/kg. It is an eye irritant. The NOEL (2 year) in rats is 17.5 mg/kg/day.

In animals (rats), guazatine is eliminated after degrading completely in 3 days. It is not registered in the USA, but is registered in Australia, South Africa, Finland and Hungary.

The MRL value of guazatine on crops like garlic, onions and tomatoes is 0.1 mg/kg.

Gummosis of orange

See **Crop – orange orchard pest control**

GY-81 fungicide

See **Sodium tetrathiocarbonate fungicide**

Gymnosperms

Gymnosperms are a group of plants (with four distinct orders) that reproduce via an exposed seed or an ovule. Gymnosperms, which include conifers, do not have fruits and flowers and are thus different from **angiosperms**.

H

2-Hydrazinoethanol as a PGR

2-Hydrazinoethanol, $C_2H_8N_2O$ (2-hydrazinoethanol; 2-hydroxyethylhydrazine), is a colourless liquid that is soluble in water. Also called BOH, it is stable in cool and dark conditions. Dilute solutions are easily oxidized. The CAS Registry Number is 109-84-2.

$HOCH_2CH_2NHNH_2$

2-Hydrazinoethanol

The PGR promotes flowering of the **bromeliad** group (family Bromeliaceae) that includes the pineapple. It is formulated as a soluble concentrate (SL). But it is toxic to humans and is a carcinogen and a fetal toxin.

4-Hydroxyphenylpyruvate dioxygenase (HPPD)

4-Hydroxyphenylpyruvate dioxygenase (HPPD) is an enzyme that is one of the components in the carotenoid biosynthesis pathway of plants. It directly plays a part in the production of plastoquinone, which is a **cofactor** of phytoene desaturase, an early enzyme in carotenoid biosynthesis. Inhibition of HPPD results in bleaching of plant cells thus affecting physiochemical systems of the plant. The actual action of HPPD involves complex reactions with tyrosine (hydroxyphenyl pyruvate) to yield homogentistic acid (2,5-dihydroxyphenyl acetic acid).

Herbicides have been developed to target this enzyme and achieve the necessary weed control. These herbicides belong to the general class, benzoylcyclohexane-1,3-diones, known as the **triketones**. Some examples of HPPD-inhibiting herbicides are given below.

Isoxaflutole, $C_{15}H_{12}F_3NO_4S$ (5-cyclopropyl-1,2-oxazol-4-yl) (α,α,α-trifluoro-2-mesyl-*p*-tolyl)methanone, is a cyclopropylisoxazole or oxazole herbicide. Isoxaflutole is an example of an HPPD inhibitor. It effectively controls weeds in maize. Sulcotrione, $C_{14}H_{13}ClO_5S$ (2-(2-chloro-4-mesylbenzoyl) cyclohexane-1,3-dione), a 2-benzoyl cyclohexanedione molecule, is another HPPD inhibitor. A pre- and post-emergence herbicide, sulcotrione controls broadleaved weeds and grasses in maize, either by itself or in combination with **atrazine**, **bromoxynil** and many others. Its CAS Registry Number is 99105-77-8. Pyrazolynate, $C_{19}H_{16}Cl_2N_2O_4S$ (4-(2,4-dichlorobenzoyl)-1,3-dimethylpyrazol-5-yl toluene-4-sulfonate), is a benzoyl pyrazole compound. It is used for the selective control of annual and some perennial grass and broadleaved weeds in rice. A mixture of pyrazolynate with bromobutide and **pretilachlor** are used to achieve a broader spectrum. The compound, in flooded rice water, is hydrolysed to the same hydroxypyrazole.

Pyrazoxyfen, $C_{20}H_{16}Cl_2N_2O_3$, a selective compound effective on annual and perennial grass and broadleaved weeds, also acts by inhibiting the enzyme HPPD. Its CAS Registry Number is 71561-11-0. It is used in combination with bromobutide and **pretilachlor** on rice fields. Similarly, benzofenap ($C_{22}H_{20}Cl_2N_2O_3$) with a CAS Registry Number of 82692-44-2, is also a benzopyrazole rice herbicide which causes bleaching in the target plant and the accumulation of phytoene. HPPD inhibition takes place because of the metabolite hydroxypyrazole. **Mesotrione,** $C_{14}H_{13}NO_7S$ (2-(4-mesyl-2-nitrobenzoyl) cyclohexane-1, 3-dione), is another HPPD inhibitor. Being a benzoyl cyclohexanedione compound, mesotrion is used on maize crops to protect them against a range of weeds.

8-Hydroxyquinoline sulfate fungicide

8-Hydroxyquinoline sulfate, $C_{18}H_{16}N_2O_6S$ (bis(8-hydroxyquinolinium) sulfate), is a yellow-coloured, highly hygroscopic crystalline fungicidal powder. It is soluble in water (300 g/l) and slightly soluble in glycerine and alcohols. 8-Hydroxyquinoline sulfate can be prepared by the distillation of coal tar.

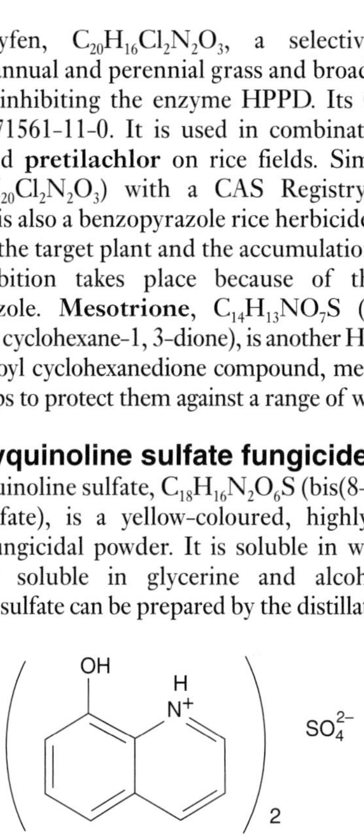

8-Hydroxyquinoline sulfate

This systemic PGR is a fungicide, bactericide, bacteriostat, fungistat and a soil sterilant. It controls grey mould (*Botrytis cinerea*) in vines and many soil-borne diseases.

Some of the common formulations of 8-hydroxyquinoline sulfate are soluble concentrate (SL) and water-soluble powder (SP). The bacteriostatic/**fungistatic** property makes it suitable in the pharmaceutical industry and as an ingredient in deodorants.

8-Hydroxyquinoline sulfate is a Class III toxin (slightly hazardous), with an acute oral LD_{50} in rats and mice ranging from 500 to 1250 mg/kg. The ADI is fixed at 0.15 mg/kg b.w./day. The compound is not toxic to birds, fish and bees.

In mammals, 8-hydroxyquinoline sulfate is eliminated within 36 h. It is registered for use in South Africa, New Zealand and Hungary.

Half-life

The half-life period of a substance is defined as the time required for reducing it via its reaction, to half of its original value. This time is independent of the original amount of the substance. For example, the half-life of radioactive compounds varies from 1590 years for radium, to a few seconds for some other radioactive substances. The half-life is given by the equation Eqn H.1. The half-life is dependent on the disintegration constant of the element.

$$t_{1/2} = \frac{\ln(2)}{\lambda} = \text{constant} \qquad \textbf{(H.1)}$$

where
λ = the disintegration constant of the element
ln (2) = natural logarithm of 2

Phosphorus ^{32}P, which has a half-life of 14.3 days is used in mechanistic studies of the phosphate fertilizer pathway in plants.

The half-life of a pesticide depends on: (i) its chemical nature; (ii) the amount applied to a crop; and (iii) the physical nature of the plant surface. An active ingredient or its metabolite with a long half-life is a matter of concern, since it is a sign of the product's persistence in the environment. Such persistent residues can be plotted on linear, semi-logarithmic curves, by relating degradation of residue concentration with respect to time, to arrive at the duration required for its degradation to safe levels. Formulated products have to indicate this time interval, for an accurate estimation of the pre-harvest interval.

Organochlorine insecticides have the greatest impact on environmental quality because of: (i) their relatively long persistence; (ii) their very high lipid-water partition coefficients; and (iii) their relative resistance to biological degradation. On the other hand, most organophosphate insecticides readily degrade in the presence of light, air and moisture.

The average half-life of some persistent pesticides are **DDT** 2.8 years, **dieldrin** 2.5 years, **chlordane** 1.0 year, **malathion** up to 2 weeks and **parathion** 3–6 months.

Half-life also indicates the time taken to weaken a disease inoculum (e.g. a plant virus) to estimate the spread of infection.

Halosulfuron herbicide

See **Halosulfuron-methyl herbicide**

Halosulfuron-methyl herbicide

Halosulfuron-methyl, $C_{13}H_{15}ClN_6O_7S$ (methyl 3-chloro-5-(4,6-dimethoxypyrimidin-2-ylcarbamoylsulfamoyl)-1-methylpyrazole-4-carboxylate), is a pyrimidinylsulfonylurea herbicide. A derivative of halosulfuron, halosulfuron-methyl is a white-coloured powder, soluble in water at 1650 mg/l. It is stable under normal handling and storage conditions. It is produced by reaction of *N*-1-methyl-3-chloro-4-carboxymethyl-5-sulfonylisocyanate pyrazole with 2-amino-4,6-dimethoxypyrimidine. The CAS Registry Number of halosulfuron-methyl is 100784-20-1.

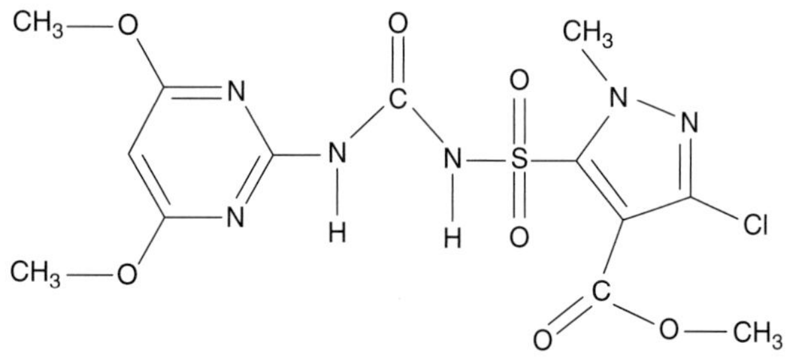

Halosulfuron-methyl

Halosulfuron-methyl, a selective, pre- and post-emergence herbicide, acts by interfering with acetolactate synthase enzyme, thus rapidly stopping growth and cell division in both roots and shoots. It is used in the control of annual broadleaved weeds in maize, sugarcane, rice and lawns. If applied with water or furilazole safener, the spectrum of weed control in maize becomes broader.

Halosulfuron-methyl, available as wettable powder (WP) or water-dispersible granules (WG), is applied at a rate of 45–115 g/ha and repeated as necessary. Trade names include Sandea, Permit and Manage. Reducing agents and acids reduce the efficacy of the compound, while **chemigation**, leaching and spray drift can cause injury to non-target plants and crops. Aquatic plants are also adversely affected.

The acute oral LD_{50} of halosulfuron-methyl among rats is >1287 mg/kg, while the same among birds is >2250 mg/kg. Its acute dermal LD_{50} in rats is >5000 g/kg. The compound may cause irritation to the eyes and skin. A 96 h acute LC_{50} among fish species tested, ranges from >118 to >131 mg/l. But, this compound is a potential groundwater contaminant and hence care has to be taken to prevent spillage or leakage near sources of water. Applicators and handlers are advised to follow label instructions and be equipped with appropriate protective garments and eye gear. The EPA rates this herbicide as slightly toxic.

The aerobic soil half-life of halosulfuron-methyl is 55 days, the compound being degraded by microbes (although this is slow in anaerobic conditions). Soil hydrolysis is rapid. The aquatic half-life ranges from 13 to 27 days.

Halosulfuron-methyl is registered for use in the USA, Australia, New Zealand and Greece.

Halosulfuron, $C_{12}H_{13}ClN_6O_7S$ (3-chloro-5-(4,6-dimethoxypyrimidin-2-ylcarbamoylsulfamoyl)-1-methylpyrazole-4-carboxylic acid), is a pyrimidinylsulfonylurea herbicide. When this substance is used as an ester or a salt, its identity would be modified, as in the case of halosulfuron-methyl.

Haloxyfop-p herbicide

Haloxyfop-p, $C_{15}H_{11}ClF_3NO_4$ ((*R*)-2-{4-[3-chloro-5-(trifluoromethyl)-2-pyridyloxy] phenoxy} propionic acid), is an aryloxyphenoxypropionic herbicide. It is a colourless crystalline substance, soluble in water (at 43.4 mg/l at pH 2.6), acetone, methanol, isopropanol, dichloromethane, toluene and hexane. It remains stable in its airtight, original container. The CAS Registry Number is 95977-29-0.

The aquatic half-life of haloxyfop-p is 78 days at pH 7. When this substance is used as an ester or a salt, its identity may change, for example in the case of haloxyfop-p-ethyl and haloxyfop-p-methyl ($C_{16}H_{13}ClF_3NO_4$ with a CAS Registry Number of 72619-32-0). The unresolved isomeric mixture of this substance has the ISO common name haloxyfop.

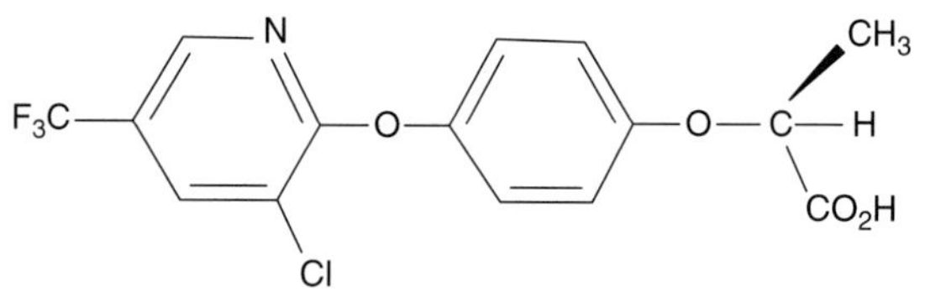

Haloxyfop-p

Haloxyfop-p is a selective pre- and post-emergence herbicide that is absorbed by the target plants and hydrolysed to haloxyfop. Haloxyfop inhibits the normal functioning of meristematic growth tissues. Haloxyfop-p controls annual and perennial grasses in sugarbeet, fodder beet, canola, potato, leaf vegetables, onion, sunflower, soybean and vines.

Haloxyfop-p is compatible with many other herbicidal compounds. It is available as an emulsifiable concentrate (EC).

The acute oral LD_{50} values of haloxyfop-p among rats and mallard ducks are 337 mg/kg and >2150 mg/kg, respectively. The EC_{50} of haloxyfop-p for *Daphnia* is 96.4 mg/l. The NOEL of haloxyfop-p in rats (up to 2 years) is 0.065 mg/kg/day. The use of protective garments and goggles is strongly advised since contact of the compound with skin, eyes and food and drink can cause irritation and injury. Spillage of haloxyfop-p (e.g. via washing of equipment) into water sources is to be avoided.

In rats, following oral administration, haloxyfop-p undergoes metabolism to haloxyfop which is excreted in the faeces and urine. The DT_{50} of haloxyfop-p in soil and water is in the range of 55–100 days at pH 7 and between 20 and 25°C. (See also **ACCase inhibitors**.)

Haloxyfop-p-methyl herbicide
See **Haloxyfop-p herbicide**

Hand dusters
See **Dusting machinery**

Harpin
See **Harpin aβ protein for biocontrol**

Harpin aβ protein for biocontrol
Harpins are a class of acidic, heat-stable proteins that are produced by the Gram-negative pathogenic bacteria *Erwinia amylovora* and may also be produced by modifying other strains. Harpins induce resistance among crops against pathogens, without any direct interaction with the pathogen. Harpins have evolved as a safe alternative to chemical pesticides that induce resistance in pathogens. These proteins activate a natural defence system within plants called **systemic acquired resistance (SAR)**. Harpins are used against fungal and viral pathogens.

Harpin is a non-toxic biopesticide with the acute oral LD_{50} of harpin in rats being >2000 mg/kg. It is safe for handlers, as well as non-target creatures like birds, fish and honey bees. An example of a trade name is Messenger.

Harpin aβ protein (amyloid beta, Aβ or Abeta) is a peptide of several amino acids, and is sourced from bacterial species. When applied to crops it elicits SAR among the crops. Commercially, this substance is produced by a strain of *Escherichia coli* (containing a DNA sequence from *E. amylovora*).

Harpin aβ protein is effective against a variety of fungal, bacterial, viral and soil-borne plant pathogens. It reduces infestation by insect and nematode pests and also enhances plant growth such as germination, flowering and fruit set on crops.

Water-dispersible granules (WG) (trade name ProAct) of this compound may be used as a spray, plant dip or as a seed treatment. As a drench, the application rate of 0.35–10 g a.i/ha at a 14–21 day interval is common. The protein is compatible with compounds except chlorinated water.

This compound, being required at such low rates, is classified as a Category IV toxin by the EPA. The acute oral LD_{50} of harpin aβ protein in rats is 5050 mg/kg, while in rainbow trout the acute LC_{50} is >3270 mg/l and in honey bees (48 h) it is >13 μg per bee.

Harpin aβ protein rapidly degrades in soil and does not enter the food chain. Both harpins and harpin aβ protein are registered in the USA.

Harrows
Harrows are implements used for light, shallow or secondary cultivation. The disc harrow, spike-tooth harrow and spring-tooth are among the most common types of harrows.

Harrows are ideal for breaking clods, levelling surfaces, sowing seeds, incorporating organic matter into the soil and destroying weeds. Harrows can penetrate deep into the soil and help in cultivation with their heavy prongs. Harrows have a significant role in **integrated pest management (IPM)**.

Modern-day harrows can be zigzag or spring-tined and are powered. (See also **Weed management by mechanical methods**.)

HCB insecticide
HCB is the synonym of **hexachlorobenzene**, which is a persistent organochlorine insecticide.

HCH insecticide
HCH is the synonym of **hexachlorocyclohexane**, which is an organochlorine insecticide.

Helicoverpa nucleopolyhedrovirus (HNPV)
See **Biocontrol agents**

Hemiptera
See **Pest order – Hemiptera**

Heptachlor epoxide
See **Heptachlor insecticide**

Heptachlor insecticide
Heptachlor, $C_{10}H_5Cl_7$ (1,4,5,6,7,8,8-heptachloro-3a,4,7,7a-tetrahydro-4,7-methano-1*H*-indene), is a cyclodiene organochlorine insecticide. A whitish powder with a CAS Registry Number 76-44-8, heptachlor is almost insoluble in water (0.056 mg/l) and soluble in acetone, benzene, xylene and ethanol. It is stable in normal storage conditions (e.g. temperatures up to 160°C). Heptachlor is obtained by the treatment of chlordane with *N*-bromosuccinimide, followed by chlorination with hydrogen chloride in nitromethane in the presence of aluminum trichloride or with monochloro iodide in carbon tetrachloride. It is even more active than **chlordane**.

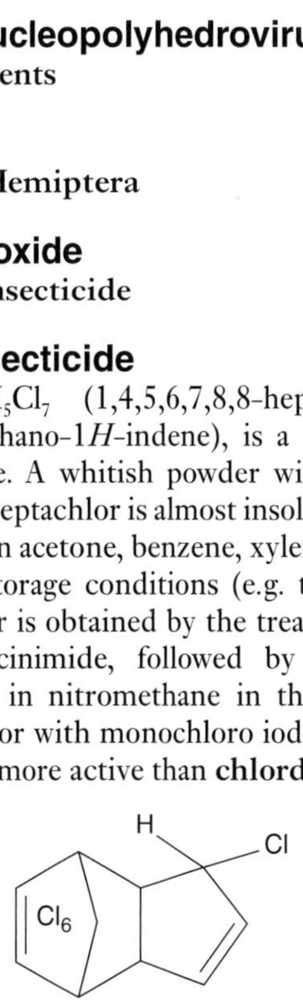

Heptachlor

Heptachlor, a non-systemic insecticide with stomach and respiratory action, has been used against ants, mosquitoes, fire ants, termites, weevils, thrips and other soil insects, a few of which affect crops like maize in many countries. Heptachlor is commonly formulated as emulsifiable concentrates (EC), dusts and wettable powders (WP).

However, the toxicity and the chronic effects of heptachlor on humans, birds and other non-target organisms have prompted countries to ban or severely restrict the use of this compound. It is declared a Class II toxin (moderately hazardous) by the WHO. The acute oral LD_{50} among rats, mice, pigs and chickens range from 30 mg/kg to 320 mg/kg. It is a suspected endocrine disruptor, developmental toxic and a cholinesterase inhibitor. These attributes make heptachlor a PAN Bad Actor chemical. Heptachlor is also toxic to birds, fish and bees.

The principal metabolites of heptachlor in animals are heptachlor epoxide (the CAS Registry Number is 1024-57-3) and 1-exo-hydroxychlordene epoxide. Both heptachlor and the metabolites persist for years in soil and water (they adsorb to sediments in the water). This compound also accumulates in the fat tissues of organisms. Hydrogenation of heptachlor leads to β-dihydroheptachlor that is very active as an insecticide, but is relatively safe to mammals (rat oral LD_{50} >5000 mg/kg). Both heptachlor and heptachlor epoxide are also **persistent organic pollutants** (POP). Symptoms of poisoning include hyper-excitability, tremors, convulsions and paralysis.

Heptachlor is not registered in the USA and is very restricted in its use in other countries. It is registered for use in Cambodia. In many countries, it has been used only as a termiticide. India has banned the use or import of heptachlor.

Heptenophos insecticide

Heptenophos, $C_9H_{12}ClO_4P$ (7-chlorobicyclo [3.2.0] hepta-2,6-dien-6-yl *O,O*-dimethyl phosphate), is an organophosphorus insecticide. With a CAS Registry Number of 23560-59-0, the moderately volatile compound is a pale amber liquid, soluble in water (2.2 g/l), and miscible with acetone, methanol and xylene. It is hydrolysed in acidic and alkaline media.

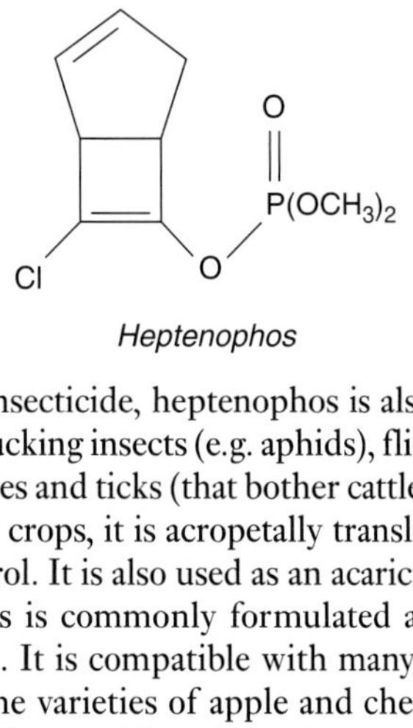

Heptenophos

A systemic insecticide, heptenophos is also an ectoparasiticide. It controls sucking insects (e.g. aphids), flies (order Diptera) and fleas, lice, mites and ticks (that bother cattle, pigs and sheep). When sprayed on crops, it is acropetally translocated and brings about insect control. It is also used as an acaricide.

Heptenophos is commonly formulated as an emulsifiable concentrate (EC). It is compatible with many other pesticides. It may injure some varieties of apple and cherry. It is available commercially as Hostaquick and Ragdan.

The WHO has categorized this chemical as a highly hazardous chemical (Class Ib). It may even get absorbed via the skin. The acute oral LD_{50} among rats is in the range of 96–121 mg/kg. The symptoms of poisoning include dizziness, hypersensitivity of the skin, nervous system disfunctioning, and may include coma and death. Due to its high acute toxicity rating, and its being a cholinesterase inhibitor, it is classified as a PAN Bad Actor chemical.

Mammals eliminate most of heptenophos via urine. In plants, heptenophos evaporates quickly, leaving no residues after 5 days. In soil, it is rapidly degraded by microorganisms, with the DT_{50} being between 4 and 36 h.

Heptenophos is not registered in the USA or throughout the EU. It is registered in Cambodia.

Herbaceous plants

A herbaceous plant, or simply a herb, is one which sheds leaves or stems (completely or partly) at the end of its growing season. Depending on its life cycle, plants can be divided into **annuals**, **biennials** and **perennials**. Annuals die down completely in 1 year or a season after flowering and fruiting. Biennials shed their parts and survive between seasons, and finally die at the end of the next growing season. Examples of biennials are carrots and ragwort. Herbaceous perennials survive by means of roots, or a hardened, modified stem (caudex), tubers or bulbs. Herbaceous perennials include most grasses and ferns. Thistle, bindweed, dandelion and ivies are examples of herbaceous plants that are weeds.

Herbicidal carryover effect

See **Residual pesticides**

Herbicidal inhibition of ALS

Some herbicides inhibit the synthesis of vital amino acids in plants. Such inhibitors primarily affect three different enzymes, 5-enolpyruvylshikimate 3-phosphate synthase (**EPSPS**), acetolactate synthase (ALS) and glutamine synthetase (GS), and achieve breakdown of normal plant functions.

Glutamine synthetase (GS) is an enzyme that aids the formation of glutamine from ammonium or glutamate ions. Thus it aids in the supply of nitrogen for many enzymatic pathways to enable synthesis of amino acids. GS inhibitors block the conversion of glutamic acid and ammonia to glutamine.

Acetolactate synthase (ALS) is an enzyme capable of synthesizing branched-chain amino acids. ALS prompts pyruvate molecules to form alpha-acetolactate, with the help of thiamine pyrophosphate. ALS also synthesizes acetohydroxybutyrate and ketobutyrate in the final production of leucine, valine and isoleucine. Since ALS is present in fungi and bacteria, chemical classes that inhibit ALS may be used to control all these pathogens. The enzyme acetohydroxy acid synthase (AHAS) is the other term for acetolactate synthase (ALS).

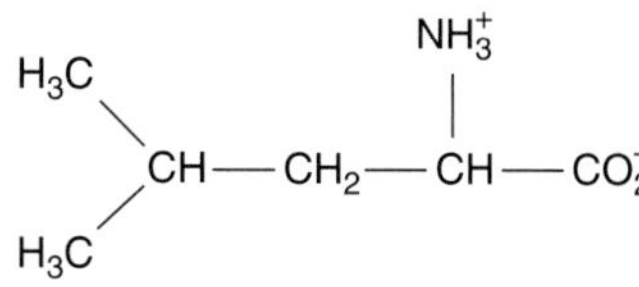

L-leucine

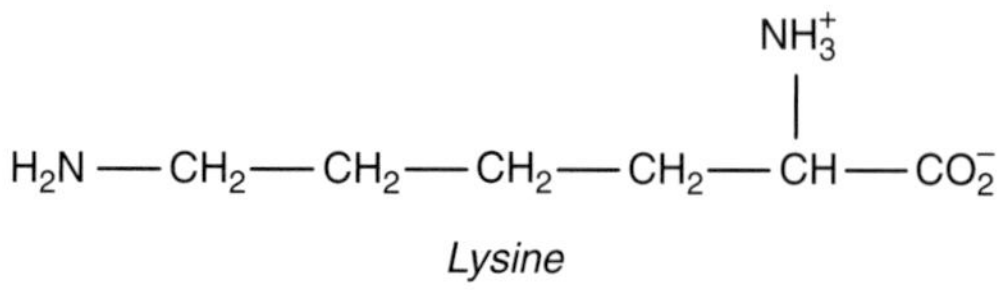

Lysine

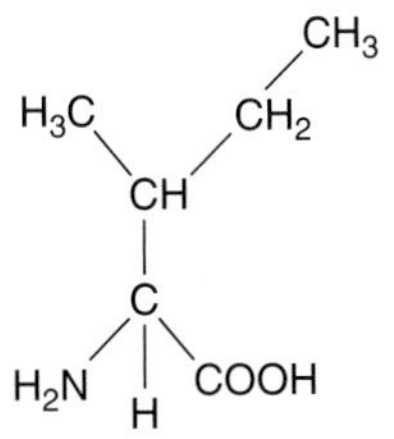

Isoleucine

Imidazolinones, sulfonylureas, triazolopyrimidines, pyrimidinyl oxybenzoates and sulfonylamino carbonyl triazolinones are known to be ALS inhibitors. ALS inhibitors block branched-chain amino acids leucine, lysine and isoleucine, resulting in the inhibition of DNA synthesis.

ALS inhibitors are selective herbicides, needed in very low doses (0.15 kg/ha). Foliar and soil applications of such compounds are effective on annual weeds. **Cloransulam-methyl, diclosulam, florasulam** and **metosulam** are some examples.

Formulations with these components have sometimes developed resistance in weeds.

Herbicidal inhibition of EPSPS

Some herbicides inhibit the synthesis of vital amino acids in plants. Such inhibitors primarily affect three different enzymes, 5-enolpyruvylshikimate 3-phosphate synthase (EPSPS), acetolactate synthase (ALS) and glutamine synthetase (GS), and achieve breakdown of normal plant functions.

EPSPS is a critical enzyme in the shikimate pathway. It converts shikimate 3-phosphate to 5-enolpyruvyl-3-shikimate phosphate. This end product acts as a precursor to many of the aromatic amino acids and other chromatic compounds like tyrosine and phenylalanine. Amino acids help in proper cell wall formation and in the development of the defence system. EPSPS inhibition effectively blocks the biosynthesis of amino acids in the plant cells of dicots as well as grasses.

Glyphosate, a very popular foliar herbicide, is an EPSPS inhibitor. An analogue of amino acid glycine, glyphosate is a post-emergence, **non-selective**, systemic herbicide, effective against all types of broadleaved grasses. Resistance has been observed in some ryegrass species.

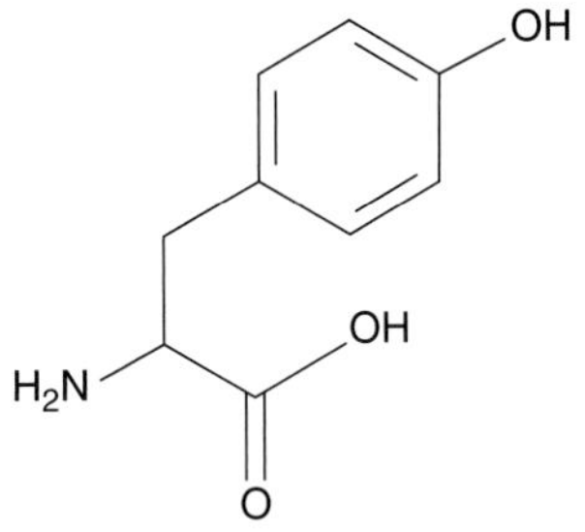

Tyrosine

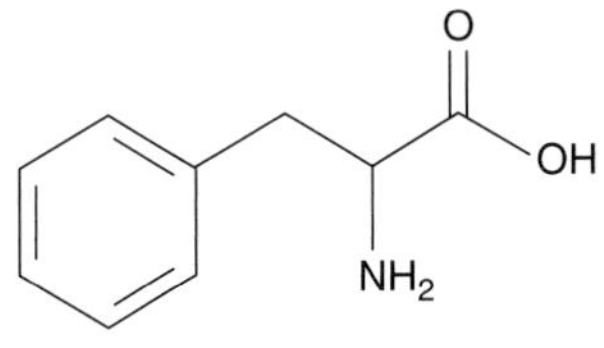

Phenylalanine

Herbicidal inhibition of glutamine synthetase (GS)

Some herbicides inhibit the synthesis of vital amino acids in plants. Such inhibitors primarily affect three different enzymes, 5-enolpyruvylshikimate 3-phosphate synthase (EPSPS), acetolactate synthase (ALS) and glutamine synthetase (GS), and achieve breakdown of normal plant functions.

GS is a very important enzyme that controls the metabolism of nitrogen by prompting the formation of glutamine from ammonium or glutamate ions. The activity of GS helps in the supply of nitrogen to many enzymatic pathways in order for the proper synthesis of amino acids. GS-inhibiting compounds block the conversion of glutamic acid and ammonia to glutamine.

The inhibition of the enzyme GS by amino-acid-inhibiting herbicides leads to the accumulation of high levels of ammonia. **Glufosinate** and **bialaphos** are two commercial herbicides acting on GS. Glufosinate is **non-selective**, non-leaching GS inhibitor that degrades rapidly in soil. It is used for controlling a variety of weeds in non-agricultural crops.

Maize has been made resistant to glufosinate by cloning the gene coding for phosphinothricin acetyl transferase from the bacterium *Streptomyces hygroscopicus*. Such resistant maize converts glufosinate to non-phytotoxic acetylated metabolites. **Bialaphos** is a fermentation product of *Streptomcyces* spp. It is a pro-herbicide which only converts into the desired phosphinothricin in the target plant. Bialaphos independently, is almost inactive on the GS system.

Herbicide

A compound, either organic or inorganic, used to destroy unwanted vegetation like weeds, grasses, woody plants, and sometimes even crop plants, is known as a herbicide or a weedicide. The word herbicide has been derived from the Latin word 'herba' (meaning 'plant') and 'caedere' (meaning 'to kill'). The term weedicide is no longer used, since the chemical/natural agents can kill non-weed plants as well (as in non-selective herbicides).

Until the beginning of the 1920s inorganic herbicides such as **sodium chlorate** (NaCl), ammonium sulfate, arsenic and boron compounds and ash and factory wastes were used. At that time, cultural methods for weed control like burning, flooding and increasing salinity were common. Since then, many specific organic herbicides have been introduced. Herbicides work by inhibiting various enzymes that disturb critical biochemical processes in target plants. Some modes of action include inhibition of acetolactate synthase, acetyl-CoA carboxylase, carotenoid biosyntheses, amino acid synthesis and photosynthesis at photosystem I or II.

Herbicides may be applied to the soil or on the foliage. They may be applied:

- before weeds are visible (pre-emergence stage); or
- at specific stages after emergence (post-emergence) either before crops are planted (pre-plant stage) or after crops have been planted (post-plant stage).

Some chemical classes of herbicides include phenylpyrazole, **triazines**, **thiadiazole**, bipyridylium, amides, urea, uracid and aryloxyphenoxypropionates. Some examples of herbicide active ingredients include **2,4-D**, **MCPA**, **MCPB**, **diquat**, **paraquat**, **pyrazoxyfen**, refenacet, **butylate**, **molinate**, **dicamba**, **picloram** and **cinmethylin**.

While most herbicide usage is a reaction to the nuisance of weeds, some chemicals are applied for agricultural convenience. For example some crops like cotton are sprayed with chemicals to kill outwardly growing leaves, thereby making harvesting easy. Such herbicides are described as **desiccants** and **defoliants**. On other occasions, herbicides are applied in minute quantities to help regulate growth (e.g. to increase yields, hasten flowering, delay ripening or increase leaf drop). These chemicals are called **plant growth regulators (PGRs)**.

Repeated use of a few components of the same chemical group can cause the development of resistance in weeds. Many ryegrass varieties, for instance, have developed resistance to diuron, chlorosulfuron, **atrazine** and diclofop. In India, the most serious case of herbicide resistance and later cross-resistance has been reported in littleseed canarygrass (*Phalaris minor*) in wheat. **Resistance management** is therefore an important feature of herbicide application.

Herbicides may be of two major types: (i) **selective herbicides** such as 2,4-D, phenols, carbamates and urea derivatives, permitting elimination of weeds without injury to crops; and (ii) **non-selective herbicides** comprising soil sterilants (sodium compounds) and silvicides (ammonium sulfamate), which can kill woody plants and trees.

Weeds are plants and they are often closely related to crop plants, having similar morphology/architecture and physiology (cellular and vascular organization and processes). Many weeds even belong to the same genus as the crop plants and resemble them so closely that they can hardly be distinguished. Hence controlling weeds in crop fields (i.e. killing undesirable plants, weeds, among desirable plants, the crop), is a huge challenge for weed scientists. Proper technical know-how and sensitivity in herbicide usage is a prerequisite for successful adoption of herbicide technology.

Herbicide action – by inhibition of amino acid synthesis

There are 20 amino acids essential for the growth of plants. These are variously called **proteinogenic amino acids** or primary amino acids. These amino acids are coded for by the DNA and form the subunits of all the proteins synthesized by a plant. Enzymes and intermediates are required for the synthesis of amino acids.

Inhibition of amino acid synthesis is the target of some herbicides. Specifically, they inhibit: (i) aromatic amino acid synthesis, by blocking the enzyme 5-enolpyruvylshikimate 3-phosphate synthase (**EPSPS**); (ii) branched-chain amino acid synthesis, especially the enzymes **acetolactate synthase (ALS)** or acetohydroxyacid synthase; and (iii) enzyme **glutamine synthetase (GS)**.

The herbicidal action of amino acid synthesis inhibitors is evidenced by a gradual depletion of proteins vital for plant growth, seen as stunted growth, bleached leaves and necrotic plant tips.

Herbicide action – by inhibition of fatty acid elongation

Inhibition of fatty acid elongation in plants results in the inhibition of lipid synthesis. The **thiocarbamate**/thiolcarbamate or **carbamothioate** group is an inhibitor of fatty acid elongation, which is mainly responsible for production of plant surface wax. Apart from fatty acid elongation, this group of herbicides also inhibits gibberellin biosynthesis. **Butylate**, **triallate**, **cycloate**, **pebulate** and **EPTC** are some members of this class of herbicides, with common trade names like Sutan, Avadex and Avadex B-W.

Thiocarbamates, applied at the pre-emergence stage into the soil, are volatile by nature and are not persistent in soil (except for butylate). Effective against annual grass weeds, they are recommended for use with safeners because they cause injury to the crop. **Dichlormid** is a commonly used safener. Most thiocarbamate herbicides are degraded by microorganisms.

Diallate and **triallate** are similar except for one chlorine atom. Both are soil applied at pre- or post-emergence stages and are effective against grasses such as wild oats in crops of lentil, sugarbeet and pea. Soil incorporation at the depth of around 15 cm gives best results. Triallate is more widely used because of its selectivity.

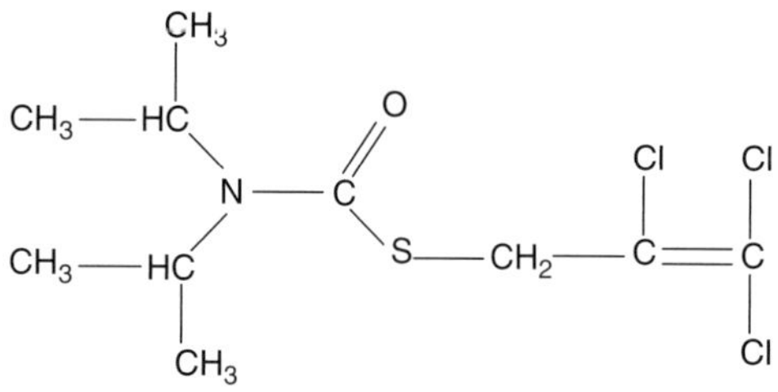

Triallate

EPTC (*S*-ethyl dipropyl carbamothioate) is another thiocarbamate, used at the pre-plant stages in maize, lucerne (alfalfa), flax, potato and sugarbeet to effectively control **yellow nutsedge** and **common lambsquarters**. EPTC is converted into EPTC sulfoxide, which inhibits plant processes. To avoid injury, EPTC is used along with safeners on crops like lucerne, flax, potato and sugarbeet. The symptoms are inhibition of shoot growth at the germinating stage and abnormal growth of leaves.

Cycloate is also mixed into the soil at the pre-plant stage for the protection of sugarbeet and spinach.

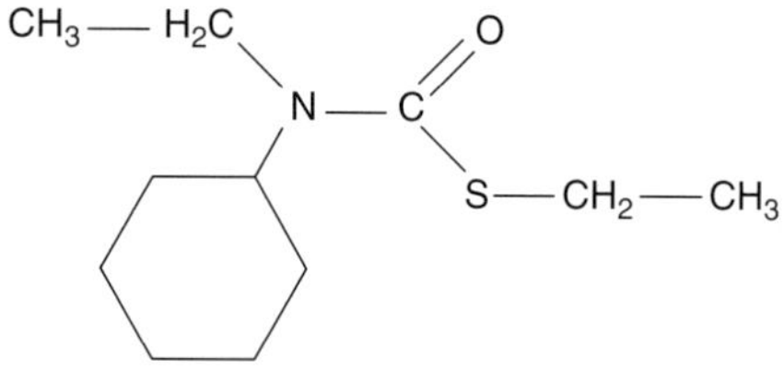

Cycloate

Butylate is the only thiocarbamate which may be used without a safener. It is effective against several annual grasses and nutsedges and some broadleaved weeds in maize. Its soil half-life is short. The effects of thiocarbamates include inhibition of shoot growth and the abnormal and unhealthy growth of grass leaves.

Herbicide action – carotenoid synthesis inhibition

Carotenoids are organic pigments that have a dual function of absorbing light energy during photosynthesis and making it available to chlorophyll, while preventing chlorophyll from photo damage. Carotenoids are present in chloroplasts and chromoplasts of the plant. These substances, around 400 in number, are responsible for the bright colour of fruits and flowers.

Herbicides have been developed that inhibit the phytoene desaturase enzyme that is responsible for carotenoid synthesis. The HRAC has coded such herbicides under F1-12, F2-28 and so on. This herbicide group belongs to the **pyridazinone** family.

Chemical groups of **triketones**, **isoxazolidinones** and triazoles are also considered inhibitors in the carotenoid synthesis pathway. The effect of this inhibition is disruption of plant processes leading to bleached and chlorotic foliage, tissue necrosis and ultimately, death. **Norflurazon**, **sulcotrione** and **amitrole** are some of the herbicidal active ingredients with this mode of action.

Herbicide action – contact membrane disrupters

There are some herbicides that inhibit the photosynthetic function of the plant by destroying cell membranes. These are classified as membrane disruptors with the site of action being photosystem I electron diversion (commonly called photosynthesis inhibitors). The bipyridiliums are included in this category.

Bipyridilium compounds are a group of light-dependent, membrane-disrupting contact herbicides that are effective as foliar sprays. These compounds work by accepting electrons from photosystem I and undergoing a series of reduction reactions to produce hydrogen peroxide and hydroxyl radicals which in turn react with and destroy lipid, fatty acids and chlorophyll in the cell. The cell wall disintegrates, finally killing the plant. **Paraquat** and **diquat** are the two compounds in this class.

Bipyridilium herbicides are contact, non-selective and quick-acting herbicides and need a **surfactant** to spread. Broadleaved grasses and plants are controlled by paraquat and diquat. Their rapid action makes them good **desiccants**.

Bipyridiliums get adsorbed and thus de-activated by organic soils. They are thus unsuitable as a soil herbicide. But the tendency to generate free radicals poses a problem to mammals, in case the compound is consumed. This has led to paraquat being banned throughout the EU.

Herbicide action – lipid synthesis inhibitors

Fatty acid synthesis in plants is an essential process in the development of plant lipids. Lipids include fatty acids, neutral fats and steroids. Plant lipids play an important part in the structure and function of cell membranes. The enzyme Acetyl-CoA carboxylase is involved in plant fatty acid biosynthesis and in the production of, for example, cell membranes, cuticular wax and flavonoids.

Aryloxyphenoxypropionic acids, cyclohexenediones, thiocarbamates, **chloroacetamides** and substituted pyridazinones are the major herbicide families which inhibit lipid synthesis. Each herbicide in these families acts by inhibiting acetyl-CoA carboxylase (**ACCase inhibitors**) and fatty acid elongation (fatty acid elongation inhibitors). Many of these herbicides also inhibit other biochemical mechanisms such as photosynthesis and carotenoid biosynthesis.

Lipid biosynthesis inhibitors commonly have a selective action and are used on perennial and annual grasses. They act on the meristematic tissue and vegetative parts. Some of them have short soil activity. Effects of these herbicides are seen as necrosis, small leaf size, stunting and translucent banding of new leaves. HRAC codes for herbicides that inhibit lipid synthesis are 1, 8 and 16.

Herbicide action – meristematic mitotic inhibitors

Meristems are regions in plants where cell division is the most frequent, like the tips of shoots and roots. Meristematic tissues constitute undifferentiated cells, which can divide continuously thus allowing growth of the plant. Mitosis is the process by which a cell divides into two daughter cells.

Herbicides that inhibit this process of cell division or mitosis, cause death of the plant and are called meristematic mitotic inhibitors. Most of these herbicides are applied to the soil or germinating seedlings. The general effects include: (i) stunted shoots and roots; (ii) non-emergence of seedlings; and (iii) unopened or improperly opened leaf buds. Small seeded grains and broadleaved weeds are effectively controlled by such herbicides.

There are two known areas of inhibition, namely, shoot meristem inhibition and root meristem inhibition.

The root meristem inhibition is effected by the dinitroanilines or toluidine group of herbicides which are based on *para*-toluidine. These chemicals attach to the protein tubulin and affect the polymerization and production of microtubules. Microtubules are protein strands that are an important constituent of the cytoskeleton. Microtubules (composed of tubulin) are involved in cell division, cell wall development and elongation.

Dinitroanilines are volatile, pre-emergence, pre-plant herbicides that affect root growth. This herbicide group controls small-seeded, broadleaved weeds in cotton, soybean, potato, groundnut, sunflowers, peas and fruits. However, some of the dinitroaniline herbicides persist in soil and can harm the environment, mainly due to their low solubility in water. Benefin, ethalfluralin, isopropalin, **oryzalin**, **pendimethalin**, prodiamine and **trifluralin** are some examples of dinitroanilines, a few trade names being Paarlan, Surflan and Prowl. Other herbicidal groups grouped together with dinitroanilines are **benzoic acid (DCPA)**, phosphoramidates and pyridine herbicides. These are grouped under 3 of the HRAC classification.

The shoot meristem inhibition is brought about by acetamides, carbamothioates and *N*-phenyl carbamates. Carbamothioates are responsible for injured and puckered leaves and buds that do not open. A notched distal leaf margin is the symptom of the action of acetamides.

Carbamate herbicides inhibit cell division, besides interfering with microtubule formation. Carbamates, not widely known as herbicides, can selectively control certain annual grasses. **Propham** and chlorpropham, which are arylcarbamate esters, are examples. These are soil-applied, non-leaching and volatile agents which have a short soil life. They are effective on annual grasses, especially dodder (*Cuscuta europaea*).

Pronamide is a substituted amide that works selectively in fruit crops like raspberry, blackberry and cherry. This agent prevents microtubules from forming, by attaching itself to tubulin. **Dithiopyr** (pyridine) is another agent that inhibits mitosis in a variety of grass weeds in turf and rice.

Herbicide action – others

Chloroacetamides (2-chloro-*N*,*N*-diallylacetamide) are a group of pre-emergence herbicides which include allidochlor and chloroacetic acid diallylamide. These are soil-applied herbicides and are highly toxic to many grasses (e.g. crabgrass, annual bluegrass and wild oats) and broadleaved weeds (e.g. pigweed, carpet weed, **barnyard grass** and foxtail varieties). Most of the formulations are tolerated by maize, sorghum, beans and cabbage. Randall is a known trade name.

Sodium chlorate, with contact action, is used as a herbicide (along with other chemicals). Toxicity is due to the reduction of chlorate to hypochlorite and chlorite. Plant deficiency of **molybdenum** also increases the toxic action of chlorate.

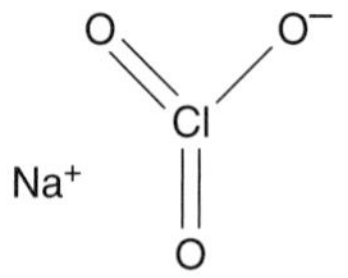

Sodium chlorate

Herbicide action – photosynthesis inhibitors

Photosynthesis is a metabolic pathway that uses light energy to convert carbon dioxide into carbohydrates, especially sugars. There are two major stages in photosynthesis: (i) the light-dependent stage; and (ii) the light-independent stage. The light-dependent stage captures the energy of sunlight and stores it in the form of energy-giving molecules ATP and NADPH. In the second light-independent stage (dark reactions) the ATP and NADPH produced in the light-dependent stage are used to convert atmospheric carbon dioxide into sugar molecules.

The light-dependent reactions involve two protein complexes, namely, photosystem I and photosystem II. Photosystem II uses photons from sunlight and excites a pair of electrons which are transferred through a chain of electron carriers. A water molecule is split in the process and the lost electrons are replaced by the electrons from water. The energy of the excited electrons is used to create a proton gradient with the hydrogen ions generated during the splitting of water. This gradient is used to generate ATP. In photosystem I electrons are again excited by light and these electrons are transferred through an electron transport chain to $NADP^+$ where hydrogen ions too are added to form NADPH. The transferred electrons in photosystem I are replaced by the excited electrons from photosystem II. The ATP and NADPH thus produced are used further in the light-independent stage to synthesize sugar molecules using atmospheric carbon dioxide.

Herbicides that inhibit photosynthesis, act by blocking light reactions, thus preventing the conversion of this light energy into its chemical form. More herbicides act on the photosynthetic process than on any other biochemical process. The common symptoms of inhibition of photosynthesis include: (i) **chlorosis**; (ii) rapid bleaching of leaf tissue; and (iii) cell membrane disruption.

Herbicides that act on photosystem I keep the electrons from reaching the electron transport system, thus inhibiting the overall photosynthesis process. They are called photosystem I electron acceptors. The specific action is the reduction of molecular oxygen to a toxic superoxide radical. Photosystem I herbicides are generally contact herbicides which disrupt membranes.

Bipyridylium compounds are the major group of chemicals that inhibit photosystem I. They cause photo-oxidative stress by diverting energy from photosystem I. **Diquat** and **paraquat**, two compounds in this class, are fast-acting, **non-selective**, contact herbicides and act only when absorbed by leaves. Translocation is poor and they have almost no soil action.

Herbicides that inhibit photosystem II block electron transport by binding to adjacent sites on the D-1 quinone proteins that help to form the transport chain.

Photosynthesis inhibitors, acting on photosystem II, include **triazines** (***s*-triazines**, *as*-triazines), **phenylureas**, **uracils**, hydrobenzonitriles, **acylanilides**, **pyridazinones**, ***bis*-carbamates** and benzothiadiazine betazon (Table H.1).

Table H.1. Photosystem II electron transport inhibitors. (Table courtesy of Shubhangi Borse.)

Common name[a]		Substituents		
Family	Pesticide (or a.i.)	R_1	R_2	R_3
s-Triazine				
Chlorotriazine	Atrazine	Cl	$NH\text{-}CH_2\text{-}CH_3$	$NH\text{-}CH(CH_3)_2$
Methoxytriazine	Prometon	$O\text{-}CH_3$	$NH\text{-}CH(CH_3)_2$	$NH\text{-}CH(CH_3)_2$
Methylthiotriazine	Ametryne	SCH_3	$NH\text{-}CH_2\text{-}CH_3$	$NH\text{-}CH(CH_3)_2$
as-Triazine	Metribuzin	$C(CH_3)_3$	NH_2	SCH_3
Uracil	Terbacil	$C(CH_3)_3$	Cl	CH_3
Cynophenol or hydroxynitrile	Dichlobenil	Cl	Cl	
Acylanilide	Propanil	$\text{-}CH_2\text{-}CH_3$	Cl	
Bis-carbamate	Desmedipham	H	$CH_3\text{-}CH_2$	H
Phenylurea	Diuron	$3,4\text{-}Cl_2\text{-}C_6H_3$	CH_3	

[a]The common name refers to the family name of the pesticide and the names of the individual pesticides within that family. For example *s*-triazine is a family of pesticides, chlorotriazine is a class in that family and atrazine is the pesticide (or a.i.) under consideration which has three substituents (R_1, R_2 and R_3).

The triazine family includes **chlorotriazines (ametryn, simazine, atrazine**, cyanazine, **metribuzin** and **hexazinone**), **methoxytriazines** and **methylthiotriazines**. These are effective in controlling weeds in maize, sorghum and soybean. The effect is seen in the browning of leaves. The site of action is the D-1 quinone-binding protein of photosynthetic electron transport.

Phenylureas include **linuron** and **tebuthiuron**, which are effective on soybean and on pastures. Their site of action is the same as that of triazine. It is similar in the case of **uracils** (terbacil), benzothiadiazoles (bentazon), nitriles (**bromoxynil**), **carbamates** (**desmedipham**) and dicarboxylic acids (**endothal**). The symptoms are chlorosis and browning of leaves and the mode of action is the D-1 quinone-binding protein of the photosynthetic electron transport.

Monuron (3-(p-chlorophenyl)-1,1-dimethyl urea) (CMU) is a commonly available phenyl-substituted urea. The herbicides in this group are very toxic and persistent. They are generally soil-applied, non-selective chemicals, absorbed by the roots.

Amino triazole (**amitrole**) is a very widely known herbicide for perennial weeds. It is absorbed and translocated easily. The mode of action is its accumulation in meristematic tissues and leaves, forming metabolites like carbohydrate derivatives with glucose and amino acid derivatives. It forms chelates with essential metals like Mg and Fe, which inhibit growth.

Herbicide action – protoporphyrinogen oxidase (PROTOX) inhibition

Protoporphyrinogen oxidase (PROTOX) is an enzyme that is responsible for the biosynthesis of chlorophyll and haem in plants. PROTOX catalyses the oxidation of protoporphyrinogen IX to protoporphyrin IX. PROTOX is also called PPG oxidase. Any disturbance in the synthesis pathway causes irreversible damage to plant cell components including the cell wall.

Diphenyl ether is a herbicide that inhibits PROTOX. This inhibition results in a considerable build up of the chlorophyll precursors that are extremely light sensitive. This, through photoreactions, results in formation of destructive free radicals that oxidize lipids (lipid peroxidation) and other oxidizable cell components. This results in the destruction of the cell wall, seen as the desiccation of mature foliage.

Chemically, diphenyl ether is classified either as a PROTOX inhibitor or as a carotenoid biosynthesis inhibitor, its activity peaking under high temperature and intense light.

N-phenylphthalimides, oxazolidinediones, phenylpyrazoles, pyrimidindiones, **triazolinones** and others are also believed to be PROTOX inhibitors.

Bifenox, **pyraflufen-ethyl**, **sulfentrazone**, butafenacil, oxyfluorfen are some herbicides that share this mode of action.

Herbicide action through hormone-like substances

Plant hormones such as **gibberellins**, **auxins**, **ethylene**, **abscisic acid** and **cytokinins** are associated with plant growth and regulation. Auxins are a very important group of plant hormones. The most important of them is indoleacetic acid (**indole-3-acetic acid**). Other naturally occurring auxins are 4-chloro-indole acetic acid, phenylacetic acid (PAA) and indole-3-butyric acid (IBA). These are responsible for the growth and development of the fruit and the meristematic region, root initiation, organ differentiation and abscission. Auxins, which control the organized growth in plant parts, depend upon the polarized division of meristematic cells for their successful action. The effect of an auxin is generally on a site different from where it is produced.

Synthetic herbicides have been developed that mimic these natural hormones. They inhibit the action of natural growth hormones. Hormone-like herbicidal groups include **phenoxy acids, benzoic acid** and 2-pyridine carboxylic acids. Their action is on nucleic acid metabolism, causing hormonal imbalance (especially seen in the rapid biosynthesis of RNA), leading to aberrant growth of plant parts and tissues.

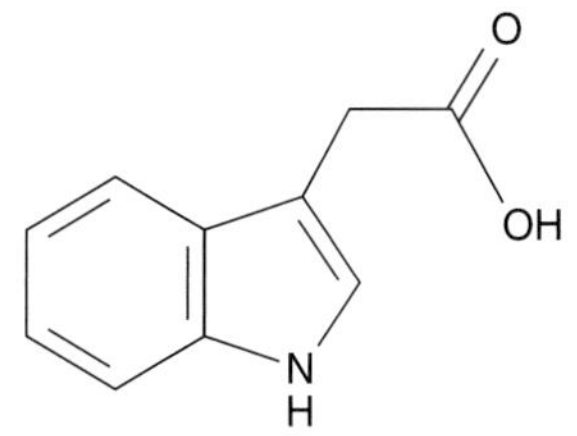

Indoleacetic acid

These herbicides are highly systemic and translocate via the apoplast and symplast to the meristematic regions and vegetative propagules (not so much in some grasses). Most of them are phytotoxic in nature and can have selective action. Some of the symptoms include malformed new leaves, leaf cupping, crinkling or elongation with near parallel veins and stem brittleness.

Certain organic herbicides like substituted phenoxyacetic acids, compare with natural plant hormones like auxins. These include **2,4-D** (2,4-dichlorophenoxyacetic acid), 2,4,5-T (2,4,5 trichlorophenoxyacetic acid) and **MCPA** (4-chloro-2-methylphenoxy) acetic acid) and their derivatives. For instance, phenoxypropionic acid herbicides (e.g. 4-CPP, **dichlorprop**, 3,4-DP, fenoprop) have a selective action on target species. Their toxicity to plants (or even mammals) arises from minor modifications to the aliphatic side chain of the molecule. These molecular modifications determine the retention and penetration of the herbicide in plants.

Several alterations to the herbicidal molecule (especially to the number of carbons in the aliphatic chain of a chlorinated phenoxyacid molecule) are made to increase its efficiency.

A number of derivatives of chlorinated benzoic acid have come into extensive use for the control of perennial weeds in non-crop or uncultivated areas. Substituted benzoic acids have a growth-regulating activity which is milder than that of 2,4-D. Such derivatives are useful in cases where plants have developed resistance to 2,4-D. They are effective on the Canada thistle, smilax, honeysuckle, persimmons and sumac. Nitrobenzene and amino benzoic acid derivatives are effective as pre-emergence herbicides for fruits like strawberries.

Herbicide antagonism

Herbicide antagonism is the reduction in control of some weed species, brought about as a result of applying mixtures of two or more herbicides. Antagonistic interactions take place when two or more herbicides are mixed or when a herbicide formulation is sprayed alongside a fertilizer or another pesticide.

The practice of mixing pesticidal and fertilizer chemicals in the same tank may also cause antagonism. Even water can interfere with a herbicide function, for instance, clay inactivates **paraquat** and **glyphosate**, while Na, Ca, K and P ions interfere with **MCPA**, **2,4-D**, for example. Herbicides may adsorb on organic matter in water. Commonly, antagonism is evidenced when post-emergence grass herbicides are applied along with post-emergence broadleaved herbicides.

To counter antagonism more of the herbicide is commonly added to the mixture to achieve the intended objective. Other measures include knowledge of the compounds and the solvents. Label instructions should be followed. Addition of adjuvants and additives to the tank mix alleviates the problem (e.g. ammonium sulfate).

Herbicide classification – by mode of action

Herbicides are classified in several ways for ease of understanding, such as by: (i) application; (ii) penetration; (iii) persistence; and (iv) mechanisms (or modes) of action (MOA). The action of herbicides is to be understood as interference in normal plant metabolism, which inhibits plant growth or functions. Seven broadly seen modes of action are: (i) amino acid synthesis inhibitors (which includes **glyphosate** and its analogues, **imidazolinones**, sulfonylureas and triazopyrimidine sulphonamides; (ii) contact membrane disrupters (bipyradiniums and **diphenyl ethers**); (iii) hormone-like inhibitors (**benzoic acids**, quinoline carboxylic acids); (iv) lipid synthesis inhibitors (cyclohexanediones, aryloxyphenoxy-propionates); (v) photosynthesis inhibitors (bicarbamates, hydroxybenzonitriles, **triazines**, **uracils**, ureas); (vi) pigment synthesis inhibitors (aminotriazole, isoxazolidiones, pridaziones); and (vii) meristem mitotic inhibitors (**dinitroanilines**, *N*-phenylcarbamates, carbomothiates, **chloroacetamides**).

The Herbicide Resistance Action Committee (HRAC), an international body, is a platform that offers, among others, a dialogue for measures to manage resistance to herbicides. More information may be accessed from the website (http://www.hracglobal.com).

HRAC has classified herbicides based on their MOA and allotted codes for the same, for ease of reference. Please refer to Appendix 12 for the HRAC MOA classification scheme for herbicides (reproduced with the permission of Dr John Soteres, Chairman, HRAC).

Herbicide dazomet

See **Dazomet fungicide**

Herbicide mefluidide

See **Mefluidide as a PGR**

Herbicide Resistance Action Committee (HRAC)

See **Herbicide classification – by mode of action**

Herbicide safeners

Herbicide safeners are diverse chemical compounds that are designed to reduce the phytotoxicity of herbicides to crops by favourable activity at the molecular level. Structurally similar to herbicides (e.g. safener dichlormid and herbicide **EPTC**), these compounds selectively target weeds that grow among crops, that are sometimes botanically similar to them.

While the specific activity of each herbicide safener might be unique, all of them work by inducing the action of several enzymes (especially cytochrome P450) that metabolize or detoxify chemicals or groups such as **chloroacetanilides**, anyloxyphenoxy propionates, sulfonylurea, **imidazolinone** and sulfonamide.

Examples of herbicide safeners are **dichlormid**, **benoxacor**, furilazole (dichloroacetamide derivatives), fenchlorazole-ethyl (phenylpyrazoles), **cloquintocet-mexyl** (quinoline carboxylic acid esters) and compounds belonging to the thiocarbamates, phenylpyrimidines and methylbenzyltolylureas.

Mostly used in grass crops such as maize, wheat, sorghum and rice, herbicide safeners are applied either along with the respective herbicides at the pre-plant stage or the post-weed-emergence stage. For example, naphthalic anhydride is commonly used with maize herbicides. Safeners have a soil and a seed application too. The amount of a safener needed per herbicide application is very minute (up to a ratio of 1:30).

The selectivity element of herbicide safeners is illustrated in the protective action of fenchlorazole-ethyl on wheat when the herbicide fenoxaprop-ethyl is applied to the wheat crop to eliminate weeds like crabgrass. The safener enhances and speeds up the de-esterification and detoxification of fenoxaprop-ethyl in wheat, thus preventing damage to the plant, while not protecting the crabgrass, thus allowing the weed to be killed.

Naphthalic anhydride protects maize from **thifensulfuron-methyl** (a sulfonylurea compound) by inducing its de-esterification to its parent acid thifensulfuron. Similarly, the dichloracetamide safener furilazole speeds up the de-esterification of **halosulfuron-methyl** to halosulfuron acid in maize seedlings by enhancing the activity of P450 monooxygenases and other esterases.

Safeners prompt the glutathione conjugation of chloroacetanilide and thiocarbamate herbicides either by elevating the levels of reduced glutathione (glutathione *S*-transferase) or by inducing the activity of glutathione-dependent enzymes. Elevating the glutathione levels may be achieved by means such as regulating the assimilatory sulfate reduction to cysteine and activating the key enzymes in the glutathione biosynthesis pathway.

The biological action of safeners is normally determined by substitutions of nitrogen atoms.

Herbicide *S*-metolachlor

See ***S*-metolachlor herbicide**

Herbicide sodium trichloroacetate

See **Trichloroaceticacid sodium (TCA sodium) herbicide**

Herbicide spray coverage

See **Spray application concepts**

Herbicide synergists

See **Pesticide synergists**

Hexachlorobenzene insecticide

Hexachlorobenzene, C_6Cl_6 (hexachlorobenzene or perchlorobenzene), is an organochlorine insecticide and a chlorocarbon.

With a CAS Registry Number of 118-74-1, it is a white crystalline, persistent compound that is almost insoluble in water. It does not react with acids or alkalis, and exhibits very slow photolysis. It is also referred to as **HCB**.

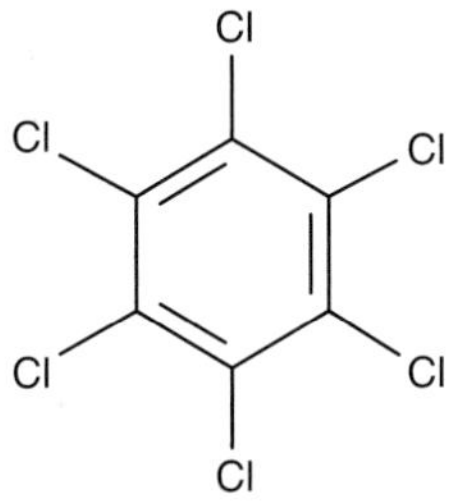

HCB

HCB is also used as a seed protectant and a selective fungicide, while fumigating spores. It is an extremely hazardous compound (Class Ia, according to the WHO) in the acute toxicity rating. The oral LD_{50} among rats is 4000 mg/kg. It is moderately to very highly toxic to aquatic organisms. It also irritates the skin and the eyes. HCB, being an organochlorine compound, is a cholinesterase inhibitor, a carcinogen, a developmental/reproductive toxic and a suspected endocrine disruptor. It is thus a PAN Bad Actor chemical. It also causes porphyria (an inherited disease of the blood haem) as tested among rats.

Symptoms of poisoning include hypersensitivity to touch and sound, dizziness, tremors and convulsions. Caution must be taken to avoid skin contact with the compound, including for example the donning of proper full-sleeved clothes and chemical-resistant goggles.

The principal metabolites of HCB in mammals include **pentachlorophenol**, tetrachlorohydroquinone, pentachlorobenzene, tetrachlorobenzene and 2,4,5- and 2,4,6-trichlorophenols. The soil half-life of HCB is around 6 years. This compound is very persistent in the environment and also finds its way into the food chain.

Hexachlorobenzene is not registered by the USA.

Hexachlorocyclohexane insecticide

Hexachlorocyclohexane, $C_6H_6Cl_6$ (1,2,3,4,5,6 hexachlorocyclohexane), is an organochlorine insecticide. With a CAS Registry Number of 58-89-9, this compound is soluble in water, acetone, methanol, ethanol, benzene, xylene and many other solvents. It is extremely stable to light, air, oxidation, acids and temperatures up to 180°C. It is more soluble in water than **DDT** (to which it is similar). It is also referred to as **HCH**.

HCH is moderately hazardous according to the WHO classification. It is a carcinogenic compound and a suspected endocrine disruptor. It is thus a PAN Bad Actor chemical. While HCH is toxic if consumed in large quantities, it is readily absorbed through skin and can cause acute toxicity via the dermal route.

The most toxic isomer of HCH, called **gamma-HCH**, has insecticidal properties. It is a product of the catalytic chlorination of benzene. Gamma-HCH (also called lindane) is a white crystalline solid that is insoluble in water. Lindane is used to control soil-inhabiting insects, stored grain pests and also as a seed treatment agent. It is also used against animal ectoparasites and as a rodenticide. Insect groups controlled include members of the Aphididae, Muscidae, Lepidoptera and Coleoptera. The CAS Registry Number of lindane is 58-89-9 and the molecular formula is $C_6H_6Cl_6$.

The EPA and WHO both classify lindane as moderately toxic, as per the acute toxicity rating. The oral LD_{50} in rats is 88 mg/kg while the dermal LD_{50} is 1000 mg/kg. It is found to harm its handlers, where symptoms of nervous system disorders are common (dizziness, hypersensitivity to sound and touch, nausea, tremors). Lindane is also a carcinogen and a suspected endocrine disruptor. It is a PAN Bad Actor chemical.

Since lindane is very stable to the action of heat, light and oxidation, it persists in the environment. It is not registered in many countries for example the USA. It is banned for use and manufacture in India. HCH is allowed as a restricted compound in Australia and many African countries like Mali, Burkina Faso, South Africa and Tanzania.

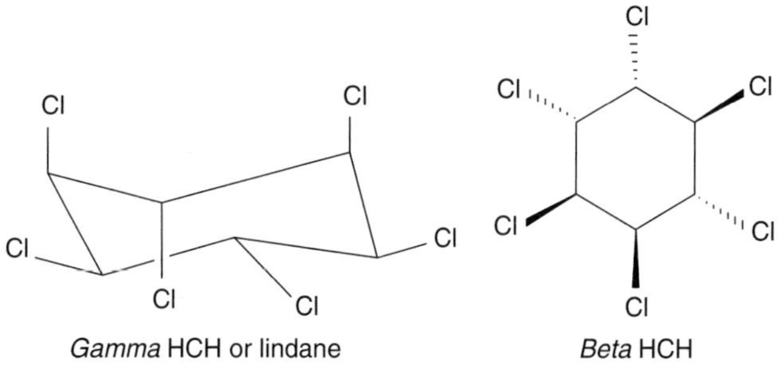

Gamma HCH and beta HCH

Hexaconazole fungicide

Hexaconazole, $C_{14}H_{17}Cl_2N_3O$ ((*RS*)-2-(2,4-dichlorophenyl)-1-(1*H*-1, 2,-triazol-1-yl) hexan-2-ol), is a triazole fungicide. It is a white crystalline solid, soluble in water (0.017 g/l), dichloromethane (336 g/l), methanol (246 g/l) and toluene (59 g/l). The compound is stable to hydrolysis, to photolysis in water and to normal temperatures. Hexaconazole is produced by the reaction of 1-(2,4-dichlorophenyl)pentan-2-one and trimethyloxosulfonium iodide, followed by opening of the resulting epoxide with triazole in the presence of a base. The CAS Registry Number of hexaconazole is 79983-71-4.

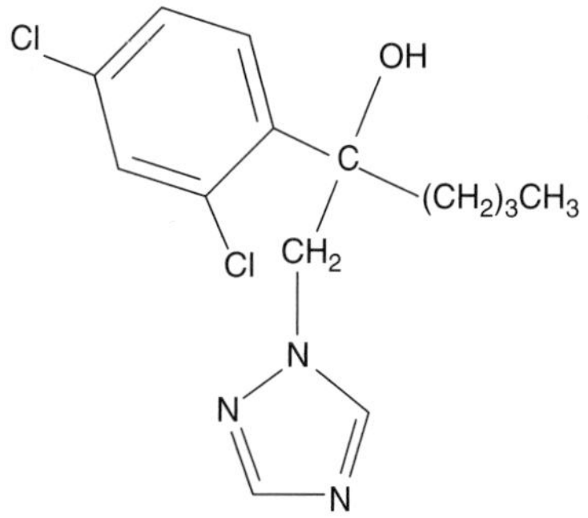

Hexaconazole

Hexaconazole is a broad-spectrum, systemic compound with protective and curative action. Many infections such as **powdery mildew**, blight, **scab** and rust affecting crops like apples, grapes, groundnuts, bananas, rice, cucurbits, tea and coffee are controlled by hexaconazole. Ascomycetes, basidiomycetes and deuteromycetes such as *Podosphaera leucotricha* and *Venturia inaequalis*, *Exobasidium vexans*, *Hemileia vastatrix* and *Cercospora* spp. are some of the pathogens controlled by this compound. The action of the compound inhibits ergosterol biosynthesis.

The fungicide is formulated as an oil-miscible liquid (OL) and a water-soluble granule (SG), with common doses being 15–250 g/ha. Anvil, Force and Silicon are some trade names.

Hexaconazole is classified by the WHO as a Class U toxin (unlikely to be hazardous) and by the EPA as a Class IV toxin. The acute oral LD_{50} in rats ranges from 2189 to 6071 mg/kg and in mallard ducks is >4000 mg/kg. It may pose a slight hazard to the eyes and skin, if not applied with proper care and precaution. The NOEL (2 year period) in rats and mice is 10 mg/kg diet and 40 mg/kg diet, respectively. The ADI is 0.005 mg/kg b.w./day.

Hexaconazole is extensively metabolized via multiple reactions, to yield hexaconazole acid as one of the degradation products. It is rapidly degraded in soils.

The chemical is registered in India, South Africa, Tanzania, Australia, New Zealand, Canada and many other countries, but not in the USA The MRL for apples and pears in Australia is 0.1 mg/kg in each case and in Japan is 0.5 mg/kg in each case.

Hexaflumuron insecticide

Hexaflumuron, $C_{16}H_8Cl_2F_6N_2O_2$ (1-[3,5-dichloro-4(1,1,2,2-tetrafluroethoxy) phenyl]-3-(2,6-difluorobenzoyl) urea), is an insecticide and acaricide. It is a colourless solid which is slightly soluble in water and moderately soluble in methanol and xylene. Its CAS Registry Number is 86479-06-3.

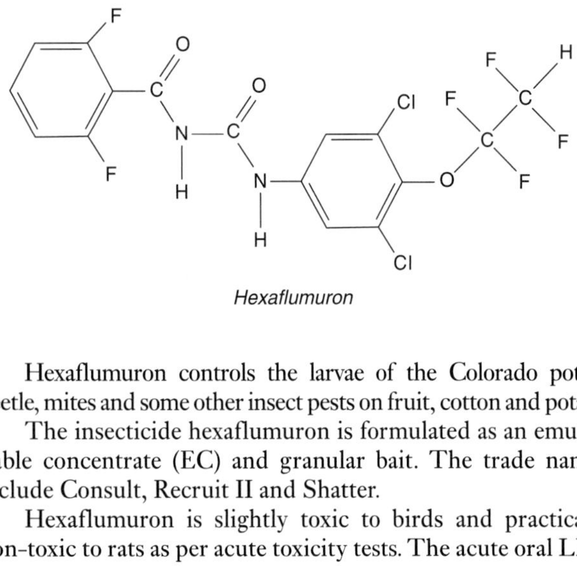

Hexaflumuron

Hexaflumuron controls the larvae of the Colorado potato beetle, mites and some other insect pests on fruit, cotton and potato.

The insecticide hexaflumuron is formulated as an emulsifiable concentrate (EC) and granular bait. The trade names include Consult, Recruit II and Shatter.

Hexaflumuron is slightly toxic to birds and practically non-toxic to rats as per acute toxicity tests. The acute oral LD_{50} in rats is >5000 mg/kg. It is rated as a Class U toxin by the WHO (unlikely to be hazardous). But the insecticide is highly toxic to fish, other aquatic creatures and bees with the 96 h LC_{50} in rainbow trout being <0.5 mg/l and the 48 h LD_{50} in bees being <0.1 mg per bee (oral and contact). Its soil half-life is 100–280 days depending on the soil type.

The insecticide is registered in the USA, Australia, Vietnam and some other countries.

Hexamethylphosphoramide chemosterilant

Hexamethylphosphoramide, $C_6H_{18}N_3OP$ (*N*-[bis(dimethylamino)-phosphoryl]-N-methylmethanamine), is a **chemosterilant**, which is a phosphoramide (i.e. an amide of phosphoric acid). Commonly referred to as **HMPA**, this is a colourless liquid that is miscible in water and many organic solvents. Besides being a chemosterilant, HMPA is an ingredient of industrial processes, being used as a solvent for polymers and organometallic compounds, and an additive in organic synthesis. It has a CAS Registry Number of 680-31-9.

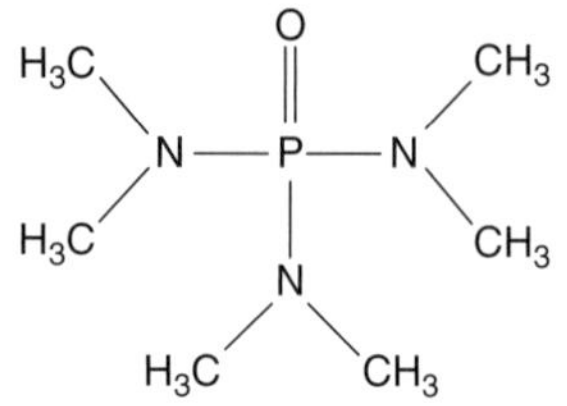

HMPA

HMPA causes sterility in male houseflies and both sexes of the American cockroach (*Periplanata americana*).

The acute oral LD_{50} among rats is 2525 mg/kg. The compound irritates the skin and the eyes. Since it is a carcinogen, it is classified as a PAN Bad Actor chemical.

When HMPA comes in contact with hydrochloric acid, it degrades to less toxic compounds.

HMPA is not registered for use in the USA.

Hexazinone herbicide

Hexazinone, $C_{12}H_{20}N_4O_2$ (3-cyclohexyl-6-dimethylamino-1-methyl-1,3,5-triazine-2,4(1*H*,3*H*)-dione), is a triazinone herbicide. It is a colourless crystalline substance, soluble in water (>30,000 mg/l), acetone 70.9 g/100 g and hexane 0.3 g/100 g. It is stable in aqueous media between pH 5 and pH 9 and below 37°C. It is prone to photodegradation and biodegradation. It is synthesized by the reaction of cyanoamine with methyl chloroformate, followed by methylation and treatment with dimethylamine and cyclohexylisocyanate. The CAS Registry Number is 51235-04-2.

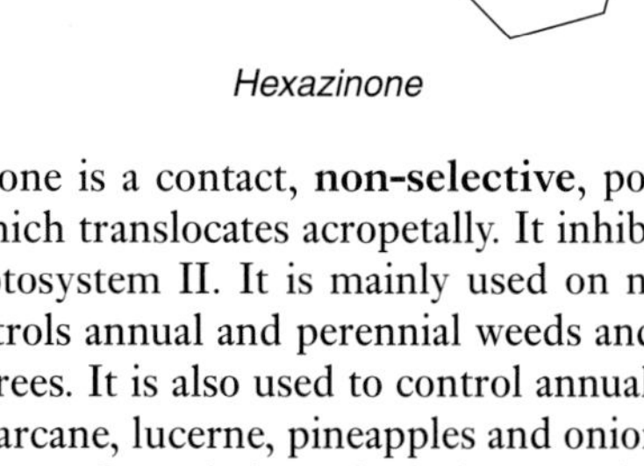

Hexazinone

Hexazinone is a contact, **non-selective**, post-emergence herbicide, which translocates acropetally. It inhibits photosynthesis at photosystem II. It is mainly used on non-crop sites where it controls annual and perennial weeds and broadleaved bushes and trees. It is also used to control annual and biennial weeds in sugarcane, lucerne, pineapples and onions.

Hexazinone formulations include granules (GR) and water-soluble powders (SP), while Velpar is a common trade name. Hexazinone can be combined with herbicides like diuron and **bromacil**.

According to the WHO and EPA, hexazinone is a slightly hazardous Class III toxin. However, since this compound is a groundwater contaminant, it is classified as a PAN Bad Actor chemical. It is necessary for labels to display the signal word 'CAUTION', and for instructions to be strictly followed.

Plants and animals degrade hexazinone rapidly. Microbial action and photodegradation are the two routes of breakdown. In a cold and dry climate and in soil with a high organic content, this compound does not degrade quickly. The DT_{50} of hexazinone in soil is 1–6 months, depending on climate and soil type. The final degradation product is carbon dioxide, although others include monomethyl hexazinone.

Hexazinone is registered for use in a large number of countries including India, the USA, Canada, Australia and South Africa. Japan Food Chemical Research Foundation has fixed 0.1 ppm as the MRL for sugarcane and 0.8 ppm for pineapples.

High performance liquid chromatography (HPLC)

High performance liquid chromatography (HPLC), popularly known as high pressure liquid chromatography, is chromatography that separates the components of a mixture. The basic components of HPLC are the stationary and mobile phases, a pressure pump and a detector (Fig. H.1).

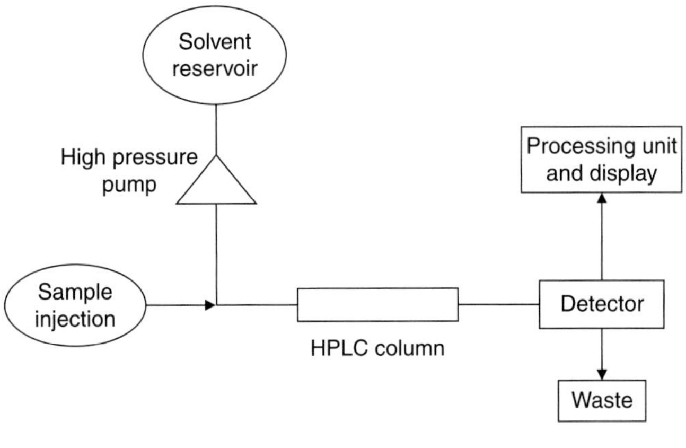

Fig. H.1. Flow diagram of high performance liquid chromatography (HPLC).

The stationary phase constitutes fine particles of sorbent material that is packed densely in the columns of the HPLC. The mixture is dissolved in a suitable solvent and forced through the columns by the pump along with the mobile phase. In the column, the constituents of the mixture are separated and detected with a suitable detector.

HPLCs are of two types depending on the relative polarity of the solvent and stationary phase. The two types are: (i) normal phase HPLC; and (ii) reversed phase HPLC. The normal phase is no longer used as much as the reversed phase. In the normal phase, the column is filled with a stationary phase of small silica particles, while the solvent is hexane and iso-octane. The column is up to 200 mm long, with an internal diameter of 4.6 mm or less. The least polar analyte elutes first, then moderately polar and finally highly polar. Polar compounds in the mixture, on passing through the column, stick longer to the polar slice than non-polar compounds do. Hence, the non-polar compounds will pass through the column more quickly. By decreasing the polarity of the mobile phase, analyte retention can be increased. The limitations of normal phase HPLC are: (i) irreversible retention; (ii) long column regeneration; and (iii) deactivation of silica by water.

In reversed phase HPLC, the stationary phase is non-polar. In this case, the column size is the same, but the silica is modified to make it non-polar by attaching long hydrocarbon chains (of eight or 18 carbon atoms) to the surface. The solvent is polar in nature (e.g. a mixture of water and alcohol). As the solvent and the sample move through the column, polar molecules mostly move with the solvent, while non-polar compounds in the mixture will tend to be attracted to the hydrocarbon groups because of van der Waals' dispersion forces. They will also be less soluble in the solvent. This means that it is the polar molecules that will travel through the column more quickly. Increasing the polarity of the mobile phase increases analytic retention.

HPLC separations are achieved using five basic operational modes, namely: (i) liquid-solid chromatography; (ii) liquid-liquid chromatography; (iii) bonded phase chromatography – ion suppression and ion pair; (iv) low exchange chromatography – strong base anion exchange and strong acid anion exchange; and (v) size exclusion chromatography – gel permeation chromatography and gel filtration chromatography.

The solvent reservoir capacity varies from a simple 1 l flask to a very sophisticated solvent storage system that ensures uniform compositions with the help of filters and magnetic stirrers. Degassing is necessary in gradient elution where concentrations of the dissolved air in the two solvents differ and when one solvent is water.

The HPLC pump is designed to deliver a precise, pulse-free flow of the mobile phase and the solvent, through the column at high pressure (500–5000 psi). The pump also needs to be resistant to the chemical action of the mobile phases. The sample is introduced into the chromatography system via a typical sample injector, a six-port injection valve which has two portions, the load and the inject portion. The injection valve is the most widely used injection system for sample introduction.

The success of an analysis depends on the column and the operating conditions. The column material, its dimensions and the packaging material are of importance to performance. Column packing is made up of polystyrene and silica-based ion exchange materials which include octadecylsilyl-bonded (ODS) silica, phenyl-bonded silica (PBS) and cynopropyl-bonded silica. Silica-based material is structurally strong, suitable for a variety of solvents and can withstand higher temperature to improve exchange. The limiting factor is the pH range of 2–7.5.

Any change in column effluent is converted into an electrical signal by a detector, of which the HPLC has many, including the refractive index detector (RI), fixed wavelength UV detector, variable wavelength UV detector and photoconductivity detector.

The time taken for a particular compound to travel through the column to the detector is known as the retention time. Different compounds have different retention times. For a particular compound, the retention time will vary depending on:

(i) the pressure used; (ii) the nature and the composition of the two phases; and (iii) the temperature of the column.

Highly Hazardous Pesticides of PAN (HHPs)
See **PAN International**

High-pressure sprayers
See **Application equipment – larger power-driven sprayers**

Hilling operation
See **Weed management by mechanical methods**

HMPA chemosterilant
HMPA is the synonym for **hexamethylphosphoramide**, which is a **chemosterilant**.

Hoe
A hoe is a long-handled tool with a thin horizontal metal blade (30 cm × 24 cm), used for weeding and cultivating soil between rows of crops with a view to controlling weeds.

A hoe blade can be modified to suit farm activity. A large blade suits digging and a small one suits weeding. Today, hoes are attached to tractors and moved hydraulically.

These implements can be innovatively designed to suit the land in question and the convenience of the farmer. Gardeners use hand hoes (Fig. H.2) made of a thin metal blade attached to a long handle for loosening the soil and chopping and removing weeds.

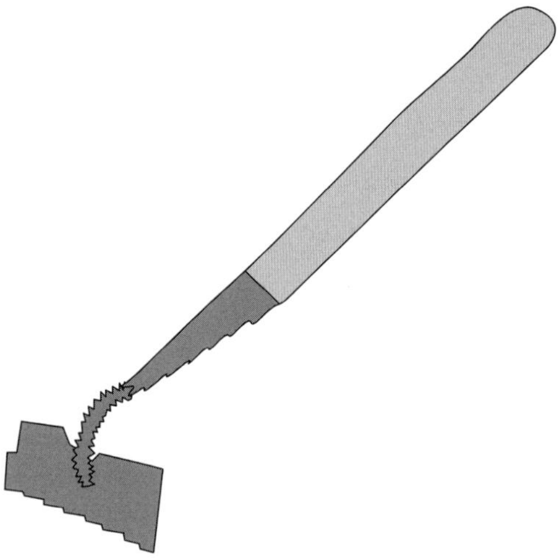

Fig. H.2. A simple hoe. Courtesy of Leena Achari, with permission.

Hoeing
See **Weed management by mechanical methods**

Honey bees
Honey bees are the species of bees that belong to the genus *Apis* in the insect order Hymenoptera (Fig. H.3). Honey bees produce and store honey, and are important pollinators of flowering plants. These creatures deserve a mention among these pages for two main reasons. The first is they play a critical role in the ecosystem of regions, aiding crop production and contributing to the economic progress of a region. The second is that the use of pesticides can cause an imbalance in the life cycle of honey bees. Most toxicity studies of pesticides take cognizance of the effect of a chemical on honey bees.

Fig. H.3. Honey bee. Sketch courtesy of Leena Achari.

Honey bees build hives out of wax and hundreds of bees live in the hive together. They feed on the nectar and pollen of thousands of flowers, and in the process, pollinate plants. The honey they produce is the result of regurgitation of the consumed nectar. The distinctive flavour of different types of honey is a result of the diversity in the nectar that the colony feeds on. Honey has both culinary and medicinal uses and honey production is the reason why many bee keepers keep bees.

Out of the seven recognized species of honey bees (e.g. the dwarf and the giant honey bees), only two are exploited commercially, namely, the eastern honey bee (*Apis cerana*) and the western or European honey bee (*Apis mellifera*). The eastern honey bee is found in South and South-east Asia. It is reared on a small scale for honey production. The western honey bee is thought to have originated in eastern Africa. From there, this honey bee has moved to North America, Europe, Asia and other parts of Africa. This species is used extensively for pollination of various crops. Commercially, pollination has now become the major use of honey bees, rather than production of honey.

In recent times, there has been a large decline in honey bee populations all over the world. Although there are several reasons for this decline, pesticide toxicity is proposed as a major cause, resulting in the disappearance of entire colonies.

Honeydew
See **Pest – aphid**

Hopper burn
See **Pest – plant hopper**

Hormone
Hormones (also called chemical messengers) are chemical molecules, which at very low concentrations (e.g. 10 mol/l) influence growth, cell differentiation, cell division and gene expression in all living beings. The term has its origins in Greek which means 'to set in motion'. Hormones regulate physiological processes such as metabolism, growth, reproduction, **moulting**, pigmentation and many others.

Plant hormones are variously called plant growth regulators or phytohormones. Excess secretion is usually phytotoxic, leading to some herbicides being synthesized based on these chemical substances.

Plant hormones, required at specific times during plant growth, move easily through plant tissues, towards sites of active growth. Plant hormones are classified into five major classes based on their structural similarities and their effects on plant physiology. They are **abscisic acid**, **auxins**, **cytokinins**, **ethylene** and **gibberellins**. There are other plant hormone classes called, for example, brassinolides and jasmonates.

In animals, hormones are produced in glandular systems (of which there are two types, the endocrine system and the exocrine system) which secrete the fluid, thus helping the regulation of biochemical processes in performing routine and critical functions. Examples of glands are thyroid, adrenal, pituitary and the ovaries. Amines, peptides, neurohormones, sex hormones and other steroids are some types of hormones in the animal body.

Horticultural oils

Oils extracted from natural sources or from the oil refining industry, aimed at providing pest control, are called horticultural oils (**pesticidal oils**).

Previously, petroleum-based products, extracted from crude oil, consisted of traces of nitrogen and sulfur-linked compounds. But, since these products lead to phytotoxicity, a new class of highly refined, sulfur-free, light, low viscosity oils was developed and these are now prevalent in many pest management situations. Oils from natural sources include vegetable oil, mint oil, castor oil and those extracted from maize, safflower and **neem**.

Horticultural oils are sprayed to control some types of insects, fungi, mites and weeds. These oils kill or inhibit all stages of the pest life cycle, with the mode of action being to penetrate shells and skin, while interfering with metabolic processes or hindering respiration. When applied to weeds, the oil causes water stress since the stomata and lenticels become clogged. Targeted insects include aphids, beetles, caterpillars, leaf miners, **mealybugs**, mites, scales and whiteflies. When large tracks of land have to be cleared of weeds, crude oil with a high content of aromatic, unsaturated sulfonated compounds is used.

These pesticidal oils are applied like any other pesticides, with common formulations being emulsifiable concentrates (EC). These oils can be mixed with soap-water for ease of application.

Pesticidal oils do not persist in the environment, and hence are considered safe for use. These compounds may cause irritation to the skin and eye and so handlers are advised to be armed with PPE. In warm weather, the spray application must be restricted to the early morning or late evening hours. Crops must also be irrigated before the application can begin.

There are various types of oils used in different plant situations. For instance, dormant oil is sprayed during the autumn (fall) on the bark of leafless trees infested with insects and mites. Dormant oils are heavier and contain sulfur compounds and hence are phytotoxic. Then, there are summer oils which are normally applied to the foliage of plants.

Many types of horticultural oils are registered in many countries. They are permitted under organic farming.

Hose-end sprayer

See **Application equipment – hand sprayers**

Hydrogen cyanide rodenticide

Hydrogen cyanide, HCN (formonitrile), is a rodenticide and also an insecticide. Also called hydrocyanic acid or prussic acid, HCN is a colourless, flammable, weakly acidic liquid that is completely miscible with water and soluble in ethanol and diethyl ether. Its CAS Registry Number is 74-90-8. HCN also exists as salts, formed out of sodium, potassium and calcium cyanide.

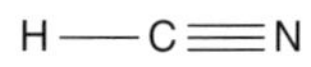

Hydrogen cyanide

Hydrogen cyanide, with a respiratory action, acts on rodents and insects by preventing oxygen transfer from haemoglobin into tissue cells. It serves as a good fumigant in warehouses and grain stores. HCN acts by conjugation with ferricytochrome oxidase.

The formulations of hydrogen cyanide are as a gas or gas-generating product.

Hydrogen cyanide is an extremely toxic pesticide and is accordingly classified as a Category I toxin by the EPA. The acute oral LD_{50} in rats is 10–15 mg/kg. It irritates the eyes and the skin. Hydrogen cyanide is toxic to mammals, birds, fish and bees. Exposure to 0.35 mg/l air for 30 min can be fatal to humans. In animals, the cyanide ion is converted to thiocyanate. HCN is phytotoxic and damages moist leaves of plants.

HCN is generated during the production of calcium cyanide (C_2CaN_2) and sodium cyanide (CAS Registry Number of 143-33-9 and a molecular formula of NaCN). These two are also prominent insecticides and rodenticides. Sodium cyanide is very soluble in water (58 g/l) and slightly soluble in ethanol. This salt hydrolyses readily, releasing hydrogen cyanide.

Calcium cyanide (with a CAS Registry Number 592-01-8 and a molecular formula C_2CaN_2) produces toxic gases on contact with water or damp air. Calcium and sodium cyanide are used to fumigate rabbit burrows and rat runs. The acute oral LD_{50} of sodium cyanide in rats is 6.44 mg/kg. India has banned the use, manufacture and export of calcium cyanide.

Hydrogen phosphide

See **Fumigants**

Hydrophytes

See **Aquatic weeds**

Hydroprene as an IGR

Hydroprene, $C_{17}H_{30}O_2$ (ethyl (*E*,*E*)-(*RS*)-3,7,11-trimethyldodeca-2,4-dienoate), is an insect growth regulator, commonly used for non-crop situations. The (*E*,*E*)-(*S*)-isomer is more effective against insects than is the (*E*,*E*)-(*R*)-isomer. Its CAS Registry Number is 41096-46-2. Hydroprene is a yellowish-brown liquid soluble in water (2 mg/l) and common organic solvents. It is stable under normal storage conditions for more than 3 years.

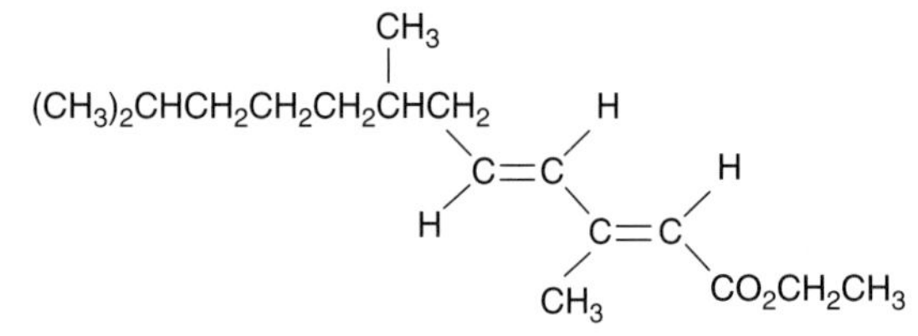

Hydroprene

Hydroprene mimics **insect juvenile hormones** and prevents **moulting** and adulthood in insects, leading to dehydration, body changes and death. This compound controls household insects like cockroaches, beetles and moths.

The IGR may be combined with many insecticides such as **cypermethrin**, **acephate** or **deltamethrin**. Formulations include **aerosols**, liquids and baits. It is available in the market under trade names like Gencor.

Hydroprene is classified as slightly toxic by the EPA. The WHO classifies this compound as a Class U toxin (unlikely to be hazardous). It may affect normal development of young aquatic organisms and bees. It is not a hazard to humans as long as label instructions are strictly followed. Labels need to display the signal words 'CAUTION' and 'WARNING'.

Hydroprene is rapidly degraded in soil with a half-life of a few days. It is degraded by plants. The IGR is registered for use in the USA.

Hydroxybenzonitrile herbicides

Hydroxybenzonitriles are a herbicidal group which inhibits photosystem II electron transport. Among the other photosystem electron transport inhibitors are **phenylureas**, ***s*-triazines**, triazinones, **uracils**, hydroxybenzonitriles, **acylanilides**, pyridazinones, ***bis*-carbamates**, benzothiadiazine and bentazon. Hydroxybenzonitriles comprise **bromoxynil**, **ioxynil** and **dichlobenil**.

Bromoxynil and ioxynil are contact herbicides which selectively target grasses, broadleaved weeds and annuals (e.g. chickweed and mayweed) without harming cereal crops such as wheat or barley. There is no soil activity. The herbicide dichlobenil is a soil active agent which is a cell-wall synthesis inhibitor.

Hydroxypyrimidine

See **Fungicidal modes of action**

Hyperaccumulator plants

Plants that absorb disproportionately large amounts of toxic metals or metalloids throughout their lives are called hyperaccumulators. This process is believed to reflect a deviation from normal and this plant attribute represents a breakdown of the homeostatic mechanism.

Hyperaccumulators exist throughout the plant kingdom, but are most common among the family Brassiceae. These plants accumulate metals such as nickel, copper, cobalt, lead, zinc and **manganese**. The concentration levels of these metals are found to be around 1000 mg metal/g dry weight in the leaves for nickel, copper, cobalt and lead and 100 mg cadmium/g dry weight. However, plants can absorb such metals only as free ions or as soil soluble complexes.

Phyto-mining with hyperaccumulators and indicator plants can help to restore soils with heavy metal content and pesticidal compounds, without employing traditional decontamination techniques. Nitrogen compounds, chlorinated solvents, BTEX compounds (benzene, toluene, ethylbenzene and xylenes) are some examples of compounds that can be removed by hyperaccumulators.

It has to be noted that in its natural course, phytoremediation can take many, many years to accomplish what conventional methods (i.e. man-made methods) can do in a much shorter duration.

Hyperplasia

See **Common viral plant diseases**

Hypoplasia

See **Common viral plant diseases**

I

IAA as a bioPGR

See **Indole-3-acetic acid (IAA or AIA) as a bioPGR**

IGR chlorfluazuron

See **Chlorfluazuron insecticide**

IGR cyromazine

See **Cyromazine insecticide**

IGR fluazuron

See **Fluazuron acaricide**

IGR pyriproxyfen

See **Pyriproxyfen as an IGR**

Imazamethabenz acid

See **Imazamethabenz-methyl herbicide**

Imazamethabenz-methyl herbicide

Imazamethabenz-methyl, $C_{16}H_{20}N_2O_3$, is a mixture of methyl (±)-6-(4-isopropyl-4-methyl-5-oxo-2-imidazolin-2-yl)-*m*-toluate and methyl (±)-2-(4-isopropyl-4-methyl-5-oxo-2-imidazolin-2yl)-*p*-toluate. An **imidazolinone** herbicide, imazamethabenz-methyl is a whitish crystalline solid, soluble in water, methylene chloride and dimethyl sulfoxide. It rapidly hydrolyses at higher pH levels. It is produced by the reaction of 3,4-(4,5)-dicarboxymethyltoluene with 2-aminoamide-3-methylbutane in the presence of two equivalents of potassium *tert*-butoxide. The CAS Registry Number of imazamethabenz-methyl is 81405-85-8 (see Imazamethabenz-methyl at bottom of the page).

Imazamethabenz-methyl is a pre-emergence pro-herbicide (**pro-pesticide**), which has to be converted, in the target weed, to imazamethabenz acid ($C_{15}H_{18}N_2O_3$) to be an active compound. The mode of action is the inhibition of the enzyme acetolactate synthase (ALS), critical in the synthesis of branched-chain amino acids. It is effective on weeds in wheat, rye and barley crops. Imazamethabenz-methyl controls a range of weeds including loose bluegrass, meadow grass, silkybent (*Apera spica-venti*) and oat grass (*Arrhenatherum elatius*).

Low doses of this herbicide are usually sufficient to control weeds. For wheat, 0.35–0.53 kg a.i./ha are sufficient and for barley, between 0.2 and 0.43 kg a.i./ha. The herbicide may be mixed with a surfactant for better effect. Formulations include liquid concentrates and water-soluble granules (SG).

Imazamethabenz-methyl is not a very toxic herbicide, with the acute oral LD_{50} in rats and mallard ducks being >5000 mg/kg and 2150 mg/kg, respectively. It is not very hazardous via the inhalation route. It is also not toxic to honey bees, birds and earthworms. The WHO classifies this chemical as a Class U toxin (unlikely to be hazardous). The ADI is 0.0625 mg/kg b.w./day.

In animals, this herbicide is eliminated rapidly and there is no accumulation in any major organ. The **bioconcentration factor** in fish is 1.6. The major metabolite is carboxylic acid, with imazamethabenz-methyl acid being produced too. However, imazamethabenz-methyl persists in soils with a half-life of up to 180 days. It adsorbs to sandy and loamy soils.

The herbicide is registered in the USA but is not approved throughout the EU. The generally accepted MRL of this compound is 0.01 mg/kg.

Imazamox herbicide

Imazamox, $C_{15}H_{19}N_3O_4$ ((*RS*)-2-(4-isopropyl-4 methyl-5-oxo-2-imidazolin-2-yl)-5-methoxymethyl-nicotinic acid), is an imidazolinone herbicide. It is a white odourless powder, miscible in water, and soluble in acetone, acetonitrile and dichloromethane. It remains stable for about 1 month in the pH range of 5–9. It is produced by the reaction of 2,3-dicarboxymethyl-5-methoxymethylpyridine with 2-aminoamide-3-methylbutane in the presence of two equivalents of potassium *tert*-butoxide. The CAS Registry Number is 114311-32-9.

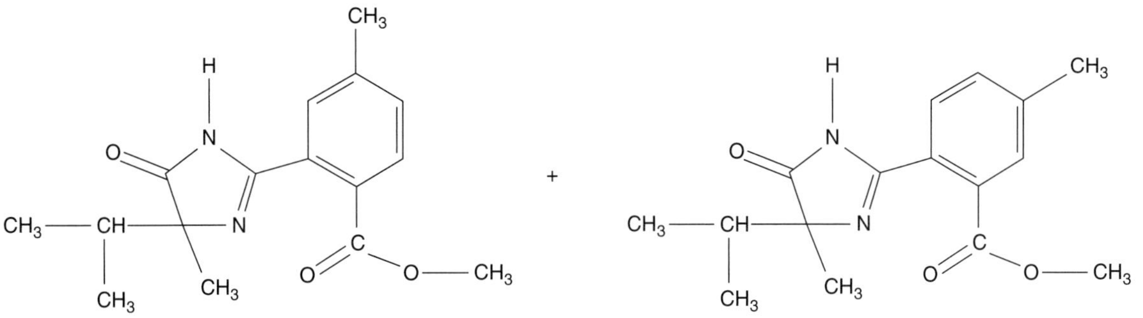

Imazamethabenz-methyl

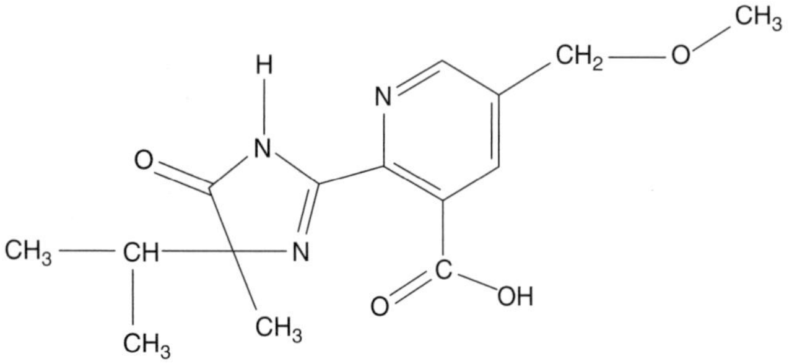

Imazamox

Imazamox is a pre-emergence herbicide, suitable for leguminous crops (e.g. soybean, beans, lucerne). It controls the growth of a wide range of weeds including velvetleaf, annual ragweed, wild mustard and wild oats. The herbicide reaches the growing tips and acts by inhibiting the biosynthesis of branched-chain aliphatic amino acids like leucine, isoleucine and valine. The herbicide is formulated as dispersible granules and applied with an oil **adjuvant**.

The herbicide is a slightly toxic chemical with practically safe oral and dermal LD_{50} values among rats, mice and rabbits. It is toxic to fish (with an acute LC_{50} in rainbow trout of >122 mg/l). The acute LD_{50} in honey bees is >25 μg per bee. It causes irritation to the skin and eyes.

Imazamox degrades most rapidly among all the imidazolinones, the main route being hydroxylation followed by conjugation to glucose. This helps reduce the pre-harvest time. In sandy and silt soils, imazamox is adsorbed with a half-life of up to 30 days. The soil metabolite is not mobile. The photodegradation half-life is 6.8 h.

Imazamox is registered in India, the USA, Australia, Canada, South Africa and a few EU nations such as the UK, France, Spain, Greece and Hungary.

Imazapic herbicide

Imazapic, $C_{14}H_{17}N_3O_3$ ((*RS*)-2-(4-isopropyl-4-methyl-5-oxo-2-imidazolin-2-yl)-5-methylnicotic acid), is a member of the **imidazolinone** class of herbicides. It is soluble in water at 2230 mg/l and in acetone at 18.9 mg/l. It degrades rapidly in water solution by the action of sunlight. It stores well in airtight containers in dark and ventilated areas. It is produced by the reaction of 2,3-dicarboxymethyl-5-methylpyridine with 2-aminoamide-3-methylbutane in the presence of two equivalents of *tert*-butoxide. The CAS Registry Number is 104098-48-8.

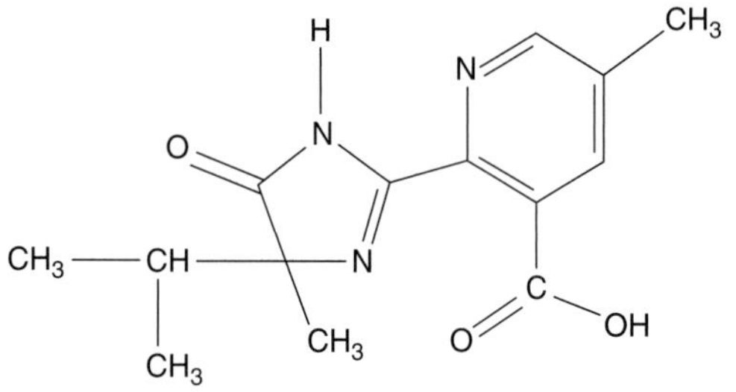

Imazapic

Imazapic, a selective, pre- and post-emergence herbicide, controls annual and perennial grasses and broadleaved weeds in groundnuts, sugarcane and canola. The mode of action is the inhibition of the action of the enzyme acetolactate synthase (ALS) (also called **acetohydroxy acid synthase** (AHAS)). This affects protein synthesis and cell growth. Some examples of weeds that can be controlled are wild oats, millets, mustards, cocklebur, **Johnson grass** and **Bermuda grass**.

The herbicide is formulated as dispersible granules and soluble concentrates (SL). Commercially, it is available as Plateau and Cadre, among others. The herbicide is applied along with **adjuvants** like vegetable oil concentrate and methylated seed oil.

The compound is not toxic by the oral route to rats. In normal use, it is slightly toxic to birds. It does not irritate the skin in normal handling. It is non-toxic to honey bees and fish. The ADI for imazapic is 0.3 mg/kg/day.

Imazapic adsorbs to acidic soils and sandy and loamy soils, with a half-life of >100 days. It has low leaching potential and seldom goes deeper than 30 cm into the soil. Upon ingestion, it is rapidly excreted by animals and does not bioaccumulate. It degrades by exposure to sunlight.

Tolerance to the compound occurs when a plant alters the route of metabolism of the chemical, thus not inhibiting the enzyme ALS.

Imazapic is registered in the USA, Australia and Vietnam, but is not allowed for use in the EU. This compound is not permitted for use in aquatic environments.

Imazapyr herbicide

Imazapyr, $C_{13}H_{15}N_3O_3$ (2-(4-isopropyl-4-methyl-5-oxo-2-imidazolin-2-Yl) nicotinic acid), is a member of the **imidazolinone** class of herbicides. It is soluble in water, acetone and dimethyl sulfoxide. It rapidly degrades in water via photolysis with a half-life of up to 5.3 days. It is produced by the reaction of 2,3-dicarboxymethylpyridine with 2-aminoamide-3-methylbutane in the presence of two equivalents of *tert*-butoxide. The CAS Registry Number is 81334-34-1.

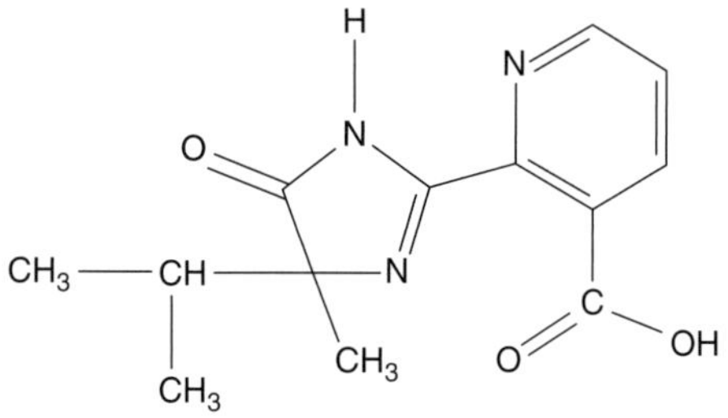

Imazapyr

A broad-spectrum compound used in non-crop situations, imazapyr affects the growing points of plants by inhibiting the enzyme acetolactate synthase, critical in the synthesis of the amino acids leucine, isoleucine and valine. Many annual and perennial grasses and broadleaved weeds, shrubs, vines and woody plants are controlled in forest areas, meadows and public places. Beech varieties, stickywilly (or cleavers) and daisy fleabanes are some examples of weeds controlled.

The common formulations for this herbicide are granules (GR) and soluble concentrates (SL). The herbicide is applied as a low-volume spray or applied on the basal bark.

Commercially, it is available under trade names like Arsenal, Chopper and Stalker.

Imazapyr is not toxic either orally or dermally, to rats or birds (with an oral LD_{50} and an 8 day dietary LD_{50} in rats and mallard ducks of >5000 mg). It is also not toxic to fish or to honey bees. The RfD for imazapyr is 2.50 mg/kg/day. However, it has caused malignancy during laboratory testing and is also feared to contaminate groundwater. It is a persistent chemical, with the half-life in dry soils being 149 days. Hence, it is a PAN Bad Actor chemical.

Imazapyr metabolizes into immobile imidazolyrrolo-pyridine. Upon ingestion by mammals, it is rapidly eliminated via the urine, with almost no **bioaccumulation**.

The herbicide is registered for use in the USA, Canada, Australia and South Africa.

Imazaquin herbicide

Imazaquin, $C_{17}H_{17}N_3O_3$ ((*RS*)-2-(4-isopropyl-4-methyl-5-oxo-2-imidazolin-2-yl) quinoline-3-carboxylic acid), is a herbicide that belongs to the **imidazolinone** family of herbicides. It exists as a liquid at room temperature, soluble in water at 0.06 g/100 ml and in dimethyl sulfoxide, at 15.9 g/100 ml. It decomposes between 219°C and 224°C, and also under intense sunlight in aqueous solution (with a half-life in water of 15 h). Imazaquin is produced by the reaction of 2,3-dicarboxymethylquiniline with 2-aminoamide-3-methylbutane in the presence of two equivalents of *tert*-butoxide. Its CAS Registry Number is 81335-37-7.

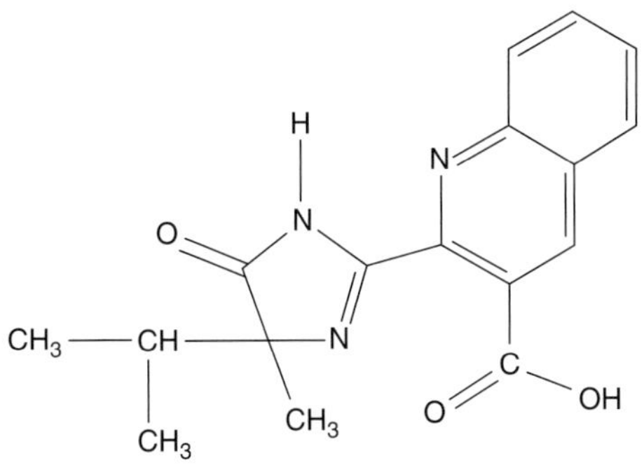

Imazaquin

Imazaquin, a selective, pre- and post-emergence herbicide, is mainly used on soybean crops to control grasses and broad-leaved weeds like pigweeds, foxtails, nightshade, mustards, **Johnson grass**, **common lambsquarters** and many others. The mode of action is inhibition of enzyme acetolactate synthase, critical to the synthesis of branched-chain amino acids, leucine, isoleucine and valine. The compound's action causes **chlorosis** and **necrosis**, killing them within 1 month.

At the time of pre-emergent application, rainfall or irrigation is necessary to activate the chemical. At the post-emergence stage, the herbicide is to be mixed with a non-ionic surfactant before application. It is formulated as an emulsifiable concentrate (EC) and as dispersible granules. It may be applied at a rate of 100–140 g/ha. Scepter, Squadron and Partner are some known trade names.

The oral LD_{50} and acute dermal LD_{50} in rats and rabbits is >5000 and 2000 mg/kg, respectively. It is also not toxic to fish. It is rated as a general use pesticide (GUP) according to the EPA. The herbicide does not leach much, but gets adsorbed on silt, acidic and sandy soils with a half-life of up to 90 days. Microbial decomposition is also known to happen, with the chemical decarboxylating into CO_2 and many other metabolites.

Imazaquin is registered in the USA and EU countries such as the UK, Belgium, France and Hungary.

Imazethapyr herbicide

Imazethapyr, $C_{15}H_{19}N_3O_3$ ((*RS*)-5-ethyl-2-(4-isopropyl-4-methyl-5-oxo-2-imidazolin-2-yl) nicotinic acid), is a herbicide that belongs to the **imidazolinone** family of herbicides. It is a colourless, crystalline substance, soluble in water to the tune of 0.14 g/100 ml. Its solubility in dimethyl sulfoxide is 42.25 g/100 ml and in methylene chloride, 18.48 g/100 ml. It decomposes at 180°C and degrades on photolysis with a DT_{50} of 3 days. It is produced by the reaction of 2,3-dicarboxymethyl-5-ethylpyridine with 2-aminoamide-3-methylbutane in the presence of two equivalents of *tert*-butoxide. The CAS Registry Number is 81335-77-5.

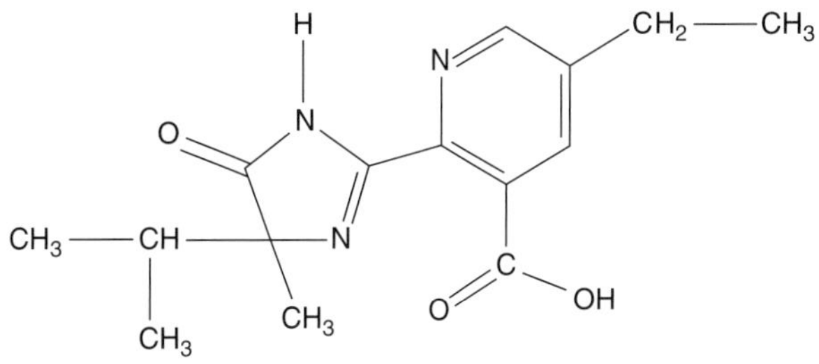

Imazethapyr

Imazethapyr, a broad-spectrum, pre-plant, pre-emergence/early post-emergence herbicide, is a compound suitable for soybeans and other leguminous crops. It controls a wide variety of weeds like pigweeds, **common lambsquarters**, morning glory, crabgrass, **Johnson grass** and many more. It acts by inhibiting the enzyme **acetolactate synthase** (ALS), causing **chlorosis** and necrosis and finally death of the plant.

The herbicide imazethapyr is applied to the foliage or incorporated into the soil and is formulated as a soluble concentrate (SL). A few examples of trade names include Hammer, Overtop, Pursuit and Resolve. Soybeans, groundnuts, beans, peas, lucerne and imidazolinone-resistant/tolerant maize are tolerant to imazethapyr. The chemical is applied at rates from 40 g a.i./ha to 140 g a.i./ha.

Imazethapyr is classified as a general use pesticide (GUP) according to the EPA and is slightly toxic to non-toxic to mammals, birds and fish. It is safe on normal use via oral, dermal and inhalation routes. It has low **bioaccumulation** in fish. The oral LD_{50} in rats, mice and rabbits is >5000 mg/kg b.w., while in bobwhite quails and mallard ducks, it is >2150 mg/kg. The acute inhalation LC_{50} of imazethapyr in rats is 3.27 mg/l.

In tolerant crops, imazethapyr is inactivated because it is metabolized into a hydroxyl-ethyl metabolite which is conjugated to glucose. In animals, α-hydroxyethyl-imazethapyr is formed, which is excreted via the urine. In soils, the herbicide has a low leaching potential and it is adsorbed on sandy loam, silt loam and acidic soils. The half-life is up to 90 days.

Imazethapyr is registered in India, Australia, Canada, Tanzania and the USA, but not in the EU.

Imazosulfuron herbicide

Imazosulfuron, $C_{14}H_{13}ClN_6O_5S$ (1-(2-chloroimidazo [1,2-*a*] pyridin-3-yl-sulfonyl)-3-(4,6-dimethoxypyrimidin-2-yl) urea), is a pyrimidinylsulfonylurea herbicide. The oxidative degradation DT_{50} is >1 h. It is produced by the reaction of 2-chloroimidazo [1,2-a]pyridine sulfonylisocyanate with 2-amino-4,6-dimethoxypyrimidine. The CAS Registry Number is 122548-33-8.

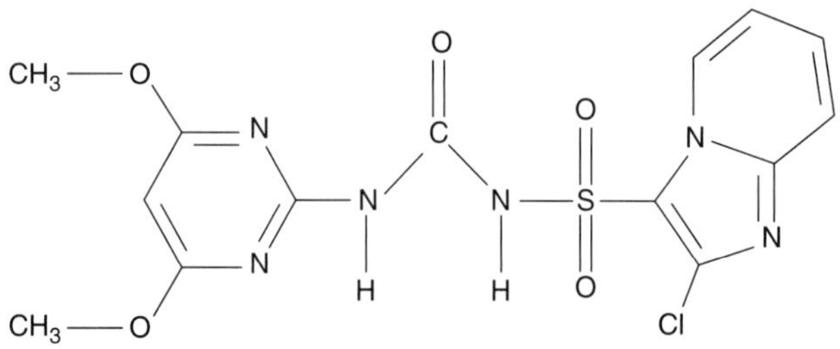

Imazosulfuron

The insecticide, a predominantly pre-emergence herbicide, inhibits **acetolactate synthase** (ALS), an enzyme crucial to the production of branched-chain amino acids. Imazosulfuron is used for a range of weeds and sedges (especially **yellow nutsedge**) in rice, potato and turf.

Imazosulfuron is available in granular (GR) and suspension concentrate (SC) forms. It may be combined with many insecticides, fungicides and fertilizers, especially ***s*-metolachlor** and **rimsulfuron**. The application rate of this chemical is 50–2000 g a.i./ha, the higher dosage being suitable for rice fields. Applications post-emergence are also carried out, depending on the severity of the problem.

The insecticide imazosulfuron is slightly toxic (the acute oral LD_{50} among rats >5000 mg/kg b.w. and the acute percutaneous LD_{50} >2000 mg/kg b.w.). It is practically non-toxic to birds and honey bees. Irritation is not caused to the skin, except in cases of extreme exposure.

The compound produces metabolites like HMS (which is 1-(2-chloroimidazo [1,2-*a*]pyridin-3-yl-sulfonyl)-3-(4-hydroxy-6-methoxy-pyrimidin-2-yl)urea). In plants, the half-life is 1.6 days, while in soils and water it can be as high as 75 days, the higher range being due to aerobic conditions.

Imazosulfuron is registered by the USA, the Philippines, Vietnam and a few EU members like Germany and Italy. The MRL established for fruits and vegetables by the UK (Office of Public Sector Information) is 0.01 mg/kg.

Imidacloprid insecticide

Imidacloprid, $C_9H_{10}ClN_5O_2$ ((*E*)-1-(6-chloro-3-pyridylmethyl)-*N*-nitroimidazolidin-2-ylideneamine), is a chlorinated analogue of **nicotine** and belongs to the chloronicotinyl group of neonicotinoid insecticides. It is soluble in water (0.5 mg/l) and in dichloromethane (50–100 mg/l). Neonicotinoid insecticides are a new class of insecticides and include **nitenpyram, acetamiprid, thiacloprid, thiamethoxam, clothianidin** and dinotefuran. Imidacloprid is produced by the reaction of 2-chloro-5-chloromethylpyridin with ethylene diamine followed by cyclization with nitroguanidin. Its CAS Registry Number is 138261-41-3.

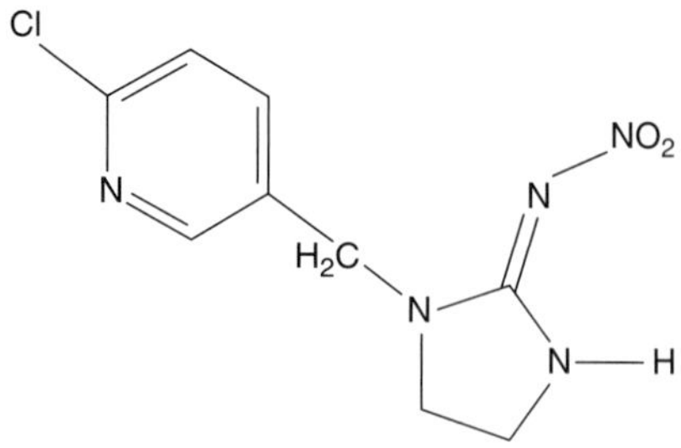

Imidacloprid

Imidacloprid, with a unique structure and new modes of action, fulfilled the need for an alternative to organophosphates, carbamates and pyrethroids. It is highly effective in the control of hemipteran pests (bugs, aphids, leafhoppers, plant hoppers and whiteflies), while it is not so effective on nematodes and mites.

As a systemic neonicotinoid, imidacloprid binds to the **nicotinic acetylcholine receptor (nAChR)** of sucking, hemipteran pests (bugs, **aphids, leafhoppers, plant hoppers** and **whiteflies**) leading to the accumulation of the acetylcholine neurotransmitter. It blocks the nicotinergic neuronal pathway and thus interferes with stimulus transmission. It is absorbed from the roots and moves upwards and has residual activity. This attribute lends it the anti-feeding effect on insects. The crops on which imidacloprid is effective are potatoes, lettuce, cotton and cereal crops like rice and maize. A large number of viral pathogens are also controlled, since the pests that imidacloprid controls are **vectors** of viruses.

The insecticide imidacloprid is applied to leaves and soil and is used for treating seeds. It is available under trade names like Merit and Admire. Formulations include dustable powder (DP), granules (GR), wettable powder (WP), flowable slurry concentrate (for seed dressing) and soluble concentrate (SL). It is compatible with fungicides during seed treatment. Its systemic properties make it effective for seed treatment and as a soil insecticide.

The WHO has categorized this insecticide as a moderately hazardous compound (Class II toxin). The acute oral LD_{50} among male and female rats is 424–475 mg/kg, while among mice, it is 131–168 mg/kg. The dermal LD_{50} in rats is >5000 mg/kg. It is not irritating to the skin and eyes, under normal usage.

The photolytic half-life of imidacloprid is 57 min. After oral administration of [$_{14}$C]imidacloprid in rats, most of the compound was rapidly absorbed and eliminated via urine and faeces. Metabolic reactions yield 6-chloronicotinic acid, 6-hydroxynicotinic acid and derivatives.

Imidacloprid is registered for use in many countries like India, Australia, the USA, Canada and all the countries of the EU.

Imidazolinone herbicides

The imidazolinone family of herbicides was aimed at soybean protection. Imidazolinones inhibit the biosynthesis of branched-chain amino acids like leucine, isoleucine and valine, by preventing the catalytic action of enzyme **acetolactate synthase (ALS)** (also called acetohydroxy acid synthase (AHAS)). Thus, death occurs due to non-availability of amino acids for protein synthesis. Growing tips are affected. The other branched-chain amino acid inhibitors include **sulfonylureas** and **triazolopyrimidines**.

Imidazolinones are a rare group of herbicides that have high specific activity, requiring low dosage per unit area. The members of this family include **imazamethabenz-methyl**, **imazamox**, **imazapic**, **imazapyr**, **imazaquin** and **imazethapyr**.

Imidazolinones readily move in plants (both apoplastically and symplastically) and act on the plant meristem, stopping growth. They appear to be more active against dicot than monocot species.

Persistence in soil is commonly high among imidazolinones, making sensitive crops more susceptible to injury. Most weeds that are resistant to sulfonylurea herbicides are also resistant to imidazolinone herbicides. They tolerate the herbicide by altering the routes of metabolizing the herbicide, thus disallowing inhibition of the enzyme ALS.

Imiprothrin insecticide

Imiprothrin, $C_{17}H_{22}N_2O_4$ (mixture of 20% 2,5-dioxo-3-prop-2-ynylimidazolidin-1-ylmethyl (1*R*,3*S*)-2,2-dimethyl-3-(2-methylprop-1-enyl)cyclopropane carboxylae and 80% 2,5-dioxo-3-prop-2-ynylimidazolidin-1-ylmethyl(1*R*,3*R*)-2,2-dimethyl-3-(2-methylprop-1-enyl)cyclopropane carboxylate), is a pyrethroid ester insecticide that has an imidazolidinyl ring in its alcohol moiety. It is a golden-yellow liquid, soluble in water (up to 93.5 mg/l) and soluble in hexane and methanol. It is stable to metals and sunlight, but not in alkaline media. The CAS Registry Number of imiprothrin is 72963-72-5.

Imiprothrin, a non-crop insecticide, targets household pests like cockroaches, silverfish, spiders and crickets. It is used in warehouses, schools, homes and hospitals.

Imiprothrin is commonly formulated as an aerosol. It can be combined with other pesticides like *d*-phenothrin or cypermethrin, but not with alkaline compounds.

The acute oral LD_{50} of imiprothrin in rats ranges from 2400 to 4500 mg/kg. The acute dermal LD_{50} in rats is >2000 mg/kg. The acute inhalation LC_{50} is >2 mg/l among male and female rats. It is toxic to fish and other aquatic species. It may be harmful to inhalation and to contact on mammals, including humans. Symptoms of excessive exposure include irritation to skin and eyes, sensitivity to pricking, dizziness, nausea, vomiting, diarrhoea, excessive salivation and fatigue. In severe cases fluid in the lungs and muscle twitching may develop. Seizures may occur and are more common with more toxic cyano-pyrethroids. As a precaution, applicators are advised to wear appropriate PPE, while handling the chemical.

Imiprothrin degrades in aerobic soils, the primary metabolic pathway being ester cleavage followed by the elimination of a hydroxylmethyl group to form PGH (1-propargylimidazolidine-2,4-dione).

Imiprothrin is registered for use in India, South Africa, Australia, the USA, Canada and Finland (see Imiprothrin at bottom of the page).

Immersed weeds

See **Aquatic weeds**

Inabenfide as a PGR

Inabenfide, $C_{19}H_{15}ClN_2O_2$ (4′-chloro-2′-(α-hydroxybenzyl) isonicotinanilide), is a plant growth regulator. It is a pale yellow crystalline substance, soluble in water, acetone and ethyl acetate. Its CAS Registry Number is 82211-24-3. Although stable to heat and sunlight, it hydrolyses in alkaline media.

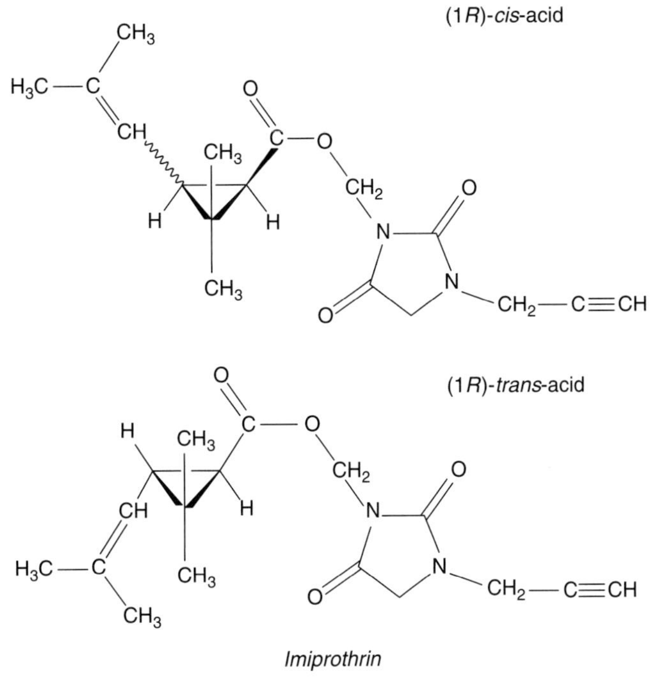

Imiprothrin

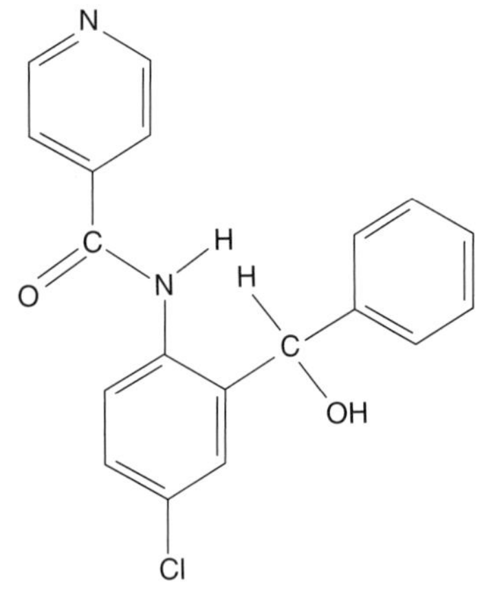

Inabenfide

Inabenfide, as a PGR, is mainly a herbicide used on rice. By inhibiting gibberellin biosynthesis, it keeps the length of internodes and leaf blades short, thus preventing lodging.

The compound is formulated as granules (GR) and as a wettable powder (WP). It is applied to the soil surface under submerged conditions, the dose range being 1.5–2.4 kg/ha. A common trade name is Seritard.

This PGR is safe on mammals, fish and birds according to both oral and dermal acute toxicity tests, although the acute inhalation LC_{50}, for a 4 h period, among rats is >0.46 mg/l air. It does not irritate the skin or eye. It thus has been classified by the EPA, as a Class IV toxin (practically non-toxic).

Inabenfide metabolizes into hydroxylated inabenfide, dihydrodiol inabenfide, methylated catechol inabenfide, inabenfide ketone, inabenfide *N*-oxide and an amine derivative.

Indanofan herbicide

Indanofan, $C_{20}H_{17}ClO_3$ ((*RS*)-2-[2-(3-chlorophenyl)-2,3-epoxypropyl]-2-ethylindan-1,3-dione), is a herbicide. It is a colourless, crystalline substance, soluble in water (17 mg/l) and very soluble in acetone and hexane. Indanofan is made by the reaction of 2-[2-(3-chlorophenyl) propen-3-yl]-2-ethylindan-1, 3-dione with acetic acid. The CAS Registry Number is 133220-30-1.

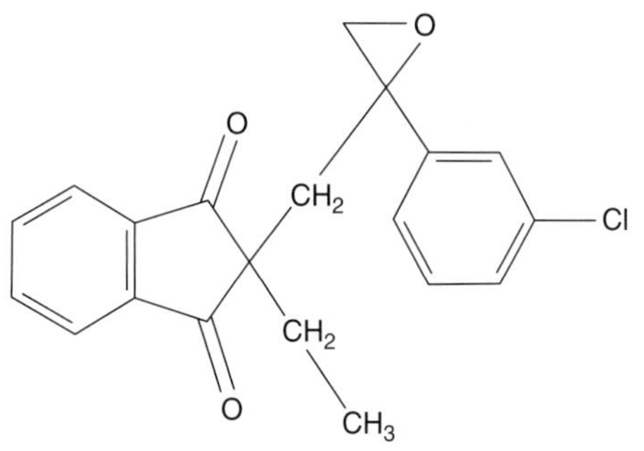

Indanofan

A selective, systemic pre- and post-emergence compound, indanofan is a herbicide suitable for rice. Indanofan acts by inhibiting the biosynthesis of very long chain fatty acids.

Indanofan is moderately toxic to rats (the acute oral LD_{50} is 460 mg/kg). The acute inhalation LC_{50} in rats is 1.57 mg/l. But it is not toxic to birds, honey bees and fish in normal use. The ADI of indanofan is 0.0035 mg/kg b.w./day.

Indole-3-acetic acid (IAA or AIA) as a bioPGR

Indole-3-acetic acid, $C_{10}H_9NO_2$ (1*H*-indol-3-ylacetic acid), also called indol-3-ylacetic acid or β-indoleacetic acid, is a plant-derived plant growth regulator and a very important plant auxin. It is also known as IAA and AIA. It is an off-white crystalline solid sparingly soluble in water and sensitive to light. It stores well in dark, cool, dry, well-ventilated areas. Indol-3-ylacetic acid is synthesized by the reaction of indole with glycolic acid in the presence of a base, at 250°C. The CAS Registry Number is 87-51-4.

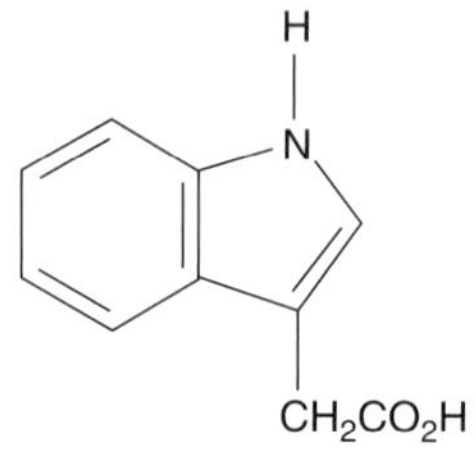

Indol-3-ylacetic acid

IAA is produced in the cells of the apical bud and young leaves, mainly from tryptophan. IAA is responsible for cell division and elongation in the cambial tissue, thus encouraging rooting in cuttings. It is this property that is exploited by agriculturists as both a herbicide and a growth promoter. As a herbicide, IAA induces such rapid and uncontrolled growth in broadleaved dicot plants that they are killed. The discovery of IAA led to the development of the phenoxy herbicides 2,4-dichlorophenoxyacetic acid (**2,4-D**) and 2,4,5-trichlorophenoxyacetic acid (2,4,5-T). These are metabolically and environmentally more stable analogues of IAA.

IAA is incompatible with strong oxidizing agents. It is available in the market as a tablet (TB) and water-dispersible powder for slurry seed treatment (WS). A known trade name is Rhizopon A.

IAA naturally occurs in plants and does not show any adverse effect on mammals, the environment and non-target organisms. It is rapidly degraded in soil. It is approved for use in organic farming. Excessive exposure causes irritation to the eyes and skin. Any exposure should be followed by removal of contaminated protective equipment and thorough washing of the affected part. If the patient is conscious, two to four cups of milk should be offered to the patient.

IAA degrades into hazardous products of carbon monoxide and oxides of nitrogen.

Indol-3-ylacetic acid is registered in the USA, Australia, South Africa and Vietnam.

Induced systemic resistance (ISR)

A pathogen attack provokes in a plant a series of complex defence responses which enables it to fight back. Such responses are triggered, in varying proportions, when the plant is attacked by fungi, insects, bacteria or viruses of a pathogenic or non-pathogenic nature. It is the response time to an infection that determines if the plant would succumb to the disease or survive.

Out of the many ways in which plants activate their defence systems, the two that are well studied include **systemic acquired resistance** (**SAR**) and induced systemic resistance

(ISR). ISR is a series of plant physiological defence responses elicited by biotic factors in the environment, wherein the symptoms are not necessarily pathogenic in nature. ISR has distinct signalling pathways, commonly regulated by **jasmonic acid** and **ethylene**. Root-colonizing bacteria, also called plant growth-promoting rhizobacteria (PGPR), activate elicitors (molecules that stimulate a certain action) to produce an ISR response. Such groups of PGPRs are *Azospirillum* spp., *Bacillus* spp., *Arthrobacter* spp. and *Pseudomonas* spp. (e.g. *Pseudomonas fluorescens* and *Pseudomonas putida*).

PGPRs are any group of root-colonizing bacteria that do not cause pathogenic symptoms in plants. PGPRs are capable of promoting growth directly or indirectly. Production of phytohormones and metabolites, and the improvement in nutrition status in plants are direct actions of PGPRs. Indirectly, PGPR can have a biocontrol property, controlling diseases by producing antibiotics and antifungal by-products. They also have a role in degrading soil pollutants, as rhizo-remediators.

Inhibition of fungal respiration

See **Fungicidal modes of action**

Inorganic pesticides

Chemical pesticides are grouped into organic and inorganic pesticides. Inorganic pesticides are derivatives of metals like arsenic, tin, lead, mercury, cyanide, fluorine, sulfur and copper. Most of the inorganic pest control agents are very highly toxic and are used sparingly. Recently inorganic compounds have been synthesized into pesticides with low toxicity levels and less persistence. Some examples include minerals such as silicate and borate.

Most inorganic pesticides act by forming complexes with the enzyme in the target pest, or coagulating with a specific metal group or the protein and interfering with the pest's normal biochemical functioning. Arsenical compounds inhibit ATP production, while mercury has a strong affinity for sulfhydryl groups (SH) in haemoglobin and proteins. The organometallic compound methylmercury, used as a fungicide, is a volatile and fat-soluble compound that has distinct symptoms of toxicity.

Inorganic pesticides do not degrade easily and persist for a long time in the ecosystem. Most of them score high on acute toxicity and are linked with the observable fall in populations of some fish and bee species over the last few years. Many inorganic pesticides leach into groundwater sources and enter the food chain. There are also other problems such as residues on crops and accumulation in animal tissues.

Thus the use of inorganic pesticides is being rapidly phased out the world over, and if used, they are restricted to non-crop applications.

Insect growth regulators (IGRs)

The concept of growth regulation became popular with the synthesis of chemical analogues that interfere with insect biochemical processes and hence, their life cycle. Insect growth regulators (IGRs) are not insecticidal or persistent in nature and normally do not cause resistance in the target pests.

The life cycle of insects includes prominent stages such as the egg, young larva, mature larva, nymph, pupa and the adult. There may be many sub-stages (called instars) in the larval stage. The growth phase is marked by the process of moulting, a process by which the organism sheds its rigid outer covering or exoskeleton and replaces it with a larger one to accommodate a larger body. The exoskeleton is made up of, among other compounds, a carbohydrate known as **chitin**. Normal synthesis of chitin is responsible for successful moulting. Chitin also plays a role at the insect embryonic stage (the egg stage).

Hormones that initiate the growth of the insect from egg to adult stage are known as insect juvenile hormones (JH). There are distinct JHs identified such as JH 0, JH I, JH II and JH III. Juvenile hormones are acyclic sesquiterpenoids that activate the process of reproduction in females and synthesize yolk proteins. The production of the hormone gradually decreases with every instar, allowing the next moult to proceed towards adulthood. JHs are produced by the *corpora allata* near the insect brain.

Another insect growth hormone, produced in the insect prothoracic gland is ecdysone, a steroidal pro-hormone (α-ecdysone and β-ecdysone), also known as ecdysterone or crustecdysone. Ecdysone inhibitors do not allow the hormone to signal moulting, hence killing the insect in the current stage.

IGRs act on the moulting process and the growth stages of insects. **Chitin synthesis inhibitors** block chitin production, which results in the new exoskeleton not being formed. Anti-juvenile hormones interfere with the production of the juvenile hormones, thus preventing the insect life cycle from progressing. Some IGRs mimic juvenile hormones, leading to abnormal reactions, like more larval stages in the life cycle. JH analogues also lead to sterilization, disrupted diapause and other unnatural behaviour patterns, finally causing death.

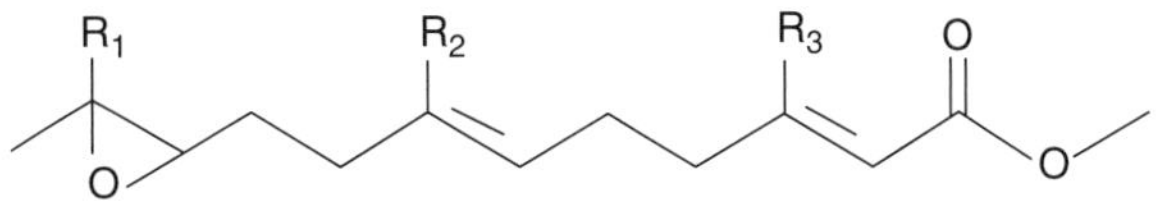

JH 0, $R_1 = R_2 = R_3 =$ ethyl (Et)

JH I, $R_1 = R_2 =$ Et, $R_3 =$ methyl (Me)

JH II, $R_1 =$ Et, $R_2 = R_3 =$ Me

JH III, $R_1 = R_2 = R_3 =$ Me

JH 0, JH I, JH II and JH III

Some examples of IGR compounds synthesized and used as active ingredients are **dicyclanil**, **fluazuron**, **methoprene**, **triflumuron** and fenoxycarb and these generally target fleas, flies, roundworms, cockroaches, ticks, lice and many other insects.

Some plant compounds have properties that regulate insect growth (and resemble the hormone JH III, a widely prevalent juvenile hormone). Such natural compounds have helped in the synthesis of a wide range of IGRs. Such beneficial botanical growth regulators are referred to as plant-derived IGRs. *Serratula coronata*, a plant from family Asteraceae (that includes asters and sunflowers), found in some Far East Asian regions, is said to produce ecdysteroids, as is the common bugle or carpetweed, *Ajuga reptans*, found in Europe. Ecdysteroids are a group of insect moulting hormones that include ecdysone and 20-hydroxyecdysone. Their presence in some plant species is responsible for repellent/anti-feedant properties. In plants, they are polyhydroxylated compounds that have three hexane rings and one pentane ring. Certain polyhydroxylated sterols that resemble ecdysteroids are called phytoecdysones.

Some naturally occurring plant compounds inhibit the gland that produces the juvenile hormones, while others have sterilant properties. The seeds and many other parts of the **neem** tree (*Azadirachta indica*) contain triterpenoid limonoids, with multiple biological effects against many insect pests including growth inhibition and anti-feedant and oviposition-deterring properties. The major component responsible is **azadirachtin**.

Insect juvenile hormones (JH)
See **Insect growth regulators (IGRs)**

Insect pathology
The study of pathology as specifically applied to insects is insect pathology. A very large variety of insects suffer from disorders originating from pathogen infections, metabolic imbalance, nutritional deprivation, physical or pesticidal injury or any other reason. Insects are also carriers of many pathogens. Insect pathology studies the biological relationship between an insect, especially a crop pest, and the microorganisms that parasitize it. The use of this study to man's advantage is the reason that insect pathology is popular.

Insect (pheromone) traps
See **Pheromones**

Insecticide aluminium phosphide
See **Aluminium phosphide rodenticide**

Insecticide borax
See **Borax herbicide**

Insecticide carbodiimide CGA 140408
See **Diafenthiuron acaricide**

Insecticide carbophenothion
See **Carbophenothion acaricide**

Insecticide chlorethoxyfos
See **Chlorethoxyfos insecticide**

Insecticide chlorfenapyr
See **Chlorfenapyr acaricide**

Insecticide chloropicrin
See **Chloropicrin insecticide**

Insecticide dazomet
See **Dazomet fungicide**

Insecticide enestroburin
See **Enestroburin fungicide**

Insecticide fosthiazate
See **Fosthiazate nematicide**

Insecticide gamma-HCH
See **Hexachlorocyclohexane insecticide**

Insecticide garlic oil
See **Garlic oil as fungicide**

Insecticide heptachlor epoxide
See **Heptachlor insecticide**

Insecticide hydrogen cyanide
See **Hydrogen cyanide rodenticide**

Insecticide jojoba oil
See **Jojoba oil fungicide**

Insecticide lindane
See **Hexachlorocyclohexane insecticide**

Insecticide oleic acid
See **Oleic acid herbicide**

Insecticide pyridaben
See **Pyridaben acaricide**

Insecticide Resistance Action Committee (IRAC)
The Insecticide Resistance Action Committee (IRAC) began in 1984, as a specialist technical group of CropLife to provide a conscious industry response to delay or prevent the development of resistance among insects and mites to chemical insecticides and chemical classes. The mission of IRAC is to facilitate communication and awareness on insecticide resistance and to promote steps to manage insecticide resistance, thereby encouraging sustainable agriculture. The IRAC has classified insecticides based on their mode of action and this is in Appendix 11 (reproduced with permission from IRAC; http://www.irac-online.org/).

Insecticide thiram
See **Thiuram disulfide fungicide**

Insecticides
Any chemical or natural substance used to kill insects is an insecticide. Insecticides are used when insects pose a threat to the health of man, animal or the economy.

Natural substances used as insecticides have a plant/fungal/viral origin. They include **rotenone**, **pyrethrum** and **nicotine**. Chemical toxins can be divided into inorganic and organic types. Inorganic insecticides, now limited in their use, include compounds of metals like copper, tin, mercury, arsenic and lead. Synthetic organic compounds include chlorinated hydrocarbons (**DDT**, **lindane** and **chlordane**), esters of phosphorus and many others. They have a specific mode of action, thus guaranteeing good results (while inviting resistance from insects). For instance, **organochlorines** and **organophosphorus** pesticides are nerve toxins. They either work by opening the sodium channels in the nerve cells of the insect, or bind to **acetylcholinesterase** and other cholinesterases, resulting in disruption of nerve impulses, killing the insect or interfering with its ability to carry on normal functions. Carbamate insecticides work like organophosphates, but have a short life and are less toxic. **Pyrethroids** mimic the natural insecticide, pyrethrum. Neonicotinoids are nicotinic **acetylcholine receptor** agonists and remain active for a long time.

Insecticides are grouped based on how they act on the pest. Based on the mechanism of action, there are: (i) stomach insecticides (ingested with food and suitable for chewing and sucking insects); (ii) **contact** insecticides which penetrate the cuticle; and (iii) **fumigant** insecticides, which are inhaled.

Insecticides are either applied directly on to the crop or are soil drenched. The chemicals are absorbed (**systemic**) either by the leaves or the roots, from where they are translocated to the target site.

Insecticides are formulated in various ways including emulsifiable concentrates (EC), wettable powders (WP) and dustable powders (DP), with every formulation suited for a specific type of insect infestation. It is often pertinent to apply insecticides with **adjuvants** and spreaders for better retention, effectiveness, rainfastness and persistence. Examples of such materials are polychlorinated biphenyls (PCBs).

Insecticides must be used with considerable caution on food, plants or animal forage. Many chemical compounds break down easily, but might persist on plants. Some chemical classes do not break down easily, and hence persist in the food chain. Arsenic compounds, **fonophos, phorate, aldrin, chlordane, DDT, dieldrin, malathion, prothiofos** and many chlorinated hydrocarbons, organophosphates and carbamates are examples of persistent compounds. This property causes harmful effects (reproductive, respiratory, endocrine and carcinogenic) on surrounding wildlife, **beneficial organisms**, fish, birds and mammals including man. They also pose a serious threat to groundwater sources.

Examples of some common insecticides are malathion (a contact insecticides that controls aphids, mites, thrips, cockroaches, **leafhoppers**, sawflies, caterpillars, fruit flies, leaf rollers, blister beetles, whiteflies and many others), **diazinon** (a contact insecticide that controls wire worms, cut worms and maggots, aphids, beetles, cabbage moths, thrips, sawflies, spider mites and caterpillars) and **dimethoate** (a systemic insecticide effective on some leaf miners and **scale insects**). Table I.1 lists a huge number of chemical insecticides that have or had been in use over a long period of time to curb insect pest problems.

Botanical/natural insecticides are also useful against aphids, beetles, caterpillars, fleas, leafhoppers and mosquitoes. Similarly, there are plant-derived insecticides that are being exploited for pest management. **Neem**, **nicotine**, **pyrethrum**, **rotenone**, **ryania** and sabadilla are some examples. Most natural products are non-toxic to non-target species and tend to degrade quickly, making them a safer alternate. A large number of plants produce volatile substances with a strong odour. These are essential oils, many of which have insecticidal and repellent effects. These are broad-spectrum, commonly monoterpenoid, antiseptic components that include simple hydrocarbons (limonene), alcohols (**geraniol**, menthol, terpineol), cyclic ethers (1,8-cineole), aldehydes (**citronella**), phenols (carvaiol) and thymol as well as esters (linalyl acetate). Some are phenol derivatives (asarone, eugenol) and aldehyde derivatives (**cinnamaldehyde**). Marigold contains a compound called α-terthienyl that is nematicidal and larvicidal in nature. It is also used to control mosquitoes. Many cereal crops have reactive cyclic hydroxamic acids, **1,4-benzoxazin-3-ones** that inhibit insect gut proteases, and hence are able to resist fungal and bacterial infections. Sesamin (also called **fagarol**), $C_{20}H_{18}O_6$, a component of sesame oil, has significant insecticidal properties.

Insectigation

Chemigation is the application of any chemical through irrigation waters and includes fertigation, herbigation, insectigation and **fungigation**. When insecticides are applied in this manner, the operation is called insectigation.

Drip irrigation and micro-irrigation are the commonly preferred vehicles of chemigation of any kind. Insectigation in particular, is ideal for row crops like vegetables and strawberries.

Table I.1. Major insecticidal active ingredients with their chemical group. (Table courtesy of Kalyani Paranjape.)

Chemical group	Active ingredients (a.i.)
Organophosphate compounds	Chlorfenvinphos, dichlorvos, dicrotophos, dimethylvinphos, hetenophos, mevinphos, monocrotophos, naled, phosphamidon, propaphos, tetrachlorvinphos, chlorethoxyphos, chlorpyrifos, chlorpyrifos-methyl, coumaphos, cyanophos, diazinon, ethrimfos, famphur, fenitrothion, fenthion, flupyrazofos, isazofos, isoxathion, methacrifos, parathion, parathion-methyl, phoxim, pirimiphos-methyl, pyridaphenthion, quinalphos, sulfotep, tebupirimfos, temephos, triazophos, azinphos-ethyl, azinphos-methyl, chlormephos, dimethoate, disulfoton, ethion, formothion, malathion, mecarbam, methidathion, phenthoate, phorate, phosalone, phosmet, prothiofos, sulprofos, terbufos, thiometon, azamethiphos demeton-*S*-methyl, omethoate, oxydementon-methyl, profenofos, pyraclofos, vamidothion, cadusafos, ethoprophos, phenamiphos, isofenphos, propetamphos, acephate, fosthiazate, methamidophos, EPN, fonofos
Carbamate compounds	Bendiocarb, carbofuran, carbosulfan, carbaryl, ethiofencarb, fenobucarb, formetanate, furathiocarb, isoprocarb, methiocarb, metolcarb, pirimicarb, propoxur, trimethacarb, xylylcarb, alanycarb, aldicarb, aldoxycarb, butocarboxim, butoxycarboxim, methomyl, oxamyl, thiodicarb, thiofanox, cartap, thiocyclam, bensultap
Pyrethroids	Pyrethrin I, cinerin I, jasmolin I, pyrethrin II, cinerin II, jasmolin II, allethrin, bioallethrin, empenthrin, imiprothrin, prallethrin, resmethrin, cyphenothrin, bioresmethrin, kadethrin, tetramethrin, phenothrin, d-phenothrin, permethrin, cypermethrin, alpha-cypermethrin, beta-cypermethrin, theta-cypermethrin, zeta-cypermethrin, deltamethrin, fenvalerate, esfenvalerate, flucythrinate, fluvalinate, tau-fluvalinate, ZX 8901, cyfluthrin, beta-cyfluthrin, flumethrin, acrinathrin, cycloprothrin, cyhalothrin lambda-cyhalothrin, fenpropathrin, tralomethrin, bifenthrin, tefluthrin, transfluthrin, etofenprox, flufenprox, halfenprox, silafluofen
Organochlorine compounds	Cyclodieners, aldrin, chlordane, dieldrin, endosulfan, endrin heptachlor, isodrin, kepon, mirex, DDD, DDE, DDT, dicofol, hexachlorocyclohexane, methoxychlor, perthane, toxaphene, hexachlorobenzene, pentachlorophenol

Integrated pest management (IPM)

The widespread application of a cheap and effective insecticide was once considered a quick solution to every insect pest problem. However, no eradication programme ever met with sustained success because of the development of resistance in the pest to the toxin. Newer chemicals and more innovative application methods also failed to bring pest infestation under control. To make matters worse, this chemical attack began to disrupt the ecological balance and increase pollution.

The concept of integrated pest management (IPM) emerged from the realization that sensitivity to ecology is crucial if any natural system has to be altered. According to the US EPA, IPM is an effective and environmentally sensitive approach to pest management that relies on a combination of common sense practices. According to the Jawaharlal Nehru Krushi Vishwa Vidyalay, Jabalpur, India 'IPM is a sustainable approach to managing pests by combining biological, cultural, mechanical and chemical tools in a way that minimizes economic, health and environmental risks'.

IPM aims at rationalizing the use of chemical pesticides, such that pest populations are managed to acceptable levels, while not posing a risk to human and animal health. The approach is far more promising than simple biological control programmes. The components of IPM should be seen as a part of the umbrella of **good agricultural practices (GAP)**.

IPM involves the continuous and comprehensive assessment of pest problems, so that an appropriate combination of techniques such as chemical insecticides, **parasitoids**, predators, **semiochemicals**, cultural practices, physical/mechanical methods and regulatory measures can be utilized to manage the situation. Of paramount importance is the protection of the ecological balance in the area under pest influence.

Specifically, IPM is carried out, on the one hand, by studying pest life cycles and pest interaction with the environment. On the other, it involves the use of biotechnology to bring about modifications in pest behaviour (via plant breeding to incorporate genetic factors to deter insect pest attack), and also pest-resistant hybrid crops.

Knowledge of pest life cycles is important since planting time and harvest time can be adjusted accordingly, to escape intense damage. Over generations, farming families become experts at predicting pest attacks and methods to bypass severe infestations and infections. For example, planting peas by the middle of spring helps avoid the cool and wet weather, conditions which suit damping-off and root rot. Sowing tomatoes, onions and any of the cruciferous crops (e.g. cabbage, cauliflowers, broccoli) earlier than conventionally accepted helps them escape the onslaught of **late blight** and maggots. Planting sorghum in the early part of the season helps the crop escape attacks from some insects like cutworms.

Some other tools include the selective use of the refugia technique (refugia being areas in which a population of organisms can survive through a period of unsuitable or harsh conditions). Use of fungi such as *Verticillium lecanii*, *Paecilomyces variotii* and ***Beauveria bassiana*** for insect control is also common. Use of biological products and **biorationals** is equally relevant in IPM.

The use of pesticides is important in any IPM programme, with specific attention given to the timing of use and quantity of the chemical. Their use has to be short lived, so that they do not accumulate in the food chain. They must not be mutagenic and carcinogenic, must be safe to handle and yet effective in controlling pests. At the same time, the application should be monitored such that resistance among pests, to the chemical, is kept at a minimum. A major advantage of IPM is the cost saving that results from not having to be solely dependent on pesticides.

Integrated weed management (IWM)

Integrated weed management (IWM) is a set of tools used to maintain weeds at an acceptable level of safety. IWM encompasses a combination of manual, chemical and biological methods.

Weed control and selective prevention are the hallmarks of a successful weed management strategy. Although specific prescriptions for overall weed management may differ with crop, climate and soil types, there are a few activities that form an integrated weed control strategy. These include: (i) improving soil quality; (ii) rotating crops; (iii) planting smother crops; (iv) mulching; (v) mechanical weed control; and (vi) herbicide intervention.

The key in soil building is to lend support to crops so that they stay ahead of weeds. Soil building is a basic tenet of weed control. A compacted soil is considered more favourable for weed seeds than loose, aerated soil. Certain weed species (e.g. foxtail) grow well in low calcium and sulfur soils. Shallow incorporation of green manure, organic matter and soil nutrients (e.g. phosphorus, calcium, sulfur and a few trace elements) help crops to grow vigorously and develop strong roots. However, weed seeds should not be present in the manure.

Crop rotation has always proved to be helpful in breaking weed cycles. Growing crops rich in soil nutrients (i.e. those crops that leave behind some nutrient value to the soil) in rotation with a season in hay will help the subsequent crop to be less challenged by weeds because of a compulsory break from fresh weed seed. Shifts in weed population can be achieved by rotating crops. Some commonly rotated crops are maize after soybean and barley after wheat. It is important to rotate crops from different genera.

Crop rotation successfully controls bacterial and insect populations too. Many diseases like bacterial blights that attack wheat, barley, rye and maize, or **smut** and **scab** on wheat, barley and grasses, can be significantly controlled by rotating them with other crops.

Smother crops are those grain or vegetable crops that grow fast and thus suppress weed growth in that area. Such crops bear fruit quickly, are economically useful, while being a natural weed control agent. Common smother crops are vegetables like pumpkins, beans, cucumbers, peas, potatoes, tomatoes and other crops like Sudan grass, rye, lucerne, barley and oats. Their planting time should be monitored to prevent them from being counter-productive. They are best planted in the autumn and removed in the spring just in time for the desired crop plantation. (Autumn or fall months are from September to November in the northern hemisphere, and March to May in the southern hemisphere.) In some cases, certain smother crops may even be inter-seeded or over-seeded into the desired crop. Smother crops help improve the soil, prevent erosion and provide abundant soil nutrition.

The use of machines for weed control is very critical to manage weeds on a farm. Tilling operations can be either deep in the soil or shallow. It is ideal to use different sets of tools for different weeding and tilling needs. For instance, **disking**, harrowing and ploughing can go to greater soil depths and are very effective.

Mulches are materials that are spread over the surface of soil. Mulches are made either of organic or artificial materials. Grass clippings, dry leaves, newspaper and plastic are used as mulches. Mulches retain soil moisture, suppress weeds and keep soil temperature low. Organic mulches also help improve the soil's fertility, as they decompose easily, adding nutrients to the soil. Mulches prevent or delay weed germination by blocking or limiting sunlight to the weeds. Organic mulches include manure, lawn clippings, sawdust, leaves and straw that decompose and further nourish the soil. Mulches also prevent rapid evaporation around the crop, thus encouraging their growth.

Organic mulches are best applied on warm soil in late spring. Inorganic mulches are also in use and are made of plastic sheets (commonly black in colour). Since they do not decompose, they have to be removed after use. They are best used on cool soil in the early spring so that they can give the soil the benefit of the greenhouse effect.

Row crop cultivation is an important weed management tool. Multiple cultivations can help give more yields, for instance in maize three instead of two cultivations give better yields. Weed control in grain sorghum depends upon three or four tillage operations to prepare a seedbed and encourage weed germination. Inter-row crop cultivation may be carried out after sorghum emerges. Here, to save time, intervention of herbicides can allow sorghum to be planted earlier and lend the benefit of a longer growing season.

The use of herbicides in the right quantities and at the right time depending on the crop in question is critical for weed management. There are specific recommendations about the use of herbicides throughout this volume.

International Agency for Research on Cancer (IARC)

See **World Health Organization (WHO)**

International legislation

See our **Pesticide legislation** series under 'P' for an outline of legislation for a few countries.

Invert emulsions

See **Emulsion**

Iodofenphos insecticide

Iodofenphos is another term used for **jodfenphos insecticide**.

Iodosulfuron herbicide

See **Iodosulfuron-methyl-sodium herbicide**

Iodosulfuron-methyl-sodium herbicide

Iodosulfuron-methyl-sodium, $C_{14}H_{13}IN_5NaO_6S$ (methyl-4-iodo-2-(4-methoxy-1,3,5-triazine-2-ylcarbamosulfamoyl) benzoate), is a triazinylsulfonylurea herbicide. It is a crystalline powder soluble in water, *n*-heptane, *n*-hexane, toluene and acetonitrile. It is reasonably stable in dry, cool and well-ventilated conditions as well as photostable in water (DT_{50} >50 days). It may be a fire hazard in direct sunlight. The CAS Registry Number is 144550-36-7. This substance is a derivative of iodosulfuron $C_{13}H_{12}IN_5O_6S$ (4-iodo-2-[(4-methoxy-6-methyl-1,3,5-triazin-2-yl) carbamoylsulfamoyl] benzoic acid), with a CAS Registry Number of 185119-76-0.

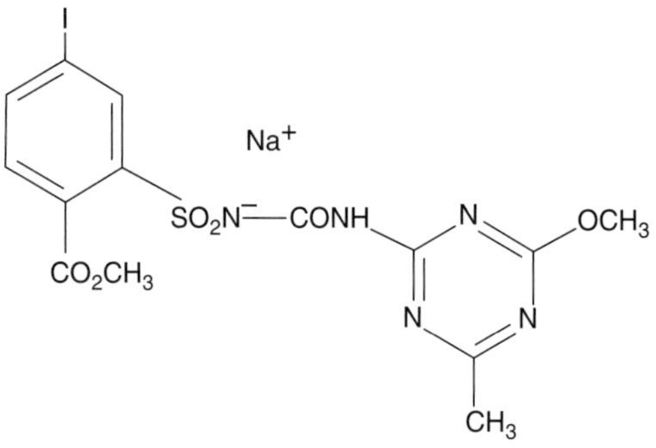

Iodosulfuron-methyl-sodium

Iodosulfuron-methyl-sodium, a selective, post-emergence cereal herbicide, inhibits the synthesis of amino acids like valine and isoleucine, thus stopping cell division in a number of weeds in wheat, rye and barley.

The above-mentioned herbicide is available as emulsifiable concentrates (EC) or as oil dispersions (OD). The compound is applied on cereals with a maximum application rate of 0.01 kg a.i./ha. It should not be mixed with oxidizing agents, acids and bases.

This herbicide is moderately toxic orally or dermally with most acute LD_{50} values in rats and birds being above 2000 mg/kg. The oral **NOAEL/NOEL** for a 12 month period among rats is 200 ppm. The ADI of iodosulfuron-methyl-sodium is 0.03 mg/kg b.w./day. The chemical is slightly to moderately toxic to earthworms, honey bees and fish. Extreme exposure to the herbicide may cause irritation to eyes and skin, necessitating the use of PPE during application, including a face shield, full cotton overalls and elbow-length PVC or nitrile gloves.

Iodosulfuron-methyl-sodium is rapidly absorbed, metabolized and excreted in animals. In soils, degradation takes place by microbial action, taking a maximum of 10 days in the case of low soil moisture. The chemical or its metabolites do not leach and threaten groundwater sources. Photochemical oxidative degradation takes place in 6.3 days.

The herbicide is registered in India (along with methosulfuron methyl), the USA, Australia, Canada and all but one of the members of the EU. Australia has established an MRL value of 0.01 mg/kg for barley and 0.05 mg/kg for forage crops.

Ioxynil herbicide

Ioxynil, $C_7H_3I_2NO$ (4-hydroxy-3,5-diiodo benzonitrile), is a benzonitrile herbicide. The colourless solid is moderately soluble in water and fairly soluble in organic solvents. Ioxynil exists as esters or salts, and thus there are ioxynil-lithium, ioxynil octanoate and ioxynil-sodium. The herbicide is produced by iodination of 4-hydroxybenzonitrile with iodine monochloride. On burning, it produces toxic fumes and nitrogen oxide. The CAS Registry Number is 1689-83-4.

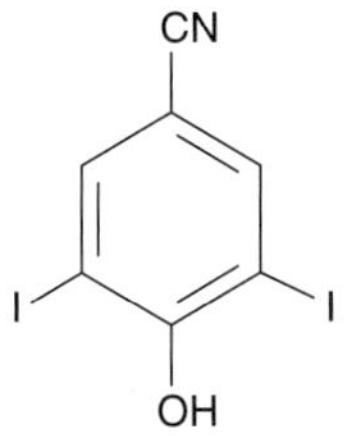

Ioxynil

A foliar, post-emergence herbicide, ioxynil controls broad-leaved weeds on cereal crops, onions and garlic. It acts by inhibiting photosynthesis at photosystem II.

The herbicide ioxynil is formulated as emulsifiable concentrate (EC). It is available in the market under trade names like Actril, Bentrol, Certol and Iotril.

The herbicide is moderately to highly toxic through all routes. The acute oral LD_{50} values in rats and bobwhite quails are 114 and 62 mg/kg, respectively, while the inhalation LC_{50} of ioxynil in rats is 0.38 mg/l. In normal use, ioxynil does not irritate the skin and the eyes (the dermal LD_{50} is 1050 mg/kg). It is toxic to fish and other aquatic organisms. The ADI is 0.005 mg/kg/day. It is also suspected to disrupt endocrine functioning. Depending on the formulation, labels should have the signal words 'WARNING' or 'DANGER'. Symptoms of toxicity to exposure include dizziness, vomiting, weakness and fever. In the case of a spill, contaminated clothes and accessories should be removed and kept aside and the patient given a thorough wash, before medical attention can be given.

In animals ioxynil is almost completely eliminated via urine. The two major degradation products are 3-iodo-4-hydroxybenzonitrile and an isomer of ioxynil.

Ioxynil is registered in Australia, New Zealand, South Africa and many other African countries, and many members of the EU. But ioxynil or its derived products are not registered in the USA. European MRLs for onions and carrots are 0.2 mg/kg, while for other vegetables, fruit and cereal grains, the value is 0.05 mg/kg. (See also **Hydroxybenzonitrile herbicides**.)

Iprobenfos fungicide

Iprobenfos, $C_{13}H_{21}O_3PS$ (*S*-benzyl *O*,*O*-diisopropyl phosphorothioate), is an organophosphorus fungicide. It is a colourless liquid, soluble in water (430 mg/l at 20°C), acetone, acetonitrile, methanol and xylene. The photolysis half-life in aqueous medium is up to 11 days, while the aqueous hydrolysis half-life at pH 7 is 276 days. Its CAS Registry Number is 26087-47-8.

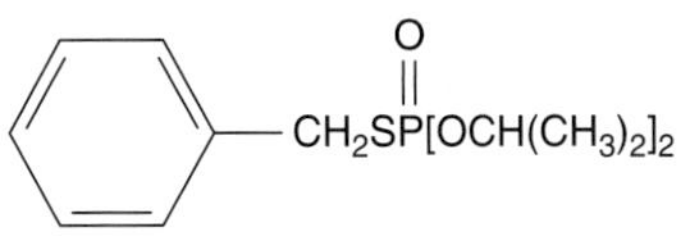

Iprobenfos

The fungicide iprobenfos, mainly used on rice, is a systemic compound, absorbed through the roots. It acts as an inhibitor of phospholipid biosynthesis. It provides both curative and protective properties. It effectively controls leaf blast, ear blast, stem rot and sheath blight. *Pyricularia oryzae*, which is the causative agent of blast, is controlled by iprobenfos.

Iprobenfos is formulated as emulsifiable concentrate (EC), granules (GR) and dustable powder (DP). It is commercially sold as Kitazin.

This fungicide is considered slightly hazardous according to the WHO. The acute oral LD_{50} in rats is 680 mg/kg and the dermal LD_{50} >4000 mg/kg. The inhalation LC_{50} in rats is 0.34 mg/l. Since the compound is a cholinesterase inhibitor, it is considered to be a PAN Bad Actor chemical. Symptoms of toxicity include excessive salivation, dizziness, nausea, vomiting, blurred vision and respiratory depression. Applicators and handlers are advised to wear shields for face and eyes. Respirators and gloves are also recommended.

The fungicide iprobenfos undergoes metabolism in many steps including P-S-$CH_2C_6H_5$ and P-O-C_3H_7 cleavage, isomerisation and desulfurization.

Iprobenfos is registered in India, Cambodia and Vietnam.

Iprodione fungicide

Iprodione, $C_{13}H_{13}Cl_2N_3O_3$ (3-(3, 5-dichlorophenyl)-*N*-isopropyl-2,4-dioxoimidazolidine-1-carboxamide), is a dichlorophenyl **dicarboximide** fungicide (imidazole fungicide). It is soluble in water (at 13 mg/l), ethanol and methanol (at 25 g/l) as well as dimethylformamide (at 500 g/l). It is also known as glycophene or promidione.

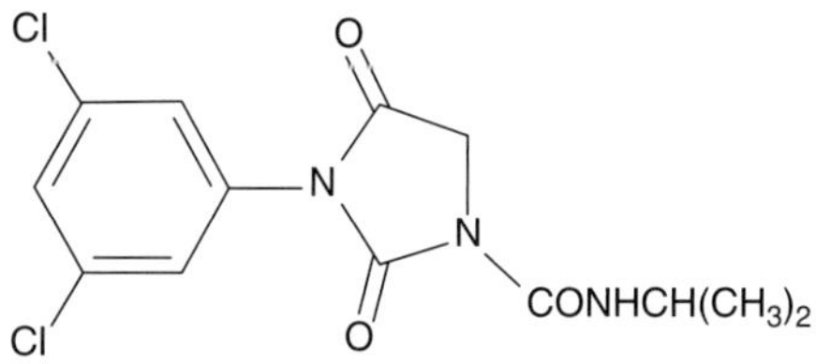

Iprodione

Iprodione, a contact fungicide, is used to control diseases on vegetables, pome and stone fruit and cotton. It inhibits the germination of spores and mycelium growth.

The fungicide iprodione is used as a post-harvest fungicide and during seed treatment. Some trade names are Kidon, Rovral and Chipco 26019. The fungicide may be combined with **thiabendazole**, **carbendazim** and other pesticides. Some formulations have been listed as restricted use pesticides (RUP) according to the EPA.

Iprodione is classified as a Class U toxin (unlikely to be hazardous) by the WHO. The oral LD_{50} is 3500 mg/kg in rats and 4000 mg/kg in mice. It is not toxic via the dermal or inhalation routes in normal use and is non-toxic to bees. However, it is moderately toxic to some bird species and to fish. It is a carcinogen, and feared to be a groundwater contaminant and an endocrine disruptor, and hence it is a PAN Bad Actor chemical. The NOEL of iprodione in rats is 4.2 mg/kg and the ADI in humans is optimized to be 0.3 mg/kg/day.

Iprodione is metabolized to 3,5-DCA (3, 5-dichloroaniline) and eliminated via urine in 24 h. In soil, it rapidly degrades in alkaline soil (half-life of 27 min in alkaline soils). Degradation also takes place by microbial action.

Iprodione is registered in India, the USA, Canada, Australia, Cambodia, New Zealand, South Africa, Tanzania and most EU nations (except five members).

Iron as a plant nutrient

See **Plant nutrient excesses and deficiencies**

Irradiation

Exposure to radiation of wavelengths shorter than those of visible light (gamma, X-ray or UV) either for medical purposes (e.g. cancer therapy, removal of skin blemishes), destruction of bacteria in milk and other foodstuffs or for inducing polymerization of monomers or vulcanization of rubber is called irradiation. UV irradiation was formerly used to induce activation of vitamin D in milk and was sometimes used to sterilize the air in operating rooms.

Isazofos insecticide

Isazofos, $C_9H_{17}ClN_3O_3PS$ (*O*-5-chloro-1-isopropyl-1*H*-1,2,4-triazol-3-yl *O,O*-diethyl phosphorothioate), is a triazole organothiophosphate insecticide, with nematicidal property (organothiophosphate nematicide). It is a yellow liquid, readily soluble in water (168 mg/l), and miscible with chloroform, methanol and benzene. It is hydrolysed more rapidly in alkaline media and moist conditions. It has a DT_{50} of 19 days at pH 9. When heated above 200°C, it emits very toxic fumes of Cl^-, NO_x, PO_x and SO_x. Isazofos is synthesized by reacting ethyl 3-isopropylcarbamate with cyanogen chloride followed by cyclization with phosgene. It has a CAS Registry Number of 42509-80-8.

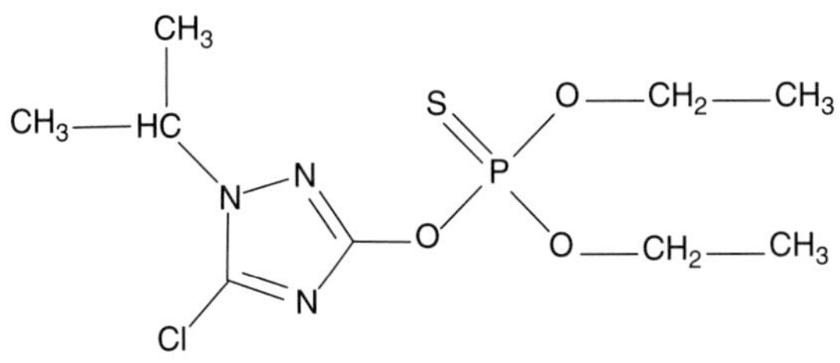

Isazofos

Isazofos, a systemic, broad-spectrum, soil-applied compound, is a cholinesterase inhibitor that has stomach action. It protects a range of crops including bananas, papaya, citrus fruits, sugarcane, cotton, groundnut, rice, maize and vegetables. Banana root nematodes and banana weevil borers, reniform nematodes and some other root-knot infestations are effectively controlled by isazofos. Some of the insects that are controlled by isazofos are beetles that attack citrus and ornamentals, leaf folders and gall midges.

Insecticide isazofos is formulated for foliar spraying and for seed treatment. The application dose is accepted in the range of 0.5–2 kg/ha. It is available in the market as Miral and Triumph. However, it is now an obsolete compound.

Isazofos is toxic by ingestion, inhalation and skin contact. The acute oral LD_{50} is 34 mg/kg in mice, 175 mg/kg in rabbits and 27 mg/kg in rats. The dermal LD_{50} of isazofos in rats is 118 mg/kg. It is toxic to many bird species and is very toxic to fish and aquatic organisms. It is classified as a Class Ib toxin and is a PAN Bad Actor chemical. Symptoms of toxicity include salivation, tremors, incoordination, headache, dizziness, nausea, vomiting, abdominal cramps, respiratory depression, tightness in the chest, a productive cough and many others. An exposed person is strongly advised to remove and dispose of clothing and accessories, wash the affected part with plenty of water and move into open well-ventilated areas. Suitable protective clothing, respiratory equipment are a must. Labels should clearly display the signal word 'DANGER'.

The EU, the USA and many other countries do not allow the use of isazofos, while its use is highly restricted in Bangladesh, Indonesia, Malaysia, Pakistan and Sri Lanka. Antigua has banned some formulations. It is registered in Cambodia and Tanzania.

Isobardac slimicide

Isobardac, $C_{22}H_{48}ClN$ (*N*-decyl-*N*,*N*-dimethyl-1-decanaminium chloride), is a slimicide, bactericide and an algicide. It is a solid substance, soluble in propanol and benzene. It is also called dialkyl dimethyl ammonium chloride. The CAS Registry Number is 7173-57-5. Isobardac is used in swimming pools and industrial water treatment plants.

Isofenphos insecticide

Isofenphos, $C_{15}H_{24}NO_4PS$ (*O*, ethyl *O*-2-isopropoxycarbonylphenyl isopropylphosphoramidothioate), is an organophosphorus insecticide. It is colourless oil, soluble in water (at 18 mg/l), isopropanol, hexane, acetone, kerosene, xylene and toluene. It has an aqueous hydrolysis of up to 2.8 years at varying pH levels. In the presence of strong reducing or oxidizing agents, isofenphos can produce toxic gases like **phosphine**. The CAS Registry Number is 25311-71-1.

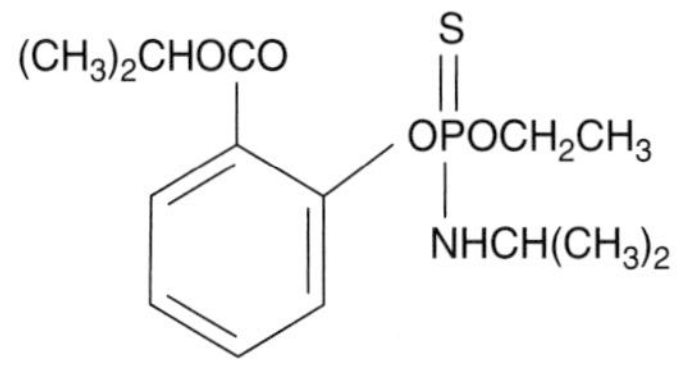

Isofenphos

A selective, systemic stomach poison that inhibits cholinesterase, isofenphos is applied at the pre-plant or pre-emergence stage and is used in the control of soil insects like cabbage root flies, carrot flies, corn rootworm, white grubs and wireworms, commonly in maize, carrots, bananas and turf grass. It is also used as a termiticide. It is a nerve toxic and inhibits cholinesterase.

Isofenphos is formulated as emulsifiable concentrates (EC), wettable powders (WP) and granules (GR). It is available in the market under trade names like Amaze, Oftanol and Pryfon.

This insecticide is highly toxic via the oral and inhalation routes and moderately toxic via the dermal route as observed in experiments on rats, mice and rabbits. Birds are highly susceptible to isofenphos with an LD_{50} ranging between a few ppm and 20 mg/kg. The LD_{50} in fish is a maximum of 2 mg/kg. The acute oral LD_{50} of isofenphos in rats is 28 mg/kg and the dermal LD_{50} in rats is 188 mg/kg. In rabbits, it ranges from 162 to 315 mg/kg. It is highly toxic via the inhalation route. The ADI of isofenphos is 0.0001 mg/kg/day. The chronic effect on animals and humans is on the nervous system. The WHO classifies it as a Class Ib toxin (highly hazardous). Since isofenphos is a cholinesterase inhibitor, it is considered a PAN Bad Actor chemical. According to the EPA, it is a restricted use pesticide (RUP). Thus labels must bear the signal words, 'DANGER – POISON'. Toxicity symptoms include excessive salivation, pin-point pupils, muscle twitching, weakness, tremors, incoordination and tightness in chest.

Isofenphos is highly persistent in soil with a half-life of up to 300 days. It leaches and breaks down slowly, producing isopropyl salicylate and cyclic isofenphos. Plants metabolize the chemical to isofenphos and an isofenphos-oxygen analogue.

Isofenphos is registered for use in Australia and Cambodia.

Isoprocarb insecticide

Isoprocarb, $C_{11}H_{15}NO_2$ (*O*-cumenyl methylcarbamate), is a carbamate insecticide. It is a colourless crystalline substance, sparingly soluble in water (0.265 g/l) and readily soluble in acetone (400 g/l) and methanol (125 g/l). It is hydrolysed in alkaline media. Isoprocarb is produced by the reaction of 2-isopropylphenol with methylisocyanate. It has a CAS Registry Number of 2631-40-5.

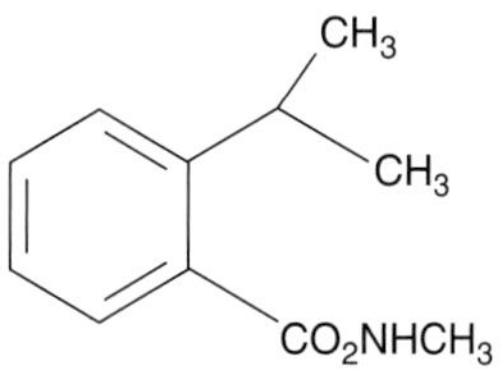

Isoprocarb

The insecticide, with contact and stomach action, effectively controls **leafhoppers**, plant hoppers and **aphids** on a variety of crops like rice, cocoa, sugarcane and vegetables. Its residual activity gives effective pest control. It acts by inhibiting cholinesterase.

The insecticide is formulated as a dustable powder (DP), emulsifiable concentrate (EC) and granules (GR). It is compatible with many other pesticides, but incompatible with alkaline material. It is applied at the rate of 0.5–0.75 kg a.i./ha. It is available under brand names such as Etrofolan and Mipcin.

The insecticide isoprocarb is a Class II toxin (moderately hazardous), with the acute oral LD_{50} among rats being 403–485 mg/kg, while the percutaneous LD_{50} among male rats and mice is >500 mg. It is a PAN Bad Actor chemical, since it is a cholinesterase inhibitor. Symptoms of poisoning with this insecticide include muscle weakness, dizziness, salivation, nausea, vomiting, sweating, headache, effects on the CNS and depression. PPE such as splash-resistant safety goggles and chemical-resistant clothing, gloves and masks are advised.

In animals and plants, isoprocarb metabolizes into 2-isopropylphenol and 2-(1-hydroxy-1-methylethyl)-phenyl-*N*-methylcarbamate. Isoprocarb has low mobility in soil. Its DT_{50} is from 3 to 20 days under flooded conditions.

Isoprocarb is registered for use in Cambodia, the Philippines and Vietnam.

Isoproturon herbicide

Isoproturon, $C_{12}H_{18}N_2O$ (3-(4-isopropylphenyl)-1,1-dimethylurea or 3-*p*-cumenyl-1,1-dimethylurea), is a herbicide of the **phenylurea** class. It is a colourless, odourless crystalline substance. It is soluble in water (72 mg/l at 20°C) and polar organic solvents. It hydrolyses in strong alkaline media and when heated. It is produced by the reaction of 4-isopropylphenyl isocyanate with dimethylamine. The CAS Registry Number is 34123-59-6.

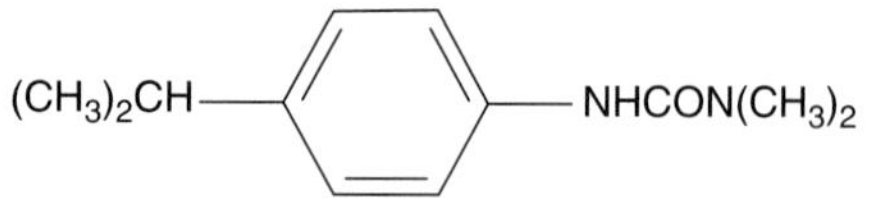

Isoproturon

Isoproturon, a selective, systemic, pre- and post-emergence herbicide, is absorbed by the roots and leaves of broadleaved weeds and grasses in cereals. It acts by inhibiting photosynthesis, thus killing weed pests. Examples of target weeds are slender meadow foxtails (*Alopecurus myosuroides*) and wild oats (*Avena fatua*).

The compound is formulated as a suspension concentrate (SC) and a wettable powder (WP). Common brand names include Belgram, Tolkan and Narilon.

The acute oral LD_{50} of isoproturon in rats ranges from 1826 to 3600 mg/kg b.w. It is not a skin or eye irritant. The NOEL in a 90 day study in dogs was approximately 3 mg/kg b.w./day. However, it is very toxic to aquatic organisms. It can become a possible source of surface/groundwater pollution if it is not used as per the label instructions. It is phytotoxic to some varieties of wheat (durum). The WHO classifies it as a Class III toxin (slightly hazardous).

In water, isoproturon is quite persistent, with a half-life of about 1 month. In soil, with a half-life of up to 40 days, isoproturon undergoes microbial, photochemical and enzymatic degradation. Metabolites include 4-(2-hydroxyisopropyl) aniline, 3-(4-isopropylphenyl)-1-methylurea, 3-(4-isopropylphenyl) urea, 4,4′-diisopropylazobenzene and 4,4′-diisopropylazoxybenzene. The metabolite 4,4′-diisopropylazobenzene is persistent in soil.

This herbicide is registered for use in India, by most EU members, Australia and Tanzania.

Isouron herbicide

Isouron, $C_{10}H_{17}N_3O_2$ (3-(5-*tert*-butyl-1,2-oxazol-3-yl)-1,1-dimethylurea), is a herbicide in the oxazole and urea class. It is produced by the reaction of 3-*tert*-butyl-isocyanate with dimethylamine. The CAS Registry Number is 55861-78-4.

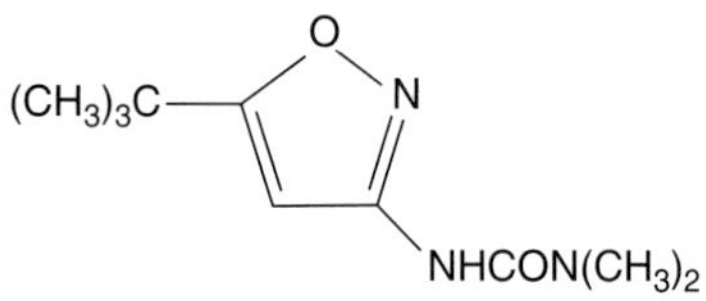

Isouron

Isouron is a selective, systemic, pre- and post-emergence herbicide. Once absorbed through the roots, the compound translocates and inhibits photosynthetic electron transport at the photosystem II receptor site. It effectively controls a variety of weeds in crops (e.g. sugarcane and pineapples) and also at non-crop sites.

Common formulations of isouron are granules (GR), suspension concentrates and wettable powder (WP). Application rates range from 0.5 to 1.5 kg/ha for crops while high doses are needed for controlling weeds on non-crop lands.

The WHO has classified isouron as a slightly hazardous chemical (Class III toxin). While the acute oral LD_{50} among rats and mice ranges between 520 and 760 mg/kg, the acute percutaneous LD_{50} among rats is >5000 mg/kg. It is only very slightly toxic to birds and fish.

Herbicide isouron is eliminated upon oral administration within 48 h. In plants too, it is metabolized quickly. The soil half-life is 22 days, degradation taking place via microbial action.

Isoxaben herbicide

Isoxaben, $C_{18}H_{24}N_2O_4$ (*N*-[3-(1-ethyl-1-methylpropyl)-1,2-oxazol-5-yl]-2,6-dimethoxybenzamide), is a herbicide of the amide, oxazole and benzamide class. It is soluble in water at 1.0 mg/l. The CAS Registry Number is 82558-50-7.

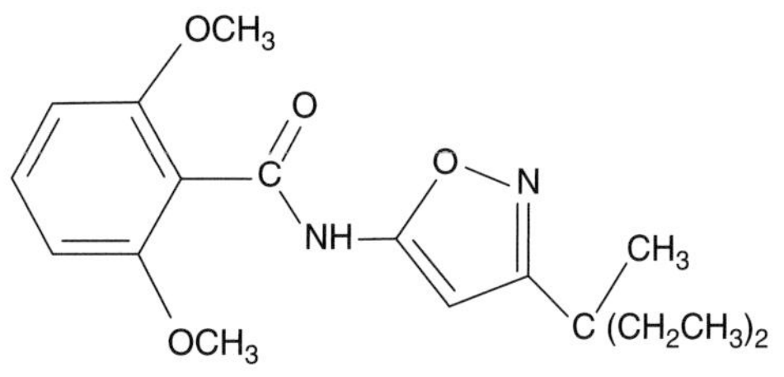

Isoxaben

Isoxaben, a pre/early-emergence herbicide, inhibits cell wall biosynthesis of broadleaved weeds in turf and other non-crop areas like industrial sites and highways. The added advantage is its long residual life. It stores well in a dry and cool place.

Common formulations of the herbicide are granules (GR) and suspension concentrates. In most cases, it is mixed with water, avoiding spray drift, and applied from >700 g to nearly 2 kg/ha.

The WHO categorizes isoxaben as a Class U toxin (unlikely to be hazardous) since the acute oral LD_{50} among rats, mice and birds is >5000 mg/kg, >10,000 mg/kg and >2000 mg/kg, respectively. It is moderately toxic to some fish species, but only slightly toxic to honey bees. However, it is a possible carcinogen and a potential groundwater contaminant and can harm the aquatic environment. It is also moderately toxic by inhalation and can harm the skin and eyes. Applicators are thus advised to wear appropriate PPE including safety glasses, gloves and long-sleeved, full-body garments.

Isoxaben is registered in the USA, Canada, Australia and many countries of the EU including the UK, Belgium, Germany, Finland, Spain and Portugal. The MRL of isoxaben for wheat is established at 0.05 mg/kg. The MRL for grapes in Belgium is 0.05 mg/kg and Australia, 0.01 mg/kg. India, Canada and Brazil do not permit any detectable residues on grapes.

Isoxadifen herbicide safener

See **Isoxadifen-ethyl herbicide safener**

Isoxadifen-ethyl herbicide safener

Isoxadifen-ethyl, $C_{18}H_{17}NO_3$ (ethyl 4,5-dihydro-5, 5-diphenyl-1, 2-oxazole-3-carboxylate), is a herbicide safener. It is a solid, brownish granular substance with a slight odour. It is stable in dry, cool, airtight containers and normal temperatures and should be kept away from food and feed. The CAS Registry Number of isoxadifen-ethyl is 163520-33-0.

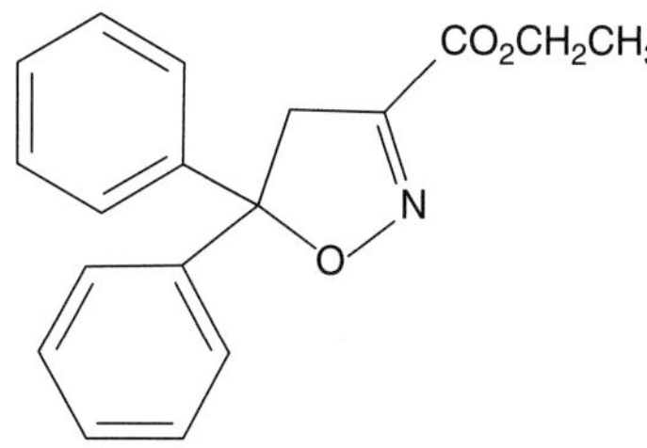

Isoxadifen-ethyl

Isoxadifen-ethyl, as a safener, is used at the post-emergence stage to protect maize from various weeds. It can be used along with **foramsulfuron**, **rimsulfuron** and **dicamba**.

Isoxadifen-ethyl is a derivative of isoxadifen ($C_{16}H_{13}NO_3$), with an IUPAC name of 4,5-dihydro-5,5-diphenyl-1,2-oxazole-3-carboxylic acid. The CAS Registry Number of this parent compound is 209866-92-2.

Isoxadifen is observed to cause ill effects on organs on chronic exposure. Neither isoxadifen-ethyl nor isoxadifen is registered in the USA.

Isoxaflutole herbicide

Isoxaflutole, $C_{15}H_{12}F_3NO_4S$ ((5-cyclopropyl-1, 2-oxazol-4-yl) (α, α, α-trifluoro-2-mesyl-*p*-tolyl) ketone), is an oxazole or cyclopropylisoxazole herbicide. This 4-benzoyl isoxazole molecule is a colourless-to-pale yellow solid, sparingly soluble in water, stable to light and heat for 14 days at 54°C. The DT_{50} in water is 1 day at pH 7. The CAS Registry Number is 141112-29-0.

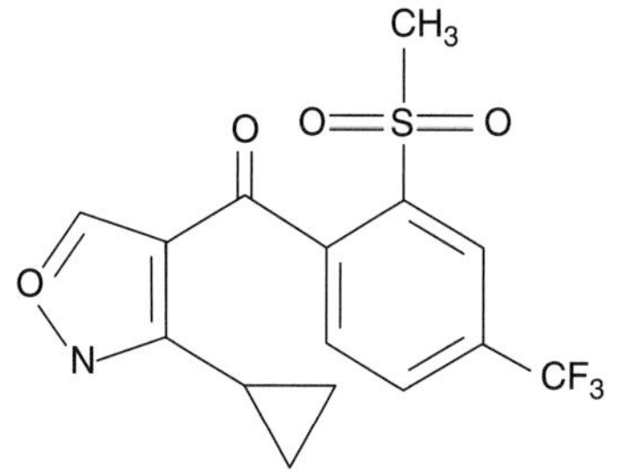

Isoxaflutole

A systemic, broad-spectrum herbicide, used at pre-emergence or pre-plant stage, isoxaflutole controls annual grass weeds and broadleaved weeds in maize, rice and sugarcane. It is a **4-hydroxyphenylpyruvate dioxygenase** inhibitor. It thus disrupts pigment biosynthesis.

The herbicide isoxaflutole is formulated as suspension concentrates (SC) and water-dispersible granules (WG). It can be applied at 75–140 g/ha. It may be used in place of **metolachlor**, **acetochlor** and **pendimethalin** for the same effect.

Isoxaflutole is a slightly toxic compound (with an acute oral LD_{50} among rats of >5000 mg/kg and a dermal LD_{50} of >2000 mg/kg b.w.). It is slightly toxic to bird species and to bees. It may cause irritation to the eyes and skin. In the case of extreme exposure, contaminated clothing should be removed and the patient should be made to wash himself clean with soap and water, after which the patient should be moved into fresh air. This chemical has been shown to be carcinogenic and has thus been classified as a PAN Bad Actor chemical.

The herbicide isoxaflutole is mobile in soil, but degrades easily. One of the breakdown products is mobile, persistent and phytotoxic. The chemical is rapidly metabolized in plants.

Isoxaflutole is registered in many countries of the EU, the USA, Canada, Australia, New Zealand and many countries of Africa such as Cameroon, Chad and South Africa. (See also **4-Hydroxyphenylpyruvate dioxygenase (HPPD)**.)

Isoxapyrifop herbicide

Isoxapyrifop, $C_{17}H_{16}Cl_2N_2O_4$ ((*RS*)-2-[2-[4-(3, 5-dichloro-2-pyridyloxy) phenoxy]propionyl]-isoxazolidine), is an aryloxyphenoxypropionic herbicide. Its CAS Registry Number is 87757-18-4. Isoxapyrifop is listed as an obsolete pesticide by the WHO. It is not registered by the USA (see Isoxapyrifop at bottom of the page).

Isoxathion insecticide

Isoxathion, $C_{13}H_{16}NO_4PS$ (*O,O*-diethyl *O*-5-phenyl-1,2-oxazol-3-yl phosphorothioate), is an isoxazole organothiophosphate insecticide. It is a pale yellow liquid, soluble in water at 1.9 mg/l and readily soluble in organic solvents. It is unstable to alkaline materials and decomposes at 160°C. Isoxathion has a CAS Registry Number of 18854-01-8.

Isoxathion, a broad-spectrum, cholinesterase-inhibiting compound, has contact and stomach action. It is effective against a variety of aphids and **scale insects** (in the families Diaspididae and Margarodidae). The compound thus protects crops like cabbage, citrus and rice.

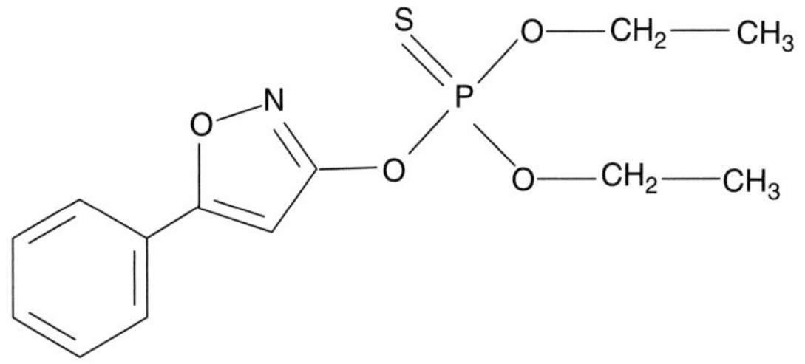

Isoxathion

The insecticide is formulated as emulsifiable concentrates (EC), wettable powders (WP), dustable powders (DP) and micro-granules. It is commonly available under the trade name of Karphos. Application rates vary from 500 to 1000 g a.i./ha.

The compound is a highly hazardous chemical according to the WHO (Class Ib toxin). The acute oral LD_{50} among rats is 112 mg/kg, while the dermal LD_{50} is >2000 mg/kg b.w. It is very toxic to honey bees (the acute 48 h contact LD_{50} being 0.082 μg per bee) and to fish and birds. The ADI of isoxathion is 0.003 mg/kg b.w./day. Being a cholinesterase inhibitor and a highly toxic chemical, isoxathion is considered a PAN Bad Actor chemical. Symptoms of poisoning can include excessive salivation and sweating, headache, muscle twitching, pin-point pupils, weakness, tremor, incoordination, dizziness, nausea, vomiting, diarrhoea and seizures.

The DT_{50} in soil is up to 40 days. In animals, the chemical metabolizes, after a series of reactions, to form 3-hydroxy-5-phenylisoxazole, before being eliminated. There is a similar degradation route of the insecticide in plants.

Isoxathion is registered in Vietnam, but is not allowed in the USA or throughout the EU.

Isoxazolidinone herbicides

The isoxazolidinone group of herbicides includes pigment-production inhibitors, also called carotenoid inhibitors (the other group of pigment-production inhibitors being called protoporphyrinogen oxidase or **PROTOX inhibitors**). Isoxazolidinone herbicides are pre-emergence herbicides used in the soybean crop. They are used for selective control over a wide range of weeds. The main active ingredient in this group is **clomazone**.

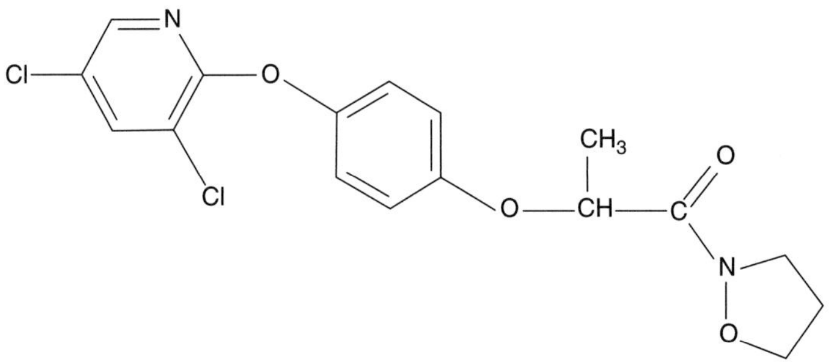

Isoxapyrifop

J

Jasmonic acid (JA) as a PGR

Jasmonic acid (JA), $C_{12}H_{18}O_3$ ((1*R*,2*R*)-3-oxo-2-(2*Z*)-2-pentenyl-cyclopentaneacetic acid), is a methyl ester-based plant growth regulator, with a CAS Registry Number of 6894-38-8. It is a hormone-like signalling substance found in jasmine and many other flowers and fruits. It is a volatile substance derived from the oxygenation of the octadecanoid α-linolenic acid by the enzyme lipoxygenase. While JA prompts and signals senescence and fruit ripening, it can inhibit growth, germination and photosynthesis. The chloroplast contains many of the enzymes needed for JA biosynthesis. JA is also found in some fungal species.

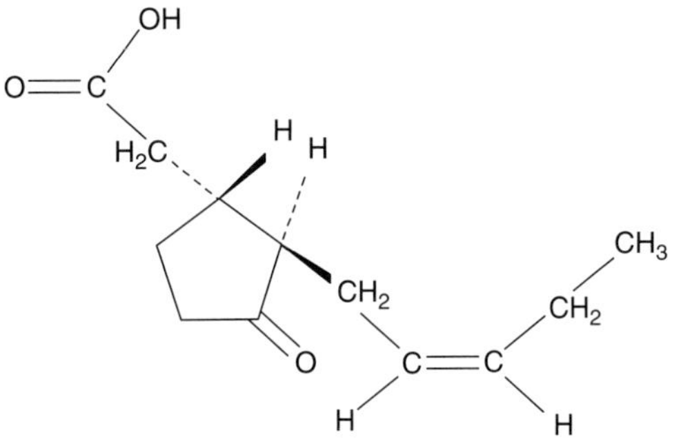

Jasmonic acid

The role of JA is in signalling a defensive response in the plant, which is seen in a heightened response to a pest attack, an injury or wound (**systemic acquired resistance**). JA also signals a defensive response to unfavourable weather conditions. JA also produces, in the plant, attractively fragranced nectar, to attract natural enemies of its pests. JA and its derivatives are used on potatoes, strawberries, grapes and bananas.

Jasmonate and its derivatives, such as methyl jasmonate (MeJA), can be synthesized from the fatty acid linolenic acid in the oxylipin octadecanoid pathway. These two are used to prevent fruit (e.g. banana) from ripening, and as microbial agents in stored crops.

JH

See **Insect growth regulators (IGRs)**

Jodfenphos insecticide

Jodfenphos, $C_8H_8Cl_2IO_3PS$ (*O*-2, 5-dichloro-4-iodophenyl *O*,*O*-dimethyl phosphorothioate), belongs to the phenyl organothiophosphate insecticide group. It is a colourless crystal with a mild odour, soluble in water (2 mg/l), ketones, aromatic hydrocarbons, alcohols and aliphatic hydrocarbons, in varying degrees. It is relatively stable in weak acids and alkalis, but unstable in strongly acidic and alkaline media. It has a CAS Registry Number of 18181-70-9.

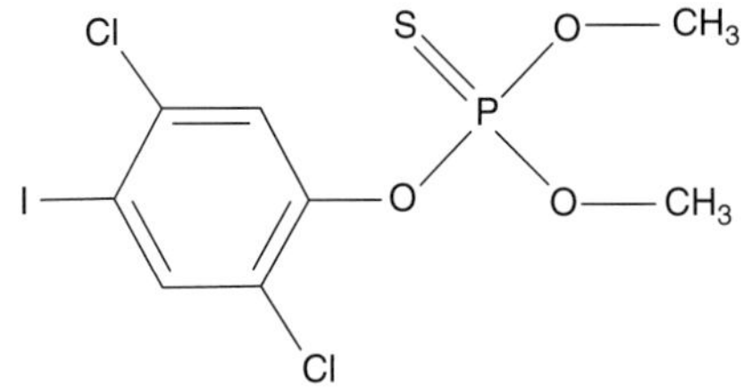

Jodfenphos

Jodfenphos (also known as **iodofenphos**), an organophosphorus cholinesterase inhibitor, is mainly used to protect stored grains, livestock (against ectoparasites) and public places against a host of insect pests like flies, cockroaches, moths, mosquitoes, weevils, mites, ticks and lice. It is thus an insecticide-acaricide of low mammalian toxicity.

The formulation types are emulsifiable concentrate (EC), wettable power (WP), liquid suspension and flowable concentrate for seed treatment (FS).

Jodfenphos has a high toxicity to fish and water fleas. It is moderately toxic to birds and bees. Product labels are required to display the signal word 'CAUTION'. It is classified as a Class III toxin (slightly hazardous) by the WHO. The acute oral LD_{50} in rats ranges from 2100 to 3000 mg/kg and in dogs is 3000 mg/kg. The dermal LD_{50} in rats is 2150 mg/kg.

The compound may be absorbed from the gastrointestinal tract, skin and by inhalation. In mammals, jodfenphos is rapidly eliminated, mainly in the urine, with metabolites being phosphoric acid, dimethyl phosphoric acid, monomethyl phosphoric acid and desmethyl-jodfenoxon. It does not accumulate in body tissues.

This insecticide, considered obsolete by the WHO, is not registered in the USA or throughout the EU.

Johnson grass

See **Weed – Johnson grass**

Jojoba oil fungicide

Jojoba oil, derived from jojoba seeds (*Simmondsia californica* Nutt), is used as a fungicide, an insecticide and a repellent. It is a stable golden liquid with a faint odour. When ignited, jojoba oil produces carbon monoxide, carbon dioxide and thick smoke, all of which can cause irritation. It stores well in cool, dark, dry and ventilated areas in airtight containers (below 35°C). Jojoba oil has a CAS Registry Number 61789-91-1.

Jojoba oil acts by forming a layer between the leaf surface and the pest. Its application on foliage kills fungal spores and the mycelium of **powdery mildew** pathogens. It suffocates whitefly eggs and larvae and kills them. Cotton and fruits like melons and grapes are protected by jojoba oil. As a repellent, it keeps insects, dogs and cats away.

Very low concentrations of jojoba oil are sufficient for the oil to be effective. Spray formulations are common, with trade names including E-erase and Permatrol. This oil may be used with other crop protection agents for a wider reach.

Jojoba oil is not toxic in normal use. It may cause irritation to the skin and the eyes in cases of prolonged exposure. But since oils are toxic to aquatic organisms, jojoba oil should not be disposed of in water bodies. However, due to its low toxicity, it is safe for humans, the environment and for non-target organisms.

Upon ingestion, jojoba oil is eliminated unchanged in the faeces, so accumulation in body tissues is rare. It is rapidly decomposed in the environment and has a US Organic Materials Review Institute (OMRI) approval for its use in organic farming. Jojoba oil is registered in the USA.

Juglone fungicide

Juglone, $C_{10}H_6O_3$ (5-hydroxy-1, 4-naphthalenedione or 5-hydroxynaphthoquinone), is a naturally occurring orange, solid, organic compound. Juglone has a CAS Registry Number of 481-39-0. It is insoluble in benzene, poorly soluble in water, but soluble in dioxane. It exists as a glycoside in parts of black walnut, pecan and some other plants (particularly the roots, nuts, leaves and shells of the black walnut).

Juglone is synthesized from isochorismic acid and 2-oxoglutaric acid. Juglone is obtained by the oxidation of 1,4,5-trihydroxynapthalene, derived from glucoside.

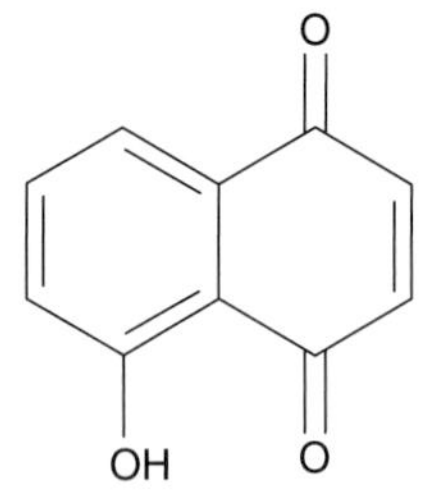

Juglone

The compound juglone inhibits the growth of plants around trees that contain it, especially the black walnut (*Juglans nigra*); such an inhibitory property is called **allelopathy**. It inhibits respiration of sensitive plants such as tomato, potato, apples, berries and azalea.

Juglone has fungicidal/**fungistatic** and bacteristatic effects, with the mode of action being blockage of transcription by RNA polymerase II at the cell level. Juglone is a highly selective, irreversible inhibitor of peptidyl-prolyl isomerases which facilitate the isomerization of peptide bonds. Anthracnose and **scab** pathogens are controlled in the presence of juglone.

Juglone is toxic upon ingestion. It causes irritation to the eyes, respiratory system and skin.

Some insects and plants metabolize juglone converting it into non-toxic substances. Tolerant crops are onions, maize, soybeans, wheat and snap beans. Insects detoxify the compound into 1,4,5-trihydroxynaphthalene.

Juvenile hormones (JH)

See **Insect growth regulators (IGRs)**

K

Kadethrin insecticide

Kadethrin, $C_{23}H_{24}O_4S$ (5-benzyl-3-furylmethyl (1*R*, 3S)-3-[(*E*)-(dihydro-2-oxo-3(2*H*)-thienylidene) methyl]-2, 2-dimethylcyclopropanecarboxylate), is a pyrethroid ester insecticide. Its CAS Registry Number is 58769-20-3. It is a yellow, viscous oil, insoluble in water but soluble in organic solvents. It is extremely volatile and unstable to light and heat. Similarly, it is incompatible with alkaline and lime-based materials, and also with strong oxidizers. It is to be stored in aluminium cans. It is produced by esterification of 2,2-dimethyl-3-(2,3,4,5-tetrahydro-2-oxothien-3-ylidene methylcyclopropanecarboxylate with 4-(2-benzyl) furfuryl alcohol (see Kadethrin at bottom of the page).

This insecticide mainly has household applications. Due to its potency, it has to be applied with appropriate PPE, like dust- and splash-proof safety goggles and with gloves.

Kadethrin is toxic to bees and fish and it is moderately toxic to rats with an LD_{50} of 1324 mg/kg. Direct contact with skin and eyes can cause injury. If consumed, the compound causes nausea, vomiting and abdominal pain, cough, chest pain and respiratory failure.

It is commonly formulated as a spray. Trade names include AgroEvo.

Kadethrin breaks down rapidly in soil or in sunlight, due to its photo-instability.

Kairomones

Kairomones are volatile compounds or **semiochemicals** released by organisms for the specific purpose of self-protection, mating or added nutrition. These chemical messengers specifically benefit individuals of another species and not the emitter. Sometimes, it may actually harm the emitter. Floral scents are kairomones for the prospective honey bee. The scales of *Heliothis armigera* act as a kairomone for the parasitic wasp ***Trichogramma** chilonis*, while *Heliothis zea* emits kairomones that attract *Trichogramma evanescens*.

Hexatriacontane, hexadecane, pentacosane, docosane and mixtures of eugenol with **geraniol** or phenyl ethyl are examples of kairomones.

Kairomones (methyl eugenol, cucurbitacins, phenols and 1-octen-3-ol) are used for inter-specific communication.

With the growing need for more specific and less environment-polluting pest control measures, kairomones are used to attract known crop pests to a desired location, where they can be trapped or killed. These chemicals may also be used to attract **beneficial organisms** for monitoring their rehabilitation or for mating.

Kairomones are used along with simple, sticky traps for control purposes. Chemical attractants can be used in conjunction with inexpensive traps for monitoring insect pest populations.

Karanjin bioinsecticide

Karanjin, $C_{18}H_{12}O_4$ (3-methoxy-2-phenylfuro [2, 3-h] chromen-4-one), is a plant-derived flavonoid (furanoflavonol) insecticide and acaricide. It is soluble in methanol, ethanol, chloroform, benzene, nitric acid, acetic acid and sulfuric acid. It is obtained from the seeds of the *Karanja* tree (*Millettia pinnata*, also known as *Pongamia glabra* Vent.) or from the Indian beech (*Derris indica* (Lam)) a tree growing wild in south India. Karanjin has a CAS Registry Number of 521-88-0.

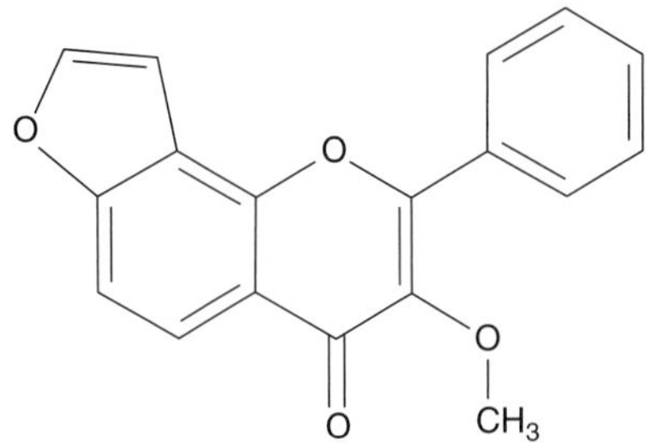

Karanjin

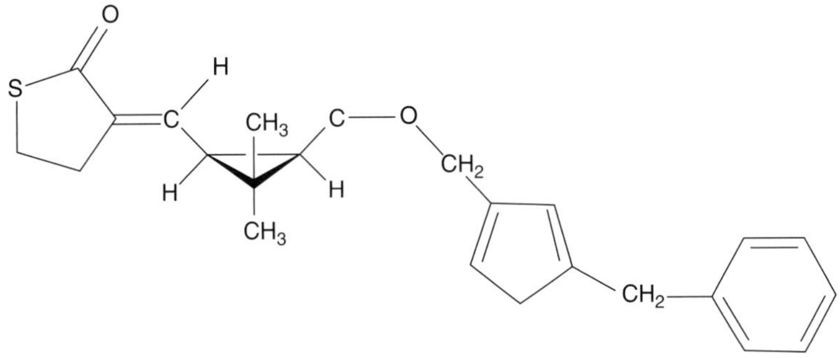

Kadethrin

Karanjin acts as an anti-feedant or repellent on insects by inhibiting nitrification and by affecting the nervous system. It also has IGR properties since it suppresses the hormone **ecdysone**. Pests such as mites, **scale insects, whitefly, thrips** and leaf miners stop eating and die. Vegetables, cotton, tea and tobacco are protected with the application of karanjin.

Karanjin is formulated as an emulsifiable concentrate (EC) and a known trade name is Escort. Normal dosage ranges from 100 to 500 g/ha with frequent applications being recommended. Karanjin is incompatible with sulfur.

Karanjin has no adverse effect on handlers, non-target organisms or the environment.

Karbutilate herbicide

Karbutilate, $C_{14}H_{21}N_3O_3$ (3-(3,3-dimethylureido) phenyl *tert*-butylcarbamate), is a carbamate herbicide. It is a white crystalline solid, soluble in water at 94.4 mg/l and stable under storage.

Karbutilate controls annual and perennial broadleaved weeds, grasses and vines. It has mainly non-crop uses, for example along the banks of drainage ditches and industrial plant sites.

The formulation types of karbutilate are granules (GR) and wettable powder (WP). It can be combined with other pesticides like **simazine**. Trade names include Tandex.

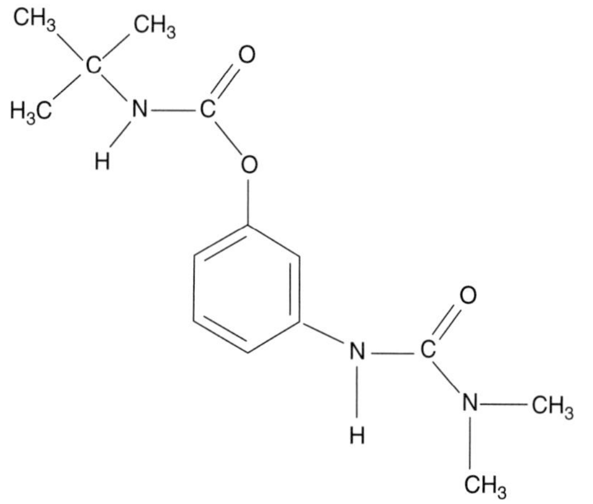

Karbutilate

The acute oral LD_{50} of karbutilate in albino rats is 3000 mg/kg b.w. The herbicide is non-toxic to wildlife and fish. However, it may be slightly irritating to the skin and harmful upon ingestion or inhalation, hence the ADI is 0.005 mg/kg b.w.

Karbutilate is registered in the USA.

Karpa disease in cotton

See **Crop – cotton pest control**

Kasugamycin biofungicide

Kasugamycin, $C_{14}H_{25}N_3O_9$ ([5-amino-2-methyl-6-(2,3,4,5,6-pentahydroxycyclohexyloxy) tetrahydropyran-3-yl] amino-α-iminoacetic acid), is a microorganism-derived systemic bactericide and antibiotic fungicide. It is isolated from the soil actinomycete *Streptomyces kasugaensis* by fermentation. Kasugamycin is a whitish solid and has a CAS Registry Number of 6980-18-3. It is available as kasugamycin hydrochloride hydrate (kasugamycin HCl). Kasugamycin hydrochloride hydrate has a CAS Registry Number of 19408-46-9 and a molecular formula of $C_{14}H_{28}ClN_3O_{10}$. It is highly soluble in water.

Kasugamycin is an aminoglycoside antibiotic that inhibits both Gram-positive and Gram-negative bacteria and blocks translation initiation. It also controls the infection caused by *Pseudomonas* spp., *Erwinia atroseptica*, *Cladosporium fulvum* and *Xanthomonas campestris* pv. *vesicatoria*. The crops thus protected by kasugamycin are beans, tomatoes, potatoes, carrots, onions and citrus fruit.

Kasugamycin hydrochloride hydrate, a systemic compound, inhibits protein biosynthesis, preventing incorporation of amino acids into protein. Its curative and protective properties controls and prevents rice blast, leaf spot, grey mould and black rot (e.g. caused by *Cercospora* and *Pyricularia grisea*). It is also a bacteriostat and a wood preservative.

Kasugamycin hydrochloride hydrate is available as wettable powder (WP), dustable powder (DP) and granules (GR). Trade names include Kasumin and Kasai. It is applied as a foliar spray, a dust or a seed treatment agent at rates of 20 g a.i./l, or more. It should not be combined with alkaline compounds (see Kasugamycin hydrochloride hydrate at top of the next page).

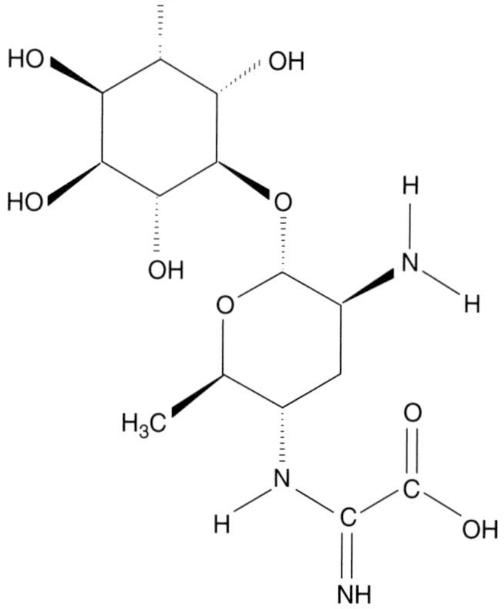

Kasugamycin

Kasugamycin hydrochloride hydrate is not highly toxic to mammals. Its toxicity is Class U according to the WHO, and IV according to the EPA. The acute oral LD_{50} of hydrochloride hydrate in male rats is >5000 mg/kg and the dermal LD_{50} in rabbits is >2000 mg/kg. It is non-irritating to eyes and skin and excreted with no residues in the organs. In plants and soils it is degraded into kasugamycinic acid and finally to ammonia, oxalic acid, carbon dioxide and water.

Neither of these compounds are permitted in the USA or the EU. India has registered the use of kasugamycin.

Kasugamycin hydrochloride hydrate

See **Kasugamycin biofungicide**

Kawadi disease in cotton

See **Crop – cotton pest control**

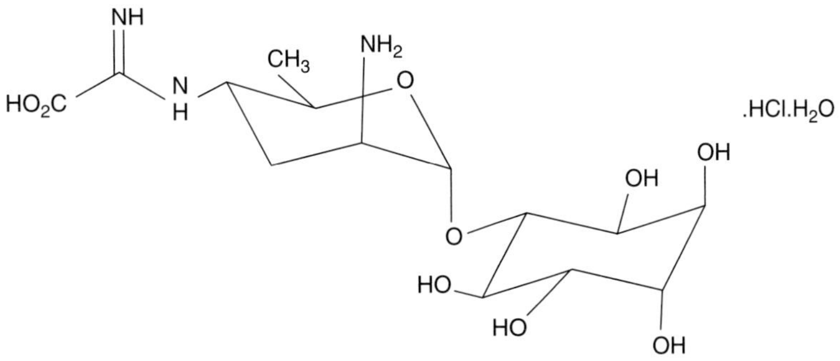

Kasugamycin hydrochloride hydrate

Khaira disease of rice

Khaira disease of rice, common among transplanted rice, is a non-pathogenic condition and may be caused by non-availability of zinc to the plants. The disease occurs particularly during the period of decomposition of the previous season's stubble in the flooded field. Symptoms of *khaira* are similar to ***pansukh* disease**, and begin with **chlorosis**, seen in the basal leaves with brown-bronze spots on the lamina, leading to the drying up of the entire leaf. Roots are unhealthy and pruned. Yields are severely affected.

Khaira may be controlled by spraying a solution made of zinc sulfate (2.5 kg) and slaked lime or calcium hydroxide (1.25 kg) in 500 l water. The first application may be done as soon as the early symptoms appear followed by a second spraying 10 days thereafter.

Kharif crops

Crops that are planted at the onset of rains are called *kharif* crops (*kharif* meaning autumn). Such a distinction is possible in regions which have a sustained rainy season or monsoons. The *kharif* season is prevalent in the vast region of Asia, which characteristically has a long spell of hot summer followed by monsoon. So, these crops can be described as summer to autumn crops. The *kharif* season normally begins by June and ends in October, although the months may change depending on location and seasonal fluctuation.

Common *kharif* crops are rice, sorghum, millet, green gram, pigeon pea, groundnut and soybean.

There are a few crops that are sown in the autumn season and harvested in winter. This seasonal cycle is known as ***rabi***, meaning spring. The *rabi* season is roughly between September and April, and begins at the end of the monsoon season. Wheat, gram, peas and barley are examples of *rabi* crops.

KIH 2023

See **Bispyribac-sodium herbicide**

KIH 2031

See **Pyrithiobac-sodium herbicide**

Kittleson's killer

See **Captan fungicide**

Kresoxim-methyl fungicide

Kresoxim-methyl, $C_{18}H_{19}NO_4$ (methyl (*E*)-2-methoxyimino-[2-(*O*-tolyloxymethyl) phenyl] acetate), is a fungicide. With a CAS Registry Number of 143390-89-0, it is a light-brown crystalline substance, insoluble in water (2 mg/l). It is produced by the Grignard reaction of 1-(2-methylphenoxy)-2-bromobenzene in tetrahydrofuran (THF) with methyloxalylimidazole to give 2-(2-methyloxalylimidazole)phenylglyoxylic acid methyl ester, which upon reaction with methylhydroxylamine hydrochloride in methanol containing sodium carbonate yields the desired product.

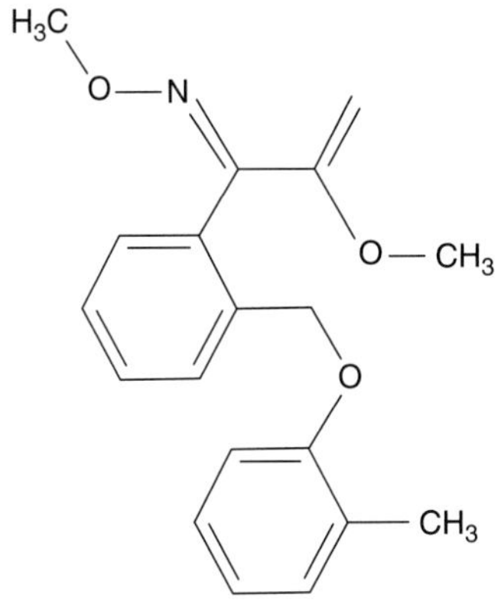

Kresoxim-methyl

Kresoxim-methyl, a systemic compound with protectant properties, acts by inhibiting mitochondrial respiration, spore production and germination and mycelial growth. Its major action is in the control of **scab** and **powdery mildews** on vines, apples, cucurbits, oak trees and ornamentals like dahlias, geraniums, marigolds and roses. The pathogens controlled are *Venturia inaequalis*, *Podospheara leucotricha* and *Uncinula necator*. Kresoxim-methyl has a long residual action, a gas phase and high lipophilicity. Thus curative and protective activity is seen on leaves and within the plant.

The compound is formulated commonly as a suspension emulsion (SE) and a suspension concentrate (SC).

Kresoxim-methyl has an acute oral LD_{50} of >5000 mg/kg b.w. and a dermal LD_{50} of >2000 mg/kg b.w. in rats. It is not a skin or eye irritant. It is not permanently toxic to fish, earthworms or bees. The ADI for kresoxim-methyl is 0.4 mg/kg b.w./day. However, kresoxin-methyl is considered to be a carcinogen and has a high potential to contaminate underground water sources (with an adsorption coefficient, K_{OC}, of 437).

In mammals, kresoxim-methyl is rapidly metabolized to many products including the inactive carboxylic acid. These are quickly eliminated via the urine and faeces. There are no cases of **bioaccumulation** recorded. The soil DT_{50} of kresoxim-methyl ranges from <5 days to 35 days, depending on the soil type, while its acid metabolite is considered persistent. In water, this fungicide dissipates in <2 days.

This compound is registered throughout the EU (except one country), India, Australia, the USA, South Africa, Tanzania and many other countries. According to the EU Pesticide Database, the MRL for soft fruit and sweet peppers is 1.0 mg/kg, for aubergine and tomatoes 0.5 mg/kg, for pumpkins, winter squash and pome fruits 0.2 mg/kg and for cereal grains 0.05 mg/kg.

L

Lacewing *Chrysoperla* sp.

See ***Chrysoperla* as biocontrol agents**

Lacrimatory agent

A chemical agent that stimulates the causing of tears is a lacrimatory agent. Many chemical pesticides (e.g. **chloropicrin**) have lachrymatory action. The term originates from Latin '*Lacrima*' meaning tear. Benzyl bromide, ethyl bromoacetate, xylyl bromide, chloroacetophenone (CN) and *o*-chlorobenzylidenemalononitrile (CS) are some substances used as lacrimatory agents. Many pesticide active ingredients are lacrimatory and may cause irritation, redness, watering and even injury on exposure.

Lady beetle

See ***Adalia bipunctata* (ladybug) as a biocontrol agent**; **Ladybird beetles as biocontrol agents**

Ladybird beetles as biocontrol agents

Ladybird beetles, belonging to the family Coccinellidae and order Coleoptera, are a large group of small, brightly coloured and spotted insects that are widely present throughout the world. Ladybird beetle adults, which grow up to 10 mm in length and have mandibles to chew and bite, are variously called lady bugs and lady beetles. They are important to agriculture since they are predatory in nature and can voraciously feed on insect crop pests like aphids, scale insects, mites and whiteflies. Thus these beetles are exploited for their biocontrol property, to ensure biological control of pests and for pest management, which might help minimize pesticide application.

The general life cycle of ladybird beetles includes stages including eggs, which hatch into larvae, having three to four instars, which in turn pupate to form adults. The larvae and adults cause the highest casualty to pest populations.

Stethorus utilis (subfamily Scymninae) and *Stethorus punctillum* feed on tetranychid mites, while *Cryptolaemus montrouzieri* feeds on citrus **mealybugs**. *Menochilus sexmaculatus* feeds on aphids, scale insects and whiteflies, while many species of *Delphastus* are known to feed on whiteflies. *Rhyzobius lophanthae* feeds on scale insects. ***Adalia bipunctata*** and other species are commonly found in many parts of India and Asia. **Convergent lady beetles** *Hippodamia convergens* and many other species of ladybird beetles are commercially available for use in vegetable crops and orchards.

Ladybird beetles are threatened by various wasps and **tachinid flies** (e.g. *Epilachna* spp. and *Aplomyiopsis* spp.), although these beetles are known to spray a toxin on the predator that repels or even kills it.

Ladybird beetle *Menochilus sexmaculatus* (Coleoptera) is a predatory insect on many crop pests including **aphids, scale insects** and **whitefly**. These form an important tool in the overall biological pest control in farm situations afflicted by aphids. Ladybeetles become particularly precious as natural enemies, as they have the ability to be sustained by other prey when aphids become scarce in number.

Adults are oval to semi-circular in shape and yellow to red in colour with black spots or stripes (Figs L.1 and L.2). The adult female lays 10–50 yellow-coloured eggs on the lower side of the leaves or near an aphid colony. The eggs hatch in about 1 week to give greyish coloured 4 mm long larvae with three pairs of legs and pointed mandibles with which they feed on smaller insects. These larvae prey on the leaves and stems of plants and then pupate, to emerge as adult beetles in about 3–12 days depending on temperature. *M. sexmaculatus* are voracious predators of aphids as they are capable of feeding on about 5000 aphids in their whole life time. Once the supply of aphids is exhausted, they search for other insects.

Ladybird beetles can minimally tolerate a few chemical insecticides applied in fields at recommended rates, except for aphid insecticides such as **carbaryl** and **methamidophos**.

Ladybug

See ***Adalia bipunctata* (ladybug) as a biocontrol agent**; **Ladybird beetles as biocontrol agents**

Fig. L1. Ladybird beetle. Sketch courtesy of Leena Achari.

Fig. L.2. Adult ladybird beetle on a soybean leaf. Courtesy of John Obermeyer, Purdue Extension Entomology. With permission.

Laminarine biofungicide

Laminarine, $(C_6H_{10}O_5)_n$ (where n = 20–30), is a polysaccharide carbohydrate fungicide. It is extracted from kelp (seaweeds) and the marine brown algae, *Laminaria digitata* (Hudson). Laminarine has a CAS Registry number of 9008-22-4 and an IUPAC name of (1→3)-β-D-glucan.

Laminarine, a preventive systemic, pre-emergence fungicide, stimulates the plant's natural defence mechanism and thus is a **systemic acquired resistance** (**SAR**) inducer. This fungicide is ideal for treatment of wheat against *Septoria* sp. and **powdery mildews**.

Laminarine is formulated as a wettable powder (WP) which is sprayed on leaves with ground equipment. A common trade name is Iodus 40. It may be used with other crop protective agents (see Laminarine at bottom of the page).

The acute oral LD_{50} of laminarine among rats is >2000 mg/kg, while the dermal LD_{50} is >5000 mg/kg b.w. The acute LD_{50} in bobwhite quails is >2000 mg/kg. In short, laminarine is non-toxic to mammals and does not have any adverse effect on non-target organisms or on the environment. It is approved for organic farming.

Laminarine is registered for use in EU countries, including Greece and the UK and also in the USA.

Larvicidal oil analysis

A larvicidal oil is a mixture of mineral oils, free from any extraneous impurities including water and dirt, but excluding additives to improve physical properties. A larvicide kills 90% of *Anopheles stephensi* and 75% of *Aedes aegypti*. An environmentally safe pesticide, it can be analysed by the following experiment. The sample, without any insecticidal chemical, should have a relative density of a maximum of 0.940 at around 30°C. Its **flash point** should be 65.6°C. The spreading pressure should be 4.6×10^{-2}N/m for Grade I, 2.5×10^{-2}N/m for Grade II and 1.8×10^{-2}N/m for Grade III. These form the three standard solutions.

The analytical method for spreading pressure determination involves the preparation of the three standard solutions, with different spreading pressures. Solution 1 is a 100 g/l of the solution of oleyl alcohol in medicinal paraffin (spreading pressure 4.6×10^{-2} N/m), solution 2 is a 10 g/l of oleyl alcohol in medicinal paraffin (spreading pressure 2.5×10^{-2} N/m) and solution 3 is 10 g/l of terpineol in medicinal paraffin oil (spreading pressure 1.8×10^{-2} N/m).

A glass funnel (of at least 20 cm diameter) is cleaned with chromic acid solution and rinsed with water to remove all acid traces. Fix the funnel in a vertical position in a retort stand and connect the stem to the water supply with a rubber tube. Fill the funnel with water until it is overflowing to give a clean surface for testing. Tilt the funnel slightly to bring the water level to about 3 mm below the rim of the funnel.

Take a clean glass rod in each hand, dip one rod into the sample and the other into the standard solution corresponding to the grade of the larvicidal oil and deposit a drop from each of the rods simultaneously on the water surface. Observe the spreading of the oil on the water surface, by observing the water surface at eye level.

If the sample occupies more than half the surface, the spreading pressure is greater than that of the standard. If the sample and the standard solution occupy about equal areas, the spreading pressures are almost equal. In this case, the sample is acceptable from the spreading pressure point of view. If the standard solution occupies more than half the surface, the spreading pressure of the sample is lower than that of the standard. In this case the sample is unacceptable. Only initial observations should be recorded.

Late blight

See **Plant disease late blight**

Lead hydrogen arsenate

See **Arsenical inorganic insecticides**

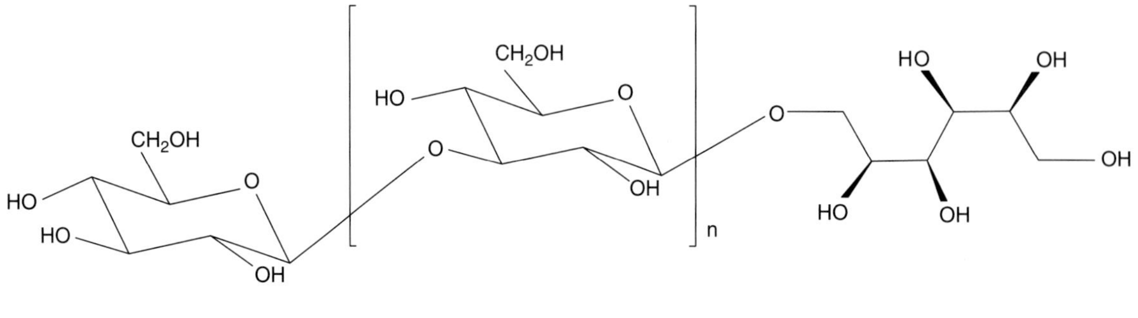

M series n = 20–30

Laminarine

Leaf spot
See **Plant disease *Alternaria* leaf spot of cole crops**

Leafhopper
See **Pest – leafhopper**

Legislation
See our **Pesticide legislation** series under 'P' for an outline of legislation for a few countries.

Lemonol
See **Geraniol bioinsecticide**

Lenacil herbicide
Lenacil, $C_{13}H_{18}N_2O_2$ (3-cyclohexyl-1, 5,6,7-tetrahydrocyclo-pentapyrimidine-2, 4(3*H*)-dione), is a **uracil** herbicide. It is available as colourless to tan-coloured crystals, slightly soluble in water and readily soluble in organic solvents. Lenacil has a CAS Registry Number of 2164-08-1. Lenacil is made by condensation of cyclohexylurea and ethyl-2-oxocyclopentane-1-carboxylate with H_3PO_4.

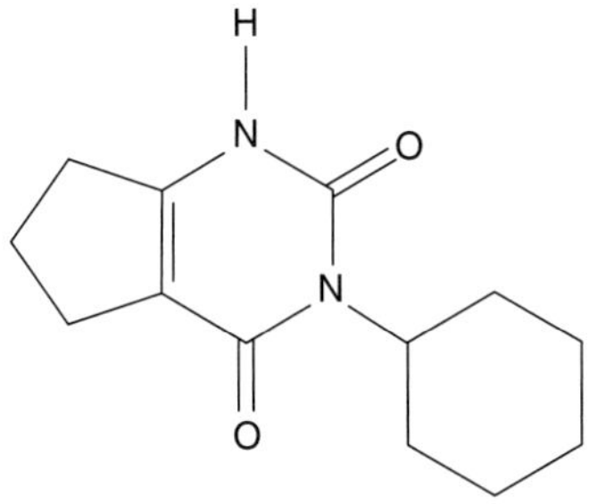

Lenacil

Lenacil is a selective, pre-plant, pre-emergence and systemic herbicide, absorbed by the roots and affecting the process of photosynthesis. It controls annual grasses and broadleaved weeds in sugarcane, sugarbeet, fodder beet and ornamental plants.

Available formulations of lenacil are wettable powder (WP) and soluble concentrates (SL) that are commonly sprayed on soil. It may be formulated with most herbicides. Trade names include Venzar and Adol.

According to the EPA lenacil is unlikely to present acute hazard in normal use. Irritation to skin, eyes and the respiratory tract may be caused in cases of acute exposure. In soil, complete microbial degradation of lenacil occurs within 5–6 months.

According to the EU Pesticide Database, the MRL of lenacil for sugarbeet is 0.2 mg/kg. Lenacil is registered for use in Cyprus, Greece, Hungary, Portugal and the UK.

Leptospermone bioherbicide
Leptospermone, $C_{15}H_{22}O_4$ (2,2,4,4-tetramethyl-6-(3-methylbutonoyl)cyclohexane-1,3,5-trione), the primary phytotoxic component from the bottlebrush plant (*Calistemon* spp.) was the basis for a synthetic, systemic, pre-emergence herbicide **mesotrione**.

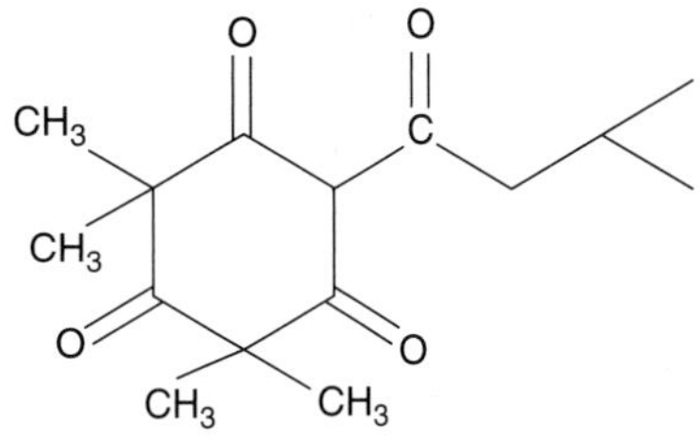

Leptospermone

Leptospermone is also the basis of a maize herbicide called **sulcotrione**. It is very effective on **common lambsquarters** and velvetleaf, especially when combined with **atrazine**. Leptospermone is also a bactericide that inhibits *Clostridium* species.

The herbicide-cum-antibacterial is considered non-toxic to the environment.

Lethal concentration$_{50}$ (LC$_{50}$)
See **Acute toxicity**

Lethal dose$_{50}$ (LD$_{50}$)
See **Acute toxicity**

Lethal dose low (LC$_{Lo}$)
See **Acute toxicity**

L-glutamic acid-γ-aminobutyric acid biofungicide
The combined use of naturally occurring compounds L-glutamic acid and gamma-aminobutyric acid has proved to be biofungicidal in nature and a yield booster in fruits, vegetables, fodder crops, grapes and ornamentals. It is seen to be effective against **powdery mildews** on grapes (see Gamma-aminobutyric acid and L-glutamic acid at bottom of the page).

The above combination, available as a wettable powder (WP), may be sprayed from the ground or soil drenched.

The US FDA gives the status of GRAS (generally recognized as safe) for L-glutamic acid. This combination product is considered safe and does not persist in the environment. Except for restriction of its direct use in surface waters, the combination of these two products is approved for organic farming.

Lime sulfur fungicide
Lime sulfur, or **calcium polysulfide** ($CaS{\cdot}S_x$), has been widely used since before the 1900s, against mildew, mites and **scale insects**. Boiling varying quantities of sulfur and lime

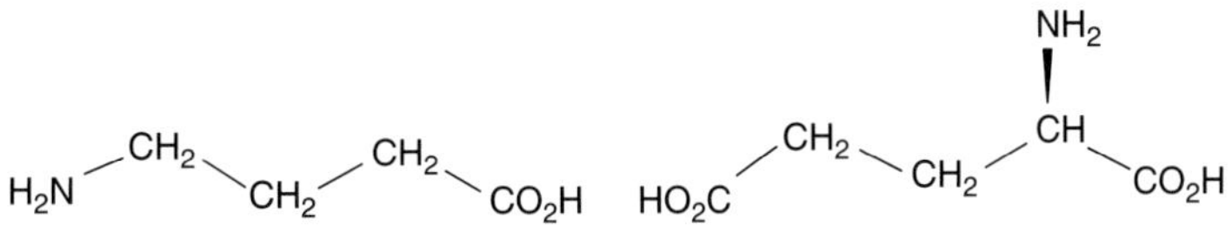

Gamma-aminobutyric acid and L-glutamic acid

water to different degrees and filtering out the solution gives lime sulfur and calcium thiosulfate.

Lime sulfur has contact and vapour action as a fungicide. Sulfur competes with oxygen as a hydrogen acceptor and disrupts hydrogenation and dehydrogenation in fungal cells, forming hydrogen sulfide. Such abnormal changes disturb normal functioning of the fungal cells.

High concentrations of lime sulfur can be phytotoxic. Dusting applications during the summer (particularly in bright sunny weather) are to be avoided as they may cause burning symptoms and prove to be phytotoxic, especially among cucurbits. The best time for sulfur dust applications are the early morning or after sunset. Being alkaline in reaction, lime sulfur is incompatible with a large number of pesticides.

While lime sulfur has been used as a contact fungicide, **elemental sulfur** is used as a residual fungicide.

Lindane insecticide

See **Hexachlorocyclohexane insecticide**

Linuron herbicide

Linuron, $C_9H_{10}Cl_2N_2O_2$ (3-(3,4-dichlorophenyl)-1-methoxy-1-methylurea), is a **phenylurea** herbicide (a substituted urea herbicide). It is an odourless, white crystalline solid, soluble in water at 75 ml/l (at 25°C), but more soluble in ethanol and acetone. It is incompatible with strong oxidizing agents. It is produced by reaction of 3,4-dichlorophenyl isocyanate with methoxymethylamine. Linuron has a CAS Registry Number of 330-55-2. It is a restricted use pesticide (RUP) according to the EPA.

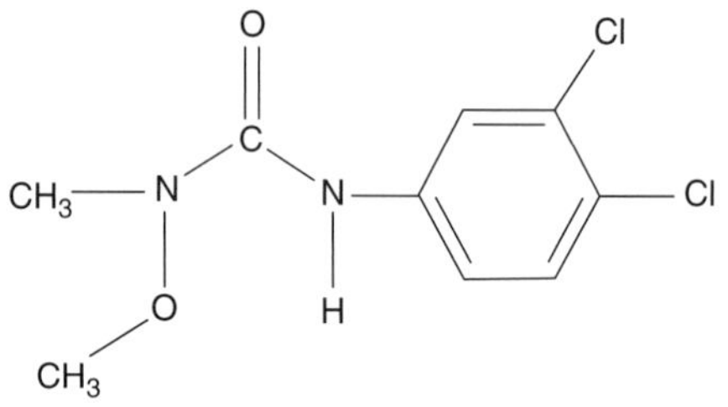

Linuron

Linuron, a pre- and a post-emergence selective compound, controls a wide range of broadleaved and grassy weeds on soybean, cotton, potato, maize, onion, winter wheat, legumes, vegetables and fruits. It has a post-harvest application. It is also effective on tulips, lilies, daffodils and on forestry plantations. It works by inhibiting photosynthesis.

Linuron is available as wettable powder (WP), dispersible granules (GR) and liquid concentrate. Trade names include Afalon, Linex and Linorox.

The acute oral LD_{50} is 1500 mg/kg in rats and 2400 mg/kg in mice. Linuron is slightly toxic to birds (the 5–8 day dietary LC_{50} is >5000 ppm in Japanese quails). It is toxic to fish and shellfish, but non-toxic to bees. Linuron formulations are required to have the signal word 'CAUTION' on their labels. It is mildly irritating to eyes and skin. The EPA classifies linuron as a possible human carcinogen and a potential groundwater contaminant and thus it is a PAN Bad Actor chemical. The ADI is 0.002 mg/kg/day. According to the WHO, linuron is a Class U toxic (unlikely to be hazardous).

Many substituted ureas like linuron are irritating to the eyes, skin and mucous membranes. They can cause coughing and shortness of breath, nausea, vomiting, diarrhoea, headache and confusion.

In the environment, linuron is broken down mainly by microbes, into a much less active metabolite 3,4-dichloroaniline, apart from carbon dioxide. Tolerant plants metabolize linuron into inactive products. In crops, linuron has low residual action and persistence. Linuron is moderately persistent in soils, with a field half-life of up to 150 days in various soils.

It is registered for use in Australia, Canada, the EU, India, South Africa and the USA.

Liquid baits

When active ingredients are mixed with sweetened water or with food-flavoured liquids so as to attract pests, they are called liquid baits. Some insecticides and rodenticides are available as liquid baits.

Liquid rodenticides are effective in areas where rodents cannot find water and where ready availability of food renders traditional baits ineffective.

Liquid insecticide baits, packaged as ready-to-use, sugar-based liquids are used for controlling ants and to a lesser extent, cockroaches.

Liquid formulations

A pesticide formulation in which the active ingredient (a.i.) is mixed with water, or an organic or oil-based solvent, is called a liquid formulation. Emulsifiable concentrates, soluble concentrates and flowables are examples of liquid formulations.

Pesticides that are insoluble in water are formulated in such a way that they form a stable suspension or emulsion when diluted with water. This enables the a.i. to be uniformly distributed over the target. But pesticides that are soluble in water or other solvents are formulated as solutions or soluble concentrates.

The solvent is mixed with a suitable wettable powder, dispersing agents or **wetting agents**, to form a formulation that can then be sprayed. Liquid formulations pose little inhalation hazard since the powder is suspended in water, permitting it to be poured.

However, liquid formulations and flowables share a common weakness with suspensions. The a.i. may be insoluble, so that agitation may be required before spraying. There may also be difficulty in mixing the suspension or rinsing out the container due to heavy sedimentation, which may also clog nozzles and pumps.

A solution or soluble concentrate (SL) is made of water-soluble a.i.s dissolved in water or a petroleum-based solvent for dilution prior to field application. Such formulations that may contain the a.i. and a carrier are easy to make, mix and apply and do not need any agitation.

Solutions may be used in any type of sprayer and are not abrasive to the equipment. Besides, phytotoxicity is not very common and except in cases of some dissolved salts, most are not harmful to the applicator. Typical examples are formulations with **paraquat** and **glyphosate**.

Solutions and soluble concentrates are common among seed-treatment formulations. The a.i. should either be soluble in water or be a seed-compatible solvent (a non-phytotoxic one). Seeds are commonly treated in a tumbling device. Seeds must be treated and dried so as to prevent germination, yet retain seed viability.

Liquid nitrogen as a termiticide

Liquid nitrogen, a colourless clear liquid, is nitrogen at very low temperatures. This cryogenic liquid is produced by fractional distillation of liquid air. Liquid nitrogen, referred to as LN_2, is soluble in ethanol but insoluble in water. It has to be stored in closed, non-corrosive containers, away from moisture. It has a CAS Registry Number of 7727-37-9.

Liquid nitrogen

The ability to maintain temperatures far below 0°C makes liquid nitrogen useful in storage and laboratory work which involves potassium nitrate or saltpetre (used in gunpowder) and ammonium nitrate (an important fertilizer and explosive).

Liquid nitrogen is approved for use as a pesticide for dry wood termites. The termites die when exposed to -29°C for 5 min.

The mode of application of liquid nitrogen is to inject it into spaces between walls and into wood where pests like carpenter ants and a variety of termites freeze to death.

However, since liquid nitrogen is extremely corrosive to the eyes and skin and as it can cause asphyxia, only certified applicators are allowed to handle it. The use of a waterproof coat, rubber gloves, protective glasses and head gear is highly recommended, as is washing with soap after the job is done. Its toxic concentration in humans is 90 ppm and in mice, 250 ppm.

Liquid nitrogen is considered safe to the environment upon application. Once it gets outdoors, it turns into a gas and disperses.

Low-concentrate solutions

Low-concentrate solutions are easy-to-apply and ready-to-use liquid formulations that are used mainly in households against home pests. They consist of a small amount of active ingredient (often 1% or less per unit volume) dissolved in an organic solvent like petroleum-based solvents. They come in small sizes, to avoid post-application storage problems. These formulations are used in spraying and fogging.

However, drift potential and inhalation problems may pose a risk. Such products are not widely available and are highly priced.

Low-input sustainable agriculture (LISA)

Any system of farming which attempts to maximize output by keeping the cost low is called low-input sustainable agriculture or LISA. Organic inputs are abundantly used under LISA and hence the two systems, LISA and organic farming, are similar.

But LISA differs slightly from organic farming. Although the two farming systems depend predominantly on natural and biologically available inputs, LISA does so to keep the costs of production low, whereas the organic farming system focuses on avoidance of synthetic inputs.

Hence, LISA will use easily available environmentally friendly material such as compost, biofertilizers, rock phosphates and green manure. Crop rotation with legumes is routinely practised for nitrogen supplies. Farmers advocating this low-input farming generally grow certain crops to break pest cycles. LISA does not exclude synthetic fertilizers and pesticides. However, they are used in moderation because the focus is to keep the energy cost low. LISA avoids the use of expensive mechanized tillage, major irrigation systems and aerial spraying of pesticides and fertilizers, knowing full well that although yields may be moderate, a low-cost structure can still ensure greater profits.

LISA attempts to synthesize the best management practices carried out in organic farming and select practices of modern agriculture.

Low-pressure sprayers

See **Application equipment – larger power-driven sprayers**

Lufenuron as an IGR

Lufenuron, $C_{17}H_8Cl_2F_8N_2O_3$((RS)-1-[2,5-dichloro-4-(1,1,2,3,3,3-hexafluoropropoxy)phenyl]-3-(2,6-difluorobenzoyl)urea), is an insect growth regulator of the benzoylurea class. Its colourless crystals are partially soluble in water (<0.06 mg/l) and soluble in ethanol, acetone, toluene, and *n*-hexane. It is stable in air and light. The CAS Registry Number is 103055-07-8 (see Lufenuron at bottom of the page).

Lufenuron, a **chitin synthesis inhibitor**, is an effective agent against lepidopteran members in crops, eriophid mites and **western flower thrips**. In non-crop situations too, lufenuron is effective against fleas on animals and on cockroaches in households. This chemical also acts as an anti-fungal agent in plants.

Lufenuron is formulated as an emulsifiable concentrate (EC) and is commercially available as Match. According to the WHO, lufenuron is a Class III toxin (slightly hazardous). It is safe on mammals, since it is not broken down by the liver or kidneys.

Lycosa spp. as biocontrol agent

Spiders of the genus *Lycosa* (Fig. L.3) belong to the order Araneae, which is the largest order of the class Arachnida. These arthropods are predatory in nature, and feed on many crop pests like brown plant hoppers, stem borers, **leafhoppers**,

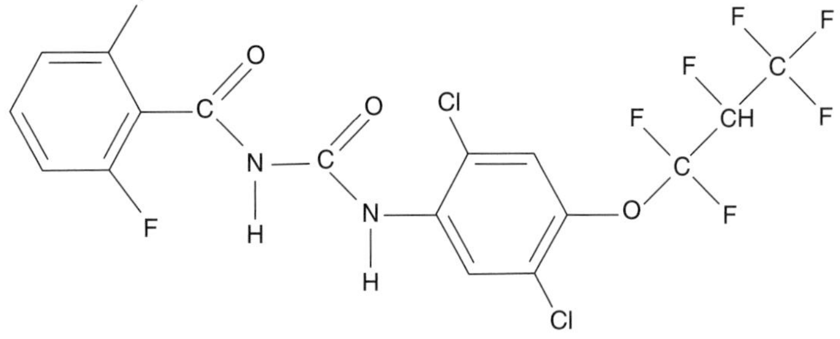

Lufenuron

Fig. L.3. *Lycosa* spp. Sketch courtesy of Leena Achari.

moths and flies. This makes them important from the viewpoint of pest management.

The adult is greyish with yellow stripes on the thorax. It has strong jaws with an orange elevated spot on both the sides. With eight eyes, these creatures grow to about 25 mm in length. *Lycosa* spiders live on the ground and are poor climbers. They pounce upon their prey, inject venom into it and suck out the body fluids.

The female spider protects her eggs in a silky egg sac. On hatching, the female carries the spiderlings on her back till they grow and alight.

Although not exploited fully, these spiders can be used to control plant hoppers and leafhoppers in rice fields and to protect other crops. The fact that they are not toxic to humans is an added benefit.

Toxic insecticides (especially **carbofuran**) should be avoided when planning to use *Lycosa* spp. for pest control.

M

MACE

See **Acetylcholinesterase**

Macronutrients

See **Plant nutrient excesses and deficiencies**

Magnesium as a plant nutrient

See **Plant nutrient excesses and deficiencies**

Malathion analysis

A widely used insecticide, malathion is a clear amber liquid that is slightly toxic. Technical analysis of this insecticide requires a sample with a minimum malathion purity of 92%. Acidity (as H_2SO_4) should be a maximum of 5 g/kg and water content, 1 g/kg (maximum).

Gas-liquid chromatography apparatus is needed for this analysis, with a glass column made of borosilicate (182 cm long, 2 mm in internal diameter and 6 mm in external diameter). The column packing material is chromosorb W-HP (100–120 mesh) treated with OV-210. Other equipment includes a flame-ionization detector. The temperature should be 180°C in the oven, 190°C at the injection port and 250°C in the flame-ionization detector. The gas flow rates of hydrogen, air and nitrogen (carrier gas) should be 30, 350 and 30 ml/min, respectively. The retention time of the malathion peak should be 10 min and of the internal standard peak, 7 min.

The internal standard solution chosen is 1,3-diphenoxy benzene. It is prepared by dissolving 30 g of internal standard in 1 l chloroform. Stable in sealed containers and under refrigeration, this solution is brought to room temperature before use. Analytical grade malathion is used for calibration. Weigh about 425, 500 and 575 mg (to the nearest 0.1 mg) quantities of standard malathion directly into 50 ml volumetric flasks. Pipette exactly 5 ml internal standard solution into each flask. Dilute to the mark and mark the flasks as A, B and C, respectively, with solution B being the working calibration solution for gas chromatography and the others being used for checking the linearity of the gas chromatography. These solutions also remain stable in sealed containers and under refrigeration, but must be brought to room temperature before use in experiments.

The experimental procedure highlights the need to regularly check the linearity of the gas-liquid chromatography equipment. Peak area measurements using a digital integrator determine the appropriate attenuation setting and the quantity of calibration solution B (between 2 and 4 μl) that must be injected to yield an area count of at least 100,000. The attenuation setting should be used for all samples and calibrations in the set. Inject an equal volume of the three calibration solutions in the gas-liquid chromatography system and record the response ratio for each injection and average these ratios for each solution. Divide the average response ratio for each solution by the corresponding malathion content in milligrams, which should lie within ±2%.

A sample containing about 500 mg malathion, along with 5 ml internal standard solution should be poured into a 50 ml volumetric flask, with the level being made up by chloroform. The contents of this flask should be mixed well. Inject duplicate aliquots of calibration solution B and record the response ratios (ensure they lie within 2%). Inject duplicate aliquots of the sample solution, record the response ratios for each sample solution and take the average. Use the average to calculate the malathion content of the sample.

The response ratio (r) for each injection is given by Eqn M.1 and the malathion content is calculated using Eqn M.2.

$$r = \frac{\text{Area of the malathion peak}}{\text{Area of internal standard peak}} \qquad \textbf{(M.1)}$$

$$\text{Malathion content} = \frac{r_2 \times m_1 \times P}{r_1 \times m_2} \qquad \textbf{(M.2)}$$

where

r_1 = Average ratio for calibration solution B

r_2 = Average ratio for sample solution

m_1 = Mass (g) of malathion standard in the calibration solution

m_2 = Mass (g) of sample taken

P = Purity of malathion standard (g/kg)

The water-dispersible powder (WS) of malathion is a homogenous mixture of technical grade malathion with fillers and other additives and can be easily stirred with water. Malathion is extracted from the sample with chloroform and malathion estimated by gas-liquid chromatography as explained above.

An impurity found in malathion, namely, isomalathion, is a very toxic ingredient. It is possible to determine the presence of isomalathion from the water-dispersible powder of malathion. About 1.5 g of the powder is dissolved in 25 ml chloroform and 2.0 ml internal standard solution is added and shaken to dissolve the malathion, allowing the phases to separate and for the supernatant solution to filter out. This may be used for analysis by gas-liquid chromatography. The response ratio for the injected aliquot is determined and compared with the response of a standard containing a known quantity of isomalathion to get the percentage of isomalathion in the sample. The retention time for isomalathion peak is 26 min compared with 7 min for the internal standard peak.

The isomalathion content can be calculated using Eqn M.3.

$$\text{Isomalathion content (g/kg) of sample} = \frac{r_2 \times m_1}{r_1 \times m_2} \qquad \textbf{(M.3)}$$

where
r_1 = Average of response ratio for calibration isomalathion solution
r_2 = Average of response ratio for sample solution
m_1 = Mass (mg) of isomalathion in the calibration solution
m_2 = Mass (g) of sample taken

The emulsifiable concentrate (EC) of malathion is a stable liquid and may also be analysed for malathion content. The sample, dissolved in chloroform, is added to the internal standard. This sample solution is estimated by gas-liquid chromatography according to the procedure described above by comparing the response ratio of the sample with that of the internal standard.

A similar procedure is applied for analysing malathion dustable powder (DP), which is a fine, free-flowing powder.

Malathion insecticide

Malathion, $C_{10}H_{19}O_6PS_2$ (*S*-1,2-bis (ethoxycarbonyl) ethyl *O,O*-di-methyl phosphorodithioate), is an organophosphorus insecticide and an acaricide. It is a clear amber liquid, soluble in water (145 mg/l) and miscible with solvents such as alcohols and esters. This compound is decomposed by acids and alkalis. Malathion has a CAS Registry Number of 121-75-5.

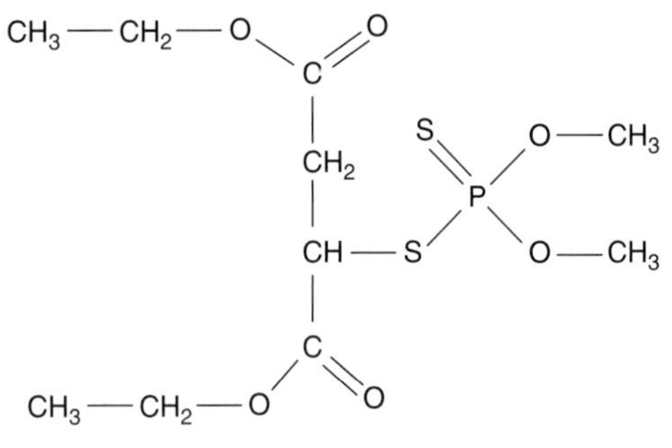

Malathion

Malathion is a non-systemic insecticide and a very widely used organophosphorus compound that finds application in agriculture, households and public places. It is also an ectoparasiticide. The pathogens controlled by malathion are members of Diptera, Lepidoptera, Hymenoptera, Hemiptera, Coleoptera and Acari. The most notable applications of malathion include the control of the Mediterranean fruit fly and the West Nile virus. Common crops protected are cotton, a variety of fruit (pome and stone), potatoes, rice and vegetables. It is also used in the control of mosquitoes, lice and ticks.

Malathion is formulated as emulsifiable concentrate (EC), wettable powder (WP), dustable powder (DP) and as a ULV liquid (UL). A mixture of malathion and corn (maize) syrup forms a toxic bait for fruit flies. The maximum application rate for the control of mosquito larvae is 250 g/ha. Common trade names include Calmathion, Celthion and Emmatos extra.

The LD_{50} of malathion in rats ranges from 1000 to 10,000 mg/kg and in mice, from 400 to 4000 mg/kg. Its dermal LD_{50} among rats is >4000 mg/kg. The acute oral LD_{50} of malathion in birds is between 167 mg/kg (blackbirds and starlings) and 1485 mg/kg (pheasants), making it a potential toxin. It is highly toxic to many fish species (with the LC_{50} being as low as 0.06 mg/l in some cases) and to honey bees. The ADI of malathion is 0.02 mg/kg/day. While it may be combined with many insecticides and fungicides, malathion can harm some varieties of cucurbits, beans, apples, pears and grapes. The WHO classifies malathion as a slightly hazardous chemical (Class III toxin). It is considered a PAN Bad Actor chemical, since it is a cholinesterase inhibitor. Being an organophosphate compound, malathion may cause toxic symptoms of excessive salivation, sweating, muscle twitching, dizziness, nausea, vomiting, pin-point pupils, blurred or dark vision, tightness in the chest and even seizures. Removal of contaminated clothes, thorough washing of skin, eyes and body, exposure to fresh air and resting in a half-upright position, may give relief, before any medical intervention.

Malathion is registered in India, Australia, the Philippines, the USA and in many countries of Africa. According to the USDA Foreign Agricultural Service's Pesticide MRL database, the indicative MRL for cottonseed in the USA and South Africa is 20 ppm and in Japan it is 2 ppm. The EU maintains an MRL of 0.02 ppm for its 'Oilseeds' group. Similarly, for rice grain, the MRL is fixed at 8 ppm.

Male sterilization

The process of sterilization of male insect pests and their subsequent release as a means of biocontrol is called male sterilization. Such treated males, when released, cannot produce offspring with normal females, thereby bringing about a reduction of population. Populations of the screw worm fly (*Cochliomyia hominivorax*) and of the Mexican fruit fly (*Anastrepha ludens*) have been reduced with this technique. Exposure to gamma radiation is a method of sterilization.

Maleic hydrazide as a PGR

Maleic hydrazide, $C_4H_4N_2O_2$ (6-hydroxy-2*H*-pyridazin-3-one; 1,2-dihydropyridazine-3,6-dione), is a plant growth regulator and a herbicide. It is a white crystalline substance, soluble in water, methanol, hexane and toluene and degraded by light, alkaline material, oxidizing agents and strong acids. The starting materials for the synthesis of maleic hydrazide are maleic anhydride, hydrochloric acid and hydrazine. The PGR is allotted three CAS Registry Numbers, namely, 10071-13-3, 123-33-1 and 51542-52-0.

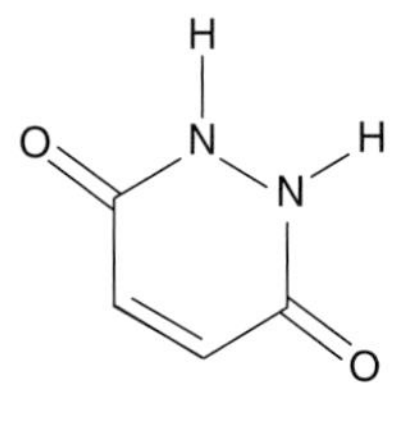

Maleic hydrazide

Maleic hydrazide shows good translocation once absorbed by the leaves and roots. As a PGR, it inhibits cell division in the meristematic regions. It inhibits post-harvest sprouting or growth in potatoes, onions, beets and tobacco. As a herbicide, it suppresses growth of grass, shrubs and trees.

The PGR is formulated as a suspension emulsion/suspoemulsion (SE), emulsifiable concentrate (EC) and soluble concentrate (SL). The ADI is 0.19 mg/kg/day. Trade names include Mazide 25 and Regulox.

According to the WHO, maleic hydrazide is unlikely to present acute hazard in normal use (Class U), while the EPA classifies it as slightly toxic (Class III).

The acute dermal LD_{50} of maleic hydrazide in rabbits is >2000 mg/kg. However, it severely irritates the eyes of rabbits. The oral LD_{50} in rats is 6950 mg/kg. It is not acutely toxic to fish, birds or even **beneficial organisms** in the soil. Symptoms of exposure are irritation of the skin, eyes, nose and throat. Chronic exposure leads to liver damage.

This PGR is excreted by animals. It breaks down into succinic acid, fumaric acid and maleic acid in plants. It is mobile in sandy soils and photodegrades in water.

Maleic hydrazide is banned in India, Austria and Denmark, but registered in Australia, the USA and Canada. According to the The Japan Food Chemical Research Foundation, the indicative MRL of maleic hydrazide for soybeans and groundnuts in Japan is 0.02 ppm.

Management of pesticide poisoning

The subject of managing a victim of pesticide poisoning is vast. A brief note on this topic may be seen under **Worker safety norms and first-aid techniques**.

Mancopper fungicide

Mancopper, $C_8H_{12}CuMnN_4S_8$ (ethylenebis(dithiocarbamate)), is a mixed-metal complex containing about 13.7% manganese and about 4% copper. It is an alkylenebis dithiocarbamate fungicide. The CAS Registry Number of mancopper is 53988-93-5 (see Mancopper at bottom of the page).

Copper fungicide mancopper, which is a non-specific thiol reactant, acts as a protective fungicide by inhibiting respiration and the germination of fungal spores. It is used to control **downy mildew** on vines and in cereal seed treatment to control *Septoria* and *Fusarium* spp.

Some of the common formulations of mancopper are wettable powder (WP) and seed treatment (SD). A well-known trade name is Dithane C 90.

The acute oral LD_{50} of mancopper in rats is 9600 mg/kg. Mancopper is not listed as a PAN Bad Actor chemical. It is also not EPA registered. The Pesticide Properties Database (PPDB) of the Agriculture and Environment Research Unit (AERU), UK, does not approve the use of mancopper in the UK. It is considered obsolete by the WHO.

Mancozeb fungicide

Mancozeb, $(C_4H_6MnN_2S_4)_x$ $(Zn)_y$ (manganese ethylenebis-(dithiocarbamate) (polymeric) complex with zinc salt), is an alkylenebis dithiocarbamate compound. It is a greyish-yellow powder with a slight hydrogen sulfide odour, soluble in water (6.2 ppm at pH 7.5, 25°C) and insoluble in most organic solvents. It stores well under normal storage conditions, away from moisture and heat. Mancozeb is also called manzeb. It is produced by the reaction of a concentrated solution of zinc sulfate with a suspension of **maneb** in water. The CAS Registry Number of mancozeb is 8018-01-7.

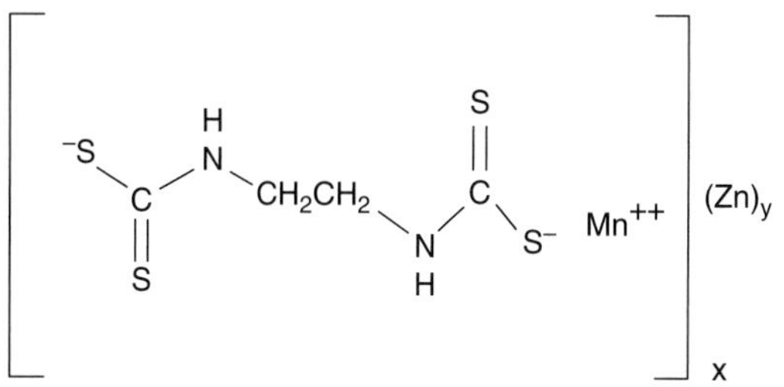

x:y = 1:0.091

Mancozeb

Mancozeb, a carbamate compound, is a broad-spectrum, non-systemic, contact, preventive fungicide, and a cholinesterase inhibitor. It inactivates the sulfhydryl groups of amino acids and enzymes of the fungal cells that result in the disruption of lipid metabolism, ATP production and respiration. This is done by forming a complex with metal-containing enzymes. It controls fungal diseases like **early blights**, **late blights**, blast, **scab**, rusts, leaf spot, leaf **smut**, damping-off in fruit, nut and field crops. Specifically, mancozeb protects potatoes and tomatoes from early and late blight, bananas from sigatoka (caused by *Mycosphaerella* spp.), apples and pears from scab, maize and millet from leaf smut and **downy mildew**, and rice from blast.

This fungicide is commonly formulated as dustable powder (DP), oil dispersion (OD) and wettable powder (WP). Some common trade names include Dithane M-45, Matilda WG, Fore, Nemispot and Manzanc. This fungicide may be combined as a foliar application or as a seed treatment agent, with other compounds like **copper oxychloride**, cymoxanil,

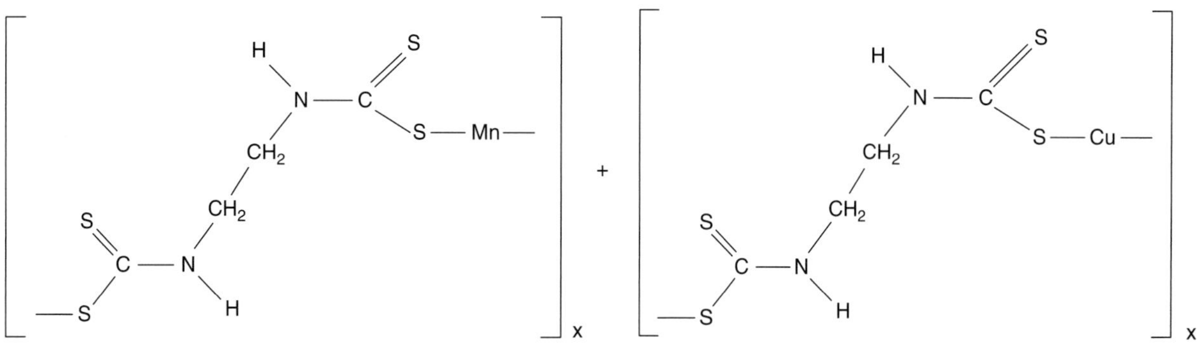

Mancopper

zineb, maneb, **benthiavalicarb** and **zoxamide**. The application rates range from 1500 to 2000 g/ha. Mancozeb is incompatible with oxidizing agents or acids.

The fungicide mancozeb is classified as a Class IV toxin according to the EPA and as Class U (unlikely to be hazardous) by the WHO. It is an eye irritant but not a skin irritant. Its acute oral LD_{50} in rats is >5000 mg/kg. In the bird species tested, the acute oral LD_{50} was found to be >5500 mg/kg. Mancozeb has low toxicity to non-target creatures and beneficial arthropods. It is moderately to highly toxic to fish and aquatic organisms (with 48 h LC_{50} ranging from 2.2 to 9 mg/l in many species). It is a potential groundwater contaminant. Since mancozeb is a cholinesterase inhibitor, it can affect the nervous system and is a suspected endocrine disruptor. Extended exposure can result in symptoms of poisoning like nervous system disorders, fatigue, headache, blurred vision, nausea, convulsions, slurred speech, confusion and slowed heartbeat. A major toxicological concern with respect to mancozeb and other **dithiocarbamates** is its primary metabolite, **ethylenethiourea** (ETU), shown to cause damage to the thyroid gland, placenta and increased tumour incidence (carcinogenic) in test animals. ETU (CAS Registry Number 96-45-7 and molecular formula $C_3H_6N_2S$) is a mobile impurity that persists for up to 10 weeks. Mancozeb is thus a PAN Bad Actor chemical. The RfD of mancozeb is 0.003 mg/kg/day. In cases of exposure the person should wash the affected skin with water and soap. Moving the person into a well-ventilated area can also provide relief in cases of inhalation. Vomiting must be induced in cases of oral exposure. Appropriate antidotes and medical attention are critical. All labels on mancozeb formulations should carry the signal word 'CAUTION'.

Mancozeb breaks down rapidly in soil and the soil aerobic half-life is 2 days. However, the hydrolysis half-life is 166 days, since it tends to adsorb to sediment and suspended solids. Mancozeb degrades in water with a half-life of 1–2 days in slightly acidic to slightly alkaline conditions. ETU has the potential to be mobile in soils and persist for up to 5–10 weeks. In plants, it is metabolized to form intermediates like ethylenethiourea, ethylenethiuram monosulfide, ethylenethiuram disulfide and sulfur and some natural products derived from glycine. Adherence to label instructions can ensure safe application of mancozeb to the target crops.

Mancozeb is registered for use in India, the USA, all of the EU, Madagascar, South Africa, Australia and New Zealand. The MRL value for mancozeb on citrus fruit like oranges and lemons is 5 mg/kg.

According to the USDA Foreign Agricultural Service's Pesticide MRL database, the indicative MRL of mancozeb for apples is 7 ppm in the USA, 5 ppm in Japan and South Africa, 5 ppm in the EU (for its 'Pome fruits' group) and 3 ppm in India for its 'Other fruits' group. The MRL for most cucurbit crops in the USA, South Africa, Japan and the EU is 2 ppm.

Maneb fungicide

Maneb, $C_4H_6MnN_2S_4$ (manganeseethylenebis(dithiocarbamate)), is a polymeric fungicide (a polymer of ethylene(bis)dithiocarbamate units linked with manganese). This alkylenebis dithiocarbamate compound is commonly available as a yellow powder with a faint odour, that is soluble in water, common organic solvents and in chelating agents such as sodium salts of ethylenediaminetetraacetic acid. It is stable to light but decomposes in the presence of moisture, oxygen, acidic conditions and in temperatures above 30°C. Maneb should thus be stored in airtight, water-tight containers, placed in well-ventilated locations. Its CAS Registry Number is 12427-38-2.

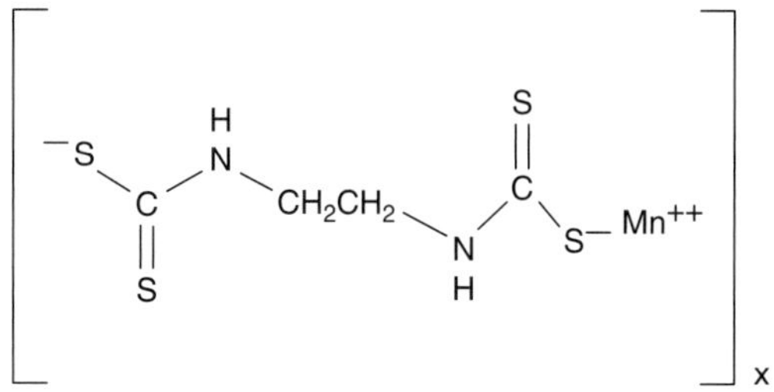

Maneb

Maneb, a non-specific thiol reactant, has a protective action and inhibits respiration. It controls many fungal diseases in a wide range of crops (e.g. **early** and **late blights** of potatoes and tomatoes), sigatoka disease (*Cercospora musae*) on bananas and many seed-borne diseases of cereals (e.g. rice, wheat and maize). Maneb also controls fungal diseases including **downy mildew**, leaf spot and leaf blight that attack pepper, tobacco, tea and fruits such as grapes, apples and pears.

Some of the common formulation types of maneb are wettable powder (WP), water-dispersible granules (WG) and seed-treatment formulations. Trade names include Farmaneb, Manesan, Manex, Manzate, Nereb and Newspor. Maneb is compatible with many systemic fungicides which help to increase efficacy.

Maneb is classified as a Class IV toxin according to the EPA, and is registered as a general use pesticide (GUP). It is classified as a Class U toxin by the WHO (unlikely to be hazardous). The acute oral LD_{50} in rats is >5000 mg/kg and a dermal LD_{50} in rats is >5000 mg/kg. While maneb is practically non-toxic to birds, it is highly toxic to fish with the 48 h LC_{50} being around 1.9 ppm. This fungicide is also phytotoxic to some varieties of apples, tomatoes, tobacco and cucurbit members, at seedling or early growth stages. The ADI is 0.03 mg/kg b.w./day. However, since this compound is a cholinesterase inhibitor, a reproductive toxin and a suspected endocrine disruptor, it is classified as a PAN Bad Actor chemical. Accidental exposure to large quantities of maneb may result in hyperactivity and incoordination, loss of muscular tone, nausea, vomiting, diarrhoea, confusion, coma and death. It causes irritation to the eye, nose and throat, even at moderate levels of exposure. Maneb residues are also observed to be present on some crops such as beans and tomatoes, necessitating a strict post-application time interval of >15 days (referred to as a **restricted entry interval** or REI). Most **ethylenebisdithiocarbamates** (EBDC) metabolize into **ethylenethiourea** (ETU), a known teratogen. The generation of ETU also occurs when EBDCs are subjected to heat. Containers of maneb must bear the signal word 'CAUTION'.

Upon ingestion, rats metabolize maneb to ethylenediamine, ethylenebis (thiuram) monosulfide and ethylenebis (thiourea). In plants, the main metabolite is ETU, which is then rapidly metabolized further.

Maneb is degraded in the environment by hydrolysis, oxidation, photolysis and metabolism. Like mancozeb, maneb breaks down in soil to ETU. While maneb has medium persistence (with a half-life of up to 36 days), ETU is more persistent and mobile. Maneb is degraded completely within 1 h under anaerobic aquatic conditions. The MRL for tomatoes is 3 mg/kg and onion is 1 mg/kg.

Maneb is registered in the USA, Canada, Cambodia, South Africa and EU members including the UK, Germany, Poland, France and Italy.

Manganese as a plant nutrient

See **Plant nutrient excesses and deficiencies**

Manzeb fungicide

See **Mancozeb fungicide**

Maximum residue level or limit (MRL)

The concept of a maximum residue level (MRL) has gained popularity with the increased awareness that chemicals often leave behind at least a trace of their residues on crops, and their consumption via food could lead to abnormality or even death in living systems.

The MRL denotes the maximum concentration of pesticide residue that is allowable on food or feed, in order to avoid unacceptable risks. These MRL values are set, assuming the implementation of **good agricultural practices (GAP)** and the passage of reasonable time after pesticide application, before harvest. MRL values are usually less than a milligram (mg) per kilogram (kg) of food/feed and are based on a number of field trials. A national regulatory authority sets the MRL values for that country. The **acceptable daily intake (ADI)** values are also considered while setting MRL, specifically since ADI values take into account the amount of the food/feed that is likely to be consumed. The MRL is normally expressed in milligrams per kilogram (mg/kg) of fresh weight.

Several tests are carried out to determine the consumption of pesticide residues, such as short-term and long-term exposure to pesticides via inhalation, ingestion, skin and eyes. These lead to the establishment of ADI and acute **reference dose** (aRfD) values for each pesticide. It is only when the resultant values are accepted, that a pesticide is approved for use. So, MRL values prompt wise and optimal use of farm pesticides and may not necessarily be a unique tool to decide safety limits. Residues above the safety levels may attract review/prohibition of the use of such pesticides.

MRL values are fixed based on field trials. In case an estimate cannot be worked out, the level is set at the limit of determination (LOD) (normally in the range of 0.01–0.05 mg/kg). The LOD reflects the lowest level at which analysis can be done quantitatively, which in turn, depends on the sensitivity of the equipment carrying out the test.

In many countries extensive analysis of food, fruit and vegetables from the market is carried out to validate the recommendations of the regulating agency and confirm that they are in fact safe. Such analysis of market samples also checks if the recommendations are indeed followed by growers.

MCPA herbicide

MCPA, $C_9H_9ClO_3$ ((4-chloro-2-methylphenoxy) acetic acid), is a phenoxyacetic herbicide. MCPA stands for 2-methyl-4-chlorophenoxyacetic acid. It is a weak acid that exists as a colourless crystalline substance, readily soluble in water and organic solvents. Normally stable to light and storage, MCPA decomposes on heating to produce toxic fumes. It forms water-soluble alkali-metal and amine salts in water, making it essential to store MCPA away from moisture, strong alkaline media, food and feed. MCPA is made by chlorination of *o*-cresol and coupling with monochloroacetic acid. Its CAS Registry Number is 94-74-6.

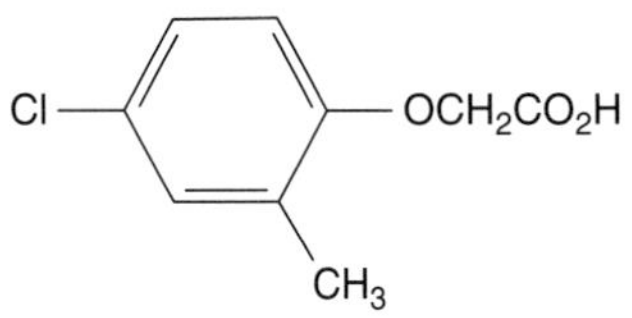

MCPA

MCPA is a post-emergence, selective, systemic herbicide compound that mimics growth hormones. It translocates to the meristematic regions of plants to inhibit growth. MCPA is used to control a wide variety of annual and perennial broadleaved weeds in cereals (e.g. wheat, rice, sorghum, barley and oats), vegetables, turf, lucerne and forestry crops.

MCPA is often formulated as a soluble concentrate (SL), water-soluble powder (SP) and emulsifiable concentrate (EC). It can be combined with **2,4-D**, **dicamba**, **mecoprop** and bromoxynil, and sprayed aerially, by ground boom sprayers, hand-held sprayers or other types of equipment. Trade names include Agritox, Chiptox, Cornox, Methoxone and Tigrex. The period from the four-leaf stage to the boot stage is ideal for MCPA application (the boot stage is the growth stage just before the emergence of the seedhead).

MCPA is an acutely toxic compound, which upon ingestion by rats, is rapidly excreted. The oral LD_{50} of MCPA in rats is 765 mg/kg. It is toxic to birds (the oral LC_{50} of MCPA is 377 mg/kg). But this herbicide is practically non-toxic to honey bees and slightly toxic to fish. It may be phytotoxic to vines, vegetables, onion, cabbage, lettuce, turnip and cotton. If inhaled, MCPA can cause nausea and a headache. Contact with the eyes and skin may cause redness. Other toxicity symptoms (of oral exposure) include stomach pain, vomiting, weakness and related ailments. Wearing a face shield or goggles, chemical-resistant gloves, boots made of material such as neoprene or polyvinyl chloride and cotton overalls buttoned to the neck and wrist offer protection. All product labels are required to carry the signal word 'DANGER' because it is a severe eye irritant. According to the WHO, MCPA is a Class III toxin (slightly hazardous). It is a possible carcinogen.

MCPA is rapidly absorbed and excreted via the urine in an almost unchanged form. In plants, MCPA degrades to form 2-hydroxymethyl-4-chlorophenoxyacetic acid, while in soil, it degrades to 4-chloro-2-methylphenol, followed by ring hydroxylation and ring opening. Photolytic metabolism also leads to 4-chloro-2-methylphenol. Although insoluble in water, MCPA can have high mobility in soil. MCPA is also subject to microbial degradation.

MCPA is registered in India, the USA, Canada, Australia, New Zealand, the Philippines and many EU members.

According to the USDA Foreign Agricultural Service's Pesticide MRL database, the indicative MRL of MCPA for barley grain and wheat in USA is 1 ppm, while in South Africa it is 0.05 ppm. Australia maintains an MRL of 0.02 ppm for its 'Cereal grains' group, while the EU has fixed it at 0.05 ppm for its 'Cereal grains' group.

MCPA-thioethyl herbicide

MCPA-thioethyl, $C_{11}H_{13}ClO_2S$ (*S*-ethyl-4-chloro-*o*-tolyoxythioacetate), is a phenoxyacetic herbicide. It is a colourless crystalline substance, which is moderately soluble in organic solvents. It is unstable in alkaline media (with a DT_{50} in water of 2 days at pH 9). MCPA-thioethyl is produced by the reaction of ethylmercaptan with 4-chloro-2-methylphenoxy methyl carboxychloride. Its CAS Registry Number is 25319-90-8.

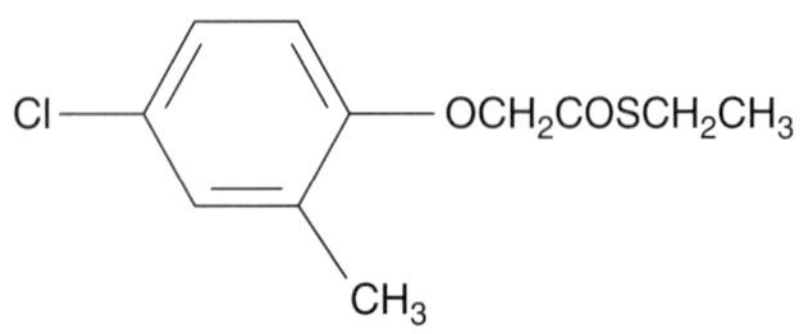

MCPA-thioethyl

MCPA-thioethyl is a selective, systemic, post-emergence, hormone-like herbicide, which inhibits growth in the meristematic region. Annual and perennial broadleaved weeds growing among cereals are controlled by this herbicide. **Common lambsquarters** (*Chenopodium album*) and false pickerelweed (*Monochoria vaginalis*) are some examples of weeds controlled by this herbicide.

The herbicide is available as granules (GR) and emulsifiable concentrate (EC). With a widely accepted dosage of 210 g/ha, this herbicide is best sprayed with water. Trade names of MCPA-thioethyl include Fenobit and Herbit.

The herbicide MCPA-thioethyl is toxic to rats with an acute oral LD_{50} in rats ranging from 749 to 877 mg/kg. The dermal LD_{50} of MCPA-thioethyl in rats is >5000 mg/kg b.w. It is slightly toxic to birds and fish. The ADI in rats is 0.013 mg/kg b.w./day. The aRfD of MCPA-thioethyl in rabbits is 0.15 mg/kg b.w./day. Irritation to skin, eyes and the respiratory tract are symptoms of poisoning, while headache, vomiting, diarrhoea, confusion, kidney failure and increased heart rate may occur with severe exposure. The WHO categorizes this herbicide as slightly toxic (Class III toxin). It is classified as a PAN Bad Actor chemical due its acute toxicity and because it is a possible carcinogen.

MCPA-thioethyl is registered for use in Portugal and Tanzania but not in the USA.

MCPB herbicide

MCPB, $C_{11}H_{13}ClO_2$ (4-(4-chloro-*o*-tolyloxy) butyric acid), is a phenoxybutyric herbicide. It is an acid that exists as a colourless crystalline substance, moderately soluble in water and readily soluble in organic solvents. This compound forms water-soluble alkali-metal and amine salts in water. It must be stored in a tightly closed container in a cool and dry ventilated area, away from heat sources. MCPB is produced by chlorination of *o*-cresol and coupling with γ-chlorobutyric acid. MCPB has a CAS Registry Number of 94-81-5.

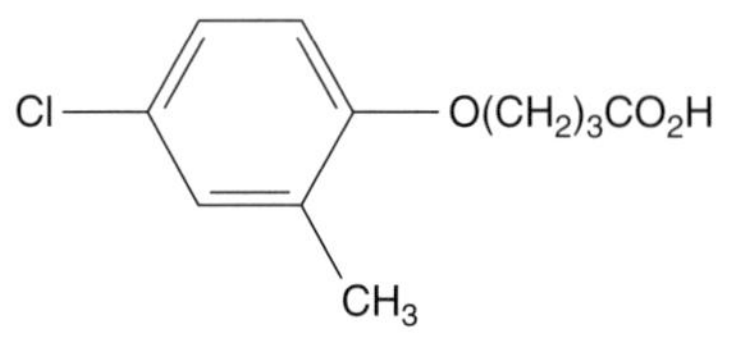

MCPB

MCPB is a selective, systemic, pre-emergence, hormone-type herbicide that must undergo β-oxidation of the butyric acid side chain to form the phytotoxic **MCPA**. Due to this dependency, MCPB is more selective than MCPA, and is effective on sensitive annual and perennial broadleaved weeds like Canadian thistle, mustard, purslane, ragweed, **common lambsquarters** and pigweed in cereals, peas and potatoes.

Maximum application rates of MCPB are 1.7 kg/ha, for the control of Canada thistle at the bud stage. It is available in the market under trade names like Bellmac Straight. To avoid spray drift and toxicity to non-target organisms, it is sprayed with medium or coarse droplets.

The acute oral LD_{50} in rats is 680 mg/kg. MCPB is moderately toxic to birds, and it is classified as practically non-toxic to birds on an acute dietary basis. It is slightly toxic to freshwater invertebrates. Symptoms of poisoning with MCPB are irritation of the skin, eyes and respiratory tract, headache, vomiting, diarrhoea, aggressive behaviour and kidney failure. The REI after application is 12 h. According to the WHO, MCPB comes under toxicity Class III (slightly hazardous). Product labels should clearly display the signal word 'CAUTION'.

MCPB undergoes β-oxidation to MCPA, which is subsequently degraded to 4-chloro-2-methylphenol, followed by ring hydroxylation and ring opening. MCPB is not volatile, not persistent and not likely to bioconcentrate. It is also prone to leach and runoff.

The MRL of MCPB permitted on peas is 0.1 ppm. MCPB is registered in Australia, Canada, Hungary, New Zealand, the UK and the USA.

Mealybug

See **Pest – mealybug**

Mecarbam insecticide

Mecarbam, $C_{10}H_{20}NO_5PS_2$ (*S*-(*N*-ethoxycarbonyl-*N*-methylcarbamoylmethyl) *O,O*-diethyl phosphorodithioate), is an organophosphorus insecticide. It is a brownish-yellow oil, soluble in organic hydrocarbons and in water at 1 g/l. It is miscible with alcohols, esters, ketones and aromatic and chlorinated hydrocarbons at room temperature. It is readily hydrolysed in aqueous media below pH 3. Mecarbam has a CAS Registry Number of 2595-54-2.

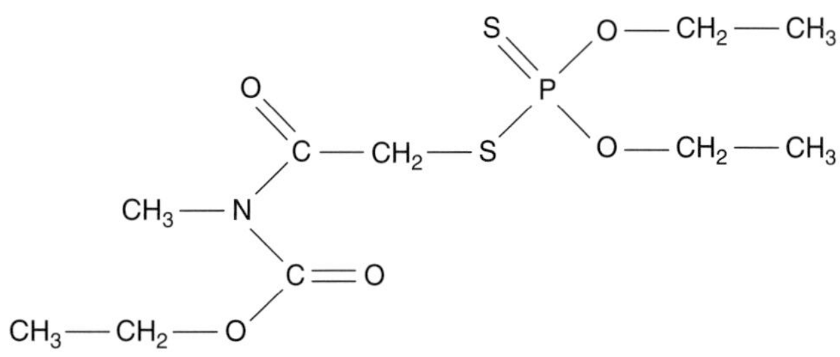

Mecarbam

Mecarbam is a contact (and slightly systemic) insecticide and acaricide with a long residual activity. It is used in the control of **aphids**, suckers, **whitefly**, **scale insects**, **mealybugs**, red spider mites, **leafhoppers**, plant hoppers and **thrips** on crops such as fruit orchards, rice, onion, carrots, brassicas and cotton.

This insecticide is commonly formulated as an emulsifiable concentrate (EC). It is compatible with most pesticides, except if they are highly alkaline. It is available under trade names like Murfotox and Pestan.

The acute oral LD_{50} of mecarbam among rats is 36 mg/kg. The inhalation LC_{50} in rats is 0.7 mg/l. The acute 96 h LC_{50} in fish is 0.004 mg/l. The ADI is 0.002 mg/kg b.w./day. The aRfD is 0.008 mg/kg b.w./day. Mecarbam is highly phytotoxic to aubergines. It is also toxic to bees. The WHO has categorized it as a Class Ib toxin (highly hazardous chemical). Mecarbam is a PAN Bad Actor chemical. The compound is a cholinesterase inhibitor. Symptoms of poisoning with mecarbam are excessive salivation, sweating, rhinorrhea and tearing, muscle twitching, tremor, headache, nausea, abdominal cramps, diarrhoea, respiratory depression, tightness in the chest, fluid in the lungs and pin-point pupils. In severe cases it causes seizures, incontinence, respiratory depression and unconsciousness.

In mammals, mecarbam is rapidly metabolized by hydrolysis, oxidative desulfuration and degradation of the carbamoyl moiety. *O*-Deethylation also takes place to a minor extent.

Mecarbam persists in the soil for 4–6 weeks. Belize has banned the use of mecarbam, while Cambodia has registered its use.

Mecoprop herbicide

Mecoprop, $C_{10}H_{11}ClO_3$ ((*RS*)-2-(4-chloro-*o*-tolyloxy) propionic acid), is a phenoxypropionic herbicide. It is an acidic colourless, crystalline substance, moderately soluble in water and readily so in acetone, alcohol, diethyl ether and ethanol. It is stable to heat and to hydrolysis, and atmospheric oxidation, and it forms salts that are often water soluble. Mecoprop is produced by chlorination of *o*-cresol and coupling with 2-chloropropionic acid. The two CAS Registry Numbers of mecoprop are 7085-19-0 and 93-65-2.

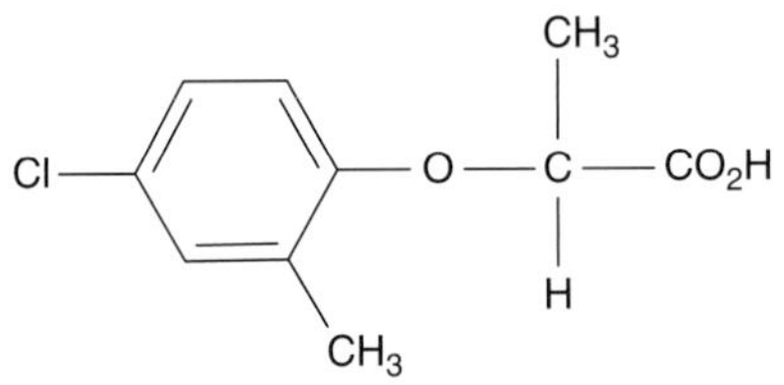

Mecoprop

A selective, systemic, post-emergence, hormone-like herbicide, mecoprop can control broadleaved weeds (especially cleavers, ivy, chickweed, clovers and suface-creeping plants) in wheat, barley, oats, fruit trees, turf and meadows. It acts in a matter of 3–4 weeks.

Mecoprop is available as granules (GR) and liquid concentrate and is marketed under trade names such as Kilprop, Actril M, Atlas and Mecopar. This herbicide is often used in combination with other herbicides such as **2,4-D**, **dicamba** and **MCPA**.

The oral LD_{50} values in rats and mice range from 650 to 1210 mg/kg. The dermal LD_{50} of mecoprop in rats is >4000 mg/kg. Mecoprop is irritating to the eyes, causing redness and swelling. It is slightly, to moderately toxic to birds, but practically non-toxic to honey bees and fish. It is observed to be a possible carcinogen and groundwater contaminant. The WHO has classified this compound as slightly hazardous (Class III toxin).

Symptoms of poisoning with mecoprop include irritation of skin, eyes and respiratory tract, vomiting, a burning sensation in the chest, headache, aggressive behaviour, kidney failure and increased heart rate.

In plants, mecoprop is hydroxylated at the methyl group with the formation of 2-hydroxymethyl-4-chlorophenoxypropionic acid. In soil, mecoprop is degraded predominantly by microorganisms to 4-chloro-2-methylphenol. Its DT_{50} in soil is 7–13 days. Adsorption increases in the presence of organic matter. In spite of high solubility and mobility, the herbicide does not persist and has a half-life of less than 1 month in water.

Mecoprop is registered in Australia, Cambodia, the EU, New Zealand, the USA and the UK.

Mecoprop-P herbicide

Mecoprop-P, $C_{10}H_{11}ClO_3$ ((*R*)-2-(4-chloro-*o*-tolyloxy) propionic acid), is a phenoxypropionic herbicide. It forms colourless odourless crystals which are moderately soluble in water and readily soluble in organic solvents. But, only the (*R*)-isomer of mecoprop is herbicidally active. Mecoprop-P is susceptible to photolysis, but stable to atmospheric oxidation and hydrolysis. It has a CAS Registry Number of 16484-77-8.

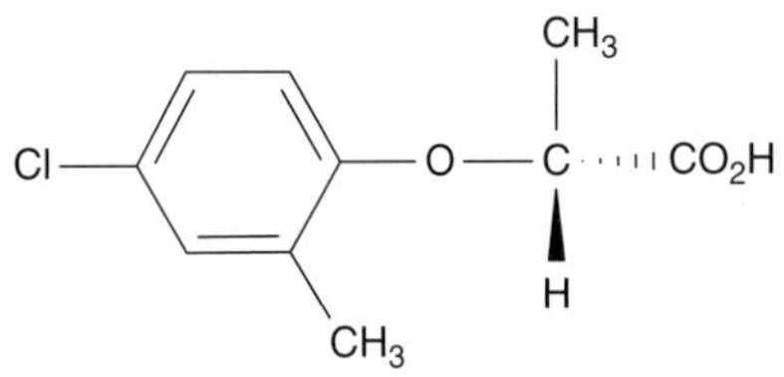

Mecoprop-P

Mecoprop-P is a selective, systemic post-emergence herbicide that translocates to the roots. It is used for the control of broadleaved weeds (especially cleavers, knotweed, chickweed, stitchwort and clovers) in wheat, barley, oats, turf and lawn.

Mecoprop-P is available as a soluble concentrate (SL), emulsifiable concentrate (EC), a ready-to-use solution and as granules (GR). It is compatible with many herbicides.

The acute oral LD_{50} of mecoprop-P is 431 mg/kg in rats and 602 mg/kg in birds. The acute 96 h LC_{50} in rainbow trout is >100 mg/l. The ADI in rats is 0.01 mg/kg b.w./day. Symptoms of poisoning are irritation of the skin, eyes and respiratory tract, dizziness, headache, kidney failure and increased heart rate. This herbicide is a potential groundwater contaminant and carcinogen according to the PAN Pesticide Database and hence it is listed as a PAN Bad Actor chemical. Mecoprop-P is classified by the WHO as a Class III toxin (slightly hazardous).

In mammals, mecoprop-P is predominantly excreted unmodified in the urine. In plants, mecoprop-P is metabolized rapidly via hydroxylation at the methyl group forming 2-hydroxymethyl-4-chlorophenoxypropionic acid. The compound has an average field half-life of 7–21 days. In soil, mecoprop-P is readily degraded by microorganisms to yield

4-chloro-2-methylphenol, although mecoprop-P can have residual activity for up to 60 days. It does not bond strongly to soil and may leach out of the soil.

Mecoprop-P is registered in Australia, Canada, some EU members and the USA.

Median effective concentration

The statistically obtained concentration of a pesticide in an environment, which is expected to produce a certain effect in 50% of the test organisms in a given population under defined conditions, is known as the median effective concentration (also referred to as the EC_{50}). The expected concentration of extremely toxic chemicals is in the range of 1–50 mg/kg b.w. in oral tests, while in dermal exposure acute toxicity is denoted by concentration levels between 1 and 200 mg/kg.

Median lethal concentration

The statistically derived concentration of a pesticide in an environment expected to kill 50% of the test organisms in a given population under defined conditions is known as the median lethal concentration. It is normally expressed as the weight of a substance per unit weight or volume of water, air or feed (mg/l or ppm).

Median lethal dose

The statistically derived pesticide dose expected to kill 50% of the test organisms in a given population under a defined set of conditions is known as median lethal dose. It is normally expressed as milligrams of test material per kilogram of body weight of the organism per day (mg/kg b.w./day).

Mefenacet herbicide

Mefenacet, $C_{16}H_{14}N_2O_2S$ (2-benzothiazol-2-yloxy-*N*-methylacetanilide), is a benzothiazole herbicide. It is a colourless crystalline substance, sparingly soluble in water and soluble in dichloromethane. It is stable to light, acid and alkali. Mefenacet is made by the reaction of 2-chlorobenzthiazol with *N*-methylbenzene hydroxyacetamide. It has a CAS Registry Number of 73250-68-7.

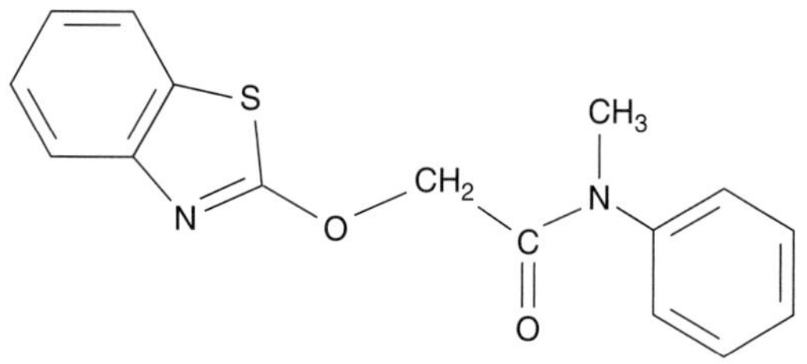

Mefenacet

Mefenacet is a selective, pre-emergence soil herbicide, used mainly in transplanted rice. It inhibits cell growth, cell division and the very-long-chain-fatty-acid (VLCFA). It is used to control annual grasses and some dicotyledon weeds (e.g. *Cyperus* spp. and *Echinochloa* spp.) on wheat, maize and beans.

Mefenacet is available in granular formulations and wettable powder (WP) under trade names like Hinochloa.

According to the WHO and EPA, mefenacet is a practically safe compound. In plants, mefenacet is metabolized to 4-aminophenol via oxidation of an intermediate *N*-methylaniline. Other metabolites of mefenacet are benzothiazolone and benzothiazolyloxy acetic acid; both are degraded by hydroxylation.

Mefenacet is registered in countries like Egypt, Japan, Korea, Vietnam and Taiwan for agricultural use. According to the The Japan Food Chemical Research Foundation, the indicative MRL of mefenacet for rice in Japan is 0.1 ppm.

Mefenpyr-diethyl herbicide safener

Mefenpyr-diethyl, $C_{16}H_{18}Cl_2N_2O_4$ (diethyl (*RS*)-1-(2,4-dichlorophenyl)-5-methyl-2-pyrazoline-3,5-dicarboxylate), a derivative of mefenpyr, is a herbicide safener. It is a whitish crystalline substance, soluble in water (20 mg/kg), acetone, toluene and ethyl acetate. It is hydrolysed by both acids and alkalis. The CAS Registry Number is 135590-91-9.

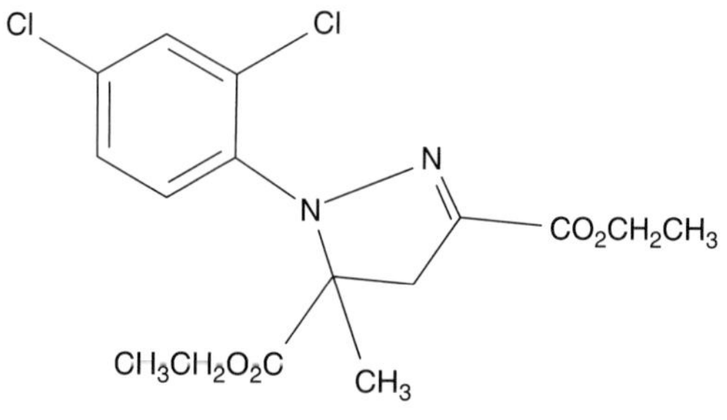

Mefenpyr-diethyl

As a safener, mefenpyr-diethyl enhances the metabolism of herbicides like **mesosulfuron-methyl, fenoxaprop-p-ethyl** and **iodosulfuron-methyl-sodium** in cereal crops such as wheat, rye and barley, for selective weed control.

Mefenpyr-diethyl is formulated as an emulsion, oil in water (EW), a suspension concentrate (SC) and a suspension emulsion/suspo-emulsion (SE). The safener is commercially available under trade names such as Husar, Huskie, Infinity, Mesomaxx, Precept and Puma Super.

Mefenpyr-diethyl is slightly toxic to rats and is moderately toxic to birds (with the acute oral LD_{50} being >2000 mg/kg). It is toxic to fish and bees. The LC_{50} values among fish species tested range between 2.4 and 4.2 mg/l. The 48 h oral LD_{50} in honey bees is >900 µg per bee. Mefenpyr-diethyl is a PAN Bad Actor chemical. The NOEL of this chemical in rats (based on a 2 year study) is 49 mg/kg b.w./day. The ADI is 0.03 mg/kg/day.

The herbicide safener is completely mineralized in soil by hydrolysis and by microbial and photolytic degradation. The soil DT_{50} is <10 days. The abiotic hydrolysis DT_{50} is >1 year at pH 5. It photodegrades in 29 days.

Mefenpyr-diethyl is registered for use in Australia, Hungary and Portugal, but not in the USA. Italy has fixed an MRL of 0.05 ppm for wheat.

Mefluidide as a PGR

Mefluidide, $C_{11}H_{13}F_3N_2O_3S$ (5′-(1,1,1-trifluoromethanesulfonamido)acet-2′, 4′-xylidide), is a colourless and odourless crystalline substance that regulates plant growth. It is soluble in water, acetone, methanol, acetonitrile, ethyl acetate, *n*-octanol, diethyl ether, dichloromethane, benzene and xylene. An aqueous solution of mefluidide degrades when exposed to UV radiation. The compound must be stored at temperatures above 4°C.

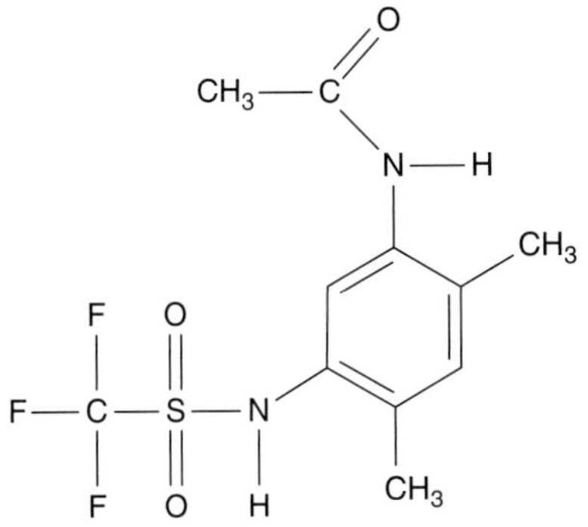

Mefluidide

Mefluidide is a post-emergence plant growth regulator and a herbicide. It suppresses the production of seeds, elongation of stems, and growth and division of meristematic cells in perennial grasses in turf, lawns and public places. It also controls volunteer maize and wheat, **Johnson grass** and other weeds in soybeans. It also enhances the sucrose content of sugarcane.

Mefluidide is compatible with growth regulator herbicides, except those liquid formulations which are acidic in nature. Mefluidide is available in the market as Embark and Vistar. The compound is formulated as a soluble concentrate (SL) that may be sprayed at any time of the year, by ground equipment or boom sprayer. Mefluidide may be applied at any time of the year when the plant is healthy and actively growing. While it is ideal to spray woody plants just before trimming, application on soybean is done at the two to six trifoliate stage.

The PGR is practically non-toxic to rabbit skin (with dermal LD_{50} being >20,000 mg/kg). The oral LD_{50} in rabbits is >4000 mg/kg, while that in mice and rats is between 1920 and 4000 mg/kg. The substance is practically non-toxic to birds, fish and bees. The WHO has categorized this chemical as a slightly hazardous toxin (Class III toxin). Contamination of water sources and spray drift should be strictly avoided. Applicators should be equipped with PPE while applying the chemical and during the clean-up activities to avoid exposure.

In animals, mefluidide residues are excreted intact. In soil, 5-amino-2, 4-dimethyltrifluoromethanesulfonanillide is formed as a metabolite. Mefluidide is susceptible to photodecomposition in aqueous solutions. This compound has a soil half-life of up to 2 weeks, but may remain in soil in its unchanged form in unfavourable soil conditions. Microbial breakdown also takes place.

Mepanipyrim fungicide

Mepanipyrim, $C_{14}H_{13}N_3$ (*N*-(4-methyl-6-prop-1-ynylpyrimidin-2-yl)aniline), is an anilinopyrimidine fungicidal compound. It is a pale yellow, odourless powder soluble in water (2.08 mg/l), acetone (139 g/l), toluene (55.4 g/l) and dichloromethane (277 g/l). It is stable up to 180°C. Mepanipyrim is produced by the reaction of 4-methyl-6-prop-1-ynylpyrimidine with a mixture of NaOH and formanilide at room temperature, followed by the addition of 4N HCl. The mixture is then refluxed. The CAS Registry Number of this fungicide is 110235-47-7.

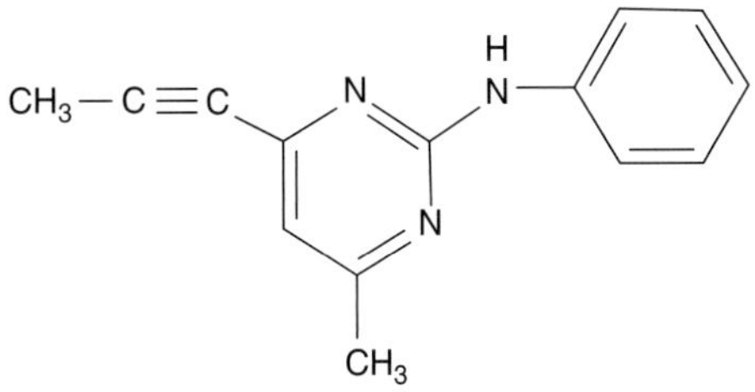

Mepanipyrim

Mepanipyrim is a contact, preventive fungicide, used to control grey mould on strawberries, tomatoes and cucumbers, **scab** on apples and pears and **powdery mildew** on cucumbers and strawberries. It also works as a bactericide and wood preservative. Mepanipyrim inhibits the elongation of the germ tube and the formation of the appressorium, which in turn inhibits penetration into the host plant. It also affects methionine biosynthesis, thus making it difficult for fungi to produce the protein required for degrading the plant cell wall.

Common formulations of mepanipyrim are wettable powder (WP) and suspension concentrate. Mepanipyrim is diluted before spraying, with application rates ranging from 0.2 to 0.75 kg/ha. Some trade names of mepanipyrim are Frupica, Japica and Cockpit.

The acute oral LD_{50} values in rats and mice are >5000 mg/kg, while in mallard ducks and bobwhite quails the values are >2250 mg/kg. Mepanipyrim is a slight eye irritant but not a skin irritant. The inhalation LC_{50} of mepanipyrim in rats is 0.59 mg/l. The acute 96 h LC_{50} of mepanipyrim in rainbow trout is > 0.74 mg/l. It is not toxic to honey bees and earthworms. The ADI is 0.02 mg/kg b.w./day. The WHO has categorized mepanipyrim as unlikely to be hazardous (Class U toxin), while it is classified as a PAN Bad Actor chemical since it is carcinogen.

In rats, mepanipyrim is almost completely excreted through faeces and urine within 96 h. It is degraded into several metabolites by the hydroxylation of aniline and pyrimidine methyl groups. In plants, the main metabolites are 2-hydroxypropyl-, 2,3-dihydroxypropyl- and 4-hydroxyphenyl-derivatives of mepanipyrim. This fungicide is moderately persistent in soil with the DT_{50} being 56.5 days. It is stable in water (DT_{50} >1 year at various pH levels).

Mepanipyrim is registered in many countries of the EU including Denmark, the UK, Germany, Greece and Portugal. It is not registered in the USA.

The commonly accepted MRL value on apples and pears is 0.01 mg/kg. According to the The Japan Food Chemical Research Foundation, the indicative MRL value of mepanipyrim fixed in Japan for tomato is 5 ppm and for cucumber is 2 ppm.

Mephosfolan insecticide

Mephosfolan, $C_8H_{16}NO_3PS_2$ (diethyl[(*EZ*)-4-methyl-1,3-dithiolan-2-ylidene] phosphoramidate), is a phosphoramidate insecticide. It is a yellow-to-amber coloured liquid soluble in water (57 mg/l). It works as an insecticide and acaricide. It is a moderately volatile substance. It is hydrolysed by acids and alkalis.

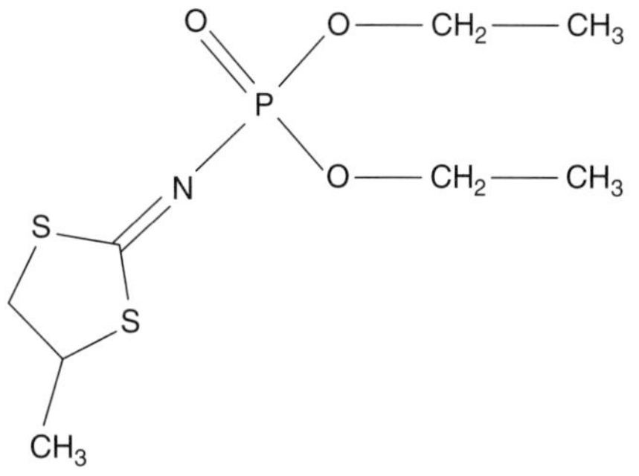

Mephosfolan

Mephosfolan, a cholinesterase inhibitor, can control many pests in aquatic environments. It is a highly toxic compound that is now obsolete. It is extremely toxic to rats, birds and to most aquatic organisms. The acute LD_{50} in rats is 9 mg/kg.

Exposure to mephosfolan may produce symptoms like sweating, headache, dizziness, profound weakness, muscle spasms, seizures, pin-point pupils with blurred vision, abdominal pain and many others. Respiratory depression and respiratory paralysis may also occur. Mephosfolan is a PAN Bad Actor chemical, since it inhibits cholinesterase.

This insecticide is not registered in the USA or the EU. It has also been refused registration in India, among other countries.

Mepiquat as a PGR

Mepiquat, $C_7H_{16}N$ (1,1-dimethylpiperidinium), is a colourless, odourless, hygroscopic crystalline substance designated as a plant growth regulator. It is soluble in water, ethanol, acetone, chloroform, benzene, ethyl acetate and cyclohexane. The compound is stable in an aqueous medium and artificial sunlight, but it decomposes at 285°C. Mepiquat forms salts like mepiquat pentaborate and mepiquat chloride, which are also active ingredients. The CAS Registry Number of mepiquat is 15302-91-7.

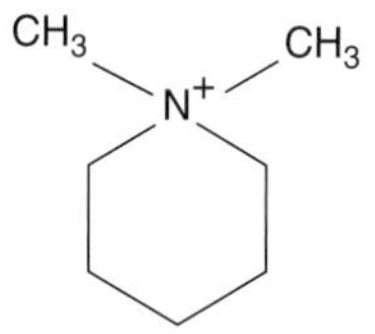

Mepiquat

Mepiquat inhibits the production of gibberellic acid after translocating throughout the plant. This PGR is used to slow down sprouting, prevent lodging (by keeping stems short) and reduce vegetative growth. At the same time, it also hastens boll maturation in cotton. It protects onions and garlic (from sprouting) and cereals (from lodging).

This PGR is formulated as a soluble concentrate (SL) and a ULV liquid (UL). It may be combined with **ethephon** to enhance its effect. One of the trade names of mepiquat is Pix.

The acute oral LD_{50} of mepiquat in rats is 1490 mg/kg. The ADI is 1.5 mg/kg/day. According to the WHO, it is a Class III toxin (i.e. it is slightly hazardous). The EPA has also classified it as a slightly toxic chemical. It is non-toxic to bees.

In animals, mepiquat is excreted via urine and faeces. It is registered in the UK and some countries of the EU, whereas only mepiquat chloride ($C_7H_{16}ClN$) is registered in India and the USA.

Meptyldinocap fungicide

See **Dinocap fungicide**

Mercuric chloride fungicide

See **Mercuric compounds as fungicides**

Mercuric compounds as fungicides

Mercury compounds such as mercuric chloride, mercuric oxide and mercurous chloride are toxic and can be used as fungicides and wound protectants. They can also effectively control root maggots. Many pathogens including *Fusarium* sp., *Sclerotium cepivorum* and *Delia* spp. on turf, onions and brassicas are controlled by these agents. Canker on fruit trees, rubber trees, vines and shrubs, and bark injuries and pruning wounds are also taken care of by the application of mercury compounds. Given the toxicity of mercury compounds, all products are expected to clearly display signal words like 'DANGER', 'POISON' or 'WARNING'.

Mercuric chloride, Cl_2Hg (mercury (II) chloride; mercury dichloride), is a well known fungicide. It is a colourless, crystalline powder that is soluble in water (69 g/l at 20°C), ethanol, diethyl ether and pyridine. Mercuric chloride is reduced to mercury (I) chloride and to metallic mercury chemically or by sunlight. Due to the hazardous characteristic of mercury compounds (especially mercuric chloride) much care is advised during its storage and use. Mercuric chloride should be kept in a tightly closed container and stored in a cool, dry, ventilated area, protected from direct sunlight and incompatible substances (e.g. strong oxidizing agents, strong bases, ammonia, copper, iron, potassium). The CAS Registry Number of mercuric chloride is 7487-94-7.

Mercuric chloride is incompatible with sodium, lead, sulfites, phosphates, albumin, gelatin, alkalis, lime water, arsenic, bromides, **borax** and carbonates while mercurous chloride is not compatible with bromides, iodides, ammonia, cyanides and hydrogen peroxide.

Mercuric chloride is classified by the WHO as a Class Ia (extremely hazardous) toxin. It is hazardous in the case of contact with skin (causing burning and redness), eyes (causing burning and injury) and via inhalation (it is corrosive and causes a burning sensation, shortness of breath and a cough). The acute oral LD_{50} in rats ranges from 1 to 5 mg/kg. The compound is toxic to fish (the acute LC_{50} species ranging from 0.037 to 0.903 mg/l). It is phytotoxic to many crops. The ADI is 0.0003 mg/kg b.w./day. Mercuric chloride is toxic to the brain, nervous system (both central and peripheral) and reproductive system.

Oral exposure to mercuric chloride causes abdominal cramps, abdominal pain, a burning sensation, diarrhoea, nausea, shock or collapse, sore throat, vomiting and a metallic taste. Rinsing the mouth and administering a drink of activated charcoal and antidotes such as dimercaprol may help in reviving the victim. Impervious protective clothing (including boots, gloves, lab coat, goggles (a full face shield, if necessary) and apron or coveralls) is recommended to prevent skin/eye contact.

A self-contained breathing apparatus (approved by NIOSH) and fully protective gear may also be required. Label instructions should be referred to for mixing, loading and cleaning up dry and wet spills.

Mercuric oxide, HgO (mercury oxide; mercury (II) oxide), is an inorganic fungicide. It is an orange-red powder soluble in water (53 mg/l), and insoluble in organic solvents. The CAS Registry Number of this compound is 21908-53-2.

Mercuric oxide is a fungicidal wound protectant. It is also a very toxic substance which can be absorbed via the dermal, inhalation and oral routes. The WHO has classified mercuric oxide as a Class Ib toxin (highly hazardous) and the EPA, as a Class I toxin. The acute oral LD_{50} in rats is 18 mg/kg. The compound is toxic to fish. It can affect the eyes, skin, respiratory tract and kidneys of humans. Mercuric oxide bioaccumulates in the food chain.

Mercurous chloride, Hg_2Cl_2 (dimercury dichloride), is a white solid soluble in water (0.2 mg/100 ml) and insoluble in ethanol and ether. It slowly decomposes in sunlight into mercuric chloride and metallic mercury. Mercurous chloride is not compatible with bromides, iodides, ammonia, cyanides, silver salts, iodoform and hydrogen peroxide. It has the additional properties of being a purgative (laxative) and a de-worming agent. Mercurous chloride is formulated as dustable powder (DP).

Mercurous chloride is toxic, although due to its low solubility in water, it is generally less dangerous than mercuric chloride. Mercurous chloride causes irritation to the respiratory tract, eye and skin. It also causes skin sensitization. This compound is a moderately hazardous compound according to the WHO.

Mercuric chloride was formerly used more widely. However, it is one of the most toxic forms of mercury because of its rapid solubility in water. The substance is banned as a pesticide in the EU and the USA. It is not registered in India.

Methoxyethyl mercuric chloride (MEMC), C_3H_7ClHgO ((2-methoxyethyl) mercury (II) chloride or (2-methoxyethyl) mercuric chloride), is an organomercury fungicide. A colourless, crystalline powder, MEMC is soluble in acetone and decomposes in hydrochloric acid. The CAS Registry Number of this compound is 123-88-6.

MEMC acts by reacting with the sulfydryl group in fungal pathogens and inhibits growth. The compound has been used in soil and seed treatment against fungal pathogens including *Ustilago scitaminea*, *Rhizobium solani* and *Fusarium* sp.

But MEMC is a highly acute toxin with the acute oral LD_{50} among rats being 30 mg/kg. It is also a persistent chemical which is a reproductive toxin, a carcinogen and a suspected endocrine disruptor, according to the PAN Pesticide Database. Hence it is not used or its use is severely restricted in many countries.

India allows restricted use of MEMC formulations.

Mercuric oxide fungicide

See **Mercuric compounds as fungicides**

Mercurous chloride fungicide

See **Mercuric compounds as fungicides**

Mercury fungicides – technical analysis

Organic and inorganic mercuric compounds are a major group of fungicidal agents that protect crops, paints and wood products. Mercuric chloride, mercuric oxide and methoxyethyl mercuric chloride (MEMC) are some of the major fungicides used, albeit conservatively. The following passage is a brief analytical method for a sample containing an organic mercury compound.

Take 0.5–1 g of the compound in a ground-glass, double-necked flask and attach to it, a Friedrich condenser and a 50 ml dropping funnel. Add 10 ml concentrated sulfuric acid, followed by about 25 ml 50% hydrogen peroxide through the dropping funnel. These additions should be done drop by drop with intermittent cooling to prevent hazard. Allow the sample to dissolve naturally. Then heat the digest under a full flame for 30 min. Allow the clear digest to cool and then transfer quantitatively with water to a volumetric flask such that 5 ml aliquots contain microgram or milligram quantities of mercury.

Select a standard Hg^{2+} solution so that the aliquot measured contains 0.1–10 times the estimated amount of mercury in the sample solution. To the sample and standard solution, add an amount of ^{203}Hg tracer up to 0.01 times the Hg^{2+} content and enough concentrated sulfuric acid to make the acid concentration 1 N. The total volume should be 5–6 ml. Add 5 ml carbon tetrachloride free of organic mercury content. Stir vigorously for 10 min and separate by centrifugation. Measure the ^{203}Hg in an aliquot of each phase by gamma scintillation counting. Normalize the count rate of each aliquot of the aqueous phase to a volume of 5 ml and calculate the isotope dilution.

Meristems

See **Herbicide action – meristematic mitotic inhibitors**

Mesosulfuron-methyl herbicide

Mesosulfuron-methyl, $C_{17}H_{21}N_5O_9S_2$ (methyl 2-[(4, 6-dimethoxypyrimidin-2-ylcarbamoyl)sulfamoyl]-α-(methanesulfonamido)-*p*-toluate), is a pyrimidinyl sulfonylurea herbicide. It is a cream-coloured powder soluble in water at 21 mg/kg. The herbicide has a CAS Registry Number of 208465-21-8 (see Mesosulfuron-methyl at top of the next page).

Mesosulfuron-methyl is an acetolactase herbicide, and acts by inhibiting **acetohydroxy acid synthase** (AHAS). It is used to control a variety of weeds in wheat and rye (e.g. perennial ryegrass, foxtail, wild oats and many broadleaved weeds).

The formulation types of mesosulfuron-methyl are oil dispersion (OD) and water-dispersible granule (WG). The trade names associated with mesosulfuron-methyl are Alister, Atlantis, Cossack, Osprey, Othello, Pacifica and Sigma.

Mesosulfuron-methyl has low acute oral, dermal and inhalation toxicity. It is not a skin and eye irritant. However, it is considered as a potential groundwater contaminant, according to the PAN Pesticide Database.

The herbicide is predominantly recovered in the faeces within 24 h. It degrades in the environment easily, although it binds weakly to the soil.

Mesosulfuron-methyl is registered for use in India, Australia, Canada, South Africa, the UK and the USA. In India, the registered formulation contains mesosulfuron-methyl in combination with iodosulfuron methyl sodium.

Mesotrione herbicide

Mesotrione, $C_{14}H_{13}NO_7S$ (2-(4-mesyl-2-nitrobenzoyl)cyclohexane-1,3-dione), is a benzoyl cyclohexanedione herbicide. It is a beige-to-tan coloured liquid with a faint odour. It can decompose

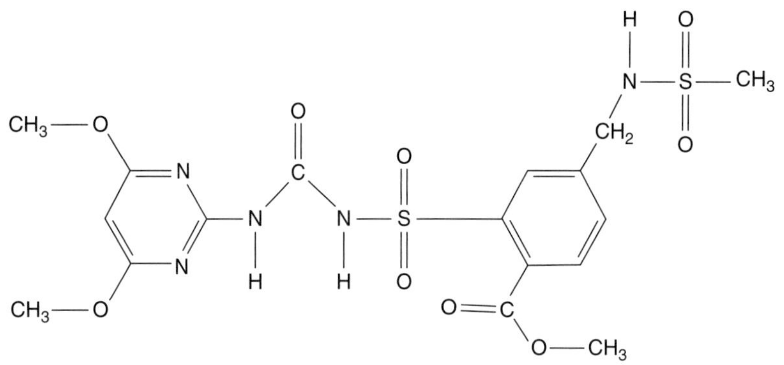

Mesosulfuron-methyl

at high temperatures forming toxic gases. It is soluble in water (160 mg/l at 20°C) and highly soluble in acetone and ethyl acetate. It is stable to aqueous photolysis, with a DT_{50} of >80 days. It is stable under normal use and storage conditions, as long as contact with mild steel, cast iron or aluminum is avoided. Its CAS Registry Number is 104206-82-8.

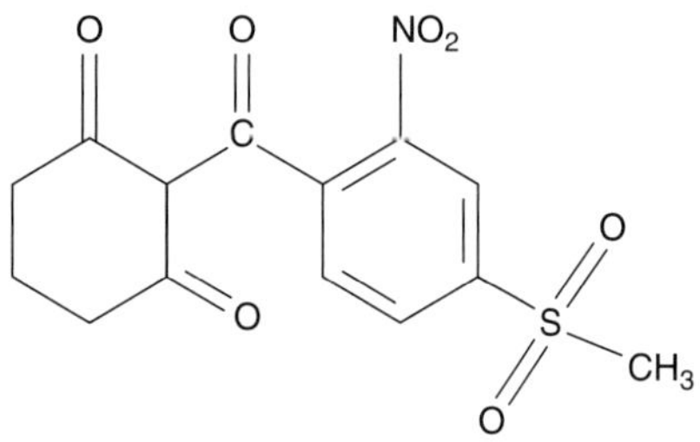

Mesotrione

This pre- and post-emergence herbicide is effective in controlling broadleaved weeds in maize. It acts by inhibiting the enzyme **4-hydroxyphenylpyruvate dioxygenase (HPPD)** in target plants. It also inhibits carotenoid biosynthesis, resulting in a bleaching effect.

Mesotrione is applied through ground or aerial application equipment. Trade names include Callisto and Marvel. This is a relatively new class of herbicide and if used in Resistance Management Programmes it can mitigate the increase in resistant biotypes.

The herbicide may cause mild eye and skin irritation. It is not a skin sensitizer. Prolonged or frequently repeated skin contact may cause allergic reactions in some individuals. The acute oral LD_{50} in rats is >5000 mg/kg b.w. The acute dermal LD_{50} in rats is >5000 mg/kg b.w. It is unlikely to be hazardous to aquatic life. The acute LC_{50} is >120 mg/l for rainbow trout (96 h). The acute LC_{50} is >120 mg/l in bluegill sunfish.

Mesotrione is moderately persistent in soil with a half-life of over 30 days. It has moderate mobility in soil. It is registered for use in Canada, the USA, South Africa and many EU members including Germany, the UK, Italy and France.

Metabolism

The transformation that occurs to a substance, inside an organism (plant, animal or man) or in the environment, is called metabolism. In an organism, metabolism may be witnessed from the time a substance enters it up to the time the waste product is eliminated.

The process of this transformation, which includes steps such as degradation, synthesis, hydrolysis, reduction and oxidation, involves the participation of enzymes and acids. These transformations are often localized with respect to the organs, tissues and types of cells involved.

Basal metabolism is the rate of heat produced by an individual, who is awake, but in complete mental and physical repose for at least 12 h. Under these conditions, oxidation of stored nutrients provides the sole source of energy expended. The heat thus produced is measured by calorimetry.

A metabolite is an intermediate substance produced and utilized in the processes of a living cell. Metabolites bring about the growth of living tissues and are also a source of energy in the body. Examples of metabolites are nucleic acids, glucose, cholesterol and many types of enzymes. In the case of some organophosphates (OP), their breakdown leads to a metabolite chemical (an oxon) that is toxic to target pests. **Parathion** (metabolite 4-nitrophenol) and **chlorpyrifos** (metabolite TCP, i.e. **3,5,6-trichloro-2-pyridinol** (3,5,6-TCP)), are some examples of OPs with hazardous metabolites. An organochloride such as **heptachlor** metabolizes into heptachlor epoxide, which is moderately toxic.

There are several methods of studying the metabolism of agrochemicals, the method using isotopes being considered one of the most effective. The study of uptake, distribution and metabolism of an applied chemical is made possible with the use of radio-labelled material. Stable isotopes are also used for this study.

Metabolic activity for the breakdown of chemical pesticides takes place mainly in leaves and roots. Photodegradation is one of the major routes of degradation. Detoxifying enzymes are responsible for the metabolic processes in tissues. Young plants have high levels of detoxifying enzymes, and hence they metabolize pesticides at higher rates. In the rhizosphere, microbial breakdown is a major step of degradation.

The study of metabolism is important for better regulation of pesticide use and for environmental safety. It has become very important to assess the activity of metabolites in treated plants, to estimate total residue content, their composition and distribution. It is also necessary to gauge the presence and activity of metabolites in the environment.

Metabolite

See **Metabolism**

Metalaxyl fungicide

Metalaxyl, $C_{15}H_{21}NO_4$ (methyl *N*-(methoxyacetyl)-*N*-(2-6-xylyl)-DL-alaninate), is an acylalanine **phenylamide** fungicide. It is a fine, white powder soluble in ethanol (400 g/l), toluene (340 g/l) and water (8.4 g/l). This compound is stable in neutral and acidic media as well as at temperatures up to 300°C. Metalaxyl is produced by the reaction of 2,6-dimethylaniline with an excess of methyl 2-bromopropionate in the presence of sodium bicarbonate, followed by acetylation with methoxyacetyl chloride.

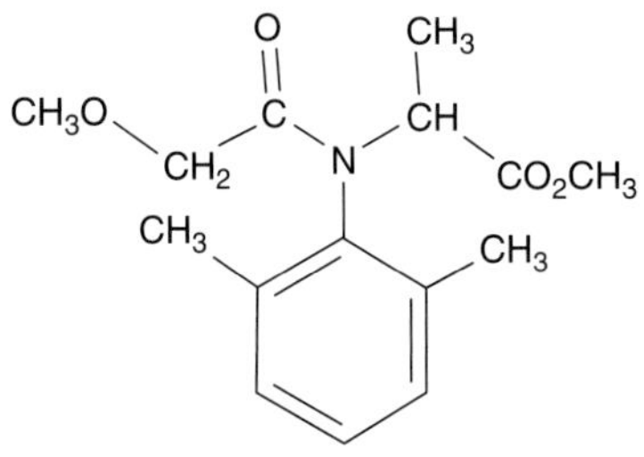

Metalaxyl

Metalaxyl, a known systemic fungicide with curative and protective action, works by inhibiting the process of RNA synthesis in fungi, specifically that of ribosomal RNA. Metalaxyl is especially effective on the order Peronosporales and covers a wide variety of crops including potatoes, peas and tomatoes (against *Phytophthora infestans*), tobacco (against *Peronospora tabacina*), vines (against *Plasmopara viticola*) and **downy mildew** on certain vegetables. Many other crops such as maize, peas, sorghum, cotton, onions, cucurbits and citrus are also protected by mixing metalaxyl with other fungicides.

Some common metalaxyl formulations include powder for dry seed treatment (DS), flowable concentrate for seed treatment (FS), wettable powder (WP) and granules. Application rates range from 150 to 250 g/ha, and applications are done at both pre- and post- harvest stages. Seed and soil treatments and foliar sprays are common pre-harvest applications. This fungicide is also applied on stored crops for protection against rot. Trade names include Apron, Ridomil and Subdue. Other protective fungicides like **copper oxychloride**, **mancozeb** and ***Bacillus subtilis*** can be combined with metalaxyl.

Metalaxyl is classified by the WHO as a Class III toxin (slightly hazardous). The oral acute LD_{50} in rats is 633 mg/kg. The compound does not cause any skin irritation when tested on rabbits and guinea pigs. The NOEL in dogs is found to be 7.8 mg/kg b.w./day for 6 months. Metalaxyl is not toxic to bees or birds, if used as directed (the LD_{50} in mallard ducks for an 8 day period is 1466 mg/kg and an 8 day LC_{50} in many other bird species is >10,000 mg/kg). The RfD of metalaxyl is 0.06 mg/kg/day.

When ingested by animals, metalaxyl is almost entirely eliminated via urine and the bowels. Metabolism proceeds by hydrolysis of the ester bond followed by oxidation of the 2-(6)-methyl group and of the phenyl ring and *N*-dealkylation. The compound does not seem to accumulate in the system. In plants, metalaxyl is taken up via leaves, stems and roots, and metabolism takes place via oxidation of the phenyl ring and methyl group and cleavage of the methyl ester and *N*-dealkylation. Metalaxyl degrades in soil and water within 50 days, with the DT_{50} in water ranging from 22 to 48 days. At high pH levels, the DT_{50} is 12 days.

Metalaxyl is registered in many countries including India, the USA, Australia, Canada, the EU, South Africa, Uganda and Senegal.

According to the USDA Foreign Agricultural Service's Pesticide MRL database, the indicative MRL of metalaxyl for cottonseed in the USA is 0.01 ppm. The EU maintains a specific MRL of 0.1 ppm for its 'Oilseeds' group.

Metalaxyl-M fungicide

Metalaxyl-M, $C_{15}H_{21}NO_4$ (methyl *N*-(methoxyacetyl)-*N*-(2,6,-xylyl)-D-alaninate), is a pale yellowish-brown viscous liquid. It is soluble in water (26 gm/l) and *n*-hexane (59 gm/l) and miscible with acetone, ethyl acetate, toluene and dichloromethane.

The fungicide is hydrolytically stable under neutral to acidic conditions (with a DT_{50} of >200 days), while in alkaline media, the DT_{50} is 116 days. Metalaxyl-M is stable under normal use and storage conditions (i.e. a dry, well-ventilated, secure area, away from heat). The CAS Registry Number of metalaxyl-M is 70630-17-0.

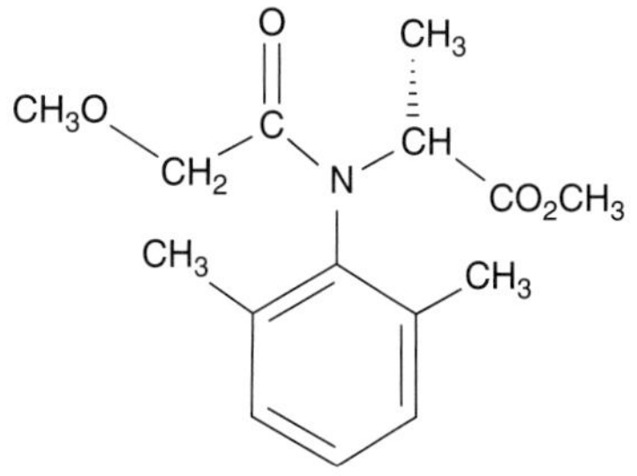

Metalaxyl-M

Metalaxyl-M is a known systemic fungicide, used to control diseases caused by *Phytophthora*, *Peronospora* and *Pythium* that affect crops like potatoes, tomatoes, tobacco, maize, peas, sorghum, vegetables and citrus. Its mode of action is the inhibition of RNA synthesis in the organism.

This fungicide is generally used in a mixture with other protectant fungicides and is used as foliar sprays and as soil applications. It can be applied in the range of 250–1000 g/ha and in the range of 35–300 g/100 kg seed for seed treatment. Formulations that are common among these mixtures are emulsifiable concentrate (EC), granules (GR) and seed treatment formulations.

Metalaxyl-M's acute oral LD_{50} is >953 mg/kg in male rats and 375 mg/kg in female rats. The acute 48 h oral LD_{50} in honey bees is >127 μg per bee. The LD_{50} in birds is up to 1419 mg/kg. The acute 14 day LC_{50} in earthworms is 830 mg/kg. The ADI is 0.08 mg/kg b.w./day.

Due to its acute toxicity (mammals) and its potential to contaminate groundwater, metalaxyl-M is classified as a PAN Bad Actor chemical. Thus care has to be taken to avoid rinse water flowing into groundwater supplies. Disposal of empty containers, consumables and excess chemical formulations should be as per the guidelines of local health and environment regulating agencies.

Metalaxyl-M is registered for use in India and the UK (Category B). The MRL value of metalaxyl-M is 2.0 ppm for table grapes and lettuce; 1.0 ppm for wine grapes, pome fruit and cabbages; 0.5 ppm for strawberries, citrus, sweet peppers, onions and garlic; 0.1 ppm for carrots, cauliflower and broccoli; 0.05 ppm for other vegetables and fruit; and 0.02 ppm for cereal grains.

Metaldehyde molluscicide

Metaldehyde, $C_8H_{16}O_4$ (r-2,c-4,c-6,c-8-tetramethyl-1,3,5,7-tetoxocane or 2,4,6,8-tetramethyl-1,3,5,7-tetraoxacyclo-octane), a crystalline powder, is a cyclic tetramer of acetaldehyde and a known molluscicide. It is made by the polymerization of acetaldehyde in an alcohol–water mixture in the presence of HBr (hydrogen bromide) and other alkaline earth bromides. Metaldehyde is soluble in water (190 mg/l) and methanol. Its CAS Registry Number is 108-62-3 (polymeric form).

The ingestion of metaldehyde causes the gut walls of slugs and snails to become permeable, leading to their death. Contact with mollusc feet induces excessive secretion of mucous, again leading to death by dehydration, especially in hot and dry conditions. Gastropod eggs are killed by metaldehyde vapour.

Metaldehyde (4–6%) is mixed with wheat bran or barley flour, shaped into very small pellets and spread evenly. The pellet formulation may contain concentrations of up to 10 kg a.i./ha. Animal repellents such as vanillylamide are added to protect mammals from consuming these baits. However, dustable powders and sprays are more effective. Pastes are also applied around plants. Common trade names are Slug-tox, Slug-Death, Slugit pellets and Ariotox.

Metaldehyde is slightly to moderately toxic by the oral route with LD_{50} values in the range of 100–1270 mg/kg in cats, rats, mice, guinea pigs and rabbits. By the dermal route, the LD_{50} in rats is between 2275 and 5000 mg/kg, thus making it moderately toxic. Symptoms of high acute exposure include increased heart rate, high blood pressure, drowsiness, salivation and tremors. Metaldehyde is also suspected to be a mutagen. Humans show symptoms of toxicity even with minute quantities of metaldehyde. Baits (4% a.i.) can be a threat to certain wildlife species, but they are non-toxic to bees and fish. Metaldehyde is considered a potential groundwater contaminant and a possible carcinogen according to the PAN Pesticide Database. The WHO has categorized it as a Class II toxin (moderately hazardous).

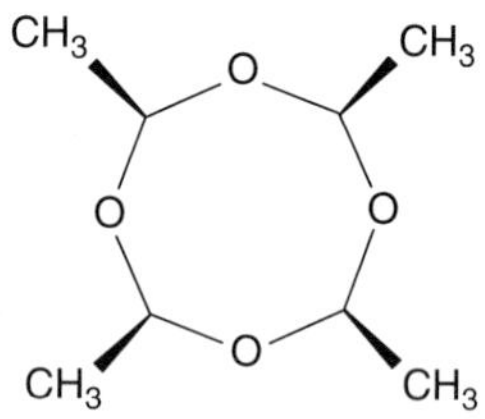

Metaldehyde

In soil and water, metaldehyde gradually depolymerizes into acetaldehyde. But, it is not believed to persist for long in the environment. This chemical is registered in India, the USA, Australia, the Netherlands, Canada, New Zealand and Portugal.

Metam salts

Methyldithiocarbamic salt, metam ($C_2H_5NS_2$, methyldithiocarbamic acid), and its salts like metam-sodium and metam-potassium are **dithiocarbamate** fungicides as well as soil fumigants, nematicides, insecticides and herbicides.

Metam-sodium, $C_2H_4NNaS_2$ (sodium methyldithiocarbamate), is a soil sterilant that controls soil-borne fungi, insects, weeds and nematodes. It is a colourless, crystalline substance that is highly soluble in water (>700 g/l) and much less so in acetone and xylene. It decomposes in the presence of heavy metals and acidic medium (with a DT_{50} of <24 h). The compound's activity is initiated when it is converted to **methyl isothiocyanate**.

Metam-sodium is applied at a pre-plant stage and is generally formulated as a soluble concentrate (SL). This soil fumigant is an extremely toxic compound with an acute oral LD_{50} of 895 mg/kg in rats and 285 mg/kg in mice. It is a highly phytotoxic compound that also causes irritation to the skin and eyes. The ADI/RfD is 0.001 mg/kg b.w.

Metamifop herbicide

Metamifop, $C_{23}H_{18}ClFN_2O_4$ ((*R*)-2-[4-(6-chloro-1,3-benzoxazol-2-yloxy) phenoxy]-2′-fluoro-*N*-methylpropionanilide), is a new aryloxyphenoxypropionate (AOPP) herbicide. It is a light brown, odourless substance that is very soluble in acetone and

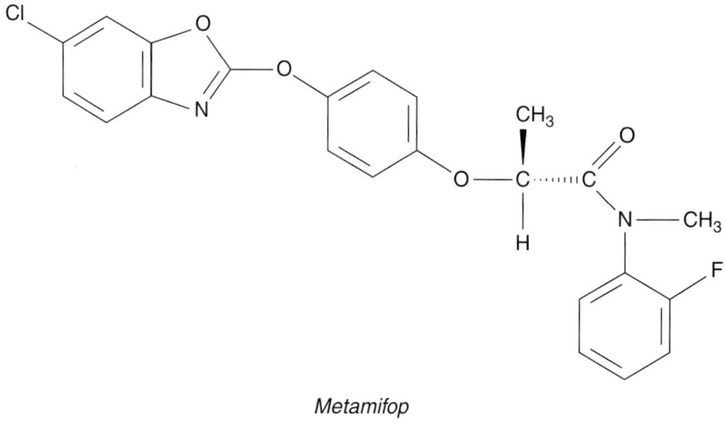

Metamifop

methanol, and in water at 6.87 mg/l. It is stable at 54°C. The CAS Registry Number of metamifop is 256412-89-2 (see Metamifop at bottom of the page).

Metamifop is a post-emergence, acetyl-CoA carboxylase inhibitor, causing **chlorosis** and stunting growth in susceptible annual grass weeds like *Echinochloa* spp., *Leptochloa chinensis*, *Digitaria* spp. and *Eleusine indica*. It is used mainly in rice cultivation.

This chemical is formulated as an emulsifying concentrate (EC) and as granules (GR). It is applied post-emergence, in paddy and directly-sown rice fields, at 90–200 g a.i./ha.

Metamifop may cause sensitization by skin contact. It is a slight eye irritant. Metamifop's acute oral LD_{50} for rats is >2000 mg/kg and for bees is >100 µg per bee. The acute percutaneous LD_{50} in rats is >2000 mg/kg.

The soil degradation of metamifop is both chemical and microbial. The herbicide's DT_{50} is up to 60 days (25°C).

Metamitron herbicide

Metamitron, $C_{10}H_{10}N_4O$ (4-amino-4,5-dihydro-3-methyl-6-phenyl-1,2,4-triazin-5-one), is a triazinone herbicide. It is a yellowish crystalline solid, fairly soluble in water and organic solvents. It decomposes in alkaline media, in the presence of strong oxidizers and on heating (the latter producing an explosion and toxic fumes, e.g. nitrogen oxides). Hence, this chemical must be stored in its original container in a dry, well-ventilated secure store room. Metamitron is made by the reaction of *N'*-acetyl-benzylhydrazone with hydroxyl amine in pyridine, containing potassium hydroxide. It has a CAS Registry Number of 41394-05-2.

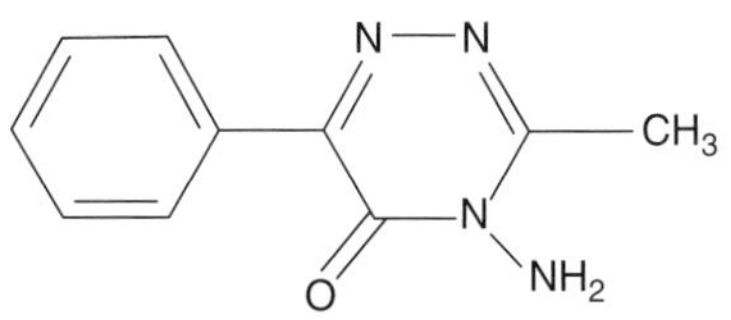

Metamitron

Metamitron is a selective, systemic, pre- and post-emergence herbicide, well absorbed by all plant parts and translocated acropetally. It controls grasses and broadleaved weeds by inhibiting photosynthesis in crops such as sugar and fodder beets, as well as in non-crop areas like industrial sites, lawns and irrigation canals.

Metamitron is available in wettable powder (WP) formulations that can be diluted and sprayed on target weeds, along with **phenmedipham** and chloridazon. Application dosage ranges from 0.35 to 4.2 kg/ha. Goltix is one of the trade names containing metamitron.

The acute oral LD_{50} of metamitron in rats is 1183 mg/kg, while the 4 h inhalation LC_{50} is 3.17 mg/l. Toxicity of this chemical to bird species varies from 190 to 1302 mg/l (making it moderately toxic). The acute 48 h oral LD_{50} of metamitron in honey bees is >97.2 µg per bee. The aRfD of metamitron in rats is 0.1 mg/kg b.w./day. According to the WHO and EPA, metamitron is a Class III toxin (slightly hazardous). Wearing of PPE (including dust-resistant safety goggles, gloves, overalls, etc.) is recommended while handling the chemical, to keep the skin and eyes safe.

The principal metabolite of metamitron in plants is 3-methyl-6-phenyl-1,2,4-triazin-5(4*H*)-one. In soil, metamitron degrades to form stable metabolites.

Metamitron is registered in many countries of the EU, but not in the USA.

Metam-sodium

See **Metam salts**; **Ethylenebisdithiocarbamate (EBDC) fungicides**

Metazachlor herbicide

Metazachlor, $C_{14}H_{16}ClN_3O$ (2-chloro-*N*-(pyrazol-1-ylmethyl) acet-2′,6′-xylidide), is a pyrazole herbicide. It is a yellowish crystalline substance, soluble in water (430 mg/l), chloroform (>1000 g/kg) and ethyl acetate (590 g/kg). It is stable for 2 years at up to 40°C. It is produced by the reaction of 2*N*,6-dimethylaniline with formaldehyde, followed by treatment with chloroacetyl chloride, and then with potassium pyrazole.

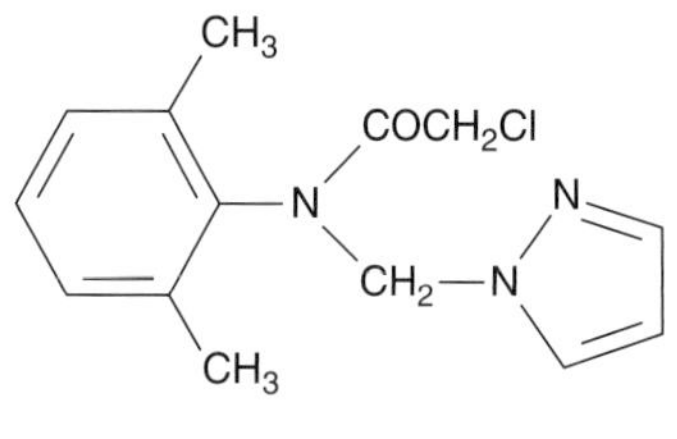

Metazachlor

Metazachlor, a selective, pre-emergence and early post-emergence compound, is absorbed by the roots and acts by inhibiting cell division and blocking protein synthesis. It is used on members of the genus *Brassica*, onions, groundnuts, potatoes, sugarcane, sunflowers, tobacco, pome fruits and many others. It controls annual grasses and broadleaved weeds.

This herbicide is formulated as a soluble concentrate (SL), and applied at the rate of 1.0–1.5 kg a.i./ha.

The acute oral LD_{50} of metazachlor is 2150 mg/kg in rats and >2000 mg/kg in bobwhite quails. The acute percutaneous LD_{50} in rats is >6810 mg/kg and it causes no irritation to the mucous membranes in rabbits. The LC_{50} (4 h) in rats is >34.5 mg/l. The LC_{50} (96 h) in rainbow trout is 4 mg/l and in carp is 15 mg/l. It is not toxic to bees. In long-term feeding trials, the NOEL in rats is 3.6 mg/kg. The ADI is 0.036 mg/kg/day. The WHO classifies metazachlor as a Class U toxin (unlikely to be hazardous).

After oral administration in rats, metazachlor is seen to be metabolized and eliminated rapidly in the form of polar conjugates via the kidneys. In plants, the residues contain the unchanged 2,6-dimethylaniline moiety. Microbial degradation in aerobic soil is rapid. The main metabolites are metazachlor oxalic acid and metazachlor sulfonic acid.

Metazachlor is registered in much of EU and South Africa but not in the USA.

Metconazole fungicide

Metconazole, $C_{17}H_{22}ClN_3O$ (1 *RS*,5 *RS*:1 *RS*,5 *SR*)-5-(4-chlorobenzyl)-2,2-dimethyl-1(1 *H*-1,2,4-triazole-1-ylmethyl) cyclopentanol), is an off-white, odourless crystalline fungicidal substance of the azole type. It is soluble in water (15 mg/l), methanol 235 mg/l and acetone (238.9 g/l). It is thermally and hydrolytically stable. Metconazole is produced by the reaction of 3-dimethyl-2-oxocyclopentanecarboxylate in benzene with sodium hydride, followed by the addition of 4-chlorobenzoyl chloride. The product thus obtained is decarboxylated by heating with hydrogen bromide. The CAS Registry Number of metconazole is 125116-23-6.

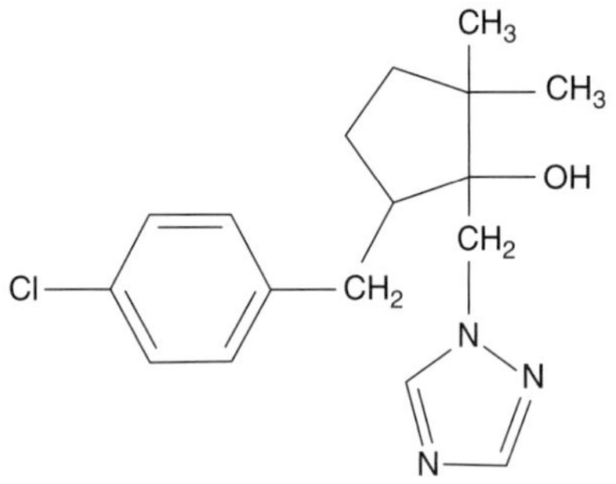

Metaconazole

This fungicide is a systemic, **ergosterol** biosynthesis inhibitor that interferes with synthesis of fungal cell membranes. It is most commonly used on cereal crops like wheat and barley for protection from rust diseases (*Septoria* spp.). Metconazole also controls black sigatoka disease (*Mycosphaerella fijiensis*) on bananas.

Metconazole is commonly formulated as a soluble concentrate (SL) and a suspension concentrate that can be sprayed via foliar broadcast spray and as ground or aerial equipment. The soluble concentrate generally contains 9% a.i. It is commercially available as Caramba.

The acute oral LD_{50} of metconazole in rats is up to 1459 mg/kg, and it causes moderate irritation in the eyes. It is not phytotoxic if applied according to the instructions on the label. The compound is practically non-toxic to bees and earthworms. According to the WHO, metconazole is a Class III toxin (slightly hazardous).

The fungicide metconazole is registered in many countries of the EU, the USA and Canada. According to the USDA Foreign Agricultural Service's Pesticide MRL database, the MRL value in the USA for wheat is 0.15 mg/kg, for barley, 2.5 ppm and for soybean, 0.05 ppm. This fungicide is relatively to very persistent in soil with a DT_{50} ranging from 84 to 546 days.

METEPA chemosterilant

METEPA, $C_9H_{18}N_3OP$ (1-[*bis*(2-methyl-1-aziridinyl)phosphoryl]-2-methyl aziridine), is a **chemosterilant**. It is a straw-coloured liquid that dissolves completely in water.

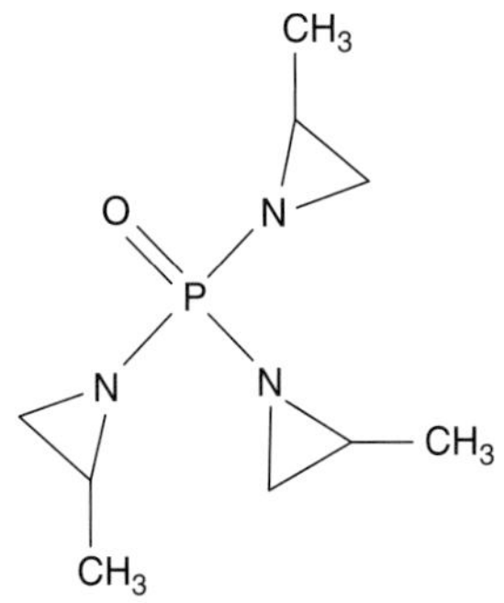

METEPA

METEPA causes sterility in both sexes of pests like the house fly, stable fly and screw worm fly. It is also effective on the defoliating insect *Pyrausta machearalis*, which attacks plantations of teak (*Tectona grandis*), especially after monsoons. It is phytotoxic on cotton leaves.

The acute oral LD_{50} in 11-day-old cockerels is 329 mg/kg.

Methabenzthiazuron herbicide

Methabenzthiazuron, $C_{10}H_{11}N_3OS$ (1-benzothiazol-2-yl-1,3-dimethylurea), is a urea (dimethylurea) herbicide. It is a colourless, odourless crystalline substance, soluble in water (59 mg/l), acetone (115.9 g/l) and methanol (65.9 g/l). It is unstable in strong acids and alkalis. Methabenzthiazuron is produced by the reaction of 2-methylaminobenzthiazole with methylisocyanate. Its CAS Registry Number is 18691-97-9.

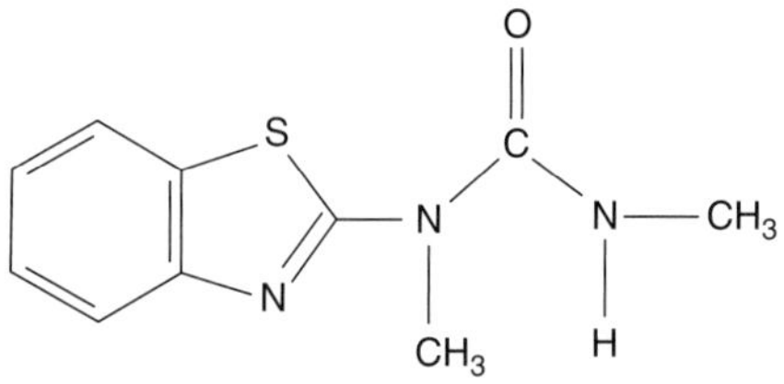

Methabenzthiazuron

Methabenzthiazuron, a persistent, broad-spectrum, systemic compound, acts by inhibiting photosynthetic electron transport (photosystem II). It is absorbed through the roots and protects against broadleaved weeds and grasses in vineyards and in fields of maize, legumes, garlic and onions.

The common formulation type of methabenzthiazuron is the wettable powder (WP). The herbicide may be combined with other herbicides (except urea-based, strongly alkaline/acidic or liquid fertilizers) for increased efficiency.

The WHO classifies this compound as a Class U toxin (unlikely to be hazardous). The NOEL (2 year) in rats and mice is 150 mg/kg diet, while the ADI is 0.05 mg/kg/day. The compound is not toxic to bees.

Methabenzthiazuron degrades slowly in soils. Its major metabolites in plants are 1-hydroxymethyl-1-methyl-3-(benzothiazol-2-yl) urea (its glucoside) and 3-methyl-3-(benzothiazol-2-yl) urea.

The herbicide is registered in India, Australia and New Zealand, but not in the USA.

Methacrifos insecticide

Methacrifos, $C_7H_{13}O_5PS$ ((E)-*O*-(2-methoxycarbonylprop-1-enyl) *O,O*-dimethyl phosphorothioate), is an organophosphorus insecticide. It is a colourless liquid, soluble in water (400 mg/l). It is miscible with many organic solvents such as methanol, benzene, hexane and dichloromethane. Methacrifos is relatively unstable in alkaline conditions and to hydrolysis (the DT_{50} is 29 days at pH 7 and 9.5 days at pH 9). This chemical is not in use/not manufactured in most countries (so it is considered obsolete). Methacrifos is produced by the reaction of 2-hydroxymethylenepropionic acid methyl ester with dimethylthiophosphoryl chloride. It has a CAS Registry Number of 62610-77-9.

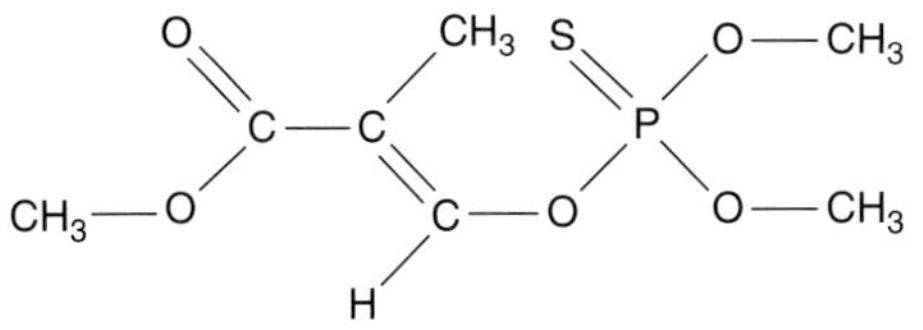

Methacrifos

Methacrifos, a cholinesterase inhibitor, is a broad-spectrum insecticide and acaricide with added vapour action and a long residual activity. It is also used to protect stored products.

This insecticide is formulated in forms such as dustable powder (DP) and emulsifiable concentrate (EC). A known trade name is Damfin.

The acute oral LD_{50} of methacrifos in rats is 678 mg/kg while the dermal LD_{50} is >3100 mg/kg and the inhalation LC_{50} is 2.2 mg/l. It is a mild skin irritant. Methacrifos can injure plant foliage. The ADI of methacrifos in humans is up to 0.006 mg/kg b.w. Methacrifos has been listed as a Class II toxin (moderately hazardous) according to the WHO. Since it is a cholinesterase inhibitor and because of its toxicity, this insecticide is categorized as a PAN Bad Actor chemical. Symptoms of poisoning are excessive salivation, sweating, rhinorrhea and tearing, nausea, vomiting, abdominal cramps, diarrhoea, respiratory depression, incontinence, tremors, dizziness, tightness in the chest, wheezing, a productive cough and pin-point pupils, sometimes with blurred or dark vision.

In animals, rapid metabolism and elimination occur via hydrolysis of the methyl esters as well as cleavage of the phosphoric and vinyl ester bonds. Mineralization to CO_2 and oxidative desulfuration also occur.

This insecticide is not in use in many countries such as India, the USA or the UK. Commonly established MRLs for citrus fruit and stone fruit are 0.05 mg/kg.

Methamidophos insecticide

Methamidophos, $C_2H_8NO_2PS$ (*O,S*-dimethyl phosphoramidothioate), is an organophosphorus insecticide. It is a colourless crystalline substance with a mercaptan-like odour. It is soluble in water (>200 g/l) and in organic solvents such as isopropanol, dichloromethane, hexane and toluene. Methamidophos decomposes on heating and in alkaline media (the DT_{50} is 70 h at pH 9). All formulations should be stored in their original labelled containers in secure, locked, cool, well-ventilated storage areas. Methamidophos is prepared by the reaction of *O,O*-dimethyl phosphoramidothioate and methyl iodide. The thiono-thiolo rearrangement reaction takes place more generally by the action of alkyl halides on phosphorothionates (Pistschimuka reaction). Its CAS Registry Number is 10265-92-6.

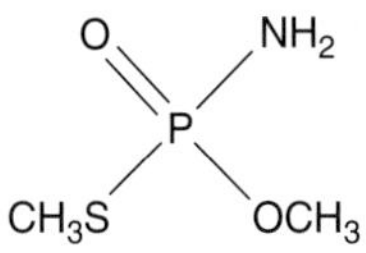

Methamidophos

Methamidophos, a cholinesterase inhibitor, is a systemic, broad-spectrum insecticide and acaricide with some contact toxicity and a residual activity of up to 10 days. It is absorbed by the roots and leaves. Methamidophos is used in the control of spider mites and chewing and sucking insects such as aphids, Colorado potato beetles, armyworms, whiteflies, **leafhoppers**, thrips and cabbage loopers. Potatoes, citrus and non-citrus fruit, hops, brassicas, beet, cotton, maize, tobacco and other crops are protected by this chemical.

Methamidophos is formulated commonly as a soluble concentrate (SL) and sprayed prior to harvesting (at 0.5–1.5 kg/ha). It is compatible with many other pesticides. Common trade names include Monitor, Tamaron, Morithion and Filitox.

The acute oral LD_{50} of methamidophos is 21 mg/kg and 16 mg/kg in male and female rats, respectively, and from 10 to 30 mg/kg in rabbits. The dermal LD_{50} is 50 mg/kg in rats. The acute 96 h LC_{50} values in various fish species tested are 25–100 mg/l. The acute oral LD_{50} values in mallard ducks, Japanese quails and hens range from 8 to 50 mg/kg b.w. Methamidophos is toxic to bees. The ADI is 0.004 mg/kg diet/day. The WHO has listed the pesticide as a Class Ib toxin (extremely hazardous). It is a PAN Bad Actor chemical, since it is an acutely poisonous cholinesterase inhibitor and a potential groundwater contaminant. Symptoms of poisoning include pin-point pupils with blurred or dark vision, tightness in the chest, respiratory depression, excessive salivation, headache, dizziness, nausea, vomiting, abdominal cramps, diarrhoea, sweating, seizures, incontinence and loss of consciousness. This chemical appears to cause delayed neuropathy. Given the toxicity of this chemical, it is mandatory for only authorized and registered personnel to handle it for application. Utmost care must be taken to wear proper PPE and to strictly follow label instructions. All product labels must display the signal word ‘DANGER’ or ‘POISON’.

In mammals, this insecticide is absorbed through the skin, stomach and lungs and eliminated via urine as deaminated methamidophos and phosphoric acid. Metabolism proceeds mainly by deamination and demethylation. In plants, degradation is by hydrolysis with loss of the amino, *S*-methyl or *O*-methyl groups. Degradation time is 6.1 days in soil, and 309 days in water at pH 5. Sunlight and microbial action also contribute towards degradation.

A few countries such as New Zealand, Niger, Nigeria, Sudan, Vanatu, Brazil, China, Japan, Liberia, Malaysia, Mauritius, Suriname, Thailand and Sri Lanka have permitted restricted use of methamidophos. It has been banned in Indonesia, Kenya, Kuwait, Jamaica and Libya. Many countries have banned the import of methamidophos formulations.

The pesticide methamidophos is registered in Australia, South Africa, Canada and Cameroon but not in the USA. According to the USDA Foreign Agricultural Service’s Pesticide MRL database, the indicative MRL of methamidophos for cottonseed in the EU, India and South Africa is 0.02 ppm and in the USA, it is 0.01 ppm. The MRL for carcass meat, milk and fat of cattle, goats and sheep is 0.01 mg/kg. The MRL for fruits and vegetables in Finland is 0.2 mg/kg.

Methasulfocarb fungicide

Methasulfocarb, $C_9H_{11}NO_4S_2$ (*S*-4-methylsulfonyloxyphenyl methylthiocarbamate), is a fungicide and plant growth regulator of the thiocarbamate class. It is a colourless, crystalline substance stable to light and soluble in water (480 mg/l), acetone, alcohol and benzene. Methasulfocarb has a CAS Registry Number of 66952-49-6.

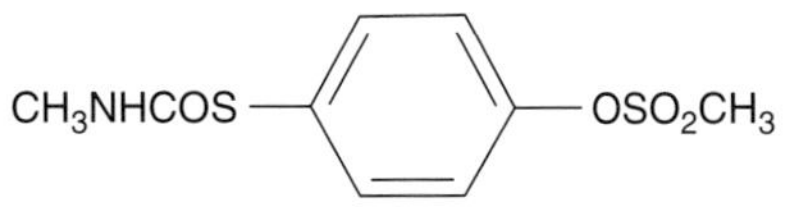

Methasulfocarb

Methasulfocarb is a soil fungicide with a long residual activity. It is used to control *Corticium*, *Fusarium*, *Mucor*, *Pseudomonas*, *Pythium*, *Rhizoctonia* and *Trichoderma* spp. in rice. This fungicide is especially effective in the control of seedling blight and bacterial seedling rot of rice, where a single application controls much of the infection.

The fungicide is formulated as a dustable powder (DP), and Kayabest is a common trade name.

The WHO classifies methasulfocarb as a Class II toxin. It is non-mutagenic to mice and non-teratogenic to rats. The acute oral LD_{50} in rats is 112–119 mg/kg, while the values in male and female mice are 342 mg/kg and 262 mg/kg, respectively. Acute percutaneous LD_{50} values in rats and mice are >5000 mg/kg. The inhalation LC_{50} of methasulfocarb in rats is 0.44 mg/l. The acute 96 h LC_{50} of methasulfocarb in carp is 1.95 mg/l.

Methasulfocarb is not registered in the EU or the USA.

Methemoglobin

See **4-Aminopyridine (4-AP) avicide**

Methidathion insecticide

Methidathion, $C_6H_{11}N_2O_4PS_3$ (*S*-(2,3-dihydro-5-methoxy-2-oxo-1,3,4-thiadiazol-3-ylmethyl)*O*,*O*-dimethyl phosphorodithioate), is an organophosphorus insecticide. It is a colourless crystalline substance, soluble in water (200 mg/l), ethanol, acetone, toluene, hexane and *n*-acetone. Methidathion is relatively unstable in both extremely acidic and extremely alkaline media, with a half-life of 30 min at pH 13. It is produced by the reaction of 2-methoxy-1,3,4-thiadiazol-5(4H)-one with dimethyldithiophosphoric acid via the *N*-chloromethyl derivative or in a one-step reaction in the presence of formaldehyde. Its CAS Registry Number is 950-37-8.

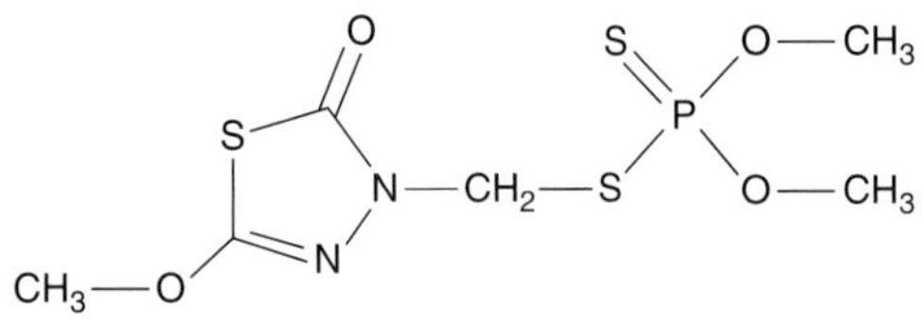

Methidathion

Methidathion is a broad-spectrum and non-systemic insecticide and acaricide but is strictly for non-residential use. It acts by inhibiting cholinesterase, which affects the nervous system of pests.

The insecticide methidathion is used to control a wide range of sucking and chewing insects and spider mites, on a variety of crops such as nuts (e.g. walnuts, almonds), citrus, artichokes, pears, plums, peaches, apples, cotton, lucerne, safflower and olives. Some specific insect pests controlled are weevils, **scale insects**, bollworms, beetles and whiteflies.

This insecticide is formulated as emulsifiable concentrate (EC), wettable powder (WP) and ULV liquid (UL). Some common trade names are Supracide, Ultracide and Suprathion. Methidathion is compatible with many pesticides, except those with alkaline materials. The application dosage normally ranges from 0.4 to 0.6 kg/ha. It is sprayed by ground methods which can put birds, bees and other non-target creatures at risk. This risk can be countered by applying a highly diluted formulation.

Methidathion is a highly toxic compound according to the EPA (Class I toxin). The WHO lists it as a Class Ib toxin (extremely hazardous). The compound is listed as a PAN Bad Actor chemical because it is a cholinesterase inhibitor, an acutely toxic poison, a potential groundwater contaminant and a possible carcinogen. All formulation labels must bear the signal word 'DANGER'. The **restricted entry interval** or REI for methidathion is 3 days for all crops. The acute oral LD_{50} ranges from 18 to 80 mg/kg among rats, mice, rabbits and guinea pigs. It is a severe skin irritant (on the basis of tests on rats). It is toxic to fish and birds. Its ADI is 0–0.005 mg/kg b.w./day. Being an organophosphorus compound, symptoms of poisoning include headache, giddiness, blurred vision, weakness, nausea, diarrhoea, cramps and discomfort in the chest. Wearing chemical-resistant gloves, an apron and a vapour-removing respirator are strongly recommended, especially for ground boom applications.

In mammals and plants, methidathion is rapidly metabolized and excreted. The overall metabolic pattern follows many processes such as hydrolysis of the ester bond and cleavage of the heterocyclic moieties into fragments. Methidathion and its metabolites are rapidly degraded in soil and water by chemical, photolytic and biological processes (the DT_{50} being 3–18 days). Microbial action also contributes to degradation. But, since metabolites do not adsorb, they may contaminate groundwater. Residues have also been found on some food samples.

Methidathion is registered for use in the USA, Australia, Madagascar and Vietnam.

According to the USDA Foreign Agricultural Service's Pesticide MRL database, the indicative MRL of methidathion for apple in South Africa is 0.5 ppm, while the MRL for the whole 'Pome fruits' group in the EU and the USA is 0.05 ppm.

Methiocarb insecticide

Methiocarb, $C_{11}H_{15}NO_2S$ (4-methylthio-3,5-xylyl methylcarbamate), is a carbamate insecticide. It is a colourless, crystalline substance which is sparingly soluble in water, but soluble in hexane, isopropanol and dichloromethane. It is unstable in alkaline media, with the hydrolysis DT_{50} being 6 h at pH 9. Methiocarb is produced by the reaction of 3,5-dimethyl-4-methylthiophenol and methylisocyanate. Its CAS Registry Number is 2032-65-7.

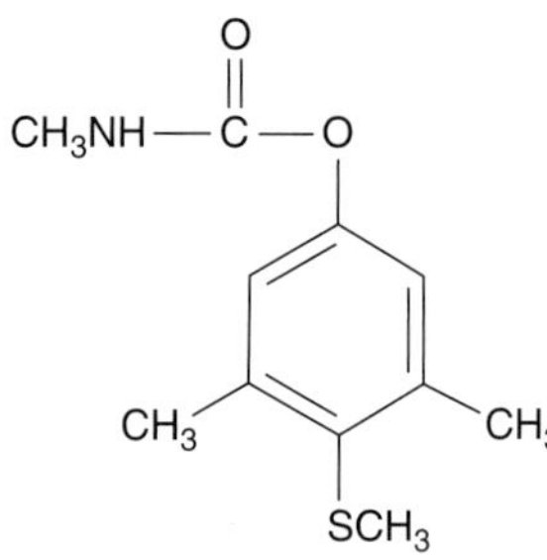

Methiocarb

Methiocarb is a contact, broad-spectrum, residual insecticide that also acts as a molluscicide, acaricide and a bird repellent. Its neurotoxic action (via cholinesterase inhibition) controls members of orders such as the Lepidoptera, Coleoptera, Diptera, **Thysanoptera** and Hemiptera (including soil and seed-borne insects). Insect pests targeted include fruit flies, aphids, thrips, **grasshoppers**, mites, leaf miners and beetles. Spider mites, snails and slugs are also controlled by methiocarb.

Methiocarb is formulated as wettable powder (WP), suspension concentrate (SC), bait (RB), granular bait (GB) and seed treatment, under trade names like Mesurol, Decoy and Draza. It is compatible with many pesticides, except those made of strong alkalis or acids and oil-based formulations. The application time should be adhered to strictly in cases of apple orchards to avoid yield loss (fruit thinning). Slug baits are applied in the form of pellets at a rate of 20 pellets/m^2 corresponding to 3–5 kg/ha. Slugs and snails are rapidly paralysed on contact. This toxic effect is independent of temperature, but better under humid conditions.

Methiocarb is a highly toxic chemical. The WHO lists it as a Class Ib toxin (extremely hazardous). The acute oral LD_{50} in rats is 19 mg/kg, while its dermal LD_{50} is >5000 mg/kg b.w. It is toxic to birds, fish and honey bees. It is also toxic by the inhalation route among rats (the 4 h inhalation LC_{50} being 0.433 mg/l). The acute LD_{50} in Japanese quails is 5 mg/kg. The acute 96 h LC_{50} of methiocarb in rainbow trout is 0.65 mg/l, while the acute 48 h EC_{50} for *Daphnia magna* is 0.008 mg/l. The acute 48 h contact LD_{50} of methiocarb in honey bees is 0.23 μg per bee. The ADI of methiocarb is 0.013 mg/kg/b.w./day. Signs of poisoning in humans include dizziness, muscle weakness, stomach cramping, diarrhoea, excessive sweating, nausea, vomiting, blurred vision and convulsions. Applicators and people involved in manufacturing, handling and loading the pesticide are strongly advised to wear special protective clothing. Since it is a cholinesterase inhibitor and a potential groundwater contaminant, methiocarb is considered a PAN Bad Actor chemical.

In mammals, methiocarb is rapidly absorbed and excreted via urine, with only a small proportion in the faeces. Its metabolism involves hydrolysis, oxidation and hydroxylation, followed by excretion in free or conjugated form. In plants, the methylthio group of methiocarb is oxidized to sulfoxide and sulfone, with hydrolysis to the corresponding thiophenol, methylsulfoxide-phenol and methylsulfonyl-phenol which are the metabolite products. Methiocarb degradation in soil is rapid.

Methiocarb is registered in many countries of the EU, South Africa, the USA, Australia, New Zealand and the Philippines, though its use is restricted. According to The Japan Food Chemical Research Foundation, the indicative MRL of methiocarb for rice, wheat and barley in Japan is 0.05 ppm.

Methomyl insecticide

Methomyl, $C_5H_{10}N_2O_2S$ (*S*-methyl *N*-(methylcarbamoyloxy) thioacetimidate), is a carbamate insecticide. Methomyl, also known as metomil and mesomile, is a colourless crystalline solid, highly soluble in water and in most organic solvents. Methomyl is stable in the solid form but flammable as liquid formulations. It rapidly decomposes in alkaline solutions, on heating and in moist soils. Heating produces toxic fumes of nitrogen oxides, sulfur oxides and hydrogen cyanide, making it necessary to store it very carefully as per label instructions. Methomyl is produced by the chlorination of acetaldoxime and conversion of the resulting α-chlorooxime with sodium methylmercaptide. Its CAS Registry Number is 16752-77-5.

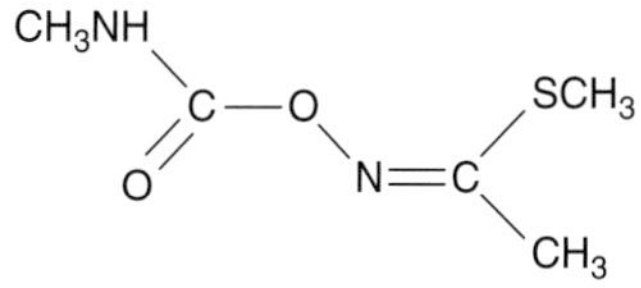

Methomyl

The pesticide methomyl is available as water-soluble powder (SP), wettable powder (WP), water-soluble liquid and as bait (in fly bait). Trade names of methomyl include Acinate, Methavin, Kipsin and Agrinate.

The WHO has categorized methomyl as a Class Ib toxin (extremely hazardous). It is toxic to many beneficial insects like honey bees. The acute oral LD_{50} values in rats, mice and guinea pigs range from 15 to 24 mg/kg. It is also highly toxic to birds (the acute oral LD_{50} ranging from 24.2 to 34 mg/kg) and aquatic organisms (the 96 h LC_{50} ranging from 0.8 to 3.4 mg/l). It can also cause human deaths in cases of acute exposure, making it necessary for handlers to wear protective gloves, clean protective clothing and a respirator of the organic-vapour type when handling this material. Symptoms of poisoning with methomyl include malaise, dizziness, muscle weakness, sweating, headache, salivation, nausea, abdominal pain, diarrhoea, blurred vision, muscle twitching, confusion and CNS depression (coma and seizures). Used containers should be crushed or burned and all disposable material should be disposed according to the local laws of disposal of harmful wastes. The ADI is 0.03 mg/kg/day. Since methomyl is a cholinesterase inhibitor, an acute toxin, a possible groundwater contaminant and a reproductive toxic, it is classified as a PAN Bad Actor chemical.

Methomyl is very mobile in sandy loam and silty clay loam soils. It has a soil half-life of up to 33 days, mainly resulting in carbon dioxide. The chemical degrades in plants within a matter of 3–5 days, reducing its potential to accumulate. In water, degradation is rapid in alkaline conditions and in the presence of sunlight. Upon ingestion by mammals, it is broken down in the liver and is readily excreted mainly via urine.

Methomyl is registered for use in India, Australia, New Zealand, the USA, Canada, Madagascar, the Philippines and South Africa. According to the USDA Foreign Agricultural Service's Pesticide MRL database, the indicative MRL for wheat grain in India and South Africa is 2 ppm, and in the USA, it is 1 ppm.

Methoprene as an IGR

Methoprene, $C_{19}H_{34}O_3$ (isopropyl (*E*,*E*)-(*R*,*S*)-11-methoxy-3,7,11-trimethyldodeca-2,4-dienoate), an amber-yellow liquid with a faint fruity odour (in its technical grade form) is an insect growth regulator and an insecticide. It is slightly soluble (1.4 mg/l) in water and miscible with common organic solvents. It is sensitive to UV light. Its CAS Registry Number is 40596-69-8.

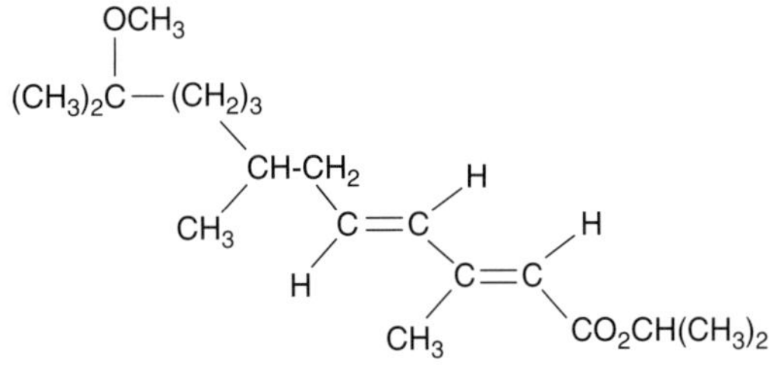

Methoprene

Methoprene, a **juvenile hormone (JH)** analogue, is an IGR and a larvicide. As an IGR, it interferes with the normal developmental life stages of insects, preventing them from maturing and reproducing. The shortage of JH supply is a **moulting** signal for an insect. Since methoprene mimics a JH, the insect remains in its pupal or larval stage, thus disrupting its biological cycle and killing it. Methoprene may be used as an animal ectoparasiticide for flea control, and a larvicide for mosquito control.

Methoprene is applied on a wide variety of crops such as cereals and vegetables. It is available as emulsifiable concentrate (EC) and soluble concentrate (SL) formulations, as well as in briquette, aerosol and bait forms. It is available in the market as Altosid, Apex, Diacan, Dianex and Kabat.

The IGR methoprene is not considered to be hazardous to the surrounding living systems or the environment. The oral LD_{50} for methoprene is >34,600 mg/kg in rats and >5000 mg/kg in dogs. It is practically non-toxic to birds and fish. Methoprene residues may have a slight potential for bioconcentration in bluegill sunfish and crayfish. The ADI is 0.1 mg/kg/day. The WHO has categorized methoprene as a Class U toxin (unlikely to be hazardous). Special protective gear might be needed only in case of prolonged exposure (in which case a NIOSH-approved organic vapour/pesticide respirator, impervious gloves, chemical goggles or safety glasses with side shields may be used). Labels of products containing methoprene must bear the signal word 'CAUTION'.

In mammals, methoprene rapidly and completely breaks down and is excreted in the urine and faeces. In plants, methoprene degradation principally involves ester hydrolysis, *O*-demethylation, and oxidative cleavage. This compound adsorbs to most soils, but breaks down by the influence of sunlight and microbial action, leading to a half-life of around 10 days. Hence, persistence and mobility in soil are not issues to contend with. In water too, its half life is of up to 2 days.

Methoprene is registered for use in Australia, South Africa and the USA. According to The Japan Food Chemical Research Foundation, the indicative MRL of methoprene for rice, wheat and barley in Japan is 0.05 ppm.

Methoxychlor analysis

Methoxychlor insecticide is a widely used compound, popular because of its low toxicity and low persistence. It is formulated as granules, flakes or powder. Its chemical composition is similar to **DDT**.

The special features of methoxychlor analysis are that the total organic chlorine content in the sample should be 295–315 g/kg. The hydrolysable chlorine content is required to be 97–117 g/kg; the maximum chloral hydrate content, 0.25g/kg; the maximum H_2SO_4 for acidity, 3 g/kg; and the water content, a maximum of 10 g/kg. The solid material that is insoluble in acetone can be up to 10 g/kg.

Methoxychlor insecticide

Methoxychlor, $C_{16}H_{15}Cl_3O_2$ (1,1,1-trichloro-2-2-bis(4-methoxyphenyl)ethane), is an organochlorine insecticide. It does not dissolve readily in water. Methoxychlor decomposes on heating and burning, to produce toxic and corrosive gases like hydrogen chloride. It can be made by reacting CH_3Cl or dimethyl sulfate with sodium phenate to produce anisole which is then reacted with chloral. The CAS Registry Number of o,p′-methoxychlor is 30667-99-3 and of p,p′-methoxychlor, 72-43-5.

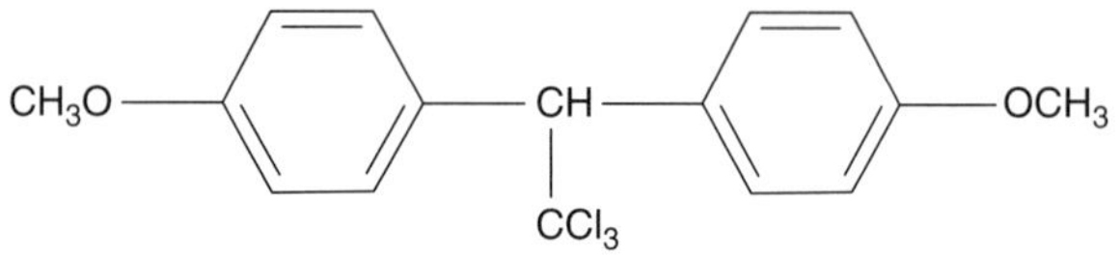

Methoxychlor

Methoxychlor is a broad-spectrum insecticide used in many situations. Field crops, vegetables, livestock and homes are protected from pests such as mosquitoes, cockroaches and fleas. The significant advantage of methoxychlor is its rapid metabolism, low toxicity and its property to not bioaccumulate.

Methoxychlor is available under trade names like Chemform, Methoxo, Metox and Methoxcide. Due to its low solubility in water, methoxychlor is mixed with a petroleum-based fluid and sprayed, or it is used as dust.

This compound is not a hazard to most mammals and organisms that may come into contact with it. The acute oral LD_{50} is 5000–6000 mg/kg in rats, the dermal LD_{50} is >2000 mg/kg in rabbits and the acute oral LD_{50} is >2000 mg/kg in mallard ducks. The ADI is 0.1 mg/kg/day. Acute exposure to this chemical may cause a prickling, tingling or creeping sensation on skin, headache, dizziness, vomiting, tremors, convulsions and coma.

Most methoxychlor was excreted by animals that were tested within 24 h, with monophenol, bisphenol and dihydroxybenzophenone being a few of the metabolites that exist in a conjugated form. The insecticide is persistent in soil since it adsorbs to soil colloids. The soil half-life in anaerobic soil can be as high as 4 months. However, degradation can occur in 1 week's time in aerobic soils. Strong adsorption and insolubility in water reduce the chances of methoxycarb contaminating groundwater, except in cases of very high application rates, rainfall and organic matter. According to the WHO, methoxychlor comes under toxicity Class U (unlikely to be hazardous), while according to the EPA, it is a Class IV toxin (practically non-toxic) and is listed as a general use pesticide (GUP). But, it is a suspected endocrine disruptor according to the PAN Pesticide Database.

Methoxyethyl mercuric chloride (MEMC) fungicide

See **Mercuric compounds as fungicides**

Methoxytriazine herbicides

Methoxytriazines are a type of *s*-triazine, which are a part of the triazine family. Methoxytriazines have a methoxy substituent in their structure.

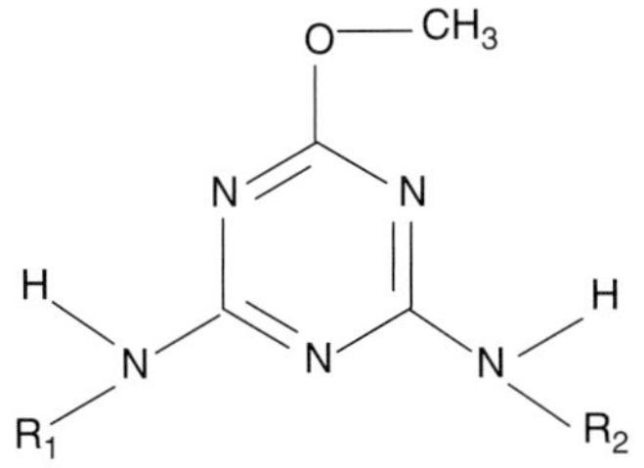

R_1	R_2
H	H
HiC_3H_7	HiC_3H_7
HET	HiC_3H_7
$(Et)_2$	HiC_3H_7

Methoxytriazine

Methoxytriazines are water-soluble herbicides that have a distinct foliar activity. Since members of this group are toxic to crops, especially maize, these are used as soil sterilants on wastelands and non-farm lands, for the control of weeds. This is especially true in the case of **prometon**.

Methyl bromide fumigant

Methyl bromide, CH_3Br (bromomethane), is a colourless, odourless gas, soluble in water (13.4 g/l at 25°C). It hydrolyses rapidly in alkaline media and readily dissolves in organic solvents like ethers, esters, ketones and aromatic hydrocarbons. The CAS Registry Number of methyl bromide is 74-83-9.

Methyl bromide is a multi-purpose fumigant best used in grain warehouses, mills, ships and soil. In addition, it acts like a **soil sterilant**, killing or inhibiting soil-borne pathogens, nematodes, mites and rodents, besides weeds and weed seeds. Methyl bromide is commonly used in combination with **chloropicrin** in varying proportions and is formulated as a gas (GA).

This chemical is highly toxic to man with an inhalation threshold limit valued at 0.065 mg/l air. It is a Category II toxin according to the EPA (moderately toxic). It is also extremely phytotoxic and so its use has to be strictly as directed. Handlers of CH_3Br have to wear respirators, masks and protective garments during fumigation. Since it is an acutely toxic chemical, a reproductive toxin and a suspected endocrine disruptor, methyl bromide is classified as a PAN Bad Actor chemical. It is registered for use in Australia, Cambodia, Canada, India, New Zealand, the Philippines, South Africa, Vietnam and the USA. India allows only restrictive use of this chemical.

Methyl iodide

See **Soil fumigants**

Methyl isothiocyanate analysis

Methyl isothiocyanate is a soil fumigant, fungicide, nematicide and weedicide. The estimation of this moderately toxic compound can be determined by: (i) reaction with ammonia to form *N*-methylthiourea which is then allowed to react with pentacyanoamine ferrate to produce a blue colour which is estimated colorimetrically; (ii) extraction from soil samples with carbon tetrachloride and measuring the methyl isothiocyanate content from its infrared spectrum over the region of 3.5–5.5 μm with the help of a calibration curve; and (iii) using thin layer chromatography.

Place a cooled mixture of 10 ml each of concentrated ammonium hydroxide and absolute methanol in a 50 ml receiver tube. Plug the mouth of the tube lightly to reduce loss of ammonia and partially immerse the tube in an ice bath. Weigh 50 g of soil sample and quickly transfer to a 100 ml conical distillation flask. Add 25 ml of 1:9 methanol + ethyl ether and connect the flask to a water-cooled condenser and to the receiver tube. Heat the flask and distil until 25 ml of distillate is collected. Remove the receiver tube and stopper, and mix. Allow to stand at room temperature for approximately 20 min.

Transfer all the contents of the receiver to a 250 ml beaker, rinsing with a small quantity of methanol, if required. Evaporate until the solution is reduced to about 4 ml. Transfer the concentrated solution to a 25 ml graduated cylinder. Wash the beaker with several small portions of water and add these to the cylinder until a volume of 10 ml is obtained. To this solution, add 2.0 ml phosphate buffer (pH 7.4) and 1.0 ml freshly prepared 0.1% sodium pentacyanoamine ferrate solution. Dilute to 15 ml with water. Mix and measure the absorbance after 15 min at 610 nm. Determine the amount of *N*-methylthiourea in the sample from a standard curve similarly prepared from a range of solutions containing 2–50 μg *N*-methylthiourea in 10 ml water.

To use the infrared method, add a solution of 50 ml each of water and carbon tetrachloride to 50 g of soil sample and shake vigorously. Centrifuge for 10 min. Decant the supernatant liquid into a 250 ml separatory funnel, allowing the phases to separate. Filter 25 ml of the carbon tetrachloride layer into a 50 ml flask with a glass stopper and add 2 g anhydrous sodium sulfate to it. After standing for up to 3 h, filter the portion of the dried extract through Whatman No. 1 filter paper and run the infrared spectra in the region 3.5–5.5 μm. The methyl isothiocyanate content can be determined from a standard curve using the net absorbance at 4.7 μm.

In the thin layer chromatography method, a sample containing carbon tetrachloride and methyl isothiocyanate is used. The solvent systems used can be 1:1 ethylacetate-methanol, 6:3 ethylacetate-benzene, 19:1 chloroform-methanol and 7:2 ethylacetate-pyridine. The separate spot is visualized by exposing the dried chromatograms to gaseous chlorine and spraying with a starch-iodine reagent. Methyl isothiocyanate shows up a blue-black spot which gradually fades to brown. As little as 0.5 μg can be detected.

Methyl isothiocyanate nematicide

Methyl isothiocyanate (CH_3NCS) is a colourless crystalline substance with a horseradish-like odour. It is an unstable and reactive substance which is sensitive to light and oxygen and hydrolysed in alkaline medium. It is easily soluble in organic solvents such as ethanol, acetone and cyclohexanone and soluble in water at 8.2 g/l (at 20°C). The CAS Registry Number of methyl isothiocyanate is 556-61-6.

Methyl isothiocyanate is a known nematicide and a **soil fumigant**. The fumigant property controls soil insects, soil fungi and weed seeds. It is phytotoxic to all green plants and must be used as directed. Methyl isothiocyanate is formulated as an emulsifiable concentrate (EC) and the common trade names include Trapex. The compound is generally used in mixtures of **1,3-dichloropropene** and **chloropicrin**.

The acute oral LD_{50} in mice ranges from 90 to 104 mg/kg and the inhalation LD_{50} in rats is 1.9 mg/l air. In soil and water, the bulk of methyl isothiocyanate is degraded and evaporated within 3 weeks at around 20°C. It is registered in the USA. (See also **Nematicides**.)

Methylarsonic acid herbicide

Methylarsonic acid, CH_5AsO_3 (methylarsonic acid), is an organo-arsenic compound. It exists in the form of white hydroscopic crystals, soluble in ethanol and in water (2.56×10^5 mg/l at 20°C).

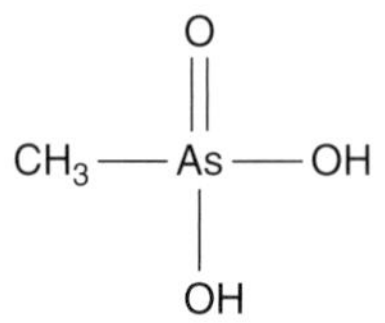

Methylarsonic acid

Methylarsonic acid is a selective, contact herbicide with some systemic properties. It is readily absorbed by leaves and translocated throughout the plant. Methylarsonic acid is used on turf to control summer grass. It is usually formulated as a soluble concentrate (SL). This compound is now a largely obsolete herbicide.

The WHO classifies methylarsonic acid as a Class III toxin, (slightly hazardous). It is mildly irritating to the eyes, skin and respiratory tract. Accidental ingestion may result in gastroenteritis. The substance may result in impaired functioning of the kidneys and liver. This substance is carcinogenic to humans and toxic to aquatic organisms. Its acute oral LD_{50} in rats is 2833 mg/kg. The NOEL (2 years) in dogs is 2 mg/kg/day. The ADI is 0.03 mg/kg/day.

Methylarsonic acid is not registered in the USA or in the EU.

Methyldymron herbicide

Methyldymron, $C_{17}H_{20}N_2O$ (1-methyl-3-(1-methyl-1-phenylethyl)-1-phenylurea), is a phenyl urea herbicide. It is a colourless, odourless crystalline substance, stable to heat and light. The solubility in water is 120 mg/l (at 20°C), while in acetone, it is 913 g/l, hexane, 8.2 g/l and methanol, 637 g/l.

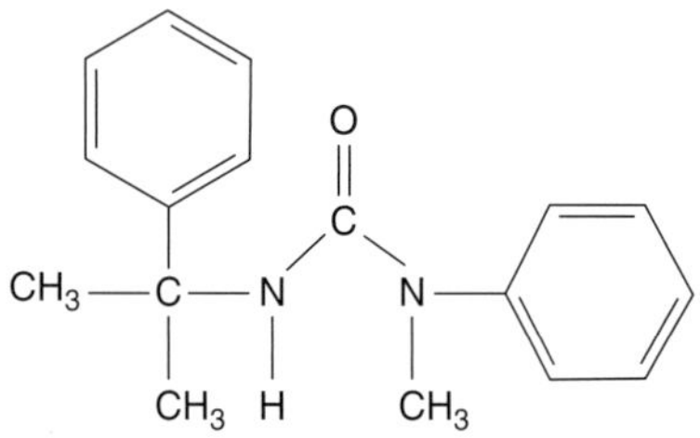

Methyldymron

The WHO has classified methyldymron as a Class U toxin (unlikely to be hazardous). The chemical is not registered in the USA.

Methylene bis (thiocyanate) slimicide

Methylene bis (thiocyanate), $C_3H_2N_2S_2$ (IUPAC name thiocyanatomethyl thiocyanate), is a slimicide that decomposes in water and reacts with strongly acidic and alkaline media. The CAS Registry Number of methylene bis (thiocyanate) is 6317-18-6.

Methylene bis(thiocyanate)

The slimicide is used as a micro-biocide, fungicide and algicide. The toxic activity of this slimicide is believed to be brought about by the inhibition of cell respiration. The thiocyanate moiety inactivates the electron transport system by forming complexes with the ferric ion of cytochrome, thus inhibiting cell respiration.

Methylene bis (thiocyanate) is used as a **biocide** in a number of applications. Its major use is in water cooling systems and paper mills as an inhibitor of algae, fungi and bacteria. It has also been used as a preservative in latex emulsions and oils. The use of the compound is as a slimicide in the production of paper and paperboard.

The substance irritates the eyes and the skin and prolonged contact may cause skin sensitization. Methylene bis (thiocyanate) is registered by the USA as an antimicrobial and slimicide.

Methyllycaconitine

See **Acetylcholine receptors**

Methyl- or ethyl-thiotriazine herbicides

Methyl- or ethyl-thiotriazines are triazines with methyl- or ethyl-thiol substituents in their structure.

The methylthiotriazine group (SCH_3) shows variable selective action. The selectivity is attributed to their ability to adsorb to soil particles. The same property makes this group less persistent than **chlorotriazines**. These herbicides are not very mobile. **Prometryn** and ametryn are two methyl-thiotriazine herbicides.

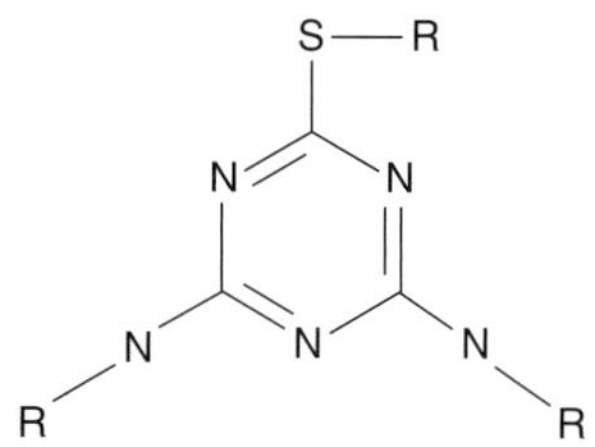

Where R = CH_3 or CH_2-CH_3

Methyl- or ethyl-thiotriazine

Ametryn, $C_9H_{17}N_5S$ (N^2-ethyl-N^4-isopropyl-6-methylthio-1,3,5-triazine-2,4-diamine), a white powder soluble in water (200 mg/l), is a systemic herbicide, with selectivity in sugarcane, banana and pineapple. Though it degrades in UV light

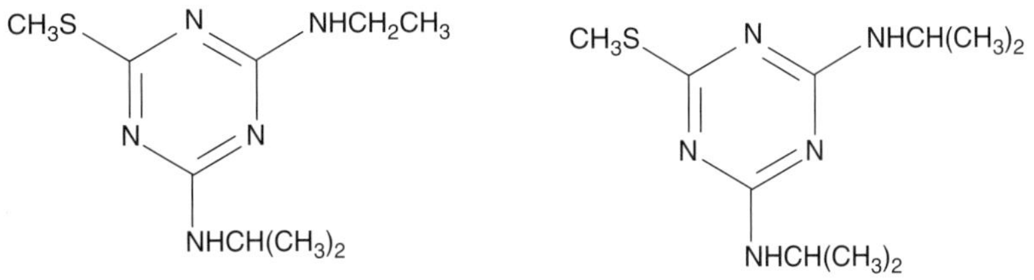

Ametryn and Prometryn

and in highly alkaline medium, ametryn is very persistent in water and moderately persistent in soil (with a half-life of up to 37 days). It has a CAS Registry Number of 834-12-8.

The herbicide ametryn inhibits photosynthesis and other enzymatic processes. It is a selective, systemic herbicide, absorbed by the leaves and roots, with acropetal translocation in the xylem, and accumulation in the apical meristems. It controls broadleaved and grass weeds in pineapple, sugarcane, bananas and plantains. It can also be used as a growth regulator, in the form of a **desiccant** for maize and potato. Ametryn is available as emulsifiable concentrates (EC) and wettable powders (WP).

Ametryn is toxic to mammals and moderately toxic to fish and birds. Due to its caustic nature, it is used for chemical fallow. The WHO's Acute Hazard Classification categorizes ametryn as a slightly toxic chemical (Class III toxin). It is not registered in the EU.

Prometryn, $C_{10}H_{19}N_5S$ (N^2,N^4-diisopropyl-6-methylthio-1,3,5-triazine-2,4-diamine), is a herbicide that is selective on cotton and celery. The WHO has classified it as a Class U toxin (unlikely to be hazardous). However, because it is a potential groundwater contaminant (solubility in water is 33 mg/l), a suspected endocrine disruptor and a confirmed reproductive toxin, this herbicide is a PAN Bad Actor chemical. Its aerobic soil half-life is 274 days and its anaerobic half-life is >300 days (see Ametryn and Prometryn at top of the page).

Methylthiotriazine herbicides

See **Methyl- or ethyl-thiotriazines herbicides**

Metiram fungicide

Metiram, $[C_{16}H_{33}N_{11}S_{16}Zn_3]_x$ (zinc ammoniate ethylenebis (dithiocarbamate)-poly(ethylenethiuram disulfide)), is an alkylenebis dithiocarbamate compound. It is a yellow powder, almost insoluble in water but soluble in pyridine. It is stable at 30°C. Metiram slowly decomposes in light and highly alkaline materials. The manufacturing process involves a reaction of ammonium ethylenebis(dithiocarbamate) with zinc chloride and then with hydrogen peroxide. The CAS Registry Number of metiram is 9006-42-2 (see Metiram at bottom of the page).

Metiram is a non-systemic, protective, broad-spectrum foliar fungicide. It is a non-specific thiol reactant that inhibits respiration by binding to SH-containing enzyme systems (i.e. enzymes that contain thiol groups (-SH), the analogue of hydroxyl groups (-OH)). Metiram controls a wide range of diseases like **scab**, shot hole, rust, **leaf spot** and **downy mildews**. It is also used to keep crops well during storage or transport. The crops that are protected are pome fruit, cereals, tobacco, groundnuts, vegetables and ornamentals. The pathogens controlled are species of *Puccinia*, *Peronospora*, *Septoria* and *Venturia*.

Metiram is formulated as water-dispersible granules (WG) and sold under trade names like Polyram, Arbatene and Polyram-Combi. Recommended application rates range from 1.5 to 4.0 kg/ha (it is used generously since resistance is not expected to develop). Metiram is used with compounds like cymoxanil.

Metiram is classified by the EPA as a Class IV toxin and a general use pesticide (GUP). Acute percutaneous LD_{50} in rats is >2000 mg/kg. The acute oral LD_{50} in rats is >5000 mg/kg. The compound is practically safe to birds and beneficial organisms, but is toxic to fish and *Daphnia magna* (the acute 96 h LC_{50} in rainbow trout is 0.33 mg/l, while the acute 48 h EC_{50} in *D. magna* is 0.77 mg/l). The ADI is 0.03 mg/kg/day. Formulations are expected to bear the signal word 'CAUTION' on their labels. Acute exposure may result in symptoms of poisoning like irritation to the skin, eyes and the respiratory tract. The product is also a proven carcinogen and a reproductive toxin. It is also believed to be an endocrine disruptor and a groundwater contaminant. It has been listed as a PAN Bad Actor chemical.

The fungicide is readily absorbed and rapidly degraded in rats, but has been known to leave residues in body parts such as the kidneys and gastrointestinal tract. The toxicologically significant metabolite of all ethylenebisdithiocarbamates is **ethylenethiourea (ETU)**. Metabolism in animals and plants proceeds via hydrolysis

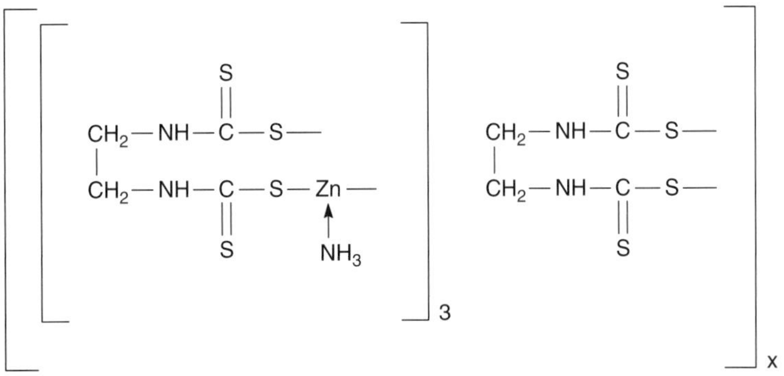

Metiram

and ring formation. Metiram binds to most soils, but due to its low solubility in water and rapid degradation in the presence of moisture, sunlight and microorganisms, metiram does not persist in soil and underground water. But, ETU and other metabolites are mobile and persistent.

Metiram is registered in many countries like India, South Africa, Australia, New Zealand, Tanzania, Cambodia, the EU, the USA and Canada. But, some of the applications have been restricted or cancelled – such as the use of metiram on tobacco seedlings and roses. The MRL value of metiram is 0.05 mg/kg on rice, 1 mg/kg on wheat and 2 mg/kg on barley.

Metobenzuron herbicide

Metobenzuron, $C_{22}H_{28}N_2O_5$ ((*RS*)-1-methoxy-3-[4-(2-methoxy-2,4,4-trimethylchroman-7-yloxy) phenyl]-1-methylurea), is a phenyl urea herbicide. It is a white-coloured powder, soluble in water to the extent of 0.4 mg/l (25°C). It is produced by the reaction of 2-(2-methoxy-2,4,4-trimethylchroman-7-yloxy)phenyl isocyanate with dimethylamine. The CAS Registry Number of metobenzuron is 111578-32-6 (see Metobenzuron at bottom of the page).

Metobenzuron is used in maize cultivation. A common trade name of this herbicide is Mitsui. It is not registered for use in the USA.

Metobromuron herbicide

Metobromuron, $C_9H_{11}BrN_2O_2$(3-(4-bromophenyl)-1-methoxy-1-methylurea), is a phenyl urea herbicide. It is a colourless crystal. It is soluble in water (330 mg/l at 20°C), in acetone (500 g/l), dichloromethane (550 g/l), methanol (240 g/l), toluene (100 g/l), octanol (70 g/l) and hexane (2.6 g/l). The compound is hydrolysed by strong acids and alkalis. It is produced by bromination of 3-phenyl-1-methoxy-1-methylurea. The CAS Registry Number of metobromuron is 3060-89-7.

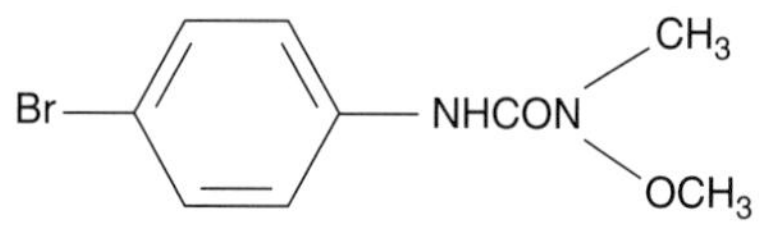

Metobromuron

Metobromuron, a selective, systemic, pre-emergence soil-applied herbicide, is absorbed by the roots and leaves of annual broadleaved weeds and grasses. It acts by inhibiting photosynthetic electron transport at the photosystem II receptor site. It is applied in fields of common beans (*Phaseolus vulgaris*), potatoes, tomatoes, tobacco, maize and sugarbeet.

This herbicide is formulated as soluble concentrate (SL) and wettable powder (WP).

The WHO has classified metobromuron as a Class U toxin (unlikely to be hazardous). It is not a skin or eye irritant. The NOEL (2 year) in rats is 250 mg/kg diet. The ADI is 0.008 mg/kg b.w./day. It is not toxic to bees and earthworms.

In animals, the metabolism proceeds mainly via *N*-demethylation, *N*-dealkoxylation and phenyl ring hydroxylation, followed by cleavage and conjugation processes. In plants, the methyl and methoxy groups are split off and the urea moiety is further degraded to 4-bromoaniline. In soil, the compound binds strongly to soil, indicating low leaching potential. It is stable to abiotic degradation in water.

Metolachlor herbicide

Metolachlor, $C_{15}H_{22}ClNO_2$ (2-chloro-*N*-(6-ethyl-*o*-tolyl)-*N*-[(1*RS*)-2-methoxy-1-methylethyl] acetamide or 2-chloro-6′-ethyl-*N*-[(1*RS*)-2-methoxy-1-methylethyl]acet-*o*-toluidide), is a **chloroacetanilide** herbicide. It is an off-white liquid, soluble in water (530 ppm/l) and very soluble in benzene, hexane and dichloromethane. It is produced by the reaction of 2-methyl-6-ethylanilin with 2-chloroisopropyl methyl ether, followed by treatment with chloroacetyl chloride. The CAS Registry Number of metolachlor is 51218-45-2.

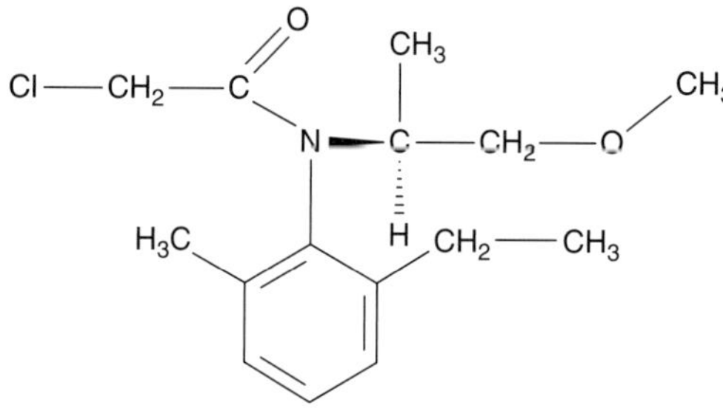

Metolachlor

This herbicide is used to control broadleaved and grassy weeds in cereals (e.g. maize), sorghum, groundnuts, cotton and ornamental plants. It acts by inhibiting protein synthesis and photosynthesis.

Metolachlor may be combined with **atrazine** and cyanazine. The WHO classifies metolachlor as a Class III toxin (slightly hazardous). Metolachlor is toxic to fish, with 96 h LC_{50} values ranging from 2 to 15 mg/l. It can become toxic to sorghum and some protein-rich crops. It is a proven groundwater contaminant, a possible carcinogen and endocrine disruptor according to the PAN Pesticide Database and hence it is a PAN Bad Actor chemical. Metolachlor is persistent in the soil with an anaerobic soil DT_{50} of >60 days, and an hydrolytic half-life of 200 days.

Metolachlor is registered in India, the USA, Canada, Australia and many African countries.

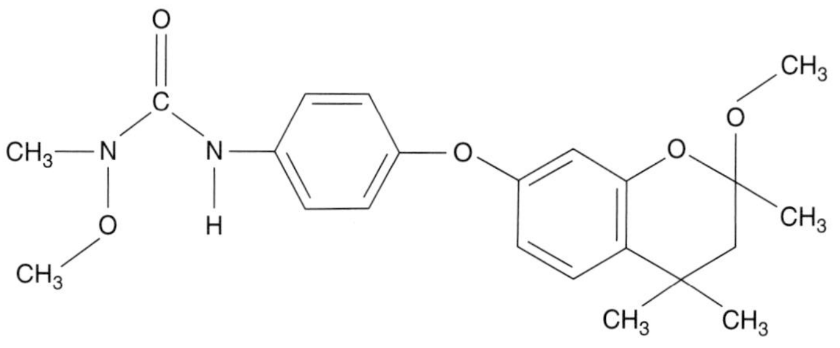

Metobenzuron

Metolcarb insecticide

Metolcarb, $C_9H_{11}NO_2$ (*m*-toyl methylcarbamte), is a carbamate insecticide. It is a colourless solid, moderately soluble in water and readily soluble in xylene and methanol. The compound is unstable in strong acids, alkalis and oxidizing agents, and a fire/explosion hazard in the presence of heat (with the production of toxic gases). It must be stored in a cool, dry, dark location in tightly sealed containers or cylinders. Metolcarb is produced by the reaction between 3-methylphenol and metolcarb. It has a CAS Registry Number of 1129-41-5.

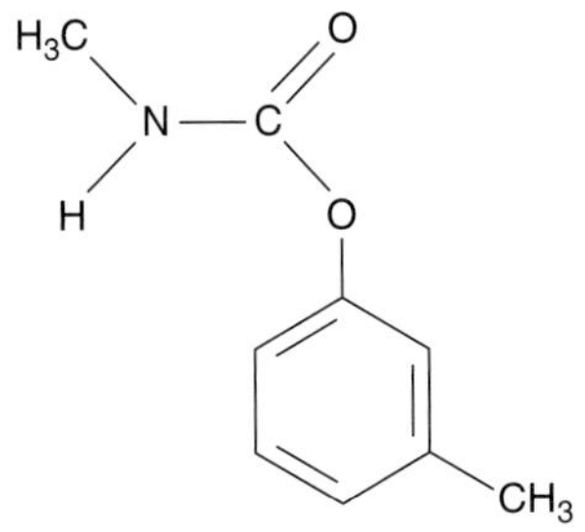

Metolcarb

Metolcarb, a cholinesterase inhibitor, is a systemic insecticide with contact and respiratory action. It is used in the control of plant hoppers and **leafhoppers** on rice, besides citrus **mealybugs, pink bollworms**, cotton aphids and onion thrips. Formulations include dustable powder (DP), emulsifiable concentrate (EC) and wettable powder (WP). Metolcarb is compatible with many non-alkaline and non-oxidizing pesticides. Metacrate is a common brand name.

According to the WHO, metolcarb is a Class II (moderately hazardous) toxin. As a nerve toxin, metolcarb can enter the body via the skin and inhalation routes. Symptoms of exposure are varied and chronic and include headache, sweating, nausea and vomiting, diarrhoea, memory problems and loss of both appetite and weight, followed by death. Hence, a respirator with gas-absorbent properties should be used, along with protective clothing, gloves, gas mask and goggles. Metolcarb is a cholinesterase inhibitor and is listed as a PAN Bad Actor chemical.

The main pathways of metabolism of metolcarb in plants, mammals and insects include phenyl ring hydroxylation, carbamate ester hydrolysis and conjugation.

Metolcarb is not registered in the USA.

Metominostrobin fungicide

Metominostrobin, $C_{16}H_{16}N_2O_3$ ((*E*)-2-methoxyimina-*N*-methyl-2(2-phenoxyphynyl) acetamide), is an oximinoacetamide of the **strobilurin** type. It is a white crystalline substance that is slightly unstable to light. This compound is slightly soluble in water (0.128 g/l) and highly soluble in dichloromethane and chloroform (1380 and 1280 g/l, respectively). The CAS Registry Number of metominostrobin is 133408-50-1.

Metominostrobin, a systemic fungicide with protective properties, acts as a **QoI** (quinone outside inhibitor), inhibiting respiratory electron transport functioning. It has both curative and protective properties. This compound controls *Pyricularia oryzae* on rice, even after infection has set in. It also protects rice crops against many other diseases like brown spot, rice blast and sheath blight.

Commonly formulated as granules (GR), the recommended application is 1.5–1.8 kg/ha. A popular trade name for this compound is Oribright.

The acute oral LD_{50} of metominostrobin in rats is between 708 and 776 mg/kg. Tests on experimental animals have shown that the compound does not irritate the skin. The NOEL observed in both male and female rats, mice and dogs is between 1.6 and 2.7 mg/kg/day. Its ADI is 0.016 mg/kg/day. In birds, fish, bees and earthworms, the acute oral LC_{50} values are >5200 ppm (mallard duck), 18.1 mg/l (96 h period in carp), >100 μg per bee (in a 48 h contact dose) and 114 mg/kg (for *Eisenia foetida* for a 14 day period).

Rats rapidly eliminate the absorbed dose of metominostrobin. Metabolic pathways in animals include hydroxylation of the phenoxyphenyl group, followed by demethylation and subsequent conjugation of the *N*-methyl amide group. In plants too, the *N*-methyl amide group undergoes hydroxylation and demethylation. Metominostrobin degrades in aerobic soil with a DT_{50} of 98 days.

The fungicide metominostrobin is registered in Vietnam, although not in the USA or the EU.

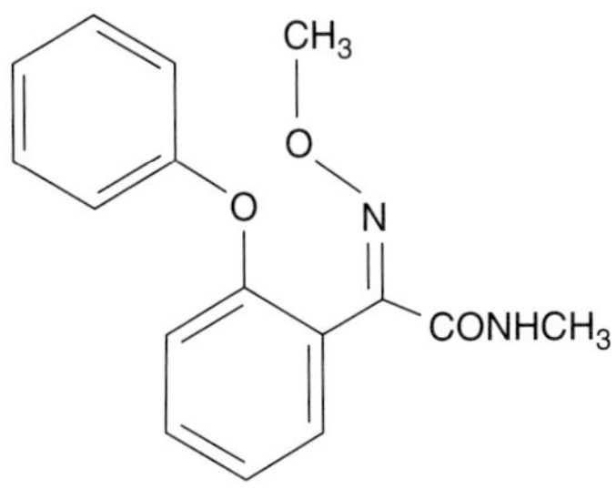

Metominostrobin

Metosulam herbicide

Metosulam, $C_{14}H_{13}Cl_2N_5O_4S$ (2′,6′-dichloro-5,7-dimethoxy-3′-methyl [1,2,4] triazolo [1,5-α] pyrimidine-2-sulfonanilide), is a **triazolopyrimidine** herbicide. It is a cream to tan-coloured powder, moderately soluble in water and fairly soluble in organic solvents. It is stable under normal storage conditions, and it decomposes above its melting point of 211.5°C. Metosulam is made by a reaction of 4,6-dimethoxy-(-2sulfonyl)-2,3-triazolyl pyrimidine chloride with 2,6-dichloro-5-methylaniline. Metosulam has a CAS Registry Number of 139528-85-1.

Metosulam, a selective, pre- and post-emergence herbicide with a residual life, effectively controls broadleaved weeds in wheat, barley, rye and maize. Once it is taken up by roots and foliage, this herbicide inhibits acetolactate synthase or ALS (acetohydroxy acid synthase or AHAS). Weeds such as scarlet pimpernel (*Anagallis arvensis*), **common lambsquarters** (*Chenopodium album*), *Galium aparine*, *Lamium pupureum*, *Polygonum* spp.,

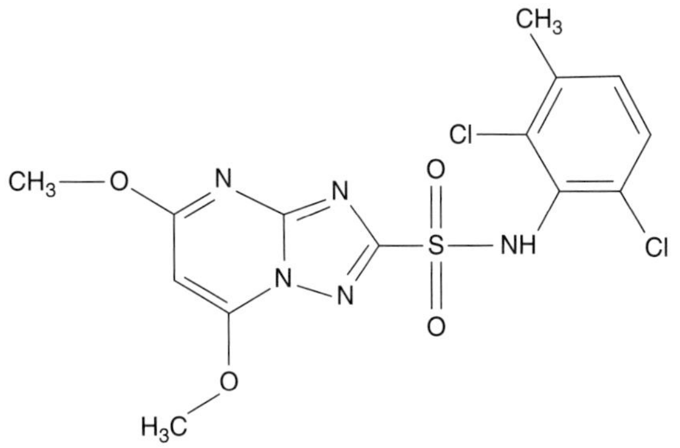

Metosulam

Raphanus raphanistrum, *Sinapis arvensis*, *Senecio vulgaris*, *Solanum nigrum* and *Stellaria media* are controlled by metosulam.

Metosulam is available as suspension concentrates (SC) and water-dispersible granules (WG). It is available under the trade names of Pronto, among others.

The acute oral LD_{50} of metosulam in rats is >5000 mg/kg, while the acute LD_{50} in mallard ducks is >2000 mg/kg. The acute 96 h LC_{50} in rainbow trout is >29.3 mg/l, and the acute 48 h EC_{50} in *Daphnia magna* is >100 mg/l. The inhalation LC_{50} of metosulam in rats is 1.9 mg/l. The ADI of metosulam in rats is 0.05 mg/kg b.w./day. The European MRL of metosulam for wheat and maize grains is 0.01 mg/kg. The WHO lists this herbicide as a Class U toxin (unlikely to be hazardous).

Metosulam is registered in Australia, Greece and Germany.

Metoxuron herbicide

Metoxuron, $C_{10}H_{13}ClN_2O_2$ (3-(3-chloro-4-methoxyphenyl)-1,1-dimethylurea), is a phenyl urea herbicide and a plant growth regulator. It is a colourless, crystalline substance, soluble in water at 20°C at 678 mg/l. It is also soluble in solvents like acetone, cyclohexanone, acetonitrile, diethyl ether, benzene and toluene. Though stable otherwise, metoxuron is hydrolysed by strong acids and alkalis. Metoxuron is produced by the reaction of 3-chloro-4-methoxyphenyl isocyanate with dimethylamine. Its CAS Registry Number is 19937-59-8.

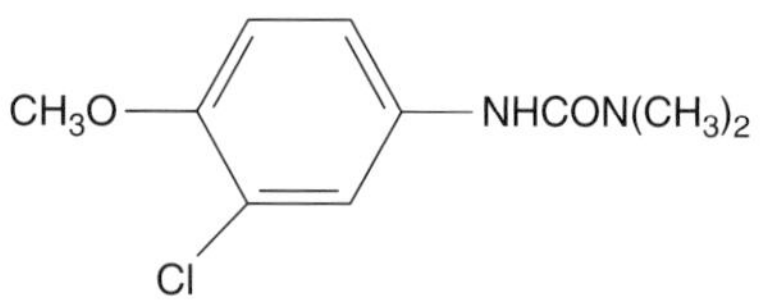

Metoxuron

A selective, pre- and post-emergence herbicide, metoxuron is absorbed by leaves and roots, after which it acts by inhibiting photosynthetic electron transport. It is used to control grasses and broadleaved weeds in carrots and cereals (e.g. wheat, barley and rye).

Formulation types for metoxuron are granules (GR), wettable powder (WP) and water-dispersible granules (WG), with the application dosage usually ranging from 2.4 to 4 kg/ha. Trade names include Dosanex and Investt.

The WHO has classified metoxuron as a Class U toxin (unlikely to be hazardous). It is not toxic to bees.

In plants, the degradation of metoxuron involves demethylation of the terminal nitrogen and hydrolysis of the urea moiety. In soil, breakdown involves degradation to 3-chloro-4-methoxyaniline, ring hydroxylation and cleavage. Metoxuron is banned in India and not registered in the USA.

Metrafenone fungicide

Metrafenone, $C_{19}H_{21}BrO_5$ (3′-bromo-2,3,4, 6′-tetramethoxy-2′, 6-dimethylbenzophenone), is a benzophenone compound. It is a white crystalline solid substance, barely soluble in water (0.552 mg/l) and easily soluble in dichloromethane (1950 g/l), acetone (403 g/l) and acetonitrile. It degrades after irradiation by simulated sunlight (DT_{50} of 3.1 days). The CAS Registry Number of metrafenone is 220899-03-6.

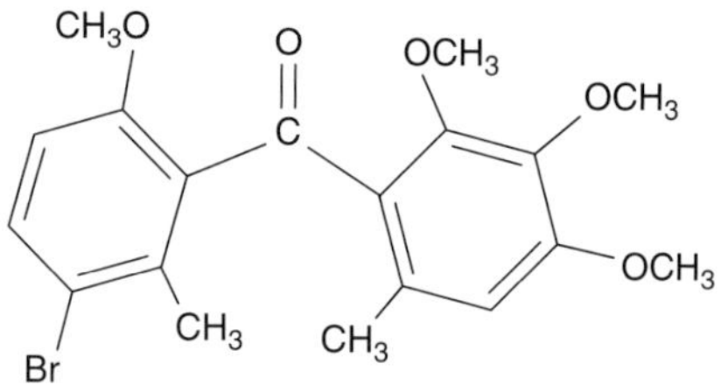

Metrafenone

Metrafenone, a preventive and curative compound, inhibits the growth of hyphae and spores of fungal pathogens such as *Pseudocercosporella herpotrichoides* and *Erysiphe graminis* that cause **powdery mildew** and eyespot in wheat and barley.

Metrafenone is commonly formulated as a suspension concentrate (SC), and trade names include Flexity and Vivando. Applications may be repeated as required, without risk to non-target organisms.

The acute oral LD_{50} values in rats and bobwhite quails are >5000 mg/kg and >2025 mg/kg, respectively. The inhalation LC_{50} of metrafenone in rats is >5.0 mg/l. The ADI is determined at 0.25 mg/kg b.w. Metrafenone is slightly to moderately toxic to fish with an acute 96 h LC_{50} of >0.82 mg/l in rainbow trout. It is considered safe for beneficial creatures such as predatory beetles, wasps, bees and earthworms.

Metrafenone is extensively metabolized in animals which rapidly eliminate the absorbed dose as glucuronic acid conjugates via urine and bile. The metabolic route in animals includes ortho-dealkylation, aliphatic oxidation, debromination, ring hydroxylation and conjugation. In soil and water, the DT_{50} is between 32 and 124 days and between 3.2 and 4.6 days, respectively. However the compound is considered a possible carcinogen by the PAN Pesticide Database.

Metrafenone is registered for use in Canada, Denmark, Germany, Greece, Hungary, the Netherlands, New Zealand and the UK.

Metribuzin herbicide

Metribuzin, $C_8H_{14}N_4OS$ (4-amino-6-*tert*-butyl-4,5dihydro-3-methylthio-1,2,4 triazin-5-one), is a triazinone herbicide. It is produced by the reaction of 4-amino-5-oxo-3-thioxo-6-*tert*-butyl-2,3,4,5-tetrahydro-1,2,4-triazine with methyliodide, in the presence of methoxide. Metribuzin exists as colourless crystals, highly soluble in water (1030 mg/l) and moderately soluble in hexane and other organic solvents. It has a CAS Registry Number of 21087-64-9.

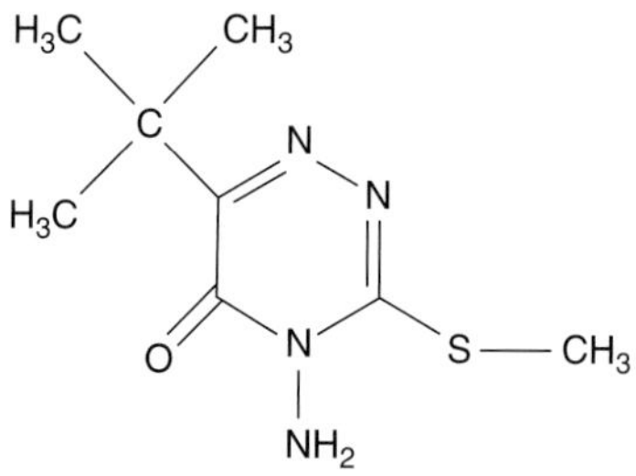

Metribuzin

Metribuzin is a selective, systemic, pre- and post-emergence soil and foliar herbicide, which translocates acropetally, once absorbed by the roots and leaves. It acts as a photosynthesis electron-transport inhibitor (photosystem II) and is used on susceptible grasses and broadleaved weeds in fields of soybean, potatoes, tomatoes, sugarcane, asparagus, maize and turf grass, as well as on fallow lands.

The herbicide is available as water-dispersible granules (WG), suspension concentrate (SC) and wettable powder (WP). It is applied using low-pressure ground sprayers, aerial sprayers or sprinklers. It is available in the market under trade names such as Sensor and Lexone.

The acute oral LD_{50} of metribuzin is 1090–2300 mg/kg in rats and 245–274 mg/kg in guinea pigs. The 4 h inhalation LC_{50} for metribuzin in rats is >0.65 mg/l, indicating moderate toxicity via the inhalation route. It is practically non-toxic by the dermal route. Metribuzin is toxic to birds, with an acute oral LD_{50} of 169.2 mg/kg in bobwhite quails, indicating moderate toxicity. This herbicide is slightly toxic to fish. The WHO lists metribuzin as a moderately toxic chemical. It has been listed as a PAN Bad Actor chemical since it is a proven reproductive toxin and a possible source of endocrine damage.

In plants, metribuzin undergoes oxidative deamination and further degradation to water-soluble conjugates. In soil and water, microbial breakdown of metribuzin is the major mechanism of loss. However, since the chemical is soluble in water and has a half-life of up to 120 days, it is mobile in water and has a high potential to contaminate groundwater.

The herbicide metribuzin is registered in a large number of countries like India, the USA, EU countries, Canada, Australia and New Zealand.

According to the Japan Food Chemical Research Foundation, the indicative MRL of metribuzin for rice, wheat and barley in Japan is 0.75 ppm. According to the USDA Foreign Agricultural Service's Pesticide MRL database, the indicative MRL for soybean is 0.3 ppm in the USA and 0.1 ppm in South Africa. The EU maintains an MRL of 0.1 ppm for its 'Oilseeds and oil fruits' group. As for wheat grain, the MRL is 0.75 ppm in the USA and 0.1 ppm in South Africa. (See also ***as*-Triazines**.)

Metsulfuron herbicide

Metsulfuron, $C_{13}H_{13}N_5O_6S$ (2-(4-methoxy-6-methyl-1, 3,5-triazin-2-ylcarbamoylsulfamoyl) benzoic acid), is a sulfonylurea herbicide. This substance is used as an ester or a salt, such as metsulfuron-methyl. The CAS Registry Number of metsulfuron is 79510-48-8.

Metsulfuron is mainly used to control many types of weeds in cereal crops. This compound is registered in Cambodia and many EU members, but is not registered in the USA.

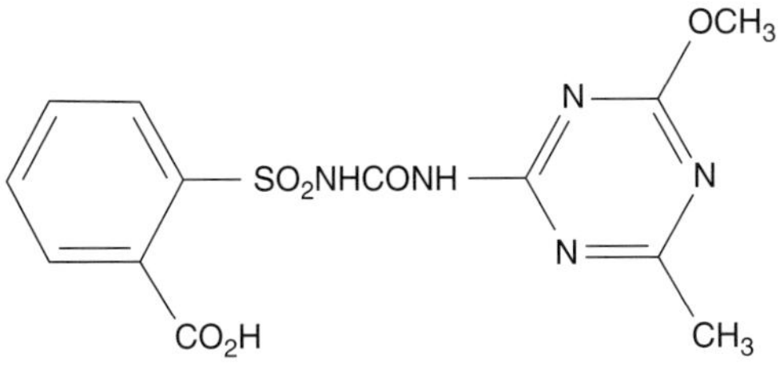

Metsulfuron

Metsulfuron-methyl herbicide

Metsulfuron-methyl, $C_{14}H_{15}N_5O_6S$ (methyl 2-(4-methoxy-6-methyl-1, 3,5-triazin-2-ylcarbamoylsulfamoyl) benzoate), is a triazinylsulfonylurea herbicide. It is a whitish solid, soluble in water at 3.3 mg/l (25°C). It is stable to photolysis, but breaks down in UV light. The CAS Registry Number of metsulfuron-methyl is 74223-64-6.

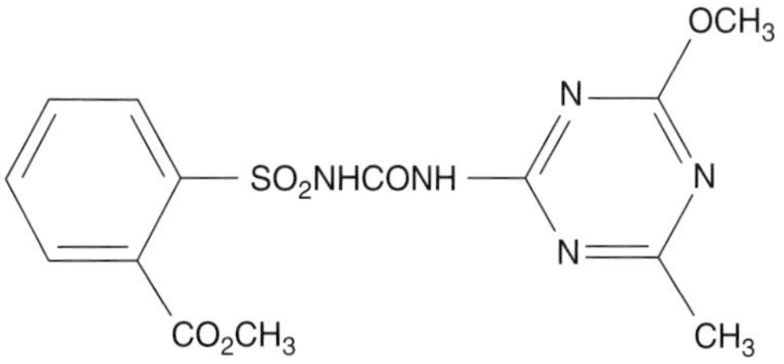

Metsulfuron-methyl

The herbicide metsulfuron-methyl is a systemic, selective, pre- and post-emergence herbicide that has a residual effect on leaves and in the soil. This compound kills broadleaved weeds and some annual grasses that grow among wheat, rye and barley crops. The herbicide is absorbed rapidly by susceptible plants, after which it inhibits cell division.

Metsulfuron-methyl can be used with other foliar herbicides, and may be applied at the rate of up to 7.84 g/ha. It is commonly manufactured as a dry flowable formulation, and sold under trade names such as Ally, Allie, Gropper and Escort.

This herbicide is classified by the EPA in Acute Toxicity Category III, and all labels must bear the signal word 'CAUTION'. Metsulfuron-methyl is a moderate eye irritant and mild skin irritant. It has a very low toxicity to mammals, birds, fish and insects. Its ADI is 0.25 mg/kg/day. The NOEL is observed to be 25.0 mg/kg/day. Due to its residual activity in soils, it can be injurious to certain crops such as sunflower and maize if such crops are planted in fields exposed to metsulfuron-methyl, less than 1 year ago. Metsulfuron-methyl is broken down quickly and eliminated from the body. In tests with radiolabelled metsulfuron-methyl in rats, the excretion half-life ranged from 9 to 29 h. It did not bioaccumulate in fish.

Metsulfuron-methyl is registered in countries such as India, the USA, Australia and Canada.

Mevinphos insecticide

Mevinphos, $C_7H_{13}O_6P$ ((*EZ*)-2-methoxycarbonyl-1-methylvinyl dimethyl phosphate or methyl (*EZ*)-3-(dimethoxyphosphinoyloxy) but-2-enoate), is an organophosphorus insecticide. It is a colourless liquid, completely miscible with water, alcohols, ketones and chlorinated hydrocarbons and slightly soluble in aliphatic hydrocarbons, petroleum ether and carbon disulfide. It is hydrolysed in an aqueous alkaline solution, with a DT_{50} of 1.4 h at pH 11. When mevinphos is manufactured (by the Perkow reaction) from trimethyl phosphite and methyl α-chloroacetoacetate, the main product, (*E*)-form is more potent than another isomer, the (*Z*)-form. Technical products consist of the (*E*)- and (*Z*)-isomers in a ratio of about 2:1. It has a CAS Registry Number of 7786-34-7.

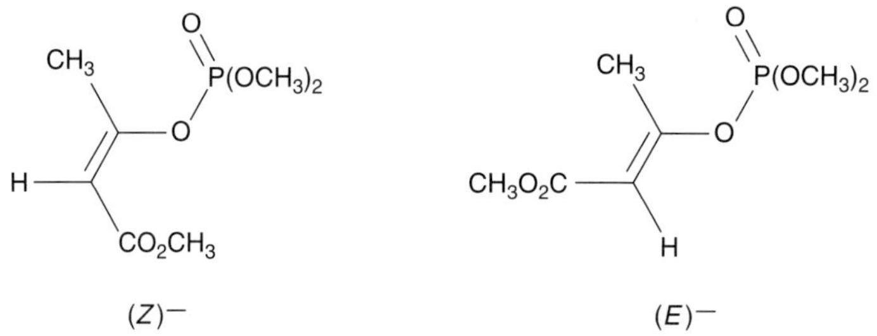

Mevinphos

Mevinphos, a cholinesterase inhibitor, is a systemic insecticide and acaricide with contact and respiratory action. It is used in the control of chewing and sucking insects and spider mites, on a wide range of crops including melons, vines, cucurbits, aubergines, beet, potatoes, cotton, canola and sorghum. Mevinphos has a short residual activity.

The compound is formulated in forms such as emulsifiable concentrate (EC) and soluble concentrate (SL). It is compatible with many pesticides (except alkaline substances). It is available under the trade names of Phosdrin, Mevinox and Duraphos.

The oral LD_{50} of mevinphos is 3–12 mg/kg in rats, indicating extreme toxicity. The acute oral LD_{50} is from 1.34 to 4.6 mg a.i./kg b.w. in many of the bird species tested. It is toxic to bees (the LD_{50} is 0.027 μg per bee). The compound is highly toxic by the dermal route as well, with a dermal LD_{50} of 4.2 mg/kg in rats, and 40 mg/kg in mice. Symptoms of poisoning are excessive salivation, sweating, muscle twitching, weakness, tremors, incoordination, headache, dizziness, nausea, vomiting, abdominal cramps, diarrhoea, tightness in chest, wheezing, productive cough, fluid in lungs and pin-point pupils, sometimes with blurred or dark vision. It may cause seizures, incontinence and respiratory depression. The WHO lists mevinphos as an extremely hazardous toxin (Class Ia toxin). Since cholinesterase is inhibited by mevinphos, and since endocrine disruption and groundwater contamination is suspected, it is considered to be a PAN Bad Actor chemical.

In mammals, following oral administration, metabolites are eliminated in 3–4 days via urine and faeces. Demethylation by glutathion-*S*-alkyl transferase is an important degradation route of *E*-mevinphos in mammals. In plants, mevinphos is hydrolysed rapidly to less toxic products, including phosphoric acid dimethyl ester and phosphoric acid. In comparison with the (*Z*)-isomer, the (*E*)-isomer is a more effective insecticide and is detoxified quickly in plants compared with animals.

Mevinphos has a soil half-life of 2–3 days. Hence, in spite of its poor adsorption, its mobility may not contaminate groundwater. Mevinphos is readily broken down by water, losing its insecticidal activity within 2–4 weeks.

Many countries have banned or restricted the use of mevinphos. The USA has cancelled its use and India has refused registration of specific compositions. It is registered in South Africa, Thailand and Australia.

The MRL for mevinphos is 1 ppm in broccoli, Brussel sprouts, cabbage, cauliflower, cherries and strawberries; 0.5 ppm in apples, grapes, peaches, lettuce, spinach; and 0.2 ppm in cucumbers, tomatoes, citrus fruit and pears.

Microbial pesticides

When microbes or their proteins and enzymes are used for insect control, they form products called microbial insecticides. Microbial insecticides are now considered important, from the environmental safety perspective. Beneficial microorganisms and natural enemies of insects are identified, isolated, multiplied and formulated into various products as microbial insecticides. These are also known as **biopesticides**. This volume contains descriptions of many bioinsecticides, biofungicides and **bioherbicides**.

Micro-emulsions

Micro-emulsion formulations are thermodynamically stable, isotropic dispersions of active ingredients (a.i.), hydrocarbons, surface active agents and water. Micro-emulsion products appear as clear fluids, in which micro-crystals of the a.i. remain suspended indefinitely. These formulations rapidly disperse in spray water.

Since these products require a high concentration of surfactants, they usually contain a low proportion of a.i. Such products are also fairly costly.

Micro-encapsulated formulations

Encapsulation is the process of enclosing or coating an active ingredient (a.i.) in a film or sheath to control its dissolution rate. When very small capsules (e.g. 20 μm diameter) are used to enclose dry or wet material, so that the material cannot readily permeate through the capsule, it is termed micro-encapsulation. There are examples of micro-encapsulation in living systems too, with enzymes being micro-encapsulated until they reach the desired site of action.

Fertilizers, pesticides, pharmaceuticals and **pheromones** are micro-encapsulated. The covering layer is made of, for example, polyamides, polyvinyl plastic coating, elastomer or gelatine.

Micro-encapsulated formulations are mixed with water and sprayed on the target. The release of the a.i. begins gradually after the cover breaks down, enabling its availability in a controlled manner. Micro-encapsulation is also advantageous when the a.i. is harmful to **beneficial organisms** in the vicinity. These formulations can also be programmed to act when the right conditions set in. This is called 'trigger release' technology, and is used when an a.i. is required to act at a certain temperature range. Some herbicide encapsulations, for example, are programmed to be released at temperatures that prompt the germination of weed seeds.

However, slower-than-expected release of the a.i. in a micro-encapsulated formulation can cause harmful effects.

Pest control could be adversely affected, while residues could persist, increasing the **restricted entry interval (REI)** and harvest intervals. The chemicals may be ingested by nearby organisms, including bees, leading to a reduction in their population.

Micro-encapsulation

See **Micro-encapsulated formulations; Encapsulation**

Microsporidia

See **Biocontrol of insects – entomofungi**

Microtubules

See **Herbicide action – meristematic mitotic inhibitors**

Mildew preventive

A mildew preventative is a compound used to prevent the growth of parasitic fungi (that are usually stain-producing) on such organic materials as textiles, leather and paper. Compounds used widely as mildew preventives include phenols, benzoic acid, **formaldehyde** and organic salts of copper, zinc and mercury.

Mildiomycin biofungicide

Mildiomycin, $C_{19}H_{30}N_8O_9$ ((2*R*, 4R)-2-{(2*R*, 5S, 6S)-2-[4-amino-1, 2-dihydro-5-(hydroxymethyl)-20oxopyrimidin-1-yl]-5,6-dihydro-5-(L-serylamino)-2*H*-pyran-6-yl}-2,4-dihydroxy-5-guanidinopentanoic acid), is a peptidyl nucleoside antibiotic fungicide. It is a hygroscopic solid, readily soluble in water but sparingly soluble in dioxane, dimethyl sulfoxide and pyridine. It decomposes at pH values greater than 9 and less than 2. This biofungicide is produced by the fermentation of *Streptoverticillium rimafaciens*.

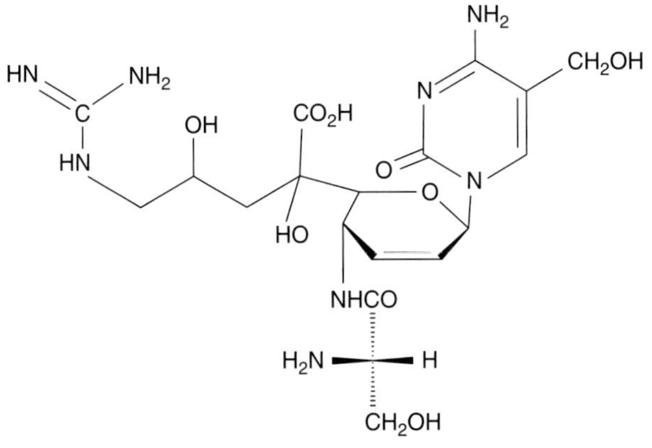

Mildiomycin

Mildiomycin, a systemic fungicide with a curative property, is a common agent used against **powdery mildews** in cucumbers, grapes, strawberry and tobacco. It acts by inhibiting protein biosynthesis by blocking peptidyl-transferase. This compound controls pathogens such as *Erysiphe* spp., *Podosphaera* spp. and *Sphaerotheca* spp.

This fungicide is formulated as a wettable powder (WP). It is applied to the foliage as a spray at concentration of 40–80 ppm.

Mildiomycin has an acute oral LD_{50} in the range of 4120–4300 mg/kg in rats. The acute 72 h LC_{50} in carp is >40 mg/l. It is found to be non-irritating to the skin and eyes. The 3 month NOEL is 3.24 mg/kg in male mice, 40.6 mg/kg in female mice, and 42 mg/kg in rats.

The low toxicity of mildiomycin allows this biofungicide to be used extensively for pest management.

Mint oil biorepellent

Mint oil is obtained from members of the genus *Mentha* in the mint family Lamiaceae. The compound that is extracted from mint oil is menthol. Menthol, $C_{10}H_{20}O$ ((1*R*,2*S*,5*R*)-2-isopropyl-5-methylcyclohexanol), is a whitish solid that is slightly soluble in water.

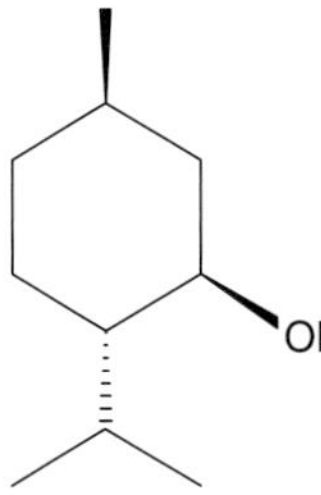

Mint oil

Mint oil is used to effectively repel mosquitoes and ants. It is additionally used as a remedy for indigestion and breathing problems.

A combination of **citric acid** and mint oil is an effective fungicide and bactericide. Fungi such as *Phytophthora* spp., *Sclerotinia* spp. and *Botrytis* spp., and bacteria (e.g. *Pseudomonas* sp.) are effectively controlled with this mixture, since the oil and acid combination inhibits respiration. **Downy mildew** is also brought under control by this combination. A known trade name is Fungastop.

Menthol has low toxicity to mammals and is not known to cause ill effects on the environment. The oral LD_{50} of menthol in rats is 3300 mg/kg.

The USA does not register the use of mint oil, peppermint oil and spearmint oil as insecticides.

Minute pirate bugs (*Orius tristicolor*) as biocontrol agents

The minute pirate bug, or *Orius tristicolor* from the order Hemiptera and family Anthocoridae, is a predatory insect on a wide range of host insect pests on crops. Also called the flower bug, this insect is important from the biocontrol point of view, due to its predatory nature.

Adult pirate bugs are 2 mm long, oval-shaped, black insects with white wing patches (Fig. M.1). They have a beak-like mouthpart that stretches up to the basal portions of the fore-legs. The adult female lays over 100 small eggs on petioles or on the midrib of leaves. Eggs hatch within 1 week to produce wingless orangish nymphs. Nymphs go through five instar stages in about 10 days to emerge as adults, which live for about 1 month. These pirate bugs have varied habitats, including flowers, and are most active at temperatures from 20 to 30°C.

Nymphs and adults are both primarily predatory in nature, and can feed on 30 or more spider mites in a day. They also feed on eggs and larvae of thrips, aphids, **scale insects**, **leafhoppers**, corn borers and armyworms. These insects pierce their prey with their mouthparts and suck out the body fluids.

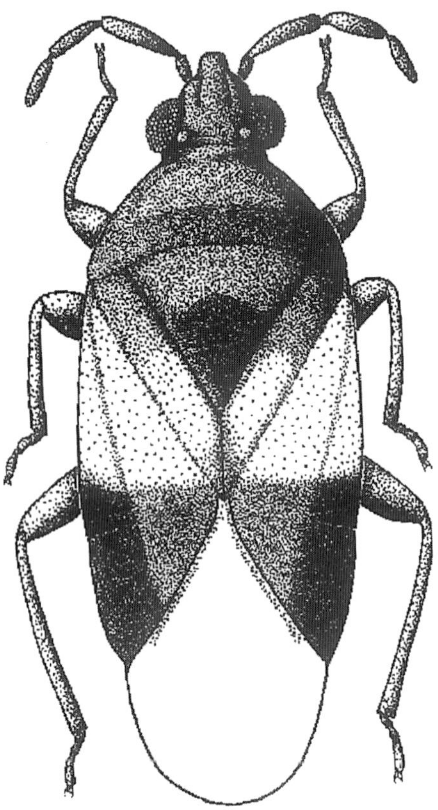

Fig. M.1. Pirate bug. Sketch courtesy of Leena Achari, with permission.

In the absence of prey, minute pirate bugs can feed on pollen and plant sap for sustenance.

Minute pirate bugs are supplied commercially as adults in vermiculite or rice hull, although they cannot survive in bottles for long periods. These predatory bugs are successfully used as biological control agents on cotton, groundnut, lucerne, maize, pea, strawberry and tea. Releasing about 100–4000 bugs/ha proves to be effective.

Pirate bugs are sensitive to foliar- and soil-applied chemical insecticides. Planting flowering plants provides enough pollen and nectar for the bugs to survive harsh periods of prey scarcity.

Mirex insecticide

Mirex, $C_{10}Cl_{12}$ (dodecachloropentacyclo[5.3.0.0^{2,6}.0^{3,9}.0^{4,8}]decane or perchloropentacyclo[5.3.0.0^{2,5}.0^{3,9}.0^{4,8}]decane), is a chlorinated hydrocarbon that is an insecticide. It is a white, crystalline, odourless solid, almost insoluble in water but soluble in tetrahydrofuran, carbon disulfide and chloroform. It is an extremely stable substance and degrades only in sunlight to form the toxic photomirex. It is also resistant to microbial degradation. Mirex is made by the dimerization of hexachlorocyclopentadiene in the presence of aluminium chloride. Mirex has a CAS Registry Number of 2385-85-5.

Although mirex had been used on several insect pests (e.g. fire ants), its stability, toxicity and persistence has prompted most countries to ban its use. According to the PAN Pesticide Database, mirex is a moderately toxic chemical which is a carcinogen and a suspected endocrine disruptor. It is a persistent organic pollutant (POP), which remains in the environment for up to 10 years. It is toxic to non-target and **beneficial organisms**.

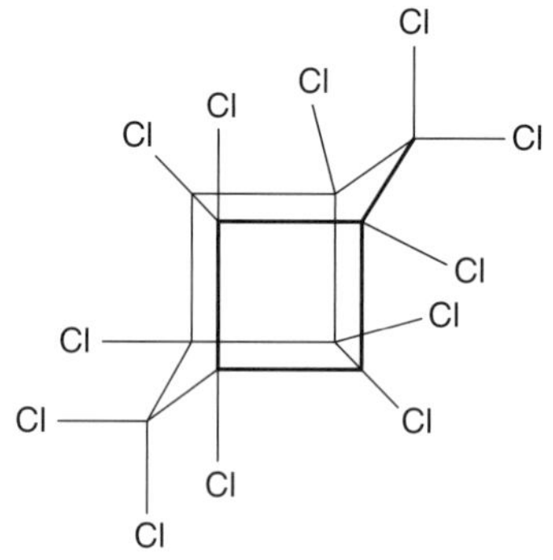

Mirex

Miticide

See **Acaricides (miticides)**

Mitosis

See **Herbicide action – meristematic mitotic inhibitors**

Molinate herbicide

Molinate, $C_9H_{17}NOS$ (*S*-ethyl azepane-1-carbothioate), is a thiocarbamate herbicide. It is a colourless liquid, moderately soluble in water and miscible with organic solvents. Molinate is unstable to light, but stable at room temperature. The compound is made by the reaction of chlorothiolformate with hexamethyleneimine. Its CAS Registry Number is 2212-67-1.

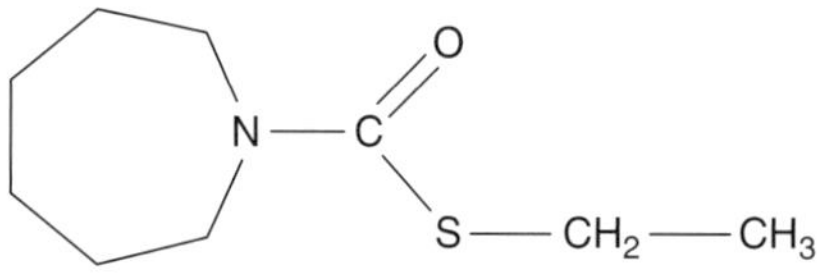

Molinate

Molinate is a selective, pre-plant, soil-applied, systemic herbicide, with good acropetal translocation (thus reaching the leaves). Mainly used on rice crops, it controls germinating broadleaved weeds and grasses by inhibiting lipid metabolism.

Molinate is available as granules (GR) and as an emulsifiable concentrate (EC), under trade names such as Hydram and Sakkimol. It is used in combination with **propanil**.

The acute oral LD_{50} of molinate in rats is between 501 and 720 mg/kg, while the acute dermal LD_{50} is 2000 mg/kg. The 4 h inhalation LC_{50} in rats is 1.36 mg/l. Molinate is practically safe on birds, but is toxic to some fish species (the 96 h LC_{50} is 0.21–0.32 mg/l in rainbow trout and bluegill sunfish). Molinate is irritating to the skin and eyes. According to the WHO, molinate is a Class II toxin (moderately toxic). It is a PAN Bad Actor chemical.

In plants, molinate is rapidly metabolized to carbon dioxide and naturally occurring plant constituents. In soil, its microbial breakdown involves hydrolysis to ethyl mercaptan, carbon dioxide and dialkylamine. Molinate does not persist in soil and has a half-life of up to 21 days. Microorganisms also contribute to its breakdown in the soil. To avoid degradation of molinate by sunlight, it is necessary to incorporate the chemical into the soil.

Molinate is registered in Australia, Vietnam, Cambodia and some EU members.

Molluscicide cloethocarb

See **Cloethocarb insecticide**

Molluscicide methiocarb

See **Methiocarb insecticide**

Molluscicide thiodicarb

See **Thiodicarb insecticide**

Molluscicides

Molluscicides are chemical or non-chemical substances used to kill molluscs. Commonly, molluscicides are either metal salts or toxic contact agents that inhibit biochemical processes in the creatures. Molluscicides of plant origin are also gaining popularity.

Snails and slugs, which belong to the **class Gastropoda**, are major mollusc pests. An effective strategy to control infestation is a combination of chemical and natural products. It is ideal to also exploit their natural enemies. Baits and traps may be used wherever possible.

Chemical compounds used include **copper sulfate, calcium cyanamide, organotin** compounds, sodium ferric EDTA, **metaldehyde**, some carbamates, iron phosphate, **endosulfan** and **niclosamide**.

Carbamate insecticides (cholinesterase inhibitors) and phosphate esters are also used to control molluscs.

Thiodicarb, $C_{10}H_{18}N_4O_4S_3$ (3,7,9,13-tetramethyl-5,11-dioxa-2,8,14-trithia-4,7,9,12-tetraazapentadeca-3,12-diene-6,10-dione), a carbamate insecticide is also a molluscicide. It is a colourless crystal, sparingly soluble in water and readily soluble in organic solvents. It is made by the reaction of *N,N*-thiobis (methylcarbamic acid fluoride) with 2-methylthioacetaldoxim in the presence of a base. The recommended application rate for pellets is 5 kg/ha.

Trimethacarb, $C_{11}H_{15}NO_2$ (a mixture 3,4,5- and 2,3,5-trimethylphenyl methylcarbamate), is a cholinesterase inhibitor and a known molluscicide, besides being a bird and mammal repellent. Trimethacarb is obtained by the reaction of methylisocyanate with 3,4,5-trimethylphenol.

Phosphate esters like **phorate, azinphos-methyl**, have been observed to have molluscicidal properties in some instances.

Fentin acetate, $C_{20}H_{18}O_2Sn$ (triphenyltin (IV) acetate), is primarily a non-systemic fungicide effective on **blights** (potatoes), leaf spots (celery, onions, sugarbeet and groundnuts) and rusts (coffee) but is also used to control algae and snails in rice fields and water snails in fish ponds. Fentin acetate is classified as moderately toxic (Category II) by both the WHO and the EPA. Fentin acetate as well as fentin hydroxide have been refused registration in India and the EU.

Calcium oxide (as a molluscicide), commonly known as lime, has historically been used to control snails and slugs. It is a white, crystalline, alkaline solid, insoluble in water. With a molecular formula of CaO, lime is toxic to molluscs, but is not harmful to plants and animals.

Sodium pentachlorophenate has molluscicidal properties and continuous use of this compound at a high concentration can almost eliminate snail populations from the field. The compound is hazardous if inhaled and is also absorbed by the skin. It has a molecular formula of C_6Cl_5NaO and a CAS Registry Number of 131-52-2.

Other molluscicides such as yurimin and organotin compounds are known. Yurimin exhibits acute toxicity and is not commonly used.

The perennial climbing plant, **endod** (*Phytolacca dodecandra*), a native of Ethiopia, grows soapy berries that, when dried, powdered and mixed with water, exhibit molluscicidal effects, while being a low-cost biodegradable agent. The compounds in the berries include triterpenoid saponin and oleanolic acid glucoside. The components degrade rapidly, probably by bacterial action, neutralizing their molluscicidal property quickly. Many varieties of this plant grow in many Asian countries.

The endod berries are proven to control disease-transmitting snails (with a LC_{50} of 1.85 ppm). Endod exhibits adequate control against *Biomphalaris pfeifferi*. Besides being a molluscicide, the soapy solution formed by endod berries can also control larvae of mosquitoes and other insects. At concentrations of up to 30 mg/l, endod can control certain snails that are intermediate hosts to pathogens. The berries also have antifungal properties.

Molluscs

Molluscs are animals belonging to the phyllum Mollusca. These unsegmented animals, with body cavities and highly developed blood and nervous systems, live on land and in both fresh and sea water. The body is divided into a head, a muscular foot and a curvaceous back that has a chalky shell in many species. The back helps in respiration in some species. Molluscs do not have a standard, fixed shape, but their bodies can mould according to need.

Most molluscs are slow moving. Reproduction is sexual, with some species having separate sexes and others being hermaphrodites. There are six classes of molluscs: (i) Monoplacophora; (ii) Amphineura (chitons); (iii) Gastropoda (slugs, snails and limpets); (iv) Scaphopoda (tusk shells); (v) Lamellibranchia (bivalve molluscs); and (vi) Cephalopoda (squids and octopuses).

Gastropods are the only class of molluscs with terrestrial species. Some species of gastropods are agricultural pests (Figs M.2 and M.3).

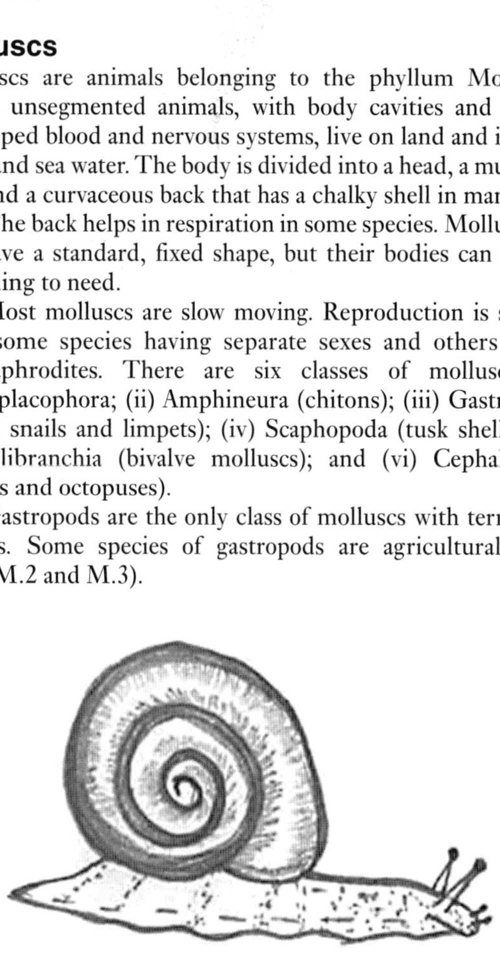

Fig. M.2. Sketch of a snail. Sketch courtesy of Leena Achari.

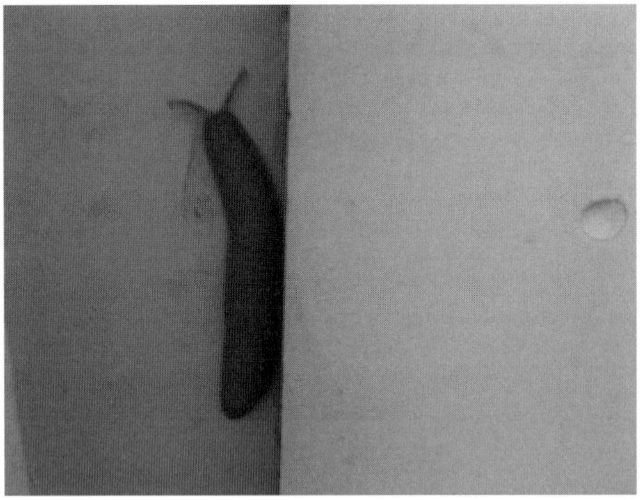

Fig. M.3. Photo of a slug. Courtesy of Leena Achari.

Molybdenum as a plant nutrient

See **Plant nutrient excesses and deficiencies**

Monocrotophos insecticide

Monocrotophos, $C_7H_{14}NO_5P$ (dimethyl (*E*)-1-methyl-2-(methylcarbamoyl) vinyl phosphate), is an organophosphorus insecticide. It is a colourless hygroscopic crystalline substance, highly soluble in water and methanol (100%), moderately soluble in acetone, *n*-actanol and toluene, and sparingly soluble in kerosene and diesel oil. It is relatively stable in aqueous media and hydrolysed in alkaline solutions. DT_{50} values (20°C) at pH 3, 7 and 9 are 131, 66 and 17 days, respectively. Monocrotophos decomposes at less than 38°C. It is produced by the reaction of trimethylphosphite with chloroacetoacetic acid methylamide in the presence of a base. Its CAS Registry Number is 6923-22-4.

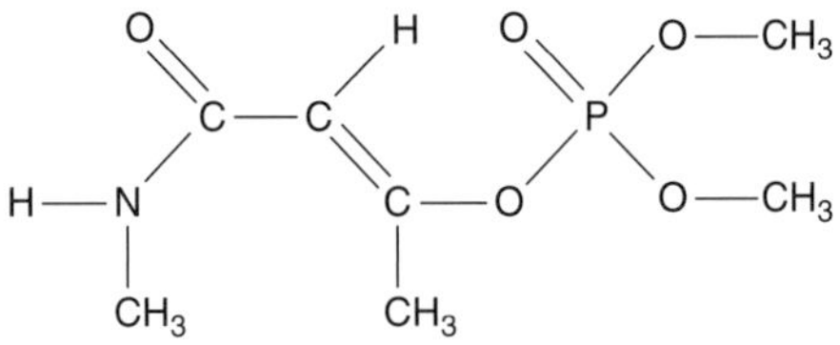

Monocrotophos

Monocrotophos is a systemic, broad-spectrum insecticide and acaricide that penetrates plant tissues rapidly. It acts by inhibiting **acetylcholinesterase**, an enzyme essential for normal nerve impulse transmission and is thus considered a nerve toxin. Monocrotophos is used in the control of spider mites and a host of insects. The main crops that this insecticide is used on are cotton and maize. However, it is also used on citrus, olives, rice, sorghum, sugarcane, sugarbeet, groundnuts (peanuts), potatoes, soybeans and tobacco.

Monocrotophos is formulated as soluble concentrates (SL) and ULV liquids (UL). It is compatible with most insecticides, **acaricides** and fungicides, as long as they do not comprise alkaline material. Trade names for monocrotophos include Azodrin, Bilobran, Monocil 40, Monocron, Nuvacron and Pillardrin.

Monocrotophos is toxic via all routes, oral, dermal and inhalation, and is categorized as a Class Ib toxin by the WHO. The oral LD_{50} in mice is <20 mg/kg. This compound is highly toxic to birds and bees and moderately toxic to fish (the LD_{50} is 0.028–0.033 mg per bee). Monocrotophos can mildly irritate the skin and eyes. It is listed as a PAN Bad Actor chemical based on its property of cholinesterase inhibition and acute toxicity. The ADI is 0.0006 mg/kg b.w. Applicators should use PPE including respirators and should treat all contaminated equipment and garments according to the instructions on the label. The **restricted entry interval** (REI) for unprotected workers is 48 h.

In mammals, following oral administration, more than half of the insecticide is excreted within a day, predominantly via the urine. In plants, demethyl hydrogen phosphate and *N*-methylacetoacetamide are formed. Monocrotophos is rapidly degraded, for example the soil DT_{50} is <7 days.

India and many African countries such as Madagascar, Senegal and Burkina Faso have registered the use of monocrotophos. Several other countries have restricted it and some (e.g. the USA) have banned its use. The MRL of monocrotophos is 0.02–1 mg/kg.

Monolinuron herbicide

Monolinuron, $C_9H_{11}ClN_2O_2$ (3-(4-chlorophenyl)-1-methoxy-1-methylurea), is a **phenylurea** herbicide. It is a colourless, crystalline substance, soluble in water (0.735 g/l), and readily soluble in solvents such as alcohols, acetone, dioxane and xylene. This herbicide decomposes in both acidic and alkaline media, as well as by UV light. When liquid formulations of monolinuron are heated, they can become combustible, and release toxic fumes. It is produced by the reaction of 4-chlorophenyl isocyanate with *N*-methylhydroxylamine, followed by treatment with dimethyl sulfate. The CAS Registry Number of monolinuron is 1746-81-2.

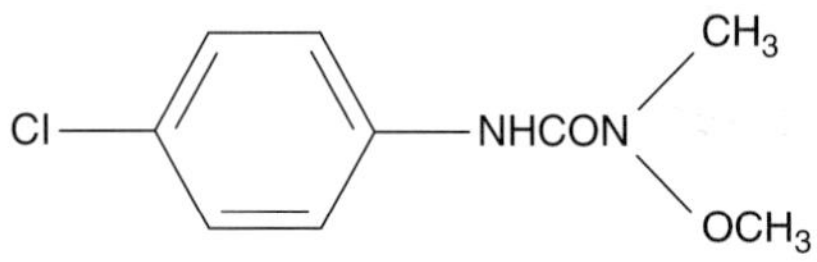

Monolinuron

Monolinuron is a selective, broad-spectrum, systemic, pre- or post-emergence herbicide, with additional algicidal properties. It acts by inhibiting photosynthesis at the photosystem II level, of susceptible broadleaved weeds and annual grasses in asparagus, maize, field beans, vines, onions, potatoes and ornamental plants. As an algicide, it controls blanket weed (*Galenia pubescens*) and hair algae.

Monolinuron is formulated as an emulsifiable concentrate (EC) or wettable powder (WP). It may be used in combination with other herbicides for increased weed control. A common trade name is AgriGuard.

The WHO classifies the herbicide monolinuron as a Class U toxin (unlikely to be hazardous). It can harm aquatic organisms. Since the compound produces toxic fumes like hydrogen chloride and nitrogen oxides, care has to be taken while spraying to prevent the drift from spreading into surrounding areas. Dust formulations too should be carefully applied.

The chemical must be kept away from fire, food and sewer water to avoid contamination. Handlers are advised to use breathing apparatus, protective gloves and safety spectacles. The NOEL in rats (for a 2 year period) is 0.5 mg/kg/day. According to ECCO (the European Community Co-ordination project), the ADI of monolinuron is 0.003 mg/kg b.w. Though monolinuron is not a PAN Bad Actor chemical, its parent compound, **paraquat dichloride**, is.

The DT_{50} of monolinuron in soil is 45–60 days. It breaks down to form 3-(2-hydroxy-4-chlorophenyl) urea and the corresponding 3-hydroxy compound (3-(4-chlorophenyl)-1-methyl urea and *N*-(4-hydroxyphenyl)-*N'*-methoxy-*N'*-methylurea. Degradation in plants is similar. The USA has not registered the use of monolinuron.

Monuron herbicide

See **Herbicide action – photosynthesis inhibitors**

Mosaic disease

Mosaic disease is used to describe a viral condition where the leaf develops a pattern of discoloration. Such patterns have a defined boundary, delineated by veins, and may be green and angular, while the background may be yellowish. If such distinct areas later coalesce, a mottle symptom may result.

The mosaic disease is common among crops such as squash, cucumber, pumpkin, watermelon and gourd. The plant affected by the mosaic virus gets a mottled appearance and becomes distorted. Often the leaves turn upwards, resulting in stunted growth and discoloration of the infected plant part. This disease can be transmitted through **vectors** such as **aphids** or beetles. Examples of mosaic diseases are rose mosaic, sugarcane mosaic, cowpea mosaic, soil-borne wheat mosaic and wheat streak mosaic.

There is no specific chemical available to control the spread of mosaic disease. However, control of pest vectors help control the spread of this virus to a large extent. (See also **Common viral plant diseases**.)

Mottling disease

Mottle is a leaf symptom in which numerous, small areas of chlorotic and irregularly shaped discoloration develop. These areas do not have defined boundaries transcending veins. There is more than one reason for mottling.

Leaf mottles could indicate deficiency of one of more nutrients (e.g. a zinc deficiency). Besides malnutrition, plant viruses can also cause mottling. Chlorotic mottle in cowpea, impatiens necrotic spot, maize dwarf mosaic on maize, leaf roll and bud necrosis disease are some examples of viral diseases. The symptoms of the infection vary from necrosis to formation of spots to stunted growth leading to dwarfism.

Soil that undergoes periods of inadequate aeration because of waterlogged conditions can also result in spots of rust colour on leaves that are also called mottles. (See also **Common viral plant diseases**.)

Mould (sooty mould)

Moulds are fungi that exist as several hyphae, interconnected in a network called a colony. Moulds reproduce, either sexually (meiosis) or asexually (mitosis), through small spores that may have one or many nuclei.

Sooty mould is a black mould seen mainly on leaves, but also on branches and twigs. It is a dark grey-black, sticky substance that forms a velvety crust-like coating. Sooty mould does not kill the plant, but its build-up blocks sunlight, slowing the plant's photosynthesis and reducing the growth of healthy new shoots, leaves or fruit. These factors affect yield and cause plants to droop and **wilt** faster in dry weather than normal plants would.

Besides being abundantly present on vegetation, moulds are seen on damp corners of homes and storehouses, and also on stale cooked food and over-ripened fruits.

The causal agents of sooty mould (or **black sooty mould**) are *Cladosporium* spp., *Aureobasidium* spp., *Antennariella* spp., *Limacinula* spp, *Aspergillus niger*, *Capnodium* spp., *Scorias* spp. and *Capnodium* spp.

Moulds are also present in other colours. Blue or green mould is caused by *Penicillium* spp. and a grey mould by *Botrytis cinerea*.

Wind-blown spores of these pathogens germinate on the **honeydew** excreted by **aphids**, **mealybugs**, **whitefly**, **leafhoppers**, plant hoppers, psyllids and **scale insects**. These spores then grow as long, black strands or a powdery mass over the leaf surface. Although the coating can be removed by rubbing, severe infections can penetrate the plant and rejuvenate each year from the wood of the plants.

For effective control of sooty moulds, bulbs should be stored in moisture-free conditions. There has to be an effective and long-term plan to control insects that produce honeydew. Pesticides that are effective against specific insects and horticultural oils effectively control infestation. Use of **neem** oil, insecticidal soap and appropriate insecticides, fungicides and **acaricides** lend some measure of control.

Moulting

See **Insect growth regulators (IGRs)**

Mulches

See **Weed management by mechanical methods**

Myclobutanil fungicide

Myclobutanil, $C_{15}H_{17}ClN_4$ (2-*p*-chlorophenyl-2-(1 *H*-1,2,4-triazon-1-ylmethyl)hexanenitrile;2-(4-chlorophenyl)-2-(1 *H*-1,2,4-triazol-1-ylmethyl)hexanenitrile), is a fungicide of the azole type. A pale yellow solid, myclobutanil is soluble in water at 142 mg/l and also in ketones, esters and alcohols. It decomposes in sunlight with a DT_{50} of 0.8 days. Burning of this fungicide can create an explosive hazard, with the release of toxic gases. Storage is recommended in airtight containers, in dry, cool, well-ventilated rooms, at temperatures between 1 and 49°C. The CAS Registry Number is 88671-89-0.

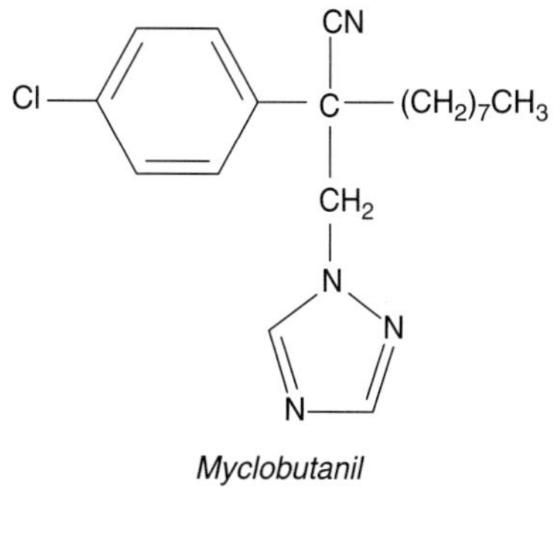

Myclobutanil

Myclobutanil is a systemic compound that acts by inhibiting **ergosterol** biosynthesis (via inhibition of steroid demethylation). Ergosterol is a critical part of fungal cell membranes. Myclobutanil controls ascomycetes, fungi imperfecti and basidiomycetes in a range of crops that include apples, strawberries, pears, vines, barley, cotton, rice and wheat. Diseases such as rust, shot hole, blight, **powdery mildew**, **scab** and seed-borne diseases are controlled with myclobutanil.

The fungicide is commonly formulated as wettable powder (WP), emulsifiable concentrate (EC) and as a seed treatment. Principal trade names include Systhane. Combination with **captan**, **sulfur**, **mancozeb** and **ziram** increases the efficiency of myclobutanil. Seed, soil, foliage, drench and dip applications are common.

The acute oral LD_{50} is 1600 mg/kg in male rats and 2290 mg/kg in female rats. The acute oral LD_{50} in bobwhite quails is 510 mg/kg, making it moderately toxic to birds. It is non-toxic to bees and non-phytotoxic if used as directed. Myclobutanil causes irritation to the eyes and skin. It is a reproductive toxin, based on tests on rats. The WHO lists this chemical as a Class III toxin (slightly hazardous). Symptoms of exposure to this fungicide are skin rash, nausea, headache, diarrhoea, vomiting, nose bleed and eye irritation. Handling, application, loading and disposal should be done with utmost care, to prevent drift and contamination.

While myclobutanil is eliminated by animals in the faeces and urine, decomposition in both animals and plants is via oxidation of the butyl group to a ketone and an alcohol and partial conjugation to a glucoside. The primary metabolite is 1,2,4-triazole. Myclobutanil is environmentally mobile and has been found in groundwater. In anaerobic soil conditions, this fungicide is very stable, and in silt loamy soil, the DT_{50} is 66 days, which means that it is persistent. Myclobutanil is registered for use in India, the USA, Canada, Germany, the UK and many other countries.

According to the USDA Foreign Agricultural Service's Pesticide MRL database, the indicative MRL of myclobutanil for apples in the USA and South Africa is 0.5 ppm. The EU maintains an MRL of 0.5 ppm for its 'Pome fruit' group. For cucumber, it is 0.1 ppm in South Africa. The MRL value on many berry varieties is accepted as 0.02 mg/kg. According to the Japan Food Chemical Research Foundation, the indicative MRL of myclobutanil for wheat in Japan is 0.03 ppm.

Mycoherbicides

See **Bioherbicides**

Mycoplasmas

Mycoplasmas are extremely small organisms that are known to cause disease in crops. Their size varies from that of a virus to that of a bacterium. They lack a cell wall (unlike bacteria) but respond to antibiotics (like bacteria). They are transmitted by insects (like viruses).

Aster yellows is the name of a disease brought on by mycoplasmas.

Mycotoxicosis

See **Mycotoxin**

Mycotoxin

A toxic metabolite produced during the break down of substrates by fungal species is called a mycotoxin. Different fungal species produce different mycotoxins. The reason for toxin production is self-proliferation and host subjugation. Many of the fungal species producing mycotoxins are plant pathogens and are present on crop yields.

Mycotoxins enter the food chain through human consumption of grains and other yields, or via animal feed. They do not degrade easily in the digestive systems of humans, poultry and pigs. They do not easily break down even upon cooking.

Mycotoxins differ in their intensity of toxicity and cause allergic reactions, weakened immune systems, liver and/or kidney damage, haemorrhaging and sometimes, death. Poisoning by mycotoxins is known as mycotoxicosis.

There are many types of mycotoxins, which are produced by different fungal species. Species of the genus *Aspergillus*, a common genus, produce mycotoxins. *Aspergillus flavus* produces **aflatoxin** that is observed to be carcinogenic. *Cladisporium*, *Fusarium*, *Penicillium* and *Stachybotrys* species are other examples of fungi that produce mycotoxins. Ochratoxin, penicillic acid (*Penicillium* spp.) and **trichothecene** (*Stachybotrys* spp.) are other examples of mycotoxins.

N

1-Naphthylacetic acid as a PGR (α-naphthaleneacetic acid)

1-Naphthylacetic acid, $C_{12}H_{10}O_2$ (1-naphthylacetic acid), is a colourless, crystalline substance that is used as a plant growth regulator. This PGR, also referred to as NAA, is soluble in water, xylene, carbon tetrachloride, acetone and diethyl ether. It is stable and is compatible with salts of acids. It is a metabolite of 1-naphthylacetamide. The CAS Registry Number of 1-naphthylacetic acid is 86-87-3.

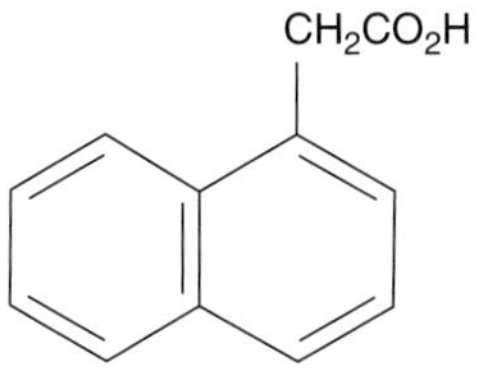

1-Naphthylacetic acid

The PGR 1-naphthylacetic acid promotes good growth of roots in fruit crops including apples, pears, grapes, guavas and mangoes. It also prevents premature fruit and flower drop. However, it can harm the plant when applied at high doses.

1-Naphthylacetic acid is variously formulated as a soluble concentrate (SL), emulsifiable concentrate (EC) and a water-soluble powder (SP). It is commercially available as Fruitone and Celmone.

According to the WHO, it is a product unlikely to present acute hazard in normal use, and according to the EPA, it is a Class III toxin. It is non-toxic to bees when used as directed. The acute oral LD_{50} in rats is between 1000 and 5900 mg/kg while that in mallard ducks is >10,000 mg/kg.

The PGR is registered for use in many EU countries including the UK.

2-(1-Naphthyl) acetamide as a PGR

2-(1-Naphthyl) acetamide, $C_{12}H_{11}NO$, is a colourless, crystalline substance, soluble in water, acetone, ethanol and isopropanol, but insoluble in kerosene. With plant regulating properties, it is stable under normal storage conditions. The CAS Registry Number of 2-(1-naphthyl) acetamide is 86-86-2.

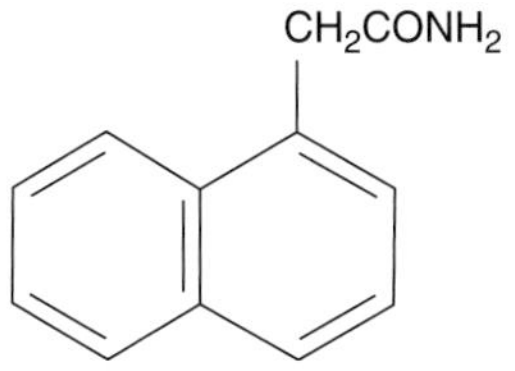

2-(1-Naphthyl) acetamide

2-(1-Naphthyl) acetamide is a plant growth regulator, controlling foliage growth in apple and pear trees and preventing their premature fruit fall. It acts by inducing formation of an abscission zone in the peduncle.

The PGR is commonly available as a wettable powder (WP) under the trade name Amid-Thin.

According to the WHO and EPA (USA), 2-(1-naphthyl) acetamide is classified as a chemical unlikely to present acute hazard in normal use (Class U and Category III toxin, respectively). The acute oral LD_{50} in rats is 1690 mg/kg, while the contact LD_{50} in rabbits is >2000 mg/kg. It causes irritation to the eyes.

(2-Naphthyloxy) acetic acid as a PGR

(2-Naphthyloxy) acetic acid, $C_{12}H_{10}O_3$ ((2-naphthyloxy) acetic acid), is a plant growth regulator. It is soluble in water, ethanol, acetic acid and diethyl ether. It is soluble in water at 203 mg/l and very soluble in acetate. The CAS Registry Number of (2-naphthyloxy) acetic acid is 120-23-0. It forms water-soluble alkali-metal and amine salts.

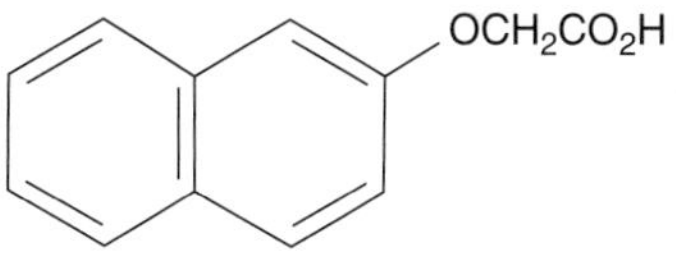

(2-Naphthyloxy) acetic acid

As a synthetic auxin, this compound is absorbed well by the plant. It is used on tomatoes, strawberries, capsicum, grapes and pineapples for better root development. It also prevents premature ripening of fruits, a property being exploited for better storage.

(2-Naphthyloxy) acetic acid is best used as soluble concentrates (SL) and emulsifiable concentrates (EC) and applied as a spray. It is available in the market under the trade name of Betapal.

According to the WHO, this PGR is a Class III toxin (slightly hazardous). It is non-toxic to bees. The acute oral and dermal LD_{50} is 1470 mg/kg and >2000 mg/kg, respectively. Exposure to it can cause irritation to the eyes and the respiratory tract.

In plants, (2-naphthyloxy) acetic acid undergoes degradation to 2-naphthol, followed by ring hydroxylation and ring opening.

This PGR is registered in Cyprus, Vietnam and Hungary, but not in the USA and most nations in the EU.

3-[*N*-butyl-*N*-acetyl]-aminopropionic acid, ethyl ester biorepellent

3-[*N*-Butyl-*N*-acetyl]-aminopropionic acid, ethyl ester is a plant-derived insect repellent. Its approved name is ethyl

3-[*N*-butyl-*N*-acetyl]-aminopropionate. It is similar to the naturally occurring β-alanine. The CAS Registry Number of 3-[*N*-butyl-*N*-acetyl]-aminopropionic acid, ethyl ester is 52304-36-6.

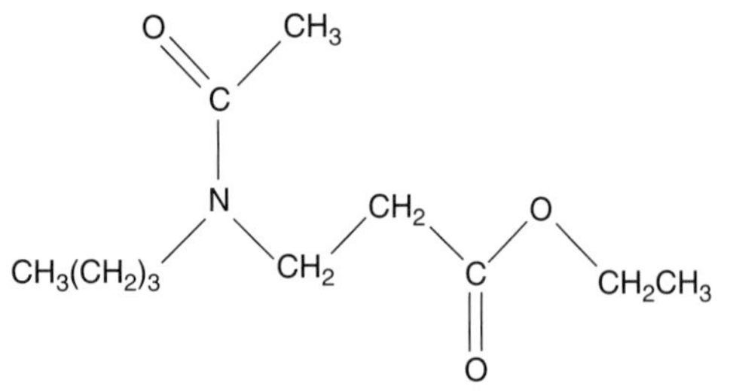

3-[N-Butyl-N-acetyl]-aminopropionic acid, ethyl ester

Production of this insect repellent is mainly for commercial uses. The target pests of this repellent are mosquitoes, deer ticks, body lice and biting flies. Trade names include IR353.

This insect repellent does not have any adverse effects on mammals, birds or the environment, except upon direct contact. It is registered in Australia, Greece and the USA.

Nabam fungicide

Nabam, $C_4H_6N_2Na_2S_4$ (disodium ethylenebis(dithiocarbamate), is an alkylenebis dithiocarbamate fungicide. It is a colourless crystalline substance with a slight odour, soluble in water at 200 g/l. It is stable as an aqueous solution but decomposes by light, moisture and heat. The CAS Registry Number of nabam is 142-59-6.

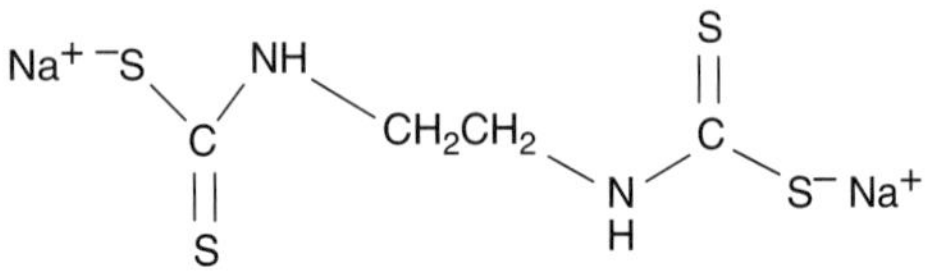

Nabam

Nabam is a broad-spectrum fungicide with protective action. It also has nematicidal, algicidal and bactericidal properties. It is a non-specific thiol reactant that inhibits respiration. Slime-forming microorganisms are among the common targets. Nabam controls some soil-borne fungal diseases of cotton, capsicums and onions. It controls algae in rice fields. *Aspergillus niger*, *Fusarium oxysporum* and *Pseudomonas fluorescens* are some targets of nabam. Nabam also has non-crop uses like in water cooling systems, oil recovery drilling systems and paper/pulp systems.

Nabam is available commercially as Dithane D-14, AMA-31, Nabasan, Alcotreat MOS and many others. Nabam is formulated in the form of soluble concentrate (SL) or a powder. It is applied by hand and by pump. It is also combined with zinc sulfate to form **zineb** or manganese sulfate to form **maneb**. Nabam is incompatible with **lime sulfur** and soluble spray oils. Non-crop application doses range widely from 1.5 to 210 ppm.

Nabam is classified by both the WHO and the EPA (USA) as a Class II toxin (moderately hazardous). It is irritating to the skin and mucous membrane. It is phytotoxic to plant foliage. The acute oral LD_{50} in rats and mice is in the range of 395–580 mg/kg. The NOEL in rats receiving 1000–2500 mg/kg diet for 10 days shows a goitrogenic effect (a substance causing disorders in the functioning of the thyroid gland). It is moderately toxic to fish and *Daphnia magna*. It is practically non-toxic to birds and to bees. It is a carcinogen, a reproductive toxin and a suspected endocrine disruptor, according to the PAN Pesticide Database. Toxicity symptoms include irritation of skin, eyes and the respiratory tract.

In plants, the main metabolite is **ethylenethiourea (ETU)** and it is persistent. It is stable to chemical degradation (hydrolysis and photodegradation) and is highly soluble in water, and is mobile and thus may reach groundwaters. Because ETU is the most stable, toxic and mobile degradate of nabam, it is a subject for environmental concern.

Nabam is registered for use in Canada and the USA. In the USA, it is not allowed on lettuce, mustard greens and spinach. The tolerance limit of nabam for most fruits and vegetables is set at 7 ppm.

Naled insecticide

Naled, $C_4H_7Br_2Cl_2O_4P$ (1,2-dibromo-2,2-dichloroethyl dimethyl phosphate), is an organophosphorus insecticide. It is a colourless liquid with a slightly pungent odour, practically insoluble in water, but soluble in aromatic and chlorinated solvents, aliphatic solvents and mineral oils. It rapidly hydrolyses in aqueous medium and by sunlight and even at room temperature. In the presence of metals and reducing agents, bromine is given out and **dichlorvos** is formed. It may pose a fire and explosion hazard in the presence of oxidizers, producing toxic oxides of phosphorus, hydrogen bromide and hydrogen chloride. Hence, it is stored in airtight containers, in secure, cool, dark, dry rooms. Naled is synthesized by bromination of dichlorvos. It has a CAS Registry Number of 300-76-5.

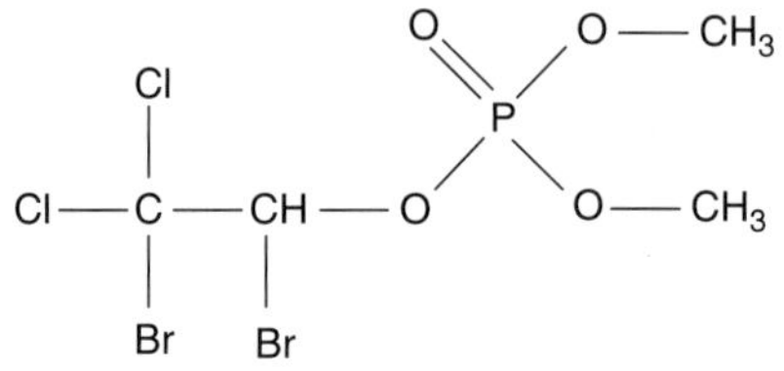

Naled

Naled, a non-systemic insecticide and an acaricide, is a cholinesterase inhibitor that has respiratory and stomach actions. Like all organophosphate compounds, naled acts by interfering with the activity of cholinesterase, an enzyme that is essential for the proper working of the nervous systems of both humans and insects. Spider mites, aphids and a vast number of insects affecting cotton, rice, lucerne, soybean, tobacco, mushrooms and fruits and vegetables are controlled. It is also used to control pests like flies, ants, fleas, cockroaches, silverfish and mosquitoes in animal shelters, homes and public places. Of all these pests, adult mosquitoes are most commonly controlled by naled. It also acts as a fumigant of limited efficacy, and as an ectoparasiticide.

Naled is formulated as emulsifiable concentrate (EC), dustable powder (DP) and ULV liquid (UL). The vapour action can be harnessed to control insects in greenhouses by applying liquid formulations through heating pipes. Very small quantities of naled can be applied by aircraft- or truck-mounted sprayers or by ULV sprayers. Naled is sold in the market as Dibrom, Bromex and Fly Killer-D.

According to the WHO, naled is categorized as a Class II toxin (moderately hazardous). The oral LD_{50} for naled in rats and mice ranges from 50 to 430 mg/kg, while the inhalation LC_{50} is 7.7 mg/kg and the dermal LD_{50} is 800 mg/kg. Naled is highly to moderately toxic to birds (with the LD_{50} being as low as 37 mg/kg in some species). Naled is highly toxic to bees and highly to moderately toxic to fish. It may be very highly toxic to aquatic invertebrate species. Naled causes severe irritation to skin and causes eyes to burn. In addition, high exposure to naled can cause toxicity and even death. Naled is also a suspected carcinogen and a reproductive and development-stage toxin, according to the PAN Pesticide Database. It is thus a PAN Bad Actor chemical. Naled can injure some varieties of apples, pears, melons, plums, peaches and cotton.

Naled breaks down into **dichlorvos** in animals, plants and the environment. In the presence of metals and reducing agents, bromine is lost and dichlorvos is formed. This insecticide is not persistent in soil as it adsorbs only weakly to soil particles, but is nearly insoluble in water. Soil microorganisms break down most of naled in the soil. It is rapidly broken down in water with a half-life of 2 days. Naled is practically non-persistent in the environment, with a reported field half-life of less than 1 day. Symptoms of poisoning include excessive salivation, sweating, rhinorrhoea, muscle twitching, weakness, headache, dizziness, incoordination, respiratory depression and tightness in the chest, wheezing, nausea, vomiting, abdominal cramps and many others. In cases of exposure, contaminated clothing should be removed and the skin washed with plenty of water and soap. Since it may be absorbed through the skin and by inhalation, handlers must be equipped with protective gloves, clean protective clothing and a respirator. If it has been ingested, vomiting should be induced in a conscious person.

Naled is registered by the EPA (USA) as an insecticide. However, the residential uses of naled have been severely restricted or banned in many countries including the USA, Canada, Australia and Vietnam.

Naphthenic acid fungicide

Naphthenic acid is a combination of several carboxylic acids that act as fungicides on wood. It also has animal repellent properties. Naphthenic acid and its zinc and copper salts are soluble in most organic solvents but almost insoluble in water. They are slightly volatile and are unstable in concentrated acidic or alkaline media. The CAS Registry Number of naphthenic acid is 1338-24-5.

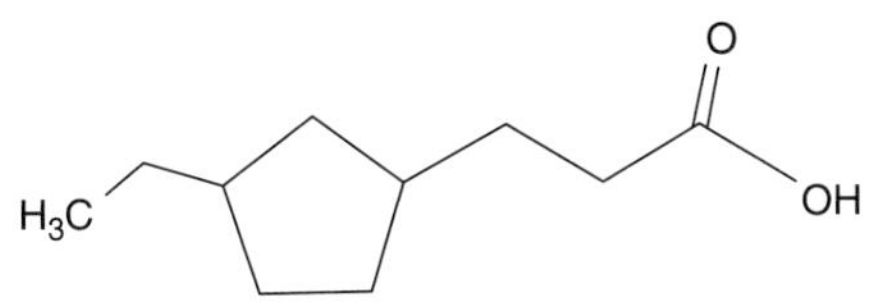

Naphthenic acid

Naphthenic acid salts protect wood from wood-decaying fungi that cause soft rot and white rot. It is also applied on tree-wounds.

This fungicide is available as copper and zinc naphthenate, and a common formulation is emulsifiable concentrate (EC). Exposure via inhalation and skin may cause irritation and coughing.

Napropamide herbicide

Napropamide, $C_{17}H_{21}NO_2$ ((*RS*)-*N*,*N*-diethyl-2-(1-naphthyloxy) propionamide), is an amide herbicide. It is a brown or colourless crystalline solid, soluble in water at 73 mg/l, acetone, ethanol, xylene and hexane. Its CAS Registry Number is 15299-99-7.

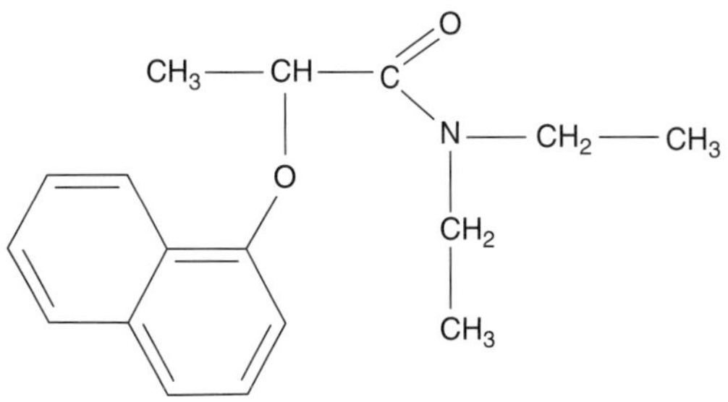

Napropamide

Napropamide is a selective systemic amide herbicide, that is normally applied to the soil and used to control a number of annual grasses and broadleaved weeds that take up the chemical through the roots and translocate it throughout the plant. A number of crops are protected by napropamide including tobacco, sunflowers, vegetables, vines and fruit trees.

The herbicide napropamide is available as emulsifiable concentrate (EC), wettable powder (WP), granules (GR) and suspension concentrates (SC). Trade names containing napropamide include Devrinol and R-7465. Napropamide may be combined with a variety of compounds like **monolinuron**, nitralin, **simazine**, **trifluralin** and **tebutam**.

Napropamide is a slightly toxic compound and is rated as a Class III toxin by the EPA (USA) and a Class U toxin by the WHO and is considered a general use pesticide (GUP) by the EPA. The signal word 'CAUTION' is required on the label of products containing the compound. The acute oral LD_{50} is 5000 mg/kg in male rats and 4680 mg/kg in female rats. Dermal toxicity, as tested on rabbits and guinea pigs, show acute dermal LD_{50} values of >2000 mg/kg. This herbicide is not toxic to bees (the acute oral LD_{50} is 121 μg per bee). The ADI is 0.1 mg/kg b.w./day.

The elimination of napropamide is rapid following oral administration (with 99% being excreted within 4 days) due to extensive metabolism. It is quite persistent in the soil environment, with the aerobic soil half-life being over 450 days. The hydrolysis half-life is over 1 month. Photodegradation is an important mechanism of breakdown. This, in addition to its solubility, makes napropamide a potential groundwater contaminant, especially in soils with low organic matter and clay. In plants, it rapidly metabolizes via ring hydroxylation and N-dealkylation followed by conjugation.

Napropamide is registered for use in Australia, Canada and the USA.

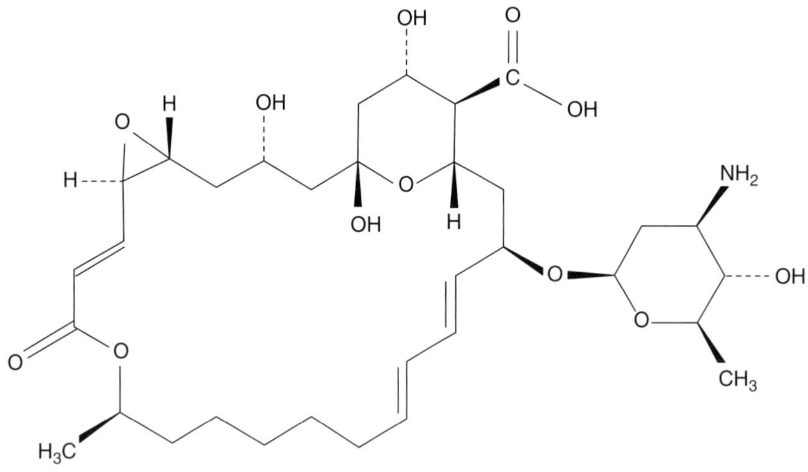

Natamycin

Natamycin biofungicide (pimaricin or myprozine)

Natamycin, $C_{33}H_{47}NO_{13}$ (8*E*,14*E*,16*E*,18*E*,20*E*)-(1*R*,3*S*,5*R*,7*R*12*R*,22*R*,24*S*,25*R*,26*S*)-22-(3-amino-3,6-dideoxy-β-D-mannopyranosyloxy)-1,326-trihydroxy-12-methyl-10-oxo-6, 11,28-trioxatricyclo[22.3.1.$0^{5,7}$]octacosa-8,14,16,18,20-pentaene-25-carboxylic acid), a pale yellow crystalline substance, is a fungicidal antibiotic compound. It has a CAS Registry Number of 7681-93-8. It is stable in dry air, but not to light. It reacts with acidic and alkaline solvents to form water-soluble salts. It is not very soluble in water but is soluble in methanol, ethanol and dimethyl sulfoxide. Natamycin is produced by the aerobic fermentation of bacteria *Streptomyces natalensis* and *Streptomyces chattanoogensis*, commonly found in soil. The final product is derived by the extraction process (see Natamycin at top of the page).

The antibiotic natamycin is a relatively non-toxic agent that controls basal rot (caused by *Fusarium oxysporum*) and moulds. Ornamental and food crops are thus protected. The fungicide acts by binding to **ergosterol** and inhibiting growth. This compound also controls yeast.

This compound is not acutely toxic. In rats, the LD_{50} is ≥2300 mg/kg, and doses of 500 mg/kg/day over 2 years did not cause any significant change in the growth of test rats. In humans, large daily doses of natamycin caused toxicity symptoms including nausea and vomiting. The WHO lists this fungicide as a Class III toxin (slightly hazardous).

Natamycin is registered in Canada, but not in the USA.

Neburon herbicide

Neburon, $C_{12}H_{16}C_{12}N_2O$ (1-butyl-3-(3, 4-dichlorophenyl)-1-methylurea), is a phenyl urea herbicide. It is a colourless crystalline substance soluble in water at 4.8 mg/l (at 20°C) and even less soluble in hydrocarbon solvents. Neburon is produced by the reaction of 3,4-dichlorophenyl isocyanate with *N*-butyl-*N*-methylamine (see Neburon at bottom of the page).

Neburon is a selective herbicide that inhibits photosynthesis at photosystem II. Once absorbed through the roots, it controls annual grasses and broadleaved weeds growing in legume crops, cereals, garlic, pear, beans and woody ornamentals.

The herbicide neburon is available as a flowable concentrate (SC) and a wettable powder (WP). The WHO classifies neburon as a Class U toxin (unlikely to be hazardous). However, at higher exposure levels, it can irritate the skin, eye and the respiratory tract. The acute oral LD_{50} in rats is >11,000 mg/kg. It is also not toxic to bees. Formulation labels should carry the signal word 'CAUTION'.

Neburon undergoes dealkylation of the terminal nitrogen atom, ring hydroxylation and degradation to dichlorodihydroxyaniline. The residual activity in soil is limited to approximately 3–4 months. It hydrolyses under basic conditions forming aniline derivatives.

Necrosis

The condition of an organism or a plant where cells or tissues die or degenerate for no apparent reason is called necrosis. Necrosis is associated with plant parts becoming darkened and wilted and is a reflection of nutrient deficiency or a hypersensitive response in defence of a pathogen attack. Necrosis is thus a

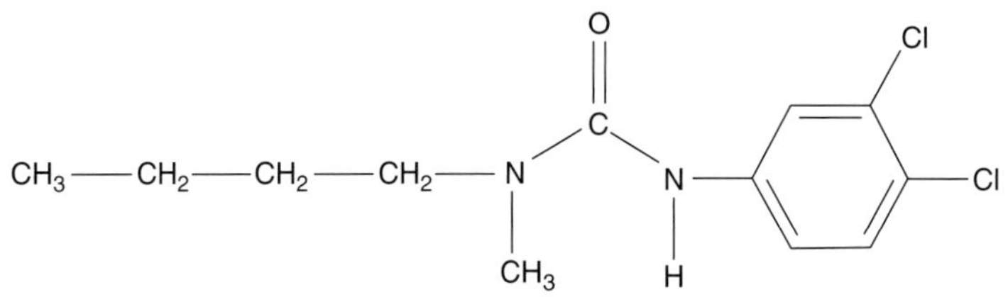

Neburon

symptom and not a disease. Such a condition makes the plant vulnerable to pathogen attacks.

Depleted soil, unsuitable soil pH and malnutrition lead to necrosis. For instance, insufficient levels of phosphorus, nitrogen, potassium, iron, **boron** and **magnesium** cause leaf necrosis (generally in older leaves) along with other symptoms like scorching, **chlorosis** and stunting in many plants (e.g. sugarbeet, apple, potato, raspberry, cabbage).

During a pathogen attack, a lot of energy of the plant is diverted towards developing resistance to the enemy, resulting in energy depletion, leading to necrosis of affected parts. Commonly, fungal or bacterial infections are visible as leaf lesions or grey/yellow/purple spots or streaks, which, if left unchecked, can cause necrosis. Since most pathogens, especially viruses, need a healthy plant to complete their life cycle, the plant often responds with a self-kill strategy, known as a hypersensitive response, where it triggers the death of host cells, thus restricting pathogen spread. Such an action is referred as programmed cell death.

Selecting disease-resistant plants and avoiding bacterial and fungal diseases controls necrosis. Rotating crops helps break pest cycles and consequently controls necrosis.

Net necrosis is an example of an aphid-transmitted virus disease called potato leaf roll virus (PLRV) disease. It affects the potato tubers by causing browning of the vascular system and leaf and stem necrosis occur. Other examples are the papaya droopy necrosis virus and papaya apical necrosis virus, where necrotic symptoms include plant drooping, yellowing and wilting of the leaf.

Neem as a biopesticide

Neem (*Azadirachta indica*) is a tree in the family Meliaceae and order Sapindales. Growing abundantly in tropical and semi-tropical regions, it is believed to be a native of the Indian subcontinent. The neem tree is known to have numerous pesticidal and medicinal properties. What makes it a valuable asset in crop protection are its selective anti-fungal, antibacterial, anti-feedant, antiseptic and insecticidal properties. All parts of the tree are said to have useful properties. Neem is also considered a natural fertilizer.

Neem derivatives control hundreds of pests worldwide, mainly by repelling and inhibiting their growth. Examples of pests include **whitefly**, **aphids**, **scale insects**, **mealybugs**, soil grubs, spider mites, locusts, **thrips**, beetle varieties, moth larvae, **nematodes** and the fungi that cause **powdery mildew** and black spot.

The various plant parts of neem contain nimbin, nimbinin, limonoids, azadirachtin, salanin and azadiradione. **Azadirachtin**, $C_{35}H_{44}O_{16}$, works as an insect growth regulator, thus preventing the insect larval stage from **moulting** and becoming an adult. The extremely bitter taste of neem also works as an anti-feedant thus repelling insects that suck sap or eat foliage. In households, it controls ants, bedbugs, cockroaches, houseflies, termites and mosquitoes. The CAS Registry Number of azadirachtin is 11141-17-6.

Neem cake, the resultant product of cold-pressing neem seed and kernels (Fig. N.1), is applied as organic manure on rice, cotton and sugarcane. The cake reduces alkalinity in soil, and produces organic acids on decomposition. It is observed to improve soil fertility, texture, aeration and water-holding capacity, while enhancing the rhizosphere microflora. Neem cake is commonly used as a fertilizer for its supply of nitrogen, phosphorus and potassium (NPK), although in limited quantities.

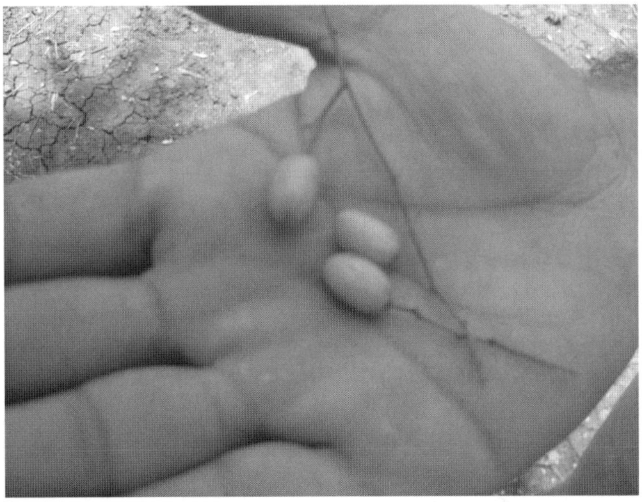

Fig. N.1. Neem seeds are used in making neem oil and cake. Courtesy of Kalyani Paranjape.

Neem oil (CAS Registry Number 8002-65-1) is the yellowish-brown oil pressed from the neem seed. It has a repulsive odour. Being biodegradable, it breaks down easily and quickly. It has proven to be non-toxic to mammals, birds and adult beneficial insects. However, care is recommended while spraying neem oil solution on **bees**, ladybugs and plant foliage, especially in direct sunlight. Neem oil is considered slightly toxic. It is registered in the USA and New Zealand.

Neem often works slower than other pesticides, and effectiveness is reduced in cooler climates. Neem does not persist in the environment and should be re-applied after rain.

Neem products include Azatin, Align and Turplex. Azadirachtin formulations (all neem products) are registered in India. (See also **Azadirachtin bioinsecticide**.)

Neem cake

See **Neem as a biopesticide**

Nematicide abamectin

See **Abamectin bioinsecticide**

Nematicide aldicarb

See **Aldicarb insecticide**

Nematicide aldoxycarb

See **Aldoxycarb insecticide**

Nematicide azadirachtin

See **Azadirachtin bioinsecticide**

Nematicide cadusafos

See **Cadusafos insecticide**

Nematicide carbofuron

See **Carbofuron insecticide**

Nematicide cloethocarb
See **Cloethocarb insecticide**

Nematicide copper hydroxide
See **Copper hydroxide fungicide**

Nematicide cytokinins (mixed)
See **Cytokinins as PGRs**

Nematicide dazomet
See **Dazomet fungicide**

Nematicide emamectin benzoate
See **Emamectin benzoate bioinsecticide**

Nematicide ethoprophos
See **Ethoprophos insecticide**

Nematicide ethylene dibromide
See **Ethylene dibromide insecticide**

Nematicide isazofos
See **Isazofos insecticide**

Nematicide nabam
See **Nabam fungicide**

Nematicide oxamyl
See **Oxamyl insecticide**

Nematicide pyraclofos
See **Pyraclofos insecticide**

Nematicides
Nematodes are soil-borne, worm-like creatures, which are considered serious pests of crops, although there are beneficial varieties as well. Nematodes belong to phylum Nematoda. Nematodes can cause severe damage to crops by colonizing their roots causing knots and galls. Examples of nematodes considered as crop pests include the *Meloidogyne* sp.

Chemical nematicides are substances used to kill or control nematodes in many ways. These belong to several chemical groups including carbamates, organophosphates, organothiophosphates, phosphonothioates and oximes. Nematicides are available as emulsifiable concentrates (EC), gases (GA) (fumigants) and as granules (GR). Most of these control agents are highly toxic compounds that have very low LD_{50} values. Examples of nematicide active ingredients (a.i.) are **fenamiphos, ethoprophos, methyl bromide, 1,3-dichloropropene, metam-sodium, dazomet, methyl isothiocyanate, terbufos** and **aldicarb**. A few (e.g. **chloropicrin** and methyl bromide) can only be used in a restricted way due to their **acute** and **chronic toxicity**. Their dosage rates range from 0.6 to 10 kg/ha.

Fumigant nematicides effectively act against nematodes and can control them far better than other formulations. Most fumigants (including fumigant nematicides) are considered toxic to the environment and may contaminate groundwater. Their application has to be carried out with a lot of care. Although the ideal conditions for most fumigants are dry weather, warm temperature and soils that are not rich in organic matter, nematicide granules work best when their incorporation into soil is followed by rain or irrigation, since moisture helps to activate the a.i.

There are nematicidal products that have a natural origin too, for example **avermectins** or **neem cake.** Nematophagous fungi such as *Paecilomyces* sp. provide biocontrol. Soil solarization and steaming also control nematodes. Crops protected by nematicides include bananas, grapes, tomatoes, potatoes and cucumber.

However, when specific nematode infestations become severe, the nematicidal effect becomes limited, because by then, the pathogen is well entrenched in the root system, or safely inside a gall. Higher doses of the applied chemical, can injure the plant root, or prove phytotoxic. To be effective, nematicides have to persist long enough for nematodes to be exposed to lethal concentrations. Nematode control may thus become an uphill task. (See also **Cultural practices for nematode management.**)

Nematodes
Nematodes belong to the phylum Nematoda and are tiny, usually unsegmented, multi-cellular, worm-like organisms that commonly live in all types of soil, throughout the world. The elongated, cuticled body and undulatory movements of nematodes make them popularly known as eelworms. They do not have special respiratory organs. The life cycle of nematodes starts with a single-celled egg. It goes through four juvenile stages (j1 to j4). While both females and males are required for reproduction in many species, many other species are hermaphrodites.

The nematode population is largest in the warm and moist season as well as in sandy and sandy-loamy soils. Nematodes can have a long dormant phase, helping them to survive harsh conditions and without a host plant. They reproduce when environmental factors such as soil, temperature and host are favourable.

Most of them are beneficial to the farmer owing to their nutrient-releasing property. However, parasitic nematodes cause great harm to crops and thus are a focus of attention for pesticide manufacturers and users. They can also affect animals and humans. Most nematode species live in the top few centimetres of the soil and require thin films of water to move about. They can move very slowly through the soil by themselves and are often transported by agricultural tools, erosion, landslides and flowing water.

Nematodes can be classified based on what they feed on: (i) those that eat decaying matter are called saprophytic nematodes; (ii) those that live on smaller nematodes, bacteria, algae or fungi are called predator nematodes; and (iii) those that attack crops and cause yield losses and sometimes even fungal infections are called parasitic nematodes. The mouthparts of nematodes are suited to their food habits, and hence they have either a tube-like structure that can suck up the inner parts of microorganisms, or a piercing needle-like structure called a stylet or spear that punctures the prey. Some are entomopathogenic in nature and can control some species of grubs, thrips and beetles (Fig. N.2).

Parasitic nematodes that attack roots from outside and eat root tips and cortical tissues are called ectoparasitic nematodes.

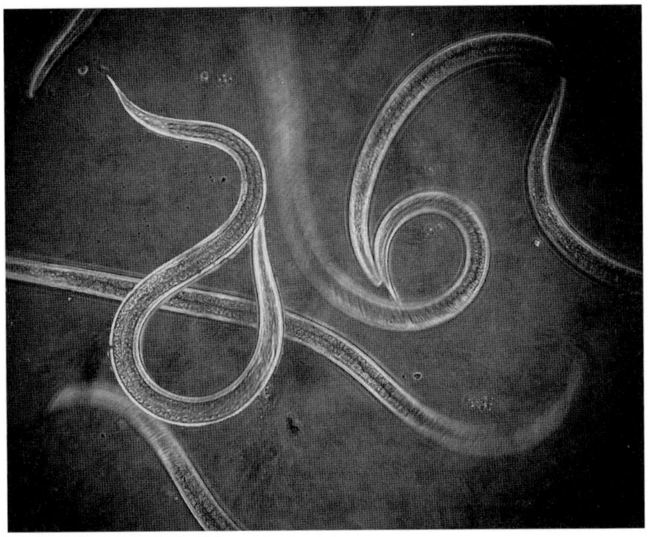

Fig. N.2. Entomopathogen of scarab larvae in infective juvenile stage. Courtesy of David Cappaert, Michigan State University (Bugwood.org), reproduced with permission from D. Cappaert.

Fig. N.3. Golden yellow cyst of *Globodera rostochiensis* on potato root. Courtesy of Theirry Vrain, Agriculture and Agri-Food Canada (Bugwood.org), with permission from Jonathan D. Eisenback.

Fig. N.4. Nematode on *Hosta* spp. Courtesy of Department of Plant Pathology Archive, North Carolina State University (Bugwood.org), with permission from Lane P. Tredway.

Endoparasitic nematodes enter the root tissues of a plant by piercing it and feeding on the root cortex where the root cells are killed. They move through the cortex, the injured and exposed areas forming lesions on the root and increasing the chances of a fungal infection. A nematode problem can be diagnosed by actually estimating the nematode population in the soil.

Parasitic nematodes may either be migratory (moving through soil) or sedentary, depending on the stage in their life cycle. Some sedentary female nematodes attach themselves to the host root, injecting a growth promoting substance into it. They then feed off the nutritious source, and swell into a bean-shaped structure. Sting nematodes (*Belonolaimus* spp.), stubby root nematodes (*Trichodorus* spp.) and awl nematodes (*Dolichodorus* spp.) are examples of ectoparasitic nematodes. The sedentary endoparasite, in its second juvenile (j2) stage is most dangerous to plant health since it has to feed itself vigorously to develop beyond this j2 stage.

The parasitic behaviour of plant nematodes varies widely. There are many species that attack plant roots (e.g. the root knot/gall-forming *Meloidogyne* and the cyst-forming *Globodera* and *Heterodera*, which are sedentary endoparasites) (Fig. N.3).

There are a few nematodes that damage plant parts above the ground. For example *Anguina tritici*, *Aphelenchoides* sp. and *Ditylenchus* sp., affect buds, seeds and leaves by forming galls and lesions on ornamentals and strawberries. Figure N.4 shows a plant affected by a foliar nematode.

Some other common types include citrus nematodes and ring, lesion and burrowing nematodes. *Radopholus citrophilus*, a burrowing nematode, is a migratory endoparasite and affects citrus crops. The effects of nematode attack are reflected in altered root growth, root tissue injury and fungal infection.

Many species of ring nematodes, which have a ring-like appearance, are ectoparasites that suck plant nutrients from plant roots with their piercing mouthparts. Some species in the genera *Criconemoides* and *Macroposthonia* are major pests of crops. Crop infestation by ring nematodes causes intense pruning of feeder roots and retardation of growth. For example *Macroposthonia xenoplax* is a serious pest on almond, apricot, plum, walnut and peach. Figure N.5 is a microscopic view of *M. xenoplax*.

Substances that are used to kill nematodes are called **nematicides**. They can be chemical or biological in nature. Chemical nematicides are commonly classified on the basis of their chemical structure. They can be halogenated hydrocarbons, methylisocyanate derivatives, carbamates, oxime carbamates, antibiotic, organophosphorus, phosphonothioates and miscellaneous nematicides.

Halogenated hydrocarbons and methylisocyanate derivatives act as **fumigants**, while organophosphate and carbamate compounds are commonly non-fumigants. However, given the extreme toxicity of methylisocyanate and hydrocarbons, many of these compounds are not registered in most countries. Some examples of chemical nematicides are **methyl bromide, chloropicrin, DCIP, GY-81, methyl isothiocyanate, dazomet, carbon disulfide, cloethocarb, ethylene dibromide, metam, cadusafos, fosthiazate, ethoprophos, fenamiphos, terbufos** and **1,3-dichloropropene**.

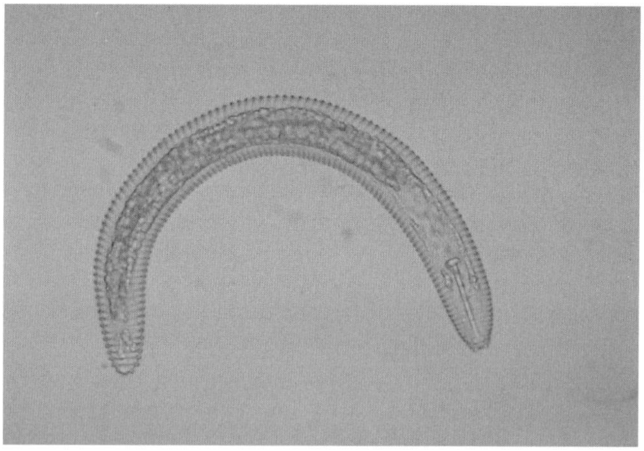

Fig. N.5. Microscopic view of ring nematode *Macroposthonia xenoplax*, a pest species of several crops. Courtesy of H.J. Larsen (Bugwood.org), with permission.

Interestingly, endoparasites may easily escape the effects of fumigants and also some nematophagous microorganisms in the soil (for want of physical contact). (See also **Cultural practices for nematode management**.)

Neonicotinoid insecticides

See **Acetylcholine receptors**

Nereistoxin

Nereistoxin (NTX) is a naturally occurring toxin, which can be sourced from *Lumbriconereis heteropoda* and some other species of sea annelids. It is a compound of significance since a few insecticides have been synthesized from it. Its principle site of action is the cholinergic synapses. The IUPAC name is 4-*N*,*N*-dimethylamino-1,2-dithiolane.

Insecticides developed from nereistoxin include **cartap** and **bensultap**.

Neuropeptides and insect control

Neuropeptides are molecules that are released and emitted by neurons, to perform specific tasks and to help communicate with other neurons. Most vertebrates and invertebrates have several genes that encode precursors of neuropeptides. These chemicals are present in the insect nervous system and have a vital role in regulating almost every aspect of insect life (e.g. growth, metamorphosis, pairing, mating, reproduction, lipid metabolism). If exploited wisely, these neuropeptides can be successfully used for insect control.

Prothoracicotropic hormone (PTTH) and adipokinetic hormone (AKH) are examples of two groups of neuropeptides. PTTH, the first of such hormones discovered by man, is critical in insect development stages like **moulting** and pupation. The AKH, present in a desert locust, is said to interact with the flight muscle, regulate metabolism of lipids and sugars, and help release diglycerides. This hormone is also harnessed for use against pest vectors that cause disease to man and animals.

Proctolin, present in many insect orders, has a role in skeletal muscle contraction and relaxation, apart from acting on other insect systems. A gut peptide, proctolin has diverse actions and can be used as an agent to control pests.

The use of insect neuropeptides has a high potential in IPM programmes due to their high selectivity and low dosage requirements.

Neutral Bordeaux

Neutral Bordeaux is another term for **Bordeaux mixture fungicide**.

Niclosamide molluscicide

Niclosamide, $C_{13}H_8Cl_2N_2O_4$ (2′,5-dichloro-4′-nitrosalicylanilide), is a yellow crystalline solid substance, soluble in water at 110 mg/l (at pH 9.1), ethanol and diethyl ether. It is hydrolysed by concentrated acids and alkalis, and also in aqueous solutions, with a DT_{50} of 1 week. It is produced by reaction of 2-chloro-4-nitroaniline with 2-hydroxy-5-nitrobenzyl chloride in the presence of a base. The CAS Registry Number is 50-65-7.

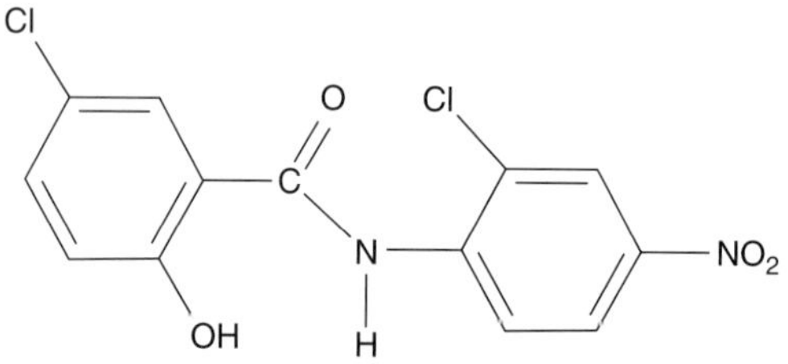

Niclosamide

Niclosamide, used against slugs and snails (especially the green apple snail) in paddy fields, is most effective between pH 5 and 9, subject to the concentration of Ca and Mg. A derivative of niclosamide, niclosamide-olamine, with molecular formula of $C_{15}H_{15}Cl_2N_3O_5$, and solubility in water of 0.1 g/l, is also a molluscicide. Niclosamide can be used to kill freshwater snails that are intermediate hosts for the human pathogens that cause schistosomiasis and fascioliasis.

This molluscicide is classified as a Class U toxin (unlikely to be hazardous) by the WHO. It is highly toxic to fish (the 96 h LC_{50} being as low as 0.1mg/l in some species). It is also toxic to frogs, salamanders and plankton. An acute oral LD_{50} in rats is >5000 mg/kg.

Niclosamide rapidly degrades in pond sediment, with a DT_{50} of between 1.1 and 2.9 days. It is registered for use in Canada, New Zealand, the Philippines and Vietnam.

Nicobifen fungicide

Nicobifen is another term for **boscalid** fungicide.

Nicosulfuron herbicide

Nicosulfuron, $C_{15}H_{18}N_6O_6S$ (1-(4, 6-dimethoxypyrimidin-2-yl)-3-(3-dimethylcarbamoyl-2-pyridylsulfonyl) urea), is a pyrimidinylsulfonylurea herbicide. Its solubility in water is 0.04 g/100 g at 25°C. Nicosulfuron is soluble in acetone, acetonitrile, chloroform, hexane, dimethylformamide, dichloromethane, ethanol and toluene. The herbicide is stable under normal temperatures and pressures. It is produced by the reaction of 2-sulfamoylchloride-3-*N*-dimethylaminocarboxypyridin with 2-amino-4, 6-dimethylpyrimidin. The CAS Registry Number is 111997-09-4.

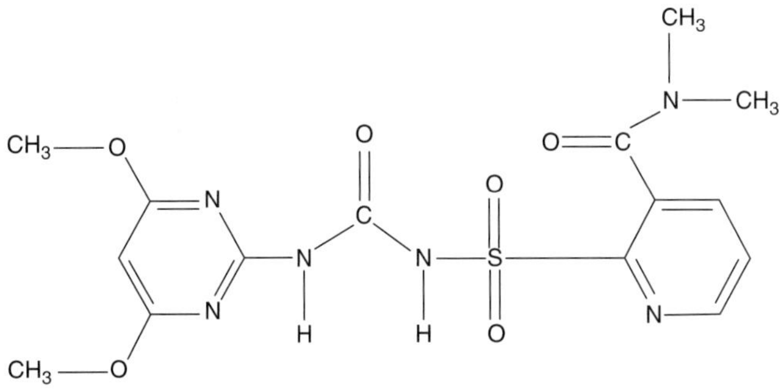

Nicosulfuron

Nicosulfuron is a member of the sulfonylurea family of herbicides. It controls weeds by inhibiting the plant enzyme acetolactate synthase (ALS). This enzyme blocks the production of the amino acids, valine and isoleucine, essential building blocks of proteins and other plant components. Susceptible weeds include **Johnson grass**, quack grass, foxtails, **barnyard grass**, pigweed and many others. The main crop protected is maize.

The herbicide nicosulfuron comes in a 75% water-dispersible granule (WG) formulation. It may be tank mixed with a limited number of other compounds. The trade names for products containing nicosulfuron include Accent, Challenger, Milagro, Mistral and Motivel.

The WHO classifies nicosulfuron as a Class U toxin (unlikely to be hazardous) and it is a general use pesticide (GUP) according to the EPA. The products containing nicosulfuron must bear the signal word 'CAUTION' on their label.

The acute oral LD_{50} for technical nicosulfuron is >5000 mg/kg b.w. in rats. Nicosulfuron does not irritate the skin (the acute dermal LD_{50} is >2000 mg/kg in rats), but may moderately irritate the eye. The acute inhalation LC_{50} is >5.9 mg/l. It is slightly toxic to birds, and practically non-toxic to freshwater fish, invertebrates and honey bees. The NOEL is 125 mg/kg b.w./day. The ADI is 1.25 mg/kg b.w./day. It is a potential groundwater contaminant.

Nicosulfuron is registered in Canada, the UK, the USA and South Africa.

Nicotine bioinsecticide

Nicotine, $C_{10}H_{14}N_2$ ((*S*)-3(1-methylpyrrolidin-2-yl) pyridine), a hygroscopic, oily, colourless, alkaloid liquid, is a botanical insecticide. Miscible with water and organic solvents, nicotine forms salts with acids that are usually solid and water soluble. It has a CAS Registry Number of 54-11-5.

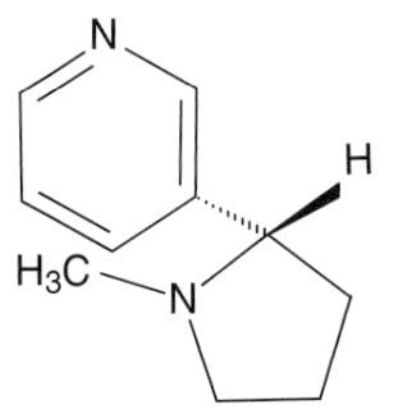

Nicotine

Nicotine is a naturally occurring, non-systemic insecticide, derived mainly from the tobacco plant (*Nicotiana tabacum*). Other plants containing nicotine include *Nicotiana rustica* and common milkweed (*Aesclepias syriaca*). With a respiratory action, nicotine acts as an inhibitor of the **acetylcholine receptor** (nicotinic acetylcholine receptor or nAChR) by interfering with the transmitters between the muscles and the nerves. Another similar insecticidal alkaloid is nornicotine. Nicotine bioinsecticide primarily targets piercing and sucking insects such as aphids, thrips and whiteflies that attack vines and a wide range of vegetables and flowering plants. Being a nerve toxin, it causes tremors and disorientation in the pests.

This bioinsecticide is formulated as a liquid concentrate of nicotine sulfate, which is diluted before use. It is available as Nicabate, Black leaf, Nicobrevin and Nicorette. Warm weather is preferred for application. Applications may be frequent and as required.

Nicotine causes irritation to the skin, especially in the presence of nicotine dust. Nicotine can penetrate the skin and enter the bloodstream and may prove fatal, even to humans (with around 60 mg of nicotine being lethal to a person, especially in the form of tobacco patches and sprays). The acute oral LD_{50} and acute dermal LD_{50} of nicotine in rabbits and rats is 50–60 mg/kg, while in mice, the acute oral LD_{50} is 3 mg/kg. The WHO has classified nicotine as a Class Ib toxin (extremely hazardous). Nicotine is not hazardous to birds and fish. It can be toxic to bees, but it repels them, so deaths are not significant. Symptoms of nicotine poisoning include salivation, vomiting and convulsions. Wearing appropriate PPE is highly recommended. Based on acute toxicity studies, nicotine has been considered a PAN Bad Actor chemical.

Nicotine is registered in many countries including the USA, Canada, Cambodia and Finland. Given its rapid disintegration, its use is permitted in organic farming.

The MRL of nicotine in the USA for mushrooms is 0.01 mg/kg.

Nicotinic acetylcholine receptor (nAChR)

See **Acetylcholine receptors**

Nitenpyram insecticide

Nitenpyram, $C_{11}H_{15}ClN_4O_2$ ((*E*)-*N*-(6-chloro-3-pyridylmethyl)-*N*-ethyl-*N'*-methyl-2-nitrovinylidenediamine), is a kind of neonicotinoid insecticide with veterinary applications. It acts on the nicotinic **acetylcholine receptor** (nAChR) of insect nervous systems. It is also considered to be a part of the nitromethylene and pyridylmethylamine group of insecticides. This yellow crystalline powder is soluble in water and incompatible with strong oxidizing agents. It has a CAS Registry Number of 150824-47-8.

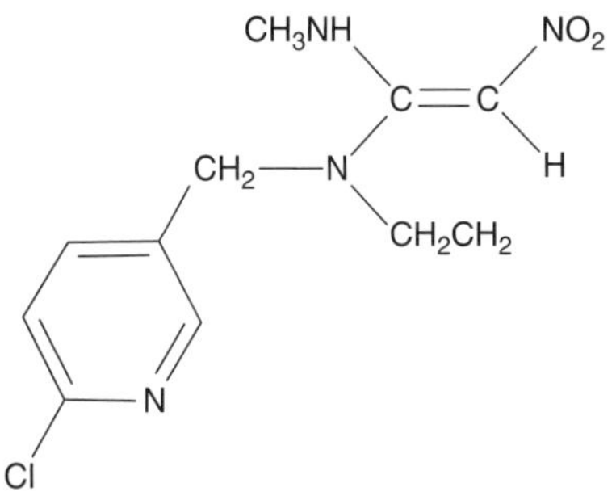

Nitenpyram

The insecticide nitenpyram, an effective agent against ectoparasites, is an nAChR agonist with systemic translaminar activity. Nitenpyram blocks neural transmission through the CNS, effectively killing the pest. It acts on fleas on animals such as cats, dogs and livestock and this is the main use of nitenpyram. It has some limited additional activity on sucking insects on rice and is sometimes used on other crops.

A common brand name of nitenpyram is Capstar. The insecticide is usually supplied as a dustable powder (DP), granules (GR), drops and impregnated collars and tablets.

The effect of nitenpyram is not long lasting since it has insignificant residual action. However, symptoms of acute exposure include facial swelling, diarrhoea and vomiting.

The oral LD_{50} of nitenpyram is 867 mg/kg in mice, while in rabbits it is >3 g/kg. The acute LD_{50} in mallard ducks is 1124 mg/kg. It is not toxic to fish or to *Daphnia magna* (the acute 96 h LC_{50} in rainbow trout is 10 mg/l). The acute 48 h LD_{50} in honey bees 0.138 µg per bee.

Nitenpyram is registered for use in Australia and New Zealand but not in the USA.

Nithiazine insecticide

Nithiazine, $C_5H_8N_2O_2S$ ((*E*2)-2-nitromethylene-1,3-thiazinane-tetrahydro-2-(nitromethylene)-2H-1,3thiazine), is a nitromethylene nicotinoid insecticide. It has a CAS Registry Number of 58842-20-9.

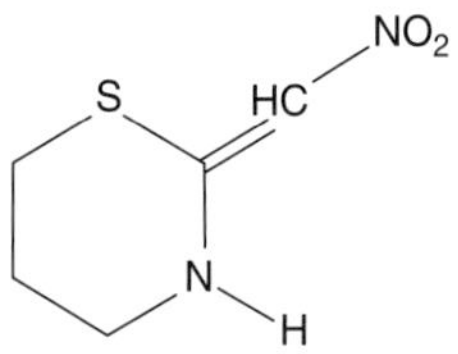

Nithiazine

Insecticide nithiazine is used to control ectoparasites on poultry, farm animals and pets in stables and sheds. It is also used in industrial areas, rubbish pits and rest rooms.

Nithiazine was not very acutely toxic by the dermal route of exposure in laboratory animal studies, but was moderately toxic by the oral exposure route. It is observed to cause loss in weight and temperature in rats at doses of 30 mg/kg b.w. It causes severe eye irritation and thus care is recommended during application/handling.

Nithiazine is registered for use in the USA.

Nitrapyrin bactericide

Nitrapyrin, $C_6H_3Cl_4N$ (2-chloro-6-trichloromethylpyridine), is a bactericide and bacteriostatic agent. It is a white crystalline solid, soluble in water at 40 mg/kg (22°C), propanone, methyl dichloride and dimethyl benzene. It is incompatible with aluminium, magnesium or nitrogen and nitrogen fertilizers. Its CAS Registry Number is 1929-82-4.

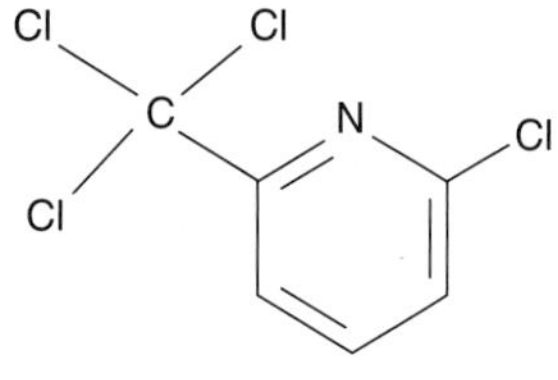

Nitrapyrin

Nitrapyrin, a pre- and post-plant emergence, and a post-harvest compound, acts by inhibiting *Nitrosomonas* spp. that convert ammonium nitrogen to nitrite nitrogen. It makes nitrogen available to maize, cotton, rice, sorghum, strawberries and wheat.

This bactericide is available as an emulsifiable concentrate (EC) formulation. It is broadcast in the soil, applied in bands or as top dressing. The maximum single application rate is up to 1 kg/ha on sorghum, wheat and maize.

According to the WHO, nitrapyrin comes under toxicity Class III (slightly hazardous). It is slightly to moderately toxic to rats and mice, with the acute oral LD_{50} ranging from 713 to 1231 mg/kg. Its acute percutaneous LD_{50} in rabbits is 2830 mg/kg. However, it is a carcinogen and a developmental toxin. It is also a potential groundwater contaminant. It is thus classified as a PAN Bad Actor chemical, according to the PAN Pesticide Database.

Nitrapyrin is hydrolysed in soil to 6-chloropyridine-2-carboxylic acid (6-chloropicolinic acid or 6-CPA), which is absorbed by plants. 6-CPA degrades via hydroxylation and microbial mineralization. The two are mobile in mineral soils.

Nitrogen as a macronutrient

See **Plant nutrient excesses and deficiencies**

Nitrothal-isopropyl fungicide

Nitrothal-isopropyl, $C_{14}H_{17}NO_6$ (diisopropyl 5-nitroisophthalate), is a yellow crystalline substance, soluble in water (2.7 mg/l), chloroform and acetone (both above 100 g/l). The fungicide hydrolyses under strong alkaline conditions. Its CAS Registry Number is 10552-74-6. It is produced by the reaction of 5-nitroisophthaloyl chloride with isopropanol in an inert solvent.

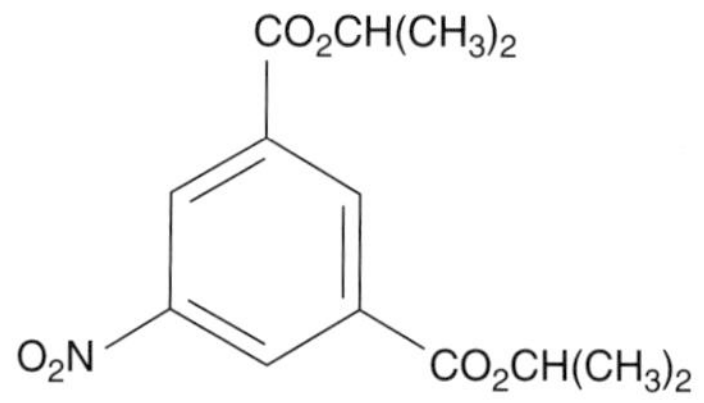

Nitrothal-isopropyl

Nitrothal-isopropyl, a non-systemic, contact fungicide, is used for the control of **powdery mildews** and **scab** which affect vines, vegetables and apples.

The fungicide is commonly formulated as a wettable powder (WP) and may be combined with other fungicidal compounds.

The acute oral dose of nitrothal-isopropyl in rats is >6400 mg/kg (slightly hazardous), while that in bees is >100 µg per bee (moderately toxic). The EPA has classified this compound as a Class IV toxin while the WHO lists it under Class U (unlikely to be hazardous). It is moderately toxic to aquatic creatures with the acute 96 h LC_{50} in rainbow trout being 0.56 mg/l. It is not toxic to earthworms.

In rats, nitrothal-isopropyl is eliminated within 1 week of ingestion, while in plants the metabolites are 3-carbo-isopropoxy-5-nitrobenzoic acid and 5-nitroisophthalic acid. The DT_{50} in soil is up to 12 days.

Nitrothal-isopropyl is registered for use in South Africa, Australia and New Zealand but it is not registered by the USA.

NOAEL (NOEL)

NOAEL stands for no observed adverse effect level and denotes the quantity of a toxic substance that an organism is exposed to, which does not cause any adverse effects, compared with the control population. NOAEL is also commonly referred as NOEL (no observed effect level).

Experimental animals used for such toxicity assessment are rats, mice and dogs. To arrive at a dose–response relationship and risk assessment, these animals are exposed to large doses of chemicals.

Risk determination involves: (i) identification of the hazard; (ii) quantification of the exposure; and (iii) extrapolation of laboratory results to field situations. However, all risk assessment techniques are meant to determine if the risk–benefit ratio is favourable to the public.

Pesticide residues detected on crops at the time of harvest ought to be lower than or equal to the tolerance level. Residual levels commonly decrease further during storage and transit and then further by washing, peeling and cooking.

Nomenclature for easy reference

For quick reference, plant pathologists classify most disease-causing organisms using their Latin binomial, which consist of two scientific names. The first is the genus, while the second, the species. These words, although in Latin, help in describing the type of disease, based on the pathogen causing it. Once named, it makes disease knowledge easy to document and control. This system is followed for fungal and bacterial pathogens.

Besides the scientific names, there are common names that describe the disease by its symptoms. Agriculturists normally use common names when describing diseases. For example, the common name '**powdery mildew**' is a disease that is caused by various pathogens and affects many plants. They are all so called because of the similarity in symptoms. However, the pathogen causing powdery mildew on apple is *Podosphaera leucotricha* and the one affecting cucumber is *Sphaerotheca fuliginea*. A scientist is assured that *P. leucotricha* will not normally affect cucumber. Other examples of common names are **wilt**, rot and **scab**.

In the case of naming viral pathogens, the naming system followed includes the first-identified host of the pathogen and the disease symptom. The name is usually in English. For instance, the cucumber mosaic virus was so named because it was first identified on cucumber and forms mosaic patterns. At present, this naming system does not seem to be very reliable since the same cucumber mosaic virus is known to attack other plants (e.g. beans). A more precise classification system based on viral chemical make up, shape and infection is needed for better categorization of these pathogens.

Non-ionic surfactants

See **Surfactants**

Non-parasitic diseases

A disease caused by an abiotic factor or a non-parasitic organism is called a non-parasitic disease. Such causal agents are found in soil, the environment and in the chemicals applied to the plant or the farm. Non-parasitic diseases are not infectious and the symptoms reflect some physiological disorders in plants. Some examples of causes and symptoms are described below.

The most important factor of non-pathogenic disease is the soil. Since plants depend so much on soil, its physical and chemical conditions affect plant growth. Soil can be for example sandy, clayey or loamy and any unfavourable soil type can affect plant growth negatively. The aeration of a soil also affects the growth of vegetation and the rhizosphere. The capacity of a soil to hold water also affects plants. Soil pH regulates the availability of plant nutrients and too much alkalinity or acidity can hamper growth.

Nutrient deficiency or excess is equally harmful to plants. Nutritional balance is important for effective metabolism in plants and it is provided by **macronutrients** (i.e. major elements) and micronutrients (i.e. minor or trace elements). Macronutrients are required in large quantities and micronutrients, in minute quantities, which does not mean that one is more important than the other. Deficiencies and excesses of these elements affect plant growth adversely and even cause toxicity. For example, copper may become toxic at a concentration above 0.5 ppm. Soil oxygen should be at least of 2.85% for optimum root growth, below this level leaves can become chlorotic and growth of the plant is hampered. Excess potassium induces a deficiency of calcium, magnesium or iron. The diseases caused by such abiotic factors are referred to as non-parasitic diseases. Nutrient imbalance reduces the resistance of plants against bacterial and fungal diseases.

Water requirements of different crops vary with growth stages. Water stress results from deficits or excesses of water supply. This affects the normal process of water absorption, transpiration, photosynthesis and nutrient uptake by plants. Blossom end rot of tomato is caused by water deficit. Highly compact soil hampers aeration and water-holding capacity, resulting in slow plant growth. Excess water or a waterlogged area affects plant growth. Tip brown disease of rice is caused by waterlogged conditions, which reduce the amount of oxygen available to the roots.

Unfavourable temperatures cause physiological problems in plant life. Very low temperature produces freezing injury in potato crops and frost injury to a number of winter crops. Low temperatures lead to **chlorosis** in sugarcane and some cereals. Hailstorms damage many crops (e.g. papaya). High temperatures can injure fruits and vegetables, and scorch leaves and burn their tips. For shade-loving plants like pineapple, even excess and direct light is injurious. On the other hand, plants that grow in the absence of light cause the stems and leaves to grow much more rapidly than normal, which is called etiolation. Vegetable crops such as potato, tomato, cabbage, onion, banana and soybean are very sensitive to extreme temperatures.

Toxins in the environment can have a negative effect on plant life. Poisonous gases in the form of industrial fumes, smoke, CO, SO_2, ethylene, Cl_2, F_2, NH_3, O_3, NO_2 and peroxyacetyl nitrate cause irreparable damage to plant parts. Liquid and solid effluents from factories and from pesticide handling sites are taken up by plants and cause abnormal stunting, necrosis, discoloration and even death. Similar effects are seen in the depletion of the ozone layer. Mango orchards are affected if they are close to brick kilns.

Incomplete decomposition of crop residues produces toxic products such as CH_4, H_2S and organic acids (e.g. acetic, lactic, formic and butyric acids). Roots are commonly injured, especially when soils are poorly aerated, clayey and waterlogged.

Many pesticide active ingredients (a.i.) are phytotoxic. They enter the food chain when applied to crops. Phytotoxicity increases with persistence. Many chlorinated and arsenical compounds as well as auxin-type herbicides can be injurious to crops. Some examples of strong pesticide a.i. are **2,4-D**, **MCPA**, simazine and **atrazine**. Symptoms of toxicity include leaf thickening, crinkling, in-rolling of leaves, chlorosis and dieback symptoms.

A few non-parasitic diseases occur in post-harvest storage and transit of vegetables and fruits. These diseases are caused by poor ventilation and insufficient storage conditions. The black heart of potato tubers, brown heart of apples and apple scald diseases, may be cited as examples of this, where no pathogen can be isolated (Figs N.6 and N.7). Some volatile substances of mild toxicity are produced that accumulate in the harvested materials during storage.

Fig. N.6. The brown heart of apple. Courtesy of Kalyani Paranjape.

Fig. N.7. The black heart of potato. Courtesy of Kalyani Paranjape.

Non-selective herbicides

Chemicals used for controlling or destroying weeds or herbs are called herbicides. There are two types of herbicides: (i) selective; and (ii) non-selective. The selective herbicide eliminates weeds without injuring the crop. Non-selective herbicides are toxic to all the plants that are in contact with it. These herbicides are to be applied with care, to avoid crop injury. Examples include **glyphosate** and **paraquat**. Such compounds are used as **soil sterilants** and as silvicides. Selective herbicides are selective to a kind of target weed, say grasses, or annuals or perennials on rice crops. An example of a selective herbicide is **2,4-D** for broadleaved weeds.

Non-systemic fungicides

Non-systemic fungicides (also called contact fungicides or residual protective fungicides) are compounds that do not penetrate into the plant tissues but form a protective barrier on the plant surface, and thus prevent the fungus from entering the host. Most non-systemic fungicides react with the thiol groups present in the enzymes of the fungus but may also have other sites of action. Non-systemic fungicidal groups include inorganic sulfur or copper and organic mercury compounds, examples being **captan**, dithane and **fludioxonil**. Systemic fungicides are absorbed and move within the plant tissues. Compounds might differ in the degree of systemicity. **Benzimidazole** and **dicarboximide** fungicides are systemic in nature (e.g. **benomyl**, **carbendazim**, mecarbinzid, **fluoroimide**). The movement is mostly upwards, except **fosetyl-aluminium**, which is bidirectional.

Norflurazon herbicide

Norflurazon, $C_{12}H_9ClF_3N_3O$ (4-chloro-5-methylamino-2-(α,α,α-trifluoro-*m*-tolyl)pyridazin-3(2*H*)-one), is a pyridazinone herbicide. The herbicide norflurazon is a member of the carotenoid inhibitors (there are two pigment-production inhibiting groups, namely carotenoid inhibitors and protoporphyrinogen oxidase (**PROTOX**) inhibitors). Norflurazon is a buff, non-corrosive odourless, crystalline solid, soluble in water at 28 ppm, acetone (at 5 g/100 ml) and ethyl alcohol (14.2 g/100 ml). It has a CAS Registry Number of 27314-13-2.

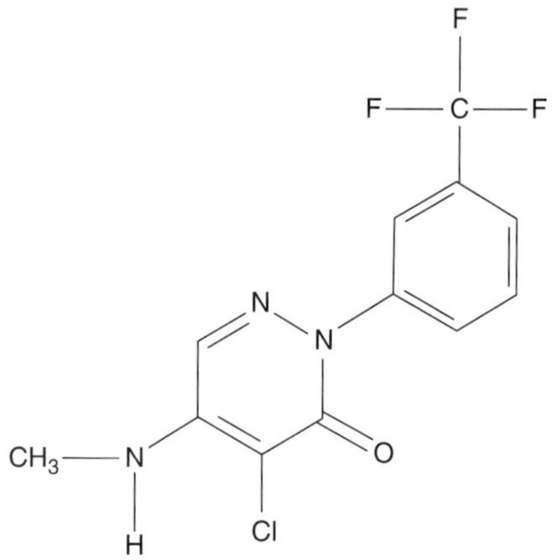

Norflurazon

Norflurazon is registered as a selective pre-emergence herbicide used to control annual grasses and broadleaved weeds in cotton, soybeans, nuts (e.g. almonds and walnuts) and a variety of citrus and non-citrus fruits. Susceptible weeds absorb the herbicide through the roots and inhibit carotenoid biosynthesis and chlorophyll photodegradation. This leads to necrosis and death of the target pests.

The herbicide is diluted in water and broadcast, aerially sprayed or band applied. Application doses range from 2.5 to 9 kg/ha, according to the severity of the problem.

Norflurazon has an acute oral LD_{50} of 9000 mg/kg in rats. It does not irritate the eye or skin in normal use. It is also not toxic to birds. The ADI for norflurazon is 0.02 mg/kg b.w./day. The WHO has listed this chemical as a Class U toxin (unlikely to be hazardous).

A major portion of the herbicide is eliminated in rat urine and faeces. Degradation involves demethylation and hydroxylation processes, involving the replacement of chlorine on carbon-4 of the pyridazinone ring, and conjugation, through sulfur introduced at carbon-4 of the pyridazinone ring. Norflurazon persists in soil, with a half-life ranging from

38 to 731 days, depending on the soil content. Norflurazon presents a potential for groundwater contamination.

The MRL of norflurazon for almonds, apricots, apples and soybeans is 0.1 ppm, while for soybean hay, it is 1.0 ppm and for groundnuts, it is 0.2 ppm.

Norflurazon is registered for use in Australia, the USA and New Zealand. However, it is not allowed to be used in most of the EU.

N-Phenylphthalamic acid as a PGR

N-Phenylphthalamic acid, $C_{14}H_{11}NO_3$ (*N*-phenylphthalamic acid), is a greyish white powder, soluble in water, methanol, ethanol, acetone and acetonitrile. It has a CAS Registry Number of 4727-29-1. It hydrolyses in strong acids and bases and decomposes slowly above 100°C and in UV light.

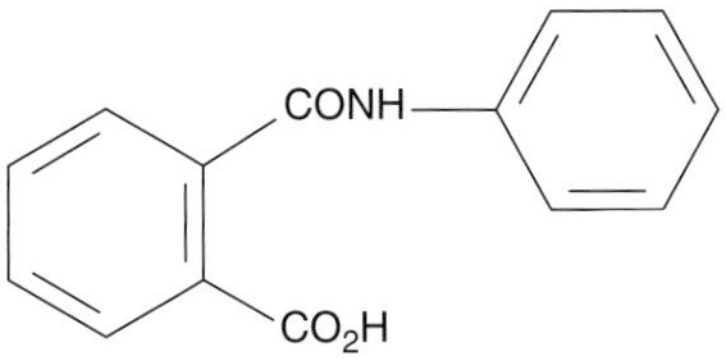

N-Phenylphthalamic acid

N-Phenylphthalamic acid, as a plant growth regulator, helps in effective pollination, by keeping the stigma healthy and viable. Thus, crops have better fruit setting, and higher yield. Plants also become more resistant to frost. This PGR is used on tomatoes, peppers, beans, peas, soybeans, lucerne, rape, lupin, sunflower and many fruits.

The PGR is available as wettable powder (WP), with a common trade name being Nevirol. The acute oral LD_{50} in male rats is 8837 mg/kg. It is non-toxic to bees, fish and other aquatic organisms and other non-target organisms.

Nuarimol fungicide

Nuarimol, $C_{17}H_{12}ClFN_2O$ ((*RS*)-2-chloro-4′-fluoro-α-(pyrimidin-5-yl)benzhydryl alcohol), is a pyrimidinyl carbinol fungicide. It has a CAS Registry Number of 63284-71-9. It is a colourless, crystalline substance, soluble in water (26 mg/l), acetone (170 g/l), methanol, acetonitrile, benzene and chloroform. It is stable at 52°C, but quickly decomposes in UV light. It is produced by the reaction of 5-bromopyrimidine with 4-fluorobenzoyl-2-chlorobenzene in the presence of butyllithium.

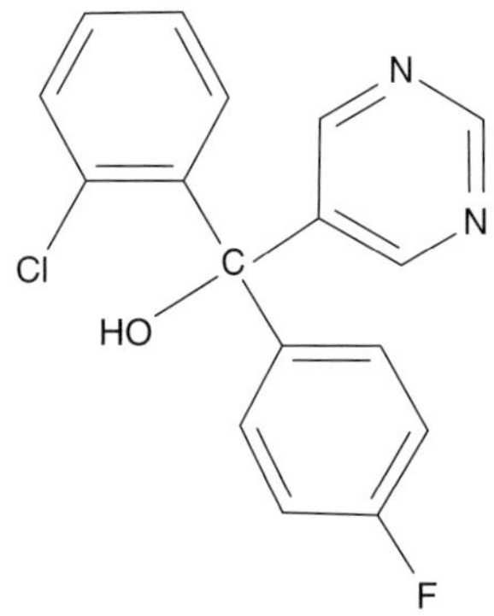

Nuarimol

Nuarimol, a systemic foliar and seed fungicide, has curative and protective properties. It inhibits the biosynthesis of **ergosterol**. It effectively controls *Cercosporella* spp., *Septoria* spp., *Ustilago* spp., **scab**, leaf spot and **powdery mildews** in a range of crops such as cereals, stone fruit, cucurbits and apples.

Fungicide nuarimol is usually combined with other compounds like **maneb**, **sulfur**, imazalil and **mancozeb**, the combination resulting in additional insecticidal and acaricidal properties.

Nuarimol is formulated as an emulsifiable concentrate (EC), suspension concentrate (SC), wettable powder (WP) and seed treatment. Trade names include Trimidal and Gauntlet.

Nuarimol is categorized as a Class III toxin (slightly toxic), since toxicologically, the acute oral LD_{50} in rats is in the range of 1250–2500 mg/kg, while in mice, it is much higher. The value in bobwhite quails is 2000 mg/kg. It is practically non-toxic to bees and fish and is quite safe on earthworms and other beneficial microorganisms. The NOEL for a 2 year period in rats and mice is 50 mg/kg. Exposure to nuarimol causes both acute and chronic effects. Residues on food have been disallowed in many countries since nuarimol is observed to cause birth defects and cancer.

Nuarimol is eliminated rapidly after ingestion in the case of rats. In plants, soil and water, photodegradation products and microbial products (that are photodegraded) are produced.

Use of nuarimol is severely restricted across the world and it is not registered in the EU or the USA.

Nuclear polyhedroviruses (NPVs)

See **Biocontrol of insects – baculoviruses**

O

Occupational Safety and Health Administration (OSHA)

The Occupational Safety and Health Administration (OSHA), an agency under the jurisdiction of the USA, is responsible for establishing and enforcing standards for exposure to harmful materials in industrial environments affecting the health and well-being of industrial personnel. Its standards are based on the **threshold limit values (TLV)** recommended by the American Conference of Governmental Industrial Hygienists. The website (http://www.osha.gov) provides more information.

Odorant

See **Odour**

Odour

Odour is the property of a substance, perceived when its molecules reach the olfactory nervous system of the recipient, in the form of a physiological sensation. Living things can perceive even minute amounts (less than 1 μg) of volatile compounds. A concentration of 1.7×10^7 molecules/cm^3 is necessary for detection by the olfactory system. The nose is said to be 100 times more sensitive than the best analytical apparatus for detecting threshold odour values.

Among the most powerful unpleasant odours are those of organic sulfur compounds, especially ethyl merchantman. Organic substances having a pleasant odour (e.g. **essential oils**) are broadly designated as aromatic, regardless of the chemical nature.

Characteristic odours serve to identify or caution industrial workers and chemical handlers about toxic and noxious gases. A substance with a distinctive odour, deliberately added to give a particular odour to a product is called an **odorant**. Odorants are added to harmful but odourless materials (e.g. CO or natural gas), to warn users. Odorants are also added to perfumes and toiletries. Flavours and odours are closely related and both are profoundly affected by sub-microgram amounts of volatile compounds.

Ofurace fungicide

Ofurace, $C_{14}H_{16}ClNO_3$ ((±)-α-2-chloro-*N*-(2,6-dimethylphenyl)-*N*-(tetrahydro-2-oxo-3-furanyl)=acetamide), is an anilide fungicide. Its CAS Registry Number is 58810-48-3. It is a colourless crystalline substance, soluble in water (146 mg/l), 1,2-dichloroethane (300–600 g/l), acetone, ethyl acetate and methanol. While it is stable at high temperatures and acidic media, it hydrolyses rapidly in alkali (DT_{50} of 7 h) and undergoes photochemical degradation in water (DT_{50} of 7 days). It is produced by refluxing 3-*N*-(2,6-dimethylphenyl) amino-γ-butyrolactone with chloroacetyl chloride in the presence of pyridine.

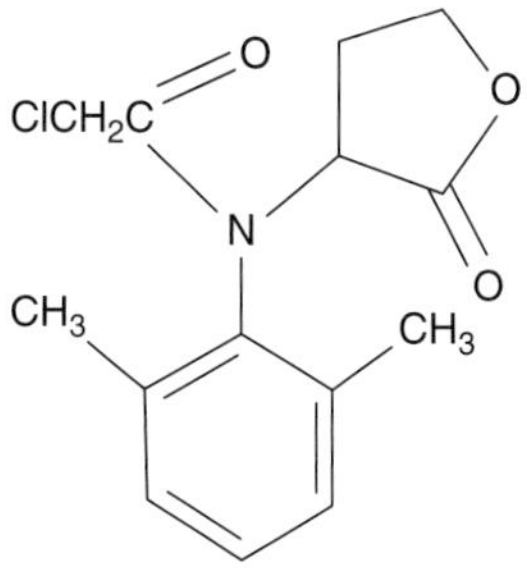

Ofurace

The fungicide is a systemic, protective compound with curative properties. Once absorbed by roots and leaves, it interferes with the synthesis of ribosomal RNA thus retarding spore germination and mycelium formation. Diseases controlled are **powdery mildews** and **late blights**, while crops protected are potatoes, tomatoes and vines.

Ofurace is formulated as a suspension concentrate (SC) or a wettable powder (WP).

The acute oral LD_{50} of ofurace in mice, rats and rabbits is in the range of 2600–>5000 mg/kg. The acute LD_{50} of ofurace in birds is 5000 mg/kg. It is non-toxic to bees and practically non-toxic to fish. It is classified as a Class III toxin by the EPA and a Class U toxin (unlikely to be hazardous) by the WHO. The short-term dietary NOEL of ofurace in rats is shown to be 2.5 mg/kg. The ADI for ofurace is 0.03 mg/kg b.w./day.

Animals rapidly eliminate ingested ofurace after extensively transforming it. In plants, the absorbed ofurace is metabolized by hydroxylation and conjugation. In soil, it degrades with a DT_{50} of around 6 days, with a small amount being adsorbed on the soil.

Ofurace is not permitted for use throughout the EU or in the USA.

Oil-miscible liquids

Some pesticide formulations are made by mixing active ingredients with vegetable or mineral oils. Such compositions are called oil-miscible liquid formulations. For better results and improved spraying, surface-active agents are added to the formulation.

Oleic acid herbicide

Oleic acid, $C_{18}H_{34}O_2$ ((9*Z*)-octadec-9-enoic acid), occurs naturally in animal and plant fat. It is abundant in all oil crops, in poultry and meat and also in human adipose tissue. Sources of oleic acid include the **neem** plant, dead bees, olive oil and groundnut oil. Oleic acid is one of the many fatty acids that are present in nature. Fatty acids are any long-chained carboxylic acids that have an even number of carbon atoms in the chain (or tail). They are a source of adenosine-5′-triphosphate (ATP). Fatty acids differ with the length of the chain and the number of bonds. Palmitic acid, linolenic acid, oleic acid, butyric acid and stearic acid are some examples of fatty acids.

Oleic acid is used as an insecticide, herbicide and fungicide. It interferes with the normal functioning of the cell membranes and kills pathogens. Specifically, the lipophilic carbon chains penetrate and disrupt the lipoprotein matrix and destroy the cuticle and cellular membranes. Such fatty acid-based insecticides are effective against aphids, thrips and **scale insects**. As a herbicide, it is used to control moss. It can protect grapes from fungal pathogens. It is also an ingredient in making insect **pheromones**.

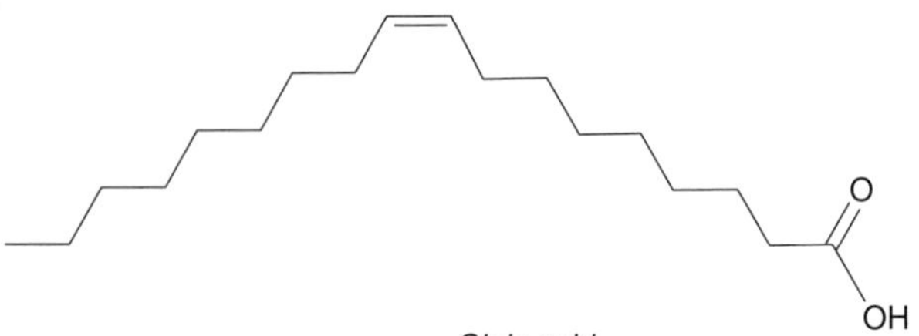

Oleic acid

Fatty acids are available as liquid concentrates and are ideal for foliar sprays or as a soil drench. Trade names include Savona and M-Pede.

Oleic acid is a non-toxic substance, as are most other fatty acids. It has an acute oral LD_{50} in rats and mice of >5000 mg/kg and an acute dermal LD_{50} of >2000 mg/kg. It is safe for fish and bees, and is fully degradable. The fatty acid is approved by the Organic Materials Review Institute of the USA (US OMRI) for use in organic farming. It is registered for use in New Zealand.

Omethoate acaricide

Omethoate, $C_5H_{12}NO_4PS$ (2-dimethoxyphosphinoylthio-*N*-methyl-acetamide), is an organothiophosphate acaricide. It is a colourless liquid, readily soluble in water, alcohol and acetone and slightly soluble in diethyl ether. The CAS Registry Number of omethoate is 1113-02-6.

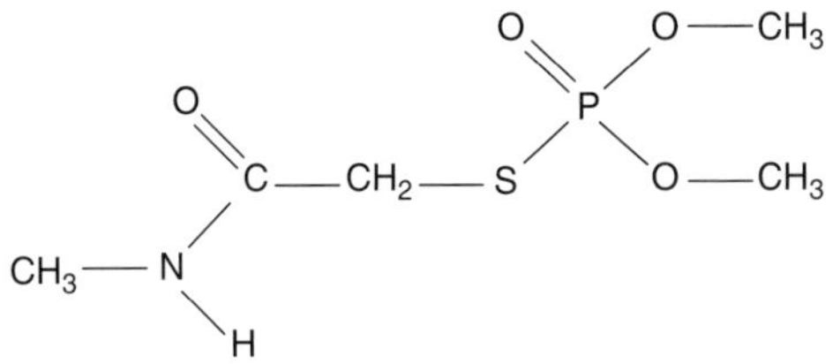

Omethoate

Omethoate is a systemic acaricide and insecticide. It controls spider mites, aphids, beetles, caterpillars, scale insects and other insect pests on fruit crops, hops, rice and other cereals. It is produced by the reaction of *O*,*O*-dimethylphosphorylmercapto acetic acid with methyl isocyanate. It is commonly available as a soluble concentrate (SL). A trade name is Folimat.

The insecticide is an acute toxin and is classified as a Class Ib toxin (highly hazardous) by the WHO. It causes severe irritation to the skin and mucous membranes. The acute oral LD_{50} in rats ranges from 28 to 68 mg/kg. It is also toxic to cats and chickens. The ADI is 0.0003 mg/kg/day. The PAN Pesticide Database has classified it as a PAN Bad Actor chemical.

In plants, omethoate is relatively persistent. In soil, its half-life values range from 3.8 to 25 days. It is also observed to leach from the soils and may contaminate groundwater sources owing to its high solubility in water.

Omethoate is registered for use in Australia, South Africa, Tanzania and Vietnam. It is not registered in the USA and is banned throughout the EU.

Oral exposure

Oral exposure of a substance is the amount that has been ingested/swallowed by an individual. In the present context, exposure to a pesticide may occur in any of three ways – oral, dermal and inhalation.

Oral exposure may occur accidentally, but often occurs because of carelessness. Smoking or eating without washing hands, after using a pesticide is an example of exposure. Consuming food that has been recently sprayed with a pesticide is another.

The seriousness of the exposure depends upon the oral toxicity of the material and the amount swallowed.

Orbencarb herbicide

Orbencarb, $C_{12}H_{16}ClNOS$ (*S*-2-chlorobenzyl diethyl thiocarbamate), is a thiocarbamate herbicide. It is a colourless liquid, which is soluble in water and very soluble in xylene, alcohol and acetone. It is slightly decomposed by sunlight in aqueous solution (DT_{50} of 1.7 days). It is stable to hydrolysis and has a half-life of >1 year. Orbencarb is synthesized by the reaction of 2-chloro-mercaptotoluene with *N*-diethyl carbamoyl chloride. Its CAS Registry Number is 34622-58-7.

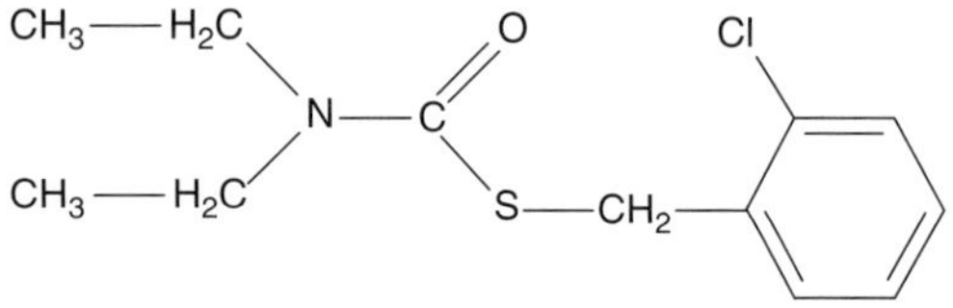

Orbencarb

Orbencarb is a selective, pre-emergence, systemic herbicide, which, once absorbed, is translocated to growth points. Here it inhibits growth of leaves of germinated weeds by interfering with lipid synthesis. It controls broadleaved weeds in barley, wheat, rye, maize, soybeans, cotton and turf.

The herbicide orbencarb is available as an emulsifiable concentrate (EC) and is commonly available in the market as Lanray.

The acute oral LD_{50} of orbencarb in rats is >800 mg/kg and the dermal LD_{50} is >3000 mg/kg. The inhalation LC_{50} in rats is 2.94 mg/l. It is practically non-toxic to birds (the acute LD_{50} in mallard ducks is 2000 mg/kg), fish and honey bees. It may be irritating to the skin, eyes and the respiratory tract. Since it is a cholinesterase inhibitor, it causes headache, dizziness and nausea. The EPA and WHO categorize orbencarb as a slightly toxic compound (Class III).

In plants, orbencarb undergoes rapid degradation. It is not permitted in the USA or in the EU.

Organic pesticides

Pesticides are grouped in two categories, organic and inorganic. Organic pesticides, containing carbon, can be further divided into three groups: (i) synthetic pesticides (synthesized in laboratories and manufactured); (ii) natural pesticides produced by larger organisms including plants; and (iii) natural pesticides produced by microorganisms.

Organic pesticides such as **rotenone**, **pyrethrum**, **nicotine**, **ryania**, sabadilla, **neem** oil, fruit- and seed-based oils and extracts, rosemary oils, clove oils and cinnamon oils are products of living organisms. Most of these substances are used by plants to protect themselves against attack from parasites and pathogens. Natural organic pesticides are often low in toxicity and are environmentally safer than synthetic organic pesticides, although exceptions exist. **Nicotine**, for example, from the genus *Nicotiana*, is highly toxic to mammals, although it is a natural pesticide.

The four main classes of organic insecticides are: (i) organophosphates; (ii) organocarbamates; (iii) organochlorides; and (iv) organic **pyrethroids**. Organophosphates produce their effects by inhibiting the action of acetyl cholinesterase enzyme, which leads to increase in acetylcholine in pre-ganglionic, post-ganglionic and other receptors. Acetylcholine is also a transmitter in the CNS.

Organic pesticides are just as strong as other forms of insect- and weed-controlling pesticides. However, since organic pesticides have high specificity, they are less prone to harming non-targeted life forms. The apparent low toxicity could be due to this specificity. For instance, if the octopamine receptor is targeted, then mammals and plants are safe since these receptors are absent in these non-target species.

However, organic insecticides are common agents of accidental poisoning due to their easy accessibility and improper handling. A WHO estimate suggests that the majority of acute poisoning is caused by organophosphates.

Organic sulfur compounds

See **Sulfur as a pesticide**

Organochlorine pesticides

An organochlorine pesticide is composed primarily of carbon, hydrogen and chlorine. These chemically stable neurotoxins have high lipophilicity and high hydrophobicity. Their structural variety leads to a broad range of applications. While some of these compounds are isolated from natural sources, their mode of action is severe and includes the persistent opening of the sodium channels in neurons, resulting in repetitive firing of action potentials. K^+ permeability is reduced, and that of Na, Ca ATPase are inhibited. These effects slow nerve re-polarization. Their persistence in the environment is the reason behind most of the active ingredients being restricted for use, or banned.

Organochlorines are commonly prepared as insecticides, most often belonging to chemical classes such as dichlorodiphenylethanes, chlorinated cyclodienes, chlorinated benzenes and cyclohexanes. **DDT**, 1,1-dichloro-2,2-bis-(4-chlorphenyl) ethane (DDD, also known by the generic name TDE), methoxychlor and dicofol are dichlorodiphenylethanes. **Aldrin**, **dieldrin**, **endrin**, **heptachlor** and **endosulfan** are part of the chlorinated cyclodiene group. **Lindane**, **toxaphene**, **mirex**, **hexachlorobenzene** (HCB) and **hexachlorocyclohexane** (HCH) are chlorinated benzenes and cyclohexanes.

Several studies have established acute and chronic illnesses in all those exposed to organochlorines. Most of these organochlorines are banned and even targeted for global elimination under the recently signed Stockholm Convention on Persistent Organic Pollutants (POPs). However, some of these pesticides are still used by a few countries, on a wide array of crops including vegetables, fruits, vines, cereal grains, cashew, cotton and ornamental trees, although their use is said to be strictly monitored.

Organomercurial fungicides

Organomercurial fungicides were discovered in the 1910s. Their popularity stemmed from their effective control over seed-borne and soil-borne fungal diseases in cereal crops such as wheat, barley and oats. These compounds also have multi-site modes of action.

Chloro(2-hydroxyphenyl) mercury and **phenyl mercury acetate** have been shown to provide effective protection in seed treatment. Inhibition of mycelia of *Fusarium* spp. and *Dreschlera* spp. is well established. In addition, organomercurial compounds control bunt disease.

However, environmental toxicity and persistence of mercury-based compounds led to their being used for strictly restricted purposes or banned from use in many countries.

Organophosphorus (OP) fungicides

While most organophosphorus (OP) compounds are suitable as insecticides, a few compounds that have a thiol ester linkage (P-S-R) in the molecule or with a thiono-ester (having nitrogen-containing heterocycles) can target some common fungal diseases among crops. Organophosphate fungicides have slightly lower **acute toxicity** than the strictly insecticidal OPs. **Powdery mildews**, blast in rice, **downy mildew** and soil-borne fungal diseases are some of the common diseases that are controlled by certain OP fungicides. **Pyrazophos**, **tolclofos–methyl**, **fosetyl-aluminium** and **iprobenfos** are a few examples of active ingredients that control *Rhizoctonia solani*, *Corticium rolfsii* and *Phycomycete* spp.

Organophosphorus (OP) insecticides

Organophosphorus (OP) insecticides are neutral esters of phosphoric, phosphoramidic and **phosphonic acids** and their mono- or di-thio derivatives. The credit of development of this class of insecticides goes to Gerhard Schrader who devised tetraethylpyrophosphate.

The dominant mode of action of organophosphate compounds is the inhibition of **acetylcholinesterase** (AChE). AChE breaks down the neurotransmitter chemical acetylcholine, which is released from the nerve ending. The structural variation among the many OP compounds is reflected in properties like solubility, **biodegradability**, activity in soil, as well as their action on targeted organisms.

OP insecticides have high efficacy and low persistence in the natural environment. They also have a low cost of production. Although newer synthetic insecticides are increasing in popularity, OP insecticides are still the largest insecticide group sold and consumed. There are over 70 OP insecticides registered so far.

The general structure of OP insecticides is shown by the formula below. This structure may carry various constituent groups (e.g. the electron-withdrawing groups). The group indicated by X is called a 'leaving group' because it may leave by disruption of the P–X bond when the OP molecule reacts with the hydroxide ion or serine enzymes of which AChE is the primary site of action.

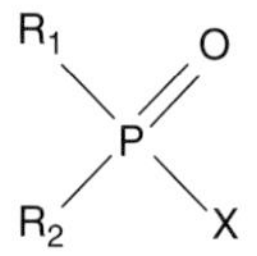

R = Alkyl, Aryl, O⁻Alkyl, NH_2, NHAcyl, etc., X = Leaving group

OP insecticide

The majority of OP insecticides are made up of compounds that belong to the thionate (P=S) type, due to their higher stability, better penetrability and lower mammalian toxicity when compared with the P=O analogues (oxons). Some optically active OP insecticides have the phosphorus atom as an asymmetric centre.

The common property of OP insecticides is that of slightly low translocation in spite of good systemicity. Sufficient water solubility and a proper lipophilic–hydrophilic balance for cuticle penetration (via foliar application) prove beneficial for systemic OP insecticides, which then can become broad spectrum in nature. A typical example is the water-soluble acephate, which is a broad-spectrum insecticide, readily distributed throughout the plant.

There are three subclasses of OP insecticides based on the type of constituent compounds or esters. They are: (i) the vinyl phosphate group; (ii) the aliphatic phosphorothioates; and (iii) the miscellaneous group.

The vinyl phosphate group has three –OR (ester) groups and one P=O (oxon) group. This group has ten constituent types of insecticides, including **chlorfenvinphos** and **dichlorvos**. The second subclass of OP insecticides is called the aliphatic phosphorothioate esters, which have 13 constituent types, which include **demeton-*S*-methyl, phorate, terbufos, phenthoate, malathion, cadusafos** and **mecarbam**. This category can be further divided into three subdivisions: (i) esters of phenol (e.g. **parathion, methyl-parathion**, fenthion, **temephos**); (ii) esters of hetrocyclicenols (including **diazinon, chlorpyrifos** and chlorpyrifos-methyl); and (iii) esters of s-methyl heterocycles (a.i. such as **azinphos-methyl, phosalone** and **azamethiphos**). Aliphatic phosphorothioate insecticides have three –OR (ester) groups and one P=S (thioate) group.

The third subclass of OP insecticides is referred to as miscellaneous. It includes **acephate, dimethoate, dimethylvinphos, ethoprophos, fenamiphos**, flupyrozofos, **formothion**, fosthiazate, **heptenophos, methamidophos**, omenthoate, phenthoate and **vamidothion**.

Organotin fungicides

Organotin compounds are compounds of tin with hydrocarbon substituents. They can be prepared by reacting a Grignard reagent with tin halides. Diethyltin diiodide, tributyltin oxide, *n*-butyltin trichloride, tri-*n*-alkyltins, triphenyltin, fentin chloride and fentin acetate are some of the well-known organotin compounds. Most of these compounds are used as heat stabilizers in preparing **biocides** (e.g. fungicides, miticides and molluscicides), wood preservatives and as marine anti-fouling paints and agents. Fentin chloride ($C_{18}H_{15}ClSn$; triphenyltin chloride; CAS Registry Number of 639-58-7) and **fentin acetate** are often used in rice fields.

The mode of action of these organotin compounds is the inhibition of oxidative phosphorylation and ATP synthase. They carry the **Fungicide Resistance Action Committee (FRAC)** Code C6. However, high toxicity and environment hazards have restricted the use of some of these compounds, especially in agriculture and marine applications. For instance, triorganotins, such as tributyltins and triphenyltins, are very toxic and have caused adverse effects on the environment when used as anti-fungal agents (e.g. in paper mills and in the textile industry), anti-fungal paints and agricultural fungicides. This toxicity is evident in anti-fouling activities. Fouling is the tendency of marine organisms and other biological material to stick and grow on man-made structures like vessels, pipelines and oil rigs. Ships that are not protected by anti-fouling systems, may gather a large amount of these organisms (i.e. many kilograms per square metre) in less than a year, leading to an increase in fuel costs. However, hazardous environmental effects of organotin compounds have prompted countries and organizations to ban organotin compounds, especially tributyltin (TBT)-related compounds. TBT belongs to the trialkyl organotin class, with other members being tributyltin benzoate, tributyltin chloride and tributyltin oxide. Pollution caused by TBT and triphenyltin have caused adverse effects on plankton, crustaceans and gastropods, even at very low concentrations. The compounds are known to cause irreversible sexual abnormality.

The Philippines have banned the use of organotins on rice fields in the country. The EU has also placed restrictions on the use of many organotins, especially TBT. Agencies such as the International Maritime Organization have also proposed a ban on the use of TBT in anti-fouling paints. Tri-*n*-alkyltins are phytotoxic and are banned in some countries.

Ortho dichlorobenzene termiticide

Ortho dichlorobenzene, $C_6H_4Cl_2$ (*o*-dichlorobenzene or 1,2-dichlorobenzene), is a termiticide. It is a liquid with a water solubility of 155.8 mg/l. It has a CAS Registry Number of 95-50-1. It is also called *o*-dichlorobenzene or 1,2-dichlorobenzene.

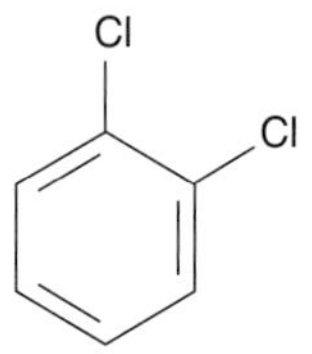

Ortho dichlorobenzene

Ortho dichlorobenzene is a toxic and persistent compound that mainly has industrial applications, such as an automotive and marine degreaser/decarbonizer and as an industrial deodorant. The LD_{50} in rats and mice of the termiticide is 840 and 1228 mg/kg b.w., respectively. It causes irritation to the eye, skin and the respiratory tract. It is toxic to aquatic invertebrates.

Since ortho dichlorobenzene is persistent in the environment, its use is very restricted. The chemical is currently registered for use in Australia.

Orthophenylphenol (OPP)

See **2-Phenylphenol fungicide**

Orthoptera

See **Pest order – Orthoptera**

Orthosulfamuron herbicide

Orthosulfamuron, $C_{16}H_{20}N_6O_6S$ (1-(4,6-dimethoxypyrimidin-2-yl)-3-[2-(dimethylcarbamoyl) phenylsulfamoyl] urea), is a pyrimidinylsulfonyl urea herbicide. It is a white, odourless, fine powder, soluble in water (0.062 g/l), *n*-heptane (0.23 mg/l), xylene (130 mg/l), acetone (20 g/l), dichloromethane (56 g/l) and methanol (8.3 g/l). Its CAS Registry Number is 213464-77-8.

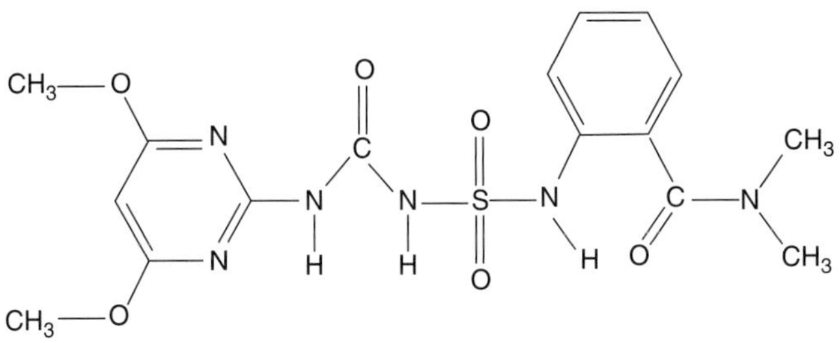

Orthosulfamuron

Orthosulfamuron is a pre- and post-emergence, systemic herbicide whose pesticidal mode of action is the inhibition of the plant enzyme **acetolactate synthase** (also known as acetohydroxy acid synthase). The inhibition of this enzyme blocks branched-chain amino acid biosynthesis of valine, leucine and isoleucine involved in plant growth, leading to plant death. The compound is a rice herbicide, active against sedges, grasses and some broadleaved weeds.

It is formulated as a wettable powder (WP) and as granules (GR). The proposed application rate is up to 74 g/ha, applied by ground boom and aerial equipment.

Orthosulfamuron is moderately irritating to the eyes but is neither a skin irritant nor a dermal sensitizer. The acute oral and acute dermal LD_{50} in rats is >5000 mg/kg. The acute inhalation LC_{50} in rats is >2.19 mg/l. It is practically non-toxic to birds, mammals and honey bees under acute exposure conditions and only slightly toxic to fish and freshwater aquatic invertebrates. However, it is phytotoxic to aquatic plants and terrestrial dicotyledonous plants, following acute exposure. It may also be carcinogenic in nature.

The herbicide orthosulfamuron is registered for use in the USA.

Oryzalin herbicide

Oryzalin, $C_{12}H_{18}N_4O_6S$ (3,5-dinitro-N^4,N^4-dipropylsulfanilamide), is a **dinitroaniline** herbicide. It is a yellow-orange crystalline substance, soluble in water at 2.5 ppm at pH 7 (at 25°C) and readily soluble in acetone, ethanol, methanol and acetonitrile. Though it is stable under normal temperatures and pressures, the dust form poses an explosive or fire hazard. It is produced by the reaction of 4-dipropylamino-3,5-dinitrobenzene sulfonylchloride with ammonium hydroxide. The CAS Registry Number of oryzalin is 19044-88-3.

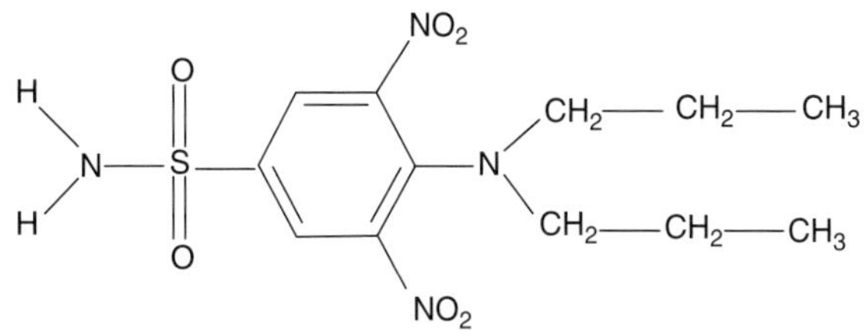

Oryzalin

Oryzalin is a selective pre-emergence, surface-applied herbicide used for the control of annual grasses and broadleaved weeds in fruit trees, nut trees, vineyards and flowering plants. It targets seeds, which have not germinated yet, and kills them.

The herbicide oryzalin is available in aqueous suspension, dry flowable form and wettable powder (WP) formulations. The trade names include Dirimal, Rycelan and Surflan.

Oryzalin has low acute toxicity in mammals and may cause mild irritation to the skin or eye in some individuals. It is not hazardous to birds. All products containing oryzalin must bear the signal word 'CAUTION'. The NOEL in a 1 year feeding study with dogs is 5 mg/kg/day. The ADI is 0.0094 mg/kg/day. The EPA has classified oryzalin as a possible human carcinogen and also a groundwater contaminant.

About 80% of oryzalin is eliminated via the urine and the faeces within 3 days, when administered to rats. Its soil half-life is 20 days. Microbial degradation has a role in its breakdown in soils. It is subject to photodecomposition at the soil surface.

Oryzalin is registered for use in the USA, Australia, New Zealand, Spain and France.

Oxabetrinil herbicide safener

Oxabetrinil, $C_{12}H_{12}N_2O_3$ ((*Z*)-1, 3-dioxolan-2-ylmethoxyimino (phenyl) acetonitrile), is a herbicide safener. It is a colourless crystalline substance, soluble in water (20 mg/l), acetone, cyclohexanone, toluene, methanol and dichloromethane. It is stable up to 240°C and stable to hydrolysis at pH 5–9 for 1 month. Its CAS Registry Number is 94593-79-0.

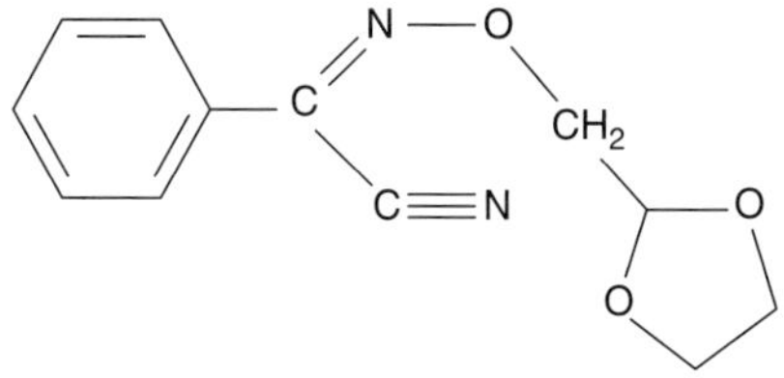

Oxabetrinil

Oxabetrinil stimulates the metabolism of **metolachlor** by inducing glutathione *S*-transferase and is used to protect the sorghum crop (including the sweet and forage varieties) from injury by metolachlor. Metolachlor is used for the control of broadleaved weeds in sorghum. Oxabetrinil can be used in combination with **triazines** (1,3,5-triazines e.g. **atrazine** and **terbutryn**) to increase tolerance to metolachlor. The ideal use of oxabetrinil is during seed treatment to protect crop seedling damage by hastening metabolism and detoxification.

The safener is commonly used as a wettable powder (WP). A common trade name is Concep II. It is applied on sorghum seeds in the form of a smooth slurry, at the rate of 45 g/25 kg of seeds.

The WHO categorizes oxabetrinil as a Class U toxin (unlikely to be hazardous). The **NOAEL** of oxabetrinil in rats (for a 90 day period) is 1500 ppm. The ADI is 0.0047 mg/kg b.w./day. However, it is important to keep seeds, that have been treated with herbicides and safeners, away from animals, poultry, food and feed. The active ingredient, container and spray tank should be washed thoroughly and the rinse water disposed of according to the regulations of the local or national regulatory agency.

In rats, oxabetrinil is rapidly excreted. In plants, it is excreted through the opening of the dioxolane ring by hydroxylation and glycosylation to water-soluble conjugation products.

Oxabetrinil is registered in Australia, but is not registered for use by the USA or the EU. The MRL of oxabetrinil set for most foods is 0.1 mg/kg.

Oxadiargyl herbicide

Oxadiargyl, $C_{15}H_{14}Cl_2N_2O_3$ (5-*tert*-butyl-3-[2,4-dichloro-5-(prop-2-ynyloxy) phenyl]-1,3,4-oxadiazol-2(3*H*)-one), is an oxadiazolone herbicide. It is a beige powder, soluble in water at 0.37 mg/l at 20°C and in acetone at 250 g/l. Its CAS Registry Number is 39807-15-3.

Oxadiargyl, a pre-emergence, selective soil herbicide, is used for the control of broadleaved weeds and sedges in rice, vegetables and sugarcane. The mode of action is the inhibition of protoporphyrinogen oxidase (PPO or **PROTOX**).

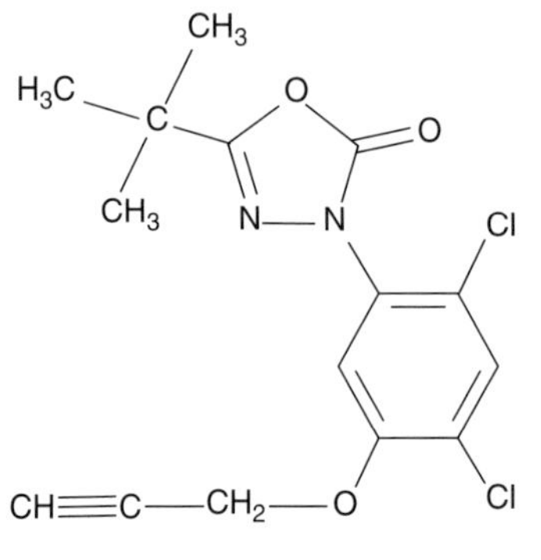

Oxadiargyl

The herbicide oxadiargyl is formulated as emulsifiable concentrates (EC), suspension concentrates (SC), water-dispersible granules (WG) and wettable powders (WP). Trade names include Topstar and Carioca.

The acute oral LD_{50} of oxadiargyl in rats is >5000 mg/kg b.w. The acute dermal LD_{50} in rats is >2000 mg/kg b.w. The ADI in rats is 0.008 mg/kg b.w./day.

The rate and extent of excretion of this herbicide is rapid and most of it is eliminated within 24 h. It is extensively metabolized in animals. Oxadiargyl is not readily biodegradable.

Oxadiargyl is registered in India, Australia, France, Italy, Poland, Vietnam, Mauritania and Gambia.

Oxadiazole herbicides

Oxadiazoles are a type of protoporphyrinogen oxidase (**PROTOX**) **inhibitor**, which is a type of pigment production inhibitor. The other types of PROTOX inhibitors are *N*-phenylthalamides, ***p*-nitrodiphenyl ethers** and **triazolinones**. **Oxadiazon** and **oxadiargyl** are two main herbicides in this category.

PROTOX is the enzyme that converts protoporphyrinogen IX to protoporphyrin IX (proto). Proto is a chlorophyll precursor and is critical in the porphyrin pathway (porphyrins are organic compounds relating to pigments and include chlorophyll and haem). Proto is also photodynamic in nature (i.e. producing singlet oxygen), causing membrane peroxidation and death of cells if produced in an uncontrolled manner. Inhibition of PROTOX leads to accumulation of the substrate, which slowly oxidises to proto, leading to its accumulation. This leads to a drop in the production of chlorophyll and haem. Since haem regulates the porphyrin pathway, its inhibition disturbs the pathway. Such disruption of normal plant processes lead to high photosensitive damage (due to excess proto), necrosis and death of the plant, characterized by discoloration of leaves (**chlorosis**).

Oxadiazon herbicide

Oxadiazon, $C_{15}H_{18}C_{12}N_2O_3$ (5-tert-butyl-3-(2,4-dichloro-5-isopropoxyphenyl)-1,3,4-oxadiazol-2(3*H*)-one), is a pre- and post-emergence **oxadiazole** herbicide, soluble in water at 0.57 mg/kg but highly soluble in acetone and benzene. Its CAS Registry Number is 19666-30-9.

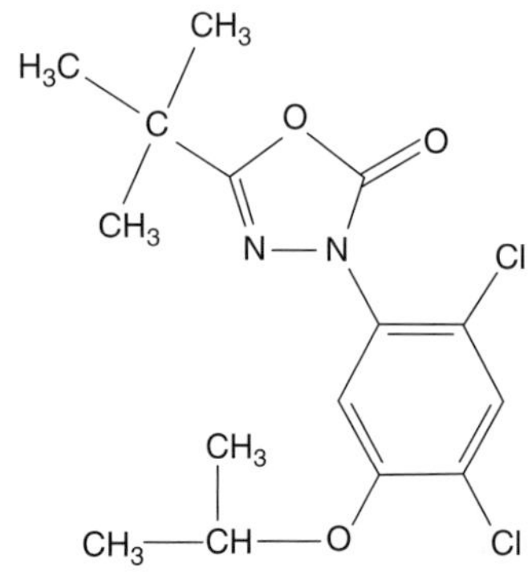

Oxadiazon

The herbicide oxadiazon, a protoporphyrinogen oxidase (**PROTOX**) inhibitor, is effective on several weeds in ornamental crops. It is commonly sprayed as an emulsifiable concentrate (EC) mixed with water.

It persists in soil due to its strong adsorption properties (the soil DT_{50} being 502 days). It is also known to be a reproductive toxin, and a carcinogen, although it is categorized as a Class U toxin (according to acute toxicity ratings).

Oxadiazon is registered for use in India, Australia, the UK, France, Greece, the USA and a few African countries.

Oxadixyl fungicide

Oxadixyl, $C_{14}H_{18}N_2O_4$ (2-methoxy-*N*-(2-oxo-1, 3-oxazolidin-3-yl) aceto-2′, 6′-xylidide), is a colourless, odourless crystalline substance of the **phenylamide** (oxazolidinone) group. Oxadixyl is soluble in acetone (344 g/l), methanol, ethanol and water (3.4 g/l) and is stable in normal storage conditions, with a DT_{50}

of around 4 years. It is produced by intra-molecular condensation of 1-chloroethyl-2-(methoxyacetyl)-2-(2,6-xylyl)-hydrazine carboxylate with sodium hydride. Its CAS Registry Number is 77732-09-3.

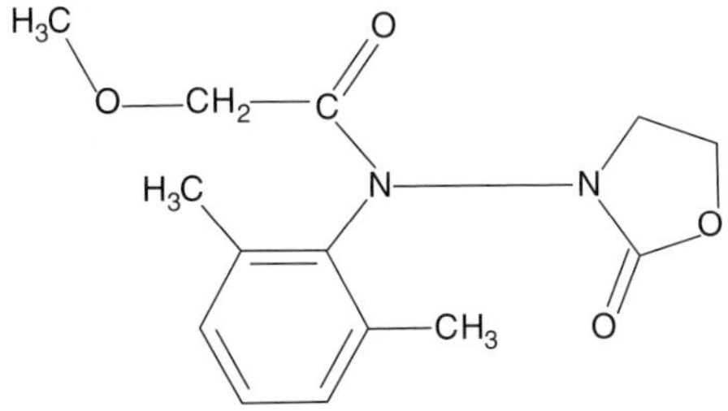

Oxadixyl

Oxadixyl is a systemic fungicide that acts by blocking the synthesis of fungal protein (especially ribosomal RNA). It is absorbed by the leaves and roots, and has a curative and protective action. Oxadixyl is used along with other fungicides to control fungi in the order Peronosporales that cause **downy mildews, late blights** and rusts. Foliar and seed applications (seed dressing) give protection to wheat, rye, sorghum, oats, potatoes, citrus fruits, vegetables (e.g. cabbage, carrot, beans, cucumber), cotton, tobacco and sunflowers. This fungicide also protects lawns and turf. The pathogens controlled include members of genera *Pythium* and *Phytophthora*.

The foliar application rate of oxadixyl is in the range of 200–300 g/ha. Common formulations are wettable powders (WP), water-dispersible granules (WG) and granules (GR). Some popular brands include Recoil and Sandofan.

The fungicide is classified by both the WHO and the EPA as slightly toxic (Class III toxin). The acute oral LD_{50} in rats is 1860 mg/kg in female rats and 3480 mg/kg in male rats. In the case of birds, it is >2510 mg/kg in mallard ducks, while the oral LD_{50} in bees is >200 µg per bee. The compound is not a skin sensitizer and is non-mutagenic and non-teratogenic. The NOEL for 1 year in dogs is 500 mg/kg diet. The ADI in humans is 0.05 mg/kg b.w.

The fungicidal compound is absorbed rapidly and eliminated in 144 h, via urine and faeces, after metabolism occurs. In plants, most of the compound remains unchanged, and out of the minute amount that enters the plant system, less than half is metabolized. Oxadixyl has a DT_{50} of 2–3 months in field conditions, with oxadixyl acid being the main metabolite. However, it has shown a possible trend towards being carcinogenic, based on tests on laboratory animals. Oxadixyl is included in the list of **Highly Hazardous Pesticides of PAN**.

Oxadixyl is registered in Australia and Tanzania, among others. However, this chemical is banned in the whole of EU.

Oxamyl insecticide

Oxamyl, $C_7H_{13}N_3O_3S$ (*N,N*-dimethyl-2-methylcarbamoyloxyimino-2-(methylthio) acetamide), is an oxime carbamate insecticide. It is a colourless, crystalline substance with a garlic-like odour, soluble in water at 280 g/l (25°C), methanol, ethanol, acetone and toluene. Oxamyl decomposes slowly at pH 5 (with a DT_{50} of >31 days in aqueous solutions), while at higher pH levels, decomposition is very quick. Oxamyl is produced by a chain of reactions, including chlorination of the oxime of methylglycolate, reaction with methanethiol and alkali, and conversion to the carbamate with methyl isocyanate. The CAS Registry Number of oxamyl is 23135-22-0.

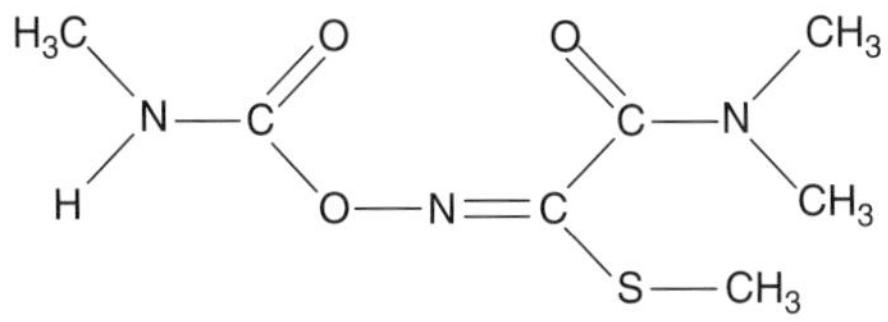

Oxamyl

Oxamyl, a systemic, broad-spectrum, cholinesterase-inhibiting compound with contact action, has additional acaricidal and nematicidal properties. This chemical controls sucking insects, spider mites, ticks and several soil pests such as roundworms. Crops that are protected include ornamentals, cucurbits, beet, potatoes, bananas, pineapples, groundnuts, soybeans, cotton and tobacco. The chemical moves downwards to the plant roots after being applied. Cotton cultivation accounts for a major proportion of oxamyl formulations.

The herbicide is formulated as granules (GR) for soil application. It may also be foliar sprayed. It is available in the market as Blade, Pratt, Oxamil, Tevday and Vydate. Oxamyl is applied up to twice in a season, at between 250 g/ha and 1.2 kg/ha (commonly at 450 g/ha). This pesticide may be applied at the pre-plant stage, planting stage or at the time of germination. Seed dips are also sometimes undertaken.

The insecticide-acaricide-nematicide oxamyl is a highly toxic chemical, categorized as a Class Ib toxin by the WHO and as a Category I pesticide by the EPA. The oral LD_{50} is 5.4 mg/kg in rats and the dermal LD_{50} is 2960 mg/kg in rabbits. The 4 h inhalation LC_{50} in rats is 0.12–0.17 mg/l. The acute oral LD_{50} in quails is 4.18 mg/kg. It is slightly to moderately toxic to fish and highly toxic to bees. Oxamyl causes irritation to the eyes and skin. All product labels must carry the signal words 'DANGER – POISON'. The ADI in oxamyl is 0.03 mg/kg/day. Mixers and loaders are advised to follow label instructions strictly and be prepared with appropriate PPE. After application of oxamyl, there is a **restricted entry interval (REI)** of 48 h for all crops except citrus, for which the REI is 4 days. Symptoms of poisoning with oxamyl are muscle weakness, dizziness and sweating, headache, salivation, nausea, abdominal pain, blurred vision, muscle twitching and slurred speech, CNS depression and seizures.

When oxamyl, a carbamate insecticide, is administered to rats, most of the dose is rapidly eliminated via urine as breakdown products. Oxamyl is of low persistence in soil, the break down being brought about by aerobic and anaerobic bacteria. The compound is also hydrolysed rapidly in neutral and alkaline soils. In the absence of organic matter, oxamyl readily leaches out of the soil. It is shown to contaminate groundwater sources, since microbes are not active at those depths. An increase in temperature causes a decrease in adsorption.

In plants, oxamyl hydrolyses to the corresponding oximino compound which in turn conjugates with glucose, the metabolism taking place rapidly.

Oxamyl is registered for use in Australia, Burkina Faso, Canada, Cape Verde, Chad, Cyprus, more than half of the EU including the UK, France, Belgium, Italy and Hungary,

Gambia, Mali, Mauritania, New Zealand, Niger, the Philippines, Senegal, South Africa and the USA.

Oxasulfuron herbicide

Oxasulfuron, $C_{17}H_{18}N_4O_6S$ (oxetan-3-yl 2-[(4, 6-dimethylpyrimidin-2-yl) carbamoylsulfamoyl] benzoate), is a pyrimidinylsulfonylurea herbicide. It is a white, odourless powder, soluble in water (at 1700 mg/l), hexane (2.2 mg/l) and acetone (9300 mg/l). Oxasulfuron is produced by the reaction of 2-[[(3-oxetanyl) oxy] carbonyl]benzenesulfonamide with (4,6-dimethylpyrimidine-2yl) carbamate. Its CAS Registry Number is 144651-06-9.

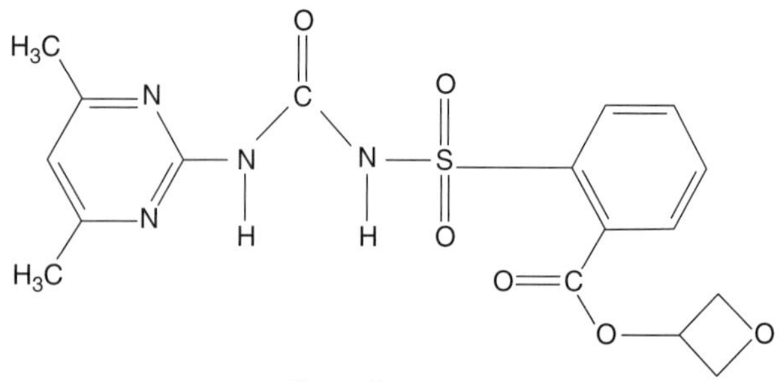

Oxasulfuron

The herbicide is used mainly on soybean to control broad-leaved weeds and grasses at the post-emergence stage. It is an acetolactate synthase (ALS) inhibitor absorbed by shoots and roots and translocated. The compound is formulated as water-dispersible granules (WG). A known trade name is Chart.

Oxasulfuron is a Class III toxin. It is not a skin and eye irritant. The ADI for a 1 year period in dogs is 0.013 mg/kg b.w. It is moderately toxic to aquatic organisms and earthworms.

Most oxasulfuron (up to 94%) is excreted in 48 h at the low dose level. The metabolism in animals is limited (with almost 85% excreted as the parent compound). It is classified as 'not readily biodegradable'. It is rapidly and highly absorbed, especially at lower pH. The soil DT_{50} is in the range of 10–33 days. The aqueous photolysis DT_{50} of 3.5 days suggests that this herbicide is not persistent. It is also considered mobile.

Oxasulfuron is registered for use in Italy.

Oxaziclomefone herbicide

Oxaziclomefone, $C_{20}H_{19}Cl_2NO_2$ (3-[1-(3,5-dichlorophenyl)-1-methylethyl]-3,4-dihydro-6-methyl-5-phenyl-2H-1, 3-oxazin-4-one), is a herbicide. It is a white crystalline substance, soluble in water (0.18 ppm). The CAS Registry Number is 153197-14-9.

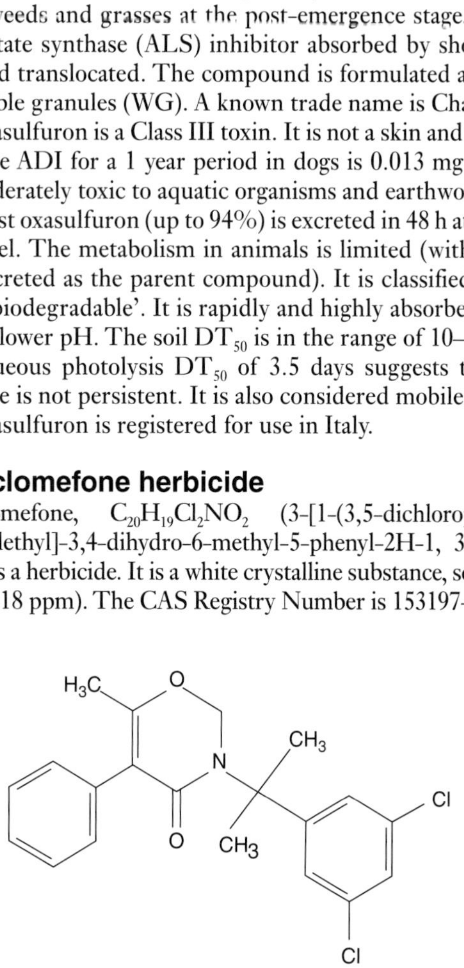

Oxaziclomefone

Oxaziclomefone is effective in controlling cockspur and annual sedges in rice crops. It blocks plant growth by inhibiting cell expansion in grass roots. It shows some residual activity. It is available as granules (GR), wettable powders (WP) and soluble concentrates (SL).

The herbicide oxaziclomefone is a slightly toxic chemical, and can irritate the eyes. It is toxic to fish with a 48 h LC_{50} in carp being around 5 ppm.

This chemical is not permitted for use in the USA or in the EU.

Oxine-copper fungicide

Oxine-copper, $C_{18}H_{12}CuN_2O_2$ (bis(quinolin-8-olato-*O,N*)copper (II) or cupric 8-quinolinoxide), is a copper fungicide. It is a complex of copper with quinolin-8-ol. Oxine-copper is an olive-green powder, soluble in dichloromethane (410 mg/l), acetone, ethanol and water (1.04 mg/l). It remains stable up to 270°C, as well as in acidic and alkaline media (pH 5–9). This fungicide is also known as oxine-Cu and **copper 8-quinolinoleate**. The CAS Registry Number of oxine-copper is 10380-28-6.

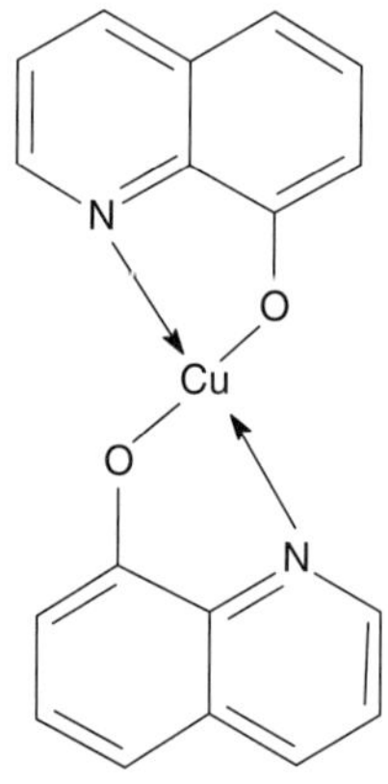

Oxine-copper

Oxine-copper, a non-systemic, broad-spectrum, protective fungicide, is used to control blotch, bunt, leaf spot, canker and **scab** in a variety of crops including wheat, beet, flax, canola, sunflowers, beans, celery, peas and pome fruits. Proliferation of species belonging to the genera *Cercospora*, *Phoma*, *Pythium*, *Alternaria*, *Botrytis* and *Sclerotinia* is controlled. Copper ions react with sulfur-containing enzymes in the plant and thus inhibit fungal metabolism. Copper prevents entry of pathogens into the plants, hence it is applied on parts recently pruned or injured.

Commonly known trade names include Dokrin, Oxin-doh and Quinondo. Oxine-copper is formulated as a powder for dry seed treatment (DS), paste (PA), wettable powder (WP) and suspension concentrate (SC).

The acute oral LD_{50} of oxine-copper among rats is 585 and 500 mg/kg (male and female, respectively). It is highly toxic to fish (the LC_{50} in rainbow trout is 8.94 μg/l). The 8 day LD_{50} in bobwhite quails is 618 mg/kg. However, it does not pose a great hazard, if used as directed. This fungicide is also non-carcinogenic, non-mutagenic and non-teratogenic, if used as directed. It is non-toxic to bees. The NOEL in rats is up to 1.11 mg/kg b.w. for 2 years.

Oxine-copper is also used as an anti-sap stain chemical, and is used in wood preservation. Sap stain is the discoloration of freshly cut rubber wood and other sapwoods, caused by the blue-stain organism *Botryodiplodia theobromae* and other deuteromycete-mould fungi.

Animals eliminate almost all the oxine-copper that was ingested within 72 h, with the main metabolites being the parent, **8-hydroxyquinoline** and glucuronide and sulfate conjugates. In soil, the DT_{50} is 2 days, and tests reveal that oxine-copper remains in the top 6 cm of the soil.

Oxolinic acid bactericide

Oxolinic acid, $C_{13}H_{11}NO_5$ (5-ethyl-5, 8-dihydro-8-oxo [1,3] dioxolo[4,5-g]quinoline-7-carboxylic acid), is a bactericide. It is a white crystalline powder, soluble in hexane, xylene and methanol (<10 g/kg at 20°C). It reacts with oxidizing agents like nitrates, chlorine, strong acids and bases, leading to a fire hazard. Hence, it should be stored in airtight glass containers. It has a CAS Registry Number of 14698-29-4.

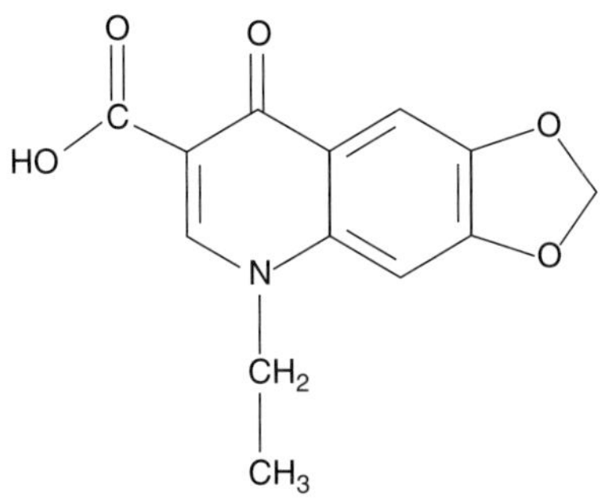

Oxolinic acid

Oxolinic acid, a systemic, quinolone antibiotic, provides preventive and curative control of Gram-negative bacteria in rice crops. The mode of action is the inhibition of the enzyme DNA-gyrase, with target pests including *Escherichia coli*, *Proteus* spp. and *Klebsiella* spp. It also has fungicidal properties.

Oxolinic acid is available in dry powder (DP) and wettable powder (WP) formulations.

The bactericide is toxic to mammals with an oral LD_{50} among male rats being 630 mg/kg and among female rats being 570 mg/kg. The dermal LD_{50} among rats is >2000 mg/kg. It is slightly to moderately toxic to fish species, with a 48 h LC_{50} in carp being 10 mg/l. Toxicity symptoms include nausea and vomiting, gastrointestinal bleeding and weakness, thus requiring the use of PPE during the time of handling the chemical. It is a skin and eye irritant.

The bactericide is registered in Vietnam and Cambodia, but not in the USA and the EU.

Oxpoconazole fumarate fungicide

Oxpoconazole fumarate, $C_{42}H_{52}Cl_2N_6O_8$ (bis[(*RS*)-1-{2-[3-(4-chlorophenyl)propyl]-2,4,4-trimethyl-1,3-oxazolidin-3-ylcarbonyl}imidazolium] fumarate), is a conazole fungicide. It is a colourless crystalline substance, hardly soluble in water (0.0895 g/l). It is stable in neutral and alkali media. It photodegrades in about 10.6 h. The CAS Registry Number of oxpoconazole fumarate is 174212-12-5.

Oxpoconazole fumarate is a fungicide which acts as an inhibitor of **ergosterol** biosynthesis. It has a curative and residual action and controls *Venturia*, *Alternaria*, *Botrytis*, *Sclerotinia* and *Monilinia* spp. by obstructing the life cycles of these fungal pathogens.

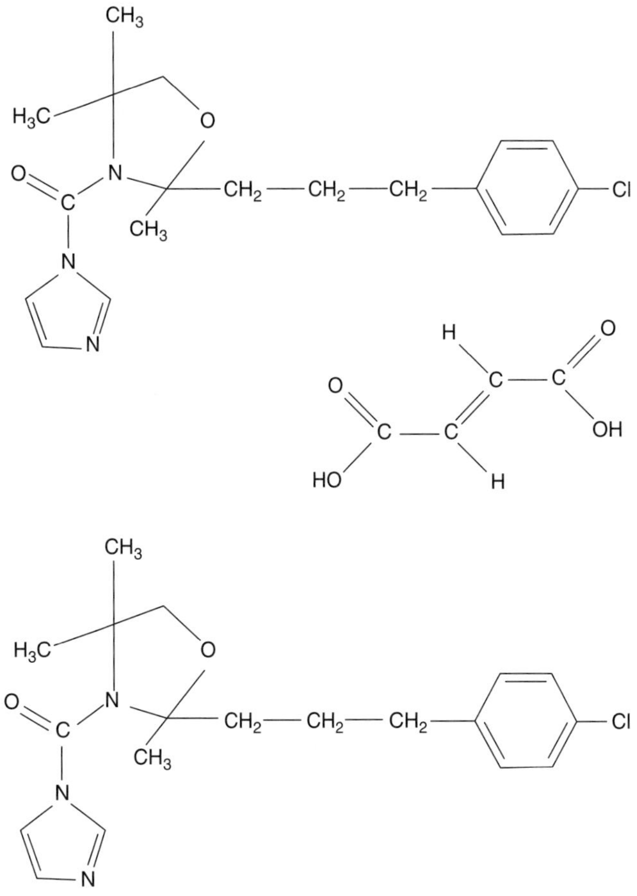

Oxpoconazole fumarate

The acute oral LD_{50} in rats is in the range of 1035–1424 mg/kg. While it is observed to be slightly irritating to eyes, it is practically non-irritating to the skin (according to tests on rabbits and rats). The acute oral LD_{50} for bird species tested is up to 1731.3 mg/kg, whereas fish species show LC_{50} values of up to 15.8 mg/l. The DT_{50} on field soil is from 23 to 34 days.

Oxpoconazole fumarate is a derivative of oxpoconazole. Oxpoconazole, $C_{19}H_{24}ClN_3O_2$ ({(*RS*)-2-[3-(4-chlorophenyl) propyl]-2,4,4-trimethyl-1,3-oxazolidin-3-yl}(imidazol-1-yl) methanone), is a fungicide. It is generally available as its fumarate salt. This fungicide is an imidazole derivative with an *N*-acylimidazole group, used mainly on rice seedlings but also on some fruit trees (e.g. apples and peaches). This compound controls *Botrytis cinerea*, *Venturia* spp., *Monilinia* spp., *Rhizoctonia solani* and *Magnaporthe oryzae*. The application dose on fruit trees is commonly 0.01% and a known trade name is All-shine.

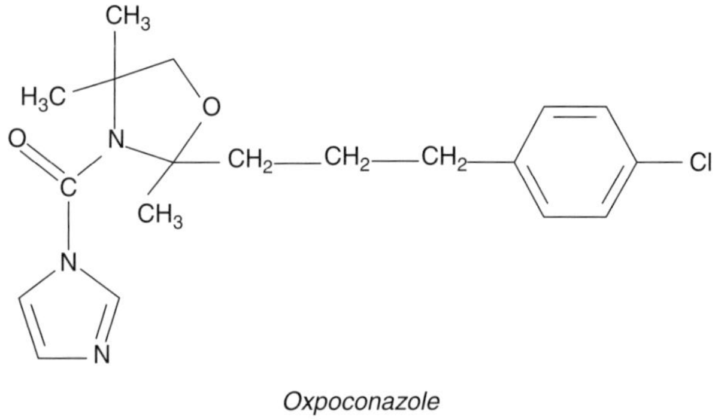

Oxpoconazole

Oxycarboxin fungicide

Oxycarboxin, $C_{12}H_{13}NO_4S$ (5,6-dihydro-2-methyl-1,4-oxathiine-3-carboxanilide 4,4-dioxide), is a fungicide of the oxathiin carboxamide class. It is a white solid, soluble in water (1.4 g/l), acetone (83.7 g/l) and hexane (8.8 mg/l). It remains stable at 55°C for over 2 weeks. Oxycarboxin hydrolyses in 44 days at a pH of 6 and is unstable in highly acidic or alkaline media. It is produced by oxidation of oxathiin, which is prepared by the reaction of acetoacetanilide with sulfonyl chloride, followed by treatment with a potassium hydroxide solution and then with 2-mercaptoethanol. It has a CAS Registry Number of 5259-88-1.

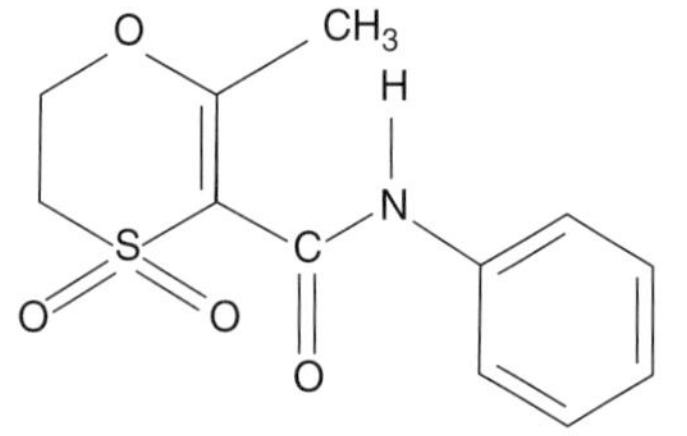

Oxycarboxin

Oxycarboxin, a systemic fungicide, has a curative action and is effective in controlling rust diseases in cereals. However, it is commonly used in ornamental plants and turf. It acts by disrupting complex II in the respiratory electron transport chain, thus inhibiting mitochondrial function.

Since it is commonly applied to the leaves, mixing oxycarboxin with other pesticide formulations (especially insecticides and **acaricides**) is to be avoided.

Oxycarboxin is commonly formulated as an emulsifiable concentrate (EC) or a wettable powder (WP). It is applied via **chemigation**. A trade name is Plantvax.

The acute oral LD_{50} of oxycarboxin in male rats is much higher (5816 mg/kg) than in the case of female rats (1632 mg/kg). The LD_{50} in mallard ducks is 1250 mg/kg. The contact LD_{50} in bees is >181 µg per bee. It is moderately toxic to some fish species. The 2 year NOEL value in rats is 15 mg/kg b.w. and the ADI in humans is 0.15 mg/kg/day. The WHO and EPA classify oxycarboxin as a pesticide that is unlikely to be hazardous. Irritation and allergic reactions to the eyes and skin may be expected, with prolonged exposure causing lung damage. In the case of dermal contact, contaminated clothes should be removed and the exposed part washed thoroughly with soap and water. In the case of oral exposure, vomiting may be induced if the victim is conscious.

The DT_{50} for oxycarboxin in sandy loamy soil is up to 8 weeks. Oxycarboxin forms four distinct metabolites, with residues binding tightly to the soil.

Oxycarboxin is registered in Australia, South Africa, the USA and Canada. Since oxycarboxin is commonly used on non-food categories, there are no MRLs established for it.

Oxydemeton-methyl insecticide

Oxydemeton-methyl, $C_6H_{15}O_4PS_2$ (*S*-(2-ethylsulfinylethyl) *O*,*O*-dimethyl phosphorothioate), is an aliphatic organothiophosphate insecticide. It is a colourless liquid, completely miscible with water and soluble in common organic solvents. It rapidly hydrolyses in alkaline media, and the DT_{50} values substantiate the same (at pH 4 and 9, the DT_{50} is 107 and 2 days, respectively). Oxydemeton-methyl is produced by treatment of demeton-*S*-methyl with hydrogen peroxide or with bromine. It has a CAS Registry Number of 301-12-2.

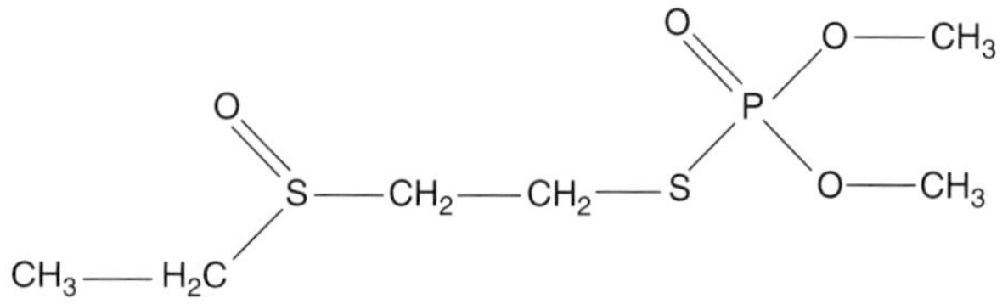

Oxydemeton-methyl

The insecticide, a systemic, cholinesterase-inhibiting compound, has contact action. It is effective against aphids, sawflies and thrips on fruits, vegetables, cereals and ornamentals.

The compound is commonly available under the trade name of Metasystox-R. It is formulated as emulsifiable concentrate (EC) and soluble concentrate (SL). Oxydemeton-methyl is compatible with insecticides and fungicides so long as they are not alkaline materials. Caution is advised against handling treated ornamentals until the spray has dried.

Considering the potential harm oxydemeton-methyl can cause to bird species (the LD_{50} is from 7 to 60 mg/kg) and some **beneficial organisms** like honey bees, its use has been restricted in many countries. It is also toxic to fish (the 96 h LC_{50} range in freshwater fish is 0.7–0.12 ppm). Product combinations with this chemical may cause irritation to the skin. Its oral LD_{50} in rats is 65–75 mg/kg and the dermal LD_{50} is 250 mg/kg. The ADI in rats is 0.0003 mg/kg b.w./day. The inhalation LC_{50} in rats is 0.471 mg/l. The acute 48 h contact LD_{50} in honey bees is 0.175 µg per bee. According to the WHO Acute Hazard classification, oxydemeton-methyl comes under toxicity Class Ib (highly hazardous). It is a PAN Bad Actor chemical.

Symptoms of poisoning with oxydemeton-methyl are excessive salivation and sweating, muscle twitching, tremor, headache, dizziness, nausea, abdominal cramps, diarrhoea, respiratory depression, tightness in the chest, wheezing, cough, pin-point pupils, seizures and incontinence.

The risks involved in the application of oxydemeton-methyl are moderately high for mixers, loaders, applicators and handlers. Hence, the **restricted entry interval (REI)** is up to 30 days, depending on the amounts applied. Reduced rates of application, safer application methods (e.g. tree injections) and insistence on protective equipment can help mitigate risks of exposure.

In both animals and plants, elimination of this insecticide is very quick. The metabolized product, demeton-*S*-methylsulfon, is biologically active and is of concern to the environment. Oxydemeton-methyl degrades in a variety of soils (the DT_{50} is up to 65 days), where the sulfoxide oxidizes to a sulfone group followed by an oxidative and hydrolytic cleavage of the side chain, with the formation of dimethylphosphoric acid and phosphoric acid.

Oxydemeton-methyl is banned in the EU, with many countries imposing restrictions on the type and method of application. It is registered in Australia, Cambodia, India, South Africa and the USA.

Oxyfluorfen herbicide

Oxyfluorfen, $C_{15}H_{11}ClF_3NO_4$ (2-chloro-α, α, α-trifluoro-p-tolyl 3-ethoxy-4-nitrophenyl ether), is a *p*-nitrophenyl ether herbicide, and it is used as an algicide too. It is soluble in water (0.116 mg/l at 25°C), acetone, cyclohexanone and isophorone. It decomposes rapidly by UV radiation, though it remains stable up to 50°C. Oxyfluorfen is a member of the ***p*-nitrodiphenyl ether** group, a type of protoporphyrinogen oxidase (**PROTOX**) **inhibitor** that inhibits pigment production in plants. It is produced by the reaction of 3,4-dichlorobenzotrifluoride with resorcinol, the resulting product being treated with ethanol in the presence of potassium hydroxide. It has a CAS Registry Number of 42874-03-3.

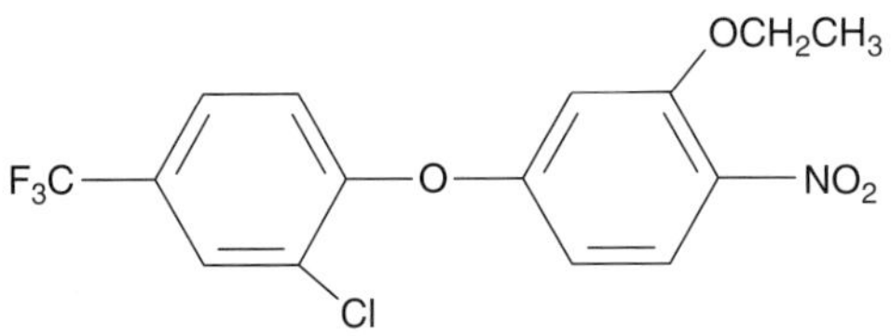

Oxyfluorfen

Oxyfluorfen, a selective, pre- and post-emergence herbicide, is effective against broadleaved weeds and annual grass species in cotton, soybean, cereals, maize, onions, vegetables and all types of fruits, nut trees, turf and forestry. Its main uses are on wine grapes and almonds. This herbicide is absorbed, via the leaves, but is not well translocated.

Herbicide oxyfluorfen is formulated as an emulsifiable concentrate (EC) and as granules (GR). Trade names include Goal and Koltar. Application rates range between 0.25 and 2.0 kg/ha.

The oral LD_{50} of oxyfluorfen ranges from 2700 to 5000 mg/kg in rats, mice and dogs, while the same in birds ranges between 2200 and 4000 mg/kg. The dermal LD_{50} is >5000 mg/kg in both rats and rabbits. The 96 h LC_{50} among a wide range of aquatic organisms varies from 32 to 400 μg/l. Oxyfluorfen may cause severe irritation to the skin and the eye. Its ADI is 0.003 mg/kg/day. Soybean and cotton may be injured by contact with oxyfluorfen. According to the WHO Acute Hazard studies, oxyfluorfen is a Class U toxin (unlikely to be hazardous).

The herbicide is moderately persistent, with a field half-life of up to 40 days, due partly to its relative stability to hydrolysis and microbial action. In moist soils and water, its mode of degradation is photodegradation, evaporation/co-distillation. The herbicide adsorbs strongly to soil particles (especially to clay and organic matter). This property, as well as its very low solubility in water, makes it practically impossible for the herbicide to leach or be mobile. In plants, metabolism is slow and residues have been found in carrots and oats during field tests.

Oxyfluorfen is registered in India, the USA, Australia, Canada, New Zealand, the Philippines, South Africa and many EU countries including Cyprus, France, Greece, Hungary, Poland, Italy and Spain.

Oxythioquinox acaricide

Oxythioquinox, $C_{10}H_6N_2OS_2$ (6-methyl-1, 3-dithiolo (4, 5-b) quinoxalin-2-one), is a dithiocarbonate acaricide, fungicide and insecticide. Its CAS Registry Number is 2439-01-2.

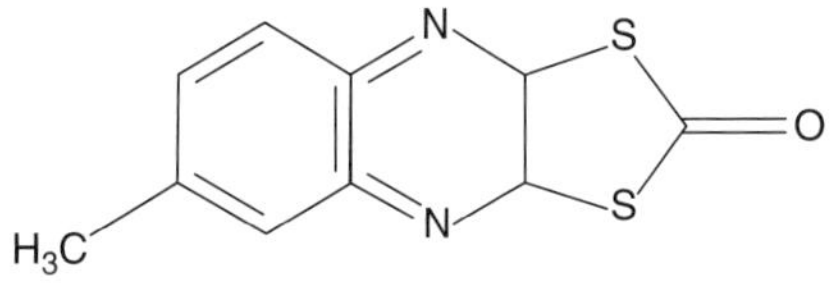

Oxythioquinox

The acaricide oxythioquinox is also registered as an insecticide and fungicide and it is used to control mites, mite eggs and **powdery mildew**. The current agricultural application is for citrus crops. The other use of this chemical is on ornamentals.

Oxythioquinox is formulated mostly as emulsifiable concentrates (EC) and applied through low-volume concentrated spray by high-volume sprayers or by hand-held sprayers. One of the trade names of this pesticide is Morestan.

The WHO has classified oxythioquinox as a Class III toxin (slightly hazardous). The acute oral LD_{50} in rats is 1100 mg/kg and in mice is 1070 mg/kg. The acute percutaneous LD_{50} in rats is 500 mg/kg. It is moderately to highly toxic to fish and other aquatic organisms. It can cause permanent eye damage, and is toxic on contact and ingestion. It has also proven to be a carcinogen and a reproductive toxin, based on studies on laboratory animals. It is thus classified as a PAN Bad Actor chemical.

Oxythioquinox is registered for use in Australia.

P

2-Phenylphenol fungicide

2-Phenylphenol, $C_{12}H_{10}O$ (biphenyl-2-ol), is a pinkish crystalline substance belonging to the phenol family. It is also called orthophenylphenol (OPP). 2-Phenylphenol is soluble in water (0.7 g/l), ethanol and glycol ethers.

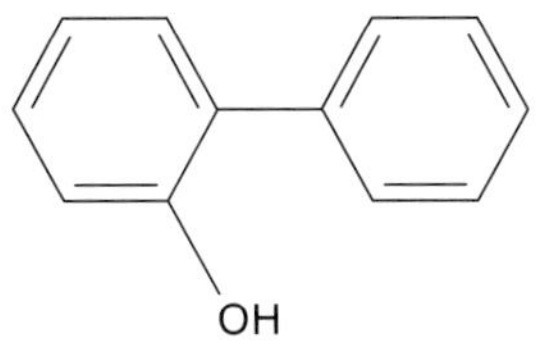

2-Phenylphenol

Besides being a fungicide, 2-phenylphenol is also a nematicide, fumigant and bactericide. Some common trade names are Dowcidel and Preventol O Extra. The main use of 2-phenylphenol is as a deodorizer in agricultural areas, hospitals and other public utilities that need sanitation. It is also used as a fungal preservative in paint and paper industries. A sodium salt of 2-phenylphenol is sodium 2-phenylphenoxide, with molecular formula $C_{12}H_9NaO$. This compound is soluble in water (1.1 kg/kg water) and forms a solution with a pH of 12–13, which is used as an inert ingredient in many formulations that act as insecticides and herbicides.

The compound is slightly toxic to mammals, and care has to be taken during application.

OPP is registered for use in Australia, Canada, the USA and the Philippines. The Netherlands and the UK have established an MRL of 12 mg/kg for citrus, while Denmark has fixed a value of 5 mg/kg for the same. The accepted MRLs for OPP are in the range of 0.1–125 mg/kg for a wide range of fruit and fruit products.

Paclobutrazol as a PGR

Paclobutrazol, $C_{15}H_{20}ClN_3O$ ((2*RS*,3*RS*)-1-(4-chlorophenyl)-4,4-dimethyl-2-(1*H*-1,2,4-trizol-1-yl)penten-3-ol), is a plant growth regulator. A white crystalline substance, paclobutrazol is soluble in water, acetone, cyclohexane, dichloromethane, hexane, xylene, methanol and propylene glycol. It remains relatively stable under normal conditions, and also to hydrolysis and UV light.

Paclobutrazol is absorbed in the xylem and growth tips and inhibits the production of gibberellin. Thus, plant size remains small, while blossoms and fruits are enhanced, to increase yields. Paclobutrazol also has some fungicidal activity against mildew and rusts, and helps builds plant resistance to frost and unfavourable environmental conditions. Crops on which paclobutrazol is used include rice, turf and ornamentals.

Paclobutrazol is formulated as a suspension concentrate (SC) and wettable powder (WP) and applied as foliar spray, soil drench or as a trunk injection. It is available in the market under trade names such as Bonzi, Clipper, Cultar, Celstar and Paelay.

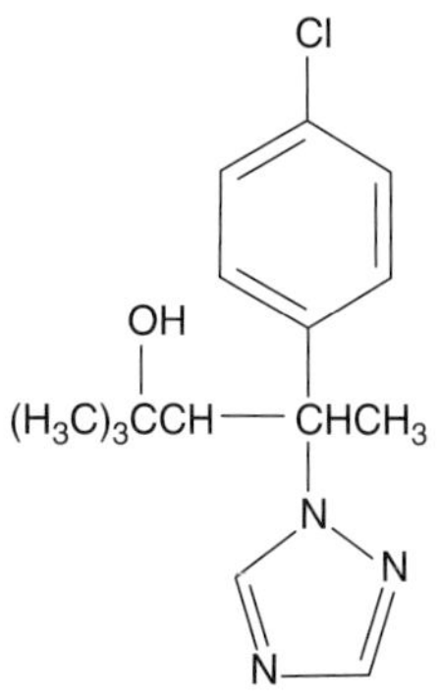

Paclobutrazol

The ADI for paclobutrazol is 0.1 mg/kg. The WHO classifies it as a Class III toxin (i.e. slightly hazardous). According to the EPA, it is a Class IV toxin. This PGR is registered in many countries of the EU and in the USA.

Paenibacillus popilliae as a biocontrol agent

See **Biocontrol of crop pests – entomopathogenic microorganisms**

PAN Bad Actor chemicals

See **PAN International**

PAN International

The Pesticide Action Network (PAN) was founded in 1982 and is a network of over 600 non-governmental organizations, institutions and individuals in over 90 countries worldwide, working to replace the use of hazardous pesticides with ecologically sound and socially just alternatives. Its projects and campaigns are coordinated by five autonomous regional centres: (i) PAN North America, with headquarters based in California, USA; (ii) PAN Europe, with headquarters based in London, UK and Hamburg, Germany; (iii) PAN Asia and the Pacific, with headquarters based in Malaysia; (iv) PAN Africa, with headquarters based in Dakar, Senegal; and (v) PAN Latin America, with headquarters based in Santiago, Chile and Peru.

The work of PAN International is guided by five common strategic objectives, which were collectively developed by representatives from all regions. These are to: (i) protect health and the environment by eliminating highly hazardous pesticides from the market and replacing them with sustainable solutions; (ii) resist development and stop the introduction and use of genetic engineering into agriculture; (iii) promote empowerment of grass roots movements and citizens to fight agrochemical and seed corporations and challenge corporate globalization; (iv) increase public investment, development, adoption and implementation of non-chemical alternative pest management systems; and (v) develop further PAN International structures.

The rationale of organizations like PAN is to provide a parallel voice to highlight the dangers of pesticide use and its role in reducing biodiversity and damaging ecological systems, by providing information on the effects of pesticide use.

PAN has classified pesticides on the basis of the toxicity or ill effects of different pesticides. Hence, there are PAN Bad Actor chemicals which indicate those chemicals that are either carcinogenic in nature, cholinesterase inhibitors, endocrine disruptors, groundwater contaminants, reproductive/developmental toxins or are acutely toxic. There is a list of Dirty Dozen chemicals that include organochlorine pesticides and persistent organic pollutants (POPs) such as **aldicarb**, texaphene, **chlordane** and **heptachlor**, **chlordimeform**, chlorobenzilate, DBCP (1,2-dibromo-3-chloropropane), **DDT**, **aldrin**, **dieldrin**, **mirex**, **endrin**, EDB (**ethylene dibromide** or 1,2 dibromoethane), HCH and **lindane**, paraquat, **parathion** and **methyl parathion**, **pentachlorophenol** and 2,4,5-T (2,4,5-trichlorophenoxyacetic acid). Polychlorinated biphenyls, dibenzo-*p*-dioxins are included in this list. They are known to persist in the environment and are carcinogens and mutagens. Highly Hazardous Pesticides of PAN (HHPs) is another list of chemicals that indicate high toxicity and environmental concern. A chemical that may have at least one of the toxicity characteristics (e.g. high acute toxicity, chronic effects such as developmental abnormality or mutagenicity, or causing irreversible adverse damage to human health or the environment) is an HHP. The Stockholm Convention (for elimination of POPs), the Rotterdam Convention on the Prior Informed Consent Procedure for Certain Hazardous Chemicals and Pesticides in International Trade (for better information exchange) and the Montreal Protocol (on substances that deplete the ozone layer) regulate the use, trade and production of certain chemicals. All these chemicals are on the PAN list of HHP.

More information about general and specific activities of PAN worldwide may be found at http://www.pan-international.org. This volume refers to the PAN terminology in the spirit of increasing the reader's awareness.

Pansukh disease of rice (*Dakhina* or *Chatra*)

A non-pathogenic disease of transplanted rice, common in India, is called *Pansukh* (translated as 'dried-up leaf'). Symptoms of this condition are lack-lustre leaves that dry up and turn bronze, progressing along the margins to the leaf tip. In a matter of a month, affected plants (clumps) dry up. Shortage of oxygen to the roots and the presence of H_2S, nitrite of nitrate-nitrogen (nitrogen present in the form of nitrate), sulfite formed by reduction of sulfate or oxidation of H_2S, or even methane in the field may be the cause of this disease.

Keeping the field well drained at the early stages of plant growth helps prevent the development of the disease. Addition of ammonium sulfate at the rate of 30–50 kg/ha also helps to keep this condition away, while providing sufficient nutrition. Green manuring also reduces the incidence of the disease.

Paraclox slimicide

Paraclox, C_8H_6ClNO (2-chloro-4′-hydroxy-2-isonitrosoacetophenone), is a slimicide. It has a CAS Registry Number of 34911-46-1. Paraclox is registered by the EPA as an antimicrobial and slimicide.

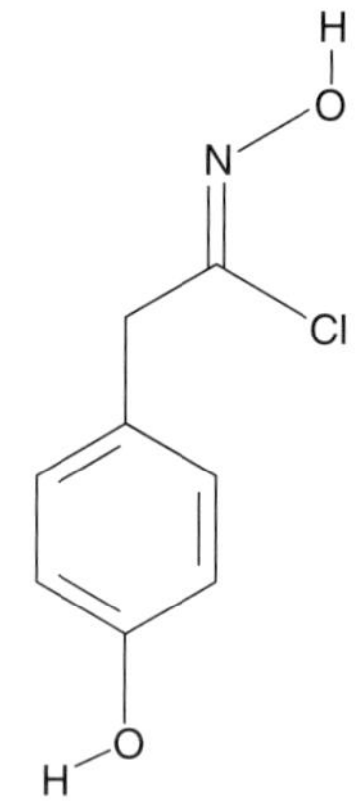

Paraclox

Paraquat dichloride herbicide

See **Paraquat herbicide**

Paraquat herbicide

Paraquat, $C_{12}H_{14}Cl_2N_2$ (1,1′-dimethyl-4,4′-bipyridinium dichloride), is a herbicide. It is a colourless crystalline quaternary ammonium compound, with high solubility in water. Paraquat is made by reacting pyridine coupled with sodium in anhydrous ammonia to give 4,4′-bipyridine, which is then methylated with chloromethane. Its CAS Registry Number is 1910-42-5.

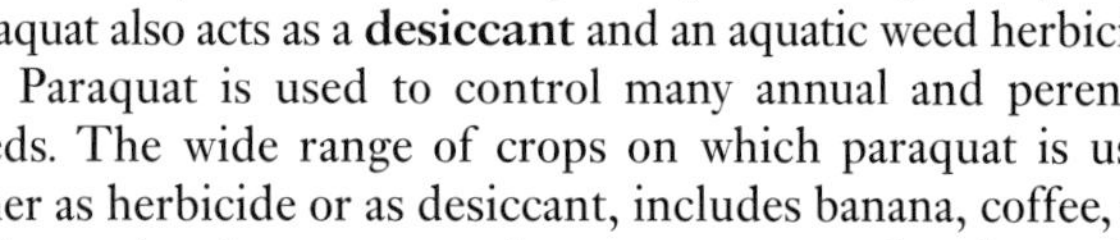

Paraquat

Paraquat is a quick-acting, **non-selective**, contact herbicide with significant translocation throughout the plant. The mode of action is by the inhibition of photosynthesis at photosystem I. Paraquat also acts as a **desiccant** and an aquatic weed herbicide.

Paraquat is used to control many annual and perennial weeds. The wide range of crops on which paraquat is used, either as herbicide or as desiccant, includes banana, coffee, tea, rubber, onion, lucerne, sugarbeet, sugarcane and soybeans. The compound is also used to clear weeds from forestry land, barren land and land that is being prepared for cropping. Paraquat is used near harvest time as a desiccant, to enable reduced foliage and effective harvesting of the crop.

Paraquat is a highly toxic herbicide classified as a Class I (most toxic) chemical by the EPA. It is acutely toxic when ingested, with an LD_{50} of between 110 and 180 mg/kg in rats, and much lower values in other mammals like cats, cows and dogs. In humans, the LD_{50} is 35 mg/kg. The compound is slightly to moderately toxic to birds and many aquatic species, and non-toxic to honey bees. Contact with paraquat leads to skin irritation, respiratory ailments and abnormal foetal development. It is toxic if absorbed through the skin and can cause nose bleeding if inhaled. Symptoms of poisoning include excitability, lung congestion, burning of the mouth and throat, convulsions, incoordination and abdominal pain. Other toxic effects include rapid heart rate, kidney failure and liver injury. It is also linked to the development of Parkinson's disease.

The ADI of paraquat is 0.004 mg/kg b.w./day. It is classified as a restricted use pesticide (RUP) according to the EPA, making it necessary for only trained and licensed personnel to handle it.

Since the half-life of paraquat in soil is over 3 years, it is considered highly persistent. UV radiation, sunlight and microbial activity degrade paraquat to some extent. The decomposition products are less toxic than paraquat. Since it is adsorbed on soil particles, it is not very mobile and thus does not contaminate groundwater, even though it is highly soluble in water. However, studies show that paraquat could be a groundwater contaminant. It is an endocrine disruptor.

Paraquat and paraquat dibromide have been banned in the EU and their use is highly restricted in the USA.

Paraquat dichloride, $C_{12}H_{14}Cl_2N_2$ (1,1′-dimethyl-4, 4′-bipyridinium dichloride), is a quaternary ammonium herbicide. A derivative of paraquat, it is one of the most widely used herbicides in the world. This compound is a colourless, hygroscopic crystalline substance, soluble in water at 700 g/l (20°C) but practically insoluble in most other organic solvents.

The WHO has classified paraquat dichloride as a Class II toxin (moderately hazardous). The substance is irritating to the eyes, skin and the respiratory tract. It is also toxic to human beings when swallowed. It is very toxic to aquatic organisms and may cause long-term effects in the aquatic environment. The acute oral LD_{50} of paraquat dichloride is 129–157 mg/kg and the acute percutaneous LD_{50} is >911 mg/kg in rats. The NOEL in dogs (1 year) is 0.65 mg/kg b.w./day. The ADI is 0.005 mg/kg b.w./day. Paraquat dichloride becomes biologically inactive upon contact with soil. It is a PAN Bad Actor chemical and also a member of the PAN **Dirty Dozen**. Paraquat dichloride is registered for use in India and many other countries.

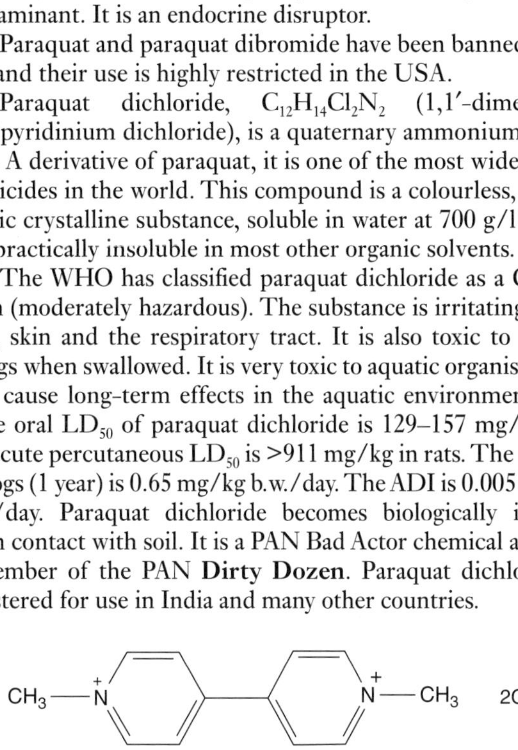

Paraquat dichloride

Parasitic plants as pests

Plants that live off other plants and cause harm to the host are called parasitic plants. These plants do not make their own food, instead attaching themselves to a host plant, depriving it of nutrients and space. Two common parasitic plants are mistletoe and dodder.

Mistletoe varieties parasitize pine trees and conifers, for example, producing short greenish stems, and causing cankers, swellings and injury to the stems of the host plant. Some parasites attack deciduous trees and grow thickly on them, causing leaf drop in the host.

There are many herbicides that can be used on parasitic plants. However, the best way to control these plant pests is to uproot and destroy them, and to weed out any new growth in the surrounding areas. Affected parts of the host should be severed and clean crop seeds should be bought and sown.

Parasitoids as biocontrol agents

See **Biocontrol of crop pests – parasitoids**

Parathion insecticide

Parathion, $C_{10}H_{14}NO_5PS$ (*O,O*-diethyl *O*-4-nitrophenyl phosphorothioate), is a phenyl organophosphorus insecticide. It is a pale yellow liquid with a phenol-like odour. Parathion is soluble in water (11 mg/l) and completely miscible with most organic solvents such as dichloromethane, isopropanol, toluene and hexane. It is hydrolysed rapidly in alkaline media and isomerizes on heating above 130°C to the *O, S*-diethyl isomer. Parathion is produced by the reaction between sodium *p*-nitrophenolate and diethylthiophosphoryl chloride. It has a CAS Registry Number of 56-38-2.

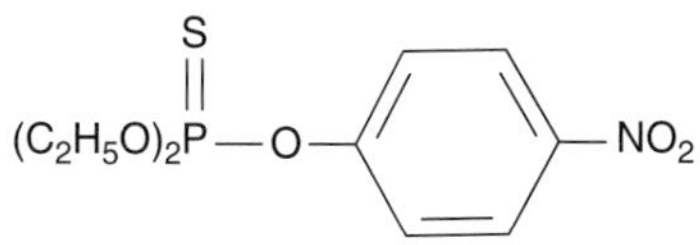

Parathion

Parathion, a cholinesterase inhibitor, is a non-systemic insecticide and acaricide with limited respiratory action. It is used to control soil insects and mites in cereals, citrus fruits, vines, hops, vegetables, ornamentals and cotton.

Formulations of parathion include wettable powder (WP), dustable powder (DP), emulsifiable concentrates (EC), granules (GR) and aerosols (AD). Common trade names are Alkron, Aphamite, Folidol, Fosfermo, Fosferno, Gearphos, Penphos and Thiofos. It is compatible with other insecticides and fungicides, but incompatible with alkaline materials. Some species of ornamentals, cucurbits, sorghum, apple, pear and tomato might be injured by the use of parathion.

The WHO classifies parathion as a Class Ia toxin (extremely hazardous). Parathion has an oral LD_{50} in rats of 2–30 mg/kg. It is extremely toxic to birds like mallard ducks and pigeons (the LD_{50} is up to 6 mg/kg), moderately toxic to fish and aquatic invertebrates (the 96 h LC_{50} is 4.3 mg/l) and toxic to honey bees (the 24 h LD_{50} is 0.07–0.10 μg per bee). Parathion is listed as a PAN Bad Actor chemical and is also a PAN **Dirty Dozen** chemical. It is highly toxic to humans and is readily absorbed into the bloodstream from the skin, lungs or gut. Exposed parts should be washed immediately with soap and water and all contaminated clothing removed. Persons with cardiovascular, liver or kidney diseases, glaucoma or CNS abnormalities may be at increased risk from exposure to parathion. High temperatures or exposure to UV light may increase its toxicity. Labels of parathion products must bear the signal word 'DANGER'.

South Africa has registered parathion. Many other countries have restricted its use.

Parathion-methyl insecticide

Parathion-methyl, $C_8H_{10}NO_5PS$ (*O,O*-dimethyl *O*-4-nitrophenyl phosphorothioate), is an organophosphorus insecticide. It is also called methyl-parathion. A dimethyl ester of parathion, this compound is a colourless, odourless crystalline substance. It is soluble in water (55 mg/l) and readily soluble in common organic solvents such as dichloromethane, toluene and hexane. It is hydrolysed in alkaline and acidic media; the DT_{50} values (25°C) at pH 5, 7 and 9 are 68, 40 and 33 days, respectively.

Parathion-methyl photodegrades in water (with a half-life of 8 days in summer and 38 days in winter). Its CAS Registry Number is 298-00-0.

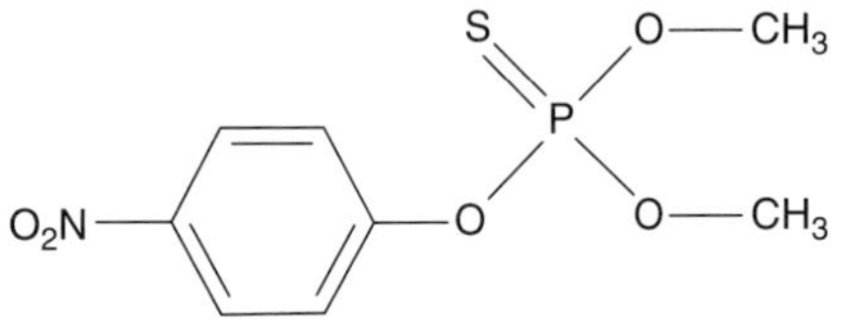

Parathion-methyl

Parathion-methyl, a cholinesterase inhibitor, is a non-systemic insecticide and acaricide with some respiratory action. It is used in the control of chewing and sucking insects in a wide range of crops, including cereals, fruits, vines, vegetables, ornamentals and cotton.

The toxicity and degradation patterns of parathion-methyl are similar to those of parathion. It has oral LD_{50} values of 6–50 mg/kg in rats, 14.5–19.5 mg/kg in mice and 420 mg/kg in rabbits. It is highly toxic via the dermal route as well, and has dermal LD_{50} values of 67 mg/kg in rats, 1200 mg/kg in mice and 300 mg/kg in rabbits. In view of the extreme toxicity of this insecticide, agricultural treatments on broccoli, Brussels sprouts, cauliflower, celery, lettuce and spinach have been stopped in many countries. Non-food uses such as the control of mosquitoes and treatment on lawns have been severely restricted. The **restricted entry interval (REI)** for workers, mixers and handlers has also been increased to up to 4 days. India, Australia and the USA are some countries that have allowed the use of this chemical. The EU has banned its use.

Paris Green rodenticide

Paris Green is another term for **copper acetoarsenite**. It is prepared from copper (II) acetate and arsenic trioxide, and used as a rodenticide and insecticide. The name Paris Green comes from the fact that copper acetoarsenite was first used to kill rats in Paris in France. (See also **Arsenical inorganic insecticides**.)

Paste and paint as formulations

Fungicides are normally mixed with water and occasionally with alcohol or other carriers, and applied as paint on plant surfaces. For example, **Bordeaux mixture** can be made in to a paint-like consistency with linseed oil or water for painting tree wounds. Most wood products are preserved with a paste/paint formulation.

Pasteurization

Pasteurization is a processing technique for beverages, wherein food is heated below its boiling point to inactivate or kill bacteria, and ultimately, to extend food shelf life. The term pasteurization was coined after Louis Pasteur, the inventor of this process.

A soil sample may also be pasteurized by heating it up to 180°C for about 30 min, and then cooling it. This process kills most insects, both those that are pests and those that are beneficial, as well as weed seeds.

Pathfinder element

A pathfinder element helps locate the main ore deposit. Pathfinder elements are usually metallic, are present in small proportions (less than 1%) and are associated with ore deposits at the time of formation. Zinc acts as the 'pathfinder' for lead (Pb), copper (Cu) and silver (Ag) ores, while **molybdenum** (Mo) is found in Cu deposits.

Pathway

A pathway is a sequence of reactions, usually of a biochemical nature, in which complex substances are converted to simple end products. The course of a pathway is largely determined by preferential factors involving coenzymes and other catalysts. An example is the tricarboxylic acid (TCA) cycle, which is the pathway in the degradation of foodstuffs and cell constituents into CO_2 and H_2O.

Pebulate herbicide

Pebulate, $C_{10}H_{21}NOS$ (*S*-propyl butyl (ethyl) thiocarbamate), is a thiocarbamate herbicide. It is a colourless liquid, moderately soluble in water and miscible with organic solvents. It is stable up to 200°C. Pebulate is made by the reaction of ethylbutylamine with phosgene, followed by treatment with propylmercaptan. It has a CAS Registry Number of 1114-71-2.

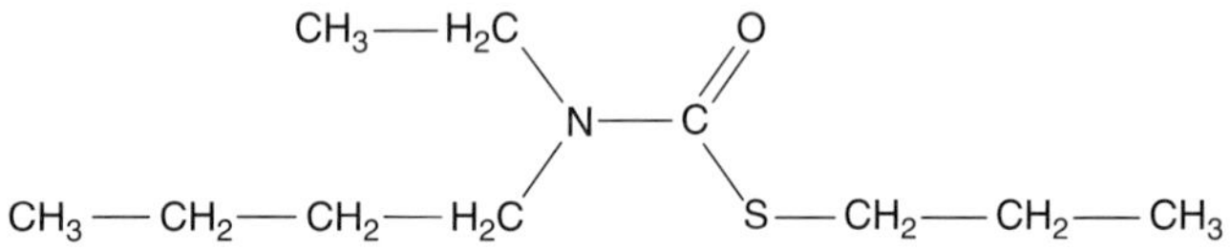

Pebulate

Pebulate, a selective, systemic pre-emergence herbicide, is absorbed by plant roots and translocated throughout the plant. It is an inhibitor of long-chain fatty acid synthesis. When incorporated into the soil, pebulate inhibits the germination of susceptible annual grasses and broadleaved weeds in plants such as sugarbeet and tomatoes.

Formulations of pebulate include granules (GR) and emulsifiable concentrates (EC), and trade names include Tillam. Application to a depth of 5–7.6 cm in normal soil conditions is ideal for weed control. In the case of very wet soils, loss by volatilization has been recorded. The compound is compatible with other herbicides.

Pebulate is moderately toxic to rats (the acute oral LD_{50} is 1120 mg/kg and the dermal LD_{50} is >4640 mg/kg b.w.). The inhalation LC_{50} in rats is 3.5 mg/l. The acute 48 h LD_{50} in honey bees is 11 μg per bee, while the acute 48 h EC_{50} in *Daphnia magna* is 5.9 mg/l. Pebulate is practically non-toxic to birds (with the acute LD_{50} for mallard ducks being 5206 mg/kg.). Symptoms of poisoning include irritation to the skin, eyes and the respiratory tract and possibly headaches, dizziness and nausea. The WHO lists pebulate as a Class II toxin (moderately hazardous). The herbicide may pollute groundwater.

In plants, pebulate is rapidly metabolized to carbon dioxide and naturally occurring plant constituents. In soil, it breaks down mainly by microbial degradation to form mercaptan, ethylbutylamine and carbon dioxide (the soil DT_{50} is 2–3 weeks). It adsorbs to dry soil and may volatilize in wet soil.

As per information available at the time of writing, pebulate is registered for use in Australia.

Penconazole fungicide

Penconazole, $C_{13}H_{15}Cl_2N_3$ ((*RS*)-1-[2-(2,4-dichlorophenyl) pentyl]-1*H*-1,2,4-triazole), is a triazole fungicide. This fine white powder is reasonably stable to hydrolysis (up to pH 13) and heat (up to 350°C). It can be produced by the alkylation of 2,4-dichlorophenylacetonitrile with bromopropane, followed by conversion of the nitrile to the acetate. Reduction with lithium aluminum hydride gives the corresponding alcohol, which is first reacted with methanesulfonyl chloride and then with sodium triazolide. The CAS Registry Number is 66246-88-6.

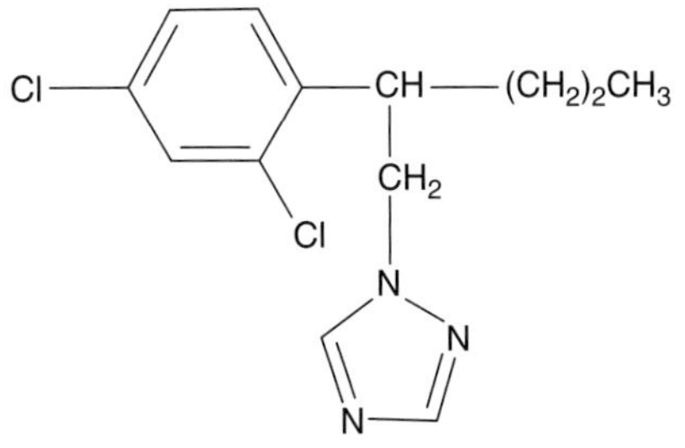

Penconazole

Penconazole is a systemic fungicide that acts by inhibiting the biosynthesis of the cell membrane ergosterol. It is also known as a sterol **demethylation inhibitor** (SBI). Penconazole is absorbed by the foliage, and with its protective and curative action, disrupts the growth of fungi (ascomycetes, deuteromycetes and basidiomycetes) on vines, vegetables and fruit (stone/pome). Penconazole may effectively control **powdery mildew** and fruit **scab**.

Commonly formulated as an emulsion, emulsifiable concentrate (EC) or as a wettable powder (WP), this fungicide is available under trade names such as Topas, Pentos and Dallas. Application rates range from 25 to 75 g/ha.

As for mammalian toxicology, the acute oral LD_{50} of penconazole is 2125 mg/kg in rats, 2444 mg/kg in mice, 1590–3000 mg/kg across many bird species and >5 µg per bee in bees. Penconazole is not a skin irritant or a skin sensitizer. It is found to be a non-mutagenic compound, when tested on guinea pigs and rabbits. With a NOEL of 3.8 mg/kg in rats, and an ADI in humans of 0.03 mg/kg b.w./day, penconazole is considered a Class U toxin by the WHO (unlikely to be hazardous). It is a possible endocrine disruptor.

Penconazole is rapidly eliminated from animals via urine and faeces, leaving insignificant residues in tissues. In plants, penconazole is metabolized to triazolylalanine and triazolylacetic acid. It is degenerated in soil with an aerobic DT_{50} of 61–188 days, although it is immobile in soil. In water, penconazole undergoes insignificant photolytic degradation, finally resulting in 1,2,4-triazole.

According to an FAO study, the MRL for penconazole for pome fruit is established at 0.5 mg/kg, although in the UK, the MRL is established at 0.2 mg/kg for apples and pears and 0.05 mg/kg for blueberries and cranberries.

Penconazole is registered for use in several countries including India, South Africa, Australia, New Zealand and all EU countries except Denmark and Poland.

Pencycuron fungicide

Pencycuron, $C_{19}H_{21}ClN_2O$ ((4-chlorobenzyl)-1-cyclopentyl-3-phenylurea), is a **phenylurea** fungicide. It is a colourless, odourless, crystalline substance, soluble in water at 0.3 mg/l, but readily soluble in dichloromethane (>250 g/l), *n*-octanol and *n*-heptane. Pencycuron has a hydrolysis DT_{50} of up to 302 days. It degrades in water and on soil surfaces in the presence of light. This compound is produced by the reaction of *N*-(4-chlorobenzyl)cyclopentylamine with phenyl isocyanate. It has a CAS Registry Number of 66063-05-6.

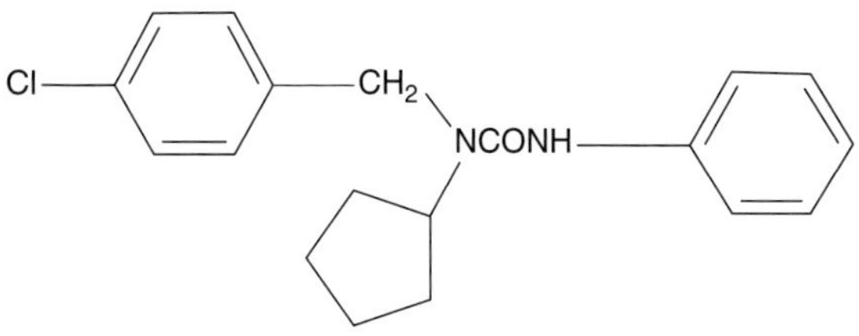

Pencycuron

Pencycuron is a non-systemic fungicide that acts by inhibiting cell division. It also has a protective action. Pathogens such as *Rhizoctonia solani* and *Pellicularia* spp., which cause black scurf of potatoes, sheath blight of rice and damping-off of seedlings (especially of ornamentals) are controlled by pencycuron. Sugarbeet and cotton are other crops that are protected from some fungal pathogens. Pencycuron has a wide variety of applications, namely, as a foliar spray, for soil incorporation, in seed treatment and as a dust formulation. Common trade names include Prestige and Monceren G.

The fungicide may be combined with many compounds like **imidacloprid**, **edifenphos** and **thiram**. Some common formulations are driftless formulations (DL), dustable powders (DP) and dry seed treatment powder (DS). Commonly recommended dosages include 15–20 g/100 kg potato seeds or 3–5 kg/ha in seed furrow for potatoes, 500 g/ha foliar application for sugarbeet and 1250–1500 g/ha of drench or soil incorporation for vegetables.

Pencycuron is non-teratogenic, non-carcinogenic and non-mutagenic, and it is non-irritating to the skin and eyes in normal use. Its acute oral LD_{50} is more than 5000 mg/kg among rats and over 2000 mg/kg in bobwhite quails. The NOEL in rats is 1.8 mg/kg b.w. and the ADI is 0.018 mg/kg/day. The EPA has classified pencycuron under Class IV, while the WHO classifies it as Class U (unlikely to be hazardous). Pencycuron is safe for bees, several beetle species and soil microorganisms. Excessive exposure causes symptoms like coughing, nausea, weight loss and irritation to the eyes and mucous membranes.

Most of pencycuron is eliminated in rats within 3 days, either unchanged or as sulfate and glucuronic acid conjugates. Pencycuron degrades only marginally and is seen in the form of hydroxylated metabolites. In soil, pencycuron is only slightly mobile, with the main metabolites being the derivatives of p-chlorobenzylamine and p-chlorobenzylformamide.

The MRL of pencycuron has been established in Europe as 0.1 mg/kg on potatoes and 0.05 mg/kg for sweet potatoes and sugarcane.

Pencycuron is registered for use in India, South Africa, New Zealand, the UK, Australia, Denmark and Germany.

Pendimethalin herbicide

Pendimethalin, $C_{13}H_{19}N_3O_4$ (*N*-(1-ethylpropyl)-2, 6-dinitro-3, 4-xylidine), is a **dinitroaniline** herbicide. It is an orange-yellow crystalline solid with a fruity flavour. It is hardly soluble in water (0.275 ppm at 25°C) but soluble in acetone (161.0 g/100 ml), xylene, corn (maize) oil, heptane, benzene, toluene, chloroform and dichloromethane. The compound is slowly decomposed by light, but may pose a fire hazard when exposed to heat or flame. Pendimethalin is produced by the reaction of 2,6-dichloro-4,5-dimethyl chlorobenzene and ethylpropylamine. Its CAS Registry Number is 40487-42-1.

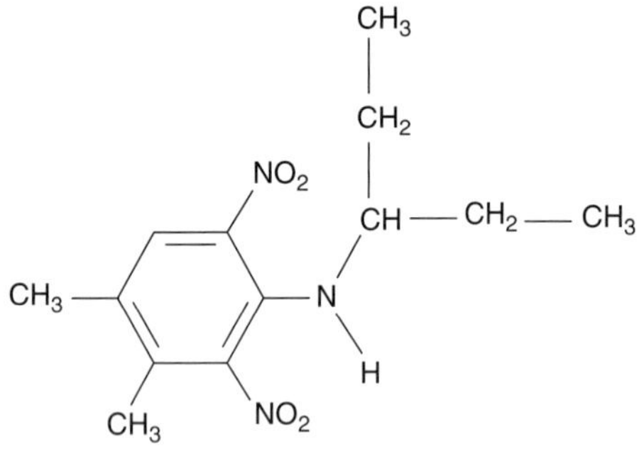

Pendimethalin

Pendimethalin is a selective, pre-emergence herbicide, used to control most annual grasses and certain broadleaved weeds. It acts by inhibiting cell division and cell elongation. This herbicide is used to control weeds like crabgrass in crops such as maize, rice, cotton, soybean, sunflowers, tobacco, potatoes and groundnuts.

Available formulations of pendimethalin include emulsifiable concentrates (EC), wettable powders (WP) and water-dispersible granules (WG), and trade names include Penoxalin, Trump, Blazer, Prowl and Way-Up. The compound may be incorporated into the soil, followed by irrigation to ensure it reaches deep into the soil.

According to the WHO, pendimethalin is a Class III toxin (slightly hazardous). Acute exposure may cause irritation to the skin, eyes and the respiratory tract. The acute oral LD_{50} for technical pendimethalin in rats is >5000 mg/kg, while the acute dermal LD_{50} in rabbits is >2000 mg/kg. The ADI is 0.125 mg/kg/day. In a 90 day feeding study of rats, the NOEL was 500 ppm (40 mg/kg/day). The compound is not toxic to bees or mammals. Pendimethalin products must carry the signal word 'CAUTION' or 'WARNING'. However, there are indications that this herbicide may prove carcinogenic or harm endocrine functioning.

Most of the pendimethalin, orally administered to rats, is eliminated within 24 h, via faeces and urine. It adsorbs strongly to soil organic matter and clay and does not leach through the soil so does not contaminate groundwater. It is stable to hydrolysis, but is degraded by light in aquatic systems. It also binds to water sediments and gradually degrades in around 2 weeks.

Pendimethalin is registered in countries such as India Australia, most of the EU and the USA.

Penoxsulam herbicide

Penoxsulam, $C_{16}H_{14}F_5N_5O_5S$ (3-(2,2-difluoroethoxy)-N-(5,8-dimethoxy[1,2,4]triazolo[1,5-c]pyrimidin-2-yl)-α,α,α-trifluorotoluene-2-sulfonamide), is a sulfonamide herbicide. It is an off-white-coloured solid with a musty odour. This compound is soluble in distilled water (0.0049 g/l), acetone (20.3 g/l) and methanol (1.48 g/l). It is stable to hydrolysis and not flammable.

Penoxsulam is an acetolactate synthase (ALS) inhibitor herbicide. Used mainly as a post-emergence foliar spray or a granular formulation on water-seeded rice crops, it helps control broadleaved weeds, sedges, aquatic plants and certain grasses. Penoxsulam is also used on fruit orchards and turf to control broadleaved weeds and sedges. Some examples of weeds are white clover (*Trifolium repens*), plantain varieties (*Plantago* spp.), dandelions and pennywort. Penoxsulam comes in liquid and granular formulations.

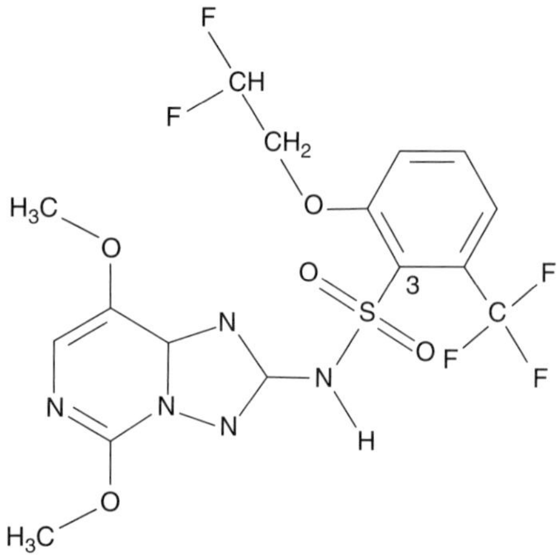

Penoxsulam

The WHO classifies penoxsulam as a Class U toxin (unlikely to be hazardous). It is a mild eye irritant and very slight skin irritant. The NOEL for rats is 500 mg/kg b.w./day. The ADI is 0.147 mg/kg b.w./day. This herbicide poses a risk of being carcinogenic.

In laboratory studies, more than 90% of the dose administered to rats was excreted within 36–48 h, and many metabolites were produced in the urine, faeces and bile. The parent chemical and a 2-hydroxyphenyl derivative were the major compounds. Penoxsulam may be a groundwater contaminant as it does not bind strongly to either soil or sediment, thus tending to be very mobile in both aqueous and terrestrial environments.

Penoxsulam is registered for use in Germany, France, Portugal and the USA.

Pentachlorophenol insecticide

Pentachlorophenol (PCP), C_6HCl_5O (2,3,4,5,6-pentachlorophenol), is an organochlorine insecticide, fungicide and **non-selective** contact herbicide. It has a CAS Registry Number of 87-86-5.

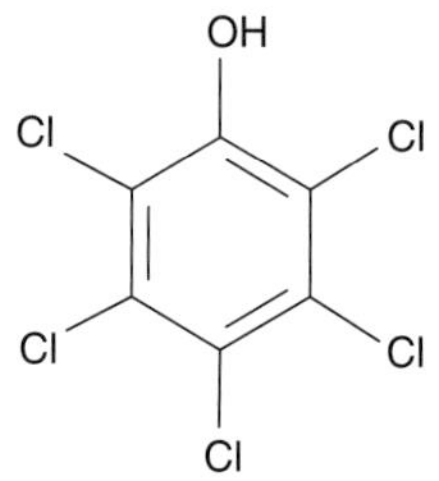

Pentachlorophenol

PCP acts as a **biocide** through its ability to uncouple mitochondrial oxidative phosphorylation. In this form of activity, it has molluscicidal, fungicidal, termiticidal, insecticidal, herbicidal and plant growth regulating (defoliating) features. It is used mainly against fungal rot on wood, wood-boring insects and as a pre-emergence herbicide and defoliant. It may also be used in industrial water systems.

PCP is a moderately toxic compound, with oral LD_{50} values ranging from 27 to 211 mg/kg in rats, 74–130 mg/kg in mice and about 70–300 mg/kg in rabbits. The inhalation LC_{50} in rats is 0.2–2.1 mg/l. It is slightly toxic to practically non-toxic to bird species. This biocide irritates the mucous membranes, skin and eyes. The WHO has listed PCP as a moderately toxic compound (Class II toxin) and it is a restricted use pesticide (RUP) according to the EPA. Labels are required to bear the signal word 'DANGER'.

The biocide PCP decomposes on exposure to sunlight. Tetra-, tri- and dichlorophenols and *m*-chlorophenol are formed in anaerobic soils, resulting in pentachloroanisole and lesser chlorinated phenols. In anaerobic soils, PCP can be degraded within 14 days to 5 years, with help from microbial action and alkaline soils.

PCP is metabolized in rats as the glucuronide. PCP is suspected to be a probable human carcinogen. PCP is banned in India, but registered in the USA, Canada, South Africa and Cameroon.

Pentanochlor herbicide

Pentanochlor, $C_{13}H_{18}ClNO$ ((*RS*)-3′-chloro-2-methylvaler-*p*-toluidide), is an anilide herbicide. Also called solan, it is a colourless solid, soluble in water (8.5 mg/l at 20°C) and xylene (0.25 kg/l). Pentanochlor is stable in aqueous solutions in the pH range 7–9 at room temperature. It also produces toxic chloride and nitrogen oxide gas. The storage environment should be ventilated, cool and dry and away from food.

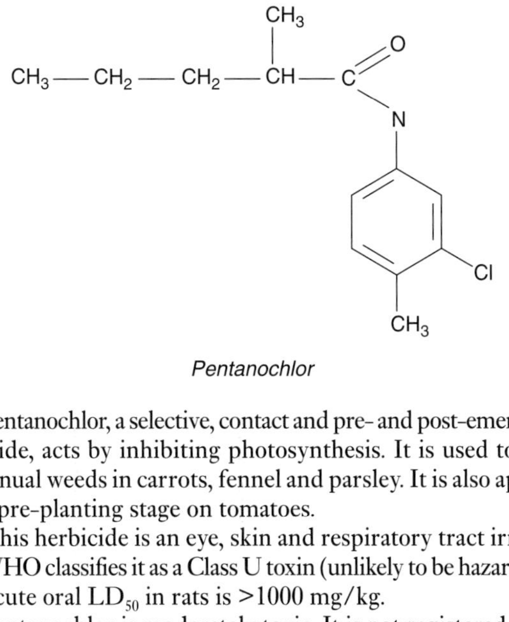

Pentanochlor

Pentanochlor, a selective, contact and pre- and post-emergence herbicide, acts by inhibiting photosynthesis. It is used to control annual weeds in carrots, fennel and parsley. It is also applied at the pre-planting stage on tomatoes.

This herbicide is an eye, skin and respiratory tract irritant. The WHO classifies it as a Class U toxin (unlikely to be hazardous). The acute oral LD_{50} in rats is >1000 mg/kg.

Pentanochlor is moderately toxic. It is not registered in the USA.

Penthiopyrad fungicide

Penthiopyrad, $C_{16}H_{20}F_3N_3OS$ ((*RS*)-*N*-[2-(1,3-dimethylbutyl)-3-thienyl]-1-methyl-3-(trifluoromethyl)-pyrazole-4-carboxamide), is a pyrazolecarboxamide fungicide. It is a white powder, soluble in water to 7.53 mg/l. The CAS Registry Number is 183675-82-3.

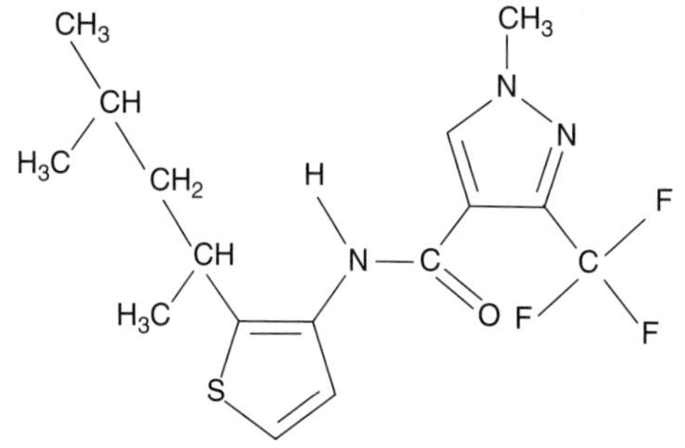

Penthiopyrad

Penthiopyrad acts by disrupting complex II in the respiratory electron transport chain. It inhibits mitochondrial function by affecting the succinate dehydrogenase. Penthiopyrad effectively controls rust, **powdery mildew**, apple **scab**, grey mould and leaf and stem rot on groundnuts, cereals, fruits and vegetables. The pathogens that are kept in check by penthiopyrad are species of *Rhizoctonia* spp., *Podosphaera* spp. and *Venturia* spp.

A common formulation of penthiopyrad is suspension concentrate (SC). This fungicide may be combined with chlorothalonil, **tebuconazole** and **azoxystrobin**. It also complements **sterol biosynthesis inhibitors** (**SBI**) and quinone outside inhibitors (**QoI**).

The acute oral LD_{50} of penthiopyrad in rats is >2000 mg/kg, while a 96 h LC_{50} among carp fish is 1.17 ppm. The inhalation LC_{50} is 5.67 mg/l among rats. The acute 48 h EC_{50} in *Daphnia magna* is 40.0 mg/l.

Penthiopyrad is registered for use in Canada, but it is not registered in the USA.

Pentoxazone herbicide

Pentoxazone, $C_{17}H_{17}ClFNO_4$ (3-(4-chloro-5-cyclopentyloxy-2-fluorophenyl)-5-isopropylidene-1,3-oxazolidine-2, 4-dione), is a herbicide. It is a colourless, crystalline powder soluble in water (0.216 mg/l at 20°C), methanol (24,800 mg/l) and hexane (5100 mg/l at 20°C). Pentoxazone is stable to heat, light and acid.

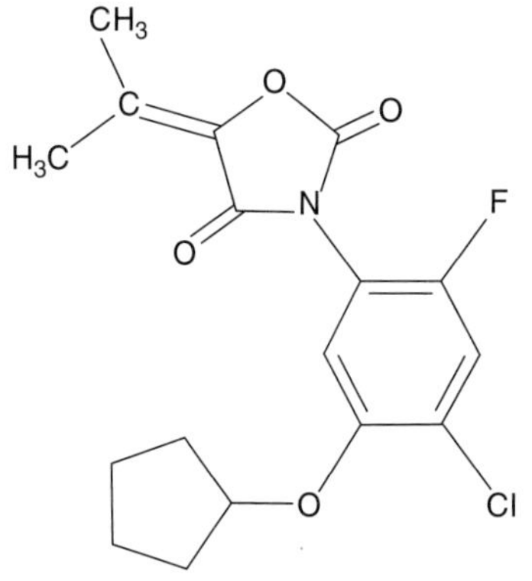

Pentoxazone

Pentoxazone, an oxazolidinedione compound, is an inhibitor of protoporphyrinogen oxidase (**PROTOX**). It is a new oxazolidinedione compound used to control annual broadleaved barnyard grass (*Echinochloa crus-galli*) and pickerel weed (*Monocholia* sp.)

in paddy. Pentoxazone is the only member of the oxazolidinedione class, at present.

Useful formulations include granules (GR) and soluble concentrates (SC).

Pentoxazone is neither a skin nor an eye irritant. It is also not a skin sensitizer. Its acute oral LD_{50} values are >5000 mg/kg in rats and 2250 mg/kg in birds. The acute dermal LD_{50} in rats is >2000 mg/kg. The NOEL in male and female rats is 6.92 mg/kg and 43.8 mg/kg, respectively. This compound is not carcinogenic or teratogenic. About 95% of administered pentoxazone dose is excreted within 48 h. Pentoxazone is not registered in the USA.

Perennial crops

Plants that live for more than 2 years are called perennials. Such plants grow with the help of vegetative structures and root systems.

Perennials survive unfavourable seasons and tillage practices of farmers. A portion of stem or a root (rhizome, bulb or crown) remains dormant until conditions are right for growth.

There are broadly three types of perennials: (i) simple; (ii) creeping; and (iii) woody. **Woody perennials**, which include shrubs and trees, undergo secondary thickening. Most perennials normally propagate by seed but can also be effectively propagated by stems, bulbs and buds.

Perennial weeds are very aggressive and very difficult to control. Both vegetative parts and seeds call for different strategies of weed management. The timing of the management practice is also crucial: for instance weedicides have to be applied at the seedling stage. What makes perennial weed management more complex is the weeds' rapid adaptation to the environment. Although perennials are seen growing in annually cropped land, they are also commonly found in non-cultivated areas such as roadsides, pastures and wasteland.

Woody plants, which include semi-woody vegetation (e.g. bitter nightshade) and near-woody types of weeds, compete with range grasses, which may be utilized as fodder or straw. Examples of some woody species are blackberries, palmettos and members of the Cactaceae family. Woody perennials are also often noxious: for example plants such as poison-ivy of the genus *Rhus* cause skin irritation. Woody plants tend to have a deep root system because of their xeric characteristics (ability to survive in semi-arid regions).

Permethrin insecticide

Permethrin, $C_{21}H_{20}Cl_2O_3$ (3-phenoxyben2yl (1*RS*, 3*RS*; 1*RS*, 3*SR*)-3-(2, 2-dichlorovinyl)-2, 2-dimethylcyclopropanecarboxylate), belongs to the pyrethroid class of synthetic insecticides and **acaricides**. It is an odourless, colourless crystalline solid, stable in light and air and during storage under cool, dry conditions. Permethrin is almost insoluble in water (0.2 mg/l) but highly soluble in xylene, hexane and methanol. It remains stable at higher temperatures (of 50°C) and is more stable in acidic conditions than alkaline. This compound contains halo-substituted chrysanthemate as the acidic moiety. It is produced by the reaction of 3-phenoxybenzyl alcohol with permethrin acid. Permethrin has a CAS Registry Number of 52645-53-1.

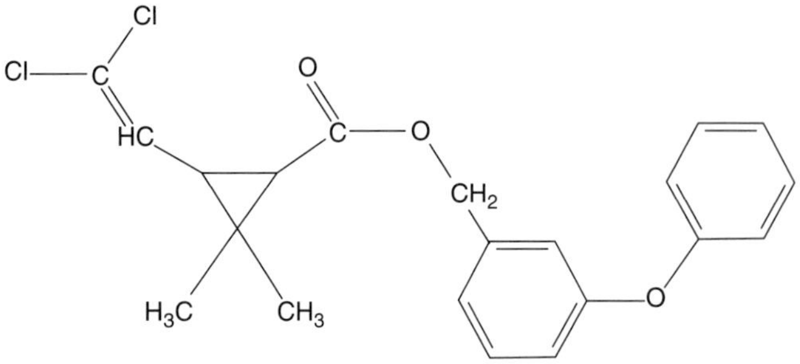

Permethrin

Permethrin is a non-systemic, broad-spectrum insecticide with acaricidal and repellent properties. Being a neurotoxin, this compound binds to sodium channels and thus affects the nervous system of susceptible pests. Its residual action makes this compound more effective. Permethrin is applied to wood-boring insects, beetles, ticks, fleas, lice, ants and termites. In agriculture, it is mainly used on crops such as cotton, wheat, maize and lucerne. A common use of permethrin and other pyrethroids like **resmethrin** and theta-cypermethrin is for protection from household and animal shelter pests.

Formulations of permethrin include dusts (DP), emulsifiable concentrates (EC), smokes (FU) and ULV liquid (UL). Common brand names are Rentokil woodworm killer, Lyclear and Dragnet. Permethrin is applied at the rate of 150 g a.i./ha in cotton, and 25–50 g a.i./ha for fruits. It mixes well with many insecticides and fungicides. Permethrin could injure some ornamental plants.

The WHO lists permethrin as a Class II toxin (moderately hazardous). While this compound is considered of low toxicity for dogs, birds, rats and humans, it is toxic to bees, fish and other aquatic life and beneficial insects. The LC_{50} (at 96 h) in fish species is 2.5 μg/l. In rats, the oral LD_{50} of permethrin is 430 mg/kg and the dermal LD_{50} is over 4000 mg/kg. The ADI is 0.05 mg/kg b.w./day. Permethrin may be a carcinogen and may pose a threat to the endocrine system. Symptoms of poisoning are irritation to skin and eyes, headache, dizziness, nausea, vomiting and sometimes coma and seizures (especially with cyano-pyrethroids). It is necessary to dispose of unused permethrin according to the instructions on the label, and to ensure that liquid formulations and water used for rinsing mixing and loading containers do not leach into groundwater. Permethrin is a PAN Bad Actor chemical.

Mammals eliminate permethrin rapidly via hydrolysis of the ester bond, hydroxylation and elimination as the glucoside conjugate. Both in soil and in water, permethrin degrades rapidly. The DT_{50} in soil is a little over 1 month.

Permethrin is registered by India, Australia, Cambodia, the Philippines, the USA, Canada, South Africa and many African countries.

Persistence

See **Soil persistence of pesticides**

Persistent organic pollutants

Persistent organic pollutants (POPs) are toxic organic compounds that do not degrade through chemical, biological and photolytic processes. POPs have come into sharp focus because of their **chronic toxicity**, accumulation in body fat, and thus in the food chain (biomagnification).

POPs are major environmental contaminants. They are of two major types: (i) organochlorines (OC); and (ii) polycyclic aromatic hydrocarbons (PAHs). PAHs are the result of incomplete combustion of fossil fuels, producing for example asphalt and coal tar. Some POPs have been used in industry or generated as by-products. These include PCBs (polychlorinated biphenyls) and chlorinated dibenzo-*p*-dioxins and dibenzofurans, which may be generated during the manufacture of some chlorinated compounds or by pyrolysis.

Some of the chemical characteristics of POPs include high lipid solubility, semi-volatility and high molecular masses, all of which make them persistent in nature. Semi-volatility allows these chemicals to travel through the air, contaminating the of pops atmosphere. Exposure to such pollutants leads to developmental abnormalities, cell mutation, cancer, allergies and death.

Some POPs have been regularly used as pesticides in the past. However, greater awareness has alerted many countries to the dangers of POPs (although there are exceptions). **Aldrin**, **chlordane**, **DDT**, **dieldrin**, **endrin**, **heptachlor**, hexachlorobenzene, **mirex**, polychlorinated biphenyls, polychlorinated dibenzo-*p*-dioxins, polychlorinated dibenzofurans, and toxaphen are organic pollutants.

There is increasing global awareness about immediately banning POPs in agriculture and their systematic phase-out from other industries. This effort is seen in many deliberations on policy matters, including at the Stockholm Convention and the Rotterdam Convention.

Personal protective equipment (PPE)

Personal protective equipment (PPE) is a group of protective body garments and apparatus that is aimed at preventing direct contact with toxic chemicals like pesticides. Exposure to pesticides has various routes – oral, dermal and by inhalation. Hands, face and the neck are most vulnerable to exposure and hence their protection is an integral part of PPE.

The scope of PPE is very vast due to the variation in the intensity and type of pesticide applications (e.g. mists, vapours, sprays, dusts, corrosive liquids). The rationale behind PPE is that although accidents cannot be prevented, the toll caused by them can certainly be minimized.

Following are examples of a few items of PPE with their characteristic properties.

Gloves, boots, face masks, goggles, respirators, aprons and full body coats are examples of commonly used PPE (Figs P.1, P.2, P.3 and P.4). Gloves and boots are used to protect the head and feet, while goggles and face masks are used for the eyes and nose. Respirators are used to avoid breathing in dust, mist or vapour. Hats and helmets protect the head, hair, eyes and nose from spills, dust and mist. A wide-brimmed, rubber, rain hat is also commonly used. Long, waterproof rubber or plastic aprons are effective especially when mixing or handling concentrated pesticides.

Helmets must have a face shield to protect the face from spills or droplets. Safety goggles are designed in many ways to ensure prevention of the toxin entering the eyes. Some come with full side-shields to protect eyes from stray droplets coming unnoticed. Others come with an attachment for respirators. Still others are stylized as a transparent face shield for better air circulation and greater vision. Safety goggles must meet ANSI Z87.1-1989 standards.

To prevent the entry of pesticides through the ear canal, the use of ear plugs (made of material like polyurethane foam or

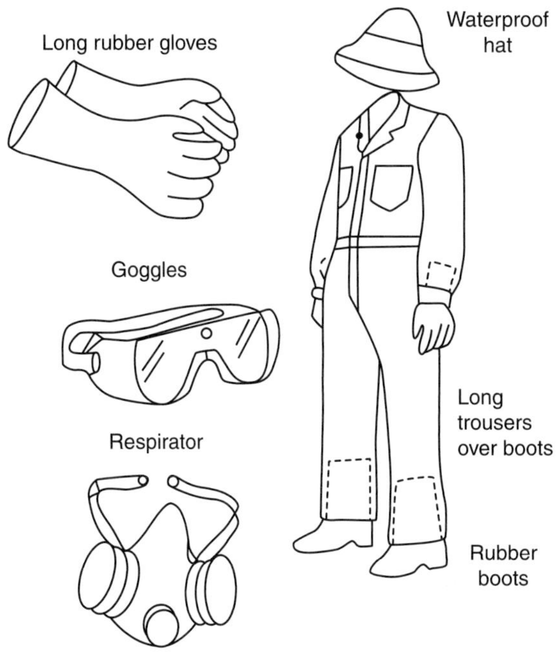

Fig. P.1. Protective clothing and equipment. Copyright of Province of British Columbia. All rights reserved. Reproduced with permission of the Province of British Columbia (Ilona Ugro, Copyright Officer, Intellectual Property Program, Ministry of Citizens' Services) (www.ipp.gov.bc.ca).

Fig. P.2. Apron. Copyright of Province of British Columbia. All rights reserved. Reproduced with permission of the Province of British Columbia (Ilona Ugro, Copyright Officer, Intellectual Property Program, Ministry of Citizens' Services) (www.ipp.gov.bc.ca).

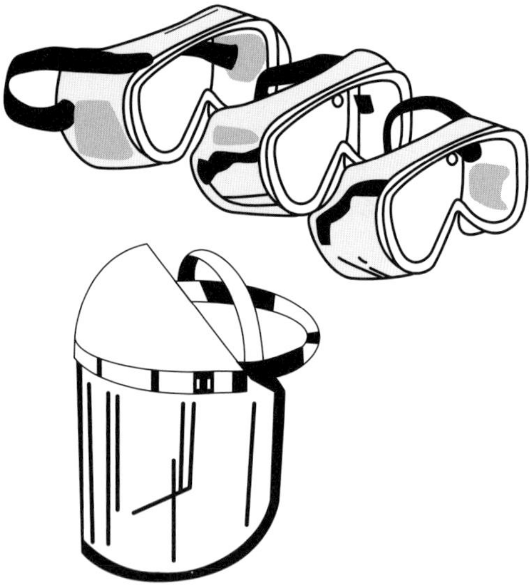

Fig. P.3. Goggles and face shields. Copyright of Province of British Columbia. All rights reserved. Reproduced with permission of the Province of British Columbia (Ilona Ugro, Copyright Officer, Intellectual Property Program, Ministry of Citizens' Services) (www.ipp.gov.bc.ca).

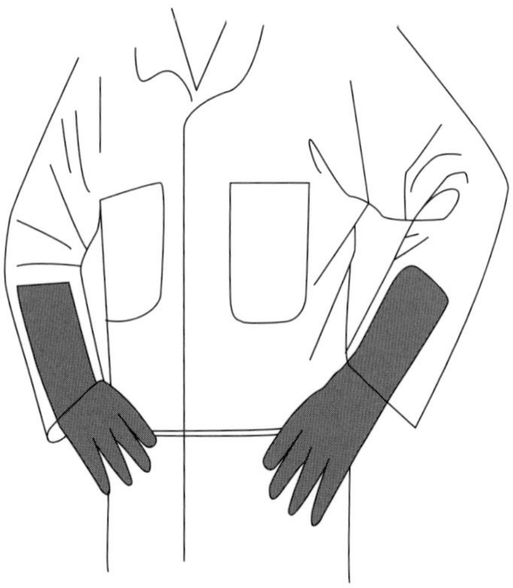

Fig. P.4. Gloves. Copyright of Province of British Columbia. All rights reserved. Reproduced with permission of the Province of British Columbia (Ilona Ugro, Copyright Officer, Intellectual Property Program, Ministry of Citizens' Services) (www.ipp.gov.bc.ca).

cylindrical foam) or muffling is recommended. These ear protectors also reduce noise levels.

Gloves, meant to protect the applicator from dermal exposure, should extend up to the mid-arm. Materials such as cotton or other natural fibres and leather are to be avoided since these absorb chemicals. Rubber, latex, PVC, butyl rubber (>14 ml) or nitrile rubber (≥14 ml) may be used. Similar precautions should be exercised in the case of shoes/boots (Fig. P.5). In the case of both gloves and shoes/boots, they should be checked for areas of damage where leakage may occur before they are worn.

A respirator is a device that offers protection to the lungs and respiratory tract. A respirator filters out particles like dust,

Fig. P.5. Boots. Copyright of Province of British Columbia. All rights reserved. Reproduced with permission of the Province of British Columbia (Ilona Ugro, Copyright Officer, Intellectual Property Program, Ministry of Citizens' Services) (www.ipp.gov.bc.ca).

mist and vapour. Dusts and mists are easy to filter while fumes and vapours are more difficult and require specific protection. Respirators may be in the form of nose filters/disposable masks, cartridge respirators, canister-type respirators/gas masks or self-contained breathing apparatus (SCBA). The structure of a respirator depends on factors such as: (i) the type of filter required; (ii) the toxicity of pesticide; and (iii) the extent of possible exposure. Respirators need approval by The National Institute for Occupational Safety and Health (NIOSH).

A simple filter is a two-strap dust mask which fits conveniently while retaining its shape. Such masks may be equipped with exhale valves. Two-strap dust masks are commonly used while applying dusts and granules. Canister and cartridge respirators are used to filter out vapour (Figs P.6 and P.7); the canister and the cartridge have to be replaced periodically. Most respirators can be attached to the face mask and head gear.

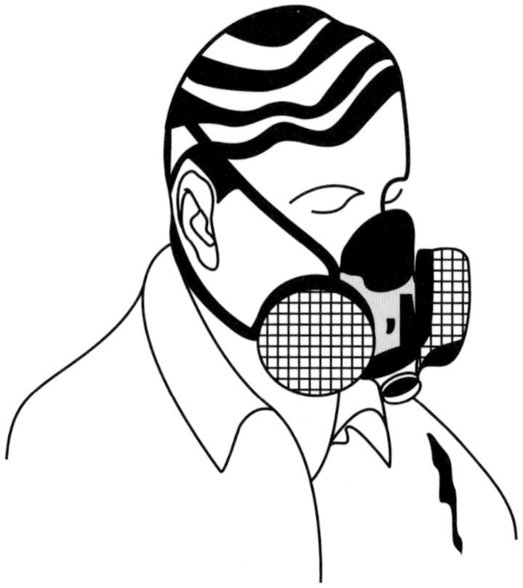

Fig. P.6. Chemical cartridge respirators. Copyright of Province of British Columbia. All rights reserved. Reproduced with permission of the Province of British Columbia (Ilona Ugro, Copyright Officer, Intellectual Property Program, Ministry of Citizens' Services) (www.ipp.gov.bc.ca).

Fig. P.7. Canister respirators. Copyright of Province of British Columbia. All rights reserved. Reproduced with permission of the Province of British Columbia (Ilona Ugro, Copyright Officer, Intellectual Property Program, Ministry of Citizens' Services) (www.ipp.gov.bc.ca).

A powered respirator supplies filtered air to the user and runs on a battery-powered blower. Called powered air protection respirator (PAPR), it makes breathing easier, which is an advantage over the canister/cartridge respirator (Figs P.8 and P.9).

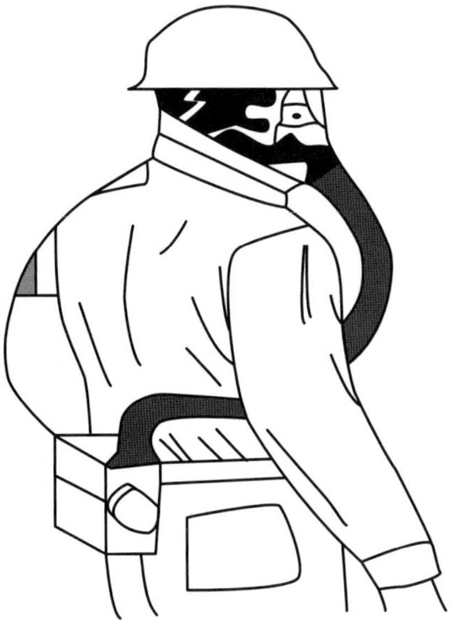

Fig. P.8. Powered air purifiers. Copyright of Province of British Columbia. All rights reserved. Reproduced with permission of the Province of British Columbia (Ilona Ugro, Copyright Officer, Intellectual Property Program, Ministry of Citizens' Services) (www. pp.gov.bc.ca).

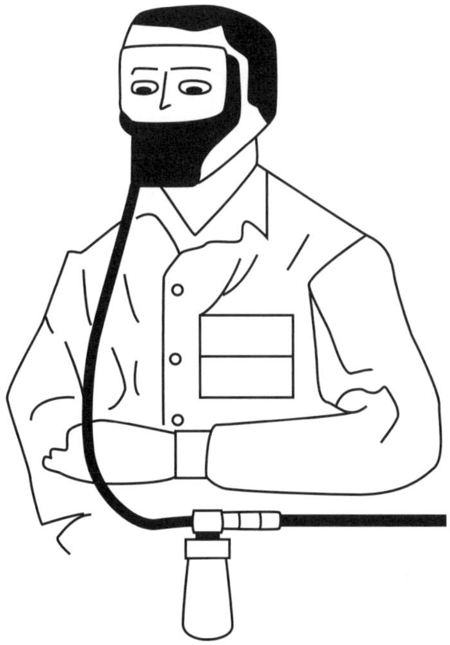

Fig. P.9. Supplied-air respirators. Copyright of Province of British Columbia. All rights reserved. Reproduced with permission of the Province of British Columbia (Ilona Ugro, Copyright Officer, Intellectual Property Program, Ministry of Citizens' Services) (www.ipp.gov.bc.ca).

Other equipment, like SCBA, is useful when toxic pollutants are present in high density and when oxygen may be insufficient. The SCBA includes a full-face gas mask, metering devices and a cylinder of compressed air. Only trained personnel are allowed to use SCBA.

Any fumigation procedure has to be undertaken only with specialized PPE, since many fumigating compounds can penetrate rubber, neoprene and leather and cause severe injury. Full-face masks and complete protective clothing (hat, jacket, trousers or coveralls, rubber gloves and an airtight, full-face mask) are required to be worn, with specially made, freshly opened canisters. All pesticide formulations are required to be used strictly according to label instructions.

Care, maintenance and disposal of PPE are important to note. To begin with, only such PPE that fit properly and comfortably should be worn. Caring for PPE begins with following instructions on the label. Contaminated PPE items that are re-usable should be washed separately, wearing chemically-resistant gloves. Boots, helmets, goggles, respirators and other bulky items may be washed by hand (again wearing gloves), while others may be washed in a washing machine with a strong detergent and hot water. After every wash, the machine should be run empty for an entire cycle. PPE can be dried outdoors in the sunlight or in a dryer. Again, the dryer must be cleaned of pesticide residues. Disposables should be discarded as per disposal norms.

Environmental factors for PPE are to be kept in mind while choosing PPE. The choice of PPE is influenced by environmental conditions, such as the location of the site, the temperature during application and other factors. A practical and safe approach to PPE is necessary during pesticide application. Hot and humid conditions make it very cumbersome to wear plastic or rubberized clothing and hence there is reluctance to use such clothing. For the safety of workers, application during early morning/late evening hours is recommended. In highly windy sites, face masks, gloves and shoes may be more important than respirators. On the other hand, in forestry applications (truck injections, weed control, etc.), a thick coverall and gloves with sturdy boots and a simple face mask might be more important than chemical resistant goggles and gloves, and rubberized clothing, since the bigger danger may emerge from insects, mites and reptiles.

Pest – alfalfa (lucerne) weevil

See **Crop – lucerne (alfalfa) pest control**

Pest – American bollworm

See **Crop – cotton pest control**

Pest – aphid

Aphids are sap-sucking insects that belong to order Hemiptera and the superfamily Aphidoidea. These are 1–10 mm in length and have soft bodies that are green, brown or pink (Fig. P.10). They have a segmented antenna and a small protrusion in the region of the rectum. Aphids feed with the help of a stylet, which is a hard and sharp structure at the mouth. Reproduction happens sexually and asexually, resulting in eggs and live young nymphs.

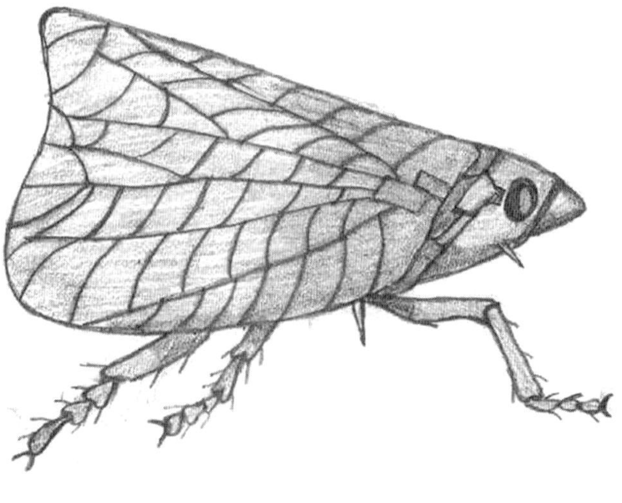

Fig. P.10. Aphid. Sketch courtesy of Leena Achari, with permission.

Known by many names such as greenfly, lice or blackfly, aphids are widely studied because of the destruction they cause to crops. From over 4000 species from ten known families, around 250 species are serious pests of crops and forest trees. Some aphid species are monophagous (those that feed on only one plant species), while others are polyphagous (e.g. the green peach aphid *Myzus persicae*). Many species are also known to inject a triglycerol substance, for example in defence against a predator, and this may be toxic to the predator. Some aphids also excrete a dark and shiny substance called **honeydew**, which attracts fungal pathogens (e.g. causing **black sooty mould**) and ants.

Common symptoms of aphid infestation include: (i) leaf curling and yellowing; (ii) curling of leaves and stunting of the plant; (iii) development of galls inside plants; and (iv) growth of a cottony substance on plant surfaces. The injection of the toxin exudate stunts plant growth. Aphids also act as **vectors** of viruses. Peanut stripe virus and the groundnut rosette virus complex are two major viruses spread by aphids, causing mottling, yellowing or curling of leaves. Crops like cucumbers, squashes, pumpkins, beans, potatoes, lettuces and beets are affected by aphid attack. As well as affecting plant leaves, aphids can also affect plant shoots and flowers and sometimes roots (e.g. lettuce root aphid).

The feeding pattern of aphids (i.e. puncturing plant phloem vessel and feeding on the sap which is rich in sugars and amino acids) and the huge population they produce (Fig. P.11 shows their life cycle), make their management and control important in today's age of world food shortage. Following are a few strategies for their management.

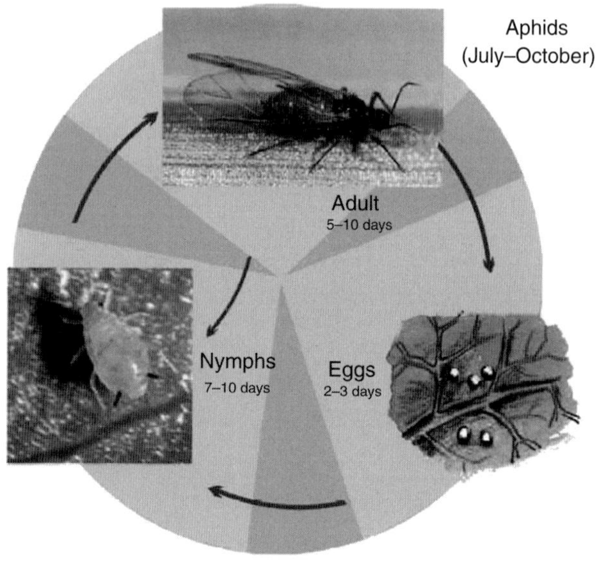

Fig. P.11. Life cycle of aphids. Courtesy of: http://takingroots.in/ipm_aphids (licensed under creative commons) with permission from Ramanjaneyulu GV, Executive Director, Centre for Sustainable Agriculture, Secunderabad, India.

A strong spray of water on plants washes off aphids. Over-fertilization by nitrogen has to be avoided to prevent excessively succulent plant growth. The deliberate use of natural enemies of aphids is an effective aphid control method. Natural enemies include predatory **ladybirds**, parasitic wasps, syrphid fly aphid midge larvae, crab spiders, common **lacewings** and entomopathogenic fungi (e.g. *Lecanicillium lecanii*). Various species of parasitic wasps lay their eggs inside an aphid, making its skin turn crusty and brown, a form called the **aphid mummy**. Usually, within 14 days of sighting a mummy, aphid populations decrease, especially in warm and humid conditions. Aphids are also susceptible to certain bacteria and fungi. Entire colonies may be wiped out in humid conditions by a fungal infection, after causing the aphids to turn reddish brown, with a shrivelled texture.

Hot and sunny days and extremely cold temperatures can limit populations of many aphid species. Chemical pesticides used for aphid control are **dimethoate**, methyl parathion, **monocrotophos** and **chlorpyrifos** (used at a concentration of 0.05%). Dust formulations are used at higher concentrations (1.5% and above). A spray of **neem** seed kernel extract (5%) a week after the first incidence, followed by an insecticidal spray 2 weeks later can control aphids.

Pest – brinjal/aubergine shoot and fruit borer

The shoot and fruit borer *Leucinodes orbonalis* (family Pyraustidae, order Lepidoptera) is a borer pest. The adult is a moth with brown and white patches and a wingspan of about 24 mm. The moth becomes active at sunset.

L. orbonalis is believed to have originated in the tropics and is widespread in India, and many other countries in Asia and Africa. This pest is significant in our present context since its favoured hosts are plants of the genus *Solanum* (e.g. aubergine/brinjal (*Solanum indicum*), potato, tomato and some medicinal herbs such as *Solanum xanthocarpum*); it also affects other plants such as capsicum, beetroot, green peas and tobacco. The shoot and fruit borer is popularly known as the brinjal/aubergine borer because of the pest's affinity to this crop.

The adult moth of *L. orbonalis* lays its eggs on the shoot tips and flower buds. The pinkish white larvae that hatch out of the eggs cause the damage to the plant by boring into the fruit and shoot, leaving their faecal matter in and on the plant. Subsequently, the larvae pupate in the soil.

IPM techniques for the borer include: (i) practising crop rotation; (ii) collection and destruction of dried shoot tips and damaged fruits; (iii) use of pheromone traps for controlling and monitoring the pest; and (iv) spraying repellents made of **neem** seed kernels.

Spray application of **phosalone** (0.07%), **carbaryl** (0.1%), **profenofos** (0.05%) or **cypermethrin** (0.025%) normally renders borer control.

Pest – brown plant hopper (BPH) of rice

See **Pest – plant hopper; Pest – rice brown plant hopper (BPH)**

Pest – castor semilooper

The adult castor semilooper, *Achaea janata* (order Lepidoptera) is a defoliating type of pest. It is a dark brown moth with black and white patterns on the rear wings. The adult lays blue-green eggs on both sides of the leaf, which hatch in 3–5 days to produce brown-coloured larvae with black lines running along the sides, or black-coloured larvae with white spots. These feed extensively on castor leaves, severely defoliating the plant and leaving behind only skeletal venation. After feeding for almost 1 month, the larvae pupate in silky cocoons among the fallen leaves. Within the next 10–15 days, the pupae are transformed into dark-brown-coloured adult moths. The period of pest intensity is high from February to October throughout India (Fig. P.12).

IPM techniques to control the castor semilooper include: (i) deep summer ploughing, which helps destroy the pupae; (ii) avoiding the use of chemical pesticides in order to encourage natural control agents such as ***Trichogramma***, *Apanteles*, *Bracon*, ***Chrysopa*** and **ladybird beetles**; (iii) spraying a suspension of 5% neem seed kernel to repel the egg-laying adults; and (iv) spraying a chilli-garlic suspension to destroy the young larvae.

Ecological methods effective for pest control include: (i) using light traps to capture and destroy adult insects; (ii) placing bird perches in the crop to attract birds that control the larval population; and (iii) releasing eggs of *Trichogramma evanescens minutum* at 100,000/ha.

Fig. P.12. Life cycle of castor semilooper. Courtesy of: http://takingroots.in/ipm_castor_semilooper (licensed under creative commons) with permission from Ramanjaneyulu GV, Executive Director, Centre for Sustainable Agriculture, Secunderabad, India.

Chemical measures to control pests include spraying of **quinalphos** (2 ml/l water), **chlorpyrifos** (2 ml/l water), **carbaryl** (3 g/l water), **monocrotophos** (1.5 ml/l water) or **endosulfan** (2 ml/l water) if four to five semiloopers are observed per plant after about 1 month of sowing.

Pest – castor stem and capsule borer

The adult stem and capsule borer of castor, *Dichocrosis punctiferalis* (Pyralidae, order Lepidoptera), is a borer type of a pest. The adult, a yellow moth with black spots, lays pink eggs on flower buds or on young capsules. These hatch within 1 week to produce pale reddish-brown coloured larvae with outgrowths all over the body. The larvae bore into the stem and capsules for 2–3 months, causing the stem to dry up. Then they pupate there for about 1 week after which the adult yellow moths emerge. The period of pest intensity is high during September to March in India (Fig. P.13).

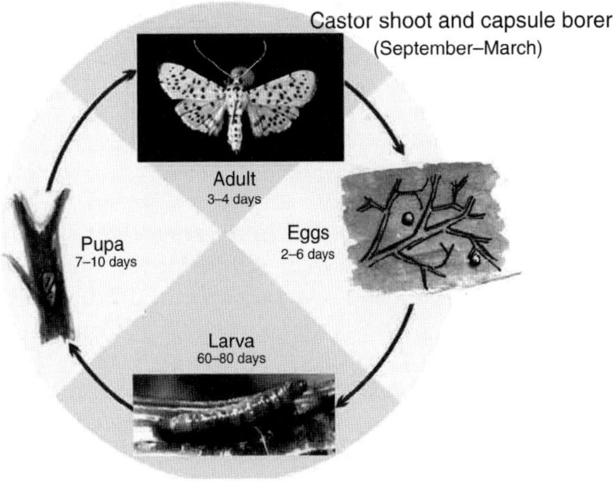

Fig. P.13. Life cycle of castor shoot and capsule borer. Courtesy of: http://takingroots.in/ipm_cscb (licensed under creative commons) with permission from Ramanjaneyulu GV, Executive Director, Centre for Sustainable Agriculture, Secunderabad, India.

IPM measures to control castor stem and capsule borer include: (i) collecting and destroying the dried capsules and branches, thus killing the larval and pupal stages; and (ii) spraying 5% **neem** seed kernel suspension to prevent the adults from laying eggs.

Chemical control measures include spraying with pesticides such as **monocrotophos** (1.6 ml/l water) or **dimethoate** (2 ml/l), **quinalphos** (1.5%) or **parathion-methyl** (2%) at the flowering stage and a second spraying after 15–20 days.

Pest – chilli thrips

Chilli thrip, *Scirtothrips dorsalis* (also called yellow tea thrip), is an insect of the order **Thysanoptera**. It is an insect pest, believed to have originated in tropical countries which subsequently spread to other regions. This pest affects a range of crops such as asparagus, bananas, castor beans, grapes, strawberries, tomatoes, tobacco, pepper, cotton, tea, groundnuts and roses.

Chilli thrips are yellow coloured with dark antennae and dark stripes on the lower abdomen (Fig. P.14). The females lay eggs (up to 40), usually in buds and young leaves. There are two nymphal and two pupal stages. Pupation takes place in the soil or the biomass around the host plant. They develop within 2 weeks to reach adulthood. Chilli thrips feed on new growth, especially at the larval stage, leading to the wrinkling and curling of leaves, damage to leaves and flower buds, and sometimes plant death. Chilli thrips are also thought to be **vectors** of three **tospoviruses**.

Fig. P.14. Adult chilli thrip. Courtesy of L. Osborne, University of Florida, Professional Landscape Management Recommendations for Control of Chilli Thrips. With permission.

Chilli thrips are controlled by spraying the foliage with pesticides such as chlorofenapyr, **imidacloprid**, **abamectin**, **acephate**, spinosyn A and B, **azadirachtin**, **thiamethoxam** or **cyfluthrin**.

Pest – Colorado potato beetle

The Colorado potato beetle (*Leptinotarsa decemlineata*) is an important pest of potato crops. Also known as the Colorado beetle or the potato bug, this leaf beetle belongs to family Chrysomelidae. It is 10 mm long, with a yellow/orange body and five brown stripes along each of the front pairs of wings (Fig. P.15). Besides potato, the pest attacks pepper, aubergine, tomato and tobacco.

Fig. P.15. Colorado potato beetle. Courtesy of John Obermeyer, Purdue Extension Entomology. With permission.

The Colorado potato beetle is a defoliating pest. Both adults and larvae feed on foliage and can completely defoliate entire potato fields. Severe infestation can significantly reduce yield. Larvae (especially late instar larvae) tend to cause more damage than adults. The presence of yellow eggs under the leaves and scattered eaten leaves are initial signs of infestation.

The practice of mulching is considered a control measure for this pest. Mulching done up to 30 cm makes it difficult for adult beetles to climb up the stems. Chemical pesticides produce resistance in the Colorado potato beetle. **Rotenone** (plant-derived toxin) and **pyrethrum** (extracted from chrysanthemum) are effective control measures. Two applications within 1 week have a controlling effect on the beetle.

Beauveria bassiana is a pathogenic fungus that infects the Colorado potato beetle. Commercial formulations of the fungus are applied as a spray. **Azadirachtin** in **neem** oil can disturb the **moulting** of beetles at the larval stage. Neem oil is commonly sprayed on the leaves for best results. Timing is of the essence here, since neem oil might be ineffective after the larval stage. ***Bacillus thuringiensis***, parasitic nematodes, lady beetles and spined soldier bugs are all effective against the larvae. Parasitic nematode species feed on the larvae on the ground.

Horseradish, flax, coriander and nasturtium, grown around susceptible crops, can repel the potato beetle. The use of resistant plant varieties and nylon netting can prevent infestation.

Pest – corn earworm

The corn earworm is another term for the American bollworm (*Helicoverpa armigera*). (See also **Crop – cotton pest control**.)

Pest – corn plant hopper

See **Pest – plant hopper**

Pest – *Daktulosphaera vitifoliae*

A major pest of grapevines, *Daktulosphaera vitifoliae*, also known as grape phylloxera, is a member of the order Hemiptera and family Phylloxeridae and is categorized as a louse, although it is related to aphids. The sap-sucking adult females are normally tiny, winged, pale yellow to greenish brown and are similar to aphids. However, they have a very long drawn-out life cycle with root, leaf, winged (Fig. P.16) and sexual forms. Phylloxeras prefer thick, clayey soil, making their control difficult.

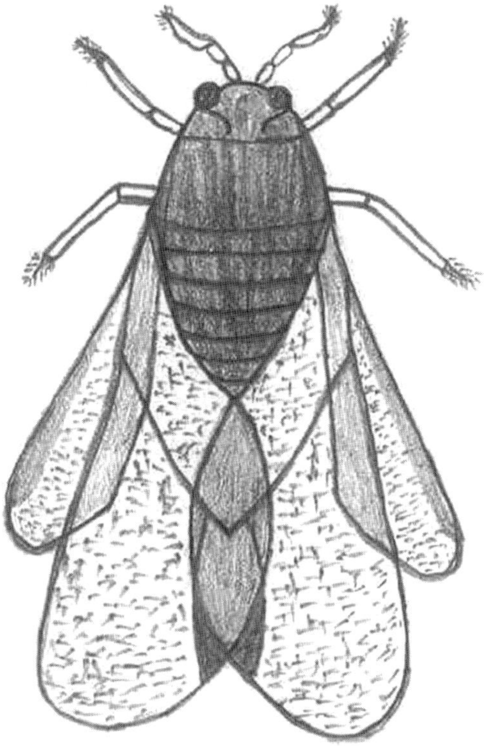

Fig. P.16. Phylloxera. Sketch courtesy of Leena Achari, with permission.

The nymphs of each of these forms suck plant sap, damaging leaves or roots and finally killing the plant. When they feed on the roots of grapevines, they form knot-like galls on them. Such deformed roots can no longer transport nutrients to the rest of the plant, and they become vulnerable to secondary fungal infections. Nymphs also form protective wart-like galls on leaves.

Using chemicals such as **endosulfan** may help kill a part of the population, but since there are many generations inhabiting the soil, feeder roots and leaves, effective control cannot be achieved. The use of resistant rootstock is a major step in the prevention of phylloxera infestation. Thoroughly clean agricultural equipment prevents the spread of infestation. Soil solarization, incorporation of organic matter in the soil and water management may be other cultural measures to prevent vine damage. The crop should be inspected regularly so that action can be taken at the first sight of weakly growing or necrotic patches among the crop.

Pest – grasshopper

Grasshoppers are insects belonging to the suborder Caelifera in the order **Orthoptera** (this order also includes cockroaches). They are herbivorous insects that have short ovipositors and two antennae. They have strong and large hind legs, with which they leap (Fig. P.17). While the forewings are unsuitable for flight, the hind wings are fan-like and membranous and are used for flying. They are folded under the forewings when at rest. Some species make prominent sounds by rubbing their hind legs on their forewings.

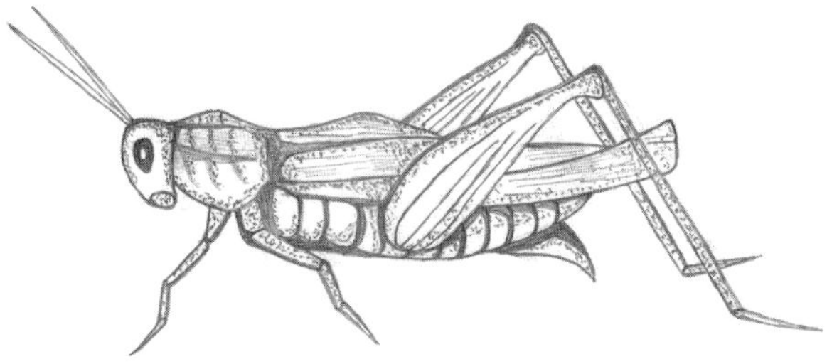

Fig. P.17. Grasshopper. Sketch courtesy of Leena Achari, with permission.

Female grasshoppers lay their eggs in the soil and these eggs overwinter until they hatch in warm weather. The wingless nymphs that emerge begin to feed on leaves from the first day. They then go through five **moulting** stages until they turn into adults.

Although grasshoppers are voracious eaters, they are not choosy about their plant food. Crops are thus not directly threatened by these insects, unless there is nothing else available.

Species that form swarms are short-horned grasshoppers called locusts and during such group movement, they display a distinct behaviour. They also change colour when in groups. Locusts are known to have destroyed entire crops in very short periods of time.

Grasshoppers are not considered to be a major crop pest, with only a few species being a potential threat to fields (as shown in Table P.1). Pesticides such as **malathion**, **carbaryl** and **diflubenzuron** are used to control grasshoppers. Common formulations include emulsifiable concentrates and wettable powders, sprayed on fruit trees, pines and any crop with severe infestation. However, since some grasshopper species are part of the diet in some parts of the world, application of chemicals should be done with caution. (See also **Pest – locust**.)

Pest – groundnut leaf miner

The adult groundnut leaf miner or *Aproaerema modicella* (family Gelechiidae and order Lepidoptera) is a defoliating pest. Besides groundnut, it affects crops such as pigeon pea or *Cajanus cajan* (called *toor* or *arhar* in India), soybean (*Glycine max*) and some other minor crops.

The groundnut leaf miner is a brownish-grey coloured moth (Fig. P.18). The female lays about 200 shiny white eggs along the midrib on the lower side of leaflets. These hatch to produce pale green-coloured larvae with black heads. Young larvae mine between the epidermal layers of the leaf. This leads to the drying of leaves. Larvae bring the edges of the leaf together and pupate inside it before growing into the adult moth. This major pest of groundnut is found all over India, its intensity being the highest in September and October. The groundnut leaf miner is also a serious pest in the groundnut and soybean growing areas of South and South-east Asia.

IPM techniques for the groundnut leaf miner include: (i) simple rotation of crops, as the pest cannot survive on other crops; (ii) avoiding chemical pesticides, to help increase the number

Table P.1. Some important grasshopper pests worldwide. (Table courtesy of Kedar Champhekar.)

Species	Distribution	Crops affected
Zonocerus variegatus (variegated grasshopper)	Africa	Cassava, finger millet, cocoa, castor, cashew, coffee, cotton, sweet potato
Melanoplus bivittatus, M. differentialis	North America	Lucerne, small grains, maize
Camnula pellucida (clear-winged grasshopper)	North America	Small grains, cereals, forage grasses, vegetable crops
Brachystola magna	USA and Mexico	Cotton
Oedalus senegalensis (Senegalese grasshopper)	Semi-arid grasslands in Africa, Middle East and India	Millets

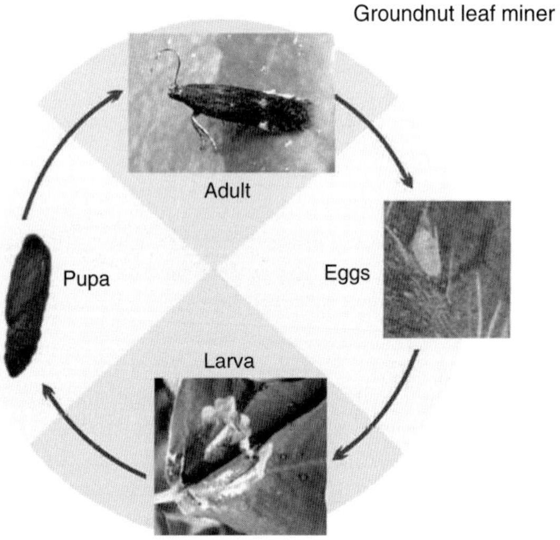

Fig. P.18. Groundnut leaf miner. Courtesy of: http://takingroots.in/ipm_glm (licensed under creative commons) with permission from Ramanjaneyulu GV, Executive Director, Centre for Sustainable Agriculture, Secunderabad, India.

of parasites that can naturally control the pest; (iii) spraying 5% **neem** seed kernel suspension to prevent adult moths from mating and laying eggs; and (iv) setting up light traps to capture and kill adult insects.

Chemical control measures include dusting with **phosalone** (4%) or **carbaryl** (10%), or spraying **fenitrothion** (0.025%), **phosalone** (0.05%) or **monocrotophos** (0.05%). **Chlorpyrifos** is also commonly applied to control the leaf miner. (See also **Crops – groundnut pest control**.)

Pest – grub

Grub worms or **white grubs** are terms for the larvae of several beetle species in the Scarabaeidae family. Beetles in this family are commonly called scarabs and include Japanese beetles, dung beetles and June bugs. Grubs are considered crop pests due to their feeding tendency.

Grubs are roundish, soft, whitish worms with a brown head. Their bodies become C-shaped when feeding. Adults mate in trees and lay their eggs in the soil in May or June, from which larvae emerge in about 2 months. The larvae start feeding on the rotting matter near trees or crops and gradually move downwards towards the plant roots, as the second instar. Here, they may overwinter under the soil. After spring, the grubs start coming to the surface (from more than 40 cm underground) and continue feeding until the next winter, when as third instars, they overwinter again. By the following May, pupae and adults are found in the upper surface of the soil, and then, the cycle is repeated. The greatest damage is caused in the second year, when feeding is at its maximum. Grub worms prefer well-drained, loose but moist soil.

Since all grub species are not crop pests, care should be taken to target only those that cause harm to crops. Many scavenge on dung and organic matter.

Phyllophaga implicita (June beetles), *Cyclodephala* spp. and *Popillia japonica* (Japanese beetles) are some important grub species. Areca nut, coconut, tapioca, sugarcane, maize, wheat, oats, barley, sugarbeet, soybeans, sweet potatoes, potatoes, turf and ornamental plants are some of the crops affected by grubs. Affected plants wilt and droop, and often die. The others have a stunted growth. Common signs of infestations in lawns are brownish patches and brittle roots. The presence of around five grubs per grass clump is considered a severe attack.

It is important that any chemical intervention be done either just before or after the beetles lay their eggs, or when the grubs have just hatched. **Imidacloprid** may be applied at the pre-egg stage, while trichlorfen and **carbaryl** may be used on young larvae. **Diazinon** may be used, if it is legally allowed in the region. Non-chemical methods include the use of milky spores, a slow-acting bacterium that infects the grubs of Japanese beetles. This bacterium is said to continue its life cycle for a long time, keeping a control on grub populations. The entomopathogenic nematode, *Heterorhabditis bacteriophora*, which is an insect control agent and a host to a beneficial bacterium *Photorhabdus luminescens*, also controls grubs.

Pest – *Helicoverpa armigera*

See **Crop – cotton pest control**

Pest – leafhopper

Leafhoppers are small jumping insects from the family Cicadellidae in the order Hemiptera. They are about 2–30 mm long. The life cycle includes eggs, followed by a nymphal stage (with brightly coloured nymphs) and this is directly followed by the emergence of the adult (Fig. P.19). The adults have sharp, piercing and sucking mouthparts, with which they feed on the plant sap of a variety of plants. Leafhoppers produce a protective granular substance called brochosome, with which they groom themselves.

Leafhoppers are a widespread and diverse group of insects and many species are known to be parasitic on different plants all over the world. They attack rice, maize, potato, tomato, fruit trees, sugarcane, lucerne and ornamentals (e.g. roses).

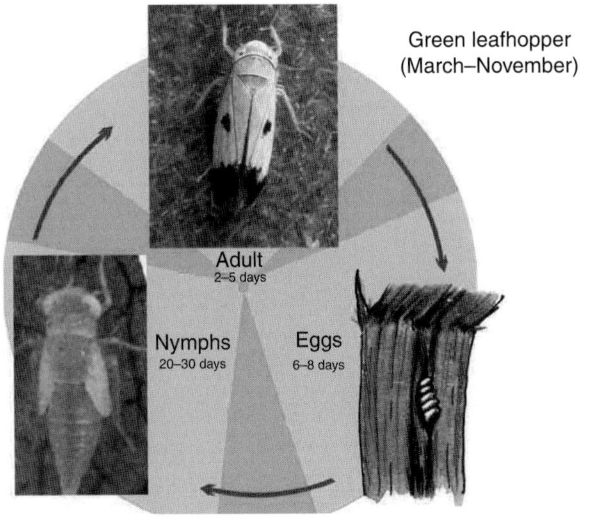

Fig. P.19. Green leafhopper. Courtesy of: http://takingroots.in/ipm_rglh (licensed under creative commons) with permission from Ramanjaneyulu GV, Executive Director, Centre for Sustainable Agriculture, Secunderabad, India.

Maize leafhoppers (*Cicadulina bimaculata*) are common in fields of maize, sorghum and sugarcane. Apart from the direct damage to crops, they transmit the maize streak virus. *Homalodisca vitripennis*, a serious pest of grapes, citrus and almonds, transmits the bacterial plant pathogen *Xylella fastidiosa* (which causes citrus-X disease). Some of the important pest species among leafhoppers are the green rice leafhopper (*Nephotettix* spp.) which attacks rice and the beet leafhopper (*Circulifer tenellus*) which transmits the curly top virus to many plants in the aubergine (brinjal) family. The use of pesticides, **parasitoid** wasps (from the family Myrmaridae) and viruses are some control measures.

Pest – locust

Locusts belong to the insect order **Orthoptera** and are a kind of **grasshopper**. However, they differ from grasshoppers in their behaviour. Locusts show two distinct phases in their life cycle, the solitary phase and the gregarious phase, the transition being indicated by a change in colour and behaviour.

In warm and wet conditions (e.g. during rainfall), females deposit eggs a few centimetres below the soil surface with their ovipositors (Fig. P.20). The eggs hatch into nymphs in 2 weeks and crawl out of the soil. Nymphs, also called hoppers, go through a series of instars or development stages before becoming winged adults. With millions of such nymphs growing together, they form aggregates, called bands. These bands fly across large distances, at a low height, devastating any vegetation on the way. Adult locusts also form aggregates called swarms in a similar fashion and migrate in search of fresh vegetation.

Fig. P.20. Female locust depositing eggs in sand. Sketch courtesy of Kedar Champhekar, with permission.

Probably the most serious locust pest is the desert locust (*Schistocera gregaria*) whose nymphs change colour from green to yellow and black and whose adults change from brown to red to yellow. This dramatic change in colour occurs when they transform from the solitary to the gregarious phase. Locusts can produce up to four generations in a year. They are said to voraciously feed on plants and usually travel up to 200 km/day.

Swarms of up to 80 million locusts/km^2 travel in search of food and converge on vegetation of almost any type. These pests can devastate an area of about 1200 km^2 in a very short time. Plagues of these locusts have been occurring since 1926 and continue almost every alternate year in their favoured regions. The desert locust has a wide distribution affecting crops across North Africa (including the Sahara desert and the Arabian Peninsula), north-western India and Australia. When such plagues occur in African nations, they worsen the famine situation there.

A few species of locusts that are a threat to crops are listed in Table P.2.

Table P.2. Important pest species of locusts worldwide. (Table courtesy of Kedar Champhekar.)

Species	Distribution	Crops affected
Schistocera gregaria (desert locust)	North Africa, the Sahara, the Arabian Peninsula, north-western India and Australia	Highly polyphagous, prefers cereals
Locusta migratoria (migratory locust)	Four subspecies covering Africa, Asia and Australasia	Cereals
Nomadacris septemfasciata (red locust)	Central and southern Africa	Cereals, other grass crops
Chortoicetes terminifera (Australian plague locust)	Australia	Damage to pasture and young crops
Locustana pardalina (brown locust)	Southern Africa	Cereals
Patanga succincta (Bombay locust)	South-east Asia, India, South China	Maize and other cereals

Pest – mealybug

The *Solenopsis* mealybug or *Phenacoccus solenopsis* (in the order Hemiptera and family Pseudococcidae) is a sucking pest with a varied host range that includes cotton, okra, sugarcane, grape, coffee, tomato, aubergine (brinjal) and ornamental plants. Mealybugs, believed to have originated in the USA, are included in the group of **scale insects**.

The adults generally reproduce parthenogenically. The wingless females retain nymphal characteristics, this phenomenon being called neoteny. Females lay a group of up to 600 small eggs in ovisacs, most of which hatch as female nymphs within 10 days, and grow into crawling adults within 3 weeks. Mealybugs, which can have 15 generations in a year, can survive harsh cold conditions in the egg stage.

Both nymphs and adults suck sap from leaves, twigs, stems, roots and fruits, thus weakening the plant, distorting shoots and twisting leaves. They secrete a powdery wax layer around them for protection. They also produce a sugary by-product called **honeydew** that attracts ants and aids the development of black **sooty mould**. The mould slows photosynthesis in the plant reducing growth of healthy new shoots, leaves or fruit.

Infestation spreads as the nymphs are transported by wind, rain, birds, farm tools and by the ants that feed on the honeydew.

Important pest species are the pink hibiscus mealybug, the grape mealybug, the obscure mealybug and the citrus mealybug (Fig. P.21).

Strategies for IPM of the *Solenopsis* mealybug include: (i) release of predators such as the wasp *Aenasius bambawalei* (order Hymenoptera) and **coccinellid** beetles, popularly known as **ladybugs** or lady beetles (family Coccinellidae, order Coleoptera); (ii) removal of alternate host plants such as *Parthenium* sp.; and (iii) destroying ant colonies during land preparation.

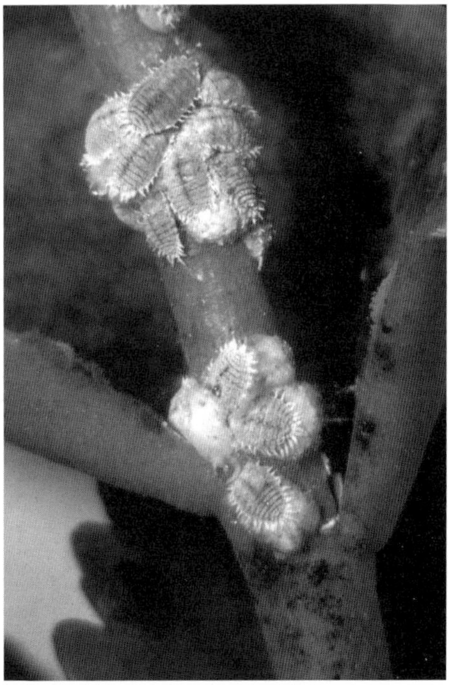

Fig. P.21. Citrus mealybugs (*Planococcus citri*). Courtesy of United States National Collection of Scale Insects Photographs Archive, USDA Agricultural Research Service, Bugwood.org, with permission.

The threshold for pesticide intervention is 40 plants with more than one stem completely colonized by the bug. Organophosphate and carbamate insecticides are commonly preferred for the control of the mealybug. Spraying of a wettable powder (WP) or an emulsifiable concentrate (EC) of **carbaryl** (2500 g/ha), **thiodicarb** (625 g/ha) and **quinalphos** (2000 ml/ha) is very effective. Similarly, **chlorpyrifos**, **diazinon**, **profenofos**, and a spray of fish oil, **neem** oil and **buprofezin**, are all also common. Most of these applications may be repeated up to three times in a season.

Pest – mite

Mites and ticks are the two small arthropod members of the subclass Acari (or Acarina) and the **class Arachnida**. The study of ticks and mites is called acarology. Mites are a predominant group of this subclass.

Mites are extremely small-sized, invertebrate, free-living or parasitic organisms that can adapt to several habitats. With over 48,200 species, mites can have several hosts including plants, honey bees, mammals and humans. Low relative humidity favours the multiplication of mites. The life cycle of mites, which is completed in less than 1 month, includes eggs, larvae, two nymphal stages (protonymph and deutonymph) and adult (Fig. P.22).

Commonly known mite species are spider mites, gall mites, house dust mites, strawberry mites and red spider mites. They belong to families such as the Tetranychidae, Eriophyidae and Pyroglyphidae. Plant parasitic mites include all species of the genus *Tetranychus* (including red spider mite, two spotted spider mite (*Tetranychus urticae*), strawberry mite and Pacific mite) (Fig. P.23). Mites are commonly a problem on tomato, beans, aubergine and cucumber, in orchards and on ornamentals.

Most *Tetranycus* species range from yellow to greenish and red to brownish in colour. They feed on the leaf-cell contents using their piercing mouthparts. They prefer to feed on the underside of the leaf. Since mite populations can be very large, with overlapping

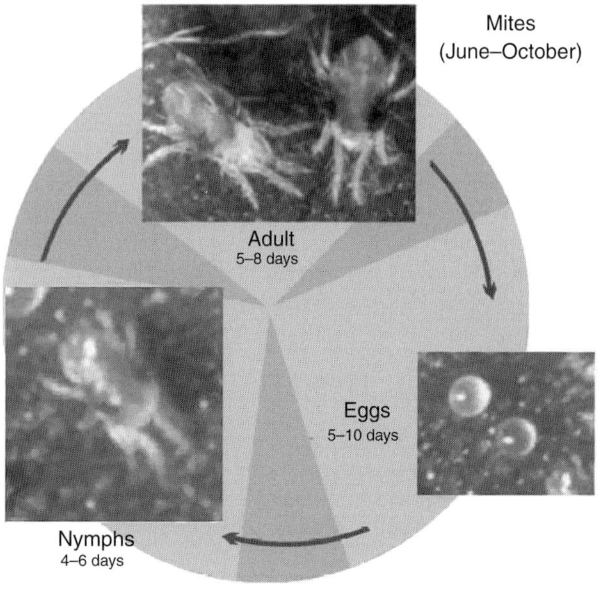

Fig. P.22. Life cycle of mites. Courtesy of: http://takingroots.in/ipm_mites (licensed under creative commons) with permission from Ramanjaneyulu GV, Executive Director, Centre for Sustainable Agriculture, Secunderabad, India.

Fig. P.23. Two-spotted spider mite (*Tetranychus urticae* Koch). Courtesy of Eugene E. Nelson, Bugwood.org. With permission.

generations, the damage to plants is significant if infestation is not controlled. The symptoms are white or yellow specks on leaves, due to reduction in chlorophyll, and defoliation.

Permethrin is perhaps the most widely used pesticide in the treatment of fleas, ticks, itch mites, head lice and other human pests, although it is carcinogenic, besides being extremely toxic to fish and cats. Insecticidal soaps, diluted with water may be sprayed on infested plants. This application disrupts the cellular functions of the mite and kills it. **Horticultural oils, neem** oil and hot water may also be used to control mites. The application should cover the undersides of the foliage. Other measures that help to control mites include: (i) conserving natural enemies; (ii) providing optimum irrigation; (iii) regular washing of leaves; and (iv) reducing dust.

Predatory mites such *as Parasitus* sp., *Phytoseiulus persimilis* and several species in the genus ***Amblyseius*** (especially *Amblyseius womersleyi* and *Amblyseius fallacis*) control spider mites. **Green lacewings** are also effective, with the third instar grub of **Chrysoperla** *carnea* consuming up to 30 adult spider mites/day.

Pest – pink bollworm of cotton

The pink bollworm *Pectinophora gossypiella* (Gelichidae: Lepidoptera) is a borer pest of cotton. The adult is a dark-coloured moth with greyish patches on the wings. The adult moth lays about 100 eggs on floral parts, which hatch within 3–7 days into cream-coloured young larvae that change colour from pink to reddish brown. These larvae feed on pollen and seeds, and bore into cotton bolls, causing the bolls to fail to fully open. They cause large-scale damage by feeding for over 1 month, before pupating inside the cotton boll or in the soil, to grow into adult moths 3 weeks later. The period of pest intensity in India is highest in the months May–November (Fig. P.24). Other countries affected by this pest are Pakistan, China, South Africa and other cotton-growing countries. They also damage the fruits of lady's finger (*Abelmoschus esculentus*).

There are a few effective IPM techniques for pink bollworm control. Deep summer ploughing kills soil pupae and hibernating larvae, while destroying infested bolls and stubble, eliminating new populations. Spraying a 5% **neem** seed kernel suspension repels egg-laying adults. Pheromone traps can confuse adult moths, the result being that they fail to mate. Intercropping cotton with groundnut, green gram and soybean may encourage natural enemies such as **ladybird beetles,** ***Chrysopa*** **spp.** and **syrphids. Trap plants** like marigold and lady's finger should be planted on field bunds if cotton is the main crop. Applying bacterial formulations of ***Bacillus thuringiensis*** (at the rate of

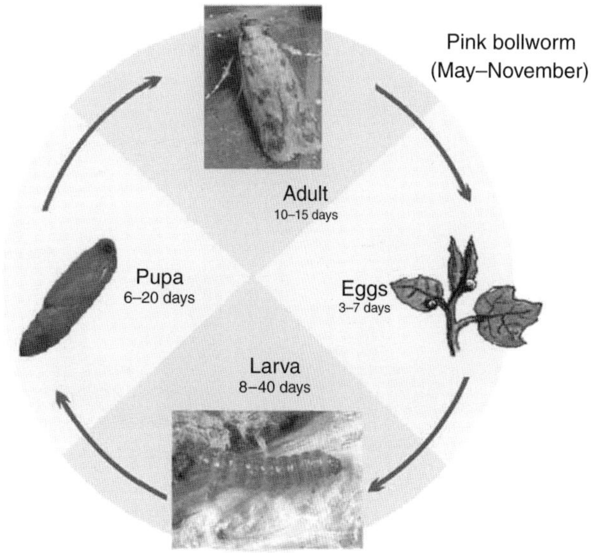

Fig. P.24. Life cycle of the pink bollworm. Courtesy of: http://takingroots.in/sites/default/files/Pink_Bollworm_02.jpg (licensed under creative commons) with permission from Ramanjaneyulu GV, Executive Director, Centre for Sustainable Agriculture, Secunderabad, India.

1 kg/ha) and ***Trichogramma*** **spp.** are also helpful. Selection of resistant seed varieties is highly recommended.

Chemical control includes different processes of seed treatment with ingredients like sulfuric acid, hot water and **aluminium phosphide**. Insecticides such as **chlorpyrifos** or **endosulfan** of varying concentration (up to 40%), sprayed at rates of 2.5 l/ha are effective.

Pest – plant hopper

All plant hoppers are insects belonging to the single superfamily Fulgoroidea and order Hemiptera. Most of the species are known for jumping or hopping, although on plants they walk slowly to avoid being spotted. Plant hoppers grow up to 5 cm in the tropics, and are characterized by an enlarged head, a thick base segment of the antenna and wax plates in females for protecting the eggs. These insects also produce a sort of **honeydew** that attracts fungal and bacterial infections.

While plant hoppers are not considered very serious crop pests, they are known to cause crop damage by sucking the sap out of plants, and for being **vectors** for virus diseases such as grassy stunt virus, ragged stunt virus in rice, wilted stunt diseases and diseases of phloem phytoplasmas. The main targets of these pests are cereal crops like rice, maize, sorghum and sugarcane.

Important plant hopper species include: (i) *Nilaparvata lugens* or the brown plant hopper (BPH) of rice; (ii) *Sogatella furcifera* (white-backed plant hopper or WBPH); (iii) *Perkinsiella maidis* (corn plant hopper), which attacks maize, sorghum and sugarcane; (iv) the small brown plant hopper, a pest of cereals and sugarcane; and (v) *Perkinsiella saccharicida* (sugarcane plant hopper) a pest on sugarcane. All these species are common in the tropics, Australia and Hawaii. Both nymphs and adults feed on plants by sucking sap (Fig. P.25). The symptoms of a plant hopper attack are yellowing, browning and drying of leaves, a condition called hopper burn (Fig. P.26). Plant hopper attacks reduce the leaf area, chlorophyll production and finally, photosynthesis.

Fig. P.25. Life cycle of the brown plant hopper. Courtesy of: http://takingroots.in/ipm_rbph (licensed under creative commons) with permission from Ramanjaneyulu GV, Executive Director, Centre for Sustainable Agriculture, Secunderabad, India.

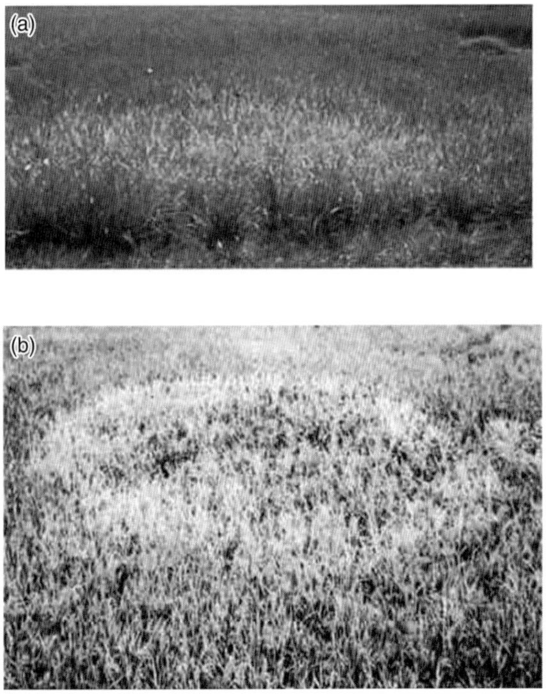

Fig. P.26. Damage caused by brown plant hopper. (a) Drying field; (b) hopper burn. Courtesy of: http://takingroots.in/ipm_rbph (licensed under creative commons) with permission from Ramanjaneyulu GV, Executive Director, Centre for Sustainable Agriculture, Secunderabad, India.

The females lay eggs in the rice seedbed and get carried to the rice field. Nitrogen fertilization and water help in the growth of the pests. The corn plant hopper (*P. maidis*) sucks sap from leaves, shoots and sheath (producing honeydew) and causes galls under leaf surfaces, on the veins. This leads to stunted growth of the plant and loss of yields. WBPHs are another common rice pest. Again, nymphs and adults both feed on the phloem with the help of a feeding tube, causing the tillers to dry and turn brown. These pests can attack the rice plant at any growth stage.

Neem oil and neem kernel extract are two biological measures to control plant hoppers. Natural enemies like lady bugs, water bugs, spiders, **coccinellid** beetles and mirid bugs can be used to manage pest nymphs and adults, while some mites eat eggs. Maintaining sufficient row space and optimum use of nitrogen are two cultural measures to prevent serious infestation.

Pest – psyllid

Psyllids or jumping plant lice are small insects resembling cicadas. They belong to the family Psyllidae in the order Hemiptera. Psyllids have strong jumping legs and short antennae and most species are host specific. The larvae and adults suck sap out of the shoot and new leaves, leading to lack of vigour in the plant and defoliation.

Although psyllids are not a very major pest group, some species such as *Trioza* spp. and *Psylla* spp. may cause considerable damage. Apples, pears and citrus fruits are under risk of yield loss. Some psyllids are **vectors** for disease-causing organisms, for example the Asian citrus psyllid (*Diaphorina citri*) is a vector of a bacterial pathogen that causes the Huanglongbing disease of citrus. Another example of a psyllid pest is *Heteropsylla cubana*.

Pest – red flour beetle and confused flour beetle

The red flour beetle (*Tribolium castaneum*) and the confused flour beetle (*Tribolium confusum*), both pests of stored grain products, are quite similar in appearance, both being reddish-brown, flat and elongated with a shiny appearance (Figs P.27 and P.28). The main difference is in their antennae. The last three segments of the antennae of the red flour beetle are abruptly larger than the preceding ones, forming a club, while in the confused flour beetle, the segments increase in size gradually. The thorax of the red flour beetle has curved sides while that of the confused flour beetle has straight sides. The red flour beetle can fly short distances while the confused flour beetle is unable to fly.

The adults and larvae of both beetles are pests of processed or stored agricultural and dairy products. They attack flour, milled grain products (e.g. cereals) and food (e.g. milk powder, chocolate, dry fruits and legume seeds). These beetles infest storage houses as well as homes, and can survive on food accumulated in crevices and cracks. The two pests have similar life cycles, comprising stages of egg (around 400), larva, pupa and adult. (See also **Stored grain pest control**.)

Pest – red hairy caterpillar

The red hairy caterpillar denotes the larval stage of the Lepidopteran moth *Amsacta albistriga* of the family Arctiidae. It is characterized by long red hair (Fig. P.29). The caterpillar is a defoliating pest, and a common threat to crops of groundnuts and maize in India, South-east Asia and parts of Africa and America.

The adult moth, which is white with black spots, lays clusters of cream-coloured eggs on and around plants. The eggs hatch within 3 days to produce young larvae that feed on the crop leaves, defoliating the crop. Within 3 weeks, the larvae pupate in soil, and then emerge as adult moths. The period of pest intensity is highest from the months of June–August in India (Fig. P.30).

IPM techniques include deep ploughing in summer which helps kill pupae. Killing adult moths by lighting a fire at night is an effective method to reduce the pest population. Trap crops such as jatropha can help in collecting many larvae in one

Fig. P.27. Red flour beetle. Courtesy of Natasha Wright, Florida Department of Agriculture and Consumer Services, Bugwood.org. With permission.

Fig. P.28. Confused flour beetle. Courtesy of Pest and Diseases Image Library, Bugwood.org. With permission.

location. The use of pest-resistant seeds is very important for pest prevention. An emulsifiable concentrate of **dichlorvos** (of around 1 kg in up to 2000 l water) is commonly used.

Pest – rice brown plant hopper (BPH)

The brown plant hopper (BPH) *Nilaparvata lugens*, belonging to the family Delphacidae in the order Hemiptera, is a very important sucking pest of rice crops (Fig. P.31). These pests suck the sap from the base of rice stems close to the water, causing the plants to turn yellow and then dry up.

Fig. P.29. The red hairy caterpillar, *Amsacta albistriga*. Courtesy of *Satrends*, the monthly newsletter of International Crops Research Institute for the Semi-Arid Tropics (ICRISAT). With permission from Lydia Flynn, Editor in Chief.

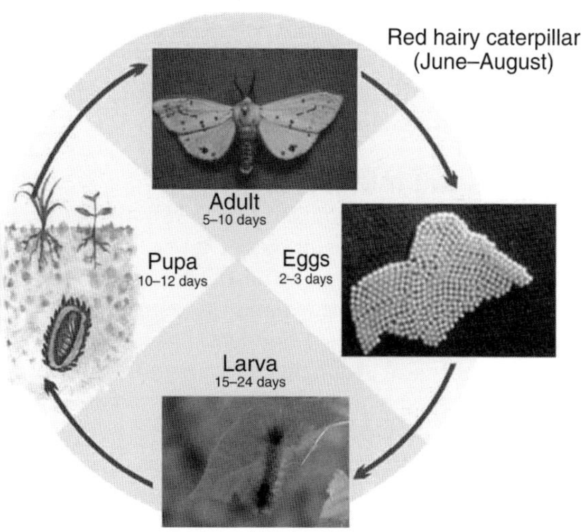

Fig. P.30. Life cycle of the red hairy caterpillar. Courtesy of: http://takingroots.in/ipm_rhc (licensed under creative commons) with permission from Ramanjaneyulu GV, Executive Director, Centre for Sustainable Agriculture, Secunderabad, India.

The golden-winged adult lays eggs into the sheath or the midrib of the leaf and these hatch into greenish wingless nymphs. The nymphs grow into adults in a month and prefer humid conditions.

Both nymphs and adults are a nuisance to the crop. BPH infestation is very serious in India, Bangladesh, Japan, Taiwan, Korea, Malaysia, the Philippines and other rice-growing countries.

IPM techniques include reducing humidity in the field by maintaining a good distance between crop rows and intermittently draining out water. Natural predators such as the mirid bug, *Cyrtorhinus* spp. (e.g. *Cyrtorhinus lividipennis*) and spider, ***Lycosa* spp.** have proved to be successful in controlling the plant hopper. *C. lividipennis* feeds on the insides of BPH eggs

Fig. P.31. Brown plant hoppers (*Nilaparvata lugens*). Courtesy of Sylvia Villareal/IRRI (http://www.flickr.com/photos/ricephotos/4270847124/ IRRI's Creative Commons policy).

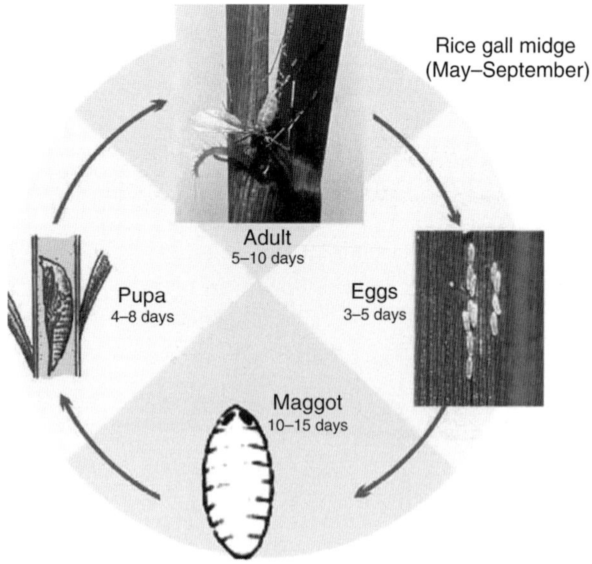

Fig. P.32. Life cycle of rice gall midge. Courtesy of: http://takingroots.in/ipm_rgm (licensed under creative commons) with permission from Ramanjaneyulu GV, Executive Director, Centre for Sustainable Agriculture, Secunderabad, India.

and also on nymphs and adults. A mixture of urea and **neem** oil (in a ratio of 5:1) is also an effective biological control measure.

Chemical insecticides include a dustable powder (DP) formulation or soluble concentrate (SC) of **carbaryl**, granular (GR) formulations of **carbofuran** and emulsifiable concentrates (EC) of **fenthion**. Applications of **imidacloprid** and **monocrotophos** are also common.

Pest – rice gall midge

The adult rice gall midge, a mosquito-like fly with an orange-coloured body and long legs, belongs to the order Diptera. This midge species is considered a pest of rice crops because of its tendency to defoliate.

The adult midge lays eggs along the veins of the lower part of rice leaves. Maggots (larvae of midges) hatch out within 5 days and feed on the apical shoot, which leads to tillers that do not form panicles. Maggots pupate in the hollow top-most leaf after about 14 days. Adults emerge 1 week later.

The rice gall midge causes most damage to crops in India from May to September. The pest is also present in some African countries such as Nigeria, Mali and Sierra Leone.

IPM techniques for rice gall midge control include removing grassy weed hosts of the midge and keeping nitrogen fertilization to a minimum. Introducing phytoseiid mites (***Amblyseius*** *imbricatus*) and some wasp and spider species (e.g. lynx and wolf spiders) into the crop controls the pest population, as they feed on midge eggs and maggots. Spraying a suspension of 5% neem kernel repels adult insects.

Since most of the pest's life cycle is inside the plant, chemical control is not effective (Fig. P.32).

Pest – rice green leafhopper

The leafhopper *Nephotettix virescens* (order Hemiptera) is a sucking pest of rice. It is a wedge-shaped, green insect with a black spot in the central part and a black patch in the posterior part of its wings.

The life cycle of the hopper includes eggs (laid under the leaf by the adult), green nymphs (with incompletely developed wings) and adults. Both nymphs and adults cause damage to the crop by sucking sap from the leaves. These pests are also vectors of the rice tungro virus. The rice green leafhopper is present in Thailand, India, Bangladesh and Sri Lanka.

IPM techniques for rice green leafhoppers are similar to those of rice brown plant hoppers (BPH). In addition, light traps over cans of kerosene help in trapping and killing leafhoppers. **Imidacloprid** and **thiamethoxam** (25 g/ha) and **clothianidin** (12.5 g/ha) provide chemical protection.

Pest – rice gundhi bug

The hemipteran rice gundhi bug (*Leptocorisa varicornis*) is a sucking pest. It is considered a minor pest. The thin, greenish-brown adult lays eggs in rows, which hatch into green-coloured winged nymphs, which grow into adults (Fig. P.33). Nymphs and adults cause damage to rice plants by sucking on the grain during the milky stage of grain development, thus degrading grain quality. They also emit an unpleasant odour.

The green gundhi bug is found in India, Japan and the Philippines, with pest intensity being highest from July to November.

Spraying a 5% neem seed kernel suspension repels adult bugs from laying eggs, and application of **carbaryl** and **malathion** provide chemical control. Removing host weeds deprives the pest of a safe hideout. Heavy rainfall adversely affects the pest population.

Pest – rice leaf folder

The rice leaf folder *Cnaphalocrosis medinalis*, belonging to the order Lepidoptera, is a defoliating moth. Although a minor pest, it assumes importance since it affects important crops like rice, wheat, barley and sugarcane.

The adult is a brownish-orange moth with two dark wavy lines on its wings. The female lays round eggs along the midrib of leaves, which hatch into green larvae within 1 week. The larvae fold a leaf around themselves and feed on the leaf tissue. After feeding, they pupate inside the leaf fold. The pupae are transformed into adult moths within a period that lasts from

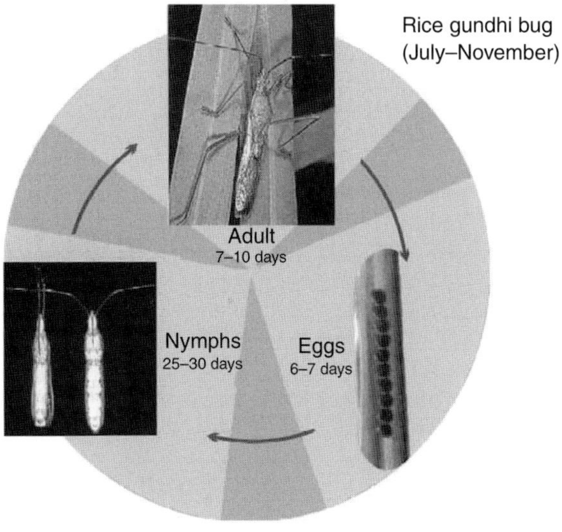

Fig. P.33. Life cycle of rice gundhi bug. Courtesy of: http://takingroots.in/ipm_rgb (licensed under creative commons) with permission from Ramanjaneyulu GV, Executive Director, Centre for Sustainable Agriculture, Secunderabad, India.

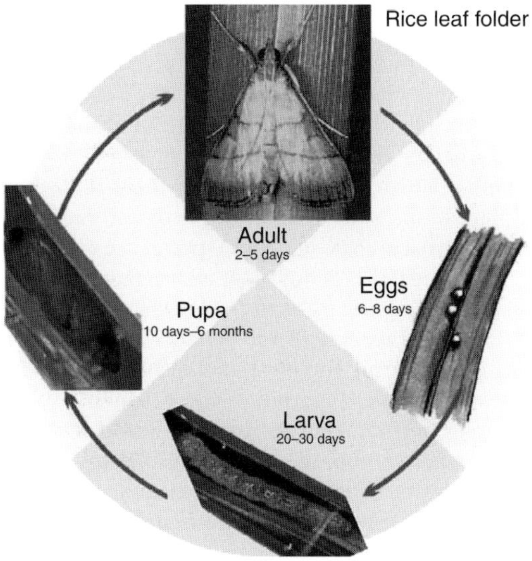

Fig. P.34. Life cycle of the rice leaf folder. Courtesy of: http://takingroots.in/ipm_rlf (licensed under creative commons) with permission from Ramanjaneyulu GV, Executive Director, Centre for Sustainable Agriculture, Secunderabad, India.

10 days to 6 months (Fig. P.34). India, Bangladesh, Pakistan, China, Taiwan, Indonesia, Korea, the Philippines and Thailand are affected by rice leaf folder infestations.

IPM techniques for rice leaf folder control include spraying a 5% **neem** seed kernel suspension to prevent adults from laying eggs, and spraying chilli-garlic solution to kill young larvae. ***Trichogramma** chilonis* (100,000 adult parasites/ha) at the seedling stage, and products of ***Bacillus thuringiensis*** are considered useful. A granulosis viral infection can kill larvae. Use of **malathion, parathion-methyl, diazinon, chlorpyrifos, phasolone** and **quinolphos** is recommended, depending on the severity of the infestation. Most of these pesticides are used as emulsifiable concentrates.

Pest – scale insect

Scale insects are small insects of the order Hemiptera and superfamily Coccoidea. With over 8000 species, many are considered crop parasites because they suck sap from the plant.

Scale insects vary in size from 1 mm to 5 mm; they secrete a waxy protective coating that looks like fish or reptilian scales. Most females are immobile once they attach themselves to the plant (except in the case of the **mealybug**). Males may or may not have wings, but die in 1 day. The waxy coating on these insects protects them from pesticides.

Horticultural oils and soapy water help to control scale insects. The use of specific pesticides is also common in the control of these pests.

Pest – sorghum shoot borer

The dark brown-coloured sorghum shoot borer *Chilo partellus* is a moth belonging to the order Lepidoptera. It is a borer pest on sorghum, and crops like millet, maize and sugarcane. Most sorghum-growing regions in Asia and Africa face a threat from the shoot borer. In India, April to October is the most damaging period.

The female moth, which is significantly larger than the male, lays about 500 flat, scaly eggs under the leaves, which hatch within 1 week to produce yellow-coloured, spotted larvae.

The larvae bore into the shoot tips and stems, which results in dead hearts and the death of the plant. A month later, the larvae pupate for up to 14 days, to emerge as adult moths.

IPM techniques for the control of the sorghum shoot borer include the sowing of fast-growing seeds earlier than the normal schedule. Destruction of crop stubble prevents the pest from any active feeding. Natural enemies like the egg parasitoid ***Trichogramma* spp.** and egg parasites such as *Apanteles* spp. and *Bracon* spp. keep the pest population under control.

Chemical control includes spraying pesticides such as **carbaryl** or **endosulfan** a month after sowing. Spraying is repeated three times, with a gap of 15 days. Application of 10% **carbaryl** or 4% **cartap hydrochloride** granules after the emergence of the crop is also effective. This treatment may have to be repeated in cases of severe infestation.

Pest – sorghum shoot bug

The sorghum shoot bug, *Peregrinus maidis*, belonging to the order Homoptera, is a sucking pest on sorghum crops. It is a yellow- to dark brown-coloured winged bug. The female, which is much larger than the male, lays about 100 elongated and tapering eggs into the leaf. The eggs hatch within 1 week to produce yellow-coloured nymphs that develop wings as they grow. Nymphs develop into adults within 3 weeks.

Both nymphs and adults suck sap from the shoot, leading to drying up of the plant. India, the West Indies and the Philippines are among the countries that face the threat of this bug, with the period from September to January considered as most damaging.

The sorghum shoot bug may be managed by spraying a 5% **neem** seed kernel suspension to repel adults, and by spraying a 5% extract of *Vitex* leaves (known as the *Nirgundi* tree in Hindi) to repel nymphs and adults.

Chemical control measures include spraying monocrotophos or dimethioate at 0.1%. Another method is spraying **endosulfan** (0.1%) 15 days after sowing.

Pest – sorghum shoot fly

The sorghum shoot fly or *Atherigona soccata*, belonging to the order Diptera, is a borer pest of sorghum. This small, grey-coloured fly lays white elongated eggs on the lower surface of the leaf which hatch to produce yellowish grubs or maggots about 6 mm long. The sorghum shoot fly also infests other crops such as wheat, maize, small millets and grasses.

The grub bores within shoot tips and feeds for around 2 weeks, turning the central leaf into a dead heart. The grubs pupate on the plant or in soil, to grow into adult grey flies in just over 1 week.

Shoot fly infestations occur in Asian and African countries. June and July are the months when there is a potential for damage in India.

IPM measures for sorghum shoot fly control include: (i) advancing sowing prior to mid-July so as to avoid the pest; (ii) disease-resistant seeds; (iii) spraying 5% **neem** seed kernel suspension to repel adults from laying eggs; and (iv) planting resistant varieties.

Chemical control measures include the application of **phorate** or **carbofuran** during sowing at the rate of 2.5 kg a.i./ha, or spraying pesticides like **cypermethrin**.

Pest – spotted bollworm

See **Crop – cotton pest control**

Pest – spotted borer/spotted bollworm

The spotted borer *Earias vittella* is an important borer pest. Belonging to the order Lepidoptera and family Noctuidae, this pest of cotton and lady's fingers (okra, *Abelmoschus esculentus*) is a robust yellowish-green moth with a dark green stripe on the forewings.

The female adult lays blue-coloured eggs on shoot tips and flowers, which hatch into larvae that bore into the terminal shoot, flowers and fruits. Crop yield drops substantially as a result. In cotton, the boll quality is severely affected. After 15 days, the larvae pupate to emerge as adults in 10 days (Fig. P.35).

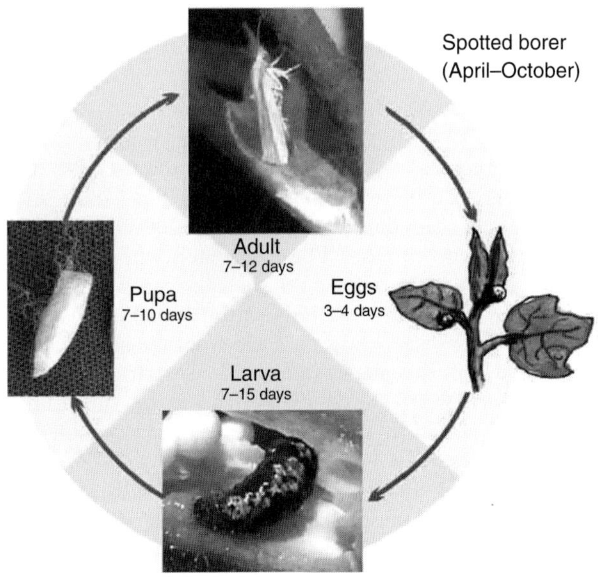

Fig. P.35. Life cycle of the spotted borer. Courtesy of: http://takingroots.in/sites/default/files/Spotted_Borer_02.jpg (licensed under creative commons) with permission from Ramanjaneyulu GV, Executive Director, Centre for Sustainable Agriculture, Secunderabad, India.

Most okra- and cotton-producing nations bear the brunt of the spotted bollworm, with April to October being the worst months for potential damage.

Destroying damaged plant parts and cutting shoot tips every 45 days to remove eggs are two simple methods to prevent infestation. Deep ploughing to expose pupae to predators and intercropping with groundnut and soybean are other methods to reduce the spread of infestation. Introducing **ladybird beetles, syrphids,** ***Trichogramma*** **spp.** (*Trichogramma chilonis, Trichogramma brasiliensis*), ***Chrysopa*** **sp.** and spider species of the genus *Thomisus* are some biocontrol methods to protect cotton and okra crops from the spotted borer. Interchangeably planting a few okra plants as a trap crop among cotton crops and vice versa, helps to trap larvae which can then be killed.

Cypermethrin, **triazophos** and **endosulfan** are some chemical insecticides that can help control the growth of this pest. Emulsifiable concentrates are commonly used and dosage rates range from 600 ml to 2.5 l/ha.

Pest – thrip

Thrips are insects belonging to the order **Thysanoptera** (which has two suborders Terebrantia and Tubulifera). Thrips are agricultural pests that are normally 1 mm long (although a few species are up to 14 mm in length), slender and with fringed wings. Most species are poor fliers. Thrips feed on plant tissues by piercing them with the stylet on the left mandible. The left mandible is larger than the right mandible. Thrips reproduce rapidly and frequently (either parthenogenetically or by mating), when weather and food supply are favourable. Thrips are known to inject a toxic substance through their saliva that reacts with the plant tissue, causing mottling or discoloration. Besides directly causing harm to plants, some species are **vectors**, transmitting viruses in the genus *Tospovirus* that cause disease (e.g. tomato spotted wilt virus and *Impatiens* necrotic spot virus). A few thrip species are considered beneficial since they feed on mites and some fungal spores. Examples of important thrip pest species are *Thrips tabaci* and *Thrips palmi*.

T. tabaci (Thysanoptera: Thripidae) is one of the most damaging insect pests on onion, garlic, cotton, cabbage, celery, tomato, beans, cucumber and pineapple. Brownish adult females, up to 2 mm in length, make many slits in plant tissues and lay each of the more than 50 bean-shaped eggs singly in individual slits. In about 10 days, the eggs hatch into yellowish slender nymphs that are miniature but wingless versions of the adults. The nymphs feed on leaves, while developing through two instars. Then they enter the pupal stage, and gradually develop into adults.

Damage caused by *T. tabaci* includes silvery patches on leaves which affect photosynthesis. There is also excessive transpiration, affecting plant growth and yields. *T. tabaci* also acts as a vector of tomato spotted wilt virus in different crops.

Releasing beneficial predators such as black hunter thrips (*Aelothrips* sp.), **minute pirate bugs** (*Orius* spp.), **green lacewing** (***Chrysoperla*** spp.) and bugs of the genus *Geocoris* can control thrips effectively. Cultural methods such as mulching (to reduce moisture loss and increase organic matter) and planting trap crops (e.g. carrots, cucurbits and yellow or white flowering plants) help in keeping the pest numbers low. Using clean and pest-resistant varieties and physically washing out thrips from leaves are two other methods of pest control. **Imidacloprid**, spinosad and **abamectin** are three chemical agents which may be used sparingly, in order to kill thrips while protecting their natural enemies.

The western flower thrip (*Frankliniella occidentalis*) is another example of a thrip pest (Fig. P.36). It is a native of North America, but has spread to South America, Australia and Europe via infested plant material. There are over 500 species of the genus *Frankliniella* that cause plant damage.

Fig. P.36. Western flower thrip (*Frankliniella occidentalis*). Courtesy of Dr Mark S. Hoddle, Dept. of Entomology, University of California Riverside, with permission.

Adult female thrips, which are slightly larger than males, commonly reproduce by parthenogenesis. They lay extremely small, bean-shaped eggs in flower tissues. Eggs hatch into yellowish nymphs with red eyes. The nymphs complete the first two instars on flowers, feeding on the tissues and the fruits developing from within the flower. Nymphs develop through the next two stages in soil. The life cycle of the western flower thrip may depend on the duration of the nymphal stages. Symptoms of damage are areas of discoloration on the plant.

Western flower thrips are vectors of plant viral diseases like the tomato spotted wilt disease caused by **tomato spotted wilt virus** (genus Tospovirus, family **Bunyaviridae**).

Infestations of western flower thrips are less serious during wet weather. These thrips can be managed by growing barriers of non-host plants around crops and eliminating plants that the thrips are especially attracted to (called reservoir plants), such as Jimson weed (*Datura stramonium*) and other plants with brightly coloured flowers. Bugs of the genus *Orius* and agents like the fungus *Metarhizium anisopliae* show promise as biological control agents.

Pest – *Thrips tabaci*

See **Pest – thrip**

Pest – tobacco caterpillar

The tobacco caterpillar *Spodoptera litura*, also known as the cotton leafworm or tobacco cutworm, is the larval stage of a dark brown moth belonging to the order Lepidoptera. This defoliating pest is important in agriculture as it attacks cotton, castor, groundnut, cabbage, sorghum, tomato, sugarcane and many other crops.

The adult female lays hundreds of eggs in groups, which hatch within 5 days into larvae. The larvae are initially greenish but keep changing colour as they develop through instars. Larvae feed actively at night for about 1 month, and then gradually retire into the soil to pupate. The adult emerges after about 10 days (Fig. P.37).

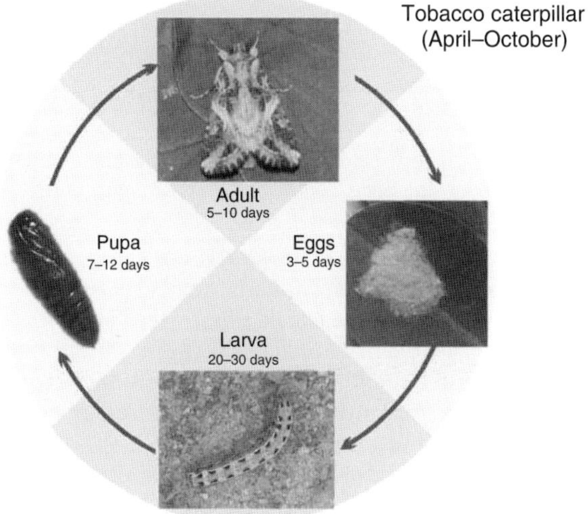

Fig. P.37. Tobacco caterpillar. Courtesy of: http://takingroots.in/sites/default/files/Tobacco_Caterpiller_02.jpg (licensed under creative commons) with permission from Ramanjaneyulu GV, Executive Director, Centre for Sustainable Agriculture, Secunderabad, India.

Intercropping with cowpea and soybean, growing trap crops like castor and border crops such as maize and sorghum together help trap the pest, as does nurturing birds and other natural enemies of the tobacco caterpillar. Eggs and larvae can be manually removed from leaf surfaces. Applying a suspension of **neem** seed kernel, chilli and garlic can kill young larvae. Other useful measures include introducing an inoculum of nucleopolyhedrovirus (NPV), releasing eggs of the **parasitoids** ***Trichogramma*** sp. (150,000/ha), or larval parasitoids *Telenomus remus* or *Bracon* sp., or predators such as ***Chrysoperla carnea*** (at the rate 50,000/ha). ***Bacillus thuringiensis*** **(*Bt*)** products are also useful. **Endosulfan** (600–750 ml/ha), **fenvalerate** (400–500 ml/ha) and **cypermethrin** (0.025%) also control the caterpillar.

Pest – western flower thrip

See **Pest – thrip**

Pest – wheat and rice weevils

The wheat weevil (*Sitophilus granarius*) and rice weevil (*Sitophilus oryzae*) are very destructive pests of stored grains the world over, causing large losses.

The adults of these weevils are about 3–5 mm in length, the size varying according to the size of the grain infested. The reddish-brown adults of the wheat weevil have long snouts and chewing mouthparts and they cannot fly (Fig. P.38). Rice weevils are dull reddish brown with four light spots on the thorax. Rice weevils can fly. Both the weevils have a four-stage life cycle (i.e. egg, larva, pupa and adult).

Fig. P.38. Wheat weevil. Courtesy of: Clemson University, USDA Cooperative Extension Slide Series, Bugwood.org. With permission.

Weevils lay eggs inside grain kernels, one egg in each kernel, by boring a hole in the grain, laying the egg and covering it with a gel-like secretion. While the female wheat weevil lays between 36 and 254 eggs in all, the rice weevil lays between 300 and 400 eggs.

The adults of rice and wheat weevils live from 6 to 8 months and so damage is substantial, if control measures are not put in place. Besides rice and wheat, other grains such as oat and maize are also attacked. (See also **Stored grain pest control**.)

Pest – whitefly

Whiteflies are hemipteran creatures that belong to the single family Aleyrodidae. There are over 1000 species of whiteflies, of which many are considered pests. Crops that are affected include fruits, vegetables (e.g. tomatoes, beans, cucurbits, potatoes and sweet potatoes) and cotton.

Around 1.5 mm in length, whitefly adults and nymphs secrete an abdominal, white, waxy substance on themselves, which makes them look like tiny white moths. The nymphal instars are wingless, while the adults have wings. Warm, humid days are favourable for growth and reproduction, which is frequent, resulting in many short generations.

Whiteflies feed by puncturing the veins and sucking out the phloem sap. Such attacks by armies of whiteflies make plants lose their turgor. In addition, they excrete **honeydew** which attracts **black sooty mould**. The mould covers the surface of afflicted plant leaves, making photosynthesis difficult.

Besides inflicting direct damage, whiteflies are responsible for transmitting various viral diseases including mosaic, tomato yellow leaf curl and mottle. Many species in the genus *Bemisia* are **vectors** for virus diseases. For example the sweet potato whitefly (*Bemisia tabaci*) and the silverleaf whitefly (known as *Bemisia tabaci* biotype B and sometimes referred to as *Bemisia argentifolii*) transmit the tomato mottle virus (a geminivirus). Tomato mottle virus infection leads to the mottling, **chlorosis** and curling of leaves, and reduced plant height. The cabbage whitefly (*Aleyrodes proletella*) is another whitefly pest species.

Since whiteflies quickly develop resistance to chemical pesticides, IPM strategies are more effective. **Clothianidin**, **thiamethoxam**, dinotefuran and **imidacloprid** are some of the chemical compounds deemed necessary to control severe infestations of whiteflies. To manage resistance, chemical rotation with insecticides of different classes is alternated with cultural practices and the use of biological methods. Spraying of water or a solution of a mild soap on affected plants, the use of yellow sticky paper and manual removal of affected leaves are some cultural practices.

There are many whitefly predators and **parasitoids** that may be used in the control of whiteflies. **Green lacewings**, **ladybirds**, predatory mites (*Euseius* spp.), ***Trichogramma*** wasps and **minute pirate bugs** are some of them. Many birds also feed on whiteflies. Some biofungicides such as ***Beauveria bassiana*** or *Paecilomyces fumosoroseus* have been used to secure whitefly control.

Pest management

See **Integrated pest management (IPM)**

Pest Management Regulatory Agency (PMRA)

See **Pesticide legislation in Canada**

Pest order – Blattaria

See **Phylum Arthropoda**

Pest order – Coleoptera

See **Phylum Arthropoda**

Pest order – Diptera

See **Phylum Arthropoda**

Pest order – Hemiptera

The order Hemiptera is a large and diverse classification of insects, commonly called true bugs. ('Hemiptera' comes from '*hemi*' in Greek meaning 'half' and '*pteron*' meaning 'wing'.) The two former orders, Homoptera and Heteroptera, were initially suborders of the order Hemiptera. This has subsequently been changed and the new suborders of Hemiptera are Auchenorrhyncha, Coleorrhyncha, Heteroptera and Sternorrhyncha.

The order Hemiptera comprises around 80,000 species of varied sizes and shapes (with the largest number of hemimetabolous insects), and includes a large number of crop pests. Most hemipteran insects are sucking creatures with specialized sharp mouthparts that can pierce plant tissues or animal skin. They may be winged or wingless, with some possessing both forms during their life cycle.

The order Hemiptera contains a large proportion of very damaging agricultural insect pests such as **aphids, mealybugs**, leaf bugs, **whitefly**, plant hoppers (superfamily Fulgoroidea), **leafhoppers** (family Cicadellidae) and **scale insects**. Besides being plant pests, many of them are also known to be predators of **beneficial organisms** like bees, butterflies and wasps. Further, many of them are **vectors** of viruses. For example, aphids are responsible for the transmission of over 100 viral infections in plants. The green peach aphid (*Myzus persicae*) transmits the turnip mosaic virus. Leafhoppers transmit diseases such as beet curlytop and dwarf disease of rice. A few species are parasites on humans, a well-known example of this being the bed bug. Some hemipterans are beneficial as they predate upon crop pests. For example, *Geocoris* spp. predate on various insect and mite pests. Some species are also commercially important – for instance, the lac bug (*Kerria lacca*) that yields lac, a scarlet substance that is used for dyeing wool and silk, as a cosmetic and as a medicinal drug.

Pest order – Hymenoptera
See **Phylum Arthropoda**

Pest order – Isoptera
See **Phylum Arthropoda**

Pest order – Lepidoptera
See **Phylum Arthropoda**

Pest order – Mallophaga
See **Phylum Arthropoda**

Pest order – Mantodea
See **Phylum Arthropoda**

Pest order – Mecoptera
See **Phylum Arthropoda**

Pest order – Neuroptera
See **Phylum Arthropoda**

Pest order – Odonata
See **Phylum Arthropoda**

Pest order – Orthoptera
This insect order comprises approximately 10,000 species worldwide, and includes **grasshoppers**, crickets, locusts and related insects, many of which are crop pests.

Most insects of this order have long hind legs with powerful thighs which enables them to make long jumps. Leathery forewings protect the fan-like hind wings. Many of them make sounds by rubbing hard parts of the body together. Insects of the order Orthoptera have hearing organs and biting mouthparts, used for eating foliage.

Pest order – Phasmatodea
See **Phylum Arthropoda**

Pest order – Siphonaptera
See **Phylum Arthropoda**

Pest order – Thysanoptera
Members of the order Thysanoptera are called thrips, of which there are 5000 species worldwide. Thrips (mostly 1 mm long) are tiny, slender insects with piercing mouthparts and narrow hairy wings. Thrips are not good fliers and are transported long distances by wind. Thrips are hemimetabolous insects and young thrips look like small, wingless adults. Adult thrips cut into the plant surface and lay eggs. This makes thrip control difficult. Eggs mostly hatch in 3 days. The thrip life cycle includes egg, larval, pre-pupal, pupal and adult stages, of which, in the pre-pupal and pupal stages, thrips remain immobile and hide in the plant or in the soil. In most cases, males are smaller than females and parthenogenesis is common in this order.

Most thrips feed on plants and cause damage to flowers, pollen, buds and plant tissues of young shoots and leaves. Most thrips are considered crop pests, although there are exceptions. Some thrip species are known to feed on fungal spores. Plant-feeding thrips puncture the surface of the plant and suck out the contents, while injecting toxic digestive juices into the plant. Wilting, growth deformity, bleaching and scarring are the effects of thrip damage. Thrips also act as **vectors** for plant viruses, for example **western flower thrips** (*Frankliniella occidentalis*) are vectors of **tomato spotted wilt virus**.

Some thrip species are predators of mites and other pests. For example, banded thrips, black hunter thrips, six-spotted thrips and species belonging to the genus *Franklinothrips* in the family Aeolothripidae are obligate predators. Two of these ant-like *Franklinothrips* species, *Franklinothrips orizabensis* and *Franklinothrips vespiformes*, are used as control agents against pest species of thrips, mites, whitefly nymphs and leaf miner larvae. Thus some thrip species are biocontrol agents.

Eulophidae and Trichogrammatidae are two Hymenoptera families that parasitize eggs. Other biocontrol agents of pest thrips in their adult and larval stages are aphid wasps, **anthocorid** bugs and predatory mites such as *Neoseiulus mackenziei* and ***Amblyseius*** *cucumeris*.

Use of sticky traps is a common tool for thrip control. **Malathion**, **acetamiprid** and **thiamethoxam** are some of the chemical insecticides used to control thrips. Some examples of pest thrips include *Thrips tabaci* and *Stenchaetothrips biformis* on rice, *Scirtothrips dorsalis* or chilli thrips and *F. occidentalis*, which has a broad host range.

Pesticidal oils
See **Horticultural oils**

Pesticide antagonism
Pesticide antagonism is the reduction in efficacy of a pesticide composition as the result of mixing two or more compounds. Applying mixtures of some pesticide active ingredients reduces their toxicity towards pest species. For example, herbicide antagonistic interactions take place when two or more herbicides are mixed or when a herbicide is mixed with a fertilizer.

The practice of tank mixing of pesticidal and fertilizer chemicals may also cause antagonism. Even water can interfere with a pesticide action, for instance, clay inactivates **paraquat** and **glyphosate**, while Na, Ca, K and P ions interfere with **MCPA** and **2,4-D**. Commonly, antagonism is seen when post-emergence grass herbicides are applied along with post-emergence broadleaved herbicides.

To counter antagonism, more of the herbicide is commonly added to the mixture to achieve the intended objective. This may lead to incorporation of excess chemicals into the environment. Knowledge of the properties of pesticide compounds and the solvents is a viable alternative. Addition of **adjuvants** and additives to the tank mix alleviates the problem. Label instructions regarding mixing and chemical interactions should be followed.

Pesticide deposits and residues
When pesticides are applied to crops, they settle on targeted parts, like the foliage. The remnants of the active ingredients of pesticides are termed deposits if they remain on the surface for a brief period of time, before they undergo chemical change. If, however, a pesticide remains on the surface of the plant for a reasonably long time, without being broken down by atmospheric factors or microbial action, it is called a residue. A pesticide

residue is the concentration of a substance that remains in or on a food commodity, soil, air or water after its application. It includes the unchanged parent compound and any degradation product and toxic impurity. The life of a residue depends on the chemical itself and also, on how and where it is used.

The distinction between deposits and residues is not very uniform, but it is important from the point of view of **chronic toxicity**. On the face of it, a residual pesticide helps control pest attacks for a long time, since the pesticidal effect is still active. But by the same token, such a pesticide may enter the food chain, and cause long-term effects on man, animals and the environment. Residues can persist in the soil or on the plant surface, interfering with the normal growth of the non-target soil microorganisms, including **beneficial organisms**. Hence, pesticide residues are a subject of continued research.

Pesticide residues in food are under strong legislation in most countries. Usually a **maximum residue level** or limit (MRL) or tolerance limit based on toxicological data is arrived at, when good agricultural practices (GAP) have been followed. GAP is defined as a set of practices recommended by the agricultural authorities within the country, to ensure safe and healthy use of chemicals in agriculture.

If an approved use of pesticide causes a certain residue level, the legal MRL should not be higher than this level, even if higher levels may be harmless. The **acceptable daily intake** (**ADI**) is the key concept for determining tolerance levels of residues. Monitoring agencies include the **Environmental Protection Agency** (**EPA**), Food and Drug Administration (FDA) and United States Department of Agriculture (USDA), all belonging to the USA.

Pesticide disposal

An important aspect of pesticide application is their disposal. The responsibility of disposing of excess, unused, mixed or spilt pesticides lies with handlers and mixers. The personnel involved in handling, mixing, loading and disposing of pesticides should be well trained and proficient in this field. Knowledge of the chemical properties of the pesticides used can be an invaluable asset.

First, it should be ascertained if the excess pesticide in a container can be used, after the seal is opened. The option of returning excess or unused pesticide to the dealer should be explored. Pesticides should be disposed of only if the options of storage (after opening the seal) or return are not open. Improper disposal of pesticides can cause serious harm to man, animals and the environment.

Disposal of unused pesticides and empty containers should be done according to instructions on the label. Commonly, containers, nozzles and other tools are triple-rinsed or pressure-rinsed and set aside for disposal or reuse, as required. When dry, disposable material is packed in a designated container, sealed and marked for disposal.

Most rinsates from containers and equipment cleaning or spill clean-up may be added to a tank mixture of the same pesticide, if this is permissible on the label, for reuse in a short time. However, rinsates that contain dirt or cleaning agents cannot be reused.

Pesticides may be disposed of as landfill, as long as it is done without contaminating groundwater, surface water or sewage treatment facilities. Pesticide wastes that cannot be disposed of immediately should be kept aside safely and marked, until disposal is possible. If label instructions are unclear or insufficient, the country's pesticide or environmental control agency should be contacted for clarity. Burial pits for disposal should be lined with 5–10 cm of clay and coated with 2–3 cm of lime. Waste should be added to the pit in layers not more than 10–15 cm deep and inter-mixed with lime and biodegradable household waste to assist in biological degradation. Specific instructions for disposal of certain pesticide formulations should be carried out according to the norms of the regulating agency.

As far as **personal protective equipment** (**PPE**) is concerned, it has to be replaced regularly as its chemical resistance decreases each time it is used, hence disposal is also an issue. Absorbent clothing material such as cotton, denim and canvas are disposable and must be cleaned well before discarding. Gloves, footwear and aprons, which might be reusable, must be discarded at the first signs of wearing out or damage. Label instructions must be followed for PPE disposal.

To prevent having to dispose of drums, cans and containers after each use, it should be established if the same container can be refilled at the dealer's end. This would avoid the need to rinse out the container at the user's end. Containers may even be designed in small convenient packages, enough for even a single application. Alternatively, containers, along with consumables like paper, cotton and plastics can be burned or incinerated, as is permissible under local laws. Recycling or reconditioning may also be carried out, if this is permissible.

Pesticide formulations

A pesticide is a substance used for preventing pests such as insects, fungi, nematodes, rodents or weeds from causing harm to man, his crops and other belongings. A pesticide, either singly, or in combination, can inhibit, repel or kill the pest. The chemical or active ingredient (a.i.) in its pure form can be very toxic. Converting a concentrated a.i. into a relatively dilute or otherwise acceptable form is called formulation. It involves selecting, mixing, blending or dilution of one or more a.i. to obtain an end-use product. All measures are taken to ensure that the ingredients do not react chemically with each other.

As explained, the creation or putting together of any registered product is sometimes referred to as formulation. The term may be used to refer either to the product sold by the manufacturer or the process used for preparing the a.i.s for practical use. The list of the ingredients and their amounts or percentages required in such an industrial product constitutes a product formula. The materials added in a formulation (except the biologically a.i.s) are known as formulants.

A formulation improves the properties of the chemicals used, and makes them suitable for handling, storage and application, which consequently influence their effectiveness and safety. Formulations are made so that: (i) larger areas can be covered; (ii) overdosing is avoided; and (iii) damage to plants, animals, applicators and non-target pests is minimized.

Some formulations are to be used direct from the package (ready to use), while others need to be diluted with water, oil or other carriers. A single pesticide is often processed into different formulations. Different formulations of the same a.i. often behave differently.

Pesticide formulations can be in dry or liquid forms. These are available for example as emulsifiable concentrates (EC), solutions (S), flowables (F), dusts (D), baits (B), soluble powders (SP), granules (GR) or pellets.

Solid or dry formulations can be either ready-to-use or concentrates to be diluted before spraying. Some examples of solid formulations are:

- Dusting powders are ready-to-use preparations that are commonly used in seed treatment. They are also widely used for non-agricultural insects and pests on livestock. Powders have two disadvantages: (i) they tend to stray with the wind and hit non-target organisms; and (ii) they cake with moisture, which makes them cumbersome for application. Powders also come in wettable (WP), water-dispersible and water-soluble forms.
- Baits, another formulation, usually come in ready-to-use forms. Baits attract pests like rats, but may be accidentally consumed by non-target animals or children.
- Granules (GR) are usually ready-to-use forms that have a low risk of drift. Granules are used for insect, weed and nematode control in soil and aquatic environments. Granules also come in water-soluble and water-dispersible forms.
- Pellets are another pesticide formulation. These are very similar to granular formulations. The uniformity of the particles helps usage with precision application equipment.
- Tablets constitute another formulation type of pesticides, available in water-dispersible or water-soluble forms. These are pre-measured dosage delivery systems, and are usually applicable to low-dose a.i. Tablets are more suitable for small areas. However, because of the cost of the manufacturing process, tablets are a costly form of application.

Some examples of liquid formulations are:

- Solution or soluble concentrate (SC) is a term for a liquid that consists of a.i.s dissolved in water or petroleum-based solvents. Such a formulation is required to be diluted prior to application, and is commonly used for seed treatment.
- Emulsifiable concentrate (EC) is one of the most commonly used formulations. This composition is made of a.i.s that are soluble in water-immiscible, organic solvents. ECs are added to water and sprayed on target crops. They are economical and easy to use. However, since several solvents are flammable and phytotoxic, ECs may be harmful to the environment and the applicator.
- Emulsion (EW) is an oil-in-water formulation. This is a good option for oil-soluble a.i.s, because the oil component aids the a.i., while the water component helps lower the volatility of the organic solvent. Emulsions may be flammable and may be toxic to the skin. There are two types of emulsions: **micro-emulsions** and **invert emulsions**.
- ULV liquid (UL) formulation is designed to spray very small droplets evenly over a large area. These drops are suitable for a.i.s that are extremely soluble in crop-compatible solvents. Small quantities of ULVs are adequate for large areas, when administered with specialized equipment. ULV products often contain surface active agents and drift control substances, because of their high potential to drift off the target. Such formulations may also be absorbed through skin.
- Formulations described as flowable or liquid are similar to a wettable powder and require to be diluted before application.
- An **aerosol**, another popular formulation, is made up of fine solid and liquid particles stored in a gas under pressure. Aerosols are made of one or more a.i.s and a solvent. They are available in a ready-to-use form, and sold in pressurized, sealed containers. An aerosol releases smoke or fog through electrical- or gasoline-powered aerosol generators.

Other formulations are **micro-encapsulated** products and fumigants, which cannot be said to be either liquid or dry/solid formulation. Their special properties make them suitable for certain pest control situations.

Pesticide label

A pesticide label is an indication that the chemicals in a formulation have been approved by the environmental protection agency of the country for the purpose intended. The label provides, for example: (i) concentrations of the active ingredients in the composition; (ii) directions of how to use in a safe manner; (iii) the toxicity rating; and (iv) indication of any hazard(s) during application. Pictorial representation via the hazard diamond may also be included on labels. The number and colour code denote hazardous characteristics such as being corrosive, explosive, flammable, poisonous or radioactive.

The use of any pesticide contrary to the instructions provided on the label is prohibited and the label may be regarded as a legal document.

Pesticide legislation

In view of the fact that thousands of active ingredients are used as pesticides, stringent rules and regulations have been laid down in each country to ensure the safe use of these toxic chemicals. Regulations are closely implemented since the majority of these chemicals are used in close proximity to grain, fodder, animals and water. Depending on the use and exposure to pesticide toxins, each country has its own agency/department/ministry to implement legislation that allows, restricts or bans the use of farm/non-farm pesticides, to ensure damage to individuals and the environment is minimized.

The labels and leaflets approved by a country's regulating agency normally indicate the necessary precautions while using pesticides. Labels contain information on: (i) the name of the product; (ii) the chemical composition; (iii) the name of the manufacturer; (iv) symptoms of poisoning; (v) first-aid measures; (vi) cautionary statements; (vii) directions concerning usage; (viii) restrictions (if any); (ix) instructions for storage; (x) information regarding disposal of used packages; (xi), application equipment; and (xii) waiting period (the time interval between purchase of a pesticide and its actual use).

Reproduced in Appendix 8 of this volume is a list of pesticides that are registered for manufacture and use in India. Appendix 9 lists those pesticides that are banned from being manufactured, imported and used in India, the pesticides refused registration and pesticides that are restricted in use in India. This volume also contains brief notes on the legislation in some randomly selected countries (see entries below). The information collected at the time of writing may have changed since.

Pesticide legislation in Bangladesh

The Ministry of Agriculture, through the Plant Protection Wing of the Department of Agricultural Extension, administers the Pesticide Ordinance, promulgated in 1971. Upon an amendment to the Ordinance, all matters are undertaken by the Bangladesh Crop Protection Association. Decisions are based on the basic rules laid down in the Ordinance, with compliance being implemented by the FAO.

Pesticide legislation in Canada

The Pest Control Product Act (PCPA) lays out the guidelines for pesticide registration and use in Canada. The Act is administered by Health Canada's Pest Management Regulatory Agency (PMRA), which reports to the Minister of Health (of the Ministry of Health). At different levels of the government (e.g. territorial and provincial legislation) there are agencies that share responsibility for regulating pesticide products and monitoring their use, sale and disposal, while complying with the overall guidelines and objectives of the federal legislation. Within each province, the provincial legislative authority has the autonomy to regulate all aspects of the pesticide, including a ban on the use of a compound in that province (which might not be recommended by the federal legislation). More information may be accessed from http://hc-sc.gc.ca.

Pesticide legislation in China

In China, pesticide registration and management are undertaken by the Evaluation and Adjudication Committee of Pesticide Registration, which comprises designated members of the Ministry of Agriculture, the Ministry of Health, the Customs Office and some other agencies. There is a proper demarcation of responsibilities, namely, pesticide registration, pesticide production, pesticide standards and pesticide market supervision, which are handled by designated members from this group of agencies.

The Institute for the Control of Agrochemicals (ICA) and the Ministry of Agriculture undertake the task of detailed analysis of each chemical, to evaluate its use. At the provincial levels too, there are ICAs, which do the preliminary work of evaluation.

Pesticide legislation in Denmark

The Danish Ministry of the Environment is the pivotal body that synchronizes pesticide policies with those of the EU. In Denmark, like most of the other EU member states, no chemical or biocide may be used or sold unless its ramifications on the environment are assessed.

The Ministry of the Environment deals with chemical inspection, while the Plant Directorate of the Ministry of Food, Agriculture and Fisheries looks out for cross-compliance rules. Information on which trademarks are to be inspected is shared, so as to monitor the type of consignments that come in. (See also **Pesticide legislation in the European Union (EU)**.)

Pesticide legislation in France

The principal authorities involved in pesticide legislation are the Ministry of Agriculture (*Brigade Nationale Des Controles Phytosanitaires*) and the Minister of Economy and Finances. The government drafted a plan for 'the reduction of risks linked to pesticides' (for 2006–2009) to review its use of pesticides, with a view to reduce the hazards by such use. The new plan was to have streamlined timelines and procedures for the pesticide registration process. The French Food Safety Agency is responsible for evaluating the risk:benefit ratio of chemicals. EU directive 91/414 enables better information sharing, for the benefit of all concerned. For a detailed discussion on the French system, visit http://www.crop-protection-monthly.co.uk/Copy%20of%20kynetec.htm (See also **Pesticide legislation in the European Union (EU)**.)

Pesticide legislation in Germany

In November 1998, the German Federal Ministry of Agriculture constitutionally defined principles for good plant protection practice. There have been discussions on measures to ensure that no product, banned in Germany, enters Germany through another EU member country.

Pesticide legislation in India

Some of the ministries and bodies responsible for ensuring the safe use of registered plant protection products in India are: (i) the Ministry of Agriculture; (ii) the Secretariat of the Central Insecticide Board and Registration Committee; (iii) the Directorate of Plant Protection, Quarantine and Storage; (iv) the Indian Agricultural Research Institute; and (v) the Ministry of Environment and Forests.

The Insecticides Act, 1968, regulates the import, manufacture, sale, transport, distribution and use of 'insecticides' (this term being used to include herbicides, fungicides and rodenticides), for the sake of prevention of risk to human beings, animals and the environment. The Ministry of Agriculture administers this Act. The Insecticides Act, 1968 and the Insecticides Rules, 1971 are administered by the Ministry of Agriculture, the Department of Agriculture and Cooperation; the Environment Protection Act, 1986 is administered by the Ministry of Environment and Forests, while the Prevention of Food Adulteration Act, 1954 is administered by the Ministry of Health and Family Welfare. (See also **Central Insecticide Board and Registration Committee (CIBRC)**.)

Pesticide legislation in Indonesia

The Minister of Agriculture of the Republic of Indonesia is empowered to issue permits for pesticides via Government Stipulation No. 7 (1973). The Ministry of Agricultural Decree No. 434.1 is an important piece of legislation since it combines the various aspects of the registration process into a uniform system, which includes technical and administrative requirements for enabling registration, review and deregulation of a pesticide. The standard adopted for evaluation is in line with those of international institutions such as the FAO/WHO and the EPA of the USA.

Pesticide legislation in Malaysia

The main piece of legislation is the Pesticides Act of 1974, which is implemented by the Pesticides Board, with members from various government agencies. The board is under the Department of Agriculture. The Department of Agriculture is responsible for evaluation, registration and review of all pesticides, which it does in keeping with the Code of Conduct on the

Distribution and Use of Pesticides. There are other Acts which pertain specifically to fumigation activities, waste disposal, MRL and safety of workers.

Pesticide legislation in Nepal

The Pesticide Act of 1991 established a Pesticide Registration and Management Division, and a Pesticide Committee for the task of proper evaluation of pesticides and their registration.

Consumer groups, scientists and the government are well represented on these bodies.

Pesticide legislation in Pakistan

Agricultural Pesticides Ordinance of 1971, supported by the Agricultural Pesticides Rules 1973 is responsible for the import, manufacture, formulation, sale, distribution and use of pesticides in Pakistan. Rules regarding registration and deregulation of pesticides are modified based on inputs by the Agricultural Pesticides Technical Advisory Committee.

Pesticide legislation in Republic of Korea

The Ministry of Agriculture and Forestry were given the responsibility for registration of pesticides until 1996, after which the responsibility was given to the Administrator of Rural Development Administration.

Pesticide legislation in South Pacific nations

Many South Pacific nations have designed their pesticide regulation policies based on the International Code of Conduct on the Distribution and Use of Pesticides (FAO Code). Hence although most countries have their ministries/departments that implement Acts, their concerns and measures revolve around those highlighted by FAO.

Pesticide legislation in Sri Lanka

According to the Control of Pesticides Act No. 33 (1980), the Department of Agriculture regulates the import, manufacture and sale of pesticides, irrespective of the end use of the active ingredient. The Registrar of Pesticides is the national authority for implementing the Control of Pesticides Act. The Registrar is assisted by a committee (the Pesticide Technical and Advisory Committee or PeTAC) and members and institutions under the Department of Agriculture. There are laws and institutions in place to keep a watch on any wrong-doing and illegal usage. Conformation to international conventions such as POP (persistent organic pollutant) and PIC (prior informed consent) is carried out as a routine measure.

Pesticide legislation in the European Union (EU)

European legislation aims to harmonize the registration of plant protection products across the European Union (EU). EU legislation places more stress on the safety of the environment than the quantity of food produced. Thus, every chemical is thoroughly tested before being approved for use according to standards set out in the relevant EU directives, with member-level implementation being carried out by a designated body in the member state (e.g. the Plant Protection Products Regulations implements it in the UK). Every chemical comes up for re-evaluation, where it might have restrictions placed on its use or it may be banned.

The EU's two-tier registration system first assesses active ingredients for inclusion at the level of the community. Products containing these ingredients are then separately assessed by member states for any other aspect, before they are registered.

Accordingly, the European Commission Directorate General for Health and Consumers states:

> In the European Union, no plant protection product can be used unless it has first been scientifically established that: they have no harmful effects on consumers, farmers and local residents and passers-by; they do not cause unacceptable effects on the environment; they are sufficiently effective against pests.
>
> European Commission Directorate General for Health and Consumers DG SANCO (2009)

Pesticide legislation in the UK

National policies pertaining to the environment in the UK comprise areas of plant protection, health hazards, food and rural affairs. The spirit of the regulations reflects that of the European Union (EU).

A number of government organizations and departments share the responsibility of putting in place the right laws required to monitor the production, sale, export and purchase of products that may cause damage to any aspect of the environment. Most legislation has the whole of the UK (i.e. England, Wales, Scotland and Northern Ireland) in its purview.

The bodies responsible for pesticide approval include: (i) the Department for Environment, Food and Rural Affairs (Defra), which has strategic policy responsibility for pesticides, chemicals and detergents; (ii) the Department of Health; (iii) the Health and Safety Executive (HSE); and (iv) the Chemicals Regulation Directorate (CRD). Relevant legislation includes: (i) Plant Protection Products Regulations (PPPR); (ii) Control of Substances Hazardous to Health (COSHH); and (iii) the Food and Environment Protection Act (FEPA).

The CRD and the HSE represent the UK in the European Community in activities pertaining to active ingredients. Since every pesticide used in the UK must be formally approved, the CRD has the task of approving each chemical biocide, detergent and pesticide, whether for crop use or non-crop use. UK legislation recognizes four different types of approvals: (i) on-label approvals; (ii) off-label approvals; (iii) parallel import; and (iv) commodity substance approvals. Additionally, the CRD is responsible for fulfilling the objectives of Defra and the HSE, namely, prevention of death or injury from unsafe use of such chemicals, and spreading awareness of safe use and disposal of such substances.

The PPPR is a recent piece of legislation and implements an EU Directive (91/414/EEC) which regulates 'Plant Protection Products'. Plant protection products include agricultural pesticides and growth regulators. (See also **Pesticide legislation in the European Union (EU)**.)

Pesticide legislation in the USA

See **Environmental Protection Agency (EPA) of the USA**

Pesticide legislation in Vietnam

The Plant Protection Department of Vietnam, under the administration of the Ministry of Agriculture and Rural

Development, was established in 1961. These bodies have formulated their policies of, for example, pesticide registration and waste disposal based on the 1989 policy framework of the International Code of Conduct on the Distribution and Use of Pesticides (approved by the FAO and WHO).

Pesticide mixing and loading

Mixing and loading pesticides involves a high risk of exposure to the chemicals. It is thus necessary to fully understand the operation and to take all necessary precautions during the task. The following steps should be adhered to.

Operations should not be carried out by a single person but by at least two people. Label instructions must be placed in a visible area and strictly followed. Mixing or loading should be carried out outdoors, or in a well-lit and ventilated room.

Measuring and mixing of quantities and calculations for the dilution of chemicals should be carried out with care and accuracy. Tanks should be checked for leaks before a tank mix is loaded with water. Loading should be carried out away from a water source to prevent contamination of the water source and precautions should be taken to avoid backflow to the water source. Pesticide containers should be placed on a stable platform before they are opened.

Dry pesticidal compounds should remain in watertight containers before they are mixed. Mixing should be done evenly to avoid lumps or an excessively watery solution. Use of calibrated and clean pesticide applicators is recommended. Nozzles, valves and hoses must be checked before they are used. Closed mixing and loading systems are preferable for applying large quantities of pesticides. To avoid contamination, food, utensils and feed must be kept away so that they are not exposed to the chemicals.

After use, all the mixing and loading equipment should be cleaned, and the cleaning site should be decontaminated. Cleaning should be carried out with suitable and appropriate PPE.

Pesticide nomenclature

Any pesticide has three different names, namely: (i) the chemical name; (ii) a trade name; and (iii) a standard or common name.

Common or standard names usually consist of ordinary nouns given to the pesticides. Standardization organizations such as the International Organization of Standardization approve such common or standard names.

The chemical names of a pesticide are based on the chemical formulation of that pesticide. International bodies such as the International Union for Pure and Applied Chemistry (IUPAC) or Chemical Abstracts (CA) finalize a name for each active ingredient (a.i.) according to its chemical formula. CA also assigns a Chemical Abstracts Service (CAS) Registry Number to each pesticide. Such chemical names and numbers make it easier to refer to each a.i.

A pesticide can have one or more proprietary trade names, each with a different concentration of the pesticide, a specific formulation and a unique mix with other pesticides.

Pesticide resistance management

According to the Insecticide Resistance Action Committee (IRAC) (2007), insecticide resistance is 'a heritable change in the sensitivity of a pest population that is reflected in the repeated failure of a product to achieve the expected level of control when used according to the label recommendation for that pest species'. The following natural cycles and biological processes contribute towards the development of resistance.

Organisms of a species that survive a pesticide attack, do so possibly due to a behavioural trait (avoidance) or a biochemical trait (detoxification) not present in their susceptible brethren. When such resistant organisms produce the next generation, they pass on these traits so the next generation has fewer susceptible individuals. Also, the huge number of offspring produced in each short pest generation, with multiple generations being produced per season, increase the probability of random mutation and thus dominance of the resistant gene. (In comparison, it is observed that most biocontrol agents such as insect predators and parasites do not produce a vast number of offspring, thereby reducing the possibility of random mutation and the development of resistance in their population.) Resistance as a trait is autosomal (reflected in either sex) although it is an incompletely dominant trait (the offspring will carry a less susceptible gene).

Sometimes, a pest organism modifies its biochemical behaviour by producing more enzymes (e.g. esterases and glutathione transferases) that can break down toxins. Biochemical traits in organisms sometimes reduce sensitivity to a particular toxin or increase the rate of excretion.

As an evolutionary trend, just as insects can damage plants, plants also produce synthesized substances that have toxic, repellent and anti-feedant properties. Hence, gradual development of tolerance towards poisons has developed in plants.

Every occurrence of resistance leads to the need for more frequent and higher rates of application of pesticides. This, in effect, leads to increased selection pressure, thus increasing the number of resistant genes.

In 1914, A.L. Melander documented, for the first time, resistance in scale insects. Gradually, pests such as the diamondback moth, Colorado potato beetle, housefly, rat and mosquito and the weed *Amaranthus palmeri* have also been seen to develop resistance towards many pest control agents. Compounds belonging to the organophosphate and carbamate groups, and specifically **glyphosate**, **DDT** and ***Bacillus thuringiensis***, are now becoming ineffective on pests.

It is very important to manage pesticide resistance, since the pest–pesticide battle is an ongoing one. Pesticide applications are inevitable, in view of the losses caused by pests on crops. An extreme step would be to apply so much pesticide in one application that no pest would survive, eliminating selection pressure and the inheritance of dominant traits. However, this is not an acceptable strategy because of the persistence of many pesticides and their toxicity to non-target organisms, handlers, consumers and the environment.

Often, a single line of attack is chosen by farmers or advocated by manufacturers for the successful control of a certain type of pest. Repeated use of the same compound or a.i. from the same chemical class, however, invites challenges on two fronts. One is cross resistance. Cross resistance is the change at the targeted pest biochemical site, which makes the pest resist a compound from any chemical class, as long as it acts on the same target site. An example is sulfonylurea herbicides (weed control agents) that inhibit acetolactate synthase (ALS). The development of resistance to sulfonylurea herbicides brings about resistance to **imidazolinone** and **triazolopyrimidine** herbicides too, because of the changes brought about at the site of the enzyme activity. The other challenge is multiple resistance (i.e. multiple resistance mechanisms in a population, or in a few individuals in the population), where the resistance may be either target-site oriented or non-target-site oriented. Simply, if a pesticide is used continuously to control a pest species until resistance is developed in the new generation, then another class of pesticides is introduced and used continuously until the pest develops resistance to that class too. This way multiple resistance develops. This can be very complicated and difficult to control and manage.

If used in moderation, resistance to pesticides may be kept under control for long periods. If chemical intervention is used only in times of crisis, it helps keep the susceptible population alive and keeps selection pressure under control, while holding the pest population within threshold levels. Often, pesticides with different properties are used for managing resistance, especially with fungal pathogens where a mixture of systemic and contact fungicides is used so that there isn't over-dependence on either. Rotating pesticides with varying modes of action and alternating the use of pesticides belonging to different chemical classes helps manage the development of resistance. Mixing two (or more) pesticides to accentuate either compound's efficacy is another way to postpone resistance to the chemical. Genetic modification of plants to create a pest-resistant variety is yet another strategy for combating resistance. For instance, sowing smut-resistant sugarcane frees the farmer from anxieties of over-use of smut fungicides.

Cultural practices (e.g. crop rotation, tilling, solarization and seed bed preparation), manual intervention, use of biological control agents and pheromone traps are critical in postponing the development of resistance.

Pesticide spill management

When a pesticidal compound is accidentally released or spilt from its container, it is called a pesticide spill. Spill management involves the quick, effective clean-up and proper disposal of the spilt compound, along with appropriate protection of the personnel involved. A transport accident causes huge environmental losses since it affects the soil, people and wildlife in the neighbouring areas. A small spill, resulting from human misjudgement or from a leak or crack in the container may be hazardous too, since the exposure may go unnoticed for a long time.

Small liquid spills should be absorbed by sand, mud or any other absorbent material. If a small container is leaking, it must be placed in a larger chemical-resistant container. If a tank or drum has developed a leak, the site has to be cordoned off and people have to be alerted to stay out of the area. Large quantities of absorbent material like sand, sawdust, clay, amorphous silica, diatomite or newspaper might be spread on the affected area. It has to be ensured that the absorbent material does not cause a fire hazard (e.g. sawdust in the presence of hypochlorite-containing herbicides).

Dry pesticides should be swept up and collected in a disposable bag or drum, without allowing the dust to fly around. The collected substance can be reused, if it has not disintegrated. If, however, clumps have formed or impurities have been mixed with the pesticide as a result of the spill then disposal of the collected substance is necessary. Inadvertent adulteration may cause nozzles to clog.

A spill clean-up kit should always to be carried in transit, storage areas and application sites. Easy to assemble and widely available, the kit must include clean personal protective clothing and chemically resistant equipment (gloves, footwear, apron, disposable coveralls, protective eyewear and a respirator). Spill pillows, absorbent clay, cat litter, activated charcoal, sawdust, commercial absorbents, newspapers and vermiculite are required for liquid spills, along with contaminant tubes to prevent spills from spreading. A diaphragm pump must also be included. Duct tape is critical in a spill kit, for closing any leaks. A plastic cover of polyethylene or plastic tarpaulin, a broom and shovel or dust pan are some important tools needed in the kit for managing dry spills. A spray bottle filled with water to mist dry spills, sufficient cotton tape, ropes to cordon people off, contact telephone numbers of emergency assistance agencies, a large sturdy, chemical-resistant drum to collect spills and a permanent marker for labelling cans, are some other common items for the kit.

Pesticide storage management

Pesticides, being active chemicals, may affect the environment around them. They may react to changes in temperature, pressure and moisture levels, causing them to degrade or volatilize. Some pesticides may react in the presence of certain other pesticides. So such chemicals need to be stored carefully.

A suitably located, correctly designed pesticide storage site is a must, irrespective of its size. The following few important features may be noted for proper storage.

First, it is important to decide and plan which pesticides are actually going to be needed, and in what quantity. Storing more than 1 year's supply might be an unwise decision since the passage of time may degrade some chemicals. Once the quantities of various pesticides are determined and the pesticides stored, the storage room, rack, shelf or pallet must be physically secured by a lock. This puts a check on unauthorized use or theft.

It is necessary to ensure that the storeroom, shelves or racks are not at the floor level where they can come in contact with water. All original containers made of cardboard or metal should be kept in strong plastic drums, cans and bags, so that they may be safeguarded from moisture. The floors of storage rooms should be made of unabsorbent material such as cement or tiles, instead of carpet or wood. Racks, cupboards and shelves that hold pesticides in small storage containers should be coated with epoxy paint or covered with polyurethane. As far as possible, pesticides should be retained in their original containers and in a sealed condition. The storage site should be kept cool, dry and well ventilated. Volatile pesticides should be stored separately from other pesticides, fertilizers and seeds. Vapours from opened containers can contaminate nearby pesticides and chemicals.

It is very important to preserve labels of pesticide formulations. Labels carry details of chemicals, first-aid measures and storage recommendations. Moisture in the air or regular handling can remove the label, fading the printed matter. Commonly, labels should be covered with a transparent tape or lacquer so that they remain stuck on the container. It is critical that label instructions are followed in case of concerns over storage and disposal and in an emergency.

The storage site should be well lit to enable handlers to clearly see labels and detect early symptoms of degradation, discoloration, spills or leaks.

There has to be some provision for accidental exposure, such as a well-covered drum of water, eye-wash dispenser, a clean set of clothes, gloves and face mask. However, these should be stored away from the pesticide storage area to avoid contamination from vapours and fumes leaking out of the containers. Food, feed and seeds should not be kept at the site of pesticide storage. All containers should be kept tightly closed so that they are not inadvertently opened by inexperienced people.

A tight watch has to be maintained for signs of leakage, spillage and discoloration in containers, and lumps and clogs in dry formulations. Containers that are kept for disposal should be stored separately, preferably after rinsing.

An up-to-date inventory of stored formulations is critical. A record has also to be maintained about: (i) the shelf life; (ii) the vendor for each chemical; and (iii) any unusual event in the storage site.

Pesticide synergists

Piperonyl butoxide: See Pesticide synergists

Synergists are chemical substances that are added to pesticide products to enhance the potency of active ingredients. Since they increase the efficacy of the pesticide, synergists are credited with successfully reducing the amounts of pesticide needed for an application. A pest may be able to detoxify certain pesticides because its enzymes are able to breakdown chemicals into harmless products. However, synergists reduce the action of pesticide deactivation, thus enabling the pesticide to kill a larger population of pests.

A large group of enzymes, called cytochrome P450, is present in most living organisms (e.g. insects, animals, plants, fungi). The enzyme complex and its substrates function by reacting with organic compounds and forming simpler substances. This enzymatic action is relevant for the study of pesticide toxicity and drug interaction. In the case of insects, any compound within a pesticidal product that can inhibit the metabolic enzyme system of an insect, or help increase the penetration of the insecticide, would be a synergist for that pesticide product. Sesamex, for instance, can inhibit insect P450 and is thus an insecticide synergist.

Sometimes, insecticides work as herbicide synergists. Organophosphates like **malathion** and **terbufos** have been shown to synergize the effects of a broad range of sulfonylureas and other herbicides by inhibiting the activity of microsomal P450s that prompt the oxidation of these herbicides. The activity of **metribuzin** on some weeds in soybean is enhanced by picotinic acid tert-butylamide.

Insecticide synergists, including piperonyl butoxide (discussed below) and diethyl maleate, inhibit the metabolic action of glutathione *S*-transferase, by producing conjugates. The detoxifying activity of several esterases is also inhibited by several synergists like triphenyl phosphate.

The enhanced activity of atrazine on giant foxtails and large crabgrass in the presence of the herbicide tridiphane is an example of herbicide synergism. Tridiphane inhibits atrazine metabolism by the enzyme glutathione *S*-transferase in the weeds, thus enhancing weed control. The maize crop can still metabolize the **atrazine** dose, and is thus spared the ill effects of atrazine, while the weeds are killed.

However, sometimes such enhanced action of a pesticide may have adverse effects. For instance, the contact herbicide **propanil**, which is well tolerated by rice plants, can actually kill the plant if insecticides such as **carbaryl** or **diazinon** are applied either simultaneously or within a fortnight of the herbicide application.

Synergists are distinct from inerts or additives which improve the physical characteristics of the product. By themselves, most synergists do not pose toxic hazards, but when present in the formulation, they may increase toxicity to non-target organisms and even to man.

Some other popular synergists are **tribufos**, piperonyl butoxide, sulfoxide and bucarpolate. Many pesticide ingredients are combined with synergists, for example **rotenone**, citrus oils, some carbamates and many pyrethroids are mixed with piperonyl butoxide for increased efficacy.

Tribufos, a plant growth regulator with a molecular formula $C_{12}H_{27}OPS_3$, acts as a synergist in insecticides. Tribufos is observed to make **fenvalerate** effective on some resistant insect species.

Piperonyl butoxide, $C_{19}H_{30}O_5$ (5-[2-(2-butoxyethoxy) ethoxymethyl]-6-propyl-1,3-benzodioxole), is an insecticide synergist. It is a very effective inhibitor of the cytochrome P450-dependent polysubstrate monooxygenases. Polysubstrate monooxygenase commonly enables detoxification of toxic plant **allelochemicals** into simpler products that can be excreted.

With a CAS Registry Number of 51-03-6, piperonyl butoxide is found to be quite toxic to aquatic organisms and amphibians. It significantly reduces the degradation and detoxification in human systems, thus becoming hazardous to applicators and handlers of this synergist. It is observed to affect the liver and the nervous system (see Piperonyl butoxide at top of next page).

Pesticide tolerance limit

The tolerance limit refers to the amount of toxic residue allowable in edible plant parts. It is the limit of a pesticide/chemical in the plant or animal product certified by legal authorities as safe for human consumption. (See also **Tolerance**; **Maximum residue level or limit (MRL)**.)

Pesticides

A pesticide is any chemical, organic or biological formulation aimed at controlling, preventing or eliminating plant or animal pests harmful to human health and the economy. Such compounds are either toxic or extremely repelling by nature. Pesticides are used in various places such as homes, gardens, hospitals, commercial buildings, public places, crops and stored grain sites. As there may be different kinds of pests in each situation, customized pesticidal measures are needed in each situation.

Pests include insects, nematodes, weeds, fungal and bacterial microorganisms, mites, molluscs, algae, rodents, pest vectors and such other life forms that cause a nuisance to the welfare of animals and human beings.

Pesticides can be classified on the basis of: (i) the pests they target; (ii) the chemical structure of the compounds used; and (iii) the degree and type of health hazards involved. Hence, there are many specific subgroups of pesticides such as herbicides, insecticides, miticides, molluscicides,

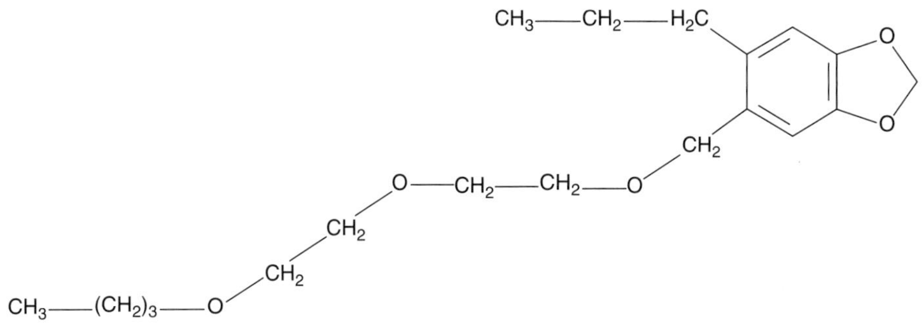

Piperonyl butoxide

nematicides, repellents and rodenticides. Categories of agents like **defoliants, desiccants** and attractants are also commonly included in the broad group of pest control agents, although these may not cause toxicity to the target. Similarly, vector control agents that inhibit transmission of disease to man, animal or crop are also included in the category of pesticides.

Most pesticides are products of a chemical nature (**DDT, 2,4-D**). However, there are many compounds obtained from plants (e.g. **azadirachtin** from the **neem** tree, the alkaloid veracivine ($C_{27}H_{43}NO_8$) from lily species, **nicotine** from tobacco, **rotenone** from yam and **pyrethrins** from chrysanthemum). Minerals (e.g. **cryolite, sulfur** and **copper**) are also sometimes pesticidal in nature, while some microorganisms also have pesticidal properties such as ***Bacillus thuringiensis***, ***Helicoverpa*** **nucleopolyhedrovirus** (HNPV) for controlling *Helicoverpa* moths, wasp species from the genus ***Trichogramma*** and fungal species, for example *Trichoderma harzianum*.

Chemical pesticides are the predominant choice of pest control. However, since these agents vary in their degree of toxicity, there is risk of adverse effects on non-target plants, animals, beneficial insects, as well as handlers and applicators. Pesticide residues present in the food chain can also cause much damage to human life. To avoid or prevent such calamities, pesticides must be selected and applied with care.

Pesticide regulating agencies exist in each country to monitor proper use of chemicals. These include agencies such as the **Environmental Protection Agency (EPA)** based in the USA and the **Pest Management Regulatory Agency (PMRA)** based in Canada. Organizations such as the **World Health Organization (WHO)** and the Pesticide Action Network (**PAN**) also play an important role in monitoring the safety of farmers. In India, the **Central Insecticide Board and Registration Committee (CIBRC)** under the Ministry of Agriculture is responsible for decisions concerned with registering pesticides in the country for specific uses. The main task of such agencies is to issue guidelines on hazards caused by specific chemicals. They also classify pesticides on the basis of toxicity, which is a useful tool for everyone concerned.

Overuse or even regular use of a pesticide can bring about changes in the pest system by which resistance is developed to that chemical. Agencies such as the **Fungicide Resistance Action Committee (FRAC)**, the **Herbicide Resistance Action Committee (HRAC)** and the **Insecticide Resistance Action Committee (IRAC)** work towards managing resistance such that minimum applications may lead to maximum effects on the pest.

Pesticides and the environment

In the context of pesticides and the hazards associated with them, the term 'environment' not only includes large, natural and geographical features such as mountains and rivers in a certain location, but also the micro-environment (e.g. home gardens, recreation parks, offices, industries, restaurants, hedges along the roads, ponds, man-made lakes and dams). It also includes the plants, animals, farms, soil, water and the air around us.

The use of pesticides can affect all aspects of this environment – misuse causing definite and visible harm. To avoid misuse, people and institutions engaged in the manufacture, sourcing, selling and handling of pesticides have to be made aware of the risks to the environment. Since this task involves a very large number of people from various walks of life, and since the frequency of pesticide use is very high, accidents involving pesticide spills, runoffs, leaks, overuse, drifts and related hazards affecting man and beast are common and frequent. The safe use of pesticides involves care while using, mixing, loading, selling and storing them. The following passage lists some of the many possible hazards and precautions that may be taken.

At the national level, specific characteristics of an active ingredient (a.i.) (e.g. its reactivity and combustibility) along with its reaction with soil and water form the basis of registering a chemical pesticide in that country. Such a pesticide-registering agency of the country establishes the toxicity of that chemical and decides the formulations and uses permissible for that chemical. This way, basic information is made available about chemicals which are harmful to the environment (for instance to bees and other pollinators). This is especially true with regards to newer molecules synthesized into pesticides. As newer studies reveal potential hazards of existing pesticides, their registration status is also reviewed. It is necessary for users to stay up to date on the pesticides they are using.

Pesticide labels contain specific information regarding, for example, mixing, dosage details, fire hazard and storage parameters. Adherence to label instructions is critical to safety.

Pesticide pollution of soil, water and air is either broad-based or topical. An accidental spill and leak constitutes topical pollution, where soil and animal life around the accident site is harmed. Other examples include spills during mixing and loading, and the improper disposal, rinsing and clean-up of spills and leaks. Fire hazards, specific to the pesticide, may also cause damage. Smoke and toxic fumes can add to environmental pollution.

Broad-based pesticide pollution takes place when a pesticidal compound enters surface water bodies or underground water sources (water beneath the earth's surface that moves along depending on the structure of rock and stones above it). Occasions for such broad-based pollution include rinsing out containers and PPE in streams, porous soils or near channels made to rejuvenate bore wells.

If the wrong location is selected for storing, mixing and loading pesticides, this can also lead to environmental pollution and health hazards. The ideal location must be far from schools, hospitals, old-age homes, parks and animal shelters, to avoid chronic exposure (i.e. long duration exposure to small quantities of toxins). Attention to label instructions is critical when applying chemicals near parks, gardens, farms or grain store houses.

Spills near parks, forest areas and animal sanctuaries cause a great deal of damage to wildlife, insect populations and plant growth in the area. Bird life can also be disturbed. Unintentional exposure to pesticides via contact or upon ingestion can kill populations of non-target species of birds and small mammals. Water contaminated with pesticide drift or a spill-off/runoff of a pesticide, may kill or inhibit aquatic plants, invertebrates and fish, while simultaneously being a threat to predatory birds and small mammals. Depending on the persistence of such chemicals in soil, colonies of microorganisms and earthworms and beneficial beetles can be wiped out. Accumulation in body tissues allows chemicals to enter the food chain.

There are cases of pesticides being moved away from the intended site of application by strong winds and vapour (air), via leaching and runoff (water) and via living systems (as accumulated residues in tissues of plants and animals). Various forms of pesticide formulations have different potential for polluting the environment. For instance, formulations like dust, wettable powders and spray droplets have a greater tendency to move away from the intended site than other formulations. Such a movement of particles in the air, away from the intended target site is known as a drift. Drift is caused by the force of the wind or the ventilation system at the time of application. It is important to note that large droplets produced with low pressure by coarse nozzles do not drift and may be ideal for use in sensitive areas. Granules and pellets are also good formulations when environment pollution is a concern.

Pollution via the air is commonly caused during fumigation. To conserve the environment, such an application should be made when the wind speed is slow. To avoid the entry of pesticides into the food chain, fumigating near fodder crops or fruit crops should be strictly avoided.

Pollution via water is very widespread and is seen in the form of leaching and runoff. When liquid formulations (e.g. soil drench, emulsifiable concentrate and soluble concentrates) are applied excessively on the plant, the excess seeps into the soil, contaminating groundwater sources. If the soil is not porous, these chemicals runoff and may accumulate in a seepage tank, well or local reservoir. In a domestic garden situation, liquid formulations may flow down a slope and collect in a corner, to cause harm to passers by. Even if non-sewage water is channelled through an underground pipe, its contamination can harm sensitive plants and **beneficial organisms**. Unscrupulous rinsing and washing of tools, containers and **PPE** can also contribute to seepage and runoff, thus increasing risk to pets, birds and plants. Label instructions clearly mention a **buffer zone** from water bodies, which is the distance between the application site and the nearest water body.

It is very important to ensure that contamination of the food chain by the use of pesticides is avoided. A strict **restricted entry interval (REI)** is recommended. At the same time, the recommended pre-harvest interval is also to be adhered to. As mentioned before, pesticides should not be applied on forage, pasture, fruit crops or any other target directly consumed by people or animals.

Treated fodder should not immediately be offered to cattle. Meat animals should not be allowed to consume treated hay until the prescribed time period, usually 21 days.

Fishing should not be carried out for at least 3 days after the application of an aquatic herbicide or after treating an area for mosquitoes or flies. Pesticide residues are a serious issue among persistent pesticides. The knowledge of **maximum residue limits** (MRLs) of pesticides for each crop or crop group, makes manufacturers, users and inspectors alert to any deviation from safe use.

Factors that facilitate contamination of groundwater are important to note. The chemical characteristic of the pesticide is an important factor. Mobile chemicals tend to move rapidly through soil layers. A highly water-soluble pesticide that is not very persistent (meaning that it breaks down easily by reacting with soil pH, minerals and microorganisms) will contaminate groundwater quite easily, compared with water-insoluble persistent chemicals. The property of adsorption also contributes to contamination. The less adsorbable the chemical, the more potential it has to contaminate groundwater.

Coarse and sandy soil carries the chemicals downwards easily; clayey soil does not. Some chemicals adsorb to organic matter, slowing their downward movement.

Knowledge of the a.i. and the soil are critical for minimizing environmental pollution. Pesticide application should be planned after studying irrigation patterns or after reading the forecast for rain. Some labels state that applications should be avoided immediately after rain or irrigation; advice for other pesticides state that irrigation should follow immediately after the pesticide has been applied. Non-adherence to label instructions may either lead to wastage of pesticide (via runoff) or plant burn and phytotoxicity and death of soil organisms.

Environmental contamination because of pesticide use can be controlled by the sincere efforts of all people concerned with pesticides, the most basic components being the employment of trained and conscientious personnel and adherence to label instructions.

Pethoxamid herbicide

Pethoxamid, $C_{16}H_{22}ClNO_2$ (2-chloro-*N*-(2-ethoxyethyl)-*N*-(2-methyl-1-phenylprop-1-enyl) acetamide), is an amide herbicide. It is an off-white, odourless, crystalline solid soluble in water (400 mg/l at 20°C) and highly soluble in organic solvents (117 g/l in heptane and 250 g/l in hexane at 20°C) (see Pethoxamid at top of next page).

Pethoxamid is a systemic, pre-emergence and early post-emergence herbicide that is used to control several grass and broadleaved weeds.

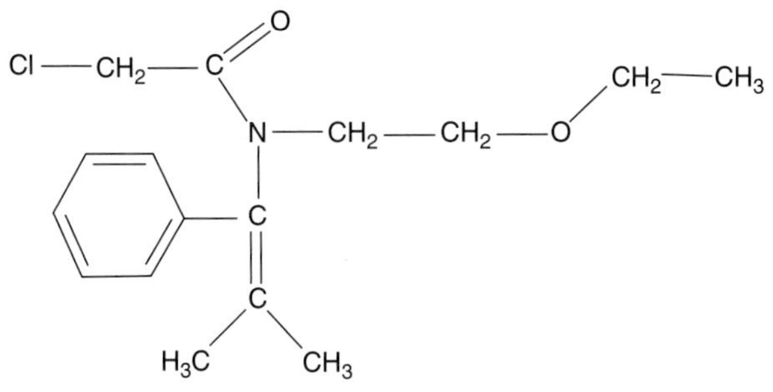

Pethoxamid

The acute oral LD_{50} in rats is 983 mg/kg while its acute dermal LD_{50} is >2000 mg/kg b.w. The ADI is 0.01 mg/kg b.w./day. The NOEL (1 year) in dogs is 2 mg/kg b.w./day. The acute oral LD_{50} in honey bees is 200 µg per bee. This herbicide sensitizes the skin.

Most (90%) of this herbicide is absorbed when administered orally in rats and most of it is excreted within 96 h. The metabolism in animals is extensive with the production of sulfoxides and sulfones. *N*-(2-ethoxyethyl)-*N*-(2-methyl-1-phenyl-1-propenyl)-2-sulfoacetamide is a key metabolite in soil and it is known to be mobile.

Pethoxamid is registered in Hungary, Denmark, France, Germany, but not in the UK or the USA.

PGR AIA
See **Indole-3-acetic acid (IAA or AIA) as a bioPGR**

PGR benzoic acid
See **Benzoic acid herbicides**

PGR carbaryl
See **Carbaryl insecticide**

PGR clodinafop-propargyl herbicide
See **Clodinafop-propargyl herbicide**

PGR dalapon
See **Dalapon herbicide**

PGR diquat
See **Diquat herbicide**

PGR diquat dibromide
See **Diquat dibromide herbicide**

PGR fosamine
See **Fosamine herbicide**

PGR gibberellin A_4 and gibberellin A_7
See **Gibberellins**

PGR IAA
See **Indole-3-acetic acid (IAA or AIA) as a bioPGR**

PGR methasulfocarb
See **Methasulfocarb fungicide**

PGR propham
See **Propham herbicide**

PGR prynachlor
See **Chloroacetanilide herbicides**

Phenmedipham herbicide
Phenmedipham, $C_{16}H_{16}N_2O_4$ (methyl 3-(3-methyl carbaniloyloxy) carbamate), is a carbanilate herbicide. It is a colourless, crystalline substance, moderately soluble in water and soluble in polar organic solvents. Although stable up to 200°C, it is important to store phenmedipham in controlled temperatures and in airtight containers, away from unsuspecting people or food. Phenmedipham is made by the reaction of 3-aminotoluene with 3-methylcarbamic benzonyl chloride. It has a CAS Registry Number of 13684-63-4.

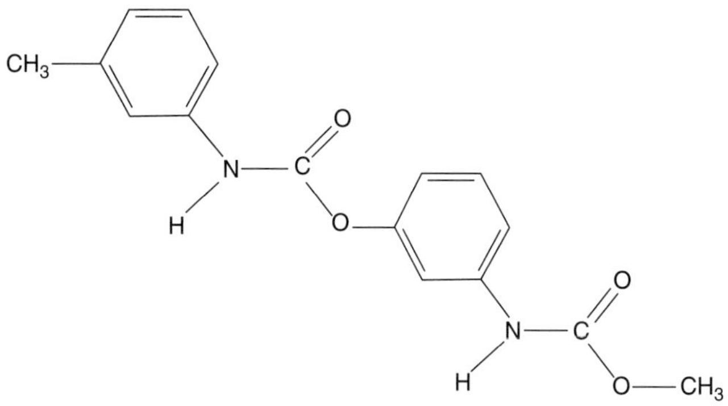

Phenmedipham

Phenmedipham is a selective, systemic, post-emergence foliar herbicide, absorbed through the leaves. It acts as a photosynthesis electron-transport inhibitor, and is used to control annual broadleaved weeds and grass weeds in beet crops, strawberries, peas and spinach.

This herbicide is available as emulsifiable concentrates (EC), soluble concentrates (SL) and water-dispersible granules (WG), under trade names like Betanal, Alegro and Beetup. It is best applied as a foliar spray, since soil application is not effective. Phenmedipham is compatible with other herbicides.

The WHO and EPA classify phenmedipham under the toxicity Class U (product unlikely to present acute hazard in normal use) and Class III (slightly toxic), respectively.

The dermal LD_{50} of phenmedipham is >4000 mg/kg in rats and >10,000 mg/kg in rabbits. The acute oral LD_{50} among rats and mice is >8000 mg/kg. The ADI of phenmedipham in rats is 0.03 mg/kg b.w./day. It causes dermatitis of hands, if spray mist is inhaled.

In mammals and plants, phenmedipham is hydrolysed to methyl *N*-(3-hydroxyphenyl) carbamate and conjugates. In soil, methyl *N*-(3-hydroxyphenyl) carbamate and *m*-aminophenol are major metabolites of phenmedipham.

Phenmedipham is registered in the USA, most of the EU, Australia, Canada and several other countries. According to the EU Pesticide Database, the MRL of phenmedipham is 0.1 mg/kg for strawberries and 0.05 mg/kg for vegetables, fruit and cereal grains. (See also ***Bis*-carbamate herbicides**.)

Phenothrin insecticide
Phenothrin, $C_{23}H_{26}O_3$ (3-phenoxybenzyl (1*RS*) *cis- trans*-2,2 dimethyl-3-(2-methyl-prop-1-enyl) cyclopropanecarboxylate), is a non-systemic, pyrethroid insecticide and an acaricide. It is a

yellowish-brown liquid, totally soluble in water and moderately soluble in methanol and hexane. Though stable under UV radiation, it is hydrolysed by alkalis. It is also light sensitive and combustible. This insecticide is produced by esterification of chrysanthemic acid with 3-phenoxy phenyl alcohols. It has a CAS Registry Number of 26002-80-2 (see Phenothrin at top of the next page).

Phenothrin acts by disrupting the life cycle of insects (including killing eggs). It is mainly used to control insects (e.g. fleas, ticks, lice) in homes, industries and public places and stored grains. It is also used in mosquito control and as an ingredient in aerosol insecticides.

The insecticide phenothrin is formulated as an aerosol, emulsifiable concentrate (EC) or oil-miscible liquid (OL). A common trade name is Sumithrin. Phenothrin is often used with **methoprene**, an insect growth regulator that interrupts the insect's biological life cycle by killing the eggs.

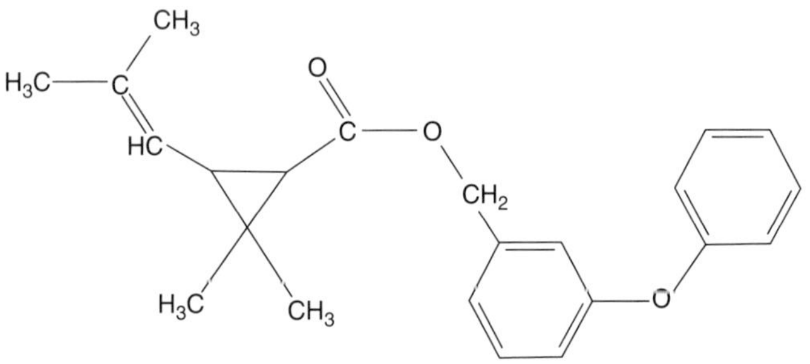

Phenothrin

According to the WHO, phenothrin is a Class U toxin (a product unlikely to present acute hazard in normal use). The oral LD_{50} in rats and mice ranges from 500 to >5000 mg/kg. However, it is toxic to humans, bees and fish and is probably a carcinogen. The compound also irritates the skin. Phenothrin product labels are required to carry the signal word 'WARNING', due to the possibility of deaths of pets such as cats and dogs. The ADI for phenothrin is 0.07 mg/kg/day. Symptoms of poisoning include irritability to sound or touch, sensation of prickling, tingling or creeping on skin, numbness, headache, weakness, dizziness, nausea, vomiting, diarrhoea, excessive salivation and sometimes fluid in the lungs and seizures.

Phenothrin is degraded by photo-oxidation in soil and water. Due to its sensitivity to sunlight, it does not accumulate in the environment.

Phenothrin is registered by the USA, South Africa, Australia and Canada, but it is banned or restricted in most EU countries.

Phenoxy acid herbicides

Phenoxy acid herbicides were the first selective, organic, auxinic herbicides produced. These have growth-promoting properties, which can be exploited to bring about abnormal growth in targeted plants, to eventually kill them. Phenoxy acid herbicides are particularly effective on broadleaved weeds, causing twisting of leaves and stems, and finally death. These herbicides are used to protect grassy crops, like cereals and turf.

Auxinic herbicides can be categorized into four classes: (i) phenoxy acids; (ii) **benzoic acids**; (iii) pyridine carboxylic acids; and (iv) quinolinecarboxylic acids.

Commercially developed, widely used phenoxy acid herbicides include **2,4-D**, **2,4-DB**, **dichlorprop**, **MCPA**, **MCPB**, **MCPA-thioethyl**, 2,4,5-T and **mecoprop-P**. Besides protecting crops against broadleaved plants, these herbicides also increase crop yields. Most are foliar applied since they are absorbed by the leaves.

Phenoxy acid compounds are composed of a substituted phenoxy ring and a carboxylic acid bound to an oxygen molecule. Variation in the acid (acetic or butyric acid) can change the degree of selectivity and persistence. Generally, the more substitution, the longer is the persistence. These herbicides are water soluble and may leach into the soil. Degradation is rapid via microbial action and the degradation of the side chain.

Phenthoate insecticide

Phenthoate, $C_{12}H_{17}O_4PS_2$ (*S*-α-ethoxycarbonylbenzyl *O,O*-dimethyl phosphorodithioate), is an organophosphorus insecticide. It is a colourless, crystalline substance soluble in water at 10 mg/l, methanol, ethanol, acetone, xylene, cyclohexanol and carbon tetrachloride. This compound gradually decomposes under alkaline conditions. It slightly degrades after 20 days at pH 9.7. Phenthoate is produced by the reaction of the sodium salt of *O,O*-dimethyldithiophosphonic acid with phenylbromoethyl acetate. It has a CAS Registry Number of 2597-03-7.

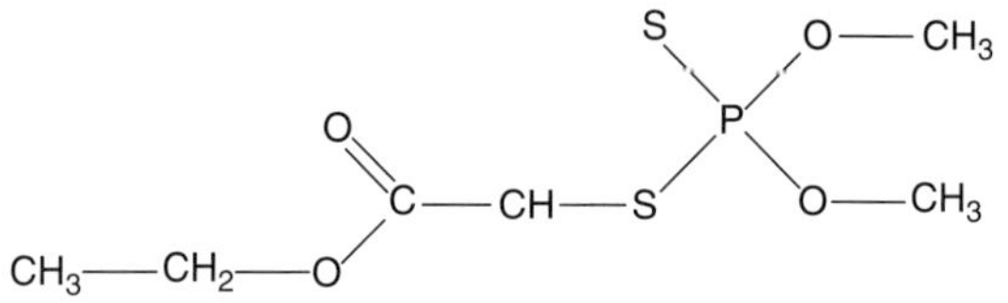

Phenthoate

Phenthoate, an **acetylcholinesterase** (AChE) inhibitor, is a non-systemic insecticide and acaricide with stomach action. It is used to control aphids, **scale insects**, jassids, lepidopteran larvae, bollworms, borers, **leafhoppers**, **mealybugs**, thrips, whitefly and spider mites in fruit (citrus/pome), olives, chestnuts, mulberries, cotton, maize, rice, coffee, tea, sugarcane, tobacco, vegetables and ornamentals. A non-crop use is mosquito control.

The formulations of phenthoate include wettable powder (WP), dustable powder (DP), emulsifiable concentrate (EC), ULV liquid (UL) and granules (GR). Trade names include Cidial, Elsan and Papthion. The compound is compatible with most other pesticides, except those which are alkaline in reaction. It is phytotoxic to some varieties of grape, fig, peach and apple.

Phenthoate has an oral LD_{50} of 240 and 400 mg/kg in rats and mice, respectively, and 300 mg/kg in quail and birds. The dermal LD_{50} in rats is >4000 mg/kg. The acute 96 h LC_{50} for carp is 2.5 mg/l. Phenthoate is a moderately toxic compound, absorbed through the gastrointestinal tract and skin, and via inhalation. It is toxic to fish and bees (the acute 48 h LD_{50} being 0.306 μg per bee). It also has the potential of being carcinogenic. Symptoms of poisoning are excessive salivation, muscle twitching, tightness in the chest, wheezing, productive cough, fluid in lungs and pin-point pupils. The ADI for phenthoate is 0.003 mg/kg b.w./day.

In animals, phenthoate is degraded by hydrolysis of the carboethoxy moiety, cleavage of the P–S or C–S bond and removal of the methoxy group by either direct demethylation or hydrolytic cleavage of the P=O bond. The main metabolites are demethyl phenthoate acid, demethyl phenthoate oxon acid, *O,O*-dimethyl phosphorothioic and phosphorothioic acids. In plants, oxidation to the phosphorothiote and hydrolysis lead to phosphoric acid dimethyl and monomethyl phosphate. Phenthoate is rapidly degraded in soil (with the DT_{50} of <1 day in most soils except dry sandy loam soil). Microbial action is the major means of degradation. Polar products, CO_2 and phenthoate acid are some of the breakdown products.

Thailand has banned phenthoate. The USA, Canada and many countries of the EU have not registered its use. India, the Philippines, South Africa, Cambodia, Mali and many African countries have registered the use of phenthoate.

Phenyl mercury acetate fungicide

Phenyl mercury acetate, $C_8H_8HgO_2$ (phenylmercury (II) acetate), is an organomercurial pesticide. It is used as a fungicide, slimicide and herbicide. The compound is a coarse yellowish, hygroscopic crystalline powder with an odour of acetic acid. It is soluble in water (0.16 g/100 g) and extremely soluble in ethanol and methanol (95%). The product is not compatible with strong oxidizing agents, sulfur and ammonia. Carbon dioxide and carbon monoxide may be produced when phenyl mercury acetate is heated to decomposition. The compound has a CAS Registry Number of 55-68-5.

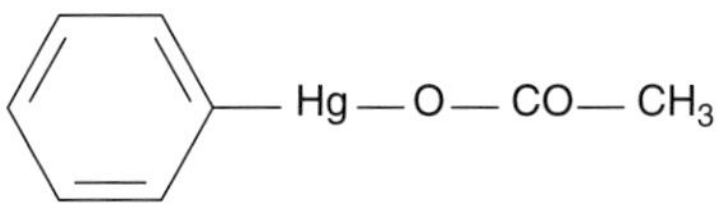

Phenyl mercury acetate

Phenyl mercury acetate is used in controlling root-borne fungal diseases in rice and orchards. It is also used to control external seed-borne diseases in crops like wheat, barley, rice, sorghum, oats, beet, cotton and turf. When applied to the leaves of plants, it acts as an anti-transpirant. Phenyl mercury acetate is also used as a preservative and in the wood pulp and paper industries.

Phenyl mercury acetate is extremely hazardous on ingestion, inhalation and eye contact, and irritating to the skin. Inflammation of the eye is characterized by redness, watering and itching. Severe over-exposure can result in death.

The acute oral LD_{50} of phenyl mercury acetate is 41 mg/kg among rats and 13.25 mg/kg in mice. The LC_{50} at 96 h in fish is less than 1 mg/l. The compound is not toxic to bees. When released into the soil, phenyl mercury acetate may not evaporate significantly, but may leach into groundwater. It does not bioaccumulate significantly. The use of certified dust respirators, gloves, splash goggles and laboratory coats is necessary while applying the pesticide.

Phenyl mercury acetate, like most mercury-based compounds, are either banned or severely restricted for use in most countries. It is registered for use in Australia. In India, it is banned for use, manufacture and import, for agricultural purposes. Norway allows the use of this compound only as a seed dressing. Most other countries have banned the use of phenyl mercury acetate in farm-related activities.

Phenylamide fungicides (PAFs)

The development of phenylamide fungicides (PAFs) is an important landmark in the development of effective and safe systemic fungicides. Formerly known as acylalanines, phenylamides have proved to be better than **dithiocarbamates**, copper fungicides and **chlorothalonil** in their control of oomycetes (i.e. members of the orders Peronosporales, Sclerosporales, Pythiales and Saproleginales). The main advantages of these compounds are their preventive and curative properties, acropetal translocation, long-lasting action, high specificity and crop tolerance.

Metalaxyl, **metalaxyl-M**, furalaxyl, **benalaxyl**, **metolachlor** and **ofurace** are some members of the **phenylamide** class of fungicides, with the first two being the most popular. Pathogens targeted are species in the genera *Peronospora*, *Albugo*, *Peronosclerospora*, *Pseudoperonospora*, *Sclerospora*, *Pythium* and *Phytophthora*. Diseases such as **late blight** and **downy mildew** are effectively controlled in crops like potatoes, grapes, vegetables, tobacco and avocadoes. Demand for phenylamides is in foliar, soil and seed applications, in that order.

The mode of action of phenylamides is by the inhibition of RNA synthesis. This class of compounds covers a range of disease complexes when mixed with multi-site fungicides (commonly, **mancozeb**, which is a dithiocarbamate).

The effectiveness and safety of PAFs have led to extensive commercialization. However, this has resulted in development of resistance among some species, such as *Phytophthora infestans* and *Pseudoperonospora cubensis*. This has led to restriction on the application per season.

Phenylbenzene fungicide

Biphenyl fungicide is also known as phenylbenzene fungicide.

Phenylpyrrole fungicides

Phenylpyrroles are a new class of fungicides (e.g. **strobilurins** and spiroketalamines) that have been developed in recent years. They are based on and related to the structure of pyrrolnitrin, which is a secondary metabolite of the bacterium *Pseudomonas pyrrocinia*. Synthesized out of the structure of pyrrolnitrin are two chemicals – **fenpiclonil** and **fludioxonil**.

Phenylpyrroles are effective agents that control seed-borne fungal pathogens and foliar diseases. Crops like rice, wheat, maize, potatoes and peas are well protected from fungal seed pathogens such as *Fusarium* spp., *Botryotinia fuckeliana*, *Botrytis cinerea*, *Monilinia fructicola* and *Venturia inaequalis*. The advantage of these fungicides is that only limited amounts are taken up in the seed and seedling, most of it remaining on the surface and in the rhizosphere.

Interference in glycerol synthesis involving protein kinase (PK-III) and inhibition of glucose metabolism are considered to be two modes of action linked to phenylpyrroles.

Stability to light, safety to applicators and the environment and degradation into simpler compounds are some important features of this class of fungicides. The degradation route is via oxidation and opening of the pyrrole ring.

Phenylurea herbicides

Phenylurea herbicides, originating from urea, are post-emergence, soil-based, broad-spectrum herbicides, which are absorbed by plant roots and easily translocated through the plant. They act by inhibiting photosystem II in the photosynthesis pathway. Anisuron, chloreturon, chlorotoluron, chloroxuron, **daimuron**, difenoxuron, dimefuron, diuron, **fenuron**, **fluometuron**, fluothiuron, isoproturon, **linuron**, **metobromuron**, **metoxuron**, **monolinuron**, **siduron**, tetrafluron and **thidiazuron** are some urea-based herbicides.

Phenylurea herbicides act as soil sterilants and cause **chlorosis** and necrosis to small-seeded broadleaved weeds at doses above 2 kg/ha. Their low mammalian toxicity and residual activity (up to 6 months) make them a popular choice among farmers growing sugarcane, cotton, lucerne, soybean, maize, sorghum, potatoes, citrus fruits and turf. Trade names of herbicides belonging to the phenylurea family include Karmex, Lorox and Tupersan.

Herbicide selectivity is based on the way plants metabolize the herbicide, by reactions such as *N*-demethylation or *N*-demethoxylation. (See also **Herbicide action – photosynthesis inhibitors**.)

Pheromone confusion techniques

See **Pheromones**

Pheromones

Pheromones are chemicals that are produced by individuals of a species (generally insects) and are used for signalling to other members of the same species, to induce social response from them. Pheromones control many insect activities like aggregation, alarm signals, mating and migration.

Pheromones, along with other chemicals like allomones, **kairomones** and synomones, have behaviour-modifying properties, although the target for each one is different. The latter three chemicals act on individuals of species other than the ones producing them. Allomones benefit the producer, kairomones benefit the receiver and synomones benefit both the producer and the receiver. All these chemicals are together called **semiochemicals**, the messenger chemicals.

It is possible to use the properties of pheromones to influence the behaviour of insect pests. The major advantages of pheromones are that they are highly specific, and hence do not significantly affect non-target organisms. The pheromones most commonly employed in pest control are sex pheromones of female origin. Such chemicals were first discovered in moths (Lepidoptera), but sex pheromones exist in other animals including humans.

Most sex pheromones are esters of alcohols, and are highly specific. Specificity also depends on factors such as the rate and time of release and ratios of the various components in the pheromone complex.

Man commonly uses pheromones as a pest management technique. These chemicals are widely used to mislead the pest species to a location away from the crop. Pest populations may also be roughly monitored, which can lead to a more or less accurate estimate of the amount and time of pesticide application (to ensure optimum usage and minimum wastage of pesticides). Insect pheromone traps are simple tools used for pest management. A trap is made up of a simple funnel with a polythene bag attached to it or it can be a paper trap with a sticky floor. A small quantity of the pheromone is placed in a capsule at the mouth of the trap. Insects are attracted to the pheromone, and slide down the funnel into the bag or get stuck to the floor of a paper trap. Such pests may be killed or counted for pest population monitoring.

Since pheromone traps usually use chemicals used by female insects, only male insects are trapped. Hence, the estimation of the female population in a given area may be difficult with this method. Pheromone traps may not reduce the pest population, since a very significant number of males would have to be trapped for the desired result. Another drawback is that eggs and larvae of pests, which cause major plant casualty, are not inhibited by pheromones. In spite of these problems, pheromones are commonly accepted as an effective pest management technique. Pests controlled by pheromone traps include the codling moth (*Cydia pomonella*) on apple and the **pink bollworm** (*Pectinophora gossypiella*) on cotton. Other compounds such as aggregation pheromones can also be used in some cases, for example for male boll weevils (*Anthonomus grandis*) on cotton.

Crop damage may also be controlled by pheromone confusion techniques. Saturating the crop environment or laying artificial trails confuses male pests about the exact location of the females, thus jeopardizing mating. This reduces the frequency of fertilization. Hollow capillary polymer fibres containing encapsulated pheromone droplets are broadcast across fields using spray equipment. In the ensuing confusion, crop damage is minimized, while optimizing the use of pesticide, and in turn reducing the risk to natural predators.

The use of semiochemicals in pest management is gaining wider acceptance by farmers, although these techniques always play a complementary role in pest control.

Pheromones as a pest management technique

See **Pheromones**

Phorate insecticide

Phorate, $C_7H_{17}O_2PS_3$ (*O,O*-diethyl *S*-ethylthiomethyl phosphorodithioate), is an organophosphorus insecticide, an organothiophosphate nematicide and an acaricide. It is a colourless liquid, soluble in water (50 mg/l) and miscible with alcohols, ketones, ethers and vegetable oils. Its hydrolysis DT_{50} at pH 5–7 is 3.2 days. This compound, in its aqueous form, is degraded by light. Phorate is produced by the reaction of *O,O*-diethyldithiophosphoric acid with ethanethiol and formaldehyde. Phorate has a CAS Registry Number of 298-02-2.

Phorate, a systemic, cholinesterase inhibitor, has good contact and vapour action, making it effective against sucking plant pests. Examples of pests controlled include **leafhoppers**, leaf miners, mites, nematodes (especially *Meloidogyne* spp.) and rootworms. Crops protected include plants of the brassica group, many bean varieties, carrots, maize, sorghum, potatoes, tomatoes, sugarcane, sugarbeet, lucerne, cotton, rice, groundnuts, beetroot, sugarbeet and fodder beet. The compound's persistence in soil makes it ideal as a soil pesticide against nematodes and rootworms.

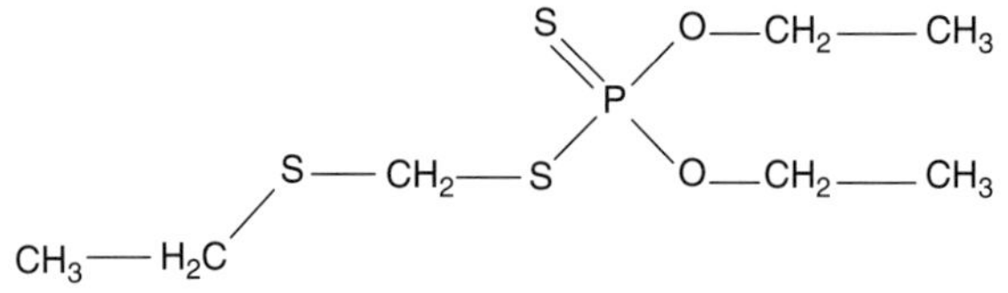

Phorate

Phorate is formulated as granules (GR) and emulsifiable concentrates (EC), and is available commercially under trade names such as Foraat, Gramitox, Rampart and Thimate. It is compatible with some fungicides and fertilizers, but not with preparations containing water.

The oral LD_{50} is 1.0–3.7 mg/kg in rats, and 3.5–6.59 mg/kg in mice. Phorate can severely injure the skin, according to tests on rats and rabbits. The acute 1 h inhalation LC_{50} in rats is reported to be 0.06 mg/l. Symptoms of acute exposure to phorate in humans occur at very low doses and include numbness, tingling sensation, incoordination, headache, tremor, nausea, abdominal cramps, sweating, blurred vision, difficulty breathing or respiratory depression, and slow heartbeat. Higher doses can cause death. The ADI for phorate is 0.0002 mg/kg/day. The granular form of phorate is phytotoxic to some varieties of apples, tobacco, beet, carrots, beans, maize and tomatoes. Phorate is highly to very highly toxic to birds, fish and bees. Labels must carry the signal words 'HIGHLY TOXIC – POISON'. Phorate comes under the toxicity Class Ia (extremely hazardous) according to the WHO, and is a PAN Bad Actor chemical as it is a cholinesterase inhibitor, an acute toxin and a potential groundwater pollutant.

In animals and plants, phorate is metabolized to the sulfoxide and sulfone and their phosphorothioate analogues, followed by hydrolysis to dithio-, thio- and orthophosphoric acids. In soil, the sulfoxide, sulfone and phosphorothioate analogues undergo hydrolysis with a half-life of 2–14 days. The sulfoxide metabolite is persistent in plants and soils.

The insecticide phorate is severely restricted or banned in many countries, for example the whole of the EU, Belize, Thailand and Kuwait. It is registered in India, Australia, the USA and Canada.

Phosalone insecticide

Phosalone, $C_{12}H_{15}ClNO_4PS_2$ (*S*-6-chloro-2,3-dihydro-2-oxobenzoxazol-3-ylmethyl *O*,*O*-diethyl), is an organothiophosphate insecticide and acaricide. It is a colourless crystalline substance with a garlic odour. Phosalone is soluble in water (1.7 mg/l) and extremely soluble in ethyl acetate, acetone, toluene, acetonitrile, benzene and xylene. It is hydrolysed by alkaline media, such as calcium arsenate and **lime sulfur** (the DT_{50} at pH 9 being 9 days). Phosalone emits toxic fumes of chlorine, phosphorus, nitrogen and sulfur oxides when heated to decomposition. This compound is produced by treating *O*,*O*-dimethyldithiophosphoric acid with *N*-chloromethyl-5-chlorobenzoxazolone. It has a CAS Registry Number of 2310-17-0.

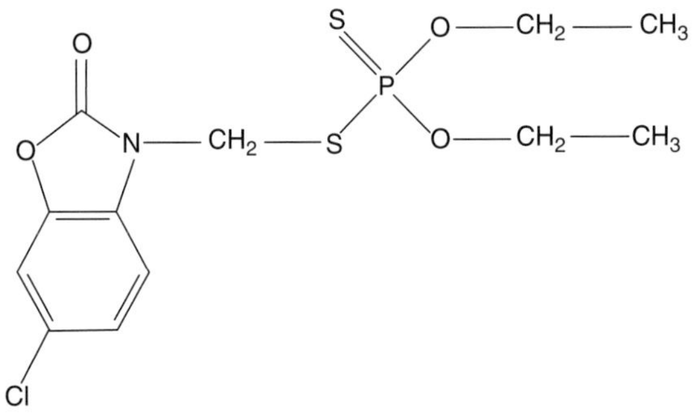

Phosalone

Phosalone, a cholinesterase inhibitor, is a non-systemic insecticide and acaricide that penetrates plant cuticle. It is effective against members of the orders Coleoptera (beetles), Hemiptera and Homoptera (aphids, **leafhoppers**, plant hoppers and miners), Lepidoptera (moths) and **Thysanoptera** (thrips), and **class Arachnida** (mites and ticks). Phosalone is used to protect crops like apples, pears, grapes, potatoes, vegetables and flowering plants.

The insecticide and acaricide phosalone is formulated as emulsifiable concentrate (EC), wettable powder (WP) and dustable powder (DP). It is compatible with many pesticides and is available under trade names like Zolone and Rubitox.

The acute oral LD_{50} of phosalone is in the range of 82–205 mg/kg in rats, and between 290 and 2150 mg/kg in birds. This chemical is toxic to fish (the acute LC_{50} being between 0.05 and 3.4 ppm among tested species). Phosalone is phytotoxic at higher doses to apples. Its ADI is 0.006 mg/kg b.w./day. The chemical is moderately toxic and product labels are required to carry the signal word 'WARNING'. Symptoms of poisoning include salivation, sweating, headache, dizziness, nausea, abdominal cramps and pin-point pupils (sometimes with blurred or dark vision). Symptoms may include seizures, incontinence and respiratory depression. It is listed as a PAN Bad Actor chemical.

In mammals, phosalone is oxidized to the phosphorothiate, and then hydrolysed to water-soluble metabolites. On plants it degrades to chlorbenzoxazolone in 14 h. It degrades in soil with rapid microbial action, and its DT_{50} is 2–3 days.

Phosalone is registered in India, Cambodia and Vietnam but not in the EU or the USA.

Phosdiphen fungicide

Phosdiphen, $C_{14}H_{11}Cl_4O_4P$ (bis (2,4-dichlorophenyl) ethyl phosphate), is an organophosphate ester fungicide. It is a yellowish liquid, soluble in water (0.7 mg/l), acetone, xylene and hexane. Although stable at high temperatures and under storage conditions, it is unstable in alkaline media. Phosdiphen has a CAS Registry Number of 36519-00-3.

Phosdiphen, applied mostly as an emulsifiable concentrate (EC), is used to control *Pyricularia oryzae*, the causative agent

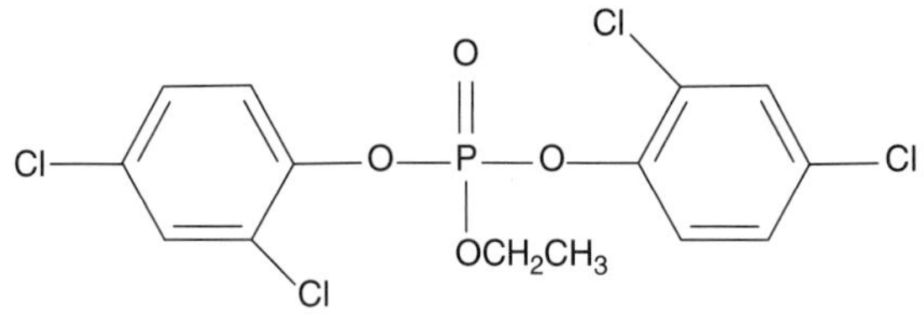

Phosdiphen

of rice blast. It is normally used in combination with **kasugamycin**. The hydrolysis DT_{50} is 4 days at pH 9.

Toxicologically, phosdiphen has an acute oral LD_{50} in rats and mice in the range of 5300–6200 mg/kg, indicating low mammalian toxicity. The LC_{50} in carp fish is 1.7 mg/l. Tests on rabbits and mice indicate that the compound is non-teratogenic and non-mutagenic and does not irritate the skin or eye. The WHO lists phosdiphen as a Class III toxin (slightly hazardous). The compound has a NOEL of 200 mg/kg in rats. Phosdiphen is not registered in the USA.

Phosmet insecticide

Phosmet, $C_{11}H_{12}NO_4PS_2$ (*O*,*O*-dimethyl S-phthalimidomethyl phosphorodithioate), is an isoindole organothiophosphate insecticide and an organothiophosphate acaricide. It is a pinkish white, waxy crystalline substance, soluble in water (25 mg/l) and readily soluble in acetone, benzene, toluene, xylene, methanol and kerosene. It decomposes in sunlight and at temperatures above 100°C. It rapidly hydrolyses in alkaline media (with the DT_{50} at pH 8.3 being <4 h). Phosmet is produced by a reaction of *N*-chloromethylphthalimide with dimethyldithiophosphoric acid. Its CAS Registry Number is 732-11-6.

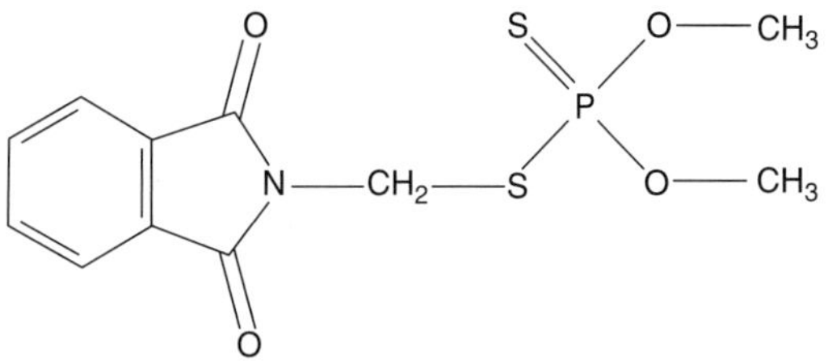

Phosmet

Phosmet, a cholinesterase inhibitor, is a non-systemic insecticide, ectoparasiticide and acaricide with contact action. It is used in the control of sucking and chewing insects, lepidopterous larvae, aphids, suckers, fruit flies, beetles, weevils, thrips and mites that attack crops like apples, olives, cotton, sorghum, maize and lucerne. Specific pests include spider mites, codling moths, Colorado potato beetles and boll weevils. This compound, when applied as an aqueous spray, controls cattle grubs, ticks and flies on farm animals.

The insecticide phosmet is formulated as wettable powder (WP) and dustable powder (DP) (at application rates of 2–40 mg/kg b.w.), emulsifiable concentrates (EC), soluble concentrates (SL) and as aqueous sprays. It is available in the market under the trade names of APPA, Decemthion, Imidan, Percolate, Phthalophos and Prolate. The commonly recommended dose on cotton is 3–5 kg/ha and 0.75 1 kg/ha on forage crops. Although phosmet is moderately toxic via the oral route (the acute oral LD_{50} being in the range of 113–160 mg/kg in rats), it is practically non-hazardous if used in small doses. It is slightly toxic via the inhalation route. Phosmet is highly toxic to bees. The EPA and WHO have classified phosmet as moderately toxic, requiring all product labels to clearly display the signal word 'WARNING'. The compound is not very irritating to the eyes or the skin in normal use. It is considered safe for birds, mammals and fish except if exposed to severely high doses. Symptoms of poisoning include the common symptoms of organophosphate chemical poisoning, such as excessive salivation, muscle twitching and weakness, sweating, pin-point pupils (sometimes with blurred or dark vision), tremors, headache, dizziness, vomiting, diarrhoea, respiratory depression, tightness in the chest, wheezing, cough and fluid in the lungs. In severe cases, poisoning can cause seizures, incontinence, unconsciousness and cholinesterase inhibition. The ADI of phosmet is 0.01 mg/kg b.w./day. Phosmet is a PAN Bad Actor chemical.

Animals metabolize phosmet to phthalamic acid, phthalic acid and derivatives, which are eliminated via urine. In plants and soil, phosmet is rapidly broken down to non-toxic metabolites, even though residues have been found in some fruits and forage. Degradation is faster in the presence of alkaline and moist soil and via microbial action.

Phosmet is registered by the USA, Australia, Canada, France, Germany, Italy and Spain.

Phosphamidon insecticide

Phosphamidon, $C_{10}H_{19}ClNO_5P$ ((*EZ*)-2-chloro-2-diethylcarbamoyl-1-methylvinyl dimethyl phosphate), is an organophosphorus insecticide and nematicide. It is a pale yellow liquid, miscible with water, acetone and toluene, and soluble in aromatic hydrocarbons. It is rapidly hydrolysed by alkalis (with a half-life of 13.8 days at pH 7 at 23°C). Phosphamidon decomposes above 160°C. It is produced by the reaction of trimethylphosphite with α, α-dichloroacetic acid diethylamide. Phosphamidon has a CAS Registry Number of 13171-21-6.

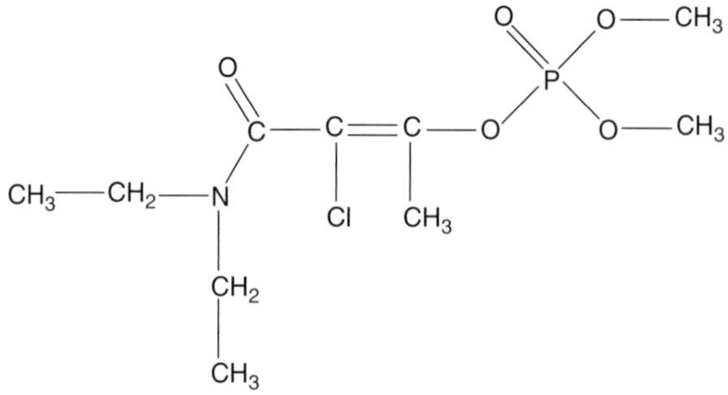

Phosphamidon

Phosphamidon is a systemic insecticide and acaricide with limited contact action. It is absorbed by the leaves and roots and acts as a cholinesterase inhibitor. It is used in the control of leaf beetles and miners, stem borers, fruit flies, moth and beetle larvae and spider mites in rice, sugarcane, fruit, vines, olives, vegetables, beet, sorghum, maize, lucerne and forestry.

The insecticide phosphamidon is formulated as a soluble/suspension/emulsifiable concentrate and an ULV liquid (UL). It is compatible with many pesticides, except for **copper oxychloride**, **captan**, **folpet** and **sulfur**, besides alkaline materials. The application dose on citrus and cotton crops is in the range of 300–1000 g/ha. This insecticide is available as Dimecron.

Phosphamidon is highly toxic to mammals and is listed as a Class Ia toxin (extremely hazardous) by the WHO. The oral LD_{50} is 17 mg/kg b.w. in rats and 10 mg/kg b.w. in mice. It is also very toxic via dermal exposure (the dermal LD_{50} in rats is 374 mg/kg b.w). The compound is highly toxic to birds and bees, and moderately toxic to fish. It causes slight skin irritation and moderate eye irritation in some animals.

The ADI for phosphamidon is 0.0005 mg/kg b.w./day. As is the case with any organophosphorus compound, the symptoms of poisoning with phosphamidon are excessive salivation, sweating, muscle twitching, tremors, headache, dizziness, vomiting, diarrhoea, respiratory depression, tightness in chest, productive cough, fluid in the lungs and pin-point pupils. Symptoms like unconsciousness and seizures may occur in severe cases.

Mammals eliminate almost all of phosphamidon within 24 h, via urine. It is absorbed from the gastrointestinal tract, through the skin and by inhalation of spray mists and dusts. Metabolism occurs during the passage by oxidative de-alkylation of the amide group and hydrolysis of the phosphorus-ester bond. In soils, the DT_{50} is 7–25 days depending upon the soil type. Higher doses may harm some varieties of cherry, plum, peach, maple and sorghum.

Phosphamidon is registered in India, Madagascar and Tanzania. As many as 46 countries such as Sudan, Libya, Thailand, Belize and Kuwait have banned or restricted the use of phosphamidon formulations. Those that have not registered it include the USA, Denmark, Finland, Netherlands and the UK.

Phosphine gas

Phosphine (PH_3) is a toxic, odourless, flammable gas that is produced as a by-product of decomposition or combustion of many pesticidal compounds. It has an IUPAC name of phosphane and is also known as phosphorus trihydride. Phosphine is soluble in water at 31 mg/100 ml. The CAS Registry Number of phosphine is 7803-51-2.

Phosphine is produced when white phosphorus reacts with sodium hydroxide.

Phosphinothricin

See **Glufosinate-ammonium herbicide**

Phosphonic acid fungicide

Phosphonic acid, with molecular formula H_3O_3P, is a hygroscopic white crystalline substance commonly used as monopotassium (H_2KO_3P), dipotassium (HK_2OP_3) or potassium dihydrogen phosphite. It is soluble in water at 45 g/l. It is stable in normal, dry storage conditions, but it turns into toxic **phosphine gas** at temperatures above 100°C. Phosphonic acid has a CAS Registry Number of 13598-36-2.

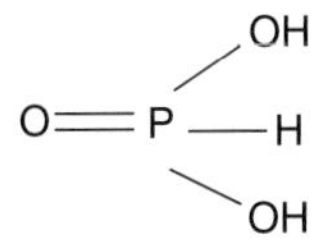

Phosphonic acid

Phosphonic acid is a systemic fungicide that also renders immunity in the plant. Its action controls the infection of plant-damaging species of the genera *Phytophthora* (e.g. *Phytophthora capsici* and *Phytophthora palmivora* that cause rot diseases) and *Pythium*. Blights of leaves and stems are also controlled by phosphonic acid. Crops protected include citrus, cucurbits, coconuts, stone fruits and turf.

The common formulation of phosphonic acid is soluble concentrate (SL) and common trade names include Agri-Fos.

Toxicologically, the WHO has classified phosphonic acid under Class U (unlikely to be hazardous). The LC_{50} among fish is greater than 100 mg/l, although the compound is not harmful to bees. The ADI in humans is 2.9 mg/kg/day. Phosphonic acid degrades in soil in 16 weeks.

Most countries have made phosphonic acid residues exempt from MRL values. Phosphonic acid is registered for use in the USA, South Africa, Australia and New Zealand.

Phosphorus as a macronutrient

See **Plant nutrient excesses and deficiencies**

Phoston as a PGR

See **Chlorphonium as a PGR**

Photosystem I

See **Herbicide action – photosynthesis inhibitors**

Photosystem II

See **Herbicide action – photosynthesis inhibitors**

Phoxim insecticide

Phoxim, $C_{12}H_{15}N_2O_3PS$ (*O,O*-diethyl α-cyanobenzylideneamino-oxyphosphonothioate), is an oxime organothiophosphate insecticide and an acaricide. This yellow-coloured liquid is soluble in water (1.5 mg/l), toluene, *n*-hexane, dichloromethane, isopropanol, vegetable and mineral oils. Phoxim slowly hydrolyses in aqueous media (the DT_{50} at 22°C at pH 4 and 9 are 26.7 and 3.1 days, respectively) and gradually decomposes under UV radiation. It is synthesized by treating α-cyanobenzaldoxim with diethylthiophosphoryl chloride. Phoxim has a CAS Registry Number of 14816-18-3.

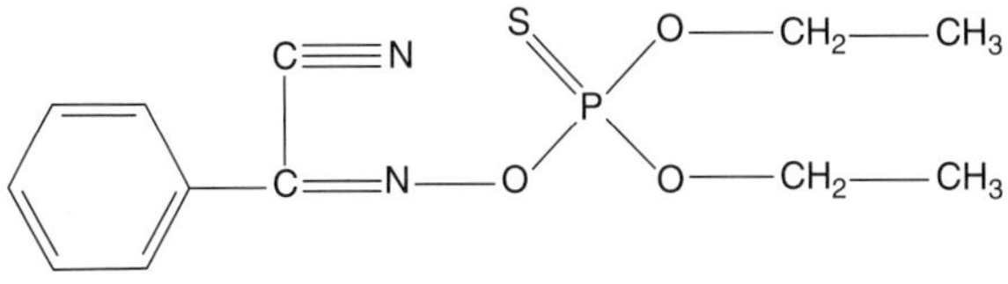

Phoxim

Phoxim, a moderate cholinesterase inhibitor, is a non-systemic, selective compound with both crop and non-crop applications. It controls soil insects and mites in maize, potatoes and vegetables. It is also used in seed treatment. In non-crop applications, phoxim controls ticks, mites, lice, locusts, ants and soil insects in storehouses, granaries, homes, buildings and on animals.

The insecticide phoxim is formulated as an emulsifiable concentrate (EC), dustable powder (DP), ULV liquid (UL) and as granules (GR). It is available in the market under the trade names of Baythion and Volaton.

The acute oral LD_{50} in male and female rats is 10,000 mg/kg b.w. Phoxim is toxic to bees. The acute 48 h EC_{50} for *Daphnia magna* is 0.00081 mg/l. The ADI for phoxim is 0.004 mg/kg b.w./day. The compound may exhibit some phytotoxicity to cotton. According to the WHO, phoxim comes under toxicity Class II (moderately hazardous).

Symptoms of phoxim exposure are excessive sweating, headache, weakness, giddiness, nausea, vomiting, stomach pain, blurred vision, slurred speech and muscle twitching. Swallowing massive doses may cause convulsions and loss of reflexes and sphincter control.

Mammals rapidly absorb, metabolize and eliminate phoxim, with degradation products such as diethylphosphoric acid, desethyl phoxim, phoxim carboxylic acid and the oxo-analogue PO-phoxim.

Phoxim is severely restricted for use in most countries and is a PAN Bad Actor chemical. It is not registered in the USA. The compound is registered in Denmark, Vietnam and South Africa.

Phreatophytes

See **Woody perennials**

Phthalide fungicide

Phthalide, $C_8H_2Cl_4O_2$ (4,5,6,7-tetrachlorophthalide), is a colourless crystalline substance and is a foliar fungicide. It is soluble in water at 2.5 mg/l and relatively more soluble in solvents like acetone, benzene and tetrahydrofuran. Phthalide is stable to heat, light and very low pH (pH 2), while it is unstable in alkaline medium with a DT_{50} of 10 days at pH 6.8. It has a CAS Registry Number of 27355-22-2.

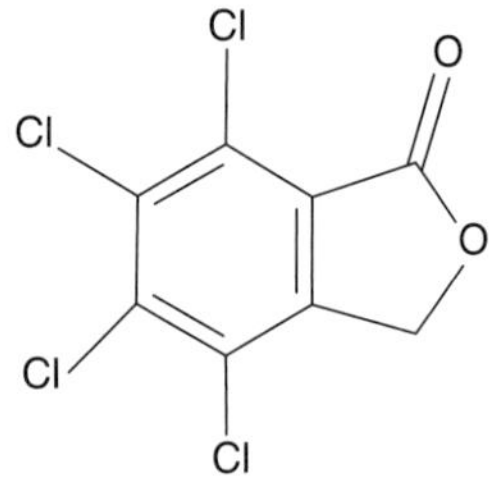

Phthalide

As a fungicide, phthalide has a protective action, inhibiting melanin biosynthesis in the pathogen, thus disabling it from penetrating the plant. It is used to control *Pyricularia oryzae* which causes rice blast.

Common formulations that phthalide is available in are dustable powder (DP), wettable powder (WP) and suspension concentrate (SC). Trade names include Rabcide. Phthalide may be combined with many non-alkaline pesticidal chemicals.

Toxicologically, phthalide has an acute oral and dermal LD_{50} of >10,000 mg/kg in rats and mice. It is non-irritating to the skin and eyes and is practically non-toxic to fish, *Daphnia magna*, hens and **beneficial organisms** like bees and earthworms. The WHO lists phthalide as a Class U toxin (unlikely to be hazardous).

In animals (rats) the main metabolite is 2-hydroxymethy-3, 4,5,6-tetrachlorobenzoic acid, while in plants (rice) the main metabolites are 4,7-dichlorophthalide and 4,6,7-trichlorophthalide. Phthalide metabolizes in soil into 2-hydroxymethyl-3, 4,5,6-tetrachlorobenzoic acid.

Phthalide is registered for use in Vietnam and Cambodia.

Phthalimide fungicides

Phthalimide is a class of compounds that include broad-spectrum fungicides with contact multi-site action. **Captan**, **folpet** and **captafol** are derivatives of tetrahydrophthalimide or phthalimide.

Reaction with the sulfydryl group of fungal cells and oxidation of glutathione are some of the ways in which phthalimides act on, and thus inhibit fungal pathogens. These surface protectants have fungitoxic gaseous degradation products. Phthalimide fungicides are used to effectively control various species of *Alternaria*, *Botrytis*, *Pythium*, *Cercospora*, *Cladosporium*, *Monilinia*, *Rhizoctonia*, *Phytophthora*, *Venturia* and *Septoria* on crops like cereals, citrus and non-citrus fruits, vegetables, potatoes and grapes.

Phylum Arthropoda

Arthropods form the largest phylum in the **Animal Kingdom**, with more than a million species described. The term is derived from the Greek words for 'jointed feet'. The bodies of arthropods are divided into a series of segments which bear organs such as legs or wings. The head bears sense organs and mouthparts. Arthropods also have a horny, rigid cuticle made up of the protein **chitin**. In order for the organism to go through its growth stages, this hard covering has to be broken out of in a process called **moulting**.

Phylum Arthropoda has several classes.

Insects belong to the class Insecta of the phylum Arthropoda. It is the most diverse group of animals, constituting more than half of Earth's living species. Insects typically have three segments, a head, a thorax and an abdomen. The head holds a pair of antennae and the mouth, which is made of four parts. The thorax has three segments with a pair of legs on each segment and usually a pair of wings each on the two rear segments. The abdomen has multiple segments. Class Insecta has 29 orders. The orders relevant to the subject of this work are briefly discussed here.

Order **Thysanoptera**, with 5000 species of **thrips**, constitutes well-known crop pests. These small, slender-bodied insects have piercing mouthparts and tiny narrow wings fringed with hair. Thrips commonly feed on flower heads.

Order **Hemiptera** includes 55,000 species of diverse types of bugs including bed bugs and **aphids** (Fig. P.39). Members of this order have beak-like mouthparts and thread-like stylets used for piercing and sucking. The thickened membranous section on the forewings overlaps the abdomen when the wings are at rest. Many members are pests of crops and homes.

Order Lepidoptera includes over 100,000 species of butterflies and moths. All members of the order have two pairs of wings covered with powdery scales and a long sucking proboscis formed from a pair of mouthparts. This is coiled when not in use. Out of the 80 families in this order, most are of moths which fold their wings horizontally and are mostly nocturnal. Many members of this order pose a serious threat to crops in one or more stages of their life cycle.

Order Diptera is the order of true flies, and includes thousands of species. True flies have only one pair of functional wings, the hind pair being reduced to a pair of knobs used to maintain balance. There are distinct groups in the order and these include midges, mosquitoes, houseflies, hoverflies, fruit flies and maggots. Members of this order either feed by sucking fluids like sap, or blood or fluids emanating from decaying matter.

Order Hymenoptera constitutes about 100,000 species and includes ants, **bees** and wasps. Several members of this group have two pairs of glossy, membranous wings, with the forewings linked to the hind wings. They have slender, waist-like portions between the thorax and abdomen. The mouthparts are adapted for biting and sucking. Many species live in developed social organization. This order includes pests like sawflies, wood wasps and gall wasps, many of which are parasitic in nature.

Fig. P.39. *Aphis craccivora* Koch. Courtesy of J. Poorani, with permission. National Bureau of Agriculturally Important Insects (formerly PDBC), Karnataka, India (www.aphidweb.com).

Order Coleoptera is a very large and diverse order of beetles with over 300,000 species. A distinguishing feature of the members of this order is that the forewings are modified into horny sheaths that completely conceal the hind wings when the insect is resting. These organisms range in size from 0.7 mm to 15.2 cm. Beetle larvae vary from legless grubs to caterpillar-like forms preying on other insects. Borers and **weevils** are common members of this order and are pests of crops.

Order Isoptera includes about 2000 species of termites. Termites are small insects that develop fragile wings that are shed before mating. They have a highly developed social organization. These are pests of houses and granaries. The king and the queen build intricate colonies and grow thousands of offspring. Strong wingless sterile worker termites maintain and defend the colony that may exist for up to 100 years.

Order **Orthoptera** includes **grasshoppers** and crickets. Most members of this order have long hind legs with powerful thighs to help them jump. They make noises by rubbing hard parts of the body together. The locust, an agricultural pest, belongs to this order.

Order Mallophaga constitutes many types of bird louse. These wingless flat insects live as parasites on birds and some mammals, feeding on blood and skin. The chicken louse is an example. Order Anoplura with around 250 species includes sucking lice. These insect are parasites on mammals, and feed on host blood by biting or piercing. The human louse is a common example.

Order Siphonaptera includes 1800 species of fleas that are parasitic on birds and mammals. They are brown, wingless insects with bristly, vertically compressed bodies. They have piercing mouthparts for feeding and have well-developed hind legs used for jumping.

Order Neuroptera includes about 4000 species of predatory insects, with biting mouthparts and net-like wings which form a roof over the abdomen when at rest. **Lacewings** are a significant biocontrol agent belonging to the order Neuroptera.

Order Mecoptera has 300 species described and includes scorpion flies. The elongated heads of these insects point vertically downwards. The prominent male reproductive organs curl up over the back like a scorpion's tail. Order Odonata includes dragonflies and damselflies. These are strong fliers with a long, narrow abdomen, very short antennae and huge eyes. They feed on smaller insects. These flies do not pose a threat to crops in normal circumstances. Order Mantodea includes insects with extended forelegs that assume a praying position. They are quick to grasp their prey. The female eats the male after mating is over. These organisms do not threaten crops. Order Phasmatodea includes stick-insects and leaf-insects. These insects have thin, twig-like bodies. They feed at night. The hard-shelled eggs of these insects resemble seeds.

Order Blattaria includes over 3500 species of cockroaches. These have long antennae, short oval bodies and thick leathery forewings that protect the hind wings. When at rest, the wings are folded back. These are house pests but do not threaten crops.

Phylum Mollusca

Members of the phylum Mollusca, also called molluscs, are unsegmented animals with body cavities and highly developed circulatory and nervous systems. The body is divided into a head, a muscular foot and a humped back covered by a mantle of skin. This mantle usually forms the animal's chalky shell. Molluscs do not have a standard shape and the body shape is altered, depending on the need in some species. The classes in this phylum include Monoplacophora, Amphineura, Gastropoda (slugs, snails and limpets), Scaphopoda, Lamellibranchia and Cephalopoda (squids, cuttlefish and octopuses). Molluscs are found on land, in freshwater and salty water.

Phylum Nematoda

Phylum Nematoda comprises over 10,000 species, and includes roundworms. All roundworms have long bodies, pointed at each end, covered by a thick cuticle. There are parasitic species and species which live in soil, freshwater and the sea. Their sizes range from microscopic to about 1 m long. The sexes are usually separate. Order Tylenchida is an important order that include plant pathogenic species belonging to the genera *Meloidogyne* and *Belonolaimus*.

Phytoalexins

Phytoalexins are a group of antimicrobial compounds produced and accumulated by plants in response to external infection or environmental stress. These compounds have low molecular weights and may belong to chemical classes like terpenoids, alkaloids, diterpene, isoflavonoids, phenolic acid, indole, biphenyl or glycosteroids. Phenylpropanoid phenols, which are important constituents of the cell wall and of lignin, are excellent examples of these phytoalexins. Some examples of phytoalexins are allixin found in garlic and the antibiotic stilbene in groundnuts. Antimicrobial compounds are found in all plants; their identification is made possible by high performance liquid chromatographic analysis and mass spectrometry.

When plant cells recognize a pathogen cell or its own dead tissue particle, the plant puts up two types of defence responses: (i) **induced systemic resistance (ISR)**; and (ii) long-term **systemic acquired resistance (SAR)**. In either case, the plant induces hormones such as **abscisic acid**, ethylene or **jasmonic acid**, which send signals of aggression to the gene, leading to the death of the invading cells or the host's own cells, in and around the site of infection. Resistance is also evident from the inhibition of hormone production by pathogens. Such changes brought about by the plant are also sensed by potential predators that then stay away from it. Synthesis of phytoalexins occurs as a result of diversion of primary metabolic precursors into secondary metabolic pathways.

In this context, it is important to note that the compounds that exist in the plant (pre-existing/pre-formed) even before an infection sets in are called phytoanticipins. They may also be produced from existing cell constituents. Phytoanticipins belong to chemical classes such as phenols, thioglucosides and triterpenoids and also have a role to play in the defence mechanism of plants (if, for instance, there is a bacterial or fungal attack). Simple phenolic compounds like catechol in the dead outer scales of some onion species are phytoanticipins that can counter *Colletotrichum circinans*. Chlorogenic acids are believed to inhibit the synthesis of cutinase in *Monilinia fructicola* (the causal agent of brown rots in peaches and cherries), making it difficult to penetrate the plant cuticle. However, in tomatoes and some cereals, several *Septoria* spp. can detoxify these defence agents, penetrate the cell and cause fungal diseases.

Phytoanticipins

See **Phytoalexins**

Phytoecdysones

See **Insect growth regulators (IGRs)**

Phytoremediation

Certain plant species can reduce, remove or degrade such contaminants as pesticidal toxins and hazardous heavy metals present in the soil. The process by which these plants do so is called phytoremediation. Phytoremediation is understood, in common parlance, as **bioremediation** through green plants. Contaminated soils are redeemed and restored by cultivating such plant species.

Phytoremediation occurs through the action of internal enzymes, phytoanticipins, plant roots and leaves. There are many aspects of phytoremediation including: (i) phytoextraction; (ii) phytodegradation; and (iii) phytovolatilization.

The phenomenon of plants absorbing toxins through the roots and transporting them to the aerial parts is called phytoextraction. Plants that absorb large quantities of such toxins (especially a metal or metalloid) throughout their lives are called hyperaccumulator plants, displaying a property called continuous phytoextraction. Soils can be restored by cultivating **hyperaccumulator plants**. Continuous phytoextraction can be complemented with a technique called induced phytoextraction, in which accelerants or chelators are added to speed up the mobilization of such metals as cadmium, lead, copper and chromium. Plants like mustard and sunflower are known to respond to induced phytoextraction. Soils are redeemed of these unfavourable elements when such plants are harvested.

Another technique of phytoremediation is called phytodegradation, in which plants convert organic pollutants into less toxic substances by secreting enzymes that act upon them. Hybrid poplar trees and *Datura innoxia* are known to convert the explosive RDX ('Research Department explosive') and trichloroethylene (TCE) into their metabolic components through enzymes like laccase, nitrilase and peroxidase. Such enzymatic action is believed to make these components less potent.

Some plants absorb toxins through their roots, transport them to the leaves and volatilize them into the atmosphere through the stomata. Such a phytoremediation technique is called phytovolatilization. Yellow poplars and lucerne are found to remove mercury by this method. Similar volatilization has also been observed in certain microbes, which, for instance, volatilize selenium into dimethylselenide.

Lucerne and some deep-rooted grasses and legumes contribute to phytoremediation by encouraging microbial activity that breaks down organic pollutants. Microbes oxidize the chemical residues which get volatilized or degraded. The practice of growing grasses and forage plants for ground cover on polluted sites reduces water and wind erosion and is a healthy strategy for mine sites and arid regions. Planting deep-rooted hyperaccumulator trees and perennials is also ideal for reducing pollution from groundwater.

Picloram herbicide

Picloram, $C_6H_3Cl_3N_2O_2$ (4-amino-3, 5,6-trichloropyridine-2-carboxylic acid), is a pyridine herbicide. It is a crystalline organic solid with a chlorine odour. Picloram is moderately soluble in water and soluble in polar organic solvents. It decomposes in hot concentrated alkalis. Picloram is made from α-picoline by chlorination, followed by hydrolysis of the trichloro-picoline derivative and amidation. It has a CAS Registry Number of 1918-02-1.

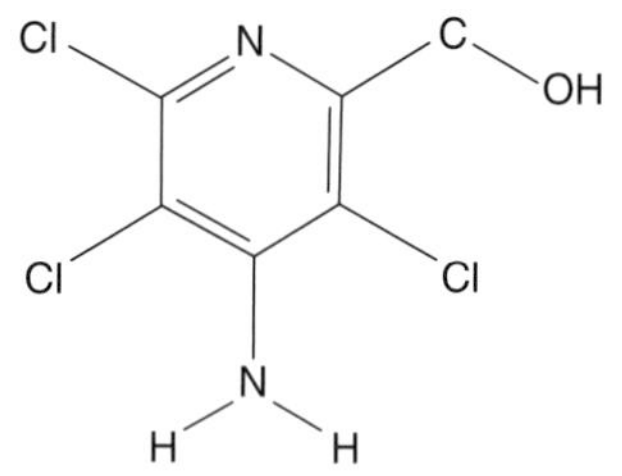

Picloram

Picloram is a selective, systemic herbicide with good translocation properties. It is used in both crop and non-crop situations. Once absorbed by the roots and leaves, picloram acts by mimicking auxins (e.g. indoleacetic acid) to control woody plants and many annual and perennial broadleaved plants (e.g. Canada thistle and field bindweed, knapweed and leafy spurge). Grasses tolerate picloram and so cannot be controlled by this compound.

Commonly available formulations of picloram are granules (GR) and soluble concentrate (SL). These are available in the market under the trade name of Tordon and Grazon.

Picloram is applied on foliage, injected into woody perennials or very carefully incorporated into the soil around targeted plants, so that it leaches to the roots. Picloram may be used with **MCPA**, **triclopyr**, diuron, **2,4-D**, as well as with fertilizers. It is effective at very low doses (i.e. from 500 g to 1.7 kg/ha). **Chemigation** should be strictly avoided to guard against contamination of groundwater. It is also necessary to observe a pre-grazing and a pre-harvest interval of 7–30 days.

According to the WHO and EPA, picloram comes under toxicity Class U (product unlikely to present acute hazard in normal use) and Class IV (practically non-toxic), respectively. Its oral LD_{50} is 8200 mg/kg in rats and 1061–4000 mg/kg in mice. The compound is non-toxic to bees and slightly toxic to practically non-toxic to fish and birds (the acute oral LD_{50} being 2000–5000 mg/kg in ducks). This herbicide may hurt the eyes in case of exposure. Studies indicate that it may disturb the functioning of the endocrine system. Due to this and it being a groundwater contaminator, picloram is considered a PAN Bad Actor chemical. Product labels are required to carry the signal word 'CAUTION'.

The herbicide picloram is the most persistent of its family of herbicides, with gradual microbial action as the main source of breakdown. Although it does not accumulate in body tissues, picloram does not adhere to soil either, which causes it to leach to groundwater sources. On plant surfaces, soil and water, picloram photodecomposes with a possible cleavage of the pyridine ring. When subject to photolysis, its soil DT_{50} is 60–90 days. It is not hydrolysed rapidly either.

Picloram is registered by the USA, Australia, Canada, the Philippines and many members of the EU including the UK, France, Italy and Spain.

Picolinafen herbicide

Picolinafen, $C_{19}H_{12}F_4N_2O_2$ (4′-fluoro-6-(α,α,α-trifluoro-m-tolyloxy) pyridine-2-carboxanilide), is an anilide and pyridine herbicide. A fine crystalline solid with a musty smell, picolinafen is soluble in distilled water (3.9×10^{-5} g/l at 20°C), acetone (557 g/l) and dichloromethane (764 g/l). It is not highly flammable or explosive.

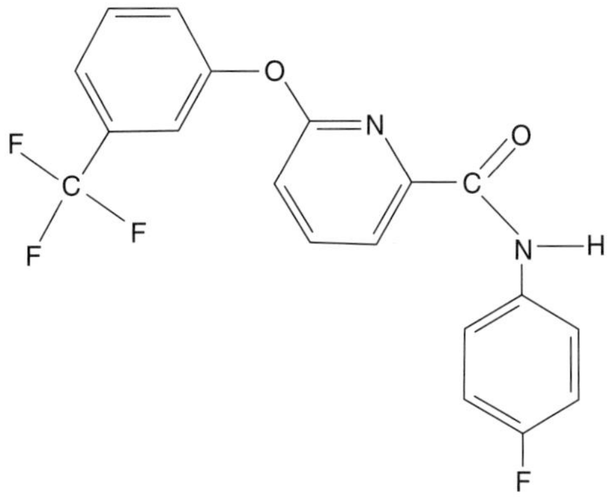

Picolinafen

Picolinafen is a post-emergence, selective herbicide, richly absorbed by the leaves. It is formulated as a soluble concentrate (SL) or wettable granules, and used mainly as a cereal herbicide to control broadleaved weeds.

Picolinafen is not a highly toxic compound. It is slightly toxic to rats and birds but is not harmful to **beneficial organisms** like bees. It is not carcinogenic, nor does it show significant signs of accumulation. The NOEL (1 year) of icolinafen in dogs is 1.4 mg/kg b.w./day. Exposure to eyes must be safeguarded against due to the possibility of irritation.

Picolinafen is rapidly absorbed and almost entirely excreted in 48 h via urine and faeces. It is registered in Australia, Canada and some EU countries, but not in the USA.

Picolinic acid herbicides

Picolinic acid, $C_6H_5NO_2$ (pyridine-2-carboxylic acid), made by adding a carboxyl group to the pyridine ring, belongs to a herbicidal group of indoleacetic acid (IAA) mimics, which also lend these herbicides growth-regulating properties. It is a stable, off-white powder, soluble in water. It has a CAS Registry Number of 98-98-6. The other IAA mimic members are **phenoxy acid**, **benzoic acid** and quinolinecarboxylic acids (**quinclorac** and **quinmerac**).

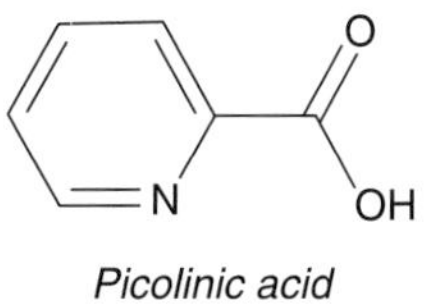

Picolinic acid

Herbicides based on picolinic acid are **picloram**, **clopyralid**, **fluroxypyr** and **triclopyr**. Broadleaved and perennial weeds are effectively controlled by these compounds, which may be applied both pre- and post-emergence. As a PGR, picolinic acid produces epinastic effects.

Phytotoxicity is one of the properties of these active ingredients, even at very low doses. Applicators must take special care to wear splash goggles, a laboratory coat and approved/certified dust respirators.

Picoxystrobin fungicide

Picoxystrobin, $C_{18}H_{16}F_3NO_4$ (methyl (2*E*)-3-methoxy-2-{2-[6-(trifluoromethyl)-2-pyridyloxymethyl] phenyl} acrylate), is a **strobilurin** fungicide (methoxyacrylate). It is a cream-coloured, volatile solid substance soluble in water (3.1 mg /l), methanol (96 g/l) and ethyl acetate (250 g/l). Picoxystrobin has a CAS Registry Number of 117428-22-5.

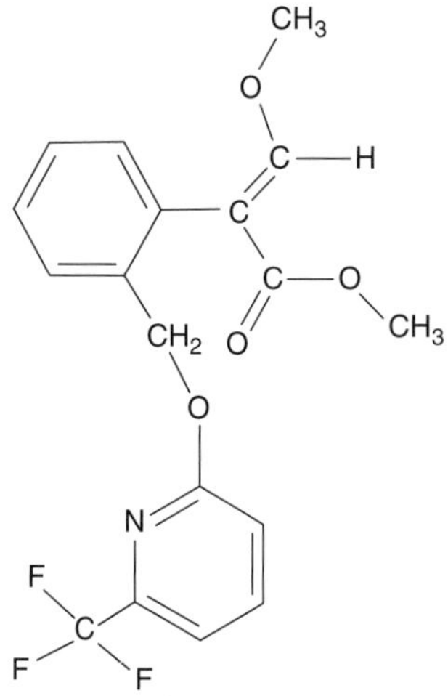

Picoxystrobin

Picoxystrobin, mainly a cereal fungicide, has a preventive and eradicative action. Picoxystrobin inhibits mitochondrial respiration by blocking electron transfer at quinol outer binding sites of cytochrome bc1. After being absorbed into the foliage of plants, it controls **powdery mildew**, net blotch and brown rust on wheat, barley and oats. Other crops protected include grapes, apples and vegetables.

The fungicide picoxystrobin is usually formulated as a soluble concentrate (SL) and used as a spray solution. It may be combined with **chlorothalonil** and cyprodinil. Known trade names include Acanta, Credo, Furlong and Galileo.

Toxicologically, picoxystrobin has low toxicity to rats and rabbits, with an acute oral LD_{50} of >5000 mg/kg in rats and >2250 mg/kg in birds. Although the fungicide is not harmful to bees, it is toxic to fish, algae and earthworms (the 96 h LD_{50} in fish ranges from 75 to 160 μg/l). The ADI of picoxystrobin in dogs is 0.043 mg/kg b.w./day. Picoxystrobin is moderately persistent in aerobic soils with a field DT_{50} of 20 days, while the aqueous photolysis DT_{50} at pH 7 is 3 weeks.

According to the EPA, picoxystrobin has a low-risk status. This fungicide is not registered by the USA, but it is registered in New Zealand and also in Germany and France (mainly for barley).

The MRL of picoxystrobin is 0.2 ppm for barley and oats and 0.05 ppm for wheat.

Pindone rodenticide

Pindone, $C_{14}H_{14}O_3$ (2-(2,2-dimethyl-1-oxopropyl) indane-1, 3-dione), is an indandione rodenticide. It is a yellowish-brown crystalline solid, slightly soluble in water but readily soluble in organic solvents. This rodenticide is incompatible with strong oxidizing agents. Pindone is made by condensing dimethylphthalate with pinacolone in the presence of metallic sodium. It has a CAS Registry Number of 83-26-1. Pindone is a first generation rodenticide. Other types of indandione compounds include **diphacinone** and **chlorophacinone**.

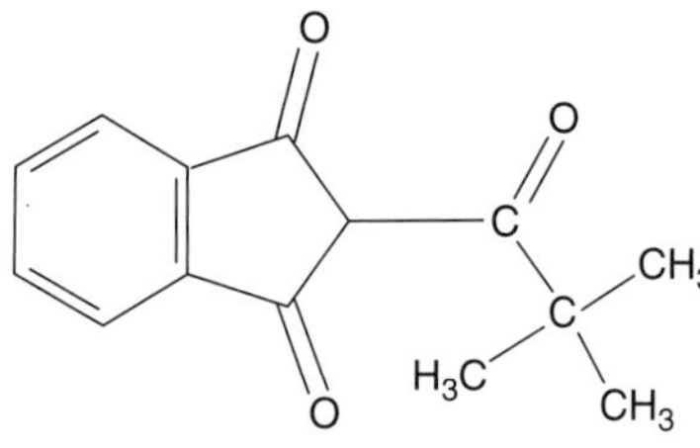

Pindone

Pindone is an anticoagulant agent that works by blocking the blood clotting system of target animals (by inhibiting the synthesis of vitamin K-dependent clotting factors). It needs to be consumed continually for over a 15 day period, for the desired effect. Pindone is used to control rabbits as well as rats and mice. It is similar to **warfarin**.

The rodenticide is commonly used as a 0.005–0.05% bait concentrate for controlling rats and mice. Commonly known trade names are Pival, RABBIT and Pivalyn. It is also formulated as an aqueous concentrate. Pindone, available as pindone sodium (salt) and pindone acid may be mixed with food to form the bait.

Pindone is a restricted use pesticide (RUP) and may be handled, used and applied only by authorized personnel. It is commonly used in place of sodium fluroacetate, which is toxic to humans and non-target animals. The oral LD_{50} of pindone in male rats is 280 mg/kg. It is very toxic to rabbits and to fish and aquatic organisms. It is thus necessary to apply baits carefully, to avoid non-target species from consuming them. Pindone is toxic via inhalation, contact or oral modes. Symptoms of over-exposure of pindone are excessive bleeding from cuts and bruises, black tarry stools and abdominal and back pain. Pindone is a PAN Bad Actor chemical.

This rodencide is registered for use in Australia and New Zealand but not in the USA.

Pink bollworm

See **Crop – cotton pest control**

Pinoxaden herbicide

Pinoxaden, $C_{23}H_{32}N_2O_4$ (8-(2,6-diethyl-p-tolyl)-1,2,4,5-tetrahydro-7-oxo-7H-pyrazolo[1,2-d][1,4,5]oxadiazepin-9-yl 2,2-dimethylpropionate), is a phenylpyrazole herbicide. It is a fine beige-coloured powder, soluble in water (200 mg/l at 20°C), acetone (250 g/l) and dichloromethane (500 g/l). Pinoxaden is stable to hydrolysis (see Pinoxaden at top of next column).

Pinoxaden, a systemic, post-emergence herbicide, is used to control annual grass weeds in cereal crops like wheat and barley. It acts as an acetyl-coenzyme A carboxylase (**ACCase**) inhibitor, inhibiting fatty acid synthesis.

The herbicide under study is recommended to be applied only once per season with a typical application rate of up to 70 g/ha. Ground boom or aerial spraying equipments may be used.

The acute oral LD_{50} of pinoxaden in rats is >5000 mg/kg, while its acute dermal LD_{50} is >2000 mg/kg. It is irritating to the eyes. It is a mobile chemical and may reach water sources via runoff or rinse waters. Non-target **beneficial organisms** may be harmed because of drift from sprays and so caution should be exercised during application. The ADI is 0.1 mg/kg b.w./day.

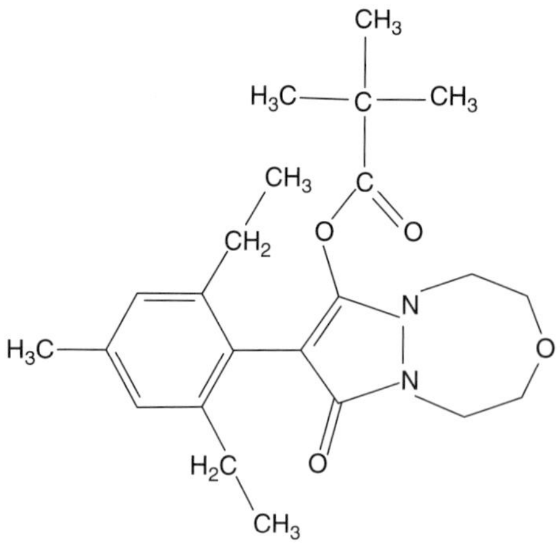

Pinoxaden

Most of the pinoxaden administered orally is excreted via the urine and the faeces. It degrades rapidly in aerobic soil, with the half-life ranging from 2 to 3 days.

Pinoxaden is registered for use in Australia, the Netherlands, New Zealand and the USA.

Piperalin fungicide

Piperalin, $C_{16}H_{21}Cl_2NO_2$ (3-[(*RS*)-2-methylpiperidino] propyl 3,4-dichlorobenzoate), is an unclassified fungicide. It is a viscous amber liquid, soluble in water (20 mg/l) and miscible in solvents like acetone, chloroform and dichloromethane. Piperalin is stable to heat and in normal conditions for up to about 7 years. It has a CAS Registry Number of 3478-94-2.

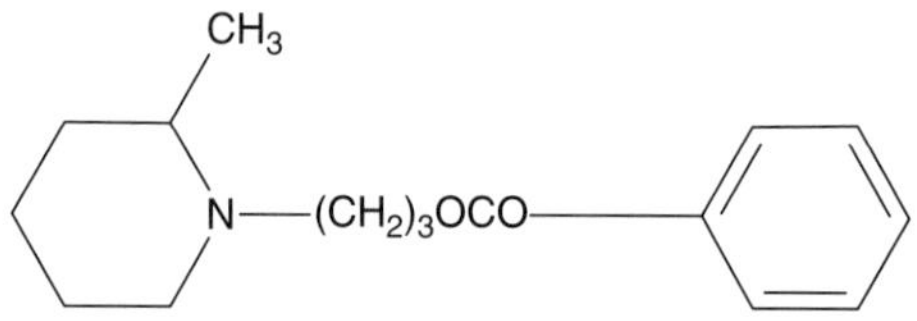

Piperalin

Piperalin, an inhibitor of growth and **ergosterol** biosynthesis, is commonly used as a fungicide on vines and on flowering plants (e.g. dahlias, roses and zinnia). It is mainly used against **powdery mildews**, and is particularly useful against *Ustilago maydis*.

The fungicide piperalin is formulated as a soluble concentrate (SL) or liquid, and applied as a foliar spray using a high-volume high-pressure sprayer. It is often necessary for this fungicide to be used with surfactants like Triton B-1956. The **restricted entry interval (REI)** of piperalin is 12 h. **Chemigation** of fields or direct application to wetlands is to be avoided to prevent the possible contamination of groundwater sources.

The acute oral LD_{50} of piperalin in rats is 2500 mg/kg, while that in bobwhite quails is >5380 ppm. It is toxic to some fish species (the acute LC_{50} bluegill sunfish is 0.77 mg/l). Piperalin is not carcinogenic, mutagenic or teratogenic in nature and does not cause sensitization of the skin. It is categorized as a Class II toxin (moderately hazardous). Endangered species are not at risk from the use of piperalin, except those affected by groundwater contamination.

Piperalin hydrolyses very rapidly at pH 9, forming two degradates, DCBA or dichlorobenzoic acid and 3-(2-methylpiperidino) propyl alcohol. Degradation also occurs microbially. Piperalin is immobile in several types of soil.

Fungicide piperalin is registered for use in the USA.

Piperine

See **Black pepper oil as a repellent**

Piperonyl butoxide

See **Pesticide synergists**

Piperophos herbicide

Piperophos, $C_{14}H_{28}NO_3PS_2$ (*S*-2-methylpiperidinocarbonyl methyl *O*, *O*-dipropyl phosphorodithioate), is an organophosphate herbicide. It is a pale yellow, slightly viscous liquid, moderately soluble in water and miscible with organic solvents. It is stable under normal storage conditions, but incompatible with strong oxidizing agents. Piperophos is made by the reaction of *O*,*O*-di-isopropyl phosphorodi-thioate with 1-methyl-piperidinylcarboxy chloride. The CAS Registry Number of of piperophos is 24151-93-7.

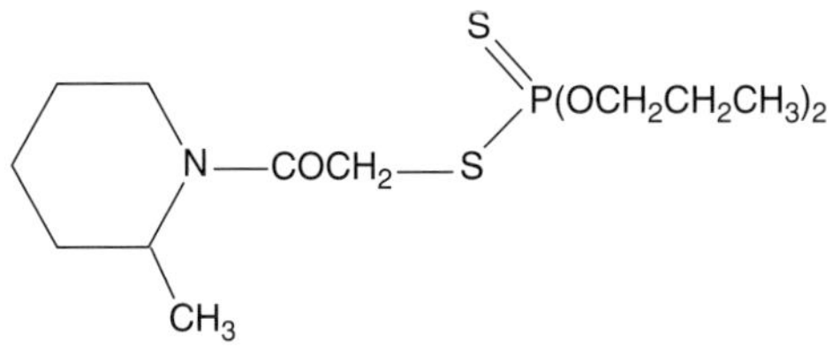

Piperophos

Piperophos is a selective, systemic herbicide that controls annual grasses and sedges in directly seeded or transplanted rice. It acts by inhibiting cell division, after it has been absorbed by roots and leaves.

Piperophos is available as an emulsifiable concentrate (EC) under trade names like Rilof.

The oral LD_{50} of piperophos among rats is 324 mg/kg, while its dermal LD_{50} among rabbits is >2150 mg/kg. The compound is slightly toxic to fish (the 96 h LC_{50} for rainbow trout is >6 mg/l), and non-toxic to bees. The WHO and EPA have listed piperophos as moderately hazardous.

Symptoms of poisoning include excessive salivation, sweating, muscle twitching, tremors, incoordination, dizziness, vomiting, diarrhoea, tightness in the chest, fluid in the lungs and pin-point pupils. In severe cases, it causes seizures and unconsciousness. Thus, handlers of this fungicide must be armed with appropriate PPE like positive pressure self-contained breathing apparatus (SCBA), and product labels should clearly display the signal word 'DANGER'. This fungicide is a PAN Bad Actor chemical because of its cholinesterase-inhibiting properties.

In animals, piperophos degrades by a series of reactions, including hydrolysis of the thiolophosphate and hydroxylation of the piperidine moiety. It degrades rapidly in plants. The compound is relatively stable to hydrolysis and photolysis.

As per information available, piperophos is registered in Tanzania but not in the USA.

Pirimicarb insecticide

Pirimicarb, $C_{11}H_{18}N_4O_2$(2-dimethylamino-5,6-dimethyl-pyrimidin-4-yl dimethylcarbamate), is a carbamate insecticide. It is a colourless solid, moderately soluble in water, acetone, ethanol, xylene and chloroform. It is decomposed by boiling with strong acids and alkalis and is unstable to UV light in its aqueous form. Pirimicarb undergoes photochemical degradation within 24 h at pH 5–9. This insecticide is produced by a reaction of 2-dimethylamino-5,6-dimethyl-4-pyrimidone with dimethylcarbamic acid chloride, in the presence of a base or phosgene and dimethylamine. Pirimicarb has a CAS Registry Number of 23103-98-2.

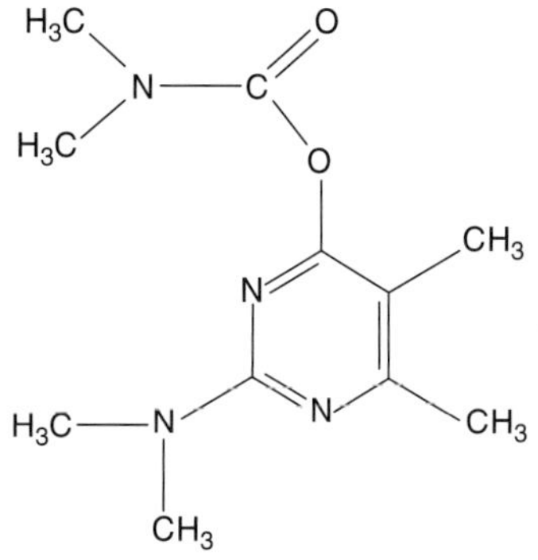

Pirimicarb

Pirimicarb, a cholinesterase inhibitor, is a selective, systemic insecticide with contact and respiratory action. It also may be used as a fumigant. This compound is mainly absorbed by the roots and translocated through the xylem. Its action targets all types of aphids (e.g. the grain aphid (*Macrosiphum avenae*), the black margined aphid, the yellow aphid (*Monellia* sp.) and the cotton aphid (*Aphis gossypii*)). Crops protected by pirimicarb include potatoes, cereals, cotton, tobacco and sugarbeet, besides many ornamental crops.

Pirimicarb is produced in forms such as an aerosol dispenser (AD), wettable powder (WP), emulsifiable concentrate (EC), dustable powder (DP), water-dispersible granules (WG) and smoke (FU). It is available under the brand names of Aphox and Pirimor. It is compatible with neutral insecticides, fungicides and **acaricides**. Pirimicarb is required in very small quantities (from 0.0125 to 0.5 kg a.i./ha). It may be used on aphid species resistant to organophosphorus insecticides.

The WHO has classified pirimicarb as a moderately hazardous chemical. The acute oral LD_{50} values among rats, mice and dogs range from 100 to 200 mg/kg, while the same for birds is in the range of 20–50 mg/kg. Pirimicarb is also of moderate toxicity to bees, and is slightly toxic to **beneficial organisms** such as pest parasites and predators. It mildly irritates the eyes. The temporary ADI for human beings is 0.004 mg/kg b.w./day. Since this insecticide is a proven carcinogen and a cholinesterase inhibitor, it has been listed as a PAN Bad Actor chemical by the PAN Pesticide Database.

In mammals, the major metabolites of pirimicarb include 2-dimethylamino-5,6-dimethyl-4-hydroxypyrimidine and 2-methylamino-5,6-dimethyl-4-hydroxypyrimidine. In alkaline soils, pirimicarb has a half-life of up to 7 weeks under aerobic conditions. The principal route of degradation in soil is by hydrolysis of the carbamate moiety from the pyrimidine ring. The degradation products have low soil mobility.

Pirimicarb is registered for use in Australia, Canada, New Zealand, South Africa, Vietnam and EU members like the UK, Denmark, Finland, Germany, France and Italy.

Pirimiphos-ethyl insecticide

Pirimiphos-ethyl, $C_{13}H_{24}N_3O_3PS$ ((*O*-2-diethylamino-6-methyl-pyrimidin-4-yl) *O,O*-diethyl phosphorothioate), is an obsolete pyrimidine organothiophosphate insecticide. It is a straw-coloured liquid, soluble in water (2.3 mg/l) and miscible with most organic solvents. This compound is relatively stable in neutral to slightly alkaline pH at a range of temperatures. It is hydrolysed by acidic and alkaline media, and produces toxic and flammable **phosphine gas** when it comes in contact with reducing agents such as hydrides. Heat and partial oxidation produce toxic oxides of phosphorus, nitrogen and sulfur. Pirimiphos-ethyl is obtained by the condensation of diethylguanidine with ethyl acetate to give the 4-hydroxypyrimidine derivative which is then treated with diethyl thiophosphoryl chloride. The CAS Registry Number of pirimiphos-ethyl is 23505-41-1.

Pirimiphos-ethyl is a broad-spectrum insecticide with contact and respiratory action that acts as an acetylcholin-esterase (AChE) inhibitor. It controls soil pests from among members of orders Diptera and Coleoptera (i.e. flies and beetles). Foliar and seed-borne pests are also controlled by this insecticide. Crops protected include vegetables, bananas, mushrooms and turf.

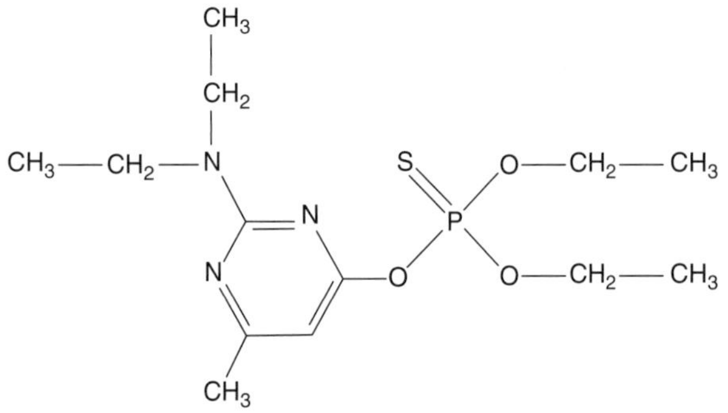

Pirimiphos-ethyl

Pirimiphos-ethyl is formulated as a powder, emulsifiable concentrate (EC) and as encapsulated granules and is available under trade names including Primicid. It may be used in seed treatment along with fungicides.

Pirimiphos-ethyl is a highly toxic insecticide and is classified as a Class Ib toxin (highly hazardous). As an organophosphorus pesticide, it inhibits cholinesterase and thus the nervous system. It is absorbed by the skin and the respiratory and gastrointestinal tracts. It is highly toxic to honey bees. It is phytotoxic in cases of excessive seed dressing. The half-life in soil is up to 70 days (in aerobic soil with a pH of 6–7.5). Some tests indicate that this chemical does not cause untoward geno-toxicity to fish and mammals. Symptoms of poisoning include incoordination, headache, dizziness, vomiting, abdominal cramps, diarrhoea, respiratory depression, productive cough and pin-point pupils, sometimes with blurred vision. It may cause seizures, incontinence and unconsciousness.

The EPA of the USA as well as the EU have cancelled the registration of pirimiphos-ethyl as an insecticide. It is registered for use in Cambodia.

Pirimiphos-methyl insecticide

Pirimiphos-methyl, $C_{11}H_{20}N_3O_3PS$ (*O*-(2-diethylamino-6-methylpyrimidin-4-yl)*O,O*-dimethyl phosphorothioate), is a pyrimidine organothiophosphate insecticide that is also used as an acaricide. It is a straw-coloured liquid that is soluble in water (9.9 mg/l at pH 2.2) and miscible with alcohols, ketones and halogenated hydrocarbons. This compound is hydrolysed by concentrated acidic and alkaline media. The DT_{50} at 22°C and at pH 4, 7 and 9 are 26.7, 7.2, and 3.1 days, respectively. It is produced by condensation of diethylguanidine with ethyl acetate, followed by the reaction with dimethylphosphoryl chloride. The CAS Registry Number is 29232-93-7.

Pirimiphos-methyl is a broad-spectrum insecticide and acaricide with a contact and respiratory action. It has a post-harvest application and is used in the control of chewing, sucking, boring insects (especially whitefly, thrips, aphids, onion fly, carrot fly and many types of grain beetles) and mites on tomatoes, cucumbers, capsicums, aubergines, carrots, onions, sugarcane, maize, sorghum and rice. Specific examples include the confused flour beetle, red flour beetle, corn (maize) sap beetle, rice weevil and the Indian meal moth. Pirimiphos-methyl is also used in warehouses, industrial premises, gardens and animal houses (to control ectoparasites such as horn and face flies on cattle).

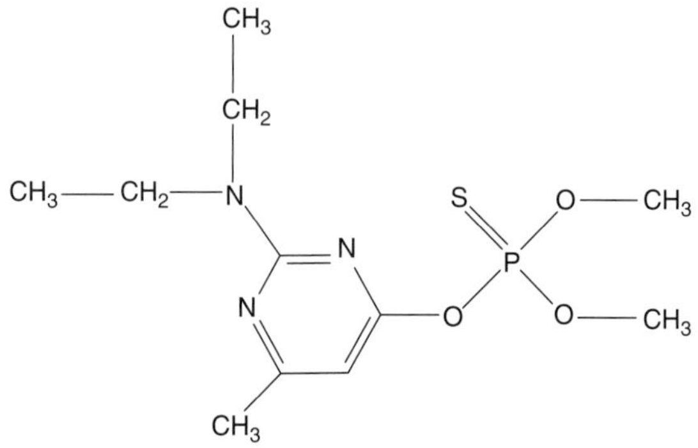

Pirimiphos-methyl

Pirimiphos-methyl is formulated as a dustable powder (DP), emulsifiable concentrate (EC), smoke generator (FU), aerosol (AD) and as an ULV liquid (UL). It may be used in combination with common insecticides and fungicides. Adding pirimiphos-methyl as an interior surface paint additive renders a residual effect, which is otherwise not an attribute of the chemical. This compound is available in the market under the trade name of Actellic.

Acute oral toxicity of pirimiphos-methyl in rats and mice is in the range of 1108–2050 mg/kg, while in quail it is 140 mg/kg, indicating high toxicity to birds. Pirimiphos-methyl is highly toxic to fish too. Being an organo phosphorus compound, it causes cholinesterase inhibition in humans too, with symptoms like nausea, dizziness, confusion and at very high exposures, death. Mixers, handlers and farm workers are also at risk of being exposed to the application of pirimiphos-methyl, and as a result, the wearing of masks, gloves and long garments is recommended. The WHO categorizes this insecticide as a

Class III toxin (slightly hazardous), while the PAN Pesticide Database lists it as a PAN Bad Actor chemical.

Pirimiphos-methyl degrades quite easily in both mammals and plants. Most of the compound evaporates in plants within 3 days, the degradation product being *O*-2-ethylamino-6-methylpyrimidin-4-yl, *O*-dimethyl phosphorothioate. In mammals, the P–O bond is cleaved extensively, leading to *N*-dealkylation of the pyrimidine moiety. The DT_{50} in soil is less than 30 days.

Owing to the type of application of pirimiphos-methyl, there is no danger of it polluting water. So drinking water is considered safe in the presence of pirimiphos-methyl. Although pirimphos-methyl is highly toxic to birds and fish, these risks are not of concern based on the use pattern of pirimiphos-methyl.

Pirimiphos- methyl is registered for use in India, Australia, the EU, the USA and South Africa.

Plant disease *Alternaria* leaf spot of cole crops

The fungal genus *Alternaria* is considered one of very damaging pathogens of crops and humans, since it is widely prevalent in almost all parts of the world. *Alternaria*, an ascomycete fungal plant pathogen, belongs to the family Pleosporaceae. It is found in both soil and water. The various species of *Alternaria* are responsible for a wide range of plant infections known as **leaf spot**, rot diseases and **blight** diseases. The crops that are easily vulnerable are most cole crops (i.e. members of the genus *Brassica*), potatoes, tomatoes, carrots and many more. Some of the many phytopathogenic species include *Alternaria alternata*, *Alternaria brassicae*, *Alternaria solani* and *Alternaria citri*. Only one example of leaf spot disease of brassicae caused by one species is discussed below.

A. brassicae (along with *Alternaria brassicicola*) is one of the main *Alternaria* species that causes the common leaf spot disease in all cole crops (broccoli, cabbage, cauliflower, radish, turnip, broccoli, etc.), cotton, canola and some grasses.

The large multi-celled, asexual spores (conidia) germinate only under high moisture conditions and their germ tubes enter the leaf. The sources of the inoculum are infected crop residues and the weed host (e.g. *Sida* sp.). Symptoms of infection are small, brown, rounded spots with purple margins on young leaves, and necrotic spots on older leaves. In cotton, the boll turns blackish brown after being infected by the pathogen. This is accompanied by premature defoliation and yield loss. The vulnerability of this infection increases with potassium deficiency.

A. brassicae, which can survive dry winter conditions, is spread to healthy crops by wind and tools. Leaf spot occurs in almost all cotton-growing countries of the world and hence is considered a disease of economic significance.

Disease-resistant seed usually keeps the leaf spot under check. IPM techniques include: (i) sufficient doses of potassium; (ii) maintenance of proper crop rows; and (iii) ensuring ideal land preparation, stubble and irrigation management. Biocontrol agents like *Aspergillus fumigatus*, *Aspergillus niger*, *Fusarium culmorum*, *Fusarium moniliforme* and *Trichoderma viride* also help to keep the disease under control. Repeated spraying of **mancozeb** 2.5 g/l or **copper oxychloride** 3 g/l or **captafol** with a 15–20 day interval is very useful in the control of the disease. Application of **fludioxonil**, imazalil, iprodione, **maneb** and **chlorothalonil** in varying concentrations is also recommended for the control of the leaf spot pathogen.

Alternaria species are known to cause allergic reactions and disease conditions in people with reduced immunity. (See also **Plant disease bacterial leaf spot**).

Plant disease antagonists

Antagonists of a plant disease are microorganisms, present in the environment during a pathogenic infection, that hinder the progress of the infection. The means of hindrance are known to be via competition, predation and hyperparasitism. Examples of plant disease antagonists include several species of the fungal genera *Trichoderma*, *Laetisaria* and *Gliocladium* and bacterial genera *Pseudomonas*, *Bacillus* and *Streptomyces*. Their strains are available to protect crops such as sugarbeet, lentil, chickpea, sugarcane, pea, pearl millet, groundnut, sunflower, cotton, pulses and pigeon pea against a range of diseases caused by pathogenic fungi and bacteria as well as by insect pests.

Trichoderma harzianum, *Trichoderma viride*, *Trichoderma virens* and some *Gliocladium* species can control soil-borne, seed-borne and root diseases caused by *Pythium* spp., *Fusarium oxysporum*, *Sclerotium rolfsii*, *Rhizoctonia solani* and *Phytophthora* spp. Similarly, *Pseudomonas cepacia*, *Pseudomonas putida* and *Pseudomonas fluorescens* are successful in controlling several wilt and root rot diseases.

The discovery of disease antagonism and the scope of disease control have made most of these microorganisms an excellent choice for biocontrol of pests.

Plant disease bacterial blight

Bacterial blight is one of the most prominent of bacterial diseases that affects many crops such as rice, wheat, barley and ornamentals. The pathogen responsible for bacterial blight is *Xanthomonas* sp. The disease is also sometimes called leaf spot.

The symptoms of blight are the appearance of watery spots or lesions on the foliage that become bigger, leading to the wilting of the leaf. The bacterial fluids can spread by wind or water to other leaves and plants, thus spreading the infection. Subsequently, the spots turn black or brown and dry up, leading to the death of the plant. In the case of rice, the spots begin at the leaf tip. On cotton, the symptoms are seen as angular red and brown leaf spots. Other crops that are infected by the blight are soybeans, tomatoes and peppers.

Examples of blight pathogens include *Xanthomonas campestris* pv. *malvacearum* causing blights in cotton, *Xanthomonas oryzae* pv. *oryzae* causing blights in rice and *Xanthomonas campestris* pv. *phaseoli* that affects common beans. There is another bacterial pathogen, *Erwinia chrysanthemi* that causes the *Erwina* blight on philodendrons. The symptoms include yellowing of new leaves and a rotting smell from the stem. Fire blight of pear and apple are caused by *Erwinia amylovora*.

Hot and humid weather is favourable for blight infection to spread, with strong winds and rains helping to spread the infection. Removal of crop residues and also infected plants is crucial to disease management.

Plant disease bacterial leaf spot

The bacterial leaf spot is caused by *Pseudomonas* species and affects cole crops, cucurbit crops, beans and maize. The important feature of the pathogen is that it can affect any part of the plant at any stage of its growth.

Symptoms of the infection are brown or yellowish spots on leaves which are watery and translucent. Sometimes, there are angular spots. They then turn dry and may fall off. In some cases, the leaves develop holes in them. The affected plant bears unhealthy fruit, often with lesions. The main strains include *Pseudomonas syringae* pv. *maculicola* and *P. syringae* pv. *lachrymans*. There are minor differences in the symptoms in each crop.

Plant disease blast

Blast disease is an economically important affliction of the rice crop. The causative agent is the ascomycete fungus *Magnaporthe grisea*, also called *Pyricularia grisea*. The disease is variously known as **rice leaf blast** and rice rotten neck. The pathogen is also responsible for blast and blight infections of other cereals such as wheat, rye and barley. The name of the blast changes according to the crop affected, such as wheat blast, rye blast, and so on.

The infection begins with the pathogens producing spores that easily spread with wind and water, to neighbouring plants. The symptoms of the infection are greyish lesions (which may be spindle-shaped) with dark borders on all the above-ground parts. These lesions expand, finally killing the whole leaf. Spores are produced sexually and asexually and hyphae penetrate the root and vascular tissues. The neck region becomes dark and weak. The panicle breaks, with the whole plant getting a burnt appearance. The plant is killed and the fungus is ready to spread.

Disease-resistant seed, elimination of crop residues and nitrogen supplies to the crops can help keep the crop healthy. Seed treatment with **captan** or **carbendazim** (around 2.5 g/kg of seeds) prevents infection from setting in. Applications of **edifenphos**, pyroquilon and triclyclazole are useful to control blast disease.

Plant disease blotch

Blotch is a term used to describe a distinct plant disease or a disease complex, where common symptoms are brown or black dead areas on leaves or fruit. Many types of blotch affect different crops. *Septoria* blotch, spot blotch, net blotch and sooty blotch are common types of blotch diseases.

Spot blotch of wheat, caused by *Cochliobolus sativus*, affects leaves and makes them prone to common root rot and other diseases.

Septoria blotch, also called *Septoria* leaf blotch or *Septoria* leaf spot, is a very common wheat disease. The causal agent is *Septoria tritici*.

Sooty blotch is a disease complex caused by fungi such as *Peltaster fructicola, Geastrumia polystigmatis* and *Leptodontium elatius*. It affects apples, causing greenish blotches on the surface of the fruit.

Net blotch, a foliar disease of barley, causes light-brown elongated spots with distinctive dark-brown net-like patterns on the leaves and other parts of young plants. This symptom is followed by leaf **chlorosis** and tissue death. The loss of yield is proportional to the amount of leaf area destroyed on the upper two leaves. Grain yield, brewing quality and carbohydrate content of kernels suffer. The causal agent of net blotch is *Drechslera teres*.

Rotating barley crops with any other crop, removal of barley residue and volunteer barley are some initial control measures. The use of resistant cultivars and the avoidance of early planting also help control net blotch.

The *Septoria* complex that targets wheat, barley, oats, rye and grasses is caused by *Septoria nodorum* and the disease is known as glume blotch. The symptoms are small, oblong, dark-brown blotches, with minute black dots on floral bracts. Crop rotation with non-cereal crops, removal of the stubble and volunteer wheat can help to control blotch.

Most blotch diseases may be controlled by the application of copper-based fungicides like **Bordeaux mixture** as well as **phenyl mercury acetate**, in the recommended dosage.

Plant disease corn (maize) smut

See **Crop – maize pest control**

Plant disease damping-off

Damping-off is the general term used for a set of symptoms of fungal infections that affect seeds or seedlings of a number of plants. This disease is associated with several soil-borne diseases of roots and seeds, known as rot diseases. Pathogens causing damping-off include *Phytophthora* sp., *Pythium* sp., *Rhizoctonia solani*, *Sclerotium rolfsii*, *Fusarium* sp. and *Botrytis* sp. The infections of these pathogens are variously called root/crown rot and grey/white moulds.

The soil fungus enters the roots of a seedling, after which spore-containing sporangia are produced. Infection proceeds with the germination of spores. The fungus absorbs root nutrients, giving the roots a rotten, water-soaked, blackish appearance. The infection immediately affects the seedling, causing wilting, discoloration and falling of leaves (especially older leaves) and finally the death of the sapling.

In the case of damping-off of a seed, infection begins when moisture enters the seed cover for germination. Damping-off results in abnormal germination and a weak and rotting seedling. Lesions appear along the soil line.

Damping-off is a serious problem in tobacco, cabbage, soybean, cucumber, pepper and tomato crops. Often, damping-off is confused with a nematode attack or excessive fertilization, thus leading to the wrong diagnosis and treatment.

There are a number of ways in which damping-off can be prevented. For example: (i) germinating seeds in sterile soils before transplanting them; (ii) providing good ventilation at the time of germination; (iii) sowing clean and disease-resistant seeds in soils of low pH; (iv) preparing seedbeds with preventive fungicides; and (v) incorporation of anti-fungal biological ingredients (e.g. the extracts of **neem** seed, garlic and camomile).

Seed treatment with biocontrol agents such as *Trichoderma viride* and *Trichoderma harzianum*, along with intervention with fungicides like **carbendazim** 50 WP, **quintozene**, **iprodione**, **fosetyl-aluminium** and **captan** (via soil drench, dust or seed-dip) also control damping-off.

Plant disease downy mildew

Downy mildew is a common term used to describe a particular set of disease symptoms on foliage, caused by several types of oomycetes (e.g. *Peronospora*, *Sclerospora*, *Sclerophthora* and *Plasmopara*) that belong to the orders Peronosporales and Sclerosporales. Spinach, sugarbeet, grapes and members of the cucurbit family are commonly affected by this infection.

The life cycle begins in the spring when the overwintering oospores and the mycelia penetrate the plant cells. They take up nutrients, producing sporangia which exit via the stomata on the undersides of leaves. Wind transports the sporangia, leading to secondary infections. The pathogens need wet and cool weather conditions to propagate. The symptoms are first seen on older leaves. A few mildew infections are briefly discussed here.

Spinach downy mildew is caused by *Peronospora farinosa*. Initial symptoms of infection are greenish spots on cotyledons, followed by yellow lesions on leaves, especially the older ones. The fungus produces asexual sporangia in the presence of moisture, which get carried in the wind, spreading the infection. This species of the fungus can overwinter and can infect **common lambsquarters**, a noxious weed. Maintaining row space and keeping foliage dry may help prevent the disease. Mefanoxam, **fosetyl-aluminium** and **copper hydroxide** may be used to control the disease.

In the case of sugarbeet (*Beta vulgaris*), the fungus *Peronospora farinosa* f. sp. *betae* infects the plant at the seedling stage. The cotyledons become distorted and discoloured, affecting plant growth. Oospores can overwinter on volunteer plants, or in the soil. Disease management involves stubble management, crop rotation and soil application of pesticides.

Cucurbit downy mildew (caused by *Pseudoperonospora cubensis*) is specific to members of the gourd family, such as cantaloupe, cucumber, pumpkin, squash and watermelon. The disease is one of the most important diseases of cucurbits worldwide; it generally directly affects the foliage, and affects yields. Symptoms include irregular pale green spots on the upper part of the leaves that turn into angular, yellow lesions. As the infection advances, the leaves become necrotic, and the lesions become blackish, purple or red. Sporangia are produced on the underside of leaves. Cool and wet conditions favour the disease, which can cause the whole plant to die. Normally the infection spreads from the base of the leaf outwards.

Downy mildew on maize is caused by agents such as *Peronosclerospora sorghi* (sorghum downy mildew), *Sclerospora graminicola* (graminicola downy mildew or green ear) and *Sclerophthora macrospora* (crazy top), besides many others. Symptoms vary, depending on the pathogen. If infection occurs at the seedling stage, the plant dies of **chlorosis** and stunted growth. Older leaves develop streaks and lesions. Leaves of infected plants are narrower and more erect, and tillering is excessive. The pathogen may either infect leaves or seeds.

Grape downy mildew is another important disease, caused by *Plasmopara viticola*. Warm, humid and wet weather favour the oospores, overwintering on leaf litter. The fungus germinates on healthy plants and releases zoospores that infect the leaves causing lesions. This affects the stem and shoots, affecting yields. Sometimes, berries also display a greyish coating of sporangia.

Well-drained soils, management of debris and rotation of cucurbits with cereal crops can help keep infection under control. Copper sprays, sulfur dusts and active ingredients from the maneb-ziram family (**dithiocarbamates**) provide protection (through contact, broad-spectrum fungicides). **Fosetyl-aluminium**, **captan** and **metalaxyl** are also useful. The application of any fungicide might have to be followed with adherence to the **restricted entry interval (REI)**.

Downy mildew is believed to have originated in North America, from where it entered France. Downy mildew infection has caused severe damage to several French vineyards. After observing farmers in the Bordeaux region use a solution of copper sulfate and lime, Dr Pierre-Marie-Alexis Millardet developed a preparation, now called **Bordeaux mixture**, which is still used today for controlling downy mildew and some other fungal diseases.

Plant disease early blight

Early blight is a fungal infection which is a major foliar disease on tomatoes and potatoes.

The causative agent is *Alternaria solani*. The pathogen can remain dormant in the soil or in the debris until conditions like early season rain and excessive irrigation present themselves, when they then produce spores. The spores germinate forming small, dark lesions on leaves that expand into black spots. In fruits, the lesions look unpleasant, while in leaves they are circular.

Early blight may also be a seed-borne disease, if seeds of an infected plant are used in the next season. Such use can cause damping-off disease, blight or leaf spots in the new seedlings, causing defoliation and death of the plant.

Use of resistant seeds is always recommended. Injury to plants should be avoided. Crop residues should be disposed of properly. Pesticidal intervention is also required, and includes application of **azoxystrobin**, pyraclostrobin, **mancozeb**, **maneb**, **ziram** or **chlorothalonil**. ***Bacillus subtilis*** may also be used to control early blight.

Plant disease gummosis of orange

See **Crop – orange orchard pest control**

Plant disease late blight

Late blight is a fungal disease common among potatoes, tomatoes, capsicum, aubergine and some ornamental crops.

The main causal agent is the pathogen *Phytophthora infestans*. A full-blown infection can be seen on all plant parts, especially in potato and tomato plants. Infection sets in during wet conditions with the spores being transported to neighboring plants in the field. The symptoms are brownish lesions on the fruit, and dark lesions on the underside of leaves, with the spores appearing as whitish granules around the lesions. The lesions dry up and turn dark, drying up and killing the leaf. On the stems, greyish-brown lesions appear on the junction of leaves and stems.

Late blight infections can attract bacterial infections such as soft rot, which has its own infectious stage, thus multiplying the farmer's worries.

Cultural measures are very useful in preventing late blight, and these include: (i) use of resistant varieties; (ii) removal of infected tubers; (iii) elimination of volunteer plants and crop remains; and (iv) crop rotation. Pesticide intervention is also called for, at times. Commonly used chemical compounds are **chlorothalonil**, **mancozeb**, **maneb**, **metiram** (in combination with triphenyltin hydroxide), dimethomorph and **propamocarb**.

The large-scale destruction of the potato crop caused by a severe potato late blight infection in Ireland in 1845–1846 is of historical importance. The disease deprived Ireland of its staple food and was referred to as the Irish potato famine. Scientists, who were already probing into blight cases in and around Europe, intensified their investigations, with a young scientist, Heinrich Anton DeBary finally being able to establish the reason for the disease in 1853.

Plant disease powdery mildew

Powdery mildew is a widely known fungal disease caused by many different species of fungi in the order Erysiphales. Since the symptoms are quite distinctive, it can be spotted easily. This disease is relevant for this volume because of its wide host range that includes grapes, wheat, onions, strawberries, members of the gourd family and many other fruits and vegetables.

Typical symptoms of powdery mildew are white powdery spots on leaves (mainly lower leaves) and stems of infected plants. Affected leaves become yellow and drop off. As the disease progresses, the spots get larger and denser with a large number of asexual spores being formed. Stunted plant growth and distortion of buds and fruit are the characteristics of powdery mildew. Optimum use of fertilizer nitrogen and disciplined irrigation are two measures that can reduce the incidence of powdery mildew. Stubble management can also minimize infection.

On grapevines, the causal agent of powdery mildew is *Erysiphe necator* (or *Uncinula necator*). The infection starts with the onset of warm weather, when the overwintering fruiting bodies produce asexual spores, which are transported to healthy plants by the wind. The plant tissues become infected in the primary stage. This leads to whitish-grey patches on the undersides of lower leaves that further produce several thousand spores. Secondary infection occurs when these spores blow onto other healthy plants.

If infection occurs within the first month of bloom, the berries and shoots are most affected. Grape yields are greatly reduced, although plant death is not very common.

Disciplined irrigation and fertilizer management are key factors that help prevent powdery mildew. Leaf wetness must be monitored. Vines should be planned to allow as much sunlight as possible. Use of pesticides, from the very early growth stage until flowering, weakens the disease potential. Chemicals such as **Bordeaux mixture** (copper sulfate and hydrated lime), **kresoxim-methyl**, **quinoxyfen**, myclobutanil and pyraclostrobin are commonly used, with the full knowledge that repetitions of the same chemical group cause resistance to develop. Fungicides should be combined to ensure control and reduce the chance of the causal organism developing resistance.

In wheat, powdery mildew is caused by *Blumeria graminis* f. sp. *tritici*. Overwintering fruiting bodies (cleistothecia) are wind-blown onto healthy plants, where they produce spores that appear like white spots on leaves and stems. This way the mycelia and conidia successfully establish themselves in the plant cell, resulting in the plant becoming chlorotic. The infection is seen most commonly on the lower leaves. Humid and cool weather favour this disease.

Excessive nitrogen fertilization, absence of sufficient plant row space and over-irrigation are ideal conditions for powdery mildew. Infections are rampant in the early growing stage (especially, stem elongation). Use of mildew-resistant seeds and the application of **propiconazole** and triademefon are recommended solutions to controlling the disease.

Similarly, powdery mildew affects the cucurbit family with *Podosphaera xanthii* and *Erysiphe cichoracearum* being the main causal agents. The fungal infection affects the quality, taste and colour of pumpkins, squashes and cucumbers. The symptoms are whitish-yellow powdery mats on petioles and leaf surfaces. The pathogen favours humidity but not watery conditions. **Azoxystrobin**, **trifloxystrobin**, **myclobutanil** and ***Bacillus subtilis*** products are usually used for mildew control.

The fungus that causes onion powdery mildew is *Leveillula taurica*. This pathogen has a broad host range and can also infect cucumbers, tomatoes and garlic. Lesions on older leaves are the initial signs of infection. With the growth of mycelium, the plant becomes chlorotic. The use of potassium salts and baking soda helps halt disease progression.

Besides the control agents mentioned above, various combinations of substances such as potassium bicarbonate, **neem** oil, sulfur and copper protect the plant against further infection. **Triforine** is commonly used on ornamental plants.

Plant disease scab

Scab, both common and powdery, is a well-known disease that is most commonly seen in tuber crops like potatoes, radish, beet and carrot. However, scab can also affect apples, cereal crops, capsicum, tomatoes and the cucurbit group. Scab can be caused by bacterial or fungal (mould) pathogens. Common scab of potato is caused by soil-living bacterial *Streptomyces* species, for example *Streptomyces scabies*, while powdery scab is caused by the mould *Spongospora subterranea*. Apple scab is caused by *Venturia inaequalis*. Cucumber scab is caused by *Cladosporium cucumerinum*.

The general symptoms of scab are crusty, rough patches on the surface of the fruit, tuber, leaf or stem, or dull greyish-black lesions on the plant surface. The disease cycle begins with dormant fungal spores in leaf litter that become active in warmer or moist weather. These settle and germinate on the surface of the susceptible tree, and the germ tube penetrates the plant cuticle. The mycelium that forms develops into yellow spots that then form black lesions bearing conidia. The conidia are released and germinate on fresh hosts, continuing the cycle. The plant loses its leaves and fruit, leading to poor yields.

Although not totally effective, early application of dithiocarbamates (e.g. **maneb**, **captafol**, **metiram**) in formulations like flowable concentrates and wettable powders (WP) can help control scab in the cucurbit family. **Benzimidazole fungicides**, copper compounds, sulfur products and Bordeaux mixture are other pesticides that help to control scab on various crops. Cultural practices like removing and destroying affected plant parts can inhibit infection effectively. Using disease-resistant seed and treating seeds with appropriate fungicides before sowing are other preventive methods. In the case of tuber crops, irrigating tubers sufficiently and regularly during its development stage is helpful. Enriching the soil with organic matter and avoiding liming in the early stages can help prevent the common scab in potato. On the other hand if there is a history of powdery scab in a field, over-irrigation is to be avoided and the soil should be kept dry. Cucurbit crops should be planted after a gap of 2 years.

Plant disease shot hole

Shot hole is a disease caused by fungal and bacterial pathogens. It is seen often among stone fruits like peach, plum, apricots, cherries and olives. The bacterial species *Xanthomonas arboricola* is the main causative agent of shot hole. *Stigmina carpophila*, *Blumeriella gaapi*, *Wilsonomyces carpophilus* and *Cercospora* sp. are some fungal pathogens that cause shot hole in peaches and plums.

Small, circular, deep purple spots and lesions appear on fruit, twigs and leaves, and as the disease progresses, the spots may become brown, raised, gummy or corky, depending on the fruit type. Shot hole may easily be confused with other leaf and fruit disease, or with an insect attack, leading to the wrong diagnosis and delayed pesticidal action.

Spores overwinter in the leaf litter and become active in moist conditions. The pathogen needs a continuous period of wetness to germinate, penetrate and cause infection in the plant.

Remedial measures include a spray of **copper oxychloride** at the early bud formation stage. Copper can inhibit the growth of most bacteria and fungal pathogens. The removal and destruction of affected twigs and fallen leaves and avoidance of overhead irrigation reduce the occurrence of shot hole.

Plant disease smut

Smuts are plant diseases caused by dark, thick-walled, multicellular fungi with a large number of seed or soil-borne teliospores. These fungi belong to phylum Basidiomycota and class Teliomycetae. All smuts are crop pathogens that target members of the grass family (Graminaceae), such as maize, barley, wheat, oats, sugarcane, rice and forage grasses. The fungi attack the plant's reproductive system, forming galls that darken and burst, and releasing fungal spores that infect other plants nearby.

The disease is initiated when teliospores germinate in moist conditions, producing hyphae that infect germinating seeds by penetrating the plant coleoptile. Sporulation takes place in the endosperm tissue, until the kernel is filled with a mass of sporangia, consisting of a dark brown to black mass of teliospores.

Loose smut of barley, oats and wheat, semi-loose smut of barley and oats, covered smut and the common bunt or stink smut of wheat are important smut diseases. Diseased heads emerge from the shoot a few days earlier than healthy heads and are filled with black, powdery masses of smut teliospores, which are released.

Smut-free seeds are a strongly recommended preventive measure. Seed treatment with **carboxin** is the most important control method.

In loose smut of barley, caused by *Ustilago nuda* or *Ustilago tritici*, the entire head becomes a dusty mass of spores. The smutted head emerges a couple of days before the healthy heads do. Loose smut is the most destructive and most common barley smut known. Covered smut, caused by *Ustilago hordei*, attacks barley, oats and rye.

The sugarcane whip smut is caused by *Ustilago scitaminea*. The disease takes its name from the whip-like, black, irregularly curved structure, emerging from the upper leaves at the terminal bud. Sometimes, it emerges from lateral shoots. The whip contains millions of spores that are released on maturity, and subsequently, carried by the wind. It takes around 3 months from initiation for the inoculum to spread. Besides the whip, the other symptoms of whip smut are excessive tillering and spindly shoots with narrow leaves.

Younger sugarcane buds just coming out of the soil are most susceptible to whip smut infection. Because water is necessary for spore germination, prudent irrigation can prevent the spread of smut. Besides, use of disease-free seeds and rouging of cane can prevent smut.

Black smut of rice is caused by *Tilletia horrida* (Fig. P.40). Rice smut occurs during the ripening of rice and attacks several spikelets. Grain heads turn black and burst open to infect healthy kernels.

Fig. P.40. Black smut of rice caused by *Tilletia horrida*. Courtesy of Elena Gasich, Interactive Agricultural Ecological Atlas of Russia and Neighbouring Countries (www.agroatlas.ru). With permission.

Corn (maize) smut, occurring throughout the world and caused by *Ustilago maydis*, is easily distinguished by the presence of large conspicuous galls or smut heads instead of grain kernels. The action of the fungus converts the kernels into black, powdery fungal tissues.

Common bunt, also known as stinking smut, is a disease of both spring and winter wheat. The disease is caused by fungi such as *Tilletia tritici* and *Tilletia laevis*. It stunts the growth of the plants and forms smut balls in the grain heads. When the grain is milled, the smut balls emit a fishy odour that lowers the quality of the flour. Plants with common bunt may be moderately stunted, with brownish black grain heads. Also, a dark brown to black mass of teliospores is hidden under the modified periderm. Millions of spores are released at harvest, which continue the cycle.

Plant disease sterility mosaic

See **Crop – pigeon pea pest control**

Plant disease *Tikka*

See **Crop – groundnut pest control**

Plant disease wheat leaf rust

See **Crop – wheat pest control**

Plant disease wheat loose smut

See **Plant disease smut**; **Crop – wheat pest control**

Plant disease wheat stem rust

See **Crop – wheat pest control**

Plant disease wilt

Wilt disease is a disease that affects the plant vascular system, blocking the flow of fluid to the leaves. It is mostly seen as a disease of non-woody parts of plants.

Fusarium and *Verticillium* wilt diseases are two soil-borne fungal infections. Root-knot nematodes (*Meloidogyne* spp.) also cause wilt diseases. Tomato spotted wilt virus is responsible for the spread of wilt diseases in tomato.

The symptoms of this disease include downward curling and spotting of leaves and stunted growth. It diminishes the plant's ability to transpire. Fruits turn rough and become distorted. Wilt disease can be spread by **thrips**. Thus to control the spread of wilt, infestation by thrips should be controlled by spraying pesticides. (See also **Common viral plant diseases**.)

Plant disorder diagnosis checklist

Diagnosing a plant disorder is the key to ideal crop management. Plant ailments can arise due to one or more of three factors: (i) pests, including insects, mites, nematodes, **grasshoppers** and mice; (ii) diseases, including infections caused by bacteria, viruses or fungi; and (iii) physiological conditions including a wide range of environmental and physical conditions such as soil type, water content and climatic conditions. Diagnosis often becomes difficult since each of these three groups of factors can negatively affect the plant in more ways than one, and many of the disorders show common symptoms. It is essential to differentiate between contagious (infective) disorders and diseases from topical or one-off factors. For example, a nutrient deficiency of the soil may cause necrosis to the entire field, whereas a nematode attack may kill only a portion of the field. The following measures and tips prove helpful in diagnosing common plant disorders.

Keen inspection of plants or a sample of the crop is the most important step in arresting major disorders at an early stage. It is more easily done in smaller plant rows. The first warning is the sighting of any abnormal symptoms (e.g. wilting of plant parts, yellowing of leaves, holes in leaves, waterlogging in patches in the field or dried-up soil surface). Symptoms of one disorder can easily be confused with another disorder: for instance **chlorosis** of leaves could be due to either a nutrient deficiency or an insect attack. It is very important to realize that potential ailments and management practices pertaining to any particular crop or field are unique and general guidelines may have to be treated simply as guidelines.

To correctly diagnose a disorder that has been observed, it is recommended that the affected plant be uprooted carefully and examined closely (in a laboratory) for peculiarities in the roots, leaves and the rhizosphere. Insect eggs, root maggots and root knots may be clearly indicated by close examination. A study of the rhizosphere can indicate excesses and deficiencies in the soil. The nature of the soil plays an important part in plant growth. Besides the pH value, other factors such as the soil composition (saline, sodic or clayey), mineral content, level of organic matter and soil porosity are critical to plant health. Poorly aerated soil encourages fungal rot diseases. It is important to study the surrounding area of a field or garden taking into account, for example: (i) the prevailing direction of the wind; (ii) the amount of sunlight; (iii) the presence or absence of trees; (iv) knowledge of vulnerability of the crop to particular pests and deficiencies in that location; (v) knowledge of the ideal course of crop rotation in that area; and (vi) the population of natural enemies of common pests in the field. Fertilization and irrigation patterns should be adjusted to suit the crop and soil conditions of the location under study.

Once peculiar conditions and the presence of pathogens have been recorded, the analyst can confidently suggest and implement corrective methods and thus prevent potential threat to crops.

Plant growth regulators (PGRs)

An organic compound, either natural or synthetic, that modifies or controls one or more specific physiological processes within the plant is a plant growth regulator (PGR). PGRs form a major class of herbicidal agents. They are designed on the structure of plant hormones that regulate cell responses in plants. Man turns the principles of plant hormones on to undesirable plants (weeds) in order to control them and achieve better crop yields. High concentrations of a growth regulating chemical lead to abnormalities in the weed plant, thus killing or inhibiting its growth. This volume includes many notes on various growth regulating compounds that help achieve weed control.

Plant cells naturally produce substances called plant hormones that are signal molecules with specialized functions that prompt formation of flowers, fruits and seeds, as well as shedding of leaves and fruits. Cell elongation, cell division and growth are also activated by these hormones. The death of a group of cells at the time of pathogen attack is also initiated by hormones. Plant hormones promote some activities while inhibiting others. These chemicals give plants their distinct character. Plant hormones produced in one cell may or may not be used locally. These chemicals are slowly transported, as required, through the plant's vascular system (e.g. xylem) and cytoplasm. Plant cells may also defer the use of hormones for a later period. Plant hormones are required in extremely low concentrations, since even these quantities are sufficient to initiate critical events in plant growth.

Plant hormones can be grouped into five major classes: (i) **auxin**; (ii) **ethylene**; (iii) **abscisic acid**; (iv) **cytokinins**; and (v) **gibberellins**. Jasmonates and peptides are some of the other hormones that regulate, promote and control various stages of plant growth, ensuring good health of plants.

As mentioned above, PGRs synthesized by man can be herbicidal (to weeds) but others can be used as growth promoters for crops. Most PGRs are anti-auxins, **defoliants** or growth inhibitors. Some active ingredients include **carbaryl**, **flurprimidol**, **paclobutrazol**, **jasmonic acid**, **propham**, 4-CPA, **2,4-D**, **endothal**, fenoprop and **dichlorprop**. The chemical classes to which PGRs belong include phenoxyacetic acids, phenoxypropionic acids, carbamates and conazoles. Some of them keep plant height at an optimum, accommodate fuller grain heads and prevent lodging while others lead to improved rooting of cuttings, increased rate of ripening and easier scission (separation of the fruit from stem).

Plant growth-promoting rhizobacteria (PGPR)

See **Induced systemic resistance (ISR)**

Plant nutrient excesses and deficiencies

Plant nutrients are elements necessary for plant growth and these are needed in varying degrees. The availability of these inputs is crucial for the plant to complete its healthy life cycle. It is observed that in the absence of nutrients in sufficient quantities, plants are more susceptible to infections and insect attacks.

Nutrients are of two types: (i) macronutrients; and (ii) micronutrients. Nitrogen, phosphorus, potassium, calcium, magnesium and sulfur are macronutrients. They are required in large quantities. A brief about each of them is given in the following passage.

Nitrogen (N) is a vital component of proteins, amino acids, chlorophyll, purines, pyrimidines and coenzymes and is needed for vigorous growth, branching, leaf production, size enlargement and the green colour of leaves and stems. The element also controls the efficient utilization of potassium (K) and phosphorus (P). N is taken up in the form of ammonium or nitrate ions. Deficiency of N or absence of amino acids inhibits chlorophyll synthesis, causes derangement of plastids and many other processes. It also retards growth, causes foliage to turn yellow or pale and results in shrivelling grains and low yield. Older leaves are affected first.

Excess N produces leathery, crinkled, dark green leaves with succulent growth. It delays maturity and impairs quality of the crop. This situation increases susceptibility to diseases, as well as lodging, since the stem internodes become very long.

Phosphorus (P) is a constituent of nucleoproteins and phospholipids such as lecithin, ADP, ATP and pyridoxal phosphate. It is absorbed by plants mainly as $H_2PO_4^-$. Phosphate is a mobile nutrient and accumulates mostly in tissues and cells. It plays a key role in energy transfer, respiration and photosynthesis.

P influences vigour and improves the quality of the crop. It encourages cell formation, root development and leaf growth. P strengthens the stem and increases resistance to diseases. It facilitates the action of N-fixing bacteria in legumes.

P-deficient plants fail to grow healthily and rapidly, and have a weak root system. Stems and foliage become reddish or pinkish and the plant remains stunted. High P concentration in soil causes zinc deficiency and low concentration may induce increased uptake of Zn, Fe, Mn and Cu.

Potassium (K) is found bound to proteins. It plays a specific role in the activation of many enzymes in living organisms. K enhances resistance to diseases, insect attacks and adverse weather conditions. It helps in the formation of starch and has a role to play in stomatal opening.

Deficiency of K causes accumulation of soluble carbohydrates, amino acids and amines. It produces reddish-brown discoloration, drying and scorching of leaf margins. K deficiency produces an abnormal concentration of organic acids which affects protein synthesis and translocation of photosynthates. Such plants show necrosis on old leaves. Leaf tips become curly, leaf margins turn upward and growth is stunted. Excess of potassium delays maturity of plants.

Calcium (Ca) in combination with pectin in plants forms calcium pectate, which is an essential constituent of cell walls. It promotes soil bacterial action and helps to fix free nitrogen as nitrates. Ca helps to develop a good root system. Deficiency of Ca produces mottled discoloration of leaves or distorted, cup-shaped and chlorotic leaves (a condition similar to 'dieback'). Plant roots die at the tips or remain short and stubby. Ca deficiency affects many crops such as tobacco, banana and tomato. Ca deficiency is associated with accumulation of soluble toxic salts of iron, aluminium and manganese. Excess Ca causes **scab** disease in potato.

Magnesium (Mg) is a constituent of chlorophyll and plays an important role in some enzyme activity related to carbohydrate metabolism. Mg deficiency causes yellowing of older leaves, intraveinal white stripes, chlorosis and premature defoliation. Coffee, banana, grapevine and groundnut crops suffer due to Mg deficiency in the soil.

Sulfur (S) is present in S-containing amino acids, tripeptides, proteins and vitamins like thiamine and biotin. Sulfur forms an important part of straw and plant stalks. Onion, garlic, cabbage, radish, tomato, coconut and groundnut grow well in soils that contain high levels of S.

S deficiency causes soluble carbohydrates and nitrogen to accumulate, thus negatively affecting photosynthesis and protein synthesis. S fertilization enhances root and nodule development in legumes. Tea is susceptible to S deficiency. Plants grown in S-deficient soils have new leaves turning yellow; roots and stems growing abnormally long; fruits with thin skin and less juice. Chlorosis appears on young leaves. In cereal crops and oil seed crops maturity is delayed, hence oil production is significantly reduced.

Iron, manganese, copper, zinc, molybdenum, **boron** and chlorine are micronutrients, which are required in small quantities by plants for growth. They are as important as macronutrients.

Iron (Fe) is mainly concerned with photosynthesis. In iron-deficient soils, acute chlorotic symptoms develop in some plants. The synthesis of RNA in chloroplasts is seriously affected by Fe deficiency. Legume root nodules show a close correlation between the symbiotic nitrogen-fixing capacity and the concentration of the iron-sulfur protein ferredoxin in the nodules. Fe deficiency causes severe chlorosis in young leaves whereas veins remain green. Excess Mn induces Fe deficiency. In calcareous soils, ferrous salts are converted into unavailable ferric salts. In acid soils, iron is available, but excess soluble phosphate makes Fe unavailable. Soil or foliar application of **ferrous sulfate** corrects the Fe deficiency in plants. Chelated iron compounds are recommended in acid soils.

Manganese (Mn), a component of chloroplasts and a catalyst in photosystem II (oxygen evolution) has a direct role in photosynthesis. Excess Mn induces Fe chlorosis due to competition between the two during absorption. Mn plays a big role in nitrate assimilation as an activator for hydroxylamine and nitrate reductase. Mn deficiency increases the IAA-oxidase activity. Interveinal chlorosis of spinach, potato, oats, banana, citrus and beans, and spots on potato are caused by Mn deficiency. The deficiencies are corrected by foliar spraying of manganese sulfate. Mn deficiency symptoms are green leaf veins, grey lesions on the leaves and yellow stems that are mostly hard and woody. Most symptoms appear first on young leaves.

Copper (Cu) is required for the photosynthetic process and the site of action is the Hill reaction. Cu deficiency interferes with protein synthesis. As a component of plant enzymes such as cytochrome oxidaze, Cu is directly involved in the oxidation process. Cu deficiency initiates multiple bud formation in the leaf axils, chlorosis, splitting of fruits, dieback of shoots, plant wilting, necrosis on young leaves and rosetting.

Zinc (Zn) is an essential component of various plant enzymes and is crucial for protein synthesis and growth regulation. Zn is needed for the synthesis of tryptophan, which is the precursor of IAA. Zn is a component of certain metalloenzymes (e.g. dehydrogenase, carbonic anhydrase, tripeptides) and Zn nutrition relates to Ca, P and Mg metabolism.

Zn salts can easily be absorbed through leaves and hence a foliar application is useful. Zinc-deficiency in soils can be corrected by the incorporation of zinc salts at the pre-plant stage. Reduction in phosphorus application makes zinc more available. Zn deficiency symptoms appear in younger leaves as interveinal chlorosis leading to reduction in shoot growth and shortening of internodes in tomato, apple, tea and coffee. Mottle leaf of citrus, rosette of apple and little leaf of grape vine are Zn deficiency diseases. Crinkling and stunting of leaves are also common. In soils with high calcium and phosphorus content, Zn deficiency is common. Zn is available at between pH 5.5 and pH 7.0 but availability increases at lower pH. Excess rainfall may leach Zn from the soil surface.

Molybdenum (Mo) functions in turning nitrates (NO_3) into amino acids. It also plays a complicated role in the metabolism of ascorbic acid, P and chloroplast disorganization. Sandy and gravelly soils are low in Mo, because it leaches out of these soils. Low pH and high iron oxide content reduce Mo availability. Mo is associated with N fixation by non-symbiotic bacteria (*Azotobacter* spp., *Anabina* spp.).

Mo deficiency in plants is expressed as stunted roots that are thick and swollen at the tips. Yellow, water-soaked patches appear on leaves and stems which turn into interveinal necrotic yellow spots. Leaves become mottled and cupped on the dorsal side. Leaf tips burn and dieback symptoms appear. Fruits have less bright colours, with brown spots on the rinds.

Boron (B) helps in active salt absorption, and metabolism of N and P. Plants take up soluble B from the soil. B deficiency causes: (i) distortion of young plants (e.g. in tomato seedlings); and (ii) degeneration of cambium cells in meristematic tissues of some vegetables and fruits resulting in necrosis and splitting of fruits (pear) and swelling of roots (beet root). Heart rot may occur in beet plants. Break down of cauliflower stem may also take place. Application of **borax** before sowing supplies the required dose. B may become toxic to plants in Ca-deficient acid soils.

Chlorine (Cl) is present in plants as chloride and is a trace element. It is essential for the production of oxygen in the Hill reaction of photosynthesis. Cl acts as an electron-transporting agent between two cytochrome pigments and the vitamin K bypass chloride-dependent step.

Cl-deficient plants develop symptoms of wilting in young leaf tips, leading to chlorosis, followed by bronzing and necrosis. Sometimes leaves remain rolled up. Roots also grow abnormally with club tips. Tomato, sugarbeet, cabbage and carrot are often found to be Cl deficient. Excess Cl may cause thickening of leaf tissues resulting in brittleness.

Resistance to pathogen attacks in a plant may be increased by incorporation of the right amounts of essential nutrients. Many symptoms of nutrient excesses and deficiencies are similar to symptoms of pest infection. Thus a sound knowledge of the role of plant nutrients is crucial.

Plant pathology

The branch of science that deals with the various biotic and abiotic factors causing diseases in plants is called plant pathology (phytopathology). Biotic factors causing plant disease include microorganisms such as fungi, viruses, bacteria and protozoans. These are commonly infectious in nature. Examples of abiotic factors are anaerobic conditions, pollution, toxic waters and highly acidic, sodic or alkaline soils. These can cause plant diseases, although they are non-infectious.

The scope of a plant pathologist includes the ways and means of controlling and managing disease factors to ensure a balance in the environment. To that extent, plant pathology integrates many biological disciplines.

To trace the history of the development of plant pathology, necroses, rot, **wilt**, spots and **chlorosis** were some of the first abnormal signs observed and documented. Thus, pathology was studied on the basis of symptoms. However, since many diseases have similar symptoms, it was difficult to predict or control loss of yields due to a specific set of symptoms. As microorganisms began to be studied, the disease agent started to become the focus of plant pathology. With modern technology and an in-depth knowledge of the structures of both plants and pathogens, the role of the plant pathologist is now critical in controlling economic disasters for farmers.

Plant protection with fungicides

See **Fungicides**

Plant-derived IGRs

See **Insect growth regulators (IGRs)**

Planting time

See **Integrated pest management (IPM)**

Plough

The plough is one of the oldest agricultural tillage implements. It is used to: (i) break up the land; (ii) turn over the soil into ridges and furrows; and (iii) bury vegetation. Ploughing exposes new soil to the air in preparation for sowing. The basic plough has a horizontal blade to cut the furrow and a projecting mouldboard to turn the soil over. Ploughs have been used since the Bronze Age. A traditional wooden plough is shown in Fig. P.41.

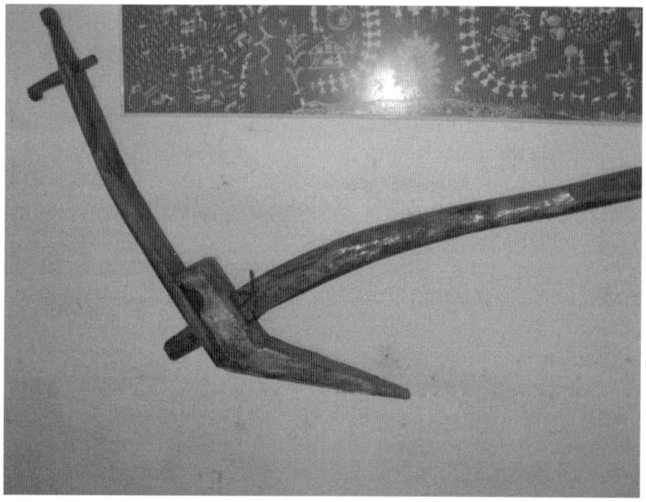

Fig. P.41. Plough. Courtesy of Mrs Madhuri Shirgaonkar of Pawankhind Resorts, Maharashtra. With permission.

There are two types of ploughs, the mouldboard plough and the disc plough. These ploughs are mechanically drawn. The modern plough has a steel frame to which a number of objects or accessories are fixed in order to achieve various tasks.

p-Nitrodiphenyl ether herbicides

p-Nitrodiphenyl ethers are a herbicide group which include **acifluorfen**, **bifenox**, **oxyfluorfen**, **fomesafen** and lactofen. These are part of the large family of protoporphyrinogen oxidase (**PROTOX**) inhibitors. The other chemical classes of PROTOX inhibitors are *N*-phenylthalamides, **triazolinones** and **oxadiazoles**.

p-Nitrodiphenyl ethers are used to control weeds in crops such as soybean, groundnut, rice, maize and sorghum. Lactofen (trade name Cobra) protects cotton and soybean from broad-leaved and grass weeds, whereas fomesafen protects only soybean. Bifenox can be applied at the post-emergence stage on foliage and at the pre-emergence stage to soil. The soil life of *p*-nitrodiphenyl ethers is up to 60 days.

Poison

See **Toxicity**

Pollution

Introduction of potentially toxic substances into any environment is called pollution of that environment. Pollution, which may exist in many forms, is relevant in this book, since pesticides may become a major cause of pollution in the environment. Pesticide application carried out by untrained and unauthorized hands may pollute groundwater and soil.

Water pollution due to discharge of toxic chemical wastes is closely regulated by environment protection agencies in each country, for example the EPA and FDA in the USA.

The most commonly known atmospheric contaminants are: (i) SO_2 emitted from fuels used in electric power production and industrial processing; and (ii) automobile exhaust rich in CO and tetraethyl-lead residues. Mandatory use of low-sulfur fuels and catalytic converters has addressed these conditions.

There have been many global efforts to reduce sources of pollution in the air, water and soil, but these efforts are commonly neutralized with poor governance, weak policies and unabated commercialization.

Polyoxin B

See **Polyoxin biofungicide**

Polyoxin biofungicide

Polyoxin, produced by fermentation with *Streptomyces cacaoi* var. *asoensis*, and similar substances such as Polyoxin B and Polyoxin D are classified as antibiotic fungicides. Polyoxin has a CAS Registry Number of 11113-80-7.

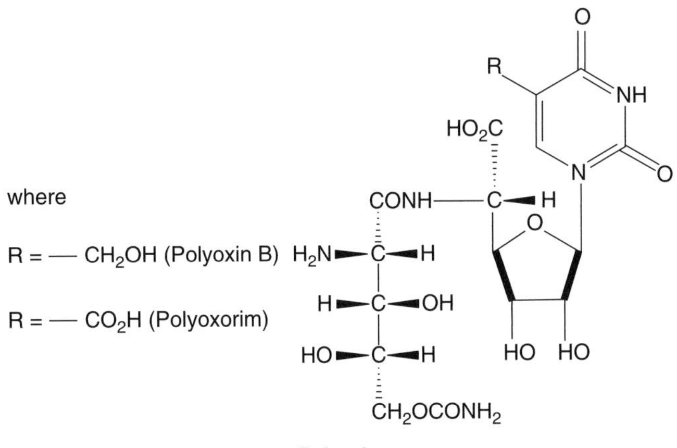

Polyoxins

Polyoxin is a nitrogenous antibiotic, composed of heterocyclic moieties. Polyoxins have both protective and systemic action. They act by inhibiting the synthesis of **chitin**, an essential component of the fungal cell wall. This leads to suppression of spores and controls fungal pathogens such as *Alternaria* spp., *Pseudoperonospora* spp., *Fusarium* spp., *Pyricularia* spp. and *Botrytis* spp. Thus, many types of mildews, rots and leaf spots on grapes, strawberries, apples and wheat are controlled. Polyoxin is also sometimes used as a bactericide.

Polyoxin may be mixed with non-alkaline substances like captafol, **edifenphos** or **oxine-copper** and is formulated as a wettable powder (WP), emulsifiable concentrate (EC) and dustable powder (DP). Polyoxin is registered for use in Vietnam.

Polyoxin B, $C_{17}H_{25}N_5O_{13}$ (5-(2-amino-5-*O*-carbamoyl-2-deoxy-L-xylonamido)-1,5-dideoxy-1-(1,2,3,4-tetrahydro-5-hydroxymethyl-2,4-dioxopyrimidin-1-yl)-β-D-allofuranuronic acid), is an amorphous and hygroscopic powder, and is very soluble in water at 1 kg/l. Its CAS Registry Number is 19396-06-6. Polyoxin B is stable up to pH 8.

Polyoxin B, with both preventive and curative properties, is rapidly absorbed and translocated in the plant. It helps control *Alternaria* spp., *Botrytis cinerea* and *Sclerotinia sclerotiorum*, which are causal agents for **powdery mildews**, grey moulds, blights and other fungal infections. This fungicide thus helps protect apples, pears, vines, melons, tobacco, capsicums, tomatoes, carrots, roses and carnations.

Being a biofungicide, it may be applied as regularly as fortnightly, with each dose being up to 12 kg/ha diluted with water. However, care should be taken with repeating applications and to ensure that polyoxin B is combined with other fungicidal compounds, to prevent the development of resistance in the disease-causing organisms.

Polyoxin B is safe for mammals (EPA Class IV) with the acute oral LD_{50} in rats being up to 21,200 mg/kg and a 2 year NOEL in rats of >48000 mg/kg. This fungicide is not harmful to fish, honey bees and other **beneficial organisms**. Polyoxin B is registered for use in Vietnam, but not in the USA.

Polyoxin D, $C_{17}H_{23}N_5O_{14}$ (5-(2-amino-5-*O*-carbamoyl-2-deoxy-L-xylonamido)-1-(5-carboxy-1,2,3,4-tetrahydro-2,4-dioxopyrimidin-1-yl)-1,5-dideoxy-β-D-allofuranuronic acid), is another polyoxine fungicide. It has a CAS Registry Number of 22976-86-9. Being hygroscopic in nature, it is soluble in water (>200 mg/kg). Polyoxin D is also known as polyoxorim.

Polyoxin D is used for controlling *Rhizoctonia solani* (the cause of sheath blight in rice) and members of the genera *Drechslera* and *Bipolaris*, which cause turf diseases.

Polyoxin D is safe for mammals (it is a Class IV toxin according to the EPA) with the acute oral LD_{50} in rats being >15,000 mg/kg and a 2 year NOEL in rats of >50 g/kg. However, since it has been tested as a reproductive toxin it is listed as a PAN Bad Actor chemical (according to the PAN Pesticide Database).

Polyoxin D is generally available as a zinc salt to enable better plant retention. Polyoxin D-zinc ($C_{17}H_{21}N_5O_{14}Zn$), also a fungicide, is mainly used on turf and in parks, as a spray. Although it is safe for animals and birds, this compound is not intended for seed treatment.

Polyoxin D

See **Polyoxin biofungicide**

Polyoxorim

See **Polyoxin biofungicide**

Polyquat

See **Algicides (algaecides)**

Post-harvest technology/operations

Post-harvest technology refers to various operations carried out on harvested crops or processed foods, to ensure preservation and retention of quality and taste, before such goods are consumed. The utilization of by-products and residues obtained during processing is integral to post-harvest technology. Globally, up to 30% of total produce is lost annually due to non-availability of appropriate processing and preservation technologies or storage facilities. Most losses are due to pest infection and infestation. This great economic loss has made post-harvest operations very critical in today's world.

The precise set of steps necessary in post-harvest operations varies with: (i) the commodity to be stored; (ii) the expected standards of taste and quality; and (iii) the facilities available at the farm. The nature of processing produce will vary according to the final desired product, as in the case of rice grains being converted into fortified rice or white, polished rice. The typical unit operations for food crops are cleaning, grading, drying, curing, shelling, decortication, parboiling, dehusking, polishing, extracting, mixing, fortification, freezing and storing. Rapidly degrading substances such as fruits, vegetables, milk, milk products and other animal products may require additional operations (e.g. washing, canning, bottling, pickling, addition of preservatives and additives, and deep-freezing). There is a unique set of operations required for handling by-products and residues. All these operations are time consuming and need specialized machinery and experienced labour. Preservation of produce includes the application of pesticides, which in turn needs specific skill sets and awareness of toxicity.

In the case of grain storage, a critical element is to dry the grain to less than 10% moisture content. This is needed since a very large number of insects and microorganisms (e.g. weevils and moulds) attack stored grains. Storage bins and packing material should be moisture-proof. Warehouses must be well-ventilated and kept dry and well lit.

Efficient post-harvest technology helps in creating added value to the produce. Extraction of **essential oils**, vitamins, proteins and flavours from specific produce has a specialized value in the pharmaceuticals and health-care industries. Instant foods or ready-to-eat meals are other examples of value-added processing that fetch a high price in the food market.

Post-harvest treatment

See **Fungicide application methods**

Potassium as a macronutrient

See **Plant nutrient excesses and deficiencies**

Powder in water-soluble packaging

Powder in water-soluble packages is a ready-to-use formulation used occasionally for treating crops. Such a powder is one of dispersible dry formulations of pesticides. The predetermined quantity of active ingredient (a.i.) is packaged in pouches or sachets made with a soluble film. Such packaging is made from polyvinyl alcohol (PVA). After the PVA film dissolves, the resultant suspended or dissolved powder is applied to the field as usual.

Such a formulation does not involve mixing and hence is safe to the user. But the water-soluble packaging (mainly film) adds to the cost of the product. It also requires protection from moisture during manufacture, transportation and storage, to prevent change in the physical properties of the compound. Hence, the outer packaging is made of aluminium or polyethylene foil, increasing the cost even further.

Rolls or sheets impregnated with the a.i. in a polymer film are another form of this formulation. The film may be added to a spray tank so the impregnated rolls/films release the compound ready for use.

Powders for seed treatment

It is crucial to protect seeds and seedlings from insects and microbial pathogens, throughout crop growth and after harvest. Seed treatment methods are widely practised in agriculture. A seed treatment agent can be a preventive, contact active ingredient (a.i.) used either on the seed or in the soil. It could alternatively be a systemic agent that targets seed-borne diseases. Seeds are treated with powders, water-dispersible powders, slurries or suspensions.

Although seed-treatment powders are formulated in a manner similar to other powder formulations, special attention is given to the adhesive agents to enable efficient seed coating, and also to non-phytotoxicity to the seed or embryotic plants. Sometimes the a.i. is applied as a film coating, while at other times the seeds are mixed in a **wettable or water-dispersible powder**. Such powders, along with good adhesive agents, are formulated for suspension in water to form a slurry.

An example of a soluble dry formulation is a **water-soluble powder** for seed treatment. Here, the a.i. in powder form is dissolved in water and applied to the seeds in a tumbling device. Here again, the ingredients need to be non-phytotoxic to seeds or emerging seedlings. Care should be taken to ensure that seeds do not germinate while being treated.

Emulsion formulations are also used for treating seeds. Oil-in-water emulsions of the a.i. are mixed with seeds for protection against pathogens. These are called **emulsions for seed treatment**. (See also **Seed protection devices**; **Wettable or water-dispersible powders**.)

Powdery mildew

See **Plant disease powdery mildew**

Powdery mildew of grapes

See **Crop – grape pest control**

Power dusters

See **Dusting machinery**

Power wheelbarrow sprayer

See **Application equipment – hand sprayers**

Prallethrin insecticide

Prallethrin, $C_{19}H_{24}O_3$ ((*RS*)-2-methyl-4-oxo-3-prop-2-ynylcyclopent-2-enyl(1*RS*,3*RS*;1*RS*,3*SR*)-2,2-dimethyl-3-(2-methylprop-1-enyl)cyclopropanecarboxylate), belongs to the pyrethroid class of insecticides. It has a CAS Registry Number of 23031-36-9. It is soluble in water at 8 mg/l but highly soluble in hexane, xylene and methanol. It is stable under normal storage conditions but is incompatible with strong oxidizing agents and alkaline media.

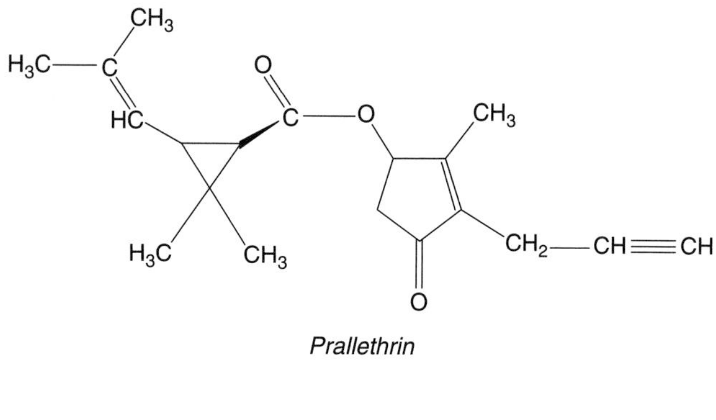

Prallethrin

Prallethrin is an insecticide with contact action. Being a pyrethroid, it acts on the insect's nervous system. It mainly has public health applications and controls members of the classes Blattodea, Culicidae and Muscidae, which include cockroaches, mosquitoes and flies.

The insecticide prallethrin is formulated as an emulsifiable concentrate (EC) and aerosol dispenser (AD) (as vaporizing mats and liquid vaporizers) and it may be mixed with other pesticides. A commonly known trade name is Etoc.

Prallethrin is classified as a Class II toxic by the WHO. The oral LD_{50} is 640 mg/kg in rats and the dermal LD_{50} is >5000 mg/ kg. It is not carcinogenic in nature. However, it is very toxic to honey bees (on contact, with a 48 h LD_{50} of 0.028 μg per bee) and to fish (with a 96 h LC_{50} ranging between 360 ppm and 22 μg/l). Prallethrin is moderately toxic to birds. Exposure to applicators may cause symptoms such as irritation to the skin and eyes, numbness, dizziness, vomiting, fluid in the lungs and muscle twitching. The use of safety glasses and gloves is thus recommended.

Prallethrin is registered in India, the USA, Australia and Canada. This insecticide is not registered by countries in the EU.

Pre-baiting

Pre-baiting is the process of regulating the time and place of bird feed. It involves feeding target bird pests with safe food at first, so that once the birds are accustomed to the food at a certain time and place, when the feed is replaced with treated bait they consume the maximum possible. Pre-baiting is required before the use of **avicides**.

Pre-baiting may also be carried out for rodents and other small mammalian pests.

Preservatives

Any agent that prolongs the useful life of a material is called a preservative. Many substances are preserved by: (i) particular additives; (ii) low temperature; (iii) ionizing radiation; and (iv) antioxidants. Antioxidants, fungicides, aldehydes and paints are some preservatives used to protect lube oils, textiles, biological specimens, wood and metals.

Most preservatives protect the material against microbial decomposition or oxidation. Wood is protected by coating with arsenic or copper compounds. Food is preserved by a host of non-toxic additives such as salt, vinegar, sugar, sodium bisulfate and sodium nitrate. The safety of preservatives has to be verified before such usage, especially if there is an exposure hazard.

Pressurized cans

See **Application equipment – hand sprayers**

Pretilachlor herbicide

Pretilachlor, $C_{17}H_{26}ClNO_2$ (2-chloro-2′, 6′-diethyl-N-(2-propoxyethyl) acetanilide), is a **chloroacetanilide herbicide**. This substance is soluble in water (0.5 g/l) and very soluble in benzene, hexane, methanol and dichloromethane. Pretilachlor is relatively stable to hydrolysis, with a DT_{50} of >200 days (up to pH 9). Pretilachlor is produced by the reaction of 2, 6-diethylaniline with α-chloroethyl propyl ether, followed by reaction with chloroacetyl chloride. It has a CAS Registry Number of 51218-49-6.

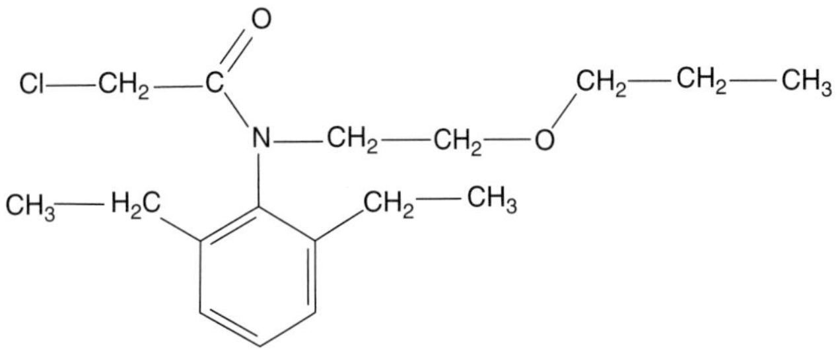

Pretilachlor

A rice herbicide, pretilachlor selectively acts by inhibiting mitosis and cell division. It controls annual grass weeds and sedges in rice. It may be used with **fenclorim**. Pretilachlor is formulated as an emulsifiable concentrate (EC). It may injure direct seeded rice, if applied alone.

According to the WHO, pretilachlor comes under toxicity Class U (unlikely to be hazardous). It is registered for use in India, Vietnam, Cambodia, the Philippines and African countries such as Guinea-Bissau, Madagascar, Mali, Niger and Senegal.

Primary tillage

See **Weed management by mechanical methods**

Primisulfuron-methyl herbicide

Primisulfuron-methyl, $C_{15}H_{12}F_4N_4O_7S$ (methyl 2-[4, 6-bis (difluoromethoxy) pyrimidin-2-ylcarbamoylsulfamoyl] benzoate), is a pyrimidinyl sulfonylurea herbicide. Under normal conditions, primisulfuron-methyl is a colourless, crystalline solid. The water solubility is 70 mg/l at 20°C. It is soluble in acetone (36 g/l), cyclohexane (27 g/l), isopropanol (0.4 g/l), methanol (3.5 g/l) and xylene (0.3 g/l). It is produced by the reaction of 2-sulfamoylchloride benzoic acid with 2-amino-4,6-difluoromethoxypyrimidin. The CAS Registry Number of this herbicide is 86209-51-0.

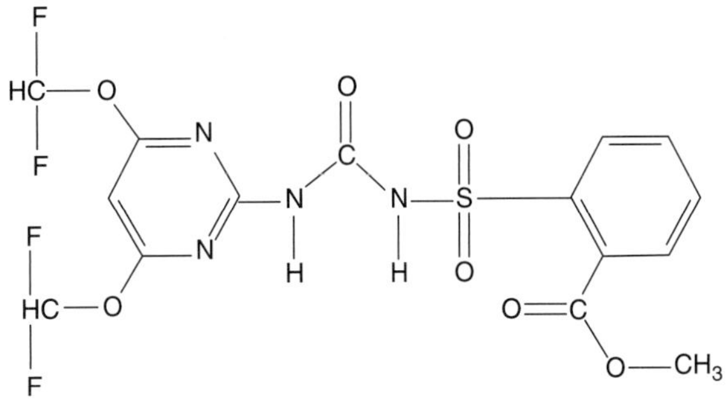

Primisulfuron-methyl

Primisulfuron-methyl is a selective, systemic, post-emergence herbicide used to control broadleaved weeds in crop and non-crop fields. Translocated throughout the plant, this herbicide acts by blocking cell growth in the actively growing regions of the plant and by inhibiting photosynthesis.

The formulations of primisulfuron-methyl are wettable powder (WP) and water-dispersible granules (WG) in water-soluble packets. It may be found in mixtures of **2,4-D**, **dicamba**, cyanazine, **bromoxynil** and **atrazine**. Some trade names include Beacon, Rifle and Tell.

Primisulfuron-methyl is listed as a practically non-toxic chemical by the EPA. The labels for products containing this chemical must bear the signal word 'CAUTION'. It is a general use pesticide (GUP) according to the EPA. It is a slight eye and skin irritant. The acute dermal LD_{50} is >2010 mg/kg in rats and rabbits. Primisulfuron-methyl is not teratogenic, carcinogenic or mutagenic. It is slightly toxic to freshwater fish, aquatic organisms and shrimp. It is non-toxic to honey bees.

Most of primisulfuron-methyl is excreted by rats within 10 days, mostly in the unchanged form, indicating little metabolic activity. The breakdown of the compound is slightly quicker in acidic soils and water. It is of low to moderate persistence in the soil environment, with a field half-life of 4 to 60 days.

Primisulfuron-methyl is registered in countries like Canada, New Zealand and the USA.

Probenazole fungicide

Probenazole, $C_{10}H_9NO_3S$ (3-allyloxy-1, 2-benz[*d*]isothiazole 1, 1-dioxide), is a benzothiazole fungicide, and also a bactericide. It is a colourless crystalline substance soluble in water (150 mg/l), hexane, benzene, ethanol and acetone. Probenazole is produced by the reaction of 3-chloro-1, 2-benzoisothiazole 1, 1-dioxide with allyl alcohol. The CAS Registry Number is 27605-76-1.

The compound probenazole is systemic in nature and moves upwards from the roots. It is effective on select diseases in rice and cucumbers (e.g. blast in rice and leaf spot in cucumbers). As a bactericide, it controls rice leaf blight and rice grain rot. Probenazole is also believed to systemically activate the natural defence system of the host plant by stimulating the salicylic acid-mediated defence-signalling pathway.

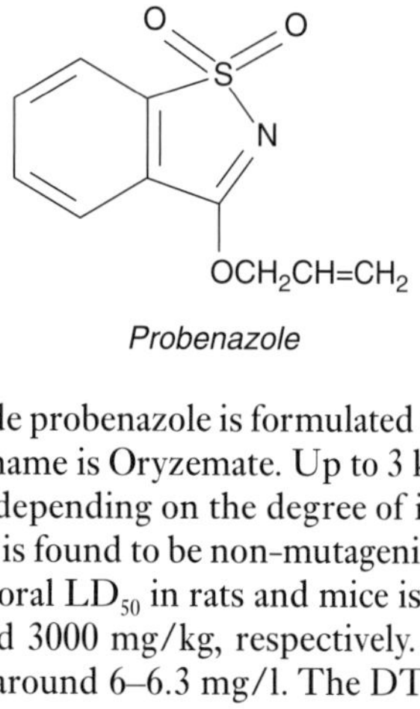

Probenazole

The fungicide probenazole is formulated as granules (GR). A popular trade name is Oryzemate. Up to 3 kg/ha of this fungicide is applied depending on the degree of infection.

Probenazole is found to be non-mutagenic and non-teratogenic. The acute oral LD_{50} in rats and mice is 2030 mg/kg and between 2750 and 3000 mg/kg, respectively. Among fish species, the LD_{50} is around 6–6.3 mg/l. The DT_{50} of probenazole in soil and water is less than 24 h. The WHO has classified probenazole as a Class U toxin that is unlikely to present acute hazard in normal use.

Probenazole is registered for use in India.

Prochloraz fungicide

Prochloraz, $C_{15}H_{16}Cl_3N_3O_2$ (*N*-propyl-*N*-[2-(2,4,6-trichlorophenoxy) ethyl] imidazole-1-carboxamide), is a colourless crystalline substance. It is a conazole (imidazole) fungicide of the amide class, stable in water (pH 7) and soluble in diethyl ether, chloroform and xylene. It photodegrades and decomposes at higher temperatures and in both concentrated acids and alkaline solvents. Prochloraz is produced by the reaction of 2-(2,4,6-trichlorophenoxy)ethyl bromide with propylamine, followed by phosgenation and formation of the substituted urea with imidazole. It has a CAS Registry Number of 67747-09-5.

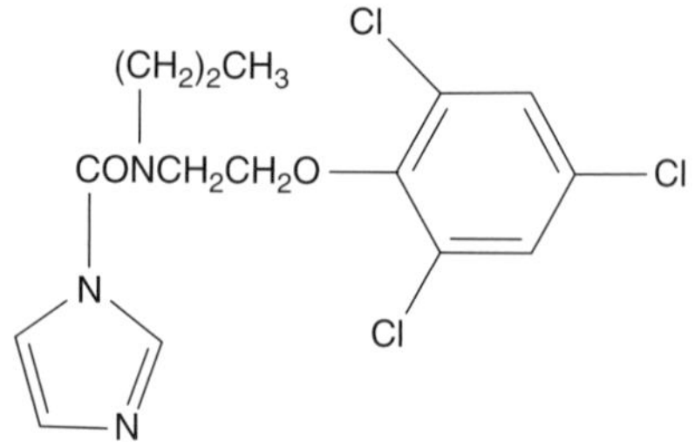

Prochloraz

Prochloraz, an **ergosterol** biosynthesis-inhibiting, broad-spectrum, protective and curative fungicide, is effective against *Botrytis* spp., *Fusarium* spp., *Alternaria* spp., *Pseudocercosporella* spp., *Pyrenophora* spp. and *Septoria* spp. The wide range of crops covered by prochloraz includes cereal crops, coffee, canola, legumes, citrus fruits, vegetables and turf.

Prochloraz is used as a dip treatment and as a seed treatment. Normally, it is formulated as an emulsifiable concentrate (EC) and a wettable powder (WP). The known trade names include Sportak. Prochloraz is usually combined with any one or more pesticides like **mancozeb**, cyproconazole, **anthraquinone**, **oxine-copper** or **carbendazim**. Seed treatment (0.2–0.5 g/kg) will control cereal diseases caused by *Cochliobolus* spp., *Fusarium* spp., *Pyrenophora* spp. and *Septoria* spp.

The acute oral LD_{50} in rats is 1600 mg/kg and in mice is 2400 mg/kg. Similarly, the acute oral LD_{50} in mallard ducks is 3132 mg/kg. It is non-toxic to bees and most beneficial insects. The ADI in dogs is found to be 0.01 mg/kg b.w./day. The fungicide shows indications of being a carcinogen.

Prochloraz is rapidly metabolized as a result of the imidazole ring being cleaved. In plants, the final degradation product is *N*-propyl-*N*-(2-(2,4,6-trichlorophenoxy)ethyl)urea. In soil and water prochloraz does not leach much and adsorbs to soil particles. The DT_{50} of prochloraz is up to 1 month (5–37 days). According to the WHO, it is a Class III toxin (slightly hazardous).

South Africa, Australia, Cambodia, New Zealand, the Philippines and most EU countries have registered prochloraz for use for various applications. The USA has not registered it. According to the EU Pesticide Database, the MRL for lettuce is 5.0 mg/kg; for barley and oat grains, it is 1.0 mg/kg; for garlic, rye and wheat grains, it is 0.5 mg/kg; and for most vegetables, it is 0.05 mg/kg.

Procymidone fungicide

Procymidone, $C_{13}H_{11}C_{12}NO_2$ (*N*-(3,5-dichlorophenyl)-1,2-dimethylcyclopropane-1,2-dicarboximide), is a colourless crystalline substance and a known **dicarboximide** fungicide. It is soluble in water at 4.5 mg/l, acetone (180 g/l), xylene (43 g/l) and dimethylformamide (230 g/l). Procymidone is stable to heat, light and moisture, but not to alkaline materials. It is produced by the reaction of 1,2-dimethylcyclopropanedicarboxylic anhydride and 3,5-dichloroaniline, the resulting monoamide is then cyclized with acetic anhydride and a catalytic amount of anhydrous sodium acetate at 100°C. Its CAS Registry Number is 32809-16-8.

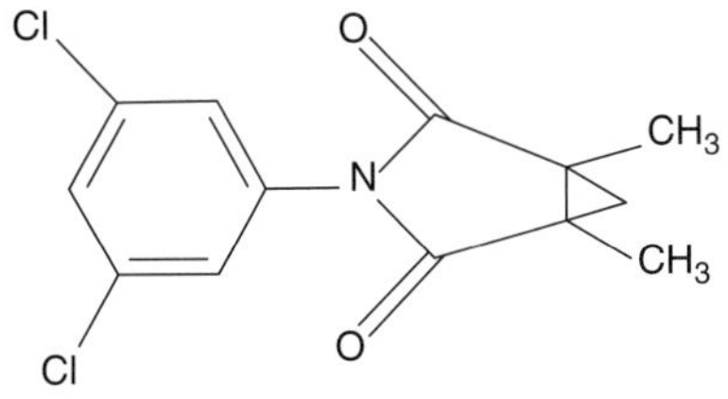

Procymidone

Procymidone is a systemic, protective compound that is applied on various vegetables, fruits, cereals, sunflowers, tomatoes, tobacco, beans, rape and soybeans to protect against species of *Botrytis*, *Sclerotinia* and *Helminthosporium*. The mode of action of procymidone on fungal pathogens is the interference with the functioning and assembly of microtubules. It is effective against strains of *Botrytis* that have acquired resistance to **benzimidazole** compounds.

Procymidone is normally formulated as wettable powder (WP), suspension concentrate (SC), dustable powder (DP) and water-dispersible granules (WG). The fungicide may also be combined with **maneb**, **zineb** and **chlorothalonil**. Procymidone is available under trade names such as Cyon 500 SC, Fortress 500, Kimono, Prodone, Progress and Prolex. It is commonly applied on the aerial parts of fruits and vegetables as well as on the soil (to control soil-borne pathogens).

Toxicologically, the acute oral LD_{50} values in rats and mice are between 6800 and 9100 mg/kg, while the LD_{50} in fish is up to 22.9 mg/l. It is not toxic to bees. Although this fungicide is classified as a Class U toxin (unlikely to be hazardous) in acute toxicity rating, it is a carcinogen and may disturb the functioning of the endocrine system. It is thus considered to be a PAN Bad Actor chemical. PPE (e.g. safety glasses or goggles, rubber gloves, shoes plus socks, long-sleeved shirt and long trousers) must be used when handling procymidone.

On ingestion, procymidone is eliminated rapidly, while in soils, procymidone persists for over 3 months. It is registered in Australia, Cambodia, New Zealand and South Africa.

Profenofos insecticide

Profenofos, $C_{11}H_{15}BrClO_3PS$ (*O*-(4-bromo-2-chlorophenyl) *O*-ethyl *S*-propyl phosphorothioate), is an organophosphorus insecticide. It is a pale yellow liquid with a garlic-like odour that is soluble in water at 28 mg/l. It hydrolyses in alkaline media (the DT_{50} at pH 9 is 5.7 h). Profenofos is produced by the reaction of *O*-(4-bromo-2-chlorophenyl) *O,O*-diethyl thiophosphate with propylbromide and triethylenediamine in acetonitrole. It has a CAS Registry Number of 41198-08-7.

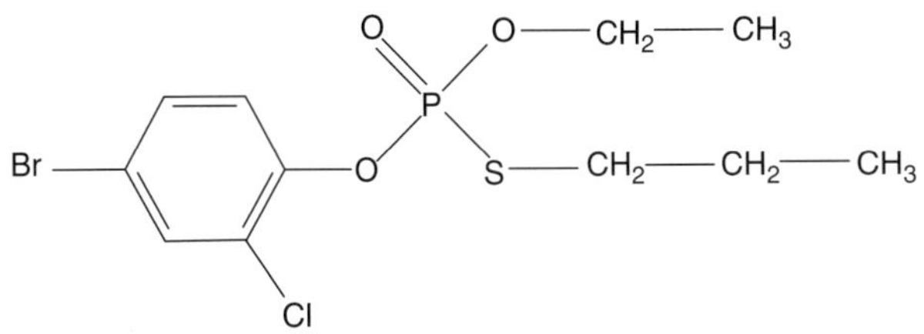

Profenofos

Profenofos, a cholinesterase inhibitor, is a non-systemic insecticide and acaricide. It is considered as a restricted use pesticide (RUP) according to the EPA. It is used to control lepidopteran insects, aphids, whiteflies, spider mites and hoppers, specifically on cotton, maize (corn), sugarbeet, soybeans, potatoes, vegetables and tobacco. There are no residential uses of profenofos.

The insecticide profenofos is formulated as emulsifiable concentrates (EC). It is available under the trade name Curacron. The maximum recommended application rate is 2.5 kg a.i./ha to control lepidopteran pests in cotton crops, to be applied at a maximum of twice per season.

The acute oral LD_{50} of profenofos in rats and rabbits is between 358 and 700 mg/kg, while the acute percutaneous LD_{50} in rats and rabbits is 3300 and 472 mg/kg, respectively. The NOEL of profenofos in rats (2 years) is 0.3 mg a.i./kg diet and the ADI is 0.01 mg/kg b.w./day. It is highly toxic to birds (the acute 8 h LC_{50} in bobwhite quails is up to 200 ppm, while in mallard ducks it is 150–612 ppm). It is also toxic to fish (the acute 96 h LC_{50} for rainbow trout is 0.08 mg/l), bees and crustaceans. Slight reddening of cotton may occur on application. Profenofos can cause dizziness, respiratory paralysis and death of humans and animals in cases of extreme exposure. It is a cholinesterase inhibitor and hence a PAN Bad Actor chemical. The WHO has listed this chemical as moderately toxic (a Class II toxin).

In rats and plants, profenofos is metabolized at various rates. The principal metabolic routes of profenofos are oxidation of the sulfide group to the sulfoxide and sulfone, and hydrolysis of the phenyl phosphate ester bond in both animals and plants. Profenofos breaks down rapidly in aerobic and anaerobic conditions and dissipates in neutral to alkaline soils with a mean half-life in soil of 1 week. The mobility of profenofos in soils is reduced because of its low solubility in water and its affinity for soil organic matter.

Although there is no serious risk to water sources, for workers handling and mixing profenofos formulations, safety recommendations include reducing the application rates to 900 g/ha, and using protective clothing and equipment for workers re-entering treated fields.

Profenofos is registered for use in the USA, Australia, Vietnam, the Philippines, South Africa and many other African countries including Burkina Faso, Mali, Mauritania and Tanzania. However, it is not registered for use in the EU region.

Prohexadione calcium as a PGR

Prohexadione calcium, $C_{10}H_{10}CaO_5$ (calcium 3-oxido-5-oxo-4-propionylcyclohex-3-enecarboxylate), is a plant growth regulator. It is a white-yellow-brown solid substance, soluble in water, acetone, toluene, hexane, methanol and ethyl acetate. Prohexadione calcium can remain stable for at least 2 years when stored away from direct sunlight, under normal conditions, in high density polyethylene containers. Prohexadione calcium is produced by the treatment of diethyl-(1-propylketone) succinate with metallic sodium in ethanol. Its CAS Registry Number is 127277-53-6.

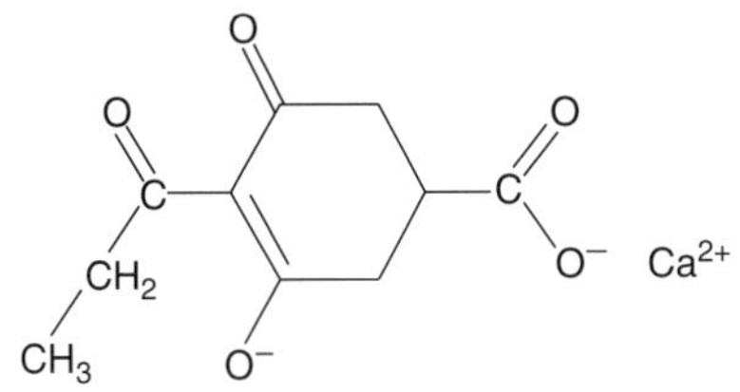

Prohexadione calcium

Prohexadione calcium, as a PGR and retardant, is absorbed by the foliage and translocated throughout the plant. It is used as an anti-lodging agent in small grain cereals, and as a growth retardant in groundnuts and flowers.

Prohexadione calcium is formulated as water-dispersible granules (WG) or as a soluble concentrate (SL), sprayed after dilution in water. It is available in the market under the trade name of Viviful.

The EPA finds prohexadione unlikely to be carcinogenic to humans. Its ADI is 0.2 mg/kg/day.

The degradation pathway of this PGR involves protonation followed by cleavage of the propionyl side chain. In animals, prohexadione calcium is rapidly absorbed from the gastrointestinal tract leading to high concentrations of radioactivity in blood and plasma.

According to US EPA, prohexadione calcium residues in or on pome fruit are tolerated at 3.0 ppm. According to the EU Pesticide Database, the MRL of prohexadione calcium for barley grain and wheat grain is 0.2 mg/kg, while for vegetables, fruit and other cereal grains, it is 0.05 mg/kg.

Prohexadione calcium is registered in the USA, Australia, New Zealand, the UK, France, Germany, Poland, Italy and South Africa.

Prokaryotes

Organisms that do not possess a distinct nucleus and whose cell contents are not bound in membranes are called prokaryotes. Bacteria are examples of prokaryotes. A bacterial cell binds itself in a cell membrane or a wall, that keeps all the constituents bound together. But, within the cell, the structures are not compartmentalized.

Prometon herbicides

See **Triazine herbicides**

Prometryn herbicides

See **Triazine herbicides**

Pronamide

See **Herbicide action – meristematic mitotic inhibitors**

Propachlor herbicide

Propachlor, $C_{11}H_{14}ClNO$ (2-chloro-*N*-isopropylacetanilide), is a **chloroacetanilide herbicide**. The CAS Registry Number is 1918-16-7.

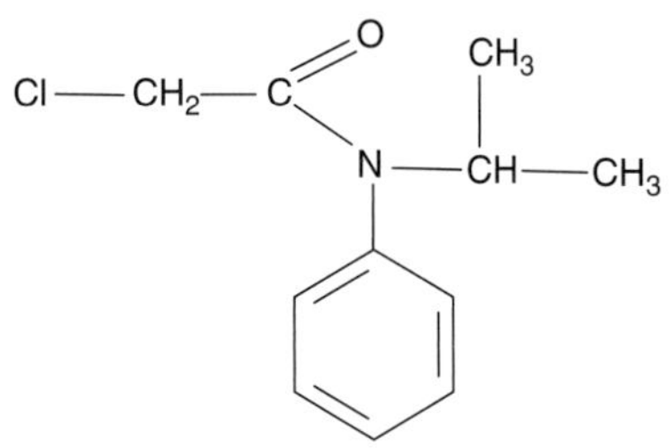

Propachlor

The herbicide propachlor is a selective, systemic compound which affects the cell formation and protein synthesis of grasses and broadleaved weeds in sorghum and maize. According to the WHO, propachlor is a Class III toxin (slightly hazardous). But, it is categorized as a PAN Bad Actor chemical since it is a cholinesterase inhibitor and a reproductive toxin. While the herbicide is non-toxic to mammals and honey bees, it is moderately toxic to birds and fish and other aquatic creatures.

Propachlor is registered for use in Australia, Cyprus, New Zealand and South Africa.

Propamocarb hydrochloride fungicide

Propamocarb, $C_9H_{20}N_2O_2$ (propyl 3-(dimethylamino) propamocarbamate), is a carbamate fungicide. The active ingredient propamocarb hydrochloride, with a molecular formula $C_9H_{21}ClN_2O_2$, is a colourless crystalline, hygroscopic substance. It is soluble in water at 867 g/l and in methanol (over 500 g/l). Propamocarb hydrochloride is photostable and stable to temperatures up to 400°C. It is produced by the reaction of *N,N*-dimethylpropane-1,3-diamine with propyl chloroformate in the presence of a base, followed by treatment with hydrogen chloride. Propamocarb hydrochloride has a CAS Registry Number of 25606-41-1.

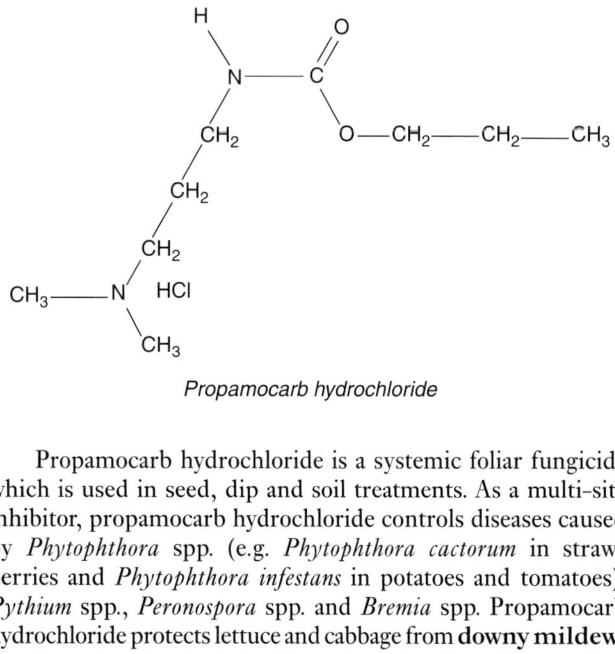

Propamocarb hydrochloride

Propamocarb hydrochloride is a systemic foliar fungicide which is used in seed, dip and soil treatments. As a multi-site inhibitor, propamocarb hydrochloride controls diseases caused by *Phytophthora* spp. (e.g. *Phytophthora cactorum* in strawberries and *Phytophthora infestans* in potatoes and tomatoes), *Pythium* spp., *Peronospora* spp. and *Bremia* spp. Propamocarb hydrochloride protects lettuce and cabbage from **downy mildew**. Its sites of action include disruption of cell wall formation, interference with synthesis of phospholipids and fatty acids and inhibition of mycelial growth and spore production.

Used as a suspension concentrate (SC), propamocarb hydrochloride is commonly available as trade names like Banol and Previcur. This fungicide also has drench and root-dip applications.

The acute oral LD_{50} of propamocarb hydrochloride among test animals is as follows: (i) rats, between 2000 and 8550 mg/kg; (ii) dogs 1450 mg/kg; and (iii) mice between 1960 and 2800 mg/kg. Similarly, the acute oral LD_{50} among several bird species is in the range of 3050–25,000 mg/kg, while the same in bees is >0.1 mg per bee. The 96 h LC_{50} in fish species is 235–616 mg/l. The NOEL (2 year period) of this fungicidal compound in rats is 1000 mg/kg. Propamocarb hydrochloride is not phytotoxic when used as recommended. Propamocarb hydrochloride does not inhibit cholinesterase since it is a methyl carbamate compound.

It is a potential groundwater contaminant. Disposal of wastes and spills should be carried out strictly according to the disposal regulations of the region. In addition, farmers are advised against feeding livestock with grass that is freshly treated with the fungicide.

Propamocarb hydrochloride is almost entirely eliminated via the urine and faeces. It is immobile in soil. The soil DT_{50} is 1 month, with microbial action contributing to its degradation.

Fungicide propamocarb hydrochloride is registered for use in Australia, the USA, the UK, Canada, Vietnam, South Africa and all members of the EU.

The MRL values established for potatoes and cucumbers are 0.5 mg/kg and 2.0 mg/kg, respectively.

Propanil herbicide

See **Acylanilide herbicides**

Propaphos insecticide

Propaphos, $C_{13}H_{21}O_4PS$ (4-(methylthio) phenyl dipropyl phosphate), is an organophosphorus insecticide. It is a colourless liquid soluble in water (125 mg/l) and organic solvents. It is stable at <230°C. It decomposes in alkaline medium. It is obtained by reacting *p*-thiomethylphenol with *O,O*-di-isopropyl phosphoryl chloride. Its CAS Registry Number is 7292-16-2.

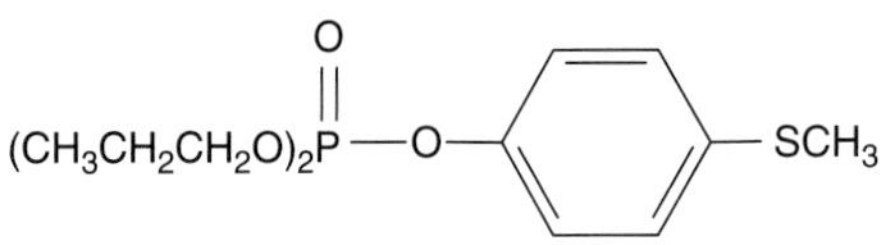

Propaphos

Propaphos, a cholinesterase inhibitor, is a systemic insecticide with contact action. It is mainly used in the control of the **rice stem borer** (*Chilo suppressalis*), the small brown plant hopper (*Laodelphax striatella*) and the rice leaf beetle (*Oulema oryzae*). Some varieties of these insects are resistant to many organophosphorus and carbamate insecticides.

Insecticide propaphos is formulated as emulsifiable concentrates (EC), granules (GR) and dustable powders (DP). It is available in the market under the trade name of Kayaphos.

This insecticide is highly toxic and may be fatal if inhaled, swallowed or absorbed through the skin. It is thus necessary to wear protective gloves, clothing, goggles, respirators and positive-pressure self-contained breathing apparatus (SCBA) while applying the chemical. It is toxic to bees. According to the WHO, propaphos is a Class Ib toxin (i.e. a highly hazardous toxin). Due to its cholinesterase inhibiting property, it is considered to be a PAN Bad Actor chemical.

Propaphos is rapidly excreted by mammals mainly via the urine. Oxidation of the sulfide group to the sulfoxide and sulfone, and hydrolysis of the phenyl phosphate ester bond occurs with propaphos in both animals and plants.

Propaphos is registered for use in Cambodia but not in the USA.

Propaquizafop herbicide

Propaquizafop, $C_{22}H_{22}ClN_3O_5$ (2-isopropylideneamino-oxyethyl(*R*)-2-[4-(6-chloroquinoxalin-2-ylodxy) phenoxy] propionate), is a herbicide (of miscellaneous class). It is a colourless crystalline substance, sparingly soluble in water and readily soluble in organic solvents. It is rapidly hydrolysed under alkaline conditions. Propaquizafop is produced by the reaction of 2-(6-chloroaminoxalinyl) phenoxyproponic acid ethoxy carbonyl chloride with isopropylidene hydroxylamine. Its CAS Registry Number is 111479-05-1 (see Propaquizafop at bottom of the page).

Propaquizafop is a systemic, selective, post-emergence herbicide translocated after being absorbed by foliage and roots. It is used in the control of a range of annual and perennial grasses in soybeans, sugarcane, cotton and vegetables. It inhibits fatty acid synthesis.

The herbicide propaquizafop is available as an emulsifiable concentrate (EC). It is available in the market under the trade name of Agil.

According to the WHO, propaquizafop is a Class U toxin (unlikely to be hazardous). The acute dermal LD_{50} of propaquizafop for rabbits is >2000 mg/kg. The acute oral LD_{50} in rats is 5000 mg/kg. The ADI of propaquizafop in mice is 0.015 mg/kg b.w./day.

In plants, propaquizafop is rapidly broken down to the free acid, followed by further metabolism to quinoxaline oxyphenol.

Propaquizafop is registered by the USA, Australia, the Philippines and EU members such as Germany, Hungary and the UK. According to the EU Pesticide Database, the MRL for sugarbeet is 0.05 mg/kg.

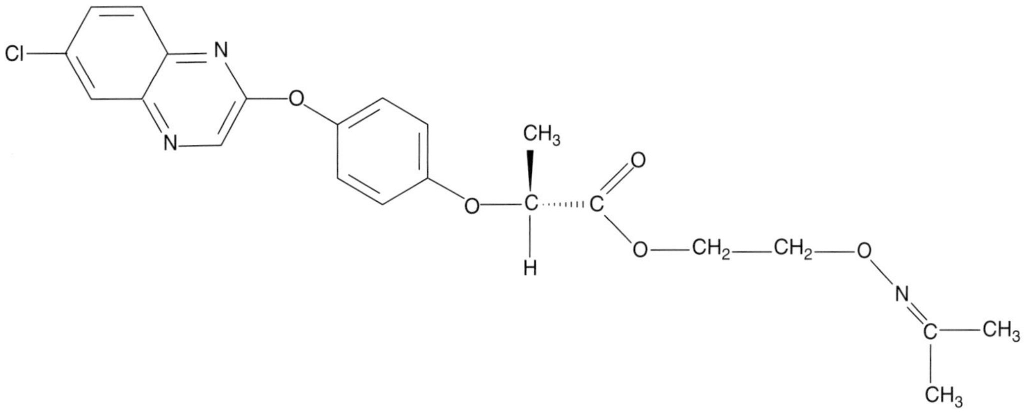

Propaquizafop

Propargite acaricide

Propargite, $C_{19}H_{26}O_4S$ ((1*RS*, 2*RS*; 1*RS*, 2*SR*)-2-(4-*tert*-butylphenoxy) cyclohexyl prop-2-ynyl sulfite), is a sulfite ester acaricide. It is a dark viscous oil, almost insoluble in water (0.5 ppm at 25°C) but soluble in most organic solvents. Propargite is produced by the reaction of 4-*tert*-butylphenol with 1,2-epoxycyclohexane, followed by treatment with thionyl chloride to give the chlorosulfiinate, and then with 2-propynol (propargyl alcohol) in the presence of a base. The CAS Registry Number of propargite is 2312-35-8.

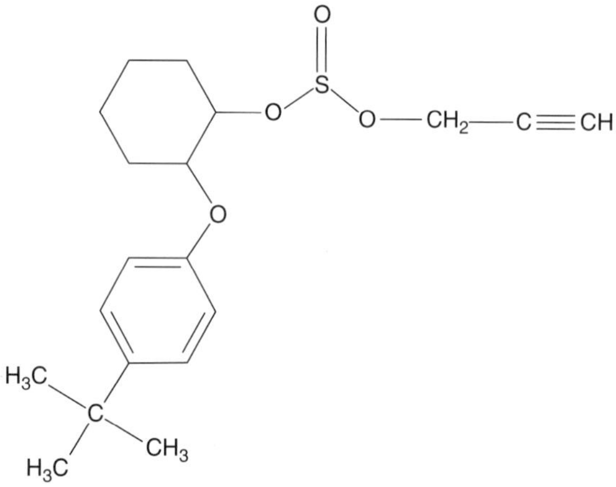

Propargite

Propargite is a non-systemic acaricide that controls mites on fruit crops, nuts, vegetables and ornamentals. It also protects cotton, maize, potato and groundnuts.

This acaricide is formulated as an emulsifiable concentrate (EC) and a wettable powder (WP), and is sprayed on foliage. The trade names of propargite include Omite and Redador.

The WHO classifies propargite as a Class III toxin (slightly hazardous). According to the EPA's acute toxicity rating for products, propargite is listed as highly toxic. The acaricide sensitizes and irritates the eye and skin. It is highly toxic to amphibians and fish. The acute oral LD_{50} is about 2800 mg/kg in rats and 311 mg/kg in rabbits. The acute dermal LD_{50} among rabbits is 4000 mg/kg. The NOEL in rats (2 year feed) is 15 mg/kg/day. The ADI is 0.15 mg/kg/day. Propargite is a carcinogen and also a reproductive toxin and hence it is considered to be a PAN Bad Actor chemical.

Propargite is observed to remain as a residue on crops and has a half-life from 11 to 53 days.

Propargite is registered for use in France, Germany, Estonia, Poland, Belgium, Australia, South Africa, New Zealand, the USA and some other countries.

Propargyl bromide

See **Soil fumigants**

Propazine herbicide

Propazine is a selective systemic *s*-triazine herbicide. Chemically, it is 6-chloro-N^2,N^4-diisopropyl-1,3,5-triazine-2,4-diamine, with molecular formula $C_9H_{16}ClN_5$. It is a colourless crystalline powder having solubility in water (5 mg/l) and organic solvents like benzene, toluene and diethyl ether (up to 6 g/kg). It is stable in neutral, slightly acidic and slightly alkaline media. Propazine is hydrolysed by acids and alkalis on heating to form hydroxypropazine.

Propazine, a pre- and post-emergence, selective compound, is a photosynthetic electron transport inhibitor at the photosystem II receptor site and is metabolized in tolerant plants to the corresponding 6-hydroxy compounds. On application, propazine is absorbed by plant roots with acropetal translocation in the xylem and accumulation in the apical meristem and leaves. It is used for the control of grasses and broadleaved weeds in crops such as sorghum, carrot and chervil. As a post-emergence compound, it controls annual broadleaved weeds in ornamentals.

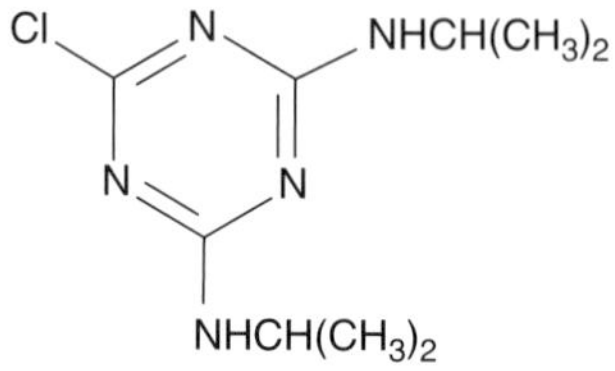

Propazine

The herbicide is available as liquid, wettable powder (WP) and water-dispersible granules (WG). It is known by the trade names Prozinex, Propasin and Primatol. Pre-emergence doses range from 0.5 to 3.0 kg/ha.

Propazine is classified under Class U by the WHO and Category IV by the EPA. The acute oral LD_{50} of the herbicide is >7000 mg/kg in rats. It is a mild skin and eye irritant. The dietary LC_{50} for birds, bobwhite quails and mallard ducks is >10,000 mg/kg. The LC_{50} values (96 h) in fish species range from 17.5 to 32 mg/l. It is slightly toxic to earthworms (with the LC_{50} being >1000 mg/kg) and non-toxic to bees. The NOEL in rats is 200 mg a.i./kg diet. The ADI is optimized at 0.017 mg/kg b.w./day. Sugarbeet and many vegetables are sensitive to this herbicide.

Around 46% of propazine is eliminated in metabolized form within 24 h in mammals. In the case of susceptible plants, propazine accumulates and causes death. However, in tolerant plants, it hydrolyses to 6-hydroxypropazine. The dealkylated amino groups are further decomposed. Propazine is degraded in soil by microbial activity. The soil half-life of the herbicide is 80–100 days.

This compound is registered for use in the USA, Canada, Australia and New Zealand. However, the EU has banned its use recently.

Pro-pesticides

A pro-pesticide (e.g. a pro-herbicide or any other '-cide') is a biologically inactive compound that undergoes structural transformations upon application, to become active. In the target site, such a compound may be activated enzymatically, chemically or physically. This property of the precursor ensures higher selectivity and hence safety to non-target organisms, due to different metabolism routes of the parent and the metabolite. Altered physicochemical properties can lead to improved stability and solubility, influencing distribution within an organism. Slow release of the compound from its derivative can lead to longer protection.

For example, **imazamethabenz-methyl** is a pro-herbicide that has to be converted into imazamethabenz acid in the plant to be an active herbicide. Similarly, the insecticide **carbosulfan** is a pro-insecticide of **carbofuran**, which is the active pesticidal agent.

Propetamphos insecticide

Propetamphos, $C_{10}H_{20}NO_4PS$ ((*RS*)-[(*E*)-*O*-2-isopropoxycarbonyl-1-methylvinyl *O*-methyl ethylphosphoramidothioate]), is a phosphoramidothioate insecticide and an acaricide. It is a yellowish oily liquid with solubility in water being 110 mg/l. It is readily miscible with acetone, ethanol, methanol, hexane, dimethyl sulfoxide and xylene. It is stable to light and in aqueous solution. Its DT_{50} values (at 20°C) at pH 3, 6 and 9 are 11 days, 1 year and 41 days, respectively. Propetamphos is made by the reaction of thiophosphorus trichloride with 2-hydroxymesitylic acid and isopropylate, followed by treatment with ethylamine and methanol. It has a CAS Registry Number of 31218-83-4.

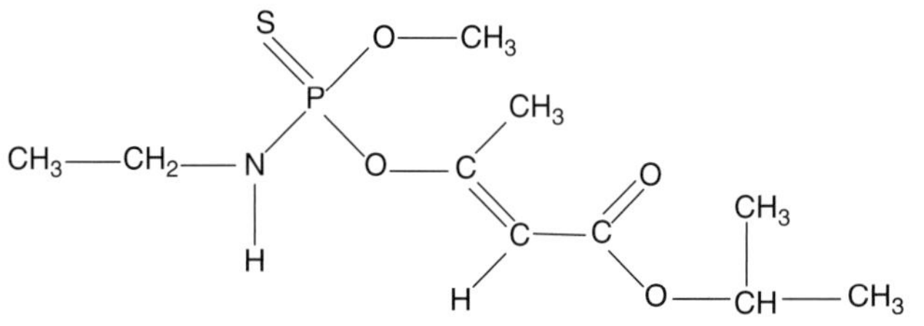

Propetamphos

Propetamphos, a cholinesterase inhibitor, is an insecticide with contact and stomach action and a long residual activity. It is used to control pests such as cockroaches, flies, fleas, mosquitoes, cloth moths, ants and animal ectoparasites (e.g. ticks, lice and mites).

The insecticide propetamphos is formulated as a wettable powder (WP), dustable powder (DP) and an emulsifiable concentrate (EC), which is sprayed by a compressed air sprayer or a hand-held pump, after dilution in water (with a maximum of 0.5% of a.i.). It is available under the trade name Safrotin.

Rats completely metabolize and excrete propetamphos via urine and exhaled air. Carbon dioxide is formed at lower doses. Propetamphos is detoxified through hydrolytic reactions involving the phosphorus and carboxylic ester bonds. Carbon dioxide is finally produced although the residual activity in soil may be up to 3 months.

Propetamphos is very toxic to rats (the oral LD_{50} is 75–119 mg/kg). It is toxic to some species of birds, with the acute oral LD_{50} being between 45 and 200 mg/kg. The LC_{50} among fish species tested is as low as 0.13 mg/l to 0.36 mg/l. It is classified as a moderately toxic compound by the EPA, with many formulations being declared as restricted use pesticides (RUPs). Acute exposure to propetamphos results in symptoms such as incoordination, headache, dizziness, tremors, nausea, abdominal cramps, sweating, blurred vision, breathing problems, unconsciousness, convulsions and even death. Handlers/applicators are advised to use PPE and prohibit all broadcast/spot applications. If used as directed, there is no ecological risk attached to using this compound. The WHO rates propetamphos as a highly hazardous pesticide (Class Ia toxin).

India, Australia, the USA and a few other countries have registered the use of propetamphos.

Propham herbicide

Propham, $C_{10}H_{13}NO_2$ (isopropyl phenylcarbamate; isopropyl carbanilate), is a carbanilate herbicide and also a plant growth regulator. It is a colourless, crystalline substance, stable in water, esters, alcohols, acetone and xylene. It is stable up to 100°C, and also in light. It has a CAS Registry Number of 122-42-9. It can be hydrolysed slowly in acidic and alkaline media.

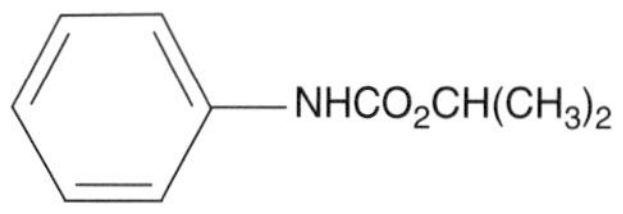

Propham

Propham is a selective systemic compound, absorbed by the roots and translocated acropetally. It may be applied at all three planting stages – pre-plant, pre-emergence or post-emergence. As a PGR, propham inhibits root growth, normal cell division, the activity of messenger RNA and protein and amylase synthesis. It controls weeds (e.g. wild oats, volunteer barley and bluegrass) in lucerne, sugarbeet, spinach, lettuce, legumes, flax, safflowers and lentils.

The herbicide propham is compatible with many other pesticides for better weed control and also with liquid fertilizers. Propham is formulated as wettable powder (WP), emulsifiable concentrate (EC), suspension concentrate (SC), granules (GR) and dustable powder (DP). Trade names include Birgin and Triherbide.

Propham is considered to be relatively safe to mammals, fish, bees and other **beneficial organisms**, when used according to the label instructions. The WHO acute hazard studies list propham as a Class U toxin (unlikely to be hazardous). The EPA classifies propham as a Class IV toxin (i.e. non-irritating to skin). However, since it is a cholinesterase inhibitor, it is a PAN Bad Actor chemical. Acute exposure could lead to dizziness, breathlessness, incoordination and nausea.

In animals, the major metabolic route of propham is *N*-oxidation to hydroxy-isopropyl phenylcarbamate. In plants, the major metabolite found is isopropyl *N*-2-hydroxycarbanilate. In soil, propham is readily degraded by microorganisms, with the formation of aniline and carbon dioxide.

Propham is registered for use in New Zealand but not in the USA. According to the EU Pesticide Database, the MRL of propham for vegetables, fruit and cereal grains is 0.05 mg/kg. (See also **Herbicide action – meristematic mitotic inhibitors**.)

Propiconazole fungicide

Propiconazole, $C_{15}H_{17}Cl_2N_3O_2$ ((±)-1-[2-(2,4-dichlorophenyl)-4-propyl-1,3-dioxolan-2-ylmethyl]-1*H*-1,2,4-triazole), is a yellowish viscous liquid and is a known conazole fungicide (triazole class). It is soluble in water (100 mg/l) and hexane (47 g/l) and is stable up to 320°C. It is incompatible with strong oxidizing agents and can produce toxic gases when decomposed on heating. Propiconazole can be produced by ketalization of 2-bromo-2′,4′-dichloroacetophenone with 1,2-pentanediol, followed by nucleophilic substitution with sodium 1,2,4-triazolide in dimethylformamide at 160°C. Its CAS Registry Number is 60207-90-1.

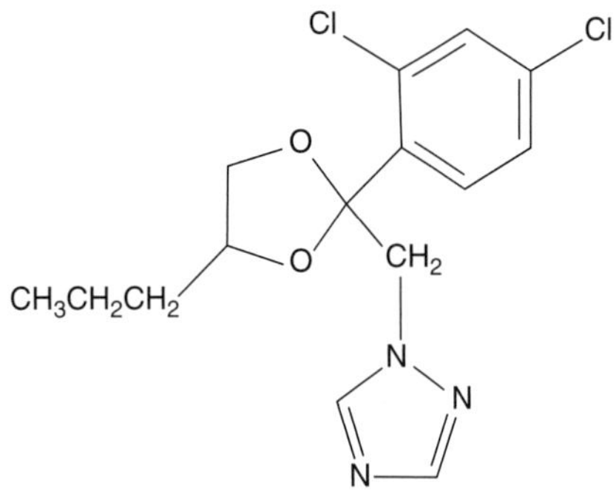

Propiconazole

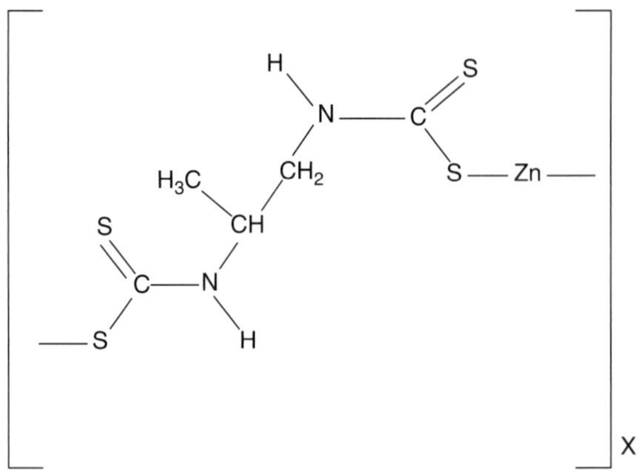

Propineb

Propiconazole, a systemic, broad-spectrum, eradicative, protective, foliar fungicide, acts as an **ergosterol** biosynthesis inhibitor. Specific pathogens controlled are *Mycosphaerella musicola* and *Mycosphaerella fijiensis*, *Sclerotinia* spp., *Septoria* spp., *Puccinia* spp., *Erysiphe graminis*, *Helminthosporium oryzae* and *Rhizoctonia solani*. Crops that are protected include rice, coffee, maize, sorghum, turf, bananas and some other fruits.

Propiconazole is usually formulated as soluble concentrate (SL), emulsifiable concentrate (EC) and emulsifiable gel (GL). Formulations include mixtures of propiconazole with **carbendazim, fenpropidin, pyrazophos, thiabendazole**, imazalil and **cypermethrin**. Propiconazole may be sold under trade names such as Tilt, Alamo, Banner, Orbit and Quilt.

Toxicologically, the acute oral LD_{50} in some bird species is above 2223 mg/kg, while in rats it is 1517 mg/kg and in rabbits it is 1344 mg/kg. It is not irritating to the skin and eyes in normal use. While it is not toxic to bees, it is moderately toxic to fish with the EC_{50} ranging from 0.9 to 21 ppm for a wide range of fish species that have been tested. The ADI is 0.04 mg/kg b.w./day. The WHO has classified propiconazole as a Class II toxin (moderately toxic). However, it is observed to be a possible carcinogen, potential groundwater contaminant and suspected endocrine disruptor.

After ingestion by rats and upon absorption by plants, propiconazole is broken down by the hydroxylation of the propyl side chain and the dioxolane ring, forming 1,2,4-triazole. In water bodies, the DT_{50} for propiconazole is from 3 to 12 weeks. The estimated soil half-life is 110 days. Toxic triazole compounds are formed via degradation of propiconazole.

Propiconazole is registered in India, the USA, Australia, New Zealand, the Philippines and South Africa and almost all the countries in the EU.

Propineb fungicide

Propineb, $(C_5H_8N_2S_4Zn)_x$ (polymeric zinc propylenebis (dithiocarbamate)), is a **dithiocarbamate** fungicide (alkylenebis (dithiocarbamate) or polymeric dithiocarbamate fungicide). It is a white powdery inorganic zinc compound, soluble in water (0.01 g/l) and in toluene and hexane (at less than 0.1 g/l). Propineb decomposes in acidic, alkaline and moist conditions. This fungicide is produced by the reaction of an aqueous solution of 1,2-propylenediamine with carbon disulfide in NaOH. The zinc salt is obtained by precipitation with zinc sulfate. Its CAS Registry Number is 12071-83-9.

Propineb, a multi-site, broad-spectrum, quick-acting, contact, protective compound, has a long residual activity. Common diseases such as **downy mildews**, blights, leaf spot, rots and moulds are controlled, because the fungicide inhibits the growth of species of *Phytophthora*, *Alternaria* and *Septoria* and many other genera. Crops thus protected are vines, apples, pears, tomatoes, onions, potatoes, rice, groundnuts, coffee, vegetables and tobacco.

Propineb is formulated mainly as a wettable powder (WP). Dust formulations are also known. Commonly used as a foliar application, the fungicide may be combined with other compounds including triadimefon, **oxadixyl** and triadimefon. Some of the known trade names are Melody and Antracol. Propineb is widely used in most parts of the world.

Toxicologically, the oral and dermal LD_{50} in rats is >5000 mg/kg. The NOEL (2 years) in rats is 50 mg/kg of diet, while the ADI is 0.007 mg/kg b.w./day. Propineb is safe for earthworms with the acute 14 day LC_{50} being >700 mg/kg. It is toxic to honey bees (the acute 48 h oral LD_{50} is >70 μg per bee). It is non-toxic to slightly toxic to aquatic organisms. The WHO has categorized propineb as a Class U toxin (unlikely to be hazardous). However, since propineb is a proven developmental toxin (according to the PAN Pesticide Database), it has been listed as a PAN Bad Actor chemical. Propineb degrades to propylene urea in soil.

Propineb is registered in countries like India, Australia, Cambodia, New Zealand, the Philippines, South Africa, Tanzania and some countries of the EU (e.g. Italy, Portugal and Germany). In most of Europe where propineb is registered, the MRLs are fixed at 5 mg/kg for citrus and lettuce, 3 mg/kg for tomatoes, 2 mg/kg for grapes and oat grains and 0.5 mg/kg for onions and cucurbit crops.

Propisochlor herbicide

Propisochlor, $C_{15}H_{22}ClNO_2$ (2-chloro-6′-ethyl-*N*-isopropoxymethylacet-*o*-toluidide), is a selective, **chloroacetanilide** herbicide. It is a light brown to purple oil, moderately soluble in water and soluble in most organic solvents. Propisochlor is produced by the reaction of 2-methyl-6-ethylaniline with chloromethyl isopropyl ether, followed by the reaction with chloroacetyl chloride. The CAS Registry Number is 86763-47-5.

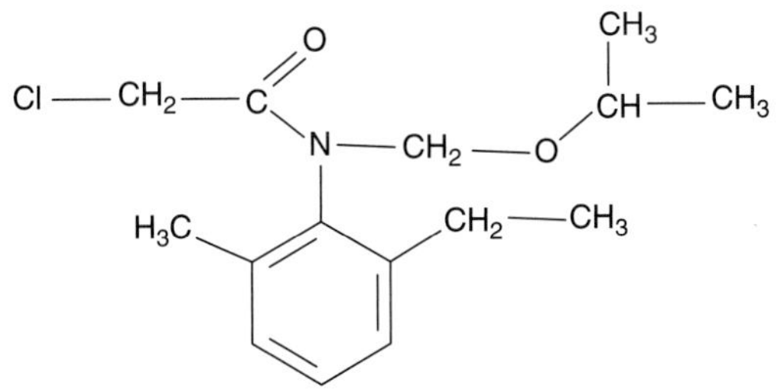

Propisochlor

Propisochlor, a pre-emergence and pre-planting selective herbicide, is absorbed by shoots of germinating plants and it inhibits protein synthesis in annual grasses and some broad-leaved weeds in maize, sunflower, soybean, potatoes and peas.

The herbicide is available as emulsifiable concentrate (EC) under the trade name Proponit.

Classified under toxicity Class III (slightly hazardous) by the WHO, propisochlor has an acute oral LD_{50} of 3433 mg/kg and 2088 mg/kg in male and female rats, respectively. The NOEL in rats is 250 mg/kg/day. The acute oral LD_{50} in mallard ducks is 2000 mg/kg. The LC_{50} (96 h) is 7.95 mg/l for carp and 0.25 mg/l for rainbow trout. The herbicide is non-toxic to bees. The ADI is optimized at 2.5 mg/kg/day.

The compound is registered for use in Belgium, Hungary, Poland and Vietnam.

Propoxur analysis

Technical propoxur, mixed with additives, is a greyish solid powder with a phenolic odour. The sample should contain propoxur with a minimum purity of 97%, with 0.1% acetone-insoluble solid material and 0.2% water.

The apparatus needed for the analysis includes: (i) a liquid chromatograph with stainless steel columns that are 25 cm long with an internal diameter of 4.6 mm, packed with ≤10 μm C-18 bonded silica gel (partisil-10 ODS-3 or equivalent); (ii) a pressure system of 11 × 10^3 kPa; and (iii) a UV spectrophotometer detector. The column must be at ambient temperature and must have a flow rate of 1.5 ml/min (~6.9 × 10^3 kPa), the detector sensitivity should be 0.16 AUFS, the wavelength, 280 nm, the injection volume, 20 μl and the chart speed, 0.5 cm/min. The retention time of propoxur should be 3.75 min and that of the internal standard, 6.50 min.

For making the internal standard solution, weigh 6 g *n*-butyrophenone into a 200 ml standard flask. Add HPLC grade acetonitrile to dissolve and make up to the level with acetonitrile. Weigh about 300 mg analytical grade propoxur of known purity (to the nearest 0.1 mg) into a 50 ml volumetric flask. Pipette 10 ml internal stand solution into the flask, dilute to the mark with acetonitrile and mix well. Pipette 10 ml of this solution into a 100 ml volumetric flask, dilute to the mark with acetonitrile and mix well.

Pump the liquid chromatography mobile phase through the column until the system is equilibrated (flat base line). Give 8 min between injections.

Adjust the operating parameters to get the propoxur to elute in 3.5–4.5 min. Adjust the injection size and attenuate to give the largest possible on-scale peaks. Repeat with an injection of the calibration solution. Calculate the response ratios by dividing the peak area or height of propoxur by that of the internal standard peak. The response ratios must be within ±1%. Average the duplicate response ratios obtained with the calibration solution injections. Inject duplicate aliquots of the sample solution and average the response ratios of propoxur and internal standard peaks.

Equations P.1 and P.2 are used to calculate the propoxur content.

$$\text{The response ratio (r)} = \frac{\text{Area or height of propoxur peak}}{\text{Area or height of internal standard peak}} \quad \textbf{(P.1)}$$

$$\text{Propoxur content (g/kg)} = \frac{r_2 \times m_1 \times P}{r_1 \times m_2} \quad \textbf{(P.2)}$$

where

r_1 = Average response ratio for standard solution
r_2 = Average response ratio for sample solution
m_1 = Mass (g) of propoxur standard in solution
m_2 = Mass (g) of sample taken
P = Purity of propoxur (g/kg) used in standard

Propoxur insecticide

Propoxur, $C_{11}H_{15}NO_3$ (2-isopropoxyphenyl methylcarbamate), is a carbamate insecticide, acaricide and a molluscicide. A white crystalline solid that is moderately soluble in most organic solvents, propoxur is hydrolysed in strongly alkaline media. When heated, propoxur emits highly toxic fumes. It has a half-life of 40 min at pH 10. The insecticide is prepared by reacting 2-isopropoxyphenol and methylisocyanate. It has a CAS Registry Number of 11426-1.

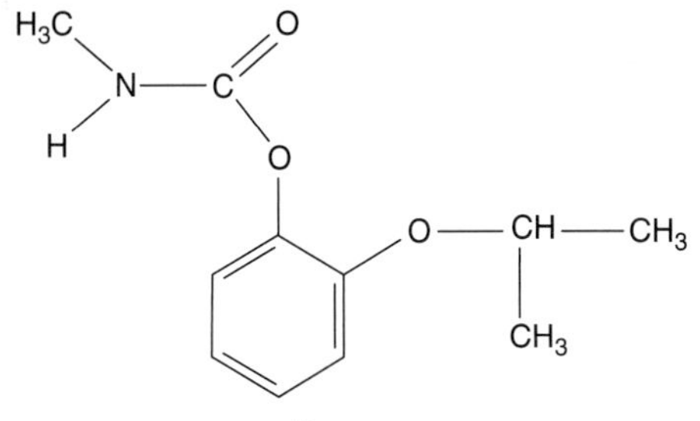

Propoxur

Propoxur, a non-systemic insecticide with a long residual activity, is a non-crop pest control agent. It controls mosquitoes, flies, cockroaches, fleas, ticks, ants and mites in homes, lawns and on farm animals. In crop situations, it is sometimes used to control aphids, **leafhoppers** and gypsy moths on sugarcane, grapes, maize, rice, vegetables and cotton.

Propoxur is compatible with insecticides and fungicides like **azinphos-methyl, chlorpyrifos, cyfluthrin, dichlorvos, disulfoton** or **methiocarb**. Propoxur is available in formulations such as emulsifiable concentrates (EC), wettable powders (WP), baits, aerosols, fumigants (FU), granules (GR) and oilsprays. Trade names of propoxur formulations include the common brand Baygon.

The LD_{50} for propoxur in rats is from 83 mg/kg to 150 mg/kg and in mice it is 23.5 mg/kg. Propoxur is highly toxic to very highly toxic to many bird species (its toxicity varying with the species). It is moderately to slightly toxic to fish and other aquatic organisms. Propoxur is highly toxic to honey bees. The ADI for propoxur is 0.02 mg/kg/day. Acute exposure to propoxur in human beings leads to cholinesterase inhibition, leading to blurred vision, nausea and sweating. However, chronic exposure has proved to be carcinogenic in laboratory animals. Pesticide products containing propoxur must bear the signal words 'CAUTION' or 'DANGER', depending on the actual combinations and concentration. Applicators or mixers should avoid skin contact and must use PPE (e.g. clean overalls, impervious shoes, clean rubber gloves and face masks).

The principal pathways of propoxur metabolism in plants and animals include deisopropylation and conjugation. Propoxur is of moderate to low persistence in the soil environment (with a reported field half-life of 14–50 days). With high solubility in water and poor soil binding, it may prove risky to groundwater.

Propoxur is registered in India, the USA, South Africa and some other African countries, but is not registered in the EU.

Propoxycarbazone-sodium herbicide

Propoxycarbazone-sodium, $C_{15}H_{17}N_4NaO_7S$ (sodium 4,5-dihydro-*N*-{[(2-methoxycarbonyl)phenyl]sulfonyl}-4-methyl-5-oxo-3-propoxy-1*H*-1,2,4-triazole-1-carboximidate), is a sulfonylaminocarbonyltriazolinone herbicide. The compound is a derivative of propoxycarbazone. It is a colourless, odourless, crystalline powder that is soluble in water (42 g/l at pH 7), dichloromethane (1.5 g/l), xylene and isopropanol (both at <0.1 g/l). It is stable to hydrolysis at pH 4–9. The CAS Registry Number is 181274-15-7.

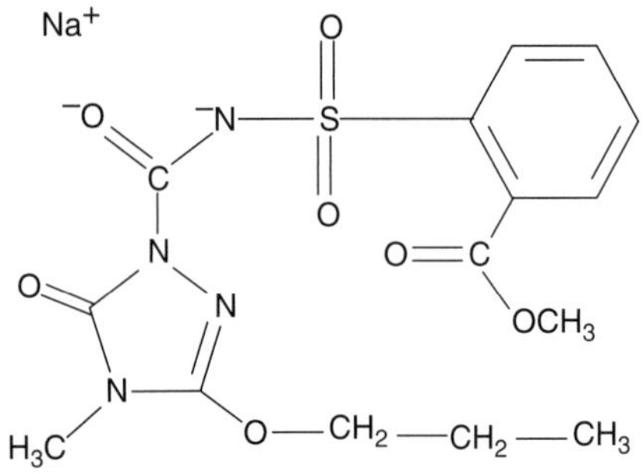

Propoxycarbazone-sodium

The compound is an inhibitor of the biosynthesis of the essential amino acids valine and isoleucine (via the inhibition of acetolactate synthase or ALS). The result is that cell division and plant growth is halted. The compound is taken up by the leaves and plant roots with both acropetal as well as basipetal translocation in xylem and phloem. The herbicide is used for post-emergence control of annual and some perennial grasses (e.g. jointed goat grass and wild oats) and some broadleaved weeds through ground or aerial application in wheat, rye and triticale.

The herbicide is available as water-soluble granules (SG) and water-dispersible granules (WG) and is sold under selected trade names such as Attribute and Olympus. Concentrations of 30–70 g/ha is considered sufficient. A maximum of two applications per season is recommended.

The EPA has classified propoxycarbazone-sodium under Category IV as it has found the herbicide to be negative in all tests of neurotoxicity, oncogenicity and genotoxicity. It is also non-irritating to the skin and eyes. The acute oral LD_{50} of the herbicide in rats is >5000 mg/kg, the main target organ being the gastrointestinal tract (it causes gastric irritation). The NOEL in male and female rats (2 years) is 43 and 49 mg/kg b.w./day, respectively. The acute oral LD_{50} in bobwhite quails is >2000 mg/kg whereas the dietary LC_{50} is >10,566 mg/kg diet. The LC_{50} (96 h) for rainbow trout is 77.2 mg/l, while in earthworms, it is >1000 mg/kg dry soil. In bees the oral lethal dose is >319 μg per bee. The ADI is optimized to be 0.4 mg/kg b.w./day.

Propoxycarbazone-sodium is expected to degrade in soil by metabolism and abiotic processes, and its hydrolysis is slow. It is incompletely absorbed in animals, the administered dose being excreted within 48 h. The excreted compound is mostly the unchanged parent active ingredient. The process of metabolism occurs by cleavage. In plants, the residues found have been observed to be the parent compound and its metabolite, 2-hydroxypropoxy. Soil metabolism experiments have shown a half-life of 77–103 days. The half-life of propoxycarbazone-sodium in European soils is found to be 9 days under field conditions and about 60 days in various soil types under laboratory conditions.

The herbicide is registered in many EU countries such as Finland, France, Germany and Poland.

Proquinazid fungicide

Proquinazid, $C_{14}H_{17}IN_2O_2$ (6-iodo-2-propoxy-3-propylquinazolin-4(3*H*)-one), is a fungicide (unclassified). It is a slightly volatile substance soluble in many organic solvents and only slightly soluble in water. Its CAS Registry Number is 189278-12-4.

With a protective, curative and long-lasting action, proquinazid acts by inhibiting the growth of the germ tube appressorium, which ultimately starves and kills the spore (since the formation of the appressorium is necessary for infection and progress of the life cycle). Vapour action also may contribute to fungicidal activity. The residual effect lasts for 3–6 weeks. The compound is effective in controlling **powdery mildews** on cereals (e.g. wheat, barley, oats and rye) and grapes.

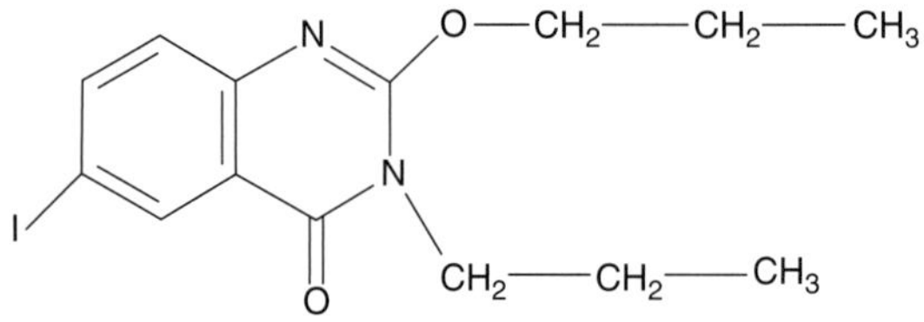

Proquinazid

Proquinazid is normally formulated as an emulsifiable concentrate (EC) that is mixed with water and used as a spray before rainfall. It is also commercially available as the trade names Talius and Talendo, among others. It is used at a rate of 0.25 l/ha. It is best applied before flowering or heading. Due to residual activity, the pre-harvest interval is 28 days. Repeat application of proquinazid, if at all, should be made only in combination with a compound with a different mode of action.

The acute oral LD_{50} of proquinazid in rats is 4846 mg/kg and the dermal LD_{50} is >5000 mg/kg b.w. The 4 h inhalation LC_{50} in rats is >5.2 mg/l. The ADI in rats is 0.01 mg/kg b.w./day. It is slightly toxic to bird species (the acute LD_{50} in bobwhite quails is >2250 mg/kg) and moderately toxic to honey bees (the 48 h acute oral LD_{50} is >125 μg per bee) and moderately toxic to fish (the acute 96 h LC_{50} for rainbow trout is 0.349 mg/l). It is practically non-toxic to earthworms.

Proquinazid is absorbed, metabolized and eliminated by test rats. Metabolism is set off via hydroxylation of the phenyl ring and propyl and propoxy side chains.

Proquinazid is registered in the EU (e.g. in Cyprus, Germany, Greece, Hungary, Portugal and the UK) but not in the USA. According to the EU Pesticide Database the MRL for grapes is 0.5 mg/kg and for cereal grains, it is 0.05 mg/kg.

Prosulfocarb herbicide

Prosulfocarb, $C_{14}H_{21}NOS$ (*S*-benzyl dipropylthiocarbamate), is a thiocarbamate herbicide. It is a colourless liquid that is moderately soluble in water and miscible with organic solvents. It is stable for at least 2 months at 52°C. Prosulfocarb is made by the reaction of *N*-dipropylcarbamoyl chloride with mercaptotoluene. It has a CAS Registry Number of 52888-80-9.

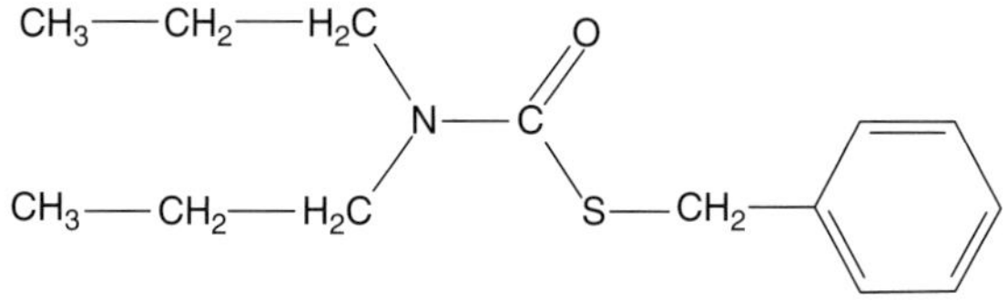

Prosulfocarb

Prosulfocarb is a selective, systemic herbicide, absorbed by the leaves and roots. It is a lipid metabolism inhibitor. It causes twisting and inhibition of shoots and roots. Leaves do not emerge and the plant turns a dark green. It is used in pre- and post-emergence for the control of a wide range of grass and broadleaved weeds in winter wheat, barley and rye.

Prosulfocarb is commonly available as an emulsifiable concentrate (EC) that is mixed with water and applied as a spray. It is available in the market as Boxer and Defi.

The acute oral LD_{50} of prosulfocarb in rats is 1820 mg/kg and the dermal LD_{50} is >2000 mg/kg/day. The inhalation LC_{50} of prosulfocarb in rats is >4.72 mg/l. The acute LD_{50} of prosulfocarb in bobwhite quails is >2250 mg/kg. It is toxic to fish and aquatic organisms (the acute 96 h LC_{50} in rainbow trout is 0.84 mg/l) and honey bees (the acute 48 h oral LD_{50} is >80 µg per bee). It is practically non-toxic to earthworms. The ADI of prosulfocarb in rats is 0.005 mg/kg/day. Symptoms of poisoning include irritation to the skin, eyes and respiratory tract. Since prosulfocarb is a weak inhibitor of cholinesterase, exposure may lead to headache, dizziness and nausea. According to the WHO acute hazard rating criteria, prosulfocarb comes under toxicity Class II (moderately hazardous).

Prosulfocarb breaks down into mercaptan, carbon dioxide and dipropylamine. In soil, prosulfocarb undergoes hydrolytic decomposition with a DT_{50} of 10–35 days, with prosulfocarb sulfoxide being a dominant metabolite.

Herbicide prosulfocarb is registered in Australia and in many EU nations, but not in the USA.

Prosulfuron herbicide

Prosulfuron, $C_{15}H_{16}F_3N_5O_4S$ (1-(4-methoxy-6-methyl-1,3,5-triazine-2-yl)-3-[2-(3,3,3-trifluropropyl)phenylsulfonyl]urea), is a sulfonylurea herbicide. It is a colourless, odourless crystalline compound having solubility in distilled water (29 mg/l), ethanol (8.4 g/l), acetone (160 g/l), toluene (6.1 g/l) and dichloromethane (180 g/l) (all at 25°C). It is not degraded photolytically but is hydrolysed rapidly and has a half-life (DT_{50}) of 5–10 days at pH 5 and >1 year at pH 7 and 9. The CAS Registry Number is 94125-34-5.

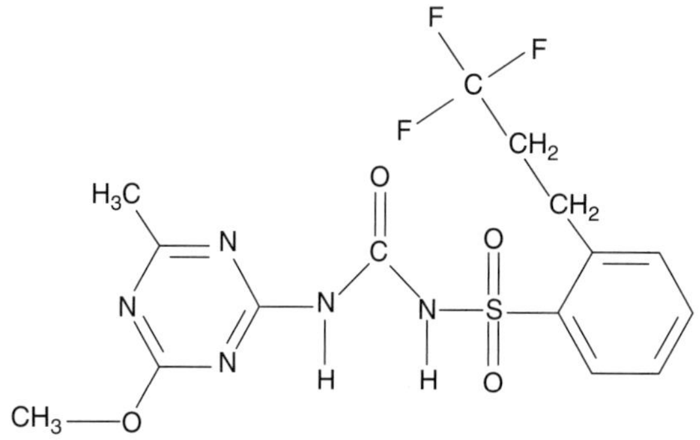

Prosulfuron

Prosulfuron is a branched-chain amino acid synthesis inhibitor and acts by inhibiting biosynthesis of the essential amino acids valine, leucine and isoleucine, stopping cell division and plant growth. The herbicide is absorbed through leaves and roots in the plant system and is translocated in the xylem and phloem to the site of action. The application kills the weed in about 3 weeks. Prosulfuron is used as a post-emergence selective herbicide in cereals, millets, pastures and turf for broad-spectrum control of species of *Amaranthus*, *Abutilon*, *Chenopodium*, *Polygonum*, *Rumex* and *Stellaria* and other annual broadleaved weeds.

The compound is available in the form of water-dispersible granules (WG) and is marketed under trade names like Peak. It may be applied at 12–30 g/ha. The formulation is not compatible with organophosphate insecticides.

Prosulfuron is classified under toxicity Class III both by the WHO as well as the EPA. The compound is found to be non-mutagenic and non-teratogenic. However, according to toxicological studies, the acute oral LD_{50} of the herbicide in rats is >986 mg/kg and in mice it is 1247 mg/kg. It is not irritating to the skin and eyes. The NOEL in rats is 200 mg a.i./kg. The dietary LC_{50} (8 days) values in bobwhite quails and mallard ducks is >5000 ppm whereas the LC_{50} (96 h) for carp, catfish and rainbow trout is >100 mg/l and in earthworms it is >1000 mg/kg. The LD_{50} (48 h) in bees is also >100 µg per bee. Predatory insects like beetles are not negatively affected up to an application rate of 30 g/ha. The ADI is optimized to be 0.017 mg/kg b.w./day.

Around 95% of prosulfuron is extensively absorbed by animals and almost all the quantity is excreted within 48 h. The metabolic pathway involved is *O*-demethylation and side-chain hydroxylation. In plants, besides hydroxylation, the metabolic pathway involves cleavage of phenyl and triazine ring systems. In soil, the degradation of the compound depends on the temperature, moisture and the organic content of the soil. The DT_{90} is 14–120 days under field conditions. Prosulfuron does not leach below 50 cm into the soil.

This compound is registered in the USA, the UK, France, Hungary and Germany.

Proteinogenic amino acids

Proteinogenic amino acid is another term for amino acids. (See also **Herbicide action – by inhibition of amino acid synthesis**.)

Prothiofos insecticide

Prothiofos, $C_{11}H_{15}Cl_2O_2PS_2$ ((*RS*)-(*O*-2,4-dichlorophenyl *O*-ethyl S-propyl phosphorodithioate)), is a phenyl organothiophosphate insecticide. It is a colourless liquid with a weak characteristic odour. It is nearly insoluble in water (0.07 mg/l) but soluble in dichloromethane, isopropanol and toluene. It has a CAS Registry Number of 34643-46-4. It is relatively stable in aqueous media. It is produced by the reaction of 2,4-dichlorophenol with *O*-ethyl *O*-thioethyl thiophosphoryl chloride in the presence of a base.

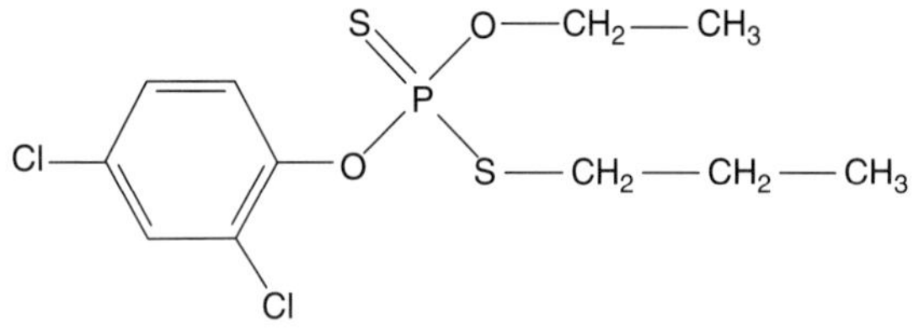

Prothiofos

Prothiofos, a cholinesterase inhibitor, is a non-systemic insecticide with contact and stomach action. It is used in the control of leaf-eating caterpillars, **mealybug** species (*Pseudococcus* spp.) and cutworms in a range of crops, including vegetables, fruits, maize, sugarcane, sugarbeet, tea, tobacco and ornamentals.

The insecticide prothiofos is formulated in forms such as wettable powder (WP) and emulsifiable concentrate (EC). It is compatible with many other non-alkaline pesticides. It is available in the market under the trade name Tokuthion.

The EPA classifies prothiofos as slightly toxic, while the WHO lists it as moderately toxic. Prothiofos is slightly to moderately toxic to fish and aquatic organisms. It is non-toxic to bees. It can cause irritation to the skin when exposed to it. Since it is an acetylcholine inhibitor, it is listed as a PAN Bad Actor chemical. Russetting may occur in some varieties of apples.

In mammals, prothiofos is rapidly absorbed and metabolized by oxidation to prothiofos-oxon which leads to the formation of 2,4-dichlorophenylethyl hydrogen phosphorothioate and 2,4-dichlorophenylethyl hydrogen phosphate. In plants, prothiofos is hydrolysed to 2,4-dichlorophenol. Prothiophos is very strongly adsorbed to soil. The DT_{50} under field conditions is 1–2 months. Here too, the final product is CO_2.

Prothiofos is permitted for use in Australia, New Zealand and South Africa, but not in the USA.

Prothrombin

Prothrombin is a coagulation protein in the blood that is needed to help clot blood. Prothrombin, produced in the liver, forms thrombin upon activation in the blood-clotting mechanism.

There are 12 factors that help blood to clot and vitamin K is necessary for the process. Prothrombin time is the time blood takes to clot. It is also referred to as Factor II.

The prefix 'pro' commonly indicates that the compound is a precursor. A precursor is an intermediate compound in a living system, which upon activation converts to a functional substance. **Ergosterol** is a precursor to vitamin D_2 and is thus known as a pro-vitamin.

PROTOX inhibitors (herbicides)

Protoporphyrinogen oxidase (PPO or PROTOX) is an enzyme that is critical in the tetrapyrrole biosynthesis pathway which produces chlorophyll and haem. Haem and chlorophyll are produced in plastids and mitochondria, and in the biosynthesis process, proto IX is produced.

PROTOX-inhibiting herbicides induce the accumulation of proto IX in the cytoplasm which, in the presence of light, produces singlet oxygen that damages cell membranes. PROTOX inhibitors include *N*-phenylthalamides, *p*-nitrodiphenyl ethers, **triazolinones** and **oxadiazoles**.

Resistance to PROTOX inhibitors is rare although the tall waterhemp (*Amaranthus tuberculatus*) has developed resistance. Currently, *A. tuberculatus* is controlled by **glyphosate**.

Prynachlor herbicide

See **Chloroacetanilide herbicides**

Purple nutsedge

See **Weed – purple nutsedge**

Push–pull hand pump sprayer

See **Application equipment – hand sprayers**

Pyraclofos insecticide

Pyraclofos, $C_{14}H_{18}ClN_2O_3PS$ ((*RS*)-[*O*-1-(4-chlorophenyl) pyrazol-4-yl *O*-ethyl S-propyl phosphorothioate]), is a heterocyclic organothiophosphate insecticide and nematicide. It is a pale yellow oil, soluble in water at 33 mg/l and miscible with most organic solvents. Although stable in original containers in a cool, dry place, it decomposes on heating, to produce toxic vapours of NO_x, SO_x, PO_x and Cl^-. It has a CAS Registry Number of 77458-01-6.

Pyraclofos, a cholinesterase inhibitor, is a non-systemic compound with respiratory, contact and stomach action. It is used to control members of the orders Lepidoptera (*Spodoptera* and *Heliothis* spp.) and Coleoptera and **class Arachnida** in crop and non-crop situations. Nematodes on fruits, vegetables, field crops, ornamentals and forestry are also controlled.

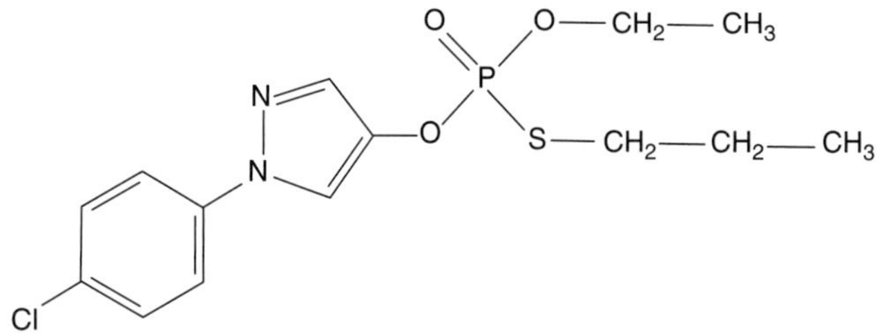

Pyraclofos

Pyraclofos is formulated in forms such as emulsifiable concentrate (EC), wettable powder (WP) and granules (GR). It is toxic to bees, but not irritating to the eyes and skin.

The acute oral LD_{50} of pyraclofos in rats is 237 mg/kg and the dermal LD_{50} is >2000 mg/kg. The acute LD_{50} of pyraclofos in bobwhite quails is 164 mg/kg. The acute 96 h LC_{50} of pyraclofos for carp is 0.028 mg/l. The acute 48 h contact LD_{50} of pyraclofos for honey bees is 0.953 μg per bee.

The inhalation LC_{50} of pyraclofos in rats is 1.46 mg/l. These studies clearly indicate that pyraclofos is moderately to acutely toxic to rats, honey bees, fish and birds. The WHO has classified it as a moderately toxic chemical (Class II toxin).

It is slightly phytotoxic to some varieties of apple, pear, peach and citrus. It is toxic to bees. Since it inhibits cholinesterase, pyraclofos is a PAN Bad Actor chemical.

In mammals, following oral administration more than 90% of the dose is eliminated within 24 h via urine. The degradation routes are cleavages of the P–S-, P–O-alkyl and P–O-aryl bonds in both animals and plants. Pyraclofos is metabolized to an active **acetylcholinesterase** (AChE) inhibitor, probably by the oxidation of the sulfur atom in the phosphorothiolate linkage. The half-life in soil is 50 days under both aerobic and anaerobic conditions.

Pyraclofos is registered for use in Australia, Cambodia and Vietnam. However, it is not allowed for use in the whole of the EU or the USA.

Pyraflufen-ethyl herbicide

Pyraflufen-ethyl, $C_{15}H_{13}Cl_2F_3N_2O_4$ (ethyl 2-chloro-5-(4-chloro-5-difluoromethoxy-1-methylpyrazole-3-yl)-4-fluorophenoxyacetate), is a phenylpyrazole herbicide. It is a fine cream-coloured powder, soluble in water at 0.082 mg/l, acetone (167–182 g/l) and methanol (7.39 g/l) (the latter two both at 20°C). It is stable in water at pH 4. Its half-life at neutral pH and at 25°C is 13 days. It is rapidly hydrolysed at pH 9. The DT_{50} for aqueous photolysis is 30 h. The CAS Registry Number is 129630-19-9.

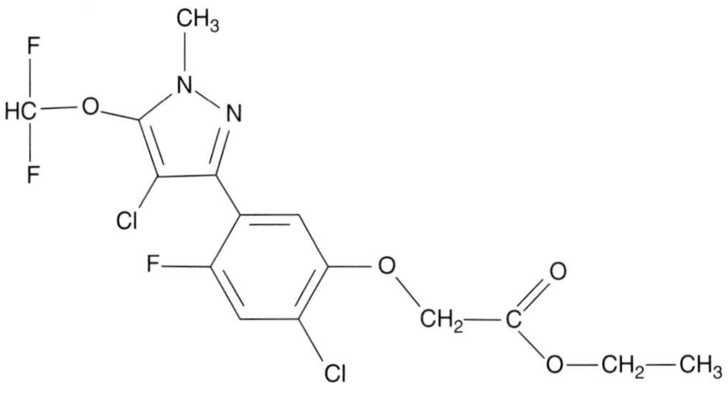

Pyraflufen-ethyl

This herbicide inhibits enzyme protoporphyrinogen-IX oxidase and acts as a contact herbicide. When applied to foliage, the herbicide is readily absorbed into the tissues and in the presence of light, rapid necrosis and desiccation of stems and leaves occur. Pyraflufen-ethyl is used as a post-emergence selective herbicide in cereals for the control of broadleaved weeds, especially *Galium aparine*, *Chenopodium album*, *Matricaria inodora*, *Lamium purpureum*, *Stellaria media*, *Senecio vulgaris*, *Veronica hederifolia* and *Veronica persica* at 10–20 g/ha.

Pyraflufen-ethyl is available as soluble concentrate (SL) as well as emulsifiable concentrate (EC) and is marketed as selected products under the trade names of Desiccan and Ecopart.

The compound is found to be non-mutagenic and non-carcinogenic. According to toxicological studies, the acute oral LD_{50} of the herbicide in rats is >5000 mg/kg and the acute percutaneous LD_{50} is >2000 mg/kg. The NOEL (2 years) in male and female rats is 86.7 mg/kg b.w. and 111.5 mg/kg b.w., respectively. The dietary LC_{50} (8 days) in birds (bobwhite quails and mallard ducks) is >5000 mg/kg, whereas the acute oral LD_{50} in bobwhite quails is >2000 mg/kg. The LC_{50} (96 h) in the fish species tested is >0.1 mg/l and in earthworms it is >1000 mg/kg dry soil. The LD_{50} (48 h) in bees is also >112 μg per bee. The ADI of pyraflufen-ethyl is optimized to be 0.2 mg/kg b.w./day.

More than 50% of the compound is rapidly absorbed in animals and almost all of it is excreted within 24 h. Ester hydrolysis and *N*-demethylation brings about the complete metabolism of the compound. In plants, metabolism proceeds via de-esterification followed by *N*-demethylation. The half-life (DT_{50}) of the compound in soil under field conditions is 1–7 days.

The compound is registered in the USA, some EU nations and Australia.

Pyrasulfotole herbicide

Pyrasulfotole, $C_{14}H_{13}F_3N_2O_4S$ ((5-hydroxy-1,3-dimethylpyrazole-4-yl)(α,α,α-trifluoro-2-mesyl-*p*-toly)methanone), is a herbicide. It is a beige-coloured powder, soluble in water at 69.1 mg/l (pH 7), ethanol (21.6 g/l), acetone (89.2 g/l), dichloromethane (120–150 g/l) and ethyl-acetate (37.2 g/l) (all at 20°C). It is stable against biotic hydrolysis and aqueous photolysis (pH 7). The CAS Registry Number is 365400-11-9.

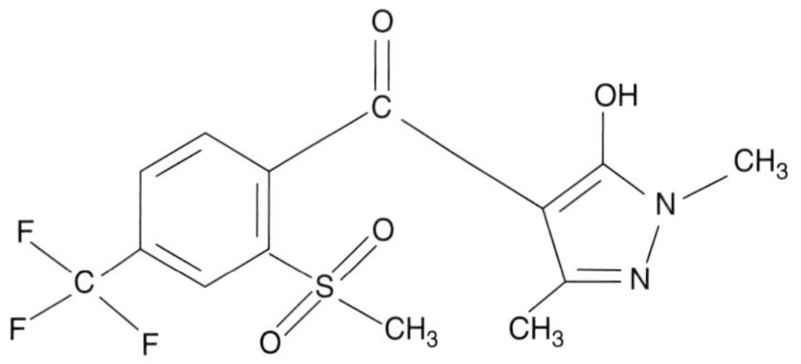

Pyrasulfotole

The herbicide inhibits the enzyme **4-hydroxyphenylpyruvate dioxygenase**. It is used as a mixture in combination with **bromoxynil**, **mefenpyr-diethyl** or **MCPA** ester for post-emergence control of broadleaved weeds in cereals.

Available as an emulsifiable concentrate (EC), the most common trade names of pyrasulfotole are Huskie, Infinity and Percept. The recommended application rate of the herbicide is 37.5 g/ha.

The acute oral as well as percutaneous LD_{50} values in rats are >2000 mg/kg. The **NOAEL** is 25 ppm. The dietary LC_{50} in bobwhite quails is >4911 mg/kg feed and the acute LD_{50} is >2000 mg/kg b.w. The LC_{50} (96 h) for fish (e.g. bluegill sunfish and rainbow trout) is >100 mg/l and in earthworms it is >1000 mg/kg soil. The oral LD_{50} in bees is >120 μg per bee. The ADI of pyrasulfotole is optimized to be 0.01 mg/kg b.w./day.

The metabolic pathways for pyrasulfotole are quite similar in mammals and plants. It involves *N*-demethylation followed by glycosylation of the parent compound. Pyrasulfotole readily degrades in soil as well as air. The half-life in soil under field conditions is about 5–31 days. The compound does not significantly volatilize from soil or water.

Pyrasulfotole is registered in the USA, Canada and Australia.

Pyrazolynate herbicide

See **4-Hydroxyphenylpyruvate dioxygenase (HPPD)**

Pyrazophos fungicide

Pyrazophos, $C_{14}H_{20}N_3O_5PS$ (ethyl 2-diethoxyphosphinothioyloxy-5-methylpyrazolo [1,5-a] pyrimidine-6-carboxylate), is an organophosphorus (OP) fungicide. This compound has a CAS Registry Number of 13457-18-6. Pyrazophos is a colourless crystal melting at 51°C. It is soluble in water and in most organic solvents, and is hydrolysed by acids and alkalis.

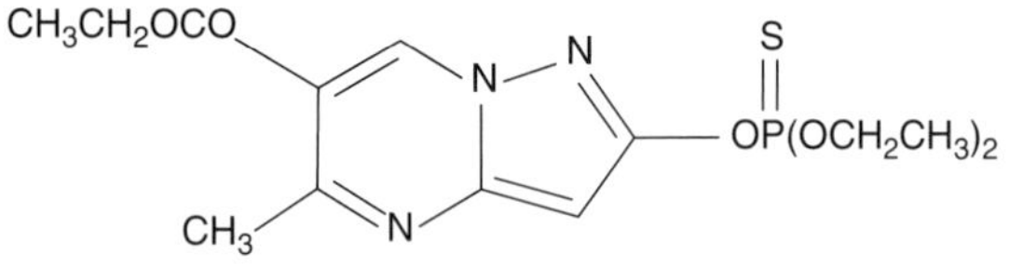

Pyrazophos

Pyrazophos is a systemic fungicide with both protective and curative action. This fungicide controls pathogens such as *Erysiphe* spp., *Uncinula* spp., *Oidium* spp., *Rhynchosporium* spp. and *Podosphaera* spp., on cucurbits, tomatoes, apples, strawberries and cereals. Its mode of action is believed to be the inhibition of melanin biosynthesis and that of appressoria by fungal conidia.

Fungicide pyrazophos is available as emulsifiable concentrate (EC), with application rates of 15–30 g a.i./100 l being typical. Furesan and Sigma are two known trade names.

The acute oral LD_{50} of pyrazophos in rats is 151 mg/kg and the dermal LD_{50} is >2000 mg/kg b.w. The LD_{50} for birds is between 118 and 480 mg/kg, while the LC_{50} in fish is between 2.8 and 6.1 mg/l. Pyrazophos is highly toxic to honey bees (the acute 48 h contact LD_{50} is 0.25 µg per bee), while it is only slightly toxic to earthworms. The ADI is fixed at 0.004 mg/kg b.w./day. The WHO has listed pyrazophos as a moderately hazardous chemical. The symptoms of poisoning are excessive salivation, sweating, muscle twitching, weakness, respiratory depression, tightness in the chest, abdominal cramps, nausea and other OP toxicity symptoms. This compound, being a cholinesterase inhibitor, is a PAN Bad Actor chemical.

When administered pyrazophos, rats metabolize and excrete the compound via the urine. Ethyl 2-hydroxy-5-methyl-6-pyrazolo [1, 5-a] pyrimidine-carboxylate is the most important metabolite. In soil too, pyrazophos degrades, at varying rates, to ultimately produce CO_2. Hydrolysis of the phosphate bond forms the β-glucoside of the pyrazolopyrimidine.

Pyrazophos is not registered in the USA or in the EU. However, an MRL value of 0.05 mg/kg appears to be common for vegetables, fruit and cereal grains.

Pyrazosulfuron-ethyl herbicide

Pyrazosulfuron-ethyl, $C_{14}H_{18}N_6O_7S$ (ethyl 5-[(4,6-dimethoxypyrimidin-2-ylcarbamoyl)sulfamoyl]-1-methylpyrazole-4-carboxylate), is a systemic herbicide of the class sulfonylurea. It is a colourless, crystalline solid that is soluble in water at 9.76 mg/l, methanol (4.32 g/l), benzene (15.6 g/l), chloroform (200 g/l) and acetone (33.7 g/l) (all at 20°C). It is stable at 50°C at pH 7 for about 6 months and unstable in acidic as well as alkaline media.

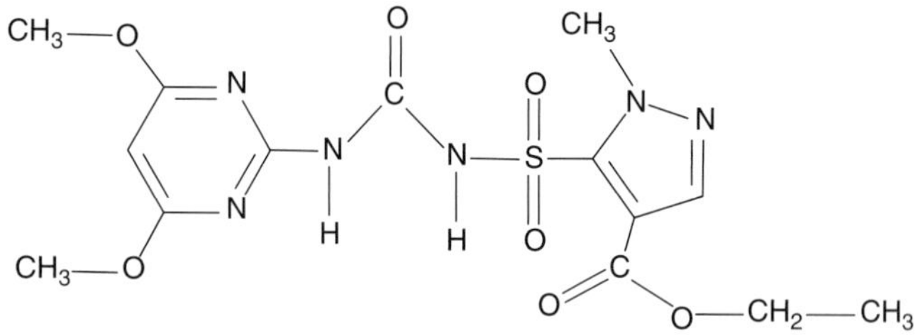

Pyrazosulfuron-ethyl

The herbicide acts by inhibiting the biosynthesis of the essential amino acids valine and isoleucine, thus stopping cell division and ultimately causing retarded plant growth. It is a systemic herbicide, which is absorbed by roots and leaves and translocated to the meristem cells.

Formulated as granules (GR), wettable powder (WP), soluble concentrate (SL) and water-dispersible granules (WG), pyrazosulfuron-ethyl herbicide is available as trade names such as Agreen, Sirius and Pyrazosun and also as mixtures. The herbicide is used in the pre- or post-emergence control of both annual and perennial broadleaved weeds usually in transplanted rice crops.

Pyrazosulfuron-ethyl is classified as being not hazardous in normal use (Class U) according to the WHO. It is not listed as a carcinogenic or mutagenic compound by the EPA and toxicological studies have shown its acute oral LD_{50} value in rats and mice to be >5000 mg/kg. The NOEL value (78 weeks) in mice is 4.3 mg/kg/day. Its ADI is 0.043 mg/kg/day. The acute oral LD_{50} in bobwhite quails is >2250 mg/kg and the LD_{50} in bees is observed to be >100 µg per bee. In fish species (e.g. rainbow trout and bluegill sunfish) the LC_{50} (96 h) is >180 mg/l.

The herbicidal dose in rats is found to be excreted in urine and faeces within 48 h. The metabolism of the compound in animals involves demethylation of the methoxy group. The half-life (DT_{50}) of pyrazosulfuron-ethyl in soil is less than 15 days, however, in paddy fields or in river water it is about 28 days. Pyrazosulfuron-ethyl is a PAN Bad Actor chemical.

It is registered for use in Vietnam but is not registered by the EPA in the USA.

Pyrazoxyfen herbicide

See **4-Hydroxyphenylpyruvate dioxygenase (HPPD)**

Pyrethrin analysis

For the estimation of pyrethrin I, the aqueous solution is filtered through a Büchner funnel lined with a 7 cm filter paper coated with diatomaceous silica and washed three times with 15 ml distilled water. The filtrate is transferred to a 500 ml separating funnel and the solution extracted with two 50 ml portions of light petroleum. The petroleum extract is washed successively with two 10 ml portions of distilled water. The petroleum is filtered through a cotton plug into a clear 250 ml separating funnel and the cotton plug washed with 5 ml petroleum. The aqueous phase from these extracts is reserved, for the determination of pyrethrin II. The petroleum layer is extracted with 5 ml 0.1 N sodium hydroxide, by shaking vigorously. The aqueous layer is drawn into a 100 ml conical flask and the petroleum layer is washed with a second portion of 5 ml 0.1 N sodium hydroxide. The aqueous extract is added to the flask. To this, 10.0 ml of Denigés' reagent is added and allowed to stand for 60 min in the dark at 25° ±0.5°C. Following this, 20 ml ethanol is added and mercurous chloride is precipitated with 2 ml saturated sodium chloride solution. The mixture is warmed to about 60°C, and set aside for several minutes until the mercurous chloride has coagulated and settled. The supernatant liquid is decanted through a 9 cm filter paper in a small glass funnel, retaining most of the precipitate in the flask. The precipitate is washed in the flask with 10 ml or more of anhydrous ethanol decanting the ethanol through the same filter paper. The precipitate is washed with three successive amounts of 10 ml hot chloroform, decanting the washings through the same filter paper. The filter paper and any precipitate in the paper are transferred to the flask containing the bulk precipitate. To this flask 30 ml concentrated hydrochloric acid and 20 ml distilled water are added and cooled. Then 6 ml chloroform or carbon tetrachloride and 1 ml iodine monochloride are added and titrated with 0.01 M potassium iodate, shaking vigorously after each addition, until the red colour disappears from the chloroform layer. Potassium iodate reacts with mercurous mercury to form mercuric mercury and iodine which gives the red colour to the chloroform layer. Further addition of potassium iodate in the presence of hydrochloric acid oxidizes iodine to iodine chloride. A blank is carried out by repeating the whole procedure without the pyrethrins.

Equation P.3 is used to calculate the pyrethrin I content.

$$\text{Pyrethrin I content (g/kg)} = (V_1 - V_2) \times \frac{7.125}{m} \quad \textbf{(P.3)}$$

where

V_1 = Volume of 0.01 M potassium iodate required for titration of pyrethrin

V_2 = Volume of 0.01 M potassium iodate required for titration of the blank solution

m = Mass of initial sample

For the estimation of pyrethrin II, the aqueous extract (kept aside after extracting with light petroleum as described above) is filtered into a Gooch crucible. The filtrate is concentrated to a volume of about 50 ml and transferred to a 500 ml separating funnel. This is then acidified with 10 ml hydrochloric acid and saturated with sodium chloride. Using 50 ml ether for the extraction, the aqueous layer is drawn off into a separating funnel; the extraction is then repeated with 50 ml ether. The aqueous phase is drained into a 100 ml conical flask and the ether extract is transferred to the first separating funnel. The aqueous phase is returned to the second separating funnel and extracted with two 35 ml portions of ether, combining all the ether extracts in the first separating funnel. The aqueous phase is discarded. The ether extract with three 10 ml portions of saturated sodium chloride solution is washed and filtered through a cotton plug into a 500 ml conical flask, followed by the cotton plug being washed with 10 ml ether. The ether extract is evaporated in a water bath, removing the last traces of vapour with a current of air. The residue is dried at 100°C for 10 min in an oven. Then 75 ml boiling water is added and the solution is filtered through a Whatman no. 42 filter paper, washing the flask and filter paper with five successive 20 ml portions of boiling water or until the washings are neutral to litmus. One or two drops of phenolphthalein indicator solution are added to the combined filtrate and titrated with 0.02 N sodium hydroxide. A blank is carried out by repeating the whole procedure on the aqueous liquid set aside in the blank determination for pyrethrin I.

Equation P.4 is used to calculate the pyrethrin II content based on 1 ml 0.02 N NaOH solution = 0.00374 g of pyrethrin.

$$\text{Pyrethrin II content (g/kg)} = (V_3 - V_4) \times \frac{4.675}{m} \quad \textbf{(P.4)}$$

where

V_3 = Volume of 0.02 N NaOH required for titration of pyrethrin II

V_4 = Volume of 0.02 N NaOH required for titration of the blank solution

m = Mass of initial sample

Pyrethroid insecticides

Pyrethroids are synthetic, stable analogues of an organic insecticide called pyrethrum, which constitutes chrysanthemic acid and pyrethrin acid (broken up into pyrethrin I and II, jasmolin I and II and cinerin I and II). Pyrethrum is naturally found in some chrysanthemum species (e.g. *Chrysanthemum cinerariaefolium*) and historically has been used in China, the Middle East and parts of Africa. These compounds are non-volatile and are very slightly soluble in water, but soluble in kerosene. Pyrethroids are not compatible with alkaline materials. Common formulations are the dry powder form (with a 25% pyrethrin I and II content).

This naturally occurring compound is a neuro-active insecticide that targets the sodium channel of the nerve membrane. Although low in mammalian toxicity and residual life, pyrethrum is photolabile, and hence cannot be applied to open fields. That is why compounds, based on the pyrethrum structure, were developed, and known as pyrethroids. Pyrethroids are used both in crop and in household situations.

Pyrethroids have a wide range of chemical structures (for instance, with or without the α-cyano group), stereo-chemical configurations and biological action. They are thus broadly distinguished on the basis of these criteria. There are 40 commonly known pyrethroids, most of them being esters of pyrethrin I (37 are esters and three are non-esters). The first synthetic pyrethroid to be developed was **allethrin**. The first generation pyrethroids include allethrin, **resmethrin** and **bioallethrin**. **Permethrin, cypermethrin, alpha-cypermethrin, deltamethrin, fenvalerate** and **esfenvalerate** were synthesized to be better versions of the first generation pyrethroids and these came to be known as second and third generation pyrethroids. Allethrin, kedethrin, **phenothrin, transfluthrin** and resmethrin mainly have household applications, but differ in toxicity.

R = CH_3 (Chrysanthemate)
CO_2CH_3 (Pyrethrate)
R_1 = $CH{=}CH_2$ (Pyrethrin)
CH_3 (Cinerin)
$CH_2{-}CH_3$ (Jasmolin)

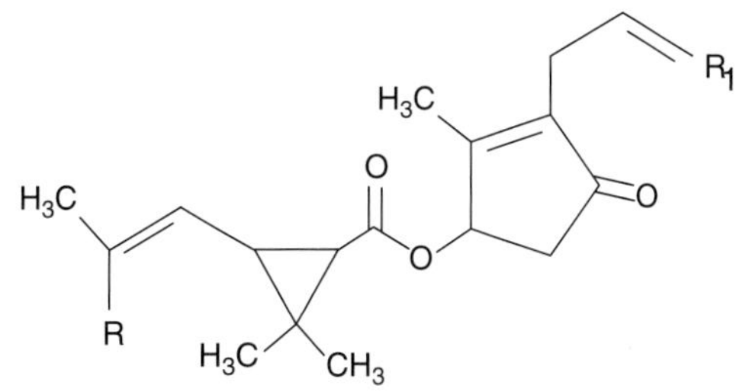

Pyrethroid

Synthetic pyrethroids (or just pyrethroids) are photostable and relatively non-volatile, broad-spectrum compounds with longer residual activity. They are effective on soil pests and mites, and relatively safe to mammals and humans. Pyrethroid toxicity is relatively lower than that of **organophosphorus** (OP) **insecticides** because pyrethroid esters are hydrolysed and oxidized leading to pyrethroid detoxification. In mammals, the central ester bond is cleaved leading to the decomposition of the pyrethroid. They are also cost-effective and hence popular.

Pyrethroid insecticides are commonly applied to leafy vegetables, cotton (e.g. lamda cyhalothrin), fruits and some crop trees against arthropods. Pyrethroid insecticides can be formulated as liquid concentrates, foliar sprays, granules (GR) and wettable powders (WP).

Neonicotinoid and pyrethroid insecticides have become popular alternatives to OP insecticides, and some directly substitute some OPs.

Pyrethroids, which are axonic poisons, are toxic to fish, other aquatic organisms and **beneficial organisms** such as bees and dragonflies, even at very small doses (in fish, it could be 2 ppt). For them to become lethal to target pests, pyrethroids are combined with **piperonyl butoxide**, an inhibitor of oxidase enzymes. The oral LD_{50} in rats can be as low as 200 mg/kg and in mice is 370 mg/kg. Pyrethroids can cause irritation to the skin. Exposure via inhalation may result in sneezing, head aches, facial flushing, asthmatic breathing and incoordination.

Pyrethrum

See **Pyrethroid insecticides**

Pyrethrum analysis

The material present in commercial pyrethrum flowers should be free from extraneous impurities or added modifying agents. The requirements are: (i) pyrethrin minimum purity of 190 g/kg; and (ii) the insoluble material in dichlorodifluoro methane is a maximum of 15 g/kg.

The pyrethrins are extracted from the sample with light petroleum and hydrolysed with alkali to chrysanthemum monocarboxylic acid and dicarboxylic acid. The acids are converted into water-soluble barium salts and filtered to remove insoluble materials and the acids are liberated by acidifying with sulfuric acid.

Chrysanthemum monocarboxylic acid which is 'pyrethrin I' is extracted with light petroleum and reacted with Denigés' reagent to form mercurous chloride, which is treated with potassium iodate. Chrysanthemum dicarboxylic acid, known as 'pyrethrin II' is extracted with ether from the water layer after removal of the monocarboxylic acid extraction and determined by titration with standard alkali.

To prepare 0.01 M potassium iodate solution, dissolve 2.14 g pure potassium iodate (dried at 105°C) in distilled water in a 1000 ml volumetric flask to make up to the mark. (1 ml of this solution = 0.0057g of pyrethrin I.)

Iodine monochloride solution is prepared by dissolving 10.00 g potassium iodide (KI) and 6.44 g potassium iodate in 75 ml concentrated HCl and 5 ml chloroform. It is adjusted to a faint iodine colour in the chloroform layer by adding dilute KI or iodate solution and kept in a dark cupboard and readjusted when necessary.

Denigés' reagent is prepared by mixing yellow mercuric oxide with 40 ml distilled water and slowly adding 20 ml concentrated sulfuric acid followed by another 40 ml distilled water until the solution is complete. The presence of Hg^+ is tested as follows: to 10 ml of the reagent, add a few drops of iodine monochloride solution, 30 ml concentrated HCl and 20 ml distilled water. After the flask is cooled, add 6 ml chloroform or carbon tetrachloride and titrate with the 0.01 M potassium iodate solution, shaking vigourously after each addition until the red colour in the chloroform layer disappears.

About 10 g of pyrethrum flowers are accurately weighed and the pyrethrins are extracted with light petroleum in a Soxhlet extractor for 7 h. When the extraction is complete, the solution is evaporated to about 40 ml, the flask stoppered and kept at 0°C for 2 h or preferably overnight.

The extract is filtered through a small cotton plug in a funnel previously wetted with light petroleum. The filtrate is collected in a conical flask. To this, 20 ml of chilled light petroleum is added and the flask shaken to dislodge the resinous material in the flask (using a rubber-tipped rod helps dislodge the material). The operation is repeated twice with 10 ml portions of chilled light petroleum.

The filtered solution is evaporated in a water bath after adding glass beads, making sure that the residue is heated long enough to remove the last traces of solvent. To the residue in the flask, 15–20 ml 0.5 N alcoholic sodium hydroxide is added and this is then refluxed gently for 45 min using a water-cooled condenser. The solution is transferred to a 600 ml beaker and distilled water is added to make the volume to 200 ml. The solution is boiled to reduce the volume to about 150 ml. The solution is transferred to a 250 ml standard flask and 1g diatomaceous silica filtering aid and 10 ml barium chloride solution (100g/l) are added without shaking. Distilled water is added to make up to the mark and the mixture is shaken well to mix thoroughly. The solution is filtered. To exactly 200 ml of the filtered solution, one drop of phenolphthalein indicator is added and neutralized with 4 N sulfuric acid and 1 ml acid excess. The solution is left aside overnight. The solution is used for pyrethrin analysis.

Alternately, to analyse an unknown sample containing pyrethrins, weigh 0.10–0.15 g accurately into a 250 ml conical flask and add 50 ml light petroleum and 1 g diatomaceous silica filtering aid. After mixing, the flask is stoppered and set aside at 0°C overnight. The solution is filtered through asbestos in a Gooch crucible into a 250 ml conical flask, washing with three 15 ml portions of cold light petroleum and combining the filtrate and washings.

The solution is evaporated in a water bath after adding glass beads, making sure that the residue is heated long enough to remove the last traces of solvent. To this 20–25 ml of 1 N alcoholic sodium hydroxide is added and refluxed for 45 min using a water-cooled reflux condenser. The solution is transferred to a 600 ml beaker and distilled water is added to make the solution volume 200 ml. The aqueous layer is reduced to 150 ml by boiling and transferred to a 500 ml separating funnel, with the aqueous layer drawn off into a 250 ml volumetric flask. The oily layer is washed with water and the washings added to the aqueous layer in the volumetric flask.

To this volumetric flask, 1 g diatomaceous silica filtering aid and 10 ml barium chloride solution (100 g/l) are added without shaking. The solution is made up to the mark with distilled water and mixed well. Finally, 200 ml of this solution is filtered, a drop of phenolphthalein indicator is added and neutralized with 4 N sulfuric acid, and 1 ml of acid excess is added. This solution is used for pyrethrin analysis. (See **Pyrethrin analysis**.)

Pyribenzoxim herbicide

Pyribenzoxim, $C_{32}H_{27}N_5O_8$ (benzophenone *O*-[2,6-bis(4,6-dimethoxypyrimidin-2-yloxy)benzoyl]oxime), is a pyrimidinyloxybenzoic broad-spectrum herbicide. It is a white solid with low solubility in water (3.5 mg/l).

Pyribenzoxim acts by inhibiting the biosynthesis of essential amino acids. It is formulated as an emulsifiable concentrate (EC) and marketed under the trade name Pyanchor. The herbicide is used in the post-emergence control of **barnyard grass**, black grass and polygonums in rice and wheat. It is applied at the rate of 30 g/ha.

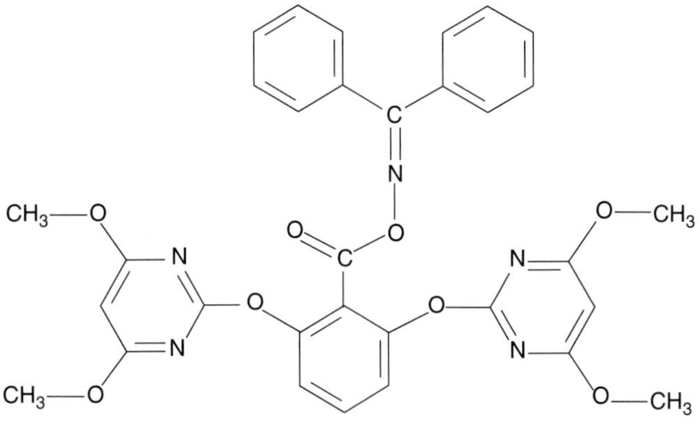

Pyribenzoxim

The compound is not carcinogenic and does not cause irritation to the skin and eye. Its toxicological studies show its low toxicity to mammals. Its acute oral LD_{50} value in rats is >5000 mg/kg. It is moderately toxic to fish (e.g. Japanese rice fish, *Orizias latipes*) and the LC_{50} (96 h) is >100 mg/l. The LD_{50} (24 h) in bees is observed to be >100 µg per bee.

Pyribenzoxim is relatively immobile and is found to rapidly degrade in soil. The half-life (DT_{50}) in soil under field conditions is 7 days (according to a study with sandy loam soil under flooded conditions), as a result of microbial and chemical reactions (mainly hydrolysis). The metabolites of the compound include 2,6-bis(4,6-dimethoxypyrimidin-2-yloxy) benzoic acid and 2-hydroxy-6-(4,6-dimethoxypyrimidin-2-yloxy)benzoic acid. The possibility of environmental contamination from the use of pyribenzoxim is expected to be low.

Pyributicarb herbicide

Pyributicarb, $C_{18}H_{22}N_2O_2S$ (*O*-3-*tert*-butylphenyl 6-methoxy-2-pyridyl (methyl) thiocarbamate), is a thiocarbamate herbicide. This compound is a colourless crystal, sparingly soluble in water and readily soluble in organic solvents. Pyributicarb is made by the reaction of 3-*tert*-butylphenol with 6-methoxy-2-*N*-methyl-*N*-thionyl chloride. When heated to decomposition it emits toxic fumes of NO_x and SO_x. It has a CAS Registry Number of 88678-67-5.

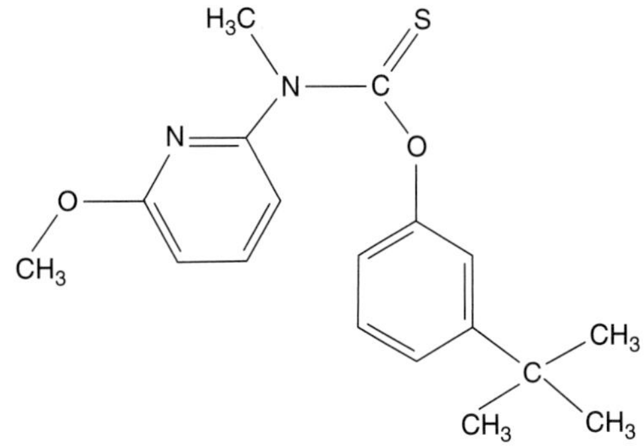

Pyributicarb

Pyributicarb is a systemic herbicide, well absorbed and translocated by the plant. It acts by inhibiting elongation of the roots and aerial parts. Being a thiocarbamate, pyributicarb particularly acts on annual grass weeds, especially post-germination, though it may be used at any time on perennial grasses in turf. Pyributicarb is available as a wettable powder (WP).

Herbicide pyributicarb is not acutely toxic by ingestion or inhalation. The oral LD_{50} in rats and mice is >5000 mg/kg. The 48 h LC_{50} in carp fish is 11 mg/l. Excessive exposure may cause irritation to the skin, eyes and the respiratory tract, and symptoms of cholinesterase inhibition such as headache, dizziness and nausea. The half-life of pyributicarb in paddy soil is up to 3 weeks.

Pyridaben acaricide

Pyridaben, $C_{19}H_{25}ClN_2OS$ (2-*tert*-butyl-5-(4-tert-butylbenzylthio)-4-chloropyridazin-3-(2H)-one), is an acaricide with insecticidal properties. It is a colourless crystalline substance, almost insoluble in water and soluble in acetone, benzene, xylene, ethanol and hexane. Pyridaben is produced by the reaction of mucochloric acid with 4-*tert*-butylmercaptotoluene, followed by cyclocondensation of the resulting malealdehydic acid derivative with *N*-*tert*-butylhydrazine. It has a CAS Registry Number of 96489-71-3.

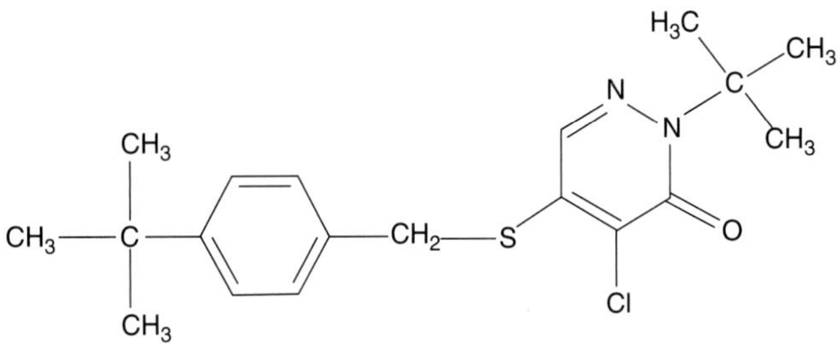

Pyridaben

Pyridaben is an inhibitor of mitochondrial electron transport at complex I. It is a very effective non-systemic insecticide and acaricide with a long residual activity. This compound is mainly active against the larval and nymph stages of mites, whiteflies and **leafhoppers** (families Acari, Aleyrodidae and Cicadellidae) that attack many fruit trees and vegetables.

The formulation types of pyridaben are emulsifiable concentrate (EC), soluble concentrate (SL) and wettable powder (WP). Trade names of pyridaben include Sanmite.

Pyridaben does not irritate the skin or eyes and is not considered to be mutagenic. The WHO classifies pyridaben as a Class III toxin (slightly hazardous) and the EPA rates it as moderately toxic (according to its acute toxicity rating of product labels). The acute oral LD_{50} of pyridaben among rats is between 820 and 1350 mg/kg. The acute dermal LD_{50} in rats is >2000 mg/kg.

Tests on rats and goats show that an orally administered dose of pyridaben is metabolized into many breakdown products which are then excreted mainly via the faeces. Photochemical degradation is seen in plants, while microbial degradation is seen in the environment, finally producing polar products and CO_2.

Pyridaben is registered for use in countries such as Australia, Canada, Vietnam and the USA. MRLs have been established under the Food and Drugs Act at 2 ppm in strawberries, 1.5 ppm in peaches/nectarines, 1 ppm in peppers, 0.75 ppm in pears, 0.5 ppm in apples, 0.3 ppm in grapes and 0.1 ppm in cucumbers.

Pyridaphenthion insecticide

Pyridaphenthion, $C_{14}H_{17}N_2O_4$ PS (*O*-(1,6-dihydro-6-oxo-1-phenylpyridazin-3-yl) *O,O*-diethyl phosphorothioate), is an organophosphorus insecticide and an acaricide. It is a pale yellow solid that is soluble in water (74 mg/kg), acetone, methanol and diethyl ether. When heated to decomposition, it emits very toxic fumes of SO_x, PO_x and NO_x. It has a CAS Registry Number of 119-12-0.

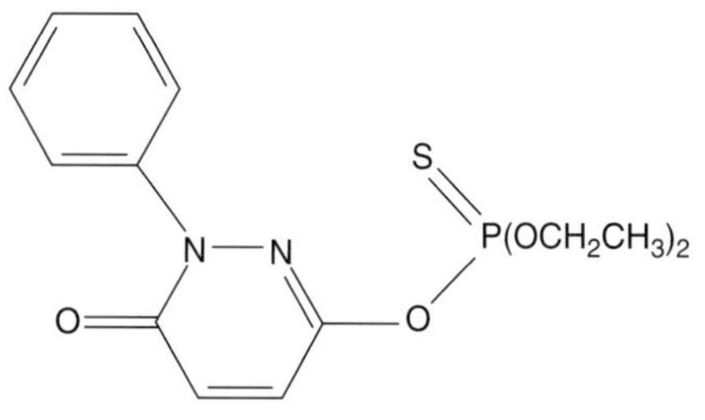

Pyridaphenthion

Pyridaphenthion, a cholinesterase inhibitor, is an insecticide and acaricide with contact action. It is used in the control of sucking and chewing insects (e.g. stem borers, leaf rollers, **leafhoppers**, plant hoppers, beetles, thrips, aphids, sawflies and onion flies), spider mites and red spiders on rice fields, vegetables, fruit and ornamentals.

Pyridaphenthion is formulated as emulsifiable concentrate (EC), wettable powder (WP) and dustable powder (DP). It is non-phytotoxic when used as directed. It is available in the market under the trade names of Ofunnack and Oreste.

The acute oral LD_{50} of pyridaphenthion in rats is between 769.4 mg/kg and 850 mg/kg, indicating toxicity, although it is not very toxic to rats and rabbits via the dermal route in normal use. It is highly toxic to honey bees (the LD_{50} is 0.08 μg per bee) and birds (the acute oral LD_{50} in Japanese quails is 68 mg/kg). Pyridafenthion is classified as a Class III toxin (slightly hazardous).

In mammals, metabolites such as *O*-ethyl-*O*-(3-oxo-2-phenyl-2H-pyridazine-6-yl) phosphorothioate and the corresponding phosphate are formed, along with some amounts of the parent compound. In rice, phenyl maleic hydrazide, *O,O*-diethyl thiophosphoric acid and phenyl maleic hydrazide are formed. The soil DT_{50} is 11–24 days.

The herbicide pyridaphenthion is registered in Vietnam and many African countries (e.g. Burkina Faso, Chad, Gambia, Guinea-Bissau, Madagascar, Mali, Mauritania, Niger and Senegal). It is not registered for use in the EU or the USA.

Pyridate herbicide

Pyridate, $C_{19}H_{23}ClN_2O_2S$ (6-chloro-3-phenylpyridazin-4-yl *S*-octyl thiocarbonate), is a pyridazine herbicide. It is a colourless crystalline substance, sparingly soluble in water, readily soluble in organic solvents and stable in neutral media. Its CAS Registry Number is 55512-33-9. Pyridate is made by the reaction of acetophenone with diethyloxalate, followed by reaction with hydrazine, phosphorus pentachloride and octylmercaptocarboxy chloride.

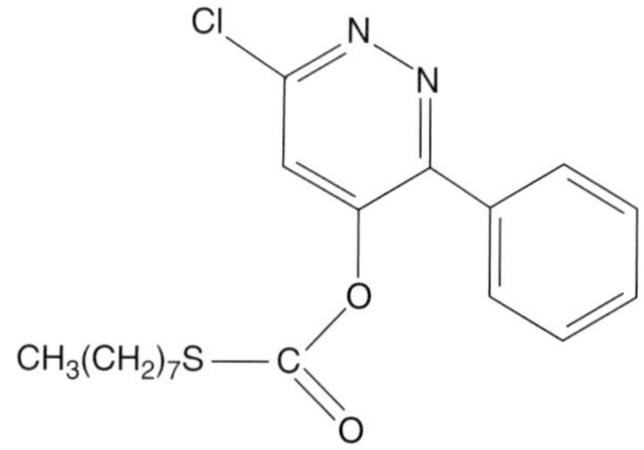

Pyridate

Pyridate, a post-emergence, selective herbicide, is used to control broadleaved weeds in groundnuts, maize, rice and other cereals. After being absorbed by the leaves, it inhibits photosynthesis by blocking photosystem II electron transport.

The formulations in which pyridate is available are emulsifiable concentrate (EC) and wettable powder (WP). It is available in the market under the trade name of Lentagran.

The acute oral and percutaneous LD_{50} of pyridate in rats and rabbits is >2000 mg/kg, respectively. Pyridate is moderately toxic to birds (the acute oral LD_{50} in bobwhite quails is from 1269 to 1502 mg/kg) and to *Daphnia magna* (the LC_{50} is 0.83 mg/l). It is practically non-toxic to honey bees and earthworms, but variously toxic to fish (with the LC_{50} for various species ranging from 1.2 to >100 mg/l). According to the WHO and EPA, pyridate is a slightly toxic compound. It has the potential to contaminate groundwater.

In animals, the principal metabolite of pyridate is 3-phenyl-4-hydroxy-6-chloropyridazine, which is further detoxified to form *O*- and *N*-glucuronides. In plants, pyridate is hydrolysed into the metabolite 3-phenyl-4-hydroxy-6-chloropyridazine. It degrades in soil (the DT_{50} is <2.8 days).

Pyridate is registered in Australia, many countries of the EU, South Africa and New Zealand, but it is not registered by the USA.

Pyridazinone herbicides

Pyridazinones are a herbicidal group which inhibits photosystem II electron transport in the photosynthesis pathway. **Pyridate** is one of the major constituents of this group. The other members include brompyrazon, chloridazon, flufenpyr, metflurazon, **norflurazon** and pydanon.

Pyrifenox fungicide

Pyrifenox, $C_{14}H_{12}Cl_2N_2O$ (2′,4′-dichloro-2-(3-pyridyl) acetophenone (*EZ*)-*O*-methyloxime), is a pyridine class fungicide. It is a viscous, sweet-odoured, yellowish liquid. Its solubility in water decreases with higher pH (e.g. 300 mg/l at pH 5.0 and 130 mg/l at pH 9). Pyrifenox is produced by condensation of an alkyl ester of 2,4-dichlorobenzoic acid with ethyl 3-pyridylacetate, followed by reaction with *O*-methylhydroxylamine. It is also soluble in *n*-hexane (210 g/l). It has a CAS Registry Number of 88283-41-4.

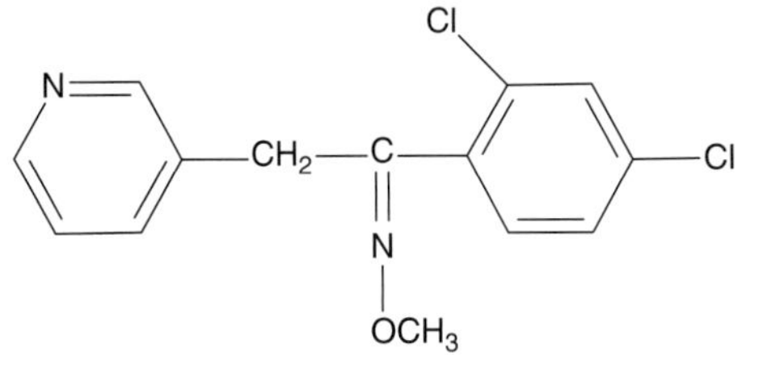

Pyrifenox

Pyrifenox is a systemic compound with curative and protective properties. It inhibits the biosynthesis of **ergosterol** after it is absorbed by the leaves and roots. It controls ascomycetes, basidiomycetes and deuteromycetes and specifically the causes of **powdery mildew**, leaf spot, blossom blight and **scab**, thus covering a wide range of crops including vines, groundnuts, sugarbeet, bananas, cucurbits, pome fruit and several vegetables.

Pyrifenox is commonly combined with **captan**, **mancozeb** or **folpet**. Common formulation types include wettable powder (WP) and emulsifiable concentrate (EC). The main trade names include Dorado, Corona and Podigrol.

Toxicologically, the acute oral LD_{50} values among rats and mice are 2912 mg/kg and >2000 mg/kg, respectively. It causes slight irritation to the skin. The LC_{50} (96 h) in fish species tested for toxicity was in the range of 6.6–7.1 mg/l. Pyrifenox is non-mutagenic and non-teratogenic. The LD_{50} in bird species tested was >2000 mg/kg, while the same in bees (48 h) is 59 μg per bee. The ADI in humans is 0.1 mg/kg/day. Applicators are advised against inhaling and accidentally touching the compound. Appropriate PPE such as waterproof gloves, long-sleeved shirt, long trousers and protective eyewear are recommended. Any accidental oral exposure should be treated with administration of medicinal charcoal with plenty of water. The WHO classifies pyrifenox as a Class III toxin (i.e. slightly hazardous).

In animals (rats), pyrifenox is rapidly absorbed and eliminated. In plants, the DT_{50} in groundnut leaves is 4 days and in apple leaves, it is 3 days, with the degradation taking place via hydrolysis and the oxime group being eliminated. In soils, pyrifenox is degraded without leaching.

The USA and many countries in the EU have not allowed use of this fungicide. South Africa has registered its use. The MRL established for tomatoes is 0.20 mg/kg while for sugarbeet, legumes and pulses it is 0.1 mg/kg.

Pyriftalid herbicide

Pyriftalid, $C_{15}H_{14}N_2O_4S$ ((*RS*)-7-(4,6-dimethoxypirimidin-2-ylthio)-3-methyl-2-benzofuran-1(*3H*)-one), is a pyrimidinyloxybenzoic herbicide. It is a white, odourless solid with low solubility in water (1.8 mg/l at 25°C). The CAS Registry Number is 135186-78-6.

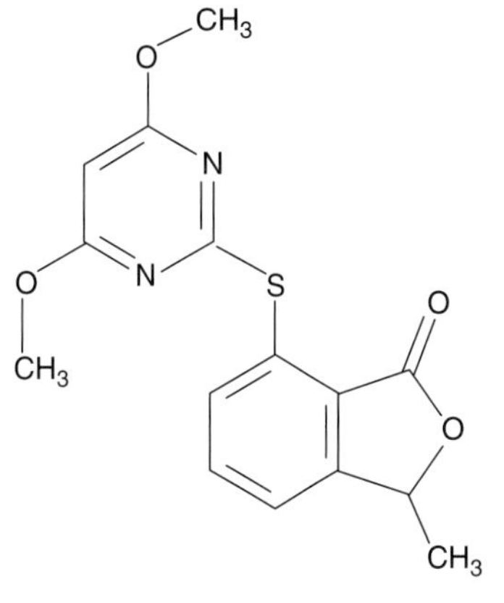

Pyriftalid

Pyriftalid acts by inhibiting the biosynthesis of amino acids. It is formulated as granules (GR), water-dispersible granules (WG) and soluble concentrate (SL) as mixtures and marketed under trade names such as Apiro Star and Apiro Max. On application under flooded conditions, pyriftalid is absorbed through the roots and the stem base of susceptible weeds. The herbicide is being developed for the control of grass weeds in rice. The rate of application is 100–300 g/ha.

The WHO classifies pyriftalid under Class III (slightly hazardous). The herbicide is found to be of low toxicity to mammals. Its acute oral LD_{50} value in rats is >5000 mg/kg, whereas the percutaneous LD_{50} is >2000 mg/kg. In birds (bobwhite quail), the oral LD_{50} is 1505 mg/kg. It is moderately toxic to rainbow trout (fish) and the LC_{50} (96 h) is 81 mg/l. The LD_{50} (contact) in bees is observed to be >100 μg per bee.

Plants do not show detectable residues of the herbicide, and in soil and the environment it degrades rapidly. The half-life (DT_{50}) of pyriftalid in soil is found to be 5–20 days.

Pyriminobac-methyl herbicide

Pyriminobac-methyl, $C_{17}H_{19}N_3O_6$ (methyl (*EZ*)-2-(4,6-dimethoxypyrimidin-2-yloxy)-6-(1-methoxyiminoethyl) benzoate), is a selective systemic herbicide belonging to the class pyrimidinyloxybenzoic. It is a white powder with low solubility in water (9.25 mg/l at 25°C). However, the solubility in organic solvents is as follows: methanol (14.6 g/l), *n*-hexane (0.456 g/l), acetone (126.0 g/l) and ethyl acetate (64.6 g/l). The compound is believed to be moderately volatile. The compound is stable to aqueous photolysis in natural water (i.e. glaciers, polar ice caps and groundwater) at pH 7 and the DT_{50} is 205 days. The CAS Registry Number is 136191-56-5 (see Pyriminobac-methyl at top of the next page).

Pyriminobac-methyl acts by inhibiting the biosynthesis of branched-chain amino acids. Pyriminobac-methyl is selectively developed for **barnyard grass** (*Echinocloa* spp.) in paddy. On early application post-emergence, it is absorbed by the foliage.

Formulated as granules (GR), soluble concentrate (SL) and wettable powder (WP) and marketed under trade names Hieclean and Topgun, the rate of application of the herbicide is 30–120 g/ha.

The WHO classifies pyriminobac-methyl under Class U (unlikely to be hazardous). The herbicide is found to be of low toxicity to mammals and moderately toxic to birds, fish and earthworms. Its acute oral LD_{50} value in rats is >5000 mg/kg, whereas the percutaneous LD_{50} is >2000 mg/kg. The NOEL (2 years) in

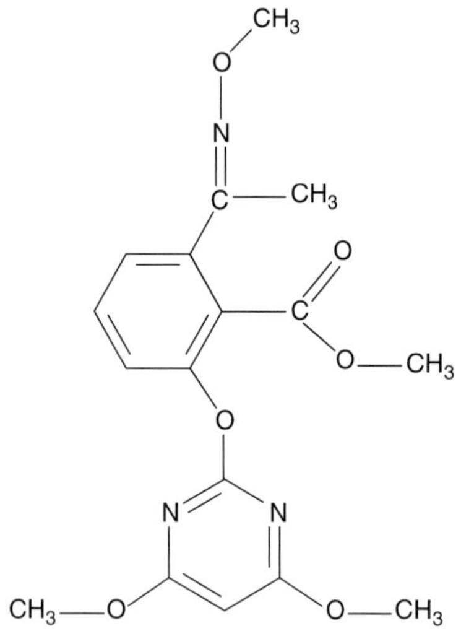

Pyriminobac-methyl

male and female rats is 0.9 and 1.2 mg/kg b.w./day, respectively. In bobwhite quails, the oral LD_{50} is >2000 mg/kg. The LC_{50} (96 h) in rainbow trout is 21.2 mg/l, while the LD_{50} in bees is >200 µg per bee. The NOEL in earthworms is >1000 mg/kg soil.

The herbicide forms many metabolites detectable in plants and animals. In animals almost all of the compound is excreted through urine and faeces and in plants, the residues are mostly found in the straw. The K_{OC} is 736 ml/g.

Pyriminobac-methyl is not approved for use in the EU. The herbicide is not registered in India.

Pyriproxyfen as an IGR

Pyriproxyfen, $C_{20}H_{19}NO_3$ (4-phenoxyphenyl(*RS*)-2-(2-pyridyloxy)propylether 2-[1-(4-phenoxyphenoxy)propan-2-yloxy] pyridine), is a pyridine-based insect growth regulator. It is a colourless crystalline substance, soluble in hexane, methanol and xylene. It has very low solubility in water, while it also does not adsorb to soil surfaces. It is moderately volatile. Its CAS Registry Number is 95737-68-1.

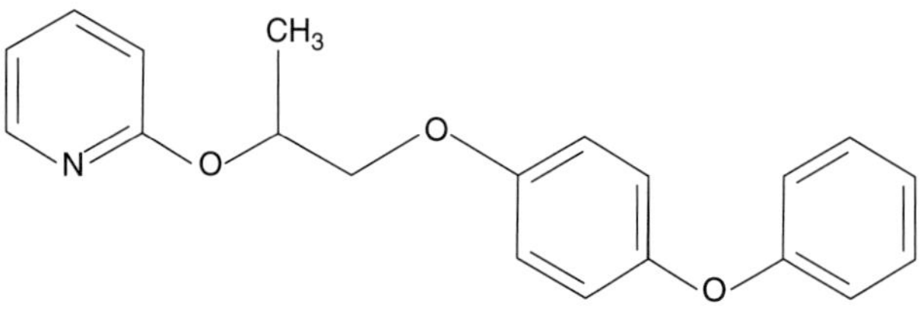

Pyriproxyfen

An IGR and an insecticide, pyriproxyfen is found to be effective against a variety of arthropods and is classified as a **juvenile hormone** analogue, preventing larvae from developing into adults, hence rendering them unable to reproduce. This IGR is used on cotton crops against whitefly. Fleas on pets and fire ants can also be controlled.

Pyriproxyfen is available in the market under the trade names of Nylar and Sumilarv. It is effective at 25–100 g a.i./ha. It is classified by the WHO as Class U (unlikely to be hazardous in normal use). The half-life of pyriproxyfen at a soil depth of up to 15 cm is 36 days.

Pyriproxyfen is registered in India, Australia, the USA, Canada, New Zealand, South Africa and many EU countries such as Denmark, Belgium, Hungary, Germany and France.

Pyrithiobac herbicide

See **Pyrithiobac-sodium herbicide**

Pyrithiobac-sodium herbicide

Pyrithiobac-sodium is a salt of pyrithiobac. **Pyrithiobac,** $C_{13}H_{11}ClN_2O_4S$ (2-chloro-6-(4, 6-dimethoxypyrimidin-2-ylthio) benzoic acid), is a pyrimidinylthiobenzoic acid herbicide. It has a CAS Registry Number of 123342-93-8.

The herbicide pyrithiobac inhibits branched-chain amino acid synthesis by inhibiting the enzyme acetolactate synthase (ALS). Pyrithiobac, a benzoate, controls a range of grass and broadleaved weeds in cotton.

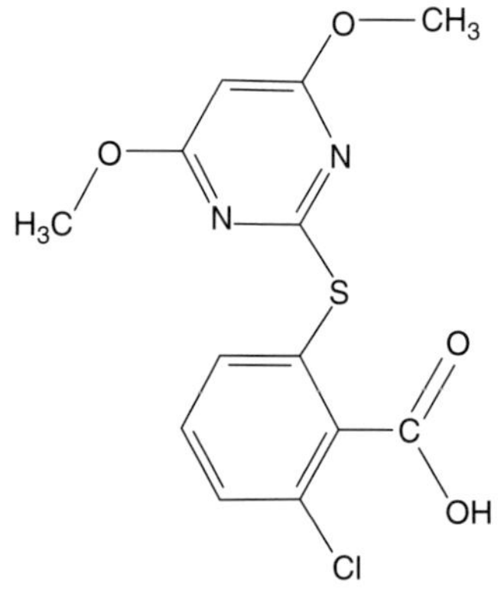

Pyrithiobac

Pyrithiobac-sodium, $C_{13}H_{10}ClN_2NaO_4S$ (sodium 2-chloro-6-(4, 6-dimethoxy pyrimidin-2-ylthio benzoate), is a pyrimidinylthiobenzoic acid herbicide. It is a yellowish-white powder, readily soluble in water and organic solvents. Pyrithiobac-sodium is made by diazotization of 2-amino-6-chlorobenzoic acid ethyl ester, followed by treatment with 4, 6 dimethoxy-2-mercaptopyrimidine, and hydrolysis of the ester to the acid. It has a CAS Registry Number of 123343-16-8. Pyrithiobac-sodium is also called KIH 2031.

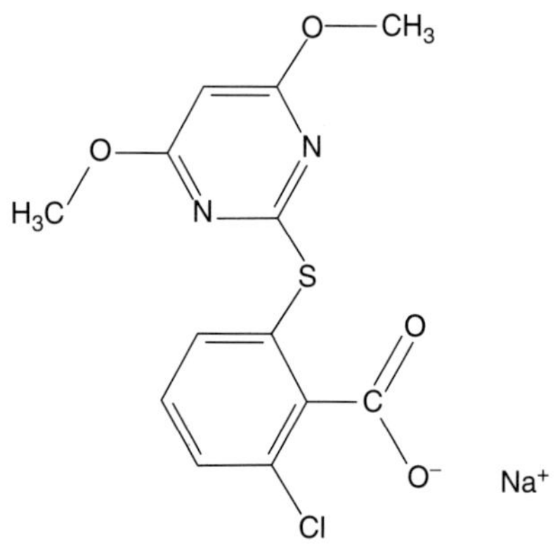

Pyrithiobac-sodium

Pyrithiobac-sodium is a pre- and post-emergence, selective herbicide which acts by inhibiting the plant enzyme ALS, thus blocking branched-chain amino acid biosynthesis. It controls a wide range of grasses and broadleaved weeds in cotton including morning glory, pigweed and cocklebur.

Pyrithiobac-sodium is usually supplied as a water-soluble powder (SP). The pre-emergence application rate is 35 g a.i./ha, while for the post-emergence stage, it is from 70 to 105 g a.i./ha. Its activity is enhanced by the addition of **adjuvants** (such as non-ionic surfactants and crop oil concentrates) and with the combination of carefully chosen herbicides like **bromoxynil** and **glyphosate**. It is available in the market under the name of Staple.

The acute oral LD_{50} of pyrithiobac-sodium in rats is >3300 mg/kg, while the acute percutaneous LD_{50} in rabbits is >2000 mg/kg. The 4 h inhalation LC_{50} in rats is >6.9 mg/l. The NOEL (2 years) in male rats is 58.7 mg/kg b.w./day, while in female rats, it is 278 mg/kg b.w./day. It is slightly toxic to birds (the acute oral LD_{50} in bob white quails is >2250 mg/kg b.w.). It is practically non-toxic to fish species and to honey bees. Exposure to skin and eyes may cause harm and hence appropriate PPE is recommended. The WHO classifies pyrithiobac-sodium as a Class U toxin (unlikely to be hazardous). It is likely to contaminate groundwater.

More than 90% of pyrithiobac-sodium, administered to rats, was excreted in urine and faeces within 48 h, with the major metabolite being the *O*-desmethyl derivative. In plants, major metabolites were the phenol formed by mono-demethylation and its glucose conjugate. Microbial and photochemical degradation play a major role in degradation in the environment, with a long half-life in silty soil (DT_{50} of 60 days). Pyrithiobac-sodium is degraded photochemically in water.

Pyrithiobac-sodium is registered in India, Australia and the USA.

Pyroxasulfone herbicide

Pyroxasulfone, $C_{12}H_{14}F_5N_3O_4S$ (3-[5-(difluoromethoxy)-1-methyl-3-(trifluoromethyl)pyrazol-4-ylmethylsulfonyl]-4,5-dihydro-5,5-dimethyl-1,2-oxazole), is a selective herbicide belonging to the pyrazole class. The CAS Registry Number is 447399-55-5.

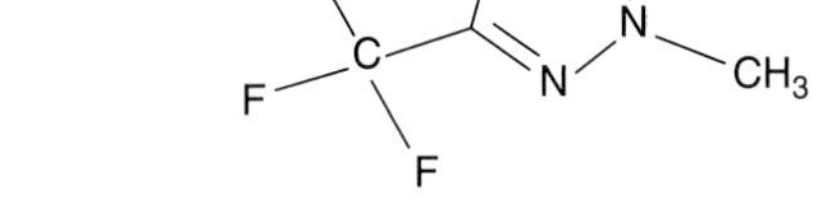

Pyroxasulfone

Pyroxasulfone is an acetolactate synthase (ALS) inhibitor. Although its mode of action is not yet fully understood, it is found to affect meristems and coleoptile development. The compound is found to have good residual action. Pyroxasulfone is used for the pre-emergence control of annual grasses and some broadleaved weeds in wheat, maize, soybean and other crops.

Pyroxasulfone is not registered by the EPA in the USA nor is it listed as a PAN Bad Actor chemical. It is not approved for use in the UK. The herbicide is not registered in India.

Pyroxsulam herbicide

Pyroxsulam, $C_{14}H_{13}F_3N_6O_5S$ (*N*-(5,7-dimethoxy[1,2,4]triazolo[1,5-a]pyrimidin-2-yl)-2-methoxy-4-(trifluoromethyl)-3-pyridinesulfonamide), belongs to class triazolopyrimidine. The compound is soluble in distilled water (0.0626 g/l at 20°C). It is soluble in acetone (2.79 g/l) and ethyl acetate (2.17 g/l) at 20°C. The herbicidal compound is stable to hydrolysis at pH 5, 7 and 9 at 25°C and the aqueous half-life is 3.2 days.

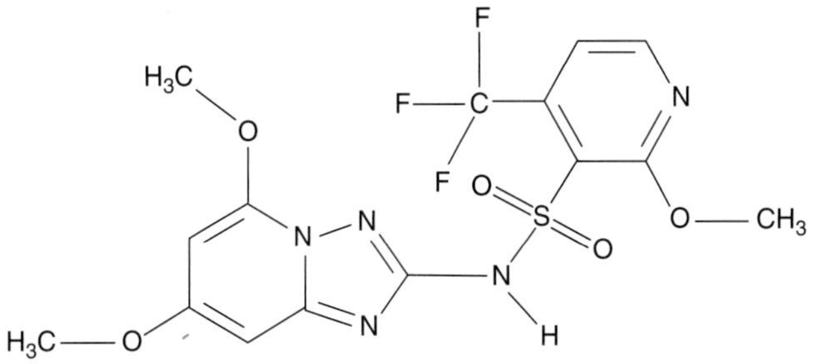

Pyroxsulam

Pyroxsulam is a broad-spectrum, systemic, post-emergence cereal herbicide that provides excellent control of tough grass weeds and a few broadleaved weeds. It inhibits the synthesis of branched-chain amino acids, leucine, isoleucine and valine. The herbicide is absorbed and translocated systemically through the phloem and xylem to meristematic tissues.

Pyroxsulam is used to control a host of weeds including hemp-nettle and species of mustard, geranium, speedwell and viola. It is commonly used on winter and spring wheat. In combination with **MCPA**, the control spectrum improves to include cleavers, chickweed and pigweed. Pyroxsulam is also known to provide short soil residual control of newly emerging annual weeds. The normal rate of application of the herbicide is 9–18.75 g/ha. Some known trade names include Admitt, Merit Gold and Poweflex.

The herbicide is found to be moderately toxic to mammals as well as to birds, fish and earthworms. Its acute oral as well as percutaneous LD_{50} value in rats is >2000 mg/kg. The **NOAEL** (2 years) in male mice is 100 mg/kg b.w. For birds (bobwhite quail and mallard duck) the oral LD_{50} is >2000 mg/kg. The LC_{50} (96 h) in rainbow trout is >87 mg/l, while the LD_{50} (48 h) in bees is >107 μg per bee. The LC_{50} (14 days) in earthworms is >10,000 mg/kg soil.

In animals and in plants, pyroxsulam is found to be the only residue of concern, although 2-desmethyl is also formed. The compound is rapidly absorbed and excreted via urine and faeces. In soil, the degradation is primarily by aerobic microbial means. The compound is weakly to moderately adsorbed on the soil particles and hence there is limited movement in the soil profile. In water, degradation is caused by photolysis and aerobic microbial degradation. The aerobic microbial DT_{50} is 18 days.

Pyroxsulam is registered by the EPA in the USA. The herbicide is not registered in India. It is registered in Canada, Australia, Finland, the UK, Germany and Hungary, among others.

Q

QoI fungicides

Quinone outside inhibitor (QoI) fungicides are fungicides that act at the quinol outer binding site of the cytochrome bc_1 complex. These fungicides act by inhibiting the electron transfer from cytochrome b to cytochrome c_1 in the mitochondrial membrane. QoI fungicides include three groups of fungicides, namely, the well-known **strobilurins** and the others represented by **fenamidone** and **famoxadone**. These are active against a broad range of ascomycetes, basidiomycetes, oomycetes and deuteromycetes on a wide range of crops such as cereals, vines, tomatoes, potatoes and cucurbits.

All these fungicides are in the same cross-resistance group and hence care must be taken to prevent resistance from developing in the target fungi. Measures that are effective at reducing the development of resistance include: (i) adhering to label instructions, with respect to dosage; (ii) maintaining a sufficient interval between two applications; and (iii) alternating these fungicides with others which have distinctly different modes of action.

QuEChERS

See **Quick, Easy, Cheap, Effective, Rugged and Safe method of residue analysis**

Quick, Easy, Cheap, Effective, Rugged and Safe method of residue analysis

In view of multiple pesticide residues present in crops, there has been a need for a method that would help determine the presence of more than one residue in a sample. A number of analytical methods can achieve high quality results for a wide range of pesticides. The **QuEChERS** method, which stands for Quick, Easy, Cheap, Effective, Rugged and Safe method for pesticide residue analysis, provides high quality results in a fast and easy way. This method has been validated in over 200 pesticides.

In this method, the sample is extracted with acetonitrile (MeCN) containing 1% acetic acid (HAc) and simultaneous liquid-liquid partitioning formed by adding anhydrous magnesium sulfate ($MgSO_4$) plus sodium acetate (NaAc) followed by a simple clean-up step called dispersive solid-phase extraction. The method is carried out by shaking a solution in a teflon centrifuge tube which contains 1 ml 1% HAc in MeCN + 0.49 anhydrous $MgSO_4$ + 0.1 g anhydrous NaAc per gram of sample. The tube is centrifuged and portions of the extract are treated with an amine sorbent and anhydrous $MgSO_4$ to remove fatty acids and water from the sample. Then the extract is centrifuged and transferred to auto sampler vials for concurrent analysis by GC/MS and liquid chromatography/tandem mass spectrometry depending on the instrumentation available.

The QuEChERS method produces high recovery rates for most pesticide types and gives very accurate results because an internal standard is used to correct for samples with different water content. Chlorinated solvents are not used in this method. There is very little equipment and space needed for this method, making it easy to carry out.

Quinalphos insecticide

Quinalphos, $C_{12}H_{15}N_2O_3PS$ (*O,O*-diethyl*O*-quinoxalin-2-yl phosphorothioate), is an organophosphorus (OP) insecticide. The insecticide is a colourless crystal with solubility in water being 17.8 mg/l. It is readily soluble in most organic solvents such as toluene, xylene, diethyl ether, ethyl acetate, acetone, acetonitrile, methanol and ethanol. It is stable for 2 weeks at room temperature and under ambient storage conditions (in non-polar organic solvents, in the presence of a stabilizing agent). Quinalphos is rather unstable in both alkaline and acidic conditions. The CAS Registry Number is 13593-09-8.

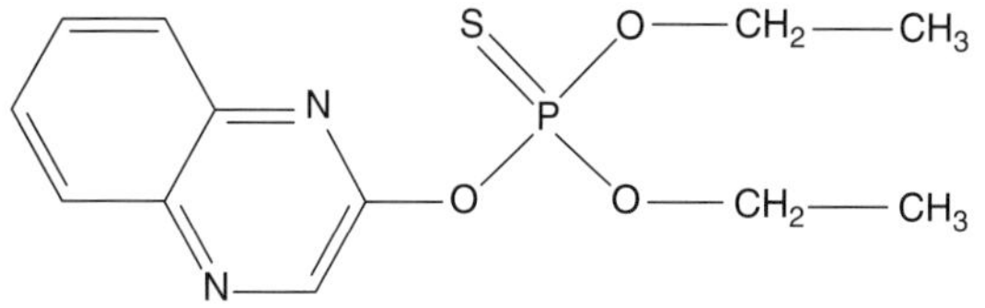

Quinalphos

Quinalphos, a cholinesterase inhibitor, is an insecticide and acaricide with contact action. It penetrates plant tissues through translaminar action. It is used in the control of many insect pests in the orders Lepidoptera, Coleoptera, Diptera and Hemiptera. Thus, it controls a wide range of pests such as caterpillars, aphids, bollworms, borers, **leafhoppers**, **mealybugs**, mites, plant hoppers and thrips. Crops that are protected are cotton, vegetables, groundnuts, soybeans, cocoa, tea, coffee, beet, vines and potatoes.

The insecticide quinalphos is formulated in forms such as emulsifiable concentrate (EC), granules (GR), wettable powder (WP), dustable powder (DP) and ULV liquid (UL). It is compatible with many pesticides, except those that are alkaline in nature. It is available under the trade names of Bayrusil and Ekalux.

In normal use, quinalphos does not irritate the skin and eyes of tested animals. But since it is an OP compound, exposure to quinalphos may cause blurred vision, nausea, headaches, confusion, excessive salivation and sweating. Quinalphos is highly toxic to bees. The insecticide may injure some varieties of fruit trees.

Mammals rapidly absorb and metabolize quinalphos into 2-hydroxyquinoxaline, which is excreted in the urine and bile within a short time. In plants too, while some of the original quinalphos is lost to evaporation, the principal metabolite is 2-hydroxyquinoxaline. In any case, the metabolite further breaks down to simpler substances. In soil, the insecticide is rapidly degraded under aerobic conditions, with the DT_{50} being 3 weeks.

India has registered the use of quinalphos, while the EU has banned its use.

Quinclorac herbicide

Quinolinecarboxylic acids have been developed into herbicides since it was found that substituted quinoline derivatives have plant activity. Among the many derivatives that were synthesized, quinclorac and **quinmerac** have been widely commercialized. Quinclorac has a Cl substitution in position 3 and quinmerac has a CH_3 substitution in position 3.

Quinclorac is a quinolinecarboxylic acid derivative, one of the chemical classes to which auxinic herbicides belong (the others being **phenoxy acid**, **benzoic acid** and **picolinic acid**). This herbicide is an auxin mimic. It is soluble in water at 72 mg/l.

Quinclorac is commonly used as a herbicide in rice, used in both the pre- and the post emergence stage. Along with controlling annual grasses in rice, it also protects against field bindweed and some other annual broadleaved weeds in crops such as winter wheat and canola.

Quinclorac, which is normally stable to heat and light, can undergo photolytic decomposition in paddy water giving 3,7-dichloroquinoine. It can break down because of the action of microorganisms too, especially in paddy waters. In tolerant plants, about half of quinclorac is metabolized, hence deactivating it. Mammals excrete the compound unchanged and the LD_{50} suggests that it is relatively safe to use. However, its solubility in water makes it a potential groundwater contaminant.

The soil DT_{50} of quinclorac ranges from 45 to over 200 days, indicating high persistence, due to adsorption to organic matter-rich soils. Also, **barnyard grass** and smooth crabgrass have developed resistance to quinclorac.

Quinclorac is registered in the USA, Canada, Vietnam and Australia.

Quinmerac herbicide

Quinmerac is a quinolinecarboxylic acid derivative, one of the chemical classes to which auxinic herbicides belong (the others being **phenoxy acid**, **benzoic acid** and **picolinic acid**). This herbicide is an auxin mimic. The CAS Registry Number is 90717-03-6.

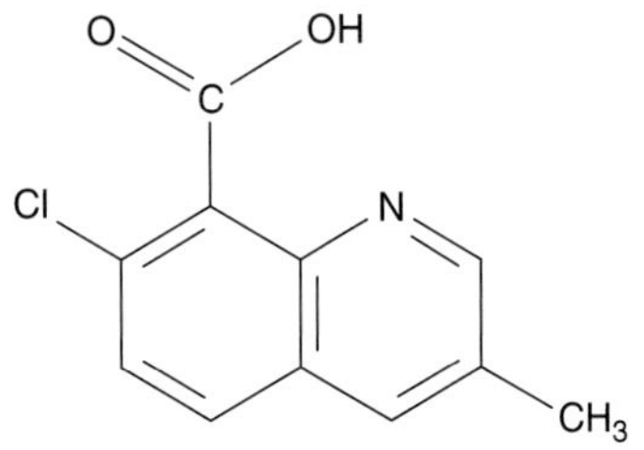

Quinmerac

Quinmerac is a pre-emergence herbicide useful to crops such as sugarbeet, cereals and canola and also to non-crop situations. It is absorbed by the roots in moist conditions.

In both plants and soils, oxidation of the 3-methyl group (forming alcohol) and hydroxylation at the 2-quinoline position take place. In mammals, the compound is excreted unchanged. The DT_{50} in the field ranges from 3 to 33 days. Persistence may be an issue, subject to soil moisture.

Quinoclamine herbicide

Quinoclamine, $C_{10}H_6ClNO_2$, (2-amino-3-chloro-1,4-naphthoquinone), is a naphthoquinone herbicide. It is a crystalline yellow solid, soluble in acetic acid, acetone, chlorobenzene and nitrobenzene at variable concentrations. The compound is stable up to 155°C in solution when stored in the dark.

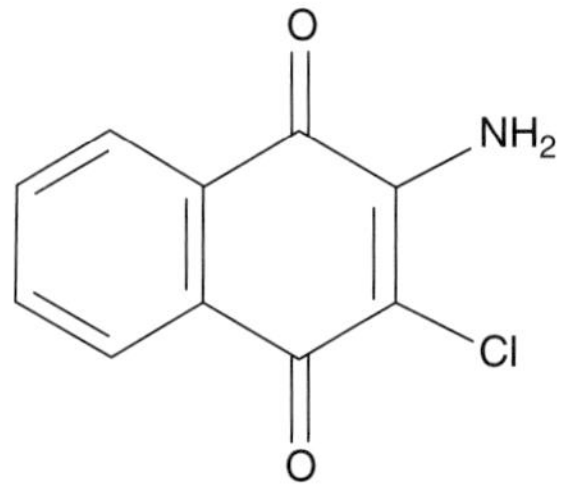

Quinoclamine

The herbicide quinoclamine is used for controlling algae, moss and liverwort in rice crops. High spray volumes (e.g. >935 l/ha) are required for quinoclamine to be effective. Quinoclamine is available as wettable powder (WP) and granule (GR) formulations and is available in the market under the trade name Mageton.

In toxicological studies, the LD_{50} of quinoclamine in rats is 1360 mg/kg. The acute percutaneous LD_{50} in rats is >500 mg/kg. The NOEL of quinoclamine (2 years) in rats is 5.7 mg/kg/day. Quinoclamine has an LC_{50} for carp of 0.7 mg/l. The WHO classifies quinoclamine as a Class III chemical (slightly hazardous).

Quinoxyfen fungicide

Quinoxyfen, $C_{15}H_8Cl_2FNO$ (5,7-dichloro-4-quinolyl 4-fluorophenyl ether), is a quinoline class of fungicide. It is an off-white flocculent solid, soluble in water (at 0.128 mg/l at pH 5), acetone (116 g/l) and dichloromethane (589 g/l). It degrades rapidly in light.

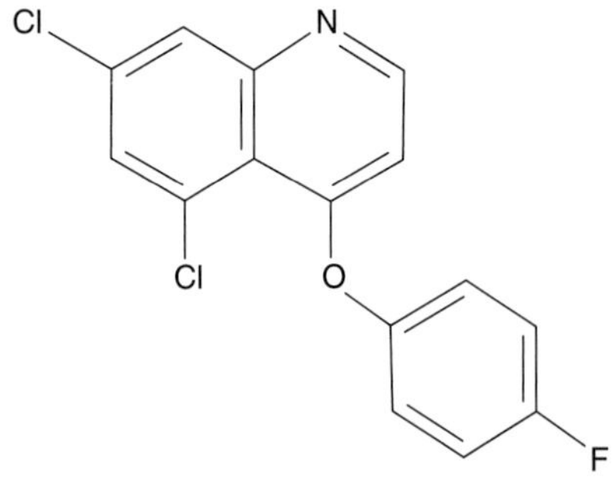

Quinoxyfen

Quinoxyfen is a systemic fungicide and has a preventive action. It is used to control the damage caused by **powdery mildew** on cereals and grapes. It has a multi-site mode of action that includes inhibition of cell signalling and the development of the appressorium.

Trade names of quinoxyfen include Fortress and Orka. The common application dose is 250 g/ha for cereals and 50–120 g/ha for grapes. This fungicide is sprayed as a soluble concentrate (SL).

The single acute oral LD_{50} dose of quinoxyfen on rats and mice is >5000 mg/kg., indicating relatively non-toxic, mammalian toxicity for the product. However, it is highly toxic to many fish species (the 96 h LC_{50} is in the range of 168–284 ppb), shrimps and oysters. Abiding by label instructions is therefore strongly recommended. Direct application to water or application in situations of possible runoff or drift is to be avoided. The WHO has classified quinoxyfen as a Class U toxin (unlikely to be a hazard).

Quinoxyfen is stable to hydrolytic degradation and is hence considered very persistent. The fungicide is also very persistent in soil (with a DT_{50} of 97 days) since it adsorbs to soil particles and organic matter. The half-life is reduced slightly with reduced pH.

Quinoxyfen is registered for use in South Africa, Australia, the USA and many EU members such as the UK, Italy, France, Hungary, Belgium, Germany and Poland.

Quintozene fungicide

Quintozene (pentachloronitrobenzene), with molecular formula $C_6Cl_{15}NO_2$, is a colourless needle-like aromatic hydrocarbon that belongs to the chlorophenyl or nitroaniline class. It is soluble in water (0.1 mg/l), toluene (1140 g/l), heptane and methanol and stable to heat and acids. It is incompatible with excessive sunlight and alkaline material.

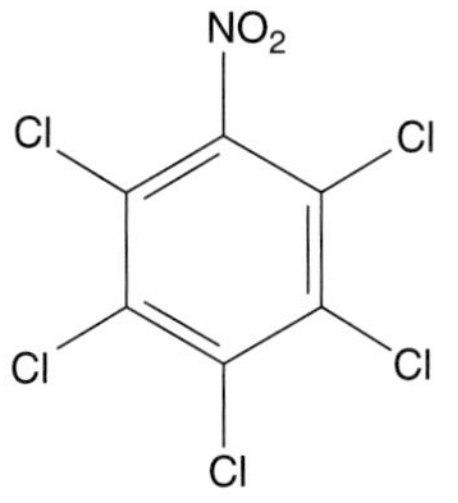

Quintozene

Quintozene, a contact fungicide suitable for soil application and seed treatment, works by lipid peroxidation. It helps control various diseases including damping-off, bulb rot, dollar spot, **smut** rot, white rot, club rot, **scab** and bunt. The crops that are protected include brassicas, lettuce, tomatoes, onions, garlic, potatoes, wheat, rice, maize, groundnuts, bananas and flowering plants. The pathogens that are effectively controlled are *Rhizoctonia solani*, *Sclerotium* sp., *Botrytis* sp. and many others. The common dosage is up to 1.5 kg/ha.

The fungicide quintozene is commonly formulated as a dustable powder (DP), emulsifiable concentrate (EC) or soluble concentrate (SL), granules (GR) and a seed treatment agent, and it is combined with agents like **metalaxyl** and ***Bacillus subtilis***. Some well-known trade names include Blocker and System 3. Quintozene also has nematicidal properties.

Toxicologically, the acute oral LD_{50} values of quintozene among rats, birds and bees are >5000 mg/kg, >2000 mg/kg and >100 µg per bee, respectively. The LC_{50} among fish species, shrimps and oysters ranges between 0.029 and 0.55 ppm, respectively. The 2 year (oncogenic) NOEL in rats is 1 mg/kg/day and the ADI for humans is 0.01 mg/kg b.w. Although quintozene is classified under Class U by the WHO (unlikely to be hazardous) and is not a PAN Bad Actor chemical, it is a possible carcinogen.

Quintozene is either eliminated in its unchanged form, or is changed into metabolites. In the case of mammals (rats and sheep), pentachloroaniline, **pentachlorophenol** and pentachlorobenzene are formed. In plants, along with pentachloroaniline, a series of chlorophenyl methyl sulfoxides and sulfones are produced. Quintozene persists in the soil for up to 10 months with gradual biodegradation and some amount of volatilization.

Quintozene is severely restricted or banned in Canada, the USA, New Zealand and Australia. It is banned throughout the EU.

Quizalofop

See **Quizalofop-ethyl herbicide**

Quizalofop-ethyl herbicide

Quizalofop-ethyl, $C_{19}H_{17}ClN_2O_4$ (ethyl (2*RS*)-2-[4-(6-chloroquinoxalin-2-yloxy)phenoxy]propionate), is a herbicide belonging to class aryloxyphenoxypropionate. It is a derivative of quizalofop ($C_{17}H_{13}ClN_2O_4$ with a CAS Registry Number of 76578-12-6). Quizalofop-ethyl, developed as a commercial herbicidal compound, is a colourless crystalline compound, soluble in water (0.3 mg/l at 20°C), in benzene (290 g/l) and ethanol (9 g/l). It is normally stable at 50°C for 90 days but is unstable in the presence of light. The (*R*)-isomer of quizalofop-ethyl is quizalofop-P-ethyl.

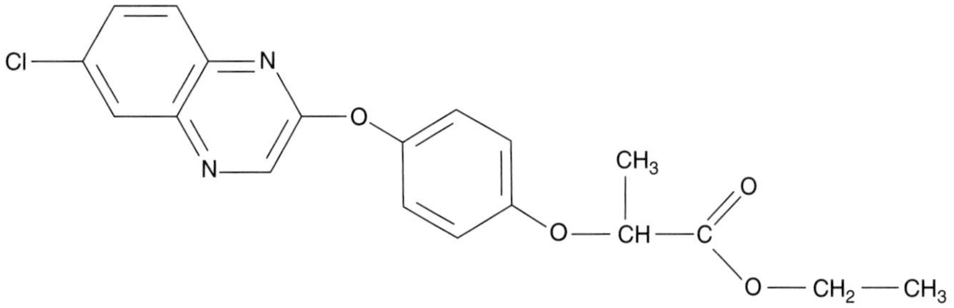

Quizalofop-ethyl

Quizalofop-ethyl is a systemic broad-spectrum herbicide that provides excellent post-emergence control of annual and perennial grass weeds in crops like potato, soybean, sugarbeet, groundnut, canola, sunflower, cotton, flax and vegetables. The systemic herbicide is absorbed by the leaves and translocated systemically through the phloem and xylem to the meristematic tissues.

Formulated as an emulsifiable concentrate (EC) as well as soluble concentrate (SL), quizalofop-ethyl inhibits synthesis of fatty acids, by inhibiting acetyl CoA carboxylase. The selected commercial product of quizalofop-ethyl includes Targa.

Classified by the WHO and the EPA as a slightly hazardous chemical (Class III), the herbicide is found to be moderately toxic to mammals and highly toxic to aquatic organisms. Its acute oral LD_{50} values in male and female rats are 1670 mg/kg and 1480 mg/kg, respectively, whereas the percutaneous LD_{50} in rats and mice is >5000 mg/kg. In birds such as bobwhite quail and mallard duck, the oral LD_{50} is >2000 mg/kg, however, the LC_{50} (96 h) in rainbow trout is 10.7 mg/l. The NOEL (104 weeks) in rats is 0.9 mg/kg/day. An ADI is standardized to be 0.01 mg/kg b.w./day, while the LD_{50} in bees is observed to be >50 µg per bee.

Quizalofop-ethyl in animals is metabolized rapidly and is completely excreted through the faeces. In plants, most of it is found unchanged on treated leaves. It rapidly degrades to quizalofop in soil and the DT_{50} is <1 day.

Quizalofop is not registered by the EPA in the USA. However, quizalofop-ethyl is registered as a herbicide in the USA and India although it is banned in the EU.

A mixture of two diastereoisomers with the same mode of action as quizalofop-P-ethyl is called quizalofop-P-tefuryl ((*RS*)-tetrahydrofurfuryl (*R*)-2-[4-(6-chloroquinoxalin-2-yloxy) phenoxy]propionate). Quizalofop-P-tefuryl ($C_{22}H_{21}ClN_2O_5$) is a white solid powder (with the technical grade compound being an orange waxy solid). The compound is stable for almost 2 years at 25°C. The herbicide is used to control annual grass weeds, specifically *Avena fatua* and *Alopecurus myosuroides* and perennial grasses such as *Sorghum halepense* and *Elymus repens* in crops like canola, sugar and fodder beet, potatoes, linseed, sunflower, peas, beans and other pulses.

Formulated as an emulsifiable concentrate (EC), the herbicidal application rate is normally between 20 and 100 g/ha. The product is available under names like Panarex, Pantera and Rango.

This herbicide is moderately toxic to animals with the oral LD_{50} in rats being 1012 mg/kg. Although less toxic to birds, the herbicidal compound is highly toxic to aquatic organisms. The LC_{50} in rainbow trout is 0.51 mg/l and in worms is >500 mg/kg dry soil. The oral LD_{50} value in bees is 16.8 µg per bee, indicating high toxicity.

Quizalofop-P-tefuryl in animals is extensively degraded by hydrolysis, hydroxylation and conjugation. In plants, it is rapidly hydrolysed to quizalofop.

While applying quizalofop compounds, care has to be taken to prevent drift. Adherence to label instructions is necessary, paying particular attention to the appropriate temperature and moisture range at which to carry out applications. In the case of oral exposure, vomiting is not to be induced.

Quizalofop-P-tefuryl is registered for use as a herbicide in India.

Quizalofop-P-ethyl

See **Quizalofop-ethyl herbicide**

Quizalofop-P-tefuryl herbicide

See **Quizalofop-ethyl herbicide**

R

Rabi season

See *Kharif* **crops**

Radiation

Radiation is the emission and uniform propagation of waves transmitting energy through a medium. Each wave has a certain wavelength and frequency. If this energy is capable of ionizing an atom, it is referred as ionizing radiation. Examples are alpha, beta, gamma rays and X-rays. Ionizing radiation interacts with chemical and physical matter in a way different from the other main type of radiation, namely, the non-ionizing radiation (radio waves, infra red and light rays). Non-ionizing radiation does not carry enough energy to dislodge an electron from the atom.

Radiation is used in medicine in the form of X-rays and radioactive isotopes (**radioisotopes**). It is used as a vitamin activator, sterilizing agent and polymerization initiator. It is also the basis of all types of spectroscopic analysis. Gamma rays and electron beams are used extensively in the field of agriculture. Two examples are sterilization of substrates and sterile insect technique.

Radioisotopes

An atom whose nucleus has excess energy that is either emitted outwards or within itself, is said to be radioactive. This energy is the result of an imbalance caused as a result of an unequal number of neutrons and protons in atoms of the same elements. Such atoms are referred to as isotopes. Such imbalance weakens the binding energy in the nucleus, leading to the nucleus giving out one or more neutrons to regain its stability. This process causes the atom to decay until its stable form is reached. Radioactive isotopes are called radioisotopes.

The energy emitted from radioisotopes can be harnessed for man's use in many ways, for example in medicine (e.g. bone imaging), archaeology (e.g. carbon dating) and agriculture (e.g. assessing chemical uptake in pest control and residue analysis). It should be noted that exposure to unmonitored radioactivity causes untold harm and abnormality to living things. Hence, it is critical to estimate when radioactive levels can return to the original level (the atom's stable/decayed form) as this helps decide if such a material is safe to use.

Radioisotopes are used to study the uptake, distribution and metabolism of a pesticide. The permitted and preferred radio-labelled isotopes include ^{14}C, ^{32}P and ^{35}S, depending on, for example, the chemical structure of the pesticide substance and the duration of the experiment.

There are regulations about entry into and use of fields that are subjected to radioactivity.

Recirculating sprayers

See **Application equipment – sprayers**

Red squill biorodenticide

Red squill is a compound extracted from the squill plant, *Urginea maritima*. The bulbs of this Mediterranean plant contain scilliroside ($C_{32}H_{44}O_{12}$), which is a cardiac glycoside and a known rodenticide. With a CAS Registry Number of 507-60-8, scilliroside is a yellow crystalline substance, very slightly soluble in water and soluble in alcohol and dioxane. It is sensitive to high temperatures and prolonged sunlight. The IUPAC name of scilliroside is 3β-(β-D-glucopyranosyloxy)-17β-(2-oxo-2*H*-pyran-5-yl)-14β-androst-4-ene-6β,8,14-triol 6-acetate (see Scilliroside at top of the next page).

Red squill has been used as an ingredient in rat poison for hundreds of years. Once ingested by rats, it creates neurotoxicity in the target. Since rats are unable to vomit, the substance remains in the body, causing cardiac complications, thus killing the animal. It is formulated as powders, granules or as a bait. Commercial products containing scilliroside include Silmine, Red Squill and Rat-o-Cide.

The LD_{50} in female rats is 1.4 mg/kg, while in males it is 5.3 mg/kg. It is 0.17 mg/kg in the house mouse but 37 mg/kg in field mice. It is considered to be a highly toxic chemical.

The scilliroside in red squill degrades with time and with exposure to air. Scillirosidin is the metabolite of scilliroside, occurring due to bacterial action. In other warm-blooded animals and man, scilliroside acts as an emetic and is thus thrown out of the system within 6 days, and hence is not so hazardous to them, unless it has been ingested in excessive quantities. In such an event, confusion and cardiac arrhythmias may occur.

Red squill is not produced any more as a rodenticide or is banned or restricted in many countries, including the UK.

Reductase-isomerase inhibiting fungicides

See **Sterol biosynthesis inhibitor fungicides (SBIs)**

Reference dose (RfD)

The **acceptable daily intake (ADI)** is often referred to as reference dose.

Repellent

A substance that causes an insect or animal to turn away from it or reject it as food is called a repellent. Repellents may be in the form of gases, liquids or solids and are commonly available as oily liquids or aerosol preparations.

Repellents are commonly used to keep away insects, household pests, birds, mice, rats and rabbits. Common targets of repellents are mosquitoes, black flies, biting midges, sand flies, stable flies, lice, ticks, cockroaches and ants.

Repellent activity is also seen in paints, electronic devices with sound waves and deodorants. Standard repellents include:

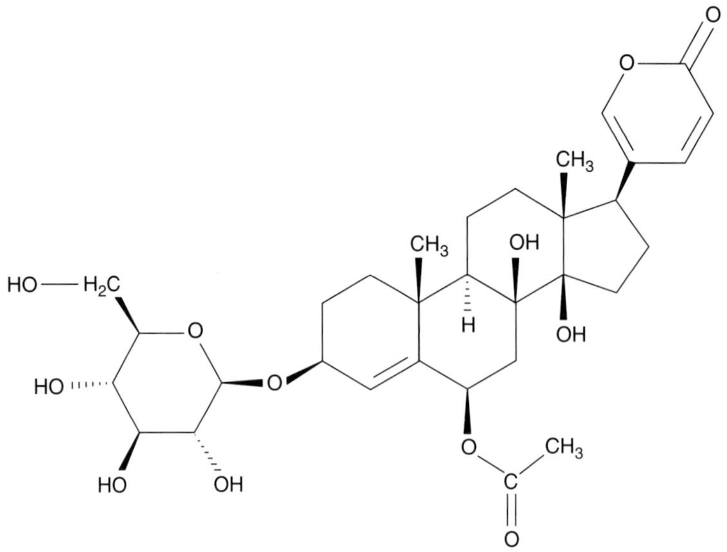

Scilliroside

(i) **citronella oil** and **dimethyl phthalate** (for mosquitoes and ticks); (ii) actidione, **thiuram disulfide** and trinitrobenzene (for rats); (iii) copper naphthenate and lime/sulfur mixtures (for rabbits); and (iv) naphthalene, quinine and sandalwood oil (for birds). Some other repellants include *N*,*N*-diethyl-*m*-toluamide **(DEET)**, **permethrin**, *N*,*N*-dipropylamino-3-cyclohexanecarboxamide,2-methyl-1-piperidyl-3-cyclohexenecarboxamide, 1-piperidyl-3-cyclohexenecarboxamide and 1-(hexahydor-1H-azepinyl) 2-methylcyclohexene carboxamide. Eucalyptus, nepetalactone (catnip oil) and **neem** oil are other natural repellents like citronella oil. Fermented eggs and sulfur compounds are some ingredients used to repel mammals.

Repellents used for insects include creams and skin patches and repellents on cloth. Repellent jackets are usually used by those who have to go into jungle tracts or forests. Repellents are chiefly based on taste. Bird repellents vary widely with the type of bird.

A repellent is registered for use on humans only when it passes toxicological tests on laboratory animals like rats and rabbits. The tests include exposure through oral, dermal and inhalation routes.

In chemistry, a repellent is a substance which does not mix or blend with another substance because of its physicochemical nature. All hydrophobic materials have water-repellent properties, largely due to differences in surface tension or electric charge. Oils, fats, waxes and certain types of plastics are examples of repellents.

Use of ultrasonic devices can help repel rodents, cockroaches, mosquitoes and other insects.

Repellent azamethiphos
See **Azamethiphos insecticide**

Repellent canola oil
See **Canola oil bioinsecticide**

Repellent cyhalothrin
See **Cyhalothrin insecticide**

Repellent jojoba oil
See **Jojoba oil fungicide**

Repellent methiocarb
See **Methiocarb insecticide**

Repellent ziram
See **Ziram fungicide**

Residual pesticides
Chemicals which do not break down into simpler forms, but persist in the soil are said to be residual. Herbicides and fungicides are commonly found to have residual effects.

The action of microbes helps in the degradation of pesticidal chemicals. Since microbes flourish in warm and moist weather such conditions encourage breakdown. Excessive rains can deprive these organisms of necessary oxygen, thus inhibiting breakdown.

When planning a pesticide programme, the possible risk to subsequent crops should be considered. However, if pest nuisance is predicted throughout the year, then persistence may be beneficial.

Soil residual life is the period of time that a pesticide is active in the soil. It is also called **soil persistence**. When the herbicidal effect is felt even after accomplishing its task of weed control, it is called the **herbicidal carryover effect**.

Resistance management
See **Pesticide resistance management**

Resistance to antibiotic streptomycin
See **Antibiotics**

Resmethrin insecticide
Resmethrin, $C_{22}H_{26}O_3$ (5-benzyl-3-furanylfuranylmethyl (1*RS*,3*RS*:1*R*,3*SR*)-2,2-dimethyl-3-(2-methylprop-1-enyl) cyclopanecarboxylate), is a synthetic pyrethroid, known mainly

for its control of mosquito populations in infested areas. It is a waxy, creamy white solid with a CAS Registry Number of 10453-86-8. It is produced by the reaction of chrysanthemic acid with 5-benzyl-3-furylmethyl alcohol (see Resmethrin at bottom of the page).

Being a neurotoxin, resmethrin interacts with the sodium channels in the peripheral nervous system and the CNS of target organisms. This leads to repetitive action in the nervous system membranes. The insecticide has applications in households, animal houses and public places, where it helps to control a range of flying and crawling insects such as mosquitoes, spiders, cockroaches, ticks and fleas. Other important applications include greenhouses, mushroom houses and industrial areas.

Ready-to-use formulations of resmethrin may be applied as aerial (truck-mounted) ULV sprays and shampoos. The chemical is available commercially as Crossfire, Respond and Scourge. Maximum single application rates range from 0.0076 kg/ha for mosquito control to 0.3 kg/ha for outdoor sites.

Resmethrin is only slightly toxic by ingestion and dermal routes in rats, mice, rabbits and birds. However, it is extremely toxic to fish and aquatic species with a 96 h LC_{50} of up to 0.001 mg/l. The insecticide is also highly toxic to bees. Exposure to resmethrin can cause irritation to eyes and skin, and a tingling and creeping feeling on the skin. Nausea, fluid in the lungs, muscle twitching and seizures may also occur. When handling or loading this chemical, it is necessary to wear clean overalls, gloves, a face shield and a face mask. A supply of fresh water is recommended on the site, to take care of any spillages. Resmethrin is classified as a Class III chemical according to the acute hazard criterion, but it is a development/reproductive toxin and a carcinogen. Hence it is a PAN Bad Actor chemical. All product labels should carry the signal word 'CAUTION'.

Resmethrin is observed to be moderately persistent in soil, with the half-life being in the range of 1–6 months depending on the soil type. Degradation products include benzaldehyde benzyl alcohol, chrysanthemic and phenylacetic acids. It binds tightly to the soil and is not very soluble in water, so the threat to groundwater is low except in runoff. In water too, degradation is quick with the half-life being about 1 month.

Resmethrin is registered for use in Canada and the USA but it is banned throughout the EU.

Restricted entry interval (REI)

The restricted entry interval (REI) of a pesticide is the duration, immediately after application, when entry into the treated area is risky and hence restricted. This interval helps to reduce incidents of cases of acute toxicity.

Needless to say, such time intervals vary according to: (i) the chemicals involved in pest control; (ii) the crop concerned; (iii) the method of application; and (iv) the climatic conditions at the time of application. The rule of thumb is 48–72 h for extremely hazardous active ingredients (i.e. Class Ia or Ib according to the WHO classification), the higher range being applicable in cases of hot and arid weather. The REI for moderately hazardous chemicals is 24 h and for slightly toxic chemicals it is 4–12 h.

There are some norms that need strict adherence. Product labels on each proprietary formulation/mixture need to display the REI, keeping in mind that if two or more chemicals are mixed to synthesize a formulated product, then the higher of the individual REIs would be considered the REI for the formulation. It is also necessary to display posters of warning signs (e.g. the skull and crossbones depiction) before and during the REI.

Handlers should be trained on using application equipment and chemicals, as well as on the proper disposal of chemicals.

Rhizosphere

The rhizosphere (the sphere of roots) is the narrow soil zone which is continuously influenced by plant roots. Living roots produce exudates that are constantly being introduced into the soil. This component influences microbial action that enriches the zone. Death of plant roots also add to the qualitative content of the rhizosphere. The rhizoplane is the actual surface of plant roots. The rhizosphere and rhizoplane form the actual **root–soil interface**. Plant growth and development are determined to a great extent by the soil environment in the root region.

The rhizosphere, created by plants, contains soluble compounds that include sugars, amino acids, organic acids and various other compounds. The quality and quantity varies with the physiological condition of the plant, plant species and several abiotic conditions such as temperature, water content and soil pH. In addition to soluble exudates, volatile compounds (e.g. alcohols, fatty acids, alkyl sulfides), plant mucilage (gel plus acidic polysaccharides) and lysates released from epidermal and root cap cells are important constituents of the rhizosphere environment.

The microbial groups found in the rhizosphere include bacteria, actinomycetes, fungi, nematodes, protozoa, algae and micro-arthropods. The rhizosphere–rhizoplane environment supports much higher densities of various microorganisms

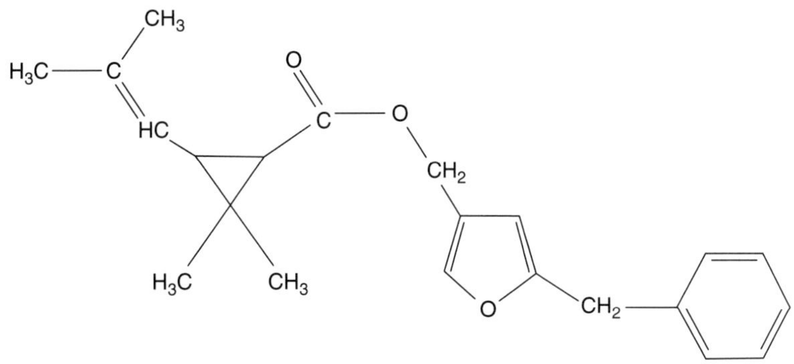

Resmethrin

than the surrounding soil. It is reflected in the RS value (rhizosphere to soil ratio). For bacteria, RS values may reach 20, since they are abundant in grooves between the epidermal cells and among root hairs, where mucilage is thicker. The ratio goes on decreasing for actinomycetes, fungi and algae in that order.

Microbial activities are often localized and are never uniform in soil, substrates and sediments. This non-uniformity may be due to differences in temperatures or substrate concentrations or any other factor. The total microbial biomass in the surface soil may range between 110 and 1940 kg C/ha in arable soil and 500–2100 kg C/ha in forest soil. The range may be wider in grassland soil. The forest litter layer has higher population densities than a root-free soil surface. The surface soil has larger microbial populations than deeper sediments. However, microbial populations are sometimes even seen in geological materials deep below the Earth's surface. Different scales are used to estimate the quantity and distribution of microorganisms even in the same geographical unit.

The ability of microorganisms to inhabit and colonize the rhizosphere–rhizoplane environment is known as the **colonization potential** or rhizosphere competence. Some of the microorganisms colonizing the rhizosphere may be harmful to plants because they spread disease (soil pathogens). However, whether pathogenic or beneficial, microorganisms play an important role in terrestrial ecosystems.

Rhizosphere competence

See **Rhizosphere**

Rice hull ash

Rice hull ash (RHA) is the result of burning rice hulls, a residue of the rice plant. RHA may be produced in an open field or by using it in a hull-fired stove.

Rice hull is a natural, renewable by-product of rice production. RHA is non-crystalline silica and hence, a natural silica fertilizer. RHA is incorporated into a seedbed, at the rate of up to 1 kg/m^2, before sowing the rice seed. The seedlings are observed to become disease resistant and capable of producing a better yield.

Rice leaf blast

See **Crop – rice pest control**

Rice stem borer

See **Crop – rice pest control**

Rice stem rot

See **Crop – rice pest control**

Rimsulfuron herbicide

Rimsulfuron, $C_{14}H_{17}N_5O_7S_2$ (1-(4,6-dimethoxypyrimidine-2-yl)-3-(3-ethylsulfonyl-2-pyridisulfonyl)urea), is a selective systemic herbicide belonging to the class sulfonylurea. The compound is a colourless, crystalline substance, soluble in water at <10 mg/l at 25°C. The hydrolysis DT_{50} is up to 4.6 days. The CAS Registry Number is 122931-48-0 (see Rimsulfuron at top of the next column).

Rimsulfuron, a post-emergence herbicide, provides effective control of most annual and perennial grass weeds and broadleaved weeds in crops such as maize, potato and tomato.

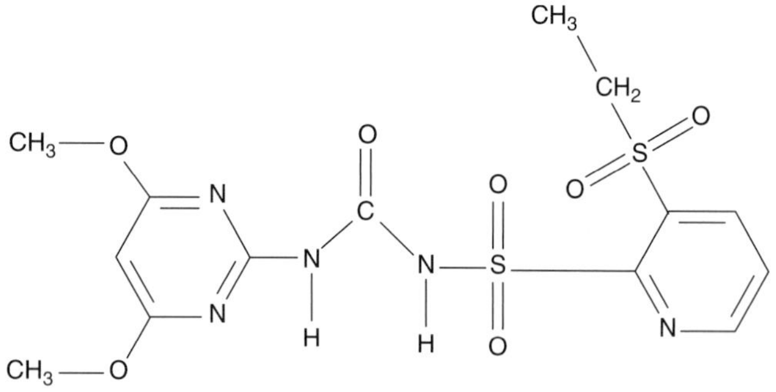

Rimsulfuron

In its mode of action, it inhibits biosynthesis of essential amino acids, leucine, isoleucine and valine, ultimately affecting cell division and plant growth. On application, the herbicide is translocated to meristematic tissues.

Formulated as water-dispersible granules (WG), the compound is marketed under trade names like Matrix, Titus and Solida.

The herbicide is found to be less toxic to mammals as well as to birds, fish and earthworms. Its acute oral LD_{50} value in rats is >5000 mg/kg. The NOEL (2 years) in male and female rats is 300 ppm and 3000 ppm, respectively. In birds, bobwhite quail and mallard duck, the oral LD_{50} is >2000 mg/kg. The LC_{50} (96 h) in rainbow trout is >390 mg/l; the LD_{50} (contact) in bees is observed to be >100 μg per bee. The LC_{50} (14 days) in earthworms is >1000 mg/kg.

Rimsulfuron is rapidly absorbed and excreted via urine and faeces in animals. It is rapidly degraded in soil by means of chemical pathways, the rate being influenced by acidic and alkaline soils. Microbial degradation plays a minor role. The DT_{50} in soil is 10–20 days. The residues in plants can be determined by HPLC.

Rimsulfuron is registered throughout the EU except in Malta.

Rodent control – miscellaneous

Besides the use of rodenticides, there are other measures that can be taken to protect places that are favoured by rodents.

Fumigation with the help of **hydrogen cyanide** (HCN), **sodium cyanide** (NaCN), **methyl bromide** (CH_3Br) or **chloropicrin** (CCl_3NO_2) is a common practice to control rats and rabbits. Fumigation becomes the ideal choice when buildings, installations and structures such as warehouses and hospitals are to be treated or when ships and aircraft are to be cleared of the menace of rodents. Fumigants are available as powders and pellets as well as gases.

Chemosterilants (e.g. α-chlorohydrin) are used as a means to control rodents, without killing them. Chemosterilants cause infertility in animals and have shown to be effective on Norway (brown) rats (*Rattus norvegicus*). α-Chlorohydrin [3-chloro-1, 2- propanediol] ingested at 150–160 mg/kg is the acute oral LD_{50} for rodent species. Commonly known trade names are Epibloc and Gametrics.

The development of a series of phenolic compounds which exist in nature show huge potential as repellents. Electronic devices produce ultrasonic sound waves and are occasionally used. The use of traps and the prevention of clutter in buildings and homes help keep rodents away. Predatory animals like cats and dogs may also be exploited.

Since rodenticides are extremely toxic substances, they must be kept away from unsuspecting people and their pets. Addition of extremely bad-tasting material (e.g. denatonium benzoate) to a rodenticide can avoid accidental consumption by humans (even 0.001% is sufficient).

Rodenticide bait selection and formulations

See **Bait selection and formulations of rodenticides**

Rodenticide sodium cyanide

See **Hydrogen cyanide rodenticide**

Rodenticides

Rodenticides are chemical or biological chemicals that kill or inhibit rats, mice, rabbits, squirrels, voles and hamsters. Rodents are well-known pests in agricultural and household situations, causing extensive damage to standing crops, stored grain and household goods, the world over. Rodents can transmit diseases either directly by bites or via infected urine and faeces, or indirectly via a vector such as a fly or mosquito. The range of diseases transmitted by rodents includes plague, salmonellosis, typhus and some haemorrhagic and relapsing fevers. The organisms causing these diseases include viruses, bacteria and protozoa.

Rodents are mammals that belong to the order Rodentia and phylum Chordata. These mammals are characterized by their constant gnawing, an action critical for preventing their incisors from growing too long.

Stored grains and processed foods are a major target of most rodents, besides standing crops in fields, greenhouses and kitchen gardens. In the pursuit of food, rodents also damage man-made material (e.g. fencing, nets, paper, plastic sheets, cloth, wood, straw) that may be used to protect foodstuffs. In addition, an adult rat produces some 30,000 droppings and 8 l of urine during its life time. The cost of cleaning soiled commodities exceeds the value of actual damage.

Rodents are the most damaging pests of growing crops such as cereals and plantation crops. Some pest species are *Sigmodon hispidus*, *Arvicanthis niloticus*, *Praomys natalensis*, *Bandicota bengalensis*, *Rattus tiomanicus*, *Rattus argentiventer*, *Rattus exulans*, *Microtus agrestis* and *Microtus arvalis*.

Rodenticides are applied as poisoned baits and placed where rodents are active. They are also used as dusts for tracking the route of rodents, such that they also ingest the poison dust. In some situations, rodenticides are used in liquid forms.

Zinc phosphide, **thallium sulfate** and **warfarin** are well-known chemical rodenticides. **Coumarin** derivatives and the ones based on 1,3-indandione are also used.

Sodium monofluoroacetate (Compound 1080) is a deadly rodenticide even to humans. Its molecular formula is $NaFC_2H_2O_2$. It is made by reacting ethylchloroacetate with potassium fluoride in an autoclave at 200°C. The resulting fluoroacetate is saponified with a methanolic solution of sodium hydroxide and the product, crystallized out.

Strychnine and **red squill** are botanical (herbal/organic/biological) rodenticides. Repeated non-lethal doses of strychnine can lead to bait shyness whereas powder and extracts of red squill kill Norway (brown) rats (*Rattus norvegicus*). A relatively new rodenticide, *N*-3-pyridylmethyl-*N*-p nitrophenyl urea, called Vacor, is comparatively harmless to humans.

The effective and safe use of rodenticides requires skill. Baits should be kept in the required quantities in such a way as to prevent their consumption by non-target animals. Non-chemical methods for rodent control may also be used, including the use of traps and ultrasonic devices. Predatory animals such as cats may be encouraged to visit areas ridden with rodents. Other methods to control rodents include changing the environment to make it less conducive to rodents (e.g. removal of clutter).

Rodents

See **Rodenticides**

Role of residue analysis

Residue analysis is the determination of the residues of farm chemicals present in fresh or processed food. Such an analysis helps in assessing the significance of plant-unique metabolites. Such metabolites can be identified under laboratory conditions or in a limited field plot. Specific methods using mass spectrometry, immuno-assays, gas chromatography and liquid chromatography have been developed for analysing residues of metabolites or the unchanged chemical. It is critical to confirm that the identified metabolite is not mutagenic, carcinogenic or in any other way, toxic or persistent.

Further action may have to be taken after determining the quantity of the metabolite. If the metabolite represents less than 10% of the total radiochemical residue and is 0.01 mg/kg or less, further action is not needed. If the concentration is more than 10%, its presence in the field has to be confirmed. A phased toxicological assessment is carried out. For this, the study of the metabolic pathway is beneficial. For example, the loss of the methyl-carbamyl group from *N*-methyl-carbamate insecticide immediately indicates that no anti-cholinesterase activity will be present in the metabolite. Even so, microbial mutation tests are recommended to confirm the assessments.

Short-term feeding studies of higher doses are commonly undertaken to assess: (i) how much residue is accumulated; (ii) how soon the compound will be cleared from the body; and (iii) in what form the compound will be cleared from the body (i.e. metabolites/unchanged).

Rollers

See **Weed management by mechanical methods**

Root–soil interface

See **Rhizosphere**

Rotary seed dresser

See **Seed protection devices**

Rotary tillers

See **Weed management by mechanical methods**

Rotary weeders

See **Weed management by mechanical methods**

Rotavators

Rotary tillers are also called rotavators. (See also **Weed management by mechanical methods**.)

Rotenone bioinsecticide

Rotenone, $C_{23}H_{22}O_6$((2*R*,6a*S*,12a*S*)-1,2,6,6a,12,12a-hexahydro-2-isopropenyl-8,9-dimethoxychromeno[3,4-*b*]furo[2,3-*h*]chromen-6-one), is a botanical insecticide. It is a colourless, odourless crystal that is sparingly soluble in water and freely soluble in acetone. Rotenone has a CAS Registry Number of 83-79-4. When exposed to air and light, rotenone undergoes hydroxylation at C-7 followed by dehydration to give dehydrorotenone, which inactivates rotenone. It is chemically prepared by condensation of the appropriate pyrrolidine amine derivative and tabu acid chloride.

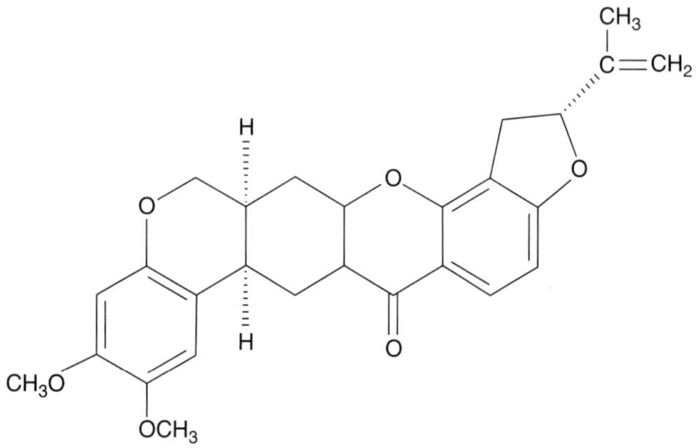

Rotenone

A broad-spectrum, non-systemic insecticide, acaricide and a fish poison (piscicide), rotenone occurs naturally in the roots and stems of plants such as the tuba plant (*Derris elliptica*) and lance pod (*Lonchocarpus utilis*), the rotenone content being between 5 and 11%. The insecticide works by inhibiting the respiratory enzyme NADH dehydrogenase, thus interfering with the electron transport chain in the mitochondria. With sufficient contact action, rotenone kills cucumber beetles, potato beetles and cabbage worms. It has a foliar application, and is preferably applied at the pre-harvest stage. Crops treated with rotenone include vegetables, legumes, fruit and nut trees and sugarcane. Rotenone may be applied to cats, dogs and cattle to protect against ticks, mites and flying and crawling insects.

Rotenone-containing products are formulated as dusts, dips, sprays, dispersible powders, liquids and emulsifiable concentrates. It is known under trade names such as Chem-fish and Tox-R.

The acute oral LD_{50} of rotenone in rats is up to 90 mg/kg. The acute oral LC_{50} in pheasants is 1680 mg/kg, while in rainbow trout, blue gill and catfish it ranges from 2.6 to 22.5 ppb.

Rotenone comes under the toxicity Class I (indicating it is a highly toxic chemical) when it is formulated as an emulsifiable concentrate (EC), with all product labels clearly displaying the signal word 'DANGER'. In other forms it is considered to be a Class III toxin, requiring the signal word 'CAUTION'. Dermal exposure causes redness, while inhalation causes a cough, sore throat, headache, breathing difficulty, nausea and unconsciousness. Accidental ingestion causes diarrhoea, cramps and vomiting. Appropriate PPE is recommended during handling of rotenone, making sure that first-aid measures include thorough washing of exposed areas and removal of contaminated clothes.

Rotenone does not leave residues since it rapidly biodegrades in warm conditions. However, studies have revealed development of Parkinson's disease in humans. The compound also breaks down in water. In both water and soil, the half-life is between 1 and 3 days. It is also sensitive to heat and higher temperatures. (In plants, many photodecomposition products form in the presence of light.)

The insecticide rotenone is registered in Australia, Cambodia, Canada and the USA.

Routes of entry and effects of exposure of a pesticide

Toxic chemicals need to be handled with extreme care to prevent exposure via accidental spillage or incorrect handling. The manifestation of **acute toxicity** depends on the amount of pesticide involved and the route of exposure. When body parts are exposed to a pesticide, the pesticide either remains localized on the skin or is absorbed into the body. There are four routes of exposure: (i) dermal; (ii) oral; (iii) inhalation; and (iv) by injection. Pesticides can thus come into contact with humans or wildlife in various different ways. Table R.1 below shows some symptoms of toxicity.

Table R.1. Symptoms of some toxic effects of pesticide exposure. (Table courtesy of Leena Achari and Shubhangi Borse.)

Effects	Symptoms
Dermatological effects	Rash, itching, redness, swelling
Respiratory effects	Irritation, cough, breathing problems
Gastrointestinal effects	Diarrhoea, vomiting, nausea
Reproductive effects	Infertility, foetal abnormalities
Neurological effects	Headache, confusion, coma, convulsions

Handling of pesticides for a considerable duration can lead to **chronic toxicity**. It may cause irreversible damage at the tissue and cell level, causing, for example, developmental abnormality, carcinogenicity and endocrine dysfunction. Frequent contact with pesticides can cause toxic effects in the body resulting in severe disorders or even death.

Ryania bioinsecticide

Ryania is a natural pesticide made from the ground stems of *Ryania speciosa*, a native plant of tropical South America. This plant stem contains the principal alkaloid ryanodine (0.2%), which is insecticidal in nature. The molecular formula of ryanodine is $C_{25}H_{35}NO_9$ and the CAS Registry Number 15662-33-6. It is more stable to light than **rotenone** or pyrethrum. It is soluble in water (100–10,000 mg/l), methanol, acetone, ether, chloroform and benzene.

Ryania, a contact poison, is used for the control of: (i) citrus thrips on citrus; (ii) codling moth on apples, pears and walnuts; and (iii) the European corn borer on maize. It is effective on pests that have developed resistance to OP pesticides.

Ryania is applied as a wettable powder (WP). It is available as trade names Rynanodine, Ryanicide and Ryan 50.

The oral LD_{50} for ryania in rats is between 750 and 1200 mg/kg, while the values are between 150 and 2500 mg/kg in guinea pigs, rabbits and dogs. Ryania is moderately toxic to birds (the LD_{50} being in the range of 1.78–13.3 mg/kg) and moderately toxic to fish (the 96 h LC_{50} being 3.2–18.5 mg/l).

Symptoms of poisoning with ryania include vomiting, weakness, diarrhoea, coma and death. It is considered to be a general use pesticide (GUP) according to the EPA and it is classified as a slightly toxic chemical. Products containing this pesticide are labelled with the signal word 'CAUTION'.

Because of the lack of persistence of ryanodine in the environment, residues do not end up in the food chain. The major degradation product of ryanodine is anhydroxyanodine.

S

Saflufenacil herbicide

Saflufenacil, $C_{17}H_{17}ClF_4N_4O_5S$ (*N′*-{2-chloro-4-fluoro-5-[1,2,3,6-tetrahydro-3-methyl-2,6-dioxo-4-(trifluoromethyl) pyrimidin-1-yl]benzoyl}-*N*-isopropyl-*N*-methylsulfamide), is a foliar contact and residual herbicide belonging to the pyrimidindione class. The compound is a white powder that at 20°C is soluble in water (2100 mg/l), acetone (275 g/l), methanol (29.8 g/l), toluene (2.3 g/l) and ethyl acetate (65.5 g/l). The compound is stable at both room temperature and at elevated temperatures in the presence of metal and metal ions. The CAS Registry Number is 372137-35-4.

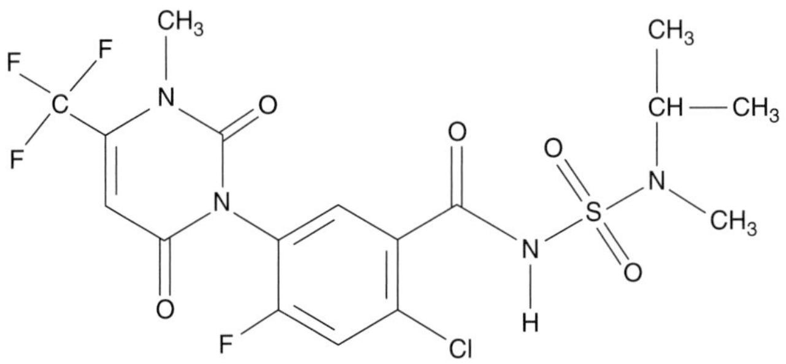

Saflufenacil

Saflufenacil is used alone or mixed with **glyphosate** for rapid control of many broadleaved weeds by pre-plant and pre-emergence applications to a wide range of food crops. Saflufenacil is used as a **desiccant** and/or defoliant on sunflowers. It is an inhibitor of protoporphyrinogen oxidase (**PROTOX**). On application, the herbicide is absorbed by leaves and roots and with translocation in the apoplast. It is available in a range of formulations including water-dispersible granules (WG), emulsifiable concentrates (EC) and suspension concentrates (SL). Trade names include Heat, Kixor, Sharpen, Optil, TreeVix and Integrity. The formulations are applied pre-emergence in maize and sorghum at 50–125 g/ha. It can be used to burn down foliage in soybeans, cereals, cotton and legumes.

The herbicide is found to be moderately toxic to mammals as well as to birds and fish. Its acute oral LD_{50} value in rats is >2000 mg/kg. The NOEL (18 months) for mice is 4.6 mg/kg b.w. The ADI is 0.046 mg/kg/day. In bobwhite quails, the oral LD_{50} is >2000 mg/kg b.w. The LC_{50} (96 h) in fish is >98 mg/l, while the contact LD_{50} in bees is observed to be 100 μg per bee. The EC_{50} (14 days) in earthworms is >1000 mg/kg soil. Given its exposure in normal use, it is considered to be a Class III toxin (slightly hazardous).

Saflufenacil is rapidly absorbed and excreted within 96 h in animals. In plants, the metabolism is rapid in non-susceptible plants and the metabolic route is via *N*-dealkylation of the sulfamide side chain and hydrolysis of the uracil ring. The half-life in soil is in the range of 1–10 weeks, the longer half-life being seen in acidic soils.

Saflufenacil is registered for use in Canada.

Sap stain

See **Oxine-copper fungicide**

S-bioallethrin insecticide

S-bioallethrin, $C_{19}H_{26}O_3$ ((*S*)-3-allyl-2-methyl-4-oxocyclopent-2-enyl(1*R*,3*R*)-2,-2-dimethyl-3-(2-methylprop-1-enyl)cyclopropanecarboxylate), is a pyrethroid ester insecticide, with a bioallethrin *S*-cyclopentyl isomer. It is an orange-yellow viscous liquid with a mild odour that is soluble in most organic solvents such as ethanol and acetone, but hardly soluble in water. *S*-bioallethrin is incompatible with strong acids and oxidizing agents and produces carbon dioxide and carbon monoxide upon decomposition. It is also unstable in the presence of light, heat, moisture and air. It may be stored in the original containers, at ambient temperatures. *S*-bioallethrin has a CAS Registry Number of 28434-00-6.

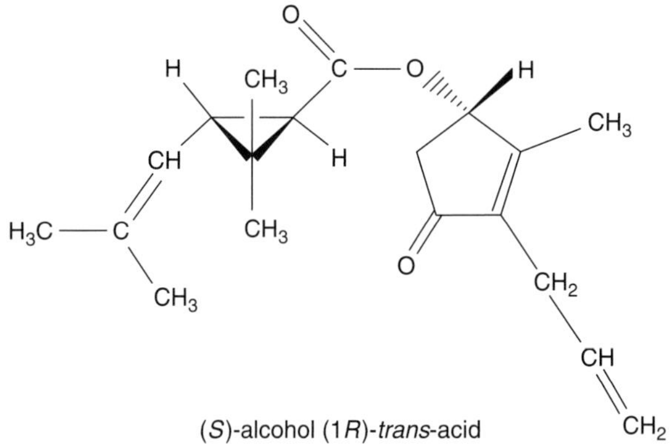

(*S*)-alcohol (1*R*)-*trans*-acid

S-bioallethrin

S-bioallethrin is effective against insects in the order Blattodea and families Culicidae and Muscidae (i.e. cockroaches, mosquitoes and houseflies). It is used in public health and domestic pest control situations. It acts on the nervous system of insects, disturbing the proper functioning of neurons by interacting with the sodium channel. This non-residual insecticide acts rapidly.

Most formulations of *S*-bioallethrin can be mixed with other pyrethroids in combination with **piperonyl butoxide**.

S-bioallethrin is classified as a moderately toxic compound by the WHO and EPA according to acute toxicity tests. The acute oral LD_{50} in male rats is 784 mg/kg. The acute contact LD_{50} in rabbits is 1545 mg/kg. It is only slightly toxic to birds. However, this insecticide is highly toxic to fish and many other aquatic organisms. The 48 h EC_{50} of *S*-bioallethrin in *Daphnia magna* is 0.016 mg/l. The 96 h LC_{50} of *S*-bioallethrin in rainbow trout is 0.0105 mg/l and in bluegill sunfish, it is 0.0332 mg/l. High levels of exposure to *S*-bioallethrin cause irritation to the eyes, irritability to sound, a prickling and tingling sensation on the skin, dizziness, nausea, headache, confusion and fluid in the lungs. The more toxic cyano-pyrethroids may cause seizures. It undergoes photolysis with its half-life in water being 19 h.

S-bioallethrin is registered in India, South Africa, the USA, Australia and many others countries.

Scab

See **Plant disease scab**

Scale insects

See **Pest – scale insect**

sec-Butylamine fungicide

sec-Butylamine, $C_4H_{11}N$ ((*RS*)-2-aminobutane), also known as 2-aminobutane, is a fumigant fungicide introduced in 1971 and classified as an aliphatic nitrogen compound. It is a stable, highly flammable, colourless liquid with an ammoniacal odour. It is miscible with water and forms water-soluble salts with acids. *sec*-Butylamine is stable at room temperature, but sensitive to light and incompatible with strong oxidizing agents and strong acids. It is produced by heating a mixture of *sec*-butylbromide potassium isocyanate and ethanol in DMF, followed by hydrolysis of the resulting urethane. The CAS Registry Number of *sec*-butylamine is 13952-84-6.

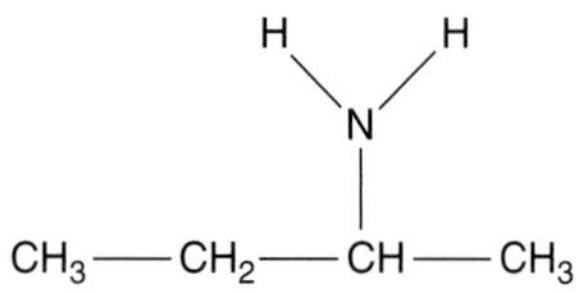

sec-Butylamine

The fungicidal fumigant is mainly used on potato (seed and ware) against skin spots, rots and moulds. This fumigant is also used during post-harvest treatment of potatoes, fruits and ornamentals. It controls diseases like stem rots in oranges.

sec-Butylamine is formulated as a soluble concentrate (SL) to be sprayed and used as a dip. It is classified as a Class II toxin (moderately hazardous) by the WHO. The acute oral LD_{50} toxicity in rats is recorded to be 380 mg/kg. The vapours are highly toxic on inhalation and the compound is irritating to eyes and skin; the percutaneous LD_{50} in rabbits is >2500 mg/kg. The **NOAEL** (2 years) in rats is recorded to be 2500 mg/kg diet. The acute oral LD_{50} toxicity in hens is observed at 250 mg/kg and the LC_{50} in fish (young bluegill sunfish) at >50 mg/l. This chemical can cause coughing, shortness of breath and throat irritation, with relief setting in only with fresh air and artificial respiration. It can cause redness, pain and severe burning on the skin. Removal of contaminated garments and thorough rinsing can provide relief. *sec*-Butylamine causes redness, pain, blurred vision and/or severe deep burns in the eyes. First aid includes rinsing the eyes with a lot of water and removing contact lenses (if easily possible) before seeking medical help.

sec-Butylamine has a recommended MRL of 50 mg/kg for dried citrus pulp and citrus molasses, 30 mg/kg for citrus fruits, and 0.5 mg/kg for citrus juice.

Secondary tillage

See **Weed management by mechanical methods**

Sedges

Sedges are monocotyledonous plants that are similar to grasses. Sedges belong to the family Cyperaceae and the order Poales. They are a diverse group of plants that are commonly seen in the tropics where the soils are not very fertile and may be poorly drained. Examples of important sedge varieties are **purple nutsedge**, yellow nutsedge and papyrus sedge.

Seed protection devices

Seeds have to be protected even before they are sown, and hence treating seeds is an important activity of crop protection. There are several methods and devices used to coat seeds with pesticides. Seeds may be coated with wettable powders (WP), suspension concentrates (SC) or emulsifiable concentrates (EC). **Stickers** such as vinyl pyrrolidone, other copolymers, Fuller's earth or clay help the pesticide to stick to the seeds. The key to seed treatment is the correct dosage: too much can injure the seed while too little may be ineffective.

Dust treaters for seeds are devices commonly used to mix pesticide dust with seeds in a mechanical mixing chamber. This equipment mixes the recommended seed quantity with the pesticide, with the help of a feed hopper and a vibrating feeder, both of which adjust the flow.

Slurry treaters for seeds coat seeds with wettable powder formulations to form a slurry. A slurry is a thick suspension of the chemical in a small volume of water or other solvent. Minimal moisture ensures neither germination nor deterioration of seeds. Slurry tanks are mechanical mixing chambers that have over 100 l capacity. Such slurries should be kept agitated before use to avoid separation of water and sedimentation of the pesticide.

Liquid or direct treaters are designed to apply a small amount of pesticide to a large quantity of seeds. Limited quantities of one or more moderately volatile liquids are mixed with the seeds in these treaters, such that each seed is not required to be entirely covered.

Mist formulations are also common among seed protection techniques. A recommended amount of liquid pesticide is delivered into the treater through a pressure tube or a moving disk, in the form of a mist, directly onto a determined dump of seed placed in the treater.

In order to maintain suspensions and emulsions, it is necessary to agitate it with power sprayers by a return flow pipe or by mechanical agitators.

A rotary seed dresser is a small drum fitted with a handle. The drum has a lid through which seeds and the pesticides may be introduced for treating and coating. Similarly, gravity seed dressers are fitted with perforated cones, such that the seeds in the drum get coated with pesticides and may be received through the small openings at the bottom. Such a treater is recommended for a volatile chemical coating. There are other machines which are fitted with a hopper and a pressure-gauged sprayer unit, which can control the pressure and the flow of spray. Grains dropped in at the top of the hopper are treated with the pesticide, gradually slide downwards and out through small slits at the bottom. These machines are called grain-treating machines.

In addition there are many other large motorized seed dressers and mixers with innovative fitments, suitable for seeds, grains and containers for grain storage.

For protecting seeds during storage, it is important that the containers and bags are devoid of pathogens. Protection of sacks and bags is achieved by passing them over a long conveyor belt during which a liquid formulation slowly laces the fabric. Impregnation is achieved by using a low-volume nozzle. The machine used for such an activity is called a sack-impregnating machine.

Seed treatment

See **Fungicide application methods**

Seed-dip method

See **Fungicide application methods**

Selective herbicides

See **Non-selective herbicides**

Selective pesticide

A selective pesticide is a major group of pesticides, classified on the basis of its penetration and selective action on the pest. Such a compound kills only certain species since it inhibits specific enzymes in those target species, not present in others.

For example, the herbicide 2,4-D is useful as a broadleaved weed killer where grassy plants are being grown and it will disrupt the growth of broadleaved plants such as clover or tomatoes. But it will be safe on grassy crops and ineffective on grass weeds.

Pesticides can be selective or non-selective (i.e. broad spectrum). Selective pesticides are generally preferred in IPM programmes, where they are combined with biological control agents.

Selectivity of a pesticide can be exploited in a large field where the target is predominantly present. However, application of selective pesticides has to be carried out at favourable temperatures with predictable drift.

Semiochemicals

Semiochemicals are chemicals produced by insects and higher organisms to communicate a message to others of their kind and elicit a sexual or any other response. Semiochemicals are volatile, fatty and acid-derived compounds.

Semiochemicals are produced within the body but are emitted externally to elicit a response. There are four groups of chemical messengers: (i) allomones; (ii) kairomones; (iii) pheromones; and (iv) synomones.

An allomone is a chemical produced and released by an individual of one species that affects the behaviour of a member of another species. Allomones are chemicals that aid the producer, so for instance, the leaves of the black walnut tree contain allelopathic **juglone**, or certain antibiotics produced by some organisms.

Kairomones induce a behavioural response in a species other than in the species producing the compound. Floral scents are kairomones to the honey bee.

Pheromones are produced by the members of one species and affect another member of the same species. Pheromones are present in bees, ants, wasps and termites. The chemical can signal, for example, alarm, aggregation or a call for mating. Pheromones persist for some time, since they are less volatile compounds.

Synomones are chemicals that modify the behaviour of different species to benefit both the emitting species and the receptive species.

Semiochemicals are isolated and synthesized for controlling insect pests and weeds. For example the Mediterranean fruit fly has been controlled to varying degrees using a terpenoid compound derived from the seed oil of a plant called wild celery or angel's herb (*Angelica archangelica*). These compounds also have a role as attractants or repellents.

Sesamin insecticide

See **Insecticides**

Sethoxydim herbicide

Sethoxydim, $C_{17}H_{29}NO_3S$ ((+)-(*EZ*)-2-(1-ethoxyiminobutyl) 5-[2-(ethylthio)propyl]-3-hydroxy- 2-cyclohexen-1-one), is a cyclohexanedione oxime herbicide. It is an oily, odourless liquid, soluble in water, methanol, hexane and acetone. Sethoxydim is made by the reaction of ethyl mercaptan with crotonal, followed by reaction with acetone, dimethylmalonate, hydrolysis reagents, butyryl chloride and hydroxy ethylamine. This herbicide stores well in normal storage conditions. Sethoxydim has a CAS Registry Number of 74051-80-2.

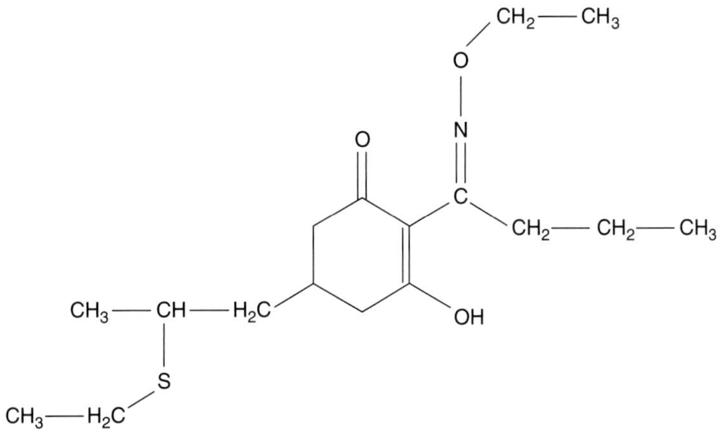

Sethoxydim

Sethoxydim, a selective, post-emergence herbicide, is translocated rapidly both acropetally and basipetally. It is an inhibitor of fatty acid synthesis in annual and perennial grasses in broadleaved crops such as soybean, lucerne, groundnut, maize, cotton and some fruits and vegetables. The grasses controlled include crabgrass, annual ryegrass, wild oats and quackgrass. It also has non-crop applications, for example in hedges and public gardens.

The herbicide sethoxydim is commonly formulated as an emulsifiable concentrate (EC). It is commercially available as Aljaden, Checkmate, Expand, Fervinal, Tritex-Extra and Vantage.

Sethoxydim is slightly toxic by ingestion as is seen in toxicity to rats (the oral LD_{50} of sethoxydim in rats is 2600–3100 mg/kg, while in birds it ranges from 2600 to 5000 mg/kg). In normal use, it is practically non-toxic by dermal absorption. Sethoxydim is slightly to moderately toxic to aquatic organisms. It is observed to be safe on **beneficial organisms**, in normal use. However, if inhaled, sethoxydim causes irritation to the nose and throat, and incoordination. In the case of accidental contact with eyes, it causes irritation and watering. According to the WHO and EPA, sethoxydim is classified as slightly hazardous (Class III). The acute percutaneous LD_{50} in rats and mice is >5000 mg/kg. It is non-toxic to earthworms.

When orally administered to rats, nearly all of the sethoxydim is eliminated within 48 h. In soil too, the herbicide does not persist for long (the DT_{50} is up to 25 days). It does not leach more than 10 cm into the soil. Photodegradation and microbial action are the main means of break down. It is stable to aqueous hydrolysis with a half-life of 40 days in neutral waters.

The herbicide sethoxydim is registered in Australia, Cambodia, Canada and the USA. According to the USDA Foreign Agricultural Service's Pesticide MRL database, the indicative MRL of sethoxydim for cottonseed in the USA is 5 ppm and in the EU it is 0.05 ppm.

Sex pheromones

See **Pheromones**; **Semiochemicals**

Siduron herbicide

Siduron, $C_{14}H_{20}N_2O$ (1-(2-methylcyclohexyl)-3-phenylurea), is a urea herbicide. The compound is a colourless, odourless crystalline substance, soluble in water at 18 mg/l, ethanol 160 g/kg and dimethylacetamide 367 g/kg (all at 25°C). It is stable in neutral aqueous solutions and up to 138°C. The compound decomposes slowly in acidic and alkaline media, with the DT_{50} being >30 days. The CAS Registry Number is 1982-49-6.

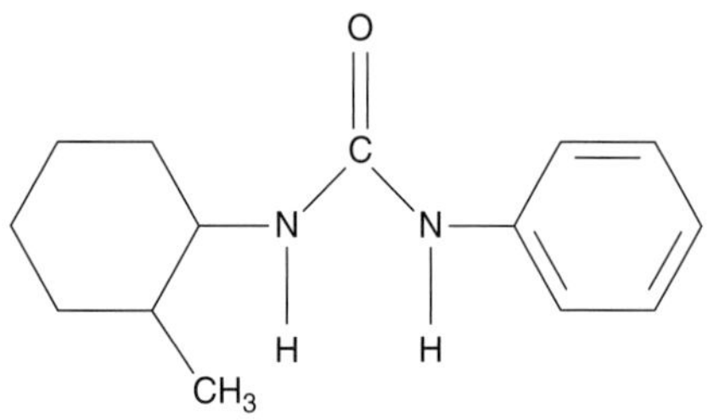

Siduron

Siduron is a selective pre-emergence herbicide. It selectively controls *Digitaria* spp. and certain germinating annual weed grasses such as crabgrass (smooth and hairy), foxtail and **barnyard grass**. A photosynthetic electron transport inhibitor at the photosystem II receptor site, siduron is absorbed by the roots and translocated in the xylem.

Formulated as a wettable powder (WP) for spray application, the product is marketed under trade names like Tupersan. Commercial products are also made (granular formulations) with siduron mixed with fertilizers and/or pesticides. It is often applied on bare soil at the pre-emergence stage (i.e. before the weeds sprout).

The WHO and EPA have classified siduron as a Class U (unlikely to be hazardous) and Class III compound (slightly toxic), respectively. The herbicide is found to be not particularly toxic to mammals and birds. Its acute oral LD_{50} in rats is >7500 mg/kg. In birds such as bobwhite quail and mallard duck, the oral LC_{50} is >10,000 ppm. The LC_{50} (48 h) in fish such as carp is 18 ppm and Japanese goldfish, 10–40 ppm. Siduron is known to cause irritation to the eyes, skin and the respiratory tract. The acute percutaneous LD_{50} in rabbits is >5500 mg/kg. The inhalation LC_{50} (4 h) in rats is >5.8 mg/l. The NOEL (2 years) in rats is 500 mg/kg diet.

In soil, siduron is degraded microbially leading to the formation of metabolites like methylcyclohexylurea, methylcyclohexylamine, **phenylurea** and aniline. Siduron resists leaching, remains in the top layer of soil, and degrades at a moderate rate. The half-life (DT_{50}) in soil is 4–5 months. Adsorption of siduron increases as the clay content and/or organic matter content of soil increases; clays of high cation exchange capacity adsorb more than those of low cation exchange capacity. The amount adsorbed on Keyport silt loam in equilibrium with 1 ppm in soil solution at 22.5°C is 2.5 ppm. A study of degradation of the herbicide in plants revealed no metabolite formation in barley.

Siduron is not registered in the EU.

Silafluofen insecticide

Silafluofen, $C_{25}H_{29}FO_2Si$ ((4-ethoxyphenyl) [3-(4-fluoro-3-phenoxyphenyl) propyl] (dimethyl) silane), is a pyrethroid insecticide. It is a liquid, soluble in water and organic solvents. It is stable for over 2 years. It has a CAS Registry Number of 105024-66-6.

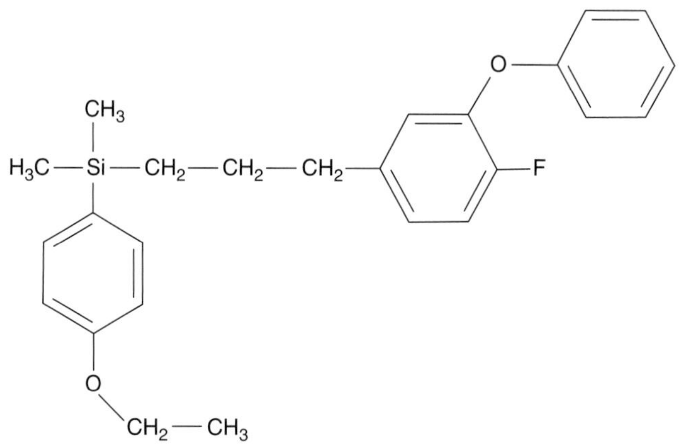

Silafluofen

Silafluofen is a broad-spectrum insecticide, which means it can be used on a wide variety of pests in the Coleoptera, Diptera, Heteroptera, Homoptera, Isoptera, Lepidoptera, **Orthoptera** and **Thysanoptera** (which includes flies, cockroaches, moths, beetles and thrips). It acts mainly as a stomach poison.

Silafluofen is formulated as an emulsifiable concentrate (EC), dustable powder (DP), wettable powder (WP) and emulsion, water in oil (EO). It is available in the market as Joker and Silatop. Its efficacy increases when mixed with **Bordeaux mixture**.

This insecticide is not phytotoxic when used as directed. Exposure to silafluofen causes irritation to the skin and eyes, headache, dizziness, numbness, excessive salivation, and many more symptoms of poisoning. This necessitates that handlers of this insecticide be well equipped with PPE.

Silafluofen is registered in India, Japan and Vietnam, but not in the USA. According to the Japan Food Chemical Research Foundation, the MRL fixed for rice is 0.5 ppm and for apple, 5 ppm.

Silicosis

See **Dust**

Silthiofam fungicide

Silthiofam, $C_{13}H_{21}NOSSi$ (*N*-allyl-4, 5-dimethyl-2-(trimethylsilyl) thiophene-3-carboxamide), is a thiophenecarboxamide fungicide. It is a white crystalline powder, soluble in water (39.9 mg/l), *n*-heptane (15.5 g/l) and ethyl acetate (>250 g/l), besides other solvents. Its degradation time changes with pH (i.e. the DT_{50} is 61 days at pH5 and 314 days at pH 9). The CAS Registry Number of silthiofam is 175217-20-6.

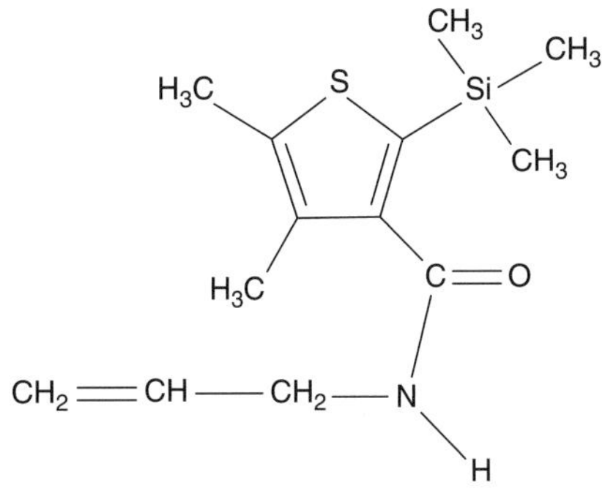

Silthiofam

Silthiofam, with long-lasting protectant properties, is used as an antifungal agent on wheat and barley seeds. It disrupts the energy-dependent metabolic activity in the mitochondria by inhibiting the production and transport of adenine nucleotide ATP. This affects growth.

Silthiofam is formulated as a flowable concentrate for seed treatment (FS). A common dose is 25 g/100 kg of seeds. Commercially available silthiofam formulations include Latitude.

Toxicologically, the acute oral LD_{50} in rats and most bird species is >5000 mg/kg. In rats, the compound's dermal LD_{50} is >5000 mg/kg b.w., while the inhalation LC_{50} is 2.8 mg/l. Silthiofam is neither irritating to the skin and eyes of rabbits, nor is it a skin sensitizer in guinea pigs. It is slightly toxic to birds (the acute LD_{50} in bobwhite quails is > 2250 mg/kg). The acute 96 h LC_{50} in fish species is between 11 and 14 mg/l. It is not toxic to earthworms and honey bees. In animals, silthiofam is absorbed, metabolized and eliminated (mainly in urine), in about 2 days. The ADI is 0.064 mg/kg b.w./day. The metabolic products of silthiofam include compounds of hydroxyl and carboxylic acid groups. In plants (wheat), extensive metabolism produces a carbohydrate conjugate and several minor metabolites. Silthiofam degrades in soil in about 1 month, the exact period depending on the type of soil.

Silthiofam is registered for use in many EU countries (e.g. the UK, Germany and Belgium) and also in South Africa, but not in the USA.

Simazine herbicide

See **Triazine herbicides**

Simeconazole fungicide

Simeconazole, $C_{14}H_{20}FN_3OSi$ ((*RS*)-2-(4-fluorophenyl)-1-(1*H*-1, 2,4-triazol-1-yl)-3-(trimethylsilyl) propan-2-ol), is a triazole fungicide. It is a white crystalline substance that is soluble in water (57.5 mg/l). The CAS Registry Number of simeconazole is 149508-90-7.

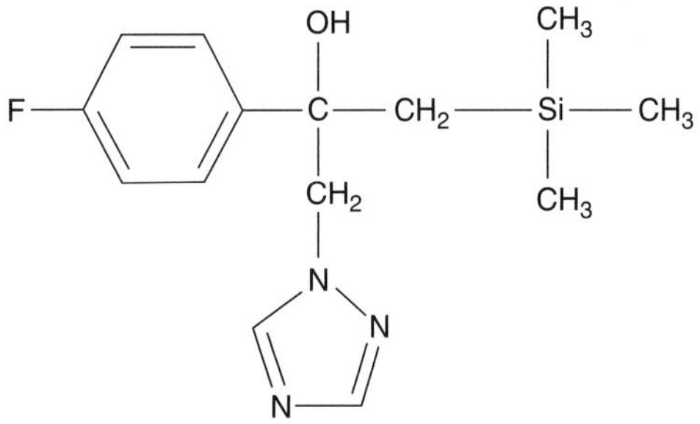

Simeconazole

Simeconazole is a systemic, broad-spectrum triazole fungicide and a sterol **demethylation inhibitor** (DMI). It works by inhibiting C-14-demethylase in sterol biosynthesis. It is absorbed by the roots, from where it translocates. Simeconazole is used for the control of sheath blight in rice (caused by *Thanatephorus cucumeris*) and other fungal disease conditions like **scab** (*Venturia inaequalis*), *Alternaria* blotch, rust, blossom blight, brown rot, blister blight, **powdery mildew** and eyespot. It is used to protect apple, rice, maize, strawberry, wheat, vegetables, soybean, turf and barley. It is very effective as a seed treatment.

Simeconazole is formulated as granules (GR) and wettable powder (WP). Common trade names are Mongarit and Sanlit. A dose of 4–10 g/100 kg of seeds can effectively control most common seed-borne infections.

The acute oral LD_{50} of simeconazole for male rats, female rats, male mice and female mice are 611 mg/kg, 682 mg/kg, 1178 mg/kg and 1018 mg/kg, respectively. The fungicide is non-irritating to the skin and eyes of rabbits and practically so in rats (the dermal LD_{50} being >5000 mg/kg b.w.) The ADI is 0.0085 mg/kg/day. The acute inhalation LC_{50} in rats is >75.17 mg/kg. Formulations with simeconazole need to have the signal word 'CAUTION' clearly displayed on the label, since it is a slightly hazardous chemical.

Simeconazole metabolizes to glucuronide, silanol and some other products (e.g. silanol, which is the simplest silicon alcohol (SiH_3OH) and is formed in rice).

Simeconazole is registered in Japan and many countries in Asia and the Middle East, but is not registered anywhere in the EU.

Simetryn herbicide

Simetryn, $C_8H_{15}N_5S$ (*N*2,*N*4-di-ethyl-6-methylthio-1,3,5-triazine-2,4-diamine), is a herbicide belonging to the triazine class. The compound is a white powder, soluble in water (400 mg/l), acetone (400 g/l) and methanol (380 g/l). It is stable in neutral, weakly acidic and weakly alkaline media. The compound is stable under normal storage conditions in the original container for at least 2 years.

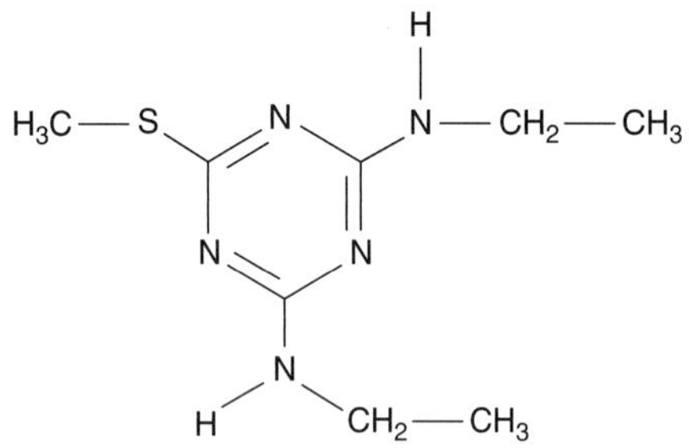

Simetryn

Simetryn is a selective herbicide that provides effective control of broadleaved weeds in rice. Its mode of action is that it inhibits photosynthetic electron transport at the photosystem II receptor site by binding to the plastoquinone-binding protein.

Formulated as water-dispersible granules (WG), the product is used in combination with **thiobencarb**. It is compatible with most herbicide formulations except very strong acid or alkaline formulations and it is marketed under the trade name Pownax M, among others.

Classified by the WHO and EPA as a slightly hazardous (Class III) compound, simetryn is found to cause no acute systemic toxicity unless large amounts have been ingested. The toxicity symptoms include irritation of the eyes, skin and respiratory tract. Its acute oral LD_{50} in rats is 750–1195 mg/kg, while its acute percutaneous LD_{50} is >3200 mg/kg. The NOEL (2 years) in rats is 1.2 mg/kg/day. The provisional value of ADI is 0.025 mg/kg b.w./day. The herbicidal compound is slightly to moderately toxic to fish (the LC_{50} for rainbow trout is 7 ppm). It is found to be non-toxic to bees.

Simetryn is degraded by two oxidative metabolic pathways and is excreted in animals via urine and faeces. Its DT_{50} in rats is 10 h.

Simetryn is neither registered in India nor included in the list of banned or restricted compounds by the **Central Insecticide Board and Registration Committee (CIBRC)**, India. It is also not registered in the EU or the USA.

Sinkers

See **Aquatic weeds**

Slimicide nabam

See **Nabam fungicide**

Slimicides

Any pesticide which is designed to kill slime-producing organisms is called a slimicide. Microorganisms produce extra-cellular polysaccharides that are referred to as slime. *Streptococcus mutans* and some strains of staphylococci are examples of slime-producing bacteria. Since these microorganisms flourish in humid, neutral conditions and in the temperature range of 38–43°C, the need for slimicides is felt in industrial cooling systems, wastewater treatment plants, paper and pulp manufacturing units, pools and tanks. Slime can clog up machines and produce a bad odour. The accumulation of slime is also associated with pulp, dirt, fibre and wet organic material. A slimicide can control most of the bad effects of slime.

Most slimicides are water soluble and are combined with surfactants such as ammonium salt or pyridium salt, for increased efficacy. These active ingredients (e.g. 4,5-dichloro-1,2-dithiol-3-one) are also bactericides and fungicides, besides being biocidal. Use of slimicides has always been restricted because of their toxicity to aquatic organisms.

Sodium hypochlorite, 4,5-dichloro-1,2-dithiol-3-one, sodium bromide, 2-(thiocyanomethylthio)benzothiazole (**TCMTB**) and methylenebisthiocyanate are examples of a few effective slimicides.

Slurry treaters for seeds

See **Seed protection devices**

Small motorized sprayers

See **Application equipment – hand sprayers**

S-metolachlor herbicide

S-metolachlor, $C_{15}H_{22}ClNO_2$ (a mixture of (α*RS*,1*S*)-2-chloro-6′-ethyl-N-(2-methoxy-1 methylethyl)aceto-o-toluidide and (α*RS*,1*R*)-2-chloro-6′-ethyl-N-(2-methoxy-1-methylethyl) aceto-o-toluidide), is a **chloroacetanilide** herbicide. It is a slightly volatile substance that is soluble in water at 480 mg/l and miscible in many organic solvents. It has a CAS Registry Number of 87392-12-9.

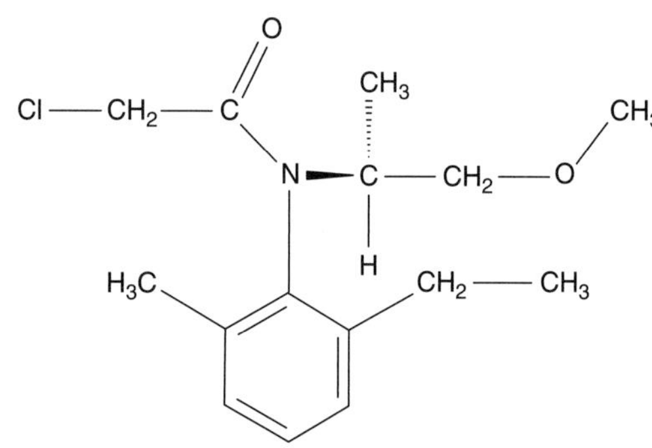

S-metolachlor

The herbicide *S*-metolachlor is a selective and pre-emergence herbicide, absorbed through the roots and shoots. It controls broadleaved weeds in soybeans, maize, groundnuts and cotton.

This herbicide is classified as a slightly toxic compound by the WHO, but it is listed as a PAN Bad Actor chemical since it is a known groundwater contaminant. It is believed to cause **chronic toxicity**, and affect the endocrine system. It is observed to lead to cancerous growth in tested species. The ADI of *S*-metolachlor in dogs is 0.1 mg/kg b.w./day. The acute oral LD_{50} in rats is 2600 mg/kg and the acute dermal LD_{50} in rabbits is >2000 mg/kg. It does not cause irritation to the eyes or the skin. It is moderately toxic to birds (the acute oral LD_{50} being 2510 mg/kg). It is extremely toxic to bees.

S-metolachlor is considered a mobile compound and it degrades into metabolites that may cause harm to the environment (including metolachlor ethylsulfonic acid and ethane sulfonic acid). *S*-metolachlor is considered to be moderately persistent in different soil types and has an average field dissipation half-life of 114 days.

According to the EU Pesticide Database the MRL values of *S*-metolachlor in cereals, fruit and vegetables is 0.05 mg/kg.

S-metolachlor is registered in Australia, South Africa, Tanzania and the USA. (See also **Metolachlor herbicide**.)

Smoke

A colloidal or microscopic dispersion of a solid in gas is known as smoke. Smoke may result from combustion of coal, wood, metals or chemicals. The particles that are emitted may be toxic to nearby life and the environment. Smoke is a major factor of air pollution. The outcome of an aerosol application is similar to smoke.

Smother crops

See **Integrated weed management (IWM)**

Smut

See **Plant disease smut**

Sodium arsenite insecticide

See **Arsenical inorganic insecticides**

Sodium chlorate

See **Herbicide action – others**

Sodium cyanide

See **Hydrogen cyanide rodenticide**

Sodium dimethyldithiocarbamate slimicide

Sodium dimethyldithiocarbamate (sodium dimethylcarbamodithioate) is a slimicide and fungicide with molecular formula $C_3H_6NNaS_2$. It is a yellow, crystalline liquid, completely soluble in water. It is incompatible with acids and peroxides. It gradually decomposes in aqueous solutions to give carbon disulfide and ethylamine.

This compound is an antimicrobial agent, **biocide**, fungicide, bactericide and algicide. It is used in the sugar, paper, pulp and chemical industries.

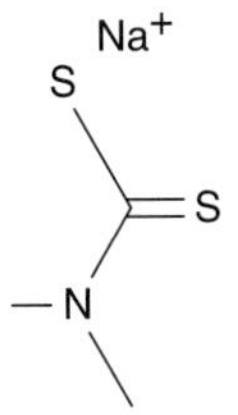

Sodium dimethyldithiocarbamate

The acute oral LD_{50} of this slimicide in rats is 1000 mg/kg. It is highly toxic to fish and *Daphnia* spp. (the static acute 96 h LC_{50} for rainbow trout is 0.85 mg/l and the static acute 48 h EC_{50} for *Daphnia magna* is 0.0715 mg/l). It can be toxic via dermal and inhalation routes.

Sodium ethylenebisdithiocarbamate fungicide

See **Ethylenebisdithiocarbamate (EBDC) fungicides**

Sodium ferric ethylenediaminetetraacetate (ferric (III) EDTA complex) as a molluscicide

Sodium ferric ethylenediaminetetraacetate (or ferric (III) EDTA complex or EDTA, iron chelate) is a complex formed with iron and sodium EDTA. Ferric (III) EDTA complex is an active ingredient used variously as a micronutrient in agriculture, as a catalyst and finally as a molluscicide.

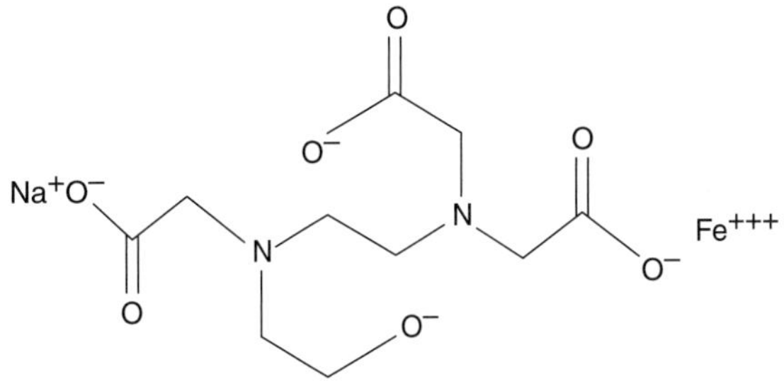

Sodium ferric ethylenediaminetetraacetate

As a molluscicide, the ferric (III) EDTA complex is broadcast on the soil around affected areas of, for example, vegetable crops, fruit orchards, greenhouses, home gardens and ornamentals. When sodium ferric ethylenediaminetetraacetate is ingested by slugs or snails, the iron in the compound interacts with the haemocyanin common to the blood of crustaceans, and eventually causes death.

Small pellets or granules (GR) are the formulations of this molluscicide. It may be spread by a rotary or ground spreader or by hand-held equipment. Slug & Snail Killer is a common trade name of this molluscicide.

Sodium ferric ethylenediaminetetraacetate is hazardous to birds or mammals. Excessive exposure may cause irritation to the eyes of applicators. *Daphnia* spp. and other crustaceans and aquatic organisms may be adversely affected. Thus, this molluscicide is not to be applied to aquatic bodies.

Sodium fluoride insecticide

See **Fluoride inorganic insecticides**

Sodium fluoroacetate rodenticide

Sodium fluoroacetate, $C_2H_2FNaO_2$ (otherwise called **sodium monofluoroacetate** or **Compound 1080**), is an organofluorine rodenticide. It is an odourless, whitish powder, soluble in water and alcohol. It is derived from fluoroacetic acid, which also exists naturally in some plant species. It can be prepared by treating sodium chloroacetate with potassium fluoride. Sodium fluoroacetate is also produced by reacting ethyl α-chloroacetate with potassium fluoride at 200°C, followed by the hydrolysis of the product in methanolic hydroxide. It has a CAS Registry Number of 62-74-8.

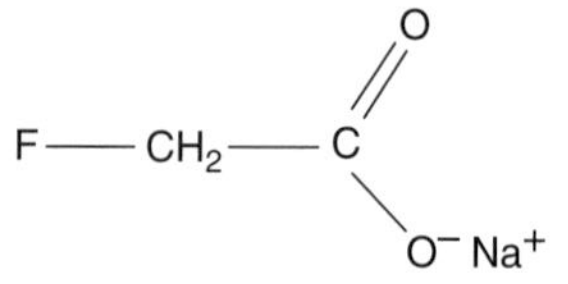

Sodium fluoroacetate

Sodium fluoroacetate disrupts the citric acid cycle which is critical to cellular metabolism. Pests like rats, coyotes, mice, squirrels and possums are effectively controlled by this rodenticide and thus many field crops are protected. This rodenticide also has repellent effects on these mammalian pests.

Normally, this rodenticide is formulated as a bait (GB/RB), with 1080 being a common trade name. Formulations are made black in colour to alert unsuspecting residents, farmers or applicators.

Sodium fluoroacetate is classified as a Class Ia compound (i.e. an extremely hazardous chemical) by the WHO. It is highly toxic to mammals (e.g. dogs, cats and even humans), birds and insects. In humans, a dose in the range of 2–10 mg/kg can be fatal. Exposure to sodium fluoroacetate may cause nausea, vomiting, sweating, confusion and cardiac and respiratory disorders.

Small doses of sodium fluoroacetate, when administered orally, are eliminated within 4 days. This rodenticide degrades into harmless compounds in soil and water. Although it is banned or restricted in many countries, it is used in New Zealand, Australia, Mexico and the USA for the control of the pests mentioned above. However, it has been banned for its acute toxicity in many countries such as Cuba and Belize.

Sodium hexafluorosilicate insecticides

See **Fluoride inorganic insecticides**

Sodium 2-mercaptobenzothiazole slimicide

Sodium 2-mercaptobenzothiazole, $C_7H_5NNaS_2$ (sodium benzothiazolethiolate), is a slimicide and fungicide. A salt of 2-mercaptobenzothiazolate, it is soluble in water at >100 mg/l at 20°C. It is unstable when exposed to air for prolonged periods or to high temperatures, and produces toxic fumes. It is reactive to oxidizing material and mineral acids.

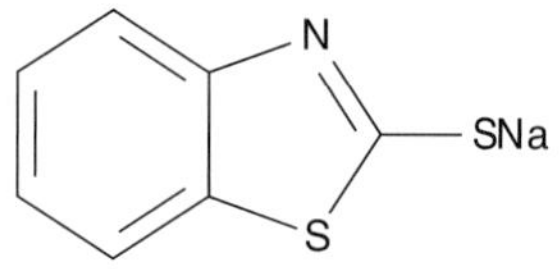

Sodium 2-mercaptobenzothiazole

The LC_{50} of sodium 2-mercaptobenzothiazole for rainbow trout is 730 μg/l, which is highly toxic. It is extremely toxic to *Daphnia magna*. Contact with the skin is risky, making appropriate PPE necessary. Mammalian toxicity occurs at higher doses, according to studies on rabbits.

Sodium 2-mercaptobenzothiazole is registered by the EPA as an antimicrobial and fungicide.

Sodium monofluoroacetate rodenticide

Sodium monofluoroacetate rodenticide is another term for **sodium fluoroacetate rodenticide**.

Sodium *N*-methyl-dithiocarbamate fungicide

Vapam or metam-sodium is the popular name of sodium *N*-methyl-dithiocarbamate. (See **Metam salts; Ethylenebisdithiocarbamate (EBDC) fungicides**)

Sodium pentachlorophenate molluscicide

See **Molluscicides**

Sodium tetrathiocarbonate fungicide

Sodium tetrathiocarbonate, CNa_2S_4 (sodium tetrathio (peroxocarbonate)), is a fungicide, fumigant, insecticide and also a nematicide. Also referred by its code GY-81, it is an orange crystalline, hygroscopic solid, which is completely soluble in water. The photolytic LT_{50} (median lethal time) of sodium tetrathiocarbonate is <1 day. It is produced by the reaction of carbon disulfide with sulfur in alkali ethoxide. The CAS Registry Number of sodium tetrathiocarbonate is 7345-69-9.

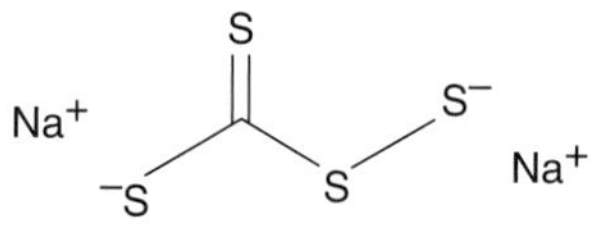

Sodium tetrathiocarbonate

The fungicide sodium tetrathiocarbonate provides good control over nematodes and also of grape aphids, *Phylloxera vitifoliae*. Root rot is also prevented on grapes, lemons and oranges.

The compound, formulated as a soluble concentrate (SL), is available as the trade name Enzone, among others.

Sodium tetrathiocarbonate has an acute oral LD_{50} in rats of 630 mg/kg and can cause severe irritation to the skin and eyes, in case of excessive exposure. The LD_{50} in bees is >25 mg per bee. The ADI is 6 mg/kg/day.

The fungicide sodium tetrathiocarbonate degrades almost fully in the soil into carbonates, sulfates and carbon disulfide gas and microbes also contribute to the break down. It is not persistent in the environment. The dissipation in soils is complete within 7 days.

The fungicide is registered for use in New Zealand and the USA.

Sodium trichloroacetate

See **Trichloroaceticacid sodium (TCA sodium) herbicide**

Soil conditioners

Soil is one of the most critical ingredients in crop growth. To ensure optimum health of the soil, it may be necessary to add various man-made inputs called soil conditioners. These substances help improve the physical characteristics of the soil and act as a plant growth medium. Soil conditioners correct modern-day problems such as excessive erosion, groundwater contamination and surface sealing.

Erosion occurs when the force of rainfall or irrigation water exceeds the cohesive forces between the soil particles. Soil aggregates break up into finely divided material that flows as runoff. Valuable top soil is lost, leading to the pollution of downstream water with fertilizer and pesticide chemicals. Also fine material is deposited on the surface of the soil, forming seals and blocking soil pores. This promotes ponding, reduces soil aeration, inhibits seedling emergence and finally lowers yields. There are other problems like surface crusting which further block leaching. The soil tends towards becoming saline or acidic.

Soil conditioners can be both natural and synthetic. Green manure, peat, gypsum and sawdust are some examples of natural conditioners. Synthetic conditioners include two types of polyacrylamides (PAMs).

The first kind of PAM is the water-soluble, high molecular weight (>10^7) not cross-linked 'anionic PAMS' with approximately 20% carboxylate groups. These PAMs are safe, effective flocculants that facilitate liquid–solid separations and may be used in erosion control and in industrial wastewater treatment.

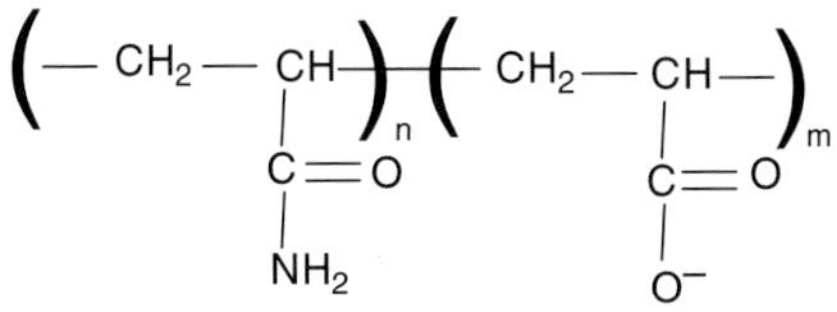

Anionic polyacrilamide

The second type of PAM, called super absorbers, consist of deliberately cross-linked, water-insoluble copolymers of acrylamide and metal salts of acrylic acid. Cross-linked systems reversibly absorb large volumes of water and are used to retain and provide water for agriculture or horticulture in arid environments.

When mixed with soil or water and sprayed, these polymers are capable of absorbing water 40–500 times their weight, hence avoiding runoff and contamination. Both hydration and swelling are reversible, with plant roots creating a sufficient pressure gradient to extract more than 90% of the water held by the polymer.

Synthetic polyelectrolytes like copolymers of maleic anhydride with vinyl acetate, polyethylene glycol, polyurea formaldehyde, as well as polyurethanes and cellulose xanthate have been used as soil conditioners. Sodium acrylate–acrylamide copolymers cross-linked with methylenebisacrylamide improve soil properties, specifically water distribution, availability and drainage characteristics. Other ways to enhance soil conditioning and protection are hydroseeding, hydroponics, mulching for trees and shrubs and increased vegetation.

Soil drenching

See **Fungicide application methods**

Soil enzymes

Every living system in almost every ecosystem contains enzymes, which are proteins that catalyse the synthesis of a range of proteins, carbohydrates, nucleic acids and other components in an organism. Enzymes have an important role to play in the degradation of and recycling of nutrients. Any alteration in the enzyme protein structure might have far-reaching consequences for living organisms. Soil remains alive and vibrant due to enzymatic reactions.

A characteristic of enzymes is their specificity, which is brought about by the property of the chemical groups attached to their susceptible bonds. Thus, trypsin is a proteolytic enzyme that can hydrolyse certain peptide bonds in protein, while the enzyme maltase hydrolyses maltose to glucose.

Microorganisms, soil animals and plants release enzymes into the soil environment. Enzymes mediate a large number of biochemical reactions in soil, especially those related to nutrient cycling (especially C, N, S and P). Soil also hosts many free enzymes, extra-cellular enzymes and mineral fractions that interact with each other, making the enzymatic activity in soil very high. The biochemical enzymatic reaction is expressed by Eqn S.1.

$$[E]+[S] \underset{K_2}{\overset{K_1}{\rightleftharpoons}} [ES] \xrightarrow{K_3} [E]+[P] \qquad (S.1)$$

where
E = Enzyme
S = Substrate
ES = Enzyme–substrate complex
P = Product released
K_1, K_2 and K_3 = rate constants
[ ] indicates concentration

The large biomass and the high metabolic activity of microorganisms make them the largest source of enzymes in the soil. The enzymes produced by them include protease, amylase, glucose isomerase, pectinase and lipase. For instance, ribonucleases and alkaline phosphatase are excreted by *Bacillus subtilis* under certain conditions.

Soil enzymatic reactions change with for example temperature, pH, ionic environment and substrate concentration. Human activities such as cultivation, cropping, fertilization, pH alteration, tillage, residue management and waste disposal affect the levels and distribution of enzyme activity in soils. The activities of some enzymes can be used as a monitoring tool to assess the effect of pollutants on the health of the ecosystem.

Soil fumigants

Fumigation of the soil, as a part of soil preparation, has been practised for many years. Soil **fumigants** control a wide range of plant pests and pathogens such as fungi, nematodes, insects, bacteria and weeds. Some well-known, but very hazardous fumigants (e.g. **methyl bromide** and **chloropicrin**) are being phased out because of the threat to the ozone layer, as well as their high toxicity. Effective alternatives to former fumigants are methyl iodide, propargyl bromide, **metam-sodium (vapam)** and **dazomet**.

Methyl iodide (CH_3I), a substitute to methyl bromide, is extremely soluble in water (4 g/l), but is a carcinogen and very toxic. Its water solubility makes it a potential water contaminant. Hence, although it is registered in the USA and Australia, its long-term use is debated.

Propargyl bromide, C_3H_3Br (IUPAC name 3-bromoprop-1-yne and CAS Registry Number 106-96-7), is similar to methyl bromide, although it has lower ozone-depleting potential. It is effective on weeds (e.g. **barnyard grass** and some nutsedges) and some nematodes.

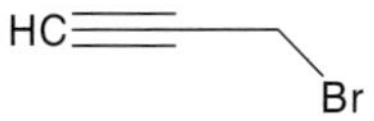

Propargyl bromide

An ideal soil fumigant ought to be a halogen and isothiocyanate compound with low molecular weight. It must be able to move with the water and into the soil, and be effective to a depth of 30 cm, even when followed by compaction. It should ideally also be compatible with herbicides, to take care of weed seeds that are lying dormant under the soil surface.

To avoid exposure of soil fumigants to non-target organisms, soil should be compacted or covered with impermeable tarpaulins to ensure the chemical does not contaminate the air.

Soil organic partition coefficient (K_{OC})

The ratio of the mass of chemical (i.e. pesticide, in this volume) adsorbed in the organic matter component (carbon) of soil (or sediment) to that in the aqueous phase at equilibrium is called the soil organic partition coefficient (K_{OC}). The K_{OC} is calculated by dividing the **distribution coefficient** value (K_d) by the fraction of organic carbon present in the soil or sediment. K_{OC} values indicate how mobile a chemical is in the environment. Higher values indicate low mobility and vice versa.

Soil persistence of pesticides

Persistence of a chemical is its property to delay breakdown into simpler metabolites. The breakdown of a chemical takes place via microbial action, volatilization, photolysis, oxidation, leaching and hydrolysis as well as runoff. Since pesticidal or fertilizer compounds directly or indirectly reach the soil (via rain, spray applications and irrigation) the soil is the main repository of farm chemicals, making chemical persistence in soil a critical factor in soil health.

Persistence of soil chemicals, leading to accumulation in the soil, can have hazardous and chronic adverse effects on the microbial population and groundwater sources. To prevent over-polluting the environment, it is important to assess the behaviour of pesticides in soil, especially the speed and pattern of breakdown. Registration of a pesticide ought to be confirmed only when the pesticide meets the standards of safe use.

Some useful criteria considered when analysing a pesticide for its safe use are: (i) its structural properties; (ii) its reaction to different pH levels and/or different levels of vapour pressure; (iii) its sorption characteristics; and (iv) its solubility in water. Its fate in soil, water, sunlight, alkaline/acidic media and its reactivity to microbial activity are also important considerations that decide its persistence. The breakdown of organic compounds is strongly influenced by microbial enzymes, breaking down or metabolizing the organic compounds into polar products, which in turn lead to simple molecules that are not as potent. The soil factors per se that influence pesticide breakdown are soil texture, organic content, water content and redox potential. The way a pesticide is applied also determines the rate and pattern of persistence. Broadcast chemicals face many more fluctuating environmental factors than those chemicals used for treating seeds or incorporated into the soil.

However, sometimes, a pesticide is valued for its persistence. The residual life of a pesticide ensures long-term protection against pathogens such as fungi, nematodes and weeds. Herbicides applied at the pre-emergence stage control the early growth of weeds, to give enough time for the crop to establish. Similarly, if soil insecticides and nematicides are present for a substantial period, early plant growth can take place unhindered. However, over-persistence or an excess residual life can result in phytotoxicity and plant injury to both current and subsequent crops.

Thus, a balance is critical between persistence and toxicity. The kinetics of pesticide loss can be very complex in the field, because more than one loss process may be involved. A photolabile compound might initially be lost very rapidly from the soil surface. But the remaining amount may leach underground and persist for a long time. Detailed studies of the degradation pathways of pesticides in soils use a radio-labelled compound, mostly ^{14}C. Use of chromatography techniques such as thin layer and high pressure liquid chromatography allows separation of extracted breakdown products which can be quantified by liquid distillation and subsequently identified by mass spectroscopy.

An example of persistence (sometimes unpredictably) is chlorosulfuron which hydrolyses at low pH levels and remains persistent in alkaline soil, causing phytotoxicity. Similarly, rapid reduction is common for nitro groups in anaerobic soils. (See also **Residual pesticides**.)

Soil residual life

See **Residual pesticides**

Soil solarization

Soil solarization is an agricultural technique used to kill weeds, nematodes, fungi and some bacteria which occur in open fields or greenhouses. Soil-borne pests are removed using the sun's energy. This process is used only when there is a serious pest or disease problem. Soil solarization works best in areas that have long stretches of clear, hot weather.

To undertake soil solarization, the site has to be prepared for planting (e.g. by carrying out tilling, adding soil amendments and watering the site) before covering the area with a clear, polyethylene sheet (4 mm), burying the edges at the ends of the field in a trench to anchor them. The field should be thus covered for a period of up to 4 weeks. The effect of solarization is evident in the top 12 cm of soil becoming free of pathogens to a large extent. Planting should be done without deeply cultivating the soil. Since **beneficial organisms** and microorganisms are also killed during solarization, supplementing the soil with microbe-rich material and compost puts nutrients back into the soil.

Solarization may also be carried out with steam and this is called soil sterilization by steam. This method also rids the soil of pathogens, while activating some beneficial bacteria to produce nutritive enzymes. There are different methods used in steam sterilization depending on the soil strata that needs to be treated. The common condition of steam sterilization is a loose soil surface. The equipment for surface steaming includes heat-resistant synthetic insulation fleece, a harrow and drainage pipes. Steam is injected underneath the surface, for a period of up to 5 h for sterilizing an area of about 400 m^2 and up to a depth of 30 cm. Sterilization at deeper levels may be achieved using drainage pipes or suction pipes.

Usually solarization is a good technique for pest management and can help substitute toxic fumigants and soil-applied pesticides. The use of bromomethane (curtailed by the Montreal Protocol), **methyl bromide** and other harmful chemicals may be thus avoided. (See also **Weed management by mechanical methods**; **Fungicide application methods**.)

Soil sterilants

Fumigants are also called soil sterilants or **soil fumigants**, since their common application is on or in the soil.

Soil sterilization by steam

See **Soil solarization**

Solarization of soil

See **Weed management by mechanical methods**

Soluble concentrates

A soluble concentrate (SL) is a liquid formulation that contains a spray concentration of a completely soluble active ingredient (a.i.), the primary carrier being water or a water-soluble solvent. These are easy to manufacture and low on cost. Once prepared, a SL does not require agitation and may be used in any sprayer without clogging the equipment. (However, depending on the a.i., corrosion may occur.)

Seed-treatment formulations are commonly soluble concentrates. Care must be taken to ensure that the seeds remain viable. (See also **Liquid formulations**.)

Sooty mould

See **Mould (sooty mould)**

Spade

See **Weed management by mechanical methods**

Spill management

See **Pesticide spill management**

Spinning disc sprayers

See **Application equipment – sprayers**

Spinosad bioinsecticide

Spinosad (a mixture of 50–95% (2*R*,3a*S*,5a*R*,5b*S*,9*S*,13*S*,14*R*,16a*S*,16b*R*)-2-(6-deoxy-2,3,4-tri-*O*-methyl-α-L-mannopyranosyloxy)-13-(4-dimethylamino-2,3,4,6-tetradeoxy-β-D-erythropyranosyloxy)-9-ethyl-2,3,3a,5a,5b,6,7,9,10,11,12,13,14,15,16a,16b-hexadecahydro-14-methyl-1*H*-as-indaceno[3,2-d]oxacyclododecine-7,15-dione and 50–5% (2*S*,3a*R*,5a*S*,5b*S*,9*S*,13*S*,14*R*,16a*S*,16b*S*)-2-(6-deoxy-2,3,4-tri-*O*-methyl-α-L-mannopyranosyloxy)-13-(4-dimethylamino-2,3,4,6-tetradeoxy-β-D-erythropyranosyloxy)-9-ethyl-2,3,3a,5a,5b,6,7,9,10,11,12,13,14,15,16a,16b-hexadecahydro-4,14-dimethyl-1*H-as*-indaceno[3,2-*d*]oxacyclododecine-7,15-dione) is a bioinsecticide. The actinomycete *Saccharopolyspora spinosa* has been the basis for the development of spinosad which belongs to a new family, macrolides. Spinosad and spinetoram are the two insecticides of the spinosyn group of insecticides. The molecular formula of spinosyn A is $C_{41}H_{65}NO_{10}$ and that of spinosyn D is $C_{42}H_{67}NO_{10}$.

Spinosads are effective in the control of chewing insects in cotton and vegetables.

The wide range of pests controlled by spinosad are caterpillars, thrips, beetle and fly pests that affect a range of fruit and vegetable crops, ornamentals, turf and stored grains. Spinosad has contact activity on all life stages of insects and is most potent when ingested. Spinosad acts by altering the function of nicotinic- and γ-aminobutyric acid (GABA)-gated ion channels of insect nervous systems.

The main formulation types available are suspension concentrates (SC), wettable powders (WP), water-dispersible granules (WG) and granules for direct application (GR). Spinosad may be co-formulated with other insecticide active ingredients. A common dose for SC is 120–480 g/l.

Spinosad formulations are registered in India. Indian MRL values established by the Ministry of Health and Family Welfare are 0.02 g for cottonseed oil, cabbage and cauliflower and 0.01 g for red gram.

Spinosyns

See **Spinosad bioinsecticide**

Spiroxamine fungicide

Spiroxamine, $C_{18}H_{35}NO_2$ (8-*tert*-butyl-1, 4-dioxaspiro [4, 5] decan-2-ylmethyl (ethyl) (propyl) amine), is an amine fungicide (spiroketalamine). It is a yellow liquid that is soluble in water, toluene, isopropanol and dichloromethane. It is stable to hydrolysis and has a DT_{50} of 5 days. It is produced by ketalization of 4-*tert*-butylcyclohexanone with 1-chloro-2,3-dihydroxypropane, followed by treatment with ethylpropylamine. Its CAS Registry Number is 118134-30-8 (see Spiroxamine at bottom of the page).

Spiroxamine, a systemic, protective and eradicative foliar fungicide, acts as a sterol biosynthesis inhibitor (SBI), thereby controlling a wide range of diseases of grapes, bananas, wheat, barley, oats and other cereals, vegetables and peas. **Powdery mildew** (caused by *Erysiphe graminis* and *Erysiphe necator*), black and yellow sigatoka (caused by *Mycosphaerella fijiensis* and *Mycosphaerella musicola*, respectively) and rusts (*Rhynchosporium* sp. and *Pyrenophora teres*) are some of the diseases that are controlled.

The fungicide spiroxamine is commonly formulated as capsule suspension (CS), emulsifiable concentrate (EC), emulsion, oil in water (EW) or suspension concentrate (SC). Some of the known trade names available are Impulse, Prosper and Falcon 460, while some formulations are prepared by combining compounds such as **tebuconazole** and **triadimenol**. The commonly accepted dosage of spiroxamine is 400 g/ha for powdery mildew and 320 g/ha for sigatoka.

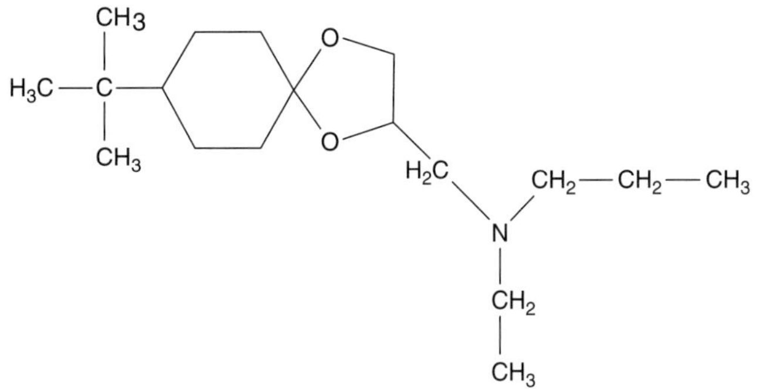

Spiroxamine

The WHO has classified spiroxamine as moderately hazardous (Class II). The acute oral LD_{50} values of spiroxamine in rats and bobwhite quails are between 500 and 595 mg/kg. Spiroxamine is a severe skin irritant. The 96 h LC_{50} values for some tested fish species are in the range of 3–18.5 mg/kg. It is not toxic to earthworms in normal use. The ADI is 0.025 mg/kg/day, while the NOEL in rats and mice is 70 and 160 mg/kg diet, respectively.

In rats, spiroxamine is absorbed and eliminated within 48 h, leaving a carboxylic acid compound as the main metabolite. In plants too, metabolism is extensive. In soils, spiroxamine degrades to carbon dioxide following oxidation on the *tert*-butyl moiety and dealkylation of the amine. The soil DT_{50} is up to 2 months, making spiroxamine moderately persistent.

Spiroxamine is registered for use in more than 60 countries, including the USA, Australia, South Africa, Columbia, Costa Rica, France, the UK, Germany and Hungary. According to the EU Pesticide Database, the MRL values of spiroxamine are for grapes 1.0 mg/kg, for barley grain and oat grain, 0.3 mg/kg and for vegetables, fruit and other cereal grains, 0.05 mg/kg.

Spotted bollworm

See **Pest – spotted borer/spotted bollworm**

Spray application concepts

The method and timing of the application of a pesticide are as important as the pesticide itself. These factors have come into sharper focus in recent times, with increasing emphasis on minimizing chemical intervention. Since spray applications are common among pesticides, there is increased focus on application equipment, especially the nozzle type, since nozzles determine droplet size, spray coverage, drift and contact angles.

In light of effective disease control with moderate intervention of chemical pesticides, the following concepts should be understood.

A pesticide may be sprayed either by a hand-held sprayer, a vehicle-mounted sprayer, by **chemigation** (pressurized sprinkling through the irrigation system) or by an aerial sprayer.

Spray coverage is a function of the amount of carrier applied per unit area and the number of droplets per unit of leaf surface. The rule of thumb is that coverage needs to be heavy in the case of non-systemic herbicides and optimal for systemic herbicides. However, the ideal coverage depends on the target plant species. It is considered beneficial/sufficient for spray droplets to be in the range of 100–300 μm to achieve the correct balance between maximum coverage and minimum spray drift.

Spray droplet size is crucial in a pesticide application (especially for herbicides or fungicides) since the size of the drop determines how much of the compound will actually be in contact with the target pest, which in turn decides its ultimate uptake. Adjusting the spray equipment as well as the physical properties of the pesticide solution can aid in a uniform spray of small droplets. A large droplet does not have a large surface spread, and so, it has a relatively limited area. Small droplets improve coverage of the target.

Spray nozzle designs cater for different droplet sizes and spray volumes. Increasing the pressure on the nozzle decreases the spray droplet size but increases spray volumes. But a pressure increase brings about only a small increase in volume output. Decreasing the droplet size improves coverage, but increases the risk of drift. Droplet sizes are measured in microns (1 μm equals 0.001 mm) and range from 'very fine' to 'extremely coarse', as per a standard established by the American Society of Agricultural and Biological Engineers (ASABE). Very fine droplets are <150 μm in size, while extremely coarse droplets are >550 μm. On a practical level, nozzles are classified according to the volume mean diameter (VMD). The VMD is defined as the diameter of a droplet, whose volume, when multiplied with the total number of droplets, equals the total volume of the sample.

As a rule of thumb, contact applications for a foliar fungicide, are most effective if the spray droplets are fine to medium, from the point of view of maximum coverage of leaf surfaces. If the soil or basal parts of plants need to be drenched, coarser droplets are needed. For example, basal rot anthracnose will require a coarser spray. If a wider coverage is desired, as in the case of treatment of dollar spot, a fine spray application is needed.

While studying the phenomenon of water droplet infiltration and spread, one must appreciate the importance of three forces: (i) gravity; (ii) cohesion (the attraction of water molecules to each other, which causes surface tension); and (iii) adhesion (the attraction of water molecules to another substance). When cohesive forces are greater than adhesive forces, a water droplet does not spread and the surface is said to be hydrophobic. A spray droplet needs to spread just enough for optimum surface coverage of herbicide. The contact angle of a droplet is a function of the surface tension of the solution and the mass of the droplet. The greater the angle, the lower the wettability and thus, the lower the uptake. Contact angles of a spray droplet without and with an **adjuvant** are shown in Fig. S.1. It is important to note that some adjuvants increase the size of each droplet by thickening their size.

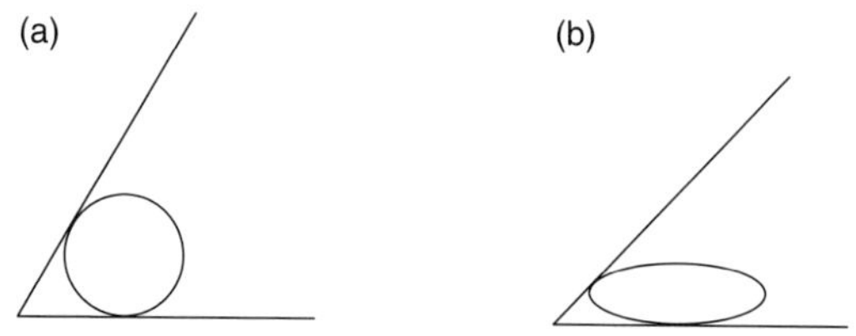

Fig. S.1. Contact angle of a spray droplet (a) without a wetting agent, and (b) with a wetting agent.

A spray droplet with a high contact angle cannot facilitate pesticide uptake at desirable levels. A high contact angle, of say a herbicide, may lead to the droplet falling off the leaf instead of spreading over the surface, affecting herbicide efficiency. As it is necessary for a droplet to spread, wetting agents (surfactants) are used to reduce the interfacial tensions inside the droplet. Liquids heavier than water form smaller spray angles, while those lighter than water form wider spray angles. However, sometimes the unique characteristics of some weed species (hairy or waxy leaf surfaces) can make it difficult for some non-systemic herbicides to be efficient. Examples of such hard-to-wet plants include giant foxtails and **common lambsquarters** (*Chenopodium album*).

Non-ionic surfactants or surface active wetting agents reduce the surface tension of water, thus enabling the water molecules to spread on the surface. Wetting agents help to reduce the interfacial tensions of the spray droplets. The effect of gravity on a droplet as it settles causes the initial spread of the droplet. This results in reduced concentration of the adjuvant at the droplet front, leading to increased surface tension. This temporarily halts the droplet from spreading until the flow from the droplet centre impacts the adjuvant concentration at the droplet front, lowering the surface tension and allowing further spread. This continues until the adjuvant concentration can no longer affect the droplet spread.

A spray angle is very important in determining coverage. Nozzle spray angles range from 65° to 110°. Two typical spray patterns are flat fan and hollow cone. In the flat fan spray nozzle, the spray forms a narrow, elliptical, inverted 'V' pattern, with some overlap of coverage. The fan angle, at 80° or 100°, ensures overlap even at lower boom heights. The nozzle pressure is around 280 kPa, but can vary depending on target pests.

The other type of nozzle, the hollow cone nozzle, can either produce a droplet that is finer than that of a fan spray nozzle, or one that produces a really coarse droplet. Flat fan nozzles are widely used and are ideal for a level field crop.

The amount of pesticide applied in a specified time by the nozzle is the nozzle flow rate. It is commonly measured in litres per minute (l/min).

Boom height is the distance between the nozzle and the target. Since the target might be a tree or a patch of broad-leaved weeds, the boom height should be adjustable. If the height is more than 0.5 m, the spray can drift.

Spray retention of a spray solution is the percentage of the solution that actually stays on the leaf. Spray retention of a herbicidal active ingredient depends on factors such as: (i) wax formation on leaf; (ii) leaf ridges, scales and hairs; (iii) dust on leaf surfaces; and (iv) adjuvant type and concentration. Hard-to-wet plants do not retain sprays on their leaves without adjuvants. Spray retention decreases as spray volume increases. The ideal spray retention can be brought about by keeping the surface tension, droplet size and contact angle low.

An accurate estimation of spray effect is crucial for environmental safety. The area between the pesticide application point and the nearest ecologically sensitive location is the buffer zone.

Spray application equipment

Pesticides are commonly applied to crops in the form of sprays. Spraying ensures a uniform and thorough coverage of the plant part. Spraying is done for pest control or for nutrient supply. Spray equipments are either hand operated or motorized.

Hand-operated equipment such as knapsack (backpack) sprayers and shoulder-mounted mist blowers are common on small farms or gardens that are dependent on semi-skilled labour. Hand spraying helps to direct most of the spray fluid to the target organism. This operation brings flexibility to the user, thus making such spray application an efficient one. However, the major disadvantages of hand spraying are its slow speed and high labour requirement. Carriage-mounted sprayers are designed for spraying large areas.

Aerial spraying employs either a nozzle boom or spinning cages to reach large tracts of fields. Such spraying results in deposition of much of the spray liquid on the upper surfaces of the leaves. The disadvantages of aerial spraying are the cost and toxins straying onto non-target areas.

A large number of herbicides (contact), insecticides and fungicides are sprayed. **Bordeaux mixture, copper oxychloride, zineb, maneb, captan, carbendazim, oxycarboxin** and **benomyl** are active ingredients sprayed on pests. **Powdery mildew** is controlled by a spray of wettable sulfur, besides that of **ethirimol, chloroneb** and **tridemorph**.

Spray coverage

See **Spray application concepts**

Spray droplet size

See **Spray application concepts**

Spray modifiers

See **Adjuvants**

Spray retention

See **Spray application concepts**

Spreaders

See **Adjuvants**

Spudding

See **Weed management by mechanical methods**

S,S,S-tributyl phosphorotrithioate as a PGR

S,S,S-tributyl phosphorotrithioate, $C_{12}H_{27}OPS_3$, is a colourless to pale yellow liquid, with a mercaptan-like odour. It is soluble in water, alcohols and aliphatic, aromatic and chlorinated hydrocarbons. It is relatively stable to acids and heat. It hydrolyses slowly under alkaline conditions.

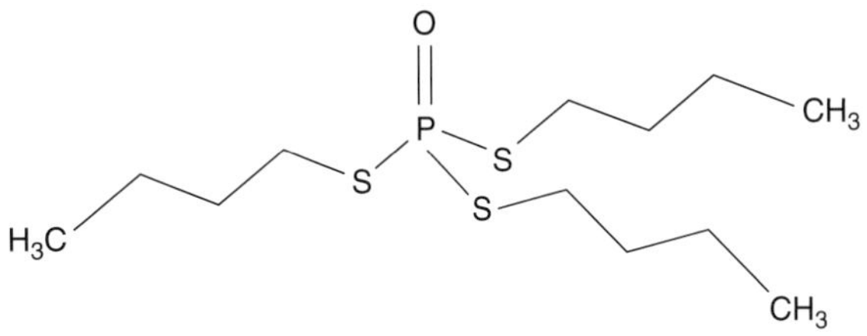

S,S,S,-tributyl phosphorotrithioate

S,S,S-tributyl phosphorotrithioate (also known as **tribufos**) is a plant growth regulator, which is absorbed by the leaves. It stimulates the formation of an abscission layer between the plant stem and the leaf petioles, causing leaves to drop. It is used only for defoliating cotton, to facilitate harvesting.

This PGR is formulated as an emulsifiable concentrate (EC). It is very phytotoxic, but relatively non-toxic to honey bees. The WHO classifies it as a Class II toxin (i.e. moderately hazardous). The ADI for *S,S,S*-tributyl phosphorotrithioate is 0.001 mg/kg/day.

Stale seedbed

The use of a stale seedbed (or false seedbed) is a form of weed management. It is a process of preparing a seedbed (well prior to sowing). When the seedbed is prepared well in advance, annual weeds start germinating and emerging from the soil, which can then be removed or killed. At this stage, the seedbed becomes ready for the crop seed to be sown.

Care should be taken not to delay planting after weeding is done. It is also important to note that the soil is almost weed free only up to a depth of around 6 cm.

Starlicide avicide

See **Avicides**

Sterilization

The complete destruction of all bacteria and other infectious organisms in an industrial, food or medical product is called sterilization. It must be followed by aseptic packaging to prevent re-contamination. The methods used involve either wet or dry heat, use of chemicals such as **formaldehyde** (HCHO) and ethylene oxide filtration (pharmaceutical products), or irradiation by UV or gamma radiation. Milk is sterilized by heating for a few seconds at 138°C.

Rendering an insect or organism incapable of reproduction by radiation or chemical treatment is also termed sterilization. (See also **Male sterilization**.)

Sterol biosynthesis inhibitor fungicides (SBIs)

Compounds that inhibit the synthesis of sterols in fungi, discovered in the 1960s, are known as sterol biosynthesis inhibitors (SBIs). Sterols are a naturally occurring class of organic molecules that are vital to cell function. **Ergosterols**, the most characteristic sterol among higher fungi, are critical in maintaining the integrity of biological membranes and in regulating membrane-bound enzymes. Thus, plant diseases caused by ascomycetes, basidiomycetes and deuteromyces are controlled to varying degrees by SBIs. However, oomycetes (*Pythium* spp. and *Phytophthora* spp.) and members of the Peronosporales (e.g. causal agent of **downy mildew**) are not inhibited. Diseases caused by *Botrytis* spp., *Alternaria* spp., *Rhizoctonia* spp., *Sclerotinia sclerotiorum* and *Pyricularia oryzae* are controlled by SBI inhibitors.

Many classes of compounds inhibit sterol biosynthesis at various stages. These are (i) demethylation inhibitors (DMIs); (ii) amines; (iii) hydroxyanilides; and (iv) squalene epoxidase inhibitors.

Sterol biosynthesis begins with the isoprenoid pathway, which includes oxidation to 2,3-oxidosqualene, ring closure and modification by methylation at the C-24 side-chain position followed by demethylation at the C-14 position (i.e. removal of the methyl group at the C-14 position). The C-14 demethylation proceeds through many reactions resulting in Δ^{14}-double bond sterol (this is the target of DMIs). The reaction is completed by the reduction of the Δ^{14}-double bond catalysed by the enzyme Δ^{14}-reductase. This reduction step is a target site for another group of SBIs, the reductase-isomerase inhibitors. The last stage of biosynthesis of ergosterol involves isomerization and formation of many double bonds and is known as Δ^{8}-Δ^{7}-isomerization. This process is also the target for some fungicides.

Antifungal demethylation inhibitors (DMIs) act by blocking the C-14 demethylation step of ergosterol biosynthesis. Under their influence, physical and structural functions of fungal membranes are disturbed. Respective sterol precursors accumulate in the membranes and fungal growth stops. **Metconazole**, **myclobutanil**, **nuarimol**, **penconazole**, **prochloraz**, **tebuconazole**, **triforine**, **tetraconazole**, flutriafol and diniconazole are DMI fungicides.

Most DMI fungicides are systemic in nature but translocation is restricted to the apoplast. Sublethal doses of DMIs cause fungistatic action. DMIs do not effectively inhibit spore germination, as they inhibit mycelial growth.

DMIs are used as foliar sprays and seed dressing. Those that are used for both foliar application and seed dressing include **prochloraz**, **bitertanol**, **triadimenol** and propiconazole. Some DMIs are also used to treat post-harvest diseases. Triforine is a DMI used mostly for foliar application, while imazalil is used in all three functions. Diseases such as **smuts**, bunts, **powdery mildew**, rots and rusts are controlled. Wheat, barley, grapes, apples and bananas are protected.

DMIs are commonly mixed with other SBIs acting at other sites in the sterol biosynthesis process. DMIs have low acute toxicity in mammals but they are toxic to fish.

The class of chemicals that inhibit reductase-isomerase are called amines. These fungicides inhibit the step after C-14 demethylation in the overall sterol biosynthesis pathway, and they are called reductase-isomerase inhibiting fungicides. Some examples of reductase-isomerase inhibitors are **tridemorph**, **fenpropimorph**, **fenpropidin** and **spiroxamine**. In the sterol biosynthesis pathway, the enzymatic steps of Δ^{14}-reduction and Δ^{8}-Δ^{7}-isomerization proceed through intermediates of carbocationic nature. The nitrogen from these inhibitors binds to the enzyme by mimicking the intermediate. This retards sterol biosynthesis and inhibits fungal growth.

Such reductase-isomerase inhibitors are broad-spectrum fungicides that are systemic in nature and are translocated within the apoplast. Powdery mildew and rusts in cereals and ornamentals as well as leaf blotch in barley and black and yellow sigatoka of banana are common diseases controlled by these inhibitors.

Yet another step in the fungal ergosterol biosynthesis pathway is the C3-keto-reductase step. This is another target site of inhibition. Hydroxyanilide fungicides (fenhexamid) act to inhibit this crucial step. They have a narrow spectrum of activity and are usually used on *Botrytis* spp.

The fungicides that inhibit squalene epoxidase have a specific mechanism, by which they act against many fungi. The action is associated with the development of intracellular squalene concentrations, which are believed to interfere with fungal membrane function and cell wall synthesis. These fungicides are not as effective on yeasts.

Many SBIs have caused resistance to develop among fungal species. Hence, it is recommended that application of SBIs be carried out according to the recommendations of the **Fungicide Resistance Action Committee (FRAC)**.

Stickers

See **Adjuvants**; **Additives**

Stockholm Convention on Persistent Organic Pollutants

The Stockholm Convention is a global treaty:

> that aims to protect human health and the environment from the effects of persistent organic pollutants (POPs).

> The Convention has a range of control measures to reduce and, where feasible, eliminate the release of POPs, including emissions of unintentionally produced POPs such as dioxins.
>
> UNEP (2004)

The Convention also aims to ensure the sound management of stockpiles and wastes that contain POPs.

This Convention came into force in 2004 with the inclusion of the first list of 12 chemicals being declared as POPs. Another list of nine chemicals was added in 2010.

More can be learnt from various internet portals (e.g. http://chm.pops.int/Countries/StatusofRatification/tabid/252/language/en-US/Default.aspx and http://www.environment.gov.au/topics/environment-protection/chemicals-management/pops).

Stored grain pest control

Stored grains are post-harvest food products and include grains, pulses, oil seeds and root crops. These are usually annual crops. They are often attacked by pathogens, throughout the world, causing great loss of food supplies and defeating the very purpose of warehousing. Stored grain pest control is generally neglected in most poor economies, leading to further food insecurity. It is important to study the pests that cause such losses in order to implement measures to prevent such loss.

Weevils, beetles and moths are the three main types of stored grain pests. Besides these, there are fungal pathogens and rodents that attack and destroy stored grain. Infestation of grain can cause allergies in people, when they eat food contaminated with insects or their excreta.

Apart from keeping pathogens away, a warehouse should be structured such that the grain in the bins remains in a dry, well-ventilated, moisture-proof and leak-proof condition. It also demands strict adherence to proper hygiene and sanitation in and around the warehouse, as well as periodic inspection of the harvest, especially with the use of sticky pheromone traps, bin probes and physical traps at the floor level for creeping insects and rodents.

Specifically, grain has to be prepared for storage after all crop residues, trash and stones are removed, and the grain cleaned and sorted. Pre-binning procedures include high-temperature drying and sorting and separating the grain of different ages. While filling a bin with new grain, it is ideal to leave a gap of at least 15 cm (6 inches) from the lid. The temperature inside the bin should be comparable to the outside temperature (+/−4°C).

Grain found infested should be immediately separated and disposed of properly. Pesticides such as **methoxychlor**, **cyfluthrin**, **chlorpyrifos-methyl**, **chloropicrin**, **carbendazim** and **carvone** should be appropriately sprayed inside and around the bins, to kill any pest that may remain among the residues. Empty bins may also need to be fumigated, after they have been physically cleaned and dried. Fumigant agents include **chinomethionat**, **chlordane**, **chloropicrin** and **aluminium phosphide**, many of which are banned for use except for fumigating empty bins. While filling the bin, there are some protectant chemicals that may be applied at 6–8 ppm on the grain. These include **pirimiphos-methyl** and chlorpyrifos-methyl. Once filled, chemicals such as **malathion** and certain pyrethroids, besides some of the above-mentioned compounds, may be sprayed to prevent further infestation. Wheat, oats, barley, rye, sorghum and sunflower are a few of the grains that are thus protected. Use of chemical pesticides is justified if grains are to be stored for over 4–5 months. Chemicals should be used only according to instructions on the label. Use of **diatomaceous earth**, ***Bacillus thuringiensis*** **(*Bt*)** and the leaves and extract of *Azadirachta indica* (**neem**) has also shown good results.

Rodent baits may be placed at many points in the warehouse, but not inside the bins.

Stored grain pests are very difficult to identify as they are very tiny and look quite similar to each other. Stored grain pests can be divided into three categories according to the way they feed on the grain, namely: (i) primary pests; (ii) secondary pests; and (iii) surface-feeding caterpillars. Primary pests lay eggs inside or outside the grain, and the larvae, on hatching, develop inside the grain, feeding on it. Secondary pests develop outside the grain and feed on the grain from outside or from within cracks developed in the grain. Surface-feeding caterpillars feed on the grain from the outside, usually on the top of the grain surface, but also sometimes from the bottom of the heaps of grains stored above aeration ducts or perforated drying floors.

Some of the important stored product pests are: the granary or wheat weevil (*Sitophilus granarius*), the saw-toothed grain beetle (*Oryzaephilus surinamensis*), the red flour beetle (*Tribolium castaneum*; see Fig. P.27), the rice weevil (*Sitophilus oryzae*), the Indian meal moth (*Plodia interpunctella*), the angoumois grain moth (*Sitotoga cerealella*), the African armyworm (*Spodoptera exempta*), the diamondback moth (*Plutella xylostella*) and the flat grain beetle (*Cryptolestes pusillus*).

These pests commonly lay eggs on or inside the grain, and feed and grow on the broken grain, flour or the dust.

Weevils are a common and serious pest in stored harvest grains. Weevils are beetles that belong to the superfamily Curculionoidea and order Coleoptera in the class Insecta. These dark brown to blackish, 1.25 cm long insect pests include rice weevils, wheat weevils, corn weevils and cotton boll weevils. Rice weevils can fly and so can spread extensively. They lay several eggs, each one being separately laid into a rice kernel, with the help of the adult's mandibles. The weevil seals the egg with a sticky secretion. The larvae that hatch start feeding inside the kernel itself.

Most weevils may be controlled by the chemical and biological compounds mentioned earlier. Besides these, the use of natural enemies may also help control the weevils. For example the histerid beetle (*Teretrius nigrescens*) predates on the large grain borer (*Prostephanus truncatus*). Both the adult and the larvae are attracted by the pheromone of the borer.

Stover

Dried and mature stalks and leaves of cereal crops are called stover. Stover of maize, sorghum and millet can be used as feed, bedding or for soil incorporation. It is also used as a surface mulch.

However, stover provides a home to pathogens that overwinter on them, only to become active (e.g. sporulation in a fungus) in spring and move to young and healthy plants.

Streptocycline biopesticide

Streptocycline, $C_{43}H_{62}ClH_9O_{20}$, with a CAS Registry Number of 39278-10-9 is an antibiotic agent that can control fungicidal and bacterial pathogens. It is a combination of streptomycin and **tetracycline**. Antibiotics, grouped under **streptomycins**, tetracyclines and **cycloheximides**, are metabolic products of certain bacterial species and have pesticidal effects.

Streptocycline inhibits the fungal enzymes that help in penetrating the host cell wall. Antibiotic effects of streptocycline on *Erwinia* spp. (e.g. *Erwinia carotovora*) have been recorded. Many rot diseases of maize, potatoes and onions may be controlled by the use of this compound. Meanwhile, seed dip treatment with a streptocycline solution (3 g/l) is also used to control bacterial leaf spot in bean varieties.

Applicators should be careful to avoid dermal exposure. Over-application of streptocycline may cause injury (e.g. stunted growth) to some plants such as members of the *Brassica* genus (see Streptocycline at bottom of the page).

Streptomyces griseus as an antibiotic

See **Cycloheximide as a bioPGR**

Streptomycin bactericide

Streptomycin, $C_{21}H_{39}N_7O_{12}$ (*O*-2-deoxy-2-methylamino-α-L-glucopyranosyl-(12)-*O*-5-deoxy-3-C-formyl-α-L-lyxofuranosyl-(14)-N[1],N[3]-diamidino-D-streptamine), is used as a bactericide as well as an antibiotic fungicide. Streptomycin is an off-white odourless and slightly bitter, hygroscopic powder, soluble in water (20 mg/ml) and insoluble in ethanol, isopropanol and carbon tetrachloride. This substance may be used as a salt, as in streptomycin sesquisulfate, streptomycin sulfate or streptomycin nitrate. It is unstable under strong acidic and alkaline conditions. It has a CAS Registry Number of 57-92-1.

Streptomycin, derived from the actinobacterium *Streptomyces griseus*, is an aminoglycoside, used as a systemic bactericide, fungicide and algicide. By inhibiting protein production, it combats the growth of these pathogens in fruits, vegetables, tobacco, ornamentals and in ponds. Gram-positive strains of bacteria are more susceptible than Gram-negative bacteria. It also helps to prevent tuber and seed rots, if cuttings of potato tubers or seeds of cotton, beans and cereals are dipped in a solution of streptomycin. Fire blights of apples and pears, walnut blight, bacterial canker and angular leaf blight are controlled by streptomycin. *Xanthomonas* spp. and *Pseudomonas* spp. are controlled by streptomycin. Trade names of streptomycin formulations include Agri-mycin-17, Plantomycin and Rimosidin. The compound is available as water-soluble powder (SP), pellets or emulsifiable concentrate (EC).

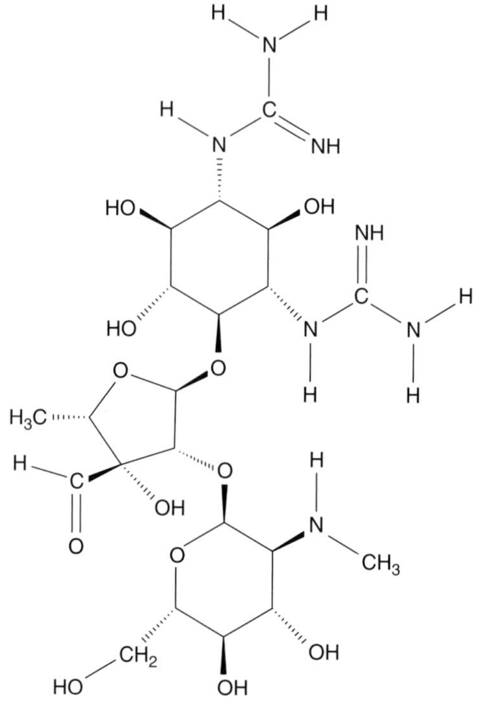

Streptomycin

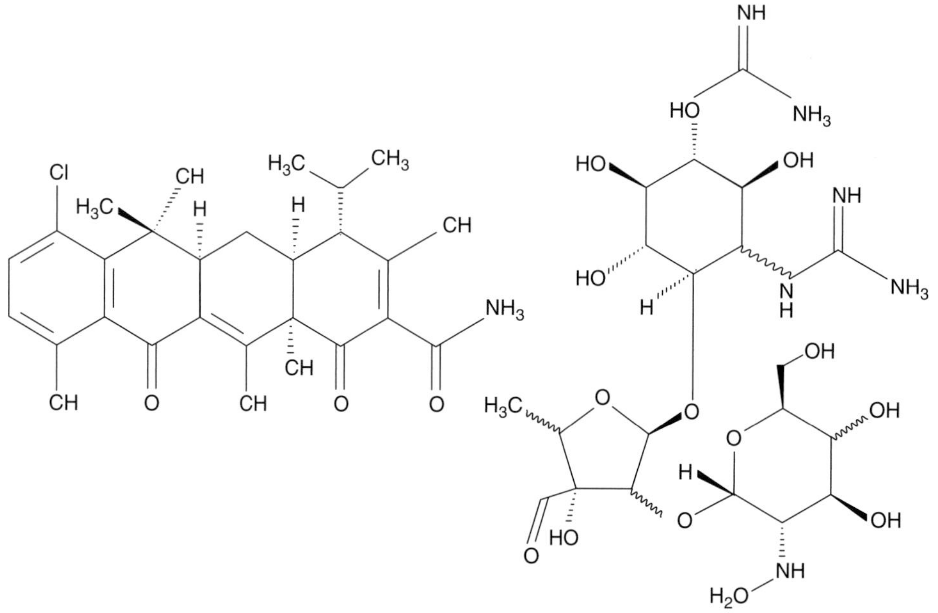

Streptocycline

The EPA has classified streptomycin as a Class IV toxin (i.e. relatively non-toxic). Since streptomycin may irritate the skin, product labels should bear the signal word 'CAUTION'. Streptomycin may cause **chlorosis** in maize, grapes, pears, peaches and certain ornamentals. It is registered as a general use pesticide (GUP) in many countries, according to the EPA. Streptomycin is practically non-toxic to rats and birds. It is slightly toxic to some species of fish and very toxic to algae. The ADI of streptomycin is 0.05 mg/kg/day.

Aminoglycosides are not extensively absorbed, and they are not metabolized by the body. Streptomycin is eliminated from the body depending on the kidney function.

s-Triazines herbicides

See **Triazine herbicides**

Strobilurin fungicide

Strobilurins are an important group of chemical compounds with a novel mode of action. They may be compared with quinone inhibiting compounds. Strobilurins, extracted from the fungus *Strobilurus tenacellus*, are the basis of a number of fungicides developed in recent years. It is also believed that the structure of strobilurin was developed from that of a group of natural β-methoxyacrylates, the simplest of which are strobilurin A and oudemansin A. Strobilurins act by inhibiting mitochondrial respiration in fungi. They bind at the quinine-outer binding site on cytochrome b and block electron transfer between cytochrome b and cytochrome c1. This disrupts the energy cycle within the fungus by halting the production of ATP.

Strobilurin A is the most important lead structure which has been modified to synthesize numerous fungicidal chemicals. Some examples of strobilurins include **trifloxystrobin**, pyraclostrobin, **azoxystrobin** and **kresoxim-methyl**. Strobilurin-based fungicides are effective on cereal crops, tomatoes and potatoes. The fungi suppressed by strobilurin-based fungicides include *Erysiphe gramminis* f.sp. *tritici* and *Septoria tritici*.

Strychnine rodenticide

Strychnine, with a molecular formula of $C_{21}H_{22}N_2O_2$, is a colourless, crystalline alkaloid and is used as a rodenticide and sometimes as an avicide. It acts by blocking the strychnine-sensitive glycine receptor situated near the spinal cord and brain.

Strychnine is a highly toxic rodenticide. Death can be brought about by an oral LD_{50} of between 10 and 50 mg/kg b.w. It is specially used for killing rodents and birds. It is absorbed through the gastrointestinal tract and stimulates the CNS. Symptoms of poisoning include nausea, vomiting, blurred vision and convulsions.

Sub-soiler

See **Weed management by mechanical methods**

Sugarcane red rot

See **Crop – sugarcane pest control**

Sugarcane whip smut

See **Plant disease smut**

Sulcotrione herbicide

See **4-Hydroxyphenylpyruvate dioxygenase (HPPD)**

Sulfentrazone herbicide

Sulfentrazone, $C_{11}H_{10}Cl_2F_2N_4O_3S$ (2′,4′-dichloro-5′-(4-difluoromethyl-4,5-dihydro-3-methyl-5-oxo-1*H*-1,2,4-triazol-1-yl) methanesulfonanilide), is a herbicide belonging to the class aryl triazolinone. The compound is a tan-coloured solid, with a faint sulfur-like odour that is soluble in water at 0.11 mg/l (pH 6) at 25°C and to some extent in acetone and other polar organic solvents. It is stable to hydrolysis and is readily photolysed in water. The CAS Registry Number is 122836-35-5.

Sulfentrazone, a pre-emergence herbicide mainly applied to soil, provides effective control of broadleaved weeds, some grasses and *Cyperus* spp. in soybeans, sugarcane and tobacco. It is a protoporphyrinogen oxidase (**PROTOX**) inhibitor of the chlorophyll biosynthesis pathway. On application, the herbicide is absorbed by the roots and foliage, and translocated primarily in the apoplasm. Plants emerging from treated soil turn necrotic and die after exposure to light. Foliar contact causes rapid desiccation and necrosis of exposed plant tissue.

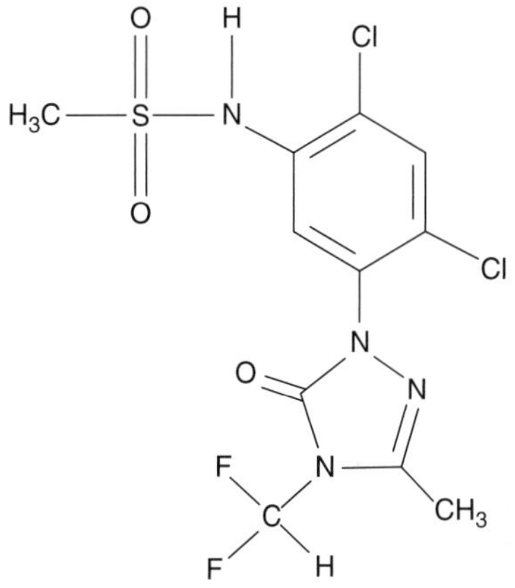

Sulfentrazone

The herbicide is formulated as a soluble concentrate (SL) and water-dispersible granules (WG) and marketed under trade names like Authority. It is not recommended for aerial application or through **chemigation**. It is to be used mainly on soybean in non-windy conditions.

Sulfentrazone is classified by the EPA as a slightly hazardous (Class III) compound. Its acute oral LD_{50} value in rats is >2855 mg/kg and its acute dermal LD_{50} in rats is >2000 mg/kg. The acute inhalation LC_{50} in rats is >4.13 mg/l. The lowest RfD value of sulfentrazone is 0.25 mg/kg.

Sulfentrazone is practically non-toxic to birds (the acute oral LD_{50} is >2250 mg/kg) and is toxic to fish (the LC_{50} in rainbow trout is >120 ppm).

Most of the administered dose is metabolized and eliminated via urine and faeces of rats within 72 h. It appears that sulfentrazone is highly mobile and persistent, and it can possibly leach into groundwater and move off-site to surface water.

Sulfentrazone is registered in the USA, Canada, the Philippines and South Africa but not in India.

Sulfometuron methyl herbicide

Sulfometuron methyl, $C_{15}H_{16}N_4O_5S$ (methyl 2-(4,6-dimethylpyrimidin-2-ylcarbamoylsulfamoyl) benzoic acid), is a urea herbicide (pyrimidinyl sulfonylurea family). It is a colourless solid which is soluble in water (244 mg/l at pH 7), in acetone (3300 mg/kg) and in acetonitrile (1800 mg/kg). Aqueous suspensions of the compound are stable to hydrolysis at pH 7 to pH 9 but unstable in alkaline media. Sulfometuron methyl is produced by reaction of 2-sulfonylisocyanate methylbenzoate with 2-amino-4,6-dimethylpyrimidin. The CAS Registry Number of sulfometuron is 74222-97-2.

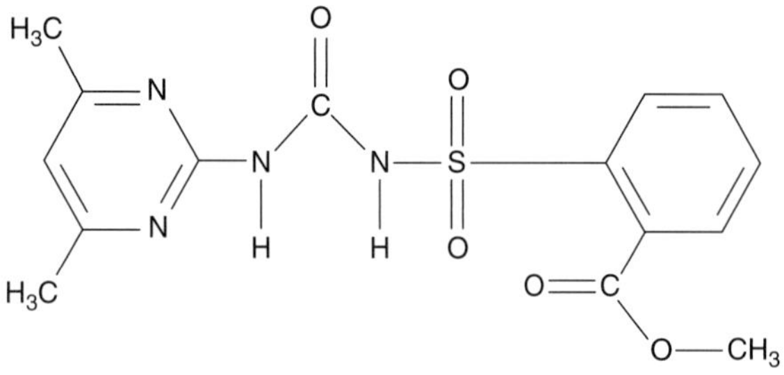

Sulfometuron methyl

The herbicide is a broad-spectrum compound, suitable for both the pre- and the post-emergence stages of plant life. It acts by inhibiting the enzyme acetolactate synthase (ALS), thus not facilitating the production of the three amino acids, leucine, isoleucine and valine. Growth of the targeted plant is affected, stopping cell division at the root tips (meristematic tissues). Sulfometuron methyl mainly has non-farm applications for example on turf, fences and in conifer forests. **Johnson grass** (*Sorghum halepense*) is well controlled with this herbicide.

The herbicide sulfometuron methyl is marketed as Oust and formulated as water-dispersible granules (WG). It is applied at concentrations ranging from 26 to 420 g/ha.

The EPA and WHO classify sulfometuron methyl as a relatively non-toxic compound (Class IV and Class U, respectively). The acute oral LD_{50} in rats is >5000 mg/kg. It is practically non-toxic to birds. The NOEL (2 years) in rats is 50 mg/kg diet, while the ADI/RfD is 0.275 mg/kg b.w./day. It is toxic to fish such as rainbow trout and bluegill sunfish, and moderately toxic to bees. It is a slight skin and eye irritant. The inhalation LC_{50} (4 h) in rats is >11 mg/l air. The LC_{50} value (96 h) in rainbow trout is >12.5 mg/l.

The half-life of sulfometuron methyl varies between 10 and 25 days. Degradation in soil is helped by microbial action and by hydrolysis. In animals, it is metabolized to hydroxylated sulfometuron methyl.

Sulfometuron methyl is registered in the USA, Canada and Australia, but not in the EU.

Sulfonylurea herbicides

Sulfonylureas are a group of herbicides that are branched-chain amino acid inhibitors, which (along with inhibitors of aromatic amino acids and **glutamine synthetase**) inhibit amino acid biosynthesis. Specifically, they act on the biosynthesis of the three aliphatic amino acids, valine, leucine and isoleucine.

The structure of sulfonylurea is a combination of urea and triazine and its inhibitory action affects photosynthesis and respiration, besides protein synthesis. The symptoms of their inhibition include **chlorosis**, necrosis and discoloration.

Many herbicides are included in this group such as bensulfuron, chlorimuron, **oxasulfuron**, **nicosulfuron**, **triasulfuron** and **thifensulfuron**. Since most sulfonylureas are extremely effective even in small doses (from 8 to 80 g/ha), over-use should be prevented to keep toxicity or resistance at bay.

Sulfosulfuron herbicide

Sulfosulfuron, $C_{16}H_{18}N_6O_7S_2$ (1-(4,6-dimethoxypyrimidin-2-yl)-3-(2-ethylsulfonylimidazo[1,2-a]pyridin-3-yl)sulfonylurea), is a sulfonylurea herbicide. It is a white, odourless solid that is soluble in water (18 ppm at pH 5 and 482 ppm at pH 9). The compound is stable under normal storage conditions. Sulfosulfuron is produced by reaction of 2-ethylsulfonylimidazophenyl-3-sulfonaamide with 2-(4,6-dimethoxy) pyrimidine isocyanate. The CAS Registry Number is 141776-32-1.

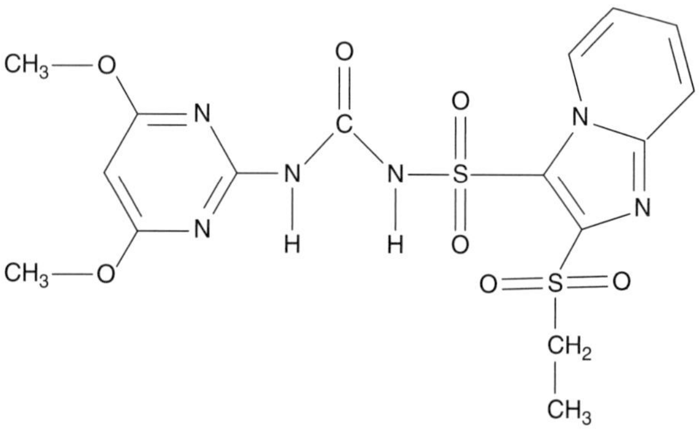

Sulfosulfuron

The herbicide is used widely throughout the world for control of broadleaved and grass weeds growing in maize and other cereal crops. The fundamental mode of action entails inhibition of acetolactate synthase (ALS), an essential enzyme in aliphatic amino acid synthesis.

The herbicide sulfosulfuron is formulated as water-dispersible granules (WG). It is not to be combined with any compound with a pH of less than 5 (as solubility can suffer and the herbicide may become less effective).

The EPA has identified sulfosulfuron as a Class III toxin. It causes irritation to the skin but may not injure the eyes. The acute oral LD_{50} in rats is >5000 mg/kg. The acute oral LD_{50} in bobwhite quails is >5620 ppm. The LC_{50} in rainbow trout is >95 mg/l. It is slightly toxic to birds and fish. It is toxic to bees (the oral LD_{50} is >30 μg per bee). Phytotoxicity occurs in some varieties of barley and oats. Crop injury may result if sulfosulfuron is applied immediately after crop emergence. It is practically non-toxic by inhalation. The NOEL (2 years) in rats is 24.4–30.4 mg/kg b.w./day. The ADI is 0.24 mg/kg/day.

In animals, sulfosulfuron is eliminated rapidly from rats, with limited metabolism. A major pathway of metabolism in soil is cleavage of the sulfonylurea linkage with oxidative demethylation. Degradation products include sulfonamide and dimethoxypyrimidinamine. It is observed that it does not bioaccumulate, although studies have revealed its presence in wheat straw and forage. In soil, both hydrolytic cleavage as well as photodegradation is observed. The compound also degrades rapidly in water with a DT_{50} between 20 and 32 days.

Sulfosulfuron is registered for use in India, Australia, Canada, South Africa, the USA and many countries of the EU including the UK, Denmark, Finland, Germany and Hungary.

Sulfotep insecticide

Sulfotep, $C_8H_{20}O_5P_2S_2$ (*O,O,O′,O′*-tetraethyl dithiopyrophosphate), is an aliphatic organothiophosphate insecticide. It is a pale yellow liquid, soluble in water at 10 mg/l and miscible with most organic solvents. Sulfotep is incompatible with strong oxidizers, producing toxic phosphorus oxides. It corrodes iron. The insecticide produces the highly toxic **phosphine gas** in the presence of hydrides and other strong reducing agents. It slowly hydrolyses in aqueous media. Sulfotep should be stored in a cool, dry, well-ventilated area in tightly sealed containers. Sulfotep is synthesized by the action of water on diethyl phosphorochloridothionate in the presence of pyridine. Its CAS Registry Number is 3689-24-5.

Sulfotep, a cholinesterase inhibitor, is a non-systemic insecticide, acaricide and fumigant, with contact and respiratory action. The insecticide is used in the control of aphids, thrips, spider mites and whiteflies on glasshouse crops, ornamentals and mushrooms.

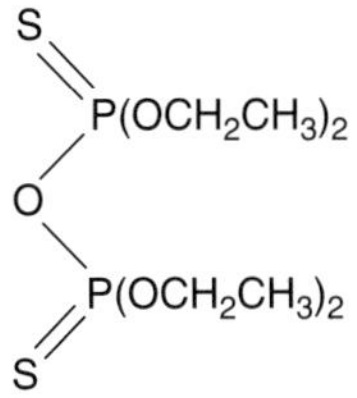

Sulfotep

The insecticide-acaricide sulfotep is extremely hazardous (Class Ia toxin). The oral LD_{50} in rats is 5 mg/kg, while the dermal LD_{50} in rats is 65 mg/kg. The inhalation LC_{50} of sulfotep in rats is 0.05 mg/l. It is extremely toxic to fish, *Daphnia* spp. and other aquatic species (the acute 96 h LC_{50} in rainbow trout is 0.00361 mg/l). The ADI for sulfotep is 0.001 mg/kg b.w./day. It is phytotoxic to some varieties of chrysanthemums, orchids and roses. Being an acute poison and a cholinesterase inhibitor, it has been listed as a PAN Bad Actor chemical. The symptoms of toxicity include excessive salivation and sweating, muscle twitching and weakness, diarrhoea, abdominal cramps, pin-point pupils and blurred and dark vision, respiratory depression and tightness, incontinence and confusion, dizziness and tremors, and many others. Eyes and skin should be well protected with appropriate PPE to ensure safety to applicators, mixers and others in the vicinity. Disposal of wasted or unused amounts of sulfotep products should be carried out with due care to ensure the chemical does not enter the soil or water sources by accident. Sulfotep is available in the market under the trade name of Bladafum.

Mammals eliminate sulfotep (via urine) in the form of diethylthiophosphate. In plants, sulfotep metabolizes into monothiotep, tetraethylpyrophosphate, diethylthiophosphonic acid and diethylphosphonic acid. Many of these rapidly degrade into simple thiophosphates or phosphates. How sulfotep is degraded varies with different soil types. Sulfotep is activated by oxidative desulfuration to tetraethyl monothiopyrophosphate and pyrophosphate in the environment.

Many countries have banned the use of sulfotep as a result of its acute toxicity, for example the USA and the EU.

Sulfur as a pesticide

Sulfur (molecular formula S_x) one of the oldest known fungicides, is an inorganic compound and also acaricide. It is a yellow-coloured powder that exists in many forms. Sulfur is soluble in water (0.063 g/m^3), hot benzene, acetone, ether and petroleum ether. Dusting sulfur, wettable sulfur, elemental sulfur and bio-sulfur (elemental sulfur with hydrophilic characteristics) are the various forms of sulfur for crop use. Sulfur is a stable compound that exists as a sulfate and a sulfide. It can form sulfides in the presence of strong alkalis.

Sulfur is an essential element in all living cells, being a part of many amino acids in plants and animals. It promotes healthy root formation and nodulation, and is thus used as a fertilizer.

Sulfur compounds are classified as organic and inorganic types. Inorganic sulfur compounds include elemental sulfur, lime sulfur, **calcium polysulfides**, ferric sulfate, calcium sulfates and other salts of sulfuric acid. Organic sulfur compounds are a result of changes in the functional group. Compounds developed out of the sulfur analogues of ethers (thioether) and alcohols (thiols) are organic in nature. Carbamates are an example.

Sulfur acts as a non-systemic, protective fungicide and an acaricide with vapour and contact action. It is a non-specific thiol reactant that inhibits respiration. It is used on livestock to keep ticks and mites away (by adding it to salt licks). As a fungicide, it effectively controls **powdery mildews** on apples, pears, peaches, apricots, grapes, beet crops, vegetables and in forest crops. It also controls black spot and **scab**. It controls eriophyid mites such as the plum rust mite (*Aculus fockeui*). When applied, sulfur lowers the pH level, making the medium acidic.

Chemically, sulfur is a very stable compound and is compatible with most other insecticides and fungicides. However, for reasons of phytotoxicity, its mixing with oil is avoided.

Organic sulfur compounds used as fungicides are referred to as **dithiocarbamate fungicides**, which are derived from dithiocarbamic acid (NH_2 CS_2 H). This acid is an unstable powder which readily forms salts upon reaction with metals. These are one of the most versatile and most widely used groups of compounds. Thiocarbamates are protectants and used in seed-dressing treatments. They are also used as preservatives in textiles and paints.

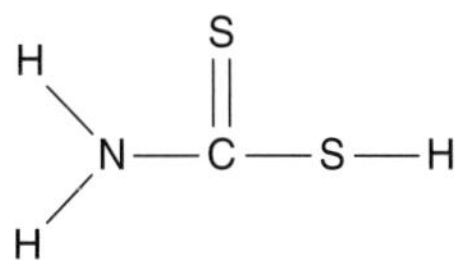

Dithiocarbamic acid

Thiocarbamates may be classified into three groups: (i) dimethyl disulfides; (ii) metallic dithiocarbamates; and (iii) ethylenebisdithiocarbamates. It is considered that the thioureide ion is responsible for the antifungal activity of **thiuram disulfides** and dialkyl dithiocarbamates. They disrupt the activity of fungal cells, besides acting as strong chelating agents. Bisdithiocarbamate groups act by production of diisothiocyanate groups, with ethylene thiuram monosulfide acting as an intermediate product. Isothiocyanate reacts with SH groups essential for functioning of SH-bearing enzymes. Isothiocyanates react with the thiol group of proteins and enzymes. N-methyl dithiocarbamate has been found to cause alteration of permeability of the mycelium of *Rhizoctonia solani*.

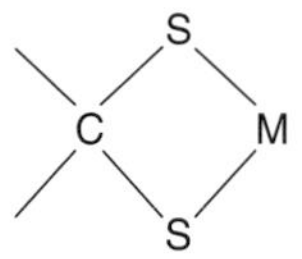

Thioureide ion

The formulations of sulfur are dustable powder (DP), water-dispersible granule (WG) and wettable powder (WP). **Sulfur dust** is applied to crops with a duster or through the aerial route. It is also common to dust clothing or limbs with sulfur powder. Sulfur is either applied alone or as **lime sulfur** (which is a combination of calcium hydroxide and **elemental sulfur**). It can be combined with other compounds such as **zineb**, copper or gypsum. The common application doses include 6 kg/ha on citrus and 4–10 kg/ha on vines. It is available commercially under the trade names, Cabritol, Sulfex and Sulphy. Mixing sulfur with oils may become phytotoxic. Dusting is not advised on hot sunny days, as it may induce burning symptoms.

The EPA classifies sulfur as a Class IV toxin. It causes irritation to the skin and mucous membranes. The acute oral and dermal LD_{50} in rats are >5000 and >2000 mg/kg, respectively. In birds, the acute oral LC_{50} in bobwhite quails is >5000 ppm. It is non-toxic to fish and bees and practically non-toxic to humans and animals. It may be phytotoxic to cucurbits and apricots.

In plants, degradation of sulfur proceeds mainly by microbial reduction. In soil, it does not leach into groundwater.

Sulfur is registered for use in India, South Africa, Madagascar, Australia, New Zealand, the UK, the USA and Canada. The MRL value of sulfur for almonds and cashew nuts is 50 mg/kg. (See also **Elemental sulfur fungicide**.)

Sulfur as a plant nutrient

See **Plant nutrient excesses and deficiencies**

Sulfur dust

See **Elemental sulfur fungicide**

Sulfur knapweed moth

See ***Agapeta zoegana* (sulfur knapweed moth) as a biocontrol agent**

Sulprofos insecticide

Sulprofos, $C_{12}H_{19}O_2PS_3$ ((RS)-*O*-ethyl *O*-4-(methylthio) phenyl *S*-propyl phosphorodithioate), is a phenyl organothiophosphate insecticide. It is a colourless, volatile oil with a mercapton-like odour. It is nearly insoluble in water (0.31 mg/l) but readily soluble in isopropanol, dichloromethane, hexane and toluene. Sulprofos undergoes photolysis in water and on soil surfaces. Sulprofos is produced by reaction of *p*-thiomethylphenol with *O*-ethyl *S*-propyl thionophosphoric acid chloride. It has a CAS Registry Number of 35400-43-2.

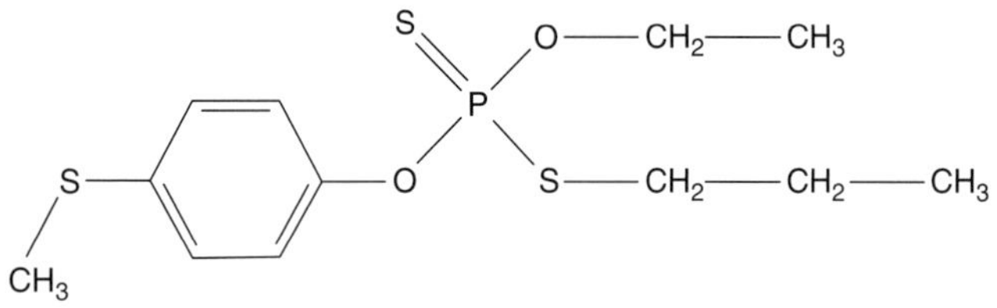

Sulprofos

Sulprofos, a cholinesterase inhibitor, is a non-systemic insecticide with contact and stomach action. It is used in the control of thrips, whiteflies and other sucking insects in cotton, soybeans, tobacco, vegetable and tomatoes.

Sulprofos is formulated as emulsifiable concentrate (EC) and ULV liquid (UL). It is available in the market under the trade name of Bolstar.

The WHO has classified sulprofos as a Class II toxin. The acute oral LD_{50} in rats is 176 mg/kg, while the same in bobwhite quails is 47 mg/kg. Sulprofos is moderately toxic to fish, with the acute 96 h LC_{50} in bluegill fish being 11 mg/l. In normal use, this insecticide does not irritate the eyes or skin. The ADI for sulprofos is 0.003 mg/kg b.w./day, while the (2 days) short-term dietary NOEL among rats is 6 ppm diet. Being an OP insecticide, exposure to sulprofos causes excessive salivation and sweating, muscle twitching and weakness, headache, dizziness, nausea, vomiting, blurred or dark vision, tremor, abdominal cramps, respiratory depression and chest tightness, seizures and incontinence.

In mammals and plants, the degradation products are oxygen analogues, sulprofos sulfoxide and sulprofos sulfone. The latter two are the main breakdown products in plants and soil. Sulprofos is readily adsorbed in soil, with the half-life being in the range of several days to several weeks, depending on the soil type.

The insecticide sulprofos is considered obsolete and is not registered in the USA or in the EU.

Super absorbers

See **Soil conditioners**

Surface active agent

See **Adjuvants**

Surface tension

See **Surfactants**

Surfactants

In the context of crop use, a surfactant is a material used in liquid formulations that modifies the physical properties of the liquid by changing its surface characteristics. Surfactants have a water-soluble (hydrophilic) polar head and an oil-soluble (lipophilic/hydrophobic) long hydrocarbon chain. This enables them to penetrate the target plant along with the pesticide active ingredient (a.i.).

Surfactants include **wetting agents**, emulsifiers and dispersing agents. Their main function is to reduce the surface tension of water (when mixed at rates of 0.25–0.5% (v/v) in the spray solution). The property of the surface of a liquid to remain pulled inwards and resist external forces is its surface tension. Each molecule of a liquid pulls other molecules towards itself with equal force. This cohesion can be seen in the formation of a drop. To make a uniform solution, the surface tension has to be brought down adequately.

Surfactants are of four types (the classification being based on the differences in the polar head groups): (i) anionic; (ii) cationic; (iii) non-ionic; and (iv) ampholytic.

Anionic surfactants have a negative charge and are excellent wetting agents and detergents. But since they ionize easily, they interact and bind with other ions of the spray solution (commonly Ca, Mg, Fe ions) forming insoluble complexes with the solution. An example of an anionic surfactant is sodium stearate ($C_{17}H_{35}COO^{-}Na^{+}$).

Cationic surfactants have a positive charge and have a basic structure of quaternary amine salts. These are not very effective surfactants, since they are phytotoxic, forming insoluble precipitates in the solution. But these surfactants have been found suitable in some cases. An example of a cationic surfactant is cetyltrimethyl ammonium bromide ($C_{16}H_{33}(CH_3)_3N^{+}Br^{-}$).

Non-ionic surfactants are the most widely used surfactants. They have a long lipophilic hydrocarbon chain, referred to as a tail. They are used in various formulations because they tend not to create foam and have low phytotoxicity and low mammalian toxicity. They do not ionize, thus do not form insoluble salts. An example of a non-ionic surfactant is poly(ethylene oxide) laurylalcohol ($C_{12}H_{25}O(CH_2CH_2O)_{23}H$). Non-ionic surfactants have a negative polarity and originate from ethylene oxide groups ($CH_2.CH_2.O$). Water solubility increases with an increase in ethylene oxide units (while oil solubility decreases), while the length of the hydrocarbon chain can determine lipophilicity. Another example of a non-ionic surfactant is organosilicones (trisiloxane surfactants). These are excellent surfactants because at a certain concentration (0.1% w/w), they reduce surface tension and allow rapid spreading and wetting on hydrophobic surfaces, leading to increased foliar uptake.

The concentration of a surfactant, in say a herbicidal solution, determines the efficiency of the herbicide activity. Continuous increase in the concentration leads to the formation of micelles or aggregates of the lipophilic tails of the surfactants. The concentration at which surfactant monomers form micelles is the critical micelle concentration (CMC).

Ampholytic surfactants contain both a positive and a negative charge on the same molecule. The pH of the solute and the a.i. determine the ionization of this surfactant, thus making its performance unpredictable. So, the polarity of the surfactant varies.

Vegetable seed oils used as **adjuvants** in a herbicide formulation, commonly originate from soybean, sunflower and canola. These oils contain fatty acids like oleic and linoleic acids. Vegetable seed oils help to maintain the physical state of liquid formulations at normal temperatures. Among these kinds of adjuvants, methylated crop oil gives the best spray retention, because it is said to give an optimum hydrophilic/lipophilic balance. Methylated crop oil is manufactured by adding methanol units to vegetable-based oil.

Fertilizers (inorganic salts) such as ammonium sulfate and urea ammonium nitrate aid in the performance of some pesticides, especially in some broadleaved plants and grasses. These chemicals also increase absorption and translocation of these pesticides. Ammonium sulfate increases the solution density and buffering capacity. It prevents the formation of complexes of cations of Ca, Mg and Fe (in hard water) with the pesticide components. Such complexes reduce herbicide uptake. The mode of prevention of such a complex is by adding the ammonium compound to the herbicide, thereby allowing the ammonium ions to form salts with the herbicide, and then adding the hard water to the solution.

Some other compounds such as **citric acid** and high-fructose corn (maize) syrup are used as adjuvants in a blended mix of 60:40, oil: surfactant ratio.

The ideal choice of an adjuvant depends on the combined criteria of lowest phytotoxicity and maximum performance. There are many adjuvants in the market and it is prudent to make a choice depending on the merit of individual compounds as well as one's particular requirements.

Suspension concentrates (SC)

Water-based compositions in which water-insoluble active ingredients (a.i.) are wet-ground to give a heterogeneous mixture of fine solid particles suspended in an aqueous medium is called a suspension concentrate (SC). Such pourable formulations require suspending agents and dispersants to ensure a stable product. An advantage of this formulation is its volumetric measurability. The very small particle size of the a.i. enables better uptake by the target site. However, this composition may sometimes need agitation. Seeds are commonly treated with a suspension concentrate (FS).

Suspensions made by mixing a dry a.i. in a liquid are referred to as flowables. The dry a.i. is a wettable powder (WP), which is stirred into a suspension. Fungicides are often formulated as flowables. Sometimes this mixture may be made thick or creamy for specific applications.

Symptoms of pesticide poisoning

Pesticide application comes with its share of hazards to man, animal and the environment. While harm to mammals, birds, fish and **beneficial organisms** are known and can be consciously controlled by adhering to label instructions, accidents can never be ruled out. Exposure to a pesticidal compound can occur by oral, dermal or inhalation routes and depending on the severity, can lead to severe injury or even death.

Hence, it is important to be aware of toxicity symptoms, so that treatment can be provided to restore the victim in time. Some of these symptoms are externally visible, while others are not. Some of the common symptoms of poisoning are burning of the eyes and pin-point pupils, rashes and blisters on the skin, nausea, vomiting, stomach cramps and loose motions, headaches, dizziness, confusion and convulsions, chest pain, breathlessness and unconsciousness. Each group of chemical compounds has a common set of symptoms.

Exposure to organophosphate (OP) and carbamate chemicals causes distinct symptoms like pin-point pupils, blurred vision, nausea, diarrhoea, headache, giddiness, convulsions, slow reflexes, breathlessness and tightness in the chest. Organochlorine (OC) compounds cause nervousness and also liver and kidney damage.

Early identification of toxicity symptoms can help in providing prompt treatment. At all times, it is necessary to be aware of the first-aid measures before the victim can receive specific medical treatment. For instance atropine, given in prescribed measures, can be an antidote for OP and carbamate poisoning, or vitamin K can help begin the process of blood clotting, in the case of accidental intake of **warfarin**. (See also **Worker safety norms and first-aid techniques**.)

Synergists

See **Pesticide synergists**

Synthetic pyrethroids

See **Pyrethroid insecticides**

SYP-Z071 fungicide

See **Enestroburin fungicide**

Syrphids

See **Biocontrol of crop pests – predators**

Systemic acquired resistance (SAR)

Whenever subjected to trauma, plants, like any other living systems, prepare themselves for a fight, in order to survive. Trauma may be due to pathogen attacks, injury or malnutrition. Plants may vary in their responses to an attack, with different corresponding elicitors associated with such responses. It is this understanding of the biochemical changes in a plant that has paved the way for developing genetically modified crop varieties that are robust and disease resistant.

A distinct defence response elicited by the attack of non-pathogenic bacteria is called **induced systemic resistance** (**ISR**). The ISR requires the activation of signalling pathways involving **jasmonic acid** and **ethylene**.

Plants activate their defence in the event of a pathogen attack from insects, bacteria and fungi. Such an attack elicits in them distinct signal transduction pathways which are crucial for defending themselves and resisting the attack. Such induced resistance, triggered by local infection, is termed systemic acquired resistance (SAR). SAR is seen to act effectively against disease symptoms caused by *Phytophthora parasitica*, *Peronospora tabacina*, the tobacco mosaic virus (TMV) and tobacco necrosis virus (TNV) affecting cucumbers, tobacco and *Arabidopsis thaliana* (mouse-ear cress or thale cress). The initial response in most plants is the hypersensitive response, a mechanism that prevents the spread of disease. The response may be the sudden death of cells in the infection zone, the behaviour expressed by a gene in some plant species.

The state of SAR brings about an overall umbrella of protection against a range of pathogens, making the SAR activity broad based. SAR also brings about a change in gene expression, with a set of genes, identified as SAR genes, being associated with the onset of SAR. Many SAR proteins belong to the same protein class as pathogenesis-related proteins (PR proteins) for instance, PR-1, β-1,3-glucanase and some forms of chitinase.

For SAR to be activated, the production and accumulation of the signal molecule salicylic acid (SA) is necessary, with the isochorismate pathway being the major source of SA during SAR. SAR seems to provide whole-plant protection, given that an infection provokes SA accumulation in many uninfected plant parts. Extensive research on SAR against various pathogen attacks on various plants has indicated the possibility of many chemicals (e.g. phosphate, silicon, nucleic acids and even a lipid molecule) being responsible for activating SAR. It is now understood that after SA is accumulated, a positive regulator protein NPRI or NIMI moves to the nucleus to activate transcription of the PR genes, the expression of which causes resistance activity.

It is considered that for an agent to activate SAR, the chemical should not exhibit direct antimicrobial activity and should resist the same pathogens as the naturally activated SAR would. Other chemicals such as 2,6-dichloro-isonicotinic acid (INA) or benzo (1,2,3) thiadiazole-7-carbothioic acid *S*-methyl ester (BTH) can also activate SAR effectively.

Systemic fungicides

See **Fungicides**

Systemic herbicides

Herbicidal action is initiated by either physical contact of the herbicide with the plant or its penetration into the plant. A herbicide that enters the plant part, translocates to plant tissues within the plant and acts on a specific plant process is called a systemic herbicide. Systemic herbicides translocate either via leaves (if a foliar application is made) downwards, or via the roots (through soil application) upwards. The site of action is almost always away from the point of application.

Systemic herbicides are often used against established perennial weeds. Examples of systemic herbicides are those based on phenoxy acids such as **2,4-D** and **MCPB**. These chemicals mimic growth hormones and so are readily absorbed by plants. Similarly, **dicamba**, a benzoic acid herbicide, is also an effective systemic foliar chemical.

Sometimes, plants have features (e.g. **epicuticular wax** or leaf hairs) that hinder the absorption of systemic herbicide formulations. These may be overcome by use of effective adjuvants that help reduce the surface tension of the solvent and increase the retention and coverage of the spray.

Systemic pesticides

Any chemical that is absorbed into the system of a target organism or plant and acts from within, by inhibiting or damaging some biochemical processes, is a systemic pesticide. At the same time, an organism or a plant treated with a systemic pesticide also becomes a more able participant in setting up conditions that are unfavourable to invaders. Some systemic chemicals are already pesticidal in nature, while others become chemically altered to toxic chemicals, once they are absorbed.

A systemic pesticide differs from **non-systemic** ones, in that it is not so easily affected by external factors such as rain or strong winds, although it should be applied in favourable temperatures, according to instructions on the label. A systemic chemical also acts like a protectant to a plant by repelling a pest from harming it, or killing an infestation inside the plant tissue. In the case of plants, systemic pesticides, taken up through the roots, stems or leaves, move through the plant's vascular system to most parts of the plant and act against the pests within the plant, as in the case of stem borers and root nematodes. Commonly, but not always, a systemic insecticide or fungicide moves upwards and outwards.

Some examples of systemic pesticides are difenoconazole and triadimefon (both fungicides), **imidacloprid**, **acephate** and **permethrin** (insecticides) and **2,4-D**, **atrazine** and **glyphosate** (which are herbicides).

There are some variations among systemic chemicals, some being locally systemic (i.e. they will not move long distances), while others also have contact activity.

Systemic pesticides should be applied before harvest to avoid the chemicals from entering the food chain. Care has also to be taken to prevent systemic insecticides that poison pollen and nectar from killing bees and other important pollinators.

The most common methods of application of systemic pesticides are: (i) spraying the leaves; (ii) drenching the soil; (iii) treating the seeds; and (iv) injecting the trunk.

T

2-(Thiocyanatomethylthio) benzothiazole fungicide

2-(Thiocyanatomethylthio) benzothiazole (**TCMTB**) with molecular formula $C_9H_6N_2S_3$ is a fungicide as well as bactericide. It is a dark orange liquid, stable under normal storage conditions. It is soluble in water at 125 mg/l. TCMTB is produced by reacting 2-mercaptobenzothiazole with a series of compounds such as sodium ethoxide or chloromethylthiocyanate.

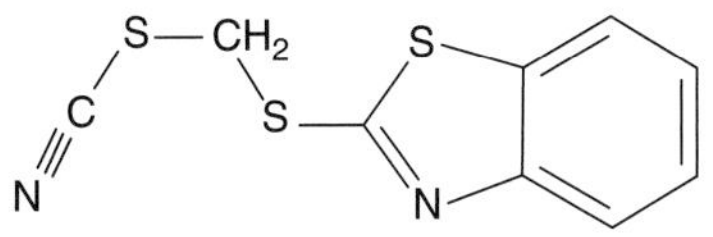

2-(Thiocyanatomethylthio) benzothiazole

2-(Thiocyanatomethylthio) benzothiazole is used to protect cereal seeds against fungal attacks. It also protects timber, paper and leather from algal, fungal and bacterial attacks.

In toxicological studies, the compound TCMTB is found to be highly toxic to fish. It is moderately toxic to birds (with an LD_{50} of around 1500 mg/kg). The NOEL in dogs is 3.8 mg/kg b.w., while the ADI in humans is 0.01 mg/kg b.w./day.

The degradation of TCMTB in soil is very fast with a DT_{50} of a little over 1 day. This chemical also hydrolyses in alkaline medium.

This fungicide is registered for use in Australia, New Zealand, the USA and Canada. However, it is not allowed for use in the EU.

2,3,6-TBA herbicide (TBA)

2,3,6-TBA, $C_7H_3Cl_3O_2$ (2,3,6-trichlorobenzoic acid), is one of the compounds that belongs to the group of auxin-like herbicides. TBA is chlorinated at the 2, 3 and 6 positions. Degradation occurs only in reductive conditions and in the laboratory, upon the action of *Brevibacterium* spp. TBA is very soluble in water (7700 mg/l) and even more so in acetone and benzene. The CAS Registry Number of TBA is 50-31-7.

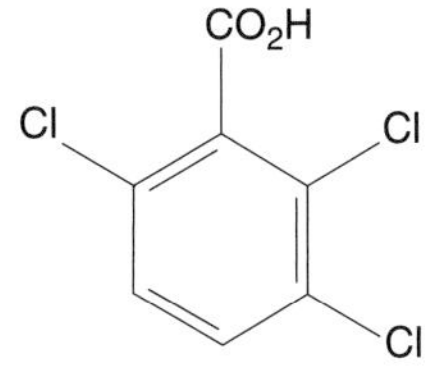

TBA

TBA, which is normally a stable compound, is a systemic benzoic-acid-based herbicide. It is used in cereal grain crops to control deep-rooted, perennial, broadleaved plants, as well as annual weeds.

The WHO categorizes TBA as a moderately toxic chemical. It is moderately toxic to rats.

TBA does not metabolize in plants or soil and hence is considered very persistent. Its solubility in water makes it a leaching hazard. However, in soils that are rich in organic matter it gets adsorbed, making those amounts unavailable, thus reducing phytotoxicity.

3,5,6-Trichloro-2-pyridinol

See **Chlorpyrifos insecticide**

Tablet formulation

A tablet is a dry formulation of a pesticide (or a fertilizer). There are two types of formulation, tablets for direct application (DT) and water-dispersible tablets (WT). These products are made by compression techniques and can be produced in large volumes. They can be produced both mechanically and by hand.

The main feature of the tablet is its slow release. Since tablets administer a low dose of the active ingredient (a.i.), they are very convenient when a slightly longer-term effect is desired. Tablets are commonly used in rice-producing regions, including Japan. Once applied to the surface of the paddy water, tablets sink and slowly break up to release the a.i.

The water-dispersible tablet (WT) is similar to the water-dispersible granule (WG). WT is designed to be added to spray water in conventional application equipment, where the tablet disintegrates by mechanical means or by effervescence. The WT compositions include surface active components, swelling agents and lubricating agents (to aid the break up of the tablet and a.i. release).

The attractive features of WT are safety in handling and convenience. However, the complex formulation, manufacturing and packing processes make tablet formulations expensive (sturdy packaging is necessary to protect it from breaking and from moisture).

Tachinid flies

See **Biocontrol of crop pests – parasitoids**

Tardigrades

See **Biological control of nematodes**

Tau-fluvalinate insecticide

Tau-fluvalinate, $C_{26}H_{22}ClF_3N_2O_3$ ((*RS*)-α-cyano-3-phenoxybenzyl *N*-(2-chloro-α,α,α-trifluoro-p-tolyl)-D-valinate), is a pyrethroid insecticide. It is a viscous amber oil that is soluble in water, toluene, acetonitrile, isopropanol, dimethylformamide and *n*-octanol. It is unstable in sunlight and incompatible in alkaline medium.

Tau-fluvalinate is produced by the reaction of 2-bromo-3-methyl butyric acid with 2-chloro-3-trifluoromethylaniline, followed by esterification with α-cyano-3-phenoxybenzyl alcohol. Its CAS Registry Number is 102851-06-9 (see Tau-fluvalinate at bottom of the page).

Tau-fluvalinate is an insecticide and acaricide with contact and stomach action. It is used to control a wide range of insects including moths, aphids, thrips, **leafhoppers**, whitefly and spider mites on ornamentals and greenhouse plants. Food use is mainly related to honey bees, where it is used to control a mite (*Varroa jacobsoni*) that parasitizes the Asian honey bee (*Apis cerana*). However, it is occasionally applied to apples, pears, peaches, carrots and cole crops (which are grown for seed). As an insecticide-acaricide, tau-fluvalinate affects the nervous system (both the CNS and the peripheral nervous system) by keeping the sodium channels in the neuronal membrane open, overstimulating the nerve cells, causing paralysis and even death.

Tau-fluvalinate is formulated as an emulsifiable concentrate (EC), ULV liquid (UL) and as vapour-releasing products. Drenching, dipping, aerial spraying and fogging are some methods of applying tau-fluvalinate. Tau-fluvalinate is available under the trade names of Apistan, Klartan, Minadox and Mavrik.

Tau-fluvalinate is moderately toxic to rats (the acute oral LD_{50} is 546 mg/kg, while the dermal LD_{50} is >2000 mg/kg b.w.). The inhalation LC_{50} in rats is >0.56 mg/l. It is slightly toxic to birds with the acute LD_{50} in bobwhite quails being > 2510 mg/kg. Tau-fluvalinate is extremely toxic to fish and to *Daphnia magna* (the acute 96 h LC_{50} for rainbow trout is 0.000794 mg/l). The acute 48 h contact LD_{50} of tau-fluvalinate in honey bees is 12 μg per bee. However, this insecticide is not toxic to earthworms. The ADI in rats is 0.005 mg/kg b.w./day. According to the WHO, it is a Class III toxin (slightly hazardous). Exposure to tau-fluvalinate results in a prickling and creeping sensation on the skin, irritability to sound and touch, headache, dizziness, nausea, fluid in the lungs and muscle twitching and other symptoms of pyrethroids. It is also a reproductive toxin.

In animals, the principal metabolites include aniline acid, 3-phenoxy benzyl alcohol, 3-phenoxybenzoic acid (3-PBA) and 3-4′-hydroxy phenoxybenzoic acid. In soil and water, it degrades rapidly, under aerobic conditions, but is persistent in anaerobic conditions.

Tau-fluvalinate is registered in Australia, Canada, South Africa, the USA and in most countries of the EU (e.g. Denmark, Finland, Germany, Greece, Portugal and the UK). According to the EU Pesticide Database, the MRL for wheat is 0.02 mg/kg and for potatoes is 0.01 mg/kg.

TCMTB fungicide

TCMTB fungicide is the abbreviation for **2-(thiocyanatomethylthio) benzothiazole fungicide**.

Tebuconazole fungicide

Tebuconazole, $C_{16}H_{22}ClN_3O$ ((*RS*)-1-*p*-chlorophenyl-4,4-dimethyl-3-(1*H*-1,2,4-triazol-1-ylmethyl)pentan-3-ol), is a triazole class of fungicide (which is a **demethylation inhibitor** (DMI)). It is also known as fenetrazole and ethyltrianol. It is a colourless, crystalline substance which is soluble in water (36 ml/l), dichloromethane (>200 gm/l), isopropanol and toluene. It is stable to high temperatures, photolysis and hydrolysis (the hydrolysis DT_{50} is >1 year). Its CAS Registry Number is 107534-96-3.

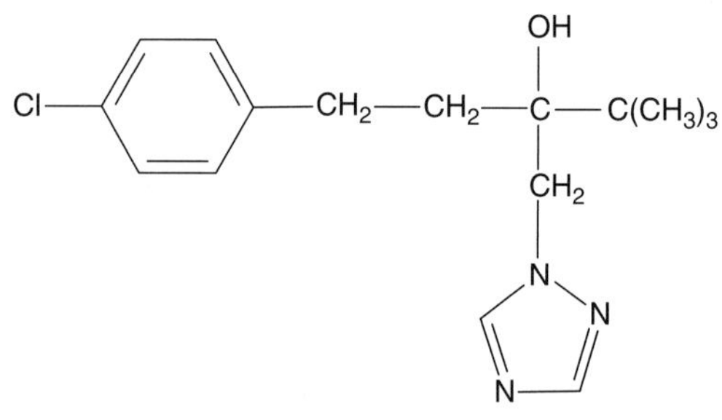

Tebuconazole

Tebuconazole is a systemic foliar fungicide that has curative and protective properties. Tebuconazole is a DMI, inhibiting steroid (ergosterol biosynthesis) demethylation. It is rapidly absorbed by growing parts of the plant and translocates acropetally. It is used to control a variety of diseases such as **smut** and bunt diseases, rust, **powdery mildew**, leaf spots, rot, blights and many others caused by *Tilletia* spp., *Ustilago* spp., *Fusarium* spp., *Erysiphe graminis*, *Mycosphaerella* spp. and many others. It is a common seed-treatment agent, especially for cereals. Thus, the various crops protected include cereals (e.g. wheat, oats and barley), canola, soybean, potato, tomato, banana, tea and pome and stone fruit.

This fungicide is formulated as emulsifiable concentrates (EC), granules, wettable powders (WP), aerosols (AD) and as a powder for seed treatment. Trade names include Elite, Folicur,

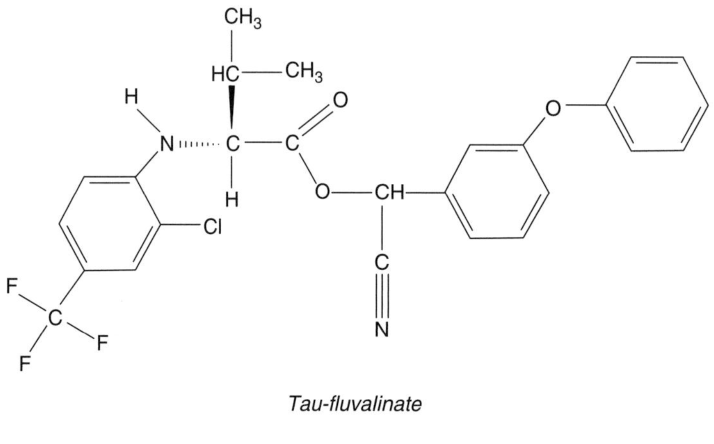

Tau-fluvalinate

Rexil and Horizon. Many products are combined with, for example, **spiroxamine, triadimenol** and fluoxastrobin.

Toxicologically, the acute oral LD_{50} in rats is between 1700 and 4000 mg/kg, while the inhalation LD_{50} in rats is 0.37 mg/l air (as an aerosol). The 2 year NOEL among rats is 300 mg/kg diet, while in mice it is only 20 mg/kg diet. The ADI is 0.03 mg/kg b.w. The acute percutaneous LD_{50} in rats is >5000 mg/kg. It is moderately toxic to birds with the acute oral LD_{50} in quails ranging from 2912 to 4438 mg/kg. It is moderately toxic to fish and aquatic creatures. It is slightly toxic to honey bees and non-toxic to earthworms.

Animals eliminate more than 99% of administered tebuconazole within 3 days via urine and faeces, with triazole-containing compounds as metabolites. Adsorption/desorption studies indicate low mobility in the soil. Soil degradation is moderate with the half-life being 12 days. Low vapour pressure and strong adsorption result in low volatilization into the air.

While the WHO classifies tebuconazole as a Class III toxin (slightly hazardous), the EPA considers it to be a moderately toxic fungicide. It has been rated by the EPA's OPP as a possible carcinogen (Rating C). It is also suspected to be harmful to endocrine functioning.

Tebuconazole is registered for use in India, Australia, the USA, Canada, Japan, Russia, South Korea, Brazil and many countries of the EU.

Tebutam herbicide

Tebutam, $C_{15}H_{23}NO$ (*N*-benzyl-*N*-isopropyl-2,2-dimethylpropionamide), is a herbicide of the benzamide class. It is a brownish liquid that is soluble in water (790 mg/l at 20°C). It is stable to aqueous hydrolysis as well as photolysis. The CAS Registry Number of tebutam is 35256-85-0. Tebutam is also known as butam.

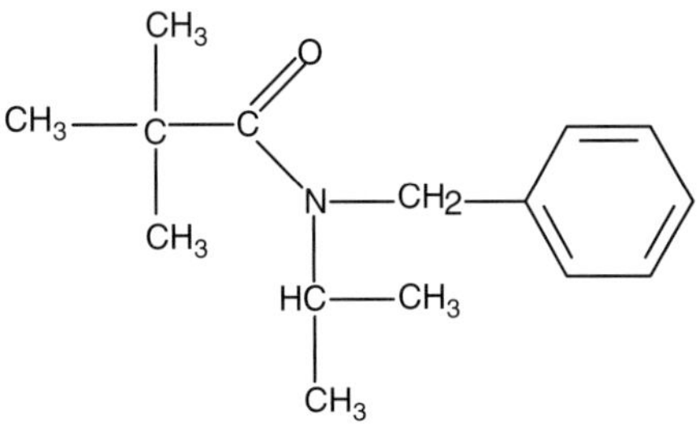

Tebutam

Tebutam, a pre-emergence herbicide, controls broad-leaved weeds and grasses in brassicas.

Tebutam has low toxicity in mammals as well as birds (with the acute oral LD_{50} among rats being 6210 mg/kg). It is moderately toxic to fish and aquatic invertebrates (the acute 96 h LC_{50} in fish is 19 mg/l). It is only slightly toxic to honey bees and earthworms. The WHO classifies tebutam in Class U (unlikely to be hazardous) and also considers the compound to be obsolete. The ADI is fixed at 0.15 mg/kg b.w./day.

Tebutam is moderately persistent in soil and has a DT_{50} of 60 days in aerobic conditions. The compound is mobile in water with the K_{OC} (organic-carbon sorption coefficient) being 25 ml/g and thus tebutam has high leachability.

Tebutam is not registered for use in the USA or by any country in the EU.

Tebuthiuron herbicide

Tebuthiuron, $C_9H_{16}N_4OS$ (1-(5-*tert*-butyl-1,3,4-thiadiazol-2-yl)-1,3-dimethylurea), is a urea-based herbicide. With a CAS Registry Number of 34014-18-1, tebuthiuron is a pungent, off-white crystalline solid, soluble in water (2.5 g/l) and in organic solvents such as methanol (170 g/l) and 2-methoxyethanol, acetone. Subject to thermal decomposition, tebuthiuron may release toxic oxides of carbon, nitrogen and sulfur. It is hydrolysed by strong alkalis and strong acids. It also presents a fire hazard.

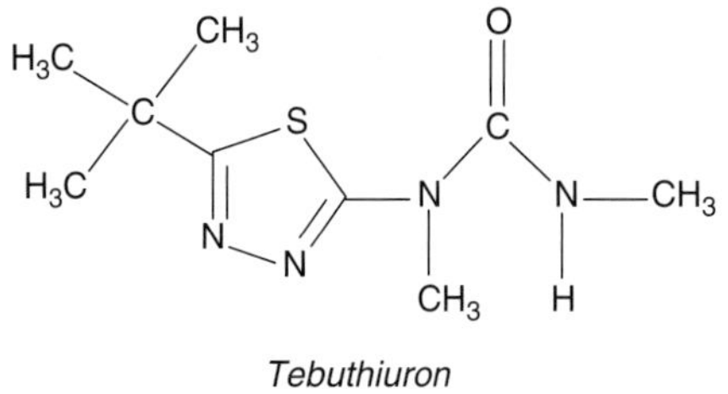

Tebuthiuron

Tebuthiuron, a systemic, broad-spectrum soil herbicide, is absorbed and translocated throughout the plant. Photosynthetic electron transport inhibition at the photosystem II receptor site is its mode of action. Perennial grasses, broadleaved weeds, wood and herbaceous plants (e.g. lucerne and clover) that grow on non-crop sites are controlled.

Tebuthiuron is commonly sprayed or broadcast directly on the ground, and so is applied at the rate of 0.6–6.8 kg/ha, depending on the type of weed and level of its infestation. Some formulations include water-dispersible granules (WG), wettable powder (WP) and pellets. It is compatible with other herbicides. Trade names include Brush, Spike, Herbec and Bullet. Since it is phytotoxic to some varieties and also non-targeted plants, its application must be done with care.

Tebuthiuron is classified under toxicity Class III by the WHO (slightly hazardous) as well as the EPA. It is, however, a chemical with moderate toxicity. The acute oral LD_{50} is 644 mg/kg in rats and 57.9 mg/kg in mice. It is also found to be a developmental and reproductive toxin. Hence the compound is listed as a PAN Bad Actor chemical. It is non-toxic to bees, birds and aquatic organisms. It is registered as a general use herbicide according to the EPA.

Tebuthiuron persists in soils with high organic matter content and low moisture content. It is also considered a potential groundwater contaminant. Its partial degradation proceeds via microbial action. The major metabolic pathways in plants and animals include *N*-demethylation of the *tert*-butyl side chain and the substituted urea side chain, respectively.

Herbicide tebuthiuron is registered for use in the USA and other countries such as Australia, South Africa and Vietnam. It is not registered in India.

Tecnazene fungicide

Tecnazene, $C_6HCl_4NO_2$ (1, 2, 4, 5-tetrachloro-3-nitrobenzene), is an aromatic hydrocarbon which is used as a fungicide. It is a colourless, odourless crystalline substance soluble in water (0.44 mg/l), but much more readily soluble in ethanol (40 g/l), benzene, carbon disulfide and chloroform. It decomposes when exposed to UV irradiation but is stable to acids, bases and to heat up to 300°C. When heated to decomposition, it emits toxic fumes. Tecnazene is produced by nitration of 1, 2, 4, 5-tetrachlorobenzene. It has a CAS Registry Number of 117-18-0.

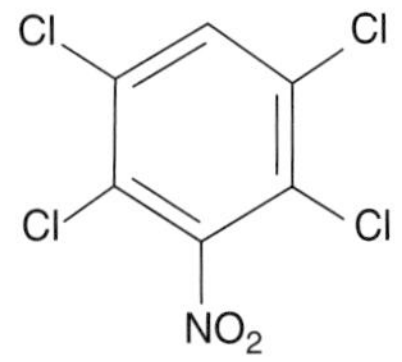

Tecnazene

Tecnazene is a protective and curative fungicide with lipid peroxidation being its mode of action, which leads to the membrane structure and function being disrupted. Rot and moulds are prevented by this compound. It protects potatoes in storage and prevents sprouting and thus is also a plant growth regulator. Smoke formulations control *Botrytis* sp. on glasshouse crops such as chrysanthemums, lettuce and tomatoes.

The fungicide tecnazene is formulated as dustable powders (DP), granules and as a fumigant. It is available under trade names like Fusarex, TCNB, Folosan and Fumite.

Tecnazene is not teratogenic, embryotoxic or carcinogenic. The WHO classifies tecnazene as Class U (unlikely to be hazardous). It is known to irritate the eyes, skin and the respiratory tract. The fungicide is moderately toxic to rats, fish, honey bees and earthworms. The acute oral LD_{50} among rats ranges from 1256 to 2047 mg/kg. It is phytotoxic to some varieties of roses. The acute 96 h LC_{50} in rainbow trout is 0.37 mg/l, while the acute 48 h EC_{50} for *Daphnia magna* is 0.2 mg/l. The NOEL is 50 mg/kg/day in rats and 15 mg/kg/day in rabbits. The ADI is 0.02 mg/kg b.w./day. Any exposure to tecnazene should be remedied by immediate medical attention. It is important to remove contaminated clothing, while thoroughly rinsing the affected skin or eyes. It is not advisable to induce vomiting, in cases of accidental ingestion.

In animals, tecnazene is rapidly absorbed, metabolized and excreted via the faeces and urine, the main metabolite being mercapturic acid conjugate. In plants, sulfhydryl groups substitute the chlorine atom. Tecnazene is rapidly lost from treated soil, possibly via evaporation.

Tecnazene is not registered by the USA or the EU. The MRL values of tecnazene for banana, mango and guava are 0.05 mg/kg in each case.

Teflubenzuron as an IGR

Teflubenzuron, $C_{14}H_6Cl_2F_4N_2O_2$ (1-(3, 5-dichloro-2, 4, difluorophenyl)-3-(2, 6-diflurobenzoyl) urea), is a yellowish crystalline substance, soluble in water (0.019 mg/l), acetone, ethanol, dimethyl sulfoxide and cyclohexane. It is stable at room temperature and under normal conditions for 2 years. It has a CAS Registry Number of 83121-18-0.

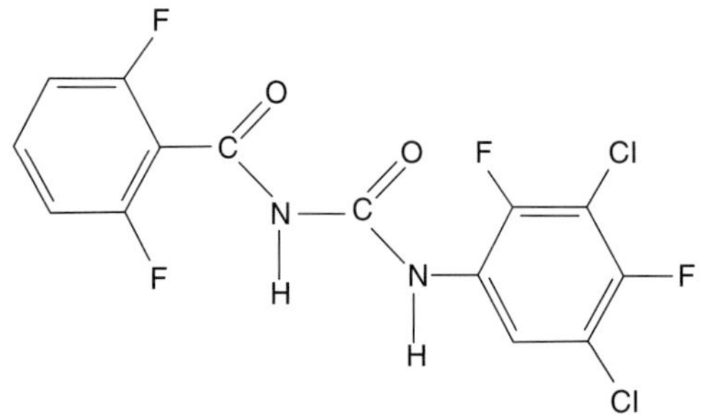

Teflubenzuron

Teflubenzuron, an acyl urea derivative, is a non-systemic insect growth regulator. It acts by interfering with **chitin** synthesis and thereby disrupting the **moulting** process of insects. It controls lice, locusts and many other pests including flies, cockroaches, whiteflies, hoppers, moths and mosquitoes by preventing moulting of the larvae. Crops on which this insecticide may be applied are potatoes, cabbage, pome, stone and citrus fruit, vines, soybean, cotton and tobacco. It controls ectoparasites on animals too.

The insecticide teflubenzuron is supplied as a suspension concentrate (SC) and ULV liquid (UL). It is sprayed after diluting with water. Trade names include Nomolt.

The WHO has classified this insecticide as not acutely toxic. According to the EPA, teflubenzuron comes under the toxicity Class IV, indicating insignificant toxicity. The acute oral LD_{50} of teflubenzuron in rats is >5038 mg/kg, indicating slight toxicity. The dermal LD_{50} is >2000 mg/kg b.w. It is moderately toxic to birds (the acute LD_{50} in the quail is >2250 mg/kg). However, it is toxic to fish and moderately toxic to honey bees. It is practically non-toxic to earthworms. The ADI in mice is 0.01 mg/kg b.w./day.

Poisoning with teflubenzuron can cause irritation to the eyes, skin and mucous membranes, coughing and shortness of breath, vomiting, diarrhoea, headache, electrolyte depletion and many other abnormal signs.

In rats, following oral administration, teflubenzuron and its metabolites are rapidly excreted in the faeces and urine. The compound adsorbs to soil particles and the DT_{50} in soil is in the range of 2–12 weeks, with microbial action being a significant breakdown agent. The major metabolite of teflubenzuron is 3,5-dichloro-2,4-difluorophenylurea. In aqueous solution, teflubenzuron has a half-life of about 10 days.

Teflubenzuron is registered for use in many countries of the EU including the UK, Italy, Belgium, Germany, Hungary and France. The Gambia, Burkina Faso, Madagascar, Mali, Senegal and South Africa are some other countries that have registered this compound. The MRLs for rice, wheat, barley, cauliflower and sugarbeet are fixed at 0.05 ppm, while for maize and soybean it is 0.1 ppm.

Tefluthrin insecticide

Tefluthrin, $C_{17}H_{14}ClF_7O_2$ (2,3,5,6-tetrafluoro-4-methylbenzyl (*Z*)-(1*S*,3*RS*)-3-(2-chloro-3,3,3-trifluroprop-1-enyl)2,2-dimethylcyclopropanecarboxylate), is a prominent soil and foliar-applied pyrethroid ester insecticide. Tefluthrin is almost insoluble in water but readily soluble in acetone, hexane, toluene and dichloromethane. It is stable to hydrolysis in acidic media for more than 30 days, if stored at a normal temperature. It is produced by the reaction of 3-(2chloro-2-trifluoromethyl)-2,2-dimethylcyclopropanecarboxylic acid with 2,3,5,6-tetrafluoro-4-methyl benzylalcohol. The CAS Registry Number of tefluthrin is 79538-32-2 (see Tefluthrin at top of the next page).

Mainly used as a foliar insecticide, tefluthrin is used on cereal crops such as wheat, oats, barley and maize. It also protects sugarcane, potato, groundnuts and strawberries from various wireworms, wheat bud fly and seed- and soil-related pathogens belonging to the orders Coleoptera, Lepidoptera and Diptera.

Tefluthrin, like many other foliar insecticides, is sprayed by dissolving it in a suitable organic solvent. Tefluthrin may

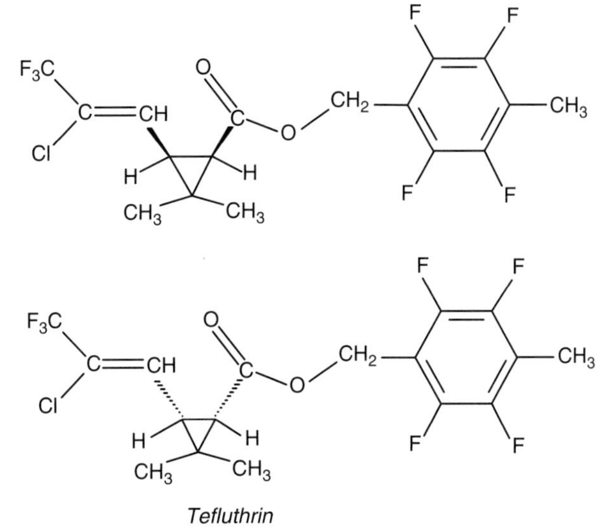

Tefluthrin

also be used along with a suitable fungicide/insecticide acaricide (e.g. **fludioxonil** or **dimethoate**). Seed treatment formulations and sprays are common. Tefluthrin is available under trade names like Force ST.

The acute oral LD_{50} of tefluthrin in rats is 21.8 mg/kg and the dermal LD_{50} is 177 mg/kg b.w. A 4 h inhalation LC_{50} in rats is 0.037 mg/l. The acute 96 h LC_{50} in rainbow trout is 0.00006 mg/l. The acute LD_{50} in sparrows is >267 mg/kg. The acute 48 h oral LD_{50} among honey bees is 0.28 μg per bee, indicating acute toxicity, while the acute 14 day LC_{50} of tefluthrin in earthworms is 1.0 mg/kg. The ADI among dogs is 0.005 mg/kg/day. The WHO classifies this insecticide as a Class Ib toxin (highly hazardous). It is also a PAN Bad Actor chemical. Exposure to tefluthrin, like any other pyrethroid insecticide, includes irritation to the skin and eyes and irritability to sound or touch, a prickling, tingling or creeping sensation on the skin, headache, dizziness, vomiting, diarrhoea, excessive salivation, fatigue, fluid in the lungs and seizures.

Although tefluthrin is a persistent soil insecticide, tefluthrin does not appear to negatively affect the soil microflora. No residues are found on major crops treated with tefluthrin.

Tefluthrin is registered for use in Australia, Canada, Cyprus, Denmark, Germany, Greece, Hungary, the Netherlands, Portugal and the USA.

In the EU, the MRL of tefluthrin is 0.02 mg/kg for sugarbeet, while it is 0.01 mg/kg for other foods of plant origin.

Tembotrione herbicide

Tembotrione, $C_{17}H_{16}ClF_3O_6S$ (2-{2-chloro-4-mesyl-3-[(2,2,2-trifluoroethoxy)methyl]benzoyl}cyclohexane-1,3-dione), is a herbicide of the benzoylcyclohexanedione class (triketone family). It is a beige powder soluble in water (0.22 g/l at pH 4 and 28.3 g/l at pH 7 and 20°C), DMSO, methylene chloride (>600 mg/l) and acetone (300–600 mg/l) at 20°C. The CAS Registry Number of this herbicide is 335104-84-2.

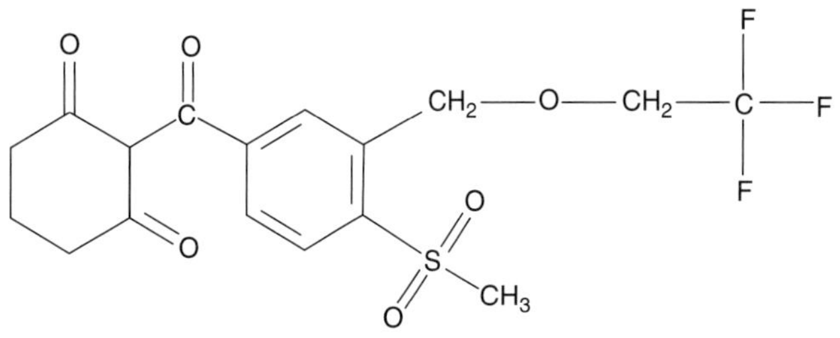

Tembotrione

Tembotrione, a post-emergence herbicide, is used for the control of broad-spectrum and grassy weeds, especially those found in maize. The mode of action is the inhibition of the enzyme **4-hydroxyphenylpyruvate dioxygenase** (HPPD) that blocks the pathway of prenylquinone biosynthesis in plants. As a result, chlorophyll is unable to protect the weed leaf from UV radiation. This leads to **chlorosis** that kills the plant.

The product is applied either in single or more doses at the rate of 100 g/ha per season. It can be combined with other chemicals such as **isoxadifen**.

The acute oral LD_{50} of the herbicide in rats is >2000 mg/kg. From tests conducted on laboratory animals, the chemical may cause irritation to the eyes and the skin. The NOEL in male rats is 0.04 mg/kg/day. The acute oral LC_{50} is >292 mg/l in mallard ducks, >1788 mg/kg in bobwhite quails and >100 mg/l in rainbow trout (96 h). In bees, the 72 h oral LD_{50} is >92.8 μg per bee. In earthworms, the 14 day LD_{50} is >1000 mg/kg soil. The compound is not a PAN Bad Actor chemical.

The herbicide is registered in the USA, Canada, Germany and Portugal, but not in India.

Temephos insecticide

Temephos, $C_{16}H_{20}O_6P_2S_3$ (*O,O,O′,O′*-tetramethyl *O′,O′*-thiodi-*p*-phenylene bis (phosphorothioate)), is a phenyl organothiophosphate insecticide. The technical product is a brown viscous liquid that is soluble in water at 0.03 mg/l, but more soluble in diethyl ether, hexane and in aromatic hydrocarbons.

It is hydrolysed by strong acids and alkalis and unstable above 49°C. It has a CAS Registry Number of 3383-96-8.

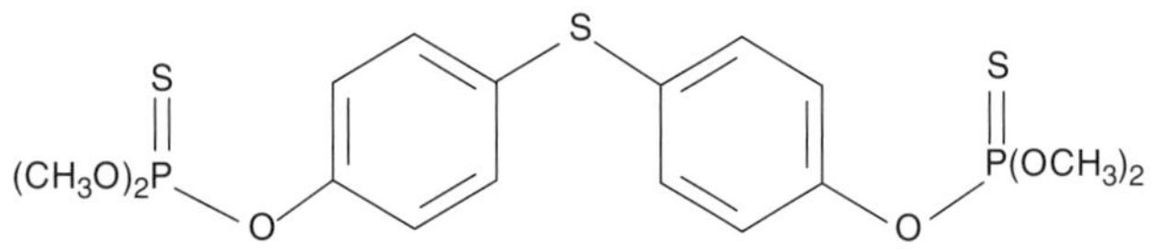

Temephos

Temephos, a cholinesterase inhibitor and an ectoparasiticide, is a non-systemic insecticide with significant larvicidal properties. Temephos finds application in areas of public health. It does not have any food-related or residential applications. It is used to control the larvae of fleas, mosquitoes, lice, gnats, midges and black flies (examples being *Anopheles*, *Aedes*, *Culex* and *Simulium* spp.). Temephos is used on animals and around swamps, lakes and tidal water to control the spread of malaria and guinea worm infestation.

Temephos is formulated as emulsifiable concentrate (EC), granules (GR), dustable powder (DP), fumigant and cold fogging concentrate. By mixing it with other insecticides (e.g. **trichlorfon**) the control becomes more broad based. It is non-phytotoxic when used at the recommended application rates. The typical application dosage is 0.1–0.5 kg/ha. It is available in the market under trade names such as Abate, Biothion, Difenthos, Ecopro and Nimitox.

The oral LD_{50} values of temephos among rats range between 1226 mg/kg and 13,000 mg/kg, while for mice, they range from 460 mg/kg to 4700 mg/kg. This chemical is thus toxic to these rodent species. Temephos is extremely toxic to birds, fish and *Daphnia* spp. The LD_{50} among quail, pheasant and dove species ranges from 35 mg/kg to 85 mg/kg. Similarly, the LD_{50} in fish ranges from 0.16 mg/l to 3.49 mg/l and this value is much lower among *Daphnia magna*. Temephos is toxic to honey bees (the acute 48 h contact LD_{50} is 1.55 µg per bee). Despite this, since exposure to temephos is rarely under common farm or household conditions (except during mosquito treatment), there is no serious threat to most mammals except those exposed to very high levels. The WHO has categorized temephos as a Class U toxin (unlikely to be hazardous). Personnel employed in the mixing and application of temephos formulations are advised to use protective equipment (e.g. closed mixing and loading systems) and long, protective clothes. Exposure may result in excessive salivation, sweating, muscle twitching, tremor, respiratory depression, productive cough, fluid in the lungs, headache, dizziness, nausea, vomiting, abdominal cramps, diarrhoea and pin-point pupils, sometimes with blurred or dark vision.

Mammals eliminate most temephos in its unchanged form. The principal metabolites in mammals are sulfate ester conjugates of 4,4′-thiodiphenol, 4,4′-sulfinyldiphenol and 4,4′-sulfonyldiphenol. In plants, it gets slowly oxidized to the sulfoxide, and to a lesser extent, to the sulfone and the mono- and di-orthophosphate. Further degradation occurs very slowly. Temephos has moderate persistence in most soils, since it adsorbs to soil particles. In water, it undergoes photolysis rapidly and breaks down within 4 days, with insignificant residues being found in runoff water or in soil water.

Temephos is registered for use in Australia, India, New Zealand, South Africa and the USA but not in the EU.

Temephos technical analysis

Temephos is an organophosphate compound considered as a Class U toxin by the WHO. For analysing the quantity of temephos, the sample should contain a minimum purity of 900 g temephos/kg of the sample, the acidity should be 3 g sulfuric acid/kg, water content 2 g/kg and the sample must be free from extraneous impurities.

For the analysis, a liquid chromatograph with stainless steel columns (30 cm in length and 3.9 mm in internal diameter, packed with 10 µm silica gel) is needed that is equipped with a pumping system capable of maintaining a pressure of 17×10^3 kPa and a UV detector capable of measuring UV absorption at 254 nm. Conditions suitable for the HPLC are: (i) ambient column temperature; (ii) a flow rate of 1 ml/min; and (iii) a wavelength of 250 nm. The retention time for the internal standard solution should be 9.6 min, for temephos, 11.50 min and for the alternative standard (dimethyl 4-nitrophthalate), 18.0 min.

The temephos calibration solution is made by weighing (to the nearest 0.1 mg) 50, 60 and 70 mg of analytical grade temephos of known purity into different 50 ml standard flasks labelled A, B and C. To each flask 5 ml internal standard solution 4-nitrophenyl 4-nitrobenzoate and 25 ml dry ethyl acetate is added, while diluting to volume with *n*-hexane. Solution B is the working solution for the HPLC, while A and C are used to check the linearity of the equipment. The internal standard solution consists of 1.5 g nitrophenyl 4-nitrobenzoate in a 250 ml volumetric flask, diluted to volume with dry ethyl acetate and stored under refrigeration, until it is time for use. (Alternatively, 2.75 g dimethyl 4-nitrophthalate in a 250 ml volumetric flask and diluted to volume with dry ethyl acetate may be used as the internal standard solution.) HPLC-grade ethyl acetate is dried over a molecular sieve (0.5 nm, 8–12 mesh beads) and filtered through a 0.45 µm millipore filter. HPLC-grade *n*-hexane is dried over a molecular sieve and filtered through a 0.45 µm millipore filter. The liquid chromatography mobile phase is a degassed mixture of 10 ml ethyl acetate and 90 ml *n*-hexane.

Anhydrous methanol (50 ml) is pumped through the liquid chromatography column, followed by 100 ml dry ethyl acetate. Sufficient mobile phase should be pumped in to equilibrate the system, at which point, 5 µl aliquots of calibration solution B should be injected until a constant response is obtained (i.e. when the response ratios (area or height of internal standard peak: area or height of temephos peak) of three consecutive injections agree to within 2%).

Linearity checks should be carried out on the liquid chromatography at least when the column is installed in the apparatus. When peak height is the criterion for the chromatographic response, the attenuation and injection volume (3–6 µl) of solution B should be determined with respect to the peak height of the internal standard. This condition is used for all samples and calibration solutions.

Triplicate aliquots of appropriate volume of solutions A, B and C are injected into the liquid chromatograph, and the response ratios for each injection and averages for each solution are determined. The response ratios for each solution are divided by the corresponding temephos content and the resulting response factors compared. These should agree to within 2%.

For sample analysis, 60 mg (to the nearest 0.1 mg) of sample is weighed in duplicate into two 50 ml volumetric flasks. Exactly 5 ml internal standard solution and 25 ml dry ethyl acetate are dissolved in each flask and diluted up to the mark with *n*-hexane.

Duplicate aliquots of both calibration solution B and of the sample solutions are injected and the response ratios, calculated. The duplicate response ratios for solution B and the sample solutions are averaged. Duplicate aliquots of calibration solution injections before and after sample solutions are also injected.

Equations T.1 and T.2 are used to calculate the temephos content.

$$\text{The response ratio (r)} = \frac{\text{Area or height of temephos peak}}{\text{Area or height of internal standard peak}} \quad (\text{T.1})$$

$$\text{Temephos (g/kg)} = \frac{r_2 \times m_1 \times P}{r_1 \times m_2} \quad (\text{T.2})$$

where

r_1 = Average response ratio for standard solution B
r_2 = Average response ratio for sample solution
m_1 = Mass (g) of temephos standard in solution B
m_2 = Mass (g) of sample taken
P = Purity of temephos standard (g/kg)

TEPA chemosterilant

TEPA, $C_6H_{12}N_3OP$ (tris(1-aziridinyl)phosphine oxide), is a **chemosterilant**. It is a colourless and odourless liquid. It has a CAS Registry Number of 545-55-1. It is highly soluble in organic solvents such as acetone, ether and alcohol. It is also called aphoxide.

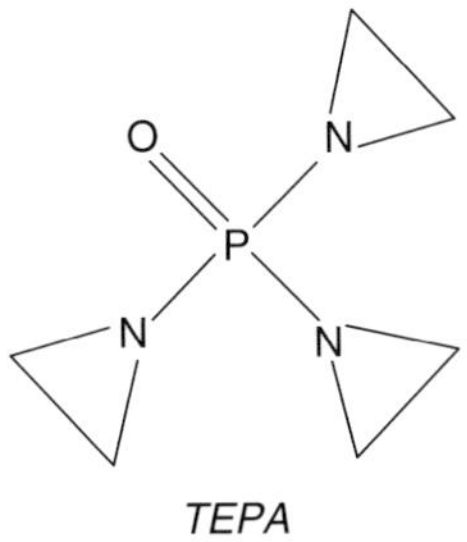

TEPA

TEPA is effective against the housefly, mosquito, Mexican fruit fly, German cockroach and citrus red mite. In its modes of action, it can inactivate sperm, degenerate ovaries and induce pycnosis of nuclei in various insect species. In crop situations, citrus red mites, maize earworms, armyworms and some other pests are effectively controlled. TEPA is more effective in causing sterility than other commonly available active ingredients.

Tepraloxydim herbicide

Tepraloxydim, $C_{17}H_{24}ClNO_4$ ((*EZ*)-(RS)-2-{1-[(2E)-3-chloroallyloxyimino]propyl}-3-hydroxy-5-perhydropyran-4-ylcyclohex-2-en-1-one), is a cyclohexanedione oxime herbicide. It is a white odourless powder, soluble in water. Tepraloxydim is made by condensation of 5-perhydropyranyl-3-hydroxy-2-carbonyl-propylcyclohex-2-ene-1-one with chloroallyloxyamine. It has a CAS Registry Number of 149979-41-9.

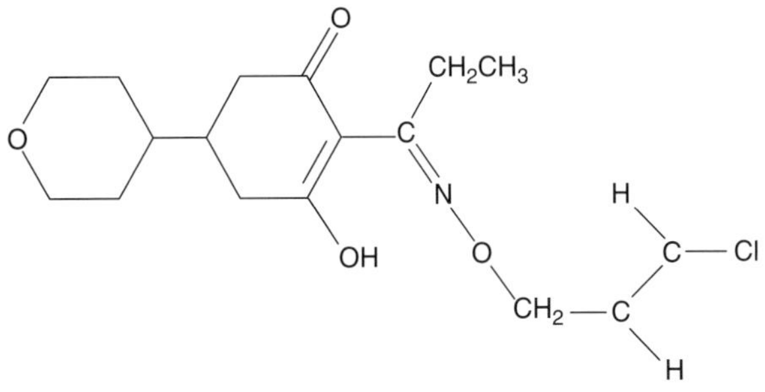

Tepraloxydim

Tepraloxydim, a post-emergence, broad-spectrum, systemic compound, is used to control annual grass weeds in flax, linseed, soybean, cotton, sugarbeet, fodder beet, beans, bulb onions, carrots, leeks and cabbages. Once absorbed and translocated, it acts as an **ACCase inhibitor**.

The herbicide is often supplied as an emulsifiable concentrate (EC), diluted in water.

The acute oral and dermal LD_{50} of tepraloxydim among rats is >2000 mg/kg b.w., while the inhalation LC_{50} is 5.1 mg/l. The ADI is 0.025 mg/kg b.w./day. It is toxic to honey bees (the acute 48 h oral LD_{50} is 25 μg per bee) and moderately toxic to earthworms (the acute 14 day LC_{50} is 100 mg/kg) and to some aquatic organisms.

Tepraloxydim is registered in Australia, Canada, New Zealand, the USA and in EU countries such as the Netherlands, the UK, Poland, Italy, Denmark and Finland.

Terbacil herbicide

Terbacil, $C_9H_{13}ClN_2O_2$ (3-*tert*-butyl-5-chloro-6-methyluracil), is a **uracil** herbicide. It forms colourless crystals that are highly soluble in water, readily soluble in organic solvents but sparingly soluble in mineral oils. It is very stable in aqueous alkaline media at normal temperatures and in the dark. Terbacil is produced by the reaction of *tert*-butyl isocyanate with 3-aminocrotonate, followed by cyclization and chlorination. Its CAS Registry Number is 5902-51-2.

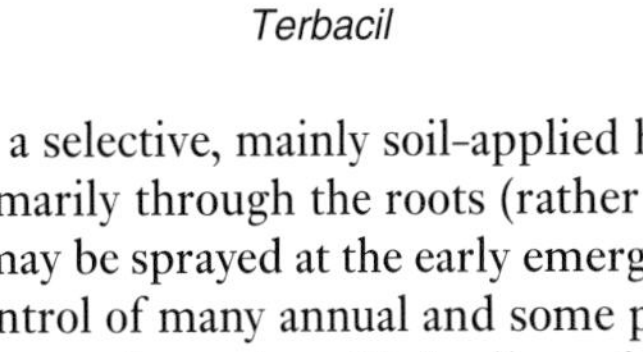

Terbacil

Terbacil is a selective, mainly soil-applied herbicide and it is absorbed primarily through the roots (rather than the leaves and stems). It may be sprayed at the early emergence stage. It is used for the control of many annual and some perennial weeds in apples, citrus and peaches. Terbacil works in plants by inhibiting photosynthesis.

Terbacil is commonly available as a wettable powder (WP). It is sprayed on the soil aerially or by a spray boom. The application rate ranges from 0.5 to 9 kg/ha. It is available in the market under the trade names of Compound 732, Sinbar and

Geonter. This compound may also be combined with other herbicides for a wider coverage.

The herbicide terbacil has low oral toxicity to rats. It is moderately toxic to birds (with the LD_{50} in quails being >2250 mg/kg.) Terbacil is slightly toxic to fish and freshwater invertebrates (with an LC_{50} of 65 mg/l in *Daphnia* spp.). It is phytotoxic to sensitive plants such as oats and lucerne. Symptoms of poisoning with terbacil are irritation of the skin, eyes and respiratory tract.

In animals, the principal biotransformation pathways of terbacil are hydroxylation of the 6-methyl group, and replacement of the 5-chloro group with a hydroxyl group. Terbacil is highly persistent in soils with a half-life between 50 and 180 days. Soil persistence varies with application rate, rainfall and soil type. The herbicide is also highly soluble in water, with poor adsorption. Thus, it poses a risk to water sources. Leaching may be slightly low in highly organic soil. Microbial degradation does occur in some soils. **Bioaccumulation** is also known with terbacil.

Terbacil is registered in Australia, Canada, New Zealand, South Africa and the USA, but it is not registered anywhere in the EU.

Terbufos insecticide

Terbufos, $C_9H_{21}O_2PS_3$ (*S-tert*-butylthiomethyl *O,O*-diethyl phosphorodithioate), is an aliphatic organothiophosphate insecticide. It is a slightly yellow liquid, soluble in water (4.5 mg/l) and readily soluble in aromatic hydrocarbons, alcohol and acetone. It remains stable for more than 2 years at room temperature. It decomposes above 120°C and is hydrolysed by strong alkalis (pH >9) and acids (pH <2). It is prepared by the reaction of *O,O*-diethyl-dithiophosphoric acid with formaldehyde and *tert*-butanethiol. It has a CAS Registry Number of 13071-79-9.

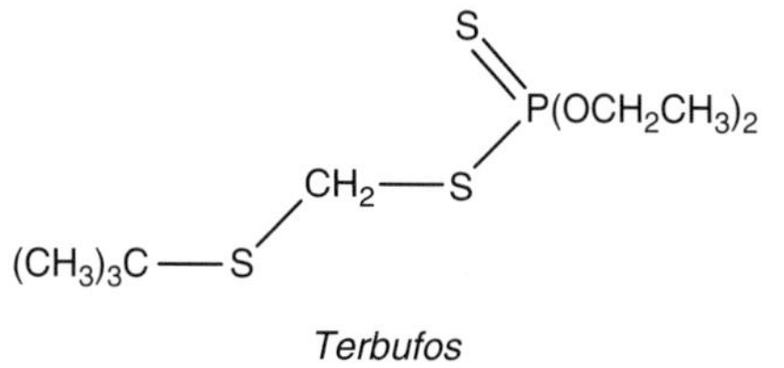

Terbufos

Terbufos, a cholinesterase inhibitor, is a soil insecticide and a nematicide with residual life and contact action. Rootworms, wireworms and maggots are the main pests controlled by terbufos. It also controls larvae of beetle *Diabrotica* spp., and maggot pests like *Tetanops myopaeformis* and *Delia* spp. Terbufos protects cabbages, onions, maize and sugarbeet. Nematodes controlled include *Radopholus similis*, *Meloidogyne incognita*, *Helicotylenchus* spp. and *Ditylenchus dipsaci* which infest crops, especially bananas and sugarbeet.

Terbufos is commonly formulated as granules (GR). It is available in the market as AC 92100, Counter, Aragran, Plydox and ST-100, among others.

Terbufos is highly toxic to rats and dogs by both the oral and the dermal routes. High doses also cause congestion of the liver, kidneys and lungs. The acute oral LD_{50} in rats is 1.3 mg/kg. The dermal LD_{50} of terbufos in rabbits is 1.0 mg/kg b.w. The inhalation LC_{50} in rats is 0.0061 mg/l. The acute LD_{50} in mallard ducks is 185 mg/kg. Terbufos is extremely toxic to *Daphnia magna* and to most fish species (the acute 96 h LC_{50} in bluegill sunfish is 0.004 mg/l). The insecticide's toxicity to honey bees is seen in its acute, contact LD_{50} dose of 4.1 μg per bee. It is also very toxic to earthworms (the acute LC_{50} being 4 mg/kg). The ADI is 0.0006 mg/kg/day. Terbufos is classified as a Class Ia toxin (i.e. extremely hazardous). Exposure causes excessive salivation, sweating, weakness, tremors, incoordination, headache, abdominal cramps, diarrhoea, respiratory depression, productive cough, pin-point pupils, blurred vision, seizures and incontinence.

Terbufos is metabolized in plants, animals and soils in much the same way. It involves the oxidation of the sulfide group into the sulfoxide, then to sulfone and finally to diethyl hydrogen phosphorodithioate and phosphate. There is no accumulation in body tissues, although terbufos is readily absorbed via all routes. The half-life in soil is 9–27 days. Almost all of the ^{14}C-labelled terbufos used during a trial was eliminated via the urine, within 1 week.

Terbufos is registered for use in Australia, Cameroon, Canada, New Zealand, the Philippines, South Africa and the USA, but it is not registered anywhere in the EU.

Terbumeton herbicide

Terbumeton, $C_{10}H_{19}N_5O$ (N^2-*tert*-butyl-N^4-ethyl-6-methoxy-1,3,5-triazine-2,4-diamine), is a herbicide of the 1,3,5-triazine group. It is a colourless, crystalline substance that is soluble in water (130 mg/l), methanol (220 g/l), acetone and toluene. At high temperatures, it hydrolyses in strong acids and alkalis, forming hydroxytriazine. It is produced by the reaction of 2-*tert*-butylamino-4-chloro-6-ethylamino-*s*-triazine with methanol and sodium hydroxide. Terbumeton has a CAS Registry Number of 33693-04-8.

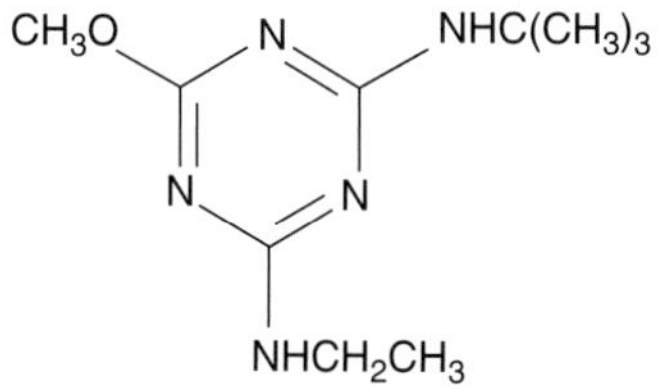

Terbumeton

Terbumeton, a selective, systemic herbicide, acts on a wide range of susceptible grasses and broadleaved weeds by inhibiting photosynthetic electron transport at the photosystem II receptor site. It is well absorbed by the leaves and roots of weed pests. The chemical is applied to fruit trees (e.g. apples and citrus) and vineyards.

Typical formulations of terbumeton are suspension concentrates (SC) and wettable powders (WP), with common dosage ranging from 3 to 6 kg a.i./ha.

The herbicide terbumeton is classified as moderately hazardous by the EPA and the WHO (Class II toxin). Its acute oral LD_{50} in rats is 651 mg/kg and in mice, 2343 mg/kg (with the NOEL in rats being 7.5 mg/kg/day). Among rabbits, it is slightly irritating to the eyes. Among fish, the dietary LC_{50} values (96 h) are between 14 and 30 mg/l. The herbicide is non-toxic to bees. The ADI is optimized to be 0.075 mg/kg b.w.

Terbumeton, once absorbed by mammals, is almost completely eliminated. In animals, plants and soil, terbumeton metabolizes via demethoxylation, dealkylation and conjugation. Microbial action also takes place in soil leading to hydroxytriazine.

Terbumeton is not registered in the USA or anywhere in the EU.

Terbuthylazine herbicide

Terbuthylazine, $C_9H_{16}ClN_5$ (6-chloro-*N*-(1,1-dimethylethyl)-*N*′-ethyl-1,3,5-triazine-2,4-diamine), is a herbicide from the 1,3,5-triazine group. It is a colourless powder that is soluble in water (9 mg/l), ethanol (14 g/l) and acetone (41 g/l). It is stable in weak acids and alkalis. It is produced by the reaction of cyanuric chloride with *tert*-butylamine, followed by treatment with ethylamine in the presence of sodium hydroxide. The CAS Registry Number is 5915-41-3.

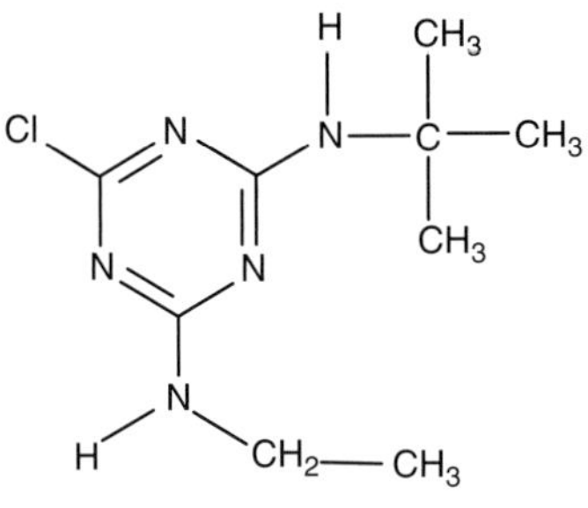

Terbuthylazine

Terbuthylazine, a broad-spectrum, pre- and post-emergence, systemic herbicide, is absorbed mainly by the roots and acts by inhibiting photosynthetic electron transport at the photosystem II receptor site. It is effective on several types of weeds in maize, sorghum, potatoes, sugarcane, peas, beans, citrus crops and coffee.

Formulated as a suspension concentrate (SC) and as water-dispersible granules (WG), terbuthylazine is applied at up to 3 kg/ha. It can be combined with other herbicides such as glyphosate, *S*-metachlor and some others. The chemical is available as Click and Folar.

Terbuthylazine is classified under toxicity Class U (unlikely to be hazardous) by the WHO and Class III by the EPA. The acute oral LD_{50} in rats is 1590 mg/kg, while in ducks and quails, it is >1000 mg/kg. The LC_{50} among fish ranges from 2.2 to 52 mg/l. Terbuthylazine is practically non-toxic to bees and beneficial soil organisms such as earthworms. The NOEL among rats is 0.35 mg/kg/day. The ADI is optimized to be 0.002 mg/kg b.w./day. The compound is not listed as a PAN Bad Actor chemical.

Terbuthylazine herbicide is registered in the USA, Australia, Germany, Greece, Hungary, the Netherlands, New Zealand, Portugal and South Africa.

Terbutryn herbicide

Terbutryn, $C_{10}H_{19}N_5S$ (*N*[2]-*tert*-butyl-*N*[4]-ethyl-6-methylthio-1,3,5-triazine-2,4-diamine), is a herbicide of the triazine class. It is an almost colourless crystalline powder that is soluble in water (22 mg/l). It is hydrolysed in strong acid or basic solutions and is decomposed by exposure to UV. It is stable in dry and cool storage conditions. The CAS Registry Number is 886-50-0.

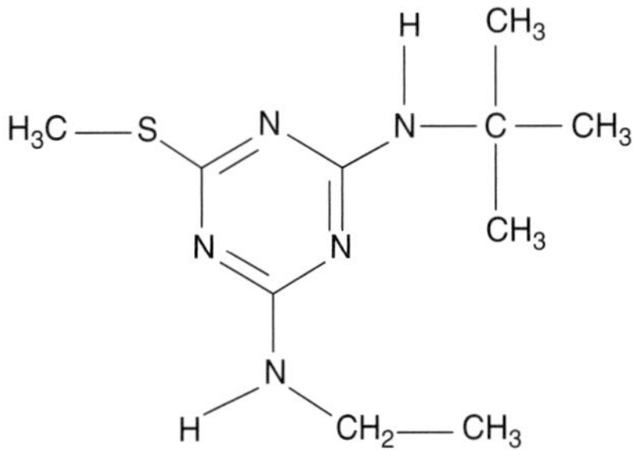

Terbutryn

Terbutryn, a selective, pre- and post-emergence herbicide, acts as an algicide and aquatic weedicide, with additional biocidal properties. After being absorbed by roots and foliage, it acts as a photosynthesis inhibitor for poppies, meadow grass and mayweed. It controls most grasses and broadleaved weeds in cereals (e.g. wheat, barley and sorghum), sugarcane, cotton, sunflowers, peas and potatoes. Aquatic weeds and algae in water bodies are also controlled. It has a residual activity of up to 10 days.

Terbutryn is available as a wettable powder (WP) or a soluble concentrate (SL). It can be combined with other herbicides such as **terbuthylazine, metolachlor, metobromuron** and **atrazine** to enhance weed control cover. On sighting of weeds in cereals, the chemical may be applied at the rate of 0.2–0.4 kg/ha, while it is applied at up to 3 kg/ha in sugarcane. Common trade names include Plantonit and Gesaprim and many formulations are 50–80% WP.

The WHO ranks terbutryn as a Class U toxin (unlikely to be hazardous). The dose, which kills half of the test animals, the LD_{50}, is 2450–2500 mg/kg in rats and 3884 mg/kg in mice. However, at higher doses, symptoms like incoordination, convulsions and swelling of the lungs are observed. It is a possible carcinogen and endocrine disruptor (it is listed in the EU list of endocrine disruptors). While the chemical is practically safe to birds (the LC_{50} (8 day dietary) is >20,000 mg/kg in bobwhite quails) and bees, it is moderately toxic to fish. The concentration which is lethal to fish in water (the 96 h LC_{50}) is between 3 and 4 mg/l. Terbutryn has an ADI of 0.01 mg/kg.

Triazine herbicides are relatively persistent in the environment and may leave residues in foods and water. Terbutryn is adsorbed to clayey or organic soil with a half-life of up to 28 days. The chemical also adsorbs to sediment and suspended matter in water with a long half-life. The major degradation product is hydroxyterbutryn, which is more mobile than the parent and is feared to leach into groundwater. In plants, it breaks down into stable products.

Terbutryn is not registered for use in the USA or throughout the EU, but is registered in Australia, Burkina Faso, Cape Verde, Chad, Gambia, Guinea-Bissau, Madagascar, Mali, Mauritiana, New Zealand, Niger, Senegal and South Africa.

Termiticide bifenthrin

See **Bifenthrin insecticide**

Termiticide isofenphos

See **Isofenphos insecticide**

Terpene

See **Chemosterilant**

Tertiary tillage

See **Weed management by mechanical methods**

Tetrachlorvinphos insecticide

Tetrachlorvinphos, $C_{10}H_9Cl_4O_4P$ ((*Z*)-2-chloro-1-(2,4,5-trichlorophenyl) vinyl dimethyl phosphate), is an organophosphorus insecticide. It is a colourless, crystalline solid, soluble in water (11 mg/l), acetone, chloroform, dichloromethane and xylene. It is hydrolysed slowly in acidic media and more in alkaline media. The CAS Registry Number is 22248-79-9 (*Z*-isomer).

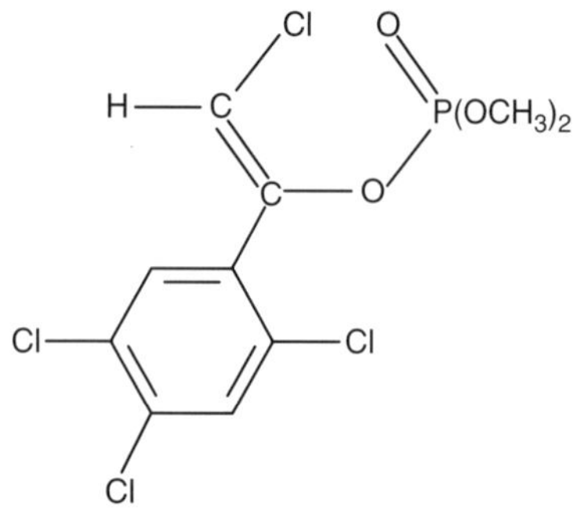

Tetrachlorvinphos

Tetrachlorvinphos, a non-systemic insecticide and acaricide, is a cholinesterase inhibitor, and has contact and stomach action. It controls members of the orders Diptera (flies) and Lepidoptera (moths) in maize, rice, tobacco, cotton, vegetables and pastures as well as on stored products. As an ectoparasite control agent, it works well against flies that infest farm animals.

This insecticide is formulated as wettable powder (WP), suspension concentrate (SC), emulsifiable concentrate (EC) and dustable powder (DP). It is compatible with many pesticides except those of an alkaline nature and dodine. It is available on the market as Gardona, Rabond, Stirofos, Gardcide and Debantic. The acute oral LD_{50} of tetrachlorvinphos is >4000 mg/kg among rats, while the dermal LD_{50} is >1500 mg/kg b.w. The inhalation LC_{50} in rats is 0.85 mg/l. It is moderately toxic to birds (the acute LD_{50} in mallard ducks is 2000 mg/kg), while it is very toxic to fish and to *Daphnia magna* (the acute LC_{50} in rainbow trout is 0.43 mg/l, while it is 0.002 mg/l in *D. magna*.). Tetrachlorvinphos is toxic to honey bees. It is also known to injure (via russeting) some apple varieties under certain conditions. Tetrachlorvinphos is a cholinesterase inhibitor and a possible endocrine disruptor. Exposure to the insecticide may cause some or all of the following symptoms: excessive salivation, muscle twitching, weakness, tremor, incoordination, headache, dizziness, vomiting, abdominal cramps, diarrhoea, tightness in the chest, wheezing, productive cough, blurred or dark vision, seizures, incontinence, respiratory depression and loss of consciousness. Protective clothing is required for personnel associated with this chemical, including long-sleeved shirt and trousers, and chemical-resistant gloves, shoes and socks.

In mammals, orally administered tetrachlorvinphos is rapidly metabolized through demethylation and hydrolysis of the vinyl phosphate linkage, forming metabolites such as 2,4,5-trichlorophenylethanediol glucuronide, 2,4,5-trichloromandelic acid and 2-chloro-1-(2,4,5-trichlorophenyl) vinyl methyl hydrogen phosphate. Some metabolites bioaccumulate.

Tetrachlorvinphos is banned throughout the EU and in Angola. Canada, the USA, South Africa and Australia have registered its use.

Tetraconazole fungicide

Tetraconazole, $C_{13}H_{11}Cl_2F_4N_3O$ ((*RS*)-2-(2, 4-dichlorophenyl)-3-(1*H*-1, 2, 4-triazol-1-yl)propyl 1,1,2,2-tetrafluoroethyl ether), is a triazole class of fungicide (which is a **demethylation inhibitor** (DMI)). Tetraconazole is a colourless viscous liquid that is soluble in water, acetone, methanol and 1, 2-dichloroethane. It is stable in sunlight (when in water) and in mild alkaline solutions. Tetraconazole has a CAS Registry Number of 112281-77-3.

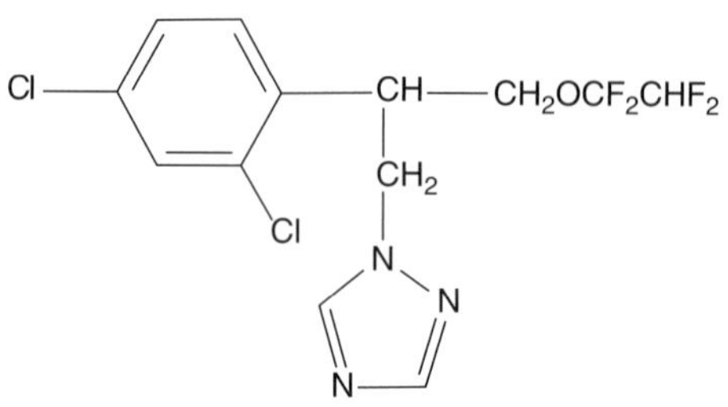

Tetraconazole

Tetraconazole, a systemic sterol C^{14} DMI, has a broad, protective and curative action upon its absorption by most parts of susceptible target plants (i.e. the roots, foliage and stems). Its action is seen against a variety of fungal diseases like **powdery mildews**, rusts, *Septoria* leaf spot, *Cercospora* leaf spot, **scab** and loose smuts. Causes of powdery mildew on different crops that are controlled by tetraconazole include *Erysiphe graminis* (cereals), *Erysiphe betae* (sugarcane), *Erysiphe necator* (grapes), *Erysiphe cichoracearum* (cucurbits), *Sphaerotheca macularis* (strawberry) and *Leveillula taurica* (tomatoes). It also acts on *Puccinia* spp., *Septoria* spp., *Uromyces betae*, *Venturia* spp. and many others that affect sugarcane, pome fruits and vegetables.

Some commonly recommended doses of this fungicide include 100–125 g/ha, 100 g/ha and 40–50 g/ha, for cereals, sugarcane and tomatoes, respectively.

Commonly used for foliar and seed treatment, tetraconazole is formulated as emulsifiable concentrate (EC), wettable powder (WP) among others, which is applied from the ground or aerially. This fungicide is also applied via **chemigation**. Trade names include Lospel, Emerald and Domark. For managing resistance to the chemical, tetraconazole applications must be alternated with non-triazole fungicides. A maximum of two applications may be made with an interval of 4 weeks.

Toxicologically, tetraconazole has an acute oral LD_{50} among rats of between 1031 mg/kg and 1248 mg/kg while in birds it ranges from 63 mg/kg to 650 mg/kg. It is toxic to bees and earthworms and also highly toxic to fish and aquatic organisms and plants (with as low an LC_{50} as 0.42 mg/l in some species). The ADI of tetraconazole in rats is 0.004 mg/kg b.w./day. It is classified as a moderately hazardous chemical (Class II toxin) by the WHO and is feared to be a carcinogen.

Tetraconazole is absorbed, metabolized and almost completely excreted upon ingestion by animals, with the main metabolite being 1,2,4-triazole, while in plants, the metabolic products are tetraconazole acid and alcohol. This fungicide is persistent in soil and water.

Tetraconazole is registered for use in the USA, South Africa and countries of the EU such as the UK, Italy, Hungary, Germany, Belgium, France and Spain. In the EU, most nations have fixed an MRL for many vegetables and grapes at 0.02 mg/kg (according to the EU Pesticide Database).

Tetracycline bactericides

Tetracyclines are a group of compounds with broad-spectrum activity against a wide range of Gram-negative and Gram-positive bacteria. Most important tetracyclines inhibit cell growth by reversibly binding to ribosomes and thus inhibiting protein synthesis. Tetracyclines are also used in the medical field for controlling respiratory and urogenital tract diseases.

Oxytetracycline is used for promoting growth and curing diseases of animals and fish.

Disease control by oxytetracycline with Agrimycin (at 50–200 ppm) against bacterial spot of tomato, bacterial spot of peach, fire blight and wild fire of tobacco is now known. The pathogen *Xanthomonas campestris* pv. *pruni* is managed well with this compound. Oxytetracycline injectibles are common formulations suitable for trunks of peach and pear trees, for controlling some bacterial infections like X disease.

The bactericide is critical in cases where pathogens have developed resistance to streptomycin (as in the case of *Erwinia amylovora*).

Tetradifon acaricide

Tetradifon, $C_{12}H_6Cl_4O_2S$ (4-chlorophenyl 2,4,5-trichlorophenyl sulfone), is a bridged diphenyl acaricide. It is a colourless crystal, soluble in water and different organic solvents (e.g. kerosene, methanol and acetone) from 10 g/l to 255 g/l. It is stable in acidic and alkaline media and on exposure to UV radiation. Tetradifon is produced by the Friedel-Crafts reaction of 2,4,5-trichlorobenzene sulfonyl chloride and chlorobenzene. Its CAS Registry Number is 116-29-0.

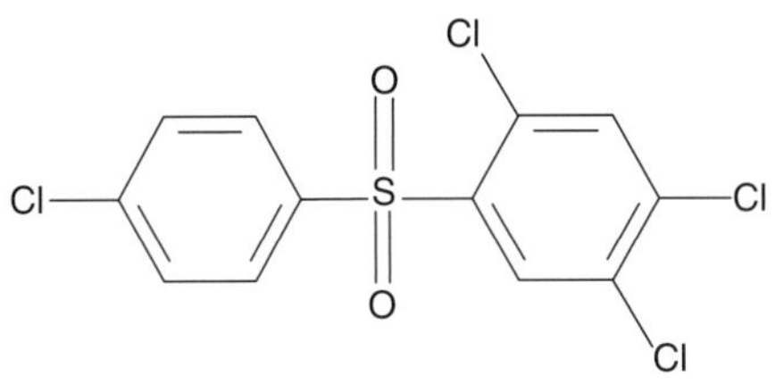

Tetradifon

This acaricide is a non-systemic compound that inhibits the growth of most mites in all life stages. Tetradifon is used on vegetables, greenhouse crops, ornamentals, cotton and sugarcane.

Known trade names are Tedion V-18, Duracide and Tedone. It is available in a variety of formulations including emulsifiable concentrates (EC), wettable powders (WP) and smoke generators (FU).

The WHO classifies tetradifon as a Class U toxin (unlikely to be hazardous). The acute oral LD_{50} in rats ranges from 5000 mg/kg b.w. to 14,700 mg/kg b.w. The NOEL in rats (2 year feed) is 15 mg/kg b.w./day. It is relatively non-toxic to bees. The acaricide can irritate the eyes.

About three-quarters of the tetradifon administered to test mammals is excreted in the faeces (via bile) within 2 days. In soil, tetradifon is persistent, since the compound and its degradates adsorb to soil particles and do not leach. It degrades in aerobic conditions.

Tetradifon is not registered in India, the USA and throughout the EU. It is registered for use in Australia and South Africa.

Tetramethrin insecticide

Tetramethrin, $C_{19}H_{25}NO_4$ (cyclohex-1-ene-1, 2-dicarboximidomethyl (1*RS*, 3*RS*; 1*RS*, 3*SR*)-2, 2-dimethyl-3-(2-methylprop-1-enyl) cyclopropanecarboxylate), is a pyrethroid insecticide. It is a yellowish crystalline substance, soluble in water, benzene, xylene, acetone, toluene, trichloroethane and cyclohexanon. Although it is stable up to 50°C, it is sensitive to alkalis and strong acids. Its CAS Registry Number is 7696-12-0. Tetramethrin is synthesized by the esterification of chrysanthemum monocarboxylic acid with tetrahydronaphthalimidomethyl alcohol.

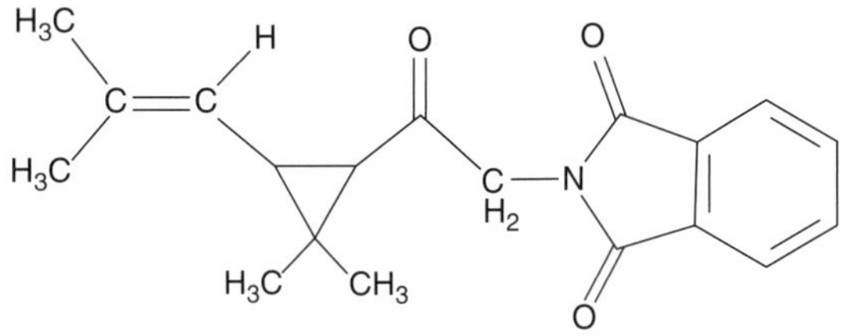

Tetramethrin

Tetramethrin is a contact insecticide, which is very effective on flies, cockroaches, mosquitoes, wasps and other insect pests in public places, homes and gardens. It is not recommended for agricultural use.

This insecticide, commercially available as Neo-Pynamin and Duracide is generally formulated as an aerosol dispenser (AD), emulsifiable concentrate (EC) and dustable powder (DP). It is compatible with **synergists** and other insecticides (e.g. **permethrin** and **resmethrin**).

According to the WHO and EPA, tetramethrin is unlikely to present acute hazard in normal use (Class U of the WHO). The acute oral LD_{50} in rats is >5000 mg/kg, while the acute percutaneous LD_{50} is >5000 mg/kg (indicating that it does not irritate the skin, for practical purposes). But it is toxic to bees, birds and fish and is also feared to be a carcinogen. It is also moderately toxic by the inhalation route. The acute oral LD_{50} in bird species tested is >1000 mg/kg, while for bluegill sunfish, the LC_{50} (96 h) is 0.021 mg/l. It is advised that tetramethrin is not brought in contact with water and food sources. Although toxicity to users is very rare, symptoms of prolonged exposure include shortness of breath, muscle twitching, skin and eye irritation.

In animals, tetramethrin is metabolized and eliminated via urine and faeces within 5 days. In soil and water, degradation involves cleavage of the ester bond and other breakdown reactions, leading to chrysanthemic acid derivatives, tetramethrin ((1*R*)-isomers) and phenoxybenzoic acid.

Tetramethrin is registered for use in the USA, Australia, South Africa and many other countries. But the derivatives of tetramethrin are not registered in the USA.

Tetramethylthiuram disulfide

See **Thiuram disulfide**

Thallium as a rodenticide

Thallium (Tl) is a soft and malleable metal occurring naturally in pyrites and lorandite and is obtained from extracting lead and zinc ores. It is an odourless and tasteless compound that is commonly considered a highly toxic rodenticide. Its salts may also be pesticidal.

Thallium reacts with water and air and forms hydroxides and oxides. Thallium sulfate is one of them. Thallium sulfate

(Tl_2SO_4), also called thallous sulfate, is also used as an ant killer. The CAS Registry Number of thallium sulfate is 7446-18-6.

The oral LD_{50} in rats is >50 mg/kg b.w. Thallium poisoning does not have an antidote. Its toxicity may produce injury to skin since it is absorbed via the skin. It also has carcinogenic tendencies.

Thallium sulfate

See **Thallium as a rodenticide**

The classical (or inoculative or importation) method of applying weed control

See **Biological weed control agents – application methods**

The Prevention of Food Adulteration Act, 1954

The Prevention of Food Adulteration Act (PFA), 1954, was enacted on 29 September 1954 in India. The PFA governs the standards of food and methods of analysis and sampling. The Act also covers the penalties for not following these set standards.

The Central Committee for Food Standards (CCFS) under the Ministry of Health and Family Welfare recommends the quality of food commodities under the PFA Act. Among its various sub-committees, the Pesticide Residue sub-committee advises the CCFS on matters including tolerance limits for pesticides in different items of food, based on use.

More information on the above may be obtained from http://fssai.gov.in/Regulations/PreventionofFoodAdulterationAct1954.aspx and related links.

Thenylchlor herbicide

See **Chloroacetanilide herbicides**

Therapeutants

See **Fungicides**

Thiabendazole fungicide

Thiabendazole, $C_{10}H_7N_3S$ (2-(thiazol-4-yl) benzimidazole; 2-(1,3-thiazol-4-yl)benzimidazole), is a **benzimidazole** type of fungicide. Thiabendazole (or tiabendazole) is an off-white powder that is soluble in water (at 0.16 g/l at pH 4), *n*-heptane, methanol (8.28 g/l) and 1,2-dichloroethane. It is hydrolytically stable with a photolysis DT_{50} of 29 h at pH 5. It has a CAS Registry Number of 148-79-8.

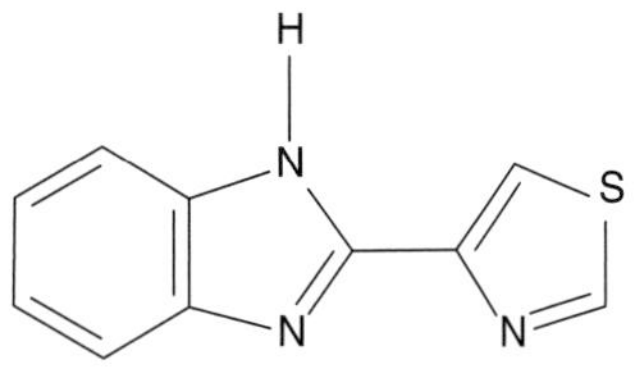

Thiabendazole

Thiabendazole, a systemic fungicide with curative action, also sticks to the outer cover of the crop, protecting it from outside too. The fungicide is a favoured post-harvest pest control agent, used as a seed treatment and as a stem injection. It also has good translocation to leaf margins. Thiabendazole acts by binding to tubulin, which inhibits mitosis and hence, growth of the fungus. It controls many fungal pathogens from the genera *Aspergillus*, *Botrytis*, *Cercospora*, *Fusarium*, *Penicillium*, *Rhizoctonia*, *Sclerotinia* and *Gibberella*. Typical diseases controlled by thiabendazole include mould and blights in oranges, mangoes, avocados, chicory, potatoes and vegetables. It also acts as an anti-parasitic agent on farms, by controlling hookworms and roundworms.

The fungicide is usually formulated as a suspension concentrate (SC), soluble concentrate (SL), tablet (TB) and wettable powder (WP). A common application dose is 2 g a.i./l. Thiabendazole may be combined with select pesticides such as **carboxin**. Some of the known trade names include Mertect, Tecto, Thiabensun, Mintezol, Tresaderm and Arbotect.

Toxicologically, the acute oral LD_{50} in mice, rats and rabbits is 3600, 3100 and >3800 mg/kg b.w., respectively, while the same for some bird species is >2250 mg/kg, indicating moderate toxicity to mammals and birds. It is toxic to fish (with the LC_{50} for rainbow trout being 0.55 mg/l). The ADI is 0.1 mg/kg and a 2 year NOEL among rats is 10 mg/kg b.w./day. The WHO has categorized thiabendazole as a Class U toxin (unlikely to be hazardous), while the EPA categorizes it in Category III (slightly toxic). Excessive exposure to this fungicide leads to dizziness, vomiting, diarrhoea, lethargy, fever, chills, rash and local oedema, headache and liver injury. Since it is a carcinogen and a developmental toxin, thiabendazole is classified as a PAN Bad Actor chemical.

Animals absorb, metabolize and eliminate thiabendazole almost completely, with hydroxylation at the 5-position and conjugation with glucuronic acid and sulfate. In plants, thiabendazole forms a major portion of the residues. This is a highly persistent fungicide, with a soil DT_{50} of 1 year. This property means that it is not readily leached from soil.

Thiabendazole is registered in South Africa, Tanzania, Hungary, Germany, the UK, the USA, Canada and some other countries. According to the Canadian Pest Control Product Act, MRL values are 10 mg/kg each for apples, citrus fruits, pears and potatoes, but only 0.4 mg/kg for bananas.

Thiacloprid insecticide

Thiacloprid, $C_{10}H_9ClN_4S$ ((*Z*)-3-(6-chloro-3-pyridylmethyl)-1,3-thiazolidin-2-ylidenecyanamide), is a neonicotinoid insecticide (a chloro-nicotinoid). It has a CAS Registry Number of 111988-49-9. It is a yellowish, odourless, crystalline powder, soluble in water to the tune of 185 mg/l. It is stable up to 50°C and to metals/metal ions.

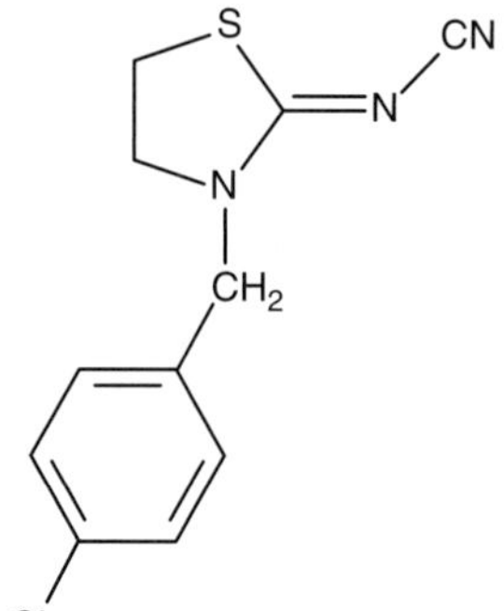

Thiacloprid

Thiacloprid, a systemic insecticide, is applied to crops like apples, pears, cotton, potatoes and rice to control whiteflies, aphids and codling moths. Its mechanism of action involves disruption of the insect nervous system by inhibiting nicotinic **acetylcholine receptors**.

The insecticide thiacloprid is formulated as oil dispersions (OD), flowable concentrates (OF and SC) or as water-dispersible granules (WG). The application rate for thiacloprid is in the range of 100–280 g/ha diluted as required. It is commercially available as Alanto, Bariard, Biscaya and Proteus.

The WHO has classified thiacloprid as a moderately hazardous chemical. The acute oral LD_{50} in rats is 444 mg/kg, while the dermal LD_{50} is >2000 mg/kg b.w. This insecticide is toxic via the inhalation route (the inhalation LC_{50} in rats is >0.481 mg/l). This chemical is toxic to birds and honey bees. The acute LD_{50} in Japanese quails is 49 mg/kg, while the acute 48 h oral LD_{50} in honey bees is 17.32 µg per bee. The ADI in rats is 0.01 mg/kg b.w./day. It is not harmful to fish but is toxic to marine invertebrates. Groundwater contamination cannot be ruled out. Thiacloprid is believed to cause cancer. It is a PAN Bad Actor chemical.

Thiacloprid degrades in soil through microbial means and has a half-life of >4 days. In aquatic conditions, the half-life varies from 2 months to >1 year. Degradation products are thiacloprid amine, thiacloprid sulfonic acid and YRC 2894 amide, all of which are persistent. The EPA recommends the use of PPE for applicators and workers, as well as following proper label instructions with regards to managing spray drift and maintaining a **buffer zone** covering aquatic and non-aquatic areas.

Thiacloprid is registered in India, Australia, Canada, the USA and throughout the EU. Local Environment Risk Assessment for Pesticides (LERAP, UK) recommends broadcast air-assisted sprayers for thiacloprid to reduce drift.

According to the EU Pesticide Database, the MRLs of thiacloprid are 2.0 mg/kg for lettuce; 1.0 mg/kg for sweet peppers and beans (with pods), 0.5 mg/kg for tomatoes and strawberries, and 0.02 mg/kg for citrus fruits and cereals.

Thiadiazole-copper fungicide

Thiadiazole-copper, $C_4H_4CuN_6S_4$, is an azole class of fungicide, also used as a bactericide. It is also registered as a herbicide. It is stable in acidic conditions, and can be stored in conditions that are cool, dry and well ventilated.

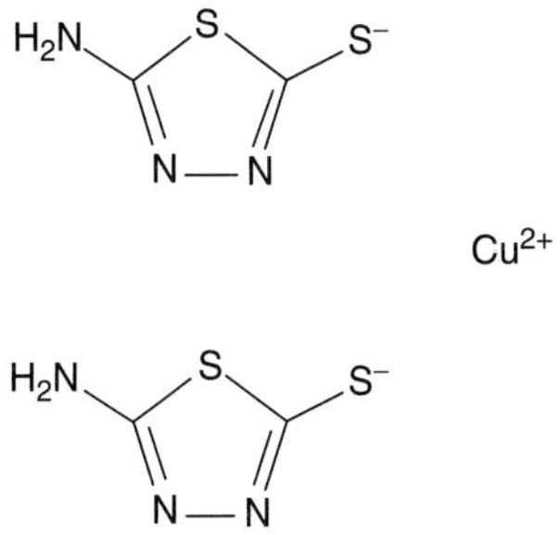

Thiodiazole-copper

Thiadiazole-copper is a systemic, broad-spectrum, protective compound that is effective against fungal and bacterial diseases of rice, orange, tobacco, cabbage and some other vegetables. The thiazolyl group and the copper element control bacterial pathogens. Cupric ions provide the protective property, while its antiseptic property halts the spread of disease. Bacterial leaf blight of rice, **scab** and canker of orange, leaf spots, *Fusarium* wilts and rots are some diseases controlled by thiadiazole-copper. Its residual activity is approximately 2 weeks.

Thiadiazole-copper is formulated as a suspension concentrate (SC). Spray applications at the onset of a disease give the best results. It is compatible with many pesticidal, non-alkaline formulations. It is commercially available as LongTM.

The fungicide-bactericide-herbicide is safe on mammals, fish, birds and bees. In case of the liquid coming in contact with the skin and eyes, both eyelids need to be rinsed with fresh water for up to 15 min. Symptomatic treatment is recommended in cases of oral exposure.

Thiadiazole-copper is registered for use in the Philippines and Vietnam.

Thiamethoxam insecticide

Thiamethoxam, $C_8H_{10}ClN_5O_3S$ ((*EZ*)-3-(2-chloro-1,3-thiazol-5-ylmethyl)-5-methyl -1,3,5-oxadiazinan-4-ylidene(nitro)amine), is a neonicotinoid insecticide belonging to the thianicotinyl subgroup of chemicals. It hydrolyses in alkaline conditions but is stable in acidic medium. It has a CAS Registry Number of 153719-23-4.

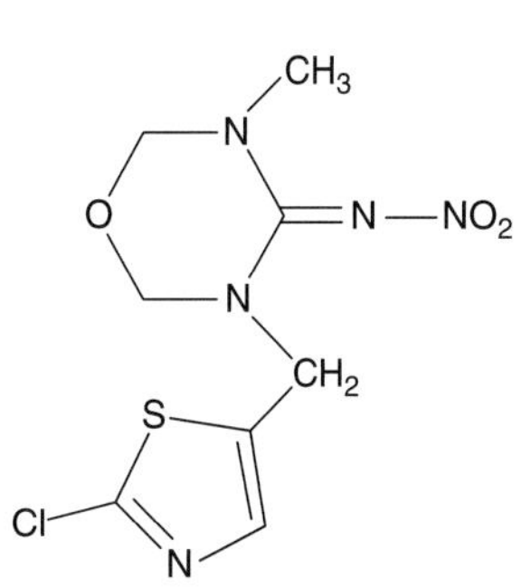

Thiamethoxam

Thiamethoxam acts as a broad-spectrum, systemic insecticide, and it is mainly used as a seed treatment. As a neonicotinoid, it paralyses the insect by blocking the electron transfer between nerve cells. It is absorbed quickly by plants and transported to all parts. It is also an anti-feedant. The range of pests controlled includes aphids, thrips, whiteflies, beetles, centipedes, millipedes, sawflies, leaf miners, stem borers, plant hoppers (white, green and brown) and termites. Crops protected are mango, citrus, aubergine (brinjal), potato, rice, wheat and cotton. Thiamethoxam is also applied to control fruit flies and house flies in homes and animal shelters.

The available formulations of thiamethoxam are suspension concentrate (SC), water-dispersible granules (WG), granules (GR), flowable concentrate (FS) and water-dispersible powder for slurry seed treatment (WS). It is applied at rates of 10–200 g/ha. Common trade names are Helix XTra, Actara, Evident and Cruiser. Besides being used as a seed treatment, it is also applied to the foliage.

The WHO classifies thiamethoxam as a Class III toxin (i.e. slightly hazardous). The acute oral LD_{50} in rats is 1563 mg/kg, while the acute percutaneous LD_{50} in rats is >2000 mg/kg. This insecticide is toxic to birds (the acute oral LD_{50} ranging from 576 to 1552 mg/kg). It is slightly toxic to fish and *Daphnia magna*.

Thiamethoxam is very toxic to honey bees and moderately toxic to earthworms and other beneficial soil organisms (the contact LD_{50} in honey bees is 0.024 mg per bee, while the 14 day LC_{50} in earthworms is >1000 mg/kg soil).

In animals, thiamethoxam is quickly and completely absorbed, rapidly distributed in the body and equally rapidly eliminated. A metabolite of thiamethoxam in soil is **clothianidin.** In soil, the DT_{50} is 51 days. Aqueous photolysis occurs rapidly.

Thiamethoxam is registered for use in India, Australia, the USA, Burkina Faso, Cambodia, Canada, Mali, Mauritania, New Zealand, Niger, the Philippines, Senegal, South Africa, Vietnam and almost throughout the EU.

Thiazopyr herbicide

Thiazopyr, $C_{16}H_{17}F_5N_2O_2S$ (methyl 2-difluoromethyl-5-(4, 5-dihydro-1, 3-thiazol-2-yl)-4-isobutyl-6-trifluoromethylnicotinate), is a pyridine herbicide. It is a light tan-coloured crystalline solid, soluble in water and organic solvents. It is not very photostable in water, with a DT_{50} of 15 days. Thiazopyr is made by amination of 3-chlorocarbonyl-5-carboxymethyl-4-(2-methylpropyl)-2-trifluoromethyl-6-difluoro-methylpyridine with ethanolamine followed by treatment with phosphorus oxychloride and cyclization. Hydrolysis of thiazopyr occurs in alkaline media. Its CAS Registry Number is 117718-60-2.

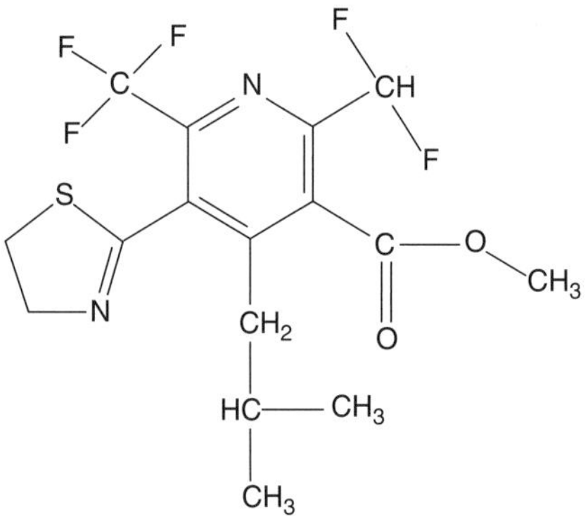

Thiazopyr

Thiazopyr, a pre-emergence herbicide, inhibits cell division in susceptible annual grasses and some broadleaved weeds (including some small-seed broadleaved weeds) growing among fruit trees, vines, citrus, sugarcane and pineapple.

The formulations in which thiazopyr is available are emulsifiable concentrate (EC) and as granules (GR). It is available in the market under the trade name of Visor.

In rats, the acute oral LD_{50} of thiazopyr is >5000 mg/kg, while the dermal LD_{50} is >5000 mg/kg b.w. and the inhalation LC_{50} is >1.2 mg/l. Thiazopyr is very highly to moderately toxic to non-target terrestrial and aquatic plants and algae, earthworms and honey bees. Thiazopyr is also moderately toxic to birds (the oral LD_{50} in bobwhite quails is 1913 mg/kg). This herbicide is also suspected to be a carcinogen and a groundwater contaminant.

Thiazopyr is persistent in soil, with a half-life of between 111 and 437 days. The major soil metabolite, monoacid, is slightly adsorbed on soils.

The European Commission (EC) does not allow thiazopyr to be used as an active ingredient in the EU. Thiazopyr is registered for use in Australia and the USA.

Thidiazuron as a PGR

Thidiazuron, $C_9H_8N_4OS$ (1-phenyl-3-(1, 2, 3-thiadiazol-5-yl) urea), is a plant growth regulator and a herbicide. It is a colourless, odourless crystalline substance that is stable in water, hexane, methanol, dichloromethane, toluene, acetone and ethyl acetate. It is rapidly converted to the photo-isomer 1-phenyl-3-(1, 2, 5-thiadiazol-3-yl) urea in the presence of light. It is stable at room temperature (between pH 5 and 9). It has a CAS Registry Number of 51707-55-2.

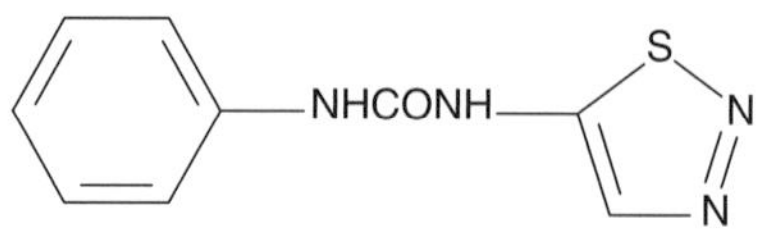

Thidiazuron

Thidiazuron is absorbed by the leaves and, as a pre-harvest defoliant, stimulates formation of an abscission layer between the plant stem and the leaf petioles, causing entire leaves to drop. It is used for defoliating cotton, resulting in removing immature leaves and fruiting structures, which can stain cotton. It is used as a growth regulator.

The herbicide thidiazuron is formulated as a wettable powder (WP), emulsifiable concentrate (EC) and an oil-miscible flowable concentrate or oil-miscible suspension (OF). An example of an application rate is 0.21 kg/ha. Thidiazuron is available in the market under the trade name, Dropp.

According to the WHO and EPA, thidiazuron is unlikely to be hazardous in normal use. Excess exposure to thidiazuran can cause irritation to the eyes, skin and mucous membranes, and cause coughing and shortness of breath, nausea, vomiting, diarrhoea, headache, confusion and electrolyte depletion.

In animals, the metabolism involves hydroxylation of the phenyl group, followed by formation of water-soluble conjugates. Thidiazuron is excreted in the urine and faeces within 96 h. It strongly adsorbs to soil.

The herbicide is registered for use in Australia, New Zealand, South Africa and the USA.

Thifensulfuron-methyl herbicide

Thifensulfuron-methyl, $C_{12}H_{13}N_5O_6S_2$ (methyl 3-[[[[(4-methoxy-6-methyl-1,3,5-triazin-2-yl)amino]carbonyl]amino]sulfonyl]-2-thiophenecarboxylate), is a herbicide of the sulfonylurea class. It is an off-white and odourless solid substance, soluble in water (ranging from 223 to 8830 mg/l, depending on pH), *o*-xylene (0.212 g/l), ethyl acetate (3.3 g/l) and dichloromethane (23.8 g/l). Thifensulfuron-methyl is produced by reaction of methyl 3-sulfamoyl-2-thiophenecarboxylate with isocyanate to give 2-(methoxycarbonyl)-3-thiophenesulfonyl isocyanate, which is then treated with 2-amino-4,6-dimethoxypyrimidine. The CAS Registry Number is 79277-27-3.

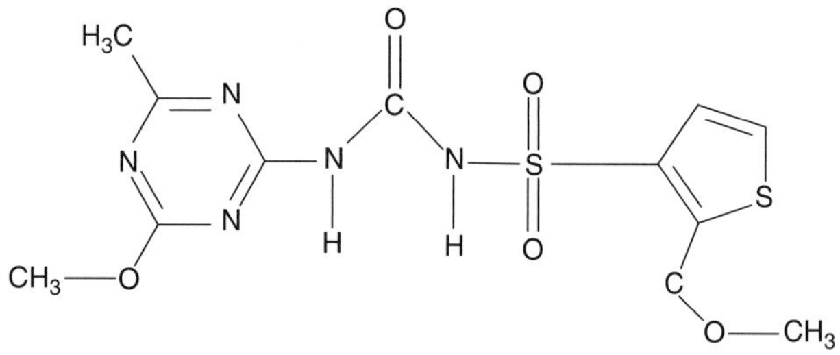

Thifensulfuron-methyl

Thifensulfuron-methyl, a selective, post-emergence, systemic herbicide, is translocated throughout the plant. It acts by inhibiting the enzyme acetolactate synthase (ALS) thus affecting cell division and growth in the plants. It shows little or no soil activity. The chemical is used to control annual and broadleaved weeds in cereal crops such as maize, wheat (spring and winter), oats and barley (winter and spring). It is also effective on soybean and pasture.

Formulated as a suspension concentrate (SC), tablet (TB) and water-dispersible granules (WG), thifensulfuron-methyl may be combined with herbicides such as **tribenuron-methyl** and **metsulfuron-methyl**. Some trade names include Harmony, Canvas and Quantum 75.

Thifensulfuron-methyl is classified under toxicity Class III (i.e. slightly hazardous) by the WHO. The acute oral LD_{50} in rats is >5000 mg/kg and in rabbits is >2000 mg/kg. The LD_{50} in birds is >2250 mg/kg. It is practically safe for honey bees and earthworms. The 2 year NOEL among rats is 25 ppm/kg diet. The ADI is optimized to be 0.008 mg/kg body weight.

The half-life of thifensulfuron-methyl in soil is 1–7 days. The hydrolysis DT_{50} is 180 days at neutral pH but is faster in acidic media. Photodecomposition is not significant.

Thifluzamide fungicide

Thifluzamide, $C_{13}H_6Br_2F_6N_2O_2S$ (2′,6′-dibromo-2-methyl-4′-trifluoromethoxy-4-trifluoromethyl-1,3-thiazole-5-carboxanilide), is a thiazole carboxamide fungicide. It is a light brown powder that is soluble in water at 1.6 mg/l at pH 5.7 (with solubility going up with increasing pH). It is stable to hydrolysis between pH 5 and 9, with a DT_{50} of <4 days. It is produced by reacting 2-chloro-3-carbonyl-4-trifluoroethyl butylate with thioacetamide in DMF, the resulting thiazolecarboxylate being converted in two steps to thifluzamide. It has a CAS Registry Number of 130000-40-7.

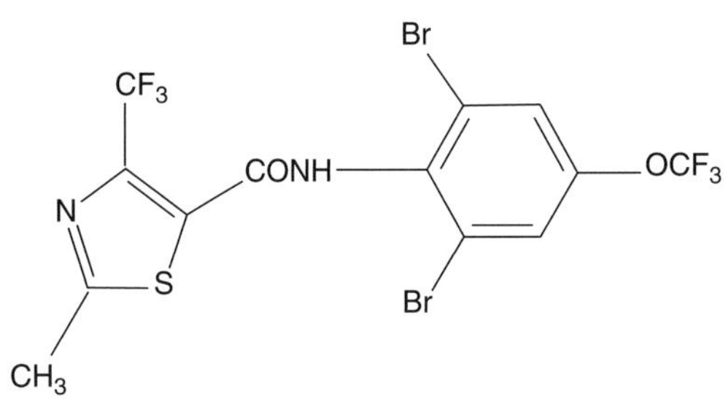

Thifluzamide

Thifluzamide acts by inhibiting succinate dehydrogenase in the tricarboxylic acid cycle, after being absorbed by the roots and leaves of the plant. Commonly used as a soil drench or a foliar spray, thifluzamide protects rice, potato, coffee, strawberry and maize against *Rhizoctonia* spp. (especially *Rhizoctonia solani* that causes sheath blight).

Thifluzamide is applied on rice foliage, in the range of 50–150 g/ha. It is formulated as granules (GR), suspension concentrates (SC) and as wettable powder (WP). Commercially known trade names are Pulsor and Greatam G.

Toxicologically, the acute oral LD_{50} values of thifluzamide in rats and rabbits are >6500 mg/kg and >5000 mg/kg, respectively. In fish, the LC_{50} is as low as 1.2 mg/l. The NOEL in rats is 1.4 mg/kg b.w./day. Thifluzamide is moderately toxic to birds (with an acute LD_{50} in bobwhite quail of 2250 mg/kg) and honey bees (with an acute 48 h contact LD_{50} of 100 μg per bee). The ADI is 0.014 mg/kg b.w./day. Thifluzamide is categorized by the WHO as a toxin that is not likely to be hazardous (Class U).

In animals, thifluzamide is extensively metabolized, while it is unchanged in plant leaves. It persists in soil (the DT_{50} is 4 years) with very insignificant microbial degradation.

Thifluzamide is registered for use in India, Vietnam, Japan, China and South Korea where it is mainly used on rice crops.

Thin-layer chromatography (TLC)

Thin-layer chromatography (TLC) is a planar chromatographic technique exclusively used as a microanalytical tool. The principle of TLC has been known for more than 100 years.

The advantages of TLC are: (i) cost-efficient optimization of separation due to easy changes of mobile or stationary phases; (ii) the possibility of a high sample throughput in a short time; and (iii) suitability for screening tests. Since the TLC layer acts as a storage medium for data, the analytical information can be stored for a long time. The separated substances can be further analysed with procedures such as infrared, NMR and mass spectrometry.

Thin-layer chromatographic separation can be divided into four basic steps: (i) sample preparation; (ii) sample application; (iii) development of a chromatogram; and (iv) qualitative and quantitative determination (TLC evaluation).

The retention of a solute in TLC is characterized by R_f, which is the distance travelled by the compound from the origin to the centre of the separated zone divided by the distance from the origin of the solvent front (Eqn T.3).

$$R_f = \frac{\text{Distance moved from the origin by the compound}}{\text{Distance moved from the origin by the solvent}} \quad \text{(T.3)}$$

The capacity factor K can be defined as the ratio of retention times of a solute in the stationary phase (t_s) and the mobile phase (t_m), respectively (Eqn T.4).

$$K = \frac{t_s}{t_m} \quad \text{(T.4)}$$

The relationship between the capacity factor and R_f is shown by Eqns T.5 and T.6.

$$R_f = \frac{t_m}{t_m + t_s} = \frac{1}{K+1} \quad \text{(T.5)}$$

$$K = \frac{1-R_f}{R_f} = \frac{1}{R_f - 1} \quad \text{(T.6)}$$

The relationship between the R_f value of the solute and the concentration of the major component in the eluent (C) is not linear (Eqns T.7 and T.8). The R_m value was introduced to obtain a linear correlation between the retention of solutes and the composition of the eluent.

$$R_m = \log\left(\frac{1}{R_f - 1}\right) \quad \text{(T.7)}$$

$$R_m = R_{mo} + bC \quad \text{(T.8)}$$

where

R_{mo} = the theoretical value

R_m = the value of a solute extrapolated to zero concentration of the major component in the mobile phase b = the change of the R_m value caused by unit change of the concentration of the major component in the mobile phase

A wide variety of organic and inorganic supports have been used for separation in TLC. Silicas with varying surface pH and pore diameter, and other sorbents like alumina, **diatomaceous earth**, celluloses and their derivatives have been used extensively in making TLC plates. The average particle size of sorbents is between 10 μm and 50 μm. The performance of the TLC layer depends on the characteristics of the sorbents. The solvent strength (elution capacity) is determined by its capacity to move a solute, with higher solute mobility indicating higher elution strength. The solvent strength ranges from *n*-heptane-hexane to water as follows: *n*-heptane-hexane >cyclohexane >toluene >dichloromethane >acetonitrile >1-butanol >2-propanol >methanol >water. There are two methods for conducting quantitative evaluation of chromatograms. In one method, a spot is scratched from the absorbent material and the quantity determined using a spectrophotometer and gas liquid chromatograph. In the second method, spot areas are measured in relation to the weight of the material or measurement of density or colour using a photometer. The method is limited to non-volatile compounds and limited to separation numbers (10–15 peaks).

TLC is widely used to determine residues of pesticides in many types of samples such as foods, water and various environmental situations.

Thiobencarb herbicide

Thiobencarb, $C_{12}H_{16}ClNOS$ (*S*-4-chlorobenzyl diethylthiocarbamate), is a thiocarbamate herbicide. It is a pale yellow to brownish yellow liquid that is stable in sunlight and moderately soluble in water and miscible with organic solvents. It is made by reacting *p*-chloromercaptotoluene with *N*-diethyl carbamoyl chloride. Its CAS Registry Number is 28249-77-6.

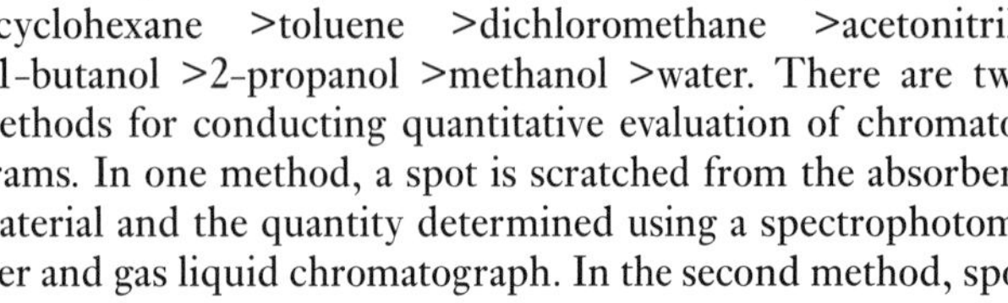

Thiobencarb

Thiobencarb, a selective, pre- and early post-emergence herbicide, is absorbed by the foliage and roots of a number of susceptible grasses and weeds in rice. It interferes with protein synthesis. It inhibits photosynthesis and **ACCase**.

The formulations in which thiobencarb is available are granules (GR) and emulsifiable concentrates (EC). It is used in soil treatment, and sprayed after dilution on rice, lettuce and celery. Typical application rates are between 2.24 kg/ha and 9 kg/ha. Trade names include Bolero, Bencarb, Saturno and Tamariz.

The acute oral LD_{50} of thiobencarb in mice is 560 mg/kg. The dermal LD_{50} in rats is >5000 mg/kg b.w., while the inhalation LC_{50} in rats is >2.43 mg/l. It is moderately toxic to birds (with the acute LD_{50} in bobwhite quails being >2000 mg/kg), *Daphnia magna* and to honey bees. However, it is not toxic to earthworms (the acute 14 day LC_{50} being 437 mg/kg). The ADI in rats is 0.01 mg/kg b.w./day. This herbicide is moderately hazardous according to the WHO and EPA. Exposure causes irritation to the skin, eyes and respiratory tract, headache, dizziness and nausea.

In soils, thiobencarb is persistent and does not leach in the water, with the half-life in aerobic conditions being up to 3 weeks and in anaerobic conditions being up to 8 months. Microbial action is the main source of breakdown, followed by photolysis upon exposure to sunlight. In water, thiobencarb is stable to hydrolysis. 4-Chlorobenzoic acid and 4-chlorobenzaldehyde are some of the breakdown products.

Thiobencarb is registered in India, Australia, Cambodia, Madagascar, the Philippines, Tanzania, Uganda, Vietnam and the USA.

Thiocarb

See **Molluscicides**

Thiocarbamate herbicides

Thiocarbamates are a group of pesticides with a wide range of uses. Some are used as fungicides, while others have herbicidal activity. Thiocarbamate fungicides are applied for protection of seeds and seedlings of ornamental plants, turf, vegetables and fruit. Two thiocarbamate fungicides are **methasulfocarb** and prothiocarb. Unlike carbamates, thiocarbamates have very little insecticidal properties and their use is limited to the control of household pests and control of vectors. An example is **cartap**. Thiocarbamates are also used as **biocides** in industrial plants and their use as herbicides is well known.

Thiocarbamates are volatile liquids or solids with low melting points. Water solubility varies with active ingredients (a.i.). Their volatility makes these compounds evaporate from soil, while their solubility in water makes them leach and move laterally in soil.

Breakdown mechanisms of thiocarbamates include photodegradation. While most thiocarbamates are commonly unstable in alkaline conditions, compounds such as EPTC, pebulate or diallate are very stable at pH 2 or 10. Their sulfoxide and sulfone derivatives are also stable at pH 2, but much less so at pH 10. Normally, the sequential oxidation of thiocarbamates to thiocarbamate sulfoxide and thiocarbamate sulfone decreases the hydrolytic stability. Soil microorganisms contribute significantly to the disappearance of thiocarbamates from the soil. In microorganisms and plants, thiocarbamates undergo many processes like hydrolysis and sulfoxidation to form carbon dioxide. Their mode of action is typically the inhibition of cholinesterase. Some symptoms of exposure to thiocarbamates include anorexia, squinting, hyper-salivation, lachrymation, laboured breathing, incoordination and depression.

Thiocarbamate herbicides include **butylate**, **cycloate**, diallate, **EPTC**, **esprocarb**, ethiolate, isopolinate, methiobencarb, **molinate**, orbencarb, **pebulate**, **prosulfocarb**, **pyributicarb**, sulfallate, **thiobencarb**, tiocarbazil, tri-allate and **vernolate**.

Diallate, $C_{10}H_{17}Cl_2NOS$ (*S*-2, 3-dichloroallyl diisopropylthiocarbamate), is a pre- and post-emergence herbicide and a thiocarbamate. Now obsolete, diallate has a CAS Registry Number of 2303-16-4. This chemical is used to control wild oats, blackgrass and some other grass weeds. Since it is a cholinesterase inhibitor, it has been classified as a PAN Bad Actor chemical. Diallate is not registered in the USA or throughout the EU.

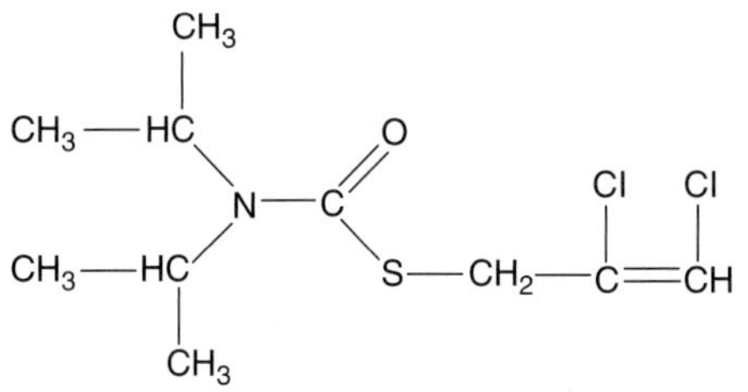

Diallate

Ethiolate herbicide, $C_7H_{15}NOS$ (*S*-ethyl diethyl (thiocarbamate)), is another thiocarbamate. It has a CAS Registry Number of 2941-55-1. Ethiolate is a PAN Bad Actor chemical since it is a cholinesterase inhibitor and is not registered in many countries like the USA.

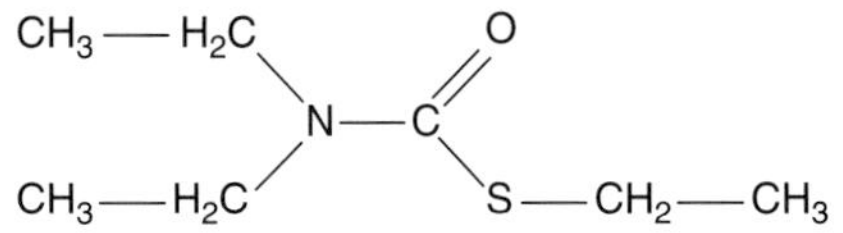

Ethiolate

Isopolinate, $C_{10}H_{19}NOS$ (*S*-isopropyl azepane-1-carbothioate), is a selective, systemic herbicide. With a CAS Registry Number of 3134-70-1, isopolinate controls broadleaved weeds and grasses, once it is taken up by the roots and translocated. Isopolinate is a cholinesterase inhibitor, and an irritant to the eyes, skin and respiratory tract. It is not widely registered.

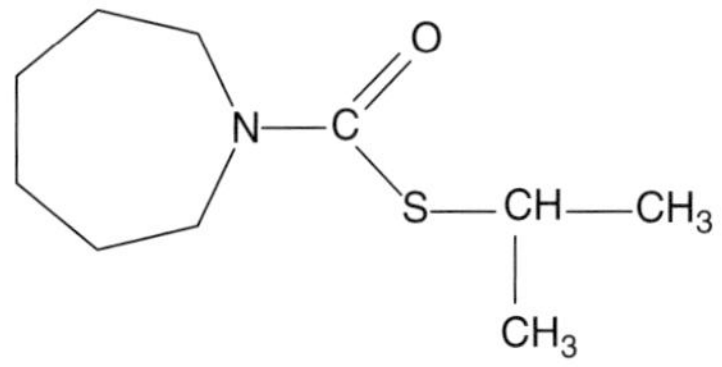

Isopolinate

Methiobencarb herbicide, $C_{13}H_{19}NO_2S$ (*S*-4-methoxybenzyl diethylthiocarbamate), is a thiocarbamate compound. It has a CAS Registry Number of 18357-78-3. It is a PAN Bad Actor chemical and not registered in many countries.

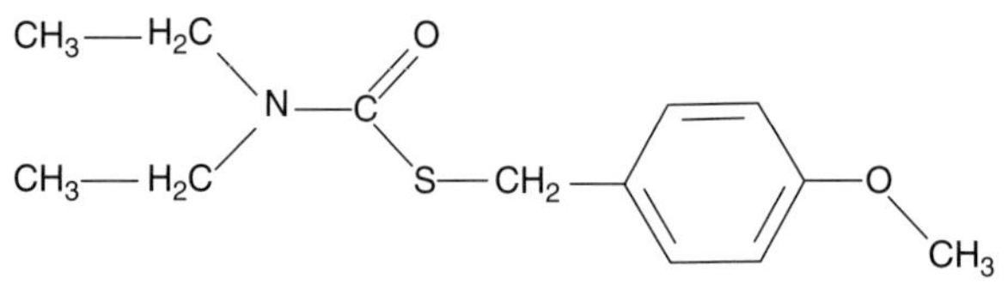

Methiobencarb

Sulfallate, $C_8H_{14}ClNS_2$ (2-chloroallyl diethyl-dithiocarbamate), is a pre-emergence thiocarbamate herbicide. It is an amber liquid that is incompatible with alkalis and strong oxidizing agents. It has a CAS Registry Number of 95-06-7. Sulfallate inhibits cholinesterase and is thus a PAN Bad Actor chemical.

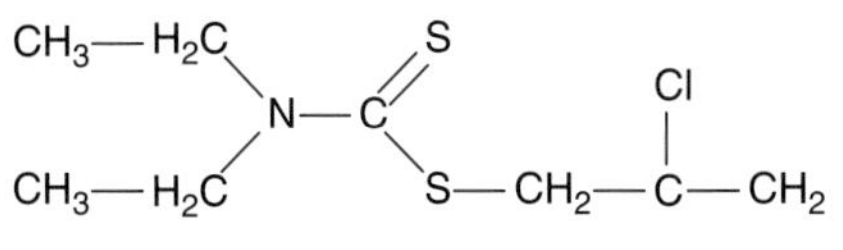

Sulfallate

Tiocarbazil, $C_{16}H_{25}NOS$ (*S*-benzyl di-sec-butylthiocarbamate), is a thiocarbamate herbicide. It has a CAS Registry Number of 36756-79-3. It is a selective herbicide that controls **barnyard grass** in submerged rice fields. It inhibits lipid synthesis and cholinesterase. The WHO Acute Hazard classification lists tiocarbazil as unlikely to be hazardous (Class U toxin). Tiocarbazil is a PAN Bad Actor chemical and is not registered in many countries.

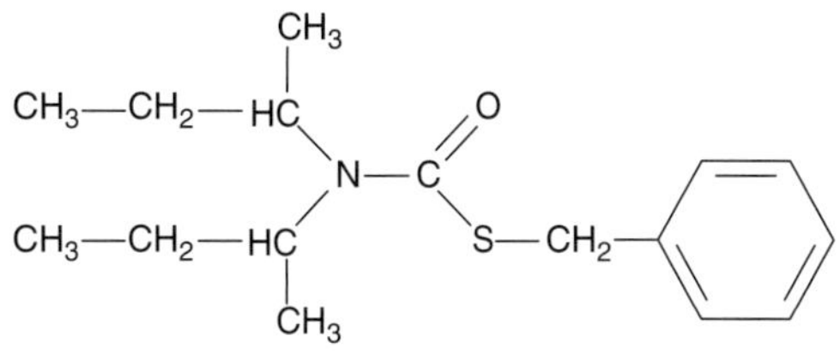

Tiocarbazil

Thiocyclam insecticide

Thiocyclam, $C_5H_{11}NS_3$ (*N,N*-dimethyl-1,2,3-trithian-5-ylamine), is a **nereistoxin** analogue insecticide. It is a colourless, odourless solid compound, soluble in water (at 84 g/l), dimethyl sulfoxide, methanol, ethanol, acetonitrile, acetone, ethyl acetate, chloroform, toluene and hexane. It is derived from the naturally produced nereistoxin of the annelid *Lumbriconereis heteropoda*. It has a CAS Registry Number of 31895-21-3. It is normally available as its salts (e.g. thiocyclam hydrochloride and thiocyclam oxalate).

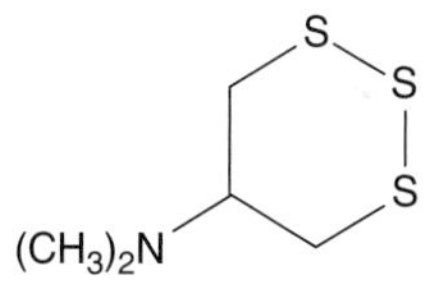

Thiocyclam

Thiocyclam, a selective compound, is a pro-insecticide nerve toxin that is an **acetylcholine receptor** (nAChR) agonist. It rapidly converts to a nereistoxin in a living system and has stomach and contact action on susceptible members of the orders Lepidoptera, Coleoptera, Diptera and **Thysanoptera**. Some examples of pests controlled are the Colorado potato beetle, stem borers, corn borers and weevils. Crops thus protected include potato, rice, maize, sugarbeet, sugarcane and fruit orchards. It acts by blocking the nervous system (via ganglionic blocking action) and paralysing the insect.

Thiocyclam is available as dustable powder (DP), granule (GR) and water-soluble powder (SP). It is available in the market under the brand name of Evisect.

According to the WHO and EPA, thiocyclam is a moderately hazardous toxin (Class II toxin). The acute oral LD_{50} in rats is 370 mg/kg, while the dermal LD_{50} in rats is >880 mg/kg b.w. It is toxic to fish and to earthworms (the acute 96 h LC_{50} in rainbow trout is 0.04 mg/l, while the acute 14 day LC_{50} in earthworms is 87 mg/kg). The ADI of thiocyclam is 0.008 mg/kg b.w./day. It is non-phytotoxic at recommended doses to most crops.

Thiocyclam is degraded in natural water via nereistoxin and its oxide, with the rate of degradation increasing with irradiation by sunlight. The DT_{50} value is up to 3 days in surface water. Thiocyclam hydrolyses at higher pH, with a DT_{50} of up to 7 days. The soil DT_{50} is 1 day. In plants, thiocyclam has limited systemic activity with acropetal translocation.

The insecticide thiocyclam is registered for use in Cameroon.

Thiodicarb insecticide

Thiodicarb, $C_{10}H_{18}N_4O_4S_3$ (3,7,9,13-tetramethyl-5,11-dioxa-2,8,14-trithia-4,7,9,12-tetraazapentadeca-3,12-diene-6,10-dione), is an oxime carbamate insecticide. It has a CAS Registry Number of 59669-26-0. It is a colourless, crystalline substance, soluble in water (35 mg/l), dichloromethane, acetone, methanol and xylene. It must be stored in a cool, dry container. It rapidly hydrolyses at pH 9 and above and decomposes in light. It degrades to dimethyl disulfide and carbon dioxide at temperatures above 60°C.

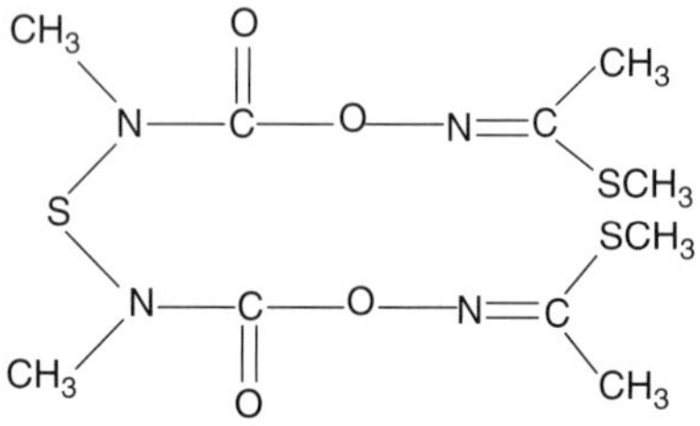

Thiodicarb

Thiodicarb is produced by the reaction of *N*,*N*′-thiobis (methylcarbamic acid fluoride) with 2-methylthioacetaldoxim in the presence of a base.

Thiodicarb is a cholinesterase inhibitor, meaning that it disrupts the action of enzymes involved in the functioning of the CNS. Symptoms of toxicity include increased salivation, headache, profuse sweating, muscle weakness and cramps, dizziness, nausea and vomiting. Thiodicarb is commonly used as an insecticide, ovicide and molluscicide to protect agricultural crops and vegetables from slugs and Lepidopteran, Coleopteran and some Hemipteran insect pests. It predominantly has a stomach action, although there is some contact action. Thiodicarb is used primarily on cotton, sweetcorn and soybeans (mostly sweetcorn). It also controls insects in leafy vegetables, cole crops and ornamentals. Common pests controlled by it are cotton bollworm, budworm, beet armyworm, corn earworm and black cutworm. As an insecticide, thiodicarb is effective against eggs as well as larvae, although the latter must feed upon treated foliage in order to be controlled.

Thiodicarb is commonly formulated as solutions and wettable powders (WP), ready-to-use baits (RB) and suspension concentrates (SC). The application rates of thiodicarb range from 560 g/ha to 8.5 kg/ha. Application may be carried out with ground and air equipment. It is available in the market under the brand names of Larvin, Judge and Toro.

The acute oral LD_{50} of thiodicarb in rats is 50 mg/kg, indicating high toxicity. This chemical is toxic to fish, earthworms and honey bees. The acute LD_{50} in bobwhite quails is 2023 mg/kg (moderate toxicity). The dermal LD_{50} in rats is >2000 mg/kg. The inhalation LC_{50} of thiodicarb in rats is 0.66 mg/l. The ADI in rats is 0.01 mg/kg b.w./day. The WHO categorizes thiodicarb as moderately hazardous (Class II). It is toxic via inhalation. Symptoms of poisoning are muscle weakness, dizziness, sweating, headache, salivation, abdominal pain, diarrhoea and nervous system depression.

In animals, a metabolic by-product called acetamide is formed. Acetamide is a potential carcinogen. However, this substance breaks down to carbon dioxide. In the environment, decomposition depends on sufficient aeration, microbial activity, temperature, pH and soil type (light-textured soils cause rapid degradation). It is not mobile in the soil and has a soil half-life of <7 days. Soil accumulation or groundwater contamination is not considered to be a threat. The major by-product of photolysis of thiodicarb is **methomyl**.

Thiodicarb is registered for use in India, Australia, Cambodia, Cameroon, Canada, Madagascar, New Zealand, the Philippines, South Africa and the USA. Some countries such as Belize have restricted the registration of thiodicarb.

According to the EU Pesticide Database, the MRL value is 1.0 mg/kg for mandarins, wine grapes, limes and lemons, 0.5 mg/kg for oranges and plums and 0.05 mg/kg for cereal grains, vegetables and fruits.

Thiofanox insecticide

Thiofanox, $C_9H_{18}N_2O_2S$ ((*EZ*)-(3,3-dimethyl-1-methylthiobutanone *O*-methylcarbamoyloxime), is an oxime carbamate insecticide. It is a colourless, pungent solid that is soluble in chlorinated and aromatic hydrocarbons, ketones and non-polar solvents, stable to heat under normal storage conditions and relatively stable to hydrolysis in the pH range of 5–9. It is decomposed by strong acids and alkalis. Thiofanox is obtained by reaction of 3,3-dimethyl-1-(methylthio)-2-butanone with hydroxylamine, followed by reaction with isocyanate. The CAS Registry Number is 39196-18-4.

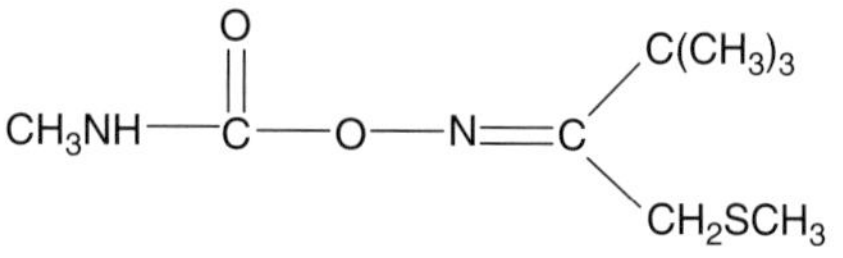

Thiofanox

The insecticide thiofanox is an extremely toxic chemical to mammals, birds, fish and honey bees. The WHO classifies it as a Class Ib toxin (highly hazardous). The acute oral LD_{50} in rats is 8.5 mg/kg and the dermal LD_{50} is 39.0 mg/kg b.w. The acute LD_{50} of thiofanox in bobwhite quails is 43 mg/kg. The acute 96 h LC_{50} of thiofanox in rainbow trout is 0.13 mg/l. The acute 48 h contact LD_{50} of thiofanox in honey bees is 0.058 μg per bee. Symptoms of poisoning with thiofanox are malaise, dizziness, sweating, headache, salivation, vomiting, diarrhoea, blurred vision, incoordination, muscle twitching and also coma, seizures and cardio-respiratory depression. The exposed person must be made to remove contaminated clothes and rinse the affected body part thoroughly with water and soap. Eyes must be rinsed with plenty of water. Medical attention should be sought. All products should carry the signal words 'DANGER – POISON'.

The methylthio group is rapidly oxidized to the sulfoxides and then to the sulfone in soil. Further degradation occurs to water-soluble metabolites. In soil, plants and animals, thiofanox undergoes a two-step oxidation to its sulfoxide 3,3-dimethyl-1-(methylsulfiny)-2-butanone *O*-[(methylamino) carbonyl] oxime and sulfone 3,3-dimethyl-1-methyl(sulfony)-2-butanone *O*-[(methylamino) carbonyl] oxime.

Thiofanox is registered for use in Cambodia.

Thiometon insecticide

Thiometon, $C_6H_{15}O_2PS_3$ (*S*-2-ethylthioethyl *O,O*-dimethyl phosphorodithiate), is an organophosphorus insecticide. It is a colourless oil, soluble in water (200 mg/l) and in common organic solvents (except alkanes), but slightly soluble in petroleum ether and mineral oils. It is hydrolysed more rapidly in alkaline media than in acidic media. It is produced by reaction of the sodium salt of *O,O*-dimethyldithiophosphoric acid with ethylmercaptochloroethanol. It has a CAS Registry Number of 640-15-3.

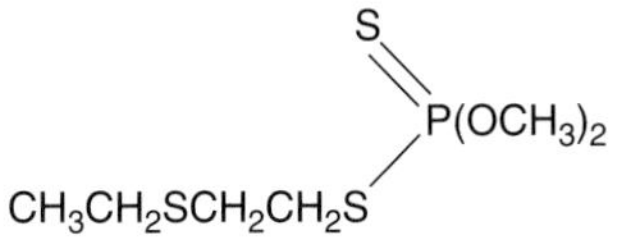

Thiometon

Thiometon, a cholinesterase inhibitor, is a systemic insecticide and acaricide with contact and stomach action. It is used in the control of sucking insects (e.g. aphids, woolly aphids, suckers, sawfly larvae and thrips), bryobia mites, tetranychid mites and fruit-tree red spider mites (non-OP resistant) on a wide range of crops, including ornamentals, strawberries, citrus and other fruit, olives, vines, beet, tobacco, cotton and cereals.

Thiometon is formulated in forms such as emulsifiable concentrates (EC) and ULV liquid (UL). Foliar spray and soil drench are the two main applications of thiometon. It is compatible with most other pesticides, except alkaline materials. It may be applied at a concentration of 0.1% when the plant is still growing. The REI should be 4 days for tree crops and 24 h for all other crops. It is available in the market under the trade name, Ekatin.

The WHO classifies thiometon as a Class Ib toxin (highly hazardous). The acute oral LD_{50} in rats is 40 mg/kg and the dermal LD_{50} is 1429 mg/kg b.w. The inhalation LC_{50} of thiometon in rats (4 h period) is 1.93 mg/l. The acute LD_{50} in birds is 61.75 mg/kg. The acute 96 h LC_{50} of thiometon in rainbow trout is 3.7 mg/l. The acute 48 h contact LD_{50} of thiometon in honey bees is 0.55 μg per bee. The ADI is 0.003 mg/kg b.w./day. It is non-phytotoxic when used as directed. Symptoms of exposure to thiometon are excessive salivation, sweating, weakness, tremor, incoordination, headache, dizziness, nausea, vomiting, abdominal cramps, diarrhoea, respiratory depression, tightness in the chest, wheezing, fluid in the lungs, pinpoint pupils, blurred or dark vision, seizures, incontinence, respiratory depression and loss of consciousness. Exposed persons should be made to remove contaminated clothes and rinse the affected part thoroughly with water and soap. Eyes must be rinsed with plenty of water. If ingested, the mouth must be rinsed and the patient offered a slurry of activated charcoal in water. Vomiting must be induced. Medical attention should be sought. All product labels should carry the signal words 'DANGER – POISON'.

In mammals, metabolism is rapid. It occurs by conversion of P=S to P=O, oxidation of the sulfide to sulfoxide and sulfone, and hydrolysis to the dimethyl thiolophosphoric acid. In plants, oxidation and hydrolytic reactions are the most important factors in degradation. The main metabolites are thiometons sulfoxide and thiometon sulfone.

Thiometon is registered for use in India, Australia and Cambodia.

Thiophanate-methyl fungicide

Thiophanate-methyl, $C_{12}H_{14}N_4O_4S_2$ (dimethyl 4,4′-(*o*-phenylene)bis(3-thioallophanate)), is a **benzimidazole** precursor fungicide and a carbamate fungicide. It exists as colourless crystals that are soluble in water (at 0.0224 g/l), acetone (58.1 g/l), methanol (29.2 g/l), chloroform (26.2 g/l) and ethyl acetate (11.9 g/l). It is stable to air, sunlight and acidic media at room temperature but unstable in alkaline media, strong oxidizing acids and peroxides. It reacts with metals to produce flammable hydrogen. It is not compatible with compounds containing copper. Thiophanate methyl is produced by reaction of *o*-phenylenediamine and methoxycarbonyl isothiocyanate. It has a CAS Registry Number of 23564-05-8.

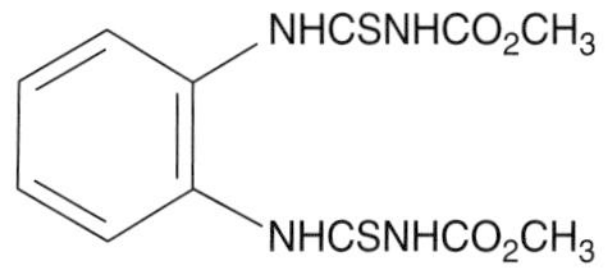

Thiophanate-methyl

Thiophanate-methyl, being a systemic fungicide acts as a **carbendazim** precursor. It has protective and curative properties. Thiophanate-methyl is also a wound protectant, used during pruning. Absorbed by the leaves and roots, it is effective on a wide range of fungal diseases such as eye spot of cereals, **scab** on pears and apples (and also apple rot caused by *Gloeosporium* spp.), canker and **powdery mildews** on fruit crops. It also protects almonds, tea, rice, coffee, groundnuts, tobacco, sugarcane, cucurbits, strawberries and many other crops from a wide range of pathogens including *Fusarium* spp., *Botrytis* spp., *Sclerotinia* spp. and *Pyricularia oryzae*.

Some of the common trade names of thiophanate-methyl include Alert, Capital, Topsin-M, Cycosin and Hilnate. The formulation types of this compound are dustable powder (DP), paste (PA), soluble concentrate (SC) and wettable powder (WP). A common dosage for most fungal control is 30–50 g a.i./ha.

The WHO and EPA classify thiophanate-methyl as unlikely to be hazardous (Class U and Class IV toxins, respectively). It is a mild skin and eye irritant. The acute oral LD_{50} in rats is in the range of 6640–7500 mg/kg, while among bird species tested, it is >4640 mg/kg. The acute percutaneous LD_{50} in male and female rats is >10,000 mg/kg. The NOEL (2 year) in rats is 160 mg/kg diet and in dogs is 50 mg/kg diet. It is not toxic to bees. The ADI is 0.08 mg/kg b.w. Excessive exposure to thiophanate-methyl may cause mild irritation to the skin (dermatitis, itching, redness and swelling) and to the eyes (congested ocular mucosa). Labels should display the signal word 'CAUTION'. Thiophanate-methyl is a PAN Bad Actor chemical, since the chemical is a carcinogen. It is a potential groundwater contaminant.

In rats, following oral administration, more than half of the fungicide is eliminated via the urine within 2 days. The principal metabolite in rats is methyl 5-hydroxybenzimidazol-2-carbamate, while in plants, cyclization occurs, leading to the formation of carbendazim. Soil persistence is up to 4 weeks, the breakdown occurring due to UV light and cyclization, leading to formation of carbendazim and then 2-aminobenzimidazole and 5-hydroxy-2-aminobenzimidazole.

The fungicide is registered in countries like India, South Africa, Australia, Canada, the USA and almost the whole of the EU. The MRL of thiophanate-methyl for tomatoes is 2 mg/kg, for lady's finger (okra) 1 mg/kg and for potatoes it is 0.1 mg/kg. According to the EU Pesticide Database, the MRL for grapes and strawberries is 3.8 mg/kg; for tomatoes, apricots, nectarines and peaches, it is 2.0 mg/kg; for plums and barley and oat grains, it is 0.3 mg/kg; and for maize grain, it is 0.01 mg/kg.

Thio-TEPA chemosterilant

Thio-TEPA, $C_6H_{12}N_3PS$ (phosphorothioic tri(ethyleneamide) or tris(1-aziridinyl)phosphine sulfide), is a **chemosterilant**. Thio-TEPA is a white, crystalline solid, soluble in water, ethanol, diethyl ether, benzene and chloroform. Thio-TEPA is stable in alkaline solution. It is unstable in acids and is sensitive to light. When heated to decomposition, thio-TEPA emits toxic fumes of phosphorus, sulfur and nitrogen oxides. Thio-TEPA, a sulfur analogue of TEPA, is derived from aziridine thiophosphoryl chloride. It has a CAS Registry Number of 52-24-4.

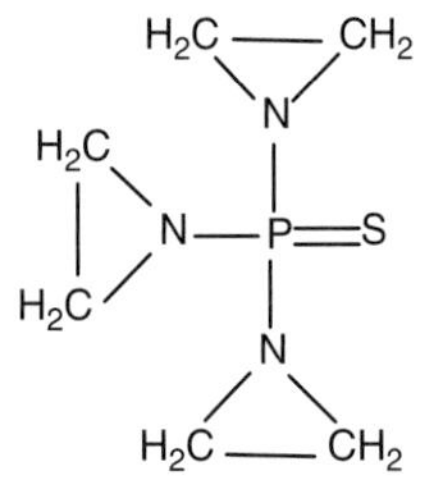

Thio-TEPA

Thio-TEPA induces mutagenic changes in oocytes, and thus controls the life cycle of insects. Thio-TEPA is effective against housefly, cockroach and screw worm fly.

Thio-TEPA is not commonly used commercially, due to its adverse effects on the environment.

Thiram fungicide

See **Thiuram disulfide fungicide**

Thiuram disulfide fungicide

Thiuram disulfide, also known as thiram or tetramethylthiuram disulfide (TMTD) or bis-(dimethylthiocarbamyl) disulfide or thiuram, is an organic sulfur compound formed by joining two molecules of dithiocarbamic acid through the 'S' atom. Thiram is an off-white crystalline powder having a characteristic odour. It is soluble in alcohol, benzene, chloroform and carbon disulfide but insoluble in water, dilute alkali and petroleum ether. The chemical formula is $[(CH_3)_2NCH]_2S_2$. The CAS Registry Number is 137-26-8.

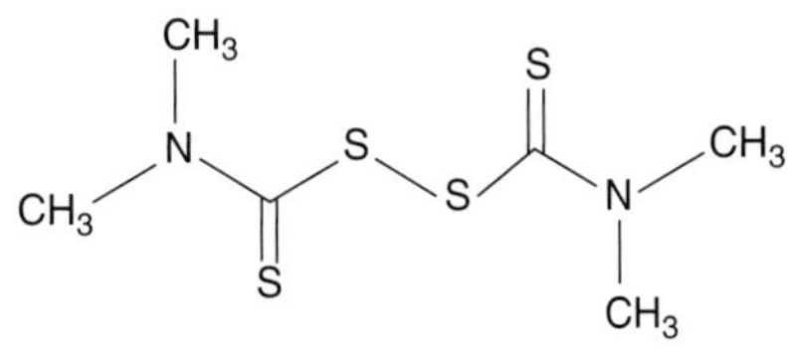

Thiuram

Thiuram disulfide, a non-volatile, protectant compound (with preventive action) has a wide variety of crop applications, which includes being a fungicide, a bird and animal repellent, a bacteriostat, a seed disinfectant and an ectoparasiticide. Its other uses are in the treatment of human skin ailments (including scabies and other fungal and bacterial infections), as a vulcanizing agent and for heat-resistant uses. As a fungicide, it is used to control damping-off, **smut**, stem gall and rot diseases in many crops (e.g. millets, onions and fruits) with particular activity against species of *Botrytis* spp., *Monilia* spp., *Colletotrichum* spp. and *Fusarium* spp. It is commonly used in seed treatment. Thiuram disulfide has multiple modes of action that include interference with various enzymes involved in the respiration process, inhibition of spore germination and growth of mycelia. This multiple-site activity makes it difficult for pests to develop resistance to thiuram disulfide, unless it is over-used.

The sulfur compound thiuram disulfide is compatible with most other pesticides. It is often used as an admixture with systemic compounds to enhance their activity and minimize the risk of resistance development. Thiuram disulfide is available as wettable powder (WP), dustable powder (DP), water-dispersible granules (WG), pastes (PA) and suspensions. For soil treatment, 10–15 kg a.i./ha is usually recommended. It is not advisable to spray fruits meant for canning. As repellent for rodents and small mammals, it is mixed with white latex paint before applying to tree trunks.

The WHO has classified thiram as a slightly toxic hazardous chemical. The acute oral LD_{50} in male rats is 3700 mg/kg and the acute percutaneous LD_{50} in rats is >2000 mg/kg. Thiram is highly toxic to fish (with the LC_{50} ranging from 0.13 mg/l to 4 mg/l in fish species). It does not bioconcentrate in aquatic organisms. The ADI is 0.015 mg/kg/day. It causes irritation to the skin, eyes and the respiratory tract. Other symptoms of toxicity include loss of muscle tone, hyperactivity and convulsions. It could be toxic by ingestion and inhalation, with a tolerance level of 5 mg/m^3 air. This chemical is not phytotoxic if used as directed. However, studies have shown that this compound can harm the reproductive and endocrine systems.

The soil half-life of thiram is up to 2 weeks. It also adsorbs to soil particles, especially clayey soils and soils with high organic matter. So there is no risk of groundwater contamination. Doses of thiuram disulfide ingested by mammals are metabolized and excreted.

The chemical is registered in India, Australia, the USA, Canada, Burkina Faso, Cambodia, Chad, Madagascar, Mali, New Zealand and most of the EU.

Threshold limit value (TLV)

A substance's threshold limit value (TLV) is a set of standards, established by the American Conference of Governmental Industrial Hygienists, that evaluate concentrations of airborne substances. They are time-weighted averages based on conditions that workers may be exposed to without adverse effect. TLVs are revised annually and provide the basis for regulations of OSHA.

TLVs serve as guiding tools for those devising measures for controlling health hazards.

Thrips

See **Pest – thrip**

Thysanoptera

See **Order Thysanoptera**

Tillage

Tillage is a mechanical cultivation system, intended to alter the soil structure, to manage crop residues and to prepare the land for fresh planting or as a seedbed or rootbed. Tilling helps remove plant debris from the soil and is a means of removing or managing weeds. The two major types of tillage are primary and secondary. Primary tillage involves breaking and loosening the soil to a depth of up to 90 cm using mouldboard and disc ploughs. In secondary tillage, soil is stirred at relatively shallow depths, using the spike-tooth harrow, spring-tooth harrow and disc harrow.

There are many types of tillage systems. Conventional tillage involves a combination of primary and **secondary tillage**. Here, soil compaction is more likely to occur on wet soil than dry soil. The practice of conservation tillage retains organic residues of crops, keeping the field surface rough. This is practised when growing grain crops. Minimum tillage calls for the use of the bare minimum of equipment for tilling the land. No tillage is a procedure whereby a crop is planted directly into the soil after the previous crop is harvested and the land is not tilled in preparation for sowing. Planting of wheat in the previous wheat stubble is an example of no-tillage planting.

Continuous cropping and tillage by tractors, heavy machinery, animal hooves and shoes cause changes in the soil porosity and increase soil compaction. (See also **Weed management by mechanical methods**.)

Tobacco caterpillar

See **Pest – tobacco caterpillar**

Tolclofos-methyl fungicide

Tolclofos-methyl, $C_9H_{11}C_{12}O_3PS$ (*O*-2,6-dichloro-p-tolyl *O,O*-dimethyl phosphorothioate), is an organophosphorus fungicide. It is a colourless crystal, soluble in water (1.1 mg/l) and in most organic solvents. While stable for over 2 years under normal conditions, tolclofos-methyl decomposes in alkaline and acidic conditions. Tolclofos-methyl is made from *O,O*-dimethyl-thiophosphorylchloride, which is made from thiophosphoryl-trichloride and sodium methanolate, and 2,6 dichloro-p-cresol in the presence of a base. Its CAS Registry Number is 57018-04-9.

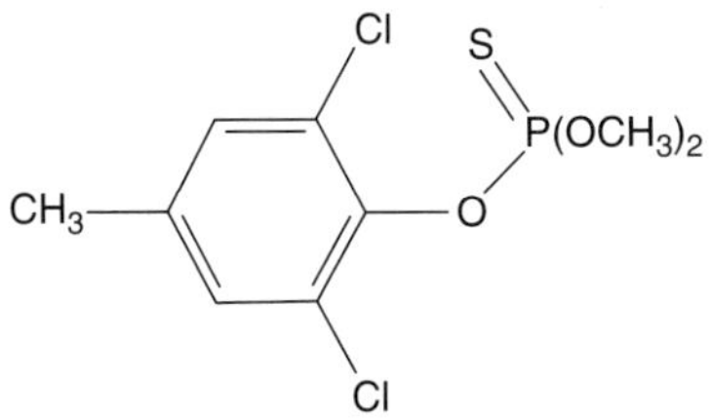

Tolclofos-methyl

Tolclofos-methyl, a non-systemic, contact fungicide with protective and curative action, is effective on a broad range of plant pathogens like *Rosellinia necatrix*, *Corticium rolfsii*, *Typhula incarnata*, *Helminthosporium sigmoideum* and *Rhizoctonia solani*. It inhibits spore germination, hyphal growth, phospholipid synthesis, multiplication of sporidia and formation of sclerotia. This fungicide is used on potato and cotton.

Available formulations are wettable powder (WP), water-dispersable granule (WG) and a flowable product. It is commonly sprayed in the furrows at planting time.

Fungicide tolclofos-methyl is considered unlikely to be hazardous in normal use by the WHO and the EPA. The acute oral LD_{50} in rats and bobwhite quails is >5000 mg/kg. The inhalation LC_{50} of tolclofos-methyl in rats is >3.32 mg/l. It is toxic to fish (the acute 96 h LC_{50} in rainbow trout is 0.69 mg/l) and moderately toxic to honey bees (the acute 48 h contact LD_{50} is >100 μg per bee). The ADI among mice is 0.064 mg/kg b.w./day. Acute exposure to tolclofos-methyl causes excessive salivation, sweating, diarrhoea, vomiting, tightness in the chest, wheezing, a productive cough and fluid in the lungs, blurred or dark vision, seizures and incontinence. Applicators and handlers are advised to be equipped with PPE. Product labels should clearly display the signal word 'CAUTION'. This fungicide, being an inhibitor of cholinesterase, is a PAN Bad Actor chemical.

The metabolic pathways of tolclofos-methyl include oxidation of P=S to P=O group, cleavage of the P–O-aryl linkage, demethylation of P–O–CH_3 and hydroxylation/oxidation of the P–CH_3 to –CH_2OH and –COOH in mammals, plants and soil. The chlorinated phenol is further degraded to carbon dioxide and mineralized products.

Tolclofos-methyl is registered for use in Australia, Madagascar, New Zealand, South Africa and many countries of the EU such as Finland, Germany, the Netherlands and the UK. According to the EU Pesticide Database, the MRL for lettuce is 1.0 mg/kg while that for potatoes is 0.1 mg/kg.

Tolerance

Often chemical residues that remain on the crops at the time of harvest enter the human or animal diet, making it necessary to determine the amount of residues that are safe for consumption.

The maximum amount of residue, which may remain on a harvested crop, is called the tolerance. The tolerance is usually set at 100 times less than the highest dose which has no adverse effects on the test population.

The terms 'tolerance' and 'resistance' both refer to a relative acceptance of a population to the effects of a toxicant. The reasons for tolerance are natural or acquired genetic traits of the population to by-pass or rapidly metabolize a toxicant. Natural tolerance can be estimated when the population is first exposed to a toxin.

Each country has agencies which are entrusted with the responsibility of fixing tolerance limits on the amount of active ingredient that may remain in our food. These limits on pesticides left on foods are called 'tolerances' or MRLs. (See also **Maximum residue level or limit (MRL)**.)

Tolylfluanid fungicide

Tolylfluanid, $C_{10}H_{13}Cl_2FN_2O_2S_2$ (*N*-dichlorofluoromethylthio-*N*′,*N*′-dimethyl-*N*-*p*-tolylsulfamide), is a phenylsulfamide fungicide. It is a solid colourless powder, soluble in water at 0.90 mg/l. It is produced by the reaction of the potassium salt of *N*′,*N*′-dimethyl-*N*-*p*-tolysulfamide with dichlorofluoromethylthiochloride in xylene. Tolylfluanid has a CAS Registry Number of 731-27-1.

Tolylfluanid is used on crops such as tomatoes, hops, grapes and apples. It is sprayed aerially or from the ground, with an application rate of up to 3 kg/ha, repeated approximately ten times, with a gap of a week, as required. It is important to maintain a pre-harvest interval (PHI) of 5–35 days. It is also available as a wettable powder (WP).

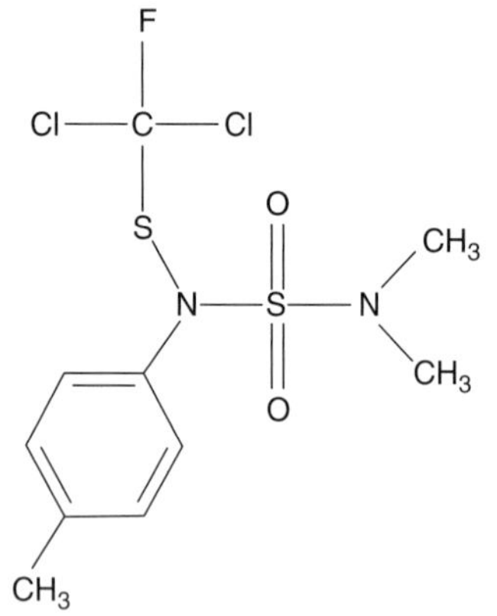

Tolylfluanid

The insecticide tolylfluanid is classified as a Class U toxin (unlikely to be hazardous) according to the WHO. The acute oral toxicity LD_{50} in rats is >5000 mg/kg. The NOEL in rats for 90 days is 20.1 mg/kg/day. However, it is a proven carcinogen and thus a PAN Bad Actor chemical.

Tolylfluanid is readily absorbed and rapidly hydrolysed within 48 h. In soil too, it degrades rapidly to produce *N′*,*N′*-dimethyl-*N′*-*p*-tolysulfamide, which is a product of environmental concern. Tolylfluanid is stable to aqueous hydrolysis.

Australia, Spain and the USA have registered the use of this fungicide.

Tomato spotted wilt virus

Tomato spotted wilt virus belongs to the genus *Tospovirus* and was the first virus to be discovered in this genus. The genus name is derived from the name of tomato spotted wilt virus (now about 20 **tospoviruses** have been described).

This virus is transmitted by **thrips** particularly the **western flower thrips** (*Frankliniella occidentalis*) and infects a very broad range of plants spread over 35 families, of both dicots and monocots. Some of the important crops affected are vegetables such as tomato, potato, aubergine, cowpea and celery, as well as a wide range of ornamentals. There are a number of weed species which act as reservoirs for this virus. The virus is transmitted by thrips from these weeds to crop plants, compelling the farmer to remove such weeds from the vicinity of the crops. The broad host range of the virus and the short life cycle of the western flower thrips have favoured the spread of this disease. Common symptoms of infection include yellow or brown ring spots, black streaks on petioles or stems, necrotic leaf spots and tip dieback.

Control of thrips, removal of reservoir weeds and immediate destruction of infected plant material are the best ways of managing this disease.

Topramezone herbicide

Topramezone, $C_{16}H_{17}N_3O_5S$ ([3-(4,5-dihydro-1,2-oxazol-3-yl)-4-mesyl-*o*-tolyl](5-hydroxy-1-methylpyrazol-4-yl)), is a herbicide of the benzoylpyrazole class (or it can be described as a methanone herbicide with a pyrazole structure). It is a white crystalline solid, soluble in water (510 mg/l at pH 3.1), isopropanol, acetone, acetonitrile, *n*-heptane and ethyl acetate. At 50°C, it is stable in aqueous solution for 5 days (at pH 4.7 and 9).

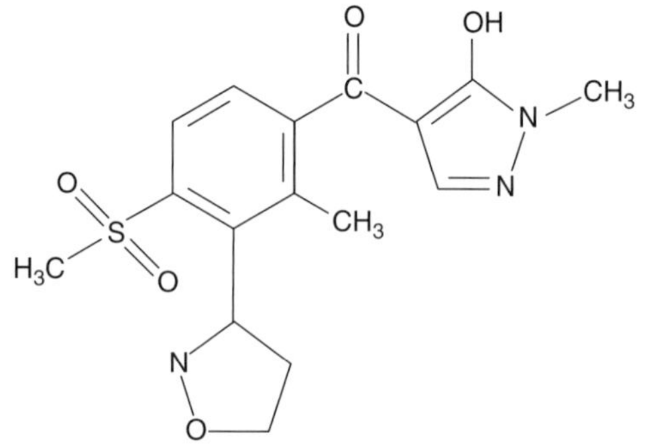

Topramezone

Topramezone, a systemic, highly selective herbicide, is translocated within the plant. It acts by inhibiting the enzyme HPPD (*p*-hydroxyphenyl pyruvate dioxigenase) in photosynthetic cells thus causing necrosis and bleaching, leading to plant death. It is effectively used to control broadleaved plants and grass weeds in maize.

The herbicide is commonly formulated as suspension concentrates (SC). It can be combined with related herbicides like **atrazine** and dimethenamid-P, and together with which, it increases soil residual activity against broadleaved weeds and grasses. Trade names include Clio and Impact.

Toxicity studies of topramezone reveal that it causes slight irritation to the eyes and skin in test animals. The acute oral toxicity of topramezone in rats is >2000 mg/kg, while the acute oral LD_{50} in birds is >2000 mg/kg b.w. and in bees is >72.05 μg per bee. The LC_{50} in fish and worms is >100 mg/l and >1000 mg/kg, respectively. The **NOAEL** value optimized in rats is 0.4 mg/kg b.w. The compound is listed as a PAN Bad Actor chemical.

Topramezone degrades in soil by the action of microorganisms, the half-life in field conditions being up to 81 days. However, in plants and animals the compound is poorly metabolized and the parent compound is the main component in residues.

Topramezone is registered in Canada, Germany, Hungary, the Netherlands and the USA.

Tospoviruses

Tospoviruses are viruses in the genus *Tospovirus* which belong to the family **Bunyaviridae**. They are negative-strand RNA viruses and it is the only group in this family which infects plants (all other members infecting animals). The genus name has been derived from the **tomato spotted wilt virus** which was the first virus of this genus to be discovered. Now about 20 tospoviruses have been described. These have a single-stranded RNA genome with a negative polarity.

Tospoviruses are arboviruses or arthropod-borne viruses. About ten species of thrips (order **Thysanoptera**) belonging to the family Thripidae have been confirmed as **vectors** for about 13 tospoviruses. Studies show that thrips are infected by the virus in the larval stage.

Tospoviruses cause spotting and wilting of the infected plant leading to reduced productivity and eventual death. There is no cure and the infected plants should be immediately removed and destroyed.

Toxaphene insecticide

Toxaphene, $C_{10}H_{10}Cl_8$, is an organochlorine insecticide. It has a CAS Registry Number of 8001-35-2. It is a yellowish, semi-crystalline gum, soluble in organic solvents but almost insoluble in water (0.4 mg/l). Toxaphene is unstable in the presence of alkalis, strong oxidizers, prolonged sunlight and at temperatures higher than 155°C.

A broad-spectrum, contact, persistent pesticide, toxaphene is widely used on cotton, maize, fruit, vegetables and small grains. It is also an ectoparasiticide used on livestock and controls lice, fleas, ticks and scab mites.

Toxaphene is highly toxic and has been shown to cause malignant tumours and birth defects in animals and humans. It is extremely toxic to fish, birds and rats. It is very toxic via inhalation and ingestion, the symptoms of poisoning being hypersensitivity to sound and touch, headache, dizziness, nausea, vomiting, incoordination, confusion, convulsions, seizures and coma. It is a PAN Bad Actor chemical and is included in the list of **Dirty Dozen** chemicals.

The half-life of toxaphene in soil can be as high as 12 years, depending on the soil type and climate. Its high partition coefficient indicates its likelihood to bioconcentrate. The use of toxaphene has been banned or severely restricted. It is currently used only for: (i) scabies control in cattle (as a dip); (ii) insect control for pineapples and bananas; and (iii) emergency treatment of cotton, maize and small grains. Toxaphene has been banned in 37 countries (at the time of writing) including Austria, Belize, Brazil, Costa Rica, Dominican Republic, Egypt, the EU, India, Ireland, Kenya, Korea, Mexico, Panama, Singapore, Thailand and Tonga. Its use has been severely restricted in 11 other countries, including Argentina, Columbia, Dominica, Honduras, Nicaragua, Pakistan, South Africa, Turkey and Venezuela.

Toxic dose low (TD_{Lo})

The toxic dose low (TD_{Lo}) value is the lowest dose (per unit of body weight) required to produce signs of toxicity in a particular animal species. The lowest concentration of a substance in air to which animals have been exposed for any given period of time that has produced any toxic effect in animals is called the toxic concentration low (TC_{Lo}).

Toxicity

Any substance that is harmful to living tissues when applied in relatively small doses is a poison or a toxin. Toxicity means the ability of a substance to cause damage to or death of a living organism, or one or more of its parts (e.g. the nervous or respiratory systems), when ingested, inhaled or absorbed by the skin.

The degree of the damage brought about by a poison depends on, for example: (i) the nature of the substance; (ii) the time of exposure to it; (iii) its particle size; (iv) its affinity for living tissues; and (v) the sensitivity of the tissue or organ. It is not always easy to specifically categorize poisons from non-poisons since this distinction depends on many variables. So, an indication on toxicity is provided for every substance.

Toxicity may be brought about by a chemical or a biological organism, for example a fungal or bacterial pathogen or a virus.

A toxin may sometimes manifest itself only in a particular form. For instance, some metals are toxic as fumes or in the powder form, and some when mixed in an acid or dissolved in water. Many substances that are intensely poisonous are actually beneficial when administered in micro-amounts. Examples include some prescription drugs such as strychnine.

There are two main types of toxicity, acute and chronic.

In the transportation industry, it is necessary to indicate if a substance is a poison (and here the term 'poison' is often used). Poisons are classified as follows:

- Poison A – to indicate a gas or liquid so toxic that even an extremely small amount of the gas or the vapour is life threatening.
- Poison B – to mean less toxic liquids and solids that are hazardous either by contact with the skin or by ingestion.
- Poison C – these are liquids or solids that release toxic or strongly irritating fumes when heated or when exposed to air (excluding Class A poisons).
- Poison D – to mean substances that are radioactive in nature.

The term poison can also mean other things. For example, any material with high capture probability for neutrons that may divert an undesirable number of neutrons from the fission chain reaction is called a poison in nuclear technology. Also a substance that reduces or destroys the activity of a catalyst is called a poison. (See also **Acute toxicity**; **Chronic toxicity**; **Toxicity testing**.)

Toxicity classifications

In order to safeguard their environment, countries around the world constitute regulatory agencies, designed to form policies and directives that control the types and amounts of chemicals that may be used in agriculture, food storage and related industries. Although the criteria for registering or banning a particular ingredient might differ with each country, a broad classification of all pesticide active ingredients, whether chemical or biological, have been provided by the **World Health Organization (WHO)** and the **Environmental Protection Agency (EPA) of the USA**, and this is normally used by all regulatory agencies as the basis of their own policies.

The EPA, in association with related agencies such as the **Federal Insecticide, Fungicide and Rodenticide Act (FIFRA)**, classifies toxic chemicals into four classes.

- Toxicity Class I means the chemical is most toxic. It requires the signal words 'DANGER – POISON', with the skull and crossbones symbol. Labels could have words such as 'Fatal if swallowed', 'Poisonous if inhaled', 'Extremely hazardous by skin contact – rapidly absorbed through skin' or 'Corrosive – causes eye damage and severe skin burns'. Generally, agents of Class I will kill an adult person at a dose of less than 5 g (less than a teaspoon).
- Toxicity Class II means the chemical is moderately toxic. The signal word is 'WARNING', possibly followed by 'Harmful or fatal if swallowed', 'Harmful or fatal if absorbed through the skin', 'Harmful or fatal if inhaled' or 'Causes skin and eye irritation'. Class II can be fatal at 5–30 g.

- Toxicity Class III means the chemical is slightly toxic and the signal word is 'CAUTION'. It is possibly followed by 'Harmful if swallowed', 'May be harmful if absorbed through the skin', 'May be harmful if inhaled' or 'May irritate eyes, nose, throat, and skin'. Class III can be fatal at >30 g.
- Toxicity Class IV means the chemical is practically non-toxic. It requires no signal word (since 2002).

Furthermore, the EPA classifies pesticides into: (i) general use pesticides (GUP), meaning those that can be applied by anybody; and (ii) restricted use pesticides (RUP), meaning those whose application needs to be carried out by or under the supervision of a certified individual. Application of RUPs requires that a record of the application be kept. (See also **World Health Organization (WHO)**; **Environmental Protection Agency of the USA (EPA)**; and our **Pesticide legislation** series under 'P' for an outline of legislation for a few countries.)

Toxicity testing

To check the effects of toxicity of a certain substance on the bodies of animals, various tests are carried out. A toxic compound may come in contact with the body through injection, inhalation and consumption or through direct skin contact. Thus, tests are designed depending upon the probable route of contact. Generally, rats or mice are used as test animals but it is also common to use rabbits and dogs in some tests.

Substances which may be inhaled are sprayed into a chamber where test animals are kept for a predetermined duration of time and later observed for toxic effects, like stomach disorders or sudden death. This is known as the inhalation test.

The dermal test involves a procedure by which the toxic substance is applied or rubbed onto the skin of the test animals. Later these animals are observed for effects such as reddening, rash or allergic response.

In injection/oral consumption testing, the concerned substance is injected or given to the test animals through food or water. The effects of toxicity are checked later. Increased body temperature, stomach disorders, rashes, allergies, sudden death or abnormal behaviour are noted and recorded as symptoms of toxicity.

To monitor long-term effects, smaller doses of a toxic substance are administered orally for a long duration of time.

The **lethal dose$_{50}$ (LD$_{50}$)** is one way by which the toxicity of chemicals is measured. The LD_{50} is that amount of a substance which is required to kill 50% of the test animals under specified test conditions. A single dose of the test substance is usually placed directly into the stomach of animals via a tube. Different groups of animals are given increasing doses of the test substance to see which dose will kill half of them. The animals are observed for 14 days, and the animals that haven't already died are then killed and dissected. Researchers record the number of deaths on each day, and various signs of illness. This test can be done for both dermal and oral routes of exposure. For example, an acute oral LD_{50} value indicates the amount of dose administered orally necessary to kill half of the given test animals under given test conditions.

The smaller the LD_{50} value, the smaller the dose required to kill half of the test animals and the more poisonous is the pesticide. So, a pesticide with a dermal LD_{50} of 25 is more poisonous than a pesticide with a dermal LD_{50} of 2000.

LD_{50} values neither tell us how a chemical acts nor how sensitive different organs within an animal or human might be to a given chemical. These values can vary depending upon the route of entry, for instance in the case of diclorvos, the dermal LD_{50} in rats is 75 mg/kg whereas the intraperitoneal LD_{50} is 15 mg/kg and the oral LD_{50} is 56mg/kg.

Some of the toxic substances are present in the atmosphere in gaseous form. They have the potential to cause injury even if they are inhaled in small quantities. To measure the extent of their toxicity, their effects on the animals are tested and the term LC_{50} is used to denote their toxicity levels.

The **lethal concentration$_{50}$ (LC$_{50}$)** is that concentration of a substance in air that is lethal to 50% of the total population of the given test animals, under the given test conditions.

Compounds which come in direct or indirect contact with human beings need to be checked for their toxicity. Usually LD_{50} values are considered while measuring toxicity. However, certain other values such as the LD_{01} also help in the measurement of the toxicity. The dose required to kill 1% of the animal test population is denoted in terms of the LD_{01}or lethal dose one.

Although conventionally LD_{50} values are used to test toxicity effects caused by the compounds, some other values like the LD_{01} and LD_{100} are also used to denote the extent of toxicity. The dose required to kill 100% of the animal test population is given in terms of the lethal dose100 (LD_{100}). However, the LD_{50} value does not give any account of the minimum dose that kills the test organism. Therefore, the **lethal dose low (LD$_{Lo}$)** value is used to indicate the lowest dose causing lethality to the specified group of animals.

The **toxic dose low (TD$_{Lo}$)** is the value used to show the lowest dose, producing signs of toxicity in a population of test organisms. The LD_{50} and LD_{Lo} values show the dose lethal for 50% of the test population and lowest dose required to kill the test animals, respectively, but the dose required to produce any kind of signs of toxicity might be different from these doses.

Pesticide LD_{50} values are measured in units of weight such as milligrams per kilogram (mg/kg). The LD_{50} value refers to the milligrams of pesticide that was needed to kill half of the test animals for each kilogram of the animal's body. For example, an acute oral LD_{50} of 5 mg/kg in rats for a pesticide indicates that it is toxic when there are 5 mg of this chemical given orally for every kilogram (or about 1 kg) of the animal. Similarly, all other toxicity doses namely LC_{50}, LD_{01}, LD_{100}, LD_{Lo} and TD_{Lo} are measured in milligrams per kilogram (mg/kg).

Toxicological tests have been carried out on many of the compounds used as repellents and this led to discarding of 2-ethyl-hexane-1, 3-diol from the market. Compounds are tested as repellents on human skin only after passing the four standard toxicological tests: (i) rabbit eye irritation; (ii) rabbit dermal; (iii) rat inhalation; and (iv) rat acute ingestion. All of these, plus EPA regulations have considerably reduced the number of compounds submitted for general screening.

Some of the tests carried out for assessing risks arising out of the use of plant pesticides are: (i) acute oral, dermal and inhalation studies on rats or other mammals; (ii) skin and eye irritation tests on rabbits; (iii) 14 day sub-acute and 90 day sub-chronic studies on rats and dogs; (iv) contact allergy on guinea pigs or rabbits; (v) 2 year and 18 month carcinogenicity tests on rats and mice, respectively; and (vi) neurotoxicity and kinetics and biotransformation tests.

Although all these may not be required, these tests help to confirm the safety of newer compounds and may help regulatory agencies to re-register a compound in the country, and establish ADI and MRL values. (See also **Acute toxicity**; **Chronic toxicity**; **Toxicity**.)

Toxicity types

The toxicity of a substance is the capacity of that substance to cause harm or injury to any living system directly or indirectly. Here, any living organism or different body parts of that organism can be considered as a living system.

Different substances cause different kinds and levels of toxicity. Therefore, based on the extent of damage caused by the toxic agent, toxicity can be divided into two main types: (i) **acute toxicity**; and (ii) **chronic toxicity**. Different tests (e.g. the LD_{50} test) designed to determine toxicity of a substance can also categorize the toxicity caused by that substance as acute or chronic. Toxicity is expressed as LC_{50} or LD_{50}. (See also **Acute toxicity**; **Chronic toxicity**; **Toxicity testing**.)

Tracer

A tracer is a chemical entity added to reacting compounds in a chemical process, which can be traced through the process by appropriate detection methods, for example using a Geiger counter. Compounds containing tracers are often said to be 'tagged' or 'labelled'. Carbon-14 and radioactive forms of iodine and sodium are the commonly used tracers. Many biochemical processes, inhibitions, absorption and metabolism can be studied using tracers. Radioactive enzymes such as ribonuclease, pepsin, trypsin and others are also available for tracer studies.

Tracking powders

See **Dustable powder**

Tralkoxydim herbicide

Tralkoxydim, $C_{20}H_{27}NO_3$ (2-[1-(ethoxyimino) propyl]-3-hydroxy-5-mesitylcyclohex-2-enone), is a cyclohexanedione oxime. It has a CAS Registry Number of 87820-88-0. It is a colourless crystalline substance that is fairly soluble in water and readily soluble in organic solvents. It remains stable for more than a year and half. Tralkoxydim is made by condensation of 2,4,6-trimethylbenzaldehyde with acetone followed by condensation with diethylmalonate, acylation with propionic anhydride and treatment with thioxyamine.

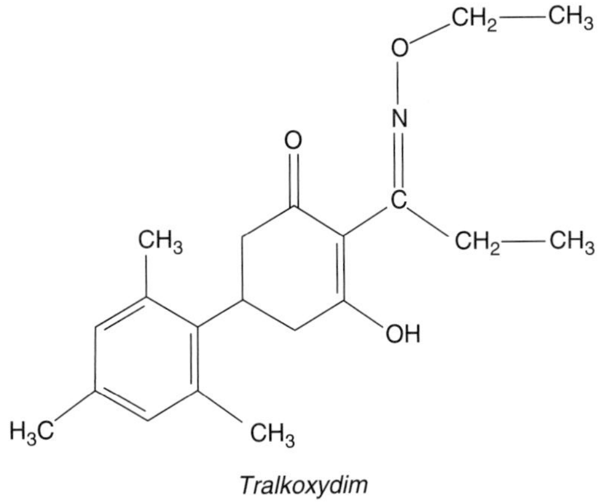

Tralkoxydim

Tralkoxydim, a selective, post-emergence herbicide, is used in the control of grass weeds in wheat and barley. It is absorbed by the leaves and reaches the growth points, where it inhibits cell division and fatty acid synthesis. Examples of weed pests include wild oats, green and yellow foxtail and annual ryegrass.

Tralkoxydim is available as soluble concentrates (SL) and emulsifiable concentrates (EC). It should be applied at 150–400 g/ha. Tralkoxydim is applied by ground and aerial application. It is available under the trade names of Achieve and Grasp.

The acute oral LD_{50} of tralkoxydim in rats is 934 mg/kg and the dermal LD_{50} is >2000 mg/kg b.w. The inhalation LC_{50} in rats at 4 h is >3.5 mg/l. Tralkoxydim is moderately toxic to birds, with an acute LD_{50} of 2024 mg/kg. It is also toxic to fish (the acute 96 h LC_{50} is >6.1 mg/l) and moderately toxic to honey bees (the acute contact LD_{50} is >100 μg per bee). The ADI of tralkoxydim in dogs is 0.005 mg/kg b.w./day. The WHO categorizes tralkoxydim as a Class III toxin. This herbicide may be phytotoxic to non-target plants via spray drift or runoff.

Tralkoxydim degrades rapidly in crops and also in aerobic soil. The DT_{50} is about 3 days. It hydrolyses quickly in acidic conditions with a half-life of 6.3 days at pH 5. The herbicide photodegrades in water, with a half-life of 19.3 days, although in anaerobic aquatic conditions, it persists for as many as 115 days. The major degradate of tralkoxydim is tralkoxydim acid, which along with the other degradation products are very mobile in soil.

Tralkoxydim is registered in Australia, Canada, some EU countries, New Zealand, Tanzania and the USA. According to the EU Pesticide Database, the MRL values for cereal grains and straw are 0.01 mg/kg.

Tralomethrin insecticide

Tralomethrin, $C_{22}H_{19}Br_4NO_3$ ((*S*)-α-cyano-3-phenoxybenzyl (1*R*,3*S*)-2,2-dimethyl-3-[(*RS*)-1,2,2,2-tetrabromoethyl]cyclopropanecarboxylate), is a light yellowish substance belonging to the pyrethroid class of insecticides. It has a CAS Registry Number of 66841-25-6. It is almost insoluble in water but soluble in most organic solvents. Tralomethrin is made by bromination of the dichlorovinyl group of cypermethrin.

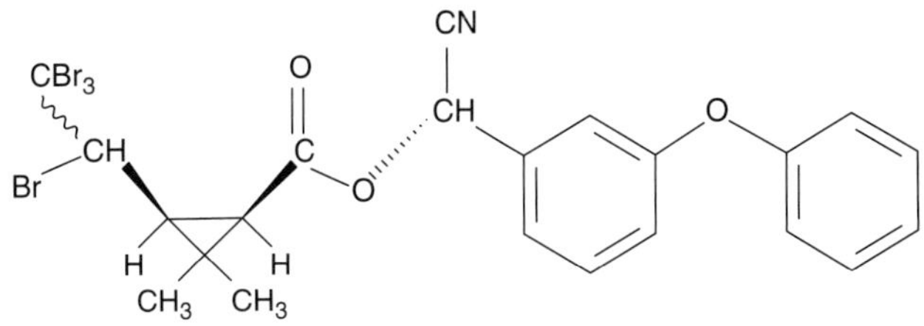

Tralomethrin

Tralomethrin, an insecticide and acaricide, is a nerve toxin, active against spiders, ticks, mites and other members of the Arachnida. The compound affects the sodium channels in insect neurons causing tremors, spasms and even death in extreme cases. Apart from cotton, it has non-crop applications. Tralomethrin is most effective during the early growth stage of insects.

Common formulations of tralomethrin are liquids, wettable powders (WP) and sprays. Scout X-TRATM, Saga and Hot Shot are common brand names. Most formulations are effective on insects of cotton crops, where an application rate of 18–27 g/ha is made depending on the severity of the infestation. Non-crop doses may be a little more depending on the conditions and the severity of the infestation.

The insecticide is very toxic to mammals with an LD_{50} in rats ranging from 85 mg/kg to 99 mg/kg. It is extremely toxic to fish and other aquatic organisms like *Daphnia magna* and also to honey bees (the acute 48 h contact LD_{50} in honey bees is 0.13 μg per bee). The inhalation LC_{50} in rats is 0.4 mg/l. The ADI of tralomethrin is 0.0075 mg/kg b.w./day. The WHO categorizes tralomethrin as a moderately hazardous toxin (Class II toxin). It is also a skin and eye irritant. Symptoms of poisoning with tralomethrin are irritation of the skin and eyes, irritability to sound or touch, abnormal facial sensation, sensation of prickling, tingling or creeping on the skin, numbness, headache, dizziness, nausea, vomiting, diarrhoea, excessive salivation and fatigue. Seizures may also occur. Product labels should display the signal word 'WARNING'.

Tralomethrin is completely degraded in soils (especially in aerobic sandy loam soils) and metabolizes into various products such as **deltamethrin**, decamethrin, phenoxybenzoic acid and finally carbon dioxide. The half-life may be a maximum of 33 days. This compound is not very mobile. It is, however, observed to bioaccumulate. This property, in addition to its poor solubility in water, ensures that it is not a threat to groundwater.

Tralomethrin is registered in African countries such as Burkina Faso, Cambodia, Cape Verde, Chad, Gambia, Guinea-Bissau, Mali, Mauritania, Niger and Senegal, and also in Canada, Vietnam and the USA.

Trans-α-ionone bioinsecticide

Trans-α-ionone, $C_{13}H_{20}O$ (4-(2,6,6-trimethyi-1-cyclohex-2-enyl) but-3-en-2-one), is a plant-derived insecticide. It is present in many tropical and subtropical plants including copaiba (*Copaifera officinalis* (Jacq) L.). Trans-α-ionone is a yellow liquid that is insoluble in water. It has a CAS Registry Number of 9066-80-2.

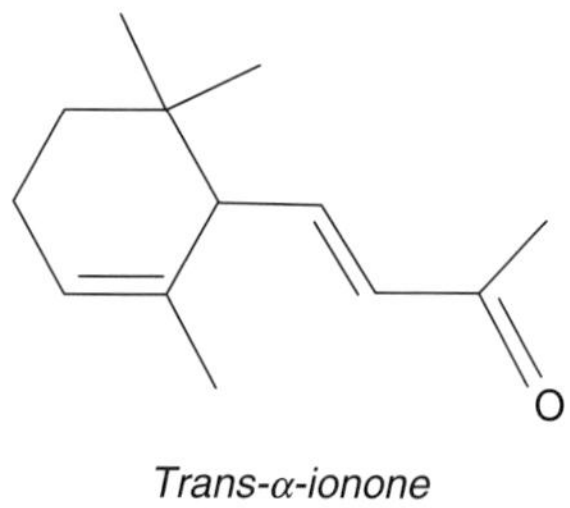

Trans-α-ionone

Trans-α-ionone is extracted as an insect repellent that is effective in controlling flying and biting insects which are a nuisance to the public and in animal shelters. Sometimes it is also used as an anti-fungal agent.

Trans-α-ionone is available in ready-to-use formulations and is best used when insects and mosquitoes are active. The trade names include Scent-Off, Twist-ons and Scent-Off Pellets.

Trans-α-ionone is considered safe for mammals, the environment and for non-target organisms. It has US OMRI approval for organic farming.

Transfluthrin insecticide

Transfluthrin, $C_{15}H_{12}Cl_2F_4O_2$ (2,3,5,6-tetrafluorobenzyl (1*R*, 3*S*)-3-(2,2-dichlorovinyl)-2, 2-dimethylcyclopropanecarboxylate), is a pyrethroid ester insecticide. It is a colourless, volatile, crystalline substance, soluble in water and readily soluble in organic solvents. It is stored away from direct sunlight and has a shelf life up to 3 years. Transfluthrin is produced by the esterification of permethrin acid with 2,3,5,6-tetrafluorobenzyl alcohol. It has a CAS Registry Number of 118712-89-3.

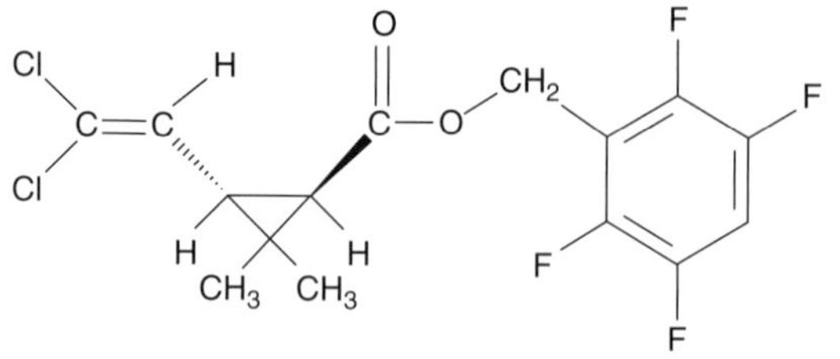

Transfluthrin

Transfluthrin is effective on pre-synaptic voltage gate channels in nerve membranes causing a knock-down effect in insects such as mosquitoes, flies, moths, cockroaches and whitefly, in all stages of their life cycle. It is active after inhalation or contact and also has repellent activity.

Formulated as sprays, mats and mosquito coils, transfluthrin is available under the trade names of Baygon and Bayothrin.

Transfluthrin is slightly toxic to rats, with an acute oral LD_{50} and dermal LD_{50} of >5000 mg/kg b.w. The inhalation LC_{50} of transfluthrin in rats is 0.513 mg/l. It is moderately toxic to birds (the acute LD_{50} in bobwhite quails is >2000 mg/kg) and extremely toxic to fish and other aquatic creatures (the acute LC_{50} in rainbow trout is >0.0007 mg/l, while the acute 48 h EC_{50} in *Daphnia magna* is >0.0017 mg/l). Exposure to transfluthrin causes irritation to the skin and eyes, sensitivity to sound and touch, headache, dizziness, nausea, vomiting, diarrhoea, excessive salivation and fatigue, fluid in the lungs, muscle twitching and seizures. According to the WHO, transfluthrin is categorized under Class U (unlikely to be hazardous).

Tetrafluorobenzoic acid and glucuronate of tetrafluorobenzyl alcohol are formed as metabolites in animals and aquatic ecosystems. It is non-mutagenic, non-carcinogenic and non-teratogenic. It has relatively low residual activity and is easily biodegradable.

Transfluthrin is registered for use in Australia, Greece, India and South Africa.

Trap plants

Plants that are used as a means of biological control, by virtue of their being attractive to insects, mites and other pests, are called trap plants. All pests have preferences towards some crops and particular varieties of those crops. So, if an identified pest can be attracted to a trap plant, then the pest infests the trap plant, thus sparing the crop plant. For example, rose plants may be planted among crops that are susceptible to mite infestation. Tomatoes or aubergine (brinjal) may be planted among poinsettia to control whitefly.

Trap plants are generally planted around crops, especially in greenhouses. Otherwise, they may even be interspersed among the crop. It is important that once infestation on trap plants is noticed, the infested plants are removed.

This intervention is a means of minimizing the use of pesticides on crops, while exercising biocontrol methods of pest control.

Triadimenol fungicide

Triadimenol, $C_{14}H_{18}ClN_3O_2$ (1-(4-chlorophenoxy)-3,3-dimethyl-1-(1*H*-1,2,4-triazole-1-yl)butan-2-ol), is a triazole fungicide (of the **demethylation inhibitor** (DMI) group). It is a colourless, odourless crystalline substance, soluble in water (at 72 mg/l), isopropanol (140 mg/l) and toluene (20–50 mg/l). It is a compound of two diastereoisomers which are stable to hydrolysis.

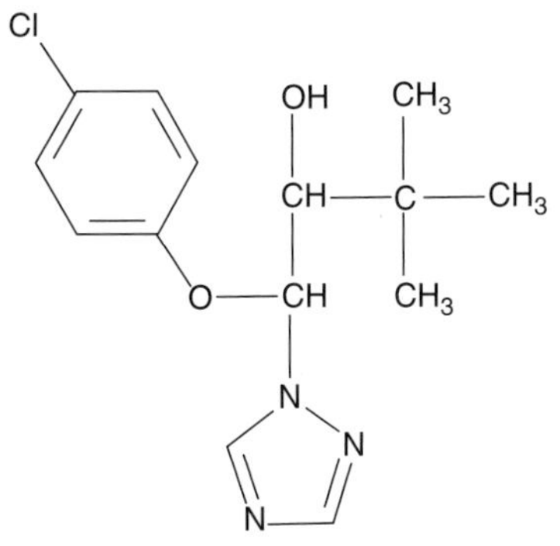

Triadimenol

Triadimenol, a systemic fungicide with protective, curative and eradicant properties, inhibits cell division since it disrupts the biosynthesis of **gibberellins** and **ergosterol**. It readily translocates to the young growing tissues. Primarily a cereal fungicide, triadimenol is also used on vegetables, fruit orchards and vines. It controls diseases like **powdery mildews**, rusts, **smuts**, bunts and blights.

The various formulations of triadimenol include powders for seed treatment (DS), suspension concentrate (SC), wettable powder (WP), emulsifiable concentrate (EC) and water-dispersible granules (WG). Some of the trade names include Bayfidan, Triadim, Baytan and Euro. Sprays apply 100–250 g a.i./ha, while seed treatment formulations apply 20–60 g/100 kg seeds.

Triadimenol is classified as slightly toxic by both the WHO and the EPA. The compound does not show mutagenic or teratogenic effects. The acute oral LD_{50} in rats is between 700 mg/kg and 1100 mg/kg whereas the acute percutaneous LD_{50} is >5000 mg/kg, indicating slight to moderate toxicity. The acute oral LD_{50} in bobwhite quails is >2000 mg/kg indicating slight toxicity. This fungicide is not toxic to fish, honey bees and earthworms. The 2 year NOEL in rats is 125 mg/kg diet and the ADI, 0.03 mg/kg b.w.

Triadimenol is metabolized in animals by oxidation of the *tert*-butyl moiety to the corresponding alcohol and then to carboxylic acid. In plants, the fungicide forms conjugates with sugar compounds. In soil, triadimefon is formed as a result of degradation by microbes.

Triallate herbicide

Triallate, $C_{10}H_{16}Cl_3NOS$ (*S*-2,3,3-trichloroallyl diisopropyl-(thiocarbamate)), is a thiocarbamate herbicide. It is a pale yellow to brown liquid, moderately soluble in water and easily miscible with organic solvents. It is stable under normal storage conditions and to UV degradation. Triallate is made by the reaction of isopropylamine with phosgene, followed by coupling with 1,1-dichloro-2-chloro-3-mercapto-propylene-1. It has a CAS Registry Number of 2303-17-5.

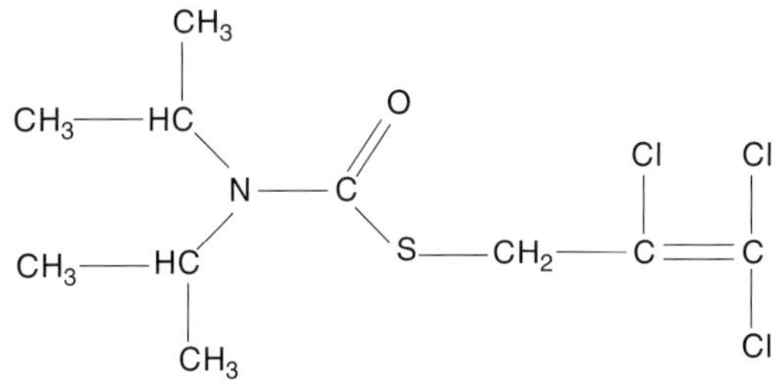

Triallate

Triallate is a selective, pre- or post-emergence herbicide that acts by inhibiting cell division, cell elongation and inhibition of fatty acid and lipid biosynthesis of wild oats and some annual grasses that bother wheat, barley, rye and field beans. Triallate is commonly applied to the soil.

Available as emulsifiable concentrates (EC) and granules (GR), triallate is available in the market as Avadex BW, Buckle, Dipthal and Showdown.

The oral LD_{50} of triallate in rats is 800–2165 mg/kg, and in mice is 930 mg/kg. The dermal LD_{50} for triallate is 8200 mg/kg in rabbits, indicating non-toxicity in the practical sense. The inhalation 4 h LC_{50} in cats is 0.4 mg/l. The herbicide is a mild eye irritant. The toxicity in birds is moderate (with the acute oral LD_{50} in bobwhite quails being 2251 mg/kg). However, triallate is toxic to fish and *Daphnia magna*. Exposure to this herbicide causes irritation to the skin, eyes and respiratory tract, headache, dizziness and nausea. But since it is a cholinesterase inhibitor, it is considered a PAN Bad Actor chemical. The WHO and EPA classify this herbicide as a slightly toxic chemical (Class III toxin).

In plants, 2,3,3-trichloropropenesulfonic acid is the major metabolite of triallate. It persists in soils, adsorbing to loam and clay particles. Metabolism proceeds via hydrolytic cleavage with the formation of dialkylamine, carbon dioxide and mercaptan moieties of triallate. Soil microbes also degrade the chemical. The average soil half-life is 82 days. When ingested by animals, triallate is rapidly absorbed into the bloodstream, readily broken down into metabolites, and then excreted by treated animals. Groundwater contamination is not a concern as far as triallate is concerned. Plants also degrade triallate, reducing the potential for soil accumulation. If applied to the surface of wet soil at high temperatures, triallate is lost through volatilization, compelling applicators to incorporate it into the soil. In cold weather, triallate persists to the next season, due to slow photodecomposition.

Triallate is registered in Australia, Canada, India, the Netherlands, New Zealand, the UK and the USA. (See also **Herbicide action – lipid synthesis inhibitors**.)

Triasulfuron herbicide

Triasulfuron, $C_{14}H_{16}ClN_5O_5S$ (1-[2-(2-chloroethoxy)phenylsulfonyl]-3-(4-methoxy-6-methyl-1,3,5-triazin-2-yl)urea), is a triazinylsulfonylurea herbicide. It is a colourless crystal that is soluble in water (at 1.5 g/l at pH 7 and at 20°C), xylene (166 mg/l), acetone and dichloroethane. It stores well in airtight, original containers under ambient conditions. Triasulfuron is prepared by the reaction of 2-chloroethoxyphenylsulfamoylchloride with 2-amino-4-methylpyrimidine. The CAS Registry Number of triasulfuron is 82097-50-5.

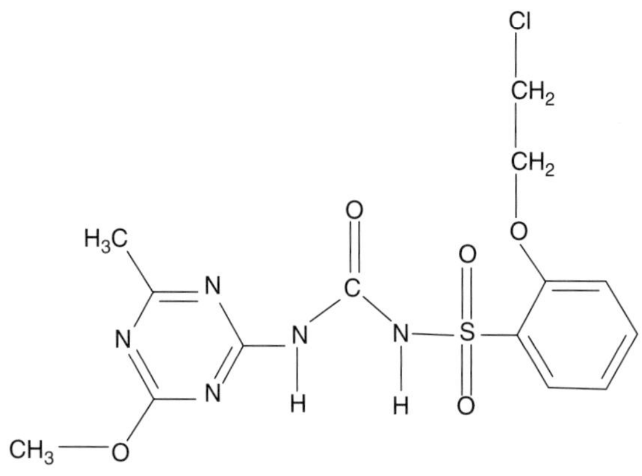

Triasulfuron

Triasulfuron, a selective, pre- and post-emergence herbicide, is absorbed by the leaves and roots of susceptible broadleaved weeds. After quickly reaching the growing tips, it controls susceptible weeds in wheat and barley.

This herbicide is formulated as water-dispersible granules (WG). A typical dose is 5–10 g a.i./ha.

The compound is classified by the WHO as being an unlikely hazard. Although a mild skin irritant, the acute oral LD_{50} among rats and mice is >5000 mg/kg. The acute dermal LD_{50} in rats is >2000 mg/kg. The NOEL (2 years) in rats is 32.1 mg/kg b.w./day. The ADI is 0.01 mg/kg/day. It is slightly toxic to honey bees (the LD_{50} is >100 μg per bee).

In plants, metabolism proceeds by hydroxylation of the phenyl ring and hydrolytic cleavage of the urea bridge. Metabolites include *O*-demethyl triasulfuron.

Triasulfuron is registered in many countries of the EU and also in Australia and the USA.

Triazine herbicides

A triazine is an organic heterocyclic compound containing a six-member ring formed from three nitrogen and three carbon atoms. Triazines are trimerized products of urea and ammonia, for example cyanuric acid or melamine. These may also be used as slow-release nitrogen fertilizers.

The triazine family of herbicides came into existence after simazine was discovered. The triazine group of herbicides characteristically has two types of structures: (i) symmetrical triazine (*s*-triazines); and (ii) asymmetrical triazine (***as*-triazines**). *s*-Triazines are characterized by a symmetrical hexameric ring, consisting of alternating carbon and nitrogen atoms.

s-Triazines (or symmetrical diamine triazine) can be divided into three different subclasses, which differ in the substitution pattern by one substituent. These are **chlorotriazines**, **methoxytriazines** and **methyl- or ethyl-thiotriazine herbicides**.

A large number of herbicides are based on the generalized symmetrical diamine triazine or *s*-triazine. Most triazines are photosynthesis inhibitors (commonly called Hill reaction inhibitors). Herbicides in this group are tailored for uptake by plant roots (although there are some that have foliar activity) and are effective in places where long-term **non-selective** weed control is desired. Soil persistence ensures long-term control. Triazines are detoxified in tolerant crops like maize, sugarcane and sorghum by the enzyme glutathione *s*-transferase, which is not present in susceptible weeds and sensitive crops such as oats and wheat. The use of triazines is restricted because they leach into groundwater and may pollute drinking water. Triazine herbicides have low mammalian toxicity. This group includes ametryne, **atrazine**, cyanazine, **desmetryn**, dimethametryne, prometon, prometryn, propazine, **simazine**, **simetryn**, **terbumeton**, **terbuthylazine**, terbutryn and trietazine.

Cyanazine, $C_9H_{13}ClN_6$ (2-(4-chloro-6-ethylamino-1,3,5-triazin-2-ylamino)-2-methylpropiononitrile), is a chlorotriazine herbicide. Its CAS Registry Number is 21725-46-2.

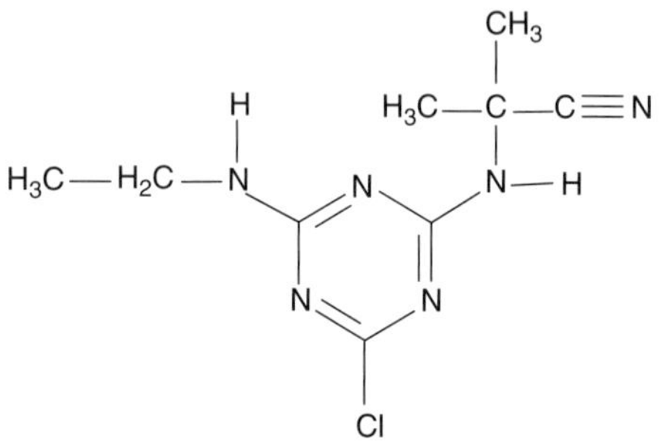

Cyanazine

A pre- and post-emergence herbicide, cyanazine controls annual grasses and broadleaved weeds and is mostly used on maize and cotton and occasionally on sorghum and wheat.

The WHO Acute Hazard classification enlists cyanazine as a Class II toxin (moderately hazardous). Since cyanazine is a groundwater contaminant and a reproductive toxin, it has been listed as a PAN Bad Actor chemical.

Cyanazine is registered in Australia, Burkina Faso, Cape Verde, Chad, Gambia, Guinea-Bissau, Mali, Mauritania, New Zealand, Niger, Senegal and South Africa.

Dimethametryne, $C_{11}H_{21}N_5S$ (N^2-(1,2-dimethylpropyl)-N^4-ethyl-6-methylthio-1,3,5-triazine-2, 4-diamine), is a methylthiotriazine herbicide. Its CAS Registry Number is 22936-75-0.

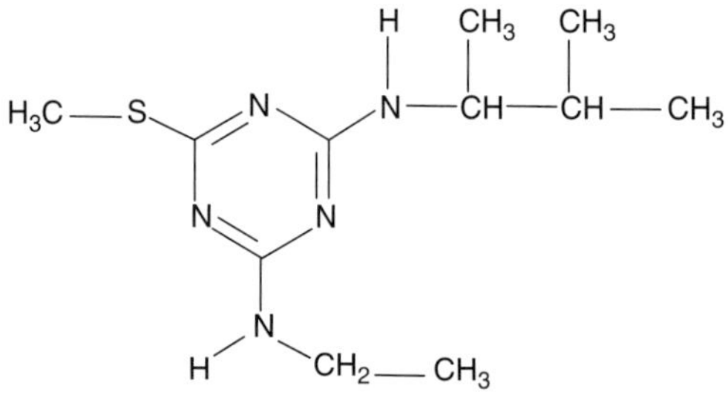

Dimethametryne

Dimethametryne, a photosynthetic electron transport inhibitor at the photosystem II receptor site, is a selective herbicide absorbed by roots and leaves. Commonly used with other herbicides, it is used to control broadleaved weeds in rice and other crops.

The acute oral LD_{50} in rats is >3000 mg/kg. It causes irritation to the skin and eyes in cases of exposure. The ADI is 0.01 mg/kg/day. The WHO lists it as a Class III toxin (slightly hazardous).

Prometon, $C_{10}H_{19}N_5O$ (N^2,N^4-diisopropyl-6-methoxy-1,3,5-triazine-2,4-diamine), is a methoxytriazine herbicide. It is a colourless powder, soluble in water (750 mg/l) and more readily soluble in benzene, acetone, methanol and dichloromethane. While stable in slightly acidic or alkaline media, prometon is unstable on warming at most pH levels, and decomposes on exposure to UV radiation. It has a CAS Registry Number of 1610-18-0.

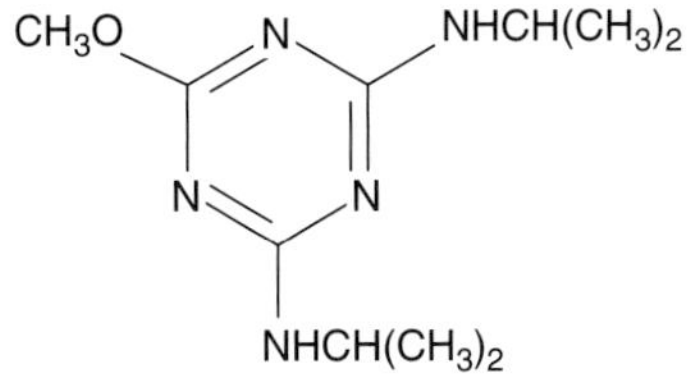

Prometon

Prometon, a **non-selective**, pre- or post-emergence herbicide, acts on photosystem II, competing with plastoquinone and modifying electron transport processes. As a photosynthesis inhibitor, it disrupts the fixation of carbon dioxide and other energy components. It has a residual activity. It controls many annual and perennial broadleaved weeds and grasses, such as **Johnson grass**, **Bermuda grass**, foxtail, mustards, ragweeed, quack grass, horsetail, crab grass, goose grass, lambsquarters, nightshade, purslane, velvetleaf and wild oats.

According to the WHO Acute Hazard classification, prometon is categorized as a Class U toxin (unlikely to be hazardous). Prometon is listed as a PAN Bad Actor chemical since it is considered a groundwater contaminant. The acute oral LD_{50} in rats and mice are 2980 mg/kg and 2160 mg/kg, respectively. The acute percutaneous LD_{50} in rabbits is >2000 mg/kg. It may mildly irritate the skin and eyes. The inhalation LC_{50} in rats is >3.26 mg/l air. The ADI is 0.05 mg/kg b.w./day. It is non-toxic to bees and slightly toxic to birds.

In soil, the methoxy group cleaves hydrolytically to give hydroxyl metabolites. Prometon also degrades by microbial action. It persists in soil up to 1 year depending on soil type, moisture and application rate (the DT_{50} is 41 days in most soils).

Prometon is registered in the USA, but is not permitted for use anywhere in the EU.

Prometryn, $C_{10}H_{19}N_5S$ (N^2,N^4-diisopropyl-6-methylthio-1,3,5-triazine-2, 4-diamine), is a triazine herbicide. It has a CAS Registry Number of 7287-19-6. It is a white powder, soluble in water (33 mg/l), acetone, alcohol and hexane. It is stable to hydrolysis in neutral and slightly acidic or alkaline media, but decomposes on exposure to UV radiation.

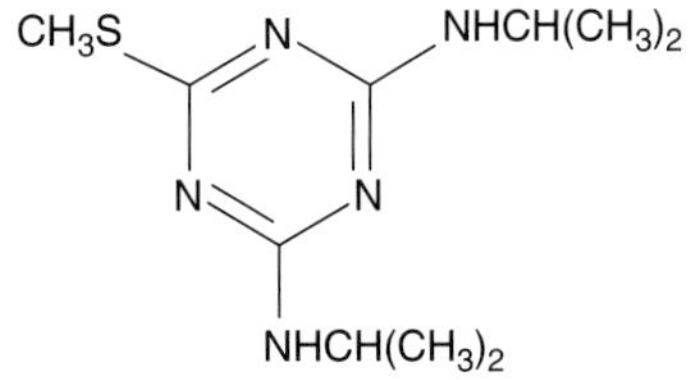

Prometryn

Prometryn is a selective, systemic, pre- and post-emergence compound with both contact and residual action. It inhibits photosynthetic electron transport at the photosystem II receptor site. It is used for most annual grasses and broadleaved weeds in cotton and celery. Spray, broadcast and band applications of prometryn are common.

The oral acute LD_{50} in rats is >2000 mg/kg. Prometryn is slightly irritating to the eyes and skin. The 4 h inhalation LC_{50} in rats is >5170 mg/m^3. It is non-toxic to bees. According to the WHO Acute Hazard classification, prometryn is categorized as a Class U toxin (unlikely to be hazardous). However, the herbicide is a PAN Bad Actor chemical since it is a reproductive toxin and is feared to inhibit normal functioning of the endocrine gland.

Plants metabolize prometryn slowly by oxidizing the methylthio group to hydroxyl metabolites and by dealkylating the side chains. In soil, microbial degradation takes place with a DT_{50} of 50 days, indicating low mobility in soil.

Prometryn is registered in Australia, Burkina Faso, Canada, Cape Verde, Chad, Gambia, Guinea-Bissau, Mali, Mauritania, New Zealand, Niger, Senegal, South Africa, Tanzania and the USA.

Propazine, $C_9H_{16}ClN_5$ (6-chloro-N^2,N^4-diisopropyl-1,3,5-triazine-2,4-diamine), is a chlorotriazine herbicide. It has a CAS Registry Number of 139-40-2. It is a white, colourless powder that is soluble in water (5.0 mg/l) and in organic solvents. It is hydrolysed in acids and alkalis on warming to form hydroxyl propazine.

It is a selective, pre-plant stage, pre-emergence, systemic herbicide, absorbed by the roots and translocated to the apical meristems and leaves of grasses and broadleaved weeds in sorghum and glasshouse plants. Propazine is available as liquid formulations and as wettable powders (WP).

The acute oral LD_{50} in rats is >7000 mg/kg. It does not harm the skin in normal use. It may irritate the eyes in cases of exposure. The inhalation LC_{50} in rabbits is >2.04 mg/l air. The NOEL in 130 days feeding trials in rats at the rate of 250 mg/kg diet showed no ill effects. The ADI is 0.017 mg/kg b.w./day. It is non-toxic to bees. It is phytotoxic to sugarbeet and many vegetables.

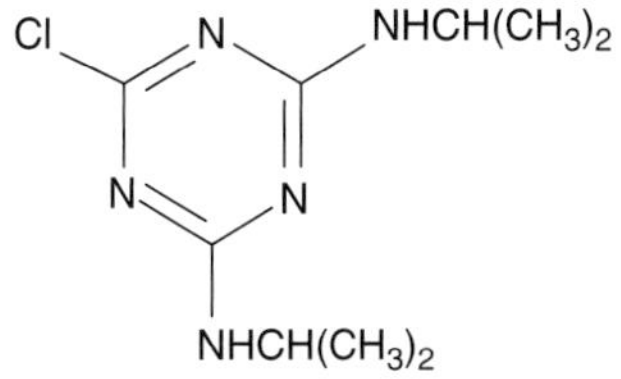

Propazine

According to the WHO Acute Hazard classification, propazine is categorized as a Class U toxin (unlikely to be hazardous). In animals, unchanged propazine is eliminated within 1 day of oral administration. In soils, microbial degradation occurs by hydrolysis of the chlorine atom and dealkylation of amino groups.

Propazine is registered in Australia, New Zealand, South Africa and the USA.

Simazine, $C_7H_{12}ClN_5$ (6-chloro-N^2,N^4-diethyl-1,3,5-triazine-2,4-diamine), is a **chlorotriazine** herbicide and is also an algicide. It has a CAS Registry Number of 122-34-9. It is a colourless powder, soluble in water, alcohol, acetone, toluene and hexane. It is hydrolysed in strongly acidic and alkaline media and degraded by UV light (See Simazine at top of the next page).

This herbicide, a selective compound, is a photosynthetic electron transport inhibitor at the photosystem II receptor site. It is used to control broadleaved weeds and annual grasses in vegetables, nuts, turf, vines and orchards. At higher rates of application, simazine control becomes broad based (i.e. the a.i. can

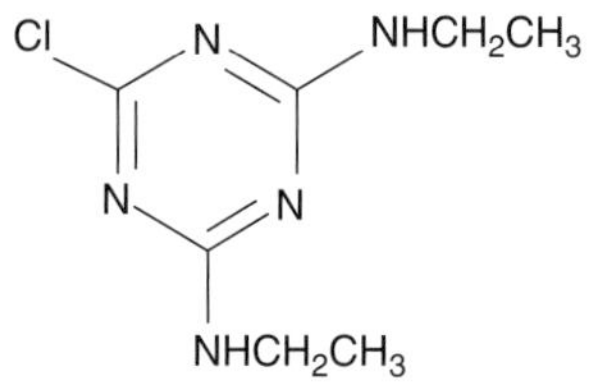

Simazine

control more pests than at normal rates). As an algicide, it controls submerged weeds and algae in cooling towers and aquaria.

The types of formulations available are granules (GR), suspension concentrates (SC), water-soluble granules (SG) and wettable powders (WP).

The WHO categorizes simazine as a Class U toxin (unlikely to be hazardous). The oral acute toxicity in rats and mice is >5000 mg/kg. The acute percutaneous LD_{50} in rats is >3100 mg/kg. It is very toxic to bees. Simazine is phytotoxic to sugarbeet, tobacco, tomatoes, rice and vegetables. It is a PAN Bad Actor chemical.

Following oral administration in mammals, almost the whole dose is eliminated within 24 h as a dealkylated metabolite. In soils and in tolerant plants, simazine is metabolized to a 6-hydroxy analogue which is further dealkylated to amino compounds.

Simazine is registered in Australia, Cambodia, Canada, Madagascar, New Zealand, South Africa, Vietnam and the USA. It is banned in the EU.

Simetryn, $C_8H_{15}N_5S$ (N^2,N^4-diethyl-6-methylthio-1,3,5-triazine-2,4-diamine), is a triazine herbicide. It has a CAS Registry Number of 1014-70-6. It is a white crystalline compound, soluble in water (400 mg/l) and very soluble in organic solvents.

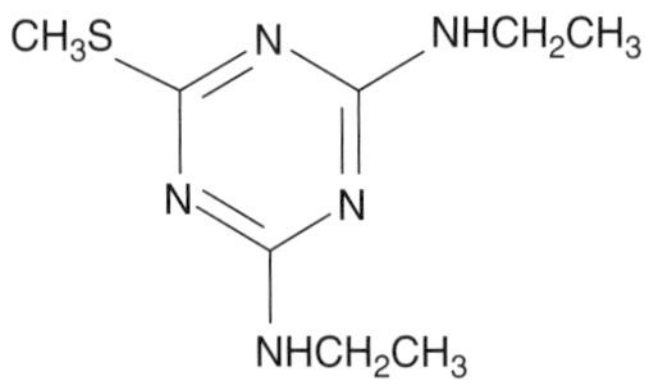

Simetryn

Simetryn is a selective systemic herbicide which inhibits photosynthesis and is mainly used on rice, maize, soybean, wheat, groundnut and cotton to control weeds such as bog pondweed, dwarf hairgrass, **barnyard grass**, dwarf arrowhead and spiral wild celery. It is commonly combined with thiobencarb to control broadleaved weeds in rice.

The acute oral LD_{50} in rats is 750–1195 mg/kg, while the acute percutaneous LD_{50} in rats is >3200 mg/kg. It does not irritate the skin and the eyes and also is not toxic to bees. The ADI is 0.025 mg/kg b.w./day. The WHO classifies it as a Class III toxin (slightly hazardous).

Animals excrete the metabolites in urine and faeces, while in plants two metabolic oxidative pathways are possible.

Trietazine, $C_9H_{16}ClN_5$ (6-chloro-N^2,N^2,N^4-triethyl-1,3,5-triazine-2,4-diamine), is a chlorotriazine herbicide. It has a CAS Registry Number of 1912-26-1. It is a colourless, crystalline substance soluble in water (20 mg/l) and very soluble in polar solvents such as acetone, chloroform, dioxane and ethanol. It is hydrolysed in strongly acidic and alkaline conditions, with the loss of chlorine.

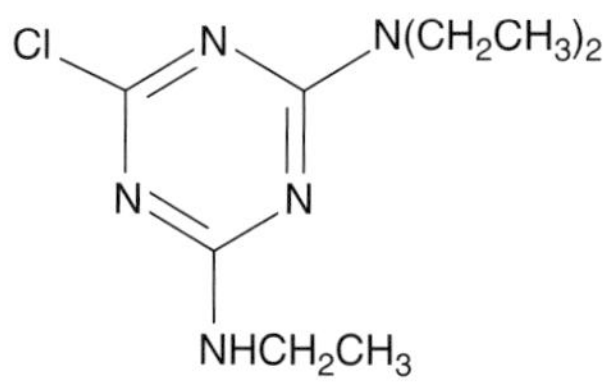

Trietazine

Trietazine is a selective herbicide, absorbed through the roots and leaves. It inhibits photosystem II Site A, and is used to control weeds in potatoes, peas and beans. It is most often used with simazine, **linuron** or **terbutryn**. It is available in formulations like suspension concentrates (SC) and wettable powder (WP).

The acute oral LD_{50} in rats is 494–841 mg/kg and the percutaneous LD_{50} in rats is >600 mg/kg. It is non-irritating to the skin and is non-toxic to honey bees. Based on 90 day feeding trials with 15 mg/kg diet, rats showed no ill effects.

Trietazine undergoes dealkylation of side chains, followed by ring cleavage and evolution of carbon dioxide. According to the WHO Acute Hazard classification, trietazine is categorized as a Class U toxin (meaning it is unlikely to be hazardous).

Triazolinone herbicides

Triazolinones are *N*-phenylthalamide herbicides which are protoporphyrinogen oxidase (PROTOX) inhibitors (chlorophyll biosynthesis pathway). Triazolinones can be soil incorporated or foliar applied.

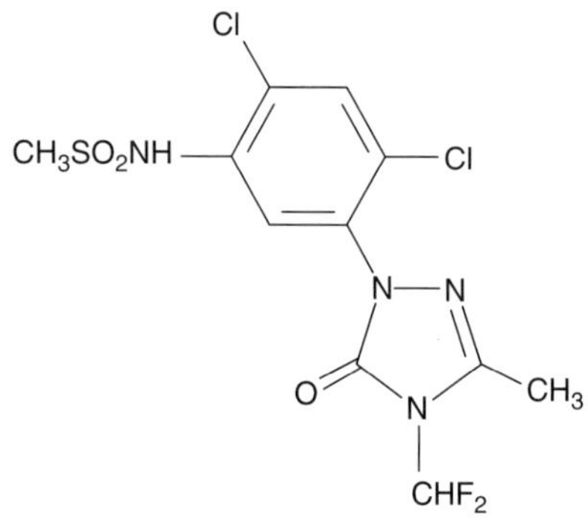

Sulfentrazone

Sulfentrazone, an aryl triazolinone, is a pre-emergence compound controlling a range of grasses and broadleaved weeds in soybean, sugarcane and tobacco. The CAS Registry Number of sulfentrazone is 122836-35-5. Sulfentrazone is highly soluble in water (at 400 mg/l) so it poses a threat to groundwater sources. This herbicide is registered in the USA, Canada and a few other countries, but not in the EU. (See **PROTOX inhibitors**.)

Triazolopyrimidine herbicides

The triazolopyrimidines are a group of herbicides that inhibit branched-chain amino acid synthesis, specifically by affecting the enzyme **acetolactate synthase** (**ALS**). The other herbicidal groups that inhibit ALS are the **sulfonylureas**, **imidazolinones** and **pyrithiobac**. The triazolopyrimidine group includes **cloransulam-methyl**, **diclosulam**, **florasulam**, **flumetsulam**, **metosulam** and **penoxsulam**.

Triazophos insecticide

Triazophos, $C_{12}H_{16}N_3O_3PS$ (*O,O*-diethyl *O*-1-phenyl-1*H*-1,2,4-trizol-3-yl phosphorothioate), is a triazole organophosphorus insecticide that is a pale yellow oil. Its solubility in water is 30–40 mg/l and it is readily soluble in organic solvents such as ethyl acetate, acetone, ethanol, toluene and hexane. It is hydrolysed by aqueous acids and alkalis. It is produced by reaction of 1-phenyl-3-hydroxy-1,2,4-triazole, suspended in acetone with diethoxythiophosphoryl chloride in the presence of triethylamine. It has a CAS Registry Number of 24017-47-8.

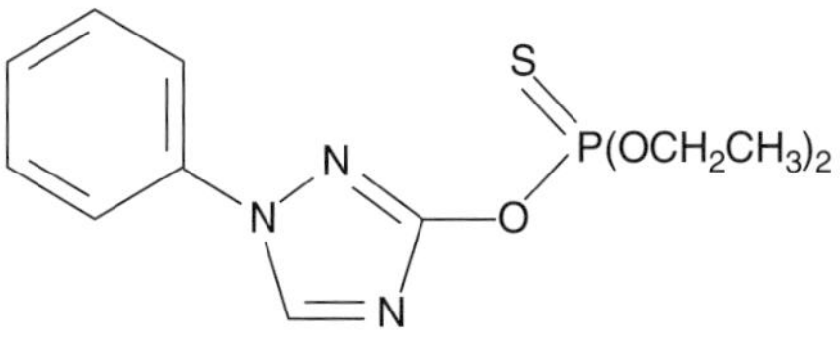

Triazophos

Triazophos, a cholinesterase inhibitor, is a non-systemic, broad-spectrum insecticide, acaricide and nematicide, with penetrating action. It is absorbed into plant tissues. With contact and stomach action, triazophos controls aphids, thrips, midges, beetles, lepidopterous larvae, cutworms and soil insects, spider mites and other species of mites that are commonly seen on ornamentals, fruit trees (including citrus), vines and a number of other crops. These include bananas, strawberries, vegetables, canola, cereals, sugarbeet, sugarcane, maize, soybean, groundnuts, guavas, mangoes, oil palms, olives, cotton, coffee and rice as well as grassland and forestry. It is also used to control some free-living nematodes, particularly among ornamentals and strawberries.

Triazophos is formulated as emulsifiable concentrate (EC), granules (GR), ULV liquid (UL), wettable powder (WP) and powder for dry seed treatment (DS).

Triazophos is a highly toxic insecticide, classified as a Class Ib toxin by the WHO (highly hazardous). It is extremely toxic to rats, via both oral and inhalation routes (the acute oral LD_{50} in rats is 66 mg/kg and the inhalation LC_{50} is 0.53 mg/l). The dermal LD_{50} is >2000 mg/kg. It is toxic to bees and fish. The short-term dietary NOEL of triazophos in rats is 1.5 mg/kg. Triazophos is also extremely toxic to birds (with the acute LD_{50} in bobwhite quails being 8.3 mg/kg) and to fish, *Daphnia magna* and many other aquatic creatures (the acute LC_{50} in rainbow trout is 0.038 mg/l, while the acute 48 h EC_{50} for *D. magna* is 0.0026 mg/l). Triazophos is toxic to both honey bees and earthworms. The ADI is 0.001 mg/kg b.w./day. Triazophos is a PAN Bad Actor chemical. Exposure to triazophos can lead to some or all of the following toxicity symptoms: excessive salivation, sweating, muscle twitching, weakness, tremors, incoordination, headache, dizziness, nausea, abdominal cramps, diarrhoea, respiratory depression, wheezing, fluid in the lungs, pin-point pupils, seizures and loss of consciousness. However, it is non-phytotoxic if used as directed.

In cotton plants, 1-phenyl-3-hydroxy-1, 2,4-trizole is formed as a metabolite. In mammals, administered triazophos is rapidly metabolized via disruption of P–O (triazolyl) bond possibly by hydrolysis of triazophos oxon. The soil half-life is 6–12 days.

Triazophos is registered in India, the Philippines, Vietnam, Mali, Mauritani, Niger, Senegal and Chad. However, it is not registered in the USA or anywhere in the EU.

Tribenuron-methyl herbicide

Tribenuron-methyl, $C_{15}H_{17}N_5O_6S$ (methyl 2-[4-methoxy-6-methyl-1,3,5-triazin-2-yl (methyl)carbamoylsulfamoyl] benzoate), is a triazinylsulfonylurea herbicide. It is a light brown, odourless solid, soluble in water (at 0.05 g/l at 20°C), acetone (43.8 mg/l), acetonitrile, carbon tetrachloride and hexane. This herbicide is produced by reaction of methyl-2-sulfamoylchloride benzoic acid with 2-methylamino-4-methoxy-6-methyl-1,3,5-triazine. It hydrolyses in acidic medium. The CAS Registry Number of tribenuron-methyl is 101200-48-0.

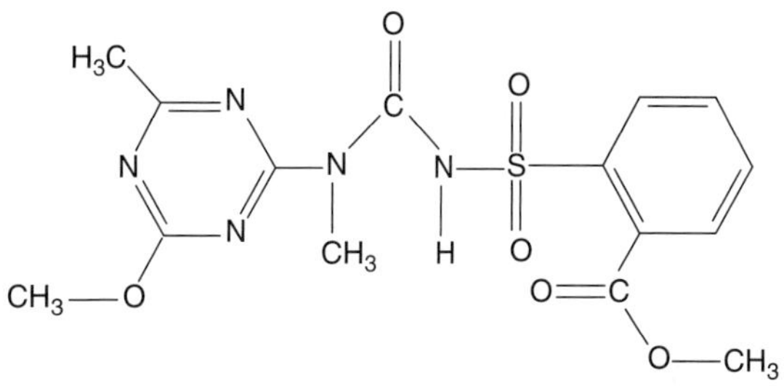

Tribenuron-methyl

Tribenuron-methyl, a post-emergence compound, is a branched-chain amino acid synthesis inhibitor (by affecting the enzyme acetolactate synthase (ALS)), acting especially on valine and isoleucine. Its action, seen on broadleaved weeds, starts after it is readily absorbed by the leaves and roots. This herbicide stops the growth of the plant, by stopping cell division, leading to death within 3 weeks. Tribenuron-methyl is used on cereals.

This herbicide is formulated either as tablets (TB) or as water-dispersible granules (WG). Up to 30 g/ha is a common dose of the a.i.

Tribenuron-methyl is classified as a Class III toxin (i.e. the toxin is slightly hazardous) according to the WHO classification. The acute oral LD_{50} in rats is >5000 mg/kg. The acute dermal LD_{50} in rabbits is >2000 mg/kg. In normal use, this herbicide does not irritate the skin or the eyes. This chemical is non-teratogenic in rats. The ADI is 0.008 mg/kg/day.

The soil half-life of tribenuron-methyl is up to 7 days, with degradation being caused by hydrolysis and microbial action. Hydrolysis is brought about faster in acidic soils.

Tribenuron-methyl is registered in Australia, Denmark, South Africa and the USA.

Tribufos as a PGR

Tribufos, $C_{12}H_{27}OPS_3$ (*S,S,S*-tributyl phosphorotrithionate), is a colourless to pale yellow liquid with a mercaptan-like odour. It has a CAS Registry Number of 78-48-8, soluble in water, alcohols and aliphatic, aromatic and chlorinated hydrocarbons. It hydrolyses slowly under alkaline conditions.

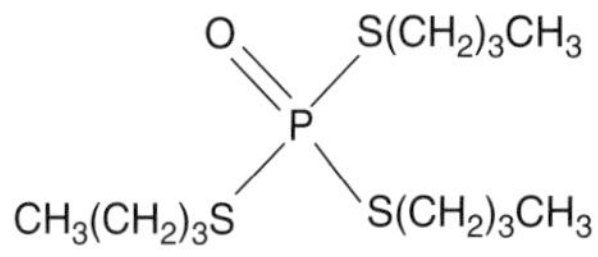

Tribufos

Tribufos is a plant growth regulator and an insecticide **synergist**. Upon absorption into leaves, tribufos stimulates the formation of an abscission layer between the plant stem and the leaf petioles, causing leaf drop. Tribufos is mainly used on cotton to facilitate harvesting.

Formulated as an emulsifiable concentrate (EC), tribufos is available under trade names such as Deleaf, Easy-Off-D and Folex. Application doses range from 560 g/ha to 2 kg/ha. This herbicide is commonly tank mixed, where the dosage is less.

The insecticide is a moderately hazardous chemical, via oral, dermal and inhalation routes. The acute oral LD_{50} in rats is 234–435 mg/kg and the dermal LD_{50} is 850 mg/kg b.w. The inhalation LC_{50} of tribufos in rats is 2.46–4.65 mg/l. It is very toxic to birds with the acute oral LD_{50} in bobwhite quails being 142–163 mg/kg. Tribufos is very toxic to fish and aquatic organisms. The acute 96 h LC_{50} in bluegill sunfish ranges from 0.72 mg/l to 0.84 mg/l. It is relatively non-toxic to bees. As an acute poison, the symptoms caused by its exposure include excessive salivation, sweating, rhinorrhea and tearing, muscle twitching, weakness, headache, dizziness, nausea, vomiting, abdominal cramps and pin-point pupils. It may also cause tremor, incoordination, respiratory depression, tightness in the chest, fluid in the lungs, seizures, incontinence and loss of consciousness. It is very phytotoxic. The ADI is 0.001 mg/kg/day.

Upon oral administration in test mammals, tribufos is absorbed, metabolized and excreted rapidly. Metabolism occurs by a series of processes such as hydrolysis, methylation and oxidation, to finally give (3-hydroxy)-butylmethylsulfone. The unchanged herbicide is also found in cotton. In soil, tribufos is strongly adsorbed and hence there is almost no leaching. The main metabolite is 1-butanesulfonic acid.

Tribufos is registered for use in the USA, Argentina and Mexico.

Tributyltin (TBT) fungicide

See **Organotin fungicides**

Trichlorfon insecticide

Trichlorfon, $C_4H_8Cl_3O_4P$ (dimethyl 2, 2, 2-trichloro-1-hydroxyethylphosphonate), is an organophosphorus insecticide. It has a CAS Registry Number of 52-68-6. It is a colourless crystalline powder, soluble in water (120 g/l), benzene, toluene and most organic solvents, except aliphatic hydrocarbons. It is rapidly converted to **dichlorvos** by alkalis and then hydrolysed; the DT_{50} values at pH 4, 7 and 9 are 510 days, 46 h and <30 min, respectively. Dimethyl phosphite reacts with chloral to give trichlorfon. The product is a racemate (i.e. the 1:1 mixture of (1*R*)- and (1*S*)-enantiomers).

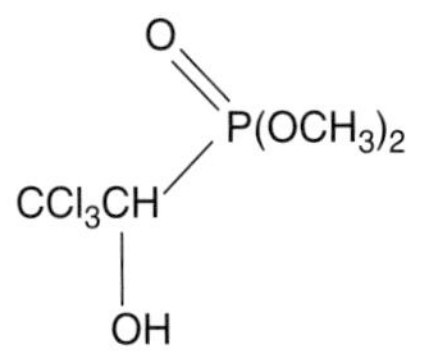

Trichlorfon

Trichlorfon, a cholinesterase inhibitor, is a non-systemic insecticide with contact and stomach action. It is used in the control of pests belonging to the orders Diptera, Lepidoptera, Hymenoptera, Hemiptera and Coleoptera. It is also used to control household pests such as flies, cockroaches, fleas, bed bugs, silver fish and ants. It is used to control ectoparasites on farm animals.

Formulated as wettable powder (WP), water-soluble powder (SP), dustable powder (DP), granules (GR), ULV liquid (UL), a coating agent and a pour-on, trichlorfon is compatible with most other pesticides, except alkaline materials and oils. It is available as trade names Dipterex and Tugon.

In mammals, trichlorfon is rapidly absorbed, metabolized and excreted almost completely in the urine within 6 h. In mammals and plants, major metabolites are dimethyl- and monomethyl-phosphoric acid and conjugates of dichloroacetic and dichloroethanol. It is rapidly broken down and has relatively high mobility in soil and is thus likely to leach into groundwater sources. In an aqueous medium at higher pH levels, it is converted to dichlorvos.

Trichlorfon is moderately to highly toxic to birds with an LC_{50} as low as 700 ppm for some species. Similarly, it is highly toxic to many species of fish, and moderately to highly toxic to many beneficial insects. The acute oral LD_{50} in rats is 212 mg/kg and the dermal LD_{50} is >5000 mg/kg b.w. The inhalation LC_{50} in rats is >0.53 mg/l, indicating toxicity. The ADI of trichlorfon in rats is 0.045 mg/kg b.w./day. It is very toxic to honey bees and to fish and *Daphnia magna*. The acute LD_{50} in birds is around 36 mg/kg. The WHO categorizes trichlorfon as moderately toxic (a Class II toxin). It is a PAN Bad Actor chemical, since it inhibits cholinesterase. There are indications of trichlorfon causing tumours, at higher exposure. Symptoms of poisoning include excessive salivation, sweating, rhinorrhea and tearing, muscle twitching and weakness, pin-point pupils with blurred or dark vision, headache, dizziness, nausea, vomiting, abdominal cramps and diarrhoea. Syptoms may also include tremors, incoordination, respiratory depression, tightness in the chest, wheezing, fluid in the lungs, seizures, incontinence and loss of consciousness.

Trichlorfon is registered in India, the USA, Canada, Australia and a few other countries, but not in the EU and Kuwait. According to the EU Pesticide Database, the MRL of trichlorfon for vegetables is 0.5 mg/kg and for cereal grains is 0.1 mg/kg.

Trichlorfon technical analysis

The analysis of trichlorfon, an organophosphate insecticide, may be done, ensuring a minimum purity of 97% per kilogram of the sample. It should be free from other extraneous materials or modifying agents. The quantity of H_2SO_4 should be 3 g/kg in order to provide the necessary acidity. The water content must be a maximum of 4 g/kg.

In a 250 ml conical flask, dissolve 10 ml 2-aminoethanol in about 1 g of the sample. Cool in ice water and add 50 ml 4 N chlorine-free nitric acid. Keep at 20°C and titrate against 0.1 N silver nitrate electrometrically or using an indicator. After obtaining a chloride correction value for all reagents, determine the amount of chlorine (both hydrolysable and inorganic chlorine).

To determine the inorganic chlorine, weigh about 1 g (to the nearest 0.1 mg) of the sample into a 250 ml conical flask and dissolve in 100 ml distilled water. Set aside for 5 min at room temperature, acidify with 5 ml 4 N nitric acid (chlorine free) and titrate against 0.1 N silver nitrate electrometrically or using an indicator.

The trichlorfon content can be calculated using Eqn T.9.

$$\text{Trichlorfon content (g/kg)} = \left(\frac{v_1}{m_1} - \frac{v_2}{m_2}\right) \times 25.75 \qquad (T.9)$$

where

v_1 = Volume of 0.1 N silver nitrate in millilitres (ml) for the total chlorine (hydrolysable plus inorganic)
v_2 = Volume (ml) of 0.1 N silver nitrate for the inorganic chlorine
m_1 = Mass (g) of sample used for the total chlorine experiment
m_2 = Mass (g) of sample used for the inorganic chlorine estimation

The emulsifiable concentrate (EC) of trichlorfon is analysed by taking 1 g of sample and hydrolysing it quantitatively in a mixture of methanol and 2-aminoethanol to give chlorine ions. The inorganic chlorine is then estimated in the same way as given above for technical grade trichlorfon. The trichlorfon content is calculated from the difference in volumes of silver nitrate. The water-soluble powder (SP) is determined by hydrolysing a mixture in methanol and aminoethanol and estimating the chlorine content.

Trichloroaceticacid sodium (TCA sodium) herbicide

Trichloroaceticacid (TCA) sodium, $C_2Cl_3NaO_2$ (6-chloro-N^2,N^4-diisopropyl-1,3,5-triazine-2,4-diamine), is a halogenated alkanoic acid, used as a herbicide. It is also called **sodium trichloroacetate**. It is a yellowish powder, highly soluble in water (1.2 kg/l), methanol (232 g/l) and acetone (7.6 g/l). It decomposes in aqueous, strongly alkaline media.

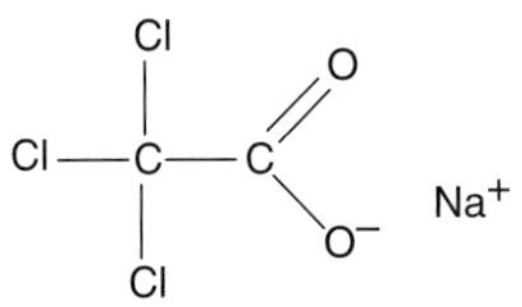

Trichloroaceticacid sodium

TCA sodium, a selective, systemic fungicide, is absorbed by plant roots and sometimes by the leaves. The mode of action is the inhibition of lipid synthesis. The herbicide is used for the control of annual and perennial grasses as well as volunteer cereals in canola, carrots, potatoes, tomatoes, peas, brassicas, sugarbeet, lucerne, sunflowers, sugarcane, cotton, flax, rice and ornamentals. It is formulated as granules (GR) and water-dispersible granules (WG).

TCA sodium is classified under toxicity Class U by the WHO and as Class III by the EPA (indicating it is unlikely to be hazardous). The acute oral LD_{50} of the herbicide is 3200–5000 mg/kg in rats and 3600–5600 mg/kg in mice. It irritates the skin, eyes and mucous membranes. The herbicide is non-toxic to bees and fish. The compound is not listed as a PAN Bad Actor chemical.

Handlers are well advised to remove contaminated clothing, wash their face, hands and eyes in cases where symptoms such as cough, sore throat, redness of eyes and skin indicate exposure to the chemical. Medical treatment should be sought in cases of severe toxicity.

Although TCA sodium is persistent in soil (up to 3 months), it degrades by microbial action, leading to the formation of chloride. Heavy rainfall causes leaching of TCA sodium.

The TCA sodium herbicide is not registered by the EPA in the USA or in the EU.

Trichloroethanol

See **Chloralose rodenticide**

Trichogramma spp. as biocontrol agent

Wasps in the genus *Trichogramma* (family Trichogrammatidae and order Hymenoptera) are some of the most widely studied agents of biological control in the field of entomology. The genus *Trichogramma* has many species and most of them are **parasitoids** of well-known crop pests. These free-living, stingless creatures measure up to 1 mm in length and have a lifespan of about 10 days. They feed on flower nectar. Males have antennae with many long hairs whereas females have antennae with a few short hairs. These wasps are abundantly present in several terrestrial and some aquatic habitats. *Trichogramma brevicapillum*, *Trichogramma nubilale* and *Trichogramma exiguum* are some species in the genus *Trichogramma*.

An adult female wasp scouts for eggs of about 200 species and insert her own eggs into the host eggs with the help of her ovipositor. Within 24 h, the eggs hatch into larvae that feed on the host egg's contents. The host eggs start turning black within 3–4 days as a result of melanin being deposited on the inner surface. After the pupal stage, the adult wasps emerge from the dead egg within 5 days, only when the temperature is warm enough and there is sufficient humidity.

The wide range of host eggs include those of pests such as cotton bollworms, sugarcane borers, corn borers, spruce budworms, tobacco budworms, cutworms, armyworms and cabbage loopers. Moth species include the diamondback moth, codling moth and gypsy moth.

Commercial rearing of *Trichogramma* spp. is common in many countries since there is a demand for these wasps to parasitize pest eggs. Pest populations can be controlled and kept at below economic levels by 12,000–50,000 *Trichogramma*/acre. Crops on which this wasp is used include maize, rice, sugarcane, cole crops, cotton, vegetables, sugarbeet, fruit trees and pine and spruce trees. *Trichogramma brassicae* and *Trichogramma pretiosum* are examples of species that are commercially exploited. These wasps thrive if crops are intercropped with nectar-producing plants.

These biocontrol agents may be killed in the presence of broad-spectrum insecticides like **parathion**, **thiodicarb** and **cyhalothrin**. Ants are a natural enemy and must be kept away from *Trichogramma* sp. in host eggs. Eggs must be protected from direct sunlight.

Trichothecene mycotoxins

Trichothecenes, a large family of sesquiterpene compounds, are a group of mycotoxins that are structurally similar to each other. These mycotoxins have a 12, 13-epoxy ring and a hydroxyl or acetyl group.

Species of *Fusarium*, *Trichoderma* and *Stachybotrys* produce trichothecenes on grains such as wheat, oats or maize, normally in damp and indoor conditions. *Stachybotrys chartarum* is an example of one such pathogen.

Trichothecenes react with rRNA and inhibit protein synthesis as soon as they are ingested. They are absorbed through the skin, causing skin irritation. Examples of trichothecenes include T-2 toxin and **vomitoxin**.

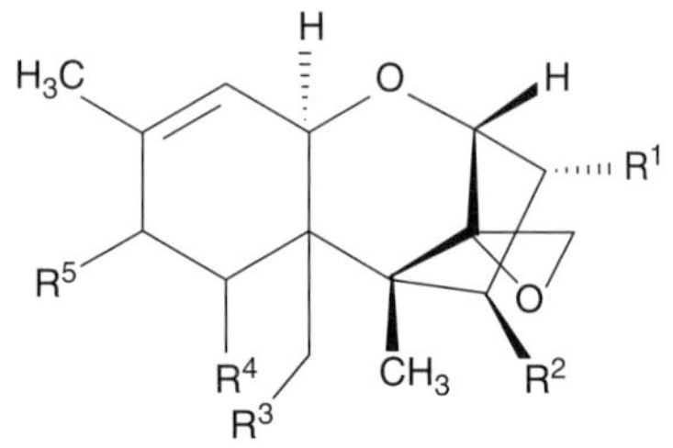

	R_1	R_2	R_3	R_4	R_5
T_2 Toxin	OH	$OCOCH_3$	$OCOCH_3$	H	$OCOCH_2CH(CH_3)_2$
HT — 2Toxin	OH	OH	$OCOCH_3$	H	$OCOCH_2CH(CH_3)_2$
Nivalenol	OH	OH	OH	OH	=O

Trichothecenes

Triclopyr herbicide

Triclopyr, $C_7H_4Cl_3NO_3$ (3,5,6-trichloro-2-pyridyloxyacetic acid), is a pyridine herbicide. It is a colourless solid that is fairly soluble in water, hydrocarbon solvents and polar organic solvents. It is stable under normal storage conditions and to hydrolysis, but subject to photodecomposition. Triclopyr exists in its salt form or as an ester, such as triclopyr-butotyl, triclopyr-ethyl and triclopyr-triethylammonium. Triclopyr is made by reacting 2,3,5,6-tetrachloropyridine with paraformaldehyde in dimethyl sulfoxide containing potassium cyanide, followed by hydrolysis of the compound to the acid. It has a CAS Registry Number of 55335-06-3.

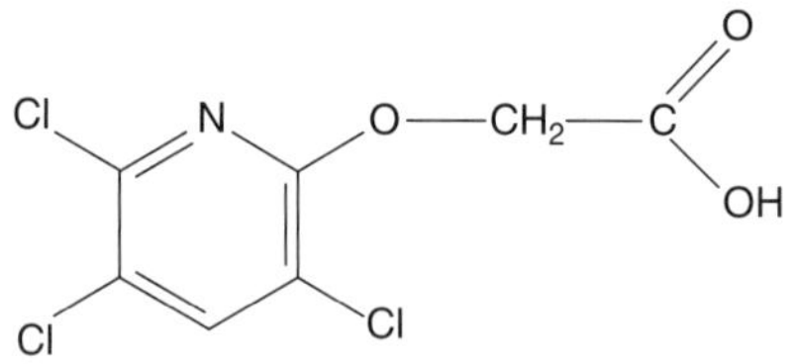

Triclopyr

Triclopyr, a selective, systemic herbicide, is rapidly absorbed by leaves and roots and it is then translocated to the meristematic tissue. It promotes growth much like an auxin (IAA) does, in broadleaved weeds, woody plants and grass weeds in rice and grasslands, turf and coniferous forests. By mimicking IAA, tryiclopyr kills these weeds.

Trade names associated with triclopyr include Access, Crossbow, Grazon, Redeem, Rely, Remedy and Turflon. Triclopyr is often mixed with herbicides like **picloram** or 2,4-D for a much broader control.

The WHO categorizes triclopyr as a slightly hazardous chemical (Class III toxin). The oral LD_{50} is 630–729 mg/kg in rats, 550 mg/kg in rabbits and 310 mg/kg in guinea pigs, indicating moderate toxicity. Triclopyr is toxic to birds with the LD_{50} in mallard ducks being 1698 mg/kg. The herbicide is slightly toxic to non-toxic to fish and invertebrates (e.g. *Daphnia magna*). Exposure to triclopyr causes irritation to the skin and eyes.

Triclopyr is susceptible to photolytic degradation (the half-life in water is 10 h). It is moderately persistent in soil, with the half-life of 46–90 days. Metabolites include **3,5,6-trichloro-2-pyridinol** and 3,5,6-trichloro-2-methoxypyridine. Microbial action aids in the breakdown. The ester and amine salt break down to form triclopyr.

Soil adsorption of triclopyr is limited, indicating its potential to leach. Tolerant plants inactivate triclopyr by decarboxylation and then by conjugation. Mammals excrete this herbicide rapidly via the urine. Studies have found the presence of triclopyr residues in some berries.

This herbicide is registered in Australia, Burkina Faso, Cape Verde, Chad, Gambia, Guinea-Bissau, Hungary, Madagascar, Mauritania, the Netherlands, New Zealand, Niger, the Philippines, South Africa, the USA and EU members (e.g. the UK, Italy, Hungary, Greece, Germany and France).

Tricyclazole fungicide

Tricyclazole, $C_9H_7N_3S$ (5-methyl-1,2,4-triazolo[3,4-b][1,3] benzothiazole), is a triazolobenzothiazole fungicide, which is in the form of a crystalline solid. It is soluble in methanol (26.5 g/l), acetone (13.8 g/l), xylene (4.9 g/l) and water (0.596 g/l). It is relatively stable to UV light and stable when stored in airtight packaging in a well-ventilated, dry room. Tricyclazole is produced by the reaction of 2-hydrazino-4-methylbenzothiazole with triethyl orthoformate. Tricyclazole has a CAS Registry Number of 41814-78-2.

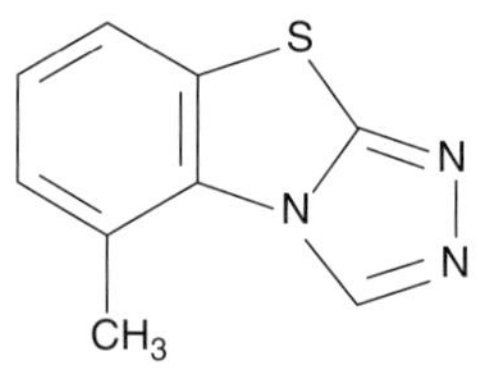

Tricyclazole

Tricyclazole, a systemic fungicide, works as a melanin biosynthesis inhibitor. The fungicide is rapidly absorbed by plant roots and foliage and is translocated in the xylem and apoplast. It is best used to control rice blast (*Pyricularia oryzae*) in transplanted and direct-seeded rice. The leaves can be drenched or roots soaked for best results. Two applications are commonly recommended for season-long control of the pathogen.

Some of the tricyclazole formulations are dustable powder (DP), granules (GR) and wettable powder (WP) and common trade names include Beam, Bim and Agni.

Tricyclazole is classified by both the WHO and the EPA as a Class II toxin (moderately hazardous). It is not irritating to the skin but may irritate the eyes. The acute oral LD_{50} among rats is 314 mg/kg, mice 245 mg/kg, and mallard ducks and bobwhite quails, >100 mg/kg. The NOEL (2 years) in rats is 9.6 mg/kg b.w. The ADI for tricyclazole is 0.03 mg/kg/day. In case of an accident, contaminated garments must be removed and the body part washed thoroughly with soap and water. If swallowed, vomiting should be induced unless the patient is unconscious. If inhaled, the patient should be moved into fresh air. It is important to wear safety glasses, rubber gloves, shoes, socks, a long-sleeved shirt and long trousers when using this fungicide.

Tricyclazole is rapidly and extensively metabolized in animals. In plants, the main metabolite is the hydroxymethyl analogue.

Tricyclazole is registered for use in countries like India, Madagascar, Vietnam, Cambodia and Portugal, but is not permitted for use in the USA and the EU. The MRL value for rice is 3 ppm and for wheat is 0.02 ppm.

Tridemorph fungicide

Tridemorph, $C_{19}H_{39}NO$, (a reaction mixture of 4-alkyl-2,6-dimethylmorpholines, where 'alkyl' is a mixture of C_{11}–C_{14} homologues of which 60–70% is tridecyl), is an amine morpholine type of compound. It is a yellow oily liquid with a slight amine-like odour that is soluble in water (1.1 mg/l) and miscible in ethanol, acetone, ethyl acetate, benzene, chloroform and olive oil. It is stable at up to 50°C for 2 years in airtight containers. Tridemorph is produced as the main product (60–70%) in the reaction of tridecylamine and propylene oxide, followed by dehydration and cyclization in the presence of concentrated sulfuric acid. It has two CAS Registry Numbers, 24602-86-6 and 81412-43-3.

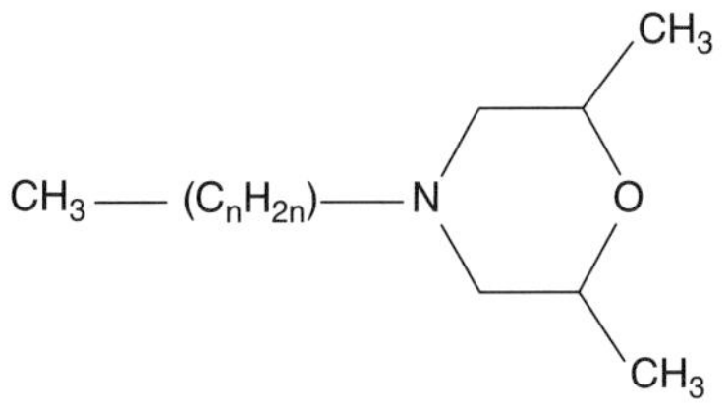

n = 10, 11, 12 (60–70%) or 13

Tridemorph

Tridemorph, a systemic fungicide with eradicant action, works by inhibiting **ergosterol** biosynthesis, by inhibition of sterol reduction and isomerization. The fungicide is absorbed by the roots and leaves to give some protective action. It is used to control *Erysiphe graminis* in cereals, *Mycosphaerella* spp. in bananas, and *Corticium salmonicolor* and *Exobasidium vexans* in tea. Application of this compound to bananas, wheat and barley is common.

Tridemorph is used in combination with cyproconazole, **fenpropimorph**, **flusilazole**, **propiconazole**, **tebuconazole**, **triadimenol** and **carbendazim** to extend the spectrum of disease control. The formulation type of tridemorph is oil-miscible liquid (OL) and a commercial product is Calixin. The maximum single dose of tridemorph must not exceed 375 g/ha. Operators must wear suitable protective gloves when handling contaminated surfaces, using closed systems to empty tridemorph products into spray tanks.

The fungicide tridemorph is classified by the WHO as a moderately hazardous toxin (Class II toxin). It is irritating to the skin and eyes (based on tests on rabbits), with prolonged or repeated exposure causing dermatitis and/or conjunctivitis. Inhalation of tridemorph vapour may cause adverse health effects. The acute oral LD_{50} in rats is 480 mg/kg. The NOEL (2 years) in female rats is 2 mg/kg b.w. and in male rats is 4.5 mg/kg b.w. It is toxic to birds (with the LD_{50} among quails being 1388 mg/kg) and to fish and other aquatic life. The 24 h LD_{50} in honey bees is >200 mg per bee. The ADI of tridemorph is 0.016 mg/kg b.w./day. Under certain climatic conditions, scorch may occur on some winter wheat varieties. Tridemorph is not a PAN Bad Actor chemical.

In animals (rats), tridemorph is rapidly absorbed after oral administration and is eliminated completely within 2 days. In plants, the metabolism proceeds via oxidation of the 4-alkyl side chain and/or by opening of the morpholine ring. Tridemorph residues in cereal grains at harvest are <0.05 mg/kg. Its half-life in soils is up to 50 days.

Tridemorph is registered for use in India, South Africa, Australia, Costa Rica and Ecuador but not in the USA or the EU.

Trifloxystrobin fungicide

Trifloxystrobin, $C_{20}H_{19}F_3N_2O_4$ (methyl (*E*)-methoxyimino-{(*E*)-α-[1-(α,α,α-trifluoro-m-tolyl)ethylideneaminooxy]-o-tolyl}acetate), is an oximinoacetate **strobilurin** type of fungicide. It is an odourless, white solid that is soluble in water (610 μg/l), methanol (76 g/l) and toluene (500 g/l). Trifloxystrobin is obtained starting from substituted acetophenone, via oxidation with potassium permanganate in alkaline solution, esterification with methanol, reaction with methoxyamine, bromization and condensation. It has a CAS Registry Number of 141517-21-7.

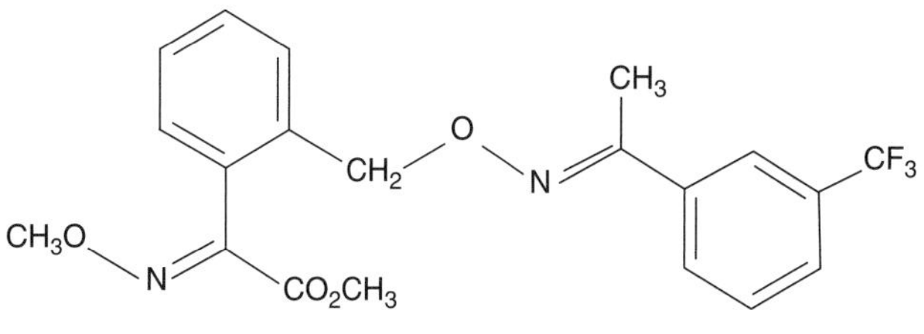

Trifloxystrobin

Trifloxystrobin is a mesostemic, broad-spectrum fungicide with mainly preventive activity. Being a strobilurin compound, it acts on the fungal respiration system since it affects the mitochondrial respiration pathway. A quinone outside inhibitor (**QoI**), trifloxystrobin inhibits mitochondrial respiration by blocking electron transfer at the Qo centre of cytochrome bc1. Thus trifloxystrobin inhibits fungal spore germination and mycelial growth. The particular characteristics of trifloxystrobin are its vapour movement, good rain fastness and residual activity. The important aspect of trifloxystrobin is its efficacy against all four fungal classes (i.e. ascomycetes, deuteromycetes, basidiomycetes and oomycetes). It is used in cereals, soybean, maize, rice, cotton, grapes, bananas, cucurbits, pome and stone fruit, ornamentals and turf.

Commonly used as a foliar fungicide, trifloxystrobin is used to control **powdery mildew**, apple **scab** and peanut (groundnut) leaf spot. The application of the fungicide ranges from 50 g a.i./ha to 550 g a.i./ha, the actual rate being subject to the crop, the disease and the method of application. It is usually combined with other fungicides such as **tebuconazole**, cyproconazole and **propiconazole**. Trifloxystrobin is commercially available as Flint, Compass and Stratego. It is formulated as a flowable concentrate for seed treatment (FS), suspension concentrate (SC) and water-dispersible granules (WG).

Trifloxystrobin is non-mutagenic, non-teratogenic and non-carcinogenic (and rated by the EPA as 'not a likely human carcinogen'). It is classified by the WHO as a Class III toxin (slightly hazardous). The acute percutaneous LD_{50} in rats is >2000 mg/kg. The acute oral LD_{50} in rats is >5000 mg/kg, while in bobwhite quails and mallard ducks it is >2000 and >2250 mg/kg, respectively. The NOEL (2 years) in rats is 9.8 mg/kg b.w./day. It is very toxic to fish (the acute 96 h LC_{50} in rainbow trout is 0.015 mg/l). It is also toxic to honey bees and moderately toxic to earthworms (the acute 48 h oral LD_{50} is >200 μg per bee and the acute 14 day LC_{50} is >1000 mg/kg in earthworms). The ADI of trifloxystrobin in rats is 0.1 mg/kg/day. Symptoms of exposure to trifloxystrobin are irritation to the eyes, while long-term skin contact may cause allergic reactions and skin sensitization.

In animals, more than half of trifloxystrobin is excreted in 48 h via urine and faeces. It is metabolized by O-demethylation, oxidation and conjugation and completely eliminated from the body. In plants, soil and the aquatic environment, the metabolic profile is the same. Trifloxystrobin dissipates rapidly from the soil and surface water. One of its metabolites, CGA-321113, is mobile (and has been detected even at 91 cm in the soil) and is persistent.

Trifloxystrobin is registered in India, the USA, South Africa, Australia, New Zealand and almost the entire EU. The MRL value for barley and grain in the USA is 0.05 ppm. According to the EU Pesticide Database, the MRL value for grapes is 5.0 mg/kg; for rye grains and wheat grains is 0.05 mg/kg; and for tomatoes, beans (with pods), pomes and strawberries it is 0.5 mg/kg.

Triflumizole fungicide

Triflumizole, $C_{15}H_{15}ClF_3N_3O$ ((*E*)-4-chloro-α,α,α-trifluoro-*N*-(1-imidazol-1-yl-2-propoxyethyidene)-o-toluidine), is an imidazole compound. It is a colourless crystal with solubility in water being 0.0102 g/l. It is highly soluble in chloroform (2220 g/l) and acetone (1440 g/l). Triflumizole is unstable in highly alkaline and acidic media. Triflumizole is produced by the reaction of 2-trifluoromethyl-4-chloropropoxyacetanilide with imidazole in the presence of phosphorus pentachloride. It has two CAS Registry Numbers, namely, 68694-11-1 and 99387-89-0.

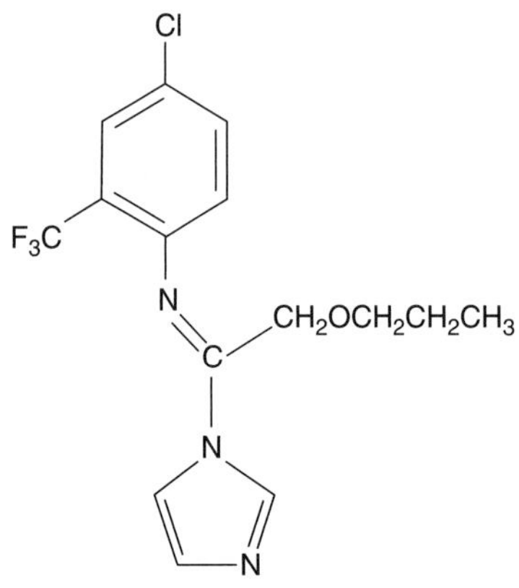

Triflumizole

Triflumizole is a systemic fungicide with protective and curative action. Absorbed by the leaves and roots, it inhibits sterol demethylation (**ergosterol** biosynthesis). It is used in the control of rust and **scab** in pome fruit (*Gymnosporangium* and *Venturia* spp.), **powdery mildew** in fruits and vegetables (*Podosphaera* sp., *Sphaerotheca* sp. and *Erysiphe* sp.) and diseases caused by *Fusarium* spp., *Fulvia* spp. and *Ustilago* spp. in cereals. Triflumizole is also used on ornamentals grown in greenhouses and shade houses and in seed treatment against *Bipolaris oryzae*, *Tilletia* spp. and *Ustilago* spp. in cereals.

Triflumizole is commonly formulated as an emulsifiable concentrate (EC) and as a smoke generator (FU) and some of the trade names include Trifmine and Procure. Other applications are as a soil drench, foliar spray or **chemigation** agent. The recommended application rate of triflumizole is at 180–300 g/ha in vegetables and 700–1000 g/ha in orchards. It is usually combined with other fungicides such as cyflufenamid.

The fungicide triflumizole is classified by the WHO and EPA as a slightly toxic chemical (Class III toxin). It is a mild eye irritant. The acute oral LD_{50} in rats is 695–715 mg/kg, while in Japanese quails it is 2467–4308 mg/kg. It is highly toxic to fish, molluscs and zooplankton. The acute 96 h LC_{50} in rainbow trout is 0.57 mg/l, while the acute 48 h oral LD_{50} in honey bees is 14 μg per bee. The ADI of triflumizole in rats is 0.05 mg/kg b.w./day.

Triflumizole rapidly metabolizes in aerobic soil by the action of microbes, producing carbon dioxide and several other products. In sandy, loamy soils, it has a half-life of 18 days. Triflumizole has a low leaching potential. In plants, the main residues remain unchanged and the major metabolite formed is the result of cleavage of the imidazole group.

The fungicide is registered in Australia, the USA and the Philippines, but not in the EU. According to the EU Pesticide Database, the MRL value for tomato is 1.0 mg/kg and for zucchini it is 0.2 mg/kg.

Triflumuron as an IGR

Triflumuron, $C_{15}H_{10}ClF_3N_2O_3$ (1-(2-chlorobenzoyl)-3-(4-trifluoromethoxyphenyl) urea), is a benzoylphenylurea **chitin synthesis inhibitor** and is thus an insect growth regulator. It is a colourless and odourless powder, soluble in water and many organic solvents. Triflumuron is hydrolysed by alkalis. It has a CAS Registry Number of 64628-44-0.

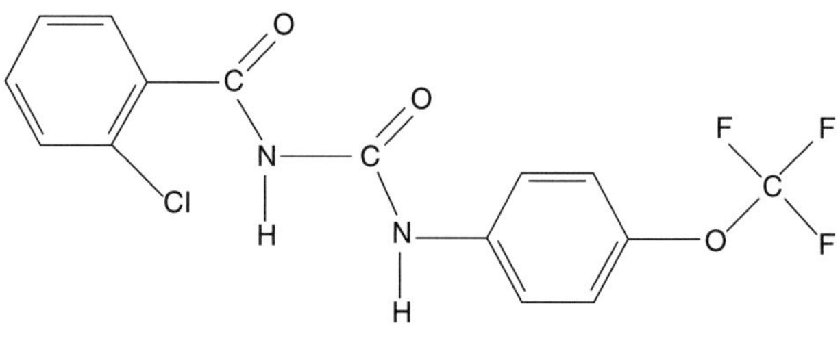

Triflumuron

Triflumuron interferes with chitin synthesis and thus disrupts the **moulting** process of insects. This IGR affects the growth cycle of leaf miners, bollworms, cabbage moths and codling moth on crops like maize, cotton, citrus fruits, stone fruits, soybean, cabbage and many other vegetables. It is also used to control ectoparasites such as flies and fleas on livestock.

This IGR is formulated as suspension concentrates (SC) and pour-on formulations that can be sprayed after diluting in water, as required.

According to the WHO Acute Hazard classification, triflumuron is categorized as a Class U toxin (unlikely to be hazardous).

The acute oral LD_{50} and dermal LD_{50} in rats is >5000 mg/kg b.w. while the same in birds is 561 mg/kg, indicating high toxicity. Triflumuron is also very toxic to fish and aquatic invertebrates. The acute 48 h contact LD_{50} of triflumuron in honey bees is > 200 µg per bee. The ADI in dogs is 0.014 mg/kg b.w./day. Exposure to triflumuron (like many substituted ureas) causes irritation to the eyes, skin and mucous membranes, coughing and shortness of breath, nausea, vomiting, diarrhoea, headache, confusion due to electrolyte depletion, poor protein metabolism and weight loss.

Triflumuron was slightly metabolized in rats by hydrolytic cleavage, forming metabolites and conjugates containing the 2-chlorophenyl ring. In plants, following spray application to apples, soybeans and potatoes, triflumuron is metabolized in the same way as in animals.

This IGR is persistent in soil with half of the applied compound degrading to carbon dioxide in over 100 days. Microbial action contributes to the degradation process.

Triflumuron is registered in Australia, New Zealand, Burkina Faso, Cape Verde, Gambia, Greece, Madagascar, Mali, Mauritania, Niger, Senegal and South Africa. According to the EU Pesticide Database, the MRL of triflumuron for pome fruit is 0.5 mg/kg and for stone fruit, 0.3 mg/kg.

Trifluralin herbicide

Trifluralin, $C_{13}H_{16}F_3N_3O_4$ (α,α,α-trifluoro-2,6-dinitro-*N*, *N*-dipropyl-p-toluidine), is a **dinitroaniline** herbicide. This widely used herbicide is a yellow, volatile (when wet), crystalline substance that is soluble in water (at 0.0024 g/100 ml), xylene and acetone. It is produced by nitration of 4-chlorotrifluorotoluene followed by treatment with dipropylamine. The CAS Registry Number of trifluralin is 1582-09-8.

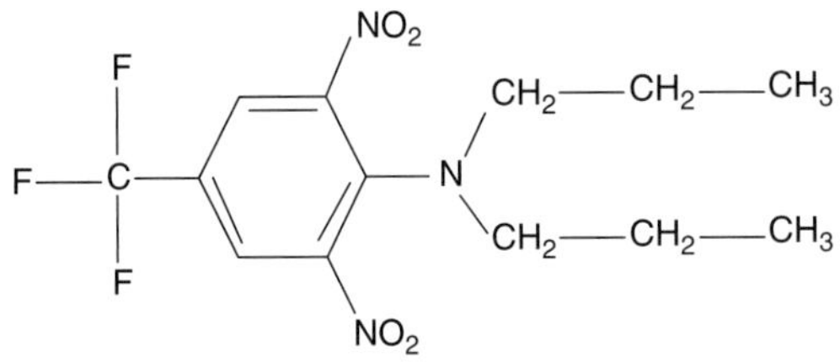

Trifluralin

A pre-emergence, selective, soil herbicide, trifluralin controls a large number of annual grass and broadleaved weeds in wheat, barley, sugarbeet, soybean, tomatoes, beans, carrots, brassicas, vines, sunflower, canola, turf, fodder, lucerne and cotton. It acts by interrupting mitosis and thus inhibits root development and cell division. So weed growth is controlled at an early stage.

The herbicide trifluralin is applied by ground equipment for soil treatment. It is available as granules (GR) and emulsifiable concentrates (EC). Trade names include Treflan and Ipersan.

The WHO has classified trifluralin as a Class U toxin (unlikely to be hazardous). The acute oral LD_{50} in rats is >5000 mg/kg. The acute dermal LD_{50} in rabbits is >5000 mg/kg. However, it is highly toxic to aquatic creatures but not toxic to birds. The NOEL in mice is 73 mg/kg/day. The ADI is 0.015 mg/kg b.w./day. Trifluralin is, however, suspected to harm the endocrine system, and to be a carcinogen. There is also a risk of **bioaccumulation**. Product labels must carry the signal words 'CAUTION' or 'WARNING' depending on the formulation. However, to avoid exposure to trifluralin, users and handlers should be equipped with coveralls, chemical-resistant gloves, shoes and socks. Inhaling vapours of this herbicide or accidentally swallowing it can cause headache, dizziness, nausea and vomiting. A 12 h REI is recommended for all applications of trifluralin.

Most of trifluralin administered to rats is eliminated via the faeces. In soil, trifluralin breaks down very slowly finally forming carbon dioxide. However, soil persistence is a problem.

Trifluralin is registered in India, Australia, New Zealand and the USA.

Triflusulfuron herbicide

Triflusulfuron, $C_{16}H_{17}F_3N_6O_6S$ (2-[4-dimethylamino-6-(2,2,2-trifluoroethoxy)-1,3,5-triazin-2-ylcarbamoylsulfamoyl]-m-toluic acid), is a triazinylsulfonylurea herbicide. It is soluble in water (1 mg/l). It is known to irritate the skin and the eyes. The CAS Registry Number of triflusulfuron is 135990-29-3.

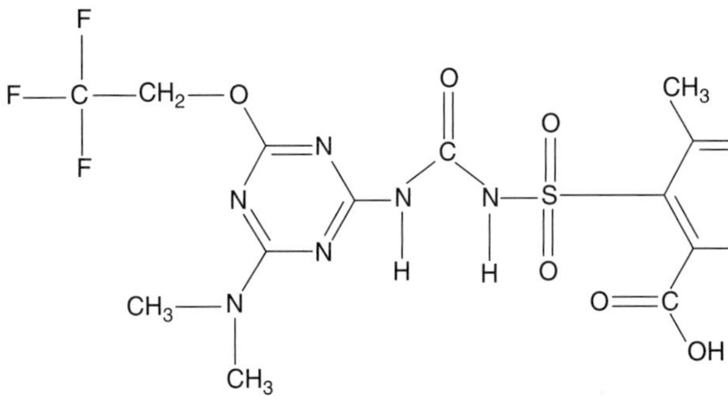

Triflusulfuron

Triflusulfuron commonly exists as a salt or ester and the active ingredient arising from it is triflusulfuron-methyl. Triflusulfuron-methyl, $C_{17}H_{19}F_3N_6O_6S$ (methyl 2-[4-dimethylamino-6-(2,2,2-trifluoroethoxy)-1,3,5-triazin-2-ylcarbamoylsulfamoyl]-m-toluate), is a derivative of triflusulfuron. It is a white solid, soluble in water at 260 mg/l. It is produced by reacting methyl 2-sulfamoylchloride-3-methyl benzoic acid with 2-amino-4-dimethylamino-6-(2,2,2-trifluoro)ethoxy-1,3,5-triazine. Its CAS Registry Number is 126535-15-7.

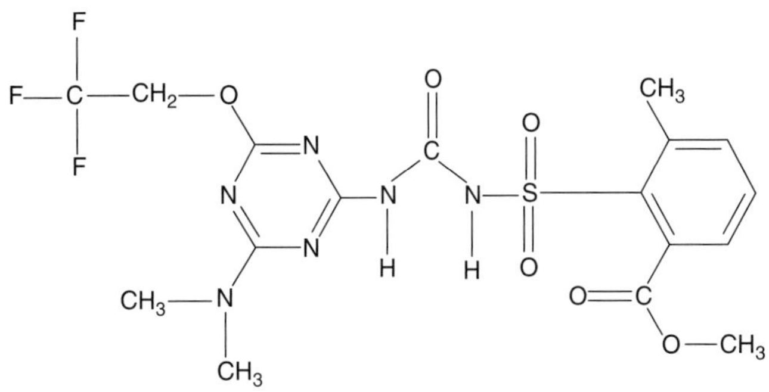

Triflusulfuron-methyl

Triflusulfuron-methyl, a selective post-emergence sulfonylurea herbicide, controls annual and perennial broadleaved weeds and grasses in sugarbeet (*Beta vulgaris*). It acts on the enzyme acetolactate synthase (ALS), critical in branched-chain amino acid biosynthesis. Velvetleaf and cocklebur are two examples of weeds that are controlled with triflusulfuron-methyl.

Triflusulfuron-methyl is formulated as a wettable powder (WP). Split applications help extend the period of weed control. Upbeat and Safari are two known trade names.

The acute oral LD_{50} in rats is 5000 mg/kg and the acute inhalation LC_{50} is 5.1 mg/l. The acute dermal LD_{50} in rabbits is 2000 mg/kg. Triflusulfuron-methyl is moderately to highly toxic to algae, water plants, bees and earthworms, necessitating proper disposal of used/unused chemical. The NOEL in rats is 2.44 mg/kg while the ADI is 0.04 mg/kg/day. Exposure to the chemical can cause respiratory distress.

When triflusulfuron-methyl is administered to small mammals, the majority of the dose is excreted rapidly in faeces and urine. The major breakdown pathway is demethylation of the dimethylamino substituent on the triazine ring. *N*-desmethyl triazine amine and *N,N*-bis-desmethyl triazine amine are two of the important soil metabolites produced, which may be hazardous to the environment. The soil half-life is very short. However, it is stable to aquatic hydrolysis. Triflusulfuron-methyl is suspected to be a carcinogen and water contaminant.

Triflusulfuron-methyl is registered in the USA, Canada and several countries of the EU.

Triflusulfuron-methyl herbicide

See **Triflusulfuron herbicide**

Triforine fungicide

Triforine, $C_{10}H_{14}Cl_6N_4O_2$ (*N,N'*-{piperazine-1,4-diylbis-[(trichloromethyl)methylene]}diformamide;1,1'-piperazine-1,4-diyldi-[*N*-(2,2,2-trichloroethyl)formamide]), is a piperazine compound. It is a whitish-brown crystalline substance that is soluble in water (at 12.5 mg/l) and DMF (330 g/l), but insoluble in benzene, petroleum ether and cyclohexane. It decomposes at 151°C to release toxic, hazardous gases. Triforine decomposes on contact with concentrated sulfuric acid, hydrochloric acid or alkaline substances. Triforine is produced by reaction of piperazine with (1,2,2,2-tetrachloroethyl) formamide in the presence of triethylamine. It has a CAS Registry Number of 26644-46-2.

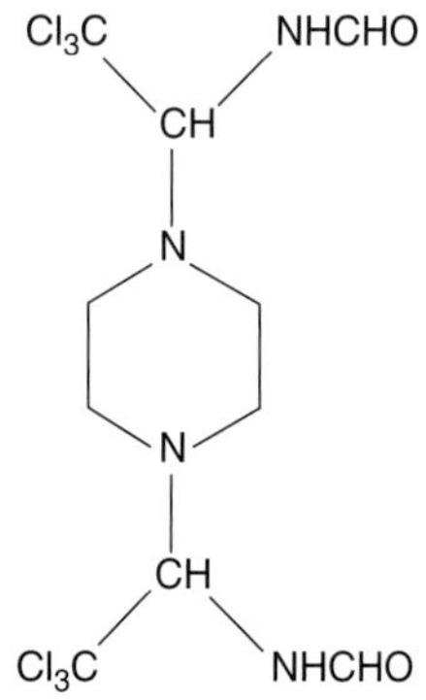

Triforine

Triforine, a piperazine derivative, is a systemic fungicide with protective and curative properties. It also has slight acaricidal activity (against red spider mites). It inhibits **ergosterol** biosynthesis and also other developmental stages of the fungal pathogen. It is used to control **powdery mildews** on cereals, vines, ornamentals and some vegetables, and rusts on cereals, fruits and beans. It also controls blight, black rot, **scab** and storage diseases on cucurbits, apples, hops, mangoes, mushrooms, sugarbeet and tobacco.

The fungicide triforine is formulated as a dispersible concentrate (DC), liquid seed treatment agent and emulsifiable concentrate (EC) and it is commercially available as Saprol, Denarin and Funginex. Triforine can be combined with **carbendazim**, **permethrin**, **mancozeb**, **bupirimate** and some other common insecticides and herbicides in the recommended manner, but it is incompatible with **wetting agents**, spreader-stickers or **adjuvants**. It is applied by hose-end sprayers, low-pressure sprayers and backpack sprayers at an application rate of 515 g/ha for sprays, with repeat application as needed.

The WHO Acute Hazard classification and EPA categorize triforine as unlikely to be hazardous. The acute oral LD_{50} in rats, mice, dogs and bobwhite quails is >16,000 mg/kg, >6000 mg/kg, >2000 mg/kg and >5000 mg/kg, respectively. It is not toxic via the dermal route, in normal use, but causes irritation to the eyes. The ADI is 0.02 mg/kg b.w./day. Because of its low hazard to beneficial insects, triforine may be used in IPM programmes. The fungicide is phytotoxic to certain varieties of pear. However, triforine is observed to cause reproductive deformities and hence is a PAN Bad Actor chemical and it has been suspected of causing malignant tumours in some studies.

In rats, triforine is metabolized and excreted after 48 h, with chloralformamide as one of the metabolites. In animals, triforine is completely absorbed and is excreted mainly via the kidneys. Triforine and its metabolites do not accumulate in the environment and are non-persistent in soil (with a half-life of 3 weeks). The metabolites in soil include piperazine.

Triforine is registered in South Africa, Australia, New Zealand, Canada, the Philippines and Tanzania, but it is banned throughout the EU. The MRL for apples and pears in Australia is 1.0 mg/kg.

Trigger pump sprayers

See **Application equipment – hand sprayers**

Triketones

See **Weed control agents – biological origin**

Trimethacarb insecticide

Trimethacarb, $C_{11}H_{15}NO_2$ (a mixture of 3,4,5- and 2,3,5-trimethylphenyl methylcarbamate in the ratio between 3.5:1 and 5:1), is a phenyl methylcarbamate insecticide. It is a brownish crystalline solid, soluble in water (58 mg/l) and not readily soluble in organic solvents. While it is stable to light, it is decomposed by strong acids and alkalis. Trimethacarb is obtained by the reaction of a mixture of 2,3,5- and 3,4,5-trimethylphenol with methylisocynate. It has a CAS Registry Number of 12407-86-2 (see Trimethacarb at top of the next page).

Trimethacarb is a cholinesterase-inhibiting insecticide with stomach and contact action and a long residual activity. It is also a molluscicide. It is used in the control of corn rootworm larvae (in maize) and a range of mollusc pests. Trimethacarb acts also as a mammal and avian repellent. Common formulations are granules (GR) and wettable powders (WP).

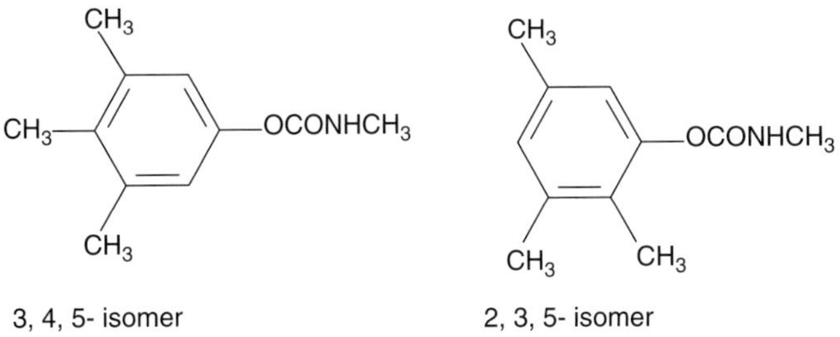

Trimethacarb

The acute oral LD_{50} of trimethacarb in rats is 566 mg/kg, indicating high toxicity. It is moderately toxic to fish (the acute LC_{50} in fish is 40 mg/l). Trimethacarb is a PAN Bad Actor chemical. It is phytotoxic to the seeds of some crops, including maize, sorghum, wheat and rice, unless used as directed. Trimethacarb is toxic to fish and honey bees when sprayed. Trimethacarb also causes rapid (but reversible) over-stimulation of the parasympathetic nervous system due to over-exposure via inhalation, ingestion or skin absorption. The symptoms of poisoning are headache, sweating, blurred vision, giddiness, nausea, vomiting, diarrhoea, salivation, lacrimation, pin-point pupils, muscle twitching, and in severe cases, convulsions and respiratory failure.

The metabolic pathways of trimethacarb in plants, mammals and insects involve mainly oxidation of the *N*-methyl or ring-methyl substituents and conjugation. Some hydroxylation of the phenyl ring and ester hydrolysis occurs. (See also **Molluscicides**.)

Trinexapac as a PGR

Trinexapac, $C_{11}H_{12}O_5$ (ethyl(*RS*)-4-cyclopropyl(hydroxy) methylene-3,5-dioxocyclohexanecarboxylic), is a plant growth regulator. It is a yellowish-brown liquid that is soluble in water, ethanol, acetone, toluene, *n*-octanol and *n*-hexane and is stable under normal environmental conditions. When this substance is used as an ester or a salt, its identity should be stated, for example trinexapac-ethyl. It is generally used as an ester. The trinexapac has a CAS Registry Number of 104273-73-6.

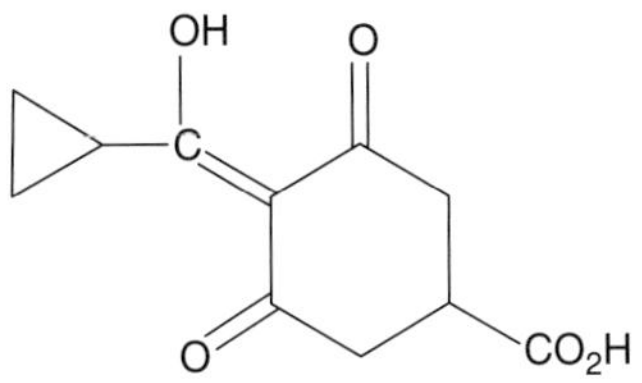

Trinexapac

This PGR reduces stem growth and inhibits internode elongation. It is absorbed by the foliage and translocates to growing shoots where it inhibits a key enzyme in the biosynthesis of gibberellic acid. Trinexapac prevents lodging in cereals.

Trinexapac is formulated as an emulsifiable concentrate (EC). Although less stable in alkaline medium, it is compatible with other pesticides. It is available in the market under the trade names of Moddus and Primo.

The WHO classifies it as a Class III toxin (slightly hazardous). It is non-toxic to earthworms.

In animals and plants, excretion of trinexapac occurs within 24 h as an acid metabolite. In soil and water, the ester undergoes rapid degradation. Further metabolism of trinexapac is rapid.

Trinexapac is registered in Australia, New Zealand, South Africa, the USA and parts of the EU.

Triticonazole fungicide

Triticonazole, $C_{17}H_{20}ClN_3O$ ((*RS*)-(*E*)-5-(4-chlorobenzylidene)-2,2-dimethyl-1-(1*H*-1,2,4-triazol-1-ylmethyl)cyclopentanol), is a triazole compound. It is an odourless, white powder that is soluble in water (9.3 mg/l). It is slightly decomposed at 180°C. It has a CAS Registry Number of 131983-72-7.

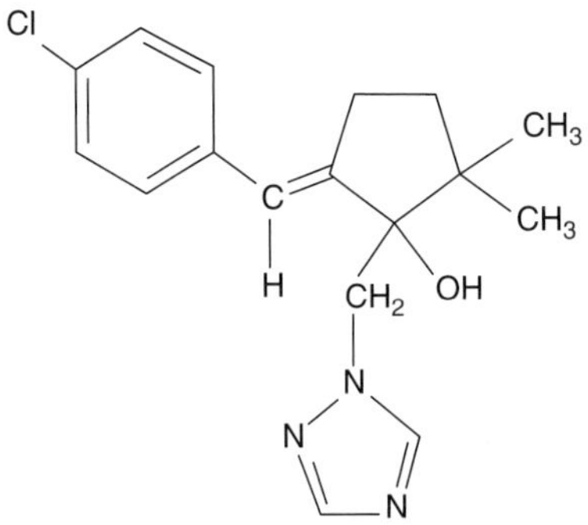

Triticonazole

Triticonazole, a specific triazole fungicide, is best known for controlling seed-borne diseases of maize and barley (caused by *Rhynchosporium secalis*). It also acts as a seed disinfectant and protects against foliar diseases (controllable by seed treatment) such as rusts, leaf blotch (*Septoria tritici*), **powdery mildew**, leaf spot and head smut of maize (*Sphacelotheca reiliana*). The compound inhibits sterol demethylation and has preventive action.

The fungicide triticonazole is commonly formulated as flowable concentrate for seed treatment (FS) and a popular trade name for this compound is Real.

Triticonazole is classified by the EPA as a slightly toxic compound, while it is categorized as unlikely to be hazardous by the WHO (Class U). It is not a skin or eye irritant. The acute percutaneous LD_{50} and oral LD_{50} in rats is >2000 mg/kg, suggesting moderate toxicity. The NOEL in rats is 29.4 mg/kg b.w./day and 38.3 mg/kg b.w./day in male and female rats, respectively. The ADI is 0.025 mg/kg b.w./day. It is moderately toxic to birds, fish and to honey bees. The acute LD_{50} in bobwhite quails is >2000 mg/kg, while the acute 96 h LC_{50} in rainbow trout is >3.6 mg/l. The acute 48 h contact LD_{50} in honey bees is >100 μg per bee. The 4 h inhalation LC_{50} in rats is >5.61 mg/l.

In rats, almost the entire triticonazole is eliminated through faeces within 1 week. In plants, a dihydroxy metabolite is formed. In soil, metabolism starts with hydroxylation.

Triticonazole is registered by the USA, Australia, Canada, South Africa and many EU members such as the UK, Belgium, France, Italy and Germany. According to the EU Pesticide Database, the MRL of triticonazole for cereals is 0.01 mg/kg.

True epiphytes

See **Epiphytes**

Trunk injections

See **Fungicide application methods**

U

Ultra-low volume (ULV) liquid formulations

Formulations made out of active ingredients (a.i.) that are highly soluble in the carrier solvent and dispersed in very small droplets over a large area are referred to as ultra-low volume (ULV) liquid formulations. The international code for this formulation is UL. The solvent is non-phytotoxic and the ULV product is ready to use.

ULVs carry highly concentrated solutions of the a.i. and hence, it is important to reach the target site accurately. Thus, equipment such as rotary nozzles and **aerosols** must be used for these formulations. Weather conditions should be considered before application. Most of these formulations are made up of surface-active agents and drift control substances.

These special-purpose formulations are used mostly in outdoor and fogging applications, such as in agriculture, forestry and mosquito control programmes. Pesticides applied with a ULV formulation can effectively penetrate the cuticle and foliage, although there can be risks to the applicator while handling. The equipment needed to apply ULV liquid formulations is costly.

Ultra-low volume (ULV) sprayers

See **Application equipment – sprayers**

Uncoupling as a mode of action

Uncouplers are chemical molecules that prevent the flow of electrons in the electron transport chain during photosynthesis. Uncoupling energy (ATP) production and respiration are two fundamental mechanisms of many pesticides. Uncouplers, which are weak organic acids and phenols, transport H^+ ions across the membrane so that energy is wasted as heat and not used to produce ATP.

The functioning of chloroplasts, mitochondrial membranes and bacterial cells are all inhibited by the action of uncouplers. Pesticides with this mode of action are dinitrophenols, **pentachlorophenol** and ioxynil.

Uniconazole as a PGR

Uniconazole, $C_{15}H_{18}ClN_3O$ ((*E*)-(*RS*)-1-(4-chlorophenyl)-4, 4-dimethyl-2-(1*H*-1, 2, 4-trizol-1-yl)-pent-1-en-3-ol), belonging to the conazole group, is a plant growth regulator, with fungicidal activity. It is a white-coloured, crystalline solid, soluble in water, acetone, ethyl acetate, methanol and xylene. It is stable under normal storage conditions in original containers. The CAS Registry Number of the racemate is 83657-22-1, while that of the (*S*) isomer (called uniconazole-P) is 83657-17-4 (see Uniconazole at top of the next column).

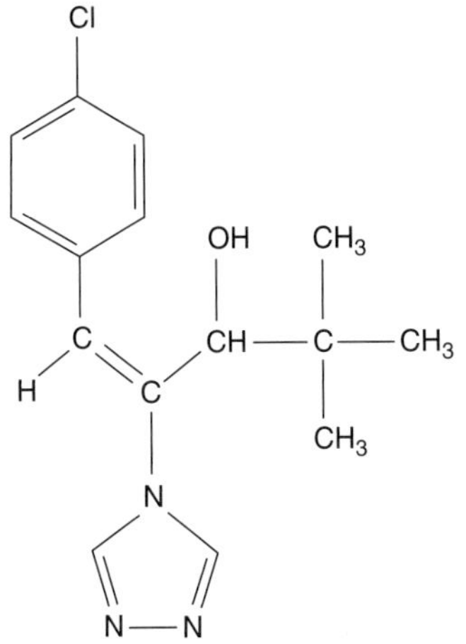

Uniconazole

Uniconazole, absorbed by the stems and roots and translocated to growing points, inhibits gibberellin synthesis and sterol biosynthesis. It is used to reduce vegetative growth and thereby reduces lodging in rice.

Uniconazole is formulated as a wettable powder (WP), granules (GR) and soluble concentrate (SL). It is available in the market under trade names such as Lomica and Sumiseven.

The PGR is non-phytotoxic and non-mutagenic when used as directed. The ADI is 0.02 mg/kg/day. The acute oral LD_{50} in rats ranges from 1790 to 2020 mg/kg, while the dermal LD_{50} is >2000 mg/kg. The short-term dietary NOEL in rats is 2.25 mg/kg. It is moderately toxic to birds (with the acute LD_{50} in bobwhite quails being 1155 mg/kg) and toxic to *Daphnia magna* and honey bees (the acute 480 h contact LD_{50} of uniconazole is >20 µg per bee). The acute 14 day LC_{50} in earthworms is 1000 mg/kg. Chronic exposure may be carcinogenic. It can injure the eyes and the skin. Protective clothing such as a long-sleeved shirt and long trousers, waterproof gloves, shoes plus socks should be used when handling and spraying this chemical.

According to the EPA and WHO uniconazole is a slightly hazardous chemical (Class III). Uniconazole-P is registered for use in Australia, Canada, South Africa, Vietnam and the USA. It is not approved in the EU.

Uracil herbicides

Uracil is a planar, unsaturated, naturally occurring pyrimidine derivative. This light-absorbing compound was first isolated by hydrolysis of yeast nuclein in wheat germ and cattle spleen. It forms the basis of herbicidal compounds that inhibit the photosystem II electron transport chain.

Uracil herbicides include **bromacil**, flupropacil, **lenacil** and **terbacil**. Most members control grasses and broadleaved weeds in pineapple, citrus and non-citrus fruit, soybean, cotton and sugarbeet.

Uracil compounds can injure plant roots as they leach into the soil. Exposure may cause irritation to the eyes, skin and the respiratory tract. At higher doses, they act like soil sterilants.

Urea sulfate herbicide

Urea sulfate ($CH_6N_2O_5S$) is a yellow-orange coloured liquid that is soluble in water and decomposes at temperatures above 80°C. The CAS Registry Number is 21351-39-3.

$NH_2CONH_2{\cdot}H_2SO_4$

Urea sulfate

Urea sulfate is used as a **desiccant** and **defoliant** and functions as a cell membrane disruptor. It is available commercially under trade names such as Cotton Quik.

The herbicide is non-toxic to animals and birds. It is a severe skin and eye irritant. The acute oral LD_{50} in male and female rats is 1200 mg/kg and 350 mg/kg, respectively. The acute dermal LD_{50} in rats is >2 g/kg, while the inhalation LC_{50} is >10.8 mg/l. The LC_{50} in mallard ducks is >5620 ppm.

V

Validamycin biofungicide

Validamycin, $C_{20}H_{35}NO_{13}$ (1*L*-(1,3,4/2,6)-2,3-dihydroxy-6-hydroxymethyl-4-[(1S,4R,5S,6S)-4,5,6-trihydroxy-3-hydroxymethylcyclohex-2-enylamino]cyclohexyl β-D-glucopyranoside), is a glucopyranosyl antibiotic fungicide. Also called validamycin A, it exists as an odourless, colourless and hygroscopic powder. It is soluble in water (>610 × 10^3 mg/l), methanol (62.3 g/l) and dimethylformamide. It is slightly unstable in acidic media. The biofungicide is produced by fermenting *Streptomyces hygroscopicus* var. *limonius*. This organism is a Gram-positive soil-borne bacterium that belongs to the Actinomycetes group. The CAS Registry Number is 37248-47-8.

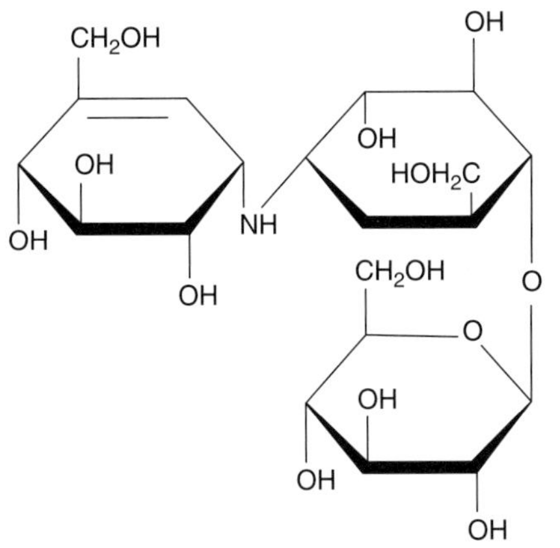

Validamycin

Validamycin is a non-systemic antibiotic with fungistatic properties. It inhibits the enzyme trehalase, which disturbs the digestion of the carbohydrate trehalose and transport of glucose to the hyphae. This inhibition leads to abnormal hyphal branching. This fungicide controls soil-borne diseases caused by *Rhizoctonia solani* and *Fusarium oxysporum*. Rice, potato, cotton, sugarbeet and tomatoes are protected against sheath and leaf blight, damping-off and root rot.

The common formulations of validamycin are liquid, driftless formulations and soluble concentrate (SL). The compound is applied as a soil drench or as seed dressing (with common doses being 1 g/kg for seed dressing). It can be combined with **etofenprox, kasugamycin** and a few other fungicides. Some popular trade names of validamycin are Validacin and Solacol.

The WHO classifies validamycin as a Class U toxin (unlikely to be hazardous), while the EPA classifies it as a Class IV toxin (practically non-toxic). The acute oral LD_{50} in rats and mice is >20,000 mg/kg. The NOEL (2 years) during feeding trials in rats is 40.4 mg/kg b.w./day. The acute percutaneous LD_{50} in rats is >5000 mg/kg, and so it is practically safe to touch. It is non-irritating to the skin and eyes. The LC_{50} (4 h) in rats is >5 mg/l air. The LC_{50} (72 h) in carp is >40 mg/l.

In animals (rats) and plants, cleavage to glucose and validoxylamine A occurs after oral administration of validamycin. It undergoes rapid microbial degradation in soil (with a soil half-life of <5 h). Validoxylamine A (CAS Registry Number 82309-75-9 and a molecular formula $C_{14}H_{25}NO_8$) also possesses pesticidal properties, especially against soil-borne *Fusarium* spp.

The countries in which validamycin is used are Cambodia, India and Vietnam. However, it is not used in the USA.

Valifenalate fungicide

Valifenalate, $C_{19}H_{27}ClN_2O_5$ (methyl *N*-(isopropoxycarbonyl)-*L*-valyl-(3RS)-3-(4-chlorophenyl)-β-alaninate), is a valinamide carbamate fungicide (acylamino acid class of fungicide or carboxylic acid amides). It is an odourless, fine, white powder that is soluble in water (at 2.41 × 10^{-2} g/l, pH 4.9 to 5.9), acetone (29.3 g/l), ethyl acetate (25.4 g/l) and methanol (28.8 g/l). The compound is stable in aqueous solutions (pH 4) but degrades rapidly in air. The CAS Registry Number of valifenalate is 283159-90-0.

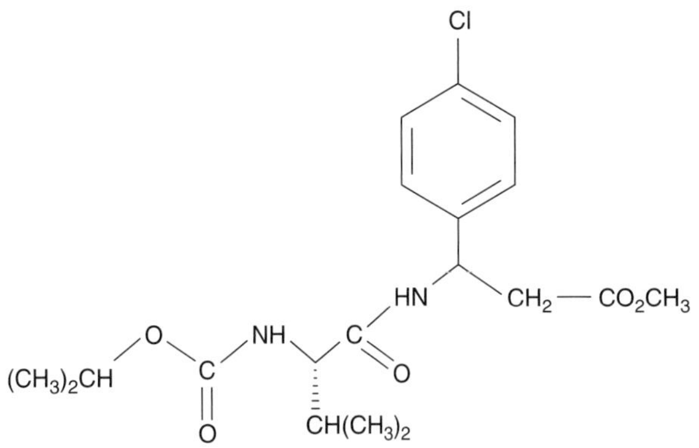

Valifenalate

Valifenalate, once absorbed and translocated throughout the plant, gives a long-term protective and curative effect. It acts on cell wall synthesis and thus affects all growth stages of the pathogen, reducing spore germination and mycelial growth. The fungicide is used to control **late blights** in tomatoes and potatoes (the main agent being *Phytophthora infestans*), **downy mildews** in vines (*Plasmopara viticola*), onions (*Peronospora destructor*) and in tobacco. Thus, the fungicide controls many pathogens belonging to the class Oomycetes, particularly members of the Peronosporaceae. It is used in combination with other fungicides and is applied at 120–150 g/ha.

The fungicide is formulated as water-dispersible granules (WG). It is available in the market under trade names such as Java and Yaba.

The acute oral LD_{50} in rats is >5000 mg/kg, while its acute percutaneous LD_{50} is >2000 mg/kg. The NOEL for male and female rats is 150 and 1000 mg/kg b.w./day, respectively. The ADI is 0.168 mg/kg b.w./day. It is non-irritating to the eyes and skin. The LD_{50} (14 days) in bobwhite quails and mallard ducks is >2250 mg/kg. The LC_{50} (5 days) is 2649 mg/kg b.w. in mallard ducks, indicating moderate toxicity. It is toxic to fish and to bees.

In animals (rats) valifenalate is extensively metabolized and completely excreted via the faeces within 3 days. In plants, it is slowly metabolized.

According to the European Food Safety Authority (EFSA, 2009), the MRL for aubergines and tomatoes could be as high as 0.1 mg/kg (as against an earlier set limit of 0.01 mg/kg). Valifenalate is not registered by the EPA in the USA, while in the EU it is only registered in Hungary.

Vamidothion insecticide

Vamidothion, $C_8H_{18}NO_4PS_2$ (*O,O*-dimethyl *S*-2-(1-methylcarbamoylethylthio) phosphorothioates), is an organophosphorus insecticide. It is a colourless needle-like substance, readily soluble in water (4 kg/l), benzene, toluene, methyl ethyl, ethyl acetate, acetonitrile and dichloromethane, where it remains stable. It is decomposed in strong acidic or alkaline media. It is produced by the reaction of *O,O*-dimethylphosphoric acid chloride with the sodium salt of *N*-methylcarbamylethylmercaptoethanethiol. Its CAS Registry Number is 2275-23-2.

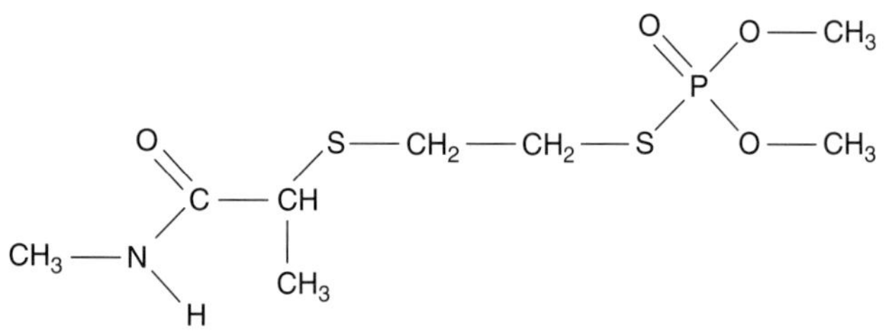

Vamidothion

Vamidothion, a cholinesterase inhibitor, is a systemic insecticide and acaricide which persists in the environment. It is effective on sucking pests like aphids (e.g. apple woolly aphid), leaf/plant hoppers and cicadas that attack cotton, sugarcane, grapes and rice.

The insecticide is formulated commonly as emulsifiable concentrates (EC) and as a paste (PA). Commonly accepted application rates are 37–50 g/ha, which may be repeated in case of severe infestations. Orchards generally call for a single treatment of 50 g/100 l solution. A pre-harvest interval, usually up to 4 weeks, is recommended for most crops. A known trade name is Kilaval.

In plants, sulfoxides are produced (the sulfide group is oxidized into the sulfoxide), while des-*O*-methyl vamidothion is also formed. Vamidothion sulfoxide is seen to persist in plants for over 20 days. However, this metabolite is less toxic than vamidothion itself. In mammals and insects, demethylation and hydrolysis occur, resulting in degradation products such as phosphoric acid, dimethyl phosphates, monomethyl phosphates and other water-soluble metabolites. The soil (aerobic) half-life of vamidothion is between 1 and 1.5 days.

The WHO has categorized vamidothion as a Class Ia toxin (highly hazardous). It is toxic to aquatic organisms and honey bees. It is a PAN Bad Actor chemical because of its high toxicity and its inhibition of cholinesterase.

Australia and Cambodia have registered the use of vamidothion. The USA and the EU region have not registered its use. India has refused registration.

Vapam fungicide

See **Ethylenebisdithiocarbamate (EBDC) fungicides**; **Metam salts**

Vectors of plant viruses

Organisms that transmit a pathogen are called vectors. Most plant viruses such as the tomato spotted wilt virus or the many mosaic viruses need a vector to enter plant cells. **Leafhoppers**, nematodes and protozoa can act as vectors of plant viruses. That is, they help in carrying a particular plant virus from one host to another. The vector of the tomato spotted wilt virus is an insect, while that of the tobacco ring spot virus is a nematode.

Out of all the vectors, insect vectors are the most important means by which plant viruses are spread. Insects in the order Homoptera (e.g. leafhoppers, plant hoppers, whiteflies and aphids) are among the most common vectors of plant viruses. Initially, most plant viruses multiply at the site of infection, giving rise to localized symptoms such as necrotic spots on the leaves or wilting of the plant part. Thereupon, the virus spreads systemically through the phloem or cytoplasmic bridges called plasmodesmata, the channels used by plant cells to transfer fluids. The virus may also modify the plasmodesmata in order to facilitate its own movement.

Sometimes viruses can also be transmitted from the sap of an infected plant to a healthy plant by contact with a wounded part. For example infected tools can carry the inoculum to a healthy plant. Viruses can also be transmitted to the next generation through seeds or pollen. The seed-borne mosaic virus is transmitted through seeds or pollen grains.

Vernolate herbicide

Vernolate, $C_{10}H_{21}NOS$ (*S*-propyl dipropyl (thiocarbamate)), is a thiocarbamate herbicide. It is a clear liquid with a faint odour. It is soluble in water at 90 mg/l at 20°C, and is miscible with common solvents such as xylene, methylisobutyl ketone and kerosene. It is stable up to 200°C and must be stored in a cool, dry place. Vernolate is made by reacting *N*-dipropylcarbamoyl chloride with propylmercaptan. Its CAS Registry Number is 1929-77-7.

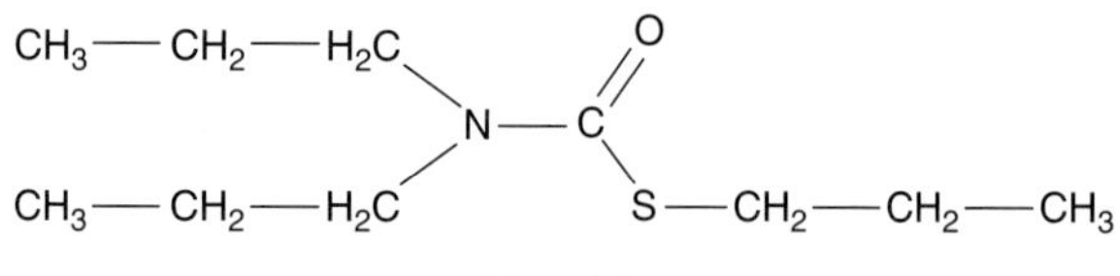

Vernolate

The selective, pre-plant or post-emergence herbicide is a lipid metabolism inhibitor, resulting in retarded growth. When incorporated in the soil, it halts the germination of broadleaved weeds and grasses in groundnuts, maize, soybeans, tobacco and sweet potatoes.

Vernolate is formulated as emulsifiable concentrates (EC) and granules (GR). It is available commercially as Reward and Surpass.

Vernolate is classified by the WHO as moderately hazardous (Class II). The acute oral LD_{50} in rats is between 1200 and 1900 mg/kg, while the acute dermal LD_{50} in rabbits is >1955 mg/kg. The compound is slightly to moderately toxic to fish and only slightly toxic to birds. It is considered to be a general use pesticide (GUP) according to the EPA, with formulations displaying the signal word 'CAUTION' on their labels. Care should be taken to avoid spills and contact with food and drink. Handlers should be equipped with chemical safety glasses or a face shield. Due to its toxicity and its potential to contaminate groundwater, vernolate is classified as a PAN Bad Actor chemical.

In moist soil, vernolate is microbially decomposed to mercaptan, amine, carbon dioxide and isopropanol, while in dry soils it is adsorbed. In plants, degradation involves the production of carbon dioxide. It is not registered for use in the USA or the EU, while it is registered for use in Australia.

Vinclozolin fungicide

Vinclozolin, $C_{12}H_9Cl_2NO_3$ ((*RS*)-3-(3, 5-dichlorophenyl)-5-methyl-5-vinyl-1, 3-oxazolidine-2, 4-dione), is a dicarboximide fungicidal compound. It is a colourless, crystalline substance with a slight odour. It is soluble in water (2.6 mg/l), methanol, acetone, toluene and dichloromethane. It is stable up to 50°C for 24 h in acidic media. It hydrolyses in dilute sodium hydroxide (the DT_{50} is <3.8 h). The fungicide is produced by the reaction of 3, 5-dichlorophenyl isocyanate with an alkyl ester of 2-hydroxy-2-vinylpropionic acid. The CAS Registry Number is 50471-44-8.

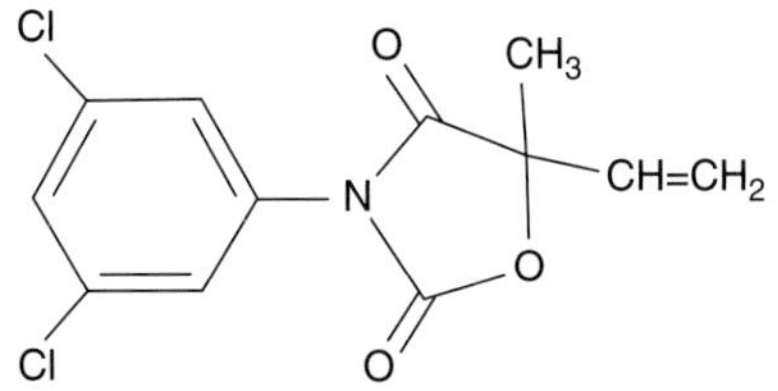

Vinclozolin

Vinclozolin is a **non-systemic**, protective fungicide. Its mode of action is signal transduction, which prevents spore germination. The use of vinclozolin helps in the control of diseases caused by *Botrytis cinerea*, *Fusarium* spp., *Sclerotinia sclerotiorum* and *Monilinia fructicola* in grapes and other vines, strawberries, vegetables and ornamental plants.

The compound is formulated as dry flowables, water-dispersible granules (WG) and wettable powder (WP). It is commercially available as Ronilan and Verlan. It may be combined with fungicides like **thiram** and **carbendazim**.

The WHO classifies vinclozolin as a Class U toxin (unlikely to be hazardous). Although the compound is low in acute toxicity, it is a developmental toxin and a carcinogen. It is also feared to contaminate underground water sources. Some of the breakdown products are shown to aid formation of tumours. The acute oral LD_{50} in rats and mice is >15,000 mg/kg. The acute percutaneous LD_{50} in rats is >5000 mg/kg, while the same in quails is >2510 mg/kg. It is not toxic to bees and earthworms. The ADI for vinclozolin is 0.01 mg/kg b.w./day.

In rats, vinclozolin is eliminated through urine and faeces, with the main metabolite being *N*-(3, 5-dichlorophenyl)-2-methyl-2, 3, 4-trihydroxybutanamide. In plants, the primary metabolites formed are (1-carboxy-1-methyl) allyl 3, 5-dichlorophenylcarbamate and *N*-(3, 5-dichlorophenyl)-2-hydroxy-2-methyl-3-butenamide. The metabolites exist as conjugates. In soil, the half-life is from 3 days to 3 weeks, with the residue binding to the soil.

Vinclozolin is registered for use in South Africa, Canada and the USA, but not in the EU.

Virus

Virus (meaning 'poison' in Latin) is an ultramicroscopic disease-causing agent which can cause infections in animals, plants and bacteria. The invention of the electron microscope has enabled the study of viruses. The study of viruses is known as virology.

Viruses are obligate, acellular parasites that are often host specific. They have rod-shaped or isometric-, helical-, spherical- or polyhedral-shaped bodies that contain nucleic acid genetic material (RNA or DNA), which are the information-carrying molecules, and a protein coat protecting the genetic material. Viruses can pass through bacterial filters and can exist in crystallized forms as well as in the form of infectious nucleic acid. Some species of viruses are enclosed by a membrane-like envelope around them. Such enveloped viruses are generally found in animals, while naked viruses are often seen in plants.

The enzymes DNA polymerase and RNA polymerase, which are responsible for the replication function, are present in all viruses. Some viruses may have small amounts of carbohydrates, a lipid membrane or an RNA-dependent DNA polymerase. However, the virus does not have the synthesis mechanism and so depends on the host organism for replication. Replication takes place via a six-step process, namely: (i) adsorption or attachment; (ii) penetration; (iii) un-coating; (iv) biosynthesis; (v) maturation; and (vi) release. Although contact between the virus and host may happen accidentally, the process of attachment is dependent on the presence of specific receptor sites on the host cell, to which the virus particle (virion) can attach. The virus protein coat undergoes several changes by which the host surface fuses with the virion. Then the virus penetrates the host and infects it by inserting its protein into the host cell. This is followed by the dissolution or dissociation of the viral capsid, releasing the nucleic acid into the host cell. Multiplication of the genome follows, by synthesis of the viral messenger RNA (mRNA), capsid proteins and enzymes. The daughter virions undergo a process of assembly where they acquire the shape and form of the original virus. Once assembled, the virus is ready to be released from the host cell, a process which usually kills the cell.

Viruses are classified based on some prominent similarities, for example similar replication process, structure, host range, size and shape, and mode of transmission. For instance, the Baltimore system classifies viruses on the mechanism of mRNA production (which recognizes seven groups). According to the International Committee on Taxonomy of Viruses (ICTV), viruses are classified as order (-virales), followed by family (-viridae), subfamily (-virinae), genus (-virus) and then species (-virus). There are five orders, 82 families, 307 genera and 2083 species, with many viruses as yet being unclassified.

The study of viruses was first demonstrated by Louis Pasteur in 1884, while other luminaries in this area include Adolf Mayer, Martinus Beijerinck and Frederick Twort. Tobacco mosaic virus (TMV) is one of the most studied viruses.

Most plant viruses belong to the positive-sense single-stranded RNA (ssRNA+) genome structures. Virus families such as the Bromoviridae and Sequiviridae are important for their pestilence on crops, since they infect plant families such as the Solanaceae (potatoes, tomatoes, aubergine and tobacco), Cucurbitaceae (cucumbers, melons and gourds) and Cruciferae (cabbage, turnip and radish). Viruses that have negative-sense single-stranded RNAs also include some plant pathogens; for example, the family **Bunyaviridae** has the prominent genus *Tospovirus* (which includes tomato spotted wilt virus).

Viruses are normally transmitted by **vectors**. For instance, most of the virus species belonging to the ssRNA+ genome mentioned above are transmitted by arthropods and insects in the orders Coleoptera, **Thysanoptera**, Hemiptera and family Aphididae. Some types of cucumber beetles, white-fringed beetles, sugarcane beetles, aphids and whiteflies are some examples of vectors.

Viruses infect bacteria (bacteriophages), fungi (mycovirus), insects and almost all living organisms. Some commonly known viral diseases that affect plants are tomato spotted wilt virus, tobacco mosaic virus, rose mosaic virus, tobacco ring spot virus and beet yellow vein virus.

Viral diseases may be difficult to identify. The symptoms can vary from one plant to the next and also with age and growing conditions. The general types of symptoms are: (i) mosaic; (ii) ring spot; (iii) russeting; and (iv) leaf curling. To control the spread of a viral infection, the unhealthy plants are pulled out and destroyed.

Volume mean diameter (VMD)

See **Spray application concepts**

Vomitoxin

Vomitoxin, $C_{15}H_{20}O_6$ ((3α,7α)-3,7,15-trihydroxy-12,13-epoxytrichothec-9-en-8-one), a member of a group of **mycotoxins** called **trichothecenes**, is a toxin produced by many species in the genus *Fusarium* under certain moist conditions. Other toxin compounds produced by various *Fusarium* spp. include nivalenol, diacetoxyscirpenol and T-2 toxin. Such fungal toxin production could be a strategy for self-preservation.

Vomitoxin, also called deoxynivalenol (DON), is a protein synthesis inhibitor. It is frequently seen in *Fusarium* spp. (e.g. *Fusarium graminearum* and *Fusarium culmorum*) that infect cereal crops such as wheat, barley, maize and oats and are responsible for head blight of wheat and ear rot of maize. Vomitoxin levels produced as a result of these diseases are particularly high in conditions of high humidity.

The toxin present in the grains of such crops enters the food chain by direct human consumption or via fodder. It results in anorexic conditions in cattle and other animals that consume the fodder or grain. Ulceration in the stomach is also observed in animals. It seriously affects pigs and poultry too. Vomitoxin is less toxic than the other trichothecene mycotoxins. It is not very toxic to humans. An MRL of 1 ppm in food has been fixed by the US FDA, while it is up to 4 ppm in animals.

W

Warfarin rodenticide

Warfarin, $C_{19}H_{16}O_4$ ((*RS*)-4-hydroxy-3-(3-oxo-1-phenylbutyl) coumarin), also known as coumaphane, is a chronic rodenticide. It is an odourless and tasteless substance, which is administered in very low dosages. It is produced by reacting 1-phenyl-3-oxo-butene-1 with 4-hydroxycoumarin. Warfarin is a synthetic derivative of a **coumarin** anticoagulant, used both as a rodenticide and in medical practice (for prevention of thrombosis and embolism). Warfarin is practically insoluble in water, soluble to very slightly soluble in acetone, benzene, ethanol, ether, toluene and xylene, and moderately soluble in methanol and ethanol.

It inhibits the enzyme vitamin K epoxide reductase whose function is to recycle vitamin K after its involvement in the carboxylation of blood coagulation proteins. General internal bleeding is induced by the reduction of the **prothrombin** content of the blood. It is an anticoagulant rodenticide and produces toxic symptoms on repeated consumption. Warfarin is highly effective on Norway (brown) rats (*Rattus norvegicus*) and house mice (*Mus domesticus*). The compound does not induce bait shyness.

Warfarin is commonly formulated as a water-soluble liquid concentrate, powder, bait concentrate and ready-to-use bait (RB). The commercial formulations known include Grey Squirrel Bait. It is compatible with other rodenticides like **pindone**, sulfaquinoxaline and calciferol. It may also be mixed with talc and used as a tracking powder, which accumulates on the animal's skin and fur, and is licked during grooming. The compound may be mixed with pindone, calciferol and sulfaquinoxaline.

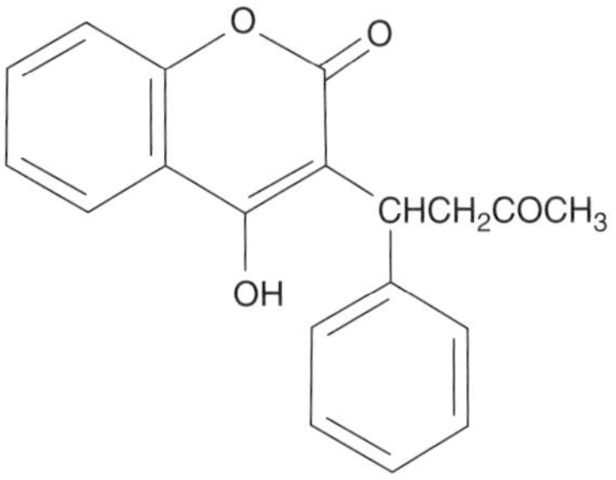

Warfarin

The acute oral LD_{50} of warfarin in rodent species ranges from 186 mg/kg to 374 mg/kg. The LD_{100} in rats is 0.2 mg/kg/day for a consecutive period of 5 days. The oral LD_{50} for cats is 2.5–20 mg/kg. The dermal LD_{50} is 1400 mg/kg. A 96 h LC_{50} in rainbow trout using a 0.54% formulation of warfarin sodium salt is >10,000 ppm. This formulation is considered non-toxic to rainbow trout. Many rodent species have developed resistance to warfarin. It is classified as a Class Ib toxin (highly hazardous) by the WHO.

The use of warfarin itself as a rat poison is now declining because of the development of resistance and the availability of more effective alternatives. Examples are 4-hydroxycoumarin compounds such as **coumatetralyl** and **brodifacoum**, referred to as 'super-warfarin'. Unlike warfarin, which is readily excreted, newer anticoagulant poisons also accumulate in the liver and kidneys after ingestion.

Water-dispersible granules (WG)

See **Granules**

Water-soluble granules (SG)

See **Granules**

Water-soluble powder formulation

The FAO describes a soluble powder as 'a powder to be applied as a true solution of the active ingredient after dissolution in water, but may contain insoluble inert ingredients'.

Dissoluble dry formulations include water-soluble formulations where the active ingredient (a.i.) dissolves in water and produces a reasonably concentrated solution. Here, particle size is important only to safeguard against any errant particle or impurity clogging the spray equipment, because most particles will dissolve anyway.

Wax bars

Wax bars are a pesticide formulation which have specialized applications. Herbicidal agents are sometimes applied to turf grass and vineyards using these bars. Wax bars do not drift and do not need calibration, but require application to specific target locations.

Weed

Weeds are plants whose nuisance value outweighs their utility. A report of the Terminology Committee of the Weed Society of America defines a weed as a plant growing where it is not desired. Some of the main characteristics of weeds include: (i) their competitive nature; (ii) their persistence; (iii) their high reproductive capacity; (iv) their rapid growth; (v) the harm they cause to man, animals and crops; and (vi) commonly, their having no use and hence being unwanted. The classification of weeds follows the lines of botanical classification and thus weeds can be classified on the basis of habitat, life form and their being either monocotyledonous or dicotyledonous. Examples of some common weeds are cogongrass, **goosegrass**, **Bermuda grass**, **purple nutsedge** and large crabgrass. Although some weeds are used as animal feed, and may help control erosion, fix nitrogen and provide shelter for wildlife, the costs of having unwanted plants are enormous.

Weeds interfere with the yield and quality of crops. For example in India, crops such as sugarcane, rice and cotton are often heavily infested with weeds. If the weeds are aquatic in nature, they can clog waterways and canals. Some weeds cause allergic reactions such as hay fever and respiratory problems. Some species act as anti-feedants and can cause harm to animals.

Terrestrial weeds include grasses, sedges or broadleaved plants. Grass seedlings have one leaf as they emerge from the seed. The leaves are narrow and upright with parallel veins. Most grasses have fibrous roots. The growing point on seedling grasses is sheathed, and located below the soil surface during the early part of their life cycle. Some grasses are annuals and others are perennials. **Sedges** are similar to grasses and have triangular stems and three rows of leaves. Generally, sedges are found in wet places, but the principal pest species are found in fertile, well-drained soils. **Yellow nutsedge** is a perennial weed species that produces rhizomes and tubers. Broadleaved seedlings have two broad leaves with netlike veins as they emerge from seeds. Broadleaved plants have a taproot and a coarse root system. Their growing points, at the end of each stem and roots, are exposed. Broadleaved plants may be **annuals**, **biennials** and **perennials**. There are many types of perennials such as bulbous perennials and creeping perennials. Bulbous perennials propagate by seeds and bulbs. Creeping perennials such as **Johnson grass**, Canada thistle, Russian knapweed, yellow nutsedge and leafy spurge spread via seeds, rhizomes or stolons.

In addition to the above groups of weeds, there are spore-bearing plants (e.g. field horsetail and bracken fern), conifers (e.g. *Juniperus* sp.) and monocots (e.g. wild garlic and *Iris* sp.).

Aquatic plants, like land plants, have stems, leaves, flowers and roots. Most of the aquatic plants are perennials. They are classified as emergent, floating and submerged plants. Cattails, bulrushes and arrowheads are emergents, which extend above the water surface. Floating plants float on the surface. Water lilies, duckweeds, water lettuce and water hyacinth are examples of this class of plants. The plants growing below the water surface are called submerged plants, and water milfoil, pond weeds and coontails are some examples of this class.

Weed plants typically prosper due to the large number of seeds (in the hundreds) produced by each plant. Most of the seeds can survive unfavourable weather conditions, and delay germination until moisture becomes available. Many weeds form seeds in response to day-length signals. Weeds germinate at different times, to avert any major seasonal catastrophe like floods or frost. Sometimes large weeds form viable seeds even if they are clipped at the bloom stage. Weed seeds are commonly transported with crop seeds. Weed seeds also travel large distances by sticking to clothes, tyres, harvest equipment and harvest sacks. Some are eaten by birds and travel with them, without being damaged by the digestive system. Winds also carry thousands of weed seeds (e.g. Russian thistle) across long distances, while water is another carrier (e.g. curly docks). Compounds, both chemical and biological, that are intended to inhibit or kill weeds are called herbicides. These were formerly called weedicides, but since these agents can equally harm valuable crops, trees and other favoured vegetation, they have been termed **herbicides**. Table W.1 lists some of the major agricultural weeds of the world and their ranking with regard to their effect on crop productivity and difficulty to control.

Weed – barnyard grass

Barnyard grass (*Echinochloa crus-galli*) is an annual grass that is native to the tropics, but is widely seen in Asia, Africa, Europe and now also in North America. Known in some countries as cockspur, this plant grows up to 2 m in height with erect stems, long leaves and a purplish or greenish inflorescence (Fig. W.1).

Table W.1. Major agricultural weeds of the world. (From Weed Science Society of America (WSSA) (1989); extracts of this list are reproduced with permission from James V. Anderson, Director of Publications, WSSA.)

Class (in *Angiosperma*)	Family	Species	Common name	Rank[a]
Monocotyledonae	Cyperaceae	*Cyperus esculentus* L.	Yellow nutsedge	16
		Cyperus rotundus L.	Purple nutsedge	1
		Avena fatua L.	Wild oat	13
		Cynodon dactylon (L.) Pers	Common Bermuda grass	2
		Digitaria sanguinalis (L.) Scop.	Large crabgrass	11
		Echinochloa colona (L.) Link	Junglerice	4
		Echinochloa crus-galli (L.) Beauv.	Barnyard grass	3
		Eleucine indica (L.) Gaertn.	Goosegrass	5
		Elytrigia repens (L.) Nevski	Quackgrass	20
		Imperata cylindrica (L.) Beauv.	Cogongrass	7
		Paspalum conjugatum Berg.	Sour paspalum	17
		Rottboellia cochinchinensis (Lour.) W. Clayton	Itchgrass	18
		Sorghum halepense (L.) Pers.	Johnson grass	6
	Pontederiaceae	*Eichhornia crassipes* (Mart.) Solms		8
Dicotyledonae	Amaranthaceae	*Amaranthus hybridus* L.	Smooth pigweed	14
		Amaranthus spinosus L.	Spiny pigweed	15
	Asteraceae	*Ageratum conyzoides* L.	Tropical ageratum	19
		Ageratum houstonianum Mill		19
	Chenopodiaceae	*Chenopodium album* L.	Common lambsquarters	10
	Convolvulaceae	*Convolvulus arvensis* L.	Field bindweed	12
	Portulacaceae	*Portulaca oleracea* L.	Common purslane	9

[a]These are 20 weed species that affect crop yields globally and are difficult to control. They are ranked from 1 to 20, with 1 being the worst.

Fig. W.1. Barnyard grass (*Echinochloa crus-galli*). Courtesy of (a) Howard F. Schwartz, Colorado State University, Bugwood.org. (b) Joseph M. DiTomaso, University of California Davis, Bugwood.com. With permission.

Barnyard grass propagates through thousands of seeds. Barnyard grass grows well in ample sunlight and wet soil.

Barnyard grass is grown as a forage crop and also converted to silage. It acts as an alternative host for maize dwarf disease and for mosaic viruses such as sugarcane mosaic virus and wheat streak mosaic virus. However, it is a serious weed in many crops like rice, maize, beet, vegetables and potato. Since it resembles rice, it is difficult to weed it out of the rice field. It has been ranked as the third worst weed in the world with regard to the effect on crop productivity and difficulty to control. Its rapid growth and propagation depletes soil nutrition and reduces crop yields significantly.

Proper seed selection, seedbed preparation (by ploughing and harrowing), hand weeding (as soon as it is spotted), close plant rows and flooding paddies are some of the cultural methods of control of this weed pest. **Sodium trichloroacetate, thiobencarb, butachlor, fenoxaprop-p-ethyl, quinclorac** and **pretilachlor** are some of the chemicals that control the growth of barnyard grass.

Two fungal species, *Exserohilum monoceras* and *Cochilobolus lunatus*, though sometimes harmful to the rice plant, can be exploited to control barnyard grass biologically.

Weed – Bermuda grass

Bermuda grass (*Cynodon dactylon*) is a perennial grass that grows widely in Asia, Australia, southern Europe, North Africa and many other countries. It is commonly known as *Durva* (in many parts of India) or couch grass. Bermuda grass is a creeping plant with long, erect and flattened green leaves with tufts of hair in between the leaf blade and sheath (Fig. W.2). The plant forms a carpet on the ground. The plant propagates via the rhizomes, seeds and stolons. It grows aggressively and survives even under unfavourable conditions such as droughts and saline conditions, with its seeds remaining viable for up to 2 years. It prefers warm climates.

Fig. W.2. Bermuda grass (*Cynodon dactylon* L. Pers). Courtesy of Steve Dewey, Utah State University, Bugwood.org. With permission.

Cultural measures to control Bermuda grass include keeping the soil dry and using **harrows** and spades to dig up rhizomes (without breaking them) from under the soil to manually destroy them. Preventing direct sunlight reaching the weed by mulching with black, polyethylene sheets can slow down its growth. Clippings may be used as manure only after the weed is totally decomposed.

Bermuda grass is grown as turf grass or as forage. However, it is a noxious weed that affects the yields of crops (it is among the top three most noxious weeds). It is difficult to eradicate the weed from sugarcane, maize and cotton.

Sethoxydim, **fluazifop**, **triclopyr**, **diquat** and **glyphosate** are some of the chemical herbicides that can help control the growth of the weed. Cultural practices include plastic mulching, depriving the weed of water and tilling practices.

Weed – cogongrass

Cogongrass (*Imperata cylindrica*) is considered an invasive weed pest in over 70 countries. The yellowish-green perennial grass has blade-like long leaves that grow out of short, insignificant stems. The leaves have a rough margin, which might be unpleasant to grazing animals. The plant can grow up to 1.22 m. The plant develops a rhizome about 15 cm below the ground and this grows like a mat. The rhizome, along with wind-blown seeds, helps the plant to propagate. The plant always grows as a group and can be spotted as patches on stretches of land (Fig. W.3).

Cogongrass spreads aggressively on disturbed sites like mines, landfill and construction sites. It is also transported to distant places through pieces of the rhizome and seed. It affects moist pine sites where it spreads like a field.

Once cogongrass is well entrenched on a piece of land, it is difficult to control due to the growth of the rhizome both into the soil and laterally. Tilling up to a depth of 15 cm, followed by harrowing from March onwards can keep the rhizomes from growing. Prior to herbicidal intervention, mowing should be carried out, injuring the plant. Foliar applications of a combination of **glyphosate** and **imazapyr**, repeated two to three times, can control the growth and proliferation of the weed rhizome, although it is to be noted that imazapyr may injure crops. **Sulfometuron** may also be applied with imazapyr.

Weed – common lambsquarters

Common lambsquarters (*Chenopodium album*) is a fast-growing annual weed belonging to the family Amaranthaceae. It is believed to have originated in both Europe and Asia, but is today prevalent throughout the world.

The leaves of common lambsquarters are hairy and have a toothed margin, with upper leaves being narrow (Fig. W.4). It bears very small greenish-yellow flowers which are wind pollinated. The lambsquarters' plant grows from 10 cm to 150 cm high and prefers soils high in nitrogen.

The young leaves and seeds of this plant are eaten as food in some regions of the world, as well as by a variety of insects and birds.

C. album is considered to be a weed and a menace to agriculture. Since each plant can produce a large number of seeds, it is important to destroy the plant before the seeds emerge. It can cause significant crop losses in maize, soybean and sugarbeet.

This weed can be controlled at a young stage by mechanical methods such as tillage at night (dark tillage, when weed emergence is believed to be slower), rotary **hoeing** or flaming. Crop rotation with small grain crops is also helpful. *C. album* attracts leaf miners to which the plant becomes vulnerable. This phenomenon can be turned to the farmer's advantage, by planting the weed as a **trap plant** for managing leaf miner infestations in crops like tomatoes, cucumbers and aubergines. *C. album* is one of the host plants of the beet **leafhopper**, which transmits the beet curly top virus to beet plants.

Weed – common purslane

Common purslane (*Portulaca oleracea*), from the family Portulacaceae, is a succulent, drought-prone, summer annual that grows horizontally on the soil. It has a heavy taproot, with a purple-red stem and small, smooth, alternate, oval leaves. Yellow flowers bloom in clusters at the end of branches. The branched stems creep on the ground, and reach up to 60 cm in length, to form a mat. It is propagated by seeds.

This weed is a pest of gardens and crops. The abundant seed production helps the weed to cover vast tracts of a field.

Fig. W.3. Cogongrass (*Imperata cylindrica* (L.) Beauv.). Courtesy of Charles T. Bryson, USDA Agricultural Research Service, Bugwood.org. With permission.

Fig. W.4. Common lambsquarters (*Chenopodium album*). Courtesy of (a) Chris Evans, River to River CWMA, Bugwood.org. (b) Lynn Sosnoskie, University of Georgia, Bugwood.com. With permission.

Besides, it serves as an alternate host for various plant diseases and nematodes. It is said to have medicinal properties and is also consumed as food in some places.

There are pre- and post-emergence herbicides used to control purslane. Benefin, **dithiopyr**, **ethofumesate**, **isoxaben**, **pendimethalin** and **trifluralin** may be used for pre-emergent control, while **2,4-D**, **dicamba**, **MCPA** and **clopyralid** are good post-emergence options.

Weed – common ragweed

Common ragweed (*Ambrosia artemisiifolia*) is a very common weed in North America. It is also commonly known as annual ragweed, hay fever weed, tassel weed and carrot weed. This 1 m tall weed, seen in many countries in Europe and the Far East, belongs to the family Asteraceae, emerging in late spring.

The deeply-lobed leaves of common ragweed are hairy, when young (Fig. W.5). It bears cylindrical inflorescences and sets seeds by late summer. It relies on wind pollination and can cause hay fever. It is a sturdy, fast-growing plant that can grow in drought-prone soil, with the seeds remaining viable for up to 10 years. Its allelopathic properties inhibit the growth and development of neighbouring plants. It can cause up to 30% losses in soybean.

Fig. W.5. Common ragweed (*Ambrosia artemisiifolia* L.). Courtesy of Steve Dewey, Utah State University, Bugwood.org.

Crop rotation with small grain crops can control the growth of this weed to some extent. Biological control can be brought about by introducing caterpillar species and the ragweed leaf beetle (*Zygogramma suturalis*). Birds may also be enticed into eating the seeds. Night tillage may reduce emergence of seedlings.

Weed – giant ragweed

Giant ragweed (*Ambrosia trifida*) is an annual plant belonging to the family Asteraceae. The plant grows up to 3 m tall, with large three- to five-lobed leaves (Fig. W.6). It bears small, green flowers from midsummer to late autumn. It is a wind-pollinated plant and the pollen is widely dispersed in the wind, causing hay fever.

Fig. W.6. Giant ragweed (*Ambrosia trifida*). Courtesy of John D. Byrd, Mississippi State University, Bugwood.org. With permission.

Giant ragweed is one of the most competitive weeds of **sugarcane** and soybean and can reduce yields by up to 50%. It prefers moist conditions and fertile loamy soils where it develops into a giant plant, although it can survive on less fertile soils too.

An effective means of control is mechanical methods such as mowing, as well as appropriate use of chemical pesticides.

Weed – goosegrass

Goosegrass (*Eleusine indica*) is a summer annual grass, from the family Poaceae. Also called the silver crabgrass, the tuft-like stems radiate outwards from the centre which is whitish. The stems spread horizontally on the ground for about 0.6 m, forming a mat (Fig. W.7). The root system is fibrous and the stems do not form roots at the nodes. It grows very well in full sunlight, and in poorly drained soil. It reproduces by seeds. The weed grows in lawns and turf areas and is not destroyed by mowing.

Goosegrass may be confused with smooth crabgrass (*Digitaria ischaemum*) which is also a prostrate plant with the leaves rolled in the bud.

Pre-emergence herbicides work well to control many stubborn grass weeds. Oxadiazone, oxadiazolin and ***S*-metolachlor** are a few compounds that are to be applied as pre-emergence herbicides (ideal for application 2 weeks before germination).

Fig. W.7. Goosegrass (*Eleusine indica*). Courtesy of John D. Byrd, Mississippi State University, Bugwood.org. With permission.

Cultural practices include aerating the ground, since this weed flourishes in compacted soil.

Weed – Johnson grass

Johnson grass (*Sorghum halepense*) is one of the most noxious perennial grass weeds that are the bane of farmers in most parts of the world.

This is an erect, tall, dense plant with many wide leaves with prominent mid-ribs (Fig. W.8). The panicles are open and the branches have several spikelets that produce and release several thousand seeds. Johnson grass propagates via seeds and the rhizome that can extend several inches underneath the surface of the earth. Thus, the weed proliferates rapidly.

This weed is a nuisance to cereal crops of sorghum, wheat and maize, legumes, soybean and cotton. The weed can easily compete with the crop, since it is sturdy, fast growing and resistant to many control agents.

Multiple control strategies are used to control Johnson grass. Prevention of the stands from establishing themselves in the soil is the first measure. Pre-emergence applications of

Fig. W.8. Johnson grass (*Sorghum halepense* (L.) Pers). Courtesy of Howard F. Schwartz, Colorado State University, Bugwood.org. With permission.

dinitroaniline compounds (e.g. benefin and **oryzalin**) help control the emergence of the weed. Spot application of post-emergence agents such as **nicosulfuron** and **primisulfuron** are also useful. **Glyphosate** sprays are also effective, although phytotoxicity to crops should be kept in mind.

Weed – junglerice

Junglerice (*Echinochloa colona*), an annual graminoid, is a sedge that belongs to the family Poaceae. It is a crop weed which mainly infests rice crops and is found in the major rice-growing areas of the world and in the USA.

Junglerice is a plant that grows up to 0.6 m, assumes a prostrate position while young, and then becomes erect. The wide and long leaf blades are folded in the buds. The leaves have a purple band, a unique feature of this weed. The plant is propagated by seeds and the rhizomes. The plant can produce thousands of seeds, which are transported into rice fields via rice seeds, seedlings, tools and birds. The plant competes with the crop for space and nitrogen. Junglerice inhabits humid and moist regions and irrigated crop fields. Other crops that are infested with this weed are sugarcane, maize, cotton, banana and sorghum.

Besides the direct damage to the crop that this weed causes, it is also the alternate host to *Magnaporthe grisea*, the fungus that causes rice blast and seedling blight in rice crops. It is also a host to mosaic virus in rice and sugarcane, and to some nematodes.

Pre-emergence treatment with **ametryn**, **hexazinone**, **glyphosate**, **terbacil**, **bensulide**, ethalfluralin, **flumetsulam** or **pendimethalin** and post-emergence applications of fenoxaprop, **fluazifop**, hexazinone, glyphosate or terbacil can render good to excellent control over junglerice.

Weed – parthenium

Parthenium (*Parthenium hysterophorus*), native to South and North America, is an annual herb with a deep taproot and an erect stem that becomes woody with age. The plant belongs to the family Asteraceae in the order Asterales. With light green and lobed hairy leaves, the plant grows small cream-coloured flowers, which bear scaly, wedged-shaped seeds.

Parthenium is a troublesome weed on farms and wasteland because of its adaptability to most soils, weather conditions and disturbed conditions such as dug up land or flooded conditions. The weed produces an extremely high number of seeds and their rapid germination and growth challenge the growth of native grasses and crops. A number of crops such as maize, sorghum, cotton, sugarcane and pulse crops are affected by parthenium. The weed is also allelopathic in nature, causing ill effects in sheep and allergies to man, the culprit being parthenin.

Mulching, crop rotation, hand pulling and the use of harrows and ploughs to remove or kill the saplings are some of the cultural practices used to control parthenium. There are also some beetle, weevil and moth species whose larvae feed on the stem, inhibiting its vigorous growth and profuse seed production. The beetle *Zygogramma bicolorata* and a leaf mining moth, *Bucculatrix parthenica*, are two examples. Herbicides like **simazine**, **atrazine**, **butachlor** and **alachlor** are effective in controlling parthenium.

Weed – purple nutsedge

Purple nutsedge (*Cyperus rotundus*) is a true sedge that belongs to the family Cyperaceae in the order Poales. It is a weed in more than 50 crops worldwide, where it infests a number of vegetables, cereals and turf.

The plant grows up to 0.75 m and is a perennial that grows from rhizomes and tubers (attached to the rhizomes). The stem is triangular, while the leaves are dark green, narrow, shiny, thick, ridged in the middle and pointed at the tips (Fig. W.9). The leaves emerge in groups of three from the base of the plant. The vegetative portions die every winter, while the tubers and rhizomes remain underground (up to 1 m) until warmer temperatures prompt propagation.

There is another nutsedge, the yellow nutsedge (*Cyperus esculentus*), that is also a perennial sedge species and similar to purple nutsedge. A characteristic difference is that the purple nutsedge is a slightly smaller plant with dark green leaves and dark brown seeds, with the tubers and rhizomes forming a chain, and one tuber is able to branch off into a new one, forming a colony. The yellow nutsedge is a larger light-green plant, with light-brown seeds, where tubers are linked to rhizomes at their end points (of rhizomes). The flowering stem in the purple nutsedge is longer than the basal leaves, while it is generally shorter in the other species.

Cultural practices for controlling both purple and yellow nutsedges include planting crops in close rows, to deprive the nutsedge of sunlight. Crops may be rotated with fast-growing crops for the same result. Fertilization of crops in bands is suggested, to deprive the weed of nutrients. Use of disc **harrows** and mowers can uproot rhizomes and injure seedlings, thus controlling growth. Continuous solarizing, weeding and tilling can disturb normal growth.

Purple nutsedge is controlled by non-selective **glyphosate** at the flowering stage, to ensure proper translocation to the tubers and rhizomes. Applications can be repeated as spot treatment, taking care to avoid drift. Treatment with selective herbicides (e.g. **butylate**, **EPTC**, **halosulfuron** and **pebulate**) is recommended at the pre-plant stage, especially for crops such as tomato, maize, potato, cucumber, melons and beans as these herbicides effectively inhibit weed emergence for both the nutsedge species. Halosulfuron and trifloxysulfuron sodium may be applied once the weed emerges, followed by no irrigation for 3–4 days.

In spite of its nuisance to farmers, purple nutsedge has been used in traditional medicine to counter fever, inflammation and some other ailments.

Weed – velvetleaf

Velvetleaf (*Abutilon theophrasti*) is an annual plant also known as buttonweed or China jute. The plant is believed to have originated in Southern Asia and is in the family Malvaceae (the hibiscus family).

Velvetleaf is a tall, thin plant with heart-shaped, hairy leaves, yellow flowers and rounded pods. The fine hair on the leaves give it a velvety feel.

This rapidly growing plant is a weed in many cultivated areas where it produces hundreds of seeds that can remain viable for several years. It is a serious pest in maize, sorghum, soybean and cotton.

The bug species *Niesthrea louisianica*, from the order Hemiptera and family Rhopalidae, has been used as a biological

(a) (b)

Fig. W.9. Purple nutsedge (*Cyperus rotundus*). Courtesy of (a) Charles T. Bryson, USDA Agricultural Research Service, Bugwood.org. With permission. (b) Virginia Tech Weed Identification Guide (College of Agriculture and Life Science, Department of Plant Pathology, Physiology and Weed Science). With permission from Elizabeth Grabau.

control agent for velvetleaf. A field inundated with the bug population can help bring about significant reduction in seed viability. Although the bug can cause damage to other plants of the same family as velvetleaf, their controlled release can control weed growth.

Weed – water hyacinth

Water hyacinth (*Eichhornia crassipes*) is a free-floating aquatic macrophyte, seen in South-east Asian regions (and believed to have originated in the Amazon basin). The plant has thick, waxy, rounded leaves (up to 20 cm) with numerous fine veins, and the leaf stalks are spongy. The stems, which sprout from the plant base, enable vegetative reproduction. The plant has a rhizome that is up to 30 cm long and commonly grows to a height of 0.5 m. Roots are produced at the node bearing the primary leaf. As much as 50% of a single water hyacinth's biomass can be roots. The fruit is encapsulated with many seeds in it. Water hyacinth bears attractively coloured flowers which appear in clusters (Fig. W.10). They are coloured dark violet to bluish or pinkish to whitish.

E. crassipes mats reduce levels of dissolved oxygen and light and increase the water temperature. It is considered as one of the most noxious aquatic weeds.

A characteristic feature of some of the herbicides used on water hyacinth is that their efficiency increases with temperature. Glyphosate and 2,4-D are examples. The **non-selective** herbicide **glyphosate** (applied at 2 kg/ha) is effective in controlling water hyacinth, without harming fish. The compound is slightly toxic to aquatic invertebrates. **2,4-D** (2,4-dichlorophenoxyacetic acid) aerially sprayed at 1–12 kg/ha is effective. However, it is slightly to moderately toxic to birds such as waterfowl, fish and aquatic invertebrates and so its salt formulations are more appropriate for use against *E. crassipes*. Brand names include Aqua-Kleen. Other herbicides that are used for the control of water hyacinth include copper sulfate and copper chelates, which may be applied at 3.5 mg/l for weed control, although they can be toxic to fish, some mammals, aquatic invertebrates and soil organisms.

Fig. W.10. Water hyacinth (*Eichhornia crassipes*). Courtesy of Wellgrow Horti Trading (www.wellgrowhorti.com). With permission.

Weed – yellow nutsedge

See **Weed – purple nutsedge**

Weed and feed method of crop management

See **Weed management by mechanical methods**

Weed control agents – biological origin

Any factor that can control weeds or regulate growth in an area is a weed control agent. The term biological control, first coined by H.S. Smith (1919), is defined as the action of parasites, predators or pathogens in maintaining another organism's population at a lower average density than would occur in their absence.

Such agents are sometimes freely available in nature. Grazing animals and insects are two such examples. However, fungi and other microbes and allelopathic plants also help control the growth of certain plants. Many herbicides have been synthesized from microbial hormones. A **mycoherbicide** is a bioherbicide based on a fungus.

Indole-3-acetic acid, terpene, 1,8-cineole, **sulcotrione** and **cytokinins** form the basis of a large number of herbicidal chemicals, all of them being derived from plants and having molecules that are phytotoxic to targeted plants. Most plant-growth retardants or herbicides act by blocking gibberellin biosynthesis, plant acetyl-CoA carboxylase and carotenoid biosynthesis. Natural phytotoxic products of microbial origin include **bilanaphos** ($C_{11}H_{22}N_3O_6P$), a tripeptide derivative isolated from a strain of *Streptomyces* and **gibberellins**, a PGR, first isolated from the fungus *Gibberella fujikuroi*, with gibberellic acid often commercially produced from this strain. Phytotoxins related to microbial organisms include pelargonic acid, glufosinate and **triketones**. The triketone family of herbicides (inhibitors of *p*-hydroxyphenylpyruvate dioxygenase) is derived from the naturally occurring **leptospermones**. Leptospermone is an allelopathic compound derived from a bottlebrush variety (*Calistemon* spp.). Sulcotrione is a herbicide that targets maize-specific weeds. **Mesotrione** is also synthesized from leptospermones. *Phytophthora palmivora* controls strangler vine in citrus crops. A sub-species of *Colletotrichum gloeosporioides* controls vetch species in soybean and rice fields; *C. gloeosporioides* f. sp. *malvae* is a leaf pathogen that can control many round-leaved plants. *Alternaria cassia* controls sicklepod. **Cornexistin**, useful as a broad-spectrum post-emergence herbicide in maize is fungal in origin.

One of the common natural means of weed control is allelopathy. Allelopathy is a complex behaviour of one plant (or a species) which harms the growth of another plant (or species) through the secretion of chemicals that are phytotoxic to the inhibited plants. Allelopathy involves biomolecules called allelochemicals, which include chemical classes such as flavonoids, terpenoids and alkaloids, many of which are secondary metabolites. Any part of an allelopathic plant may display such behaviour (e.g. the leaves or the roots). Allelopathy is widely seen in black walnut, which suppresses the growth of plants around it because of the phytotoxin **juglone** (5-hydroxyanthraquinone), it contains. Phytotoxins from the leaves and roots of the bottlebrush plant and eucalyptus suppress specific plant species.

Research has shown how the allelopathic property of some plants can be harnessed to control weeds. Some allelochemicals encourage the growth of some crops, while inhibiting some others. For instance, some species of *Leucaena leucocephala* are believed to increase the yield of rice, while they inhibit wheat yields. Some varieties of rye, clover, rice, mustard and wheat are grown as cover crops and are subsequently killed. Their allelopathic effect suppresses the growth of weeds such as common purslane and pigweeds during the next cropping season. Crops that are resistant to allelopathic chemicals are grown on the same field, especially in reduced tillage situations. For instance, *Sorghum sudanense* is grown as a cover crop to prepare for a minimum-till cultivation of soybeans. Sometimes allelochemicals are combined with herbicidal ingredients for a more sustained effect on weed control.

A challenge that is faced in the process of fully exploiting natural biocontrol sources of weed control is to expand the number of weeds that can be controlled by biocontrol measures.

Weed control by flooding

See **Weed management by mechanical methods**

Weed management by mechanical methods

The objective of weed management is to contain the growth of weeds so they do not cause economic, social, environmental or ecological loss to mankind. Although mechanical (or non-chemical) methods are the oldest means of weed management, they do have an important place, even today, in any comprehensive weed management strategy, because they combine the benefits of prevention and control of weeds.

In ancient times, the practice of 'slash and burn' was carried out, where land was cleared of wild vegetation, burned and then farmed, before a new patch was located.

Mechanical methods comprise the use of tools, machines and manual labour to remove/reduce/control weeds on a farm. Tillage operations, flooding, burning, **hoeing**, digging and mulching are examples. Hand weeding and hand hoeing have been used to control annual and biennial weeds.

Mechanical methods are suitable where there are scattered pockets of infestation on a large plot. Such physical methods are favoured, especially when weeds have developed resistance to a herbicide. The disadvantages of mechanical methods are: (i) difficulty in complete weed elimination; (ii) the futility of these methods in heavy and continuous rains; and (iii) persistence of perennial weeds.

Some common tools and techniques of weed control are summarized here.

A spud is a 5 cm chisel with a handle. Spudding is a selective technique and is effective against scattered weeds in a small area, such as dandelions in lawns.

A spade is another tool that works well on scattered biennial weeds such as burdock and musk thistle.

Hoes are bladed tools used to disturb the surface of soil, remove weeds, create ridges and furrows and mount soil around plants. **Hoeing** is mainly used for the selective uprooting of particular unwanted plants or general removal of all kinds of weeds, a process that may have to be repeated continually to prevent weeds taking hold. Hoes come with both short and long handles (now short-handled hoes are common). Hoeing has been carried out since ancient times, with hand hoeing being the standard practice for sugarbeet.

Hoe blades have various widths and weights, depending upon the job to be performed. Hoeing is faster and easier than hand pulling or spudding and thus is more widely used in gardens for high value crops such as sugarbeet.

Activities like disking, rotavating, harrowing and ploughing reach various soil depths to kill deep-rooted perennial weeds. **Rotavators** or rotary tillers are cultivators with blades attached behind tractors. A disk is a type of plough that uses blades to cut the surface of soil up to around 10 cm deep. A plough is sometimes used along with a disk. Accessories such as cut-away disks, long-ridged shields, cutters, trimmers, oscillating sickles, shearers and rotary blades are attached for performing various tasks. Rotary weeders are used for cutting weeds (even in rice fields) and for mulching the soil. Mechanical weed pullers are used to pull tall-growing weeds out of short-growing crops, especially in moist soil. They are made up of two rubber tyres that rotate together, uprooting the weeds and shearing them. Tough weeds such as crabgrass and **barnyard grass** are known to continue to grow via their rhizome.

Weed control by flooding is a common technique for both rice fields and upland crops. Flooding is seen to effectively control perennial weeds. The farm is surrounded by dykes and submerged under 25 cm of water for several weeks (sometimes up to 2 months). Wild morning glory, white horse-nettle (*Solanum elaeagnifolium*), **Johnson grass** (*Sorghum halepense*) and Russian knapweed (*Acroptilon repens*) may be controlled by this method. Many annual weeds as well as nematodes and some soil fungi may also be controlled.

Control by fire or burning of weeds is an ancient method of eliminating undesired vegetation. Small, well-targeted propane burners can effectively control velvet mesquite, burroweed and others. Controlled burning is carried out for rangelands and pastures and for destroying woody plants in forests (as in the case of some cedars and baked hazel). Burning also stimulates plant growth. Burning followed by grazing (by goats) on newly emerging weeds can effectively control weed growth.

Solarization of soil has been accomplished by covering the soil with plastic sheets. When the soil surface heats up to high temperatures due to the bright sunlight, it kills non-dormant weed seeds. However, this process requires several weeks to destroy seed viability, and hence soil sterilization is generally used in greenhouses. Solarization is carried out using a Fresnel lens (made of acrylic plastic), where its parallel grooves refract the sun's rays and converge the radiant energy to a desired point. This results in very high temperatures, which destroy seed viability in less than 1 min. Soil sterilization also controls growth of soil pests.

Cultivator blades and vehicle tyres should be kept clean to prevent weed seeds from entering fields, and reduce the transfer of mud, stones and husk. Only weed-free seed should be bought and sold to prevent weed infestation. Combine harvesters (a machine that combines harvesting, threshing and cleaning) remove many weed seeds with the help of gushes of air and sieves during harvesting. Removing weed seeds becomes easier if the size, weight, shape and colour of the crop seeds are distinct from those of weed seeds.

Seed conditioning is considered necessary for efficient weed-free seed storage. Seed conditioning is a set of processes (and tools) carried out to clean seed produce and enhance its shelf life. Mechanical cleaning devices and air screen cleaners, mills and specific gravity separators ensure uniform seed drying and seed flow.

When land is set aside for agriculture, the seedbed is prepared for the operation by ripping, turning and ploughing the soil. Such an activity is called **tillage**. Depending on the cultural practices prevalent in a given area and the type of crop rotation followed, the extent of tillage varies. Thus, there are non-till (or zero tillage), reduced tillage and conservation tillage systems. Tillage operations are a prevalent method of weed management. Tillage leads to the burial and cutting of weeds. Besides the primary objective of weed control, tillage operations benefit the soil in other ways, for example: (i) loosening and aerating it; (ii) incorporating green manure; and (iii) encouraging the growth of microorganisms. An example of a tillage sequence includes mouldboard ploughing followed by disking, harrowing and/or cultivation.

Primary tillage makes the soil loose by disrupting and roughing the surface. Primary tillage equipment such as the ripper, mouldboard plough, chisel plough, tandem disk harrow and rotary tiller are used to break and loosen the soil to depths of up to 75 cm. Fertilizers are commonly added at this stage.

Mouldboard and disk ploughs can effectively incorporate plant residues since they invert the surface soil and bury the residue. The mouldboard plough breaks up and loosens the soil and buries plant residues, weeds and their propagules. If weeds are not buried, they may recover.

Disk ploughing and tandem disk harrowing, used for incorporating plant residues, are effective in controlling vine weeds such as prostrate knotweeds.

Chisel ploughs are effective in a wide range of soils. The depth of tillage varies with the soil. Greater depths are necessary when hard pans are present.

The sweep plough, equipped with sweeps of various sizes (46–210 cm in width), helps incorporate plant residues. However, since a mouldboard plough is found to be more efficient, it is commonly preferred to the sweep plough.

The rotary tiller has a more limited use because it destroys soil structure by pulverizing the soil. It is also expensive to operate. The most common use of a rotary tiller is to till soil in a narrow strip in front of the planter unit. Fertilizers and herbicides can also be applied to the soil during planting with a rotary tiller. This is known as the weed and feed method of crop management.

Secondary tillage is meant for weed management, herbicide incorporation, erosion management and seedbed preparation. It also produces fine soil and helps shape rows. Best results are obtained when sufficient time has elapsed between tillage operations, to allow weed seeds to germinate so that they can be destroyed by the next tillage operation. The more weed seedlings that are killed with frequent tillage operations before planting, the less weed interference there is in the succeeding crop. Secondary tillage implements (e.g. field cultivator, dibbers, tandem disk harrows, rod weeders, seeded conditioners, tillers, roller harrows and combinations of these tools) are used to work the soil up to a depth of 15 cm. **Harrows** dig the surface of the soil, unlike a plough which goes much deeper. A dibber is a simple tool with a pointed end, designed to dig holes in the ground for planting seeds. Rotary tillers (also called rotary hoes, rotary ploughs or rotavators) are powered cultivators that work by rotating tines or blades. Rollers are tools used to flatten the surface of the soil after ploughing. They also break up soil mounds or heaps of organic matter. A sub-soiler is a tool that digs into the soil at depths twice that of a rotary tiller. With the means of one sharp-tipped blade, it can reach almost 15 cm inside the soil, loosening up compacted soil. Each implement leaves varying mounds of plant residue on the soil surface. The field cultivator and tandem disk harrow are the most popular secondary tillage implements in view of their fast, wide-pass operations that do not require large tractors. Both can incorporate herbicides and leave the land levelled for weed-free seeds. Rod weeders maintain plant residue on the surface although they are not suited to uneven areas (i.e. areas with stones or stumps). Seedbed conditioners are used to reduce soil clods, incorporate herbicides and level the field. The roller harrow has limited use in effectively reducing clods to give smaller soil aggregates and a firm seedbed for small-seed crops such as perennial grasses and legumes.

Another novel method of weed management during seedbed preparation is making a **stale seedbed** or a **false seedbed**.

Tertiary tillage begins after planting and includes cultivation of row crops. Most of the cultivator tools perform well in the absence of plant residues. The rotary hoe, finger weeder, round wire-teeth harrow and spike-tooth harrow are effective on weeds in the very early stage of germination. For example, spike-tooth harrowing should be done before the emergence of maize, with a second harrowing, with teeth set at a slant, occurring between maize emergence and the three- to four-leaf stage. Intra-row tillage implements used after crop emergence should be used after the early morning turgidity of plants to avoid injury.

Inter-row crop cultivation is carried out to manage weeds in row crops. In the case of a maize crop, land is disked early to encourage weed seed germination before mouldboard ploughing. Three sets of cultivations are carried out to control weeds. After surface planting maize, one or two rotary hoeings are done after the maize emerges. Cultivators are equipped with vertical fins, which move soil laterally into the maize row. Cultivator shovels next to the row may be at 4.5 cm from the row. The first cultivation is done at a speed that would not move soil over the maize. A few weeks later, the maize is cultivated again, and a third cultivation occurs when the maize is about 50 cm tall. This is called a hilling operation and soil is moved into the maize row to cover as many weeds as possible. However, in years of excessive rain, weeds are poorly controlled.

Inter-row crop cultivation equipment includes rollers, shovels (vertical and horizontal), a combination disk and power-driven rotary cultivators. Most cultivators are effective on well-emerged weeds but not effective on inter-row weeds unless sufficient soil can be removed from the row to cover the weeds. The first cultivation should be 5–7 days after the white-root stage of weed seedlings, and repeated in 7 days. Inter-row cultivation is limited by soil moisture, and it is often ineffective against narrow-leaved weeds.

Inter-row crop cultivation is thought to be essential to obtain adequate weed control without reducing grain yields. Inter-row crop cultivation generally should begin 2–3 weeks after crop emergence, but rains may prevent timely cultivation.

Furrow openers are attachments that remove the dry surface soil, clods and plant residues over the crop row, enhancing soil–seed contact, soil moisture and subsequent crop emergence. Furrow openers also remove germinating weeds and sufficient weed seeds from over the row to reduce the weed population in the row. Weed seed germination between rows may be delayed because of lack of soil water. The crop can then emerge ahead of the weeds and become taller. A furrow opener can result in most of the weed seed being removed up to a depth of 6–8 cm under the soil.

Weeds – classification

Weeds are commonly classified botanically, but may also be classified based on their morphology, growth form, life cycle and habitat.

Botanically, plants are classified into groups such as division, class, order, family, sub-family and genus. Examples of plant divisions are Pteridophyta, Angiosperma and Gynmosperma. A division such as Angiosperma has two classes, namely Monocotyledoneae (monocots) and Dicotyledoneae (dicots). Classes are sub-grouped into families such as the Cyperaceae, Apiaceae and Aquifoliaceae.

On the basis of morphology, weeds are broadly classified by the number of cotyledons (seed leaves), into monocotyledonous plants (those having one cotyledon) and dicotyledonous plants (those having two cotyledons), which are together grouped as angiosperms. Seed-producing plants (spermatophytes) are of two types – **angiosperms** and **gymnosperms**. Most weeds are angiosperms. Monocots include grasses, sedges, lilies, spiderworts and cattails. Examples of monocot weeds are *Echinochloa colona* (junglerice), *Setaria glauca* (yellow foxtail) and *Cynodon dactylon* (common **Bermuda grass**). Colloquially, monocot weeds are referred to as 'grass-like' because their leaf shape or growth form resembles that of grasses. The sedge family, Cyperaceae, has a wide number of weed species in the genera *Cyperus*, *Eleocharis*, *Scirpus* and *Bulbostylis*. Sedges can survive harsh climatic conditions and grow either as annuals producing a huge number of seeds or as perennials with tubers and rhizomes for propagation. **Purple nutsedge** (*Cyperus rotundus*) is the worst weed species among sedges. It aggressively invades agricultural land because of its ability to produce a network of underground tubers, each tuber being capable of producing a new plant. **Yellow nutsedge** is similar to **purple nutsedge** in both propagation and nuisance. Umbrella sedge and rice flat sedge are other examples.

Dicot weeds are often referred to as 'broadleaved weeds'. Among broadleaved weeds, important species include common purslane and **common lambsquarters**, morning glories and field bindweed. These adapt to new surroundings and compete with crops. Many weed species are poisonous, causing ailments that range from a runny nose and skin irritation to hay fever and even death. This classification of weeds into broadleaved, grass and sedge weeds, or annual and perennial weeds, is useful when considering treatment with broad-spectrum and **non-selective** herbicides.

Based on their growth forms, weeds are divided into **herbaceous plants** and **woody perennials**. Herbaceous species are further divided into vines, climbers and stranglers. Woody perennials are subdivided into trees, shrubs (both being called woody weeds) and woody vines.

Based on life cycle, weeds are divided into **annuals, biennials** and **perennials**.

Weeds can be divided on the basis of habitat. The characteristics of weeds are influenced by factors such as soil, local climate, dispersal methods, reproductive ability, their ability to compete with crops and other weeds. This classification of weeds is not very rigid as the same weed may be found in different habitats at the same time.

Woody species growing on pasture and non-crop land are called brush plants. Examples of brush species are the big sagebrush, blackberry, blackbrush and buckbrush. Weeds that grow in water are called **aquatic weeds**. Examples include *Impomoea* spp., *Typha* spp. and *Hydrilla verticillata*. Depending on root position, they can either be floating **hydrophytes**, suspended hydrophytes or anchored submerged hydrophytes, and their various adaptations. For easy distinction, such plants are called **floaters**, **banker weeds** and **sinkers**.

The favourable micro-climatic conditions in orchards encourage some weeds to flourish. The predominant orchard weed species are *Ageratum* spp., *Oxalis* spp. and *Paspalum* spp. Lawns and parks are commonly infested with a large number of annuals and perennials, which include *Triflorum* spp., *Oxalis* spp. and *Poa annua*. A large number of annual, biennial and perennial weeds are found on uncultivated land. These weeds are mostly hardy in nature, as in the cases of *Digitaria* spp., *Cenchrus* spp., *Lantana camara* and *Xanthum strumarium*.

Another type of classification is based on the way weeds have inhabited and invaded cultivated land. Thus, there are **apophytes** and anthropophytes, further divided into ergasiophytes, ergasiolipophytes, ergasiophygophytes, archeophytes, neophytes, epekophytes, ephemerophytes, oekiophytes and spontaneous apophytes. Ergasiophytes are plants that have been introduced into a new region for cultivation or for some specific reason. Ergasiolipophytes, epekophytes, agryophytes and neoindigenophytes are types of naturalized alien plant species that have become self-sustaining and can reproduce in large numbers (e.g. more than 100 million in 50 years) and spread in a large area, even away from their parent plants.

Epiphytes are another group that is recognized by their relationship to habitat. Lichens, orchids, ferns and mosses are commonly known examples of epiphytes. Epiphytes can either be **true epiphytes**, half-epiphytes or nest epiphytes.

Weevils
See **Pest – wheat and rice weevils**

Western flower thrips
See **Pest – thrip**

Wettable or water-dispersible powders
Wettable powders (WPs) or water-dispersible powders have assumed an important place among the dry formulations available to the farmer. These are finely divided and relatively insoluble powders combined with an inert material and a wetting/dispersing agent. The **wetting agents** help in keeping the physical properties and suspension qualities of the formulation intact, although occasionally, the formulation may have to be agitated.

While WPs look similar to dustable powders, WPs are formulated to remain suspended in a carrier, usually water, and then sprayed on the target crop. WPs can be also harnessed for slurry seed treatment, for effective pre-sowing.

The manufacture of WPs involves blending of the components, followed by mechanical milling of the mixture to reduce the particle size. Re-blending ensures uniform size of particles, as required. Active ingredients that are crystalline materials with low vapour pressure or material with high melting points are most suitable to the WP form.

WPs are economical and simple to use and store. But they cannot be accurately measured volumetrically and hence require weighing. Dust inhalation is also another risk. (See also **Powders for seed treatment**.)

Wettable sulfur
See **Elemental sulfur fungicide**

Wetting agents
See **Spray application concepts**

White grubs
See **Pest – grub**

White muscardine
See ***Beauveria bassiana* (fungus) as a biocontrol agent**

Whitefly
See **Pest – whitefly**

Wilt
See **Plant disease wilt**

Withholding period/pre-harvesting interval
The minimum time necessary to be taken between the last application of a pesticide to a crop (including pasture) and harvesting for human consumption or grazing with livestock is called the withholding period. It is also the time interval between the last application of a pesticide to an animal and the collection of eggs or milk, or slaughter for human consumption. This time period is necessary for the natural degradation of the pesticide, and is necessary for the safety of the consumer. It is also referred as the pre-harvest interval.

Woody perennials
Plants whose stems develop secondary thickening are called woody perennials. This growth makes the stems hard and resistant to breakage. Usually, these stems have an annular growth. When weeds develop such secondary thickening, they become aggressive and competitive, compared with farm crops, orchards or even forestry. Woody perennials pose a threat to farmers. Blackberry (*Rubus fruticosus*), red alder (*Alnus rubra*), big leaf maple (*Acer macrophyllum* Pursh), Scotch broom (*Cytisus scoparius*), cabbage palmetto (*Sabal palmetto*) and some members of the Cactaceae family are examples of woody perennials.

Most woody plants have a strong survival instinct, commonly seen through an efficient method of seed production and dissemination, and also through vegetative forms of reproduction. The seeds are known to postpone germination in unfavourable conditions. The plants are also known to grow tall and spread horizontally, shading out beneficial plants and range grasses.

In arid regions, most of these plants show a xeric growth form, namely through deep roots and certain crown buds.

A large number of weed varieties along river channels are able to take up most of the moisture available from the water table or the unsaturated zone just above it. These are called phreatophytes. Some examples of phreatophytes are willows, ironwood, smoke tree and tamarix (which are considered to absorb the most moisture).

Some woody plants in humid regions are poisonous and irritating to people, including poison ivy and poison sumac (members of the genus *Rhus*).

Worker safety norms and first-aid techniques

Personnel involved in handling, mixing, applying and storing pesticides are most prone to the risk of exposure to toxic ingredients. Their safety is of paramount importance. It is necessary therefore to educate these personnel about the safe use of pesticides, by training them on the correct use or disposal of equipment, containers, tools and tanks. They must also know the measures to be taken in case of accidents. It is also necessary that they abide by safety norms by using **personal protective equipment (PPE)**.

Worker safety is enforced by respective agencies that are part of a country's ministry or department responsible for pesticide registration (e.g. the **Environmental Protection Agency (EPA) of the USA** or the **Central Insecticide Board and Registration Committee** of India). A few recommendations for ensuring the safety of workers are discussed here.

Only those workers who have been trained by government-authorized agencies should be allowed to handle pesticides. Such licensed personnel should clearly display their credentials, if ever asked to do so.

Workers should be well informed about the particular task at hand, in terms of: (i) the toxicity class of the pesticide; (ii) its form and formulation (e.g. an aerosol (AD) or an emulsifiable concentrate (EC)); and (iii) the necessary protection equipment to be worn.

It is important to distribute to workers, audiovisual materials (e.g. posters and CDs) that depict the norms and directions for mixing compounds, using PPE, handling emergencies in case of accidents, cleaning of tools and equipment and storing of unused pesticides.

Every pesticide application site, such as a farm, warehouse or a industrial site, should be well equipped with washing and decontaminating provisions nearby. Such locations should be equipped with: (i) sufficient water supply; (ii) a technically approved, cleaning medium; (iii) disinfectant soap; (iv) a well-equipped washroom; (v) clean safety garments; and (vi) at least one well-ventilated room with a bed. The other important materials that ought to be stocked include clean cotton pads, common salt, skin ointments, aspirin tablets, **antidotes** and first-aid measures. At least one person should know how to restore victims to normalcy. There should be posters and leaflets available that clearly display some common emergency measures, such as how to induce vomiting.

As far as first aid is concerned, label instructions prove to be the most important in an emergency situation, since each formulation gives specific first-aid/medical advice for that specific proprietary combination.

In addition to label instructions, the following are to be treated as universally acceptable measures in case of accidents. These help prevent exposures from becoming fatal.

The first of the first-aid techniques is to remove the victim from the site of exposure. Treatment will differ with the exposure route. In cases of dermal exposure, the person's clothes should be removed immediately and the skin washed with clean water and a mild liquid soap. If possible, the victim should be immersed in a tub or tank of water. If the skin is burnt or blistered, the area should be covered with a loose clean cotton cloth. In cases where the pesticide has got into the eyes, the affected eye should be washed with eyewash or the open eye should be washed with clean running water for 15 min. In cases where the victim has inhaled a toxic pesticide, he/she should be removed to an airy, open room and if their breathing difficulty is intense the victim should be given artificial respiration.

There should be a health supervisor, especially in large industrial sites producing pesticides, whose basic task is to monitor individual handlers and mixers by recording body weight, temperature and blood pressure. There should also be a method of taking blood and urine samples from victims in the case of accidents.

In the case of a pesticide being swallowed, the first step is to help the victim rinse out their mouth, to prevent residues harming the insides of the mouth. Plenty of water or milk should be given to the victim. Vomiting should be induced only if permissible on the label and only if the poison is not a corrosive substance. If it is permissible, putting a finger deep into the mouth can help induce nausea and vomiting. The person should be kneeling forward with face down to facilitate vomiting. The victim may be served two to three glasses of water or a strong solution of mustard or common salt (two to four tablespoonfuls in a glass of water). Ipecac syrup (0.01 mg/kg b.w.) is also recommended. For infants, one to two teaspoonfuls of this syrup after a large serving of water or milk induces vomiting. It should be noted that vomiting may persist even after the poison is ejected. Ipecac syrup is not advised after activated charcoal is administered. In the case of acid poisoning, milk of magnesia is administered, while in the case of alkali poisoning, diluted vinegar (1:1) solution in water is given. Apomorphine hydrochloride 5 mg can be given subcutaneously.

However, if the patient is convulsing or unconscious, vomiting should not be induced. In such a case, administering an antidote is essential. A mixture of activated charcoal, magnesium oxide and tannic acid in a 2:1:1 ratio acts as an antidote to most poisons. Activated charcoal is carbon that has been processed to become porous and it has an increased potential for adsorption and chemical reactions. Amyl nitrite (for cyanide), liquid paraffin (for a solvent with benzene or gasoline), atropine (for organophosphorus/carbamate poisoning), calcium gluconate, milk of magnesia and mustard powder are some antidotes or medicines recommended to be included in the first-aid kit.

Gastric lavage is also recommended for unconscious victims or those that have consumed long-chain hydrocarbons or anti-emetic compounds. The lavage should be performed in the operating theatre to prevent aspiration.

Symptomatic treatment should be administered as soon as possible, since many cases of chemical poisoning are reversible. The major supportive care normally includes maintaining the oxygen inflow, correction of electrolyte imbalance, correction of metabolic imbalance and treatment of seizures. Some important highlights to be kept in mind are given below.

Table W.2. Pesticides are classified by the WHO according to their acute toxicities. (From WHO, 2004.)

		LD_{50} for rat (mg/kg b.w.)			
		Oral toxicity		Dermal toxicity	
Class	Description	Solid	Liquid	Solid	Liquid
Ia	Extremely hazardous	≤5	≤20	≤10	≤40
Ib	Highly hazardous	5–50	20–200	10–100	40–400
II	Moderately hazardous	50–500	200–2000	100–1000	400–4000
III	Slightly hazardous	≥501	≥2001	≥1001	≥4001
Table 5 or U	Unlikely to present acute hazard in normal use	≥2000	≥3000	–	–
Table 6	Not classified, obsolete				
Table 7	Fumigants not classified				

Oxygen is both a supportive treatment and a curative antidote (especially against poisoning by cyanide gas and carbon monoxide vapour). External oxygen support restores the normal oxygen concentration at the tissue level and increases its saturation in the alveoli arterial blood. Oxygen can be provided by mechanical respiratory support, intranasal catheter or a face mask.

Pesticide poisoning results in a disturbed water–electrolyte balance (e.g. corpuscles may be destroyed). An infusion of sodium chloride, potassium or magnesium bicarbonate and glucose can help restore the balance. It is important to keep the victim's head at a lower level than the body.

Hypertension, common during acute poisoning, is remedied by maintenance of the airway, improved ventilation and correction of hypoxia, which is critical to the survival of victims.

In reality, very few farms and manufacturing sites have well-equipped decontamination sites. In fact, a supply of running water or a large tank/pool for immersing the victim is also often not present. Most set ups have a pumping system, where a nearby stream or another water source is tapped into in case water is needed for washing or bathing. Hence it is necessary for local government agencies to supervise and critically analyse the set up, perhaps insisting upon one well-equipped site for two or three small farms.

World Health Organization (WHO)

Pesticides are classified by international agencies, on the basis of acute toxicities. After several toxicity studies and estimates of pesticide residues, each of these agencies has arrived at guidelines that may be referred to when registering an active ingredient (a.i.). Two widely known such international bodies are the World Health Organization (WHO) and the **Environmental Protection Agency (EPA) of the USA**. Most toxicities are defined according to rat oral and dermal LD_{50} (e.g. see Table W.2 which shows pesticides classified by the WHO according to their acute toxicities). For ease of communication, groupings are defined to indicate levels and forms of toxicity. In general, liquids are regarded as more dangerous than solids, dermal toxicity more dangerous than oral toxicity. A list of phrases that specify the hazard, called R-phrases or risk phrases, is also used (e.g. R22 = harmful if swallowed; R28 = very toxic when swallowed; and R62 = possible risk of impaired fertility).

The WHO is the coordinating authority for health in the United Nations. The WHO is responsible for providing leadership on global health matters, assessing trends in health, helping shape global policies on health and many other matters. More information can be accessed from the WHO website (http://www.who.int).

Carcinogenicity is also classified according to well-defined terms according to the strength of evidence of potency. A substance with extremely low but definite potency as a carcinogen (e.g. amitrole) will be classified, together with very high potency substances (e.g. aflatoxin). Carcinogenicity is classified according to the **International Agency for Research on Cancer (IARC)**, which is a part of the WHO. Their classification is as follows:

- Group I – The agent is carcinogenic to humans.
- Group 2A – The agent is probably carcinogenic to humans.
- Group 2B – The agent is possbily carcinogenic to humans.
- Group 3 – The agent is not classifiable as to its carcinogenicity to humans.
- Group 4 – The agent is probably not carcinogenic to humans.

More information is available from the IRAC website (http://www.iarc.fr/).

X

Xylylcarb insecticide

Xylylcarb, $C_{10}H_{13}NO_2$ (3,4-xylyl methylcarbamate), is a colourless solid carbamate insecticide. Soluble in water (580 mg/l) and moderately soluble in acetonitrile, cyclohexane and xylene, it is hydrolysed in alkaline media. Xylylcarb is obtained by treating 3,4-dimethylphenol with methylisocyanate. The CAS Registry Number of xylylcarb is 2425-10-7.

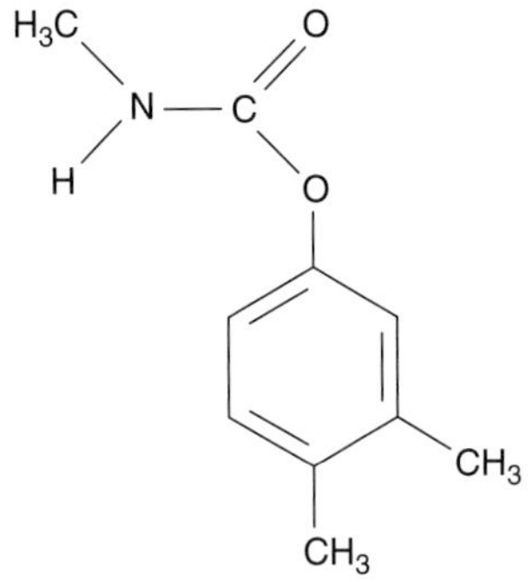

Xylylcarb

The non-systemic insecticide xylylcarb is a cholinesterase inhibitor. It is used to control sucking insects like **leafhoppers** and plant hoppers on rice and **scale insects** on fruit.

It is available in the market under brand names such as Meobal.

The compound is formulated as an emulsifiable concentrate (EC), a wettable powder (WP), a dustable powder (DP) or as micro-granules.

Xylylcarb is classified as moderately toxic (Class II toxin) by the WHO and is considered as a PAN Bad Actor chemical, since it inhibits cholinesterase. Product labels should carry the signal word 'WARNING'.

The principal pathways of xylylcarb metabolism in plants and animals are oxidation of the 4-methyl substituent, hydroxylation of the *N*-methyl group and hydrolysis of the carbamate ester forming phenols and conjugates.

Xylylcarb is not registered in the USA.

Y

Yellow nutsedge

See **Weed – purple nutsedge**

Yellow phosphorus rodenticide

Phosphorus (P_4), an element that is found in almost all living organisms, is mostly used in the manufacture of fertilizers and pesticides. Phosphorus has many allotropes, including white and red phosphorus. White phosphorus, being unstable, starts turning red, resulting in yellow phosphorus.

Yellow phosphorus is a highly toxic rodenticide. The oral LD_{50} in rats is <50 mg/kg b.w. Yellow phosphorus is made into bait by mixing it with peanut butter or molasses. It is most toxic to the gastrointestinal tract and liver. Death occurs due to cardiovascular failure.

Z

Zephiramine slimicide

Zephiramine, $C_{23}H_{42}ClN$ (benzyl-dimethyl-tetradecylazanium chloride), is a slimicide, algicide and rodenticide. It is a white solid that is soluble in water.

It is registered by the EPA as a rodenticide. It is used to control slimes and algae in swimming pools and industrial water treatment plants. It is formulated as disinfectants and sanitizers.

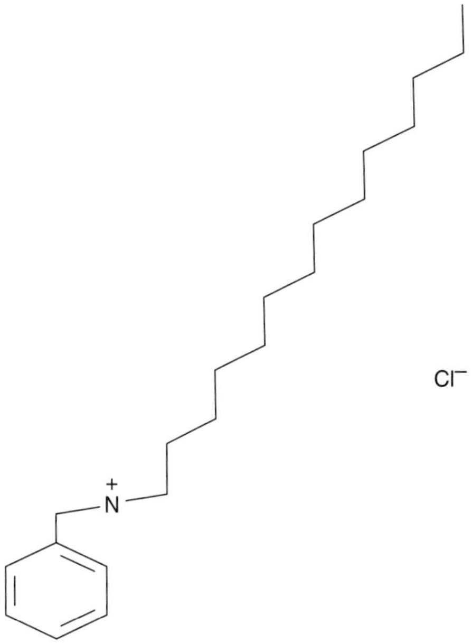

Zephiramine

Zero tolerance

The limit for a pesticide residue in food is assumed to be zero and therefore any detectable residue is deemed illegal. Zero tolerances are used by some regulatory systems where no MRLs have been established for particular pesticide or crop combinations.

Zeta-cypermethrin insecticide

Zeta-cypermethrin, $C_{22}H_{19}Cl_2NO_3$ (a mixture of the stereoisomers (*S*)-α-cyano-3-phenoxybenzyl (1*RS*,3*RS*;1*RS*,3*SR*)-3-(2,2-dichlorovinyl)-2,2-dimethylcyclopropanecarboxylate, where the ratio of the (*S*);(1*RS*,3*RS*) isomeric pair to the (*S*);(1*RS*,3*SR*) isomeric pair lies in the ratio range 45–55:55–45), is an insecticide. It has a CAS Registry Number of 52315-07-8. It is a dark brown viscous liquid which is soluble in water and miscible in most of the organic solvents. It can remain stable for 1 year at 50°C.

(*S*)-alcohol (1*R*)-*cis*-acid

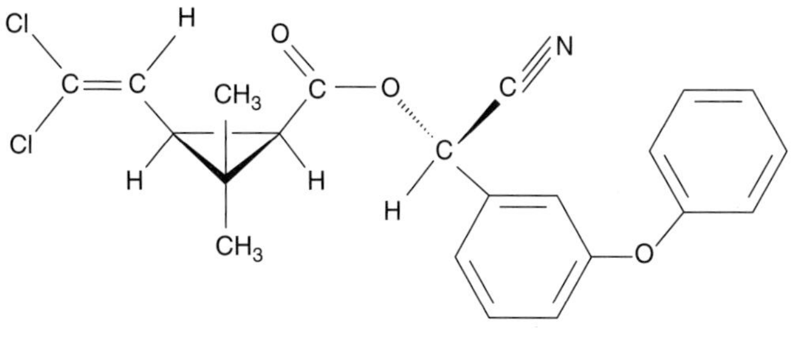

(*S*)-alcohol (1*R*)-*trans*-acid

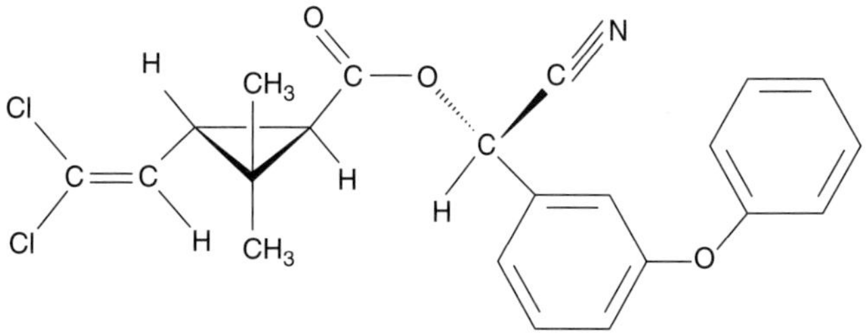

(*S*)-alcohol (1*S*)-*cis*-acid

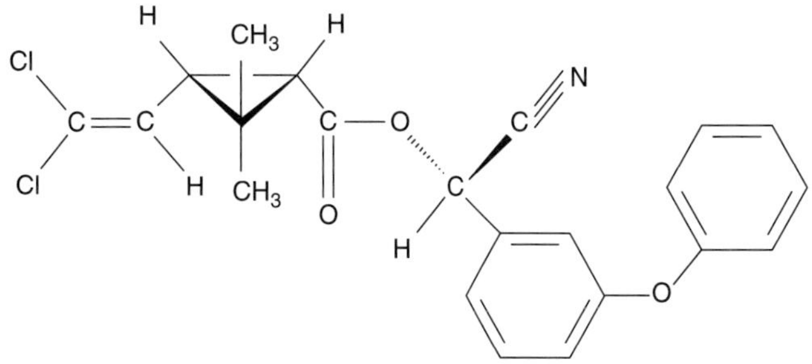

(*S*)-alcohol (1*S*)-*trans*-acid

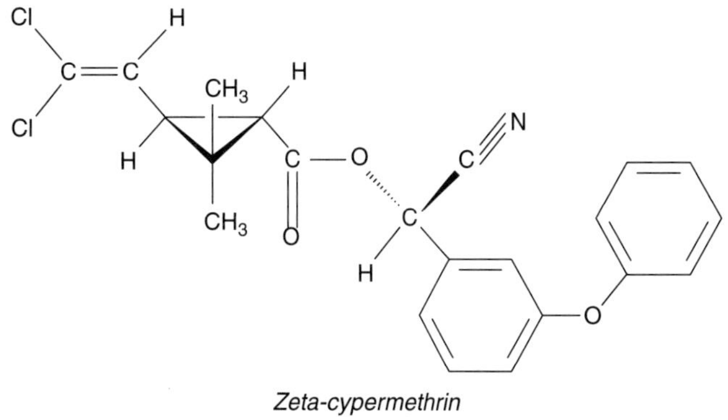

Zeta-cypermethrin

Zeta-cypermethrin is a pyrethroid chemical used as an insecticide. It has contact and stomach action. It is used in the control of beetles, aphids and members of the order Lepidoptera on cotton, fruits, vegetables, field crops and ornamentals. Also, it is used in forestry and in public health. It is available under the trade name Fury. It is formulated as an emulsifiable concentrate (EC), wettable powder (WP) and emulsion, oil in water (EW).

According to the WHO, zeta-cypermethrin comes under the toxicity Class Ib (highly hazardous). Ingestion of the compound can cause irritation to the skin and eyes, headache, dizziness, nausea, vomiting, diarrhoea, excessive salivation and fatigue. It is thus a PAN Bad Actor chemical.

In mammals, the chemical is eliminated rapidly via urine and faeces. It is non-toxic to bees.

Zeta-cypermethrin is registered in Australia, Canada, the USA, New Zealand, South Africa and many countries of the EU including the UK, Germany, Cyprus, France and Hungary.

Zinc arsenate

See **Arsenical inorganic insecticides**

Zinc as a plant nutrient

See **Plant nutrient excesses and deficiencies**

Zinc ethylenebisdithiocarbamates

See **Ethylenebisdithiocarbamate (EBDC) fungicides**

Zinc phosphide rodenticide

Zinc phosphide, Zn_3P_2 (trizinc diphosphate), is a member of the inorganic phosphide group. Zinc phosphide is a grey powder that is practically insoluble in water and ethanol but soluble in carbon disulfide and benzene. It decomposes in humid conditions. If it reacts with acids, it releases **phosphine gas** (PH_3). Zinc phosphide can be prepared by reacting finely powdered zinc with phosphorus.

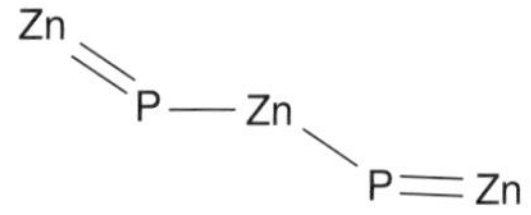

Zinc phosphide

Zinc phosphide has a strong, pungent, garlic-like odour, which attracts rodents. Zinc phosphide is used as a rodenticide as it has high mammalian toxicity. It is also effective on mice, house rats, field mice, ground squirrels and prairie dogs. When ingested, the compound comes in contact with stomach acids, releasing phosphine gas, which is absorbed into the bloodstream. Phosphine gas is toxic and inflammable. This causes damage to blood vessels and cardiovascular collapse. Death can occur anywhere within 15 min to 4 h, from heart and kidney failure.

In order to control crop loss brought about by rodents, zinc phosphide is used in doses of 1 kg/ha if the area is very badly infested. The first dose of zinc phosphide has to be sufficiently large to be effective, since bait shyness can occur. A 0.75–2% concentration in rodent baits is commonly considered an effective dose. It may be necessary to introduce the non-toxic bait food in a **pre-baiting** period while using zinc phosphide. It is quite possible that the extended offering of the same bait results in bait shyness. Zinc phosphide can be toxic to humans too, if it is absorbed into the gastrointestinal tract or if it is inhaled as phosphine gas. Even 2.8 mg/l of air can be fatal to man if care is not taken while handling the compound. Zinc phosphide is also very toxic to birds, cows, goats, sheep and pigs. Hence, rodent baits made of zinc phosphide are prepared along with antimony potassium tartrate, which can induce vomiting if the bait is consumed by non-target species. Secondary poisoning occurs rarely, unless a predator feeds on a freshly poisoned rodent. Its selectivity and slow action allows remedial options like antidotes and symptomatic treatment to be used on non-target species.

Zinc phosphide metabolizes into phosphoric acid and phosphate and phosphine gas. The compound is excreted via urine as hypophosphite and a number of reducing substances. It also does not accumulate in animal tissues. It is economical and safe for humans, if proper care is taken. It is relatively selective and slow acting, thus enhancing the chances of safety to non-target species. Zinc phosphide detoxifies in the target species and decomposes in the bait form, thus reducing long-term hazards. In soils too, zinc phosphide metabolizes into non-toxic forms and phosphine gas is liberated. In plants, it is not absorbed into roots and foliage.

The personnel administering zinc phosphide should have full protective clothing and should not eat, drink or smoke while handling the substance. It is considered as a restricted use pesticide (RUP) according to the EPA and also in many other countries, which indicates that it can be purchased and used by certified applicators only. In case of accidents, antidotes or symptomatic treatment may be administered.

The common trade names of zinc phosphide include Arvex, Azinphos, Commando, Ratol and Ridall. (See also **Anticoagulant rodenticides**.)

Zineb fungicide

Zineb, $C_4H_6N_2S_4Zn$ (zinc ethylenebis(dithiocarbamate) (polymeric)), is a polymeric foliar fungicide. It has a CAS Registry Number of 12122-67-7. It is an alkylenebis dithiocarbamate compound and is available as a pale yellow-coloured powder or crystals. It is soluble in water (10 mg/l), slightly soluble in pyridine and almost insoluble in common organic solvents. It dissolves irreversibly in some chelating agents such as EDTA. It is not compatible with alkaline substances and compounds containing mercury. It should be stored in a cool dry place but decomposes if stored for a long time. Thermal decomposition may release toxic oxides of nitrogen and sulfur. Zineb can be produced by reacting ethylenebis(dithiocarbamate) sodium salt and nabam, with zinc sulfate solution.

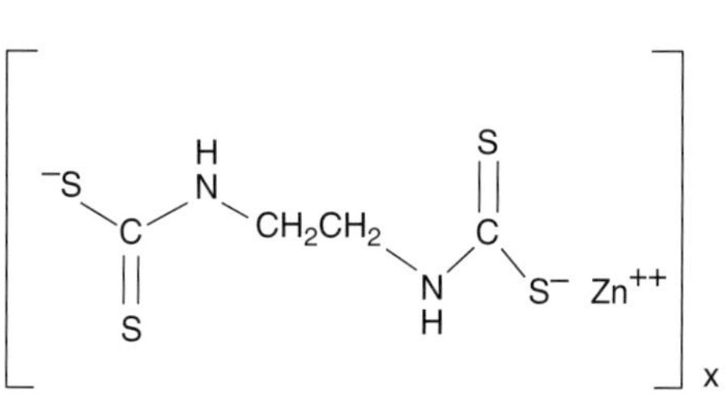

Zineb

Zineb inhibits respiration in the pathogen and thus effectively controls **downy mildews**. Its formulations (e.g. dustable powder (DP) and wettable powder (WP)) may be combined with other fungicides such as **copper oxychloride** and sulfur. It is available under trade names like Zinc ethylenebisdithiocarbamate (EBDC), Devizeb, Discon-Z, Lodaco, Mancozan, Tiezene, Zebtox, Ziden, Aspor, Dipher, Hexathane and Chem Zineb.

Zineb is a non-phytotoxic compound, except for zinc-sensitive plants such as tobacco and the cucurbit family. Zineb is a Category IV (practically non-toxic) toxin according to the EPA. It is slightly irritating to the skin and mucous membranes. The acute percutaneous LD_{50} for rats is >6000 mg/kg. The acute oral LD_{50} for rats is >5200 mg/kg. The ADI is 0.03mg/kg b.w./day. Zineb is broken down chemically to ethylene thiourea (ETU), which can cause abnormal foetal development. The teratogenic NOEL in rats was found to be 1000 mg/kg for zineb administered in single, daily oral doses. The WHO has categorized zineb as a Class U toxin (unlikely to be hazardous) according to acute toxicity criteria.

In plants, the main metabolite of zineb is **ethylenethiourea**. Ethylenethiuram monosulfide, ethylenethirum disulfide and sulfur are also formed. In rats, most of zineb is recovered unchanged in the faeces, with only 11–17% of an oral dose being absorbed into the body from the gastrointestinal tract of the rat. Zineb does not persist in soil and produces many metabolites. Its half-life in the field is 16 days and most of it is lost from a sprayed field within 4 months. It is a PAN Bad Actor chemical.

Zineb is registered for use in India, Australia, Canada and New Zealand. It is not registered in the USA or in the EU.

Ziram fungicide

Ziram, $C_6H_{12}N_2S_4Zn$ (zinc bis (dimethyldithiocarbamate)), is a dimethyldithiocarbamate fungicide. It is an off-white colour powder, which is soluble in water (0.97–18.3 mg/l), acetone (2.3 g/l), methanol (0.11g/l), toluene (2.33 g/l) and *n*-hexane (0.07 g/l). It is rapidly degraded in acidic solutions. It is not compatible with compounds containing iron, copper, mercury or lime. Ziram is produced by precipitation from an aqueous solution of sodium dimethyldithiocarbamate in the presence of a water-soluble zinc salt.

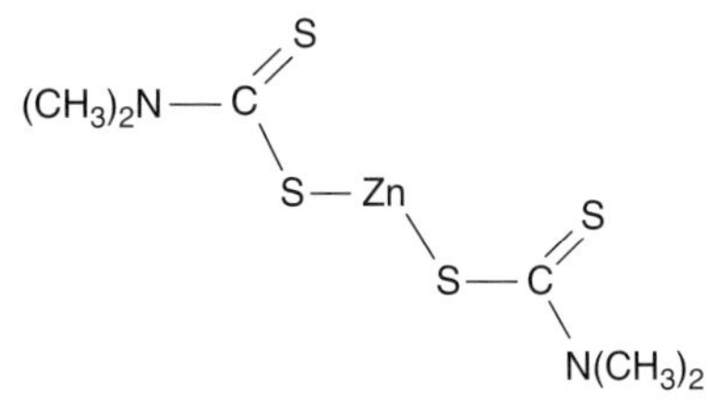

Ziram

Ziram is a foliar fungicide with protective action. Besides being a fungicide, it repels birds and rodents. It inhibits spore germination and mycelium growth by interfering with various enzymes involved in the respiration process. This property of acting on many sites of the biochemical process ensures that the product is free from resistance problems and effective against many fungi. It controls **scab**, rusts, leaf curl and anthracnose in fruits, nuts, vines, vegetables and ornamentals. It is well tolerated by all crops. It may be applied to the soil or for treating seeds. The presence of **dithiocarbamate** and zinc makes ziram very effective against *Taphrina* spp., *Phytophtora* spp. and *Septoria* spp. In a non-crop application, it is used as an accelerator in rubber manufacturing, adhesives and textiles. As a repellent, it is applied as a paste on tree trunks or sprayed onto ornamentals, dormant fruit trees and other crops. The application rate as a repellent on seeds is 0.05–0.20 kg/100 kg seed.

Some of the formulation types of ziram are liquid, wettable powder (WP) and paste (PA). Trade names of ziram include Crittam, Cuman, Miram and Attivar. Fungicide products of chemicals such as **bitertanol**, dodine and **myclobutanil** often contain ziram.

Both the WHO and the EPA classify ziram as a slightly toxic compound (Class III). It is irritating to the mucous membranes and eyes. Ziram carries the signal word 'DANGER' on its label due to severe eye irritation and possible irreversible eye damage. The acute percutaneous LD_{50} for rabbits is >2000 mg/kg. The acute oral LD_{50} for rats is 2068 mg/kg, while the LD_{50} in guinea pigs is between 100 and 150 mg/kg. The **NOAEL** (52 weeks) in dogs is 1.6 mg/kg b.w./day. In birds, the LD_{50} in bobwhite quails is 97 mg/kg. It is not toxic to bees. Ziram is non-phytotoxic except to zinc-sensitive crops such as tobacco and cucurbits. The ADI is 0.03 mg/kg b.w.

Almost all of ziram is eliminated within 2 days in rats. In plants, metabolites include the dimethylamine salt of dimethyldithiocarbamic acid and tetramethylthiourea, and sulfur. In soil it does not leach.

Ziram is a PAN Bad Actor chemical. It is registered for use in Australia, India, the USA and some EU members including Spain, France, Italy, Greece and the UK. The MRL value for ziram on fruits like bananas is 2 mg/kg, guava, 0.05 mg/kg and papaya, 7 mg/kg.

Zoxamide fungicide

Zoxamide, $C_{14}H_{16}Cl_3NO_2$ ((*RS*)-3, 5-dichloro-*N*-(3-chloro-1-ethyl-1-methyl-2-oxopropyl)-*p*-toluamide), is a benzamide fungicide. It is a white powder that is soluble in water (0.681 mg/l) and acetone (55.7 g/l). The CAS Registry Number is 156052-68-5.

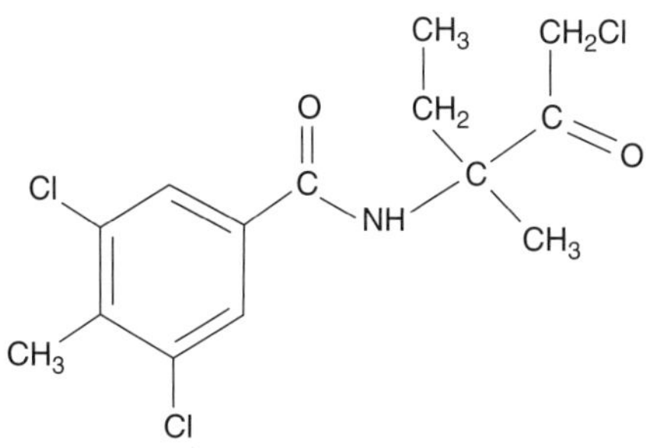

Zoxamide

Zoxamide is a preventive fungicide with rainfast and residual properties. It prevents nuclear division by binding to the β-subunit of tubulin and causing disruption of the microtubule cytoskeleton. When applied at the rate of 120–200 g/ha, it controls oomycete fungi on potatoes and tomatoes, *Plasmopara viticola* on vines and *Pseudoperonospora cubensis* on cucurbits. Zoxamide is not cross-resistant to current oomycete fungicides and **QoI** fungicides.

Zoxamide is used as a pre-formulated mixture with either **mancozeb** or cymoxanil. The formulation types of the fungicide are water-dispersible granule (WG) and wettable powder (WP).

Zoxamide is a non-teratogenic fungicide. It is a skin sensitizer and a moderate eye irritant but not a skin irritant. The acute oral LD_{50} for rats is >5000 mg/kg. The acute percutaneous LD_{50} in rats is >2000 mg/kg. The NOEL (1 year) in dogs is found to be 50 mg/kg b.w./day. The ADI is 0.1 mg/kg b.w./day.

Zoxamide is transient in the environment with low potential for leaching or persistence. In animals, both absorption and elimination are rapid, elimination taking place through urine and faeces. Several metabolites are formed as a result of hydrolysis, oxidation, reductive de-halogenation and conjugation. In plants, zoxamide penetrates surface layers with limited translocation. In soil, it has low mobility with no leaching potential. Over or constant use might make it a potential groundwater contaminant. However, it is not a PAN Bad Actor chemical.

Zoxamide is registered for use in South Africa, the UK and the USA.

Appendix 1: Common Chemical Compounds

Active ingredient	Class/group	Molecular formula	CAS Registry No.
Acephate	Insecticide	$C_4H_{10}NO_3PS$	30560-19-1
Acetamiprid	Insecticide	$C_{10}H_{11}ClN_4$	
Acifluorfen	Pesticide	$C_{14}H_7ClF_3NO_5$	50594-66-6
Acrinathrin	Insecticide	$C_{26}H_{21}F_6NO_5$	101007-06-1
Aldicarb	Insecticide	$C_7H_{14}N_2O_2S$	116-06-3
Aldoxycarb	Insecticide	$C_7H_{14}N_2O_4S$	1646-88-4
Aldrin	Insecticide	$C_{12}H_8Cl_6$	309-00-2
Allethrin	Insecticide	$C_{19}H_{26}O_3$	584-79-2
Alloxydim-sodium	Cyclohexanedione oxime	$C_{17}H_{24}NNaO_5$	55635-13-7
Alpha-cypermethrin	Insecticide	$C_{22}H_{19}Cl_2NO_3$	67375-30-8
Aminoethoxyvinylglycine	Plant growth regulator		49669-74-1
Amisulbrom	Fungicide	$C_{13}H_{13}BrFN_5O_4S_2$	348635-87-0
Amitrole	Herbicide	$C_2H_4N_4$	
Ancymidol	Plant growth regulator	$C_{15}H_{16}N_2O_2$	12771-68-5
Anilofos	Herbicide	$C_{13}H_{19}ClNO_3PS_2$	64249-01-0
Antraquinone	Bird repellent	$C_{14}H_8O_2$	84-65-1
Apholate	Chemosterilant	$C_{12}H_{24}N_9P_3$	52-46-0
Atrazine	Herbicide	$C_8H_{14}ClN_5$	1912-24-9
Azafenidin	Herbicide	$C_{15}H_{13}Cl_2N_3O_2$	68049-83-2
Azamethiphos	Insecticide	$C_9H_{10}ClN_2O_5PS$	35575-96-3
Azinphos-ethyl	Insecticide	$C_{12}H_{16}N_3O_3PS_2$	2642-71-9
Azinphos-methyl	Insecticide	$C_{10}H_{12}N_3O_3PS_2$	86-50-0
6-Benzylaminopurine (BAP)	Plant growth regulator	$C_{12}H_{11}N_5$	1214-39-7
Benalaxyl	Fungicide	$C_{20}H_{23}NO_3$	71626-11-4
Bendiocarb	Insecticide	$C_{11}H_{13}NO_4$	22781-23-3
Benomyl	Fungicide	$C_{14}H_{18}N_4O_3$	17804-35-2
Benoxacor	Herbicide safener	$C_{11}H_{11}Cl_2NO_2$	98730-04-2
Bensulide	Herbicide	$C_{14}H_{24}NO_4PS_3$	741-58-2
Bensultap	Insecticide	$C_{17}H_{21}NO_4S_4$	17606-31-4
Benthiavalicarb-isopropyl	Fungicide	$C_{18}H_{24}FN_3O_3S$	177406-68-7
Benzobicyclon	Herbicide	$C_{22}H_{19}ClO_4S_2$	156963-66-5
Benzoic acid	Herbicide	$C_7H_6O_2$	65-85-0
Beta-cypermethrin	Insecticide	$C_{22}H_{19}Cl_2NO_3$	65731-84-2
Bifenox	Herbicide	$C_{14}H_9Cl_2NO_5$	42576-02-3
Bifenthrin	Insecticide	$C_{23}H_{22}ClF_3O_2$	82657-04-3
Bilanafos	Herbicide	$C_{11}H_{22}N_3O_6P$	35597-43-4
Bioallethrin	Insecticide	$C_{19}H_{26}O_3$	28057-48-9
Bioresmethrin	Insecticide	$C_{22}H_{26}O_3$	28434-01-7
Biphenyl	Fungicide	$C_{12}H_{10}$	92-52-4
Bis (trichloromethyl) sulfone	Slimicide, fungicide	$C_2C_{l6}O_2S$	3064-70-8
Bismerthiazol	Bactericide	$C_5H_6N_6S_4$	79319-85-0
Bispyribac-sodium	Herbicide	$C_{19}H_{17}N_4NaO_8$	125401-92-5
Bitertanol	Fungicide	$C_{20}H_{23}N_3O_2$	55179-31-2
Blasticidin-S	Fungicide	$C_{17}H_{26}N_8O_5$	2079-00-7
Borax	Herbicide, insecticide	$Na_2B_4O_7.10H_2O$	
Bordeaux mixture	Fungicide, bactericide	$Ca_3Cu_4H_6O_{19}S_4$	
Boscalid	Fungicide	$C_{18}H_{12}Cl_2N_2O$	188425-85-6
Brodifacoum	Rodenticide	$C_{31}H_{23}BrO_3$	56073-10-0

Continued

Continued.

Active ingredient	Class/group	Molecular formula	CAS Registry No.
Bromacil	Herbicide	$C_9H_{13}BrN_2O_2$	314-40-9
Bromadiolone	Rodenticide	$C_{30}H_{23}BrO_4$	28772-56-7
Bromethalin	Rodenticide	$C_{14}H_7Br_3F_3N_3O_4$	63333-35-7
Bromofenoxim	Herbicide	$C_{13}H_7Br_2N_3O_6$	13181-17-4
Bromoxynil	Herbicide	$C_7H_3Br_2NO$	1689-84-5
Bronopol	Bactericide	$C_3H_6BrNO_4$	52-51-7
Bupirimate	Fungicide	$C_{13}H_{24}N_4O_3S$	41483-43-6
Burgundy mixture	Fungicide		11125-96-5
Butachlor	Herbicide	$C_{17}H_{26}ClNO_2$	23184-66-9
Butamifos	Herbicide	$C_{13}H_{21}N_2O_4PS$	36335-67-8
Butene-1,4-diyl,bis(bromoacetate)	Slimicide, bactericide, fungicide	$C_8H_{10}Br_2O_4$	
Butocarboxim	Insecticide	$C_7H_{14}N_2O_2S$	34681-10-2
Butoxycarboxim	Insecticide	$C_7H_{14}N_2O_4S$	34681-23-7
Butroxydim	Herbicide	$C_{24}H_{33}NO_4$	138164-12-2
Butylate	Herbicide	$C_{11}H_{23}NOS$	2008-41-5
Cadusafos	Insecticide	$C_{10}H_{23}O_2PS_2$	95465-99-9
Cafenstrole	Herbicide	$C_{16}H_{22}N_4O_3S$	125306-83-4
Captafol	Fungicide	$C_{10}H_9Cl_4NO_2S$	24-25-06-1
Captan (Kittleson's killer)	Fungicide	$C_9H_8Cl_3NO_2S$	133-06-2
Carbaryl	Insecticide	$C_{12}H_{11}NO_2$	63-25-2
Carbendazim	Fungicide	$C_9H_9N_3O_2$	10605-21-7
Carbetamide	Herbicide	$C_{12}H_{16}N_2O_3$	16118-49-3
Carbofuran	Insecticide	$C_{12}H_{15}NO_3$	1563-66-2
Carbosulfon	Insecticide	$C_{20}H_{32}N_2O_3S$	55285-14-8
Carboxin	Fungicide	$C_{12}H_{13}NO_2S$	5234-68-4
Carpropamid	Fungicide	$C_{15}H_{18}Cl_3NO$	104030-54-8
Cartap	Insecticide	$C_7H_{15}ClN_3O_2S_2$	15263-53-3
Carvone	Fungicide	$C_{10}H_{14}O$	99-49-0
Chinomethionat	Fungicide	$C_{10}H_6N_2OS_2$	2439-01-2
Chloralose	Avicide	$C_8H_{11}Cl_3O_6$	15879-93-3
Chloramben	Herbicide	$C_7H_5Cl_2NO_2$	28283-88-7
Chlordane	Insecticide	$C_{10}H_6Cl_8$	57-74-9
Chlordene	Insecticide		56534-02-2
Chlorethoxyfos	Insecticide	$C_6H_{11}Cl_4O_3PS$	54593-83-8
Chlorfenvinphos	Insecticide, acaricide	$C_{12}H_{14}Cl_3O_4P$	470-90-6
Chlorfluazuron	Insecticide	$C_{20}H_9Cl_3F_5N_3O_3$	71422-67-8
Chlorflurenol	Plant growth regulator	$C_{15}H_9ClO_3$	2464-37-1
Chlormephos	Insecticide	$C_5H_{12}ClO_2PS_2$	24934-91-6
Chlormequat	Growth retardant	$C_5H_{13}ClN$	7003-89-6
Chloroacetamides	Herbicide		
Chloroneb	Fungicide	$C_8H_8Cl_2O_2$	2675-77-6
Chlorophacinone	Rodenticide	$C_{23}H_{15}ClO_3$	3691-35-8
Chloropicrin	Fungicide	CCl_3NO_2	76-06-2
Chlorothalonil	Fungicide	$C_8Cl_4N_2$	1897-45-6
Chlorotriazines	Herbicide		
Chlorphonium	Plant growth regulator	$C_{19}H_{32}Cl_2P$	115-78-6
Chlorpyrifos	Insecticide	$C_9H_{11}Cl_3NO_3PS$	2921-88-2
Chlorpyrifos-methyl	Insecticide	$C_7H_7Cl_3NO_3PS$	5598-13-0
Cholecalciferol	Rodenticide	$C_{27}H_{44}O$	67-97-0
Clethodim	Herbicide	$C_{17}H_{26}ClNO_3S$	99129-21-2
Cloethocarb	Insecticide	$C_{11}H_{14}ClNO_4$	82496-76-2
Cloprop	Plant growth regulator	$C_9H_9ClO_3$	101-10-0
Clopyralid	Herbicide	$C_6H_3Cl_2NO_2$	1702-17-6
Cloransulam-methyl	Herbicide	$C_{15}H_{13}ClFN_5O_5S$	147150-35-4
Clothianidin	Insecticide	$C_6H_8ClN_5O_2S$	210880-92-5
Cloxyfonac	Plant growth regulator	$C_9H_9ClO_4$	6386-63-6
Copper hydroxide	Fungicide	CuH_2O_2	20427-59-2
Copper octanoate	Fungicide	$C_{16}H_{30}CuO_4$	20543-04-8
Copper oxychloride	Fungicide	$ClCu_2H_3O_3$	1332-40-7
Copper sulfate	Fungicide	$CuSO_4$	7758-98-7
Coumaphos	Insecticide	$C_{14}H_{16}ClO_5PS$	56-72-4
Coumarin	Rodenticide	$C_9H_6O_2$	91-64-5
Coumetetralyl	Rodenticide	$C_{19}H_{16}O_3$	5836-29-3

Continued

Continued.

Active ingredient	Class/group	Molecular formula	CAS Registry No.
Cyanophos	Insecticide	$C_9H_{10}NO_3PS$	2636-26-2
Cycloprothrin	Insecticide	$C_{26}H_{21}Cl_2NO_4$	63935-38-6
Cyfluthrin	Insecticide	$C_{22}H_{18}C_{l2}FNO_3$	68359-37-5
Cyhalothrin	Insecticide	$C_{23}H_{19}ClF_3NO_3$	68085-85-8
Cynamide	Herbicide	H_2NCN	
Cypermethrin	Insecticide	$C_{22}H_{19}Cl_2NO_3$	52315-07-8
Cyromazine	Insecticide	$C_6H_{10}N_6$	66215-27-8
2,4-D	Herbicide	$C_8H_6Cl_2O_3$	
2,4-DB	Herbicide	$C_{10}H_{10}Cl_2O_3$	94-82-6
1,3-Dichloropropene	Nematicide	$C_3H_4Cl_2$	99614-02-5
Daimuron	Phenylurea herbicide	$C_{17}H_{20}N_2O$	42609-52-9
Dalapon	Herbicide	$C_3H_4Cl_2O_2$	75-99-0
d-Allethrin	Insecticide	$C_{19}H_{26}O_3$	42534-61-2
Daminozide	Plant growth regulator	$C_6H_{12}N_2O_3$	1596-84-5
Dazomet	Fungicide, herbicide, insecticide, nematicide	$C_5H_{10}N_2S_2$	533-74-4
DCIP	Nematicide	$C_6H_{12}Cl_2O$	108-60-1
DDT	Insecticide	$C_{14}H_9Cl_5$	50-29-3 (for p,p'-DDT), 789-02-6 (for o,p'-DDT)
DEET	Insect repellent	$C_{12}H_{17}NO$	134-62-3
Deltamethrin	Insecticide	$C_{22}H_{19}Br_2NO_3$	52918-63-5
Demeton-*S*-methyl	Insecticide	$C_6H_{15}O_3PS_2$	919-86-8
Desmedipham	Herbicide	$C_{16}H_{16}N_2O_4$	13684-56-5
Desmetryn	Herbicide	$C_8H_{15}N_5S$	1014-69-3
Di-allate	Herbicide	$C_{10}H_{17}C_{l2}NOS$	2303-16-4
Diazinon	Insecticide	$C_{12}H_{21}N_2O_3PS$	333-41-5
Dicamba	Herbicide	$C_8H_6Cl_2O_3$	1918-00-9
Dichlobenil	Herbicide	$C_7H_3Cl_2N$	1194-65-6
Dichlone	Fungicide	$C_{10}H_4Cl_2O_2$	117-80-6
Dichlormid	Herbicide safener	$C_8H_{11}C_{l2}NO$	3-7764-25-3
Dichlorprop	Herbicide	$C_9H_8Cl_2O_3$	120-36-5
Dichlorprop-P	Herbicide	$C_9H_8Cl_2O_3$	15165-67-0
Dichlorvos	Insecticide	$C_4H_7Cl_2O_4P$	62-73-7
Diclofop-methyl	Herbicide	$C_{16}H_{14}Cl_2O_4$	51338-27-3
Diclosulam	Herbicide	$C_{13}H_{10}Cl_2FN_5O_3S$	145701-21-9
Dicrotophos	Insecticide	$C_8H_{16}NO_5P$	141-66-2
Dicyclanil	Insect growth regulator	$C_8H_{10}N_6$	112636-83-6
Dieldrin	Insecticide	$C_{12}H_{10}OC_{l6}$	60-57-1
Dienochlor	Insecticide, acaricide	$C_{10}Cl_{10}$	2227-17-0
Difenacoum	Rodenticide	$C_{13}H_{24}O_3$	56073-07-5
Difethialone	Rodenticide	$C_{31}H_{23}BrO_2S$	104653-34-1
Diflubenzuron	Insecticide	$C_{14}H_9ClF_2N_2O_2$	35367-38-5
Diflufenican	Herbicide	$C_{19}H_{11}F_5N_2O_2$	83164-33-4
Diflufenzopyr	Herbicide	$C_{15}H_{12}F_2N_4O_3$	109293-97-2
Di-halo	Bactericide, fungicide, slimicide	$C_5H_6BrClN_2O_2$	16079-88-2
Dimepiperate	Herbicide	$C_{15}H_{21}NOS$	61432-55-1
Dimethenamid	Herbicide	$C_{12}H_{18}ClNO_2S$	87674-68-8
Dimethipin	Plant growth regulator	$C_6H_{10}O_4S_2$	55290-64-7
Dimethoate	Insecticide	$C_5H_{12}NO_3PS_2$	60-51-5
Dimethyl phthalate	Insect repellent	$C_{10}H_{10}O_4$	131-11-3
Dimethylvinphos	Insecticide	$C_{10}H_{10}Cl_3O_4P$	2274-67-1
Dinitramin	Herbicide	$C_{11}H_{13}F_3N_4O_4$	29091-05-2
Dinocap	Fungicide	$C_{18}H_{24}N_2O_6$	39300-45-3
Dinoterb	Herbicide	$C_{10}H_{12}N_2O_5$	89396-94-1
Diofenolan	Insect growth regulator	$C_{18}H_{20}O_4$	63837-33-2
Diphacinone	Indandione rodenticide	$C_{25}H_{16}O_3$	82-66-6
Diquat dibromide	Herbicide	$C_{12}H_{12}Br_2N_2$	85-00-7
Disulfoton	Insecticide	$C_8H_{19}O_2PS_3$	298-04-4
Dithiopyr	Herbicide	$C_{15}H_{16}F_5NO_2S_2$	97886-45-8
Edifenphos	Fungicide	$C_{14}H_{15}O_2PS_2$	17109-49-8
Endosulfan	Insecticide	$C_9H_6Cl_6O_3S$	115-29-7
Endothall	Herbicide	$C_8H_{10}O_5$	145-73-3

Continued

Continued.

Active ingredient	Class/group	Molecular formula	CAS Registry No.
Endrin	Insecticide	$C_{12}H_8Cl_6O$	72-20-8
EPN	Insecticide	$C_{14}H_{14}NO_4PS$	2104-64-5
EPTC	Herbicide	$C_9H_{19}NOS$	759-94-4
Esfenvalerate	Insecticide	$C_{25}H_{22}ClNO_3$	66230-04-4
Esprocarb	Herbicide	$C_{15}H_{23}NOS$	85785-20-2
Ethephon	Plant growth regulator	$C_2H_6ClO_3P$	16672-87-0
Ethiofencarb	Insecticide	$C_{11}H_{15}NO_2S$	29973-13-5
Ethiolate	Herbicide	$C_7H_{15}NOS$	2941-55-1
Ethion	Insecticide	$C_9H_{22}O_4P_2S_4$	563-12-2
Ethirimol	Fungicide	$C_{11}H_{19}N_3O$	23947-60-6
Ethoprophos	Insecticide	$C_8H_{19}O_2PS_2$	13194-48-4
Ethychlozate	Plant growth regulator	$C_{11}H_{11}ClN_2O_2$	27512-72-7
Ethylene dibromide	Insecticide	$C_2H_4Br_2$	106-93-4
Etofenprox	Insecticide	$C_{25}H_{28}O_3$	80844-07-1
Etrimfos	Insecticide	$C_{10}H_{17}N_2O_4PS$	38260-54-7
Famoxadone	Fungicide	$C_{22}H_{18}N_2O_4$	131807-57-3
Fatty acid	Herbicide	$C_{18}H_{34}O_2$	
Fenamidone	Fungicide	$C_{17}H_{17}N_3OS$	161326-34-7
Fenamiphos	Insecticide	$C_{13}H_{22}NO_3PS$	22224-92-6
Fenarimol	Fungicide	$C_{17}H_{12}Cl_2N_2O$	60168-88-9
Fenbuconazole	Fungicide	$C_{19}H_{17}ClN_4$	114369-43-6
Fenchlorazole	Herbicide safener	$C_{10}H_4Cl_5N_3O_2$	103112-36-3
Fenclorim	Herbicide safener	$C_{10}H_6Cl_2N_2$	3740-92-9
Fenfuram	Fungicide	$C_{12}H_{11}NO_2$	24691-80-3
Fenobucarb	Insecticide	$C_{12}H_{17}NO_2$	3766-81-2
Fenoxaprop-P-ethyl,	Herbicide	$C_{18}H_{16}ClNO_5$	71283-80-2
Fenpropathrin	Insecticide, acarcide	$C_{22}H_{23}NO_3$	39515-41-8
Fenpropidin	Fungicide	$C_{19}H_{31}N$	67306-00-7
Fenpropimorph	Fungicide	$C_{20}H_{33}NO$	67564-91-4
Fenthion	Insecticide	$C_{10}H_{15}O_3PS_2$	73981-34-7
Fentin	Fungicide	$C_{18}H_{15}Sn$	668-34-8
Fentrazamide	Herbicide	$C_{16}H_{20}ClN_5O_2$	158237-07-1
Fenuron	Herbicide	$C_9H_{12}N_2O$	101-42-8
Ferbam	Fungicide	$C_9H_{18}FeN_3S_6$	14484-64-1
Ferimzone	Fungicide	$C_{15}H_{18}N_4$	89269-64-7
Ferrous sulfate	Herbicide	FeO_4S	7720-78-7
Flamprop	Herbicide	$C_{16}H_{13}ClFNO_3$	58667-63-3
Flazasulfuron	Herbicide	$C_{13}H_{12}F_3N_5O_5S$	104040-78-0
Flocoumafen	Rodenticide	$C_{33}H_{25}F_3O_4$	90035-08-8
Florasulam	Herbicide	$C_{12}H_8F_3N_5O_3S$	145701-23-1
Fluazifop	Herbicide	$C_{15}H_{12}F_3NO_4$	69335-91-7
Fluazinam	Fungicide	$C_{13}H_4Cl_2F_6N_4O_4$	79622-59-6
Fluazuron	Acaricide	$C_{20}H_{10}Cl_2F_5N_3O_3$	86811-58-7
Flucarbazone-sodium	Herbicide	$C_{12}H_{10}F_3N_4NaO_6S$	181274-17-9
Flucetosulfuron	Herbicide	$C_{18}H_{22}FN_5O_8S$	412928-75-7
Fluchloralin	Herbicide	$C_{12}H_{13}ClF_3N_3O_4$	33245-39-5
Flucythrinate	Insecticide	$C_{26}H_{23}F_2NO_4$	70124-77-5
Flufenacet	Herbicide	$C_{14}H_{13}F_4N_3O_2S$	142459-58-3
Flufenpyr-ethyl	Herbicide	$C_{16}H_{13}ClF_4N_2O_4$	188489-07-8
Flumetralin	Plant growth regulator	$C_{16}H_{12}ClF_4N_3O_4$	62924-70-3
Flumetsulam	Herbicide	$C_{12}H_9F_2N_5O_2S$	98967-40-9
Flumiclorac-pentyl	Herbicide	$C_{21}H_{23}ClFNO_5$	87546-18-7
Fluometuron	Herbicide	$C_{10}H_{11}F_3N_2O$	2164-17-2
Fluoroacetamide	Rodenticide	C_2H_4FNO	640-19-7
Fluoroglycofen-ethyl	Herbicide	$C_{18}H_{13}ClF_3NO_7$	77501-90-7
Fluoromide	Fungicide	$C_{10}H_4Cl_2FNO_2$	41205-21-4
Flupoxam	Herbicide	$C_{19}H_{14}ClF_5N_4O_2$	119126-15-7
Flupropanate	Herbicide	$C_3H_2F_4O_2$	756-09-2
Flupyrazofos	Insecticide		122431-24-7
Fluquinconazole	Fungicide	$C_{16}H_8Cl_2FN_5O$	136426-54-5
Fluridon	Herbicide	$C_{19}H_{14}F_3NO$	59763-91-6

Continued

Continued.

Active ingredient	Class/group	Molecular formula	CAS Registry No.
Fluridone	Herbicide	$C_{19}H_{14}F_3NO$	59756-60-4
Flurochloridone	Herbicide	$C_{12}H_{10}Cl_2F_3$	89286-81-7
Fluroxypyr	Herbicide	$C_7H_5Cl_2FN_2O_3$	69377-81-7
Flurprimidol	Plant growth regulator	$C_{15}H_{15}F_3N_2O_2$	56425-91-3
Flurtamone	Herbicide	$C_{18}H_{14}F_3NO_2$	96525-23-4
Flusilazole	Fungicide	$C_{16}H_{15}F_2N_3Si$	85509-19-9
Fluthiacet-methyl	Herbicide	$C_{15}H_{15}ClFN_3O_3S_2$	117337-19-6
Flutolanil	Fungicide	$C_{17}H_{16}F_3NO_2$	66332-96-5
Folpet	Fungicide	$C_9H_4Cl_3NO_2S$	133-07-3
Fomesafen	Herbicide	$C_{15}H_{10}ClF_3N_2O_6S$	72178-02-0
Fonofos	Insecticide	$C_{10}H_{15}OPS_2$	944-22-9
Foramsulfuron	Herbicide	$C_{17}H_{20}N_6O_7S$	173159-57-4
Forchlorfenuron	Plant growth regulator	$C_{12}H_{10}ClN_3O$	68157-60-8
Formaldehyde	Fumigant	CH_2O	50-00-0
Formetanate	Insecticide	$C_{11}H_{15}N_3O_2$	22259-30-9
Formic acid	Acaricide	CH_2O_2	64-18-6
Formothion	Insecticide	$C_6H_{12}NO_4PS_2$	2540-82-1
Fosamine	Herbicide	$C_3H_8NO_4P$	59682-52-9
Fosetyl-aluminium	Fungicide	$C_6H_{18}AlO_9P_3$	39148-24-8
Fosthiazate	Nematicide	$C_9H_{18}NO_3PS_2$	98886-44-3
Fuberidazole	Fungicide	$C_{11}H_8N_2O$	3878-19-1
Furathiocarb	Insecticide	$C_{18}H_{26}N_2O_5S$	65907-30-4
Gibberellic acid	Plant growth regulator	$C_{19}H_{22}O_6$	77-06-05
Glufosinate-ammonium	Herbicide	$C_5H_{15}N_2O_4P$	77182-82-2
Glyphosate	Herbicide	$C_3H_8NO_5P$	1071-83-6
Guanidine	Fungicide	CH_5N_3	646-34-4
Guazatine	Fungicide	$C_{18}H_{41}N_7$	108173-90-6
GY-81	Fungicide	CNa_2S_4	7345-69-9
2-Hydrazinoethanol	Plant growth regulator	$C_2H_8N_2O$	109-84-2
8-Hydroxyquinoline sulfate	Fungicide	$C_{18}H_{16}N_2O_6S$	134-31-6
Halosulfuron-methyl	Herbicide	$C_{13}H_{15}ClN_6O_7S$	100784-20-1
Haloxyfop-p	Herbicide	$C_{15}H_{11}ClF_3NO_4$	95977-29-0
Heptachlor	Insecticide	$C_{10}H_8Cl_6$	76-44-8
Heptenophos	Insecticide	$C_9H_{12}ClO_4P$	23560-59-0
Hexachlorobenzene	Insecticide	C_6Cl_6	118-74-1
Hexachlorocyclohexane	Insecticide	$C_6H_6Cl_6$	58-89-9
Hexaconazole	Fungicide	$C_{14}H_{17}Cl_2N_3O$	79983-71-4
Hexaflumuron	Insecticide	$C_{16}H_8Cl_2F_6N_2O_3$	86479-06-3
Hexamethylphosphoramide	Chemosterilant	$C_6H_{18}N_3OP$	680-31-9
Hexazinone	Herbicide	$C_{12}H_{20}N_4O_2$	51235-04-2
Hydrogen cyanide	Rodenticide	CHN	74-90-8
Hydroprene	Insect growth regulator	$C_{17}H_{30}O_2$	41096-46-2
Imazamethabenz-methyl	Herbicide	$C_{16}H_{20}N_2O_3$	81405-85-8
Imazamox	Herbicide	$C_{15}H_{19}N_3O_4$	114311-32-9
Imazapic	Herbicide	$C_{14}H_{17}N_3O_3$	104098-48-8
Imazapyr	Herbicide	$C_{13}H_{15}N_3O_3$	81334-34-1
Imazaquin	Herbicide	$C_{17}H_{17}N_3O_3$	81335-37-7
Imazethapyr	Herbicide	$C_{15}H_{19}N_3O_3$	81335-77-5
Imazosulfuron	Herbicide	$C_{14}H_{13}ClN_6O_5S$	122548-33-8
Imidacloprid	Insecticide	$C_9H_{10}ClN_5O_2$	138261-41-3
Imiprothrin	Insecticide	$C_{17}H_{22}N_2O_4$	72963-72-5
Inabenfide	Plant growth regulator	$C_{19}H_{15}ClN_2O_2$	82211-24-3
Indanofan	Herbicide	$C_{20}H_{17}ClO_3$	133220-30-1
Iodosulfuron-methyl-sodium	Herbicide	$C_{14}H_{13}IN_5NaO_6S$	144550-36-7
Ioxynil	Herbicide	$C_7H_3I_2NO$	1689-83-4
Iprobenfos	Fungicide	$C_{13}H_{21}O_3PS$	26087-47-8
Iprodione	Fungicide	$C_{13}H_{13}Cl_2N_3O_3$	61840-50-4
Isazofos	Insecticide	$C_9H_{17}ClN_3O_3PS$	42509-80-8
Isobardac	Slimicide	$C_{22}H_{48}ClN$	138698-36-9
Isofenphos	Insecticide	$C_{15}H_{24}NO_4PS$	25311-71
Isopolinate	Herbicide	$C_{10}H_{19}NOS$	3134-70-1

Continued

Continued.

Active ingredient	Class/group	Molecular formula	CAS Registry No.
Isoprocarb	Insecticide	$C_{11}H_{15}NO_2$	2631-40-5
Isoproturon	Herbicide	$C_{12}H_{18}N_2O$	34123-59-6
Isouron	Herbicide	$C_{10}H_{17}N_3O_2$	55861-78-4
Isoxaben	Herbicide	$C_{18}H_{24}N_2O_4$	82558-50-7
Isoxadifen-ethyl	Herbicide safener	$C_{18}H_{17}NO_3$	163520-33-0
Isoxaflutole	Herbicide	$C_{15}H_{12}F_3NO_4S$	141112-29-0
Isoxapyrifop	Herbicide	$C_{17}H_{16}Cl_2N_2O_4$	87757-18-4
Isoxathion	Insecticide	$C_{13}H_{16}NO_4PS$	18854-01-8
Juglone	Fungicide	$C_{10}H_6O_3$	481-39-0
Kadethrin	Insecticide	$C_{23}H_{24}O_4S$	58769-20-3
Karbutilate	Herbicide	$C_{14}H_{21}N_3O_3$	4849-32-5
KIH 2023	Herbicide	$C_{19}H_{17}N_4NaO_8$	
Kittleson's killer	See Captan		
Lenacil	Herbicide	$C_{13}H_{18}N_2O_2$	2164-08-1
Linuron	Herbicide	$C_9H_{10}Cl_2N_2O_2$	330-55-2
Lufenuron	Insect growth regulator	$C_{17}H_8Cl_2F_8N_2O_3$	103055-07-8
Malathion	Insecticide	$C_{10}H_{19}O_6PS_2$	121-75-5
Maleic hydrazide	Plant growth regulator	$C_4H_4N_2O_2$	10071-13-3, 123-33-1, 51542-52-0
Mancopper	Fungicide	$C_8H_{12}CuMnN_4S_8$	53988-93-5
Mancozeb	Fungicide	$[C_4H_6MnN_2S_4]_xZn_y$	8018-01-7
Maneb	Fungicide	$C_4H_6MnN_2S_4$	12427-38-2
MCPA	Herbicide	$C_9H_9ClO_3$	94-74-6
MCPA-thioethyl	Herbicide	$C_{11}H_{13}ClO_2S$	25319-90-8
MCPB	Herbicide	$C_{11}H_{13}ClO_2$	94-81-5
Mecarbam	Insecticide	$C_{10}H_{20}NO_5PS_2$	2595-54-2
Mecoprop	Herbicide	$C_{10}H_{11}ClO_3$	7085-19-0, 93-65-2
Mecoprop-P	Herbicide	$C_{10}H_{11}ClO_3$	16484-77-8
Mefenacet	Herbicide	$C_{16}H_{14}N_2O_2S$	73250-68-7
Mefenpyr-diethyl	Herbicide safener	$C_{16}H_{18}Cl_2N_2O_4$	
Mefluidide	Plant growth regulator	$C_{11}H_{13}F_3N_2O_3S$	55599-99-0
MEMC	Fungicide	C_3H_7ClHgO	123-88-6
Mepanipyrim	Fungicide	$C_{14}H_{13}N_3$	110235-47-7
Mepiquat	Plant growth regulator	$C_7H_{16}N$	
Mercuric chloride	Fungicide	Cl_2Hg	7487-94-7
Mercurous chloride	Fungicide	Hg_2Cl_2	39448-57-2
Mesosulfuron-methyl	Herbicide	$C_{17}H_{21}N_5O_9S_2$	208465-21-8
Metalaxyl	Fungicide	$C_{15}H_{21}NO_4$	57837-19-1
Metalaxyl-M	Fungicide	$C_{15}H_{21}NO_4$	
Metallic dithiocarbamates	Fungicide	$C_9H_{18}FeN_3S_6$ or $[(CH_3)_2NCSS]_3Fe$	14484-64-1
Metamifop	Herbicide	$C_{23}H_{18}ClFN_2O_4$	
Metamitron	Herbicide	$C_{10}H_{10}N_4O$	41394-05-2
Metazachlor	Herbicide	$C_{14}H_{16}ClN_3O$	
Metconazole	Fungicide	$C_{12}H_{22}ClN_3O$	125116-23-6
METEPA	Chemosterilant	$C_9H_{18}N_3OP$	57-39-6
Methabenzthiazuron	Herbicide	$C_{10}H_{11}N_3OS$	18691-97-9
Methacrifos	Insecticide	$C_7H_{13}O_5PS$	62610-77-9
Methamidophos	Insecticide	$C_2H_8NO_2PS$	10265-92-6
Methasulfocarb	Fungicide	$C_9H_{11}NO_4S_2$	66952-49-6
Methidathion	Insecticide	$C_6H_{11}N_2O_4PS_3$	950-37-8
Methiobencarb	Herbicide	$C_{13}H_{19}NO_2S$	18357-78-3
Methiocarb	Insecticide	$C_{11}H_{15}NO_2S$	2032-65-7
Methiocarb	Molluscicide	$C_{11}H_{15}NO_2S$	716-16-5
Methomyl	Insecticide	$C_5H_{10}N_2O_2S$	16752-77-5
Methoprene	Insecticide	$C_{19}H_{34}O_3$	40596-69-8
Methoxychlor	Insecticide	$C_{16}H_{15}Cl_3O_2$	30667-99-3, 72-43-5
Methyl bromide	Fumigant	CH_3Br	74-83-9
Methyl isothiocyanate	Nematicide	CH_3NCS	556-61-6
Methylarsonic acid	Herbicide	CH_5AsO_3	65513-69-1
Methyldymron	Herbicide	$C_{17}H_{20}N_2O$	42609-73-4
Methylene bis (thiocyanate)	Slimicide	$C_3H_2N_2S_2$	

Continued

Continued.

Active ingredient	Class/group	Molecular formula	CAS Registry No.
Metiram	Fungicide	$[C_{16}H_{33}N_{11}S_{16}Zn_3]_x$	9006-42-2
Metobenzuron	Herbicide	$C_{22}H_{28}N_2O_5$	
Metobromuron	Herbicide	$C_9H_{11}BrN_2O_2$	3060-89-7
Metolachlor	Herbicide	$C_{15}H_{22}ClNO_2$	51218-45-2
Metolcarb	Insecticide	$C_9H_{11}NO_2$	1129-41-5
Metominostrobin	Fungicide	$C_{16}H_{16}N_2O_3$	133408-50-1
Metosulam	Herbicide	$C_{14}H_{13}Cl_2N_5O_4S$	139528-85-1
Metoxuron	Herbicide, plant growth regulator	$C_{10}H_{13}ClN_2O_2$	19937-59-8
Metrafenone	Fungicide	$C_{19}H_{21}BrO_5$	220899-03-6
Metribuzin	Herbicide	$C_8H_{14}N_4OS$	21087-64-9
Metsulfuron	Herbicide	$C_{13}H_{13}N_5O_6S$	79510-48-8
Metsulfuron-methyl	Herbicide	$C_{14}H_{15}N_5O_6S$	82197-07-7
Mevinphos	Insecticide	$C_7H_{13}O_6P$	7786-34-7, 26718-65-0
Molinate	Herbicide	$C_9H_{17}NOS$	2212-67-1
Monocrotophos	Insecticide	$C_7H_{14}NO_5P$	6923-22-4
Monolinuron	Herbicide	$C_9H_{11}ClN_2O_2$	56449-36-6
Myclobutanil	Fungicide	$C_{15}H_{17}ClN_4$	88671-89-0
(2-Naphthyloxy) acetic acid	Plant growth regulator	$C_{12}H_{10}O_3$	120-23-0
2-(1-Naphthyl) acetamide	Plant growth regulator	$C_{12}H_{11}NO$	
2-(1-Naphthyl) acetic acid	Plant growth regulator	$C_{12}H_{10}O_2$	
Nabam	Fungicide	$C_4H_6N_2Na_2S_4$	142-59-6
Naled	Insecticide	$C_4H_7Br_2Cl_2O_4P$	300-76-5
Naphthenic acid	Fungicide		1338-24-5
Napropamide	Herbicide	$C_{17}H_{21}NO_2$	
Neburon	Herbicide	$C_{12}H_{16}Cl_2N_2O$	555-37-3
Netamycin	Fungicide	$C_{33}H_{47}NO_{13}$	
Nicosulfuron	Herbicide	$C_{15}H_{18}N_6O_6S$	111991-09-4
Nitempyram	Insecticide	$C_{11}H_{15}ClN_4O_2$	150824-47-8
Nithiazine	Insecticide	$C_5H_8N_2O_2S$	58842-20-9
Nitrapyrin	Bactericide	$C_6H_3Cl_4N$	1929-82-4
Nitrothal-isopropyl	Fungicide	$C_{14}H_{17}NO_6$	10552-74-6
Norflurazon	Herbicide	$C_{12}H_9ClF_3N_3O$	27314-13-2
N-phenylphthalamic acid	Plant growth regulator	$C_{14}H_{11}NO_3$	4727-29-1
Nuarimol	Fungicide	$C_{17}H_{12}ClFN_2O$	63284-71-9
Ofurace	Fungicide	$C_{14}H_{16}ClNO_3$	58810-48-3
Orbencarb	Herbicide	$C_{12}H_{16}ClNOS$	34622-58-7
Ortho dichlorobenzene	Termiticide	$C_6H_4Cl_2$	
Orthosulfamuron	Herbicide	$C_{16}H_{20}N_6O_6S$	
Oryzalin	Herbicide	$C_{12}H_{18}N_4O_6S$	19044-88-3
Oxabetrinil	Herbicide safener	$C_{12}H_{12}N_2O_3$	94593-79-0
Oxadiargyl	Herbicide	$C_{15}H_{14}Cl_2N_2O_3$	39807-15-3
Oxadixyl	Fungicide	$C_{14}H_{18}N_2O_4$	77732-09-3
Oxamyl	Insecticide	$C_7H_{13}N_3O_3S$	23135-22-0
Oxasulfuron	Herbicide	$C_{17}H_{18}N_4O_6S$	144651-06-9
Oxaziclomefone	Herbicide	$C_{20}H_{19}Cl_2NO_2$	
Oxidiazoles	Herbicide group		
Oxine-copper	Fungicide	$C_{18}H_{12}CuN_2O_2$	10380-28-6
Oxolinic acid	Bactericide	$C_{13}H_{11}NO_5$	14698-29-4
Oxpoconazole fumarate	Fungicide	$C_{42}H_{52}Cl_2N_6O_8$	
Oxycarboxin	Fungicide	$C_{12}H_{13}NO_4S$	5259-88-1
Oxydemeton-methyl	Insecticide	$C_6H_{15}O_4PS_2$	301-12-2
Oxyfluorfen	Herbicide	$C_{15}H_{11}ClF_3NO_4$	42874-03-3
2-Phenylphenol	Fungicide	$C_{12}H_{10}O$	
Paclobutrazol	Plant growth retardant	$C_{15}H_{20}ClN_3O$	76738-62-0
Paraquat	Herbicide	$C_{12}H_{14}Cl_2N_2$	1910-42-5
Parathion	Insecticide	$C_{10}H_{14}NO_5PS$	56-38-2
Parathion-methyl	Insecticide	$C_8H_{10}NO_5PS$	298-00-0
Pebulate	Herbicide	$C_{10}H_{21}NOS$	1114-71-2
Penconazole	Fungicide	$C_{13}H_{15}C_{l2}N_3$	66246-88-6

Continued

Continued.

Active ingredient	Class/group	Molecular formula	CAS Registry No.
Pencycuron	Fungicide	$C_{19}H_{21}ClN_2O$	66063-05-6
Pendimethalin	Herbicide	$C_{13}H_{19}N_3O_4$	64719-41-1
Pentachlorophenol (PCP)	Insecticide	C_6HCl_5O	87-86-5
Pentanochlor	Herbicide	$C_{13}H_{18}ClNO$	2307-68-8
Penthiopyrad	Fungicide	$C_{16}H_{20}F_3N_3OS$	183675-82-3
Pentoxazone	Herbicide	$C_{17}H_{17}ClFNO_4$	
Permethrin	Insecticide	$C_{21}H_{20}Cl_2O_3$	52645-53-1
Pethoxamid	Herbicide	$C_{16}H_{22}ClNO_2$	106700-29-2
Phenmedipham	Herbicide	$C_{16}H_{16}N_2O_4$	13684-63-4
Phenothrin	Insecticide	$C_{23}H_{26}O_3$	26002-80-2
Phenthoate	Insecticide	$C_{12}H_{17}O_4PS_2$	2597-03-7
Phenylmercury acetate	Fungicide	$C_8H_8HgO_2$	62-38-4
Phorate	Insecticide	$C_7H_{17}O_2PS_3$	298-02-2
Phosalone	Insecticide	$C_{12}H_{15}ClNO_4PS_2$	2310-17-0
Phosdiphen	Fungicide	$C_{14}H_{11}Cl_4O_4P$	36519-00-3
Phosmet	Insecticide	$C_{11}H_{12}NO_4PS_2$	732-11-6
Phosphamidon	Insecticide	$C_{10}H_{19}ClNO_5P$	13171-21-6
Phosphonic acid	Fungicide	H_3O_3P	13598-36-2
Phoxim	Insecticide	$C_{12}H_{15}N_2O_3PS$	14816-18-3
Phthalide	Fungicide	$C_8H_2Cl_4O_2$	27355-22-2
Picloram	Herbicide	$C_6H_3Cl_3N_2O_2$	1918-02-1
Picolinafen	Herbicide	$C_{19}H_{12}F_4N_2O_2$	
Picolinic acid	Herbicide	$C_6H_5NO_2$	98-98-6
Picoxystrobin	Fungicide	$C_{18}H_{16}F_3NO_4$	117428-22-5
Pindone	Rodenticide	$C_{14}H_{14}O_3$	83-26-1
Pinoxaden	Herbicide	$C_{23}H_{32}N_2O_4$	243973-20-8
Piperalin	Fungicide	$C_{16}H_{21}Cl_2NO_2$	3478-94-2
Piperophos	Herbicide	$C_{14}H_{28}NO_3PS_2$	24151-93-7
Pirimicarb	Insecticide	$C_{11}H_{18}N_4O_2$	23103-98-2
Pirimiphos-ethyl	Insecticide	$C_{13}H_{24}N_3O_3PS$	23505-41-1
Pirimiphos-methyl	Insecticide	$C_{11}H_{20}N_3O_3PS$	29232-93-7
Prallethrin	Insecticide	$C_{19}H_{24}O_3$	23031-36-9
Pretilachlor	Herbicide	$C_{17}H_{26}ClNO_2$	51218-49-6
Primisulfuron-methyl	Herbicide	$C_{15}H_{12}F_4N_4O_7S$	86209-51-0
Probenazole	Fungicide	$C_{10}H_9NO_3S$	27605-76-1
Prochloraz	Fungicide	$C_{15}H_{16}Cl_3N_3O_2$	67747-09-5
Procymidone	Fungicide	$C_{13}H_{11}Cl_2NO_2$	32809-16-8
Profenophos	Insecticide	$C_{11}H_{15}BrClO_3PS$	41198-08-7
Prohexadione calcium	Plant growth regulator	$C_{10}H_{10}CaO_5$	127277-53-6
Propachlor	Herbicide	$C_{11}H_{14}ClNO$	1918-16-7
Propamocarb	Fungicide	$C_9H_{21}ClN_2O_2$	25606-41-1
Propaphos	Insecticide	$C_{13}H_{21}O_4PS$	7292-16-2
Propaquizafop	Herbicide	$C_{22}H_{22}ClN_3O_5$	111479-05-1
Propetamphos	Insecticide	$C_{10}H_{20}NO_4PS$	31218-83-4
Propham	Herbicide	$C_{10}H_{13}NO_2$	122-42-9
Propiconazole	Fungicide	$C_{15}H_{17}Cl_2N_3O_2$	60207-90-1
Propineb	Fungicide	$(C_5H_8N_2S_4Zn)_x$	12071-83-9
Propisochlor	Herbicide	$C_{15}H_{22}ClNO_2$	86763-47-5
Propoxur	Insecticide	$C_{11}H_{15}NO_3$	114-26-1
Proquinazid	Fungicide	$C_{14}H_{17}IN_2O_2$	189278-12-4
Prosulfocarb	Herbicide	$C_{14}H_{21}NOS$	52888-80-9
Prothiofos	Insecticide	$C_{11}H_{15}Cl_2O_2PS_2$	34643-46-4
Pyraclofos	Insecticide	$C_{14}H_{18}ClN_2O_3PS$	77458-01-6
Pyrazophos	Fungicide	$C_{14}H_{20}N_3O_5PS$	13457-18-6
Pyributicarb	Herbicide	$C_{18}H_{22}N_2O_2S$	88678-67-5
Pyridaphenthion	Insecticide	$C_{14}H_{17}N_2O_4$	119-12-0
Pyridate	Herbicide	$C_{19}H_{23}ClN_2O_2S$	55512-33-9
Pyrifenox	Fungicide	$C_{14}H_{12}Cl_2N_2O$	88283-41-4.
Pyriproxyfen	Insecticide	$C_{20}H_{19}NO_3$	95737-68-1
Pyrithiobac	Herbicide	$C_{13}H_{11}ClN_2O_4S$	123342-93-8
Pyrithiobac-sodium	Herbicide	$C_{13}H_{10}ClN_2NaO_4S$	123343-16-8

Continued

Continued.

Active ingredient	Class/group	Molecular formula	CAS Registry No.
Quinalphos	Insecticide	$C_{12}H_{15}N_2O_3PS$	13593-03-8
Quinclorac	Herbicide	$C_{10}H_5Cl_2NO_2$	84087-01-4
Quinmerac	Herbicide	$C_{11}H_8ClNO_2$	90717-03-6
Quinoclamine	Herbicide	$C_{10}H_6ClNO_2$	2797-51-5
Quinoxyfen	Fungicide	$C_{15}H_8Cl_2FNO$	124495-18-7
Quintozene	Fungicide	$C_6Cl_5NO_2$	82-68-8
Resmethrin	Insecticide	$C_{22}H_{26}O_3$	10453-86-8
S, S, S-tributyl phosphorotrithionate	Plant growth regulator	$C_{12}H_{27}OPS_3$	
S-bioallethrin	Insecticide	$C_{19}H_{26}O_3$	28434-00-6
sec-Butylamine	Fungicide	$C_4H_{11}N$ or $CH_3CH_2CH(CH_3)NH_2$	13952-84-6
Sethoxydim	Herbicide	$C_{17}H_{29}NO_3S$	74051-80-2
Silafluofen	Insecticide	$C_{25}H_{29}FO_2Si$	105024-66-6
Silthiofam	Fungicide	$C_{13}H_{21}NOSSi$	175217-20-6
Simeconazole	Fungicide	$C_{14}H_{20}FN_3OSi$	149508-90-7
S-metolachlor	Herbicide	$C_{15}H_{22}ClNO_2$	
Sodium 2-mercaptibenzothiazolate	Slimicide, fungicide	$C_7H_5NNaS_2$	
Sodium dimethyldithiocarbamate	Slimicide, fungicide	$C_3H_{10}NNaO_2S_2$	72140-17-1
Sodium fluoroacetate	Rodenticide	$C_2H_2FNaO_2$	62-74-8
Sodium tetrathiocarbonate	Fungicide	CNa_2S_4	7345-69-9
Spiroxamine	Fungicide	$C_{18}H_{35}NO_2$	118134-30-8
Strychnine	Rodenticide, avicide	$C_{21}H_{22}N_2O_2$	
Sulfallate	Herbicide	$C_8H_{14}ClNS_2$	95-06-7
Sulfotep	Insecticide	$C_8H_{20}O_5P_2S_2$	3689-24-5
Sulfur	Fungicide	S_x	
Sulprofos	Insecticide	$C_{12}H_{19}O_2PS_3$	35400-43-2
SYP-Z071	Fungicide	$C_{22}H_{22}ClNO_4$	
2-(Thiocyanatomethylthio)	Fungicide, bactericide	$C_9H_6N_2S_3$	
2-(Thiocyanatomethylthio) benzothiazole	Fungicide	$C_9H_6N_2S_3$	
Tau-fluvalinate	Insecticide	$C_{26}H_{23}ClF_3N_2O_3$	102851-06-9
Tebuconazole	Fungicide	$C_{16}H_{22}ClN_3O$	107534-96-3
Tebupirimfos	Insecticide	$C_{13}H_{23}N_2O_3PS$	96182-53-5
Tebutam	Herbicide	$C_{15}H_{23}NO$	35256-85-0
Tebuthiuron	Herbicide	$C_9H_{16}N_4OS$	34014-18-1
Tecnazene	Fungicide, plant growth regulator	$C_6HCl_4NO_2$	117-18-0
Teflubenzuron	Insect growth regulator	$C_{14}H_6Cl_2F_4N_2O_2$	83121-18-0
Tefluthrin	Insecticide	$C_{17}H_{14}ClP_7O_2$	79538-32-2
Tembotrione	Herbicide	$C_{17}H_{16}ClF_3O_6S$	
Temephos	Insecticide	$C_{16}H_{20}O_6P_2S_3$	3383-96-8
TEPA	Chemosterilant	$C_6H_{12}N_3OP$	545-55-1
Tepraloxydim	Herbicide	$C_{17}H_{24}ClNO_4$	149979-41-9
Terbacil	Herbicide	$C_9H_{13}ClN_2O_2$	5902-51-2
Terbufos	Insecticide	$C_9H_{21}O_2PS_3$	13071-79-9
Terbumeton	Herbicide	$C_{10}H_{19}N_5O$	33693-04-8
Terbuthylazine	Herbicide	$C_9H_{16}ClN_5$	63026-57-3
Terbutryn	Herbicide	$C_{10}H_{19}N_5S$	886-50-0
Tetrachlorvinphos	Insecticide	$C_{10}H_9Cl_4P$	22248-79-9
Tetraconazole	Fungicide	$C_{13}H_{11}Cl_2F_4N_3O$	112281-77-3
Tetramethrin	Insecticide	$C_{19}H_{25}NO_4$	7696-12-0
Thiabendazole	Fungicide	$C_{10}H_7N_3S$	148-79-8
Thiacloprid	Insecticide	$C_{10}H_9ClN_4S$	111988-49-9
Thiadiazuran	Plant growth regulator	$C_9H_8N_4OS$	51707-55-2
Thiamethoxam	Insecticide	$C_8H_{10}ClN_5O_3S$	153719-23-4
Thiazopyr	Herbicide	$C_{16}H_{17}F_5N_2O_2S$	117718-60-2
Thifensulfuron-methyl	Herbicide	$C_{12}H_{13}N_5O_6S_2$	79277-27-3
Thifluzamide	Fungicide	$C_{13}H_6Br_2F_6N_2O_2S$	130000-40-7
Thiobencarb	Herbicide	$C_{12}H_{16}ClNOS$	28249-77-6
Thiocyclam	Insecticide	$C_5H_{11}NS_3$	31895-21-3
Thiodiazole-copper	Fungicide	$C_4H_4CuN_6S_4$	
Thiodicarb	Insecticide	$C_{10}H_{18}N_4O_4S_3$	59669-26-0

Continued

Continued.

Active ingredient	Class/group	Molecular formula	CAS Registry No.
Thiofanox	Insecticide	$C_9H_{18}N_2O_2S$	39196-18-4
Thiometon	Insecticide	$C_6H_{15}O_2PS_3$	640-15-3
Thiophanate-methyl	Fungicide	$C_{12}H_{14}N_4O_4S_2$	23564-05-8
Thio-TEPA	Chemosterilant	$C_6H_{12}N_3PS$	52-24-4
Thiuram disulphide	Fungicide	$[(CH_3)_2NCH]_2S_2$	137-26-8
Tiocarbazil	Herbicide	$C_{16}H_{25}NOS$	36756-79-3
Tolclofos-methyl	Fungicide	$C_9H_{11}C_{l2}O_3PS$	57018-04-9
Topramezone	Herbicide	$C_{16}H_{17}N_3O_5S$	
Toxaphene	Insecticide	$C_{10}H_{10}Cl_8$	8001-35-2
Tralkoxydim	Herbicide	$C_{20}H_{27}NO_3$	87820-88-0
Tralomethrin	Insecticide	$C_{22}H_{19}Br_4NO_3$	66841-25-6
Transfluthrin	Insecticide	$C_{15}H_{12}Cl_2F_4O_2$	118712-89-3
Triallate	Herbicide	$C_{10}H_{16}Cl_3NOS$	2303-17-5
Triazophos	Insecticide	$C_{12}H_{16}N_3O_3PS$	24017-47-8
Tribufos	Plant growth regulator	$C_{12}H_{27}OPS_3$	78-48-8
Trichlorfon	Insecticide	$C_4H_8Cl_3O_4P$	52-68-6
Trichloroaceticacid sodium	Herbicide	$C_2Cl_3NaO_2$	
Triclopyr	Herbicide	$C_7H_4Cl_3NO_3$	55335-06-3
Tricyclazole	Fungicide	$C_9H_7N_3S$	41814-78-2
Tridemorph	Fungicide	$C_{19}H_{39}NO$	24602-86-6, 81412-43-3
Triflumizole	Fungicide	$C_{15}H_{15}ClF_3N_3O$	68694-11-1, 99387-89-0
Triflumuron	Insect growth regulator	$C_{15}H_{10}ClF_3N_2O_3$	64628-44-0
Triforine	Fungicide	$C_{10}H_{14}Cl_6N_4O_2$	26644-46-2
Trimethacarb	Insecticide	$C_{11}H_{15}NO_2$	12407-86-2
Trinexapac	Plant growth regulator	$C_{11}H_{12}O_5$	104273-73-6
Triticonazole	Fungicide	$C_{17}H_{20}ClN_3O$	131983-72-7
Valifenalate	Fungicide	$C_{19}H_{27}ClN_2O_5$	
Vamidothion	Insecticide	$C_8H_{18}NO_4PS_2$	2275-23-2
Vernolate	Herbicide	$C_{10}H_{21}NOS$	1929-77-7
Vinclozolin	Fungicide	$C_{12}H_9Cl_2NO_3$	9002-04-4
Vomitoxin	Mycotoxins	$C_{15}H_{20}O_6$	
Warfarin	Rodenticide	$C_{19}H_{16}O_4$	81-81-2
Xylylcarb	Insecticide	$C_{10}H_{13}NO_2$	2425-10-7
Zeta-cypermethrin	Insecticide	$C_{22}H_{19}Cl_2NO_3$	52315-07-8
Zinc naphthenate	Fungicide	$(2(C_{11}H_7O_2).Zn)$	12001-85-3
Zinc phosphide	Rodenticide	Zn_3P_2	
Zineb	Fungicide	$C_4H_6N_2S_4Zn$	12122-67-7
Ziram	Fungicide	$C_6H_{12}N_2S_4Zn$	98391-07-2
Zoxamide	Fungicide	$C_{14}H_{16}Cl_3NO_2$	160171-18-6

Appendix 2: International Codes for Pesticide Formulations

Code	Details
AD	Aerosol dispenser
CS	Capsule suspension
DC	Dispersible concentrate
DL	Driftless formulations
DP	Dustable powder
DS	Powder for dry seed treatment
DT	Tablet for direct application
EC	Emulsifiable concentrate
EG	Emulsifiable granule
EO	Emulsion, water in oil
EP	Emulsifiable powder
EW	Emulsion, oil in water
FS	Flowable concentrate for seed treatment
FU	Smoke generator
GA	Gas
GB/RB	Granular bait/bait (ready for use)
GL	Emulsifiable gel
GR	Granule
GW	Water-soluble gel
LS	Solution for seed treatment
ME	Micro-emulsion
OD	Oil dispersion
OF	Oil-miscible flowable concentrate or oil-miscible suspension
OL	Oil-miscible liquid
OP	Oil dispersible powder
PA	Paste
PB/RB	Plate bait/bait (ready for use)
PS	Seed coated with a pesticide
RB	Bait (ready for use)
SC	Suspension concentrate (= flowable concentrate)
SD	Suspension concentrate for direct application
SE	Suspension emulsion/suspo-emulsion
SG	Water-soluble granule
SL	Soluble concentrate
SO	Spreading oil
SP	Water-soluble powder
ST	Water-soluble tablet
SU	Ultra-low volume (ULV) suspension
TB	Tablet
UL	Ultra-low volume (ULV) liquid
WG	Water-dispersible granule
WP	Wettable powder
WS	Water-dispersible powder for slurry seed treatment
WT	Water-dispersible tablet

Appendix 3: Common Biocontrol Agents and Biopesticides

Biocontrol Agents

Biocontrol agent	Identification	Target pests/application
Adalia bipunctata	Two-spotted ladybug or lady beetle or coccinellid beetle	Used on aphids
Adoxophyes orana	Granulosis virus (class Baculoviridae)	Used on summer fruit tortrix moths
Agapeta zoegana	Root-boring moth or sulfur knapweed moth	Used on knapweeds, spotted knapweed (*Centaurea maculosa*), Russian knapweed and diffuse knapweed (*Centaurea diffusa*)
Agasicles hygrophila	Alligator weed flea beetle	Used on weed *Alternanthera philoxeroides*
Agrobacterium radiobacter	Gram-negative rod bacterium	Used on soil-borne pathogen *Agrobacterium tumefaciens*
A. radiobacter K84	Gram-negative rod bacterium	It inhibits pathogenic strains and species (e.g. *A. tumefaciens*)
A. radiobacter K1026	Gram-negative rod bacterium	It is used to treat germinating seeds or roots and stems of many crops and effective control against *A. tumefaciens* and *Agrobacterium rhizogenes*
Amblyseius mites	Predatory mites	Used on mites and thrips
Amblyseius californicus	Predatory mites	Used to control common red spider mite or two-spot mite (*Tetranychus urticae*), carmine mite (*Tetranychus cinnabarinus*), broad mite (*Polyphagotarsonemus latus*) and the cyclamen mite (*Tarsonemus pallidus*)
Amblyseius newsami	Predatory mites	Used to control tea mites
Amblyseius andersoni	Predatory mites	Used to control red spider mite, fruit-tree red spider mite, gall mite, russet mite, two-spotted spider mite, European red mite and apple rust mite
Amblyseius barkeri (or *Amblyseius mckenziei*)	Predatory mites	Used to control thrip species including western flower thrips (*Frankliniella occidentalis*) and onion thrips (*Thrips tabaci*). Feeds on pollen, fungal spores and spider mite eggs when main targets are scarce
Amblyseius cucumeris	Predatory mites	Used to control thrip *F. occidentalis* and also feeds on spider mites, strawberry mites, rust mites and pollen
Amblyseius degenerans	Predatory mites	Used to control western flower thrips (*F. occidentalis*) and onion thrips (*T. tabaci*) and to some extent tarsonemid mites
Amitus spiniferus	Wasp	It parasitizes whitefly nymphs, predominantly the cottony white fly *Aleurothrixus flocossus*
Aphidius matricariae	Wasp (genus *Aphidius*)	Used on nymph and adult stages of green peach aphid on chrysanthemums
Aphidius colemani	Wasp (genus *Aphidius*)	It is most prominently used to control the the green peach aphid (*Myzus persicae*) and peach potato aphid (*M. persicae* var. *persicae*). However, it cannot control the cotton aphid (*Aphis gossypii*), melon aphid and tobacco aphid (*M. persicae* var. *nicotianae*) and potato aphid, effectively
Aphidius ervi	Wasp (genus *Aphidius*)	Used to control potato aphid and foxglove aphid

Continued

Continued.

Biocontrol agent	Identification	Target pests/application
Aphidoletes aphidimyza	Aphid midge	Used on aphids
Arbuscular mycorrhizal fungi (AMF)	Symbiotic fungi (genus *Glomus*)	Used on root rot
Aspergillus flavus	Soil-borne fungus (order Eurotiales)	Competes with other mycotoxin-producing strains and dominates the field, thus preventing their growth
Bacillus licheniformis	Gram-positive thermophilic soil bacterium	It prevents the growth of fungal pathogens (e.g. *Alternaria* spp., *Botrytis* spp., *Fusarium* spp. and *Erysiphe* spp.) and diseases (e.g. leaf spot and powdery mildews)
Bacillus mycoides	Gram-positive soil bacterium (family Bacillaceae and order Bacillales)	It controls the target fungal pest *Cercospora beticola* Sacc. that causes Cercospora leaf spot disease
Bacillus pumilus isolate GB34	Gram-positive, rod-shaped, aerobic, endospore-forming, motile bacterium	Used on *Pythium* spp., *Rhizoctonia* spp. and *Fusarium* spp.
B. pumilus isolate QST 2808	Gram-positive, rod-shaped, aerobic, endospore-forming, motile bacterium	Used on *Alternaria* spp., *Erysiphae* spp., *Sclerotinia* spp. and *Botrytis* spp.
Bacillus sphaericus	Aerobic, obligate, motile, Gram-positive, endospore-forming bacterium	Used on *Culex* spp., *Aedes* spp., *Psorophora* spp. and *Anopheles* spp.
Bacillus subtilis	Gram-positive, rod-shaped, aerobic, endospore-producing bacterium (order Bacillales and family Bacillaceae)	Used on *Rhizoctonia solani*, *Fusarium* spp., *Alternaria* spp., *Aspergillus* spp. and *Botrytis cinerea*
Bacillus thuringiensis (*Bt*)	Gram-positive, motile, soil microorganism	Used on coleopterans (beetles) and dipterans (flies) (e.g. Colorado potato beetle)
Baculoviruses	Nucleopolyhedroviruses (NPVs) and granuloviruses (GVs)	Used to control moths and insects
Beauveria bassiana	Fungus	Used on Colorado potato beetle, black vine weevil, coffee borer, stem borers, codling moths, whiteflies, thrips, aphids, termites and mosquitoes
Enchytracidae	Potworms	Controls nematode pests
Paenibacillus popilliae	Gram-positive, rod-shaped bacterium	Used on white grubs (grub species include *Leucopholis coneophora* on groundnut and *Leucopholis lepidophora*)
Granulovirus	Codling moth granulosis virus	Protects against many types of worms and moths that attack apples and pears
Braconid	Parasitic wasp (order Hymenoptera, family Braconidae)	Used on ants, aphids, bollworms, caterpillars, leaf miners, leafhoppers and maggots
Chrysopa sp. and *Chrysoperla* sp.	Lacewings (family Chrysopidae in the order Neuroptera)	Used on aphids, thrips, mealy bugs, whitefly, bollworms, beetle larvae, spider mites, leaf miner nymphs and eggs of leafhoppers and scale insects
Epiricania melanoleuca	A moth larvae	Used to control a leafhopper on sugar cane
Fusarium spp.	Filamentous soil fungi (family Nectriaceae)	Used against root rots, damping-off and seed blights
Gliocladium catenulatum	Soil-borne fungal saprophyte (phylum Ascomyota, order Hyprocreales)	It inhibits plant pathogenic fungal strains
Gliocladium virens	Soil-borne fungus	Used on *Pythium* spp., *R. solani* and *Sclerotium rolfsii*
Helicoverpa nucleopolyhedrovirus (HNPV)	Virus	Kills the larvae of *Helicoverpa*, which destroy sorghum, gram, pigeon pea, cotton, sunflower, maize, tomato and safflower
Hippodamia convergens	Convergent lady beetles	Used on aphids on crops such as vegetables, grains, legumes, strawberries and tree crops
Ladybird beetles (*Menochilus sexmaculatus*)	Predatory insect (Coleoptera: Coccinellidae)	Used on aphids, scale insects and whiteflies
Lycosa spp.	Predatory arachnid (wolf spider) (Arachnida: Araneae)	Used on brown plant hoppers, stem borers, leafhoppers, moths, flies and even small lizards and frogs
Pirate bugs (*Orius tristicolor* and *Anthocoris nemoralis*)	Predatory bug (Hemiptera: Anthocoridae)	Used on aphids, bollworm, leafhopper, scale insects, thrips, whiteflies and other small caterpillars
Microsporidia	Unicellular parasite	Control lepidopteran pests in storage
Syrphids	Flies	Controls aphids
Tachinid flies	Parasitoid	Used to control lepidopteran pests
Trichogramma spp.	Wasps of the genus *Trichogramma*, (Hymenoptera: Trichogrammatidae)	Used on cotton bollworm, codling moth and corn borer
Tardigrades	Nematode predators	Control nematode pests
Alternaria cassia	Ascomycete fungal pathogen	Used on sicklepod (*Senna obtusifolis*) and coffee senna (*Cassia occidentalis*)
Alternaria destruens	Ascomycete fungus	Used on dodders (Cuscuta spp.) and it is hyperparasitic on Erysiphae spp. (agents of powdery mildew diseases)

Biopesticides

Chemical name	Type of pesticide	Derived from	Target pests/application
4-Allyl-2-methoxyphenol insecticide	Bioinsecticide	Derived from plants such as cloves, cinnamon, nutmeg, holy basil/ tulsi (*Ocimum sanctum*)	Used on aphids, armyworms, beetles, cutworms and mites
Abamectin bioinsecticide	Microorganism-derived insecticide	Fermentation product of *Streptomyces avermitilis*	Used on mites, leaf miners, suckers, beetles and fire ants in crops such as cotton, citrus, pome fruit and vegetables
Aminoethoxyvinylglycine	Actinobacteria	*Streptomyces* sp.	It is used to protect apples and pears from premature ripening and dropping
Amphotericin B	Anti-fungal antibiotic	Fermentation with a strain of *Streptomyces nodosus*	Used against deep mycotic infections
Ampelomyces quisqualis	Biofungicide	Mitosporic fungus	Used against powdery mildews
Anagrapha falcifera nucleopolyhedrovirus	Bioinsecticide	Alfalfa looper (*Anagrapha falcifera* (Kirby))	Used on *Heliothis* and *Helicoverpa* and species of Lepidoptera such as the tobacco budworm and cotton bollworm
Antibiotics	Biopesticide	Derived from strains of *Penicillium notatum* and *Penicillium chrysogenum*	It inhibits the growth of, and even destroys, bacteria and rickettsiae
Anticarsia gemmatalis nucleopolyhedrovirus (MNPV or AgMNPV)	Bioinsecticide	Culture is prepared on the velvet bean caterpillar (*Anticarsia gemmatalis* Hubner)	Used on sugarcane borer and the velvet bean caterpillar
Anthraquinone biorepellent	Biorepellent	Rye grass species and teak	Keeps birds away
Aureofungin	Biofungicide	Synthesized by a soil-inhabiting *Streptomycete* species	It control seed-borne diseases in many crops
Autographa californica nucleopolyhedrovirus	Bioinsecticide	First isolated from alfalfa looper *A. californica*	Used on maize, vegetables, fruit crops and ornamentals and any other crop the looper infests
Azadirachtin	Plant-derived insecticide	Obtained from seeds of the neem plant (*Azadirachta indica*)	Used on whitefly, leaf miners, caterpillars, aphids, thrips, beetles and soil pathogens
Bilanaphos	Bioherbicide	Isolated from *Streptomyces hygroscopicus* by fermentation	Used on for example cucurbits and apple against many weed species
Black pepper oil	Animal repellent	Obtained from seeds of black pepper plant (*Piper nigrum*)	Used to repel small mammals (e.g. dogs, cats, groundhogs)
Blasticidin-S	Biofungicide antibiotic	Fermentated by soil actinomycete, *Streptomyces griseochromogenes*	Used to control rice blast caused by *Pyricularia oryzae* Cavara
Canola oil	Plant-derived insecticide	Obtained from seeds of two species of canola plants (*Brassica napus* and *Brassica campestris*) in the family Cruciferae	Used on mites and thrips
Capsaicin	Biorepellent/ bioinsecticide	Hot pepper extract	Controls small mammals
Carvone	BioPGR/biofungicide	Caraway oil	Anti-sprouting and rot diseases of potatoes
Chenopodium ambrosioides	Bioinsecticide	Extracted from *C. ambrosioides* (American wormseed or Mexican tea)	Used on aphids, mites, whitefly, thrips and leaf miners
Cinnamaldehyde	Plant-derived fungicide	Derived from the bark of the cinnamon tree (*Cinnamomum zeylanicum*)	Used on dry bubble (*Lecanicillium fungicola*), pitch canker disease and diseases caused by *Verticillium*, *Rhizoctonia* and *Pythium*
Citric acid	Plant-derived insecticide	Derived from citrus fruits	Used on ants, beetles, aphids, caterpillars, mites, whiteflies and leafhoppers
Citronella oil	Plant-derived insect repellent	Extracted by distillation process from citronella (family Cardiopteridaceae)	Used on mosquitoes, gnats, flies and fleas

Continued

Continued.

Chemical name	Type of pesticide	Derived from	Target pests/application
Cornexistin	Bioherbicide	Derived from *Paecilomyces variotii* Sank	Used on weeds in maize fields
Corn oil	Plant-derived insecticide	Derived from maize plant	Used to control insects. Also used as an adjuvant in herbicide sprays to enable better penetration of the a.i.
Cottonseed oil	Plant-derived insecticide	Extracted by crushing cottonseeds	Used on mites, aphids and whitefly
Cottonseed oil plus maize oil plus garlic oil	Plant-derived fungicide	Extracted from respective plants	Used to control powdery mildew
Cycloheximide	BioPGR	It is an antibiotic produced by *Streptomyces griseus*	It is a plant growth regulator (PGR)
Cytokinins	Plant-derived PGRs and nematicides	They are naturally present in plants	They are PGRs and they are used as a nematicide on avocados, berries, citrus, coffee, herbs, kiwi, tree nuts, tropical fruit and vegetables
Diatomaceous earth	Biopesticide	Fossils of certain algae	Helps to detach cuticles from insects
Dihydroazadirachtin	Bioinsecticide	Derived from azadirachtin from the seeds and kernels of *A. indica* (neem)	Used to control insects and soil-borne pests and also acts like an anti-feedant
Dioxapyrrolomycin	Bioinsecticide/ bionematicide	It is an antibiotic isolated from *Streptomyces fumanus* Sveshnikova	It is an insecticide and acaricide. An anti-parasitic for cattle, sheep and poultry
DMDP	Bionematicide	It is isolated from tropical legume plants called lancepods (*Lonchocarpus* spp.) and Indian beech or honge (*Derris indica*)	Used to control soil insects and nematodes
(*E*)-3, 7-dimethyl-2, 6-octadien-1-ol	Biorepellent and insecticide	Extract of citronella plant (*Cymbopogon nardus*) plus geraniol	It repels mosquitoes, house flies, stable flies, horn flies, head lice, ants and lone star ticks
Emamectin benzoate	Microorganism-derived insecticide	Produced by the fungus *Streptomyces avermitilis*; belongs to the avermectin family	It paralyses and kills pests such as nematodes, insects and mites, leaf miners, moths, caterpillars, mites, bollworms, cutworms and armyworms
Ergocalciferol	Biorodenticide	Obtained from yeast	Used for rodent control
Garlic oil	Biofungicide, antibacterial and insecticide	Obtained from crushing garlic	Used to control aphids, bugs, leafhoppers and whiteflies
Geraniol bioinsecticide geranyl alcohol/guaniol	Bioinsecticide and insect repellent	It is an essential oil made by the combination of geraniol and citronellol	Used against mosquitoes
Gibberellins	PGR	Naturally present in plants, fungi and bacteria	It acts as a PGR
Mixture of gibberellin A_4 and gibberellin A_7	PGR	Naturally present in plants, fungi and bacteria	It reduces russetting in apples, improves the length and quality of fruit and increases seed germination
Harpin aβ protein	Biopesticide	Produced with a weak strain of *Escherichia coli* (with DNA sequence from *Erwinia amylovora*)	Effective against a variety of fungal, bacterial, viral and soil-borne plant pathogens and also reduces infestation by insect and nematode pests
Indole-3-acetic acid	PGR	Naturally present in plants	It acts as a PGR
Jasmonic acid (JA)	PGR	It is found in jasmine and many other flowers and fruits and it is also found in some fungal species	It signals senescence and fruit ripening. It can inhibit growth, germination and photosynthesis
Jojoba oil	Biofungicide	Derived from jojoba seeds (*Simmondsia californica* Nutt)	Used to control fungal spores and the mycelium of powdery mildew pathogens, whitefly eggs and larvae. Also used as a repellent for insects, dogs and cats
Karanjin	Bioinsecticide/ bioacaricide/bioPGR	Obtained from the seeds of the Karanja tree (*Millettia pinnata*, aka *Pongamia glabra* Vent.)	Used to control mites, scales, whiteflies, thrips and leaf miners

Continued

Continued.

Chemical name	Type of pesticide	Derived from	Target pests/application
Kasugamycin	Antibiotic fungicide and microorganism-derived bactericide	Isolated from the soil actinomycete *Streptomyces kasugaensis* by a fermentation process	It inhibits both Gram-positive and Gram-negative bacteria, prevents rice blasts, leaf spot, grey mould, black rot (caused by *Cercospora* and *Pyricularia grisea*) and controls infection caused by *Pseudomonas* spp., *Erwinia atroseptica*, *Cladosporium fulvum* and *Xanthomonas campestris* pv. *vesicatoria*
Laminarine	Biofungicide	Extracted from kelp (i.e. seaweed)	Effective against fungal pathogens of cereals such as *Septoria* and powdery mildews
L-glutamic acid plus γ-aminobutyric acid	Biofungicide	Both are found naturally in all living organisms as metabolic primers	Prevention of powdery mildew
Mildiomycin	Biofungicide	Produced by the fermentation of *Streptoverticillium rimafaciens*	Used for control of *Erysiphe* spp., *Podosphaera* spp. and *Sphaerotheca* spp. which cause powdery mildews
Mint oil	Biorepellant	Extracted from mint	It repels mosquitoes and ants
Combination of citric acid and mint oil	Biofungicide and biobactericide	Extracted from citrus fruit and mint leaves	It controls fungi such as *Phytophthora* spp., downy mildews, *Sclerotinia* spp. and *Botrytis* spp. and many bacteria (e.g. *Pseudomonas* spp.)
Natamycin	Antibiotic, biofungicide	It is produced by the fermentation of *Streptomyces natalensis* and *Streptomyces chattanoogensis*	Used against *Fusarium oxysporum*
Nicotine	Bioinsecticide	It is derived from the tobacco plant (*Nicotiana tabacum* L. and other *Nicotiana* spp., e.g. *Nicotiana rustica*), pituri (*Duboisia hopwoodii*) and common milkweed (*Aesclepias syriaca*)	Used to control aphids, thrips and whitefly
3-[*N*-butyl-*N*-acetyl]-aminopropionic acid, ethyl ester	Biorepellent	It is similar to the naturally occurring β-alanine	Used to control mosquitoes, deer ticks, body lice and biting flies
Neem	Biopesticide	Obtained from *A. indica* (neem)	Used to control whiteflies, aphids, scale insects, mealy bugs, soil grubs, spider mites, locusts, thrips, beetle varieties, moth larvae, nematodes, powdery mildew and black spot
Oleic acid	Bioherbicide	Naturally occurring fatty acid found in animal and plant fat, abundant in all oil crops and in poultry and meat that man consumes and also present in human adipose tissue	It is effective against aphids, thrips and scale insects. As a herbicide, it is used for moss control. It can protect grapes from fungal pathogens. It is also used as an ingredient in making insect pheromones used in IPM techniques
Polyoxins	Antibiotic, biofungicide	It is produced by fermentation with *Streptomyces cacaoi* var. *asoensis*	Controls *Alternaria* leaf spot, *Alternaria longipes*, *Pseudoperonospora cubensis*, Fusarium wilt, *Alternaria panax*, *Pyricularia oryzae*, powdery mildew and *B. cinerea*
Polyoxin B	Antibiotic, biofungicide	It is produced by fermentation with *S. cacaoi* var. *asoensis*	Controls *Alternaria* spp., *B. cinerea*, *Sclerotinia sclerotiorum*, powdery mildews, grey moulds and blights
Polyoxin D	Antibiotic, biofungicide	It is produced by fermentation with *S. cacaoi* var. *asoensis*	Used for controlling *R. solani* (sheath blight in rice), *Drechslera* spp. and *Bipolaris* spp.

Continued

Continued.

Chemical name	Type of pesticide	Derived from	Target pests/application
Ryania	Bioinsecticide	Made from the ground stems of *Ryanis speciosa*	Used on citrus thrips, citrus codling moth and European corn borer
Rotenone	Bioinsecticide	It is extracted from *Derris elliptica*, *Derris malaccensis*, *Lonchocarpus utilis* and *Lonchocarpus urucu*	Used to control potato beetles, cucumber beetles, flea beetles, cabbage worms and lice
Streptomycin	Antibiotic, biobactericide	Derived from the actinobacterium *Streptomyces griseus*	Used on seed rots, fire blights of apples and pears, walnut blight and bacterial canker. Also against *Xanthomonas* spp. and *Pseudomonas* spp.
Streptocycline	Antibiotic, fungicide	A combination of streptomycin and tetracycline	Used on maize, potato and onion
Trans-α-ionone	Bioinsecticide	Naturally present in many tropical and subtropical plants including copaiba (*Copaifera officinalis* (Jacq) L.)	Used as an insect repellent on flying and biting insects
Validamycin	Antibiotic, biofungicide	It is produced by fermenting *Streptomyces hygroscopicus*	Used to control soil-borne diseases caused by *R. solani* and *F. oxysporum*
Strobilurin	Biofungicide	It is extracted from the fungus *Strobilurus tenacellus*	Used to control *Erysiphe graminis* f.sp. *tritici* and *Septoria tritici*
Trifloxystrobin	Biofungicide	It is extracted from the fungus *S. tenacellus*	Used on fungal ascomycetes, deuteromycetes, basidiomycetes and oomycetes
Azoxystrobin	Biofungicide	Experimental synthesizing from *Oudemansiella mucida* and *S. tenacellus*	Controls powdery mildews, rust, cereal leaf blotch, blast, damping off, downy mildew and early/late blight
Kresoxim-methyl	Biofungicide	It is a naturally occurring anti-fungal compound in certain wood-decaying mushrooms	Used to control *Venturia inaequalis*, *Podosphaera leucotricha* and *Uncinula necator*
Fenpiclonil	Biofungicide	Originates from *Pseudomonas pyrrocinia* (pyrrolnitrin)	Used to control seed-borne fungal diseases caused by *Fusarium* spp. and *Tilletia* spp. and many robust isolates of *Fusarium nivale*
Fludioxonil	Biofungicide	Originates from *P. pyrrocinia* (pyrrolnitrin)	Controls *Gibberella*, *Fusarium*, *Tilletia*, *Helminthosporium*, *Septoria*, *Botrytis*, *Sclerotinia*, *Rhizoctonia* and *Alternaria*

Appendix 4: Periodic Table

Periodic Table of Elements

1	2	3	4	5	6	7	8	9	10	11	12	13	14	15	16	17	18
1 H																	2 He
3 Li	4 Be											5 B	6 C	7 N	8 O	9 F	10 Ne
11 Na	12 Mg											13 Al	14 Si	15 P	16 S	17 Cl	18 Ar
19 K	20 Ca	21 Sc	22 Ti	23 V	24 Cr	25 Mn	26 Fe	27 Co	28 Ni	29 Cu	30 Zn	31 Ga	32 Ge	33 As	34 Se	35 Br	36 Kr
37 Rb	38 Sr	39 Y	40 Zr	41 Nb	42 Mo	43 Tc	44 Ru	45 Rh	46 Pd	47 Ag	48 Cd	49 In	50 Sn	51 Sb	52 Te	53 I	54 Xe
55 Cs	56 Ba	57–71 *	72 Hf	73 Ta	74 W	75 Re	76 Os	77 Ir	78 Pt	79 Au	80 Hg	81 Ti	82 Pb	83 Bi	84 Po	85 At	86 Rn
87 Fr	88 Ra	89–103 #	104 Rf	105 Db	106 Sg	107 Bh	108 Hs	109 Mt	110 Ds	111 Uuu	112 Uub	113 Uut	114 Uuq	115 Uup			

Metals

Non-metals

Transition metals

Elements of the f-series column labelling

* Lanthenides	57 La	58 Ce	59 Pr	60 Nd	61 Pm	62 Sm	63 Eu	64 Gd	65 Tb	66 Dy	67 Ho	68 Er	69 Tm	70 Yb	71 Lu
# Actinides	89 Ac	90 Th	91 Pa	92 U	93 Np	94 Pu	95 Am	96 Cm	97 Bk	98 Cf	99 Es	100 Em	101 Md	102 No	103 Lr

Periodic Table of Plant Nutrients

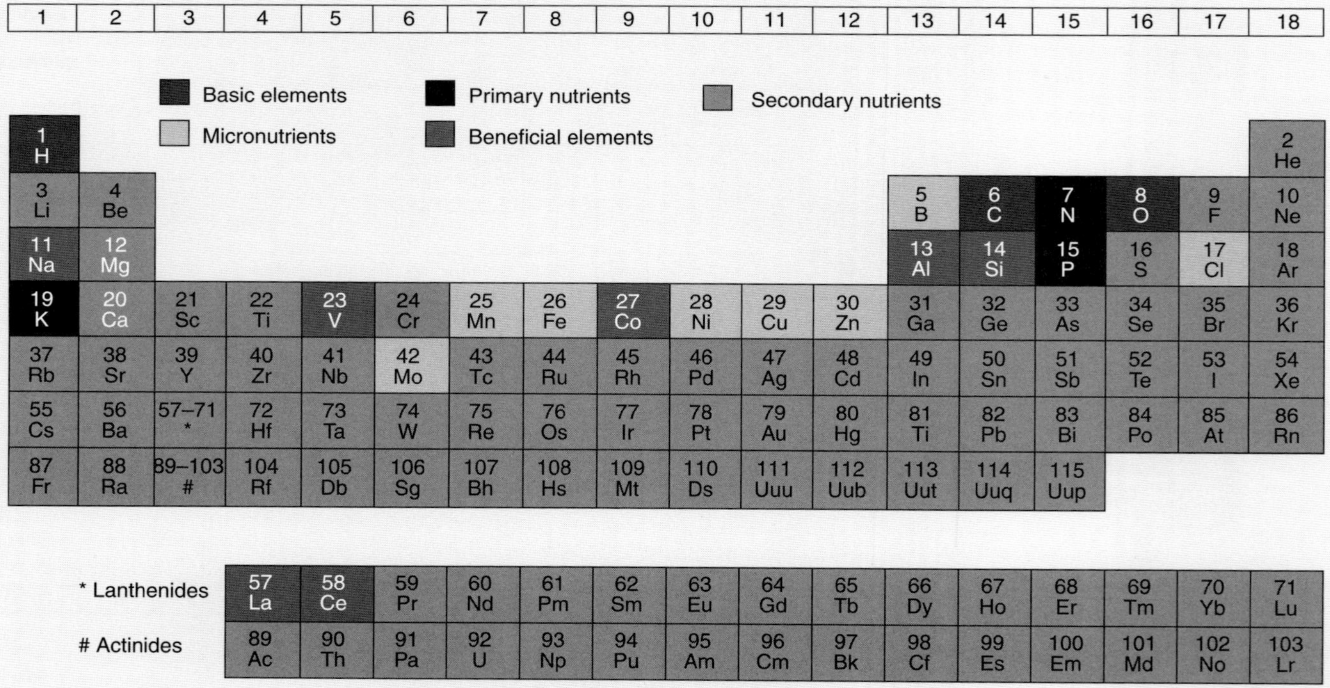

Position of plant nutrient elements in the Periodic Table

Appendix 5: Units of Measurement

Metric and Imperial Units

Area measurements

Metric unit	Equivalent	Imperial unit	Equivalent
1 square centimetre (cm^2)	100 square millimetres (mm^2)	1 square inch (in^2)	6.45 cm^2
1 square metre (m^2)	10,000 cm^2	1 square foot (ft^2)	144 in^2
1 acre	100 m^2	1 square yard (yd^2)	9 ft^2 = 0.836 m^2
1 hectare (ha)	10,000 m^2	1 acre	4840 yd^2 = 0.405 hectare (ha)
1 square kilometre (km^2)	1,000,000 m^2	1 square mile	259 ha

Capacity measurements

Metric unit	Equivalent	Imperial unit	Equivalent
1 centilitre (cl)	10 millilitres (ml)	1 fluid ounce (fl oz)	0.0296 l
1 decilitre (dl)	10 centilitres	1 gill (gi)	5 fluid oz = 0.1480 l
1 litre (l)	10 decilitres	1 pint (pt)	20 fluid oz = 0.568 l
1 decalitre (dal)	10 l	1 quart (qt)	2 pints = 1.136 l
1 hectolitre (hl)	10 decalitres	1 gallon (gal)	4 quarts = 4.546 l
1 kilolitre (kl)	10 hectolitres	1 peck (pk)	2 gallons = 9.092 l
		1 bushel (bu)	4 pecks = 36.4 l

Linear measurements

Metric unit	Equivalent	Imperial unit	Equivalent
1 centimetre (cm)	10 millimetres (mm)	1 inch (in)	25.4 mm
1 metre (m)	100 cm	1 foot (ft)	12 inches = 0.3048 m
1 kilometre (km)	1000 m	1 yard (yd)	3 feet = 0.9144 m
		1 rod (rd)	5.5 yards
		1 mile (mi)	1,760 yards = 1.609 km
		1 nautical mile	1.852 km

Weight

Metric unit	Equivalent	Imperial unit	Equivalent
1 centigram (cg)	10 milligrams (mg)	1 grain (gr)	0.065 g
1 decigram (dg)	100 mg	1 dram (dr)	1.772 g
1 gram (g)	1000 mg	1 ounce (oz)	28.35 g
1 decagram (dag)	10 g	1 pound (lb)	0.4536 kg
1 hectogram (hg)	100 g	1 stone	6.35 kg
1 kilogram (kg)	1000 g	1 hundred weight (cwt)	50.80 kg
1 tonne (t) or metric ton	1000 kg	1 t (short)	0.907 t
		1 t (long)	1.016 t

Volume measurements

Metric unit	Equivalent	Imperial unit	Equivalent
1 cubic centimetre (cc or cm^3)	1 millilitre (ml)	1 cubic inch (in^3)	16.4 cm^3
1 litre (l)	1000 ml	1 cubic foot (ft^3)	1,728 in^3 = 0.0283 m^3
1 cubic metre (m^3)	1000 l	1 cubic yard (yd^3)	27 ft^3 = 0.765 m^3

Temperature Scales

Fahrenheit (F): under standard conditions, water boils at 212°F and freezes at 32°F.
Celsius or Centigrade (C): water boils at 100°C and freezes at 0°C.
Kelvin (K): water boils at 373.15K and freezes at 273.15K.

To convert from Celsius (C) to Fahrenheit (F): $F = (C \times 9/5) + 32$
To convert from Fahrenheit (F) to Celsius (C): $C = (F - 32) \times 5/9$
To convert from Celsius (C) to Kelvin (K): $K = C + 273$
To convert from Kelvin (K) to Celsius (C): $C = K - 273$

Fahrenheit (°F)	Celsius/Centigrade (°C)	Kelvin (K)
−4	−20	253
14	−10	263
32	0	273
50	10	283
68	20	293
86	30	303
104	40	313
122	50	323
140	60	333
158	70	343
176	80	353
194	90	363
212	100	373

Appendix 6: SI Units

Type of unit	Quantity	Name of unit	Symbol
Base units	Amount of substance	mole	mol
	Electric current	ampere	A
	Length	metre	m
	Luminous intensity	candela	Cd
	Mass	kilogram	kg
	Temperature	kelvin	K
	Time	second	s
Supplementary units	Plane angle	radian	rad
	Solid angle	steradian	sr
Derived units with special names	Electric capacitance	farad	F
	Electric change	coulomb	C
	Electric conductance	siemens	S
	Electric resistance	ohm	O
	Electromotive force	volt	V
	Energy	joule	J
	Force	newton	N
	Frequency	hertz	Hz
	Illumination	lux	lx
	Inductance	henry	H
	Luminous flux	lumen	lm
	Magnetic flux	weber	Wb
	Magnetic flux density	tesla	T
	Power	watt	W
	Pressure	pascal	Pa
Units permitted for use	Time	minute	min
	Time	hour	h
	Time	day	day
	Time	year	a
	Area	hectare	ha
	Volume	litre	l
	Volume	hectolitre	hl
	Mass	metric ton	t

Prefixes for Units in the SI System

Name	Prefix	Abbreviation/symbol	Factor	Example
Quintillion	exa	E	10^{18}	
Quadrillion	peta	P	10^{15}	
Trillion	tera	T	10^{12}	
Billion	giga	G	10^{9}	Gigawatt (GW)
Million	mega	M	10^{6}	Megahertz (MHz)
Thousand	kilo	k	10^{3}	Kilogram (kg)
Hundred	hector	h	10^{2}	
Ten	deka	Da	10	
One-tenth	deci	d	10^{-1}	
One-hundredth	centi	c	10^{-2}	Centimetre (cm)
One-thousandth	milli	m	10^{-3}	Milligram (mg)
One-millionth	micro	μ	10^{-6}	Microgram (μg)
One billionth	nano	n	10^{-9}	Nanosecond (ns)
One-trillionth	pico	p	10^{-12}	Picofarad (pf)
One-quadrillionth	femto	f	10^{-15}	
One-quintillionth	atto	a	10^{-18}	

Some Fundamental Physical Constants

Constants	Symbol	Value
Speed of light in vacuum	c	3×10^{8}
Gravitational constant	G	6.63×10^{-11} Js
Planck's constant	h	6.63×10^{-34}
Universal gas constant	R	8.31 J/mol K
Boltzman constant	k	1.38×10^{-23} J/K
Stefan-Boltzman constant	o	5.67×10^{-8} W/m^{2} K^{4}
Avogadro's number	NA	6.02×10^{23} mol^{-1}
Faraday constant	F	9.65×10^{4} C/mol
Solar constant	S	1.38 kW/m^{2}
Bohr radius	ao	5.29×10^{-11} m

Appendix 7: Greek Alphabet and Roman Numbers

The Greek Alphabet

Name	Capital	Lower case
alpha	Α	α
beta	Β	β
gamma	Γ	γ
delta	Δ	δ
epsilon	Ε	ε
zeta	Ζ	ζ
eta	Η	η
theta	Θ	θ
iota	Ι	ι
kappa	Κ	κ
lambda	Λ	λ
mu	Μ	μ
nu	Ν	ν
xi	Ξ	ξ
omicron	Ο	ο
pi	Π	π
rho	Ρ	ρ
sigma	Σ	σ
tau	Τ	τ
upsilon	Υ	υ
phi	Φ	φ
chi	χ	χ
psi	Ψ	ψ
omega	Ω	ω

Arabic and Roman Numericals

Arabic	Roman
1	I
2	II
3	III
4	IV
5	V
6	VI
7	VII
8	VIII
9	IX
10	X
11	XI
12	XII
13	XIII
14	XIV
15	XV
16	XVI
17	XVII
18	XVIII
19	XIX
20	XX
21	XXI
30	XXX
40	XL
49	XLIX
50	L
60	LX
70	LXX
80	LXXX
90	XC
99	XCIX
100	C
101	CI
144	CXLIV
200	CC
400	CD or CCCC
500	D
900	CM or DCCCC
1000	M
1900	MCM or MDCCCC
1995	MCMXCV
1999	MCMXCIX
2000	MM

Appendix 8: Pesticides Registered in India

This is a list of insecticides/pesticides that are registered under section 9(3) of the Indian Insecticides Act, 1968 (http://cibrc.nic.in/reg_products.htm). The list is correct as of 30 November 2012.

Serial number	Name of the pesticide
1	2,4-Dichlorophenoxy acetic acid
2	Acephate
3	Acetamiprid
4	Alachlor
5	Allethrin
6	Alphacypermethrin
7	Alphanaphthyl acetic acid
8	Aluminium phosphide
9	*Ampelomyces quisqualis*
10	Anilophos
11	Atrazine
12	Aureofungin
13	Azadirachtin (neem products)
14	Azimsulfuron
15	Azoxystrobin
16	*Bacillus sphaericus*
17	*Bacillus thuringiensis* var. *israelensis*
18	*Bacillus thuringiensis* var. *kurstaki*
19	*Bacillus thuringiensis* var. *galleriae*
20	Barium carbonate
21	*Beauveria bassiana*
22	Bendiocarb
23	Benfuracarb
24	Benomyl
25	Bensulfuron methyl
26	Beta cyfluthrin
27	Bifenazate
28	Bifenthrin
29	Bispyribac sodium
30	Bitertanol
31	Bromadiolone
32	Buprofezin
33	Butachlor
34	Captan
35	Carbaryl
36	Carbendazim
37	Carbofuran
38	Carbosulfan
39	Carboxin
40	Carfentazone ethyl
41	Carpropamid
42	Cartap hydrochloride
43	Chlorantraniliprole
44	Chlorfenapyr

Continued

Continued.

Serial number	Name of the pesticide
45	Chlorfluazuron
46	Chlorimuron ethyl
47	Chlormequat chloride (CCC)
48	Chlorothalonil
49	Chlorpropham
50	Chlorpyriphos
51	Chlorpyriphos methyl
52	Chlothianidin
53	Chromafenozide
54	Cinmethylene
55	Clodinafop-propargyl
56	Clomazone
57	Copper hydroxide
58	Copper oxychloride
59	Copper sulfate
60	Coumachlor
61	Coumatetralyl
62	Cuprous oxide
63	Cyfluthrin
64	Cyhalofop-butyl
65	Cymoxanil
66	Cypermethrin
67	Cyphenothrin
68	Dazomet
69	Deltamethrin (Decamethrin)
70	Diafenthiuron
71	Diazinon
72	Dichloro diphenyl trichloroethane (DDT)
73	Dichloropropene and dichloropropane mixture (DD mixture)
74	Diclorvos (DDVP)
75	Diclofop-methyl
76	Dicofol
77	Difenoconazole
78	Diflubenzuron
79	Dimethoate
80	Dimethomorph
81	Dinocap
82	Dinotefuron
83	Dithianon
84	Diuron
85	Dodine
86	D-trans allethrin
87	Edifenphos
88	Emamectin benzoate
89	Endosulfan*
90	Ethephon
91	Ethiprole
92	Ethion
93	Ethofenprox (Etofenprox)
94	Ethoxysulfuron
95	Ethylene dichloride and carbon tetrachloride mixture (EDCT mixture 3:1)
96	Famoxadone
97	Fenamidone
98	Fenarimol
99	Fenazaquin
100	Fenitrothion
101	Fenobucarb (BPMC)
102	Fenoxaprop-p-ethyl
103	Fenpropathrin
104	Fenpyroximate
105	Fenthion
106	Fenvalerate

Continued

Continued.

Serial number	Name of the pesticide
107	Fipronil
108	Flonicamid
109	Fluazifop-p-butyl
110	Flubendiamide
111	Fluchloralin
112	Flufenacet
113	Flufenoxuron
114	Flufenzine
115	Flusilazole
116	Fluvalinate
117	Forchlorfenuron
118	Fosetyl-Al
119	Gibberellic acid
120	Glufosinate ammonium
121	Glyphosate
122	Hexaconazole
123	Hexazinone
124	Hexythiazox
125	Hydrogen cyanamide
126	Imazamox
127	Imazethapyr
128	Imidacloprid
129	Imiprothrin
130	Indoxacarb
131	Iprobenfos (Kitazin)
132	Iprodione
133	Iprovalicarb
134	Isoprothiolane
135	Isoproturon
136	Kasugamycin
137	Kresoxim methyl
138	Lambdacyhalothrin
139	Lime sulfur
140	Linuron
141	Lufenuron
142	Magnesium phosphide plates
143	Malathion
144	Mancozeb
145	Mandipropamid
146	Mepiquate chloride
147	Mesosulfuron methyl + iodosulfuron methyl sodium
148	Metaflumizone
149	Metalaxyl
150	Metalaxyl-M
151	Metaldehyde
152	*Metarhizium anisopliae*
153	Methabenzthiazuron
154	Methomyl
155	Methoxy ethyl mercury chloride (MEMC)
156	Methyl bromide
157	Methyl chlorophenoxy acetic acid (MCPA)
158	Methyl parathion
159	Metiram
160	Metolachlor
161	Metribuzin
162	Metsulfuron methyl
163	Milbemectin
164	Monocrotophos
165	Myclobutanil
166	Novaluron
167	Nuclear polyhedrosis virus of *Helicoverpa armigera*
168	Nuclear polyhedrosis virus of *Spodoptera litura*

Continued

Continued.

Serial number	Name of the pesticide
169	Orthosulfamuron
170	Oxadiargyl
171	Oxadiazon
172	Oxycarboxin
173	Oxydemeton-methyl
174	Oxyfluorfen
175	Paclobutrazol
176	Paraquat dichloride
177	Penconazole
178	Pencycuron
179	Pendimethalin
180	Permethrin
181	Phenthoate
182	Phorate
183	Phosalone
184	Phosphamidon
185	Pinoxaden
186	Prallethrin
187	Pretilachlor
188	Primiphos-methyl
189	Profenophos
190	Propanil
191	Propaquizafop
192	Propergite
193	Propetamphos
194	Propiconazole
195	Propineb
196	Propoxur
197	*Pseudomonas fluorescens*
198	Pyrachlostrobin
199	Pyrazosulfuron ethyl
200	Pyrethrins (pyrethrum)
201	Pyridalyl
202	Pyriproxyfen
203	Pyrithiobac sodium
204	Quinalphos
205	Quizalofop ethyl
206	Quizalofop-P-tefuryl
207	S-bioallethrin
208	Sirmate
209	Sodium cyanide
210	Spinosad
211	Spiromesifen
212	Streptomycin + tetracycline
213	Sulfosulfuron
214	Sulfur
215	Tebuconazole
216	Temephos
217	Thiacloprid
218	Thifluzamide
219	Thiobencarb (Benthiocarb)
220	Thiodicarb
221	Thiomethoxain
222	Thiometon
223	Thiophanate-methyl
224	Thiram
225	Transfluthrin
226	Triacontanol
227	Triadimefon
228	Triallate
229	Triazophos
230	Tricholorofon

Continued

Continued.

Serial number	Name of the pesticide
231	*Trichoderma harzianum*
232	*Trichoderma viride*
233	Tricyclazole
234	Tridemorph
235	Trifloxistrobin
236	Trifluralin
237	Validamycin
238	*Verticillium lecanii*
239	Zinc phosphide
240	Zineb
241	Ziram

*Endosulfan has been banned by the Supreme Court of India with effect from 13 May 2011 for production, use and sale, all over India, until further orders vide ad-Interim order in the Writ Petition (Civil) No. 213 of 2011.

Appendix 9: Pesticides Banned in India

This is a list of pesticides that are banned, refused registration and restricted in use (as of 31 December 2012).

I. Pesticides/Formulations Banned in India

A. Pesticides banned for manufacture, import and use

Serial number	Pesticide
1	Aldicarb
2	Aldrin
3	Benzene hexachloride
4	Calcium cyanide
5	Chlorbenzilate
6	Chlordane
7	Chlorofenvinphos
8	Copper acetoarsenite
9	Dibromochloropropane
10	Dieldrin
11	Endrin
12	Ethyl mercury chloride
13	Ethyl parathion
14	Ethylene dibromide
15	Heptachlor
16	Lindane (Gamma-HCH) (Banned vide Gazette Notification No. S.O. 637(E) dated 25 March 2011: banned for manufacture, import or formulate with effect from 25 March 2011 and banned for use with effect from 25 March 2013.)
17	Maleic hydrazide
18	Menazon
19	Metoxuron
20	Nitrofen
21	Paraquat dimethyl sulfate
22	Pentachloro nitrobenzene
23	Pentachlorophenol
24	Phenyl mercury acetate
25	Sodium methane arsonate
26	TCA (trichloroacetic acid)
27	Tetradifon
28	Toxaphene (camphechlor)

B. Pesticide formulations banned for import, manufacture and use

Serial number	Pesticide
1	Carbofuron 50% SP
2	Methomyl 12.5% L
3	Methomyl 24% formulation
4	Phosphamidon 85% SL

C. Pesticide/pesticide formulations banned for use but continued to be manufactured for export

Serial number	Pesticide
1	Captafol 80% powder
2	Nicotin sulfate

D. Pesticides withdrawn

Withdrawal may become operative as soon as the required data (as per the guidelines) is generated and submitted by the pesticides industry to the Government and accepted by the Registration Committee (S.O. 915(E) dated 15 June 2006).

Serial number	Pesticide
1	Dalapon
2	Ferbam
3	Formothion
4	Nickel chloride
5	Paradichlorobenzene (PDCB)
6	Simazine
7	Warfarin

II. Pesticides Refused Registration

Serial number	Pesticide
1	Ammonium sulfamate
2	Azinphos ethyl
3	Azinphos methyl
4	Binapacryl
5	Calcium arsenate
6	Carbophenothion
7	Chinomethionate (Morestan)
8	Dicrotophos
9	EPN
10	Fentin acetate
11	Fentin hydroxide
12	Lead arsenate
13	Leptophos (Phosvel)
14	Mephosfolan
15	Mevinphos (Phosdrin)
16	2,4,5-T
17	Thiodemeton/disulfoton
18	Vamidothion

III. Pesticides Restricted for Use in India

Name of pesticide	Details of restrictions
Aluminium phosphide	Pest control operations with aluminium phosphide may be undertaken only by government/ government undertakings/government organizations/pest control operators under the strict supervision of government experts or experts whose expertise is approved by the Plant Protection Advisor to the Government of India, except: [1]aluminium phosphide 15% 12 g tablet and [2]aluminum phosphide 6% tablet. (Registration Committee (RC) decision circular F No. 14-11(2)-CIR-II (Vol. II) dated 21 September 1984 and G.S.R. 371(E) dated 20 May 1999) [1]Decision of 282nd RC held on 2 November 2007 and [2]Decision of 326th RC held on 15 February 2012. The production, marketing and use of aluminium phosphide tube packs with a capacity of 10 and 20 tablets of 3 g each of aluminium phosphide are banned completely. (S.O. 677(E) dated 17 July 2001)
Captafol	The use of captafol as a foliar spray is banned. Captafol can only be used as a seed dresser. (S.O. 569(E) dated 25 July 1989) The manufacture of captafol 80% powder for dry seed treatment (DS) is banned for use in the country except manufacture for export. (S.O. 679(E) dated 17 July 2001)
Cypermethrin	Cypermethrin 3% smoke generator is to be used only through pest control operators and is not allowed to be used by the general public. (Order of the Honourable High Court of Delhi in WP(C) 10052 of 2009 dated 14 July 2009 and LPA-429/2009 dated 08 September 2009)
Dazomet	The use of dazomet is not permitted on tea. (S.O. 3006(E) dated 31 December 2008)
Diazinon	Diazinon is banned for use in agriculture except for household use. (S.O. 45(E) dated 8 January 2008)
Dichlorodiphenyltrichloroethane (DDT)	The use of DDT for the domestic Public Health Programme is restricted up to 10,000 t/year, except in case of any major outbreak or epidemic. M/s Hindustan Insecticides Ltd, the sole manufacturer of DDT in the country, may manufacture DDT for export to other countries for use in vector control for public health purposes. The export of DDT to parties and state non-parties shall be strictly in accordance with the paragraph 2(b) article 3 of the Stockholm Convention on Persistent Organic Pollutants (POPs). (S.O. 295(E) dated 8 March 2006) Use of DDT in agriculture is withdrawn. In very special circumstances warranting the use of DDT for plant protection work, the state or central Government may purchase it directly from M/s Hindustan Insecticides Ltd to be used under expert governmental supervision. (S.O. 378(E) dated 26 May 1989)
Fenitrothion	The use of fenitrothion is banned in agriculture except for locust control in scheduled desert areas and for public health purposes. (S.O. 706(E) dated 3 May 2007)
Fenthion	The use of fenthion is banned in agriculture except for locust control, household and public health purposes. (S.O. 46(E) dated 8 January 2008)
Lindane (gamma-HCH)	Lindane is banned for manufacture, import or formulate. However, it was allowed for use up to 24 March 2013 for termite control in buildings including wood, and termite control in agriculture as per approved label claims by the Registration Committee and for exports. (S.O. 637(E) dated 25 March 2011 and S.O. 1472(E) dated 29 August 2007)
Methoxy ethyl mercuric chloride (MEMC)	The use of MEMC is banned completely except for seed treatment of potato and sugarcane. (S.O. 681(E) dated 17 July 2001)
Methyl bromide	Methyl bromide may be used only by government/government undertakings/government organizations/ pest control operators under the strict supervision of government experts or experts whose expertise is approved by the Plant Protection Advisor to the Government of India. (G.S.R. 371 (E) dated 20 May 1999 and an earlier RC decision)
Methyl parathion	Methyl parathion 50% EC and 2% DP formulations are banned for use on fruits and vegetables. (S.O. 680(E) dated 17 July 2001) The use of methyl parathion is permitted only on those crops approved by the Registration Committee where honey bees are not acting as pollinators. (S.O. 658 (E) dated 4 September 1992)
Monocrotophos	Monocrotophos is banned for use on vegetables. (S.O. 1482 (E) dated 10 October 2005)
Sodium cyanide	The use of sodium cyanide shall be restricted for fumigation of cotton bales under expert supervision approved by the Plant Protection Advisor to the Government of India. (S.O. 569 (E) dated 25 July 1989)

Appendix 10: FRAC Code List©

FRAC
FUNGICIDE RESISTANCE
ACTION COMMITTEE

FRAC Code List©: Fungicides sorted by mode of action (including FRAC Code numbering)

MOA	TARGET SITE AND CODE	GROUP NAME	CHEMICAL GROUP	COMMON NAME	COMMENTS	FRAC CODE
A: nucleic acids synthesis	**A1:** NA polymerase **I**	**PA** – fungicides (**P**henyl**A**mides)	acylalanines	benalaxyl furalaxyl metalaxyl metalaxyl-M (=mefenoxam)	Resistance and cross resistance well known in various Oomycetes but mechanism unknown. **High risk. See FRAC Phenylamide Guidelines for resistance management**	4
			oxazolidinones	oxadixyl		
			butyrolactones	ofurace		
	A2: adenosin-deaminase	hydroxy-(2-amino-) pyrimidines	hydroxy-(2-amino-) pyrimidines	bupirimate dimethirimol ethirimol	Medium risk Resistance and cross resistance known in powdery mildews. Resistance management required.	8
	A3: DNA/RNA synthesis (proposed)	heteroaromatics	isoxazoles	hymexazole	Resistance not known.	32
			isothiazolones	octhilinone		
	A4: DNA topoisomerase type II (gyrase)	carboxylic acids	carboxylic acids	oxolinic acid	Bactericide. Resistance known. Risk in fungi unknown. Resistance management required.	31
B: mitosis and cell division	**B1:** ß-tubuline assembly in mitosis	**MBC** - fungicides (**M**ethyl **B**enzimidazole **C**arbamates)	benzimidazoles	benomyl carbendazim fuberidazole thiabendazole	Resistance common in many fungal species. Several target site mutations, mostly E198A/G/K, F200Y in β-tubulin gene. Positive cross resistance between the group members. Negative cross resistance to N-Phenylcarbamates. **High risk. See FRAC Benzimidazole Guidelines for resistance management.**	1
			thiophanates	thiophanate thiophanate-methyl		
	B2: ß-tubulin assembly in mitosis	N-phenyl carbamates	N-phenyl carbamates	diethofencarb	Resistance known. Target site mutation E198K. Negative cross resistance to benzimidazoles. High risk. Resistance management required.	10
	B3: ß-tubulin assembly in mitosis	benzamides	toluamides	zoxamide	Low to medium risk. Resistance management required.	22
	B4: cell division (proposed)	phenylureas	phenylureas	pencycuron	Resistance not known	20
	B5: delocalisation of spectrin-like proteins	benzamides	pyridinylmethyl-benzamides	fluopicolide	Resistance not known	43

Continued

Continued.

MOA	TARGET SITE AND CODE	GROUP NAME	CHEMICAL GROUP	COMMON NAME	COMMENTS	FRAC CODE
C. respiration	**C1:** complex I NADH Oxido-reductase	pyrimidinamines	pyrimidinamines	diflumetorim	Resistance not known.	**39**
	C2: complex II: succinate-dehydrogenase	SDHI (**S**uccinate **deh**ydrogenase **i**nhibitors)	phenyl-benzamides	benodanil flutolanil mepronil	Resistance known for several fungal species in field populations and lab mutants. Target site mutations in sdh gene, e.g. H/Y (or H/L) at 257, 267, 272 or P225L, dependent on fungal species. Resistance management required. **Medium to high risk.** **See FRAC SDHI Guidelines for resistance management.**	**7**
			pyridinyl-ethyl-benzamides	fluopyram		
			furan- carboxamides	fenfuram		
			oxathiin-carboxamides	carboxin oxycarboxin		
			thiazole-carboxamides	thifluzamide		
			pyrazole-carboxamides	bixafen furametpyr isopyrazam penflufen penthiopyrad sedaxane		
			pyridine-carboxamides	boscalid		
	C3: complex III: cytochrome bc1 (ubiquinol oxidase) at Qo site *(cyt b gene)*	**QoI**-fungicides (**Q**uinone **o**utside **I**nhibitors)	methoxy-acrylates	azoxystrobin enestrobin picoxystrobin pyraoxystrobin	Resistance known in various fungal species. Target site mutations in cyt b gene (G143A, F129L) and additional mechanisms. Cross resistance shown between all members of the QoI group. **High risk.** **See FRAC QoI Guidelines for resistance management.**	**11**
			methoxy-carbamates	pyraclostrobin pyrametostrobin		
			oximino acetates	kresoxim-methyl trifloxystrobin		
			oximino-acetamides	dimoxystrobin metominostrobin orysastrobin		
			oxazolidine-diones	Famoxadone		
			dihydro-dioxazines	Fluoxastrobin		
			Imidazolinones	Fenamidone		
			benzyl-carbamates	Pyribencarb		
	C4: complex III: cytochrome bc1(ubiquino-ne reductase) at Qi site	**QiI** - fungicides (**Q**uinone **i**nside **I**nhibitors)	cyano- imidazole	cyazofamid	Resistance risk unknown butassumed to be medium to high(mutations at target site knownin model organisms).Resistance management required.	**21**
			sulfamoyl-triazole	Amisulbrom		
	C5: uncouplers of oxidative phosphorylation		dinitrophenyl crotonates	Binapacryl meptyldinocap dinocap	Resistance not known. Also acaricidal activity.	**29**
			2,6-dinitro-Anilines	Fluazinam	Low risk. However, resistance claimed in *Botrytis* in Japan.	
			pyrimidinone-hydrazones	Ferimzone	Resistance not known.	

Continued

Continued.

MOA	TARGET SITE AND CODE	GROUP NAME	CHEMICAL GROUP	COMMON NAME	COMMENTS	FRAC CODE
C. respiration (continued.)	**C6:** inhibitors of oxidative phosphorylation, ATP synthase	organo tin compounds	tri phenyl tin compounds	fentin acetate fentin chloride fentin hydroxide	Some resistance cases known. Low to medium risk.	**30**
	C7: ATP production (proposed)	thiophene-carboxamides	thiophene-carboxamides	Silthiofam	Resistance reported. Risk low.	**38**
	C8: complex III: cytochrome bc1 (ubiquinone reductase) at Q × (unknown) site	Q × I – fungicide (Quinone × Inhibitor)	triazolo-pyrimidylamine	ametoctradin	Resistance risk assumed to bemedium to high (single siteinhibitor). Resistance management required.	**45**
D: amino acids and protein synthesis	**D1:** methionine biosynthesis (proposed) *(cgs gene)*	**AP** - fungicides (**A**nilino-**P**yrimidines)	anilino-pyrimidines	cyprodinil mepanipyrim pyrimethanil	Resistance known in *Botrytis*and *Venturia*, sporadically in *Oculimacula*. **Medium risk. See FRAC Anilinopyrimidine } Guidelines for resistance management.**	**9**
	D2: protein synthesis	enopyranuronic acid antibiotic	enopyranuronic acid antibiotic	blasticidin-S	Low to medium risk. Resistance management required.	**23**
	D3: protein synthesis	hexopyranosyl antibiotic	hexopyranosyl antibiotic	kasugamycin	Resistance known in fungal and bacterial (*P. glumae*) pathogens. Medium risk. Resistance management required.	**24**
	D4: protein synthesis	glucopyranosyl antibiotic	glucopyranosyl antibiotic	streptomycin	Bactericide. Resistance known. High risk. Resistance management required.	**25**
	D5: protein synthesis	tetracycline antibiotic	tetracycline antibiotic	oxytetracycline	Bactericide. Resistance known. High risk. Resistance management required.	**41**

Continued

Continued.

MOA	TARGET SITE AND CODE	GROUP NAME	CHEMICAL GROUP	COMMON NAME	COMMENTS	FRAC CODE
E: signal transduction	**E1:** Signal transduction (mechanism unknown)	aza-naphthalenes	quinolines	quinoxyfen	Resistance to quinoxyfen known. Medium risk. Resistance management required. Cross resistance found in *Erysiphe (Uncinula) necator* but not in *Blumeria graminis*.	13
			quinazolinone	proquinazid		
	E2: MAP/Histidine-Kinase in osmotic signal transduction *(os-2, HOG1)*	**PP**-fungicides (**P**henyl**P**yrroles)	phenylpyrroles	fenpiclonil fludioxonil	Resistance found sporadically, mechanism speculative. Low to medium risk. Resistance management required.	12
	E3: MAP/Histidine-Kinase in osmotic signal transduction *(os-1, Daf1)*	dicarboximides	dicarboximides	chlozolinate iprodione procymi-done vinclozolin	Resistance common in *Botrytis* and some other pathogens. Several mutations in OS-1, mostly I365S. Cross resistance common between the group members. **Medium to high risk. See FRAC Dicarboximide Guidelines for resistance management.**	2
F: lipids and membrane synthesis	F1	formerly dicarboximides				
	F2: phospholipid biosynthesis, methyltrans-ferase	phosphoro-thiolates	phosphoro-thiolates	edifenphos iprobenfos (IBP) pyrazophos	Resistance known in specific fungi. Low to medium risk. Resistance management required if used for risky pathogens.	6
		dithiolanes	dithiolanes	isoprothiolane		
	F3: lipid peroxidation (proposed)	**AH**-fungicides (**A**romatic **H**ydrocarbons) (chlorophenyls, nitroanilines)	aromatic hydrocarbons	biphenyl chloroneb dicloran quintozene (PCNB) tecnazene (TCNB) tolclofos-methyl	Resistance known in some fungi. Low to medium risk. Cross resistance patterns complex due to different activity spectra.	14
		heteroaromatics	1,2,4-thiadiazoles	etridiazole		
	F4: cell membrane permeability, fatty acids (proposed)	carbamates	carbamates	iodocarb propa-mocarb prothiocarb	Low to medium risk. Resistance management required.	28
	F5: phospholipid biosynthesis and cell wall deposition (proposed)	**CAA**-fungicides (**C**arboxylic **A**cid **A**mides)	cinnamic acid amides	dimethomorph flumorph	Resistance known in *Plasmopara viticola* but not in *Phytophthora infestans.* Cross resistance between all members of the CAA group. **Low to medium risk. See FRAC CAA Guidelines for resistance management**	40
			valinamide carbamates	benthiavalicarb iprovalicarb valifenalate		
			mandelic acid amides	mandipropamid		
	F6: microbial disrupters of pathogen cell membranes	Microbial (*Bacillus* sp.)	*Bacillus subtilis* and the fungicidal lipopeptides they produce	*Bacillus subtilis* strain QST 713	No resistance reported. Assumed to be low risk based on mode of action of non-specific membrane disruption	44

Continued

Continued.

MOA	TARGET SITE AND CODE	GROUP NAME	CHEMICAL GROUP	COMMON NAME	COMMENTS	FRAC CODE
G: sterol biosynthesis in membranes	**G1:** C14- demethylase in sterol biosynthesis *(erg11/cyp51)*	**DMI**-fungicides (**De****M**ethylation **I**nhibitors) (SBI: Class I)	piperazines	triforine	There are big differences in the activity spectra of DMI fungicides. Resistance is known in various fungal species. Several resistance mechanisms are known incl. target site mutations in cyp51 (erg 11) gene, e.g. V136A, Y137F, A379G, I381V; cyp51 promoto; ABC transporters and others. Generally wise to accept that cross resistance is present between DMI fungicides active against the same fungus. DMI fungicides are Sterol Biosynthesis Inhibitors (SBIs), but show no cross resistance to other SBI classes. **Medium risk.** **See FRAC SBI Guidelines for resistance management.**	**3**
			pyridines	pyrifenox		
			pyrimidines	fenarimol nuarimol		
			imidazoles	imazalil oxpoconazole pefurazoate prochloraz triflumizole		
			triazoles	azaconazole bitertanol bromuconazole cyproconazole difenoconazole diniconazole epoxiconazole etaconazole fenbuconazole fluquinconazole flusilazole flutriafol hexaconazole imibenconazole ipconazole metconazole myclobutanil penconazole propiconazole prothioconazole simeconazole tebuconazole tetraconazole triadimefon triadimenol triticonazole		
	G2: Δ^{14}-reductase and $\Delta^{8}\rightarrow\Delta^{7}$-isomerase in sterol biosynthesis *(erg24, erg2)*	Amines ("Morpholines") (SBI: Class II)	morpholines	aldimorph dodemorph fenpropimorph tridemorph	Decreased sensitivity for powdery mildews. Cross resistance within the group generally found but not to other SBI classes. **Low to medium risk.** **See FRAC SBI Guidelines for resistance management.**	**5**
			piperidines	fenpropidin piperalin		
			spiroketal-amines	spiroxamine		
	G3: 3-keto reduc-tase, C4- de-methylation *(erg27)*	hydroxyanilides (SBI: Class III)	hydroxyanilides	fenhexamid	Low to medium risk. Resistance management required.	**17**
	G4: squalene-epoxidase in sterol biosyn-thesis *(erg1)*	(SBI class IV)	thiocarbamates	pyributicarb	Resistance not known, fungicidal and herbicidal activity	**18**
			allylamines	naftifine terbinafine	Medical fungicides only	

Continued

Continued.

MOA	TARGET SITE AND CODE	GROUP NAME	CHEMICAL GROUP	COMMON NAME	COMMENTS	FRAC CODE
H: glucan synthesis	**H3:** trehalase and inositol-biosynthesis	glucopyranosyl antibiotic	glucopyranosyl antibiotic	validamycin	Resistance not known	**26**
H: glucan synthesis	**H4:** chitin synthase	polyoxins	peptidyl pyrimidine nucleoside	polyoxin	Resistance known. Medium risk. Resistance management required.	**19**
I: melanin synthesis in cell wall	**I1:** reductase in melanin biosynthesis	**MBI-R** (**M**elanin **B**iosynthesis **I**nhibitors – **R**eductase)	isobenzo-furanone	fthalide	Resistance not known	**16.1**
			pyrrolo-quinolinone	pyroquilon		
			triazolobenzo-thiazole	tricyclazole		
I: melanin synthesis in cell wall	**I2:** dehydratase in melanin biosynthesis	**MBI-D** (**M**elanin **B**iosynthesis **I**nhibitors – **D**ehydratase)	cyclopropane-carboxamide	carpropamid	Resistance known. Medium risk. Resistance management required.	**16.2**
			carboxamide	diclocymet		
			propionamide	fenoxanil		
P: host plant defence induction	**P1:** salicylic acid pathway	benzo-thiadiazole BTH	benzo-thiadiazole BTH	acibenzolar-S-methyl	Resistance not known	**P**
P: host plant defence induction	**P2**	benzisothiazole	benzisothiazole	probenazole (also antibacterial and antifungal activity)	Resistance not known	**P**
P: host plant defence induction	**P3**	thiadiazole-carboxamide	thiadiazole-carboxamide	tiadinil isotianil	Resistance not known	**P**
P: host plant defence induction	**P4** (proposed)	natural compound		laminarin	Resistance not known	**P**
Unknown mode of action (U numbers not appearing in the list derive from reclassified fungicides)	unknown	cyanoacetamide-oxime	cyanoacetamide-oxime	cymoxanil	Resistance claims described. Low to medium risk. Resistance management required.	**27**
	unknown	phosphonates	ethyl phosphonates	fosetyl-Al	Few resistance cases reported in few pathogens. Low risk	**33**
				phophorous acid and salts		
	unknown	phthalamic acids	phthalamic acids	tecloftalam (Bactericide)	Resistance not known	**34**
	unknown	benzotriazines	benzotriazines	triazoxide	Resistance not known	**35**
	unknown	benzene-sulfona-mides	benzene-sulfona-mides	flusulfamide	Resistance not known	**36**
	unknown	pyridazinones	pyridazinones	diclomezine	Resistance not known	**37**
	unknown	thiocarbamate	thiocarbamate	methasulfocarb	Resistance not known	**42**

Continued

Continued.

MOA	TARGET SITE AND CODE	GROUP NAME	CHEMICAL GROUP	COMMON NAME	COMMENTS	FRAC CODE
Unknown mode of action (Continued).	microtubule disruption (proposed)	thiazole carboxamide	ethylamino-thiazole carboxamide	ethaboxam	Resistance not known	U5
	unknown	phenyl-acetamide	phenyl-acetamide	cyflufenamid	Resistance in *Sphaerotheca*. Resistance management required	U6
	actin disruption (proposed)	benzophenone	benzophenone	metrafenone	Less sensitive isolates detected in wheat powdery mildew. Medium risk. Resistance management required.	U8
	Cell membrane disruption (proposed)	guanidines	guanidines	dodine	Resistance known in *Venturia inaequalis*. Low to medium risk. Resistance management recommended.	U12
not classified	unknown	diverse	diverse	mineral oils, organic oils, potassium bicarbonate, material of biological origin	Resistance not known	NC
Multi-site contact activity	multi-site contact activity	inorganic	inorganic	copper (different salts)	Generally considered as a low risk group without any signs of resistance developing to the fungicides No cross resistance between group members M1 to M9	M1
		inorganic	inorganic	sulphur		M2
		dithiocarbamates and relatives	dithio-carbamates and relatives	ferbam mancozeb maneb metiram propineb thiram zineb ziram		M3
		phthalimides	phthalimides	captan captafol folpet		M4
		chloronitriles (phthalonitriles)	chloronitriles (phthalonitriles)	chlorothalonil		M5
		sulfamides	sulfamides	dichlofluanid tolylfluanid		M6
		guanidines	guanidines	guazatine iminoctadine		M7
		triazines	triazines	anilazine		M8
		quinones (anthraquinones)	quinones (anthra-quinones)	dithianon		M9

Appendix 11: IRAC MoA Classification Scheme

6. The MoA Classification Scheme

The MOA classification scheme developed and endorsed by IRAC is based on the best available evidence of the MoA of available insecticides. Details of the listing have been agreed by IRAC companies and approved by internationally recognized industrial and academic insect toxicologists and biochemists.

It is our aim to ensure that insecticide and acaricide users are aware of MoA groups and that they have a sound basis on which to implement season-long, sustainable resistance management through the effective use of alternations, sequences or rotations of insecticides with different modes of action. To help delay resistance, it is strongly recommended that growers also integrate other control methods into insect or mite control programmes. Further advice is given in Appendix 2.

Note: Inclusion of a compound in the MoA list does not necessarily signify regulatory approval.

6.1. Rules for inclusion of a compound in the MoA list

- Chemical nomenclature is based on that appearing in *The Pesticide Manual*, 15th edition, November 2009, Ed. C.D.S. Tomlin, published by The British Crop Protection Council. ISBN 9781901396188
- To be included in the active list, compounds must have, or be very close to having, a minimum of one registered use in at least one country.
- In any one MoA classification sub-group, where more than one active ingredient in that chemical sub-group is registered for use, the chemical sub-group name is used.
- In any one MoA classification sub-group, where only one active ingredient is registered for use, the name of that exemplifying active ingredient is used
- Where more than one chemical sub-group or exemplifying active ingredient appears in a single MoA group, each is named according to the above rules, with chemical subgroups having precedence over single active ingredients.

6.2. Classification Table

IRAC MoA Classification v 7.1, June 2011[1]		
Main Group and Primary Site of Action	Chemical Sub-group or exemplifying Active Ingredient	Active Ingredients
1* **Acetylcholinesterase (AChE) inhibitors** Nerve action {Strong evidence that action at this protein is responsible for insecticidal effects} * *Please see footnotes for further information on the use of compounds between sub-groups*	**1A** Carbamates	Alanycarb, Aldicarb, Bendiocarb, Benfuracarb, Butocarboxim, Butoxycarboxim, Carbaryl, Carbofuran, Carbosulfan, Ethiofencarb, Fenobucarb, Formetanate, Furathiocarb, Isoprocarb, Methiocarb, Methomyl, Metolcarb, Oxamyl, Pirimicarb, Propoxur, Thiodicarb, Thiofanox, Triazamate, Trimethacarb, XMC, Xylylcarb
	1B Organophosphates	Acephate, Azamethiphos, Azinphos-ethyl, Azinphos-methyl, Cadusafos, Chlorethoxyfos, Chlorfenvinphos, Chlormephos, Chlorpyrifos, Chlorpyrifos-methyl, Coumaphos, Cyanophos, Demeton-S-methyl, Diazinon, Dichlorvos/ DDVP, Dicrotophos, Dimethoate, Dimethylvinphos, Disulfoton, EPN, Ethion, Ethoprophos, Famphur, Fenamiphos, Fenitrothion, Fenthion, Fosthiazate, Heptenophos, Imicyafos, Isofenphos, Isopropyl *O*- (methoxyaminothio-phosphoryl) salicylate, Isoxathion, Malathion, Mecarbam, Methamidophos, Methidathion, Mevinphos, Monocrotophos, Naled, Omethoate, Oxydemeton-methyl, Parathion, Parathion-methyl, Phenthoate, Phorate, Phosalone, Phosmet, Phosphamidon, Phoxim, Pirimiphos-methyl, Profenofos, Propetamphos, Prothiofos, Pyraclofos, Pyridaphenthion, Quinalphos, Sulfotep, Tebupirimfos, Temephos, Terbufos, Tetrachlorvinphos, Thiometon, Triazophos, Trichlorfon, Vamidothion
2 **GABA-gated chloride channel antagonists** Nerve action {Strong evidence that action at this protein is responsible for insecticidal effects}	**2A** Cyclodiene organochlorines	Chlordane, Endosulfan
	2B Phenylpyrazoles (Fiproles)	Ethiprole, Fipronil
3* **Sodium channel modulators** Nerve action {Strong evidence that action at this protein is responsible for insecticidal effects} * *Please see footnotes for further information on the use of compounds between sub-groups*	**3A** Pyrethroids Pyrethrins	Acrinathrin, Allethrin, d-*cis-trans* Allethrin, d-*trans* Allethrin, Bifenthrin, Bioallethrin, Bioallethrin S-cyclopentenyl isomer, Bioresmethrin, Cycloprothrin, Cyfluthrin, *beta*-Cyfluthrin, Cyhalothrin, *lambda*-Cyhalothrin, *gamma*-Cyhalothrin, Cypermethrin, *alpha*-Cypermethrin, *beta*-Cypermethrin, *theta*-cypermethrin, *zeta*-Cypermethrin, Cyphenothrin, (1*R*)-*trans*- isomers], Deltamethrin, Empenthrin (*EZ*)- (1*R*)-isomers], Esfenvalerate, Etofenprox, Fenpropathrin, Fenvalerate, Flucythrinate, Flumethrin, *tau*-Fluvalinate, Halfenprox, Imiprothrin, Kadethrin, Permethrin, Phenothrin [(1*R*)-*trans*- isomer], Prallethrin, Pyrethrins (pyrethrum), Resmethrin, Silafluofen, Tefluthrin, Tetramethrin, Tetramethrin [(1*R*)-isomers], Tralomethrin, Transfluthrin,
	3B DDT Methoxychlor	DDT Methoxychlor
4 **Nicotinic acetylcholine receptor (nAChR) agonists** Nerve action {Strong evidence that action at one or more of this class of protein is responsible for insecticidal effects}	**4A** Neonicotinoids	Acetamiprid, Clothianidin, Dinotefuran, Imidacloprid, Nitenpyram, Thiacloprid, Thiamethoxam,
	4B Nicotine	Nicotine

Continued

Continued.

Main Group and Primary Site of Action	Chemical Sub-group or exemplifying Active Ingredient	Active Ingredients
5 **Nicotinic acetylcholine receptor (nAChR) allosteric activators** Nerve action {Strong evidence that action at one or more of this class of protein is responsible for insecticidal effects}	Spinosyns	Spinetoram, Spinosad
6 **Chloride channel activators** Nerve and muscle action {Strong evidence that action at one or more of this class of protein is responsible for insecticidal effects}	Avermectins, Milbemycins	Abamectin, Emamectin benzoate, Lepimectin, Milbemectin
7 **Juvenile hormone mimics** Growth regulation {Target protein responsible for biological activity is unknown, or uncharacterized}	**7A** Juvenile hormone analogues	Hydroprene, Kinoprene, Methoprene
	7B Fenoxycarb	Fenoxycarb
	7C Pyriproxyfen	Pyriproxyfen
8 **Miscellaneous non-specific (multi-site) inhibitors**	**8A** Alkyl halides	Methyl bromide and other alkyl halides
	8B Chloropicrin	Chloropicrin
	8C Sulfuryl fluoride	Sulfuryl fluoride
	8D Borax	Borax
	8E Tartar emetic	Tartar emetic
9 **Selective homopteran feeding blockers** Nerve action {Target protein responsible for biological activity is unknown, or uncharacterized}	**9B** Pymetrozine	Pymetrozine
	9C Flonicamid	Flonicamid
10 **Mite growth inhibitors** Growth regulation {Target protein responsible for biological activity is unknown, or uncharacterized}	**10A*** Clofentezine Hexythiazox Diflovidazin **Please see footnotes for further information on this sub-grouping*	Clofentezine, Hexythiazox, Diflovidazin
	10B Etoxazole	Etoxazole

Continued

Continued.

Main Group and Primary Site of Action	Chemical Sub-group or exemplifying Active Ingredient	Active Ingredients
11 **Microbial disruptors of insect midgut membranes** (includes transgenic crops expressing *Bacillus thuringiensis* toxins, however specific guidance for resistance management of transgenic crops is not based on rotation of modes of action)	*Bacillus thuringiensis* or *Bacillus sphaericus* and the insecticidal proteins they produce	*Bacillus thuringiensis* subsp. *israelensis Bacillus sphaericus Bacillus thuringiensis* subsp. *aizawai Bacillus thuringiensis* subsp. *kurstaki Bacillus thuringiensis* subsp. *tenebrionis* Bt crop proteins: Cry1Ab, Cry1Ac, Cry1Fa, Cry2Ab, mCry3A, Cry3Ab, Cry3Bb, Cry34/35Ab1
12 **Inhibitors of mitochondrial ATP synthase** Energy metabolism {Compounds affect the function of this protein, but it is not clear that this is what leads to biological activity}	**12A** Diafenthiuron	Diafenthiuron
	12B Organotin miticides	Azocyclotin, Cyhexatin, Fenbutatin oxide
	12C Propargite	Propargite
	12D Tetradifon	Tetradifon
13 **Uncouplers of oxidative phosphorylation via disruption of the proton gradient** Energy metabolism	Chlorfenapyr DNOC Sulfluramid	Chlorfenapyr DNOC Sulfluramid
14 **Nicotinic acetylcholine receptor (nAChR) channel blockers** Nerve action {Compounds affect the function of this protein, but it is not clear that this is what leads to biological activity}	Nereistoxin analogues	Bensultap, Cartap hydrochloride, Thiocyclam, Thiosultap-sodium
15 **Inhibitors of chitin biosynthesis, type 0** Growth regulation {Target protein responsible for biological activity is unknown, or uncharacterized}	Benzoylureas	Bistrifluron, Chlorfluazuron, Diflubenzuron, Flucycloxuron, Flufenoxuron, Hexaflumuron, Lufenuron, Novaluron, Noviflumuron, Teflubenzuron, Triflumuron
16 **Inhibitors of chitin biosynthesis, type 1** Growth regulation {Target protein responsible for biological activity is unknown, or uncharacterized}	Buprofezin	Buprofezin
17 **Moulting disruptor, Dipteran** Growth regulation {Target protein responsible for biological activity is unknown, or uncharacterized}	Cyromazine	Cyromazine

Continued

Continued.

Main Group and Primary Site of Action	Chemical Sub-group or exemplifying Active Ingredient	Active Ingredients
18 **Ecdysone receptor agonists** Growth regulation {Strong evidence that action at this protein is responsible for insecticidal effects}	Diacylhydrazines	Chromafenozide, Halofenozide, Methoxyfenozide, Tebufenozide
19 **Octopamine receptor agonists** Nerve action {Good evidence that action at one or more of this class of protein is responsible for insecticidal effects}	Amitraz	Amitraz
20 **Mitochondrial complex III electron transport inhibitors** Energy metabolism {Good evidence that action at this protein complex is responsible for insecticidal effects}	**20A** Hydramethylnon	Hydramethylnon
	20B Acequinocyl	Acequinocyl
	20C Fluacrypyrim	Fluacrypyrim
21 **Mitochondrial complex I electron transport inhibitors** Energy metabolism {Good evidence that action at this protein complex is responsible for insecticidal effects}	**21A** METI acaricides and insecticides	Fenazaquin, Fenpyroximate, Pyrimidifen, Pyridaben, Tebufenpyrad, Tolfenpyrad
	21B Rotenone	Rotenone (Derris)
22* **Voltage-dependent sodium channel blockers** Nerve action {Good evidence that action at this protein complex is responsible for insecticidal effects} **Please see footnotes for further information on sub-grouping*	**22A** Indoxacarb	Indoxacarb
	22B Metaflumizone	Metaflumizone
23 **Inhibitors of acetyl CoA carboxylase.** Lipid synthesis, growth regulation {Good evidence that action at this protein is responsible for insecticidal effects}	Tetronic and Tetramic acid derivatives	Spirodiclofen, Spiromesifen, Spirotetramat

Continued

Continued.

Main Group and Primary Site of Action	Chemical Sub-group or exemplifying Active Ingredient	Active Ingredients
24 **Mitochondrial complex IV electron transport inhibitors** Energy metabolism {Good evidence that action at this protein complex is responsible for insecticidal effects}	**24A** Phosphine	Aluminium phosphide, Calcium phosphide, Phosphine, Zinc phosphide
	24B Cyanide	Cyanide
25 **Mitochondrial complex II electron transport inhibitors** Energy metabolism {Good evidence that action at this protein complex is responsible for insecticidal effects}	Beta-ketonitrile derivatives	Cyenopyrafen, Cyflumetofen
28 **Ryanodine receptor modulators** Nerve and muscle action {Good evidence that action at this protein complex is responsible for insecticidal effects}	Diamides	Chlorantraniliprole, Cyantraniliprole, Flubendiamide
un Compounds of unknown or uncertain MoA[2] {Target protein responsible for biological activity is unknown, or uncharacterized}	Azadirachtin	Azadirachtin
	Benzoximate	Benzoximate
	Bifenazate	Bifenazate
	Bromopropylate	Bromopropylate
	Chinomethionat	Chinomethionat
	Cryolite	Cryolite
	Dicofol	Dicofol
	Pyridalyl	Pyridalyl
	Pyrifluquinazon	Pyrifluquinazon

Table Notes:
MoA assignments will usually involve identification of the target protein responsible for the biological effect, although groupings can be made where compounds share distinctive physiological effects and have related chemical structures.
Groups 26 and 27 are unassigned at this time and have therefore been omitted from the table.

[1] Inclusion of a compound in the classification above does not necessarily signify regulatory approval
[2] A compound with an unknown or controversial MoA or an unknown mode of toxicity will be held in category 'un' until evidence becomes available to enable that compound to be assigned to a more appropriate MoA class

6.3. Criteria for descriptors of the quality of MoA information

{Strong evidence that action at this protein (or protein complex) is responsible for insecticidal effects}	Potent effects on the function of the target protein and either resistance due to mutation / overexpression / removal of this protein or correlation of potency between effects on the protein and biological activity for a set of analogues.
{Good evidence that action at this protein (or protein complex) is responsible for insecticidal effects}	Highly potent effects on the function of the protein combined with clearly consistent physiological effects
{Compounds affect the function of this protein, but it is not clear that this is what leads to biological activity}	Compounds (or their metabolites) have moderate or low potency on the function of the protein, and there is little or no evidence associating this effect with biological activity. Compounds may be grouped because of similarity of structure and distinctive physiological effect.
{Target protein responsible for biological activity is unknown, or uncharacterized}	Compounds may be grouped because of similarity of structure and distinctive physiological effect.

6.4. Notes regarding sub-groups

Sub-groups represent distinct structural classes believed to have the same MoA. In principle, they provide a useful level of differentiation between compounds that may bind at the same target site but are nevertheless structurally different enough that the risk of metabolic cross-resistance is lower than for close chemical analogs. Subgroups are likely to be metabolized by different enzymes and may bind differently enough within the target site that the chance of selection for either metabolic or target-site resistance is reduced compared to close analogs. In the absence of other alternatives, it may be possible to rotate compounds between subgroups if it is clear that cross resistance mechanisms do not exist in the target populations. By definition, subgroups are established to represent distinct chemical classes with a common MoA. Whether they should be rotated or not will depend on knowledge and experience of cross-resistance patterns, resistance mechanisms, and furthermore on the pest, crop and region considered.

Sub-group number	Notes
1A & B	If there are no other alternatives, compounds from groups 1A and 1B may be rotated in situations where cross-resistance mechanisms are known to be absent in the insect populations to be treated.
3A & B	If there are no other alternatives, compounds from groups 3A and 3B may be rotated in situations where cross-resistance mechanisms (e.g., kdr) are known to be absent in the insect populations to be treated. Because DDT is no longer used in agriculture, this is only applicable for the control of human disease vectors such as mosquitoes, because of a lack of alternatives.
10A	Hexythiazox is grouped with clofentezine because they commonly exhibit cross-resistance, even though they are structurally distinct, and the target site for neither compound is known. Diflovidazin has been added to this group because it is a close analogue of Clofentezine and expected to have the same mode of action.
22A & B	Although these compounds are believed to have the same target site, they have been sub-grouped because they are chemically distinct, and current evidence indicates that the risk of metabolic cross-resistance is low.

6.5. General notes & MoA Classification Scheme Updates

- Further details on the MoA Group Descriptors are given in Appendix 3.
- A list of active ingredients in alphabetical order with their respective MoA classification is given in Appendix 5.
- The Classification Scheme has been prepared using the most up-to-date information available to IRAC. It is provided to user groups, grower organisations, extension personnel, regulatory authorities such as the US EPA and all those involved in resistance management, as an agreed definitive statement by the agrochemical industry on the MoA of insecticides currently in use.
- The IRAC MoA classification is reviewed and reissued at intervals as required. The latest version is always available for reference via the IRAC website (www.irac-online.org).
- Submissions for new active ingredients together with recommendations for their inclusion in specific new or existing MoA classes, together with citations or evidence for classification should be made to IRAC through the website.
- IRAC member companies review draft versions before an agreed final version of any update is published. In addition, a number of internationally well-known insect toxicologists and biochemists can be consulted regarding additions, deletions or other changes to the list. Details of the procedures followed for allocation of new insecticidal materials to the MoA classification are given in Appendix 4.
- Changes to the listing may have serious consequences. In those countries where insecticide labels display the IRAC MoA number or class name as an aid to good IRM (see Appendix 1), changes may be especially costly to implement. In general, changes are therefore only endorsed when the scientific evidence supporting the change is compelling.

- Superseded, obsolete or withdrawn compounds for which no current registration exists, and that are no longer in common usage, are not listed.
- In a continued effort to refine the list, readers are kindly asked to inform IRAC of factual errors or omissions, citing definitive evidence wherever possible. Such submissions should be directed to IRAC via the website. Suggestions for improvements are likewise welcome.

Appendix 12: HRAC Mode of Action Classification Scheme

Classification of herbicides is based on their mode of action (MOA). (Courtesy of Herbicide Resistance Action Committee (HRAC). Reproduced with the permission of Dr John Soteres, Chairman, HRAC.)

HRAC group	MOA	Chemical family	Active ingredient	Weed Science Society of America classification[a]
A	Inhibition of acetyl CoA carboxylase (ACCase)	Aryloxyphenoxy-propionates 'FOPs'	Clodinafop-propargyl, Cyhalofop-butyl, Diclofop-methyl, Fenoxaprop-P-ethyl, Fluazifop-P-butyl, Propaquizafop, Quizalofop-P-ethyl	1
		Cyclohexanediones 'DIMs'	Alloxydim, Butroxydim (clefoxydim proposed), Clethodim, Cycloxydim, Sethoxydim, Tepraloxydin, Tralkoxydim	
B	Inhibition of acetolactate synthase (ALS) (acetohydroxyacid synthase, AHAS)	Sulfonylureas	Amidosulfuron, Azimsulfuron, Bensulfuron, Bensulfuron-methyl, Chlorimuron-ethyl, Chlorsulfuron, Cinosulfuron cyclosulfamuron, Ethametsulfuron-methyl, Ethoxysulfuron, Flazasulfuron, Flupyrsulfuron-methyl-Na, Foramsulfuron, Halosulfuron-methyl, Imazosulfuron, Iodosulfuron, Metsulfuron-methyl, Nicosulfuron, Oxasulfuron, Primisulfuron-methyl, Prosulfuron, Pyrazosulfuron-ethyl, Rimsulfuron, Sulfometuron-methyl, Sulfosulfuron, Thifensulfuron-methyl, Triasulfuron, Tribenuron-methyl, Trifloxysulfuron, Triflusulfuron-methyl, Tritosulfuron	2
		Imidazolinones	Imazapic, Imazamethabenz-methyl, Imazamox imazapyr imazaquin, Imazethapyr	
		Triazolopyrimidines	Cloransulam-methyl, Diclosulam, Florasulam, Flumetsulam, Metosulam	
		Pyrimidiny(thio) benzoates	Bispyribac-Na, Pyribenzoxim, Pyriftalid, Pyrithiobac-Na, Pyriminobac-methyl	
		Sulfonylaminocarbonyl-triazolinones	Flucarbazone-Na, Procarbazone-Na	
C1	Inhibition of photosynthesis at photosystem II	Triazines	Ametryne, Atrazine, Cyanazine, Desmetryne, Dimethametryne, Prometon, Prometryne, Propazine, Simazine, Simetryne, Terbumeton, Terbuthylazine, Terbutryne, Trietazine	5
		Triazinones	Hexazinone, Metamitron, Metribuzin	
		Triazolinone uracils	Amicarbazone, Bromacil, Lenacil, Terbacil	
		Pyridazinones	Pyrazon=chloridazon	
		Phenyl-carbamates	Desmedipham, Phenmedipham	
C2	Inhibition of photosynthesis at photosystem II	Ureas	Chlorobromuron, Chlorotoluron, Chloroxuron, Dimefuron, Diuron, Ethidimuron, Fenuron, Fluometuron (see F3), Isoproturon, Isouron, Linuron, Methabenzthiazuron, Metobromuron, Metoxuron, Monolinuron, Neburon, Sinduron, Tebuthiuron	7
		Amides	Propanil, Pentanochlor	
C3	Inhibition of photosynthesis at photosystem II	Nitriles	Bromofenoxim (also M), Bromoxynil (also group M), Ioxynil (also group M)	
D	Photosystem I electron diversion	Bipyridyliums	Diquat, Paraquat	22

Continued

Continued.

HRAC group	MOA	Chemical family	Active ingredient	Weed Science Society of America classification[a]
E	Inhibition of protoporphyrinogen oxidase (PPO)	Diphenylether	Acifluorfen-Na, Bifenox, Chlomethoxyfen, Fluoroglycofen-ethyl, Fomesafen, Halosafen, Lactofen, Oxyfluorfen	14
		Phenylpyroazonles	Fluazolate, Pyraflufen-ethyl	
		N-phenylphthalimides	Cinidon-ethyl, Flumioxazin, Flumiclorac-pentyl	
		Thiadiazoles	Fluthiacet-methyl, Thidiazimin	
		Oxadiazoles	Oxadiazon, Oxadiargyl	
		Triazolinones	Azafenidin, Carfentrazone-ethyl, Sulfentrazone	
		Oxazolidinediones	Pentoxazone	
		Pyrimidindiones	Benzfendizone, Butafenacil	
		Others	Pyrazogyl, Profluazol	
F2	Bleaching: inhibition of 4-hydroxyphenyl-pyruvate-dioxygenase (4-HPPD)	Triketones	Mesotrione, Sulcotrione	28
		Isoxazoles	Isoxachlortole, Isoxaflutole	
		Pyrazoles	Benzofenap, Pyrazolynate, Pyrazoxyfen	
F3	Bleaching: inhibition of carotenoid biosynthesis (unknown target)	Triazoles	Amitrole (*in vivo* inhibition of lycopene cyclase)	11
		Isoxazolidinones	Clomazone	13
		Ureas	Fluometuron (see C2)	
		Diphenylether	Aclonifen	
G	Inhibition of EPSP synthase (EPSP = 5-enolpyrovylshikimate-3-phosphate)	Clycines	Glyphosate, Sulfosate	9
H	Inhibiiton of glutamine synthetase	Phosphinic acids	Glufosinate-ammonium bialaphos=bilanaphos	10
I	Inhibition of DHP (dihydropteroate) synthase	Carbamates	Asulam	18
K1	Microtubule assembly inhibition	Dinitroanilines	Benefin = benfluralin, Butralin, Dinitramine ethalfluralin, Oryzalin, Pendimethalin, Trifluralin	3
		Phosphoroamidates	Amiprophos-methyl, Butamiphos	
		Pyridines	Dithiopyr, Thiazopyr	
		Benzenedicarboxylic acids	DCPA = chlorthal-dimethyl	3
K2	Inhibition of mitosis/ microtubule organization	Carbamates	Chlorpropham, Propham, Carbetamide	23
K3	Inhibition of cell division (inhibition of very long chain fatty acids)	Chloroacetamides	Acetochlor, Alachor, Butachlor, Dimethachlor, Metazachlor, Metolachlor, Pethoxamid, Pretilachlor, Propachlor, Propisochlor, Thenylchlor	
		Acetamides	Diphenamid, Napropamide, Naproanilide	
		Oxyacetamides	Flufenacet, Mefenacet	
		Tetrazolinones	Fentrazamide	
		Others	Anilofos, Cafenstrole, Indanofan, Piperophos	
L	Inhibition of cell wall (cellulose) synthesis	Nitriles	Dichlobenil, Chlorthiamid	20
		Benzamides	Isoxaben	21
		Triazolocarboxamides	Flupoxam	
M	Uncoupling (membrane disruption)	Dinitrophenols	DNOC, Dinoseb, Dinoterb	24
N	Inhibition of lipid synthesis – not ACCase inhibition	Thiocarbamates	Butylate, Cycloate, Dimepiperate, EPTC, Esprocarb, Molinate, Orbencarb, Pebulate, Prosulfocarb, Thiobencarb = benthiocarb, Tiocarbazil, Triallate, Vernolate	
		Phosphorodithioates	Bensulide	
		Benzofuranes	Benfuresate, Ethofumesate	
		Chloro-carbonic acids	TCA, Dalapon, Flupropanate	26

Continued

Continued.

HRAC group	MOA	Chemical family	Active ingredient	Weed Science Society of America classification[a]
O	Action like indole acetic acid (synthetic auxins)	Phenoxy-carboxylic-acids	Clomeprop, 2,4-D, 2,4-DB dichlorprop = 2,4-DP, MCPA, MCPB, Mecoprop = MCPP = CMPP	
		Benzoic acids	Chloramben, Dicamba, TBA	
		Pyridine carboxylic acids	Clopyralid, Fluroxypyr, Picloram, Triclopry	
		Quinoline carboxylic acids	Quinclorac (also group L), Quinmerac	
		Others	Benazolin-ethyl	
P	Inhibition of auxin transport	Phthalamates semicarbazones	Naptalam, Diflufenzopyr-Na	19
R				
S				
Z	Unknown	Arylaminopropionic acids	Flamprop-M-methyl/isopropyl	25
		Pyrazolium	Difenzoquat	8
		Organoarsenicals	DSMA, MSMA	17
		Others	Bromobutide (chloro)-flurenol, Cinmethylin, Cumyluron, Dazomet, dymron = daimuron, Methyl-dimuron = methyl-dymron, Etobenzanid, Fosamine, Metam, Oxaziclomefone, Oleic acid, Pelargonic acid, Pyributicarb	

[a]This column shows the equivalent codes used by the Weed Science Society of America for classifying each particular MOA.

Appendix 13: Chemicals Evaluated for Carcinogenic Potential

August 2010

Chemicals Evaluated for Carcinogenic Potential

Office of Pesticide Programs
U.S. Environmental Protection Agency

BACKGROUND

What is this list?

The following list provides an overview of pesticide chemicals evaluated for carcinogenic potential by EPA's Pesticide Program through August 2010. The evaluation of many of these chemicals is an ongoing process. Therefore, the information in this list may be subject to change as new and/or additional data are submitted to EPA. This list will be updated annually.

How should the information provided in this list be used?

Although this list is available to the public, note that the list represents only the potential carcinogenicity hazard for the chemical with no consideration of exposure information. This list is not intended to be used independent of the full risk assessment for the chemical. When EPA completes a risk assessment on a pesticide, a variety of toxicity information, including potential for noncancer effects (e.g., neurotoxicity, developmental and reproductive toxicity, immunotoxicity, etc) and carcinogenicity, are considered in determining whether to register a pesticide and what requirements for use of the pesticide need to be in place to protect human health. The simple fact of being listed here does not imply that the pesticide poses a significant cancer hazard to the public from use

What does the report date mean?

The date included in the list for each chemical is the date of the most recent review of potential carcinogenicity hazard for that chemical. This date provides a key as to which set of cancer guidelines were used in the review. Note that the classification of potential carcinogenicity generally is not reevaluated unless new data are submitted.

How does EPA review pesticides for potential carcinogenicity?

In evaluating and describing the potential carcinogenicity of a pesticide, EPA's Pesticide Program follows the Agency's Guidelines for Carcinogen Risk Assessment (see http://epa.gov/cancerguidelines/ for more information). The Health Effects Division of the Pesticide Program performs an independent review of all the available evidence to determine the carcinogenic potential of pesticides.

The results of the independent review are peer-reviewed by the Cancer Assessment Review Committee. This committee recommends a "descriptor" (e.g., likely to be carcinogenic to humans, not likely to be carcinogenic to humans, suggestive evidence of carcinogenic potential) to convey the cancer hazard potential of the compound. This descriptor is also referred to as the cancer classification. The evidence for the human cancer potential and the extent to which a person might be exposed (how much time and to what quantity of the pesticide) will determine how the Agency regulates the pesticide. In some cases, EPA may request a review by the FIFRA Scientific Advisory Panel.

What does EPA consider in its review of cancer risk?

In determining the cancer-causing potential of a chemical, EPA considers the full range of available evidence. This information includes

- laboratory animal findings,
- metabolism studies,
- structural relationships with other carcinogens, and
- if available, mode of carcinogenic action information and epidemiologic findings in humans.

All of the information is considered in a weight-of-the-evidence approach. In this weight-of-evidence evaluation, EPA undertakes a critical analysis of each available study to determine its quality and reliability. Then the entire body of evidence is integrated and examined for consistency (repeatability of findings in studies), cohesiveness (a logical pattern of responses), and for biological plausibility (i.e., are the observed findings consistent with current understanding of carcinogenic processes). How the Agency determines the cancer potential of a compound can be found at: http://epa.gov/cancerguidelines/

Most of the cancer determinations for pesticides are largely based on laboratory animal studies because, under its statutory authority, EPA requires registrants to submit an extensive range of laboratory studies on pesticides including long-term rodent cancer studies (http://www.epa.gov/pesticides/regulating/data_requirements.htm). The findings in laboratory animals are generally assumed to be relevant to humans unless there is evidence to the contrary. When human information is available, EPA would consider that information (for example see http://www.epa.gov/pesticides/health/ag-health.html.)

When does EPA review pesticides for potential carcinogenicity?

EPA reviews studies submitted when a pesticide is proposed for registration. Studies are required in two species (mice and rats) and two sexes (males and females), as well as a battery of mutagenicity assays. These studies are required for all pesticides used on food and some non-food pesticides that could lead to long-term exposures in humans. In future reviews of the pesticide, the cancer classification may be reevaluated if new studies have been submitted.

All existing pesticide tolerances that were in place as of August 1996 were reassessed for their human health and environmental risks by August 2006 as required by the Food Quality Protection Act of 1996. However, if there was no new information on carcinogenicity, the compound was not re-evaluated simply to determine how it would be described under the 2005 cancer guidelines.

How have the Agency's cancer assessment guidelines changed?

There have been a number of different documents issued by the Agency for cancer evaluation.

- In 1976, EPA issued its first set of principles and interim procedures to guide evaluation of human cancer potential.
- In 1986, EPA issued guidance, which included a letter system (A-E) for designating degree of carcinogenic potential.
  - In the 1986 guidelines, hazard identification and the weight-of evidence process focused on tumor findings.
  - See http://epa.gov/cancerguidelines/guidelines-carcinogen-risk-assessment-1986.htm for a detailed description of this classification system.
- In 1996, EPA released "Proposed Guidelines for Carcinogen Risk Assessment," which used a weight of evidence narrative and standardized descriptors to replace the 1986 alphanumeric classification to classify carcinogenic potential.
  - In the 1996 proposal, emphasis was placed on available mode of carcinogenic action information and discussing characterization of hazard, dose-response, and exposure assessments.
  - The hazard and weight of evidence process embraced an analysis of all relevant biological information and emphasized understanding the agent's mode of action in producing tumors to reduce the uncertainty in describing the likelihood of harm.
  - See 1996 -http://www.epa.gov/raf/publications/pdfs/propcra_1996.pdf
- In 1999, EPA issued a revised version of its 1996 proposal that responded to public and peer review comments, which included:
  - a framework approach to evaluate the mode of action in the assessment of potential carcinogenesis.
  - See-http://epa.gov/raf/publications/pdfs/CANCER_GLS.PDF
- In March, 2005, EPA released its final *Guidelines for Carcinogen Risk Assessment* (EPA/630/P03/001B). These guidelines represent the culmination of a long development process, replacing EPA's original cancer risk assessment guidelines (1986) and its interim final guidelines (1999). Key changes included:
  - Five weight of evidence descriptors chosen for use in narratives ("*Carcinogenic to Humans*," "*Likely to Be Carcinogenic to Humans*," "*Suggestive Evidence of Carcinogenic Potential*," "*Inadequate Information to Assess Carcinogenic Potential*," and "*Not Likely to Be Carcinogenic to Humans*.")
  - Emphasis on analysis of data will precede use of defaults
  - Improved guidance on modeling and expanded discussions of sensitive subpopulations including children (supplemental guidance)
  - See http://www.epa.gov/raf/publications/pdfs/CANCER_GUIDELINES_FINAL_3-25-05.PDF

Why are there several different cancer classifications in the list?

As discussed above, EPA's guidelines for evaluating the potential carcinogenicity of chemicals have been updated over the years to reflect increase transparency in describing the cancer potential of a compound and to reflect the understanding of ways chemicals may cause cancer.

Not all pesticides have been evaluated under EPA's 2005 Cancer Guidelines. Agency policy states that for risk assessments that were completed before issuance of the 2005 Guidelines, the need for a re-assessment should be determined on a case-by-case basis by the program and that the existing assessment should continue to be considered scientifically sound based on the guidance used when the assessment was completed (http://www.epa.gov/raf/publications/pdfs/2005_cancer_guidelines_mem.pdf)

How can I find the basis of EPA's decision regarding the carcinogenicity of a pesticide?

The Reregistration Eligibility Decision for each pesticide contains a discussion of the available data and information used in the human health and environmental risk assessments, which includes a description of the evidence used to determine the cancer potential of the chemical. Also, the Cancer Assessment Review Committee report is available by requesting it through the Freedom of Information Act (FOIA) (link to http://www.epa.gov/pesticides/foia/).

Chemicals Evaluated for Carcinogenic Potential

Office of Pesticide Programs
U.S. Environmental Protection Agency

CHEMICAL	CAS NO.*	PC CODE**	CANCER CLASSIFICATION***	REPORT DATE****
1,3-Dibromo-5,5-dimethylhydantoin	77-48-5	006317	Not Likely to Be Carcinogenic to Humans	8/28/2000
1,3-dichloro-5-methylhydantoin	89415-87-2	128826	Not Likely to Be Carcinogenic to Humans	8/28/2000
2, 4 - DBA	94-82-6	030801	Not Likely to Be Carcinogenic to Humans	6/13/2003
2,4-D + Salts & Esters	94-75-7	030001	Group D–Not Classifiable as to Human Carcinogenicity	1/29/1997
2-Benzyl-4-chlorophenol	120-32-1	062201	Group C–Possible Human Carcinogen	9/5/1995
4-aminopyridine	504-24-5	069201	Group D–Not Classifiable As To Human Carcinogenicity	8/6/2007
Acephate	30560-19-1	103301	Group C–Possible Human Carcinogen	5/8/1985
Acequinocyl	57960-19-7	006329	Not Likely to Be Carcinogenic to Humans	11/13/2003
Acetamide	60-35-5	111101	Group C–Possible Human Carcinogen	5/29/1990
Acetamiprid	135410-20-7	099050	Not Likely to Be Carcinogenic to Humans	12/11/2001
Acetochlor	34256-82-1	121601	Suggestive Evidence of Carcinogenic Potential	1/3/2007
Acibenzolar-S-methyl	135158-54-2	061402	Not Likely to Be Carcinogenic to Humans	12/9/1999
Acifluorfen sodium	62476-59-9	114402	Multiple Descriptors: Likely to be Carcinogenic to Humans at High Doses Not Likely to be Carcinogenic to Humans at Low Doses	7/9/2003
Acrinathrin	101007-06-1	129141	Group D–Not Classifiable as to Human Carcinogenicity	7/15/1996
ADBAC	68424-85-1	069105	Not Likely to Be Carcinogenic to Humans	12/8/1999
Alachlor	15972-60-8	090501	Multiple Descriptors: Likely to be Carcinogenic to Humans (High Doses); Not Likely to be Carcinogenic to Humans (Low Doses)	6/27/1997
Aldicarb	116-06-3	098301	Group E–Evidence of Non-carcinogenicity for Humans	7/17/2002
Ametryn	834-12-8	080801	Data Are Inadequate for an Assessment of Human Carcinogenic Potential	9/17/2004
Amicarbazone	129909-90-6	114004	Not Likely To Be Carcinogenic To Humans	8/10/2005
Aminocyclopyrachlor	858956-08-8, 858956-35-1, 858954-83-3, 124423-84-3, 1759-53-1	288008	Not Likely To Be Carcinogenic To Humans	3/17/2010
Aminopyralid	150114-71-9	005100	Not Likely To Be Carcinogenic To Humans	7/12/2005
Amitraz	33089-61-1	106201	Suggestive Evidence of Carcinogenic Potential	7/18/2006
Amitrole	61-82-5	004401	Multiple Descriptors: Not Likely To Be Carcinogenic To Humans At Doses That Do Not Alter Rat Thyroid Hormone Homeostasis	5/11/2006
Aquashade	2650-18-2	110301	Not Likely To Be Carcinogenic To Humans	9/27/2005
Asulam	3337-71-1	106901	Group C–Possible Human Carcinogen	12/6/2001

Continued

Continued.

CHEMICAL	CAS NO.*	PC CODE**	CANCER CLASSIFICATION***	REPORT DATE****
Atrazine	1912-24-9	080803	Not Likely to be Carcinogenic to Humans	12/13/2000
Avermectin (see Emamectin Benzoate)	65195-55-3	122804	Group E–Evidence of Non-carcinogenicity for humans	6/27/1996
Azafenidin	68049-83-2	119016	Data Are Inadequate for an Assessment of Human Carcinogenic Potential	10/18/1999
Azinphos-methyl	86-50-0	058001	Not Likely to Be Carcinogenic to Humans	12/7/1993
Azoxystrobin	131860-33-8	128810	Not Likely to Be Carcinogenic to Humans	1/14/1997
Bendiocarb	22781-23-3	105201	Group E–Evidence of Non-carcinogenicity for Humans	12/16/1997
Benfluralin	1861-40-1	084301	Suggestive Evidence of Carcinogenicity, but Not Sufficient to Assess Human Carcinogenic Potential	12/27/2001
Benomyl	17804-35-2	099101	Group C–Possible Human Carcinogen	9/21/2000
Bensulide	741-58-2	009801	Not Likely to Be Carcinogenic to Humans	6/10/1999
Bentazon	25057-89-0	275200	Group E–Evidence of Non-carcinogenicity for Humans	1/14/1992
Benthiavalicarb-isopropyl	177406-68-7	098379	Likely to be Carcinogenic to Humans	10/18/2005
Beta Cyfluthrin	68359-37-5	118831	Not Likely To Be Carcinogenic To Humans	1/27/2010
Bifenazate	149877-41-8	000586	Not Likely to Be Carcinogenic to Humans	8/28/2001
Bifenthrin	82657-04-3	128825	Group C–Possible Human Carcinogen	2/19/2003
Bioallethrin	584-79-2	004003	Suggestive Evidence of Carcinogenicity, but Not Sufficient to Assess Human Carcinogenic Potential	10/29/2003
Bispyrabac Sodium	125401-92-5	078906	Not Likely to Be Carcinogenic to Humans	8/2/2001
Bitertanol	55179-31-2	117801	Not Likely To Be Carcinogenic To Humans	11/30/2005
Borax	1303-96-4	011102	Group E–Evidence of Non-carcinogenicity for humans	11/24/1993
Boric acid	10043-35-3	011001	Group E–Evidence of Non-carcinogenicity for humans	11/24/1993
Boron	7440-42-8	128945	Group E–Evidence of Non-carcinogenicity for humans	11/24/1993
Boscolid	188425-85-6	128008	Suggestive Evidence of Carcinogenicity, but Not Sufficient to Assess Human Carcinogenic Potential	11/14/2002
Bromacil	314-40-9	012301	Group C–Possible Human Carcinogen	1/13/1993
Bromoxynil	1689-84-5	035301	Group C–Possible Human Carcinogen	6/24/1998
Bromuconazole	116255-48-2	120503	Group E–Evidence of Non-carcinogenicity for humans	4/24/1995
Bronopol	52-51-7	216400	Group E–Evidence of Non-carcinogenicity for humans	6/12/1995
Buprofezin	69327-76-0	275100	Suggestive Evidence of Carcinogenicity, but Not Sufficient to Assess Human Carcinogenic Potential	3/15/2000
Butachlor	23184-66-9	112301	Likely to be Carcinogenic to Humans	2/24/1999
Butafenacil	134605-64-4	122004	Not Likely to Be Carcinogenic to Humans	7/11/2003

Continued

Continued.

CHEMICAL	CAS NO.*	PC CODE**	CANCER CLASSIFICATION***	REPORT DATE****
Butylate	2008-41-5	041405	Group E–Evidence of Non-carcinogenicity for humans	11/25/1992
Cacodylic acid	75-60-5	012501	Group B–Probable Human Carcinogen	12/14/1999
Cadusafos	95465-99-9	128864	Group E–Evidence of Non-carcinogenicity for humans	5/28/1992
Captafol	2939-80-2	081701	Group B–Probable Human Carcinogen	5/19/1987
Captan	133-06-2	081301	Multiple Descriptors: Likely at prolonged, high-level exposures, but not likely at dose levels that do not cause cytotoxicity and regenerative cell hyperplasia	9/22/2004
Carbaryl	63-25-2	056801	Likely to be Carcinogenic to Humans	2/12/2002
Carbendazim (MBC)	10605-21-7	128872	Group C–Possible Human Carcinogen	4/7/1989
Carbofuran	1563-66-2	090601	Not Likely to Be Carcinogenic to Humans	6/17/1997
Carboxin	5234-68-4	090201	Not Likely to Be Carcinogenic to Humans	6/5/2003
Carfentrazone-ethyl	128639-02-1	128712	Not Likely to Be Carcinogenic to Humans	5/16/2001
Chlorantraniliprole	500008-45-7	090100	Not Likely To Be Carcinogenic To Humans	3/4/2009
Chlordimeform	6164-98-3	059701	Group B–Probable Human Carcinogen	12/20/1985
Chlorethoxyfos	54593-83-8	129006	Group D–Not Classifiable as to Human Carcinogenicity	3/9/1995
Chlorfenapyr	122453-73-0	129093	Suggestive Evidence of Carcinogenicity, but Not Sufficient to Assess Human Carcinogenic Potential	3/18/2003
Chlorimuron-ethyl	90982-32-4	128901	Not Likely To Be Carcinogenic To Humans	2/5/2009
Chloroaniline, p-	106-47-8	017203	Group B–Probable Human Carcinogen	4/27/1995
Chloroneb	2675-77-6	027301	Data Are Inadequate for an Assessment of Human Carcinogenic Potential	12/18/2003
Chloropicrin	76-06-2	081501	Not Likely To Be Carcinogenic To Humans	6/30/2010
Chlorothalonil	1897-45-6	081901	Likely To Be Carcinogenic To Humans	10/20/1997
Chlorpropham	101-21-3	018301	Group E–Evidence of Non-carcinogenicity for humans	10/11/1994
Chlorpyrifos	2921-88-2	059101	Group E–Evidence of Non-carcinogenicity for humans	11/23/1993
Chlorpyrifos methyl	1351032	059102	Not Likely to Be Carcinogenic to Humans	5/17/1999
Chlorsulfuron	64902-72-3	118601	Group E–Evidence of Non-carcinogenicity for humans	7/17/2002
Chlorthal-dimethyl (DCPA)	1861-32-1	078701	Group C–Possible Human Carcinogen	2/10/1995
Clethodim	99129-21-2	121011	Not Likely To Be Carcinogenic To Humans	9/28/2007
Clodinafop-propargyl	105512-06-9	125203	Suggestive Evidence of Carcinogenic Potential	2/8/2006
Clofencet (MON 21200)	82697-71-0	128726	Group C–Possible Human Carcinogen	7/23/1996
Clofentezine	74115-24-5	125501	Group C–Possible Human Carcinogen	4/3/1990
Clomazone	81777-89-1	125401	Not Likely to Be Carcinogenic to Humans	1/31/2001
Clopyralid	1702-17-6	117403	Not Likely to Be Carcinogenic to Humans	12/20/1999
Cloquintocet-mexyl	99607-70-2	700099	Not Likely to Be Carcinogenic to Humans	8/31/1999

Continued

Continued.

CHEMICAL	CAS NO.*	PC CODE**	CANCER CLASSIFICATION***	REPORT DATE****
Cloransulam-methyl	147150-35-4	129116	Group E–Evidence of Non-carcinogenicity for humans	9/30/1997
Clothianidin	210880-92-5	044309	Not Likely to Be Carcinogenic to Humans	1/6/2003
Cocamide Diethanolamine	68603-42-9	224600	Likely to be Carcinogenic to Humans	7/25/2001
Coumaphos	56-72-4	036501	Not Likely to Be Carcinogenic to Humans	6/25/1999
Cresol, p-Chloro-m-	59-50-7	064206	Group D–Not Classifiable as to Human Carcinogenicity	11/28/1995
Cryolite	15096-52-3	075101	Group D–Not Classifiable as to Human Carcinogenicity	12/22/1995
Cumyluron	99485-76-4	027902	Suggestive Evidence Of Carcinogenic Potential	6/11/2008
Cyanazine	21725-46-2	100101	Group C–Possible Human Carcinogen	7/30/1991
Cyazofamid	120116-88-3	085651	Not Likely To Be Carcinogenic To Humans	6/3/2009
Cyclanilide	113136-77-9	026201	Not Likely to Be Carcinogenic to Humans	4/9/1997
Cycloate	1134-23-2	041301	Not Likely to Be Carcinogenic to Humans	9/25/2003
Cyflufenamid	180409-60-3	555550	Likely To Be Carcinogenic To Humans	6/22/2010
Cyfluthrin	68359-37-5	128831	Not Likely to Be Carcinogenic to Humans	5/21/2002
Cyhalofop-butyl	122008-85-9	082583	Not Likely To Be Carcinogenic To Humans	12/20/2007
Cyhalothrin	68085-85-8	128867	Group D–Not Classifiable as to Human Carcinogenicity	9/15/1994
Cyhexatin	13121-70-5	101601	Data Are Inadequate for an Assessment of Human Carcinogenic Potential	4/7/2005
Cymoxanil	57966-95-7	129106	Not Likely to Be Carcinogenic to Humans	1/2/2003
Cypermethrin	52315-07-8	109702	Group C–Possible Human Carcinogen	9/27/1988
Cyproconazole	94361-06-5	128993	Not Likely To Be Carcinogenic To Humans at doses that do not cause a mitogenic response in the liver	12/4/2007
Cyprodinil	121552-61-2	288202	Not Likely to Be Carcinogenic to Humans	1/14/1998
Cyprosulfamide	221667-31-8	877400	Not Likely To Be Carcinogenic To Humans	2/29/2008
Cyromazine	66215-27-8	121301	Group E–Evidence of Non-carcinogenicity for humans	1/6/1995
Daminozide	1596-84-5	035101	Group B–Probable Human Carcinogen	7/26/1991
Dantochlor (BCDMH)	118-52-5	028501	Not Likely to Be Carcinogenic to Humans	8/14/2000
Dazomet	533-74-4	035602	Group D–Not Classifiable as to Human Carcinogenicity	12/7/1993
DEET	134-62-3	080301	Group D–Not Classifiable as to Human Carcinogenicity	1/4/1996
Deltamethrin	52918-63-5	097805	Not Likely to Be Carcinogenic to Humans	9/9/2003
Desmedipham	13684-56-5	104801	Group E–Evidence of Non-carcinogenicity for humans	11/20/1995
Diazinon	333-41-5	057801	Not Likely to Be Carcinogenic to Humans	6/17/1997
Dicamba	1918-00-9	029801	Group D–Not Classifiable as to Human Carcinogenicity	7/29/1996
Dichlobenil	1194-65-6	027401	Group C–Possible Human Carcinogen	7/18/1995

Continued

Continued.

CHEMICAL	CAS NO.*	PC CODE**	CANCER CLASSIFICATION***	REPORT DATE****
Dichloran	99-30-9	031301	Suggestive Evidence Of Carcinogenic Potential	5/11/2006
Dichlormid	37764-25-3	900497	Not Likely To Be Carcinogenic To Humans	11/15/2005
Dichlorobenzamide, 2,6-	2008-88-4	027402	Group D–Not Classifiable as to Human Carcinogenicity	11/28/1995
Dichlorvos	62-73-7	084001	Suggestive Evidence of Carcinogenicity, but Not Sufficient to Assess Human Carcinogenic Potential	3/1/2000
Diclofop-methyl	51338-27-3	110902	Likely to be Carcinogenic to Humans	5/24/2000
Diclosulam	145701-21-9	129122	Not Likely to Be Carcinogenic to Humans	11/9/1999
Dicofol	115-32-2	010501	Group C–Possible Human Carcinogen	6/24/1992
Dicrotophos	141-66-2	035201	Suggestive Evidence of Carcinogenicity, but Not Sufficient to Assess Human Carcinogenic Potential	10/18/1999
Didecyl dimethyl ammonium chloride (DDAC)	7173-51-5	069149	Group E–Evidence of Non-carcinogenicity for Humans	4/11/2000
Difenoconazole	119446-68-3	128847	Group C–Possible Human Carcinogen	7/27/1994
Difenzoquat methyl sulfate	43222-48-6	106401	Group E–Evidence of Non-carcinogenicity for humans	5/24/1994
Diflubenzuron	35367-38-5	108201	Group E–Evidence of Non-carcinogenicity for humans	4/27/1995
Diflufenzopyr Sodiium	109293-98-3	005107	Not Likely to Be Carcinogenic to Humans	10/6/1998
Dimethenamid	87674-68-8	129051	Group C–Possible Human Carcinogen	9/15/1995
Dimethipin	55290-64-7	118901	Group C–Possible Human Carcinogen	1/5/1990
Dimethoate	60-51-5	035001	Group C–Possible Human Carcinogen	3/26/2002
Dimethomorph	110488-70-5	268800	Not Likely to Be Carcinogenic to Humans	5/13/1998
Dimethoxane	828-00-2	001001	Suggestive Evidence of Carcinogenic Potential	12/21/2000
Dimethyl ether	115-10-6	900382	Group D–Not Classifiable as to Human Carcinogenicity	1/12/1994
Dimethylhydantoin	16079-88-2	006315	Not Likely to Be Carcinogenic to Humans	8/28/2000
Dinocap	39300-45-3	036001	Group E–Evidence of Non-carcinogenicity for Humans	6/22/1994
Dinoseb	88-85-7	037505	Group C–Possible Human Carcinogen	6/19/1986
Dinotefuran	165252-70-0	044312	Not Likely to Be Carcinogenic to Humans	3/5/2004
Diphenylamine	122-39-4	038501	Not Likely to Be Carcinogenic to Humans	4/1/1997
Diquat dibromide	85-00-7	032201	Group E–Evidence of Non-carcinogenicity for Humans	5/12/1994
Disodium methanearsonate	144-21-8	013802	Not Likely to Be Carcinogenic to Humans	7/26/2000
Disulfoton	298-04-4	032501	Group E–Evidence of Non-carcinogenicity for Humans	4/21/1997
Dithianon	3347-22-6	099201	Suggestive Evidence of Carcinogenic Potential	2/23/2006
Dithiopyr (MON 7200)	97886-45-8	128994	Group E–Evidence of Non-carcinogenicity for Humans	5/29/1997
Diuron	330-54-1	035505	Known/Likely	5/8/1997
Dodine	2439-10-3	044301	Not Likely To Be Carcinogenic To Humans	1/24/2008

Continued

Continued.

CHEMICAL	CAS NO.*	PC CODE**	CANCER CLASSIFICATION***	REPORT DATE****
Ecolyst		069089	Not Likely to Be Carcinogenic to Humans	10/19/1999
Emamectin Benzoate (Deoxy Avermectin)	137512-74-4	122806	Not Likely to Be Carcinogenic to Humans	3/19/1998
Endosulfan	115-29-7	079401	Not Likely to Be Carcinogenic to Humans	1/31/2000
Endothall	145-73-3	038901	Not Likely To Be Carcinogenic To Humans	10/23/2008
Epoxiconazole	106325-08-0, 133855-98-8	123909	Likely to be Carcinogenic to Humans	1/24/2001
Esbiothrin	28434-00-6	004007	Suggestive Evidence of Carcinogenicity, but Not Sufficient to Assess Human Carcinogenic Potential	12/2/2003
Esfenvalerate	66230-04-4	109303	Group E–Evidence of Non-carcinogenicity for Humans	7/1/1996
Ethaboxam	162650-77-03	090205	Suggestive Evidence of Carcinogenic Potential	3/23/2006
Ethalfluralin	55283-68-6	113101	Group C–Possible Human Carcinogen	9/14/1994
Ethephon	16672-87-0	099801	Group D–Not Classifiable as to Human Carcinogenicity	8/15/1994
Ethion	563-12-2	058401	Group E–Evidence of Non-carcinogenicity for humans	1/26/1994
Ethiprole	181587-01-9	005550	Suggestive Evidence Of Carcinogenic Potential	10/28/2010
Ethofumesate	26225-79-6	110601	Group D–Not Classifiable as to Human Carcinogenicity	2/24/1994
Ethoprop	13194-48-4	041101	Likely to be Carcinogenic to Humans	10/7/1998
Ethyl dipropylthiocarbamate (EPTC)	759-94-4	041401	Not Likely to Be Carcinogenic to Humans	8/31/1999
Ethylene thiourea (ETU)	96-45-7	600016	Group B–Probable Human Carcinogen	7/7/1999
Etofenprox	80844-07-1	128965	Multiple Descriptors: Not Likely Below a Defined Dose Range	2/8/2006
Etoxazole	153233-91-1	107091	Not Likely to Be Carcinogenic to Humans	8/7/2003
Famoxadone	131807-57-3	113202	Not Likely to Be Carcinogenic to Humans	4/16/2003
Fenamidone	161326-34-7	046679	Not Likely to Be Carcinogenic to Humans	7/12/2002
Fenamiphos	22224-92-6	100601	Group E–Evidence of Non-carcinogenicity for Humans	11/23/1993
Fenarimol	60168-88-9	206600	Not Likely to Be Carcinogenic to Humans	9/5/2001
Fenazaquin	120928-09-8	044501	Not Likely To Be Carcinogenic To Humans	5/15/2007
Fenbuconazole	114369-43-6	129011	Group C–Possible Human Carcinogen	4/15/1996
Fenbutatin-oxide	13356-08-6	104601	Group E–Evidence of Non-carcinogenicity for Humans	3/2/1993
Fenhexamide	126833-17-8	090209	Not Likely to Be Carcinogenic to Humans	3/4/1999
Fenitrothion	122-14-5	105901	Group E–Evidence of Non-carcinogenicity for Humans	7/13/1993
Fenoxycarb	72490-01-8	125301	Likely to be Carcinogenic to Humans	4/1/1996
Fenpropathrin	39515-41-8	127901	Not Likely to be Carcinogenic to Humans	12/22/2003
Fenpropidin	67306-00-7	012305	Suggestive Evidence Of Carcinogenic Potential	6/9/2009
Fenpropimorph	67564-91-4	121402	Not Likely To Be Carcinogenic To Humans	10/19/2005

Continued

Continued.

CHEMICAL	CAS NO.*	PC CODE**	CANCER CLASSIFICATION***	REPORT DATE****
Fenpyroximate	134098-61-6	129131	Not Likely to Be Carcinogenic to Humans	2/19/1997
Fenthion	55-38-9	053301	Group E–Evidence of Non-carcinogenicity for Humans	3/11/1996
Fenvalerate	51630-58-1	109301	Group E–Evidence of Non-carcinogenicity for Humans	2/10/2003
Ferbam	128-04-1	034801	Likely to be Carcinogenic to Humans	4/6/2000
Fipronil	120068-37-3	129121	Group C–Possible Human Carcinogen	7/18/1995
Flazasulfuron	104040-78-0	119011	Not Likely To Be Carcinogenic To Humans	11/16/2005
Flonicamid	158062-67-0	128016	Suggestive Evidence of Carcinogenicity, but Not Sufficient to Assess Human Carcinogenic Potential	2/24/2005
Florasulam	145701-23-1	129108	Not Likely To Be Carcinogenic To Humans	5/31/2007
Fluazifop-P-Butyl	79241-46-6	122809	Not Likely To Be Carcinogenic To Humans	9/19/2008
Fluazinam	79622-59-6	129098	Suggestive Evidence of Carcinogenicity, but Not Sufficient to Assess Human Carcinogenic Potential	3/29/2001
Flubendiamide	272451-65-7	027602	Not Likely To Be Carcinogenic To Humans	4/3/2008
Flucarbazone-sodium	181274-17-9	114009	Not Likely to Be Carcinogenic to Humans	7/19/2000
Fludioxonil	131341-86-1	071503	Group D–Not Classifiable as to Human Carcinogenicity	9/19/1996
Flufenacet (Thiaflumide)	142459-58-3	121903	Not Likely to Be Carcinogenic to Humans	7/16/1997
Flufenoxuron	101463-69-8	108203	Not Likely To Be Carcinogenic To Humans	8/15/2006
Flufenpyr-ethyl	188489-07-8	108853	Not Likely to Be Carcinogenic to Humans	6/8/2003
Flumetralin	62924-70-3	123001	Not Likely To Be Carcinogenic To Humans	6/21/2007
Flumetsulam (XRD-498)	98967-40-9	129016	Group E–Evidence of Non-carcinogenicity for Humans	3/24/1993
Flumiclorac pentyl	87546-18-7	128724	Group E–Evidence of Non-carcinogenicity for Humans	9/7/1994
Flumioxazin	103361-09-7, 141490-50-8	129034	Not Likely to Be Carcinogenic to Humans	2/22/2001
Fluometuron	2164-17-2	035503	Group C–Possible Human Carcinogen	8/28/1996
Fluopicolide	239110-15-7	027412	Not Likely to Be Carcinogenic to Humans	12/12/2006
Fluopyram	658066-35-4	080302	Likely To Be Carcinogenic To Humans	11/25/09
Fluoxastrobin	361377-29-9	028869	Not Likely To Be Carcinogenic To Humans	1/24/2005
Fluridone	59756-60-4	112900	Group E–Evidence of Non-carcinogenicity for Humans	7/1/1985
Fluroxypyr	81406-37-3	128968	Not Likely To Be Carcinogenic To Humans	6/26/2003
Fluroxypyr acid (see also PC Code 128968)	69377-81-7	128959	Not Likely to Be Carcinogenic To Humans	6/26/2003
Flurprimidol	56425-91-3	125701	Not Likely To Be Carcinogenic To Humans	9/29/2005
Fluthiacet methyl	117337-19-6	108803	Likely to be Carcinogenic to Humans	11/20/1998
Flutolanil	66332-96-5	128975	Group E–Evidence of Non-carcinogenicity for Humans	6/9/1994

Continued

Continued.

CHEMICAL	CAS NO.*	PC CODE**	CANCER CLASSIFICATION***	REPORT DATE****
Flutriafol	76674-21-0	128940	Not Likely To Be Carcinogenic To Humans	6/1/2009
Folpet	133-07-3	081601	Not likely to be carcinogenic to humans at doses that do not cause an irritation response in the mucosal epithelium	10/13/2010
Fomesafen	108731-70-0	123802	Not Likely to Be Carcinogenic to Humans	11/3/2005
Fonofos	944-22-9	041701	Group E–Evidence of Non-carcinogenicity for Humans	11/10/1993
Forchlorfenuron	68157-60-8	128819	Not Likely To Be Carcinogenic To Humans	3/11/2008
Formasulfuron	173159-57-4	122020	Not Likely to Be Carcinogenic to Humans	9/19/2001
Formetanate hydrochloride	23422-53-9	097301	Group E–Evidence of Non-carcinogenicity for Humans	5/20/1996
Fosetyl-Al	39148-24-8	123301	Not Likely To Be Carcinogenic To Humans	4/22/1999
Fosthiazate	98886-44-3	129022	Not Likely to Be Carcinogenic to Humans	9/15/2003
Furiazole (MON 13900)	121776-33-8	911596	Likely to be Carcinogenic to Humans	10/15/1999
Furmecyclox	60568-05-0	122601	Group B–Probable Human Carcinogen	7/3/1985
Gamma Cyhalothrin	76703-62-3	128807	Not Likely to Be Carcinogenic to Humans	3/1/2004
Glufosinate-ammonium	77182-82-2	128850	Not Likely to Be Carcinogenic to Humans	5/17/1999
Glutaraldehyde	111-30-8	043901	Not Likely to Be Carcinogenic to Humans	5/18/2006
Glyphosate	1071-83-6	417300	Group E–Evidence of Non-carcinogenicity for Humans	12/31/1991
Halosulfuron methyl (MON 1200)	100784-20-1	128721	Not Likely to Be Carcinogenic to Humans	2/26/1998
Haloxyfop-methyl	690806-40-2	125201	Group B–Probable Human Carcinogen	9/18/1989
Hexaconazole	79983-71-4	128925	Group C–Possible Human Carcinogen	1/21/1999
Hexazinone	51235-04-2	107201	Group D–Not Classifiable as to Human Carcinogenicity	7/27/1994
Hexythiazox	78587-05-0	128849	Likely To Be Carcinogenic To Humans	8/12/09
HOE107892	135590-91-9	811800	Not Likely to Be Carcinogenic to Humans	11/24/1998
Hydramethylnon	67485-29-4	118401	Group C–Possible Human Carcinogen	3/28/1991
Hydrogen cyanamide	420-04-2	014002	Group C–Possible Human Carcinogen	9/15/1993
Hydroprene	41096-46-2	486300	Group D–Not Classifiable as to Human Carcinogenicity	6/8/1995
Imazalil	35554-44-0	111901	Likely to be Carcinogenic to Humans	12/7/1999
Imazamethabenz	81405-85-8	128842	Group D–Not Classifiable as to Human Carcinogenicity	6/11/1987
Imazamox	114311-32-9	129171	Not Likely to Be Carcinogenic to Humans	2/27/1997
Imazapic	81334-60-3	129041	Group E–Evidence of Non-carcinogenicity for Humans	9/27/1995
Imazapyr	81334-34-1	128821	Group E–Evidence of Non-carcinogenicity for Humans	10/5/1995
Imazaquin Acid	81335-37-7	128848	Not Likely To Be Carcinogenic To Humans	10/31/2005
Imazethapyr	81335-77-5	128922	Not Likely to Be Carcinogenic to Humans	1/31/2002
Imazosulfuron	122548-33-8	118602	Not Likely To Be Carcinogenic To Humans	3/13/2009

Continued

Continued.

CHEMICAL	CAS NO.*	PC CODE**	CANCER CLASSIFICATION***	REPORT DATE****
Imidacloprid	105827-78-9	129099	Group E–Evidence of Non-carcinogenicity for Humans	11/10/1993
Indaziflam	950782-86-2	080818	Not Likely To Be Carcinogenic To Humans	4/22/2010
Indoxacarb	173584-44-6	067710	Not Likely to Be Carcinogenic to Humans	7/17/2000
Iodomethane	74-88-4	000011	Multiple Descriptors: Not Likely to be Carcinogenic to Humans at doses that do not alter rat thyroid hormone homeostasis	11/10/2005
Iodosulfuran	144550-36-7	122021	Not Likely to Be Carcinogenic to Humans	1/5/2004
Ipoconazole	125225-28-7	125618	Not Likely To Be Carcinogenic To Humans	5/28/2008
Iprodione	36734-19-7	109801	Likely to be Carcinogenic to Humans	2/26/1998
Iprovalicarb	140923-17-7	098359	Likely to be Carcinogenic to Humans	4/11/2002
Isofenphos	25311-71-1	109401	Group E–Evidence of Non-carcinogenicity for Humans	1/13/1998
Isophorone	78-59-1	047401	Group C–Possible Human Carcinogen	9/2/1999
Isoxaben	82558-50-7	125851	Suggestive Evidence of Carcinogenic Potential	10/7/2008
Isoxadifen-ethyl	163520-33-0	823000	Not Likely to Be Carcinogenic to Humans	1/29/2001
Isoxaflutole	141112-29-0	123000	Likely to be Carcinogenic to Humans	8/6/1997
Kasugamycin	6980-18-3	230001	Not Likely To Be Carcinogenic To Humans	8/17/2005
Kathon 886	55965-84-9	107106	Group D–Not Classifiable as to Human Carcinogenicity	5/18/1995
KBR 3023	119515-38-7	070705	Not Likely to Be Carcinogenic to Humans	6/9/1999
Kresoxim-methyl	143390-89-0	129111	Likely to be Carcinogenic to Humans	8/19/1999
Lactofen	77501-63-4	128888	Multiple Descriptors: Likely to be Carcinogenic in Humans at High Doses. Not Likely to be Carcinogenic to Humans at Low Doses	10/17/2006
Lambda cyhalothrin	91465-08-6	128897	Group D–Not classifiable as to Human Carcinogenicity	9/12/2002
Lindane	58-89-9	009001	Suggestive Evidence of Carcinogenicity, but Not Sufficient to Assess Human Carcinogenic Potential	11/29/2001
Linuron	330-55-2	035506	Group C–Possible Human Carcinogen	11/20/2001
Malathion	121-75-5	057701	Suggestive Evidence of Carcinogenicity, but Not Sufficient to Assess Human Carcinogenic Potential	4/28/2000
Maleic hydrazide	123-33-1	051501	Group E–Evidence of Non-carcinogenicity for Humans	11/10/1993
Mancozeb	8018-01-7	014504	Group B–Probable Human Carcinogen	7/7/1999
Mandipropamid	374726-62-2	036602	Not Likely To Be Carcinogenic To Humans	1/21/2009
Maneb	12427-38-2	014505	Group B–Probable Human Carcinogen	7/7/1999
MB46513 (photodegradate of Fipronil)	120067-83-6	600050	Not Likely to Be Carcinogenic to Humans	12/6/2000
MCPA + Salts	94-74-6	030501	Not Likely to Be Carcinogenic to Humans	10/29/2003
MCPB Acid	94-81-5	019201	Not Likely To Be Carcinogenic To Humans	10/1/2008

Continued

Continued.

CHEMICAL	CAS NO.*	PC CODE**	CANCER CLASSIFICATION***	REPORT DATE****
Mecoprop-P	16484-77-8	129046	Suggestive Evidence of Carcinogenicity, but Not Sufficient to Assess Human Carcinogenic Potential	3/13/2003
Mefenoxam	70630-17-0	113502	Not Likely to Be Carcinogenic to Humans	5/17/2000
Mefluidide	53780-34-0	114001	Not Likely To Be Carcinogenic To Humans	5/30/2007
Melamine	108-78-1	777201	Group D–Not Classifiable as to Human Carcinogenicity	7/21/1993
Mepanipyrim	110235-97-7	288203	Likely to be Carcinogenic to Humans	4/20/2004
Mepiquat Chloride	24307-26-4	109101	Not Likely To Be Carcinogenic To Humans	2/19/2003
Meptyldinocap (DE-126/ Dinocap II)	131-72-6	036000	Group E–Evidence Of Non-Carcinogenicity For Humans	3/17/2009
Mercaptobenzothiazole, 2-	149-30-4	051701	Group C–Possible Human Carcinogen	11/19/1992
Mesosulfuron methyl	208465-21-8	122009	Not Likely to Be Carcinogenic to Humans	3/4/2004
Mesotrione	104206-82-8	122990	Not Likely to Be Carcinogenic to Humans	4/12/2001
Metaflumizone	139968-49-3	281250	Not Likely To Be Carcinogenic To Humans	1/24/2006
Metalaxyl	57837-19-1	113501	Group E–Evidence of Non-carcinogenicity for Humans	4/20/1994
Metaldehyde	108-62-3	053001	Suggestive Evidence of Carcinogenic Potential	6/23/2005
Metam sodium	137-42-8	039003	Likely To Be Carcinogenic To Humans	5/14/2009
Metconazole	125116-23-6	125619	Not Likely to Be Carcinogenic to Humans	4/14/2006
Methamidophos	10265-92-6	101201	Not Likely to Be Carcinogenic to Humans	10/6/1997
Methidathion	950-37-8	100301	Group C–Possible Human Carcinogen	2/19/1988
Methiocarb	2032-65-7	100501	Group D–Not Classifiable as to Human Carcinogenicity	3/2/1993
Methomyl	16752-77-5	090301	Group E–Evidence of Non-carcinogenicity for Humans	10/25/1996
Methoxyfenozide	161050-58-4	121027	Not Likely to Be Carcinogenic to Humans	7/1/1999
Methyl bromide	74-83-9	053201	Not Likely To Be Carcinogenic To Humans	8/4/1992
Methyl isothiocyanate	6317-18-6	068103	There are insufficient data to characterize the cancer risk of MITC.	4/30/2009
Methyl parathion	298-00-0	053501	Not Likely to Be Carcinogenic to Humans	12/1/1997
Metiram	9006-42-2	014601	Group B–Probable Human Carcinogen	7/7/1999
Metofluthrin	240444-70-6	109709	Not Likely to Be Carcinogenic to Humans at doses that do not result in a mitogenic response.	7/26/2007
Metolachlor	51218-45-2	108801	Group C–Possible Human Carcinogen	11/16/1994
Metrafenone	220899-03-6	000325	Suggestive Evidence of Carcinogenic Potential	7/6/2006
Metribuzin	21087-64-9	101101	Group D–Not Classifiable as to Human Carcinogenicity	5/16/1995
Metsulfuron methyl	74223-64-6	122010	Not Likely to Be Carcinogenic to Humans	3/14/2002
Mevinphos	7786-34-7	015801	Not Likely To Be Carcinogenic To Humans	5/17/2000
MGK 264	113-48-4	057001	Group C–Possible Human Carcinogen	6/7/1995
MGK Replellent 326	136-45-8	047201	Group B–Probable Human Carcinogen	11/12/2002

Continued

Continued.

CHEMICAL	CAS NO.*	PC CODE**	CANCER CLASSIFICATION***	REPORT DATE****
Molinate	2212-67-1	041402	Suggestive Evidence of Carcinogenicity, but Not Sufficient to Assess Human Carcinogenic Potential	12/14/2000
MON 4660	71526-07-3	600046	Likely to be Carcinogenic to Humans	12/9/1999
Monosodium acid methanearsonate (MMA)	2163-80-6	013803	Not Likely to Be Carcinogenic to Humans	7/26/2000
MSMA-calcium salt	5902-95-4	013806	Not Likely to Be Carcinogenic to Humans	12/14/2000
Myclobutanil	88671-89-0	128857	Group E–Evidence of Non-carcinogenicity for Humans	6/16/1994
Naled	300-76-5	034401	Group E–Evidence of Non-carcinogenicity for Humans	8/31/1994
Napropamide	15299-99-7	103001	Not Likely To Be Carcinogenic To Humans	7/7/2005
Naptalam Sodium Salt	132-67-2	030703	Group D–Not Classifiable as to Human Carcinogenicity	9/7/1994
Napthalene Acetates	2122-70-5	056008	Not Likely To Be Carcinogenic To Humans	3/5/2009
Nicosulfuron	111991-09-4	129008	Group E–Evidence of Non-carcinogenicity for Humans	9/1/1998
Nitrapyrin	1929-82-4	069203	Likely to be Carcinogenic to Humans	3/26/2005
Norflurazon	27314-13-2	105801	Group C–Possible Human Carcinogen	11/2/1990
Novaluron	116714-46-6	124002	Not Likely to Be Carcinogenic to Humans	2/4/2004
Orthophenylphenol (see also PC 064104)	90-43-7	064103	Not Likely to Be Carcinogenic to Humans	10/12/2005
Orthophenylphenol, Sodium salt (see also PC 064103)	132-27-4	064104	Not Likely to Be Carcinogenic to Humans	10/12/2005
Orthosulfamuron	213464-77-3	108209	Suggestive Evidence of Carcinogenic Potential	10/26/2006
Oryzalin	19044-88-3	104201	Likely to be Carcinogenic to Humans	6/25/2003
Oxadiazon	19666-30-9	109001	Likely To Be Carcinogenic To Humans	5/1/2001
Oxadixyl	77732-09-3	126701	Group C–Possible Human Carcinogen	1/4/1989
Oxamyl	23135-22-0	103801	Group E–Evidence of Non-carcinogenicity for Humans	11/5/1996
Oxydemeton-methyl	301-12-2	058702	Not Likely to Be Carcinogenic to Humans	7/24/1997
Oxyfluorfen	42874-03-3	111601	Likely To Be Carcinogenic To Humans	4/20/2010
Oxytetracycline	2058-46-0	006308	Group D–Not Classifiable as to Human Carcinogenicity	12/18/1992
Oxythioquinox	2439-01-2	054101	Group B–Probable Human Carcinogen	2/15/1996
Paclobutrazol	76738-62-0	125601	Group D–Not Classifiable as to Human Carcinogenicity	6/23/1994
Paradichlorobenzene	106-46-7	061501	Not Likely To Be Carcinogenic To Humans	6/5/2007
Paranitrophenol	100-02-7	056301	Group D–Not Classifiable as to Human Carcinogenicity	5/14/1996
Paraquat dichloride	1910-42-5	061601	Group E–Evidence of Non-carcinogenicity for Humans	4/19/2000
Parathion, ethyl-	56-38-2	057501	Group C–Possible Human Carcinogen	9/11/1991
Pebulate	1114-71-2	041403	Not Likely to Be Carcinogenic to Humans	12/7/1998

Continued

Continued.

CHEMICAL	CAS NO.*	PC CODE**	CANCER CLASSIFICATION***	REPORT DATE****
Pendimethalin	40487-42-1	108501	Group C–Possible Human Carcinogen	7/24/1992
Penoxulam	219714-96-2	119031	Suggestive Evidence of Carcinogenicity, but Not Sufficient to Assess Human Carcinogenic Potential	3/24/2004
Pentachloronitrobenzene (PCNB)	82-68-8	056502	Group C–Possible Human Carcinogen	12/18/1992
Pentachlorophenol	87-86-5	063001	Group B–Probable Human Carcinogen	1/3/1991
Permethrin	52645-53-1	109701	Likely to be Carcinogenic to Humans	10/23/2002
Phenmedipham	13684-63-4	098701	Group D–Not Classifiable as to Human Carcinogenicity	4/28/1993
PHMB	32289-58-0	111801	Suggestive Evidence of Carcinogenicity, but Not Sufficient to Assess Human Carcinogenic Potential	4/9/2003
Phorate	298-02-2	057201	Group E–Evidence of Non-carcinogenicity for Humans	12/30/1993
Phosalone	2310-17-0	097701	Not Likely to Be Carcinogenic to Humans	8/12/1999
Phosmet	732-11-6	059201	Suggestive Evidence of Carcinogenicity, but Not Sufficient to Assess Human Carcinogenic Potential	10/27/1999
Phosphamidon	13171-21-6	018201	Group C–Possible Human Carcinogen	5/31/1989
Phostebupirim	96182-53-5	129086	Group E–Evidence of Non-carcinogenicity for Humans	4/27/1993
Picloram Acid	1918-02-1	005101	Group E–Evidence of Non-carcinogenicity for Humans	4/1/1994
Picloram Acid Ethylhexyl Ester	2545-60-0	005103	Group E–Evidence of Non-carcinogenicity for Humans	4/1/1994
Picloram Acid Potassium Salt	35832-11-2	005104	Group E–Evidence of Non-carcinogenicity for Humans	4/1/1994
Picloram Acid Triisopropanolamine Salt	6753-47-5	005102	Group E–Evidence of Non-carcinogenicity for Humans	4/1/1994
Pinoxaden	293973-20-8	147500	Data Are Inadequate for an Assessment of Human Carcinogenic Potential	5/18/2005
Piperonyl butoxide	51-03-6	067501	Group C–Possible Human Carcinogen	6/7/1995
Pirimicarb	23103-98-2	106101	Likely to be Carcinogenic to Humans	7/13/2005
Pirimiphos-methyl	29232-93-7	108102	Cannot Be Determined	1/29/1998
Polymeric Betaine		103679	Inadequate Information to Assess Carcinogenic Potential	10/3/2006
Potassium dichromate	7778-50-9	068302	Not Likely to Be Carcinogenic to Humans	8/28/2001
Prallethrin	23031-36-9	128722	Not Likely to Be Carcinogenic to Humans	6/27/2003
Primisulfuron-methyl	86209-51-0	128973	Group D–Not Classifiable as to Human Carcinogenicity	5/3/1990
Prochloraz	67747-09-5	128851	Group C–Possible Human Carcinogen	7/1/1988
Procymidone	32809-16-8	129044	Group B–Probable Human Carcinogen	4/5/1991
Prodiamine	29091-21-2	110201	Group C–Possible Human Carcinogen	6/10/1991
Profenofos	41198-08-7	111401	Group E–Evidence of Non-carcinogenicity for Humans	2/6/1995

Continued

Continued.

CHEMICAL	CAS NO.*	PC CODE**	CANCER CLASSIFICATION***	REPORT DATE****
Prohexadione	127277-53-6	112600	Not Likely to Be Carcinogenic to Humans	4/14/2000
Prometon	1610-18-0	080804	Group D–Not Classifiable as to Human Carcinogenicity	11/25/1992
Prometryn	7287-19-6	080805	Group E–Evidence of Non-carcinogenicity for Humans	7/26/1994
Pronamide	23950-58-5	101701	Group B–Probable Human Carcinogen	12/10/2001
Propachlor	1918-16-7	019101	Likely to be Carcinogenic to Humans	10/16/1997
Propamocarb hydrochloride	25606-41-1	119302	Not Likely To Be Carcinogenic To Humans	5/31/2000
Propanil	709-98-8	028201	Suggestive Evidence of Carcinogenicity, but Not Sufficient to Assess Human Carcinogenic Potential	6/19/2001
Propargite	2312-35-8	097601	Group B–Probable Human Carcinogen	7/23/1992
Propazine	139-40-2	080808	Not Likely to Be Carcinogenic to Humans	12/8/2005
Propetamphos	31218-83-4	113601	Not Likely to Be Carcinogenic to Humans	10/31/1998
Propiconazole	60207-90-1	122101	Group C–Possible Human Carcinogen	9/14/1992
Propoxur	114-26-1	047802	Group B–Probable Human Carcinogen	6/17/1996
Propoxycarbazone-Sodium	181274-15-7	122019	Not Likely to Be Carcinogenic to Humans	4/6/2004
Prosulfuron	94125-34-5	129031	Data Are Inadequate for an Assessment of Human Carcinogenic Potential	1/24/2000
Prothioconazole	178928-70-6	113961	Not Likely To Be Carcinogenic To Humans	12/31/2007
Pymetrozine	123312-89-0	101103	Likely to be Carcinogenic to Humans	9/22/1999
Pyraclostrobin	175013-18-0	099100	Not Likely to Be Carcinogenic to Humans	2/15/2007
Pyraflufen ethyl	129630-19-9	030090	Likely to be Carcinogenic to Humans	10/8/2002
Pyrasulfotole	365400-11-9	000692	Suggestive Evidence Of Carcinogenic Potential	5/17/2007
Pyrazon	1698-60-8	069601	Not Likely To Be Carcinogenic To Humans	7/28/2005
Pyrethrins	8003-34-7	069001	Not Likely To Be Carcinogenic To Humans at doses that do not cause mitogenic repsonse in the liver cell proliferation	2/14/2008
Pyridaben	96489-71-3	129105	Group E–Evidence of Non-carcinogenicity for Humans	5/11/1994
Pyridalyl	179101-81-6	295149	Not Likely To Be Carcinogenic To Humans	8/26/2004
Pyridate	55512-33-9	128834	Not Likely To Be Carcinogenic To Humans	1/24/2000
Pyrimethanil	53112-28-0	288201	Group C–Possible Human Carcinogen	2/11/1997
Pyriproxyfen	95737-68-1	129032	Group E–Evidence of Non-carcinogenicity for Humans	8/15/1995
Pyrithiobac-sodium	123343-16-8	078905	Group C–Possible Human Carcinogen	9/5/1995
Pyroxasulfone	447399-55-5	090099	Likely To Be Carcinogenic To Humans	05/11/2010
Pyroxsulam	422556-08-9	108702	Not Likely To Be Carcinogenic To Humans	7/12/2007
Quinchlorac	84087-01-4	128974	Group D–Not Classifiable as to Human Carcinogenicity	8/26/1992
Quinoxyfen	124495-18-7	055459	Not Likely to Be Carcinogenic to Humans	1/28/2003
Quizalofop ethyl	76578-14-8	128711	Group D–Not Classifiable as to Human Carcinogenicity	3/17/1988

Continued

Continued.

CHEMICAL	CAS NO.*	PC CODE**	CANCER CLASSIFICATION***	REPORT DATE****
Resmethrin	10453-86-8	097801	Likely to be Carcinogenic to Humans	5/25/2005
Rimsulfuron	122931-48-0	129009	Not Likely to Be Carcinogenic to Humans	2/19/1998
RoteNone	83-79-4	071003	Group E–Evidence of Non-carcinogenicity for Humans	10/5/1988
Saflufenacil (BAS 800 H)	372137-35-4	118203	Not Likely To Be Carcinogenic To Humans	7/22/2009
S-Bioallethrin	28434-00-6	004004	Suggestive Evidence of Carcinogenicity, but Not Sufficient to Assess Human Carcinogenic Potential	12/2/2003
Sethoxydim	74051-80-2	121001	Not Likely to Be Carcinogenic to Humans	3/19/2003
Simazine	122-34-9	080807	Not Likely to be Carcinogenic to Humans	4/14/2005
s-Metolachlor	87392-12-9	108800	Group C–Possible Human Carcinogen	9/28/2001
Sodium bentazon	50723-80-3	103901	Group E–Evidence Of Non-Carcinogenicity For Humans	3/8/2006
Sodium omadine	15922-78-8	088004	Group D–Not Classifiable as to Human Carcinogenicity	5/16/1995
Spinetoram	187166-40-1 + 187166-15-0	110008	Not Likely To Be Carcinogenic To Humans	9/20/2007
Spinosad	131929-60-7	110003	Not Likely to Be Carcinogenic to Humans	7/18/2002
Spirodiclofen	148477-71-8	124871	Likely to be Carcinogenic to Humans	6/10/2004
Spiromesifen	283594-90-1	024875	Not Likely To Be Carcinogenic To Humans	5/21/2008
Spirotetramat	203313-25-1	392201	Not Likely To Be Carcinogenic To Humans	3/26/2009
Spiroxamine	118134-30-8	120759	Not Likely to Be Carcinogenic to Humans	11/14/2003
Sulfentrazone	122836-35-5	129081	Group E–Evidence of Non-carcinogenicity for Humans	5/7/1996
Sulfosate	81591-81-3	128501	Group E–Evidence of Non-carcinogenicity for Humans	7/26/1994
Sulfosulfuron	141776-32-1	085601	Not Likely to be Carcinogenic to Humans	12/16/2008
Sulfuryl fluoride	2699-79-8	078003	Not Likely to Be Carcinogenic to Humans	5/24/2001
Sulprofos	35400-43-2	111501	Group E–Evidence of Non-carcinogenicity for Humans	3/26/1996
Sumithrin	26002-80-2	069005	Not Likely to Be Carcinogenic to Humans	5/30/2006
Tau-fluvalinate	102851-06-9	109302	Not Likely To Be Carcinogenic To Humans	9/29/2005
TCMTB (Busan 72)	21564-17-0	035603	Group C–Possible Human Carcinogen	8/28/1996
Tebuconazole	107534-96-3	128997	Group C–Possible Human Carcinogen	9/15/1993
Tebufenozide	112410-23-8	129026	Group E–Evidence of Non-carcinogenicity for Humans	8/29/1994
Tebufenpyrad	119168-77-3	090102	Suggestive Evidence of Carcinogenicity, but Not Sufficient to Assess Human Carcinogenic Potential	7/15/2002
Tebuthiuron	34014-18-1	105501	Group D–Not Classifiable as to Human Carcinogenicity	3/1/1993
Telone	542-75-6	029001	Group B–Probable Human Carcinogen	3/19/2002
Tembotrione	335104-84-2	012801	Suggestive Evidence of Carcinogenic Potential	5/22/2007
Tepraloxydim	149979-41-9	121005	Data Are Inadequate for an Assessment of Human Carcinogenic Potential	2/27/2001

Continued

Continued.

CHEMICAL	CAS NO.*	PC CODE**	CANCER CLASSIFICATION***	REPORT DATE****
Terbacil	5902-51-2	012701	Group E–Evidence of Non-carcinogenicity for Humans	9/30/1994
Terbufos	13071-79-9	105001	Group E–Evidence of Non-carcinogenicity for Humans	3/9/1994
Terbuthylazine	5915-41-3	080814	Group D–Not Classifiable as to Human Carcinogenicity	8/24/1994
Terbutryn	886-50-0	080813	Group C–Possible Human Carcinogen	3/3/1988
Terrazole	2593-15-9	084701	Group B–Probable Human Carcinogen	6/29/1999
Tetrachlorvinphos	961-11-5	083701	Likely to be Carcinogenic to Humans	3/7/2002
Tetraconazole	112281-77-3	120603	Likely to be Carcinogenic to Humans	1/11/2000
Tetramethrin	7696-12-0	069003	Group C–Possible Human Carcinogen	12/11/1989
Thiabendazole	148-79-8	060101	Multiple Descriptors: Likely to be Carcinogenic to Humans at High Does; Not Likely to be Carcinogenic to Humans at Low Doses	3/8/2002
Thiacloprid	111988-49-9	014019	Likely to be Carcinogenic to Humans	3/26/2003
Thiamethoxam	153719-23-4	060109	Not Likely to Be Carcinogenic to Humans	6/13/2005
Thiazopyr (MON 13200)	117718-60-2	129100	Suggestive Evidence Of Carcinogenic Potential	12/6/2007
Thidiazuron	51707-55-2	120301	Not Likely To Be Carcinogenic To Humans	8/31/2005
Thiencarbazone-methyl	317815-83-1	015804	Not Likely To Be Carcinogenic To Humans at doses that do not cause urothelium cytotoxicity	2/29/2008
Thifensulfuron methyl	79277-27-1	128845	Not Likely To Be Carcinogenic To Humans	12/12/2006
Thiobencarb (Bolero)	28249-77-6	108401	Group D–Not Classifiable as to Human Carcinogenicity	6/10/1996
Thiocyclam hydrogen oxalate	31895-22-4	128868	Group D–Not Classifiable as to Human Carcinogenicity	9/15/1994
Thiodicarb	59669-26-0	114501	Group B–Probable Human Carcinogen	6/10/1996
Thiophanate-methyl	23564-05-8	102001	Likely to be Carcinogenic to Humans	8/24/1999
Thiram	137-26-8	079801	Not Likely to Be Carcinogenic to Humans	4/14/2003
Tolyfluanid	731-27-1	309200	Likely to be Carcinogenic to Humans	6/18/2002
Topramezone	210631-68-8	123009	Multiple Descriptors: Not Likely to be Carcinogenic to Humans at Doses that Do Not Alter Rat Thyroid Hormone Homeostasis	5/19/2005
Tralkoxydim	87820-88-0	121000	Suggestive Evidence of Carcinogenicity, but Not Sufficient to Assess Human Carcinogenic Potential	6/30/2004
Triadimefon	43121-43-3	109901	Group C–Possible Human Carcinogen	12/4/1996
Triadimenol	55219-65-3	127201	Group C–Possible Human Carcinogen	1/29/1988
Triallate	2303-17-5	078802	Group C–Possible Human Carcinogen	1/12/1994
Triasulfuron	82097-50-5	128969	Group E–Evidence of Non-carcinogenicity for Humans	2/27/1991
Triazamate	112143-82-5	128100	Not Likely to Be Carcinogenic to Humans	12/1/1997
Tribenuron methyl	101200-48-0	128887	Group C–Possible Human Carcinogen	7/14/1989
Tribufos	78-48-8	074801	Multiple Descriptors: Likely to be Carcinogenic to Humans (High Doses); Not Likely to be	5/22/1997

Continued

Continued.

CHEMICAL	CAS NO.*	PC CODE**	CANCER CLASSIFICATION***	REPORT DATE****
Tributyltin maleate	14275-57-1	083118	Group D–Not Classifiable As To Human Carcinogenicity	3/31/2005
Trichlorfon	52-68-6	057901	Multiple Descriptors: Likely to be Carcinogenic to Humans (High Doses), Not Likely to be Carcinogenic to Humans (Low Doses)	7/15/1999
Triclopyr	55335-06-3	116001	Group D–Not Classifiable as to Human Carcinogenicity	5/9/1996
Triclosan	3380-34-5	054901	Not Likely To Be Carcinogenic To Humans	1/4/2008
Tridiphane	58138-08-2	123901	Group C–Possible Human Carcinogen	4/22/1986
Trifloxystrobin	141517-21-7	129112	Not Likely to Be Carcinogenic to Humans	6/16/1999
Trifloxysulfuron	290332-10-4	119009	Not Likely to Be Carcinogenic to Humans	7/22/2003
Triflumizole	68694-11-1	128879	Group E–Evidence of Non-carcinogenicity for Humans	8/10/1993
Trifluralin	1582-09-8	036101	Group C–Possible Human Carcinogen	4/11/1986
Triflusulfuron-methyl	126535-15-7	129002	Group C–Possible Human Carcinogen	5/28/1996
Triforine	26644-46-2	107901	Suggestive Evidence of Carcinogenicity, but Not Sufficient to Assess Human Carcinogenic Potential	6/29/2004
Trinexapac-Ethyl	95266-40-3	112602	Not Likely To Be Carcinogenic To Humans	9/5/2008
Triphenyltin hydroxide (TPTH)	76-87-9	083601	Group B–Probable Human Carcinogen	5/24/1990
Triticonazole	131983-72-7	125620	Not Likely to be Carcinogenic to Humans	6/15/2006
Troysan polyphase (IPBC)	55406-53-6	107801	Not Likely to Be Carcinogenic to Humans	12/4/1996
UDMH	57-14-7	600018	Group B–Probable Human Carcinogen	7/26/1991
UMP-488 (PAL 6000)	111578-32-6	129025	Group E–Evidence of Non-carcinogenicity for Humans	5/6/1994
Uniconazole	83657-22-1	128976	Group C–Possible Human Carcinogen	10/11/1990
Vinclozolin	50471-44-8	113201	Group C–Possible Human Carcinogen	6/20/2000
Zeta-Cypermethrin	52315-07-8	129064	Group C–Possible Human Carcinogen	9/27/1988
Ziram	137-30-4	034805	Suggestive Evidence of Carcinogenicity, but Not Sufficient to Assess Human Carcinogenic Potential	2/6/2003
Zoxamide	156052-68-5	101702	Not Likely to Be Carcinogenic to Humans	2/7/2001

For information about future review of these pesticides see the Registration Review schedule/status at http://www.epa.gov/oppsrrd1/registration_review/

***CAS No.:** a chemical identifier designated by the Chemical Abstracts Service

****PC Code:** a unique chemical identifier used by the Office of Pesticide Programs

*****Cancer Classification:** Simple descriptors used to express conclusions regarding the carcinogenic hazard potential of a chemical based on all relevant information, which is usually laboratory studies. These phases changed over time, as the Agency's guidance for carcinogenicity assessment was updated and revised.

******Report:** Date of the Carcinogen Assessment Review Committee (CARC) report. The CARC report date does not necessarily reflect the date of the pesticide risk assessment. For example, all existing pesticide tolerances that were in place as of August 1996 were re-assessed for their health risks by August 2006 as required by the Food Quality Protection Act, and if there was no new information on carcinogenicity, the compound was not re-evaluated by the CARC.

For information about future review of these pesticides see the Registration Review schedule/status at http://www.epa.gov/oppsrrd1/registration_review/

Appendix 14: Glossary

This is a brief list of terms/key words relating to the subject of pesticides, with explanations that may assist the reader.

A

Abiotic: The non-living factors in the environment (e.g. climate, soil, air and light).
Abiotic degradation: Degradation of a substance by purely physical or chemical mechanisms. Examples include hydrolysis and photolysis.
Abrasion: The process of wearing away by rubbing or grinding.
Abscission: The natural detachment or shedding of fruits, leaves or stems from the parent plant. The abscission zone is the region at the base of some plant organs consisting of a layer of cells which protects the wound when the organ falls off.
Absorption: The process by which one substance takes up a liquid with which it is in contact. Plants absorb nutrients and pesticides from the soil through roots or foliage (through the stomata or the cuticle). There is a special type of clay used to soak up spills of chemicals/pesticides.
Acaricide: A pesticide used to control mites and ticks. It is also called a miticide.
ACCase: Acetyl-coenzyme A (CoA) carboxylase, responsible for production of fatty acids.
Acceptable daily intake (ADI): The ADI is the highest daily intake level of a chemical, which may be continued over the whole life of a person, without appearing to pose a health risk and is thus a life-time exposure. The ADI is usually expressed in milligrams per kilogram of body weight (mg/kg b.w.).
Acceptable operator exposure level (AOEL): The acceptable operator exposure level is the maximum amount of active substance to which the operator may be exposed without any adverse health effects. It is expressed in milligrams per kilogram of body weight per day (mg/kg b.w./day).
Acidophile weeds: Weed species that are characteristically acid plants (e.g. *Cynodon dactylon*, *Digitaria sanguinalis*).
Acetylcholine (ACh): A neurotransmitter in the nervous system.
Acidic herbicides: Herbicides that possess carboxyl or phenolic functional groups and ionize in aqueous systems to yield anionic species are acidic herbicides. TCA, dalapon, phenols and nitriles are examples of this class.
Acidity: Quality of being acid or sour; having a pH of less than 7.
Actinomycetales: This is an order of actinobacteria, the cells of which form branching filaments (<1.5 μm in diameter) that develop into mycelia.
Action level (regulatory): (i) For food commodities, an administrative maximum residue limit (MRL) used by regulatory authorities to initiate action where no legally defined MRL has been established. (ii) For the environment, the concentration of a pesticide in air, soil or water at which emergency measures or preventative actions are to be taken.
Activated charcoal: A high quality charcoal that has been treated with oxygen and which adsorbs liquids and gases easily. It is used to protect a crop from herbicide toxicity.
Active: A designation by the EPA indicating that a pesticide may be marketed in the USA.
Active ingredient (a.i.): In any pesticide product, the component that kills, or otherwise controls, target pests is the active ingredient.
Active site: The location on an enzyme (protein) to which the substrate molecules bind and get converted to their products. The shape of this site must be maintained for the enzyme to remain functional.
Active substance: Any substance or microorganism that can act against harmful organisms on plants. It is usually combined in a pesticide product.
Active transport: The transport of substances, usually polar molecules or ions, across a membrane against a concentration gradient. Active transport is mediated by specific carrier proteins that selectively bind a substrate and transport it.
Actual dosage: The amount of active ingredient (not formulated product) that is applied to an area or other target.
Acute dermal toxicity: The toxicity of a single dose of pesticide when exposed to the skin.
Acute effects: Symptoms appearing immediately after exposure to a pesticide (24 h).
Acute inhalation toxicity: The toxicity of a single dose when inhaled into the lungs.
Acute oral toxicity: The toxicity of a single dose when swallowed or licked.

Acute reference dose (aRfD): This is defined, based on the facts at the time of evaluation, as an estimate of the amount of a chemical substance in food (or drinking water), expressed on a body weight basis, that can be ingested over a short period of time (usually during one meal or 1 day) without appreciable health risk to the consumer.
Acute toxicity: Ability of a substance to cause adverse effects within a short period (24 h) following a single dose or single exposure.
Additive: Substances added to a pesticide to improve its activity (e.g. oils).
Additive resistance: Resistance governed by more than one gene, each of which can be expressed independently, but which is reinforced by the expression of each of the additional genes.
Adherence: Sticking to a surface.
ADI: Abbreviation for acceptable daily intake.
Adjuvant: A substance other than water that enhances the effectiveness of a pesticide product is an adjuvant (e.g. oils or surfactants).
Adsorb: To hold on a surface (as opposed to absorb) as a thin film on the outside surface or on internal surfaces within the material.
Advection: The bulk movement of groundwater or solutes at approximately the mean linear groundwater velocity parallel to the direction of groundwater flow.
Adventitious: Accidental or casual occurrence of buds or root.
AEA: Abbreviation for Agricultural Engineers' Association.
Aerobic: Organisms that need air or oxygen for living. The opposite is anaerobic.
Aerosol: An extremely fine mist or fog consisting of solid or liquid particles (<30 μm) suspended in air.
Agitation: The process of stirring or mixing a liquid or solution in a sprayer or tank with a mechanical device, air or hydraulic action. The tank is called an agitator.
Agricultural chemical: A general term used to refer to pesticides, soil conditioners, adjuvants, additives and even fertilizers and veterinary drugs, which are used to grow crops and raise farm animals for production and protect them from pests and diseases. Such agricultural chemicals are also called agrochemicals.
Air blast: A type of pesticide sprayer that can deliver high and low volumes of spray.
Algae: Simple, rootless plants that grow in open water bodies and lit aquaria. Algae can affect water quality adversely by lowering the dissolved oxygen content.
Algicide: Chemical used to control or kill algae and aquatic weeds.
Alien weeds (or Anthrophytes): Weeds introduced from some other part of the world. Without their natural enemies, these weeds become difficult to control. Water hyacinth, Mexican poppy and carrot grass are examples.
Aliphatic materials: Aliphatic materials have open chain or linear molecular structures. As herbicides, these are less toxic to plants than aromatic compounds.
Alkaline: Substances having a pH greater than 7 (the opposite of acidic).
Alkaloids: A class of nitrogen-containing plant products, that are commonly alkaline and toxic. Examples are morphine and quinine.
Allele: One of the two forms of a gene (or marker) that can exist at a single locus.
Allelochemicals: The chemical inhibition of one plant by another.
ALS: Abbreviation for acetolactate synthase, also known as acetohydroxy acid synthase (AHAS), an enzyme essential for amino acid production.
Amine: A class of compounds derived from ammonia by replacing the hydrogen molecules with organic radicals. Amine salts are soluble in water. MCPA and 2,4-D are examples.
Amino acid: An organic molecule containing a carboxyl group and an amino group. It is the basic building block of proteins.
Anaerobic: Conditions in which there is no or less oxygen.
Anionic surfactant: Salt of an organic acid (sulfonates of sodium or ammonium).
Annidation: The sharing of resources by plant species in a given area.
Annual: A plant that completes its life cycle in 1 year.
Antagonism: Decreased activity arising from the effect of one chemical on another.
Antibiotic: Pesticidal substance produced by microorganisms.
Anticoagulant: A chemical that interferes with clotting of blood.
Antidote: A substance that neutralizes the effect of a poison.
Apical dominance: The inhibition of the development of some or all of the lateral buds by the terminal (apical) bud of a shoot.
Apical meristem: The meristem at the tip of the stem or root that gives rise to primary tissues and is responsible for increase in length rather than girth of the axis.
Apophytes: Robust weed species living independently that enter cultivated fields.
Apoplast: The free diffusional space outside the plasma.
Aquatic ecosystem: A strictly aqueous ecosystem with no terrestrial elements.
Aqueous: A term used to indicate the presence of water in a solution.
Arable crops: Annual crops that need farming and tillage activities (e.g. seedbed preparation, ploughing and cultivation). Arable land is land that is fit for tillage.

Arachnid: A wingless arthropod with two main body parts and four pairs of jointed legs. Spiders, ticks and mites belong to the class Arachnida.
Arboricide: Chemicals used for killing trees.
Aromatics: Chemicals characterized by the presence of one or more benzene (C_6H_6) ring, and usually with a pleasant odour.
Arsenicals: Pesticides containing arsenic.
Arthropod: An invertebrate animal characterized by segmented body parts and a life cycle that involves moulting of the hard body covering. Insects and mites belong to the phylum Arthropoda.
ATP: Abbreviation for adenosine triphosphate. It is present in all living organisms.
Auxanometer: An apparatus designed to measure the length of plant parts.
Auxin: A hormone-like substance found in plants, which stimulates cell growth in plant tissue. Indole-3-acetic acid (IAA) is an example.
Auxin-binding protein (ABP): A 22 kDa protein that is thought to recognize auxin and act as a receptor at the cell surface.
Auxotrophy: Auxotrophy is the inability of a living system to synthesize an essential compound (e.g. an amino acid or a fatty acid) required for its normal functioning. This may occur due to a mutant gene that inhibits the synthesis.
Avian: Related to birds.
Avicide: A pesticide used to kill or repel birds. Birds belong to the class Aves.
Awn: A stiff bristle-like projection, usually at the tip of an organ. The glumes and lemmas of grasses (family Gramineae) commonly possess awns.
Axillary bud: A bud formed in the upper angle between a twig (or petiole) and the stem.
Axis: The main stem of a flower or panicle.

B

Back-siphoning: The movement of liquid or pesticide solution mixture back into the water source, through the filling hose.
Bacteria: Microscopic organisms that may be either pathogenic or beneficial to man and his crops.
Bactericide: An agent or chemical that destroys or controls bacteria.
Bacteriostat: Material used to prevent growth or multiplication of bacteria.
Band application: Application of a pesticide in a strip alongside or around a structure or any object.
Basipetal: Movement or translocation from apex to base, the opposite of acropetal.
Basophile weeds: Weed species that are characteristically alkaline are basophile weeds. Alkali grass and quack grass are basophile weeds.
BCF: Abbreviation for bioconcentration factor, a measure of the tendency for a substance (chemical) to accumulate in the tissues of an organism.
Bioaccumulation: Progressive increase in the amount of a substance in an organism because the rate of intake exceeds the organism's ability to remove it from its system.
Bioactivation: Transformation of a pesticide within an organism into a more biochemically active metabolite.
Bioagent/biocontrol agent: A living organism employed to control a pest.
Bioassay: Determination of the concentration or potency of a chemical by its effect on living cells or tissues.
Bioavailability: Extent to which a pesticide residue can be taken up into an organism from its food and environment, and the rate at which this occurs.
Biocidal product: Non-crop pesticidal agents used to protect against wood rot, algae and slime.
Biodegradation: Conversion or breakdown of the chemical structure of a pesticide catalysed by enzymes *in vitro* or *in vivo*, resulting in loss of biological activity.
Biological assays: Measurement of the potency of any stimulus (physical, chemical or biological) by means of the reactions that it produces in living matter.
Biomagnification: Increase in concentration of a toxic/pollutant substance (e.g. a pesticide residue) in a particular food chain is known as biomagnification.
Biomarker: A biomarker is an indicator (molecular, biochemical, cellular or an organism) signalling an event or condition in a biological system.
Biomass: The total quantity (in weight) of organisms in a given area or volume.
Biorational pesticides: Pesticides derived from natural materials. Microbial pesticides, plant-incorporated protectants and biochemical pesticides are the three types.
Bioremediation or biotreatment: Treating contaminated land with naturally occurring or genetically modified microbes that degrade complex organic compounds (e.g. pesticides, insecticides, polychlorobenzenes, radioactive substances and heavy metals) which are thus made less harmful, less mobile or even useful and render the soil suitable for cultivation. The process of using metal-accumulating plants to remediate soil and water contaminated with heavy metal compounds is termed bioremediation through plants. Plants such as common ragweed (*Ambrosia artemisifolia*), hemp dogbane (*Apocynum cannabinum*), nodding thistle (*Carduus nutans*), Asiatic dayflower (*Commelina communis*), water hyacinth (*Eichornia crassipes*), water lettuce (*Pistia stratiotes*), zebra rush, hemlock water dropwort (*Oenanthe crocata*), taro and shrub violet (*Hybanthus floribundus*) are known to take up lead, mercury, nickel, cobalt and even salt.
Biostatic effect: The microbial decomposition of pesticides is sometimes reduced in the presence of some chemicals, which is called a biostatic effect. Such a chemical is also called an extender. Boron has a biostatic effect on 2,4-D.

Biota: Animals and plants of a given habitat.
Biotechnology: Any technique that uses material from living organisms to make or modify a product, for specific uses.
Biotic insecticide: A microorganism that is known to be an insect pathogen is applied in the same manner as a conventional insecticide to control a pest species.
Biotic potential: The inherent characteristic of an organism to survive and reproduce.
Biotransformation pathway: Sequence of the changes occurring in the structure of a pesticide when it is introduced into a specific biological test system.
Blank material (sample): Laboratory-simulated test material, free of the pesticide being analysed, used to test the method, apparatus and reagents of a study.
Blight: Common name for a number of different diseases of plants, especially when collapse is sudden.
Boom: A section of pipe to which pesticide sprayer nozzles can be attached to increase the area that can be treated in a single pass of a vehicle.
Bound residue: Chemicals in soil, plant or animal tissue originating from a pesticide, which cannot be extracted by a standard method (e.g. Soxhlet solvent extraction).
Branched-chain amino acids: Amino acids that have a chemical structure that includes side chains that are branched. Examples include leucine, isoleucine and valine.
Brand/trade name: The name or designation of a specific pesticide product or device made by a manufacturer or formulator for commercial purposes.
Broadcast rate equivalent: For band treatments, it is the amount of herbicide applied per unit when only the band area is considered. All rates for band treatment should be expressed as the broadcast rate equivalent.
Broadcasting: The process of uniformly scattering agricultural inputs (seed, fertilizer, manure or pesticides) on the soil of a field, manually or mechanically.
Broadleaved plants: Plants (e.g. dandelions and roses) with wide, rounded or flattened leaves and netted veins. Generally, these do not include mosses or grass-like plants.
Broad-spectrum pesticide: A pesticide toxic to a wide range of pests (i.e. non-selective).
Brush plants: Woody plant species growing on unfertilized land are called 'brush'. Big sagebrush (*Artemisia tridentata*), blackberry, blackbrush (*Coleogyne ramosissima*) and buckbrush are some brush species. Their growth and survival pattern prevent any beneficial or aesthetic use of the area. Brush control can be brought about by the use of chemical weedicides, but other practices like cattle grazing, mowing and tilling the land, and mulching are also useful. Specific chemical herbicides such as 2,4-D, atrazine, fluroxypyr, picloram, imazapyr, triclopyr and glyphosate may be used in varying concentrations to control these weeds.
Buffer zone: A safe distance between the edge of a permitted site of pesticide application and a sensitive non-target area (e.g. a watercourse).
Buffering agents: Agents that increase dispersion or solubility of pesticides in water.
b.w.: Body weight.

C

C_3 plant: A plant in which the first product of carbon dioxide fixation is the three-carbon compound. C_3 plants are photosynthetically less efficient than C_4 plants. Examples of C_3 plants are wheat, rice and barley.
C_4 plant: A plant in which the first product of carbon dioxide fixation is the four-carbon compound, oxaloacetic acid. Examples of C_4 plants are sugarcane and maize.
Calibration: The measurement and adjustment of pesticide application equipment to apply a pesticide formulation at a desired application rate.
Callus: Tissue overgrowth around a wound or canker.
Cambium: A layer of actively dividing cells in plants producing new tissues.
Canister: A container with absorbent material that filters fumes and vapours from the air before they are inhaled by an applicator.
Canker: A lesion on a stem.
Carbamate insecticide: A synthetic organic pesticide derived from carbamic acid that contains carbon, hydrogen, nitrogen and sulfur.
Carbohydrates: A major source of energy for organisms that include starch, sugar, cellulose and woody materials. Carbohydrate is made up of carbon, hydrogen and oxygen.
Carbon cycle: The processes by which carbon compounds are incorporated into living tissues via photosynthesis, only to return to the atmosphere through respiration, decay of dead organisms and the burning of fossil fuels.
Carcinogen: A substance that can cause cancer in living animals.
Carotenoid: A plant pigment that protects chlorophyll from being photodegraded and helps in the absorption of light during photosynthesis. Two major types of carotenoids are xanthophylls and carotene.
Carrier: An inert liquid, solid or gas added to an active ingredient to make a pesticide disperse effectively. Water or oil, used to dilute a formulation, is also a carrier.
Caryopsis: The seed (grain) or fruit of grasses, including that of cereals.
Catalyst: A substance that speeds up the rate of a reaction but is not used up in the reaction. It reduces the activation energy.

Cell: An individual biological unit, capable of independent function and division to form two new identical cells.
Cell membrane: A thin envelope surrounding the cell through which food materials pass in and waste materials and other metabolic products pass out.
Cell nucleus: A part of the cell responsible for functional activity and the continued life of the cell.
Cell wall: A thin, mechanically strong structure surrounding all plant cells. It is made up of a complex mixture of polysaccharides and other polymers. It regulates cell volume and determines cell shape.
Cellulose: A carbohydrate composed of continuous units of the sugar glucose linked in a specific way (β-1,4 linkage). It is a part of the cell walls of plants and algae.
Cereal: A grass grown for its edible seed or grain, such as maize, wheat and rice.
Chelating agent: A substance (usually an organic chemical, e.g. ethylenediaminetetraacetic acid) that combines with metal ions in solution to form insoluble compounds.
Chemical oxygen demand (COD): The amount of dissolved oxygen in water that will be consumed during oxidation of organic and inorganic substances. A high COD is undesirable.
Chemosterilant: A chemical compound capable of preventing animal reproduction.
Chemosynthesis: The process of carbon dioxide assimilation by the chemical oxidation of simple inorganic compounds.
Chiral molecule: A molecule containing one or more asymmetric carbon atoms. An asymmetric carbon atom has four different substituents and lacks a centre of symmetry.
Chlorinated hydrocarbon: A persistent pesticide containing chlorine, carbon and hydrogen. Examples are chlordane, DDT and methoxychlor.
Chloroplast: The organelle in plant cells that contains chlorophyll and is the site of photosynthesis.
Chromatography: A method of separation and identification of closely related chemical substances by their differential movement on or through certain materials, such as paper and resins, and by their specific colour reactions with certain reagents.
Chromosome: A thread-like body found in cell nuclei, comprised of genes arranged in linear order. While genes are the units of heredity, chromosomes are the units of transmission from one generation to the next.
Chronic toxicity: Ability of a substance to cause adverse (toxic) effects on a living organism when that organism is exposed to the substance continuously or in small repeated doses.
Clastogen: A chemical that is able to cause structural changes in chromosomes, primarily by breaking. Such effects are referred to as clastogenic.
Codex Alimentarius Commission (CAC): The organization charged with the development of the Codex standards and texts. The Codex Alimentarius Commission is an intergovernmental body jointly sponsored by the Food and Agriculture Organization of the United Nations (FAO) and the World Health Organization (WHO).
Coding: The process by which the sequence of nucleotides within a certain area of RNA determines the sequence of amino acids in the synthesis of the particular protein.
Codon: A coding unit. A sequence of three purine and pyrimidine bases on DNA or messenger RNA that code for a specific amino acid.
Coenzyme: A low molecular weight chemical activating enzyme action.
Cofactor: Compound or element that is required for efficient enzyme activity.
Coleoptile: The sheath covering the first leaf of a grass seedling as it emerges.
Commodity substance: A chemical with pesticidal properties, used mainly as a non-pesticidal substance. Its use as a pesticide is subject to formal approval. Examples are sulfuric acid and carbon dioxide.
Common name: A name given to a pesticide's active ingredient by a recognized committee on pesticide nomenclature. Many pesticides are known by a number of trade or brand names but the active ingredient(s) has only one recognized common name. An example is the common name for Sevin insecticide is carbaryl.
Compatibility: The degree to which two or more pesticides can be combined or mixed without significant reduction in effectiveness or safety.
Competitive inhibitor: A molecule that is similar in structure to a natural substrate that an enzyme acts upon. It competes with the substrate in binding to the active site.
Complete metamorphosis: The process of insect development, which includes the egg, larva, pupa and adult stages.
Composite sample: A term for combined samples from replicate trials.
Compound: A specific association of atoms held together by strong chemical bonds (e.g. water, H_2O).
Concentration: The amount of active ingredient in a given volume or weight of formulation.
Conjugate: A water-soluble derivative of a chemical formed by its combination with for example glucuronic acid, gluthathione, sulfate, acetate or glycine.
Correlation coefficient: A measure of the degree of the linear relationship between two variables.
Cotyledon: The first leaf or leaves produced (cotyledon leaves) by the embryo of a flowering plant. Grasses are monocotyledonous, while legumes are dicotyledonous.
Cover crop: A crop grown to cover the land to prevent erosion and leaching. It may also become a green-manure crop.
Critical concentration: The lowest concentration of a pesticide where adverse effects are felt on organisms or in the environment.
Crop diversification: A diverse cropping system with crop rotation year on year.

Cross-contamination: The inadvertent mixing of two or more pesticides.
Cross-pollination: Deposition of pollen from one flower of a plant to the stigma of another.
Culm: A monocotyledonous stem that is hollow except at the nodes or joints.
Cultivar: A variety that is propagated for its unique and distinguishing features.
Cumulative pesticides: Those chemicals that tend to accumulate or build up in the tissues of animals or in the environment (e.g. soil, water).
Cuticle: Waxy layer on plant epidermis that retains water and resists pest invasion.
Cytochrome P450: (Also known as cytochrome P-450 monooxygenases or CytP450s.) Multifunctional enzymes widely distributed among living organisms responsible for binding and activating molecular oxygen, incorporating one of its atoms into an organic substrate, and reducing the second atom to form water.
Cytokinins: Growth-regulating chemicals that promote cell division in plant cells.
Cytology: The science of the structure, function and life history of the cell.
Cytoplasm: The material enclosed by a cell wall in which the nucleus, endoplasmic reticulum, mitochondria and other organelles are situated.
Cytosol: The liquid substance of protoplasm, excluding all the organelles such as nuclei, plastids and ribosomes.

D

Danger: A signal word, used with or without the word 'Poison' and the skull and crossbones symbol, on pesticide labels to indicate a highly toxic substance (toxicity Class I).
Days-to-harvest: The least number of days between the last pesticide application and the harvest date, as set by law. It is the same as 'harvest intervals'.
DC_{50}: The median effective concentration (ppm or ppb) of the toxicant in the environment (usually in water) that produces a designated effect in 50% of the test organisms exposed.
Deciduous: Plants or trees that shed leaves or awns at a particular season or stage (leaf loss may also be induced by drought). Perennial plants lose their leaves during the winter.
Deflocculating agent: A material added to a suspension to prevent settling.
Defoliant: An agent that causes foliage to drop from a plant.
Degradation: The process by which a chemical compound is reduced to simpler compounds by the action of microorganisms, water, air, sunlight or other agents. Degradation products are usually less toxic than the original compound.
Dehydrogenase: A class of enzymes that oxidize substrates in the process of removing electrons and hydrogen atoms.
Denatured protein: Protein whose properties have been altered by treatment with physical or chemical agents.
Deoxyribonucleic acid (DNA): A polymer of nucleotides connected via a phosphate-deoxyribose sugar backbone. DNA carries the genetic material of a cell. The enzyme that catalyses the reaction of adding new DNA bases to a growing DNA strand is known as the DNA polymerase. The natural process in which a DNA molecule is copied is called DNA replication.
Dermal absorption: The dermal absorption of a pesticide is a measure of the amount that enters the body via the skin and is systemically available.
Desiccant: A pesticide or a hygroscopic chemical used to draw moisture from or dry up a plant, plant part or insect.
Detoxification: A process of making a chemical less toxic.
Diastereoisomers: Molecules that have non-superimposable structures that are not mirror images.
Diatomaceous earth: A whitish powder prepared from deposits formed by the silicified skeletons of diatoms.
Dicotyledon: Having two cotyledons in each seed. Most annual weeds and many perennial weeds are dicotyledonous.
Dinitro: A common designation for compounds containing two nitrogen atoms, such as dinitrophenols.
Dioxins and furans: Chemicals produced as by-products of heating chlorine-containing compounds in the presence of organic (carbon-containing) materials. They are among the most chronic toxic chemicals known so far. Major sources are incinerators, power plants, pulp and paper mills, diesel engines and refineries.
Diploid: Having a double set of chromosomes (i.e. pairs of homologous chromosomes) in the nucleus so that twice the haploid number is present, usually written as 2n.
Disease: A condition causing damage to a plant.
Dispersal: Distribution of small particles in a medium.
Dissipation and persistence: Residues can disappear in two steps. The first is the initial phase which is fast and is called dissipation. The slower second phase is known as persistence. Dissipation follows the law of 'first-order kinetics'. Chemicals persist because of different rates of translocation and degradation of various materials.
Dissipation time 50% (DT_{50}): Time required for half the initial quantity or concentration of a pesticide to dissipate from a system (e.g. water or soil).
Dose (dosage): A measure of something that is applied or given with a certain purpose.
Double helix: The spiral shape that two linear strands of DNA assume when bonded together.
Drift: The movement of pesticide droplets or particles by wind and air currents from the target area to a non-targeted area.
Dry weight basis: Residue concentration reported assuming the residue is wholly contained in the dry matter of the sample.
DT_{50}: The time taken for an active substance to degrade/dissipate by 50%.

E

Ear: The heavy spike or spike-like inflorescence of a plant (e.g. ear of wheat).
EC_{50}: The median effective concentration. Statistically derived concentration of a substance (pesticide) in an environmental medium expected to produce a certain effect in 50% of the test organisms in a given population under a defined set of conditions.
Ecology: A division of biology concerned with organisms and their relation to the environment.
Ecosystem: A system formed by the interaction of a community of organisms with their environment.
Ecotoxicological: The effects of environmental toxicants on populations of organisms originating, growing or living naturally in a particular region or environment. The concentration of a pesticide (active ingredient, formulations and relevant metabolites) that is likely to affect a determinable ecological characteristic of an exposed system is the ecotoxicologically relevant concentration (ERC). It is related to the toxicity characteristics.
Ectoparasite: A parasite feeding on a host from the exterior or outside the body.
ED_{50}: The median effective dose, expressed as milligrams per kilogram of body weight (mg/kg b.w.), which produces a designated effect in 50% of the test organisms exposed.
Electron-transport phosphorylation (or oxidative phosphorylation): Synthesis of ATP involving a membrane-associated electron transport chain and the creation of a proton-motive force.
Electrophoresis: A technique used to separate chemical substances according to the electrical charge on their molecules which migrate in response to the application of an electric field.
Emulsifiable concentrate (EC): An oil solution containing a high concentration of active ingredient and an emulsifying agent.
Emulsion: A mixture of two liquids that are not soluble in one another.
Enantiomer: One of two stereoisomers that are mirror images of each other that are 'non-superposable'.
Enantiotopic: The stereochemical term 'enantiotopic' refers to the relationship between two groups in a molecule which, if one or the other were replaced, would generate a chiral compound. The two possible compounds resulting from that replacement would be enantiomers.
Encapsulated formulation: A pesticide formulation with the active ingredient enclosed in capsules of polyvinyl or other materials.
Endocrine disruptors: Chemicals that interfere with glands and the hormones produced by the glands.
Endoparasite: A parasite that enters host tissue and feeds from within the body.
Endosperm: The part of the seed that acts as a food store for the growing plant.
Enterohepatic circulation: Cyclical process in which a pesticide residue is absorbed and transported to the liver, metabolized (often including conjugation), transported to the intestine by the bile, reabsorbed (often after deconjugation), and transported to the liver for further metabolism.
Entomopathogen: An organism capable of creating a disease condition in insects. Many fungi, bacteria, nematodes and viruses infect insect species to complete their life cycle, but in the process kill the host. This property is exploited in pest management.
Entomophagy: The consumption of insects by other organisms.
Environment: Air, land, water, all plants, man and other animals and the interrelationships that exist among them.
Enzyme: A protein that catalyses, or speeds up, a specific biochemical reaction without changing the nature of the reaction and without being changed itself.
Enzyme-linked immunosorbant assay (ELISA): A serological test in which the sensitivity of the antibody–antigen reaction is increased by attaching an enzyme to one of the two reactants.
Ephemeral: Plant with a short life cycle, producing several generations in a year.
Epidemiology: The study of the incidence, distribution and control of diseases.
Epinasty: Downward curling of a leaf blade as a result of cell growth on the upper side of the petiole being more rapid than that on the lower side.
Essential fatty acids: Fatty acids (e.g. linolenic and linoleic acids) that are necessary for growth of the human body, but which cannot be synthesized by the body.
Etiolation: Excessive spindliness in plants characterized by long, weak stems (long internodes) and small, pale yellow leaves, owing to insufficient light or to disease.
European and Mediterranean Plant Protection Organization (EPPO): EPPO is an intergovernmental organization. Its objectives are to protect plants, to develop international strategies against the introduction and spread of dangerous pests and to promote safe and effective control methods.
Exoskeleton: The rigid external skeleton for the body of some invertebrate animals (e.g. the segmented external skeleton of an arthropod).
Explosive limits: The range of concentrations (percentage by volume in air) of a flammable gas or vapour that can result in an explosion for ignition in a confined space.
Extractability: Degree to which a pesticide residue may be removed from a matrix (e.g. soil) through use of appropriate extraction techniques.
Extraneous residue limit (ERL): The maximum concentration of a pesticide residue or contaminant that is recommended by the Codex Alimentarius Commission to be legally permitted or recognized as acceptable in or on a food, agricultural commodity or animal feed. The concentration is expressed in milligrams of pesticide residues or contaminant per kilogram of the commodity.

F

Fallow: The practice of leaving land either uncropped or with volunteer vegetation during an otherwise normal growing season. Land could be left fallow to allow water to accumulate, for the soil to breathe or to control weeds.
FAO: Food and Agriculture Organization of the United Nations.
Fatty acids: A class of compounds containing a long hydrocarbon chain and a terminal carboxyl group. Fatty acids can be saturated (containing no carbon–carbon double bond), mono-unsaturated (containing one carbon–carbon double bond) or polyunsaturated (containing multiple carbon–carbon double bonds).
FDA: The US Food and Drug Administration, the federal agency that monitors pesticide residues on food products.
Fecundity: The reproductive capacity of a weed.
FIFRA: The Federal Insecticide, Fungicide, and Rodenticide Act, the federal law pertaining to pesticide regulation and use in the USA. The original act was enacted in 1947. It has been amended many times since then.
Filler: Diluent in powder form.
Floret: One of the small flowers making up a composite flower cluster.
Flowable: A formulation in which a large amount of very finely ground solid particles is suspended in a liquid carrier.
Fogger: An aerosol generator that produces very fine droplets (aerosol or smoke) to cause a fog.
Foliage: The leaves, needles, stems and blades of plants.
Formulation: A preparation of a pesticide for practical use. Formulations include preparations such as wettable powder, granules and emulsifiable concentrate. The added material in a pesticide formulation other than the active ingredient is called the formulant.
Free radical: Any molecule with one unshared electron giving the molecule a partial negative charge. It is highly reactive and can cause permanent damage to cellular membranes.
Fume: A smoke, vapour or gas that smells strongly or is dangerous to inhale.
Fumigant: A pesticide formulation that volatilizes, forming a toxic gas that kills pests by penetrating the voids in their body.
Fungus (plural, fungi): Non-photosynthetic organisms such as moulds, mildews, yeasts, mushrooms and puffballs. Fungi are usually non-mobile, filamentous and multi-cellular.

G

Gamete: A mature reproductive cell (sperm or ovum) containing half of the total number of chromosomes in a cell.
Gene cloning: Making copies of a specific gene, the fundamental unit of heredity.
Gene expression: The production of a protein encoded by a gene.
Gene mapping: Determination of the relative positions of genes on a DNA molecule or chromosome and the distance between them.
General use pesticide: That which can be used by the general public without undue hazard to the applicator or environment, as long as label instructions are followed.
Genome: The total genetic composition of the chromosomes in the nucleus of a gamete.
Genus: A taxonomic rank used in the biological classification of living and fossil organisms. It is the principal subdivision of a family above species.
Germicide: A substance that kills germs (microorganisms).
Gibberellins: A group of plant growth-regulating substances.
Good agricultural practice (GAP): GAP includes nationally authorized safe uses of pesticides and other additives on crops (standing and in storage) necessary for effective and reliable pest control, keeping in mind public and occupational health and environmental safety considerations.
Gram (g): A metric weight measurement equal to 1/1000th of a kilogram; approximately 28.5 g equals 1 oz (one ounce) and 454 kg equals to 1 lb (one pound).
Graminicide: A herbicide that targets the grass family (Graminae).
Granule (G): A small compact particle of a substance mixed with or coated onto an inert carrier to form a low concentrate particle which has a very low drift hazard.
Guard cells: Specialized epidermal cells that form the boundaries of stomata, the pores in plant leaf surfaces.
Gymnosperms: A class of plants whose seeds are not enclosed in an ovule. This taxonomic class includes pines, spruces and firs. The leaves of gymnosperms are needle-like and often do not shed. These plants are called softwoods. The four phyla of gymnosperms are cycads, ginkgo, gnetophytes and conifers.

H

Half-life ($t_{½}$): The time in which half the amount of the initial deposit of a substance is eliminated by reacting or otherwise getting dissipated. Half-life values can serve as a general quick reference to the rate of pesticide disappearance and can be described by first order kinetics with the rate constant k, $t^{½} = 0.693/k$.
Haploid: Having one set of chromosomes in a cell or an individual.
Harvest interval: The period between the last application of a pesticide to a crop and the harvest, as permitted by law.
Hazardous distance for the most sensitive effect (HDSE): Statistically determined safety margin corresponding to a distance from treated areas at which protection of the terrestrial environment can be adequately achieved as measured by the most sensitive non-target species.

Health advisory level (HAL): Estimate of the upper concentration limit for a pesticide in drinking water that can be consumed for a lifetime without adverse effects. HALs generally do not have formal legal significance but have been used, particularly in the USA, for preliminary risk assessment.
Hectare (ha): A metric area equal to 10,000 m^2 or approximately 2.47 acres.
Hemimetabolous insects: Insects that have a shortened or an incomplete metamorphosis in their life cycle. These insects do not have a pupal stage and such insects include crickets, cockroaches and bugs, many of which are in the order Hemiptera. Here, the stage between the eggs and the adult is usually that of a nymph, which roughly resembles the adult. Insects with four stages (eggs, larvae, pupae and adults) in their life cycle are called holometabolous (e.g. beetles, butterflies and flies).
Herbaceous plant: A plant that does not develop woody tissue.
Herbicide: A substance used to kill or inhibit the growth of plants, especially weeds.
Herbicide resistance: The inherited ability of a plant to survive and reproduce following exposure to a dose of herbicide that is normally lethal to the wild type. In a plant, resistance may be naturally occurring or induced by such techniques as genetic engineering or selection of variants produced by tissue culture or mutagenesis.
Heterosis: Increased growth, vigour or yield frequently shown in the F_1 generation of a cross.
Host: A living plant or animal that a pest depends on for its survival.
HPLC/MS: High performance liquid chromatography apparatus coupled to a mass spectrometer.
Hybrid: The first generation offspring of a sexual cross-fertilization between two individuals differing in one or more genes (i.e. a heterozygote).
Hydraulic sprayer: A machine that delivers large volumes of spray at high pressures (up to several hundred psi) to the target. They are used primarily for treating trees. It is the same as a high pressure sprayer.
Hydrocarbon: A chemical whose molecules consist entirely of carbon and hydrogen atoms. Methane (natural gas) is the simplest hydrocarbon, with each molecule containing one atom of carbon and four atoms of hydrogen. Crude oil is a mixture largely of hydrocarbons.
Hydrolysis: Reaction in which a chemical bond is cleaved and a new bond formed with the oxygen atom of a molecule of water.
Hydrophilic: Having chemical properties relating to water (water-like). Compounds are polar or ionizable and can be soluble in water and are generally lipophobic (which is a property of insolubility in lipids or other non-polar solvents).
Hydrophobic: Water-'fearing' compounds that are relatively insoluble in water and other polar solvents and are soluble in lipids (lipophilic) and other non-polar solvents.
Hyperparasite: A parasite that attacks another parasite.

I

Immobilization: Restricted mobility of a pesticide in a plant or soil due to strong binding, or the formation of complex organic forms in microbial or plant tissue.
Immunoassay: Ligand-binding assay based on antibodies capable of specific binding to the pesticide analyte. Most commonly used in a competitive binding format where analyte molecules compete with a specific antigen complex labelled for detection using a radioisotope (radioimmunoassay or RIA) or an enzyme (enzyme-linked immunosorbant assay or ELISA).
Impurity: By-product of the manufacture of a pesticide. They need to be regulated if found to be toxic or otherwise persistent.
In situ: In the original situation.
In vitro: A term used to describe effects in biological material outside the living animal.
Inbred: A normally cross-pollinated plant that has been self-fertilized.
Inert ingredients: Inactive components of a formulation used for making it safer, more effective, easier to mix and apply.
Infestation: Pests that are found widely in an area or location.
Inflorescence: The arrangement of flowers on a stem. Types of inflorescence include corymb, cyme, solitary, raceme and umbel.
Inhalation: Breathing into the lungs.
Inhalation toxicity: The toxicity of a pesticide when breathed in through the lungs.
Inhibitor: A chemical used to prevent physiological processes in an organism/plant.
Inorganic compound: A compound that does not contain carbon atoms (the only exception being carbon dioxide) (e.g. copper sulfate, sodium arsenite and sulfuric acid).
Insect growth regulator: An agent that disrupts the life processes of insects.
Insecta: A class in the phylum Arthropoda characterized by a body composed of three segments and three pairs of legs.
Insecticide: A chemical agent used to kill insects or control unwanted insects.
Insoluble: Does not dissolve in liquid.
Integrated pest management (IPM): A management system that uses all suitable techniques and methods in as compatible a manner as possible to maintain pest populations at levels below those causing economic injury. Pests include weeds too.
Intercropping: Growing two crops simultaneously in the same field.
Internode: The portion of the stem between two nodes or joints.
Interphase: The stage of cell division when DNA is replicated.
Interval: The time between two pesticide applications, or between the final pesticide application and harvest.
Invertebrate: The very large group of animals that lack backbones. Insects, jellyfish, starfish, shellfish, squid and spiders are examples.

Isoenzyme: Forms of an enzyme having catalytic properties ascribed to that enzyme.
Isomer: Briefly, two compounds are said to be isomeric if they are composed of the same elements in the same proportion by weight but have different chemical and physical properties.

J

Juvenile hormone: A hormone produced by an insect that inhibits moulting.

K

Kernel: Whole grain of a cereal; the mature ovule.
kg: Kilogram. The metric weight measurement equal to 1000 g (approximatey 2.2 lb).
K_{OC}: The organic-carbon sorption partition coefficient, which is the tendency of a chemical to be adsorbed by soil, expressed as K_{OC} = concentration adsorbed/concentration dissolved/percentage of organic carbon in the soil.
Krebs cycle: The sequence of reactions by which most living cells generate energy during the process of aerobic respiration. It takes place in the mitochondria, using up oxygen and producing carbon dioxide and water as waste products, and ADP is converted to energy-rich ATP.

L

Label: All printed material attached to, or part of, the container or wrapper of the package of a pesticide.
Laboratory sample: Sample or subsample(s) sent to or received by the laboratory.
Lamina: Leaf blade.
Larva (plural, larvae): The developmental stage of insects that hatch from an egg with complete metamorphosis. A mature larva becomes a pupa. (Some other invertebrates have larvae, including crustaceans, mites and ticks.) The pesticide used to kill larvae is called a larvicide.
LC_{50}: Lethal concentration 50% or median lethal concentration – the concentration of an active ingredient in air/liquid expected to cause death in 50% of the test animals treated. The LC_{50} is measured in milligrams per kilogram body weight (mg/kg b.w.) or in parts per million (ppm).
LD_{50}: Lethal dose 50% or median lethal dose – the dose (quantity) of the active ingredient that will kill 50% of the test animals.
Leaching: The movement of a substance downwards through the soil as a result of water movement or via its dissolution in water is leaching. The moving aqueous phase through the soil column is called the leachate.
Ligase: An enzyme that can rejoin the cut strand of DNA.
Lignin: The complex phenolic polymer that endows the xylem and other tissues in plants with resistance to compression and decay.
Lipid: Group of organic compounds which are esters of fatty acids and which are characterized by being insoluble in water but soluble in many organic solvents.
Lipophilic: Having chemical properties relating to lipids (lipid-like).
Lipophilicity: Affinity for fat as described by partitioning behaviour between water and an immiscible organic solvent, favouring the latter.
Litre (L): A metric measurement of volume which is 1000 dm^3 or, when dealing with liquid measurement, a little more than 1 quart.
Locus: The site on a chromosome where a particular gene (or a sequence) is located.
Low concentrate solution (S): A solution that contains a low concentration of active ingredient in a highly refined oil or solvent.
Lowest observed adverse effect level (LOAEL): The lowest dose in a study at which an adverse effect occurs.
Low-pressure boom sprayer: A machine that delivers low to moderate volumes of pesticide at pressures of 30–60 psi. Used on field, forage and pasture crops.
Lysine: A non-essential amino acid found in eggs, milk, meat, legumes, grains and nuts.
Lysis: The dissolution or destruction of cells by the disruption of the cell membrane.

M

Marker gene: The gene used to control or detect a transgenic organism from a non-transgenic organism.
Maximum permissible intake (MPI): This is derived by multiplying the acceptable daily intake (ADI) with the average body weight.
Maximum residue level (or limit) (MRL): Maximum concentration of a residue that is legally permitted in or on food and feed, as set by the Codex Alimentarius Commission or a national regulatory authority. It is normally expressed in milligrams per kilogram (mg/kg) fresh weight.
Median effective concentration (EC_{50}): Statistically derived concentration of a substance (pesticide) in an environmental medium expected to produce a certain effect on 50% of the test organisms in a given population under a defined set of conditions.
Median lethal concentration (LC_{50}): Statistically derived concentration of a pesticide in an environmental medium expected to kill 50% of test organisms in a given population under defined conditions.

Median lethal dose (LD_{50}): Statistically derived dose of a pesticide expected to kill 50% of test organisms in a given population under a defined set of conditions. Normally expressed as milligrams of test material per kilogram of body weight (mg/kg b.w.) of the organism.
Meiosis: The process of division of sexual cells in which the number of chromosomes in each nucleus is reduced to half the normal number found in normal somatic cells. When two sexual cells fuse, each contributes its half of the chromosomes. The resulting embryo contains the full chromosome complement. Cells with half the chromosome complement are called haploid cells, while those with the full complement are diploid cells.
Mesophyll: Layer of chloroplast-containing cells (parenchyma or chlorenchyma cells) between the epidermal layers of a leaf.
Messenger RNA (mRNA): A chain of ribonucleotides that codes for a specific protein. mRNA, a molecule which is synthesized from the DNA template (a reaction catalysed by the enzyme RNA polymerase) carries the genetic code for a protein from the DNA in the nucleus to the ribosomes in the cytoplasm where the code is read and the protein synthesized.
Metabolism: A general term that with reference to pesticides refers to the change of a pesticide from an active to an inactive state within an organism and the product formed is called a metabolite.
Metamorphosis: Changes in the shape, form and size of insects during their life cycle.
Microbial pesticide: Bacteria, viruses, fungi and other microorganisms used for controlling pests such as insects or other microorganisms. Microbial pesticides are also called biorational products.
Microgram (μg): A metric weight measurement equal to 1 millionth of a gram.
Mineralization: Conversion of an element from an organic form to an inorganic form. Mineralization of pesticides most commonly refers to the microbial degradation to carbon dioxide as a terminal metabolite.
Miscible liquids: Two or more liquids capable of being mixed in any proportion that will remain mixed under normal conditions.
Mites: Tiny animals closely related to ticks. A mite has six legs during the larval and nymphal stages, but eight legs as adults. A miticide is a pesticide that is used to control mites.
Mitochondria: Minute bodies (organelles) in the cytoplasm of most eukaryotic cells which act as a source of chemical energy (in the form of adenosine triphosphate, ATP) necessary for the organism to carry out important tasks.
Mitosis: The normal process of division of somatic cells (i.e. non-sexual cells) in which the chromosomes are duplicated to give two daughter nuclei each having a chromosome complement equal to that of the original (parent) nucleus.
Mode of action: A term used to describe the method by which a pesticide may alter living processes in the target organism, leading to death/inhibition.
Molluscicide: A chemical used to kill or control snails and slugs.
Monocotyledon: Any seed plant (angiosperm) having a single cotyledon or seed leaf. Leaves are mostly parallel veined (e.g. cereals, grasses, rushes and sedges).
Monoculture: The growing of a single crop in successive seasons over a wide area.
Mosaic: Yellow/green leaf patterns produced by certain virus infections.
mRNA: Messenger RNA, a molecule which is synthesized from the DNA template by the enzyme RNA polymerase.
MTD: Maximum tolerated dose, the highest dose of a chemical that does not alter the lifespan or severely affect the health of an animal.
Mutagenicity: Ability to cause permanent genetic changes in an organism.
Mutant: An individual having an abnormal structure, property or behaviour in which it differs distinctly from its parent. This is an inheritable change and may be caused or induced by high energy irradiation or by certain chemical substances.
Mycoplasmas: The smallest known living organisms that can reproduce and exist apart from other living organisms. They obtain their food from plants.

N

National estimate of dietary intake (NEDI): An estimate of the long-term intake by consumers, of crops that contain pesticide/toxic residues at the median level (i.e. supervised trials median residue, STMR). The NEDI is compared against the acceptable daily intake (ADI).
Necrosis: Death of plant or animal tissues, resulting in discoloured, sunken or necrotic (dead) areas.
Negligible residue: A tolerance that is set on a food or feed crop permitting an ultra-small amount of pesticide at harvest as a result of indirect contact with the chemical.
Nematicide: A pesticide used to control unwanted nematodes (eelworms).
Neoprene: A synthetic, protective rubber often used to make gloves and boots.
Nereistoxin: (NTX; 4-*N*,*N*-dimethylamino-1,2-dithiolane). A naturally occurring toxin with redox effects on neuronal nicotinic acetylcholine receptors in chick retina. Found in marine annelid or segmented worm, *Lumbrineris brevicirra*. Bensultap, cartap, polythialan, thiocyclam and thiosultap contain nereistoxins.
Neurotoxicity: The poisonous effect of a pesticide on nerve tissue.
Non-ionic surfactant: A surfactant that does not ionize in solution and is therefore compatible with both anionic and cationic substances.
Non-selective pesticide: A pesticide that is toxic to a wide range of pests.
No-observable effect concentration or level (NOEC/NOEL): Highest concentration of pesticide in the test system that causes no biological effect to the target organism.

No-observed adverse effect level (NOAEL): The highest dose of substance that does not cause any detectable toxic effect in the experimental animal studied.
Nozzles: Devices that determine the flow rate, droplet size and discharge pattern of a pesticide solution during application. Nozzles are mounted in various ways, usually at the ends of wands or along spray booms. Types of nozzles include flat fan, even flat fan, cone, flooding, atomizing, broadcast and solid stream.
Nucleic acid: A compound of high molecular weight that consists of pentose (ribose (RNA) or deoxyribose (DNA)), phosphoric acid, and nitrogen bases (purines and pyrimidines), present in all living things including the infectious parts of plant viruses.
Nucleotide: A subunit of DNA or RNA consisting of a nitrogenous base (adenine, guanine, thymine, uracil or cytosine) a phosphate molecule and a sugar molecule. Thousands of nucleotides are linked together to form a DNA strand. A nucleotide sequence is the order of nucleic acid bases in a DNA or RNA strand.
Nucleus: the membrane-enclosed protoplasmic body structure within eukaryotic cells containing the genetic material (DNA) organized in chromosomes.
Nymph: The developmental stage of insects that hatch from an egg with gradual metamorphosis. Nymphs often resemble the adult insect but are smaller.

O

Obligate: Having a specific requirement for a particular environmental factor (e.g. a specific host) and being unable to survive if it is not available.
OC: Organochlorine pesticides or those containing chlorine but the term is used to refer to persistent ones such as aldrin, chlordane, DDT, dieldrin, heptachlor and lindane.
Octanol/water partition coefficient (POW): Partition coefficient for a pesticide in the two-phase system octanol/water. It is a distribution coefficient reflecting the relative lipophilicity of a pesticide and its potential for bioconcentration. The value is often expressed in logarithmic (base 10) form (log POW).
Oils: Liquids used to carry or dilute an active ingredient in a pesticide formulation.
OP: Organophosphorus pesticides or those containing phosphorus. OP commonly refers to the ones with cholinesterase-inhibiting esters of phosphate or thiophosphate.
Oral toxicity: The ability of a pesticide to cause injury or acute illness when taken by mouth.
Organic pesticide: Pesticides that contain carbon. The two major types are petroleum oils and synthetic organic pesticides.
Ovicide: A chemical that destroys an organism's (insects and mites) eggs.
Ovisorption: The process in which ripe eggs are not deposited but are absorbed back into the body. This occurs when the female does not get enough proteins or when the female fails to locate a suitable host for laying the egg.
Ovule: A female germ cell which develops into a seed after fertilization.
Oxidation: Removal or exchange of electrons usually by combining with oxygen.
Oxon: A degradation product of a chemical that shows the toxic potential of the original chemical. Oxons occur in OP insecticides, where a phosphorus–oxygen bond replaces a phosphorus–sulfur bond. This change is aided by the animal liver. Diazinon and chlorpyrifos are two examples of oxons.

P

Parasite: A plant, animal or microorganism living in, on or with another living organism for the purpose of obtaining all or part of its food.
Parenchyma: The plant tissue consisting of cells concerned with physiological activities (i.e. capable of photosynthesis or storage).
Parthenogenesis: The development of an egg cell into an embryo without fertilization.
Partition coefficient: Ratio of the concentrations of a substance in solution in two phases which are in equilibrium.
Parts per billion (ppb): Ratio of amounts expressed as parts pesticide per billion (10^9) parts of sample.
Parts per million (ppm): Ratio of amounts expressed as parts pesticide per million (10^6) parts of sample.
Pathogen: A disease-producing microorganism such as a bacterium or virus.
Pathway: A sequence of interconnecting enzyme reactions.
Peat: Partially decomposed plant material.
Pedalfer: A soil in which the soluble lime has been leached from the surface layers by heavy persistent rain.
Pedocal: A soil in which lime accumulates in the surface layer, which consequently becomes alkaline. Pedocals occur in regions with light rainfall and are usually fertile.
Pellet: Solid formulation of pesticide that is larger than a granule.
Penetrant: A kind of additive or adjuvant which aids the pesticide to get through the outer surface and into the plant.
Penumbra: The extent of shading of one leaf by another over the total photoperiod.
Persistent, bioaccumulative and toxic (PBTs) chemicals: Toxic chemicals that persist in the environment and bioaccumulate in food chains.
Persistent organic pollutants (POPs): Chemical substances that persist in the environment and bioaccumulate through the food web. They threaten human health, as well as the health of ecosystems. POPs are subject to international negotiations with the aim being moving towards global elimination.

Persistent pesticide: A pesticide that does not break down chemically or breaks down very slowly, and remains in the environment after a growing season.
Personal protective equipment (PPE): Devices and clothing intended to protect a person from exposure to pesticides.
Pest: An insect, rodent, nematode, fungus, weed or other life form that is injurious to health or the environment.
Pesticide: Any substance or mixture intended for preventing, destroying, attracting, repelling or controlling any organism, animal or plant that interferes in the production, storage, transport, distribution and processing of food produce, animal feeds and the surroundings of man and his stock.
Pesticide chemical name: Scientific name of a pesticide following the recommendations of IUPAC for naming of chemical compounds or another accepted naming convention such as Chemical Abstracts.
Pesticide common name: Simple name assigned to a pesticide active ingredient by the International Organization for Standardization (ISO) to be used as a generic or non-proprietary name.
Pesticide formulation: The form of a pesticide product in which the active ingredient (a.i.) is offered for sale. Along with the a.i., a formulation includes adjuvants and additives that improve some useful property of the final product.
Pesticide residue: Any specified substances in food or feed, including derivatives of a pesticide, metabolites, reaction products and impurities resulting from the use of a pesticide.
Pesticide trade name: Proprietary name assigned to a pesticide or its formulations by the company manufacturing or selling it.
Petroleum oils: Pesticides refined from crude oil (e.g. xylene, refined oil and kerosene).
pH: Negative log of the hydrogen ion concentration ($pH = -\log [H^+]$). A measure of the acidity/alkalinity of a liquid.
Phenology: The study of the changes taking place in living systems as a result of periodic changes in nature. Bird migration and flowering are seasonal changes.
Phenotype: The physical characteristics of an organism, determined by a combination of the genetic composition (genotype) and the environment of the individual.
Pheromone: A volatile hormone secreted by an animal that influences the sexual or developmental behaviour of others of the same species.
Phloem: Part of the plant's vascular system adapted to transport photosynthetic products such as transporting sugar from where it is synthesized in the leaves to the rest of the plant.
Photoautotroph: An organism that uses the energy from the sun to synthesize its own organic food from inorganic carbon dioxide (CO_2) and water (H_2O). Photoautotrophs include most green plants.
Photolysis: Chemical reaction caused by light in which a bond is cleaved. The splitting of water by light into gaseous oxygen, hydrogen ions and electrons is called photolysis of water. Breakdown of a chemical due to light is called photodegradation.
Photonasty: A movement of plant parts induced by external light stimuli.
Photoperiod: The optimum duration of light and darkness for the normal growth and maturity of a plant in nature.
Photophosphorylation (photosynthetic phosphorylation): The production of ATP from ADP and inorganic phosphate, utilizing light-induced photosynthetic electron transport as a source of energy.
Photoreceptor: Any light-sensitive region of a plant.
Photorespiration: Light-dependent oxygen uptake and carbon dioxide production (distinct from mitochondrial respiration), seen primarily in C_3 plants.
Photosynthesis: The enzymatic conversion of light energy into chemical energy and use of this to form carbohydrates and oxygen from carbon dioxide (CO_2) and water (H_2O) in plant chloroplasts.
Photosystem: An array of pigment–protein complexes and electron transfer components that function together to harvest light energy, transfer the energy to photochemical reaction centres, and move the excited electrons in a controlled fashion to produce usable biochemical energy. This complex needs chlorophyll and carotenoid molecules in order to function.
Phototrophs: Organisms that derive the necessary energy for the synthesis of organic compounds from sunlight (include photoautotrophs and photoheterotrophs).
Physiological drought: A drought condition suffered by plants that occurs despite there being sufficient water in the soil. For example this may happen because of exclusive loss of water through transpiration due to increased light intensity and temperature or because of a lack of roots to absorb soil moisture. Alternatively it may be that the soil water is temporarily unavailable, as when water is frozen.
Phytotoxic: A substance that is toxic to plants.
Piscicide: A chemical used to kill fish.
Plant growth regulator: A substance (not fertilizers/nutrients) that alters the expected growth, flowering or reproduction rate of plants.
Plasmalemma: The cytoplasmic membrane found on the outside of the protoplast adjacent to the cell wall.
Plasmodesmata: Fine protoplasmic connections between cells which extend through the cell wall.
Plastic mulch: Thin polythene film used as a mulch for field crops.
Plastid: Specialized cytoplasmic structures in a cell where chemical activity occurs.
Ploidy: The degree of replication of chromosomes (e.g. haploid and diploid) in a plant.
Plough in: To bury a crop or weeds by turning them under the soil with a plough.
Podosol (podzol): A type of zonal acid infertile soil (a pedalfer) found in regions of heavy rainfall and long cool or cold winters.

Poison: Any chemical or agent that can cause illness or death when eaten, absorbed through the skin or inhaled, by man or animals.
Pollen: In plants, the male germ cells produced in the anther. When ripe, pollen sacs at the end of the anthers split open to release the pollen.
Pollutant: A harmful chemical or waste material discharged into water, soil or air.
Post-emergence: Period after a crop or pest has appeared.
ppm: Parts per million.
Predator: An animal or an organism that attacks, kills and feeds on other animals or organisms (e.g. hawks, owls, snakes and many insects, fungi and nematodes).
Pre-emergence herbicides: A treatment of a fully cultivated field after sowing of the crop seed but before it has emerged.
Pre-harvest interval: The time between the last pesticide application and the harvest of treated crops.
Preoviposition: The interval between emergence of the adult female and when it lays the first egg.
Pre-planting treatment: Application of an agent before the crop is planted.
Protease: An enzyme that degrades protein into its component amino acids.
Protectant: A fungicide applied to the plant's surface before pathogen attack to prevent penetration and subsequent infection.
Protein: A large molecule composed of one or more chains of amino acids in a specific order. Proteins are necessary for the structure, function and regulation of an organism's cells, tissues and organs. Each protein has a unique function.
Protein synthesis: The production of proteins in a cell.
Protopam chloride (2-PAM): An antidote for certain organophosphate pesticide poisoning, but not for carbamate poisoning.
Protoplasm: The colourless material comprising the living part of a cell including the cytoplasm, nucleus and other organelles except the cell wall.
Protoplast: A plant cell whose cell wall has been removed.
Protoporphyrin: An intermediate in the biosynthetic pathways for chlorophyll and haem. This molecule is a strong photosensitizer and can cause cellular damage.

Q

QoI: Quinone outside inhibitor.
Quenching: The process of causing the de-excitation of a molecule from an excited singlet or triplet state back down to the ground state. Quenching of chlorophyll is an important protective role of carotenoids. If a triplet molecule is not rapidly quenched, it can react with oxygen to generate singlet oxygen. Singlet oxygen can cause much cellular damage.

R

Racemate: Diastereoisomers differ in their physical and chemical properties, whereas enantiomers have identical physical and chemical properties and differ only in their ability to rotate the plane of polarized light clockwise or counterclockwise. A mixture containing equal amounts of a pair of enantiomers is called a racemic mixture or racemate.
Receptor: Special proteins in the cell membrane that bind to a signal or chemical messenger from the environment and transmit signals to the cell to initiate a response.
Recombinant DNA (rDNA): A strand of DNA synthesized in the laboratory by splicing together selected parts of DNA strands from different organisms or by adding a selected part to an existing DNA strand.
Recovery, analytical: Fraction or percentage of a pesticide residue recoverable following extraction and analysis of a matrix containing the pesticide.
Redox potential: Electrical potential indicating the relative activity of oxidized and reduced species. The redox potential of an environmental matrix is a measure of the extent to which oxidizing species can act as terminal electron acceptors in respiration.
Re-entry interval: Minimum time between pesticide application and human re-entry to a treated area.
Reference dose: Expected dose resulting from human exposure to a pesticide at the level at which it is regulated in the environment.
Registration: An approval by a responsible government authority for the sale and use of a pesticide upon evaluation of scientific tests demonstrating that it is effective for the purposes intended and not unduly hazardous to life and the environment.
Relevant metabolite: A metabolite of an active substance that is subject to risk assessment because of its toxicity and the levels found.
Repellent: A compound that keeps insects, rodents, birds or other pests away from plants, domestic animals, buildings or other treated areas.
Residential buffer zones: Area of land close to homes, where spraying is disallowed.
Residual pesticide: A pesticide that remains in the environment for a long time.
Residue: Pesticide that remains on or in a crop/soil/animals.
Resistance: Ability of a population to survive and reproduce following repeated exposure to a pesticide. Resistance may also be induced by genetic engineering, tissue culture or mutagenesis.
Restricted use pesticide: A pesticide that can be purchased and used only by certified applicators or persons under their direct supervision. A pesticide classified for restricted use is listed under FIFRA, Section 3(d)(1)(C).

Rhizomes: A specialized horizontal stem that gives rise to shoots and roots at its nodes; usually found underground (as in e.g. iris and rhubarb).
Ribosome: Small, almost spherical structures found in all cells. They are the site of translation of mRNA into protein and each ribosome contains enzymes essential for protein formation.
Rinsate: Water or other liquid used for rinsing a pesticide container or equipment.
Risk assessment: A risk assessment determines whether pesticides at specific levels present a concern for consumer health.
RNA polymerase: An enzyme that catalyses the synthesis of mRNA, in which the nucleotide sequence of DNA is copied.
Rodent: Any animal of the order Rodentia (e.g. mice, rats, squirrels and gophers).
Rodenticide: Substances used to kill or repel rodents.
Rotation: The practice of growing different crops on the same land in a regular, recurring sequence. It is a good agricultural practice.
Runoff: The movement of water and associated materials on the soil surface. Runoff usually proceeds to bodies of surface water.

S

Safener: A substance added to a pesticide formulation to eliminate or reduce phytotoxic effects of the pesticide to certain crops.
Sample: Portion of material selected from a larger quantity so that it is representative of the whole. Examples are aggregate sample, aliquot, composite sample, control sample, laboratory sample, primary sample, random sample and retention sample.
Seed protectant: A chemical applied to seed before planting to protect seeds and new seedlings from disease and insects.
Selective pesticide: A substance that kills one kind of pest and spares much of the other fauna or flora, including beneficial species, either through differential toxic action or by virtue of other means such as the formulation, dosage and timing of application.
Self-pollination: Pollination with pollen from the flower(s) of the same plant.
Senescence: Process or state of growing old.
Sensitive areas: Locations that are easily and severely affected by a pesticide application. Examples include streams, ponds, parks, houses and barns.
Sequencing: A means of identifying the order of nucleotides in DNA, RNA or amino acids in a polypeptide.
Sequential application: Two or more chemicals applied at different times to the same crop.
Shikimate pathway: Important chemical pathway in plants and bacteria for producing amino acids and other aromatic products.
Signal words: Mandatory words on pesticide labels that indicate their toxicity class, for example 'CAUTION', 'WARNING' and 'DANGER – POISON' (this last with the skull and crossbones symbol).
Silvicides: Herbicidal compounds used for killing or inhibiting the growth of trees, shrubs, brushes and weed trees.
Slimicide: A substance intended to prevent, inhibit or destroy biological slimes produced by a combination of algae, bacteria or fungi.
Slurry: A thick suspension of a pesticide made from a wettable powder and water.
Smut: A fungus with sooty spore masses, belonging to the fungal group basidiomycetes.
Soil organic partition coefficient (K_{OC}): Ratio of a pesticide concentration adsorbed in the organic matter component of soil or sediment to that in the aqueous phase at equilibrium. The K_{OC} is calculated by dividing the K_d value by the fraction of organic carbon present in the soil or sediment.
Soil partition coefficient (K_d): Experimental ratio of a pesticide concentration in the soil to that in the aqueous (dissolved) phase at equilibrium. It is valid only for the specific concentration and solid:solution ratio of the test. The K_d is a distribution coefficient reflecting the relative affinity of a pesticide for adsorption by soil solids and its potential for leaching movement through the soil.
Soil sterilant: A chemical that prevents the growth of weeds. They have residual effects and can damage the soil significantly.
Solution: The mixture of a substance in another substance (usually a liquid) in which all ingredients are completely soluble without changing their chemical characteristics.
Solvent: A liquid that will dissolve another substance (or substances) to form a solution.
Species: One kind of plant or animal. It is abbreviated as sp. (singular) and spp. (plural). If spp. follows the name of the genus (e.g. *Rhizoctonia* spp.) this means that a number of species of that genus are indicated.
Specific leaf area (SLA): Leaf area per unit of leaf dry weight.
Specimens: Samples collected from a system for examination, analysis or storage.
Spectrophotometer: An instrument that analyses characteristics of organic compounds through transmittance of light waves through solutions. Ultraviolet and infrared wavelengths are commonly employed. Ordinary light can be used in certain types of instruments and electromagnetic fields are employed in some instruments.
Split application: The total amount of pesticide or fertilizer divided between two or more applications made at different times to the same crop.
Spore: An inactive form of a microorganism that is capable of becoming active again.
Spot treatment: A pesticide application directed at a small area or a part of a target.
Spreader: Ingredient (adjuvant) added to a spray mixture to improve contact between the pesticide and the plant surface.
Spudding: Removal of weeds by cutting below the soil surface.

Stale seedbed: A seedbed prepared in advance of crop planting to ensure germination of many weed seeds which are then destroyed mechanically or by use of non-residual herbicides.
Stamen: The male part of a flower which consists of an anther and a filament.
STEL: Short-term exposure limit, usually a continuous 15 min exposure period.
Sterilize: To treat with an agent to kill every living thing in a certain area.
Sticker: Substance that increases the adhesiveness of a product applied to a surface.
Stoma (plural, stomata): A pore in the epidermis of aerial plant parts providing a means for gaseous exchange between the internal tissues and the atmosphere.
Stomach poison: A pesticide that must be eaten in order to become fatal.
Storage stability test: A test that measures the chemical and physical stability of the product stored under defined (often worst) conditions.
Stratification: Exposing seeds to a cold period to encourage germination or break seed dormancy.
Stubble: That part of a crop left in the ground after harvesting, including the roots and short stem remains, particularly the dry stalks of cereals.
Surface active agent: Adjuvant/surfactant that reduces surface tension between two different substances (such as water agent and oil).
Surface water: Water on the earth's surface (e.g. rivers, lakes, ponds and streams).
Surfactant: A chemical that increases the dispersing, spreading and wetting properties of a pesticide by reducing its surface tension.
Susceptibility: The inability to withstand a pesticide at normal rates of use (sensitivity).
Swath: The width of the area covered by a sprayer or duster making one sweep.
Synergism: Increased activity resulting from the effect of one chemical on another.
Synthetic organic pesticides: Man-made pesticides that contain carbon, hydrogen and other elements.
Systemic: Compound that is absorbed and translocated throughout the target.

T

Tank mix: Two or more pesticides mixed in the spray tank before spraying.
Target: The site or pest towards which control measures are being directed.
Temporary ADI: A dose (measured in milligrams per kilogram of body weight, i.e. mg/kg b.w.) of a pesticide that can be consumed daily for a short period of time, without any significant adverse/toxic effects. It is established as temporary, since the risks involved in a lifetime of consumption of the specific pesticide cannot be stated with certainty. To ensure safe limits of daily doses, a temporary ADI is always set at a level that is higher than the normal safe limit for the given pesticide. See entry for Acceptable daily intake (ADI).
Termiticide (termicide): A chemical agent used to kill termites.
Test portion (analytical portion): Subsample of proper size for a chemical analysis or other test, removed from the test sample.
Test sample (analytical sample): Homogenous sample, prepared from the laboratory sample by mixing, grinding, blending and fine chopping, from which test portions are removed for analysis with minimal sampling error.
Threshold: Concentration of a pesticide in an organism or environmental compartment below which an adverse effect is not expected.
Thylakoid membranes: A series of sealed membrane compartments found inside the chloroplast of plant cells.
Ticks: Small blood-sucking, eight-legged arthropods belonging to the class Arachnida (which also includes spiders and mites). They are vectors of diseases.
Tillage: The use of implements to cultivate the soil in the preparation of seedbeds and root-beds for crops.
TLV: The threshold limit value of a chemical is a level that is allowed for repeated exposure that does not result in adverse health effects. It is set by a competent authority.
Tolerance limit: The maximum concentration (in ppm) of pesticide residue that is permitted in or on food at a specified stage of harvesting, transport, marketing or preparation of food up to the final point of consumption. The term now has been replaced by maximum residue level or limit (MRL).
Topical application: Treatment of a localized surface site.
Toxicity: The degree to which a substance can harm humans or animals.
Transgenic plant: A plant that has a new genetically engineered DNA sequence present in every one of its cells. Genetically engineered plants are transgenic plants.
Translocation: Movement of a substance within the system of the organism or plant.
Transpiration: Loss of water from plant surfaces, primarily through stomata.
TWA: The time weighted average is the average concentration of a chemical in air over the total exposure time – usually an 8 h work day.
Tyrosine: A non-essential amino acid.

U

Ultra low volume (ULV) spray: Spray with the total volume rate of spray being very low (5 l/ha or less). The formulations in a ULV spray are applied undiluted.

Use site: The immediate environment where a pesticide (or its apparatus/equipment) is being mixed, loaded, applied, transported, stored or rinsed off or disposed of.

V

Vacuole: A membrane-bound (tonoplast) space, which is larger than a vesicle, that stores material, either dissolved in water or as a crystalline or flocculent mass. Vacuoles are key organelles for storage of compounds and osmoregulation in plants.
Vapour pressure: The property of a chemical to evaporate. The higher the vapour pressure the more volatile the chemical.
Vector: A vehicle for transporting a disease-producing organism (pathogen) from one host to another. In vector ecology, the most common vectors are insects and other arthropods.
Vermin: Pests; usually rats, mice or insects.
Vertebrate: An animal with a backbone or spinal column. Mammals, fish, birds, snakes and frogs are vertebrates.
Viricide: A substance that inactivates a virus completely and permanently.
Virus: A microorganism that can grow and reproduce only in living cells of other organisms. Often, viruses cause diseases in their hosts.
Viscosity: This is a measure of resistance of a fluid to flow. Viscosity usually increases as temperature decreases.
Volatile: The property of a solid/liquid substance to evaporate when exposed to air.
Volunteer weeds: Unwanted plants of a crop that grow from seed that remains on the field from a previous crop.

W

Warning: A signal word used on pesticide labels to designate a pesticide that is moderately toxic (toxicity Class II).
Water-based pesticides: Pesticides that use water as the only diluent or carrier.
Water-dispersible granule (WG): Formulation containing granules which readily disperse in water to form a suspension.
Water-dispersible powder (WP): Pesticide in a dry form with surfactant (mixed or coated), a fine solid carrier, for dispersion in water to form a suspension.
Water-soluble powder (SP): Powder formulation to be applied as a true solution of active ingredient after mixing with water but which may contain insoluble inert ingredients.
Water table: The upper level of the water-saturated zone in the ground.
Weed: A pest plant.
Weed control efficiency (WCE): The percentage reduction in the weed population in a plot that has been treated compared with an untreated plot. WCE denotes the efficiency of the applied treatment.
Wettable powder (WP): A dry (powder) preparation that is mixed with water to form a suspension that is used for spraying.
Wetting agent: A surfactant used in spray formulations to assist dispersion of a powder in the diluent or spreading of spray droplets on surfaces.
Winter annuals: Plants that germinate in the autumn or winter, do most of their growing in the spring, produce flower and seeds, and then die by early summer.
Withholding period: Minimum permissible time between the last application of a pesticide to a crop (including pasture) and harvesting for human consumption or grazing with livestock.
wt: Weight.

X

Xenobiotic substance: Compound that is present in an organism but is not a natural component of that organism (e.g. man-made contaminants in organisms).
Xylem: Part of the plant's vascular system adapted to the transport of water and solutes from the roots to aerial parts.

Z

Zero tolerance: Limit for a pesticide residue in food or feed which is assumed to be zero and therefore any detectable residue is deemed illegal.
Zygote: The cell resulting from the fusion of the male and female (haploid) gametes; the fertilized ovum.

Bibliography

Abel, S. and Theologis, A. (1996) Early genes and auxin action. *Plant Physiology* 111, 9–17.

Abeles, F.B. (1973) *Ethylene in Plant Biology*. Academic Press, New York.

Adams, A.J. and Lindquist, R.K. (1991) A review of the performance of air-assisted sprayers for use in glasshouses. *British Crop Production Council (BCPC) Monograph* 46, 227–235.

Adenle, V.O. and Cardwell, K.F. (2000) Seed transmission of *Peronosclerospora sorghi*, causal agent of maize downy mildew in Nigeria. *Plant Pathology* 49, 628–635.

Adrian, N.R., Robinson, J.A. and Suflita, J.M. (1994) Spatial variability in biodegradation rates as evidenced by methane production from an aquifer. *Applied and Environmental Microbiology* 60, 3632–3639.

Aelion, C.M., Dobbins, D.C. and Pfaender, F.K. (1989) Adaptation of aquifer microbial communities to the biodegradation of xenobiotic compounds: influence of substrate concentration and preexposure. *Environmental Toxicology and Chemistry* 8, 75–86.

Aga, D.S. (ed.) (2000) *Fate of Pharmaceuticals in the Environment and in Water Treatment Systems.* 219th American Chemical Society (ACS) National Meeting. ACS, San Francisco, California.

Agrios, G.N. (1997) *Plant Pathology*. Academic Press, San Diego, California.

Ahrens, W.H. (ed.) (1994) *Herbicide Handbook*, 7th edn. Weed Science Society of America, Champaign, Illinois, pp. 173–175, 313–318, 344.

Alabouvette, C. and Lemanceau, P. (1999) Joint action of microbials for disease control. In: Hall, F.R. and Menn, J.J. (eds) *Biopesticides – Use and Delivery*. Humana Press, Totowa, New Jersey, pp. 117–135.

Alberts, B., Bray, D., Johnson, A., Lewis, J., Raff, M., Roberts, K. and Walter, P. (1998) *Essential Cell Biology: an Introduction to the Molecular Biology of the Cell.* Garland Publications, New York, 630 pp.

Alberts, B., Johnson, A., Lewis, J., Raff, M., Roberts, K. and Walter, P. (2002) *Molecular Biology of the Cell.* Garland Science, Taylor & Francis, London, 1463 pp.

Albro, P.W., Lister, M.I., Cahe, K. *et al.* (1979) A radio-immunuassay for chlorinated dibenzo-p-dioxins. *Toxicology and Applied Pharmacology* 50, 137–146.

Aldridge, W.N. and Reiner, E. (1972) *Enzyme Inhibitors as Substrates: Interactions of Esterases with Esters of Organophosphorus and Carbamic Acids*, Vol XVI. North-Holland Publishing, Amsterdam, 328 pp.

Allan, G.G., Chopra, C.S., Friedhoff, J.F., Gara, R.I., Maggi, M.W., Neogi, A.N., Roberts, S.C. and Wilkens, R.M. (1973) Pesticides, pollution, and polymers. *ChemTech* 3, 171–178.

Allard, A.S. and Neilson, A.H. (1997) Bioremediation of organic wastes sites: a critical review of microbiological aspects. *International Biodeterioration and Biodegradation* 39, 253–285.

Ames, B.N., Profet, M. and Gold, L.S. (1990b) Nature's chemicals and synthetic chemicals: comparative toxicology. 3. *Proceedings of the National Academy of Sciences USA* 87, 7782–7786.

Ammerman, E., Lorenz, G., Schelberger, K., Wenderoth, B., Sauter, H. and Rentzea, C. (1992) BAS 490F – a broad-spectrum fungicide with a new mode of action. In: *Proceedings of the Brighton Crop Protection Conference – Pests and Diseases*. BCPC Publications, Croydon, UK, pp. 403–410.

Amsellem, Z., Zidack, N.K., Quimby, P.C. and Gressel, J. (1999) Long term dry preservation of viable mycelia of two mycoherbicidal organisms. *Crop Protection* 18, 643–649.

Anderson, M.E. (1995) What do we mean by 'dose'? *Inhalation Toxicology* 7, 909–915.

Andrews, J.H. (1990) Biological control in the phyllosphere: realistic goal or false hope? *Canadian Journal of Plant Pathology* 12, 300–307.

Anjaiah, V., Koedam, N., Nowak-Thompson, B., Loper, J.E., Hofte, M., Tambong, J.T. and Cornelis, P. (1998) Involvement of phenazines and anthranilate in the antagonism of *Pseudomonas aeruginosa* PNAI and Tn5 derivatives toward *Fusarium* spp. and *Pythium* spp. *Molecular Plant–Microbe Interactions* 11, 847–854.

Ankley, G.T., Tietge, J.E., DeFoe, D.L., Jensen, K.M., Holcombe, G.W., Durhan. E.J. and Diamond, S.A. (1997) Effects of ultraviolet light and methoprene on survival and development of *Rana pipiens*. *Environmental Toxicology and Chemistry* 17, 2530–2542.

Anonymous (1971) Borlaug's warning. *Nature* 233, 444.

Anonymous (1992) *Neem, a Tree for Solving Global Problems*. National Academy Press, Washington, DC, pp. 1–139.

Anonymous (1997) Code of conduct for the import and release of exotic biological control agents. *Biocontrol News and Information* 18, 119N–124N.

Anonymous (1999) Strategy of developing the pesticide industry or news in brief. *Pesticide Information* 25(2), 50–52.

Anonymous (2008) Average crop losses due to pests and cost-benefit ratio of pesticides use in different crops. In: Kishore, P. (ed.) *Crop Care, Official Magazine of Crop Care Federation of India*, 75pp.

Anonymous (2009) *Directorate of Plant Protection, Quarantine and Storage.* Faridabad, Haryana, India.

Anthony, N.M., Harrison, J.B. and Sattelle, D.B. (1993) *GABA Receptor Molecules of Insects.* Birkhauser Verlag, Basel, Switzerland, pp. 172–209.

Antolic, S., Salopek, B., Kojic-Prodic, B., Magnus, V. and Cohen, J.D. (1999) Structural characterization and auxin properties of dichlorinated indole-3-acetic acids. *Plant Growth Regulation* 27, 21–31.

Arias, H.R. (1997) Topology of ligand binding sites on the nicotinic acetylcholine receptor. *Brain Research Reviews* 25, 133–191.

Arima, K., Imanaka, H., Kousaka, M., Fukuda, A. and Tamura, G. (1965) Studies of pyrrolnitrin, a new antibiotic. I. Isolation and properties of pyrrolnitrin. *Journal of Antibiotics* 18, 201–204.

Ariyoshi, T., Shiiba, S., Hasegawa, H. and Arizono, K. (1990) Profile of metal-binding proteins and heme oxygenase in red carp treated with heavy metals, pesticides and surfactants. *Bulletin of Environmental Contamination and Toxicology* 44(4), 643–649.

Assaf, N.A. and Turco, R.F. (1994) Influence of carbon and nitrogen application on the mineralization of atrazine and its metabolites in soil. *Pesticide Science* 41, 47.

Atallah, Y.H., Whitacre, D.M. and Butz, R.G. (1981) Fate of hexachlorocyclopentadiene in the environment. In: Khan, M.A.Q. and Stanton, R.H. (eds) *Toxicology of Halogenated Hydrocarbons: Health and Ecological Effects.* Pergamon Press, Elmsford, New York, p. 344–355.

Atwal, A.S. and Dhaliwal, G.S. (2005) *Agricultural Pests of South Asia and Their Management.* Kalyani Publishers, New Delhi, India.

Austerweil, M. and Grinstein, A. (1997) Automatic pesticide application in greenhouses. *Phytopararasitica* 25, 71S–80S.

Babij, J., Zhu, Q., Brain, P. and Hollomon, D.W. (2000) Resistance risk assessment of cereal eyespot, *Tapesia yallundae* and *Tapesia acuformis*, to the anilinopyrimidine fungicide, cyprodinil. *European Journal of Plant Pathology* 106, 895–905.

Backman, P.A., Brannen, P.M. and Mahaffee, W.F. (1994) Plant response and disease control following seed inoculation with *Bacillus subtilis.* In: Ryder, M.H., Stephens, P.M. and Bowen, G.D. (eds) *Improving Plant Productivity with Rhizosphere Bacteria.* International Workshop on Plant Growth-promoting Bacteria. CSIRO, Adelaide, South Australia, pp. 3–8.

Backman, P.A., Wilson, M. and Murphy, J.F. (1997) Bacteria for biological control of plant diseases. In: Rechcigl, N.A. and Rechcigl, J.E. (eds) *Environmentally Safe Approaches to Crop Disease Control.* CRC Press, Boca Raton, Florida, pp. 95–109.

Baerselman, F. (1993) A long-term policy plan for crop protection: the Dutch Multi-year Crop Protection Plan (MJP-G), a contribution towards sustainable agriculture. In: Zadoks, J.C. (ed.) *Modern Crop Protection: Developments and Perspectives.* Wageningen Academic Publishers, Wageningen, The Netherlands, pp. 237–241.

Bai, D., Lummis, S.C.R., Leicht, W., Breer, H. and Sattelle, D.B. (1991) Actions of imadacloprid and related nitromethylene on cholinergic receptors of an identified insect motor neurone. *Pesticide Science* 33, 197–204.

Bailey, A.M., Burden, R.S., James, C.S., Keon, J.P.R., Croxen, R., Bard, M. and Hargreaves, J.A. (1994) Isolation of the erg2 gene, encoding sterol 8-7 isomerase from the maize smut pathogen *Ustilago maydis. Experimental Mycology* 18, 87–92.

Baker, K.F. and Cook, R.J. (1974) *Biological Control of Plant Pathogens.* W.H. Freeman, San Francisco, California.

Baker, R. (1987) Mycoparasitism: ecology and physiology. *Journal of Plant Pathology* 9, 370–379.

Baker, R.W. and Lonsdale, H.K. (1974) Controlled release: mechanisms and rates. In: Tanquary, A.C. and Lacy, R.E. (eds) *Controlled Release of Biologically Active Agents.* Plenum Press, New York, pp. 15–71.

Baldwin, F.L. (2000) Transgenic crops: a view from the US Extension Service. *Pest Management Science* 56, 584–585.

Ball, W.E. and French, O.C. (1935) Sulfuric acid for control of weeds. *California Agricultural Experiment Station Report* 596.

Ballschmiter, K., Buchert, H. and Bihler, S. (1981) Baseline studies of the global pollution. VI. The pattern of pollution by chlorinated hydrocarbons in the Caspian Sea. *Fresenius' Zeitschrift für Analytische Chemie* 306, 323.

Ballschmiter, K., Rappe, C. and Buser, H.R. (1989) Chemical properties, analytical methods and environmental levels of PCBs, PCTs, PCNs and PBBs. In: Kimbrough, R.D. and Jensen, A.A. (eds) *Halogenated Biphenyls, Terphenyls, Naphthalenes, Dibenzodioxins and Related Products*, 2nd edn. Elsevier, Amsterdam, pp. 47–70.

Bals, E.J. (1970) Rotary atomization. *Agricultural Aviation* 12, 85–90.

Bals, E.J. (1975) The importance of controlled droplet application (CDA) in pesticide applications. In: *Proceedings of the 8th British Insecticide and Fungicide Conference.* BCPC Publications, Croydon, UK, pp. 153–160.

Bandurski, R.S., Schulze, A., Dayanandan, P. and Kaufman, P. (1984) Response to gravity by *Zea mays* seedlings. I. Time course of the response. *Plant Physiology* 74, 284–288.

Barbier-Brygoo, H. (1995) Tracking auxin receptors using functional approaches. *Critical Reviews in Plant Sciences* 14, 1–25.

Barlow, F., Elliott, M., Farnham, A.W., Hadaway, A.B., Janes, N.F., Needham, P.H. and Wickham, J.C. (1971) Insecticidal activity of the pyrethrins and related compounds. IV. Essential features for insecticidal activity in chrysanthemates and related cyclopropane esters. *Pesticide Science* 2, 115–118.

Barnett, G.S. and Matthews, G.A. (1992) Effect of different fan nozzles and spray liquids on droplet spectra with special reference to drift control. *International Pest Control* 34, 81–83.

Barnett, H.L. and Lilly, V.G. (1962) A destructive mycoparasite: *Gliocladium roseum. Mycologia* 54, 72–77.

Bateman, R.P. (1997) The development of a mycoinsecticide for the control of locusts and grasshoppers. *Outlook on Agriculture* 26, 13–18.

Batie, S.S. and Healy, R.G. (1983) The future of American Agriculture. *Scientific American* 248(2), 45–53.

Baum, J.A., Johnson, T.B. and Carlton, B.C. (1999) *Bacillus thuringiensis* natural and recombinant bioinsecticide products. In: Hall, F.R. and Menn, J.J. (eds) *Biopesticides – Use and Delivery.* Humana Press, Totowa, New Jersey, pp. 189–209.

Bausmith, D.S. and Neufeld, R.D. (1999) Soil biodegradation of propylene glycol based aircraft de-icing fluids. *Water Environment Research* 71, 459–464.

Bear, J. (1979) *Hydraulics of Groundwater.* McGraw-Hill, New York.

Beilstein Information (2013) Beilstein Data Base. Beilstein Information Systems, GmbH. Available at: http://www.elsevier.com/about/press-releases/science-and-technology/elsevier-acquires-the-beilstein-database,-worlds-leading-archive-of-chemical-data (accessed 17 December 2013).

Belden, J. and Lady, M. (2000) Impact of atrazine on organophosphate insecticide toxicity. *Environmental Toxicology and Chemistry* 19, 2266–2274.

Benezet, H.J. and Matsumura, F. (1973) Isomerization of gamma-BHC to alpha-BHC in the environment. *Nature* 243, 480.

Benn, W and Lange, R. (1988) Cold atomization from the viewpoint of new technology for effective control of pathogens in glassshouse installations. *Gartenbau* 35, 1. (in German)

Benson, J.A. (1992) Electrophysiological pharmacology of the nicotinic and muscarinic cholinergic responses of isolated neuronal somata from locust thoracic ganglia. *Journal of Experimental Biology* 170, 203–233.

Benveniste, P. and Rahier, A. (1992) Target sites of sterol biosynthesis inhibitors in plants. In: Koller, W. (ed.) (1992) *Target Sites of Fungicide Action*. CRC Press, Boca Raton, Florida, pp. 207–226.

Beraud, J.M., Claument, J. and Montury, A. (1991) ICIA0051, a new herbicide for the control of annual weeds in maize. In: *Proceedings of the Brighton Crop Protection Conference – Weeds*. BCPC Publications, Croydon, UK, pp. 51–56.

Berezkin, V.G. and Buzaev, V.V. (1997) New thin layer chromatography plate with closed sorbent layer and details of the application. *Journal of Chromatography A* 758, 125–134.

Bergman, F., Wilson, I.B. and Nachmansohn, D. (1950) Acetylcholinesterase. IX. Structural features determining the inhibition by amino acids and related compounds. *Journal of Biological Chemistry* 186, 693–703.

Berlin, M. (1986) Mercury. In: Friberg, L., Nordberg, G. and Nordman, C. (eds) *Specific Metals* (Vol. 2), 2nd edn. Elsevier, Amsterdam, pp. 386–445.

Best, G.A and Ruthven, A.D. (1995) *Pesticides – Developments, Impacts and Controls*. Woodhead Publishing, Sawston, Cambridge, UK.

Bhowmik, P.C. (1988) Cinmethylin for weed control in soybeans, *Glycine max*. *Weed Science* 36, 678–682.

Bhupinder, P.S., Khambay, B.B. and Bromilow, R.H. (2000) Discovery of pesticides. In: Hall, N. (ed.) *The New Chemistry*. Cambridge University Press, Cambridge, UK, pp. 232–258.

Bjorkqvist, S. (2008) *Tracking Pesticides and Promoting Alternatives in Laos*. Pesticide Action Network (PAN) North America, Oakland, California.

Bjugstad, N. and Torgrimsen, T. (1996) Operator safety and plant deposits when using pesticides in greenhouses. *Journal of Agricultural Engineering Research* 65, 205–212.

Blanchflower, W.J., Hughes, P.J. and Kennedy, D.G. (1997) Determination of nicarbazin in eggs by liquid chromatography-atmospheric pressure chemical ionization mass spectrometry. *Journal of AOAC (Association of Official Analytical Chemists) International* 80, 1177.

Blechert, S., Brodschelm, W., Holder, S., Kammerer, S., Kutchan, T.M., Mueller, M.J., Xia, Z. and Zenk, M.H. (1995) The octadecanoic pathway: signal molecules for the regulation of secondary pathways. *Proceedings of the National Academy of Sciences USA* 92, 4099–4105.

Bloomquist, J.R. (1993) Toxicology, mode of action and target site-mediated resistance to insecticides acting on chloride channels. *Comparative Biochemistry and Physiology* 106, 301–314.

Bloomquist, J.R. (1996) Ion channels as targets for insecticides. *Annual Review of Entomology* 41, 163–190.

Bloomquist, J.R. (2003) Chloride channels as tools for developing selective insecticides. *Archives of Insect Biochemistry and Physiology* 54, 145–146.

Blumton, A.K., Fraser, J.D., Young, R.W., Goodbred, S., Porter, S.L. and Luukkonen, D.L. (1990) Pesticide and PCB residues for loggerhead shrikes in the Shenandoah Valley, Virginia, 1985–1988. *Bulletin of Environmental Contamination and Toxicology* 45, 697.

Blus, L.J. (1995) Organochlorine pesticides. In: Hoffman, D.J., Rattner B.A., Burton, G.A. and Cairns, J. (eds) *Handbook of Ecotoxicology*. Lewis Publishers, Boca Raton, Florida, pp. 275–300.

Boettner, E.A., Ball, G.L., Hollingsworth, Z. and Aquino, R. (1981) Organic and organotin compounds leached from PVC and CPVC pipes. *US Environmental Protection Agency (EPA) Report* EPA-600/1-81-062. EPA, Washington, DC.

Bonning, B.C. and Hammock, B.D. (1996) Development of recombinant baculoviruses for insect control. *Annual Review of Entomology* 41, 191–210.

Borazjani, A., Graves, C.H. Jr and Hedin, P.A. (1985) Occurence of juglonge in various tissues of pecan and related species. *Phytopathology* 75, 1419–1421.

Borchers, J.G. and Perry, D.A. (1992) The influence of soil texture and aggregation on carbon and nitrogen dynamics in Southwest Oregon forests and clearcuts. *Canadian Journal of Forest Research* 22, 298–305.

Bottger, M., Engvild, K.C. and Soll, H. (1978) Growth of *Avena* coleoptiles and pH drop of protoplast suspensions induced by chlorinated indoleacetic acids. *Planta* 140, 89–92.

Bouchard, D.C., Lavy, T.L. and Marx, D.B. (1982) Fate of metribuzin, metolachlor, and fluometuron in soil. *Weed Science* 30, 629–632.

Boude, M.R. (1982) Epidemiology of downy mildew disease of maize, sorghum and pearl millet. *Tropical Pest Management* 28, 220–223.

Boyette, C.D. (2000) The bioherbicide approach: using plant pathogens to control weeds. In: Cobb, A.H. and Kirkwood, R.C. (eds) *Herbicides and their Mechanisms of Action*. Sheffield Academic Press, Sheffield, UK, pp. 134–152.

Boyette, C.D. and Abbas, H.K. (1994) Host range alteration of the bioherbicidal fungus *Alternaria crassa* with fruit pectin and plant filtrates. *Weed Science* 42, 487–491.

Boyette, C.D., Quimby, P.C. Jr, Caesar, A.J., Birdsall, J.L., Connick, W.J. Jr and Daigle, D.J. (1996) Adjuvants, formulations and spraying systems for improvement of mycoherbicides. *Weed Technology* 10, 637–644.

Brandon, D.L., Binder, R.G., Bates, A.H. and Montague, W.C. (1992) Monoclonal antibody-based ELISA for thiabendazole in liver. *Journal of Agricultural and Food Chemistry* 40, 1722–1726.

Braverman, M.P., Lavy, T.L. and Barnes, C.J. (1986) The degradation and bioactivity of metolachlor in the soil. *Weed Science* 34, 479–484.

Bravo, P. (2007) *GMO Crops Mean More Pesticides*. Red de Acción sobre Plaguicidas (RAP) y sus Alternativas en América Latina. Pesticide Action Network, Dakar Fann, Sénégal.

Breeze, V.G. and West, C.J. (1987) Long- and short-term effects of vapour of the herbicide 2,4-D butyl on the growth of tomato plants. *Weed Research* 27, 13–21.

Breitenbach, J., Zhu, C. and Sandmann, G. (2001) Bleaching herbicide norflurazon inhibits phytoene desaturase by competition with the cofactors. *Journal of Agricultural and Food Chemistry* 49, 5270–5272.

Brinkman, U.A.T. and De Kok, A. (1980) Production, properties and usage. In: Kimbrough, R.D. (ed.) *Halogenated Biphenyls, Terphenyls, Naphthalenes, Dibenzodioxins and Related Products*. Elsevier/North-Holland Biomedical Press, Amsterdam, p. 1–40.

Briskin, D.P. and Hanson, J.B. (1992) How does the plant plasma membrane H^+-ATPase pump protons? *Journal of Experimental Botany* 43, 269–289.

Bromilow, R.H. and Chamberlain, K. (1991) Pathways and mechanisms of transport of herbicides in plants. In: Kirkwood R.C. (ed.) *Herbicides and their Mechanisms of Action*. Plenum Press, New York, pp. 245–284.

Bromilow, R.H. and Chamberlain, K. (1995) Principles governing uptake and transport of chemicals. In: Trapp, S. and McFarlane, J.C. (eds) *Plant Contamination: Modeling a Simulation of Organic Chemical Processes*. CRC Press, Boca Raton, Florida, pp. 37–68.

Brooker, N.L., Mischke, C.F., Patterson, C.D., Mischke, S., Bruckart, W.L. and Lydon, J. (1996) Pathogenicity of *bar*-transformed *Colletotrichum gloeosporioides* f. sp. *aeschynomene*. *Biological Control* 7, 159–166.

Brooks, W.M. (1988) Entomogenous protozoa. In: Ignoffo, C.M. (ed.) *Handbook of Natural Pesticides*, Vol. V, *Microbial Insecticides*. CRC Press, Boca Raton, Florida, pp. 1–149.

Brown, A.W.A. (1958) *Insecticide Resistance in Arthropods*. World Health Organization, Geneva.

Brown, A.W.A. (1978) *Ecology of Pesticides*. Wiley, New York.

Brown, D.E., Rashotte, A.M., Murphy, A.S., Normanly, J., Tague, B.W., Peer, W.A., Taiz, L. and Muday, G.K. (2001) Flavonoids act as negative regulators of auxin transport *in vivo* in *Arabidopsis*. *Plant Physiology* 126, 524–535.

Brown, D.P. (1987) Mortality of workers exposed to polychlorinated biphenyls – an update. *Archives of Environmental Health* 43, 333.

Brown, D.P. and Jones, M. (1981) Mortality and industrial hygiene study of workers exposed to polychlorinated biphenyls. *Archives of Environmental Health* 36, 120.

Brown, V. (1988) *Acute and Sub-acute Toxicology*. Hodder & Stoughton, London, 125 pp.

Bucha, H.C. and Todd, C.W. (1951) 3-(p-Chlorophenyl)-1,1-dimethylurea; a new herbicide. *Science* 114, 493–494.

Buckingham, S.D., Balk, M.L., Lummis, S.C.R., Jewess, P. and Sattelle, D.B. (1995) Actions of nitromethylenes on an or-bungarotoxin-sensitive neuronal nicotinic acetylcholine receptor. *Neuropharmacology* 34, 591–597.

Buckingham, S.D., Lapied, B., Corronc, H.L., Grolleau, F. and Sattelle, D.B. (1997) Imidacloprid actions on insect neuronal acetylcholine receptors. *Journal of Experimental Biology* 200, 2685–2692.

Bull, C.T. (1999) Allelopathic interactions in the biological control of postharvest plant pathogens. In: Inderjit, Dakshini, K.M.M. and Foy, C.L. (eds) *Principles and Practices in Chemical Ecology*. CRC Press, Boca Raton, Florida, pp. 517–530.

Burg, R.W., Miller, B.M., Baker, E.E., Birnbaum, J., Currie, S.A., Hartman, R., Kong, Y.L., Monaghan, R.L., Olson, G., Putter, I., Tunac, J.B., Wallick, H., Stapley, E.O., Oiwa, R. and Omura, S. (1979) Avermectins, new family of potent anthelmintic agents: producing organism and fermentation. *Antimicrobial Agents and Chemotherapy* 15(3), 361–367.

Burkhard, N. and Guth, J.A. (1981) Rate of volatilisation of pesticides from soil surfaces. *Pesticide Science* 12, 37–44.

Burnet, W.M.W., Christopher, J.T., Holtum, J.A.M. and Powles, S.B. (1994a) Identification of two mechanisms of sulfonylurea resistance within one population of rigid ryegrass (*Lolium rigidum*) using a selective germination medium. *Weed Science* 42, 468–473.

Burström, H. (1950) Studies on growth and metabolism of roots: IV. Positive and negative auxin effects on cell elongation. *Journal of Plant Physiology* 3, 277–292.

Buta, J.G. and Moline, H.E. (1998) Methyl jasmonate extends shelf life and reduces microbial contamination of fresh-cut celery and peppers. *Journal of Agricultural and Food Chemistry* 46, 1253–1256.

Byrne, F.J. and Devonshire, A.L. (1993) Insensitive acetylcholinesterase and esterase polymorphism in susceptible and resistant populations of the tobacco whitefly *Bemisia tabaci* (Genn.). *Pesticide Biochemistry and Physiology* 45, 34–42.

Cabanne, F., LeFebvre, A. and Scalla, R. (1987) Behaviour of the herbicide EL-107 in wheat and rape grown under controlled conditions. *Weed Research* 27, 135–142.

Cadogan, B.L., Sundaram, A., Mickle, R., Robinson, A.G., Knowles, K.R. and Scharbach, R.D. (1998) Efficacy of tebufenozide sprayed aerially with an anti-drift technique that uses only upwind atomizers. *Crop Protection* 17, 315–321.

Cahill, M., Gorman, K., Day, S. and Denholm, I. (1996) Baseline determination and detection of resistance to imidacloprid in *Bemisia tabaci* (Homoptera: Aleyrodidae). *Bulletin of Entomological Research* 86, 343–349.

Camara, M. and Sylla C.H. (2011) A big step forward for pesticide dealers in Mali. *Pesticides News* 92 (June).

Campbell, M.A., Bandiera, S., Robertson, L., Parkinson, A. and Safe, S. (1981) Octachloronapthalene induction of hepatic microsomal aryl hydrocarbon hydroxylase activity in the immature male rat. *Toxicology* 22, 123–132.

Cannon, R.J.C. (1996) *Bacillus thuringiensis* use in agriculture: a molecular perspective. *Biological Reviews* 71, 561–636.

Canny, M.J. (1995) Apoplastic water and solute movement: new rules for an old space. *Annual Review of Plant Physiology and Plant Molecular Biology* 46, 215–236.

Capinera, J.L. (ed.) (2008) *Encyclopedia of Entomology*, 2nd edn. Springer Science+Business Media, Dordrecht, The Netherlands.

Capping, L.G. (2009) *The Manual of Biocontrol Agents*. British Crop Production Council (BCPC), Farnham, Surrey, UK.

Carde, R.M. and Minks, A.L. (1997) *Insect Pheromone Research*. Champman & Hall, London, 864 pp.

Carlton, B.C. (1993) Genetics of BT insecticidal crystal proteins and strategies for the construction of improved strains. In: Duke, S.O., Menn, J.J. and Plimmer, J. (eds) *Pest Control With Enhanced Environmental Safety*. ACS Symposium Series 524. American Chemical Society (ACS), Washington, DC, pp. 326–337.

Carr, C.S. and Hughes, J.B. (1998) Enrichment of high rate PCE dechlorination and comparative study of lactate, methanol, and hydrogen as electron donors to sustain activity. *Environmental Science & Technology* 32, 1817–1824.

Casida, J.E. (ed.) (1973) *Pyrethrum, the Natural Insecticide*. Academic Press, London, 329 pp.

Casida, J.E. and Quistad, G.B. (1998) Golden age of insecticide research: past, present, or future? *Annual Review of Entomology* 43, 1–16.

Castelli, M.G., Martelli, G.P., Spagone, C., Capellini, L. and Fanelli, R. (1983) Quantitative determination of PCBs in marine organisms analyzed by high resolution gas chromatography selected ion monitoring. *Chemosphere* 12, 291–298.

Castilla, N. and Hernandez, J. (1995) Protected cultivation in the Mediterranean area. *Plasticulture* 107, 13–20.

Cecil, A. (1997) Modified spray nozzle design reduces drift whilst maintaining chemical coverages. In: *Proceedings of the Brighton Crop Protection Conference*. BCPC Publications, Croydon, UK, pp. 543–548.

Chadwick, R.W., Copeland, M.F. and Froelich, R. (1983) Chlorobenzene-impaired lindane metabolism and the effect of pretreatment with chlorobenzene, lindane, or chlorobenzene plus lindane. *Journal of Toxicology and Environmental Health* 12, 599–610.

Chalfie, M., Tu, Y., Euskirchen, G., Ward, W.W. and Prasher, D.C. (1994) Green fluorescent protein as a marker for gene expression. *Science* 263, 802–805.

Chamberlain, P. and Symes, K.C. in Karsa, D.R. and Stephenson, R.A. (eds) (1993) *Encapsulation and Controlled Release*. Royal Society of Chemistry, Cambridge, UK, pp. 131–140.

Chambers, J.E. and Levi, P.E. (eds) (1992) *Organophosphates Chemistry, Fate and Effects*. Academic Press, San Diego, California.

Chambers, J.E. and Yardbrough, J. (1982) *Effects of Chronic Exposures to Pesticides*. Raven Press, New York, 250 pp.

Champagnat, P. (1961) Dominance apicale. Topisme. Epinastie. In: Ruhland, W. (ed.) *Encyclopedia of Plant Physiology*, Vol. 14. Springer, Berlin, pp. 872–908.

Chandran, R.S. and Singh, M. (1999) Reduction of norflurazon leaching in a sandy soil by adjuvants. *Bulletin of Environmental Contamination and Toxicology* 62(3), 315–323.

Chang, F.Y., Stephenson, G.R. and Smith, L.W. (1971) Influence of herbicides on insecticide metabolism in leaf tissue. *Journal of Agricultural and Food Chemistry* 19(6), 1187–1190.

Chao, S.L., Dennehy, T.J. and Casida, J.E. (1997) Whitefly (Hemiptera: Aleyrididae) binding site for imidacloprid and related insecticides: a putative nicotine acetylcholine receptor. *Journal of Economic Entomology* 90, 879–882.

Chatonnet, A. and Lockridge, O. (1989) Comparison of butyrylcholinesterase and acetylcholinesterase. *Biochemical Journal* 260, 625–634.

Chattopadhyay, S.B. and Mustafee, T. (2008) *Plant Diseases and Their Management*. Aditya Books, New Delhi, India.

Chen, J.-G., Ullah, H., Young, J.C., Sussman, M.R. & Jones, A.M. (2001) The role of auxin-binding protein 1 in the expansion of tobacco leaf cells. *Genes & Development* 15, 902–911.

Chen, L.H., Long, G.W. and Tien, Y.S. (1992) Development of labour-saving and low-cost auto-fogsprayer for plastic house cluture. *Bulletin of the Taichung District Agricultural Improvement Station* 35, 11–24. (in Chinese)

Chen, P.H., Chang, K.T. and Lu, Y.D. (1981) Polychlorinated biphenyls and polychlorinated dibenzofurans in the toxic rice-bran oil that caused PCB poisoning in Taichung. *Bulletin of Environmental Contamination and Toxicology* 26, 489–495.

Chet, I. and Inbar, J. (1994) Biological control of fungal pathogens. *Applied Biochemistry and Biotechnology* 48, 37–43.

Chisholm, C.J., Bottoms, D.J., Dwyer, M.J., Lines, J.A. and Whyte, R.T. (1992) Safety, health and hygiene in agriculture. *Safety Science* 15(4–6), 225–248.

Chism, W.J., Bingham, S.W. and Shaver, R.L. (1991) Uptake, translocation and metabolism of quinclorac in two grass species. *Weed Technology* 5, 771–775.

Choi, K.W., Han, O.H., Lee, H.J., Yun, Y.H., Moon, Y.H., Kim, M., Kuk, Y.I., Han, S.U. and Guh, J.O. (1998) Generation of resistance to the diphenyl ether herbicide, oxyfluorfen, via expression of the *Bacillus subtilis* protoporphyrinogen oxidase gene in transgenic tobacco plants. *Bioscience, Biotechnology, and Biochemistry* 62, 558–560.

Christensen, C.M. (ed.) (1982) *Storage of Cereal Grains and Their Products*, 3rd edn. American Association of Cereal Chemists, St Paul, Minnesota.

Ciesielski, T. (1872) Untersuchungen über die Abwärtskrümmung der Wurzel. *Beitrage zur Biologie der Pflanzen* 1, 1–30.

Clare, B.G. (1993) *Agrobacterium*: biological plant disease control. In: Kim L. (ed.) *Advanced Engineered Pesticides*. Marcel Dekker, New York, pp. 129–146.

Clark, D.R. Jr (1990) Dicofol (Kelthane) as an environmental contaminant: a review. *United States Fish and Wildlife Service Technical Report* No. 29, pp. 1–37.

Clark, J.M. with Scott, J.G., Campos, F. and Bloomquist, J.R. (1995) Resistance to avermectins: extent, mechanisms, and management implications. *Annual Review of Entomology* 40, 1–30.

Clark, L. (1998) Review of bird repellents. In: Baker, R.O. and Crabb, A.C. (eds) *Proceedings of the 18th Vertebrate Pest Conference*. University of California, Davis, California, pp. 330–337.

Clarke, B.S. and Jewess, P.J. (1990) Inhibition of chitin synthesis in *Spodoptera littoralis* larvae by flufenoxuron, teflubenzuron and diflubenzuron. *Pesticide Science* 28, 377–388.

Clarke, J.R., Marquardt, R.R., Oosterveld, A., Frohlich, A.A., Madrid, F.J. and Dawood, M. (1993) Development of a quantitative and sensitive enzyme-linked immunosorbent assay for ochratoxin A using antibodies from the yolk of the laying hen. *Journal of Agricultural and Food Chemistry* 41, 1784–1789.

Clein, J.S. and Schimel, J.P. (1994) Reduction in microbial activity in birch litter due to drying and rewetting events. *Soil Biology and Biochemistry* 26, 403–406.

Clemons, G.P. and Sisler, H.D. (1969) Formation of a fungitoxic derivative from Benlate. *Phytopathology* 59, 705–706.

Cline, S. and Neely, D. (1984) Relationship between juvenile-leaf resistance to anthracnose and the presence of juglone and hydrojuglone glucoside in black walnut. *Phytopathology* 74, 185–188.

Coats, J.R. (1993) What happens to degradable pesticides? *ChemTech* 23, 25–29.

Cockerline, R., Shilling, M. and Safe, S. (1981) Polychlorinated naphthalenes as hepatic microsomal enzyme inducers in the immature male rat. *General Pharmacology* 12, 83.

Cohen, E. (1987) Chitin biochemistry: synthesis and inhibition. *Annual Review of Entomology* 322, 71–93.

Cohen, E. and Casida, J.E. (1980a) Properties of *Tribolium* gut chitin synthetase. *Pesticide Biochemistry and Physiology* 13, 121–128.

Cohen, E. and Casida, J.E. (1980b) Inhibition of *Tribolium* gut chitin synthetase. *Pesticide Biochemistry and Physiology* 13, 129–136.

Combellack, J.H., Western, N.M. and Richardson, R.G. (1996) A comparison of the drift potential of a novel twin fluid nozzle with conventional low volume flat fan nozzles when using a range of adjuvants. *Crop Protection* 15, 147–152.

Cook, R.J. and Baker, K.F. (1983) *The Nature and Practice of Biological Control of Plant Pathogens*. APS Press, St Paul, Minnesota.

Cooke, M., Roberts, D.J. and Tillett, M.E. (1980) Polychlorinated naphthalenes, polychlorinated biphenyls and DDT residues in British birds of prey. *The Science of the Total Environment* 15, 237–246.

Copping, L.G. (ed.) (1998) *The Biopesticide Manual*, 1st edn. British Crop Production Council, Farnham, UK, pp. 1–333.

Corbell, N. and Loper, J.E. (1995) A global regulator of secondary metabolite production in *Pseudomonas fluorescens* Pf5. *Journal of Bacteriology* 177, 6230–6236.

Corcia Di, A., Crescenzi, C., Guerriero, E. and Samperi, R. (1997) Ultratrace determination of atrazine and its six major degradation products in water by solid-phase extraction and liquid chromatography–electrospray/mass spectrometry. *Environmental Science & Technology* 31, 1658–1663.

Cosgrove, D.J. (1999) Enzymes and other agents that enhance cell wall extensibility. *Annual Review of Plant Biology* 50, 391–417.

Cottrell, H.J. and Heywood, B.J. (1965) Benzenesulphonylcarbamates: new herbicides. *Nature* (London) 207, 655–656.

Coulson, J.R., Vail, P.V., Dix, M.E. *et al.* (2001) 100 years of biological control research and development in the United States Department of Agriculture, 1883–1993. *USDA/ARS Admin Report 2000–1*. United States Department of Agriculture/Agricultural Research Service (USDA/ARS), Washington, DC.

Council on Agricultural Science and Technology (CAST) (1989) Mycotoxins: economics and health risks. *Task Force Report* no. 116. CAST, Ames, Iowa.

Cox, C. (1997) Herbicide fact sheet – dichlobenil. *Journal of Pesticide Reform* 17(1).

Cramp, S. (1973) The effects of pesticides on British wildlife. *British Veterinary Journal* 129, 315–323.

Crombie, L. (1980) Chemistry and biosynthesis of natural pyrethrins. *Pesticide Science* 11, 102–118.

Crouse, G.D., Sparks, T.C., DeAmicis, C.V. and Kirst, H.A. (1999) Chemistry and insecticidal activity of the spinosyns. In: *Pesticide Chemistry: the Food–Environment Challenge*. Special Publication Royal Society of Chemistry 233. Royal Society of Chemistry, Cambridge, UK, pp. 155–166.

Cummings, A.M. and Gray, L.E. (1989) Antifertility effect of methoxychlor in female rats: dose- and time-dependent blockade of pregnancy. *Toxicology and Applied Pharmacology* 94, 454–462.

Curle, P.D., Emmerson, C.D., Gregory, A.H., Hartmann, J. and Nixon, P. (1998) Packaging of agrochemicals. In: Knowles, D.A. (ed.) *Chemistry and Technology of Agrochemical Formulations*. Kluwer, Dordrecht, The Netherlands, pp. 264–301.

Darvas, B. and Polgar, L.A. (1998) Novel-type insecticides: specificity and effects on non-target organisms. In: Ishaaya, I. and Degheele, D. (eds) *Insecticides with Novel Modes of Action*. Springer, Berlin, pp. 198–208.

Darwin, C. (1881) *The Power of Movement in Plants*. D. Appleton and Company, New York.

Datta, Surajit K. De (1981) *Principles and Practices of Rice Production*. A Wiley-Interscience Publication. Wiley, New York.

Davies, J.H. (1985) The pyrethroids: a historical introduction. In: Leahey, J.P. (ed.) *The Pyrethroid Insecticides*. Taylor & Francis, London, pp. 1–41.

Davis, D. and Safe, S. (1990) Immunosuppressive activities of polychlorinated biphenyls in C57BL/6N mice: structure-activity relationships as Ah receptor agonists and partial antagonists. *Toxicology* 63, 97–111.

Davis, J. (1986) *Statistics and Data Analysis in Geology*, 2nd edn. Wiley, New York.

Davis, P.J. (ed.) (1995) *Plant Hormones: Physiology, Biochemistry and Molecular Biology*. Kluwer Academic Publishers, Dordrecht.

De Cock, A. and Degheele, D. (1998) Buprofezin: a novel chitin synthesis inhibitor affecting specifically planthoppers, whiteflies and scale insects. In: Ishaaya, I. and Degheele, D. (eds) *Insecticides with Novel Modes of Action*. Springer, Berlin, pp. 84–87.

De Freitas, P.L., Zobel, R.W. and Snyder, V.A. (1999) Corn root growth in soil columns with artificially constructed aggregates. *Crop Science* 39, 725–730.

Dekker, J. (1972) Resistance. In: March, E. (ed.) *Systemic Fungicides*. Wiley, New York, pp. 156–714.

DeSavingny, D. and Ivy, E.E. (1974) Microencapsulated pesticides. In: Vandegaer, J.E. (ed.) *Microencapsulation Processes and Applications*. Plenum Press, New York, pp. 89–95.

Devine, G.L., Harling, Z.K., Scarr, A.W. and Devonshire, A.W. (1996) Lethal and sublethal effects of imidacloprid on nicotine-tolerant *Myzus nicotianae* and *Myzus persicae*. *Pesticide Science* 48, 57–62.

Devonshire, A.L. (1989) Insecticide resistance in *Myzus persicae*: from field to gene and back again. *Pesticide Science* 26, 375–382.

Devonshire, A.L. and Field, L.M. (1991) Gene amplification and insecticide resistance. *Annual Review of Entomology* 36, 1–23.

Dey, P.M. and Harborne, J.B. (eds) (1997) *Plant Biochemistry*. Academic Press, New York, p. 415.

Dhawan, S., Erickson, L.E. and Fan, L.T. (1993) Model development and simulation of bioremediation in soil beds with aggregates. *Groundwater* 31, 271–284.

Dibble, J.T. and Bartha, R. (1979) Effect of environmental parameters on the biodegradation of oil sludge. *Applied and Environmental Microbiology* 37, 729–739.

Diehl, S.V., Graves, C.H. Jr and Hedin, P.A. (1992) Cytochemical responses of pecan to *Cladosporium caryigenum*: *in situ* localization and quantification of fungitoxic phenols. *Phytopathology* 82, 1037–1041.

Diel, F., Horr, B., Borck, H., Savtchenko, H., Mitsche, T. and Diel, E. (1999) Pyrethroids and piperonyl-butoxide affect human T-lymphocytes *in vitro*. *Toxicology Letters* 107, 65–74.

Doble, S.J., Matthews, G.A., Rutherford, I. and Southcombe, E.S.E. (1985) A system for classifying hydraulic and other atomizers into categories of spray quality. In: *Proceedings of the British Crop Protection Conference – Weed*, 3. BCPC Publications, Croydon, UK, pp. 1125–1133.

Dobrat, W. and Martijn, A. (eds) (1995) 437/Glufosinate ammonium. In: *CIPAC Handbook G, Analysis of Technical and Formulated Pesticides*. Black Bear Press, Cambridge, UK, pp. 89–93.

Donham, K.J. (1990) Prologue: agricultural occupational and environmental health: policy strategies for the future. *American Journal of Industrial Medicine* 18(2), 107–119.

Donham, K.J. and Horvath, E. (1988) Agricultural occupational medicine. In: Zenz, C. (ed.) *Occupational Medicine*, 2nd edn. Yearbook Medical Publishers, Chicago, pp. 933–957.

Donzel, B. and Dorn, E. (1992) Appraisal of the fate of agrochemicals in plants and soil; a cost effective integrated approach. *Pure and Applied Chemistry* 64, 1965–1976.

Dowing, A.L. and Truesdale, G.A. (1955) Some factors affecting the rate of solution of oxygen in water. *Journal of Applied Chemistry* 5, 570–581.

Duggan, R.E. and Weatherwax, J.R. (1967) Dietary intake of pesticide chemicals. *Science* 157, 1006–1010.

Duke, S.O. and Kenyon, W.H. (1986) Effects of dimethazone (FMC 57020) on chloroplast development: II. Pigment synthesis and photosynthetic function in cowpea (*Vigna unguiculata* L.) primary leaves. *Pesticide Biochemistry and Physiology* 25, 11–18.

Duke, S.O., Kenyon, W.H. and Paul, R.N. (1985) Clomazone effects on chloroplast development in pitted morning glory (*Ipomoea lacunosa*) cotyledons. *Weed Science* 33, 786–794.

Duke, S.O., Dayan, F.E., Romagni, J.G. and Rimando, A.M. (2000) Natural products as sources of herbicides: current status and future trends. *Weed Research* 40, 99–111.

Dukes, J., Zhong, H., Greer, M., Hester, P., Hogan, D. and Barber, J.A.S. (2004a) A comparison of two spray nozzle systems used to aerially apply the ultra-low-volume adulticide fenthion. *Journal of the American Mosquito Control Association* 20(1), 27–35.

Dukes, J., Zong, H., Greer, M., Hester, P., Hogan, D. and Barber, J.A.S. (2004b) A comparison of two ultra-low-volume adulticide fenthion. *Journal of the American Mosquito Control Association* 20(1), 36–44.

Dunne, C., Crowley, J.J., Moënne-Loccoz, Y., Dowling, D.N., de Bruijn, F.J. and O'Gara, F. (1998) Biological control of *Pythium ultimum* by *Stenotrophomonas maltophilia* W81. *Plant Pathology* 47, 299–307.

Dyke, M.I. van and Prosser, J.I. (2000) Enhanced survival of *Pseudomonas fluorescens* in soil. *Soil Biology and Biochemistry* 32, 1377–1382.

Eastham, H.M., Lind, R.J., Clarke, B.S., Towner P., Reynolds S.E., Wolstenholme, A.J. and Wonnacott, S. (1998) Characterization of a nicotinic acetylcholine receptor from the insect *Manduca sexta*. *European Journal of Neuroscience* 10, 879–889.

Eaton, D.L. and Groopman, J.D. (eds) (1994) *The Toxicology of Aflatoxins: Human Health, Veterinary and Agricultural Significance*. Academic Press, San Diego, California.

Eder, G., Sturm, R. and Ernst, W. (1987) *Chemosphere* 16, 2487.

Edwards, C. (1973) *Environmental Pollution by Pesticides*. Plenum Press, New York, 542 pp.

Egan, A.R., Michelotti, E.L., Young, D.H., Wilson, W.J. and Mattioda, H. (1998) RH-7281: a novel fungicide for control of downy mildew and late blight. In: *Proceedings of the Brighton Crop Protection Conference – Pests and Diseases*. Vol. 2. BCPC Publications, Croydon, UK, pp. 335–342.

Elad, Y. (1996) Mechanisms involved in the biological control of *Botrytis cinerea* incited diseases. *European Journal of Plant Pathology* 102, 719–732.

Elbert, A. and Nauen, R. (2000) Resistance of *Bemisia tabaci* (Homoptera: Aleyrodidae) to insecticides in southern Spain with special reference to neonicotinoids. *Pest Management Science* 56, 60–64.

Elliott, M. (1977) Synthetic pyrethroids. In: Elliott, M. (ed.) *Synthetic Pyrethroids*. ACS Symposium Series 42. American Chemical Society (ACS), Washington, DC, pp. 1–28.

Elliott, M. (1996) Synthetic insecticides related to natural pyrethrins. In: Copping, L.C. (ed.) *Crop Protection Agents from Nature: Natural Products and Analogues*. Royal Society of Chemistry, Cambridge, UK, pp. 254–300.

Elliott, M., Farnham, A.W., Janes, N.F., Needham, P.H., Pulman, D.A. and Stevenson, J.H. (1973) A photostable pyrethroid. *Nature* 246, 169–170.

Elliott, M., Janes, N.F. and Potter, C. (1978) The future of pyrethoids in insect control. *Annual Review of Entomology* 23, 443–469.

Ellis, M.K., Whitfield, A.C., Gowans, L.A., Auton, T.R., Provan, W.M., Lock, E.A. and Smith, L.L. (1995) Inhibition of 4-hydroxyphenylpyruvate dioxygenase by 2-(2-nitro-4-trifluoromethylbenzoyl)-cyclohexane-1, 3-dione and 2-(2-chloro-4-methanesulfonylbenzoyl)-cyclohexane-1, 3-dione. *Toxicology and Applied Pharmacology* 133, 12–19.

El Naggar, S.F., Creekmore, R.W., Cornell, S.S., Dunn, C. and McLaughlin, T. (2000) Metabolism of benzfendizone herbicide in rat. In: *Proceedings of the American Chemical Society (ACS) 220th Agro-114*. ACS, San Francisco, California, (Abstract).

Emden, H. and Peaball, D.B. (1996) *Beyond Silent Spring, Integrated Pest Management and Chemical Safety*. Chapman & Hall, London, 322 pp.

Emden, H.F. van and Servie, M.W. (2004) *Pest and Vector Control*. Cambridge University Press, Cambridge, UK.

Endo, S. and Tsurumachi, M. (2001) Insecticide susceptibility of the brown planthopper and the white-backed planthopper collected from southeast Asia. *Journal of Pesticide Science* 26, 82–86.

Engelhart, E. and Loewi, O. (1930) Fermentative azetylcholispaltung im blut und ihre hemming. *Archiv for Experimentelle Pathologie und Pharmakologie* 150, 1.

Enright, W.J. (1989) *Use of Somatotropin in Livestock Production*. Elsevier, London, pp. 132–156.

Epstein, E. and Ludwig-Muller, J. (1993) Indole-3-butyric acid in plants: occurrence, synthesis, metabolism and transport. *Journal of Plant Physiology* 88, 382–389.

Epstein, E. and Wiesman, Z. (1987) Improved vegetative propagation of olive cultivars with TBA-alanine. *Olea* 18, 35–38.

Erlanger, B.F. (1980) The preparation of antigenic hapten-carrier conjugates: a survey. *Methods in Enzymology* 70, 85–104.

Ernstsen, A. and Sandberg, G. (1986) Identification of 4-chloroindole-3-acetic acid and indole-3-aldehyde in seeds of *Pinus sylvestris*. *Journal of Plant Physiology* 68, 511–518.

Erskine, A.J. (1992) *Atlas of Breeding Birds of the Maritime Provinces*. Nimbus/Nova Scotia Museum Program, Halifax, Canada, pp. 1–270.

Eshel, D., Gamliel, A., Katan, J. and Grinstein, A. (1999) Evaluation of soil fumigants on soilborne fungal pathogens in a controlled-environment system and in soil. *Crop Protection* 18, 437–443.

Eto, M. (1974) *Organophosphorus Pesticides: Organic and Biological Chemistry*. CRC Press, Cleveland, Ohio, p. 387.

Eto, M. and Casida, J.E. (eds) (1995) *Progress and Prospects of Organophosphorus Agrochemicals*. Kyushu University Press, Fukuoka, Japan, p. 190.

Eto, M. and Kuano, K. (1992) Prenyl imidazoles and related compounds controlling hormonal development processes of insects. In: Otto, D. and Weber, B. (eds) *Insecticides: Mechanism of Action and Resistance*. Intercept Ltd, Andover, UK, pp. 155–165.

European Commission Directorate General for Health and Consumers (DG SANCO) (2009) European Crop Protection – Regulatory Overview. European Crop Protection Association (ECA). Available at: http://www.ecpa.eu/page/regulatory-overview (accessed 14 June 2009).

European Food Safety Authority (EFSA) (2009a) Peer review of the pesticide risk assessment of the active substance cadusafos. *EFSA Journal* 262, 6.

European Food Safety Authority (EFSA) (2009b) Reasoned opinion of EFSA: modification of existing MRLs for valifenalate (valipenal) in tomatoes and aubergines. *EFSA Journal* 7(11), 1388.

European Union (EU) Pesticide Database (2013) EU Pesticide Database. Available at: http://ec.europa.eu/sanco_pesticides/public/?event=homepage&CFID=744483&CFTOKEN=41666187&jsessionid=09045ff4bf62927e55b74b4117d667c54651TR (accessed 12 June 2013).

Evans, M.L. (1985) The action of auxin on plant cell elongation. *Critical Reviews in Plant Sciences* 2, 317.

Farmer, E.E. and Ryan, C.A. (1990) Interplant communication: airborne methyl jasmonate induces synthesis of proteinase inhibitors in plant leaves. *Proceedings of the National Academy of Sciences USA* 87, 7713–7716.

Faustman, E.M. and Omenn, G.S. (2001) *Risk assessment*. In: Klaassen, C.D. (ed.) *Cassarett and Doull's Toxicology. The Basic Science of Poisons*. McGraw-Hill, New York, pp. 83–104.

Federici, B.A. (1999) Naturally occurring baculoviruses for insect pest control. In: Hall, F.R. and Menn, J.J. (eds) *Biopesticides – Use and Delivery*. Humana Press, Totowa, New Jersey, pp. 301–320.

Fedtke, C. (1982) *Biochemistry and Physiology of Herbicide Action*. Springer, New York.

Fenton, A.M., Stephens, P.M., Crowley, J., O'Callaghan, M. and O'Gara, F. (1992) Exploitation of a gene(s) involved in 2,4-diacetylphloroglucinol biosynthesis to confer a new biocontrol capability to a *Pseudomonas* strain. *Applied and Environmental Microbiology* 58, 3873–3878.

Feung, C.S., Hamilton, R.H. and Mumma, R.O. (1977) Metabolism of indole-3-acetic acid. IV. Biological properties of amino acid conjugates. *Plant Physiology* 59, 91–93.

Fimreite, N. (1970) Mercury uses in Canada and their possible hazards as sources of mercury contamination. *Environmental Pollution* 1, 119–131.

Fimreite, N. (1974) Mercury contamination of aquatic birds in northwest Ontario. *Journal of Wildlife Management* 38, 120–131.

Finklea, H.C. and Fontenot, M.F. Jr (1995) Accelerated bioremediation of triazine contaminated soils: a practical case study. In: Skipper, H.D. and Turco, R.F. (eds) *Bioremediation: Science and Applications*. Soil Science Society of America (SSSA) Special Publication 43. American Society of Agronomy (ASA), Madison, Wisconsin, pp. 221–235.

Fischer, E. (1894) Influence of configuration on the action of enzymes. *Berichte der deutschen chemischen Gesellschaft* 27, 2985–2993.

Fischler, H.M. and Korte, F. (1969) Sensibilisierte und unsensibilisierte photoisomerisierung von cylclodieninsektiziden. *Tetrahedron Letters* 15, 888.

Fishbein, L. (1972) Chromatographic and biological aspects of polychlorinated biphenyls. *Journal of Chromatography* 68, 345.

Fisher, M.H. (1993) Recent progress in avermectim research. In: Duke, S.O., Menn, J.J. and Plimmer, J. (eds) *Pest Control With Enhanced Environmental Safety*. ACS Symposium Series 524. American Chemical Society (ACS), Washington, DC, pp. 169–182.

Fleming, W.J., Clark, D.R. Jr and Henny, C.J. (1984) Organochlorine pesticides and PCBs: a continuing problem for the 1980s. *Transactions of the North American Wildlife and Natural Resources Conference* 48, 186–199.

Focardi, S., Fossi, C., Leonzio, C. and Romei, R. (1986) PCB congeners, hexachlorobenzene, and organochlorine insectidices in human fat in Italy. *Bulletin of Environmental Contamination and Toxicology* 36, 644–650.

Fogg, P. and Carter, A.D. (1998) Biobeds: the development and evaluation of a biological system for pesticide waste and washings. *British Crop Production Council (BCPC) Symposium Series* 70, 49–58.

Food and Agriculture Organization of the United Nations (FAO) (1990) *FAO Sampling Guidelines. Guidelines on Producing Pesticide Residues Data from Supervised Trails*. FAO, Rome.

Ford, M.G. (1988) Molecular basis of drug and pesticide action. In: Lunt, G.G. (ed.) *Neurotox '88*. Excerpta Medica, Amsterdam, pp. 509–527.

Forlani, F., Arnoldi, A. and Pagani, S. (1992) Development of an enzyme-linked immunosorbent assay for triazole fungicides. *Journal of Agricultural and Food Chemistry* 40, 328–331.

Foster, R.J., Mc.Rae, D.H. and Bonner, J. (1952) Auxin induced growth inhibition a natural consequence of two-point attachment. *Proceedings of the National Academy of Sciences USA* 38, 1014–1022.

Fournier, D., Karch, F., Bride, J.M., Hall, L.M.C., Bergé, J.B. and Spierer, P. (1989) *Drosophila melanogaster* acetylcholinesterase gene. Structure, evolution and mutations. *Journal of Molecular Biology* 210, 15–22.

Fraley, R.T. (1992) Sustaining the food supply. *Biotechnology* 10, 40–43.

Fravel, D.R. (1988) Role of antibiosis in the biocontrol of plant diseases. *Annual Review of Phytopathology* 26, 75–91.

Frederiksen, R.A. and Renfro, B.L. (1977) Global status of maize downy mildew. *Annual Review of Phytopathology* 15, 249–275.

Freeze, R.A. and Cherry, J.A. (1979) *Groundwater*. Prentice-Hall, Englewood Cliffs, New Jersey.

Frehse, H. and Anderson, J.P.E. (1983) Pesticide residues in soil – problems between concept and concern. In: Greenhaigh, R. and Drescher, N. (eds) *IUPAC Residue Chemistry: Human Welfare and the Environment*. Vol. 4, *Pesticide Residues and Formulation Chemistry*. Pergamon Press, Oxford, pp. 23–32.

French-Constant, R.H. and Bonning, B.C. (1989) Rapid microtitre plate test distinguishes insecticide resistant acetylcholinesterase genotype in the mosquitoes *Anopheles albimanus*, *An. nigerrimus* and *Culex pipiens*. *Medical and Veterinary Entomology* 3, 9–16.

Frias, I., Caldeira, M.T., Perez-Castineira, J.R., Navarro-Avinio, J.P., Culianez-Macia, F.A., Kuppinger, O., Stransky, H., Pages, M., Hager, A. and Serrano, R. (1996) A major isoform of the maize plasma membrane H+-ATPase: characterization and induction by auxin in coleoptiles. *The Plant Cell* 8, 1533–1544.

Friedman, H.B. (1992) DDT (Dichlorodiphenyltrichloroethane): a chemist's tale. *Journal of Chemical Education* 69, 362–365.

Fry, S.C. (1989) Cellulases, hemicelluloses and auxin-stimulated growth: a possible relationship. *Journal of Plant Physiology* 75, 532–536.

Fuerst, E.P. and Gronwald, J.W. (1986) Induction of rapid metabolism of metolachlor in sorghum shoots by CGA-92194 and other antidotes. *Weed Science* 34, 354–361.

Fujimura, M., Ochiai, N., Ichiishi, A., Usami, R., Horikoshi, K. and Yamaguchi, I. (2000) Fungicide resistance and osmotic stress sensitivity in *os* mutants of *Neurospora crassa*. *Pesticide Biochemistry and Physiology* 67, 125–133.

Fukui, R., Fukui, H. and Alvarez, A.M. (1999) Comparisons of single versus multiple bacterial species on biological control of anthurium blight. *Phytopathology* 89, 366–373.

Fuller, M.S. and Gisi, U. (1985) Comparative studies of the *in vitro* activity of the fungicides oxadixyl and metalaxyl. *Mycologia* 77, 424–432.

Gahukar, R.T. (1999) *Agro-Medical Guide of Synthetic Pesticides*. Agri-Horticultural Publishing House, Nagpur, India.

Gallardo, F., Fu, J.M., Canton, F.R., Garcia-Gutierrez, A., Canovas, F.M. and Kirby, E.G. (1999) Expression of a conifer glutamine synthetase gene in transgenic poplar. *Planta* 210, 19–26.

Galweiler, L., Guan, C., Muller, A., Wisman, E. and Mendgen, K. (1998) Regulation of polar auxin transport by AtPINl in *Arabidopsis* vascular tissue. *Science* 282, 2226–2230.

Gamliel, A., Hadar, E. and Katan, J. (1993) Improvement of growth and yield of *Gypsophila paniculata* by solarization or fumigation of soil or container medium in continuous cropping system. *Plant Disease* 77, 933–938.

Gamliel, A., Grinstein, A., Peretz, Y., Klein, L., Nachmias, A., Tsror, L., Livescu, L. and Katan, J. (1997) Reduced dosage of methylbromide for controlling verticillium wilt of potato in experimental and commercial plots. *Plant Disease* 81, 469–474.

Gamliel, A., Grinstein, A., Klein, L., Cohen, Y. and Katan, J. (1998) Permeability of plastic films to methyl bromide: field study. *Crop Protection* 17, 241–248.

Gandar, J.C. and Nitsch, C. (1967) Isolement de l'ester méthylique d'un acide chloro-3-indolylacétique a partir de graines immatures de pois, *Pisum sativum*. *Comptes Rendus de l'Académie des Sciences* 265, 1795–1798.

Garbers, C., DeLong, A., Deruère, J., Bernasconi, P. and Soll, D. (1996) A mutation in protein phosphatase 2A regulatory subunit A affects auxin transport in *Arabidopsis*. *EMBO Journal* 15, 2115–2124.

Gardner, D.E., Crapo, J.D. and McClellan, R.O. (1993) *Toxicology of the Lung*, 2nd edn. Raven Press, New York.

Garry, V.F., Burroughs, B., Tarone, R. and Kesner, J.S. (1999) Herbicides and adjuvants: an evolving view. *Toxicology and Industrial Health* 15(1–2), 159–167.

Gaskell, S.J. (1997) Electrospray: principles and practice. *Journal of Mass Spectrometry* 32, 677–688.

Gasztonyi, M. and Lyr, H. (1995) Miscellaneous fungicides. In: Lyr, H. (ed.) *Modern Selective Fungicides: Properties, Applications, Mechanisms of Action*, 2nd edn. Fischer Verlag, New York, pp. 389–414.

Gazit, Y., Ishaaya, I. and Perry, A.S. (1989) Detoxification and synergism of diflubenzuron and chlorfluazuron in the red flour beetle *Tribolium castaneum*. *Pesticide Biochemistry and Physiology* 34, 103–110.

Gehmann, K., Nyfeler, R., Leadbeater, A.J., Nevell, D. and Sozzi, D. (1990) CGA 173 506: a new phenylpyrrole fungicide for broad-spectrum disease control. In: *Proceedings of the Brighton Crop Protection Conference – Pests and Diseases*. Vol. 2. BCPC Publications, Croydon, UK, pp. 399–406.

Geyer, H., Scheunert, I. and Korte, F. (1986) Bioconcentration potential of organic environmental chemicals in humans. *Regulatory Toxicology and Pharmacology* 6, 313.

Gilbert, A.J. (1998) Design guidelines, features and performance characteristics and development of current pesticide containers. *British Crop Production Council (BCPC) Symposium Series* 70, 9–16.

Gimeno, M. (1996) An overview of the latest development of microencapsulation for agricultural products. *Journal of Environmental Science and Health, Part B* 31, 407–420.

Gisi, U. and Staehle-Csech, U. (1988) Resistance risk evaluation of phenylamide and EBI fungicides. In: *Proceedings of the Brighton Crop Protection Conference – Pests and Diseases*. Vol. 1. BCPC Publications, Croydon, UK, pp. 359–366.

Gladen, B.C., Rogan, W.J., Hardy, P., Thullen, J., Tingelstad, J. and Tully, M. (1988) Development after exposure to polychlorinated biphenyls and dichlorodiphenyl dichloroethene transplacentally and through human milk. *Journal of Pediatrics* 113, 991.

Goksøyr, J. (1955) The effect of some dithiocarbamyl compounds on the metabolism of fungi. *Physiologia Plantarum* 8, 719–835.

Goldstein, J.A. and Safe, S. (1989) Mechanism of action and structure-activity relationships for the chlorinated dibenzo-p-dioxins and related compounds. In: *Halogenated Biphenyls, Naphthalenes, Dibenzodioxins and Related Compounds*, 2nd edn. Elsevier, Amsterdam, pp. 239–293.

Goldstein, J.A., McKinney, J.D., Lucier, G.W., Hickman, P., Bergman, H. and Moore, J.A. (1976) Toxicological assessment of hexachlorobiphenyl isomers and 2,3,7,8,-tetrachlorodibenzofuran in chicks. II. Effects on drug metabolism and porphyrin accumulation. *Toxicology and Applied Pharmacology* 36, 81–92.

Goot, H. van der and Timmerman, H. (1988) Sterical isomerism: one of the determinant factors for chemical and biological properties. In: Ariens, E.J., van Rensen, J.J.S. and Welling, W. (eds) *Stereoselectivity of Pesticides, Biological and Chemical Problems*. Elsevier, Amsterdam, pp. 11–38.

Gopferich, A. (1997) Polymer bulk erosion. *Macromolecules* 30, 2598–2604.

Gorbach, S., Their, W., Kellner, H.M., Schulze, E.F., Kuenzler, K. and Fischer, H. (1974) Environmental impact of pyrazophos. 1. Contribution, degradation in plants and in the rat. *Environmental Quality and Safety* 3(Suppl.), 840–844.

Gorski, S.F. (1993) Slow release delivery system for herbicides in container grown stock. *Weed Technology* 7, 894–899.

Gosselin, R.E., Smith, R.P., Hodge, H.C. and Braddock, J.E. (1984) *Clinical Toxicology of Commercial Products*, 5th edn. Lippincott Williams & Wilkins, Philadelphia, Philadelphia.

Gowariker, V., Krishnamurthy, V.N., Gowariker, S., Dhanorkar, M. and Paranjape, K. (2007) *The Fertilizer Encyclopedia*. Vasudha Research and Publications, Pune, India.

Gowariker, V., Krishnamurthy, V.N., Gowariker, S., Dhanorkar, M. and Paranjape, K. (2009) *The Fertilizer Encyclopedia*. Wiley, Hoboken, New Jersey.

Graves, J.B., Leonard, B.R., Micinski, S., Burris, E., Martin, S.H., White, C.A. and Baldwin, J.L. (1993) Monitoring insecticide resistance in tobacco budworm and bollworm in Louisiana. In: *Proceedings of the Beltwide Cotton Research Conference, 10–14 January, New Orleans*. National Cotton Council of America, Memphis, Tennesee, pp. 788–794.

Gray, A.J. and Soderlund, D.M. (1985) Mammalian toxicity of pyrethroids. In: Hutson, D.H. and Roberts, T.R. (eds) *Insecticides, Progress in Pesticide Biochemistry and Toxicology*, Vol. 5. Wiley, Chichester, UK, pp. 193–348.

Grayson, B.T., Williams, K.S., Freehauf, P.A., Pease, R.R., Ziesel, W.T., Sereno, R.L. and Reinsfelder, R.E. (1987) The physical and chemical properties of the herbicide cinmethylin. *Pesticide Science* 21, 143–153.

Green, E.A., Bernhard, U. and Bacci, L. (1998) Control of powdery mildews with quinoxyfen in horticultural crops. In: *Proceedings of the Brighton Crop Protection Conference – Pests and Diseases*. Vol. 3. BCPC Publications, Croydon, UK, pp. 857–862.

Grewal, P. and Georgis, R. (1999) Entomopathogenic nematodes. In: Hall, F.R. and Menn, J.J. (eds) *Biopesticides – Use and Delivery*. Humana Press, Totowa, New Jersey, pp. 271–299.

Gueldner, R.C., Reilly, C.C., Pusey, P.L., Costello, C.E., Arrendale, R.F., Cox, R., Himmelsbach, D., Crumley, G. and Cutler, H. (1988) Isolation and identification of iturins as antifungal peptides in biological control of peach brown rot with *Bacillus subtilis*. *Journal of Agricultural and Food Chemistry* 36, 366–370.

Gueldner, R.C., Yates, I.F., Reilly, C.C., Wood, B.W. and Smith, M.T. (1994) Levels of hydrojuglone glucoside in developing pecan leaves in relation to scab susceptibility. *Journal of the American Society for Horticultural Science* 119, 498–504.

Guilfoyle, T.J., Hagen, G., Ulmasov, T. and Murfett, J. (1998) How does auxin turn on genes? *Plant Physiology* 118, 341–347.

Gullickson, W.D. (1994) History of pyrethrum in the 1970s and 1980s. In: Casida, J.E. and Quistad, B.B. (eds) *Pyrethrum Flowers: Production, Chemistry, Toxicology and Uses*. Oxford University Press, Oxford, UK, pp. 1–32.

Gunning, R.V., Moores, G.D. and Devonshire, A.L. (1996) Insensitive acetylcholinesterase and resistance to thiodicarb in Australian *Helicoverpa armigera* Hubner (Lepidoptera: Noctuidae). *Pesticide Biochemistry and Physiology* 55, 21–28.

Gunning, R.V., Moores, G.D. and Devonshire, A.L. (1997) Esterases and fenvalerate resistance in a field population of *Helicoverpa punctigera* (Lepidoptera: Noctuidae) in Australia. *Pesticide Biochemistry and Physiology* 58, 155–162.

Gustafson, D.I. (1990) Controlled delivery and soil application. In: Wilkins, R.M. (ed.) *Controlled Delivery of Crop-Protection Agents*. Taylor & Francis, London, pp. 23–42.

Hager, A., Menzel, H. and Krauss, A. (1971) Versuche und hypothese zur primärwirkung des auxins beim streckungswachtum. *Planta* 100, 47–75.

Hajjar, N.P. and Casida, J.E. (1979) Structure activity relationship of benzoylphenyl ureas as toxicants and chitin synthesis inhibitors in *Oncopeltus fasciatus*. *Pesticide Biochemistry and Physiology* 11, 33–45.

Hall, F.R. and Menn, J.J. (eds) (1999) *Biopesticides – Use and Delivery*. Humana Press, Totowa, New Jersey, pp. 1–626.

Hama, H. and Iwata, S. (1971) Insensitive cholinesterase in the Nakagawara strain of the green rice leafhopper, *Nephotetrix cincticeps* Uhler (Hemiptera: Cicadellidae) as a cause of resistance to carbamate insecticides. *Japanese Journal of Applied Entomology and Zoology* 6, 183–191.

Hance, R.J. (1987) Herbicide behaviour in the soil, with particular reference to the potential for groundwater contamination. In: Huston, D.H. and Roberts, T.R. (eds) *Herbicides*. Wiley, New York, pp. 223–247.

Handa, S.K. (1999) *Principles of Pesticide Chemistry*. Agrobios Publishers, New Delhi, India.

Handelsman, J. and Parke, J.L. (1989) Mechanisms in biocontrol of soilborne plant pathogens. In: Kosuge, T. and Nester, E.W. (eds) *Plant–Microbe Interactions*, Vol. 3, *Molecular and Genetic Perspectives*. McGraw-Hill, New York, pp. 27–61.

Handelsman, J. and Stabb, E.V. (1996) Biocontrol of soilborne plant pathogens. *The Plant Cell* 8, 1855–1869.

Hangarter, R.P., Peterson, M.D. and Good, N.E. (1980) Biological activities of indoleacetic acids and their use as auxins in tissue culture. *Plant Physiology* 65, 761–767.

Hanrahan, J.P. (1987) *Beta-Agonists and Their Effects on Animal Growth and Carcass Quality*. Elsevier Applied Science, London, pp. 106–118.

Hansel, W. and McEntee, K. (1955) Bovine hyper-keratosis (X-disease): a review. *Journal of Dairy Science* 38, 875.

Hansen, H. and Grossmann, K. (2000) Auxin-induced ethylene triggers abscisic acid biosynthesis and growth inhibition. *Plant Physiology* 124, 1437.

Hansen, J.L. and Spiegel, M.H. (1983) Hydrolysis studies of aldicarb, aldicarb sulfoxide and aldicarb sulfone. *Environmental Toxicology and Chemistry* 2, 147–153.

Hara, I. (1985) Health status and PCBs in blood of workers exposed to PCBs and of their children. *Environmental Health Perspectives* 59, 85–90.

Harborne, J.B. (1993) *Introduction to Ecological Biochemistry*. Academic Press, London, 318 pp.

Harkin, J.M., Garner, W.Y., Honeycutt, R.C. and Nigg, H.N. (1986) *Evaluation of Pesticides in Ground Water*. ACS Symposium Series 315. American Chemical Society (ACS), Washington, DC, pp. 219–254.

Harms, H.H. (1992) *In-vitro* systems for studying phytotoxicity and metabolic fate of pesticides and xenobiotics in plants. *Pesticide Science* 35, 277–281.

Harris, D.A., Johnson, K.S. and Ogilvy, J.M.E. (1991) A system for the treatment of waste water from agrochemical production and field use. In: *Proceedings of the Brighton Crop Protection Conference – Weeds*. BCPC Publications, Croydon, UK, pp. 715–722.

Harris, R.F. (1981) Effect of water potential on microbial growth and activity. In: Parr, J.F., Gardner, W.R., Elliot, L.F. and Madison, W.I. (eds) *Water Potential Relations in Soil Microbiology*. American Society of Agronomy (ASA), Madison, Wisconsin, pp. 23–95.

Harris, R.F. and Arnold, S.M. (1995) Redox and energy aspects of soil bioremediation. In: Skipper, H.D. and Turco, R.F. (eds) *Bioremediation: Science and Applications*. Soil Science Society of America (SSSA) Special Publication 43. American Society of Agronomy (ASA), Madison, Wisconsin, pp. 54–87.

Harrison, R.O., Goodrow, M.H., Gee, S.J. and Hammock, B.D. (1991) Hapten synthesis for pesticide immunoassay development. In: Vanderlaan, M., Stanker, L.S., Watkins, B.E. and Roberts, D.W. (eds) *Immunoassays for Trace Chemical Analysis*. ACS Symposium Series 451. American Chemical Society (ACS), Washington, DC, pp. 14–27.

Hartley, C.S. and Graham-Bryce, B.J. (1980) *Physical Principles of Pesticide Behaviour*, Volumes 1 and 2. Academic Press, New York.

Harvey, G.R. and Steinhauer, W.G. (1974) Transport of polychlorobiphenyls to the North Atlantic. *Atmospheric Environment* 8, 777–782.

Harvey, I.C., Bourdot, G.W., Saville, D.J. and Sands, D.C. (1998) A comparison of auxotrophic and wild strains of *Sclerotinia sclerotiorum* used as a mycoherbicide against Californian thistle (*Cirsium arvense*). *Biocontrol, Science and Technology* 8, 73–81.

Hatano, T., Katayama, M. and Marumo, S. (1987) 5,6 Dichloroindole 3 acetic acid as a potent auxin: its synthesis and biological activity. *Experientia* 43, 1237–1239.

Hatano, T., Kato, Y., Katayama, M. and Marumo S. (1989) A new indolic antiauxin, α-(5,7-dichloroindole-3-)isobutyric acid: its chemical synthesis and biological activity. *Experientia* 45, 400–402.

Hatzios, K.K. (ed.) (1997) *Regulation of Enzymatic Systems Detoxifying Xenobiotics in Plants*. NATO ASI Series, High Technology, Vol. 37. Kluwer Academic Publishers, Dordrecht, p. 385.

Hayes, A.W. (2001) *Principles and Methods of Toxicology*, Vol. XIX. Taylor & Francis, Philadelphia, 1887 pp.

Hayes, W.J. Jr (1991) Dosage and other factors influencing toxicity. In: Hayes, W.J. Jr and Laws, E.R. Jr (eds) *Handbook of Pesticide Toxicology: General Principles*, Vol. 1. Academic Press, New York, pp. 39–105.

Heap, I. (2013) The International Survey of Herbicide Resistant Weeds. Available at http://www.weedscience.com (accessed 11 October 2013).

Heath, R.G., Spann, J.W. and Dreitzer, J.F. (1969) Marked DDE impairment of mallard reproduction in controlled studies. *Nature* 224, 47–47.

Henderson, J., Macdonald, H., Lazarus, C.M., Napier, R.M. and Hawes, C.R. (1996) Protein retention in the endoplasmic reticulum is not compromised by baculovirus infection. *Cell Biology International* 6, 413–422.

Hendricks, S. (2007) *DDT? Again?* Pesticide Action Network (PAN) North America, Oakland, California.

Hermsen, B., Stetzer, E., Thees, H.R, Schrattenholz, A., Ebbinghaus, U., Kretschmer, A., Methfessel, C., Reinhardt, S. and Maelicke, A. (1998) Neuronal nicotinic receptors in the locust *Locusta migratoria*: cloning and expression. *Journal of Biological Chemistry* 273, 18394–19404.

Herve, J.J. (1985) Agricultural and public health usage. In: Leahey, J.P. (ed.) *The Pyrethroid Insecticides*. Taylor & Francis, London, pp. 343–417.

Hess, T.F. and Schmidt, S.K. (1995) Improved procedure for obtaining statistically valid parameter estimates from soil respiration data. *Soil Biology and Biochemistry* 27, 1–7.

Hesse, T., Feldwisch, J., Balshusemann, D., Bauw, G., Puype, M., Vandekerckhove, J., Lobler, M., Klambt, D., Schell, J. and Palme, K. (1989) Molecular cloning and structural analysis of a gene from *Zea mays* (L.) coding for a putative receptor for the plant hormone auxin. *EMBO Journal* 8, 2453–2461.

Hewitt, H.G. (1998) *Fungicides in Crop Protection*. CAB International, Wallingford, UK.

Hickey, J.J. and Anderson, D.W. (1968) Chlorinated hydrocarbons and eggshell changes in raptorial and fish-eating birds. *Science* 162, 271–273.

Hiei, Y., Ohta, S., Komari, T. and Kumashiro, T. (1994) Efficient transformation of rice (*Oryza sativa* L.) mediated by *Agrobacterium* and sequence analysis of the boundaries of the T-DNA. *The Plant Journal* 6, 271–282.

Higuchi, K. (1976) *PCB Poisoning and Pollution*. Kondansha Ltd and Academic Press, Tokyo.

Higuchi, T. (1961) Rate of release of medicaments from ointment bases containing drugs in suspension. *Journal of Pharmaceutical Sciences* 50, 874–883.

Hill, D.S. (1996) *The Economic Importance of Insects*. Chapman & Hall, London.

Hirshberg, J. and McIntosh, L. (1983) Molecular basis of herbicide resistance in *Amaranthus hybridus*. *Science* 222, 1346–1349.

Hjelm, O. (1996) *Organohalogens in Forest Soil*. Department of Water and Environmental Studies, Linkoping University, Linkoping, Sweden.

Hobbie, S.E. and Chapin, F.S. III (1996) Winter regulation of tundra litter carbon and nitrogen dynamics. *Biogeochemistry* 35, 327–338.

Hofer, I., Beck, T. and Wallnöfer, P. (1971) The influence of the fungicide benomyl on soil microflora. *Z Pflanzenkr Pflanzenschutz* 7, 398–405 (in German).

Hoffman, J.C. and Vaughn, K.C. (1996) Flupoxam induces classic club morphology but is not a mitotic disrupter herbicide. *Pesticide Biochemistry and Physiology* 55, 49–53.

Holden, J.S. (1979) Absorption and metabolism of permethrin and cypermethrin in the cockroach and the cotton-leafworm larvae. *Pesticide Science* 10, 295–307.

Holden, L.R. and Jeffery, A.G. (1992) Results of the National Alachlor Well Water Survey. *Environmental Science and Technology* 26, 935–943.

Holt, J.S. and LeBaron, H.M. (1990) Significance and distribution of herbicide resistance. *Weed Technology* 4, 141–149.

Holt, J.S., Powles, S.B. and Holtum, J.A.M. (1993) Mechanisms and agronomic aspects of herbicide resistance. *Annual Review of Plant Physiology and Plant Molecular Biology* 44, 203–229.

Hong, S. and Smith, A.E. (1996) Abiotic and biotic degradation of dithiopyr in golf course greens. *Journal of Agricultural and Food Chemistry* 44, 3393–3398.

Hornby, D. (1983) Suppressive soils. *Annual Review of Phytopathology* 21, 65–85.

Horst, M.N. (1981) The biosynthesis of crustacean chitin by a microsomal enzyme from larval brine shrimp. *Journal of Biological Chemistry* 256, 1412–1419.

Hotson, I.K. June (1982) The avermectins: a new family of antiparasitic agents. *Journal of the South African Veterinary Association* 53(2), 87–90.

Houle, G. and Babeux, P. (1998) The effects of collection date, IBA, plant gender, nutrient availability, and rooting volume on adventitious root and lateral shoot formation by *Salix planifolia* stem cuttings from the Ungava Bay area (Quebec, Canada). *Canadian Journal of Botany* 76, 1687–1692.

Hsu, S.T., Ma, C.I., Hsu, S.K., Wu, S.S., Hsu, N.H. and Yeh, C.C. (1985) Discovery and epidemiology of PCB poisoning in Taiwan: a four-year followup. *Environmental Health Perspectives* 59, 5–10.

Huang, Y., Williamson, M.S., Devonshire, A.L., Windass, J.D., Lansdell, S.J. and Millar, N.S. (1999) Molecular characterization and imidacloprid selectivity of nicotinic acetylcholine receptor subunits from the peach-potato aphid *Myzus persicae*. *Journal of Neurochemistry* 73, 380–389.

Hunkin, G.G., Reed, T.A. and Brand, G.N. (1984) Some observations on field experiences with monitoring wells. *Groundwater Monitoring & Remediation* 4(1), 43–45.

Hutchins, S.R. (1991) Biodegradation of monoaromatic hydrocarbons by aquifer microorganisms using oxygen, nitrate, or nitrous oxide as the terminal electron acceptor. *Applied and Environmental Microbiology* 57, 2403–2407.

Hutzinger, O., Safe, S. and Zitko, V. (1974) *The Chemistry of PCBs*. CRC Press, Boca Raton, Florida.

Ikonomou, M.G., Blades, A.T. and Kerbarale, A.T. (1990) Investigations of the electrospray interface for liquid chromatography mass spectrometry. *Analytical Chemistry* 62, 957–967.

Insecticide Resistance Action Committee (2007) Resistance Definition. Available at: http://en.wikipedia.org/wiki/Pesticide_resistance (accessed 25 October 2013).

International Agency for Research on Cancer (IARC) (2013) Available at: http://www.iarc.fr/ (accessed 23 August 2013).

Ishaaya, I. (1992) 13 selective insect control agents: mechanisms and agricultural importance. In: Otto, D. and Weber, B. (eds) *Insecticides: Mechanism of Action and Resistance*. Intercepts Ltd, Andover, UK, pp. 127–134.

Ishii, H. and Yamaguchi, A. (1977) Tolerance of *Venturia nashicola* to thiophanate-methyl and benomyl in Japan. *Annals of the Phytopathological Society of Japan* 43, 557–561.

Jackson, J.L., Jacobson, S.W. and Humphrey, H.E.B. (1990) Effects of *in utero* exposure to polychlorinated biphenyls and related contaminants on cognitive functioning in young children. *Journal of Pediatrics* 116, 38–45.

Jackson, M.A., Schisler, D.A., Slininger, P.J. and Boyette, C.D. (1996) Fermentation strategies for improving the fitness of a bioherbicide. *Weed Technology* 10, 645–650.

Jacobs, M. and Gilbert, S.F. (1983) Basal localization of the presumptive auxin transport carrier in pea stem cells. *Science* 220, 1297–1300.

Jacobs, M. and Rubery, P.H. (1988) Naturally occurring auxin transport regulators. *Science* 241, 346–349.

Jacobson, S.W., Fein, G.G., Jacobson, J.L., Schwartz, P.M. and Dowler, J.K. (1985) The effect of intrauterine PCB exposure on visual recognition memory. *Child Development* 56, 853.

James, D.W. and Gutterson, N.I. (1986) Multiple antibiotics produced by *Pseudomonas fluorescens* HV37a and their differential regulation by glucose. *Applied and Environmental Microbiology* 52, 1183–1189.

Janisiewicz, W. (1996) Ecological diversity, niche overlap, and coexistence of antagonists used in developing mixtures for biocontrol of postharvest diseases of apples. *Phytopathology* 86, 473–479.

Jarvis, N.J. (1994) The MARCO model (version 3.1). Technical description and sample simulations. Dissertation 19. Department of Soil Science, Swedish University of Agricultural Science, Uppsala, Sweden.

Jarvis, N.J. (1998) Modelling the impact of preferential flow on nonpoint source pollution. In: Magdi Selim, H. and Wa, L. (eds) *Physical Nonequilibrium in Soils: Modelling and Application*. Ann Arbor Press, Chelsea, Michigan, pp. 195–217.

Jarvis, N.J., Stahli, M., Bergstrom, L. and Johnsson, H. (1994) Simulation of dichlorprop and bentazon leaching in soils of contrasting texture using the MACRO model. *Journal of Environmental Science and Health, Part A* 29, 1255–1277.

Jastrow, J.D., Miller, R.M. and Lussenhop, J. (1998) Contributions of interacting biological mechanisms to soil aggregate stabilization in restored prairie. *Soil Biology and Biochemistry* 30, 905–916.

Jauhianen, A., Kangas, J., Laitinen, S. and Savolainen, K. (1992) Biological monitoring of workers exposed to mevinphos in greenhouses. *Bulletin of Environmental Contamination and Toxicology* 49, 37–43.

Jeger, M.J., Gilijamse, E., Bock, C.H. and Frinking, H.D. (1998) The epidemiology, variability and control of the downy mildews of pearl millet and sorghum, with particular reference to Africa. *Plant Pathology* 47, 544–569.

Jelaska, S., Magnus, V., Seretin, M. and Lacan, G. (1985) Induction of embryogenic callus in *Cucurbita pepo* hypocotyl explants by indole-3-ethanol and its sugar conjugates. *Journal of Plant Physiology* 64, 237–242.

Jensen, R.K., Sleight, S.D., Goodman, J.I., Aust, S.D. and Trosko, J.E. (1982) Polybrominated biphenyls as promoters in experimental hepatocarcinogenesis in rats. *Carcinogenesis* 3, 1183–1186.

Jensen, W.B. (1980) *The Lewis Acid–Base Concepts: an Overview*. Wiley, New York.

Jim, C.Y. (1998) Soil characteristics and management in an urban park in Hong Kong. *Environmental Management* 22, 683–695.

Johal, G.S. and Briggs, S.P. (1992) Reductase activity encoded by the HM1 disease resistance gene in maize. *Science* 258, 985–987.

Johnson, B.J. (1998) Influence of spring-applied herbicides on Bermudagrass (*Cynodon dactylon*) greens overseeded with roughstalk bluegrass (*Poa trivialis*). *Weed Technology* 12, 1–6.

Jones, R.G. (1990) Cinmethylin – a new herbicide developed for use in rice. In: Grayson, B.T., Green, M.B. and Copping, L.C. (eds) *Pest Management in Rice*. Elsevier, London, pp. 349–357.

Joseph-Horne, T., Hollomon, D.W. and Wood, P.M. (2001) Fungal respiration: a fusion of standard and alternative components. *Biochimica et Biophysica Acta* 1504, 179–195.

Kaethner, J.M. (1977) Conformational change theory for auxin structure–activity relationships. *Nature* 267, 19–23.

Kagabu, S. (1997) Chloronicotinyl insecticides-discovery, application and future perspective. *Reviews in Toxicology* 1, 75–129.

Kagabu, S. (1999) Discovery of chloronicotinylinsecticides. In: Yamamoto, I. and Casida, J.E. (eds) *Nicotinoid Insecticides and the Nicotinic Acetylcholine Receptor*. Springer, New York, pp. 91–106.

Kagabu, S. and Medej, S. (1995) Stability comparison of imidacloprid and related compounds under simulated sunlight, hydrolysis conditions, and to oxygen. *Bioscience, Biotechnology, and Biochemistry* 59, 980–985.

Kaiser, J.P. and Bollag, J.M. (1992) Influence of soil inoculum and redox potential on the degradation of several pyridine derivatives. *Soil Biology and Biochemistry* 24, 351–357.

Kaldorf, M. and Ludwig-Müller, J. (2000) AM fungi might affect the root morphology of maize by increasing indole- 3-butyric acid biosynthesis. *Journal of Plant Physiology* 109, 58–67.

Kaniewski, W.K., Lawson, C., Sammons, B., Haley, L., Hart, J., Delannay, X. and Tumer, N. (1990) Field resistance of transgenic Russet Burbank potato to effects of infection by potato virus X and potato virus Y. *Biotechnology* 8, 750–754.

Kanis, E., Nieuwhof, G.J., de Greef, K.H., van der Hel, W., Vestergen, M.W.A., Huisman, J. and van der Wal, P. (1990) Effect of recombinant porcine somatotropin on growth and carcass quality in growing pigs: interactions with genotype, gender and slaughter weight. *Journal of Animal Science* 68, 1193–1200.

Karalliedde, L., Feldman, S., Henry, J. and Marrs, T. (2001) *Organophosphates and Health*. IC Press, London.

Karlson, P. and Butenandt, A. (1959) Pheromones (ectohormones) in insects. *Annual Review of Entomology* 4, 39–58.

Karlson, P. and Luscher, M. (1959) 'Pheromones': a new term for a class of biological active substances. *Nature*, 183, 55–56.

Kas, K.A., Sarafinas, A. and Stephens, R.W. (1999) A method to prepare 2-(3-pyridyl)-4,5-dihydrothiazoles. European Patent EP0936222. European Patent Office, Munich, Germany.

Kashimoto, T., Miyata, H., Kunita, S., Tung, T.-C., Hsu, S.T., Chang, K.J. and Tang, S.Y. (1981) The role of polychlorinated diben2ofuran in Yusho (PCB poisoning). *Archives of Environmental Health* 36, 321–326.

Katayama, M. and Gautam, R.K. (1996) Synthesis and biological activities of substituted 4,4,4-trifluoro-3-(indole-3-)butyric acids, novel fluorinated plant growth regulators. *Bioscience, Biotechnology, and Biochemistry* 60, 755–759.

Katayama, M. and Gautam, R.K. (1997) Synthesis and biological activities of fluorinated plant growth regulators, 4,4,4-trifluoro-3-(3-indolyl) butyric acids and 4,4,4-trifluoro-3-(2-indolyl) butyric acid bearing a methyl group on the indole nucleus. *Journal of Pesticide Science* 22, 331–337.

Katayama, M., Kato, Y., Hatano, T., Hatori, M. and Marumo, S. (1998) Synthesis and biological activities of 5,6-difluoroindole-3-acetic acid; a new fluoroindole auxin. *Journal of Pesticide Science* 23, 289–295.

Katekar, G.F. (1979) Auxins: on the nature of the receptor site and molecular requirements for auxin activity. *Phytochemistry* 18, 223–233.

Keel, C., Schnider, U., Maurhofer, M., Voisard, C., Laville, J., Burger, U., Wirthner, P., Haas, D. and Défago, G. (1992) Suppression of root diseases by *Pseudomonas fluorescens* CHA0 – importance of the bacterial seconday metabolite 2,4-diacetylphloroglucinol. *Molecular Plant–Microbe Interactions* 5, 4–13.

Keiding, J. (1967) Persistance of resistance populations after the relaxation of selection pressure. *World Review of Pest Control* 6, 115–130.

Kelsey, T.W. (1994) The agrarian myth and policy responses to farm safety. *American Journal of Public Health* 84(7), 1171–1177.

Kendall, S.J., Hollomon, D.W., Cooke, L.R. and Jones, D.R. (1993) Changes in sensitivity to DMI fungicides in *Rhynchosporium secalis*. *Crop Protection* 12, 357–362.

Keon, J.P.R., Broomfield, P.L.E., Hargreaves, J.A. and White, G.A. (1992) Molecular genetic analysis of carboxin resistance in *Ustilago maydis*. In: *Proceedings of the Brighton Crop Protection Conference – Pests and Diseases*. BCPC Publications, Croydon, UK, pp. 221–226.

Key, J.L. (1989) Modulation of gene expression by auxin. *BioEssays* 11, 52–58.

Khambay, B.P.S., Farnham, A.W. and Beddie, D.G. (1994) Relationship between pyrethroid structure and level of resistance in houseflies. In: Briggs, G.G. (ed.) *Advances in the Chemistry of Insect Control III*. Royal Society of Chemistry, Cambridge, UK, pp. 117–126.

Khan, Z., Amudavi, D. and Pickett, J. (2008) *Push–Pull Technology Transforms Small Farms in Kenya*. Pesticide Action Network (PAN) North America, Oakland, California.

Kirk, D.A., Evenden, M.D. and Mineau, P. (1996) Past and current attempts to evaluate the role of birds as predators of insect pests in temperate agriculture. *Current Ornithology* 13, 175–269.

Klein, M., Müller, M., Dust, M., Görlitz, G., Gottesbüren, B., Hassink, J., Kloskowski, R., Kubiak, R., Resseler, H., Schüfer, H., Stein, B. and Vereecken, H. (1997) Validation of the pesticide leaching model pelmo using lysimeter studies performed for registration. *Chemosphere* 35, 2563–2587.

Kleschick, W.A., Parker, M.A. and Turner, J.A. (1999–2103) *Encyclopedia of AgroChemicals*. Wiley, New York.

Knapp, L.W. (1966) Occupational and rural accidents. *Archives of Environmental Health* 13(4), 501–506.

Knight, T.J., Bush, D.R. and Langston-Unkerfer, P.J. (1988) Oats tolerant of *Pseudomonas syringae* pv. *tabaci* contain tabtoxinine-beta-lactam-insensitive leaf glutamine synthetases. *Plant Physiology* 88, 333–339.

Knudson, T.A. (1984) A Harvest of Harm. An award-winning series of six feature articles in the *Des Moines Register*, 16–30 September. Des Moines, Iowa.

Kobayashi, H. and Rittman, B.E. (1982) Microbial removal of hazardous organic compounds. *Environmental Science & Technology* 16, 170A–183A.

Koeman, J.H., van Velzen-Blad, H.C., de Vries, R. and Vos, J.G. (1973) Effects of PCB and DDE in cormorants and evaluation of PCB residues from an experimental study. *Journal of Reproduction and Fertility* (Supplement) 19, 353.

Kogure, K., Simidu, U. and Taga, N. (1979) A tentative direct microscopic method for counting living marine bacteria. *Canadian Journal of Microbiology* 25, 415–420.

Kollmeyer, W.D., Flattum, R.F., Foster, J.P., Powell, J.E., Schroeder, M.E. and Soloway, S.B. (1999) Discovery of the nitromethylene heterocycle insecticides. In: Yamamoto, I. and Casida, J.E. (eds) *Nicotinoid Insecticides and the Nicotinic Acetylcholine Receptor*. Springer, New York, pp. 71–90.

Kolpin, D.W., Barbash, J.E. and Gilliom, R.J. (1998) Occurrence of pesticides in shallow groundwater of the United States: initial results from the National Water-Quality Assessment Program. *Environmental Science & Technology* 32, 558–566.

Kondo, A. (1979) *Microcapsule Processing and Technology*. Marcel Dekker, New York.

Koo, S.J., Neal, J.C. and DiTomaso, J.M. (1997) Mechanism of action of quinclorac in grass roots. *Pesticide Biochemistry and Physiology* 57, 44–53.

Kortenhoff, A. (1993) Developments in the use of pesticides. In: Zadoks, J.C. (ed.) *Modern Crop Protection: Developments and Perspectives*. Wageningen Academic Publishers, Wageningen, The Netherlands, pp. 3–10.

Koshland, D. Jr (1994) The key–lock theory and the induced fit theory. *Angewandte Chemie International Edition* 33, 2375–2378.

Kostov, O. and Cleemput, O.V. (2001) Nitrogen transformations in copper-contaminated soils and effects of lime and compost application on soil resiliencey. *Biology and Fertility of Soils* 33, 10–16.

Krause, A., Hancock, W.G., Minard, R.D., Freyer, A.J., Honeycutt, R.C., LeBaron, H.M., Paulson, D.L., Liu, S. and Bollag, J.M. (1985) Microbial transformation of the herbicide metolachlor by a soil actinomycete. *Journal of Agricultural and Food Chemistry* 33, 545–589.

Krause, R.T. and August, E.M. (1983) Applicability of a carbamate insecticide multiresidue method for determining additional types of pesticides in fruits and vegetables. *Journal of AOAC (Association of Official Analytical Chemists) International* 66, 234–240.

Kremer, R.J. and Kennedy, A.C. (1996) Rhizobacteria as biocontrol agents of weeds. *Weed Technology* 10, 601–609.

Krieger, R. (2001) *Handbook of Pesticide Toxicology (a) Principles (b) Agents*. Academic Press, San Diego, California.

Kroes, J.G., Van Dam, J.C., Huygen, J. and Vervoort, R.W. (1999) *User's Guide of SWAP version 2.0; Simulation of Water Flow, Solute Transport and Plant Growth in the Soil-Water-Atmosphere-Plant Environment, Report 81*. Wageningen Agricultural University, DLO Winand Staring Centre, Technical Document 53.

Krynitsky, A.J. (1997) Determination of sulfonylurea herbicides in water by capillary electrophoresis and by liquid chromatography/mass spectrometry. *Journal of AOAC (Association of Official Analytical Chemists) International* 80, 392–400.

Kuck, K.H. and Thielert, W. (1987) On the systemic properties of HWG 1608, the active ingredient of the fungicides Folicur and Raxil. *Pflanzenschutz-Nachrichten Bayer* 40, 133–152.

Kumari, B. and Kathpal, T.S. (2010) *Pesticides: Methods of their Residue Estimation*. New India Publishing Agency, New Delhi, India.

Kume, R., Tashima, S., Matsumoto, K., Ando, I. and Shiraishi, T. (1996) Release of active ingredient from controlled relase granules of SSF-126 and its rice blast control activity in pot trial. *Journal of Pesticide Science* 21, 438–440.

Kuratsune, M. and Shapiro, R.E. (1984) *PCB Poisoning in Japan and Taiwan*. Alan R. Liss, New York.

Kuratsune, M., Yoshimura, T., Matsuzaka, J. and Yamaguchi, A. (1972) Epidemiologic study of polychlorinated biphenyls. *Environmental Health Perspectives* 1, 119–128.

Kuthiala, A., Gupta, R.K. and Davis, E.E. (1992) Effect of the repellent deet on the anternal chemoreceptors for oviposition in *Aedes aegypti* (Diptera, Culicidae). *Journal of Medical Entomology* 29, 639–643.

Kydonieus, A.F. (ed.) (1980) *Controlled Release Technologies: Methods, Theory and Applications*, Volumes 1 and 2. CRC Press, Boca Raton, Florida.

Kydonieus, A.G. (1977) The effect of some variables on the controlled release of chemicals from polymeric membranes. In: Scher, H.B. (ed.) *Controlled Release Pesticides*. American Chemical Society (ACS), Washington, DC, pp. 152–167.

Lam, B.S., Strobel, G.A., Harrison, L.A. and Lam, S.T. (1987) Transposon mutagenesis and tagging of fluorescent *Pseudomonas*: antimycotic production is necessary for control of Dutch elm disease. *Proceedings of the National Academy of Sciences USA* 84, 6447–6451.

Landrigan, P.J. (1992) Environmental pollution and health. *Lancet* 340(8829), 1220.

Langenberg, W.G., Zhang, L., Court, D., Giunchedi, L. and Mirtra, A. (1997) Transgenic tobacco plants expressing the bacterial *rnc* gene resist virus infection. *Molecular Breeding* 3, 391–399.

Larsen, P. (1961) Biological determination of natural auxin. In: Ruhland, W. (ed.) *Encyclopedia of Plant Physiology*, Vol. 14. Springer, Berlin, pp. 521–582.

Law, L.M. and Goerlitz, D.F. (1974) Selected chlorinated hydrocarbons in bottom material from streams tributary to San Francisco Bay. *Pesticide Monitoring Journal* 8, 33.

Lawhorne, L. (1976) The health of farmers. *Journal of Iowa Medical Society* 66(10), 409–418.

Lawton, R.W., Ross, M.R., Feingold, J. and Brown, J.F. (1985) Effects of PCB exposure on biochemical and hematological findings in capacitor workers. *Environmental Health Perspectives* 60, 165.

LeBaron, H.M. (1992) *Proceedings of the 3rd Tropical Weed Science Conference*, Kuala Lumpur, Malaysia.

Lee, D.L., Prisbylla, M.P., McLean Provan, W., Frase, T. and Mutter, L.C. (1997) The discovery and structural requirements of inhibitors of p-hydroxyphenylpyruvate dioxygenase. *Weed Science* 45, 601–609.

Lee, L.J. and Ngim, J. (2000) A first report of glyphosate-resistant goosegrass (*Eleusine indica* (L) Gaertn) in Malaysia. *Pest Management Science* 56, 336–339.

Lee, S.A., Clarke, B.S., Jenner, D.W. and Williamson, F.A. (1990) Cytochemical demonstration of the effects of the acylureas flufenoxuron and diflubenzuron on the incorporation of chitin into insect cuticle. *Pesticide Science* 28, 367–375.

Leece, B., Denomme, M.A., Towner, R., Li, S.M.A. and Safe, S. (1985) Polychlorinated biphenyls: correlation between *in vivo* and *in vitro* quantitative structure-activity relationships (QSARs). *Journal of Toxicology and Environmental Health* 16, 379–388.

Leech, C.A., Jewess, P., Marshall, J. and Satelle, D.B. (1991) Nitromethylene actions on *in situ* and expressed insect nicotinic acetylcholine receptors. *FEBS Letters* 290, 90–94.

Leighton, T., Marks, E.P. and Leighton, F. (1981) Pesticides: insecticides and fungicides are chitin synthesis inhibitors. *Science* 213, 905–907.

Leonhardt, B.A. (1990) Controlled-release systems for the delivery of insect pheromones. In: Wilkins, R.M. (ed.) *Controlled Delivery of Crop-Protection Agents*. Taylor & Francis, London, pp. 169–190.

Leusen, van A.M., Siderius, H., Hoogenboom, B.E. and Leusen, van D. (1972) A new and simple synthesis of the pyrrole ring system from Michael acceptors and tosylmethylisocyanides. *Tetrahedron Letters* 52, 5337–5340.

Lewis, D.L., Hodson, R.E. and Freeman, L.F. (1985) Multiphasic kinetics for transformation of methyl parathion by *Flavobacterium* species. *Applied and Environmental Microbiology* 50, 553–557.

Liesivouri, J., Liukkonen, S. and Pirhonen, P. (1988) Re-entry intervals after pesticide application in greenhouses. *Scandinavian Journal of Work Environment and Health* 14(Suppl. 1), 35–36.

Ligon, J.M., Hill, D.S., Hammer, P., Torkewitz, N., Hofmann, D., Kempf, H.J. and Van Pee, K.-H. (1999) Natural products with antimicrobial activity from *Pseudomanas* biocontrol bacteria. In: Brooks, F.T. and Roberts, T.R. (eds) *Pesticide Chemistry and Bioscience, the Food–Environment Challenge*. Royal Society of Chemistry, Cambridge, UK, pp. 179–189.

Lin, N.V.F. (2000) *In vitro* studies of cellular and molecular developmental toxicity of adjuvants, herbicides, and fungicides commonly used in Red River valley, Minnesota. *Journal of Toxicology and Environmental Health* 60(6), 423–439.

Lindquist, R.K., Adams, A.J., Hall, F.R. and Adams, I.H.H. (1988) Biological efficacy of bifenthrin, applied, using an air-assisted electrostatic sprayer, to control *Aphis gossypii* Glover (Homoptera: Aphididae) on potted chrysanthemums. In: *Proceedings of the Brighton Crop Protection Conference*. BCPC Publications, Croydon, UK, pp. 187–192.

Lindquist, R.K., Powell, C.C. and Hall, F.R. (1993) Glasshouse treatment. In: Matthews, G.A. and Hislop, E.C. (eds) *Application Technology for Crop Protection*. CAB International, Wallingford, UK, pp. 275–290.

Lindstedt, S., Holme, E., Lock, E.A., Hjalmarsson, O. and Strandvik, B. (1992) Treatment of hereditary tyrosinemia type 1 by inhibition of 4-hydroxy phenyl pyruvate dioxygenase. *Lancet* 340, 813–817.

Lindstrom, J.E., Prince, R.C., Clark, J.C., Grossman, M.J., Yeager, T.R., Braddock, J.F. and Brown, E.J. (1991) Microbial populations and hydrocarbon biodegradation potentials in fertilized shoreline sediments affected by the T/V Exxon-Valdez oil-spill. *Applied and Environmental Microbiology* 57, 2514–2522.

Lipson, D.A., Raab, T.K., Schmidt, S.K. and Monson, R.K. (2001) An empirical model of amino acid transformations in an alpine soil. *Soil Biology and Biochemistry* 33, 189–198.

Liu, M.Y., Tzeng, Y.J. and Sun, C.N. (1981) Diamondback moth resistance to several synthetic pyrethroids. *Journal of Economic Entomology* 74, 393–396.

Liu, S.Y., Zhang, R. and Bollag, J.M. (1988) Biodegradation of metolachlor in a soil perfusion experiment. *Biology and Fertility of Soils* 5, 276–281.

Löbler, M. and Klämbt, D. (1985) Auxin-binding protein from coleoptile membranes of corn (*Zea mays* L.). II. Localization of a putative auxin receptor. *Journal of Biological Chemistry* 260, 9854–9859.

Lock, E.A. and Wilks, M.F. (2001) Paraquat. In: Kriger, R.I. (ed.) *Handbook of Pesticides Toxicology*, 2nd edn. Academic Press, San Diego, California, pp. 1559–1603.

Loon, van L.C., Bakker, P.A.H.M. and Pieterse, C.M.J. (1998) Systemic resistance induced by rhizosphere bacteria. *Annual Review of Phytopathology* 36, 453–483.

Loper, J.E. and Stowell, L.J. (1992) *Biological Control of Plant Diseases*. The American Phytopathological Society (APS) Press Slide Collection. APS Press, St Paul, Minnesota.

Lopez-Martinez, N., Shimabukuro, R.H. and Prado, R. De (1998) Effect of quinclorac on auxin-induced growth, transmembrane proton gradient and ethylene biosynthesis in *Echinochloa* spp. *Australian Journal of Plant Physiology* 25, 851–857.

Lübbe, C., Pée, K.H. van, Salcher, O. and Lingens, F. (1983) The metabolism of tryptophan and 7-chlorotryptophan in *Pseudomonas pyrrocinia* and *Pseudomonas aureofaciens*. *Hoppe Seyler's Zeitschrift fur Physiologische Chemie* 364, 447–453.

Lukens, H.R., Williams, C.B., Levinson, S.A., Dendliker, W.B., Murayama, D. and Barn, R.L. (1997) Fluorescence immunoassay technique for detecting organic environmental contaminants. *Environmental Science & Technology* 11, 292–297.

Lumsden, R.D., Walter, J.R. and Baker, C.P. (1996) Development of *Gliocladium virens* for damping-off disease control. *Canadian Journal of Plant Pathology* 18, 463–468.

Macdonald, H., Henderson, J., Napier, R.M., Venis, M.A., Hawes, C. and Lazarus, C.M. (1994) Authentic processing and targeting of active maize auxin-binding protein in the baculovirus expression system. *Plant Physiology* 105, 1049–1057.

Mackay, D., Wan-Ying, S. and Kuo-Ching, M. (1997) *Illustrated Handbook of Physical-Chemical Properties and Environmental Fate for Organic Chemicals: Pesticide Chemicals*. Lewis Publishers, London.

Macrae, I.C. and Alexander, M. (1965) Microbial degradation of selected herbicides in soil. *Journal of Agricultural and Food Chemistry* 33, 72–76.

Magnus, A.H. (1970) Mercury-tolerance in *Pyrenophora avenae* in Norway. *Meldinger Norges Landbrukshoyskle* 49, 1–8.

Magnus, V., Nigovic, B., Hangarter, R.P. and Good, N.E. (1992) N-(Indoleyl-3-acetyl) amino acids as sources of auxin in plant tissue culture. *Plant Growth Regulation* 11, 19–28.

Malik, R.K. and Singh, S. (1995) Littleseed canarygrass (*Phalaris minor*) resistance to isoproturon in India. *Weed Technology* 9, 419–425.

Mangels, G. (1991) Behavior of the imidazolinone herbicides in the aquatic environment. In: Shaner, D. and O'Connor, S. (eds) *The Imidazolinone Herbicides*. CRC Press, Boca Raton, Florida, pp. 184–190.

Marek, I.J. and Koskinen, W.C. (1996) LC/MS analysis of 4-methoxy-6-methyl-1,3,5-triazine-2-yl-containing sulfonylurea herbicides in soil. *Journal of Agricultural and Food Chemistry* 44, 3878–3881.

Mark, E.M. (1969) *Report of the Secretary's Commission on Pesticides and their Relationship to Environmental Health*. US Department of Health, Education and Welfare, Washington, DC, pp. 1–677.

Marrs, G.J. and Scher, H.B. (1990) Development and uses of microencapsulation. In: Wilkins, R.M. (ed.) *Controlled Delivery of Crop-Protection Agents*. Taylor & Francis, London, pp. 65–90.

Marshall, J., Buckingham, S.D., Shingai, R., Lunt, G.G., Goosey, M.W., Darlison, M.G., Sattelle, D.B. and Barnard, E.A. (1990) Sequence and functional expression of a single alpha subunit of an insect nicotinic acetylcholine receptor. *EMBO Journal* 9, 4391–4398.

Martin, F.N. and Loper, J.E. (1999) Soilborne plant diseases caused by *Pythium* spp.: ecology, epidemiology, and prospects for biological control. *Critical Reviews in Plant Sciences* 18, 111–181.

Martin, H. (1972) *Pesticide Manual*, 3rd edn. British Crop Production Council, Farnham, Surrey, UK.

Martin, N.A., Beresford, R.M. and Harrington, K.C. (eds) (2005) *Pesticide Resistance: Prevention and Management Strategies 2005*. New Zealand Plant Protection Society Incorporated, Hastings, New Zealand. 166 pp. (ISBN 0-473-10441-5). Available at: http://www.nzpps.org/resistance/benzimidazole.php (accessed 26 August 2013).

Marumo, S., Hattori, H., Abe, H. and Munakata, K. (1968) Isolation of 4-chloroindolyl-3-acetic acid from immature seeds of *Pisum sativum*. *Nature* 219, 959–960.

Masson, F., Szponarski, W. and Rossignol, M. (1996) The heterogeneity of the plasma membrane H^+-ATPase response to auxin. *Plant Growth Regulation* 18, 15–21.

Matoba, Y., Ohnishi, J.I. and Matsuo, M. (1993) A simulation of insecticides in indoor aerosol space spraying. *Chemosphere* 26, 1167–1186.

Matsson, M., Ackrell, B.A.C., Cochran, B. and Hederstedt, L. (1998) Carboxin resistance in *Paracoccus denitrificans* conferred by a mutation in the membrane-anchor domain of succinate-quinone reductase (complex II). *Archives of Microbiology* 170, 27–37.

Matsumura, F. (1985) *Toxicology of Insecticides*, 2nd edn. Plenum Press, New York, pp. 1–598.

Matsumura, F. and Clark, J.M. (1982) ATP-utilizing system in the squid axons: a review on the biochemical aspects of ion-transport. *Progress in Neurobiology* 18, 231–255.

Matsuno, A.-Y. and Hatefi, Y. (1997) Ubiquinol: cytochrome *c* oxidoreductase: the redox reactions of the bis-heme cytochrome *b* in unenergized and energized submitochondrial particles. *Journal of Biological Chemistry* 272, 16928–16933.

Matthews, G.A. (1999) *Application of Pesticides to Crops*. Imperial College Press, London.

Mayerhoff, A. (2009) Chemicals: Our Champions, Our Killers. *Los Angeles Times* dated 28 December 2008. Reproduced in *PAN North America*, Spring issue.

Mayer, M.S. and McLaughlin, J.R. (1991) *Handbook of Insect Pheromones and Sex Attractants*. CRC Press, Ann Arbor, Boston, Massachussets, pp. 1–1083.

Mayer, R.T., Chen, A.C. and DeLoach, J.R. (1981) Chitin synthesis inhibiting insect growth regulators do not inhibit chitin synthase. *Experientia* 37, 337–338.

Mazzola, M., Fujimoto, D.K., Thomashow, L.S. and Cook, R.J. (1995) Variation in sensitivity of *Gaeumannomyces graminis* to antibiotics produced by fluorescent *Pseudomonas* spp. and effect on biological control of take-all of wheat. *Applied and Environmental Microbiology* 61, 2554–2559.

McCaffery, A.R. (1998) Resistance to insecticides in Heliothine Lepidoptera: a global view. *The Royal Society* 353, 1735–1750.

McCarthy, J.F. (1992) Regulatory issues and adjuvants. In: *Adjuvants for Agrichemicals*. CRC Press, Boca Raton, Florida, pp. 245–246.

McClanahan, R.H., Solsten, R.T., Feng, P.C.C. and Logusch, E.W. (1995) *In vitro* metabolism of thiazopyr plant metabolites by rat liver enzymes. *Pesticide Biochemistry and Physiology* 51, 90–98.

McConn, M., Creelman, R.A., Bell, E., Mullet, J.E. and Browse, J. (1997) Jasmonate is essential for insect defense in *Arabidopsis*. *Proceedings of the National Academy of Sciences USA* 94, 5473–5477.

McCracken, T., Jonasson, K., Bennett, A., Dean, R. and Johansen, A. (1996) *Spray Dynamics for an Improved Agricultural Spray Nozzle System*. ILASS Americas (Institute for Liquid Atomization and Spray Systems, North and South America), Irvine, California.

McCutchen, W.F. and Flexner, L. (1999) Joint actions of baculoviruses and other control agents. In: Hall, F.R. and Menn, J.J. (eds) *Biopesticides – Use and Delivery*. Humana Press, Totowa, New Jersey, pp. 341–355.

McKirdy, S.J. and Jones, R.A.C. (1996) Use of Imidacloprid and newer generation synthetic pyrethroids to control the spread of barley yellow dwarf luteovirus in cereals. *Plant Disease* 80, 895–901.

McRae, D.H. and Bonner, J. (1952) Diortho substituted phenoxyacetic acids as antiauxins. *Plant Physiology* 27, 834–838.

Mehlhorn, H. (2008) *Encyclopedia of Parasitology*, 3rd edn. Springer, Berlin.

Meister, R.T. (ed.) (1992) *Farm Chemical Handbook*. Meister Publishing, Willoughby, Ohio.

Mendes, I.C., Bandick, A.K., Dick, R.P. and Bottomley, P.J. (1999) Microbial biomass and activities in soil aggregates affected by winter cover crops. *Soil Science Society of America Journal* 63, 873–881.

Menn, J.J. and Hall, F.R. (1999) Biopesticides present status and future prospects. In: Hall, F.R. and Menn, J.J. (eds) *Biopesticides – Use and Delivery*. Humana Press, Totowa, New Jersey, pp. 1–10.

Meola, S.M. and Mayer, R.T. (1980) Inhibition of cellular proliferation of imaginal epidermal cells by diflubenzuron in pupae of the stable fly. *Science* 207, 985–987.

Mes, J., Marchand, L. and Davies, D.J. (1990) Organochlorine residues in adipose tissue of Canadians. *Bulletin of Environmental Contamination and Toxicology* 45, 681.

Metcalf, R.L. (1989) Insect resistance to insecticides. *Pesticide Science* 26, 333–358.

Michell, R.H., Jafferji, S.S. and Jones, L.M. (1979) The possible involvement of phosphatidylinositol breakdown in the mechanism of stimulus-response coupling at receptors which control cell-surface calcium gates. *Advances in Experimental Medicine and Biology* 83, 447–464.

Mihara, K., Ohkawa, H. and Miyamoto, J. (1981) Metabolism of tolclofos-methyl in rats and mice. *Journal of Pesticide Science* 6, 65–74.

Mikami, N., Imanishi, K., Yamada, H. and Miyamoto, J. (1984) Photodegradation of the fungicide tolclofos-methyl in water and on soil surface. *Journal of Pesticide Science* 9, 215–222.

Millard, C.B., Kryger, G., Ordentlich, A., Greenblatt, H.M., Harel, M., Raves, M.L. and Segall, Y. (1999) Crystal structures of aged phosphonylated acetylcholinesterase: nerve agent reaction products at the atomic level. *Biochemistry* 38, 7032–7039.

Miller, G.H., Doukas, P.H. and Seydel, J.K. (1972) Sulfonamide structure-activity relationship in a cell-free system. Correlation of inhibition of folate synthesis with antibacterial activity and physicochemical parameters. *Journal of Medical Chemistry* 15, 700–706.

Miller, P., Fung, C. and Gingher, B. (1991) Animal metabolism. In: Shaner, D. and O'Connor, S. (eds) *The Imidazolinone Herbicides*. CRC Press, Boca Raton, Florida, pp. 168–172.

Miller, T.A. and Salgado, V.I. (1985) The mode of action of pyrethroids on insects. In: Leahey, J.P. (ed.) *Pyrethroid Insecticides*. Taylor & Francis, London, pp. 43–97.

Mills, S.A. and Frankenberger, W.T. Jr (1994) Evaluation of phosphorus sources promoting bioremediation of diesel fuel in soil. *Bulletin of Environmental Contamination and Toxicology* 53, 280–284.

Milne, G.W.A. (1995) *CRC Handbook of Pesticides*. Wiley-Interscience. Wiley, New York.

Ministry of Agriculture, Fisheries and Food (1997) *Assessment of Humaneness of Vertebrate Control Agents*. Ministry of Agriculture, Fisheries and Food, Pesticides Safety Directorate, York, UK.

Mishustin, E.N. and Naumova, A.N. (1962) Bacterial fertilizers, their effectiveness and mode of action. *Mikrobiologiya* 31, 543–555 (in Russian with English abstract).

Mitchell, J.W. and Livingston, G.A. (1968) *Methods of Studying Plant Hormones and Growth-Regulating Substances*. Agriculture Handbook No. 336. Agricultural Research Service, United States Department of Agriculture (USDA), Washington, DC.

Mitsui, T., Nobusawa, C. and Fukami, J. (1984) Mode of inhibition of chitin synthesis by diflubenzuron in the cabbage armyworm, *Mamestra brassicae* L. *Journal of Pesticide Science* 9, 19–26.

Mitsui, T., Tada, M., Nobusawa, C. and Yamaguchi, I. (1985) Inhibition of UDP-N-acetylglucosamine transport by diflubenzuron across biomembranes of the midgut epithelial cells in the cabbage armyworm, *Mamestra brassicae* L. *Journal of Pesticide Science* 10, 55–60.

Miyata, H., Kashimoto, T. and Kunita, N. (1977) Detection and determination of polychlorodibenzofurans in normal human tissues and Kanemi rice oil caused 'Kanemi Yusho'. *Journal of the Food Hygiene Society of Japan* 19, 260.

Monaco, T.J., Weller, S.C. and Ashton, F.M. (2002) *Weed Science: Principles and Practices*, 4th edn. Wiley, New York.

Montville, T.J. and Bruno, M.E.C. (1993) Evidence that dissipation of proton motive force is a common mechanism of action for bacteriocins and other antimicrobial proteins. *Journal of Food Microbiology* 24, 53–74.

Moores, G.D., Devine, G.J. and Devonshire, A.L. (1994) Insecticide-insensitive acetylcholinesterase can enhance esterase-based resistance in *Myzus persicae* and *Myzus nicotianae*. *Pesticide Biochemistry and Physiology* 49, 114–120.

Moores, G.D., Gao, X.W., Denholm, I. and Devonshire A.L. (1996) Characterisation of insensitive acetylcholinesterase in insecticide-resistant cotton aphids, *Aphis gossypii* Glover (Homoptera: Aphididae). *Pesticide Biochemistry and Physiology* 56, 102–110.

Moreland, D.E. and Hilton, J.L. (1976) Actions on photosynthetic systems. In: Audus, L.J. (ed.) *Herbicides, Physiology, Biochemistry, Ecology*, Vol. 1, 2nd edn. Academic Press, London, pp. 493–523.

Morgan, P. and Watkinson, R.J. (1992) Factors limiting the supply and efficiency of nutrient and oxygen supplements for the *in situ* biotreatment of contaminated soil and groundwater. *Water Research* 26, 73–78.

Morgan, P.W. (1976) Effects on ethylene physiology. In: Audus L.J. (ed.) *Herbicides*, Vol. 1. Academic Press, New York.

Morgan, W.M. (1981) The distribution and persistence of iprodione applied by thermal fogging in a glasshouse tomato crop. *Annals of Applied Biology* 98, 93–99.

Moriarty, F., Bell, A.A. and Hanson, H. (1986) Does p,p'-DDE thin eggshells? *Environmental Pollution Series A* 40, 257–286.

Morita, M., Nakagawa, J., Akiyama, K., Mimura, S. and Isono, N. (1977) Detailed examination of polychlorinated dibenzofurans in PCB preparations and Kanemi Yusho oil. *Bulletin of Environmental Contamination and Toxicology* 18, 67–73.

Moscardi, F. (1988) Production and use of entomopathogens in Brazil. In: Roberts, D.W. and Granados, R.R. (eds) *Biological Pesticides and Novel Plant – Pest Resistance for Insect Pest Management*. Cornell University Press, Ithaca, New York, pp. 53–60.

Moseley, W.M., Paulissen, J.B. and Goodwin, M.C. (1992) Recombinant bovine somatotropin improves growth performance in finishing beef steers. *Journal of Animal Science* 70, 412–425.

Mueller, F. (2000) *Agrochemicals*. Wiley-VCH, Weinheim, Germany.

Mueller, M.J. (1997) Enzymes involved in jasmonic acid biosynthesis. *Journal of Plant Physiology*. 100, 653–663.

Müller, B.P., Zumdick, A., Schuphan, I. and Schmidt, B. (2001) Metabolism of the herbicide glusinateammonium in plant cell cultures of transgenic (rhizomania-resistant) and non-transgenic sugarbeet (*Beta vulgaris*), carrot (*Daucus carota*), purple foxglove (*Digitalis purpurea*) and thorn apple (*Datura stramonium*). *Pest Management Science* 57, 46–56.

Muller-Rober, B.T., Kobmann, R.G., Hannah, L.C., Willmitzer, L. and Sonnewald, U. (1990) One of the two ADP-glucose pyrophosphorylase genes from potato responds strongly to elevated levels of sucrose. *Molecular and General Genetics* 224, 136–146.

Mumma, R.O. and Brady, J.F. (1987) Immunological assays for agrochemicals. In: Greenhalgh, R. and Roberts, T.R. (eds) *Pesticide Science and Biotechnology*. Proceedings of the 6th International Congress of Pesticide Chemistry. Blackwell Scientific Publications, Oxford, pp. 341–348.

Mutel, C.F. and Donham, K.J. (1983) *Medical Practices in Rural Communities*. Springer, New York, pp. 77–78.

Nakagawa, Y. and Matsumura, F. (1994) Diflubenzuron affects gammathioGTP stimulated $Ca2^{+}$ transport *in vitro* in intracellular vesicles from the integument of the newly molted American cockroach *Periplaneta americana* L. *Insect Biochemistry and Molecular Biology* 24, 1009–1015.

Nakagawa, Y., Ishii, S. and Matsumura, F. (1996) Diflubenzuron stimulates phosphorylation of a 39 kDa integumental protein from newly molted American cockroach (*Periplaneta americana*). *Insect Biochemistry and Molecular Biology* 26, 891–898.

Nakajima, H., Muranaka, T., Ishige, F., Akutsu, K. and Oeda, K. (1997) Fungal and bacterial resistance in transgenic plants expressing human lysozyme. *Plant Cell Reproduction* 16, 674–679.

Nakajima, M., Itoi, K., Takamatsu, Y., Sadao, S., Furukawa, Y., Furuya, K., Honma, T., Kadotani, J., Kozasa, M. and Haneishi, T. (1991) Cornexitin: a new fungal metabolite with herbicidal activity. *Journal of Antibiotics* 44, 1065–1072.

Napier, R.M. and Venis, M.A. (1995) Tansley review No.79 – auxin action and auxin-binding proteins. *New Phytologist* 129, 167–201.

National Safety Council (NSC) (1950–1998) *Accident Facts*. NSC, Chicago, Illinois.

Naxine, M. (2010) *Pesticides in India: Recent Facts and Figures*. Agri-horticultural Publishing House, Nagpur, India.

Nedwell, D.B. and Rutter, M. (1994) Influence of temperature on growth rate and competition between two psychrotolerant Antarctic bacteria: low temperature diminishes affinity for substrate uptake. *Applied and Environmental Microbiology* 60, 1984–1992.

Nelson, O.E. and Rines, H.W. (1962) The enzymatic deficiency in the waxy mutant of maize. *Biochemical and Biophysical Research Communications* 9, 297–300.

Neumann, R. and Guyer, W. (1987) Biochemical and toxicological differences in the modes of reaction of the benzoylureas. *Pesticide Science* 20, 147–156.

Newhouse, K., Wang, T. and Anderson, P.C. (1991) Imidazolinone-resistant crops. In: Shaner, D. and O'Connor, S. (eds) *The Imidazolinone Herbicides*. CRC Press, Boca Raton, Florida, pp. 139–150.

New Zealand Food Safety Authority (NZFSA) (2013) NZFSA Pesticide MRL Database. Available at: https://pxmrl.maf.govt.nz/ (accessed 30 July 2013).

Nicolas, G. and Sillans, D. (1989) Immediate and latent effects of carbon dioxide on insects. *Annual Review of Entomology* 34, 97–116.

Nigovic, B., Kojic-Prodic, B., Antolic, S., Tomic, S., Puntarec, V. and Cohen, J.D. (1996) Structural studies on monohalogenated derivatives of the phytohormone indole-3-acetic acid (auxin). *Acta Crystallographica Section B: Structural Science* 52, 332–343.

Nimrod, A.C. and Benson, W.H. (1996) Environmental estrogenic effects of alkylphenol ethoxylates. *Critical Reviews in Toxicology* 26(3), 335–364.

Noppun, V., Miyataand, T. and Saito, T. (1987) Insensitivity of acetylcholinesterase in phenthoate resistant diamondback moth, *Plutella xylostella*. *Applied Entomology and Zoology* 22, 116–18.

Norris, S.R., Barrette, T.R. and DellaPenna, D. (1995) Genetic dissection of carotenogenesis in *Arabidopsis thaliana* defines plastoquinone as the electron acceptor for phytoene desaturation. *The Plant Cell* 7, 2139–2149.

Okada, K., Ueda, J., Komaki, M.K., Bell, C.J. and Shimura, Y. (1991) Requirement of the auxin polar transport system in early stages of *Arabidopsis* floral bud formation. *The Plant Cell* 3, 677–684.

Olsen, P., Settle, H. and Swift, R. (1980) Organochlorine residues in wings of ducks in South-Eastern Australia. *Australian Wildlife Research* 7, 139.

Omura, S. and Shiomi, K. (2007) Discovery, chemistry, and chemical biology of microbial products. *Pure and Applied Chemistry* 79 (4), 581–591.

Organization for Economic Cooperation and Development (OECD) (1981) *Acute Inhalation Toxicity*. Guideline for Testing of Chemicals No. 403 (adopted 12 May 1981). OECD, Washington, DC.

Osteen, C.D. and Szmedra, P.I. (1989) *Agricultural Pesticide Use Trends and Policy Issues*. United States Department of Agriculture (USDA) Economic Research Service, Washington, DC, 622 pp.

Owen, M.D.K. (1997) North American developments in herbicide-tolerant crops. In: *Proceedings of the Brighton Crop Protection Conference*. Vol. 3. BCPC Publications, Croydon, UK, pp. 955–963.

Padgette, S.R., Huynh, Q.K., Borgmeyer, J., Shah, D.M., Brand, L.A., Re, D.B., Bishop, B.F., Rogers, S.G., Fraley, R.T. and Kishore, G.M. (1988) Bacterial expression and isolation of *Petunia* hybrid 5-enolpyruvylshikimate-3-phosphate synthase. *Biochimica et Biophysica Acta* 258, 564–573.

Pallett, K.E., Cramp, S.M., Little, J.P., Veerasekaran, P., Crudace, A.J. and Slater, A.E. (2001) Isoxaflutole: the background to its discovery and the basis of its herbicidal properties. *Pest Management Science* 57, 133–142.

Panda, H. (2008) *Pesticides, Insecticides, Fungicides, Herbicides with Formulae and Processes*. National Institute of Industrial Research, Karachi, Pakistan.

Parkin, T.B. and Shelton, D.R. (1992) Spatial and temporal variability of carbofuran degradation in soil. *Journal of Environmental Quality* 21, 672–678.

Parking, C.S. (1993) Methods for measuring spray droplet sizes. In: Matthews, G.A. and Hislop, E.C. (eds) *Application Technology for Crop Protection*. CAB International, Wallingford, UK, pp. 57–84.

Parkinson, A. and Safe, S. (1987) Mammalian biologic and toxic effects of PCBs. In: Safe, S. and Hutzinger, O. (eds) *Polychlorinated Biphenyls (PCBs): Mammalian and Environmental Toxicology*, Vol. 1. Springer, Heidelberg, Germany, pp. 49–75.

Parkinson, A., Cockerline, R. and Safe, S. (1980a) Polychlorinated biphenyl isomers and congeners as inducers of both 3-methylcholanthrene- and phenobarbitone-type microsomal activity. *Chemico-Biological Interactions* 29, 277–289.

Parkinson, A., Robertson, L., Safe, L. and Safe, S. (1980b) Polychlorinated biphenyls as inducers of hepatic microsomal enzymes: structure-activity rules. *Chemico-Biological Interactions* 30, 271–285.

Parkinson, A., Safe, S.H. and Robertson, L.W. (1983) Immunochemical quantitation of cytochrome P-450 isozymes and epoxide hydrolase in liver microsomes from polychlorinated or polybrominated biphenyl-treated rats. A study of structure-activity relationships. *Journal of Biological Chemistry* 258, 5967–5976.

Paul, E. and Clark, F.E. (1996) *Soil Microbiology and Biochemistry*. Academic Press, New York.

Paulitz, T.C. and Fernando, W.G.D. (1996) Biological seed treatments for the control of soil-borne plant pathogens. In: Utkhede, R.S. and Gupta, V.K. (eds) *Management of Soilborne Diseases*. Kalyani Publishers, New Delhi, India. pp. 185–217.

Payne, G.A. and Widstrom, N.W. (1992) Aflatoxin in maize. *Critical Reviews in Plant Sciences* 10, 423–440.

Peakall, D.B. and Bart, J.R. (1983) Impacts of aerial application of insecticides on forest birds. *Critical Reviews in Environmental Control* 13, 117–165.

Pearce, P.A. and Baird, J.C. (1970) DDT closes New Brunswick woodcock season. *Canadian Field-Naturalist* 85, 82.

Penninckx, I.A.M.A., Eggermont, K., Terras, F.R.G., Thomma, B.P.H.J., De Samblanx, G.W., Bchala, A., Metraux, J.-P., Manners, J.M. and Broekaert, W.F. (1996) Pathogen-induced systemic activation of a plant defensin gene in *Arabidopsis* follows a salicylic acid-independent pathway. *The Plant Cell* 8, 2309–2323.

Perlak, F.J. and Fischhoff, D.A. (1993) *Advances in Engineered Pesticides*. Marcel Dekker, New York, pp. 199–211.

Perry, A.S., Yamamoto, I., Ishaaya, I. and Perry, R.Y. (1999) *Insecticides in Agriculture and Environment: Retrospects and Prospects*. Springer, Berlin, pp. 1–261.

Pesticide Action Network (PAN) North America (2009a) Agriculture @ the Crossroads. *PAN North America*, Spring issue.

Pesticide Action Network (PAN) North America (2009b) Campaigns. Lindane's Last Stand? *PAN North America*, Spring issue.

Pesticide Action Network (PAN) North America (2009c) Campaigns. No Methyl Iodide for New York. *PAN North America*, Spring issue.

Pesticide Action Network (PAN) North America (2013) PAN Pesticide Database. Pesticide Action Network, North America. Available at: http://www.pesticideinfo.org (accessed 26 August 2013).

Pesticide and Toxic Chemical News (1987) Review Board Urges Agriculture Canada to Reinstate Alachlor. *Pesticide and Toxic Chemical News* 25 November, p. 15.

Peterson, G.E. (1967) The discovery and development of 2,4-D. *Agricultural History* 41, 243–253.

Pfister, G. and Bahadir, M. (1990) Approaches to evaluation of controlled release pesticides in the laboratory and field. In: Wilkins, R.M. (ed.) *Controlled Delivery of Crop-Protection Agents*. Taylor & Francis, London, pp. 279–309.

Phalen, R.F. (1984) *Inhalation Studies: Foundations and Techniques, Animal Models*. CRC Press, Boca Raton, Florida, pp. 211–241.

Pierson, L.S. III, Wood, D.W. and Pierson, E.A. (1998) Homoserine lactone-mediated gene regulation in plant-associated bacteria. *Annual Review of Phytopathology* 36, 207–226.

Pillonel, C. and Meyer, T. (1997) Effect of phenylpyrroles on glycerol accumulation and protein kinase activity of *Neurospora crassa*. *Pesticide Science* 49, 229–238.

Pimentel, D. (1971) *Ecological Effects of Pesticides on Non-Target Species*. Office of Science and Technology, Executive Office of the President, Washington, DC, pp. 1–220.

Pimental, D. (2007) *Encyclopedia of Pest Management*. CRC Press, Boca Raton, Florida.

Pitterna, T., Cassayre, J., Hüter, O.F., Jung, P.M.J., Maienfisch, P., Kessabi, F.M., Quaranta, L. and Tobler, H. (2009) New ventures in the chemistry of avermectins. *Bioorganic & Medicinal Chemistry Letters* 17, 4085–4095.

Plimmer, J.R. and Klingebiel, U.I. (1976) Photolysis of hexachlorobenzene. *Journal of Agricultural and Food Chemistry* 24, 721–723.

Pohland, A.E. (1993) Mycotoxins in review. *Food Additives and Contaminants* 10, 17–28.

Poland, A. and Knutson, J.C. (1982) 2,3,7,8-Tetrachlorodibenzo-thorn-dioxin and related halogenated aromatic hydrocarbons: examination of the mechanism of toxicity. *Annual Review of Pharmacology and Toxicology* 22, 517–554.

Poland, A., Greenlee, W.F. and Kende, A.S. (1979) Studies on the mechanism of action of the chlorinated dibenzo-*p*-dioxins and related compounds. *Annals of the New York Academy of Sciences* 320, 214–230.

Pomerantz, I., Burke, J., Firestone, D., McKinney, J., Roach, J. and Trotter, W. (1978) Chemistry of PCBs and PBBs. *Environmental Health Perspectives* 24, 133–149.

Pope, D.F. and Mathews, J.E. (1993) *Bioremediation of Hazardous Waste Sites: Practical Approaches to Implementation*. EPA/600/K-93/002. United States Environmental Protection Agency (EPA), Washington, DC, pp. 7.1–7.5.

Post, L.C., De Jong, B.J. and Vincent, W.R. (1974) 1-(2,6-Disubstituted benzoyl)-3-phenylurea insecticide: inhibitors of chitin synthesis. *Pesticide Biochemistry and Physiology* 4, 473–483.

Postma, J., van Elsas, J.D., Govaert, J.M. and van Veen, J.A. (1988) The dynamics of *Rhizobium leguminosarum* biovar. *trifolii* introduced into soil as determined by immunofluorescence and selective plating techniques. *FEMS Microbiology Ecology* 53, 251–260.

Postma, J., Hok-A-Hin, C.H. and van Veen, J.A. (1990) Role of microniches in protecting introduced *Rhizobium leguminosarum* biovar. *trifolii* against competition and predation in soil. *Applied and Environmental Microbiology* 56, 495–502.

Pothuluri, J.V., Moorman, T.B., Obenhuber, D.C. and Wauchope, R.D. (1990) Aerobic and anaerobic degradation of alachlor in samples from a surface-to-groundwater profile. *Journal of Environmental Quality* 19, 525–530.

Powles, S.B., Lorraine-Colwill, D.F., Dellow, J.J. and Preston, C. (1998) Evolved resistance to glyphosate in rigid ryegrass (*Lolium rigidum*) in Australia. *Weed Science* 46, 604–607.

Powlson, D.S. (1980) The effects of grinding on microbial and non-microbial organic matter in soil. *Journal of Soil Science* 31, 77–85.

Pozueta-Romero, J., Frehner, M., Viale, A. and Akazawa, T. (1991) Direct transport of ADPglucose by an adenylate translocator is linked to starch biosynthesis in amyloplasts. *Proceedings of the National Academy of Sciences USA* 88, 5769.

Pratt, D. and May, J. (1994) Agricultural occupational medicine. In: Zenz, C., Dickerson, O.B. and Horvath, E.P. Jr (eds) *Occupational Medicine*, 3rd edn. Mosby-Year Book, St Louis, Missouri, pp. 883–902.

Preiss, J. (1988) *The Biochemistry of Plants*. Academic Press, Orlando, Florida, pp. 184–249.

Prosen, B.J., Korreck, W.M. and Armstrong, J.M. (1991) Design and preliminary performance results of a full scale bioremediation system utilizing an on-site oxygen generation system. In: Hinchee, R.E. and Olfenbuttel, R.F. (eds) In situ *Bioreclamation: Application and Investigation for Hydrocarbons and Contaminated Site Remediation*. Butterworth-Heinemann, Woburn, Massachusetts, pp. 523–528.

Purschwitz, M.A. and Field, W.E. (1990) Scope and magnitude of injuries in the agricultural workplace. *American Journal of Industrial Medicine* 18(2), 179–192.

Quimby, P.C., Fulghum, F.E., Boyette, C.D. and Connick, W.J. (1988) An invert emulsion replaces dew in biocontrol of sicklepod – a preliminary study. In: Hovde, D.A. and Beestman, G.B. (eds) *Pesticide Formulations and Application Systems*. Vol. 8. ASTMSTP980. American Society Testing and Materials (ASTM), Philadelphia, pp. 264–270.

Rakhaminova, A.B., Khavkin, E.E. and Yaguzhinsky, L.S. (1978) [Model for auxin receptor.] *Biokhimiya* 43, 806–823 (in Russian with English abstract).

Ramek, M. and Tomic, S. (1998a) Quantum chemical conformational analysis of the auxin phytohormone 4-methyl-3-indole acetic acid. *International Journal of Quantum Chemistry* 70, 1169–1175.

Ramek, M. and Tomic, S. (1998b) *Ab initio* RHF investigation of mono- and dichlorinated indole-3-acetic acid (IAA) phytohormones. *Journal of Molecular Structure (Theochem)* 454, 167–173.

Rao, V.S. (2000) *Principles of Weed Science*, 2nd edn. Science Publishers, Inc., Enfield, New Hampshire.

Rappe, C. and Buser, H.R. (1989) Chemical and physical properties, analytical methods, sources and environmental levels of halogenated dibenzodioxins and dibenzofurans. In: Kimbrough, R.D. and Jensen, A.A. (eds) *Halogenated Biphenyls, Terphenyls, Naphthalenes, Dibenzodioxins and Related Products*, 2nd edn. Elsevier, Amsterdam, pp. 71–103.

Rappe, C., Buser, H.R. and Bosshardt, H.P. (1979) Health effects of halogenated aromatic hydrocarbons. *Annals of the New York Academy of Sciences* 320, 1–18.

Rappe, C., Buser, H.R., Stalling, D.I., Smith, L.M and Dougherty, R.C. (1981) Identification of polychlorinated dibenzofurans in environmental samples. *Nature* 292, 524–526.

Rasiah, V., Vononey, R.P. and Kachanoski, R.G. (1992) Biodegradation of an oily waste as influenced by nitrogen forms and sources. *Water, Air, Soil Pollution* 65, 143–151.

Ratcliff, D.A. (1967) Decrease in eggshell weight in certain birds of prey. *Nature* 215, 208–210.

Ray, L.E., Murray, H.E. and Giam, C.S. (1983) Analysis of water and sediment from the Nueces Estuary/Corpus Christi Bay (Texas) for selected organic pollutants. *Chemosphere* 12, 1039–1045.

Rayle, D.L. and Cleland, R.E. (1992) The Acid Growth Theory of auxin-induced cell elongation is alive and well. *Plant Physiology* 99, 1271–1274.

Registry of Toxic Effects of Chemical Substances (RTECS) Quarterly Microfiche (1987) Metolachlor. AN3430000. *RTECS Quarterly Microfiche*, April.

Reinecke, D.M. (1999) 4-Chloroindole-3-acetic acid and plant growth. *Plant Growth Regulation* 27, 3–13.

Reinecke, D.M., Ozga, J.A., Ilić, N., Magnus, V., Kojić-Prodić, B. (1999) Molecular properties of 4-substituted indole-3-acetic acids affecting pea pericarp elongation. *Plant Growth Regulation* 27, 39–48.

Rengam, S. (2007a) The People's Caravan for Food Sovereignty. *PAN AP* (Pesticide Action Network Asia Pacific), November issue.

Rengam, S. (2007b) Women and Pesticides in Agriculture. *PAN AP* (Pesticide Action Network Asia Pacific), November issue.

Restifo, L.L. and White, K. (1990) Molecular and genetic approaches to neurotransmitters and neuromodulator systems in *Drosophila*. *Advances in Insect Physiology* 22, 115–219.

Reyes, Z. (1965) Methods of producing microcapsules by chemical methods. US Patent No. 3173878. Issued 16 March 1965. US Patent Office, Alexandria, Virginia.

Ridley, W.P., Fujiwara, H., Cheng, T. and Honegger, J.L. (1998) Metabolism of thiazopyr in laying hens. *Journal of Agricultural and Food Chemistry* 46, 4398–4405.

Ringer, R.K., Aulerich, R.J. and Bleavins, M.R. (1981) Biological effects of PCBs and PBBs on mink and ferrets: a review. In: Khan, M.A.Q. and Stanton, R.H. (eds) *Toxicology of Halogenated Hydrocarbons: Health and Ecological Effects*. Pergamon Press, Elmsford, New York, pp. 329–343.

Risebrough, R.W., Rieche, P., Herman, S.G., Peakall, D.B. and Kirven, M.N. (1968) Polychlorinated biphenyls in the global ecosystem. *Nature* 220, 1098–1102.

Ritger, P.L. and Peppas, N.A. (1987a) A simple equation for description of solute release. I. Fickian and non-Fickian release from non-swellable devices in the form of slabs, spheres, cylinders or discs. *Journal of Controlled Release* 5, 23–26.

Ritger, P.L. and Peppas, N.A. (1987b) A simple equation for description of solute release. II. Fickian and anomalous release from swellable devices. *Journal of Controlled Release* 5, 37–42.

Roberts, P.V., Hopkins, G.D., Mackay, D.M. and Semprini, L. (1990) A field evaluation of *in-situ* biodegradation of chlorinated ethenes. Part 1: Methodology and field site characterization. *Groundwater* 28, 591–604.

Roberts, T. (2000) *Metabolism of Agrochemicals in Plants*. Wiley, New York.

Roberts, T.R. and Hutson, D.H. (eds) (1999a) *Metabolic Pathways of Agrochemicals*, Vol. 2. Royal Society of Chemistry, Cambridge, UK, pp. 17–24, 544, 1195.

Roberts, T.R. and Hutson, D.H. (eds) (1999b) *Metabolic Pathways of Agrochemicals. Part 2: Insecticides and Fungicides*. Royal Society of Chemistry, Cambridge, UK, pp. 187–522, 795–816.

Robinson, R.C. and Wilson, C.W. (1979) Control of *Botrytis cinerea* in glasshouse crops with prodione applied through termal fogging equipment. In: *Proceedings of the Brighton Crop Protection Conference*. BCPC Publications, Croydon, UK, pp. 425–432.

Rodricks, J.V., Hesseltine, C.W. and Mehlman, M.E. (eds) (1977) *Mycotoxins in Human and Animal Health*. Pathotox Publishing, Park Forest South, Illinois.

Rogan, W.J., Gladen, B.C., McKinney, J.D., Carreras, N., Hardy, P., Thullen, J., Tinglestad, J. and Tully, M. (1986) Neonatal effects of transplacental exposure to PCBs and DDE. *Journal of Pediatrics* 109, 335–341.

Roitman, J.N., Mahoney, N.E., Janisievicz, W.J. and Benson, M. (1990) A new chlorinated phenylpyrrole antibiotic produced by the antifungal bacterium *Pseudomonas cepacia*. *Journal of Agricultural and Food Chemistry* 38, 538–541.

Ross, J.H., Driver, J.H., Cochran, R.C., Thongsinthusak, T. and Krieger, R.I. (2001) Could pesticide toxicology studies be more relevant to occupational risk assessment? *Annals of Occupational Hygiene* 45, 5–17.

Rossi, F.S., Di Tomaso, J.M. and Neal, J.C. (1993) Fate of fenoxyaprop-ethyl applied to moisture-stressed smooth crabgrass (*Digitaria ischaemum*). *Weed Science* 41, 335–340.

Rosskopf, E.N., Charudattan, R. and Kadir, J.B. (1999) Use of plant pathogens in weed control. In: Bellows, T.S. and Fisher, T.W. (eds) *Handbook of Biological Control: Principles and Applications of Biological Control*. Academic Press, San Diego, California, pp. 891–918.

Rouchaud, J., Neus, R., Bulcke, K. and Cools, H. (1998) Sulcotrione metabolism in summer corn crops. *Bulletin of Environmental Contamination and Toxicology* 61, 669–676.

Rougerie, I. and Leroux, P. (1997) Quinoxyfen, a fungicide against powdery mildew in wheats and barleys. *Phytoma – La Défense des végétaux* 499, 57–58.

Roulston, W.J., Schnitzerling, H.J. and Schuntner, C.A. (1968) Acetylcholinesterase insensitivity in the Biarra strain of the cattle tick *Boophilus microplus*, as a cause of resistance to organophosphorus and carbamate acaricides. *Australian Journal of Biological Sciences* 21, 759–767.

Rowinski, R.S. (1996) The use of aeroplanes for plant protection in and out of Europe. *EPPO Bulletin* 26, 123–130.

Rudd, R.L. (1964) *Pesticides and the Living Landscape*. University of Wisconsin Press, Madison, Wisconsin, pp. 1–320.

Ruegger, M., Dewey, E., Hobbie, L., Brown, D., Bernasconi, P., Turner, J., Muday, G. and Estelle, M. (1997) Reduced NPA-binding in the tir3 mutant of *Arabidopsis* is associated with a reduction in polar auxin transport and diverse morphological defects. *The Plant Cell* 9, 745–757.

Russell, P.E. (2003) Fungicide Resistance Action Committee (FRAC) Pathogen Risk List, December 2005, adapted from EPPO 2002. FRAC Monograph No. 3. Available from: http://www.frac.info. (accessed 25 June 2010).

Russell, S.G., Monaco, T.J. and Weber, J.B. (1991) Influence of soil moisture on phytotoxicity of cinmethylin to various crops. *Weed Science* 39, 402–407.

Safe, S. (1984) Polychlorinated biphenyls (PCBs) and polybrominated biphenyls (PBBs): biochemistry, toxicology and mechanism of action. *Critical Reviews in Toxicology* 13, 319–394.

Safe, S. (1990) Polychlorinated biphenyls (PCBs), dibenzo-p-dioxins (PCDDs), dibenzofurans (PCDFs) and related compounds: environmental and mechanistic considerations which support the development of toxic equivalency factors (TEFs). *Critical Reviews in Toxicology* 21, 51–88.

Safe, S., Safe, L. and Mullin, M. (1985) Polychlorinated biphenyls (PCBs) – congener-specific analysis of a commercial mixture and a human milk extract. *Journal of Agricultural and Food Chemistry* 33, 24–29.

Safe, S., Safe, L. and Mullin, M. (1987) Polychlorinated biphenyls: environmental occurrence and analysis. In: Safe, S. and Hutzinger, O. (eds) *Polychlorinated Biphenyls (PCBs): Mammalian and Environmental Toxicology*, Vol. 1. Springer, Heidelberg, Germany, pp. 133–145.

Salcher, O. and Lingens, F. (1978) Biosynthese von pyrrolnitrin. Nachweis von 3-chloranthranilsäure und 7-chlorindolessigsäure im kulturmedium von *Pseudomonas aureofaciens*. *Tetrahedron Letters* 34, 3101–3102.

Salisbury, F.B. and Ross, C.W. (1992) *Plant Physiology*, 4th edn. Wadsworth Publishing Company, Belmont, California.

Salminen, J. and Haimi, J. (1996) Effects of pentachlorophenol in forest soil: a microcosm experiment for testing ecosystem responses to anthropogenic stress. *Biology and Fertility of Soils* 23, 182–188.

Sandmann, G. and Boger, P. (1986) Interference of dimethazone with formation of terpenoid compounds. *Zeitschrift für Naturforschung B* 41c, 729–732.

Sands, D.C. and Miller, R.V. (1993a) Altering the host range of mycoherbicides by genetic manipulation. In: Duke, S.O., Menn, J.J. and Plimmer, J. (eds) *Pest Control With Enhanced Environmental Safety*. ACS Symposium Series 524. American Chemical Society (ACS), Washington, DC, pp. 101–110.

Sands, D.C. and Miller, R.V. (1993b) Evolving strategies for biological control of weeds with plant pathogens. *Pesticide Science* 37, 399–403.

Sarniguet, A., Kraus, J., Henkels, M.D., Muehlchen, A.M. and Loper, J.E.E. (1995) The sigma factor σ^s affects antibiotic production and biological control activity of *Pseudomonas fluoroscens* Pf-5. *Proceedings of the National Academy of Sciences USA* 92, 12255–12259.

Sasaki, M., Ooishi, T., Takayama, C. and Mukai, K. (1984) Synthesis and antifungal activity of *O*, *O*-dialkyl *O*-aryl phosphorothioates and related compounds. *Journal of Pesticide Science* 9, 737–744.

Sathe, T.V. (2003) *Agrochemicals and Pest Management*. Daya Publishing House, Delhi, India.

Sattelle, D.B. (1980) Acetylcholine receptors of insects. *Advances in Insect Physiology* 15, 215–315.

Sawhney, B.L., Pignatello, J.J. and Steinberg, S.M. (1988) Determination of 1,2-dibromoethane (EDB) in field soils: implications for volatile organic compounds. *Journal of Environmental Quality* 17, 149–152.

Sawicki, R.M. (1985) Resistance to pyrethroid insecticides in arthropods. In: Hutson, D.H. and Roberts, T.R. (eds) *Insecticides, Progress in Pesticide Biochemistry and Toxicology*, Vol. 5. Wiley, Chichester, UK, pp. 143–192.

Sawyer, T. and Safe, S. (1982) PCB isomers and congeners: induction of aryl hydrocarbon hydroxylase and ethoxyresorufin *O*-deethylase enzyme activities in rat hepatoma cells. *Toxicology Letters* 13, 87–94.

Schafer, E.W. Jr (1991) Bird control chemicals – nature, modes of action, and toxicity. In: Pimentel, D. (ed.) *CRC Handbook of Pest Management in Agriculture* (Vol. 2), 2nd edn. CRC Press, Boca Raton, Florida, pp. 599–610.

Scher, H.B. (1977) Microencapsulated pesticides. In: Scher, H.B. (ed.) *Controlled Release Pesticides*. American Chemical Society (ACS), Washington, DC, pp. 127–144.

Scher, H.B. (ed.) (1999) *Controlled-Release Delivery Systems for Pesticides*. Marcel Dekker, New York.

Scher, H.B., Rodson, M. and Lee, K.S. (1998) Microencapsulation of pesticides by interfacial polymerization utilizing isocyanate or aminoplast chemistry. *Journal of Pesticide Science* 54, 394–400.

Scherff, R.H. (1973) Control of bacterial blight of soybean by *Bdellovibrio bacteriovorus*. *Phytopathology* 63, 400–402.

Schlamann, J.E., ten Hoopen, H.J.G. and Heijnen, J.J. (1996) Large scale production of secondary metabolites by plant cell cultures. In: DiCosmo, F. and Misawa, M. (eds) *Plant Cell Culture Secondary Metabolism Toward Industrial Application*. CRC Press, Boca Raton, Florida, pp. 11–25.

Schmitt, C.J., Zajicek, J.L. and Ribick, M.A. (1985) National pesticide monitoring program – residues of organochlorine chemicals in freshwater fish, 1980–81. *Archives of Environmental Contamination and Toxicology* 14, 225.

Schmitzer, P.R., Eilers, R.J. and Cseke, C. (1993) Lack of cross-resistance of imazaquin-resistant *Xanthium strumarium* acetolactate synthase to flumetsulam and chlorimuron. *Plant Physiology* 103, 281–283.

Schneider, E.A., Kazakoff, C.W. and Wighman, F. (1985) Gas chromatography-mass spectrometry evidence for several endogenous auxins in pea seedling organs. *Planta* 165, 232–241.

Schneider, G. (1970) Morphactins: physiology and performance. *Annual Review of Plant Physiology* 21, 499–536.

Schnider, U., Keel, C., Blumer, C., Troxler, J., Defago, G. and Haas, D. (1995) Amplification of the housekeeping sigma factor in *Pseudomonas fluorescens* CHA0 enhances antibiotic production and improves biocontrol abilities. *Journal of Bacteriology* 177, 5387–5392.

Schrader, G. (1963) *Die Entwicklung neuer insektizider Phosphorsaure Ester*. Verlag Chemie, Weinheim, Germany, p. 444.

Schroeder, M.E. and Flattum, R.F. (1984) The mode of action and neurotoxic properties of the nitromethylene heterocycle insecticides. *Pesticide Biochemistry and Physiology* 22, 148–160.

Schulten, H.R. and Leinweber, P. (2000) New insights into organic-mineral particles: composition, properties and models of molecular structure. *Biology and Fertility of Soils* 30, 399–432.

Schultz, A., Wengenmayer, F. and Goodman, H.M. (1990) Genetic engineering of herbicide resistance in higher plants. *Critical Reviews in Plant Sciences* 9, 1–15.

Schwarzenbach, R.P., Gschwend, P.M. and Imboden, D.M. (1993) *Environmental Organic Chemistry*. Wiley, New York.

Schwinn, F.J. and Staub, T. (1995) Oomycete fungicides: phenylamides and other fungicides against oomycetes. In: Lyr, H. (ed.) *Modern Selective Fungicides: Properties, Applications, Mechanisms of Action*, 2nd edn. Fischer Verlag, New York, pp. 323–346.

Scott, J.G. (1989) Cross-resistance to the biological insecticide abamectin in pyrethroid resistant strains of house flies. *Pesticide Biochemistry and Physiology* 34, 27–31.

Scott, J.G. (1995) Effects of temperature on insecticide toxicity. In: Roe, R.M. and Kuhr, R.J. (eds) *Reviews in Pesticide Toxicology*. North Carolina State University, Raleigh, North Carolina, pp. 111–135.

Scott, P.M. (1993) Mycotoxins. *Journal of the Association of Official Analytical Chemists* 76, 112–119.

Scow, K.M. and Alexander, M. (1992) Effect of diffusion and sorption on the kinetics of biodegradation: experimental results with synthetic aggregates. *Soil Science Society of America Journal* 56, 128–134.

Secor, J. (1994) Inhibition of barnyardgrass 4-hydroxyphenylpyruvate dioxygenase by sulcotrione. *Plant Physiology* 106, 1429–1433.

Segal, L.M. and Wighman, F. (1982) Gas chromatographic and GC-MS evidence for the occurrence of 3-indolylpropionic acid and 3-indolylacetic acid in seedlings of *Cucurbita pepo*. *Journal of Plant Physiology* 56, 367–370.

Semprini, L. and McCarty, P.L. (1991) Comparison between model simulations and field results for *in situ* biorestoration of chlorinated aliphatics. Part 1, Biostimulation of methanotrophic bacteria. *Groundwater* 29, 365–374.

Semprini, L., Roberts, P.V., Hopkins, G.D. and McCarty, P.L. (1990) A field-evaluation of *in situ* biodegradation of chlorinated ethenes. Part 2, Results of biostimulation and biotransformation experiments. *Groundwater* 28, 715–727.

Semprini, L., Hopkins, G.D., Roberts, P.V., Grbic-Galic, D. and McCarty, P.L. (1991) A field evaluation of *in situ* biodegradation of chlorinated ethenes. Part 3, Studies of competitive inhibition. *Groundwater* 29, 239–250.

Sgard, F., Fraser, S.P., Katkowska, M.J., Djamgoz, M.B.A., Dunbar, S.J. and Windass, J.D. (1998) Cloning and functional characterisation of two novel nicotinic acetylcholine receptor α subunits from the insect pest *Myzus persicae*. *Journal of Neurochemistry* 71, 903–912.

Shah, D., Horsch, R.B., Klee, K.J., Kishore, G., Winther, J., Turner, N.E., Hironake, C.M., Sanders, P.R., Gaser, C.S., Aybert, S., Siegel, N.R., Rogers, S.G. and Fraley, R.T. (1986) Engineering herbicide tolerance in transgenic plants. *Science* 233, 478–481.

Shaner, D.L. and Mallipudi, N.M. (1991) Mechanisms of selectivity of the imidazolinones. In: Shaner, D. and O'Connor, S. (eds) *The Imidazolinone Herbicides*. CRC Press, Boca Raton, Florida, pp. 92–102.

Shardlow, M. (2006) Sheep dip banned to save rivers. *Pesticides News* No. 72 (June issue).

Sharpe, R.M., Fisher, J.S., Millar, M.M., Jobling, S. and Sumpter, J.P. (1995) Gestational and lactational exposure of rats to xenoestrogens results in reduced testicular size and sperm production. *Environmental Health Perspectives* 103(12), 1136–1143.

Sherwood, P. (2007) Mustard Can Drive Away Pests. *PAN North America*, Spring issue.

Shibuya, S. and Shimazaki, I. (1999) *Shibuya Index: a Complete Index of Pesticides*, 8th edn. Zen-Noh, Tokyo, p. 873.

Shiva, V. (2008) We Need to Reclaim the Food System. *PAN North America*, Spring issue.

Shivanandappa, T., Joseph, P. and Krishnakumari, M.K. (1988) Response of blood and brain cholinesterase to dermal exposure of Bromophos in the rat. *Toxicology* 48, 199–208.

Siegel, M. (1981) Sterol-binding fungicides: effect on sterol biosynthesis and site of action. *Plant Disease* 65, 986–989.

Silman, I. and Futerman, A.H. (1987) Modes of attachment of acetylcholinesterase to the surface membrane. *European Journal of Biochemistry* 170, 11–22.

Sims, R.C. (1993) *Bioremediation of Hazardous Waste Sites: Practical Approaches to Implementation.* EPA/600/K-93/002. US Environmental Protection Agency (EPA), Washington, DC, pp. 3.1–3.8.

Sims, S.B., Greenplate, J.T., Stone, T.B., Caprio, M.A. and Gould, F.L. (1996) *Molecular Genetics and Evolution of Pesticide Resistance.* ACS Symposium Series 645. American Chemical Society (ACS), Washington, DC, pp. 229–242.

Singh, R.S. (2002) *Introduction to Principles of Plant Pathology.* Oxford & IBH Publishing, New Delhi, India.

Singh, S. and Singh, J.S. (1996) Water-stable aggregates and associated organic matter in forest, savanna and cropland soils of a seasonally dry tropical region, India. *Biology and Fertility of Soils* 22, 76–82.

Six, J., Elliott, E.T. and Paustian, K. (1999) Aggregate and soil organic matter dynamics under conventional and no-tillage systems. *Soil Science Society of America Journal* 63, 1350–1358.

Six, J., Paustian, K., Elliott, E.T. and Combrink, C. (2000) Soil structure and soil organic matter: I. Distribution of aggregate size classes and aggregate associated carbon. *Soil Science Society of America Journal* 64, 681–689.

Skogland, T., Lomeland, S. and Goksoyr, J. (1988) Respiratory burst after freezing and thawing of soil: experiments with soil bacteria. *Soil Biology and Biochemistry* 20, 851–856.

Slovin, J. (1997) Phytotoxic conjugates of indole-3-acetic acid: potential agents for biochemical selection of mutants in conjugate hydrolysis. *Plant Growth Regulation* 21, 215–221.

Slusarenko, A.J., Fraser, R.S.S. and van Loon, L.C. (2000) *Mechanisms of Resistance to Plant Diseases.* Kluwer Academic Publishers, Dordrecht, The Netherlands.

Smart, G.C. (1995) Entomopathogenic nematodes for the biological control of insects. *Journal of Nematology* 27, 529–534.

Smeda, R.J., Vaughn, K.C. and Morrison, I.N. (1992) A novel pattern of herbicide cross-resistance in a trifluralin-resistant biotype of green foxtail (*Setaria viridis* (L.) Beauv.). *Pesticide Biochemistry and Physiology* 42, 227–241.

Smissaert, H.R. (1964) Cholinesterase inhibition in spider mites susceptible and resistant to organophosphate. *Science* 143, 129–131.

Smith, A.G. (1991) Chlorinated hydrocarbon insecticides. In: Hayes, W.J. Jr and Laws, E.R. Jr (eds) *Handbook of Pesticide Toxicology.* Academic Press, San Diego, California, pp. 731–915.

Smith, D.R. and Renfro, B.L. (1999) Downy mildews. In: White, D.G. (ed.) *Compendium of Corn Diseases.* The American Phytopathology Society, St Paul, Minnesota, pp. 25–32.

Smith, H.S. (1919) On some phases of insect control by the biological method. *Journal of Economic Entomology* 12, 288–292.

Smith, K.L. (1999) Dispersible microparticles. In: Scher, H.B. (ed.) *Controlled-Release Delivery Systems for Pesticides.* Marcel Dekker, New York, pp. 137–149.

Solano-Serena, F., Marchal, R., Blanchet, D. and Vandecasteele, J.P. (1998) Intrinsic capacities of soil microflora for gasoline degradation. *Biodegradation* 9, 319–326.

Song, H.G. and Bartha, R. (1990) Effects of jet fuel spills on the microbial community of soil. *Applied and Environmental Microbiology* 56, 646–651.

Soto, A.M., Sonnenschein, C., Chung, K.L., Fernandez, M.F., Olea, N. and Serrano, F.O. (1995) The E-screen assay as a tool to identify estrogens: an update on estrogenic environmental pollutants. *Environmental Health Perspectives* 103 (Suppl. 7), 113–122.

Sparks, T.C., Thompson, G.D., Kirst, H.A., Hertlein, M.B., Mynderse, J.S., Turner, J.R. and Worden, T.V. (1998) Fermentation-derived insect control agents, the spinosyns. In: Hall, F.R. and Menn J.J. (eds) *Biopesticides: Use and Delivery.* Springer Science+ Business Media, New York, pp. 171–188.

Srivastava, M.P. (2003) Transfer of plant pathology knowledge for rural prosperity – an Asian perspective. *Australasian Plant Pathology* 32, 187–194.

Srivastava, U.K. and Patel, N.T. (1990) *Pesticides Industry in India.* Oxford & IBH Publishing, New Delhi, India.

Staswick, P.E., Yuen, G.Y. and Lehman, C.C. (1998) Jasmonate signaling mutants of *Arabidopsis* are susceptible to the soil fungus *Pythium irregulare. The Plant Journal* 16, 747–754.

Stedman, E. and Barger, G. (1925) Physostigmine (eserine). Part III. *Journal of the Chemical Society* 127, 247–258.

Steinberg, S.M., Pignatello, J.J. and Sawhney, B.L. (1987) Persistence of 1,2-dibromoethane in soils: entrapment in intraparticle micropores. *Environmental Science and Technology* 21, 1201–1208.

Steinfield, U., Sierotski, H., Parisi, S., Poirey, S. and Gisi, U. (2001) Sensitivity of mitochondrial respiration to different inhibitors in *Venturia inaequalis. Pest Management Science* 57, 787–796.

Steinmann, T., Geldner, N., Grebe, M., Mangold, S., Jackson, C.L., Paris, S. Gälweiler, L., Palme, K. and Jürgens, G. (1999) Coodinated polar localization of auxin efflux carrier PINI1 by GNOM ARF GEF. *Science* 286, 316–318.

Stenersen, J. (2004) *Chemical Pesticides – Mode of Action and Toxicology.* CRC Press, Boca Raton, Florida.

Stenersen, J. and Somme, L. (1963) Notes on cross-resistance and genetics of resistance to the DDT-group insecticides in the stable fly (*Stomoxys calcitrans*). *Norsk Entomologisk Tidsskrift* 12, 113–117.

Sterling, T.M. and Hall, J.C. (1997) Mechanism of action of natural auxins and the auxinic herbicides. In: Roe, R.M., Burton, J.D. and Kuhr, R.J. (eds) *Herbicide Activity: Toxicology, Biochemistry and Molecular Biology.* IOS Press, Burke, Virginia, pp. 111–141.

Stout, S.J., daCunha, A.R., Picard, G.L. and Safarpour, M.M. (1996) Microwave-assisted extraction coupled with liquid chromatography/electrospray ionization mass spectrometry for the simplified determination of imidazolinone herbicides and their metabolites in plant tissues. *Journal of Agricultural and Food Chemistry* 44, 3548–3553.

Strauch, E., Wohlleben, W. and Pühler, A. (1988) Cloning of a phosphinothricin N-acetyltransferase gene from *Streptomyces viridochromogenes* Tü494 and its expression in *Streptomyces lividans* and *Escherichia coli*. *Gene* 63, 65–75.

Sullivan, J.B. Jr, Gonzales, M., Krieger, G.R. and Runge, C.F. (1992) Health-related hazards of agriculture. In: Sullivan, J.B. Jr and Krieger, G.R. (eds) *Hazardous Materials Toxicology, Clinical Principles of Environmental Health*. Williams and Wilkins, Baltimore, Maryland, pp. 642–666.

Sussman, J.L., Harel, M., Frolow, F., Oefner, C., Goldman, A., Toker, L. and Silman, I. (1991) Atomic structure of acetylcholinesterase from *Torpedo californica*: a prototypic acetylcholine-binding protein. *Science* 253, 872–879.

Swindoll, C.M., AeHon, C.M. and Pfaender, F.K. (1988) Influence of inorganic and organic nutrients on aerobic biodegradation and on the adaptation response of subsurface microbial communities. *Applied and Environmental Microbiology* 54, 212–217.

Takahashi, H., Yoshida, M., Murao, N. and Maita, K. (1994) Different lethality between micron-sized and submicron-sized aerosols of organophosphorus insecticide, chlorfenvinphos, in rats. *Toxicology Letters* 73, 103–111.

Takahashi, Y., Matsumoto, A., Seino, A., Ueno, J., Iwai, Y. and Omura, S. (2002) *Streptomyces avermectinius* sp. nov., an avermectin-producing strain. *International Journal of Systematic and Evolutionary Microbiology* 52, 2163–2168.

Tanabe, S., Kanna, N., Subramanian, A., Watanabe, S. and Tatsukawa, R. (1987) Highly toxic coplanar PCBs: occurrence, source, persistency and toxic implications to wildlife and humans. *Environmental Pollution* 47, 147–163.

Taylor, W.A. and Andersen, P. (1997) A review of the benefits of air assisted spraying in arable crops. *Protection* 1, 473–482.

Tecle, B., Shaner, D.L., Cunha, A.D., Devine, P.J. and van Ellis, M.R. (1997) Comparative metabolism of imidazolinone herbicides. In: *Proceedings of the Brighton Crop Protection Conference*. Vol. 2. BCPC Publications, Croydon, UK, pp. 605–610.

Terras, F.R.G., Penninckx, I.A.M.A., Goderis, I.J. and Broekaert, W.F. (1998) Evidence that the role of plant defensins in radish defense responses is independent of salicylic acid. *Planta* 206, 117–124.

Teske, M.E., Bowers, J.F., Rafferty, J.E. and Barry, J.W. (1993) FSCBG: an aerial spray dispersion model for predicting the fate of released material behind aircraft. *Environmental Toxicology and Chemistry* 12, 453–464.

Thakur, R.P. and Mathur, K. (2002) Downy mildews of India. *Crop Protection* 21, 333–345.

Theologis, A. (1998) Ethylene signaling: redundant receptors all have their say. *Current Biology* 8, R875–R878.

Thimann, K.V. (1977) *Hormone Action in the Whole Life of Plants*. A general overview written by one of the founders of auxin biochemistry. The University of Massachusetts Press, Amherst, Massachusetts.

Thimann, K.V. and Behnke-Rogers, J. (1950) *The Use of Auxins in the Rooting of Woody Cuttings*. Maria Moors Cabot Foundation, Publication no. 1. Harvard Forest, Petersham, Massachusetts.

Thomashow, L.S., Weller, D.M., Bonsall, R.F. and Pierson, L.S. III (1990) Production of the antibiotic phenazine-1-carboxylic acid by fluorescent pseudomonas species in the rhizosphere of wheat. *Applied Environmental Microbiology* 56, 908–912.

Thompson, C.J., Rao Movva, N., Tizard, R., Crameri, R., Davies, J.E., Lauwereys, M. and Botterman, J. (1987) Characterization of the herbicide-resistance gene bar from *Streptomyces hygroscopicus*. *EMBO Journal* 6, 2519–2523.

Thompson, I.P., Bailey, M.J., Ellis, R.J., Boyd, E.M, Maguire, N. and Meharg, A.A. (1999) Concentration dependent response of soil microbial communities to 1,2-dichlorobenzene. *Environmental Toxicology and Chemistry* 18, 1891–1898.

Thomson, D.R., Gut, L.J. and Jenkins, J.W. (1999) Pheromones for insect control, strategies and successes. In: Hall, F.R. and Menn, J.J. (eds) *Biopesticides – Use and Delivery*. Humana Press, Totowa, New Jersey, pp. 385–412.

Thorn, G.D. and Richardson, L.T. (1962) Exudate produced by *Glomerella cingulata* spores in the presence of copper and dithiocarbamate ions. *Canadian Journal of Botany* 40, 25–33.

Thornhill, E.W., Matthews, G.A. and Clayton, J.S. (1995) Potential operator exposure to herbicides: a comparison between knapsack and CDA hand sprayers. In: *Proceedings of the Brighton Crop Protection Conference*. Vol. 3. BCPC Publications, Croydon, UK, pp. 507–512.

Thornhill, E.W., Matthews, G.A. and Clayton, J.S. (1996) Potential operator exposure to insecticides: a comparison between knapsack and CDA spinning disc sprayers. In: *Proceedings of the Brighton Crop Protection Conference*. Vol. 3. BCPC Publications, Croydon, UK, pp. 1175–1180.

Thurman, E.M. and Batian, C. (1998) Determination of atrazine and atrazine mercapture in drinking water samples and in urine using immunoaffinity SPE with positive ion spray HPLC/MS. Paper presented at the 15th Symposium on Liquid Chromatography/Mass Spectrometry, Montrux, Switzerland, 9–10 November.

Thurman, E.M., Ferrer, I. and Barcelo, D. (2001) Choosing between atmospheric pressure chemical ionization and electrospray interfaces for the HPLC/MS analysis of pesticides. *Annals of Chemistry* 73, 5441–5449.

Tiedje, J.M. (1993) *In Situ Bioremdiation: When Does It Work?* National Academy Press, Washington, DC.

Tillman, R.W., Siegel, M.R. and Long, J.W. (1973) Mechanism of action and fate of the fungicide chlorothalonil (2,4,5,6-tetrachloroisophthalonitrile) in biological systems. 1. Reactions with cells and subcellular components of *Saccharomyces pastorianus*. *Pesticide Biochemistry and Physiology* 3, 160–167.

Tisdale, W.H. and Flenner, A.L. (1942) Derivatives of dithiocarbamic acid as pesticides. *Industrial and Engineering Chemistry Research* 34, 501–502.

Tomic, S., Gabdoulline, R.R., Kojic-Prodic, B. and Wade, R.C. (1998a) Classification of auxin plant hormones by interaction property similarity indices. *Journal of Computer-Aided Molecular Design* 12, 63–79.

Tomic, S., Ramek, M. and Kojic-Prodic, B. (1998b) Combined ab-initio SCF and molecular mechanics studies of propionic and isobutryric acids and their indole derivates related to the phytohormone auxin (indole-3-acetic acid). *Croatica Chemica Acta* 71, 511–525.

Tomizawa, M., Latli, B. and Casida, J.E. (1996) Novel neonicotinoid-agarose affinity column for *Drosphila* and *Musca* nicotinic acetylcholine receptors. *Journal of Neurochemistry* 67, 1669–1676.

Tomizawa, M., Latli, M. and Casida, J.E. (1999) Structure and function of insect nicotinic acetylcholine receptors studied with nicotinoid insecticide affinity probes. In: Yamamoto, I. and Casida, J.E. (eds) *Nicotinoid Insecticides and the Nicotinic Acetylcholine Receptor*. Springer, New York, pp. 271–292.

Tomlin, C.D.S. (ed.) (1997) *The Pesticide Manual*, 11th edn. British Crop Production Council, Farnham, Surrey, UK, pp. 195, 206, 643 and 813.

Tomlin, C.D.S. (ed.) (2000) *The Pesticide Manual*, 12th edn. British Crop Production Council (BCPC), Farnham Surrey, UK, p. 1250.

Tomlinson, A.J., Benson, L.M., Guzman, N.A. and Naylor, S. (1996) Preconcentration and microreaction technology on-line with capillary electrophoresis. *Journal of Chromatography A* 744, 3–15.

Torstensson, L., Olson, G., Norup, S. and Stenberg, B. (1994) Biobeds minimize environmental risks when filling agricultural spraying equipment. Paper presented at the 35th Swedish Crop Protection Conference, pp. 223–233.

Toth, L.A., Oberbeck, C., Straign, C.M., Frazier, S. and Rehg, J.E. (2000) Toxicity evaluation of prophylactic treatments for mites and pinworms in mice. *Contemporary Topics in Laboratory Animal Science* 39(2), 18–21.

Treacy, M.F. (1999) Recombinant baculoviruses. In: Hall, F.R. and Menn, J.J. (eds) *Biopesticides – Use and Delivery*. Humana Press, Totowa, New Jersey, pp. 321–340.

Tucker, S.B. and Flannigan, S.A. (1983) Cutaneous effects from occupational exposure to fenvalerate. *Archives of Toxicology* 54, 195–202.

Tuominen, H., Puech, L., Fink, S. and Sundberg, B. (1997) A radial concentration gradient of indole-3-acetic acid is related to secondary xylem development in hybrid aspen. *Plant Physiology* 115, 577–585.

Tupper, K.A. (2009) Endosulphan Faces Growing Bans Around the World. *PAN North America*, Spring issue.

Uchida, M., Asai, T. and Sugimoto, T. (1985) Inhibition of cuticle deposition and chitin biosynthesis by a new insect growth regulator, buprofezin, in *Nilaparvata lugens* Stål. *Agricultural and Biological Chemistry* 49, 1233–1234.

Uggla, C., Moritz, T., Sandberg, G. and Sundberg, B. (1996) Auxin as a positional signal in pattern formation in plants. *Proceedings of the National Academy of Sciences USA* 93, 9282–9286.

Uggla, C., Mellerowicz, E.J. and Sundberg, B. (1998) Indole-3-acetic acid controls cambial growth in scots pine by positional signalling. *Plant Physiology* 117, 113–121.

Ulmasov, T., Hagen, G. and Guilfoyle, T.J. (1997) ARF1, a transcription factor that binds to auxin response elements. *Science* 276, 1865–1868.

United Nations Environment Programme (UNEP) (2004) Stockholm Convention. Protecting Human Health and the Environment from Persistant Organic Pollutants. Available at: http://chm.pops.int/Countries/StatusofRatification/tabid/252/language/en-US/Default.aspx (accessed 4 November 2013).

University of Hertfordshire (2013) The Pesticide Properties Data Base (PPDB). Available at: http://sitem.herts.ac.uk/aeru/footprint/en/index.htm (accessed 31 July 2013).

University of Michigan, USA (2008) Organic Crops Outperform Chemical Farming. *PAN North America*, Spring 2008.

Urech, P.A. and Staub, T. (1985) The resistance strategy for acylalanine fungicides. *EPPO Bulletin* 15, 539–543.

US Environmental Protection Agency (EPA) (1980) *Pesticide Regulation Standard for Metolachlor*. Issued September 1980. EPA, Washington, DC.

US Environmental Protection Agency (EPA) (1987a) *Metolachlor*. Chemical Fact Sheet No. 106. Issued January 1987. Office of Pesticide Programs, EPA, Washington, DC.

US Environmental Protection Agency (EPA) (1987b) *Metolachlor Health Advisory Draft Report*. Issued August 1987. Office of Drinking Water, EPA, Washington, DC.

US Environmental Protection Agency (EPA) (1996a) *Health Effects Test Guidelines*. OPPTS 870.1000, Acute Oral Toxicity, EPA 712-C96-189. EPA, Washington, DC.

US Environmental Protection Agency (EPA) (1996b) *Health Effects Test Guidelines*. OPPTS 870.1100, Acute Oral Toxicity, EPA 712-C96-190. EPA, Washington, DC.

US Environmental Protection Agency (EPA) (1996c) *Health Effects Test Guidelines*. OPPTS 870.1200, Acute Oral Toxicity, EPA 712-C96-192. EPA, Washington, DC.

US Environmental Protection Agency (EPA) (1996d) *Health Effects Test Guidelines*. OPPTS 870.1300, Acute Oral Toxicity, EPA 712-C96-193. EPA, Washington, DC.

US Environmental Protection Agency (EPA) (2013) Available at: http://www.epa.gov (accessed 17 August 2013).

US Food and Drug Administration (US FDA) (1994a) [Updated October 1999] Method HRAV-5A. In: *Pesticide Analytical Manual (PAM)*. US FDA (Centre for Food Safety and Applied Nutrition, Office of Plant and Diary Foods and Beverages), Washington, DC.

US Food and Drug Administration (US FDA) (1994b) [Updated October 1999] Method HRAV-12A. In: *Pesticide Analytical Manual (PAM)*. US FDA (Centre for Food Safety and Applied Nutrition, Office of Plant and Diary Foods and Beverages), Washington, DC.

van Daalen, J., Meltzer, J.J., Mulder, R. and Wellinga, K. (1972) A selective insecticide with a novel mode of action. *Naturwissenschaften* 59, 312–313.

van Emon, J.M. and Lopez-Avila, V. (1992) Immunochemical methods for environmental analysis. *Analytical Chemistry* 64, 79A–88A.

van Gestel, M., Merckx, R. and Vlassak, K. (1993) Microbial biomass and activity in soils with fluctuating water contents. *Geoderma* 56, 617–626.

van Huizen, R., Ozga, J.A., Reinecke, D.M., Twitchin, B. and Mander, L.N. (1995) Seed and 4-chloroindole-3-acetic acid regulation of gibberellin metabolism in pea pericarp. *Plant Physiology* 109, 1213–1217.

van Overbeek, L.S., van Veen, J.A. and van Elsas, J.D. (1997) Induced reporter gene activity, enhanced stress resistance, and competitive ability of a genetically modified *Pseudomonas fluorescens* strain released into the field plot planted with wheat. *Applied and Environmental Microbiology* 63, 1965–1973.

van Valkenburg, J.W. (1999) *Adjuvants for Herbicides*. Weed Science Society of America, Champaign, Illinois, pp. 1–8.

van Zuydam, R.P. and van de Zande, J.C. (1996) Application technology, emission and safety in glasshouse spraying. *EPPO Bulletin* 26, 95–101.

Vasil, V., Brown, S.M., Re, D., Fromm, M.E. and Vasil, I.K. (1991) Stably transformed callus lines from microprojectile bombardment of cell suspension cultures of wheat. *Journal of Biotechnology* 9, 743–747.

Vasquez, C. (2008) Farmworkers Demand EPA Protection from Fumigant Pesticides. *PAN North America*, Summer issue.

Veerasekaran, P., Kirkwood, R.C. and Parnell, E.W. (1981) Studies of the mechanism of action of asulam in plants. Part 2: Effect of asulam on the biosynthesis of folic acid. *Pesticide Science* 12, 330–338.

Vijayan, P., Shockey, J.C., Lévesque, A., James, C.R. and John, B. (1998) A role for jasmonate in pathogen defence of *Arabidopsis*. *Proceedings of the National Academy of Sciences USA* 95, 7209–7214.

Vijverberg, H.P.M. and Oortigiesen, M. (1988) Steric structure and action of pyrethroids. In: Ariens, E.J., van Rensen, J.J.S. and Welling, W. (eds) *Stereoselectivity of Pesticides, Biological and Chemical Problems*. Elsevier, Amsterdam, pp. 151–182.

Viktorov, A.V. with Yurkiv, V.A. (2003) Effect of ivermectin on function of liver macrophages. *Bulletin of Experimental Biology and Medicine* 136(6), 569–571.

Villaume, J.F. (1985) Investigations at sites contaminated with DNAPLs. *Groundwater Monitoring & Remediation* 5(2), 60–74.

Vincent, P.G. and Sisler, H.D. (1968) Mechanism of antifungal action of 2,4,5,6-tetrachloroisophthalonitrile. *Physiologia Plantarum* 21, 1249–1264.

Vincent, R., Fraiser, V., Chaillou, S., Limami, M.A., Deleens, E., Phillipson, B., Douat, C., Boutin, J.P. and Hirel, B. (1997) Overexpression of a soybean gene encoding cytosolic glutamine synthetase in shoots of transgenic *Lotus corniculatus* L. plants triggers changes in ammonium assimilation and plant development. *Planta* 201, 424–433.

Vogt, E. (1936) *Die Chemischen Pflanzenschutzmittel. Ihre Anwendung und Wirkung*. Sammlung Goschen, Walter de Gryter & Co., Berlin, p. 76.

Voss, G. and Ramos, G. (2003) *Chemistry of Crop Protection*. Wiley-VCH, Weinheim, Germany.

Voytek, J.E. (1983) Considerations in the design and installation of monitoring wells. *Groundwater Monitoring & Remediation* 3(1), 70–71.

Waksman, S.A. (1927) *Principles of Soil Microbiology*. Williams & Wilkins, Baltimore, Maryland.

Walter, J.F. (1999) Commercial experience with neem products. In: Hall, F.R. and Menn, J.J. (eds) *Biopesticides – Use and Delivery*. Humana Press, Totowa, New Jersey, pp. 155–170.

Wang, Y.S., Madsen, E.L. and Alexander, M. (1985) Microbial degradation by mineralization or cometabolism determined by chemical concentration and environment. *Journal of Agricultural and Food Chemistry* 33, 495–499.

Ward, E.R., Voltrath, S.L., Johnson, M.A. and Potter, S.L. (2001) Herbicide-tolerant protoporphyrinogen oxidase produced by DNA shuffling of plant genes. US Patent Publication Application No. 20010016956 A1 20010823. US Patent Office, Alexandria, Virginia.

Ware, G.W. and Whitcare, D.M. (2004) *The Pesticide Book*, 6th edn. Messterpro Information Resources, Willoughby, Ohio.

Warshaw, R., Fischbein, A., Thornton, J., Miller, A. and Selikoff, I.J. (1979) Decrease in vital capacity in PCB-exposed workers in a capacitor manufacturing facility. *Annals of the New York Academy of Sciences* 320, 277–283.

Wasserman, M., Wasserman, D., Cucos, S. and Miller, H.J. (1979) World PCB's map: storage and effects in man and his biologic environment in the 1970s. *Annals of the New York Academy of Sciences* 320, 69.

Watts, M. (2008) Pesticides and Breast Cancer. *PAN North America*, Spring issue.

Waxman, M.F. (1998) *Agrochemical and Pesticide Safety Handbook*. Lewis Publishers, Boca Raton, Florida.

Weed Science Society of America (WSSA) (1989) *The Composite List of Weeds*, revised edn. WSSA, Champaign, Illinois, p. 112.

Weindling, R. (1932) *Trichoderma lignorum* as a parasite of other soil fungi. *Phytopathology* 22, 837–845.

Welhouse, G.J. and Bleam, W.F. (1992) NMR spectroscopic investigation of hydrogen bonding in atrazine. *Environmental Science & Technology* 26, 959–964.

Welhouse, G.J. and Bleam, W.F. (1993a) Atrazine hydrogen-bonding potentials. *Environmental Science & Technology* 27, 494–500.

Welhouse, G.J. and Bleam, W.F. (1993b) Cooperative hydrogen bonding of atrazine. *Environmental Science & Technology* 27, 500–505.

Weller, D.M. (1988) Biological control of soilborne plant pathogens in the rhizosphere with bacteria. *Annual Review of Phytopathology* 26, 379–407.

Went, F.W. (1929) Wuchsstoff und wachstum. *Recueil des Travaux Botaniques Néerlandais* 25, 1–116.

Wessels, J.S.C. and van der Veen, R. (1956) The action of some derivatives of phenylurethan and of 3-phenyl-1,1-dimethylurea on the Hill reaction. *Biochimica et Biophysica Acta* 19, 548–549.

West, A.E., Brooks, P.D., Fisk, M.C., Smith, L., Holland, E., Jaeger, C.H., Babcock, S., Kai, R. and Schmidt, S.K. (1999) Landscape patterns of CH_4 fluxes in an alpine tundra ecosystem. *Biogeochemistry* 45, 243–264.

Wharfe, J.R. and van den Broek, W.L.F. (1978) Chlorinated hydrocarbons in macroinvertebrates and fish from the Lower Medway Estuary, Kent. *Marine Pollution Bulletin* 9, 76–79.

White, D.G. (ed.) (1999) *Compendium of Corn Diseases*, 3rd edn. APS Press, St Paul, Minnesota.

Whitehead, C.W. and Switzer, C.M. (1963) The differential response of strains of wild carrot to 2,4-D and related herbicides. *Canadian Journal of Plant Science* 43, 255–262.

Whitiker, T.G. (1990) Problems associated with testing agricultural commodities for aflatoxin: errors in sampling, sample preparation and analysis. In: *A Perspective on Aflatoxin in Field Crops and Animal Food Products in the United States*. United States Department of Agriculture, Washington, DC, pp. 1–11.

Wickstrom, K., Pyysalo, H. and Perttila, M. (1981) Organochlorine compounds in the liver of cod (*Gadus morhua*) in the northern Baltic. *Chemosphere* 10, 999–1004.

Widholm, J.M. (1992) Properties and uses of photoautotrophic plant cell culture. *International Review of Cytology* 132, 109–175.

Wiemeyer, S.N. and Porter, R.D. (1970) DDE thins eggshells of captive American kestrels. *Nature* 227, 737–738.
Wierenga, J.M. and Hollingworth, R.M. (1993) Inhibition of altered acetyl cholinesterases from insecticide-resistant Colorado potato beetles (Coleoptera, Chrysomelidae). *Journal of Economic Entomology* 86, 673–679.
Wiesman, Z., Riov, J. and Epstein, E. (1989) Characterization and rooting ability of indole-3-butyric acid conjugates formed during rooting of mung bean cuttings. *Plant Physiology* 91, 1080–1084.
Wilder, L.I. (1933/1971) *Farmer Boy*. Harper and Row, New York.
Wiles, J.A. and Jepson, P.C. (1993) The dietary effects of deltamethrin upon *Nebria brevicollis* (F.) (Coleoptera: Carabidae). *Pesticide Science* 38, 329–334.
Wilkinson, R.E. (1981) Metolachlor influence on growth and terpenoid synthesis. *Pesticide Biochemistry and Physiology* 16, 63–71.
Williams, D.T., LeBel, G.L. and Junkins, E. (1988) Organohalogen residues in human adipose autopsy samples from six Ontario municipalities. *Journal of the Association of Official Analytical Chemists* 71, 410–414.
Williamson, M.S., Martinez-Torres, D., Hick, C.A. and Devonshire, A.L. (1996) Identification of mutations in the housefly para-type sodium channel gene associated with knockdown resistance (kdr) to pyrethroid insecticides. *Molecular and General Genetics* 252, 51–60.
Wilson, B.F. (2000) Apical control of branch growth and angle in woody plants. *American Journal of Botany* 87, 601–607.
Wilson, I.B. and Bergman, F. (1950) Studies on cholinesterase: VII. The active surface of acetylcholine esterase derived from effects of pH on inhibitors. *Journal of Biological Chemistry* 185, 479–489.
Wilson, I.B., Bergman, F. and Nachmansohn, D. (1950) Acetylcholinesterase. X. Mechanism of the catalysis of acylation reactions. *Journal of Biological Chemistry* 186, 781–790.
Windholz, M., Budavari, S., Blumetti, R.F. and Otterbein, E.S. (eds) (1983) Metolachlor. In: *The Merck Index*, 10th edn. Merck & Co., Rahway, New Jersey, p. 880.
Winding, A., Binnerup, S.J. and Sorensen, J. (1994) Viability of indigenous soil bacteria assayed by respiratory activity and growth. *Applied and Environmental Microbiology* 60, 2869–2875.
Winnie, M. (2009) Ending Hunger Means Ending Income Disparity. *PAN North America*, Spring issue.
Wolf, C.N., Schuldt, P.H. and Baldwin, M.M. (1955) S-triazine derivatives – a new class of fungicides. *Science* 121, 61–62.
Wollweber, D. and Tietjen, K. (1999) Chloronicotinyl insecticides: a success of the new chemistry. In: Yamamoto, I. and Casida, J.E. (eds) *Nicotinoid Insecticides and the Nicotinic Acetylcholine Receptor*. Springer, New York, pp. 109–125.
Wood, G.E. (1992) Mycotoxins in foods and feeds in the United States. *Journal of Animal Science* 70, 3941–3949.
World Health Organization (WHO) (1986) *Organophosphorus Insecticides: a General Introduction*. Environmental Health Criteria 63. WHO, Geneva, p. 181.
World Health Organization (WHO) (1992a) *Trichlorfon*. Environmental Health Criteria 132. WHO, Geneva, p. 162.
World Health Organization (WHO) (1992b) *Fenitrothion*. Environmental Health Criteria 133. WHO, Geneva, p. 184.
World Health Organization (WHO) (2004) *The WHO Recommended Classification of Pesticides by Hazards and Guidelines to Classification*. WHO, Geneva.
Worthing, C.R. (ed.) (1983) Metolachlor. In: *The Pesticide Manual*, 7th edn. British Crop Production Council, Farnham, Surrey, UK, p. 377.
Wright, S.F., Starr, J.L. and Paltineau, I.C. (1999) Changes in aggregate stability and concentration of glomalin during tillage management transition. *Soil Science Society of America Journal* 63, 1825–1829.
Wygoda, H.J. and Rietz, S. (1996) Plant protection equipment in glasshouses. *EPPO Bulletin* 26, 87–93.
Wyllie, T.D. and Morehouse, L.G. (eds) (1977–1978) *Mycotoxic Fungi Mycotoxins, Mycotoxicosis: an Encyclopedic Handbook*, Vols. 1–3. Marcel Dekker, New York.
Wymore, L.A., Poirier, C., Watson, A.K. and Gotlief, A.R. (1988) *Colletotrichum coccodes*, a potential bioherbicide for control of velvetleaf (*Abutilon theophrasti*). *Plant Disease* 72, 534–538.
Yaduraju, N.T. and Ahuja, K.N. (1995) *Illustrated Dictionary of Weed Science*. Venus Publications, New Delhi.
Yamamoto, I. (1965) Nicotinoids as insecticides. *Advances in Pest Control Research* 6, 231–260.
Yang, S.F. and Hoffman, N.E. (1984) Ethylene biosynthesis and its regulation in higher plants. *Annual Review of Plant Physiology* 35, 155–189.
Yen, J.T., Mersmann, H.J., Hill, D.A. and Pond, W.G. (1990) Effects of ractopamine on genetically obese and lean pigs. *Journal of Animal Science* 68, 3705–3712.
Yeoh, C.L., Kuwano, E. and Eto, M. (1983) *In vitro* metabolism of the fungicide IBP [diisopropyl s-benzyl phosphorothioate] by insects and rat liver. *Pesticide Science* 8, 213–219.
Yoshimatsu, K., Hatano, T., Katayama, M., Marumo, S., Kamada, H. and Shimomura, K (1990) IAA derivative induced tropane alkaloid production in root cultures of a *Duboisia* hybrid. *Phytochemistry* 29, 3525–3528.
Young, B.G. and Hart, S.E. (1998) Optimizing foliar activity of isoxaflutole on giant foxtail (*Setaria faberi*) with various adjuvants. *Weed Science* 46, 397–402.
Young, B.W. (1991) A method for assessing the drift potential of hydraulic spray clouds and the effect of air assistance. *British Crop Production Council (BCPC) Monograph* 46, 77–86.
Young, D.H. and Slawecki, R.A. (2001) Mode of action of zoxamide (RH-7281), a new oomycete fungicide. *Pesticide Biochemistry and Physiology* 69, 100–111.
Zahner, H., Drautz, H. and Weber, W. (1982) Novel approaches to metabolite screening. In: Bu'Lock, J.D., Nisbet, L.J. and Winstanley, D.J. (eds) *Bioactive Secondary Metabolites. Search and Discovery*. Academic Press, London, pp. 51–70.
Zainur, A.S., Tassell, R., Kellaway, R.C. and Dodemaide, W.R. (1989) Recombinant growth hormone in growing lambs: effects on growth, feed utilization, body and carcass characteristics and on wool growth. *Australian Journal of Agricultural Research* 40, 195–206.

Zak, D.R., Tilman, D., Parmenter, R.R., Rice, C.W., Fisher, F.M., Vose, J., Milchunas, D. and Martin, C.W. (1994) Plant production and soil microorganisms in later-successional ecosystems: a continental-scale study. *Ecology* 75, 2333–2347.

Zama, P. and Hatzios, K.K. (1987) Interactions between the bioicide metolachlor and the safener CGA-92194 in sorghum leaf protoplasts. *Pesticide Biochemistry and Physiology* 27, 86–96.

Zejda, J.E., McDuffie, H.H. and Dosman, J.A. (1993) Epidemiology of health and safety risks in agriculture and related industries. Practical applications for rural physicians. *Western Journal of Medicine* 158(1), 56–63.

Zettl, R., Feldwisch, J., Boland, W., Schell, J. and Palme, K. (1992) 5' Azido-3 6-tritiated-1-naphthylphthalamic acid, a photoactivable probe for naphthylphthalamic acid receptor proteins for higher plants: identification of a 23 kDa protein from maize coleoptile plasma membranes. *Proceedings of the National Academy of Sciences USA* 89, 480–484.

Zhong, H., Latham, M., Hester, P.G., Frommer, R.L. and Brock, C. (2003) Impact of aerial applied ultra-low-volume insecticide naled with flat-fan nozzle system on honeybee survival and productivity. *Archives of Environmental Contamination and Toxicology* 45, 216–220.

Zhong, H., Latham, M., Payne, S. and Brock, C. (2004) Minimizing the impact of mosquito adulticide naled on honeybees *Apis mellifera* (Hymenoptera: Apidae): aerial ultra-low volume application using a high-pressure nozzle system. *Journal of Economic Entomology* 97, 1–7.

Zimdahl, R.L. and Clark, S.K. (1982) Degradation of three acetanilide herbicides in soil. *Weed Science* 30, 545–548.

Zimmerlin, A. and Durst, F. (1990) Xenobiotics metabolism in plants: aryl hydroxylation of dilofop by a cytochrome P-450 enzyme from wheat. *Phytochemistry* 29, 1729–1732.

Zimmerman, P.W. and Wilcoxon, F. (1935) Several chemical growth substances which cause initiation of roots and other responses in plants. *Contributions from the Boyce Thompson Institute* 7, 209–229.

Zimmerman, P.W. and Wilcoxon, F. (1942) Flowering habit and correlation of organs as modified by TIBA. *Contributions from the Boyce Thompson Institute* 12, 321–343.

Zlotkin, E., Rochat, H., Kopeyan, F.M. and Lissitzky, S. (1971) Purification and properties of the insect toxin from the venom of the scorpion *Androctonus australis* Hector. *Biochimie* 53, 1073–1078.

Zong, H., Dukes, J., Creer, M., Hester, P., Shirley, M. and Anderson, B. (2003) Ground deposition impact of aerially applied fenthion on the fiddler crabs, Uca pugilator. *Journal of the American Mosquito Control Association* 19, 47–52.

Website References

dbtbiosafety.nic.in. Indian Biosafety Rules and Regulations. Date last visited: 14 January 2014.

http://69.59.152.188/Detail_Chemical.jsp?Rec_Id=PC33780. 2-Mercaptobenzothiazole, Sodium Salt – Identification, Toxicity, Use, Water Pollution Potential, Ecological Toxicity and Regulatory Information. PAN Pesticide Database, Pesticide Action Network, North America. Date last visited: 12 July 2013.

http://69.59.152.188/Detail_Chemical.jsp?Rec_Id=PC37445. Cloethocarb – Identification, Toxicity, Use, Water Pollution Potential, Ecological Toxicity and Regulatory Information. PAN Pesticide Database, Pesticide Action Network, North America. Date last visited: 12 July 2013.

http://ag.arizona.edu/PLP/plpext/diseases/fruits/citrus/citrusgum.html. Extension Plant Pathology. Date last visited: 12 July 2013.

http://ag.udel.edu/enwc/research/delphacid/specialconcern.html. Species of Special Concern. College of Agricultural and Natural Resources, University of Delaware. Date last visited: 12 July 2013.

http://agebb.missouri.edu/storage/pests/insect.htm. Stored Grain Insects. State Extension Entomologist, University of Missouri, Columbia. Date last visited: 12 July 2013.

http://agr.wa.gov/FoodAnimal/Organic. Organic Food Program. Washington State Department of Agriculture. Date last visited: 12 July 2013.

http://agritech.tnau.ac.in/org_farm/orgfarm_agripest.html. Organic Farming. TNAU Agritech Portal. Tamil Nadu Agricultural University, Coimbatore. Date last visited: 12 July 2013.

http://agropedia.iitk.ac.in/sites/default/files/JPEG/20081218/Agropedia_sorghum.jpg. Agropedia, ICAR, NAIP. Date last visited: 27 August 2013.

http://apsjournals.apsnet.org/doi/pdf/10.1094/PHYTO-95-1209. Ishikawa, R., Shirouzu, K., Nakashita, H., Lee, H.-Y., Motoyama, T., Yamaguchi, I., Teraoka, T., and Arie, T. (2005) Foliar spray of validamycin A or validoxylamine A controls tomato fusarium wilt. *The American Phytopathological Society* 95 (10). Date last visited: 12 July 2013.

http://baroda.indiabizclub.com/catalog/414956'biocontrol'baroda. Baroda Catalog. Gujarat Life Sciences Pvt. Ltd. Date last visited: 12 July 2013.

http://bayercropscience.com. Date last visited: 12 July 2013.

http://biology.suite101.com/article.cfm/five_kingdoms_of_life. Five Kingdoms of Science. Nature and Science. Suite 101. Date last visited: 12 July 2013.

http://books.google.co.in/books?id=CdBpwokWL08C&pg=PA331&lpg=PA331&dq=fentrazamide+as+a+herbicide&source=bl&ots=G_0txhNYty&sig=aOXBP6WF_t0C4jTJTnCi4qcp7vU&hl=en&ei=pKTiS4jEJNKxrAeNz-. Krämer, W. and Schirmer, U. (eds) Modern Crop Protection Compounds, Volume I. Wiley, Weinheim. Date last visited: 12 July 2013.

http://books.google.co.in/books?id=ib8Qhju9EQEC&pg=PA1660&lpg=PA1660&dq=Flumetsul am+target+weeds&source=bl&ots=DeHvs6Jrrp&sig=Z0LCrAj1vAF2UG7lRYDd9s9D2Cs&hl=en&ei=pjksTIrsE5CGrQfQ4e3EDQ&sa=X&oi=book_result&ct=result&resnum=5&ved=0CBkQ6AEwBA#v=onepage&q=Flumetsulam%20target%20weeds&f=false. Irving Krieger, R. (ed.) (2001) *Handbook of Pesticide Toxicology*, Volume 1. Academic Press. Date last visited: 12 July 2013.

http://botanydictionary.org/chlorosis.html. Dictionary of Botany. Date last visited: 8 July 2013.

http://btny.agriculture.purdue.edu/turfcast/. Turf Cast 2013. Department of Botany and Plant Pathology, Purdue University. Date last visited: 7 August 2013.

http://cameochemicals.noaa.gov/chemical/4909. Chlormephos Chemical Datasheet. CAMEO Chemicals version 2.4. Date last visited: 8 July 2013.

http://chm.pops.int/Countries/StatusofRatification/tabid/252/language/en-US/Default.aspx. Protecting Human Health and the Environment from Persistant Organic Pollutants. United Nations Environment Programme (UNEP) (2004) Stockholm Convention. Date last visited: 4 November 2013.

http://ci.nii.ac.jp/naid/110001867889/en. Herbicidal Property of a New Paddy Bleaching Herbicide, Benzobicyclon and Herbicidal Activity against Invaded Weeds from Levee. The Japanese Society for Chemical Regulation of Plants. Date last visited: 8 July 2013.

http://cibrc.nic.in/reg_products.htm. Insecticides Registered Under Section 9 (3) of the Insecticide Act, 1968. Date last visited: 12 July 2013.

http://circa.europa.eu/Public/irc/env/bio_reports/library?l=/assessement_directive/chlorophacinonepdf/_EN_1.0_&a=d. Directive 98/8/EC Concerning the Placing of Biocidal Products on the Market. Inclusion of Active Substances in Annex I or IA to Directive 98/8/EC. Assessment Report. Chlorophacinone. Date last visited: 12 July 2013.

http://citrusindia.blogspot.com/. Citrus Cultivation in India. Date last visited: 12 July 2013.

http://compendium.bayercropscience.com/BAYER/CropScience/CropCompendium/BCSCropComp.nsf/id/carpropamid.htm. Carpropamid. Bayer Crop Science. Global Internet Portal. Date last visited: 13 July 2013.

http://compendium.bayercropscience.com/BAYER/CropScience/CropCompendium/BCSCropComp.nsf/id/EN_Ethephon?Open&ccm=300010. Ethephon. Bayer Crop Science. Global Internet Portal. Date last visited: 13 July 2013.

http://compendium.bayercropscience.com/BAYER/CropScience/CropCompendium/BCSCropComp.nsf/id/ethoprophos.htm. Ethoprophos. Bayer Crop Science. Global Internet Portal. Date last visited: 13 July 2013.
http://compendium.bayercropscience.com/BAYER/CropScience/CropCompendium/BCSCropComp.nsf/id/fenamidone.htm. Fenamidone. Bayer Crop Science. Global Internet Portal. Date last visited: 13 July 2013.
http://compendium.bayercropscience.com/BAYER/CropScience/CropCompendium/BCSCropComp.nsf/id/Oulema_oryzae.htm?open&ccm=200010. *Oulema oryzae*, Rice Leaf Beetle. Bayer Crop Science. Global Internet Portal. Date last visited: 13 July 2013.
http://compendium.bayercropscience.com/BAYER/CropScience/CropCompendium/BCSCropComp.nsf/id/thiodicarb.htm#. Thiodicarb. Bayer Crop Science. Global Internet Portal. Date last visited: 13 July 2013.
http://compendium.bayercropscience.com/BAYER/CropScience/CropCompendium/BCSCropComp.nsf/id/EN_Tribufos?open&ccm=300050. Tribufos. Bayer Crop Science. Global Internet Portal. Date last visited: 13 July 2013.
http://compendium.bayercropscience.com/BAYER/CropScience/CropCompendium/BCSCropComp.nsf/id/triflumuron.htm. Triflumuron. Bayer Crop Science. Global Internet Portal. Date last visited: 13 July 2013.
http://croptechnology.unl.edu/pages/informationmodule.php?idinformationmodule=1130447045&topicorder=7&maxto=7&minto=1. Micronutrient Fertilizers. Soils – Part 8: Characteristics of Fertilizer Materials. Plant & Soil Sciences eLibrary. Date last visited: 13 July 2013.
http://ddr.nal.usda.gov/bitstream/10113/13274/1/IND43835945.pdf. Toews, M.D., Campbell, J.F., Arthur, F.H. and Ramaswamy, S.B. Outdoor flight activity and immigration of Rhyzopertha dominica into seed wheat warehouses. National Agricultural Library Digital Collection. USDA National Agricultural Library. Date last visited: 13 July 2013.
http://drs.nio.org/drs/bitstream/2264/3364/1/Brazil_J_Microbiol_40_269p.pdf. Parvathi, A., Krishna, K., Jose, J., Joseph, N. and Nair, S. (2009) Biochemical and molecular characterization of *Bacillus pumilus* isolated from coastal environment in Cochin, India. *Brazilian Journal of Microbiology* 40, 269–275. Date last visited: 13 July 2013.
http://ec.europa.eu/sanco_pesticides/public/?event=homepage&CFID=744483&CFTOKEN=41666187&jsessionid=09045ff4bf62927e55b74b4117d667c54651TR. EU Pesticide Database. Date last visited: 12 June 2013.
http://edis.ifas.ufl.edu/ch081. Post-harvest Decay Control Recommendations for Florida Citrus Fruits. University of Florida. Date last visited: 13 July 2013.
http://edis.ifas.ufl.edu/in589. Parasitoid of the Citrus Leaf Miner, *Semielacher petiolatus* (Girault). (Insecta: Hymenoptera: Eulophidae). University of Florida. Date last visited: 13 July 2013.
http://edis.ifas.ufl.edu/in638. Chilli Thrips (Castor Thrips, Assam Thrips, Yellow Tea Thrips and Strawberry Thrips), *Scirtothrips dorsalis* Hood, Provisional Management Guidelines. University of Florida. Date last visited: 13 July 2013.
http://edis.ifas.ufl.edu/PI088. Pesticide Toxicity Profile: Carbamate Pesticides. University of Florida. Date last visited: 13 July 2013.
http://edis.ifas.ufl.edu/pi125. Pesticide Toxicity Profile: Bipyridylium herbicides. University of Florida. Date last visited: 13 July 2013.
http://efw.bpa.gov/environmental_services/Document_Library/Vegetation_Management/sheets/Azafenidin.pdf. Azafenidin Herbicide Fact Sheet. US Department of Energy, Bonneville Power Administration. Date last visited: 13 July 2013.
http://en.wikipedia.org/wiki/Abamectin#column-one. Abamectin. Wikipedia, the free encyclopedia. Date last visited: 13 July 2013.
http://en.wikipedia.org/wiki/Acetochlor. Acetochlor. Wikipedia, the free encyclopedia
http://en.wikipedia.org/wiki/Adalia_bipunctata. *Adalia bipunctata*. Wikipedia, the free encyclopedia. Date last visited: 13 July 2013.
http://en.wikipedia.org/wiki/Agapeta_zoegana. *Agapeta zoegana*. Wikipedia, the free encyclopedia. Date last visited: 13 July 2013.
http://en.wikipedia.org/wiki/Anthraquinone#column-one. Anthraquinone. Wikipedia, the free encyclopedia. Date last visited: 13 July 2013.
http://en.wikipedia.org/wiki/Aphidoletes_aphidomyza. *Aphidoletes aphidomyza*. Wikipedia, the free encyclopedia. Date last visited: 13 July 2013.
http://en.wikipedia.org/wiki/Azadirachtin#column-one. Azadirachtin. Wikipedia, the free encyclopedia. Date last visited: 13 July 2013.
http://en.wikipedia.org/wiki/Beauveria_bassiana. *Beauveria bassiana*. Wikipedia, the free encyclopedia. Date last visited: 13 July 2013.
http://en.wikipedia.org/wiki/6-Benzylaminopurine#column-one. 6-Benzylaminopurine. Wikipedia, the free encyclopedia. Date last visited: 13 July 2013.
http://en.wikipedia.org/wiki/Blasticidin_S#column-one. Blasticidin S. Wikipedia, the free encyclopedia. Date last visited: 13 July 2013.
http://en.wikipedia.org/wiki/Brown_planthopper. Brown Planthopper. Wikipedia, the free encyclopedia. Date last visited: 13 July 2013.
http://en.wikipedia.org/wiki/Capsaicin#column-one. Capsaicin. Wikipedia, the free encyclopedia. Date last visited: 15 July 2013.
http://en.wikipedia.org/wiki/Category:Carbamate_insecticides. Carbamate Insecticides. Wikipedia, the free encyclopedia. Date last visited: 15 July 2013.
http://en.wikipedia.org/wiki/Carbofuran. Carbofuran. Wikipedia, the free encyclopedia. Date last visited: 15 July 2013.
http://en.wikipedia.org/wiki/Carvone. Carvone. Wikipedia, the free encyclopedia. Date last visited: 15 July 2013.
http://en.wikipedia.org/wiki/Chemosterilant. Chemosterilant. Wikipedia, the free encyclopedia. Date last visited: 15 July 2013.
http://en.wikipedia.org/wiki/Chlorfenvinphos. Chlorfenvinphos. Wikipedia, the free encyclopedia. Date last visited: 15 July 2013.
http://en.wikipedia.org/wiki/Chlorosis. Chlorosis. Wikipedia, the free encyclopedia. Date last visited: 15 July 2013.
http://en.wikipedia.org/wiki/Cinnamaldehyde#As_an_agrichemical. Cinnamaldehyde. Wikipedia, the free encyclopedia. Date last visited: 15 July 2013.
http://en.wikipedia.org/wiki/Citric_acid#Others. Citric Acid. Wikipedia, the free encyclopedia. Date last visited: 15 July 2013.
http://en.wikipedia.org/wiki/Citronella_oil#column-one. Citronella. Wikipedia, the free encyclopedia. Date last visited: 15 July 2013.
http://en.wikipedia.org/wiki/Clopyralid Clopyralid. Wikipedia, the free encyclopedia. Date last visited: 16 July 2013.
http://en.wikipedia.org/wiki/Coumarin. Coumarin. Wikipedia, the free encyclopedia. Date last visited: 16 July 2013.

http://en.wikipedia.org/wiki/Coumatetralyl. Coumatetralyl. Wikipedia, the free encyclopedia. Date last visited: 16 July 2013.
http://en.wikipedia.org/wiki/Cover_crop. Cover Crop. Wikipedia, the free encyclopedia. Date last visited: 16 July 2013.
http://en.wikipedia.org/wiki/Cymbopogon. Cymbopogon. Wikipedia, the free encyclopedia. Date last visited: 16 July 2013.
http://en.wikipedia.org/wiki/Cytokinin#column-one. Cytokinin. Wikipedia, the free encyclopedia. Date last visited: 16 July 2013.
http://en.wikipedia.org/wiki/1,2-Dibromoethane. 1,2-Dibromoethane. Wikipedia, the free encyclopedia. Date last visited: 13 July 2013.
http://en.wikipedia.org/wiki/2,4-Dichlorophenoxyacetic_acid#column-one. 2,4-Dichlorophenoxyacetic Acid. Wikipedia, the free encyclopedia. Date last visited: 13 July 2013.
http://en.wikipedia.org/wiki/Endotoxin. Endotoxin. Wikipedia, the free encyclopedia. Date last visited: 16 July 2013.
http://en.wikipedia.org/wiki/Environmental_impact_assessment. Environmental Impact Assessment. Wikipedia, the free encyclopedia. Date last visited: 16 July 2013.
http://en.wikipedia.org/wiki/Ergosterol. Ergosterol. Wikipedia, the free encyclopedia. Date last visited: 16 July 2013.
http://en.wikipedia.org/wiki/Ergot. Ergot. Wikipedia, the free encyclopedia. Date last visited: 16 July 2013.
http://en.wikipedia.org/wiki/Fluvalinate. Fluvalinate. Wikipedia, the free encyclopedia. Date last visited: 16 July 2013.
http://en.wikipedia.org/wiki/Formaldehyde. Formaldehyde. Wikipedia, the free encyclopedia. Date last visited: 16 July 2013.
http://en.wikipedia.org/wiki/Formic_acid. Formic Acid. Wikipedia, the free encyclopedia. Date last visited: 16 July 2013.
http://en.wikipedia.org/wiki/Garlic#cite_note-58. Garlic. Wikipedia, the free encyclopedia. Date last visited: 16 July 2013.
http://en.wikipedia.org/wiki/Geraniol. Geraniol. Wikipedia, the free encyclopedia. Date last visited: 16 July 2013.
http://en.wikipedia.org/wiki/Gibberellic_acid. Gibberellic Acid. Wikipedia, the free encyclopedia. Date last visited: 16 July 2013.
http://en.wikipedia.org/wiki/Guanidine. Guanidine.Wikipedia, the free encyclopedia. Date last visited: 16 July 2013.
http://en.wikipedia.org/wiki/Heptachlor. Heptachlor. Wikipedia, the free encyclopedia. Date last visited: 16 July 2013.
http://en.wikipedia.org/wiki/Hexachlorobenzene. Hexachlorobenzene. Wikipedia, the free encyclopedia. Date last visited: 16 July 2013.
http://en.wikipedia.org/wiki/Hexachlorocyclohexane. Hexachlorocyclohexane. Wikipedia, the free encyclopedia. Date last visited: 16 July 2013.
http://en.wikipedia.org/wiki/Hexamethylphosphoramide. Hexamethylphosphoramide. Wikipedia, the free encyclopedia. Date last visited: 16 July 2013.
http://en.wikipedia.org/wiki/Indole-3-acetic_acid#column-one. Indole-3-acetic Acid. Wikipedia, the free encyclopedia. Date last visited: 16 July 2013.
http://en.wikipedia.org/wiki/Karanjin. Karanjin. Wikipedia, the free encyclopedia. Date last visited: 16 July 2013.
http://en.wikipedia.org/wiki/Kasugamycin. Kasugamycin.Wikipedia, the free encyclopedia. Date last visited: 16 July 2013.
http://en.wikipedia.org/wiki/Lewis_acids_and_bases. Lewis Acids and Bases. Wikipedia, the free encyclopedia. Date last visited: August 10, 2013
http://en.wikipedia.org/wiki/Lindane. Lindane. Wikipedia, the free encyclopedia. Date last visited: 16 July 2013.
http://en.wikipedia.org/wiki/Menthol. Menthol. Wikipedia, the free encyclopedia. Date last visited: 16 July 2013.
http://en.wikipedia.org/wiki/Methomyl. Methomyl. Wikipedia, the free encyclopedia. Date last visited: 16 July 2013.
http://en.wikipedia.org/wiki/Nuclear_Polyhedrosis_Virus. Nuclear Polyhedrosis Virus. Wikipedia, the free encyclopedia. Date last visited: 16 July 2013.
http://en.wikipedia.org/wiki/Oleic_acid. Oleic Acid. Wikipedia, the free encyclopedia. Date last visited: 16 July 2013.
http://en.wikipedia.org/wiki/Pesticide_resistance. Pesticide Resistance. Insecticide Resistance Action Committee. Wikipedia, the free encyclopedia. Date last visited: 16 July 2013.
http://en.wikipedia.org/wiki/Pesticide_toxicity_to_bees. Pesticide Toxicity to Bees. Wikipedia, the free encyclopedia. Date last visited: 16 July 2013.
http://en.wikipedia.org/wiki/Phthorimaea_operculella. *Phthorimaea operculella*. Wikipedia, the free encyclopedia. Date last visited: 17 July 2013.
http://en.wikipedia.org/wiki/Planthopper. Plant Hopper. Wikipedia, the free encyclopedia. Date last visited: 17 July 2013.
http://en.wikipedia.org/wiki/Plant_hormone#Gibberellins. Plant Hormone – Gibberellins. Wikipedia, the free encyclopedia. Date last visited: July 17. 2013.
http://en.wikipedia.org/wiki/Plant_nutrition. Plant Nutrition. Wikipedia, the free encyclopedia. Date last visited: .17 July 2013.
http://en.wikipedia.org/wiki/Polyoxins. Polyoxins. Wikipedia, the free encyclopedia. Date last visited: 17 July 2013.
http://en.wikipedia.org/wiki/Potato. Potato. Wikipedia, the free encyclopedia. Date last visited: 17 July 2013.
http://en.wikipedia.org/wiki/Procymidone. Procymidone. Wikipedia, the free encyclopedia. Date last visited: 17 July 2013.
http://en.wikipedia.org/wiki/Propoxur. Propoxur. Wikipedia, the free encyclopedia. Date last visited: 17 July 2013.
http://en.wikipedia.org/wiki/Pseudomonas_fluorescens. *Pseudomonas fluorescens*. Wikipedia, the free encyclopedia. Date last visited: 13 August 2013.
http://en.wikipedia.org/wiki/Pyriproxyfen. Pyriproxyfen. Wikipedia, the free encyclopedia. Date last visited: 17 July 2013.
http://en.wikipedia.org/wiki/Rotenone. Rotenone.Wikipedia, the free encyclopedia. Date last visited: 17 July 2013.
http://en.wikipedia.org/wiki/Simazine. Simazine. Wikipedia, the free encyclopedia. Date last visited: 17 July 2013.
http://en.wikipedia.org/wiki/Sodium fluoroacetate. Sodium Fluoroacetate. Wikipedia, the free encyclopedia. Date last visited: 14 July 2013.
http://en.wikipedia.org/wiki/Streptomycin. Streptomycin. Wikipedia, the free encyclopedia. Date last visited: 17 July 2013.
http://en.wikipedia.org/wiki/Tebuconazole. Tebuconazole. Wikipedia, the free encyclopedia. Date last visited: 17 July 2013.
http://en.wikipedia.org/wiki/Temefos. Temefos. Wikipedia, the free encyclopedia. Date last visited: 17 July 2013.
http://en.wikipedia.org/wiki/Thiacloprid. Thiacloprid. Wikipedia, the free encyclopedia. Date last visited: 17 July 2013.
http://en.wikipedia.org/wiki/Thiamethoxam. Thiamethoxam. Wikipedia, the free encyclopedia. Date last visited: 17 July 2013.
http://en.wikipedia.org/wiki/Thiram. Thiram. Wikipedia, the free encyclopedia. Date last visited: 17 July 2013.
http://en.wikipedia.org/wiki/Tiabendazole. Tiabendazole. Wikipedia, the free encyclopedia. Date last visited: 17 July 2013.
http://en.wikipedia.org/wiki/Toxaphene. Toxaphene. Wikipedia, the free encyclopedia. Date last visited: 17 July 2013.

http://en.wikipedia.org/wiki/Transfluthrin. Transfluthrin. Wikipedia, the free encyclopedia. Date last visited: 17 July 2013.

http://en.wikipedia.org/wiki/Trichothecene. Trichothecane. Wikipedia, the free encyclopedia. Date last visited: 17 July 2013.

http://en.wikipedia.org/wiki/Whitefly. Whitefly. Wikipedia, the free encyclopedia. Date last visited: 17 July 2013.

http://en.wiktionary.org/wiki/chloroacetanilide. Chloroacetanilide. Wiktionary, a wiki-based open content dictionary. Date last visited: 17 July 2013.

http://expasy.org/sprot/hamap/AGRRK.html. Swiss Institute of Bioinformatics. HAMP: *Agrobacterium radiobacter* (strain K84/ATCC BAA-868) Complete Proteome. Date last visited: 23 July 2013.

http://extension.missouri.edu/publications/DisplayPub.aspx?P=G7275. Managing Whiteflies on Indoor and Outdoor Plants. University of Missouri Extension. Date last visited: 23 July 2013.

http://extoxnet.orst.edu/pips/acetochl.htm. Extension Toxicology Network. Pesticide Information Profiles. Acetochlor. Date last visited: 23 July 2013.

http://extoxnet.orst.edu/pips/carbofur.htm. Extension Toxicology Network. Pesticide Information Profiles. Carbofuran. Date last visited: 23 July 2013.

http://extoxnet.orst.edu/pips/clomazon.htm. Extension Toxicology Network. Pesticide Information Profiles. Clomazone. Date last visited: 23 July 2013.

http://extoxnet.orst.edu/pips/cyanazin.htm. Extension Toxicology Network. Pesticide Information Profiles. Cyanazin. Date last visited: 23 July 2013.

http://extoxnet.orst.edu/pips/esfenval.htm. Extension Toxicology Network. Pesticide Information Profiles. Esfenvalerate. Date last visited: 23 July 2013.

http://extoxnet.orst.edu/pips/ethephon.htm. Extension Toxicology Network. Pesticide Information Profiles. Ethephon. Date last visited: 23 July 2013.

http://extoxnet.orst.edu/pips/ethion.htm. Extension Toxicology Network. Pesticide Information Profiles. Ethion. Date last visited: 23 July 2013.

http://extoxnet.orst.edu/pips/fluazifo.htm. Extension Toxicology Network. Pesticide Information Profiles. Fluazifop-p-butyl. Date last visited: 23 July 2013.

http://extoxnet.orst.edu/pips/fluometu.htm. Extension Toxicology Network. Pesticide Information Profiles. Fluometuron. Date last visited: 23 July 2013.

http://extoxnet.orst.edu/pips/hexachlo.htm. Extension Toxicology Network. Pesticide Information Profiles. Hexachlorobenzene. Date last visited: 23 July 2013.

http://extoxnet.orst.edu/pips/napropam.htm. Extension Toxicology Network. Pesticide Information Profiles. Napropamide. Date last visited: 23 July 2013.

http://extoxnet.orst.edu/pips/oxamyl.htm. Extension Toxicology Network. Pesticide Information Profiles. Oxamyl. Date last visited: 23 July 2013.

http://extoxnet.orst.edu/pips/oxyfluor.htm. Extension Toxicology Network. Pesticide Information Profiles. Oxyfluorfen. Date last visited: 23 July 2013.

http://extoxnet.orst.edu/pips/propazin.htm. Extension Toxicology Network. Pesticide Information Profiles. Propazine. Date last visited: 23 July 2013.

http://extoxnet.orst.edu/pips/propetam.htm. Extension Toxicology Network. Pesticide Information Profiles. Propetamphos. Date last visited: 23 July 2013.

http://extoxnet.orst.edu/pips/propoxur.htm. Extension Toxicology Network. Pesticide Information Profiles. Propoxur. Date last visited: 23 July 2013.

http://extoxnet.orst.edu/pips/resmethr.htm. Extension Toxicology Network. Pesticide Information Profiles. Resmethrin. Date last visited: 23 July 2013.

http://extoxnet.orst.edu/pips/rotenone.htm. Extension Toxicology Network. Pesticide Information Profiles. Rotenone. Date last visited: 23 July 2013.

http://extoxnet.orst.edu/pips/ryania.htm. Extension Toxicology Network. Pesticide Information Profiles. Ryania. Date last visited: 23 July 2013.

http://extoxnet.orst.edu/pips/sethoxyd.htm. Extension Toxicology Network. Pesticide Information Profiles. Sethoxydim. Date last visited: 23 July 2013.

http://extoxnet.orst.edu/pips/simazine.htm. Extension Toxicology Network. Pesticide Information Profiles. Simazine. Date last visited: 23 July 2013.

http://extoxnet.orst.edu/pips/streptom.htm. Extension Toxicology Network. Pesticide Information Profiles. Streptomycin. Date last visited: 23 July 2013.

http://extoxnet.orst.edu/pips/temephos.htm. Extension Toxicology Network. Pesticide Information Profiles. Temephos. Date last visited: 23 July 2013.

http://extoxnet.orst.edu/pips/terbacil.htm. Extension Toxicology Network. Pesticide Information Profiles. Terbacil. Date last visited: 23 July 2013.

http://extoxnet.orst.edu/pips/thiabend.htm. Extension Toxicology Network. Pesticide Information Profiles. Thiabendazole. Date last visited: 23 July 2013.

http://extoxnet.orst.edu/pips/thiram.htm. Extension Toxicology Network. Pesticide Information Profiles. Thiram. Date last visited: 26 July 2013.

http://extoxnet.orst.edu/pips/trialla.htm. Extension Toxicology Network. Pesticide Information Profiles. Triallate. Date last visited: 26 July 2013.

http://extoxnet.orst.edu/pips/triclopy.htm. Extension Toxicology Network. Pesticide Information Profiles. Triclopyr. Date last visited: 26 July 2013.

http://extoxnet.orst.edu/pips/triforin.htm. Extension Toxicology Network. Pesticide Information Profiles. Triforine. Date last visited: 26 July 2013.

http://faculty.ucr.edu/~legneref/botany/medicine.htm. Plants of Economic or Aesthetic Importance – Medicinal Plants. University of California, Riverside. Date last visited: 26 July 2013.

http://fieldguide.mt.gov/displaySpecies.aspx?family=Chenopodiaceae. Montana Field Guide. Montana's Official State Website. Date last visited: 26 July 2013.

http://forestpathology.cfans.umn.edu/microbes.htm. Blanchette R.A. Microorganisms Causing Decay in Trees and Wood. Research Projects, University of Minnesota. Date last visited: 28 July 2013.

http://gain.fas.usda.gov/Pages/Default.aspx. Global Agricultural Information Network. United States Department of Agriculture. Foreign Agricultural Service. Date last visited: 26 July 2013.

http://govpulse.us/entries/2007/09/19/E7-18496/chloroneb-cypermethrin-methidathion-nitrapyrin-oxyfluorfen-pirimiphos-methyl-sulfosate-tebuthiuron-t. Chloroneb, Cypermethrin, Methidathion, Nitrapyrin, Oxyfluorfen, Pirimiphos-methyl, Sulfosate, Tebuthiuron, Thiabendazole, Thidiazuron, and Tribuphos: Tolerance Actions. Govpulse. Date last visited: 28 July 2013.

http://horizon.documentation.ird.fr/exl-doc/pleins_textes/pleins_textes_5/pt5/nemato/36411.pdf. Hartwig, J. and Sikora, R.A. (1991) Mode-of-action of the carbamate nematicides cloethocarb, aldicarb and carbofuran on *Heteroderu schachtii*. *Revue Nématologie* 14 (4), 525–530. Date last visited: 28 July 2013.

http://hortweeds.tennessee.edu/webapp/test/moa.htm. Mode of Action Definitions. The University of Tennessee, Knoxville. Date last visited: 28 July 2013.

http://ipm.illinois.edu/pubs/iapmh/12chapter.pdf. *Illinois Agricultural Pest Management Handbook* (2008) (Chapter Weed Resistance to Herbicides). College of Agricultural, Consumer and Environmental Sciences, University of Illinois Extension. Date last visited: 28 July 2013.

http://ipm.ncsu.edu/safety/factsheets/resistan.pdf. Bellinger, R.G. (1996) *Pest Resistance to Pesticides*. Clemson University. Date last visited: 28 July 2013.

http://ipmguidelines.org/Turfgrass/Chapters/CH01/default.aspx. Crop and Pest Management Guidelines. Cornell University. Cornell Cooperative Extension Publication. Date last visited: 28 July 2013.

http://jxb.oxfordjournals.org/content/55/404/1947.full. Cooper, R.M. and Williams, J.S. (2004) Elemental sulphur as an induced antifungal substance in plant defence. *Journal of Experimental Botany*. Date last visited: 13 August 2013.

http://krishiworld.com/field-crops-part-2/. Field Crops (Part 2). Krishi World. Date last visited: 23 August 2013.

http://landscaping.about.com/cs/lazylandscaping/g/covercrop.htm. Cover Crops. Date last visited: 28 July 2013.

http://monographs.iarc.fr/ENG/Monographs/vol73/mono73-11.pdf. Chlorothalonil. Date last visited: 28 July 2013.

http://mrec.ifas.ufl.edu/lso/THRIPS/CHILLIWEB2/chilli-doc/CHILLI%20THRIPS%20Management.pdf. Chilli Thrips. Texas AgriLife Extension Service. Date last visited: 29 July 2013.

http://nabc.cals.cornell.edu/Publications/Reports/nabc_10/nabc_10_complete.pdf. Agricultural Biotechnology and Environmental Quality: Gene Escape and Pest Resistance. National Agricultural Biotechnology Council Report. Date last visited: 24 August 2013.

http://nematodeinformation.com/kill-black-cutworms-agrotis-ipsilon-hufnagel-with-entomopathogenic-nematodes. Kill Black Cutworms (*Agrotis ipsilon* Hufnagel) with Entomopathogenic Nematodes. Nematode Information. Date last visited: 29 July 2013.

http://nematodeinformation.com/tag/predatory-insects. Use Good Bugs to Control Bad Bugs: Predatory Insects. Nematode Information. Date last visited: 29 July 2013.

http://nhb.gov.in/bulletin_files/fruits/grape/gra008.pdF. Grapes. Date last visited: 28 July 2013.

http://npic.orst.edu/factsheets/captantech.pdf. Captan (Technical Fact Sheet). National Pesticide Information Center (NPIC). Date last visited: 29 July 2013.

http://ntp.niehs.nih.gov. National Toxicology Program, Department of Health and Human Services. National Institute of Environmental Health Sciences (part of National Institutes of Health, USA). Date last visited: 29 July 2013.

http://ntp.niehs.nih.gov/?objectid=7E6CAEBD-BDB5-82F8-F8C29152153B80B1. Evaluation of the Potential Use of Toxicogenomics Technology in Hazard Identification. Current Areas of Research, National Toxicology Program. Date last visited: 29 July 2013.

http://media.wiley.com/product_data/excerpt/42/04714919/0471491942-1.pdf. Famoxadone. In: Handbook of Residue Analytical Methods for Agrochemicals. 2003 John Wiley and Sons. Date last visited: 28 August 2013.

http://oehha.ca.gov/water/phg/pdf/thioben.pdf. Public Health Goal for Thiobencarb in Drinking Water. Office of Environmental Health Hazard Assessment, California Environmental Protection Agency. Date last visited: 29 July 2013.

http://okeechobee.ifas.ufl.edu/News%20columns/Selective%20Pesticides.htm. Selective Pesticides. UF/IFAS Okeechobee County Extension Service. Date last visited: 29 July 2013.

http://onlinelibrary.wiley.com/doi/10.1002/ps.2780060406/abstract. Suett, D.L. (2006) Persistence and degradation of chlorfenvinphos, chlormephos, disulfoton, phorate and pirimiphos-ethyl following spring and late-summer soil application. *Pest Management Science*. Date last visited: 29 July 2013.

http://phx.corporate-ir.net/phoenix.zhtml?c=117919&p=irol-newsArticle&ID=970461&highlightregistrations. FMC Corporation Acquires Exclusive Access to Sankyo Agro Fungicide. FMC Corporation. Date last visited: 29 July 2013.

http://plant-disease.ippc.orst.edu/ShowDisease.aspx. *Pacific Northwest Plant Disease Management Handbook*. Date last visited: 29 July 2013.

http://plants.usda.gov/java/profile?symbol=SEVI4. *Setaria viridis* (L.) P. Beauv. Green Bristlegrass. Plants Database. United States Department of Agriculture (USDA), Natural Resources Conservation Service. Date last visited: 29 July 2013.

http://pmep.cce.cornell.edu/profiles/extoxnet/24d-captan/bendiocarb-ext.html. Bendiocarb. Pesticide Information Profiles. Extension Toxicology Network. Date last visited: 29 July 2013.

http://pmep.cce.cornell.edu/profiles/extoxnet/carbaryl-dicrotophos/carbaryl-ext.html. Carbaryl. Pesticide Information Profiles. Extension Toxicology Network. Date last visited: 29 July 2013.

http://pmep.cce.cornell.edu/profiles/extoxnet/carbaryl-dicrotophos/carbofuran-ext.html. Carbofuran. Pesticide Information Profiles. Extension Toxicology Network. Date last visited: 29 July 2013.

http://pmep.cce.cornell.edu/profiles/extoxnet/carbaryl-dicrotophos/carboxin-ext.html. Carboxin. Pesticide Information Profiles. Extension Toxicology Network. Date last visited: 29 July 2013.

http://pmep.cce.cornell.edu/profiles/extoxnet/carbaryl-dicrotophos/chloramben-ext.html. Chloramben. Pesticide Information Profiles. Extension Toxicology Network. Date last visited: 29 July 2013.

http://pmep.cce.cornell.edu/profiles/extoxnet/carbaryl-dicrotophos/chlorothalonil-ext.html. Chlorothalonil. Pesticide Information Profiles. Extension Toxicology Network. Date last visited: 29 July 2013.

http://pmep.cce.cornell.edu/profiles/extoxnet/carbaryl-dicrotophos/cyfluthrin-ext.html. Cyfluthrin. Pesticide Information Profiles. Extension Toxicology Network. Date last visited: 29 July 2013.

http://pmep.cce.cornell.edu/profiles/extoxnet/dienochlor-glyphosate/eptc-ext.html. EPTC. Pesticide Information Profiles. Extension Toxicology Network. Date last visited: 28 August 2013.

http://pmep.cce.cornell.edu/profiles/extoxnet/dienochlor-glyphosate/ethylene-dibromide-ext.html. Ethylene Dibromide. Pesticide Information Profiles. Extension Toxicology Network. Date last visited: 29 July 2013.

http://pmep.cce.cornell.edu/profiles/extoxnet/dienochlor-glyphosate/fluazifop-p-butyl-ext.html. Fluazifop-p-butyl. Pesticide Information Profiles. Extension Toxicology Network. Date last visited: 29 July 2013.

http://pmep.cce.cornell.edu/profiles/extoxnet/dienochlor-glyphosate/fluometuron-ext.html. Fluometuron. Pesticide Information Profiles. Extension Toxicology Network. Date last visited: 28 August 2013.

http://pmep.cce.cornell.edu/profiles/extoxnet/haloxyfop-methylparathion/hexazinone-ext.html. Hexazinone. Pesticide Information Profiles. Extension Toxicology Network. Date last visited: 29 July 2013.

http://pmep.cce.cornell.edu/profiles/extoxnet/metiram-propoxur/propiconazole-ext.html. Propiconazole. Pesticide Information Profiles. Extension Toxicology Network. Date last visited: 28 August 2013.

http://pmep.cce.cornell.edu/profiles/extoxnet/metiram-propoxur/propoxur-ext.html. Propoxur. Pesticide Information Profiles. Extension Toxicology Network. Date last visited: 28 August 2013.

http://pmep.cce.cornell.edu/profiles/extoxnet/pyrethrins-ziram/scilliroside-ext.html. Scilliroside. Pesticide Information Profiles. Extension Toxicology Network. Date last visited: 29 July 2013.

http://pmep.cce.cornell.edu/profiles/extoxnet/pyrethrins-ziram/vernolate-ext.html. Vernolate. Pesticide Information Profiles. Extension Toxicology Network. Date last visited: 29 July 2013.

http://pmep.cce.cornell.edu/profiles/extoxnet/pyrethrins-ziram/vinclozolin-ext.html. Vinclozolin. Pesticide Information Profiles. Extension Toxicology Network. Date last visited: 29 July 2013.

http://pmep.cce.cornell.edu/profiles/fung-nemat/febuconazole-sulfur/oxycarboxin/fung-prof-oxycarboxin.html. Oxycarboxin. Pesticide Information Profiles. Extension Toxicology Network. Date last visited: 29 July 2013.

http://pmep.cce.cornell.edu/profiles/herb-growthreg/24-d-butylate/ametryn/herb-prof-ametryn.html. Ametryn. Pesticide Information Profiles. Extension Toxicology Network. Date last visited: 29 July 2013.

http://pmep.cce.cornell.edu/profiles/herb-growthreg/fatty-alcohol-monuron/fluridone/herb-prof-fluridone.html. Fluridone (Sonar, Brake, Pride) Herbicide Profile 3/86. Chemical Fact Sheet For Fluridone. Cornell University. Date last visited: 28 August 2013.

http://pmep.cce.cornell.edu/profiles/herb-growthreg/naa-rimsulfuron/prohexadione-calcium/prohexadione_pet_301.html. Prohexadione-calcium (Viviful) Pesticide Petition Filing 3/01. Cornell University. Date last visited: 28 August 2013.

http://pmep.cce.cornell.edu/profiles/herb-growthreg/naa-rimsulfuron/prometon/herb-prof-prometon.html. Prometon. Pesticide Information Profiles. Extension Toxicology Network. Date last visited: 29 July 2013.

http://pmep.cce.cornell.edu/profiles/herb-growthreg/naa-rimsulfuron/propham/herb-prof-propham.html. Propham (Chem Hoe) Herbicide Profile 2/85. Cornell University. Date last visited: 28 August 2013.

http://pmep.cce.cornell.edu/profiles/insect-mite/cadusafos-cyromazine/chlorethoxyfos/chloreth1197.pdf. New Active Ingredient Chlorethoxyfos Contained in the Pesticide Product Fortress 5G Insecticide. Date last visited: 29 July 2013.

http://pmep.cce.cornell.edu/profiles/insect-mite/cadusafos-cyromazine/cyromazine/insect-prof-cyromazine.html. Cyromazine (Larvadex, Trigard) Chemical Fact Sheet 12/86. Cornell University, New York. Date last visited: 29 July 2013.

http://pmep.cce.cornell.edu/profiles/insect-mite/mevinphos-propargite/oxydemeton-methyl/insect-prof-oxydem-methyl.html. Oxydemeton-methyl (Metasystox-R) Chemical Profile 4/85. Cornell University, New York. Date last visited: 29 July 2013.

http://pmep.cce.cornell.edu/profiles/insect-mite/propetamphos-zetacyperm/rotenone/insect-prof-rotenone.html. Rotenone EPA Pesticide Fact Sheet 10/88. Cornell University, New York. Date last visited: 29 July 2013.

http://pmep.cce.cornell.edu/profiles/insect-mite/propetamphos-zetacyperm/tetrachlorvinphos/prof-tetrachlorvinphos.html. Tetrachlorvinphos (Rabon, Gardona) EPA Pesticide Fact Sheet 10/88. Cornell University, New York. Date last visited: 29 July 2013.

http://pmep.cce.cornell.edu/profiles/insect-mite/propetamphos-zetacyperm/thiodicarb/insect-prof-thiodicarb.html. Thiodicarb (Larvin) Chemical Fact Sheet 8/84. Cornell University, New York. Date last visited: 29 July 2013.

http://pmep.cce.cornell.edu/profiles/rodent/rodent_M_Z/scilliroside/scillir_prf_0185.html. Scilliroside (Red Squill, Rodine) – Chemical Profile 1/85. Cornell University, New York. Date last visited: 29 July 2013.

http://pmep.cce.cornell.edu/regulation/sec18/2009/agrimek_s18_0609.pdf. Agri-Mek 0.15EC. 1995. Syngenta FIFRA. Date last visited: 28 August 2013.

http://pubchem.ncbi.nlm.nih.gov/summary/summary.cgi?cid=41554. CTK1H5547 – Compound Summary (CID 41554). PubChem Compound. Date last visited: 29 July 2013.

http://pubs.acs.org/doi/abs/10.1021/bk-1994-0551.ch006. Hydantocidin and Cornexistin – New Herbicidal Antibiotics. American Chemical Society (ACS) Symposium Series. Date last visited: 30 July 2013.

http://pubs.acs.org/doi/abs/10.1021/jf60161a615. Hock, W. and Sisler, H. (1969) Metabolism of chloroneb by *Rhizoctonia solani* and other fungi. *Journal of Agricultural and Food Chemistry* 17 (1), 123–128. Date last visited: 30 July 2013.

http://pubs.cas.psu.edu/FreePubs/pdfs/uo229.pdf. How to Handle Chemical Spills. Pesticide Safety Fact Sheet. The Pennsylvania State University. Date last visited: 30 July 2013.

http://purdueturftips.blogspot.in/ Turf Tips. Purdue Extension. Date last visited: 7 August 2013.

http://pxmrl.nzfsa.govt.nz/Residues/Country/Australia/Pesticide/CHLOROPICRIN/Default.aspx. Chloropicrin. New Zealand Food Safety Authority – Pesticide MRL Database. Date last visited: 30 July 2013.

http://pxmrl.nzfsa.govt.nz/Residues/Country/India/Pesticide/CHLORPYRIFOS/Default.aspx. Chlorpyrifos. New Zealand Food Safety Authority – Pesticide MRL Database. Date last visited: 30 July 2013.

http://pxmrl.nzfsa.govt.nz/Residues/Country/India/Pesticide/CYPERMETHRIN/Default.aspx. Cypermethrin. New Zealand Food Safety Authority – Pesticide MRL Database. Date last visited: 30 July 2013.

http://resources.cas.psu.edu/ipm/POP/croprotat.pdf. Healthy Plants Start Here – Crop Rotation. Date last visited: 30 July 2013.

http://resources.ciheam.org/om/pdf/c15-3/CI011017.pdf. Biological and Chemical Control of Rice Blast Disease (*Pyricularia oryzae*) in Northern Greece. National Articultual Research Foundation, Plant Protection Institute of Thessaloniki (Greece). Date last visited: 30 July 2013.

http://scialert.net/abstract/?doi=ppj.2010.31.38. Tahat, M.M., Kamaruzaman, S. and Othman, R. (2010) Mycorrhizal fungi as a biocontrol agent. *Plant Pathology Journal* 9 (9), 198–207. Date last visited: 30 July 2013.

http://sciencelinks.jp/j-east/article/200112/000020011201A0497579.php. Mitsuru, K., Hiromi, O., Norio, S., Hitoshi, K., Yasutomo, T. and Makotu, K. Herbicidal properties of cafenstrole in turfgrasses. *Journal of Japanese Society of Turfgrass Science* 29 (2), 115–120. Date last visited: 30 July 2013.

https://secure.virtuality.net/panukcom/PDFs/Banned%20In%20The%20EU.pdf. Which Pesticides Are Banned in Europe? Pesticide Action Network UK, PAN Africa and The Rufford Maurice Laing Foundation. Date last visited: 28 August 2013.

http://services.indiabizclub.com/catalog/119240 amblyseius+cucumeris tucson. *Amblyseius cucumeris*. Buglogical Beneficial Insect Catalog, Tucson, USA. Date last visited: 30 July 2013.

http://sitem.herts.ac.uk/aeru/footprint/en/index.htm. The Pesticide Properties Data Base (PPDB). University of Hertfordshire. Date last visited: 31 July 2013.

http://sitem.herts.ac.uk/aeru/footprint/en/Reports/1.htm#none. 1,3-Dichloropropene. General Information. Pesticide Properties Data Base (PPDB). University of Hertfordshire. Date last visited: 31 July 2013.

http://sitem.herts.ac.uk/aeru/footprint/en/Reports/12.htm. Acetochlor. General Information. Pesticide Properties Data Base (PPDB). University of Hertfordshire. Date last visited: 31 July 2013.

http://sitem.herts.ac.uk/aeru/footprint/en/Reports/47.htm. Azafenidin. General Information. Pesticide Properties Data Base (PPDB). University of Hertfordshire. Date last visited: 31 July 2013.

http://sitem.herts.ac.uk/aeru/footprint/en/Reports/92.htm. Bromofenoxim. General Information. Pesticide Properties Data Base (PPDB). University of Hertfordshire. Date last visited: 1 August 2013.

http://sitem.herts.ac.uk/aeru/footprint/en/Reports/105.htm#none. Butylate. General Information. Pesticide Properties Data Base (PPDB). University of Hertfordshire. Date last visited: 31 July 2013.

http://sitem.herts.ac.uk/aeru/footprint/en/Reports/117.htm. Carbetamide. General Information. Pesticide Properties Data Base (PPDB). University of Hertfordshire. Date last visited: 31 July 2013.

http://sitem.herts.ac.uk/aeru/footprint/en/Reports/123.htm#none. Carfentrazone-ethyl. General Information. Pesticide Properties Data Base (PPDB). University of Hertfordshire. Date last visited: 31 July 2013.

http://sitem.herts.ac.uk/aeru/footprint/en/Reports/145.htm. Chloroneb. General Information. Pesticide Properties Data Base (PPDB). University of Hertfordshire. Date last visited: 31 July 2013.

http://sitem.herts.ac.uk/aeru/footprint/en/Reports/162.htm. Cinidon-ethyl. General Information. Pesticide Properties Data Base (PPDB). University of Hertfordshire. Date last visited: 31 July 2013.

http://sitem.herts.ac.uk/aeru/footprint/en/Reports/168.htm. Clomazone. General Information. Pesticide Properties Data Base (PPDB). University of Hertfordshire. Date last visited: 31 July 2013.

http://sitem.herts.ac.uk/aeru/footprint/en/Reports/188.htm. Cycloate. General Information. Pesticide Properties Data Base (PPDB). University of Hertfordshire. Date last visited: 31 July 2013.

http://sitem.herts.ac.uk/aeru/footprint/en/Reports/193.htm. Cyhalofop-butyl. General Information. Pesticide Properties Data Base (PPDB). University of Hertfordshire. Date last visited: 31 July 2013.

http://sitem.herts.ac.uk/aeru/footprint/en/Reports/211.htm. Di-allate. General Information. Pesticide Properties Data Base (PPDB). University of Hertfordshire. Date last visited: 31 July 2013.

http://sitem.herts.ac.uk/aeru/footprint/en/Reports/277.htm. Ethirimol. General Information. Pesticide Properties Data Base (PPDB). University of Hertfordshire. Date last visited: 31 July 2013.

http://sitem.herts.ac.uk/aeru/footprint/en/Reports/281.htm. Ethoxysulfuron. General Information. Pesticide Properties Data Base (PPDB). University of Hertfordshire. Date last visited: 31 July 2013.

http://sitem.herts.ac.uk/aeru/footprint/en/Reports/296.htm. Fenclorim. General Information. Pesticide Properties Data Base (PPDB). University of Hertfordshire. Date last visited: 31 July 2013.

http://sitem.herts.ac.uk/aeru/footprint/en/Reports/305.htm. Fenpiclonil. General Information. Pesticide Properties Data Base (PPDB). University of Hertfordshire. Date last visited: 31 July 2013.

http://sitem.herts.ac.uk/aeru/footprint/en/Reports/308.htm. Fenpropimorph. General Information. Pesticide Properties Data Base (PPDB). University of Hertfordshire. Date last visited: 31 July 2013.

http://sitem.herts.ac.uk/aeru/footprint/en/Reports/330.htm. Fludioxonil. General Information. Pesticide Properties Data Base (PPDB). University of Hertfordshire. Date last visited: 31 July 2013.

http://sitem.herts.ac.uk/aeru/footprint/en/Reports/342.htm. Fluquinconazole. General Information. Pesticide Properties Data Base (PPDB). University of Hertfordshire. Date last visited: 31 July 2013.

http://sitem.herts.ac.uk/aeru/footprint/en/Reports/349.htm. Flurtamone. General Information. Pesticide Properties Data Base (PPDB). University of Hertfordshire. Date last visited: 31 July 2013.

http://sitem.herts.ac.uk/aeru/footprint/en/Reports/361.htm. Formothion. General Information. Pesticide Properties Data Base (PPDB). University of Hertfordshire. Date last visited: 31 July 2013.

http://sitem.herts.ac.uk/aeru/footprint/en/Reports/363.htm. Fosetyl-aluminium. General Information. Pesticide Properties Data Base (PPDB). University of Hertfordshire. Date last visited: 31 July 2013.

http://sitem.herts.ac.uk/aeru/footprint/en/Reports/365.htm. Fuberidazole. General Information. Pesticide Properties Data Base (PPDB). University of Hertfordshire. Date last visited: 31 July 2013.

http://sitem.herts.ac.uk/aeru/footprint/en/Reports/405.htm. Isazofos. General Information. Pesticide Properties Data Base (PPDB). University of Hertfordshire. Date last visited: 31 July 2013.

http://sitem.herts.ac.uk/aeru/footprint/en/Reports/412.htm. Isoxaflutole. General Information. Pesticide Properties Data Base (PPDB). University of Hertfordshire. Date last visited: 31 July 2013.

http://sitem.herts.ac.uk/aeru/footprint/en/Reports/413.htm. Isoxathion. General Information. Pesticide Properties Data Base (PPDB). University of Hertfordshire. Date last visited: 31 July 2013.

http://sitem.herts.ac.uk/aeru/footprint/en/Reports/490.htm. Ofurace. General Information. Pesticide Properties Data Base (PPDB). University of Hertfordshire. Date last visited: 31 July 2013.

http://sitem.herts.ac.uk/aeru/footprint/en/Reports/493.htm. Orbencarb. General Information. Pesticide Properties Data Base (PPDB). University of Hertfordshire. Date last visited: 31 July 2013.

http://sitem.herts.ac.uk/aeru/footprint/en/Reports/497.htm. Oxadixyl. General Information. Pesticide Properties Data Base (PPDB). University of Hertfordshire. Date last visited: 31 July 2013.

http://sitem.herts.ac.uk/aeru/footprint/en/Reports/501.htm. Oxydemeton-methyl. General Information. Pesticide Properties Data Base (PPDB). University of Hertfordshire. Date last visited: 31 July 2013.

http://sitem.herts.ac.uk/aeru/footprint/en/Reports/533.htm. Polyoxins. General Information. Pesticide Properties Data Base (PPDB). University of Hertfordshire. Date last visited: 31 July 2013.

http://sitem.herts.ac.uk/aeru/footprint/en/Reports/543.htm. Propachlor. General Information. Pesticide Properties Data Base (PPDB). University of Hertfordshire. Date last visited: 31 July 2013.

http://sitem.herts.ac.uk/aeru/footprint/en/Reports/557.htm. Prosulfocarb. General Information. Pesticide Properties Data Base (PPDB). University of Hertfordshire. Date last visited: 31 July 2013.

http://sitem.herts.ac.uk/aeru/footprint/en/Reports/560.htm. Prothiofos. General Information. Pesticide Properties Data Base (PPDB). University of Hertfordshire. Date last visited: 31 July 2013.

http://sitem.herts.ac.uk/aeru/footprint/en/Reports/563.htm. Pyraclofos. General Information. Pesticide Properties Data Base (PPDB). University of Hertfordshire. Date last visited: 31 July 2013.

http://sitem.herts.ac.uk/aeru/footprint/en/Reports/604.htm. Sulfotep. General Information. Pesticide Properties Data Base (PPDB). University of Hertfordshire. Date last visited: 31 July 2013.

http://sitem.herts.ac.uk/aeru/footprint/en/Reports/607.htm. Sulprofos. General Information. Pesticide Properties Data Base (PPDB). University of Hertfordshire. Date last visited: 31 July 2013.

http://sitem.herts.ac.uk/aeru/footprint/en/Reports/608.htm. Tau-fluvalinate. General Information. Pesticide Properties Data Base (PPDB). University of Hertfordshire. Date last visited: 31 July 2013.

http://sitem.herts.ac.uk/aeru/footprint/en/Reports/615.htm. Tecnazene. General Information. Pesticide Properties Data Base (PPDB). University of Hertfordshire. Date last visited: 31 July 2013.

http://sitem.herts.ac.uk/aeru/footprint/en/Reports/616.htm. Teflubenzuron. General Information. Pesticide Properties Data Base (PPDB). University of Hertfordshire. Date last visited: 31 July 2013.

http://sitem.herts.ac.uk/aeru/footprint/en/Reports/618.htm. Temephos. General Information. Pesticide Properties Data Base (PPDB). University of Hertfordshire. Date last visited: 31 July 2013.

http://sitem.herts.ac.uk/aeru/footprint/en/Reports/621.htm. Terbufos. General Information. Pesticide Properties Data Base (PPDB). University of Hertfordshire. Date last visited: 31 July 2013.

http://sitem.herts.ac.uk/aeru/footprint/en/Reports/625.htm. Tetrachlorvinphos. General Information. Pesticide Properties Data Base (PPDB). University of Hertfordshire. Date last visited: 31 July 2013.

http://sitem.herts.ac.uk/aeru/footprint/en/Reports/630.htm. Thiacloprid. General Information. Pesticide Properties Data Base (PPDB). University of Hertfordshire. Date last visited: 31 July 2013.

http://sitem.herts.ac.uk/aeru/footprint/en/Reports/633.htm. Thiazopyr. General Information. Pesticide Properties Data Base (PPDB). University of Hertfordshire. Date last visited: 31 July 2013.

http://sitem.herts.ac.uk/aeru/footprint/en/Reports/636.htm. Thiobencarb. General Information. Pesticide Properties Data Base (PPDB). University of Hertfordshire. Date last visited: 31 July 2013.

http://sitem.herts.ac.uk/aeru/footprint/en/Reports/638.htm. Thiofanox. General Information. Pesticide Properties Data Base (PPDB). University of Hertfordshire. Date last visited: 31 July 2013.
http://sitem.herts.ac.uk/aeru/footprint/en/Reports/644.htm. Tolclofos-methyl. General Information. Pesticide Properties Data Base (PPDB). University of Hertfordshire. Date last visited: 31 July 2013.
http://sitem.herts.ac.uk/aeru/footprint/en/Reports/653.htm. Triazophos. General Information. Pesticide Properties Data Base (PPDB). University of Hertfordshire. Date last visited: 31 July 2013.
http://sitem.herts.ac.uk/aeru/footprint/en/Reports/656.htm. Tribufos. General Information. Pesticide Properties Data Base (PPDB). University of Hertfordshire. Date last visited: 31 July 2013.
http://sitem.herts.ac.uk/aeru/footprint/en/Reports/661.htm. Tridemorph. General Information. Pesticide Properties Data Base (PPDB). University of Hertfordshire. Date last visited: 31 July 2013.
http://sitem.herts.ac.uk/aeru/footprint/en/Reports/663.htm. Trietazine. General Information. Pesticide Properties Data Base (PPDB). University of Hertfordshire. Date last visited: 31 July 2013.
http://sitem.herts.ac.uk/aeru/footprint/en/Reports/665.htm. Triflumizole. General Information. Pesticide Properties Data Base (PPDB). University of Hertfordshire. Date last visited: 31 July 2013.
http://sitem.herts.ac.uk/aeru/footprint/en/Reports/671.htm. Trimethacarb. General Information. Pesticide Properties Data Base (PPDB). University of Hertfordshire. Date last visited: 1 August 2013.
http://sitem.herts.ac.uk/aeru/footprint/en/Reports/711.htm. Diisobutylamine. General Information. Pesticide Properties Data Base (PPDB). University of Hertfordshire. Date last visited: 1 August 2013.
http://sitem.herts.ac.uk/aeru/footprint/en/Reports/1021.htm. Cinmethylin. General Information. Pesticide Properties Data Base (PPDB). University of Hertfordshire. Date last visited: 31 July 2013.
http://sitem.herts.ac.uk/aeru/footprint/en/Reports/1065.htm. Thiocyclam. General Information. Pesticide Properties Data Base (PPDB). University of Hertfordshire. Date last visited: 31 July 2013.
http://sitem.herts.ac.uk/aeru/footprint/en/Reports/1087.htm. Butroxydim. General Information. Pesticide Properties Data Base (PPDB). University of Hertfordshire. Date last visited: 31 July 2013.
http://sitem.herts.ac.uk/aeru/footprint/en/Reports/1158.htm#none. Dimethametryn. General Information. Pesticide Properties Data Base (PPDB). University of Hertfordshire. Date last visited: 31 July 2013.
http://sitem.herts.ac.uk/aeru/footprint/en/Reports/1171.htm. Thenylchlor. General Information. Pesticide Properties Data Base (PPDB). University of Hertfordshire. Date last visited: 31 July 2013.
http://sitem.herts.ac.uk/aeru/footprint/en/Reports/1204.htm. Simeconazole. General Information. Pesticide Properties Data Base (PPDB). University of Hertfordshire. Date last visited: 31 July 2013.
http://sitem.herts.ac.uk/aeru/footprint/en/Reports/1209.htm. Thifluzamide. General Information. Pesticide Properties Data Base (PPDB). University of Hertfordshire. Date last visited: 31 July 2013.
http://sitem.herts.ac.uk/aeru/footprint/en/Reports/1282.htm. Transfluthrin. General Information. Pesticide Properties Data Base (PPDB). University of Hertfordshire. Date last visited: 31 July 2013.
http://sitem.herts.ac.uk/aeru/footprint/en/Reports/1295.htm. Simetryn. General Information. Pesticide Properties Data Base (PPDB). University of Hertfordshire. Date last visited: 31 July 2013.
http://sitem.herts.ac.uk/aeru/footprint/en/Reports/1380.htm. Propisochlor. General Information. Pesticide Properties Data Base (PPDB). University of Hertfordshire. Date last visited: 31 July 2013.
http://sitem.herts.ac.uk/aeru/footprint/en/Reports/1463.htm. Citronella oil. General Information. Pesticide Properties Data Base (PPDB). University of Hertfordshire. Date last visited: 31 July 2013.
http://sitem.herts.ac.uk/aeru/footprint/en/Reports/1497.htm. Flucetosulfuron. General Information. Pesticide Properties Data Base (PPDB). University of Hertfordshire. Date last visited: 31 July 2013.
http://sitem.herts.ac.uk/aeru/footprint/en/Reports/1562.htm. Chlorphonium Chloride. General Information. Pesticide Properties Data Base (PPDB). University of Hertfordshire. Date last visited: 31 July 2013.
http://sitem.herts.ac.uk/aeru/footprint/en/Reports/1677.htm. Propaphos. General Information. Pesticide Properties Data Base (PPDB). University of Hertfordshire. Date last visited: 31 July 2013.
http://sitem.herts.ac.uk/aeru/footprint/en/Reports/1682.htm. Butamifos. General Information. Pesticide Properties Data Base (PPDB). University of Hertfordshire. Date last visited: 31 July 2013.
http://sitem.herts.ac.uk/aeru/iupac/Reports/163.htm. Cinosulfuron. General Information. Pesticide Properties Data Base (PPDB). University of Hertfordshire. Date last visited: 1 August 2013.
http://sitem.herts.ac.uk/aeru/iupac/331.htm. Flufenacet. General Information. Pesticide Properties Data Base (PPDB). University of Hertfordshire. Date last visited: 1 August 2013.
http://sitem.herts.ac.uk/aeru/iupac/Reports/448.htm. Metamitron. General Information. Pesticide Properties Data Base (PPDB). University of Hertfordshire. Date last visited: 1 August 2013.
http://sitem.herts.ac.uk/aeru/iupac/542.htm. Prometryn. General Information. Pesticide Properties Data Base (PPDB). University of Hertfordshire. Date last visited: 1 August 2013.
http://sitem.herts.ac.uk/aeru/iupac/Reports/599.htm. Spiroxamine. General Information. Pesticide Properties Data Base (PPDB). University of Hertfordshire. Date last visited: 1 August 2013.
http://sitem.herts.ac.uk/aeru/iupac/Reports/617.htm. Tefluthrin. General Information. Pesticide Properties Data Base (PPDB). University of Hertfordshire. Date last visited: 1 August 2013.

http://sitem.herts.ac.uk/aeru/iupac/Reports/619.htm. Tepraloxydim. General Information. Pesticide Properties Data Base (PPDB). University of Hertfordshire. Date last visited: 1 August 2013.

http://sitem.herts.ac.uk/aeru/iupac/Reports/637.htm. Thiodicarb. General Information. Pesticide Properties Data Base (PPDB). University of Hertfordshire. Date last visited: 1 August 2013.

http://sitem.herts.ac.uk/aeru/iupac/Reports/640.htm. Thiophanate-methyl. General Information. Pesticide Properties Data Base (PPDB). University of Hertfordshire. Date last visited: 1 August 2013.

http://sitem.herts.ac.uk/aeru/iupac/Reports/647.htm. Tralomethrin. General Information. Pesticide Properties Data Base (PPDB). University of Hertfordshire. Date last visited: 1 August 2013.

http://sitem.herts.ac.uk/aeru/iupac/Reports/657.htm. Trichlorfon. General Information. Pesticide Properties Data Base (PPDB). University of Hertfordshire. Date last visited: 1 August 2013.

http://sitem.herts.ac.uk/aeru/iupac/Reports/664.htm. Trifloxystrobin. General Information. Pesticide Properties Data Base (PPDB). University of Hertfordshire. Date last visited: 1 August 2013.

http://sitem.herts.ac.uk/aeru/iupac/Reports/666.htm. Triflumuron. General Information. Pesticide Properties Data Base (PPDB). University of Hertfordshire. Date last visited: 1 August 2013.

http://sitem.herts.ac.uk/aeru/iupac/Reports/673.htm. Triticonazole. General Information. Pesticide Properties Data Base (PPDB). University of Hertfordshire. Date last visited: 1 August 2013.

http://sitem.herts.ac.uk/aeru/iupac/Reports/731.htm. 3,5,6-trichloro-2-pyridinol. General Information. Pesticide Properties Data Base (PPDB). University of Hertfordshire. Date last visited: 1 August 2013.

http://sitem.herts.ac.uk/aeru/iupac/Reports/1027.htm. S-metolachlor. General Information. Pesticide Properties Data Base (PPDB). University of Hertfordshire. Date last visited: 1 August 2013.

http://sitem.herts.ac.uk/aeru/iupac/Reports/1157.htm. Flucarbazone-sodium. General Information. Pesticide Properties Data Base (PPDB). University of Hertfordshire. Date last visited: 1 August 2013.

http://sitem.herts.ac.uk/aeru/iupac/1206.htm. Oxpoconazole fumarate. General Information. Pesticide Properties Data Base (PPDB). University of Hertfordshire. Date last visited: 1 August 2013.

http://sitem.herts.ac.uk/aeru/iupac/Reports/1576.htm. Isoxadifen ethyl. General Information. Pesticide Properties Data Base (PPDB). University of Hertfordshire. Date last visited: 1 August 2013.

http://sitem.herts.ac.uk/aeru/iupac/Reports/1582.htm. Fluxofenim. General Information. Pesticide Properties Data Base (PPDB). University of Hertfordshire. Date last visited: 1 August 2013.

http://sitem.herts.ac.uk/aeru/vsdb/Reports/1879.htm. Oxolinic acid. General Information. Pesticide Properties Data Base (PPDB). University of Hertfordshire. Date last visited: 1 August 2013.

http://strawberry.ifas.ufl.edu/plantpathfiles/Fung%20label/fung%202009/helena%20prophyt.pdf. Helena® Prophyt®. Helena Holding Company, Luxembourg Industries. Date last visited: 1 August 2013.

http://takingroots.in/ipm_aphids. Aphids [*Aphis craccivora*, *Myzus persicae*, *Aphis gossypii*]. Centre for Sustainable Agriculture and Krishi Media. Date last visited: 1 August 2013.

http://takingroots.in/ipm_castor_semilooper. Castor Semilooper [*Achaea janata*, *Parallelia algira*]. Centre for Sustainable Agriculture and Krishi Media. Date last visited: 1 August 2013.

http://takingroots.in/ipm_cscb. Castor Stem and Capsule Borer [*Dichocrosis punctiferalis*]. Centre for Sustainable Agriculture and Krishi Media. Date last visited: 1 August 2013.

http://takingroots.in/ipm_glm. Groundnut Leaf Miner [*Aproarema modicella*]. Centre for Sustainable Agriculture and Krishi Media. Date last visited: 1 August 2013.

http://takingroots.in/ipm_rbph. Rice Brown Plant Hopper [*Nilparvata lugens*]. Centre for Sustainable Agriculture and Krishi Media. Date last visited: 1 August 2013.

http://takingroots.in/ipm_rgb. Rice Gundhy Bug [*Leptocorisa varicornis*]. Centre for Sustainable Agriculture and Krishi Media. Date last visited: 1 August 2013.

http://takingroots.in/ipm_rglh. Rice Green Leaf Hopper [*Nephotettix virens*]. Centre for Sustainable Agriculture and Krishi Media. Date last visited: 1 August 2013.

http://takingroots.in/ipm_rgm. Rice Gall Midge [*Orseolia oryzae*]. Centre for Sustainable Agriculture and Krishi Media. Date last visited: 1 August 2013.

http://takingroots.in/ipm_rhc. Red Hairy Caterpillar [*Amsacta albistriga*]. Centre for Sustainable Agriculture and Krishi Media. Date last visited: 1 August 2013.

http://takingroots.in/ipm_rlf. Rice Leaf Folder [*Cnephalocrosis medinalis*]. Centre for Sustainable Agriculture and Krishi Media. Date last visited: 1 August 2013.

http://takingroots.in/sites/default/files/Pink_Bollworm_02.jpg. Pink Bollworm. Centre for Sustainable Agriculture and Krishi Media. Date last visited: 1 August 2013.

http://takingroots.in/sites/default/files/Spotted_Borer_02.jpg. Spotted Borer. Centre for Sustainable Agriculture and Krishi Media.

http://takingroots.in/sites/default/files/Tobacco_Caterpiller_02.jpg. Tobacco Caterpillar. Centre for Sustainable Agriculture and Krishi Media. Date last visited: 1 August 2013.

http://tomclothier.hort.net/page13.html. Damping-off. Tom Clothier's Garden Walk and Talk. Date last visited: 1 August 2013.

http://toxipedia.org/display/toxipedia/Chlorfenvinphos. Chlorfenvinphos. Toxipedia. Date last visited: 1 August 2013.

http://toxipedia.org/display/toxipedia/Ethiofencarb#Ethiofencarb-PesticideDatabaseEthiofencarb. Ethiofencarb. Toxipedia. Date last visited: 1 August 2013.
http://toxipedia.org/display/toxipedia/Ethion. Ethion. Toxipedia. Date last visited: 1 August 2013.
http://toxipedia.org/display/toxipedia/Ethylene+Dibromide. Ethylene Dibromide. Toxipedia. Date last visited: 1 August 2013.
http://toxipedia.org/display/toxipedia/Tetramethrin. Tetramethrin. Toxipedia. Date last visited: 1 August 2013.
http://toxipedia.org/display/toxipedia/Toxaphene. Toxaphene. Toxipedia. Date last visited: 1 August 2013.
http://toxipedia.org/display/toxipedia/Trichlorfon. Trichlorfon. Toxipedia. Date last visited: 1 August 2013.
http://toxipedia.org/display/toxipedia/Triclopyr. Triclopyr. Toxipedia. Date last visited: 1 August 2013.
http://toxnet.nlm.nih.gov/cgi-bin/sis/search/a?dbs+hsdb:@term+@DOCNO+668. Tribufos. Toxicology Data Network. Date last visited: 1 August 2013.
http://toxnet.nlm.nih.gov/cgi-bin/sis/search/a?dbs+hsdb:@term+@DOCNO+2800. Chlordene. Toxicology Data Network. Date last visited: 1 August 2013.
http://toxnet.nlm.nih.gov/cgi-bin/sis/search/a?dbs+hsdb:@term+@DOCNO+6684. Isazofos. Toxnet. Toxicology Data Network. Date last visited: 1 August 2013.
http://toxnet.nlm.nih.gov/cgi-bin/sis/search/r?dbs+hsdb:@term+@rn+6317-18-6. Methylene Thiocyanate. Hazardous Substances Data Bank. Toxicology Data Network. Date last visited: 1 August 2013.
http://translate.google.co.in/translate?hl=en&sl=es&u=http://gipcitricos.ivia.es/amitus-spiniferus.html&ei=ksV1TYGOCYnorQeS-Lm_Cg&sa=X&oi=translate&ct=result&resnum=6&ved=0CDcQ7gEwBTgU&prev=/search%3Fq%3DAmitus%2Bspiniferus%26start%3D20%26hl%3Den%26sa%3DN%26biw%3D1024%26bih%3D599%26prmd%3Divnsb. Gestión Integrada De Plagas De Cítricos. GIP Citrus Instituto valenciano de investigaciones agrarias (IVIA). Date last visited: 2 August 2013.
http://umbbd.msi.umn.edu/ipt/ipt_map.html. Chi Ieng Lai. Isoproturon Pathway Map. University of Minnesota. Date last visited: 2 August 2013.
http://universalium.academic.ru/91376/chemosterilant. Academic Dictionaries and Encyclopedia. Date last visited: 2 August 2013.
http://webbook.nist.gov/cgi/cbook.cgi?ID=C2104645&Mask=4#Thermo-Phase.Phosphonothioic Acid, Phenyl-, O-ethyl O-(4-nitrophenyl) Ester. National Institute of Standards and Technology. Date last visited: 2 August 2013.
http://webcache.googleusercontent.com/search?q=cache:http://sitem.herts.ac.uk/aeru/footprint/en/Reports/546.htm. Propaquizafop (Ref: CGA 233380). Date last visited: 28 August 2013.
http://wssa.net/weed/biological-control/. Biological Control. Weed Science Society of America (WSSA) Position Statement on Biological Control of Weeds. Date last visited: 23 August 2013.
http://www.abcbirds.org/abcprograms/policy/toxins/aims/aims/toxicitytable.cfm. American Bird Conservancy. Date last visited: 2 August 2013.
http://www.academicjournals.org/article/article1380538490_Siddique%20et%20al.pdf. Siddique, S., Parveen, Z. and Chaudhry, M.N. (2013) Repellency, fumigant and contact toxicities of essential oils of selected species from family Myrtaceae against *Tribolium castaneum*. *Journal of Medicinal Plants Research* 7 (3), 126–139. Date last visited: 2 August 2013.
http://www.academicjournals.org/article/article1380784245_Khan%20et%20al.pdf. Khan, A., Haque, E., Rahman, M. and Nessa, F. (2012) Antibacterial activity and cytytoxicity of rhizomes of *Dryneria quercifolia*. *Journal of Medicinal Plants Research* 6 (13), 2576–2580. Date last visited: 2 August 2013.
http://www.aces.edu/pubs/docs/A/ANR-1073/ANR-1073.pdf. Nutsedge Control in Commercial Vegetables. Alabama Cooperative Extension System. ANR-1073. Date last visited: 2 August 2013.
http://www.adavalue.com/en/biological-control.html. Biological Control (or Integrated Control). Adavalue. Date last visited: 2 August 2013.
http://www.afpmb.org/sites/default/files/pubs/techguides/tg15.pdf. Pesticide Spill Prevention and Management – Technical Guide No. 15. 2009. Armed Forces Pest Management Board Information Services Division. Date last visited: 2 August 2013.
http://www.ag.ndsu.edu/procrop/env/envebl05.htm. Emergence of Broadleaf Crops. North Dakota State University (NDSU) Extension Service. Date last visited: 2 August 2013.
http://www.ag.ndsu.edu/pubs/plantsci/hortcrop/pp697-3.htm. North Dakota State University (NDSU). Date last visited: 2 August 2013.
http://www.ag.ndsu.edu/smgrains/Best%20of%20Best%20West%202010/Sawfly%20Stem%20Maggot%20Bottineau.pdf. Small Grains. North Dakota State University (NDSU) Extension Service. Date last visited: 2 August 2013.
http://www.agf.gov.bc.ca/cropprot/grapeipm/mildew.htm. Grape Diseases. (*Uncinula necator*). Ministry of Agriculture, British Columbia. Date last visited: 2 August 2013.
http://www.agf.gov.bc.ca/pesticides/a_3.htm. About Pesticides. Ministry of Agriculture, British Columbia. Date last visited: 2 August 2013.
http://www.agf.gov.bc.ca/pesticides/g_4.htm. Pesticide Emergencies. Ministry of Agriculture, British Columbia. Date last visited: 2 August 2013.
http://www.agf.gov.bc.ca/pesticides/infosheets/abamectin.pdf. Abamectin. Pesticide Info. British Columbia. Ministry of Agriculture, Food and Fisheries. Date last visited: 28 August 2013.
http://www.agnet.org/view.php?id=20110421140203_982650. Food & Fertilizer Technology Center. Taiwan. Date last visited: 2 August 2013.
http://www.agra-trading.com/page/organic-insecticide.php. Insecticide/Fungicide Agra Marketing Group. Date last visited: 2 August 2013.
http://www.agribusinessweek.com/the-other-insect-pests-corn-planthoppers-and-leaf-aphids/. The Other Insect Pests: Corn Planthoppers and Leaf Aphids. Agriculture Business Week. Date Last Visited: 2 August 2013.
http://www.agric.gov.ab.ca/app23/getproductinfo?id=8&view=dressed. Amitrol 240. Herbicide Selector. Alberta Agriculture and Rural Development. Date Last Visited: 2 August 2013.

http://www.agroatlas.ru/en/content/diseases/Oryzae/Oryzae_Tilletia_horrida/. Diseases. *Tilletia horrida* Takah. – Black Smut of Rice. Interactive Agricultural Ecological Atlas of Russia and Neighbouring Countries.

http://www.agrocare.com.cn/Products/Acetochlor.htm. Acetochlor. Agro-care Chemical Industry Group. Date last visited: 2 August 2013.

http://www.agrocare.com.cn/Products/Tebuconazole.htm. Tebuconazole. Agro-care Chemical Industry Group. Date last visited: 2 August 2013.

http://www.agropages.com/AgroData/Detail-1188.htm. Pyributicarb. Crop Protection Database. Agro data. Agropages.com. Date last visited: 2 August 2013.

http://www.agropages.com/AgroData/Detail-1716.htm. Pyrithiobac-sodium. Crop Protection Database. Agro data. Agropages.com. Date last visited: 2 August 2013.

http://www.agry.purdue.edu/turf/. Department of Agronomy, Purdue University. Date last visited: 7 August 2013.

http://www.agscientific.com/cycloheximide-streptomyces-griseus.html. AG Scientific, Inc. Date last visited: 7 August 2013.

http://www.alanwood.net/pesticides/1,3-dichloropropene.html. 1,3-Dichloropropene. Compendium of Pesticide Common Names. Date last visited: 7 August 2013.

http://www.alanwood.net/pesticides/acetochlor.html. Acetochlor. Compendium of Pesticide Common Names. Date last visited: 7 August 2013.

http://www.alanwood.net/pesticides/alachlor.html. Alachlor. Compendium of Pesticide Common Names. Date last visited: 7 August 2013.

http://www.alanwood.net/pesticides/aluminium%20phosphide.html. Aluminium Phosphide. Compendium of Pesticide Common Names. Date last visited: 7 August 2013.

http://www.alanwood.net/pesticides/amisulbrom.html. Amisulbrom. Compendium of Pesticide Common Names. Date last visited: 7 August 2013.

http://www.alanwood.net/pesticides/atrazine.html. Atrazine. Compendium of Pesticide Common Names. Date last visited: 7 August 2013.

http://www.alanwood.net/pesticides/benzobicyclon.html. Benzobicyclon. Compendium of Pesticide Common Names. Date last visited: 7 August 2013.

http://www.alanwood.net/pesticides/butachlor.html. Butachlor. Compendium of Pesticide Common Names. Date last visited: 7 August 2013.

http://www.alanwood.net/pesticides/butenachlor.html. Butenachlor. Compendium of Pesticide Common Names. Date last visited: 7 August 2013.

http://www.alanwood.net/pesticides/butylamine.html. Butylamine. Compendium of Pesticide Common Names. Date last visited: 7 August 2013.

http://www.alanwood.net/pesticides/carbetamide.html. Carbetamide. Compendium of Pesticide Common Names. Date last visited: 28 August 2013.

http://www.alanwood.net/pesticides/carbofuran.html. Carbofuron. Compendium of Pesticide Common Names. Date last visited: 7 August 2013.

http://www.alanwood.net/pesticides/carpropamid.html. Carpropamid. Compendium of Pesticide Common Names. Date last visited: 7 August 2013.

http://www.alanwood.net/pesticides/chlorethoxyfos.html. Chlorethoxyfos. Compendium of Pesticide Common Names. Date last visited: 7 August 2013.

http://www.alanwood.net/pesticides/chlorfluazuron.html. Chlorfluazuron. Compendium of Pesticide Common Names. Date last visited: 7 August 2013.

http://www.alanwood.net/pesticides/chlorflurenol.html. Chlorflurenol. Compendium of Pesticide Common Names. Date last visited: 7 August 2013.

http://www.alanwood.net/pesticides/chlormephos.html. Chlormephos. Compendium of Pesticide Common Names. Date last visited: 7 August 2013.

http://www.alanwood.net/pesticides/chlormequat.html. Chlormequat. Compendium of Pesticide Common Names. Date last visited: 7 August 2013.

http://www.alanwood.net/pesticides/chloroneb.html. Chloroneb. Compendium of Pesticide Common Names. Date last visited: 7 August 2013.

http://www.alanwood.net/pesticides/chlorophacinone.html. Chlorophacinone. Compendium of Pesticide Common Names. Date last visited: 7 August 2013.

http://www.alanwood.net/pesticides/chloropicrin.html. Chloropicrin. Compendium of Pesticide Common Names. Date last visited: 7 August 2013.

http://www.alanwood.net/pesticides/chlorothalonil.html. Chlorothalonil. Compendium of Pesticide Common Names. Date last visited: 7 August 2013.

http://www.alanwood.net/pesticides/chlorphonium.html. Chlorphonium. Compendium of Pesticide Common Names. Date last visited: 7 August 2013.

http://www.alanwood.net/pesticides/chlorpyrifos-methyl.html. Chlorpyrifos-methyl. Compendium of Pesticide Common Names. Date last visited: 7 August 2013.

http://www.alanwood.net/pesticides/clethodim.html. Clethodim. Compendium of Pesticide Common Names. Date last visited: 7 August 2013.

http://www.alanwood.net/pesticides/cloethocarb.html. Cloethocarb. Compendium of Pesticide Common Names. Date last visited: 7 August 2013.

http://www.alanwood.net/pesticides/cloprop.html Cloprop. Compendium of Pesticide Common Names. Date last visited: 7 August 2013.

http://www.alanwood.net/pesticides/cloquintocet.html. Cloquintocet. Compendium of Pesticide Common Names. Date last visited: 7 August 2013.

http://www.alanwood.net/pesticides/cloransulam.html. Cloransulam. Compendium of Pesticide Common Names. Date last visited: 7 August 2013.

http://www.alanwood.net/pesticides/clothianidin.html Clothianidin. Compendium of Pesticide Common Names. Date last visited: 7 August 2013.

http://www.alanwood.net/pesticides/cloxyfonac.html Cloxyfonac. Compendium of Pesticide Common Names. Date last visited: 7 August 2013.

http://www.alanwood.net/pesticides/coumaphos.html. Coumaphos. Compendium of Pesticide Common Names. Date last visited: 7 August 2013.

http://www.alanwood.net/pesticides/cyanazine.html. Cyanazine. Compendium of Pesticide Common Names. Date last visited: 7 August 2013.

http://www.alanwood.net/pesticides/cyanophos.html. Cyanophos. Compendium of Pesticide Common Names. Date last visited: 7 August 2013.

http://www.alanwood.net/pesticides/cycloheximide.html. Cycloheximide. Compendium of Pesticide Common Names. Date last visited: 7 August 2013.

http://www.alanwood.net/pesticides/cycloprothrin.html. Cycloprothrin. Compendium of Pesticide Common Names. Date last visited: 7 August 2013.

http://www.alanwood.net/pesticides/cyclosulfamuron.html. Cyclosulfamuron. Compendium of Pesticide Common Names. Date last visited: 7 August 2013.

http://www.alanwood.net/pesticides/cyfluthrin.html. Cyfluthrin. Compendium of Pesticide Common Names. Date last visited: 7 August 2013.
http://www.alanwood.net/pesticides/cyhalofop.html. Cyhalofop. Compendium of Pesticide Common Names. Date last visited: 7 August 2013.
http://www.alanwood.net/pesticides/cyhalothrin.html. Cyhalothrin. Compendium of Pesticide Common Names. Date last visited: 7 August 2013.
http://www.alanwood.net/pesticides/cyromazine.html. Cyromazine. Compendium of Pesticide Common Names. Date last visited: 7 August 2013.
http://www.alanwood.net/pesticides/dcip.html. DCIP. Compendium of Pesticide Common Names. Date last visited: 7 August 2013.
http://www.alanwood.net/pesticides/delachlor.html. Delachlor. Compendium of Pesticide Common Names. Date last visited: 7 August 2013.
http://www.alanwood.net/pesticides/derivatives/cyhalofop-butyl.html. Cyhalofop-butyl. Compendium of Pesticide Common Names. Date last visited: 7 August 2013.
http://www.alanwood.net/pesticides/derivatives/dinoterb%20acetate.html. Dinoterb Acetate. Compendium of Pesticide Common Names. Date last visited: 7 August 2013.
http://www.alanwood.net/pesticides/derivatives/emamectin%20benzoate.html. Emamectin Benzoate. Compendium of Pesticide Common Names. Date last visited: 7 August 2013.
http://www.alanwood.net/pesticides/derivatives/fenchlorazole-ethyl.html Fenchlorazole-ethyl. Compendium of Pesticide Common Names. Date last visited: 7 August 2013.
http://www.alanwood.net/pesticides/derivatives/fenuron%20tca.html Fenuron TCA. Compendium of Pesticide Common Names. Date last visited: 7 August 2013.
http://www.alanwood.net/pesticides/derivatives/flamprop-isopropyl.html. Flamprop-isopropyl. Compendium of Pesticide Common Names. Date last visited: 7 August 2013.
http://www.alanwood.net/pesticides/derivatives/flamprop-methyl.html. Flamprop-methyl. Compendium of Pesticide Common Names. Date last visited: 7 August 2013.
http://www.alanwood.net/pesticides/derivatives/fluazifop-butyl.html. Fluazifop-butyl. Compendium of Pesticide Common Names. Date last visited: 7 August 2013.
http://www.alanwood.net/pesticides/derivatives/fluazifop-methyl.html. Fluazifop-methyl. Compendium of Pesticide Common Names. Date last visited: 7 August 2013.
http://www.alanwood.net/pesticides/derivatives/flucarbazone-sodium.html. Flucarbazone-sodium. Compendium of Pesticide Common Names. Date last visited: 7 August 2013.
http://www.alanwood.net/pesticides/derivatives/flumiclorac-pentyl.html. Flumiclorac-pentyl. Compendium of Pesticide Common Names. Date last visited: 7 August 2013.
http://www.alanwood.net/pesticides/derivatives/fosetyl-aluminium.html. Fosetyl-aluminium. Compendium of Pesticide Common Names. Date last visited: 7 August 2013.
http://www.alanwood.net/pesticides/derivatives/index-derivatives-frame.html. Bromacil-sodium. Compendium of Pesticide Common Names. Index of Salt and Ester Derivatives. Date last visited: 28 August 2013.
http://www.alanwood.net/pesticides/derivatives/iodosulfuron-methyl-sodium.html. Iodosulfuron-methyl-sodium. Compendium of Pesticide Common Names. Date last visited: 7 August 2013.
http://www.alanwood.net/pesticides/derivatives/oxpoconazole%20fumarate.html. Oxpoconazole Fumarate. Compendium of Pesticide Common Names. Date last visited: 7 August 2013.
http://www.alanwood.net/pesticides/derivatives/prohexadione-calcium.html. Prohexadione-calcium. Compendium of Pesticide Common Names. Date last visited: 28 August 2013.
http://www.alanwood.net/pesticides/derivatives/propamocarb%20hydrochloride.html. Propamocarb Hydrochloride. Compendium of Pesticide Common Names. Date last visited: 28 August 2013.
http://www.alanwood.net/pesticides/derivatives/pyrithiobac-sodium.html. Pyrithiobac-sodium. Compendium of Pesticide Common Names. Date last visited: 28 August 2013.
http://www.alanwood.net/pesticides/di-allate.html. Di-allate. Compendium of Pesticide Common Names. Date last visited: 7 August 2013.
http://www.alanwood.net/pesticides/diethatyl.html. Diethatyl. Compendium of Pesticide Common Names. Date last visited: 7 August 2013.
http://www.alanwood.net/pesticides/dimethachlor.html. Dimethachlor. Compendium of Pesticide Common Names. Date last visited: 7 August 2013.
http://www.alanwood.net/pesticides/epn.html. EPN. Compendium of Pesticide Common Names. Date last visited: 12 August 2013.
http://www.alanwood.net/pesticides/eptc.html. EPTC. Compendium of Pesticide Common Names. Date last visited: 12 August 2013.
http://www.alanwood.net/pesticides/ergocalciferol.html. Ergocalciferol. Compendium of Pesticide Common Names. Date last visited: 12 August 2013.
http://www.alanwood.net/pesticides/esfenvalerate.html. Esfenvalerate. Compendium of Pesticide Common Names. Date last visited: 12 August 2013.
http://www.alanwood.net/pesticides/esprocarb.html. Esprocarb. Compendium of Pesticide Common Names. Date last visited: 12 August 2013.
http://www.alanwood.net/pesticides/ethephon.html. Ethephon. Compendium of Pesticide Common Names. Date last visited: 12 August 2013.
http://www.alanwood.net/pesticides/ethiofencarb.html. Ethiofencarb. Compendium of Pesticide Common Names. Date last visited: 12 August 2013.
http://www.alanwood.net/pesticides/ethiolate.html. Ethiolate. Compendium of Pesticide Common Names. Date last visited: 12 August 2013.
http://www.alanwood.net/pesticides/ethion.html. Ethion. Compendium of Pesticide Common Names. Date last visited: 12 August 2013.
http://www.alanwood.net/pesticides/ethirimol.html. Ethirimol. Compendium of Pesticide Common Names. Date last visited: 12 August 2013.

http://www.alanwood.net/pesticides/ethofumesate.html. Ethofumesate. Compendium of Pesticide Common Names. Date last visited: 12 August 2013.

http://www.alanwood.net/pesticides/ethoprophos.html. Ethoprophos. Compendium of Pesticide Common Names. Date last visited: 12 August 2013.

http://www.alanwood.net/pesticides/ethoxysulfuron.html. Ethoxysulfuron. Compendium of Pesticide Common Names. Date last visited: 12 August 2013.

http://www.alanwood.net/pesticides/ethychlozate.html. Ethychlozate. Compendium of Pesticide Common Names. Date last visited: 12 August 2013.

http://www.alanwood.net/pesticides/ethylene%20dibromide.html. Ethylene Dibromide. Compendium of Pesticide Common Names. Date last visited: 28 August 2013.

http://www.alanwood.net/pesticides/etobenzanid.html. Etobenzanid. Compendium of Pesticide Common Names. Date last visited: 12 August 2013.

http://www.alanwood.net/pesticides/etofenprox.html. Etofenprox. Compendium of Pesticide Common Names. Date last visited: 12 August 2013.

http://www.alanwood.net/pesticides/etrimfos.html. Etrimfos. Compendium of Pesticide Common Names. Date last visited: 12 August 2013.

http://www.alanwood.net/pesticides/fenclorim.html. Fenclorim. Compendium of Pesticide Common Names. Date last visited: 12 August 2013.

http://www.alanwood.net/pesticides/ferbam.html. Ferbam. Compendium of Pesticide Common Names. Date last visited: 28 August 2013.

http://www.alanwood.net/pesticides/ferimzone.html. Ferimzone. Compendium of Pesticide Common Names. Date last visited: 28 August 2013.

http://www.alanwood.net/pesticides/flamprop-m.html. Flamprop-m. Compendium of Pesticide Common Names. Date last visited: 12 August 2013.

http://www.alanwood.net/pesticides/fluazifop-p.htm. Fluazifop-p. Compendium of Pesticide Common Names. Date last visited: 12 August 2013.

http://www.alanwood.net/pesticides/flucetosulfuron.html. Flucetosulfuron. Compendium of Pesticide Common Names. Date last visited: 12 August 2013.

http://www.alanwood.net/pesticides/flufenpyr.html. Flufenpyr. Compendium of Pesticide Common Names. Date last visited: 12 August 2013.

http://www.alanwood.net/pesticides/flupropanate.html. Flupropanate. Compendium of Pesticide Common Names. Date last visited: 28 August 2013.

http://www.alanwood.net/pesticides/flupyrsulfuron.html. Flupyrsulfuron. Compendium of Pesticide Common Names. Date last visited: 12 August 2013.

http://www.alanwood.net/pesticides/flurazole.html. Flurazole. Compendium of Pesticide Common Names. Date last visited: 12 August 2013.

http://www.alanwood.net/pesticides/fluroxypyr.html. Fluroxypyr. Compendium of Pesticide Common Names. Date last visited: 12 August 2013.

http://www.alanwood.net/pesticides/fluthiacet.html. Fluthiacet. Compendium of Pesticide Common Names. Date last visited: 12 August 2013.

http://www.alanwood.net/pesticides/fluxofenim.html. Fluxofenim. Compendium of Pesticide Common Names. Date last visited: 12 August 2013.

http://www.alanwood.net/pesticides/formaldehyde.html. Formaldehyde. Compendium of Pesticide Common Names. Date last visited: 12 August 2013.

http://www.alanwood.net/pesticides/formetanate.html. Formetanate. Compendium of Pesticide Common Names. Date last visited: 12 August 2013.

http://www.alanwood.net/pesticides/fuberidazole.html. Fuberidazole. Compendium of Pesticide Common Names. Date last visited: 12 August 2013.

http://www.alanwood.net/pesticides/glufosinate.html. Glufosinate. Compendium of Pesticide Common Names. Date last visited: 12 August 2013.

http://www.alanwood.net/pesticides/glyphosate.html. Glyphosate. Compendium of Pesticide Common Names. Date last visited: 28 August 2013.

http://www.alanwood.net/pesticides/guazatine.html. Guazatine. Compendium of Pesticide Common Names. Date last visited: 28 August 2013.

http://www.alanwood.net/pesticides/halosulfuron.html. Halosulfuron. Compendium of Pesticide Common Names. Date last visited: 12 August 2013.

http://www.alanwood.net/pesticides/hempa.html. Hempa. Compendium of Pesticide Common Names. Date last visited: 28 August 2013.

http://www.alanwood.net/pesticides/heptenophos.html. Heptenophos. Compendium of Pesticide Common Names. Date last visited: 28 August 2013.

http://www.alanwood.net/pesticides/hexachlorobenzene.html. Hexachlorobenzene. Compendium of Pesticide Common Names. Date last visited: 28 August 2013.

http://www.alanwood.net/pesticides/hydroprene.html. Hydroprene. Compendium of Pesticide Common Names. Date last visited: 28 August 2013.

http://www.alanwood.net/pesticides/index_cn_frame.html. Index of Common Names. Compendium of Pesticide Common Names. Date last visited: 12 August 2013.

http://www.alanwood.net/pesticides/index_cn_frame.html. Fenpropimorph. Compendium of Pesticide Common Names. Index of Common Names. Date last visited: 28 August 2013.

http://www.alanwood.net/pesticides/index_cn_frame.html. Fenpropinid. Compendium of Pesticide Common Names. Index of Common Names. Date last visited: 28 August 2013.

http://www.alanwood.net/pesticides/index-iupac-i.html. Index of IUPAC Systematic Names – 1. Compendium of Pesticide Common Names. Date last visited: 12 August 2013.

http://www.alanwood.net/pesticides/iodosulfuron.html. Iodosulfuron. Compendium of Pesticide Common Names. Date last visited: 12 August 2013.

http://www.alanwood.net/pesticides/iodosulfuron-methyl-sodium.html. Indosulfuron-methyl-sodium. Compendium of Pesticide Common Names. Date last visited: 28 August 2013.

http://www.alanwood.net/pesticides/isazofos.html. Isazofos. Compendium of Pesticide Common Names. Date last visited: 12 August 2013.

http://www.alanwood.net/pesticides/isopolinate.html. Isopolinate. Compendium of Pesticide Common Names. Date last visited: 12 August 2013.

http://www.alanwood.net/pesticides/isoproturon.html. Isoproturon. Compendium of Pesticide Common Names. Date last visited: 12 August 2013.
http://www.alanwood.net/pesticides/isoxaflutole.html. Isoxaflutole. Compendium of Pesticide Common Names. Date last visited: 12 August 2013.
http://www.alanwood.net/pesticides/isoxathion.html. Isoxathion. Compendium of Pesticide Common Names. Date last visited: 12 August 2013.
http://www.alanwood.net/pesticides/kasugamycin.html. Kasugamycin. Compendium of Pesticide Common Names. Date last visited: 12 August 2013.
http://www.alanwood.net/pesticides/methiobencarb.html. Methiobencarb. Compendium of Pesticide Common Names. Date last visited: 12 August 2013.
http://www.alanwood.net/pesticides/oxabetrinil.html. Oxabetrinil. Compendium of Pesticide Common Names. Date last visited: 12 August 2013.
http://www.alanwood.net/pesticides/oxine-copper.html. Oxine-copper. Compendium of Pesticide Common Names. Date last visited: 12 August 2013.
http://www.alanwood.net/pesticides/oxolinic%20acid.html. Oxolinic Acid. Compendium of Pesticide Common Names. Date last visited: 12 August 2013.
http://www.alanwood.net/pesticides/oxyfluorfen.html. Oxyfluorfen. Compendium of Pesticide Common Names. Date last visited: 12 August 2013.
http://www.alanwood.net/pesticides/penoxsulam.html. Penoxsulam. Compendium of Pesticide Common Names. Date last visited: 12 August 2013.
http://www.alanwood.net/pesticides/piperalin.html. Piperalin. Compendium of Pesticide Common Names. Date last visited: 12 August 2013.
http://www.alanwood.net/pesticides/polyoxins.html. Polyoxins. Compendium of Pesticide Common Names. Date last visited: 28 August 2013.
http://www.alanwood.net/pesticides/polyoxorim.html. Polyoxorim. Compendium of Pesticide Common Names. Date last visited: 28 August 2013.
http://www.alanwood.net/pesticides/prallethrin.html. Prallethrin. Compendium of Pesticide Common Names. Date last visited: 28 August 2013.
http://www.alanwood.net/pesticides/pretilachlor.html. Pretilachlor. Compendium of Pesticide Common Names. Date last visited: 12 August 2013.
http://www.alanwood.net/pesticides/probenazole.html. Probenazole. Compendium of Pesticide Common Names. Date last visited: 28 August 2013.
http://www.alanwood.net/pesticides/prochloraz.html. Prochloraz. Compendium of Pesticide Common Names. Date last visited: 28 August 2013.
http://www.alanwood.net/pesticides/prometon.html. Prometon. Compendium of Pesticide Common Names. Date last visited: 12 August 2013.
http://www.alanwood.net/pesticides/prometryn.html. Prometryn. Compendium of Pesticide Common Names. Date last visited: 12 August 2013.
http://www.alanwood.net/pesticides/propachlor.html. Propachlor. Compendium of Pesticide Common Names. Date last visited: 12 August 2013.
http://www.alanwood.net/pesticides/propaquizafop.html. Propaquizafop. Compendium of Pesticide Common Names. Date last visited: 28 August 2013.
http://www.alanwood.net/pesticides/propazine.html. Propazine. Compendium of Pesticide Common Names. Date last visited: 12 August 2013.
http://www.alanwood.net/pesticides/propham.html. Propham. Compendium of Pesticide Common Names. Date last visited: 28 August 2013.
http://www.alanwood.net/pesticides/propiconazole.html. Propiconazole. Compendium of Pesticide Common Names. Date last visited: 28 August 2013.
http://www.alanwood.net/pesticides/propineb.html. Propineb. Compendium of Pesticide Common Names. Date last visited: 28 August 2013.
http://www.alanwood.net/pesticides/propisochlor.html. Propisochlor. Compendium of Pesticide Common Names. Date last visited: 12 August 2013.
http://www.alanwood.net/pesticides/propetamphos.html. Propetamphos. Compendium of Pesticide Common Names. Date last visited: 28 August 2013.
http://www.alanwood.net/pesticides/proquinazid.html. Proquinazid. Compendium of Pesticide Common Names. Date last visited: 28 August 2013.
http://www.alanwood.net/pesticides/prosulfocarb.html. Prosulfocarb. Compendium of Pesticide Common Names. Date last visited: 28 August 2013.
http://www.alanwood.net/pesticides/prynachlor.html. Prynachlor. Compendium of Pesticide Common Names. Date last visited: 12 August 2013.
http://www.alanwood.net/pesticides/rotenone.html. Rotenone. Compendium of Pesticide Common Names. Date last visited: 12 August 2013.
http://www.alanwood.net/pesticides/sethoxydim.html. Sethoxydim. Compendium of Pesticide Common Names. Date last visited: 12 August 2013.
http://www.alanwood.net/pesticides/silafluofen.html. Silafluofen. Compendium of Pesticide Common Names. Date last visited: 12 August 2013.
http://www.alanwood.net/pesticides/simazine.html. Simazine. Compendium of Pesticide Common Names. Date last visited: 12 August 2013.
http://www.alanwood.net/pesticides/simeconazole.html. Simeconazole. Compendium of Pesticide Common Names. Date last visited: 12 August 2013.
http://www.alanwood.net/pesticides/s-metolachlor.html. S-metolachlor. Compendium of Pesticide Common Names. Date last visited: 12 August 2013.
http://www.alanwood.net/pesticides/sodium%20fluoroacetate.html. Sodium Fluoroacetate. Compendium of Pesticide Common Names. Date last visited: 12 August 2013.
http://www.alanwood.net/pesticides/spiroxamine.html. Spiroxamine. Compendium of Pesticide Common Names. Date last visited: 12 August 2013.
http://www.alanwood.net/pesticides/streptomycin.html. Streptomycin. Compendium of Pesticide Common Names. Date last visited: 12 August 2013.
http://www.alanwood.net/pesticides/sulfallate.html. Sulfallate. Compendium of Pesticide Common Names. Date last visited: 12 August 2013.
http://www.alanwood.net/pesticides/sulfotep.html. Sulfotep. Compendium of Pesticide Common Names. Date last visited: 12 August 2013.
http://www.alanwood.net/pesticides/sulprofos.html. Sulprofos. Compendium of Pesticide Common Names. Date last visited: 12 August 2013.
http://www.alanwood.net/pesticides/tebuthiuron.html. Tebuthiuron. Compendium of Pesticide Common Names. Date last visited: 12 August 2013.
http://www.alanwood.net/pesticides/tefluthrin.html. Tefluthrin. Compendium of Pesticide Common Names. Date last visited: 12 August 2013.
http://www.alanwood.net/pesticides/tembotrione.html. Tembotrione. Compendium of Pesticide Common Names. Date last visited: 12 August 2013.

http://www.alanwood.net/pesticides/temephos.html. Temephos. Compendium of Pesticide Common Names. Date last visited: 12 August 2013.

http://www.alanwood.net/pesticides/tepa.html. Tepa. Compendium of Pesticide Common Names. Date last visited: 12 August 2013.

http://www.alanwood.net/pesticides/terbufos.html. Terbufos. Compendium of Pesticide Common Names. Date last visited: 12 August 2013.

http://www.alanwood.net/pesticides/thenylchlor.html. Thenylchlor. Compendium of Pesticide Common Names. Date last visited: 12 August 2013.

http://www.alanwood.net/pesticides/thiabendazole.html. Thiabendazole. Compendium of Pesticide Common Names. Date last visited: 12 August 2013.

http://www.alanwood.net/pesticides/thiacloprid.html. Thiacloprid. Compendium of Pesticide Common Names. Date last visited: 12 August 2013.

http://www.alanwood.net/pesticides/thiamethoxam.html. Thiamethoxam. Compendium of Pesticide Common Names. Date last visited: 12 August 2013.

http://www.alanwood.net/pesticides/thidiazuron.html. Thidiazuron. Compendium of Pesticide Common Names. Date last visited: 12 August 2013.

http://www.alanwood.net/pesticides/thifluzamide.html. Thifluzamide. Compendium of Pesticide Common Names. Date last visited: 12 August 2013.

http://www.alanwood.net/pesticides/thiobencarb.html. Thiobencarb. Compendium of Pesticide Common Names. Date last visited: 12 August 2013.

http://www.alanwood.net/pesticides/thiocyclam.html. Thiocyclam. Compendium of Pesticide Common Names. Date last visited: 12 August 2013.

http://www.alanwood.net/pesticides/thiodicarb.html. Thiodicarb. Compendium of Pesticide Common Names. Date last visited: 12 August 2013.

http://www.alanwood.net/pesticides/thiofanox.html. Thiofanox. Compendium of Pesticide Common Names. Date last visited: 12 August 2013.

http://www.alanwood.net/pesticides/thiophanate-methyl.html.Thiophanate-methyl. Compendium of Pesticide Common Names. Date last visited: 12 August 2013.

http://www.alanwood.net/pesticides/thiotepa.html. Thiotepa. Compendium of Pesticide Common Names. Date last visited: 12 August 2013.

http://www.alanwood.net/pesticides/tiocarbazil.html. Tiocarbazil. Compendium of Pesticide Common Names. Date last visited: 12 August 2013.

http://www.alanwood.net/pesticides/tolclofos-methyl.html. Tolclofos-methyl. Compendium of Pesticide Common Names. Date last visited: 12 August 2013.

http://www.alanwood.net/pesticides/tralkoxydim.html. Tralkoxydim. Compendium of Pesticide Common Names. Date last visited: 12 August 2013.

http://www.alanwood.net/pesticides/tralomethrin.html. Tralomethrin. Compendium of Pesticide Common Names. Date last visited: 12 August 2013.

http://www.alanwood.net/pesticides/tri-allate.html. Tri-allate. Compendium of Pesticide Common Names. Date last visited: 12 August 2013.

http://www.alanwood.net/pesticides/tribufos.html. Tribufos. Compendium of Pesticide Common Names. Date last visited: 12 August 2013.

http://www.alanwood.net/pesticides/trichlorfon.html. Trichlorfon. Compendium of Pesticide Common Names. Date last visited: 12 August 2013.

http://www.alanwood.net/pesticides/triclopyr.html. Triclopyr. Compendium of Pesticide Common Names. Date last visited: 12 August 2013.

http://www.alanwood.net/pesticides/tridemorph.html. Tridemorph. Compendium of Pesticide Common Names. Date last visited: 12 August 2013.

http://www.alanwood.net/pesticides/trietazine.html. Trietazine. Compendium of Pesticide Common Names. Date last visited: 12 August 2013.

http://www.alanwood.net/pesticides/trifloxystrobin.html. Trifloxystrobin. Compendium of Pesticide Common Names. Date last visited: 12 August 2013.

http://www.alanwood.net/pesticides/triflumizole.html. Triflumizole. Compendium of Pesticide Common Names. Date last visited: 12 August 2013.

http://www.alanwood.net/pesticides/triflumuron.html. Triflumuron. Compendium of Pesticide Common Names. Date last visited: 12 August 2013.

http://www.alanwood.net/pesticides/triforine.html. Triforine. Compendium of Pesticide Common Names. Date last visited: 12 August 2013.

http://www.alanwood.net/pesticides/trinexapac.html. Trinexapac. Compendium of Pesticide Common Names. Date last visited: 12 August 2013.

http://www.alanwood.net/pesticides/triticonazole.html. Triticonazole. Compendium of Pesticide Common Names. Date last visited: 12 August 2013.

http://www.alanwood.net/pesticides/valifenalate.html. Valifenalate. Compendium of Pesticide Common Names. Date last visited: 12 August 2013.

http://www.alanwood.net/pesticides/xylachlor.html. Xylachlor. Compendium of Pesticide Common Names. Date last visited: 12 August 2013.

http://www.al.gov.bc.ca/pesticides/d_1.htm. Protective Clothing and Equipment for Pesticide Applicators. Safety Precautions. The Ministry of Agriculture. Date last visited: 28 August 2013.

http://www.apo-tokyo.org/00e-books/AG-15_NonPesticide/AG-15_NonPesticide.pdf. Nonpesticide Methods for Controlling Diseases and Insect Pests. Asian Productivity Organization. Date last visited: 13 August 2013.

http://www.apsnet.org/edcenter/intropp/topics/Mycotoxins/Pages/Trichothecenes.aspx. Mycotoxins in Crops: a Threat to Human and Domestic Animal Health. The American Phytopathological Society. Date last visited: 13 August 2013.

http://www.apsnet.org/online/common/comment/onion.asp. Diseases of Onion (*Allium cepa* L.) and Garlic (*A. sativum* L.). The American Phytopathological Society. Date last visited: 13 August 2013.

http://www.apsnet.org/publications/phytopathology/backissues/Documents/1977Articles/Phyto67n09_1155.PDF. Chemical Control of Stem Rot of Rice in California (Disease Control and Pest Management). The American Phytopathological Society. Date last visited: 13 August 2013.

http://www.apsnet.org/publications/PlantDisease/BackIssues/Documents/1986Articles/PlantDisease70n05_448.PDF. Niblack, T.L. and Hussey, R.S. (1986) Evaluation of *Arthrobotrys amerospora* as a biocontrol agent for *Heterodera glycines* on soybean. *Plant Disease* 70 (5). The American Phytopathological Society. Date last visited: 13 August 2013.

http://www.apvma.gov.au/products/review/completed/bromoxynil.php. Changes to the Withholding Period on Labels of Products Containing Bromoxynil. 2009. Australian Pesticides and Veterinary Medicines Authority, Australia. Date last visited: 13 August 2013.

http://www.apvma.gov.au/registration/assessment/docs/tan_iodosulfuron.pdf. Trade Advice Notice on Iodosulfuron-Methyl-Sodium and Mefenpyr-Diethyl in the Product Hussar OD Selective Herbicide. 2010. Australian Pesticides and Veterinary Medicines Authority. Date last visited: 13 August 2013.

http://www.ars.usda.gov/research/publications/publications.htm?seq_no_115=231570. Yellow Nutsedge Control in Potato with Imazosulfuron (Sustainable Potato Cropping Systems for Irrigated Agriculture in the Pacific Northwest). United States Department of Agriculture – Agricultural Research Service. Date last visited: 13 August 2013.

http://www.ars.usda.gov/research/publications/publications.htm?seq_no_115=236044. Systemic insecticides for control of black vine weevil, (Coleoptera: Curculionidae). In: Container- and Field-Grown Nursery Crops. Management of Invasive Scarabs, Root Weevils and Other Beetles of Quarantine Significance in Horticultural, Turf and Nursery Crops. United States Department of Agriculture – Agricultural Research Service. Date last visited: 13 August 2013.

http://www.ars.usda.gov/SP2UserFiles/person/990/Chitwood2003NematicideReview.pdf. Nematicides. United States Department of Agriculture – Agricultural Research Service. Date last visited: 13 August 2013.

http://www.asia-chem.com/insecticide/etofenprox.htm. Etofenprox. Asiachem Chemical Co. Ltd. Date last visited: 13 August 2013.

http://www.atsdr.cdc.gov/mmg/mmg.asp?id=1062&tid=131. Toxic Substances Portal – 1,2-Dibromoethane. Agency for Toxic Substances and Disease Registry. Date last visited: 13 August 2013.

http://www.atsdr.cdc.gov/substances/toxsubstance.asp?toxid=62. Toxic Substances Portal – Chlordane. Agency for Toxic Substances and Disease Registry. Date last visited: 13 August 2013.

http://www.awri.com.au/industry_support/viticulture/agrochemicals/mrls. Maximum Residue Limits (MRLs). The Australian Wine Research Institute. Date last visited: 13 August 2013.

http://www.bayercropscience.co.uk/assets/Uploads/Triumph-Version-4-Guideline-version-4.pdf. Environmental Information Sheet Triumph. Bayer CropScience. 2011. Date last visited: 14 August 2013.

http://www.bayercropscience.co.za/DATA/MSDS/Mocap%20%20GR%205929830.pdf. Material Safety Data Sheet. Mocap GR. 2007. Bayer CropScience. Date last visited: 14 August 2013.

http://www.bellona.org/aquaculture/artikler/Teflubenzuron. Teflubenzuron. Aqua Web. Bellona. Date last visited: 14 August 2013.

http://www.beyondpesticides.org/infoservices/pesticidefactsheets/toxic/trichlorfon.htm. Trichlorfon. Beyond Pesticides. Date last visited: 14 August 2013.

http://www.biconet.com/biocontrol/cucumeris.html. Thrip Free Crops Always Have Greater Yields. Beneficial Insects and Organisms. Biconet. Date last visited: 14 August 2013.

http://www.biobest.be/producten/109/3/0/0/. Biological Control: Beneficial Insects and Mites: *Amblyseius* – Breeding System. Biconet. Date last visited: 14 August 2013.

http://www.biobest.be/producten/110/3/0/0/. Biological Control: Beneficial Insects and Mites: *Degenerans* System. Biobest. Date last visited: 14 August 2013.

http://www.biobest.be/producten/114/3/0/0/. Biological Control: Beneficial Insects and Mites: *Aphidius* System. Biconet. Date last visited: 14 August 2013.

http://www.biocontrol.entomology.cornell.edu/pathogens/ampelomyces.html. *Ampelomyces quisqualis*. Biological Control. College of Agriculture and Life Sciences, Cornell University. Date last visited: 14 August 2013.

http://www.biocontrol.entomology.cornell.edu//pathogens/baculoviruses.html. Baculoviruses. Biological Control. College of Agriculture and Life Sciences, Cornell University. Date last visited: 14 August 2013.

http://www.biocontrol.entomology.cornell.edu/predators/Aphidoletes.html. *Aphidoletes aphidimyza*. Biological Control. College of Agriculture and Life Sciences, Cornell University. Date last visited: 14 August 2013.

http://www.biocontrol.entomology.cornell.edu/weedfeed/Agapeta.html. *Agapeta zoegana*. Biological Control. College of Agriculture and Life Sciences, Cornell University. Date last visited: 14 August 2013.

http://www.biocontrol.entomology.cornell.edu/weedfeed/Agasicles.html. *Agasicles hygrophila*. Biological Control. College of Agriculture and Life Sciences, Cornell University. Date last visited: 14 August 2013.

http://www.biofac.com/Urban___Greenhouse/Biological_Agents/body_biological_agents.html. Using Biocontrol Agents in Commercial Greenhouses. Oklahoma Cooperative Extension Services, Oklahama State University. Date last visited: 14 August 2013.

http://www.biological-research.com/Fungi/fungi.html. Fungi. Biological Research and Investment Corporation, 'BRIC'. Date last visited: 14 August 2013.

http://www.biotus.fi/DowebEasyCMS/?Page=Amblyseiusbarkeri. *Amblyseius barkeri* – predator mite. Biotus. Date last visited: 14 August 2013.

http://www.brisbaneinsects.com/brisbane_planthoppers/index.html. Leafhoppers, Treehopper and Planthoppers – Suborder Auchenorrhyncha (Order Hemiptera). Date last visited: 14 August 2013.

http://www.bugladyconsulting.com/Neoseiulus%20%28=Amblyseius%29%20californicus.htm. Predatory Mite – *Neoseiulus* (*Amblyseius*) *californicus*. Buglady Consulting. Date last visited: 14 August 2013.

http://www.bugwood.org. The Bugwood Network. Center for Invasive Species and Ecosytem Health. The University of Georgia. Date last visited: 14 August 2013.

http://www.bugwood.org/arthropod/day5/Borgemeister.pdf. Impact Assessment of *Teretrius nigrescens* Lewis (Col.: Histeridae) in West Africa, a Predator of the Larger Grain Borer, *Prostephanus truncatus* (Horn) (Col.: Bostrichidae). 1st International Symposium on Biological Control of Arthropods. Date last visited: 14 August 2013.

http://www.cabdirect.org/abstracts/20000809555.html. Sanyal, P.K. (2000) Nematophagous activity of *Arthrobotrys oligospora* against pre-parasitic stages of *Strongyloides papillosus*. CABDirect. Date last visited: 14 August 2013.

http://www.cals.ncsu.edu/course/ent425/tutorial/resistance.html. Resistance to Pesticides. General Entomology. North Carolina State University. Date last visited: 14 August 2013.

http://www.capl.sci.eg/ActiveIngredient/Oxycarboxin.html#topPage. Oxycarboxin. Central Agricultural Pesticides Laboratory. Date last visited: 14 August 2013.

http://www.carrb.com/75rpt/Disease.htm. Cause and Control of Rice Diseases. Date last visited: 14 August 2013.

http://www.cdc.gov/niosh/docs/81-123/pdfs/0098.pdf. Occupational Safety and Health Guideline for Captafol. US Department of Human and Health Services. Date last visited: 14 August 2013.

http://www.cdc.gov/niosh/topics/heatstress/ Heat Stress. The National Institute for Occupational Safety and Health (NIOSH) Workplace Safety and Health Topics. Centers for Disease Control and Prevention, USA. Date last visited: 26 August 2013.

http://www.cdph.ca.gov/programs/hesis/Documents/cyclohex.pdf. Cycloheximide. Hazard Evaluation System and Information Service. Date last visited: 14 August 2013.

http://www.cdpr.ca.gov/docs/emon/pubs/fatememo/carbofuran.pdf. Environmental Fate of Carbofuran. Environmental Monitoring Branch, Department of Pesticide Regulation. Date last visited: 14 August 2013.

http://www.cdpr.ca.gov/docs/emon/pubs/fatememo/cyflthrn.pdf. Environmental Fate of Cyfluthrin. Environmental Monitoring Branch, Department of Pesticide Regulation. Date last visited: 14 August 2013.

http://www.cdpr.ca.gov/docs/emon/pubs/fatememo/pyrprxfn.pdf. Environmental Fate of Pyriproxyfen. Environmental Monitoring Branch, Department of Pesticide Regulation. Date last visited: 16 August 2013.

http://www.cdpr.ca.gov/docs/emon/pubs/reviews/em9003.pdf. Literature Review of the Environmental Fate of Chloropicrin. Environmental Monitoring Branch, Department of Pesticide Regulation. Date last visited: 16 August 2013.

http://www.cdpr.ca.gov/docs/registration/ais/publicreports/5788.pdf. Polyoxin-D. Department of Pesticide Regulation. Date last visited: 16 August 2013.

http://www.cdpr.ca.gov/docs/registration/ais/publicreports/5789.pdf. Quinoxyfen. Department of Pesticide Regulation. Date last visited: 16 August 2013.

http://www.chemblink.com/products/64-02-8.htm. Sodium Edetate. ChemBlink. Online Database of Chemicals from Around the World. Date last visited: 16 August 2013.

http://www.chemcas.com/msds/cas/msds83/105-86-2.asp. 2,6-Octadien-1-ol, 3,7-Dimethyl-, Formate, (E)-. ChemCAS. Date last visited: 16 August 2013.

http://www.chemcas.com/msds/cas/msds91/115-78-6.asp. Phosphonium, Tributyl(2,4-dichlorobenzyl)-, Chloride. ChemCAS. Date last visited: 16 August 2013.

http://www.chemcas.net/chemical_property/cas/msds5/59756-60-4.asp. Fluridone. Chemical Information. Date last visited: 28 August 2013.

http://www.chemicalbook.com/ChemicalProductProperty_EN_CB0271910.htm. Cis-Chlordene. Chemical Book. Date last visited: 16 August 2013.

http://www.chemicalbook.com/ChemicalProductProperty_EN_CB3489413.htm. Sulfotep. Chemical Book. Date last visited: 16 August 2013.

http://www.chemicalbook.com/ChemicalProductProperty_EN_CB7249734.htm. Thiophanate-methyl. Chemical Book. Date last visited: 16 August 2013.

http://www.chemicalbook.com/ChemicalProductProperty_EN_CB7475748.htm. EPN. Chemical Book. Date last visited: 16 August 2013.

http://www.chemicalbook.com/ChemicalProductProperty_EN_CB9288793.htm. Garlic Oil. Chemical Book. Date last visited: 16 August 2013.

http://www.chemicalbook.com/ChemicalProductProperty_EN_CB9754559.htm. Tecnazene. Chemical Book. Date last visited: 16 August 2013.

http://www.chemicalland21.com/specialtychem/perchem/CINNAMALDEHYDE.htm. Cinnamaldehyde. Chemical Book. Date last visited: 16 August 2013.

http://www.chemicalland21.com/specialtychem/perchem/SODIUM%20MERCAPTOBENZOTHIAZOLE.htm. Sodium Mercaptobenzothiazole. Date last visited: 8 July 2013.

http://www.chem.unep.ch/gpa_trial/15hept.htm. Heptachlor. Global Programme of Action (GPA) for the Protection of the Marine Environment from Land-Based Activities. United Nations Environment Programme (UNEP)-GPA Coordination Office. Date last visited: 28 August 2013.

http://www.chemyq.com/En/xz/xz13/128178ygdlt.htm. Isazofos.ChemYQ. Date last visited: 16 August 2013.

http://www.chinese-pesticide.com/insecticides/etofenprox.htm. Etofenprox. Bessen Chemical Ltd. Date last visited: 16 August 2013.

http://www.chlorpyrifos.com/pdf/ChlorpyrifosUseBenefits_18May2009.pdf. Use and Benefits of Chlorpyrifos in US Agriculture. Date last visited: 16 August 2013.

http://www.cnsforum.com/educationalresources/imagebank/. Image Bank. CNS Forum. Lundbeck Institute. Date last visited: 16 August 2013.

http://www.coastalwiki.org/coastalwiki/Quinoxyfen. Quinoxyfen. CoastalWiki. Date last visited: 16 August 2013.

http://www.codexalimentarius.net/pestres/data/pesticides/details.html?id=8. Maximum Residue Limits for Carbaryl. Pesticide Residues in Food and Feed. CODEX Alimentarius. Date last visited: 16 August 2013.

http://www.codexalimentarius.net/pestres/data/pesticides/details.html?id=194. Maximum Residue Limits for Haloxyfop. Pesticide Residues in Food and Feed. CODEX Alimentarius. Date last visited: 16 August 2013.

http://www.cottonseed.com/publications/facts.asp. Twenty Facts about Cottonseed Oil. National Cottonseed Products Association. Date last visited: 16 August 2013.

http://www.crl-pesticides.eu/docs/public/tmplt_article.asp?CntID=745&LabID=400&Lang=EN. Pesticides Authorised for Use in Cereals. EU Reference Laboratories for Residues of Pesticides. Date last visited: 16 August 2013.

http://www.cropinfo.in/infobase/webpage/id/50. Thiacloprid Insecticide. Cropinfo.com. Date last visited: 16 August 2013.
http://www.croplife.co.za/Portals/13/Croplife_Documents/Insecticides.pdf. Insecticides July 2013. CropLife South Africa. Date last visited: 28 August 2013.
http://www.cropnutrition.com/nutrient-knowledge. Periodic Table of Crop Nutrients. Crop Nutrition. Date last visited: 13 August 2013.
http://www.cropnutrition.com/phosphorus-stratified-soil-management. Phosphorus Stratified Soil Management. Crop Nutrition. Date last visited: 14 August 2013.
http://www.cropscience.bayer.com/en/Commitment/Managing-Weed-Resistance.aspx. Managing Weed Resistance. Bayer CropScience. Date last visited: 14 August 2013.
http://www.cropscience.bayer.com/en/Crop-Compendium/Pests-Diseases-Weeds/Diseases/Alternaria-spp-Cotton.aspx. *Alternaria* spp. (Cotton). Bayer CropScience. Date last visited: 14 August 2013.
http://www.currentbiotica.com/curl.aspx?url=Journals3-IssueIII/CB-Short_notes_2.pdf. Firake, D.M. and Pande, R. (2009) Relative toxicity of Proclaim 5% SG (Emamectin benzoate) and Dipel 8L. (*Bacillus thuringiensis* var. *kurstaki*) against *Spodoptera litura* (Fab.) by leaf roll method. *Current Biotica* 3 (3), 445–449. Date last visited: 16 August 2013.
http://www.currentbiotica.com/curl.aspx?url=Journals5-IssueIII/CB5(3)-Short-notes1.pdf. Khedkar, A.A., Patel, M.G. and Bharpoda, T.M. (2011) Screening of different varieties/genotypes for their susceptibility against mustard aphid, *Lipaphis erysimi* (Kaltenbach). *Current Biotica* 5 (3), 359–363. Date last visited: 16 August 2013.
http://www.deepdyve.com/lp/taylor-francis/beauveria-brongniartii-subjected-to-spray-drying-in-a-composite-03y8Mn22Zm. Horaczek, A. and Viernstein, H. (2004) *Beauveria brongniartii* subjected to spray-drying in a composite carrier matrix system. *Journal of Microencapsulation*. Date last visited: 16 August 2013.
http://www.dias.kvl.dk/Plantvirology/esymptoms/symp-color.html. Leaf Colors. Ole Søgaard Lund. Date last visited: 16 August 2013.
http://www.docstoc.com/docs/19246532/German-MRLS-for-Table-Grapes. German MRLS for Table Grapes. Atlanta MRL Data base – Version.2. Docstoc. Date last visited: 16 August 2013.
http://www.druglead.com/cds/Karanjin.html. Karanjin. Chemicals Pesticides. DrugLead. Date last visited: 16 August 2013.
http://www.druglead.com/cds/simeconazole.html. Simeconazole. Chemicals Pesticides. DrugLead. Date last visited: 16 August 2013.
http://www.drugs.com/search.php?searchterm=streptomycin. Streptomycin. Drugs.com. Date last visited: 16 August 2013.
http://www.ecpa.eu/page/regulatory-overview. European Crop Protection – Regulatory Overview. European Crop Protection Association (ECA). European Commission Directorate General for Health and Consumers (DG SANCO). Date last visited: 14 June 2009.
http://www.efsa.europa.eu/fr/efsajournal/doc/1388.pdf. Reasoned Opinion of EFSA: Modification of Existing MRLs for Valifenalate (Valiphenal) in Tomatoes and Aubergines. 2009. European Food Safety (EFSA). Date last visited: 16 August 2013.
http://www.ehow.com/about_5076807_gibberellic-acid.html#ixzz1CgYWxU44. What is Gibberellic Acid? eHow. Date last visited: 16 August 2013.
http://www.ehow.com/facts_8210499_chloroacetanilide-herbicides.html. Chloroacetanilide Herbicides. eHow. Date last visited: 16 August 2013.
http://www.ehow.com/info_8422352_definition-necrosis-plants.html. The Definition of Necrosis in Plants. Date last visited: 16 August 2013.
http://www.elsevier.com/about/press-releases/science-and-technology/elsevier-acquires-the-beilstein-database,-worlds-leading-archive-of-chemical-data. Beilstein Data Base. Beilstein Information Systems, GmbH. Date last visited: 17 December 2013.
http://www.encyclo.co.uk/define/Selective%20pesticide. Selective Pesticide. Encyclo the on-line Encyclopedia. Date last visited: 16 August 2013.
http://www.entomology.wisc.edu/mbcn/kyf410.html. Know Your Friends – the Entomopathogen *Beauveria bassiana*. University of Wisconsin–Madison. Date last visited: 16 August 2013.
http://www.enviro-news.com/glossary/selective_pesticide.html. Selective Pesticide. Enviro News & Business. Date last visited: 16 August 2013.
http://www.environment-agency.gov.uk/business/topics/hazsubstances/default.aspx. Hazardous Substances. The Environment Agency. Date last visited: 17 August 2013.
http://www.environment-agency.gov.uk/business/topics/pollution/39411.aspx. Climate Change Agreements Scheme. The Environment Agency. Date last visited: 17 August 2013.
http://www.environment-agency.gov.uk/business/topics/pollution/129647.aspx. The Kyoto Mechanisms. The Environment Agency. Date last visited: 17 August 2013.
http://www.enzolifesciences.com/BML-A239/kasugamycin/. Kasugamycin. Enzo Life Sciences Inc. Date last visited: 17 August 2013.
http://www.epa.gov. United States Environmental Protection Agency. Date last visited: 17 August 2013.
http://www.epa.gov/opp00001/reregistration/REDs/chlorflurenol_red.pdf. Reregistration Eligibility Decision (RED) for Chlorflurenol Methyl Ester. 2007. United States Environmental Protection Agency. Date last visited: 17 August 2013.
http://www.epa.gov/opp00001/reregistration/REDs/oxamyl_red.pdf. Reregistration Eligibility Decision (RED) for Oxamyl. 2000. United States Environmental Protection Agency. Date last visited: 17 August 2013.
http://www.epa.gov/oppsrrd1/REDs/0012red_carboxin.pdf. Reregistration Eligibility Decision (RED) for Carboxin. 2004. United States Environmental Protection Agency. Date last visited: 17 August 2013.
http://www.epa.gov/oppsrrd1/REDs/2665red.pdf. Reregistration Eligibility Decision (RED) for Thiobencarb. 1997. United States Environmental Protection Agency. Date last visited: 17 August 2013.
http://www.epa.gov/oppsrrd1/REDs/cycloate_red.pdf. Reregistration Eligibility Decision (RED) for Cycloate (S-ethyl cyclohexyl(ethyl)thiocarbamate). United States Environmental Protection Agency. Date last visited: 17 August 2013.
http://www.epa.gov/oppsrrd1/REDs/factsheets/2380fact.pdf. Reregistration Eligibility Decision (RED) Facts for Sodium and Zinc Salts of 2-Mercaptobenzothiazole. United States Environmental Protection Agency. Date last visited: 17 August 2013.
http://www.epa.gov/oppsrrd1/REDs/factsheets/chlorethfact.pdf. Reregistration Eligibility Decision (RED) Facts for Chlorethoxyfos. United States Environmental Protection Agency. Date last visited: 17 August 2013.

http://www.epa.gov/oppsrrd1/REDs/factsheets/chloroneb_factsheet.pdf. Reregistration Eligibility Decision (RED) Facts for Chloroneb. United States Environmental Protection Agency. Date last visited: 17 August 2013.

http://www.epa.gov/oppsrrd1/REDs/factsheets/oxadixyl_fs.htm. Reregistration Eligibility Decision (RED) Facts for Oxadixyl. United States Environmental Protection Agency. Date last visited: 17 August 2013.

http://www.epa.gov/oppsrrd1/REDs/methoxychlor_red.htm. Methoxychlor Reregistration Eligibility Decision (RED). 2004. United States Environmental Protection Agency. Date last visited: 17 August 2013.

http://www.epa.gov/oppsrrd1/REDs/phenylphenol_red.pdf. Reregistration Eligibility Decision for 2-Phenylphenol and Salts (Orthophenylphenol or OPP). United States Environmental Protection Agency. Date last visited: 17 August 2013.

http://www.epa.gov/oppsrrd1/REDs/taufluvalinate_red.pdf. Reregistration Eligibility Decision for Tau-fluvalinate. 2005. United States Environmental Protection Agency. Date last visited: 17 August 2013.

http://www.epa.gov/oppsrrd1/REDs/triforine_red.pdf. Reregistration Eligibility Decision for Triforine. 2008. United States Environmental Protection Agency. Date last visited: 17 August 2013.

http://www.epa.gov/pesticides/biopesticides/ingredients/factsheets/factsheet_000669.htm#contact. Black Pepper Oil (000669) Fact Sheet. United States Environmental Protection Agency. Date last visited: 19 August 2013.

http://www.epa.gov/pesticides/biopesticides/ingredients/factsheets/factsheet_011332.htm. Canola Oil (011332) Fact Sheet. United States Environmental Protection Agency. Date last visited: 19 August 2013.

http://www.epa.gov/pesticides/biopesticides/ingredients/factsheets/factsheet_230000.htm. Polyoxin D Zinc Salt (230000) Fact Sheet. United States Environmental Protection Agency. Date last visited: 19 August 2013.

http://www.epa.gov/pesticides/biopesticides/ingredients/tech_docs/brad_000669.pdf. Oil of Black Pepper. Biopesticides Registration Action Document. United States Environmental Protection Agency. Date last visited: 19 August 2013.

http://www.epa.gov/pesticides/chem_search/ppls/034704-00937-20060711.pdf. Dyna-Shield Mefonoxam Fungicide. United States Environmental Protection Agency. Date last visited: 17 August 2013.

http://www.epa.gov/pesticides/chemical/foia/cleared-reviews/reviews/021901/021901.htm. Index of Cleared Science Reviews. United States Environmental Protection Agency. Date last visited: 17 August 2013.

http://www.epa.gov/pesticides/factsheets/chemicals/deet.htm. The Insect Repellent DEET. United States Environmental Protection Agency. Date last visited: 17 August 2013.

http://www.epa.gov/pesticides/factsheets/ipm.htm. Integrated Pest Management (IPM) Principles. Date last visited: 17 August 2013.

http://www.epa.gov/pesticides/insect/choose.htm. Choosing an Insect Repellent: How Do I Know Which Repellent is Right for Me? Date last visited: 17 August 2013.

http://www.epa.gov/pestwise/grants/regionalaggrants.html. Regional IPM Agricultural Grants. PestWise, an EPA Partnership Program. United States Environmental Protection Agency (EPA). Date last visited: 17 August 2013.

http://www.epa.gov/ttn/atw/hlthef/ethyl-di.html. Ethylene Dibromide (Dibromoethane). 2000. Technology Transfer Network –Air Toxics Web Site. United States Environmental Protection Agency. Date last visited: 19 August 2013.

http://www.epa.nsw.gov.au/pesticides/pesticides.htm. Pesticides. NSW Environment Protection Authority. Date last visited: 16 August 2013.

http://www.eppo.int/MEETINGS/2013_meetings/wp_pr_sarajevo.htm. 51st meeting of the Working Party on Phytosanitary Regulations. European and Mediterranean Plant Protection Organization. Date last visited: 19 August 2013.

http://www.ext.colostate.edu/mg/gardennotes/244.html. Cover Crops and Green Manure Crops. 2011. Department of Horticulture & Landscape Architecture. Colorado State University Extension. Date last visited: 19 August 2013.

http://www.ext.colostate.edu/ptlk/2115.html. Iron Chlorosis. 2013. Plantalk. Colorado State University Extension. Date last visited: 19 August 2013.

http://www.extension.org/pages/43627/what-is-necrosis-on-a-plant-and-what-does-it-look-like. What Is Necrosis on a Plant and What Does It Look Like? 2011. Extension. Date last visited: 19 August 2013.

http://www.extension.umn.edu/distribution/horticulture/dg1168.html. Tomato-Tobacco Mosaic Virus Disease. University of Minnesota Extension. Date last visited: 19 August 2013.

http://www.fao.org/agriculture/crops/core-themes/theme/pests/en/. AGP-Pest and Pesticide Management. Food and Agriculture Organization of the United Nations (FAO). Date last visited: 19 August 2013.

http://www.fao.org/docrep/w8338e/w8338e06.htm. Cyfluthrin. Food and Agriculture Organization of the United Nations (FAO) Corporate Document Repository. Date last visited: 19 August 2013.

http://www.fao.org/docrep/x2230e/x2230e10.htm. The Use of Spices and Medicinals as Bioactive Protectants for Grains. Food and Agriculture Organization of the United Nations (FAO) Corporate Document Repository. Date last visited: 19 August 2013.

http://www.fao.org/fileadmin/templates/agphome/documents/Pests_Pesticides/JMPR/Report09/Cadusafos.pdf. Cadusafos. Food and Agriculture Organization of the United Nations (FAO). Date last visited: 19 August 2013.

http://www.fargro.co.uk/content/croptalk/croptalk_spring2011.pdf. Fargro Launch Two New Biopesticide Products for the UK market. CropTalk. Spring 2011. Date last visited: 19 August 2013.

http://www.farmchemicalsinternational.com/article/33147/restrictions-on-key-insecticides-overturned. Restrictions on Key Insecticides Overturned. 22 February 2013. FarmChemicalsInternational. Date last visited: 19 August 2013.

http://www.farmchemicalsinternational.com/crop-protection-database/product/brief/51780. Azoxystrobin. FarmChemicalsInternational. Date last visited: 19 August 2013.

http://www.farmchemicalsinternational.com/crop-protection-database/product/brief/91680. Chlorpyrifos. 2013. FarmChemicalsInternational. Date last visited: 19 August 2013.

http://www.farmchemicalsinternational.com/crop-protection-database/product/brief/260220. Metalaxyl. 2013. FarmChemicalsInternational. Date last visited: 19 August 2013.

http://www.farmchemicalsinternational.com/crop-protection-database/product/brief/327660. Pyraclostrobin. FarmChemicalsInternational. Date last visited: 19 August 2013.

http://www.farmchemicalsinternational.com/crop-protection-database/product/brief/368100. Sulphur. FarmChemicalsInternational. Date last visited: 19 August 2013.

http://www.fertilizersandpesticides.com/articles/pesticide-residues.html. Decontamination of Pesticide Residues on Fruits and Vegetables. Fertilizers and Pesticides. Date last visited: 19 August 2013.

http://www.fertilizersandpesticides.com/bio-vitrioll.html. Bio Vitrioll – Organic Fungicide. Fertilizers and Pesticides. Date last visited: 19 August 2013.

http://www.ffcr.or.jp/zaidan/FFCRHOME.nsf/pages/e-AFS06. Administration of Food Safety '06. 2013. Date last visited: 19 August 2013.

http://www.ffcr.or.jp/zaidan/FFCRHOME.nsf/pages/guideline-mrls. Guideline for Application for Establishment and Revision of Maximum Residue Limits for Agricultural Chemicals used Outside Japan. Guideline for Establishment of MRLs. Date last visited: 19 August 2013.

http://www.ffcr.or.jp/zaidan/psl.nsf/88e391f9074a0ca54925663d001a0caa/9447a4416dd3e1ea49256644001a6ae8?OpenDocument. Pyrifenox. MRLs of Agricultural Chemicals. The Japan Food Chemical Research Foundation. Date last visited: 27 August 2013.

http://www.flickr.com/photos/ricephotos/4270847124/. International Rice Research Institute (IRRI)'s Creative Commons Policy. Date last visited: 1 August 2013.

http://www.floridata.com/ref/P/phas_vul.cfm. Phaseolus vulgaris. Floridata. Date last visited: 22 August 2013.

http://www.fluoridealert.org. Fluoride Action Network. Date last visited: 22 August 2013.

http://www.fluoridealert.org/issues/pollution/. Pollution. Fluoride Action Network. Date last visited: 22 August 2013.

http://www.fluoridealert.org/pesticides/epage.clodinafop-propargyl.htm. Adverse Effects. Clodinafop-propargyl. Fluoride Action Network. Date last visited: 22 August 2013.

http://www.fluoridealert.org/pesticides/flamprop-m-page.htm. Flamprop-M. Fluoride Action Network. Date last visited: 22 August 2013.

http://www.fluoridealert.org/pesticides/flazasulfuron-page.htm. Flazasulfuron. Fluoride Action Network. Date last visited: 22 August 2013.

http://www.fluoridealert.org/pesticides/flufenpyr.page.htm. Flufenpyr. Fluoride Action Network. Date last visited: 22 August 2013.

http://www.fluoridealert.org/pesticides/flurazole-page.htm. Flurazole. Fluoride Action Network. Date last visited: 22 August 2013.

http://www.fluoridealert.org/pesticides/fluxofenim.msds.ciba.1996.htm. Fluxofenim. Material Safety Data Sheet for Concep(R) III. Ciba-Geigy. 1996. Fluoride Action Network. Date last visited: 22 August 2013.

http://www.fluoridealert.org/pesticides/oxyfluorfen-page.htm. Oxyfluorfen. Fluoride Action Network. Date last visited: 22 August 2013.

http://www.fluoridealert.org/pesticides/silafluofen-page.htm. Silafluofen. Fluoride Action Network. Date last visited: 22 August 2013.

http://www.fluoridealert.org/pesticides/sodium.fluoroacetate-page.htm. Sodium Fluoroacetate (Also Known As 1080). Fluoride Action Network. Date last visited: 22 August 2013.

http://www.fluoridealert.org/pesticides/tau-fluvalinate-page.htm. Tau-fluvalinate. Fluoride Action Network. Date last visited: 22 August 2013.

http://www.fluoridealert.org/pesticides/tetraconazole.japan.limits.htm. Japan: Maximum Residue Levels for Agricultural Chemicals. Tetraconazole. Fluoride Action Network. Date last visited: 22 August 2013.

http://www.fluoridealert.org/pesticides/tetraconazole–page.htm. Tetraconazole. Fluoride Action Network. Date last visited: 22 August 2013.

http://www.fluoridealert.org/pesticides/thiazopyr-page.htm. Thiazopyr. Fluoride Action Network. Date last visited: 22 August 2013.

http://www.fluoridealert.org/pesticides/trifloxystrobin–page.htm. Trifloxystrobin. Fluoride Action Network. Date last visited: 22 August 2013.

http://www.foodsafety.govt.nz/elibrary/industry/proposed-amendment-nz-food-standard-2010/index.htm. Proposed Amendment to the New Zealand Food (Supplemented Food) Standard 2010. Ministry of Primary Industries. Date last visited: 30 July 2013.

http://www.foodstandards.gov.au/Pages/default.aspx. Food Standards Australia New Zealand. Date last visited: 22 August 2013.

http://www.freepatentsonline.com/5093124.html. Fatty Acid-based Pesticide with Reduced Phytotoxicity. FPO. Date last visited: 22 August 2013.

http://www.fs.fed.us/foresthealth/pesticide/pdfs/122205_Oxyfluorfen.pdf. Oxyfluorfen – Human Health and Ecological Risk Assessment – Final Report. United States Department of Agriculture, Forest Service. Date last visited: 22 August 2013.

http://www.gdrc.org/uem/eia.html. Man and Anshu. Environmental Impact Assessment. UEMRI-India. Date last visited: 22 August 2013.

http://www.ghorganics.com/page14.html. Insecticide Treatment. Golden Harvest Organics. Date last visited: 23 August 2013.

http://www.ghorganics.com/whiteflies.html. Dealing with Those Darn Whiteflies. Golden Harvest Organics. Date last visited: 23 August 2013.

http://www.google.co.in/url?sa=t&source=web&cd=12&ved=0CCEQFjABOAo&url=http%3A%2F%2Fdatasheets.scbt.com%2Fsc-212488.pdf&rct=j&q=oxolinic%20acid%20bactericide&ei=fgr7TbShJsaIrAfikMThDw&usg=AFQjCNENsgFTnNgME8C8ahGMMfcUMpSkcA&cad=rja. Oxolinic Acid. 2010. Santa Cruz Biotechnology, Inc. Date last visited: 23 August 2013.

http://www.grapes.msu.edu/pesticideResist.htm. How Pesticide Resistance Develops. Excerpt from Fruit Crop Ecology and Management. Michigan State University Extension. Date last visited: 23 August 2013.

http://www.greenhousemanagementonline.com/greenhouse-0611-quick-tips-pesticide-resistance.aspx. 3 Quick Tips: Managing Pesticide Resistance. Greenhouse Management. Date last visited: 23 August 2013.

http://www.growingproduce.com/article/35555/understand-blossom-end-rot. Understand Blossom End Rot. Alabama Cooperative Extension System; Purdue Cooperative Extension Service, Jefferson County, August 2013 Newsletter. Date last visited: 23 August 2013.

http://www.hc-sc.gc.ca/home-accueil/important-eng.php. Health Canada. Date last visited: 23 August 2013.

http://www.health.gov.au/internet/publications/publishing.nsf/Content/ohp-enhealth-manual-atsi-cnt-l'ohp-enhealth-manual-atsi-cnt-l-ch5'ohp-enhealth-manual-atsi-cnt-l-ch5.1. Environmental Health Practitioner Manual. 2010. Australian Government. Department of Health and Ageing. Date last visited: 23 August 2013.

http://www.herbiguide.com.au/InformationWeeds.aspx. Weeds. HerbiGuide Pty Ltd. Date last visited: 23 August 2013.

http://www.hindu.com/thehindu/seta/2002/06/06/stories/2002060600160300.htm. Managing Rice Brown Plant Hopper. 6 June 2006. Sci Tech. The Hindu online edition. Date last visited: 23 August 2013.

http://www.hmdb.ca/metabolites/HMDB05812. Showing Metabocard for Geraniol (HMDB05812). Human Metabolome Database version 3.5. Genome Alberta. Date last visited: 23 August 2013.

http://www.hort.purdue.edu/newcrop/morton/orange.html. Morton, J.F. (1987) Orange. In: *Fruits of Warm Climates*, pp. 134–142. Date last visited: 23 August 2013.

http://www.hracglobal.com/Portals/5/HeapCriteria.pdf. Criteria for Confirmation of Herbicide-Resistant Weeds – With Specific Emphasis on Confirming Low Level Resistance. Date last visited: 23 August 2013.

http://www.iaia.org. International Association for Impact Assessment. Date last visited: 23 August 2013.

http://www.iaia.org/iaia10/documents/reviewed_papers/Urban%20Pesticide%20Management%20in%20Quebec.pdf. Urban Pesticide Management in Quebec. Date last visited: 23 August 2013.

http://www.iarc.fr/ International Agency for Research on Cancer. Date last visited: 23 August 2013.

http://www.icrisat.org/Images/Why-grain-legumes.pdf. Why Grain Legumes. International Crops Research Institute for the Semi-Arid Tropics. Date last visited: 28 August 2013.

http://www.icrisat.org/what-we-do/satrends/dec2004.htm. Consorting to Correct. *SATrends* Issue 49 (December 2004). Monthly Newsletter for International Crops Research Institute for the Semi-Arid Tropics (ICRISAT). Date last visited: 23 August 2013.

http://www.ienica.net/crops/castor.htm. Castor. 2002. Date last visited: 23 August 2013.

http://www.ikisan.com/basics%20of%20agriculture/Soils.htm. Soil. Date last visited: 23 August 2013.

http://www.ilo.org/legacy/english/protection/safework/cis/products/icsc/dtasht/_icsc09/icsc0985.htm. Sulfotep. International Programme on Chemical Safety. Date last visited: 23 August 2013.

http://www.inchem.org. Chemical Safety Information from Intergovernmental Organizations. International Programme on Chemical Safety (IPCS). Date last visited: 23 August 2013.

http://www.inchem.org/documents/hsg/hsg/hsg82_e.htm. Carbendazim. International Programme on Chemical Safety (IPCS). Health and Safety Guide No. 82. Date last visited: 23 August 2013.

http://www.inchem.org/documents/jmpr/jmpmono/v065pr07.htm. Carbaryl. Evaluation of the Toxicity of Pesticide Residues in Food. International Programme on Chemical Safety (IPCS). Date last visited: 28 August 2013.

http://www.inchem.org/documents/jmpr/jmpmono/v065pr12.htm. Chlorpropham. Evaluation of the Toxicity of Pesticide Residues in Food. International Programme on Chemical Safety (IPCS). Date last visited: 28 August 2013.

http://www.inchem.org/documents/jmpr/jmpmono/v065pr17.htm. Diazinon. Evaluation of the Toxicity of Pesticide Residues in Food. International Programme on Chemical Safety (IPCS). Date last visited: 28 August 2013.

http://www.inchem.org/documents/jmpr/jmpmono/v065pr21.htm. Dimethrin. Evaluation of the Toxicity of Pesticide Residues in Food. International Programme on Chemical Safety (IPCS). Date last visited: 28 August 2013.

http://www.inchem.org/documents/jmpr/jmpmono/v065pr28.htm. Lindane. Evaluation of the Toxicity of Pesticide Residues in Food. International Programme on Chemical Safety (IPCS). Date last visited: 28 August 2013.

http://www.inchem.org/documents/jmpr/jmpmono/v065pr32.htm. Mevinphos. Evaluation of the Toxicity of Pesticide Residues in Food. International Programme on Chemical Safety (IPCS). Date last visited: 28 August 2013.

http://www.inchem.org/documents/jmpr/jmpmono/v068pr07.htm. Carbon tetrachloride. Evaluation of the Toxicity of Pesticide Residues in Food. International Programme on Chemical Safety (IPCS). Date last visited: 28 August 2013.

http://www.inchem.org/documents/jmpr/jmpmono/v068pr21.htm. Fenchlorphos. Evaluation of the Toxicity of Pesticide Residues in Food. International Programme on Chemical Safety (IPCS). Date last visited: 28 August 2013.

http://www.inchem.org/documents/jmpr/jmpmono/v069pr17.htm. Folpet. Evaluation of the Toxicity of Pesticide Residues in Food. International Programme on Chemical Safety (IPCS). Date last visited: 28 August 2013.

http://www.inchem.org/documents/jmpr/jmpmono/v069pr25.htm. 2-Phenylphenol and its Sodium Salt. Evaluation of the Toxicity of Pesticide Residues in Food. International Programme on Chemical Safety (IPCS). Date last visited: 28 August 2013.

http://www.inchem.org/documents/jmpr/jmpmono/v074pr08.htm. Chinomethionat. International Programme on Chemical Safety (IPCS). Date last visited: 23 August 2013.

http://www.inchem.org/documents/jmpr/jmpmono/v075pr07.htm. sec-Butylamine. International Programme on Chemical Safety (IPCS). Date last visited: 23 August 2013.

http://www.inchem.org/documents/jmpr/jmpmono/v076pr08.htm. Cartap. International Programme on Chemical Safety (IPCS). Date last visited: 23 August 2013.

http://www.inchem.org/documents/jmpr/jmpmono/v080pr17.htm. Pesticide Residues in Food – 1980. International Programme on Chemical Safety (IPCS). Date last visited: 23 August 2013.

http://www.inchem.org/documents/pds/pds/pest16_e.htm. Sodium Fluoroacetate. International Programme on Chemical Safety (IPCS). Date last visited: 23 August 2013.

http://www.inchem.org/documents/pds/pds/pest3_e.htm. Carbaryl. WHO and FAO. International Programme on Chemical Safety (IPCS). Date last visited: 23 August 2013.
http://www.inchem.org/documents/pds/pds/pest36_e.htm. Chlordane. WHO and FAO. International Programme on Chemical Safety (IPCS). Date last visited: 23 August 2013.
http://www.inchem.org/documents/pds/pds/pest62_e.htm. Chlorophacinone. WHO and FAO. International Programme on Chemical Safety (IPCS). Date last visited: 23 August 2013.
http://www.inchem.org/documents/pds/pds/pest67_e.htm. Thiometon. WHO and FAO. International Programme on Chemical Safety (IPCS). Date last visited: 23 August 2013.
http://www.inchem.org/documents/pds/pds/pest70_e.htm. Ethoprophos. WHO and FAO. International Programme on Chemical Safety (IPCS). Date last visited: 23 August 2013.
http://www.inchem.org/documents/pds/pds/pest8_e.htm. Temephos. WHO and FAO. International Programme on Chemical Safety (IPCS). Date last visited: 23 August 2013.
http://www.inchem.org/documents/pds/pds/pest83_e.htm. Resmethrin. WHO and FAO. International Programme on Chemical Safety (IPCS). Date last visited: 23 August 2013.
http://www.inra.fr/hyppz/RAVAGEUR/6diacit.htm. Citrus Whitefly. HYPP Zoology. Date last visited: 23 August 2013.
http://www.inra.fr/hyppz/ZGLOSS/6g—087.htm. Coccinellidae, Ladybirds. HYPP Zoology. Date last visited: 23 August 2013.
http://www.interchim.fr/MSDS/5/52005A.pdf. Carbodiimide (EDCA). Material Safety and Data Sheet. Uptima Interchim. Date last visited: 23 August 2013.
http://www.interfarm.co.uk/ProductGuide/UK/Default.aspx?FilterGroupId=7&FilterAttributeId=102. United Kingdom Product Guide. InterFarm (UK) Ltd. Date last visited: 23 August 2013.
http://www.interfarm.co.uk/ProductGuide/UK/Default.aspx?FilterGroupId=7&FilterAttributeId=113. United Kingdom Product Guide. InterFarm (UK) Ltd. Date last visited: 23 August 2013.
http://www.invasive.org/gist/products/handbook/12.Fluazifop.pdf. Fluazifop-P-Butyl. *Weed Control Methods Handbook*. The Nature Conservancy. Date last visited: 23 August 2013.
http://www.ipm.ucdavis.edu/PMG/NE/convergent_lady_beetle.html. Convergent Lady Beetle. Statewide IPM Program, Agriculture and Natural Resources, University of California. Date last visited: 23 August 2013.
http://www.ipm.ucdavis.edu/PMG/r735301211.html. Sugarbeet Cutworms. Statewide IPM Program, Agriculture and Natural Resources, University of California. Date last visited: 23 August 2013.
http://www.ipmimages.org/browse/subimages.cfm?sub=836. *Cercospora* Fungi. IPM Images. Date last visited: 23 August 2013.
http://www.jnkvv.nic.in. Jawaharlal Nehru Krushi Vishwa Vidyalay, Jabalpur, India. Date last visited: 28 July 2013.
http://www.jnkvv.nic.in/ipm%20project/natural_enemy1.htm. Natural Enemy Information. Integrated Pest Management. Jawaharlal Nehru Krushi Vishwa Vidyalay, Jabalpur, India. Date last visited: 23 August 2013.
http://www.jstor.org/pss/3431032. Fukuto, T.R. (1990) Mechanism of action of organophosphorus and carbamate insecticides. *Environmental Health Perspectives* 87. Date last visited: 23 August 2013.
http://www.jstor.org/pss/3987336. Gorski, S.F., Reiners, S. and Hassell, R. (1988) Chloramben for weed control on muck-grown lettuce, *Lactuca sativa*, and endive, *Cichorium endivia*. *Weed Technology* 2 (3). Date last visited: 23 August 2013.
http://www.jstor.org/pss/4044980. O'Dke, S., Lydon, J., Becerril, J.M., Sherman, T.D., Lehnen, L.P. Jr and Matsumotom, H. (1991) Protoporphyrinogen oxidase-inhibiting herbicides. *Weed Science* 39 (3). Date last visited: 23 August 2013.
http://www.kangmei.com/tefluthrin.htm. Tefluthrin. Changzhou Kangmei Chemical Industry Co Ltd. Date last visited: 23 August 2013.
http://www.knowledgebank.irri.org/RiceDoctor/component/content/article/691.html. Pests and their Typical Effects. International Rice Research Institute. Date last visited: 23 August 2013.
http://www.koppert.com/pests/aphid/products-against-aphids/detail/aphipar-2/. Aphipar. Koppert Biological Systems. Koppert B.V. Date last visited: 23 August 2013.
http://www.koppert.com/products/products-pests-diseases/products/detail/thripex-1/. Thripex. Koppert Biological Systems. Koppert B.V. Date last visited: 23 August 2013.
http://www.krishisewa.com/cms/articles/production-technology/85-castor-cultivation.html. Cultivation Technology of Castor (*Ricinus communis* L.). Krishisewa. Agriculture Information Hub. Date last visited: 23 August 2013.
http://www.ksre.ksu.edu/bookstore/Category.aspx?id=521. Kansas State University. Date last visited: 23 August 2013.
http://www.lonyetech.com/ProductsView.asp?id=83. Product Name : Thiocyclam. Lonye Technology. Date last visited: 23 August 2013.
http://www.m5.ws001.squarestart.ne.jp/foundation/agrdtl.php?a_inq=15900. Cafenstrole. Table of MRLs for Agricultural Chemicals. The Japan Food Chemical Research Foundation. Date last visited: 23 August 2013.
http://www.m5.ws001.squarestart.ne.jp/foundation/agrdtl.php?a_inq=68600. Hexazinone. Table of MRLs for Agricultural Chemicals. The Japan Food Chemical Research Foundation. Date last visited: 23 August 2013.
http://www.micotoxinas.com.br/aflafacts.pdf. The Aflatoxins. Leatherhead Food Research Association. Date last visited: 23 August 2013.
https://www.microbelibrary.org/index.php?option=com_resource&controller=article&article=2596&Itemid=0&category_id=15. Nematophagous Fungi. American Society for Microbiology. ASM Microbe Library. Date last visited: 28 August 2013.
http://www.mortonarb.org. Morton Arboratum. Date last visited: 23 August 2013.
http://www.mortonarb.org/search/node/chlorosis. Chlorosis. Morton Arboratum. Date last visited: 23 August 2013.
http://www.mortonarb.org/trees-plants/plant-clinic/help-pests/mites. Mites. Morton Arboratum. Date last visited: 23 August 2013.

http://www.mrldatabase.com. The International Maximum Residue Level Database. FAS Online. Date last visited: 23 August 2013.

http://www.ncbi.nlm.nih.gov/pubmed/10330314. Mullié, W.C., Diallo, A.O., Gadji, B. and Ndiaye, M.D. (1999) Environmental hazards of mobile ground spraying with cyanophos and fenthion for quelea control in Senegal. *Ecotoxicology and Environmental Safety* 43 (1), 1–10. Date last visited: 24 August 2013.

http://www.ncbi.nlm.nih.gov/pubmed/16039310. Jackson, T.A., Crawford, A.M. and Glare, T.R. (2005) Oryctes virus – time for a new look at a useful biocontrol agent. *Journal of Invertebrate Pathology* 89 (1), 91–94. Date last visited: 24 August 2013.

http://www.ncbi.nlm.nih.gov/pubmed/17288493. Proudfoot, A.T., Bradberry, S.M. and Vale, J.A. (2006) Sodium fluoroacetate poisoning. *Toxicology Review* 25 (4), 213–219. Date last visited: 24 August 2013.

http://www.ncbi.nlm.nih.gov/pubmed/17385511. Hadapad, A.B. and Zebitz, C.P. (2006) Mass production, survival and evaluation of solid substrate inocula of *Beauveria brongniartii* (Saccardo) Petch against *Holotrichia serrata* (Coleoptera: Scarabaeidae). *Communications in Agricultural and Applied Biological Sciences* 71 (2), 433–441. Date last visited: 24 August 2013.

http://www.ncbi.nlm.nih.gov/pubmed/18944639. Larkin, R.P. and Fravel, D.R. (1999) Mechanisms of action and dose-response relationships governing biological control of Fusarium wilt of tomato by nonpathogenic *Fusarium* spp. *Phytopathology* 89 (12), 1152–1161. Date last visited: 24 August 2013.

http://www.ncbi.nlm.nih.gov/pubmed/1955388. Nakajima, M., Itoi, K., Takamatsu, Y., Sato, S., Furukawa, Y., Furuya, K., Honma, T., Kadotani, J., Kozasa, M., Haneishi, T. (1991) Cornexistin: a new fungal metabolite with herbicidal activity. *Journal of Antibiotics (Tokyo)* 44 (10), 1065–1072. Date last visited: 24 August 2013.

http://www.ncbi.nlm.nih.gov/pubmed/20121450. Komatsu, K., Hashimoto, M., Ozeki, J., Yamaji, Y., Maejima, K., Senshu, H., Okano, Y., Kagiwada, S. and Namba, S. (2010) Viral-induced systemic necrosis in plants involves both programmed cell death and the inhibition of viral multiplication, which are regulated by independent pathways. *Molecular–Plant Microbe Interactions* 23 (3), 283–293. Date last visited: 24 August 2013.

http://www.ncipm.org.in/bacillus-SPHERICUS.htm. *Bacillus sphaericus*. National Centre for Integrated Pest Management, Indian Council of Agricultural Research. Date last visited: 24 August 2013.

http://www.ncipm.org.in/Beauveria%20bassiana.htm. *Beauveria bassiana*. National Centre for Integrated Pest Management, Indian Council of Agricultural Research. Date last visited: 24 August 2013.

http://www.nrdc.org/breastmilk/solv.asp. Healthy Milk, Healthy Baby Chemical Pollution and Mother's Milk. Date last visited: 24 August 2013.

http://www.nri.org/research-consultancy/agricultural-biosecurity-group/background. Agricultural Biosecurity Group. Natural Resources Institute, University of Greenwich. Date last visited: 24 August 2013.

http://www.nzpps.org. New Zealand Plant Protection Society. Date last visited: 26 August 2013.

http://www.nzpps.org/journal/53/nzpp_530710.pdf. Boyd-Wilson, K.S.H., Magee, L.J., Hackett, J.K. and Walter, M. (2000) Testing bacterial and fungal isolates for biological control of *Fusarium culmorum*. *New Zealand Plant Protection* 53, 71–77. Date last visited: 26 August 2013.

http://www.nzpps.org/journal/61/nzpp_611340.pdf. Moore, M.S. and Follas, G.B., Hagerty, G.C. and Beresford, R.S. (2008) Carboxylic acid amide (Caa) fungicide – resistance prevention strategy. *New Zealand Plant Protection* 61, 134–136. Date last visited: 26 August 2013.

http://www.nzpps.org/resistance/benzimidazole.php. Martin, N.A., Beresford, R.M. and Harrington, K.C. (eds) (2005) *Pesticide Resistance: Prevention and Management Strategies 2005*. New Zealand Plant Protection Society, Hastings, New Zealand. 166 pp. (ISBN 0-473-10441-5). Date last visited: 26 August 2013.

http://www.nzpps.org/_resistance/crops/pdfs/apple_black_spot.pdf. Fungicide Resistance Management Strategy for Apple Black Spot (*Venturia inaequalis*) in New Zealand. 2013. Date last visited: 26 August 2013.

http://www.omafra.gov.on.ca/english/crops/facts/95-063.htm. Identification of Diseases and Disorders of Onions. Fact Sheet. 1995. Ontario Ministry of Agriculture and Food. Date last visited: 26 August 2013.

http://www.omri.org/ Organic Materials Review Institute. Date last visited: 26 August 2013.

http://www.oniongardening.com/onion-storage. Onion Storage. Onion Gardening Tips. Date last visited: 26 August 2013.

http://www.oniongardening.com/onion-troubleshooting-guide. Onion Troubleshooting Guide. Onion Gardening Tips. Date last visited: 26 August 2013.

http://www.opsi.gov.uk/legislation/scotland/ssi2007/ssi_20070306_en_2. The Pesticides Maximum Residue Levels in Crops, Food and Feeding Stuff (Scotland). Amendment (No. 2) Regulations 2007. Date last visited: 26 August 2013.

http://www.osha.gov. Occupational Safety and Health Administration, US Department of Labor. Date last visited: 26 August 2013.

http://www.osha.gov/dts/chemicalsampling/data/CH_267600.html. Sodium Fluoroacetate. Occupational Safety and Health Administration, US Department of Labor. Date last visited: 26 August 2013.

https://www.osha.gov/pls/oshaweb/owadisp.show_document?p_table=NEWS_RELEASES&p_id=24456. OSHA and NIOSH Issue Hazard Alert on 1-Bromopropane, Urge Efforts to Safeguard Workers from Exposure to Toxic Chemical. Occupational Safety and Health Administration, US Department of Labor. Date last visited: 26 August 2013.

http://www.pan-uk.org/pestnews/Actives/24d.htm. 2,4-D Fact Sheet. *Pesticides News* No. 37, September 1997. Date last visited: 26 August 2013.

http://www.pan-uk.org/pestnews/Actives/atrazine.htm. Atrazine. *Pesticides News* No. 56, June 2002. Date last visited: 26 August 2013.

http://www.pan-uk.org/pestnews/Actives/Carbenda.htm. Carbendazim. *Pesticides News* No. 57, September 2002. Date last visited: 26 August 2013.

http://www.pan-uk.org/pestnews/Actives/cypermet.htm. Cypermethrin – a Synthetic Pyrethroid. *Pesticides News* No. 30, December 1995. Date last visited: 26 August 2013.

http://www.pan-uk.org/pestnews/Actives/rotenone.htm. Rotenone. *Pesticides News* No. 54, December 2001. Date last visited: 26 August 2013.

http://www.pan-uk.org/pestnews/Actives/tridemor.htm Tridemorph. *Pesticides News* No. 44, June 1999. Date last visited: 26 August 2013.

http://www.pan-uk.org/pestnews/Free%20Articles/PN93/PN93pp12-14.pdf. Bueno, V.H.P. (2011) Biological control in Brazil. *Pesticides News* No. 93. Date last visited: 28 August 2013.

http://www.pan-uk.org/pestnews/Issue/pn60/pn60p8.htm. Cleaning Up Europe – a Message for Developing Countries. *Pesticides News* No. 60, June 2003. Date last visited: 26 August 2013.

http://www.pan-uk.org/pestnews/Issue/pn76/pn76%20p16-17.pdf. Bio-intensive Pest Management Reduces Pesticide Use in India. Integrated Pest Management. *Pesticides News* No.76, June 2007. Date last visited: 26 August 2013.

http://www.peanutscience.com/doi/pdf/10.3146. Peanut Cultivar Response to S-metolachlor and Paraquat Alone and in Combination. American Peanut Research and Education Society, Inc. Date last visited: 26 August 2013.

http://www.pestcontrolindia.com/know-your-pest/pest/insect-agricultural-pests-brown-plant-hopper-13-21-127.aspx. Brown Plant Hopper (*Nilaparvata lugens*). Pest Control of India. Date last visited: 26 August 2013.

http://www.pesticideinfo.org. PAN Pesticide Database. Pesticide Action Network, North America. Date last visited: 26 August 2013.

http://www.pesticideinfo.org/Detail_Chemical.jsp. 2,4-D – Identification, Toxicity, Use, Water Pollution Potential, Ecological Toxicity and Regulatory Information. PAN Pesticide Database, Pesticide Action Network, North America. Date last visited: 26 August 2013.

http://www.pesticideinfo.org/Detail_Chemical.jsp?Rec_Id=PC16. Oxadixyl – Identification, Toxicity, Use, Water Pollution Potential, Ecological Toxicity and Regulatory Information. PAN Pesticide Database, Pesticide Action Network, North America. Date last visited: 26 August 2013.

http://www.pesticideinfo.org/Detail_Chemical.jsp?Rec_Id=PC49. Cottonseed Oil – Identification, Toxicity, Use, Water Pollution Potential, Ecological Toxicity and Regulatory Information. PAN Pesticide Database, Pesticide Action Network, North America. Date last visited: 26 August 2013.

http://www.pesticideinfo.org/Detail_Chemical.jsp?Rec_Id=PC32788. Miral – Identification, Toxicity, Use, Water Pollution Potential, Ecological Toxicity and Regulatory Information. PAN Pesticide Database, Pesticide Action Network, North America. Date last visited: 26 August 2013.

http://www.pesticideinfo.org/Detail_Chemical.jsp?Rec_Id=PC32816. Carbaryl – Identification, Toxicity, Use, Water Pollution Potential, Ecological Toxicity and Regulatory Information. PAN Pesticide Database, Pesticide Action Network, North America. Date last visited: 26 August 2013.

http://www.pesticideinfo.org/Detail_Chemical.jsp?Rec_Id=PC32843. Oxythioquinox – Identification, Toxicity, Use, Water Pollution Potential, Ecological Toxicity and Regulatory Information. PAN Pesticide Database, Pesticide Action Network, North America. Date last visited: 26 August 2013.

http://www.pesticideinfo.org/Detail_Chemical.jsp?Rec_Id=PC32862. Carbendazim – Identification, Toxicity, Use, Water Pollution Potential, Ecological Toxicity and Regulatory Information. PAN Pesticide Database, Pesticide Action Network, North America. Date last visited: 26 August 2013.

http://www.pesticideinfo.org/Detail_Chemical.jsp?Rec_Id=PC32991. Bendiocarb – Identification, Toxicity, Use, Water Pollution Potential, Ecological Toxicity and Regulatory Information. PAN Pesticide Database, Pesticide Action Network, North America. Date last visited: 26 August 2013.

http://www.pesticideinfo.org/Detail_Chemical.jsp?Rec_Id=PC33097. Tetramethrin – Identification, Toxicity, Use, Water Pollution Potential, Ecological Toxicity and Regulatory Information. PAN Pesticide Database, Pesticide Action Network, North America. Date last visited: 26 August 2013.

http://www.pesticideinfo.org/Detail_Chemical.jsp?Rec_Id=PC33102. Hexachlorobenzene – Identification, Toxicity, Use, Water Pollution Potential, Ecological Toxicity and Regulatory Information. PAN Pesticide Database, Pesticide Action Network, North America. Date last visited: 26 August 2013.

http://www.pesticideinfo.org/Detail_Chemical.jsp?Rec_Id=PC33170. Fonofos – Identification, Toxicity, Use, Water Pollution Potential, Ecological Toxicity and Regulatory Information. PAN Pesticide Database, Pesticide Action Network, North America. Date last visited: 26 August 2013.

http://www.pesticideinfo.org/Detail_Chemical.jsp?Rec_Id=PC33172. Esfenvalerate – Identification, Toxicity, Use, Water Pollution Potential, Ecological Toxicity and Regulatory Information. PAN Pesticide Database, Pesticide Action Network, North America. Date last visited: 26 August 2013.

http://www.pesticideinfo.org/Detail_Chemical.jsp?Rec_Id=PC33220. Butylate – Identification, Toxicity, Use, Water Pollution Potential, Ecological Toxicity and Regulatory Information. PAN Pesticide Database, Pesticide Action Network, North America. Date last visited: 26 August 2013.

http://www.pesticideinfo.org/Detail_Chemical.jsp?Rec_Id=PC33241. Ethiolate – Identification, Toxicity, Use, Water Pollution Potential, Ecological Toxicity and Regulatory Information. PAN Pesticide Database, Pesticide Action Network, North America. Date last visited: 26 August 2013.

http://www.pesticideinfo.org/Detail_Chemical.jsp?Rec_Id=PC33346. Trichlorfon – Identification, Toxicity, Use, Water Pollution Potential, Ecological Toxicity and Regulatory Information. PAN Pesticide Database, Pesticide Action Network, North America. Date last visited: 26 August 2013.

http://www.pesticideinfo.org/Detail_Chemical.jsp?Rec_Id=PC33383. Bensulide – Identification, Toxicity, Use, Water Pollution Potential, Ecological Toxicity and Regulatory Information. PAN Pesticide Database, Pesticide Action Network, North America. Date last visited: 26 August 2013.

http://www.pesticideinfo.org/Detail_Chemical.jsp?Rec_Id=PC33396. Carbosulfan – Identification, Toxicity, Use, Water Pollution Potential, Ecological Toxicity and Regulatory Information. PAN Pesticide Database, Pesticide Action Network, North America. Date last visited: 26 August 2013.

http://www.pesticideinfo.org/Detail_Chemical.jsp?Rec_Id=PC33447. Diallate – Identification, Toxicity, Use, Water Pollution Potential, Ecological Toxicity and Regulatory Information. PAN Pesticide Database, Pesticide Action Network, North America. Date last visited: 26 August 2013.

http://www.pesticideinfo.org/Detail_Chemical.jsp?Rec_Id=PC33503. Trinexapac-ethyl – Identification, Toxicity, Use, Water Pollution Potential, Ecological Toxicity and Regulatory Information. PAN Pesticide Database, Pesticide Action Network, North America. Date last visited: 26 August 2013.

http://www.pesticideinfo.org/Detail_Chemical.jsp?Rec_Id=PC33510. Cyromazine – Identification, Toxicity, Use, Water Pollution Potential, Ecological Toxicity and Regulatory Information. PAN Pesticide Database, Pesticide Action Network, North America. Date last visited: 26 August 2013.

http://www.pesticideinfo.org/Detail_Chemical.jsp?Rec_Id=PC33537. Copper 8-Quinolinoleate – Identification, Toxicity, Use, Water Pollution Potential, Ecological Toxicity and Regulatory Information. PAN Pesticide Database, Pesticide Action Network, North America. Date last visited: 26 August 2013.

http://www.pesticideinfo.org/Detail_Chemical.jsp?Rec_Id=PC33588. Clethodim – Identification, Toxicity, Use, Water Pollution Potential, Ecological Toxicity and Regulatory Information. PAN Pesticide Database, Pesticide Action Network, North America. Date last visited: 26 August 2013.

http://www.pesticideinfo.org/Detail_Chemical.jsp?Rec_Id=PC33596. Cinnamaldehyde – Identification, Toxicity, Use, Water Pollution Potential, Ecological Toxicity and Regulatory Information. PAN Pesticide Database, Pesticide Action Network, North America. Date last visited: 26 August 2013.

http://www.pesticideinfo.org/Detail_Chemical.jsp?Rec_Id=PC33601. Oxyfluorfen – Identification, Toxicity, Use, Water Pollution Potential, Ecological Toxicity and Regulatory Information. PAN Pesticide Database, Pesticide Action Network, North America. Date last visited: 26 August 2013.

http://www.pesticideinfo.org/Detail_Chemical.jsp?Rec_Id=PC33610. 1,3-Dichloropropene – Identification, Toxicity, Use, Water Pollution Potential, Ecological Toxicity and Regulatory Information. PAN Pesticide Database, Pesticide Action Network, North America. Date last visited: 26 August 2013.

http://www.pesticideinfo.org/Detail_Chemical.jsp?Rec_Id=PC33641. Sulfallate – Identification, Toxicity, Use, Water Pollution Potential, Ecological Toxicity and Regulatory Information. PAN Pesticide Database, Pesticide Action Network, North America. Date last visited: 26 August 2013.

http://www.pesticideinfo.org/Detail_Chemical.jsp?Rec_Id=PC33671. Captafol (Isomer Unspecified) – Identification, Toxicity, Use, Water Pollution Potential, Ecological Toxicity and Regulatory Information. PAN Pesticide Database, Pesticide Action Network, North America. Date last visited: 26 August 2013.

http://www.pesticideinfo.org/Detail_Chemical.jsp?Rec_Id=PC33722. Sec-butylamine – Identification, Toxicity, Use, Water Pollution Potential, Ecological Toxicity and Regulatory Information. PAN Pesticide Database, Pesticide Action Network, North America. Date last visited: 26 August 2013.

http://www.pesticideinfo.org/Detail_Chemical.jsp?Rec_Id=PC33744. Brodifacoum – Identification, Toxicity, Use, Water Pollution Potential, Ecological Toxicity and Regulatory Information. PAN Pesticide Database, Pesticide Action Network, North America. Date last visited: 26 August 2013.

http://www.pesticideinfo.org/Detail_Chemical.jsp?Rec_Id=PC33842. Ampelomyces Quisqualis – Identification, Toxicity, Use, Water Pollution Potential, Ecological Toxicity and Regulatory Information. PAN Pesticide Database, Pesticide Action Network, North America. Date last visited: 26 August 2013.

http://www.pesticideinfo.org/Detail_Chemical.jsp?Rec_Id=PC34114. Acetochlor – Identification, Toxicity, Use, Water Pollution Potential, Ecological Toxicity and Regulatory Information. PAN Pesticide Database, Pesticide Action Network, North America. Date last visited: 26 August 2013.

http://www.pesticideinfo.org/Detail_Chemical.jsp?Rec_Id=PC34260. Propachlor – Identification, Toxicity, Use, Water Pollution Potential, Ecological Toxicity and Regulatory Information. PAN Pesticide Database, Pesticide Action Network, North America. Date last visited: 26 August 2013.

http://www.pesticideinfo.org/Detail_Chemical.jsp?Rec_Id=PC34277. Vernolate – Identification, Toxicity, Use, Water Pollution Potential, Ecological Toxicity and Regulatory Information. PAN Pesticide Database, Pesticide Action Network, North America. Date last visited: 26 August 2013.

http://www.pesticideinfo.org/Detail_Chemical.jsp?Rec_Id=PC34303. Resmethrin – Identification, Toxicity, Use, Water Pollution Potential, Ecological Toxicity and Regulatory Information. PAN Pesticide Database, Pesticide Action Network, North America. Date last visited: 26 August 2013.

http://www.pesticideinfo.org/Detail_Chemical.jsp?Rec_Id=PC34318. Red Squill Glycoside – Identification, Toxicity, Use, Water Pollution Potential, Ecological Toxicity and Regulatory Information. PAN Pesticide Database, Pesticide Action Network, North America. Date last visited: 26 August 2013.

http://www.pesticideinfo.org/Detail_Chemical.jsp?Rec_Id=PC34340. Simazine – Identification, Toxicity, Use, Water Pollution Potential, Ecological Toxicity and Regulatory Information. PAN Pesticide Database, Pesticide Action Network, North America. Date last visited: 26 August 2013.

http://www.pesticideinfo.org/Detail_Chemical.jsp?Rec_Id=PC34531. Tecnazene – Identification, Toxicity, Use, Water Pollution Potential, Ecological Toxicity and Regulatory Information. PAN Pesticide Database, Pesticide Action Network, North America. Date last visited: 26 August 2013.

http://www.pesticideinfo.org/Detail_Chemical.jsp?Rec_Id=PC34532. Tefluthrin – Identification, Toxicity, Use, Water Pollution Potential, Ecological Toxicity and Regulatory Information. PAN Pesticide Database, Pesticide Action Network, North America. Date last visited: 26 August 2013.

http://www.pesticideinfo.org/Detail_Chemical.jsp?Rec_Id=PC34539. Terbacil – Identification, Toxicity, Use, Water Pollution Potential, Ecological Toxicity and Regulatory Information. PAN Pesticide Database, Pesticide Action Network, North America. Date last visited: 26 August 2013.

http://www.pesticideinfo.org/Detail_Chemical.jsp?Rec_Id=PC34550. Chlorothalonil – Identification, Toxicity, Use, Water Pollution Potential, Ecological Toxicity and Regulatory Information. PAN Pesticide Database, Pesticide Action Network, North America. Date last visited: 26 August 2013.

http://www.pesticideinfo.org/Detail_Chemical.jsp?Rec_Id=PC34561. Ethion – Identification, Toxicity, Use, Water Pollution Potential, Ecological Toxicity and Regulatory Information. PAN Pesticide Database, Pesticide Action Network, North America. Date last visited: 26 August 2013.

http://www.pesticideinfo.org/Detail_Chemical.jsp?Rec_Id=PC34569. Captan – Identification, Toxicity, Use, Water Pollution Potential, Ecological Toxicity and Regulatory Information. PAN Pesticide Database, Pesticide Action Network, North America. Date last visited: 26 August 2013.

http://www.pesticideinfo.org/Detail_Chemical.jsp?Rec_Id=PC34574. Thiram – Identification, Toxicity, Use, Water Pollution Potential, Ecological Toxicity and Regulatory Information. PAN Pesticide Database, Pesticide Action Network, North America. Date last visited: 26 August 2013.

http://www.pesticideinfo.org/Detail_Chemical.jsp?Rec_Id=PC34581. Thiabendazole – Identification, Toxicity, Use, Water Pollution Potential, Ecological Toxicity and Regulatory Information. PAN Pesticide Database, Pesticide Action Network, North America. Date last visited: 26 August 2013.

http://www.pesticideinfo.org/Detail_Chemical.jsp?Rec_Id=PC34583. Thidiazuron – Identification, Toxicity, Use, Water Pollution Potential, Ecological Toxicity and Regulatory Information. PAN Pesticide Database, Pesticide Action Network, North America. Date last visited: 26 August 2013.

http://www.pesticideinfo.org/Detail_Chemical.jsp?Rec_Id=PC34584 Thiobencarb – Identification, Toxicity, Use, Water Pollution Potential, Ecological Toxicity and Regulatory Information. PAN Pesticide Database, Pesticide Action Network, North America. Date last visited: 26 August 2013.

http://www.pesticideinfo.org/Detail_Chemical.jsp?Rec_Id=PC34586. Thiodicarb – Identification, Toxicity, Use, Water Pollution Potential, Ecological Toxicity and Regulatory Information. PAN Pesticide Database, Pesticide Action Network, North America. Date last visited: 26 August 2013.

http://www.pesticideinfo.org/Detail_Chemical.jsp?Rec_Id=PC34587. Thiofanox – Identification, Toxicity, Use, Water Pollution Potential, Ecological Toxicity and Regulatory Information. PAN Pesticide Database, Pesticide Action Network, North America. Date last visited: 26 August 2013.

http://www.pesticideinfo.org/Detail_Chemical.jsp?Rec_Id=PC34588. Thiophanate-methyl – Identification, Toxicity, Use, Water Pollution Potential, Ecological Toxicity and Regulatory Information. PAN Pesticide Database, Pesticide Action Network, North America. Date last visited: 26 August 2013.

http://www.pesticideinfo.org/Detail_Chemical.jsp?Rec_Id=PC34603. Toxaphene – Identification, Toxicity, Use, Water Pollution Potential, Ecological Toxicity and Regulatory Information. PAN Pesticide Database, Pesticide Action Network, North America. Date last visited: 26 August 2013.

http://www.pesticideinfo.org/Detail_Chemical.jsp?Rec_Id=PC34613. Triazophos – Identification, Toxicity, Use, Water Pollution Potential, Ecological Toxicity and Regulatory Information. PAN Pesticide Database, Pesticide Action Network, North America. Date last visited: 26 August 2013.

http://www.pesticideinfo.org/Detail_Chemical.jsp?Rec_Id=PC34627. Triallate – Identification, Toxicity, Use, Water Pollution Potential, Ecological Toxicity and Regulatory Information. PAN Pesticide Database, Pesticide Action Network, North America. Date last visited: 26 August 2013.

http://www.pesticideinfo.org/Detail_Chemical.jsp?Rec_Id=PC34658. Triflumuron – Identification, Toxicity, Use, Water Pollution Potential, Ecological Toxicity and Regulatory Information. PAN Pesticide Database, Pesticide Action Network, North America. Date last visited: 26 August 2013.

http://www.pesticideinfo.org/Detail_Chemical.jsp?Rec_Id=PC34668. Chlormequat chloride – Identification, Toxicity, Use, Water Pollution Potential, Ecological Toxicity and Regulatory Information. PAN Pesticide Database, Pesticide Action Network, North America. Date last visited: 26 August 2013.

http://www.pesticideinfo.org/Detail_Chemical.jsp?Rec_Id=PC34702. Vamidothion – Identification, Toxicity, Use, Water Pollution Potential, Ecological Toxicity and Regulatory Information. PAN Pesticide Database, Pesticide Action Network, North America. Date last visited: 26 August 2013.

http://www.pesticideinfo.org/Detail_Chemical.jsp?Rec_Id=PC34709. Cholecalciferol – Identification, Toxicity, Use, Water Pollution Potential, Ecological Toxicity and Regulatory Information. PAN Pesticide Database, Pesticide Action Network, North America. Date last visited: 26 August 2013.

http://www.pesticideinfo.org/Detail_Chemical.jsp?Rec_Id=PC34790. Oxydemeton-methyl – Identification, Toxicity, Use, Water Pollution Potential, Ecological Toxicity and Regulatory Information. PAN Pesticide Database, Pesticide Action Network, North America. Date last visited: 26 August 2013.

http://www.pesticideinfo.org/Detail_Chemical.jsp?Rec_Id=PC34806. Acephate – Identification, Toxicity, Use, Water Pollution Potential, Ecological Toxicity and Regulatory Information. PAN Pesticide Database, Pesticide Action Network, North America. Date last visited: 26 August 2013.

http://www.pesticideinfo.org/Detail_Chemical.jsp?Rec_Id=PC34851. Aluminum Phosphide – Identification, Toxicity, Use, Water Pollution Potential, Ecological Toxicity and Regulatory Information. PAN Pesticide Database, Pesticide Action Network, North America. Date last visited: 26 August 2013.

http://www.pesticideinfo.org/Detail_Chemical.jsp?Rec_Id=PC34911. Tralomethrin – Identification, Toxicity, Use, Water Pollution Potential, Ecological Toxicity and Regulatory Information. PAN Pesticide Database, Pesticide Action Network, North America. Date last visited: 26 August 2013.

http://www.pesticideinfo.org/Detail_Chemical.jsp?Rec_Id=PC35001. Corn Oil – Identification, Toxicity, Use, Water Pollution Potential, Ecological Toxicity and Regulatory Information. PAN Pesticide Database, Pesticide Action Network, North America. Date last visited: 26 August 2013.

http://www.pesticideinfo.org/Detail_Chemical.jsp?Rec_Id=PC35036. Temephos – Identification, Toxicity, Use, Water Pollution Potential, Ecological Toxicity and Regulatory Information. PAN Pesticide Database, Pesticide Action Network, North America. Date last visited: 26 August 2013.

http://www.pesticideinfo.org/Detail_Chemical.jsp?Rec_Id=PC35038. Cycloheximide – Identification, Toxicity, Use, Water Pollution Potential, Ecological Toxicity and Regulatory Information. PAN Pesticide Database, Pesticide Action Network, North America. Date last visited: 26 August 2013.

http://www.pesticideinfo.org/Detail_Chemical.jsp?Rec_Id=PC35042. Atrazine – Identification, Toxicity, Use, Water Pollution Potential, Ecological Toxicity and Regulatory Information. PAN Pesticide Database, Pesticide Action Network, North America. Date last visited: 26 August 2013.

http://www.pesticideinfo.org/Detail_Chemical.jsp?Rec_Id=PC35060. Chloroneb – Identification, Toxicity, Use, Water Pollution Potential, Ecological Toxicity and Regulatory Information. PAN Pesticide Database, Pesticide Action Network, North America. Date last visited: 26 August 2013.

http://www.pesticideinfo.org/Detail_Chemical.jsp?Rec_Id=PC35061. Chloropicrin – Identification, Toxicity, Use, Water Pollution Potential, Ecological Toxicity and Regulatory Information. PAN Pesticide Database, Pesticide Action Network, North America. Date last visited: 26 August 2013.

http://www.pesticideinfo.org/Detail_Chemical.jsp?Rec_Id=PC35065. Citric Acid – Identification, Toxicity, Use, Water Pollution Potential, Ecological Toxicity and Regulatory Information. PAN Pesticide Database, Pesticide Action Network, North America. Date last visited: 26 August 2013.

http://www.pesticideinfo.org/Detail_Chemical.jsp?Rec_Id=PC35066. Oil of Citronella – Identification, Toxicity, Use, Water Pollution Potential, Ecological Toxicity and Regulatory Information. PAN Pesticide Database, Pesticide Action Network, North America. Date last visited: 26 August 2013.

http://www.pesticideinfo.org/Detail_Chemical.jsp?Rec_Id=PC35087. EPN – Identification, Toxicity, Use, Water Pollution Potential, Ecological Toxicity and Regulatory Information. PAN Pesticide Database, Pesticide Action Network, North America. Date last visited: 26 August 2013.

http://www.pesticideinfo.org/Detail_Chemical.jsp?Rec_Id=PC35088. EPTC – Identification, Toxicity, Use, Water Pollution Potential, Ecological Toxicity and Regulatory Information. PAN Pesticide Database, Pesticide Action Network, North America. Date last visited: 26 August 2013.

http://www.pesticideinfo.org/Detail_Chemical.jsp?Rec_Id=PC35097. Chlorfenvinphos – Identification, Toxicity, Use, Water Pollution Potential, Ecological Toxicity and Regulatory Information. PAN Pesticide Database, Pesticide Action Network, North America. Date last visited: 26 August 2013.

http://www.pesticideinfo.org/Detail_Chemical.jsp?Rec_Id=PC35130. Prometon – Identification, Toxicity, Use, Water Pollution Potential, Ecological Toxicity and Regulatory Information. PAN Pesticide Database, Pesticide Action Network, North America. Date last visited: 26 August 2013.

http://www.pesticideinfo.org/Detail_Chemical.jsp?Rec_Id=PC35131. Propazine – Identification, Toxicity, Use, Water Pollution Potential, Ecological Toxicity and Regulatory Information. PAN Pesticide Database, Pesticide Action Network, North America. Date last visited: 26 August 2013.

http://www.pesticideinfo.org/Detail_Chemical.jsp?Rec_Id=PC35133. Rotenone – Identification, Toxicity, Use, Water Pollution Potential, Ecological Toxicity and Regulatory Information. PAN Pesticide Database, Pesticide Action Network, North America. Date last visited: 26 August 2013.

http://www.pesticideinfo.org/Detail_Chemical.jsp?Rec_Id=PC35143. Sulfotep – Identification, Toxicity, Use, Water Pollution Potential, Ecological Toxicity and Regulatory Information. PAN Pesticide Database, Pesticide Action Network, North America. Date last visited: 26 August 2013.

http://www.pesticideinfo.org/Detail_Chemical.jsp?Rec_Id=PC35155. 1080 – Identification, Toxicity, Use, Water Pollution Potential, Ecological Toxicity and Regulatory Information. PAN Pesticide Database, Pesticide Action Network, North America. Date last visited: 26 August 2013.

http://www.pesticideinfo.org/Detail_Chemical.jsp?Rec_Id=PC35160. Alachlor – Identification, Toxicity, Use, Water Pollution Potential, Ecological Toxicity and Regulatory Information. PAN Pesticide Database, Pesticide Action Network, North America. Date last visited: 26 August 2013.

http://www.pesticideinfo.org/Detail_Chemical.jsp?Rec_Id=PC35170. Chlorflurenol, Methyl Ester – Identification, Toxicity, Use, Water Pollution Potential, Ecological Toxicity and Regulatory Information. PAN Pesticide Database, Pesticide Action Network, North America. Date last visited: 26 August 2013.
http://www.pesticideinfo.org/Detail_Chemical.jsp?Rec_Id=PC35309. Ethephon – Identification, Toxicity, Use, Water Pollution Potential, Ecological Toxicity and Regulatory Information. PAN Pesticide Database, Pesticide Action Network, North America. Date last visited: 26 August 2013.
http://www.pesticideinfo.org/Detail_Chemical.jsp?Rec_Id=PC35335. Carboxin – Identification, Toxicity, Use, Water Pollution Potential, Ecological Toxicity and Regulatory Information. PAN Pesticide Database, Pesticide Action Network, North America. Date last visited: 26 August 2013.
http://www.pesticideinfo.org/Detail_Chemical.jsp?Rec_Id=PC35345. Tebuthiuron – Identification, Toxicity, Use, Water Pollution Potential, Ecological Toxicity and Regulatory Information. PAN Pesticide Database, Pesticide Action Network, North America. Date last visited: 26 August 2013.
http://www.pesticideinfo.org/Detail_Chemical.jsp?Rec_Id=PC35367. Oxamyl – Identification, Toxicity, Use, Water Pollution Potential, Ecological Toxicity and Regulatory Information. PAN Pesticide Database, Pesticide Action Network, North America. Date last visited: 26 August 2013.
http://www.pesticideinfo.org/Detail_Chemical.jsp?Rec_Id=PC35388. *Agrobacterium radiobacter* – Identification, Toxicity, Use, Water Pollution Potential, Ecological Toxicity and Regulatory Information. PAN Pesticide Database, Pesticide Action Network, North America. Date last visited: 26 August 2013.
http://www.pesticideinfo.org/Detail_Chemical.jsp?Rec_Id=PC35396. Sulprofos – Identification, Toxicity, Use, Water Pollution Potential, Ecological Toxicity and Regulatory Information. PAN Pesticide Database, Pesticide Action Network, North America. Date last visited: 26 August 2013.
http://www.pesticideinfo.org/Detail_Chemical.jsp?Rec_Id=PC35425. Vinclozolin – Identification, Toxicity, Use, Water Pollution Potential, Ecological Toxicity and Regulatory Information. PAN Pesticide Database, Pesticide Action Network, North America. Date last visited: 26 August 2013.
http://www.pesticideinfo.org/Detail_Chemical.jsp?Rec_Id=PC35437. Sethoxydim – Identification, Toxicity, Use, Water Pollution Potential, Ecological Toxicity and Regulatory Information. PAN Pesticide Database, Pesticide Action Network, North America. Date last visited: 26 August 2013.
http://www.pesticideinfo.org/Detail_Chemical.jsp?Rec_Id=PC35454. *Bacillus sphaericus* – Identification, Toxicity, Use, Water Pollution Potential, Ecological Toxicity and Regulatory Information. PAN Pesticide Database, Pesticide Action Network, North America. Date last visited: 26 August 2013.
http://www.pesticideinfo.org/Detail_Chemical.jsp?Rec_Id=PC35460. Isoxaben – Identification, Toxicity, Use, Water Pollution Potential, Ecological Toxicity and Regulatory Information. PAN Pesticide Database, Pesticide Action Network, North America. Date last visited: 26 August 2013.
http://www.pesticideinfo.org/Detail_Chemical.jsp?Rec_Id=PC35467. Azadirachtin – Identification, Toxicity, Use, Water Pollution Potential, Ecological Toxicity and Regulatory Information. PAN Pesticide Database, Pesticide Action Network, North America. Date last visited: 26 August 2013.
http://www.pesticideinfo.org/Detail_Chemical.jsp?Rec_Id=PC35494. Chlorpyrifos-methyl – Identification, Toxicity, Use, Water Pollution Potential, Ecological Toxicity and Regulatory Information. PAN Pesticide Database, Pesticide Action Network, North America. Date last visited: 26 August 2013.
http://www.pesticideinfo.org/Detail_Chemical.jsp?Rec_Id=PC35580. Terbufos – Identification, Toxicity, Use, Water Pollution Potential, Ecological Toxicity and Regulatory Information. PAN Pesticide Database, Pesticide Action Network, North America. Date last visited: 26 August 2013.
http://www.pesticideinfo.org/Detail_Chemical.jsp?Rec_Id=PC35587. Trimethacarb – Identification, Toxicity, Use, Water Pollution Potential, Ecological Toxicity and Regulatory Information. PAN Pesticide Database, Pesticide Action Network, North America. Date last visited: 26 August 2013.
http://www.pesticideinfo.org/Detail_Chemical.jsp?Rec_Id=PC35674. Cinmethylin – Identification, Toxicity, Use, Water Pollution Potential, Ecological Toxicity and Regulatory Information. PAN Pesticide Database, Pesticide Action Network, North America. Date last visited: 26 August 2013.
http://www.pesticideinfo.org/Detail_Chemical.jsp?Rec_Id=PC35675. Clomazone – Identification, Toxicity, Use, Water Pollution Potential, Ecological Toxicity and Regulatory Information. PAN Pesticide Database, Pesticide Action Network, North America. Date last visited: 26 August 2013.
http://www.pesticideinfo.org/Detail_Chemical.jsp?Rec_Id=PC35735. Cypermethrin – Identification, Toxicity, Use, Water Pollution Potential, Ecological Toxicity and Regulatory Information. PAN Pesticide Database, Pesticide Action Network, North America. Date last visited: 26 August 2013.
http://www.pesticideinfo.org/Detail_Chemical.jsp?Rec_Id=PC35754. Thiazopyr – Identification, Toxicity, Use, Water Pollution Potential, Ecological Toxicity and Regulatory Information. PAN Pesticide Database, Pesticide Action Network, North America. Date last visited: 26 August 2013.
http://www.pesticideinfo.org/Detail_Chemical.jsp?Rec_Id=PC35819. S-Bioallethrin – Identification, Toxicity, Use, Water Pollution Potential, Ecological Toxicity and Regulatory Information. PAN Pesticide Database, Pesticide Action Network, North America. Date last visited: 26 August 2013.

http://www.pesticideinfo.org/Detail_Chemical.jsp?Rec_Id=PC35823. Tricyclazole – Identification, Toxicity, Use, Water Pollution Potential, Ecological Toxicity and Regulatory Information. PAN Pesticide Database, Pesticide Action Network, North America. Date last visited: 26 August 2013.

http://www.pesticideinfo.org/Detail_Chemical.jsp?Rec_Id=PC35869. Butachlor – Identification, Toxicity, Use, Water Pollution Potential, Ecological Toxicity and Regulatory Information. PAN Pesticide Database, Pesticide Action Network, North America. Date last visited: 26 August 2013.

http://www.pesticideinfo.org/Detail_Chemical.jsp?Rec_Id=PC35906. Azafenidin – Identification, Toxicity, Use, Water Pollution Potential, Ecological Toxicity and Regulatory Information. PAN Pesticide Database, Pesticide Action Network, North America. Date last visited: 26 August 2013.

http://www.pesticideinfo.org/Detail_Chemical.jsp?Rec_Id=PC36335. Fomesafen – Identification, Toxicity, Use, Water Pollution Potential, Ecological Toxicity and Regulatory Information. PAN Pesticide Database, Pesticide Action Network, North America. Date last visited: 26 August 2013.

http://www.pesticideinfo.org/Detail_Chemical.jsp?Rec_Id=PC36357. Tetrachlorvinphos, *Z*-isomer – Identification, Toxicity, Use, Water Pollution Potential, Ecological Toxicity and Regulatory Information. PAN Pesticide Database, Pesticide Action Network, North America. Date last visited: 26 August 2013.

http://www.pesticideinfo.org/Detail_Chemical.jsp?Rec_Id=PC36365. Fluroxypyr – Identification, Toxicity, Use, Water Pollution Potential, Ecological Toxicity and Regulatory Information. PAN Pesticide Database, Pesticide Action Network, North America. Date last visited: 26 August 2013.

http://www.pesticideinfo.org/Detail_Chemical.jsp?Rec_Id=PC36385. Butocarboxim – Identification, Toxicity, Use, Water Pollution Potential, Ecological Toxicity and Regulatory Information. PAN Pesticide Database, Pesticide Action Network, North America. Date last visited: 26 August 2013.

http://www.pesticideinfo.org/Detail_Chemical.jsp?Rec_Id=PC36389. Ethiofencarb – Identification, Toxicity, Use, Water Pollution Potential, Ecological Toxicity and Regulatory Information. PAN Pesticide Database, Pesticide Action Network, North America. Date last visited: 26 August 2013.

http://www.pesticideinfo.org/Detail_Chemical.jsp?Rec_Id=PC37412. Tepraloxydim – Identification, Toxicity, Use, Water Pollution Potential, Ecological Toxicity and Regulatory Information. PAN Pesticide Database, Pesticide Action Network, North America. Date last visited: 26 August 2013.

http://www.pesticideinfo.org/Detail_Chemical.jsp?Rec_Id=PC37430. Emamectin, Benzoate – Identification, Toxicity, Use, Water Pollution Potential, Ecological Toxicity and Regulatory Information. PAN Pesticide Database, Pesticide Action Network, North America. Date last visited: 26 August 2013.

http://www.pesticideinfo.org/Detail_Chemical.jsp?Rec_Id=PC37491. Cyhalothrin – Identification, Toxicity, Use, Water Pollution Potential, Ecological Toxicity and Regulatory Information. PAN Pesticide Database, Pesticide Action Network, North America. Date last visited: 26 August 2013.

http://www.pesticideinfo.org/Detail_Chemical.jsp?Rec_Id=PC37505. Tolclofos-methyl – Identification, Toxicity, Use, Water Pollution Potential, Ecological Toxicity and Regulatory Information. PAN Pesticide Database, Pesticide Action Network, North America. Date last visited: 26 August 2013.

http://www.pesticideinfo.org/Detail_Chemical.jsp?Rec_Id=PC37571. Cloransulam-methyl – Identification, Toxicity, Use, Water Pollution Potential, Ecological Toxicity and Regulatory Information. PAN Pesticide Database, Pesticide Action Network, North America. Date last visited: 26 August 2013.

http://www.pesticideinfo.org/Detail_Chemical.jsp?Rec_Id=PC37647. 3,5,6-Trichloro-2-pyridinol – Identification, Toxicity, Use, Water Pollution Potential, Ecological Toxicity and Regulatory Information. PAN Pesticide Database, Pesticide Action Network, North America. Date last visited: 26 August 2013.

http://www.pesticideinfo.org/Detail_Chemical.jsp?Rec_Id=PC37714. Terbuchlor – Identification, Toxicity, Use, Water Pollution Potential, Ecological Toxicity and Regulatory Information. PAN Pesticide Database, Pesticide Action Network, North America. Date last visited: 26 August 2013.

http://www.pesticideinfo.org/Detail_Chemical.jsp?Rec_Id=PC37752. Ethirimol – Identification, Toxicity, Use, Water Pollution Potential, Ecological Toxicity and Regulatory Information. PAN Pesticide Database, Pesticide Action Network, North America. Date last visited: 26 August 2013.

http://www.pesticideinfo.org/Detail_Chemical.jsp?Rec_Id=PC37757. Kasugamycin – Identification, Toxicity, Use, Water Pollution Potential, Ecological Toxicity and Regulatory Information. PAN Pesticide Database, Pesticide Action Network, North America. Date last visited: 26 August 2013.

http://www.pesticideinfo.org/Detail_Chemical.jsp?Rec_Id=PC37780. Cyanophos – Identification, Toxicity, Use, Water Pollution Potential, Ecological Toxicity and Regulatory Information. PAN Pesticide Database, Pesticide Action Network, North America. Date last visited: 26 August 2013.

http://www.pesticideinfo.org/Detail_Chemical.jsp?Rec_Id=PC37795. Delachlor – Identification, Toxicity, Use, Water Pollution Potential, Ecological Toxicity and Regulatory Information. PAN Pesticide Database, Pesticide Action Network, North America. Date last visited: 26 August 2013.

http://www.pesticideinfo.org/Detail_Chemical.jsp?Rec_Id=PC38027. Coumatetralyl – Identification, Toxicity, Use, Water Pollution Potential, Ecological Toxicity and Regulatory Information. PAN Pesticide Database, Pesticide Action Network, North America. Date last visited: 26 August 2013.

http://www.pesticideinfo.org/Detail_Chemical.jsp?Rec_Id=PC38045. Isoproturon – Identification, Toxicity, Use, Water Pollution Potential, Ecological Toxicity and Regulatory Information. PAN Pesticide Database, Pesticide Action Network, North America. Date last visited: 26 August 2013.

http://www.pesticideinfo.org/Detail_Chemical.jsp?Rec_Id=PC38494. Pretilachlor – Identification, Toxicity, Use, Water Pollution Potential, Ecological Toxicity and Regulatory Information. PAN Pesticide Database, Pesticide Action Network, North America. Date last visited: 26 August 2013.

http://www.pesticideinfo.org/Detail_Chemical.jsp?Rec_Id=PC38877. Carpropamid – Identification, Toxicity, Use, Water Pollution Potential, Ecological Toxicity and Regulatory Information. PAN Pesticide Database, Pesticide Action Network, North America. Date last visited: 26 August 2013.

http://www.pesticideinfo.org/Detail_Chemical.jsp?Rec_Id=PC38879. Clomeprop – Identification, Toxicity, Use, Water Pollution Potential, Ecological Toxicity and Regulatory Information. PAN Pesticide Database, Pesticide Action Network, North America. Date last visited: 26 August 2013.

http://www.pesticideinfo.org/Detail_Chemical.jsp?Rec_Id=PC38893. Oxabetrinil – Identification, Toxicity, Use, Water Pollution Potential, Ecological Toxicity and Regulatory Information. PAN Pesticide Database, Pesticide Action Network, North America. Date last visited: 26 August 2013.

http://www.pesticideinfo.org/Detail_Chemical.jsp?Rec_Id=PC39006. Cyhalofop Butyl – Identification, Toxicity, Use, Water Pollution Potential, Ecological Toxicity and Regulatory Information. PAN Pesticide Database, Pesticide Action Network, North America. Date last visited: 26 August 2013.

http://www.pesticideinfo.org/Detail_Chemical.jsp?Rec_Id=PC39015. Cinidon-ethyl – Identification, Toxicity, Use, Water Pollution Potential, Ecological Toxicity and Regulatory Information. PAN Pesticide Database, Pesticide Action Network, North America. Date last visited: 26 August 2013.

http://www.pesticideinfo.org/Detail_Chemical.jsp?Rec_Id=PC39208. Thenylchlor – Identification, Toxicity, Use, Water Pollution Potential, Ecological Toxicity and Regulatory Information. PAN Pesticide Database, Pesticide Action Network, North America. Date last visited: 26 August 2013.

http://www.pesticideinfo.org/Detail_Chemical.jsp?Rec_Id=PC39237. Isopolinate – Identification, Toxicity, Use, Water Pollution Potential, Ecological Toxicity and Regulatory Information. PAN Pesticide Database, Pesticide Action Network, North America. Date last visited: 26 August 2013.

http://www.pesticideinfo.org/Detail_Chemical.jsp?Rec_Id=PC39240. Methiobencarb – Identification, Toxicity, Use, Water Pollution Potential, Ecological Toxicity and Regulatory Information. PAN Pesticide Database, Pesticide Action Network, North America. Date last visited: 26 August 2013.

www.pesticideinfo.org/Detail_Chemical.jsp?Rec_Id=PC42599. Hexamethylphosphoramide – Identification, Toxicity, Use, Water Pollution Potential, Ecological Toxicity and Regulatory Information. PAN Pesticide Database, Pesticide Action Network, North America. Date last visited: 28 August 2013.

http://www.pesticideinfo.org/Detail_ChemReg.jsp?Rec_Id=PC33169. Folpet – Identification, Toxicity, Use, Water Pollution Potential, Ecological Toxicity and Regulatory Information. PAN Pesticide Database, Pesticide Action Network, North America. Date last visited: 26 August 2013.

http://www.pesticideinfo.org/Detail_ChemReg.jsp?Rec_Id=PC33412. Coumaphos – Identification, Toxicity, Use, Water Pollution Potential, Ecological Toxicity and Regulatory Information. PAN Pesticide Database, Pesticide Action Network, North America. Date last visited: 26 August 2013.

http://www.pesticideinfo.org/Detail_ChemReg.jsp?Rec_Id=PC33516. Cyanazine – Identification, Toxicity, Use, Water Pollution Potential, Ecological Toxicity and Regulatory Information. PAN Pesticide Database, Pesticide Action Network, North America. Date last visited: 26 August 2013.

http://www.pesticideinfo.org/Detail_ChemReg.jsp?Rec_Id=PC34271. Propiconazole – Identification, Toxicity, Use, Water Pollution Potential, Ecological Toxicity and Regulatory Information. PAN Pesticide Database, Pesticide Action Network, North America. Date last visited: 26 August 2013.

http://www.pesticideinfo.org/Detail_ChemReg.jsp?Rec_Id=PC34923. Triflumizole – Identification, Toxicity, Use, Water Pollution Potential, Ecological Toxicity and Regulatory Information. PAN Pesticide Database, Pesticide Action Network, North America. Date last visited: 26 August 2013.

http://www.pesticideinfo.org/Detail_ChemReg.jsp?Rec_Id=PC38042. Flamprop-methyl – Identification, Toxicity, Use, Water Pollution Potential, Ecological Toxicity and Regulatory Information. PAN Pesticide Database, Pesticide Action Network, North America. Date last visited: 26 August 2013.

http://www.pesticideinfo.org/Detail_ChemReg.jsp?Rec_Id=PC40144. Thifluzamide – Identification, Toxicity, Use, Water Pollution Potential, Ecological Toxicity and Regulatory Information. PAN Pesticide Database, Pesticide Action Network, North America. Date last visited: 26 August 2013.

http://www.pesticideinfo.org/Detail_Country.jsp?Country=Cambodia. Cambodia – Pesticide Registration, Import Consent and Bans. PAN Pesticide Database, Pesticide Action Network, North America. Date last visited: 26 August 2013.

http://www.pesticide-residue-testing.com/residue-testing.asp?docid=2245&testfor=Tau-fluvalinate. Tau-fluvalinate. Pesticide Residue Testing. Date last visited: 27 August 2013.

http://www.pesticides.gov.uk. Health and Safety Executive, UK. Date last visited: July 20, 2012.

http://www.pesticides.gov.uk/guidance/industries/pesticides/advisory-groups/Resistance-Action-Groups/Resistance-Highlight/fungicide-resistance-management-in-oilseed-rape. Fungicide Resistance Management in Oilseed Rape. Resistance Action Group. Date last visited: 27 August 2013.

http://www.pesticides.gov.uk/prc.asp?id=1925. How Will the PRC Report Carbendazim Residues in Apples or Pears after 15 September 2006? Health and Safety Executive, UK. Date last visited: 27 August 2013.

http://www.plasmaneem.com/neemoil-brown-planthopper.html. Brown Planthopper Control Using Biological Insecticide – Neem Oil: Plasma Neem. Date last visited: 27 August 2013.

http://www.popstoolkit.com/about/chemical/toxaphene.aspx. Toxaphene. Persistent Organic Pollutants Toolkit. Adapted from Persistent Organic Pollutants: Information on POPs, Their Alternatives and Alternative Approaches. United Nations Environmental Program (UNEP), 1995. Date last visited: 27 August 2013.

http://www.potato2008.org/en/potato/cultivation.html. Cultivation. International Year of the Potato 2008. Food and Agriculture Organization of the United Nations (FAO). Date last visited: 27 August 2013.

http://www.potatodiseases.org/foliardiseases.html. Foliar Diseases/Conditions of Potato. Michigan State University. Date last visited: 27 August 2013.

http://www.potatodiseases.org/tuberdiseases.html.Tuber and Stem Diseases/Conditions of Potato. Michigan State University. Date last visited: 27 August 2013.

http://www.rainforestinfo.org.au/good_wood/pres_not.htm. Preservatives Not Recommended. The RIC Good Wood Guide. Date last visited: 27 August 2013.

http://www.rci.rutgers.edu/~insects/bottsum.htm. Internationally – Where/Why Biopesticides Work. Alternative Paradigms for Commercializing Biological Control Workshop. Date last visited: 27 August 2013.

http://www.rtvanderbilt.com/documents/MSDS/US/29409.pdf. Material Safety Data Sheet. Vanderbilt Chemicals, LLC. Date last visited: 27 August 2013.

http://www.sare.org/Learning-Center/SARE-Program-Materials/National-Program-Materials/What-is-Sustainable-Agriculture. What is Sustainable Agriculture? Sustainable Agriculture Research and Education and United States Department of Agriculture. Date last visited: 28 August 2013.

http://www.sciencedaily.com. Science Daily. Year last visited: 2011 and 2012.

http://www.sciencelab.com. ScienceLab.com: Chemicals and Laboratory Equipment. Date last visited: 11 July 2012.

http://www.scribd.com/doc/35962558/AD29-Pesticides-Compounds-Use-and-Hazards. Boland, J., Koomen, I., van Lidth de Jeude, J. and Oudejans, J. (2004) *Pesticides: Compounds, Use and Hazards*. Agromisa Foundation, Wageningen. Date last visited: 27 August 2013.

http://www.sera17.ext.vt.edu/Documents/BMP_Cover_Crops.pdf. Cover Crops. Phosphorus Best Management Practices – Protecting Water Quality. Date last visited: 27 August 2013.

http://www.sinoharvest.com/products/Hexazinone.shtml. Hexazinone. SinoHarvest Products. Date last visited: 27 August 2013.

http://www.sinoharvest.com/products/Imidacloprid.shtml. Imidacloprid. SinoHarvest Products. Date last visited: 27 August 2013.

http://www.sms.si.edu/irlspec/agasicles_hygrophila.htm. Agasicles hygrophila. Smithsonian Marine Station, Fort Pierce. Date last visited: 27 August 2013.

http://www.speclab.com/compound/c106934.htm. 1,2-Dibromoethane. Spectrum Chemical Fact Sheet. Florida Spectrum Environmental Services, Inc. Date last visited: 27 August 2013.

http://www.speclab.com/compound/c470906.htm. Chlorfenvinphos. Spectrum Chemical Fact Sheet. Florida Spectrum Environmental Services, Inc. Date last visited: 27 August 2013.

http://www.springerlink.com/content/hvp10v0176534381/. Sahoo, A., Sahu, S.K., Sharmila, M. and Sethunathan, N. (1980) Persistence of carbamate insecticides, carbosulfan and carbofuran in soils as influenced by temperature and microbial activity. *Bulletin of Environmental Contamination and Toxicology* 44 (6). Date last visited: 27 August 2013.

http://www.syngenta.com/global/bioline/en/icm/Pages/IntegratedCropManagement.aspx. Guide to Syngenta Bioline and to Integrated Crop Management (ICM). Date last visited: 27 August 2013.

http://www.syngenta.com/global/bioline/en/products/allproducts/Pages/Amblylinebar.aspx. Amblyline bar. *Amblyseius* (*Neoseiulus*) *barkeri*, Tarsonemid Control. Syngenta. Date last visited: 27 August 2013.

http://www.syngenta.com/global/bioline/en/products/allproducts/Pages/Amblyline-cu.aspx. Amblyline cu CRS, *Amblyseius cucumeris*, Thrips and Tarsonemid Control. Syngenta. Date last visited: 27 August 2013.

http://www.syngentacropprotection.com/prodrender/index.aspx?nav=MOA&ProdID=649&ProdNM=Proclaim. Proclaim® Insecticide. Syngenta. Date last visited: 27 August 2013.

http://www.taminco.com/products/crop-protection/thiram.html?product_id=37. Thiram. Taminco People and Molecules. Date last visited: 27 August 2013.

http://www.textbookofbacteriology.net/endotoxin.html. Bacterial Endotoxin. Todar's Online Textbook of Bacteriology. Date last visited: 27 August 2013.

http://www.thegoodscentscompany.com/data/rw1011951.html#topropr. Alpha-ionone. The Good Scent Company. Date last visited: 27 August 2013.

http://www.the-piedpiper.co.uk/th15(g).htm. Alpha Chloralose. PiedPiper Northern Limited. Date last visited: 27 August 2013.

http://www.uwlax.edu/biology/volk/fungi3/sld042.htm. Ascomycota. Date last visited: 27 August 2013..

http://www.washingtonpost.com/wp-dyn/content/story/2008/07/24/ST2008072403523.html?ref=us In Surprise Move, EPA Bans Carbofuran Residue on Food. Washington Post 25 July 2008. Date last visited: 27 August 2013.

http://www.weedscience.com. The International Survey of Herbicide Resistant Weeds. Date last visited: 11 October 2013.

http://www.weedscience.org/summary/Country.aspx. Herbicide Resistant Weeds in Canada. International Survey of Herbicide Resistant Weeds. Date last visited: 27 August 2013.

http://www.weedscience.org/summary/Country.aspx. Herbicide Resistant Weeds in India. International Survey of Herbicide Resistant Weeds. Date last visited: 27 August 2013.

http://www.weedscience.org/summary/Crop.aspx. Herbicide Resistant Weeds in Corn (Maize) Globally. International Survey of Herbicide Resistant Weeds. Date last visited: 27 August 2013.

http://www.who.int/water_sanitation_health/dwq/chemicals/wsh0304_69/en/index7.html. 2-Phenylphenol in Drinking Water. Water Sanitation Health. World Health Organization (WHO). Date last visited: 27 August 2013.

http://www.wisegeek.com/what-is-thiodicarb.html. What is Thiodicarb? wiseGEEK. Date last visited: 28 August 2013.

http://www.workcover.nsw.gov.au/formspublications/publications/Documents/chloropicrin_fact_sheet_1371.pdf. Factsheet Chloropicrin. WorkCover Publications, NSW. Date last visited: 27 August 2013.

http://www.worldofmolecules.com/pesticides/rotenone.htm. Rotenone. World of Molecules. Date last visited: 27 August 2013.

http://www.wrongdiagnosis.com/c/chemical_poisoning_esfenvalerate/intro.htm. Chemical Poisoning – Esfenvalerate. Health Guide Inc. Date last visited: 28 August 2013.

http://www.wynnsusa.com/pdf/cwc200.pdf. Wynn's Microbinator. Material Safety Data Sheet. Wynn's, a Division of Illinois Tool Works Inc. Date last visited: 28 August 2013.

http://www.xsgrowth.com/index.php/fungicides/64-quinoneoutsideinhibitor.html. An Insight into Quinone Outside Inhibitor (QoI) Fungicides. Date last visited: 28 August 2013.

http://www.xsgrowth.com/index.php/fungicides/75-starfungicides.html. Star Fungicides – a Fascinating Group of Fungicides in Plant Protection. Date last visited: 28 August 2013.

http://www1.agric.gov.ab.ca/$department/deptdocs.nsf/all/opp9595. Pest Management – Diseases – Viral. Government of Alberta. Date last visited: 28 August 2013.

http://www1.agric.gov.ab.ca/$department/deptdocs.nsf/all/prm2859. Pith Necrosis of Tomato. Government of Alberta. Date last visited: 28 August 2013.

http://www1.agric.gov.ab.ca/$department/deptdocs.nsf/all/prm7744. Diseases of Vegetables. Government of Alberta. Date last visited: 28 August 2013.

http://yosemite1.epa.gov/ee/epa/ria.nsf/vwDT/32605D1D0E52EC1A8525736700686B 9C. Florasulam: Human Health Risk Assessment for Proposed Use on Cereal Grains (Wheat, Oats, Barley, Rye, and Triticale). United States Environmental Protection Agency. Date last visited: 28 August 2013.

https://fscimage.fishersci.com/msds/03672.htm. Eugenol, 99%. Material Safety Data Sheet. Date last visited: 28 August 2013.

https://pxmrl.maf.govt.nz/. New Zealand Food Safety Authority (NZFSA) Pesticide MRL Database. NZFSA. Date last visited: 30 July 2013.

users.rcn.com/jkimball.ma.ultranet/BiologyPages/I/Insecticides.html. Insecticides. 2011. Kimball's Biology Pages. Date last visited: 28 August 2013.

www.frac.info. Fungicide Resistance Action Committee (FRAC) Pathogen Risk List, December 2005, adapted from EPPO 2002. FRAC Monograph No. 3. Date last visited: 25 June 2010.

www.ipp.gov.bc.ca. Province of British Columbia (Ilona Ugro, Copyright Officer, Intellectual Property Program, Ministry of Citizens' Services). Date last visited: 25 June 2010.

www.jnkvv.nic.in. General Tips of Integrated Pest Management. Jawaharlal Nehru Krishi Vishwa Vidyalaya (JNKVV), Jabalpur, Madhya Pradesh; Ministry of Science & Technology, Department of Scientific & Industrial Research, Government of India. Date last visited: 28 August 2013.

www.kingtaichem.com/pro_h_PYRITHIOBAC-SODIUM.htm. Pyrithiobac-sodium. Date last visited: 28 August 2013.

www.lookchem.com.Hot Product_List. LookChem. Date last visited: 28 August 2013.

www.lookchem.com/Isoxaben/. Benzamide,N-[3-(1-ethyl-1-methylpropyl)-5-isoxazolyl]-2,6-dimethoxy-. LookChem. Date last visited: 28 August 2013.

www.panna.org. Pesticide Action Network. Date last visited: 28 August 2013.

www.pesticides.gov.uk. Pesticides Safety Directorate (UK). Date last visited: 28 August 2013.

www.pestproducts.com/hydroprene.htm. Hydroprene. Insect Growth Regulator (IGR). Date last visited: 28 August 2013.

www.sigma-aldrich.com. Sigma-Aldrich. Date last visited: 28 August 2013.

www.stewardshipcommunity.com. Stewardship Community, promoting safe, effective use of pesticides/crop protection products. Date last visited: 28 August 2013.

www.wellgrowhorti.com. Wellgrow, flower seeds distributor in Malaysia. Date last visited: 28 August 2013.

Index

Note: Page numbers in **bold** refer to figures, tables and chemical structures.